Complete Solutions Manual

for

SINGLE VARIABLE CALCULUS

SIXTH EDITION

DANIEL ANDERSON
University of Iowa

JEFFERY A. COLE
Anoka-Ramsey Community College

DANIEL DRUCKER
Wayne State University

THOMSON

BROOKS/COLE

Australia · Brazil · Canada · Mexico · Singapore · Spain · United Kingdom · United States

Printed in the United States of America

1   2   3   4   5   6   7   11   10   09   08   07

Printer: Thomson/West
Cover Image: Amelie Fear, Folkmusician violins

ISBN-13: 978-0-495-01232-0
ISBN-10: 0-495-01232-7

Thomson Higher Education
10 Davis Drive
Belmont, CA 94002-3098
USA

For more information about our products,
contact us at:
Thomson Learning Academic Resource Center
1-800-423-0563

For permission to use material from this text or
product, submit a request online at
http://www.thomsonrights.com.

Any additional questions about permissions can be
submitted by email to **thomsonrights@thomson.com.**

PREFACE

This *Complete Solutions Manual* contains solutions to all exercises in the text *Single Variable Calculus,* Sixth Edition, by James Stewart. A student version of this manual is also available; it contains solutions to the odd-numbered exercises in each section, the review sections, the True-False Quizzes, and the Problem Solving sections, as well as solutions to all the exercises in the Concept Checks. No solutions to the projects appear in the student version. It is our hope that by browsing through the solutions, professors will save time in determining appropriate assignments for their particular class.

We use some non-standard notation in order to save space. If you see a symbol which you don't recognize, refer to the Table of Abbreviations and Symbols on page v.

We appreciate feedback concerning errors, solution correctness or style, and manual style. Any comments may be sent directly to jeff.cole@anokaramsey.edu, or in care of the publisher: Thomson Brooks/Cole, 10 Davis Drive, Belmont CA 94002-3098.

We would like to thank Brian Betsill, Stephanie Kuhns, and Kathi Townes, of TECHarts, for their production services; and Stacy Green, of Thomson Brooks/Cole Publishing Company, for her patience and support. All of these people have provided invaluable help in creating this manual.

Jeffery A. Cole
Anoka-Ramsey Community College

James Stewart
McMaster University

Daniel Drucker
Wayne State University

Daniel Anderson
University of Iowa

ABREVIATIONS AND SYMBOLS

CD   concave downward

CU   concave upward

D   the domain of f

FDT   First Derivative Test

HA   horizontal asymptote(s)

I   interval of convergence

I/D   Increasing/Decreasing Test

IP   inflection point(s)

R   radius of convergence

VA   vertical asymptote(s)

$\overset{\text{CAS}}{=}$ indicates the use of a computer algebra system.

$\overset{\text{H}}{=}$ indicates the use of l'Hospital's Rule.

$\overset{j}{=}$ indicates the use of Formula j in the Table of Integrals in the back endpapers.

$\overset{s}{=}$ indicates the use of the substitution $\{u = \sin x, du = \cos x\, dx\}$.

$\overset{c}{=}$ indicates the use of the substitution $\{u = \cos x, du = -\sin x\, dx\}$.

CONTENTS

4 □ APPLICATIONS OF DIFFERENTIATION   213

5 □ INTEGRALS   365

9 □ FURTHER APPLICATIONS OF INTEGRATION   731

10 □ DIFFERENTIAL EQUATIONS   783

xii   □   CONTENTS

□ **APPENDIXES   1059**

☐ DIAGNOSTIC TESTS

Test A   Algebra

1. (a) $(-3)^4 = (-3)(-3)(-3)(-3) = 81$

(b) $-3^4 = -(3)(3)(3)(3) = -81$

(c) $3^{-4} = \dfrac{1}{3^4} = \dfrac{1}{81}$

(d) $\dfrac{5^{23}}{5^{21}} = 5^{23-21} = 5^2 = 25$

(e) $\left(\dfrac{2}{3}\right)^{-2} = \left(\dfrac{3}{2}\right)^2 = \dfrac{9}{4}$

(f) $16^{-3/4} = \dfrac{1}{16^{3/4}} = \dfrac{1}{\left(\sqrt[4]{16}\right)^3} = \dfrac{1}{2^3} = \dfrac{1}{8}$

2. (a) Note that $\sqrt{200} = \sqrt{100 \cdot 2} = 10\sqrt{2}$ and $\sqrt{32} = \sqrt{16 \cdot 2} = 4\sqrt{2}$. Thus $\sqrt{200} - \sqrt{32} = 10\sqrt{2} - 4\sqrt{2} = 6\sqrt{2}$.

(b) $(3a^3b^3)(4ab^2)^2 = 3a^3b^3 16a^2b^4 = 48a^5b^7$

(c) $\left(\dfrac{3x^{3/2}y^3}{x^2y^{-1/2}}\right)^{-2} = \left(\dfrac{x^2y^{-1/2}}{3x^{3/2}y^3}\right)^2 = \dfrac{(x^2y^{-1/2})^2}{(3x^{3/2}y^3)^2} = \dfrac{x^4y^{-1}}{9x^3y^6} = \dfrac{x^4}{9x^3y^6y} = \dfrac{x}{9y^7}$

3. (a) $3(x+6) + 4(2x-5) = 3x + 18 + 8x - 20 = 11x - 2$

(b) $(x+3)(4x-5) = 4x^2 - 5x + 12x - 15 = 4x^2 + 7x - 15$

(c) $\left(\sqrt{a} + \sqrt{b}\right)\left(\sqrt{a} - \sqrt{b}\right) = \left(\sqrt{a}\right)^2 - \sqrt{a}\sqrt{b} + \sqrt{a}\sqrt{b} - \left(\sqrt{b}\right)^2 = a - b$

Or: Use the formula for the difference of two squares to see that $\left(\sqrt{a} + \sqrt{b}\right)\left(\sqrt{a} - \sqrt{b}\right) = \left(\sqrt{a}\right)^2 - \left(\sqrt{b}\right)^2 = a - b$.

(d) $(2x+3)^2 = (2x+3)(2x+3) = 4x^2 + 6x + 6x + 9 = 4x^2 + 12x + 9$.

Note: A quicker way to expand this binomial is to use the formula $(a+b)^2 = a^2 + 2ab + b^2$ with $a = 2x$ and $b = 3$:

$(2x+3)^2 = (2x)^2 + 2(2x)(3) + 3^2 = 4x^2 + 12x + 9$

(e) See Reference Page 1 for the binomial formula $(a+b)^3 = a^3 + 3a^2b + 3ab^2 + b^3$. Using it, we get

$(x+2)^3 = x^3 + 3x^2(2) + 3x(2^2) + 2^3 = x^3 + 6x^2 + 12x + 8$.

4. (a) Using the difference of two squares formula, $a^2 - b^2 = (a+b)(a-b)$, we have

$4x^2 - 25 = (2x)^2 - 5^2 = (2x+5)(2x-5)$.

(b) Factoring by trial and error, we get $2x^2 + 5x - 12 = (2x-3)(x+4)$.

(c) Using factoring by grouping and the difference of two squares formula, we have

$x^3 - 3x^2 - 4x + 12 = x^2(x-3) - 4(x-3) = (x^2-4)(x-3) = (x-2)(x+2)(x-3)$.

(d) $x^4 + 27x = x(x^3 + 27) = x(x+3)(x^2 - 3x + 9)$

This last expression was obtained using the sum of two cubes formula, $a^3 + b^3 = (a+b)(a^2 - ab + b^2)$ with $a = x$ and $b = 3$. [See Reference Page 1 in the textbook.]

(e) The smallest exponent on x is $-\frac{1}{2}$, so we will factor out $x^{-1/2}$.

$3x^{3/2} - 9x^{1/2} + 6x^{-1/2} = 3x^{-1/2}(x^2 - 3x + 2) = 3x^{-1/2}(x-1)(x-2)$

(f) $x^3y - 4xy = xy(x^2 - 4) = xy(x-2)(x+2)$

5. (a) $\dfrac{x^2 + 3x + 2}{x^2 - x - 2} = \dfrac{(x+1)(x+2)}{(x+1)(x-2)} = \dfrac{x+2}{x-2}$

(b) $\dfrac{2x^2 - x - 1}{x^2 - 9} \cdot \dfrac{x+3}{2x+1} = \dfrac{(2x+1)(x-1)}{(x-3)(x+3)} \cdot \dfrac{x+3}{2x+1} = \dfrac{x-1}{x-3}$

(c) $\dfrac{x^2}{x^2 - 4} - \dfrac{x+1}{x+2} = \dfrac{x^2}{(x-2)(x+2)} - \dfrac{x+1}{x+2} = \dfrac{x^2}{(x-2)(x+2)} - \dfrac{x+1}{x+2} \cdot \dfrac{x-2}{x-2} = \dfrac{x^2 - (x+1)(x-2)}{(x-2)(x+2)}$

$= \dfrac{x^2 - (x^2 - x - 2)}{(x+2)(x-2)} = \dfrac{x+2}{(x+2)(x-2)} = \dfrac{1}{x-2}$

(d) $\dfrac{\dfrac{y}{x} - \dfrac{x}{y}}{\dfrac{1}{y} - \dfrac{1}{x}} = \dfrac{\dfrac{y}{x} - \dfrac{x}{y}}{\dfrac{1}{y} - \dfrac{1}{x}} \cdot \dfrac{xy}{xy} = \dfrac{y^2 - x^2}{x - y} = \dfrac{(y-x)(y+x)}{-(y-x)} = \dfrac{y+x}{-1} = -(x+y)$

6. (a) $\dfrac{\sqrt{10}}{\sqrt{5}-2} = \dfrac{\sqrt{10}}{\sqrt{5}-2} \cdot \dfrac{\sqrt{5}+2}{\sqrt{5}+2} = \dfrac{\sqrt{50}+2\sqrt{10}}{(\sqrt{5})^2 - 2^2} = \dfrac{5\sqrt{2}+2\sqrt{10}}{5-4} = 5\sqrt{2}+2\sqrt{10}$

(b) $\dfrac{\sqrt{4+h}-2}{h} = \dfrac{\sqrt{4+h}-2}{h} \cdot \dfrac{\sqrt{4+h}+2}{\sqrt{4+h}+2} = \dfrac{4+h-4}{h(\sqrt{4+h}+2)} = \dfrac{h}{h(\sqrt{4+h}+2)} = \dfrac{1}{\sqrt{4+h}+2}$

7. (a) $x^2 + x + 1 = \left(x^2 + x + \frac{1}{4}\right) + 1 - \frac{1}{4} = \left(x + \frac{1}{2}\right)^2 + \frac{3}{4}$

(b) $2x^2 - 12x + 11 = 2(x^2 - 6x) + 11 = 2(x^2 - 6x + 9 - 9) + 11 = 2(x^2 - 6x + 9) - 18 + 11 = 2(x-3)^2 - 7$

8. (a) $x + 5 = 14 - \frac{1}{2}x \Leftrightarrow x + \frac{1}{2}x = 14 - 5 \Leftrightarrow \frac{3}{2}x = 9 \Leftrightarrow x = \frac{2}{3}\cdot 9 \Leftrightarrow x = 6$

(b) $\dfrac{2x}{x+1} = \dfrac{2x-1}{x} \Rightarrow 2x^2 = (2x-1)(x+1) \Leftrightarrow 2x^2 = 2x^2 + x - 1 \Leftrightarrow x = 1$

(c) $x^2 - x - 12 = 0 \Leftrightarrow (x+3)(x-4) = 0 \Leftrightarrow x+3 = 0 \text{ or } x-4 = 0 \Leftrightarrow x = -3 \text{ or } x = 4$

(d) By the quadratic formula, $2x^2 + 4x + 1 = 0 \Leftrightarrow$

$x = \dfrac{-4 \pm \sqrt{4^2 - 4(2)(1)}}{2(2)} = \dfrac{-4 \pm \sqrt{8}}{4} = \dfrac{-4 \pm 2\sqrt{2}}{4} = \dfrac{2(-2 \pm \sqrt{2})}{4} = \dfrac{-2 \pm \sqrt{2}}{2} = -1 \pm \frac{1}{2}\sqrt{2}.$

(e) $x^4 - 3x^2 + 2 = 0 \Leftrightarrow (x^2 - 1)(x^2 - 2) = 0 \Leftrightarrow x^2 - 1 = 0 \text{ or } x^2 - 2 = 0 \Leftrightarrow x^2 = 1 \text{ or } x^2 = 2 \Leftrightarrow$
$x = \pm 1 \text{ or } x = \pm\sqrt{2}$

(f) $3|x-4| = 10 \Leftrightarrow |x-4| = \frac{10}{3} \Leftrightarrow x-4 = -\frac{10}{3} \text{ or } x-4 = \frac{10}{3} \Leftrightarrow x = \frac{2}{3} \text{ or } x = \frac{22}{3}$

(g) Multiplying through $2x(4-x)^{-1/2} - 3\sqrt{4-x} = 0$ by $(4-x)^{1/2}$ gives $2x - 3(4-x) = 0 \Leftrightarrow$
$2x - 12 + 3x = 0 \Leftrightarrow 5x - 12 = 0 \Leftrightarrow 5x = 12 \Leftrightarrow x = \frac{12}{5}.$

9. (a) $-4 < 5 - 3x \leq 17 \Leftrightarrow -9 < -3x \leq 12 \Leftrightarrow 3 > x \geq -4 \text{ or } -4 \leq x < 3.$
In interval notation, the answer is $[-4, 3)$.

(b) $x^2 < 2x + 8 \Leftrightarrow x^2 - 2x - 8 < 0 \Leftrightarrow (x+2)(x-4) < 0$. Now, $(x+2)(x-4)$ will change sign at the critical values $x = -2$ and $x = 4$. Thus the possible intervals of solution are $(-\infty, -2)$, $(-2, 4)$, and $(4, \infty)$. By choosing a single test value from each interval, we see that $(-2, 4)$ is the only interval that satisfies the inequality.

(c) The inequality $x(x - 1)(x + 2) > 0$ has critical values of $-2, 0$, and 1. The corresponding possible intervals of solution are $(-\infty, -2)$, $(-2, 0)$, $(0, 1)$ and $(1, \infty)$. By choosing a single test value from each interval, we see that both intervals $(-2, 0)$ and $(1, \infty)$ satisfy the inequality. Thus, the solution is the union of these two intervals: $(-2, 0) \cup (1, \infty)$.

(d) $|x - 4| < 3$ $\Leftrightarrow$ $-3 < x - 4 < 3$ $\Leftrightarrow$ $1 < x < 7$. In interval notation, the answer is $(1, 7)$.

(e) $\dfrac{2x - 3}{x + 1} \le 1$ $\Leftrightarrow$ $\dfrac{2x - 3}{x + 1} - 1 \le 0$ $\Leftrightarrow$ $\dfrac{2x - 3}{x + 1} - \dfrac{x + 1}{x + 1} \le 0$ $\Leftrightarrow$ $\dfrac{2x - 3 - x - 1}{x + 1} \le 0$ $\Leftrightarrow$ $\dfrac{x - 4}{x + 1} \le 0$.

Now, the expression $\dfrac{x - 4}{x + 1}$ may change signs at the critical values $x = -1$ and $x = 4$, so the possible intervals of solution are $(-\infty, -1)$, $(-1, 4]$, and $[4, \infty)$. By choosing a single test value from each interval, we see that $(-1, 4]$ is the only interval that satisfies the inequality.

10. (a) False. In order for the statement to be true, it must hold for all real numbers, so, to show that the statement is false, pick $p = 1$ and $q = 2$ and observe that $(1 + 2)^2 \ne 1^2 + 2^2$. In general, $(p + q)^2 = p^2 + 2pq + q^2$.

(b) True as long as a and b are positive real numbers. To see this, think in terms of the laws of exponents:
$$\sqrt{ab} = (ab)^{1/2} = a^{1/2}b^{1/2} = \sqrt{a}\,\sqrt{b}.$$

(c) False. To see this, let $p = 1$ and $q = 2$, then $\sqrt{1^2 + 2^2} \ne 1 + 2$.

(d) False. To see this, let $T = 1$ and $C = 2$, then $\dfrac{1 + 1(2)}{2} \ne 1 + 1$.

(e) False. To see this, let $x = 2$ and $y = 3$, then $\dfrac{1}{2 - 3} \ne \dfrac{1}{2} - \dfrac{1}{3}$.

(f) True since $\dfrac{1/x}{a/x - b/x} \cdot \dfrac{x}{x} = \dfrac{1}{a - b}$, as long as $x \ne 0$ and $a - b \ne 0$.

Test B   Analytic Geometry

1. (a) Using the point $(2, -5)$ and $m = -3$ in the point-slope equation of a line, $y - y_1 = m(x - x_1)$, we get
$$y - (-5) = -3(x - 2) \quad \Rightarrow \quad y + 5 = -3x + 6 \quad \Rightarrow \quad y = -3x + 1.$$

(b) A line parallel to the x-axis must be horizontal and thus have a slope of 0. Since the line passes through the point $(2, -5)$, the y-coordinate of every point on the line is -5, so the equation is $y = -5$.

(c) A line parallel to the y-axis is vertical with undefined slope. So the x-coordinate of every point on the line is 2 and so the equation is $x = 2$.

(d) Note that $2x - 4y = 3$ $\Rightarrow$ $-4y = -2x + 3$ $\Rightarrow$ $y = \frac{1}{2}x - \frac{3}{4}$. Thus the slope of the given line is $m = \frac{1}{2}$. Hence, the slope of the line we're looking for is also $\frac{1}{2}$ (since the line we're looking for is required to be parallel to the given line). So the equation of the line is $y - (-5) = \frac{1}{2}(x - 2)$ $\Rightarrow$ $y + 5 = \frac{1}{2}x - 1$ $\Rightarrow$ $y = \frac{1}{2}x - 6$.

2. First we'll find the distance between the two given points in order to obtain the radius, r, of the circle:
$r = \sqrt{[3 - (-1)]^2 + (-2 - 4)^2} = \sqrt{4^2 + (-6)^2} = \sqrt{52}$. Next use the standard equation of a circle, $(x - h)^2 + (y - k)^2 = r^2$, where (h, k) is the center, to get $(x + 1)^2 + (y - 4)^2 = 52$.

3. We must rewrite the equation in standard form in order to identify the center and radius. Note that

$x^2 + y^2 - 6x + 10y + 9 = 0 \Rightarrow x^2 - 6x + 9 + y^2 + 10y = 0$. For the left-hand side of the latter equation, we

factor the first three terms and complete the square on the last two terms as follows: $x^2 - 6x + 9 + y^2 + 10y = 0 \Rightarrow$

$(x - 3)^2 + y^2 + 10y + 25 = 25 \Rightarrow (x - 3)^2 + (y + 5)^2 = 25$. Thus, the center of the circle is $(3, -5)$ and the radius is 5.

4. (a) $A(-7, 4)$ and $B(5, -12) \Rightarrow m_{AB} = \dfrac{-12 - 4}{5 - (-7)} = \dfrac{-16}{12} = -\dfrac{4}{3}$

(b) $y - 4 = -\frac{4}{3}[x - (-7)] \Rightarrow y - 4 = -\frac{4}{3}x - \frac{28}{3} \Rightarrow 3y - 12 = -4x - 28 \Rightarrow 4x + 3y + 16 = 0$. Putting $y = 0$,

we get $4x + 16 = 0$, so the x-intercept is -4, and substituting 0 for x results in a y-intercept of $-\frac{16}{3}$.

(c) The midpoint is obtained by averaging the corresponding coordinates of both points: $\left(\frac{-7+5}{2}, \frac{4+(-12)}{2}\right) = (-1, -4)$.

(d) $d = \sqrt{[5 - (-7)]^2 + (-12 - 4)^2} = \sqrt{12^2 + (-16)^2} = \sqrt{144 + 256} = \sqrt{400} = 20$

(e) The perpendicular bisector is the line that intersects the line segment $\overline{AB}$ at a right angle through its midpoint. Thus the

perpendicular bisector passes through $(-1, -4)$ and has slope $\frac{3}{4}$ [the slope is obtained by taking the negative reciprocal of

the answer from part (a)]. So the perpendicular bisector is given by $y + 4 = \frac{3}{4}[x - (-1)]$ or $3x - 4y = 13$.

(f) The center of the required circle is the midpoint of $\overline{AB}$, and the radius is half the length of $\overline{AB}$, which is 10. Thus, the

equation is $(x + 1)^2 + (y + 4)^2 = 100$.

5. (a) Graph the corresponding horizontal lines (given by the equations $y = -1$ and

$y = 3$) as solid lines. The inequality $y \geq -1$ describes the points (x, y) that lie

on or *above* the line $y = -1$. The inequality $y \leq 3$ describes the points (x, y)

that lie on or *below* the line $y = 3$. So the pair of inequalities $-1 \leq y \leq 3$

describes the points that lie on or *between* the lines $y = -1$ and $y = 3$.

(b) Note that the given inequalities can be written as $-4 < x < 4$ and $-2 < y < 2$,

respectively. So the region lies between the vertical lines $x = -4$ and $x = 4$ and

between the horizontal lines $y = -2$ and $y = 2$. As shown in the graph, the

region common to both graphs is a rectangle (minus its edges) centered at the

origin.

(c) We first graph $y = 1 - \frac{1}{2}x$ as a dotted line. Since $y < 1 - \frac{1}{2}x$, the points in the

region lie *below* this line.

(d) We first graph the parabola $y = x^2 - 1$ using a solid curve. Since $y \geq x^2 - 1$, the points in the region lie on or *above* the parabola.

(e) We graph the circle $x^2 + y^2 = 4$ using a dotted curve. Since $\sqrt{x^2 + y^2} < 2$, the region consists of points whose distance from the origin is less than 2, that is, the points that lie *inside* the circle.

(f) The equation $9x^2 + 16y^2 = 144$ is an ellipse centered at $(0, 0)$. We put it in standard form by dividing by 144 and get $\dfrac{x^2}{16} + \dfrac{y^2}{9} = 1$. The x-intercepts are located at a distance of $\sqrt{16} = 4$ from the center while the y-intercepts are a distance of $\sqrt{9} = 3$ from the center (see the graph).

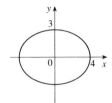

Test C Functions

1. (a) Locate -1 on the x-axis and then go down to the point on the graph with an x-coordinate of -1. The corresponding y-coordinate is the value of the function at $x = -1$, which is -2. So, $f(-1) = -2$.

   (b) Using the same technique as in part (a), we get $f(2) \approx 2.8$.

   (c) Locate 2 on the y-axis and then go left and right to find all points on the graph with a y-coordinate of 2. The corresponding x-coordinates are the x-values we are searching for. So $x = -3$ and $x = 1$.

   (d) Using the same technique as in part (c), we get $x \approx -2.5$ and $x \approx 0.3$.

   (e) The domain is all the x-values for which the graph exists, and the range is all the y-values for which the graph exists. Thus, the domain is $[-3, 3]$, and the range is $[-2, 3]$.

2. Note that $f(2 + h) = (2 + h)^3$ and $f(2) = 2^3 = 8$. So the difference quotient becomes
$$\frac{f(2+h) - f(2)}{h} = \frac{(2+h)^3 - 8}{h} = \frac{8 + 12h + 6h^2 + h^3 - 8}{h} = \frac{12h + 6h^2 + h^3}{h} = \frac{h(12 + 6h + h^2)}{h} = 12 + 6h + h^2.$$

3. (a) Set the denominator equal to 0 and solve to find restrictions on the domain: $x^2 + x - 2 = 0 \Rightarrow$ $(x - 1)(x + 2) = 0 \Rightarrow x = 1$ or $x = -2$. Thus, the domain is all real numbers except 1 or -2 or, in interval notation, $(-\infty, -2) \cup (-2, 1) \cup (1, \infty)$.

   (b) Note that the denominator is always greater than or equal to 1, and the numerator is defined for all real numbers. Thus, the domain is $(-\infty, \infty)$.

   (c) Note that the function h is the sum of two root functions. So h is defined on the intersection of the domains of these two root functions. The domain of a square root function is found by setting its radicand greater than or equal to 0. Now,

$4 - x \geq 0 \quad \Rightarrow \quad x \leq 4$ and $x^2 - 1 \geq 0 \quad \Rightarrow \quad (x-1)(x+1) \geq 0 \quad \Rightarrow \quad x \leq -1$ or $x \geq 1$. Thus, the domain of h is $(-\infty, -1] \cup [1, 4]$.

4. (a) Reflect the graph of f about the x-axis.

(b) Stretch the graph of f vertically by a factor of 2, then shift 1 unit downward.

(c) Shift the graph of f right 3 units, then up 2 units.

5. (a) Make a table and then connect the points with a smooth curve:

x	-2	-1	0	1	2
y	-8	-1	0	1	8

(b) Shift the graph from part (a) left 1 unit.

(c) Shift the graph from part (a) right 2 units and up 3 units.

(d) First plot $y = x^2$. Next, to get the graph of $f(x) = 4 - x^2$, reflect f about the x-axis and then shift it upward 4 units.

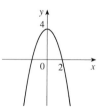

(e) Make a table and then connect the points with a smooth curve:

x	0	1	4	9
y	0	1	2	3

(f) Stretch the graph from part (e) vertically by a factor of two.

(g) First plot $y = 2^x$. Next, get the graph of $y = -2^x$ by reflecting the graph of
$y = 2^x$ across the x-axis.

(h) Note that $y = 1 + x^{-1} = 1 + 1/x$. So first plot $y = 1/x$ and then shift it
upward 1 unit.

6. (a) $f(-2) = 1 - (-2)^2 = -3$ and $f(1) = 2(1) + 1 = 3$

(b) For $x \le 0$ plot $f(x) = 1 - x^2$ and, on the same plane, for $x > 0$ plot the graph
of $f(x) = 2x + 1$.

7. (a) $(f \circ g)(x) = f(g(x)) = f(2x - 3) = (2x - 3)^2 + 2(2x - 3) - 1 = 4x^2 - 12x + 9 + 4x - 6 - 1 = 4x^2 - 8x + 2$

(b) $(g \circ f)(x) = g(f(x)) = g(x^2 + 2x - 1) = 2(x^2 + 2x - 1) - 3 = 2x^2 + 4x - 2 - 3 = 2x^2 + 4x - 5$

(c) $(g \circ g \circ g)(x) = g(g(g(x))) = g(g(2x - 3)) = g(2(2x - 3) - 3) = g(4x - 9) = 2(4x - 9) - 3$
$= 8x - 18 - 3 = 8x - 21$

Test D Trigonometry

1. (a) $300° = 300° \left(\dfrac{\pi}{180°} \right) = \dfrac{300\pi}{180} = \dfrac{5\pi}{3}$ (b) $-18° = -18° \left(\dfrac{\pi}{180°} \right) = -\dfrac{18\pi}{180} = -\dfrac{\pi}{10}$

2. (a) $\dfrac{5\pi}{6} = \dfrac{5\pi}{6} \left(\dfrac{180°}{\pi} \right) = 150°$ (b) $2 = 2 \left(\dfrac{180°}{\pi} \right) = \dfrac{360°}{\pi} \approx 114.6°$

3. We will use the arc length formula, $s = r\theta$, where s is arc length, r is the radius of the circle, and θ is the measure of the
central angle in radians. First, note that $30° = 30° \left(\dfrac{\pi}{180°} \right) = \dfrac{\pi}{6}$. So $s = (12) \left(\dfrac{\pi}{6} \right) = 2\pi$ cm.

4. (a) $\tan(\pi/3) = \sqrt{3}$ [You can read the value from a right triangle with sides 1, 2, and $\sqrt{3}$.]

(b) Note that $7\pi/6$ can be thought of as an angle in the third quadrant with reference angle $\pi/6$. Thus, $\sin(7\pi/6) = -\frac{1}{2}$,
since the sine function is negative in the third quadrant.

(c) Note that $5\pi/3$ can be thought of as an angle in the fourth quadrant with reference angle $\pi/3$. Thus,
$$\sec(5\pi/3) = \dfrac{1}{\cos(5\pi/3)} = \dfrac{1}{1/2} = 2,$$ since the cosine function is positive in the fourth quadrant.

5. $\sin\theta = a/24 \;\Rightarrow\; a = 24\sin\theta$ and $\cos\theta = b/24 \;\Rightarrow\; b = 24\cos\theta$

6. $\sin x = \frac{1}{3}$ and $\sin^2 x + \cos^2 x = 1$ $\Rightarrow$ $\cos x = \sqrt{1 - \frac{1}{9}} = \frac{2\sqrt{2}}{3}$. Also, $\cos y = \frac{4}{5}$ $\Rightarrow$ $\sin y = \sqrt{1 - \frac{16}{25}} = \frac{3}{5}$.

So, using the sum identity for the sine, we have

$$\sin(x+y) = \sin x \cos y + \cos x \sin y = \frac{1}{3} \cdot \frac{4}{5} + \frac{2\sqrt{2}}{3} \cdot \frac{3}{5} = \frac{4 + 6\sqrt{2}}{15} = \frac{1}{15}\left(4 + 6\sqrt{2}\right)$$

7. (a) $\tan\theta \sin\theta + \cos\theta = \dfrac{\sin\theta}{\cos\theta}\sin\theta + \cos\theta = \dfrac{\sin^2\theta}{\cos\theta} + \dfrac{\cos^2\theta}{\cos\theta} = \dfrac{1}{\cos\theta} = \sec\theta$

(b) $\dfrac{2\tan x}{1 + \tan^2 x} = \dfrac{2\sin x/(\cos x)}{\sec^2 x} = 2\,\dfrac{\sin x}{\cos x}\,\cos^2 x = 2\sin x\,\cos x = \sin 2x$

8. $\sin 2x = \sin x$ $\Leftrightarrow$ $2\sin x \cos x = \sin x$ $\Leftrightarrow$ $2\sin x \cos x - \sin x = 0$ $\Leftrightarrow$ $\sin x\,(2\cos x - 1) = 0$ $\Leftrightarrow$
$\sin x = 0$ or $\cos x = \frac{1}{2}$ $\Rightarrow$ $x = 0, \frac{\pi}{3}, \pi, \frac{5\pi}{3}, 2\pi$.

9. We first graph $y = \sin 2x$ (by compressing the graph of $\sin x$ by a factor of 2) and then shift it upward 1 unit.

1 □ FUNCTIONS AND MODELS

1.1 Four Ways to Represent a Function

In exercises requiring estimations or approximations, your answers may vary slightly from the answers given here.

1. (a) The point $(-1, -2)$ is on the graph of f, so $f(-1) = -2$.

(b) When $x = 2$, y is about 2.8, so $f(2) \approx 2.8$.

(c) $f(x) = 2$ is equivalent to $y = 2$. When $y = 2$, we have $x = -3$ and $x = 1$.

(d) Reasonable estimates for x when $y = 0$ are $x = -2.5$ and $x = 0.3$.

(e) The domain of f consists of all x-values on the graph of f. For this function, the domain is $-3 \le x \le 3$, or $[-3, 3]$.

The range of f consists of all y-values on the graph of f. For this function, the range is $-2 \le y \le 3$, or $[-2, 3]$.

(f) As x increases from -1 to 3, y increases from -2 to 3. Thus, f is increasing on the interval $[-1, 3]$.

2. (a) The point $(-4, -2)$ is on the graph of f, so $f(-4) = -2$. The point $(3, 4)$ is on the graph of g, so $g(3) = 4$.

(b) We are looking for the values of x for which the y-values are equal. The y-values for f and g are equal at the points $(-2, 1)$ and $(2, 2)$, so the desired values of x are -2 and 2.

(c) $f(x) = -1$ is equivalent to $y = -1$. When $y = -1$, we have $x = -3$ and $x = 4$.

(d) As x increases from 0 to 4, y decreases from 3 to -1. Thus, f is decreasing on the interval $[0, 4]$.

(e) The domain of f consists of all x-values on the graph of f. For this function, the domain is $-4 \le x \le 4$, or $[-4, 4]$.

The range of f consists of all y-values on the graph of f. For this function, the range is $-2 \le y \le 3$, or $[-2, 3]$.

(f) The domain of g is $[-4, 3]$ and the range is $[0.5, 4]$.

3. From Figure 1 in the text, the lowest point occurs at about $(t, a) = (12, -85)$. The highest point occurs at about $(17, 115)$. Thus, the range of the vertical ground acceleration is $-85 \le a \le 115$. Written in interval notation, we get $[-85, 115]$.

4. *Example 1:* A car is driven at 60 mi/h for 2 hours. The distance d traveled by the car is a function of the time t. The domain of the function is $\{t \mid 0 \le t \le 2\}$, where t is measured in hours. The range of the function is $\{d \mid 0 \le d \le 120\}$, where d is measured in miles.

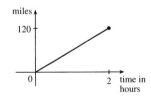

Example 2: At a certain university, the number of students N on campus at any time on a particular day is a function of the time t after midnight. The domain of the function is $\{t \mid 0 \le t \le 24\}$, where t is measured in hours. The range of the function is $\{N \mid 0 \le N \le k\}$, where N is an integer and k is the largest number of students on campus at once.

Example 3: A certain employee is paid $8.00 per hour and works a maximum of 30 hours per week. The number of hours worked is rounded down to the nearest quarter of an hour. This employee's gross weekly pay P is a function of the number of hours worked h. The domain of the function is $[0, 30]$ and the range of the function is $\{0, 2.00, 4.00, \ldots, 238.00, 240.00\}$.

5. No, the curve is not the graph of a function because a vertical line intersects the curve more than once. Hence, the curve fails the Vertical Line Test.

6. Yes, the curve is the graph of a function because it passes the Vertical Line Test. The domain is $[-2, 2]$ and the range is $[-1, 2]$.

7. Yes, the curve is the graph of a function because it passes the Vertical Line Test. The domain is $[-3, 2]$ and the range is $[-3, -2) \cup [-1, 3]$.

8. No, the curve is not the graph of a function since for $x = 0, \pm 1$, and ± 2, there are infinitely many points on the curve.

9. The person's weight increased to about 160 pounds at age 20 and stayed fairly steady for 10 years. The person's weight dropped to about 120 pounds for the next 5 years, then increased rapidly to about 170 pounds. The next 30 years saw a gradual increase to 190 pounds. Possible reasons for the drop in weight at 30 years of age: diet, exercise, health problems.

10. The salesman travels away from home from 8 to 9 AM and is then stationary until 10:00. The salesman travels farther away from 10 until noon. There is no change in his distance from home until 1:00, at which time the distance from home decreases until 3:00. Then the distance starts increasing again, reaching the maximum distance away from home at 5:00. There is no change from 5 until 6, and then the distance decreases rapidly until 7:00 PM, at which time the salesman reaches home.

11. The water will cool down almost to freezing as the ice melts. Then, when the ice has melted, the water will slowly warm up to room temperature.

12. The summer solstice (the longest day of the year) is around June 21, and the winter solstice (the shortest day) is around December 22.

13. Of course, this graph depends strongly on the geographical location!

14. The value of the car decreases fairly rapidly initially, then somewhat less rapidly.

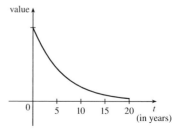

15. As the price increases, the amount sold decreases.

16. The temperature of the pie would increase rapidly, level off to oven temperature, decrease rapidly, and then level off to room temperature.

17.

18. (a)

(b)

(c)

(d)

19. (a)

(b) From the graph, we estimate the number of cell-phone subscribers worldwide to be about 92 million in 1995 and 485 million in 1999.

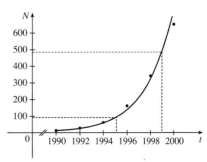

20. (a)

(b) From the graph in part (a), we estimate the temperature at 11:00 AM to be about 84.5°C.

21. $f(x) = 3x^2 - x + 2$.

$f(2) = 3(2)^2 - 2 + 2 = 12 - 2 + 2 = 12$.

$f(-2) = 3(-2)^2 - (-2) + 2 = 12 + 2 + 2 = 16$.

$f(a) = 3a^2 - a + 2$.

$f(-a) = 3(-a)^2 - (-a) + 2 = 3a^2 + a + 2$.

$f(a+1) = 3(a+1)^2 - (a+1) + 2 = 3(a^2 + 2a + 1) - a - 1 + 2 = 3a^2 + 6a + 3 - a + 1 = 3a^2 + 5a + 4$.

$2f(a) = 2 \cdot f(a) = 2(3a^2 - a + 2) = 6a^2 - 2a + 4$.

$f(2a) = 3(2a)^2 - (2a) + 2 = 3(4a^2) - 2a + 2 = 12a^2 - 2a + 2$.

$f(a^2) = 3(a^2)^2 - (a^2) + 2 = 3(a^4) - a^2 + 2 = 3a^4 - a^2 + 2$.

$[f(a)]^2 = [3a^2 - a + 2]^2 = (3a^2 - a + 2)(3a^2 - a + 2)$
$\quad = 9a^4 - 3a^3 + 6a^2 - 3a^3 + a^2 - 2a + 6a^2 - 2a + 4 = 9a^4 - 6a^3 + 13a^2 - 4a + 4$.

$f(a+h) = 3(a+h)^2 - (a+h) + 2 = 3(a^2 + 2ah + h^2) - a - h + 2 = 3a^2 + 6ah + 3h^2 - a - h + 2$.

22. A spherical balloon with radius $r + 1$ has volume $V(r+1) = \frac{4}{3}\pi(r+1)^3 = \frac{4}{3}\pi(r^3 + 3r^2 + 3r + 1)$. We wish to find the amount of air needed to inflate the balloon from a radius of r to $r + 1$. Hence, we need to find the difference

$V(r+1) - V(r) = \frac{4}{3}\pi(r^3 + 3r^2 + 3r + 1) - \frac{4}{3}\pi r^3 = \frac{4}{3}\pi(3r^2 + 3r + 1)$.

23. $f(x) = 4 + 3x - x^2$, so $f(3+h) = 4 + 3(3+h) - (3+h)^2 = 4 + 9 + 3h - (9 + 6h + h^2) = 4 - 3h - h^2$,

and $\dfrac{f(3+h) - f(3)}{h} = \dfrac{(4 - 3h - h^2) - 4}{h} = \dfrac{h(-3-h)}{h} = -3 - h$.

24. $f(x) = x^3$, so $f(a+h) = (a+h)^3 = a^3 + 3a^2h + 3ah^2 + h^3$,

and $\dfrac{f(a+h) - f(a)}{h} = \dfrac{(a^3 + 3a^2h + 3ah^2 + h^3) - a^3}{h} = \dfrac{h(3a^2 + 3ah + h^2)}{h} = 3a^2 + 3ah + h^2$.

25. $\dfrac{f(x) - f(a)}{x - a} = \dfrac{\dfrac{1}{x} - \dfrac{1}{a}}{x - a} = \dfrac{\dfrac{a - x}{xa}}{x - a} = \dfrac{a - x}{xa(x - a)} = \dfrac{-1(x - a)}{xa(x - a)} = -\dfrac{1}{ax}$

26. $\dfrac{f(x) - f(1)}{x - 1} = \dfrac{\dfrac{x + 3}{x + 1} - 2}{x - 1} = \dfrac{\dfrac{x + 3 - 2(x + 1)}{x + 1}}{x - 1} = \dfrac{x + 3 - 2x - 2}{(x + 1)(x - 1)}$

$\quad\quad = \dfrac{-x + 1}{(x + 1)(x - 1)} = \dfrac{-(x - 1)}{(x + 1)(x - 1)} = -\dfrac{1}{x + 1}$

27. $f(x) = x/(3x - 1)$ is defined for all x except when $0 = 3x - 1 \iff x = \frac{1}{3}$, so the domain
is $\{x \in \mathbb{R} \mid x \neq \frac{1}{3}\} = (-\infty, \frac{1}{3}) \cup (\frac{1}{3}, \infty)$.

28. $f(x) = (5x + 4)/(x^2 + 3x + 2)$ is defined for all x except when $0 = x^2 + 3x + 2 \iff 0 = (x + 2)(x + 1) \iff x = -2$
or -1, so the domain is $\{x \in \mathbb{R} \mid x \neq -2, -1\} = (-\infty, -2) \cup (-2, -1) \cup (-1, \infty)$.

29. $f(t) = \sqrt{t} + \sqrt[3]{t}$ is defined when $t \geq 0$. These values of t give real number results for $\sqrt{t}$, whereas any value of t gives a real
number result for $\sqrt[3]{t}$. The domain is $[0, \infty)$.

30. $g(u) = \sqrt{u} + \sqrt{4 - u}$ is defined when $u \geq 0$ and $4 - u \geq 0 \iff u \leq 4$. Thus, the domain is $0 \leq u \leq 4 = [0, 4]$.

31. $h(x) = 1 / \sqrt[4]{x^2 - 5x}$ is defined when $x^2 - 5x > 0 \iff x(x - 5) > 0$. Note that $x^2 - 5x \neq 0$ since that would result in
division by zero. The expression $x(x - 5)$ is positive if $x < 0$ or $x > 5$. (See Appendix A for methods for solving
inequalities.) Thus, the domain is $(-\infty, 0) \cup (5, \infty)$.

32. $h(x) = \sqrt{4 - x^2}$. Now $y = \sqrt{4 - x^2} \Rightarrow y^2 = 4 - x^2 \iff x^2 + y^2 = 4$, so
the graph is the top half of a circle of radius 2 with center at the origin. The domain
is $\{x \mid 4 - x^2 \geq 0\} = \{x \mid 4 \geq x^2\} = \{x \mid 2 \geq |x|\} = [-2, 2]$. From the graph,
the range is $0 \leq y \leq 2$, or $[0, 2]$.

33. $f(x) = 5$ is defined for all real numbers, so the domain is $\mathbb{R}$, or $(-\infty, \infty)$.
The graph of f is a horizontal line with y-intercept 5.

34. $F(x) = \frac{1}{2}(x + 3)$ is defined for all real numbers, so the domain is $\mathbb{R}$, or
$(-\infty, \infty)$. The graph of F is a line with x-intercept -3 and y-intercept $\frac{3}{2}$.

35. $f(t) = t^2 - 6t$ is defined for all real numbers, so the domain is $\mathbb{R}$, or
$(-\infty, \infty)$. The graph of f is a parabola opening upward since the coefficient
of t^2 is positive. To find the t-intercepts, let $y = 0$ and solve for t.
$0 = t^2 - 6t = t(t - 6) \Rightarrow t = 0$ and $t = 6$. The t-coordinate of the
vertex is halfway between the t-intercepts, that is, at $t = 3$. Since
$f(3) = 3^2 - 6 \cdot 3 = -9$, the vertex is $(3, -9)$.

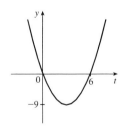

36. $H(t) = \dfrac{4 - t^2}{2 - t} = \dfrac{(2 + t)(2 - t)}{2 - t}$, so for $t \neq 2$, $H(t) = 2 + t$. The domain
is $\{t \mid t \neq 2\}$. So the graph of H is the same as the graph of the function
$f(t) = t + 2$ (a line) except for the hole at $(2, 4)$.

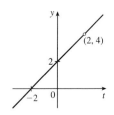

37. $g(x) = \sqrt{x - 5}$ is defined when $x - 5 \geq 0$ or $x \geq 5$, so the domain is $[5, \infty)$.

Since $y = \sqrt{x - 5}$ $\Rightarrow$ $y^2 = x - 5$ $\Rightarrow$ $x = y^2 + 5$, we see that g is the

top half of a parabola.

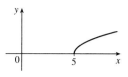

38. $F(x) = |2x + 1| = \begin{cases} 2x + 1 & \text{if } 2x + 1 \geq 0 \\ -(2x + 1) & \text{if } 2x + 1 < 1 \end{cases}$

$= \begin{cases} 2x + 1 & \text{if } x \geq -\frac{1}{2} \\ -2x - 1 & \text{if } x < -\frac{1}{2} \end{cases}$

The domain is $\mathbb{R}$, or $(-\infty, \infty)$.

39. $G(x) = \dfrac{3x + |x|}{x}$. Since $|x| = \begin{cases} x & \text{if } x \geq 0 \\ -x & \text{if } x < 0 \end{cases}$, we have

$$G(x) = \begin{cases} \dfrac{3x + x}{x} & \text{if } x > 0 \\ \dfrac{3x - x}{x} & \text{if } x < 0 \end{cases} = \begin{cases} \dfrac{4x}{x} & \text{if } x > 0 \\ \dfrac{2x}{x} & \text{if } x < 0 \end{cases} = \begin{cases} 4 & \text{if } x > 0 \\ 2 & \text{if } x < 0 \end{cases}$$

Note that G is not defined for $x = 0$. The domain is $(-\infty, 0) \cup (0, \infty)$.

40. $g(x) = \dfrac{|x|}{x^2}$. Since $|x| = \begin{cases} x & \text{if } x \geq 0 \\ -x & \text{if } x < 0 \end{cases}$, we have

$$g(x) = \begin{cases} \dfrac{x}{x^2} & \text{if } x > 0 \\ \dfrac{-x}{x^2} & \text{if } x < 0 \end{cases} = \begin{cases} \dfrac{1}{x} & \text{if } x > 0 \\ \dfrac{1}{-x} & \text{if } x < 0 \end{cases}$$

Note that g is not defined for $x = 0$. The domain is $(-\infty, 0) \cup (0, \infty)$.

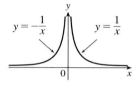

41. $f(x) = \begin{cases} x + 2 & \text{if } x < 0 \\ 1 - x & \text{if } x \geq 0 \end{cases}$

The domain is $\mathbb{R}$.

42. $f(x) = \begin{cases} 3 - \frac{1}{2}x & \text{if } x \leq 2 \\ 2x - 5 & \text{if } x > 2 \end{cases}$

The domain is $\mathbb{R}$.

43. $f(x) = \begin{cases} x + 2 & \text{if } x \leq -1 \\ x^2 & \text{if } x > -1 \end{cases}$

Note that for $x = -1$, both $x + 2$ and x^2 are equal to 1. The domain is $\mathbb{R}$.

44. $f(x) = \begin{cases} x+9 & \text{if} \quad x < -3 \\ -2x & \text{if} \quad |x| \leq 3 \\ -6 & \text{if} \quad x > 3 \end{cases}$

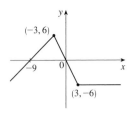

Note that for $x = -3$, both $x+9$ and $-2x$ are equal to 6; and for $x = 3$, both $-2x$

and -6 are equal to -6. Domain is $\mathbb{R}$.

45. Recall that the slope m of a line between the two points (x_1, y_1) and (x_2, y_2) is $m = \dfrac{y_2 - y_1}{x_2 - x_1}$ and an equation of the line

connecting those two points is $y - y_1 = m(x - x_1)$. The slope of this line segment is $\dfrac{7 - (-3)}{5 - 1} = \dfrac{5}{2}$, so an equation is

$y - (-3) = \frac{5}{2}(x - 1)$. The function is $f(x) = \frac{5}{2}x - \frac{11}{2}$, $1 \leq x \leq 5$.

46. The slope of this line segment is $\dfrac{-10 - 10}{7 - (-5)} = -\dfrac{5}{3}$, so an equation is $y - 10 = -\frac{5}{3}[x - (-5)]$.

The function is $f(x) = -\frac{5}{3}x + \frac{5}{3}$, $-5 \leq x \leq 7$.

47. We need to solve the given equation for y. $x + (y-1)^2 = 0 \ \Leftrightarrow \ (y-1)^2 = -x \ \Leftrightarrow \ y - 1 = \pm\sqrt{-x} \ \Leftrightarrow$

$y = 1 \pm \sqrt{-x}$. The expression with the positive radical represents the top half of the parabola, and the one with the negative

radical represents the bottom half. Hence, we want $f(x) = 1 - \sqrt{-x}$. Note that the domain is $x \leq 0$.

48. $x^2 + (y-2)^2 = 4 \ \Leftrightarrow \ (y-2)^2 = 4 - x^2 \ \Leftrightarrow \ y - 2 = \pm\sqrt{4 - x^2} \ \Leftrightarrow \ y = 2 \pm \sqrt{4 - x^2}$. The top half is given by

the function $f(x) = 2 + \sqrt{4 - x^2}$, $-2 \leq x \leq 2$.

49. For $0 \leq x \leq 3$, the graph is the line with slope -1 and y-intercept 3, that is, $y = -x + 3$. For $3 < x \leq 5$, the graph is the line

with slope 2 passing through $(3, 0)$; that is, $y - 0 = 2(x - 3)$, or $y = 2x - 6$. So the function is

$f(x) = \begin{cases} -x + 3 & \text{if } 0 \leq x \leq 3 \\ 2x - 6 & \text{if } 3 < x \leq 5 \end{cases}$

50. For $-4 \leq x \leq -2$, the graph is the line with slope $-\frac{3}{2}$ passing through $(-2, 0)$; that is, $y - 0 = -\frac{3}{2}[x - (-2)]$, or

$y = -\frac{3}{2}x - 3$. For $-2 < x < 2$, the graph is the top half of the circle with center $(0, 0)$ and radius 2. An equation of the circle

is $x^2 + y^2 = 4$, so an equation of the top half is $y = \sqrt{4 - x^2}$. For $2 \leq x \leq 4$, the graph is the line with slope $\frac{3}{2}$ passing

through $(2, 0)$; that is, $y - 0 = \frac{3}{2}(x - 2)$, or $y = \frac{3}{2}x - 3$. So the function is

$f(x) = \begin{cases} -\frac{3}{2}x - 3 & \text{if } -4 \leq x \leq -2 \\ \sqrt{4 - x^2} & \text{if } -2 < x < 2 \\ \frac{3}{2}x - 3 & \text{if } \ \ 2 \leq x \leq 4 \end{cases}$

51. Let the length and width of the rectangle be L and W. Then the perimeter is $2L + 2W = 20$ and the area is $A = LW$.

Solving the first equation for W in terms of L gives $W = \dfrac{20 - 2L}{2} = 10 - L$. Thus, $A(L) = L(10 - L) = 10L - L^2$. Since

lengths are positive, the domain of A is $0 < L < 10$. If we further restrict L to be larger than W, then $5 < L < 10$ would be

the domain.

52. Let the length and width of the rectangle be L and W. Then the area is $LW = 16$, so that $W = 16/L$. The perimeter is

$P = 2L + 2W$, so $P(L) = 2L + 2(16/L) = 2L + 32/L$, and the domain of P is $L > 0$, since lengths must be positive

quantities. If we further restrict L to be larger than W, then $L > 4$ would be the domain.

53. Let the length of a side of the equilateral triangle be x. Then by the Pythagorean Theorem, the height y of the triangle satisfies

$y^2 + \left(\frac{1}{2}x\right)^2 = x^2$, so that $y^2 = x^2 - \frac{1}{4}x^2 = \frac{3}{4}x^2$ and $y = \frac{\sqrt{3}}{2}x$. Using the formula for the area A of a triangle,

$A = \frac{1}{2}(\text{base})(\text{height})$, we obtain $A(x) = \frac{1}{2}(x)\left(\frac{\sqrt{3}}{2}x\right) = \frac{\sqrt{3}}{4}x^2$, with domain $x > 0$.

54. Let the volume of the cube be V and the length of an edge be L. Then $V = L^3$ so $L = \sqrt[3]{V}$, and the surface area is

$S(V) = 6\left(\sqrt[3]{V}\right)^2 = 6V^{2/3}$, with domain $V > 0$.

55. Let each side of the base of the box have length x, and let the height of the box be h. Since the volume is 2, we know that

$2 = hx^2$, so that $h = 2/x^2$, and the surface area is $S = x^2 + 4xh$. Thus, $S(x) = x^2 + 4x(2/x^2) = x^2 + (8/x)$, with

domain $x > 0$.

56. The area of the window is $A = xh + \frac{1}{2}\pi\left(\frac{1}{2}x\right)^2 = xh + \dfrac{\pi x^2}{8}$, where h is the height of the rectangular portion of the window.

The perimeter is $P = 2h + x + \frac{1}{2}\pi x = 30 \quad \Leftrightarrow \quad 2h = 30 - x - \frac{1}{2}\pi x \quad \Leftrightarrow \quad h = \frac{1}{4}(60 - 2x - \pi x)$. Thus,

$$A(x) = x\,\frac{60 - 2x - \pi x}{4} + \frac{\pi x^2}{8} = 15x - \frac{1}{2}x^2 - \frac{\pi}{4}x^2 + \frac{\pi}{8}x^2 = 15x - \frac{4}{8}x^2 - \frac{\pi}{8}x^2 = 15x - x^2\left(\frac{\pi + 4}{8}\right).$$

Since the lengths x and h must be positive quantities, we have $x > 0$ and $h > 0$. For $h > 0$, we have $2h > 0 \quad \Leftrightarrow$

$30 - x - \frac{1}{2}\pi x > 0 \quad \Leftrightarrow \quad 60 > 2x + \pi x \quad \Leftrightarrow \quad x < \dfrac{60}{2 + \pi}$. Hence, the domain of A is $0 < x < \dfrac{60}{2 + \pi}$.

57. The height of the box is x and the length and width are $L = 20 - 2x$, $W = 12 - 2x$. Then $V = LWx$ and so

$V(x) = (20 - 2x)(12 - 2x)(x) = 4(10 - x)(6 - x)(x) = 4x(60 - 16x + x^2) = 4x^3 - 64x^2 + 240x$.

The sides L, W, and x must be positive. Thus, $L > 0 \quad \Leftrightarrow \quad 20 - 2x > 0 \quad \Leftrightarrow \quad x < 10$;

$W > 0 \quad \Leftrightarrow \quad 12 - 2x > 0 \quad \Leftrightarrow \quad x < 6$; and $x > 0$. Combining these restrictions gives us the domain $0 < x < 6$.

58.

$$C(x) = \begin{cases} \$2.00 & \text{if } 0.0 < x \le 1.0 \\ 2.20 & \text{if } 1.0 < x \le 1.1 \\ 2.40 & \text{if } 1.1 < x \le 1.2 \\ 2.60 & \text{if } 1.2 < x \le 1.3 \\ 2.80 & \text{if } 1.3 < x \le 1.4 \\ 3.00 & \text{if } 1.4 < x \le 1.5 \\ 3.20 & \text{if } 1.5 < x \le 1.6 \\ 3.40 & \text{if } 1.6 < x \le 1.7 \\ 3.60 & \text{if } 1.7 < x \le 1.8 \\ 3.80 & \text{if } 1.8 < x \le 1.9 \\ 4.00 & \text{if } 1.9 < x < 2.0 \end{cases}$$

59. (a)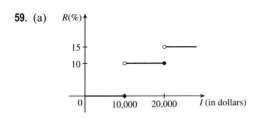

(b) On \$14,000, tax is assessed on \$4000, and $10\%(\$4000) = \$400$.

On \$26,000, tax is assessed on \$16,000, and

$10\%(\$10,000) + 15\%(\$6000) = \$1000 + \$900 = \$1900$.

(c) As in part (b), there is \$1000 tax assessed on \$20,000 of income, so the graph of T is a line segment from $(10{,}000, 0)$ to $(20{,}000, 1000)$. The tax on \$30,000 is \$2500, so the graph of T for $x > 20{,}000$ is the ray with initial point $(20{,}000, 1000)$ that passes through $(30{,}000, 2500)$.

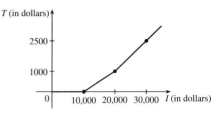

60. One example is the amount paid for cable or telephone system repair in the home, usually measured to the nearest quarter hour. Another example is the amount paid by a student in tuition fees, if the fees vary according to the number of credits for which the student has registered.

61. f is an odd function because its graph is symmetric about the origin. g is an even function because its graph is symmetric with respect to the y-axis.

62. f is not an even function since it is not symmetric with respect to the y-axis. f is not an odd function since it is not symmetric about the origin. Hence, f is *neither* even nor odd. g is an even function because its graph is symmetric with respect to the y-axis.

63. (a) Because an even function is symmetric with respect to the y-axis, and the point $(5, 3)$ is on the graph of this even function, the point $(-5, 3)$ must also be on its graph.

(b) Because an odd function is symmetric with respect to the origin, and the point $(5, 3)$ is on the graph of this odd function, the point $(-5, -3)$ must also be on its graph.

64. (a) If f is even, we get the rest of the graph by reflecting about the y-axis.

(b) If f is odd, we get the rest of the graph by rotating $180°$ about the origin.

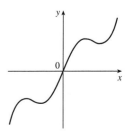

65. $f(x) = \dfrac{x}{x^2 + 1}$.

$f(-x) = \dfrac{-x}{(-x)^2 + 1} = \dfrac{-x}{x^2 + 1} = -\dfrac{x}{x^2 + 1} = -f(x)$.

So f is an odd function.

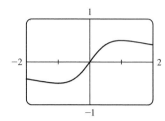

66. $f(x) = \dfrac{x^2}{x^4 + 1}$.

$f(-x) = \dfrac{(-x)^2}{(-x)^4 + 1} = \dfrac{x^2}{x^4 + 1} = f(x)$.

So f is an even function.

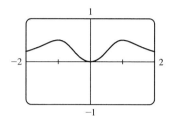

67. $f(x) = \dfrac{x}{x + 1}$, so $f(-x) = \dfrac{-x}{-x + 1} = \dfrac{x}{x - 1}$.

Since this is neither $f(x)$ nor $-f(x)$, the function f is

neither even nor odd.

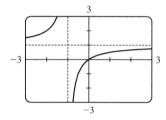

68. $f(x) = x\,|x|$.

$f(-x) = (-x)\,|-x| = (-x)\,|x| = -(x\,|x|)$
$\quad = -f(x)$

So f is an odd function.

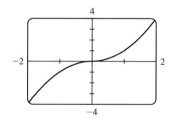

69. $f(x) = 1 + 3x^2 - x^4$.

$f(-x) = 1 + 3(-x)^2 - (-x)^4 = 1 + 3x^2 - x^4 = f(x)$.

So f is an even function.

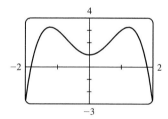

70. $f(x) = 1 + 3x^3 - x^5$, so

$f(-x) = 1 + 3(-x)^3 - (-x)^5 = 1 + 3(-x^3) - (-x^5)$
$\quad = 1 - 3x^3 + x^5$

Since this is neither $f(x)$ nor $-f(x)$, the function f is neither even nor odd.

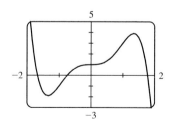

1.2   Mathematical Models: A Catalog of Essential Functions

1. (a) $f(x) = \sqrt[5]{x}$ is a root function with $n = 5$.

(b) $g(x) = \sqrt{1 - x^2}$ is an algebraic function because it is a root of a polynomial.

(c) $h(x) = x^9 + x^4$ is a polynomial of degree 9.

(d) $r(x) = \dfrac{x^2 + 1}{x^3 + x}$ is a rational function because it is a ratio of polynomials.

(e) $s(x) = \tan 2x$ is a trigonometric function.

(f) $t(x) = \log_{10} x$ is a logarithmic function.

2. (a) $y = (x - 6)/(x + 6)$ is a rational function because it is a ratio of polynomials.

(b) $y = x + x^2/\sqrt{x - 1}$ is an algebraic function because it involves polynomials and roots of polynomials.

(c) $y = 10^x$ is an exponential function (notice that x is the *exponent*).

(d) $y = x^{10}$ is a power function (notice that x is the *base*).

(e) $y = 2t^6 + t^4 - \pi$ is a polynomial of degree 6.

(f) $y = \cos \theta + \sin \theta$ is a trigonometric function.

3. We notice from the figure that g and h are even functions (symmetric with respect to the y-axis) and that f is an odd function

(symmetric with respect to the origin). So (b) $\left[y = x^5\right]$ must be f. Since g is flatter than h near the origin, we must have

(c) $\left[y = x^8\right]$ matched with g and (a) $\left[y = x^2\right]$ matched with h.

4. (a) The graph of $y = 3x$ is a line (choice G).

(b) $y = 3^x$ is an exponential function (choice f).

(c) $y = x^3$ is an odd polynomial function or power function (choice F).

(d) $y = \sqrt[3]{x} = x^{1/3}$ is a root function (choice g).

5. (a) An equation for the family of linear functions with slope 2

is $y = f(x) = 2x + b$, where b is the y-intercept.

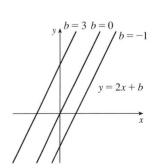

(b) $f(2) = 1$ means that the point $(2, 1)$ is on the graph of f. We can use the
point-slope form of a line to obtain an equation for the family of linear
functions through the point $(2, 1)$. $y - 1 = m(x - 2)$, which is equivalent
to $y = mx + (1 - 2m)$ in slope-intercept form.

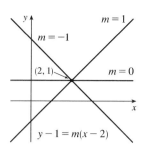

(c) To belong to both families, an equation must have slope $m = 2$, so the equation in part (b), $y = mx + (1 - 2m)$,
becomes $y = 2x - 3$. It is the *only* function that belongs to both families.

6. All members of the family of linear functions $f(x) = 1 + m(x + 3)$ have
graphs that are lines passing through the point $(-3, 1)$.

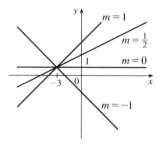

7. All members of the family of linear functions $f(x) = c - x$ have graphs
that are lines with slope -1. The y-intercept is c.

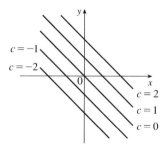

8. The vertex of the parabola on the left is $(3, 0)$, so an equation is $y = a(x - 3)^2 + 0$. Since the point $(4, 2)$ is on the
parabola, we'll substitute 4 for x and 2 for y to find a. $2 = a(4 - 3)^2 \Rightarrow a = 2$, so an equation is $f(x) = 2(x - 3)^2$.

The y-intercept of the parabola on the right is $(0, 1)$, so an equation is $y = ax^2 + bx + 1$. Since the points $(-2, 2)$ and
$(1, -2.5)$ are on the parabola, we'll substitute -2 for x and 2 for y as well as 1 for x and -2.5 for y to obtain two equations
with the unknowns a and b.

$$(-2, 2): \quad 2 = 4a - 2b + 1 \quad \Rightarrow \quad 4a - 2b = 1 \qquad \textbf{(1)}$$
$$(1, -2.5): \quad -2.5 = a + b + 1 \quad \Rightarrow \quad a + b = -3.5 \qquad \textbf{(2)}$$

$2 \cdot \textbf{(2)} + \textbf{(1)}$ gives us $6a = -6 \quad \Rightarrow \quad a = -1$. From **(2)**, $-1 + b = -3.5 \quad \Rightarrow \quad b = -2.5$, so an equation
is $g(x) = -x^2 - 2.5x + 1$.

9. Since $f(-1) = f(0) = f(2) = 0$, f has zeros of -1, 0, and 2, so an equation for f is $f(x) = a[x - (-1)](x - 0)(x - 2)$, or $f(x) = ax(x + 1)(x - 2)$. Because $f(1) = 6$, we'll substitute 1 for x and 6 for $f(x)$.

$6 = a(1)(2)(-1) \;\Rightarrow\; -2a = 6 \;\Rightarrow\; a = -3$, so an equation for f is $f(x) = -3x(x + 1)(x - 2)$.

10. (a) For $T = 0.02t + 8.50$, the slope is 0.02, which means that the average surface temperature of the world is increasing at a rate of $0.02\,°\text{C}$ per year. The T-intercept is 8.50, which represents the average surface temperature in $°\text{C}$ in the year 1900.

(b) $t = 2100 - 1900 = 200 \;\Rightarrow\; T = 0.02(200) + 8.50 = 12.50\,°\text{C}$

11. (a) $D = 200$, so $c = 0.0417D(a + 1) = 0.0417(200)(a + 1) = 8.34a + 8.34$. The slope is 8.34, which represents the change in mg of the dosage for a child for each change of 1 year in age.

(b) For a newborn, $a = 0$, so $c = 8.34$ mg.

12. (a)

(b) The slope of -4 means that for each increase of 1 dollar for a rental space, the number of spaces rented *decreases* by 4. The y-intercept of 200 is the number of spaces that would be occupied if there were no charge for each space. The x-intercept of 50 is the smallest rental fee that results in no spaces rented.

13. (a)

(b) The slope of $\frac{9}{5}$ means that F increases $\frac{9}{5}$ degrees for each increase of $1°\text{C}$. (Equivalently, F increases by 9 when C increases by 5 and F decreases by 9 when C decreases by 5.) The F-intercept of 32 is the Fahrenheit temperature corresponding to a Celsius temperature of 0.

14. (a) Let $d =$ distance traveled (in miles) and $t =$ time elapsed (in hours). At $t = 0$, $d = 0$ and at $t = 50 \text{ minutes} = 50 \cdot \frac{1}{60} = \frac{5}{6}$ h, $d = 40$. Thus we have two points: $(0, 0)$ and $\left(\frac{5}{6}, 40\right)$, so $m = \dfrac{40 - 0}{\frac{5}{6} - 0} = 48$ and so $d = 48t$.

(b)

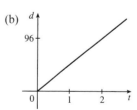

(c) The slope is 48 and represents the car's speed in mi/h.

15. (a) Using N in place of x and T in place of y, we find the slope to be $\dfrac{T_2 - T_1}{N_2 - N_1} = \dfrac{80 - 70}{173 - 113} = \dfrac{10}{60} = \dfrac{1}{6}$. So a linear equation is $T - 80 = \frac{1}{6}(N - 173) \;\Leftrightarrow\; T - 80 = \frac{1}{6}N - \frac{173}{6} \;\Leftrightarrow\; T = \frac{1}{6}N + \frac{307}{6} \;\left[\frac{307}{6} = 51.1\overline{6}\right]$.

(b) The slope of $\frac{1}{6}$ means that the temperature in Fahrenheit degrees increases one-sixth as rapidly as the number of cricket chirps per minute. Said differently, each increase of 6 cricket chirps per minute corresponds to an increase of $1°\text{F}$.

(c) When $N = 150$, the temperature is given approximately by $T = \frac{1}{6}(150) + \frac{307}{6} = 76.1\overline{6}\,°\text{F} \approx 76\,°\text{F}$.

16. (a) Let x denote the number of chairs produced in one day and y the associated

cost. Using the points $(100, 2200)$ and $(300, 4800)$, we get the slope

$\frac{4800-2200}{300-100} = \frac{2600}{200} = 13$. So $y - 2200 = 13(x - 100)$ ⇔

$y = 13x + 900$.

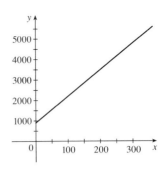

(b) The slope of the line in part (a) is 13 and it represents the cost (in dollars)

of producing each additional chair.

(c) The y-intercept is 900 and it represents the fixed daily costs of operating

the factory.

17. (a) We are given $\dfrac{\text{change in pressure}}{\text{10 feet change in depth}} = \dfrac{4.34}{10} = 0.434$. Using P for pressure and d for depth with the point

$(d, P) = (0, 15)$, we have the slope-intercept form of the line, $P = 0.434d + 15$.

(b) When $P = 100$, then $100 = 0.434d + 15$ ⇔ $0.434d = 85$ ⇔ $d = \frac{85}{0.434} \approx 195.85$ feet. Thus, the pressure is

100 lb/in^2 at a depth of approximately 196 feet.

18. (a) Using d in place of x and C in place of y, we find the slope to be $\dfrac{C_2 - C_1}{d_2 - d_1} = \dfrac{460 - 380}{800 - 480} = \dfrac{80}{320} = \dfrac{1}{4}$.

So a linear equation is $C - 460 = \frac{1}{4}(d - 800)$ ⇔ $C - 460 = \frac{1}{4}d - 200$ ⇔ $C = \frac{1}{4}d + 260$.

(b) Letting $d = 1500$ we get $C = \frac{1}{4}(1500) + 260 = 635$.

The cost of driving 1500 miles is \$635.

(c)

The slope of the line represents the cost per mile, \$0.25.

(d) The y-intercept represents the fixed cost, \$260.

(e) A linear function gives a suitable model in this situation because you have fixed monthly costs such as insurance and car

payments, as well as costs that increase as you drive, such as gasoline, oil, and tires, and the cost of these for each

additional mile driven is a constant.

19. (a) The data appear to be periodic and a sine or cosine function would make the best model. A model of the form

$f(x) = a\cos(bx) + c$ seems appropriate.

(b) The data appear to be decreasing in a linear fashion. A model of the form $f(x) = mx + b$ seems appropriate.

20. (a) The data appear to be increasing exponentially. A model of the form $f(x) = a \cdot b^x$ or $f(x) = a \cdot b^x + c$ seems appropriate.

(b) The data appear to be decreasing similarly to the values of the reciprocal function. A model of the form $f(x) = a/x$ seems

appropriate.

Some values are given to many decimal places. These are the results given by several computer algebra systems — rounding is left to the reader.

21. (a)

A linear model does seem appropriate.

(b) Using the points $(4000, 14.1)$ and $(60,000, 8.2)$, we obtain

$$y - 14.1 = \frac{8.2 - 14.1}{60,000 - 4000}(x - 4000) \text{ or, equivalently,}$$

$$y \approx -0.000105357x + 14.521429.$$

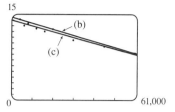

(c) Using a computing device, we obtain the least squares regression line $y = -0.0000997855x + 13.950764$.
The following commands and screens illustrate how to find the least squares regression line on a TI-83 Plus.
Enter the data into list one (L1) and list two (L2). Press $\boxed{\text{STAT}}$ $\boxed{1}$ to enter the editor.

Find the regession line and store it in Y_1. Press $\boxed{\text{2nd}}$ $\boxed{\text{QUIT}}$ $\boxed{\text{STAT}}$ $\boxed{\blacktriangleright}$ $\boxed{4}$ $\boxed{\text{VARS}}$ $\boxed{\blacktriangleright}$ $\boxed{1}$ $\boxed{1}$ $\boxed{\text{ENTER}}$.

Note from the last figure that the regression line has been stored in Y_1 and that Plot1 has been turned on (Plot1 is highlighted). You can turn on Plot1 from the Y= menu by placing the cursor on Plot1 and pressing $\boxed{\text{ENTER}}$ or by pressing $\boxed{\text{2nd}}$ $\boxed{\text{STAT PLOT}}$ $\boxed{1}$ $\boxed{\text{ENTER}}$.

Now press $\boxed{\text{ZOOM}}$ $\boxed{9}$ to produce a graph of the data and the regression line. Note that choice 9 of the ZOOM menu automatically selects a window that displays all of the data.

(d) When $x = 25,000$, $y \approx 11.456$; or about 11.5 per 100 population.

(e) When $x = 80,000$, $y \approx 5.968$; or about a 6% chance.

(f) When $x = 200,000$, y is negative, so the model does not apply.

22. (a)

(b)

Using a computing device, we obtain the least squares regression line $y = 4.85\overline{6}x - 220.9\overline{6}$.

(c) When $x = 100°\text{F}$, $y = 264.7 \approx 265$ chirps/min.

23. (a)

A linear model does seem appropriate.

(b)

Using a computing device, we obtain the least squares regression line $y = 0.089119747x - 158.2403249$, where x is the year and y is the height in feet.

(c) When $x = 2000$, the model gives $y \approx 20.00$ ft. Note that the actual winning height for the 2000 Olympics is *less than* the winning height for 1996—so much for that prediction.

(d) When $x = 2100$, $y \approx 28.91$ ft. This would be an increase of 9.49 ft from 1996 to 2100. Even though there was an increase of 8.59 ft from 1900 to 1996, it is unlikely that a similar increase will occur over the next 100 years.

24. By looking at the scatter plot of the data, we rule out the linear and logarithmic models.

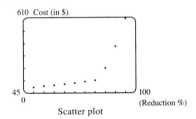

Scatter plot

We try various models:

Quadratic: $\quad y = 0.49\overline{6}x^2 - 62.28\overline{93}x + 1970.6\overline{39}$

Cubic: $\quad\;\; y = 0.0201243201x^3 - 3.88037296x^2 + 247.6754468x - 5163.935198$

Quartic: $\quad y = 0.0002951049x^4 - 0.0654560995x^3 + 5.27525641x^2 - 180.2266511x + 2203.210956$

Exponential: $\; y = 2.41422994\,(1.054516914)^x$

Power: $\quad\; y = 0.000022854971x^{3.616078251}$

Quadratic model

Cubic model

Quartic model

Exponential model

Power model

After examining the graphs of these models, we see that the cubic and quartic models are clearly the best.

25.

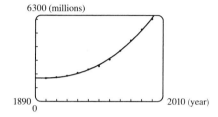

Using a computing device, we obtain the cubic function $y = ax^3 + bx^2 + cx + d$ with $a = 0.0012937$, $b = -7.06142$, $c = 12{,}823$, and $d = -7{,}743{,}770$. When $x = 1925$, $y \approx 1914$ (million).

26. (a) $T = 1.000\,431\,227d^{\,1.499\,528\,750}$

(b) The power model in part (a) is approximately $T = d^{1.5}$. Squaring both sides gives us $T^2 = d^3$, so the model matches Kepler's Third Law, $T^2 = kd^3$.

1.3   New Functions from Old Functions

1. (a) If the graph of f is shifted 3 units upward, its equation becomes $y = f(x) + 3$.

(b) If the graph of f is shifted 3 units downward, its equation becomes $y = f(x) - 3$.

(c) If the graph of f is shifted 3 units to the right, its equation becomes $y = f(x - 3)$.

(d) If the graph of f is shifted 3 units to the left, its equation becomes $y = f(x + 3)$.

(e) If the graph of f is reflected about the x-axis, its equation becomes $y = -f(x)$.

(f) If the graph of f is reflected about the y-axis, its equation becomes $y = f(-x)$.

(g) If the graph of f is stretched vertically by a factor of 3, its equation becomes $y = 3f(x)$.

(h) If the graph of f is shrunk vertically by a factor of 3, its equation becomes $y = \frac{1}{3}f(x)$.

2. (a) To obtain the graph of $y = 5f(x)$ from the graph of $y = f(x)$, stretch the graph vertically by a factor of 5.

(b) To obtain the graph of $y = f(x - 5)$ from the graph of $y = f(x)$, shift the graph 5 units to the right.

(c) To obtain the graph of $y = -f(x)$ from the graph of $y = f(x)$, reflect the graph about the x-axis.

(d) To obtain the graph of $y = -5f(x)$ from the graph of $y = f(x)$, stretch the graph vertically by a factor of 5 and reflect it about the x-axis.

(e) To obtain the graph of $y = f(5x)$ from the graph of $y = f(x)$, shrink the graph horizontally by a factor of 5.

(f) To obtain the graph of $y = 5f(x) - 3$ from the graph of $y = f(x)$, stretch the graph vertically by a factor of 5 and shift it 3 units downward.

3. (a) (graph 3) The graph of f is shifted 4 units to the right and has equation $y = f(x - 4)$.

(b) (graph 1) The graph of f is shifted 3 units upward and has equation $y = f(x) + 3$.

(c) (graph 4) The graph of f is shrunk vertically by a factor of 3 and has equation $y = \frac{1}{3}f(x)$.

(d) (graph 5) The graph of f is shifted 4 units to the left and reflected about the x-axis. Its equation is $y = -f(x + 4)$.

(e) (graph 2) The graph of f is shifted 6 units to the left and stretched vertically by a factor of 2. Its equation is

$y = 2f(x + 6)$.

4. (a) To graph $y = f(x + 4)$ we shift the graph of f,
4 units to the left.

The point $(2, 1)$ on the graph of f corresponds to the point $(2 - 4, 1) = (-2, 1)$.

(b) To graph $y = f(x) + 4$ we shift the graph of f,
4 units upward.

The point $(2, 1)$ on the graph of f corresponds to the point $(2, 1 + 4) = (2, 5)$.

(c) To graph $y = 2f(x)$ we stretch the graph of f
vertically by a factor of 2.

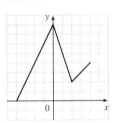

The point $(2, 1)$ on the graph of f corresponds to the point $(2, 2 \cdot 1) = (2, 2)$.

(d) To graph $y = -\frac{1}{2}f(x) + 3$, we shrink the graph of f
vertically by a factor of 2, then reflect the resulting
graph about the x-axis, then shift the resulting graph
3 units upward.

The point $(2, 1)$ on the graph of f corresponds to the point $\left(2, -\frac{1}{2} \cdot 1 + 3\right) = (2, 2.5)$.

5. (a) To graph $y = f(2x)$ we shrink the graph of f
horizontally by a factor of 2.

The point $(4, -1)$ on the graph of f corresponds to the point $\left(\frac{1}{2} \cdot 4, -1\right) = (2, -1)$.

(b) To graph $y = f\left(\frac{1}{2}x\right)$ we stretch the graph of f
horizontally by a factor of 2.

The point $(4, -1)$ on the graph of f corresponds to the point $(2 \cdot 4, -1) = (8, -1)$.

(c) To graph $y = f(-x)$ we reflect the graph of f about the y-axis.

The point $(4, -1)$ on the graph of f corresponds to the point $(-1 \cdot 4, -1) = (-4, -1)$.

(d) To graph $y = -f(-x)$ we reflect the graph of f about the y-axis, then about the x-axis.

The point $(4, -1)$ on the graph of f corresponds to the point $(-1 \cdot 4, -1 \cdot -1) = (-4, 1)$.

6. The graph of $y = f(x) = \sqrt{3x - x^2}$ has been shifted 2 units to the right and stretched vertically by a factor of 2.

Thus, a function describing the graph is

$$y = 2f(x - 2) = 2\sqrt{3(x - 2) - (x - 2)^2} = 2\sqrt{3x - 6 - (x^2 - 4x + 4)} = 2\sqrt{-x^2 + 7x - 10}$$

7. The graph of $y = f(x) = \sqrt{3x - x^2}$ has been shifted 4 units to the left, reflected about the x-axis, and shifted downward 1 unit. Thus, a function describing the graph is

$$y = \underbrace{-1 \cdot}_{\substack{\text{reflect} \\ \text{about } x\text{-axis}}} \underbrace{f\ (x + 4)}_{\substack{\text{shift} \\ 4 \text{ units left}}} \underbrace{-\ 1}_{\substack{\text{shift} \\ 1 \text{ unit left}}}$$

This function can be written as

$$y = -f(x + 4) - 1 = -\sqrt{3(x + 4) - (x + 4)^2} - 1 = -\sqrt{3x + 12 - (x^2 + 8x + 16)} - 1 = -\sqrt{-x^2 - 5x - 4} - 1$$

8. (a) The graph of $y = 2\sin x$ can be obtained from the graph of $y = \sin x$ by stretching it vertically by a factor of 2.

(b) The graph of $y = 1 + \sqrt{x}$ can be obtained from the graph of $y = \sqrt{x}$ by shifting it upward 1 unit.

9. $y = -x^3$: Start with the graph of $y = x^3$ and reflect about the x-axis. Note: Reflecting about the y-axis gives the same result since substituting $-x$ for x gives us $y = (-x)^3 = -x^3$.

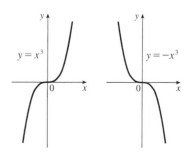

10. $y = 1 - x^2 = -x^2 + 1$: Start with the graph of $y = x^2$, reflect about the x-axis, and then shift 1 unit upward.

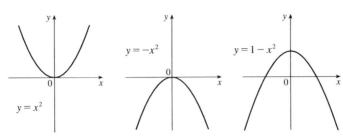

11. $y = (x + 1)^2$: Start with the graph of $y = x^2$ and shift 1 unit to the left.

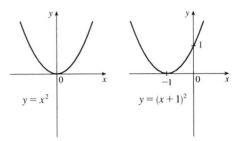

12. $y = x^2 - 4x + 3 = (x^2 - 4x + 4) - 1 = (x - 2)^2 - 1$: Start with the graph of $y = x^2$, shift 2 units to the right, and then shift 1 unit downward.

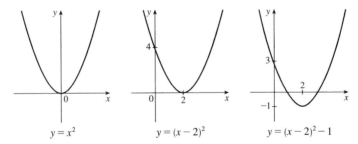

13. $y = 1 + 2\cos x$: Start with the graph of $y = \cos x$, stretch vertically by a factor of 2, and then shift 1 unit upward.

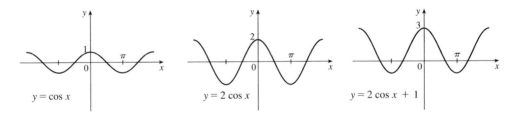

14. $y = 4\sin 3x$: Start with the graph of $y = \sin x$, compress horizontally by a factor of 3, and then stretch vertically by a factor of 4.

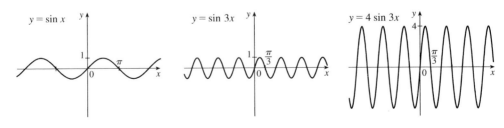

15. $y = \sin(x/2)$: Start with the graph of $y = \sin x$ and stretch horizontally by a factor of 2.

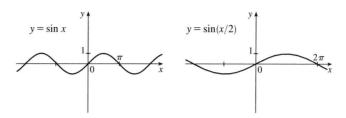

16. $y = 1/(x - 4)$: Start with the graph of $y = 1/x$ and shift 4 units to the right.

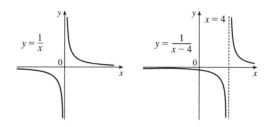

17. $y = \sqrt{x + 3}$: Start with the graph of $y = \sqrt{x}$ and shift 3 units to the left.

18. $y = (x + 2)^4 + 3$: Start with the graph of $y = x^4$, shift 2 units to the left, and then shift 3 units upward.

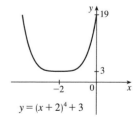

19. $y = \frac{1}{2}(x^2 + 8x) = \frac{1}{2}(x^2 + 8x + 16) - 8 = \frac{1}{2}(x + 4)^2 - 8$: Start with the graph of $y = x^2$, compress vertically by a factor of 2, shift 4 units to the left, and then shift 8 units downward.

20. $y = 1 + \sqrt[3]{x - 1}$: Start with the graph of $y = \sqrt[3]{x}$, shift 1 unit to the right, and then shift 1 unit upward.

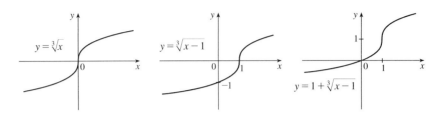

21. $y = 2/(x + 1)$: Start with the graph of $y = 1/x$, shift 1 unit to the left, and then stretch vertically by a factor of 2.

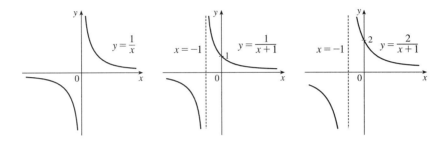

22. $y = \frac{1}{4}\tan(x - \frac{\pi}{4})$: Start with the graph of $y = \tan x$, shift $\frac{\pi}{4}$ units to the right, and then compress vertically by a factor of 4.

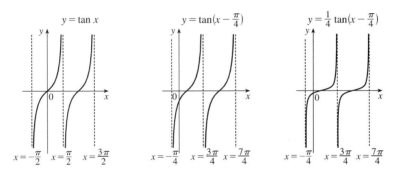

23. $y = |\sin x|$: Start with the graph of $y = \sin x$ and reflect all the parts of the graph below the x-axis about the x-axis.

24. $y = |x^2 - 2x| = |x^2 - 2x + 1 - 1| = |(x - 1)^2 - 1|$: Start with the graph of $y = x^2$, shift 1 unit right, shift 1 unit

downward, and reflect the portion of the graph below the x-axis about the x-axis.

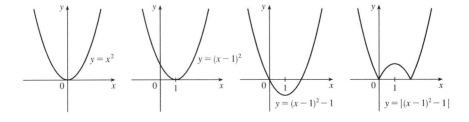

25. This is just like the solution to Example 4 except the amplitude of the curve (the $30°$N curve in Figure 9 on June 21) is

$14 - 12 = 2$. So the function is $L(t) = 12 + 2\sin\left[\frac{2\pi}{365}(t - 80)\right]$. March 31 is the 90th day of the year, so the model gives

$L(90) \approx 12.34$ h. The daylight time (5:51 AM to 6:18 PM) is 12 hours and 27 minutes, or 12.45 h. The model value differs

from the actual value by $\frac{12.45 - 12.34}{12.45} \approx 0.009$, less than 1%.

26. Using a sine function to model the brightness of Delta Cephei as a function of time, we take its period to be 5.4 days, its

amplitude to be 0.35 (on the scale of magnitude), and its average magnitude to be 4.0. If we take $t = 0$ at a time of average

brightness, then the magnitude (brightness) as a function of time t in days can be modeled by the formula

$M(t) = 4.0 + 0.35\sin\left(\frac{2\pi}{5.4}t\right)$.

27. (a) To obtain $y = f(|x|)$, the portion of the graph of $y = f(x)$ to the right of the y-axis is reflected about the y-axis.

(b) $y = \sin|x|$

(c) $y = \sqrt{|x|}$

28. The most important features of the given graph are the x-intercepts and the maximum

and minimum points. The graph of $y = 1/f(x)$ has vertical asymptotes at the x-values

where there are x-intercepts on the graph of $y = f(x)$. The maximum of 1 on the graph

of $y = f(x)$ corresponds to a minimum of $1/1 = 1$ on $y = 1/f(x)$. Similarly, the

minimum on the graph of $y = f(x)$ corresponds to a maximum on the graph of

$y = 1/f(x)$. As the values of y get large (positively or negatively) on the graph of

$y = f(x)$, the values of y get close to zero on the graph of $y = 1/f(x)$.

29. $f(x) = x^3 + 2x^2$; $g(x) = 3x^2 - 1$. $D = \mathbb{R}$ for both f and g.

$(f + g)(x) = (x^3 + 2x^2) + (3x^2 - 1) = x^3 + 5x^2 - 1$, $D = \mathbb{R}$.

$(f - g)(x) = (x^3 + 2x^2) - (3x^2 - 1) = x^3 - x^2 + 1$, $D = \mathbb{R}$.

$(fg)(x) = (x^3 + 2x^2)(3x^2 - 1) = 3x^5 + 6x^4 - x^3 - 2x^2$, $D = \mathbb{R}$.

$\left(\dfrac{f}{g}\right)(x) = \dfrac{x^3 + 2x^2}{3x^2 - 1}$, $D = \left\{x \mid x \neq \pm\dfrac{1}{\sqrt{3}}\right\}$ since $3x^2 - 1 \neq 0$.

30. $f(x) = \sqrt{3 - x}$, $D = (-\infty, 3]$; $g(x) = \sqrt{x^2 - 1}$, $D = (-\infty, -1] \cup [1, \infty)$.

$(f + g)(x) = \sqrt{3 - x} + \sqrt{x^2 - 1}$, $D = (-\infty, -1] \cup [1, 3]$, which is the intersection of the domains of f and g.

$(f - g)(x) = \sqrt{3 - x} - \sqrt{x^2 - 1}$, $D = (-\infty, -1] \cup [1, 3]$.

$(fg)(x) = \sqrt{3 - x} \cdot \sqrt{x^2 - 1}$, $D = (-\infty, -1] \cup [1, 3]$.

$\left(\dfrac{f}{g}\right)(x) = \dfrac{\sqrt{3 - x}}{\sqrt{x^2 - 1}}$, $D = (-\infty, -1) \cup (1, 3]$. We must exclude $x = \pm 1$ since these values would make $\dfrac{f}{g}$ undefined.

31. $f(x) = x^2 - 1$, $D = \mathbb{R}$; $g(x) = 2x + 1$, $D = \mathbb{R}$.

   (a) $(f \circ g)(x) = f(g(x)) = f(2x + 1) = (2x + 1)^2 - 1 = (4x^2 + 4x + 1) - 1 = 4x^2 + 4x$, $D = \mathbb{R}$.

   (b) $(g \circ f)(x) = g(f(x)) = g(x^2 - 1) = 2(x^2 - 1) + 1 = (2x^2 - 2) + 1 = 2x^2 - 1$, $D = \mathbb{R}$.

   (c) $(f \circ f)(x) = f(f(x)) = f(x^2 - 1) = (x^2 - 1)^2 - 1 = (x^4 - 2x^2 + 1) - 1 = x^4 - 2x^2$, $D = \mathbb{R}$.

   (d) $(g \circ g)(x) = g(g(x)) = g(2x + 1) = 2(2x + 1) + 1 = (4x + 2) + 1 = 4x + 3$, $D = \mathbb{R}$.

32. $f(x) = x - 2$; $g(x) = x^2 + 3x + 4$. $D = \mathbb{R}$ for both f and g, and hence for their composites.

   (a) $(f \circ g)(x) = f(g(x)) = f(x^2 + 3x + 4) = (x^2 + 3x + 4) - 2 = x^2 + 3x + 2$.

   (b) $(g \circ f)(x) = g(f(x)) = g(x - 2) = (x - 2)^2 + 3(x - 2) + 4 = x^2 - 4x + 4 + 3x - 6 + 4 = x^2 - x + 2$.

   (c) $(f \circ f)(x) = f(f(x)) = f(x - 2) = (x - 2) - 2 = x - 4$.

   (d) $(g \circ g)(x) = g(g(x)) = g(x^2 + 3x + 4) = (x^2 + 3x + 4)^2 + 3(x^2 + 3x + 4) + 4$

$$= (x^4 + 9x^2 + 16 + 6x^3 + 8x^2 + 24x) + 3x^2 + 9x + 12 + 4$$

$$= x^4 + 6x^3 + 20x^2 + 33x + 32$$

33. $f(x) = 1 - 3x$; $g(x) = \cos x$. $D = \mathbb{R}$ for both f and g, and hence for their composites.

   (a) $(f \circ g)(x) = f(g(x)) = f(\cos x) = 1 - 3 \cos x$.

   (b) $(g \circ f)(x) = g(f(x)) = g(1 - 3x) = \cos(1 - 3x)$.

   (c) $(f \circ f)(x) = f(f(x)) = f(1 - 3x) = 1 - 3(1 - 3x) = 1 - 3 + 9x = 9x - 2$.

   (d) $(g \circ g)(x) = g(g(x)) = g(\cos x) = \cos(\cos x)$ [Note that this is *not* $\cos x \cdot \cos x$.]

34. $f(x) = \sqrt{x}$, $D = [0, \infty)$; $g(x) = \sqrt[3]{1 - x}$, $D = \mathbb{R}$.

   (a) $(f \circ g)(x) = f(g(x)) = f\left(\sqrt[3]{1 - x}\right) = \sqrt{\sqrt[3]{1 - x}} = \sqrt[6]{1 - x}$.

     The domain of $f \circ g$ is $\{x \mid \sqrt[3]{1 - x} \geq 0\} = \{x \mid 1 - x \geq 0\} = \{x \mid x \leq 1\} = (-\infty, 1]$.

   (b) $(g \circ f)(x) = g(f(x)) = g(\sqrt{x}) = \sqrt[3]{1 - \sqrt{x}}$.

     The domain of $g \circ f$ is $\{x \mid x$ is in the domain of f and $f(x)$ is in the domain of $g\}$. This is the domain of f, that is, $[0, \infty)$.

   (c) $(f \circ f)(x) = f(f(x)) = f(\sqrt{x}) = \sqrt{\sqrt{x}} = \sqrt[4]{x}$. The domain of $f \circ f$ is $\{x \mid x \geq 0 \text{ and } \sqrt{x} \geq 0\} = [0, \infty)$.

   (d) $(g \circ g)(x) = g(g(x)) = g\left(\sqrt[3]{1 - x}\right) = \sqrt[3]{1 - \sqrt[3]{1 - x}}$, and the domain is $(-\infty, \infty)$.

35. $f(x) = x + \dfrac{1}{x}$, $D = \{x \mid x \neq 0\}$; $g(x) = \dfrac{x + 1}{x + 2}$, $D = \{x \mid x \neq -2\}$

   (a) $(f \circ g)(x) = f(g(x)) = f\left(\dfrac{x + 1}{x + 2}\right) = \dfrac{x + 1}{x + 2} + \dfrac{1}{\dfrac{x + 1}{x + 2}} = \dfrac{x + 1}{x + 2} + \dfrac{x + 2}{x + 1}$

$$= \dfrac{(x + 1)(x + 1) + (x + 2)(x + 2)}{(x + 2)(x + 1)} = \dfrac{(x^2 + 2x + 1) + (x^2 + 4x + 4)}{(x + 2)(x + 1)} = \dfrac{2x^2 + 6x + 5}{(x + 2)(x + 1)}$$

     Since $g(x)$ is not defined for $x = -2$ and $f(g(x))$ is not defined for $x = -2$ and $x = -1$,

     the domain of $(f \circ g)(x)$ is $D = \{x \mid x \neq -2, -1\}$.

   (b) $(g \circ f)(x) = g(f(x)) = g\left(x + \dfrac{1}{x}\right) = \dfrac{\left(x + \dfrac{1}{x}\right) + 1}{\left(x + \dfrac{1}{x}\right) + 2} = \dfrac{\dfrac{x^2 + 1 + x}{x}}{\dfrac{x^2 + 1 + 2x}{x}} = \dfrac{x^2 + x + 1}{x^2 + 2x + 1} = \dfrac{x^2 + x + 1}{(x + 1)^2}$

     Since $f(x)$ is not defined for $x = 0$ and $g(f(x))$ is not defined for $x = -1$,

     the domain of $(g \circ f)(x)$ is $D = \{x \mid x \neq -1, 0\}$.

(c) $(f \circ f)(x) = f(f(x)) = f\left(x + \dfrac{1}{x}\right) = \left(x + \dfrac{1}{x}\right) + \dfrac{1}{x + \frac{1}{x}} = x + \dfrac{1}{x} + \dfrac{1}{\frac{x^2+1}{x}} = x + \dfrac{1}{x} + \dfrac{x}{x^2 + 1}$

$$= \dfrac{x(x)(x^2 + 1) + 1(x^2 + 1) + x(x)}{x(x^2 + 1)} = \dfrac{x^4 + x^2 + x^2 + 1 + x^2}{x(x^2 + 1)}$$

$$= \dfrac{x^4 + 3x^2 + 1}{x(x^2 + 1)}, \quad D = \{x \mid x \neq 0\}$$

(d) $(g \circ g)(x) = g(g(x)) = g\left(\dfrac{x + 1}{x + 2}\right) = \dfrac{\frac{x+1}{x+2} + 1}{\frac{x+1}{x+2} + 2} = \dfrac{\frac{x + 1 + 1(x + 2)}{x + 2}}{\frac{x + 1 + 2(x + 2)}{x + 2}} = \dfrac{x + 1 + x + 2}{x + 1 + 2x + 4} = \dfrac{2x + 3}{3x + 5}$

Since $g(x)$ is not defined for $x = -2$ and $g(g(x))$ is not defined for $x = -\frac{5}{3}$,

the domain of $(g \circ g)(x)$ is $D = \left\{x \mid x \neq -2, -\frac{5}{3}\right\}$.

36. $f(x) = \dfrac{x}{1 + x}, \ D = \{x \mid x \neq -1\}; \quad g(x) = \sin 2x, \ D = \mathbb{R}.$

(a) $(f \circ g)(x) = f(g(x)) = f(\sin 2x) = \dfrac{\sin 2x}{1 + \sin 2x}$

Domain: $1 + \sin 2x \neq 0 \ \Rightarrow \ \sin 2x \neq -1 \ \Rightarrow \ 2x \neq \dfrac{3\pi}{2} + 2\pi n \ \Rightarrow \ x \neq \dfrac{3\pi}{4} + \pi n \quad (n \text{ an integer}).$

(b) $(g \circ f)(x) = g(f(x)) = g\left(\dfrac{x}{1 + x}\right) = \sin\left(\dfrac{2x}{1 + x}\right).$ Domain: $\{x \mid x \neq -1\}$

(c) $(f \circ f)(x) = f(f(x)) = f\left(\dfrac{x}{1 + x}\right) = \dfrac{\frac{x}{1+x}}{1 + \frac{x}{1+x}} = \dfrac{\left(\frac{x}{1+x}\right) \cdot (1 + x)}{\left(1 + \frac{x}{1+x}\right) \cdot (1 + x)} = \dfrac{x}{1 + x + x} = \dfrac{x}{2x + 1}$

Since $f(x)$ is not defined for $x = -1$, and $f(f(x))$ is not defined for $x = -\frac{1}{2}$,

the domain of $(f \circ f)(x)$ is $D = \left\{x \mid x \neq -1, -\frac{1}{2}\right\}$.

(d) $(g \circ g)(g) = g(g(x)) = g(\sin 2x) = \sin(2 \sin 2x).$ Domain: $\mathbb{R}$

37. $(f \circ g \circ h)(x) = f(g(h(x))) = f(g(x - 1)) = f(2(x - 1)) = 2(x - 1) + 1 = 2x - 1$

38. $(f \circ g \circ h)(x) = f(g(h(x))) = f(g(1 - x)) = f((1 - x)^2) = 2(1 - x)^2 - 1 = 2x^2 - 4x + 1$

39. $(f \circ g \circ h)(x) = f(g(h(x))) = f(g(x^3 + 2)) = f[(x^3 + 2)^2]$

$= f(x^6 + 4x^3 + 4) = \sqrt{(x^6 + 4x^3 + 4) - 3} = \sqrt{x^6 + 4x^3 + 1}$

40. $(f \circ g \circ h)(x) = f(g(h(x))) = f(g(\sqrt[3]{x})) = f\left(\dfrac{\sqrt[3]{x}}{\sqrt[3]{x} - 1}\right) = \tan\left(\dfrac{\sqrt[3]{x}}{\sqrt[3]{x} - 1}\right)$

41. Let $g(x) = x^2 + 1$ and $f(x) = x^{10}$. Then $(f \circ g)(x) = f(g(x)) = f(x^2 + 1) = (x^2 + 1)^{10} = F(x)$.

42. Let $g(x) = \sqrt{x}$ and $f(x) = \sin x$. Then $(f \circ g)(x) = f(g(x)) = f(\sqrt{x}) = \sin(\sqrt{x}) = F(x)$.

43. Let $g(x) = \sqrt[3]{x}$ and $f(x) = \dfrac{x}{1 + x}$. Then $(f \circ g)(x) = f(g(x)) = f(\sqrt[3]{x}) = \dfrac{\sqrt[3]{x}}{1 + \sqrt[3]{x}} = F(x)$.

44. Let $g(x) = \dfrac{x}{1 + x}$ and $f(x) = \sqrt[3]{x}$. Then $(f \circ g)(x) = f(g(x)) = f\left(\dfrac{x}{1 + x}\right) = \sqrt[3]{\dfrac{x}{1 + x}} = G(x)$.

45. Let $g(t) = \cos t$ and $f(t) = \sqrt{t}$. Then $(f \circ g)(t) = f(g(t)) = f(\cos t) = \sqrt{\cos t} = u(t)$.

46. Let $g(t) = \tan t$ and $f(t) = \dfrac{t}{1+t}$. Then $(f \circ g)(t) = f(g(t)) = f(\tan t) = \dfrac{\tan t}{1+\tan t} = u(t)$.

47. Let $h(x) = x^2$, $g(x) = 3^x$, and $f(x) = 1 - x$. Then

$$(f \circ g \circ h)(x) = f(g(h(x))) = f(g(x^2)) = f\left(3^{x^2}\right) = 1 - 3^{x^2} = H(x).$$

48. Let $h(x) = |x|$, $g(x) = 2 + x$, and $f(x) = \sqrt[8]{x}$. Then

$$(f \circ g \circ h)(x) = f(g(h(x))) = f(g(|x|)) = f(2 + |x|) = \sqrt[8]{2 + |x|} = H(x).$$

49. Let $h(x) = \sqrt{x}$, $g(x) = \sec x$, and $f(x) = x^4$. Then

$$(f \circ g \circ h)(x) = f(g(h(x))) = f(g(\sqrt{x})) = f(\sec\sqrt{x}) = (\sec\sqrt{x})^4 = \sec^4(\sqrt{x}) = H(x).$$

50. (a) $f(g(1)) = f(6) = 5$ (b) $g(f(1)) = g(3) = 2$

(c) $f(f(1)) = f(3) = 4$ (d) $g(g(1)) = g(6) = 3$

(e) $(g \circ f)(3) = g(f(3)) = g(4) = 1$ (f) $(f \circ g)(6) = f(g(6)) = f(3) = 4$

51. (a) $g(2) = 5$, because the point $(2, 5)$ is on the graph of g. Thus, $f(g(2)) = f(5) = 4$, because the point $(5, 4)$ is on the graph of f.

(b) $g(f(0)) = g(0) = 3$

(c) $(f \circ g)(0) = f(g(0)) = f(3) = 0$

(d) $(g \circ f)(6) = g(f(6)) = g(6)$. This value is not defined, because there is no point on the graph of g that has x-coordinate 6.

(e) $(g \circ g)(-2) = g(g(-2)) = g(1) = 4$

(f) $(f \circ f)(4) = f(f(4)) = f(2) = -2$

52. To find a particular value of $f(g(x))$, say for $x = 0$, we note from the graph that $g(0) \approx 2.8$ and $f(2.8) \approx -0.5$. Thus,

$f(g(0)) \approx f(2.8) \approx -0.5$. The other values listed in the table were obtained in a similar fashion.

x	$g(x)$	$f(g(x))$
-5	-0.2	-4
-4	1.2	-3.3
-3	2.2	-1.7
-2	2.8	-0.5
-1	3	-0.2

x	$g(x)$	$f(g(x))$
0	2.8	-0.5
1	2.2	-1.7
2	1.2	-3.3
3	-0.2	-4
4	-1.9	-2.2
5	-4.1	1.9

53. (a) Using the relationship *distance* = *rate* · *time* with the radius r as the distance, we have $r(t) = 60t$.

(b) $A = \pi r^2 \;\Rightarrow\; (A \circ r)(t) = A(r(t)) = \pi(60t)^2 = 3600\pi t^2$. This formula gives us the extent of the rippled area (in cm^2) at any time t.

54. (a) The radius r of the balloon is increasing at a rate of 2 cm/s, so $r(t) = (2 \text{ cm/s})(t \text{ s}) = 2t$ (in cm).

(b) Using $V = \frac{4}{3}\pi r^3$, we get $(V \circ r)(t) = V(r(t)) = V(2t) = \frac{4}{3}\pi(2t)^3 = \frac{32}{3}\pi t^3$.

The result, $V = \frac{32}{3}\pi t^3$, gives the volume of the balloon (in cm^3) as a function of time (in s).

55. (a) From the figure, we have a right triangle with legs 6 and d, and hypotenuse s.

By the Pythagorean Theorem, $d^2 + 6^2 = s^2 \implies s = f(d) = \sqrt{d^2 + 36}$.

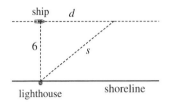

(b) Using $d = rt$, we get $d = (30 \text{ km/hr})(t \text{ hr}) = 30t$ (in km). Thus,

$d = g(t) = 30t$.

(c) $(f \circ g)(t) = f(g(t)) = f(30t) = \sqrt{(30t)^2 + 36} = \sqrt{900t^2 + 36}$. This function represents the distance between the lighthouse and the ship as a function of the time elapsed since noon.

56. (a) $d = rt \implies d(t) = 350t$

(b) There is a Pythagorean relationship involving the legs with lengths d and 1 and the hypotenuse with length s:

$d^2 + 1^2 = s^2$. Thus, $s(d) = \sqrt{d^2 + 1}$.

(c) $(s \circ d)(t) = s(d(t)) = s(350t) = \sqrt{(350t)^2 + 1}$

57. (a)

(b)

$$H(t) = \begin{cases} 0 & \text{if } t < 0 \\ 1 & \text{if } t \geq 0 \end{cases}$$

$$V(t) = \begin{cases} 0 & \text{if } t < 0 \\ 120 & \text{if } t \geq 0 \end{cases} \quad \text{so } V(t) = 120H(t).$$

(c)

Starting with the formula in part (b), we replace 120 with 240 to reflect the different voltage. Also, because we are starting 5 units to the right of $t = 0$, we replace t with $t - 5$. Thus, the formula is $V(t) = 240H(t - 5)$.

58. (a) $R(t) = tH(t)$

$$= \begin{cases} 0 & \text{if } t < 0 \\ t & \text{if } t \geq 0 \end{cases}$$

(b) $V(t) = \begin{cases} 0 & \text{if } t < 0 \\ 2t & \text{if } 0 \leq t \leq 60 \end{cases}$

so $V(t) = 2tH(t), t \leq 60$.

(c) $V(t) = \begin{cases} 0 & \text{if } t < 7 \\ 4(t-7) & \text{if } 7 \leq t \leq 32 \end{cases}$

so $V(t) = 4(t-7)H(t-7), t \leq 32$.

59. If $f(x) = m_1 x + b_1$ and $g(x) = m_2 x + b_2$, then

$(f \circ g)(x) = f(g(x)) = f(m_2 x + b_2) = m_1(m_2 x + b_2) + b_1 = m_1 m_2 x + m_1 b_2 + b_1$.

So $f \circ g$ is a linear function with slope $m_1 m_2$.

60. If $A(x) = 1.04x$, then

$$(A \circ A)(x) = A(A(x)) = A(1.04x) = 1.04(1.04x) = (1.04)^2 x,$$

$$(A \circ A \circ A)(x) = A((A \circ A)(x)) = A((1.04)^2 x) = 1.04(1.04)^2 x = (1.04)^3 x, \text{ and}$$

$$(A \circ A \circ A \circ A)(x) = A((A \circ A \circ A)(x)) = A((1.04)^3 x) = 1.04(1.04)^3 x, = (1.04)^4 x.$$

These compositions represent the amount of the investment after 2, 3, and 4 years.

Based on this pattern, when we compose n copies of A, we get the formula $\underbrace{(A \circ A \circ \cdots \circ A)}_{n \ A's}(x) = (1.04)^n x.$

61. (a) By examining the variable terms in g and h, we deduce that we must square g to get the terms $4x^2$ and $4x$ in h. If we let

$f(x) = x^2 + c$, then $(f \circ g)(x) = f(g(x)) = f(2x + 1) = (2x + 1)^2 + c = 4x^2 + 4x + (1 + c)$. Since

$h(x) = 4x^2 + 4x + 7$, we must have $1 + c = 7$. So $c = 6$ and $f(x) = x^2 + 6$.

(b) We need a function g so that $f(g(x)) = 3(g(x)) + 5 = h(x)$. But

$h(x) = 3x^2 + 3x + 2 = 3(x^2 + x) + 2 = 3(x^2 + x - 1) + 5$, so we see that $g(x) = x^2 + x - 1$.

62. We need a function g so that $g(f(x)) = g(x + 4) = h(x) = 4x - 1 = 4(x + 4) - 17$. So we see that the function g must be

$g(x) = 4x - 17$.

63. (a) If f and g are even functions, then $f(-x) = f(x)$ and $g(-x) = g(x)$.

    (i) $(f + g)(-x) = f(-x) + g(-x) = f(x) + g(x) = (f + g)(x)$, so $f + g$ is an *even* function.

    (ii) $(fg)(-x) = f(-x) \cdot g(-x) = f(x) \cdot g(x) = (fg)(x)$, so fg is an *even* function.

(b) If f and g are odd functions, then $f(-x) = -f(x)$ and $g(-x) = -g(x)$.

    (i) $(f + g)(-x) = f(-x) + g(-x) = -f(x) + [-g(x)] = -[f(x) + g(x)] = -(f + g)(x)$,

    so $f + g$ is an *odd* function.

    (ii) $(fg)(-x) = f(-x) \cdot g(-x) = -f(x) \cdot [-g(x)] = f(x) \cdot g(x) = (fg)(x)$, so fg is an *even* function.

64. If f is even and g is odd, then $f(-x) = f(x)$ and $g(-x) = -g(x)$. Now

$(fg)(-x) = f(-x) \cdot g(-x) = f(x) \cdot [-g(x)] = -[f(x) \cdot g(x)] = -(fg)(x)$, so fg is an *odd* function.

65. We need to examine $h(-x)$.

$$h(-x) = (f \circ g)(-x) = f(g(-x)) = f(g(x)) \quad \text{[because } g \text{ is even]} \quad = h(x)$$

Because $h(-x) = h(x)$, h is an even function.

66. $h(-x) = f(g(-x)) = f(-g(x))$. At this point, we can't simplify the expression, so we might try to find a counterexample to

show that h is not an odd function. Let $g(x) = x$, an odd function, and $f(x) = x^2 + x$. Then $h(x) = x^2 + x$, which is neither

even nor odd.

    Now suppose f is an odd function. Then $f(-g(x)) = -f(g(x)) = -h(x)$. Hence, $h(-x) = -h(x)$, and so h is odd if

both f and g are odd.

    Now suppose f is an even function. Then $f(-g(x)) = f(g(x)) = h(x)$. Hence, $h(-x) = h(x)$, and so h is even if g is

odd and f is even.

1.4   Graphing Calculators and Computers

1. $f(x) = \sqrt{x^3 - 5x^2}$

(a) $[-5, 5]$ by $[-5, 5]$ (b) $[0, 10]$ by $[0, 2]$ (c) $[0, 10]$ by $[0, 10]$

(There is no graph shown.)

       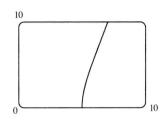

The most appropriate graph is produced in viewing rectangle (c).

2. $f(x) = x^4 - 16x^2 + 20$

(a) $[-3, 3]$ by $[-3, 3]$ (b) $[-10, 10]$ by $[-10, 10]$

        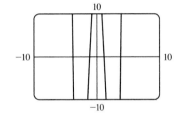

(c) $[-50, 50]$ by $[-50, 50]$ (d) $[-5, 5]$ by $[-50, 50]$

        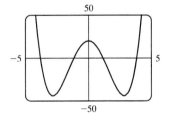

The most appropriate graph is produced in viewing rectangle (d).

3. Since the graph of $f(x) = 5 + 20x - x^2$ is a parabola opening downward, an appropriate viewing rectangle should include the maximum point.

4. An appropriate viewing rectangle for $f(x) = x^3 + 30x^2 + 200x$ should include the high and low points.

5. $f(x) = \sqrt[4]{81 - x^4}$ is defined when $81 - x^4 \geq 0 \quad \Leftrightarrow$ $x^4 \leq 81 \quad \Leftrightarrow \quad |x| \leq 3$, so the domain of f is $[-3, 3]$. Also $0 \leq \sqrt[4]{81 - x^4} \leq \sqrt[4]{81} = 3$, so the range is $[0, 3]$.

6. $f(x) = \sqrt{0.1x + 20}$ is defined when $0.1x + 20 \geq 0 \quad \Leftrightarrow$ $x \geq -200$, so the domain of f is $[-200, \infty)$.

7. The graph of $f(x) = x^3 - 225x$ is symmetric with respect to the origin. Since $f(x) = x^3 - 225x = x(x^2 - 225) = x(x + 15)(x - 15)$, there are x-intercepts at 0, -15, and 15. $f(20) = 3500$.

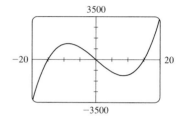

8. The graph of $f(x) = x/(x^2 + 100)$ is symmetric with respect to the origin.

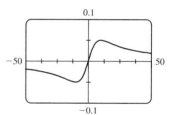

9. The period of $g(x) = \sin(1000x)$ is $\frac{2\pi}{1000} \approx 0.0063$ and its range is $[-1, 1]$. Since $f(x) = \sin^2(1000x)$ is the square of g, its range is $[0, 1]$ and a viewing rectangle of $[-0.01, 0.01]$ by $[0, 1.1]$ seems appropriate.

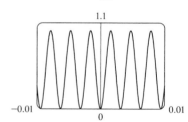

10. The period of $f(x) = \cos(0.001x)$ is $\frac{2\pi}{0.001} \approx 6300$ and its range is $[-1, 1]$, so a viewing rectangle of $[-10{,}000, 10{,}000]$ by $[-1.5, 1.5]$ seems appropriate.

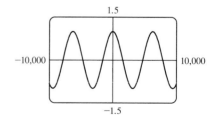

11. The domain of $y = \sqrt{x}$ is $x \geq 0$, so the domain of $f(x) = \sin\sqrt{x}$ is $[0, \infty)$ and the range is $[-1, 1]$. With a little trial-and-error experimentation, we find that an Xmax of 100 illustrates the general shape of f, so an appropriate viewing rectangle is $[0, 100]$ by $[-1.5, 1.5]$.

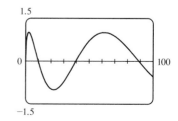

12. One period of $y = \sec x$ occurs on the interval $\left(-\frac{\pi}{2}, \frac{\pi}{2}\right) \cup \left(\frac{\pi}{2}, \frac{3\pi}{2}\right)$.

$-\frac{\pi}{2} < 20\pi x < \frac{3\pi}{2}$ ⟹ $-\frac{1}{40} < x < \frac{3}{40}$, or equivalently,

$-0.025 < x < 0.075$.

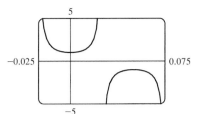

13. The first term, $10\sin x$, has period 2π and range $[-10, 10]$. It will be the dominant term in any "large" graph of $y = 10\sin x + \sin 100x$, as shown in the first figure. The second term, $\sin 100x$, has period $\frac{2\pi}{100} = \frac{\pi}{50}$ and range $[-1, 1]$. It causes the bumps in the first figure and will be the dominant term in any "small" graph, as shown in the view near the origin in the second figure.

14. $y = x^2 + 0.02\sin(50x)$

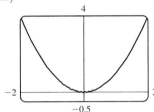

15. We must solve the given equation for y to obtain equations for the upper and lower halves of the ellipse.

$4x^2 + 2y^2 = 1$ ⟺ $2y^2 = 1 - 4x^2$ ⟺ $y^2 = \dfrac{1 - 4x^2}{2}$ ⟺

$y = \pm\sqrt{\dfrac{1 - 4x^2}{2}}$

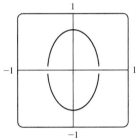

16. $y^2 - 9x^2 = 1$ ⟺ $y^2 = 1 + 9x^2$ ⟺ $y = \pm\sqrt{1 + 9x^2}$

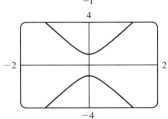

17. From the graph of $y = 3x^2 - 6x + 1$ and $y = 0.23x - 2.25$ in the viewing rectangle $[-1, 3]$ by $[-2.5, 1.5]$, it is difficult to see if the graphs intersect. If we zoom in on the fourth quadrant, we see the graphs do not intersect.

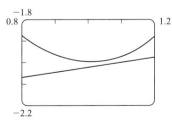

18. From the graph of $y = 6 - 4x - x^2$ and $y = 3x + 18$ in the viewing rectangle $[-6, 2]$ by $[-5, 20]$, we see that the graphs intersect twice. The points of intersection are $(-4, 6)$ and $(-3, 9)$.

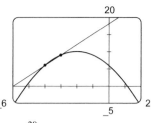

19. From the graph of $f(x) = x^3 - 9x^2 - 4$, we see that there is one solution of the equation $f(x) = 0$ and it is slightly larger than 9. By zooming in or using a root or zero feature, we obtain $x \approx 9.05$.

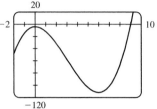

20. We see that the graphs of $f(x) = x^3$ and $g(x) = 4x - 1$ intersect three times. The x-coordinates of these points (which are the solutions of the equation) are approximately $-2.11, 0.25$, and 1.86. Alternatively, we could find these values by finding the zeros of $h(x) = x^3 - 4x + 1$.

21. We see that the graphs of $f(x) = x^2$ and $g(x) = \sin x$ intersect twice. One solution is $x = 0$. The other solution of $f = g$ is the x-coordinate of the point of intersection in the first quadrant. Using an intersect feature or zooming in, we find this value to be approximately 0.88. Alternatively, we could find that value by finding the positive zero of $h(x) = x^2 - \sin x$.

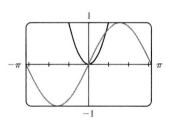

Note: After producing the graph on a TI-83 Plus, we can find the approximate value 0.88 by using the following keystrokes:

2nd CALC 5 ENTER ENTER 1 ENTER . The "1" is just a guess for 0.88.

22. (a)

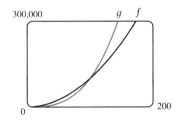

The x-coordinates of the three points of intersection are
$x \approx -3.29, -2.36$ and 1.20.

(b) Using trial and error, we find that $m \approx 0.3365$. Note that m could also be negative.

23. $g(x) = x^3/10$ is larger than $f(x) = 10x^2$ whenever $x > 100$.

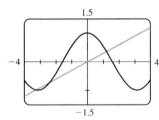

24. $f(x) = x^4 - 100x^3$ is larger than $g(x) = x^3$ whenever $x > 101$.

25.

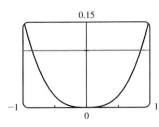

We see from the graphs of $y = |\sin x - x|$ and $y = 0.1$ that there are two solutions to the equation $|\sin x - x| = 0.1$: $x \approx -0.85$ and $x \approx 0.85$. The condition $|\sin x - x| < 0.1$ holds for any x lying between these two values, that is, $-0.85 < x < 0.85$.

26. $P(x) = 3x^5 - 5x^3 + 2x$, $Q(x) = 3x^5$. These graphs are significantly different only in the region close to the origin. The larger a viewing rectangle one chooses, the more similar the two graphs look.

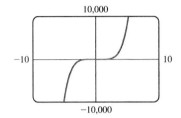

27. (a) The root functions $y = \sqrt{x}$, $y = \sqrt[4]{x}$ and $y = \sqrt[6]{x}$

(b) The root functions $y = x$, $y = \sqrt[3]{x}$ and $y = \sqrt[5]{x}$

(c) The root functions $y = \sqrt{x}$, $y = \sqrt[3]{x}$, $y = \sqrt[4]{x}$ and $y = \sqrt[5]{x}$

(d) • For any n, the nth root of 0 is 0 and the nth root of 1 is 1; that is, all nth root functions pass through the points $(0,0)$ and $(1,1)$.

 • For odd n, the domain of the nth root function is $\mathbb{R}$, while for even n, it is $\{x \in \mathbb{R} \mid x \geq 0\}$.

 • Graphs of even root functions look similar to that of $\sqrt{x}$, while those of odd root functions resemble that of $\sqrt[3]{x}$.

 • As n increases, the graph of $\sqrt[n]{x}$ becomes steeper near 0 and flatter for $x > 1$.

28. (a) The functions $y = 1/x$ and $y = 1/x^3$

(b) The functions $y = 1/x^2$ and $y = 1/x^4$

(c) The functions $y = 1/x$, $y = 1/x^2$, $y = 1/x^3$ and $y = 1/x^4$

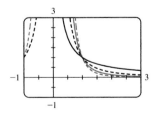

(d) • The graphs of all functions of the form $y = 1/x^n$ pass through the point $(1,1)$.

 • If n is even, the graph of the function is entirely above the x-axis. The graphs of $1/x^n$ for n even are similar to one another.

 • If n is odd, the function is positive for positive x and negative for negative x. The graphs of $1/x^n$ for n odd are similar to one another.

 • As n increases, the graphs of $1/x^n$ approach 0 faster as $x \to \infty$.

29. $f(x) = x^4 + cx^2 + x$. If $c < -1.5$, there are three humps: two minimum points and a maximum point. These humps get flatter as c increases, until at $c = -1.5$ two of the humps disappear and there is only one minimum point. This single hump then moves to the right and approaches the origin as c increases.

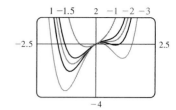

30. $f(x) = \sqrt{1 + cx^2}$. If $c < 0$, the function is only defined on $\left[-1/\sqrt{-c}, 1/\sqrt{-c}\right]$, and its graph is the top half of an ellipse. If $c = 0$, the graph is the line $y = 1$. If $c > 0$, the graph is the top half of a hyperbola. As c approaches 0, these curves become flatter and approach the line $y = 1$.

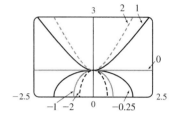

31. $y = x^n 2^{-x}$. As n increases, the maximum of the function moves further from the origin, and gets larger. Note, however, that regardless of n, the function approaches 0 as $x \to \infty$.

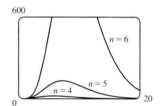

32. $y = \dfrac{|x|}{\sqrt{c - x^2}}$. The "bullet" becomes broader as c increases.

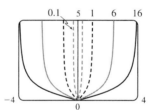

33. $y^2 = cx^3 + x^2$. If $c < 0$, the loop is to the right of the origin, and if c is positive, it is to the left. In both cases, the closer c is to 0, the larger the loop is. (In the limiting case, $c = 0$, the loop is "infinite," that is, it doesn't close.) Also, the larger $|c|$ is, the steeper the slope is on the loopless side of the origin.

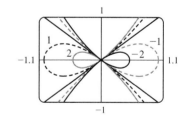

34. (a) $y = \sin(\sqrt{x})$

This function is not periodic; it oscillates less frequently as x increases.

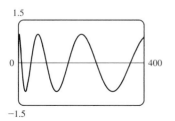

(b) $y = \sin(x^2)$

This function oscillates more frequently as $|x|$ increases. Note also that this function is even, whereas $\sin x$ is odd.

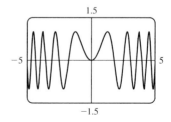

35. The graphing window is 95 pixels wide and we want to start with $x = 0$ and end with $x = 2\pi$. Since there are 94 "gaps" between pixels, the distance between pixels is $\frac{2\pi - 0}{94}$. Thus, the x-values that the calculator actually plots are $x = 0 + \frac{2\pi}{94} \cdot n$, where $n = 0, 1, 2, \ldots, 93, 94$. For $y = \sin 2x$, the actual points plotted by the calculator are $\left(\frac{2\pi}{94} \cdot n, \sin\left(2 \cdot \frac{2\pi}{94} \cdot n\right)\right)$ for $n = 0, 1, \ldots, 94$. For $y = \sin 96x$, the points plotted are $\left(\frac{2\pi}{94} \cdot n, \sin\left(96 \cdot \frac{2\pi}{94} \cdot n\right)\right)$ for $n = 0, 1, \ldots, 94$. But

$$\sin\left(96 \cdot \tfrac{2\pi}{94} \cdot n\right) = \sin\left(94 \cdot \tfrac{2\pi}{94} \cdot n + 2 \cdot \tfrac{2\pi}{94} \cdot n\right) = \sin\left(2\pi n + 2 \cdot \tfrac{2\pi}{94} \cdot n\right)$$
$$= \sin\left(2 \cdot \tfrac{2\pi}{94} \cdot n\right) \quad \text{[by periodicity of sine]}, \quad n = 0, 1, \ldots, 94$$

So the y-values, and hence the points, plotted for $y = \sin 96x$ are identical to those plotted for $y = \sin 2x$.

Note: Try graphing $y = \sin 94x$. Can you see why all the y-values are zero?

36. As in Exercise 35, we know that the points being plotted for $y = \sin 45x$ are $\left(\frac{2\pi}{94} \cdot n, \sin\left(45 \cdot \frac{2\pi}{94} \cdot n\right)\right)$ for $n = 0, 1, \ldots, 94$. But

$$\sin\left(45 \cdot \tfrac{2\pi}{94} \cdot n\right) = \sin\left(47 \cdot \tfrac{2\pi}{94} \cdot n - 2 \cdot \tfrac{2\pi}{94} \cdot n\right) = \sin\left(n\pi - 2 \cdot \tfrac{2\pi}{94} \cdot n\right)$$
$$= \sin(n\pi)\cos\left(2 \cdot \tfrac{2\pi}{94} \cdot n\right) - \cos(n\pi)\sin\left(2 \cdot \tfrac{2\pi}{94} \cdot n\right) \quad \text{[Subtraction formula for the sine]}$$
$$= 0 \cdot \cos\left(2 \cdot \tfrac{2\pi}{94} \cdot n\right) - (\pm 1)\sin\left(2 \cdot \tfrac{2\pi}{94} \cdot n\right) = \pm \sin\left(2 \cdot \tfrac{2\pi}{94} \cdot n\right), \quad n = 0, 1, \ldots, 94$$

So the y-values, and hence the points, plotted for $y = \sin 45x$ lie on either $y = \sin 2x$ or $y = -\sin 2x$.

1 Review

CONCEPT CHECK

1. (a) A **function** f is a rule that assigns to each element x in a set A exactly one element, called $f(x)$, in a set B. The set A is called the **domain** of the function. The **range** of f is the set of all possible values of $f(x)$ as x varies throughout the domain.

(b) If f is a function with domain A, then its **graph** is the set of ordered pairs $\{(x, f(x)) \mid x \in A\}$.

(c) Use the Vertical Line Test on page 16.

2. The four ways to represent a function are: verbally, numerically, visually, and algebraically. An example of each is given below.

Verbally: An assignment of students to chairs in a classroom (a description in words)

Numerically: A tax table that assigns an amount of tax to an income (a table of values)

Visually: A graphical history of the Dow Jones average (a graph)

Algebraically: A relationship between distance, rate, and time: $d = rt$ (an explicit formula)

3. (a) An **even function** f satisfies $f(-x) = f(x)$ for every number x in its domain. It is symmetric with respect to the y-axis.

(b) An **odd function** g satisfies $g(-x) = -g(x)$ for every number x in its domain. It is symmetric with respect to the origin.

4. A function f is called **increasing** on an interval I if $f(x_1) < f(x_2)$ whenever $x_1 < x_2$ in I.

5. A **mathematical model** is a mathematical description (often by means of a function or an equation) of a real-world phenomenon.

6. (a) Linear function: $f(x) = 2x + 1$, $f(x) = ax + b$

   (b) Power function: $f(x) = x^2$, $f(x) = x^a$

   (c) Exponential function: $f(x) = 2^x$, $f(x) = a^x$

   (d) Quadratic function: $f(x) = x^2 + x + 1$, $f(x) = ax^2 + bx + c$

   (e) Polynomial of degree 5: $f(x) = x^5 + 2$

   (f) Rational function: $f(x) = \dfrac{x}{x+2}$, $f(x) = \dfrac{P(x)}{Q(x)}$ where $P(x)$ and $Q(x)$ are polynomials

7.

8. (a)

   (b)

   (c)

   (d)

   (e)
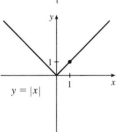

   (f)

9. (a) The domain of $f + g$ is the intersection of the domain of f and the domain of g; that is, $A \cap B$.

   (b) The domain of fg is also $A \cap B$.

   (c) The domain of f/g must exclude values of x that make g equal to 0; that is, $\{x \in A \cap B \mid g(x) \neq 0\}$.

10. Given two functions f and g, the **composite** function $f \circ g$ is defined by $(f \circ g)(x) = f(g(x))$. The domain of $f \circ g$ is the set of all x in the domain of g such that $g(x)$ is in the domain of f.

11. (a) If the graph of f is shifted 2 units upward, its equation becomes $y = f(x) + 2$.

   (b) If the graph of f is shifted 2 units downward, its equation becomes $y = f(x) - 2$.

(c) If the graph of f is shifted 2 units to the right, its equation becomes $y = f(x - 2)$.

(d) If the graph of f is shifted 2 units to the left, its equation becomes $y = f(x + 2)$.

(e) If the graph of f is reflected about the x-axis, its equation becomes $y = -f(x)$.

(f) If the graph of f is reflected about the y-axis, its equation becomes $y = f(-x)$.

(g) If the graph of f is stretched vertically by a factor of 2, its equation becomes $y = 2f(x)$.

(h) If the graph of f is shrunk vertically by a factor of 2, its equation becomes $y = \frac{1}{2}f(x)$.

(i) If the graph of f is stretched horizontally by a factor of 2, its equation becomes $y = f\left(\frac{1}{2}x\right)$.

(j) If the graph of f is shrunk horizontally by a factor of 2, its equation becomes $y = f(2x)$.

TRUE-FALSE QUIZ

1. False. Let $f(x) = x^2$, $s = -1$, and $t = 1$. Then $f(s + t) = (-1 + 1)^2 = 0^2 = 0$, but $f(s) + f(t) = (-1)^2 + 1^2 = 2 \neq 0 = f(s + t)$.

2. False. Let $f(x) = x^2$. Then $f(-2) = 4 = f(2)$, but $-2 \neq 2$.

3. False. Let $f(x) = x^2$. Then $f(3x) = (3x)^2 = 9x^2$ and $3f(x) = 3x^2$. So $f(3x) \neq 3f(x)$.

4. True. If $x_1 < x_2$ and f is a decreasing function, then the y-values get smaller as we move from left to right. Thus, $f(x_1) > f(x_2)$.

5. True. See the Vertical Line Test.

6. False. Let $f(x) = x^2$ and $g(x) = 2x$. Then $(f \circ g)(x) = f(g(x)) = f(2x) = (2x)^2 = 4x^2$ and $(g \circ f)(x) = g(f(x)) = g(x^2) = 2x^2$. So $f \circ g \neq g \circ f$.

EXERCISES

1. (a) When $x = 2$, $y \approx 2.7$. Thus, $f(2) \approx 2.7$.

(b) $f(x) = 3 \ \Rightarrow \ x \approx 2.3, 5.6$

(c) The domain of f is $-6 \le x \le 6$, or $[-6, 6]$.

(d) The range of f is $-4 \le y \le 4$, or $[-4, 4]$.

(e) f is increasing on $[-4, 4]$, that is, on $-4 \le x \le 4$.

(f) f is odd since its graph is symmetric about the origin.

2. (a) This curve *is not* the graph of a function of x since it *fails* the Vertical Line Test.

(b) This curve *is* the graph of a function of x since it *passes* the Vertical Line Test. The domain is $[-3, 3]$ and the range is $[-2, 3]$.

3. $f(x) = x^2 - 2x + 3$, so $f(a + h) = (a + h)^2 - 2(a + h) + 3 = a^2 + 2ah + h^2 - 2a - 2h + 3$, and
$$\frac{f(a + h) - f(a)}{h} = \frac{(a^2 + 2ah + h^2 - 2a - 2h + 3) - (a^2 - 2a + 3)}{h} = \frac{h(2a + h - 2)}{h} = 2a + h - 2.$$

4. There will be some yield with no fertilizer, increasing yields with increasing fertilizer use, a leveling-off of yields at some point, and disaster with too much fertilizer use.

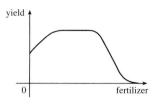

5. $f(x) = 2/(3x - 1)$. Domain: $3x - 1 \neq 0 \Rightarrow 3x \neq 1 \Rightarrow x \neq \frac{1}{3}$. $D = \left(-\infty, \frac{1}{3}\right) \cup \left(\frac{1}{3}, \infty\right)$

Range: all reals except 0 ($y = 0$ is the horizontal asymptote for f.) $R = (-\infty, 0) \cup (0, \infty)$

6. $g(x) = \sqrt{16 - x^4}$. Domain: $16 - x^4 \geq 0 \Rightarrow x^4 \leq 16 \Rightarrow |x| \leq \sqrt[4]{16} \Rightarrow |x| \leq 2$. $D = [-2, 2]$

Range: $y \geq 0$ and $y \leq \sqrt{16} \Rightarrow 0 \leq y \leq 4$. $R = [0, 4]$

7. $y = 1 + \sin x$. Domain: $\mathbb{R}$.

Range: $-1 \leq \sin x \leq 1 \Rightarrow 0 \leq 1 + \sin x \leq 2 \Rightarrow 0 \leq y \leq 2$. $R = [0, 2]$

8. $y = F(t) = 3 + \cos 2t$. Domain: $\mathbb{R}$. $D = (-\infty, \infty)$

Range: $-1 \leq \cos 2t \leq 1 \Rightarrow 2 \leq 3 + \cos 2t \leq 4 \Rightarrow 2 \leq y \leq 4$. $R = [2, 4]$

9. (a) To obtain the graph of $y = f(x) + 8$, we shift the graph of $y = f(x)$ up 8 units.

(b) To obtain the graph of $y = f(x + 8)$, we shift the graph of $y = f(x)$ left 8 units.

(c) To obtain the graph of $y = 1 + 2f(x)$, we stretch the graph of $y = f(x)$ vertically by a factor of 2, and then shift the resulting graph 1 unit upward.

(d) To obtain the graph of $y = f(x - 2) - 2$, we shift the graph of $y = f(x)$ right 2 units (for the "-2" inside the parentheses), and then shift the resulting graph 2 units downward.

(e) To obtain the graph of $y = -f(x)$, we reflect the graph of $y = f(x)$ about the x-axis.

(f) To obtain the graph of $y = 3 - f(x)$, we reflect the graph of $y = f(x)$ about the x-axis, and then shift the resulting graph 3 units upward.

10. (a) To obtain the graph of $y = f(x - 8)$, we shift the graph of $y = f(x)$ right 8 units.

(b) To obtain the graph of $y = -f(x)$, we reflect the graph of $y = f(x)$ about the x-axis.

(c) To obtain the graph of $y = 2 - f(x)$, we reflect the graph of $y = f(x)$ about the x-axis, and then shift the resulting graph 2 units upward.

(d) To obtain the graph of $y = \frac{1}{2}f(x) - 1$, we shrink the graph of $y = f(x)$ by a factor of 2, and then shift the resulting graph 1 unit downward.

11. $y = -\sin 2x$: Start with the graph of $y = \sin x$, compress horizontally by a factor of 2, and reflect about the x-axis.

12. $y = (x-2)^2$: Start with the graph of

$y = x^2$ and shift 2 units to the right.

13. $y = 1 + \frac{1}{2}x^3$: Start with the

graph of $y = x^3$, compress

vertically by a factor of 2, and

shift 1 unit upward.

14. $y = 2 - \sqrt{x}$:

Start with the graph of $y = \sqrt{x}$,

reflect about the x-axis, and shift

2 units upward.

15. $f(x) = \dfrac{1}{x+2}$:

Start with the graph of $f(x) = 1/x$

and shift 2 units to the left.

16. $f(x) = \begin{cases} 1+x & \text{if } x < 0 \\ 1+x^2 & \text{if } x \ge 0 \end{cases}$

On $(-\infty, 0)$, graph $y = 1 + x$ (the line with slope 1 and y-intercept 1)

with open endpoint $(0, 1)$.

On $[0, \infty)$, graph $y = 1 + x^2$ (the rightmost half of the parabola $y = x^2$

shifted 1 unit upward) with closed endpoint $(0, 1)$.

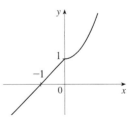

17. (a) The terms of f are a mixture of odd and even powers of x, so f is neither even nor odd.

(b) The terms of f are all odd powers of x, so f is odd.

(c) $f(-x) = \cos\big((-x)^2\big) = \cos(x^2) = f(x)$, so f is even.

(d) $f(-x) = 1 + \sin(-x) = 1 - \sin x$. Now $f(-x) \ne f(x)$ and $f(-x) \ne -f(x)$, so f is neither even nor odd.

18. For the line segment from $(-2, 2)$ to $(-1, 0)$, the slope is $\dfrac{0 - 2}{-1 + 2} = -2$, and an equation is $y - 0 = -2(x + 1)$ or,

equivalently, $y = -2x - 2$. The circle has equation $x^2 + y^2 = 1$; the top half has equation $y = \sqrt{1 - x^2}$ (we have solved for

positive y). Thus, $f(x) = \begin{cases} -2x - 2 & \text{if } -2 \le x \le -1 \\ \sqrt{1 - x^2} & \text{if } -1 < x \le 1 \end{cases}$

19. $f(x) = \sqrt{x}, \quad D = [0, \infty); \quad g(x) = \sin x, \quad D = \mathbb{R}$.

    (a) $(f \circ g)(x) = f(g(x)) = f(\sin x) = \sqrt{\sin x}$. For $\sqrt{\sin x}$ to be defined, we must have $\sin x \ge 0$ ⟺

        $x \in [0, \pi], [2\pi, 3\pi], [-2\pi, -\pi], [4\pi, 5\pi], [-4\pi, -3\pi], \dots$, so $D = \{x \mid x \in [2n\pi, \pi + 2n\pi]\,\}$, where n is an integer}.

    (b) $(g \circ f)(x) = g(f(x)) = g(\sqrt{x}) = \sin \sqrt{x}$. x must be greater than or equal to 0 for $\sqrt{x}$ to be defined, so $D = [0, \infty)$.

    (c) $(f \circ f)(x) = f(f(x)) = f(\sqrt{x}) = \sqrt{\sqrt{x}} = \sqrt[4]{x}. \quad D = [0, \infty)$.

    (d) $(g \circ g)(x) = g(g(x)) = g(\sin x) = \sin(\sin x). \quad D = \mathbb{R}$.

20. Let $h(x) = x + \sqrt{x}$, $g(x) = \sqrt{x}$, and $f(x) = 1/x$. Then $(f \circ g \circ h)(x) = \dfrac{1}{\sqrt{x + \sqrt{x}}} = F(x)$.

21. The graphs of $f(x) = \sin^n x$, where n is a positive integer, all have domain $\mathbb{R}$. For odd n, the range is $[-1, 1]$ and for even n,

the range is $[0, 1]$. For odd n, the functions are odd and symmetric with respect to the origin. For even n, the functions are

even and symmetric with respect to the y-axis. As n becomes large, the graphs become less rounded and more "spiky."

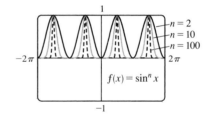

22. (a) Let x denote the number of toaster ovens produced in one week and

    y the associated cost. Using the points $(1000, 9000)$ and

    $(1500, 12{,}000)$, we get an equation of a line:

    $y - 9000 = \dfrac{12{,}000 - 9000}{1500 - 1000}(x - 1000) \quad \Rightarrow$

    $y = 6(x - 1000) + 9000 \quad \Rightarrow \quad y = 6x + 3000$.

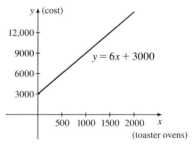

    (b) The slope of 6 means that each additional toaster oven produced adds \$6 to the weekly production cost.

    (c) The y-intercept of 3000 represents the overhead cost—the cost incurred without producing anything.

23.

Many models appear to be plausible. Your choice depends on whether you think medical advances will keep increasing life expectancy, or if there is bound to be a natural leveling-off of life expectancy. A linear model, $y = 0.2493x - 423.4818$, gives us an estimate of 77.6 years for the year 2010.

◻ PRINCIPLES OF PROBLEM SOLVING

1.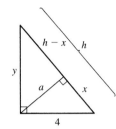

By using the area formula for a triangle, $\frac{1}{2}$ (base) (height), in two ways, we see that

$\frac{1}{2}\,(4)\,(y) = \frac{1}{2}\,(h)\,(a)$, so $a = \dfrac{4y}{h}$. Since $4^2 + y^2 = h^2$, $y = \sqrt{h^2 - 16}$, and

$$a = \frac{4\sqrt{h^2 - 16}}{h}.$$

2.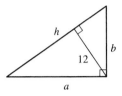

Refer to Example 1, where we obtained $h = \dfrac{P^2 - 100}{2P}$. The 100 came from

4 times the area of the triangle. In this case, the area of the triangle is

$\frac{1}{2}(h)(12) = 6h$. Thus, $h = \dfrac{P^2 - 4\,(6h)}{2P} \quad \Rightarrow \quad 2Ph = P^2 - 24h \quad \Rightarrow$

$2Ph + 24h = P^2 \quad \Rightarrow \quad h\,(2P + 24) = P^2 \quad \Rightarrow \quad h = \dfrac{P^2}{2P + 24}.$

3. $|2x - 1| = \begin{cases} 2x - 1 & \text{if } x \ge \frac{1}{2} \\ 1 - 2x & \text{if } x < \frac{1}{2} \end{cases}$ and $|x + 5| = \begin{cases} x + 5 & \text{if } x \ge -5 \\ -x - 5 & \text{if } x < -5 \end{cases}$

Therefore, we consider the three cases $x < -5$, $-5 \le x < \frac{1}{2}$, and $x \ge \frac{1}{2}$.

If $x < -5$, we must have $1 - 2x - (-x - 5) = 3 \quad \Leftrightarrow \quad x = 3$, which is false, since we are considering $x < -5$.

If $-5 \le x < \frac{1}{2}$, we must have $1 - 2x - (x + 5) = 3 \quad \Leftrightarrow \quad x = -\frac{7}{3}$.

If $x \ge \frac{1}{2}$, we must have $2x - 1 - (x + 5) = 3 \quad \Leftrightarrow \quad x = 9$.

So the two solutions of the equation are $x = -\frac{7}{3}$ and $x = 9$.

4. $|x - 1| = \begin{cases} x - 1 & \text{if } x \ge 1 \\ 1 - x & \text{if } x < 1 \end{cases}$ and $|x - 3| = \begin{cases} x - 3 & \text{if } x \ge 3 \\ 3 - x & \text{if } x < 3 \end{cases}$

Therefore, we consider the three cases $x < 1$, $1 \le x < 3$, and $x \ge 3$.

If $x < 1$, we must have $1 - x - (3 - x) \ge 5 \quad \Leftrightarrow \quad 0 \ge 7$, which is false.

If $1 \le x < 3$, we must have $x - 1 - (3 - x) \ge 5 \quad \Leftrightarrow \quad x \ge \frac{9}{2}$, which is false because $x < 3$.

If $x \ge 3$, we must have $x - 1 - (x - 3) \ge 5 \quad \Leftrightarrow \quad 2 \ge 5$, which is false.

All three cases lead to falsehoods, so the inequality has no solution.

5. $f(x) = |x^2 - 4|x| + 3|$. If $x \geq 0$, then $f(x) = |x^2 - 4x + 3| = |(x - 1)(x - 3)|$.

    Case (i):   If $0 < x \leq 1$, then $f(x) = x^2 - 4x + 3$.

    Case (ii): If $1 < x \leq 3$, then $f(x) = -(x^2 - 4x + 3) = -x^2 + 4x - 3$.

    Case (iii): If $x > 3$, then $f(x) = x^2 - 4x + 3$.

This enables us to sketch the graph for $x \geq 0$. Then we use the fact that f is an even

function to reflect this part of the graph about the y-axis to obtain the entire graph. Or, we

could consider also the cases $x < -3$, $-3 \leq x < -1$, and $-1 \leq x < 0$.

6. $g(x) = |x^2 - 1| - |x^2 - 4|$.

$$|x^2 - 1| = \begin{cases} x^2 - 1 & \text{if } |x| \geq 1 \\ 1 - x^2 & \text{if } |x| < 1 \end{cases} \quad \text{and} \quad |x^2 - 4| = \begin{cases} x^2 - 4 & \text{if } |x| \geq 2 \\ 4 - x^2 & \text{if } |x| < 2 \end{cases}$$

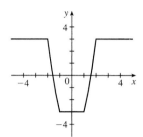

So for $0 \leq |x| < 1$, $g(x) = 1 - x^2 - (4 - x^2) = -3$, for

$1 \leq |x| < 2$, $g(x) = x^2 - 1 - (4 - x^2) = 2x^2 - 5$, and for

$|x| \geq 2$, $g(x) = x^2 - 1 - (x^2 - 4) = 3$.

7. Remember that $|a| = a$ if $a \geq 0$ and that $|a| = -a$ if $a < 0$. Thus,

$$x + |x| = \begin{cases} 2x & \text{if } x \geq 0 \\ 0 & \text{if } x < 0 \end{cases} \quad \text{and} \quad y + |y| = \begin{cases} 2y & \text{if } y \geq 0 \\ 0 & \text{if } y < 0 \end{cases}$$

We will consider the equation $x + |x| = y + |y|$ in four cases.

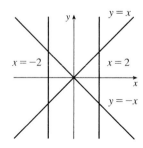

(1) $x \geq 0, y \geq 0$	(2) $x \geq 0, y < 0$	(3) $x < 0, y \geq 0$	(4) $x < 0, y < 0$
$2x = 2y$	$2x = 0$	$0 = 2y$	$0 = 0$
$x = y$	$x = 0$	$0 = y$	

Case 1 gives us the line $y = x$ with nonnegative x and y.

Case 2 gives us the portion of the y-axis with y negative.

Case 3 gives us the portion of the x-axis with x negative.

Case 4 gives us the entire third quadrant.

8. $x^4 - 4x^2 - x^2y^2 + 4y^2 = 0 \quad \Leftrightarrow \quad x^2(x^2 - 4) - y^2(x^2 - 4) = 0 \quad \Leftrightarrow$

$(x^2 - y^2)(x^2 - 4) = 0 \quad \Leftrightarrow \quad (x + y)(x - y)(x + 2)(x - 2) = 0$.

So the graph of the equation consists of the graphs of the four lines $y = -x$,

$y = x$, $x = -2$, and $x = 2$.

9. $|x| + |y| \leq 1$. The boundary of the region has equation $|x| + |y| = 1$. In quadrants

I, II, III, and IV, this becomes the lines $x + y = 1$, $-x + y = 1$, $-x - y = 1$, and

$x - y = 1$ respectively.

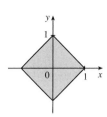

10. $|x - y| + |x| - |y| \leq 2$

Case (i):	$x > y > 0$	$\Leftrightarrow$	$x - y + x - y \leq 2$	$\Leftrightarrow$	$x - y \leq 1$	$\Leftrightarrow$	$y \geq x - 1$
Case (ii):	$y > x > 0$	$\Leftrightarrow$	$y - x + x - y \leq 2$	$\Leftrightarrow$	$0 \leq 2$ (true)		
Case (iii):	$x > 0$ and $y < 0$	$\Leftrightarrow$	$x - y + x + y \leq 2$	$\Leftrightarrow$	$2x \leq 2$	$\Leftrightarrow$	$x \leq 1$
Case (iv):	$x < 0$ and $y > 0$	$\Leftrightarrow$	$y - x - x - y \leq 2$	$\Leftrightarrow$	$-2x \leq 2$	$\Leftrightarrow$	$x \geq -1$
Case (v):	$y < x < 0$	$\Leftrightarrow$	$x - y - x + y \leq 2$	$\Leftrightarrow$	$0 \leq 2$ (true)		
Case (vi):	$x < y < 0$	$\Leftrightarrow$	$y - x - x + y \leq 2$	$\Leftrightarrow$	$y - x \leq 1$	$\Leftrightarrow$	$y \leq x + 1$

Note: Instead of considering cases (iv), (v), and (vi), we could have noted that
the region is unchanged if x and y are replaced by $-x$ and $-y$, so the region is
symmetric about the origin. Therefore, we need only draw cases (i), (ii), and
(iii), and rotate through $180°$ about the origin.

11. Let d be the distance traveled on each half of the trip. Let t_1 and t_2 be the times taken for the first and second halves of the trip.
For the first half of the trip we have $t_1 = d/30$ and for the second half we have $t_2 = d/60$. Thus, the average speed for the

entire trip is $\dfrac{\text{total distance}}{\text{total time}} = \dfrac{2d}{t_1 + t_2} = \dfrac{2d}{\dfrac{d}{30} + \dfrac{d}{60}} \cdot \dfrac{60}{60} = \dfrac{120d}{2d + d} = \dfrac{120d}{3d} = 40$. The average speed for the entire trip

is 40 mi/h.

12. Let $f = \sin$, $g = x$, and $h = x$. Then the left-hand side of the equation is $f \circ (g + h) = \sin(x + x) = \sin 2x = 2 \sin x \cos x$;
and the right-hand side is $f \circ g + f \circ h = \sin x + \sin x = 2 \sin x$. The two sides are not equal, so the given statement is false.

13. Let S_n be the statement that $7^n - 1$ is divisible by 6.
- S_1 is true because $7^1 - 1 = 6$ is divisible by 6.
- Assume S_k is true, that is, $7^k - 1$ is divisible by 6. In other words, $7^k - 1 = 6m$ for some positive integer m. Then
  $7^{k+1} - 1 = 7^k \cdot 7 - 1 = (6m + 1) \cdot 7 - 1 = 42m + 6 = 6(7m + 1)$, which is divisible by 6, so S_{k+1} is true.
- Therefore, by mathematical induction, $7^n - 1$ is divisible by 6 for every positive integer n.

14. Let S_n be the statement that $1 + 3 + 5 + \cdots + (2n - 1) = n^2$.
- S_1 is true because $[2(1) - 1] = 1 = 1^2$.
- Assume S_k is true, that is, $1 + 3 + 5 + \cdots + (2k - 1) = k^2$. Then
  $1 + 3 + 5 + \cdots + (2k - 1) + [2(k + 1) - 1] = 1 + 3 + 5 + \cdots + (2k - 1) + (2k + 1) = k^2 + (2k + 1) = (k + 1)^2$
  which shows that S_{k+1} is true.
- Therefore, by mathematical induction, $1 + 3 + 5 + \cdots + (2n - 1) = n^2$ for every positive integer n.

15. $f_0(x) = x^2$ and $f_{n+1}(x) = f_0(f_n(x))$ for $n = 0, 1, 2, \ldots$.

$f_1(x) = f_0(f_0(x)) = f_0\left(x^2\right) = \left(x^2\right)^2 = x^4$, $f_2(x) = f_0(f_1(x)) = f_0(x^4) = (x^4)^2 = x^8$,

$f_3(x) = f_0(f_2(x)) = f_0(x^8) = (x^8)^2 = x^{16}, \ldots$. Thus, a general formula is $f_n(x) = x^{2^{n+1}}$.

16. (a) $f_0(x) = 1/(2 - x)$ and $f_{n+1} = f_0 \circ f_n$ for $n = 0, 1, 2, \ldots$.

$$f_1(x) = f_0\left(\frac{1}{2-x}\right) = \frac{1}{2 - \dfrac{1}{2-x}} = \frac{2-x}{2(2-x)-1} = \frac{2-x}{3-2x},$$

$$f_2(x) = f_0\left(\frac{2-x}{3-2x}\right) = \frac{1}{2 - \dfrac{2-x}{3-2x}} = \frac{3-2x}{2(3-2x)-(2-x)} = \frac{3-2x}{4-3x},$$

$$f_3(x) = f_0\left(\frac{3-2x}{4-3x}\right) = \frac{1}{2 - \dfrac{3-2x}{4-3x}} = \frac{4-3x}{2(4-3x)-(3-2x)} = \frac{4-3x}{5-4x}, \ldots$$

Thus, we conjecture that the general formula is $f_n(x) = \dfrac{n+1-nx}{n+2-(n+1)x}$.

To prove this, we use the Principle of Mathematical Induction. We have already verified that f_n is true for $n = 1$.

Assume that the formula is true for $n = k$; that is, $f_k(x) = \dfrac{k+1-kx}{k+2-(k+1)x}$. Then

$$f_{k+1}(x) = (f_0 \circ f_k)(x) = f_0(f_k(x)) = f_0\left(\frac{k+1-kx}{k+2-(k+1)x}\right) = \frac{1}{2 - \dfrac{k+1-kx}{k+2-(k+1)x}}$$

$$= \frac{k+2-(k+1)x}{2\left[k+2-(k+1)x\right]-(k+1-kx)} = \frac{k+2-(k+1)x}{k+3-(k+2)x}$$

This shows that the formula for f_n is true for $n = k + 1$. Therefore, by mathematical induction, the formula is true for all positive integers n.

(b) From the graph, we can make several observations:

- The values at each fixed $x = a$ keep increasing as n increases.

- The vertical asymptote gets closer to $x = 1$ as n increases.

- The horizontal asymptote gets closer to $y = 1$ as n increases.

- The x-intercept for f_{n+1} is the value of the vertical asymptote for f_n.

- The y-intercept for f_n is the value of the horizontal asymptote for f_{n+1}.

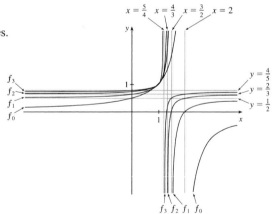

2 □ LIMITS

2.1 The Tangent and Velocity Problems

1. (a) Using $P(15, 250)$, we construct the following table:

t	Q	slope $= m_{PQ}$
5	$(5, 694)$	$\frac{694-250}{5-15} = -\frac{444}{10} = -44.4$
10	$(10, 444)$	$\frac{444-250}{10-15} = -\frac{194}{5} = -38.8$
20	$(20, 111)$	$\frac{111-250}{20-15} = -\frac{139}{5} = -27.8$
25	$(25, 28)$	$\frac{28-250}{25-15} = -\frac{222}{10} = -22.2$
30	$(30, 0)$	$\frac{0-250}{30-15} = -\frac{250}{15} = -16.\overline{6}$

(b) Using the values of t that correspond to the points closest to P ($t = 10$ and $t = 20$), we have

$$\frac{-38.8 + (-27.8)}{2} = -33.3$$

(c) From the graph, we can estimate the slope of the tangent line at P to be $\frac{-300}{9} = -33.\overline{3}$.

2. (a) Slope $= \frac{2948 - 2530}{42 - 36} = \frac{418}{6} \approx 69.67$

(b) Slope $= \frac{2948 - 2661}{42 - 38} = \frac{287}{4} = 71.75$

(c) Slope $= \frac{2948 - 2806}{42 - 40} = \frac{142}{2} = 71$

(d) Slope $= \frac{3080 - 2948}{44 - 42} = \frac{132}{2} = 66$

From the data, we see that the patient's heart rate is decreasing from 71 to 66 heartbeats/minute after 42 minutes. After being stable for a while, the patient's heart rate is dropping.

3. (a)

	x	Q	m_{PQ}
(i)	0.5	$(0.5, 0.333333)$	0.333333
(ii)	0.9	$(0.9, 0.473684)$	0.263158
(iii)	0.99	$(0.99, 0.497487)$	0.251256
(iv)	0.999	$(0.999, 0.499750)$	0.250125
(v)	1.5	$(1.5, 0.6)$	0.2
(vi)	1.1	$(1.1, 0.523810)$	0.238095
(vii)	1.01	$(1.01, 0.502488)$	0.248756
(viii)	1.001	$(1.001, 0.500250)$	0.249875

(b) The slope appears to be $\frac{1}{4}$.

(c) $y - \frac{1}{2} = \frac{1}{4}(x - 1)$ or $y = \frac{1}{4}x + \frac{1}{4}$.

4. (a)

	x	Q	m_{PQ}
(i)	2.5	$(2.5, 0.707107)$	0.585786
(ii)	2.9	$(2.9, 0.948683)$	0.513167
(iii)	2.99	$(2.99, 0.994987)$	0.501256
(iv)	2.999	$(2.999, 0.999500)$	0.500125
(v)	3.5	$(3.5, 1.224745)$	0.449490
(vi)	3.1	$(3.1, 1.048809)$	0.488088
(vii)	3.01	$(3.01, 1.004988)$	0.498756
(viii)	3.001	$(3.001, 1.000500)$	0.499875

(b) The slope appears to be $\frac{1}{2}$.

(c) $y - 1 = \frac{1}{2}(x - 3)$ or $y = \frac{1}{2}x - \frac{1}{2}$.

(d)

5. (a) $y = y(t) = 40t - 16t^2$. At $t = 2$, $y = 40(2) - 16(2)^2 = 16$. The average velocity between times 2 and $2 + h$ is
$$v_{\text{ave}} = \frac{y(2 + h) - y(2)}{(2 + h) - 2} = \frac{\left[40(2 + h) - 16(2 + h)^2\right] - 16}{h} = \frac{-24h - 16h^2}{h} = -24 - 16h, \text{ if } h \neq 0.$$

  (i) $[2, 2.5]$: $h = 0.5$, $v_{\text{ave}} = -32$ ft/s   (ii) $[2, 2.1]$: $h = 0.1$, $v_{\text{ave}} = -25.6$ ft/s

  (iii) $[2, 2.05]$: $h = 0.05$, $v_{\text{ave}} = -24.8$ ft/s   (iv) $[2, 2.01]$: $h = 0.01$, $v_{\text{ave}} = -24.16$ ft/s

(b) The instantaneous velocity when $t = 2$ (h approaches 0) is -24 ft/s.

6. (a) $y = y(t) = 10t - 1.86t^2$. At $t = 1$, $y = 10(1) - 1.86(1)^2 = 8.14$. The average velocity between times 1 and $1 + h$ is
$$v_{\text{ave}} = \frac{y(1 + h) - y(1)}{(1 + h) - 1} = \frac{\left[10(1 + h) - 1.86(1 + h)^2\right] - 8.14}{h} = \frac{6.28h - 1.86h^2}{h} = 6.28 - 1.86h, \text{ if } h \neq 0.$$

  (i) $[1, 2]$: $h = 1$, $v_{\text{ave}} = 4.42$ m/s   (ii) $[1, 1.5]$: $h = 0.5$, $v_{\text{ave}} = 5.35$ m/s

  (iii) $[1, 1.1]$: $h = 0.1$, $v_{\text{ave}} = 6.094$ m/s   (iv) $[1, 1.01]$: $h = 0.01$, $v_{\text{ave}} = 6.2614$ m/s

  (v) $[1, 1.001]$: $h = 0.001$, $v_{\text{ave}} = 6.27814$ m/s

(b) The instantaneous velocity when $t = 1$ (h approaches 0) is 6.28 m/s.

7. (a)   (i) On the interval $[1, 3]$, $v_{\text{ave}} = \dfrac{s(3) - s(1)}{3 - 1} = \dfrac{10.7 - 1.4}{2} = \dfrac{9.3}{2} = 4.65$ m/s.

   (ii) On the interval $[2, 3]$, $v_{\text{ave}} = \dfrac{s(3) - s(2)}{3 - 2} = \dfrac{10.7 - 5.1}{1} = 5.6$ m/s.

   (iii) On the interval $[3, 5]$, $v_{\text{ave}} = \dfrac{s(5) - s(3)}{5 - 3} = \dfrac{25.8 - 10.7}{2} = \dfrac{15.1}{2} = 7.55$ m/s.

   (iv) On the interval $[3, 4]$, $v_{\text{ave}} = \dfrac{s(4) - s(3)}{4 - 3} = \dfrac{17.7 - 10.7}{1} = 7$ m/s.

(b)

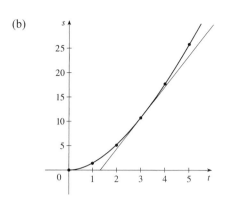

Using the points $(2, 4)$ and $(5, 23)$ from the approximate tangent

line, the instantaneous velocity at $t = 3$ is about $\dfrac{23 - 4}{5 - 2} \approx 6.3 \text{ m/s}$.

8. (a) (i) $s = s(t) = 2 \sin \pi t + 3 \cos \pi t$. On the interval $[1, 2]$, $v_{\text{ave}} = \dfrac{s(2) - s(1)}{2 - 1} = \dfrac{3 - (-3)}{1} = 6 \text{ cm/s}$.

(ii) On the interval $[1, 1.1]$, $v_{\text{ave}} = \dfrac{s(1.1) - s(1)}{1.1 - 1} \approx \dfrac{-3.471 - (-3)}{0.1} = -4.71 \text{ cm/s}$.

(iii) On the interval $[1, 1.01]$, $v_{\text{ave}} = \dfrac{s(1.01) - s(1)}{1.01 - 1} \approx \dfrac{-3.0613 - (-3)}{0.01} = -6.13 \text{ cm/s}$.

(iv) On the interval $[1, 1.001]$, $v_{\text{ave}} = \dfrac{s(1.001) - s(1)}{1.001 - 1} \approx \dfrac{-3.00627 - (-3)}{1.001 - 1} = -6.27 \text{ cm/s}$.

(b) The instantaneous velocity of the particle when $t = 1$ appears to be about -6.3 cm/s.

9. (a) For the curve $y = \sin(10\pi/x)$ and the point $P(1, 0)$:

x	Q	m_{PQ}
2	$(2, 0)$	0
1.5	$(1.5, 0.8660)$	1.7321
1.4	$(1.4, -0.4339)$	-1.0847
1.3	$(1.3, -0.8230)$	-2.7433
1.2	$(1.2, 0.8660)$	4.3301
1.1	$(1.1, -0.2817)$	-2.8173

x	Q	m_{PQ}
0.5	$(0.5, 0)$	0
0.6	$(0.6, 0.8660)$	-2.1651
0.7	$(0.7, 0.7818)$	-2.6061
0.8	$(0.8, 1)$	-5
0.9	$(0.9, -0.3420)$	3.4202

As x approaches 1, the slopes do not appear to be approaching any particular value.

(b)

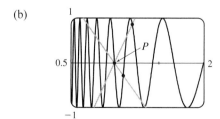

We see that problems with estimation are caused by the frequent oscillations of the graph. The tangent is so steep at P that we need to take x-values much closer to 1 in order to get accurate estimates of its slope.

(c) If we choose $x = 1.001$, then the point Q is $(1.001, -0.0314)$ and $m_{PQ} \approx -31.3794$. If $x = 0.999$, then Q is $(0.999, 0.0314)$ and $m_{PQ} = -31.4422$. The average of these slopes is -31.4108. So we estimate that the slope of the tangent line at P is about -31.4.

2.2 The Limit of a Function

1. As x approaches 2, $f(x)$ approaches 5. [Or, the values of $f(x)$ can be made as close to 5 as we like by taking x sufficiently close to 2 (but $x \neq 2$).] Yes, the graph could have a hole at $(2, 5)$ and be defined such that $f(2) = 3$.

2. As x approaches 1 from the left, $f(x)$ approaches 3; and as x approaches 1 from the right, $f(x)$ approaches 7. No, the limit does not exist because the left- and right-hand limits are different.

3. (a) $\lim\limits_{x \to -3} f(x) = \infty$ means that the values of $f(x)$ can be made arbitrarily large (as large as we please) by taking x sufficiently close to -3 (but not equal to -3).

   (b) $\lim\limits_{x \to 4^+} f(x) = -\infty$ means that the values of $f(x)$ can be made arbitrarily large negative by taking x sufficiently close to 4 through values larger than 4.

4. (a) $\lim\limits_{x \to 0} f(x) = 3$ \qquad (b) $\lim\limits_{x \to 3^-} f(x) = 4$ \qquad (c) $\lim\limits_{x \to 3^+} f(x) = 2$

   (d) $\lim\limits_{x \to 3} f(x)$ does not exist because the limits in part (b) and part (c) are not equal.

   (e) $f(3) = 3$

5. (a) $f(x)$ approaches 2 as x approaches 1 from the left, so $\lim\limits_{x \to 1^-} f(x) = 2$.

   (b) $f(x)$ approaches 3 as x approaches 1 from the right, so $\lim\limits_{x \to 1^+} f(x) = 3$.

   (c) $\lim\limits_{x \to 1} f(x)$ does not exist because the limits in part (a) and part (b) are not equal.

   (d) $f(x)$ approaches 4 as x approaches 5 from the left and from the right, so $\lim\limits_{x \to 5} f(x) = 4$.

   (e) $f(5)$ is not defined, so it doesn't exist.

6. (a) $h(x)$ approaches 4 as x approaches -3 from the left, so $\lim\limits_{x \to -3^-} h(x) = 4$.

   (b) $h(x)$ approaches 4 as x approaches -3 from the right, so $\lim\limits_{x \to -3^+} h(x) = 4$.

   (c) $\lim\limits_{x \to -3} h(x) = 4$ because the limits in part (a) and part (b) are equal.

   (d) $h(-3)$ is not defined, so it doesn't exist.

   (e) $h(x)$ approaches 1 as x approaches 0 from the left, so $\lim\limits_{x \to 0^-} h(x) = 1$.

   (f) $h(x)$ approaches -1 as x approaches 0 from the right, so $\lim\limits_{x \to 0^+} h(x) = -1$.

   (g) $\lim\limits_{x \to 0} h(x)$ does not exist because the limits in part (e) and part (f) are not equal.

   (h) $h(0) = 1$ since the point $(0, 1)$ is on the graph of h.

(i) Since $\lim\limits_{x \to 2^-} h(x) = 2$ and $\lim\limits_{x \to 2^+} h(x) = 2$, we have $\lim\limits_{x \to 2} h(x) = 2$.

(j) $h(2)$ is not defined, so it doesn't exist.

(k) $h(x)$ approaches 3 as x approaches 5 from the right, so $\lim\limits_{x \to 5^+} h(x) = 3$.

(l) $h(x)$ does not approach any one number as x approaches 5 from the left, so $\lim\limits_{x \to 5^-} h(x)$ does not exist.

7. (a) $\lim\limits_{t \to 0^-} g(t) = -1$ (b) $\lim\limits_{t \to 0^+} g(t) = -2$

(c) $\lim\limits_{t \to 0} g(t)$ does not exist because the limits in part (a) and part (b) are not equal.

(d) $\lim\limits_{t \to 2^-} g(t) = 2$ (e) $\lim\limits_{t \to 2^+} g(t) = 0$

(f) $\lim\limits_{t \to 2} g(t)$ does not exist because the limits in part (d) and part (e) are not equal.

(g) $g(2) = 1$ (h) $\lim\limits_{t \to 4} g(t) = 3$

8. (a) $\lim\limits_{x \to 2} R(x) = -\infty$ (b) $\lim\limits_{x \to 5} R(x) = \infty$ (c) $\lim\limits_{x \to -3^-} R(x) = -\infty$

(d) $\lim\limits_{x \to -3^+} R(x) = \infty$

(e) The equations of the vertical asymptotes are $x = -3$, $x = 2$, and $x = 5$.

9. (a) $\lim\limits_{x \to -7} f(x) = -\infty$ (b) $\lim\limits_{x \to -3} f(x) = \infty$ (c) $\lim\limits_{x \to 0} f(x) = \infty$

(d) $\lim\limits_{x \to 6^-} f(x) = -\infty$ (e) $\lim\limits_{x \to 6^+} f(x) = \infty$

(f) The equations of the vertical asymptotes are $x = -7$, $x = -3$, $x = 0$, and $x = 6$.

10. $\lim\limits_{t \to 12^-} f(t) = 150$ mg and $\lim\limits_{t \to 12^+} f(t) = 300$ mg. These limits show that there is an abrupt change in the amount of drug in

the patient's bloodstream at $t = 12$ h. The left-hand limit represents the amount of the drug just before the fourth injection.

The right-hand limit represents the amount of the drug just after the fourth injection.

11. (a) $\lim\limits_{x \to 0^-} f(x) = 1$

(b) $\lim\limits_{x \to 0^+} f(x) = 0$

(c) $\lim\limits_{x \to 0} f(x)$ does not exist because the limits in

part (a) and part (b) are not equal.

12. $\lim\limits_{x \to a} f(x)$ exists for all a except $a = \pm 1$.

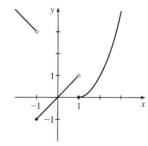

13. $\lim_{x \to 1^-} f(x) = 2, \quad \lim_{x \to 1^+} f(x) = -2, \quad f(1) = 2$

14. $\lim_{x \to 0^-} f(x) = 1, \quad \lim_{x \to 0^+} f(x) = -1, \quad \lim_{x \to 2^-} f(x) = 0,$

$\lim_{x \to 2^+} f(x) = 1, \quad f(2) = 1, \quad f(0)$ is undefined

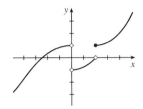

15. $\lim_{x \to 3^+} f(x) = 4, \quad \lim_{x \to 3^-} f(x) = 2, \quad \lim_{x \to -2} f(x) = 2,$

$f(3) = 3, \quad f(-2) = 1$

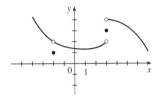

16. $\lim_{x \to 1} f(x) = 3, \quad \lim_{x \to 4^-} f(x) = 3, \quad \lim_{x \to 4^+} f(x) = -3,$

$f(1) = 1, \quad f(4) = -1$

17. For $f(x) = \dfrac{x^2 - 2x}{x^2 - x - 2}$:

x	$f(x)$
2.5	0.714286
2.1	0.677419
2.05	0.672131
2.01	0.667774
2.005	0.667221
2.001	0.666778

x	$f(x)$
1.9	0.655172
1.95	0.661017
1.99	0.665552
1.995	0.666110
1.999	0.666556

It appears that $\lim_{x \to 2} \dfrac{x^2 - 2x}{x^2 - x - 2} = 0.\bar{6} = \frac{2}{3}$.

18. For $f(x) = \dfrac{x^2 - 2x}{x^2 - x - 2}$:

x	$f(x)$
0	0
-0.5	-1
-0.9	-9
-0.95	-19
-0.99	-99
-0.999	-999

x	$f(x)$
-2	2
-1.5	3
-1.1	11
-1.01	101
-1.001	1001

It appears that $\lim_{x \to -1} \dfrac{x^2 - 2x}{x^2 - x - 2}$ *does not exist* since

$f(x) \to \infty$ as $x \to -1^-$ and $f(x) \to -\infty$ as $x \to -1^+$.

19. For $f(x) = \dfrac{\sin x}{x + \tan x}$:

x	$f(x)$
± 1	0.329033
± 0.5	0.458209
± 0.2	0.493331
± 0.1	0.498333
± 0.05	0.499583
± 0.01	0.499983

It appears that $\lim_{x \to 0} \dfrac{\sin x}{x + \tan x} = 0.5 = \dfrac{1}{2}$.

20. For $f(x) = \dfrac{\sqrt{x} - 4}{x - 16}$:

x	$f(x)$
17	0.123106
16.5	0.124038
16.1	0.124805
16.05	0.124902
16.01	0.124980

x	$f(x)$
15	0.127017
15.5	0.125992
15.9	0.125196
15.95	0.125098
15.99	0.125020

It appears that $\lim_{x \to 16} \dfrac{\sqrt{x} - 4}{x - 16} = 0.125 = \dfrac{1}{8}$.

21. For $f(x) = \dfrac{\sqrt{x+4}-2}{x}$:

x	$f(x)$
1	0.236068
0.5	0.242641
0.1	0.248457
0.05	0.249224
0.01	0.249844

x	$f(x)$
-1	0.267949
-0.5	0.258343
-0.1	0.251582
-0.05	0.250786
-0.01	0.250156

It appears that $\lim\limits_{x\to0} \dfrac{\sqrt{x+4}-2}{x} = 0.25 = \frac{1}{4}$.

22. For $f(x) = \dfrac{\tan 3x}{\tan 5x}$:

x	$f(x)$
±0.2	0.439279
±0.1	0.566236
±0.05	0.591893
±0.01	0.599680
±0.001	0.599997

It appears that $\lim\limits_{x\to0} \dfrac{\tan 3x}{\tan 5x} = 0.6 = \frac{3}{5}$.

23. For $f(x) = \dfrac{x^6-1}{x^{10}-1}$:

x	$f(x)$
0.5	0.985337
0.9	0.719397
0.95	0.660186
0.99	0.612018
0.999	0.601200

x	$f(x)$
1.5	0.183369
1.1	0.484119
1.05	0.540783
1.01	0.588022
1.001	0.598800

It appears that $\lim\limits_{x\to1} \dfrac{x^6-1}{x^{10}-1} = 0.6 = \frac{3}{5}$.

24. For $f(x) = \dfrac{9^x-5^x}{x}$:

x	$f(x)$
0.5	1.527864
0.1	0.711120
0.05	0.646496
0.01	0.599082
0.001	0.588906

x	$f(x)$
-0.5	0.227761
-0.1	0.485984
-0.05	0.534447
-0.01	0.576706
-0.001	0.586669

It appears that $\lim\limits_{x\to0} \dfrac{9^x-5^x}{x} = 0.59$. Later we will be able to show that the exact value is $\ln(9/5)$.

25. $\lim\limits_{x\to-3^+} \dfrac{x+2}{x+3} = -\infty$ since the numerator is negative and the denominator approaches 0 from the positive side as $x \to -3^+$.

26. $\lim\limits_{x\to-3^-} \dfrac{x+2}{x+3} = \infty$ since the numerator is negative and the denominator approaches 0 from the negative side as $x \to -3^-$.

27. $\lim\limits_{x\to1} \dfrac{2-x}{(x-1)^2} = \infty$ since the numerator is positive and the denominator approaches 0 through positive values as $x \to 1$.

28. $\lim\limits_{x\to0} \dfrac{x-1}{x^2(x+2)} = -\infty$ since $x^2 \to 0$ as $x \to 0$ and $\dfrac{x-1}{x^2(x+2)} < 0$ for $0 < x < 1$ and for $-2 < x < 0$.

29. $\lim\limits_{x\to-2^+} \dfrac{x-1}{x^2(x+2)} = -\infty$ since $(x+2) \to 0$ as $x \to -2^+$ and $\dfrac{x-1}{x^2(x+2)} < 0$ for $-2 < x < 0$.

30. $\lim\limits_{x\to\pi^-} \cot x = \lim\limits_{x\to\pi^-} \dfrac{\cos x}{\sin x} = -\infty$ since the numerator is negative and the denominator approaches 0 through positive values as $x \to \pi^-$.

31. $\lim\limits_{x\to2\pi^-} x \csc x = \lim\limits_{x\to2\pi^-} \dfrac{x}{\sin x} = -\infty$ since the numerator is positive and the denominator approaches 0 through negative values as $x \to 2\pi^-$.

32. $\lim\limits_{x \to 2^-} \dfrac{x^2 - 2x}{x^2 - 4x + 4} = \lim\limits_{x \to 2^-} \dfrac{x(x-2)}{(x-2)^2} = \lim\limits_{x \to 2^-} \dfrac{x}{x-2} = -\infty$ since the numerator is positive and the denominator

approaches 0 through negative values as $x \to 2^-$.

33. (a) $f(x) = \dfrac{1}{x^3 - 1}$.

From these calculations, it seems that

$\lim\limits_{x \to 1^-} f(x) = -\infty$ and $\lim\limits_{x \to 1^+} f(x) = \infty$.

x	$f(x)$
0.5	-1.14
0.9	-3.69
0.99	-33.7
0.999	-333.7
0.9999	-3333.7
0.99999	$-33{,}333.7$

x	$f(x)$
1.5	0.42
1.1	3.02
1.01	33.0
1.001	333.0
1.0001	3333.0
1.00001	33{,}333.3

(b) If x is slightly smaller than 1, then $x^3 - 1$ will be a negative number close to 0, and the reciprocal of $x^3 - 1$, that is, $f(x)$, will be a negative number with large absolute value. So $\lim\limits_{x \to 1^-} f(x) = -\infty$.

If x is slightly larger than 1, then $x^3 - 1$ will be a small positive number, and its reciprocal, $f(x)$, will be a large positive number. So $\lim\limits_{x \to 1^+} f(x) = \infty$.

(c) It appears from the graph of f that

$\lim\limits_{x \to 1^-} f(x) = -\infty$ and $\lim\limits_{x \to 1^+} f(x) = \infty$.

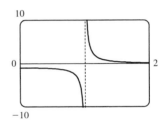

34. (a) The denominator of $y = \dfrac{x^2 + 1}{3x - 2x^2} = \dfrac{x^2 + 1}{x(3 - 2x)}$ is equal to zero when

$x = 0$ and $x = \frac{3}{2}$ (and the numerator is not), so $x = 0$ and $x = 1.5$ are

vertical asymptotes of the function.

(b)

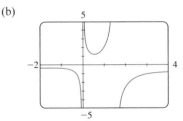

35. (a) Let $h(x) = (1 + x)^{1/x}$.

x	$h(x)$
-0.001	2.71964
-0.0001	2.71842
-0.00001	2.71830
-0.000001	2.71828
0.000001	2.71828
0.00001	2.71827
0.0001	2.71815
0.001	2.71692

(b)

It appears that $\lim\limits_{x \to 0} (1 + x)^{1/x} \approx 2.71828$.

In Chapter 7 we will identify the limit as the famous number e.

36. (a) From the graphs, it seems that $\lim\limits_{x\to 0} \dfrac{\tan 4x}{x} = 4$.

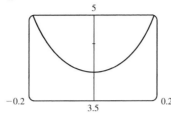

(b)

x	$f(x)$
± 0.1	4.227932
± 0.01	4.002135
± 0.001	4.000021
± 0.0001	4.000000

37. For $f(x) = x^2 - (2^x/1000)$:

(a)

x	$f(x)$
1	0.998000
0.8	0.638259
0.6	0.358484
0.4	0.158680
0.2	0.038851
0.1	0.008928
0.05	0.001465

It appears that $\lim\limits_{x\to 0} f(x) = 0$.

(b)

x	$f(x)$
0.04	0.000572
0.02	−0.000614
0.01	−0.000907
0.005	−0.000978
0.003	−0.000993
0.001	−0.001000

It appears that $\lim\limits_{x\to 0} f(x) = -0.001$.

38. For $h(x) = \dfrac{\tan x - x}{x^3}$:

(a)

x	$h(x)$
1.0	0.55740773
0.5	0.37041992
0.1	0.33467209
0.05	0.33366700
0.01	0.33334667
0.005	0.33333667

(b) It seems that $\lim\limits_{x\to 0} h(x) = \frac{1}{3}$.

(c)

x	$h(x)$
0.001	0.33333350
0.0005	0.33333344
0.0001	0.33333000
0.00005	0.33333600
0.00001	0.33300000
0.000001	0.00000000

Here the values will vary from one calculator to another. Every calculator will eventually give *false values*.

(d) As in part (c), when we take a small enough viewing rectangle we get incorrect output.

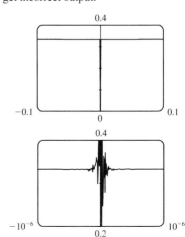

39. No matter how many times we zoom in toward the origin, the graphs of $f(x) = \sin(\pi/x)$ appear to consist of almost-vertical

lines. This indicates more and more frequent oscillations as $x \to 0$.

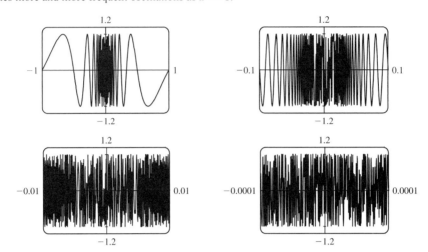

40. $\displaystyle \lim_{v \to c^-} m = \lim_{v \to c^-} \frac{m_0}{\sqrt{1 - v^2/c^2}}$. As $v \to c^-$, $\sqrt{1 - v^2/c^2} \to 0^+$, and $m \to \infty$.

41.

There appear to be vertical asymptotes of the curve $y = \tan(2\sin x)$ at $x \approx \pm 0.90$ and $x \approx \pm 2.24$. To find the exact equations of these asymptotes, we note that the graph of the tangent function has vertical asymptotes at $x = \frac{\pi}{2} + \pi n$. Thus, we must have $2\sin x = \frac{\pi}{2} + \pi n$, or equivalently, $\sin x = \frac{\pi}{4} + \frac{\pi}{2} n$. Since $-1 \le \sin x \le 1$, we must have $\sin x = \pm \frac{\pi}{4}$ and so $x = \pm \sin^{-1} \frac{\pi}{4}$ (corresponding to $x \approx \pm 0.90$). Just as $150°$ is the reference angle for $30°$, $\pi - \sin^{-1} \frac{\pi}{4}$ is the reference angle for $\sin^{-1} \frac{\pi}{4}$. So $x = \pm\left(\pi - \sin^{-1} \frac{\pi}{4}\right)$ are also equations of vertical asymptotes (corresponding to $x \approx \pm 2.24$).

42. (a) Let $y = \dfrac{x^3 - 1}{\sqrt{x} - 1}$.

From the table and the graph, we guess that the limit of y as x approaches 1 is 6.

x	y
0.99	5.92531
0.999	5.99250
0.9999	5.99925
1.01	6.07531
1.001	6.00750
1.0001	6.00075

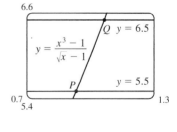

(b) We need to have $5.5 < \dfrac{x^3 - 1}{\sqrt{x} - 1} < 6.5$. From the graph we obtain the approximate points of intersection

$P(0.9313853, 5.5)$ and $Q(1.0649004, 6.5)$. Now $1 - 0.9313853 \approx 0.0686$ and $1.0649004 - 1 \approx 0.0649$, so by

requiring that x be within 0.0649 of 1, we ensure that y is within 0.5 of 6.

2.3   Calculating Limits Using the Limit Laws

1. (a) $\lim\limits_{x\to2}[f(x)+5g(x)] = \lim\limits_{x\to2}f(x) + \lim\limits_{x\to2}[5g(x)]$ [Limit Law 1]

$= \lim\limits_{x\to2}f(x) + 5\lim\limits_{x\to2}g(x)$ [Limit Law 3]

$= 4 + 5(-2) = -6$

(b) $\lim\limits_{x\to2}[g(x)]^3 = \left[\lim\limits_{x\to2}g(x)\right]^3$ [Limit Law 6]

$= (-2)^3 = -8$

(c) $\lim\limits_{x\to2}\sqrt{f(x)} = \sqrt{\lim\limits_{x\to2}f(x)}$ [Limit Law 11]

$= \sqrt{4} = 2$

(d) $\lim\limits_{x\to2}\dfrac{3f(x)}{g(x)} = \dfrac{\lim\limits_{x\to2}[3f(x)]}{\lim\limits_{x\to2}g(x)}$ [Limit Law 5]

$= \dfrac{3\lim\limits_{x\to2}f(x)}{\lim\limits_{x\to2}g(x)}$ [Limit Law 3]

$= \dfrac{3(4)}{-2} = -6$

(e) Because the limit of the denominator is 0, we can't use Limit Law 5. The given limit, $\lim\limits_{x\to2}\dfrac{g(x)}{h(x)}$, does not exist because the denominator approaches 0 while the numerator approaches a nonzero number.

(f) $\lim\limits_{x\to2}\dfrac{g(x)\,h(x)}{f(x)} = \dfrac{\lim\limits_{x\to2}[g(x)\,h(x)]}{\lim\limits_{x\to2}f(x)}$ [Limit Law 5]

$= \dfrac{\lim\limits_{x\to2}g(x)\cdot\lim\limits_{x\to2}h(x)}{\lim\limits_{x\to2}f(x)}$ [Limit Law 4]

$= \dfrac{-2\cdot0}{4} = 0$

2. (a) $\lim\limits_{x\to2}[f(x)+g(x)] = \lim\limits_{x\to2}f(x) + \lim\limits_{x\to2}g(x) = 2+0 = 2$

(b) $\lim\limits_{x\to1}g(x)$ does not exist since its left- and right-hand limits are not equal, so the given limit does not exist.

(c) $\lim\limits_{x\to0}[f(x)g(x)] = \lim\limits_{x\to0}f(x)\cdot\lim\limits_{x\to0}g(x) = 0\cdot1.3 = 0$

(d) Since $\lim\limits_{x\to-1}g(x) = 0$ and g is in the denominator, but $\lim\limits_{x\to-1}f(x) = -1 \neq 0$, the given limit does not exist.

(e) $\lim\limits_{x\to2}x^3f(x) = \left[\lim\limits_{x\to2}x^3\right]\left[\lim\limits_{x\to2}f(x)\right] = 2^3\cdot2 = 16$

(f) $\lim\limits_{x\to1}\sqrt{3+f(x)} = \sqrt{3+\lim\limits_{x\to1}f(x)} = \sqrt{3+1} = 2$

3. $\lim\limits_{x \to -2} (3x^4 + 2x^2 - x + 1) = \lim\limits_{x \to -2} 3x^4 + \lim\limits_{x \to -2} 2x^2 - \lim\limits_{x \to -2} x + \lim\limits_{x \to -2} 1$ [Limit Laws 1 and 2]

$$= 3 \lim_{x \to -2} x^4 + 2 \lim_{x \to -2} x^2 - \lim_{x \to -2} x + \lim_{x \to -2} 1 \qquad [3]$$

$$= 3(-2)^4 + 2(-2)^2 - (-2) + (1) \qquad [9, 8, \text{ and } 7]$$

$$= 48 + 8 + 2 + 1 = 59$$

4. $\lim\limits_{x \to 2} \dfrac{2x^2 + 1}{x^2 + 6x - 4} = \dfrac{\lim\limits_{x \to 2} (2x^2 + 1)}{\lim\limits_{x \to 2} (x^2 + 6x - 4)}$ [Limit Law 5]

$$= \frac{2 \lim\limits_{x \to 2} x^2 + \lim\limits_{x \to 2} 1}{\lim\limits_{x \to 2} x^2 + 6 \lim\limits_{x \to 2} x - \lim\limits_{x \to 2} 4} \qquad [2, 1, \text{ and } 3]$$

$$= \frac{2(2)^2 + 1}{(2)^2 + 6(2) - 4} = \frac{9}{12} = \frac{3}{4} \qquad [9, 7, \text{ and } 8]$$

5. $\lim\limits_{x \to 8} \left(1 + \sqrt[3]{x}\right)\left(2 - 6x^2 + x^3\right) = \lim\limits_{x \to 8} \left(1 + \sqrt[3]{x}\right) \cdot \lim\limits_{x \to 8} \left(2 - 6x^2 + x^3\right)$ [Limit Law 4]

$$= \left(\lim_{x \to 8} 1 + \lim_{x \to 8} \sqrt[3]{x}\right) \cdot \left(\lim_{x \to 8} 2 - 6 \lim_{x \to 8} x^2 + \lim_{x \to 8} x^3\right) \qquad [1, 2, \text{ and } 3]$$

$$= \left(1 + \sqrt[3]{8}\right) \cdot \left(2 - 6 \cdot 8^2 + 8^3\right) \qquad [7, 10, 9]$$

$$= (3)(130) = 390$$

6. $\lim\limits_{t \to -1} (t^2 + 1)^3 (t + 3)^5 = \lim\limits_{t \to -1} (t^2 + 1)^3 \cdot \lim\limits_{t \to -1} (t + 3)^5$ [Limit Law 4]

$$= \left[\lim_{t \to -1} (t^2 + 1)\right]^3 \cdot \left[\lim_{t \to -1} (t + 3)\right]^5 \qquad [6]$$

$$= \left[\lim_{t \to -1} t^2 + \lim_{t \to -1} 1\right]^3 \cdot \left[\lim_{t \to -1} t + \lim_{t \to -1} 3\right]^5 \qquad [1]$$

$$= \left[(-1)^2 + 1\right]^3 \cdot [-1 + 3]^5 = 8 \cdot 32 = 256 \qquad [9, 7, \text{ and } 8]$$

7. $\lim\limits_{x \to 1} \left(\dfrac{1 + 3x}{1 + 4x^2 + 3x^4}\right)^3 = \left(\lim\limits_{x \to 1} \dfrac{1 + 3x}{1 + 4x^2 + 3x^4}\right)^3$ [6]

$$= \left[\frac{\lim\limits_{x \to 1} (1 + 3x)}{\lim\limits_{x \to 1} (1 + 4x^2 + 3x^4)}\right]^3 \qquad [5]$$

$$= \left[\frac{\lim\limits_{x \to 1} 1 + 3 \lim\limits_{x \to 1} x}{\lim\limits_{x \to 1} 1 + 4 \lim\limits_{x \to 1} x^2 + 3 \lim\limits_{x \to 1} x^4}\right]^3 \qquad [2, 1, \text{ and } 3]$$

$$= \left[\frac{1 + 3(1)}{1 + 4(1)^2 + 3(1)^4}\right]^3 = \left[\frac{4}{8}\right]^3 = \left(\frac{1}{2}\right)^3 = \frac{1}{8} \qquad [7, 8, \text{ and } 9]$$

8. $\lim\limits_{u \to -2} \sqrt{u^4 + 3u + 6} = \sqrt{\lim\limits_{u \to -2} (u^4 + 3u + 6)}$ [11]

$$= \sqrt{\lim_{u \to -2} u^4 + 3 \lim_{u \to -2} u + \lim_{u \to -2} 6} \qquad [1, 2, \text{ and } 3]$$

$$= \sqrt{(-2)^4 + 3(-2) + 6} \qquad [9, 8, \text{ and } 7]$$

$$= \sqrt{16 - 6 + 6} = \sqrt{16} = 4$$

9. $\lim\limits_{x \to 4^-} \sqrt{16 - x^2} = \sqrt{\lim\limits_{x \to 4^-} (16 - x^2)}$ [11]

$= \sqrt{\lim\limits_{x \to 4^-} 16 - \lim\limits_{x \to 4^-} x^2}$ [2]

$= \sqrt{16 - (4)^2} = 0$ [7 and 9]

10. (a) The left-hand side of the equation is not defined for $x = 2$, but the right-hand side is.

(b) Since the equation holds for all $x \neq 2$, it follows that both sides of the equation approach the same limit as $x \to 2$, just as in Example 3. Remember that in finding $\lim\limits_{x \to a} f(x)$, we never consider $x = a$.

11. $\lim\limits_{x \to 2} \dfrac{x^2 + x - 6}{x - 2} = \lim\limits_{x \to 2} \dfrac{(x + 3)(x - 2)}{x - 2} = \lim\limits_{x \to 2} (x + 3) = 2 + 3 = 5$

12. $\lim\limits_{x \to -4} \dfrac{x^2 + 5x + 4}{x^2 + 3x - 4} = \lim\limits_{x \to -4} \dfrac{(x + 4)(x + 1)}{(x + 4)(x - 1)} = \lim\limits_{x \to -4} \dfrac{x + 1}{x - 1} = \dfrac{-4 + 1}{-4 - 1} = \dfrac{-3}{-5} = \dfrac{3}{5}$

13. $\lim\limits_{x \to 2} \dfrac{x^2 - x + 6}{x - 2}$ does not exist since $x - 2 \to 0$ but $x^2 - x + 6 \to 8$ as $x \to 2$.

14. $\lim\limits_{x \to 4} \dfrac{x^2 - 4x}{x^2 - 3x - 4} = \lim\limits_{x \to 4} \dfrac{x(x - 4)}{(x - 4)(x + 1)} = \lim\limits_{x \to 4} \dfrac{x}{x + 1} = \dfrac{4}{4 + 1} = \dfrac{4}{5}$

15. $\lim\limits_{t \to -3} \dfrac{t^2 - 9}{2t^2 + 7t + 3} = \lim\limits_{t \to -3} \dfrac{(t + 3)(t - 3)}{(2t + 1)(t + 3)} = \lim\limits_{t \to -3} \dfrac{t - 3}{2t + 1} = \dfrac{-3 - 3}{2(-3) + 1} = \dfrac{-6}{-5} = \dfrac{6}{5}$

16. $\lim\limits_{x \to -1} \dfrac{x^2 - 4x}{x^2 - 3x - 4}$ does not exist since $x^2 - 3x - 4 \to 0$ but $x^2 - 4x \to 5$ as $x \to -1$.

17. $\lim\limits_{h \to 0} \dfrac{(4 + h)^2 - 16}{h} = \lim\limits_{h \to 0} \dfrac{(16 + 8h + h^2) - 16}{h} = \lim\limits_{h \to 0} \dfrac{8h + h^2}{h} = \lim\limits_{h \to 0} \dfrac{h(8 + h)}{h} = \lim\limits_{h \to 0} (8 + h) = 8 + 0 = 8$

18. $\lim\limits_{x \to 1} \dfrac{x^3 - 1}{x^2 - 1} = \lim\limits_{x \to 1} \dfrac{(x - 1)(x^2 + x + 1)}{(x - 1)(x + 1)} = \lim\limits_{x \to 1} \dfrac{x^2 + x + 1}{x + 1} = \dfrac{1^2 + 1 + 1}{1 + 1} = \dfrac{3}{2}$

19. By the formula for the sum of cubes, we have

$\lim\limits_{x \to -2} \dfrac{x + 2}{x^3 + 8} = \lim\limits_{x \to -2} \dfrac{x + 2}{(x + 2)(x^2 - 2x + 4)} = \lim\limits_{x \to -2} \dfrac{1}{x^2 - 2x + 4} = \dfrac{1}{4 + 4 + 4} = \dfrac{1}{12}.$

20. $\lim\limits_{h \to 0} \dfrac{(2 + h)^3 - 8}{h} = \lim\limits_{h \to 0} \dfrac{(8 + 12h + 6h^2 + h^3) - 8}{h} = \lim\limits_{h \to 0} \dfrac{12h + 6h^2 + h^3}{h}$

$= \lim\limits_{h \to 0} (12 + 6h + h^2) = 12 + 0 + 0 = 12$

21. $\lim\limits_{t \to 9} \dfrac{9 - t}{3 - \sqrt{t}} = \lim\limits_{t \to 9} \dfrac{\left(3 + \sqrt{t}\right)\left(3 - \sqrt{t}\right)}{3 - \sqrt{t}} = \lim\limits_{t \to 9} \left(3 + \sqrt{t}\right) = 3 + \sqrt{9} = 6$

22. $\lim\limits_{h \to 0} \dfrac{\sqrt{1 + h} - 1}{h} = \lim\limits_{h \to 0} \dfrac{\sqrt{1 + h} - 1}{h} \cdot \dfrac{\sqrt{1 + h} + 1}{\sqrt{1 + h} + 1} = \lim\limits_{h \to 0} \dfrac{(1 + h) - 1}{h\left(\sqrt{1 + h} + 1\right)} = \lim\limits_{h \to 0} \dfrac{h}{h\left(\sqrt{1 + h} + 1\right)}$

$= \lim\limits_{h \to 0} \dfrac{1}{\sqrt{1 + h} + 1} = \dfrac{1}{\sqrt{1} + 1} = \dfrac{1}{2}$

23. $\displaystyle\lim_{x\to 7}\frac{\sqrt{x+2}-3}{x-7}=\lim_{x\to 7}\frac{\sqrt{x+2}-3}{x-7}\cdot\frac{\sqrt{x+2}+3}{\sqrt{x+2}+3}=\lim_{x\to 7}\frac{(x+2)-9}{(x-7)\left(\sqrt{x+2}+3\right)}$

$\displaystyle\qquad\qquad =\lim_{x\to 7}\frac{x-7}{(x-7)\left(\sqrt{x+2}+3\right)}=\lim_{x\to 7}\frac{1}{\sqrt{x+2}+3}=\frac{1}{\sqrt{9}+3}=\frac{1}{6}$

24. $\displaystyle\lim_{x\to -1}\frac{x^2+2x+1}{x^4-1}=\lim_{x\to -1}\frac{(x+1)^2}{(x^2+1)(x^2-1)}=\lim_{x\to -1}\frac{(x+1)^2}{(x^2+1)(x+1)(x-1)}=\lim_{x\to -1}\frac{x+1}{(x^2+1)(x-1)}=\frac{0}{2(-2)}=0$

25. $\displaystyle\lim_{x\to -4}\frac{\frac{1}{4}+\frac{1}{x}}{4+x}=\lim_{x\to -4}\frac{\frac{x+4}{4x}}{4+x}=\lim_{x\to -4}\frac{x+4}{4x(4+x)}=\lim_{x\to -4}\frac{1}{4x}=\frac{1}{4(-4)}=-\frac{1}{16}$

26. $\displaystyle\lim_{t\to 0}\left(\frac{1}{t}-\frac{1}{t^2+t}\right)=\lim_{t\to 0}\frac{(t^2+t)-t}{t(t^2+t)}=\lim_{t\to 0}\frac{t^2}{t\cdot t(t+1)}=\lim_{t\to 0}\frac{1}{t+1}=\frac{1}{0+1}=1$

27. $\displaystyle\lim_{x\to 16}\frac{4-\sqrt{x}}{16x-x^2}=\lim_{x\to 16}\frac{(4-\sqrt{x})(4+\sqrt{x})}{(16x-x^2)(4+\sqrt{x})}=\lim_{x\to 16}\frac{16-x}{x(16-x)(4+\sqrt{x})}$

$\displaystyle\qquad\qquad =\lim_{x\to 16}\frac{1}{x(4+\sqrt{x})}=\frac{1}{16\left(4+\sqrt{16}\right)}=\frac{1}{16(8)}=\frac{1}{128}$

28. $\displaystyle\lim_{h\to 0}\frac{(3+h)^{-1}-3^{-1}}{h}=\lim_{h\to 0}\frac{\frac{1}{3+h}-\frac{1}{3}}{h}=\lim_{h\to 0}\frac{3-(3+h)}{h(3+h)3}=\lim_{h\to 0}\frac{-h}{h(3+h)3}$

$\displaystyle\qquad\qquad =\lim_{h\to 0}\left[-\frac{1}{3(3+h)}\right]=-\frac{1}{\lim_{h\to 0}\left[3(3+h)\right]}=-\frac{1}{3(3+0)}=-\frac{1}{9}$

29. $\displaystyle\lim_{t\to 0}\left(\frac{1}{t\sqrt{1+t}}-\frac{1}{t}\right)=\lim_{t\to 0}\frac{1-\sqrt{1+t}}{t\sqrt{1+t}}=\lim_{t\to 0}\frac{\left(1-\sqrt{1+t}\right)\left(1+\sqrt{1+t}\right)}{t\sqrt{t+1}\left(1+\sqrt{1+t}\right)}=\lim_{t\to 0}\frac{-t}{t\sqrt{1+t}\left(1+\sqrt{1+t}\right)}$

$\displaystyle\qquad\qquad =\lim_{t\to 0}\frac{-1}{\sqrt{1+t}\left(1+\sqrt{1+t}\right)}=\frac{-1}{\sqrt{1+0}\left(1+\sqrt{1+0}\right)}=-\frac{1}{2}$

30. $\displaystyle\lim_{x\to -4}\frac{\sqrt{x^2+9}-5}{x+4}=\lim_{x\to -4}\frac{\left(\sqrt{x^2+9}-5\right)\left(\sqrt{x^2+9}+5\right)}{(x+4)\left(\sqrt{x^2+9}+5\right)}=\lim_{x\to -4}\frac{(x^2+9)-25}{(x+4)\left(\sqrt{x^2+9}+5\right)}$

$\displaystyle\qquad\qquad =\lim_{x\to -4}\frac{x^2-16}{(x+4)\left(\sqrt{x^2+9}+5\right)}=\lim_{x\to -4}\frac{(x+4)(x-4)}{(x+4)\left(\sqrt{x^2+9}+5\right)}$

$\displaystyle\qquad\qquad =\lim_{x\to -4}\frac{x-4}{\sqrt{x^2+9}+5}=\frac{-4-4}{\sqrt{16+9}+5}=\frac{-8}{5+5}=-\frac{4}{5}$

31. (a)

$\displaystyle\lim_{x\to 0}\frac{x}{\sqrt{1+3x}-1}\approx\frac{2}{3}$

(b)

x	$f(x)$
-0.001	0.6661663
-0.0001	0.6666167
-0.00001	0.6666617
-0.000001	0.6666662
0.000001	0.6666672
0.00001	0.6666717
0.0001	0.6667167
0.001	0.6671663

The limit appears to be $\dfrac{2}{3}$.

(c) $\displaystyle\lim_{x\to0}\left(\frac{x}{\sqrt{1+3x}-1}\cdot\frac{\sqrt{1+3x}+1}{\sqrt{1+3x}+1}\right)=\lim_{x\to0}\frac{x\left(\sqrt{1+3x}+1\right)}{(1+3x)-1}=\lim_{x\to0}\frac{x\left(\sqrt{1+3x}+1\right)}{3x}$

$$=\frac{1}{3}\lim_{x\to0}\left(\sqrt{1+3x}+1\right) \qquad\qquad \text{[Limit Law 3]}$$

$$=\frac{1}{3}\left[\sqrt{\lim_{x\to0}(1+3x)}+\lim_{x\to0}1\right] \qquad \text{[1 and 11]}$$

$$=\frac{1}{3}\left(\sqrt{\lim_{x\to0}1+3\lim_{x\to0}x}+1\right) \qquad \text{[1, 3, and 7]}$$

$$=\frac{1}{3}\left(\sqrt{1+3\cdot0}+1\right) \qquad\qquad \text{[7 and 8]}$$

$$=\frac{1}{3}(1+1)=\frac{2}{3}$$

32. (a)

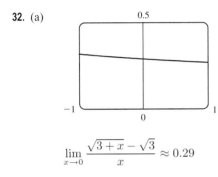

$$\lim_{x\to0}\frac{\sqrt{3+x}-\sqrt{3}}{x}\approx0.29$$

(b)

x	$f(x)$
-0.001	0.2886992
-0.0001	0.2886775
-0.00001	0.2886754
-0.000001	0.2886752
0.000001	0.2886751
0.00001	0.2886749
0.0001	0.2886727
0.001	0.2886511

The limit appears to be approximately 0.2887.

(c) $\displaystyle\lim_{x\to0}\left(\frac{\sqrt{3+x}-\sqrt{3}}{x}\cdot\frac{\sqrt{3+x}+\sqrt{3}}{\sqrt{3+x}+\sqrt{3}}\right)=\lim_{x\to0}\frac{(3+x)-3}{x\left(\sqrt{3+x}+\sqrt{3}\right)}=\lim_{x\to0}\frac{1}{\sqrt{3+x}+\sqrt{3}}$

$$=\frac{\displaystyle\lim_{x\to0}1}{\displaystyle\lim_{x\to0}\sqrt{3+x}+\lim_{x\to0}\sqrt{3}} \qquad \text{[Limit Laws 5 and 1]}$$

$$=\frac{1}{\sqrt{\displaystyle\lim_{x\to0}(3+x)}+\sqrt{3}} \qquad\qquad \text{[7 and 11]}$$

$$=\frac{1}{\sqrt{3+0}+\sqrt{3}} \qquad\qquad \text{[1, 7, and 8]}$$

$$=\frac{1}{2\sqrt{3}}$$

33. Let $f(x)=-x^2$, $g(x)=x^2\cos20\pi x$ and $h(x)=x^2$. Then

$-1\leq\cos20\pi x\leq1 \quad\Rightarrow\quad -x^2\leq x^2\cos20\pi x\leq x^2 \quad\Rightarrow\quad f(x)\leq g(x)\leq h(x)$.

So since $\displaystyle\lim_{x\to0}f(x)=\lim_{x\to0}h(x)=0$, by the Squeeze Theorem we have

$\displaystyle\lim_{x\to0}g(x)=0$.

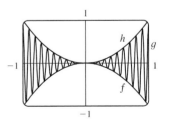

34. Let $f(x)=-\sqrt{x^3+x^2}$, $g(x)=\sqrt{x^3+x^2}\,\sin(\pi/x)$, and $h(x)=\sqrt{x^3+x^2}$. Then

$-1\leq\sin(\pi/x)\leq1 \quad\Rightarrow\quad -\sqrt{x^3+x^2}\leq\sqrt{x^3+x^2}\sin(\pi/x)\leq\sqrt{x^3+x^2} \quad\Rightarrow$

$f(x)\leq g(x)\leq h(x)$. So since $\displaystyle\lim_{x\to0}f(x)=\lim_{x\to0}h(x)=0$, by the Squeeze Theorem

we have $\displaystyle\lim_{x\to0}g(x)=0$.

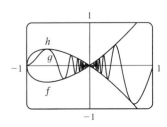

35. We have $\lim_{x \to 4} (4x - 9) = 4(4) - 9 = 7$ and $\lim_{x \to 4} (x^2 - 4x + 7) = 4^2 - 4(4) + 7 = 7$. Since $4x - 9 \le f(x) \le x^2 - 4x + 7$

for $x \ge 0$, $\lim_{x \to 4} f(x) = 7$ by the Squeeze Theorem.

36. We have $\lim_{x \to 1} (2x) = 2(1) = 2$ and $\lim_{x \to 1} (x^4 - x^2 + 2) = 1^4 - 1^2 + 2 = 2$. Since $2x \le g(x) \le x^4 - x^2 + 2$ for all x,

$\lim_{x \to 1} g(x) = 2$ by the Squeeze Theorem.

37. $-1 \le \cos(2/x) \le 1 \quad \Rightarrow \quad -x^4 \le x^4 \cos(2/x) \le x^4$. Since $\lim_{x \to 0} (-x^4) = 0$ and $\lim_{x \to 0} x^4 = 0$, we have

$\lim_{x \to 0} [x^4 \cos(2/x)] = 0$ by the Squeeze Theorem.

38. $-1 \le \sin(2\pi/x) \le 1 \quad \Rightarrow \quad 0 \le \sin^2(2\pi/x) \le 1 \quad \Rightarrow \quad 1 \le 1 + \sin^2(2\pi/x) \le 2 \quad \Rightarrow$

$\sqrt{x} \le \sqrt{x}\left[1 + \sin^2(2\pi/x)\right] \le 2\sqrt{x}$. Since $\lim_{x \to 0^+} \sqrt{x} = 0$ and $\lim_{x \to 0^+} 2\sqrt{x} = 0$, we have

$\lim_{x \to 0^+} \left[\sqrt{x}\left(1 + \sin^2(2\pi/x)\right)\right] = 0$ by the Squeeze Theorem.

39. $|x - 3| = \begin{cases} x - 3 & \text{if } x - 3 \ge 0 \\ -(x - 3) & \text{if } x - 3 < 0 \end{cases} = \begin{cases} x - 3 & \text{if } x \ge 3 \\ 3 - x & \text{if } x < 3 \end{cases}$

Thus, $\lim_{x \to 3^+} (2x + |x - 3|) = \lim_{x \to 3^+} (2x + x - 3) = \lim_{x \to 3^+} (3x - 3) = 3(3) - 3 = 6$ and

$\lim_{x \to 3^-} (2x + |x - 3|) = \lim_{x \to 3^-} (2x + 3 - x) = \lim_{x \to 3^-} (x + 3) = 3 + 3 = 6$. Since the left and right limits are equal,

$\lim_{x \to 3} (2x + |x - 3|) = 6$.

40. $|x + 6| = \begin{cases} x + 6 & \text{if } x + 6 \ge 0 \\ -(x + 6) & \text{if } x + 6 < 0 \end{cases} = \begin{cases} x + 6 & \text{if } x \ge -6 \\ -(x + 6) & \text{if } x < -6 \end{cases}$

We'll look at the one-sided limits.

$\lim_{x \to -6^+} \dfrac{2x + 12}{|x + 6|} = \lim_{x \to -6^+} \dfrac{2(x + 6)}{x + 6} = 2$ and $\lim_{x \to -6^-} \dfrac{2x + 12}{|x + 6|} = \lim_{x \to -6^-} \dfrac{2(x + 6)}{-(x + 6)} = -2$

The left and right limits are different, so $\lim_{x \to -6} \dfrac{2x + 12}{|x + 6|}$ does not exist.

41. $\left|2x^3 - x^2\right| = \left|x^2(2x - 1)\right| = \left|x^2\right| \cdot |2x - 1| = x^2 |2x - 1|$

$|2x - 1| = \begin{cases} 2x - 1 & \text{if } 2x - 1 \ge 0 \\ -(2x - 1) & \text{if } 2x - 1 < 0 \end{cases} = \begin{cases} 2x - 1 & \text{if } x \ge 0.5 \\ -(2x - 1) & \text{if } x < 0.5 \end{cases}$

So $\left|2x^3 - x^2\right| = x^2[-(2x - 1)]$ for $x < 0.5$.

Thus, $\lim_{x \to 0.5^-} \dfrac{2x - 1}{|2x^3 - x^2|} = \lim_{x \to 0.5^-} \dfrac{2x - 1}{x^2[-(2x - 1)]} = \lim_{x \to 0.5^-} \dfrac{-1}{x^2} = \dfrac{-1}{(0.5)^2} = \dfrac{-1}{0.25} = -4$.

42. Since $|x| = -x$ for $x < 0$, we have $\lim_{x \to -2} \dfrac{2 - |x|}{2 + x} = \lim_{x \to -2} \dfrac{2 - (-x)}{2 + x} = \lim_{x \to -2} \dfrac{2 + x}{2 + x} = \lim_{x \to -2} 1 = 1$.

43. Since $|x| = -x$ for $x < 0$, we have $\lim_{x \to 0^-} \left(\dfrac{1}{x} - \dfrac{1}{|x|}\right) = \lim_{x \to 0^-} \left(\dfrac{1}{x} - \dfrac{1}{-x}\right) = \lim_{x \to 0^-} \dfrac{2}{x}$, which does not exist since the

denominator approaches 0 and the numerator does not.

44. Since $|x| = x$ for $x > 0$, we have $\displaystyle\lim_{x \to 0^+} \left(\frac{1}{x} - \frac{1}{|x|} \right) = \lim_{x \to 0^+} \left(\frac{1}{x} - \frac{1}{x} \right) = \lim_{x \to 0^+} 0 = 0.$

45. (a)

(b) (i) Since $\operatorname{sgn} x = 1$ for $x > 0$, $\displaystyle\lim_{x \to 0^+} \operatorname{sgn} x = \lim_{x \to 0^+} 1 = 1.$

(ii) Since $\operatorname{sgn} x = -1$ for $x < 0$, $\displaystyle\lim_{x \to 0^-} \operatorname{sgn} x = \lim_{x \to 0^-} -1 = -1.$

(iii) Since $\displaystyle\lim_{x \to 0^-} \operatorname{sgn} x \neq \lim_{x \to 0^+} \operatorname{sgn} x$, $\displaystyle\lim_{x \to 0} \operatorname{sgn} x$ does not exist.

(iv) Since $|\operatorname{sgn} x| = 1$ for $x \neq 0$, $\displaystyle\lim_{x \to 0} |\operatorname{sgn} x| = \lim_{x \to 0} 1 = 1.$

46. (a) $\displaystyle\lim_{x \to 2^-} f(x) = \lim_{x \to 2^-} \left(4 - x^2 \right) = \lim_{x \to 2^-} 4 - \lim_{x \to 2^-} x^2 = 4 - 4 = 0$

(c)

$\displaystyle\lim_{x \to 2^+} f(x) = \lim_{x \to 2^+} (x - 1) = \lim_{x \to 2^+} x - \lim_{x \to 2^+} 1 = 2 - 1 = 1$

(b) No, $\displaystyle\lim_{x \to 2} f(x)$ does not exist since $\displaystyle\lim_{x \to 2^-} f(x) \neq \lim_{x \to 2^+} f(x).$

47. (a) (i) $\displaystyle\lim_{x \to 1^+} F(x) = \lim_{x \to 1^+} \frac{x^2 - 1}{|x - 1|} = \lim_{x \to 1^+} \frac{x^2 - 1}{x - 1} = \lim_{x \to 1^+} (x + 1) = 2$

(c)

(ii) $\displaystyle\lim_{x \to 1^-} F(x) = \lim_{x \to 1^-} \frac{x^2 - 1}{|x - 1|} = \lim_{x \to 1^-} \frac{x^2 - 1}{-(x - 1)} = \lim_{x \to 1^-} -(x + 1) = -2$

(b) No, $\displaystyle\lim_{x \to 1} F(x)$ does not exist since $\displaystyle\lim_{x \to 1^+} F(x) \neq \lim_{x \to 1^-} F(x).$

48. (a) (i) $\displaystyle\lim_{x \to 1^-} g(x) = \lim_{x \to 1^-} x = 1$

(ii) $\displaystyle\lim_{x \to 1^+} g(x) = \lim_{x \to 1^+} \left(2 - x^2 \right) = 2 - 1^2 = 1.$ Since $\displaystyle\lim_{x \to 1^-} g(x) = 1$ and $\displaystyle\lim_{x \to 1^+} g(x) = 1$, we have $\displaystyle\lim_{x \to 1} g(x) = 1.$

Note that the fact $g(1) = 3$ does not affect the value of the limit.

(iii) When $x = 1$, $g(x) = 3$, so $g(1) = 3$.

(iv) $\displaystyle\lim_{x \to 2^-} g(x) = \lim_{x \to 2^-} \left(2 - x^2 \right) = 2 - 2^2 = 2 - 4 = -2$

(v) $\displaystyle\lim_{x \to 2^+} g(x) = \lim_{x \to 2^+} (x - 3) = 2 - 3 = -1$

(vi) $\displaystyle\lim_{x \to 2} g(x)$ does not exist since $\displaystyle\lim_{x \to 2^-} g(x) \neq \lim_{x \to 2^+} g(x).$

(b)

$$g(x) = \begin{cases} x & \text{if } x < 1 \\ 3 & \text{if } x = 1 \\ 2 - x^2 & \text{if } 1 < x \leq 2 \\ x - 3 & \text{if } x > 2 \end{cases}$$

49. (a) (i) $[\![x]\!] = -2$ for $-2 \le x < -1$, so $\lim\limits_{x \to -2+} [\![x]\!] = \lim\limits_{x \to -2+} (-2) = -2$

(ii) $[\![x]\!] = -3$ for $-3 \le x < -2$, so $\lim\limits_{x \to -2^-} [\![x]\!] = \lim\limits_{x \to -2^-} (-3) = -3$.

The right and left limits are different, so $\lim\limits_{x \to -2} [\![x]\!]$ does not exist.

(iii) $[\![x]\!] = -3$ for $-3 \le x < -2$, so $\lim\limits_{x \to -2.4} [\![x]\!] = \lim\limits_{x \to -2.4} (-3) = -3$.

(b) (i) $[\![x]\!] = n - 1$ for $n - 1 \le x < n$, so $\lim\limits_{x \to n^-} [\![x]\!] = \lim\limits_{x \to n^-} (n-1) = n - 1$.

(ii) $[\![x]\!] = n$ for $n \le x < n + 1$, so $\lim\limits_{x \to n^+} [\![x]\!] = \lim\limits_{x \to n^+} n = n$.

(c) $\lim\limits_{x \to a} [\![x]\!]$ exists $\Leftrightarrow$ a is not an integer.

50. (a) See the graph of $y = \cos x$.

Since $-1 \le \cos x < 0$ on $[-\pi, -\pi/2)$, we have $y = f(x) = [\![\cos x]\!] = -1$

on $[-\pi, -\pi/2)$.

Since $0 \le \cos x < 1$ on $[-\pi/2, 0) \cup (0, \pi/2]$, we have $f(x) = 0$

on $[-\pi/2, 0) \cup (0, \pi/2]$.

Since $-1 \le \cos x < 0$ on $(\pi/2, \pi]$, we have $f(x) = -1$ on $(\pi/2, \pi]$.

Note that $f(0) = 1$.

(b) (i) $\lim\limits_{x \to 0^-} f(x) = 0$ and $\lim\limits_{x \to 0^+} f(x) = 0$, so $\lim\limits_{x \to 0} f(x) = 0$.

(ii) As $x \to (\pi/2)^-$, $f(x) \to 0$, so $\lim\limits_{x \to (\pi/2)^-} f(x) = 0$.

(iii) As $x \to (\pi/2)^+$, $f(x) \to -1$, so $\lim\limits_{x \to (\pi/2)^+} f(x) = -1$.

(iv) Since the answers in parts (ii) and (iii) are not equal, $\lim\limits_{x \to \pi/2} f(x)$ does not exist.

(c) $\lim\limits_{x \to a} f(x)$ exists for all a in the open interval $(-\pi, \pi)$ except $a = -\pi/2$ and $a = \pi/2$.

51. The graph of $f(x) = [\![x]\!] + [\![-x]\!]$ is the same as the graph of $g(x) = -1$ with holes at each integer, since $f(a) = 0$ for any

integer a. Thus, $\lim\limits_{x \to 2^-} f(x) = -1$ and $\lim\limits_{x \to 2^+} f(x) = -1$, so $\lim\limits_{x \to 2} f(x) = -1$. However,

$f(2) = [\![2]\!] + [\![-2]\!] = 2 + (-2) = 0$, so $\lim\limits_{x \to 2} f(x) \ne f(2)$.

52. $\lim\limits_{v \to c^-} \left(L_0 \sqrt{1 - \dfrac{v^2}{c^2}} \right) = L_0 \sqrt{1 - 1} = 0$. As the velocity approaches the speed of light, the length approaches 0.

A left-hand limit is necessary since L is not defined for $v > c$.

53. Since $p(x)$ is a polynomial, $p(x) = a_0 + a_1 x + a_2 x^2 + \cdots + a_n x^n$. Thus, by the Limit Laws,

$$\lim\limits_{x \to a} p(x) = \lim\limits_{x \to a} \left(a_0 + a_1 x + a_2 x^2 + \cdots + a_n x^n \right) = a_0 + a_1 \lim\limits_{x \to a} x + a_2 \lim\limits_{x \to a} x^2 + \cdots + a_n \lim\limits_{x \to a} x^n$$

$$= a_0 + a_1 a + a_2 a^2 + \cdots + a_n a^n = p(a)$$

Thus, for any polynomial p, $\lim\limits_{x \to a} p(x) = p(a)$.

54. Let $r(x) = \dfrac{p(x)}{q(x)}$ where $p(x)$ and $q(x)$ are any polynomials, and suppose that $q(a) \neq 0$. Thus,

$$\lim_{x \to a} r(x) = \lim_{x \to a} \frac{p(x)}{q(x)} = \frac{\lim\limits_{x \to a} p(x)}{\lim\limits_{x \to a} q(x)} \quad \text{[Limit Law 5]} \quad = \frac{p(a)}{q(a)} \quad \text{[Exercise 53]} \quad = r(a).$$

55. $\lim\limits_{x \to 1} [f(x) - 8] = \lim\limits_{x \to 1} \left[\dfrac{f(x) - 8}{x - 1} \cdot (x - 1) \right] = \lim\limits_{x \to 1} \dfrac{f(x) - 8}{x - 1} \cdot \lim\limits_{x \to 1} (x - 1) = 10 \cdot 0 = 0.$

Thus, $\lim\limits_{x \to 1} f(x) = \lim\limits_{x \to 1} \{ [f(x) - 8] + 8 \} = \lim\limits_{x \to 1} [f(x) - 8] + \lim\limits_{x \to 1} 8 = 0 + 8 = 8.$

Note: The value of $\lim\limits_{x \to 1} \dfrac{f(x) - 8}{x - 1}$ does not affect the answer since it's multiplied by 0. What's important is that $\lim\limits_{x \to 1} \dfrac{f(x) - 8}{x - 1}$

exists.

56. (a) $\lim\limits_{x \to 0} f(x) = \lim\limits_{x \to 0} \left[\dfrac{f(x)}{x^2} \cdot x^2 \right] = \lim\limits_{x \to 0} \dfrac{f(x)}{x^2} \cdot \lim\limits_{x \to 0} x^2 = 5 \cdot 0 = 0$

(b) $\lim\limits_{x \to 0} \dfrac{f(x)}{x} = \lim\limits_{x \to 0} \left[\dfrac{f(x)}{x^2} \cdot x \right] = \lim\limits_{x \to 0} \dfrac{f(x)}{x^2} \cdot \lim\limits_{x \to 0} x = 5 \cdot 0 = 0$

57. Observe that $0 \leq f(x) \leq x^2$ for all x, and $\lim\limits_{x \to 0} 0 = 0 = \lim\limits_{x \to 0} x^2$. So, by the Squeeze Theorem, $\lim\limits_{x \to 0} f(x) = 0.$

58. Let $f(x) = [\![x]\!]$ and $g(x) = -[\![x]\!]$. Then $\lim\limits_{x \to 3} f(x)$ and $\lim\limits_{x \to 3} g(x)$ do not exist   [Example 10]

but $\lim\limits_{x \to 3} [f(x) + g(x)] = \lim\limits_{x \to 3} ([\![x]\!] - [\![x]\!]) = \lim\limits_{x \to 3} 0 = 0.$

59. Let $f(x) = H(x)$ and $g(x) = 1 - H(x)$, where H is the Heaviside function defined in Exercise 1.3.57.

Thus, either f or g is 0 for any value of x. Then $\lim\limits_{x \to 0} f(x)$ and $\lim\limits_{x \to 0} g(x)$ do not exist, but $\lim\limits_{x \to 0} [f(x)g(x)] = \lim\limits_{x \to 0} 0 = 0.$

60. $\lim\limits_{x \to 2} \dfrac{\sqrt{6 - x} - 2}{\sqrt{3 - x} - 1} = \lim\limits_{x \to 2} \left(\dfrac{\sqrt{6 - x} - 2}{\sqrt{3 - x} - 1} \cdot \dfrac{\sqrt{6 - x} + 2}{\sqrt{6 - x} + 2} \cdot \dfrac{\sqrt{3 - x} + 1}{\sqrt{3 - x} + 1} \right)$

$$= \lim\limits_{x \to 2} \left[\dfrac{\left(\sqrt{6 - x}\right)^2 - 2^2}{\left(\sqrt{3 - x}\right)^2 - 1^2} \cdot \dfrac{\sqrt{3 - x} + 1}{\sqrt{6 - x} + 2} \right] = \lim\limits_{x \to 2} \left(\dfrac{6 - x - 4}{3 - x - 1} \cdot \dfrac{\sqrt{3 - x} + 1}{\sqrt{6 - x} + 2} \right)$$

$$= \lim\limits_{x \to 2} \dfrac{(2 - x)\left(\sqrt{3 - x} + 1\right)}{(2 - x)\left(\sqrt{6 - x} + 2\right)} = \lim\limits_{x \to 2} \dfrac{\sqrt{3 - x} + 1}{\sqrt{6 - x} + 2} = \dfrac{1}{2}$$

61. Since the denominator approaches 0 as $x \to -2$, the limit will exist only if the numerator also approaches

0 as $x \to -2$. In order for this to happen, we need $\lim\limits_{x \to -2} \left(3x^2 + ax + a + 3 \right) = 0 \iff$

$3(-2)^2 + a(-2) + a + 3 = 0 \iff 12 - 2a + a + 3 = 0 \iff a = 15$. With $a = 15$, the limit becomes

$$\lim\limits_{x \to -2} \dfrac{3x^2 + 15x + 18}{x^2 + x - 2} = \lim\limits_{x \to -2} \dfrac{3(x + 2)(x + 3)}{(x - 1)(x + 2)} = \lim\limits_{x \to -2} \dfrac{3(x + 3)}{x - 1} = \dfrac{3(-2 + 3)}{-2 - 1} = \dfrac{3}{-3} = -1.$$

62. *Solution 1:* First, we find the coordinates of P and Q as functions of r. Then we can find the equation of the line determined by these two points, and thus find the x-intercept (the point R), and take the limit as $r \to 0$. The coordinates of P are $(0, r)$. The point Q is the point of intersection of the two circles $x^2 + y^2 = r^2$ and $(x - 1)^2 + y^2 = 1$. Eliminating y from these equations, we get $r^2 - x^2 = 1 - (x - 1)^2 \quad \Leftrightarrow \quad r^2 = 1 + 2x - 1 \quad \Leftrightarrow \quad x = \frac{1}{2}r^2$. Substituting back into the equation of the shrinking circle to find the y-coordinate, we get $\left(\frac{1}{2}r^2\right)^2 + y^2 = r^2 \quad \Leftrightarrow \quad y^2 = r^2\left(1 - \frac{1}{4}r^2\right) \quad \Leftrightarrow \quad y = r\sqrt{1 - \frac{1}{4}r^2}$ (the positive y-value). So the coordinates of Q are $\left(\frac{1}{2}r^2, r\sqrt{1 - \frac{1}{4}r^2}\right)$. The equation of the line joining P and Q is thus

$$y - r = \frac{r\sqrt{1 - \frac{1}{4}r^2} - r}{\frac{1}{2}r^2 - 0}(x - 0).$$ We set $y = 0$ in order to find the x-intercept, and get

$$x = -r\frac{\frac{1}{2}r^2}{r\left(\sqrt{1 - \frac{1}{4}r^2} - 1\right)} = \frac{-\frac{1}{2}r^2\left(\sqrt{1 - \frac{1}{4}r^2} + 1\right)}{1 - \frac{1}{4}r^2 - 1} = 2\left(\sqrt{1 - \frac{1}{4}r^2} + 1\right)$$

Now we take the limit as $r \to 0^+$: $\displaystyle\lim_{r \to 0^+} x = \lim_{r \to 0^+} 2\left(\sqrt{1 - \frac{1}{4}r^2} + 1\right) = \lim_{r \to 0^+} 2\left(\sqrt{1} + 1\right) = 4$.

So the limiting position of R is the point $(4, 0)$.

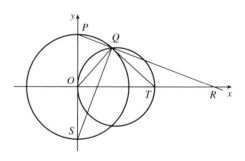

Solution 2: We add a few lines to the diagram, as shown. Note that $\angle PQS = 90°$ (subtended by diameter PS). So $\angle SQR = 90° = \angle OQT$ (subtended by diameter OT). It follows that $\angle OQS = \angle TQR$. Also $\angle PSQ = 90° - \angle SPQ = \angle ORP$. Since $\triangle QOS$ is isosceles, so is $\triangle QTR$, implying that $QT = TR$. As the circle C_2 shrinks, the point Q plainly approaches the origin, so the point R must approach a point twice as far from the origin as T, that is, the point $(4, 0)$, as above.

2.4   The Precise Definition of a Limit

1. On the left side of $x = 2$, we need $|x - 2| < \left|\frac{10}{7} - 2\right| = \frac{4}{7}$. On the right side, we need $|x - 2| < \left|\frac{10}{3} - 2\right| = \frac{4}{3}$. For both of these conditions to be satisfied at once, we need the more restrictive of the two to hold, that is, $|x - 2| < \frac{4}{7}$. So we can choose $\delta = \frac{4}{7}$, or any smaller positive number.

2. On the left side, we need $|x - 5| < |4 - 5| = 1$. On the right side, we need $|x - 5| < |5.7 - 5| = 0.7$. For both conditions to be satisfied at once, we need the more restrictive condition to hold; that is, $|x - 5| < 0.7$. So we can choose $\delta = 0.7$, or any smaller positive number.

3. The leftmost question mark is the solution of $\sqrt{x} = 1.6$ and the rightmost, $\sqrt{x} = 2.4$. So the values are $1.6^2 = 2.56$ and $2.4^2 = 5.76$. On the left side, we need $|x - 4| < |2.56 - 4| = 1.44$. On the right side, we need $|x - 4| < |5.76 - 4| = 1.76$. To satisfy both conditions, we need the more restrictive condition to hold — namely, $|x - 4| < 1.44$. Thus, we can choose $\delta = 1.44$, or any smaller positive number.

4. The leftmost question mark is the positive solution of $x^2 = \frac{1}{2}$, that is, $x = \frac{1}{\sqrt{2}}$, and the rightmost question mark is the positive solution of $x^2 = \frac{3}{2}$, that is, $x = \sqrt{\frac{3}{2}}$. On the left side, we need $|x - 1| < \left|\frac{1}{\sqrt{2}} - 1\right| \approx 0.292$ (rounding down to be safe). On the right side, we need $|x - 1| < \left|\sqrt{\frac{3}{2}} - 1\right| \approx 0.224$. The more restrictive of these two conditions must apply, so we choose $\delta = 0.224$ (or any smaller positive number).

5.

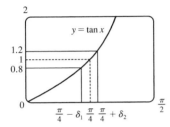

From the graph, we find that $\tan x = 0.8$ when $x \approx 0.675$, so

$\frac{\pi}{4} - \delta_1 \approx 0.675 \Rightarrow \delta_1 \approx \frac{\pi}{4} - 0.675 \approx 0.1106$. Also, $\tan x = 1.2$

when $x \approx 0.876$, so $\frac{\pi}{4} + \delta_2 \approx 0.876 \Rightarrow \delta_2 = 0.876 - \frac{\pi}{4} \approx 0.0906$.

Thus, we choose $\delta = 0.0906$ (or any smaller positive number) since this is the smaller of δ_1 and δ_2.

6.

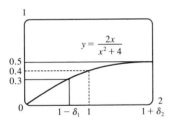

From the graph, we find that $y = 2x/(x^2 + 4) = 0.3$ when $x = \frac{2}{3}$, so

$1 - \delta_1 = \frac{2}{3} \Rightarrow \delta_1 = \frac{1}{3}$. Also, $y = 2x/(x^2 + 4) = 0.5$ when $x = 2$, so

$1 + \delta_2 = 2 \Rightarrow \delta_2 = 1$. Thus, we choose $\delta = \frac{1}{3}$ (or any smaller positive number) since this is the smaller of δ_1 and δ_2.

7. For $\varepsilon = 1$, the definition of a limit requires that we find δ such that $\left|(4 + x - 3x^3) - 2\right| < 1 \quad \Leftrightarrow \quad 1 < 4 + x - 3x^3 < 3$ whenever $0 < |x - 1| < \delta$. If we plot the graphs of $y = 1$, $y = 4 + x - 3x^3$ and $y = 3$ on the same screen, we see that we need $0.86 \leq x \leq 1.11$. So since $|1 - 0.86| = 0.14$ and $|1 - 1.11| = 0.11$, we choose $\delta = 0.11$ (or any smaller positive number). For $\varepsilon = 0.1$, we must find δ such that $\left|(4 + x - 3x^3) - 2\right| < 0.1 \quad \Leftrightarrow \quad 1.9 < 4 + x - 3x^3 < 2.1$ whenever $0 < |x - 1| < \delta$. From the graph, we see that we need $0.988 \leq x \leq 1.012$. So since $|1 - 0.988| = 0.012$ and $|1 - 1.012| = 0.012$, we choose $\delta = 0.012$ (or any smaller positive number) for the inequality to hold.

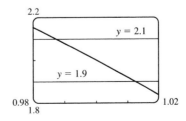

8. For $y = (4x + 1)/(3x - 4)$ and $\varepsilon = 0.5$, we need $1.91 \le x \le 2.125$. So since $|2 - 1.91| = 0.09$ and $|2 - 2.125| = 0.125$, we can take $0 < \delta \le 0.09$. For $\varepsilon = 0.1$, we need $1.980 \le 2.021$. So since $|2 - 1.980| = 0.02$ and $|2 - 2.021| = 0.021$, we can take $\delta = 0.02$ (or any smaller positive number).

9. (a)

From the graph, we find that $y = \tan^2 x = 1000$ when $x \approx 1.539$ and $x \approx 1.602$ for x near $\frac{\pi}{2}$. Thus, we get $\delta \approx 1.602 - \frac{\pi}{2} \approx 0.031$ for $M = 1000$.

(b)

From the graph, we find that $y = \tan^2 x = 10,000$ when $x \approx 1.561$ and $x \approx 1.581$ for x near $\frac{\pi}{2}$. Thus, we get $\delta \approx 1.581 - \frac{\pi}{2} \approx 0.010$ for $M = 10,000$.

10.

From the graph, we find that $x^2/\sqrt{x-5} = 100 \implies x \approx 5.066$. Thus, $5 + \delta \approx 5.0659$ and $\delta \approx 0.065$.

11. (a) $A = \pi r^2$ and $A = 1000$ cm^2 $\implies$ $\pi r^2 = 1000$ $\implies$ $r^2 = \frac{1000}{\pi}$ $\implies$ $r = \sqrt{\frac{1000}{\pi}}$ $(r > 0)$ ≈ 17.8412 cm.

(b) $|A - 1000| \le 5$ $\implies$ $-5 \le \pi r^2 - 1000 \le 5$ $\implies$ $1000 - 5 \le \pi r^2 \le 1000 + 5$ $\implies$

$\sqrt{\frac{995}{\pi}} \le r \le \sqrt{\frac{1005}{\pi}}$ $\implies$ $17.7966 \le r \le 17.8858$. $\sqrt{\frac{1000}{\pi}} - \sqrt{\frac{995}{\pi}} \approx 0.04466$ and $\sqrt{\frac{1005}{\pi}} - \sqrt{\frac{1000}{\pi}} \approx 0.04455$. So if the machinist gets the radius within 0.0445 cm of 17.8412, the area will be within 5 cm^2 of 1000.

(c) x is the radius, $f(x)$ is the area, a is the target radius given in part (a), L is the target area (1000), ε is the tolerance in the area (5), and δ is the tolerance in the radius given in part (b).

12. (a) $T = 0.1w^2 + 2.155w + 20$ and $T = 200$ $\Rightarrow$

$0.1w^2 + 2.155w + 20 = 200$ $\Rightarrow$ [by the quadratic formula or

from the graph] $w \approx 33.0$ watts $(w > 0)$

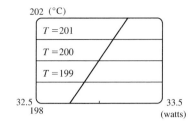

(b) From the graph, $199 \le T \le 201$ $\Rightarrow$ $32.89 < w < 33.11$.

(c) x is the input power, $f(x)$ is the temperature, a is the target input power given in part (a), L is the target temperature (200),

ε is the tolerance in the temperature (1), and δ is the tolerance in the power input in watts indicated in part (b) (0.11 watts).

13. (a) $|4x - 8| = 4|x - 2| < 0.1$ $\Leftrightarrow$ $|x - 2| < \dfrac{0.1}{4}$, so $\delta = \dfrac{0.1}{4} = 0.025$.

(b) $|4x - 8| = 4|x - 2| < 0.01$ $\Leftrightarrow$ $|x - 2| < \dfrac{0.01}{4}$, so $\delta = \dfrac{0.01}{4} = 0.0025$.

14. $|(5x - 7) - 3| = |5x - 10| = |5(x - 2)| = 5|x - 2|$. We must have $|f(x) - L| < \varepsilon$, so $5|x - 2| < \varepsilon$ $\Leftrightarrow$

$|x - 2| < \varepsilon/5$. Thus, choose $\delta = \varepsilon/5$. For $\varepsilon = 0.1$, $\delta = 0.02$; for $\varepsilon = 0.05$, $\delta = 0.01$; for $\varepsilon = 0.01$, $\delta = 0.002$.

15. Given $\varepsilon > 0$, we need $\delta > 0$ such that if $0 < |x - 1| < \delta$, then

$|(2x + 3) - 5| < \varepsilon$. But $|(2x + 3) - 5| < \varepsilon$ $\Leftrightarrow$

$|2x - 2| < \varepsilon$ $\Leftrightarrow$ $2|x - 1| < \varepsilon$ $\Leftrightarrow$ $|x - 1| < \varepsilon/2$.

So if we choose $\delta = \varepsilon/2$, then $0 < |x - 1| < \delta$ $\Rightarrow$

$|(2x + 3) - 5| < \varepsilon$. Thus, $\lim\limits_{x \to 1}(2x + 3) = 5$ by the definition of a limit.

16. Given $\varepsilon > 0$, we need $\delta > 0$ such that if $0 < |x - (-2)| < \delta$, then

$\left|\left(\frac{1}{2}x + 3\right) - 2\right| < \varepsilon$. But $\left|\left(\frac{1}{2}x + 3\right) - 2\right| < \varepsilon$ $\Leftrightarrow$

$\left|\frac{1}{2}x + 1\right| < \varepsilon$ $\Leftrightarrow$ $\frac{1}{2}|x + 2| < \varepsilon$ $\Leftrightarrow$ $|x - (-2)| < 2\varepsilon$.

So if we choose $\delta = 2\varepsilon$, then $0 < |x - (-2)| < \delta$ $\Rightarrow$

$\left|\left(\frac{1}{2}x + 3\right) - 2\right| < \varepsilon$. Thus, $\lim\limits_{x \to -2}\left(\frac{1}{2}x + 3\right) = 2$ by the definition of a

limit.

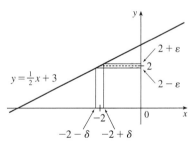

17. Given $\varepsilon > 0$, we need $\delta > 0$ such that if $0 < |x - (-3)| < \delta$, then

$|(1 - 4x) - 13| < \varepsilon$. But $|(1 - 4x) - 13| < \varepsilon$ $\Leftrightarrow$

$|-4x - 12| < \varepsilon$ $\Leftrightarrow$ $|-4||x + 3| < \varepsilon$ $\Leftrightarrow$ $|x - (-3)| < \varepsilon/4$.

So if we choose $\delta = \varepsilon/4$, then $0 < |x - (-3)| < \delta$ $\Rightarrow$

$|(1 - 4x) - 13| < \varepsilon$. Thus, $\lim\limits_{x \to -3}(1 - 4x) = 13$ by the definition of

a limit.

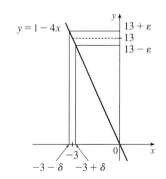

18. Given $\varepsilon > 0$, we need $\delta > 0$ such that if $0 < |x - 4| < \delta$, then

$|(7 - 3x) - (-5)| < \varepsilon$. But $|(7 - 3x) - (-5)| < \varepsilon \Leftrightarrow$

$|-3x + 12| < \varepsilon \Leftrightarrow |-3|\,|x - 4| < \varepsilon \Leftrightarrow |x - 4| < \varepsilon/3$. So

if we choose $\delta = \varepsilon/3$, then $0 < |x - 4| < \delta \Rightarrow$

$|(7 - 3x) - (-5)| < \varepsilon$. Thus, $\lim\limits_{x \to 4}(7 - 3x) = -5$ by the definition

of a limit.

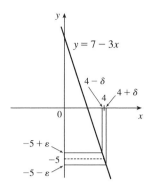

19. Given $\varepsilon > 0$, we need $\delta > 0$ such that if $0 < |x - 3| < \delta$, then $\left|\dfrac{x}{5} - \dfrac{3}{5}\right| < \varepsilon \Leftrightarrow \dfrac{1}{5}|x - 3| < \varepsilon \Leftrightarrow |x - 3| < 5\varepsilon$.

So choose $\delta = 5\varepsilon$. Then $0 < |x - 3| < \delta \Rightarrow |x - 3| < 5\varepsilon \Rightarrow \dfrac{|x - 3|}{5} < \varepsilon \Rightarrow \left|\dfrac{x}{5} - \dfrac{3}{5}\right| < \varepsilon$. By the definition

of a limit, $\lim\limits_{x \to 3}\dfrac{x}{5} = \dfrac{3}{5}$.

20. Given $\varepsilon > 0$, we need $\delta > 0$ such that if $0 < |x - 6| < \delta$, then $\left|\left(\dfrac{x}{4} + 3\right) - \dfrac{9}{2}\right| < \varepsilon \Leftrightarrow \left|\dfrac{x}{4} - \dfrac{3}{2}\right| < \varepsilon \Leftrightarrow$

$\dfrac{1}{4}|x - 6| < \varepsilon \Leftrightarrow |x - 6| < 4\varepsilon$. So choose $\delta = 4\varepsilon$. Then $0 < |x - 6| < \delta \Rightarrow |x - 6| < 4\varepsilon \Rightarrow \dfrac{|x - 6|}{4} < \varepsilon \Rightarrow$

$\left|\dfrac{x}{4} - \dfrac{6}{4}\right| < \varepsilon \Rightarrow \left|\left(\dfrac{x}{4} + 3\right) - \dfrac{9}{2}\right| < \varepsilon$. By the definition of a limit, $\lim\limits_{x \to 6}\left(\dfrac{x}{4} + 3\right) = \dfrac{9}{2}$.

21. Given $\varepsilon > 0$, we need $\delta > 0$ such that if $0 < |x - 2| < \delta$, then $\left|\dfrac{x^2 + x - 6}{x - 2} - 5\right| < \varepsilon \Leftrightarrow$

$\left|\dfrac{(x + 3)(x - 2)}{x - 2} - 5\right| < \varepsilon \Leftrightarrow |x + 3 - 5| < \varepsilon \ [x \neq 2] \Leftrightarrow |x - 2| < \varepsilon$. So choose $\delta = \varepsilon$.

Then $0 < |x - 2| < \delta \Rightarrow |x - 2| < \varepsilon \Rightarrow |x + 3 - 5| < \varepsilon \Rightarrow \left|\dfrac{(x + 3)(x - 2)}{x - 2} - 5\right| < \varepsilon \ [x \neq 2] \Rightarrow$

$\left|\dfrac{x^2 + x - 6}{x - 2} - 5\right| < \varepsilon$. By the definition of a limit, $\lim\limits_{x \to 2}\dfrac{x^2 + x - 6}{x - 2} = 5$.

22. Given $\varepsilon > 0$, we need $\delta > 0$ such that if $0 < |x + 1.5| < \delta$, then $\left|\dfrac{9 - 4x^2}{3 + 2x} - 6\right| < \varepsilon \Leftrightarrow$

$\left|\dfrac{(3 + 2x)(3 - 2x)}{3 + 2x} - 6\right| < \varepsilon \Leftrightarrow |3 - 2x - 6| < \varepsilon \ [x \neq -1.5] \Leftrightarrow |-2x - 3| < \varepsilon \Leftrightarrow |-2|\,|x + 1.5| < \varepsilon \Leftrightarrow$

$|x + 1.5| < \varepsilon/2$. So choose $\delta = \varepsilon/2$. Then $0 < |x + 1.5| < \delta \Rightarrow |x + 1.5| < \varepsilon/2 \Rightarrow |-2|\,|x + 1.5| < \varepsilon \Rightarrow$

$|-2x - 3| < \varepsilon \Rightarrow |3 - 2x - 6| < \varepsilon \Rightarrow \left|\dfrac{(3 + 2x)(3 - 2x)}{3 + 2x} - 6\right| < \varepsilon \ [x \neq -1.5] \Rightarrow \left|\dfrac{9 - 4x^2}{3 + 2x} - 6\right| < \varepsilon$.

By the definition of a limit, $\lim\limits_{x \to -1.5}\dfrac{9 - 4x^2}{3 + 2x} = 6$.

23. Given $\varepsilon > 0$, we need $\delta > 0$ such that if $0 < |x - a| < \delta$, then $|x - a| < \varepsilon$. So $\delta = \varepsilon$ will work.

24. Given $\varepsilon > 0$, we need $\delta > 0$ such that if $0 < |x - a| < \delta$, then $|c - c| < \varepsilon$. But $|c - c| = 0$, so this will be true no matter what δ we pick.

25. Given $\varepsilon > 0$, we need $\delta > 0$ such that if $0 < |x - 0| < \delta$, then $|x^2 - 0| < \varepsilon$ $\Leftrightarrow$ $x^2 < \varepsilon$ $\Leftrightarrow$ $|x| < \sqrt{\varepsilon}$. Take $\delta = \sqrt{\varepsilon}$. Then $0 < |x - 0| < \delta$ $\Rightarrow$ $|x^2 - 0| < \varepsilon$. Thus, $\lim\limits_{x \to 0} x^2 = 0$ by the definition of a limit.

26. Given $\varepsilon > 0$, we need $\delta > 0$ such that if $0 < |x - 0| < \delta$, then $|x^3 - 0| < \varepsilon$ $\Leftrightarrow$ $|x|^3 < \varepsilon$ $\Leftrightarrow$ $|x| < \sqrt[3]{\varepsilon}$. Take $\delta = \sqrt[3]{\varepsilon}$. Then $0 < |x - 0| < \delta$ $\Rightarrow$ $|x^3 - 0| < \delta^3 = \varepsilon$. Thus, $\lim\limits_{x \to 0} x^3 = 0$ by the definition of a limit.

27. Given $\varepsilon > 0$, we need $\delta > 0$ such that if $0 < |x - 0| < \delta$, then $\big||x| - 0\big| < \varepsilon$. But $\big||x|\big| = |x|$. So this is true if we pick $\delta = \varepsilon$. Thus, $\lim\limits_{x \to 0} |x| = 0$ by the definition of a limit.

28. Given $\varepsilon > 0$, we need $\delta > 0$ such that if $9 - \delta < x < 9$, then $\big|\sqrt[4]{9 - x} - 0\big| < \varepsilon$ $\Leftrightarrow$ $\sqrt[4]{9 - x} < \varepsilon$ $\Leftrightarrow$ $0 < 9 - x < \varepsilon^4$ $\Leftrightarrow$ $9 - \varepsilon^4 < x < 9$. So take $\delta = \varepsilon^4$. Then $9 - \delta < x < 9$ $\Rightarrow$ $\big|\sqrt[4]{9 - x} - 0\big| < \varepsilon$. Thus, $\lim\limits_{x \to 9^-} \sqrt[4]{9 - x} = 0$ by the definition of a limit.

29. Given $\varepsilon > 0$, we need $\delta > 0$ such that if $0 < |x - 2| < \delta$, then $\big|(x^2 - 4x + 5) - 1\big| < \varepsilon$ $\Leftrightarrow$ $\big|x^2 - 4x + 4\big| < \varepsilon$ $\Leftrightarrow$ $\big|(x - 2)^2\big| < \varepsilon$. So take $\delta = \sqrt{\varepsilon}$. Then $0 < |x - 2| < \delta$ $\Leftrightarrow$ $|x - 2| < \sqrt{\varepsilon}$ $\Leftrightarrow$ $\big|(x - 2)^2\big| < \varepsilon$. Thus, $\lim\limits_{x \to 2} (x^2 - 4x + 5) = 1$ by the definition of a limit.

30. Given $\varepsilon > 0$, we need $\delta > 0$ such that if $0 < |x - 3| < \delta$, then $\big|(x^2 + x - 4) - 8\big| < \varepsilon$ $\Leftrightarrow$ $\big|x^2 + x - 12\big| < \varepsilon$ $\Leftrightarrow$ $|(x - 3)(x + 4)| < \varepsilon$. Notice that if $|x - 3| < 1$, then $-1 < x - 3 < 1$ $\Rightarrow$ $6 < x + 4 < 8$ $\Rightarrow$ $|x + 4| < 8$. So take $\delta = \min\{1, \varepsilon/8\}$. Then $0 < |x - 3| < \delta$ $\Leftrightarrow$ $|(x - 3)(x + 4)| \leq |8(x - 3)| = 8 \cdot |x - 3| < 8\delta \leq \varepsilon$. Thus, $\lim\limits_{x \to 3} (x^2 + x - 4) = 8$ by the definition of a limit.

31. Given $\varepsilon > 0$, we need $\delta > 0$ such that if $0 < |x - (-2)| < \delta$, then $\big|(x^2 - 1) - 3\big| < \varepsilon$ or upon simplifying we need $|x^2 - 4| < \varepsilon$ whenever $0 < |x + 2| < \delta$. Notice that if $|x + 2| < 1$, then $-1 < x + 2 < 1$ $\Rightarrow$ $-5 < x - 2 < -3$ $\Rightarrow$ $|x - 2| < 5$. So take $\delta = \min\{\varepsilon/5, 1\}$. Then $0 < |x + 2| < \delta$ $\Rightarrow$ $|x - 2| < 5$ and $|x + 2| < \varepsilon/5$, so $\big|(x^2 - 1) - 3\big| = |(x + 2)(x - 2)| = |x + 2|\,|x - 2| < (\varepsilon/5)(5) = \varepsilon$. Thus, by the definition of a limit, $\lim\limits_{x \to -2} (x^2 - 1) = 3$.

32. Given $\varepsilon > 0$, we need $\delta > 0$ such that if $0 < |x - 2| < \delta$, then $|x^3 - 8| < \varepsilon$. Now $|x^3 - 8| = \big|(x - 2)(x^2 + 2x + 4)\big|$. If $|x - 2| < 1$, that is, $1 < x < 3$, then $x^2 + 2x + 4 < 3^2 + 2(3) + 4 = 19$ and so $|x^3 - 8| = |x - 2|\,(x^2 + 2x + 4) < 19\,|x - 2|$. So if we take $\delta = \min\left\{1, \frac{\varepsilon}{19}\right\}$, then $0 < |x - 2| < \delta$ $\Rightarrow$ $|x^3 - 8| = |x - 2|\,(x^2 + 2x + 4) < \frac{\varepsilon}{19} \cdot 19 = \varepsilon$. Thus, by the definition of a limit, $\lim\limits_{x \to 2} x^3 = 8$.

33. Given $\varepsilon > 0$, we let $\delta = \min\left\{2, \frac{\varepsilon}{8}\right\}$. If $0 < |x - 3| < \delta$, then $|x - 3| < 2 \quad \Rightarrow \quad -2 < x - 3 < 2 \quad \Rightarrow$

$4 < x + 3 < 8 \quad \Rightarrow \quad |x + 3| < 8$. Also $|x - 3| < \frac{\varepsilon}{8}$, so $\left|x^2 - 9\right| = |x + 3|\,|x - 3| < 8 \cdot \frac{\varepsilon}{8} = \varepsilon$. Thus, $\lim\limits_{x \to 3} x^2 = 9$.

34. From the figure, our choices for δ are $\delta_1 = 3 - \sqrt{9 - \varepsilon}$ and

$\delta_2 = \sqrt{9 + \varepsilon} - 3$. The *largest* possible choice for δ is the minimum

value of $\{\delta_1, \delta_2\}$; that is, $\delta = \min\{\delta_1, \delta_2\} = \delta_2 = \sqrt{9 + \varepsilon} - 3$.

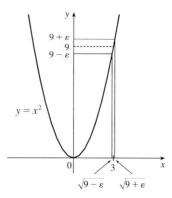

35. (a) The points of intersection in the graph are $(x_1, 2.6)$ and $(x_2, 3.4)$

with $x_1 \approx 0.891$ and $x_2 \approx 1.093$. Thus, we can take δ to be the

smaller of $1 - x_1$ and $x_2 - 1$. So $\delta = x_2 - 1 \approx 0.093$.

(b) Solving $x^3 + x + 1 = 3 + \varepsilon$ gives us two nonreal complex roots and one real root, which is

$$x(\varepsilon) = \frac{\left(216 + 108\varepsilon + 12\sqrt{336 + 324\varepsilon + 81\varepsilon^2}\,\right)^{2/3} - 12}{6\left(216 + 108\varepsilon + 12\sqrt{336 + 324\varepsilon + 81\varepsilon^2}\,\right)^{1/3}}. \text{ Thus, } \delta = x(\varepsilon) - 1.$$

(c) If $\varepsilon = 0.4$, then $x(\varepsilon) \approx 1.093\,272\,342$ and $\delta = x(\varepsilon) - 1 \approx 0.093$, which agrees with our answer in part (a).

36. *1. Guessing a value for δ* Let $\varepsilon > 0$ be given. We have to find a number $\delta > 0$ such that $\left|\dfrac{1}{x} - \dfrac{1}{2}\right| < \varepsilon$ whenever

$0 < |x - 2| < \delta$. But $\left|\dfrac{1}{x} - \dfrac{1}{2}\right| = \left|\dfrac{2 - x}{2x}\right| = \dfrac{|x - 2|}{|2x|} < \varepsilon$. We find a positive constant C such that $\dfrac{1}{|2x|} < C \quad \Rightarrow$

$\dfrac{|x - 2|}{|2x|} < C\,|x - 2|$ and we can make $C\,|x - 2| < \varepsilon$ by taking $|x - 2| < \dfrac{\varepsilon}{C} = \delta$. We restrict x to lie in the interval

$|x - 2| < 1 \quad \Rightarrow \quad 1 < x < 3$ so $1 > \dfrac{1}{x} > \dfrac{1}{3} \quad \Rightarrow \quad \dfrac{1}{6} < \dfrac{1}{2x} < \dfrac{1}{2} \quad \Rightarrow \quad \dfrac{1}{|2x|} < \dfrac{1}{2}$. So $C = \dfrac{1}{2}$ is suitable. Thus, we should

choose $\delta = \min\{1, 2\varepsilon\}$.

2. Showing that δ works Given $\varepsilon > 0$ we let $\delta = \min\{1, 2\varepsilon\}$. If $0 < |x - 2| < \delta$, then $|x - 2| < 1 \quad \Rightarrow \quad 1 < x < 3 \quad \Rightarrow$

$\dfrac{1}{|2x|} < \dfrac{1}{2}$ (as in part 1). Also $|x - 2| < 2\varepsilon$, so $\left|\dfrac{1}{x} - \dfrac{1}{2}\right| = \dfrac{|x - 2|}{|2x|} < \dfrac{1}{2} \cdot 2\varepsilon = \varepsilon$. This shows that $\lim\limits_{x \to 2}(1/x) = \frac{1}{2}$.

37. *1. Guessing a value for δ*   Given $\varepsilon > 0$, we must find $\delta > 0$ such that $|\sqrt{x} - \sqrt{a}| < \varepsilon$ whenever $0 < |x - a| < \delta$. But

$|\sqrt{x} - \sqrt{a}| = \dfrac{|x - a|}{\sqrt{x} + \sqrt{a}} < \varepsilon$ (from the hint). Now if we can find a positive constant C such that $\sqrt{x} + \sqrt{a} > C$ then

$\dfrac{|x - a|}{\sqrt{x} + \sqrt{a}} < \dfrac{|x - a|}{C} < \varepsilon$, and we take $|x - a| < C\varepsilon$. We can find this number by restricting x to lie in some interval

centered at a. If $|x - a| < \frac{1}{2}a$, then $-\frac{1}{2}a < x - a < \frac{1}{2}a$ $\Rightarrow$ $\frac{1}{2}a < x < \frac{3}{2}a$ $\Rightarrow$ $\sqrt{x} + \sqrt{a} > \sqrt{\frac{1}{2}a} + \sqrt{a}$, and so

$C = \sqrt{\frac{1}{2}a} + \sqrt{a}$ is a suitable choice for the constant. So $|x - a| < \left(\sqrt{\frac{1}{2}a} + \sqrt{a}\right)\varepsilon$. This suggests that we let

$\delta = \min\left\{\frac{1}{2}a, \left(\sqrt{\frac{1}{2}a} + \sqrt{a}\right)\varepsilon\right\}$.

2. Showing that δ works   Given $\varepsilon > 0$, we let $\delta = \min\left\{\frac{1}{2}a, \left(\sqrt{\frac{1}{2}a} + \sqrt{a}\right)\varepsilon\right\}$. If $0 < |x - a| < \delta$, then

$|x - a| < \frac{1}{2}a$ $\Rightarrow$ $\sqrt{x} + \sqrt{a} > \sqrt{\frac{1}{2}a} + \sqrt{a}$ (as in part 1). Also $|x - a| < \left(\sqrt{\frac{1}{2}a} + \sqrt{a}\right)\varepsilon$, so

$|\sqrt{x} - \sqrt{a}| = \dfrac{|x - a|}{\sqrt{x} + \sqrt{a}} < \dfrac{\left(\sqrt{a/2} + \sqrt{a}\right)\varepsilon}{\left(\sqrt{a/2} + \sqrt{a}\right)} = \varepsilon$. Therefore, $\displaystyle\lim_{x \to a} \sqrt{x} = \sqrt{a}$ by the definition of a limit.

38. Suppose that $\displaystyle\lim_{t \to 0} H(t) = L$. Given $\varepsilon = \frac{1}{2}$, there exists $\delta > 0$ such that $0 < |t| < \delta$ $\Rightarrow$ $|H(t) - L| < \frac{1}{2}$ $\Leftrightarrow$

$L - \frac{1}{2} < H(t) < L + \frac{1}{2}$. For $0 < t < \delta$, $H(t) = 1$, so $1 < L + \frac{1}{2}$ $\Rightarrow$ $L > \frac{1}{2}$. For $-\delta < t < 0$, $H(t) = 0$,

so $L - \frac{1}{2} < 0$ $\Rightarrow$ $L < \frac{1}{2}$. This contradicts $L > \frac{1}{2}$. Therefore, $\displaystyle\lim_{t \to 0} H(t)$ does not exist.

39. Suppose that $\displaystyle\lim_{x \to 0} f(x) = L$. Given $\varepsilon = \frac{1}{2}$, there exists $\delta > 0$ such that $0 < |x| < \delta$ $\Rightarrow$ $|f(x) - L| < \frac{1}{2}$. Take any rational

number r with $0 < |r| < \delta$. Then $f(r) = 0$, so $|0 - L| < \frac{1}{2}$, so $L \leq |L| < \frac{1}{2}$. Now take any irrational number s with

$0 < |s| < \delta$. Then $f(s) = 1$, so $|1 - L| < \frac{1}{2}$. Hence, $1 - L < \frac{1}{2}$, so $L > \frac{1}{2}$. This contradicts $L < \frac{1}{2}$, so $\displaystyle\lim_{x \to 0} f(x)$ does not

exist.

40. First suppose that $\displaystyle\lim_{x \to a} f(x) = L$. Then, given $\varepsilon > 0$ there exists $\delta > 0$ so that $0 < |x - a| < \delta$ $\Rightarrow$ $|f(x) - L| < \varepsilon$.

Then $a - \delta < x < a$ $\Rightarrow$ $0 < |x - a| < \delta$ so $|f(x) - L| < \varepsilon$. Thus, $\displaystyle\lim_{x \to a^-} f(x) = L$. Also $a < x < a + \delta$ $\Rightarrow$

$0 < |x - a| < \delta$ so $|f(x) - L| < \varepsilon$. Hence, $\displaystyle\lim_{x \to a^+} f(x) = L$.

   Now suppose $\displaystyle\lim_{x \to a^-} f(x) = L = \lim_{x \to a^+} f(x)$. Let $\varepsilon > 0$ be given. Since $\displaystyle\lim_{x \to a^-} f(x) = L$, there exists $\delta_1 > 0$ so that

$a - \delta_1 < x < a$ $\Rightarrow$ $|f(x) - L| < \varepsilon$. Since $\displaystyle\lim_{x \to a^+} f(x) = L$, there exists $\delta_2 > 0$ so that $a < x < a + \delta_2$ $\Rightarrow$

$|f(x) - L| < \varepsilon$. Let δ be the smaller of δ_1 and δ_2. Then $0 < |x - a| < \delta$ $\Rightarrow$ $a - \delta_1 < x < a$ or $a < x < a + \delta_2$ so

$|f(x) - L| < \varepsilon$. Hence, $\displaystyle\lim_{x \to a} f(x) = L$. So we have proved that $\displaystyle\lim_{x \to a} f(x) = L$ $\Leftrightarrow$ $\displaystyle\lim_{x \to a^-} f(x) = L = \lim_{x \to a^+} f(x)$.

41. $\dfrac{1}{(x+3)^4} > 10{,}000 \Leftrightarrow (x+3)^4 < \dfrac{1}{10{,}000} \Leftrightarrow |x+3| < \dfrac{1}{\sqrt[4]{10{,}000}} \Leftrightarrow |x-(-3)| < \dfrac{1}{10}$

42. Given $M > 0$, we need $\delta > 0$ such that $0 < |x+3| < \delta \Rightarrow 1/(x+3)^4 > M$. Now $\dfrac{1}{(x+3)^4} > M \Leftrightarrow$

$(x+3)^4 < \dfrac{1}{M} \Leftrightarrow |x+3| < \dfrac{1}{\sqrt[4]{M}}$. So take $\delta = \dfrac{1}{\sqrt[4]{M}}$. Then $0 < |x+3| < \delta = \dfrac{1}{\sqrt[4]{M}} \Rightarrow \dfrac{1}{(x+3)^4} > M$, so

$\lim\limits_{x\to -3} \dfrac{1}{(x+3)^4} = \infty$.

43. Let $N < 0$ be given. Then, for $x < -1$, we have $\dfrac{5}{(x+1)^3} < N \Leftrightarrow \dfrac{5}{N} < (x+1)^3 \Leftrightarrow \sqrt[3]{\dfrac{5}{N}} < x+1$.

Let $\delta = -\sqrt[3]{\dfrac{5}{N}}$. Then $-1-\delta < x < -1 \Rightarrow \sqrt[3]{\dfrac{5}{N}} < x+1 < 0 \Rightarrow \dfrac{5}{(x+1)^3} < N$, so $\lim\limits_{x\to -1^-} \dfrac{5}{(x+1)^3} = -\infty$.

44. (a) Let M be given. Since $\lim\limits_{x\to a} f(x) = \infty$, there exists $\delta_1 > 0$ such that $0 < |x-a| < \delta_1 \Rightarrow f(x) > M+1-c$. Since

$\lim\limits_{x\to a} g(x) = c$, there exists $\delta_2 > 0$ such that $0 < |x-a| < \delta_2 \Rightarrow |g(x)-c| < 1 \Rightarrow g(x) > c-1$. Let δ be the

smaller of δ_1 and δ_2. Then $0 < |x-a| < \delta \Rightarrow f(x)+g(x) > (M+1-c)+(c-1) = M$. Thus,

$\lim\limits_{x\to a} [f(x)+g(x)] = \infty$.

(b) Let $M > 0$ be given. Since $\lim\limits_{x\to a} g(x) = c > 0$, there exists $\delta_1 > 0$ such that $0 < |x-a| < \delta_1 \Rightarrow$

$|g(x)-c| < c/2 \Rightarrow g(x) > c/2$. Since $\lim\limits_{x\to a} f(x) = \infty$, there exists $\delta_2 > 0$ such that $0 < |x-a| < \delta_2 \Rightarrow$

$f(x) > 2M/c$. Let $\delta = \min\{\delta_1, \delta_2\}$. Then $0 < |x-a| < \delta \Rightarrow f(x)g(x) > \dfrac{2M}{c}\dfrac{c}{2} = M$, so $\lim\limits_{x\to a} f(x)g(x) = \infty$.

(c) Let $N < 0$ be given. Since $\lim\limits_{x\to a} g(x) = c < 0$, there exists $\delta_1 > 0$ such that $0 < |x-a| < \delta_1 \Rightarrow$

$|g(x)-c| < -c/2 \Rightarrow g(x) < c/2$. Since $\lim\limits_{x\to a} f(x) = \infty$, there exists $\delta_2 > 0$ such that $0 < |x-a| < \delta_2 \Rightarrow$

$f(x) > 2N/c$. (Note that $c < 0$ and $N < 0 \Rightarrow 2N/c > 0$.) Let $\delta = \min\{\delta_1, \delta_2\}$. Then $0 < |x-a| < \delta \Rightarrow$

$f(x) > 2N/c \Rightarrow f(x)g(x) < \dfrac{2N}{c}\cdot\dfrac{c}{2} = N$, so $\lim\limits_{x\to a} f(x)g(x) = -\infty$.

2.5 Continuity

1. From Definition 1, $\lim\limits_{x\to 4} f(x) = f(4)$.

2. The graph of f has no hole, jump, or vertical asymptote.

3. (a) The following are the numbers at which f is discontinuous and the type of discontinuity at that number: -4 (removable),

-2 (jump), 2 (jump), 4 (infinite).

(b) f is continuous from the left at -2 since $\lim\limits_{x\to -2^-} f(x) = f(-2)$. f is continuous from the right at 2 and 4 since

$\lim\limits_{x\to 2^+} f(x) = f(2)$ and $\lim\limits_{x\to 4^+} f(x) = f(4)$. It is continuous from neither side at -4 since $f(-4)$ is undefined.

4. g is continuous on $[-4, -2), (-2, 2), [2, 4), (4, 6)$, and $(6, 8)$.

5. The graph of $y = f(x)$ must have a discontinuity at $x = 3$ and must show that $\lim\limits_{x \to 3^-} f(x) = f(3)$.

6.

7. (a)

(b) There are discontinuities at times $t = 1, 2, 3$, and 4. A person parking in the lot would want to keep in mind that the charge will jump at the beginning of each hour.

8. (a) Continuous; at the location in question, the temperature changes smoothly as time passes, without any instantaneous jumps from one temperature to another.

(b) Continuous; the temperature at a specific time changes smoothly as the distance due west from New York City increases, without any instantaneous jumps.

(c) Discontinuous; as the distance due west from New York City increases, the altitude above sea level may jump from one height to another without going through all of the intermediate values — at a cliff, for example.

(d) Discontinuous; as the distance traveled increases, the cost of the ride jumps in small increments.

(e) Discontinuous; when the lights are switched on (or off), the current suddenly changes between 0 and some nonzero value, without passing through all of the intermediate values. This is debatable, though, depending on your definition of current.

9. Since f and g are continuous functions,

$$\lim_{x \to 3} [2f(x) - g(x)] = 2 \lim_{x \to 3} f(x) - \lim_{x \to 3} g(x) \qquad \text{[by Limit Laws 2 and 3]}$$

$$= 2f(3) - g(3) \qquad \text{[by continuity of } f \text{ and } g \text{ at } x = 3]$$

$$= 2 \cdot 5 - g(3) = 10 - g(3)$$

Since it is given that $\lim\limits_{x \to 3} [2f(x) - g(x)] = 4$, we have $10 - g(3) = 4$, so $g(3) = 6$.

10. $\lim\limits_{x \to 4} f(x) = \lim\limits_{x \to 4} \left(x^2 + \sqrt{7 - x} \right) = \lim\limits_{x \to 4} x^2 + \sqrt{\lim\limits_{x \to 4} 7 - \lim\limits_{x \to 4} x} = 4^2 + \sqrt{7 - 4} = 16 + \sqrt{3} = f(4)$.

By the definition of continuity, f is continuous at $a = 4$.

11. $\lim\limits_{x \to -1} f(x) = \lim\limits_{x \to -1} \left(x + 2x^3 \right)^4 = \left(\lim\limits_{x \to -1} x + 2 \lim\limits_{x \to -1} x^3 \right)^4 = \left[-1 + 2(-1)^3 \right]^4 = (-3)^4 = 81 = f(-1)$.

By the definition of continuity, f is continuous at $a = -1$.

12. $\lim\limits_{t\to1} h(t) = \lim\limits_{t\to1} \dfrac{2t - 3t^2}{1 + t^3} = \dfrac{\lim\limits_{t\to1}(2t - 3t^2)}{\lim\limits_{t\to1}(1 + t^3)} = \dfrac{2\lim\limits_{t\to1} t - 3\lim\limits_{t\to1} t^2}{\lim\limits_{t\to1} 1 + \lim\limits_{t\to1} t^3} = \dfrac{2(1) - 3(1)^2}{1 + (1)^3} = \dfrac{-1}{2} = h(1).$

By the definition of continuity, h is continuous at $a = 1$.

13. For $a > 2$, we have

$$\lim_{x\to a} f(x) = \lim_{x\to a} \frac{2x + 3}{x - 2} = \frac{\lim\limits_{x\to a}(2x + 3)}{\lim\limits_{x\to a}(x - 2)} \qquad \text{[Limit Law 5]}$$

$$= \frac{2\lim\limits_{x\to a} x + \lim\limits_{x\to a} 3}{\lim\limits_{x\to a} x - \lim\limits_{x\to a} 2} \qquad \text{[1, 2, and 3]}$$

$$= \frac{2a + 3}{a - 2} \qquad \text{[7 and 8]}$$

$$= f(a)$$

Thus, f is continuous at $x = a$ for every a in $(2, \infty)$; that is, f is continuous on $(2, \infty)$.

14. For $a < 3$, we have $\lim\limits_{x\to a} g(x) = \lim\limits_{x\to a} 2\sqrt{3 - x} = 2\lim\limits_{x\to a} \sqrt{3 - x}$ [Limit Law 3] $= 2\sqrt{\lim\limits_{x\to a}(3 - x)}$ [11]

$= 2\sqrt{\lim\limits_{x\to a} 3 - \lim\limits_{x\to a} x}$ [2] $= 2\sqrt{3 - a}$ [7 and 8] $= g(a)$, so g is continuous at $x = a$ for every a in $(-\infty, 3)$.

Also, $\lim\limits_{x\to3^-} g(x) = 0 = g(3)$, so g is continuous from the left at 3. Thus, g is continuous on $(-\infty, 3]$.

15. $f(x) = -\dfrac{1}{(x - 1)^2}$ is discontinuous at 1 since $f(1)$ is not defined.

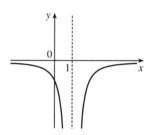

16. $f(x) = \begin{cases} 1/(x - 1) & \text{if } x \neq 1 \\ 2 & \text{if } x = 1 \end{cases}$ is discontinuous at 1 because $\lim\limits_{x\to1} f(x)$

does not exist.

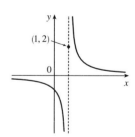

17. $f(x) = \begin{cases} 1 - x^2 & \text{if } x < 1 \\ 1/x & \text{if } x \geq 1 \end{cases}$

The left-hand limit of f at $a = 1$ is

$\lim\limits_{x\to1^-} f(x) = \lim\limits_{x\to1^-}(1 - x^2) = 0$. The right-hand limit of f at $a = 1$ is

$\lim\limits_{x\to1^+} f(x) = \lim\limits_{x\to1^+}(1/x) = 1$. Since these limits are not equal, $\lim\limits_{x\to1} f(x)$

does not exist and f is discontinuous at 1.

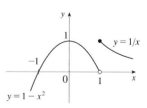

18. $f(x) = \begin{cases} \dfrac{x^2 - x}{x^2 - 1} & \text{if } x \neq 1 \\ 1 & \text{if } x = 1 \end{cases}$

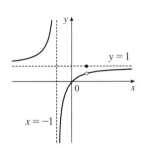

$\lim\limits_{x \to 1} f(x) = \lim\limits_{x \to 1} \dfrac{x^2 - x}{x^2 - 1} = \lim\limits_{x \to 1} \dfrac{x(x-1)}{(x+1)(x-1)} = \lim\limits_{x \to 1} \dfrac{x}{x+1} = \dfrac{1}{2}$,

but $f(1) = 1$, so f is discontinous at 1.

19. $f(x) = \begin{cases} \cos x & \text{if } x < 0 \\ 0 & \text{if } x = 0 \\ 1 - x^2 & \text{if } x > 0 \end{cases}$

$\lim\limits_{x \to 0} f(x) = 1$, but $f(0) = 0 \neq 1$, so f is discontinuous at 0.

20. $f(x) = \begin{cases} \dfrac{2x^2 - 5x - 3}{x - 3} & \text{if } x \neq 3 \\ 6 & \text{if } x = 3 \end{cases}$

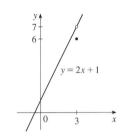

$\lim\limits_{x \to 3} f(x) = \lim\limits_{x \to 3} \dfrac{2x^2 - 5x - 3}{x - 3} = \lim\limits_{x \to 3} \dfrac{(2x+1)(x-3)}{x-3} = \lim\limits_{x \to 3}(2x+1) = 7$,

but $f(3) = 6$, so f is discontinuous at 3.

21. $F(x) = \dfrac{x}{x^2 + 5x + 6}$ is a rational function. So by Theorem 5 (or Theorem 7), F is continuous at every number in its domain,

$\{x \mid x^2 + 5x + 6 \neq 0\} = \{x \mid (x+3)(x+2) \neq 0\} = \{x \mid x \neq -3, \, -2\}$ or $(-\infty, -3) \cup (-3, -2) \cup (-2, \infty)$.

22. By Theorem 7, the root function $\sqrt[3]{x}$ and the polynomial function $1 + x^3$ are continuous on $\mathbb{R}$. By part 4 of Theorem 4, the product $G(x) = \sqrt[3]{x}\left(1 + x^3\right)$ is continuous on its domain, $\mathbb{R}$.

23. By Theorem 5, the polynomials x^2 and $2x - 1$ are continuous on $(-\infty, \infty)$. By Theorem 7, the root function $\sqrt{x}$ is continuous on $[0, \infty)$. By Theorem 9, the composite function $\sqrt{2x - 1}$ is continuous on its domain, $[\frac{1}{2}, \infty)$.

By part 1 of Theorem 4, the sum $R(x) = x^2 + \sqrt{2x - 1}$ is continuous on $[\frac{1}{2}, \infty)$.

24. By Theorem 7, the trigonometric function $\sin x$ and the polynomial function $x + 1$ are continuous on $\mathbb{R}$.

By part 5 of Theorem 4, $h(x) = \dfrac{\sin x}{x + 1}$ is continuous on its domain, $\{x \mid x \neq -1\}$.

25. By Theorem 5, the polynomial $1 - x^2$ is continuous on $(-\infty, \infty)$. By Theorem 7, $\cos$ is continuous on its domain, $\mathbb{R}$. By Theorem 9, $\cos\left(1 - x^2\right)$ is continuous on its domain, which is $\mathbb{R}$.

26. By Theorem 5, the polynomial $2x$ is continuous on $(-\infty, \infty)$. By Theorem 7, $\tan$ is continuous at every number in its domain; that is, $\{x \mid x \neq \frac{\pi}{2} + \pi n\}$. By Theorem 9, $\tan 2x$ is continuous on its domain, which is

$\{x \mid 2x \neq \frac{\pi}{2} + \pi n\} = \{x \mid x \neq \frac{\pi}{4} + \frac{\pi}{2} n\}$ (the odd multiples of $\frac{\pi}{4}$).

27. By Theorem 7, the root function $\sqrt{x}$ and the trigonometric function $\sin x$ are continuous on their domains, $[0, \infty)$ and $(-\infty, \infty)$, respectively. Thus, the product $F(x) = \sqrt{x} \sin x$ is continuous on the intersection of those domains, $[0, \infty)$, by part 4 of Theorem 4.

28. The sine and cosine functions are continuous everywhere by Theorem 7, so $F(x) = \sin(\cos(\sin x))$, which is the composite of sine, cosine, and (once again) sine, is continuous everywhere by Theorem 9.

29. $y = \dfrac{1}{1 + \sin x}$ is undefined and hence discontinuous when

$1 + \sin x = 0 \quad \Leftrightarrow \quad \sin x = -1 \quad \Leftrightarrow \quad x = -\frac{\pi}{2} + 2\pi n$, n an integer. The figure shows discontinuities for $n = -1, 0$, and 1; that

is, $-\dfrac{5\pi}{2} \approx -7.85$, $-\dfrac{\pi}{2} \approx -1.57$, and $\dfrac{3\pi}{2} \approx 4.71$.

30.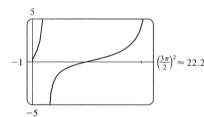

$\left(\frac{3\pi}{2}\right)^2 \approx 22.2$ $\left(\frac{9\pi}{2}\right)^2 \approx 200$

The function $y = f(x) = \tan \sqrt{x}$ is continuous throughout its domain because it is the composite of a trigonometric function and a root function. The square root function has domain $[0, \infty)$ and the tangent function has domain $\{x \mid x \neq \frac{\pi}{2} + \pi n\}$. So f is discontinuous when $x < 0$ and when $\sqrt{x} = \frac{\pi}{2} + \pi n \quad \Rightarrow \quad x = \left(\frac{\pi}{2} + \pi n\right)^2$, where n is a nonnegative integer. Note that as x increases, the distance between discontinuities increases.

31. Because we are dealing with root functions, $5 + \sqrt{x}$ is continuous on $[0, \infty)$, $\sqrt{x + 5}$ is continuous on $[-5, \infty)$, so the quotient $f(x) = \dfrac{5 + \sqrt{x}}{\sqrt{5 + x}}$ is continuous on $[0, \infty)$. Since f is continuous at $x = 4$, $\lim\limits_{x \to 4} f(x) = f(4) = \frac{7}{3}$.

32. Because x is continuous on $\mathbb{R}$, $\sin x$ is continuous on $\mathbb{R}$, and $x + \sin x$ is continuous on $\mathbb{R}$, the composite function $f(x) = \sin(x + \sin x)$ is continuous on $\mathbb{R}$, so $\lim\limits_{x \to \pi} f(x) = f(\pi) = \sin(\pi + \sin \pi) = \sin \pi = 0$.

33. Because x and $\cos x$ are continuous on $\mathbb{R}$, so is $f(x) = x \cos^2 x$. Since f is continuous at $x = \frac{\pi}{4}$,

$$\lim_{x \to \pi/4} f(x) = f\left(\frac{\pi}{4}\right) = \frac{\pi}{4}\left(\frac{\sqrt{2}}{2}\right)^2 = \frac{\pi}{4} \cdot \frac{1}{2} = \frac{\pi}{8}.$$

34. $x^3 - 3x + 1 = 0$ for three values of x, but 2 is not one of them. Thus, $f(x) = (x^3 - 3x + 1)^{-3}$ is continuous at $x = 2$ and $\lim\limits_{x \to 2} f(x) = f(2) = (8 - 6 + 1)^{-3} = 3^{-3} = \frac{1}{27}$.

35. $f(x) = \begin{cases} x^2 & \text{if } x < 1 \\ \sqrt{x} & \text{if } x \geq 1 \end{cases}$

By Theorem 5, since $f(x)$ equals the polynomial x^2 on $(-\infty, 1)$, f is continuous on $(-\infty, 1)$. By Theorem 7, since $f(x)$

equals the root function $\sqrt{x}$ on $(1, \infty)$, f is continuous on $(1, \infty)$. At $x = 1$, $\lim\limits_{x \to 1^-} f(x) = \lim\limits_{x \to 1^-} x^2 = 1$ and

$\lim\limits_{x \to 1^+} f(x) = \lim\limits_{x \to 1^+} \sqrt{x} = 1$. Thus, $\lim\limits_{x \to 1} f(x)$ exists and equals 1. Also, $f(1) = \sqrt{1} = 1$. Thus, f is continuous at $x = 1$.

We conclude that f is continuous on $(-\infty, \infty)$.

36. $f(x) = \begin{cases} \sin x & \text{if } x < \pi/4 \\ \cos x & \text{if } x \geq \pi/4 \end{cases}$

By Theorem 7, the trigonometric functions are continuous. Since $f(x) = \sin x$ on $(-\infty, \pi/4)$ and $f(x) = \cos x$ on

$(\pi/4, \infty)$, f is continuous on $(-\infty, \pi/4) \cup (\pi/4, \infty)$. $\lim\limits_{x \to (\pi/4)^-} f(x) = \lim\limits_{x \to (\pi/4)^-} \sin x = \sin \frac{\pi}{4} = 1/\sqrt{2}$ since the sine

function is continuous at $\pi/4$. Similarly, $\lim\limits_{x \to (\pi/4)^+} f(x) = \lim\limits_{x \to (\pi/4)^+} \cos x = 1/\sqrt{2}$ by continuity of the cosine function at

$\pi/4$. Thus, $\lim\limits_{x \to (\pi/4)} f(x)$ exists and equals $1/\sqrt{2}$, which agrees with the value $f(\pi/4)$. Therefore, f is continuous at $\pi/4$, so

f is continuous on $(-\infty, \infty)$.

37. $f(x) = \begin{cases} 1 + x^2 & \text{if } x \leq 0 \\ 2 - x & \text{if } 0 < x \leq 2 \\ (x-2)^2 & \text{if } x > 2 \end{cases}$

f is continuous on $(-\infty, 0)$, $(0, 2)$, and $(2, \infty)$ since it is a polynomial on

each of these intervals. Now $\lim\limits_{x \to 0^-} f(x) = \lim\limits_{x \to 0^-} (1 + x^2) = 1$ and $\lim\limits_{x \to 0^+} f(x) = \lim\limits_{x \to 0^+} (2 - x) = 2$, so f is

discontinuous at 0. Since $f(0) = 1$, f is continuous from the left at 0. Also, $\lim\limits_{x \to 2^-} f(x) = \lim\limits_{x \to 2^-} (2 - x) = 0$,

$\lim\limits_{x \to 2^+} f(x) = \lim\limits_{x \to 2^+} (x - 2)^2 = 0$, and $f(2) = 0$, so f is continuous at 2. The only number at which f is discontinuous is 0.

38. $f(x) = \begin{cases} x + 1 & \text{if } x \leq 1 \\ 1/x & \text{if } 1 < x < 3 \\ \sqrt{x - 3} & \text{if } x \geq 3 \end{cases}$

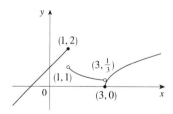

f is continuous on $(-\infty, 1)$, $(1, 3)$, and $(3, \infty)$, where it is a polynomial,

a rational function, and a composite of a root function with a polynomial,

respectively. Now $\lim\limits_{x \to 1^-} f(x) = \lim\limits_{x \to 1^-} (x + 1) = 2$ and $\lim\limits_{x \to 1^+} f(x) = \lim\limits_{x \to 1^+} (1/x) = 1$, so f is discontinuous at 1.

Since $f(1) = 2$, f is continuous from the left at 1. Also, $\lim\limits_{x \to 3^-} f(x) = \lim\limits_{x \to 3^-} (1/x) = 1/3$, and

$\lim\limits_{x \to 3^+} f(x) = \lim\limits_{x \to 3^+} \sqrt{x - 3} = 0 = f(3)$, so f is discontinuous at 3, but it is continuous from the right at 3.

39. $f(x) = \begin{cases} x+2 & \text{if } x < 0 \\ 2x^2 & \text{if } 0 \le x \le 1 \\ 2-x & \text{if } x > 1 \end{cases}$

f is continuous on $(-\infty, 0)$, $(0, 1)$, and $(1, \infty)$ since on each of

these intervals it is a polynomial. Now $\lim\limits_{x \to 0^-} f(x) = \lim\limits_{x \to 0^-} (x+2) = 2$ and

$\lim\limits_{x \to 0^+} f(x) = \lim\limits_{x \to 0^+} 2x^2 = 0$, so f is discontinuous at 0. Since $f(0) = 0$, f is continuous from the right at 0. Also

$\lim\limits_{x \to 1^-} f(x) = \lim\limits_{x \to 1^-} 2x^2 = 2$ and $\lim\limits_{x \to 1^+} f(x) = \lim\limits_{x \to 1^+} (2-x) = 1$, so f is discontinuous at 1. Since $f(1) = 2$,

f is continuous from the left at 1.

40. By Theorem 5, each piece of F is continuous on its domain. We need to check for continuity at $r = R$.

$\lim\limits_{r \to R^-} F(r) = \lim\limits_{r \to R^-} \dfrac{GMr}{R^3} = \dfrac{GM}{R^2}$ and $\lim\limits_{r \to R^+} F(r) = \lim\limits_{r \to R^+} \dfrac{GM}{r^2} = \dfrac{GM}{R^2}$, so $\lim\limits_{r \to R} F(r) = \dfrac{GM}{R^2}$. Since $F(R) = \dfrac{GM}{R^2}$,

F is continuous at R. Therefore, F is a continuous function of r.

41. $f(x) = \begin{cases} cx^2 + 2x & \text{if } x < 2 \\ x^3 - cx & \text{if } x \ge 2 \end{cases}$

f is continuous on $(-\infty, 2)$ and $(2, \infty)$. Now $\lim\limits_{x \to 2^-} f(x) = \lim\limits_{x \to 2^-} \left(cx^2 + 2x \right) = 4c + 4$ and

$\lim\limits_{x \to 2^+} f(x) = \lim\limits_{x \to 2^+} \left(x^3 - cx \right) = 8 - 2c$. So f is continuous $\Leftrightarrow$ $4c + 4 = 8 - 2c$ $\Leftrightarrow$ $6c = 4$ $\Leftrightarrow$ $c = \frac{2}{3}$. Thus, for f

to be continuous on $(-\infty, \infty)$, $c = \frac{2}{3}$.

42. $f(x) = \begin{cases} \dfrac{x^2 - 4}{x - 2} & \text{if } x < 2 \\ ax^2 - bx + 3 & \text{if } 2 < x < 3 \\ 2x - a + b & \text{if } x \ge 3 \end{cases}$

At $x = 2$: $\lim\limits_{x \to 2^-} f(x) = \lim\limits_{x \to 2^-} \dfrac{x^2 - 4}{x - 2} = \lim\limits_{x \to 2^-} \dfrac{(x+2)(x-2)}{x-2} = \lim\limits_{x \to 2^-} (x+2) = 2 + 2 = 4$

$\lim\limits_{x \to 2^+} f(x) = \lim\limits_{x \to 2^+} \left(ax^2 - bx + 3 \right) = 4a - 2b + 3$

We must have $4a - b + 3 = 4$, or $\mathbf{4a - 2b = 1}$ **(1)**.

At $x = 3$: $\lim\limits_{x \to 3^-} f(x) = \lim\limits_{x \to 3^-} \left(ax^2 - bx + 3 \right) = 9a - 3b + 3$

$\lim\limits_{x \to 3^+} f(x) = \lim\limits_{x \to 3^+} \left(2x - a + b \right) = 6 - a + b$

We must have $9a - 3b + 3 = 6 - a + b$, or $\mathbf{10a - 4b = 3}$ **(2)**.

Now solve the system of equations by adding -2 times equation **(1)** to equation **(2)**.

$$-8a + 4b = -2$$
$$\underline{10a - 4b = 3}$$
$$2a = 1$$

So $a = \frac{1}{2}$. Substituting $\frac{1}{2}$ for a in **(1)** gives us $-2b = -1$, so $b = \frac{1}{2}$ as well. Thus, for f to be continuous on $(-\infty, \infty)$,

$a = b = \frac{1}{2}$.

43. (a) $f(x) = \dfrac{x^4 - 1}{x - 1} = \dfrac{(x^2 + 1)(x^2 - 1)}{x - 1} = \dfrac{(x^2 + 1)(x + 1)(x - 1)}{x - 1} = (x^2 + 1)(x + 1)$ [or $x^3 + x^2 + x + 1$]

for $x \neq 1$. The discontinuity is removable and $g(x) = x^3 + x^2 + x + 1$ agrees with f for $x \neq 1$ and is continuous on $\mathbb{R}$.

(b) $f(x) = \dfrac{x^3 - x^2 - 2x}{x - 2} = \dfrac{x(x^2 - x - 2)}{x - 2} = \dfrac{x(x - 2)(x + 1)}{x - 2} = x(x + 1)$ [or $x^2 + x$]   for $x \neq 2$. The discontinuity

is removable and $g(x) = x^2 + x$ agrees with f for $x \neq 2$ and is continuous on $\mathbb{R}$.

(c) $\lim\limits_{x \to \pi^-} f(x) = \lim\limits_{x \to \pi^-} [\![\sin x]\!] = \lim\limits_{x \to \pi^-} 0 = 0$ and $\lim\limits_{x \to \pi^+} f(x) = \lim\limits_{x \to \pi^+} [\![\sin x]\!] = \lim\limits_{x \to \pi^+} (-1) = -1$, so $\lim\limits_{x \to \pi} f(x)$ does not

exist. The discontinuity at $x = \pi$ is a jump discontinuity.

44.

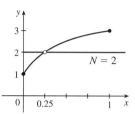

f does not satisfy the conclusion of the
Intermediate Value Theorem.

f does satisfy the conclusion of the
Intermediate Value Theorem.

45. $f(x) = x^2 + 10 \sin x$ is continuous on the interval $[31, 32]$, $f(31) \approx 957$, and $f(32) \approx 1030$. Since $957 < 1000 < 1030$,

there is a number c in $(31, 32)$ such that $f(c) = 1000$ by the Intermediate Value Theorem. *Note:* There is also a number c in

$(-32, -31)$ such that $f(c) = 1000$.

46. Suppose that $f(3) < 6$. By the Intermediate Value Theorem applied to the continuous function f on the closed interval $[2, 3]$,

the fact that $f(2) = 8 > 6$ and $f(3) < 6$ implies that there is a number c in $(2, 3)$ such that $f(c) = 6$. This contradicts the fact

that the only solutions of the equation $f(x) = 6$ are $x = 1$ and $x = 4$. Hence, our supposition that $f(3) < 6$ was incorrect. It

follows that $f(3) \geq 6$. But $f(3) \neq 6$ because the only solutions of $f(x) = 6$ are $x = 1$ and $x = 4$. Therefore, $f(3) > 6$.

47. $f(x) = x^4 + x - 3$ is continuous on the interval $[1, 2]$, $f(1) = -1$, and $f(2) = 15$. Since $-1 < 0 < 15$, there is a number c

in $(1, 2)$ such that $f(c) = 0$ by the Intermediate Value Theorem. Thus, there is a root of the equation $x^4 + x - 3 = 0$ in the

interval $(1, 2)$.

48. $f(x) = \sqrt[3]{x} + x - 1$ is continuous on the interval $[0, 1]$, $f(0) = -1$, and $f(1) = 1$. Since $-1 < 0 < 1$, there is a number c in

$(0, 1)$ such that $f(c) = 0$ by the Intermediate Value Theorem. Thus, there is a root of the equation $\sqrt[3]{x} + x - 1 = 0$, or

$\sqrt[3]{x} = 1 - x$, in the interval $(0, 1)$.

49. $f(x) = \cos x - x$ is continuous on the interval $[0, 1]$, $f(0) = 1$, and $f(1) = \cos 1 - 1 \approx -0.46$. Since $-0.46 < 0 < 1$, there

is a number c in $(0, 1)$ such that $f(c) = 0$ by the Intermediate Value Theorem. Thus, there is a root of the equation

$\cos x - x = 0$, or $\cos x = x$, in the interval $(0, 1)$.

50. $f(x) = \tan x - 2x$ is continuous on the interval $[0, 1.4]$, $f(1) = \tan 1 - 2 \approx -0.44$, and $f(1.4) = \tan 1.4 - 2.8 \approx 3.00$.

Since $-0.44 < 0 < 3.00$, there is a number c in $(0, 1.4)$ such that $f(c) = 0$ by the Intermediate Value Theorem. Thus, there is

a root of the equation $\tan x - 2x = 0$, or $\tan x = 2x$, in the interval $(0, 1.4)$.

51. (a) $f(x) = \cos x - x^3$ is continuous on the interval $[0, 1]$, $f(0) = 1 > 0$, and $f(1) = \cos 1 - 1 \approx -0.46 < 0$. Since

$1 > 0 > -0.46$, there is a number c in $(0, 1)$ such that $f(c) = 0$ by the Intermediate Value Theorem. Thus, there is a root

of the equation $\cos x - x^3 = 0$, or $\cos x = x^3$, in the interval $(0, 1)$.

(b) $f(0.86) \approx 0.016 > 0$ and $f(0.87) \approx -0.014 < 0$, so there is a root between 0.86 and 0.87, that is, in the interval

$(0.86, 0.87)$.

52. (a) $f(x) = x^5 - x^2 + 2x + 3$ is continuous on $[-1, 0]$, $f(-1) = -1 < 0$, and $f(0) = 3 > 0$. Since $-1 < 0 < 3$, there

is a number c in $(-1, 0)$ such that $f(c) = 0$ by the Intermediate Value Theorem. Thus, there is a root of the equation

$x^5 - x^2 + 2x + 3 = 0$ in the interval $(-1, 0)$.

(b) $f(-0.88) \approx -0.062 < 0$ and $f(-0.87) \approx 0.0047 > 0$, so there is a root between -0.88 and -0.87.

53. (a) Let $f(x) = x^5 - x^2 - 4$. Then $f(1) = 1^5 - 1^2 - 4 = -4 < 0$ and $f(2) = 2^5 - 2^2 - 4 = 24 > 0$. So by the

Intermediate Value Theorem, there is a number c in $(1, 2)$ such that $f(c) = c^5 - c^2 - 4 = 0$.

(b) We can see from the graphs that, correct to three decimal places, the root is $x \approx 1.434$.

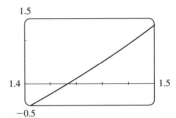

54. (a) Let $f(x) = \sqrt{x - 5} - \dfrac{1}{x + 3}$. Then $f(5) = -\frac{1}{8} < 0$ and $f(6) = \frac{8}{9} > 0$, and f is continuous on $[5, \infty)$. So by the

Intermediate Value Theorem, there is a number c in $(5, 6)$ such that $f(c) = 0$. This implies that $\dfrac{1}{c + 3} = \sqrt{c - 5}$.

(b) Using the intersect feature of the graphing device, we find

that the root of the equation is $x = 5.016$, correct to three

decimal places.

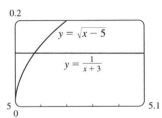

55. ($\Rightarrow$) If f is continuous at a, then by Theorem 8 with $g(h) = a + h$, we have

$$\lim_{h \to 0} f(a + h) = f\left(\lim_{h \to 0} (a + h)\right) = f(a).$$

($\Leftarrow$) Let $\varepsilon > 0$. Since $\lim_{h \to 0} f(a + h) = f(a)$, there exists $\delta > 0$ such that $0 < |h| < \delta \quad \Rightarrow$

$|f(a + h) - f(a)| < \varepsilon$. So if $0 < |x - a| < \delta$, then $|f(x) - f(a)| = |f(a + (x - a)) - f(a)| < \varepsilon$.

Thus, $\lim_{x \to a} f(x) = f(a)$ and so f is continuous at a.

56. $\lim_{h \to 0} \sin(a + h) = \lim_{h \to 0} (\sin a \cos h + \cos a \sin h) = \lim_{h \to 0} (\sin a \cos h) + \lim_{h \to 0} (\cos a \sin h)$

$$= \left(\lim_{h \to 0} \sin a\right)\left(\lim_{h \to 0} \cos h\right) + \left(\lim_{h \to 0} \cos a\right)\left(\lim_{h \to 0} \sin h\right) = (\sin a)(1) + (\cos a)(0) = \sin a$$

57. As in the previous exercise, we must show that $\lim_{h \to 0} \cos(a + h) = \cos a$ to prove that the cosine function is continuous.

$$\lim_{h \to 0} \cos(a + h) = \lim_{h \to 0} (\cos a \cos h - \sin a \sin h) = \lim_{h \to 0} (\cos a \cos h) - \lim_{h \to 0} (\sin a \sin h)$$

$$= \left(\lim_{h \to 0} \cos a\right)\left(\lim_{h \to 0} \cos h\right) - \left(\lim_{h \to 0} \sin a\right)\left(\lim_{h \to 0} \sin h\right) = (\cos a)(1) - (\sin a)(0) = \cos a$$

58. (a) Since f is continuous at a, $\lim_{x \to a} f(x) = f(a)$. Thus, using the Constant Multiple Law of Limits, we have

$$\lim_{x \to a} (cf)(x) = \lim_{x \to a} cf(x) = c \lim_{x \to a} f(x) = cf(a) = (cf)(a). \text{ Therefore, } cf \text{ is continuous at } a.$$

(b) Since f and g are continuous at a, $\lim_{x \to a} f(x) = f(a)$ and $\lim_{x \to a} g(x) = g(a)$. Since $g(a) \neq 0$, we can use the Quotient Law

of Limits: $\lim_{x \to a} \left(\dfrac{f}{g}\right)(x) = \lim_{x \to a} \dfrac{f(x)}{g(x)} = \dfrac{\lim_{x \to a} f(x)}{\lim_{x \to a} g(x)} = \dfrac{f(a)}{g(a)} = \left(\dfrac{f}{g}\right)(a).$ Thus, $\dfrac{f}{g}$ is continuous at a.

59. $f(x) = \begin{cases} 0 & \text{if } x \text{ is rational} \\ 1 & \text{if } x \text{ is irrational} \end{cases}$ is continuous nowhere. For, given any number a and any $\delta > 0$, the interval $(a - \delta, a + \delta)$

contains both infinitely many rational and infinitely many irrational numbers. Since $f(a) = 0$ or 1, there are infinitely many

numbers x with $0 < |x - a| < \delta$ and $|f(x) - f(a)| = 1$. Thus, $\lim_{x \to a} f(x) \neq f(a)$. [In fact, $\lim_{x \to a} f(x)$ does not even exist.]

60. $g(x) = \begin{cases} 0 & \text{if } x \text{ is rational} \\ x & \text{if } x \text{ is irrational} \end{cases}$ is continuous at 0. To see why, note that $-|x| \leq g(x) \leq |x|$, so by the Squeeze Theorem

$\lim_{x \to 0} g(x) = 0 = g(0)$. But g is continuous nowhere else. For if $a \neq 0$ and $\delta > 0$, the interval $(a - \delta, a + \delta)$ contains both

infinitely many rational and infinitely many irrational numbers. Since $g(a) = 0$ or a, there are infinitely many numbers x with

$0 < |x - a| < \delta$ and $|g(x) - g(a)| > |a|/2$. Thus, $\lim_{x \to a} g(x) \neq g(a)$.

61. If there is such a number, it satisfies the equation $x^3 + 1 = x \iff x^3 - x + 1 = 0$. Let the left-hand side of this equation be

called $f(x)$. Now $f(-2) = -5 < 0$, and $f(-1) = 1 > 0$. Note also that $f(x)$ is a polynomial, and thus continuous. So by the

Intermediate Value Theorem, there is a number c between -2 and -1 such that $f(c) = 0$, so that $c = c^3 + 1$.

62. $\dfrac{a}{x^3 + 2x^2 - 1} + \dfrac{b}{x^3 + x - 2} = 0 \implies a(x^3 + x - 2) + b(x^3 + 2x^2 - 1) = 0.$ Let $p(x)$ denote the left side of the last

equation. Since p is continuous on $[-1, 1]$, $p(-1) = -4a < 0$, and $p(1) = 2b > 0$, there exists a c in $(-1, 1)$ such that

$p(c) = 0$ by the Intermediate Value Theorem. Note that the only root of either denominator that is in $(-1, 1)$ is

$(-1 + \sqrt{5})/2 = r$, but $p(r) = (3\sqrt{5} - 9)a/2 \neq 0$. Thus, c is not a root of either denominator, so $p(c) = 0 \implies$

$x = c$ is a root of the given equation.

63. $f(x) = x^4 \sin(1/x)$ is continuous on $(-\infty, 0) \cup (0, \infty)$ since it is the product of a polynomial and a composite of a

trigonometric function and a rational function. Now since $-1 \leq \sin(1/x) \leq 1$, we have $-x^4 \leq x^4 \sin(1/x) \leq x^4$. Because

$\lim_{x \to 0} (-x^4) = 0$ and $\lim_{x \to 0} x^4 = 0$, the Squeeze Theorem gives us $\lim_{x \to 0} (x^4 \sin(1/x)) = 0$, which equals $f(0)$. Thus, f is

continuous at 0 and, hence, on $(-\infty, \infty)$.

64. (a) $\lim\limits_{x\to 0^+} F(x) = 0$ and $\lim\limits_{x\to 0^-} F(x) = 0$, so $\lim\limits_{x\to 0} F(x) = 0$, which is $F(0)$, and hence F is continuous at $x = a$ if $a = 0$. For

$a > 0$, $\lim\limits_{x\to a} F(x) = \lim\limits_{x\to a} x = a = F(a)$. For $a < 0$, $\lim\limits_{x\to a} F(x) = \lim\limits_{x\to a} (-x) = -a = F(a)$. Thus, F is continuous at

$x = a$; that is, continuous everywhere.

(b) Assume that f is continuous on the interval I. Then for $a \in I$, $\lim\limits_{x\to a} |f(x)| = \left|\lim\limits_{x\to a} f(x)\right| = |f(a)|$ by Theorem 8. (If a is

an endpoint of I, use the appropriate one-sided limit.) So $|f|$ is continuous on I.

(c) No, the converse is false. For example, the function $f(x) = \begin{cases} 1 & \text{if } x \geq 0 \\ -1 & \text{if } x < 0 \end{cases}$ is not continuous at $x = 0$, but $|f(x)| = 1$ is

continuous on $\mathbb{R}$.

65. Define $u(t)$ to be the monk's distance from the monastery, as a function of time, on the first day, and define $d(t)$ to be his

distance from the monastery, as a function of time, on the second day. Let D be the distance from the monastery to the top of

the mountain. From the given information we know that $u(0) = 0$, $u(12) = D$, $d(0) = D$ and $d(12) = 0$. Now consider the

function $u - d$, which is clearly continuous. We calculate that $(u - d)(0) = -D$ and $(u - d)(12) = D$. So by the

Intermediate Value Theorem, there must be some time t_0 between 0 and 12 such that $(u - d)(t_0) = 0 \iff u(t_0) = d(t_0)$.

So at time t_0 after 7:00 AM, the monk will be at the same place on both days.

2 Review

CONCEPT CHECK

1. (a) $\lim\limits_{x\to a} f(x) = L$: See Definition 2.2.1 and Figures 1 and 2 in Section 2.2.

(b) $\lim\limits_{x\to a^+} f(x) = L$: See the paragraph after Definition 2.2.2 and Figure 9(b) in Section 2.2.

(c) $\lim\limits_{x\to a^-} f(x) = L$: See Definition 2.2.2 and Figure 9(a) in Section 2.2.

(d) $\lim\limits_{x\to a} f(x) = \infty$: See Definition 2.2.4 and Figure 12 in Section 2.2.

(e) $\lim\limits_{x\to a} f(x) = -\infty$: See Definition 2.2.5 and Figure 13 in Section 2.2.

2. In general, the limit of a function fails to exist when the function does not approach a fixed number. For each of the following

functions, the limit fails to exist at $x = 2$.

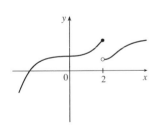

The left- and right-hand
limits are not equal.

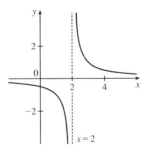

There is an
infinite discontinuity.

There are an infinite
number of oscillations.

3. See Definition 2.2.6 and Figure 14 in Section 2.2.

4. (a)–(g) See the statements of Limit Laws 1–6 and 11 in Section 2.3.

5. See Theorem 3 in Section 2.3.

6. (a) A function f is continuous at a number a if $f(x)$ approaches $f(a)$ as x approaches a; that is, $\lim_{x \to a} f(x) = f(a)$.

    (b) A function f is continuous on the interval $(-\infty, \infty)$ if f is continuous at every real number a. The graph of such a function has no breaks and every vertical line crosses it.

7. See Theorem 2.5.10.

TRUE-FALSE QUIZ

1. False.   Limit Law 2 applies only if the individual limits exist (these don't).

2. False.   Limit Law 5 cannot be applied if the limit of the denominator is 0 (it is).

3. True.   Limit Law 5 applies.

4. True.   The limit doesn't exist since $f(x)/g(x)$ doesn't approach any real number as x approaches 5. (The denominator approaches 0 and the numerator doesn't.)

5. False.   Consider $\lim_{x \to 5} \dfrac{x(x-5)}{x-5}$ or $\lim_{x \to 5} \dfrac{\sin(x-5)}{x-5}$. The first limit exists and is equal to 5. By Example 3 in Section 2.2, we know that the latter limit exists (and it is equal to 1).

6. False.   Consider $\lim_{x \to 6} [f(x)g(x)] = \lim_{x \to 6} \left[(x-6)\dfrac{1}{x-6}\right]$. It exists (its value is 1) but $f(6) = 0$ and $g(6)$ does not exist, so $f(6)g(6) \neq 1$.

7. True.   A polynomial is continuous everywhere, so $\lim_{x \to b} p(x)$ exists and is equal to $p(b)$.

8. False.   Consider $\lim_{x \to 0} [f(x) - g(x)] = \lim_{x \to 0} \left(\dfrac{1}{x^2} - \dfrac{1}{x^4}\right)$. This limit is $-\infty$ (not 0), but each of the individual functions approaches ∞.

9. False.   Consider $f(x) = \begin{cases} 1/(x-1) & \text{if } x \neq 1 \\ 2 & \text{if } x = 1 \end{cases}$

10. False.   The function f must be *continuous* in order to use the Intermediate Value Theorem. For example, let $f(x) = \begin{cases} 1 & \text{if } 0 \leq x < 3 \\ -1 & \text{if } x = 3 \end{cases}$ There is no number $c \in [0, 3]$ with $f(c) = 0$.

11. True.   Use Theorem 2.5.8 with $a = 2$, $b = 5$, and $g(x) = 4x^2 - 11$. Note that $f(4) = 3$ is not needed.

12. True.   Use the Intermediate Value Theorem with $a = -1$, $b = 1$, and $N = \pi$, since $3 < \pi < 4$.

13. True, by the definition of a limit with $\varepsilon = 1$.

14. False. For example, let $f(x) = \begin{cases} x^2 + 1 & \text{if } x \neq 0 \\ 2 & \text{if } x = 0 \end{cases}$

Then $f(x) > 1$ for all x, but $\lim\limits_{x \to 0} f(x) = \lim\limits_{x \to 0} (x^2 + 1) = 1$.

15. True. $f(x) = x^{10} - 10x^2 + 5$ is continuous on the interval $[0, 2]$, $f(0) = 5$, $f(1) = -4$, and $f(2) = 989$. Since $-4 < 0 < 5$, there is a number c in $(0, 1)$ such that $f(c) = 0$ by the Intermediate Value Theorem. Thus, there is a root of the equation $x^{10} - 10x^2 + 5 = 0$ in the interval $(0, 1)$. Similarly, there is a root in $(1, 2)$.

EXERCISES

1. (a) (i) $\lim\limits_{x \to 2^+} f(x) = 3$ **(ii)** $\lim\limits_{x \to -3^+} f(x) = 0$

(iii) $\lim\limits_{x \to -3} f(x)$ does not exist since the left and right limits are not equal. (The left limit is -2.)

(iv) $\lim\limits_{x \to 4} f(x) = 2$

(v) $\lim\limits_{x \to 0} f(x) = \infty$ **(vi)** $\lim\limits_{x \to 2^-} f(x) = -\infty$

(b) The equations of the vertical asymptotes are $x = 0$ and $x = 2$.

(c) f is discontinuous at $x = -3, 0, 2,$ and 4. The discontinuities are jump, infinite, infinite, and removable, respectively.

2. $\lim\limits_{x \to -0^+} f(x) = -2, \quad \lim\limits_{x \to 0^-} f(x) = 1, \quad f(0) = -1,$

$\lim\limits_{x \to 2^-} f(x) = \infty, \quad \lim\limits_{x \to 2^+} f(x) = -\infty$

3. $\lim\limits_{x \to 0} \cos(x + \sin x) = \cos \left[\lim\limits_{x \to 0} (x + \sin x) \right]$ [by Theorem 2.5.8] $= \cos 0 = 1$

4. Since rational functions are continuous, $\lim\limits_{x \to 3} \dfrac{x^2 - 9}{x^2 + 2x - 3} = \dfrac{3^2 - 9}{3^2 + 2(3) - 3} = \dfrac{0}{12} = 0.$

5. $\lim\limits_{x \to -3} \dfrac{x^2 - 9}{x^2 + 2x - 3} = \lim\limits_{x \to -3} \dfrac{(x + 3)(x - 3)}{(x + 3)(x - 1)} = \lim\limits_{x \to -3} \dfrac{x - 3}{x - 1} = \dfrac{-3 - 3}{-3 - 1} = \dfrac{-6}{-4} = \dfrac{3}{2}$

6. $\lim\limits_{x \to 1^+} \dfrac{x^2 - 9}{x^2 + 2x - 3} = -\infty$ since $x^2 + 2x - 3 \to 0$ as $x \to 1^+$ and $\dfrac{x^2 - 9}{x^2 + 2x - 3} < 0$ for $1 < x < 3$.

7. $\lim\limits_{h \to 0} \dfrac{(h - 1)^3 + 1}{h} = \lim\limits_{h \to 0} \dfrac{(h^3 - 3h^2 + 3h - 1) + 1}{h} = \lim\limits_{h \to 0} \dfrac{h^3 - 3h^2 + 3h}{h} = \lim\limits_{h \to 0} (h^2 - 3h + 3) = 3$

Another solution: Factor the numerator as a sum of two cubes and then simplify.

$\lim\limits_{h \to 0} \dfrac{(h - 1)^3 + 1}{h} = \lim\limits_{h \to 0} \dfrac{(h - 1)^3 + 1^3}{h} = \lim\limits_{h \to 0} \dfrac{[(h - 1) + 1]\left[(h - 1)^2 - 1(h - 1) + 1^2\right]}{h}$

$= \lim\limits_{h \to 0} \left[(h - 1)^2 - h + 2\right] = 1 - 0 + 2 = 3$

8. $\lim\limits_{t \to 2} \dfrac{t^2 - 4}{t^3 - 8} = \lim\limits_{t \to 2} \dfrac{(t + 2)(t - 2)}{(t - 2)(t^2 + 2t + 4)} = \lim\limits_{t \to 2} \dfrac{t + 2}{t^2 + 2t + 4} = \dfrac{2 + 2}{4 + 4 + 4} = \dfrac{4}{12} = \dfrac{1}{3}$

9. $\lim\limits_{r \to 9} \dfrac{\sqrt{r}}{(r - 9)^4} = \infty$ since $(r - 9)^4 \to 0$ as $r \to 9$ and $\dfrac{\sqrt{r}}{(r - 9)^4} > 0$ for $r \neq 9$.

10. $\lim\limits_{v \to 4^+} \dfrac{4 - v}{|4 - v|} = \lim\limits_{v \to 4^+} \dfrac{4 - v}{-(4 - v)} = \lim\limits_{v \to 4^+} \dfrac{1}{-1} = -1$

11. $\lim\limits_{u \to 1} \dfrac{u^4 - 1}{u^3 + 5u^2 - 6u} = \lim\limits_{u \to 1} \dfrac{(u^2 + 1)(u^2 - 1)}{u(u^2 + 5u - 6)} = \lim\limits_{u \to 1} \dfrac{(u^2 + 1)(u + 1)(u - 1)}{u(u + 6)(u - 1)} = \lim\limits_{u \to 1} \dfrac{(u^2 + 1)(u + 1)}{u(u + 6)} = \dfrac{2(2)}{1(7)} = \dfrac{4}{7}$

12. $\lim\limits_{x \to 3} \dfrac{\sqrt{x + 6} - x}{x^3 - 3x^2} = \lim\limits_{x \to 3} \left[\dfrac{\sqrt{x + 6} - x}{x^2(x - 3)} \cdot \dfrac{\sqrt{x + 6} + x}{\sqrt{x + 6} + x} \right] = \lim\limits_{x \to 3} \dfrac{(\sqrt{x + 6})^2 - x^2}{x^2(x - 3)(\sqrt{x + 6} + x)}$

$= \lim\limits_{x \to 3} \dfrac{x + 6 - x^2}{x^2(x - 3)\left(\sqrt{x + 6} + x\right)} = \lim\limits_{x \to 3} \dfrac{-(x^2 - x - 6)}{x^2(x - 3)\left(\sqrt{x + 6} + x\right)} = \lim\limits_{x \to 3} \dfrac{-(x - 3)(x + 2)}{x^2(x - 3)(\sqrt{x + 6} + x)}$

$= \lim\limits_{x \to 3} \dfrac{-(x + 2)}{x^2\left(\sqrt{x + 6} + x\right)} = -\dfrac{5}{9(3 + 3)} = -\dfrac{5}{54}$

13. $\lim\limits_{s \to 16} \dfrac{4 - \sqrt{s}}{s - 16} = \lim\limits_{s \to 16} \dfrac{4 - \sqrt{s}}{(\sqrt{s} + 4)(\sqrt{s} - 4)} = \lim\limits_{s \to 16} \dfrac{-1}{\sqrt{s} + 4} = \dfrac{-1}{\sqrt{16} + 4} = -\dfrac{1}{8}$

14. $\lim\limits_{v \to 2} \dfrac{v^2 + 2v - 8}{v^4 - 16} = \lim\limits_{v \to 2} \dfrac{(v + 4)(v - 2)}{(v + 2)(v - 2)(v^2 + 4)} = \lim\limits_{v \to 2} \dfrac{v + 4}{(v + 2)(v^2 + 4)} = \dfrac{2 + 4}{(2 + 2)(2^2 + 4)} = \dfrac{3}{16}$

15. $\lim\limits_{x \to 0} \dfrac{1 - \sqrt{1 - x^2}}{x} \cdot \dfrac{1 + \sqrt{1 - x^2}}{1 + \sqrt{1 - x^2}} = \lim\limits_{x \to 0} \dfrac{1 - (1 - x^2)}{x\left(1 + \sqrt{1 - x^2}\right)} = \lim\limits_{x \to 0} \dfrac{x^2}{x\left(1 + \sqrt{1 - x^2}\right)} = \lim\limits_{x \to 0} \dfrac{x}{1 + \sqrt{1 - x^2}} = 0$

16. $\lim\limits_{x \to 1} \left(\dfrac{1}{x - 1} + \dfrac{1}{x^2 - 3x + 2} \right) = \lim\limits_{x \to 1} \left[\dfrac{1}{x - 1} + \dfrac{1}{(x - 1)(x - 2)} \right] = \lim\limits_{x \to 1} \left[\dfrac{x - 2}{(x - 1)(x - 2)} + \dfrac{1}{(x - 1)(x - 2)} \right]$

$= \lim\limits_{x \to 1} \left[\dfrac{x - 1}{(x - 1)(x - 2)} \right] = \lim\limits_{x \to 1} \dfrac{1}{x - 2} = \dfrac{1}{1 - 2} = -1$

17. Since $2x - 1 \le f(x) \le x^2$ for $0 < x < 3$ and $\lim\limits_{x \to 1} (2x - 1) = 1 = \lim\limits_{x \to 1} x^2$, we have $\lim\limits_{x \to 1} f(x) = 1$ by the Squeeze Theorem.

18. Let $f(x) = -x^2$, $g(x) = x^2 \cos(1/x^2)$ and $h(x) = x^2$. Then since $\left| \cos(1/x^2) \right| \le 1$ for $x \neq 0$, we have

$f(x) \le g(x) \le h(x)$ for $x \neq 0$, and so $\lim\limits_{x \to 0} f(x) = \lim\limits_{x \to 0} h(x) = 0 \;\Rightarrow\; \lim\limits_{x \to 0} g(x) = 0$ by the Squeeze Theorem.

19. Given $\varepsilon > 0$, we need $\delta > 0$ such that if $0 < |x - 2| < \delta$, then $|(14 - 5x) - 4| < \varepsilon$. But $|(14 - 5x) - 4| < \varepsilon \;\Leftrightarrow\;$
$|-5x + 10| < \varepsilon \;\Leftrightarrow\; |-5|\,|x - 2| < \varepsilon \;\Leftrightarrow\; |x - 2| < \varepsilon/5$. So if we choose $\delta = \varepsilon/5$, then $0 < |x - 2| < \delta \;\Rightarrow\;$
$|(14 - 5x) - 4| < \varepsilon$. Thus, $\lim\limits_{x \to 2} (14 - 5x) = 4$ by the definition of a limit.

20. Given $\varepsilon > 0$ we must find $\delta > 0$ so that if $0 < |x - 0| < \delta$, then $|\sqrt[3]{x} - 0| < \varepsilon$. Now $|\sqrt[3]{x} - 0| = |\sqrt[3]{x}| < \varepsilon \;\Rightarrow\;$
$|x| = |\sqrt[3]{x}|^3 < \varepsilon^3$. So take $\delta = \varepsilon^3$. Then $0 < |x - 0| = |x| < \varepsilon^3 \;\Rightarrow\; |\sqrt[3]{x} - 0| = |\sqrt[3]{x}| = \sqrt[3]{|x|} < \sqrt[3]{\varepsilon^3} = \varepsilon$.
Therefore, by the definition of a limit, $\lim\limits_{x \to 0} \sqrt[3]{x} = 0$.

21. Given $\varepsilon > 0$, we need $\delta > 0$ so that if $0 < |x - 2| < \delta$, then $\left|x^2 - 3x - (-2)\right| < \varepsilon$. First, note that if $|x - 2| < 1$, then

$-1 < x - 2 < 1$, so $0 < x - 1 < 2$ $\Rightarrow$ $|x - 1| < 2$. Now let $\delta = \min\{\varepsilon/2, 1\}$. Then $0 < |x - 2| < \delta$ $\Rightarrow$

$\left|x^2 - 3x - (-2)\right| = |(x - 2)(x - 1)| = |x - 2|\,|x - 1| < (\varepsilon/2)(2) = \varepsilon$.

Thus, $\lim\limits_{x \to 2}\left(x^2 - 3x\right) = -2$ by the definition of a limit.

22. Given $M > 0$, we need $\delta > 0$ such that if $0 < x - 4 < \delta$, then $2/\sqrt{x - 4} > M$. This is true $\Leftrightarrow$ $\sqrt{x - 4} < 2/M$ $\Leftrightarrow$

$x - 4 < 4/M^2$. So if we choose $\delta = 4/M^2$, then $0 < x - 4 < \delta$ $\Rightarrow$ $2/\sqrt{x - 4} > M$. So by the definition of a limit,

$\lim\limits_{x \to 4^+}\left(2/\sqrt{x - 4}\right) = \infty$.

23. (a) $f(x) = \sqrt{-x}$ if $x < 0$, $f(x) = 3 - x$ if $0 \le x < 3$, $f(x) = (x - 3)^2$ if $x > 3$.

(i) $\lim\limits_{x \to 0^+} f(x) = \lim\limits_{x \to 0^+} (3 - x) = 3$ (ii) $\lim\limits_{x \to 0^-} f(x) = \lim\limits_{x \to 0^-} \sqrt{-x} = 0$

(iii) Because of (i) and (ii), $\lim\limits_{x \to 0} f(x)$ does not exist.   (iv) $\lim\limits_{x \to 3^-} f(x) = \lim\limits_{x \to 3^-} (3 - x) = 0$

(v) $\lim\limits_{x \to 3^+} f(x) = \lim\limits_{x \to 3^+} (x - 3)^2 = 0$ (vi) Because of (iv) and (v), $\lim\limits_{x \to 3} f(x) = 0$.

(b) f is discontinuous at 0 since $\lim\limits_{x \to 0} f(x)$ does not exist.   **(c)**

f is discontinuous at 3 since $f(3)$ does not exist.

24. (a) $g(x) = 2x - x^2$ if $0 \le x \le 2$, $g(x) = 2 - x$ if $2 < x \le 3$, $g(x) = x - 4$ if $3 < x < 4$, $g(x) = \pi$ if $x \ge 4$.

Therefore, $\lim\limits_{x \to 2^-} g(x) = \lim\limits_{x \to 2^-}\left(2x - x^2\right) = 0$ and $\lim\limits_{x \to 2^+} g(x) = \lim\limits_{x \to 2^+}(2 - x) = 0$. Thus, $\lim\limits_{x \to 2} g(x) = 0 = g(2)$,

so g is continuous at 2. $\lim\limits_{x \to 3^-} g(x) = \lim\limits_{x \to 3^-}(2 - x) = -1$ and $\lim\limits_{x \to 3^+} g(x) = \lim\limits_{x \to 3^+}(x - 4) = -1$. Thus,

$\lim\limits_{x \to 3} g(x) = -1 = g(3)$, so g is continuous at 3. $\lim\limits_{x \to 4^-} g(x) = \lim\limits_{x \to 4^-}(x - 4) = 0$ and $\lim\limits_{x \to 4^+} g(x) = \lim\limits_{x \to 4^+} \pi = \pi$.

Thus, $\lim\limits_{x \to 4} g(x)$ does not exist, so g is discontinuous at 4. But $\lim\limits_{x \to 4^+} g(x) = \pi = g(4)$, so g is continuous from the

right at 4.

(b)

25. x^3 is continuous on $\mathbb{R}$ since it is a polynomial and $\cos x$ is also continuous on $\mathbb{R}$, so the product $x^3 \cos x$ is continuous on $\mathbb{R}$.

The root function $\sqrt[4]{x}$ is continuous on its domain, $[0, \infty)$, and so the sum, $h(x) = \sqrt[4]{x} + x^3 \cos x$, is continuous on its

domain, $[0, \infty)$.

26. $x^2 - 9$ is continuous on $\mathbb{R}$ since it is a polynomial and $\sqrt{x}$ is continuous on $[0, \infty)$, so the composition $\sqrt{x^2 - 9}$ is continuous

on $\{x \mid x^2 - 9 \geq 0\} = (-\infty, -3] \cup [3, \infty)$. Note that $x^2 - 2 \neq 0$ on this set and so the quotient function $g(x) = \dfrac{\sqrt{x^2 - 9}}{x^2 - 2}$

is continuous on its domain, $(-\infty, -3] \cup [3, \infty)$.

27. $f(x) = 2x^3 + x^2 + 2$ is a polynomial, so it is continuous on $[-2, -1]$ and $f(-2) = -10 < 0 < 1 = f(-1)$. So by the

Intermediate Value Theorem there is a number c in $(-2, -1)$ such that $f(c) = 0$, that is, the equation $2x^3 + x^2 + 2 = 0$ has a

root in $(-2, -1)$.

28. Let $f(x) = 2\sin x - 3 + 2x$. Now f is continuous on $[0, 1]$ and $f(0) = -3 < 0$ and $f(1) = 2\sin 1 - 1 \approx 0.68 > 0$. So by

the Intermediate Value Theorem there is a number c in $(0, 1)$ such that $f(c) = 0$, that is, the equation $2\sin x = 3 - 2x$ has a

root in $(0, 1)$.

29. $|f(x)| \leq g(x) \quad \Leftrightarrow \quad -g(x) \leq f(x) \leq g(x)$ and $\lim\limits_{x \to a} g(x) = 0 = \lim\limits_{x \to a} -g(x)$.

Thus, by the Squeeze Theorem, $\lim\limits_{x \to a} f(x) = 0$.

30. (a) Note that f is an even function since $f(x) = f(-x)$. Now for any integer n,

$[\![n]\!] + [\![-n]\!] = n - n = 0$, and for any real number k which is not an integer,

$[\![k]\!] + [\![-k]\!] = [\![k]\!] + (-[\![k]\!] - 1) = -1$. So $\lim\limits_{x \to a} f(x)$ exists (and is equal to -1)

for all values of a.

(b) f is discontinuous at all integers.

☐ PROBLEMS PLUS

1. Let $t = \sqrt[6]{x}$, so $x = t^6$. Then $t \to 1$ as $x \to 1$, so

$$\lim_{x \to 1} \frac{\sqrt[3]{x} - 1}{\sqrt{x} - 1} = \lim_{t \to 1} \frac{t^2 - 1}{t^3 - 1} = \lim_{t \to 1} \frac{(t-1)(t+1)}{(t-1)\left(t^2 + t + 1\right)} = \lim_{t \to 1} \frac{t+1}{t^2 + t + 1} = \frac{1+1}{1^2 + 1 + 1} = \frac{2}{3}.$$

Another method: Multiply both the numerator and the denominator by $(\sqrt{x} + 1)\left(\sqrt[3]{x^2} + \sqrt[3]{x} + 1\right)$.

2. First rationalize the numerator: $\lim_{x \to 0} \dfrac{\sqrt{ax + b} - 2}{x} \cdot \dfrac{\sqrt{ax + b} + 2}{\sqrt{ax + b} + 2} = \lim_{x \to 0} \dfrac{ax + b - 4}{x\left(\sqrt{ax + b} + 2\right)}$. Now since the denominator

approaches 0 as $x \to 0$, the limit will exist only if the numerator also approaches 0 as $x \to 0$. So we require that

$a(0) + b - 4 = 0 \quad \Rightarrow \quad b = 4$. So the equation becomes $\lim_{x \to 0} \dfrac{a}{\sqrt{ax + 4} + 2} = 1 \quad \Rightarrow \quad \dfrac{a}{\sqrt{4} + 2} = 1 \quad \Rightarrow \quad a = 4$.

Therefore, $a = b = 4$.

3. For $-\frac{1}{2} < x < \frac{1}{2}$, we have $2x - 1 < 0$ and $2x + 1 > 0$, so $|2x - 1| = -(2x - 1)$ and $|2x + 1| = 2x + 1$.

Therefore, $\lim_{x \to 0} \dfrac{|2x - 1| - |2x + 1|}{x} = \lim_{x \to 0} \dfrac{-(2x - 1) - (2x + 1)}{x} = \lim_{x \to 0} \dfrac{-4x}{x} = \lim_{x \to 0} (-4) = -4$.

4. Let R be the midpoint of OP, so the coordinates of R are $\left(\frac{1}{2}x, \frac{1}{2}x^2\right)$ since the coordinates of P are $\left(x, x^2\right)$. Let $Q = (0, a)$.

Since the slope $m_{OP} = \dfrac{x^2}{x} = x$, $m_{QR} = -\dfrac{1}{x}$ (negative reciprocal). But $m_{QR} = \dfrac{\frac{1}{2}x^2 - a}{\frac{1}{2}x - 0} = \dfrac{x^2 - 2a}{x}$, so we conclude that

$-1 = x^2 - 2a \quad \Rightarrow \quad 2a = x^2 + 1 \quad \Rightarrow \quad a = \frac{1}{2}x^2 + \frac{1}{2}$. As $x \to 0$, $a \to \frac{1}{2}$, and the limiting position of Q is $\left(0, \frac{1}{2}\right)$.

5. (a) For $0 < x < 1$, $[\![x]\!] = 0$, so $\dfrac{[\![x]\!]}{x} = 0$, and $\lim_{x \to 0^+} \dfrac{[\![x]\!]}{x} = 0$. For $-1 < x < 0$, $[\![x]\!] = -1$, so $\dfrac{[\![x]\!]}{x} = \dfrac{-1}{x}$, and

$\lim_{x \to 0^-} \dfrac{[\![x]\!]}{x} = \lim_{x \to 0^-} \left(\dfrac{-1}{x}\right) = \infty$. Since the one-sided limits are not equal, $\lim_{x \to 0} \dfrac{[\![x]\!]}{x}$ does not exist.

(b) For $x > 0$, $1/x - 1 \le [\![1/x]\!] \le 1/x \quad \Rightarrow \quad x(1/x - 1) \le x[\![1/x]\!] \le x(1/x) \quad \Rightarrow \quad 1 - x \le x[\![1/x]\!] \le 1$.

As $x \to 0^+$, $1 - x \to 1$, so by the Squeeze Theorem, $\lim_{x \to 0^+} x[\![1/x]\!] = 1$.

For $x < 0$, $1/x - 1 \le [\![1/x]\!] \le 1/x \quad \Rightarrow \quad x(1/x - 1) \ge x[\![1/x]\!] \ge x(1/x) \quad \Rightarrow \quad 1 - x \ge x[\![1/x]\!] \ge 1$.

As $x \to 0^-$, $1 - x \to 1$, so by the Squeeze Theorem, $\lim_{x \to 0^-} x[\![1/x]\!] = 1$.

Since the one-sided limits are equal, $\lim_{x \to 0} x[\![1/x]\!] = 1$.

6. (a) $[\![x]\!]^2 + [\![y]\!]^2 = 1$. Since $[\![x]\!]^2$ and $[\![y]\!]^2$ are positive integers or 0, there are

only 4 cases:

   Case (i): $[\![x]\!] = 1,\ [\![y]\!] = 0 \ \Rightarrow 1 \le x < 2$ and $0 \le y < 1$

   Case (ii): $[\![x]\!] = -1,\ [\![y]\!] = 0 \Rightarrow -1 \le x < 0$ and $0 \le y < 1$

   Case (iii): $[\![x]\!] = 0,\ [\![y]\!] = 1 \ \Rightarrow 0 \le x < 1$ and $1 \le y < 2$

   Case (iv): $[\![x]\!] = 0,\ [\![y]\!] = -1 \Rightarrow 0 \le x < 1$ and $-1 \le y < 0$

(b) $[\![x]\!]^2 - [\![y]\!]^2 = 3$. The only integral solution of $n^2 - m^2 = 3$ is $n = \pm 2$

and $m = \pm 1$. So the graph is

$$\{(x,y) \mid [\![x]\!] = \pm 2,\ [\![y]\!] = \pm 1\} = \left\{(x,y) \,\middle|\, \begin{matrix} 2 \le x \le 3 \ \text{ or } \ -2 \le x < 1, \\ 1 \le y < 2 \ \text{ or } \ -1 \le y < 0 \end{matrix}\right\}.$$

(c) $[\![x+y]\!]^2 = 1 \ \Rightarrow \ [\![x+y]\!] = \pm 1 \ \Rightarrow \ 1 \le x + y < 2$

or $-1 \le x + y < 0$

(d) For $n \le x < n + 1$, $[\![x]\!] = n$. Then $[\![x]\!] + [\![y]\!] = 1 \ \Rightarrow \ [\![y]\!] = 1 - n \ \Rightarrow$

$1 - n \le y < 2 - n$. Choosing integer values for n produces the graph.

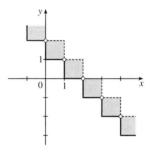

7. f is continuous on $(-\infty, a)$ and (a, ∞). To make f continuous on $\mathbb{R}$, we must have continuity at a. Thus,

$$\lim_{x \to a^+} f(x) = \lim_{x \to a^-} f(x) \ \Rightarrow \ \lim_{x \to a^+} x^2 = \lim_{x \to a^-} (x+1) \ \Rightarrow \ a^2 = a + 1 \ \Rightarrow \ a^2 - a - 1 = 0 \ \Rightarrow$$

[by the quadratic formula] $a = \left(1 \pm \sqrt{5}\right)/2 \approx 1.618$ or -0.618.

8. (a) Here are a few possibilities:

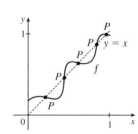

(b) The "obstacle" is the line $x = y$ (see diagram). Any intersection of the graph of f with the line $y = x$ constitutes a fixed point, and if the graph of the function does not cross the line somewhere in $(0, 1)$, then it must either start at $(0, 0)$ (in which case 0 is a fixed point) or finish at $(1, 1)$ (in which case 1 is a fixed point).

(c) Consider the function $F(x) = f(x) - x$, where f is any continuous function with domain $[0, 1]$ and range in $[0, 1]$. We shall prove that f has a fixed point. Now if $f(0) = 0$ then we are done: f has a fixed point (the number 0), which is what we are trying to prove. So assume $f(0) \neq 0$. For the same reason we can assume that $f(1) \neq 1$. Then $F(0) = f(0) > 0$ and $F(1) = f(1) - 1 < 0$. So by the Intermediate Value Theorem, there exists some number c in the interval $(0, 1)$ such that $F(c) = f(c) - c = 0$. So $f(c) = c$, and therefore f has a fixed point.

9. $\lim_{x \to a} f(x) = \lim_{x \to a} \left(\frac{1}{2} \left[f(x) + g(x) \right] + \frac{1}{2} \left[f(x) - g(x) \right] \right) = \frac{1}{2} \lim_{x \to a} \left[f(x) + g(x) \right] + \frac{1}{2} \lim_{x \to a} \left[f(x) - g(x) \right]$

$\qquad = \frac{1}{2} \cdot 2 + \frac{1}{2} \cdot 1 = \frac{3}{2}$,

and $\lim_{x \to a} g(x) = \lim_{x \to a} \left(\left[f(x) + g(x) \right] - f(x) \right) = \lim_{x \to a} \left[f(x) + g(x) \right] - \lim_{x \to a} f(x) = 2 - \frac{3}{2} = \frac{1}{2}$.

So $\lim_{x \to a} \left[f(x) g(x) \right] = \left[\lim_{x \to a} f(x) \right] \left[\lim_{x \to a} g(x) \right] = \frac{3}{2} \cdot \frac{1}{2} = \frac{3}{4}$.

Another solution: Since $\lim_{x \to a} \left[f(x) + g(x) \right]$ and $\lim_{x \to a} \left[f(x) - g(x) \right]$ exist, we must have

$\lim_{x \to a} \left[f(x) + g(x) \right]^2 = \left(\lim_{x \to a} \left[f(x) + g(x) \right] \right)^2$ and $\lim_{x \to a} \left[f(x) - g(x) \right]^2 = \left(\lim_{x \to a} \left[f(x) - g(x) \right] \right)^2$, so

$\lim_{x \to a} \left[f(x) g(x) \right] = \lim_{x \to a} \frac{1}{4} \left(\left[f(x) + g(x) \right]^2 - \left[f(x) - g(x) \right]^2 \right) \qquad$ [because all of the f^2 and g^2 cancel]

$\qquad = \frac{1}{4} \left(\lim_{x \to a} \left[f(x) + g(x) \right]^2 - \lim_{x \to a} \left[f(x) - g(x) \right]^2 \right) = \frac{1}{4} \left(2^2 - 1^2 \right) = \frac{3}{4}$.

10. (a) *Solution 1:* We introduce a coordinate system and drop a perpendicular from P, as shown. We see from $\angle NCP$ that $\tan 2\theta = \dfrac{y}{1 - x}$, and from $\angle NBP$ that $\tan \theta = y/x$. Using the double-angle formula for tangents, we get $\dfrac{y}{1 - x} = \tan 2\theta = \dfrac{2 \tan \theta}{1 - \tan^2 \theta} = \dfrac{2 \left(y/x \right)}{1 - \left(y/x \right)^2}$. After a bit of simplification, this becomes $\dfrac{1}{1 - x} = \dfrac{2x}{x^2 - y^2} \quad \Leftrightarrow \quad y^2 = x \left(3x - 2 \right)$.

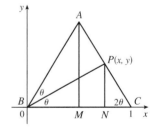

As the altitude AM decreases in length, the point P will approach the x-axis, that is, $y \to 0$, so the limiting location of P must be one of the roots of the equation $x(3x - 2) = 0$. Obviously it is not $x = 0$ (the point P can never be to the left of the altitude AM, which it would have to be in order to approach 0) so it must be $3x - 2 = 0$, that is, $x = \frac{2}{3}$.

Solution 2: We add a few lines to the original diagram, as shown. Now note that $\angle BPQ = \angle PBC$ (alternate angles; $QP \parallel BC$ by symmetry) and similarly $\angle CQP = \angle QCB$. So $\triangle BPQ$ and $\triangle CQP$ are isosceles, and the line segments BQ, QP and PC are all of equal length. As $|AM| \to 0$, P and Q approach points on the base, and the point P is seen to approach a position two-thirds of the way between B and C, as above.

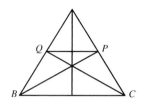

(b) The equation $y^2 = x(3x - 2)$ calculated in part (a) is the equation of the curve traced out by P. Now as $|AM| \to \infty$, $2\theta \to \frac{\pi}{2}$, $\theta \to \frac{\pi}{4}$, $x \to 1$, and since $\tan\theta = y/x$, $y \to 1$. Thus, P only traces out the part of the curve with $0 \le y < 1$.

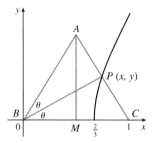

11. (a) Consider $G(x) = T(x + 180°) - T(x)$. Fix any number a. If $G(a) = 0$, we are done: Temperature at $a =$ Temperature at $a + 180°$. If $G(a) > 0$, then $G(a + 180°) = T(a + 360°) - T(a + 180°) = T(a) - T(a + 180°) = -G(a) < 0$. Also, G is continuous since temperature varies continuously. So, by the Intermediate Value Theorem, G has a zero on the interval $[a, a + 180°]$. If $G(a) < 0$, then a similar argument applies.

(b) Yes. The same argument applies.

(c) The same argument applies for quantities that vary continuously, such as barometric pressure. But one could argue that altitude above sea level is sometimes discontinuous, so the result might not always hold for that quantity.

3 □ DERIVATIVES

3.1 Derivatives and Rates of Change

1. (a) This is just the slope of the line through two points: $m_{PQ} = \dfrac{\Delta y}{\Delta x} = \dfrac{f(x) - f(3)}{x - 3}$.

(b) This is the limit of the slope of the secant line PQ as Q approaches P: $m = \lim\limits_{x \to 3} \dfrac{f(x) - f(3)}{x - 3}$.

2. The curve looks more like a line as the viewing rectangle gets smaller.

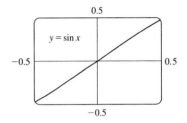

3. (a) (i) Using Definition 1 with $f(x) = 4x - x^2$ and $P(1, 3)$,

$$m = \lim_{x \to a} \frac{f(x) - f(a)}{x - a} = \lim_{x \to 1} \frac{(4x - x^2) - 3}{x - 1} = \lim_{x \to 1} \frac{-(x^2 - 4x + 3)}{x - 1} = \lim_{x \to 1} \frac{-(x - 1)(x - 3)}{x - 1}$$

$$= \lim_{x \to 1} (3 - x) = 3 - 1 = 2$$

(ii) Using Equation 2 with $f(x) = 4x - x^2$ and $P(1, 3)$,

$$m = \lim_{h \to 0} \frac{f(a + h) - f(a)}{h} = \lim_{h \to 0} \frac{f(1 + h) - f(1)}{h} = \lim_{h \to 0} \frac{\left[4(1 + h) - (1 + h)^2\right] - 3}{h}$$

$$= \lim_{h \to 0} \frac{4 + 4h - 1 - 2h - h^2 - 3}{h} = \lim_{h \to 0} \frac{-h^2 + 2h}{h} = \lim_{h \to 0} \frac{h(-h + 2)}{h} = \lim_{h \to 0} (-h + 2) = 2$$

(b) An equation of the tangent line is $y - f(a) = f'(a)(x - a)$ $\Rightarrow$ $y - f(1) = f'(1)(x - 1)$ $\Rightarrow$ $y - 3 = 2(x - 1)$, or $y = 2x + 1$.

(c)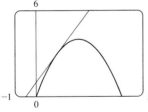

The graph of $y = 2x + 1$ is tangent to the graph of $y = 4x - x^2$ at the point $(1, 3)$. Now zoom in toward the point $(1, 3)$ until the parabola and the tangent line are indistiguishable.

4. (a) (i) Using Definition 1 with $f(x) = x - x^3$ and $P(1, 0)$,

$$m = \lim_{x \to 1} \frac{f(x) - 0}{x - 1} = \lim_{x \to 1} \frac{x - x^3}{x - 1} = \lim_{x \to 1} \frac{x(1 - x^2)}{x - 1} = \lim_{x \to 1} \frac{x(1 + x)(1 - x)}{x - 1}$$

$$= \lim_{x \to 1} \left[-x(1 + x)\right] = -1(2) = -2$$

(ii) Using Equation 2 with $f(x) = x - x^3$ and $P(1, 0)$,

$$m = \lim_{h \to 0} \frac{f(a+h) - f(a)}{h} = \lim_{h \to 0} \frac{f(1+h) - f(1)}{h} = \lim_{h \to 0} \frac{[(1+h) - (1+h)^3] - 0}{h}$$

$$= \lim_{h \to 0} \frac{1 + h - (1 + 3h + 3h^2 + h^3)}{h} = \lim_{h \to 0} \frac{-h^3 - 3h^2 - 2h}{h} = \lim_{h \to 0} \frac{h(-h^2 - 3h - 2)}{h}$$

$$= \lim_{h \to 0} (-h^2 - 3h - 2) = -2$$

(b) An equation of the tangent line is $y - f(a) = f'(a)(x - a) \quad \Rightarrow \quad y - f(1) = f'(1)(x - 1) \quad \Rightarrow \quad y - 0 = -2(x - 1)$,

or $y = -2x + 2$.

(c)

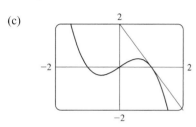

The graph of $y = -2x + 2$ is tangent to the graph of $y = x - x^3$ at the point $(1, 0)$. Now zoom in toward the point $(1, 0)$ until the cubic and the tangent line are indistinguishable.

5. Using (1) with $f(x) = \dfrac{x - 1}{x - 2}$ and $P(3, 2)$,

$$m = \lim_{x \to a} \frac{f(x) - f(a)}{x - a} = \lim_{x \to 3} \frac{\dfrac{x - 1}{x - 2} - 2}{x - 3} = \lim_{x \to 3} \frac{\dfrac{x - 1 - 2(x - 2)}{x - 2}}{x - 3}$$

$$= \lim_{x \to 3} \frac{3 - x}{(x - 2)(x - 3)} = \lim_{x \to 3} \frac{-1}{x - 2} = \frac{-1}{1} = -1$$

Tangent line: $y - 2 = -1(x - 3) \quad \Leftrightarrow \quad y - 2 = -x + 3 \quad \Leftrightarrow \quad y = -x + 5$

6. Using (1),

$$m = \lim_{x \to -1} \frac{(2x^3 - 5x) - 3}{x - (-1)} = \lim_{x \to -1} \frac{2x^3 - 5x - 3}{x + 1} = \lim_{x \to -1} \frac{(2x^2 - 2x - 3)(x + 1)}{x + 1} = \lim_{x \to -1} (2x^2 - 2x - 3) = 1.$$

Tangent line: $y - 3 = 1[x - (-1)] \quad \Leftrightarrow \quad y = x + 4$

7. Using (1), $m = \lim_{x \to 1} \dfrac{\sqrt{x} - \sqrt{1}}{x - 1} = \lim_{x \to 1} \dfrac{(\sqrt{x} - 1)(\sqrt{x} + 1)}{(x - 1)(\sqrt{x} + 1)} = \lim_{x \to 1} \dfrac{x - 1}{(x - 1)(\sqrt{x} + 1)} = \lim_{x \to 1} \dfrac{1}{\sqrt{x} + 1} = \dfrac{1}{2}.$

Tangent line: $y - 1 = \frac{1}{2}(x - 1) \quad \Leftrightarrow \quad y = \frac{1}{2}x + \frac{1}{2}$

8. Using (1), $m = \lim_{x \to 0} \dfrac{\dfrac{2x}{(x+1)^2} - 0}{x - 0} = \lim_{x \to 0} \dfrac{2x}{x(x+1)^2} = \lim_{x \to 0} \dfrac{2}{(x+1)^2} = \dfrac{2}{1^2} = 2.$

Tangent line: $y - 0 = 2(x - 0) \quad \Leftrightarrow \quad y = 2x$

9. (a) Using (2) with $y = f(x) = 3 + 4x^2 - 2x^3$,

$$m = \lim_{h \to 0} \frac{f(a+h) - f(a)}{h} = \lim_{h \to 0} \frac{3 + 4(a+h)^2 - 2(a+h)^3 - (3 + 4a^2 - 2a^3)}{h}$$

$$= \lim_{h \to 0} \frac{3 + 4(a^2 + 2ah + h^2) - 2(a^3 + 3a^2h + 3ah^2 + h^3) - 3 - 4a^2 + 2a^3}{h}$$

$$= \lim_{h \to 0} \frac{3 + 4a^2 + 8ah + 4h^2 - 2a^3 - 6a^2h - 6ah^2 - 2h^3 - 3 - 4a^2 + 2a^3}{h}$$

$$= \lim_{h \to 0} \frac{8ah + 4h^2 - 6a^2h - 6ah^2 - 2h^3}{h} = \lim_{h \to 0} \frac{h(8a + 4h - 6a^2 - 6ah - 2h^2)}{h}$$

$$= \lim_{h \to 0} (8a + 4h - 6a^2 - 6ah - 2h^2) = 8a - 6a^2$$

(b) At $(1, 5)$: $m = 8(1) - 6(1)^2 = 2$, so an equation of the tangent line

is $y - 5 = 2(x - 1)$ ⇔ $y = 2x + 3$.

At $(2, 3)$: $m = 8(2) - 6(2)^2 = -8$, so an equation of the tangent

line is $y - 3 = -8(x - 2)$ ⇔ $y = -8x + 19$.

(c)
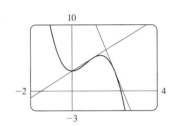

10. (a) Using (1),

$$m = \lim_{x \to a} \frac{\dfrac{1}{\sqrt{x}} - \dfrac{1}{\sqrt{a}}}{x - a} = \lim_{x \to a} \frac{\dfrac{\sqrt{a} - \sqrt{x}}{\sqrt{ax}}}{x - a} = \lim_{x \to a} \frac{(\sqrt{a} - \sqrt{x})(\sqrt{a} + \sqrt{x})}{\sqrt{ax}(x - a)(\sqrt{a} + \sqrt{x})} = \lim_{x \to a} \frac{a - x}{\sqrt{ax}(x - a)(\sqrt{a} + \sqrt{x})}$$

$$= \lim_{x \to a} \frac{-1}{\sqrt{ax}(\sqrt{a} + \sqrt{x})} = \frac{-1}{\sqrt{a^2}(2\sqrt{a})} = -\frac{1}{2a^{3/2}} \text{ or } -\frac{1}{2}a^{-3/2}$$

(b) At $(1, 1)$: $m = -\frac{1}{2}$, so an equation of the tangent line

is $y - 1 = -\frac{1}{2}(x - 1)$ ⇔ $y = -\frac{1}{2}x + \frac{3}{2}$.

At $\left(4, \frac{1}{2}\right)$: $m = -\frac{1}{16}$, so an equation of the tangent line

is $y - \frac{1}{2} = -\frac{1}{16}(x - 4)$ ⇔ $y = -\frac{1}{16}x + \frac{3}{4}$.

(c)
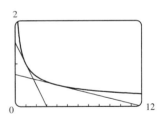

11. (a) The particle is moving to the right when s is increasing; that is, on the intervals $(0, 1)$ and $(4, 6)$. The particle is moving to

the left when s is decreasing; that is, on the interval $(2, 3)$. The particle is standing still when s is constant; that is, on the

intervals $(1, 2)$ and $(3, 4)$.

(b) The velocity of the particle is equal to the slope of the tangent line of the

graph. Note that there is no slope at the corner points on the graph. On the

interval $(0, 1)$, the slope is $\dfrac{3 - 0}{1 - 0} = 3$. On the interval $(2, 3)$, the slope is

$\dfrac{1 - 3}{3 - 2} = -2$. On the interval $(4, 6)$, the slope is $\dfrac{3 - 1}{6 - 4} = 1$.

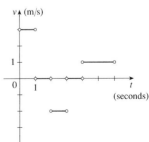

12. (a) **Runner A** runs the entire 100-meter race at the same velocity since the slope of the position function is constant.

Runner B starts the race at a slower velocity than runner A, but finishes the race at a faster velocity.

(b) The distance between the runners is the greatest at the time when the largest vertical line segment fits between the two graphs—this appears to be somewhere between 9 and 10 seconds.

(c) The runners had the same velocity when the slopes of their respective position functions are equal—this also appears to be at about 9.5 s. Note that the answers for parts (b) and (c) must be the same for these graphs because as soon as the velocity for runner B overtakes the velocity for runner A, the distance between the runners starts to decrease.

13. Let $s(t) = 40t - 16t^2$.

$$v(2) = \lim_{t \to 2} \frac{s(t) - s(2)}{t - 2} = \lim_{t \to 2} \frac{(40t - 16t^2) - 16}{t - 2} = \lim_{t \to 2} \frac{-16t^2 + 40t - 16}{t - 2} = \lim_{t \to 2} \frac{-8(2t^2 - 5t + 2)}{t - 2}$$

$$= \lim_{t \to 2} \frac{-8(t - 2)(2t - 1)}{t - 2} = -8 \lim_{t \to 2} (2t - 1) = -8(3) = -24$$

Thus, the instantaneous velocity when $t = 2$ is -24 ft/s.

14. (a) Let $H(t) = 10t - 1.86t^2$.

$$v(1) = \lim_{h \to 0} \frac{H(1 + h) - H(1)}{h} = \lim_{h \to 0} \frac{[10(1 + h) - 1.86(1 + h)^2] - (10 - 1.86)}{h}$$

$$= \lim_{h \to 0} \frac{10 + 10h - 1.86(1 + 2h + h^2) - 10 + 1.86}{h}$$

$$= \lim_{h \to 0} \frac{10 + 10h - 1.86 - 3.72h - 1.86h^2 - 10 + 1.86}{h}$$

$$= \lim_{h \to 0} \frac{6.28h - 1.86h^2}{h} = \lim_{h \to 0} (6.28 - 1.86h) = 6.28$$

The velocity of the rock after one second is 6.28 m/s.

(b) $v(a) = \lim_{h \to 0} \dfrac{H(a + h) - H(a)}{h} = \lim_{h \to 0} \dfrac{[10(a + h) - 1.86(a + h)^2] - (10a - 1.86a^2)}{h}$

$$= \lim_{h \to 0} \frac{10a + 10h - 1.86(a^2 + 2ah + h^2) - 10a + 1.86a^2}{h}$$

$$= \lim_{h \to 0} \frac{10a + 10h - 1.86a^2 - 3.72ah - 1.86h^2 - 10a + 1.86a^2}{h} = \lim_{h \to 0} \frac{10h - 3.72ah - 1.86h^2}{h}$$

$$= \lim_{h \to 0} \frac{h(10 - 3.72a - 1.86h)}{h} = \lim_{h \to 0} (10 - 3.72a - 1.86h) = 10 - 3.72a$$

The velocity of the rock when $t = a$ is $(10 - 3.72a)$ m/s.

(c) The rock will hit the surface when $H = 0 \Leftrightarrow 10t - 1.86t^2 = 0 \Leftrightarrow t(10 - 1.86t) = 0 \Leftrightarrow t = 0$ or $1.86t = 10$.
The rock hits the surface when $t = 10/1.86 \approx 5.4$ s.

(d) The velocity of the rock when it hits the surface is $v\left(\frac{10}{1.86}\right) = 10 - 3.72\left(\frac{10}{1.86}\right) = 10 - 20 = -10$ m/s.

15. $v(a) = \lim_{h \to 0} \dfrac{s(a + h) - s(a)}{h} = \lim_{h \to 0} \dfrac{\dfrac{1}{(a + h)^2} - \dfrac{1}{a^2}}{h} = \lim_{h \to 0} \dfrac{\dfrac{a^2 - (a + h)^2}{a^2(a + h)^2}}{h} = \lim_{h \to 0} \dfrac{a^2 - (a^2 + 2ah + h^2)}{ha^2(a + h)^2}$

$$= \lim_{h \to 0} \frac{-(2ah + h^2)}{ha^2(a + h)^2} = \lim_{h \to 0} \frac{-h(2a + h)}{ha^2(a + h)^2} = \lim_{h \to 0} \frac{-(2a + h)}{a^2(a + h)^2} = \frac{-2a}{a^2 \cdot a^2} = \frac{-2}{a^3} \text{ m/s}$$

So $v(1) = \dfrac{-2}{1^3} = -2$ m/s, $v(2) = \dfrac{-2}{2^3} = -\dfrac{1}{4}$ m/s, and $v(3) = \dfrac{-2}{3^3} = -\dfrac{2}{27}$ m/s.

16. (a) The average velocity between times t and $t + h$ is

$$\frac{s(t+h) - s(t)}{(t+h) - t} = \frac{(t+h)^2 - 8(t+h) + 18 - (t^2 - 8t + 18)}{h} = \frac{t^2 + 2th + h^2 - 8t - 8h + 18 - t^2 + 8t - 18}{h}$$

$$= \frac{2th + h^2 - 8h}{h} = (2t + h - 8) \text{ m/s}.$$

    (i) $[3, 4]$: $t = 3$, $h = 4 - 3 = 1$, so the average velocity is $2(3) + 1 - 8 = -1$ m/s.

                (ii) $[3.5, 4]$: $t = 3.5$, $h = 0.5$, so the average velocity is $2(3.5) + 0.5 - 8 = -0.5$ m/s.

    (iii) $[4, 5]$: $t = 4$, $h = 1$, so the average velocity is $2(4) + 1 - 8 = 1$ m/s.

                (iv) $[4, 4.5]$: $t = 4$, $h = 0.5$, so the average velocity is $2(4) + 0.5 - 8 = 0.5$ m/s.

(b) $v(t) = \lim\limits_{h \to 0} \dfrac{s(t+h) - s(t)}{h} = \lim\limits_{h \to 0} (2t + h - 8) = 2t - 8$,   (c)

so $v(4) = 0$.

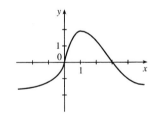

17. $g'(0)$ is the only negative value. The slope at $x = 4$ is smaller than the slope at $x = 2$ and both are smaller than the slope at $x = -2$. Thus, $g'(0) < 0 < g'(4) < g'(2) < g'(-2)$.

18. (a) Since $g(5) = -3$, the point $(5, -3)$ is on the graph of g. Since $g'(5) = 4$, the slope of the tangent line at $x = 5$ is 4. Using the point-slope form of a line gives us $y - (-3) = 4(x - 5)$, or $y = 4x - 23$.

(b) Since $(4, 3)$ is on $y = f(x)$, $f(4) = 3$. The slope of the tangent line between $(0, 2)$ and $(4, 3)$ is $\frac{1}{4}$, so $f'(4) = \frac{1}{4}$.

19. We begin by drawing a curve through the origin with a slope of 3 to satisfy $f(0) = 0$ and $f'(0) = 3$. Since $f'(1) = 0$, we will round off our figure so that there is a horizontal tangent directly over $x = 1$. Last, we make sure that the curve has a slope of -1 as we pass over $x = 2$. Two of the many possibilities are shown.

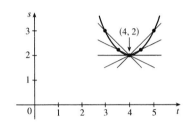

20. We begin by drawing a curve through the origin with a slope of 0 to satisfy $g(0) = 0$ and $g'(0) = 0$. The curve should have a slope of $-1, 3$, and 1 as we pass over $x = -1, 1$, and 2, respectively.

Note: In the figure, $y' = 0$ when $x \approx -1.27$ or 2.13.

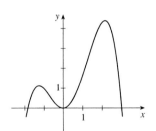

21. Using Definition 2 with $f(x) = 3x^2 - 5x$ and the point $(2, 2)$, we have

$$f'(2) = \lim\limits_{h \to 0} \frac{f(2+h) - f(2)}{h} = \lim\limits_{h \to 0} \frac{[3(2+h)^2 - 5(2+h)] - 2}{h} = \lim\limits_{h \to 0} \frac{(12 + 12h + 3h^2 - 10 - 5h) - 2}{h}$$

$$= \lim\limits_{h \to 0} \frac{3h^2 + 7h}{h} = \lim\limits_{h \to 0} (3h + 7) = 7$$

So an equation of the tangent line at $(2, 2)$ is $y - 2 = 7(x - 2)$ or $y = 7x - 12$.

22. Using Definition 2 with $g(x) = 1 - x^3$ and the point $(0, 1)$, we have

$$g'(0) = \lim_{h \to 0} \frac{g(0 + h) - g(0)}{h} = \lim_{h \to 0} \frac{\left[1 - (0 + h)^3\right] - 1}{h} = \lim_{h \to 0} \frac{(1 - h^3) - 1}{h} = \lim_{h \to 0} (-h^2) = 0$$

So an equation of the tangent line is $y - 1 = 0(x - 0)$ or $y = 1$.

23. (a) Using Definition 2 with $F(x) = 5x/(1 + x^2)$ and the point $(2, 2)$, we have
(b)

$$F'(2) = \lim_{h \to 0} \frac{F(2 + h) - F(2)}{h} = \lim_{h \to 0} \frac{\dfrac{5(2 + h)}{1 + (2 + h)^2} - 2}{h}$$

$$= \lim_{h \to 0} \frac{\dfrac{5h + 10}{h^2 + 4h + 5} - 2}{h} = \lim_{h \to 0} \frac{\dfrac{5h + 10 - 2(h^2 + 4h + 5)}{h^2 + 4h + 5}}{h}$$

$$= \lim_{h \to 0} \frac{-2h^2 - 3h}{h(h^2 + 4h + 5)} = \lim_{h \to 0} \frac{h(-2h - 3)}{h(h^2 + 4h + 5)} = \lim_{h \to 0} \frac{-2h - 3}{h^2 + 4h + 5} = \frac{-3}{5}$$

So an equation of the tangent line at $(2, 2)$ is $y - 2 = -\frac{3}{5}(x - 2)$ or $y = -\frac{3}{5}x + \frac{16}{5}$.

24. (a) Using Definition 2 with $G(x) = 4x^2 - x^3$, we have

$$G'(a) = \lim_{h \to 0} \frac{G(a + h) - G(a)}{h} = \lim_{h \to 0} \frac{\left[4(a + h)^2 - (a + h)^3\right] - (4a^2 - a^3)}{h}$$

$$= \lim_{h \to 0} \frac{4a^2 + 8ah + 4h^2 - (a^3 + 3a^2h + 3ah^2 + h^3) - 4a^2 + a^3}{h} = \lim_{h \to 0} \frac{8ah + 4h^2 - 3a^2h - 3ah^2 - h^3}{h}$$

$$= \lim_{h \to 0} \frac{h(8a + 4h - 3a^2 - 3ah - h^2)}{h} = \lim_{h \to 0} (8a + 4h - 3a^2 - 3ah - h^2) = 8a - 3a^2$$

At the point $(2, 8)$, $G'(2) = 16 - 12 = 4$, and an equation of the
tangent line is $y - 8 = 4(x - 2)$, or $y = 4x$. At the point $(3, 9)$,
$G'(3) = 24 - 27 = -3$, and an equation of the tangent line is
$y - 9 = -3(x - 3)$, or $y = -3x + 18$.

(b)

25. Use Definition 2 with $f(x) = 3 - 2x + 4x^2$.

$$f'(a) = \lim_{h \to 0} \frac{f(a + h) - f(a)}{h} = \lim_{h \to 0} \frac{\left[3 - 2(a + h) + 4(a + h)^2\right] - (3 - 2a + 4a^2)}{h}$$

$$= \lim_{h \to 0} \frac{(3 - 2a - 2h + 4a^2 + 8ah + 4h^2) - (3 - 2a + 4a^2)}{h}$$

$$= \lim_{h \to 0} \frac{-2h + 8ah + 4h^2}{h} = \lim_{h \to 0} \frac{h(-2 + 8a + 4h)}{h} = \lim_{h \to 0} (-2 + 8a + 4h) = -2 + 8a$$

26. $f'(a) = \lim_{h \to 0} \dfrac{f(a + h) - f(a)}{h} = \lim_{h \to 0} \dfrac{\left[(a + h)^4 - 5(a + h)\right] - (a^4 - 5a)}{h}$

$$= \lim_{h \to 0} \frac{(a^4 + 4a^3h + 6a^2h^2 + 4ah^3 + h^4 - 5a - 5h) - (a^4 - 5a)}{h} = \lim_{h \to 0} \frac{4a^3h + 6a^2h^2 + 4ah^3 + h^4 - 5h}{h}$$

$$= \lim_{h \to 0} \frac{h(4a^3 + 6a^2h + 4ah^2 + h^3 - 5)}{h} = \lim_{h \to 0} (4a^3 + 6a^2h + 4ah^2 + h^3 - 5) = 4a^3 - 5$$

27. Use Definition 2 with $f(t) = (2t + 1)/(t + 3)$.

$$f'(a) = \lim_{h \to 0} \frac{f(a+h) - f(a)}{h} = \lim_{h \to 0} \frac{\dfrac{2(a+h)+1}{(a+h)+3} - \dfrac{2a+1}{a+3}}{h} = \lim_{h \to 0} \frac{(2a+2h+1)(a+3) - (2a+1)(a+h+3)}{h(a+h+3)(a+3)}$$

$$= \lim_{h \to 0} \frac{(2a^2 + 6a + 2ah + 6h + a + 3) - (2a^2 + 2ah + 6a + a + h + 3)}{h(a+h+3)(a+3)}$$

$$= \lim_{h \to 0} \frac{5h}{h(a+h+3)(a+3)} = \lim_{h \to 0} \frac{5}{(a+h+3)(a+3)} = \frac{5}{(a+3)^2}$$

28. $f'(a) = \lim_{h \to 0} \dfrac{f(a+h) - f(a)}{h} = \lim_{h \to 0} \dfrac{\dfrac{(a+h)^2 + 1}{(a+h) - 2} - \dfrac{a^2 + 1}{a - 2}}{h}$

$$= \lim_{h \to 0} \frac{(a^2 + 2ah + h^2 + 1)(a-2) - (a^2 + 1)(a+h-2)}{h(a+h-2)(a-2)}$$

$$= \lim_{h \to 0} \frac{(a^3 - 2a^2 + 2a^2h - 4ah + ah^2 - 2h^2 + a - 2) - (a^3 + a^2h - 2a^2 + a + h - 2)}{h(a+h-2)(a-2)}$$

$$= \lim_{h \to 0} \frac{a^2h - 4ah + ah^2 - 2h^2 - h}{h(a+h-2)(a-2)} = \lim_{h \to 0} \frac{h(a^2 - 4a + ah - 2h - 1)}{h(a+h-2)(a-2)} = \lim_{h \to 0} \frac{a^2 - 4a + ah - 2h - 1}{(a+h-2)(a-2)}$$

$$= \frac{a^2 - 4a - 1}{(a-2)^2}$$

29. Use Definition 2 with $f(x) = 1/\sqrt{x + 2}$.

$$f'(a) = \lim_{h \to 0} \frac{f(a+h) - f(a)}{h} = \lim_{h \to 0} \frac{\dfrac{1}{\sqrt{(a+h)+2}} - \dfrac{1}{\sqrt{a+2}}}{h} = \lim_{h \to 0} \frac{\dfrac{\sqrt{a+2} - \sqrt{a+h+2}}{\sqrt{a+h+2}\sqrt{a+2}}}{h}$$

$$= \lim_{h \to 0} \left[\frac{\sqrt{a+2} - \sqrt{a+h+2}}{h\sqrt{a+h+2}\sqrt{a+2}} \cdot \frac{\sqrt{a+2} + \sqrt{a+h+2}}{\sqrt{a+2} + \sqrt{a+h+2}} \right] = \lim_{h \to 0} \frac{(a+2) - (a+h+2)}{h\sqrt{a+h+2}\sqrt{a+2}\left(\sqrt{a+2} + \sqrt{a+h+2}\right)}$$

$$= \lim_{h \to 0} \frac{-h}{h\sqrt{a+h+2}\sqrt{a+2}\left(\sqrt{a+2} + \sqrt{a+h+2}\right)} = \lim_{h \to 0} \frac{-1}{\sqrt{a+h+2}\sqrt{a+2}\left(\sqrt{a+2} + \sqrt{a+h+2}\right)}$$

$$= \frac{-1}{\left(\sqrt{a+2}\right)^2 \left(2\sqrt{a+2}\right)} = -\frac{1}{2(a+2)^{3/2}}$$

30. $f'(a) = \lim_{h \to 0} \dfrac{f(a+h) - f(a)}{h} = \lim_{h \to 0} \dfrac{\sqrt{3(a+h)+1} - \sqrt{3a+1}}{h}$

$$= \lim_{h \to 0} \frac{\left(\sqrt{3a+3h+1} - \sqrt{3a+1}\right)\left(\sqrt{3a+3h+1} + \sqrt{3a+1}\right)}{h\left(\sqrt{3a+3h+1} + \sqrt{3a+1}\right)} = \lim_{h \to 0} \frac{(3a+3h+1) - (3a+1)}{h\left(\sqrt{3a+3h+1} + \sqrt{3a+1}\right)}$$

$$= \lim_{h \to 0} \frac{3h}{h\left(\sqrt{3a+3h+1} + \sqrt{3a+1}\right)} = \lim_{h \to 0} \frac{3}{\sqrt{3a+3h+1} + \sqrt{3a+1}} = \frac{3}{2\sqrt{3a+1}}$$

Note that the answers to Exercises 31 – 36 are not unique.

31. By Definition 2, $\displaystyle\lim_{h \to 0} \frac{(1+h)^{10} - 1}{h} = f'(1)$, where $f(x) = x^{10}$ and $a = 1$.

Or: By Definition 2, $\displaystyle\lim_{h \to 0} \frac{(1+h)^{10} - 1}{h} = f'(0)$, where $f(x) = (1+x)^{10}$ and $a = 0$.

32. By Definition 2, $\displaystyle\lim_{h\to 0}\frac{\sqrt[4]{16+h}-2}{h}=f'(16)$, where $f(x)=\sqrt[4]{x}$ and $a=16$.

Or: By Definition 2, $\displaystyle\lim_{h\to 0}\frac{\sqrt[4]{16+h}-2}{h}=f'(0)$, where $f(x)=\sqrt[4]{16+x}$ and $a=0$.

33. By Equation 3, $\displaystyle\lim_{x\to 5}\frac{2^x-32}{x-5}=f'(5)$, where $f(x)=2^x$ and $a=5$.

34. By Equation 3, $\displaystyle\lim_{x\to\pi/4}\frac{\tan x-1}{x-\pi/4}=f'(\pi/4)$, where $f(x)=\tan x$ and $a=\pi/4$.

35. By Definition 2, $\displaystyle\lim_{h\to 0}\frac{\cos(\pi+h)+1}{h}=f'(\pi)$, where $f(x)=\cos x$ and $a=\pi$.

Or: By Definition 2, $\displaystyle\lim_{h\to 0}\frac{\cos(\pi+h)+1}{h}=f'(0)$, where $f(x)=\cos(\pi+x)$ and $a=0$.

36. By Equation 3, $\displaystyle\lim_{t\to 1}\frac{t^4+t-2}{t-1}=f'(1)$, where $f(t)=t^4+t$ and $a=1$.

37. $v(5)=f'(5)=\displaystyle\lim_{h\to 0}\frac{f(5+h)-f(5)}{h}=\lim_{h\to 0}\frac{[100+50(5+h)-4.9(5+h)^2]-[100+50(5)-4.9(5)^2]}{h}$

$=\displaystyle\lim_{h\to 0}\frac{(100+250+50h-4.9h^2-49h-122.5)-(100+250-122.5)}{h}=\lim_{h\to 0}\frac{-4.9h^2+h}{h}$

$=\displaystyle\lim_{h\to 0}\frac{h(-4.9h+1)}{h}=\lim_{h\to 0}(-4.9h+1)=1\text{ m/s}$

The speed when $t=5$ is $|1|=1$ m/s.

38. $v(5)=f'(5)=\displaystyle\lim_{h\to 0}\frac{f(5+h)-f(5)}{h}=\lim_{h\to 0}\frac{[(5+h)^{-1}-(5+h)]-(5^{-1}-5)}{h}$

$=\displaystyle\lim_{h\to 0}\frac{\dfrac{1}{5+h}-5-h-\dfrac{1}{5}+5}{h}=\lim_{h\to 0}\frac{\dfrac{1}{5+h}-h-\dfrac{1}{5}}{h}=\lim_{h\to 0}\frac{\dfrac{5-5h(5+h)-(5+h)}{5(5+h)}}{h}$

$=\displaystyle\lim_{h\to 0}\frac{5-25h-5h^2-5-h}{5h(5+h)}=\lim_{h\to 0}\frac{-5h^2-26h}{5h(5+h)}=\lim_{h\to 0}\frac{h(-5h-26)}{5h(5+h)}=\lim_{h\to 0}\frac{-5h-26}{5(5+h)}=\frac{-26}{25}\text{ m/s}$

The speed when $t=5$ is $\left|-\frac{26}{25}\right|=\frac{26}{25}=1.04$ m/s.

39. The sketch shows the graph for a room temperature of $72°$ and a refrigerator temperature of $38°$. The initial rate of change is greater in magnitude than the rate of change after an hour.

40. The slope of the tangent (that is, the rate of change of temperature with respect to time) at $t=1$ h seems to be about $\dfrac{75-168}{132-0}\approx-0.7\,°\text{F/min}$.

41. (a) (i) $[2000, 2002]$: $\dfrac{P(2002) - P(2000)}{2002 - 2000} = \dfrac{77 - 55}{2} = \dfrac{22}{2} = 11$ percent/year

(ii) $[2000, 2001]$: $\dfrac{P(2001) - P(2000)}{2001 - 2000} = \dfrac{68 - 55}{1} = 13$ percent/year

(iii) $[1999, 2000]$: $\dfrac{P(2000) - P(1999)}{2000 - 1999} = \dfrac{55 - 39}{1} = 16$ percent/year

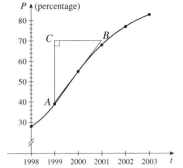

(b) Using the values from (ii) and (iii), we have $\dfrac{13 + 16}{2} = 14.5$ percent/year.

(c) Estimating A as $(1999, 40)$ and B as $(2001, 70)$, the slope at 2000 is

$\dfrac{70 - 40}{2001 - 1999} = \dfrac{30}{2} = 15$ percent/year.

42. (a) (i) $[2000, 2002]$: $\dfrac{P(2002) - P(2000)}{2002 - 2000} = \dfrac{5886 - 3501}{2} = \dfrac{2385}{2} = 1192.5$ locations/year

(ii) $[2000, 2001]$: $\dfrac{P(2001) - P(2000)}{2001 - 2000} = \dfrac{4709 - 3501}{1} = 1208$ locations/year

(iii) $[1999, 2000]$: $\dfrac{P(2000) - P(1999)}{2000 - 1999} = \dfrac{3501 - 2135}{1} = 1366$ locations/year

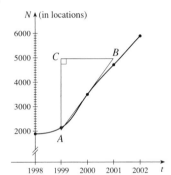

(b) Using the values from (ii) and (iii), we have

$\dfrac{1208 + 1366}{2} = 1287$ locations/year.

(c) Estimating A as $(1999, 2035)$ and B as $(2001, 4960)$, the slope at 2000 is

$\dfrac{4960 - 2035}{2001 - 1999} = \dfrac{2925}{2} = 1462.5$ locations/year.

43. (a) (i) $\dfrac{\Delta C}{\Delta x} = \dfrac{C(105) - C(100)}{105 - 100} = \dfrac{6601.25 - 6500}{5} = \$20.25/\text{unit.}$

(ii) $\dfrac{\Delta C}{\Delta x} = \dfrac{C(101) - C(100)}{101 - 100} = \dfrac{6520.05 - 6500}{1} = \$20.05/\text{unit.}$

(b) $\dfrac{C(100 + h) - C(100)}{h} = \dfrac{\left[5000 + 10(100 + h) + 0.05(100 + h)^2\right] - 6500}{h} = \dfrac{20h + 0.05h^2}{h}$

$= 20 + 0.05h, \ h \neq 0$

So the instantaneous rate of change is $\displaystyle\lim_{h \to 0} \dfrac{C(100 + h) - C(100)}{h} = \lim_{h \to 0}(20 + 0.05h) = \$20/\text{unit.}$

44. $\Delta V = V(t + h) - V(t) = 100{,}000\left(1 - \dfrac{t + h}{60}\right)^2 - 100{,}000\left(1 - \dfrac{t}{60}\right)^2$

$= 100{,}000\left[\left(1 - \dfrac{t + h}{30} + \dfrac{(t + h)^2}{3600}\right) - \left(1 - \dfrac{t}{30} + \dfrac{t^2}{3600}\right)\right] = 100{,}000\left(-\dfrac{h}{30} + \dfrac{2th}{3600} + \dfrac{h^2}{3600}\right)$

$= \dfrac{100{,}000}{3600}h\left(-120 + 2t + h\right) = \dfrac{250}{9}h\left(-120 + 2t + h\right)$

[continued]

Dividing ΔV by h and then letting $h \to 0$, we see that the instantaneous rate of change is $\frac{500}{9}(t-60)$ gal/min.

t	Flow rate (gal/min)	Water remaining $V(t)$ (gal)
0	$-3333.\overline{3}$	$100,000$
10	$-2777.\overline{7}$	$69,444.\overline{4}$
20	$-2222.\overline{2}$	$44,444.\overline{4}$
30	$-1666.\overline{6}$	$25,000$
40	$-1111.\overline{1}$	$11,111.\overline{1}$
50	$-555.\overline{5}$	$2,777.\overline{7}$
60	0	0

The magnitude of the flow rate is greatest at the beginning and gradually decreases to 0.

45. (a) $f'(x)$ is the rate of change of the production cost with respect to the number of ounces of gold produced. Its units are dollars per ounce.

(b) After 800 ounces of gold have been produced, the rate at which the production cost is increasing is $17/ounce. So the cost of producing the 800th (or 801st) ounce is about $17.

(c) In the short term, the values of $f'(x)$ will decrease because more efficient use is made of start-up costs as x increases. But eventually $f'(x)$ might increase due to large-scale operations.

46. (a) $f'(5)$ is the rate of growth of the bacteria population when $t=5$ hours. Its units are bacteria per hour.

(b) With unlimited space and nutrients, f' should increase as t increases; so $f'(5) < f'(10)$. If the supply of nutrients is limited, the growth rate slows down at some point in time, and the opposite may be true.

47. $T'(10)$ is the rate at which the temperature is changing at 10:00 AM. To estimate the value of $T'(10)$, we will average the difference quotients obtained using the times $t=8$ and $t=12$. Let $A = \frac{T(8)-T(10)}{8-10} = \frac{72-81}{-2} = 4.5$ and

$B = \frac{T(12)-T(10)}{12-10} = \frac{88-81}{2} = 3.5$. Then $T'(10) = \lim\limits_{t\to 10} \frac{T(t)-T(10)}{t-10} \approx \frac{A+B}{2} = \frac{4.5+3.5}{2} = 4°\text{F/h}$.

48. (a) $f'(8)$ is the rate of change of the quantity of coffee sold with respect to the price per pound when the price is $8 per pound. The units for $f'(8)$ are pounds/(dollars/pound).

(b) $f'(8)$ is negative since the quantity of coffee sold will decrease as the price charged for it increases. People are generally less willing to buy a product when its price increases.

49. (a) $S'(T)$ is the rate at which the oxygen solubility changes with respect to the water temperature. Its units are $(\text{mg/L})/°\text{C}$.

(b) For $T=16°\text{C}$, it appears that the tangent line to the curve goes through the points $(0,14)$ and $(32,6)$. So

$S'(16) \approx \frac{6-14}{32-0} = -\frac{8}{32} = -0.25 \ (\text{mg/L})/°\text{C}$. This means that as the temperature increases past $16°\text{C}$, the oxygen solubility is decreasing at a rate of $0.25 \ (\text{mg/L})/°\text{C}$.

50. (a) $S'(T)$ is the rate of change of the maximum sustainable speed of Coho salmon with respect to the temperature. Its units are $(\text{cm/s})/°\text{C}$.

(b) For $T = 15°C$, it appears the tangent line to the curve goes through the points $(10, 25)$ and $(20, 32)$. So

$S'(15) \approx \dfrac{32 - 25}{20 - 10} = 0.7$ (cm/s)/$°C$. This tells us that at $T = 15°C$, the maximum sustainable speed of Coho salmon is

changing at a rate of 0.7 (cm/s)/$°C$. In a similar fashion for $T = 25°C$, we can use the points $(20, 35)$ and $(25, 25)$ to

obtain $S'(25) \approx \dfrac{25 - 35}{25 - 20} = -2$ (cm/s)/$°C$. As it gets warmer than $20°C$, the maximum sustainable speed decreases

rapidly.

51. Since $f(x) = x \sin(1/x)$ when $x \neq 0$ and $f(0) = 0$, we have

$f'(0) = \lim\limits_{h \to 0} \dfrac{f(0 + h) - f(0)}{h} = \lim\limits_{h \to 0} \dfrac{h \sin(1/h) - 0}{h} = \lim\limits_{h \to 0} \sin(1/h)$. This limit does not exist since $\sin(1/h)$ takes the

values -1 and 1 on any interval containing 0. (Compare with Example 4 in Section 2.2.)

52. Since $f(x) = x^2 \sin(1/x)$ when $x \neq 0$ and $f(0) = 0$, we have

$f'(0) = \lim\limits_{h \to 0} \dfrac{f(0 + h) - f(0)}{h} = \lim\limits_{h \to 0} \dfrac{h^2 \sin(1/h) - 0}{h} = \lim\limits_{h \to 0} h \sin(1/h)$. Since $-1 \leq \sin \dfrac{1}{h} \leq 1$, we have

$-|h| \leq |h| \sin \dfrac{1}{h} \leq |h| \quad \Rightarrow \quad -|h| \leq h \sin \dfrac{1}{h} \leq |h|$. Because $\lim\limits_{h \to 0} (-|h|) = 0$ and $\lim\limits_{h \to 0} |h| = 0$, we know that

$\lim\limits_{h \to 0} \left(h \sin \dfrac{1}{h} \right) = 0$ by the Squeeze Theorem. Thus, $f'(0) = 0$.

3.2   The Derivative as a Function

1. It appears that f is an odd function, so f' will be an even

function—that is, $f'(-a) = f'(a)$.

(a) $f'(-3) \approx 1.5$ (b) $f'(-2) \approx 1$

(c) $f'(-1) \approx 0$ (d) $f'(0) \approx -4$

(e) $f'(1) \approx 0$ (f) $f'(2) \approx 1$

(g) $f'(3) \approx 1.5$

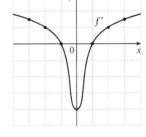

2. *Note:* Your answers may vary depending on your estimates. By

estimating the slopes of tangent lines on the graph of f, it appears that

(a) $f'(0) \approx -3$ (b) $f'(1) \approx 0$

(c) $f'(2) \approx 1.5$ (d) $f'(3) \approx 2$

(e) $f'(4) \approx 0$ (f) $f'(5) \approx -1.2$

3. (a)$' = $ II, since from left to right, the slopes of the tangents to graph (a) start out negative, become 0, then positive, then 0, then

negative again. The actual function values in graph II follow the same pattern.

(b)$' = $ IV, since from left to right, the slopes of the tangents to graph (b) start out at a fixed positive quantity, then suddenly

become negative, then positive again. The discontinuities in graph IV indicate sudden changes in the slopes of the tangents.

(c)$' =$ I, since the slopes of the tangents to graph (c) are negative for $x < 0$ and positive for $x > 0$, as are the function values of graph I.

(d)$' =$ III, since from left to right, the slopes of the tangents to graph (d) are positive, then 0, then negative, then 0, then positive, then 0, then negative again, and the function values in graph III follow the same pattern.

Hints for Exercises 4–11: First plot x-intercepts on the graph of f' for any horizontal tangents on the graph of f. Look for any corners on the graph of f—there will be a discontinuity on the graph of f'. On any interval where f has a tangent with positive (or negative) slope, the graph of f' will be positive (or negative). If the graph of the function is linear, the graph of f' will be a horizontal line.

4.

5.

6.

7.

8.

9.

10.

11.

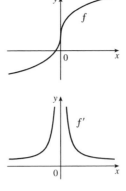

12. The slopes of the tangent lines on the graph of $y = P(t)$ are always positive, so the y-values of $y = P'(t)$ are always positive. These values start out relatively small and keep increasing, reaching a maximum at about $t = 6$. Then the y-values of $y = P'(t)$ decrease and get close to zero. The graph of P' tells us that the yeast culture grows most rapidly after 6 hours and then the growth rate declines.

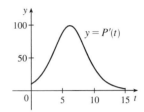

13. It appears that there are horizontal tangents on the graph of M for $t = 1963$ and $t = 1971$. Thus, there are zeros for those values of t on the graph of M'. The derivative is negative for the years 1963 to 1971.

14.

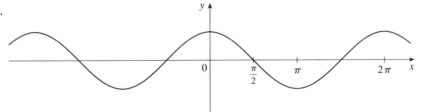

The graph of the derivative looks like the graph of the cosine function.

15. (a) By zooming in, we estimate that $f'(0) = 0$, $f'\left(\frac{1}{2}\right) = 1$, $f'(1) = 2$, and $f'(2) = 4$.

(b) By symmetry, $f'(-x) = -f'(x)$. So $f'\left(-\frac{1}{2}\right) = -1$, $f'(-1) = -2$, and $f'(-2) = -4$.

(c) It appears that $f'(x)$ is twice the value of x, so we guess that $f'(x) = 2x$.

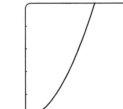

(d) $f'(x) = \lim\limits_{h \to 0} \dfrac{f(x+h) - f(x)}{h} = \lim\limits_{h \to 0} \dfrac{(x+h)^2 - x^2}{h}$

$\qquad = \lim\limits_{h \to 0} \dfrac{(x^2 + 2hx + h^2) - x^2}{h} = \lim\limits_{h \to 0} \dfrac{2hx + h^2}{h} = \lim\limits_{h \to 0} \dfrac{h(2x + h)}{h} = \lim\limits_{h \to 0} (2x + h) = 2x$

16. (a) By zooming in, we estimate that $f'(0) = 0$, $f'\left(\frac{1}{2}\right) \approx 0.75$,   (b) By symmetry, $f'(-x) = f'(x)$. So $f'\left(-\frac{1}{2}\right) \approx 0.75$,

$f'(1) \approx 3$, $f'(2) \approx 12$, and $f'(3) \approx 27$. $f'(-1) \approx 3$, $f'(-2) \approx 12$, and $f'(-3) \approx 27$.

(c)

(d) Since $f'(0) = 0$, it appears that f' may have the form $f'(x) = ax^2$. Using $f'(1) = 3$, we have $a = 3$, so $f'(x) = 3x^2$.

(e) $f'(x) = \lim\limits_{h \to 0} \dfrac{f(x+h) - f(x)}{h} = \lim\limits_{h \to 0} \dfrac{(x+h)^3 - x^3}{h} = \lim\limits_{h \to 0} \dfrac{(x^3 + 3x^2h + 3xh^2 + h^3) - x^3}{h}$

$= \lim\limits_{h \to 0} \dfrac{3x^2h + 3xh^2 + h^3}{h} = \lim\limits_{h \to 0} \dfrac{h(3x^2 + 3xh + h^2)}{h} = \lim\limits_{h \to 0} (3x^2 + 3xh + h^2) = 3x^2$

17. $f'(x) = \lim\limits_{h \to 0} \dfrac{f(x+h) - f(x)}{h} = \lim\limits_{h \to 0} \dfrac{\left[\frac{1}{2}(x+h) - \frac{1}{3}\right] - \left(\frac{1}{2}x - \frac{1}{3}\right)}{h} = \lim\limits_{h \to 0} \dfrac{\frac{1}{2}x + \frac{1}{2}h - \frac{1}{3} - \frac{1}{2}x + \frac{1}{3}}{h}$

$= \lim\limits_{h \to 0} \dfrac{\frac{1}{2}h}{h} = \lim\limits_{h \to 0} \dfrac{1}{2} = \dfrac{1}{2}$

Domain of f = domain of f' = $\mathbb{R}$.

18. $f'(x) = \lim\limits_{h \to 0} \dfrac{f(x+h) - f(x)}{h} = \lim\limits_{h \to 0} \dfrac{[m(x+h) + b] - (mx + b)}{h} = \lim\limits_{h \to 0} \dfrac{mx + mh + b - mx - b}{h}$

$= \lim\limits_{h \to 0} \dfrac{mh}{h} = \lim\limits_{h \to 0} m = m$

Domain of f = domain of f' = $\mathbb{R}$.

19. $f'(t) = \lim\limits_{h \to 0} \dfrac{f(t+h) - f(t)}{h} = \lim\limits_{h \to 0} \dfrac{\left[5(t+h) - 9(t+h)^2\right] - (5t - 9t^2)}{h}$

$= \lim\limits_{h \to 0} \dfrac{5t + 5h - 9(t^2 + 2th + h^2) - 5t + 9t^2}{h} = \lim\limits_{h \to 0} \dfrac{5t + 5h - 9t^2 - 18th - 9h^2 - 5t + 9t^2}{h}$

$= \lim\limits_{h \to 0} \dfrac{5h - 18th - 9h^2}{h} = \lim\limits_{h \to 0} \dfrac{h(5 - 18t - 9h)}{h} = \lim\limits_{h \to 0} (5 - 18t - 9h) = 5 - 18t$

Domain of f = domain of f' = $\mathbb{R}$.

20. $f'(x) = \lim\limits_{h \to 0} \dfrac{f(x+h) - f(x)}{h} = \lim\limits_{h \to 0} \dfrac{\left[1.5(x+h)^2 - (x+h) + 3.7\right] - \left(1.5x^2 - x + 3.7\right)}{h}$

$= \lim\limits_{h \to 0} \dfrac{1.5x^2 + 3xh + 1.5h^2 - x - h + 3.7 - 1.5x^2 + x - 3.7}{h} = \lim\limits_{h \to 0} \dfrac{3xh + 1.5h^2 - h}{h}$

$= \lim\limits_{h \to 0} (3x + 1.5h - 1) = 3x - 1$

Domain of f = domain of f' = $\mathbb{R}$.

21. $f'(x) = \lim\limits_{h \to 0} \dfrac{f(x+h) - f(x)}{h} = \lim\limits_{h \to 0} \dfrac{\left[(x+h)^3 - 3(x+h) + 5\right] - (x^3 - 3x + 5)}{h}$

$= \lim\limits_{h \to 0} \dfrac{(x^3 + 3x^2h + 3xh^2 + h^3 - 3x - 3h + 5) - (x^3 - 3x + 5)}{h} = \lim\limits_{h \to 0} \dfrac{3x^2h + 3xh^2 + h^3 - 3h}{h}$

$= \lim\limits_{h \to 0} \dfrac{h(3x^2 + 3xh + h^2 - 3)}{h} = \lim\limits_{h \to 0} (3x^2 + 3xh + h^2 - 3) = 3x^2 - 3$

Domain of f = domain of $f' = \mathbb{R}$.

22. $f'(x) = \lim\limits_{h \to 0} \dfrac{f(x+h) - f(x)}{h} = \lim\limits_{h \to 0} \dfrac{\left(x + h + \sqrt{x+h}\,\right) - (x + \sqrt{x}\,)}{h}$

$= \lim\limits_{h \to 0} \left(\dfrac{h}{h} + \dfrac{\sqrt{x+h} - \sqrt{x}}{h} \cdot \dfrac{\sqrt{x+h} + \sqrt{x}}{\sqrt{x+h} + \sqrt{x}} \right) = \lim\limits_{h \to 0} \left[1 + \dfrac{(x+h) - x}{h\left(\sqrt{x+h} + \sqrt{x}\,\right)} \right]$

$= \lim\limits_{h \to 0} \left(1 + \dfrac{1}{\sqrt{x+h} + \sqrt{x}} \right) = 1 + \dfrac{1}{\sqrt{x} + \sqrt{x}} = 1 + \dfrac{1}{2\sqrt{x}}$

Domain of $f = [0, \infty)$, domain of $f' = (0, \infty)$.

23. $g'(x) = \lim\limits_{h \to 0} \dfrac{g(x+h) - g(x)}{h} = \lim\limits_{h \to 0} \dfrac{\sqrt{1 + 2(x+h)} - \sqrt{1 + 2x}}{h} \left[\dfrac{\sqrt{1 + 2(x+h)} + \sqrt{1 + 2x}}{\sqrt{1 + 2(x+h)} + \sqrt{1 + 2x}} \right]$

$= \lim\limits_{h \to 0} \dfrac{(1 + 2x + 2h) - (1 + 2x)}{h\left[\sqrt{1 + 2(x+h)} + \sqrt{1 + 2x}\,\right]} = \lim\limits_{h \to 0} \dfrac{2}{\sqrt{1 + 2x + 2h} + \sqrt{1 + 2x}} = \dfrac{2}{2\sqrt{1 + 2x}} = \dfrac{1}{\sqrt{1 + 2x}}$

Domain of $g = \left[-\frac{1}{2}, \infty\right)$, domain of $g' = \left(-\frac{1}{2}, \infty\right)$.

24. $f'(x) = \lim\limits_{h \to 0} \dfrac{f(x+h) - f(x)}{h} = \lim\limits_{h \to 0} \dfrac{\dfrac{3 + (x+h)}{1 - 3(x+h)} - \dfrac{3 + x}{1 - 3x}}{h} = \lim\limits_{h \to 0} \dfrac{(3 + x + h)(1 - 3x) - (3 + x)(1 - 3x - 3h)}{h(1 - 3x - 3h)(1 - 3x)}$

$= \lim\limits_{h \to 0} \dfrac{\left(3 - 9x + x - 3x^2 + h - 3hx\right) - \left(3 - 9x - 9h + x - 3x^2 - 3hx\right)}{h(1 - 3x - 3h)(1 - 3x)}$

$= \lim\limits_{h \to 0} \dfrac{10h}{h(1 - 3x - 3h)(1 - 3x)} = \lim\limits_{h \to 0} \dfrac{10}{(1 - 3x - 3h)(1 - 3x)} = \dfrac{10}{(1 - 3x)^2}$

Domain of f = domain of $f' = \left(-\infty, \frac{1}{3}\right) \cup \left(\frac{1}{3}, \infty\right)$.

25. $G'(t) = \lim\limits_{h \to 0} \dfrac{G(t+h) - G(t)}{h} = \lim\limits_{h \to 0} \dfrac{\dfrac{4(t+h)}{(t+h) + 1} - \dfrac{4t}{t+1}}{h} = \lim\limits_{h \to 0} \dfrac{\dfrac{4(t+h)(t+1) - 4t(t+h+1)}{(t+h+1)(t+1)}}{h}$

$= \lim\limits_{h \to 0} \dfrac{\left(4t^2 + 4ht + 4t + 4h\right) - \left(4t^2 + 4ht + 4t\right)}{h(t+h+1)(t+1)} = \lim\limits_{h \to 0} \dfrac{4h}{h(t+h+1)(t+1)}$

$= \lim\limits_{h \to 0} \dfrac{4}{(t+h+1)(t+1)} = \dfrac{4}{(t+1)^2}$

Domain of G = domain of $G' = (-\infty, -1) \cup (-1, \infty)$.

26. $g'(t) = \lim\limits_{h \to 0} \dfrac{g(t+h) - g(t)}{h} = \lim\limits_{h \to 0} \dfrac{\frac{1}{\sqrt{t+h}} - \frac{1}{\sqrt{t}}}{h} = \lim\limits_{h \to 0} \dfrac{\frac{\sqrt{t} - \sqrt{t+h}}{\sqrt{t+h}\sqrt{t}}}{h} = \lim\limits_{h \to 0} \left(\dfrac{\sqrt{t} - \sqrt{t+h}}{h\,\sqrt{t+h}\,\sqrt{t}} \cdot \dfrac{\sqrt{t} + \sqrt{t+h}}{\sqrt{t} + \sqrt{t+h}} \right)$

$\quad = \lim\limits_{h \to 0} \dfrac{t - (t+h)}{h\,\sqrt{t+h}\,\sqrt{t}\,(\sqrt{t} + \sqrt{t+h})} = \lim\limits_{h \to 0} \dfrac{-h}{h\,\sqrt{t+h}\,\sqrt{t}\,(\sqrt{t} + \sqrt{t+h})} = \lim\limits_{h \to 0} \dfrac{-1}{\sqrt{t+h}\,\sqrt{t}\,(\sqrt{t} + \sqrt{t+h})}$

$\quad = \dfrac{-1}{\sqrt{t}\,\sqrt{t}\,(\sqrt{t} + \sqrt{t})} = \dfrac{-1}{t\,(2\sqrt{t})} = -\dfrac{1}{2t^{3/2}}$

Domain of g = domain of $g' = (0, \infty)$.

27. $f'(x) = \lim\limits_{h \to 0} \dfrac{f(x+h) - f(x)}{h} = \lim\limits_{h \to 0} \dfrac{(x+h)^4 - x^4}{h} = \lim\limits_{h \to 0} \dfrac{(x^4 + 4x^3h + 6x^2h^2 + 4xh^3 + h^4) - x^4}{h}$

$\quad = \lim\limits_{h \to 0} \dfrac{4x^3h + 6x^2h^2 + 4xh^3 + h^4}{h} = \lim\limits_{h \to 0} (4x^3 + 6x^2h + 4xh^2 + h^3) = 4x^3$

Domain of f = domain of $f' = \mathbb{R}$.

28. (a)

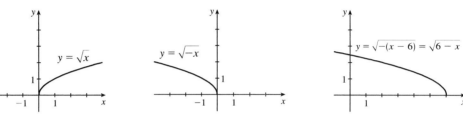

(b) Note that the third graph in part (a) has small negative values for its slope, f'; but as $x \to 6^-$, $f' \to -\infty$.

See the graph in part (d).

(c) $f'(x) = \lim\limits_{h \to 0} \dfrac{f(x+h) - f(x)}{h}$ **(d)**

$\quad = \lim\limits_{h \to 0} \dfrac{\sqrt{6 - (x+h)} - \sqrt{6 - x}}{h} \left[\dfrac{\sqrt{6 - (x+h)} + \sqrt{6 - x}}{\sqrt{6 - (x+h)} + \sqrt{6 - x}} \right]$

$\quad = \lim\limits_{h \to 0} \dfrac{[6 - (x+h)] - (6 - x)}{h\left[\sqrt{6 - (x+h)} + \sqrt{6 - x}\right]} = \lim\limits_{h \to 0} \dfrac{-h}{h\left(\sqrt{6 - x - h} + \sqrt{6 - x}\right)}$

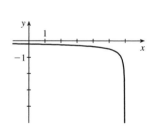

$\quad = \lim\limits_{h \to 0} \dfrac{-1}{\sqrt{6 - x - h} + \sqrt{6 - x}} = \dfrac{-1}{2\sqrt{6 - x}}$

Domain of $f = (-\infty, 6]$, domain of $f' = (-\infty, 6)$.

29. (a) $f'(x) = \lim\limits_{h \to 0} \dfrac{f(x+h) - f(x)}{h} = \lim\limits_{h \to 0} \dfrac{[(x+h)^4 + 2(x+h)] - (x^4 + 2x)}{h}$

$\quad = \lim\limits_{h \to 0} \dfrac{x^4 + 4x^3h + 6x^2h^2 + 4xh^3 + h^4 + 2x + 2h - x^4 - 2x}{h}$

$\quad = \lim\limits_{h \to 0} \dfrac{4x^3h + 6x^2h^2 + 4xh^3 + h^4 + 2h}{h} = \lim\limits_{h \to 0} \dfrac{h(4x^3 + 6x^2h + 4xh^2 + h^3 + 2)}{h}$

$\quad = \lim\limits_{h \to 0} (4x^3 + 6x^2h + 4xh^2 + h^3 + 2) = 4x^3 + 2$

(b) Notice that $f'(x) = 0$ when f has a horizontal tangent, $f'(x)$ is

positive when the tangents have positive slope, and $f'(x)$ is

negative when the tangents have negative slope.

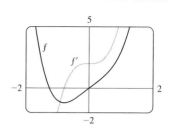

30. (a) $f'(t) = \lim\limits_{h \to 0} \dfrac{f(t+h) - f(t)}{h} = \lim\limits_{h \to 0} \dfrac{\left[(t+h)^2 - \sqrt{t+h}\,\right] - \left(t^2 - \sqrt{t}\,\right)}{h}$

$= \lim\limits_{h \to 0} \dfrac{t^2 + 2ht + h^2 - \sqrt{t+h} - t^2 + \sqrt{t}}{h} = \lim\limits_{h \to 0} \left(\dfrac{2ht + h^2}{h} + \dfrac{\sqrt{t} - \sqrt{t+h}}{h} \right)$

$= \lim\limits_{h \to 0} \left(\dfrac{h(2t+h)}{h} + \dfrac{\sqrt{t} - \sqrt{t+h}}{h} \cdot \dfrac{\sqrt{t} + \sqrt{t+h}}{\sqrt{t} + \sqrt{t+h}} \right)$

$= \lim\limits_{h \to 0} \left(2t + h + \dfrac{t - (t+h)}{h(\sqrt{t} + \sqrt{t+h}\,)} \right) = \lim\limits_{h \to 0} \left(2t + h + \dfrac{-h}{h(\sqrt{t} + \sqrt{t+h}\,)} \right)$

$= \lim\limits_{h \to 0} \left(2t + h + \dfrac{-1}{\sqrt{t} + \sqrt{t+h}} \right) = 2t - \dfrac{1}{2\sqrt{t}}$

(b) Notice that $f'(t) = 0$ when f has a horizontal tangent, $f'(t)$ is

positive when the tangents have positive slope, and $f'(t)$ is

negative when the tangents have negative slope.

31. (a) $U'(t)$ is the rate at which the unemployment rate is changing with respect to time. Its units are percent per year.

(b) To find $U'(t)$, we use $\lim\limits_{h \to 0} \dfrac{U(t+h) - U(t)}{h} \approx \dfrac{U(t+h) - U(t)}{h}$ for small values of h.

For 1993: $U'(1993) \approx \dfrac{U(1994) - U(1993)}{1994 - 1993} = \dfrac{6.1 - 6.9}{1} = -0.80$

For 1994: We estimate $U'(1994)$ by using $h = -1$ and $h = 1$, and then average the two results to obtain a final estimate.

$h = -1 \quad \Rightarrow \quad U'(1994) \approx \dfrac{U(1993) - U(1994)}{1993 - 1994} = \dfrac{6.9 - 6.1}{-1} = -0.80;$

$h = 1 \quad \Rightarrow \quad U'(1994) \approx \dfrac{U(1995) - U(1994)}{1995 - 1994} = \dfrac{5.6 - 6.1}{1} = -0.50.$

So we estimate that $U'(1994) \approx \frac{1}{2}[(-0.80) + (-0.50)] = -0.65.$

t	1993	1994	1995	1996	1997	1998	1999	2000	2001	2002
$U'(t)$	-0.80	-0.65	-0.35	-0.35	-0.45	-0.35	-0.25	0.25	0.90	1.10

32. (a) $P'(t)$ is the rate at which the percentage of Americans under the age of 18 is changing with respect to time. Its units are percent per year (%/yr).

(b) To find $P'(t)$, we use $\displaystyle\lim_{h \to 0} \frac{P(t+h) - P(t)}{h} \approx \frac{P(t+h) - P(t)}{h}$ for small values of h.

For 1950: $P'(1950) \approx \dfrac{P(1960) - P(1950)}{1960 - 1950} = \dfrac{35.7 - 31.1}{10} = 0.46$

For 1960: We estimate $P'(1960)$ by using $h = -10$ and $h = 10$, and then average the two results to obtain a final

estimate.

$h = -10 \quad \Rightarrow \quad P'(1960) \approx \dfrac{P(1950) - P(1960)}{1950 - 1960} = \dfrac{31.1 - 35.7}{-10} = 0.46$

$h = 10 \quad \Rightarrow \quad P'(1960) \approx \dfrac{P(1970) - P(1960)}{1970 - 1960} = \dfrac{34.0 - 35.7}{10} = -0.17$

So we estimate that $P'(1960) \approx \frac{1}{2}[0.46 + (-0.17)] = 0.145$.

t	1950	1960	1970	1980	1990	2000
$P'(t)$	0.460	0.145	−0.385	−0.415	−0.115	0.000

(c)

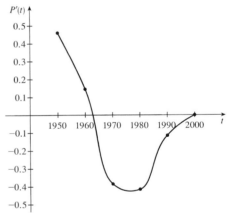

(d) We could get more accurate values for $P'(t)$ by obtaining data for the mid-decade years 1955, 1965, 1975, 1985, and 1995.

33. f is not differentiable at $x = -4$, because the graph has a corner there, and at $x = 0$, because there is a discontinuity there.

34. f is not differentiable at $x = 0$, because there is a discontinuity there, and at $x = 3$, because the graph has a vertical tangent there.

35. f is not differentiable at $x = -1$, because the graph has a vertical tangent there, and at $x = 4$, because the graph has a corner there.

36. f is not differentiable at $x = -1$, because there is a discontinuity there, and at $x = 2$, because the graph has a corner there.

37. As we zoom in toward $(-1, 0)$, the curve appears more and more like a
straight line, so $f(x) = x + \sqrt{|x|}$ is differentiable at $x = -1$. But no
matter how much we zoom in toward the origin, the curve doesn't straighten
out—we can't eliminate the sharp point (a cusp). So f is not differentiable
at $x = 0$.

38. As we zoom in toward $(0, 1)$, the curve appears more and more like a
straight line, so f is differentiable at $x = 0$. But no matter how much we
zoom in toward $(1, 0)$ or $(-1, 0)$, the curve doesn't straighten out—we can't
eliminate the sharp point (a cusp). So f is not differentiable at $x = \pm 1$.

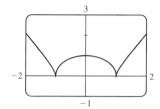

39. $a = f$, $b = f'$, $c = f''$. We can see this because where a has a horizontal tangent, $b = 0$, and where b has a horizontal tangent,
$c = 0$. We can immediately see that c can be neither f nor f', since at the points where c has a horizontal tangent, neither a
nor b is equal to 0.

40. Where d has horizontal tangents, only c is 0, so $d' = c$. c has negative tangents for $x < 0$ and b is the only graph that is
negative for $x < 0$, so $c' = b$. b has positive tangents on $\mathbb{R}$ (except at $x = 0$), and the only graph that is positive on the same
domain is a, so $b' = a$. We conclude that $d = f$, $c = f'$, $b = f''$, and $a = f'''$.

41. We can immediately see that a is the graph of the acceleration function, since at the points where a has a horizontal tangent,
neither c nor b is equal to 0. Next, we note that $a = 0$ at the point where b has a horizontal tangent, so b must be the graph of
the velocity function, and hence, $b' = a$. We conclude that c is the graph of the position function.

42. a must be the jerk since none of the graphs are 0 at its high and low points. a is 0 where b has a maximum, so $b' = a$. b is 0
where c has a maximum, so $c' = b$. We conclude that d is the position function, c is the velocity, b is the acceleration, and a is
the jerk.

43. $f'(x) = \lim\limits_{h \to 0} \dfrac{f(x+h) - f(x)}{h} = \lim\limits_{h \to 0} \dfrac{\left[1 + 4(x+h) - (x+h)^2\right] - (1 + 4x - x^2)}{h}$

$= \lim\limits_{h \to 0} \dfrac{(1 + 4x + 4h - x^2 - 2xh - h^2) - (1 + 4x - x^2)}{h} = \lim\limits_{h \to 0} \dfrac{4h - 2xh - h^2}{h} = \lim\limits_{h \to 0} (4 - 2x - h) = 4 - 2x$

$f''(x) = \lim\limits_{h \to 0} \dfrac{f'(x+h) - f'(x)}{h} = \lim\limits_{h \to 0} \dfrac{[4 - 2(x+h)] - (4 - 2x)}{h} = \lim\limits_{h \to 0} \dfrac{-2h}{h} = \lim\limits_{h \to 0} (-2) = -2$

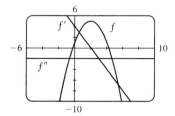

We see from the graph that our answers are reasonable because the graph of
f' is that of a linear function and the graph of f'' is that of a constant
function.

44. $f'(x) = \lim\limits_{h \to 0} \dfrac{f(x+h) - f(x)}{h} = \lim\limits_{h \to 0} \dfrac{\frac{1}{x+h} - \frac{1}{x}}{h} = \lim\limits_{h \to 0} \dfrac{x - (x+h)}{hx(x+h)} = \lim\limits_{h \to 0} \dfrac{-h}{hx(x+h)} = \lim\limits_{h \to 0} \dfrac{-1}{x(x+h)} = -\dfrac{1}{x^2}$

$f''(x) = \lim\limits_{h \to 0} \dfrac{f'(x+h) - f'(x)}{h} = \lim\limits_{h \to 0} \dfrac{-\frac{1}{(x+h)^2} - \left(-\frac{1}{x^2}\right)}{h} = \lim\limits_{h \to 0} \dfrac{-x^2 + (x+h)^2}{hx^2(x+h)^2} = \lim\limits_{h \to 0} \dfrac{2hx + h^2}{hx^2(x+h)^2}$

$\qquad = \lim\limits_{h \to 0} \dfrac{2x + h}{x^2(x+h)^2} = \dfrac{2x}{x^4} = \dfrac{2}{x^3}$

We see from the graph that our answers are reasonable because the graph of f' is that of an even function and is negative for all $x \neq 0$, and the graph of f'' is that of an odd function (negative for $x < 0$ and positive for $x > 0$).

45. $f'(x) = \lim\limits_{h \to 0} \dfrac{f(x+h) - f(x)}{h} = \lim\limits_{h \to 0} \dfrac{\left[2(x+h)^2 - (x+h)^3\right] - (2x^2 - x^3)}{h}$

$\qquad = \lim\limits_{h \to 0} \dfrac{h(4x + 2h - 3x^2 - 3xh - h^2)}{h} = \lim\limits_{h \to 0} (4x + 2h - 3x^2 - 3xh - h^2) = 4x - 3x^2$

$f''(x) = \lim\limits_{h \to 0} \dfrac{f'(x+h) - f'(x)}{h} = \lim\limits_{h \to 0} \dfrac{\left[4(x+h) - 3(x+h)^2\right] - (4x - 3x^2)}{h} = \lim\limits_{h \to 0} \dfrac{h(4 - 6x - 3h)}{h}$

$\qquad = \lim\limits_{h \to 0} (4 - 6x - 3h) = 4 - 6x$

$f'''(x) = \lim\limits_{h \to 0} \dfrac{f''(x+h) - f''(x)}{h} = \lim\limits_{h \to 0} \dfrac{[4 - 6(x+h)] - (4 - 6x)}{h} = \lim\limits_{h \to 0} \dfrac{-6h}{h} = \lim\limits_{h \to 0} (-6) = -6$

$f^{(4)}(x) = \lim\limits_{h \to 0} \dfrac{f'''(x+h) - f'''(x)}{h} = \lim\limits_{h \to 0} \dfrac{-6 - (-6)}{h} = \lim\limits_{h \to 0} \dfrac{0}{h} = \lim\limits_{h \to 0} (0) = 0$

The graphs are consistent with the geometric interpretations of the derivatives because f' has zeros where f has a local minimum and a local maximum, f'' has a zero where f' has a local maximum, and f''' is a constant function equal to the slope of f''.

46. (a) Since we estimate the velocity to be a maximum at $t = 10$, the acceleration is 0 at $t = 10$.

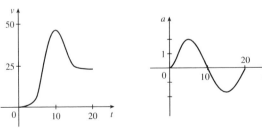

(b) Drawing a tangent line at $t = 10$ on the graph of a, a appears to decrease by 10 ft/s^2 over a period of 20 s.

So at $t = 10$ s, the jerk is approximately $-10/20 = -0.5 \ (\text{ft/s}^2)/\text{s}$ or ft/s^3.

47. (a) Note that we have factored $x - a$ as the difference of two cubes in the third step.

$$f'(a) = \lim_{x \to a} \frac{f(x) - f(a)}{x - a} = \lim_{x \to a} \frac{x^{1/3} - a^{1/3}}{x - a} = \lim_{x \to a} \frac{x^{1/3} - a^{1/3}}{(x^{1/3} - a^{1/3})(x^{2/3} + x^{1/3}a^{1/3} + a^{2/3})}$$

$$= \lim_{x \to a} \frac{1}{x^{2/3} + x^{1/3}a^{1/3} + a^{2/3}} = \frac{1}{3a^{2/3}} \text{ or } \tfrac{1}{3}a^{-2/3}$$

(b) $f'(0) = \lim\limits_{h \to 0} \dfrac{f(0 + h) - f(0)}{h} = \lim\limits_{h \to 0} \dfrac{\sqrt[3]{h} - 0}{h} = \lim\limits_{h \to 0} \dfrac{1}{h^{2/3}}$. This function increases without bound, so the limit does not

exist, and therefore $f'(0)$ does not exist.

(c) $\lim\limits_{x \to 0} |f'(x)| = \lim\limits_{x \to 0} \dfrac{1}{3x^{2/3}} = \infty$ and f is continuous at $x = 0$ (root function), so f has a vertical tangent at $x = 0$.

48. (a) $g'(0) = \lim\limits_{x \to 0} \dfrac{g(x) - g(0)}{x - 0} = \lim\limits_{x \to 0} \dfrac{x^{2/3} - 0}{x} = \lim\limits_{x \to 0} \dfrac{1}{x^{1/3}}$, which does not exist.

(b) $g'(a) = \lim\limits_{x \to a} \dfrac{g(x) - g(a)}{x - a} = \lim\limits_{x \to a} \dfrac{x^{2/3} - a^{2/3}}{x - a} = \lim\limits_{x \to a} \dfrac{(x^{1/3} - a^{1/3})(x^{1/3} + a^{1/3})}{(x^{1/3} - a^{1/3})(x^{2/3} + x^{1/3}a^{1/3} + a^{2/3})}$

$$= \lim_{x \to a} \frac{x^{1/3} + a^{1/3}}{x^{2/3} + x^{1/3}a^{1/3} + a^{2/3}} = \frac{2a^{1/3}}{3a^{2/3}} = \frac{2}{3a^{1/3}} \text{ or } \tfrac{2}{3}a^{-1/3}$$

(c) $g(x) = x^{2/3}$ is continuous at $x = 0$ and

$\lim\limits_{x \to 0} |g'(x)| = \lim\limits_{x \to 0} \dfrac{2}{3\,|x|^{1/3}} = \infty$. This shows that

g has a vertical tangent line at $x = 0$.

(d)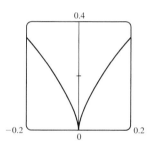

49. $f(x) = |x - 6| = \begin{cases} x - 6 & \text{if } x - 6 \geq 6 \\ -(x - 6) & \text{if } x - 6 < 0 \end{cases} = \begin{cases} x - 6 & \text{if } x \geq 6 \\ 6 - x & \text{if } x < 6 \end{cases}$

So the right-hand limit is $\lim\limits_{x \to 6^+} \dfrac{f(x) - f(6)}{x - 6} = \lim\limits_{x \to 6^+} \dfrac{|x - 6| - 0}{x - 6} = \lim\limits_{x \to 6^+} \dfrac{x - 6}{x - 6} = \lim\limits_{x \to 6^+} 1 = 1$, and the left-hand limit

is $\lim\limits_{x \to 6^-} \dfrac{f(x) - f(6)}{x - 6} = \lim\limits_{x \to 6^-} \dfrac{|x - 6| - 0}{x - 6} = \lim\limits_{x \to 6^-} \dfrac{6 - x}{x - 6} = \lim\limits_{x \to 6^-} (-1) = -1$. Since these limits are not equal,

$f'(6) = \lim\limits_{x \to 6} \dfrac{f(x) - f(6)}{x - 6}$ does not exist and f is not differentiable at 6.

However, a formula for f' is $f'(x) = \begin{cases} 1 & \text{if } x > 6 \\ -1 & \text{if } x < 6 \end{cases}$

Another way of writing the formula is $f'(x) = \dfrac{x - 6}{|x - 6|}$.

50. $f(x) = [\![x]\!]$ is not continuous at any integer n, so f is not differentiable

at n by the contrapositive of Theorem 4. If a is not an integer, then f is

constant on an open interval containing a, so $f'(a) = 0$. Thus,

$f'(x) = 0$, x not an integer.

51. (a) $f(x) = x\,|x| = \begin{cases} x^2 & \text{if } x \geq 0 \\ -x^2 & \text{if } x < 0 \end{cases}$

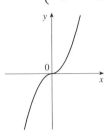

(b) Since $f(x) = x^2$ for $x \geq 0$, we have $f'(x) = 2x$ for $x > 0$.

[See Exercise 15(d).] Similarly, since $f(x) = -x^2$ for $x < 0$,

we have $f'(x) = -2x$ for $x < 0$. At $x = 0$, we have

$$f'(0) = \lim_{x \to 0} \frac{f(x) - f(0)}{x - 0} = \lim_{x \to 0} \frac{x\,|x|}{x} = \lim_{x \to 0} |x| = 0.$$

So f is differentiable at 0. Thus, f is differentiable for all x.

(c) From part (b), we have $f'(x) = \begin{cases} 2x & \text{if } x \geq 0 \\ -2x & \text{if } x < 0 \end{cases} = 2\,|x|$.

52. (a) $f'_-(4) = \lim\limits_{h \to 0^-} \dfrac{f(4 + h) - f(4)}{h} = \lim\limits_{h \to 0^-} \dfrac{5 - (4 + h) - 1}{h} = \lim\limits_{h \to 0^-} \dfrac{-h}{h} = -1$ and

$f'_+(4) = \lim\limits_{h \to 0^+} \dfrac{f(4 + h) - f(4)}{h} = \lim\limits_{h \to 0^+} \dfrac{\dfrac{1}{5 - (4 + h)} - 1}{h} = \lim\limits_{h \to 0^+} \dfrac{1 - (1 - h)}{h(1 - h)} = \lim\limits_{h \to 0^+} \dfrac{1}{1 - h} = 1.$

(b)

At 4 we have $\lim\limits_{x \to 4^-} f(x) = \lim\limits_{x \to 4^-} (5 - x) = 1$ and

$\lim\limits_{x \to 4^+} f(x) = \lim\limits_{x \to 4^+} \dfrac{1}{5 - x} = 1$, so $\lim\limits_{x \to 4} f(x) = 1 = f(4)$ and f is

continuous at 4. Since $f(5)$ is not defined, f is discontinuous at 5.

(c) $f(x) = \begin{cases} 0 & \text{if } x \leq 0 \\ 5 - x & \text{if } 0 < x < 4 \\ 1/(5 - x) & \text{if } x \geq 4 \end{cases}$

These expressions show that f is continuous on the intervals $(-\infty, 0)$, $(0, 4)$, $(4, 5)$ and $(5, \infty)$. Since

$\lim\limits_{x \to 0^+} f(x) = \lim\limits_{x \to 0^+} (5 - x) = 5 \neq 0 = \lim\limits_{x \to 0^-} f(x)$, $\lim\limits_{x \to 0} f(x)$ does not exist, so f is discontinuous (and therefore not

differentiable) at 0.

(d) From (a), f is not differentiable at 4 since $f'_-(4) \neq f'_+(4)$, and from (c), f is not differentiable at 0 or 5.

53. (a) If f is even, then

$$f'(-x) = \lim_{h \to 0} \frac{f(-x + h) - f(-x)}{h} = \lim_{h \to 0} \frac{f[-(x - h)] - f(-x)}{h}$$

$$= \lim_{h \to 0} \frac{f(x - h) - f(x)}{h} = -\lim_{h \to 0} \frac{f(x - h) - f(x)}{-h} \qquad [\text{let } \Delta x = -h]$$

$$= -\lim_{\Delta x \to 0} \frac{f(x + \Delta x) - f(x)}{\Delta x} = -f'(x)$$

Therefore, f' is odd.

(b) If f is odd, then

$$f'(-x) = \lim_{h \to 0} \frac{f(-x+h) - f(-x)}{h} = \lim_{h \to 0} \frac{f[-(x-h)] - f(-x)}{h}$$

$$= \lim_{h \to 0} \frac{-f(x-h) + f(x)}{h} = \lim_{h \to 0} \frac{f(x-h) - f(x)}{-h} \qquad [\text{let } \Delta x = -h]$$

$$= \lim_{\Delta x \to 0} \frac{f(x + \Delta x) - f(x)}{\Delta x} = f'(x)$$

Therefore, f' is even.

54. (a)

(b) The initial temperature of the water is close to room temperature because of the water that was in the pipes. When the water from the hot water tank starts coming out, dT/dt is large and positive as T increases to the temperature of the water in the tank. In the next phase, $dT/dt = 0$ as the water comes out at a constant, high temperature. After some time, dT/dt becomes small and negative as the contents of the hot water tank are exhausted. Finally, when the hot water has run out, dT/dt is once again 0 as the water maintains its (cold) temperature.

(c)

55.

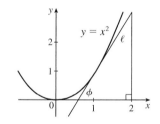

In the right triangle in the diagram, let Δy be the side opposite angle ϕ and Δx the side adjacent angle ϕ. Then the slope of the tangent line ℓ is $m = \Delta y/\Delta x = \tan \phi$. Note that $0 < \phi < \frac{\pi}{2}$. We know (see Exercise 15) that the derivative of $f(x) = x^2$ is $f'(x) = 2x$. So the slope of the tangent to the curve at the point $(1, 1)$ is 2. Thus, ϕ is the angle between 0 and $\frac{\pi}{2}$ whose tangent is 2; that is, $\phi = \tan^{-1} 2 \approx 63°$.

3.3   Differentiation Formulas

1. $f(x) = 186.5$ is a constant function, so its derivative is 0, that is, $f'(x) = 0$.

2. $f(x) = \sqrt{30}$ is a constant function, so its derivative is 0, that is, $f'(x) = 0$.

3. $f(t) = 2 - \frac{2}{3}t \quad \Rightarrow \quad f'(t) = 0 - \frac{2}{3} = -\frac{2}{3}$

4. $F(x) = \frac{3}{4}x^8 \quad \Rightarrow \quad F'(x) = \frac{3}{4}(8x^7) = 6x^7$

5. $f(x) = x^3 - 4x + 6 \quad \Rightarrow \quad f'(x) = 3x^2 - 4(1) + 0 = 3x^2 - 4$

6. $h(x) = (x-2)(2x+3) = 2x^2 - x - 6 \;\Rightarrow\; h'(x) = 2(2x) - 1 - 0 = 4x - 1$

7. $f(t) = \frac{1}{4}(t^4 + 8) \;\Rightarrow\; f'(t) = \frac{1}{4}(t^4 + 8)' = \frac{1}{4}(4t^{4-1} + 0) = t^3$

8. $f(t) = \frac{1}{2}t^6 - 3t^4 + t \;\Rightarrow\; f'(t) = \frac{1}{2}(6t^5) - 3(4t^3) + 1 = 3t^5 - 12t^3 + 1$

9. $V(r) = \frac{4}{3}\pi r^3 \;\Rightarrow\; V'(r) = \frac{4}{3}\pi(3r^2) = 4\pi r^2$

10. $R(t) = 5t^{-3/5} \;\Rightarrow\; R'(t) = 5\left[-\frac{3}{5}t^{(-3/5)-1}\right] = -3t^{-8/5}$

11. $Y(t) = 6t^{-9} \;\Rightarrow\; Y'(t) = 6(-9)t^{-10} = -54t^{-10}$

12. $R(x) = \dfrac{\sqrt{10}}{x^7} = \sqrt{10}\,x^{-7} \;\Rightarrow\; R'(x) = -7\sqrt{10}\,x^{-8} = -\dfrac{7\sqrt{10}}{x^8}$

13. $F(x) = (\frac{1}{2}x)^5 = \left(\frac{1}{2}\right)^5 x^5 = \frac{1}{32}x^5 \;\Rightarrow\; F'(x) = \frac{1}{32}(5x^4) = \frac{5}{32}x^4$

14. $f(t) = \sqrt{t} - \dfrac{1}{\sqrt{t}} = t^{1/2} - t^{-1/2} \;\Rightarrow\; f'(t) = \frac{1}{2}t^{-1/2} - \left(-\frac{1}{2}t^{-3/2}\right) = \dfrac{1}{2\sqrt{t}} + \dfrac{1}{2t\sqrt{t}} \;\left[\text{or } \dfrac{t+1}{2t\sqrt{t}}\right]$

15. $A(s) = -\dfrac{12}{s^5} = -12s^{-5} \;\Rightarrow\; A'(s) = -12(-5s^{-6}) = 60s^{-6} \quad\text{or}\quad 60/s^6$

16. $B(y) = cy^{-6} \;\Rightarrow\; B'(y) = c(-6y^{-7}) = -6cy^{-7}$

17. $y = 4\pi^2 \;\Rightarrow\; y' = 0$ since $4\pi^2$ is a constant.

18. $g(u) = \sqrt{2}\,u + \sqrt{3u} = \sqrt{2}\,u + \sqrt{3}\,\sqrt{u} \;\Rightarrow\; g'(u) = \sqrt{2}\,(1) + \sqrt{3}\left(\frac{1}{2}u^{-1/2}\right) = \sqrt{2} + \dfrac{\sqrt{3}}{2\sqrt{u}}$

19. $u = \sqrt[5]{t} + 4\sqrt{t^5} = t^{1/5} + 4t^{5/2} \;\Rightarrow\; u' = \frac{1}{5}t^{-4/5} + 4\left(\frac{5}{2}t^{3/2}\right) = \frac{1}{5}t^{-4/5} + 10t^{3/2} \quad\text{or}\quad 1/\left(5\sqrt[5]{t^4}\right) + 10\sqrt{t^3}$

20. $v = \left(\sqrt{x} + \dfrac{1}{\sqrt[3]{x}}\right)^2 = \left(\sqrt{x}\right)^2 + 2\sqrt{x}\cdot\dfrac{1}{\sqrt[3]{x}} + \left(\dfrac{1}{\sqrt[3]{x}}\right)^2 = x + 2x^{1/2 - 1/3} + 1/x^{2/3} = x + 2x^{1/6} + x^{-2/3} \;\Rightarrow\;$

$v' = 1 + 2\left(\frac{1}{6}x^{-5/6}\right) - \frac{2}{3}x^{-5/3} = 1 + \frac{1}{3}x^{-5/6} - \frac{2}{3}x^{-5/3} \quad\text{or}\quad 1 + \dfrac{1}{3\sqrt[6]{x^5}} - \dfrac{2}{3\sqrt[3]{x^5}}$

21. Product Rule: $y = (x^2 + 1)(x^3 + 1) \;\Rightarrow\;$

$$y' = (x^2 + 1)(3x^2) + (x^3 + 1)(2x) = 3x^4 + 3x^2 + 2x^4 + 2x = 5x^4 + 3x^2 + 2x.$$

Multiplying first: $y = (x^2 + 1)(x^3 + 1) = x^5 + x^3 + x^2 + 1 \;\Rightarrow\; y' = 5x^4 + 3x^2 + 2x$ (equivalent).

22. Quotient Rule: $F(x) = \dfrac{x - 3x\sqrt{x}}{\sqrt{x}} = \dfrac{x - 3x^{3/2}}{x^{1/2}} \;\Rightarrow\;$

$$F'(x) = \dfrac{x^{1/2}\left(1 - \frac{9}{2}x^{1/2}\right) - \left(x - 3x^{3/2}\right)\left(\frac{1}{2}x^{-1/2}\right)}{(x^{1/2})^2} = \dfrac{x^{1/2} - \frac{9}{2}x - \frac{1}{2}x^{1/2} + \frac{3}{2}x}{x} = \dfrac{\frac{1}{2}x^{1/2} - 3x}{x} = \frac{1}{2}x^{-1/2} - 3$$

Simplifying first: $F(x) = \dfrac{x - 3x\sqrt{x}}{\sqrt{x}} = \sqrt{x} - 3x = x^{1/2} - 3x \;\Rightarrow\; F'(x) = \frac{1}{2}x^{-1/2} - 3$ (equivalent).

For this problem, simplifying first seems to be the better method.

The notations $\overset{\text{PR}}{\Rightarrow}$ and $\overset{\text{QR}}{\Rightarrow}$ indicate the use of the Product and Quotient Rules, respectively.

23. $V(x) = (2x^3 + 3)(x^4 - 2x)$ $\overset{\text{PR}}{\Rightarrow}$

$V'(x) = (2x^3 + 3)(4x^3 - 2) + (x^4 - 2x)(6x^2) = (8x^6 + 8x^3 - 6) + (6x^6 - 12x^3) = 14x^6 - 4x^3 - 6$

24. $Y(u) = (u^{-2} + u^{-3})(u^5 - 2u^2)$ $\overset{\text{PR}}{\Rightarrow}$

$Y'(u) = (u^{-2} + u^{-3})(5u^4 - 4u) + (u^5 - 2u^2)(-2u^{-3} - 3u^{-4})$

$\quad = (5u^2 - 4u^{-1} + 5u - 4u^{-2}) + (-2u^2 - 3u + 4u^{-1} + 6u^{-2}) = 3u^2 + 2u + 2u^{-2}$

25. $F(y) = \left(\dfrac{1}{y^2} - \dfrac{3}{y^4}\right)(y + 5y^3) = \left(y^{-2} - 3y^{-4}\right)\left(y + 5y^3\right)$ $\overset{\text{PR}}{\Rightarrow}$

$F'(y) = (y^{-2} - 3y^{-4})(1 + 15y^2) + (y + 5y^3)(-2y^{-3} + 12y^{-5})$

$\quad = (y^{-2} + 15 - 3y^{-4} - 45y^{-2}) + (-2y^{-2} + 12y^{-4} - 10 + 60y^{-2})$

$\quad = 5 + 14y^{-2} + 9y^{-4}$ or $5 + 14/y^2 + 9/y^4$

26. $y = \sqrt{x}\,(x - 1) = x^{3/2} - x^{1/2}$ $\Rightarrow$ $y' = \frac{3}{2}x^{1/2} - \frac{1}{2}x^{-1/2} = \frac{1}{2}x^{-1/2}(3x - 1)$ [factor out $\frac{1}{2}x^{-1/2}$]

or $y' = \dfrac{3x - 1}{2\sqrt{x}}$.

27. $g(x) = \dfrac{3x - 1}{2x + 1}$ $\overset{\text{QR}}{\Rightarrow}$ $g'(x) = \dfrac{(2x + 1)(3) - (3x - 1)(2)}{(2x + 1)^2} = \dfrac{6x + 3 - 6x + 2}{(2x + 1)^2} = \dfrac{5}{(2x + 1)^2}$

28. $f(t) = \dfrac{2t}{4 + t^2}$ $\overset{\text{QR}}{\Rightarrow}$ $f'(t) = \dfrac{(4 + t^2)(2) - (2t)(2t)}{(4 + t^2)^2} = \dfrac{8 + 2t^2 - 4t^2}{(4 + t^2)^2} = \dfrac{8 - 2t^2}{(4 + t^2)^2}$

29. $y = \dfrac{x^3}{1 - x^2}$ $\overset{\text{QR}}{\Rightarrow}$ $y' = \dfrac{(1 - x^2)\,(3x^2) - x^3(-2x)}{(1 - x^2)^2} = \dfrac{x^2(3 - 3x^2 + 2x^2)}{(1 - x^2)^2} = \dfrac{x^2(3 - x^2)}{(1 - x^2)^2}$

30. $y = \dfrac{x + 1}{x^3 + x - 2}$ $\overset{\text{QR}}{\Rightarrow}$

$y' = \dfrac{(x^3 + x - 2)(1) - (x + 1)(3x^2 + 1)}{(x^3 + x - 2)^2} = \dfrac{x^3 + x - 2 - 3x^3 - 3x^2 - x - 1}{(x^3 + x - 2)^2} = \dfrac{-2x^3 - 3x^2 - 3}{(x^3 + x - 2)^2}$

or $-\dfrac{2x^3 + 3x^2 + 3}{(x - 1)^2(x^2 + x + 2)^2}$

31. $y = \dfrac{v^3 - 2v\sqrt{v}}{v} = v^2 - 2\sqrt{v} = v^2 - 2v^{1/2}$ $\Rightarrow$ $y' = 2v - 2\left(\frac{1}{2}\right)v^{-1/2} = 2v - v^{-1/2}$.

We can change the form of the answer as follows: $2v - v^{-1/2} = 2v - \dfrac{1}{\sqrt{v}} = \dfrac{2v\sqrt{v} - 1}{\sqrt{v}} = \dfrac{2v^{3/2} - 1}{\sqrt{v}}$

32. $y = \dfrac{t}{(t - 1)^2} = \dfrac{t}{t^2 - 2t + 1}$ $\overset{\text{QR}}{\Rightarrow}$

$y' = \dfrac{(t^2 - 2t + 1)(1) - t(2t - 2)}{[(t - 1)^2]^2} = \dfrac{(t - 1)^2 - 2t(t - 1)}{(t - 1)^4} = \dfrac{(t - 1)[(t - 1) - 2t]}{(t - 1)^4} = \dfrac{-t - 1}{(t - 1)^3}$

33. $y = \dfrac{t^2 + 2}{t^4 - 3t^2 + 1}$ $\quad\overset{\text{QR}}{\Rightarrow}$

$$y' = \frac{(t^4 - 3t^2 + 1)(2t) - (t^2 + 2)(4t^3 - 6t)}{(t^4 - 3t^2 + 1)^2} = \frac{2t[(t^4 - 3t^2 + 1) - (t^2 + 2)(2t^2 - 3)]}{(t^4 - 3t^2 + 1)^2}$$

$$= \frac{2t(t^4 - 3t^2 + 1 - 2t^4 - 4t^2 + 3t^2 + 6)}{(t^4 - 3t^2 + 1)^2} = \frac{2t(-t^4 - 4t^2 + 7)}{(t^4 - 3t^2 + 1)^2}$$

34. $g(t) = \dfrac{t - \sqrt{t}}{t^{1/3}} = \dfrac{t}{t^{1/3}} - \dfrac{t^{1/2}}{t^{1/3}} = t^{2/3} - t^{1/6}$ $\Rightarrow$ $g'(t) = \frac{2}{3}t^{-1/3} - \frac{1}{6}t^{-5/6}$

35. $y = ax^2 + bx + c$ $\Rightarrow$ $y' = 2ax + b$

36. $y = A + \dfrac{B}{x} + \dfrac{C}{x^2} = A + Bx^{-1} + Cx^{-2}$ $\Rightarrow$ $y' = -Bx^{-2} - 2Cx^{-3} = -\dfrac{B}{x^2} - 2\dfrac{C}{x^3}$

37. $y = \dfrac{r^2}{1 + \sqrt{r}}$ $\Rightarrow$

$$y' = \frac{(1 + \sqrt{r})(2r) - r^2\left(\frac{1}{2}r^{-1/2}\right)}{(1 + \sqrt{r})^2} = \frac{2r + 2r^{3/2} - \frac{1}{2}r^{3/2}}{(1 + \sqrt{r})^2} = \frac{2r + \frac{3}{2}r^{3/2}}{(1 + \sqrt{r})^2} = \frac{\frac{1}{2}r(4 + 3r^{1/2})}{(1 + \sqrt{r})^2} = \frac{r(4 + 3\sqrt{r})}{2(1 + \sqrt{r})^2}$$

38. $y = \dfrac{cx}{1 + cx}$ $\Rightarrow$ $y' = \dfrac{(1 + cx)(c) - (cx)(c)}{(1 + cx)^2} = \dfrac{c + c^2x - c^2x}{(1 + cx)^2} = \dfrac{c}{(1 + cx)^2}$

39. $y = \sqrt[3]{t}\,(t^2 + t + t^{-1}) = t^{1/3}(t^2 + t + t^{-1}) = t^{7/3} + t^{4/3} + t^{-2/3}$ $\Rightarrow$

$$y' = \frac{7}{3}t^{4/3} + \frac{4}{3}t^{1/3} - \frac{2}{3}t^{-5/3} = \frac{1}{3}t^{-5/3}(7t^{9/3} + 4t^{6/3} - 2) = (7t^3 + 4t^2 - 2)/(3t^{5/3})$$

40. $y = \dfrac{u^6 - 2u^3 + 5}{u^2} = u^4 - 2u + 5u^{-2}$ $\Rightarrow$ $y' = 4u^3 - 2 - 10u^{-3} = 2u^{-3}(2u^6 - u^3 - 5) = 2(2u^6 - u^3 - 5)/u^3$

41. $f(x) = \dfrac{x}{x + c/x}$ $\Rightarrow$ $f'(x) = \dfrac{(x + c/x)(1) - x(1 - c/x^2)}{\left(x + \frac{c}{x}\right)^2} = \dfrac{x + c/x - x + c/x}{\left(\frac{x^2 + c}{x}\right)^2} = \dfrac{2c/x}{\frac{(x^2 + c)^2}{x^2}} \cdot \dfrac{x^2}{x^2} = \dfrac{2cx}{(x^2 + c)^2}$

42. $f(x) = \dfrac{ax + b}{cx + d}$ $\Rightarrow$ $f'(x) = \dfrac{(cx + d)(a) - (ax + b)(c)}{(cx + d)^2} = \dfrac{acx + ad - acx - bc}{(cx + d)^2} = \dfrac{ad - bc}{(cx + d)^2}$

43. $P(x) = a_nx^n + a_{n-1}x^{n-1} + \cdots + a_2x^2 + a_1x + a_0$ $\Rightarrow$ $P'(x) = na_nx^{n-1} + (n - 1)a_{n-1}x^{n-2} + \cdots + 2a_2x + a_1$

44. $f(x) = \dfrac{x}{x^2 - 1}$ $\Rightarrow$

$$f'(x) = \frac{(x^2 - 1)1 - x(2x)}{(x^2 - 1)^2} = \frac{-x^2 - 1}{(x^2 - 1)^2} = -\frac{x^2 + 1}{(x^2 - 1)^2}$$

Notice that the slopes of all tangents to f are negative and $f'(x) < 0$ always.

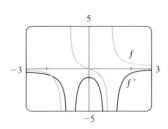

45. $f(x) = 3x^{15} - 5x^3 + 3 \Rightarrow f'(x) = 45x^{14} - 15x^2$.

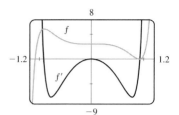

Notice that $f'(x) = 0$ when f has a horizontal tangent, f' is positive when f is increasing, and f' is negative when f is decreasing.

46. $f(x) = x + 1/x = x + x^{-1} \Rightarrow f'(x) = 1 - x^{-2} = 1 - 1/x^2$.

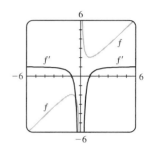

Notice that $f'(x) = 0$ when f has a horizontal tangent, f' is positive when f is increasing, and f' is negative when f is decreasing.

47. (a)

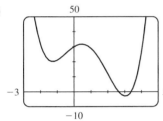

(b) From the graph in part (a), it appears that f' is zero at $x_1 \approx -1.25$, $x_2 \approx 0.5$, and $x_3 \approx 3$. The slopes are negative (so f' is negative) on $(-\infty, x_1)$ and (x_2, x_3). The slopes are positive (so f' is positive) on (x_1, x_2) and (x_3, ∞).

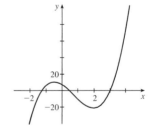

(c) $f(x) = x^4 - 3x^3 - 6x^2 + 7x + 30 \Rightarrow$

$f'(x) = 4x^3 - 9x^2 - 12x + 7$

48. (a)

(b)

From the graph in part (a), it appears that g' is zero at $x = 0$. The slopes are negative (so g' is negative) on $(-\infty, 0)$. The slopes are positive (so g' is positive) on $(0, \infty)$.

(c) $g(x) = \dfrac{x^2}{x^2 + 1}$ $\Rightarrow$

$$g'(x) = \dfrac{(x^2 + 1)(2x) - x^2(2x)}{(x^2 + 1)^2} = \dfrac{2x}{(x^2 + 1)^2}$$

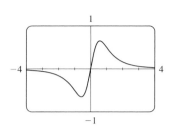

49. $y = \dfrac{2x}{x + 1}$ $\Rightarrow$ $y' = \dfrac{(x + 1)(2) - (2x)(1)}{(x + 1)^2} = \dfrac{2}{(x + 1)^2}$.

At $(1, 1)$, $y' = \frac{1}{2}$, and an equation of the tangent line is $y - 1 = \frac{1}{2}(x - 1)$, or $y = \frac{1}{2}x + \frac{1}{2}$.

50. $y = x^4 + 2x^2 - x$ $\Rightarrow$ $y' = 4x^3 + 4x - 1$. At $(1, 2)$, $y' = 7$ and an equation of the tangent line is

$y - 2 = 7(x - 1)$ or $y = 7x - 5$.

51. (a) $y = f(x) = \dfrac{1}{1 + x^2}$ $\Rightarrow$

$f'(x) = \dfrac{(1 + x^2)(0) - 1(2x)}{(1 + x^2)^2} = \dfrac{-2x}{(1 + x^2)^2}$. So the slope of the

tangent line at the point $\left(-1, \frac{1}{2}\right)$ is $f'(-1) = \dfrac{2}{2^2} = \frac{1}{2}$ and its

equation is $y - \frac{1}{2} = \frac{1}{2}(x + 1)$ or $y = \frac{1}{2}x + 1$.

(b)

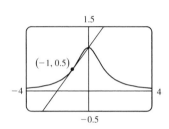

52. (a) $y = f(x) = \dfrac{x}{1 + x^2}$ $\Rightarrow$

$f'(x) = \dfrac{(1 + x^2)1 - x(2x)}{(1 + x^2)^2} = \dfrac{1 - x^2}{(1 + x^2)^2}$. So the slope of the

tangent line at the point $(3, 0.3)$ is $f'(3) = \frac{-8}{100}$ and its equation is

$y - 0.3 = -0.08(x - 3)$ or $y = -0.08x + 0.54$.

(b)

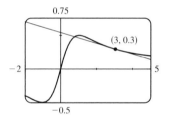

53. $y = x + \sqrt{x}$ $\Rightarrow$ $y' = 1 + \frac{1}{2}x^{-1/2} = 1 + 1/(2\sqrt{x})$. At $(1, 2)$, $y' = \frac{3}{2}$, and an equation of the tangent line is

$y - 2 = \frac{3}{2}(x - 1)$, or $y = \frac{3}{2}x + \frac{1}{2}$. The slope of the normal line is $-\frac{2}{3}$, so an equation of the normal line is

$y - 2 = -\frac{2}{3}(x - 1)$, or $y = -\frac{2}{3}x + \frac{8}{3}$.

54. $y = (1 + 2x)^2 = 1 + 4x + 4x^2$ $\Rightarrow$ $y' = 4 + 8x$. At $(1, 9)$, $y' = 12$ and an equation of the tangent line is

$y - 9 = 12(x - 1)$ or $y = 12x - 3$. The slope of the normal line is $-\frac{1}{12}$ (the negative reciprocal of 12) and an equation of the

normal line is $y - 9 = -\frac{1}{12}(x - 1)$ or $y = -\frac{1}{12}x + \frac{109}{12}$.

55. $y = \dfrac{3x + 1}{x^2 + 1}$ $\Rightarrow$ $y' = \dfrac{(x^2 + 1)(3) - (3x + 1)(2x)}{(x^2 + 1)^2}$. At $(1, 2)$, $y' = \dfrac{6 - 8}{2^2} = -\dfrac{1}{2}$, and an equation of the tangent line is

$y - 2 = -\frac{1}{2}(x - 1)$, or $y = -\frac{1}{2}x + \frac{5}{2}$. The slope of the normal line is 2, so an equation of the normal line is

$y - 2 = 2(x - 1)$, or $y = 2x$.

56. $y = \dfrac{\sqrt{x}}{x+1}$ $\Rightarrow$ $y' = \dfrac{(x+1)\left(\dfrac{1}{2\sqrt{x}}\right) - \sqrt{x}\,(1)}{(x+1)^2} = \dfrac{(x+1) - (2x)}{2\sqrt{x}\,(x+1)^2} = \dfrac{1-x}{2\sqrt{x}\,(x+1)^2}.$

At $(4, 0.4)$, $y' = \frac{-3}{100} = -0.03$, and an equation of the tangent line is $y - 0.4 = -0.03(x - 4)$, or $y = -0.03x + 0.52$. The

slope of the normal line is $\frac{100}{3}$, so an equation of the normal line is $y - 0.4 = \frac{100}{3}(x - 4)$ $\Leftrightarrow$ $y = \frac{100}{3}x - \frac{400}{3} + \frac{2}{5}$ $\Leftrightarrow$

$y = \frac{100}{3}x - \frac{1994}{15}.$

57. $f(x) = x^4 - 3x^3 + 16x$ $\Rightarrow$ $f'(x) = 4x^3 - 9x^2 + 16$ $\Rightarrow$ $f''(x) = 12x^2 - 18x$

58. $G(r) = \sqrt{r} + \sqrt[3]{r}$ $\Rightarrow$ $G'(r) = \frac{1}{2}r^{-1/2} + \frac{1}{3}r^{-2/3}$ $\Rightarrow$ $G''(r) = -\frac{1}{4}r^{-3/2} - \frac{2}{9}r^{-5/3}$

59. $f(x) = \dfrac{x^2}{1 + 2x}$ $\Rightarrow$ $f'(x) = \dfrac{(1+2x)(2x) - x^2(2)}{(1+2x)^2} = \dfrac{2x + 4x^2 - 2x^2}{(1+2x)^2} = \dfrac{2x^2 + 2x}{(1+2x)^2}$ $\Rightarrow$

$f''(x) = \dfrac{(1+2x)^2(4x+2) - (2x^2+2x)(1 + 4x + 4x^2)'}{[(1+2x)^2]^2} = \dfrac{2(1+2x)^2(2x+1) - 2x(x+1)(4+8x)}{(1+2x)^4}$

$= \dfrac{2(1+2x)[(1+2x)^2 - 4x(x+1)]}{(1+2x)^4} = \dfrac{2(1 + 4x + 4x^2 - 4x^2 - 4x)}{(1+2x)^3} = \dfrac{2}{(1+2x)^3}$

60. Using the Reciprocal Rule, $f(x) = \dfrac{1}{3-x}$ $\Rightarrow$ $f'(x) = -\dfrac{(3-x)'}{(3-x)^2} = -\dfrac{-1}{(3-x)^2} = \dfrac{1}{(3-x)^2}$ $\Rightarrow$

$f''(x) = -\dfrac{[(3-x)^2]'}{[(3-x)^2]^2} = -\dfrac{(9 - 6x + x^2)'}{(3-x)^4} = -\dfrac{-6 + 2x}{(3-x)^4} = -\dfrac{-2(3-x)}{(3-x)^4} = \dfrac{2}{(3-x)^3}.$

61. (a) $s = t^3 - 3t$ $\Rightarrow$ $v(t) = s'(t) = 3t^2 - 3$ $\Rightarrow$ $a(t) = v'(t) = 6t$

(b) $a(2) = 6(2) = 12$ m/s^2

(c) $v(t) = 3t^2 - 3 = 0$ when $t^2 = 1$, that is, $t = 1$ and $a(1) = 6$ m/s^2.

62. (a) $s = 2t^3 - 7t^2 + 4t + 1$ $\Rightarrow$ $v(t) = s'(t) = 6t^2 - 14t + 4$ $\Rightarrow$ $a(t) = v'(t) = 12t - 14$

(b) $a(1) = 12 - 14 = -2$ m/s^2

(c)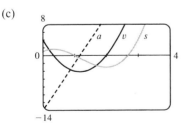

63. We are given that $f(5) = 1$, $f'(5) = 6$, $g(5) = -3$, and $g'(5) = 2$.

(a) $(fg)'(5) = f(5)g'(5) + g(5)f'(5) = (1)(2) + (-3)(6) = 2 - 18 = -16$

(b) $\left(\dfrac{f}{g}\right)'(5) = \dfrac{g(5)f'(5) - f(5)g'(5)}{[g(5)]^2} = \dfrac{(-3)(6) - (1)(2)}{(-3)^2} = -\dfrac{20}{9}$

(c) $\left(\dfrac{g}{f}\right)'(5) = \dfrac{f(5)g'(5) - g(5)f'(5)}{[f(5)]^2} = \dfrac{(1)(2) - (-3)(6)}{(1)^2} = 20$

64. We are given that $f(2) = -3$, $g(2) = 4$, $f'(2) = -2$, and $g'(2) = 7$.

(a) $h(x) = 5f(x) - 4g(x) \Rightarrow h'(x) = 5f'(x) - 4g'(x)$, so

$h'(2) = 5f'(2) - 4g'(2) = 5(-2) - 4(7) = -10 - 28 = -38.$

(b) $h(x) = f(x)g(x) \Rightarrow h'(x) = f(x)g'(x) + g(x)f'(x)$, so

$h'(2) = f(2)g'(2) + g(2)f'(2) = (-3)(7) + (4)(-2) = -21 - 8 = -29.$

(c) $h(x) = \dfrac{f(x)}{g(x)} \Rightarrow h'(x) = \dfrac{g(x)f'(x) - f(x)g'(x)}{[g(x)]^2}$, so

$h'(2) = \dfrac{g(2)f'(2) - f(2)g'(2)}{[g(2)]^2} = \dfrac{4(-2) - (-3)(7)}{4^2} = \dfrac{-8 + 21}{16} = \dfrac{13}{16}.$

(d) $h(x) = \dfrac{g(x)}{1 + f(x)} \Rightarrow h'(x) = \dfrac{[1 + f(x)]\,g'(x) - g(x)f'(x)}{[1 + f(x)]^2}$, so

$h'(2) = \dfrac{[1 + f(2)]\,g'(2) - g(2)\,f'(2)}{[1 + f(x)]^2} = \dfrac{[1 + (-3)](7) - 4(-2)}{[1 + (-3)]^2} = \dfrac{-14 + 8}{(-2)^2} = \dfrac{-6}{4} = -\dfrac{3}{2}.$

65. $f(x) = \sqrt{x}\,g(x) \Rightarrow f'(x) = \sqrt{x}\,g'(x) + g(x) \cdot \dfrac{1}{2}x^{-1/2}$, so $f'(4) = \sqrt{4}\,g'(4) + g(4) \cdot \dfrac{1}{2\sqrt{4}} = 2 \cdot 7 + 8 \cdot \dfrac{1}{4} = 16.$

66. $\dfrac{d}{dx}\left[\dfrac{h(x)}{x}\right] = \dfrac{xh'(x) - h(x) \cdot 1}{x^2} \Rightarrow \dfrac{d}{dx}\left[\dfrac{h(x)}{x}\right]_{x=2} = \dfrac{2h'(2) - h(2)}{2^2} = \dfrac{2(-3) - (4)}{4} = \dfrac{-10}{4} = -2.5$

67. (a) From the graphs of f and g, we obtain the following values: $f(1) = 2$ since the point $(1, 2)$ is on the graph of f;

$g(1) = 1$ since the point $(1, 1)$ is on the graph of g; $f'(1) = 2$ since the slope of the line segment between $(0, 0)$ and $(2, 4)$

is $\dfrac{4 - 0}{2 - 0} = 2$; $g'(1) = -1$ since the slope of the line segment between $(-2, 4)$ and $(2, 0)$ is $\dfrac{0 - 4}{2 - (-2)} = -1.$

Now $u(x) = f(x)g(x)$, so $u'(1) = f(1)g'(1) + g(1)\,f'(1) = 2 \cdot (-1) + 1 \cdot 2 = 0.$

(b) $v(x) = f(x)/g(x)$, so $v'(5) = \dfrac{g(5)f'(5) - f(5)g'(5)}{[g(5)]^2} = \dfrac{2\left(-\frac{1}{3}\right) - 3 \cdot \frac{2}{3}}{2^2} = \dfrac{-\frac{8}{3}}{4} = -\dfrac{2}{3}$

68. (a) $P(x) = F(x)\,G(x)$, so $P'(2) = F(2)\,G'(2) + G(2)\,F'(2) = 3 \cdot \frac{2}{4} + 2 \cdot 0 = \frac{3}{2}.$

(b) $Q(x) = F(x)/G(x)$, so $Q'(7) = \dfrac{G(7)\,F'(7) - F(7)\,G'(7)}{[G(7)]^2} = \dfrac{1 \cdot \frac{1}{4} - 5 \cdot \left(-\frac{2}{3}\right)}{1^2} = \dfrac{1}{4} + \dfrac{10}{3} = \dfrac{43}{12}$

69. (a) $y = xg(x) \Rightarrow y' = xg'(x) + g(x) \cdot 1 = xg'(x) + g(x)$

(b) $y = \dfrac{x}{g(x)} \Rightarrow y' = \dfrac{g(x) \cdot 1 - xg'(x)}{[g(x)]^2} = \dfrac{g(x) - xg'(x)}{[g(x)]^2}$

(c) $y = \dfrac{g(x)}{x} \Rightarrow y' = \dfrac{xg'(x) - g(x) \cdot 1}{(x)^2} = \dfrac{xg'(x) - g(x)}{x^2}$

70. (a) $y = x^2 f(x) \Rightarrow y' = x^2 f'(x) + f(x)(2x)$

(b) $y = \dfrac{f(x)}{x^2} \Rightarrow y' = \dfrac{x^2 f'(x) - f(x)(2x)}{(x^2)^2} = \dfrac{xf'(x) - 2f(x)}{x^3}$

(c) $y = \dfrac{x^2}{f(x)}$ $\Rightarrow$ $y' = \dfrac{f(x)(2x) - x^2 f'(x)}{[f(x)]^2}$

(d) $y = \dfrac{1 + x f(x)}{\sqrt{x}}$ $\Rightarrow$

$$y' = \dfrac{\sqrt{x}\,[x f'(x) + f(x)] - [1 + x f(x)]\dfrac{1}{2\sqrt{x}}}{(\sqrt{x})^2}$$

$$= \dfrac{x^{3/2} f'(x) + x^{1/2} f(x) - \frac{1}{2} x^{-1/2} - \frac{1}{2} x^{1/2} f(x)}{x} \cdot \dfrac{2x^{1/2}}{2x^{1/2}} = \dfrac{x f(x) + 2x^2 f'(x) - 1}{2x^{3/2}}$$

71. The curve $y = 2x^3 + 3x^2 - 12x + 1$ has a horizontal tangent when $y' = 6x^2 + 6x - 12 = 0$ $\Leftrightarrow$ $6(x^2 + x - 2) = 0$ $\Leftrightarrow$

$6(x + 2)(x - 1) = 0$ $\Leftrightarrow$ $x = -2$ or $x = 1$. The points on the curve are $(-2, 21)$ and $(1, -6)$.

72. $f(x) = x^3 + 3x^2 + x + 3$ has a horizontal tangent when $f'(x) = 3x^2 + 6x + 1 = 0$ $\Leftrightarrow$

$x = \dfrac{-6 \pm \sqrt{36 - 12}}{6} = -1 \pm \frac{1}{3}\sqrt{6}$.

73. $y = 6x^3 + 5x - 3$ $\Rightarrow$ $m = y' = 18x^2 + 5$, but $x^2 \geq 0$ for all x, so $m \geq 5$ for all x.

74. $y = x\sqrt{x} = x^{3/2}$ $\Rightarrow$ $y' = \frac{3}{2}x^{1/2}$. The slope of the line $y = 1 + 3x$ is 3, so the slope of any line parallel to it is also 3.

Thus, $y' = 3$ $\Rightarrow$ $\frac{3}{2}x^{1/2} = 3$ $\Rightarrow$ $\sqrt{x} = 2$ $\Rightarrow$ $x = 4$, which is the x-coordinate of the point on the curve at which the

slope is 3. The y-coordinate is $y = 4\sqrt{4} = 8$, so an equation of the tangent line is $y - 8 = 3(x - 4)$ or $y = 3x - 4$.

75. The slope of the line $12x - y = 1$ (or $y = 12x - 1$) is 12, so the slope of both lines tangent to the curve is 12.

$y = 1 + x^3$ $\Rightarrow$ $y' = 3x^2$. Thus, $3x^2 = 12$ $\Rightarrow$ $x^2 = 4$ $\Rightarrow$ $x = \pm 2$, which are the x-coordinates at which the tangent

lines have slope 12. The points on the curve are $(2, 9)$ and $(-2, -7)$, so the tangent line equations are $y - 9 = 12(x - 2)$

or $y = 12x - 15$ and $y + 7 = 12(x + 2)$ or $y = 12x + 17$.

76. $y = \dfrac{x - 1}{x + 1}$ $\Rightarrow$ $y' = \dfrac{(x + 1)(1) - (x - 1)(1)}{(x + 1)^2} = \dfrac{2}{(x + 1)^2}$. If the tangent intersects

the curve when $x = a$, then its slope is $2/(a + 1)^2$. But if the tangent is parallel to

$x - 2y = 2$, that is, $y = \frac{1}{2}x - 1$, then its slope is $\frac{1}{2}$. Thus, $\dfrac{2}{(a + 1)^2} = \dfrac{1}{2}$ $\Rightarrow$

$(a + 1)^2 = 4$ $\Rightarrow$ $a + 1 = \pm 2$ $\Rightarrow$ $a = 1$ or -3. When $a = 1$, $y = 0$ and the

equation of the tangent is $y - 0 = \frac{1}{2}(x - 1)$ or $y = \frac{1}{2}x - \frac{1}{2}$. When $a = -3$, $y = 2$ and

the equation of the tangent is $y - 2 = \frac{1}{2}(x + 3)$ or $y = \frac{1}{2}x + \frac{7}{2}$.

77. The slope of $y = x^2 - 5x + 4$ is given by $m = y' = 2x - 5$. The slope of $x - 3y = 5$ $\Leftrightarrow$ $y = \frac{1}{3}x - \frac{5}{3}$ is $\frac{1}{3}$,

so the desired normal line must have slope $\frac{1}{3}$, and hence, the tangent line to the parabola must have slope -3. This occurs if

$2x - 5 = -3$ $\Rightarrow$ $2x = 2$ $\Rightarrow$ $x = 1$. When $x = 1$, $y = 1^2 - 5(1) + 4 = 0$, and an equation of the normal line is

$y - 0 = \frac{1}{3}(x - 1)$ or $y = \frac{1}{3}x - \frac{1}{3}$.

78. $y = f(x) = x - x^2 \quad \Rightarrow \quad f'(x) = 1 - 2x$.

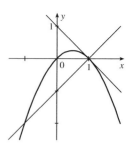

So $f'(1) = -1$, and the slope of the normal line is the negative reciprocal of that

of the tangent line, that is, $-1/(-1) = 1$. So the equation of the normal line at

$(1, 0)$ is $y - 0 = 1(x - 1) \quad \Leftrightarrow \quad y = x - 1$. Substituting this into the equation of

the parabola, we obtain $x - 1 = x - x^2 \quad \Leftrightarrow \quad x = \pm 1$. The solution $x = -1$ is

the one we require. Substituting $x = -1$ into the equation of the parabola to find

the y-coordinate, we have $y = -2$. So the point of intersection is $(-1, -2)$, as

shown in the sketch.

79.

Let (a, a^2) be a point on the parabola at which the tangent line passes through the

point $(0, -4)$. The tangent line has slope $2a$ and equation $y - (-4) = 2a(x - 0) \quad \Leftrightarrow$

$y = 2ax - 4$. Since (a, a^2) also lies on the line, $a^2 = 2a(a) - 4$, or $a^2 = 4$. So

$a = \pm 2$ and the points are $(2, 4)$ and $(-2, 4)$.

80. (a) If $y = x^2 + x$, then $y' = 2x + 1$. If the point at which a tangent meets the parabola is $(a, a^2 + a)$, then the slope of the

tangent is $2a + 1$. But since it passes through $(2, -3)$, the slope must also be $\dfrac{\Delta y}{\Delta x} = \dfrac{a^2 + a + 3}{a - 2}$.

Therefore, $2a + 1 = \dfrac{a^2 + a + 3}{a - 2}$. Solving this equation for a we get $a^2 + a + 3 = 2a^2 - 3a - 2 \quad \Leftrightarrow$

$a^2 - 4a - 5 = (a - 5)(a + 1) = 0 \quad \Leftrightarrow \quad a = 5$ or -1. If $a = -1$, the point is $(-1, 0)$ and the slope is -1, so the

equation is $y - 0 = (-1)(x + 1)$ or $y = -x - 1$. If $a = 5$, the point is $(5, 30)$ and the slope is 11, so the equation is

$y - 30 = 11(x - 5)$ or $y = 11x - 25$.

(b) As in part (a), but using the point $(2, 7)$, we get the equation

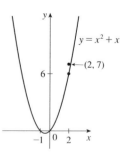

$2a + 1 = \dfrac{a^2 + a - 7}{a - 2} \quad \Rightarrow \quad 2a^2 - 3a - 2 = a^2 + a - 7 \quad \Leftrightarrow \quad a^2 - 4a + 5 = 0$.

The last equation has no real solution (discriminant $= -16 < 0$), so there is no line

through the point $(2, 7)$ that is tangent to the parabola. The diagram shows that the

point $(2, 7)$ is "inside" the parabola, but tangent lines to the parabola do not pass

through points inside the parabola.

We will sometimes use the form $f'g + fg'$ rather than the form $fg' + gf'$ for the Product Rule.

81. (a) $(fgh)' = [(fg)h]' = (fg)'h + (fg)h' = (f'g + fg')h + (fg)h' = f'gh + fg'h + fgh'$

(b) Putting $f = g = h$ in part (a), we have $\dfrac{d}{dx}[f(x)]^3 = (fff)' = f'ff + ff'f + fff' = 3fff' = 3[f(x)]^2 f'(x)$.

(c) $y = (x^4 + 3x^3 + 17x + 82)^3 \quad \Rightarrow \quad y' = 3(x^4 + 3x^3 + 17x + 82)^2(4x^3 + 9x^2 + 17)$

82. (a) $f(x) = x^n \Rightarrow f'(x) = nx^{n-1} \Rightarrow f''(x) = n(n-1)x^{n-2} \Rightarrow \cdots \Rightarrow$

$f^{(n)}(x) = n(n-1)(n-2)\cdots 2\cdot 1x^{n-n} = n!$

(b) $f(x) = x^{-1} \Rightarrow f'(x) = (-1)x^{-2} \Rightarrow f''(x) = (-1)(-2)x^{-3} \Rightarrow \cdots \Rightarrow$

$f^{(n)}(x) = (-1)(-2)(-3)\cdots(-n)x^{-(n+1)} = (-1)^n n! x^{-(n+1)}$ or $\dfrac{(-1)^n n!}{x^{n+1}}$

83. Let $P(x) = ax^2 + bx + c$. Then $P'(x) = 2ax + b$ and $P''(x) = 2a$. $P''(2) = 2 \Rightarrow 2a = 2 \Rightarrow a = 1$.

$P'(2) = 3 \Rightarrow 2(1)(2) + b = 3 \Rightarrow 4 + b = 3 \Rightarrow b = -1$.

$P(2) = 5 \Rightarrow 1(2)^2 + (-1)(2) + c = 5 \Rightarrow 2 + c = 5 \Rightarrow c = 3$. So $P(x) = x^2 - x + 3$.

84. $y = Ax^2 + Bx + C \Rightarrow y' = 2Ax + B \Rightarrow y'' = 2A$. We substitute these expressions into the equation $y'' + y' - 2y = x^2$ to get

$$(2A) + (2Ax + B) - 2(Ax^2 + Bx + C) = x^2$$

$$2A + 2Ax + B - 2Ax^2 - 2Bx - 2C = x^2$$

$$(-2A)x^2 + (2A - 2B)x + (2A + B - 2C) = (1)x^2 + (0)x + (0)$$

The coefficients of x^2 on each side must be equal, so $-2A = 1 \Rightarrow A = -\frac{1}{2}$. Similarly, $2A - 2B = 0 \Rightarrow$ $A = B = -\frac{1}{2}$ and $2A + B - 2C = 0 \Rightarrow -1 - \frac{1}{2} - 2C = 0 \Rightarrow C = -\frac{3}{4}$.

85. $y = f(x) = ax^3 + bx^2 + cx + d \Rightarrow f'(x) = 3ax^2 + 2bx + c$. The point $(-2, 6)$ is on f, so $f(-2) = 6 \Rightarrow$ $-8a + 4b - 2c + d = 6$ **(1)**. The point $(2, 0)$ is on f, so $f(2) = 0 \Rightarrow 8a + 4b + 2c + d = 0$ **(2)**. Since there are horizontal tangents at $(-2, 6)$ and $(2, 0)$, $f'(\pm 2) = 0$. $f'(-2) = 0 \Rightarrow 12a - 4b + c = 0$ **(3)** and $f'(2) = 0 \Rightarrow$ $12a + 4b + c = 0$ **(4)**. Subtracting equation **(3)** from **(4)** gives $8b = 0 \Rightarrow b = 0$. Adding **(1)** and **(2)** gives $8b + 2d = 6$, so $d = 3$ since $b = 0$. From **(3)** we have $c = -12a$, so **(2)** becomes $8a + 4(0) + 2(-12a) + 3 = 0 \Rightarrow 3 = 16a \Rightarrow$ $a = \frac{3}{16}$. Now $c = -12a = -12\left(\frac{3}{16}\right) = -\frac{9}{4}$ and the desired cubic function is $y = \frac{3}{16}x^3 - \frac{9}{4}x + 3$.

86. $y = ax^2 + bx + c \Rightarrow y'(x) = 2ax + b$. The parabola has slope 4 at $x = 1$ and slope -8 at $x = -1$, so $y'(1) = 4 \Rightarrow$ $2a + b = 4$ **(1)** and $y'(-1) = -8 \Rightarrow -2a + b = -8$ **(2)**. Adding **(1)** and **(2)** gives us $2b = -4 \Leftrightarrow b = -2$. From **(1)**, $2a - 2 = 4 \Leftrightarrow a = 3$. Thus, the equation of the parabola is $y = 3x^2 - 2x + c$. Since it passes through the point $(2, 15)$, we have $15 = 3(2)^2 - 2(2) + c \Rightarrow c = 7$, so the equation is $y = 3x^2 - 2x + 7$.

87. If $P(t)$ denotes the population at time t and $A(t)$ the average annual income, then $T(t) = P(t)A(t)$ is the total personal income. The rate at which $T(t)$ is rising is given by $T'(t) = P(t)A'(t) + A(t)P'(t) \Rightarrow$

$$T'(1999) = P(1999)A'(1999) + A(1999)P'(1999) = (961{,}400)(\$1400/\text{yr}) + (\$30{,}593)(9200/\text{yr})$$

$$= \$1{,}345{,}960{,}000/\text{yr} + \$281{,}455{,}600/\text{yr} = \$1{,}627{,}415{,}600/\text{yr}$$

So the total personal income was rising by about $1.627 billion per year in 1999.

The term $P(t)A'(t) \approx \$1.346$ billion represents the portion of the rate of change of total income due to the existing population's increasing income. The term $A(t)P'(t) \approx \$281$ million represents the portion of the rate of change of total income due to increasing population.

88. (a) $f(20) = 10{,}000$ means that when the price of the fabric is $20/yard, 10,000 yards will be sold.

$f'(20) = -350$ means that as the price of the fabric increases past $20/yard, the amount of fabric which will be sold is decreasing at a rate of 350 yards per (dollar per yard).

(b) $R(p) = pf(p) \Rightarrow R'(p) = pf'(p) + f(p) \cdot 1 \Rightarrow R'(20) = 20f'(20) + f(20) \cdot 1 = 20(-350) + 10{,}000 = 3000$.

This means that as the price of the fabric increases past $20/yard, the total revenue is increasing at $3000/($/yard). Note that the Product Rule indicates that we will lose $7000/($/yard) due to selling less fabric, but this loss is more than made up for by the additional revenue due to the increase in price.

89. $f(x) = 2 - x$ if $x \le 1$ and $f(x) = x^2 - 2x + 2$ if $x > 1$. Now we compute the right- and left-hand derivatives defined in Exercise 3.2.52:

$f'_-(1) = \lim\limits_{h \to 0^-} \dfrac{f(1+h) - f(1)}{h} = \lim\limits_{h \to 0^-} \dfrac{2 - (1+h) - 1}{h} = \lim\limits_{h \to 0^-} \dfrac{-h}{h} = \lim\limits_{h \to 0^-} -1 = -1$ and

$f'_+(1) = \lim\limits_{h \to 0^+} \dfrac{f(1+h) - f(1)}{h} = \lim\limits_{h \to 0^+} \dfrac{(1+h)^2 - 2(1+h) + 2 - 1}{h} = \lim\limits_{h \to 0^+} \dfrac{h^2}{h} = \lim\limits_{h \to 0^+} h = 0$.

Thus, $f'(1)$ does not exist since $f'_-(1) \ne f'_+(1)$, so f is not differentiable at 1. But $f'(x) = -1$ for $x < 1$ and $f'(x) = 2x - 2$ if $x > 1$.

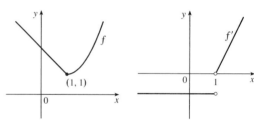

90. $g(x) = \begin{cases} -1 - 2x & \text{if } x < -1 \\ x^2 & \text{if } -1 \le x \le 1 \\ x & \text{if } x > 1 \end{cases}$

$\lim\limits_{h \to 0^-} \dfrac{g(-1+h) - g(-1)}{h} = \lim\limits_{h \to 0^-} \dfrac{[-1 - 2(-1+h)] - 1}{h} = \lim\limits_{h \to 0^-} \dfrac{-2h}{h} = \lim\limits_{h \to 0^-} (-2) = -2$ and

$\lim\limits_{h \to 0^+} \dfrac{g(-1+h) - g(-1)}{h} = \lim\limits_{h \to 0^+} \dfrac{(-1+h)^2 - 1}{h} = \lim\limits_{h \to 0^+} \dfrac{-2h + h^2}{h} = \lim\limits_{h \to 0^+} (-2 + h) = -2$,

so g is differentiable at -1 and $g'(-1) = -2$.

$\lim\limits_{h \to 0^-} \dfrac{g(1+h) - g(1)}{h} = \lim\limits_{h \to 0^-} \dfrac{(1+h)^2 - 1}{h} = \lim\limits_{h \to 0^-} \dfrac{2h + h^2}{h} = \lim\limits_{h \to 0^-} (2 + h) = 2$ and

$\lim\limits_{h \to 0^+} \dfrac{g(1+h) - g(1)}{h} = \lim\limits_{h \to 0^+} \dfrac{(1+h) - 1}{h} = \lim\limits_{h \to 0^+} \dfrac{h}{h} = \lim\limits_{h \to 0^+} 1 = 1$, so $g'(1)$ does not exist.

Thus, g is differentiable except when $x = 1$, and

$g'(x) = \begin{cases} -2 & \text{if } x < -1 \\ 2x & \text{if } -1 \le x < 1 \\ 1 & \text{if } x > 1 \end{cases}$

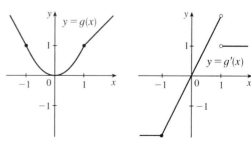

91. (a) Note that $x^2 - 9 < 0$ for $x^2 < 9$ $\Leftrightarrow$ $|x| < 3$ $\Leftrightarrow$ $-3 < x < 3$. So

$$f(x) = \begin{cases} x^2 - 9 & \text{if } x \le -3 \\ -x^2 + 9 & \text{if } -3 < x < 3 \\ x^2 - 9 & \text{if } x \ge 3 \end{cases} \Rightarrow f'(x) = \begin{cases} 2x & \text{if } x < -3 \\ -2x & \text{if } -3 < x < 3 \\ 2x & \text{if } x > 3 \end{cases} = \begin{cases} 2x & \text{if } |x| > 3 \\ -2x & \text{if } |x| < 3 \end{cases}$$

To show that $f'(3)$ does not exist we investigate $\lim\limits_{h \to 0} \dfrac{f(3+h) - f(3)}{h}$ by computing the left- and right-hand derivatives

defined in Exercise 3.2.52.

$$f'_-(3) = \lim_{h \to 0^-} \frac{f(3+h) - f(3)}{h} = \lim_{h \to 0^-} \frac{[-(3+h)^2 + 9] - 0}{h} = \lim_{h \to 0^-} (-6 - h) = -6 \quad \text{and}$$

$$f'_+(3) = \lim_{h \to 0^+} \frac{f(3+h) - f(3)}{h} = \lim_{h \to 0^+} \frac{[(3+h)^2 - 9] - 0}{h} = \lim_{h \to 0^+} \frac{6h + h^2}{h} = \lim_{h \to 0^+} (6 + h) = 6.$$

Since the left and right limits are different,

(b)

 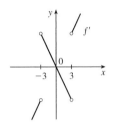

$\lim\limits_{h \to 0} \dfrac{f(3+h) - f(3)}{h}$ does not exist, that is, $f'(3)$

does not exist. Similarly, $f'(-3)$ does not exist.

Therefore, f is not differentiable at 3 or at -3.

92. If $x \ge 1$, then $h(x) = |x - 1| + |x + 2| = x - 1 + x + 2 = 2x + 1$.

If $-2 < x < 1$, then $h(x) = -(x - 1) + x + 2 = 3$.

If $x \le -2$, then $h(x) = -(x - 1) - (x + 2) = -2x - 1$. Therefore,

$$h(x) = \begin{cases} -2x - 1 & \text{if } x \le -2 \\ 3 & \text{if } -2 < x < 1 \\ 2x + 1 & \text{if } x \ge 1 \end{cases} \Rightarrow h'(x) = \begin{cases} -2 & \text{if } x < -2 \\ 0 & \text{if } -2 < x < 1 \\ 2 & \text{if } x > 1 \end{cases}$$

To see that $h'(1) = \lim\limits_{x \to 1} \dfrac{h(x) - h(1)}{x - 1}$ does not exist,

$y = h'(x)$

observe that $\lim\limits_{x \to 1^-} \dfrac{h(x) - h(1)}{x - 1} = \lim\limits_{x \to 1^-} \dfrac{3 - 3}{3 - 1} = 0$ but

$\lim\limits_{x \to 1^+} \dfrac{h(x) - h(1)}{x - 1} = \lim\limits_{x \to 1^+} \dfrac{2x - 2}{x - 1} = 2$. Similarly,

$h'(-2)$ does not exist.

93. $y = f(x) = ax^2$ $\Rightarrow$ $f'(x) = 2ax$.   So the slope of the tangent to the parabola at $x = 2$ is $m = 2a(2) = 4a$. The slope

of the given line, $2x + y = b$ $\Leftrightarrow$ $y = -2x + b$, is seen to be -2, so we must have $4a = -2$ $\Leftrightarrow$ $a = -\frac{1}{2}$. So when

$x = 2$, the point in question has y-coordinate $-\frac{1}{2} \cdot 2^2 = -2$. Now we simply require that the given line, whose equation is

$2x + y = b$, pass through the point $(2, -2)$: $2(2) + (-2) = b$ $\Leftrightarrow$ $b = 2$. So we must have $a = -\frac{1}{2}$ and $b = 2$.

94. (a) We use the Product Rule repeatedly: $F = fg \Rightarrow F' = f'g + fg' \Rightarrow$

$F'' = (f''g + f'g') + (f'g' + fg'') = f''g + 2f'g' + fg''$.

(b) $F''' = f'''g + f''g' + 2(f''g' + f'g'') + f'g'' + fg''' = f'''g + 3f''g' + 3f'g'' + fg''' \Rightarrow$

$F^{(4)} = f^{(4)}g + f'''g' + 3(f'''g' + f''g'') + 3(f''g'' + f'g''') + f'g''' + fg^{(4)}$

$= f^{(4)}g + 4f'''g' + 6f''g'' + 4f'g''' + fg^{(4)}$

(c) By analogy with the Binomial Theorem, we make the guess:

$$F^{(n)} = f^{(n)}g + nf^{(n-1)}g' + \binom{n}{2}f^{(n-2)}g'' + \cdots + \binom{n}{k}f^{(n-k)}g^{(k)} + \cdots + nf'g^{(n-1)} + fg^{(n)},$$

where $\dbinom{n}{k} = \dfrac{n!}{k!\,(n-k)!} = \dfrac{n(n-1)(n-2)\cdots(n-k+1)}{k!}$.

95. The slope of the curve $y = c\sqrt{x}$ is $y' = \dfrac{c}{2\sqrt{x}}$ and the slope of the tangent line $y = \frac{3}{2}x + 6$ is $\frac{3}{2}$. These must be equal at the

point of tangency $\left(a, c\sqrt{a}\right)$, so $\dfrac{c}{2\sqrt{a}} = \dfrac{3}{2} \Rightarrow c = 3\sqrt{a}$. The y-coordinates must be equal at $x = a$, so

$c\sqrt{a} = \frac{3}{2}a + 6 \Rightarrow \left(3\sqrt{a}\right)\sqrt{a} = \frac{3}{2}a + 6 \Rightarrow 3a = \frac{3}{2}a + 6 \Rightarrow \frac{3}{2}a = 6 \Rightarrow a = 4$. Since $c = 3\sqrt{a}$, we have

$c = 3\sqrt{4} = 6$.

96. f is clearly differentiable for $x < 2$ and for $x > 2$. For $x < 2$, $f'(x) = 2x$, so $f'_-(2) = 4$. For $x > 2$, $f'(x) = m$, so

$f'_+(2) = m$. For f to be differentiable at $x = 2$, we need $4 = f'_-(2) = f'_+(2) = m$. So $f(x) = 4x + b$. We must also have

continuity at $x = 2$, so $4 = f(2) = \lim\limits_{x \to 2^+} f(x) = \lim\limits_{x \to 2^+} (4x + b) = 8 + b$. Hence, $b = -4$.

97. $F = f/g \Rightarrow f = Fg \Rightarrow f' = F'g + Fg' \Rightarrow F' = \dfrac{f' - Fg'}{g} = \dfrac{f' - (f/g)g'}{g} = \dfrac{f'g - fg'}{g^2}$

98. (a) $xy = c \Rightarrow y = \dfrac{c}{x}$. Let $P = \left(a, \dfrac{c}{a}\right)$. The slope of the tangent line at $x = a$ is $y'(a) = -\dfrac{c}{a^2}$. Its equation is

$y - \dfrac{c}{a} = -\dfrac{c}{a^2}(x - a)$ or $y = -\dfrac{c}{a^2}x + \dfrac{2c}{a}$, so its y-intercept is $\dfrac{2c}{a}$. Setting $y = 0$ gives $x = 2a$, so the x-intercept is $2a$.

The midpoint of the line segment joining $\left(0, \dfrac{2c}{a}\right)$ and $(2a, 0)$ is $\left(a, \dfrac{c}{a}\right) = P$.

(b) We know the x- and y-intercepts of the tangent line from part (a), so the area of the triangle bounded by the axes and the

tangent is $\frac{1}{2}$(base)(height) $= \frac{1}{2}xy = \frac{1}{2}(2a)(2c/a) = 2c$, a constant.

99. *Solution 1:* Let $f(x) = x^{1000}$. Then, by the definition of a derivative, $f'(1) = \lim\limits_{x \to 1} \dfrac{f(x) - f(1)}{x - 1} = \lim\limits_{x \to 1} \dfrac{x^{1000} - 1}{x - 1}$.

But this is just the limit we want to find, and we know (from the Power Rule) that $f'(x) = 1000x^{999}$, so

$f'(1) = 1000(1)^{999} = 1000$. So $\lim\limits_{x \to 1} \dfrac{x^{1000} - 1}{x - 1} = 1000$.

Solution 2: Note that $(x^{1000} - 1) = (x - 1)(x^{999} + x^{998} + x^{997} + \cdots + x^2 + x + 1)$. So

$$\lim_{x \to 1} \frac{x^{1000} - 1}{x - 1} = \lim_{x \to 1} \frac{(x - 1)(x^{999} + x^{998} + x^{997} + \cdots + x^2 + x + 1)}{x - 1} = \lim_{x \to 1}(x^{999} + x^{998} + x^{997} + \cdots + x^2 + x + 1)$$

$$= \underbrace{1 + 1 + 1 + \cdots + 1 + 1 + 1}_{\text{1000 ones}} = 1000, \text{ as above.}$$

100.

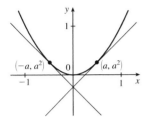

In order for the two tangents to intersect on the y-axis, the points of tangency must be at equal distances from the y-axis, since the parabola $y = x^2$ is symmetric about the y-axis. Say the points of tangency are (a, a^2) and $(-a, a^2)$, for some $a > 0$. Then since the derivative of $y = x^2$ is $dy/dx = 2x$, the left-hand tangent has slope $-2a$ and equation $y - a^2 = -2a(x + a)$, or $y = -2ax - a^2$, and similarly the right-hand tangent line has equation $y - a^2 = 2a(x - a)$, or $y = 2ax - a^2$. So the two lines intersect at $(0, -a^2)$. Now if the lines are perpendicular, then the product of their slopes is -1, so

$$(-2a)(2a) = -1 \quad \Leftrightarrow \quad a^2 = \tfrac{1}{4} \quad \Leftrightarrow \quad a = \tfrac{1}{2}. \text{ So the lines intersect at } \left(0, -\tfrac{1}{4}\right).$$

101. $y = x^2 \Rightarrow y' = 2x$, so the slope of a tangent line at the point (a, a^2) is $y' = 2a$ and the slope of a normal line is $-1/(2a)$,

for $a \neq 0$. The slope of the normal line through the points (a, a^2) and $(0, c)$ is $\dfrac{a^2 - c}{a - 0}$, so $\dfrac{a^2 - c}{a} = -\dfrac{1}{2a} \Rightarrow$

$a^2 - c = -\tfrac{1}{2} \Rightarrow a^2 = c - \tfrac{1}{2}$. The last equation has two solutions if $c > \tfrac{1}{2}$, one solution if $c = \tfrac{1}{2}$, and no solution if

$c < \tfrac{1}{2}$. Since the y-axis is normal to $y = x^2$ regardless of the value of c (this is the case for $a = 0$), we have three normal lines

if $c > \tfrac{1}{2}$ and one normal line if $c \leq \tfrac{1}{2}$.

102.

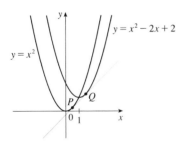

From the sketch, it appears that there may be a line that is tangent to both curves. The slope of the line through the points $P(a, a^2)$ and

$Q(b, b^2 - 2b + 2)$ is $\dfrac{b^2 - 2b + 2 - a^2}{b - a}$. The slope of the tangent line at P

is $2a \quad [y' = 2x]$ and at Q is $2b - 2 \quad [y' = 2x - 2]$. All three slopes are

equal, so $2a = 2b - 2 \quad \Leftrightarrow \quad a = b - 1$.

Also, $2b - 2 = \dfrac{b^2 - 2b + 2 - a^2}{b - a} \Rightarrow 2b - 2 = \dfrac{b^2 - 2b + 2 - (b - 1)^2}{b - (b - 1)} \Rightarrow 2b - 2 = b^2 - 2b + 2 - b^2 + 2b - 1 \Rightarrow$

$2b = 3 \Rightarrow b = \tfrac{3}{2}$ and $a = \tfrac{3}{2} - 1 = \tfrac{1}{2}$. Thus, an equation of the tangent line at P is $y - \left(\tfrac{1}{2}\right)^2 = 2\left(\tfrac{1}{2}\right)\left(x - \tfrac{1}{2}\right)$ or

$y = x - \tfrac{1}{4}$.

APPLIED PROJECT   Building a Better Roller Coaster

1. (a) $f(x) = ax^2 + bx + c$ $\Rightarrow$ $f'(x) = 2ax + b$.

The origin is at P:	$f(0) = 0$ $\Rightarrow$	$c = 0$
The slope of the ascent is 0.8:	$f'(0) = 0.8$ $\Rightarrow$	$b = 0.8$
The slope of the drop is -1.6:	$f'(100) = -1.6$ $\Rightarrow$	$200a + b = -1.6$

(b) $b = 0.8$, so $200a + b = -1.6$ $\Rightarrow$ $200a + 0.8 = -1.6$ $\Rightarrow$ $200a = -2.4$ $\Rightarrow$ $a = -\dfrac{2.4}{200} = -0.012$.

Thus, $f(x) = -0.012x^2 + 0.8x$.

(c) Since L_1 passes through the origin with slope 0.8, it has equation $y = 0.8x$.

The horizontal distance between P and Q is 100, so the y-coordinate at Q is

$f(100) = -0.012(100)^2 + 0.8(100) = -40$. Since L_2 passes through the

point $(100, -40)$ and has slope -1.6, it has equation $y + 40 = -1.6(x - 100)$

or $y = -1.6x + 120$.

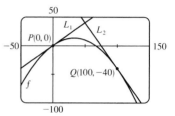

(d) The difference in elevation between $P(0, 0)$ and $Q(100, -40)$ is $0 - (-40) = 40$ feet.

2. (a)

Interval	Function	First Derivative	Second Derivative
$(-\infty, 0)$	$L_1(x) = 0.8x$	$L_1'(x) = 0.8$	$L_1''(x) = 0$
$[0, 10)$	$g(x) = kx^3 + lx^2 + mx + n$	$g'(x) = 3kx^2 + 2lx + m$	$g''(x) = 6kx + 2l$
$[10, 90]$	$q(x) = ax^2 + bx + c$	$q'(x) = 2ax + b$	$q''(x) = 2a$
$(90, 100]$	$h(x) = px^3 + qx^2 + rx + s$	$h'(x) = 3px^2 + 2qx + r$	$h''(x) = 6px + 2q$
$(100, \infty)$	$L_2(x) = -1.6x + 120$	$L_2'(x) = -1.6$	$L_2''(x) = 0$

There are 4 values of x (0, 10, 90, and 100) for which we must make sure the function values are equal, the first derivative

values are equal, and the second derivative values are equal. The third column in the following table contains the value of

each side of the condition — these are found after solving the system in part (b).

At $x =$	Condition	Value	Resulting Equation
0	$g(0) = L_1(0)$	0	$n = 0$
	$g'(0) = L_1'(0)$	$\frac{4}{5}$	$m = 0.8$
	$g''(0) = L_1''(0)$	0	$2l = 0$
10	$g(10) = q(10)$	$\frac{68}{9}$	$1000k + 100l + 10m + n = 100a + 10b + c$
	$g'(10) = q'(10)$	$\frac{2}{3}$	$300k + 20l + m = 20a + b$
	$g''(10) = q''(10)$	$-\frac{2}{75}$	$60k + 2l = 2a$
90	$h(90) = q(90)$	$-\frac{220}{9}$	$729{,}000p + 8100q + 90r + s = 8100a + 90b + c$
	$h'(90) = q'(90)$	$-\frac{22}{15}$	$24{,}300p + 180q + r = 180a + b$
	$h''(90) = q''(90)$	$-\frac{2}{75}$	$540p + 2q = 2a$
100	$h(100) = L_2(100)$	-40	$1{,}000{,}000p + 10{,}000q + 100r + s = -40$
	$h'(100) = L_2'(100)$	$-\frac{8}{5}$	$30{,}000p + 200q + r = -1.6$
	$h''(100) = L_2''(100)$	0	$600p + 2q = 0$

(b) We can arrange our work in a 12×12 matrix as follows.

a	b	c	k	l	m	n	p	q	r	s	constant
0	0	0	0	0	0	1	0	0	0	0	0
0	0	0	0	0	1	0	0	0	0	0	0.8
0	0	0	0	2	0	0	0	0	0	0	0
-100	-10	-1	1000	100	10	1	0	0	0	0	0
-20	-1	0	300	20	1	0	0	0	0	0	0
-2	0	0	60	2	0	0	0	0	0	0	0
-8100	-90	-1	0	0	0	0	729,000	8100	90	1	0
-180	-1	0	0	0	0	0	24,300	180	1	0	0
-2	0	0	0	0	0	0	540	2	0	0	0
0	0	0	0	0	0	0	1,000,000	10,000	100	1	-40
0	0	0	0	0	0	0	30,000	200	1	0	-1.6
0	0	0	0	0	0	0	600	2	0	0	0

Solving the system gives us the formulas for q, g, and h.

$$\left.\begin{array}{l} a = -0.01\overline{3} = -\frac{1}{75} \\ b = 0.9\overline{3} = \frac{14}{15} \\ c = -0.\overline{4} = -\frac{4}{9} \end{array}\right\} \quad q(x) = -\frac{1}{75}x^2 + \frac{14}{15}x - \frac{4}{9}$$

$$\left.\begin{array}{l} k = -0.000\overline{4} = -\frac{1}{2250} \\ l = 0 \\ m = 0.8 = \frac{4}{5} \\ n = 0 \end{array}\right\} \quad g(x) = -\frac{1}{2250}x^3 + \frac{4}{5}x$$

$$\left.\begin{array}{l} p = 0.000\overline{4} = \frac{1}{2250} \\ q = -0.1\overline{3} = -\frac{2}{15} \\ r = 11.7\overline{3} = \frac{176}{15} \\ s = -324.\overline{4} = -\frac{2920}{9} \end{array}\right\} \quad h(x) = \frac{1}{2250}x^3 - \frac{2}{15}x^2 + \frac{176}{15}x - \frac{2920}{9}$$

(c) Graph of L_1, q, g, h, and L_2:

This is the piecewise-defined function assignment on a TI-83 Plus calculator, where $Y_2 = L_1$, $Y_6 = g$, $Y_5 = q$, $Y_7 = h$, and $Y_3 = L_2$.

The graph of the five functions as a piecewise-defined function:

A comparison of the graphs in part 1(c) and part 2(c):

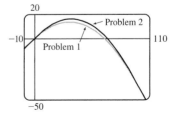

3.4   Derivatives of Trigonometric Functions

1. $f(x) = 3x^2 - 2\cos x \;\;\Rightarrow\;\; f'(x) = 6x - 2(-\sin x) = 6x + 2\sin x$

2. $f(x) = \sqrt{x}\,\sin x \;\;\Rightarrow\;\; f'(x) = \sqrt{x}\,\cos x + \sin x\left(\frac{1}{2}x^{-1/2}\right) = \sqrt{x}\,\cos x + \dfrac{\sin x}{2\sqrt{x}}$

3. $f(x) = \sin x + \frac{1}{2}\cot x \;\;\Rightarrow\;\; f'(x) = \cos x - \frac{1}{2}\csc^2 x$

4. $y = 2\csc x + 5\cos x \;\;\Rightarrow\;\; y' = -2\csc x\,\cot x - 5\sin x$

5. $g(t) = t^3\cos t \;\;\Rightarrow\;\; g'(t) = t^3(-\sin t) + (\cos t)\cdot 3t^2 = 3t^2\cos t - t^3\sin t \;\text{ or }\; t^2(3\cos t - t\sin t)$

6. $g(t) = 4\sec t + \tan t \;\;\Rightarrow\;\; g'(t) = 4\sec t\,\tan t + \sec^2 t$

7. $h(\theta) = \theta\csc\theta - \cot\theta \;\;\Rightarrow\;\; h'(\theta) = \theta(-\csc\theta\,\cot\theta) + (\csc\theta)\cdot 1 - \left(-\csc^2\theta\right) = \csc\theta - \theta\csc\theta\,\cot\theta + \csc^2\theta$

8. $y = u(a\cos u + b\cot u) \;\;\Rightarrow\;\;$

$y' = u(-a\sin u - b\csc^2 u) + (a\cos u + b\cot u)\cdot 1 = a\cos u + b\cot u - au\sin u - bu\csc^2 u$

9. $y = \dfrac{x}{2 - \tan x} \;\;\Rightarrow\;\; y' = \dfrac{(2 - \tan x)(1) - x(-\sec^2 x)}{(2 - \tan x)^2} = \dfrac{2 - \tan x + x\sec^2 x}{(2 - \tan x)^2}$

10. $y = \dfrac{1 + \sin x}{x + \cos x} \;\;\Rightarrow\;\;$

$y' = \dfrac{(x + \cos x)(\cos x) - (1 + \sin x)(1 - \sin x)}{(x + \cos x)^2} = \dfrac{x\cos x + \cos^2 x - (1 - \sin^2 x)}{(x + \cos x)^2}$

$\;\;= \dfrac{x\cos x + \cos^2 x - (\cos^2 x)}{(x + \cos x)^2} = \dfrac{x\cos x}{(x + \cos x)^2}$

11. $f(\theta) = \dfrac{\sec\theta}{1 + \sec\theta} \;\;\Rightarrow\;\;$

$f'(\theta) = \dfrac{(1 + \sec\theta)(\sec\theta\,\tan\theta) - (\sec\theta)(\sec\theta\,\tan\theta)}{(1 + \sec\theta)^2} = \dfrac{(\sec\theta\,\tan\theta)\,[(1 + \sec\theta) - \sec\theta]}{(1 + \sec\theta)^2} = \dfrac{\sec\theta\,\tan\theta}{(1 + \sec\theta)^2}$

12. $y = \dfrac{1 - \sec x}{\tan x} \;\;\Rightarrow\;\;$

$y' = \dfrac{\tan x\left(-\sec x\,\tan x\right) - (1 - \sec x)(\sec^2 x)}{(\tan x)^2} = \dfrac{\sec x\left(-\tan^2 x - \sec x + \sec^2 x\right)}{\tan^2 x} = \dfrac{\sec x\left(1 - \sec x\right)}{\tan^2 x}$

13. $y = \dfrac{\sin x}{x^2} \;\;\Rightarrow\;\; y' = \dfrac{x^2\cos x - (\sin x)(2x)}{(x^2)^2} = \dfrac{x(x\cos x - 2\sin x)}{x^4} = \dfrac{x\cos x - 2\sin x}{x^3}$

14. $y = \csc\theta\,(\theta + \cot\theta) \;\;\Rightarrow\;\;$

$y' = \csc\theta\,(1 - \csc^2\theta) + (\theta + \cot\theta)(-\csc\theta\,\cot\theta) = \csc\theta\,(1 - \csc^2\theta - \theta\cot\theta - \cot^2\theta)$

$\;\;= \csc\theta\,(-\cot^2\theta - \theta\cot\theta - \cot^2\theta) \qquad [1 + \cot^2\theta = \csc^2\theta]$

$\;\;= \csc\theta\,(-\theta\cot\theta - 2\cot^2\theta) = -\csc\theta\,\cot\theta\,(\theta + 2\cot\theta)$

15. $y = \sec\theta\,\tan\theta \;\Rightarrow\; y' = \sec\theta\,(\sec^2\theta) + \tan\theta\,(\sec\theta\,\tan\theta) = \sec\theta\,(\sec^2\theta + \tan^2\theta)$. Using the identity

$1 + \tan^2\theta = \sec^2\theta$, we can write alternative forms of the answer as $\sec\theta\,(1 + 2\tan^2\theta)$ or $\sec\theta\,(2\sec^2\theta - 1)$.

16. Using Exercise 3.3.81(a), $f(x) = x^2\sin x\,\tan x \;\Rightarrow$

$$f'(x) = (x^2)'\sin x\,\tan x + x^2(\sin x)'\tan x + x^2\sin x\,(\tan x)' = 2x\sin x\,\tan x + x^2\cos x\,\tan x + x^2\sin x\,\sec^2 x$$

$$= 2x\sin x\,\tan x + x^2\sin x + x^2\sin x\,\sec^2 x = x\sin x\,(2\tan x + x + x\sec^2 x).$$

17. $\dfrac{d}{dx}\,(\csc x) = \dfrac{d}{dx}\left(\dfrac{1}{\sin x}\right) = \dfrac{(\sin x)(0) - 1(\cos x)}{\sin^2 x} = \dfrac{-\cos x}{\sin^2 x} = -\dfrac{1}{\sin x}\cdot\dfrac{\cos x}{\sin x} = -\csc x\,\cot x$

18. $\dfrac{d}{dx}\,(\sec x) = \dfrac{d}{dx}\left(\dfrac{1}{\cos x}\right) = \dfrac{(\cos x)(0) - 1(-\sin x)}{\cos^2 x} = \dfrac{\sin x}{\cos^2 x} = \dfrac{1}{\cos x}\cdot\dfrac{\sin x}{\cos x} = \sec x\,\tan x$

19. $\dfrac{d}{dx}\,(\cot x) = \dfrac{d}{dx}\left(\dfrac{\cos x}{\sin x}\right) = \dfrac{(\sin x)(-\sin x) - (\cos x)(\cos x)}{\sin^2 x} = -\dfrac{\sin^2 x + \cos^2 x}{\sin^2 x} = -\dfrac{1}{\sin^2 x} = -\csc^2 x$

20. $f(x) = \cos x \;\Rightarrow$

$$f'(x) = \lim_{h\to 0}\frac{f(x+h) - f(x)}{h} = \lim_{h\to 0}\frac{\cos(x+h) - \cos x}{h} = \lim_{h\to 0}\frac{\cos x\cos h - \sin x\sin h - \cos x}{h}$$

$$= \lim_{h\to 0}\left(\cos x\,\frac{\cos h - 1}{h} - \sin x\,\frac{\sin h}{h}\right) = \cos x\lim_{h\to 0}\frac{\cos h - 1}{h} - \sin x\lim_{h\to 0}\frac{\sin h}{h}$$

$$= (\cos x)(0) - (\sin x)(1) = -\sin x$$

21. $y = \sec x \;\Rightarrow\; y' = \sec x\,\tan x$, so $y'\left(\frac{\pi}{3}\right) = \sec\frac{\pi}{3}\tan\frac{\pi}{3} = 2\sqrt{3}$. An equation of the tangent line to the curve $y = \sec x$

at the point $\left(\frac{\pi}{3}, 2\right)$ is $y - 2 = 2\sqrt{3}\left(x - \frac{\pi}{3}\right)$ or $y = 2\sqrt{3}\,x + 2 - \frac{2}{3}\sqrt{3}\,\pi$.

22. $y = (1 + x)\cos x \;\Rightarrow\; y' = (1 + x)(-\sin x) + \cos x \cdot 1$. At $(0, 1)$, $y' = 1$, and an equation of the tangent line is

$y - 1 = 1(x - 0)$ or $y = x + 1$.

23. $y = x + \cos x \;\Rightarrow\; y' = 1 - \sin x$. At $(0, 1)$, $y' = 1$, and an equation of the tangent line is $y - 1 = 1(x - 0)$, or $y = x + 1$.

24. $y = \dfrac{1}{\sin x + \cos x} \;\Rightarrow\; y' = -\dfrac{\cos x - \sin x}{(\sin x + \cos x)^2}$ [Reciprocal Rule].   At $(0, 1)$, $y' = -\dfrac{1 - 0}{(0 + 1)^2} = -1$, and an equation

of the tangent line is $y - 1 = -1(x - 0)$, or $y = -x + 1$.

25. (a) $y = 2x\sin x \;\Rightarrow\; y' = 2(x\cos x + \sin x \cdot 1)$.   At $\left(\frac{\pi}{2}, \pi\right)$,

$y' = 2\left(\frac{\pi}{2}\cos\frac{\pi}{2} + \sin\frac{\pi}{2}\right) = 2(0 + 1) = 2$, and an equation of the

tangent line is $y - \pi = 2\left(x - \frac{\pi}{2}\right)$, or $y = 2x$.

(b)

26. (a) $y = \sec x - 2\cos x$ $\Rightarrow$ $y' = \sec x \tan x + 2\sin x$ $\Rightarrow$

the slope of the tangent line at $\left(\frac{\pi}{3}, 1\right)$ is

$\sec\frac{\pi}{3}\tan\frac{\pi}{3} + 2\sin\frac{\pi}{3} = 2 \cdot \sqrt{3} + 2 \cdot \frac{\sqrt{3}}{2} = 3\sqrt{3}$, and an equation

is $y - 1 = 3\sqrt{3}\left(x - \frac{\pi}{3}\right)$, or $y = 3\sqrt{3}\,x + 1 - \pi\sqrt{3}$.

(b)
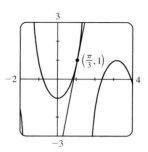

27. (a) $f(x) = \sec x - x$ $\Rightarrow$ $f'(x) = \sec x \tan x - 1$

(b)

Note that $f' = 0$ where f has a minimum. Also note that f' is negative when f is decreasing and f' is positive when f is increasing.

28. (a) $f(x) = \sqrt{x}\sin x$ $\Rightarrow$ $f'(x) = \sqrt{x}\cos x + (\sin x)\left(\frac{1}{2}x^{-1/2}\right) = \sqrt{x}\cos x + \frac{\sin x}{2\sqrt{x}}$

(b)

Notice that $f'(x) = 0$ when f has a horizontal tangent.

f' is positive when f is increasing and f' is negative when f is decreasing.

29. $H(\theta) = \theta\sin\theta$ $\Rightarrow$ $H'(\theta) = \theta\left(\cos\theta\right) + (\sin\theta)\cdot 1 = \theta\cos\theta + \sin\theta$ $\Rightarrow$

$H''(\theta) = \theta\left(-\sin\theta\right) + (\cos\theta)\cdot 1 + \cos\theta = -\theta\sin\theta + 2\cos\theta$

30. $f(x) = \sec x$ $\Rightarrow$ $f'(x) = \sec x \tan x$ $\Rightarrow$ $f''(x) = \sec x\left(\sec^2 x\right) + \tan x\left(\sec x \tan x\right) = \sec x\left(\sec^2 x + \tan^2 x\right)$.

$f''\left(\frac{\pi}{4}\right) = \sqrt{2}\left[\left(\sqrt{2}\right)^2 + 1^2\right] = \sqrt{2}\left(2 + 1\right) = 3\sqrt{2}$

31. (a) $f(x) = \dfrac{\tan x - 1}{\sec x}$ $\Rightarrow$

$f'(x) = \dfrac{\sec x\left(\sec^2 x\right) - (\tan x - 1)(\sec x \tan x)}{(\sec x)^2} = \dfrac{\sec x\left(\sec^2 x - \tan^2 x + \tan x\right)}{\sec^2 x} = \dfrac{1 + \tan x}{\sec x}$

(b) $f(x) = \dfrac{\tan x - 1}{\sec x} = \dfrac{\dfrac{\sin x}{\cos x} - 1}{\dfrac{1}{\cos x}} = \dfrac{\dfrac{\sin x - \cos x}{\cos x}}{\dfrac{1}{\cos x}} = \sin x - \cos x$ $\Rightarrow$ $f'(x) = \cos x - (-\sin x) = \cos x + \sin x$

(c) From part (a), $f'(x) = \dfrac{1 + \tan x}{\sec x} = \dfrac{1}{\sec x} + \dfrac{\tan x}{\sec x} = \cos x + \sin x$, which is the expression for $f'(x)$ in part (b).

32. (a) $g(x) = f(x)\sin x$ $\Rightarrow$ $g'(x) = f(x)\cos x + \sin x \cdot f'(x)$, so

$g'\left(\frac{\pi}{3}\right) = f\left(\frac{\pi}{3}\right)\cos\frac{\pi}{3} + \sin\frac{\pi}{3}\cdot f'\left(\frac{\pi}{3}\right) = 4\cdot\frac{1}{2} + \frac{\sqrt{3}}{2}\cdot(-2) = 2 - \sqrt{3}$

(b) $h(x) = \dfrac{\cos x}{f(x)} \;\Rightarrow\; h'(x) = \dfrac{f(x) \cdot (-\sin x) - \cos x \cdot f'(x)}{[f(x)]^2}$, so

$$h'\!\left(\tfrac{\pi}{3}\right) = \frac{f\!\left(\tfrac{\pi}{3}\right) \cdot \left(-\sin\tfrac{\pi}{3}\right) - \cos\tfrac{\pi}{3} \cdot f'\!\left(\tfrac{\pi}{3}\right)}{\left[f\!\left(\tfrac{\pi}{3}\right)\right]^2} = \frac{4\!\left(-\tfrac{\sqrt{3}}{2}\right) - \left(\tfrac12\right)(-2)}{4^2} = \frac{-2\sqrt{3}+1}{16} = \frac{1 - 2\sqrt{3}}{16}$$

33. $f(x) = x + 2\sin x$ has a horizontal tangent when $f'(x) = 0 \;\Leftrightarrow\; 1 + 2\cos x = 0 \;\Leftrightarrow\; \cos x = -\tfrac12 \;\Leftrightarrow\;$

$x = \tfrac{2\pi}{3} + 2\pi n$ or $\tfrac{4\pi}{3} + 2\pi n$, where n is an integer. Note that $\tfrac{4\pi}{3}$ and $\tfrac{2\pi}{3}$ are $\pm\tfrac{\pi}{3}$ units from π. This allows us to write the

solutions in the more compact equivalent form $(2n+1)\pi \pm \tfrac{\pi}{3}$, n an integer.

34. $y = \dfrac{\cos x}{2 + \sin x} \;\Rightarrow\; y' = \dfrac{(2 + \sin x)(-\sin x) - \cos x \cos x}{(2 + \sin x)^2} = \dfrac{-2\sin x - \sin^2 x - \cos^2 x}{(2 + \sin x)^2} = \dfrac{-2\sin x - 1}{(2 + \sin x)^2} = 0$ when

$-2\sin x - 1 = 0 \;\Leftrightarrow\; \sin x = -\tfrac12 \;\Leftrightarrow\; x = \tfrac{11\pi}{6} + 2\pi n$ or $x = \tfrac{7\pi}{6} + 2\pi n$, n an integer. So $y = \tfrac{1}{\sqrt{3}}$ or $y = -\tfrac{1}{\sqrt{3}}$ and

the points on the curve with horizontal tangents are: $\left(\tfrac{11\pi}{6} + 2\pi n,\; \tfrac{1}{\sqrt{3}}\right)$, $\left(\tfrac{7\pi}{6} + 2\pi n,\; -\tfrac{1}{\sqrt{3}}\right)$, n an integer.

35. (a) $x(t) = 8\sin t \;\Rightarrow\; v(t) = x'(t) = 8\cos t \;\Rightarrow\; a(t) = x''(t) = -8\sin t$

(b) The mass at time $t = \tfrac{2\pi}{3}$ has position $x\!\left(\tfrac{2\pi}{3}\right) = 8\sin\tfrac{2\pi}{3} = 8\!\left(\tfrac{\sqrt3}{2}\right) = 4\sqrt3$, velocity $v\!\left(\tfrac{2\pi}{3}\right) = 8\cos\tfrac{2\pi}{3} = 8\!\left(-\tfrac12\right) = -4$,

and acceleration $a\!\left(\tfrac{2\pi}{3}\right) = -8\sin\tfrac{2\pi}{3} = -8\!\left(\tfrac{\sqrt3}{2}\right) = -4\sqrt3$. Since $v\!\left(\tfrac{2\pi}{3}\right) < 0$, the particle is moving to the left.

36. (a) $s(t) = 2\cos t + 3\sin t \;\Rightarrow\; v(t) = -2\sin t + 3\cos t \;\Rightarrow\;$

  $a(t) = -2\cos t - 3\sin t$

(c) $s = 0 \;\Rightarrow\; t_2 \approx 2.55$.   So the mass passes through the equilibrium

  position for the first time when $t \approx 2.55$ s.

(b)

(d) $v = 0 \;\Rightarrow\; t_1 \approx 0.98,\; s(t_1) \approx 3.61$ cm.   So the mass travels

  a maximum of about 3.6 cm (upward and downward) from its equilibrium position.

(e) The speed $|v|$ is greatest when $s = 0$, that is, when $t = t_2 + n\pi$, n a positive integer. The mass is speeding up when v and

  a have the same sign. From the figure, we see that this is the case on the intervals $(t_1 + n\pi, t_2 + n\pi)$ where n is a whole

  number.

37.

From the diagram we can see that $\sin\theta = x/10 \;\Leftrightarrow\; x = 10\sin\theta$. We want to find the rate

of change of x with respect to θ, that is, $dx/d\theta$. Taking the derivative of $x = 10\sin\theta$, we get

$dx/d\theta = 10(\cos\theta)$. So when $\theta = \tfrac{\pi}{3}$, $\tfrac{dx}{d\theta} = 10\cos\tfrac{\pi}{3} = 10\!\left(\tfrac12\right) = 5$ ft/rad.

38. (a) $F = \dfrac{\mu W}{\mu\sin\theta + \cos\theta} \;\Rightarrow\; \dfrac{dF}{d\theta} = \dfrac{(\mu\sin\theta + \cos\theta)(0) - \mu W(\mu\cos\theta - \sin\theta)}{(\mu\sin\theta + \cos\theta)^2} = \dfrac{\mu W(\sin\theta - \mu\cos\theta)}{(\mu\sin\theta + \cos\theta)^2}$

(b) $\dfrac{dF}{d\theta} = 0 \;\Leftrightarrow\; \mu W(\sin\theta - \mu\cos\theta) = 0 \;\Leftrightarrow\; \sin\theta = \mu\cos\theta \;\Leftrightarrow\; \tan\theta = \mu \;\Leftrightarrow\; \theta = \tan^{-1}\mu$

(c)

120

0 ‾‾‾‾‾‾‾‾‾ 1

From the graph of $F = \dfrac{0.6(50)}{0.6\sin\theta + \cos\theta}$ for $0 \le \theta \le 1$, we see that

$\dfrac{dF}{d\theta} = 0 \;\Rightarrow\; \theta \approx 0.54$. Checking this with part (b) and $\mu = 0.6$, we

calculate $\theta = \tan^{-1} 0.6 \approx 0.54$. So the value from the graph is consistent

with the value in part (b).

39. $\displaystyle\lim_{x\to 0}\frac{\sin 3x}{x} = \lim_{x\to 0}\frac{3\sin 3x}{3x}$ [multiply numerator and denominator by 3]

$\displaystyle = 3\lim_{3x\to 0}\frac{\sin 3x}{3x}$ [as $x \to 0,\, 3x \to 0$]

$\displaystyle = 3\lim_{\theta\to 0}\frac{\sin\theta}{\theta}$ [let $\theta = 3x$]

$= 3(1)$ [Equation 2]

$= 3$

40. $\displaystyle\lim_{x\to 0}\frac{\sin 4x}{\sin 6x} = \lim_{x\to 0}\left(\frac{\sin 4x}{x}\cdot\frac{x}{\sin 6x}\right) = \lim_{x\to 0}\frac{4\sin 4x}{4x}\cdot\lim_{x\to 0}\frac{6x}{6\sin 6x} = 4\lim_{x\to 0}\frac{\sin 4x}{4x}\cdot\frac{1}{6}\lim_{x\to 0}\frac{6x}{\sin 6x} = 4(1)\cdot\frac{1}{6}(1) = \frac{2}{3}$

41. $\displaystyle\lim_{t\to 0}\frac{\tan 6t}{\sin 2t} = \lim_{t\to 0}\left(\frac{\sin 6t}{t}\cdot\frac{1}{\cos 6t}\cdot\frac{t}{\sin 2t}\right) = \lim_{t\to 0}\frac{6\sin 6t}{6t}\cdot\lim_{t\to 0}\frac{1}{\cos 6t}\cdot\lim_{t\to 0}\frac{2t}{2\sin 2t}$

$\displaystyle = 6\lim_{t\to 0}\frac{\sin 6t}{6t}\cdot\lim_{t\to 0}\frac{1}{\cos 6t}\cdot\frac{1}{2}\lim_{t\to 0}\frac{2t}{\sin 2t} = 6(1)\cdot\frac{1}{1}\cdot\frac{1}{2}(1) = 3$

42. $\displaystyle\lim_{\theta\to 0}\frac{\cos\theta - 1}{\sin\theta} = \lim_{\theta\to 0}\frac{\dfrac{\cos\theta - 1}{\theta}}{\dfrac{\sin\theta}{\theta}} = \frac{\displaystyle\lim_{\theta\to 0}\frac{\cos\theta - 1}{\theta}}{\displaystyle\lim_{\theta\to 0}\frac{\sin\theta}{\theta}} = \frac{0}{1} = 0$

43. $\displaystyle\lim_{\theta\to 0}\frac{\sin(\cos\theta)}{\sec\theta} = \frac{\sin\left(\displaystyle\lim_{\theta\to 0}\cos\theta\right)}{\displaystyle\lim_{\theta\to 0}\sec\theta} = \frac{\sin 1}{1} = \sin 1$

44. $\displaystyle\lim_{t\to 0}\frac{\sin^2 3t}{t^2} = \lim_{t\to 0}\left(\frac{\sin 3t}{t}\cdot\frac{\sin 3t}{t}\right) = \lim_{t\to 0}\frac{\sin 3t}{t}\cdot\lim_{t\to 0}\frac{\sin 3t}{t} = \left(\lim_{t\to 0}\frac{\sin 3t}{t}\right)^2 = \left(3\lim_{t\to 0}\frac{\sin 3t}{3t}\right)^2 = (3\cdot 1)^2 = 9$

45. Divide numerator and denominator by θ.   ($\sin\theta$ also works.)

$\displaystyle\lim_{\theta\to 0}\frac{\sin\theta}{\theta + \tan\theta} = \lim_{\theta\to 0}\frac{\dfrac{\sin\theta}{\theta}}{1 + \dfrac{\sin\theta}{\theta}\cdot\dfrac{1}{\cos\theta}} = \frac{\displaystyle\lim_{\theta\to 0}\frac{\sin\theta}{\theta}}{1 + \displaystyle\lim_{\theta\to 0}\frac{\sin\theta}{\theta}\lim_{\theta\to 0}\frac{1}{\cos\theta}} = \frac{1}{1 + 1\cdot 1} = \frac{1}{2}$

46. $\displaystyle\lim_{x\to 0}\frac{\sin(x^2)}{x} = \lim_{x\to 0}\left[x\cdot\frac{\sin(x^2)}{x\cdot x}\right] = \lim_{x\to 0}x\cdot\lim_{x\to 0}\frac{\sin(x^2)}{x^2} = 0\cdot\lim_{y\to 0^+}\frac{\sin y}{y}$ $\big[$where $y = x^2\big]$ $= 0\cdot 1 = 0$

47. $\displaystyle\lim_{x\to\pi/4}\frac{1 - \tan x}{\sin x - \cos x} = \lim_{x\to\pi/4}\frac{\left(1 - \dfrac{\sin x}{\cos x}\right)\cdot\cos x}{(\sin x - \cos x)\cdot\cos x} = \lim_{x\to\pi/4}\frac{\cos x - \sin x}{(\sin x - \cos x)\cos x} = \lim_{x\to\pi/4}\frac{-1}{\cos x} = \frac{-1}{1/\sqrt{2}} = -\sqrt{2}$

48. $\displaystyle\lim_{x\to 1}\frac{\sin(x-1)}{x^2 + x - 2} = \lim_{x\to 1}\frac{\sin(x-1)}{(x+2)(x-1)} = \lim_{x\to 1}\frac{1}{x+2}\lim_{x\to 1}\frac{\sin(x-1)}{x-1} = \frac{1}{3}\cdot 1 = \frac{1}{3}$

49. (a) $\dfrac{d}{dx}\tan x = \dfrac{d}{dx}\dfrac{\sin x}{\cos x} \;\Rightarrow\; \sec^2 x = \dfrac{\cos x \cos x - \sin x \,(-\sin x)}{\cos^2 x} = \dfrac{\cos^2 x + \sin^2 x}{\cos^2 x}.$ So $\sec^2 x = \dfrac{1}{\cos^2 x}$.

(b) $\dfrac{d}{dx}\sec x = \dfrac{d}{dx}\dfrac{1}{\cos x} \;\Rightarrow\; \sec x \tan x = \dfrac{(\cos x)(0) - 1(-\sin x)}{\cos^2 x}.$ So $\sec x \tan x = \dfrac{\sin x}{\cos^2 x}.$

(c) $\dfrac{d}{dx}(\sin x + \cos x) = \dfrac{d}{dx}\dfrac{1 + \cot x}{\csc x} \;\Rightarrow$

$\cos x - \sin x = \dfrac{\csc x\,(-\csc^2 x) - (1 + \cot x)(-\csc x \cot x)}{\csc^2 x} = \dfrac{\csc x\,[-\csc^2 x + (1 + \cot x)\cot x]}{\csc^2 x}$

$\qquad = \dfrac{-\csc^2 x + \cot^2 x + \cot x}{\csc x} = \dfrac{-1 + \cot x}{\csc x}$

So $\cos x - \sin x = \dfrac{\cot x - 1}{\csc x}.$

50. We get the following formulas for r and h in terms of θ:

$\sin\dfrac{\theta}{2} = \dfrac{r}{10} \;\Rightarrow\; r = 10\sin\dfrac{\theta}{2}$ and $\cos\dfrac{\theta}{2} = \dfrac{h}{10} \;\Rightarrow\; h = 10\cos\dfrac{\theta}{2}$

Now $A(\theta) = \frac{1}{2}\pi r^2$ and $B(\theta) = \frac{1}{2}(2r)h = rh.$ So

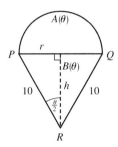

$\displaystyle\lim_{\theta\to 0^+}\dfrac{A(\theta)}{B(\theta)} = \lim_{\theta\to 0^+}\dfrac{\frac{1}{2}\pi r^2}{rh} = \frac{1}{2}\pi \lim_{\theta\to 0^+}\dfrac{r}{h} = \frac{1}{2}\pi \lim_{\theta\to 0^+}\dfrac{10\sin(\theta/2)}{10\cos(\theta/2)}$

$\qquad = \frac{1}{2}\pi \lim_{\theta\to 0^+}\tan(\theta/2) = 0$

51. By the definition of radian measure, $s = r\theta$, where r is the radius of the circle. By drawing the bisector of the angle θ, we can

see that $\sin\dfrac{\theta}{2} = \dfrac{d/2}{r} \;\Rightarrow\; d = 2r\sin\dfrac{\theta}{2}.$ So $\displaystyle\lim_{\theta\to 0^+}\dfrac{s}{d} = \lim_{\theta\to 0^+}\dfrac{r\theta}{2r\sin(\theta/2)} = \lim_{\theta\to 0^+}\dfrac{2\cdot(\theta/2)}{2\sin(\theta/2)} = \lim_{\theta\to 0}\dfrac{\theta/2}{\sin(\theta/2)} = 1.$

[This is just the reciprocal of the limit $\displaystyle\lim_{x\to 0}\dfrac{\sin x}{x} = 1$ combined with the fact that as $\theta \to 0$, $\frac{\theta}{2} \to 0$ also.]

3.5   The Chain Rule

1. Let $u = g(x) = 4x$ and $y = f(u) = \sin u$. Then $\dfrac{dy}{dx} = \dfrac{dy}{du}\dfrac{du}{dx} = (\cos u)(4) = 4\cos 4x.$

2. Let $u = g(x) = 4 + 3x$ and $y = f(u) = \sqrt{u} = u^{1/2}$. Then $\dfrac{dy}{dx} = \dfrac{dy}{du}\dfrac{du}{dx} = \frac{1}{2}u^{-1/2}(3) = \dfrac{3}{2\sqrt{u}} = \dfrac{3}{2\sqrt{4+3x}}.$

3. Let $u = g(x) = 1 - x^2$ and $y = f(u) = u^{10}$. Then $\dfrac{dy}{dx} = \dfrac{dy}{du}\dfrac{du}{dx} = (10u^9)(-2x) = -20x(1 - x^2)^9.$

4. Let $u = g(x) = \sin x$ and $y = f(u) = \tan u$. Then $\dfrac{dy}{dx} = \dfrac{dy}{du}\dfrac{du}{dx} = (\sec^2 u)(\cos x) = \sec^2(\sin x)\cdot\cos x,$

or equivalently, $[\sec(\sin x)]^2\cos x.$

5. Let $u = g(x) = \sin x$ and $y = f(u) = \sqrt{u}$. Then $\dfrac{dy}{dx} = \dfrac{dy}{du}\dfrac{du}{dx} = \frac{1}{2}u^{-1/2}\cos x = \dfrac{\cos x}{2\sqrt{u}} = \dfrac{\cos x}{2\sqrt{\sin x}}.$

6. Let $u = g(x) = \sqrt{x}$ and $y = f(u) = \sin u$. Then $\dfrac{dy}{dx} = \dfrac{dy}{du}\dfrac{du}{dx} = (\cos u)\left(\frac{1}{2}x^{-1/2}\right) = \dfrac{\cos u}{2\sqrt{x}} = \dfrac{\cos\sqrt{x}}{2\sqrt{x}}.$

7. $F(x) = (x^4 + 3x^2 - 2)^5 \Rightarrow F'(x) = 5(x^4 + 3x^2 - 2)^4 \cdot \dfrac{d}{dx}(x^4 + 3x^2 - 2) = 5(x^4 + 3x^2 - 2)^4(4x^3 + 6x)$

$\left[\text{or } 10x(x^4 + 3x^2 - 2)^4(2x^2 + 3)\right]$

8. $F(x) = (4x - x^2)^{100} \Rightarrow F'(x) = 100(4x - x^2)^{99} \cdot \dfrac{d}{dx}(4x - x^2) = 100(4x - x^2)^{99}(4 - 2x)$

$\left[\text{or } 200x^{99}(x - 2)(x - 4)^{99}\right]$

9. $F(x) = \sqrt[4]{1 + 2x + x^3} = (1 + 2x + x^3)^{1/4} \Rightarrow$

$F'(x) = \tfrac{1}{4}(1 + 2x + x^3)^{-3/4} \cdot \dfrac{d}{dx}(1 + 2x + x^3) = \dfrac{1}{4(1 + 2x + x^3)^{3/4}} \cdot (2 + 3x^2) = \dfrac{2 + 3x^2}{4(1 + 2x + x^3)^{3/4}}$

$= \dfrac{2 + 3x^2}{4\sqrt[4]{(1 + 2x + x^3)^3}}$

10. $f(x) = (1 + x^4)^{2/3} \Rightarrow f'(x) = \tfrac{2}{3}(1 + x^4)^{-1/3}(4x^3) = \dfrac{8x^3}{3\sqrt[3]{1 + x^4}}$

11. $g(t) = \dfrac{1}{(t^4 + 1)^3} = (t^4 + 1)^{-3} \Rightarrow g'(t) = -3(t^4 + 1)^{-4}(4t^3) = -12t^3(t^4 + 1)^{-4} = \dfrac{-12t^3}{(t^4 + 1)^4}$

12. $f(t) = \sqrt[3]{1 + \tan t} = (1 + \tan t)^{1/3} \Rightarrow f'(t) = \tfrac{1}{3}(1 + \tan t)^{-2/3}\sec^2 t = \dfrac{\sec^2 t}{3\sqrt[3]{(1 + \tan t)^2}}$

13. $y = \cos(a^3 + x^3) \Rightarrow y' = -\sin(a^3 + x^3) \cdot 3x^2 \quad [a^3 \text{ is just a constant}] = -3x^2\sin(a^3 + x^3)$

14. $y = a^3 + \cos^3 x \Rightarrow y' = 3(\cos x)^2(-\sin x) \quad [a^3 \text{ is just a constant}] = -3\sin x\cos^2 x$

15. Use the Product Rule. $y = x\sec kx \Rightarrow y' = x(\sec kx\tan kx \cdot k) + \sec kx \cdot 1 = \sec kx(kx\tan kx + 1)$

16. $y = 3\cot(n\theta) \Rightarrow y' = 3[-\csc^2(n\theta) \cdot n] = -3n\csc^2(n\theta)$

17. $g(x) = (1 + 4x)^5(3 + x - x^2)^8 \Rightarrow$

$g'(x) = (1 + 4x)^5 \cdot 8(3 + x - x^2)^7(1 - 2x) + (3 + x - x^2)^8 \cdot 5(1 + 4x)^4 \cdot 4$

$= 4(1 + 4x)^4(3 + x - x^2)^7\left[2(1 + 4x)(1 - 2x) + 5(3 + x - x^2)\right]$

$= 4(1 + 4x)^4(3 + x - x^2)^7\left[(2 + 4x - 16x^2) + (15 + 5x - 5x^2)\right] = 4(1 + 4x)^4(3 + x - x^2)^7(17 + 9x - 21x^2)$

18. $h(t) = (t^4 - 1)^3(t^3 + 1)^4 \Rightarrow$

$h'(t) = (t^4 - 1)^3 \cdot 4(t^3 + 1)^3(3t^2) + (t^3 + 1)^4 \cdot 3(t^4 - 1)^2(4t^3)$

$= 12t^2(t^4 - 1)^2(t^3 + 1)^3\left[(t^4 - 1) + t(t^3 + 1)\right] = 12t^2(t^4 - 1)^2(t^3 + 1)^3(2t^4 + t - 1)$

19. $y = (2x - 5)^4(8x^2 - 5)^{-3} \Rightarrow$

$y' = 4(2x - 5)^3(2)(8x^2 - 5)^{-3} + (2x - 5)^4(-3)(8x^2 - 5)^{-4}(16x)$

$= 8(2x - 5)^3(8x^2 - 5)^{-3} - 48x(2x - 5)^4(8x^2 - 5)^{-4}$

[This simplifies to $8(2x - 5)^3(8x^2 - 5)^{-4}(-4x^2 + 30x - 5)$.]

20. $y = (x^2 + 1)(x^2 + 2)^{1/3}$ $\Rightarrow$

$$y' = 2x(x^2 + 2)^{1/3} + (x^2 + 1)(\tfrac{1}{3})(x^2 + 2)^{-2/3}(2x) = 2x(x^2 + 2)^{1/3}\left[1 + \frac{x^2 + 1}{3(x^2 + 2)}\right]$$

21. $y = \left(\dfrac{x^2 + 1}{x^2 - 1}\right)^3$ $\Rightarrow$

$$y' = 3\left(\frac{x^2 + 1}{x^2 - 1}\right)^2 \cdot \frac{d}{dx}\left(\frac{x^2 + 1}{x^2 - 1}\right) = 3\left(\frac{x^2 + 1}{x^2 - 1}\right)^2 \cdot \frac{(x^2 - 1)(2x) - (x^2 + 1)(2x)}{(x^2 - 1)^2}$$

$$= 3\left(\frac{x^2 + 1}{x^2 - 1}\right)^2 \cdot \frac{2x[x^2 - 1 - (x^2 + 1)]}{(x^2 - 1)^2} = 3\left(\frac{x^2 + 1}{x^2 - 1}\right)^2 \cdot \frac{2x(-2)}{(x^2 - 1)^2} = \frac{-12x(x^2 + 1)^2}{(x^2 - 1)^4}$$

22. $y = x \sin \sqrt{x}$ $\Rightarrow$ $y' = x \cos \sqrt{x} \cdot \tfrac{1}{2}x^{-1/2} + \sin \sqrt{x} \cdot 1 = \tfrac{1}{2}\sqrt{x}\cos\sqrt{x} + \sin\sqrt{x}$

23. $y = \sin(x \cos x)$ $\Rightarrow$ $y' = \cos(x \cos x) \cdot [x(-\sin x) + \cos x \cdot 1] = (\cos x - x \sin x)\cos(x \cos x)$

24. $f(x) = \dfrac{x}{\sqrt{7 - 3x}}$ $\Rightarrow$

$$f'(x) = \frac{\sqrt{7 - 3x}(1) - x \cdot \tfrac{1}{2}(7 - 3x)^{-1/2} \cdot (-3)}{(\sqrt{7 - 3x})^2} = \frac{\sqrt{7 - 3x} + \dfrac{3x}{2\sqrt{7 - 3x}}}{(7 - 3x)^1} = \frac{2(7 - 3x) + 3x}{2(7 - 3x)^{3/2}} = \frac{14 - 3x}{2(7 - 3x)^{3/2}}$$

25. $F(z) = \sqrt{\dfrac{z - 1}{z + 1}} = \left(\dfrac{z - 1}{z + 1}\right)^{1/2}$ $\Rightarrow$

$$F'(z) = \frac{1}{2}\left(\frac{z - 1}{z + 1}\right)^{-1/2} \cdot \frac{d}{dz}\left(\frac{z - 1}{z + 1}\right) = \frac{1}{2}\left(\frac{z + 1}{z - 1}\right)^{1/2} \cdot \frac{(z + 1)(1) - (z - 1)(1)}{(z + 1)^2}$$

$$= \frac{1}{2}\frac{(z + 1)^{1/2}}{(z - 1)^{1/2}} \cdot \frac{z + 1 - z + 1}{(z + 1)^2} = \frac{1}{2}\frac{(z + 1)^{1/2}}{(z - 1)^{1/2}} \cdot \frac{2}{(z + 1)^2} = \frac{1}{(z - 1)^{1/2}(z + 1)^{3/2}}$$

26. $G(y) = \dfrac{(y - 1)^4}{(y^2 + 2y)^5}$ $\Rightarrow$

$$G'(y) = \frac{(y^2 + 2y)^5 \cdot 4(y - 1)^3 \cdot 1 - (y - 1)^4 \cdot 5(y^2 + 2y)^4(2y + 2)}{[(y^2 + 2y)^5]^2}$$

$$= \frac{2(y^2 + 2y)^4(y - 1)^3\left[2(y^2 + 2y) - 5(y - 1)(y + 1)\right]}{(y^2 + 2y)^{10}}$$

$$= \frac{2(y - 1)^3\left[(2y^2 + 4y) + (-5y^2 + 5)\right]}{(y^2 + 2y)^6} = \frac{2(y - 1)^3(-3y^2 + 4y + 5)}{(y^2 + 2y)^6}$$

27. $y = \dfrac{r}{\sqrt{r^2 + 1}}$ $\Rightarrow$

$$y' = \frac{\sqrt{r^2 + 1}\,(1) - r \cdot \tfrac{1}{2}(r^2 + 1)^{-1/2}(2r)}{(\sqrt{r^2 + 1})^2} = \frac{\sqrt{r^2 + 1} - \dfrac{r^2}{\sqrt{r^2 + 1}}}{(\sqrt{r^2 + 1})^2} = \frac{\dfrac{\sqrt{r^2 + 1}\sqrt{r^2 + 1} - r^2}{\sqrt{r^2 + 1}}}{(\sqrt{r^2 + 1})^2}$$

$$= \frac{(r^2 + 1) - r^2}{(\sqrt{r^2 + 1})^3} = \frac{1}{(r^2 + 1)^{3/2}} \text{ or } (r^2 + 1)^{-3/2}$$

Another solution: Write y as a product and make use of the Product Rule. $y = r(r^2 + 1)^{-1/2}$ $\Rightarrow$

$y' = r \cdot -\frac{1}{2}(r^2+1)^{-3/2}(2r) + (r^2+1)^{-1/2} \cdot 1 = (r^2+1)^{-3/2}[-r^2+(r^2+1)^1] = (r^2+1)^{-3/2}(1) = (r^2+1)^{-3/2}$.

The step that students usually have trouble with is factoring out $(r^2+1)^{-3/2}$. But this is no different than factoring out x^2 from x^2+x^5; that is, we are just factoring out a factor with the *smallest* exponent that appears on it. In this case, $-\frac{3}{2}$ is smaller than $-\frac{1}{2}$.

28. $y = \dfrac{\cos \pi x}{\sin \pi x + \cos \pi x}$ $\Rightarrow$

$$y' = \frac{(\sin \pi x + \cos \pi x)(-\pi \sin \pi x) - (\cos \pi x)(\pi \cos \pi x - \pi \sin \pi x)}{(\sin \pi x + \cos \pi x)^2}$$

$$= \frac{-\pi \sin^2 \pi x - \pi \sin \pi x \cos \pi x - \pi \cos^2 \pi x + \pi \sin \pi x \cos \pi x}{(\sin \pi x + \cos \pi x)^2}$$

$$= \frac{-\pi(\sin^2 \pi x + \cos^2 \pi x)}{(\sin \pi x + \cos \pi x)^2} = \frac{-\pi}{(\sin \pi x + \cos \pi x)^2} \quad \text{or} \quad \frac{-\pi}{1 + 2\sin \pi x \cos \pi x}$$

29. $y = \sin(\tan 2x)$ $\Rightarrow$ $y' = \cos(\tan 2x) \cdot \dfrac{d}{dx}(\tan 2x) = \cos(\tan 2x) \cdot \sec^2(2x) \cdot \dfrac{d}{dx}(2x) = 2\cos(\tan 2x)\sec^2(2x)$

30. $G(y) = \left(\dfrac{y^2}{y+1}\right)^5$ $\Rightarrow$ $G'(y) = 5\left(\dfrac{y^2}{y+1}\right)^4 \cdot \dfrac{(y+1)(2y) - y^2(1)}{(y+1)^2} = 5 \cdot \dfrac{y^8}{(y+1)^4} \cdot \dfrac{y(2y+2-y)}{(y+1)^2} = \dfrac{5y^9(y+2)}{(y+1)^6}$

31. $y = \sin\sqrt{1+x^2}$ $\Rightarrow$ $y' = \cos\sqrt{1+x^2} \cdot \frac{1}{2}(1+x^2)^{-1/2} \cdot 2x = \left(x \cos\sqrt{1+x^2}\right)/\sqrt{1+x^2}$

32. $y = \tan^2(3\theta) = (\tan 3\theta)^2$ $\Rightarrow$ $y' = 2(\tan 3\theta) \cdot \dfrac{d}{d\theta}(\tan 3\theta) = 2\tan 3\theta \cdot \sec^2 3\theta \cdot 3 = 6\tan 3\theta \sec^2 3\theta$

33. $y = \sec^2 x + \tan^2 x = (\sec x)^2 + (\tan x)^2$ $\Rightarrow$

$y' = 2(\sec x)(\sec x \tan x) + 2(\tan x)(\sec^2 x) = 2\sec^2 x \tan x + 2\sec^2 x \tan x = 4\sec^2 x \tan x$

34. $y = x \sin\dfrac{1}{x}$ $\Rightarrow$ $y' = \sin\dfrac{1}{x} + x\cos\dfrac{1}{x}\left(-\dfrac{1}{x^2}\right) = \sin\dfrac{1}{x} - \dfrac{1}{x}\cos\dfrac{1}{x}$

35. $y = \left(\dfrac{1 - \cos 2x}{1 + \cos 2x}\right)^4$ $\Rightarrow$

$$y' = 4\left(\frac{1-\cos 2x}{1+\cos 2x}\right)^3 \cdot \frac{(1+\cos 2x)(2\sin 2x) + (1-\cos 2x)(-2\sin 2x)}{(1+\cos 2x)^2}$$

$$= 4\left(\frac{1-\cos 2x}{1+\cos 2x}\right)^3 \cdot \frac{2\sin 2x\,(1+\cos 2x + 1 - \cos 2x)}{(1+\cos 2x)^2} = \frac{4(1-\cos 2x)^3}{(1+\cos 2x)^3}\frac{2\sin 2x\,(2)}{(1+\cos 2x)^2} = \frac{16\sin 2x\,(1-\cos 2x)^3}{(1+\cos 2x)^5}$$

36. $f(t) = \sqrt{\dfrac{t}{t^2+4}} = \left(\dfrac{t}{t^2+4}\right)^{1/2}$ $\Rightarrow$

$$f'(t) = \frac{1}{2}\left(\frac{t}{t^2+4}\right)^{-1/2} \cdot \frac{d}{dt}\left(\frac{t}{t^2+4}\right) = \frac{1}{2}\left(\frac{t^2+4}{t}\right)^{1/2} \cdot \frac{(t^2+4)(1) - t(2t)}{(t^2+4)^2}$$

$$= \frac{(t^2+4)^{1/2}}{2t^{1/2}} \cdot \frac{t^2+4-2t^2}{(t^2+4)^2} = \frac{4-t^2}{2t^{1/2}(t^2+4)^{3/2}}$$

37. $y = \cot^2(\sin\theta) = [\cot(\sin\theta)]^2 \quad\Rightarrow$

$y' = 2[\cot(\sin\theta)] \cdot \dfrac{d}{d\theta}[\cot(\sin\theta)] = 2\cot(\sin\theta) \cdot [-\csc^2(\sin\theta) \cdot \cos\theta] = -2\cos\theta\,\cot(\sin\theta)\,\csc^2(\sin\theta)$

38. $y = \left(ax + \sqrt{x^2 + b^2}\,\right)^{-2} \quad\Rightarrow\quad y' = -2\left(ax + \sqrt{x^2+b^2}\,\right)^{-3}\left(a + \tfrac{1}{2}(x^2+b^2)^{-1/2}(2x)\right) = \dfrac{-2\left(a + x/\sqrt{x^2+b^2}\,\right)}{\left(ax + \sqrt{x^2+b^2}\,\right)^3}$

39. $y = [x^2 + (1-3x)^5]^3 \quad\Rightarrow$

$y' = 3[x^2 + (1-3x)^5]^2(2x + 5(1-3x)^4(-3)) = 3[x^2 + (1-3x)^5]^2[2x - 15(1-3x)^4]$

40. $y = \sin(\sin(\sin x)) \quad\Rightarrow\quad y' = \cos(\sin(\sin x))\,\dfrac{d}{dx}(\sin(\sin x)) = \cos(\sin(\sin x))\,\cos(\sin x)\,\cos x$

41. $y = \sqrt{x + \sqrt{x}} \quad\Rightarrow\quad y' = \tfrac{1}{2}\left(x + \sqrt{x}\,\right)^{-1/2}\left(1 + \tfrac{1}{2}x^{-1/2}\right) = \dfrac{1}{2\sqrt{x + \sqrt{x}}}\left(1 + \dfrac{1}{2\sqrt{x}}\right)$

42. $y = \sqrt{x + \sqrt{x + \sqrt{x}}} \quad\Rightarrow\quad y' = \tfrac{1}{2}\left(x + \sqrt{x + \sqrt{x}}\,\right)^{-1/2}\left[1 + \tfrac{1}{2}\left(x + \sqrt{x}\,\right)^{-1/2}\left(1 + \tfrac{1}{2}x^{-1/2}\right)\right]$

43. $g(x) = (2r\sin rx + n)^p \quad\Rightarrow\quad g'(x) = p(2r\sin rx + n)^{p-1}(2r\cos rx \cdot r) = p(2r\sin rx + n)^{p-1}(2r^2\cos rx)$

44. $y = \cos^4(\sin^3 x) = [\cos(\sin^3 x)]^4 \quad\Rightarrow$

$y' = 4[\cos(\sin^3 x)]^3(-\sin(\sin^3 x))\,3\sin^2 x\,\cos x = -12\sin^2 x\,\cos x\,\cos^3(\sin^3 x)\,\sin(\sin^3 x)$

45. $y = \cos\sqrt{\sin(\tan\pi x)} = \cos(\sin(\tan\pi x))^{1/2} \quad\Rightarrow$

$y' = -\sin(\sin(\tan\pi x))^{1/2} \cdot \dfrac{d}{dx}(\sin(\tan\pi x))^{1/2} = -\sin(\sin(\tan\pi x))^{1/2} \cdot \tfrac{1}{2}(\sin(\tan\pi x))^{-1/2} \cdot \dfrac{d}{dx}(\sin(\tan\pi x))$

$= \dfrac{-\sin\sqrt{\sin(\tan\pi x)}}{2\sqrt{\sin(\tan\pi x)}} \cdot \cos(\tan\pi x) \cdot \dfrac{d}{dx}\tan\pi x = \dfrac{-\sin\sqrt{\sin(\tan\pi x)}}{2\sqrt{\sin(\tan\pi x)}} \cdot \cos(\tan\pi x) \cdot \sec^2(\pi x) \cdot \pi$

$= \dfrac{-\pi\cos(\tan\pi x)\,\sec^2(\pi x)\,\sin\sqrt{\sin(\tan\pi x)}}{2\sqrt{\sin(\tan\pi x)}}$

46. $y = \left[x + (x + \sin^2 x)^3\right]^4 \quad\Rightarrow\quad y' = 4\left[x + (x + \sin^2 x)^3\right]^3 \cdot \left[1 + 3(x + \sin^2 x)^2 \cdot (1 + 2\sin x\,\cos x)\right]$

47. $h(x) = \sqrt{x^2 + 1} \quad\Rightarrow\quad h'(x) = \dfrac{1}{2}(x^2+1)^{-1/2}(2x) = \dfrac{x}{\sqrt{x^2+1}} \quad\Rightarrow$

$h''(x) = \dfrac{\sqrt{x^2+1} \cdot 1 - x\left[\tfrac{1}{2}(x^2+1)^{-1/2}(2x)\right]}{\left(\sqrt{x^2+1}\,\right)^2} = \dfrac{(x^2+1)^{-1/2}\left[(x^2+1) - x^2\right]}{(x^2+1)^1} = \dfrac{1}{(x^2+1)^{3/2}}$

48. $y = \sin^2(\pi t) = [\sin(\pi t)]^2 \quad\Rightarrow\quad y' = 2[\sin(\pi t)]\cos(\pi t) \cdot \pi = \pi\sin(2\pi t) \quad\Rightarrow\quad y'' = \pi\cos(2\pi t) \cdot 2\pi = 2\pi^2\cos(2\pi t)$

49. $H(t) = \tan 3t \quad\Rightarrow\quad H'(t) = 3\sec^2 3t \quad\Rightarrow$

$H''(t) = 2 \cdot 3\sec 3t\,\dfrac{d}{dt}(\sec 3t) = 6\sec 3t\,(3\sec 3t\,\tan 3t) = 18\sec^2 3t\,\tan 3t$

50. $y = \dfrac{4x}{\sqrt{x+1}}$ $\Rightarrow$

$$y' = \frac{\sqrt{x+1} \cdot 4 - 4x \cdot \frac{1}{2}(x+1)^{-1/2}}{(\sqrt{x+1})^2} = \frac{4\sqrt{x+1} - 2x/\sqrt{x+1}}{x+1} = \frac{4(x+1) - 2x}{(x+1)^{3/2}} = \frac{2x+4}{(x+1)^{3/2}} \Rightarrow$$

$$y'' = \frac{(x+1)^{3/2} \cdot 2 - (2x+4) \cdot \frac{3}{2}(x+1)^{1/2}}{[(x+1)^{3/2}]^2} = \frac{(x+1)^{1/2}[2(x+1) - 3(x+2)]}{(x+1)^3} = \frac{2x+2-3x-6}{(x+1)^{5/2}} = \frac{-x-4}{(x+1)^{5/2}}$$

51. $y = (1+2x)^{10}$ $\Rightarrow$ $y' = 10(1+2x)^9 \cdot 2 = 20(1+2x)^9$.

At $(0, 1)$, $y' = 20(1+0)^9 = 20$, and an equation of the tangent line is $y - 1 = 20(x - 0)$, or $y = 20x + 1$.

52. $y = \sin x + \sin^2 x$ $\Rightarrow$ $y' = \cos x + 2 \sin x \cos x$.

At $(0, 0)$, $y' = 1$, and an equation of the tangent line is $y - 0 = 1(x - 0)$, or $y = x$.

53. $y = \sin(\sin x)$ $\Rightarrow$ $y' = \cos(\sin x) \cdot \cos x$. At $(\pi, 0)$, $y' = \cos(\sin \pi) \cdot \cos \pi = \cos(0) \cdot (-1) = 1(-1) = -1$, and an equation of the tangent line is $y - 0 = -1(x - \pi)$, or $y = -x + \pi$.

54. $y = \sqrt{5 + x^2}$ $\Rightarrow$ $y' = \frac{1}{2}(5 + x^2)^{-1/2}(2x) = x/\sqrt{5 + x^2}$. At $(2, 3)$, $y' = \frac{2}{3}$, and an equation of the tangent line is $y - 3 = \frac{2}{3}(x - 2)$, or $y = \frac{2}{3}x + \frac{5}{3}$.

55. (a) $y = f(x) = \tan\left(\frac{\pi}{4}x^2\right)$ $\Rightarrow$ $f'(x) = \sec^2\left(\frac{\pi}{4}x^2\right)\left(2 \cdot \frac{\pi}{4}x\right)$.

The slope of the tangent at $(1, 1)$ is thus

$f'(1) = \sec^2 \frac{\pi}{4}\left(\frac{\pi}{2}\right) = 2 \cdot \frac{\pi}{2} = \pi$, and its equation

is $y - 1 = \pi(x - 1)$ or $y = \pi x - \pi + 1$.

(b)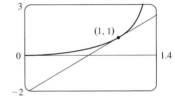

56. (a) For $x > 0$, $|x| = x$, and $y = f(x) = \dfrac{x}{\sqrt{2 - x^2}}$ $\Rightarrow$

$$f'(x) = \frac{\sqrt{2 - x^2}(1) - x\left(\frac{1}{2}\right)(2 - x^2)^{-1/2}(-2x)}{(\sqrt{2 - x^2})^2} \cdot \frac{(2 - x^2)^{1/2}}{(2 - x^2)^{1/2}}$$

$$= \frac{(2 - x^2) + x^2}{(2 - x^2)^{3/2}} = \frac{2}{(2 - x^2)^{3/2}}$$

(b)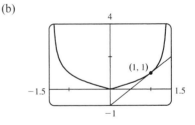

So at $(1, 1)$, the slope of the tangent line is $f'(1) = 2$ and its equation

is $y - 1 = 2(x - 1)$ or $y = 2x - 1$.

57. (a) $f(x) = x\sqrt{2 - x^2} = x(2 - x^2)^{1/2}$ $\Rightarrow$

$$f'(x) = x \cdot \frac{1}{2}(2 - x^2)^{-1/2}(-2x) + (2 - x^2)^{1/2} \cdot 1 = (2 - x^2)^{-1/2}\left[-x^2 + (2 - x^2)\right] = \frac{2 - 2x^2}{\sqrt{2 - x^2}}$$

(b)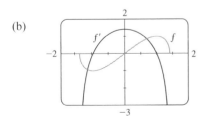

$f' = 0$ when f has a horizontal tangent line, f' is negative when f is decreasing, and f' is positive when f is increasing.

58. (a)

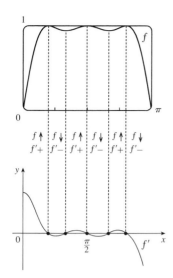

From the graph of f, we see that there are 5 horizontal tangents, so there must be 5 zeros on the graph of f'. From the symmetry of the graph of f, we must have the graph of f' as high at $x = 0$ as it is low at $x = \pi$. The intervals of increase and decrease as well as the signs of f' are indicated in the figure.

(b) $f(x) = \sin(x + \sin 2x) \quad \Rightarrow$

$$f'(x) = \cos(x + \sin 2x) \cdot \frac{d}{dx}\,(x + \sin 2x) = \cos(x + \sin 2x)(1 + 2\cos 2x)$$

59. For the tangent line to be horizontal, $f'(x) = 0$. $f(x) = 2\sin x + \sin^2 x \quad \Rightarrow \quad f'(x) = 2\cos x + 2\sin x \cos x = 0 \quad \Leftrightarrow$ $2\cos x(1 + \sin x) = 0 \quad \Leftrightarrow \quad \cos x = 0$ or $\sin x = -1$, so $x = \frac{\pi}{2} + 2n\pi$ or $\frac{3\pi}{2} + 2n\pi$, where n is any integer. Now $f\left(\frac{\pi}{2}\right) = 3$ and $f\left(\frac{3\pi}{2}\right) = -1$, so the points on the curve with a horizontal tangent are $\left(\frac{\pi}{2} + 2n\pi, 3\right)$ and $\left(\frac{3\pi}{2} + 2n\pi, -1\right)$, where n is any integer.

60. $f(x) = \sin 2x - 2\sin x \quad \Rightarrow \quad f'(x) = 2\cos 2x - 2\cos x = 4\cos^2 x - 2\cos x - 2$, and $4\cos^2 x - 2\cos x - 2 = 0 \quad \Leftrightarrow$ $(\cos x - 1)(4\cos x + 2) = 0 \quad \Leftrightarrow \quad \cos x = 1$ or $\cos x = -\frac{1}{2}$. So $x = 2n\pi$ or $(2n+1)\pi \pm \frac{\pi}{3}$, n any integer.

61. $F(x) = f(g(x)) \quad \Rightarrow \quad F'(x) = f'(g(x)) \cdot g'(x)$, so $F'(5) = f'(g(5)) \cdot g'(5) = f'(-2) \cdot 6 = 4 \cdot 6 = 24$

62. $h(x) = \sqrt{4 + 3f(x)} \quad \Rightarrow \quad h'(x) = \frac{1}{2}(4 + 3f(x))^{-1/2} \cdot 3f'(x)$, so
$h'(1) = \frac{1}{2}(4 + 3f(1))^{-1/2} \cdot 3f'(1) = \frac{1}{2}(4 + 3 \cdot 7)^{-1/2} \cdot 3 \cdot 4 = \frac{6}{\sqrt{25}} = \frac{6}{5}$

63. (a) $h(x) = f(g(x)) \quad \Rightarrow \quad h'(x) = f'(g(x)) \cdot g'(x)$, so $h'(1) = f'(g(1)) \cdot g'(1) = f'(2) \cdot 6 = 5 \cdot 6 = 30$.

(b) $H(x) = g(f(x)) \quad \Rightarrow \quad H'(x) = g'(f(x)) \cdot f'(x)$, so $H'(1) = g'(f(1)) \cdot f'(1) = g'(3) \cdot 4 = 9 \cdot 4 = 36$.

64. (a) $F(x) = f(f(x)) \quad \Rightarrow \quad F'(x) = f'(f(x)) \cdot f'(x)$, so $F'(2) = f'(f(2)) \cdot f'(2) = f'(1) \cdot 5 = 4 \cdot 5 = 20$.

(b) $G(x) = g(g(x)) \quad \Rightarrow \quad G'(x) = g'(g(x)) \cdot g'(x)$, so $G'(3) = g'(g(3)) \cdot g'(3) = g'(2) \cdot 9 = 7 \cdot 9 = 63$.

65. (a) $u(x) = f(g(x)) \quad \Rightarrow \quad u'(x) = f'(g(x))g'(x)$. So $u'(1) = f'(g(1))g'(1) = f'(3)g'(1)$. To find $f'(3)$, note that f is linear from $(2, 4)$ to $(6, 3)$, so its slope is $\dfrac{3 - 4}{6 - 2} = -\dfrac{1}{4}$. To find $g'(1)$, note that g is linear from $(0, 6)$ to $(2, 0)$, so its slope is $\dfrac{0 - 6}{2 - 0} = -3$. Thus, $f'(3)g'(1) = \left(-\frac{1}{4}\right)(-3) = \frac{3}{4}$.

(b) $v(x) = g(f(x))$ $\Rightarrow$ $v'(x) = g'(f(x))f'(x)$.   So $v'(1) = g'(f(1))f'(1) = g'(2)f'(1)$, which does not exist since

$g'(2)$ does not exist.

(c) $w(x) = g(g(x))$ $\Rightarrow$ $w'(x) = g'(g(x))g'(x)$.   So $w'(1) = g'(g(1))g'(1) = g'(3)g'(1)$. To find $g'(3)$, note that g is

linear from $(2, 0)$ to $(5, 2)$, so its slope is $\dfrac{2 - 0}{5 - 2} = \dfrac{2}{3}$. Thus, $g'(3)g'(1) = \left(\frac{2}{3}\right)(-3) = -2$.

66. (a) $h(x) = f(f(x))$ $\Rightarrow$ $h'(x) = f'(f(x))f'(x)$.   So $h'(2) = f'(f(2))f'(2) = f'(1)f'(2) \approx (-1)(-1) = 1$.

(b) $g(x) = f(x^2)$ $\Rightarrow$ $g'(x) = f'(x^2) \cdot \dfrac{d}{dx}\left(x^2\right) = f'(x^2)(2x)$.   So $g'(2) = f'(2^2)(2 \cdot 2) = 4f'(4) \approx 4(2) = 8$.

67. (a) $F(x) = f(\cos x)$ $\Rightarrow$ $F'(x) = f'(\cos x)\dfrac{d}{dx}(\cos x) = -\sin x f'(\cos x)$

(b) $G(x) = \cos(f(x))$ $\Rightarrow$ $G'(x) = -\sin(f(x))\, f'(x)$

68. (a) $F(x) = f(x^\alpha)$ $\Rightarrow$ $F'(x) = f'(x^\alpha)\dfrac{d}{dx}\left(x^\alpha\right) = f'(x^\alpha)\alpha x^{\alpha - 1}$

(b) $G(x) = [f(x)]^\alpha$ $\Rightarrow$ $G'(x) = \alpha\,[f(x)]^{\alpha - 1}\, f'(x)$

69. $r(x) = f(g(h(x)))$ $\Rightarrow$ $r'(x) = f'(g(h(x))) \cdot g'(h(x)) \cdot h'(x)$, so

$r'(1) = f'(g(h(1))) \cdot g'(h(1)) \cdot h'(1) = f'(g(2)) \cdot g'(2) \cdot 4 = f'(3) \cdot 5 \cdot 4 = 6 \cdot 5 \cdot 4 = 120$

70. $f(x) = xg(x^2)$ $\Rightarrow$ $f'(x) = xg'(x^2)\,2x + g(x^2) \cdot 1 = 2x^2 g'(x^2) + g(x^2)$ $\Rightarrow$

$f''(x) = 2x^2 g''(x^2)\,2x + g'(x^2)\,4x + g'(x^2)\,2x = 4x^3 g''(x^2) + 4xg'(x^2) + 2xg'(x^2) = 6xg'(x^2) + 4x^3 g''(x^2)$

71. $F(x) = f(3f(4f(x)))$ $\Rightarrow$

$F'(x) = f'(3f(4f(x))) \cdot \dfrac{d}{dx}(3f(4f(x))) = f'(3f(4f(x))) \cdot 3f'(4f(x)) \cdot \dfrac{d}{dx}(4f(x))$

$= f'(3f(4f(x))) \cdot 3f'(4f(x)) \cdot 4f'(x)$,   so

$F'(0) = f'(3f(4f(0))) \cdot 3f'(4f(0)) \cdot 4f'(0) = f'(3f(4 \cdot 0)) \cdot 3f'(4 \cdot 0) \cdot 4 \cdot 2 = f'(3 \cdot 0) \cdot 3 \cdot 2 \cdot 4 \cdot 2 = 2 \cdot 3 \cdot 2 \cdot 4 \cdot 2 = 96$.

72. $F(x) = f(xf(xf(x)))$ $\Rightarrow$

$F'(x) = f'(xf(xf(x))) \cdot \dfrac{d}{dx}\left(xf(xf(x))\right) = f'(xf(xf(x))) \cdot \left[x \cdot f'(xf(x)) \cdot \dfrac{d}{dx}(xf(x)) + f(xf(x)) \cdot 1\right]$

$= f'(xf(xf(x))) \cdot [xf'(xf(x)) \cdot (xf'(x) + f(x) \cdot 1) + f(xf(x))]$,   so

$F'(1) = f'(f(f(1))) \cdot [f'(f(1)) \cdot (f'(1) + f(1)) + f(f(1))] = f'(f(2)) \cdot [f'(2) \cdot (4 + 2) + f(2)]$

$= f'(3) \cdot [5 \cdot 6 + 3] = 6 \cdot 33 = 198$.

73. Let $f(x) = \cos x$. Then $Df(2x) = 2f'(2x)$, $D^2 f(2x) = 2^2 f''(2x)$, $D^3 f(2x) = 2^3 f'''(2x)$, ...,

$D^{(n)} f(2x) = 2^n f^{(n)}(2x)$. Since the derivatives of $\cos x$ occur in a cycle of four, and since $103 = 4(25) + 3$, we have

$f^{(103)}(x) = f^{(3)}(x) = \sin x$ and $D^{103} \cos 2x = 2^{103} f^{(103)}(2x) = 2^{103} \sin 2x$.

74. Let $f(x) = x \sin x$ and $h(x) = \sin x$, so $f(x) = xh(x)$. Then $f'(x) = h(x) + xh'(x)$,

$f''(x) = h'(x) + h'(x) + xh''(x) = 2h'(x) + xh''(x)$, $f'''(x) = 2h''(x) + h''(x) + xh'''(x) = 3h''(x) + xh'''(x)$, ...,

$f^{(n)}(x) = nh^{(n-1)}(x) + xh^{(n)}(x)$. Since $34 = 4(8) + 2$, we have $h^{(34)}(x) = h^{(2)}(x) = D^2 \sin x = -\sin x$ and $h^{(35)}(x) = -\cos x$. Thus, $D^{(35)} x \sin x = 35h^{(34)}(x) + xh^{(35)}(x) = -35 \sin x - x \cos x$.

75. $s(t) = 10 + \frac{1}{4} \sin(10\pi t) \quad \Rightarrow \quad$ the velocity after t seconds is $v(t) = s'(t) = \frac{1}{4} \cos(10\pi t)(10\pi) = \frac{5\pi}{2} \cos(10\pi t)$ cm/s.

76. (a) $s = A\cos(\omega t + \delta) \quad \Rightarrow \quad$ velocity $= s' = -\omega A \sin(\omega t + \delta)$.

(b) If $A \neq 0$ and $\omega \neq 0$, then $s' = 0 \quad \Leftrightarrow \quad \sin(\omega t + \delta) = 0 \quad \Leftrightarrow \quad \omega t + \delta = n\pi \quad \Leftrightarrow \quad t = \dfrac{n\pi - \delta}{\omega}$, n an integer.

77. (a) $B(t) = 4.0 + 0.35 \sin \dfrac{2\pi t}{5.4} \quad \Rightarrow \quad \dfrac{dB}{dt} = \left(0.35 \cos \dfrac{2\pi t}{5.4}\right)\left(\dfrac{2\pi}{5.4}\right) = \dfrac{0.7\pi}{5.4} \cos \dfrac{2\pi t}{5.4} = \dfrac{7\pi}{54} \cos \dfrac{2\pi t}{5.4}$

(b) At $t = 1$, $\dfrac{dB}{dt} = \dfrac{7\pi}{54} \cos \dfrac{2\pi}{5.4} \approx 0.16$.

78. $L(t) = 12 + 2.8 \sin\left(\frac{2\pi}{365}(t - 80)\right) \quad \Rightarrow \quad L'(t) = 2.8 \cos\left(\frac{2\pi}{365}(t - 80)\right)\left(\frac{2\pi}{365}\right)$.

On March 21, $t = 80$, and $L'(80) \approx 0.0482$ hours per day. On May 21, $t = 141$, and $L'(141) \approx 0.02398$, which is approximately one-half of $L'(80)$.

79. By the Chain Rule, $a(t) = \dfrac{dv}{dt} = \dfrac{dv}{ds} \dfrac{ds}{dt} = \dfrac{dv}{ds} v(t) = v(t) \dfrac{dv}{ds}$. The derivative dv/dt is the rate of change of the velocity with respect to time (in other words, the acceleration) whereas the derivative dv/ds is the rate of change of the velocity with respect to the displacement.

80. (a) The derivative dV/dr represents the rate of change of the volume with respect to the radius and the derivative dV/dt represents the rate of change of the volume with respect to time.

(b) Since $V = \dfrac{4}{3}\pi r^3$, $\dfrac{dV}{dt} = \dfrac{dV}{dr}\dfrac{dr}{dt} = 4\pi r^2 \dfrac{dr}{dt}$.

81. (a) Derive gives $g'(t) = \dfrac{45(t - 2)^8}{(2t + 1)^{10}}$ without simplifying. With either Maple or Mathematica, we first get

$g'(t) = 9\dfrac{(t - 2)^8}{(2t + 1)^9} - 18\dfrac{(t - 2)^9}{(2t + 1)^{10}}$, and the simplification command results in the expression given by Derive.

(b) Derive gives $y' = 2(x^3 - x + 1)^3(2x + 1)^4(17x^3 + 6x^2 - 9x + 3)$ without simplifying. With either Maple or Mathematica, we first get $y' = 10(2x + 1)^4(x^3 - x + 1)^4 + 4(2x + 1)^5(x^3 - x + 1)^3(3x^2 - 1)$. If we use Mathematica's Factor or Simplify, or Maple's factor, we get the above expression, but Maple's simplify gives the polynomial expansion instead. For locating horizontal tangents, the factored form is the most helpful.

82. (a) $f(x) = \left(\dfrac{x^4 - x + 1}{x^4 + x + 1}\right)^{1/2}$. Derive gives $f'(x) = \dfrac{(3x^4 - 1)\sqrt{\dfrac{x^4 - x + 1}{x^4 + x + 1}}}{(x^4 + x + 1)(x^4 - x + 1)}$ whereas either Maple or Mathematica

give $f'(x) = \dfrac{3x^4 - 1}{\sqrt{\dfrac{x^4 - x + 1}{x^4 + x + 1}}\,(x^4 + x + 1)^2}$ after simplification.

(b) $f'(x) = 0 \quad \Leftrightarrow \quad 3x^4 - 1 = 0 \quad \Leftrightarrow \quad x = \pm\sqrt[4]{\frac{1}{3}} \approx \pm 0.7598$.

(c) Yes. $f'(x) = 0$ where f has horizontal tangents. f' has two maxima and one minimum where f has inflection points.

83. (a) If f is even, then $f(x) = f(-x)$. Using the Chain Rule to differentiate this equation, we get

$f'(x) = f'(-x)\dfrac{d}{dx}(-x) = -f'(-x)$. Thus, $f'(-x) = -f'(x)$, so f' is odd.

(b) If f is odd, then $f(x) = -f(-x)$.   Differentiating this equation, we get $f'(x) = -f'(-x)(-1) = f'(-x)$, so f' is even.

84. $\left[\dfrac{f(x)}{g(x)}\right]' = \left\{f(x)\left[g(x)\right]^{-1}\right\}' = f'(x)\left[g(x)\right]^{-1} + (-1)\left[g(x)\right]^{-2}g'(x)f(x)$

$= \dfrac{f'(x)}{g(x)} - \dfrac{f(x)g'(x)}{[g(x)]^2} = \dfrac{f'(x)g(x) - f(x)g'(x)}{[g(x)]^2}$

This is an alternative derivation of the *formula* in the Quotient Rule. But part of the purpose of the Quotient Rule is to show that if f and g are differentiable, so is f/g. The proof in Section 3.2 does that; this one doesn't.

85. (a) $\dfrac{d}{dx}\left(\sin^n x \cos nx\right) = n\sin^{n-1} x \cos x \cos nx + \sin^n x\left(-n\sin nx\right)$ [Product Rule]

$= n\sin^{n-1} x \left(\cos nx \cos x - \sin nx \sin x\right)$ [factor out $n\sin^{n-1} x$]

$= n\sin^{n-1} x \cos(nx + x)$ [Addition Formula for cosine]

$= n\sin^{n-1} x \cos[(n+1)x]$ [factor out x]

(b) $\dfrac{d}{dx}\left(\cos^n x \cos nx\right) = n\cos^{n-1} x\left(-\sin x\right)\cos nx + \cos^n x\left(-n\sin nx\right)$ [Product Rule]

$= -n\cos^{n-1} x \left(\cos nx \sin x + \sin nx \cos x\right)$ [factor out $-n\cos^{n-1} x$]

$= -n\cos^{n-1} x \sin(nx + x)$ [Addition Formula for sine]

$= -n\cos^{n-1} x \sin[(n+1)x]$ [factor out x]

86. "The rate of change of y^5 with respect to x is eighty times the rate of change of y with respect to x"   ⇔

$\dfrac{d}{dx}y^5 = 80\dfrac{dy}{dx}$ ⇔ $5y^4\dfrac{dy}{dx} = 80\dfrac{dy}{dx}$ ⇔ $5y^4 = 80$ (Note that $dy/dx \neq 0$ since the curve never has a

horizontal tangent)   ⇔ $y^4 = 16$ ⇔ $y = 2$ (since $y > 0$ for all x)

87. Since $\theta° = \left(\dfrac{\pi}{180}\right)\theta$ rad, we have $\dfrac{d}{d\theta}\left(\sin\theta°\right) = \dfrac{d}{d\theta}\left(\sin\dfrac{\pi}{180}\theta\right) = \dfrac{\pi}{180}\cos\dfrac{\pi}{180}\theta = \dfrac{\pi}{180}\cos\theta°$.

88. (a) $f(x) = |x| = \sqrt{x^2} = (x^2)^{1/2}$ ⇒ $f'(x) = \frac{1}{2}(x^2)^{-1/2}(2x) = x/\sqrt{x^2} = x/|x|$ for $x \neq 0$.

f is not differentiable at $x = 0$.

(b) $f(x) = |\sin x| = \sqrt{\sin^2 x}$ ⇒

$f'(x) = \frac{1}{2}(\sin^2 x)^{-1/2}2\sin x \cos x = \dfrac{\sin x}{|\sin x|}\cos x$

$= \begin{cases} \cos x & \text{if } \sin x > 0 \\ -\cos x & \text{if } \sin x < 0 \end{cases}$

f is not differentiable when $x = n\pi$, n an integer.

(c) $g(x) = \sin |x| = \sin \sqrt{x^2}$ $\Rightarrow$

$$g'(x) = \cos |x| \cdot \frac{x}{|x|} = \frac{x}{|x|} \cos x = \begin{cases} \cos x & \text{if } x > 0 \\ -\cos x & \text{if } x < 0 \end{cases}$$

g is not differentiable at 0.

89. The Chain Rule says that $\dfrac{dy}{dx} = \dfrac{dy}{du}\dfrac{du}{dx}$, so

$$\frac{d^2y}{dx^2} = \frac{d}{dx}\left(\frac{dy}{dx}\right) = \frac{d}{dx}\left(\frac{dy}{du}\frac{du}{dx}\right) = \left[\frac{d}{dx}\left(\frac{dy}{du}\right)\right]\frac{du}{dx} + \frac{dy}{du}\frac{d}{dx}\left(\frac{du}{dx}\right) \quad \text{[Product Rule]}$$

$$= \left[\frac{d}{du}\left(\frac{dy}{du}\right)\frac{du}{dx}\right]\frac{du}{dx} + \frac{dy}{du}\frac{d^2u}{dx^2} = \frac{d^2y}{du^2}\left(\frac{du}{dx}\right)^2 + \frac{dy}{du}\frac{d^2u}{dx^2}$$

90. From Exercise 89, $\dfrac{d^2y}{dx^2} = \dfrac{d^2y}{du^2}\left(\dfrac{du}{dx}\right)^2 + \dfrac{dy}{du}\dfrac{d^2u}{dx^2}$ $\Rightarrow$

$$\frac{d^3y}{dx^3} = \frac{d}{dx}\frac{d^2y}{dx^2} = \frac{d}{dx}\left[\frac{d^2y}{du^2}\left(\frac{du}{dx}\right)^2\right] + \frac{d}{dx}\left[\frac{dy}{du}\frac{d^2u}{dx^2}\right]$$

$$= \left[\frac{d}{dx}\left(\frac{d^2y}{du^2}\right)\right]\left(\frac{du}{dx}\right)^2 + \left[\frac{d}{dx}\left(\frac{du}{dx}\right)^2\right]\frac{d^2y}{du^2} + \left[\frac{d}{dx}\left(\frac{dy}{du}\right)\right]\frac{d^2u}{dx^2} + \left[\frac{d}{dx}\left(\frac{d^2u}{dx^2}\right)\right]\frac{dy}{du}$$

$$= \left[\frac{d}{du}\left(\frac{d^2y}{du^2}\right)\frac{du}{dx}\right]\left(\frac{du}{dx}\right)^2 + 2\frac{du}{dx}\frac{d^2u}{dx^2}\frac{d^2y}{du^2} + \left[\frac{d}{du}\left(\frac{dy}{du}\right)\frac{du}{dx}\right]\left(\frac{d^2u}{dx^2}\right) + \frac{d^3u}{dx^3}\frac{dy}{du}$$

$$= \frac{d^3y}{du^3}\left(\frac{du}{dx}\right)^3 + 3\frac{du}{dx}\frac{d^2u}{dx^2}\frac{d^2y}{du^2} + \frac{dy}{du}\frac{d^3u}{dx^3}$$

APPLIED PROJECT   Where Should a Pilot Start Descent?

1. Condition (i) will hold if and only if all of the following four conditions hold:

(α) $P(0) = 0$

(β) $P'(0) = 0$ (for a smooth landing)

(γ) $P'(\ell) = 0$ (since the plane is cruising horizontally when it begins its descent)

(δ) $P(\ell) = h$.

First of all, condition α implies that $P(0) = d = 0$, so $P(x) = ax^3 + bx^2 + cx$ $\Rightarrow$ $P'(x) = 3ax^2 + 2bx + c$. But

$P'(0) = c = 0$ by condition β. So $P'(\ell) = 3a\ell^2 + 2b\ell = \ell(3a\ell + 2b)$. Now by condition γ, $3a\ell + 2b = 0$ $\Rightarrow$ $a = -\dfrac{2b}{3\ell}$.

Therefore, $P(x) = -\dfrac{2b}{3\ell}x^3 + bx^2$. Setting $P(\ell) = h$ for condition δ, we get $P(\ell) = -\dfrac{2b}{3\ell}\ell^3 + b\ell^2 = h$ $\Rightarrow$

$-\dfrac{2}{3}b\ell^2 + b\ell^2 = h$ $\Rightarrow$ $\dfrac{1}{3}b\ell^2 = h$ $\Rightarrow$ $b = \dfrac{3h}{\ell^2}$ $\Rightarrow$ $a = -\dfrac{2h}{\ell^3}$. So $y = P(x) = -\dfrac{2h}{\ell^3}x^3 + \dfrac{3h}{\ell^2}x^2$.

2. By condition (ii), $\dfrac{dx}{dt} = -v$ for all t, so $x(t) = \ell - vt$. Condition (iii) states that $\left|\dfrac{d^2y}{dt^2}\right| \le k$. By the Chain Rule,

we have $\dfrac{dy}{dt} = \dfrac{dy}{dx}\dfrac{dx}{dt} = -\dfrac{2h}{\ell^3}\left(3x^2\right)\dfrac{dx}{dt} + \dfrac{3h}{\ell^2}\left(2x\right)\dfrac{dx}{dt} = \dfrac{6hx^2v}{\ell^3} - \dfrac{6hxv}{\ell^2}$ (for $x \le \ell$) $\Rightarrow$

$\dfrac{d^2y}{dt^2} = \dfrac{6hv}{\ell^3}\left(2x\right)\dfrac{dx}{dt} - \dfrac{6hv}{\ell^2}\dfrac{dx}{dt} = -\dfrac{12hv^2}{\ell^3}x + \dfrac{6hv^2}{\ell^2}$. In particular, when $t = 0$, $x = \ell$ and so

$\dfrac{d^2y}{dt^2}\bigg|_{t=0} = -\dfrac{12hv^2}{\ell^3}\ell + \dfrac{6hv^2}{\ell^2} = -\dfrac{6hv^2}{\ell^2}$. Thus, $\left|\dfrac{d^2y}{dt^2}\right|_{t=0} = \dfrac{6hv^2}{\ell^2} \le k$. (This condition also follows from taking $x = 0$.)

3. We substitute $k = 860$ mi/h^2, $h = 35{,}000$ ft $\times \dfrac{1\text{ mi}}{5280\text{ ft}}$, and $v = 300$ mi/h into the result of part (b):

$\dfrac{6\left(35{,}000 \cdot \frac{1}{5280}\right)(300)^2}{\ell^2} \le 860 \quad \Rightarrow \quad \ell \ge 300\sqrt{6 \cdot \dfrac{35{,}000}{5280 \cdot 860}} \approx 64.5$ miles.

4. Substituting the values of h and ℓ in Problem 3 into

$P(x) = -\dfrac{2h}{\ell^3}x^3 + \dfrac{3h}{\ell^2}x^2$ gives us $P(x) = ax^3 + bx^2$,

where $a \approx -4.937 \times 10^{-5}$ and $b \approx 4.78 \times 10^{-3}$.

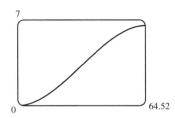

3.6 Implicit Differentiation

1. (a) $\dfrac{d}{dx}\left(xy + 2x + 3x^2\right) = \dfrac{d}{dx}(4) \quad \Rightarrow \quad \left(x \cdot y' + y \cdot 1\right) + 2 + 6x = 0 \quad \Rightarrow \quad xy' = -y - 2 - 6x \quad \Rightarrow$

$y' = \dfrac{-y - 2 - 6x}{x}$ or $y' = -6 - \dfrac{y+2}{x}$.

(b) $xy + 2x + 3x^2 = 4 \quad \Rightarrow \quad xy = 4 - 2x - 3x^2 \quad \Rightarrow \quad y = \dfrac{4 - 2x - 3x^2}{x} = \dfrac{4}{x} - 2 - 3x$, so $y' = -\dfrac{4}{x^2} - 3$.

(c) From part (a), $y' = \dfrac{-y - 2 - 6x}{x} = \dfrac{-(4/x - 2 - 3x) - 2 - 6x}{x} = \dfrac{-4/x - 3x}{x} = -\dfrac{4}{x^2} - 3$.

2. (a) $\dfrac{d}{dx}\left(4x^2 + 9y^2\right) = \dfrac{d}{dx}(36) \quad \Rightarrow \quad 8x + 18y \cdot y' = 0 \quad \Rightarrow \quad y' = -\dfrac{8x}{18y} = -\dfrac{4x}{9y}$

(b) $4x^2 + 9y^2 = 36 \quad \Rightarrow \quad 9y^2 = 36 - 4x^2 \quad \Rightarrow \quad y^2 = \frac{4}{9}(9 - x^2) \quad \Rightarrow \quad y = \pm\frac{2}{3}\sqrt{9 - x^2}$, so

$y' = \pm\frac{2}{3} \cdot \frac{1}{2}(9 - x^2)^{-1/2}(-2x) = \mp\dfrac{2x}{3\sqrt{9 - x^2}}$

(c) From part (a), $y' = -\dfrac{4x}{9y} = -\dfrac{4x}{9\left(\pm\frac{2}{3}\sqrt{9 - x^2}\right)} = \mp\dfrac{2x}{3\sqrt{9 - x^2}}$.

3. (a) $\dfrac{d}{dx}\left(\dfrac{1}{x} + \dfrac{1}{y}\right) = \dfrac{d}{dx}(1) \quad \Rightarrow \quad -\dfrac{1}{x^2} - \dfrac{1}{y^2}y' = 0 \quad \Rightarrow \quad -\dfrac{1}{y^2}y' = \dfrac{1}{x^2} \quad \Rightarrow \quad y' = -\dfrac{y^2}{x^2}$

(b) $\dfrac{1}{x} + \dfrac{1}{y} = 1 \quad \Rightarrow \quad \dfrac{1}{y} = 1 - \dfrac{1}{x} = \dfrac{x - 1}{x} \quad \Rightarrow \quad y = \dfrac{x}{x - 1}$, so $y' = \dfrac{(x - 1)(1) - (x)(1)}{(x - 1)^2} = \dfrac{-1}{(x - 1)^2}$.

(c) $y' = -\dfrac{y^2}{x^2} = -\dfrac{[x/(x - 1)]^2}{x^2} = -\dfrac{x^2}{x^2(x - 1)^2} = -\dfrac{1}{(x - 1)^2}$

4. (a) $\dfrac{d}{dx}\left(\cos x + \sqrt{y}\right) = \dfrac{d}{dx}(5) \quad \Rightarrow \quad -\sin x + \frac{1}{2}y^{-1/2} \cdot y' = 0 \quad \Rightarrow \quad \dfrac{1}{2\sqrt{y}} \cdot y' = \sin x \quad \Rightarrow \quad y' = 2\sqrt{y}\sin x$

(b) $\cos x + \sqrt{y} = 5 \;\Rightarrow\; \sqrt{y} = 5 - \cos x \;\Rightarrow\; y = (5 - \cos x)^2$, so $y' = 2(5 - \cos x)'(\sin x) = 2\sin x(5 - \cos x)$.

(c) From part (a), $y' = 2\sqrt{y}\sin x = 2\sqrt{(5 - \cos x)^2} = 2(5 - \cos x)\sin x$ [since $5 - \cos x > 0$].

5. $\dfrac{d}{dx}\left(x^3 + y^3\right) = \dfrac{d}{dx}\,(1) \;\Rightarrow\; 3x^2 + 3y^2 \cdot y' = 0 \;\Rightarrow\; 3y^2\,y' = -3x^2 \;\Rightarrow\; y' = -\dfrac{x^2}{y^2}$

6. $\dfrac{d}{dx}\left(2\sqrt{x} + \sqrt{y}\right) = \dfrac{d}{dx}\,(3) \;\Rightarrow\; 2 \cdot \dfrac{1}{2}x^{-1/2} + \dfrac{1}{2}y^{-1/2} \cdot y' = 0 \;\Rightarrow\; \dfrac{1}{\sqrt{x}} + \dfrac{y'}{2\sqrt{y}} = 0 \;\Rightarrow\;$

$\dfrac{y'}{2\sqrt{y}} = -\dfrac{1}{\sqrt{x}} \;\Rightarrow\; y' = -\dfrac{2\sqrt{y}}{\sqrt{x}}$

7. $\dfrac{d}{dx}\left(x^2 + xy - y^2\right) = \dfrac{d}{dx}\,(4) \;\Rightarrow\; 2x + x \cdot y' + y \cdot 1 - 2y\,y' = 0 \;\Rightarrow\;$

$xy' - 2y\,y' = -2x - y \;\Rightarrow\; (x - 2y)\,y' = -2x - y \;\Rightarrow\; y' = \dfrac{-2x - y}{x - 2y} = \dfrac{2x + y}{2y - x}$

8. $\dfrac{d}{dx}\left(2x^3 + x^2y - xy^3\right) = \dfrac{d}{dx}\,(2) \;\Rightarrow\; 6x^2 + x^2 \cdot y' + y \cdot 2x - (x \cdot 3y^2y' + y^3 \cdot 1) = 0 \;\Rightarrow\;$

$x^2\,y' - 3xy^2\,y' = -6x^2 - 2xy + y^3 \;\Rightarrow\; (x^2 - 3xy^2)\,y' = -6x^2 - 2xy + y^3 \;\Rightarrow\; y' = \dfrac{-6x^2 - 2xy + y^3}{x^2 - 3xy^2}$

9. $\dfrac{d}{dx}\left[x^4(x + y)\right] = \dfrac{d}{dx}\left[y^2(3x - y)\right] \;\Rightarrow\; x^4(1 + y') + (x + y) \cdot 4x^3 = y^2(3 - y') + (3x - y) \cdot 2y\,y' \;\Rightarrow\;$

$x^4 + x^4\,y' + 4x^4 + 4x^3y = 3y^2 - y^2\,y' + 6xy\,y' - 2y^2\,y' \;\Rightarrow\; x^4\,y' + 3y^2\,y' - 6xy\,y' = 3y^2 - 5x^4 - 4x^3y \;\Rightarrow\;$

$(x^4 + 3y^2 - 6xy)\,y' = 3y^2 - 5x^4 - 4x^3y \;\Rightarrow\; y' = \dfrac{3y^2 - 5x^4 - 4x^3y}{x^4 + 3y^2 - 6xy}$

10. $\dfrac{d}{dx}\left(y^5 + x^2y^3\right) = \dfrac{d}{dx}\left(1 + x^4y\right) \;\Rightarrow\; 5y^4y' + x^2 \cdot 3y^2y' + y^3 \cdot 2x = 0 + x^4y' + y \cdot 4x^3 \;\Rightarrow\;$

$y'\left(5y^4 + 3x^2y^2 - x^4\right) = 4x^3y - 2xy^3 \;\Rightarrow\; y' = \dfrac{4x^3y - 2xy^3}{5y^4 + 3x^2y^2 - x^4}$

11. $\dfrac{d}{dx}\left(x^2y^2 + x\sin y\right) = \dfrac{d}{dx}\,(4) \;\Rightarrow\; x^2 \cdot 2y\,y' + y^2 \cdot 2x + x\cos y \cdot y' + \sin y \cdot 1 = 0 \;\Rightarrow\;$

$2x^2y\,y' + x\cos y \cdot y' = -2xy^2 - \sin y \;\Rightarrow\; (2x^2y + x\cos y)y' = -2xy^2 - \sin y \;\Rightarrow\; y' = \dfrac{-2xy^2 - \sin y}{2x^2y + x\cos y}$

12. $\dfrac{d}{dx}\,(1 + x) = \dfrac{d}{dx}\left[\sin(xy^2)\right] \;\Rightarrow\; 1 = [\cos(xy^2)](x \cdot 2y\,y' + y^2 \cdot 1) \;\Rightarrow\; 1 = 2xy\cos(xy^2)\,y' + y^2\cos(xy^2) \;\Rightarrow\;$

$1 - y^2\cos(xy^2) = 2xy\cos(xy^2)\,y' \;\Rightarrow\; y' = \dfrac{1 - y^2\cos(xy^2)}{2xy\cos(xy^2)}$

13. $\dfrac{d}{dx}\,(4\cos x\,\sin y) = \dfrac{d}{dx}\,(1) \;\Rightarrow\; 4\left[\cos x \cdot \cos y \cdot y' + \sin y \cdot (-\sin x)\right] = 0 \;\Rightarrow\;$

$y'(4\cos x\,\cos y) = 4\sin x\,\sin y \;\Rightarrow\; y' = \dfrac{4\sin x\,\sin y}{4\cos x\,\cos y} = \tan x\,\tan y$

14. $\dfrac{d}{dx}\left[y\sin(x^2)\right] = \dfrac{d}{dx}\left[x\sin(y^2)\right] \;\Rightarrow\; y\cos(x^2) \cdot 2x + \sin(x^2) \cdot y' = x\cos(y^2) \cdot 2y\,y' + \sin(y^2) \cdot 1 \;\Rightarrow\;$

$y'\left[\sin(x^2) - 2xy\cos(y^2)\right] = \sin(y^2) - 2xy\cos(x^2) \;\Rightarrow\; y' = \dfrac{\sin(y^2) - 2xy\cos(x^2)}{\sin(x^2) - 2xy\cos(y^2)}$

15. $\dfrac{d}{dx}\left[\tan(x/y)\right] = \dfrac{d}{dx}(x+y) \;\;\Rightarrow\;\; \sec^2(x/y) \cdot \dfrac{y \cdot 1 - x \cdot y'}{y^2} = 1 + y' \;\;\Rightarrow$

$y\sec^2(x/y) - x\sec^2(x/y) \cdot y' = y^2 + y^2 y' \;\;\Rightarrow\;\; y\sec^2(x/y) - y^2 = y^2 y' + x\sec^2(x/y) \;\;\Rightarrow$

$y\sec^2(x/y) - y^2 = \left[y^2 + x\sec^2(x/y)\right] \cdot y' \;\;\Rightarrow\;\; y' = \dfrac{y\sec^2(x/y) - y^2}{y^2 + x\sec^2(x/y)}$

16. $\dfrac{d}{dx}\left(\sqrt{x+y}\right) = \dfrac{d}{dx}\left(1 + x^2 y^2\right) \;\;\Rightarrow\;\; \tfrac{1}{2}(x+y)^{-1/2}(1+y') = x^2 \cdot 2y\,y' + y^2 \cdot 2x \;\;\Rightarrow$

$\dfrac{1}{2\sqrt{x+y}} + \dfrac{y'}{2\sqrt{x+y}} = 2x^2 y\,y' + 2xy^2 \;\;\Rightarrow\;\; 1 + y' = 4x^2 y\sqrt{x+y}\,y' + 4xy^2\sqrt{x+y} \;\;\Rightarrow$

$y' - 4x^2 y\sqrt{x+y}\,y' = 4xy^2\sqrt{x+y} - 1 \;\;\Rightarrow\;\; y'\left(1 - 4x^2 y\sqrt{x+y}\right) = 4xy^2\sqrt{x+y} - 1 \;\;\Rightarrow$

$y' = \dfrac{4xy^2\sqrt{x+y} - 1}{1 - 4x^2 y\sqrt{x+y}}$

17. $\sqrt{xy} = 1 + x^2 y \;\;\Rightarrow\;\; \tfrac{1}{2}(xy)^{-1/2}(xy' + y \cdot 1) = 0 + x^2 y' + y \cdot 2x \;\;\Rightarrow\;\; \dfrac{x}{2\sqrt{xy}}\,y' + \dfrac{y}{2\sqrt{xy}} = x^2 y' + 2xy \;\;\Rightarrow$

$y'\left(\dfrac{x}{2\sqrt{xy}} - x^2\right) = 2xy - \dfrac{y}{2\sqrt{xy}} \;\;\Rightarrow\;\; y'\left(\dfrac{x - 2x^2\sqrt{xy}}{2\sqrt{xy}}\right) = \dfrac{4xy\sqrt{xy} - y}{2\sqrt{xy}} \;\;\Rightarrow\;\; y' = \dfrac{4xy\sqrt{xy} - y}{x - 2x^2\sqrt{xy}}$

18. $\tan(x-y) = \dfrac{y}{1+x^2} \;\;\Rightarrow\;\; (1+x^2)\tan(x-y) = y \;\;\Rightarrow\;\; (1+x^2)\sec^2(x-y) \cdot (1-y') + \tan(x-y) \cdot 2x = y' \;\;\Rightarrow$

$(1+x^2)\sec^2(x-y) - (1+x^2)\sec^2(x-y) \cdot y' + 2x\tan(x-y) = y' \;\;\Rightarrow$

$(1+x^2)\sec^2(x-y) + 2x\tan(x-y) = \left[1 + (1+x^2)\sec^2(x-y)\right] \cdot y' \;\;\Rightarrow$

$y' = \dfrac{(1+x^2)\sec^2(x-y) + 2x\tan(x-y)}{1 + (1+x^2)\sec^2(x-y)}$

19. $\dfrac{d}{dx}\left(y\cos x\right) = \dfrac{d}{dx}\left(1 + \sin(xy)\right) \;\;\Rightarrow\;\; y(-\sin x) + \cos x \cdot y' = \cos(xy) \cdot (xy' + y \cdot 1) \;\;\Rightarrow$

$\cos x \cdot y' - x\cos(xy) \cdot y' = y\sin x + y\cos(xy) \;\;\Rightarrow\;\; \left[\cos x - x\cos(xy)\right] y' = y\sin x + y\cos(xy) \;\;\Rightarrow$

$y' = \dfrac{y\sin x + y\cos(xy)}{\cos x - x\cos(xy)}$

20. $\sin x + \cos y = \sin x \cos y \;\;\Rightarrow\;\; \cos x - \sin y \cdot y' = \sin x \left(-\sin y \cdot y'\right) + \cos y \cos x \;\;\Rightarrow$

$(\sin x \sin y - \sin y)\,y' = \cos x \cos y - \cos x \;\;\Rightarrow\;\; y' = \dfrac{\cos x\,(\cos y - 1)}{\sin y\,(\sin x - 1)}$

21. $\dfrac{d}{dx}\left\{f(x) + x^2[f(x)]^3\right\} = \dfrac{d}{dx}(10) \;\;\Rightarrow\;\; f'(x) + x^2 \cdot 3[f(x)]^2 \cdot f'(x) + [f(x)]^3 \cdot 2x = 0.$ If $x = 1$, we have

$f'(1) + 1^2 \cdot 3[f(1)]^2 \cdot f'(1) + [f(1)]^3 \cdot 2(1) = 0 \;\;\Rightarrow\;\; f'(1) + 1 \cdot 3 \cdot 2^2 \cdot f'(1) + 2^3 \cdot 2 = 0 \;\;\Rightarrow$

$f'(1) + 12f'(1) = -16 \;\;\Rightarrow\;\; 13f'(1) = -16 \;\;\Rightarrow\;\; f'(1) = -\tfrac{16}{13}.$

22. $\dfrac{d}{dx}\left[g(x) + x\sin g(x)\right] = \dfrac{d}{dx}(x^2) \;\;\Rightarrow\;\; g'(x) + x\cos g(x) \cdot g'(x) + \sin g(x) \cdot 1 = 2x.$ If $x = 0$, we have

$g'(0) + 0 + \sin g(0) = 2(0) \;\;\Rightarrow\;\; g'(0) + \sin 0 = 0 \;\;\Rightarrow\;\; g'(0) + 0 = 0 \;\;\Rightarrow\;\; g'(0) = 0.$

23. $\dfrac{d}{dy}(x^4y^2 - x^3y + 2xy^3) = \dfrac{d}{dy}(0) \;\Rightarrow\; x^4 \cdot 2y + y^2 \cdot 4x^3\,x' - (x^3 \cdot 1 + y \cdot 3x^2\,x') + 2(x \cdot 3y^2 + y^3 \cdot x') = 0 \;\Rightarrow\;$

$4x^3y^2\,x' - 3x^2y\,x' + 2y^3\,x' = -2x^4y + x^3 - 6xy^2 \;\Rightarrow\; (4x^3y^2 - 3x^2y + 2y^3)\,x' = -2x^4y + x^3 - 6xy^2 \;\Rightarrow\;$

$x' = \dfrac{dx}{dy} = \dfrac{-2x^4y + x^3 - 6xy^2}{4x^3y^2 - 3x^2y + 2y^3}$

24. $\dfrac{d}{dy}(y\sec x) = \dfrac{d}{dy}(x\tan y) \;\Rightarrow\; y \cdot \sec x \tan x \cdot x' + \sec x \cdot 1 = x \cdot \sec^2 y + \tan y \cdot x' \;\Rightarrow\;$

$y\sec x \tan x \cdot x' - \tan y \cdot x' = x\sec^2 y - \sec x \;\Rightarrow\; (y\sec x \tan x - \tan y)\,x' = x\sec^2 y - \sec x \;\Rightarrow\;$

$x' = \dfrac{dx}{dy} = \dfrac{x\sec^2 y - \sec x}{y\sec x \tan x - \tan y}$

25. $x^2 + xy + y^2 = 3 \;\Rightarrow\; 2x + x\,y' + y \cdot 1 + 2yy' = 0 \;\Rightarrow\; x\,y' + 2y\,y' = -2x - y \;\Rightarrow\; y'(x + 2y) = -2x - y \;\Rightarrow\;$

$y' = \dfrac{-2x - y}{x + 2y}$. When $x = 1$ and $y = 1$, we have $y' = \dfrac{-2 - 1}{1 + 2 \cdot 1} = \dfrac{-3}{3} = -1$, so an equation of the tangent line is

$y - 1 = -1(x - 1)$ or $y = -x + 2$.

26. $x^2 + 2xy - y^2 + x = 2 \;\Rightarrow\; 2x + 2(x\,y' + y \cdot 1) - 2y\,y' + 1 = 0 \;\Rightarrow\; 2x\,y' - 2y\,y' = -2x - 2y - 1 \;\Rightarrow\;$

$y'(2x - 2y) = -2x - 2y - 1 \;\Rightarrow\; y' = \dfrac{-2x - 2y - 1}{2x - 2y}$. When $x = 1$ and $y = 2$, we have

$y' = \dfrac{-2 - 4 - 1}{2 - 4} = \dfrac{-7}{-2} = \dfrac{7}{2}$, so an equation of the tangent line is $y - 2 = \dfrac{7}{2}(x - 1)$ or $y = \dfrac{7}{2}x - \dfrac{3}{2}$.

27. $x^2 + y^2 = (2x^2 + 2y^2 - x)^2 \;\Rightarrow\; 2x + 2y\,y' = 2(2x^2 + 2y^2 - x)(4x + 4y\,y' - 1)$. When $x = 0$ and $y = \frac{1}{2}$, we have

$0 + y' = 2(\tfrac{1}{2})(2y' - 1) \;\Rightarrow\; y' = 2y' - 1 \;\Rightarrow\; y' = 1$, so an equation of the tangent line is $y - \frac{1}{2} = 1(x - 0)$

or $y = x + \frac{1}{2}$.

28. $x^{2/3} + y^{2/3} = 4 \;\Rightarrow\; \frac{2}{3}x^{-1/3} + \frac{2}{3}y^{-1/3}y' = 0 \;\Rightarrow\; \dfrac{1}{\sqrt[3]{x}} + \dfrac{y'}{\sqrt[3]{y}} = 0 \;\Rightarrow\; y' = -\dfrac{\sqrt[3]{y}}{\sqrt[3]{x}}$. When $x = -3\sqrt{3}$

and $y = 1$, we have $y' = -\dfrac{1}{\left(-3\sqrt{3}\right)^{1/3}} = -\dfrac{\left(-3\sqrt{3}\right)^{2/3}}{-3\sqrt{3}} = \dfrac{3}{3\sqrt{3}} = \dfrac{1}{\sqrt{3}}$, so an equation of the tangent line is

$y - 1 = \frac{1}{\sqrt{3}}\left(x + 3\sqrt{3}\right)$ or $y = \frac{1}{\sqrt{3}}x + 4$.

29. $2(x^2 + y^2)^2 = 25(x^2 - y^2) \;\Rightarrow\; 4(x^2 + y^2)(2x + 2y\,y') = 25(2x - 2y\,y') \;\Rightarrow\;$

$4(x + y\,y')(x^2 + y^2) = 25(x - y\,y') \;\Rightarrow\; 4y\,y'(x^2 + y^2) + 25yy' = 25x - 4x(x^2 + y^2) \;\Rightarrow\;$

$y' = \dfrac{25x - 4x(x^2 + y^2)}{25y + 4y(x^2 + y^2)}$. When $x = 3$ and $y = 1$, we have $y' = \frac{75 - 120}{25 + 40} = -\frac{45}{65} = -\frac{9}{13}$,

so an equation of the tangent line is $y - 1 = -\frac{9}{13}(x - 3)$ or $y = -\frac{9}{13}x + \frac{40}{13}$.

30. $y^2(y^2 - 4) = x^2(x^2 - 5)$ $\Rightarrow$ $y^4 - 4y^2 = x^4 - 5x^2$ $\Rightarrow$ $4y^3\, y' - 8y\, y' = 4x^3 - 10x.$

When $x = 0$ and $y = -2$, we have $-32y' + 16y' = 0$ $\Rightarrow$ $-16y' = 0$ $\Rightarrow$ $y' = 0$, so an equation of the tangent line is

$y + 2 = 0(x - 0)$ or $y = -2.$

31. (a) $y^2 = 5x^4 - x^2$ $\Rightarrow$ $2y\, y' = 5(4x^3) - 2x$ $\Rightarrow$ $y' = \dfrac{10x^3 - x}{y}.$ (b)

So at the point $(1, 2)$ we have $y' = \dfrac{10(1)^3 - 1}{2} = \dfrac{9}{2}$, and an equation

of the tangent line is $y - 2 = \frac{9}{2}(x - 1)$ or $y = \frac{9}{2}x - \frac{5}{2}.$

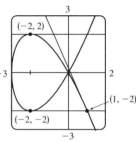

32. (a) $y^2 = x^3 + 3x^2$ $\Rightarrow$ $2y\, y' = 3x^2 + 3(2x)$ $\Rightarrow$ $y' = \dfrac{3x^2 + 6x}{2y}.$ So at the point $(1, -2)$ we have

$y' = \dfrac{3(1)^2 + 6(1)}{2(-2)} = -\dfrac{9}{4}$, and an equation of the tangent line is $y + 2 = -\frac{9}{4}(x - 1)$ or $y = -\frac{9}{4}x + \frac{1}{4}.$

(b) The curve has a horizontal tangent where $y' = 0$ $\Leftrightarrow$ (c)

$3x^2 + 6x = 0$ $\Leftrightarrow$ $3x(x + 2) = 0$ $\Leftrightarrow$ $x = 0$ or $x = -2.$

But note that at $x = 0$, $y = 0$ also, so the derivative does not exist.

At $x = -2$, $y^2 = (-2)^3 + 3(-2)^2 = -8 + 12 = 4$, so $y = \pm 2.$

So the two points at which the curve has a horizontal tangent are

$(-2, -2)$ and $(-2, 2).$

33. $9x^2 + y^2 = 9$ $\Rightarrow$ $18x + 2y\, y' = 0$ $\Rightarrow$ $2y\, y' = -18x$ $\Rightarrow$ $y' = -9x/y$ $\Rightarrow$

$y'' = -9\left(\dfrac{y \cdot 1 - x \cdot y'}{y^2}\right) = -9\left(\dfrac{y - x(-9x/y)}{y^2}\right) = -9 \cdot \dfrac{y^2 + 9x^2}{y^3} = -9 \cdot \dfrac{9}{y^3}$ [since x and y must satisfy the original

equation, $9x^2 + y^2 = 9$]. Thus, $y'' = -81/y^3.$

34. $\sqrt{x} + \sqrt{y} = 1$ $\Rightarrow$ $\dfrac{1}{2\sqrt{x}} + \dfrac{y'}{2\sqrt{y}} = 0$ $\Rightarrow$ $y' = -\dfrac{\sqrt{y}}{\sqrt{x}}$ $\Rightarrow$

$y'' = -\dfrac{\sqrt{x}\left[\dfrac{1}{2\sqrt{y}}\right]y' - \sqrt{y}\left[\dfrac{1}{2\sqrt{x}}\right]}{x} = -\dfrac{\sqrt{x}\left(\dfrac{1}{\sqrt{y}}\right)\left(-\dfrac{\sqrt{y}}{\sqrt{x}}\right) - \sqrt{y}\left(\dfrac{1}{\sqrt{x}}\right)}{2x} = \dfrac{1 + \dfrac{\sqrt{y}}{\sqrt{x}}}{2x}$

$= \dfrac{\sqrt{x} + \sqrt{y}}{2x\sqrt{x}} = \dfrac{1}{2x\sqrt{x}}$ since x and y must satisfy the original equation, $\sqrt{x} + \sqrt{y} = 1.$

35. $x^3 + y^3 = 1$ $\Rightarrow$ $3x^2 + 3y^2\, y' = 0$ $\Rightarrow$ $y' = -\dfrac{x^2}{y^2}$ $\Rightarrow$

$y'' = -\dfrac{y^2(2x) - x^2 \cdot 2y\, y'}{(y^2)^2} = -\dfrac{2xy^2 - 2x^2y(-x^2/y^2)}{y^4} = -\dfrac{2xy^4 + 2x^4y}{y^6} = -\dfrac{2xy(y^3 + x^3)}{y^6} = -\dfrac{2x}{y^5},$

since x and y must satisfy the original equation, $x^3 + y^3 = 1.$

36. $x^4 + y^4 = a^4 \;\;\Rightarrow\;\; 4x^3 + 4y^3\, y' = 0 \;\;\Rightarrow\;\; 4y^3\, y' = -4x^3 \;\;\Rightarrow\;\; y' = -x^3/y^3 \;\;\Rightarrow$

$$y'' = -\left(\frac{y^3 \cdot 3x^2 - x^3 \cdot 3y^2\, y'}{(y^3)^2}\right) = -3x^2 y^2 \cdot \frac{y - x(-x^3/y^3)}{y^6} = -3x^2 \cdot \frac{y^4 + x^4}{y^4 y^3} = -3x^2 \cdot \frac{a^4}{y^7} = \frac{-3a^4 x^2}{y^7}$$

37. (a) There are eight points with horizontal tangents: four at $x \approx 1.57735$ and

four at $x \approx 0.42265$.

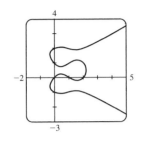

(b) $y' = \dfrac{3x^2 - 6x + 2}{2(2y^3 - 3y^2 - y + 1)} \;\;\Rightarrow\;\; y' = -1$ at $(0, 1)$ and $y' = \tfrac{1}{3}$ at $(0, 2)$.

Equations of the tangent lines are $y = -x + 1$ and $y = \tfrac{1}{3}x + 2$.

(c) $y' = 0 \;\;\Rightarrow\;\; 3x^2 - 6x + 2 = 0 \;\;\Rightarrow\;\; x = 1 \pm \tfrac{1}{3}\sqrt{3}$

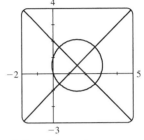

(d) By multiplying the right side of the equation by $x - 3$, we obtain the first

graph. By modifying the equation in other ways, we can generate the other

graphs.

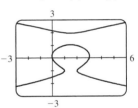

$$y(y^2 - 1)(y - 2)$$
$$= x(x - 1)(x - 2)(x - 3)$$

$$y(y^2 - 4)(y - 2)$$
$$= x(x - 1)(x - 2)$$

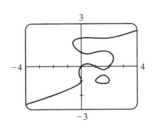

$$y(y + 1)(y^2 - 1)(y - 2)$$
$$= x(x - 1)(x - 2)$$

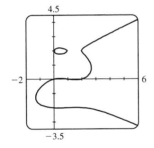

$$(y + 1)(y^2 - 1)(y - 2)$$
$$= (x - 1)(x - 2)$$

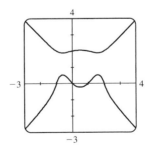

$$x(y + 1)(y^2 - 1)(y - 2)$$
$$= y(x - 1)(x - 2)$$

$$y(y^2 + 1)(y - 2)$$
$$= x(x^2 - 1)(x - 2)$$

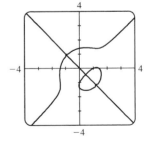

$$y(y + 1)(y^2 - 2)$$
$$= x(x - 1)(x^2 - 2)$$

38. (a)

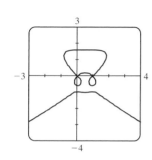

(b) There are 9 points with horizontal tangents: 3 at $x = 0$, 3 at $x = \frac{1}{2}$, and 3 at $x = 1$. The three horizontal tangents along the top of the wagon are hard to find, but by limiting the y-range of the graph (to $[1.6, 1.7]$, for example) they are distinguishable.

39. From Exercise 29, a tangent to the lemniscate will be horizontal if $y' = 0$ $\Rightarrow$ $25x - 4x(x^2 + y^2) = 0$ $\Rightarrow$

$x[25 - 4(x^2 + y^2)] = 0$ $\Rightarrow$ $x^2 + y^2 = \frac{25}{4}$ (**1**). (Note that when x is 0, y is also 0, and there is no horizontal tangent

at the origin.) Substituting $\frac{25}{4}$ for $x^2 + y^2$ in the equation of the lemniscate, $2(x^2 + y^2)^2 = 25(x^2 - y^2)$, we get

$x^2 - y^2 = \frac{25}{8}$ (**2**). Solving (**1**) and (**2**), we have $x^2 = \frac{75}{16}$ and $y^2 = \frac{25}{16}$, so the four points are $\left(\pm\frac{5\sqrt{3}}{4}, \pm\frac{5}{4}\right)$.

40. $\dfrac{x^2}{a^2} + \dfrac{y^2}{b^2} = 1$ $\Rightarrow$ $\dfrac{2x}{a^2} + \dfrac{2yy'}{b^2} = 0$ $\Rightarrow$ $y' = -\dfrac{b^2 x}{a^2 y}$ $\Rightarrow$ an equation of the tangent line at (x_0, y_0) is

$y - y_0 = \dfrac{-b^2 x_0}{a^2 y_0}(x - x_0)$. Multiplying both sides by $\dfrac{y_0}{b^2}$ gives $\dfrac{y_0 y}{b^2} - \dfrac{y_0^2}{b^2} = -\dfrac{x_0 x}{a^2} + \dfrac{x_0^2}{a^2}$. Since (x_0, y_0) lies on the ellipse,

we have $\dfrac{x_0 x}{a^2} + \dfrac{y_0 y}{b^2} = \dfrac{x_0^2}{a^2} + \dfrac{y_0^2}{b^2} = 1$.

41. $\dfrac{x^2}{a^2} - \dfrac{y^2}{b^2} = 1$ $\Rightarrow$ $\dfrac{2x}{a^2} - \dfrac{2yy'}{b^2} = 0$ $\Rightarrow$ $y' = \dfrac{b^2 x}{a^2 y}$ $\Rightarrow$ an equation of the tangent line at (x_0, y_0) is

$y - y_0 = \dfrac{b^2 x_0}{a^2 y_0}(x - x_0)$. Multiplying both sides by $\dfrac{y_0}{b^2}$ gives $\dfrac{y_0 y}{b^2} - \dfrac{y_0^2}{b^2} = \dfrac{x_0 x}{a^2} - \dfrac{x_0^2}{a^2}$. Since (x_0, y_0) lies on the hyperbola,

we have $\dfrac{x_0 x}{a^2} - \dfrac{y_0 y}{b^2} = \dfrac{x_0^2}{a^2} - \dfrac{y_0^2}{b^2} = 1$.

42. $\sqrt{x} + \sqrt{y} = \sqrt{c}$ $\Rightarrow$ $\dfrac{1}{2\sqrt{x}} + \dfrac{y'}{2\sqrt{y}} = 0$ $\Rightarrow$ $y' = -\dfrac{\sqrt{y}}{\sqrt{x}}$ $\Rightarrow$ an equation of the tangent line at (x_0, y_0)

is $y - y_0 = -\dfrac{\sqrt{y_0}}{\sqrt{x_0}}(x - x_0)$. Now $x = 0$ $\Rightarrow$ $y = y_0 - \dfrac{\sqrt{y_0}}{\sqrt{x_0}}(-x_0) = y_0 + \sqrt{x_0}\sqrt{y_0}$, so the y-intercept is

$y_0 + \sqrt{x_0}\sqrt{y_0}$. And $y = 0$ $\Rightarrow$ $-y_0 = -\dfrac{\sqrt{y_0}}{\sqrt{x_0}}(x - x_0)$ $\Rightarrow$ $x - x_0 = \dfrac{y_0\sqrt{x_0}}{\sqrt{y_0}}$ $\Rightarrow$

$x = x_0 + \sqrt{x_0}\sqrt{y_0}$, so the x-intercept is $x_0 + \sqrt{x_0}\sqrt{y_0}$. The sum of the intercepts is

$\left(y_0 + \sqrt{x_0}\sqrt{y_0}\right) + \left(x_0 + \sqrt{x_0}\sqrt{y_0}\right) = x_0 + 2\sqrt{x_0}\sqrt{y_0} + y_0 = \left(\sqrt{x_0} + \sqrt{y_0}\right)^2 = \left(\sqrt{c}\right)^2 = c$.

43. If the circle has radius r, its equation is $x^2 + y^2 = r^2$ $\Rightarrow$ $2x + 2yy' = 0$ $\Rightarrow$ $y' = -\dfrac{x}{y}$, so the slope of the tangent line

at $P(x_0, y_0)$ is $-\dfrac{x_0}{y_0}$. The negative reciprocal of that slope is $\dfrac{-1}{-x_0/y_0} = \dfrac{y_0}{x_0}$, which is the slope of OP, so the tangent line at

P is perpendicular to the radius OP.

44. $y^q = x^p$ $\Rightarrow$ $qy^{q-1}y' = px^{p-1}$ $\Rightarrow$ $y' = \dfrac{px^{p-1}}{qy^{q-1}} = \dfrac{px^{p-1}y}{qy^q} = \dfrac{px^{p-1}x^{p/q}}{qx^p} = \dfrac{p}{q}x^{(p/q)-1}$

45. $x^2 + y^2 = r^2$ is a circle with center O and $ax + by = 0$ is a line through O [assume a

and b are not both zero]. $x^2 + y^2 = r^2$ $\Rightarrow$ $2x + 2yy' = 0$ $\Rightarrow$ $y' = -x/y$, so the

slope of the tangent line at $P_0\,(x_0, y_0)$ is $-x_0/y_0$. The slope of the line OP_0 is y_0/x_0,

which is the negative reciprocal of $-x_0/y_0$. Hence, the curves are orthogonal, and the

families of curves are orthogonal trajectories of each other.

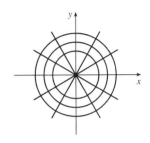

46. The circles $x^2 + y^2 = ax$ and $x^2 + y^2 = by$ intersect at the origin where the tangents are vertical and horizontal [assume a

and b are both nonzero]. If (x_0, y_0) is the other point of intersection, then $x_0^2 + y_0^2 = ax_0$ **(1)** and $x_0^2 + y_0^2 = by_0$ **(2)**.

Now $x^2 + y^2 = ax$ $\Rightarrow$ $2x + 2yy' = a$ $\Rightarrow$ $y' = \dfrac{a - 2x}{2y}$ and $x^2 + y^2 = by$ $\Rightarrow$

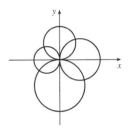

$2x + 2yy' = by'$ $\Rightarrow$ $y' = \dfrac{2x}{b - 2y}$. Thus, the curves are orthogonal at (x_0, y_0) $\Leftrightarrow$

$\dfrac{a - 2x_0}{2y_0} = -\dfrac{b - 2y_0}{2x_0}$ $\Leftrightarrow$ $2ax_0 - 4x_0^2 = 4y_0^2 - 2by_0$ $\Leftrightarrow$ $ax_0 + by_0 = 2(x_0^2 + y_0^2)$,

which is true by **(1)** and **(2)**.

47. $y = cx^2$ $\Rightarrow$ $y' = 2cx$ and $x^2 + 2y^2 = k$ [assume $k > 0$] $\Rightarrow$ $2x + 4yy' = 0$ $\Rightarrow$

$2yy' = -x$ $\Rightarrow$ $y' = -\dfrac{x}{2(y)} = -\dfrac{x}{2(cx^2)} = -\dfrac{1}{2cx}$, so the curves are orthogonal if

$c \neq 0$. If $c = 0$, then the horizontal line $y = cx^2 = 0$ intersects $x^2 + 2y^2 = k$ orthogonally

at $\left(\pm\sqrt{k}, 0\right)$, since the ellipse $x^2 + 2y^2 = k$ has vertical tangents at those two points.

48. $y = ax^3$ $\Rightarrow$ $y' = 3ax^2$ and $x^2 + 3y^2 = b$ [assume $b > 0$] $\Rightarrow$ $2x + 6yy' = 0$ $\Rightarrow$

$3yy' = -x$ $\Rightarrow$ $y' = -\dfrac{x}{3(y)} = -\dfrac{x}{3(ax^3)} = -\dfrac{1}{3ax^2}$, so the curves are orthogonal if

$a \neq 0$. If $a = 0$, then the horizontal line $y = ax^3 = 0$ intesects $x^2 + 3y^2 = b$ orthogonally

at $\left(\pm\sqrt{b}, 0\right)$, since the ellipse $x^2 + 3y^2 = b$ has vertical tangents at those two points.

49. To find the points at which the ellipse $x^2 - xy + y^2 = 3$ crosses the x-axis, let $y = 0$ and solve for x.

$y = 0 \Rightarrow x^2 - x(0) + 0^2 = 3 \Leftrightarrow x = \pm\sqrt{3}$. So the graph of the ellipse crosses the x-axis at the points $\left(\pm\sqrt{3}, 0\right)$.

Using implicit differentiation to find y', we get $2x - xy' - y + 2yy' = 0 \Rightarrow y'(2y - x) = y - 2x \Leftrightarrow y' = \dfrac{y - 2x}{2y - x}$.

So y' at $\left(\sqrt{3}, 0\right)$ is $\dfrac{0 - 2\sqrt{3}}{2(0) - \sqrt{3}} = 2$ and y' at $\left(-\sqrt{3}, 0\right)$ is $\dfrac{0 + 2\sqrt{3}}{2(0) + \sqrt{3}} = 2$. Thus, the tangent lines at these points are parallel.

50. (a) We use implicit differentiation to find $y' = \dfrac{y - 2x}{2y - x}$ as in Exercise 49. The slope

(b)

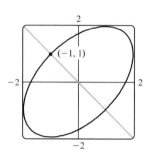

of the tangent line at $(-1, 1)$ is $m = \dfrac{1 - 2(-1)}{2(1) - (-1)} = \dfrac{3}{3} = 1$, so the slope of the

normal line is $-\dfrac{1}{m} = -1$, and its equation is $y - 1 = -1(x + 1) \Leftrightarrow$

$y = -x$. Substituting $-x$ for y in the equation of the ellipse, we get

$x^2 - x(-x) + (-x)^2 = 3 \Rightarrow 3x^2 = 3 \Leftrightarrow x = \pm 1$. So the normal line

must intersect the ellipse again at $x = 1$, and since the equation of the line is

$y = -x$, the other point of intersection must be $(1, -1)$.

51. $x^2 y^2 + xy = 2 \Rightarrow x^2 \cdot 2yy' + y^2 \cdot 2x + x \cdot y' + y \cdot 1 = 0 \Leftrightarrow y'(2x^2 y + x) = -2xy^2 - y \Leftrightarrow$

$y' = -\dfrac{2xy^2 + y}{2x^2 y + x}$. So $-\dfrac{2xy^2 + y}{2x^2 y + x} = -1 \Leftrightarrow 2xy^2 + y = 2x^2 y + x \Leftrightarrow y(2xy + 1) = x(2xy + 1) \Leftrightarrow$

$y(2xy + 1) - x(2xy + 1) = 0 \Leftrightarrow (2xy + 1)(y - x) = 0 \Leftrightarrow xy = -\frac{1}{2}$ or $y = x$. But $xy = -\frac{1}{2} \Rightarrow$

$x^2 y^2 + xy = \frac{1}{4} - \frac{1}{2} \neq 2$, so we must have $x = y$. Then $x^2 y^2 + xy = 2 \Rightarrow x^4 + x^2 = 2 \Leftrightarrow x^4 + x^2 - 2 = 0 \Leftrightarrow$

$(x^2 + 2)(x^2 - 1) = 0$. So $x^2 = -2$, which is impossible, or $x^2 = 1 \Leftrightarrow x = \pm 1$. Since $x = y$, the points on the curve

where the tangent line has a slope of -1 are $(-1, -1)$ and $(1, 1)$.

52. $x^2 + 4y^2 = 36 \Rightarrow 2x + 8yy' = 0 \Rightarrow y' = -\dfrac{x}{4y}$. Let (a, b) be a point on $x^2 + 4y^2 = 36$ whose tangent line passes

through $(12, 3)$. The tangent line is then $y - 3 = -\dfrac{a}{4b}(x - 12)$, so $b - 3 = -\dfrac{a}{4b}(a - 12)$. Multiplying both sides by $4b$

gives $4b^2 - 12b = -a^2 + 12a$, so $4b^2 + a^2 = 12(a + b)$. But $4b^2 + a^2 = 36$, so $36 = 12(a + b) \Rightarrow a + b = 3 \Rightarrow$

$b = 3 - a$. Substituting $3 - a$ for b into $a^2 + 4b^2 = 36$ gives $a^2 + 4(3 - a)^2 = 36 \Leftrightarrow a^2 + 36 - 24a + 4a^2 = 36 \Leftrightarrow$

$5a^2 - 24a = 0 \Leftrightarrow a(5a - 24) = 0$, so $a = 0$ or $a = \frac{24}{5}$. If $a = 0$, $b = 3 - 0 = 3$, and if $a = \frac{24}{5}$, $b = 3 - \frac{24}{5} = -\frac{9}{5}$.

So the two points on the ellipse are $(0, 3)$ and $\left(\frac{24}{5}, -\frac{9}{5}\right)$. Using

$y - 3 = -\dfrac{a}{4b}(x - 12)$ with $(a, b) = (0, 3)$ gives us the tangent line

$y - 3 = 0$ or $y = 3$. With $(a, b) = \left(\frac{24}{5}, -\frac{9}{5}\right)$, we have

$y - 3 = -\dfrac{24/5}{4(-9/5)}(x - 12) \Leftrightarrow y - 3 = \frac{2}{3}(x - 12) \Leftrightarrow y = \frac{2}{3}x - 5$.

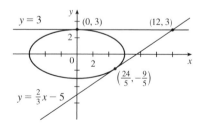

A graph of the ellipse and the tangent lines confirms our results.

53. $x^2 + 4y^2 = 5$ $\Rightarrow$ $2x + 4(2yy') = 0$ $\Rightarrow$ $y' = -\dfrac{x}{4y}$. Now let h be the height of the lamp, and let (a, b) be the point of

tangency of the line passing through the points $(3, h)$ and $(-5, 0)$. This line has slope $(h - 0)/[3 - (-5)] = \frac{1}{8}h$. But the

slope of the tangent line through the point (a, b) can be expressed as $y' = -\dfrac{a}{4b}$, or as $\dfrac{b - 0}{a - (-5)} = \dfrac{b}{a + 5}$ [since the line

passes through $(-5, 0)$ and (a, b)], so $-\dfrac{a}{4b} = \dfrac{b}{a + 5}$ $\Leftrightarrow$ $4b^2 = -a^2 - 5a$ $\Leftrightarrow$ $a^2 + 4b^2 = -5a$. But $a^2 + 4b^2 = 5$

[since (a, b) is on the ellipse], so $5 = -5a$ $\Leftrightarrow$ $a = -1$. Then $4b^2 = -a^2 - 5a = -1 - 5(-1) = 4$ $\Rightarrow$ $b = 1$, since the

point is on the top half of the ellipse. So $\dfrac{h}{8} = \dfrac{b}{a + 5} = \dfrac{1}{-1 + 5} = \dfrac{1}{4}$ $\Rightarrow$ $h = 2$. So the lamp is located 2 units above the

x-axis.

3.7   Rates of Change in the Natural and Social Sciences

1. (a) $s = f(t) = t^3 - 12t^2 + 36t$ $\Rightarrow$ $v(t) = f'(t) = 3t^2 - 24t + 36$

(b) $v(3) = 27 - 72 + 36 = -9$ ft/s

(c) The particle is at rest when $v(t) = 0$. $3t^2 - 24t + 36 = 0$ $\Leftrightarrow$ $3(t - 2)(t - 6) = 0$ $\Leftrightarrow$ $t = 2$ s or 6 s.

(d) The particle is moving in the positive direction when $v(t) > 0$. $3(t - 2)(t - 6) > 0$ $\Leftrightarrow$ $0 \le t < 2$ or $t > 6$.

(e) Since the particle is moving in the positive direction and in the

negative direction, we need to calculate the distance traveled in the

intervals $[0, 2]$, $[2, 6]$, and $[6, 8]$ separately.

$|f(2) - f(0)| = |32 - 0| = 32$.

$|f(6) - f(2)| = |0 - 32| = 32$.

$|f(8) - f(6)| = |32 - 0| = 32$.

The total distance is $32 + 32 + 32 = 96$ ft.

(f)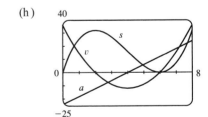

(g) $v(t) = 3t^2 - 24t + 36$ $\Rightarrow$

$a(t) = v'(t) = 6t - 24$.

$a(3) = 6(3) - 24 = -6$ (ft/s)/s or ft/s^2.

(h)

(i) The particle is speeding up when v and a have the same sign. This occurs when $2 < t < 4$ [v and a are both negative] and

when $t > 6$ [v and a are both positive]. It is slowing down when v and a have opposite signs; that is, when $0 \le t < 2$ and

when $4 < t < 6$.

2. (a) $s = f(t) = 0.01t^4 - 0.04t^3$ $\Rightarrow$ $v(t) = f'(t) = 0.04t^3 - 0.12t^2$

(b) $v(3) = 0.04(3)^3 - 0.12(3)^2 = 0$ ft/s

(c) The particle is at rest when $v(t) = 0$. $0.04t^3 - 0.12t^2 = 0$ $\Leftrightarrow$ $0.04t^2(t - 3) = 0$ $\Leftrightarrow$ $t = 0$ s or 3 s.

(d) The particle is moving in the positive direction when $v(t) > 0$. $0.04t^2(t - 3) > 0$ $\Leftrightarrow$ $t > 3$.

(e) See Exercise 1(e).

$$|f(3) - f(0)| = |-0.27 - 0| = 0.27.$$

$$|f(8) - f(3)| = |20.48 - (-0.27)| = 20.75.$$

The total distance is $0.27 + 20.75 = 21.02$ ft.

(f)

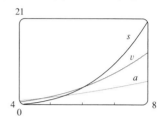

(g) $v(t) = 0.04t^3 - 0.12t^2 \Rightarrow a(t) = v'(t) = 0.12t^2 - 0.24t.$ $a(3) = 0.12(3)^2 - 0.24(3) = 0.36$ (ft/s)/s or ft/s².

(h) Here we show the graph of s, v,

and a for $0 \le t \le 4$ and

$4 \le t \le 8$.

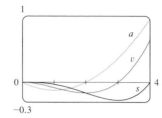

(i) The particle is speeding up when v and a have the same sign. This occurs when $0 < t < 2$ [v and a are both negative]

and when $t > 3$ [v and a are both positive]. It is slowing down when v and a have opposite signs; that is,

when $2 < t < 3$.

3. (a) $s = f(t) = \cos(\pi t/4) \Rightarrow v(t) = f'(t) = -\sin(\pi t/4) \cdot (\pi/4)$

(b) $v(3) = -\frac{\pi}{4} \sin \frac{3\pi}{4} = -\frac{\pi}{4} \cdot \frac{\sqrt{2}}{2} = -\frac{\pi\sqrt{2}}{8}$ ft/s [≈ -0.56]

(c) The particle is at rest when $v(t) = 0$. $-\frac{\pi}{4} \sin \frac{\pi t}{4} = 0 \Rightarrow \sin \frac{\pi t}{4} = 0 \Rightarrow \frac{\pi t}{4} = \pi n \Rightarrow t = 0, 4, 8$ s.

(d) The particle is moving in the positive direction when $v(t) > 0$. $-\frac{\pi}{4} \sin \frac{\pi t}{4} > 0 \Rightarrow \sin \frac{\pi t}{4} < 0 \Rightarrow 4 < t < 8$.

(e) From part (c), $v(t) = 0$ for $t = 0, 4, 8$. As in Exercise 1, we'll

find the distance traveled in the intervals $[0, 4]$ and $[4, 8]$.

$$|f(4) - f(0)| = |-1 - 1| = 2$$

$$|f(8) - f(4)| = |1 - (-1)| = 2.$$

The total distance is $2 + 2 = 4$ ft.

(f)

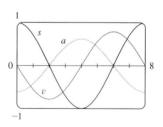

(g) $v(t) = -\frac{\pi}{4} \sin \frac{\pi t}{4} \Rightarrow$

$a(t) = v'(t) = -\frac{\pi}{4} \cos \frac{\pi t}{4} \cdot \frac{\pi}{4} = -\frac{\pi^2}{16} \cos \frac{\pi t}{4}.$

$a(3) = -\frac{\pi^2}{16} \cos \frac{3\pi}{4} = -\frac{\pi^2}{16} \left(-\frac{\sqrt{2}}{2}\right) = \frac{\pi^2\sqrt{2}}{32}$ (ft/s)/s or ft/s².

(h)

(i) The particle is speeding up when v and a have the same sign. This occurs when $0 < t < 2$ or $8 < t < 10$ [v and a are

both negative] and when $4 < t < 6$ [v and a are both positive]. It is slowing down when v and a have opposite signs;

that is, when $2 < t < 4$ and when $6 < t < 8$.

4. (a) $s = f(t) = \dfrac{t}{t^2 + 1} \Rightarrow v(t) = f'(t) = \dfrac{(t^2 + 1)(1) - t(2t)}{(t^2 + 1)^2} = \dfrac{1 - t^2}{(t^2 + 1)^2}$

(b) $v(3) = \dfrac{1 - (3)^2}{(3^2 + 1)^2} = \dfrac{1 - 9}{10^2} = \dfrac{-8}{100} = -\dfrac{2}{25}$ ft/s

(c) The particle is at rest when $v = 0 \Leftrightarrow 1 - t^2 = 0 \Leftrightarrow t = 1$ s   [$t \ne -1$ since $t \ge 0$].

(d) The particle moves in the positive direction when $v > 0 \iff 1 - t^2 > 0 \iff t^2 < 1 \iff 0 \le t < 1$.

(e) Distance in positive direction $= |s(1) - s(0)| = \left|\frac{1}{2} - 0\right| = \frac{1}{2}$ ft

(f)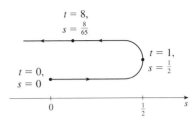

Distance in negative direction $= |s(8) - s(1)| = \left|\frac{8}{65} - \frac{1}{2}\right| = \frac{49}{130}$ ft

Total distance traveled $= \frac{1}{2} + \frac{49}{130} = \frac{57}{65}$ ft

(g) $a(t) = f''(t) = \dfrac{(1 + t^2)^2(-2t) - (1 - t^2)2(1 + t^2)(2t)}{[(1 + t^2)^2]^2}$

(h)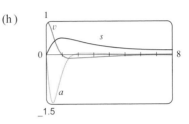

$= \dfrac{-2t(1 + t^2)[(1 + t^2) + 2(1 - t^2)]}{(1 + t^2)^4} = \dfrac{-2t(3 - t^2)}{(1 + t^2)^3}$

$a(3) = \dfrac{-6(-6)}{10^3} = \dfrac{36}{1000} = 0.036$ ft/s^2.

(i) The particle speeds up when v and a have the same sign. Both are negative for $1 < t < \sqrt{3}$. The particle slows down when v and a have opposite signs. This occurs for $0 < t < 1$ and $\sqrt{3} < t < 8$.

5. (a) From the figure, the velocity v is positive on the interval $(0, 2)$ and negative on the interval $(2, 3)$. The acceleration a is positive (negative) when the slope of the tangent line is positive (negative), so the acceleration is positive on the interval $(0, 1)$, and negative on the interval $(1, 3)$. The particle is speeding up when v and a have the same sign, that is, on the interval $(0, 1)$ when $v > 0$ and $a > 0$, and on the interval $(2, 3)$ when $v < 0$ and $a < 0$. The particle is slowing down when v and a have opposite signs, that is, on the interval $(1, 2)$ when $v > 0$ and $a < 0$.

(b) $v > 0$ on $(0, 3)$ and $v < 0$ on $(3, 4)$. $a > 0$ on $(1, 2)$ and $a < 0$ on $(0, 1)$ and $(2, 4)$. The particle is speeding up on $(1, 2)$ $[v > 0, a > 0]$ and on $(3, 4)$ $[v < 0, a < 0]$. The particle is slowing down on $(0, 1)$ and $(2, 3)$ $[v > 0, a < 0]$.

6. (a) The velocity v is positive when s is increasing, that is, on the intervals $(0, 1)$ and $(3, 4)$; and it is negative when s is decreasing, that is, on the interval $(1, 3)$. The acceleration a is positive when the graph of s is concave upward (v is increasing), that is, on the interval $(2, 4)$; and it is negative when the graph of s is concave downward (v is decreasing), that is, on the interval $(0, 2)$. The particle is speeding up on the interval $(1, 2)$ $[v < 0, a < 0]$ and on $(3, 4)$ $[v > 0, a > 0]$. The particle is slowing down on the interval $(0, 1)$ $[v > 0, a < 0]$ and on $(2, 3)$ $[v < 0, a > 0]$.

(b) The velocity v is positive on $(3, 4)$ and negative on $(0, 3)$. The acceleration a is positive on $(0, 1)$ and $(2, 4)$ and negative on $(1, 2)$. The particle is speeding up on the interval $(1, 2)$ $[v < 0, a < 0]$ and on $(3, 4)$ $[v > 0, a > 0]$. The particle is slowing down on the interval $(0, 1)$ $[v < 0, a > 0]$ and on $(2, 3)$ $[v < 0, a > 0]$.

7. (a) $s(t) = t^3 - 4.5t^2 - 7t \Rightarrow v(t) = s'(t) = 3t^2 - 9t - 7 = 5 \iff 3t^2 - 9t - 12 = 0 \iff$

$3(t - 4)(t + 1) = 0 \iff t = 4$ or -1. Since $t \ge 0$, the particle reaches a velocity of 5 m/s at $t = 4$ s.

(b) $a(t) = v'(t) = 6t - 9 = 0 \iff t = 1.5$. The acceleration changes from negative to positive, so the velocity changes from decreasing to increasing. Thus, at $t = 1.5$ s, the velocity has its minimum value.

8. (a) $s = 5t + 3t^2 \Rightarrow v(t) = \dfrac{ds}{dt} = 5 + 6t$, so $v(2) = 5 + 6(2) = 17$ m/s.

(b) $v(t) = 35 \Rightarrow 5 + 6t = 35 \Rightarrow 6t = 30 \Rightarrow t = 5$ s.

9. (a) $h = 10t - 0.83t^2 \Rightarrow v(t) = \dfrac{dh}{dt} = 10 - 1.66t$, so $v(3) = 10 - 1.66(3) = 5.02$ m/s.

(b) $h = 25 \Rightarrow 10t - 0.83t^2 = 25 \Rightarrow 0.83t^2 - 10t + 25 = 0 \Rightarrow t = \dfrac{10 \pm \sqrt{17}}{1.66} \approx 3.54$ or 8.51.

The value $t_1 = \dfrac{10 - \sqrt{17}}{1.66}$ corresponds to the time it takes for the stone to rise 25 m and $t_2 = \dfrac{10 + \sqrt{17}}{1.66}$ corresponds to the

time when the stone is 25 m high on the way down. Thus, $v(t_1) = 10 - 1.66\left(\dfrac{10 - \sqrt{17}}{1.66}\right) = \sqrt{17} \approx 4.12$ m/s.

10. (a) At maximum height the velocity of the ball is 0 ft/s. $v(t) = s'(t) = 80 - 32t = 0 \Leftrightarrow 32t = 80 \Leftrightarrow t = \frac{5}{2}$.

So the maximum height is $s\left(\frac{5}{2}\right) = 80\left(\frac{5}{2}\right) - 16\left(\frac{5}{2}\right)^2 = 200 - 100 = 100$ ft.

(b) $s(t) = 80t - 16t^2 = 96 \Leftrightarrow 16t^2 - 80t + 96 = 0 \Leftrightarrow 16(t^2 - 5t + 6) = 0 \Leftrightarrow 16(t - 3)(t - 2) = 0$.

So the ball has a height of 96 ft on the way up at $t = 2$ and on the way down at $t = 3$. At these times the velocities are

$v(2) = 80 - 32(2) = 16$ ft/s and $v(3) = 80 - 32(3) = -16$ ft/s, respectively.

11. (a) $A(x) = x^2 \Rightarrow A'(x) = 2x$. $A'(15) = 30$ mm^2/mm is the rate at which

the area is increasing with respect to the side length as x reaches 15 mm.

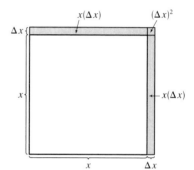

(b) The perimeter is $P(x) = 4x$, so $A'(x) = 2x = \frac{1}{2}(4x) = \frac{1}{2}P(x)$. The

figure suggests that if Δx is small, then the change in the area of the square

is approximately half of its perimeter (2 of the 4 sides) times Δx. From the

figure, $\Delta A = 2x\,(\Delta x) + (\Delta x)^2$. If Δx is small, then $\Delta A \approx 2x\,(\Delta x)$ and

so $\Delta A/\Delta x \approx 2x$.

12. (a) $V(x) = x^3 \Rightarrow \dfrac{dV}{dx} = 3x^2$. $\left.\dfrac{dV}{dx}\right|_{x=3} = 3(3)^2 = 27$ mm^3/mm is the

rate at which the volume is increasing as x increases past 3 mm.

(b) The surface area is $S(x) = 6x^2$, so $V'(x) = 3x^2 = \frac{1}{2}(6x^2) = \frac{1}{2}S(x)$.

The figure suggests that if Δx is small, then the change in the volume of the

cube is approximately half of its surface area (the area of 3 of the 6 faces)

times Δx. From the figure, $\Delta V = 3x^2(\Delta x) + 3x(\Delta x)^2 + (\Delta x)^3$.

If Δx is small, then $\Delta V \approx 3x^2(\Delta x)$ and so $\Delta V/\Delta x \approx 3x^2$.

13. (a) Using $A(r) = \pi r^2$, we find that the average rate of change is:

(i) $\dfrac{A(3) - A(2)}{3 - 2} = \dfrac{9\pi - 4\pi}{1} = 5\pi$

(ii) $\dfrac{A(2.5) - A(2)}{2.5 - 2} = \dfrac{6.25\pi - 4\pi}{0.5} = 4.5\pi$

(iii) $\dfrac{A(2.1) - A(2)}{2.1 - 2} = \dfrac{4.41\pi - 4\pi}{0.1} = 4.1\pi$

(b) $A(r) = \pi r^2 \Rightarrow A'(r) = 2\pi r$, so $A'(2) = 4\pi$.

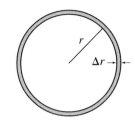

(c) The circumference is $C(r) = 2\pi r = A'(r)$. The figure suggests that if Δr is small, then the change in the area of the circle (a ring around the outside) is approximately equal to its circumference times Δr. Straightening out this ring gives us a shape that is approximately rectangular with length $2\pi r$ and width Δr, so $\Delta A \approx 2\pi r(\Delta r)$.

Algebraically, $\Delta A = A(r + \Delta r) - A(r) = \pi(r + \Delta r)^2 - \pi r^2 = 2\pi r(\Delta r) + \pi(\Delta r)^2$.

So we see that if Δr is small, then $\Delta A \approx 2\pi r(\Delta r)$ and therefore, $\Delta A/\Delta r \approx 2\pi r$.

14. After t seconds the radius is $r = 60t$, so the area is $A(t) = \pi(60t)^2 = 3600\pi t^2 \;\Rightarrow\; A'(t) = 7200\pi t \;\Rightarrow$

    (a) $A'(1) = 7200\pi$ cm^2/s
        (b) $A'(3) = 21,600\pi$ cm^2/s
        (c) $A'(5) = 36,000\pi$ cm^2/s

As time goes by, the area grows at an increasing rate. In fact, the rate of change is linear with respect to time.

15. $S(r) = 4\pi r^2 \;\Rightarrow\; S'(r) = 8\pi r \;\Rightarrow$

    (a) $S'(1) = 8\pi$ ft^2/ft
        (b) $S'(2) = 16\pi$ ft^2/ft
        (c) $S'(3) = 24\pi$ ft^2/ft

As the radius increases, the surface area grows at an increasing rate. In fact, the rate of change is linear with respect to the radius.

16. (a) Using $V(r) = \frac{4}{3}\pi r^3$, we find that the average rate of change is:

    (i) $\dfrac{V(8) - V(5)}{8 - 5} = \dfrac{\frac{4}{3}\pi(512) - \frac{4}{3}\pi(125)}{3} = 172\pi \;\mu\text{m}^3/\mu\text{m}$

    (ii) $\dfrac{V(6) - V(5)}{6 - 5} = \dfrac{\frac{4}{3}\pi(216) - \frac{4}{3}\pi(125)}{1} = 121.\overline{3}\pi \;\mu\text{m}^3/\mu\text{m}$

    (iii) $\dfrac{V(5.1) - V(5)}{5.1 - 5} = \dfrac{\frac{4}{3}\pi(5.1)^3 - \frac{4}{3}\pi(5)^3}{0.1} = 102.01\overline{3}\pi \;\mu\text{m}^3/\mu\text{m}$

  (b) $V'(r) = 4\pi r^2$, so $V'(5) = 100\pi \;\mu\text{m}^3/\mu\text{m}$.

  (c) $V(r) = \frac{4}{3}\pi r^3 \;\Rightarrow\; V'(r) = 4\pi r^2 = S(r)$. By analogy with Exercise 13(c), we can say that the change in the volume of the spherical shell, ΔV, is approximately equal to its thickness, Δr, times the surface area of the inner sphere. Thus, $\Delta V \approx 4\pi r^2(\Delta r)$ and so $\Delta V/\Delta r \approx 4\pi r^2$.

17. The mass is $f(x) = 3x^2$, so the linear density at x is $\rho(x) = f'(x) = 6x$.

    (a) $\rho(1) = 6$ kg/m
        (b) $\rho(2) = 12$ kg/m
        (c) $\rho(3) = 18$ kg/m

Since ρ is an increasing function, the density will be the highest at the right end of the rod and lowest at the left end.

18. $V(t) = 5000\left(1 - \frac{1}{40}t\right)^2 \;\Rightarrow\; V'(t) = 5000 \cdot 2\left(1 - \frac{1}{40}t\right)\left(-\frac{1}{40}\right) = -250\left(1 - \frac{1}{40}t\right)$

    (a) $V'(5) = -250\left(1 - \frac{5}{40}\right) = -218.75$ gal/min
        (b) $V'(10) = -250\left(1 - \frac{10}{40}\right) = -187.5$ gal/min

    (c) $V'(20) = -250\left(1 - \frac{20}{40}\right) = -125$ gal/min
        (d) $V'(40) = -250\left(1 - \frac{40}{40}\right) = 0$ gal/min

The water is flowing out the fastest at the beginning—when $t = 0$, $V'(t) = -250$ gal/min. The water is flowing out the slowest at the end—when $t = 40$, $V'(t) = 0$. As the tank empties, the water flows out more slowly.

19. The quantity of charge is $Q(t) = t^3 - 2t^2 + 6t + 2$, so the current is $Q'(t) = 3t^2 - 4t + 6$.

(a) $Q'(0.5) = 3(0.5)^2 - 4(0.5) + 6 = 4.75$ A   (b) $Q'(1) = 3(1)^2 - 4(1) + 6 = 5$ A

The current is lowest when Q' has a minimum. $Q''(t) = 6t - 4 < 0$ when $t < \frac{2}{3}$. So the current decreases when $t < \frac{2}{3}$ and

increases when $t > \frac{2}{3}$. Thus, the current is lowest at $t = \frac{2}{3}$ s.

20. (a) $F = \dfrac{GmM}{r^2} = (GmM)r^{-2}$ $\Rightarrow$ $\dfrac{dF}{dr} = -2(GmM)r^{-3} = -\dfrac{2GmM}{r^3}$, which is the rate of change of the force with

respect to the distance between the bodies. The minus sign indicates that as the distance r between the bodies increases,

the magnitude of the force F exerted by the body of mass m on the body of mass M is decreasing.

(b) Given $F'(20{,}000) = -2$, find $F'(10{,}000)$. $-2 = -\dfrac{2GmM}{20{,}000^3}$ $\Rightarrow$ $GmM = 20{,}000^3$.

$$F'(10{,}000) = -\dfrac{2(20{,}000^3)}{10{,}000^3} = -2 \cdot 2^3 = -16 \text{ N/km}$$

21. (a) To find the rate of change of volume with respect to pressure, we first solve for V in terms of P.

$PV = C$ $\Rightarrow$ $V = \dfrac{C}{P}$ $\Rightarrow$ $\dfrac{dV}{dP} = -\dfrac{C}{P^2}$.

(b) From the formula for dV/dP in part (a), we see that as P increases, the absolute value of dV/dP decreases.

Thus, the volume is decreasing more rapidly at the beginning.

(c) $\beta = -\dfrac{1}{V}\dfrac{dV}{dP} = -\dfrac{1}{V}\left(-\dfrac{C}{P^2}\right) = \dfrac{C}{(PV)P} = \dfrac{C}{CP} = \dfrac{1}{P}$

22. (a) $[C] = \dfrac{a^2 kt}{akt+1}$ $\Rightarrow$ rate of reaction $= \dfrac{d[C]}{dt} = \dfrac{(akt+1)(a^2 k) - (a^2 kt)(ak)}{(akt+1)^2} = \dfrac{a^2 k(akt+1-akt)}{(akt+1)^2} = \dfrac{a^2 k}{(akt+1)^2}$

(b) If $x = [C]$, then $a - x = a - \dfrac{a^2 kt}{akt+1} = \dfrac{a^2 kt + a - a^2 kt}{akt+1} = \dfrac{a}{akt+1}$.

So $k(a-x)^2 = k\left(\dfrac{a}{akt+1}\right)^2 = \dfrac{a^2 k}{(akt+1)^2} = \dfrac{d[C]}{dt}$ [from part (a)] $= \dfrac{dx}{dt}$.

23. (a) **1920:** $m_1 = \dfrac{1860 - 1750}{1920 - 1910} = \dfrac{110}{10} = 11$, $m_2 = \dfrac{2070 - 1860}{1930 - 1920} = \dfrac{210}{10} = 21$,

$(m_1 + m_2)/2 = (11 + 21)/2 = 16$ million/year

1980: $m_1 = \dfrac{4450 - 3710}{1980 - 1970} = \dfrac{740}{10} = 74$, $m_2 = \dfrac{5280 - 4450}{1990 - 1980} = \dfrac{830}{10} = 83$,

$(m_1 + m_2)/2 = (74 + 83)/2 = 78.5$ million/year

(b) $P(t) = at^3 + bt^2 + ct + d$ (in millions of people), where $a \approx 0.0012937063$, $b \approx -7.061421911$, $c \approx 12{,}822.97902$,

and $d \approx -7{,}743{,}770.396$.

(c) $P(t) = at^3 + bt^2 + ct + d$ $\Rightarrow$ $P'(t) = 3at^2 + 2bt + c$ (in millions of people per year)

(d) $P'(1920) = 3(0.0012937063)(1920)^2 + 2(-7.061421911)(1920) + 12{,}822.97902$

≈ 14.48 million/year [smaller than the answer in part (a), but close to it]

$P'(1980) \approx 75.29$ million/year (smaller, but close)

(e) $P'(1985) \approx 81.62$ million/year, so the rate of growth in 1985 was about 81.62 million/year.

24. (a) $A(t) = at^4 + bt^3 + ct^2 + dt + e$, where $a \approx -3.076923 \times 10^{-6}$,

$b \approx 0.0243620824$, $c \approx -72.33147086$, $d \approx 95442.50365$, and

$e \approx -47{,}224{,}986.6$.

(b) $A(t) = at^4 + bt^3 + ct^2 + dt + e \;\Rightarrow$

$A'(t) = 4at^3 + 3bt^2 + 2ct + d$.

(c) Part (b) gives $A'(1990) \approx 0.11$ years of age per year.

(d)

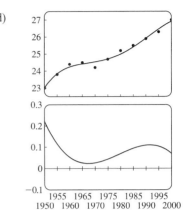

25. (a) Using $v = \dfrac{P}{4\eta l}(R^2 - r^2)$ with $R = 0.01$, $l = 3$, $P = 3000$, and $\eta = 0.027$, we have v as a function of r:

$$v(r) = \frac{3000}{4(0.027)3}(0.01^2 - r^2). \quad v(0) = 0.\overline{925}\ \text{cm/s},\ v(0.005) = 0.69\overline{4}\ \text{cm/s},\ v(0.01) = 0.$$

(b) $v(r) = \dfrac{P}{4\eta l}(R^2 - r^2) \;\Rightarrow\; v'(r) = \dfrac{P}{4\eta l}(-2r) = -\dfrac{Pr}{2\eta l}$.   When $l = 3$, $P = 3000$, and $\eta = 0.027$, we have

$$v'(r) = -\frac{3000r}{2(0.027)3}. \quad v'(0) = 0,\ v'(0.005) = -92.\overline{592}\ (\text{cm/s})/\text{cm, and}\ v'(0.01) = -185.\overline{185}\ (\text{cm/s})/\text{cm}.$$

(c) The velocity is greatest where $r = 0$ (at the center) and the velocity is changing most where $r = R = 0.01$ cm

(at the edge).

26. (a) (i) $f = \dfrac{1}{2L}\sqrt{\dfrac{T}{\rho}} = \left(\dfrac{1}{2}\sqrt{\dfrac{T}{\rho}}\right)L^{-1} \;\Rightarrow\; \dfrac{df}{dL} = -\left(\dfrac{1}{2}\sqrt{\dfrac{T}{\rho}}\right)L^{-2} = -\dfrac{1}{2L^2}\sqrt{\dfrac{T}{\rho}}$

(ii) $f = \dfrac{1}{2L}\sqrt{\dfrac{T}{\rho}} = \left(\dfrac{1}{2L\sqrt{\rho}}\right)T^{1/2} \;\Rightarrow\; \dfrac{df}{dT} = \dfrac{1}{2}\left(\dfrac{1}{2L\sqrt{\rho}}\right)T^{-1/2} = \dfrac{1}{4L\sqrt{T\rho}}$

(iii) $f = \dfrac{1}{2L}\sqrt{\dfrac{T}{\rho}} = \left(\dfrac{\sqrt{T}}{2L}\right)\rho^{-1/2} \;\Rightarrow\; \dfrac{df}{d\rho} = -\dfrac{1}{2}\left(\dfrac{\sqrt{T}}{2L}\right)\rho^{-3/2} = -\dfrac{\sqrt{T}}{4L\rho^{3/2}}$

(b) *Note:* Illustrating tangent lines on the generic figures may help to explain the results.

(i) $\dfrac{df}{dL} < 0$ and L is decreasing $\Rightarrow$ f is increasing $\Rightarrow$ higher note

(ii) $\dfrac{df}{dT} > 0$ and T is increasing $\Rightarrow$ f is increasing $\Rightarrow$ higher note

(iii) $\dfrac{df}{d\rho} < 0$ and ρ is increasing $\Rightarrow$ f is decreasing $\Rightarrow$ lower note

27. (a) $C(x) = 1200 + 12x - 0.1x^2 + 0.0005x^3$ $\Rightarrow$ $C'(x) = 12 - 0.2x + 0.0015x^2$ $/yard, which is the marginal cost function.

(b) $C'(200) = 12 - 0.2(200) + 0.0015(200)^2 = \$32/$yard, and this is the rate at which costs are increasing with respect to the production level when $x = 200$. $C'(200)$ predicts the cost of producing the 201st yard.

(c) The cost of manufacturing the 201st yard of fabric is $C(201) - C(200) = 3632.2005 - 3600 \approx \$32.20$, which is approximately $C'(200)$.

28. (a) $C(x) = 339 + 25x - 0.09x^2 + 0.0004x^3$ $\Rightarrow$ $C'(x) = 25 - 0.18x + 0.0012x^2$. $C'(100) = \$19/$item, and this is the rate at which costs are increasing with respect to the production level when $x = 100$.

(b) The cost of producing the 101st item is $C(101) - C(100) = 2358.0304 - 2339 = \$19.03$, which is approximately $C'(100)$.

29. (a) $A(x) = \dfrac{p(x)}{x}$ $\Rightarrow$ $A'(x) = \dfrac{xp'(x) - p(x) \cdot 1}{x^2} = \dfrac{xp'(x) - p(x)}{x^2}$.

$A'(x) > 0$ $\Rightarrow$ $A(x)$ is increasing; that is, the average productivity increases as the size of the workforce increases.

(b) $p'(x)$ is greater than the average productivity $\Rightarrow$ $p'(x) > A(x)$ $\Rightarrow$ $p'(x) > \dfrac{p(x)}{x}$ $\Rightarrow$ $xp'(x) > p(x)$ $\Rightarrow$

$xp'(x) - p(x) > 0$ $\Rightarrow$ $\dfrac{xp'(x) - p(x)}{x^2} > 0$ $\Rightarrow$ $A'(x) > 0$.

30. (a) $S = \dfrac{dR}{dx} = \dfrac{(1 + 4x^{0.4})(9.6x^{-0.6}) - (40 + 24x^{0.4})(1.6x^{-0.6})}{(1 + 4x^{0.4})^2}$

$= \dfrac{9.6x^{-0.6} + 38.4x^{-0.2} - 64x^{-0.6} - 38.4x^{-0.2}}{(1 + 4x^{0.4})^2} = -\dfrac{54.4x^{-0.6}}{(1 + 4x^{0.4})^2}$

(b)

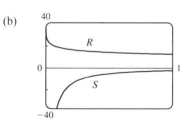

At low levels of brightness, R is quite large $[R(0) = 40]$ and is quickly decreasing, that is, S is negative with large absolute value. This is to be expected: at low levels of brightness, the eye is more sensitive to slight changes than it is at higher levels of brightness.

31. $PV = nRT \Rightarrow T = \dfrac{PV}{nR} = \dfrac{PV}{(10)(0.0821)} = \dfrac{1}{0.821}(PV)$. Using the Product Rule, we have

$\dfrac{dT}{dt} = \dfrac{1}{0.821}\left[P(t)V'(t) + V(t)P'(t)\right] = \dfrac{1}{0.821}\left[(8)(-0.15) + (10)(0.10)\right] \approx -0.2436\ \text{K/min}.$

32. (a) If $dP/dt = 0$, the population is stable (it is constant).

(b) $\dfrac{dP}{dt} = 0 \Rightarrow \beta P = r_0\left(1 - \dfrac{P}{P_c}\right)P \Rightarrow \dfrac{\beta}{r_0} = 1 - \dfrac{P}{P_c} \Rightarrow \dfrac{P}{P_c} = 1 - \dfrac{\beta}{r_0} \Rightarrow P = P_c\left(1 - \dfrac{\beta}{r_0}\right).$

If $P_c = 10{,}000$, $r_0 = 5\% = 0.05$, and $\beta = 4\% = 0.04$, then $P = 10{,}000\left(1 - \frac{4}{5}\right) = 2000$.

(c) If $\beta = 0.05$, then $P = 10{,}000\left(1 - \frac{5}{5}\right) = 0$. There is no stable population.

33. (a) If the populations are stable, then the growth rates are neither positive nor negative; that is, $\dfrac{dC}{dt} = 0$ and $\dfrac{dW}{dt} = 0$.

(b) "The caribou go extinct" means that the population is zero, or mathematically, $C = 0$.

(c) We have the equations $\dfrac{dC}{dt} = aC - bCW$ and $\dfrac{dW}{dt} = -cW + dCW$. Let $dC/dt = dW/dt = 0$, $a = 0.05$, $b = 0.001$,

$c = 0.05$, and $d = 0.0001$ to obtain $0.05C - 0.001CW = 0$ **(1)** and $-0.05W + 0.0001CW = 0$ **(2)**. Adding 10 times

(2) to **(1)** eliminates the CW-terms and gives us $0.05C - 0.5W = 0 \Rightarrow C = 10W$. Substituting $C = 10W$ into **(1)**

results in $0.05(10W) - 0.001(10W)W = 0 \Leftrightarrow 0.5W - 0.01W^2 = 0 \Leftrightarrow 50W - W^2 = 0 \Leftrightarrow$

$W(50 - W) = 0 \Leftrightarrow W = 0$ or 50. Since $C = 10W$, $C = 0$ or 500. Thus, the population pairs (C, W) that lead to

stable populations are $(0, 0)$ and $(500, 50)$. So it is possible for the two species to live in harmony.

3.8 Related Rates

1. $V = x^3 \Rightarrow \dfrac{dV}{dt} = \dfrac{dV}{dx}\dfrac{dx}{dt} = 3x^2\,\dfrac{dx}{dt}$

2. (a) $A = \pi r^2 \Rightarrow \dfrac{dA}{dt} = \dfrac{dA}{dr}\dfrac{dr}{dt} = 2\pi r\,\dfrac{dr}{dt}$ \qquad (b) $\dfrac{dA}{dt} = 2\pi r\,\dfrac{dr}{dt} = 2\pi(30\ \text{m})(1\ \text{m/s}) = 60\pi\ \text{m}^2/\text{s}$

3. Let s denote the side of a square. The square's area A is given by $A = s^2$. Differentiating with respect to t gives us

$\dfrac{dA}{dt} = 2s\,\dfrac{ds}{dt}$. When $A = 16$, $s = 4$. Substitution 4 for s and 6 for $\dfrac{ds}{dt}$ gives us $\dfrac{dA}{dt} = 2(4)(6) = 48\ \text{cm}^2/\text{s}$.

4. $A = \ell w \Rightarrow \dfrac{dA}{dt} = \ell \cdot \dfrac{dw}{dt} + w \cdot \dfrac{d\ell}{dt} = 20(3) + 10(8) = 140\ \text{cm}^2/\text{s}$.

5. $V = \pi r^2 h = \pi(5)^2 h = 25\pi h \Rightarrow \dfrac{dV}{dt} = 25\pi\,\dfrac{dh}{dt} \Rightarrow 3 = 25\pi\,\dfrac{dh}{dt} \Rightarrow \dfrac{dh}{dt} = \dfrac{3}{25\pi}\ \text{m/min}.$

6. $V = \frac{4}{3}\pi r^3 \Rightarrow \dfrac{dV}{dt} = \frac{4}{3}\pi \cdot 3r^2\,\dfrac{dr}{dt} \Rightarrow \dfrac{dV}{dt} = 4\pi\left(\frac{1}{2} \cdot 80\right)^2(4) = 25{,}600\pi\ \text{mm}^3/\text{s}.$

7. $y = x^3 + 2x \Rightarrow \dfrac{dy}{dt} = \dfrac{dy}{dx}\dfrac{dx}{dt} = (3x^2 + 2)(5) = 5(3x^2 + 2)$. When $x = 2$, $\dfrac{dy}{dt} = 5(14) = 70$.

8. $x^2 + y^2 = 25 \Rightarrow 2x\,\dfrac{dx}{dt} + 2y\,\dfrac{dy}{dt} = 0 \Rightarrow x\,\dfrac{dx}{dt} = -y\,\dfrac{dy}{dt} \Rightarrow \dfrac{dx}{dt} = -\dfrac{y}{x}\dfrac{dy}{dt}.$

When $y = 4$, $x^2 + 4^2 = 25 \Rightarrow x = \pm 3$. For $\dfrac{dy}{dt} = 6$, $\dfrac{dx}{dt} = -\dfrac{4}{\pm 3}(6) = \mp 8$.

9. $z^2 = x^2 + y^2 \quad \Rightarrow \quad 2z\dfrac{dz}{dt} = 2x\dfrac{dx}{dt} + 2y\dfrac{dy}{dt} \quad \Rightarrow \quad \dfrac{dz}{dt} = \dfrac{1}{z}\left(x\dfrac{dx}{dt} + y\dfrac{dy}{dt}\right)$. When $x = 5$ and $y = 12$,

$z^2 = 5^2 + 12^2 \quad \Rightarrow \quad z^2 = 169 \quad \Rightarrow \quad z = \pm 13$. For $\dfrac{dx}{dt} = 2$ and $\dfrac{dy}{dt} = 3$, $\dfrac{dz}{dt} = \dfrac{1}{\pm 13}(5\cdot 2 + 12\cdot 3) = \pm\dfrac{46}{13}$.

10. $y = \sqrt{1 + x^3} \quad \Rightarrow \quad \dfrac{dy}{dt} = \dfrac{dy}{dx}\dfrac{dx}{dt} = \dfrac{1}{2}(1 + x^3)^{-1/2}(3x^2)\dfrac{dx}{dt} = \dfrac{3x^2}{2\sqrt{1 + x^3}}\dfrac{dx}{dt}$. With $\dfrac{dy}{dt} = 4$ when $x = 2$ and $y = 3$,

we have $4 = \dfrac{3(4)}{2(3)}\dfrac{dx}{dt} \quad \Rightarrow \quad \dfrac{dx}{dt} = 2$ cm/s.

11. (a) Given: a plane flying horizontally at an altitude of 1 mi and a speed of 500 mi/h passes directly over a radar station.

If we let t be time (in hours) and x be the horizontal distance traveled by the plane (in mi), then we are given that $dx/dt = 500$ mi/h.

(b) Unknown: the rate at which the distance from the plane to the station is increasing   (c)

when it is 2 mi from the station. If we let y be the distance from the plane to the station,

then we want to find dy/dt when $y = 2$ mi.

(d) By the Pythagorean Theorem, $y^2 = x^2 + 1 \quad \Rightarrow \quad 2y\,(dy/dt) = 2x\,(dx/dt)$.

(e) $\dfrac{dy}{dt} = \dfrac{x}{y}\dfrac{dx}{dt} = \dfrac{x}{y}(500)$. Since $y^2 = x^2 + 1$, when $y = 2$, $x = \sqrt{3}$, so $\dfrac{dy}{dt} = \dfrac{\sqrt{3}}{2}(500) = 250\sqrt{3} \approx 433$ mi/h.

12. (a) Given: the rate of decrease of the surface area is 1 cm^2/min. If we let t be   (c)

time (in minutes) and S be the surface area (in cm^2), then we are given that

$dS/dt = -1$ cm^2/s.

(b) Unknown: the rate of decrease of the diameter when the diameter is 10 cm.

If we let x be the diameter, then we want to find dx/dt when $x = 10$ cm.

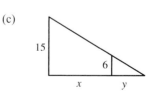

(d) If the radius is r and the diameter $x = 2r$, then $r = \frac{1}{2}x$ and

$$S = 4\pi r^2 = 4\pi\left(\tfrac{1}{2}x\right)^2 = \pi x^2 \quad \Rightarrow \quad \dfrac{dS}{dt} = \dfrac{dS}{dx}\dfrac{dx}{dt} = 2\pi x\dfrac{dx}{dt}.$$

(e) $-1 = \dfrac{dS}{dt} = 2\pi x\dfrac{dx}{dt} \quad \Rightarrow \quad \dfrac{dx}{dt} = -\dfrac{1}{2\pi x}$. When $x = 10$, $\dfrac{dx}{dt} = -\dfrac{1}{20\pi}$. So the rate of decrease is $\dfrac{1}{20\pi}$ cm/min.

13. (a) Given: a man 6 ft tall walks away from a street light mounted on a 15-ft-tall pole at a rate of 5 ft/s. If we let t be time (in s)

and x be the distance from the pole to the man (in ft), then we are given that $dx/dt = 5$ ft/s.

(b) Unknown: the rate at which the tip of his shadow is moving when he is 40 ft   (c)

from the pole. If we let y be the distance from the man to the tip of his

shadow (in ft), then we want to find $\dfrac{d}{dt}(x + y)$ when $x = 40$ ft.

(d) By similar triangles, $\dfrac{15}{6} = \dfrac{x+y}{y} \;\Rightarrow\; 15y = 6x + 6y \;\Rightarrow\; 9y = 6x \;\Rightarrow\; y = \frac{2}{3}x$.

(e) The tip of the shadow moves at a rate of $\dfrac{d}{dt}(x+y) = \dfrac{d}{dt}\left(x + \dfrac{2}{3}x\right) = \dfrac{5}{3}\dfrac{dx}{dt} = \frac{5}{3}(5) = \frac{25}{3}$ ft/s.

14. (a) Given: at noon, ship A is 150 km west of ship B; ship A is sailing east at 35 km/h, and ship B is sailing north at 25 km/h.

If we let t be time (in hours), x be the distance traveled by ship A (in km), and y be the distance traveled by ship B (in km), then we are given that $dx/dt = 35$ km/h and $dy/dt = 25$ km/h.

(b) Unknown: the rate at which the distance between the ships is changing at 4:00 PM. If we let z be the distance between the ships, then we want to find dz/dt when $t = 4$ h.

(c)

(d) $z^2 = (150 - x)^2 + y^2 \;\Rightarrow\; 2z\dfrac{dz}{dt} = 2(150 - x)\left(-\dfrac{dx}{dt}\right) + 2y\dfrac{dy}{dt}$

(e) At 4:00 PM, $x = 4(35) = 140$ and $y = 4(25) = 100 \;\Rightarrow\; z = \sqrt{(150 - 140)^2 + 100^2} = \sqrt{10{,}100}$.

So $\dfrac{dz}{dt} = \dfrac{1}{z}\left[(x - 150)\dfrac{dx}{dt} + y\dfrac{dy}{dt}\right] = \dfrac{-10(35) + 100(25)}{\sqrt{10{,}100}} = \dfrac{215}{\sqrt{101}} \approx 21.4$ km/h.

15.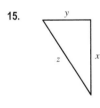

We are given that $\dfrac{dx}{dt} = 60$ mi/h and $\dfrac{dy}{dt} = 25$ mi/h. $z^2 = x^2 + y^2 \;\Rightarrow$

$2z\dfrac{dz}{dt} = 2x\dfrac{dx}{dt} + 2y\dfrac{dy}{dt} \;\Rightarrow\; z\dfrac{dz}{dt} = x\dfrac{dx}{dt} + y\dfrac{dy}{dt} \;\Rightarrow\; \dfrac{dz}{dt} = \dfrac{1}{z}\left(x\dfrac{dx}{dt} + y\dfrac{dy}{dt}\right)$.

After 2 hours, $x = 2(60) = 120$ and $y = 2(25) = 50 \;\Rightarrow\; z = \sqrt{120^2 + 50^2} = 130$,

so $\dfrac{dz}{dt} = \dfrac{1}{z}\left(x\dfrac{dx}{dt} + y\dfrac{dy}{dt}\right) = \dfrac{120(60) + 50(25)}{130} = 65$ mi/h.

16.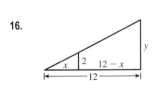

We are given that $\dfrac{dx}{dt} = 1.6$ m/s. By similar triangles, $\dfrac{y}{12} = \dfrac{2}{x} \;\Rightarrow\; y = \dfrac{24}{x} \;\Rightarrow$

$\dfrac{dy}{dt} = -\dfrac{24}{x^2}\dfrac{dx}{dt} = -\dfrac{24}{x^2}(1.6)$. When $x = 8$, $\dfrac{dy}{dt} = -\dfrac{24(1.6)}{64} = -0.6$ m/s, so the shadow is decreasing at a rate of 0.6 m/s.

17.

We are given that $\dfrac{dx}{dt} = 4$ ft/s and $\dfrac{dy}{dt} = 5$ ft/s. $z^2 = (x+y)^2 + 500^2 \;\Rightarrow$

$2z\dfrac{dz}{dt} = 2(x+y)\left(\dfrac{dx}{dt} + \dfrac{dy}{dt}\right)$. 15 minutes after the woman starts, we have

$x = (4 \text{ ft/s})(20 \text{ min})(60 \text{ s/min}) = 4800$ ft and $y = 5 \cdot 15 \cdot 60 = 4500 \;\Rightarrow$

$z = \sqrt{(4800 + 4500)^2 + 500^2} = \sqrt{86{,}740{,}000}$, so

$\dfrac{dz}{dt} = \dfrac{x+y}{z}\left(\dfrac{dx}{dt} + \dfrac{dy}{dt}\right) = \dfrac{4800 + 4500}{\sqrt{86{,}740{,}000}}(4 + 5) = \dfrac{837}{\sqrt{8674}} \approx 8.99$ ft/s.

18. We are given that $\dfrac{dx}{dt} = 24$ ft/s.

(a)

$y^2 = (90 - x)^2 + 90^2 \;\Rightarrow\; 2y\,\dfrac{dy}{dt} = 2(90 - x)\left(-\dfrac{dx}{dt}\right)$. When $x = 45$,

$y = \sqrt{45^2 + 90^2} = 45\sqrt{5}$, so $\dfrac{dy}{dt} = \dfrac{90 - x}{y}\left(-\dfrac{dx}{dt}\right) = \dfrac{45}{45\sqrt{5}}(-24) = -\dfrac{24}{\sqrt{5}}$,

so the distance from second base is decreasing at a rate of $\dfrac{24}{\sqrt{5}} \approx 10.7$ ft/s.

(b) Due to the symmetric nature of the problem in part (a), we expect to get the same answer—and we do.

$z^2 = x^2 + 90^2 \;\Rightarrow\; 2z\,\dfrac{dz}{dt} = 2x\,\dfrac{dx}{dt}$. When $x = 45$, $z = 45\sqrt{5}$, so $\dfrac{dz}{dt} = \dfrac{45}{45\sqrt{5}}(24) = \dfrac{24}{\sqrt{5}} \approx 10.7$ ft/s.

19. $A = \frac{1}{2}bh$, where b is the base and h is the altitude. We are given that $\dfrac{dh}{dt} = 1$ cm/min and $\dfrac{dA}{dt} = 2$ cm²/min. Using the

Product Rule, we have $\dfrac{dA}{dt} = \dfrac{1}{2}\left(b\,\dfrac{dh}{dt} + h\,\dfrac{db}{dt}\right)$. When $h = 10$ and $A = 100$, we have $100 = \frac{1}{2}b(10) \;\Rightarrow\; \frac{1}{2}b = 10 \;\Rightarrow$

$b = 20$, so $2 = \dfrac{1}{2}\left(20 \cdot 1 + 10\,\dfrac{db}{dt}\right) \;\Rightarrow\; 4 = 20 + 10\,\dfrac{db}{dt} \;\Rightarrow\; \dfrac{db}{dt} = \dfrac{4 - 20}{10} = -1.6$ cm/min.

20.

Given $\dfrac{dy}{dt} = -1$ m/s, find $\dfrac{dx}{dt}$ when $x = 8$ m. $y^2 = x^2 + 1 \;\Rightarrow\; 2y\,\dfrac{dy}{dt} = 2x\,\dfrac{dx}{dt} \;\Rightarrow$

$\dfrac{dx}{dt} = \dfrac{y}{x}\dfrac{dy}{dt} = -\dfrac{y}{x}$. When $x = 8$, $y = \sqrt{65}$, so $\dfrac{dx}{dt} = -\dfrac{\sqrt{65}}{8}$. Thus, the boat approaches

the dock at $\dfrac{\sqrt{65}}{8} \approx 1.01$ m/s.

21.

We are given that $\dfrac{dx}{dt} = 35$ km/h and $\dfrac{dy}{dt} = 25$ km/h. $z^2 = (x + y)^2 + 100^2 \;\Rightarrow$

$2z\,\dfrac{dz}{dt} = 2(x + y)\left(\dfrac{dx}{dt} + \dfrac{dy}{dt}\right)$. At 4:00 PM, $x = 4(35) = 140$ and $y = 4(25) = 100 \;\Rightarrow$

$z = \sqrt{(140 + 100)^2 + 100^2} = \sqrt{67{,}600} = 260$, so

$\dfrac{dz}{dt} = \dfrac{x + y}{z}\left(\dfrac{dx}{dt} + \dfrac{dy}{dt}\right) = \dfrac{140 + 100}{260}(35 + 25) = \dfrac{720}{13} \approx 55.4$ km/h.

22. Let D denote the distance from the origin $(0, 0)$ to the point on the curve $y = \sqrt{x}$.

$D = \sqrt{(x - 0)^2 + (y - 0)^2} = \sqrt{x^2 + \left(\sqrt{x}\right)^2} = \sqrt{x^2 + x} \;\Rightarrow\; \dfrac{dD}{dt} = \frac{1}{2}(x^2 + x)^{-1/2}(2x + 1)\dfrac{dx}{dt} = \dfrac{2x + 1}{2\sqrt{x^2 + x}}\dfrac{dx}{dt}$.

With $\dfrac{dx}{dt} = 3$ when $x = 4$, $\dfrac{dD}{dt} = \dfrac{9}{2\sqrt{20}}(3) = \dfrac{27}{4\sqrt{5}} \approx 3.02$ cm/s.

23. If $C =$ the rate at which water is pumped in, then $\dfrac{dV}{dt} = C - 10{,}000$, where

$V = \frac{1}{3}\pi r^2 h$ is the volume at time t. By similar triangles, $\dfrac{r}{2} = \dfrac{h}{6} \;\Rightarrow\; r = \dfrac{1}{3}h \;\Rightarrow$

$V = \frac{1}{3}\pi\left(\frac{1}{3}h\right)^2 h = \frac{\pi}{27}h^3 \;\Rightarrow\; \dfrac{dV}{dt} = \dfrac{\pi}{9}h^2\,\dfrac{dh}{dt}$. When $h = 200$ cm,

$\dfrac{dh}{dt} = 20$ cm/min, so $C - 10{,}000 = \dfrac{\pi}{9}(200)^2(20) \;\Rightarrow\; C = 10{,}000 + \dfrac{800{,}000}{9}\pi \approx 289{,}253$ cm³/min.

24. By similar triangles, $\dfrac{3}{1} = \dfrac{b}{h}$, so $b = 3h$. The trough has volume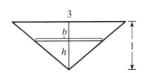

$$V = \tfrac{1}{2}bh(10) = 5(3h)h = 15h^2 \quad \Rightarrow \quad 12 = \frac{dV}{dt} = 30h\frac{dh}{dt} \quad \Rightarrow \quad \frac{dh}{dt} = \frac{2}{5h}.$$

When $h = \tfrac{1}{2}$, $\dfrac{dh}{dt} = \dfrac{2}{5 \cdot \frac{1}{2}} = \dfrac{4}{5}$ ft/min.

25. The figure is labeled in meters. The area A of a trapezoid is

$\tfrac{1}{2}(\text{base}_1 + \text{base}_2)(\text{height})$, and the volume V of the 10-meter-long trough is $10A$.

Thus, the volume of the trapezoid with height h is $V = (10)\tfrac{1}{2}[0.3 + (0.3 + 2a)]h$.

By similar triangles, $\dfrac{a}{h} = \dfrac{0.25}{0.5} = \dfrac{1}{2}$, so $2a = h \quad \Rightarrow \quad V = 5(0.6 + h)h = 3h + 5h^2$.

Now $\dfrac{dV}{dt} = \dfrac{dV}{dh}\dfrac{dh}{dt} \quad \Rightarrow \quad 0.2 = (3 + 10h)\dfrac{dh}{dt} \quad \Rightarrow \quad \dfrac{dh}{dt} = \dfrac{0.2}{3 + 10h}$. When $h = 0.3$,

$$\frac{dh}{dt} = \frac{0.2}{3 + 10(0.3)} = \frac{0.2}{6} \text{ m/min} = \frac{1}{30} \text{ m/min or } \frac{10}{3} \text{ cm/min}.$$

26. The figure is drawn without the top 3 feet.

$V = \tfrac{1}{2}(b + 12)h(20) = 10(b + 12)h$ and, from similar triangles,

$\dfrac{x}{h} = \dfrac{6}{6}$ and $\dfrac{y}{h} = \dfrac{16}{6} = \dfrac{8}{3}$, so $b = x + 12 + y = h + 12 + \dfrac{8h}{3} = 12 + \dfrac{11h}{3}$.

Thus, $V = 10\left(24 + \dfrac{11h}{3}\right)h = 240h + \dfrac{110h^2}{3}$ and so $0.8 = \dfrac{dV}{dt} = \left(240 + \dfrac{220}{3}h\right)\dfrac{dh}{dt}$.

When $h = 5$, $\dfrac{dh}{dt} = \dfrac{0.8}{240 + 5(220/3)} = \dfrac{3}{2275} \approx 0.00132$ ft/min.

27. We are given that $\dfrac{dV}{dt} = 30$ ft³/min. $V = \dfrac{1}{3}\pi r^2 h = \dfrac{1}{3}\pi\left(\dfrac{h}{2}\right)^2 h = \dfrac{\pi h^3}{12} \quad \Rightarrow$

$\dfrac{dV}{dt} = \dfrac{dV}{dh}\dfrac{dh}{dt} \quad \Rightarrow \quad 30 = \dfrac{\pi h^2}{4}\dfrac{dh}{dt} \quad \Rightarrow \quad \dfrac{dh}{dt} = \dfrac{120}{\pi h^2}$.

When $h = 10$ ft, $\dfrac{dh}{dt} = \dfrac{120}{10^2\pi} = \dfrac{6}{5\pi} \approx 0.38$ ft/min.

28. We are given $dx/dt = 8$ ft/s. $\cot\theta = \dfrac{x}{100} \quad \Rightarrow \quad x = 100\cot\theta \quad \Rightarrow$

$\dfrac{dx}{dt} = -100\csc^2\theta\dfrac{d\theta}{dt} \quad \Rightarrow \quad \dfrac{d\theta}{dt} = -\dfrac{\sin^2\theta}{100} \cdot 8$. When $y = 200$, $\sin\theta = \dfrac{100}{200} = \dfrac{1}{2} \quad \Rightarrow$

$\dfrac{d\theta}{dt} = -\dfrac{(1/2)^2}{100} \cdot 8 = -\dfrac{1}{50}$ rad/s. The angle is decreasing at a rate of $\dfrac{1}{50}$ rad/s.

29. $A = \tfrac{1}{2}bh$, but $b = 5$ m and $\sin\theta = \dfrac{h}{4} \quad \Rightarrow \quad h = 4\sin\theta$, so $A = \tfrac{1}{2}(5)(4\sin\theta) = 10\sin\theta$.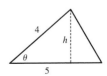

We are given $\dfrac{d\theta}{dt} = 0.06$ rad/s, so $\dfrac{dA}{dt} = \dfrac{dA}{d\theta}\dfrac{d\theta}{dt} = (10\cos\theta)(0.06) = 0.6\cos\theta$.

When $\theta = \dfrac{\pi}{3}$, $\dfrac{dA}{dt} = 0.6\left(\cos\dfrac{\pi}{3}\right) = (0.6)\left(\tfrac{1}{2}\right) = 0.3$ m²/s.

30. $\cos\theta = \dfrac{x}{10}$ $\Rightarrow$ $-\sin\theta\,\dfrac{d\theta}{dt} = \dfrac{1}{10}\dfrac{dx}{dt}$. From Example 2, $\dfrac{dx}{dt} = 1$ and

when $x = 6$, $y = 8$, so $\sin\theta = \dfrac{8}{10}$.

Thus, $-\dfrac{8}{10}\dfrac{d\theta}{dt} = \dfrac{1}{10}(1)$ $\Rightarrow$ $\dfrac{d\theta}{dt} = -\dfrac{1}{8}$ rad/s.

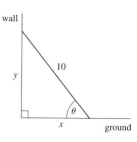

31. Differentiating both sides of $PV = C$ with respect to t and using the Product Rule gives us $P\dfrac{dV}{dt} + V\dfrac{dP}{dt} = 0$ $\Rightarrow$

$\dfrac{dV}{dt} = -\dfrac{V}{P}\dfrac{dP}{dt}$. When $V = 600$, $P = 150$ and $\dfrac{dP}{dt} = 20$, so we have $\dfrac{dV}{dt} = -\dfrac{600}{150}(20) = -80$. Thus, the volume is

decreasing at a rate of 80 cm³/min.

32. $PV^{1.4} = C$ $\Rightarrow$ $P\cdot 1.4V^{0.4}\dfrac{dV}{dt} + V^{1.4}\dfrac{dP}{dt} = 0$ $\Rightarrow$ $\dfrac{dV}{dt} = -\dfrac{V^{1.4}}{P\cdot 1.4V^{0.4}}\dfrac{dP}{dt} = -\dfrac{V}{1.4P}\dfrac{dP}{dt}$.

When $V = 400$, $P = 80$ and $\dfrac{dP}{dt} = -10$, so we have $\dfrac{dV}{dt} = -\dfrac{400}{1.4(80)}(-10) = \dfrac{250}{7}$. Thus, the volume is increasing at a

rate of $\dfrac{250}{7} \approx 36$ cm³/min.

33. With $R_1 = 80$ and $R_2 = 100$, $\dfrac{1}{R} = \dfrac{1}{R_1} + \dfrac{1}{R_2} = \dfrac{1}{80} + \dfrac{1}{100} = \dfrac{180}{8000} = \dfrac{9}{400}$, so $R = \dfrac{400}{9}$. Differentiating $\dfrac{1}{R} = \dfrac{1}{R_1} + \dfrac{1}{R_2}$

with respect to t, we have $-\dfrac{1}{R^2}\dfrac{dR}{dt} = -\dfrac{1}{R_1^2}\dfrac{dR_1}{dt} - \dfrac{1}{R_2^2}\dfrac{dR_2}{dt}$ $\Rightarrow$ $\dfrac{dR}{dt} = R^2\left(\dfrac{1}{R_1^2}\dfrac{dR_1}{dt} + \dfrac{1}{R_2^2}\dfrac{dR_2}{dt}\right)$. When $R_1 = 80$ and

$R_2 = 100$, $\dfrac{dR}{dt} = \dfrac{400^2}{9^2}\left[\dfrac{1}{80^2}(0.3) + \dfrac{1}{100^2}(0.2)\right] = \dfrac{107}{810} \approx 0.132$ Ω/s.

34. We want to find $\dfrac{dB}{dt}$ when $L = 18$ using $B = 0.007W^{2/3}$ and $W = 0.12L^{2.53}$.

$$\dfrac{dB}{dt} = \dfrac{dB}{dW}\dfrac{dW}{dL}\dfrac{dL}{dt} = \left(0.007\cdot\tfrac{2}{3}W^{-1/3}\right)(0.12\cdot 2.53\cdot L^{1.53})\left(\dfrac{20-15}{10{,}000{,}000}\right)$$

$$= \left[0.007\cdot\tfrac{2}{3}(0.12\cdot 18^{2.53})^{-1/3}\right](0.12\cdot 2.53\cdot 18^{1.53})\left(\dfrac{5}{10^7}\right) \approx 1.045\times 10^{-8} \text{ g/yr}$$

35. We are given $d\theta/dt = 2°/\text{min} = \dfrac{\pi}{90}$ rad/min. By the Law of Cosines,

$x^2 = 12^2 + 15^2 - 2(12)(15)\cos\theta = 369 - 360\cos\theta$ $\Rightarrow$

$2x\dfrac{dx}{dt} = 360\sin\theta\,\dfrac{d\theta}{dt}$ $\Rightarrow$ $\dfrac{dx}{dt} = \dfrac{180\sin\theta}{x}\dfrac{d\theta}{dt}$. When $\theta = 60°$,

$x = \sqrt{369 - 360\cos 60°} = \sqrt{189} = 3\sqrt{21}$, so $\dfrac{dx}{dt} = \dfrac{180\sin 60°}{3\sqrt{21}}\dfrac{\pi}{90} = \dfrac{\pi\sqrt{3}}{3\sqrt{21}} = \dfrac{\sqrt{7}\,\pi}{21} \approx 0.396$ m/min.

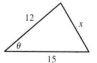

36. Using Q for the origin, we are given $\dfrac{dx}{dt} = -2$ ft/s and need to find $\dfrac{dy}{dt}$ when $x = -5$.

Using the Pythagorean Theorem twice, we have $\sqrt{x^2 + 12^2} + \sqrt{y^2 + 12^2} = 39$,

the total length of the rope. Differentiating with respect to t, we get

$$\dfrac{x}{\sqrt{x^2 + 12^2}}\dfrac{dx}{dt} + \dfrac{y}{\sqrt{y^2 + 12^2}}\dfrac{dy}{dt} = 0, \text{ so } \dfrac{dy}{dt} = -\dfrac{x\sqrt{y^2 + 12^2}}{y\sqrt{x^2 + 12^2}}\dfrac{dx}{dt}.$$

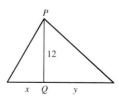

Now when $x = -5$, $39 = \sqrt{(-5)^2 + 12^2} + \sqrt{y^2 + 12^2} = 13 + \sqrt{y^2 + 12^2}$ $\Leftrightarrow$ $\sqrt{y^2 + 12^2} = 26$, and

$y = \sqrt{26^2 - 12^2} = \sqrt{532}$. So when $x = -5$, $\dfrac{dy}{dt} = -\dfrac{(-5)(26)}{\sqrt{532}\,(13)}(-2) = -\dfrac{10}{\sqrt{133}} \approx -0.87$ ft/s.

So cart B is moving towards Q at about 0.87 ft/s.

37. (a) By the Pythagorean Theorem, $4000^2 + y^2 = \ell^2$. Differentiating with respect to t,

we obtain $2y\,\dfrac{dy}{dt} = 2\ell\,\dfrac{d\ell}{dt}$. We know that $\dfrac{dy}{dt} = 600$ ft/s, so when $y = 3000$ ft,

$\ell = \sqrt{4000^2 + 3000^2} = \sqrt{25{,}000{,}000} = 5000$ ft

and $\dfrac{d\ell}{dt} = \dfrac{y}{\ell}\dfrac{dy}{dt} = \dfrac{3000}{5000}(600) = \dfrac{1800}{5} = 360$ ft/s.

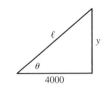

(b) Here $\tan\theta = \dfrac{y}{4000}$ $\Rightarrow$ $\dfrac{d}{dt}(\tan\theta) = \dfrac{d}{dt}\left(\dfrac{y}{4000}\right)$ $\Rightarrow$ $\sec^2\theta\,\dfrac{d\theta}{dt} = \dfrac{1}{4000}\dfrac{dy}{dt}$ $\Rightarrow$ $\dfrac{d\theta}{dt} = \dfrac{\cos^2\theta}{4000}\dfrac{dy}{dt}$. When

$y = 3000$ ft, $\dfrac{dy}{dt} = 600$ ft/s, $\ell = 5000$ and $\cos\theta = \dfrac{4000}{\ell} = \dfrac{4000}{5000} = \dfrac{4}{5}$, so $\dfrac{d\theta}{dt} = \dfrac{(4/5)^2}{4000}(600) = 0.096$ rad/s.

38. We are given that $\dfrac{d\theta}{dt} = 4(2\pi) = 8\pi$ rad/min. $x = 3\tan\theta$ $\Rightarrow$

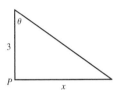

$\dfrac{dx}{dt} = 3\sec^2\theta\,\dfrac{d\theta}{dt}$. When $x = 1$, $\tan\theta = \frac{1}{3}$, so $\sec^2\theta = 1 + \left(\frac{1}{3}\right)^2 = \frac{10}{9}$

and $\dfrac{dx}{dt} = 3\left(\frac{10}{9}\right)(8\pi) = \frac{80}{3}\pi \approx 83.8$ km/min.

39. $\cot\theta = \dfrac{x}{5}$ $\Rightarrow$ $-\csc^2\theta\,\dfrac{d\theta}{dt} = \dfrac{1}{5}\dfrac{dx}{dt}$ $\Rightarrow$ $-\left(\csc\dfrac{\pi}{3}\right)^2\left(-\dfrac{\pi}{6}\right) = \dfrac{1}{5}\dfrac{dx}{dt}$ $\Rightarrow$

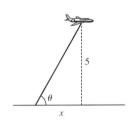

$\dfrac{dx}{dt} = \dfrac{5\pi}{6}\left(\dfrac{2}{\sqrt{3}}\right)^2 = \dfrac{10}{9}\pi$ km/min $[\approx 130$ mi/h$]$

40. We are given that $\dfrac{d\theta}{dt} = \dfrac{2\pi \text{ rad}}{2 \text{ min}} = \pi$ rad/min. By the Pythagorean Theorem, when

$h = 6$, $x = 8$, so $\sin\theta = \frac{6}{10}$ and $\cos\theta = \frac{8}{10}$. From the figure, $\sin\theta = \dfrac{h}{10}$ $\Rightarrow$

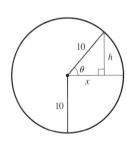

$h = 10\sin\theta$, so $\dfrac{dh}{dt} = 10\cos\theta\,\dfrac{d\theta}{dt} = 10\left(\dfrac{8}{10}\right)\pi = 8\pi$ m/min.

41. We are given that $\dfrac{dx}{dt} = 300$ km/h. By the Law of Cosines,

$y^2 = x^2 + 1^2 - 2(1)(x)\cos 120^\circ = x^2 + 1 - 2x\left(-\frac{1}{2}\right) = x^2 + x + 1$, so

$2y\,\dfrac{dy}{dt} = 2x\,\dfrac{dx}{dt} + \dfrac{dx}{dt}$ $\Rightarrow$ $\dfrac{dy}{dt} = \dfrac{2x+1}{2y}\dfrac{dx}{dt}$. After 1 minute, $x = \frac{300}{60} = 5$ km $\Rightarrow$

$y = \sqrt{5^2 + 5 + 1} = \sqrt{31}$ km $\Rightarrow$ $\dfrac{dy}{dt} = \dfrac{2(5)+1}{2\sqrt{31}}(300) = \dfrac{1650}{\sqrt{31}} \approx 296$ km/h.

42. We are given that $\frac{dx}{dt} = 3$ mi/h and $\frac{dy}{dt} = 2$ mi/h. By the Law of Cosines,

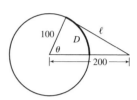

$$z^2 = x^2 + y^2 - 2xy \cos 45° = x^2 + y^2 - \sqrt{2}\,xy \quad \Rightarrow$$

$$2z \frac{dz}{dt} = 2x \frac{dx}{dt} + 2y \frac{dy}{dt} - \sqrt{2}\,x \frac{dy}{dt} - \sqrt{2}\,y \frac{dx}{dt}. \text{ After 15 minutes } \left[= \tfrac{1}{4}\text{ h}\right],$$

we have $x = \frac{3}{4}$ and $y = \frac{2}{4} = \frac{1}{2} \quad \Rightarrow \quad z^2 = \left(\frac{3}{4}\right)^2 + \left(\frac{2}{4}\right)^2 - \sqrt{2}\left(\frac{3}{4}\right)\left(\frac{2}{4}\right) \quad \Rightarrow \quad z = \frac{\sqrt{13 - 6\sqrt{2}}}{4}$ and

$$\frac{dz}{dt} = \frac{2}{\sqrt{13 - 6\sqrt{2}}}\left[2\left(\tfrac{3}{4}\right)3 + 2\left(\tfrac{1}{2}\right)2 - \sqrt{2}\left(\tfrac{3}{4}\right)2 - \sqrt{2}\left(\tfrac{1}{2}\right)3\right] = \frac{2}{\sqrt{13 - 6\sqrt{2}}} \frac{13 - 6\sqrt{2}}{2} = \sqrt{13 - 6\sqrt{2}} \approx 2.125 \text{ mi/h}.$$

43. Let the distance between the runner and the friend be ℓ. Then by the Law of Cosines,

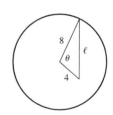

$$\ell^2 = 200^2 + 100^2 - 2 \cdot 200 \cdot 100 \cdot \cos \theta = 50{,}000 - 40{,}000 \cos \theta \quad (\star). \text{ Differentiating}$$

implicitly with respect to t, we obtain $2\ell \frac{d\ell}{dt} = -40{,}000(-\sin \theta)\frac{d\theta}{dt}$. Now if D is the

distance run when the angle is θ radians, then by the formula for the length of an arc

on a circle, $s = r\theta$, we have $D = 100\theta$, so $\theta = \frac{1}{100}D \quad \Rightarrow \quad \frac{d\theta}{dt} = \frac{1}{100}\frac{dD}{dt} = \frac{7}{100}$. To substitute into the expression for

$\frac{d\ell}{dt}$, we must know $\sin \theta$ at the time when $\ell = 200$, which we find from $(\star)$: $200^2 = 50{,}000 - 40{,}000 \cos \theta \quad \Leftrightarrow$

$\cos \theta = \frac{1}{4} \quad \Rightarrow \quad \sin \theta = \sqrt{1 - \left(\frac{1}{4}\right)^2} = \frac{\sqrt{15}}{4}$. Substituting, we get $2(200)\frac{d\ell}{dt} = 40{,}000 \frac{\sqrt{15}}{4}\left(\frac{7}{100}\right) \quad \Rightarrow$

$d\ell/dt = \frac{7\sqrt{15}}{4} \approx 6.78$ m/s. Whether the distance between them is increasing or decreasing depends on the direction in which

the runner is running.

44. The hour hand of a clock goes around once every 12 hours or, in radians per hour,

$\frac{2\pi}{12} = \frac{\pi}{6}$ rad/h. The minute hand goes around once an hour, or at the rate of 2π rad/h.

So the angle θ between them (measuring clockwise from the minute hand to the hour

hand) is changing at the rate of $d\theta/dt = \frac{\pi}{6} - 2\pi = -\frac{11\pi}{6}$ rad/h. Now, to relate θ to ℓ,

we use the Law of Cosines: $\ell^2 = 4^2 + 8^2 - 2 \cdot 4 \cdot 8 \cdot \cos \theta = 80 - 64 \cos \theta \quad (\star)$.

Differentiating implicitly with respect to t, we get $2\ell \frac{d\ell}{dt} = -64(-\sin \theta)\frac{d\theta}{dt}$. At 1:00, the angle between the two hands is

one-twelfth of the circle, that is, $\frac{2\pi}{12} = \frac{\pi}{6}$ radians. We use $(\star)$ to find ℓ at 1:00: $\ell = \sqrt{80 - 64 \cos \frac{\pi}{6}} = \sqrt{80 - 32\sqrt{3}}$.

Substituting, we get $2\ell \frac{d\ell}{dt} = 64 \sin \frac{\pi}{6}\left(-\frac{11\pi}{6}\right) \quad \Rightarrow \quad \frac{d\ell}{dt} = \frac{64\left(\frac{1}{2}\right)\left(-\frac{11\pi}{6}\right)}{2\sqrt{80 - 32\sqrt{3}}} = -\frac{88\pi}{3\sqrt{80 - 32\sqrt{3}}} \approx -18.6$.

So at 1:00, the distance between the tips of the hands is decreasing at a rate of 18.6 mm/h ≈ 0.005 mm/s.

3.9   Linear Approximations and Differentials

1. $f(x) = x^4 + 3x^2$ $\Rightarrow$ $f'(x) = 4x^3 + 6x$, so $f(-1) = 4$ and $f'(-1) = -10$.

Thus, $L(x) = f(-1) + f'(-1)(x - (-1)) = 4 + (-10)(x + 1) = -10x - 6$.

2. $f(x) = \dfrac{1}{\sqrt{2+x}} = (2+x)^{-1/2}$ $\Rightarrow$ $f'(x) = -\frac{1}{2}(2+x)^{-3/2}$ so $f(0) = \dfrac{1}{\sqrt{2}}$ and $f'(0) = -\dfrac{1}{4\sqrt{2}}$.

So $L(x) = f(0) + f'(0)(x - 0) = \dfrac{1}{\sqrt{2}} - \dfrac{1}{4\sqrt{2}}(x - 0) = \dfrac{1}{\sqrt{2}}\left(1 - \dfrac{1}{4}x\right)$.

3. $f(x) = \cos x$ $\Rightarrow$ $f'(x) = -\sin x$, so $f\left(\frac{\pi}{2}\right) = 0$ and $f'\left(\frac{\pi}{2}\right) = -1$.

Thus, $L(x) = f\left(\frac{\pi}{2}\right) + f'\left(\frac{\pi}{2}\right)\left(x - \frac{\pi}{2}\right) = 0 - 1\left(x - \frac{\pi}{2}\right) = -x + \frac{\pi}{2}$.

4. $f(x) = x^{3/4}$ $\Rightarrow$ $f'(x) = \frac{3}{4}x^{-1/4}$, so $f(16) = 8$ and $f'(16) = \frac{3}{8}$.

Thus, $L(x) = f(16) + f'(16)(x - 16) = 8 + \frac{3}{8}(x - 16) = \frac{3}{8}x + 2$.

5. $f(x) = \sqrt{1-x}$ $\Rightarrow$ $f'(x) = \dfrac{-1}{2\sqrt{1-x}}$, so $f(0) = 1$ and $f'(0) = -\frac{1}{2}$.

Therefore,

$\sqrt{1-x} = f(x) \approx f(0) + f'(0)(x - 0) = 1 + \left(-\frac{1}{2}\right)(x - 0) = 1 - \frac{1}{2}x$.

So $\sqrt{0.9} = \sqrt{1 - 0.1} \approx 1 - \frac{1}{2}(0.1) = 0.95$

and $\sqrt{0.99} = \sqrt{1 - 0.01} \approx 1 - \frac{1}{2}(0.01) = 0.995$.

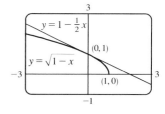

6. $g(x) = \sqrt[3]{1+x} = (1+x)^{1/3}$ $\Rightarrow$ $g'(x) = \frac{1}{3}(1+x)^{-2/3}$, so $g(0) = 1$ and $g'(0) = \frac{1}{3}$. Therefore, $\sqrt[3]{1+x} = g(x) \approx g(0) + g'(0)(x - 0) = 1 + \frac{1}{3}x$.

So $\sqrt[3]{0.95} = \sqrt[3]{1 + (-0.05)} \approx 1 + \frac{1}{3}(-0.05) = 0.98\overline{3}$,

and $\sqrt[3]{1.1} = \sqrt[3]{1 + 0.1} \approx 1 + \frac{1}{3}(0.1) = 1.0\overline{3}$.

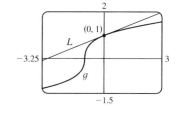

7. $f(x) = \sqrt[3]{1-x} = (1-x)^{1/3}$ $\Rightarrow$ $f'(x) = -\frac{1}{3}(1-x)^{-2/3}$, so $f(0) = 1$ and $f'(0) = -\frac{1}{3}$. Thus, $f(x) \approx f(0) + f'(0)(x - 0) = 1 - \frac{1}{3}x$. We need

$\sqrt[3]{1-x} - 0.1 < 1 - \frac{1}{3}x < \sqrt[3]{1-x} + 0.1$, which is true when

$-1.204 < x < 0.706$.

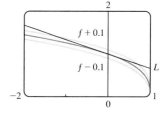

8. $f(x) = \tan x$ $\Rightarrow$ $f'(x) = \sec^2 x$, so $f(0) = 0$ and $f'(0) = 1$.

Thus, $f(x) \approx f(0) + f'(0)(x - 0) = 0 + 1(x - 0) = x$.

We need $\tan x - 0.1 < x < \tan x + 0.1$, which is true when

$-0.63 < x < 0.63$.

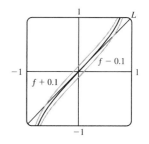

9. $f(x) = \dfrac{1}{(1+2x)^4} = (1+2x)^{-4}$ $\Rightarrow$

$f'(x) = -4(1+2x)^{-5}(2) = \dfrac{-8}{(1+2x)^5}$, so $f(0) = 1$ and $f'(0) = -8$.

Thus, $f(x) \approx f(0) + f'(0)(x-0) = 1 + (-8)(x-0) = 1 - 8x$.

We need $\dfrac{1}{(1+2x)^4} - 0.1 < 1 - 8x < \dfrac{1}{(1+2x)^4} + 0.1$, which is true

when $-0.045 < x < 0.055$.

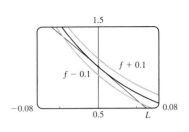

10. $f(x) = \dfrac{1}{\sqrt{4-x}}$ $\Rightarrow$ $f'(x) = \dfrac{1}{2(4-x)^{3/2}}$ so $f(0) = \frac{1}{2}$ and

$f'(0) = \frac{1}{16}$. So $f(x) \approx \frac{1}{2} + \frac{1}{16}(x-0) = \frac{1}{2} + \frac{1}{16}x$. We need

$\dfrac{1}{\sqrt{4-x}} - 0.1 < \frac{1}{2} + \frac{1}{16}x < \dfrac{1}{\sqrt{4-x}} + 0.1$, which is true when

$-3.91 < x < 2.14$.

11. (a) The differential dy is defined in terms of dx by the equation $dy = f'(x)\,dx$. For $y = f(x) = x^2 \sin 2x$,

$f'(x) = x^2 \cos 2x \cdot 2 + \sin 2x \cdot 2x = 2x(x \cos 2x + \sin 2x)$, so $dy = 2x(x \cos 2x + \sin 2x)\,dx$.

(b) $y = \sqrt{1+t^2}$ $\Rightarrow$ $dy = \frac{1}{2}(1+t^2)^{-1/2}(2t)\,dt = \dfrac{t}{\sqrt{1+t^2}}\,dt$

12. (a) For $y = f(s) = \dfrac{s}{1+2s}$, $f'(s) = \dfrac{(1+2s)(1) - s(2)}{(1+2s)^2} = \dfrac{1}{(1+2s)^2}$, so $dy = \dfrac{1}{(1+2s)^2}\,ds$.

(b) $y = u \cos u$ $\Rightarrow$ $y' = u(-\sin u) + \cos u \cdot 1$ $\Rightarrow$ $dy = (\cos u - u \sin u)\,du$

13. (a) For $y = f(u) = \dfrac{u+1}{u-1}$, $f'(u) = \dfrac{(u-1)(1) - (u+1)(1)}{(u-1)^2} = \dfrac{-2}{(u-1)^2}$, so $dy = \dfrac{-2}{(u-1)^2}\,du$.

(b) For $y = f(r) = (1+r^3)^{-2}$, $f'(r) = -2(1+r^3)^{-3}(3r^2) = \dfrac{-6r^2}{(1+r^3)^3}$, so $dy = \dfrac{-6r^2}{(1+r^3)^3}\,dr$.

14. (a) $y = (t + \tan t)^5$ $\Rightarrow$ $y' = 5(t + \tan t)^4(1 + \sec^2 t)$ $\Rightarrow$ $dy = 5(t + \tan t)^5(1 + \sec^2 t)\,dt$

(b) $y = \sqrt{z + 1/z} = (z + z^{-1})^{1/2}$ $\Rightarrow$

$y' = \frac{1}{2}(z + z^{-1})^{-1/2}(1 - z^{-2}) = \dfrac{1}{2\sqrt{z+1/z}}\left(1 - \dfrac{1}{z^2}\right) = \dfrac{z^2 - 1}{2z^2\sqrt{z+1/z}}$ $\Rightarrow$ $dy = \dfrac{z^2 - 1}{2z^2\sqrt{z+1/z}}\,dz$

15. (a) $y = \sqrt{4+5x}$ $\Rightarrow$ $dy = \frac{1}{2}(4+5x)^{-1/2} \cdot 5\,dx = \dfrac{5}{2\sqrt{4+5x}}\,dx$

(b) When $x = 0$ and $dx = 0.04$, $dy = \dfrac{5}{2\sqrt{4}}(0.04) = \frac{5}{4} \cdot \frac{1}{25} = \frac{1}{20} = 0.05$.

16. (a) $y = \dfrac{1}{x+1}$ $\Rightarrow$ $dy = -\dfrac{1}{(x+1)^2}\,dx$

(b) When $x = 1$ and $dx = -0.01$, $dy = -\dfrac{1}{2^2}(-0.01) = \frac{1}{4} \cdot \frac{1}{100} = \dfrac{1}{400} = 0.0025$.

17. (a) $y = \tan x \quad \Rightarrow \quad dy = \sec^2 x \, dx$

(b) When $x = \pi/4$ and $dx = -0.1$, $dy = [\sec(\pi/4)]^2(-0.1) = \left(\sqrt{2}\right)^2(-0.1) = -0.2$.

18. (a) $y = \cos x \quad \Rightarrow \quad dy = -\sin x \, dx$

(b) When $x = \pi/3$ and $dx = 0.05$, $dy = -\sin(\pi/3)(0.05) = -0.5\sqrt{3}\,(0.05) = -0.025\sqrt{3} \approx -0.043$.

19. $y = f(x) = 2x - x^2$, $x = 2$, $\Delta x = -0.4 \quad \Rightarrow$

$\Delta y = f(1.6) - f(2) = 0.64 - 0 = 0.64$

$dy = (2 - 2x)\,dx = (2 - 4)(-0.4) = 0.8$

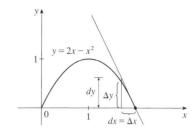

20. $y = f(x) = \sqrt{x}$, $x = 1$, $\Delta x = 1 \quad \Rightarrow$

$\Delta y = f(2) - f(1) = \sqrt{2} - \sqrt{1} = \sqrt{2} - 1 \approx 0.414$

$dy = \dfrac{1}{2\sqrt{x}}\,dx = \tfrac{1}{2}(1) = 0.5$

21. $y = f(x) = 2/x$, $x = 4$, $\Delta x = 1 \quad \Rightarrow$

$\Delta y = f(5) - f(4) = \tfrac{2}{5} - \tfrac{2}{4} = -0.1$

$dy = -\dfrac{2}{x^2}\,dx = -\dfrac{2}{4^2}(1) = -0.125$

22. $y = x^3$, $x = 1$, $\Delta x = 0.5 \quad \Rightarrow$

$\Delta y = (1.5)^3 - 1^3 = 3.375 - 1 = 2.375$.

$dy = 3x^2\,dx = 3(1)^2(0.5) = 1.5$

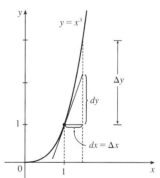

23. To estimate $(2.001)^5$, we'll find the linearization of $f(x) = x^5$ at $a = 2$. Since $f'(x) = 5x^4$, $f(2) = 32$, and $f'(2) = 80$,

we have $L(x) = 32 + 80(x - 2) = 80x - 128$. Thus, $x^5 \approx 80x - 128$ when x is near 2 , so

$(2.001)^5 \approx 80(2.001) - 128 = 160.08 - 128 = 32.08$.

24. $y = f(x) = \sin x \quad \Rightarrow \quad dy = \cos x \, dx$.   When $x = 0°$ and $dx = 1° = \frac{\pi}{180}$, $dy = \cos 0° \left(\frac{\pi}{180}\right) = 1\left(\frac{\pi}{180}\right) = \frac{\pi}{180}$, so

$\sin 1° = f(1°) \approx f(0°) + dy = 0 + \frac{\pi}{180} = \frac{\pi}{180} \approx 0.01745$.

25. To estimate $(8.06)^{2/3}$, we'll find the linearization of $f(x) = x^{2/3}$ at $a = 8$. Since $f'(x) = \frac{2}{3}x^{-1/3} = 2/\left(3\sqrt[3]{x}\right)$,

$f(8) = 4$, and $f'(8) = \frac{1}{3}$, we have $L(x) = 4 + \frac{1}{3}(x-8) = \frac{1}{3}x + \frac{4}{3}$. Thus, $x^{2/3} \approx \frac{1}{3}x + \frac{4}{3}$ when x is near 8, so

$(8.06)^{2/3} \approx \frac{1}{3}(8.06) + \frac{4}{3} = \frac{12.06}{3} = 4.02$.

26. To estimate $1/1002$, we'll find the linearization of $f(x) = 1/x$ at $a = 1000$. Since $f'(x) = -1/x^2$, $f(1000) = 0.001$,

and $f'(1000) = -0.000001$, we have $L(x) = 0.001 - 0.000001(x - 1000) = -0.000001x + 0.002$. Thus,

$1/x \approx -0.000001x + 0.002$ when x is near 1000, so $1/1002 \approx -0.000001(1002) + 0.002 = 0.000998$.

27. $y = f(x) = \tan x \Rightarrow dy = \sec^2 x\, dx$. When $x = 45°$ and $dx = -1°$,

$dy = \sec^2 45°(-\pi/180) = \left(\sqrt{2}\right)^2(-\pi/180) = -\pi/90$, so $\tan 44° = f(44°) \approx f(45°) + dy = 1 - \pi/90 \approx 0.965$.

28. $y = f(x) = \sqrt{x} \Rightarrow dy = \dfrac{1}{2\sqrt{x}}\,dx$. When $x = 100$ and $dx = -0.2$, $dy = \dfrac{1}{2\sqrt{100}}(-0.2) = -0.01$, so

$\sqrt{99.8} = f(99.8) \approx f(100) + dy = 10 - 0.01 = 9.99$.

29. $y = f(x) = \sec x \Rightarrow f'(x) = \sec x \tan x$, so $f(0) = 1$ and $f'(0) = 1 \cdot 0 = 0$. The linear approximation of f at 0 is

$f(0) + f'(0)(x - 0) = 1 + 0(x) = 1$. Since 0.08 is close to 0, approximating $\sec 0.08$ with 1 is reasonable.

30. If $y = x^6$, $y' = 6x^5$ and the tangent line approximation at $(1, 1)$ has slope 6. If the change in x is 0.01, the change in y on the

tangent line is 0.06, and approximating $(1.01)^6$ with 1.06 is reasonable.

31. (a) If x is the edge length, then $V = x^3 \Rightarrow dV = 3x^2\,dx$. When $x = 30$ and $dx = 0.1$, $dV = 3(30)^2(0.1) = 270$, so the

maximum possible error in computing the volume of the cube is about 270 cm^3. The relative error is calculated by dividing

the change in V, ΔV, by V. We approximate ΔV with dV.

Relative error $= \dfrac{\Delta V}{V} \approx \dfrac{dV}{V} = \dfrac{3x^2\,dx}{x^3} = 3\dfrac{dx}{x} = 3\left(\dfrac{0.1}{30}\right) = 0.01$.

Percentage error $=$ relative error $\times 100\% = 0.01 \times 100\% = 1\%$.

(b) $S = 6x^2 \Rightarrow dS = 12x\,dx$. When $x = 30$ and $dx = 0.1$, $dS = 12(30)(0.1) = 36$, so the maximum possible error in

computing the surface area of the cube is about 36 cm^2.

Relative error $= \dfrac{\Delta S}{S} \approx \dfrac{dS}{S} = \dfrac{12x\,dx}{6x^2} = 2\dfrac{dx}{x} = 2\left(\dfrac{0.1}{30}\right) = 0.00\overline{6}$.

Percentage error $=$ relative error $\times 100\% = 0.00\overline{6} \times 100\% = 0.\overline{6}\%$.

32. (a) $A = \pi r^2 \Rightarrow dA = 2\pi r\,dr$. When $r = 24$ and $dr = 0.2$, $dA = 2\pi(24)(0.2) = 9.6\pi$, so the maximum possible error

in the calculated area of the disk is about $9.6\pi \approx 30$ cm^2.

(b) Relative error $= \dfrac{\Delta A}{A} \approx \dfrac{dA}{A} = \dfrac{2\pi r\,dr}{\pi r^2} = \dfrac{2\,dr}{r} = \dfrac{2(0.2)}{24} = \dfrac{0.2}{12} = \dfrac{1}{60} = 0.01\overline{6}$.

Percentage error $=$ relative error $\times 100\% = 0.01\overline{6} \times 100\% = 1.\overline{6}\%$.

33. (a) For a sphere of radius r, the circumference is $C = 2\pi r$ and the surface area is $S = 4\pi r^2$, so

$$r = \frac{C}{2\pi} \quad \Rightarrow \quad S = 4\pi \left(\frac{C}{2\pi}\right)^2 = \frac{C^2}{\pi} \quad \Rightarrow \quad dS = \frac{2}{\pi} C\, dC. \text{ When } C = 84 \text{ and } dC = 0.5,\ dS = \frac{2}{\pi}(84)(0.5) = \frac{84}{\pi},$$

so the maximum error is about $\frac{84}{\pi} \approx 27 \text{ cm}^2$. Relative error $\approx \dfrac{dS}{S} = \dfrac{84/\pi}{84^2/\pi} = \dfrac{1}{84} \approx 0.012$

(b) $V = \frac{4}{3}\pi r^3 = \frac{4}{3}\pi \left(\frac{C}{2\pi}\right)^3 = \frac{C^3}{6\pi^2} \quad \Rightarrow \quad dV = \frac{1}{2\pi^2} C^2\, dC. \text{ When } C = 84 \text{ and } dC = 0.5,$

$dV = \dfrac{1}{2\pi^2}(84)^2(0.5) = \dfrac{1764}{\pi^2}$, so the maximum error is about $\dfrac{1764}{\pi^2} \approx 179 \text{ cm}^3$.

The relative error is approximately $\dfrac{dV}{V} = \dfrac{1764/\pi^2}{(84)^3/(6\pi^2)} = \dfrac{1}{56} \approx 0.018$.

34. For a hemispherical dome, $V = \frac{2}{3}\pi r^3 \quad \Rightarrow \quad dV = 2\pi r^2\, dr.$ When $r = \frac{1}{2}(50) = 25 \text{ m}$ and $dr = 0.05 \text{ cm} = 0.0005 \text{ m}$,

$dV = 2\pi(25)^2(0.0005) = \frac{5\pi}{8}$, so the amount of paint needed is about $\frac{5\pi}{8} \approx 2 \text{ m}^3$.

35. (a) $V = \pi r^2 h \quad \Rightarrow \quad \Delta V \approx dV = 2\pi rh\, dr = 2\pi rh\, \Delta r$

(b) The error is

$\Delta V - dV = [\pi(r + \Delta r)^2 h - \pi r^2 h] - 2\pi rh\, \Delta r = \pi r^2 h + 2\pi rh\, \Delta r + \pi(\Delta r)^2 h - \pi r^2 h - 2\pi rh\, \Delta r = \pi(\Delta r)^2 h.$

36. (a) $\sin\theta = \dfrac{20}{x} \quad \Rightarrow \quad x = 20\csc\theta \quad \Rightarrow$

$dx = 20(-\csc\theta \cot\theta)\, d\theta = -20\csc 30° \cot 30° \,(\pm 1°)$

$= -20(2)(\sqrt{3})\left(\pm\dfrac{\pi}{180}\right) = \pm\dfrac{2\sqrt{3}}{9}\pi$

So the maximum error is about $\pm\frac{2}{9}\sqrt{3}\,\pi \approx \pm 1.21 \text{ cm}$.

(b) The relative error is $\dfrac{\Delta x}{x} \approx \dfrac{dx}{x} = \dfrac{\pm\frac{2}{9}\sqrt{3}\,\pi}{20(2)} = \pm\dfrac{\sqrt{3}}{180}\pi \approx \pm 0.03$, so the percentage error is approximately $\pm 3\%$.

37. $V = RI \quad \Rightarrow \quad I = \dfrac{V}{R} \quad \Rightarrow \quad dI = -\dfrac{V}{R^2}\, dR.$ The relative error in calculating I is $\dfrac{\Delta I}{I} \approx \dfrac{dI}{I} = \dfrac{-(V/R^2)\, dR}{V/R} = -\dfrac{dR}{R}.$

Hence, the relative error in calculating I is approximately the same (in magnitude) as the relative error in R.

38. $F = kR^4 \quad \Rightarrow \quad dF = 4kR^3\, dR \quad \Rightarrow \quad \dfrac{dF}{F} = \dfrac{4kR^3\, dR}{kR^4} = 4\left(\dfrac{dR}{R}\right).$ Thus, the relative change in F is about 4 times the

relative change in R. So a 5% increase in the radius corresponds to a 20% increase in blood flow.

39. (a) $dc = \dfrac{dc}{dx}\, dx = 0\, dx = 0$ (b) $d(cu) = \dfrac{d}{dx}(cu)\, dx = c\dfrac{du}{dx}\, dx = c\, du$

(c) $d(u + v) = \dfrac{d}{dx}(u + v)\, dx = \left(\dfrac{du}{dx} + \dfrac{dv}{dx}\right) dx = \dfrac{du}{dx}\, dx + \dfrac{dv}{dx}\, dx = du + dv$

(d) $d(uv) = \dfrac{d}{dx}(uv)\, dx = \left(u\dfrac{dv}{dx} + v\dfrac{du}{dx}\right) dx = u\dfrac{dv}{dx}\, dx + v\dfrac{du}{dx}\, dx = u\, dv + v\, du$

(e) $d\left(\dfrac{u}{v}\right) = \dfrac{d}{dx}\left(\dfrac{u}{v}\right) dx = \dfrac{v\dfrac{du}{dx} - u\dfrac{dv}{dx}}{v^2}\, dx = \dfrac{v\dfrac{du}{dx}\, dx - u\dfrac{dv}{dx}\, dx}{v^2} = \dfrac{v\, du - u\, dv}{v^2}$

(f) $d(x^n) = \dfrac{d}{dx}(x^n)\, dx = nx^{n-1}\, dx$

40. (a) $f(x) = \sin x \;\Rightarrow\; f'(x) = \cos x$, so $f(0) = 0$ and $f'(0) = 1$. Thus, $f(x) \approx f(0) + f'(0)(x - 0) = 0 + 1(x - 0) = x$.

(b)

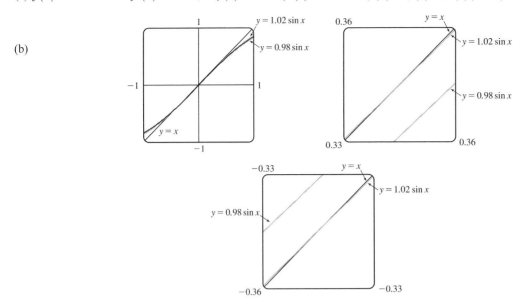

We want to know the values of x for which $y = x$ approximates $y = \sin x$ with less than a 2% difference; that is, the values of x for which

$$\left| \frac{x - \sin x}{\sin x} \right| < 0.02 \quad \Leftrightarrow \quad -0.02 < \frac{x - \sin x}{\sin x} < 0.02 \quad \Leftrightarrow$$

$$\begin{cases} -0.02 \sin x < x - \sin x < 0.02 \sin x & \text{if } \sin x > 0 \\ -0.02 \sin x > x - \sin x > 0.02 \sin x & \text{if } \sin x < 0 \end{cases} \Leftrightarrow \begin{cases} 0.98 \sin x < x < 1.02 \sin x & \text{if } \sin x > 0 \\ 1.02 \sin x < x < 0.98 \sin x & \text{if } \sin x < 0 \end{cases}$$

In the first figure, we see that the graphs are very close to each other near $x = 0$. Changing the viewing rectangle and using an intersect feature (see the second figure) we find that $y = x$ intersects $y = 1.02 \sin x$ at $x \approx 0.344$. By symmetry, they also intersect at $x \approx -0.344$ (see the third figure). Converting 0.344 radians to degrees, we get $0.344 \left(\frac{180°}{\pi} \right) \approx 19.7° \approx 20°$, which verifies the statement.

41. (a) The graph shows that $f'(1) = 2$, so $L(x) = f(1) + f'(1)(x - 1) = 5 + 2(x - 1) = 2x + 3$.

$f(0.9) \approx L(0.9) = 4.8$ and $f(1.1) \approx L(1.1) = 5.2$.

(b) From the graph, we see that $f'(x)$ is positive and decreasing. This means that the slopes of the tangent lines are positive, but the tangents are becoming less steep. So the tangent lines lie *above* the curve. Thus, the estimates in part (a) are too large.

42. (a) $g'(x) = \sqrt{x^2 + 5} \;\Rightarrow\; g'(2) = \sqrt{9} = 3$. $g(1.95) \approx g(2) + g'(2)(1.95 - 2) = -4 + 3(-0.05) = -4.15$.

$g(2.05) \approx g(2) + g'(2)(2.05 - 2) = -4 + 3(0.05) = -3.85$.

(b) The formula $g'(x) = \sqrt{x^2 + 5}$ shows that $g'(x)$ is positive and increasing. This means that the slopes of the tangent lines are positive and the tangents are getting steeper. So the tangent lines lie *below* the graph of g. Hence, the estimates in part (a) are too small.

LABORATORY PROJECT   Taylor Polynomials

1. We first write the functions described in conditions (i), (ii), and (iii):

$$P(x) = A + Bx + Cx^2 \qquad\qquad f(x) = \cos x$$
$$P'(x) = B + 2Cx \qquad\qquad f'(x) = -\sin x$$
$$P''(x) = 2C \qquad\qquad f''(x) = -\cos x$$

So, taking $a = 0$, our three conditions become

$$P(0) = f(0): \qquad A = \cos 0 = 1$$
$$P'(0) = f'(0): \qquad B = -\sin 0 = 0$$
$$P''(0) = f''(0): \quad 2C = -\cos 0 = -1 \quad \Rightarrow \quad C = -\tfrac{1}{2}$$

The desired quadratic function is $P(x) = 1 - \tfrac{1}{2}x^2$, so the quadratic approximation is $\cos x \approx 1 - \tfrac{1}{2}x^2$.

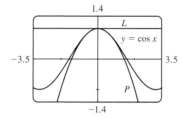

The figure shows a graph of the cosine function together with its linear approximation $L(x) = 1$ and quadratic approximation $P(x) = 1 - \tfrac{1}{2}x^2$ near 0. You can see that the quadratic approximation is much better than the linear one.

2. Accuracy to within 0.1 means that $\left|\cos x - \left(1 - \tfrac{1}{2}x^2\right)\right| < 0.1 \quad\Leftrightarrow\quad -0.1 < \cos x - \left(1 - \tfrac{1}{2}x^2\right) < 0.1 \quad\Leftrightarrow$

$$0.1 > \left(1 - \tfrac{1}{2}x^2\right) - \cos x > -0.1 \quad\Leftrightarrow\quad \cos x + 0.1 > 1 - \tfrac{1}{2}x^2 > \cos x - 0.1 \quad\Leftrightarrow\quad \cos x - 0.1 < 1 - \tfrac{1}{2}x^2 < \cos x + 0.1.$$

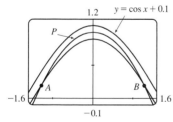

From the figure we see that this is true between A and B. Zooming in or using an intersect feature, we find that the x-coordinates of B and A are about ± 1.26. Thus, the approximation $\cos x \approx 1 - \tfrac{1}{2}x^2$ is accurate to within 0.1 when $-1.26 < x < 1.26$.

3. If $P(x) = A + B(x - a) + C(x - a)^2$, then $P'(x) = B + 2C(x - a)$ and $P''(x) = 2C$. Applying the conditions (i), (ii), and (iii), we get

$$P(a) = f(a): \qquad A = f(a)$$
$$P'(a) = f'(a): \qquad B = f'(a)$$
$$P''(a) = f''(a): \quad 2C = f''(a) \quad \Rightarrow \quad C = \tfrac{1}{2}f''(a)$$

Thus, $P(x) = A + B(x - a) + C(x - a)^2$ can be written in the form $P(x) = f(a) + f'(a)(x - a) + \tfrac{1}{2}f''(a)(x - a)^2$.

4. From Example 1 in Section 3.9, we have $f(1) = 2$, $f'(1) = \frac{1}{4}$, and

$f'(x) = \frac{1}{2}(x+3)^{-1/2}$. So $f''(x) = -\frac{1}{4}(x+3)^{-3/2}$ $\Rightarrow$ $f''(1) = -\frac{1}{32}$.

From Problem 3, the quadratic approximation $P(x)$ is

$\sqrt{x+3} \approx f(1) + f'(1)(x-1) + \frac{1}{2}f''(1)(x-1)^2 = 2 + \frac{1}{4}(x-1) - \frac{1}{64}(x-1)^2$.

The figure shows the function $f(x) = \sqrt{x+3}$ together with its linear

approximation $L(x) = \frac{1}{4}x + \frac{7}{4}$ and its quadratic approximation $P(x)$. You can see that $P(x)$ is a better

approximation than $L(x)$ and this is borne out by the numerical values in the following chart.

	from $L(x)$	actual value	from $P(x)$
$\sqrt{3.98}$	1.9950	1.99499373...	1.99499375
$\sqrt{4.05}$	2.0125	2.01246118...	2.01246094
$\sqrt{4.2}$	2.0500	2.04939015...	2.04937500

5. $T_n(x) = c_0 + c_1(x-a) + c_2(x-a)^2 + c_3(x-a)^3 + \cdots + c_n(x-a)^n$. If we put $x = a$ in this equation,

then all terms after the first are 0 and we get $T_n(a) = c_0$. Now we differentiate $T_n(x)$ and obtain

$T_n'(x) = c_1 + 2c_2(x-a) + 3c_3(x-a)^2 + 4c_4(x-a)^3 + \cdots + nc_n(x-a)^{n-1}$. Substituting $x = a$ gives $T_n'(a) = c_1$.

Differentiating again, we have $T_n''(x) = 2c_2 + 2 \cdot 3c_3(x-a) + 3 \cdot 4c_4(x-a^2) + \cdots + (n-1)nc_n(x-a)^{n-2}$ and so

$T_n''(a) = 2c_2$. Continuing in this manner, we get $T_n'''(x) = 2 \cdot 3c_3 + 2 \cdot 3 \cdot 4c_4(x-a) + \cdots + (n-2)(n-1)nc_n(x-a)^{n-3}$

and $T_n'''(a) = 2 \cdot 3c_3$. By now we see the pattern. If we continue to differentiate and substitute $x = a$, we obtain

$T_n^{(4)}(a) = 2 \cdot 3 \cdot 4c_4$ and in general, for any integer k between 1 and n, $T_n^{(k)}(a) = 2 \cdot 3 \cdot 4 \cdot 5 \cdots \cdot kc_k = k!\,c_k$ $\Rightarrow$

$c_k = \dfrac{T_n^{(k)}(a)}{k!}$. Because we want T_n and f to have the same derivatives at a, we require that $c_k = \dfrac{f^{(k)}(a)}{k!}$ for

$k = 1, 2, \ldots, n$.

6. $T_n(x) = f(a) + f'(a)(x-a) + \dfrac{f''(a)}{2!}(x-a)^2 + \cdots + \dfrac{f^{(n)}(a)}{n!}(x-a)^n$. To compute the coefficients in this equation we

need to calculate the derivatives of f at 0:

$$f(x) = \cos x \qquad\qquad f(0) = \cos 0 = 1$$
$$f'(x) = -\sin x \qquad\qquad f'(0) = -\sin 0 = 0$$
$$f''(x) = -\cos x \qquad\qquad f''(0) = -1$$
$$f'''(x) = \sin x \qquad\qquad f'''(0) = 0$$
$$f^{(4)}(x) = \cos x \qquad\qquad f^{(4)}(0) = 1$$

We see that the derivatives repeat in a cycle of length 4, so $f^{(5)}(0) = 0$, $f^{(6)}(0) = -1$, $f^{(7)}(0) = 0$, and $f^{(8)}(0) = 1$.

From the original expression for $T_n(x)$, with $n = 8$ and $a = 0$, we have

$$T_8(x) = f(0) + f'(0)(x - 0) + \frac{f''(0)}{2!}(x - 0)^2 + \frac{f'''(0)}{3!}(x - 0)^3 + \cdots + \frac{f^{(8)}(0)}{8!}(x - 0)^8$$

$$= 1 + 0 \cdot x + \frac{-1}{2!}x^2 + 0 \cdot x^3 + \frac{1}{4!}x^4 + 0 \cdot x^5 + \frac{-1}{6!}x^6 + 0 \cdot x^7 + \frac{1}{8!}x^8 = 1 - \frac{x^2}{2!} + \frac{x^4}{4!} - \frac{x^6}{6!} + \frac{x^8}{8!}$$

and the desired approximation is $\cos x \approx 1 - \frac{x^2}{2!} + \frac{x^4}{4!} - \frac{x^6}{6!} + \frac{x^8}{8!}$. The Taylor polynomials T_2, T_4, and T_6 consist of the

initial terms of T_8 up through degree 2, 4, and 6, respectively. Therefore, $T_2(x) = 1 - \frac{x^2}{2!}$, $T_4(x) = 1 - \frac{x^2}{2!} + \frac{x^4}{4!}$, and

$T_6(x) = 1 - \frac{x^2}{2!} + \frac{x^4}{4!} - \frac{x^6}{6!}$. We graph T_2, T_4, T_6, T_8, and f:

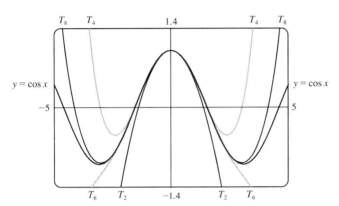

Notice that $T_2(x)$ is a good approximation to $\cos x$ near 0, $T_4(x)$ is a good approximation on a larger interval, $T_6(x)$ is a better approximation, and $T_8(x)$ is better still. Each successive Taylor polynomial is a good approximation on a larger interval than the previous one.

3 Review

CONCEPT CHECK

1. See Definition 3.1.1.

2. See the paragraph containing Formula 3 in Section 3.1.

3. (a) The average rate of change of y with respect to x over the interval $[x_1, x_2]$ is $\dfrac{f(x_2) - f(x_1)}{x_2 - x_1}$.

   (b) The instantaneous rate of change of y with respect to x at $x = x_1$ is $\displaystyle\lim_{x_2 \to x_1} \frac{f(x_2) - f(x_1)}{x_2 - x_1}$.

4. See Definition 3.1.4. The pages following the definition discuss interpretations of $f'(a)$ as the slope of a tangent line to the graph of f at $x = a$ and as an instantaneous rate of change of $f(x)$ with respect to x when $x = a$.

5. (a) A function f is differentiable at a number a if its derivative f' exists at $x = a$; that is, if $f'(a)$ exists.

   (b) See Theorem 3.2.4. This theorem also tells us that if f is *not* continuous at a, then f is *not* differentiable at a.

   (c)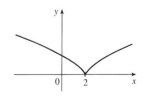

6. See the discussion and Figure 7 on page 129.

7. The second derivative of a function f is the rate of change of the first derivative f'. The third derivative is the derivative (rate of change) of the second derivative. If f is the position function of an object, f' is its velocity function, f'' is its acceleration function, and f''' is its jerk function.

8. (a) The Power Rule: If n is any real number, then $\dfrac{d}{dx}(x^n) = nx^{n-1}$. The derivative of a variable base raised to a constant power is the power times the base raised to the power minus one.

   (b) The Constant Multiple Rule: If c is a constant and f is a differentiable function, then $\dfrac{d}{dx}[cf(x)] = c\dfrac{d}{dx}f(x)$.

   The derivative of a constant times a function is the constant times the derivative of the function.

   (c) The Sum Rule: If f and g are both differentiable, then $\dfrac{d}{dx}[f(x) + g(x)] = \dfrac{d}{dx}f(x) + \dfrac{d}{dx}g(x)$. The derivative of a sum of functions is the sum of the derivatives.

   (d) The Difference Rule: If f and g are both differentiable, then $\dfrac{d}{dx}[f(x) - g(x)] = \dfrac{d}{dx}f(x) - \dfrac{d}{dx}g(x)$. The derivative of a difference of functions is the difference of the derivatives.

   (e) The Product Rule: If f and g are both differentiable, then $\dfrac{d}{dx}[f(x)\,g(x)] = f(x)\dfrac{d}{dx}g(x) + g(x)\dfrac{d}{dx}f(x)$. The derivative of a product of two functions is the first function times the derivative of the second function plus the second function times the derivative of the first function.

   (f) The Quotient Rule: If f and g are both differentiable, then $\dfrac{d}{dx}\left[\dfrac{f(x)}{g(x)}\right] = \dfrac{g(x)\dfrac{d}{dx}f(x) - f(x)\dfrac{d}{dx}g(x)}{[g(x)]^2}$.

   The derivative of a quotient of functions is the denominator times the derivative of the numerator minus the numerator times the derivative of the denominator, all divided by the square of the denominator.

   (g) The Chain Rule: If f and g are both differentiable and $F = f \circ g$ is the composite function defined by $F(x) = f(g(x))$, then F is differentiable and F' is given by the product $F'(x) = f'(g(x))\,g'(x)$. The derivative of a composite function is the derivative of the outer function evaluated at the inner function times the derivative of the inner function.

9. (a) $y = x^n \;\Rightarrow\; y' = nx^{n-1}$
   (b) $y = \sin x \;\Rightarrow\; y' = \cos x$
   (c) $y = \cos x \;\Rightarrow\; y' = -\sin x$
   (d) $y = \tan x \;\Rightarrow\; y' = \sec^2 x$
   (e) $y = \csc x \;\Rightarrow\; y' = -\csc x \cot x$
   (f) $y = \sec x \;\Rightarrow\; y' = \sec x \tan x$
   (g) $y = \cot x \;\Rightarrow\; y' = -\csc^2 x$

10. Implicit differentiation consists of differentiating both sides of an equation involving x and y with respect to x, and then solving the resulting equation for y'.

11. (a) The linearization L of f at $x = a$ is $L(x) = f(a) + f'(a)(x - a)$.

    (b) If $y = f(x)$, then the differential dy is given by $dy = f'(x)\,dx$.

    (c) See Figure 5 in Section 3.9.

TRUE-FALSE QUIZ

1. False. See the note after Theorem 4 in Section 3.2.

2. True. This is the Sum Rule.

3. False. See the warning before the Product Rule.

4. True. This is the Chain Rule.

5. True by the Chain Rule.

6. False. $\dfrac{d}{dx} f(\sqrt{x}\,) = \dfrac{f'(\sqrt{x}\,)}{2\sqrt{x}}$ by the Chain Rule.

7. False. $f(x) = |x^2 + x| = x^2 + x$ for $x \geq 0$ or $x \leq -1$ and $|x^2 + x| = -(x^2 + x)$ for $-1 < x < 0$.

So $f'(x) = 2x + 1$ for $x > 0$ or $x < -1$ and $f'(x) = -(2x + 1)$ for $-1 < x < 0$. But $|2x + 1| = 2x + 1$

for $x \geq -\frac{1}{2}$ and $|2x + 1| = -2x - 1$ for $x < -\frac{1}{2}$.

8. True. $f'(r)$ exists $\Rightarrow$ f is differentiable at r $\Rightarrow$ f is continuous at r $\Rightarrow$ $\lim\limits_{x \to r} f(x) = f(r)$.

9. True. $g(x) = x^5$ $\Rightarrow$ $g'(x) = 5x^4$ $\Rightarrow$ $g'(2) = 5(2)^4 = 80$, and by the definition of the derivative,

$\lim\limits_{x \to 2} \dfrac{g(x) - g(2)}{x - 2} = g'(2) = 80$.

10. False. $\dfrac{d^2y}{dx^2}$ is the second derivative while $\left(\dfrac{dy}{dx}\right)^2$ is the first derivative squared. For example, if $y = x$, then $\dfrac{d^2y}{dx^2} = 0$,

but $\left(\dfrac{dy}{dx}\right)^2 = 1$.

11. False. A tangent line to the parabola $y = x^2$ has slope $dy/dx = 2x$, so at $(-2, 4)$ the slope of the tangent is $2(-2) = -4$

and an equation of the tangent line is $y - 4 = -4(x + 2)$. [The given equation, $y - 4 = 2x(x + 2)$, is not even

linear!]

12. True. $\dfrac{d}{dx}(\tan^2 x) = 2 \tan x \sec^2 x$, and $\dfrac{d}{dx}(\sec^2 x) = 2 \sec x (\sec x \tan x) = 2 \tan x \sec^2 x$.

Or: $\dfrac{d}{dx}(\sec^2 x) = \dfrac{d}{dx}(1 + \tan^2 x) = \dfrac{d}{dx}(\tan^2 x)$.

EXERCISES

1. (a) $s = s(t) = 1 + 2t + t^2/4$. The average velocity over the time interval $[1, 1 + h]$ is

$$v_{\text{ave}} = \frac{s(1 + h) - s(1)}{(1 + h) - 1} = \frac{1 + 2(1 + h) + (1 + h)^2/4 - 13/4}{h} = \frac{10h + h^2}{4h} = \frac{10 + h}{4}$$

So for the following intervals the average velocities are:

(i) $[1, 3]$: $h = 2$, $v_{\text{ave}} = (10 + 2)/4 = 3$ m/s

(ii) $[1, 2]$: $h = 1$, $v_{\text{ave}} = (10 + 1)/4 = 2.75$ m/s

(iii) $[1, 1.5]$: $h = 0.5$, $v_{\text{ave}} = (10 + 0.5)/4 = 2.625$ m/s

(iv) $[1, 1.1]$: $h = 0.1$, $v_{\text{ave}} = (10 + 0.1)/4 = 2.525$ m/s

(b) When $t = 1$, the instantaneous velocity is $\lim\limits_{h \to 0} \dfrac{s(1 + h) - s(1)}{h} = \lim\limits_{h \to 0} \dfrac{10 + h}{4} = \dfrac{10}{4} = 2.5$ m/s.

2. f is not differentiable: at $x = -4$ because f is not continuous, at $x = -1$ because f has a corner, at $x = 2$ because f is not continuous, and at $x = 5$ because f has a vertical tangent.

3.

4.

5. The graph of a has tangent lines with positive slope for $x < 0$ and negative slope for $x > 0$, and the values of c fit this pattern, so c must be the graph of the derivative of the function for a. The graph of c has horizontal tangent lines to the left and right of the x-axis and b has zeros at these points. Hence, b is the graph of the derivative of the function for c. Therefore, a is the graph of f, c is the graph of f', and b is the graph of f''.

6. $2^6 = 64$, so $f(x) = x^6$ and $a = 2$.

7. (a) $f'(r)$ is the rate at which the total cost changes with respect to the interest rate. Its units are dollars/(percent per year).

(b) The total cost of paying off the loan is increasing by $1200/(percent per year) as the interest rate reaches 10%. So if the interest rate goes up from 10% to 11%, the cost goes up approximately $1200.

(c) As r increases, C increases. So $f'(r)$ will always be positive.

8. (a) Drawing slope triangles, we obtain the following estimates: $F'(1950) \approx \frac{1.1}{10} = 0.11$, $F'(1965) \approx \frac{-1.6}{10} = -0.16$, and $F'(1987) \approx \frac{0.2}{10} = 0.02$.

(b) The rate of change of the average number of children born to each woman was increasing by 0.11 in 1950, decreasing by 0.16 in 1965, and increasing by 0.02 in 1987.

(c) There are many possible reasons:

- In the baby-boom era (post-WWII), there was optimism about the economy and family size was rising.

- In the baby-bust era, there was less economic optimism, and it was considered less socially responsible to have a large family.

- In the baby-boomlet era, there was increased economic optimism and a return to more conservative attitudes.

9. $C'(1990)$ is the rate at which the total value of US currency in circulation is changing in billions of dollars per year. To estimate the value of $C'(1990)$, we will average the difference quotients obtained using the times $t = 1985$ and $t = 1995$.

Let $A = \dfrac{C(1985) - C(1990)}{1985 - 1990} = \dfrac{187.3 - 271.9}{-5} = \dfrac{-84.6}{-5} = 16.92$ and

$B = \dfrac{C(1995) - C(1990)}{1995 - 1990} = \dfrac{409.3 - 271.9}{5} = \dfrac{137.4}{5} = 27.48$. Then

$C'(1990) = \lim\limits_{t \to 1990} \dfrac{C(t) - C(1990)}{t - 1990} \approx \dfrac{A + B}{2} = \dfrac{16.92 + 27.48}{2} = \dfrac{44.4}{2} = 22.2$ billion dollars/year.

10. $f(x) = \dfrac{4 - x}{3 + x}$ $\Rightarrow$

$f'(x) = \lim\limits_{h \to 0} \dfrac{f(x+h) - f(x)}{h} = \lim\limits_{h \to 0} \dfrac{\dfrac{4 - (x+h)}{3 + (x+h)} - \dfrac{4 - x}{3 + x}}{h} = \lim\limits_{h \to 0} \dfrac{(4 - x - h)(3 + x) - (4 - x)(3 + x + h)}{h(3 + x + h)(3 + x)}$

$= \lim\limits_{h \to 0} \dfrac{-7h}{h(3 + x + h)(3 + x)} = \lim\limits_{h \to 0} \dfrac{-7}{(3 + x + h)(3 + x)} = -\dfrac{7}{(3 + x)^2}$

11. $f(x) = x^3 + 5x + 4$ $\Rightarrow$

$f'(x) = \lim\limits_{h \to 0} \dfrac{f(x+h) - f(x)}{h} = \lim\limits_{h \to 0} \dfrac{(x+h)^3 + 5(x+h) + 4 - (x^3 + 5x + 4)}{h}$

$= \lim\limits_{h \to 0} \dfrac{3x^2 h + 3xh^2 + h^3 + 5h}{h} = \lim\limits_{h \to 0} (3x^2 + 3xh + h^2 + 5) = 3x^2 + 5$

12. (a) $f'(x) = \lim\limits_{h \to 0} \dfrac{f(x+h) - f(x)}{h} = \lim\limits_{h \to 0} \dfrac{\sqrt{3 - 5(x+h)} - \sqrt{3 - 5x}}{h} \dfrac{\sqrt{3 - 5(x+h)} + \sqrt{3 - 5x}}{\sqrt{3 - 5(x+h)} + \sqrt{3 - 5x}}$

$= \lim\limits_{h \to 0} \dfrac{[3 - 5(x+h)] - (3 - 5x)}{h\left(\sqrt{3 - 5(x+h)} + \sqrt{3 - 5x}\right)} = \lim\limits_{h \to 0} \dfrac{-5}{\sqrt{3 - 5(x+h)} + \sqrt{3 - 5x}} = \dfrac{-5}{2\sqrt{3 - 5x}}$

(b) Domain of f: (the radicand must be nonnegative) $3 - 5x \geq 0$ $\Rightarrow$

$5x \leq 3$ $\Rightarrow$ $x \in \left(-\infty, \frac{3}{5}\right]$

Domain of f': exclude $\frac{3}{5}$ because it makes the denominator zero;

$x \in \left(-\infty, \frac{3}{5}\right)$

(c) Our answer to part (a) is reasonable because $f'(x)$ is always negative and f is always decreasing.

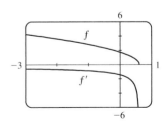

13. $y = (x^4 - 3x^2 + 5)^3 \Rightarrow$

$y' = 3(x^4 - 3x^2 + 5)^2 \dfrac{d}{dx} (x^4 - 3x^2 + 5) = 3(x^4 - 3x^2 + 5)^2 (4x^3 - 6x) = 6x(x^4 - 3x^2 + 5)^2 (2x^2 - 3)$

14. $y = \cos(\tan x) \Rightarrow y' = -\sin(\tan x) \dfrac{d}{dx} (\tan x) = -\sin(\tan x)(\sec^2 x)$

15. $y = \sqrt{x} + \dfrac{1}{\sqrt[3]{x^4}} = x^{1/2} + x^{-4/3} \Rightarrow y' = \frac{1}{2} x^{-1/2} - \frac{4}{3} x^{-7/3} = \dfrac{1}{2\sqrt{x}} - \dfrac{4}{3\sqrt[3]{x^7}}$

16. $y = \dfrac{3x - 2}{\sqrt{2x + 1}} \Rightarrow$

$y' = \dfrac{\sqrt{2x + 1}\,(3) - (3x - 2)\frac{1}{2}(2x + 1)^{-1/2}(2)}{\left(\sqrt{2x + 1}\right)^2} \cdot \dfrac{(2x + 1)^{1/2}}{(2x + 1)^{1/2}} = \dfrac{3(2x + 1) - (3x - 2)}{(2x + 1)^{3/2}} = \dfrac{3x + 5}{(2x + 1)^{3/2}}$

17. $y = 2x\sqrt{x^2 + 1} \Rightarrow$

$y' = 2x \cdot \frac{1}{2}(x^2 + 1)^{-1/2}(2x) + \sqrt{x^2 + 1}\,(2) = \dfrac{2x^2}{\sqrt{x^2 + 1}} + 2\sqrt{x^2 + 1} = \dfrac{2x^2 + 2(x^2 + 1)}{\sqrt{x^2 + 1}} = \dfrac{2(2x^2 + 1)}{\sqrt{x^2 + 1}}$

18. $y = \left(x + \dfrac{1}{x^2}\right)^{\sqrt{7}} \Rightarrow y' = \sqrt{7}\left(x + \dfrac{1}{x^2}\right)^{\sqrt{7} - 1}\left(1 - \dfrac{2}{x^3}\right)$

19. $y = \dfrac{t}{1 - t^2} \Rightarrow y' = \dfrac{(1 - t^2)(1) - t(-2t)}{(1 - t^2)^2} = \dfrac{1 - t^2 + 2t^2}{(1 - t^2)^2} = \dfrac{t^2 + 1}{(1 - t^2)^2}$

20. $y = \sin(\cos x) \Rightarrow y' = \cos(\cos x)(-\sin x) = -\sin x\,\cos(\cos x)$

21. $y = \tan\sqrt{1 - x} \Rightarrow y' = \left(\sec^2\sqrt{1 - x}\right)\left(\dfrac{1}{2\sqrt{1 - x}}\right)(-1) = -\dfrac{\sec^2\sqrt{1 - x}}{2\sqrt{1 - x}}$

22. Using the Reciprocal Rule, $g(x) = \dfrac{1}{f(x)} \Rightarrow g'(x) = -\dfrac{f'(x)}{[f(x)]^2}$, we have $y = \dfrac{1}{\sin(x - \sin x)} \Rightarrow$

$y' = -\dfrac{\cos(x - \sin x)(1 - \cos x)}{\sin^2(x - \sin x)}.$

23. $\dfrac{d}{dx}(xy^4 + x^2 y) = \dfrac{d}{dx}(x + 3y) \Rightarrow x \cdot 4y^3 y' + y^4 \cdot 1 + x^2 \cdot y' + y \cdot 2x = 1 + 3y' \Rightarrow$

$y'(4xy^3 + x^2 - 3) = 1 - y^4 - 2xy \Rightarrow y' = \dfrac{1 - y^4 - 2xy}{4xy^3 + x^2 - 3}$

24. $y = \sec(1 + x^2) \Rightarrow y' = 2x\sec(1 + x^2)\tan(1 + x^2)$

25. $y = \dfrac{\sec 2\theta}{1 + \tan 2\theta} \Rightarrow$

$y' = \dfrac{(1 + \tan 2\theta)(\sec 2\theta\,\tan 2\theta \cdot 2) - (\sec 2\theta)(\sec^2 2\theta \cdot 2)}{(1 + \tan 2\theta)^2} = \dfrac{2\sec 2\theta\,[(1 + \tan 2\theta)\tan 2\theta - \sec^2 2\theta]}{(1 + \tan 2\theta)^2}$

$= \dfrac{2\sec 2\theta\,(\tan 2\theta + \tan^2 2\theta - \sec^2 2\theta)}{(1 + \tan 2\theta)^2} = \dfrac{2\sec 2\theta\,(\tan 2\theta - 1)}{(1 + \tan 2\theta)^2} \qquad [1 + \tan^2 x = \sec^2 x]$

26. $\dfrac{d}{dx}\left(x^2\cos y + \sin 2y\right) = \dfrac{d}{dx}(xy)$ $\Rightarrow$ $x^2(-\sin y \cdot y') + (\cos y)(2x) + \cos 2y \cdot 2y' = x \cdot y' + y \cdot 1$ $\Rightarrow$

$y'(-x^2\sin y + 2\cos 2y - x) = y - 2x\cos y$ $\Rightarrow$ $y' = \dfrac{y - 2x\cos y}{2\cos 2y - x^2\sin y - x}$

27. $y = (1 - x^{-1})^{-1}$ $\Rightarrow$

$y' = -1(1 - x^{-1})^{-2}[-(-1x^{-2})] = -(1 - 1/x)^{-2}x^{-2} = -((x-1)/x)^{-2}x^{-2} = -(x-1)^{-2}$

28. $y = \left(x + \sqrt{x}\,\right)^{-1/3}$ $\Rightarrow$ $y' = -\frac{1}{3}\left(x + \sqrt{x}\,\right)^{-4/3}\left(1 + \dfrac{1}{2\sqrt{x}}\right)$

29. $\sin(xy) = x^2 - y$ $\Rightarrow$ $\cos(xy)(xy' + y \cdot 1) = 2x - y'$ $\Rightarrow$ $x\cos(xy)y' + y' = 2x - y\cos(xy)$ $\Rightarrow$

$y'[x\cos(xy) + 1] = 2x - y\cos(xy)$ $\Rightarrow$ $y' = \dfrac{2x - y\cos(xy)}{x\cos(xy) + 1}$

30. $y = \sqrt{\sin\sqrt{x}}$ $\Rightarrow$ $y' = \frac{1}{2}\left(\sin\sqrt{x}\,\right)^{-1/2}\left(\cos\sqrt{x}\,\right)\left(\dfrac{1}{2\sqrt{x}}\right) = \dfrac{\cos\sqrt{x}}{4\sqrt{x}\sin\sqrt{x}}$

31. $y = \cot(3x^2 + 5)$ $\Rightarrow$ $y' = -\csc^2(3x^2 + 5)(6x) = -6x\csc^2(3x^2 + 5)$

32. $y = \dfrac{(x + \lambda)^4}{x^4 + \lambda^4}$ $\Rightarrow$ $y' = \dfrac{(x^4 + \lambda^4)(4)(x + \lambda)^3 - (x + \lambda)^4(4x^3)}{(x^4 + \lambda^4)^2} = \dfrac{4(x + \lambda)^3(\lambda^4 - \lambda x^3)}{(x^4 + \lambda^4)^2}$

33. $y = \sqrt{x}\cos\sqrt{x}$ $\Rightarrow$

$y' = \sqrt{x}\left(\cos\sqrt{x}\,\right)' + \cos\sqrt{x}\left(\sqrt{x}\,\right)' = \sqrt{x}\left[-\sin\sqrt{x}\left(\frac{1}{2}x^{-1/2}\right)\right] + \cos\sqrt{x}\left(\frac{1}{2}x^{-1/2}\right)$

$= \frac{1}{2}x^{-1/2}\left(-\sqrt{x}\sin\sqrt{x} + \cos\sqrt{x}\,\right) = \dfrac{\cos\sqrt{x} - \sqrt{x}\sin\sqrt{x}}{2\sqrt{x}}$

34. $y = (\sin mx)/x$ $\Rightarrow$ $y' = (mx\cos mx - \sin mx)/x^2$

35. $y = \tan^2(\sin\theta) = [\tan(\sin\theta)]^2$ $\Rightarrow$ $y' = 2[\tan(\sin\theta)] \cdot \sec^2(\sin\theta) \cdot \cos\theta$

36. $x\tan y = y - 1$ $\Rightarrow$ $\tan y + (x\sec^2 y)y' = y'$ $\Rightarrow$ $y' = \dfrac{\tan y}{1 - x\sec^2 y}$

37. $y = (x\tan x)^{1/5}$ $\Rightarrow$ $y' = \frac{1}{5}(x\tan x)^{-4/5}(\tan x + x\sec^2 x)$

38. $y = \dfrac{(x-1)(x-4)}{(x-2)(x-3)} = \dfrac{x^2 - 5x + 4}{x^2 - 5x + 6}$ $\Rightarrow$ $y' = \dfrac{(x^2 - 5x + 6)(2x - 5) - (x^2 - 5x + 4)(2x - 5)}{(x^2 - 5x + 6)^2} = \dfrac{2(2x - 5)}{(x-2)^2(x-3)^2}$

39. $y = \sin\left(\tan\sqrt{1 + x^3}\,\right)$ $\Rightarrow$ $y' = \cos\left(\tan\sqrt{1 + x^3}\,\right)\left(\sec^2\sqrt{1 + x^3}\,\right)\left[3x^2/\left(2\sqrt{1 + x^3}\,\right)\right]$

40. $y = \sin^2\left(\cos\sqrt{\sin\pi x}\,\right) = \left[\sin\left(\cos\sqrt{\sin\pi x}\,\right)\right]^2 \Rightarrow$

$$y' = 2\left[\sin\left(\cos\sqrt{\sin\pi x}\,\right)\right]\left[\sin\left(\cos\sqrt{\sin\pi x}\,\right)\right]' = 2\sin\left(\cos\sqrt{\sin\pi x}\,\right)\cos\left(\cos\sqrt{\sin\pi x}\,\right)\left(\cos\sqrt{\sin\pi x}\,\right)'$$

$$= 2\sin\left(\cos\sqrt{\sin\pi x}\,\right)\cos\left(\cos\sqrt{\sin\pi x}\,\right)\left(-\sin\sqrt{\sin\pi x}\,\right)\left(\sqrt{\sin\pi x}\,\right)'$$

$$= -2\sin\left(\cos\sqrt{\sin\pi x}\,\right)\cos\left(\cos\sqrt{\sin\pi x}\,\right)\sin\sqrt{\sin\pi x}\cdot\tfrac{1}{2}(\sin\pi x)^{-1/2}(\sin\pi x)'$$

$$= \frac{-\sin\left(\cos\sqrt{\sin\pi x}\,\right)\cos\left(\cos\sqrt{\sin\pi x}\,\right)\sin\sqrt{\sin\pi x}}{\sqrt{\sin\pi x}}\cdot\cos\pi x\cdot\pi$$

$$= \frac{-\pi\sin\left(\cos\sqrt{\sin\pi x}\,\right)\cos\left(\cos\sqrt{\sin\pi x}\,\right)\sin\sqrt{\sin\pi x}\cos\pi x}{\sqrt{\sin\pi x}}$$

41. $f(t) = \sqrt{4t+1} \Rightarrow f'(t) = \tfrac{1}{2}(4t+1)^{-1/2}\cdot 4 = 2(4t+1)^{-1/2} \Rightarrow$

$f''(t) = 2(-\tfrac{1}{2})(4t+1)^{-3/2}\cdot 4 = -4/(4t+1)^{3/2}$, so $f''(2) = -4/9^{3/2} = -\tfrac{4}{27}$.

42. $g(\theta) = \theta\sin\theta \Rightarrow g'(\theta) = \theta\cos\theta + \sin\theta\cdot 1 \Rightarrow g''(\theta) = \theta(-\sin\theta) + \cos\theta\cdot 1 + \cos\theta = 2\cos\theta - \theta\sin\theta$,

so $g''(\pi/6) = 2\cos(\pi/6) - (\pi/6)\sin(\pi/6) = 2\left(\sqrt{3}/2\right) - (\pi/6)(1/2) = \sqrt{3} - \pi/12$.

43. $x^6 + y^6 = 1 \Rightarrow 6x^5 + 6y^5 y' = 0 \Rightarrow y' = -x^5/y^5 \Rightarrow$

$$y'' = -\frac{y^5(5x^4) - x^5(5y^4 y')}{(y^5)^2} = -\frac{5x^4 y^4\left[y - x(-x^5/y^5)\right]}{y^{10}} = -\frac{5x^4\left[(y^6 + x^6)/y^5\right]}{y^6} = -\frac{5x^4}{y^{11}}$$

44. $f(x) = (2-x)^{-1} \Rightarrow f'(x) = (2-x)^{-2} \Rightarrow f''(x) = 2(2-x)^{-3} \Rightarrow f'''(x) = 2\cdot 3(2-x)^{-4} \Rightarrow$

$f^{(4)}(x) = 2\cdot 3\cdot 4(2-x)^{-5}$. In general, $f^{(n)}(x) = 2\cdot 3\cdot 4\cdots n(2-x)^{-(n+1)} = \dfrac{n!}{(2-x)^{(n+1)}}$.

45. $\displaystyle\lim_{x\to 0}\frac{\sec x}{1-\sin x} = \frac{\sec 0}{1-\sin 0} = \frac{1}{1-0} = 1$

46. $\displaystyle\lim_{t\to 0}\frac{t^3}{\tan^3 2t} = \lim_{t\to 0}\frac{t^3\cos^3 2t}{\sin^3 2t} = \lim_{t\to 0}\cos^3 2t\cdot\frac{1}{\dfrac{8\sin^3 2t}{(2t)^3}} = \lim_{t\to 0}\frac{\cos^3 2t}{8\left(\displaystyle\lim_{t\to 0}\dfrac{\sin 2t}{2t}\right)^3} = \frac{1}{8\cdot 1^3} = \frac{1}{8}$

47. $y = 4\sin^2 x \Rightarrow y' = 4\cdot 2\sin x\cos x$. At $\left(\tfrac{\pi}{6}, 1\right)$, $y' = 8\cdot\tfrac{1}{2}\cdot\tfrac{\sqrt{3}}{2} = 2\sqrt{3}$, so an equation of the tangent line

is $y - 1 = 2\sqrt{3}\left(x - \tfrac{\pi}{6}\right)$, or $y = 2\sqrt{3}\,x + 1 - \pi\sqrt{3}/3$.

48. $y = \dfrac{x^2 - 1}{x^2 + 1} \Rightarrow y' = \dfrac{(x^2+1)(2x) - (x^2-1)(2x)}{(x^2+1)^2} = \dfrac{4x}{(x^2+1)^2}$.

At $(0, -1)$, $y' = 0$, so an equation of the tangent line is $y + 1 = 0(x - 0)$, or $y = -1$.

49. $y = \sqrt{1 + 4\sin x} \Rightarrow y' = \tfrac{1}{2}(1 + 4\sin x)^{-1/2}\cdot 4\cos x = \dfrac{2\cos x}{\sqrt{1 + 4\sin x}}$.

At $(0, 1)$, $y' = \dfrac{2}{\sqrt{1}} = 2$, so an equation of the tangent line is $y - 1 = 2(x - 0)$, or $y = 2x + 1$.

50. $x^2 + 4xy + y^2 = 13 \Rightarrow 2x + 4(xy' + y \cdot 1) + 2yy' = 0 \Rightarrow x + 2xy' + 2y + yy' = 0 \Rightarrow$

$2xy' + yy' = -x - 2y \Rightarrow y'(2x + y) = -x - 2y \Rightarrow y' = \dfrac{-x - 2y}{2x + y}.$

At $(2, 1)$, $y' = \dfrac{-2-2}{4+1} = -\dfrac{4}{5}$, so an equation of the tangent line is $y - 1 = -\frac{4}{5}(x - 2)$, or $y = -\frac{4}{5}x + \frac{13}{5}$.

The slope of the normal line is $\frac{5}{4}$, so an equation of the normal line is $y - 1 = \frac{5}{4}(x - 2)$, or $y = \frac{5}{4}x - \frac{3}{2}$.

51. (a) $f(x) = x\sqrt{5-x} \Rightarrow$

$$f'(x) = x\left[\frac{1}{2}(5-x)^{-1/2}(-1)\right] + \sqrt{5-x} = \frac{-x}{2\sqrt{5-x}} + \sqrt{5-x} \cdot \frac{2\sqrt{5-x}}{2\sqrt{5-x}} = \frac{-x}{2\sqrt{5-x}} + \frac{2(5-x)}{2\sqrt{5-x}}$$

$$= \frac{-x+10-2x}{2\sqrt{5-x}} = \frac{10-3x}{2\sqrt{5-x}}$$

(b) At $(1, 2)$: $f'(1) = \frac{7}{4}$.

(c)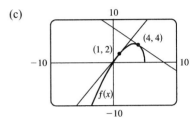

So an equation of the tangent line is $y - 2 = \frac{7}{4}(x - 1)$ or $y = \frac{7}{4}x + \frac{1}{4}$.

At $(4, 4)$: $f'(4) = -\frac{2}{2} = -1$.

So an equation of the tangent line is $y - 4 = -1(x - 4)$ or $y = -x + 8$.

(d)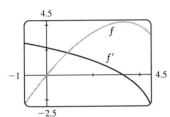

The graphs look reasonable, since f' is positive where f has tangents with positive slope, and f' is negative where f has tangents with negative slope.

52. (a) $f(x) = 4x - \tan x \Rightarrow f'(x) = 4 - \sec^2 x \Rightarrow f''(x) = -2\sec x (\sec x \tan x) = -2\sec^2 x \tan x.$

(b)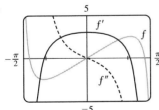

We can see that our answers are reasonable, since the graph of f' is 0 where f has a horizontal tangent, and the graph of f' is positive where f has tangents with positive slope and negative where f has tangents with negative slope. The same correspondence holds between the graphs of f' and f''.

53. $y = \sin x + \cos x \Rightarrow y' = \cos x - \sin x = 0 \Leftrightarrow \cos x = \sin x$ and $0 \le x \le 2\pi \Leftrightarrow x = \frac{\pi}{4}$ or $\frac{5\pi}{4}$, so the points are $\left(\frac{\pi}{4}, \sqrt{2}\right)$ and $\left(\frac{5\pi}{4}, -\sqrt{2}\right)$.

54. $x^2 + 2y^2 = 1 \Rightarrow 2x + 4yy' = 0 \Rightarrow y' = -x/(2y) = 1 \Leftrightarrow x = -2y$. Since the points lie on the ellipse, we have $(-2y)^2 + 2y^2 = 1 \Rightarrow 6y^2 = 1 \Rightarrow y = \pm\frac{1}{\sqrt{6}}$. The points are $\left(-\frac{2}{\sqrt{6}}, \frac{1}{\sqrt{6}}\right)$ and $\left(\frac{2}{\sqrt{6}}, -\frac{1}{\sqrt{6}}\right)$.

55. $y = f(x) = ax^2 + bx + c \Rightarrow f'(x) = 2ax + b$. We know that $f'(-1) = 6$ and $f'(5) = -2$, so $-2a + b = 6$ and

$10a + b = -2$. Subtracting the first equation from the second gives $12a = -8 \Rightarrow a = -\frac{2}{3}$. Substituting $-\frac{2}{3}$ for a in the

first equation gives $b = \frac{14}{3}$. Now $f(1) = 4 \Rightarrow 4 = a + b + c$, so $c = 4 + \frac{2}{3} - \frac{14}{3} = 0$ and hence, $f(x) = -\frac{2}{3}x^2 + \frac{14}{3}x$.

56. If $y = f(x) = \dfrac{x}{x+1}$, then $f'(x) = \dfrac{(x+1)(1) - x(1)}{(x+1)^2} = \dfrac{1}{(x+1)^2}$. When $x = a$, the equation of the tangent line is

$y - \dfrac{a}{a+1} = \dfrac{1}{(a+1)^2}(x-a)$. This line passes through $(1, 2)$ when $2 - \dfrac{a}{a+1} = \dfrac{1}{(a+1)^2}(1-a) \Leftrightarrow$

$2(a+1)^2 - a(a+1) = 1 - a \Leftrightarrow 2a^2 + 4a + 2 - a^2 - a - 1 + a = 0 \Leftrightarrow a^2 + 4a + 1 = 0$.

The quadratic formula gives the roots of this equation as $a = \dfrac{-4 \pm \sqrt{4^2 - 4(1)(1)}}{2(1)} = \dfrac{-4 \pm \sqrt{12}}{2} = -2 \pm \sqrt{3}$,

so there are two such tangent lines. Since

$$f\left(-2 \pm \sqrt{3}\right) = \dfrac{-2 \pm \sqrt{3}}{-2 \pm \sqrt{3} + 1} = \dfrac{-2 \pm \sqrt{3}}{-1 \pm \sqrt{3}} \cdot \dfrac{-1 \mp \sqrt{3}}{-1 \mp \sqrt{3}}$$

$$= \dfrac{2 \pm 2\sqrt{3} \mp \sqrt{3} - 3}{1 - 3} = \dfrac{-1 \pm \sqrt{3}}{-2} = \dfrac{1 \mp \sqrt{3}}{2},$$

the lines touch the curve at $A\left(-2 + \sqrt{3}, \frac{1 - \sqrt{3}}{2}\right) \approx (-0.27, -0.37)$

and $B\left(-2 - \sqrt{3}, \frac{1 + \sqrt{3}}{2}\right) \approx (-3.73, 1.37)$.

57. $f(x) = (x-a)(x-b)(x-c) \Rightarrow f'(x) = (x-b)(x-c) + (x-a)(x-c) + (x-a)(x-b)$.

So $\dfrac{f'(x)}{f(x)} = \dfrac{(x-b)(x-c) + (x-a)(x-c) + (x-a)(x-b)}{(x-a)(x-b)(x-c)} = \dfrac{1}{x-a} + \dfrac{1}{x-b} + \dfrac{1}{x-c}$.

Or: $f(x) = (x-a)(x-b)(x-c) \Rightarrow \ln|f(x)| = \ln|x-a| + \ln|x-b| + \ln|x-c| \Rightarrow$

$\dfrac{f'(x)}{f(x)} = \dfrac{1}{x-a} + \dfrac{1}{x-b} + \dfrac{1}{x-c}$

58. (a) $\cos 2x = \cos^2 x - \sin^2 x \Rightarrow -2\sin 2x = -2\cos x \sin x - 2\sin x \cos x \Leftrightarrow \sin 2x = 2\sin x \cos x$

(b) $\sin(x + a) = \sin x \cos a + \cos x \sin a \Rightarrow \cos(x + a) = \cos x \cos a - \sin x \sin a$.

59. (a) $h(x) = f(x)\,g(x) \Rightarrow h'(x) = f(x)\,g'(x) + g(x)\,f'(x) \Rightarrow$

$h'(2) = f(2)\,g'(2) + g(2)\,f'(2) = (3)(4) + (5)(-2) = 12 - 10 = 2$

(b) $F(x) = f(g(x)) \Rightarrow F'(x) = f'(g(x))\,g'(x) \Rightarrow F'(2) = f'(g(2))\,g'(2) = f'(5)(4) = 11 \cdot 4 = 44$

60. (a) $P(x) = f(x)\,g(x) \Rightarrow P'(x) = f(x)\,g'(x) + g(x)\,f'(x) \Rightarrow$

$P'(2) = f(2)\,g'(2) + g(2)\,f'(2) = (1)\left(\frac{6-0}{3-0}\right) + (4)\left(\frac{0-3}{3-0}\right) = (1)(2) + (4)(-1) = 2 - 4 = -2$

(b) $Q(x) = \dfrac{f(x)}{g(x)} \Rightarrow Q'(x) = \dfrac{g(x)\,f'(x) - f(x)\,g'(x)}{[g(x)]^2} \Rightarrow$

$Q'(2) = \dfrac{g(2)\,f'(2) - f(2)\,g'(2)}{[g(2)]^2} = \dfrac{(4)(-1) - (1)(2)}{4^2} = \dfrac{-6}{16} = -\dfrac{3}{8}$

(c) $C(x) = f(g(x)) \;\Rightarrow\; C'(x) = f'(g(x))g'(x) \;\Rightarrow$

$\quad C'(2) = f'(g(2))g'(2) = f'(4)g'(2) = \left(\frac{6-0}{5-3}\right)(2) = (3)(2) = 6$

61. $f(x) = x^2 g(x) \;\Rightarrow\; f'(x) = x^2 g'(x) + g(x)(2x) = x[xg'(x) + 2g(x)]$

62. $f(x) = g(x^2) \;\Rightarrow\; f'(x) = g'(x^2)(2x) = 2xg'(x^2)$

63. $f(x) = [g(x)]^2 \;\Rightarrow\; f'(x) = 2[g(x)] \cdot g'(x) = 2g(x)\, g'(x)$

64. $f(x) = x^a g(x^b) \;\Rightarrow\; f'(x) = ax^{a-1}g(x^b) + x^a g'(x^b)(bx^{b-1}) = ax^{a-1}g(x^b) + bx^{a+b-1}g'(x^b)$

65. $f(x) = g(g(x)) \;\Rightarrow\; f'(x) = g'(g(x))\,g'(x)$

66. $f(x) = \sin(g(x)) \;\Rightarrow\; f'(x) = \cos(g(x)) \cdot g'(x)$

67. $f(x) = g(\sin x) \;\Rightarrow\; f'(x) = g'(\sin x) \cdot \cos x$

68. $f(x) = g\left(\tan \sqrt{x}\right) \;\Rightarrow$

$\quad f'(x) = g'\left(\tan \sqrt{x}\right) \cdot \dfrac{d}{dx}\left(\tan \sqrt{x}\right) = g'\left(\tan \sqrt{x}\right) \cdot \sec^2 \sqrt{x} \cdot \dfrac{d}{dx}\left(\sqrt{x}\right) = \dfrac{g'\left(\tan \sqrt{x}\right)\sec^2 \sqrt{x}}{2\sqrt{x}}$

69. $h(x) = \dfrac{f(x)\,g(x)}{f(x) + g(x)} \;\Rightarrow$

$\quad h'(x) = \dfrac{[f(x) + g(x)]\,[f(x)\,g'(x) + g(x)\,f'(x)] - f(x)\,g(x)\,[f'(x) + g'(x)]}{[f(x) + g(x)]^2}$

$\quad = \dfrac{[f(x)]^2\,g'(x) + f(x)\,g(x)\,f'(x) + f(x)\,g(x)\,g'(x) + [g(x)]^2\,f'(x) - f(x)\,g(x)\,f'(x) - f(x)\,g(x)\,g'(x)}{[f(x) + g(x)]^2}$

$\quad = \dfrac{f'(x)\,[g(x)]^2 + g'(x)\,[f(x)]^2}{[f(x) + g(x)]^2}$

70. $h(x) = \sqrt{\dfrac{f(x)}{g(x)}} \;\Rightarrow\; h'(x) = \dfrac{f'(x)\,g(x) - f(x)\,g'(x)}{2\sqrt{f(x)/g(x)}\,[g(x)]^2} = \dfrac{f'(x)\,g(x) - f(x)\,g'(x)}{2[g(x)]^{3/2}\sqrt{f(x)}}$

71. Using the Chain Rule repeatedly, $h(x) = f(g(\sin 4x)) \;\Rightarrow$

$\quad h'(x) = f'(g(\sin 4x)) \cdot \dfrac{d}{dx}(g(\sin 4x)) = f'(g(\sin 4x)) \cdot g'(\sin 4x) \cdot \dfrac{d}{dx}(\sin 4x) = f'(g(\sin 4x))g'(\sin 4x)(\cos 4x)(4)$.

72. (a) $x = \sqrt{b^2 + c^2 t^2} \;\Rightarrow\; v(t) = x' = \left[1/\left(2\sqrt{b^2 + c^2 t^2}\right)\right]2c^2 t = c^2 t/\sqrt{b^2 + c^2 t^2} \;\Rightarrow$

$\quad a(t) = v'(t) = \dfrac{c^2\sqrt{b^2 + c^2 t^2} - c^2 t\left(c^2 t/\sqrt{b^2 + c^2 t^2}\right)}{b^2 + c^2 t^2} = \dfrac{b^2 c^2}{\left(b^2 + c^2 t^2\right)^{3/2}}$

(b) $v(t) > 0$ for $t > 0$, so the particle always moves in the positive direction.

73. (a) $y = t^3 - 12t + 3 \;\Rightarrow\; v(t) = y' = 3t^2 - 12 \;\Rightarrow\; a(t) = v'(t) = 6t$

(b) $v(t) = 3(t^2 - 4) > 0$ when $t > 2$, so it moves upward when $t > 2$ and downward when $0 \le t < 2$.

(c) Distance upward $= y(3) - y(2) = -6 - (-13) = 7$,

Distance downward $= y(0) - y(2) = 3 - (-13) = 16$. Total distance $= 7 + 16 = 23$.

(d)

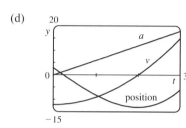

(e) The particle is speeding up when v and a have the same sign, that is, when $t > 2$. The particle is slowing down when v and a have opposite signs; that is, when $0 < t < 2$.

74. (a) $V = \frac{1}{3}\pi r^2 h \Rightarrow dV/dh = \frac{1}{3}\pi r^2$ [r constant]

(b) $V = \frac{1}{3}\pi r^2 h \Rightarrow dV/dr = \frac{2}{3}\pi rh$ [h constant]

75. The linear density ρ is the rate of change of mass m with respect to length x.

$m = x\left(1 + \sqrt{x}\right) = x + x^{3/2} \Rightarrow \rho = dm/dx = 1 + \frac{3}{2}\sqrt{x}$, so the linear density when $x = 4$ is $1 + \frac{3}{2}\sqrt{4} = 4$ kg/m.

76. (a) $C(x) = 920 + 2x - 0.02x^2 + 0.00007x^3 \Rightarrow C'(x) = 2 - 0.04x + 0.00021x^2$

(b) $C'(100) = 2 - 4 + 2.1 = \$0.10$/unit. This value represents the rate at which costs are increasing as the hundredth unit is produced, and is the approximate cost of producing the 101st unit.

(c) The cost of producing the 101st item is $C(101) - C(100) = 990.10107 - 990 = \$0.10107$, slightly larger than $C'(100)$.

77. If $x = $ edge length, then $V = x^3 \Rightarrow dV/dt = 3x^2\, dx/dt = 10 \Rightarrow dx/dt = 10/(3x^2)$ and $S = 6x^2 \Rightarrow dS/dt = (12x)\, dx/dt = 12x[10/(3x^2)] = 40/x$. When $x = 30$, $dS/dt = \frac{40}{30} = \frac{4}{3}$ cm^2/min.

78. Given $dV/dt = 2$, find dh/dt when $h = 5$. $V = \frac{1}{3}\pi r^2 h$ and, from similar

triangles, $\dfrac{r}{h} = \dfrac{3}{10} \Rightarrow V = \dfrac{\pi}{3}\left(\dfrac{3h}{10}\right)^2 h = \dfrac{3\pi}{100}h^3$, so

$2 = \dfrac{dV}{dt} = \dfrac{9\pi}{100}h^2\dfrac{dh}{dt} \Rightarrow \dfrac{dh}{dt} = \dfrac{200}{9\pi h^2} = \dfrac{200}{9\pi\,(5)^2} = \dfrac{8}{9\pi}$ cm/s

when $h = 5$.

79. Given $dh/dt = 5$ and $dx/dt = 15$, find dz/dt. $z^2 = x^2 + h^2 \Rightarrow$

$2z\dfrac{dz}{dt} = 2x\dfrac{dx}{dt} + 2h\dfrac{dh}{dt} \Rightarrow \dfrac{dz}{dt} = \dfrac{1}{z}(15x + 5h)$. When $t = 3$,

$h = 45 + 3(5) = 60$ and $x = 15(3) = 45 \Rightarrow z = \sqrt{45^2 + 60^2} = 75$,

so $\dfrac{dz}{dt} = \dfrac{1}{75}[15(45) + 5(60)] = 13$ ft/s.

80. We are given $dz/dt = 30$ ft/s. By similar triangles, $\dfrac{y}{z} = \dfrac{4}{\sqrt{241}} \Rightarrow$

$y = \dfrac{4}{\sqrt{241}}z$, so $\dfrac{dy}{dt} = \dfrac{4}{\sqrt{241}}\dfrac{dz}{dt} = \dfrac{120}{\sqrt{241}} \approx 7.7$ ft/s.

81. We are given $d\theta/dt = -0.25$ rad/h. $\tan\theta = 400/x$ $\Rightarrow$

$x = 400\cot\theta$ $\Rightarrow$ $\dfrac{dx}{dt} = -400\csc^2\theta\,\dfrac{d\theta}{dt}$. When $\theta = \frac{\pi}{6}$,

$\dfrac{dx}{dt} = -400(2)^2(-0.25) = 400$ ft/h.

82. (a) $f(x) = \sqrt{25 - x^2}$ $\Rightarrow$ $f'(x) = \dfrac{-2x}{2\sqrt{25 - x^2}} = -x(25 - x^2)^{-1/2}$.

(b)

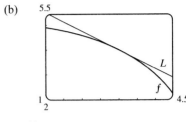

So the linear approximation to $f(x)$ near 3

is $f(x) \approx f(3) + f'(3)(x - 3) = 4 - \frac{3}{4}(x - 3)$.

(c) For the required accuracy, we want $\sqrt{25 - x^2} - 0.1 < 4 - \frac{3}{4}(x - 3)$ and

$4 - \frac{3}{4}(x - 3) < \sqrt{25 - x^2} + 0.1$. From the graph, it appears that these both

hold for $2.24 < x < 3.66$.

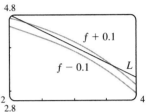

83. (a) $f(x) = \sqrt[3]{1 + 3x} = (1 + 3x)^{1/3}$ $\Rightarrow$ $f'(x) = (1 + 3x)^{-2/3}$, so the linearization of f at $a = 0$ is

$L(x) = f(0) + f'(0)(x - 0) = 1^{1/3} + 1^{-2/3}x = 1 + x$. Thus, $\sqrt[3]{1 + 3x} \approx 1 + x$ $\Rightarrow$

$\sqrt[3]{1.03} = \sqrt[3]{1 + 3(0.01)} \approx 1 + (0.01) = 1.01$.

(b) The linear approximation is $\sqrt[3]{1 + 3x} \approx 1 + x$, so for the required accuracy

we want $\sqrt[3]{1 + 3x} - 0.1 < 1 + x < \sqrt[3]{1 + 3x} + 0.1$. From the graph,

it appears that this is true when $-0.23 < x < 0.40$.

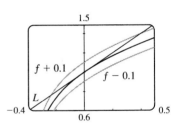

84. $y = x^3 - 2x^2 + 1$ $\Rightarrow$ $dy = (3x^2 - 4x)\,dx$. When $x = 2$ and $dx = 0.2$, $dy = [3(2)^2 - 4(2)](0.2) = 0.8$.

85. $A = x^2 + \frac{1}{2}\pi\left(\frac{1}{2}x\right)^2 = \left(1 + \frac{\pi}{8}\right)x^2$ $\Rightarrow$ $dA = \left(2 + \frac{\pi}{4}\right)x\,dx$. When $x = 60$

and $dx = 0.1$, $dA = \left(2 + \frac{\pi}{4}\right)60(0.1) = 12 + \frac{3\pi}{2}$, so the maximum error is

approximately $12 + \frac{3\pi}{2} \approx 16.7$ cm^2.

86. $\lim\limits_{x \to 1}\dfrac{x^{17} - 1}{x - 1} = \left[\dfrac{d}{dx}x^{17}\right]_{x=1} = 17(1)^{16} = 17$

87. $\lim\limits_{h \to 0}\dfrac{\sqrt[4]{16 + h} - 2}{h} = \left[\dfrac{d}{dx}\sqrt[4]{x}\right]_{x=16} = \dfrac{1}{4}x^{-3/4}\Big|_{x=16} = \dfrac{1}{4\left(\sqrt[4]{16}\right)^3} = \dfrac{1}{32}$

88. $\lim\limits_{\theta \to \pi/3}\dfrac{\cos\theta - 0.5}{\theta - \pi/3} = \left[\dfrac{d}{d\theta}\cos\theta\right]_{\theta=\pi/3} = -\sin\dfrac{\pi}{3} = -\dfrac{\sqrt{3}}{2}$

89. $\lim\limits_{x\to 0} \dfrac{\sqrt{1+\tan x}-\sqrt{1+\sin x}}{x^3} = \lim\limits_{x\to 0} \dfrac{\left(\sqrt{1+\tan x}-\sqrt{1+\sin x}\right)\left(\sqrt{1+\tan x}+\sqrt{1+\sin x}\right)}{x^3\left(\sqrt{1+\tan x}+\sqrt{1+\sin x}\right)}$

$$= \lim_{x\to 0} \frac{(1+\tan x)-(1+\sin x)}{x^3\left(\sqrt{1+\tan x}+\sqrt{1+\sin x}\right)} = \lim_{x\to 0} \frac{\sin x\,(1/\cos x-1)}{x^3\left(\sqrt{1+\tan x}+\sqrt{1+\sin x}\right)}\cdot\frac{\cos x}{\cos x}$$

$$= \lim_{x\to 0} \frac{\sin x\,(1-\cos x)}{x^3\left(\sqrt{1+\tan x}+\sqrt{1+\sin x}\right)\cos x}\cdot\frac{1+\cos x}{1+\cos x}$$

$$= \lim_{x\to 0} \frac{\sin x\cdot\sin^2 x}{x^3\left(\sqrt{1+\tan x}+\sqrt{1+\sin x}\right)\cos x\,(1+\cos x)}$$

$$= \left(\lim_{x\to 0}\frac{\sin x}{x}\right)^3 \lim_{x\to 0}\frac{1}{\left(\sqrt{1+\tan x}+\sqrt{1+\sin x}\right)\cos x\,(1+\cos x)}$$

$$= 1^3\cdot\frac{1}{\left(\sqrt{1}+\sqrt{1}\right)\cdot 1\cdot(1+1)} = \frac{1}{4}$$

90. Differentiating the first given equation implicitly with respect to x and using the Chain Rule, we obtain $f(g(x))=x \Rightarrow$

$f'(g(x))\,g'(x)=1 \Rightarrow g'(x)=\dfrac{1}{f'(g(x))}$. Using the second given equation to expand the denominator of this expression

gives $g'(x)=\dfrac{1}{1+[f(g(x))]^2}$. But the first given equation states that $f(g(x))=x$, so $g'(x)=\dfrac{1}{1+x^2}$.

91. $\dfrac{d}{dx}[f(2x)]=x^2 \Rightarrow f'(2x)\cdot 2=x^2 \Rightarrow f'(2x)=\frac{1}{2}x^2$. Let $t=2x$. Then $f'(t)=\frac{1}{2}\left(\frac{1}{2}t\right)^2=\frac{1}{8}t^2$, so $f'(x)=\frac{1}{8}x^2$.

92. Let (b,c) be on the curve, that is, $b^{2/3}+c^{2/3}=a^{2/3}$. Now $x^{2/3}+y^{2/3}=a^{2/3} \Rightarrow \frac{2}{3}x^{-1/3}+\frac{2}{3}y^{-1/3}\dfrac{dy}{dx}=0$, so

$\dfrac{dy}{dx}=-\dfrac{y^{1/3}}{x^{1/3}}=-\left(\dfrac{y}{x}\right)^{1/3}$, so at (b,c) the slope of the tangent line is $-(c/b)^{1/3}$ and an equation of the tangent line is

$y-c=-(c/b)^{1/3}(x-b)$ or $y=-(c/b)^{1/3}x+(c+b^{2/3}c^{1/3})$. Setting $y=0$, we find that the x-intercept is

$b^{1/3}c^{2/3}+b=b^{1/3}(c^{2/3}+b^{2/3})=b^{1/3}a^{2/3}$ and setting $x=0$ we find that the y-intercept is

$c+b^{2/3}c^{1/3}=c^{1/3}(c^{2/3}+b^{2/3})=c^{1/3}a^{2/3}$. So the length of the tangent line between these two points is

$$\sqrt{(b^{1/3}a^{2/3})^2+(c^{1/3}a^{2/3})^2}=\sqrt{b^{2/3}a^{4/3}+c^{2/3}a^{4/3}}=\sqrt{(b^{2/3}+c^{2/3})a^{4/3}}$$

$$=\sqrt{a^{2/3}a^{4/3}}=\sqrt{a^2}=a=\text{constant}$$

□ PROBLEMS PLUS

1. Let a be the x-coordinate of Q. Since the derivative of $y = 1 - x^2$ is $y' = -2x$, the slope at Q is $-2a$. But since the triangle is equilateral, $\overline{AO}/\overline{OC} = \sqrt{3}/1$, so the slope at Q is $-\sqrt{3}$. Therefore, we must have that $-2a = -\sqrt{3} \;\Rightarrow\; a = \frac{\sqrt{3}}{2}$.

Thus, the point Q has coordinates $\left(\frac{\sqrt{3}}{2}, 1 - \left(\frac{\sqrt{3}}{2}\right)^2\right) = \left(\frac{\sqrt{3}}{2}, \frac{1}{4}\right)$ and by symmetry, P has coordinates $\left(-\frac{\sqrt{3}}{2}, \frac{1}{4}\right)$.

2. $y = x^3 - 3x + 4 \;\Rightarrow\; y' = 3x^2 - 3$, and $y = 3(x^2 - x) \;\Rightarrow\; y' = 6x - 3$.

The slopes of the tangents of the two curves are equal when $3x^2 - 3 = 6x - 3$;

that is, when $x = 0$ or 2. At $x = 0$, both tangents have slope -3, but the curves do

not intersect. At $x = 2$, both tangents have slope 9 and the curves intersect at

$(2, 6)$. So there is a common tangent line at $(2, 6)$, $y = 9x - 12$.

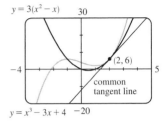

3.

We must show that r (in the figure) is halfway between p and q, that is,

$r = (p + q)/2$. For the parabola $y = ax^2 + bx + c$, the slope of the tangent line is

given by $y' = 2ax + b$. An equation of the tangent line at $x = p$ is

$y - (ap^2 + bp + c) = (2ap + b)(x - p)$. Solving for y gives us

$$y = (2ap + b)x - 2ap^2 - bp + (ap^2 + bp + c)$$

or
$$y = (2ap + b)x + c - ap^2 \qquad \textbf{(1)}$$

Similarly, an equation of the tangent line at $x = q$ is

$$y = (2aq + b)x + c - aq^2 \qquad \textbf{(2)}$$

We can eliminate y and solve for x by subtracting equation **(1)** from equation **(2)**.

$$[(2aq + b) - (2ap + b)]x - aq^2 + ap^2 = 0$$

$$(2aq - 2ap)x = aq^2 - ap^2$$

$$2a(q - p)x = a(q^2 - p^2)$$

$$x = \frac{a(q + p)(q - p)}{2a(q - p)} = \frac{p + q}{2}$$

Thus, the x-coordinate of the point of intersection of the two tangent lines, namely r, is $(p + q)/2$.

4. We could differentiate and then simplify or we can simplify and then differentiate. The latter seems to be the simpler method.

$$\frac{\sin^2 x}{1 + \cot x} + \frac{\cos^2 x}{1 + \tan x} = \frac{\sin^2 x}{1 + \frac{\cos x}{\sin x}} \cdot \frac{\sin x}{\sin x} + \frac{\cos^2 x}{1 + \frac{\sin x}{\cos x}} \cdot \frac{\cos x}{\cos x} = \frac{\sin^3 x}{\sin x + \cos x} + \frac{\cos^3 x}{\cos x + \sin x}$$

$$= \frac{\sin^3 x + \cos^3 x}{\sin x + \cos x} \quad \text{[factor sum of cubes]} \qquad = \frac{(\sin x + \cos x)(\sin^2 x - \sin x \cos x + \cos^2 x)}{\sin x + \cos x}$$

$$= \sin^2 x - \sin x \cos x + \cos^2 x = 1 - \sin x \cos x = 1 - \tfrac{1}{2}(2 \sin x \cos x) = 1 - \tfrac{1}{2}\sin 2x$$

Thus, $\dfrac{d}{dx}\left(\dfrac{\sin^2 x}{1 + \cot x} + \dfrac{\cos^2 x}{1 + \tan x}\right) = \dfrac{d}{dx}\left(1 - \tfrac{1}{2}\sin 2x\right) = -\tfrac{1}{2}\cos 2x \cdot 2 = -\cos 2x.$

5. (a) Put $x = 0$ and $y = 0$ in the equation: $f(0 + 0) = f(0) + f(0) + 0^2 \cdot 0 + 0 \cdot 0^2 \quad \Rightarrow \quad f(0) = 2f(0)$.

Subtracting $f(0)$ from each side of this equation gives $f(0) = 0$.

(b) $f'(0) = \lim\limits_{h \to 0} \dfrac{f(0 + h) - f(0)}{h} = \lim\limits_{h \to 0} \dfrac{[f(0) + f(h) + 0^2 h + 0h^2] - f(0)}{h} = \lim\limits_{h \to 0} \dfrac{f(h)}{h} = \lim\limits_{x \to 0} \dfrac{f(x)}{x} = 1$

(c) $f'(x) = \lim\limits_{h \to 0} \dfrac{f(x + h) - f(x)}{h} = \lim\limits_{h \to 0} \dfrac{[f(x) + f(h) + x^2 h + xh^2] - f(x)}{h} = \lim\limits_{h \to 0} \dfrac{f(h) + x^2 h + xh^2}{h}$

$= \lim\limits_{h \to 0} \left[\dfrac{f(h)}{h} + x^2 + xh \right] = 1 + x^2$

6. We find the equation of the parabola by substituting the point $(-100, 100)$, at which the car is situated, into the general

equation $y = ax^2$: $100 = a(-100)^2 \quad \Rightarrow \quad a = \frac{1}{100}$. Now we find the equation of a tangent to the parabola at the point

(x_0, y_0). We can show that $y' = a(2x) = \frac{1}{100}(2x) = \frac{1}{50}x$, so an equation of the tangent is $y - y_0 = \frac{1}{50}x_0(x - x_0)$.

Since the point (x_0, y_0) is on the parabola, we must have $y_0 = \frac{1}{100}x_0^2$, so our equation of the tangent can be simplified to

$y = \frac{1}{100}x_0^2 + \frac{1}{50}x_0(x - x_0)$. We want the statue to be located on the tangent line, so we substitute its coordinates $(100, 50)$

into this equation: $50 = \frac{1}{100}x_0^2 + \frac{1}{50}x_0(100 - x_0) \quad \Rightarrow \quad x_0^2 - 200x_0 + 5000 = 0 \quad \Rightarrow$

$x_0 = \frac{1}{2}\left[200 \pm \sqrt{200^2 - 4(5000)} \right] \quad \Rightarrow \quad x_0 = 100 \pm 50\sqrt{2}$. But $x_0 < 100$, so the car's headlights illuminate the statue

when it is located at the point $\left(100 - 50\sqrt{2}, 150 - 100\sqrt{2} \right) \approx (29.3, 8.6)$, that is, about 29.3 m east and 8.6 m north of

the origin.

7. We use mathematical induction. Let S_n be the statement that $\dfrac{d^n}{dx^n}(\sin^4 x + \cos^4 x) = 4^{n-1}\cos(4x + n\pi/2)$.

S_1 is true because

$$\frac{d}{dx}\left(\sin^4 x + \cos^4 x \right) = 4\sin^3 x \cos x - 4\cos^3 x \sin x = 4\sin x \cos x \left(\sin^2 x - \cos^2 x \right) x$$

$$= -4\sin x \cos x \cos 2x = -2\sin 2x \cos 2 = -\sin 4x = \sin(-4x)$$

$$= \cos\left(\tfrac{\pi}{2} - (-4x) \right) = \cos\left(\tfrac{\pi}{2} + 4x \right) = 4^{n-1}\cos\left(4x + n\tfrac{\pi}{2} \right) \text{ when } n = 1$$

Now assume S_k is true, that is, $\dfrac{d^k}{dx^k}\left(\sin^4 x + \cos^4 x \right) = 4^{k-1}\cos\left(4x + k\tfrac{\pi}{2} \right)$. Then

$$\frac{d^{k+1}}{dx^{k+1}}\left(\sin^4 x + \cos^4 x \right) = \frac{d}{dx}\left[\frac{d^k}{dx^k}\left(\sin^4 x + \cos^4 x \right) \right] = \frac{d}{dx}\left[4^{k-1}\cos\left(4x + k\tfrac{\pi}{2} \right) \right]$$

$$= -4^{k-1}\sin\left(4x + k\tfrac{\pi}{2} \right) \cdot \frac{d}{dx}\left(4x + k\tfrac{\pi}{2} \right) = -4^k \sin\left(4x + k\tfrac{\pi}{2} \right)$$

$$= 4^k \sin\left(-4x - k\tfrac{\pi}{2} \right) = 4^k \cos\left(\tfrac{\pi}{2} - \left(-4x - k\tfrac{\pi}{2} \right) \right) = 4^k \cos\left(4x + (k+1)\tfrac{\pi}{2} \right)$$

which shows that S_{k+1} is true.

Therefore, $\dfrac{d^n}{dx^n}\left(\sin^4 x + \cos^4 x \right) = 4^{n-1}\cos\left(4x + n\tfrac{\pi}{2} \right)$ for every positive integer n, by mathematical induction.

Another proof: First write

$$\sin^4 x + \cos^4 x = (\sin^2 x + \cos^2 x)^2 - 2\sin^2 x \cos^2 x = 1 - \tfrac{1}{2}\sin^2 2x = 1 - \tfrac{1}{4}(1 - \cos 4x) = \tfrac{3}{4} + \tfrac{1}{4}\cos 4x$$

Then we have $\dfrac{d^n}{dx^n}\left(\sin^4 x + \cos^4 x \right) = \dfrac{d^n}{dx^n}\left(\dfrac{3}{4} + \dfrac{1}{4}\cos 4x \right) = \dfrac{1}{4} \cdot 4^n \cos\left(4x + n\dfrac{\pi}{2} \right) = 4^{n-1}\cos\left(4x + n\dfrac{\pi}{2} \right)$.

8. If we divide $1 - x$ into x^n by long division, we find that $f(x) = \dfrac{x^n}{1-x} = -x^{n-1} - x^{n-2} - \cdots - x - 1 + \dfrac{1}{1-x}$.

This can also be seen by multiplying the last expression by $1 - x$ and canceling terms on the right-hand side. So we let

$g(x) = 1 + x + x^2 + \cdots + x^{n-1}$, so that $f(x) = \dfrac{1}{1-x} - g(x) \quad \Rightarrow \quad f^{(n)}(x) = \left(\dfrac{1}{1-x}\right)^{(n)} - g^{(n)}(x)$.

But g is a polynomial of degree $(n-1)$, so its nth derivative is 0, and therefore $f^{(n)}(x) = \left(\dfrac{1}{1-x}\right)^{(n)}$. Now

$\dfrac{d}{dx}(1-x)^{-1} = (-1)(1-x)^{-2}(-1) = (1-x)^{-2}, \ \dfrac{d^2}{dx^2}(1-x)^{-1} = (-2)(1-x)^{-3}(-1) = 2(1-x)^{-3}$,

$\dfrac{d^3}{dx^3}(1-x)^{-1} = (-3)\cdot 2(1-x)^{-4}(-1) = 3\cdot 2(1-x)^{-4}, \ \dfrac{d^4}{dx^4}(1-x)^{-1} = 4\cdot 3\cdot 2(1-x)^{-5}$, and so on.

So after n differentiations, we will have $f^{(n)}(x) = \left(\dfrac{1}{1-x}\right)^{(n)} = \dfrac{n!}{(1-x)^{n+1}}$.

9. We must find a value x_0 such that the normal lines to the parabola $y = x^2$ at $x = \pm x_0$ intersect at a point one unit from the

points $\left(\pm x_0, x_0^2\right)$. The normals to $y = x^2$ at $x = \pm x_0$ have slopes $-\dfrac{1}{\pm 2x_0}$ and pass through $\left(\pm x_0, x_0^2\right)$ respectively, so the

normals have the equations $y - x_0^2 = -\dfrac{1}{2x_0}(x - x_0)$ and $y - x_0^2 = \dfrac{1}{2x_0}(x + x_0)$. The common y-intercept is $x_0^2 + \dfrac{1}{2}$.

We want to find the value of x_0 for which the distance from $\left(0, x_0^2 + \frac{1}{2}\right)$ to $\left(x_0, x_0^2\right)$ equals 1. The square of the distance is

$(x_0 - 0)^2 + \left[x_0^2 - \left(x_0^2 + \frac{1}{2}\right)\right]^2 = x_0^2 + \frac{1}{4} = 1 \quad \Leftrightarrow \quad x_0 = \pm\frac{\sqrt{3}}{2}$. For these values of x_0, the y-intercept is $x_0^2 + \frac{1}{2} = \frac{5}{4}$, so

the center of the circle is at $\left(0, \frac{5}{4}\right)$.

Another solution: Let the center of the circle be $(0, a)$. Then the equation of the circle is $x^2 + (y - a)^2 = 1$.

Solving with the equation of the parabola, $y = x^2$, we get $x^2 + (x^2 - a)^2 = 1 \quad \Leftrightarrow \quad x^2 + x^4 - 2ax^2 + a^2 = 1 \quad \Leftrightarrow$

$x^4 + (1 - 2a)x^2 + a^2 - 1 = 0$. The parabola and the circle will be tangent to each other when this quadratic equation in x^2

has equal roots; that is, when the discriminant is 0. Thus, $(1 - 2a)^2 - 4(a^2 - 1) = 0 \quad \Leftrightarrow$

$1 - 4a + 4a^2 - 4a^2 + 4 = 0 \quad \Leftrightarrow \quad 4a = 5$, so $a = \frac{5}{4}$. The center of the circle is $\left(0, \frac{5}{4}\right)$.

10. $\displaystyle\lim_{x \to a} \dfrac{f(x) - f(a)}{\sqrt{x} - \sqrt{a}} = \lim_{x \to a}\left[\dfrac{f(x) - f(a)}{\sqrt{x} - \sqrt{a}} \cdot \dfrac{\sqrt{x} + \sqrt{a}}{\sqrt{x} + \sqrt{a}}\right] = \lim_{x \to a}\left[\dfrac{f(x) - f(a)}{x - a} \cdot \left(\sqrt{x} + \sqrt{a}\right)\right]$

$\displaystyle = \lim_{x \to a} \dfrac{f(x) - f(a)}{x - a} \cdot \lim_{x \to a}\left(\sqrt{x} + \sqrt{a}\right) = f'(a) \cdot \left(\sqrt{a} + \sqrt{a}\right) = 2\sqrt{a}\,f'(a)$

11. We can assume without loss of generality that $\theta = 0$ at time $t = 0$, so that $\theta = 12\pi t$ rad. [The angular velocity of the wheel

is 360 rpm $= 360 \cdot (2\pi \text{ rad})/(60 \text{ s}) = 12\pi$ rad/s.] Then the position of A as a function of time is

$A = (40\cos\theta, 40\sin\theta) = (40\cos 12\pi t, 40\sin 12\pi t)$, so $\sin\alpha = \dfrac{y}{1.2 \text{ m}} = \dfrac{40\sin\theta}{120} = \dfrac{\sin\theta}{3} = \dfrac{1}{3}\sin 12\pi t$.

(a) Differentiating the expression for $\sin\alpha$, we get $\cos\alpha \cdot \dfrac{d\alpha}{dt} = \dfrac{1}{3} \cdot 12\pi \cdot \cos 12\pi t = 4\pi\cos\theta$. When $\theta = \dfrac{\pi}{3}$, we have

$\sin\alpha = \dfrac{1}{3}\sin\theta = \dfrac{\sqrt{3}}{6}$, so $\cos\alpha = \sqrt{1 - \left(\dfrac{\sqrt{3}}{6}\right)^2} = \sqrt{\dfrac{11}{12}}$ and $\dfrac{d\alpha}{dt} = \dfrac{4\pi\cos\frac{\pi}{3}}{\cos\alpha} = \dfrac{2\pi}{\sqrt{11/12}} = \dfrac{4\pi\sqrt{3}}{\sqrt{11}} \approx 6.56$ rad/s.

(b) By the Law of Cosines, $|AP|^2 = |OA|^2 + |OP|^2 - 2\,|OA|\,|OP|\cos\theta \quad\Rightarrow$

$120^2 = 40^2 + |OP|^2 - 2\cdot 40\,|OP|\cos\theta \quad\Rightarrow\quad |OP|^2 - (80\cos\theta)\,|OP| - 12{,}800 = 0 \quad\Rightarrow$

$|OP| = \tfrac{1}{2}\big(80\cos\theta \pm \sqrt{6400\cos^2\theta + 51{,}200}\,\big) = 40\cos\theta \pm 40\sqrt{\cos^2\theta + 8} = 40\big(\cos\theta + \sqrt{8 + \cos^2\theta}\,\big)$ cm

[since $|OP| > 0$].   As a check, note that $|OP| = 160$ cm when $\theta = 0$ and $|OP| = 80\sqrt{2}$ cm when $\theta = \frac{\pi}{2}$.

(c) By part (b), the x-coordinate of P is given by $x = 40\big(\cos\theta + \sqrt{8 + \cos^2\theta}\,\big)$, so

$$\frac{dx}{dt} = \frac{dx}{d\theta}\frac{d\theta}{dt} = 40\left(-\sin\theta - \frac{2\cos\theta\sin\theta}{2\sqrt{8 + \cos^2\theta}}\right)\cdot 12\pi = -480\pi\sin\theta\left(1 + \frac{\cos\theta}{\sqrt{8 + \cos^2\theta}}\right) \text{ cm/s.}$$

In particular, $dx/dt = 0$ cm/s when $\theta = 0$ and $dx/dt = -480\pi$ cm/s when $\theta = \frac{\pi}{2}$.

12. The equation of T_1 is $y - x_1^2 = 2x_1(x - x_1) = 2x_1x - 2x_1^2$ or $y = 2x_1x - x_1^2$.

The equation of T_2 is $y = 2x_2x - x_2^2$. Solving for the point of intersection, we

get $2x(x_1 - x_2) = x_1^2 - x_2^2 \quad\Rightarrow\quad x = \tfrac{1}{2}(x_1 + x_2)$. Therefore, the coordinates

of P are $\big(\tfrac{1}{2}(x_1 + x_2), x_1x_2\big)$. So if the point of contact of T is $\big(a, a^2\big)$, then

Q_1 is $\big(\tfrac{1}{2}(a + x_1), ax_1\big)$ and Q_2 is $\big(\tfrac{1}{2}(a + x_2), ax_2\big)$. Therefore,

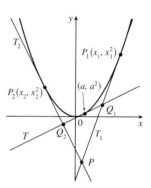

$|PQ_1|^2 = \tfrac{1}{4}(a - x_2)^2 + x_1^2(a - x_2)^2 = (a - x_2)^2\big(\tfrac{1}{4} + x_1^2\big)$ and

$|PP_1|^2 = \tfrac{1}{4}(x_1 - x_2)^2 + x_1^2(x_1 - x_2)^2 = (x_1 - x_2)^2\big(\tfrac{1}{4} + x_1^2\big)$.

So $\dfrac{|PQ_1|^2}{|PP_1|^2} = \dfrac{(a - x_2)^2}{(x_1 - x_2)^2}$, and similarly $\dfrac{|PQ_2|^2}{|PP_2|^2} = \dfrac{(x_1 - a)^2}{(x_1 - x_2)^2}$. Finally, $\dfrac{|PQ_1|}{|PP_1|} + \dfrac{|PQ_2|}{|PP_2|} = \dfrac{a - x_2}{x_1 - x_2} + \dfrac{x_1 - a}{x_1 - x_2} = 1$.

13. It seems from the figure that as P approaches the point $(0, 2)$ from the right, $x_T \to \infty$ and $y_T \to 2^+$. As P approaches the

point $(3, 0)$ from the left, it appears that $x_T \to 3^+$ and $y_T \to \infty$. So we guess that $x_T \in (3, \infty)$ and $y_T \in (2, \infty)$. It is

more difficult to estimate the range of values for x_N and y_N. We might perhaps guess that $x_N \in (0, 3)$,

and $y_N \in (-\infty, 0)$ or $(-2, 0)$.

In order to actually solve the problem, we implicitly differentiate the equation of the ellipse to find the equation of the

tangent line: $\dfrac{x^2}{9} + \dfrac{y^2}{4} = 1 \quad\Rightarrow\quad \dfrac{2x}{9} + \dfrac{2y}{4}y' = 0$, so $y' = -\dfrac{4}{9}\dfrac{x}{y}$. So at the point (x_0, y_0) on the ellipse, an equation of the

tangent line is $y - y_0 = -\dfrac{4}{9}\dfrac{x_0}{y_0}(x - x_0)$ or $4x_0x + 9y_0y = 4x_0^2 + 9y_0^2$. This can be written as $\dfrac{x_0x}{9} + \dfrac{y_0y}{4} = \dfrac{x_0^2}{9} + \dfrac{y_0^2}{4} = 1$,

because (x_0, y_0) lies on the ellipse. So an equation of the tangent line is $\dfrac{x_0x}{9} + \dfrac{y_0y}{4} = 1$.

Therefore, the x-intercept x_T for the tangent line is given by $\dfrac{x_0x_T}{9} = 1 \quad\Leftrightarrow\quad x_T = \dfrac{9}{x_0}$, and the y-intercept y_T is given

by $\dfrac{y_0y_T}{4} = 1 \quad\Leftrightarrow\quad y_T = \dfrac{4}{y_0}$.

So as x_0 takes on all values in $(0, 3)$, x_T takes on all values in $(3, \infty)$, and as y_0 takes on all values in $(0, 2)$, y_T takes on

all values in $(2, \infty)$.   At the point (x_0, y_0) on the ellipse, the slope of the normal line is $-\dfrac{1}{y'(x_0, y_0)} = \dfrac{9}{4}\dfrac{y_0}{x_0}$, and its equation is $y - y_0 = \dfrac{9}{4}\dfrac{y_0}{x_0}(x - x_0)$. So the x-intercept x_N for the normal line is given by $0 - y_0 = \dfrac{9}{4}\dfrac{y_0}{x_0}(x_N - x_0)$ ⟹

$x_N = -\dfrac{4x_0}{9} + x_0 = \dfrac{5x_0}{9}$, and the y-intercept y_N is given by $y_N - y_0 = \dfrac{9}{4}\dfrac{y_0}{x_0}(0 - x_0)$ ⟹ $y_N = -\dfrac{9y_0}{4} + y_0 = -\dfrac{5y_0}{4}$.

So as x_0 takes on all values in $(0, 3)$, x_N takes on all values in $\left(0, \frac{5}{3}\right)$, and as y_0 takes on all values in $(0, 2)$, y_N takes on all values in $\left(-\frac{5}{2}, 0\right)$.

14. $\displaystyle\lim_{x \to 0} \dfrac{\sin(3 + x)^2 - \sin 9}{x} = f'(3)$ where $f(x) = \sin x^2$. Now $f'(x) = (\cos x^2)(2x)$, so $f'(3) = 6\cos 9$.

15. (a) If the two lines L_1 and L_2 have slopes m_1 and m_2 and angles of

inclination ϕ_1 and ϕ_2, then $m_1 = \tan \phi_1$ and $m_2 = \tan \phi_2$. The triangle

in the figure shows that $\phi_1 + \alpha + (180° - \phi_2) = 180°$ and so

$\alpha = \phi_2 - \phi_1$. Therefore, using the identity for $\tan(x - y)$, we have

$\tan \alpha = \tan(\phi_2 - \phi_1) = \dfrac{\tan \phi_2 - \tan \phi_1}{1 + \tan \phi_2 \tan \phi_1}$ and so $\tan \alpha = \dfrac{m_2 - m_1}{1 + m_1 m_2}$.

(b) (i) The parabolas intersect when $x^2 = (x - 2)^2$ ⟹ $x = 1$. If $y = x^2$, then $y' = 2x$, so the slope of the tangent

to $y = x^2$ at $(1, 1)$ is $m_1 = 2(1) = 2$. If $y = (x - 2)^2$, then $y' = 2(x - 2)$, so the slope of the tangent to

$y = (x - 2)^2$ at $(1, 1)$ is $m_2 = 2(1 - 2) = -2$. Therefore, $\tan \alpha = \dfrac{m_2 - m_1}{1 + m_1 m_2} = \dfrac{-2 - 2}{1 + 2(-2)} = \dfrac{4}{3}$ and

so $\alpha = \tan^{-1}\left(\frac{4}{3}\right) \approx 53°$ [or $127°$].

(ii) $x^2 - y^2 = 3$ and $x^2 - 4x + y^2 + 3 = 0$ intersect when $x^2 - 4x + (x^2 - 3) + 3 = 0$ ⟺ $2x(x - 2) = 0$ ⟹

$x = 0$ or 2, but 0 is extraneous. If $x = 2$, then $y = \pm 1$. If $x^2 - y^2 = 3$ then $2x - 2yy' = 0$ ⟹ $y' = x/y$ and

$x^2 - 4x + y^2 + 3 = 0$ ⟹ $2x - 4 + 2yy' = 0$ ⟹ $y' = \dfrac{2 - x}{y}$. At $(2, 1)$ the slopes are $m_1 = 2$ and

$m_2 = 0$, so $\tan \alpha = \dfrac{0 - 2}{1 + 2 \cdot 0} = -2$ ⟹ $\alpha \approx 117°$. At $(2, -1)$ the slopes are $m_1 = -2$ and $m_2 = 0$,

so $\tan \alpha = \dfrac{0 - (-2)}{1 + (-2)(0)} = 2$ ⟹ $\alpha \approx 63°$ [or $117°$].

16. $y^2 = 4px$ ⟹ $2yy' = 4p$ ⟹ $y' = 2p/y$ ⟹   slope of tangent at $P(x_1, y_1)$ is $m_1 = 2p/y_1$. The slope of FP is

$m_2 = \dfrac{y_1}{x_1 - p}$, so by the formula from Problem 15(a),

$$\tan \alpha = \dfrac{\dfrac{y_1}{x_1 - p} - \dfrac{2p}{y_1}}{1 + \left(\dfrac{2p}{y_1}\right)\left(\dfrac{y_1}{x_1 - p}\right)} \cdot \dfrac{y_1(x_1 - p)}{y_1(x_1 - p)} = \dfrac{y_1^2 - 2p(x_1 - p)}{y_1(x_1 - p) + 2py_1}$$

$$= \dfrac{4px_1 - 2px_1 + 2p^2}{x_1 y_1 - py_1 + 2py_1} = \dfrac{2p(p + x_1)}{y_1(p + x_1)} = \dfrac{2p}{y_1}$$

$$= \text{slope of tangent at } P = \tan \beta$$

Since $0 \le \alpha, \beta \le \frac{\pi}{2}$, this proves that $\alpha = \beta$.

17. Since $\angle ROQ = \angle OQP = \theta$, the triangle QOR is isosceles, so

$|QR| = |RO| = x$. By the Law of Cosines, $x^2 = x^2 + r^2 - 2rx\cos\theta$. Hence,

$2rx\cos\theta = r^2$, so $x = \dfrac{r^2}{2r\cos\theta} = \dfrac{r}{2\cos\theta}$. Note that as $y \to 0^+, \theta \to 0^+$ (since

$\sin\theta = y/r$), and hence $x \to \dfrac{r}{2\cos 0} = \dfrac{r}{2}$. Thus, as P is taken closer and closer

to the x-axis, the point R approaches the midpoint of the radius AO.

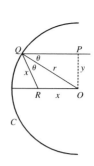

18. $\displaystyle\lim_{x\to 0}\frac{f(x)}{g(x)} = \lim_{x\to 0}\frac{f(x)-0}{g(x)-0} = \lim_{x\to 0}\frac{f(x)-f(0)}{g(x)-g(0)} = \lim_{x\to 0}\frac{\dfrac{f(x)-f(0)}{x-0}}{\dfrac{g(x)-g(0)}{x-0}} = \frac{\displaystyle\lim_{x\to 0}\dfrac{f(x)-f(0)}{x-0}}{\displaystyle\lim_{x\to 0}\dfrac{g(x)-g(0)}{x-0}} = \frac{f'(0)}{g'(0)}$

19. $\displaystyle\lim_{x\to 0}\frac{\sin(a+2x) - 2\sin(a+x) + \sin a}{x^2}$

$$= \lim_{x\to 0}\frac{\sin a\,\cos 2x + \cos a\,\sin 2x - 2\sin a\,\cos x - 2\cos a\,\sin x + \sin a}{x^2}$$

$$= \lim_{x\to 0}\frac{\sin a\,(\cos 2x - 2\cos x + 1) + \cos a\,(\sin 2x - 2\sin x)}{x^2}$$

$$= \lim_{x\to 0}\frac{\sin a\,(2\cos^2 x - 1 - 2\cos x + 1) + \cos a\,(2\sin x\,\cos x - 2\sin x)}{x^2}$$

$$= \lim_{x\to 0}\frac{\sin a\,(2\cos x)(\cos x - 1) + \cos a\,(2\sin x)(\cos x - 1)}{x^2}$$

$$= \lim_{x\to 0}\frac{2(\cos x - 1)[\sin a\,\cos x + \cos a\,\sin x](\cos x + 1)}{x^2(\cos x + 1)}$$

$$= \lim_{x\to 0}\frac{-2\sin^2 x\,[\sin(a+x)]}{x^2(\cos x + 1)} = -2\lim_{x\to 0}\left(\frac{\sin x}{x}\right)^2\cdot\frac{\sin(a+x)}{\cos x + 1} = -2(1)^2\frac{\sin(a+0)}{\cos 0 + 1} = -\sin a$$

20. Suppose that $y = mx + c$ is a tangent line to the ellipse. Then it intersects the ellipse at only one point, so the discriminant

of the equation $\dfrac{x^2}{a^2} + \dfrac{(mx+c)^2}{b^2} = 1 \;\Leftrightarrow\; (b^2 + a^2m^2)x^2 + 2mca^2x + a^2c^2 - a^2b^2 = 0$ must be 0; that is,

$$0 = (2mca^2)^2 - 4(b^2 + a^2m^2)(a^2c^2 - a^2b^2) = 4a^4c^2m^2 - 4a^2b^2c^2 + 4a^2b^4 - 4a^4c^2m^2 + 4a^4b^2m^2$$

$$= 4a^2b^2(a^2m^2 + b^2 - c^2)$$

Therefore, $a^2m^2 + b^2 - c^2 = 0$.

Now if a point (α, β) lies on the line $y = mx + c$, then $c = \beta - m\alpha$, so from above,

$$0 = a^2m^2 + b^2 - (\beta - m\alpha)^2 = (a^2 - \alpha^2)m^2 + 2\alpha\beta m + b^2 - \beta^2 \;\Leftrightarrow\; m^2 + \frac{2\alpha\beta}{a^2 - \alpha^2}m + \frac{b^2 - \beta^2}{a^2 - \alpha^2} = 0.$$

(a) Suppose that the two tangent lines from the point (α, β) to the ellipse

have slopes m and $\dfrac{1}{m}$. Then m and $\dfrac{1}{m}$ are roots of the equation

$z^2 + \dfrac{2\alpha\beta}{a^2 - \alpha^2}z + \dfrac{b^2 - \beta^2}{a^2 - \alpha^2} = 0$. This implies that $(z - m)\left(z - \dfrac{1}{m}\right) = 0 \;\Leftrightarrow$

$z^2 - \left(m + \dfrac{1}{m}\right)z + m\left(\dfrac{1}{m}\right) = 0$, so equating the constant terms in the two

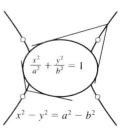

quadratic equations, we get $\dfrac{b^2 - \beta^2}{a^2 - \alpha^2} = m\left(\dfrac{1}{m}\right) = 1$, and hence $b^2 - \beta^2 = a^2 - \alpha^2$. So (α, β) lies on the

hyperbola $x^2 - y^2 = a^2 - b^2$.

(b) If the two tangent lines from the point (α, β) to the ellipse have slopes m

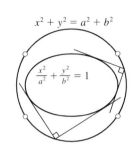

$x^2 + y^2 = a^2 + b^2$

$\dfrac{x^2}{a^2} + \dfrac{y^2}{b^2} = 1$

and $-\dfrac{1}{m}$, then m and $-\dfrac{1}{m}$ are roots of the quadratic equation, and so

$(z - m)\left(z + \dfrac{1}{m}\right) = 0$, and equating the constant terms as in part (a), we get

$\dfrac{b^2 - \beta^2}{a^2 - \alpha^2} = -1$, and hence $b^2 - \beta^2 = \alpha^2 - a^2$. So the point (α, β) lies on the

circle $x^2 + y^2 = a^2 + b^2$.

21. $y = x^4 - 2x^2 - x \ \Rightarrow \ y' = 4x^3 - 4x - 1$. The equation of the tangent line at $x = a$ is

$y - (a^4 - 2a^2 - a) = (4a^3 - 4a - 1)(x - a)$ or $y = (4a^3 - 4a - 1)x + (-3a^4 + 2a^2)$ and similarly for $x = b$. So if at

$x = a$ and $x = b$ we have the same tangent line, then $4a^3 - 4a - 1 = 4b^3 - 4b - 1$ and $-3a^4 + 2a^2 = -3b^4 + 2b^2$. The first

equation gives $a^3 - b^3 = a - b \ \Rightarrow \ (a - b)(a^2 + ab + b^2) = (a - b)$. Assuming $a \neq b$, we have $1 = a^2 + ab + b^2$.

The second equation gives $3(a^4 - b^4) = 2(a^2 - b^2) \ \Rightarrow \ 3(a^2 - b^2)(a^2 + b^2) = 2(a^2 - b^2)$ which is true if $a = -b$.

Substituting into $1 = a^2 + ab + b^2$ gives $1 = a^2 - a^2 + a^2 \ \Rightarrow \ a = \pm 1$ so that $a = 1$ and $b = -1$ or vice versa. Thus,

the points $(1, -2)$ and $(-1, 0)$ have a common tangent line.

As long as there are only two such points, we are done. So we show that these are in fact the only two such points.

Suppose that $a^2 - b^2 \neq 0$. Then $3(a^2 - b^2)(a^2 + b^2) = 2(a^2 - b^2)$ gives $3(a^2 + b^2) = 2$ or $a^2 + b^2 = \frac{2}{3}$.

Thus, $ab = (a^2 + ab + b^2) - (a^2 + b^2) = 1 - \dfrac{2}{3} = \dfrac{1}{3}$, so $b = \dfrac{1}{3a}$. Hence, $a^2 + \dfrac{1}{9a^2} = \dfrac{2}{3}$, so $9a^4 + 1 = 6a^2 \ \Rightarrow$

$0 = 9a^4 - 6a^2 + 1 = (3a^2 - 1)^2$. So $3a^2 - 1 = 0 \ \Rightarrow \ a^2 = \dfrac{1}{3} \ \Rightarrow \ b^2 = \dfrac{1}{9a^2} = \dfrac{1}{3} = a^2$, contradicting our assumption

that $a^2 \neq b^2$.

22. Suppose that the normal lines at the three points (a_1, a_1^2), (a_2, a_2^2), and (a_3, a_3^2) intersect at a common point. Now if one of

the a_i is 0 (suppose $a_1 = 0$) then by symmetry $a_2 = -a_3$, so $a_1 + a_2 + a_3 = 0$. So we can assume that none of the a_i is 0.

The slope of the tangent line at (a_i, a_i^2) is $2a_i$, so the slope of the normal line is $-\dfrac{1}{2a_i}$ and its equation is

$y - a_i^2 = -\dfrac{1}{2a_i}(x - a_i)$. We solve for the x-coordinate of the intersection of the normal lines from (a_1, a_1^2) and (a_2, a_2^2):

$y = a_1^2 - \dfrac{1}{2a_1}(x - a_1) = a_2^2 - \dfrac{1}{2a_2}(x - a_2) \ \Rightarrow \ x\left(\dfrac{1}{2a_2} - \dfrac{1}{2a_1}\right) = a_2^2 - a_1^2 \ \Rightarrow$

$x\left(\dfrac{a_1 - a_2}{2a_1 a_2}\right) = (-a_1 - a_2)(a_1 + a_2) \ \Leftrightarrow \ x = -2a_1 a_2(a_1 + a_2)$ **(1)**.   Similarly, solving for the x-coordinate of the

intersections of the normal lines from (a_1, a_1^2) and (a_3, a_3^2) gives $x = -2a_1 a_3(a_1 + a_3)$ **(2)**.

Equating **(1)** and **(2)** gives $a_2(a_1 + a_2) = a_3(a_1 + a_3) \ \Leftrightarrow \ a_1(a_2 - a_3) = a_3^2 - a_2^2 = -(a_2 + a_3)(a_2 - a_3) \ \Leftrightarrow$

$a_1 = -(a_2 + a_3) \ \Leftrightarrow \ a_1 + a_2 + a_3 = 0$.

23. Because of the periodic nature of the lattice points, it suffices to consider the points in the 5×2 grid shown. We can see that the minimum value of r occurs when there is a line with slope $\frac{2}{5}$ which touches the circle centered at $(3, 1)$ and the circles centered at $(0, 0)$ and $(5, 2)$.

To find P, the point at which the line is tangent to the circle at $(0, 0)$, we simultaneously solve $x^2 + y^2 = r^2$ and

$y = -\frac{5}{2}x \Rightarrow x^2 + \frac{25}{4}x^2 = r^2 \Rightarrow x^2 = \frac{4}{29}r^2 \Rightarrow x = \frac{2}{\sqrt{29}}r, y = -\frac{5}{\sqrt{29}}r$. To find Q, we either use symmetry or

solve $(x - 3)^2 + (y - 1)^2 = r^2$ and $y - 1 = -\frac{5}{2}(x - 3)$. As above, we get $x = 3 - \frac{2}{\sqrt{29}}r, y = 1 + \frac{5}{\sqrt{29}}r$. Now the slope of

the line PQ is $\frac{2}{5}$, so $m_{PQ} = \dfrac{1 + \frac{5}{\sqrt{29}}r - \left(-\frac{5}{\sqrt{29}}r\right)}{3 - \frac{2}{\sqrt{29}}r - \frac{2}{\sqrt{29}}r} = \dfrac{1 + \frac{10}{\sqrt{29}}r}{3 - \frac{4}{\sqrt{29}}r} = \dfrac{\sqrt{29} + 10r}{3\sqrt{29} - 4r} = \dfrac{2}{5} \Rightarrow$

$5\sqrt{29} + 50r = 6\sqrt{29} - 8r \Leftrightarrow 58r = \sqrt{29} \Leftrightarrow r = \frac{\sqrt{29}}{58}$. So the minimum value of r for which any line with slope $\frac{2}{5}$

intersects circles with radius r centered at the lattice points on the plane is $r = \frac{\sqrt{29}}{58} \approx 0.093$.

24.

Assume the axes of the cone and the cylinder are parallel. Let H denote the initial height of the water. When the cone has been dropping for t seconds, the water level has risen x centimeters, so the tip of the cone is $x + 1t$ centimeters below the water line. We want to find dx/dt when $x + t = h$ (when the cone is completely submerged).

Using similar triangles, $\dfrac{r_1}{x + t} = \dfrac{r}{h} \Rightarrow r_1 = \dfrac{r}{h}(x + t)$.

volume of water and cone at time t	$=$	original volume of water	$+$	volume of submerged part of cone
$\pi R^2(H + x)$	$=$	$\pi R^2 H$	$+$	$\frac{1}{3}\pi r_1^2(x + t)$
$\pi R^2 H + \pi R^2 x$	$=$	$\pi R^2 H$	$+$	$\frac{1}{3}\pi \dfrac{r^2}{h^2}(x + t)^3$
$3h^2 R^2 x$	$=$	$r^2(x + t)^3$		

Differentiating implicitly with respect to t gives us $3h^2 R^2 \dfrac{dx}{dt} = r^2 \left[3(x + t)^2 \dfrac{dx}{dt} + 3(x + t)^2 \dfrac{dt}{dt}\right] \Rightarrow$

$\dfrac{dx}{dt} = \dfrac{r^2(x + t)^2}{h^2 R^2 - r^2(x + t)^2} \Rightarrow \dfrac{dx}{dt}\bigg|_{x + t = h} = \dfrac{r^2 h^2}{h^2 R^2 - r^2 h^2} = \dfrac{r^2}{R^2 - r^2}$. Thus, the water level is rising at a rate of

$\dfrac{r^2}{R^2 - r^2}$ cm/s at the instant the cone is completely submerged.

25.

By similar triangles, $\dfrac{r}{5} = \dfrac{h}{16}$ $\Rightarrow$ $r = \dfrac{5h}{16}$. The volume of the cone is

$$V = \tfrac{1}{3}\pi r^2 h = \tfrac{1}{3}\pi\left(\frac{5h}{16}\right)^2 h = \frac{25\pi}{768}h^3,\ \text{so}\ \frac{dV}{dt} = \frac{25\pi}{256}h^2\,\frac{dh}{dt}.\ \text{Now the rate of}$$

change of the volume is also equal to the difference of what is being added

$(2\ \text{cm}^3/\text{min})$ and what is oozing out ($k\pi r l$, where $\pi r l$ is the area of the cone and k

is a proportionality constant). Thus, $\dfrac{dV}{dt} = 2 - k\pi r l$.

Equating the two expressions for $\dfrac{dV}{dt}$ and substituting $h = 10$, $\dfrac{dh}{dt} = -0.3$, $r = \dfrac{5(10)}{16} = \dfrac{25}{8}$, and $\dfrac{l}{\sqrt{281}} = \dfrac{10}{16}$ $\Leftrightarrow$

$l = \dfrac{5}{8}\sqrt{281}$, we get $\dfrac{25\pi}{256}(10)^2(-0.3) = 2 - k\pi\dfrac{25}{8}\cdot\dfrac{5}{8}\sqrt{281}$ $\Leftrightarrow$ $\dfrac{125k\pi\sqrt{281}}{64} = 2 + \dfrac{750\pi}{256}$. Solving for k gives us

$k = \dfrac{256 + 375\pi}{250\pi\sqrt{281}}$. To maintain a certain height, the rate of oozing, $k\pi r l$, must equal the rate of the liquid being poured in;

that is, $\dfrac{dV}{dt} = 0$. Thus, the rate at which we should pour the liquid into the container is

$$k\pi r l = \frac{256 + 375\pi}{250\pi\sqrt{281}}\cdot\pi\cdot\frac{25}{8}\cdot\frac{5\sqrt{281}}{8} = \frac{256 + 375\pi}{128} \approx 11.204\ \text{cm}^3/\text{min}$$

26. (a) $f(x) = x(x-2)(x-6) = x^3 - 8x^2 + 12x$ $\Rightarrow$

$f'(x) = 3x^2 - 16x + 12$. The average of the first pair of zeros is

$(0+2)/2 = 1$. At $x = 1$, the slope of the tangent line is $f'(1) = -1$, so an

equation of the tangent line has the form $y = -1x + b$. Since $f(1) = 5$, we

have $5 = -1 + b$ $\Rightarrow$ $b = 6$ and the tangent has equation $y = -x + 6$.

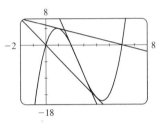

Similarly, at $x = \dfrac{0+6}{2} = 3$, $y = -9x + 18$; at $x = \dfrac{2+6}{2} = 4$, $y = -4x$. From the graph, we see that each tangent line

drawn at the average of two zeros intersects the graph of f at the third zero.

(b) A CAS gives $f'(x) = (x-b)(x-c) + (x-a)(x-c) + (x-a)(x-b)$ or

$f'(x) = 3x^2 - 2(a+b+c)x + ab + ac + bc$. Using the $\mathtt{Simplify}$ command, we get

$f'\left(\dfrac{a+b}{2}\right) = -\dfrac{(a-b)^2}{4}$ and $f\left(\dfrac{a+b}{2}\right) = -\dfrac{(a-b)^2}{8}(a+b-2c)$, so an equation of the tangent line at $x = \dfrac{a+b}{2}$ is

$y = -\dfrac{(a-b)^2}{4}\left(x - \dfrac{a+b}{2}\right) - \dfrac{(a-b)^2}{8}(a+b-2c)$. To find the x-intercept, let $y = 0$ and use the $\mathtt{Solve}$ command.

The result is $x = c$.

Using Derive, we can begin by authoring the expression $(x-a)(x-b)(x-c)$. Now load the utility file

$\mathtt{DifferentiationApplications}$. Next we author tangent $(\#1, x, (a+b)/2)$—this is the command to find an

equation of the tangent line of the function in #1 whose independent variable is x at the x-value $(a+b)/2$. We then simplify that expression and obtain the equation $y = \#4$. The form in expression #4 makes it easy to see that the x-intercept is the third zero, namely c. In a similar fashion we see that b is the x-intercept for the tangent line at $(a + c)/2$ and a is the x-intercept for the tangent line at $(b + c)/2$.

#1: (x - a)·(x - b)·(x - c)

#2: LOAD(C:\Program Files\TI Education\Derive 6\Math\DifferentiationApplications.mth)

#3: TANGENT $\left[(x - a)·(x - b)·(x - c), \ x, \ \dfrac{a + b}{2}\right]$

#4:
$$\frac{(a^2 - 2·a·b + b^2)·(c-x)}{4}$$

4 □ APPLICATIONS OF DIFFERENTIATION

4.1 Maximum and Minimum Values

1. A function f has an **absolute minimum** at $x = c$ if $f(c)$ is the smallest function value on the entire domain of f, whereas f has a **local minimum** at c if $f(c)$ is the smallest function value when x is near c.

2. (a) The Extreme Value Theorem

   (b) See the Closed Interval Method.

3. Absolute maximum at s, absolute minimum at r, local maximum at c, local minima at b and r, neither a maximum nor a minimum at a and d.

4. Absolute maximum at r; absolute minimum at a; local maxima at b and r; local minimum at d; neither a maximum nor a minimum at c and s.

5. Absolute maximum value is $f(4) = 5$; there is no absolute minimum value; local maximum values are $f(4) = 5$ and $f(6) = 4$; local minimum values are $f(2) = 2$ and $f(1) = f(5) = 3$.

6. There is no absolute maximum value; absolute minimum value is $g(4) = 1$; local maximum values are $g(3) = 4$ and $g(6) = 3$; local minimum values are $g(2) = 2$ and $g(4) = 1$.

7. Absolute minimum at 2, absolute maximum at 3, local minimum at 4

8. Absolute minimum at 1, absolute maximum at 5, local maximum at 2, local minimum at 4

9. Absolute maximum at 5, absolute minimum at 2, local maximum at 3, local minima at 2 and 4

10. f has no local maximum or minimum, but 2 and 4 are critical numbers

11. (a)

(b)

(c)

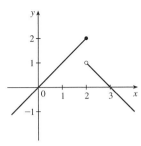

12. (a) Note that a local maximum cannot
occur at an endpoint.

(b)

Note: By the Extreme Value Theorem, f must *not* be continuous.

13. (a) *Note:* By the Extreme Value Theorem,
f must *not* be continuous; because if it
were, it would attain an absolute
minimum.

(b)

14. (a)

(b)

15. $f(x) = 8 - 3x$, $x \geq 1$. Absolute maximum $f(1) = 5$; no
local maximum. No absolute or local minimum.

16. $f(x) = 3 - 2x$, $x \leq 5$. Absolute minimum $f(5) = -7$;
no local minimum. No absolute or local maximum.

17. $f(x) = x^2$, $0 < x < 2$. No absolute or local maximum or minimum value.

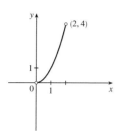

18. $f(x) = x^2$, $0 < x \le 2$. Absolute maximum $f(2) = 4$; no local maximum. No absolute or local minimum.

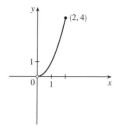

19. $f(x) = x^2$, $0 \le x < 2$. Absolute minimum $f(0) = 0$; no local minimum. No absolute or local maximum.

20. $f(x) = x^2$, $0 \le x \le 2$. Absolute maximum $f(2) = 4$. Absolute minimum $f(0) = 0$. No local maximum or minimum.

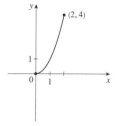

21. $f(x) = x^2$, $-3 \le x \le 2$. Absolute maximum $f(-3) = 9$. No local maximum. Absolute and local minimum $f(0) = 0$.

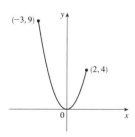

22. $f(x) = 1 + (x+1)^2$, $-2 \le x < 5$. No absolute or local maximum. Absolute and local minimum $f(-1) = 1$.

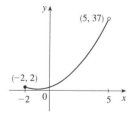

23. $f(t) = 1/t$, $0 < t < 1$. No maximum or minimum.

24. $f(t) = \cos t$, $-\frac{3\pi}{2} \le t \le \frac{3\pi}{2}$. Absolute and local maximum $f(0) = 1$; absolute and local minima $f(\pm\pi, -1)$.

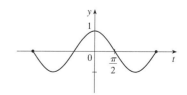

25. $f(x) = 1 - \sqrt{x}$. Absolute maximum $f(0) = 1$; no local maximum. No absolute or local minimum.

26. $f(x) = 1 - x^3$. No absolute or local extreme values.

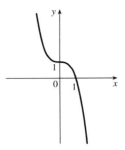

27. $f(x) = \begin{cases} 1 - x & \text{if } 0 \le x < 2 \\ 2x - 4 & \text{if } 2 \le x \le 3 \end{cases}$

Absolute maximum $f(3) = 2$; no local maximum. No absolute or local minimum.

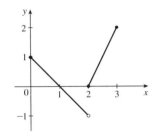

28. $f(x) = \begin{cases} 4 - x^2 & \text{if } -2 \le x < 0 \\ 2x - 1 & \text{if } 0 \le x \le 2 \end{cases}$

Abolute minimum $f(0) = -1$; no local minimim. No absolute or local maximum.

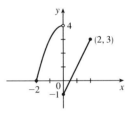

29. $f(x) = 5x^2 + 4x \quad \Rightarrow \quad f'(x) = 10x + 4$. $f'(x) = 0 \quad \Rightarrow \quad x = -\frac{2}{5}$, so $-\frac{2}{5}$ is the only critical number.

30. $f(x) = x^3 + x^2 - x \quad \Rightarrow \quad f'(x) = 3x^2 + 2x - 1$.

$f'(x) = 0 \quad \Rightarrow \quad (x + 1)(3x - 1) = 0 \quad \Rightarrow \quad x = -1, \frac{1}{3}$. These are the only critical numbers.

31. $f(x) = x^3 + 3x^2 - 24x \quad \Rightarrow \quad f'(x) = 3x^2 + 6x - 24 = 3(x^2 + 2x - 8)$.

$f'(x) = 0 \quad \Rightarrow \quad 3(x + 4)(x - 2) = 0 \quad \Rightarrow \quad x = -4, 2$. These are the only critical numbers.

32. $f(x) = x^3 + x^2 + x \quad \Rightarrow \quad f'(x) = 3x^2 + 2x + 1$. $f'(x) = 0 \quad \Rightarrow \quad 3x^2 + 2x + 1 = 0 \quad \Rightarrow \quad x = \dfrac{-2 \pm \sqrt{4 - 12}}{6}$.

Neither of these is a real number. Thus, there are no critical numbers.

33. $s(t) = 3t^4 + 4t^3 - 6t^2 \quad \Rightarrow \quad s'(t) = 12t^3 + 12t^2 - 12t$. $s'(t) = 0 \quad \Rightarrow \quad 12t(t^2 + t - 1) \quad \Rightarrow$

$t = 0$ or $t^2 + t - 1 = 0$. Using the quadratic formula to solve the latter equation gives us

$t = \dfrac{-1 \pm \sqrt{1^2 - 4(1)(-1)}}{2(1)} = \dfrac{-1 \pm \sqrt{5}}{2} \approx 0.618, -1.618$. The three critical numbers are $0, \dfrac{-1 \pm \sqrt{5}}{2}$.

34. $g(t) = |3t - 4| = \begin{cases} 3t - 4 & \text{if } 3t - 4 \geq 0 \\ -(3t - 4) & \text{if } 3t - 4 < 0 \end{cases} = \begin{cases} 3t - 4 & \text{if } t \geq \frac{4}{3} \\ 4 - 3t & \text{if } t < \frac{4}{3} \end{cases}$

$g'(t) = \begin{cases} 3 & \text{if } t > \frac{4}{3} \\ -3 & \text{if } t < \frac{4}{3} \end{cases}$ and $g'(t)$ does not exist at $t = \frac{4}{3}$, so $t = \frac{4}{3}$ is a critical number.

35. $g(y) = \dfrac{y - 1}{y^2 - y + 1}$ $\Rightarrow$

$g'(y) = \dfrac{(y^2 - y + 1)(1) - (y - 1)(2y - 1)}{(y^2 - y + 1)^2} = \dfrac{y^2 - y + 1 - (2y^2 - 3y + 1)}{(y^2 - y + 1)^2} = \dfrac{-y^2 + 2y}{(y^2 - y + 1)^2} = \dfrac{y(2 - y)}{(y^2 - y + 1)^2}$.

$g'(y) = 0$ $\Rightarrow$ $y = 0, 2$. The expression $y^2 - y + 1$ is never equal to 0, so $g'(y)$ exists for all real numbers.

The critical numbers are 0 and 2.

36. $h(p) = \dfrac{p - 1}{p^2 + 4}$ $\Rightarrow$ $h'(p) = \dfrac{(p^2 + 4)(1) - (p - 1)(2p)}{(p^2 + 4)^2} = \dfrac{p^2 + 4 - 2p^2 + 2p}{(p^2 + 4)^2} = \dfrac{-p^2 + 2p + 4}{(p^2 + 4)^2}$.

$h'(p) = 0$ $\Rightarrow$ $p = \dfrac{-2 \pm \sqrt{4 + 16}}{-2} = 1 \pm \sqrt{5}$. The critical numbers are $1 \pm \sqrt{5}$. [$h'(p)$ exists for all real numbers.]

37. $h(t) = t^{3/4} - 2t^{1/4}$ $\Rightarrow$ $h'(t) = \frac{3}{4}t^{-1/4} - \frac{2}{4}t^{-3/4} = \frac{1}{4}t^{-3/4}(3t^{1/2} - 2) = \dfrac{3\sqrt{t} - 2}{4\sqrt[4]{t^3}}$.

$h'(t) = 0$ $\Rightarrow$ $3\sqrt{t} = 2$ $\Rightarrow$ $\sqrt{t} = \frac{2}{3}$ $\Rightarrow$ $t = \frac{4}{9}$. $h'(t)$ does not exist at $t = 0$, so the critical numbers are 0 and $\frac{4}{9}$.

38. $g(x) = \sqrt{1 - x^2} = (1 - x^2)^{1/2}$ $\Rightarrow$ $g'(x) = \frac{1}{2}(1 - x^2)^{-1/2}(-2x) = \dfrac{-x}{\sqrt{1 - x^2}}$. $g'(x) = 0$ $\Rightarrow$ $x = 0$.

$g'(x)$ does not exist $\Rightarrow$ $1 - x^2 = 0$ $\Rightarrow$ $x = \pm 1$. The critical numbers are $-1, 0$, and 1.

39. $F(x) = x^{4/5}(x - 4)^2$ $\Rightarrow$

$F'(x) = x^{4/5} \cdot 2(x - 4) + (x - 4)^2 \cdot \frac{4}{5}x^{-1/5} = \frac{1}{5}x^{-1/5}(x - 4)[5 \cdot x \cdot 2 + (x - 4) \cdot 4]$

$= \dfrac{(x - 4)(14x - 16)}{5x^{1/5}} = \dfrac{2(x - 4)(7x - 8)}{5x^{1/5}}$

$F'(x) = 0$ $\Rightarrow$ $x = 4, \frac{8}{7}$. $F'(0)$ does not exist. Thus, the three critical numbers are $0, \frac{8}{7}$, and 4.

40. $g(x) = x^{1/3} - x^{-2/3}$ $\Rightarrow$ $g'(x) = \frac{1}{3}x^{-2/3} + \frac{2}{3}x^{-5/3} = \frac{1}{3}x^{-5/3}(x + 2) = \dfrac{x + 2}{3x^{5/3}}$.

$g'(-2) = 0$ and $g'(0)$ does not exist, but 0 is not in the domain of g, so the only critical number is -2.

41. $f(\theta) = 2\cos\theta + \sin^2\theta$ $\Rightarrow$ $f'(\theta) = -2\sin\theta + 2\sin\theta\cos\theta$. $f'(\theta) = 0$ $\Rightarrow$ $2\sin\theta(\cos\theta - 1) = 0$ $\Rightarrow$ $\sin\theta = 0$

or $\cos\theta = 1$ $\Rightarrow$ $\theta = n\pi$ [n an integer] or $\theta = 2n\pi$. The solutions $\theta = n\pi$ include the solutions $\theta = 2n\pi$, so the critical

numbers are $\theta = n\pi$.

42. $g(\theta) = 4\theta - \tan\theta$ $\Rightarrow$ $g'(\theta) = 4 - \sec^2\theta$. $g'(\theta) = 0$ $\Rightarrow$ $\sec^2\theta = 4$ $\Rightarrow$ $\sec\theta = \pm 2$ $\Rightarrow$ $\cos\theta = \pm\frac{1}{2}$ $\Rightarrow$

$\theta = \frac{\pi}{3} + 2n\pi, \frac{5\pi}{3} + 2n\pi, \frac{2\pi}{3} + 2n\pi$, and $\frac{4\pi}{3} + 2n\pi$ are critical numbers.

Note: The values of θ that make $g'(\theta)$ undefined are not in the domain of g.

43. A graph of $f'(x) = 1 + \dfrac{210 \sin x}{x^2 - 6x + 10}$ is shown. There are 10 zeros

between -25 and 25 (one is approximately -0.05). f' exists

everywhere, so f has 10 critical numbers.

44. A graph of $f'(x) = \dfrac{100 \cos^2 x}{10 + x^2} - 1$ is shown. There are 7 zeros

between 0 and 10, and 7 more zeros since f' is an even function.

f' exists everywhere, so f has 14 critical numbers.

45. $f(x) = 3x^2 - 12x + 5$, $[0, 3]$. $f'(x) = 6x - 12 = 0 \Leftrightarrow x = 2$. Applying the Closed Interval Method, we find that

$f(0) = 5$, $f(2) = -7$, and $f(3) = -4$. So $f(0) = 5$ is the absolute maximum value and $f(2) = -7$ is the absolute minimum

value.

46. $f(x) = x^3 - 3x + 1$, $[0, 3]$. $f'(x) = 3x^2 - 3 = 0 \Leftrightarrow x = \pm 1$, but -1 is not in $[0, 3]$. $f(0) = 1$, $f(1) = -1$, and

$f(3) = 19$. So $f(3) = 19$ is the absolute maximum value and $f(1) = -1$ is the absolute minimum value.

47. $f(x) = 2x^3 - 3x^2 - 12x + 1$, $[-2, 3]$. $f'(x) = 6x^2 - 6x - 12 = 6(x^2 - x - 2) = 6(x - 2)(x + 1) = 0 \Leftrightarrow$

$x = 2, -1$. $f(-2) = -3$, $f(-1) = 8$, $f(2) = -19$, and $f(3) = -8$. So $f(-1) = 8$ is the absolute maximum value and

$f(2) = -19$ is the absolute minimum value.

48. $f(x) = x^3 - 6x^2 + 9x + 2$, $[-1, 4]$. $f'(x) = 3x^2 - 12x + 9 = 3(x^2 - 4x + 3) = 3(x - 1)(x - 3) = 0 \Leftrightarrow x = 1, 3$.

$f(-1) = -14$, $f(1) = 6$, $f(3) = 2$, and $f(4) = 6$. So $f(1) = f(4) = 6$ is the absolute maximum value and $f(-1) = -14$

is the absolute minimum value.

49. $f(x) = x^4 - 2x^2 + 3$, $[-2, 3]$. $f'(x) = 4x^3 - 4x = 4x(x^2 - 1) = 4x(x + 1)(x - 1) = 0 \Leftrightarrow x = -1, 0, 1$.

$f(-2) = 11$, $f(-1) = 2$, $f(0) = 3$, $f(1) = 2$, $f(3) = 66$. So $f(3) = 66$ is the absolute maximum value and $f(\pm 1) = 2$ is

the absolute minimum value.

50. $f(x) = (x^2 - 1)^3$, $[-1, 2]$. $f'(x) = 3(x^2 - 1)^2(2x) = 6x(x + 1)^2(x - 1)^2 = 0 \Leftrightarrow x = -1, 0, 1$. $f(\pm 1) = 0$,

$f(0) = -1$, and $f(2) = 27$. So $f(2) = 27$ is the absolute maximum value and $f(0) = -1$ is the absolute minimum value.

51. $f(x) = \dfrac{x}{x^2 + 1}$, $[0, 2]$. $f'(x) = \dfrac{(x^2 + 1) - x(2x)}{(x^2 + 1)^2} = \dfrac{1 - x^2}{(x^2 + 1)^2} = 0 \Leftrightarrow x = \pm 1$, but -1 is not in $[0, 2]$. $f(0) = 0$,

$f(1) = \frac{1}{2}$, $f(2) = \frac{2}{5}$. So $f(1) = \frac{1}{2}$ is the absolute maximum value and $f(0) = 0$ is the absolute minimum value.

52. $f(x) = \dfrac{x^2 - 4}{x^2 + 4}$, $[-4, 4]$. $f'(x) = \dfrac{(x^2 + 4)(2x) - (x^2 - 4)(2x)}{(x^2 + 4)^2} = \dfrac{16x}{(x^2 + 4)^2} = 0$ ⇔ $x = 0$. $f(\pm 4) = \dfrac{12}{20} = \dfrac{3}{5}$ and

$f(0) = -1$. So $f(\pm 4) = \dfrac{3}{5}$ is the absolute maximum value and $f(0) = -1$ is the absolute minimum value.

53. $f(t) = t\sqrt{4 - t^2}$, $[-1, 2]$.

$f'(t) = t \cdot \dfrac{1}{2}(4 - t^2)^{-1/2}(-2t) + (4 - t^2)^{1/2} \cdot 1 = \dfrac{-t^2}{\sqrt{4 - t^2}} + \sqrt{4 - t^2} = \dfrac{-t^2 + (4 - t^2)}{\sqrt{4 - t^2}} = \dfrac{4 - 2t^2}{\sqrt{4 - t^2}}$.

$f'(t) = 0$ ⇒ $4 - 2t^2 = 0$ ⇒ $t^2 = 2$ ⇒ $t = \pm\sqrt{2}$, but $t = -\sqrt{2}$ is not in the given interval, $[-1, 2]$.

$f'(t)$ does not exist if $4 - t^2 = 0$ ⇒ $t = \pm 2$, but -2 is not in the given interval. $f(-1) = -\sqrt{3}$, $f(\sqrt{2}) = 2$, and

$f(2) = 0$. So $f(\sqrt{2}) = 2$ is the absolute maximum value and $f(-1) = -\sqrt{3}$ is the absolute minimum value.

54. $f(t) = \sqrt[3]{t}\,(8 - t)$, $[0, 8]$. $f(t) = 8t^{1/3} - t^{4/3}$ ⇒ $f'(t) = \dfrac{8}{3}t^{-2/3} - \dfrac{4}{3}t^{1/3} = \dfrac{4}{3}t^{-2/3}(2 - t) = \dfrac{4(2 - t)}{3\sqrt[3]{t^2}}$.

$f'(t) = 0$ ⇒ $t = 2$. $f'(t)$ does not exist if $t = 0$. $f(0) = 0$, $f(2) = 6\sqrt[3]{2} \approx 7.56$, and $f(8) = 0$.

So $f(2) = 6\sqrt[3]{2}$ is the absolute maximum value and $f(0) = f(8) = 0$ is the absolute minimum value.

55. $f(t) = 2\cos t + \sin 2t$, $[0, \pi/2]$.

$f'(t) = -2\sin t + \cos 2t \cdot 2 = -2\sin t + 2(1 - 2\sin^2 t) = -2(2\sin^2 t + \sin t - 1) = -2(2\sin t - 1)(\sin t + 1)$.

$f'(t) = 0$ ⇒ $\sin t = \dfrac{1}{2}$ or $\sin t = -1$ ⇒ $t = \dfrac{\pi}{6}$. $f(0) = 2$, $f(\frac{\pi}{6}) = \sqrt{3} + \dfrac{1}{2}\sqrt{3} = \dfrac{3}{2}\sqrt{3} \approx 2.60$, and $f(\frac{\pi}{2}) = 0$.

So $f(\frac{\pi}{6}) = \dfrac{3}{2}\sqrt{3}$ is the absolute maximum value and $f(\frac{\pi}{2}) = 0$ is the absolute minimum value.

56. $f(t) = t + \cot(t/2)$, $[\pi/4, 7\pi/4]$. $f'(t) = 1 - \csc^2(t/2) \cdot \dfrac{1}{2}$.

$f'(t) = 0$ ⇒ $\dfrac{1}{2}\csc^2(t/2) = 1$ ⇒ $\csc^2(t/2) = 2$ ⇒ $\csc(t/2) = \pm\sqrt{2}$ ⇒ $\dfrac{1}{2}t = \dfrac{\pi}{4}$ or $\dfrac{1}{2}t = \dfrac{3\pi}{4}$

$\left[\dfrac{\pi}{4} \le t \le \dfrac{7\pi}{4}\right.$ ⇒ $\dfrac{\pi}{8} \le \dfrac{1}{2}t \le \dfrac{7\pi}{8}$ and $\csc(t/2) \ne -\sqrt{2}$ in the last interval$]$ ⇒ $t = \dfrac{\pi}{2}$ or $t = \dfrac{3\pi}{2}$.

$f(\frac{\pi}{4}) = \dfrac{\pi}{4} + \cot\dfrac{\pi}{8} \approx 3.20$, $f(\frac{\pi}{2}) = \dfrac{\pi}{2} + \cot\dfrac{\pi}{4} = \dfrac{\pi}{2} + 1 \approx 2.57$, $f(\frac{3\pi}{2}) = \dfrac{3\pi}{2} + \cot\dfrac{3\pi}{2} = \dfrac{3\pi}{2} - 1 \approx 3.71$, and

$f(\frac{7\pi}{4}) = \dfrac{7\pi}{4} + \cot\dfrac{7\pi}{8} \approx 3.08$. So $f(\frac{3\pi}{2}) = \dfrac{3\pi}{2} - 1$ is the absolute maximum value and $f(\frac{\pi}{2}) = \dfrac{\pi}{2} + 1$ is the absolute

minimum value.

57. $f(x) = x^a(1 - x)^b$, $0 \le x \le 1$, $a > 0$, $b > 0$.

$f'(x) = x^a \cdot b(1 - x)^{b-1}(-1) + (1 - x)^b \cdot ax^{a-1} = x^{a-1}(1 - x)^{b-1}[x \cdot b(-1) + (1 - x) \cdot a]$

$= x^{a-1}(1 - x)^{b-1}(a - ax - bx)$

At the endpoints, we have $f(0) = f(1) = 0$ [the minimum value of f]. In the interval $(0, 1)$, $f'(x) = 0$ ⇔ $x = \dfrac{a}{a + b}$.

$f\left(\dfrac{a}{a + b}\right) = \left(\dfrac{a}{a + b}\right)^a\left(1 - \dfrac{a}{a + b}\right)^b = \dfrac{a^a}{(a + b)^a}\left(\dfrac{a + b - a}{a + b}\right)^b = \dfrac{a^a}{(a + b)^a} \cdot \dfrac{b^b}{(a + b)^b} = \dfrac{a^a b^b}{(a + b)^{a+b}}$.

So $f\left(\dfrac{a}{a + b}\right) = \dfrac{a^a b^b}{(a + b)^{a+b}}$ is the absolute maximum value.

58.

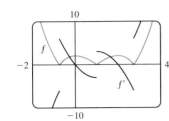

We see that $f'(x) = 0$ at about $x = 0.0$ and 2.0, and that $f'(x)$ does not

exist at about $x = -0.7$, 1.0, and 2.7, so the critical numbers of

$f(x) = \left|x^3 - 3x^2 + 2\right|$ are about -0.7, 0.0, 1.0, 2.0, and 2.7.

59. (a)

From the graph, it appears that the absolute maximum value is about

$f(-0.77) = 2.19$, and the absolute minimum value is about

$f(0.77) = 1.81$.

(b) $f(x) = x^5 - x^3 + 2 \;\Rightarrow\; f'(x) = 5x^4 - 3x^2 = x^2(5x^2 - 3)$. So $f'(x) = 0 \;\Rightarrow\; x = 0, \pm\sqrt{\frac{3}{5}}$.

$$f\left(-\sqrt{\tfrac{3}{5}}\right) = \left(-\sqrt{\tfrac{3}{5}}\right)^5 - \left(-\sqrt{\tfrac{3}{5}}\right)^3 + 2 = -\left(\tfrac{3}{5}\right)^2 \sqrt{\tfrac{3}{5}} + \tfrac{3}{5}\sqrt{\tfrac{3}{5}} + 2 = \left(\tfrac{3}{5} - \tfrac{9}{25}\right)\sqrt{\tfrac{3}{5}} + 2 = \tfrac{6}{25}\sqrt{\tfrac{3}{5}} + 2 \text{ (maximum)}$$

and similarly, $f\left(\sqrt{\tfrac{3}{5}}\right) = -\tfrac{6}{25}\sqrt{\tfrac{3}{5}} + 2$ (minimum).

60. (a)

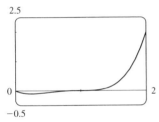

From the graph, it appears that the absolute maximum value is $f(2) = 2$,

and that the absolute minimum value is about $f(0.25) = -0.11$.

(b) $f(x) = x^4 - 3x^3 + 3x^2 - x \;\Rightarrow\; f'(x) = 4x^3 - 9x^2 + 6x - 1 = (4x - 1)(x - 1)^2$.

So $f'(x) = 0 \;\Rightarrow\; x = \tfrac{1}{4}$ or $x = 1$. Now $f(1) = 1^4 - 3\cdot1^3 + 3\cdot1^2 - 1 = 0$ (not an extremum)

and $f\left(\tfrac{1}{4}\right) = \left(\tfrac{1}{4}\right)^4 - 3\left(\tfrac{1}{4}\right)^3 + 3\left(\tfrac{1}{4}\right)^2 - \tfrac{1}{4} = -\tfrac{27}{256}$ (minimum). At the right endpoint we have

$f(2) = 2^4 - 3\cdot2^3 + 3\cdot2^2 - 2 = 2$ (maximum).

61. (a)

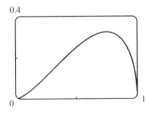

From the graph, it appears that the absolute maximum value is about

$f(0.75) = 0.32$, and the absolute minimum value is $f(0) = f(1) = 0$;

that is, at both endpoints.

(b) $f(x) = x\sqrt{x - x^2} \;\Rightarrow\; f'(x) = x \cdot \dfrac{1 - 2x}{2\sqrt{x - x^2}} + \sqrt{x - x^2} = \dfrac{(x - 2x^2) + (2x - 2x^2)}{2\sqrt{x - x^2}} = \dfrac{3x - 4x^2}{2\sqrt{x - x^2}}$.

So $f'(x) = 0 \;\Rightarrow\; 3x - 4x^2 = 0 \;\Rightarrow\; x(3 - 4x) = 0 \;\Rightarrow\; x = 0$ or $\tfrac{3}{4}$.

$f(0) = f(1) = 0$ (minimum), and $f\left(\tfrac{3}{4}\right) = \tfrac{3}{4}\sqrt{\tfrac{3}{4} - \left(\tfrac{3}{4}\right)^2} = \tfrac{3}{4}\sqrt{\tfrac{3}{16}} = \tfrac{3\sqrt{3}}{16}$ (maximum).

62. (a)

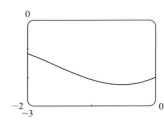

From the graph, it appears that the absolute maximum value is about $f(-2) = -1.17$, and the absolute minimum value is about $f(-0.52) = -2.26$.

(b) $f(x) = x - 2\cos x \Rightarrow f'(x) = 1 + 2\sin x$. So $f'(x) = 0 \Rightarrow \sin x = -\frac{1}{2} \Rightarrow x = -\frac{\pi}{6}$ on $[-2, 0]$.

$f(-2) = -2 - 2\cos(-2)$ (maximum) and $f\left(-\frac{\pi}{6}\right) = -\frac{\pi}{6} - 2\cos\left(-\frac{\pi}{6}\right) = -\frac{\pi}{6} - 2\left(\frac{\sqrt{3}}{2}\right) = -\frac{\pi}{6} - \sqrt{3}$ (minimum).

63. The density is defined as $\rho = \dfrac{\text{mass}}{\text{volume}} = \dfrac{1000}{V(T)}$ (in g/cm^3). But a critical point of ρ will also be a critical point of V

$\left[\text{since } \dfrac{d\rho}{dT} = -1000V^{-2}\dfrac{dV}{dT} \text{ and } V \text{ is never 0}\right]$, and V is easier to differentiate than ρ.

$V(T) = 999.87 - 0.06426T + 0.0085043T^2 - 0.0000679T^3 \Rightarrow V'(T) = -0.06426 + 0.0170086T - 0.0002037T^2$.

Setting this equal to 0 and using the quadratic formula to find T, we get

$$T = \frac{-0.0170086 \pm \sqrt{0.0170086^2 - 4 \cdot 0.0002037 \cdot 0.06426}}{2(-0.0002037)} \approx 3.9665°\text{C} \text{ or } 79.5318°\text{C}.$$ Since we are only interested

in the region $0°\text{C} \le T \le 30°\text{C}$, we check the density ρ at the endpoints and at $3.9665°\text{C}$: $\rho(0) \approx \dfrac{1000}{999.87} \approx 1.00013$;

$\rho(30) \approx \dfrac{1000}{1003.7628} \approx 0.99625$; $\rho(3.9665) \approx \dfrac{1000}{999.7447} \approx 1.000255$. So water has its maximum density at

about $3.9665°\text{C}$.

64. $F = \dfrac{\mu W}{\mu\sin\theta + \cos\theta} \Rightarrow \dfrac{dF}{d\theta} = \dfrac{(\mu\sin\theta + \cos\theta)(0) - \mu W(\mu\cos\theta - \sin\theta)}{(\mu\sin\theta + \cos\theta)^2} = \dfrac{-\mu W(\mu\cos\theta - \sin\theta)}{(\mu\sin\theta + \cos\theta)^2}$.

So $\dfrac{dF}{d\theta} = 0 \Rightarrow \mu\cos\theta - \sin\theta = 0 \Rightarrow \mu = \dfrac{\sin\theta}{\cos\theta} = \tan\theta$. Substituting $\tan\theta$ for μ in F gives us

$F = \dfrac{(\tan\theta)W}{(\tan\theta)\sin\theta + \cos\theta} = \dfrac{W\tan\theta}{\dfrac{\sin^2\theta}{\cos\theta} + \cos\theta} = \dfrac{W\tan\theta\cos\theta}{\sin^2\theta + \cos^2\theta} = \dfrac{W\sin\theta}{1} = W\sin\theta$.

If $\tan\theta = \mu$, then $\sin\theta = \dfrac{\mu}{\sqrt{\mu^2 + 1}}$ (see the figure), so $F = \dfrac{\mu}{\sqrt{\mu^2 + 1}}W$.

We compare this with the value of F at the endpoints: $F(0) = \mu W$ and $F\left(\frac{\pi}{2}\right) = W$.

Now because $\dfrac{\mu}{\sqrt{\mu^2 + 1}} \le 1$ and $\dfrac{\mu}{\sqrt{\mu^2 + 1}} \le \mu$, we have that $\dfrac{\mu}{\sqrt{\mu^2 + 1}}W$ is less than or equal to each of $F(0)$ and $F\left(\frac{\pi}{2}\right)$.

Hence, $\dfrac{\mu}{\sqrt{\mu^2 + 1}}W$ is the absolute minimum value of $F(\theta)$, and it occurs when $\tan\theta = \mu$.

65. Let $a = -0.000\,032\,37$, $b = 0.000\,903\,7$, $c = -0.008\,956$, $d = 0.03629$, $e = -0.04458$, and $f = 0.4074$.

Then $S(t) = at^5 + bt^4 + ct^3 + dt^2 + et + f$ and $S'(t) = 5at^4 + 4bt^3 + 3ct^2 + 2dt + e$.

   We now apply the Closed Interval Method to the continuous function S on the interval $0 \le t \le 10$. Since S' exists for all t,

the only critical numbers of S occur when $S'(t) = 0$. We use a rootfinder on a CAS (or a graphing device) to find that

$S'(t) = 0$ when $t_1 \approx 0.855$, $t_2 \approx 4.618$, $t_3 \approx 7.292$, and $t_4 \approx 9.570$. The values of S at these critical numbers are

$S(t_1) \approx 0.39$, $S(t_2) \approx 0.43645$, $S(t_3) \approx 0.427$, and $S(t_4) \approx 0.43641$. The values of S at the endpoints of the interval are

$S(0) \approx 0.41$ and $S(10) \approx 0.435$. Comparing the six numbers, we see that sugar was most expensive at $t_2 \approx 4.618$

(corresponding roughly to March 1998) and cheapest at $t_1 \approx 0.855$ (June 1994).

66. (a) The equation of the graph in the figure is

$v(t) = 0.00146t^3 - 0.11553t^2 + 24.98169t - 21.26872$.

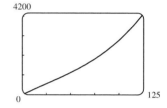

(b) $a(t) = v'(t) = 0.00438t^2 - 0.23106t + 24.98169 \quad \Rightarrow$

$a'(t) = 0.00876t - 0.23106$. $a'(t) = 0 \quad \Rightarrow \quad t_1 = \frac{0.23106}{0.00876} \approx 26.4$.

$a(0) \approx 24.98$, $a(t_1) \approx 21.93$, and $a(125) \approx 64.54$.

The maximum acceleration is about 64.5 ft/s^2 and the minimum acceleration is about 21.93 ft/s^2.

67. (a) $v(r) = k(r_0 - r)r^2 = kr_0r^2 - kr^3 \quad \Rightarrow \quad v'(r) = 2kr_0r - 3kr^2$. $v'(r) = 0 \quad \Rightarrow \quad kr(2r_0 - 3r) = 0 \quad \Rightarrow$

$r = 0$ or $\frac{2}{3}r_0$ (but 0 is not in the interval). Evaluating v at $\frac{1}{2}r_0$, $\frac{2}{3}r_0$, and r_0, we get $v\left(\frac{1}{2}r_0\right) = \frac{1}{8}kr_0^3$, $v\left(\frac{2}{3}r_0\right) = \frac{4}{27}kr_0^3$,

and $v(r_0) = 0$. Since $\frac{4}{27} > \frac{1}{8}$, v attains its maximum value at $r = \frac{2}{3}r_0$. This supports the statement in the text.

(b) From part (a), the maximum value of v is $\frac{4}{27}kr_0^3$.

(c)

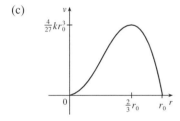

68. $g(x) = 2 + (x - 5)^3 \quad \Rightarrow \quad g'(x) = 3(x - 5)^2 \quad \Rightarrow \quad g'(5) = 0$, so 5 is a critical number. But $g(5) = 2$ and g takes on

values > 2 and values < 2 in any open interval containing 5, so g does not have a local maximum or minimum at 5.

69. $f(x) = x^{101} + x^{51} + x + 1 \quad \Rightarrow \quad f'(x) = 101x^{100} + 51x^{50} + 1 \ge 1$ for all x, so $f'(x) = 0$ has no solution. Thus, $f(x)$

has no critical number, so $f(x)$ can have no local maximum or minimum.

70. Suppose that f has a minimum value at c, so $f(x) \ge f(c)$ for all x near c. Then $g(x) = -f(x) \le -f(c) = g(c)$ for all x

near c, so $g(x)$ has a maximum value at c.

71. If f has a local minimum at c, then $g(x) = -f(x)$ has a local maximum at c, so $g'(c) = 0$ by the case of Fermat's Theorem

proved in the text. Thus, $f'(c) = -g'(c) = 0$.

72. (a) $f(x) = ax^3 + bx^2 + cx + d$, $a \neq 0$. So $f'(x) = 3ax^2 + 2bx + c$ is a quadratic and hence has either 2, 1, or 0 real roots, so $f(x)$ has either 2, 1 or 0 critical numbers.

Case (i) [2 critical numbers]:
$f(x) = x^3 - 3x \Rightarrow$
$f'(x) = 3x^2 - 3$, so $x = -1, 1$
are critical numbers.

Case (ii) [1 critical number]:
$f(x) = x^3 \Rightarrow$
$f'(x) = 3x^2$, so $x = 0$
is the only critical number.

Case (iii) [no critical number]:
$f(x) = x^3 + 3x \Rightarrow$
$f'(x) = 3x^2 + 3$,
so there is no critical number.

(b) Since there are at most two critical numbers, it can have at most two local extreme values and by (i) this can occur. By (iii) it can have no local extreme value. However, if there is only one critical number, then there is no local extreme value.

APPLIED PROJECT   The Calculus of Rainbows

1. From Snell's Law, we have $\sin \alpha = \frac{4}{3} \sin \beta \Rightarrow \frac{d}{d\alpha}(\sin \alpha) = \frac{4}{3}\frac{d}{d\alpha}(\sin \beta) \Rightarrow \cos \alpha = \frac{4}{3} \cos \beta \frac{d\beta}{d\alpha} \Rightarrow$

$\frac{d\beta}{d\alpha} = \frac{3 \cos \alpha}{4 \cos \beta}$. Now $D(\alpha) = \pi + 2\alpha - 4\beta \Rightarrow D'(\alpha) = 2 - 4\frac{d\beta}{d\alpha} = 2 - 3\frac{\cos \alpha}{\cos \beta}$. So $D'(\alpha) = 0 \Leftrightarrow$

$2 \cos \beta = 3 \cos \alpha$. Thus, $4 \cos^2 \beta = 9 \cos^2 \alpha \Rightarrow 4 - 4 \sin^2 \beta = 9 - 9 \sin^2 \alpha$. Since $3 \sin \alpha = 4 \sin \beta$,

$\sin \beta = \frac{3}{4} \sin \alpha \Rightarrow 4 - 4\left(\frac{3}{4}\sin \alpha\right)^2 = 9 - 9\sin^2 \alpha \Rightarrow \left(9 - \frac{9}{4}\right) \sin^2 \alpha = 9 - 4 = 5 \Rightarrow \sin^2 \alpha = \frac{20}{27} \Rightarrow$

$\sin \alpha = \sqrt{\frac{20}{27}}$. So $\alpha \approx 1.037$ radians or $59.4°$. We show that this α does give the minimum on $\left[0, \frac{\pi}{2}\right]$: When $\alpha = 0$,

$\sin \alpha = \frac{4}{3} \sin \beta \Rightarrow \beta = 0$, or $D(0) = \pi \approx 3.14$. When $\alpha = \frac{\pi}{2}$, $1 = \sin \frac{\pi}{2} = \frac{4}{3} \sin \beta \Rightarrow \sin \beta = \frac{3}{4} \Rightarrow$

$\beta \approx 0.85$. So $D\left(\frac{\pi}{2}\right) \approx \pi + \pi - 4(0.85) \approx 2.88$. For $\alpha \approx 1.037$, $\sin \beta = \frac{3}{4} \sin \alpha = \frac{3}{4}\sqrt{\frac{20}{27}}$, so $\beta \approx 0.702 \Rightarrow$

$D(\alpha) \approx \pi + 2(1.036) - 4(0.702) \approx 2.41$. So the minimum occurs when $\alpha \approx 1.04$ radians or $59.4°$.

2. We repeat Problem 1 with k in place of $\frac{4}{3}$. So $\sin \alpha = k \sin \beta \Rightarrow \frac{d\beta}{d\alpha} = \frac{1}{k}\frac{\cos \alpha}{\cos \beta}$. Then

$D'(\alpha) = 2 - 4\frac{d\beta}{d\alpha} = 2 - \frac{4}{k}\frac{\cos \alpha}{\cos \beta}$ and $D'(\alpha) = 0 \Leftrightarrow k \cos \beta = 2 \cos \alpha$. So $k^2 \cos^2 \beta = 4 \cos^2 \alpha \Rightarrow$

$k^2 - k^2 \sin^2 \beta = 4 - 4 \sin^2 \alpha \Rightarrow k^2 - \sin^2 \alpha = 4 - 4 \sin^2 \alpha \Rightarrow 3 \sin^2 \alpha = 4 - k^2 \Rightarrow \sin \alpha = \sqrt{\frac{4 - k^2}{3}}$.

So for $k \approx 1.3318$ (red light) the minimum occurs when $\sin \alpha_1 = \sqrt{\frac{4 - (1.3318)^2}{3}}$ or $\alpha_1 \approx 1.038$ radians, so the rainbow

angle is about $\pi - D(\alpha_1) \approx 42.3°$. For $k \approx 1.3435$ (violet light) the minimum occurs at $\alpha_2 \approx 1.026$ radians, and so the rainbow angle is about $\pi - D(\alpha_2) \approx 40.6°$.

3. At each reflection or refraction, the light is bent in a counterclockwise direction: the bend at A is $\alpha - \beta$, the bend at B is $\pi - 2\beta$, and the bend at C is again $\pi - 2\beta$, and the bend at D is $\alpha - \beta$. So the total bend is

$D(\alpha) = 2(\alpha - \beta) + 2(\pi - 2\beta) = 2\alpha - 6\beta + 2\pi$, as required. Now $\sin \alpha = k \sin \beta \quad \Rightarrow \quad \dfrac{d\beta}{d\alpha} = \dfrac{1}{k}\dfrac{\cos \alpha}{\cos \beta}$. So

$D'(\alpha) = 2 - 6\dfrac{d\beta}{d\alpha} = 2 - \dfrac{6}{k}\dfrac{\cos \alpha}{\cos \beta}$ and $D'(\alpha) = 0 \quad \Leftrightarrow \quad k \cos \beta = 3 \cos \alpha$. So $k^2 \cos^2 \beta = 9 \cos^2 \alpha \quad \Rightarrow$

$k^2 - k^2 \sin^2 \beta = 9 - 9 \sin^2 \alpha \quad \Rightarrow \quad k^2 - \sin^2 \alpha = 9 - 9 \sin^2 \alpha \quad \Rightarrow \quad \sin^2 \alpha = \dfrac{9 - k^2}{8} \quad \Rightarrow \quad \sin \alpha = \sqrt{\dfrac{9 - k^2}{8}}$.

If $k = \frac{4}{3}$, then the minimum occurs when $\sin \alpha_1 = \sqrt{\dfrac{9 - (4/3)^2}{8}}$ or $\alpha_1 \approx 1.254$ radians.

Thus, the minimum *counterclockwise* rotation is $D(\alpha_1) \approx 231°$,

which is equivalent to a *clockwise* rotation of $360° - 231° = 129°$

(see the figure). So the rainbow angle for the secondary rainbow is about

$180° - 129° = 51°$, as required. In general, the rainbow angle for the

secondary rainbow is $\pi - [2\pi - D(\alpha)] = D(\alpha) - \pi$.

4. In the primary rainbow, the rainbow angle gets smaller as k gets larger, as we found in Problem 2, so the colors appear from

top to bottom in order of increasing k. But in the secondary rainbow, the rainbow angle gets larger as k gets larger. To see this,

we find the minimum deviations for red light and for violet light in the secondary rainbow.

For $k \approx 1.3318$ (red light) the minimum occurs when $\sin \alpha_1 = \sqrt{\dfrac{9 - (1.3318)^2}{8}}$ or $\alpha_1 \approx 1.254$ radians, and so the

rainbow angle is $D(\alpha_1) - \pi \approx 50.6°$. For $k \approx 1.3435$ (violet light) the minimum occurs when $\sin \alpha_2 = \sqrt{\dfrac{9 - (1.3435)^2}{8}}$ or

$\alpha_2 \approx 1.248$ radians, and so the rainbow angle is $D(\alpha_2) - \pi \approx 53.6°$. Consequently, the rainbow angle is larger for colors

with higher indices of refraction, and the colors appear from bottom to top in order of increasing k, the reverse of their order in

the primary rainbow.

Note that our calculations above also explain why the secondary rainbow is more spread-out than the primary rainbow: in

the primary rainbow, the difference between rainbow angles for red and violet light is about $1.7°$, whereas in the secondary

rainbow it is about $3°$.

4.2   The Mean Value Theorem

1. $f(x) = 5 - 12x + 3x^2$, $[1, 3]$.   Since f is a polynomial, it is continuous and differentiable on $\mathbb{R}$, so it is continuous on $[1, 3]$ and differentiable on $(1, 3)$. Also $f(1) = -4 = f(3)$. $f'(c) = 0$ $\Leftrightarrow$ $-12 + 6c = 0$ $\Leftrightarrow$ $c = 2$, which is in the open interval $(1, 3)$, so $c = 2$ satisfies the conclusion of Rolle's Theorem.

2. $f(x) = x^3 - x^2 - 6x + 2$, $[0, 3]$.   Since f is a polynomial, it is continuous and differentiable on $\mathbb{R}$, so it is continuous on $[0, 3]$ and differentiable on $(0, 3)$. Also, $f(0) = 2 = f(3)$. $f'(c) = 0$ $\Leftrightarrow$ $3c^2 - 2c - 6 = 0$ $\Leftrightarrow$
$c = \dfrac{2 \pm \sqrt{4 + 72}}{6} = \frac{1}{3} \pm \frac{1}{3}\sqrt{19}$ $[\approx 1.79, \approx -1.12]$, so $c = \frac{1}{3} + \frac{1}{3}\sqrt{19}$ satisfies the conclusion of Rolle's Theorem.

3. $f(x) = \sqrt{x} - \frac{1}{3}x$, $[0, 9]$. f, being the difference of a root function and a polynomial, is continuous and differentiable on $[0, \infty)$, so it is continuous on $[0, 9]$ and differentiable on $(0, 9)$. Also, $f(0) = 0 = f(9)$. $f'(c) = 0$ $\Leftrightarrow$
$\dfrac{1}{2\sqrt{c}} - \dfrac{1}{3} = 0$ $\Leftrightarrow$ $2\sqrt{c} = 3$ $\Leftrightarrow$ $\sqrt{c} = \dfrac{3}{2}$ $\Rightarrow$ $c = \dfrac{9}{4}$, which is in the open interval $(0, 9)$, so $c = \dfrac{9}{4}$ satisfies the conclusion of Rolle's Theorem.

4. $f(x) = \cos 2x$, $[\pi/8, 7\pi/8]$. f, being the composite of the cosine function and the polynomial $2x$, is continuous and differentiable on $\mathbb{R}$, so it is continuous on $[\pi/8, 7\pi/8]$ and differentiable on $(\pi/8, 7\pi/8)$. Also, $f\left(\frac{\pi}{8}\right) = \frac{1}{2}\sqrt{2} = f\left(\frac{7\pi}{8}\right)$.
$f'(c) = 0$ $\Leftrightarrow$ $-2\sin 2c = 0$ $\Leftrightarrow$ $\sin 2c = 0$ $\Leftrightarrow$ $2c = \pi n$ $\Leftrightarrow$ $c = \frac{\pi}{2}n$. If $n = 1$, then $c = \frac{\pi}{2}$, which is in the open interval $(\pi/8, 7\pi/8)$, so $c = \frac{\pi}{2}$ satisfies the conclusion of Rolle's Theorem.

5. $f(x) = 1 - x^{2/3}$. $f(-1) = 1 - (-1)^{2/3} = 1 - 1 = 0 = f(1)$. $f'(x) = -\frac{2}{3}x^{-1/3}$, so $f'(c) = 0$ has no solution. This does not contradict Rolle's Theorem, since $f'(0)$ does not exist, and so f is not differentiable on $(-1, 1)$.

6. $f(x) = \tan x$. $f(0) = \tan 0 = 0 = \tan \pi = f(\pi)$. $f'(x) = \sec^2 x \geq 1$, so $f'(c) = 0$ has no solution. This does not contradict Rolle's Theorem, since $f'\left(\frac{\pi}{2}\right)$ does not exist, and so f is not differentiable on $(0, \pi)$. (Also, $f(x)$ is not continuous on $[0, \pi]$.)

7. $\dfrac{f(8) - f(0)}{8 - 0} = \dfrac{6 - 4}{8} = \dfrac{1}{4}$. The values of c which satisfy $f'(c) = \frac{1}{4}$ seem to be about $c = 0.8, 3.2, 4.4,$ and 6.1.

8. $\dfrac{f(7) - f(1)}{7 - 1} = \dfrac{2 - 5}{6} = -\dfrac{1}{2}$. The values of c which satisfy $f'(c) = -\frac{1}{2}$ seem to be about $c = 1.1, 2.8, 4.6,$ and 5.8.

9. (a), (b) The equation of the secant line is

$$y - 5 = \frac{8.5 - 5}{8 - 1}(x - 1) \iff y = \tfrac{1}{2}x + \tfrac{9}{2}.$$

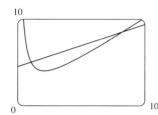

(c) $f(x) = x + 4/x \Rightarrow f'(x) = 1 - 4/x^2$.

So $f'(c) = \tfrac{1}{2} \Rightarrow c^2 = 8 \Rightarrow c = 2\sqrt{2}$, and $f(c) = 2\sqrt{2} + \frac{4}{2\sqrt{2}} = 3\sqrt{2}$. Thus, an equation of the tangent line is $y - 3\sqrt{2} = \tfrac{1}{2}\left(x - 2\sqrt{2}\right) \iff y = \tfrac{1}{2}x + 2\sqrt{2}$.

10. (a)

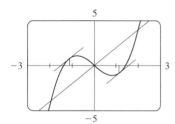

It seems that the tangent lines are parallel to the secant at $x \approx \pm 1.2$.

(b) The slope of the secant line is 2, and its equation is

$y = 2x$. $f(x) = x^3 - 2x \Rightarrow f'(x) = 3x^2 - 2$,

so we solve $f'(c) = 2 \Rightarrow 3c^2 = 4 \Rightarrow$

$c = \pm\frac{2\sqrt{3}}{3} \approx 1.155$. Our estimates were off by about 0.045 in each case.

11. $f(x) = 3x^2 + 2x + 5$, $[-1, 1]$. f is continuous on $[-1, 1]$ and differentiable on $(-1, 1)$ since polynomials are continuous and differentiable on $\mathbb{R}$. $f'(c) = \frac{f(b) - f(a)}{b - a} \iff 6c + 2 = \frac{f(1) - f(-1)}{1 - (-1)} = \frac{10 - 6}{2} = 2 \iff 6c = 0 \iff c = 0$, which is in $(-1, 1)$.

12. $f(x) = x^3 + x - 1$, $[0, 2]$. f is continuous on $[0, 2]$ and differentiable on $(0, 2)$. $f'(c) = \frac{f(2) - f(0)}{2 - 0} \iff$

$3c^2 + 1 = \frac{9 - (-1)}{2} \iff 3c^2 = 5 - 1 \iff c^2 = \tfrac{4}{3} \iff c = \pm\frac{2}{\sqrt{3}}$, but only $\frac{2}{\sqrt{3}}$ is in $(0, 2)$.

13. $f(x) = \sqrt[3]{x}$, $[0, 1]$. f is continuous on $\mathbb{R}$ and differentiable on $(-\infty, 0) \cup (0, \infty)$, so f is continuous on $[0, 1]$ and differentiable on $(0, 1)$. $f'(c) = \frac{f(b) - f(a)}{b - a} \iff \frac{1}{3c^{2/3}} = \frac{f(1) - f(0)}{1 - 0} \iff \frac{1}{3c^{2/3}} = \frac{1 - 0}{1} \iff$

$3c^{2/3} = 1 \iff c^{2/3} = \tfrac{1}{3} \iff c^2 = \left(\tfrac{1}{3}\right)^3 = \tfrac{1}{27} \iff c = \pm\sqrt{\tfrac{1}{27}} = \pm\frac{\sqrt{3}}{9}$, but only $\frac{\sqrt{3}}{9}$ is in $(0, 1)$.

14. $f(x) = \frac{x}{x + 2}$, $[1, 4]$. f is continuous on $[1, 4]$ and differentiable on $(1, 4)$. $f'(c) = \frac{f(b) - f(a)}{b - a} \iff$

$\frac{2}{(c + 2)^2} = \frac{\tfrac{2}{3} - \tfrac{1}{3}}{4 - 1} \iff (c + 2)^2 = 18 \iff c = -2 \pm 3\sqrt{2}$. $-2 + 3\sqrt{2} \approx 2.24$ is in $(1, 4)$.

15. $f(x) = (x-3)^{-2}$ $\Rightarrow$ $f'(x) = -2(x-3)^{-3}$. $f(4) - f(1) = f'(c)(4-1)$ $\Rightarrow$ $\dfrac{1}{1^2} - \dfrac{1}{(-2)^2} = \dfrac{-2}{(c-3)^3} \cdot 3$ $\Rightarrow$

$\dfrac{3}{4} = \dfrac{-6}{(c-3)^3}$ $\Rightarrow$ $(c-3)^3 = -8$ $\Rightarrow$ $c - 3 = -2$ $\Rightarrow$ $c = 1$, which is not in the open interval $(1, 4)$. This does not

contradict the Mean Value Theorem since f is not continuous at $x = 3$.

16. $f(x) = 2 - |2x - 1| = \begin{cases} 2 - (2x-1) & \text{if } 2x - 1 \geq 0 \\ 2 - [-(2x-1)] & \text{if } 2x - 1 < 0 \end{cases} = \begin{cases} 3 - 2x & \text{if } x \geq \frac{1}{2} \\ 1 + 2x & \text{if } x < \frac{1}{2} \end{cases}$ $\Rightarrow$ $f'(x) = \begin{cases} -2 & \text{if } x > \frac{1}{2} \\ 2 & \text{if } x < \frac{1}{2} \end{cases}$

$f(3) - f(0) = f'(c)(3 - 0)$ $\Rightarrow$ $-3 - 1 = f'(c) \cdot 3$ $\Rightarrow$ $f'(c) = -\frac{4}{3}$ [not ± 2]. This does not contradict the Mean

Value Theorem since f is not differentiable at $x = \frac{1}{2}$.

17. Let $f(x) = 1 + 2x + x^3 + 4x^5$. Then $f(-1) = -6 < 0$ and $f(0) = 1 > 0$. Since f is a polynomial, it is continuous, so the

Intermediate Value Theorem says that there is a number c between -1 and 0 such that $f(c) = 0$. Thus, the given equation has

a real root. Suppose the equation has distinct real roots a and b with $a < b$. Then $f(a) = f(b) = 0$. Since f is a polynomial, it

is differentiable on (a, b) and continuous on $[a, b]$. By Rolle's Theorem, there is a number r in (a, b) such that $f'(r) = 0$. But

$f'(x) = 2 + 3x^2 + 20x^4 \geq 2$ for all x, so $f'(x)$ can never be 0. This contradiction shows that the equation can't have two

distinct real roots. Hence, it has exactly one real root.

18. Let $f(x) = 2x - 1 - \sin x$. Then $f(0) = -1 < 0$ and $f(\pi/2) = \pi - 2 > 0$. f is the sum of the polynomial $2x - 1$ and the

scalar multiple $(-1) \cdot \sin x$ of the trigonometric function $\sin x$, so f is continuous (and differentiable) for all x. By the

Intermediate Value Theorem, there is a number c in $(0, \pi/2)$ such that $f(c) = 0$. Thus, the given equation has at least one real

root. If the equation has distinct real roots a and b with $a < b$, then $f(a) = f(b) = 0$. Since f is continuous on $[a, b]$ and

differentiable on (a, b), Rolle's Theorem implies that there is a number r in (a, b) such that $f'(r) = 0$. But

$f'(r) = 2 - \cos r > 0$ since $\cos r \leq 1$. This contradiction shows that the given equation can't have two distinct real roots, so

it has exactly one real root.

19. Let $f(x) = x^3 - 15x + c$ for x in $[-2, 2]$. If f has two real roots a and b in $[-2, 2]$, with $a < b$, then $f(a) = f(b) = 0$. Since

the polynomial f is continuous on $[a, b]$ and differentiable on (a, b), Rolle's Theorem implies that there is a number r in (a, b)

such that $f'(r) = 0$. Now $f'(r) = 3r^2 - 15$. Since r is in (a, b), which is contained in $[-2, 2]$, we have $|r| < 2$, so $r^2 < 4$.

It follows that $3r^2 - 15 < 3 \cdot 4 - 15 = -3 < 0$. This contradicts $f'(r) = 0$, so the given equation can't have two real roots

in $[-2, 2]$. Hence, it has at most one real root in $[-2, 2]$.

20. $f(x) = x^4 + 4x + c$. Suppose that $f(x) = 0$ has three distinct real roots a, b, d where $a < b < d$. Then

$f(a) = f(b) = f(d) = 0$. By Rolle's Theorem there are numbers c_1 and c_2 with $a < c_1 < b$ and $b < c_2 < d$

and $0 = f'(c_1) = f'(c_2)$, so $f'(x) = 0$ must have at least two real solutions. However

$0 = f'(x) = 4x^3 + 4 = 4(x^3 + 1) = 4(x+1)((x^2 - x + 1))$ has as its only real solution $x = -1$. Thus, $f(x)$ can have at

most two real roots.

21. (a) Suppose that a cubic polynomial $P(x)$ has roots $a_1 < a_2 < a_3 < a_4$, so $P(a_1) = P(a_2) = P(a_3) = P(a_4)$.

By Rolle's Theorem there are numbers c_1, c_2, c_3 with $a_1 < c_1 < a_2$, $a_2 < c_2 < a_3$ and $a_3 < c_3 < a_4$ and

$P'(c_1) = P'(c_2) = P'(c_3) = 0$. Thus, the second-degree polynomial $P'(x)$ has three distinct real roots, which is

impossible.

(b) We prove by induction that a polynomial of degree n has at most n real roots. This is certainly true for $n = 1$. Suppose

that the result is true for all polynomials of degree n and let $P(x)$ be a polynomial of degree $n + 1$. Suppose that $P(x)$ has

more than $n + 1$ real roots, say $a_1 < a_2 < a_3 < \cdots < a_{n+1} < a_{n+2}$. Then $P(a_1) = P(a_2) = \cdots = P(a_{n+2}) = 0$.

By Rolle's Theorem there are real numbers $c_1, \ldots, c_{n+1}$ with $a_1 < c_1 < a_2, \ldots, a_{n+1} < c_{n+1} < a_{n+2}$ and

$P'(c_1) = \cdots = P'(c_{n+1}) = 0$. Thus, the nth degree polynomial $P'(x)$ has at least $n + 1$ roots. This contradiction shows

that $P(x)$ has at most $n + 1$ real roots.

22. (a) Suppose that $f(a) = f(b) = 0$ where $a < b$. By Rolle's Theorem applied to f on $[a, b]$ there is a number c such that

$a < c < b$ and $f'(c) = 0$.

(b) Suppose that $f(a) = f(b) = f(c) = 0$ where $a < b < c$. By Rolle's Theorem applied to $f(x)$ on $[a, b]$ and $[b, c]$ there are

numbers $a < d < b$ and $b < e < c$ with $f'(d) = 0$ and $f'(e) = 0$. By Rolle's Theorem applied to $f'(x)$ on $[d, e]$ there is a

number g with $d < g < e$ such that $f''(g) = 0$.

(c) Suppose that f is n times differentiable on $\mathbb{R}$ and has $n + 1$ distinct real roots. Then $f^{(n)}$ has at least one real root.

23. By the Mean Value Theorem, $f(4) - f(1) = f'(c)(4 - 1)$ for some $c \in (1, 4)$. But for every $c \in (1, 4)$ we have

$f'(c) \geq 2$. Putting $f'(c) \geq 2$ into the above equation and substituting $f(1) = 10$, we get

$f(4) = f(1) + f'(c)(4 - 1) = 10 + 3f'(c) \geq 10 + 3 \cdot 2 = 16$. So the smallest possible value of $f(4)$ is 16.

24. If $3 \leq f'(x) \leq 5$ for all x, then by the Mean Value Theorem, $f(8) - f(2) = f'(c) \cdot (8 - 2)$ for some c in $[2, 8]$.

(f is differentiable for all x, so, in particular, f is differentiable on $(2, 8)$ and continuous on $[2, 8]$. Thus, the hypotheses of the

Mean Value Theorem are satisfied.) Since $f(8) - f(2) = 6f'(c)$ and $3 \leq f'(c) \leq 5$, it follows that

$6 \cdot 3 \leq 6f'(c) \leq 6 \cdot 5 \quad \Rightarrow \quad 18 \leq f(8) - f(2) \leq 30$.

25. Suppose that such a function f exists. By the Mean Value Theorem there is a number $0 < c < 2$ with

$f'(c) = \dfrac{f(2) - f(0)}{2 - 0} = \dfrac{5}{2}$. But this is impossible since $f'(x) \leq 2 < \frac{5}{2}$ for all x, so no such function can exist.

26. Let $h = f - g$. Then since f and g are continuous on $[a, b]$ and differentiable on (a, b), so is h, and thus h satisfies the

assumptions of the Mean Value Theorem. Therefore, there is a number c with $a < c < b$ such that

$h(b) = h(b) - h(a) = h'(c)(b - a)$. Since $h'(c) < 0$, $h'(c)(b - a) < 0$, so $f(b) - g(b) = h(b) < 0$ and hence $f(b) < g(b)$.

27. We use Exercise 26 with $f(x) = \sqrt{1 + x}$, $g(x) = 1 + \frac{1}{2}x$, and $a = 0$. Notice that $f(0) = 1 = g(0)$ and

$f'(x) = \dfrac{1}{2\sqrt{1 + x}} < \dfrac{1}{2} = g'(x)$ for $x > 0$. So by Exercise 26, $f(b) < g(b) \quad \Rightarrow \quad \sqrt{1 + b} < 1 + \frac{1}{2}b$ for $b > 0$.

Another method: Apply the Mean Value Theorem directly to either $f(x) = 1 + \frac{1}{2}x - \sqrt{1 + x}$ or $g(x) = \sqrt{1 + x}$ on $[0, b]$.

28. f satisfies the conditions for the Mean Value Theorem, so we use this theorem on the interval $[-b, b]$: $\dfrac{f(b) - f(-b)}{b - (-b)} = f'(c)$

for some $c \in (-b, b)$. But since f is odd, $f(-b) = -f(b)$. Substituting this into the above equation, we get

$$\frac{f(b) + f(b)}{2b} = f'(c) \quad \Rightarrow \quad \frac{f(b)}{b} = f'(c).$$

29. Let $f(x) = \sin x$ and let $b < a$. Then $f(x)$ is continuous on $[b, a]$ and differentiable on (b, a). By the Mean Value Theorem, there is a number $c \in (b, a)$ with $\sin a - \sin b = f(a) - f(b) = f'(c)(a - b) = (\cos c)(a - b)$. Thus, $|\sin a - \sin b| \le |\cos c| \, |b - a| \le |a - b|$. If $a < b$, then $|\sin a - \sin b| = |\sin b - \sin a| \le |b - a| = |a - b|$. If $a = b$, both sides of the inequality are 0.

30. Suppose that $f'(x) = c$. Let $g(x) = cx$, so $g'(x) = c$. Then, by Corollary 7, $f(x) = g(x) + d$, where d is a constant, so $f(x) = cx + d$.

31. For $x > 0$, $f(x) = g(x)$, so $f'(x) = g'(x)$. For $x < 0$, $f'(x) = (1/x)' = -1/x^2$ and $g'(x) = (1 + 1/x)' = -1/x^2$, so again $f'(x) = g'(x)$. However, the domain of $g(x)$ is not an interval [it is $(-\infty, 0) \cup (0, \infty)$] so we cannot conclude that $f - g$ is constant (in fact it is not).

32. Let $v(t)$ be the velocity of the car t hours after 2:00 PM. Then $\dfrac{v(1/6) - v(0)}{1/6 - 0} = \dfrac{50 - 30}{1/6} = 120$. By the Mean Value Theorem, there is a number c such that $0 < c < \frac{1}{6}$ with $v'(c) = 120$. Since $v'(t)$ is the acceleration at time t, the acceleration c hours after 2:00 PM is exactly 120 mi/h².

33. Let $g(t)$ and $h(t)$ be the position functions of the two runners and let $f(t) = g(t) - h(t)$. By hypothesis, $f(0) = g(0) - h(0) = 0$ and $f(b) = g(b) - h(b) = 0$, where b is the finishing time. Then by the Mean Value Theorem, there is a time c, with $0 < c < b$, such that $f'(c) = \dfrac{f(b) - f(0)}{b - 0}$. But $f(b) = f(0) = 0$, so $f'(c) = 0$. Since $f'(c) = g'(c) - h'(c) = 0$, we have $g'(c) = h'(c)$. So at time c, both runners have the same speed $g'(c) = h'(c)$.

34. Assume that f is differentiable (and hence continuous) on $\mathbb{R}$ and that $f'(x) \ne 1$ for all x. Suppose f has more than one fixed point. Then there are numbers a and b such that $a < b$, $f(a) = a$, and $f(b) = b$. Applying the Mean Value Theorem to the function f on $[a, b]$, we find that there is a number c in (a, b) such that $f'(c) = \dfrac{f(b) - f(a)}{b - a}$. But then $f'(c) = \dfrac{b - a}{b - a} = 1$, contradicting our assumption that $f'(x) \ne 1$ for every real number x. This shows that our supposition was wrong, that is, that f cannot have more than one fixed point.

4.3   How Derivatives Affect the Shape of a Graph

1. (a) f is increasing on $(1, 3)$ and $(4, 6)$.
(c) f is concave upward on $(0, 2)$.
(e) The point of inflection is $(2, 3)$.

(b) f is decreasing on $(0, 1)$ and $(3, 4)$.
(d) f is concave downward on $(2, 4)$ and $(4, 6)$.

2. (a) f is increasing on $(0, 1)$ and $(3, 7)$.
(c) f is concave upward on $(2, 4)$ and $(5, 7)$.
(e) The points of inflection are $(2, 2)$, $(4, 3)$, and $(5, 4)$.

(b) f is decreasing on $(1, 3)$.
(d) f is concave downward on $(0, 2)$ and $(4, 5)$.

3. (a) Use the Increasing/Decreasing (I/D) Test.   (b) Use the Concavity Test.

(c) At any value of x where the concavity changes, we have an inflection point at $(x, f(x))$.

4. (a) See the First Derivative Test.

(b) See the Second Derivative Test and the note that precedes Example 7.

5. (a) Since $f'(x) > 0$ on $(1, 5)$, f is increasing on this interval. Since $f'(x) < 0$ on $(0, 1)$ and $(5, 6)$, f is decreasing on these intervals.

(b) Since $f'(x) = 0$ at $x = 1$ and f' changes from negative to positive there, f changes from decreasing to increasing and has a local minimum at $x = 1$. Since $f'(x) = 0$ at $x = 5$ and f' changes from positive to negative there, f changes from increasing to decreasing and has a local maximum at $x = 5$.

6. (a) $f'(x) > 0$ and f is increasing on $(0, 1)$ and $(3, 5)$. $f'(x) < 0$ and f is decreasing on $(1, 3)$ and $(5, 6)$.

(b) Since $f'(x) = 0$ at $x = 1$ and $x = 5$ and f' changes from positive to negative at both values, f changes from increasing to decreasing and has local maxima at $x = 1$ and $x = 5$. Since $f'(x) = 0$ at $x = 3$ and f' changes from negative to positive there, f changes from decreasing to increasing and has a local minimum at $x = 3$.

7. There is an inflection point at $x = 1$ because $f''(x)$ changes from negative to positive there, and so the graph of f changes from concave downward to concave upward. There is an inflection point at $x = 7$ because $f''(x)$ changes from positive to negative there, and so the graph of f changes from concave upward to concave downward.

8. (a) f is increasing on the intervals where $f'(x) > 0$, namely, $(2, 4)$ and $(6, 9)$.

(b) f has a local maximum where it changes from increasing to decreasing, that is, where f' changes from positive to negative (at $x = 4$). Similarly, where f' changes from negative to positive, f has a local minimum (at $x = 2$ and at $x = 6$).

(c) When f' is increasing, its derivative f'' is positive and hence, f is concave upward. This happens on $(1, 3)$, $(5, 7)$, and $(8, 9)$. Similarly, f is concave downward when f' is decreasing—that is, on $(0, 1)$, $(3, 5)$, and $(7, 8)$.

(d) f has inflection points at $x = 1, 3, 5, 7$, and 8, since the direction of concavity changes at each of these values.

9. (a) $f(x) = 2x^3 + 3x^2 - 36x \implies f'(x) = 6x^2 + 6x - 36 = 6(x^2 + x - 6) = 6(x + 3)(x - 2)$.

We don't need to include the "6" in the chart to determine the sign of $f'(x)$.

Interval	$x + 3$	$x - 2$	$f'(x)$	f
$x < -3$	$-$	$-$	$+$	increasing on $(-\infty, -3)$
$-3 < x < 2$	$+$	$-$	$-$	decreasing on $(-3, 2)$
$x > 2$	$+$	$+$	$+$	increasing on $(2, \infty)$

(b) f changes from increasing to decreasing at $x = -3$ and from decreasing to increasing at $x = 2$. Thus, $f(-3) = 81$ is a local maximum value and $f(2) = -44$ is a local minimum value.

(c) $f'(x) = 6x^2 + 6x - 36 \implies f''(x) = 12x + 6$. $f''(x) = 0$ at $x = -\frac{1}{2}$, $f''(x) > 0 \iff x > -\frac{1}{2}$, and $f''(x) < 0 \iff x < -\frac{1}{2}$. Thus, f is concave upward on $\left(-\frac{1}{2}, \infty\right)$ and concave downward on $\left(-\infty, -\frac{1}{2}\right)$. There is an inflection point at $\left(-\frac{1}{2}, f\left(-\frac{1}{2}\right)\right) = \left(-\frac{1}{2}, \frac{37}{2}\right)$.

10. (a) $f(x) = 4x^3 + 3x^2 - 6x + 1 \Rightarrow f'(x) = 12x^2 + 6x - 6 = 6(2x^2 + x - 1) = 6(2x - 1)(x + 1)$. Thus, $f'(x) > 0 \Leftrightarrow x < -1$ or $x > \frac{1}{2}$ and $f'(x) < 0 \Leftrightarrow -1 < x < \frac{1}{2}$. So f is increasing on $(-\infty, -1)$ and $(\frac{1}{2}, \infty)$ and f is decreasing on $(-1, \frac{1}{2})$.

(b) f changes from increasing to decreasing at $x = -1$ and from decreasing to increasing at $x = \frac{1}{2}$. Thus, $f(-1) = 6$ is a local maximum value and $f(\frac{1}{2}) = -\frac{3}{4}$ is a local minimum value.

(c) $f''(x) = 24x + 6 = 6(4x + 1)$. $f''(x) > 0 \Leftrightarrow x > -\frac{1}{4}$ and $f''(x) < 0 \Leftrightarrow x < -\frac{1}{4}$. Thus, f is concave upward on $(-\frac{1}{4}, \infty)$ and concave downward on $(-\infty, -\frac{1}{4})$. There is an inflection point at $\left(-\frac{1}{4}, f\left(-\frac{1}{4}\right)\right) = \left(-\frac{1}{4}, \frac{21}{8}\right)$.

11. (a) $f(x) = x^4 - 2x^2 + 3 \Rightarrow f'(x) = 4x^3 - 4x = 4x(x^2 - 1) = 4x(x + 1)(x - 1)$.

Interval	$x + 1$	x	$x - 1$	$f'(x)$	f
$x < -1$	$-$	$-$	$-$	$-$	decreasing on $(-\infty, -1)$
$-1 < x < 0$	$+$	$-$	$-$	$+$	increasing on $(-1, 0)$
$0 < x < 1$	$+$	$+$	$-$	$-$	decreasing on $(0, 1)$
$x > 1$	$+$	$+$	$+$	$+$	increasing on $(1, \infty)$

(b) f changes from increasing to decreasing at $x = 0$ and from decreasing to increasing at $x = -1$ and $x = 1$. Thus, $f(0) = 3$ is a local maximum value and $f(\pm 1) = 2$ are local minimum values.

(c) $f''(x) = 12x^2 - 4 = 12\left(x^2 - \frac{1}{3}\right) = 12\left(x + 1/\sqrt{3}\right)\left(x - 1/\sqrt{3}\right)$. $f''(x) > 0 \Leftrightarrow x < -1/\sqrt{3}$ or $x > 1/\sqrt{3}$ and $f''(x) < 0 \Leftrightarrow -1/\sqrt{3} < x < 1/\sqrt{3}$. Thus, f is concave upward on $\left(-\infty, -\sqrt{3}/3\right)$ and $\left(\sqrt{3}/3, \infty\right)$ and concave downward on $\left(-\sqrt{3}/3, \sqrt{3}/3\right)$. There are inflection points at $\left(\pm\sqrt{3}/3, \frac{22}{9}\right)$.

12. (a) $f(x) = \dfrac{x^2}{x^2 + 3} \Rightarrow f'(x) = \dfrac{(x^2 + 3)(2x) - x^2(2x)}{(x^2 + 3)^2} = \dfrac{6x}{(x^2 + 3)^2}$. The denominator is positive so the sign of $f'(x)$ is determined by the sign of x. Thus, $f'(x) > 0 \Leftrightarrow x > 0$ and $f'(x) < 0 \Leftrightarrow x < 0$. So f is increasing on $(0, \infty)$ and f is decreasing on $(-\infty, 0)$.

(b) f changes from decreasing to increasing at $x = 0$. Thus, $f(0) = 0$ is a local minimum value.

(c) $f''(x) = \dfrac{(x^2 + 3)^2(6) - 6x \cdot 2(x^2 + 3)(2x)}{[(x^2 + 3)^2]^2} = \dfrac{6(x^2 + 3)[x^2 + 3 - 4x^2]}{(x^2 + 3)^4} = \dfrac{6(3 - 3x^2)}{(x^2 + 3)^3} = \dfrac{-18(x + 1)(x - 1)}{(x^2 + 3)^3}$.

$f''(x) > 0 \Leftrightarrow -1 < x < 1$ and $f''(x) < 0 \Leftrightarrow x < -1$ or $x > 1$. Thus, f is concave upward on $(-1, 1)$ and concave downward on $(-\infty, -1)$ and $(1, \infty)$. There are inflection points at $\left(\pm 1, \frac{1}{4}\right)$.

13. (a) $f(x) = \sin x + \cos x$, $0 \le x \le 2\pi$. $f'(x) = \cos x - \sin x = 0 \Rightarrow \cos x = \sin x \Rightarrow 1 = \dfrac{\sin x}{\cos x} \Rightarrow \tan x = 1 \Rightarrow x = \frac{\pi}{4}$ or $\frac{5\pi}{4}$. Thus, $f'(x) > 0 \Leftrightarrow \cos x - \sin x > 0 \Leftrightarrow \cos x > \sin x \Leftrightarrow 0 < x < \frac{\pi}{4}$ or $\frac{5\pi}{4} < x < 2\pi$ and $f'(x) < 0 \Leftrightarrow \cos x < \sin x \Leftrightarrow \frac{\pi}{4} < x < \frac{5\pi}{4}$. So f is increasing on $\left(0, \frac{\pi}{4}\right)$ and $\left(\frac{5\pi}{4}, 2\pi\right)$ and f is decreasing on $\left(\frac{\pi}{4}, \frac{5\pi}{4}\right)$.

(b) f changes from increasing to decreasing at $x = \frac{\pi}{4}$ and from decreasing to increasing at $x = \frac{5\pi}{4}$. Thus, $f\left(\frac{\pi}{4}\right) = \sqrt{2}$ is a local maximum value and $f\left(\frac{5\pi}{4}\right) = -\sqrt{2}$ is a local minimum value.

(c) $f''(x) = -\sin x - \cos x = 0 \Rightarrow -\sin x = \cos x \Rightarrow \tan x = -1 \Rightarrow x = \frac{3\pi}{4}$ or $\frac{7\pi}{4}$. Divide the interval $(0, 2\pi)$ into subintervals with these numbers as endpoints and complete a second derivative chart.

Interval	$f''(x) = -\sin x - \cos x$	Concavity
$\left(0, \frac{3\pi}{4}\right)$	$f''\left(\frac{\pi}{2}\right) = -1 < 0$	downward
$\left(\frac{3\pi}{4}, \frac{7\pi}{4}\right)$	$f''(\pi) = 1 > 0$	upward
$\left(\frac{7\pi}{4}, 2\pi\right)$	$f''\left(\frac{11\pi}{6}\right) = \frac{1}{2} - \frac{1}{2}\sqrt{3} < 0$	downward

There are inflection points at $\left(\frac{3\pi}{4}, 0\right)$ and $\left(\frac{7\pi}{4}, 0\right)$.

14. (a) $f(x) = \cos^2 x - 2\sin x$, $0 \le x \le 2\pi$. $f'(x) = -2\cos x \sin x - 2\cos x = -2\cos x (1 + \sin x)$. Note that $1 + \sin x \ge 0$ [since $\sin x \ge -1$], with equality $\Leftrightarrow \sin x = -1 \Leftrightarrow x = \frac{3\pi}{2}$ [since $0 \le x \le 2\pi$] $\Rightarrow \cos x = 0$. Thus, $f'(x) > 0 \Leftrightarrow \cos x < 0 \Leftrightarrow \frac{\pi}{2} < x < \frac{3\pi}{2}$ and $f'(x) < 0 \Leftrightarrow \cos x > 0 \Leftrightarrow 0 < x < \frac{\pi}{2}$ or $\frac{3\pi}{2} < x < 2\pi$. Thus, f is increasing on $\left(\frac{\pi}{2}, \frac{3\pi}{2}\right)$ and f is decreasing on $\left(0, \frac{\pi}{2}\right)$ and $\left(\frac{3\pi}{2}, 2\pi\right)$.

(b) f changes from decreasing to increasing at $x = \frac{\pi}{2}$ and from increasing to decreasing at $x = \frac{3\pi}{2}$. Thus, $f\left(\frac{\pi}{2}\right) = -2$ is a local minimum value and $f\left(\frac{3\pi}{2}\right) = 2$ is a local maximum value.

(c) $f''(x) = 2\sin x (1 + \sin x) - 2\cos^2 x = 2\sin x + 2\sin^2 x - 2(1 - \sin^2 x)$
$= 4\sin^2 x + 2\sin x - 2 = 2(2\sin x - 1)(\sin x + 1)$

so $f''(x) > 0 \Leftrightarrow \sin x > \frac{1}{2} \Leftrightarrow \frac{\pi}{6} < x < \frac{5\pi}{6}$, and $f''(x) < 0 \Leftrightarrow \sin x < \frac{1}{2}$ and $\sin x \ne -1 \Leftrightarrow 0 < x < \frac{\pi}{6}$ or $\frac{5\pi}{6} < x < \frac{3\pi}{2}$ or $\frac{3\pi}{2} < x < 2\pi$. Thus, f is concave upward on $\left(\frac{\pi}{6}, \frac{5\pi}{6}\right)$ and concave downward on $\left(0, \frac{\pi}{6}\right)$, $\left(\frac{5\pi}{6}, \frac{3\pi}{2}\right)$, and $\left(\frac{3\pi}{2}, 2\pi\right)$. There are inflection points at $\left(\frac{\pi}{6}, -\frac{1}{4}\right)$ and $\left(\frac{5\pi}{6}, -\frac{1}{4}\right)$.

15. $f(x) = x^5 - 5x + 3 \Rightarrow f'(x) = 5x^4 - 5 = 5(x^2 + 1)(x + 1)(x - 1)$.

First Derivative Test: $f'(x) < 0 \Rightarrow -1 < x < 1$ and $f'(x) > 0 \Rightarrow x > 1$ or $x < -1$. Since f' changes from positive to negative at $x = -1$, $f(-1) = 7$ is a local maximum value; and since f' changes from negative to positive at $x = 1$, $f(1) = -1$ is a local minimum value.

Second Derivative Test: $f''(x) = 20x^3$. $f'(x) = 0 \Leftrightarrow x = \pm 1$. $f''(-1) = -20 < 0 \Rightarrow f(-1) = 7$ is a local maximum value. $f''(1) = 20 > 0 \Rightarrow f(1) = -1$ is a local minimum value.

Preference: For this function, the two tests are equally easy.

16. $f(x) = \dfrac{x}{x^2 + 4} \Rightarrow f'(x) = \dfrac{(x^2 + 4) \cdot 1 - x(2x)}{(x^2 + 4)^2} = \dfrac{4 - x^2}{(x^2 + 4)^2} = \dfrac{(2 + x)(2 - x)}{(x^2 + 4)^2}$.

First Derivative Test: $f'(x) > 0 \Rightarrow -2 < x < 2$ and $f'(x) < 0 \Rightarrow x > 2$ or $x < -2$. Since f' changes from positive to negative at $x = 2$, $f(2) = \frac{1}{4}$ is a local maximum value; and since f' changes from negative to positive at $x = -2$, $f(-2) = -\frac{1}{4}$ is a local minimum value.

Second Derivative Test:

$$f''(x) = \frac{(x^2+4)^2(-2x) - (4-x^2)\cdot 2(x^2+4)(2x)}{[(x^2+4)^2]^2} = \frac{-2x(x^2+4)[(x^2+4)+2(4-x^2)]}{(x^2+4)^4} = \frac{-2x(12-x^2)}{(x^2+4)^3}.$$

$f'(x) = 0 \iff x = \pm 2.$ $f''(-2) = \frac{1}{16} > 0 \Rightarrow f(-2) = -\frac{1}{4}$ is a local minimum value.

$f''(2) = -\frac{1}{16} < 0 \Rightarrow f(2) = \frac{1}{4}$ is a local maximum value.

Preference: Since calculating the second derivative is fairly difficult, the First Derivative Test is easier to use for this function.

17. $f(x) = x + \sqrt{1-x} \Rightarrow f'(x) = 1 + \frac{1}{2}(1-x)^{-1/2}(-1) = 1 - \dfrac{1}{2\sqrt{1-x}}.$ Note that f is defined for $1 - x \geq 0$; that is,

for $x \leq 1$. $f'(x) = 0 \Rightarrow 2\sqrt{1-x} = 1 \Rightarrow \sqrt{1-x} = \frac{1}{2} \Rightarrow 1 - x = \frac{1}{4} \Rightarrow x = \frac{3}{4}.$ f' does not exist at $x = 1$, but we can't have a local maximum or minimum at an endpoint.

First Derivative Test: $f'(x) > 0 \Rightarrow x < \frac{3}{4}$ and $f'(x) < 0 \Rightarrow \frac{3}{4} < x < 1.$ Since f' changes from positive to negative at $x = \frac{3}{4}$, $f\left(\frac{3}{4}\right) = \frac{5}{4}$ is a local maximum value.

Second Derivative Test: $f''(x) = -\frac{1}{2}\left(-\frac{1}{2}\right)(1-x)^{-3/2}(-1) = -\dfrac{1}{4\left(\sqrt{1-x}\right)^3}.$

$f''\left(\frac{3}{4}\right) = -2 < 0 \Rightarrow f\left(\frac{3}{4}\right) = \frac{5}{4}$ is a local maximum value.

Preference: The First Derivative Test may be slightly easier to apply in this case.

18. (a) $f(x) = x^4(x-1)^3 \Rightarrow f'(x) = x^4 \cdot 3(x-1)^2 + (x-1)^3 \cdot 4x^3 = x^3(x-1)^2[3x + 4(x-1)] = x^3(x-1)^2(7x-4)$

The critical numbers are 0, 1, and $\frac{4}{7}$.

(b) $f''(x) = 3x^2(x-1)^2(7x-4) + x^3 \cdot 2(x-1)(7x-4) + x^3(x-1)^2 \cdot 7$

$\qquad = x^2(x-1)[3(x-1)(7x-4) + 2x(7x-4) + 7x(x-1)]$

Now $f''(0) = f''(1) = 0$, so the Second Derivative Test gives no information for $x = 0$ or $x = 1$.

$f''\left(\frac{4}{7}\right) = \left(\frac{4}{7}\right)^2\left(\frac{4}{7}-1\right)\left[0+0+7\left(\frac{4}{7}\right)\left(\frac{4}{7}-1\right)\right] = \left(\frac{4}{7}\right)^2\left(-\frac{3}{7}\right)(4)\left(-\frac{3}{7}\right) > 0,$ so there is a local minimum at $x = \frac{4}{7}$.

(c) f' is positive on $(-\infty, 0)$, negative on $\left(0, \frac{4}{7}\right)$, positive on $\left(\frac{4}{7}, 1\right)$, and positive on $(1, \infty)$. So f has a local maximum at $x = 0$, a local minimum at $x = \frac{4}{7}$, and no local maximum or minimum at $x = 1$.

19. (a) By the Second Derivative Test, if $f'(2) = 0$ and $f''(2) = -5 < 0$, f has a local maximum at $x = 2$.

(b) If $f'(6) = 0$, we know that f has a horizontal tangent at $x = 6$. Knowing that $f''(6) = 0$ does not provide any additional information since the Second Derivative Test fails. For example, the first and second derivatives of $y = (x-6)^4$, $y = -(x-6)^4$, and $y = (x-6)^3$ all equal zero for $x = 6$, but the first has a local minimum at $x = 6$, the second has a local maximum at $x = 6$, and the third has an inflection point at $x = 6$.

20. $f'(x) > 0$ for all $x \neq 1$ with vertical asymptote $x = 1$, so f is increasing on $(-\infty, 1)$ and $(1, \infty)$. $f''(x) > 0$ if $x < 1$ or $x > 3$, and $f''(x) < 0$ if $1 < x < 3$, so f is concave upward on $(-\infty, 1)$ and $(3, \infty)$, and concave downward on $(1, 3)$. There is an inflection point when $x = 3$.

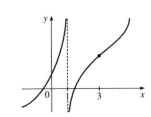

21. $f'(0) = f'(2) = f'(4) = 0$ ⇒ horizontal tangents at $x = 0, 2, 4$.

$f'(x) > 0$ if $x < 0$ or $2 < x < 4$ ⇒ f is increasing on $(-\infty, 0)$ and $(2, 4)$.

$f'(x) < 0$ if $0 < x < 2$ or $x > 4$ ⇒ f is decreasing on $(0, 2)$ and $(4, \infty)$.

$f''(x) > 0$ if $1 < x < 3$ ⇒ f is concave upward on $(1, 3)$.

$f''(x) < 0$ if $x < 1$ or $x > 3$ ⇒ f is concave downward on $(-\infty, 1)$ and $(3, \infty)$.

There are inflection points when $x = 1$ and 3.

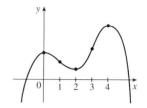

22. $f'(1) = f'(-1) = 0$ ⇒ horizontal tangents at $x = \pm 1$. $f'(x) < 0$ if $|x| < 1$ ⇒ f is decreasing on $(-1, 1)$. $f'(x) > 0$ if $1 < |x| < 2$ ⇒ f is increasing on $(-2, -1)$ and $(1, 2)$. $f'(x) = -1$ if $|x| > 2$ ⇒ the graph of f has constant slope -1 on $(-\infty, -2)$ and $(2, \infty)$. $f''(x) < 0$ if $-2 < x < 0$ ⇒ f is concave downward on $(-2, 0)$. The point $(0, 1)$ is an inflection point.

23. $f'(x) > 0$ if $|x| < 2$ ⇒ f is increasing on $(-2, 2)$.

$f'(x) < 0$ if $|x| > 2$ ⇒ f is decreasing on $(-\infty, -2)$ and $(2, \infty)$. $f'(-2) = 0$ ⇒ horizontal tangent at $x = -2$.

$\lim_{x \to 2} |f'(x)| = \infty$ ⇒ there is a vertical asymptote or vertical tangent (cusp) at $x = 2$. $f''(x) > 0$ if $x \neq 2$ ⇒ f is concave upward on $(-\infty, 2)$ and $(2, \infty)$.

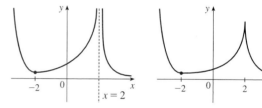

24. $f(0) = f'(0) = 0$ ⇒ the graph of f passes through the origin and has a horizontal tangent there. $f'(2) = f'(4) = f'(6) = 0$ ⇒ horizontal tangents at $x = 2, 4, 6$. $f'(x) > 0$ if $0 < x < 2$ or $4 < x < 6$ ⇒ f increasing on $(0, 2)$ and $(4, 6)$. $f'(x) < 0$ if $2 < x < 4$ or $x > 6$ ⇒ f decreasing on $(2, 4)$ and $(6, \infty)$. $f''(x) > 0$ if $0 < x < 1$ or $3 < x < 5$ ⇒ f is CU on $(0, 1)$ and $(3, 5)$.

$f''(x) < 0$ if $1 < x < 3$ or $x > 5$ ⇒ f is CD on $(1, 3)$ and $(5, \infty)$. $f(-x) = f(x)$ ⇒ f is even and the graph is symmetric about the y-axis.

25. The function must be always decreasing (since the first derivative is always negative) and concave downward (since the second derivative is always negative).

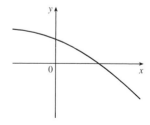

26. (a) $f(3) = 2$ $\Rightarrow$ the point $(3, 2)$ is on the graph of f. $f'(3) = \frac{1}{2}$ $\Rightarrow$ the slope of

the tangent line at $(3, 2)$ is $\frac{1}{2}$. $f'(x) > 0$ for all x $\Rightarrow$ f is increasing on $\mathbb{R}$.

$f''(x) < 0$ for all x $\Rightarrow$ f is concave downward on $\mathbb{R}$. A possible graph for f

is shown.

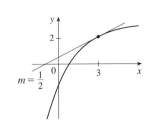

(b) The tangent line at $(3, 2)$ has equation $y - 2 = \frac{1}{2}(x - 3)$, or $y = \frac{1}{2}x + \frac{1}{2}$, and x-intercept -1. Since f is concave

downward on $\mathbb{R}$, f is below the x-axis at $x = -1$, and hence changes sign at least once. Since f is increasing on $\mathbb{R}$,

it changes sign at most once. Thus, it changes sign exactly once and there is one solution of the equation $f(x) = 0$.

(c) $f'' < 0$ $\Rightarrow$ f' is decreasing. Since $f'(3) = \frac{1}{2}$, $f'(2)$ must be greater than $\frac{1}{2}$, so no, it is not possible that $f'(2) = \frac{1}{3}$.

27. (a) f is increasing where f' is positive, that is, on $(0, 2)$, $(4, 6)$, and $(8, \infty)$; and decreasing where f' is negative, that is,

on $(2, 4)$ and $(6, 8)$.

(b) f has local maxima where f' changes from positive to negative, at $x = 2$ and at   (e)

$x = 6$, and local minima where f' changes from negative to positive, at $x = 4$

and at $x = 8$.

(c) f is concave upward (CU) where f' is increasing, that is, on $(3, 6)$ and $(6, \infty)$,

and concave downward (CD) where f' is decreasing, that is, on $(0, 3)$.

(d) There is a point of inflection where f changes from being CD to being CU, that

is, at $x = 3$.

28. (a) f is increasing where f' is positive, on $(1, 6)$ and $(8, \infty)$, and decreasing where f' is negative, on $(0, 1)$ and $(6, 8)$.

(b) f has a local maximum where f' changes from positive to negative, at $x = 6$,   (e)

and local minima where f' changes from negative to positive, at $x = 1$ and

at $x = 8$.

(c) f is concave upward where f' is increasing, that is, on $(0, 2)$, $(3, 5)$, and $(7, \infty)$,

and concave downward where f' is decreasing, that is, on $(2, 3)$ and $(5, 7)$.

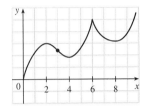

(d) There are points of inflection where f changes its direction of concavity, at

$x = 2$, $x = 3$, $x = 5$ and $x = 7$.

29. (a) $f(x) = 2x^3 - 3x^2 - 12x$ $\Rightarrow$ $f'(x) = 6x^2 - 6x - 12 = 6(x^2 - x - 2) = 6(x - 2)(x + 1)$.

$f'(x) > 0$ $\Leftrightarrow$ $x < -1$ or $x > 2$ and $f'(x) < 0$ $\Leftrightarrow$ $-1 < x < 2$. So f is increasing on $(-\infty, -1)$ and $(2, \infty)$,

and f is decreasing on $(-1, 2)$.

(b) Since f changes from increasing to decreasing at $x = -1$, $f(-1) = 7$ is a local   (d)

maximum value. Since f changes from decreasing to increasing at $x = 2$,

$f(2) = -20$ is a local minimum value.

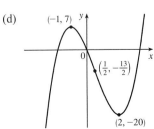

(c) $f''(x) = 6(2x - 1)$ $\Rightarrow$ $f''(x) > 0$ on $\left(\frac{1}{2}, \infty\right)$ and $f''(x) < 0$ on $\left(-\infty, \frac{1}{2}\right)$.

So f is concave upward on $\left(\frac{1}{2}, \infty\right)$ and concave downward on $\left(-\infty, \frac{1}{2}\right)$. There

is a change in concavity at $x = \frac{1}{2}$, and we have an inflection point at $\left(\frac{1}{2}, -\frac{13}{2}\right)$.

30. (a) $f(x) = 2 + 3x - x^3 \Rightarrow f'(x) = 3 - 3x^2 = -3(x^2 - 1) = -3(x+1)(x-1)$.

$f'(x) > 0 \Leftrightarrow -1 < x < 1$ and $f'(x) < 0 \Leftrightarrow x < -1$ or $x > 1$. So f is increasing on $(-1, 1)$ and f is decreasing on $(-\infty, -1)$ and $(1, \infty)$.

(b) $f(-1) = 0$ is a local minimum value and $f(1) = 4$ is a local maximum value. (d)

(c) $f''(x) = -6x \Rightarrow f''(x) > 0$ on $(-\infty, 0)$ and $f''(x) < 0$ on $(0, \infty)$.

So f is concave upward on $(-\infty, 0)$ and concave downward on $(0, \infty)$. There is an inflection point at $(0, 2)$.

31. (a) $f(x) = 2 + 2x^2 - x^4 \Rightarrow f'(x) = 4x - 4x^3 = 4x(1 - x^2) = 4x(1 + x)(1 - x)$. $f'(x) > 0 \Leftrightarrow x < -1$ or $0 < x < 1$ and $f'(x) < 0 \Leftrightarrow -1 < x < 0$ or $x > 1$. So f is increasing on $(-\infty, -1)$ and $(0, 1)$ and f is decreasing on $(-1, 0)$ and $(1, \infty)$.

(b) f changes from increasing to decreasing at $x = -1$ and $x = 1$, so $f(-1) = 3$ and $f(1) = 3$ are local maximum values.

f changes from decreasing to increasing at $x = 0$, so $f(0) = 2$ is a local minimum value.

(c) $f''(x) = 4 - 12x^2 = 4(1 - 3x^2)$. $f''(x) = 0 \Leftrightarrow 1 - 3x^2 = 0 \Leftrightarrow$ (d)

$x^2 = \frac{1}{3} \Leftrightarrow x = \pm 1/\sqrt{3}$. $f''(x) > 0$ on $\left(-1/\sqrt{3}, 1/\sqrt{3}\right)$ and $f''(x) < 0$ on $\left(-\infty, -1/\sqrt{3}\right)$ and $\left(1/\sqrt{3}, \infty\right)$. So f is concave upward on $\left(-1/\sqrt{3}, 1/\sqrt{3}\right)$ and f is concave downward on $\left(-\infty, -1/\sqrt{3}\right)$ and $\left(1/\sqrt{3}, \infty\right)$. $f\left(\pm 1/\sqrt{3}\right) = 2 + \frac{2}{3} - \frac{1}{9} = \frac{23}{9}$. There are points of inflection at $\left(\pm 1/\sqrt{3}, \frac{23}{9}\right)$.

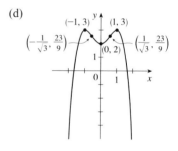

32. (a) $g(x) = 200 + 8x^3 + x^4 \Rightarrow g'(x) = 24x^2 + 4x^3 = 4x^2(6 + x) = 0$ when $x = -6$ and when $x = 0$.

$g'(x) > 0 \Leftrightarrow x > -6$ $[x \neq 0]$ and $g'(x) < 0 \Leftrightarrow x < -6$, so g is decreasing on $(-\infty, -6)$ and g is increasing on $(-6, \infty)$, with a horizontal tangent at $x = 0$.

(b) $g(-6) = -232$ is a local minimum value. (d)

There is no local maximum value.

(c) $g''(x) = 48x + 12x^2 = 12x(4 + x) = 0$ when $x = -4$ and when $x = 0$.

$g''(x) > 0 \Leftrightarrow x < -4$ or $x > 0$ and $g''(x) < 0 \Leftrightarrow -4 < x < 0$, so g is CU on $(-\infty, -4)$ and $(0, \infty)$, and g is CD on $(-4, 0)$. There are inflection points at $(-4, -56)$ and $(0, 200)$.

33. (a) $h(x) = (x+1)^5 - 5x - 2 \Rightarrow h'(x) = 5(x+1)^4 - 5$. $h'(x) = 0 \Leftrightarrow 5(x+1)^4 = 5 \Leftrightarrow (x+1)^4 = 1 \Rightarrow (x+1)^2 = 1 \Rightarrow x + 1 = 1$ or $x + 1 = -1 \Rightarrow x = 0$ or $x = -2$. $h'(x) > 0 \Leftrightarrow x < -2$ or $x > 0$ and $h'(x) < 0 \Leftrightarrow -2 < x < 0$. So h is increasing on $(-\infty, -2)$ and $(0, \infty)$ and h is decreasing on $(-2, 0)$.

(b) $h(-2) = 7$ is a local maximum value and $h(0) = -1$ is a local minimum value.

(c) $h''(x) = 20(x+1)^3 = 0$ ⇔ $x = -1$. $h''(x) > 0$ ⇔ $x > -1$ and $h''(x) < 0$ ⇔ $x < -1$, so h is CU on $(-1, \infty)$ and h is CD on $(-\infty, -1)$.

There is a point of inflection at $(-1, h(-1)) = (-1, 3)$.

(d)

34. (a) $h(x) = x^5 - 2x^3 + x$ ⇒ $h'(x) = 5x^4 - 6x^2 + 1 = (5x^2 - 1)(x^2 - 1) = (\sqrt{5}\,x + 1)(\sqrt{5}\,x - 1)(x + 1)(x - 1)$.

$h'(x) > 0$ ⇔ $x < -1$ or $-1/\sqrt{5} < x < 1/\sqrt{5}$ or $x > 1$ and $h'(x) < 0$ ⇔ $-1 < x < -1/\sqrt{5}$ or $1/\sqrt{5} < x < 1$.

So h is increasing on $(-\infty, -1)$, $(-1/\sqrt{5}, 1/\sqrt{5}\,)$, and $(1, \infty)$ and h is decreasing on $(-1, -1/\sqrt{5}\,)$ and $(1/\sqrt{5}, 1)$.

(b) $h(-1) = 0$ and $h(1/\sqrt{5}\,) = 16/(25\sqrt{5}\,)$ $[\approx 0.29]$ are local maximum values.

$h(-1/\sqrt{5}\,) = -16/(25\sqrt{5}\,)$ and $h(1) = 0$ are local minimum values.

(c) $h''(x) = 20x^3 - 12x = 4x(5x^2 - 3)$. $h''(x) = 0$ ⇔ $x = 0$ or $x = \pm\sqrt{\tfrac{3}{5}}$. $h''(x) > 0$ ⇔ $-\sqrt{\tfrac{3}{5}} < x < 0$ or

$x > \sqrt{\tfrac{3}{5}}$ and $h''(x) < 0$ ⇔ $x < -\sqrt{\tfrac{3}{5}}$ or $0 < x < \sqrt{\tfrac{3}{5}}$, so h is CU on $\left(-\sqrt{\tfrac{3}{5}}, 0\right)$ and $\left(\sqrt{\tfrac{3}{5}}, \infty\right)$ and h is

CD on $\left(-\infty, -\sqrt{\tfrac{3}{5}}\right)$ and $\left(0, \sqrt{\tfrac{3}{5}}\right)$. There are points of inflection at

$\left(-\sqrt{\tfrac{3}{5}}, -\tfrac{4}{25}\sqrt{\tfrac{3}{5}}\right) \approx (-0.77, -0.12)$, $(0, 0)$, and $\left(\sqrt{\tfrac{3}{5}}, \tfrac{4}{25}\sqrt{\tfrac{3}{5}}\right)$.

(d)

35. (a) $A(x) = x\sqrt{x+3}$ ⇒ $A'(x) = x \cdot \tfrac{1}{2}(x+3)^{-1/2} + \sqrt{x+3} \cdot 1 = \dfrac{x}{2\sqrt{x+3}} + \sqrt{x+3} = \dfrac{x + 2(x+3)}{2\sqrt{x+3}} = \dfrac{3x+6}{2\sqrt{x+3}}$.

The domain of A is $[-3, \infty)$. $A'(x) > 0$ for $x > -2$ and $A'(x) < 0$ for $-3 < x < -2$, so A is increasing on $(-2, \infty)$ and decreasing on $(-3, -2)$.

(b) $A(-2) = -2$ is a local minimum value.

(c) $A''(x) = \dfrac{2\sqrt{x+3} \cdot 3 - (3x+6) \cdot \dfrac{1}{\sqrt{x+3}}}{\left(2\sqrt{x+3}\right)^2}$

$= \dfrac{6(x+3) - (3x+6)}{4(x+3)^{3/2}} = \dfrac{3x+12}{4(x+3)^{3/2}} = \dfrac{3(x+4)}{4(x+3)^{3/2}}$

(d)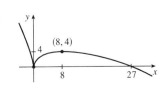

$A''(x) > 0$ for all $x > -3$, so A is concave upward on $(-3, \infty)$. There is no inflection point.

36. (a) $B(x) = 3x^{2/3} - x$ ⇒ $B'(x) = 2x^{-1/3} - 1 = \dfrac{2}{\sqrt[3]{x}} - 1 = \dfrac{2 - \sqrt[3]{x}}{\sqrt[3]{x}}$. $B'(x) > 0$ if $0 < x < 8$ and $B'(x) < 0$ if

$x < 0$ or $x > 8$, so B is decreasing on $(-\infty, 0)$ and $(8, \infty)$, and B is increasing on $(0, 8)$.

(b) $B(0) = 0$ is a local minimum value. $B(8) = 4$ is a local maximum value.

(d)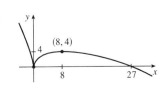

(c) $B''(x) = -\frac{2}{3}x^{-4/3} = \frac{-2}{3x^{4/3}}$, so $B''(x) < 0$ for all $x \neq 0$. B is concave

downward on $(-\infty, 0)$ and $(0, \infty)$. There is no inflection point.

37. (a) $C(x) = x^{1/3}(x+4) = x^{4/3} + 4x^{1/3}$ ⇒ $C'(x) = \frac{4}{3}x^{1/3} + \frac{4}{3}x^{-2/3} = \frac{4}{3}x^{-2/3}(x+1) = \frac{4(x+1)}{3\sqrt[3]{x^2}}$. $C'(x) > 0$ if

$-1 < x < 0$ or $x > 0$ and $C'(x) < 0$ for $x < -1$, so C is increasing on $(-1, \infty)$ and C is decreasing on $(-\infty, -1)$.

(b) $C(-1) = -3$ is a local minimum value.

(d)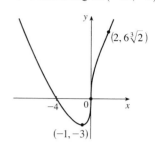

(c) $C''(x) = \frac{4}{9}x^{-2/3} - \frac{8}{9}x^{-5/3} = \frac{4}{9}x^{-5/3}(x-2) = \frac{4(x-2)}{9\sqrt[3]{x^5}}$.

$C''(x) < 0$ for $0 < x < 2$ and $C''(x) > 0$ for $x < 0$ and $x > 2$, so C is

concave downward on $(0, 2)$ and concave upward on $(-\infty, 0)$ and $(2, \infty)$.

There are inflection points at $(0, 0)$ and $\left(2, 6\sqrt[3]{2}\right) \approx (2, 7.56)$.

38. (a) $G(x) = x - 4\sqrt{x}$ ⇒ $G'(x) = 1 - \frac{2}{\sqrt{x}} = \frac{1}{\sqrt{x}}\left(\sqrt{x} - 2\right)$. $G'(x) > 0$ ⇔ $x > 4$ and $G'(x) < 0$ ⇔

$0 < x < 4$, so G is decreasing on $(0, 4)$ and increasing on $(4, \infty)$.

(b) Local minimum $G(4) = -4$. No local maximum.

(d)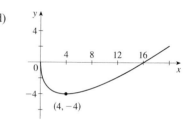

(c) $G'(x) = 1 - 2x^{-1/2}$ ⇒ $G''(x) = x^{-3/2} = 1/\sqrt{x^3}$, so

$G''(x) > 0$ for $x > 0$. Thus, G is CU on $(0, \infty)$.

G has no IPs.

39. (a) $f(\theta) = 2\cos\theta + \cos^2\theta$, $0 \leq \theta \leq 2\pi$ ⇒ $f'(\theta) = -2\sin\theta + 2\cos\theta(-\sin\theta) = -2\sin\theta(1 + \cos\theta)$.

$f'(\theta) = 0$ ⇔ $\theta = 0, \pi$, and 2π. $f'(\theta) > 0$ ⇔ $\pi < \theta < 2\pi$ and $f'(\theta) < 0$ ⇔ $0 < \theta < \pi$. So f is increasing on $(\pi, 2\pi)$ and f is decreasing on $(0, \pi)$.

(b) $f(\pi) = -1$ is a local minimum value.

(c) $f'(\theta) = -2\sin\theta(1 + \cos\theta)$ ⇒

$$f''(\theta) = -2\sin\theta(-\sin\theta) + (1 + \cos\theta)(-2\cos\theta) = 2\sin^2\theta - 2\cos\theta - 2\cos^2\theta$$

$$= 2(1 - \cos^2\theta) - 2\cos\theta - 2\cos^2\theta = -4\cos^2\theta - 2\cos\theta + 2$$

$$= -2(2\cos^2\theta + \cos\theta - 1) = -2(2\cos\theta - 1)(\cos\theta + 1)$$

Since $-2(\cos\theta + 1) < 0$ [for $\theta \neq \pi$], $f''(\theta) > 0$ ⇒ $2\cos\theta - 1 < 0$ ⇒ $\cos\theta < \frac{1}{2}$ ⇒ $\frac{\pi}{3} < \theta < \frac{5\pi}{3}$ and

$f''(\theta) < 0$ ⇒ $\cos\theta > \frac{1}{2}$ ⇒ $0 < \theta < \frac{\pi}{3}$ or $\frac{5\pi}{3} < \theta < 2\pi$. So f is CU on $\left(\frac{\pi}{3}, \frac{5\pi}{3}\right)$ and f is CD on $\left(0, \frac{\pi}{3}\right)$ and

$\left(\frac{5\pi}{3}, 2\pi\right)$. There are points of inflection at $\left(\frac{\pi}{3}, f\left(\frac{\pi}{3}\right)\right) = \left(\frac{\pi}{3}, \frac{5}{4}\right)$ and $\left(\frac{5\pi}{3}, f\left(\frac{5\pi}{3}\right)\right) = \left(\frac{5\pi}{3}, \frac{5}{4}\right)$.

(d)

40. (a) $f(t) = t + \cos t$, $-2\pi \le t \le 2\pi$ $\Rightarrow$ $f'(t) = 1 - \sin t \ge 0$ for all t and $f'(t) = 0$ when $\sin t = 1$ $\Leftrightarrow$

$t = -\frac{3\pi}{2}$ or $\frac{\pi}{2}$, so f is increasing on $(-2\pi, 2\pi)$.

(b) No maximum or minimum

(c) $f''(t) = -\cos t > 0$ $\Leftrightarrow$ $t \in \left(-\frac{3\pi}{2}, -\frac{\pi}{2}\right) \cup \left(\frac{\pi}{2}, \frac{3\pi}{2}\right)$, so f is CU on

these intervals and CD on $\left(-2\pi, -\frac{3\pi}{2}\right)$, $\left(-\frac{\pi}{2}, \frac{\pi}{2}\right)$, and $\left(\frac{3\pi}{2}, 2\pi\right)$.

Points of inflection at $\left(\pm\frac{3\pi}{2}, \pm\frac{3\pi}{2}\right)$ and $\left(\pm\frac{\pi}{2}, \pm\frac{\pi}{2}\right)$

(d)

41. The nonnegative factors $(x + 1)^2$ and $(x - 6)^4$ do not affect the sign of $f'(x) = (x + 1)^2 (x - 3)^5 (x - 6)^4$.

So $f'(x) > 0$ $\Rightarrow$ $(x - 3)^5 > 0$ $\Rightarrow$ $x - 3 > 0$ $\Rightarrow$ $x > 3$. Thus, f is increasing on the interval $(3, \infty)$.

42. $y = f(x) = x^3 - 3a^2 x + 2a^3$, $a > 0$. The y-intercept is $f(0) = 2a^3$. $y' = 3x^2 - 3a^2 = 3(x^2 - a^2) = 3(x + a)(x - a)$.

The critical numbers are $-a$ and a. $f' < 0$ on $(-a, a)$, so f is decreasing on $(-a, a)$ and f is increasing on $(-\infty, -a)$ and

(a, ∞). $f(-a) = 4a^3$ is a local maximum value and $f(a) = 0$ is a local minimum value. Since $f(a) = 0$, a is an x-intercept,

and $x - a$ is a factor of f. Synthetically dividing $y = x^3 - 3a^2 x + 2a^3$ by $x - a$ gives us the following result:

$y = x^3 - 3a^2 x + 2a^3 = (x - a)(x^2 + ax - 2a^2) = (x - a)(x - a)(x + 2a) = (x - a)^2 (x + 2a)$, which tells us

that the only x-intercepts are $-2a$ and a. $y' = 3x^2 - 3a^2$ $\Rightarrow$ $y'' = 6x$, so $y'' > 0$

on $(0, \infty)$ and $y'' < 0$ on $(-\infty, 0)$. This tells us that f is CU on $(0, \infty)$ and CD on

$(-\infty, 0)$. There is an inflection point at $(0, 2a^3)$. The graph illustrates these features.

What the curves in the family have in common is that they are all CD on $(-\infty, 0)$,

CU on $(0, \infty)$, and have the same basic shape. But as a increases, the four key points

shown in the figure move further away from the origin.

43. (a)

From the graph, we get an estimate of $f(1) \approx 1.41$ as a local maximum

value, and no local minimum value.

$f(x) = \dfrac{x + 1}{\sqrt{x^2 + 1}}$ $\Rightarrow$ $f'(x) = \dfrac{1 - x}{(x^2 + 1)^{3/2}}$.

$f'(x) = 0$ $\Leftrightarrow$ $x = 1$. $f(1) = \frac{2}{\sqrt{2}} = \sqrt{2}$ is the exact value.

(b) From the graph in part (a), f increases most rapidly somewhere between $x = -\frac{1}{2}$ and $x = -\frac{1}{4}$. To find the exact value,

we need to find the maximum value of f', which we can do by finding the critical numbers of f'.

$f''(x) = \dfrac{2x^2 - 3x - 1}{(x^2 + 1)^{5/2}} = 0$ $\Leftrightarrow$ $x = \dfrac{3 \pm \sqrt{17}}{4}$. $x = \dfrac{3 + \sqrt{17}}{4}$ corresponds to the *minimum* value of f'.

The maximum value of f' occurs at $\left(x = \dfrac{3 - \sqrt{17}}{4}\right) \approx -0.28$.

44. (a)

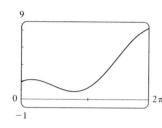

From the graph, we get estimates of $f(2.61) \approx 0.89$ as a local and absolute minimum, $f(0.53) \approx 2.26$ as a local maximum, and $f(2\pi) \approx 8.28$ as an absolute maximum. $f(x) = x + 2\cos x \ (0 \le x \le 2\pi) \quad \Rightarrow$ $f'(x) = 1 - 2\sin x. \ f'(x) = 0 \quad \Leftrightarrow \quad \sin x = \frac{1}{2} \quad \Leftrightarrow \quad x = \frac{\pi}{6}, \frac{5\pi}{6}.$ $f\left(\frac{\pi}{6}\right) = \frac{\pi}{6} + \sqrt{3}$ is the exact value of the local maximum, $f\left(\frac{5\pi}{6}\right) = \frac{5\pi}{6} - \sqrt{3}$ is the exact value of the local and absolute minimum, and $f(2\pi) = 2\pi + 2$ is the exact value of the absolute maximum.

(b) From the graph in part (a), f increases most rapidly somewhere between $x = 4.5$ and $x = 5$. Now f increases most rapidly when $f'(x) = 1 - 2\sin x$ has its maximum value. $f''(x) = -2\cos x = 0 \quad \Leftrightarrow \quad x = \frac{\pi}{2}, \frac{3\pi}{2}.$ $f'(0) = f'(2\pi) = 1, \ f'\left(\frac{\pi}{2}\right) = -1,$ and $f'\left(\frac{3\pi}{2}\right) = 3.$ The maximum value of f' occurs at $\left(\frac{3\pi}{2}, \frac{3\pi}{2}\right).$

45. $f(x) = \cos x + \frac{1}{2}\cos 2x \quad \Rightarrow \quad f'(x) = -\sin x - \sin 2x \quad \Rightarrow \quad f''(x) = -\cos x - 2\cos 2x$

(a)

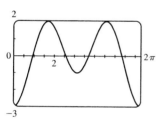

From the graph of f, it seems that f is CD on $(0, 1)$, CU on $(1, 2.5)$, CD on $(2.5, 3.7)$, CU on $(3.7, 5.3)$, and CD on $(5.3, 2\pi)$. The points of inflection appear to be at $(1, 0.4)$, $(2.5, -0.6)$, $(3.7, -0.6)$, and $(5.3, 0.4)$.

(b)

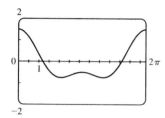

From the graph of f'' (and zooming in near the zeros), it seems that f is CD on $(0, 0.94)$, CU on $(0.94, 2.57)$, CD on $(2.57, 3.71)$, CU on $(3.71, 5.35)$, and CD on $(5.35, 2\pi)$. Refined estimates of the inflection points are $(0.94, 0.44)$, $(2.57, -0.63)$, $(3.71, -0.63)$, and $(5.35, 0.44)$.

46. $f(x) = x^3(x-2)^4 \quad \Rightarrow$

$f'(x) = x^3 \cdot 4(x-2)^3 + (x-2)^4 \cdot 3x^2 = x^2(x-2)^3[4x + 3(x-2)] = x^2(x-2)^3(7x-6) \quad \Rightarrow$

$f''(x) = (2x)(x-2)^3(7x-6) + x^2 \cdot 3(x-2)^2(7x-6) + x^2(x-2)^3(7)$

$\quad = x(x-2)^2[2(x-2)(7x-6) + 3x(7x-6) + 7x(x-2)]$

$\quad = x(x-2)^2[42x^2 - 72x + 24] = 6x(x-2)^2(7x^2 - 12x + 4)$

(a)

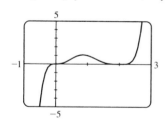

From the graph of f, it seems that f is CD on $(-\infty, 0)$, CU on $(0, 0.5)$, CD on $(0.5, 1.3)$, and CU on $(1.3, \infty)$. The points of inflection appear to be at $(0, 0)$, $(0.5, 0.5)$, and $(1.3, 0.6)$.

(b)

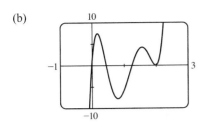

From the graph of f'' (and zooming in near the zeros), it seems that f is CD on $(-\infty, 0)$, CU on $(0, 0.45)$, CD on $(0.45, 1.26)$, and CU on $(1.26, \infty)$. Refined estimates of the inflection points are $(0, 0)$, $(0.45, 0.53)$, and $(1.26, 0.60)$.

47. In Maple, we define f and then use the command

plot(diff(diff(f,x),x),x=-2..2); . In Mathematica, we define f and then use Plot[Dt[Dt[f,x],x],{x,-2,2}]. We see that $f'' > 0$ for $x < -0.6$ and $x > 0.0$ $[\approx 0.03]$ and $f'' < 0$ for $-0.6 < x < 0.0$. So f is CU on $(-\infty, -0.6)$ and $(0.0, \infty)$ and CD on $(-0.6, 0.0)$.

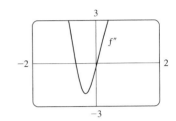

48. It appears that f'' is positive (and thus f is concave up) on $(-1.8, 0.3)$ and $(1.5, \infty)$ and negative (so f is concave down) on $(-\infty, -1.8)$ and $(0.3, 1.5)$.

49. (a) The rate of increase of the population is initially very small, then gets larger until it reaches a maximum at about $t = 8$ hours, and decreases toward 0 as the population begins to level off.

(b) The rate of increase has its maximum value at $t = 8$ hours.

(c) The population function is concave upward on $(0, 8)$ and concave downward on $(8, 18)$.

(d) At $t = 8$, the population is about 350, so the inflection point is about $(8, 350)$.

50. (a) I'm very unhappy. It's uncomfortably hot and $f'(3) = 2$ indicates that the temperature is increasing, and $f''(3) = 4$ indicates that the rate of increase is increasing. (The temperature is rapidly getting warmer.)

(b) I'm still unhappy, but not as unhappy as in part (a). It's uncomfortably hot and $f'(3) = 2$ indicates that the temperature is increasing, but $f''(3) = -4$ indicates that the rate of increase is decreasing. (The temperature is slowly getting warmer.)

(c) I'm somewhat happy. It's uncomfortably hot and $f'(3) = -2$ indicates that the temperature is decreasing, but $f''(3) = 4$ indicates that the rate of change is increasing. (The rate of change is negative but it's becoming less negative. The temperature is slowly getting cooler.)

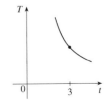

(d) I'm very happy. It's uncomfortably hot and $f'(3) = -2$ indicates that the temperature is decreasing, and $f''(3) = -4$ indicates that the rate of change is decreasing, that is, becoming more negative. (The temperature is rapidly getting cooler.)

51. Most students learn more in the third hour of studying than in the eighth hour, so $K(3) - K(2)$ is larger than $K(8) - K(7)$. In other words, as you begin studying for a test, the rate of knowledge gain is large and then starts to taper off, so $K'(t)$ decreases and the graph of K is concave downward.

52. At first the depth increases slowly because the base of the mug is wide. But as the mug narrows, the coffee rises more quickly. Thus, the depth d increases at an increasing rate and its graph is concave upward. The rate of increase of d has a maximum where the mug is narrowest; that is, when the mug is half full. It is there that the inflection point (IP) occurs. Then the rate of increase of d starts to decrease as the mug widens and the graph becomes concave down.

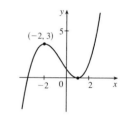

53. $f(x) = ax^3 + bx^2 + cx + d \Rightarrow f'(x) = 3ax^2 + 2bx + c$.

We are given that $f(1) = 0$ and $f(-2) = 3$, so $f(1) = a + b + c + d = 0$ and

$f(-2) = -8a + 4b - 2c + d = 3$. Also $f'(1) = 3a + 2b + c = 0$ and

$f'(-2) = 12a - 4b + c = 0$ by Fermat's Theorem. Solving these four equations, we get

$a = \frac{2}{9}, b = \frac{1}{3}, c = -\frac{4}{3}, d = \frac{7}{9}$, so the function is $f(x) = \frac{1}{9}\left(2x^3 + 3x^2 - 12x + 7\right)$.

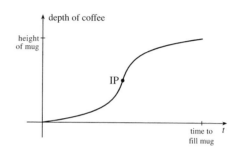

54. $y = \dfrac{1+x}{1+x^2} \Rightarrow y' = \dfrac{(1+x^2)(1) - (1+x)(2x)}{(1+x^2)^2} = \dfrac{1 - 2x - x^2}{(1+x^2)^2} \Rightarrow$

$y'' = \dfrac{(1+x^2)^2(-2-2x) - (1-2x-x^2)\cdot 2(1+x^2)(2x)}{[(1+x^2)^2]^2} = \dfrac{2(1+x^2)[(1+x^2)(-1-x) - (1-2x-x^2)(2x)]}{(1+x^2)^4}$

$= \dfrac{2(-1-x-x^2-x^3-2x+4x^2+2x^3)}{(1+x^2)^3} = \dfrac{2(x^3+3x^2-3x-1)}{(1+x^2)^3} = \dfrac{2(x-1)(x^2+4x+1)}{(1+x^2)^3}$

So $y'' = 0 \Rightarrow x = 1, -2 \pm \sqrt{3}$. Let $a = -2 - \sqrt{3}, b = -2 + \sqrt{3}$, and $c = 1$. We can show that $f(a) = \frac{1}{4}(1 - \sqrt{3})$, $f(b) = \frac{1}{4}(1 + \sqrt{3})$, and $f(c) = 1$. To show that these three points of inflection lie on one straight line, we'll show that the slopes m_{ac} and m_{bc} are equal.

$$m_{ac} = \frac{f(c) - f(a)}{c - a} = \frac{1 - \frac{1}{4}(1 - \sqrt{3})}{1 - (-2 - \sqrt{3})} = \frac{\frac{3}{4} + \frac{1}{4}\sqrt{3}}{3 + \sqrt{3}} = \frac{1}{4}$$

$$m_{bc} = \frac{f(c) - f(b)}{c - b} = \frac{1 - \frac{1}{4}(1 + \sqrt{3})}{1 - (-2 + \sqrt{3})} = \frac{\frac{3}{4} - \frac{1}{4}\sqrt{3}}{3 - \sqrt{3}} = \frac{1}{4}$$

55. Suppose that f is differentiable on an interval I and $f'(x) > 0$ for all x in I except $x = c$. To show that f is increasing on I, let x_1, x_2 be two numbers in I with $x_1 < x_2$.

    Case 1 $x_1 < x_2 < c$. Let J be the interval $\{x \in I \mid x < c\}$. By applying the Increasing/Decreasing Test to f on J, we see that f is increasing on J, so $f(x_1) < f(x_2)$.

    Case 2 $c < x_1 < x_2$. Apply the Increasing/Decreasing Test to f on $K = \{x \in I \mid x > c\}$.

    Case 3 $x_1 < x_2 = c$. Apply the proof of the Increasing/Decreasing Test, using the Mean Value Theorem (MVT) on the interval $[x_1, x_2]$ and noting that the MVT does not require f to be differentiable at the endpoints of $[x_1, x_2]$.

    Case 4 $c = x_1 < x_2$. Same proof as in Case 3.

    Case 5 $x_1 < c < x_2$. By Cases 3 and 4, f is increasing on $[x_1, c]$ and on $[c, x_2]$, so $f(x_1) < f(c) < f(x_2)$.

In all cases, we have shown that $f(x_1) < f(x_2)$. Since x_1, x_2 were any numbers in I with $x_1 < x_2$, we have shown that f is increasing on I.

56. (a) We will make use of the converse of the Concavity Test (along with the stated assumptions); that is, if f is concave upward on I, then $f'' > 0$ on I. If f and g are CU on I, then $f'' > 0$ and $g'' > 0$ on I, so $(f + g)'' = f'' + g'' > 0$ on I $\Rightarrow$ $f + g$ is CU on I.

    (b) Since f is positive and CU on I, $f > 0$ and $f'' > 0$ on I. So $g(x) = [f(x)]^2$ $\Rightarrow$ $g' = 2ff'$ $\Rightarrow$ $g'' = 2f'f' + 2ff'' = 2(f')^2 + 2ff'' > 0$ $\Rightarrow$ g is CU on I.

57. (a) Since f and g are positive, increasing, and CU on I with f'' and g'' never equal to 0, we have $f > 0$, $f' \geq 0$, $f'' > 0$, $g > 0$, $g' \geq 0$, $g'' > 0$ on I. Then $(fg)' = f'g + fg'$ $\Rightarrow$ $(fg)'' = f''g + 2f'g' + fg'' \geq f''g + fg'' > 0$ on I $\Rightarrow$ fg is CU on I.

    (b) In part (a), if f and g are both decreasing instead of increasing, then $f' \leq 0$ and $g' \leq 0$ on I, so we still have $2f'g' \geq 0$ on I. Thus, $(fg)'' = f''g + 2f'g' + fg'' \geq f''g + fg'' > 0$ on I $\Rightarrow$ fg is CU on I as in part (a).

    (c) Suppose f is increasing and g is decreasing [with f and g positive and CU]. Then $f' \geq 0$ and $g' \leq 0$ on I, so $2f'g' \leq 0$ on I and the argument in parts (a) and (b) fails.

        Example 1. $I = (0, \infty)$, $f(x) = x^3$, $g(x) = 1/x$. Then $(fg)(x) = x^2$, so $(fg)'(x) = 2x$ and $(fg)''(x) = 2 > 0$ on I. Thus, fg is CU on I.

        Example 2. $I = (0, \infty)$, $f(x) = 4x\sqrt{x}$, $g(x) = 1/x$. Then $(fg)(x) = 4\sqrt{x}$, so $(fg)'(x) = 2/\sqrt{x}$ and $(fg)''(x) = -1/\sqrt{x^3} < 0$ on I. Thus, fg is CD on I.

        Example 3. $I = (0, \infty)$, $f(x) = x^2$, $g(x) = 1/x$. Thus, $(fg)(x) = x$, so fg is linear on I.

58. Since f and g are CU on $(-\infty, \infty)$, $f'' > 0$ and $g'' > 0$ on $(-\infty, \infty)$. $h(x) = f(g(x))$ $\Rightarrow$ $h'(x) = f'(g(x))g'(x)$ $\Rightarrow$ $h''(x) = f''(g(x))g'(x)g'(x) + f'(g(x))g''(x) = f''(g(x))[g'(x)]^2 + f'(g(x))g''(x) > 0$ if $f' > 0$. So h is CU if f is increasing.

59. $f(x) = \tan x - x \;\Rightarrow\; f'(x) = \sec^2 x - 1 > 0$ for $0 < x < \frac{\pi}{2}$ since $\sec^2 x > 1$ for $0 < x < \frac{\pi}{2}$. So f is increasing

on $\left(0, \frac{\pi}{2}\right)$. Thus, $f(x) > f(0) = 0$ for $0 < x < \frac{\pi}{2} \;\Rightarrow\; \tan x - x > 0 \;\Rightarrow\; \tan x > x$ for $0 < x < \frac{\pi}{2}$.

60. Let $f(x) = 2\sqrt{x} - 3 + 1/x$. Then $f'(x) = 1/\sqrt{x} - 1/x^2 > 0$ for $x > 1$ since for $x > 1$, $x^2 > x > \sqrt{x}$. Hence, f is

increasing, so for $x > 1$, $f(x) > f(1) = 0$ or $2\sqrt{x} - 3 + 1/x > 0$ for $x > 1$. Hence, $2\sqrt{x} > 3 - 1/x$ for $x > 1$.

61. Let the cubic function be $f(x) = ax^3 + bx^2 + cx + d \;\Rightarrow\; f'(x) = 3ax^2 + 2bx + c \;\Rightarrow\; f''(x) = 6ax + 2b$.

So f is CU when $6ax + 2b > 0 \;\Leftrightarrow\; x > -b/(3a)$, CD when $x < -b/(3a)$, and so the only point of inflection occurs when

$x = -b/(3a)$. If the graph has three x-intercepts x_1, x_2 and x_3, then the expression for $f(x)$ must factor as

$f(x) = a(x - x_1)(x - x_2)(x - x_3)$. Multiplying these factors together gives us

$$f(x) = a[x^3 - (x_1 + x_2 + x_3)x^2 + (x_1x_2 + x_1x_3 + x_2x_3)x - x_1x_2x_3]$$

Equating the coefficients of the x^2-terms for the two forms of f gives us $b = -a(x_1 + x_2 + x_3)$. Hence, the x-coordinate of

the point of inflection is $-\dfrac{b}{3a} = -\dfrac{-a(x_1 + x_2 + x_3)}{3a} = \dfrac{x_1 + x_2 + x_3}{3}$.

62. $P(x) = x^4 + cx^3 + x^2 \;\Rightarrow\; P'(x) = 4x^3 + 3cx^2 + 2x \;\Rightarrow\; P''(x) = 12x^2 + 6cx + 2$. The graph of $P''(x)$ is a

parabola. If $P''(x)$ has two roots, then it changes sign twice and so has two inflection points. This happens when the

discriminant of $P''(x)$ is positive, that is, $(6c)^2 - 4 \cdot 12 \cdot 2 > 0 \;\Leftrightarrow\; 36c^2 - 96 > 0 \;\Leftrightarrow\; |c| > \frac{2\sqrt{6}}{3} \approx 1.63$. If

$36c^2 - 96 = 0 \;\Leftrightarrow\; c = \pm\frac{2\sqrt{6}}{3}$, $P''(x)$ is 0 at one point, but there is still no inflection point since $P''(x)$ never changes

sign, and if $36c^2 - 96 < 0 \;\Leftrightarrow\; |c| < \frac{2\sqrt{6}}{3}$, then $P''(x)$ never changes sign, and so there is no inflection point.

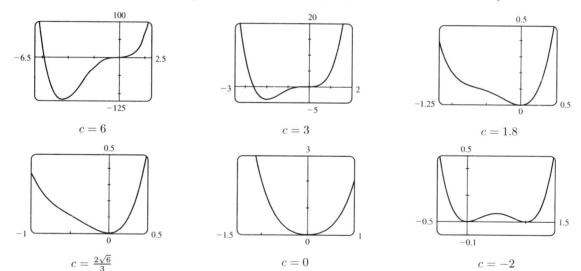

For large positive c, the graph of f has two inflection points and a large dip to the left of the y-axis. As c decreases, the graph

of f becomes flatter for $x < 0$, and eventually the dip rises above the x-axis, and then disappears entirely, along with the

inflection points. As c continues to decrease, the dip and the inflection points reappear, to the right of the origin.

63. By hypothesis $g = f'$ is differentiable on an open interval containing c. Since $(c, f(c))$ is a point of inflection, the concavity changes at $x = c$, so $f''(x)$ changes signs at $x = c$. Hence, by the First Derivative Test, f' has a local extremum at $x = c$. Thus, by Fermat's Theorem $f''(c) = 0$.

64. $f(x) = x^4 \Rightarrow f'(x) = 4x^3 \Rightarrow f''(x) = 12x^2 \Rightarrow f''(0) = 0$. For $x < 0$, $f''(x) > 0$, so f is CU on $(-\infty, 0)$; for $x > 0$, $f''(x) > 0$, so f is also CU on $(0, \infty)$. Since f does not change concavity at 0, $(0, 0)$ is not an inflection point.

65. Using the fact that $|x| = \sqrt{x^2}$, we have that $g(x) = x|x| = x\sqrt{x^2} \Rightarrow g'(x) = \sqrt{x^2} + \sqrt{x^2} = 2\sqrt{x^2} = 2|x| \Rightarrow$

$g''(x) = 2x(x^2)^{-1/2} = \dfrac{2x}{|x|} < 0$ for $x < 0$ and $g''(x) > 0$ for $x > 0$, so $(0, 0)$ is an inflection point. But $g''(0)$ does not exist.

66. There must exist some interval containing c on which f''' is positive, since $f'''(c)$ is positive and f''' is continuous. On this interval, f'' is increasing (since f''' is positive), so $f'' = (f')'$ changes from negative to positive at c. So by the First Derivative Test, f' has a local minimum at $x = c$ and thus cannot change sign there, so f has no maximum or minimum at c. But since f'' changes from negative to positive at c, f has a point of inflection at c (it changes from concave down to concave up).

67. (a) $f(x) = x^4 \sin \dfrac{1}{x} \Rightarrow f'(x) = x^4 \cos \dfrac{1}{x} \left(-\dfrac{1}{x^2}\right) + \sin \dfrac{1}{x}(4x^3) = 4x^3 \sin \dfrac{1}{x} - x^2 \cos \dfrac{1}{x}$.

$g(x) = x^4 \left(2 + \sin \dfrac{1}{x}\right) = 2x^4 + f(x) \Rightarrow g'(x) = 8x^3 + f'(x)$.

$h(x) = x^4 \left(-2 + \sin \dfrac{1}{x}\right) = -2x^4 + f(x) \Rightarrow h'(x) = -8x^3 + f'(x)$.

It is given that $f(0) = 0$, so $f'(0) = \lim\limits_{x \to 0} \dfrac{f(x) - f(0)}{x - 0} = \lim\limits_{x \to 0} \dfrac{x^4 \sin \dfrac{1}{x} - 0}{x} = \lim\limits_{x \to 0} x^3 \sin \dfrac{1}{x}$. Since

$-|x^3| \le x^3 \sin \dfrac{1}{x} \le |x^3|$ and $\lim\limits_{x \to 0} |x^3| = 0$, we see that $f'(0) = 0$ by the Squeeze Theorem. Also,

$g'(0) = 8(0)^3 + f'(0) = 0$ and $h'(0) = -8(0)^3 + f'(0) = 0$, so 0 is a critical number of f, g, and h.

For $x_{2n} = \dfrac{1}{2n\pi}$ [n a nonzero integer], $\sin \dfrac{1}{x_{2n}} = \sin 2n\pi = 0$ and $\cos \dfrac{1}{x_{2n}} = \cos 2n\pi = 1$, so $f'(x_{2n}) = -x_{2n}^2 < 0$.

For $x_{2n+1} = \dfrac{1}{(2n+1)\pi}$, $\sin \dfrac{1}{x_{2n+1}} = \sin(2n+1)\pi = 0$ and $\cos \dfrac{1}{x_{2n+1}} = \cos(2n+1)\pi = -1$, so

$f'(x_{2n+1}) = x_{2n+1}^2 > 0$. Thus, f' changes sign infinitely often on both sides of 0.

Next, $g'(x_{2n}) = 8x_{2n}^3 + f'(x_{2n}) = 8x_{2n}^3 - x_{2n}^2 = x_{2n}^2(8x_{2n} - 1) < 0$ for $x_{2n} < \frac{1}{8}$, but

$g'(x_{2n+1}) = 8x_{2n+1}^3 + x_{2n+1}^2 = x_{2n+1}^2(8x_{2n+1} + 1) > 0$ for $x_{2n+1} > -\frac{1}{8}$, so g' changes sign infinitely often on both sides of 0.

Last, $h'(x_{2n}) = -8x_{2n}^3 + f'(x_{2n}) = -8x_{2n}^3 - x_{2n}^2 = -x_{2n}^2(8x_{2n} + 1) < 0$ for $x_{2n} > -\frac{1}{8}$ and

$h'(x_{2n+1}) = -8x_{2n+1}^3 + x_{2n+1}^2 = x_{2n+1}^2(-8x_{2n+1} + 1) > 0$ for $x_{2n+1} < \frac{1}{8}$, so h' changes sign infinitely often on both sides of 0.

(b) $f(0) = 0$ and since $\sin \frac{1}{x}$ and hence $x^4 \sin \frac{1}{x}$ is both positive and negative inifinitely often on both sides of 0, and arbitrarily close to 0, f has neither a local maximum nor a local minimum at 0.

Since $2 + \sin \frac{1}{x} \geq 1$, $g(x) = x^4 \left(2 + \sin \frac{1}{x} \right) > 0$ for $x \neq 0$, so $g(0) = 0$ is a local minimum.

Since $-2 + \sin \frac{1}{x} \leq -1$, $h(x) = x^4 \left(-2 + \sin \frac{1}{x} \right) < 0$ for $x \neq 0$, so $h(0) = 0$ is a local maximum.

4.4 Limits at Infinity; Horizontal Asymptotes

1. (a) As x becomes large, the values of $f(x)$ approach 5.

(b) As x becomes large negative, the values of $f(x)$ approach 3.

2. (a) The graph of a function can intersect a vertical asymptote in the sense that it can meet but not cross it.

The graph of a function can intersect a horizontal asymptote. It can even intersect its horizontal asymptote an infinite number of times.

(b) The graph of a function can have 0, 1, or 2 horizontal asymptotes. Representative examples are shown.

No horizontal asymptote   One horizontal asymptote   Two horizontal asymptotes

3. (a) $\lim\limits_{x \to 2} f(x) = \infty$

(b) $\lim\limits_{x \to -1^-} f(x) = \infty$

(c) $\lim\limits_{x \to -1^+} f(x) = -\infty$

(d) $\lim\limits_{x \to \infty} f(x) = 1$

(e) $\lim\limits_{x \to -\infty} f(x) = 2$

(f) Vertical: $x = -1$, $x = 2$; Horizontal: $y = 1$, $y = 2$

4. (a) $\lim\limits_{x \to \infty} g(x) = 2$

(b) $\lim\limits_{x \to -\infty} g(x) = -2$

(c) $\lim\limits_{x \to 3} g(x) = \infty$

(d) $\lim\limits_{x \to 0} g(x) = -\infty$

(e) $\lim\limits_{x \to -2^+} g(x) = -\infty$

(f) Vertical: $x = -2$, $x = 0$, $x = 3$; Horizontal: $y = -2$, $y = 2$

5. If $f(x) = x^2/2^x$, then a calculator gives $f(0) = 0$, $f(1) = 0.5$, $f(2) = 1$, $f(3) = 1.125$, $f(4) = 1$, $f(5) = 0.78125$, $f(6) = 0.5625$, $f(7) = 0.3828125$, $f(8) = 0.25$, $f(9) = 0.158203125$, $f(10) = 0.09765625$, $f(20) \approx 0.00038147$, $f(50) \approx 2.2204 \times 10^{-12}$, $f(100) \approx 7.8886 \times 10^{-27}$.

It appears that $\lim\limits_{x \to \infty} \left(x^2/2^x \right) = 0$.

6. (a) From a graph of $f(x) = (1 - 2/x)^x$ in a window of $[0, 10{,}000]$ by $[0, 0.2]$, we estimate that $\lim\limits_{x \to \infty} f(x) = 0.14$

   (to two decimal places.)

(b)

x	$f(x)$
10,000	0.135308
100,000	0.135333
1,000,000	0.135335

From the table, we estimate that $\lim\limits_{x \to \infty} f(x) = 0.1353$ (to four decimal places.)

7. $\lim\limits_{x \to \infty} \dfrac{3x^2 - x + 4}{2x^2 + 5x - 8} = \lim\limits_{x \to \infty} \dfrac{(3x^2 - x + 4)/x^2}{(2x^2 + 5x - 8)/x^2}$ [divide both the numerator and denominator by x^2 (the highest power of x that appears in the denominator)]

$= \dfrac{\lim\limits_{x \to \infty} (3 - 1/x + 4/x^2)}{\lim\limits_{x \to \infty} (2 + 5/x - 8/x^2)}$ [Limit Law 5]

$= \dfrac{\lim\limits_{x \to \infty} 3 - \lim\limits_{x \to \infty} (1/x) + \lim\limits_{x \to \infty} (4/x^2)}{\lim\limits_{x \to \infty} 2 + \lim\limits_{x \to \infty} (5/x) - \lim\limits_{x \to \infty} (8/x^2)}$ [Limit Laws 1 and 2]

$= \dfrac{3 - \lim\limits_{x \to \infty} (1/x) + 4 \lim\limits_{x \to \infty} (1/x^2)}{2 + 5 \lim\limits_{x \to \infty} (1/x) - 8 \lim\limits_{x \to \infty} (1/x^2)}$ [Limit Laws 7 and 3]

$= \dfrac{3 - 0 + 4(0)}{2 + 5(0) - 8(0)}$ [Theorem 4]

$= \dfrac{3}{2}$

8. $\lim\limits_{x \to \infty} \sqrt{\dfrac{12x^3 - 5x + 2}{1 + 4x^2 + 3x^3}} = \sqrt{\lim\limits_{x \to \infty} \dfrac{12x^3 - 5x + 2}{1 + 4x^2 + 3x^3}}$ [Limit Law 11]

$= \sqrt{\lim\limits_{x \to \infty} \dfrac{12 - 5/x^2 + 2/x^3}{1/x^3 + 4/x + 3}}$ [divide by x^3]

$= \sqrt{\dfrac{\lim\limits_{x \to \infty} (12 - 5/x^2 + 2/x^3)}{\lim\limits_{x \to \infty} (1/x^3 + 4/x + 3)}}$ [Limit Law 5]

$= \sqrt{\dfrac{\lim\limits_{x \to \infty} 12 - \lim\limits_{x \to \infty} (5/x^2) + \lim\limits_{x \to \infty} (2/x^3)}{\lim\limits_{x \to \infty} (1/x^3) + \lim\limits_{x \to \infty} (4/x) + \lim\limits_{x \to \infty} 3}}$ [Limit Laws 1 and 2]

$= \sqrt{\dfrac{12 - 5 \lim\limits_{x \to \infty} (1/x^2) + 2 \lim\limits_{x \to \infty} (1/x^3)}{\lim\limits_{x \to \infty} (1/x^3) + 4 \lim\limits_{x \to \infty} (1/x) + 3}}$ [Limit Laws 7 and 3]

$= \sqrt{\dfrac{12 - 5(0) + 2(0)}{0 + 4(0) + 3}}$ [Theorem 4]

$= \sqrt{\dfrac{12}{3}} = \sqrt{4} = 2$

9. $\lim\limits_{x \to \infty} \dfrac{1}{2x + 3} = \lim\limits_{x \to \infty} \dfrac{1/x}{(2x + 3)/x} = \dfrac{\lim\limits_{x \to \infty} (1/x)}{\lim\limits_{x \to \infty} (2 + 3/x)} = \dfrac{\lim\limits_{x \to \infty} (1/x)}{\lim\limits_{x \to \infty} 2 + 3 \lim\limits_{x \to \infty} (1/x)} = \dfrac{0}{2 + 3(0)} = \dfrac{0}{2} = 0$

10. $\displaystyle\lim_{x\to\infty}\frac{3x+5}{x-4}=\lim_{x\to\infty}\frac{(3x+5)/x}{(x-4)/x}=\lim_{x\to\infty}\frac{3+5/x}{1-4/x}=\frac{\displaystyle\lim_{x\to\infty}3+5\lim_{x\to\infty}\frac{1}{x}}{\displaystyle\lim_{x\to\infty}1-4\lim_{x\to\infty}\frac{1}{x}}=\frac{3+5(0)}{1-4(0)}=3$

11. $\displaystyle\lim_{x\to-\infty}\frac{1-x-x^2}{2x^2-7}=\lim_{x\to-\infty}\frac{(1-x-x^2)/x^2}{(2x^2-7)/x^2}=\frac{\displaystyle\lim_{x\to-\infty}(1/x^2-1/x-1)}{\displaystyle\lim_{x\to-\infty}(2-7/x^2)}$

$$=\frac{\displaystyle\lim_{x\to-\infty}(1/x^2)-\lim_{x\to-\infty}(1/x)-\lim_{x\to-\infty}1}{\displaystyle\lim_{x\to-\infty}2-7\lim_{x\to-\infty}(1/x^2)}=\frac{0-0-1}{2-7(0)}=-\frac{1}{2}$$

12. $\displaystyle\lim_{y\to\infty}\frac{2-3y^2}{5y^2+4y}=\lim_{y\to\infty}\frac{(2-3y^2)/y^2}{(5y^2+4y)/y^2}=\frac{\displaystyle\lim_{y\to\infty}(2/y^2-3)}{\displaystyle\lim_{y\to\infty}(5+4/y)}=\frac{2\displaystyle\lim_{y\to\infty}(1/y^2)-\lim_{y\to\infty}3}{\displaystyle\lim_{y\to\infty}5+4\lim_{y\to\infty}(1/y)}=\frac{2(0)-3}{5+4(0)}=-\frac{3}{5}$

13. Divide both the numerator and denominator by x^3 (the highest power of x that occurs in the denominator).

$$\lim_{x\to\infty}\frac{x^3+5x}{2x^3-x^2+4}=\lim_{x\to\infty}\frac{\dfrac{x^3+5x}{x^3}}{\dfrac{2x^3-x^2+4}{x^3}}=\lim_{x\to\infty}\frac{1+\dfrac{5}{x^2}}{2-\dfrac{1}{x}+\dfrac{4}{x^3}}=\frac{\displaystyle\lim_{x\to\infty}\left(1+\frac{5}{x^2}\right)}{\displaystyle\lim_{x\to\infty}\left(2-\frac{1}{x}+\frac{4}{x^3}\right)}$$

$$=\frac{\displaystyle\lim_{x\to\infty}1+5\lim_{x\to\infty}\frac{1}{x^2}}{\displaystyle\lim_{x\to\infty}2-\lim_{x\to\infty}\frac{1}{x}+4\lim_{x\to\infty}\frac{1}{x^3}}=\frac{1+5(0)}{2-0+4(0)}=\frac{1}{2}$$

14. $\displaystyle\lim_{t\to-\infty}\frac{t^2+2}{t^3+t^2-1}=\lim_{t\to-\infty}\frac{\left(t^2+2\right)/t^3}{\left(t^3+t^2-1\right)/t^3}=\lim_{t\to-\infty}\frac{1/t+2/t^3}{1+1/t-1/t^3}=\frac{0+0}{1+0-0}=0$

15. First, multiply the factors in the denominator. Then divide both the numerator and denominator by u^4.

$$\lim_{u\to\infty}\frac{4u^4+5}{(u^2-2)(2u^2-1)}=\lim_{u\to\infty}\frac{4u^4+5}{2u^4-5u^2+2}=\lim_{u\to\infty}\frac{\dfrac{4u^4+5}{u^4}}{\dfrac{2u^4-5u^2+2}{u^4}}=\lim_{u\to\infty}\frac{4+\dfrac{5}{u^4}}{2-\dfrac{5}{u^2}+\dfrac{2}{u^4}}$$

$$=\frac{\displaystyle\lim_{u\to\infty}\left(4+\frac{5}{u^4}\right)}{\displaystyle\lim_{u\to\infty}\left(2-\frac{5}{u^2}+\frac{2}{u^4}\right)}=\frac{\displaystyle\lim_{u\to\infty}4+5\lim_{u\to\infty}\frac{1}{u^4}}{\displaystyle\lim_{u\to\infty}2-5\lim_{u\to\infty}\frac{1}{u^2}+2\lim_{u\to\infty}\frac{1}{u^4}}=\frac{4+5(0)}{2-5(0)+2(0)}=\frac{4}{2}=2$$

16. $\displaystyle\lim_{x\to\infty}\frac{x+2}{\sqrt{9x^2+1}}=\lim_{x\to\infty}\frac{(x+2)/x}{\sqrt{9x^2+1}/\sqrt{x^2}}=\lim_{x\to\infty}\frac{1+2/x}{\sqrt{9+1/x^2}}=\frac{1+0}{\sqrt{9+0}}=\frac{1}{3}$

17. $\displaystyle\lim_{x\to\infty}\frac{\sqrt{9x^6-x}}{x^3+1}=\lim_{x\to\infty}\frac{\sqrt{9x^6-x}/x^3}{(x^3+1)/x^3}=\frac{\displaystyle\lim_{x\to\infty}\sqrt{(9x^6-x)/x^6}}{\displaystyle\lim_{x\to\infty}(1+1/x^3)}$ $\qquad$ [since $x^3=\sqrt{x^6}$ for $x>0$]

$$=\frac{\displaystyle\lim_{x\to\infty}\sqrt{9-1/x^5}}{\displaystyle\lim_{x\to\infty}1+\lim_{x\to\infty}(1/x^3)}=\frac{\sqrt{\displaystyle\lim_{x\to\infty}9-\lim_{x\to\infty}(1/x^5)}}{1+0}$$

$$=\sqrt{9-0}=3$$

18. $\lim\limits_{x \to -\infty} \dfrac{\sqrt{9x^6 - x}}{x^3 + 1} = \lim\limits_{x \to -\infty} \dfrac{\sqrt{9x^6 - x}/x^3}{(x^3 + 1)/x^3} = \dfrac{\lim\limits_{x \to -\infty} -\sqrt{(9x^6 - x)/x^6}}{\lim\limits_{x \to -\infty} (1 + 1/x^3)}$ [since $x^3 = -\sqrt{x^6}$ for $x < 0$]

$\qquad = \dfrac{\lim\limits_{x \to -\infty} -\sqrt{9 - 1/x^5}}{\lim\limits_{x \to -\infty} 1 + \lim\limits_{x \to -\infty} (1/x^3)} = \dfrac{-\sqrt{\lim\limits_{x \to -\infty} 9 - \lim\limits_{x \to -\infty} (1/x^5)}}{1 + 0}$

$\qquad = -\sqrt{9 - 0} = -3$

19. $\lim\limits_{x \to \infty} \left(\sqrt{9x^2 + x} - 3x \right) = \lim\limits_{x \to \infty} \dfrac{\left(\sqrt{9x^2 + x} - 3x \right) \left(\sqrt{9x^2 + x} + 3x \right)}{\sqrt{9x^2 + x} + 3x} = \lim\limits_{x \to \infty} \dfrac{\left(\sqrt{9x^2 + x} \right)^2 - (3x)^2}{\sqrt{9x^2 + x} + 3x}$

$\qquad = \lim\limits_{x \to \infty} \dfrac{(9x^2 + x) - 9x^2}{\sqrt{9x^2 + x} + 3x} = \lim\limits_{x \to \infty} \dfrac{x}{\sqrt{9x^2 + x} + 3x} \cdot \dfrac{1/x}{1/x}$

$\qquad = \lim\limits_{x \to \infty} \dfrac{x/x}{\sqrt{9x^2/x^2 + x/x^2} + 3x/x} = \lim\limits_{x \to \infty} \dfrac{1}{\sqrt{9 + 1/x} + 3} = \dfrac{1}{\sqrt{9} + 3} = \dfrac{1}{3 + 3} = \dfrac{1}{6}$

20. $\lim\limits_{x \to -\infty} \left(x + \sqrt{x^2 + 2x} \right) = \lim\limits_{x \to -\infty} \left(x + \sqrt{x^2 + 2x} \right) \left[\dfrac{x - \sqrt{x^2 + 2x}}{x - \sqrt{x^2 + 2x}} \right] = \lim\limits_{x \to -\infty} \dfrac{x^2 - (x^2 + 2x)}{x - \sqrt{x^2 + 2x}}$

$\qquad = \lim\limits_{x \to -\infty} \dfrac{-2x}{x - \sqrt{x^2 + 2x}} = \lim\limits_{x \to -\infty} \dfrac{-2}{1 + \sqrt{1 + 2/x}} = \dfrac{-2}{1 + \sqrt{1 + 2(0)}} = -1$

Note: In dividing numerator and denominator by x, we used the fact that for $x < 0$, $x = -\sqrt{x^2}$.

21. $\lim\limits_{x \to \infty} \left(\sqrt{x^2 + ax} - \sqrt{x^2 + bx} \right) = \lim\limits_{x \to \infty} \dfrac{\left(\sqrt{x^2 + ax} - \sqrt{x^2 + bx} \right) \left(\sqrt{x^2 + ax} + \sqrt{x^2 + bx} \right)}{\sqrt{x^2 + ax} + \sqrt{x^2 + bx}}$

$\qquad = \lim\limits_{x \to \infty} \dfrac{(x^2 + ax) - (x^2 + bx)}{\sqrt{x^2 + ax} + \sqrt{x^2 + bx}} = \lim\limits_{x \to \infty} \dfrac{[(a - b)x]/x}{\left(\sqrt{x^2 + ax} + \sqrt{x^2 + bx} \right)/\sqrt{x^2}}$

$\qquad = \lim\limits_{x \to \infty} \dfrac{a - b}{\sqrt{1 + a/x} + \sqrt{1 + b/x}} = \dfrac{a - b}{\sqrt{1 + 0} + \sqrt{1 + 0}} = \dfrac{a - b}{2}$

22. $\lim\limits_{x \to \infty} \cos x$ does not exist because as x increases $\cos x$ does not approach any one value, but oscillates between 1 and -1.

23. $\lim\limits_{x \to \infty} \dfrac{x + x^3 + x^5}{1 - x^2 + x^4} = \lim\limits_{x \to \infty} \dfrac{(x + x^3 + x^5)/x^4}{(1 - x^2 + x^4)/x^4}$ [divide by the highest power of x in the denominator]

$\qquad = \lim\limits_{x \to \infty} \dfrac{1/x^3 + 1/x + x}{1/x^4 - 1/x^2 + 1} = \infty$

because $(1/x^3 + 1/x + x) \to \infty$ and $(1/x^4 - 1/x^2 + 1) \to 1$ as $x \to \infty$.

24. For $x > 0$, $\sqrt{x^2 + 1} > \sqrt{x^2} = x$. So as $x \to \infty$, we have $\sqrt{x^2 + 1} \to \infty$, that is, $\lim\limits_{x \to \infty} \sqrt{x^2 + 1} = \infty$.

25. $\lim\limits_{x \to -\infty} (x^4 + x^5) = \lim\limits_{x \to -\infty} x^5 \left(\frac{1}{x} + 1 \right)$ [factor out the largest power of x] $= -\infty$ because $x^5 \to -\infty$ and $1/x + 1 \to 1$ as $x \to -\infty$.

Or: $\lim\limits_{x \to -\infty} (x^4 + x^5) = \lim\limits_{x \to -\infty} x^4 (1 + x) = -\infty$.

26. $\lim\limits_{x \to \infty} \dfrac{x^3 - 2x + 3}{5 - 2x^2} = \lim\limits_{x \to \infty} \dfrac{(x^3 - 2x + 3)/x^2}{(5 - 2x^2)/x^2}$ [divide by the highest power of x in the denominator]

$\qquad = \lim\limits_{x \to \infty} \dfrac{x - 2/x + 3/x^2}{5/x^2 - 2} = -\infty$ because $x - 2/x + 3/x^2 \to \infty$ and $5/x^2 - 2 \to -2$ as $x \to \infty$.

27. $\lim\limits_{x \to \infty} \left(x - \sqrt{x} \right) = \lim\limits_{x \to \infty} \sqrt{x} \left(\sqrt{x} - 1 \right) = \infty$ since $\sqrt{x} \to \infty$ and $\sqrt{x} - 1 \to \infty$ as $x \to \infty$.

28. $\lim\limits_{x \to \infty} (x^2 - x^4) = \lim\limits_{x \to \infty} x^2(1 - x^2) = -\infty$ since $x^2 \to \infty$ and $1 - x^2 \to -\infty$.

29. If $t = \dfrac{1}{x}$, then $\lim\limits_{x \to \infty} x \sin \dfrac{1}{x} = \lim\limits_{t \to 0^+} \dfrac{1}{t} \sin t = \lim\limits_{t \to 0^+} \dfrac{\sin t}{t} = 1$.

30. If $t = \dfrac{1}{x}$, then $\lim\limits_{x \to \infty} \sqrt{x} \sin \dfrac{1}{x} = \lim\limits_{t \to 0^+} \dfrac{1}{\sqrt{t}} \sin t = \lim\limits_{t \to 0^+} \dfrac{t}{\sqrt{t}} \dfrac{\sin t}{t} = \lim\limits_{t \to 0^+} \sqrt{t} \cdot \lim\limits_{t \to 0^+} \dfrac{\sin t}{t} = 0 \cdot 1 = 0$.

31. (a)

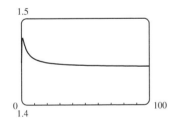

(b)

x	$f(x)$
$-10{,}000$	-0.4999625
$-100{,}000$	-0.4999962
$-1{,}000{,}000$	-0.4999996

From the table, we estimate the limit to be -0.5.

From the graph of $f(x) = \sqrt{x^2 + x + 1} + x$, we estimate the value of $\lim\limits_{x \to -\infty} f(x)$ to be -0.5.

(c) $\lim\limits_{x \to -\infty} \left(\sqrt{x^2 + x + 1} + x \right) = \lim\limits_{x \to -\infty} \left(\sqrt{x^2 + x + 1} + x \right) \left[\dfrac{\sqrt{x^2 + x + 1} - x}{\sqrt{x^2 + x + 1} - x} \right] = \lim\limits_{x \to -\infty} \dfrac{(x^2 + x + 1) - x^2}{\sqrt{x^2 + x + 1} - x}$

$= \lim\limits_{x \to -\infty} \dfrac{(x + 1)(1/x)}{\left(\sqrt{x^2 + x + 1} - x \right)(1/x)} = \lim\limits_{x \to -\infty} \dfrac{1 + (1/x)}{-\sqrt{1 + (1/x) + (1/x^2)} - 1}$

$= \dfrac{1 + 0}{-\sqrt{1 + 0 + 0} - 1} = -\dfrac{1}{2}$

Note that for $x < 0$, we have $\sqrt{x^2} = |x| = -x$, so when we divide the radical by x, with $x < 0$, we get

$\dfrac{1}{x}\sqrt{x^2 + x + 1} = -\dfrac{1}{\sqrt{x^2}}\sqrt{x^2 + x + 1} = -\sqrt{1 + (1/x) + (1/x^2)}$.

32. (a)

(b)

x	$f(x)$
$10{,}000$	1.44339
$100{,}000$	1.44338
$1{,}000{,}000$	1.44338

From the table, we estimate (to four decimal places) the limit to be 1.4434.

From the graph of $f(x) = \sqrt{3x^2 + 8x + 6} - \sqrt{3x^2 + 3x + 1}$, we estimate (to one decimal place) the value of $\lim\limits_{x \to \infty} f(x)$ to be 1.4.

(c) $\lim\limits_{x \to \infty} f(x) = \lim\limits_{x \to \infty} \dfrac{\left(\sqrt{3x^2 + 8x + 6} - \sqrt{3x^2 + 3x + 1} \right)\left(\sqrt{3x^2 + 8x + 6} + \sqrt{3x^2 + 3x + 1} \right)}{\sqrt{3x^2 + 8x + 6} + \sqrt{3x^2 + 3x + 1}}$

$= \lim\limits_{x \to \infty} \dfrac{(3x^2 + 8x + 6) - (3x^2 + 3x + 1)}{\sqrt{3x^2 + 8x + 6} + \sqrt{3x^2 + 3x + 1}} = \lim\limits_{x \to \infty} \dfrac{(5x + 5)(1/x)}{\left(\sqrt{3x^2 + 8x + 6} + \sqrt{3x^2 + 3x + 1} \right)(1/x)}$

$= \lim\limits_{x \to \infty} \dfrac{5 + 5/x}{\sqrt{3 + 8/x + 6/x^2} + \sqrt{3 + 3/x + 1/x^2}} = \dfrac{5}{\sqrt{3} + \sqrt{3}} = \dfrac{5}{2\sqrt{3}} = \dfrac{5\sqrt{3}}{6} \approx 1.443376$

33. $\lim\limits_{x\to\infty}\dfrac{2x+1}{x-2}=\lim\limits_{x\to\infty}\dfrac{\dfrac{2x+1}{x}}{\dfrac{x-2}{x}}=\lim\limits_{x\to\infty}\dfrac{2+\dfrac{1}{x}}{1-\dfrac{2}{x}}=\dfrac{\lim\limits_{x\to\infty}\left(2+\dfrac{1}{x}\right)}{\lim\limits_{x\to\infty}\left(1-\dfrac{2}{x}\right)}=\dfrac{\lim\limits_{x\to\infty}2+\lim\limits_{x\to\infty}\dfrac{1}{x}}{\lim\limits_{x\to\infty}1-\lim\limits_{x\to\infty}\dfrac{2}{x}}$

$=\dfrac{2+0}{1-0}=2,$ so $y=2$ is a horizontal asymptote.

The denominator $x-2$ is zero when $x=2$ and the numerator is not zero, so we

investigate $y=f(x)=\dfrac{2x+1}{x-2}$ as x approaches 2. $\lim\limits_{x\to2^-}f(x)=-\infty$ because as

$x\to2^-$ the numerator is positive and the denominator approaches 0 through

negative values. Similarly, $\lim\limits_{x\to2^+}f(x)=\infty$. Thus, $x=2$ is a vertical asymptote.

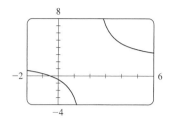

The graph confirms our work.

34. $\lim\limits_{x\to\infty}\dfrac{x^2+1}{2x^2-3x-2}=\lim\limits_{x\to\infty}\dfrac{\dfrac{x^2+1}{x^2}}{\dfrac{2x^2-3x-2}{x^2}}=\lim\limits_{x\to\infty}\dfrac{1+\dfrac{1}{x^2}}{2-\dfrac{3}{x}-\dfrac{2}{x^2}}=\dfrac{\lim\limits_{x\to\infty}\left(1+\dfrac{1}{x^2}\right)}{\lim\limits_{x\to\infty}\left(2-\dfrac{3}{x}-\dfrac{2}{x^2}\right)}$

$=\dfrac{\lim\limits_{x\to\infty}1+\lim\limits_{x\to\infty}\dfrac{1}{x^2}}{\lim\limits_{x\to\infty}2-\lim\limits_{x\to\infty}\dfrac{3}{x}-\lim\limits_{x\to\infty}\dfrac{2}{x^2}}=\dfrac{1+0}{2-0-0}=\dfrac{1}{2},$ so $y=\frac{1}{2}$ is a horizontal asymptote.

$y=f(x)=\dfrac{x^2+1}{2x^2-3x-2}=\dfrac{x^2+1}{(2x+1)(x-2)},$ so $\lim\limits_{x\to(-1/2)^-}f(x)=\infty$

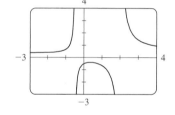

because as $x\to(-1/2)^-$ the numerator is positive while the denominator

approaches 0 through positive values. Similarly, $\lim\limits_{x\to(-1/2)^+}f(x)=-\infty,$

$\lim\limits_{x\to2^-}f(x)=-\infty,$ and $\lim\limits_{x\to2^+}f(x)=\infty.$ Thus, $x=-\frac{1}{2}$ and $x=2$ are vertical

asymptotes. The graph confirms our work.

35. $\lim\limits_{x\to\infty}\dfrac{2x^2+x-1}{x^2+x-2}=\lim\limits_{x\to\infty}\dfrac{\dfrac{2x^2+x-1}{x^2}}{\dfrac{x^2+x-2}{x^2}}=\lim\limits_{x\to\infty}\dfrac{2+\dfrac{1}{x}-\dfrac{1}{x^2}}{1+\dfrac{1}{x}-\dfrac{2}{x^2}}=\dfrac{\lim\limits_{x\to\infty}\left(2+\dfrac{1}{x}-\dfrac{1}{x^2}\right)}{\lim\limits_{x\to\infty}\left(1+\dfrac{1}{x}-\dfrac{2}{x^2}\right)}$

$=\dfrac{\lim\limits_{x\to\infty}2+\lim\limits_{x\to\infty}\dfrac{1}{x}-\lim\limits_{x\to\infty}\dfrac{1}{x^2}}{\lim\limits_{x\to\infty}1+\lim\limits_{x\to\infty}\dfrac{1}{x}-2\lim\limits_{x\to\infty}\dfrac{1}{x^2}}=\dfrac{2+0-0}{1+0-2(0)}=2,$ so $y=2$ is a horizontal asymptote.

$y=f(x)=\dfrac{2x^2+x-1}{x^2+x-2}=\dfrac{(2x-1)(x+1)}{(x+2)(x-1)},$ so $\lim\limits_{x\to-2^-}f(x)=\infty,$

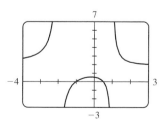

$\lim\limits_{x\to-2^+}f(x)=-\infty,$ $\lim\limits_{x\to1^-}f(x)=-\infty,$ and $\lim\limits_{x\to1^+}f(x)=\infty.$ Thus, $x=-2$

and $x=1$ are vertical asymptotes. The graph confirms our work.

36. $\lim\limits_{x \to \infty} \dfrac{1+x^4}{x^2-x^4} = \lim\limits_{x \to \infty} \dfrac{\frac{1+x^4}{x^4}}{\frac{x^2-x^4}{x^4}} = \lim\limits_{x \to \infty} \dfrac{\frac{1}{x^4}+1}{\frac{1}{x^2}-1} = \dfrac{\lim\limits_{x \to \infty}\left(\frac{1}{x^4}+1\right)}{\lim\limits_{x \to \infty}\left(\frac{1}{x^2}-1\right)} = \dfrac{\lim\limits_{x \to \infty}\frac{1}{x^4}+\lim\limits_{x \to \infty}1}{\lim\limits_{x \to \infty}\frac{1}{x^2}-\lim\limits_{x \to \infty}1}$

$\qquad = \dfrac{0+1}{0-1} = -1,$ so $y = -1$ is a horizontal asymptote.

$y = f(x) = \dfrac{1+x^4}{x^2-x^4} = \dfrac{1+x^4}{x^2(1-x^2)} = \dfrac{1+x^4}{x^2(1+x)(1-x)}.$ The denominator is

zero when $x = 0$, -1, and 1, but the numerator is nonzero, so $x = 0$, $x = -1$, and

$x = 1$ are vertical asymptotes. Notice that as $x \to 0$, the numerator and

denominator are both positive, so $\lim\limits_{x \to 0} f(x) = \infty$. The graph confirms our work.

37. $y = f(x) = \dfrac{x^3-x}{x^2-6x+5} = \dfrac{x(x^2-1)}{(x-1)(x-5)} = \dfrac{x(x+1)(x-1)}{(x-1)(x-5)} = \dfrac{x(x+1)}{x-5} = g(x)$ for $x \neq 1$.

The graph of g is the same as the graph of f with the exception of a hole in the

graph of f at $x = 1$. By long division, $g(x) = \dfrac{x^2+x}{x-5} = x+6+\dfrac{30}{x-5}.$

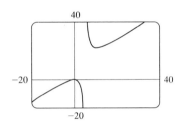

As $x \to \pm\infty$, $g(x) \to \pm\infty$, so there is no horizontal asymptote. The denominator

of g is zero when $x = 5$. $\lim\limits_{x \to 5^-} g(x) = -\infty$ and $\lim\limits_{x \to 5^+} g(x) = \infty$, so $x = 5$ is a

vertical asymptote. The graph confirms our work.

38. $\lim\limits_{x \to \infty} \dfrac{x-9}{\sqrt{4x^2+3x+2}} = \lim\limits_{x \to \infty} \dfrac{1-9/x}{\sqrt{4+(3/x)+(2/x^2)}} = \dfrac{1-0}{\sqrt{4+0+0}} = \dfrac{1}{2}.$

Using the fact that $\sqrt{x^2} = |x| = -x$ for $x < 0$, we divide the numerator by $-x$ and the denominator by $\sqrt{x^2}$.

Thus, $\lim\limits_{x \to -\infty} \dfrac{x-9}{\sqrt{4x^2+3x+2}} = \lim\limits_{x \to -\infty} \dfrac{-1+9/x}{\sqrt{4+(3/x)+(2/x^2)}} = \dfrac{-1+0}{\sqrt{4+0+0}} = -\dfrac{1}{2}.$

The horizontal asymptotes are $y = \pm\frac{1}{2}$. The

polynomial $4x^2+3x+2$ is positive for all x,

so the denominator never approaches zero,

and thus there is no vertical asymptote.

 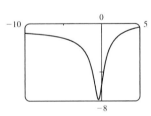

39. From the graph, it appears $y = 1$ is a horizontal asymptote.

$\lim\limits_{x \to \infty} \dfrac{3x^3+500x^2}{x^3+500x^2+100x+2000} = \lim\limits_{x \to \infty} \dfrac{\frac{3x^3+500x^2}{x^3}}{\frac{x^3+500x^2+100x+2000}{x^3}} = \lim\limits_{x \to \infty} \dfrac{3+(500/x)}{1+(500/x)+(100/x^2)+(2000/x^3)}$

$\qquad = \dfrac{3+0}{1+0+0+0} = 3,$ so $y = 3$ is a horizontal asymptote.

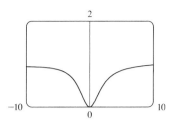

The discrepancy can be explained by the choice of the viewing window. Try $[-100{,}000,\ 100{,}000]$ by $[-1, 4]$ to get a graph that lends credibility to our calculation that $y = 3$ is a horizontal asymptote.

40. (a)

 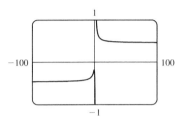

From the graph, it appears at first that there is only one horizontal asymptote, at $y \approx 0$, and a vertical asymptote at $x \approx 1.7$. However, if we graph the function with a wider viewing rectangle, we see that in fact there seem to be two horizontal asymptotes: one at $y \approx 0.5$ and one at $y \approx -0.5$. So we estimate that

$$\lim_{x \to \infty} \frac{\sqrt{2x^2 + 1}}{3x - 5} \approx 0.5 \qquad \text{and} \qquad \lim_{x \to -\infty} \frac{\sqrt{2x^2 + 1}}{3x - 5} \approx -0.5$$

(b) $f(1000) \approx 0.4722$ and $f(10{,}000) \approx 0.4715$, so we estimate that $\displaystyle\lim_{x \to \infty} \frac{\sqrt{2x^2 + 1}}{3x - 5} \approx 0.47$.

$f(-1000) \approx -0.4706$ and $f(-10{,}000) \approx -0.4713$, so we estimate that $\displaystyle\lim_{x \to -\infty} \frac{\sqrt{2x^2 + 1}}{3x - 5} \approx -0.47$.

(c) $\displaystyle\lim_{x \to \infty} \frac{\sqrt{2x^2 + 1}}{3x - 5} = \lim_{x \to \infty} \frac{\sqrt{2 + 1/x^2}}{3 - 5/x}$ [since $\sqrt{x^2} = x$ for $x > 0$] $= \dfrac{\sqrt{2}}{3} \approx 0.471404$.

For $x < 0$, we have $\sqrt{x^2} = |x| = -x$, so when we divide the numerator by x, with $x < 0$, we

get $\dfrac{1}{x}\sqrt{2x^2 + 1} = -\dfrac{1}{\sqrt{x^2}}\sqrt{2x^2 + 1} = -\sqrt{2 + 1/x^2}$. Therefore,

$$\lim_{x \to -\infty} \frac{\sqrt{2x^2 + 1}}{3x - 5} = \lim_{x \to -\infty} \frac{-\sqrt{2 + 1/x^2}}{3 - 5/x} = -\frac{\sqrt{2}}{3} \approx -0.471404.$$

41. Let's look for a rational function.

(1) $\displaystyle\lim_{x \to \pm\infty} f(x) = 0$ $\Rightarrow$ degree of numerator $<$ degree of denominator

(2) $\displaystyle\lim_{x \to 0} f(x) = -\infty$ $\Rightarrow$ there is a factor of x^2 in the denominator (not just x, since that would produce a sign change at $x = 0$), and the function is negative near $x = 0$.

(3) $\displaystyle\lim_{x \to 3^-} f(x) = \infty$ and $\displaystyle\lim_{x \to 3^+} f(x) = -\infty$ $\Rightarrow$ vertical asymptote at $x = 3$; there is a factor of $(x - 3)$ in the denominator.

(4) $f(2) = 0$ $\Rightarrow$ 2 is an x-intercept; there is at least one factor of $(x - 2)$ in the numerator.

Combining all of this information and putting in a negative sign to give us the desired left- and right-hand limits gives us

$f(x) = \dfrac{2 - x}{x^2(x - 3)}$ as one possibility.

42. Since the function has vertical asymptotes $x = 1$ and $x = 3$, the denominator of the rational function we are looking for must

have factors $(x - 1)$ and $(x - 3)$. Because the horizontal asymptote is $y = 1$, the degree of the numerator must equal the

degree of the denominator, and the ratio of the leading coefficients must be 1. One possibility is $f(x) = \dfrac{x^2}{(x - 1)(x - 3)}$.

43. $y = \dfrac{1 - x}{1 + x}$ has domain $(-\infty, -1) \cup (-1, \infty)$.

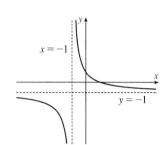

$$\lim_{x \to \pm\infty} \frac{1 - x}{1 + x} = \lim_{x \to \pm\infty} \frac{1/x - 1}{1/x + 1} = \frac{0 - 1}{0 + 1} = -1, \text{ so } y = -1 \text{ is a HA.}$$

The line $x = -1$ is a VA.

$$y' = \frac{(1 + x)(-1) - (1 - x)(1)}{(1 + x)^2} = \frac{-2}{(1 + x)^2} < 0 \text{ for } x \neq 1. \text{ Thus,}$$

$(-\infty, -1)$ and $(-1, \infty)$ are intervals of decrease.

$y'' = -2 \cdot \dfrac{-2(1 + x)}{[(1 + x)^2]^2} = \dfrac{4}{(1 + x)^3} < 0$ for $x < -1$ and $y'' > 0$ for $x > -1$, so the curve is CD on $(-\infty, -1)$ and CU on

$(-1, \infty)$. Since $x = -1$ is not in the domain, there is no IP.

44. $y = \dfrac{1 + 2x^2}{1 + x^2}$ has domain $\mathbb{R}$.

$$\lim_{x \to \pm\infty} \frac{1 + 2x^2}{1 + x^2} = \lim_{x \to \pm\infty} \frac{1/x^2 + 2}{1/x^2 + 1} = \frac{0 + 2}{0 + 1} = 2, \text{ so } y = 2 \text{ is a HA.}$$

There is no VA. $y' = \dfrac{(1 + x^2)(4x) - (1 + 2x^2)(2x)}{(1 + x^2)^2} = \dfrac{2x}{(1 + x^2)^2} > 0$

$\Leftrightarrow \quad x > 0$,

and $y' < 0 \quad \Leftrightarrow \quad x < 0$. Thus, y is increasing on $(0, \infty)$ and y is decreasing on $(-\infty, 0)$. There is a local (and absolute)

minimum at $(0, 1)$. $y'' = \dfrac{(1 + x^2)^2(2) - (2x) \cdot 2(1 + x^2)(2x)}{[(1 + x^2)^2]^2} = \dfrac{2 - 6x^2}{(1 + x^2)^3} = 0 \quad \Leftrightarrow \quad x = \pm\frac{1}{\sqrt{3}}. \quad y'' > 0 \quad \Leftrightarrow$

$-\frac{1}{\sqrt{3}} < x < \frac{1}{\sqrt{3}}$, so the curve is CU on $\left(-\frac{1}{\sqrt{3}}, \frac{1}{\sqrt{3}}\right)$ and CD on $\left(-\infty, -\frac{1}{\sqrt{3}}\right)$ and $\left(\frac{1}{\sqrt{3}}, \infty\right)$. There are IP at $\left(\pm\frac{1}{\sqrt{3}}, \frac{5}{4}\right)$.

45. $\lim_{x \to \pm\infty} \dfrac{x}{x^2 + 1} = \lim_{x \to \pm\infty} \dfrac{1/x}{1 + 1/x^2} = \dfrac{0}{1 + 0} = 0$, so $y = 0$ is a

horizontal asymptote.

$y' = \dfrac{x^2 + 1 - x(2x)}{(x^2 + 1)^2} = \dfrac{1 - x^2}{(x^2 + 1)^2} = 0$ when $x = \pm 1$ and $y' > 0 \quad \Leftrightarrow$

$x^2 < 1 \quad \Leftrightarrow \quad -1 < x < 1$, so y is increasing on $(-1, 1)$ and decreasing

on $(-\infty, -1)$ and $(1, \infty)$.

$y'' = \dfrac{(1 + x^2)^2(-2x) - (1 - x^2)2(x^2 + 1)2x}{(1 + x^2)^4} = \dfrac{2x(x^2 - 3)}{(1 + x^2)^3} > 0 \quad \Leftrightarrow \quad x > \sqrt{3} \text{ or } -\sqrt{3} < x < 0$, so y is CU on

$\left(\sqrt{3}, \infty\right)$ and $\left(-\sqrt{3}, 0\right)$ and CD on $\left(-\infty, -\sqrt{3}\right)$ and $\left(0, \sqrt{3}\right)$.

46. $y = \dfrac{x}{\sqrt{x^2 + 1}} = \dfrac{x/|x|}{\sqrt{1 + 1/x^2}}$ has domain $\mathbb{R}$. As $x \to \pm\infty$, $y \to \pm 1$, so

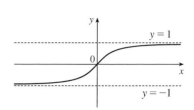

$y = \pm 1$ are HA. There is no VA. $y = x\left(x^2 + 1\right)^{-1/2}$ $\Rightarrow$

$y' = x\left(-\frac{1}{2}\right)(x^2 + 1)^{-3/2}(2x) + (x^2 + 1)^{-1/2}(1)$

$\quad = (x^2 + 1)^{-3/2}[-x^2 + (x^2 + 1)]$

$\quad = (x^2 + 1)^{-3/2} > 0 \ \text{ for all } x$

Thus, y is increasing for all x. $y'' = \left(-\frac{3}{2}\right)(x^2 + 1)^{-5/2}(2x) = \dfrac{-3x}{(x^2 + 1)^{5/2}} > 0$ for $x < 0$. So the curve is CU on $(-\infty, 0)$

and CD on $(0, \infty)$. There is an inflection point at $(0, 0)$.

47. $y = f(x) = x^4 - x^6 = x^4(1 - x^2) = x^4(1 + x)(1 - x)$. The y-intercept is

$f(0) = 0$. The x-intercepts are 0, -1, and 1 [found by solving $f(x) = 0$ for x].

Since $x^4 > 0$ for $x \neq 0$, f doesn't change sign at $x = 0$. The function does change

sign at $x = -1$ and $x = 1$. As $x \to \pm\infty$, $f(x) = x^4(1 - x^2)$ approaches $-\infty$

because $x^4 \to \infty$ and $(1 - x^2) \to -\infty$.

48. $y = f(x) = x^3(x + 2)^2(x - 1)$. The y-intercept is $f(0) = 0$. The x-intercepts are

0, -2, and 1. There are sign changes at 0 and 1 (odd exponents on x and $x - 1$).

There is no sign change at -2. Also, $f(x) \to \infty$ as $x \to \infty$ because all three

factors are large. And $f(x) \to \infty$ as $x \to -\infty$ because $x^3 \to -\infty$,

$(x + 2)^2 \to \infty$, and $(x - 1) \to -\infty$. Note that the graph of f at $x = 0$ flattens out

(looks like $y = -x^3$).

49. $y = f(x) = (3 - x)(1 + x)^2(1 - x)^4$. The y-intercept is $f(0) = 3(1)^2(1)^4 = 3$.

The x-intercepts are 3, -1, and 1. There is a sign change at 3, but not at -1 and 1.

When x is large positive, $3 - x$ is negative and the other factors are positive, so

$\lim\limits_{x \to \infty} f(x) = -\infty$. When x is large negative, $3 - x$ is positive, so

$\lim\limits_{x \to -\infty} f(x) = \infty$.

50. $y = f(x) = x^2(x^2 - 1)^2(x + 2) = x^2(x + 1)^2(x - 1)^2(x + 2)$. The y-intercept

is $f(0) = 0$. The x-intercepts are 0, -1, 1, and -2. There is a sign change at -2,

but not at 0, -1, and 1. When x is large positive, all the factors are positive, so

$\lim\limits_{x \to \infty} f(x) = \infty$. When x is large negative, only $x + 2$ is negative, so

$\lim\limits_{x \to -\infty} f(x) = -\infty$.

51. First we plot the points which are known to be on the graph: $(2, -1)$ and

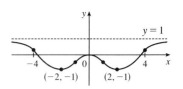

$(0, 0)$. We can also draw a short line segment of slope 0 at $x = 2$, since

we are given that $f'(2) = 0$. Now we know that $f'(x) < 0$ (that is, the

function is decreasing) on $(0, 2)$, and that $f''(x) < 0$ on $(0, 1)$ and

$f''(x) > 0$ on $(1, 2)$. So we must join the points $(0, 0)$ and $(2, -1)$ in

such a way that the curve is concave down on $(0, 1)$ and concave up on $(1, 2)$. The curve must be concave up and increasing

on $(2, 4)$ and concave down and increasing toward $y = 1$ on $(4, \infty)$. Now we just need to reflect the curve in the y-axis, since

we are given that f is an even function [the condition that $f(-x) = f(x)$ for all x].

52. The diagram shows one possible function

satisfying all of the given conditions.

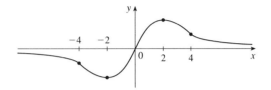

53. We are given that $f(1) = f'(1) = 0$. So we can draw a short horizontal line at the point $(1, 0)$ to represent this situation. We

are given that $x = 0$ and $x = 2$ are vertical asymptotes, with $\lim_{x \to 0} f(x) = -\infty$, $\lim_{x \to 2^+} f(x) = \infty$ and $\lim_{x \to 2^-} f(x) = -\infty$, so

we can draw the parts of the curve which approach these asymptotes.

On the interval $(-\infty, 0)$, the graph is concave down, and $f(x) \to \infty$ as

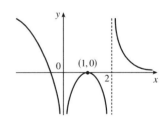

$x \to -\infty$. Between the asymptotes the graph is concave down. On the

interval $(2, \infty)$ the graph is concave up, and $f(x) \to 0$ as $x \to \infty$, so

$y = 0$ is a horizontal asymptote. The diagram shows one possible function

satisfying all of the given conditions.

54. The diagram shows one possible function

satisfying all of the given conditions.

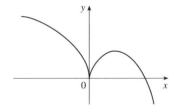

55. (a) Since $-1 \le \sin x \le 1$ for all x, $-\frac{1}{x} \le \frac{\sin x}{x} \le \frac{1}{x}$ for $x > 0$. As $x \to \infty$, $-1/x \to 0$ and $1/x \to 0$, so by the Squeeze

Theorem, $(\sin x)/x \to 0$. Thus, $\lim_{x \to \infty} \frac{\sin x}{x} = 0$.

(b) From part (a), the horizontal asymptote is $y = 0$. The function

$y = (\sin x)/x$ crosses the horizontal asymptote whenever $\sin x = 0$;

that is, at $x = \pi n$ for every integer n. Thus, the graph crosses the

asymptote *an infinite number of times.*

56. (a) In both viewing rectangles,

$$\lim_{x \to \infty} P(x) = \lim_{x \to \infty} Q(x) = \infty \text{ and}$$

$$\lim_{x \to -\infty} P(x) = \lim_{x \to -\infty} Q(x) = -\infty.$$

In the larger viewing rectangle, P and Q

become less distinguishable.

 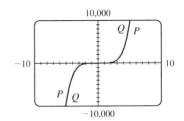

(b) $\lim_{x \to \infty} \dfrac{P(x)}{Q(x)} = \lim_{x \to \infty} \dfrac{3x^5 - 5x^3 + 2x}{3x^5} = \lim_{x \to \infty} \left(1 - \dfrac{5}{3} \cdot \dfrac{1}{x^2} + \dfrac{2}{3} \cdot \dfrac{1}{x^4} \right) = 1 - \dfrac{5}{3}(0) + \dfrac{2}{3}(0) = 1 \quad \Rightarrow$

P and Q have the same end behavior.

57. Divide the numerator and the denominator by the highest power of x in $Q(x)$.

   (a) If $\deg P < \deg Q$, then the numerator $\to 0$ but the denominator doesn't. So $\lim_{x \to \infty} [P(x)/Q(x)] = 0$.

   (b) If $\deg P > \deg Q$, then the numerator $\to \pm\infty$ but the denominator doesn't, so $\lim_{x \to \infty} [P(x)/Q(x)] = \pm\infty$

   (depending on the ratio of the leading coefficients of P and Q).

58.

   (i) $n = 0$ (ii) $n > 0$ (n odd)   (iii) $n > 0$ (n even)   (iv) $n < 0$ (n odd)   (v) $n < 0$ (n even)

From these sketches we see that

(a) $\lim_{x \to 0^+} x^n = \begin{cases} 1 & \text{if } n = 0 \\ 0 & \text{if } n > 0 \\ \infty & \text{if } n < 0 \end{cases}$

(b) $\lim_{x \to 0^-} x^n = \begin{cases} 1 & \text{if } n = 0 \\ 0 & \text{if } n > 0 \\ -\infty & \text{if } n < 0, \ n \text{ odd} \\ \infty & \text{if } n < 0, \ n \text{ even} \end{cases}$

(c) $\lim_{x \to \infty} x^n = \begin{cases} 1 & \text{if } n = 0 \\ \infty & \text{if } n > 0 \\ 0 & \text{if } n < 0 \end{cases}$

(d) $\lim_{x \to -\infty} x^n = \begin{cases} 1 & \text{if } n = 0 \\ -\infty & \text{if } n > 0, \ n \text{ odd} \\ \infty & \text{if } n > 0, \ n \text{ even} \\ 0 & \text{if } n < 0 \end{cases}$

59. $\lim_{x \to \infty} \dfrac{4x - 1}{x} = \lim_{x \to \infty} \left(4 - \dfrac{1}{x} \right) = 4$ and $\lim_{x \to \infty} \dfrac{4x^2 + 3x}{x^2} = \lim_{x \to \infty} \left(4 + \dfrac{3}{x} \right) = 4$. Therefore, by the Squeeze Theorem,

$\lim_{x \to \infty} f(x) = 4$.

60. (a) After t minutes, $25t$ liters of brine with 30 g of salt per liter has been pumped into the tank, so it contains

$(5000 + 25t)$ liters of water and $25t \cdot 30 = 750t$ grams of salt. Therefore, the salt concentration at time t will be

$$C(t) = \frac{750t}{5000 + 25t} = \frac{30t}{200 + t} \ \frac{\text{g}}{\text{L}}.$$

(b) $\lim_{t \to \infty} C(t) = \lim_{t \to \infty} \dfrac{30t}{200 + t} = \lim_{t \to \infty} \dfrac{30t/t}{200/t + t/t} = \dfrac{30}{0 + 1} = 30$. So the salt concentration approaches that of the brine

being pumped into the tank.

61. Let $g(x) = \dfrac{3x^2 + 1}{2x^2 + x + 1}$ and $f(x) = |g(x) - 1.5|$. Note that

$\lim\limits_{x \to \infty} g(x) = \frac{3}{2}$ and $\lim\limits_{x \to \infty} f(x) = 0$. We are interested in finding the

x-value at which $f(x) < 0.05$. From the graph, we find that $x \approx 14.804$,

so we choose $N = 15$ (or any larger number).

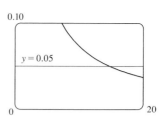

62. For $\varepsilon = 0.5$, we must find N such that whenever $x \geq N$, we have $\left|\dfrac{\sqrt{4x^2+1}}{x+1} - 2\right| < 0.5 \quad\Leftrightarrow\quad 1.5 < \dfrac{\sqrt{4x^2+1}}{x+1} < 2.5$.

We graph the three parts of this inequality on the same screen, and find that it holds whenever $x > 2.82$. So we choose $N = 3$

(or any larger number). For $\varepsilon = 0.1$, we must have $1.9 < \dfrac{\sqrt{4x^2+1}}{x+1} < 2.1$, and the graphs show that this holds whenever

$x > 18.9$. So we choose $N = 19$ (or any larger number).

63. For $\varepsilon = 0.5$, we need to find N such that $\left|\dfrac{\sqrt{4x^2+1}}{x+1} - (-2)\right| < 0.5 \quad\Leftrightarrow$

$-2.5 < \dfrac{\sqrt{4x^2+1}}{x+1} < -1.5$ whenever $x \leq N$. We graph the three parts of this

inequality on the same screen, and see that the inequality holds for $x \leq -6$. So we

choose $N = -6$ (or any smaller number).

For $\varepsilon = 0.1$, we need $-2.1 < \dfrac{\sqrt{4x^2+1}}{x+1} < -1.9$ whenever $x \leq N$. From the

graph, it seems that this inequality holds for $x \leq -22$. So we choose $N = -22$

(or any smaller number).

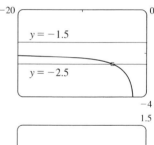

64. We need N such that $\dfrac{2x+1}{\sqrt{x+1}} > 100$ whenever $x \geq N$. From the graph, we

see that this inequality holds for $x \geq 2500$. So we choose $N = 2500$ (or any

larger number).

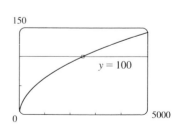

65. (a) $1/x^2 < 0.0001 \quad\Leftrightarrow\quad x^2 > 1/0.0001 = 10\,000 \quad\Leftrightarrow\quad x > 100 \quad (x > 0)$

(b) If $\varepsilon > 0$ is given, then $1/x^2 < \varepsilon \quad\Leftrightarrow\quad x^2 > 1/\varepsilon \quad\Leftrightarrow\quad x > 1/\sqrt{\varepsilon}$. Let $N = 1/\sqrt{\varepsilon}$.

Then $x > N \quad\Rightarrow\quad x > \dfrac{1}{\sqrt{\varepsilon}} \quad\Rightarrow\quad \left|\dfrac{1}{x^2} - 0\right| = \dfrac{1}{x^2} < \varepsilon$, so $\lim\limits_{x \to \infty} \dfrac{1}{x^2} = 0$.

66. (a) $1/\sqrt{x} < 0.0001$ $\Leftrightarrow$ $\sqrt{x} > 1/0.0001 = 10^4$ $\Leftrightarrow$ $x > 10^8$

(b) If $\varepsilon > 0$ is given, then $1/\sqrt{x} < \varepsilon$ $\Leftrightarrow$ $\sqrt{x} > 1/\varepsilon$ $\Leftrightarrow$ $x > 1/\varepsilon^2$. Let $N = 1/\varepsilon^2$.

Then $x > N$ $\Rightarrow$ $x > \dfrac{1}{\varepsilon^2}$ $\Rightarrow$ $\left|\dfrac{1}{\sqrt{x}} - 0\right| = \dfrac{1}{\sqrt{x}} < \varepsilon$, so $\displaystyle\lim_{x \to \infty} \dfrac{1}{\sqrt{x}} = 0$.

67. For $x < 0$, $|1/x - 0| = -1/x$. If $\varepsilon > 0$ is given, then $-1/x < \varepsilon$ $\Leftrightarrow$ $x < -1/\varepsilon$.

Take $N = -1/\varepsilon$. Then $x < N$ $\Rightarrow$ $x < -1/\varepsilon$ $\Rightarrow$ $|(1/x) - 0| = -1/x < \varepsilon$, so $\displaystyle\lim_{x \to -\infty} (1/x) = 0$.

68. Given $M > 0$, we need $N > 0$ such that $x > N$ $\Rightarrow$ $x^3 > M$. Now $x^3 > M$ $\Leftrightarrow$ $x > \sqrt[3]{M}$, so take $N = \sqrt[3]{M}$. Then

$x > N = \sqrt[3]{M}$ $\Rightarrow$ $x^3 > M$, so $\displaystyle\lim_{x \to \infty} x^3 = \infty$.

69. Suppose that $\displaystyle\lim_{x \to \infty} f(x) = L$. Then for every $\varepsilon > 0$ there is a corresponding positive number N such that $|f(x) - L| < \varepsilon$

whenever $x > N$. If $t = 1/x$, then $x > N$ $\Leftrightarrow$ $0 < 1/x < 1/N$ $\Leftrightarrow$ $0 < t < 1/N$. Thus, for every $\varepsilon > 0$ there is a

corresponding $\delta > 0$ (namely $1/N$) such that $|f(1/t) - L| < \varepsilon$ whenever $0 < t < \delta$. This proves that

$\displaystyle\lim_{t \to 0^+} f(1/t) = L = \lim_{x \to \infty} f(x)$.

Now suppose that $\displaystyle\lim_{x \to -\infty} f(x) = L$. Then for every $\varepsilon > 0$ there is a corresponding negative number N such that

$|f(x) - L| < \varepsilon$ whenever $x < N$. If $t = 1/x$, then $x < N$ $\Leftrightarrow$ $1/N < 1/x < 0$ $\Leftrightarrow$ $1/N < t < 0$. Thus, for every

$\varepsilon > 0$ there is a corresponding $\delta > 0$ (namely $-1/N$) such that $|f(1/t) - L| < \varepsilon$ whenever $-\delta < t < 0$. This proves that

$\displaystyle\lim_{t \to 0^-} f(1/t) = L = \lim_{x \to -\infty} f(x)$.

70. Definition   Let f be a function defined on some interval $(-\infty, a)$. Then $\displaystyle\lim_{x \to -\infty} f(x) = -\infty$ means that for every negative

number M there is a corresponding negative number N such that $f(x) < M$ whenever $x < N$. Now we use the definition to

prove that $\displaystyle\lim_{x \to -\infty} (1 + x^3) = -\infty$. Given a negative number M, we need a negative number N such that $x < N$ $\Rightarrow$

$1 + x^3 < M$. Now $1 + x^3 < M$ $\Leftrightarrow$ $x^3 < M - 1$ $\Leftrightarrow$ $x < \sqrt[3]{M - 1}$. Thus, we take $N = \sqrt[3]{M - 1}$ and find that

$x < N$ $\Rightarrow$ $1 + x^3 < M$. This proves that $\displaystyle\lim_{x \to -\infty} (1 + x^3) = -\infty$.

4.5   Summary of Curve Sketching

1. $y = f(x) = x^3 + x = x(x^2 + 1)$ **A.** f is a polynomial, so $D = \mathbb{R}$.

B. x-intercept $= 0$, y-intercept $= f(0) = 0$ **C.** $f(-x) = -f(x)$, so f is

odd; the curve is symmetric about the origin.   **D.** f is a polynomial, so there is

no asymptote.   **E.** $f'(x) = 3x^2 + 1 > 0$, so f is increasing on $(-\infty, \infty)$.

F. There is no critical number and hence, no local maximum or minimum value.

G. $f''(x) = 6x > 0$ on $(0, \infty)$ and $f''(x) < 0$ on $(-\infty, 0)$, so f is CU on

$(0, \infty)$ and CD on $(-\infty, 0)$. Since the concavity changes at $x = 0$, there is an

inflection point at $(0, 0)$.

H.

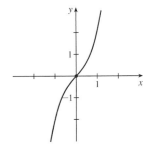

2. $y = f(x) = x^3 + 6x^2 + 9x = x(x+3)^2$ **A.** $D = \mathbb{R}$ **B.** x-intercepts are

-3 and 0, y-intercept $= 0$ **C.** No symmetry   **D.** No asymptote

E. $f'(x) = 3x^2 + 12x + 9 = 3(x+1)(x+3) < 0 \quad \Leftrightarrow \quad -3 < x < -1$,

so f is decreasing on $(-3, -1)$ and increasing on $(-\infty, -3)$ and $(-1, \infty)$.

F. Local maximum value $f(-3) = 0$, local minimum value $f(-1) = -4$

G. $f''(x) = 6x + 12 = 6(x+2) > 0 \quad \Leftrightarrow \quad x > -2$, so f is CU on $(-2, \infty)$

and CD on $(-\infty, -2)$. IP at $(-2, -2)$

H.

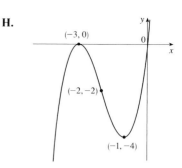

3. $y = f(x) = 2 - 15x + 9x^2 - x^3 = -(x-2)(x^2 - 7x + 1)$ **A.** $D = \mathbb{R}$ **B.** y-intercept: $f(0) = 2$; x-intercepts:

$f(x) = 0 \quad \Rightarrow \quad x = 2$ or (by the quadratic formula) $x = \frac{7 \pm \sqrt{45}}{2} \approx 0.15, 6.85$ **C.** No symmetry   **D.** No asymptote

E. $f'(x) = -15 + 18x - 3x^2 = -3(x^2 - 6x + 5)$

$\qquad = -3(x-1)(x-5) > 0 \quad \Leftrightarrow \quad 1 < x < 5$

so f is increasing on $(1, 5)$ and decreasing on $(-\infty, 1)$ and $(5, \infty)$.

F. Local maximum value $f(5) = 27$, local minimum value $f(1) = -5$

G. $f''(x) = 18 - 6x = -6(x-3) > 0 \quad \Leftrightarrow \quad x < 3$, so f is CU on $(-\infty, 3)$

and CD on $(3, \infty)$. IP at $(3, 11)$

H.

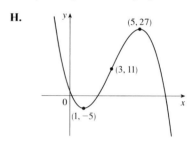

4. $y = f(x) = 8x^2 - x^4 = x^2(8 - x^2)$ **A.** $D = \mathbb{R}$ **B.** y-intercept: $f(0) = 0$; x-intercepts: $f(x) = 0 \quad \Rightarrow \quad x = 0$,

$\pm 2\sqrt{2}$ ($[\approx \pm 2.83]$ **C.** $f(-x) = f(x)$, so f is even and symmetric about the y-axis.   **D.** No asymptote

E. $f'(x) = 16x - 4x^3 = 4x(4 - x^2) = 4x(2+x)(2-x) > 0 \quad \Leftrightarrow \quad x < -2$ **H.**

or $0 < x < 2$, so f is increasing on $(-\infty, -2)$ and $(0, 2)$ and decreasing on

$(-2, 0)$ and $(2, \infty)$.   **F.** Local maximum value $f(\pm 2) = 16$, local minimum

value $f(0) = 0$ **G.** $f''(x) = 16 - 12x^2 = 4(4 - 3x^2) = 0 \quad \Leftrightarrow \quad x = \pm \frac{2}{\sqrt{3}}$.

$f''(x) > 0 \quad \Leftrightarrow \quad -\frac{2}{\sqrt{3}} < x < \frac{2}{\sqrt{3}}$, so f is CU on $\left(-\frac{2}{\sqrt{3}}, \frac{2}{\sqrt{3}}\right)$ and CD on

$\left(-\infty, -\frac{2}{\sqrt{3}}\right)$ and $\left(\frac{2}{\sqrt{3}}, \infty\right)$. IP at $\left(\pm \frac{2}{\sqrt{3}}, \frac{80}{9}\right)$

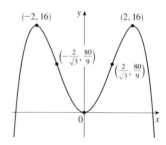

5. $y = f(x) = x^4 + 4x^3 = x^3(x+4)$ **A.** $D = \mathbb{R}$ **B.** y-intercept: $f(0) = 0$;   **H.**

x-intercepts: $f(x) = 0 \quad \Leftrightarrow \quad x = -4, 0$ **C.** No symmetry

D. No asymptote   **E.** $f'(x) = 4x^3 + 12x^2 = 4x^2(x+3) > 0 \quad \Leftrightarrow$

$x > -3$, so f is increasing on $(-3, \infty)$ and decreasing on $(-\infty, -3)$.

F. Local minimum value $f(-3) = -27$, no local maximum

G. $f''(x) = 12x^2 + 24x = 12x(x+2) < 0 \quad \Leftrightarrow \quad -2 < x < 0$, so f is CD

on $(-2, 0)$ and CU on $(-\infty, -2)$ and $(0, \infty)$. IP at $(0, 0)$ and $(-2, -16)$

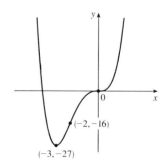

6. $y = f(x) = x(x + 2)^3$ **A.** $D = \mathbb{R}$ **B.** y-intercept: $f(0) = 0$; x-intercepts: $f(x) = 0$ $\Leftrightarrow$ $x = -2, 0$ **C.** No

symmetry   **D.** No asymptote   **E.** $f'(x) = 3x(x + 2)^2 + (x + 2)^3 = (x + 2)^2 [3x + (x + 2)] = (x + 2)^2 (4x + 2)$.

$f'(x) > 0$ $\Leftrightarrow$ $x > -\frac{1}{2}$, and $f'(x) < 0$ $\Leftrightarrow$ $x < -2$ or $-2 < x < -\frac{1}{2}$, so f is increasing on $\left(-\frac{1}{2}, \infty\right)$ and decreasing

on $(-\infty, -2)$ and $\left(-2, -\frac{1}{2}\right)$. [Hence f is decreasing on $\left(-\infty, -\frac{1}{2}\right)$ by the analogue of Exercise 4.3.53 for decreasing

functions.]   **F.** Local minimum value $f\left(-\frac{1}{2}\right) = -\frac{27}{16}$, no local maximum

G. $f''(x) = (x + 2)^2 (4) + (4x + 2)(2)(x + 2)$

$\qquad = 2(x + 2)[(x + 2)(2) + 4x + 2]$

$\qquad = 2(x + 2)(6x + 6) = 12(x + 1)(x + 2)$

$f''(x) < 0$ $\Leftrightarrow$ $-2 < x < -1$, so f is CD on $(-2, -1)$ and CU on

$(-\infty, -2)$ and $(-1, \infty)$. IP at $(-2, 0)$ and $(-1, -1)$

H.

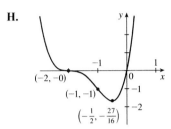

7. $y = f(x) = 2x^5 - 5x^2 + 1$ **A.** $D = \mathbb{R}$ **B.** y-intercept: $f(0) = 1$ **C.** No symmetry   **D.** No asymptote

E. $f'(x) = 10x^4 - 10x = 10x(x^3 - 1) = 10x(x - 1)(x^2 + x + 1)$, so $f'(x) < 0$ $\Leftrightarrow$ $0 < x < 1$ and $f'(x) > 0$ $\Leftrightarrow$

$x < 0$ or $x > 1$. Thus, f is increasing on $(-\infty, 0)$ and $(1, \infty)$ and decreasing on $(0, 1)$.   **F.** Local maximum value $f(0) = 1$,

local minimum value $f(1) = -2$ **G.** $f''(x) = 40x^3 - 10 = 10(4x^3 - 1)$

so $f''(x) = 0$ $\Leftrightarrow$ $x = 1/\sqrt[3]{4}$. $f''(x) > 0$ $\Leftrightarrow$ $x > 1/\sqrt[3]{4}$ and

$f''(x) < 0$ $\Leftrightarrow$ $x < 1/\sqrt[3]{4}$, so f is CD on $\left(-\infty, 1/\sqrt[3]{4}\right)$ and CU

on $\left(1/\sqrt[3]{4}, \infty\right)$. IP at $\left(\dfrac{1}{\sqrt[3]{4}}, 1 - \dfrac{9}{2\left(\sqrt[3]{4}\right)^2}\right) \approx (0.630, -0.786)$

H.

8. $y = f(x) = (4 - x^2)^5$ **A.** $D = \mathbb{R}$ **B.** y-intercept: $f(0) = 4^5 = 1024$; x-intercepts: ± 2 **C.** $f(-x) = f(x)$ $\Rightarrow$

f is even; the curve is symmetric about the y-axis.   **D.** No asymptote   **E.** $f'(x) = 5(4 - x^2)^4(-2x) = -10x(4 - x^2)^4$,

so for $x \neq \pm 2$ we have $f'(x) > 0$ $\Leftrightarrow$ $x < 0$ and $f'(x) < 0$ $\Leftrightarrow$ $x > 0$. Thus, f is increasing on $(-\infty, 0)$ and

decreasing on $(0, \infty)$.   **F.** Local maximum value $f(0) = 1024$

G. $f''(x) = -10x \cdot 4(4 - x^2)^3(-2x) + (4 - x^2)^4(-10)$

$\qquad = -10(4 - x^2)^3 [-8x^2 + 4 - x^2] = -10(4 - x^2)^3(4 - 9x^2)$

so $f''(x) = 0$ $\Leftrightarrow$ $x = \pm 2, \pm \frac{2}{3}$. $f''(x) > 0$ $\Leftrightarrow$ $-2 < x < -\frac{2}{3}$ and

$\frac{2}{3} < x < 2$ and $f''(x) < 0$ $\Leftrightarrow$ $x < -2, -\frac{2}{3} < x < \frac{2}{3}$, and $x > 2$, so f is

CU on $(-\infty, 2)$, $\left(-\frac{2}{3}, \frac{2}{3}\right)$, and $(2, \infty)$, and CD on $\left(-2, -\frac{2}{3}\right)$ and $\left(\frac{2}{3}, 2\right)$.

IP at $(\pm 2, 0)$ and $\left(\pm \frac{2}{3}, \left(\frac{32}{9}\right)^5\right) \approx (\pm 0.67, 568.25)$.

H.

9. $y = f(x) = x/(x-1)$ **A.** $D = \{x \mid x \neq 1\} = (-\infty, 1) \cup (1, \infty)$ **B.** x-intercept $= 0$, y-intercept $= f(0) = 0$

C. No symmetry   **D.** $\lim\limits_{x \to \pm\infty} \dfrac{x}{x-1} = 1$, so $y = 1$ is a HA. $\lim\limits_{x \to 1^-} \dfrac{x}{x-1} = -\infty$, $\lim\limits_{x \to 1^+} \dfrac{x}{x-1} = \infty$, so $x = 1$ is a VA.

E. $f'(x) = \dfrac{(x-1) - x}{(x-1)^2} = \dfrac{-1}{(x-1)^2} < 0$ for $x \neq 1$, so f is

decreasing on $(-\infty, 1)$ and $(1, \infty)$.   **F.** No extreme values

H.

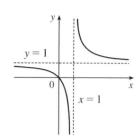

G. $f''(x) = \dfrac{2}{(x-1)^3} > 0 \iff x > 1$, so f is CU on $(1, \infty)$ and

CD on $(-\infty, 1)$. No IP

10. $y = f(x) = \dfrac{x^2 - 4}{x^2 - 2x} = \dfrac{(x+2)(x-2)}{x(x-2)} = \dfrac{x+2}{x} = 1 + \dfrac{2}{x}$ for $x \neq 2$. There is a hole in the graph at $(2, 2)$.

A. $D = \{x \mid x \neq 0, 2\} = (-\infty, 0) \cup (0, 2) \cup (2, \infty)$ **B.** y-intercept: none; x-intercept: -2 **C.** No symmetry

D. $\lim\limits_{x \to \pm\infty} \dfrac{x+2}{x} = 1$, so $y = 1$ is a HA. $\lim\limits_{x \to 0^-} \dfrac{x+2}{x} = -\infty$,

H.

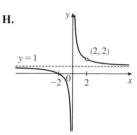

$\lim\limits_{x \to 0^+} \dfrac{x+2}{x} = \infty$, so $x = 0$ is a VA.   **E.** $f'(x) = -2/x^2 < 0$ $[x \neq 0, 2]$

so f is decreasing on $(-\infty, 0)$, $(0, 2)$, and $(2, \infty)$.   **F.** No extrema

G. $f''(x) = 4/x^3 > 0 \iff x > 0$, so f is CU on $(0, 2)$ and $(2, \infty)$ and

CD on $(-\infty, 0)$. No IP.

11. $y = f(x) = 1/(x^2 - 9)$ **A.** $D = \{x \mid x \neq \pm 3\} = (-\infty, -3) \cup (-3, 3) \cup (3, \infty)$ **B.** y-intercept $= f(0) = -\frac{1}{9}$, no

x-intercept   **C.** $f(-x) = f(x) \implies f$ is even; the curve is symmetric about the y-axis.   **D.** $\lim\limits_{x \to \pm\infty} \dfrac{1}{x^2 - 9} = 0$, so $y = 0$

is a HA. $\lim\limits_{x \to 3^-} \dfrac{1}{x^2 - 9} = -\infty$, $\lim\limits_{x \to 3^+} \dfrac{1}{x^2 - 9} = \infty$, $\lim\limits_{x \to -3^-} \dfrac{1}{x^2 - 9} = \infty$, $\lim\limits_{x \to -3^+} \dfrac{1}{x^2 - 9} = -\infty$, so $x = 3$ and $x = -3$

are VA.   **E.** $f'(x) = -\dfrac{2x}{(x^2 - 9)^2} > 0 \iff x < 0$ $(x \neq -3)$ so f is increasing on $(-\infty, -3)$ and $(-3, 0)$ and

decreasing on $(0, 3)$ and $(3, \infty)$.   **F.** Local maximum value $f(0) = -\frac{1}{9}$.

H.

G. $y'' = \dfrac{-2(x^2 - 9)^2 + (2x)2(x^2 - 9)(2x)}{(x^2 - 9)^4} = \dfrac{6(x^2 + 3)}{(x^2 - 9)^3} > 0 \iff$

$x^2 > 9 \iff x > 3$ or $x < -3$, so f is CU on $(-\infty, -3)$ and $(3, \infty)$ and

CD on $(-3, 3)$. No IP

12. $y = f(x) = x/(x^2 - 9)$ **A.** $D = \{x \mid x \neq \pm 3\} = (-\infty, -3) \cup (-3, 3) \cup (3, \infty)$ **B.** x-intercept $= 0$,

y-intercept $= f(0) = 0$.   **C.** $f(-x) = -f(x)$, so f is odd; the curve is symmetric about the origin.

D. $\lim\limits_{x \to \pm\infty} \dfrac{x}{x^2 - 9} = 0$, so $y = 0$ is a HA. $\lim\limits_{x \to 3^+} \dfrac{x}{x^2 - 9} = \infty$, $\lim\limits_{x \to 3^-} \dfrac{x}{x^2 - 9} = -\infty$, $\lim\limits_{x \to -3^+} \dfrac{x}{x^2 - 9} = \infty$,

$\lim\limits_{x \to -3^-} \dfrac{x}{x^2 - 9} = -\infty$, so $x = 3$ and $x = -3$ are VA. **E.** $f'(x) = \dfrac{(x^2 - 9) - x(2x)}{(x^2 - 9)^2} = -\dfrac{x^2 + 9}{(x^2 - 9)^2} < 0 \ [x \neq \pm 3]$

so f is decreasing on $(-\infty, -3)$, $(-3, 3)$, and $(3, \infty)$.

F. No extreme values

H.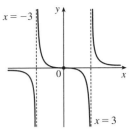

G. $f''(x) = -\dfrac{2x(x^2 - 9)^2 - (x^2 + 9) \cdot 2(x^2 - 9)(2x)}{(x^2 - 9)^4}$

$= \dfrac{2x(x^2 + 27)}{(x^2 - 9)^3} > 0$ when $-3 < x < 0$ or $x > 3$,

so f is CU on $(-3, 0)$ and $(3, \infty)$; CD on $(-\infty, -3)$ and $(0, 3)$. IP at $(0, 0)$

13. $y = f(x) = x/(x^2 + 9)$ **A.** $D = \mathbb{R}$ **B.** y-intercept: $f(0) = 0$; x-intercept: $f(x) = 0 \Leftrightarrow x = 0$

C. $f(-x) = -f(x)$, so f is odd and the curve is symmetric about the origin. **D.** $\lim\limits_{x \to \pm\infty} [x/(x^2 + 9)] = 0$, so $y = 0$ is a

HA; no VA **E.** $f'(x) = \dfrac{(x^2 + 9)(1) - x(2x)}{(x^2 + 9)^2} = \dfrac{9 - x^2}{(x^2 + 9)^2} = \dfrac{(3 + x)(3 - x)}{(x^2 + 9)^2} > 0 \Leftrightarrow -3 < x < 3$, so f is increasing

on $(-3, 3)$ and decreasing on $(-\infty, -3)$ and $(3, \infty)$. **F.** Local minimum value $f(-3) = -\frac{1}{6}$, local maximum

value $f(3) = \frac{1}{6}$

$f''(x) = \dfrac{(x^2 + 9)^2(-2x) - (9 - x^2) \cdot 2(x^2 + 9)(2x)}{[(x^2 + 9)^2]^2} = \dfrac{(2x)(x^2 + 9)[-(x^2 + 9) - 2(9 - x^2)]}{(x^2 + 9)^4} = \dfrac{2x(x^2 - 27)}{(x^2 + 9)^3}$

$= 0 \Leftrightarrow x = 0, \pm\sqrt{27} = \pm 3\sqrt{3}$

G. $f''(x) = \dfrac{(x^2 + 9)^2(-2x) - (9 - x^2) \cdot 2(x^2 + 9)(2x)}{[(x^2 + 9)^2]^2} = \dfrac{(2x)(x^2 + 9)\left[-(x^2 + 9) - 2(9 - x^2)\right]}{(x^2 + 9)^4}$

$= \dfrac{2x(x^2 - 27)}{(x^2 + 9)^3} = 0 \Leftrightarrow x = 0, \pm\sqrt{27} = \pm 3\sqrt{3}$

H.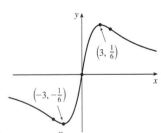

$f''(x) > 0 \Leftrightarrow -3\sqrt{3} < x < 0$ or $x > 3\sqrt{3}$, so f is CU on $\left(-3\sqrt{3}, 0\right)$

and $\left(3\sqrt{3}, \infty\right)$, and CD on $\left(-\infty, -3\sqrt{3}\right)$ and $\left(0, 3\sqrt{3}\right)$. There are three

inflection points: $(0, 0)$ and $\left(\pm 3\sqrt{3}, \pm\frac{1}{12}\sqrt{3}\right)$.

14. $y = f(x) = x^2/(x^2 + 9)$ **A.** $D = \mathbb{R}$ **B.** y-intercept: $f(0) = 0$; x-intercept: $f(x) = 0 \Leftrightarrow x = 0$

C. $f(-x) = f(x)$, so f is even and symmetric about the y-axis. **D.** $\lim\limits_{x \to \pm\infty} [x^2/(x^2 + 9)] = 1$, so $y = 1$ is a HA; no VA

E. $f'(x) = \dfrac{(x^2 + 9)(2x) - x^2(2x)}{(x^2 + 9)^2} = \dfrac{18x}{(x^2 + 9)^2} > 0 \Leftrightarrow x > 0$, so f is increasing on $(0, \infty)$

and decreasing on $(-\infty, 0)$. **F.** Local minimum value $f(0) = 0$; no local maximum

G. $f''(x) = \dfrac{(x^2 + 9)^2(18) - 18x \cdot 2(x^2 + 9) \cdot 2x}{[(x^2 + 9)^2]^2} = \dfrac{18(x^2 + 9)[(x^2 + 9) - 4x^2]}{(x^2 + 9)^4} = \dfrac{18(9 - 3x^2)}{(x^2 + 9)^3}$

$= \dfrac{-54(x + \sqrt{3})(x - \sqrt{3})}{(x^2 + 9)^3} > 0 \Leftrightarrow -\sqrt{3} < x < \sqrt{3}$

H.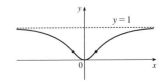

so f is CU on $\left(-\sqrt{3}, \sqrt{3}\right)$ and CD on $\left(-\infty, -\sqrt{3}\right)$ and $\left(\sqrt{3}, \infty\right)$.

There are two inflection points: $\left(\pm\sqrt{3}, \frac{1}{4}\right)$.

15. $y = f(x) = \dfrac{x-1}{x^2}$ **A.** $D = \{x \mid x \neq 0\} = (-\infty, 0) \cup (0, \infty)$ **B.** No y-intercept; x-intercept: $f(x) = 0 \Leftrightarrow x = 1$

C. No symmetry **D.** $\displaystyle\lim_{x \to \pm\infty} \dfrac{x-1}{x^2} = 0$, so $y = 0$ is a HA. $\displaystyle\lim_{x \to 0} \dfrac{x-1}{x^2} = -\infty$, so $x = 0$ is a VA.

E. $f'(x) = \dfrac{x^2 \cdot 1 - (x-1) \cdot 2x}{(x^2)^2} = \dfrac{-x^2 + 2x}{x^4} = \dfrac{-(x-2)}{x^3}$, so $f'(x) > 0 \Leftrightarrow 0 < x < 2$ and $f'(x) < 0 \Leftrightarrow$

$x < 0$ or $x > 2$. Thus, f is increasing on $(0, 2)$ and decreasing on $(-\infty, 0)$

and $(2, \infty)$. **F.** No local minimum, local maximum value $f(2) = \frac{1}{4}$.

H.

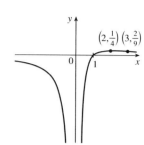

G. $f''(x) = \dfrac{x^3 \cdot (-1) - [-(x-2)] \cdot 3x^2}{(x^3)^2} = \dfrac{2x^3 - 6x^2}{x^6} = \dfrac{2(x-3)}{x^4}$.

$f''(x)$ is negative on $(-\infty, 0)$ and $(0, 3)$ and positive on $(3, \infty)$, so f is CD

on $(-\infty, 0)$ and $(0, 3)$ and CU on $(3, \infty)$. IP at $\left(3, \frac{2}{9}\right)$

16. $y = f(x) = 1 + \dfrac{1}{x} + \dfrac{1}{x^2} = \dfrac{x^2 + x + 1}{x^2}$ **A.** $D = (-\infty, 0) \cup (0, \infty)$ **B.** y-intercept: none $[x \neq 0]$;

x-intercepts: $f(x) = 0 \Leftrightarrow x^2 + x + 1 = 0$, there is no real solution, and hence, no x-intercept **C.** No symmetry

D. $\displaystyle\lim_{x \to \pm\infty} \left(1 + \dfrac{1}{x} + \dfrac{1}{x^2}\right) = 1$, so $y = 1$ is a HA. $\displaystyle\lim_{x \to 0} f(x) = \infty$, so $x = 0$ is a VA. **E.** $f'(x) = -\dfrac{1}{x^2} - \dfrac{2}{x^3} = \dfrac{-x-2}{x^3}$.

$f'(x) > 0 \Leftrightarrow -2 < x < 0$ and $f'(x) < 0 \Leftrightarrow x < -2$ or $x > 0$, so f is increasing on $(-2, 0)$ and decreasing

on $(-\infty, -2)$ and $(0, \infty)$. **F.** Local minimum value $f(-2) = \frac{3}{4}$; no local

maximum **G.** $f''(x) = \dfrac{2}{x^3} + \dfrac{6}{x^4} = \dfrac{2x+6}{x^4}$. $f''(x) < 0 \Leftrightarrow x < -3$ and

$f''(x) > 0 \Leftrightarrow -3 < x < 0$ and $x > 0$, so f is CD on $(-\infty, -3)$ and CU on

$(-3, 0)$ and $(0, \infty)$. IP at $\left(-3, \frac{7}{9}\right)$

H.

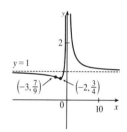

17. $y = f(x) = \dfrac{x^2}{x^2 + 3} = \dfrac{(x^2 + 3) - 3}{x^2 + 3} = 1 - \dfrac{3}{x^2 + 3}$ **A.** $D = \mathbb{R}$ **B.** y-intercept: $f(0) = 0$;

x-intercepts: $f(x) = 0 \Leftrightarrow x = 0$ **C.** $f(-x) = f(x)$, so f is even; the graph is symmetric about the y-axis.

D. $\displaystyle\lim_{x \to \pm\infty} \dfrac{x^2}{x^2 + 3} = 1$, so $y = 1$ is a HA. No VA. **E.** Using the Reciprocal Rule, $f'(x) = -3 \cdot \dfrac{-2x}{(x^2 + 3)^2} = \dfrac{6x}{(x^2 + 3)^2}$.

$f'(x) > 0 \Leftrightarrow x > 0$ and $f'(x) < 0 \Leftrightarrow x < 0$, so f is decreasing on $(-\infty, 0)$ and increasing on $(0, \infty)$.

F. Local minimum value $f(0) = 0$, no local maximum.

G. $f''(x) = \dfrac{(x^2 + 3)^2 \cdot 6 - 6x \cdot 2(x^2 + 3) \cdot 2x}{[(x^2 + 3)^2]^2}$

$= \dfrac{6(x^2 + 3)[(x^2 + 3) - 4x^2]}{(x^2 + 3)^4} = \dfrac{6(3 - 3x^2)}{(x^2 + 3)^3} = \dfrac{-18(x+1)(x-1)}{(x^2 + 3)^3}$

$f''(x)$ is negative on $(-\infty, -1)$ and $(1, \infty)$ and positive on $(-1, 1)$,

so f is CD on $(-\infty, -1)$ and $(1, \infty)$ and CU on $(-1, 1)$. IP at $\left(\pm 1, \frac{1}{4}\right)$

H.

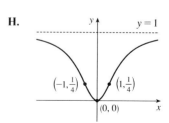

18. $y = f(x) = \dfrac{x}{x^3 - 1}$ **A.** $D = (-\infty, 1) \cup (1, \infty)$ **B.** y-intercept: $f(0) = 0$; x-intercept: $f(x) = 0$ $\Leftrightarrow$ $x = 0$

C. No symmetry **D.** $\displaystyle\lim_{x \to \pm\infty} \dfrac{x}{x^3 - 1} = 0$, so $y = 0$ is a HA. $\displaystyle\lim_{x \to 1^-} f(x) = -\infty$ and $\displaystyle\lim_{x \to 1^+} f(x) = \infty$, so $x = 1$ is a VA.

E. $f'(x) = \dfrac{(x^3 - 1)(1) - x(3x^2)}{(x^3 - 1)^2} = \dfrac{-2x^3 - 1}{(x^3 - 1)^2}$. $f'(x) = 0$ $\Rightarrow$ $x = -\sqrt[3]{1/2}$. $f'(x) > 0$ $\Leftrightarrow$ $x < -\sqrt[3]{1/2}$ and

$f'(x) < 0$ $\Leftrightarrow$ $-\sqrt[3]{1/2} < x < 1$ and $x > 1$, so f is increasing on $\left(-\infty, -\sqrt[3]{1/2}\right)$ and decreasing on $\left(-\sqrt[3]{1/2}, 1\right)$

and $(1, \infty)$. **F.** Local maximum value $f\left(-\sqrt[3]{1/2}\right) = \frac{2}{3}\sqrt[3]{1/2}$; no local minimum

G. $f''(x) = \dfrac{(x^3 - 1)^2(-6x^2) - (-2x^3 - 1)2(x^3 - 1)(3x^2)}{[(x^3 - 1)^2]^2}$

$= \dfrac{-6x^2(x^3 - 1)[(x^3 - 1) - (2x^3 + 1)]}{(x^3 - 1)^4} = \dfrac{6x^2(x^3 + 2)}{(x^3 - 1)^3}$.

$f''(x) > 0$ $\Leftrightarrow$ $x < -\sqrt[3]{2}$ and $x > 1$, $f''(x) < 0$ $\Leftrightarrow$ $-\sqrt[3]{2} < x < 0$ and

$0 < x < 1$, so f is CU on $\left(-\infty, -\sqrt[3]{2}\right)$ and $(1, \infty)$ and CD on $\left(-\sqrt[3]{2}, 1\right)$.

IP at $\left(-\sqrt[3]{2}, \frac{1}{3}\sqrt[3]{2}\right)$

H.

19. $y = f(x) = x\sqrt{5 - x}$ **A.** The domain is $\{x \mid 5 - x \geq 0\} = (-\infty, 5]$ **B.** y-intercept: $f(0) = 0$;

x-intercepts: $f(x) = 0$ $\Leftrightarrow$ $x = 0, 5$ **C.** No symmetry **D.** No asymptote

E. $f'(x) = x \cdot \frac{1}{2}(5 - x)^{-1/2}(-1) + (5 - x)^{1/2} \cdot 1 = \frac{1}{2}(5 - x)^{-1/2}[-x + 2(5 - x)] = \dfrac{10 - 3x}{2\sqrt{5 - x}} > 0$ $\Leftrightarrow$

$x < \frac{10}{3}$, so f is increasing on $\left(-\infty, \frac{10}{3}\right)$ and decreasing on $\left(\frac{10}{3}, 5\right)$.

F. Local maximum value $f\left(\frac{10}{3}\right) = \frac{10}{9}\sqrt{15} \approx 4.3$; no local minimum

G. $f''(x) = \dfrac{2(5 - x)^{1/2}(-3) - (10 - 3x) \cdot 2\left(\frac{1}{2}\right)(5 - x)^{-1/2}(-1)}{\left(2\sqrt{5 - x}\right)^2}$

$= \dfrac{(5 - x)^{-1/2}[-6(5 - x) + (10 - 3x)]}{4(5 - x)} = \dfrac{3x - 20}{4(5 - x)^{3/2}}$

H.

$f''(x) < 0$ for $x < 5$, so f is CD on $(-\infty, 5)$. No IP

20. $y = f(x) = 2\sqrt{x} - x$ **A.** $D = [0, \infty)$ **B.** y-intercept: $f(0) = 0$; x-intercepts: $f(x) = 0$ $\Rightarrow$

$2\sqrt{x} = x$ $\Rightarrow$ $4x = x^2$ $\Rightarrow$ $4x - x^2 = 0$ $\Rightarrow$ $x(4 - x) = 0$ $\Rightarrow$ $x = 0, 4$ **C.** No symmetry **D.** No asymptote

E. $f'(x) = \dfrac{1}{\sqrt{x}} - 1 = \dfrac{1}{\sqrt{x}}\left(1 - \sqrt{x}\right)$. This is positive for $x < 1$ and negative for $x > 1$, decreasing on $(1, \infty)$.

so f is increasing on $(0, 1)$ and decreasing on $(1, \infty)$.

F. Local maximum value $f(1) = 1$, no local minimum.

G. $f''(x) = (x^{-1/2} - 1)' = -\frac{1}{2}x^{-3/2} = \dfrac{-1}{2x^{3/2}} < 0$ for $x > 0$,

so f is CD on $(0, \infty)$. No IP

H.

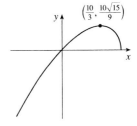

21. $y = f(x) = \sqrt{x^2 + x - 2} = \sqrt{(x+2)(x-1)}$ **A.** $D = \{x \mid (x+2)(x-1) \geq 0\} = (-\infty, -2] \cup [1, \infty)$

B. y-intercept: none; x-intercepts: -2 and 1 **C.** No symmetry **D.** No asymptote

E. $f'(x) = \frac{1}{2}(x^2 + x - 2)^{-1/2}(2x + 1) = \dfrac{2x + 1}{2\sqrt{x^2 + x - 2}}$, $f'(x) = 0$ if $x = -\frac{1}{2}$, but $-\frac{1}{2}$ is not in the domain.

$f'(x) > 0 \;\Rightarrow\; x > -\frac{1}{2}$ and $f'(x) < 0 \;\Rightarrow\; x < -\frac{1}{2}$, so (considering the domain) f is increasing on $(1, \infty)$ and

f is decreasing on $(-\infty, -2)$. **F.** No local extrema

G. $f''(x) = \dfrac{2(x^2 + x - 2)^{1/2}(2) - (2x + 1) \cdot 2 \cdot \frac{1}{2}(x^2 + x - 2)^{-1/2}(2x + 1)}{\left(2\sqrt{x^2 + x - 2}\right)^2}$

H.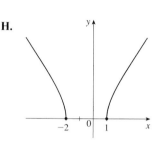

$= \dfrac{(x^2 + x - 2)^{-1/2}\left[4(x^2 + x - 2) - (4x^2 + 4x + 1)\right]}{4(x^2 + x - 2)}$

$= \dfrac{-9}{4(x^2 + x - 2)^{3/2}} < 0$

so f is CD on $(-\infty, -2)$ and $(1, \infty)$. No IP

22. $y = f(x) = \sqrt{x^2 + x} - x = \sqrt{x(x + 1)} - x$ **A.** $D = (-\infty, -1] \cup [0, \infty)$ **B.** y-intercept: $f(0) = 0$;

x-intercepts: $f(x) = 0 \;\Rightarrow\; \sqrt{x^2 + x} = x \;\Rightarrow\; x^2 + x = x^2 \;\Rightarrow\; x = 0$ **C.** No symmetry

D. $\displaystyle\lim_{x \to \infty} f(x) = \lim_{x \to \infty} \left(\sqrt{x^2 + x} - x\right) \frac{\sqrt{x^2 + x} + x}{\sqrt{x^2 + x} + x} = \lim_{x \to \infty} \frac{x^2 + x - x^2}{\sqrt{x^2 + x} + x}$

$= \displaystyle\lim_{x \to \infty} \frac{x/x}{\left(\sqrt{x^2 + x} + x\right)/x} = \lim_{x \to \infty} \frac{1}{\sqrt{1 + 1/x} + 1} = \frac{1}{2}$, so $y = \frac{1}{2}$ is a HA. No VA

E. $f'(x) = \frac{1}{2}(x^2 + x)^{-1/2}(2x + 1) - 1 = \dfrac{2x + 1}{2\sqrt{x^2 + x}} - 1 > 0 \;\Leftrightarrow\; 2x + 1 > 2\sqrt{x^2 + x} \;\Leftrightarrow$

$x + \frac{1}{2} > \sqrt{\left(x + \frac{1}{2}\right)^2 - \frac{1}{4}}$. Keep in mind that the domain excludes the interval $(-1, 0)$. When $x + \frac{1}{2}$ is positive (for $x \geq 0$),

the last inequality is *true* since the value of the radical is less than $x + \frac{1}{2}$. When $x + \frac{1}{2}$ is negative (for $x \leq -1$), the last

inequality is *false* since the value of the radical is positive. So f is increasing on $(0, \infty)$ and decreasing on $(-\infty, -1)$.

F. No local extrema

G. $f''(x) = \dfrac{2(x^2 + x)^{1/2}(2) - (2x + 1) \cdot 2 \cdot \frac{1}{2}(x^2 + x)^{-1/2}(2x + 1)}{\left(2\sqrt{x^2 + x}\right)^2}$

H.

$= \dfrac{(x^2 + x)^{-1/2}[4(x^2 + x) - (2x + 1)^2]}{4(x^2 + x)} = \dfrac{-1}{4(x^2 + x)^{3/2}}.$

$f''(x) < 0$ when it is defined, so f is CD on $(-\infty, -1)$ and $(0, \infty)$. No IP

23. $y = f(x) = x/\sqrt{x^2 + 1}$ **A.** $D = \mathbb{R}$ **B.** y-intercept: $f(0) = 0$; x-intercepts: $f(x) = 0$ $\Rightarrow$ $x = 0$

C. $f(-x) = -f(x)$, so f is odd; the graph is symmetric about the origin.

D. $\displaystyle\lim_{x \to \infty} f(x) = \lim_{x \to \infty} \frac{x}{\sqrt{x^2 + 1}} = \lim_{x \to \infty} \frac{x/x}{\sqrt{x^2 + 1}/x} = \lim_{x \to \infty} \frac{x/x}{\sqrt{x^2 + 1}/\sqrt{x^2}} = \lim_{x \to \infty} \frac{1}{\sqrt{1 + 1/x^2}} = \frac{1}{\sqrt{1 + 0}} = 1$

and

$\displaystyle\lim_{x \to -\infty} f(x) = \lim_{x \to -\infty} \frac{x}{\sqrt{x^2 + 1}} = \lim_{x \to -\infty} \frac{x/x}{\sqrt{x^2 + 1}/x} = \lim_{x \to -\infty} \frac{x/x}{\sqrt{x^2 + 1}/\left(-\sqrt{x^2}\right)} = \lim_{x \to -\infty} \frac{1}{-\sqrt{1 + 1/x^2}}$

$= \dfrac{1}{-\sqrt{1 + 0}} = -1$ so $y = \pm 1$ are HA.

No VA.

E. $f'(x) = \dfrac{\sqrt{x^2 + 1} - x \cdot \dfrac{2x}{2\sqrt{x^2 + 1}}}{[(x^2 + 1)^{1/2}]^2} = \dfrac{x^2 + 1 - x^2}{(x^2 + 1)^{3/2}} = \dfrac{1}{(x^2 + 1)^{3/2}} > 0$ for all x, so f is increasing on $\mathbb{R}$.

F. No extreme values

G. $f''(x) = -\frac{3}{2}(x^2 + 1)^{-5/2} \cdot 2x = \dfrac{-3x}{(x^2 + 1)^{5/2}}$, so $f''(x) > 0$ for $x < 0$

and $f''(x) < 0$ for $x > 0$. Thus, f is CU on $(-\infty, 0)$ and CD on $(0, \infty)$.

IP at $(0, 0)$.

H.

24. $y = f(x) = x\sqrt{2 - x^2}$ **A.** $D = \left[-\sqrt{2}, \sqrt{2}\right]$ **B.** y-intercept: $f(0) = 0$; x-intercepts: $f(x) = 0$ $\Rightarrow$

$x = 0, \pm\sqrt{2}$.   **C.** $f(-x) = -f(x)$, so f is odd; the graph is symmetric about the origin.   **D.** No asymptote

E. $f'(x) = x \cdot \dfrac{-x}{\sqrt{2 - x^2}} + \sqrt{2 - x^2} = \dfrac{-x^2 + 2 - x^2}{\sqrt{2 - x^2}} = \dfrac{2(1 + x)(1 - x)}{\sqrt{2 - x^2}}$. $f'(x)$ is negative for $-\sqrt{2} < x < -1$

and $1 < x < \sqrt{2}$, and positive for $-1 < x < 1$, so f is decreasing on $\left(-\sqrt{2}, -1\right)$ and $\left(1, \sqrt{2}\right)$ and increasing on $(-1, 1)$.

F. Local minimum value $f(-1) = -1$, local maximum value $f(1) = 1$.

G. $f''(x) = \dfrac{\sqrt{2 - x^2}(-4x) - (2 - 2x^2)\dfrac{-x}{\sqrt{2 - x^2}}}{[(2 - x^2)^{1/2}]^2}$

$= \dfrac{(2 - x^2)(-4x) + (2 - 2x^2)x}{(2 - x^2)^{3/2}} = \dfrac{2x^3 - 6x}{(2 - x^2)^{3/2}} = \dfrac{2x(x^2 - 3)}{(2 - x^2)^{3/2}}$

Since $x^2 - 3 < 0$ for x in $\left[-\sqrt{2}, \sqrt{2}\right]$, $f''(x) > 0$ for $-\sqrt{2} < x < 0$ and

$f''(x) < 0$ for $0 < x < \sqrt{2}$. Thus, f is CU on $\left(-\sqrt{2}, 0\right)$ and CD on $\left(0, \sqrt{2}\right)$.

The only IP is $(0, 0)$.

H.

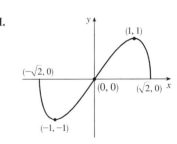

25. $y = f(x) = \sqrt{1-x^2}/x$ **A.** $D = \{x \mid |x| \le 1, x \ne 0\} = [-1,0) \cup (0,1]$ **B.** x-intercepts ± 1, no y-intercept

C. $f(-x) = -f(x)$, so the curve is symmetric about $(0,0)$.   **D.** $\lim\limits_{x \to 0^+} \dfrac{\sqrt{1-x^2}}{x} = \infty$, $\lim\limits_{x \to 0^-} \dfrac{\sqrt{1-x^2}}{x} = -\infty$,

so $x = 0$ is a VA.   **E.** $f'(x) = \dfrac{(-x^2/\sqrt{1-x^2}) - \sqrt{1-x^2}}{x^2} = -\dfrac{1}{x^2\sqrt{1-x^2}} < 0$, so f is decreasing

on $(-1,0)$ and $(0,1)$.   **F.** No extreme values

H.

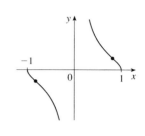

G. $f''(x) = \dfrac{2 - 3x^2}{x^3(1-x^2)^{3/2}} > 0 \iff -1 < x < -\sqrt{\tfrac{2}{3}}$ or $0 < x < \sqrt{\tfrac{2}{3}}$, so

f is CU on $\left(-1, -\sqrt{\tfrac{2}{3}}\right)$ and $\left(0, \sqrt{\tfrac{2}{3}}\right)$ and CD on $\left(-\sqrt{\tfrac{2}{3}}, 0\right)$ and $\left(\sqrt{\tfrac{2}{3}}, 1\right)$.

IP at $\left(\pm\sqrt{\tfrac{2}{3}}, \pm\tfrac{1}{\sqrt{2}}\right)$

26. $y = f(x) = x/\sqrt{x^2-1}$ **A.** $D = (-\infty, -1) \cup (1, \infty)$ **B.** No intercepts   **C.** $f(-x) = -f(x)$, so f is odd;

the graph is symmetric about the origin.   **D.** $\lim\limits_{x \to \infty} \dfrac{x}{\sqrt{x^2-1}} = 1$ and $\lim\limits_{x \to -\infty} \dfrac{x}{\sqrt{x^2-1}} = -1$, so $y = \pm 1$ are HA.

$\lim\limits_{x \to 1^+} f(x) = +\infty$ and $\lim\limits_{x \to -1^-} f(x) = -\infty$, so $x = \pm 1$ are VA.

E. $f'(x) = \dfrac{\sqrt{x^2-1} - x \cdot \dfrac{x}{\sqrt{x^2-1}}}{[(x^2-1)^{1/2}]^2} = \dfrac{x^2 - 1 - x^2}{(x^2-1)^{3/2}} = \dfrac{-1}{(x^2-1)^{3/2}} < 0$, so f is decreasing on $(-\infty, -1)$ and $(1, \infty)$.

F. No extreme values

H.

G. $f''(x) = (-1)\left(-\tfrac{3}{2}\right)(x^2-1)^{-5/2} \cdot 2x = \dfrac{3x}{(x^2-1)^{5/2}}$.

$f''(x) < 0$ on $(-\infty, -1)$ and $f''(x) > 0$ on $(1, \infty)$, so f is CD on $(-\infty, -1)$

and CU on $(1, \infty)$. No IP

27. $y = f(x) = x - 3x^{1/3}$ **A.** $D = \mathbb{R}$ **B.** y-intercept: $f(0) = 0$; x-intercepts: $f(x) = 0 \implies x = 3x^{1/3} \implies$

$x^3 = 27x \implies x^3 - 27x = 0 \implies x(x^2 - 27) = 0 \implies x = 0, \pm 3\sqrt{3}$ **C.** $f(-x) = -f(x)$, so f is odd;

the graph is symmetric about the origin.   **D.** No asymptote   **E.** $f'(x) = 1 - x^{-2/3} = 1 - \dfrac{1}{x^{2/3}} = \dfrac{x^{2/3} - 1}{x^{2/3}}$.

$f'(x) > 0$ when $|x| > 1$ and $f'(x) < 0$ when $0 < |x| < 1$, so f is increasing on $(-\infty, -1)$ and $(1, \infty)$, and

decreasing on $(-1, 0)$ and $(0, 1)$ [hence decreasing on $(-1, 1)$ since f is

continuous on $(-1, 1)$].   **F.** Local maximum value $f(-1) = 2$, local minimum

value $f(1) = -2$ **G.** $f''(x) = \tfrac{2}{3}x^{-5/3} < 0$ when $x < 0$ and $f''(x) > 0$

when $x > 0$, so f is CD on $(-\infty, 0)$ and CU on $(0, \infty)$. IP at $(0, 0)$

H.

28. $y = f(x) = x^{5/3} - 5x^{2/3} = x^{2/3}(x-5)$ **A.** $D = \mathbb{R}$ **B.** x-intercepts 0, 5; y-intercept 0   **C.** No symmetry

D. $\lim\limits_{x \to \pm\infty} x^{2/3}(x-5) = \pm\infty$, so there is no asymptote   **E.** $f'(x) = \frac{5}{3}x^{2/3} - \frac{10}{3}x^{-1/3} = \frac{5}{3}x^{-1/3}(x-2) > 0$ $\Leftrightarrow$

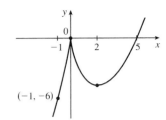

$x < 0$ or $x > 2$, so f is increasing on $(-\infty, 0)$, $(2, \infty)$ and

decreasing on $(0, 2)$.

F. Local maximum value $f(0) = 0$, local minimum value $f(2) = -3\sqrt[3]{4}$

G. $f''(x) = \frac{10}{9}x^{-1/3} + \frac{10}{9}x^{-4/3} = \frac{10}{9}x^{-4/3}(x+1) > 0$ $\Leftrightarrow$ $x > -1$, so

f is CU on $(-1, 0)$ and $(0, \infty)$, CD on $(-\infty, -1)$. IP at $(-1, -6)$

29. $y = f(x) = \sqrt[3]{x^2 - 1}$ **A.** $D = \mathbb{R}$ **B.** y-intercept: $f(0) = -1$; x-intercepts: $f(x) = 0$ $\Leftrightarrow$ $x^2 - 1 = 0$ $\Leftrightarrow$

$x = \pm 1$ **C.** $f(-x) = f(x)$, so the curve is symmetric about the y-axis.   **D.** No asymptote

E. $f'(x) = \frac{1}{3}(x^2 - 1)^{-2/3}(2x) = \dfrac{2x}{3\sqrt[3]{(x^2 - 1)^2}}$. $f'(x) > 0$ $\Leftrightarrow$ $x > 0$ and $f'(x) < 0$ $\Leftrightarrow$ $x < 0$, so f is

increasing on $(0, \infty)$ and decreasing on $(-\infty, 0)$.   **F.** Local minimum value $f(0) = -1$

G. $f''(x) = \dfrac{2}{3} \cdot \dfrac{(x^2 - 1)^{2/3}(1) - x \cdot \frac{2}{3}(x^2 - 1)^{-1/3}(2x)}{[(x^2 - 1)^{2/3}]^2}$

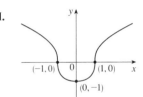

$\qquad\quad = \dfrac{2}{9} \cdot \dfrac{(x^2 - 1)^{-1/3}[3(x^2 - 1) - 4x^2]}{(x^2 - 1)^{4/3}} = -\dfrac{2(x^2 + 3)}{9(x^2 - 1)^{5/3}}$

$f''(x) > 0$ $\Leftrightarrow$ $-1 < x < 1$ and $f''(x) < 0$ $\Leftrightarrow$ $x < -1$ or $x > 1$, so

f is CU on $(-1, 1)$ and f is CD on $(-\infty, -1)$ and $(1, \infty)$. IP at $(\pm 1, 0)$

30. $y = f(x) = \sqrt[3]{x^3 + 1}$ **A.** $D = \mathbb{R}$ **B.** y-intercept: $f(0) = 1$; x-intercept: $f(x) = 0$ $\Leftrightarrow$ $x^3 + 1 = 0$ $\Rightarrow$ $x = -1$

C. No symmetry   **D.** No asymptote   **E.** $f'(x) = \frac{1}{3}(x^3 + 1)^{-2/3}(3x^2) = \dfrac{x^2}{\sqrt[3]{(x^3 + 1)^2}}$. $f'(x) > 0$ if $x < -1$,

$-1 < x < 0$, and $x > 0$, so f is increasing on $\mathbb{R}$.   **F.** No local extrema

G. $f''(x) = \dfrac{(x^3 + 1)^{2/3}(2x) - x^2 \cdot \frac{2}{3}(x^3 + 1)^{-1/3}(3x^2)}{[(x^3 + 1)^{2/3}]^2}$

$\qquad\quad = \dfrac{x(x^3 + 1)^{-1/3}[2(x^3 + 1) - 2x^3]}{(x^3 + 1)^{4/3}} = \dfrac{2x}{(x^3 + 1)^{5/3}}$

$f''(x) > 0$ $\Leftrightarrow$ $x < -1$ or $x > 0$ and $f''(x) < 0$ $\Leftrightarrow$ $-1 < x < 0$, so f is

CU on $(-\infty, -1)$ and $(0, \infty)$ and CD on $(-1, 0)$. IP at $(-1, 0)$ and $(0, 1)$

31. $y = f(x) = 3\sin x - \sin^3 x$ **A.** $D = \mathbb{R}$ **B.** y-intercept: $f(0) = 0$; x-intercepts: $f(x) = 0$ $\Rightarrow$

$\sin x\,(3 - \sin^2 x) = 0$ $\Rightarrow$ $\sin x = 0$ [since $\sin^2 x \le 1 < 3$] $\Rightarrow$ $x = n\pi$, n an integer.

C. $f(-x) = -f(x)$, so f is odd; the graph (shown for $-2\pi \le x \le 2\pi$) is symmetric about the origin and periodic with period 2π. **D.** No asymptote **E.** $f'(x) = 3\cos x - 3\sin^2 x \cos x = 3\cos x\,(1 - \sin^2 x) = 3\cos^3 x$.

$f'(x) > 0 \iff \cos x > 0 \iff x \in \left(2n\pi - \frac{\pi}{2}, 2n\pi + \frac{\pi}{2}\right)$ for each integer n, and $f'(x) < 0 \iff \cos x < 0 \iff$

$x \in \left(2n\pi + \frac{\pi}{2}, 2n\pi + \frac{3\pi}{2}\right)$ for each integer n. Thus, f is increasing on $\left(2n\pi - \frac{\pi}{2}, 2n\pi + \frac{\pi}{2}\right)$ for each integer n,

and f is decreasing on $\left(2n\pi + \frac{\pi}{2}, 2n\pi + \frac{3\pi}{2}\right)$ for each integer n.

F. f has local maximum values $f\left(2n\pi + \frac{\pi}{2}\right) = 2$ and local minimum values $f\left(2n\pi + \frac{3\pi}{2}\right) = -2$.

G. $f''(x) = -9\sin x\,\cos^2 x = -9\sin x\,(1 - \sin^2 x) = -9\sin x\,(1 - \sin x)(1 + \sin x)$.

$f''(x) < 0 \iff \sin x > 0$ and $\sin x \neq \pm 1 \iff x \in \left(2n\pi, 2n\pi + \frac{\pi}{2}\right) \cup \left(2n\pi + \frac{\pi}{2}, 2n\pi + \pi\right)$ for some integer n.

$f''(x) > 0 \iff \sin x < 0$ and $\sin x \neq \pm 1 \iff x \in \left((2n - 1)\pi, (2n - 1)\pi + \frac{\pi}{2}\right) \cup \left((2n - 1)\pi + \frac{\pi}{2}, 2n\pi\right)$

for some integer n. Thus, f is CD on the intervals $\left(2n\pi, \left(2n + \frac{1}{2}\right)\pi\right)$ and

$\left(\left(2n + \frac{1}{2}\right)\pi, (2n + 1)\,\pi\right)$ [hence CD on the intervals $(2n\pi, (2n + 1)\,\pi)$]

for each integer n, and f is CU on the intervals $\left((2n - 1)\pi, \left(2n - \frac{1}{2}\right)\pi\right)$ and

$\left(\left(2n - \frac{1}{2}\right)\pi, 2n\pi\right)$ [hence CU on the intervals $((2n - 1)\pi, 2n\pi)$]

for each integer n. f has inflection points at $(n\pi, 0)$ for each integer n.

H.

32. $y = f(x) = x + \cos x$ **A.** $D = \mathbb{R}$ **B.** y-intercept: $f(0) = 1$; the x-intercept is about -0.74 and can be found using

Newton's method **C.** No symmetry **D.** No asymptote **E.** $f'(x) = 1 - \sin x > 0$ except for $x = \frac{\pi}{2} + 2n\pi$,

so f is increasing on $\mathbb{R}$. **F.** No local extrema

G. $f''(x) = -\cos x$. $f''(x) > 0 \implies -\cos x > 0 \implies \cos x < 0 \implies$

x is in $\left(\frac{\pi}{2} + 2n\pi, \frac{3\pi}{2} + 2n\pi\right)$ and $f''(x) < 0 \implies$

x is in $\left(-\frac{\pi}{2} + 2n\pi, \frac{\pi}{2} + 2n\pi\right)$, so f is CU on $\left(\frac{\pi}{2} + 2n\pi, \frac{3\pi}{2} + 2n\pi\right)$ and CD on

$\left(-\frac{\pi}{2} + 2n\pi, \frac{\pi}{2} + 2n\pi\right)$. IP at $\left(\frac{\pi}{2} + n\pi, f\left(\frac{\pi}{2} + n\pi\right)\right) = \left(\frac{\pi}{2} + n\pi, \frac{\pi}{2} + n\pi\right)$

[on the line $y = x$]

H.

33. $y = f(x) = x\tan x$, $-\frac{\pi}{2} < x < \frac{\pi}{2}$ **A.** $D = \left(-\frac{\pi}{2}, \frac{\pi}{2}\right)$ **B.** Intercepts are 0 **C.** $f(-x) = f(x)$, so the curve is

symmetric about the y-axis. **D.** $\lim\limits_{x \to (\pi/2)^-} x\tan x = \infty$ and $\lim\limits_{x \to -(\pi/2)^+} x\tan x = \infty$, so $x = \frac{\pi}{2}$ and $x = -\frac{\pi}{2}$ are VA.

E. $f'(x) = \tan x + x\sec^2 x > 0 \iff 0 < x < \frac{\pi}{2}$, so f increases on $\left(0, \frac{\pi}{2}\right)$ **H.**

and decreases on $\left(-\frac{\pi}{2}, 0\right)$. **F.** Absolute and local minimum value $f(0) = 0$.

G. $y'' = 2\sec^2 x + 2x\tan x\,\sec^2 x > 0$ for $-\frac{\pi}{2} < x < \frac{\pi}{2}$, so f is

CU on $\left(-\frac{\pi}{2}, \frac{\pi}{2}\right)$. No IP

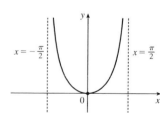

34. $y = f(x) = 2x - \tan x$, $-\frac{\pi}{2} < x < \frac{\pi}{2}$ **A.** $D = \left(-\frac{\pi}{2}, \frac{\pi}{2}\right)$ **B.** y-intercept: $f(0) = 0$; x-intercepts: $f(0) = 0 \iff$

$2x = \tan x \iff x = 0$ or $x \approx \pm 1.17$ **C.** $f(-x) = -f(x)$, so f is odd; the graph is symmetric about the origin.

D. $\displaystyle\lim_{x\to(-\pi/2)^+}(2x-\tan x)=\infty$ and $\displaystyle\lim_{x\to(\pi/2)^-}(2x-\tan x)=-\infty$, so $x=\pm\frac{\pi}{2}$ are VA. No HA.

E. $f'(x)=2-\sec^2 x<0$ $\Leftrightarrow$ $|\sec x|>\sqrt{2}$ and $f'(x)>0$ $\Leftrightarrow$ $|\sec x|<\sqrt{2}$, so f is decreasing on $\left(-\frac{\pi}{2},-\frac{\pi}{4}\right)$,

increasing on $\left(-\frac{\pi}{4},\frac{\pi}{4}\right)$, and decreasing again on $\left(\frac{\pi}{4},\frac{\pi}{2}\right)$ **F.** Local maximum value $f\left(\frac{\pi}{4}\right)=\frac{\pi}{2}-1$,

local minimum value $f\left(-\frac{\pi}{4}\right)=-\frac{\pi}{2}+1$

H.

G. $f''(x)=-2\sec x\cdot\sec x\tan x=-2\tan x\sec^2 x=-2\tan x(\tan^2 x+1)$

so $f''(x)>0$ $\Leftrightarrow$ $\tan x<0$ $\Leftrightarrow$ $-\frac{\pi}{2}<x<0$, and $f''(x)<0$ $\Leftrightarrow$

$\tan x>0$ $\Leftrightarrow$ $0<x<\frac{\pi}{2}$. Thus, f is CU on $\left(-\frac{\pi}{2},0\right)$ and CD on $\left(0,\frac{\pi}{2}\right)$.

IP at $(0,0)$

---

35. $y=f(x)=\frac{1}{2}x-\sin x,\ 0<x<3\pi$ **A.** $D=(0,3\pi)$ **B.** No y-intercept. The x-intercept, approximately 1.9, can be

found using Newton's Method. **C.** No symmetry **D.** No asymptote **E.** $f'(x)=\frac{1}{2}-\cos x>0$ $\Leftrightarrow$ $\cos x<\frac{1}{2}$ $\Leftrightarrow$

$\frac{\pi}{3}<x<\frac{5\pi}{3}$ or $\frac{7\pi}{3}<x<3\pi$, so f is increasing on $\left(\frac{\pi}{3},\frac{5\pi}{3}\right)$ and $\left(\frac{7\pi}{3},3\pi\right)$ and decreasing on $\left(0,\frac{\pi}{3}\right)$ and $\left(\frac{5\pi}{3},\frac{7\pi}{3}\right)$.

F. Local minimum value $f\left(\frac{\pi}{3}\right)=\frac{\pi}{6}-\frac{\sqrt{3}}{2}$, local maximum value

$f\left(\frac{5\pi}{3}\right)=\frac{5\pi}{6}+\frac{\sqrt{3}}{2}$, local minimum value $f\left(\frac{7\pi}{3}\right)=\frac{7\pi}{6}-\frac{\sqrt{3}}{2}$

H.

G. $f''(x)=\sin x>0$ $\Leftrightarrow$ $0<x<\pi$ or $2\pi<x<3\pi$, so f is CU on

$(0,\pi)$ and $(2\pi,3\pi)$ and CD on $(\pi,2\pi)$. IPs at $\left(\pi,\frac{\pi}{2}\right)$ and $(2\pi,\pi)$

---

36. $y=f(x)=\sec x+\tan x,\ 0<x<\pi/2$ **A.** $D=\left(0,\frac{\pi}{2}\right)$ **B.** y-intercept: none (0 not in domain); x-intercept: none,

since $\sec x$ and $\tan x$ are both positive on the domain **C.** No symmetry **D.** $\displaystyle\lim_{x\to(\pi/2)^-}f(x)=\infty$, so $x=\frac{\pi}{2}$ is a VA.

E. $f'(x)=\sec x\tan x+\sec^2 x=\sec x\,(\tan x+\sec x)>0$ on $\left(0,\frac{\pi}{2}\right)$, so f is increasing on $\left(0,\frac{\pi}{2}\right)$.

F. No local extrema

H.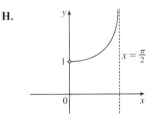

G. $f''(x)=\sec x\,(\sec^2 x+\sec x\tan x)+(\tan x+\sec x)\sec x\tan x$

$\quad\quad=\sec x\,(\sec x+\tan x)\sec x+\sec x\,(\sec x+\tan x)\tan x$

$\quad\quad=\sec x\,(\sec x+\tan x)(\sec x+\tan x)=\sec x\,(\sec x+\tan x)^2>0$

on $\left(0,\frac{\pi}{2}\right)$, so f is CU on $\left(0,\frac{\pi}{2}\right)$. No IP

---

37. $y=f(x)=\dfrac{\sin x}{1+\cos x}$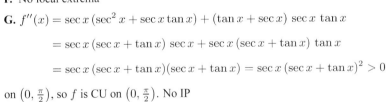

A. The domain of f is the set of all real numbers except odd integer multiples of π. **B.** y-intercept: $f(0)=0$; x-intercepts:

$x=n\pi,\ n$ an even integer. **C.** $f(-x)=-f(x)$, so f is an odd function; the graph is symmetric about the origin and has

period 2π. **D.** When n is an odd integer, $\displaystyle\lim_{x\to(n\pi)^-}f(x)=\infty$ and $\displaystyle\lim_{x\to(n\pi)^+}f(x)=-\infty$, so $x=n\pi$ is a VA for each odd integer n. No HA. **E.** $f'(x) = \dfrac{(1 + \cos x) \cdot \cos x - \sin x(-\sin x)}{(1 + \cos x)^2} = \dfrac{1 + \cos x}{(1 + \cos x)^2} = \dfrac{1}{1 + \cos x}.$ $f'(x) > 0$ for all x

except odd multiples of π, so f is increasing on $((2k - 1)\pi, (2k + 1)\pi)$ for each integer k. **F.** No extreme values

G. $f''(x) = \dfrac{\sin x}{(1 + \cos x)^2} > 0 \;\Rightarrow\; \sin x > 0 \;\Rightarrow$

$x \in (2k\pi, (2k + 1)\pi)$ and $f''(x) < 0$ on $((2k - 1)\pi, 2k\pi)$ for each

integer k. f is CU on $(2k\pi, (2k + 1)\pi)$ and CD on $((2k - 1)\pi, 2k\pi)$

for each integer k. f has IPs at $(2k\pi, 0)$ for each integer k.

H.

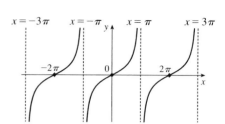

38. $y = f(x) = \dfrac{\sin x}{2 + \cos x}$ **A.** $D = \mathbb{R}$ **B.** y-intercept: $f(0) = 0$; x-intercepts: $f(x) = 0 \;\Leftrightarrow\; \sin x = 0 \;\Leftrightarrow\; x = n\pi$

C. $f(-x) = -f(x)$, so the curve is symmetric about the origin. f is periodic with period 2π, so we determine **E–G** for

$0 \leq x \leq 2\pi$. **D.** No asymptote

E. $f'(x) = \dfrac{(2 + \cos x)\cos x - \sin x(-\sin x)}{(2 + \cos x)^2} = \dfrac{2\cos x + \cos^2 x + \sin^2 x}{(2 + \cos x)^2} = \dfrac{2\cos x + 1}{(2 + \cos x)^2}.$

$f'(x) > 0 \;\Leftrightarrow\; 2\cos x + 1 > 0 \;\Leftrightarrow\; \cos x > -\tfrac{1}{2} \;\Leftrightarrow\; x$ is in $\left(0, \tfrac{2\pi}{3}\right)$ or $\left(\tfrac{4\pi}{3}, 2\pi\right)$, so f is increasing

on $\left(0, \tfrac{2\pi}{3}\right)$ and $\left(\tfrac{4\pi}{3}, 2\pi\right)$, and f is decreasing on $\left(\tfrac{2\pi}{3}, \tfrac{4\pi}{3}\right)$.

F. Local maximum value $f\left(\tfrac{2\pi}{3}\right) = \dfrac{\sqrt{3}/2}{2 - (1/2)} = \dfrac{\sqrt{3}}{3}$ and local minimum value $f\left(\tfrac{4\pi}{3}\right) = \dfrac{-\sqrt{3}/2}{2 - (1/2)} = -\dfrac{\sqrt{3}}{3}$

G. $f''(x) = \dfrac{(2 + \cos x)^2(-2\sin x) - (2\cos x + 1)2(2 + \cos x)(-\sin x)}{[(2 + \cos x)^2]^2}$

$= \dfrac{-2\sin x\,(2 + \cos x)[(2 + \cos x) - (2\cos x + 1)]}{(2 + \cos x)^4} = \dfrac{-2\sin x\,(1 - \cos x)}{(2 + \cos x)^3}$

$f''(x) > 0 \;\Leftrightarrow\; -2\sin x > 0 \;\Leftrightarrow\; \sin x < 0 \;\Leftrightarrow\; x$ is in $(\pi, 2\pi)$ [f is CU] and $f''(x) < 0 \;\Leftrightarrow$

x is in $(0, \pi)$ [f is CD]. The inflection points are $(0, 0)$, $(\pi, 0)$, and $(2\pi, 0)$.

H.

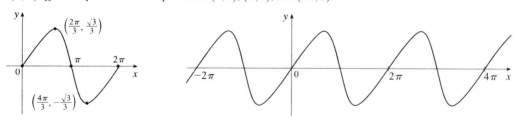

39. $m = f(v) = \dfrac{m_0}{\sqrt{1 - v^2/c^2}}$. The m-intercept is $f(0) = m_0$. There are no v-intercepts. $\lim\limits_{v \to c^-} f(v) = \infty$, so $v = c$ is a VA.

$f'(v) = -\tfrac{1}{2}m_0(1 - v^2/c^2)^{-3/2}(-2v/c^2) = \dfrac{m_0 v}{c^2(1 - v^2/c^2)^{3/2}} = \dfrac{m_0 v}{\dfrac{c^2(c^2 - v^2)^{3/2}}{c^3}} = \dfrac{m_0 c v}{(c^2 - v^2)^{3/2}} > 0$, so f is increasing on $(0, c)$. There are no local extreme values.

$$f''(v) = \frac{(c^2 - v^2)^{3/2}(m_0 c) - m_0 cv \cdot \frac{3}{2}(c^2 - v^2)^{1/2}(-2v)}{[(c^2 - v^2)^{3/2}]^2}$$

$$= \frac{m_0 c(c^2 - v^2)^{1/2}[(c^2 - v^2) + 3v^2]}{(c^2 - v^2)^3} = \frac{m_0 c(c^2 + 2v^2)}{(c^2 - v^2)^{5/2}} > 0,$$

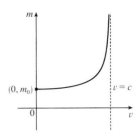

so f is CU on $(0, c)$. There are no inflection points.

40. Let $a = m_0^2 c^4$ and $b = h^2 c^2$, so the equation can be written as $E = f(\lambda) = \sqrt{a + b/\lambda^2} = \sqrt{\dfrac{a\lambda^2 + b}{\lambda^2}} = \dfrac{\sqrt{a\lambda^2 + b}}{\lambda}$.

$$\lim_{\lambda \to 0^+} \frac{\sqrt{a\lambda^2 + b}}{\lambda} = \infty, \text{ so } \lambda = 0 \text{ is a VA.}$$

$$\lim_{\lambda \to \infty} \frac{\sqrt{a\lambda^2 + b}}{\lambda} = \lim_{\lambda \to \infty} \frac{\sqrt{a\lambda^2 + b}/\lambda}{\lambda/\lambda} = \lim_{\lambda \to \infty} \frac{\sqrt{a + b/\lambda^2}}{1} = \sqrt{a}, \text{ so } E = \sqrt{a} = m_0 c^2 \text{ is a HA.}$$

$$f'(\lambda) = \frac{\lambda \cdot \frac{1}{2}(a\lambda^2 + b)^{-1/2}(2a\lambda) - (a\lambda^2 + b)^{1/2}(1)}{\lambda^2} = \frac{(a\lambda^2 + b)^{-1/2}[a\lambda^2 - (a\lambda^2 + b)]}{\lambda^2} = \frac{-b}{\lambda^2 \sqrt{a\lambda^2 + b}} < 0,$$

so f is decreasing on $(0, \infty)$. Using the Reciprocal Rule,

$$f''(\lambda) = b \cdot \frac{\lambda^2 \cdot \frac{1}{2}(a\lambda^2 + b)^{-1/2}(2a\lambda) + (a\lambda^2 + b)^{1/2}(2\lambda)}{\left(\lambda^2 \sqrt{a\lambda^2 + b}\right)^2}$$

$$= \frac{b\lambda(a\lambda^2 + b)^{-1/2}[a\lambda^2 + 2(a\lambda^2 + b)]}{\left(\lambda^2 \sqrt{a\lambda^2 + b}\right)^2} = \frac{b(3a\lambda^2 + 2b)}{\lambda^3(a\lambda^2 + b)^{3/2}} > 0,$$

so f is CU on $(0, \infty)$. There are no extrema or inflection points. The graph

shows that as λ decreases, the energy increases and as λ increases, the energy

decreases. For large wavelengths, the energy is very close to the energy at rest.

41. $y = -\dfrac{W}{24EI} x^4 + \dfrac{WL}{12EI} x^3 - \dfrac{WL^2}{24EI} x^2 = -\dfrac{W}{24EI} x^2 \left(x^2 - 2Lx + L^2\right)$

$$= \frac{-W}{24EI} x^2 (x - L)^2 = cx^2 (x - L)^2$$

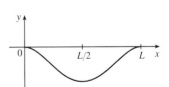

where $c = -\dfrac{W}{24EI}$ is a negative constant and $0 \le x \le L$. We sketch

$f(x) = cx^2(x - L)^2$ for $c = -1$. $f(0) = f(L) = 0$.

$f'(x) = cx^2[2(x - L)] + (x - L)^2(2cx) = 2cx(x - L)[x + (x - L)] = 2cx(x - L)(2x - L)$. So for $0 < x < L$,

$f'(x) > 0 \iff x(x-L)(2x-L) < 0$ [since $c < 0$] $\iff L/2 < x < L$ and $f'(x) < 0 \iff 0 < x < L/2$.

Thus, f is increasing on $(L/2, L)$ and decreasing on $(0, L/2)$, and there is a local and absolute

minimum at the point $(L/2, f(L/2)) = (L/2, cL^4/16)$. $f'(x) = 2c[x(x-L)(2x-L)] \implies$

$f''(x) = 2c[1 \cdot (x-L)(2x-L) + x(1)(2x-L) + x(x-L)(2)] = 2c(6x^2 - 6Lx + L^2) = 0 \iff$

$x = \dfrac{6L \pm \sqrt{12L^2}}{12} = \frac{1}{2}L \pm \frac{\sqrt{3}}{6}L$, and these are the x-coordinates of the two inflection points.

42. $F(x) = -\dfrac{k}{x^2} + \dfrac{k}{(x-2)^2}$, where $k > 0$ and $0 < x < 2$. For $0 < x < 2$, $x - 2 < 0$, so

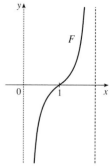

$F'(x) = \dfrac{2k}{x^3} - \dfrac{2k}{(x-2)^3} > 0$ and F is increasing. $\displaystyle\lim_{x \to 0^+} F(x) = -\infty$ and $\displaystyle\lim_{x \to 2^-}$

$F(x) = \infty$, so $x = 0$ and $x = 2$ are vertical asymptotes. Notice that when the middle

particle is at $x = 1$, the net force acting on it is 0. When $x > 1$, the net force is positive,

meaning that it acts to the right. And if the particle approaches $x = 2$, the force on it

rapidly becomes very large. When $x < 1$, the net force is negative, so it acts to the left.

If the particle approaches 0, the force becomes very large to the left.

43. $y = \dfrac{x^2 + 1}{x + 1}$. Long division gives us:

$$
\begin{array}{r}
x - 1 \\
x + 1 \overline{\smash{\big)}\ x^2 \qquad\, + 1} \\
\underline{x^2 + x} \\
-x + 1 \\
\underline{-x - 1} \\
2
\end{array}
$$

Thus, $y = f(x) = \dfrac{x^2 + 1}{x + 1} = x - 1 + \dfrac{2}{x + 1}$ and $f(x) - (x - 1) = \dfrac{2}{x + 1} = \dfrac{\frac{2}{x}}{1 + \frac{1}{x}}$ [for $x \neq 0$] $\to 0$ as $x \to \pm\infty$.

So the line $y = x - 1$ is a slant asymptote (SA).

44. $y = \dfrac{2x^3 + x^2 + x + 3}{x^2 + 2x}$. Long division gives us:

$$
\begin{array}{r}
2x - 3 \\
x^2 + 2x \overline{\smash{\big)}\ 2x^3 + \ x^2 + \ x + 3} \\
\underline{2x^3 + 4x^2} \\
-3x^2 + \ x \\
\underline{-3x^2 - 6x} \\
7x + 3
\end{array}
$$

Thus, $y = f(x) = \dfrac{2x^3 + x^2 + x + 3}{x^2 + 2x} = 2x - 3 + \dfrac{7x + 3}{x^2 + 2x}$ and $f(x) - (2x - 3) = \dfrac{7x + 3}{x^2 + 2x} = \dfrac{\frac{7}{x} + \frac{3}{x^2}}{1 + \frac{2}{x}}$

[for $x \neq 0$] $\to 0$ as $x \to \pm\infty$. So the line $y = 2x - 3$ is a SA.

45. $y = \dfrac{4x^3 - 2x^2 + 5}{2x^2 + x - 3}$. Long division gives us:

$$2x^2 + x - 3 \overline{\smash{\big)}\ \begin{array}{r} 2x - 2 \\ 4x^3 - 2x^2 \qquad + 5 \end{array}}$$
$$\underline{4x^3 + 2x^2 - 6x}$$
$$-4x^2 + 6x + 5$$
$$\underline{-4x^2 - 2x + 6}$$
$$8x - 1$$

Thus, $y = f(x) = \dfrac{4x^3 - 2x^2 + 5}{2x^2 + x - 3} = 2x - 2 + \dfrac{8x - 1}{2x^2 + x - 3}$ and $f(x) - (2x - 2) = \dfrac{8x - 1}{2x^2 + x - 3} = \dfrac{\dfrac{8}{x} - \dfrac{1}{x^2}}{2 + \dfrac{1}{x} - \dfrac{3}{x^2}}$

[for $x \neq 0$] $\to 0$ as $x \to \pm\infty$. So the line $y = 2x - 2$ is a SA.

46. $y = \dfrac{5x^4 + x^2 + x}{x^3 - x^2 + 2}$. Long division gives us:

$$x^3 - x^2 + 2 \overline{\smash{\big)}\ \begin{array}{r} 5x + 5 \\ 5x^4 \qquad + x^2 + x \end{array}}$$
$$\underline{5x^4 - 5x^3 \qquad + 10x}$$
$$5x^3 + x^2 - 9x$$
$$\underline{5x^3 - 5x^2 \qquad + 10}$$
$$6x^2 - 9x - 10$$

Thus, $y = f(x) = \dfrac{5x^4 + x^2 + x}{x^3 - x^2 + 2} = 5x + 5 + \dfrac{6x^2 - 9x - 10}{x^3 - x^2 + 2}$ and

$f(x) - (5x + 5) = \dfrac{6x^2 - 9x - 10}{x^3 - x^2 + 2} = \dfrac{\dfrac{6}{x} - \dfrac{9}{x^2} - \dfrac{10}{x^3}}{1 - \dfrac{1}{x} + \dfrac{2}{x^3}}$ [for $x \neq 0$] $\to 0$ as $x \to \pm\infty$. So the line $y = 5x + 5$ is a SA.

47. $y = f(x) = \dfrac{-2x^2 + 5x - 1}{2x - 1} = -x + 2 + \dfrac{1}{2x - 1}$ **A.** $D = \left\{ x \in \mathbb{R} \mid x \neq \tfrac{1}{2} \right\} = \left(-\infty, \tfrac{1}{2}\right) \cup \left(\tfrac{1}{2}, \infty\right)$

B. y-intercept: $f(0) = 1$; x-intercepts: $f(x) = 0$ $\Rightarrow$ $-2x^2 + 5x - 1 = 0$ $\Rightarrow$ $x = \dfrac{-5 \pm \sqrt{17}}{-4}$ $\Rightarrow$ $x \approx 0.22, 2.28$.

C. No symmetry   **D.** $\displaystyle\lim_{x \to (1/2)^-} f(x) = -\infty$ and $\displaystyle\lim_{x \to (1/2)^+} f(x) = \infty$, so $x = \tfrac{1}{2}$ is a VA.

$\displaystyle\lim_{x \to \pm\infty} [f(x) - (-x + 2)] = \lim_{x \to \pm\infty} \dfrac{1}{2x - 1} = 0$, so the line $y = -x + 2$ is a SA.

E. $f'(x) = -1 - \dfrac{2}{(2x - 1)^2} < 0$ for $x \neq \tfrac{1}{2}$, so f is decreasing on $\left(-\infty, \tfrac{1}{2}\right)$ **H.**

and $\left(\tfrac{1}{2}, \infty\right)$.   **F.** No extreme values   **G.** $f'(x) = -1 - 2(2x - 1)^{-2}$ $\Rightarrow$

$f''(x) = -2(-2)(2x - 1)^{-3}(2) = \dfrac{8}{(2x - 1)^3}$, so $f''(x) > 0$ when $x > \tfrac{1}{2}$ and

$f''(x) < 0$ when $x < \tfrac{1}{2}$. Thus, f is CU on $\left(\tfrac{1}{2}, \infty\right)$ and CD on $\left(-\infty, \tfrac{1}{2}\right)$. No IP

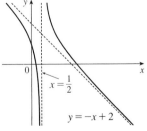

48. $y = f(x) = \dfrac{x^2 + 12}{x - 2} = x + 2 + \dfrac{16}{x - 2}$ **A.** $D = \{x \in \mathbb{R} \mid x \neq 2\} = (-\infty, 2) \cup (2, \infty)$

B. y-intercept: $f(0) = -6$; no x-intercepts. **C.** No symmetry **D.** $\displaystyle\lim_{x \to 2^-} f(x) = -\infty$ and $\displaystyle\lim_{x \to 2^+} f(x) = \infty$,

so $x = 2$ is a VA. $\displaystyle\lim_{x \to \pm\infty} [f(x) - (x + 2)] = \lim_{x \to \pm\infty} \dfrac{16}{x - 2} = 0$, so the line $y = x + 2$ is a slant asymptote.

E. $f'(x) = 1 - \dfrac{16}{(x-2)^2} = \dfrac{x^2 - 4x - 12}{(x-2)^2} = \dfrac{(x-6)(x+2)}{(x-2)^2}$, so $f'(x) > 0$ when $x < -2$ or $x > 6$ and $f'(x) < 0$

when $-2 < x < 2$ or $2 < x < 6$. Thus, f is increasing on $(-\infty, -2)$ **H.**

and $(6, \infty)$ and decreasing on $(-2, 2)$ and $(2, 6)$. **F.** Local maximum

value $f(-2) = -4$, local minimum value $f(6) = 12$

G. $f''(x) = 16(-2)(x-2)^{-3} = \dfrac{32}{(x-2)^3}$, so $f''(x) > 0$ for $x > 2$ and

$f''(x) < 0$ for $x < 2$. f is CU on $(2, \infty)$ and CD on $(-\infty, 2)$. No IP

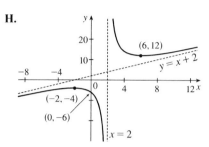

49. $y = f(x) = (x^2 + 4)/x = x + 4/x$ **A.** $D = \{x \mid x \neq 0\} = (-\infty, 0) \cup (0, \infty)$ **B.** No intercept

C. $f(-x) = -f(x)$ $\Rightarrow$ symmetry about the origin **D.** $\displaystyle\lim_{x \to \infty} (x + 4/x) = \infty$ but $f(x) - x = 4/x \to 0$ as $x \to \pm\infty$,

so $y = x$ is a slant asymptote. $\displaystyle\lim_{x \to 0^+} (x + 4/x) = \infty$ and **H.**

$\displaystyle\lim_{x \to 0^-} (x + 4/x) = -\infty$, so $x = 0$ is a VA. **E.** $f'(x) = 1 - 4/x^2 > 0$ $\Leftrightarrow$

$x^2 > 4$ $\Leftrightarrow$ $x > 2$ or $x < -2$, so f is increasing on $(-\infty, -2)$ and $(2, \infty)$ and

decreasing on $(-2, 0)$ and $(0, 2)$. **F.** Local maximum value $f(-2) = -4$, local

minimum value $f(2) = 4$ **G.** $f''(x) = 8/x^3 > 0$ $\Leftrightarrow$ $x > 0$ so f is CU on

$(0, \infty)$ and CD on $(-\infty, 0)$. No IP

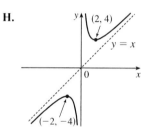

50. $y = f(x) = \dfrac{x^2 + x + 1}{x} = x + 1 + \dfrac{1}{x}$ **A.** $D = \{x \mid x \neq 0\} = (-\infty, 0) \cup (0, \infty)$ **B.** No intercept (x-intercepts would

occur when $x^2 + x + 1 = 0$ but this equation has no real roots since $b^2 - 4ac = -3 < 0$.) **C.** No symmetry

D. $\displaystyle\lim_{x \to \infty} (x + 1 + 1/x) = \pm\infty$, so no HA. But $(x + 1 + 1/x) - (x + 1) = 1/x \to 0$ as $x \to \pm\infty$, so $y = x + 1$ is a slant

asymptote. Also $\displaystyle\lim_{x \to 0^+} (x + 1 + 1/x) = \infty$, $\displaystyle\lim_{x \to 0^-} (x + 1 + 1/x) = -\infty$, so

$x = 0$ is a VA. **E.** $f'(x) = 1 - 1/x^2 > 0$ when $x^2 > 1$ $\Leftrightarrow$ $x > 1$ or **H.**

$x < -1$; $f'(x) < 0$ $\Leftrightarrow$ $-1 < x < 1$. So f is increasing on $(-\infty, -1)$,

$(1, \infty)$ and decreasing on $(-1, 0)$, $(0, 1)$. **F.** $f(1) = 3$ is a local

minimum, $f(-1) = -1$ is a local maximum. **G.** $f''(x) = 2/x^3 > 0$

$\Leftrightarrow$ $x > 0$, so f is CU on $(0, \infty)$ and CD on $(-\infty, 0)$. No IP

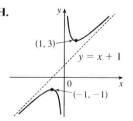

51. $y = f(x) = \dfrac{2x^3 + x^2 + 1}{x^2 + 1} = 2x + 1 + \dfrac{-2x}{x^2 + 1}$ **A.** $D = \mathbb{R}$ **B.** y-intercept: $f(0) = 1$; x-intercept: $f(x) = 0$ $\Rightarrow$

$0 = 2x^3 + x^2 + 1 = (x + 1)(2x^2 - x + 1)$ $\Rightarrow$ $x = -1$ **C.** No symmetry **D.** No VA

$\displaystyle\lim_{x \to \pm\infty} [f(x) - (2x + 1)] = \lim_{x \to \pm\infty} \frac{-2x}{x^2 + 1} = \lim_{x \to \pm\infty} \frac{-2/x}{1 + 1/x^2} = 0$, so the line $y = 2x + 1$ is a slant asymptote.

E. $f'(x) = 2 + \dfrac{(x^2 + 1)(-2) - (-2x)(2x)}{(x^2 + 1)^2} = \dfrac{2(x^4 + 2x^2 + 1) - 2x^2 - 2 + 4x^2}{(x^2 + 1)^2} = \dfrac{2x^4 + 6x^2}{(x^2 + 1)^2} = \dfrac{2x^2(x^2 + 3)}{(x^2 + 1)^2}$

so $f'(x) > 0$ if $x \neq 0$. Thus, f is increasing on $(-\infty, 0)$ and $(0, \infty)$. Since f is continuous at 0, f is increasing on $\mathbb{R}$.

F. No extreme values

G. $f''(x) = \dfrac{(x^2 + 1)^2 \cdot (8x^3 + 12x) - (2x^4 + 6x^2) \cdot 2(x^2 + 1)(2x)}{[(x^2 + 1)^2]^2}$

$= \dfrac{4x(x^2 + 1)[(x^2 + 1)(2x^2 + 3) - 2x^4 - 6x^2]}{(x^2 + 1)^4} = \dfrac{4x(-x^2 + 3)}{(x^2 + 1)^3}$

so $f''(x) > 0$ for $x < -\sqrt{3}$ and $0 < x < \sqrt{3}$, and $f''(x) < 0$ for

$-\sqrt{3} < x < 0$ and $x > \sqrt{3}$. f is CU on $\left(-\infty, -\sqrt{3}\right)$ and $\left(0, \sqrt{3}\,\right)$,

and CD on $\left(-\sqrt{3}, 0\right)$ and $\left(\sqrt{3}, \infty\right)$. There are three IPs: $(0, 1)$,

$\left(-\sqrt{3}, -\frac{3}{2}\sqrt{3} + 1\right) \approx (-1.73, -1.60)$, and

$\left(\sqrt{3}, \frac{3}{2}\sqrt{3} + 1\right) \approx (1.73, 3.60)$.

H.

52. $y = f(x) = \dfrac{(x + 1)^3}{(x - 1)^2} = \dfrac{x^3 + 3x^2 + 3x + 1}{x^2 - 2x + 1} = x + 5 + \dfrac{12x - 4}{(x - 1)^2}$

A. $D = \{x \in \mathbb{R} \mid x \neq 1\} = (-\infty, 1) \cup (1, \infty)$ **B.** y-intercept: $f(0) = 1$; x-intercept: $f(x) = 0$ $\Rightarrow$

$x = -1$ **C.** No symmetry **D.** $\displaystyle\lim_{x \to 1} f(x) = \infty$, so $x = 1$ is a VA.

$\displaystyle\lim_{x \to \pm\infty} [f(x) - (x + 5)] = \lim_{x \to \pm\infty} \frac{12x - 4}{x^2 - 2x + 1} = \lim_{x \to \pm\infty} \frac{\dfrac{12}{x} - \dfrac{4}{x^2}}{1 - \dfrac{2}{x} + \dfrac{1}{x^2}} = 0$, so the line $y = x + 5$ is a SA.

E. $f'(x) = \dfrac{(x - 1)^2 \cdot 3(x + 1)^2 - (x + 1)^3 \cdot 2(x - 1)}{[(x - 1)^2]^2}$

$= \dfrac{(x - 1)(x + 1)^2[3(x - 1) - 2(x + 1)]}{(x - 1)^4} = \dfrac{(x + 1)^2(x - 5)}{(x - 1)^3}$

so $f'(x) > 0$ when $x < -1$, $-1 < x < 1$, or $x > 5$, and $f'(x) < 0$

when $1 < x < 5$. f is increasing on $(-\infty, 1)$ and $(5, \infty)$ and decreasing

on $(1, 5)$.

F. Local minimum value $f(5) = \frac{216}{16} = \frac{27}{2}$, no local maximum

H.

G. $f''(x) = \dfrac{(x-1)^3[(x-1)^2 + (x-5)\cdot 2(x+1)] - (x+1)^2(x-5)\cdot 3(x-1)^2}{[(x-1)^3]^2}$

$= \dfrac{(x-1)^2(x+1)\{(x-1)[(x+1)+2(x-5)] - 3(x+1)(x-5)\}}{(x-1)^6}$

$= \dfrac{(x+1)\{(x-1)[3x-9] - 3(x^2-4x-5)\}}{(x-1)^4} = \dfrac{(x+1)(24)}{(x-1)^4}$

so $f''(x) > 0$ if $-1 < x < 1$ or $x > 1$, and $f''(x) < 0$ if $x < -1$. Thus, f is CU on $(-1, 1)$ and $(1, \infty)$ and

CD on $(-\infty, -1)$. IP at $(-1, 0)$

53. $y = f(x) = \sqrt{4x^2+9} \Rightarrow f'(x) = \dfrac{4x}{\sqrt{4x^2+9}} \Rightarrow$

$f''(x) = \dfrac{\sqrt{4x^2+9}\cdot 4 - 4x\cdot 4x/\sqrt{4x^2+9}}{4x^2+9} = \dfrac{4(4x^2+9) - 16x^2}{(4x^2+9)^{3/2}} = \dfrac{36}{(4x^2+9)^{3/2}}.$ f is defined on $(-\infty, \infty)$.

$f(-x) = f(x)$, so f is even, which means its graph is symmetric about the y-axis. The y-intercept is $f(0) = 3$. There are no

x-intercepts since $f(x) > 0$ for all x.

$\lim_{x\to\infty} \left(\sqrt{4x^2+9} - 2x\right) = \lim_{x\to\infty} \dfrac{\left(\sqrt{4x^2+9} - 2x\right)\left(\sqrt{4x^2+9} + 2x\right)}{\sqrt{4x^2+9} + 2x}$

$= \lim_{x\to\infty} \dfrac{(4x^2+9) - 4x^2}{\sqrt{4x^2+9} + 2x} = \lim_{x\to\infty} \dfrac{9}{\sqrt{4x^2+9} + 2x} = 0$

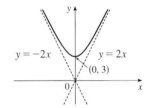

and, similarly, $\lim_{x\to-\infty} \left(\sqrt{4x^2+9} + 2x\right) = \lim_{x\to-\infty} \dfrac{9}{\sqrt{4x^2+9} - 2x} = 0,$

so $y = \pm 2x$ are slant asymptotes. f is decreasing on $(-\infty, 0)$ and increasing on $(0, \infty)$ with local minimum $f(0) = 3$.

$f''(x) > 0$ for all x, so f is CU on $\mathbb{R}$.

54. $y = f(x) = \sqrt{x^2+4x} = \sqrt{x(x+4)}$. $x(x+4) \geq 0 \Leftrightarrow x \leq -4$ or $x \geq 0$, so $D = (-\infty, -4] \cup [0, \infty)$.

y-intercept: $f(0) = 0$; x-intercepts: $f(x) = 0 \Rightarrow x = -4, 0$.

$\sqrt{x^2+4x} \mp (x+2) = \dfrac{\sqrt{x^2+4x} \mp (x+2)}{1} \cdot \dfrac{\sqrt{x^2+4x} \pm (x+2)}{\sqrt{x^2+4x} \pm (x+2)} = \dfrac{(x^2+4x) - (x^2+4x+4)}{\sqrt{x^2+4x} \pm (x+2)}$

$= \dfrac{-4}{\sqrt{x^2+4x} \pm (x+2)}$

so $\lim_{x\to\pm\infty} [f(x) \mp (x+2)] = 0$. Thus, the graph of f approaches the slant asymptote $y = x+2$ as $x \to \infty$ and it approaches

the slant asymptote $y = -(x+2)$ as $x \to -\infty$. $f'(x) = \dfrac{x+2}{\sqrt{x^2+4x}}$, so $f'(x) < 0$ for $x < -4$ and $f'(x) > 0$ for $x > 0$;

that is, f is decreasing on $(-\infty, -4)$ and increasing on $(0, \infty)$. There are no local

extreme values. $f'(x) = (x+2)(x^2+4x)^{-1/2} \Rightarrow$

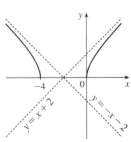

$f''(x) = (x+2)\cdot\left(-\tfrac{1}{2}\right)(x^2+4x)^{-3/2}\cdot(2x+4) + (x^2+4x)^{-1/2}$

$= (x^2+4x)^{-3/2}\left[-(x+2)^2 + (x^2+4x)\right] = -4(x^2+4x)^{-3/2} < 0$ on D

so f is CD on $(-\infty, -4)$ and $(0, \infty)$. No IP

55. $\dfrac{x^2}{a^2} - \dfrac{y^2}{b^2} = 1$ ⇒ $y = \pm\dfrac{b}{a}\sqrt{x^2 - a^2}$. Now

$$\lim_{x \to \infty}\left[\dfrac{b}{a}\sqrt{x^2 - a^2} - \dfrac{b}{a}x\right] = \dfrac{b}{a}\cdot\lim_{x \to \infty}\left(\sqrt{x^2 - a^2} - x\right)\dfrac{\sqrt{x^2 - a^2} + x}{\sqrt{x^2 - a^2} + x} = \dfrac{b}{a}\cdot\lim_{x \to \infty}\dfrac{-a^2}{\sqrt{x^2 - a^2} + x} = 0,$$

which shows that $y = \dfrac{b}{a}x$ is a slant asymptote. Similarly,

$$\lim_{x \to \infty}\left[-\dfrac{b}{a}\sqrt{x^2 - a^2} - \left(-\dfrac{b}{a}x\right)\right] = -\dfrac{b}{a}\cdot\lim_{x \to \infty}\dfrac{-a^2}{\sqrt{x^2 - a^2} + x} = 0, \text{ so } y = -\dfrac{b}{a}x \text{ is a slant asymptote.}$$

56. $f(x) - x^2 = \dfrac{x^3 + 1}{x} - x^2 = \dfrac{x^3 + 1 - x^3}{x} = \dfrac{1}{x}$, and $\lim\limits_{x \to \pm\infty}\dfrac{1}{x} = 0$. Therefore, $\lim\limits_{x \to \pm\infty}[f(x) - x^2] = 0$,

and so the graph of f is asymptotic to that of $y = x^2$. For purposes of differentiation, we will use $f(x) = x^2 + 1/x$.

A. $D = \{x \mid x \neq 0\}$ **B.** No y-intercept; to find the x-intercept, we set $y = 0$ ⇔ $x = -1$.

C. No symmetry **D.** $\lim\limits_{x \to 0^+}\dfrac{x^3 + 1}{x} = \infty$ and $\lim\limits_{x \to 0^-}\dfrac{x^3 + 1}{x} = -\infty$, **H.**

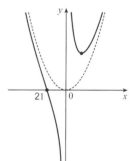

so $x = 0$ is a vertical asymptote. Also, the graph is asymptotic to the parabola

$y = x^2$, as shown above. **E.** $f'(x) = 2x - 1/x^2 > 0$ ⇔ $x > \dfrac{1}{\sqrt[3]{2}}$, so f

is increasing on $\left(\dfrac{1}{\sqrt[3]{2}}, \infty\right)$ and decreasing on $(-\infty, 0)$ and $\left(0, \dfrac{1}{\sqrt[3]{2}}\right)$.

F. Local minimum value $f\left(\dfrac{1}{\sqrt[3]{2}}\right) = \dfrac{3\sqrt[3]{3}}{2}$, no local maximum

G. $f''(x) = 2 + 2/x^3 > 0$ ⇔ $x < -1$ or $x > 0$, so f is CU on

$(-\infty, -1)$ and $(0, \infty)$, and CD on $(-1, 0)$. IP at $(-1, 0)$

57. $\lim\limits_{x \to \pm\infty}\left[f(x) - x^3\right] = \lim\limits_{x \to \pm\infty}\dfrac{x^4 + 1}{x} - \dfrac{x^4}{x} = \lim\limits_{x \to \pm\infty}\dfrac{1}{x} = 0$, so the graph of f is asymptotic to that of $y = x^3$.

A. $D = \{x \mid x \neq 0\}$ **B.** No intercept **C.** f is symmetric about the origin. **D.** $\lim\limits_{x \to 0^-}\left(x^3 + \dfrac{1}{x}\right) = -\infty$ and

$\lim\limits_{x \to 0^+}\left(x^3 + \dfrac{1}{x}\right) = \infty$, so $x = 0$ is a vertical asymptote, and as shown above, the graph of f is asymptotic to that of $y = x^3$.

E. $f'(x) = 3x^2 - 1/x^2 > 0$ ⇔ $x^4 > \dfrac{1}{3}$ ⇔ $|x| > \dfrac{1}{\sqrt[4]{3}}$, so f is increasing on $\left(-\infty, -\dfrac{1}{\sqrt[4]{3}}\right)$ and $\left(\dfrac{1}{\sqrt[4]{3}}, \infty\right)$ and

decreasing on $\left(-\dfrac{1}{\sqrt[4]{3}}, 0\right)$ and $\left(0, \dfrac{1}{\sqrt[4]{3}}\right)$. **F.** Local maximum value **H.**

$f\left(-\dfrac{1}{\sqrt[4]{3}}\right) = -4 \cdot 3^{-5/4}$, local minimum value $f\left(\dfrac{1}{\sqrt[4]{3}}\right) = 4 \cdot 3^{-5/4}$

G. $f''(x) = 6x + 2/x^3 > 0$ ⇔ $x > 0$, so f is CU on $(0, \infty)$ and CD

on $(-\infty, 0)$. No IP

58. $\lim\limits_{x \to \pm\infty} [f(x) - \cos x] = \lim\limits_{x \to \pm\infty} 1/x^2 = 0$, so the graph of f is asymptotic to

that of $\cos x$. The intercepts can only be found approximately.

$f(x) = f(-x)$, so f is even. $\lim\limits_{x \to 0} \left(\cos x + \dfrac{1}{x^2} \right) = \infty$, so $x = 0$ is a

vertical asymptote. We don't need to calculate the derivatives, since we know

the asymptotic behavior of the curve.

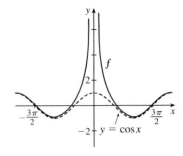

4.6 Graphing with Calculus and Calculators

1. $f(x) = 4x^4 - 32x^3 + 89x^2 - 95x + 29 \;\Rightarrow\; f'(x) = 16x^3 - 96x^2 + 178x - 95 \;\Rightarrow\; f''(x) = 48x^2 - 192x + 178$.

$f(x) = 0 \;\Leftrightarrow\; x \approx 0.5,\ 1.60;\ f'(x) = 0 \;\Leftrightarrow\; x \approx 0.92, 2.5, 2.58$ and $f''(x) = 0 \;\Leftrightarrow\; x \approx 1.46, 2.54$.

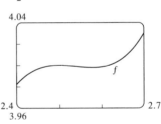

From the graphs of f', we estimate that $f' < 0$ and that f is decreasing on $(-\infty, 0.92)$ and $(2.5, 2.58)$, and that $f' > 0$ and f is increasing on $(0.92, 2.5)$ and $(2.58, \infty)$ with local minimum values $f(0.92) \approx -5.12$ and $f(2.58) \approx 3.998$ and local maximum value $f(2.5) = 4$. The graphs of f' make it clear that f has a maximum and a minimum near $x = 2.5$, shown more clearly in the fourth graph.

From the graph of f'', we estimate that $f'' > 0$ and that f is CU on

$(-\infty, 1.46)$ and $(2.54, \infty)$, and that $f'' < 0$ and f is CD on $(1.46, 2.54)$.

There are inflection points at about $(1.46, -1.40)$ and $(2.54, 3.999)$.

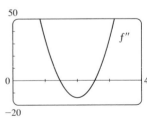

2. $f(x) = x^6 - 15x^5 + 75x^4 - 125x^3 - x \;\Rightarrow\; f'(x) = 6x^5 - 75x^4 + 300x^3 - 375x^2 - 1 \;\Rightarrow$

$f''(x) = 30x^4 - 300x^3 + 900x^2 - 750x$.

$f(x) = 0 \;\Leftrightarrow\; x = 0$ or $x \approx 5.33;\quad f'(x) = 0 \;\Leftrightarrow\; x \approx 2.50, 4.95,$ or $5.05;$

$f''(x) = 0 \;\Leftrightarrow\; x = 0, 5$ or $x \approx 1.38, 3.62$.

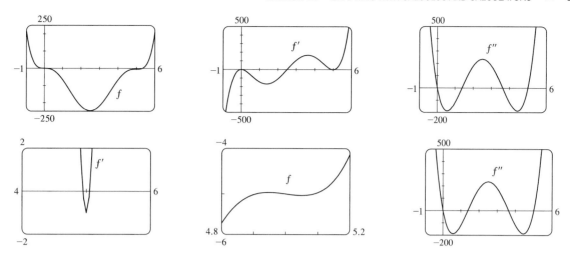

From the graphs of f', we estimate that f is decreasing on $(-\infty, 2.50)$, increasing on $(2.50, 4.95)$, decreasing on $(4.95, 5.05)$, and increasing on $(5.05, \infty)$, with local minimum values $f(2.50) \approx -246.6$ and $f(5.05) \approx -5.03$, and local maximum value $f(4.95) \approx -4.965$ (notice the second graph of f). From the graph of f'', we estimate that f is CU on $(-\infty, 0)$, CD on $(0, 1.38)$, CU on $(1.38, 3.62)$, CD on $(3.62, 5)$, and CU on $(5, \infty)$. There are inflection points at $(0, 0)$ and $(5, -5)$, and at about $(1.38, -126.38)$ and $(3.62, -128.62)$.

3. $f(x) = x^6 - 10x^5 - 400x^4 + 2500x^3 \quad\Rightarrow\quad f'(x) = 6x^5 - 50x^4 - 1600x^3 + 7500x^2 \quad\Rightarrow$
$f''(x) = 30x^4 - 200x^3 - 4800x^2 + 1500x$

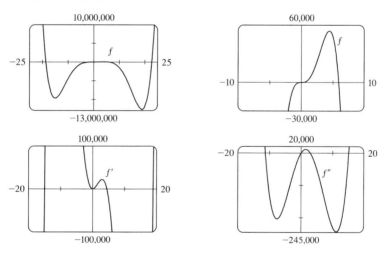

From the graph of f', we estimate that f is decreasing on $(-\infty, -15)$, increasing on $(-15, 4.40)$, decreasing on $(4.40, 18.93)$, and increasing on $(18.93, \infty)$, with local minimum values of $f(-15) \approx -9{,}700{,}000$ and $f(18.93) \approx -12{,}700{,}000$ and local maximum value $f(4.40) \approx 53{,}800$. From the graph of f'', we estimate that f is CU on $(-\infty, -11.34)$, CD on $(-11.34, 0)$, CU on $(0, 2.92)$, CD on $(2.92, 15.08)$, and CU on $(15.08, \infty)$. There is an inflection point at $(0, 0)$ and at about $(-11.34, -6{,}250{,}000)$, $(2.92, 31{,}800)$, and $(15.08, -8{,}150{,}000)$.

4. $f(x) = \dfrac{x^2 - 1}{40x^3 + x + 1}$ $\Rightarrow$ $f'(x) = \dfrac{-40x^4 + 121x^2 + 2x + 1}{(40x^3 + x + 1)^2}$ $\Rightarrow$ $f''(x) = \dfrac{80x(40x^5 - 243x^3 - 7x^2 - 3x + 3)}{(40x^3 + x + 1)^3}$

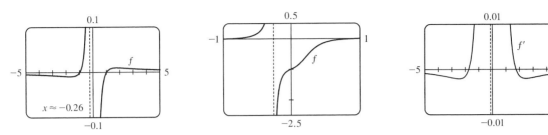

From the first graph of f, we see that there is a VA at $x \approx -0.26$. From the graph of f', we estimate that f is decreasing on $(-\infty, -1.73)$, increasing on $(-1.73, -0.26)$, increasing on $(-0.26, 1.75)$, and decreasing on $(1.75, \infty)$, with local minimum value $f(-1.73) \approx -0.01$ and local maximum value $f(1.75) \approx 0.01$.

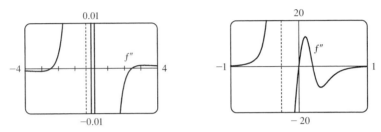

From the graphs of f'', we estimate that f is CD on $(-\infty, -2.45)$, CU on $(-2.45, -0.26)$, CD on $(-0.26, 0)$, CU on $(0, 0.21)$, CD on $(0.21, 2.48)$, and CU on $(2.48, \infty)$. There is an inflection point at $(0, -1)$ and at about $(-2.45, -0.01)$, $(0.21, -0.62)$, and $(2.48, 0.00)$.

5. $f(x) = \dfrac{x}{x^3 - x^2 - 4x + 1}$ $\Rightarrow$ $f'(x) = \dfrac{-2x^3 + x^2 + 1}{(x^3 - x^2 - 4x + 1)^2}$ $\Rightarrow$ $f''(x) = \dfrac{2(3x^5 - 3x^4 + 5x^3 - 6x^2 + 3x + 4)}{(x^3 - x^2 - 4x + 1)^3}$

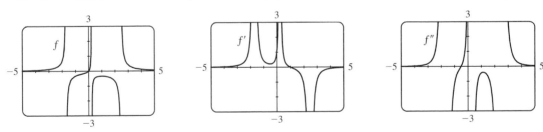

We estimate from the graph of f that $y = 0$ is a horizontal asymptote, and that there are vertical asymptotes at $x = -1.7$, $x = 0.24$, and $x = 2.46$. From the graph of f', we estimate that f is increasing on $(-\infty, -1.7)$, $(-1.7, 0.24)$, and $(0.24, 1)$, and that f is decreasing on $(1, 2.46)$ and $(2.46, \infty)$. There is a local maximum value at $f(1) = -\frac{1}{3}$. From the graph of f'', we estimate that f is CU on $(-\infty, -1.7)$, $(-0.506, 0.24)$, and $(2.46, \infty)$, and that f is CD on $(-1.7, -0.506)$ and $(0.24, 2.46)$. There is an inflection point at $(-0.506, -0.192)$.

6. $f(x) = \tan x + 5\cos x \quad \Rightarrow \quad f'(x) = \sec^2 x - 5\sin x \quad \Rightarrow \quad f''(x) = 2\sec^2 x \tan x - 5\cos x$. Since f is periodic with

period 2π, and defined for all x except odd multiples of $\frac{\pi}{2}$, we graph f and its derivatives on $\left[-\frac{\pi}{2}, \frac{3\pi}{2}\right]$.

  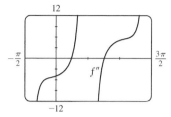

We estimate from the graph of f' that f is increasing on $\left(-\frac{\pi}{2}, 0.21\right)$, $\left(1.07, \frac{\pi}{2}\right)$, $\left(\frac{\pi}{2}, 2.07\right)$, and $\left(2.93, \frac{3\pi}{2}\right)$, and decreasing

on and $(2.07, 2.93)$.   Local minimum values: $f(1.07) \approx 4.23$, $f(2.93) \approx -5.10$.   Local maximum values:

$f(0.21) \approx 5.10$, $f(2.07) \approx -4.23$.

From the graph of f'', we estimate that f is CU on $\left(0.76, \frac{\pi}{2}\right)$ and $\left(2.38, \frac{3\pi}{2}\right)$, and CD on $\left(-\frac{\pi}{2}, 0.76\right)$ and $\left(\frac{\pi}{2}, 2.38\right)$.

f has IP at $(0.76, 4.57)$ and $(2.38, -4.57)$.

7. $f(x) = x^2 - 4x + 7\cos x$, $\ -4 \le x \le 4$. $f'(x) = 2x - 4 - 7\sin x \quad \Rightarrow \quad f''(x) = 2 - 7\cos x$.

$f(x) = 0 \ \Leftrightarrow \ x \approx 1.10$; $f'(x) = 0 \ \Leftrightarrow \ x \approx -1.49, -1.07$, or 2.89; $f''(x) = 0 \ \Leftrightarrow \ x = \pm\cos^{-1}\left(\frac{2}{7}\right) \approx \pm 1.28$.

 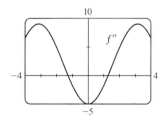

From the graphs of f', we estimate that f is decreasing ($f' < 0$) on $(-4, -1.49)$, increasing on $(-1.49, -1.07)$, decreasing

on $(-1.07, 2.89)$, and increasing on $(2.89, 4)$, with local minimum values $f(-1.49) \approx 8.75$ and $f(2.89) \approx -9.99$ and local

maximum value $f(-1.07) \approx 8.79$ (notice the second graph of f). From the graph of f'', we estimate that f is CU ($f'' > 0$)

on $(-4, -1.28)$, CD on $(-1.28, 1.28)$, and CU on $(1.28, 4)$. There are inflection points at about $(-1.28, 8.77)$

and $(1.28, -1.48)$.

8. $f(x) = \dfrac{\sin x}{x}$, $-2\pi \le x \le 2\pi$. $f'(x) = \dfrac{x \cos x - \sin x}{x^2}$ $\Rightarrow$

$f''(x) = \dfrac{x^2(\cos x - x \sin x - \cos x) - (x \cos x - \sin x)(2x)}{(x^2)^2} = \dfrac{-x^2 \sin x - 2x \cos x + 2 \sin x}{x^3}$

 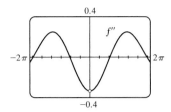

f is an even function with domain $(-\infty, 0) \cup (0, \infty)$. There is no y-intercept, but $\lim\limits_{x \to 0} f(x) = 1$. The x-intercepts are -2π,

$-\pi$, 0, π, and 2π. From the graph of f', we estimate that f is decreasing on $(-2\pi, -4.49)$, increasing on $(-4.49, 0)$,

decreasing on $(0, 4.49)$, and increasing on $(4.49, 2\pi)$. Thus, f has local minima of $f(\pm 4.49) \approx -0.22$. From the graph of

f'', we estimate that f is CD on $(-2\pi, -5.94)$, CU on $(-5.94, -2.08)$, CD on $(-2.08, 0)$ and $(0, 2.08)$, CU on $(2.08, 5.94)$,

and CD on $(5.94, 2\pi)$. f has IPs at approximately $(\pm 5.94, -0.06)$ and $(\pm 2.08, 0.42)$.

9. $f(x) = 1 + \dfrac{1}{x} + \dfrac{8}{x^2} + \dfrac{1}{x^3}$ $\Rightarrow$ $f'(x) = -\dfrac{1}{x^2} - \dfrac{16}{x^3} - \dfrac{3}{x^4} = -\dfrac{1}{x^4}(x^2 + 16x + 3)$ $\Rightarrow$

$f''(x) = \dfrac{2}{x^3} + \dfrac{48}{x^4} + \dfrac{12}{x^5} = \dfrac{2}{x^5}(x^2 + 24x + 6)$.

 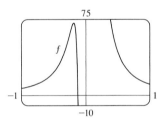

From the graphs, it appears that f increases on $(-15.8, -0.2)$ and decreases on $(-\infty, -15.8)$, $(-0.2, 0)$, and $(0, \infty)$; that f

has a local minimum value of $f(-15.8) \approx 0.97$ and a local maximum value of $f(-0.2) \approx 72$; that f is CD on $(-\infty, -24)$

and $(-0.25, 0)$ and is CU on $(-24, -0.25)$ and $(0, \infty)$; and that f has IPs at $(-24, 0.97)$ and $(-0.25, 60)$.

To find the exact values, note that $f' = 0$ $\Rightarrow$ $x = \dfrac{-16 \pm \sqrt{256 - 12}}{2} = -8 \pm \sqrt{61}$ $[\approx -0.19$ and $-15.81]$.

f' is positive (f is increasing) on $\left(-8 - \sqrt{61}, -8 + \sqrt{61}\,\right)$ and f' is negative (f is decreasing) on $\left(-\infty, -8 - \sqrt{61}\,\right)$,

$\left(-8 + \sqrt{61}, 0\right)$, and $(0, \infty)$. $f'' = 0$ $\Rightarrow$ $x = \dfrac{-24 \pm \sqrt{576 - 24}}{2} = -12 \pm \sqrt{138}$ $[\approx -0.25$ and $-23.75]$. f'' is

positive (f is CU) on $\left(-12 - \sqrt{138}, -12 + \sqrt{138}\,\right)$ and $(0, \infty)$ and f'' is negative (f is CD) on $\left(-\infty, -12 - \sqrt{138}\,\right)$

and $\left(-12 + \sqrt{138}, 0\right)$.

10. $f(x) = \dfrac{1}{x^8} - \dfrac{c}{x^4}$ $[c = 2 \times 10^8]$ $\Rightarrow$

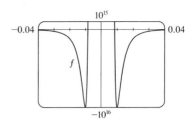

$f'(x) = -\dfrac{8}{x^9} + \dfrac{4c}{x^5} = -\dfrac{4}{x^9}(2 - cx^4)$ $\Rightarrow$

$f''(x) = \dfrac{72}{x^{10}} - \dfrac{20c}{x^6} = \dfrac{4}{x^{10}}(18 - 5cx^4).$

From the graph, it appears that f increases on $(-0.01, 0)$ and $(0.01, \infty)$ and decreases on $(-\infty, -0.01)$ and $(0, 0.01)$;

that f has a local minimum value of $f(\pm 0.01) = -10^{16}$; and that f is CU on $(-0.012, 0)$ and $(0, 0.012)$ and f is CD

on $(-\infty, -0.012)$ and $(0.012, \infty)$.

To find the exact values, note that $f' = 0$ $\Rightarrow$ $x^4 = \dfrac{2}{c}$ $\Rightarrow$ $x \pm \sqrt[4]{\dfrac{2}{c}} = \pm\dfrac{1}{100}$ $[c = 2 \times 10^8]$. f' is positive

(f is increasing) on $(-0.01, 0)$ and $(0.01, \infty)$ and f' is negative (f is decreasing) on $(-\infty, -0.01)$ and $(0, 0.01)$.

$f'' = 0$ $\Rightarrow$ $x^4 = \dfrac{18}{5c}$ $\Rightarrow$ $x = \pm\sqrt[4]{\dfrac{18}{5c}} = \pm\dfrac{1}{100}\sqrt[4]{1.8}$ $[\approx \pm 0.0116]$. f'' is positive (f is CU) on $\left(-\dfrac{1}{100}\sqrt[4]{1.8}, 0\right)$

and $\left(0, \dfrac{1}{100}\sqrt[4]{1.8}\right)$ and f'' is negative (f is CD) on $\left(-\infty, -\dfrac{1}{100}\sqrt[4]{1.8}\right)$ and $\left(\dfrac{1}{100}\sqrt[4]{1.8}, \infty\right)$.

11.

$f(x) = \dfrac{(x+4)(x-3)^2}{x^4(x-1)}$ has VA at $x = 0$ and at $x = 1$ since $\lim\limits_{x \to 0} f(x) = -\infty$,

$\lim\limits_{x \to 1^-} f(x) = -\infty$ and $\lim\limits_{x \to 1^+} f(x) = \infty$.

$f(x) = \dfrac{\dfrac{x+4}{x} \cdot \dfrac{(x-3)^2}{x^2}}{\dfrac{x^4}{x^3} \cdot (x-1)}$ $\begin{bmatrix} \text{dividing numerator} \\ \text{and denominator by } x^3 \end{bmatrix} = \dfrac{(1 + 4/x)(1 - 3/x)^2}{x(x-1)} \to 0$

as $x \to \pm\infty$, so f is asymptotic to the x-axis.

Since f is undefined at $x = 0$, it has no y-intercept. $f(x) = 0$ $\Rightarrow$ $(x+4)(x-3)^2 = 0$ $\Rightarrow$ $x = -4$ or $x = 3$, so f has

x-intercepts -4 and 3. Note, however, that the graph of f is only tangent to the x-axis and does not cross it at $x = 3$, since f is

positive as $x \to 3^-$ and as $x \to 3^+$.

From these graphs, it appears that f has three maximum values and one minimum value. The maximum values are

approximately $f(-5.6) = 0.0182$, $f(0.82) = -281.5$ and $f(5.2) = 0.0145$ and we know (since the graph is tangent to the

x-axis at $x = 3$) that the minimum value is $f(3) = 0$.

12.

$f(x) = \dfrac{(2x+3)^2(x-2)^5}{x^3(x-5)^2}$ has VAs at $x = 0$ and $x = 5$ since $\lim\limits_{x \to 0^-} f(x) = \infty$,

$\lim\limits_{x \to 0^+} f(x) = -\infty$, and $\lim\limits_{x \to 5} f(x) = \infty$. No HA since $\lim\limits_{x \to \pm\infty} f(x) = \infty$.

Since f is undefined at $x = 0$, it has no y-intercept.

$f(x) = 0 \;\Leftrightarrow\; (2x+3)^2(x-2)^5 = 0 \;\Leftrightarrow\; x = -\frac{3}{2}$ or $x = 2$, so f

has x-intercepts at $-\frac{3}{2}$ and 2. Note, however, that the graph of f is only tangent to

the x-axis and does not cross it at $x = -\frac{3}{2}$, since f is positive as $x \to \left(-\frac{3}{2}\right)^-$ and

as $x \to \left(-\frac{3}{2}\right)^+$. There is a local minimum value of $f\left(-\frac{3}{2}\right) = 0$.

The only "mystery" feature is the local minimum to the right of the VA

$x = 5$. From the graph, we see that $f(7.98) \approx 609$ is a local minimum

value.

13. $f(x) = \dfrac{x^2(x+1)^3}{(x-2)^2(x-4)^4} \;\Rightarrow\; f'(x) = -\dfrac{x(x+1)^2(x^3 + 18x^2 - 44x - 16)}{(x-2)^3(x-4)^5}$ [from CAS].

From the graphs of f', it seems that the critical points which indicate extrema occur at $x \approx -20$, -0.3, and 2.5, as estimated

in Example 3. (There is another critical point at $x = -1$, but the sign of f' does not change there.) We differentiate again,

obtaining $f''(x) = 2\dfrac{(x+1)(x^6 + 36x^5 + 6x^4 - 628x^3 + 684x^2 + 672x + 64)}{(x-2)^4(x-4)^6}$.

        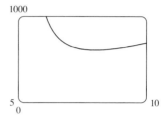

From the graphs of f'', it appears that f is CU on $(-35.3, -5.0)$, $(-1, -0.5)$, $(-0.1, 2)$, $(2, 4)$ and $(4, \infty)$ and CD

on $(-\infty, -35.3)$, $(-5.0, -1)$ and $(-0.5, -0.1)$. We check back on the graphs of f to find the y-coordinates of the

inflection points, and find that these points are approximately $(-35.3, -0.015)$, $(-5.0, -0.005)$, $(-1, 0)$, $(-0.5, 0.00001)$,

and $(-0.1, 0.0000066)$.

14. From a CAS,
$$f'(x) = \frac{2(x-2)^4(2x+3)(2x^3 - 14x^2 - 10x - 45)}{x^4(x-5)^3}$$

and
$$f''(x) = \frac{2(x-2)^3(4x^6 - 56x^5 + 216x^4 + 460x^3 + 805x^2 + 1710x + 5400)}{x^5(x-5)^4}$$

From Exercise 12 and $f'(x)$ above, we know that the zeros of f' are -1.5, 2, and 7.98. From the graph of f', we conclude that f is decreasing on $(-\infty, -1.5)$, increasing on $(-1.5, 0)$ and $(0, 5)$, decreasing on $(5, 7.98)$, and increasing on $(7.98, \infty)$.

From $f''(x)$, we know that $x = 2$ is a zero, and the graph of f'' shows us that $x = 2$ is the only zero of f''. Thus, f is CU on $(-\infty, 0)$, CD on $(0, 2)$, CU on $(2, 5)$, and CU on $(5, \infty)$.

15. $y = f(x) = \dfrac{\sqrt{x}}{x^2 + x + 1}$. From a CAS, $y' = -\dfrac{3x^2 + x - 1}{2\sqrt{x}\,(x^2 + x + 1)^2}$ and $y'' = \dfrac{15x^4 + 10x^3 - 15x^2 - 6x - 1}{4x^{3/2}(x^2 + x + 1)^3}$.

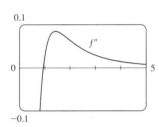

$f'(x) = 0 \iff x \approx 0.43$, so f is increasing on $(0, 0.43)$ and decreasing on $(0.43, \infty)$. There is a local maximum value of $f(0.43) \approx 0.41$. $f''(x) = 0 \iff x \approx 0.94$, so f is CD on $(0, 0.94)$ and CU on $(0.94, \infty)$. There is an inflection point at $(0.94, 0.34)$.

16. $y = f(x) = \dfrac{x^{2/3}}{1 + x + x^4}$. From a CAS, $y' = -\dfrac{10x^4 + x - 2}{3x^{1/3}(x^4 + x + 1)^2}$ and $y'' = \dfrac{2(65x^8 - 14x^5 - 80x^4 + 2x^2 - 8x - 1)}{9x^{4/3}(x^4 + x + 1)^3}$

[continued]

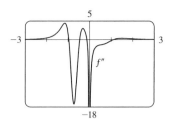

$f'(x)$ does not exist at $x = 0$ and $f'(x) = 0$ ⟺ $x \approx -0.72$ and 0.61, so f is increasing on $(-\infty, -0.72)$, decreasing on $(-0.72, 0)$, increasing on $(0, 0.61)$, and decreasing on $(0.61, \infty)$. There is a local maximum value of $f(-0.72) \approx 1.46$ and a local minimum value of $f(0.61) \approx 0.41$. $f''(x)$ does not exist at $x = 0$ and $f''(x) = 0$ ⟺ $x \approx -0.97, -0.46, -0.12,$ and 1.11, so f is CU on $(-\infty, -0.97)$, CD on $(-0.97, -0.46)$, CU on $(-0.46, -0.12)$, CD on $(-0.12, 0)$, CD on $(0, 1.11)$, and CU on $(1.11, \infty)$. There are inflection points at $(-0.97, 1.08)$, $(-0.46, 1.01)$, $(-0.12, 0.28)$, and $(1.11, 0.29)$.

17. $y = f(x) = \sqrt{x + 5 \sin x}, \ x \le 20$.

From a CAS, $y' = \dfrac{5 \cos x + 1}{2\sqrt{x + 5 \sin x}}$ and $y'' = -\dfrac{10 \cos x + 25 \sin^2 x + 10x \sin x + 26}{4(x + 5 \sin x)^{3/2}}$.

We'll start with a graph of $g(x) = x + 5 \sin x$. Note that $f(x) = \sqrt{g(x)}$ is only defined if $g(x) \ge 0$. $g(x) = 0$ ⟺ $x = 0$ or $x \approx -4.91, -4.10, 4.10,$ and 4.91. Thus, the domain of f is $[-4.91, -4.10] \cup [0, 4.10] \cup [4.91, 20]$.

From the expression for y', we see that $y' = 0$ ⟺ $5 \cos x + 1 = 0$ ⟹ $x_1 = \cos^{-1}\left(-\frac{1}{5}\right) \approx 1.77$ and $x_2 = 2\pi - x_1 \approx -4.51$ (not in the domain of f). The leftmost zero of f' is $x_1 - 2\pi \approx -4.51$. Moving to the right, the zeros of f' are x_1, $x_1 + 2\pi$, $x_2 + 2\pi$, $x_1 + 4\pi$, and $x_2 + 4\pi$. Thus, f is increasing on $(-4.91, -4.51)$, decreasing on $(-4.51, -4.10)$, increasing on $(0, 1.77)$, decreasing on $(1.77, 4.10)$, increasing on $(4.91, 8.06)$, decreasing on $(8.06, 10.79)$, increasing on $(10.79, 14.34)$, decreasing on $(14.34, 17.08)$, and increasing on $(17.08, 20)$. The local maximum values are $f(-4.51) \approx 0.62$, $f(1.77) \approx 2.58$, $f(8.06) \approx 3.60$, and $f(14.34) \approx 4.39$. The local minimum values are $f(10.79) \approx 2.43$ and $f(17.08) \approx 3.49$.

f is CD on $(-4.91, -4.10)$, $(0, 4.10)$, $(4.91, 9.60)$, CU on $(9.60, 12.25)$, CD on $(12.25, 15.81)$, CU on $(15.81, 18.65)$, and CD on $(18.65, 20)$. There are inflection points at $(9.60, 2.95)$, $(12.25, 3.27)$, $(15.81, 3.91)$, and $(18.65, 4.20)$.

18. $f(x) = \dfrac{2x - 1}{\sqrt[4]{x^4 + x + 1}}$ $\Rightarrow$ $f'(x) = \dfrac{4x^3 + 6x + 9}{4(x^4 + x + 1)^{5/4}}$ $\Rightarrow$ $f''(x) = -\dfrac{32x^6 + 96x^4 + 152x^3 - 48x^2 + 6x + 21}{16(x^4 + x + 1)^{9/4}}$

From the graph of f', f appears to be decreasing on $(-\infty, -0.94)$ and increasing on $(-0.94, \infty)$. There is a local minimum value of $f(-0.94) \approx -3.01$. From the graph of f'', f appears to be CU on $(-1.25, -0.44)$ and CD on $(-\infty, -1.25)$ and $(-0.44, \infty)$. There are inflection points at $(-1.25, -2.87)$ and $(-0.44, -2.14)$.

19.

From the graph of $f(x) = \sin(x + \sin 3x)$ in the viewing rectangle $[0, \pi]$ by $[-1.2, 1.2]$, it looks like f has two maxima and two minima. If we calculate and graph $f'(x) = [\cos(x + \sin 3x)](1 + 3\cos 3x)$ on $[0, 2\pi]$, we see that the graph of f' appears to be almost tangent to the x-axis at about $x = 0.7$. The graph of

$f'' = -[\sin(x + \sin 3x)](1 + 3\cos 3x)^2 + \cos(x + \sin 3x)(-9\sin 3x)$ is even more interesting near this x-value: it seems to just touch the x-axis.

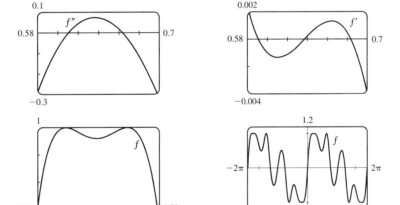

If we zoom in on this place on the graph of f'', we see that f'' actually does cross the axis twice near $x = 0.65$, indicating a change in concavity for a very short interval. If we look at the graph of f' on the same interval, we see that it changes sign three times near $x = 0.65$, indicating that what we had thought was a broad extremum at about $x = 0.7$ actually consists of three extrema (two maxima and a minimum). These maximum values are roughly $f(0.59) = 1$ and $f(0.68) = 1$, and the minimum value is roughly $f(0.64) = 0.99996$. There are also a maximum value of about $f(1.96) = 1$ and minimum values of about $f(1.46) = 0.49$ and $f(2.73) = -0.51$. The points of inflection on $(0, \pi)$ are about $(0.61, 0.99998)$, $(0.66, 0.99998)$, $(1.17, 0.72)$, $(1.75, 0.77)$, and $(2.28, 0.34)$. On $(\pi, 2\pi)$, they are about $(4.01, -0.34)$, $(4.54, -0.77)$, $(5.11, -0.72)$, $(5.62, -0.99998)$, and $(5.67, -0.99998)$. There are also IP at $(0, 0)$ and $(\pi, 0)$. Note that the function is odd and periodic with period 2π, and it is also rotationally symmetric about all points of the form $((2n + 1)\pi, 0)$, n an integer.

20. $f(x) = x^3 + cx = x(x^2 + c)$ $\Rightarrow$ $f'(x) = 3x^2 + c$ $\Rightarrow$ $f''(x) = 6x$

$c = -6$

$c = 0$

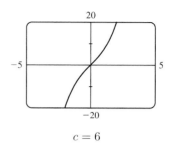
$c = 6$

x-intercepts: When $c \geq 0$, 0 is the only x-intercept. When $c < 0$, the x-intercepts are 0 and $\pm\sqrt{-c}$.

y-intercept $= f(0) = 0$. f is odd, so the graph is symmetric with respect to the origin. $f''(x) < 0$ for $x < 0$ and $f''(x) > 0$ for $x > 0$, so f is CD on $(-\infty, 0)$ and CU on $(0, \infty)$. The origin is the only inflection point.

If $c > 0$, then $f'(x) > 0$ for all x, so f is increasing and has no local maximum or minimum.

If $c = 0$, then $f'(x) \geq 0$ with equality at $x = 0$, so again f is increasing and has no local maximum or minimum.

If $c < 0$, then $f'(x) = 3[x^2 - (-c/3)] = 3\left(x + \sqrt{-c/3}\right)\left(x - \sqrt{-c/3}\right)$, so $f'(x) > 0$ on $\left(-\infty, -\sqrt{-c/3}\right)$ and $\left(\sqrt{-c/3}, \infty\right)$; $f'(x) < 0$ on $\left(-\sqrt{-c/3}, \sqrt{-c/3}\right)$. It follows that $f\left(-\sqrt{-c/3}\right) = -\frac{2}{3}c\sqrt{-c/3}$ is a local maximum value and $f\left(\sqrt{-c/3}\right) = \frac{2}{3}c\sqrt{-c/3}$ is a local minimum value. As c decreases (toward more negative values), the local maximum and minimum move further apart.

There is no absolute maximum or minimum value. The only transitional value of c corresponding to a change in character of the graph is $c = 0$.

21. $f(x) = x^4 + cx^2 = x^2(x^2 + c)$. Note that f is an even function. For $c \geq 0$, the only x-intercept is the point $(0, 0)$. We calculate $f'(x) = 4x^3 + 2cx = 4x\left(x^2 + \frac{1}{2}c\right)$ $\Rightarrow$ $f''(x) = 12x^2 + 2c$. If $c \geq 0$, $x = 0$ is the only critical point and there is no inflection point. As we can see from the examples, there is no change in the basic shape of the graph for $c \geq 0$; it merely becomes steeper as c increases. For $c = 0$, the graph is the simple curve $y = x^4$. For $c < 0$, there are x-intercepts at 0 and at $\pm\sqrt{-c}$. Also, there is a maximum at $(0,0)$, and there are

minima at $\left(\pm\sqrt{-\frac{1}{2}c}, -\frac{1}{4}c^2\right)$. As $c \to -\infty$, the x-coordinates of

these minima get larger in absolute value, and the minimum points

move downward. There are inflection points at $\left(\pm\sqrt{-\frac{1}{6}c}, -\frac{5}{36}c^2\right)$,

which also move away from the origin as $c \to -\infty$.

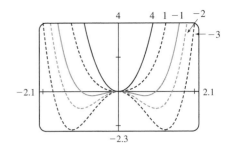

22. With $c = 0$ in $y = f(x) = x\sqrt{c^2 - x^2}$, the graph of f is just the point $(0,0)$. Since $(-c)^2 = c^2$, we only

consider $c > 0$. Since $f(-x) = -f(x)$, the graph is symmetric about the origin. The domain of f is found by

solving $c^2 - x^2 \geq 0 \iff x^2 \leq c^2 \iff |x| \leq c$, which gives us $[-c, c]$.

$$f'(x) = x \cdot \tfrac{1}{2}(c^2 - x^2)^{-1/2}(-2x) + (c^2 - x^2)^{1/2}(1) = (c^2 - x^2)^{-1/2}[-x^2 + (c^2 - x^2)] = \frac{c^2 - 2x^2}{\sqrt{c^2 - x^2}}.$$

$f'(x) > 0 \iff c^2 - 2x^2 > 0 \iff x^2 < c^2/2 \iff |x| < c/\sqrt{2}$, so f is increasing on

$\left(-c/\sqrt{2}, c/\sqrt{2}\right)$ and decreasing on $\left(-c, -c/\sqrt{2}\right)$ and $\left(c/\sqrt{2}, c\right)$. There is a local minimum value of

$f\left(-c/\sqrt{2}\right) = \left(-c/\sqrt{2}\right)\sqrt{c^2 - c^2/2} = \left(-c/\sqrt{2}\right)\left(c/\sqrt{2}\right) = -c^2/2$ and a local maximum value of $f\left(c/\sqrt{2}\right) = c^2/2$.

$$f''(x) = \frac{(c^2 - x^2)^{1/2}(-4x) - (c^2 - 2x^2)\tfrac{1}{2}\left(c^2 - x^2\right)^{-1/2}(-2x)}{[(c^2 - x^2)^{1/2}]^2}$$

$$= \frac{x(c^2 - x^2)^{-1/2}[(c^2 - x^2)(-4) + (c^2 - 2x^2)]}{(c^2 - x^2)^1} = \frac{2x\left(2x^2 - 3c^2\right)}{(c^2 - x^2)^{3/2}},$$

so $f''(x) = 0 \iff x = 0$ or $x = \pm\sqrt{\tfrac{3}{2}}\,c$, but only 0 is in the domain of f.

$f''(x) < 0$ for $0 < x < c$ and $f''(x) > 0$ for $-c < x < 0$, so f is CD on $(0, c)$

and CU on $(-c, 0)$. There is an IP at $(0, 0)$. So as $|c|$ gets larger, the maximum and

minimum values increase in magnitude. The value of c does not affect the

concavity of f.

23. Note that $c = 0$ is a transitional value at which the graph consists of the x-axis. Also, we can see that if we substitute $-c$ for c,

the function $f(x) = \dfrac{cx}{1 + c^2 x^2}$ will be reflected in the x-axis, so we investigate only positive values of c (except $c = -1$, as a

demonstration of this reflective property). Also, f is an odd function. $\displaystyle\lim_{x \to \pm\infty} f(x) = 0$, so $y = 0$ is a horizontal asymptote

for all c. We calculate $f'(x) = \dfrac{(1 + c^2 x^2)c - cx(2c^2 x)}{(1 + c^2 x^2)^2} = -\dfrac{c(c^2 x^2 - 1)}{(1 + c^2 x^2)^2}$. $f'(x) = 0 \iff c^2 x^2 - 1 = 0 \iff$

$x = \pm 1/c$. So there is an absolute maximum value of $f(1/c) = \tfrac{1}{2}$ and an absolute minimum value of $f(-1/c) = -\tfrac{1}{2}$. These

extrema

have the same value regardless of c, but the maximum points move closer to the y-axis as c increases.

$$f''(x) = \frac{(-2c^3x)(1+c^2x^2)^2 - (-c^3x^2+c)[2(1+c^2x^2)(2c^2x)]}{(1+c^2x^2)^4}$$

$$= \frac{(-2c^3x)(1+c^2x^2) + (c^3x^2-c)(4c^2x)}{(1+c^2x^2)^3} = \frac{2c^3x(c^2x^2-3)}{(1+c^2x^2)^3}$$

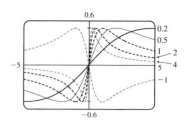

$f''(x) = 0 \iff x = 0$ or $\pm\sqrt{3}/c$, so there are inflection points at $(0,0)$ and

at $(\pm\sqrt{3}/c, \pm\sqrt{3}/4)$. Again, the y-coordinate of the inflection points does not

depend on c, but as c increases, both inflection points approach the y-axis.

24. Note that $f(x) = \dfrac{1}{(1-x^2)^2 + cx^2}$ is an even function, and also that

$\displaystyle\lim_{x\to\pm\infty} f(x) = 0$ for any value of c, so $y = 0$ is a horizontal asymptote.

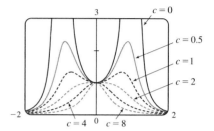

We calculate the derivatives:

$$f'(x) = \frac{-4(1-x^2)x + 2cx}{[(1-x^2)^2 + cx^2]^2} = \frac{4x\left[x^2 + \left(\frac{1}{2}c-1\right)\right]}{[(1-x^2)^2 + cx^2]^2}, \text{ and}$$

$$f''(x) = 2\frac{10x^6 + (9c-18)x^4 + (3c^2-12c+6)x^2 + 2 - c}{[x^4 + (c-2)x^2 + 1]^3}.$$

We first consider the case $c > 0$. Then the denominator of f' is positive, that is, $(1-x^2)^2 + cx^2 > 0$ for all x, so f has

domain $\mathbb{R}$ and also $f > 0$. If $\frac{1}{2}c - 1 \geq 0$; that is, $c \geq 2$, then the only critical point is $f(0) = 1$, a maximum. Graphing a few

examples for $c \geq 2$ shows that there are two IP which approach the y-axis as $c \to \infty$.

$\quad c = 2$ and $c = 0$ are transitional values of c at which the shape of the curve changes. For $0 < c < 2$, there are three critical

points: $f(0) = 1$, a minimum value, and $f\left(\pm\sqrt{1 - \frac{1}{2}c}\right) = \dfrac{1}{c(1-c/4)}$, both maximum values. As c decreases from 2 to 0,

the maximum values get larger and larger, and the x-values at which they occur go from 0 to ± 1. Graphs show that there are

four inflection points for $0 < c < 2$, and that they get farther away from the origin, both vertically and horizontally, as

$c \to 0^+$. For $c = 0$, the function is simply asymptotic to the x-axis and to the lines $x = \pm 1$, approaching $+\infty$ from both sides

of each. The y-intercept is 1, and $(0, 1)$ is a local minimum. There are no inflection points. Now if $c < 0$, we can write

$$f(x) = \frac{1}{(1-x^2)^2 + cx^2} = \frac{1}{(1-x^2)^2 - \left(\sqrt{-c}\,x\right)^2} = \frac{1}{\left(x^2 - \sqrt{-c}\,x - 1\right)\left(x^2 + \sqrt{-c}\,x - 1\right)}. \text{ So } f \text{ has vertical}$$

asymptotes where $x^2 \pm \sqrt{-c}\,x - 1 = 0 \iff x = \left(-\sqrt{-c} \pm \sqrt{4-c}\right)\big/2$ or $x = \left(\sqrt{-c} \pm \sqrt{4-c}\right)/2$. As c decreases,

the two exterior asymptotes move away from the origin, while the two interior ones move toward it. We graph a few examples

to see the behavior of the graph near the asymptotes, and the nature of the critical points $x = 0$ and $x = \pm\sqrt{1 - \frac{1}{2}c}$:

$c = -1$

$c = -2$

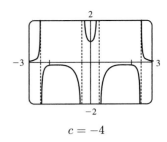

$c = -4$

We see that there is one local minimum value, $f(0) = 1$, and there are two local maximum values,

$$f\left(\pm\sqrt{1 - \tfrac{1}{2}c}\right) = \frac{1}{c(1 - c/4)}$$ as before. As c decreases, the x-values at which these maxima occur get larger, and the

maximum values themselves approach 0, though they are always negative.

25. $f(x) = cx + \sin x \quad \Rightarrow \quad f'(x) = c + \cos x \quad \Rightarrow \quad f''(x) = -\sin x$

$f(-x) = -f(x)$, so f is an odd function and its graph is symmetric with respect to the origin.

$f(x) = 0 \quad \Leftrightarrow \quad \sin x = -cx$, so 0 is always an x-intercept.

$f'(x) = 0 \quad \Leftrightarrow \quad \cos x = -c$, so there is no critical number when $|c| > 1$. If $|c| \le 1$, then there are infinitely

many critical numbers. If x_1 is the unique solution of $\cos x = -c$ in the interval $[0, \pi]$, then the critical numbers are $2n\pi \pm x_1$,

where n ranges over the integers. (Special cases: When $c = 1$, $x_1 = 0$; when $c = 0$, $x = \frac{\pi}{2}$; and when $c = -1$, $x_1 = \pi$.)

$f''(x) < 0 \quad \Leftrightarrow \quad \sin x > 0$, so f is CD on intervals of the form $(2n\pi, (2n+1)\pi)$. f is CU on intervals of the form

$((2n-1)\pi, 2n\pi)$. The inflection points of f are the points $(2n\pi, 2n\pi c)$, where n is an integer.

If $c \ge 1$, then $f'(x) \ge 0$ for all x, so f is increasing and has no extremum. If $c \le -1$, then $f'(x) \le 0$ for all x, so f is

decreasing and has no extremum. If $|c| < 1$, then $f'(x) > 0 \quad \Leftrightarrow \quad \cos x > -c \quad \Leftrightarrow \quad x$ is in an interval of the form

$(2n\pi - x_1, 2n\pi + x_1)$ for some integer n. These are the intervals on which f is increasing. Similarly, we

find that f is decreasing on the intervals of the form $(2n\pi + x_1, 2(n+1)\pi - x_1)$. Thus, f has local maxima at the points

$2n\pi + x_1$, where f has the values $c(2n\pi + x_1) + \sin x_1 = c(2n\pi + x_1) + \sqrt{1 - c^2}$, and f has local minima at the points

$2n\pi - x_1$, where we have $f(2n\pi - x_1) = c(2n\pi - x_1) - \sin x_1 = c(2n\pi - x_1) - \sqrt{1 - c^2}$.

The transitional values of c are -1 and 1. The inflection points move vertically, but not horizontally, when c changes.

When $|c| \ge 1$, there is no extremum. For $|c| < 1$, the maxima are spaced
2π apart horizontally, as are the minima. The horizontal spacing between
maxima and adjacent minima is regular (and equals π) when $c = 0$, but
the horizontal space between a local maximum and the nearest local
minimum shrinks as $|c|$ approaches 1.

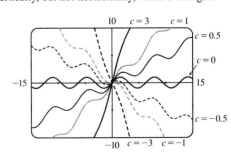

26. For $c = 0$, there is no inflection point; the curve is CU everywhere. If c increases, the curve simply becomes steeper, and there are still no inflection points. If c starts at 0 and decreases, a slight upward bulge appears near $x = 0$, so that there are two inflection points for any $c < 0$. This can be seen algebraically by calculating the second derivative:

$$f(x) = x^4 + cx^2 + x \quad \Rightarrow \quad f'(x) = 4x^3 + 2cx + 1 \quad \Rightarrow \quad f''(x) = 12x^2 + 2c. \text{ Thus, } f''(x) > 0 \text{ when } c > 0. \text{ For } c < 0,$$

there are inflection points when $x = \pm\sqrt{-\frac{1}{6}c}$. For $c = 0$, the graph has one critical number, at the absolute minimum somewhere around $x = -0.6$. As c increases, the number of critical points does not change. If c instead decreases from 0, we see that the graph eventually sprouts another local minimum, to the right of the origin, somewhere between $x = 1$ and $x = 2$. Consequently, there is also a maximum near $x = 0$.

After a bit of experimentation, we find that at $c = -1.5$, there appear to be two critical numbers: the absolute minimum at about $x = -1$, and a horizontal tangent with no extremum at about $x = 0.5$. For any c smaller than this there will be 3 critical points, as shown in the graphs with $c = -3$ and with $c = -5$.

To prove this algebraically, we calculate $f'(x) = 4x^3 + 2cx + 1$. Now if we substitute our value of $c = -1.5$, the formula for $f'(x)$ becomes

$4x^3 - 3x + 1 = (x + 1)(2x - 1)^2$. This has a double root at $x = \frac{1}{2}$, indicating that the function has two critical points: $x = -1$ and $x = \frac{1}{2}$, just as we had guessed from the graph.

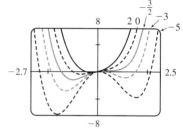

27. (a) $f(x) = cx^4 - 2x^2 + 1$. For $c = 0$, $f(x) = -2x^2 + 1$, a parabola whose vertex, $(0, 1)$, is the absolute maximum. For $c > 0$, $f(x) = cx^4 - 2x^2 + 1$ opens upward with two minimum points. As $c \to 0$, the minimum points spread apart and move downward; they are below the x-axis for $0 < c < 1$ and above for $c > 1$. For $c < 0$, the graph opens downward, and has an absolute maximum at $x = 0$ and no local minimum.

(b) $f'(x) = 4cx^3 - 4x = 4cx(x^2 - 1/c)$ $[c \neq 0]$. If $c \leq 0$, 0 is the only critical number.

$f''(x) = 12cx^2 - 4$, so $f''(0) = -4$ and there is a local maximum at $(0, f(0)) = (0, 1)$, which lies on $y = 1 - x^2$. If $c > 0$, the critical numbers are 0 and $\pm 1/\sqrt{c}$. As before, there is a local maximum at $(0, f(0)) = (0, 1)$, which lies on $y = 1 - x^2$.

$f''\left(\pm 1/\sqrt{c}\right) = 12 - 4 = 8 > 0$, so there is a local minimum at $x = \pm 1/\sqrt{c}$. Here $f\left(\pm 1/\sqrt{c}\right) = c(1/c^2) - 2/c + 1 = -1/c + 1$.

But $\left(\pm 1/\sqrt{c}, -1/c + 1\right)$ lies on $y = 1 - x^2$ since $1 - \left(\pm 1/\sqrt{c}\right)^2 = 1 - 1/c$.

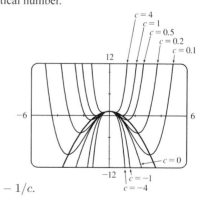

28. (a) $f(x) = 2x^3 + cx^2 + 2x \quad \Rightarrow \quad f'(x) = 6x^2 + 2cx + 2 = 2(3x^2 + cx + 1)$. $f'(x) = 0 \quad \Leftrightarrow \quad x = \dfrac{-c \pm \sqrt{c^2 - 12}}{6}$.

So f has critical points $\quad \Leftrightarrow \quad c^2 - 12 \geq 0 \quad \Leftrightarrow \quad |c| \geq 2\sqrt{3}$. For $c = \pm 2\sqrt{3}$, $f'(x) \geq 0$ on $(-\infty, \infty)$, so f' does not change signs at $-c/6$, and there is no extremum. If $c^2 - 12 > 0$, then f' changes from positive to negative at

$x = \dfrac{-c - \sqrt{c^2 - 12}}{6}$ and from negative to positive at $x = \dfrac{-c + \sqrt{c^2 - 12}}{6}$. So f has a local maximum at

$x = \dfrac{-c - \sqrt{c^2 - 12}}{6}$ and a local minimum at $x = \dfrac{-c + \sqrt{c^2 - 12}}{6}$.

(b) Let x_0 be a critical number for $f(x)$. Then $f'(x_0) = 0 \;\Rightarrow$

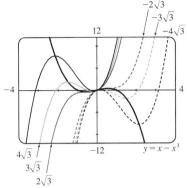

$3x_0^2 + cx_0 + 1 = 0 \;\Leftrightarrow\; c = \dfrac{-1 - 3x_0^2}{x_0}$. Now

$$f(x_0) = 2x_0^3 + cx_0^2 + 2x_0 = 2x_0^3 + x_0^2 \left(\dfrac{-1 - 3x_0^2}{x_0} \right) + 2x_0$$

$$= 2x_0^3 - x_0 - 3x_0^3 + 2x_0 = x_0 - x_0^3$$

So the point is $(x_0, y_0) = \left(x_0, x_0 - x_0^3\right)$; that is, the point lies

on the curve $y = x - x^3$.

4.7 Optimization Problems

1. (a)

First Number	Second Number	Product
1	22	22
2	21	42
3	20	60
4	19	76
5	18	90
6	17	102
7	16	112
8	15	120
9	14	126
10	13	130
11	12	132

We needn't consider pairs where the first number is larger than the second, since we can just interchange the numbers in such cases. The answer appears to be 11 and 12, but we have considered only integers in the table.

(b) Call the two numbers x and y. Then $x + y = 23$, so $y = 23 - x$. Call the product P. Then

$P = xy = x(23 - x) = 23x - x^2$, so we wish to maximize the function $P(x) = 23x - x^2$. Since $P'(x) = 23 - 2x$,

we see that $P'(x) = 0 \;\Leftrightarrow\; x = \frac{23}{2} = 11.5$. Thus, the maximum value of P is $P(11.5) = (11.5)^2 = 132.25$ and it

occurs when $x = y = 11.5$.

Or: Note that $P''(x) = -2 < 0$ for all x, so P is everywhere concave downward and the local maximum at $x = 11.5$

must be an absolute maximum.

2. The two numbers are $x + 100$ and x. Minimize $f(x) = (x + 100)x = x^2 + 100x$. $f'(x) = 2x + 100 = 0 \;\Rightarrow\; x = -50$.

Since $f''(x) = 2 > 0$, there is an absolute minimum at $x = -50$. The two numbers are 50 and -50.

3. The two numbers are x and $\dfrac{100}{x}$, where $x > 0$. Minimize $f(x) = x + \dfrac{100}{x}$. $f'(x) = 1 - \dfrac{100}{x^2} = \dfrac{x^2 - 100}{x^2}$. The critical

number is $x = 10$. Since $f'(x) < 0$ for $0 < x < 10$ and $f'(x) > 0$ for $x > 10$, there is an absolute minimum at $x = 10$.

The numbers are 10 and 10.

4. Let $x > 0$ and let $f(x) = x + 1/x$. We wish to minimize $f(x)$.

Now $f'(x) = 1 - \dfrac{1}{x^2} = \dfrac{1}{x^2}(x^2 - 1) = \dfrac{1}{x^2}(x + 1)(x - 1)$, so the only critical number in $(0, \infty)$ is 1.

$f'(x) < 0$ for $0 < x < 1$ and $f'(x) > 0$ for $x > 1$, so f has an absolute minimum at $x = 1$, and $f(1) = 2$.

Or: $f''(x) = 2/x^3 > 0$ for all $x > 0$, so f is concave upward everywhere and the critical point $(1, 2)$ must correspond to a

local minimum for f.

5. If the rectangle has dimensions x and y, then its perimeter is $2x + 2y = 100$ m, so $y = 50 - x$. Thus, the area is

$A = xy = x(50 - x)$. We wish to maximize the function $A(x) = x(50 - x) = 50x - x^2$, where $0 < x < 50$. Since

$A'(x) = 50 - 2x = -2(x - 25)$, $A'(x) > 0$ for $0 < x < 25$ and $A'(x) < 0$ for $25 < x < 50$. Thus, A has an absolute

maximum at $x = 25$, and $A(25) = 25^2 = 625$ m^2. The dimensions of the rectangle that maximize its area are $x = y = 25$ m.

(The rectangle is a square.)

6. If the rectangle has dimensions x and y, then its area is $xy = 1000$ m^2, so $y = 1000/x$. The perimeter

$P = 2x + 2y = 2x + 2000/x$. We wish to minimize the function $P(x) = 2x + 2000/x$ for $x > 0$.

$P'(x) = 2 - 2000/x^2 = (2/x^2)(x^2 - 1000)$, so the only critical number in the domain of P is $x = \sqrt{1000}$.

$P''(x) = 4000/x^3 > 0$, so P is concave upward throughout its domain and $P(\sqrt{1000}) = 4\sqrt{1000}$ is an absolute minimum

value. The dimensions of the rectangle with minimal perimeter are $x = y = \sqrt{1000} = 10\sqrt{10}$ m. (The rectangle is a square.)

7. We need to maximize Y for $N \geq 0$. $\quad Y(N) = \dfrac{kN}{1 + N^2} \quad \Rightarrow$

$Y'(N) = \dfrac{(1 + N^2)k - kN(2N)}{(1 + N^2)^2} = \dfrac{k(1 - N^2)}{(1 + N^2)^2} = \dfrac{k(1 + N)(1 - N)}{(1 + N^2)^2}$. $\quad Y'(N) > 0$ for $0 < N < 1$ and $Y'(N) < 0$

for $N > 1$. Thus, Y has an absolute maximum of $Y(1) = \frac{1}{2}k$ at $N = 1$.

8. We need to maximize P for $I \geq 0$. $\quad P(I) = \dfrac{100I}{I^2 + I + 4} \quad \Rightarrow$

$P'(I) = \dfrac{(I^2 + I + 4)(100) - 100I(2I + 1)}{(I^2 + I + 4)^2} = \dfrac{100(I^2 + I + 4 - 2I^2 - I)}{(I^2 + I + 4)^2} = \dfrac{-100(I^2 - 4)}{(I^2 + I + 4)^2} = \dfrac{-100(I + 2)(I - 2)}{(I^2 + I + 4)^2}$.

$P'(I) > 0$ for $0 < I < 2$ and $P'(I) < 0$ for $I > 2$. Thus, P has an absolute maximum of $P(2) = 20$ at $I = 2$.

9. (a)

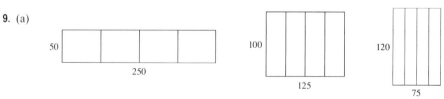

The areas of the three figures are 12,500, 12,500, and 9000 ft^2. There appears to be a maximum area of at least 12,500 ft^2.

(b) Let x denote the length of each of two sides and three dividers.

Let y denote the length of the other two sides.

(c) Area $A = $ length $\times$ width $= y \cdot x$

(d) Length of fencing $= 750 \quad \Rightarrow \quad 5x + 2y = 750$

(e) $5x + 2y = 750 \quad \Rightarrow \quad y = 375 - \frac{5}{2}x \quad \Rightarrow \quad A(x) = \left(375 - \frac{5}{2}x\right)x = 375x - \frac{5}{2}x^2$

(f) $A'(x) = 375 - 5x = 0 \quad \Rightarrow \quad x = 75$. Since $A''(x) = -5 < 0$ there is an absolute maximum when $x = 75$. Then $y = \frac{375}{2} = 187.5$. The largest area is $75\left(\frac{375}{2}\right) = 14{,}062.5$ ft^2. These values of x and y are between the values in the first and second figures in part (a). Our original estimate was low.

10. (a)

The volumes of the resulting boxes are 1, 1.6875, and 2 ft^3. There appears to be a maximum volume of at least 2 ft^3.

(b) Let x denote the length of the side of the square being cut out. Let y denote the length of the base.

(c) Volume $V = $ length $\times$ width $\times$ height $\quad \Rightarrow \quad V = y \cdot y \cdot x = xy^2$

(d) Length of cardboard $= 3 \quad \Rightarrow \quad x + y + x = 3 \quad \Rightarrow \quad y + 2x = 3$

(e) $y + 2x = 3 \quad \Rightarrow \quad y = 3 - 2x \quad \Rightarrow \quad V(x) = x(3 - 2x)^2$

(f) $V(x) = x(3 - 2x)^2 \quad \Rightarrow$

$V'(x) = x \cdot 2(3 - 2x)(-2) + (3 - 2x)^2 \cdot 1 = (3 - 2x)[-4x + (3 - 2x)] = (3 - 2x)(-6x + 3)$,

so the critical numbers are $x = \frac{3}{2}$ and $x = \frac{1}{2}$. Now $0 \leq x \leq \frac{3}{2}$ and $V(0) = V\left(\frac{3}{2}\right) = 0$, so the maximum is $V\left(\frac{1}{2}\right) = \left(\frac{1}{2}\right)(2)^2 = 2$ ft^3, which is the value found from our third figure in part (a).

11.

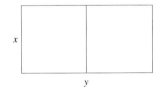

$xy = 1.5 \times 10^6$, so $y = 1.5 \times 10^6/x$. Minimize the amount of fencing, which is

$3x + 2y = 3x + 2(1.5 \times 10^6/x) = 3x + 3 \times 10^6/x = F(x)$.

$F'(x) = 3 - 3 \times 10^6/x^2 = 3(x^2 - 10^6)/x^2$. The critical number is $x = 10^3$ and $F'(x) < 0$ for $0 < x < 10^3$ and $F'(x) > 0$ if $x > 10^3$, so the absolute minimum occurs when $x = 10^3$ and $y = 1.5 \times 10^3$.

The field should be 1000 feet by 1500 feet with the middle fence parallel to the short side of the field.

12. Let b be the length of the base of the box and h the height. The volume is $32{,}000 = b^2 h \quad \Rightarrow \quad h = 32{,}000/b^2$.

The surface area of the open box is $S = b^2 + 4hb = b^2 + 4(32{,}000/b^2)b = b^2 + 4(32{,}000)/b$.

So $S'(b) = 2b - 4(32{,}000)/b^2 = 2(b^3 - 64{,}000)/b^2 = 0 \quad \Leftrightarrow \quad b = \sqrt[3]{64{,}000} = 40$. This gives an absolute minimum since $S'(b) < 0$ if $0 < b < 40$ and $S'(b) > 0$ if $b > 40$. The box should be $40 \times 40 \times 20$.

13. Let b be the length of the base of the box and h the height. The surface area is $1200 = b^2 + 4hb$ $\Rightarrow$ $h = (1200 - b^2)/(4b)$.

The volume is $V = b^2 h = b^2(1200 - b^2)/4b = 300b - b^3/4$ $\Rightarrow$ $V'(b) = 300 - \frac{3}{4}b^2$.

$V'(b) = 0$ $\Rightarrow$ $300 = \frac{3}{4}b^2$ $\Rightarrow$ $b^2 = 400$ $\Rightarrow$ $b = \sqrt{400} = 20$. Since $V'(b) > 0$ for $0 < b < 20$ and $V'(b) < 0$ for

$b > 20$, there is an absolute maximum when $b = 20$ by the First Derivative Test for Absolute Extreme Values (see page 259).

If $b = 20$, then $h = (1200 - 20^2)/(4 \cdot 20) = 10$, so the largest possible volume is $b^2 h = (20)^2(10) = 4000$ cm^3.

14.

$V = lwh$ $\Rightarrow$ $10 = (2w)(w)h = 2w^2 h$, so $h = 5/w^2$.

The cost is $10(2w^2) + 6[2(2wh) + 2(hw)] = 20w^2 + 36wh$, so

$C(w) = 20w^2 + 36w(5/w^2) = 20w^2 + 180/w$.

$C'(w) = 40w - 180/w^2 = 40\left(w^3 - \frac{9}{2}\right)/w^2$ $\Rightarrow$ $w = \sqrt[3]{\frac{9}{2}}$ is the critical number. There is an absolute minimum

for C when $w = \sqrt[3]{\frac{9}{2}}$ since $C'(w) < 0$ for $0 < w < \sqrt[3]{\frac{9}{2}}$ and $C'(w) > 0$ for $w > \sqrt[3]{\frac{9}{2}}$.

$C\left(\sqrt[3]{\frac{9}{2}}\right) = 20\left(\sqrt[3]{\frac{9}{2}}\right)^2 + \frac{180}{\sqrt[3]{9/2}} \approx \$163.54$.

15.

$10 = (2w)(w)h = 2w^2 h$, so $h = 5/w^2$. The cost is

$C(w) = 10(2w^2) + 6[2(2wh) + 2hw] + 6(2w^2)$
$= 32w^2 + 36wh = 32w^2 + 180/w$

$C'(w) = 64w - 180/w^2 = 4(16w^3 - 45)/w^2$ $\Rightarrow$ $w = \sqrt[3]{\frac{45}{16}}$ is the critical number. $C'(w) < 0$ for $0 < w < \sqrt[3]{\frac{45}{16}}$ and

$C'(w) > 0$ for $w > \sqrt[3]{\frac{45}{16}}$. The minimum cost is $C\left(\sqrt[3]{\frac{45}{16}}\right) = 32(2.8125)^{2/3} + 180/\sqrt{2.8125} \approx \$191.28$.

16. (a) Let the rectangle have sides x and y and area A, so $A = xy$ or $y = A/x$. The problem is to minimize the

perimeter $= 2x + 2y = 2x + 2A/x = P(x)$. Now $P'(x) = 2 - 2A/x^2 = 2(x^2 - A)/x^2$. So the critical number is

$x = \sqrt{A}$. Since $P'(x) < 0$ for $0 < x < \sqrt{A}$ and $P'(x) > 0$ for $x > \sqrt{A}$, there is an absolute minimum at $x = \sqrt{A}$.

The sides of the rectangle are $\sqrt{A}$ and $A/\sqrt{A} = \sqrt{A}$, so the rectangle is a square.

(b) Let p be the perimeter and x and y the lengths of the sides, so $p = 2x + 2y$ $\Rightarrow$ $2y = p - 2x$ $\Rightarrow$ $y = \frac{1}{2}p - x$.

The area is $A(x) = x\left(\frac{1}{2}p - x\right) = \frac{1}{2}px - x^2$. Now $A'(x) = 0$ $\Rightarrow$ $\frac{1}{2}p - 2x = 0$ $\Rightarrow$ $2x = \frac{1}{2}p$ $\Rightarrow$ $x = \frac{1}{4}p$. Since

$A''(x) = -2 < 0$, there is an absolute maximum for A when $x = \frac{1}{4}p$ by the Second Derivative Test. The sides of the

rectangle are $\frac{1}{4}p$ and $\frac{1}{2}p - \frac{1}{4}p = \frac{1}{4}p$, so the rectangle is a square.

17. The distance from a point (x, y) on the line $y = 4x + 7$ to the origin is $\sqrt{(x-0)^2 + (y-0)^2} = \sqrt{x^2 + y^2}$. However, it is

easier to work with the *square* of the distance; that is, $D(x) = \left(\sqrt{x^2 + y^2}\right)^2 = x^2 + y^2 = x^2 + (4x + 7)^2$. Because the distance is positive, its minimum value will occur at the same point as the minimum value of D.

$D'(x) = 2x + 2(4x + 7)(4) = 34x + 56$, so $D'(x) = 0 \iff x = -\frac{28}{17}$.

$D''(x) = 34 > 0$, so D is concave upward for all x. Thus, D has an absolute minimum at $x = -\frac{28}{17}$. The point closest to the

origin is $(x, y) = \left(-\frac{28}{17}, 4\left(-\frac{28}{17}\right) + 7\right) = \left(-\frac{28}{17}, \frac{7}{17}\right)$.

18. The square of the distance from a point (x, y) on the line $y = -6x + 9$ to the point $(-3, 1)$ is

$D(x) = (x + 3)^2 + (y - 1)^2 = (x + 3)^2 + (-6x + 8)^2 = 37x^2 - 90x + 73$. $D'(x) = 74x - 90$, so $D'(x) = 0 \iff$

$x = \frac{45}{37}$. $D''(x) = 74 > 0$, so D is concave upward for all x. Thus, D has an absolute minimum at $x = \frac{45}{37}$. The point on the

line closest to $(-3, 1)$ is $\left(\frac{45}{37}, \frac{63}{37}\right)$.

19.

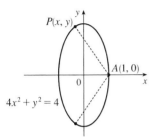

From the figure, we see that there are two points that are farthest away from

$A(1, 0)$. The distance d from A to an arbitrary point $P(x, y)$ on the ellipse is

$d = \sqrt{(x - 1)^2 + (y - 0)^2}$ and the square of the distance is

$S = d^2 = x^2 - 2x + 1 + y^2 = x^2 - 2x + 1 + (4 - 4x^2) = -3x^2 - 2x + 5$.

$S' = -6x - 2$ and $S' = 0 \implies x = -\frac{1}{3}$. Now $S'' = -6 < 0$, so we know

that S has a maximum at $x = -\frac{1}{3}$. Since $-1 \le x \le 1$, $S(-1) = 4$,

$S\left(-\frac{1}{3}\right) = \frac{16}{3}$, and $S(1) = 0$, we see that the maximum distance is $\sqrt{\frac{16}{3}}$. The corresponding y-values are

$y = \pm\sqrt{4 - 4\left(-\frac{1}{3}\right)^2} = \pm\sqrt{\frac{32}{9}} = \pm\frac{4}{3}\sqrt{2} \approx \pm 1.89$. The points are $\left(-\frac{1}{3}, \pm\frac{4}{3}\sqrt{2}\right)$.

20.

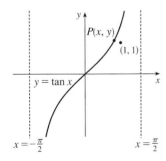

The distance d from $(1, 1)$ to an arbitrary point $P(x, y)$ on the curve

$y = \tan x$ is $d = \sqrt{(x - 1)^2 + (y - 1)^2}$ and the square of the distance is

$S = d^2 = (x - 1)^2 + (\tan x - 1)^2$. $S' = 2(x - 1) + 2(\tan x - 1)\sec^2 x$.

Graphing S' on $\left(-\frac{\pi}{2}, \frac{\pi}{2}\right)$ gives us a zero at $x \approx 0.82$, and so $\tan x \approx 1.08$.

The point on $y = \tan x$ that is closest to $(1, 1)$ is approximately $(0.82, 1.08)$.

21.

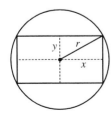

The area of the rectangle is $(2x)(2y) = 4xy$. Also $r^2 = x^2 + y^2$ so

$y = \sqrt{r^2 - x^2}$, so the area is $A(x) = 4x\sqrt{r^2 - x^2}$. Now

$A'(x) = 4\left(\sqrt{r^2 - x^2} - \frac{x^2}{\sqrt{r^2 - x^2}}\right) = 4\frac{r^2 - 2x^2}{\sqrt{r^2 - x^2}}$. The critical number is

$x = \frac{1}{\sqrt{2}}r$. Clearly this gives a maximum.

$y = \sqrt{r^2 - \left(\frac{1}{\sqrt{2}}r\right)^2} = \sqrt{\frac{1}{2}r^2} = \frac{1}{\sqrt{2}}r = x$, which tells us that the rectangle is a square. The dimensions are $2x = \sqrt{2}r$

and $2y = \sqrt{2}r$.

22.

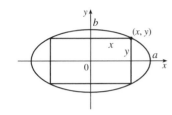

The area of the rectangle is $(2x)(2y) = 4xy$. Now $\dfrac{x^2}{a^2} + \dfrac{y^2}{b^2} = 1$ gives

$y = \dfrac{b}{a}\sqrt{a^2 - x^2}$, so we maximize $A(x) = 4\dfrac{b}{a}x\sqrt{a^2 - x^2}$.

$A'(x) = \dfrac{4b}{a}\left[x \cdot \tfrac{1}{2}(a^2 - x^2)^{-1/2}(-2x) + (a^2 - x^2)^{1/2} \cdot 1\right]$

$\qquad = \dfrac{4b}{a}(a^2 - x^{2-1/2}[-x^2 + a^2 - x^2] = \dfrac{4b}{a\sqrt{a^2 - x^2}}[a^2 - 2x^2]$

So the critical number is $x = \dfrac{1}{\sqrt{2}}a$, and this clearly gives a maximum. Then $y = \dfrac{1}{\sqrt{2}}b$, so the maximum area

is $4\left(\dfrac{1}{\sqrt{2}}a\right)\left(\dfrac{1}{\sqrt{2}}b\right) = 2ab$.

23.

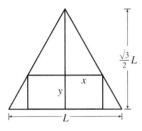

The height h of the equilateral triangle with sides of length L is $\dfrac{\sqrt{3}}{2}L$,

since $h^2 + (L/2)^2 = L^2 \;\Rightarrow\; h^2 = L^2 - \tfrac{1}{4}L^2 = \tfrac{3}{4}L^2 \;\Rightarrow$

$h = \dfrac{\sqrt{3}}{2}L$. Using similar triangles, $\dfrac{\frac{\sqrt{3}}{2}L - y}{x} = \dfrac{\frac{\sqrt{3}}{2}L}{L/2} = \sqrt{3} \;\Rightarrow$

$\sqrt{3}\,x = \dfrac{\sqrt{3}}{2}L - y \;\Rightarrow\; y = \dfrac{\sqrt{3}}{2}L - \sqrt{3}\,x \;\Rightarrow\; y = \dfrac{\sqrt{3}}{2}(L - 2x).$

The area of the inscribed rectangle is $A(x) = (2x)y = \sqrt{3}\,x(L - 2x) = \sqrt{3}\,Lx - 2\sqrt{3}\,x^2$, where $0 \le x \le L/2$. Now

$0 = A'(x) = \sqrt{3}\,L - 4\sqrt{3}\,x \;\Rightarrow\; x = \sqrt{3}\,L/(4\sqrt{3}) = L/4$. Since $A(0) = A(L/2) = 0$, the maximum occurs when

$x = L/4$, and $y = \dfrac{\sqrt{3}}{2}L - \dfrac{\sqrt{3}}{4}L = \dfrac{\sqrt{3}}{4}L$, so the dimensions are $L/2$ and $\dfrac{\sqrt{3}}{4}L$.

24.

$y = 8 - x^2$

The rectangle has area $A(x) = 2xy = 2x(8 - x^2) = 16x - 2x^3$, where

$0 \le x \le 2\sqrt{2}$. Now $A'(x) = 16 - 6x^2 = 0 \;\Rightarrow\; x = 2\sqrt{\tfrac{2}{3}}$. Since

$A(0) = A(2\sqrt{2}) = 0$, there is a maximum when $x = 2\sqrt{\tfrac{2}{3}}$. Then $y = \dfrac{16}{3}$,

so the rectangle has dimensions $4\sqrt{\tfrac{2}{3}}$ and $\dfrac{16}{3}$.

25.

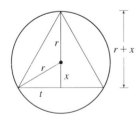

The area of the triangle is

$A(x) = \tfrac{1}{2}(2t)(r + x) = t(r + x) = \sqrt{r^2 - x^2}\,(r + x)$. Then

$0 = A'(x) = r\dfrac{-2x}{2\sqrt{r^2 - x^2}} + \sqrt{r^2 - x^2} + x\dfrac{-2x}{2\sqrt{r^2 - x^2}}$

$\qquad = -\dfrac{x^2 + rx}{\sqrt{r^2 - x^2}} + \sqrt{r^2 - x^2} \;\Rightarrow$

$\dfrac{x^2 + rx}{\sqrt{r^2 - x^2}} = \sqrt{r^2 - x^2} \;\Rightarrow\; x^2 + rx = r^2 - x^2 \;\Rightarrow\; 0 = 2x^2 + rx - r^2 = (2x - r)(x + r) \;\Rightarrow$

$x = \tfrac{1}{2}r$ or $x = -r$. Now $A(r) = 0 = A(-r) \;\Rightarrow\;$ the maximum occurs where $x = \tfrac{1}{2}r$, so the triangle has height

$r + \tfrac{1}{2}r = \tfrac{3}{2}r$ and base $2\sqrt{r^2 - (\tfrac{1}{2}r)^2} = 2\sqrt{\tfrac{3}{4}r^2} = \sqrt{3}\,r$.

26.

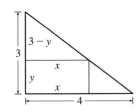

The rectangle has area xy. By similar triangles $\dfrac{3-y}{x} = \dfrac{3}{4}$ $\Rightarrow$

$-4y + 12 = 3x$ or $y = -\frac{3}{4}x + 3$. So the area is

$A(x) = x\left(-\frac{3}{4}x + 3\right) = -\frac{3}{4}x^2 + 3x$ where $0 \le x \le 4$. Now

$0 = A'(x) = -\frac{3}{2}x + 3$ $\Rightarrow$ $x = 2$ and $y = \frac{3}{2}$. Since $A(0) = A(4) = 0$,

the maximum area is $A(2) = 2\left(\frac{3}{2}\right) = 3$ cm^2.

27.

The cylinder has volume $V = \pi y^2 (2x)$. Also $x^2 + y^2 = r^2$ $\Rightarrow$ $y^2 = r^2 - x^2$, so

$V(x) = \pi(r^2 - x^2)(2x) = 2\pi(r^2 x - x^3)$, where $0 \le x \le r$.

$V'(x) = 2\pi\left(r^2 - 3x^2\right) = 0$ $\Rightarrow$ $x = r/\sqrt{3}$. Now $V(0) = V(r) = 0$, so there is a

maximum when $x = r/\sqrt{3}$ and $V\left(r/\sqrt{3}\right) = \pi(r^2 - r^2/3)\left(2r/\sqrt{3}\right) = 4\pi r^3 / \left(3\sqrt{3}\right)$.

28.

By similar triangles, $y/x = h/r$, so $y = hx/r$. The volume of the cylinder is

$\pi x^2 (h - y) = \pi h x^2 - (\pi h/r)x^3 = V(x)$. Now

$V'(x) = 2\pi h x - (3\pi h/r)x^2 = \pi h x(2 - 3x/r)$.

So $V'(x) = 0$ $\Rightarrow$ $x = 0$ or $x = \frac{2}{3}r$. The maximum clearly occurs when

$x = \frac{2}{3}r$ and then the volume is

$\pi h x^2 - (\pi h/r)x^3 = \pi h x^2 (1 - x/r) = \pi \left(\frac{2}{3}r\right)^2 h \left(1 - \frac{2}{3}\right) = \frac{4}{27}\pi r^2 h$.

29.

The cylinder has surface area

$2(\text{area of the base}) + (\text{lateral surface area}) = 2\pi(\text{radius})^2 + 2\pi(\text{radius})(\text{height})$

$$= 2\pi y^2 + 2\pi y(2x)$$

Now $x^2 + y^2 = r^2$ $\Rightarrow$ $y^2 = r^2 - x^2$ $\Rightarrow$ $y = \sqrt{r^2 - x^2}$, so the surface area is

$$S(x) = 2\pi(r^2 - x^2) + 4\pi x\sqrt{r^2 - x^2}, \ \ 0 \le x \le r$$

$$= 2\pi r^2 - 2\pi x^2 + 4\pi\left(x\sqrt{r^2 - x^2}\right)$$

Thus, $S'(x) = 0 - 4\pi x + 4\pi\left[x \cdot \frac{1}{2}(r^2 - x^2)^{-1/2}(-2x) + (r^2 - x^2)^{1/2} \cdot 1\right]$

$$= 4\pi\left[-x - \frac{x^2}{\sqrt{r^2 - x^2}} + \sqrt{r^2 - x^2}\right] = 4\pi \cdot \frac{-x\sqrt{r^2 - x^2} - x^2 + r^2 - x^2}{\sqrt{r^2 - x^2}}$$

$S'(x) = 0$ $\Rightarrow$ $x\sqrt{r^2 - x^2} = r^2 - 2x^2$ ($\star$) $\Rightarrow$ $\left(x\sqrt{r^2 - x^2}\right)^2 = (r^2 - 2x^2)^2$ $\Rightarrow$

$x^2(r^2 - x^2) = r^4 - 4r^2 x^2 + 4x^4$ $\Rightarrow$ $r^2 x^2 - x^4 = r^4 - 4r^2 x^2 + 4x^4$ $\Rightarrow$ $5x^4 - 5r^2 x^2 + r^4 = 0$.

This is a quadratic equation in x^2. By the quadratic formula, $x^2 = \frac{5 \pm \sqrt{5}}{10}r^2$, but we reject the root with the $+$ sign since it

doesn't satisfy ($\star$). [The right side is negative and the left side is positive.] So $x = \sqrt{\frac{5 - \sqrt{5}}{10}}\ r$. Since $S(0) = S(r) = 0$, the maximum surface area occurs at the critical number and $x^2 = \frac{5 - \sqrt{5}}{10}r^2 \;\Rightarrow\; y^2 = r^2 - \frac{5 - \sqrt{5}}{10}r^2 = \frac{5 + \sqrt{5}}{10}r^2 \;\Rightarrow\;$ the surface area is

$$2\pi\left(\frac{5+\sqrt{5}}{10}\right)r^2 + 4\pi\sqrt{\frac{5-\sqrt{5}}{10}}\sqrt{\frac{5+\sqrt{5}}{10}}r^2 = \pi r^2\left[2 \cdot \frac{5+\sqrt{5}}{10} + 4\frac{\sqrt{(5-\sqrt{5})(5+\sqrt{5})}}{10}\right] = \pi r^2\left[\frac{5+\sqrt{5}}{5} + \frac{2\sqrt{20}}{5}\right]$$

$$= \pi r^2\left[\frac{5+\sqrt{5}+2\cdot2\sqrt{5}}{5}\right] = \pi r^2\left[\frac{5+5\sqrt{5}}{5}\right] = \pi r^2\left(1+\sqrt{5}\right).$$

30.

Perimeter $= 30 \;\Rightarrow\; 2y + x + \pi\left(\frac{x}{2}\right) = 30 \;\Rightarrow\;$

$y = \frac{1}{2}\left(30 - x - \frac{\pi x}{2}\right) = 15 - \frac{x}{2} - \frac{\pi x}{4}$. The area is the area of the rectangle plus the area of

the semicircle, or $xy + \frac{1}{2}\pi\left(\frac{x}{2}\right)^2$, so $A(x) = x\left(15 - \frac{x}{2} - \frac{\pi x}{4}\right) + \frac{1}{8}\pi x^2 = 15x - \frac{1}{2}x^2 - \frac{\pi}{8}x^2$.

$A'(x) = 15 - \left(1 + \frac{\pi}{4}\right)x = 0 \;\Rightarrow\; x = \dfrac{15}{1 + \pi/4} = \dfrac{60}{4 + \pi}$. $A''(x) = -\left(1 + \frac{\pi}{4}\right) < 0$, so this gives a maximum.

The dimensions are $x = \dfrac{60}{4 + \pi}$ ft and $y = 15 - \dfrac{30}{4 + \pi} - \dfrac{15\pi}{4 + \pi} = \dfrac{60 + 15\pi - 30 - 15\pi}{4 + \pi} = \dfrac{30}{4 + \pi}$ ft, so the height of the

rectangle is half the base.

31.

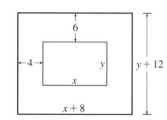

$xy = 384 \;\Rightarrow\; y = 384/x$. Total area is

$A(x) = (8 + x)(12 + 384/x) = 12(40 + x + 256/x)$, so

$A'(x) = 12(1 - 256/x^2) = 0 \;\Rightarrow\; x = 16$. There is an absolute minimum

when $x = 16$ since $A'(x) < 0$ for $0 < x < 16$ and $A'(x) > 0$ for $x > 16$.

When $x = 16$, $y = 384/16 = 24$, so the dimensions are 24 cm and 36 cm.

32.

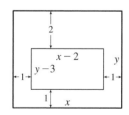

$xy = 180$, so $y = 180/x$. The printed area is

$(x - 2)(y - 3) = (x - 2)(180/x - 3) = 186 - 3x - 360/x = A(x)$.

$A'(x) = -3 + 360/x^2 = 0$ when $x^2 = 120 \;\Rightarrow\; x = 2\sqrt{30}$. This gives an absolute

maximum since $A'(x) > 0$ for $0 < x < 2\sqrt{30}$ and $A'(x) < 0$ for $x > 2\sqrt{30}$. When

$x = 2\sqrt{30}$, $y = 180/(2\sqrt{30})$, so the dimensions are $2\sqrt{30}$ in. and $90/\sqrt{30}$ in.

33.

Let x be the length of the wire used for the square. The total area is

$$A(x) = \left(\frac{x}{4}\right)^2 + \frac{1}{2}\left(\frac{10 - x}{3}\right)\frac{\sqrt{3}}{2}\left(\frac{10 - x}{3}\right)$$

$$= \frac{1}{16}x^2 + \frac{\sqrt{3}}{36}(10 - x)^2, \; 0 \le x \le 10$$

$A'(x) = \frac{1}{8}x - \frac{\sqrt{3}}{18}(10 - x) = 0 \;\Leftrightarrow\; \frac{9}{72}x + \frac{4\sqrt{3}}{72}x - \frac{40\sqrt{3}}{72} = 0 \;\Leftrightarrow\; x = \frac{40\sqrt{3}}{9 + 4\sqrt{3}}$. Now $A(0) = \left(\frac{\sqrt{3}}{36}\right)100 \approx 4.81$,

$A(10) = \frac{100}{16} = 6.25$ and $A\left(\frac{40\sqrt{3}}{9 + 4\sqrt{3}}\right) \approx 2.72$, so

(a) The maximum area occurs when $x = 10$ m, and all the wire is used for the square.

(b) The minimum area occurs when $x = \dfrac{40\sqrt{3}}{9 + 4\sqrt{3}} \approx 4.35$ m.

34.

Total area is $A(x) = \left(\dfrac{x}{4}\right)^2 + \pi\left(\dfrac{10-x}{2\pi}\right)^2 = \dfrac{x^2}{16} + \dfrac{(10-x)^2}{4\pi}$, $0 \le x \le 10$.

$A'(x) = \dfrac{x}{8} - \dfrac{10-x}{2\pi} = \left(\dfrac{1}{2\pi} + \dfrac{1}{8}\right)x - \dfrac{5}{\pi} = 0 \;\Rightarrow\; x = 40/(4+\pi)$.

$A(0) = 25/\pi \approx 7.96$, $A(10) = 6.25$, and $A(40/(4+\pi)) \approx 3.5$, so the maximum

occurs when $x = 0$ m and the minimum occurs when $x = 40/(4+\pi)$ m.

35.

The volume is $V = \pi r^2 h$ and the surface area is

$$S(r) = \pi r^2 + 2\pi rh = \pi r^2 + 2\pi r\left(\dfrac{V}{\pi r^2}\right) = \pi r^2 + \dfrac{2V}{r}.$$

$$S'(r) = 2\pi r - \dfrac{2V}{r^2} = 0 \;\Rightarrow\; 2\pi r^3 = 2V \;\Rightarrow\; r = \sqrt[3]{\dfrac{V}{\pi}} \text{ cm}.$$

This gives an absolute minimum since $S'(r) < 0$ for $0 < r < \sqrt[3]{\dfrac{V}{\pi}}$ and $S'(r) > 0$ for $r > \sqrt[3]{\dfrac{V}{\pi}}$.

When $r = \sqrt[3]{\dfrac{V}{\pi}}$, $h = \dfrac{V}{\pi r^2} = \dfrac{V}{\pi (V/\pi)^{2/3}} = \sqrt[3]{\dfrac{V}{\pi}}$ cm.

36.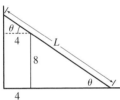

$L = 8\csc\theta + 4\sec\theta$, $0 < \theta < \dfrac{\pi}{2}$, $\dfrac{dL}{d\theta} = -8\csc\theta\cot\theta + 4\sec\theta\tan\theta = 0$ when

$\sec\theta\tan\theta = 2\csc\theta\cot\theta \;\Leftrightarrow\; \tan^3\theta = 2 \;\Leftrightarrow\; \tan\theta = \sqrt[3]{2} \;\Leftrightarrow\; \theta = \tan^{-1}\sqrt[3]{2}$.

$dL/d\theta < 0$ when $0 < \theta < \tan^{-1}\sqrt[3]{2}$, $dL/d\theta > 0$ when $\tan^{-1}\sqrt[3]{2} < \theta < \dfrac{\pi}{2}$, so L has

an absolute minimum when $\theta = \tan^{-1}\sqrt[3]{2}$, and the shortest ladder has length

$$L = 8\dfrac{\sqrt{1+2^{\,2/3}}}{2^{1/3}} + 4\sqrt{1+2^{\,2/3}} \approx 16.65 \text{ ft}.$$

Another method: Minimize $L^2 = x^2 + (4+y)^2$, where $\dfrac{x}{4+y} = \dfrac{8}{y}$.

37.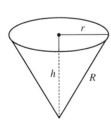

$h^2 + r^2 = R^2 \;\Rightarrow\; V = \dfrac{\pi}{3}r^2 h = \dfrac{\pi}{3}(R^2 - h^2)h = \dfrac{\pi}{3}(R^2 h - h^3)$.

$V'(h) = \dfrac{\pi}{3}(R^2 - 3h^2) = 0$ when $h = \dfrac{1}{\sqrt{3}}R$. This gives an absolute maximum, since

$V'(h) > 0$ for $0 < h < \dfrac{1}{\sqrt{3}}R$ and $V'(h) < 0$ for $h > \dfrac{1}{\sqrt{3}}R$. The maximum volume is

$$V\left(\dfrac{1}{\sqrt{3}}R\right) = \dfrac{\pi}{3}\left(\dfrac{1}{\sqrt{3}}R^3 - \dfrac{1}{3\sqrt{3}}R^3\right) = \dfrac{2}{9\sqrt{3}}\pi R^3.$$

38. The volume and surface area of a cone with radius r and height h are given by $V = \dfrac{1}{3}\pi r^2 h$ and $S = \pi r\sqrt{r^2 + h^2}$.

We'll minimize $A = S^2$ subject to $V = 27$. $V = 27 \;\Rightarrow\; \dfrac{1}{3}\pi r^2 h = 27 \;\Rightarrow\; r^2 = \dfrac{81}{\pi h}$ **(1)**.

$A = \pi^2 r^2(r^2 + h^2) = \pi^2\left(\dfrac{81}{\pi h}\right)\left(\dfrac{81}{\pi h} + h^2\right) = \dfrac{81^2}{h^2} + 81\pi h$, so $A' = 0 \;\Rightarrow\; \dfrac{-2\cdot 81^2}{h^3} + 81\pi = 0 \;\Rightarrow$

$$81\pi = \frac{2 \cdot 81^2}{h^3} \quad \Rightarrow \quad h^3 = \frac{162}{\pi} \quad \Rightarrow \quad h = \sqrt[3]{\frac{162}{\pi}} = 3\sqrt[3]{\frac{6}{\pi}} \approx 3.722. \text{ From (1), } r^2 = \frac{81}{\pi h} = \frac{81}{\pi \cdot 3\sqrt[3]{6/\pi}} = \frac{27}{\sqrt[3]{6\pi^2}} \quad \Rightarrow$$

$r = \dfrac{3\sqrt{3}}{\sqrt[6]{6\pi^2}} \approx 2.632.$ $A'' = 6 \cdot 81^2/h^4 > 0,$ so A and hence S has an absolute minimum at these values of r and h.

39.

By similar triangles, $\dfrac{H}{R} = \dfrac{H-h}{r}$ **(1)**. The volume of the inner cone is $V = \frac{1}{3}\pi r^2 h,$

so we'll solve **(1)** for h. $\dfrac{Hr}{R} = H - h \quad \Rightarrow$

$h = H - \dfrac{Hr}{R} = \dfrac{HR - Hr}{R} = \dfrac{H}{R}(R - r)$ **(2)**.

Thus, $V(r) = \dfrac{\pi}{3}r^2 \cdot \dfrac{H}{R}(R-r) = \dfrac{\pi H}{3R}(Rr^2 - r^3) \quad \Rightarrow$

$V'(r) = \dfrac{\pi H}{3R}(2Rr - 3r^2) = \dfrac{\pi H}{3R}r(2R - 3r).$

$V'(r) = 0 \quad \Rightarrow \quad r = 0 \text{ or } 2R = 3r \quad \Rightarrow \quad r = \frac{2}{3}R \text{ and from (2), } h = \dfrac{H}{R}\left(R - \frac{2}{3}R\right) = \dfrac{H}{R}\left(\frac{1}{3}R\right) = \frac{1}{3}H.$

$V'(r)$ changes from positive to negative at $r = \frac{2}{3}R,$ so the inner cone has a maximum volume of

$V = \frac{1}{3}\pi r^2 h = \frac{1}{3}\pi\left(\frac{2}{3}R\right)^2\left(\frac{1}{3}H\right) = \frac{4}{27} \cdot \frac{1}{3}\pi R^2 H,$ which is approximately 15% of the volume of the larger cone.

40. We need to minimize F for $0 \le \theta < \pi/2.$ $F(\theta) = \dfrac{\mu W}{\mu \sin\theta + \cos\theta} \quad \Rightarrow \quad F'(\theta) = \dfrac{-\mu W(\mu\cos\theta - \sin\theta)}{(\mu\sin\theta + \cos\theta)^2}$ [by the

Reciprocal Rule]. $F'(\theta) > 0 \quad \Rightarrow \quad \mu\cos\theta - \sin\theta < 0 \quad \Rightarrow \quad \mu\cos\theta < \sin\theta \quad \Rightarrow \quad \mu < \tan\theta \quad \Rightarrow \quad \theta > \tan^{-1}\mu.$

So F is decreasing on $\left(0, \tan^{-1}\mu\right)$ and increasing on $\left(\tan^{-1}\mu, \frac{\pi}{2}\right).$ Thus, F attains its minimum value at $\theta = \tan^{-1}\mu.$

This maximum value is $F(\tan^{-1}\mu) = \dfrac{\mu W}{\sqrt{\mu^2 + 1}}.$

41. $P(R) = \dfrac{E^2 R}{(R+r)^2} \quad \Rightarrow$

$$P'(R) = \frac{(R+r)^2 \cdot E^2 - E^2 R \cdot 2(R+r)}{[(R+r)^2]^2} = \frac{(R^2 + 2Rr + r^2)E^2 - 2E^2 R^2 - 2E^2 Rr}{(R+r)^4}$$

$$= \frac{E^2 r^2 - E^2 R^2}{(R+r)^4} = \frac{E^2(r^2 - R^2)}{(R+r)^4} = \frac{E^2(r+R)(r-R)}{(R+r)^4} = \frac{E^2(r-R)}{(R+r)^3}$$

$P'(R) = 0 \quad \Rightarrow \quad R = r \quad \Rightarrow \quad P(r) = \dfrac{E^2 r}{(r+r)^2} = \dfrac{E^2 r}{4r^2} = \dfrac{E^2}{4r}.$

The expression for $P'(R)$ shows that $P'(R) > 0$ for $R < r$ and $P'(R) < 0$ for $R > r.$ Thus, the maximum value of the

power is $E^2/(4r),$ and this occurs when $R = r.$

42. (a) $E(v) = \dfrac{aLv^3}{v-u} \quad \Rightarrow \quad E'(v) = aL\dfrac{(v-u)3v^2 - v^3}{(v-u)^2} = 0$ when

(b)

$2v^3 = 3uv^2 \quad \Rightarrow \quad 2v = 3u \quad \Rightarrow \quad v = \frac{3}{2}u.$

The First Derivative Test shows that this value of v gives the minimum

value of $E.$

43. $S = 6sh - \frac{3}{2}s^2 \cot\theta + 3s^2 \frac{\sqrt{3}}{2}\csc\theta$

(a) $\frac{dS}{d\theta} = \frac{3}{2}s^2\csc^2\theta - 3s^2 \frac{\sqrt{3}}{2}\csc\theta\cot\theta$ or $\frac{3}{2}s^2\csc\theta\left(\csc\theta - \sqrt{3}\cot\theta\right)$.

(b) $\frac{dS}{d\theta} = 0$ when $\csc\theta - \sqrt{3}\cot\theta = 0$ $\Rightarrow$ $\frac{1}{\sin\theta} - \sqrt{3}\frac{\cos\theta}{\sin\theta} = 0$ $\Rightarrow$ $\cos\theta = \frac{1}{\sqrt{3}}$. The First Derivative Test shows

that the minimum surface area occurs when $\theta = \cos^{-1}\left(\frac{1}{\sqrt{3}}\right) \approx 55°$.

(c)

If $\cos\theta = \frac{1}{\sqrt{3}}$, then $\cot\theta = \frac{1}{\sqrt{2}}$ and $\csc\theta = \frac{\sqrt{3}}{\sqrt{2}}$, so the surface area is

$$S = 6sh - \frac{3}{2}s^2\frac{1}{\sqrt{2}} + 3s^2\frac{\sqrt{3}}{2}\frac{\sqrt{3}}{\sqrt{2}} = 6sh - \frac{3}{2\sqrt{2}}s^2 + \frac{9}{2\sqrt{2}}s^2$$

$$= 6sh + \frac{6}{2\sqrt{2}}s^2 = 6s\left(h + \frac{1}{2\sqrt{2}}s\right)$$

44.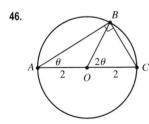

15 km/h

20 km/h

Let t be the time, in hours, after 2:00 PM. The position of the boat heading south at time t is $(0, -20t)$. The position of the boat heading east at time t is $(-15 + 15t, 0)$. If $D(t)$ is the distance between the boats at time t, we minimize $f(t) = [D(t)]^2 = 20^2 t^2 + 15^2(t-1)^2$.

$f'(t) = 800t + 450(t-1) = 1250t - 450 = 0$ when $t = \frac{450}{1250} = 0.36$ h.

0.36 h $\times \frac{60\text{ min}}{\text{h}} = 21.6$ min $= 21$ min 36 s. Since $f''(t) > 0$, this gives a minimum, so the boats are closest together

at 2:21:36 PM.

45. Here $T(x) = \frac{\sqrt{x^2+25}}{6} + \frac{5-x}{8}$, $0 \le x \le 5$ $\Rightarrow$ $T'(x) = \frac{x}{6\sqrt{x^2+25}} - \frac{1}{8} = 0$ $\Leftrightarrow$ $8x = 6\sqrt{x^2+25}$ $\Leftrightarrow$

$16x^2 = 9(x^2+25)$ $\Leftrightarrow$ $x = \frac{15}{\sqrt{7}}$. But $\frac{15}{\sqrt{7}} > 5$, so T has no critical number. Since $T(0) \approx 1.46$ and $T(5) \approx 1.18$, he

should row directly to B.

46.

In isosceles triangle AOB, $\angle O = 180° - \theta - \theta$, so $\angle BOC = 2\theta$. The distance rowed is $4\cos\theta$ while the distance walked is the length of arc $BC = 2(2\theta) = 4\theta$. The time taken is given by $T(\theta) = \frac{4\cos\theta}{2} + \frac{4\theta}{4} = 2\cos\theta + \theta$, $0 \le \theta \le \frac{\pi}{2}$.

$T'(\theta) = -2\sin\theta + 1 = 0$ $\Leftrightarrow$ $\sin\theta = \frac{1}{2}$ $\Rightarrow$ $\theta = \frac{\pi}{6}$.

Check the value of T at $\theta = \frac{\pi}{6}$ and at the endpoints of the domain of T; that is, $\theta = 0$ and $\theta = \frac{\pi}{2}$.

$T(0) = 2$, $T\left(\frac{\pi}{6}\right) = \sqrt{3} + \frac{\pi}{6} \approx 2.26$, and $T\left(\frac{\pi}{2}\right) = \frac{\pi}{2} \approx 1.57$. Therefore, the minimum value of T is $\frac{\pi}{2}$ when $\theta = \frac{\pi}{2}$; that is,

the woman should walk all the way. Note that $T''(\theta) = -2\cos\theta < 0$ for $0 \le \theta < \frac{\pi}{2}$, so $\theta = \frac{\pi}{6}$ gives a maximum time.

47. There are $(6 - x)$ km over land and $\sqrt{x^2 + 4}$ km under the river.

We need to minimize the cost C (measured in \$100,000) of the pipeline.

$C(x) = (6 - x)(4) + \left(\sqrt{x^2 + 4}\right)(8) \quad \Rightarrow$

$C'(x) = -4 + 8 \cdot \frac{1}{2}(x^2 + 4)^{-1/2}(2x) = -4 + \dfrac{8x}{\sqrt{x^2 + 4}}.$

$C'(x) = 0 \quad \Rightarrow \quad 4 = \dfrac{8x}{\sqrt{x^2 + 4}} \quad \Rightarrow \quad \sqrt{x^2 + 4} = 2x \quad \Rightarrow \quad x^2 + 4 = 4x^2 \quad \Rightarrow \quad 4 = 3x^2 \quad \Rightarrow \quad x^2 = \frac{4}{3} \quad \Rightarrow$

$x = 2/\sqrt{3}$ $[0 \le x \le 6]$. Compare the costs for $x = 0, 2/\sqrt{3}$, and 6. $C(0) = 24 + 16 = 40$,

$C\left(2/\sqrt{3}\right) = 24 - 8/\sqrt{3} + 32/\sqrt{3} = 24 + 24/\sqrt{3} \approx 37.9$, and $C(6) = 0 + 8\sqrt{40} \approx 50.6$. So the minimum cost is about

\$3.79 million when P is $6 - 2/\sqrt{3} \approx 4.85$ km east of the refinery.

48. The distance from the refinery to P is now $\sqrt{(6 - x)^2 + 1^2} = \sqrt{x^2 - 12x + 37}$.

Thus, $C(x) = 4\sqrt{x^2 - 12x + 37} + 8\sqrt{x^2 + 4} \quad \Rightarrow$

$C'(x) = 4 \cdot \frac{1}{2}(x^2 - 12x + 37)^{-1/2}(2x - 12) + 8 \cdot \frac{1}{2}(x^2 + 4)^{-1/2}(2x) = \dfrac{4(x - 6)}{\sqrt{x^2 - 12x + 37}} + \dfrac{8x}{\sqrt{x^2 + 4}}.$

$C'(x) = 0 \quad \Rightarrow \quad x \approx 1.12$ [from a graph of C' or a numerical rootfinder]. $C(0) \approx 40.3$, $C(1.12) \approx 38.3$, and

$C(6) \approx 54.6$. So the minimum cost is slightly higher (than in the previous exercise) at about \$3.83 million when P is

approximately 4.88 km from the point on the bank 1 km south of the refinery.

49.

The total illumination is $I(x) = \dfrac{3k}{x^2} + \dfrac{k}{(10 - x)^2}$, $0 < x < 10$. Then

$I'(x) = \dfrac{-6k}{x^3} + \dfrac{2k}{(10 - x)^3} = 0 \quad \Rightarrow \quad 6k(10 - x)^3 = 2kx^3 \quad \Rightarrow$

$3(10 - x)^3 = x^3 \quad \Rightarrow \quad \sqrt[3]{3}\,(10 - x) = x \quad \Rightarrow \quad 10\sqrt[3]{3} - \sqrt[3]{3}\,x = x \quad \Rightarrow \quad 10\sqrt[3]{3} = x + \sqrt[3]{3}\,x \quad \Rightarrow$

$10\sqrt[3]{3} = \left(1 + \sqrt[3]{3}\right)x \quad \Rightarrow \quad x = \dfrac{10\sqrt[3]{3}}{1 + \sqrt[3]{3}} \approx 5.9$ ft. This gives a minimum since $I''(x) > 0$ for $0 < x < 10$.

50.

The line with slope m (where $m < 0$) through $(3, 5)$ has equation $y - 5 = m(x - 3)$ or

$y = mx + (5 - 3m)$. The y-intercept is $5 - 3m$ and the x-intercept is $-5/m + 3$. So the

triangle has area $A(m) = \frac{1}{2}(5 - 3m)(-5/m + 3) = 15 - 25/(2m) - \frac{9}{2}m$. Now

$A'(m) = \dfrac{25}{2m^2} - \dfrac{9}{2} = 0 \quad \Leftrightarrow \quad m^2 = \dfrac{25}{9} \quad \Rightarrow \quad m = -\dfrac{5}{3}$ (since $m < 0$).

$A''(m) = -\dfrac{25}{m^3} > 0$, so there is an absolute minimum when $m = -\frac{5}{3}$. Thus, an equation of the line is $y - 5 = -\frac{5}{3}(x - 3)$

or $y = -\frac{5}{3}x + 10$.

51.

Every line segment in the first quadrant passing through (a, b) with endpoints on the x- and y-axes satisfies an equation of the form $y - b = m(x - a)$, where $m < 0$. By setting $x = 0$ and then $y = 0$, we find its endpoints, $A(0, b - am)$ and $B\left(a - \frac{b}{m}, 0\right)$. The distance d from A to B is given by $d = \sqrt{[(a - \frac{b}{m}) - 0]^2 + [0 - (b - am)]^2}$.

It follows that the square of the length of the line segment, as a function of m, is given by

$$S(m) = \left(a - \frac{b}{m}\right)^2 + (am - b)^2 = a^2 - \frac{2ab}{m} + \frac{b^2}{m^2} + a^2 m^2 - 2abm + b^2. \text{ Thus,}$$

$$S'(m) = \frac{2ab}{m^2} - \frac{2b^2}{m^3} + 2a^2 m - 2ab = \frac{2}{m^3}(abm - b^2 + a^2 m^4 - abm^3)$$

$$= \frac{2}{m^3}[b(am - b) + am^3(am - b)] = \frac{2}{m^3}(am - b)(b + am^3)$$

Thus, $S'(m) = 0 \iff m = b/a$ or $m = -\sqrt[3]{\frac{b}{a}}$. Since $b/a > 0$ and $m < 0$, m must equal $-\sqrt[3]{\frac{b}{a}}$. Since $\frac{2}{m^3} < 0$, we see

that $S'(m) < 0$ for $m < -\sqrt[3]{\frac{b}{a}}$ and $S'(m) > 0$ for $m > -\sqrt[3]{\frac{b}{a}}$. Thus, S has its absolute minimum value when $m = -\sqrt[3]{\frac{b}{a}}$.

That value is

$$S\left(-\sqrt[3]{\frac{b}{a}}\right) = \left(a + b\sqrt[3]{\frac{a}{b}}\right)^2 + \left(-a\sqrt[3]{\frac{b}{a}} - b\right)^2 = \left(a + \sqrt[3]{ab^2}\right)^2 + \left(\sqrt[3]{a^2 b} + b\right)^2$$

$$= a^2 + 2a^{4/3}b^{2/3} + a^{2/3}b^{4/3} + a^{4/3}b^{2/3} + 2a^{2/3}b^{4/3} + b^2 = a^2 + 3a^{4/3}b^{2/3} + 3a^{2/3}b^{4/3} + b^2$$

The last expression is of the form $x^3 + 3x^2 y + 3xy^2 + y^3 \quad [= (x + y)^3]$ with $x = a^{2/3}$ and $y = b^{2/3}$,

so we can write it as $(a^{2/3} + b^{2/3})^3$ and the shortest such line segment has length $\sqrt{S} = (a^{2/3} + b^{2/3})^{3/2}$.

52. $y = 1 + 40x^3 - 3x^5 \implies y' = 120x^2 - 15x^4$, so the tangent line to the curve at $x = a$ has slope $m(a) = 120a^2 - 15a^4$.

Now $m'(a) = 240a - 60a^3 = -60a(a^2 - 4) = -60a(a + 2)(a - 2)$, so $m'(a) > 0$ for $a < -2$, and $0 < a < 2$, and

$m'(a) < 0$ for $-2 < a < 0$ and $a > 2$. Thus, m is increasing on $(-\infty, -2)$, decreasing on $(-2, 0)$, increasing on $(0, 2)$, and

decreasing on $(2, \infty)$. Clearly, $m(a) \to -\infty$ as $a \to \pm\infty$, so the maximum value of $m(a)$ must be one of the two local

maxima, $m(-2)$ or $m(2)$. But both $m(-2)$ and $m(2)$ equal $120 \cdot 2^2 - 15 \cdot 2^4 = 480 - 240 = 240$. So 240 is the largest

slope, and it occurs at the points $(-2, -223)$ and $(2, 225)$. *Note:* $a = 0$ corresponds to a local *minimum* of m.

53. (a) If $c(x) = \dfrac{C(x)}{x}$, then, by Quotient Rule, we have $c'(x) = \dfrac{xC'(x) - C(x)}{x^2}$. Now $c'(x) = 0$ when $xC'(x) - C(x) = 0$

and this gives $C'(x) = \dfrac{C(x)}{x} = c(x)$. Therefore, the marginal cost equals the average cost.

(b) (i) $C(x) = 16{,}000 + 200x + 4x^{3/2}$, $C(1000) = 16{,}000 + 200{,}000 + 40{,}000\sqrt{10} \approx 216{,}000 + 126{,}491$, so

$C(1000) \approx \$342{,}491$. $c(x) = C(x)/x = \dfrac{16{,}000}{x} + 200 + 4x^{1/2}$, $c(1000) \approx \$342.49/\text{unit}$. $C'(x) = 200 + 6x^{1/2}$,

$C'(1000) = 200 + 60\sqrt{10} \approx \$389.74/\text{unit}$.

(ii) We must have $C'(x) = c(x)$ ⇔ $200 + 6x^{1/2} = \dfrac{16{,}000}{x} + 200 + 4x^{1/2}$ ⇔ $2x^{3/2} = 16{,}000$ ⇔

$x = (8{,}000)^{2/3} = 400$ units. To check that this is a minimum, we calculate

$c'(x) = \dfrac{-16{,}000}{x^2} + \dfrac{2}{\sqrt{x}} = \dfrac{2}{x^2}\left(x^{3/2} - 8000\right)$. This is negative for $x < (8000)^{2/3} = 400$, zero at $x = 400$,

and positive for $x > 400$, so c is decreasing on $(0, 400)$ and increasing on $(400, \infty)$. Thus, c has an absolute minimum

at $x = 400$. [*Note:* $c''(x)$ is *not* positive for all $x > 0$.]

(iii) The minimum average cost is $c(400) = 40 + 200 + 80 = \$320/\text{unit}$.

54. (a) The total profit is $P(x) = R(x) - C(x)$. In order to maximize profit we look for the critical numbers of P, that is, the

numbers where the marginal profit is 0. But if $P'(x) = R'(x) - C'(x) = 0$, then $R'(x) = C'(x)$. Therefore, if the profit

is a maximum, then the marginal revenue equals the marginal cost.

(b) $C(x) = 16{,}000 + 500x - 1.6x^2 + 0.004x^3$, $p(x) = 1700 - 7x$. Then $R(x) = xp(x) = 1700x - 7x^2$. If the profit is

maximum, then $R'(x) = C'(x)$ ⇔ $1700 - 14x = 500 - 3.2x + 0.012x^2$ ⇔ $0.012x^2 + 10.8x - 1200 = 0$ ⇔

$x^2 + 900x - 100{,}000 = 0$ ⇔ $(x + 1000)(x - 100) = 0$ ⇔ $x = 100$ (since $x > 0$). The profit is maximized if

$P''(x) < 0$, but since $P''(x) = R''(x) - C''(x)$, we can just check the condition $R''(x) < C''(x)$. Now

$R''(x) = -14 < -3.2 + 0.024x = C''(x)$ for $x > 0$, so there is a maximum at $x = 100$.

55. (a) We are given that the demand function p is linear and $p(27{,}000) = 10$, $p(33{,}000) = 8$, so the slope is

$\dfrac{10 - 8}{27{,}000 - 33{,}000} = -\dfrac{1}{3000}$ and an equation of the line is $y - 10 = \left(-\dfrac{1}{3000}\right)(x - 27{,}000)$ ⇒

$y = p(x) = -\dfrac{1}{3000}x + 19 = 19 - (x/3000)$.

(b) The revenue is $R(x) = xp(x) = 19x - (x^2/3000)$ ⇒ $R'(x) = 19 - (x/1500) = 0$ when $x = 28{,}500$. Since

$R''(x) = -1/1500 < 0$, the maximum revenue occurs when $x = 28{,}500$ ⇒ the price is $p(28{,}500) = \$9.50$.

56. (a) Let $p(x)$ be the demand function. Then $p(x)$ is linear and $y = p(x)$ passes through $(20, 10)$ and $(18, 11)$, so the slope is

$-\dfrac{1}{2}$ and an equation of the line is $y - 10 = -\dfrac{1}{2}(x - 20)$ ⇔ $y = -\dfrac{1}{2}x + 20$. Thus, the demand is $p(x) = -\dfrac{1}{2}x + 20$

and the revenue is $R(x) = xp(x) = -\dfrac{1}{2}x^2 + 20x$.

(b) The cost is $C(x) = 6x$, so the profit is $P(x) = R(x) - C(x) = -\dfrac{1}{2}x^2 + 14x$. Then $0 = P'(x) = -x + 14$ ⇒

$x = 14$. Since $P''(x) = -1 < 0$, the selling price for maximum profit is $p(14) = -\dfrac{1}{2}(14) + 20 = \$13$.

57. (a) As in Example 6, we see that the demand function p is linear. We are given that $p(1000) = 450$ and deduce that

$p(1100) = 440$, since a \$10 reduction in price increases sales by 100 per week. The slope for p is $\dfrac{440 - 450}{1100 - 1000} = -\dfrac{1}{10}$,

so an equation is $p - 450 = -\dfrac{1}{10}(x - 1000)$ or $p(x) = -\dfrac{1}{10}x + 550$.

(b) $R(x) = xp(x) = -\dfrac{1}{10}x^2 + 550x$. $R'(x) = -\dfrac{1}{5}x + 550 = 0$ when $x = 5(550) = 2750$.

$p(2750) = 275$, so the rebate should be $450 - 275 = \$175$.

(c) $C(x) = 68{,}000 + 150x$ $\Rightarrow$ $P(x) = R(x) - C(x) = -\frac{1}{10}x^2 + 550x - 68{,}000 - 150x = -\frac{1}{10}x^2 + 400x - 68{,}000$,

$P'(x) = -\frac{1}{5}x + 400 = 0$ when $x = 2000$. $p(2000) = 350$. Therefore, the rebate to maximize profits should be

$450 - 350 = \$100$.

58. Let x denote the number of \$10 increases in rent. Then the price is $p(x) = 800 + 10x$, and the number of units occupied is

$100 - x$. Now the revenue is

$$R(x) = (\text{rental price per unit}) \times (\text{number of units rented})$$
$$= (800 + 10x)(100 - x) = -10x^2 + 200x + 80{,}000 \text{ for } 0 \le x \le 100 \quad \Rightarrow$$

$R'(x) = -20x + 200 = 0$ $\Leftrightarrow$ $x = 10$. This is a maximum since $R''(x) = -20 < 0$ for all x. Now we must check the

value of $R(x) = (800 + 10x)(100 - x)$ at $x = 10$ and at the endpoints of the domain to see which value of x gives the

maximum value of R. $R(0) = 80{,}000$, $R(10) = (900)(90) = 81{,}000$, and $R(100) = (1800)(0) = 0$. Thus, the maximum

revenue of \$81,000/week occurs when 90 units are occupied at a rent of \$900/week.

59.

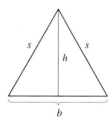

Here $s^2 = h^2 + b^2/4$, so $h^2 = s^2 - b^2/4$. The area is $A = \frac{1}{2}b\sqrt{s^2 - b^2/4}$.

Let the perimeter be p, so $2s + b = p$ or $s = (p - b)/2$ $\Rightarrow$

$A(b) = \frac{1}{2}b\sqrt{(p - b)^2/4 - b^2/4} = b\sqrt{p^2 - 2pb}/4$. Now

$$A'(b) = \frac{\sqrt{p^2 - 2pb}}{4} - \frac{bp/4}{\sqrt{p^2 - 2pb}} = \frac{-3pb + p^2}{4\sqrt{p^2 - 2pb}}.$$

Therefore, $A'(b) = 0$ $\Rightarrow$ $-3pb + p^2 = 0$ $\Rightarrow$ $b = p/3$. Since $A'(b) > 0$ for $b < p/3$ and $A'(b) < 0$ for $b > p/3$, there

is an absolute maximum when $b = p/3$. But then $2s + p/3 = p$, so $s = p/3$ $\Rightarrow$ $s = b$ $\Rightarrow$ the triangle is equilateral.

60. See the figure. The area is given by

$A(x) = \frac{1}{2}\left(2\sqrt{a^2 - x^2}\right)x + \frac{1}{2}\left(2\sqrt{a^2 - x^2}\right)\left(\sqrt{x^2 + b^2 - a^2}\right) = \sqrt{a^2 - x^2}\left(x + \sqrt{x^2 + b^2 - a^2}\right)$ for $0 \le x \le a$.

Now $A'(x) = \sqrt{a^2 - x^2}\left(1 + \dfrac{x}{\sqrt{x^2 + b^2 - a^2}}\right) + \left(x + \sqrt{x^2 + b^2 - a^2}\right)\dfrac{-x}{\sqrt{a^2 - x^2}} = 0$ $\Leftrightarrow$

$$\frac{x}{\sqrt{a^2 - x^2}}\left(x + \sqrt{x^2 + b^2 - a^2}\right) = \sqrt{a^2 - x^2}\left(\frac{x + \sqrt{x^2 + b^2 - a^2}}{\sqrt{x^2 + b^2 - a^2}}\right).$$

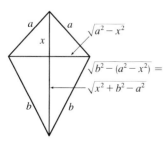

Except for the trivial case where $x = 0$, $a = b$ and $A(x) = 0$, we have

$x + \sqrt{x^2 + b^2 - a^2} > 0$. Hence, cancelling this factor gives

$$\frac{x}{\sqrt{a^2 - x^2}} = \frac{\sqrt{a^2 - x^2}}{\sqrt{x^2 + b^2 - a^2}} \quad \Rightarrow \quad x\sqrt{x^2 + b^2 - a^2} = a^2 - x^2 \quad \Rightarrow$$

$$x^2(x^2 + b^2 - a^2) = a^4 - 2a^2x^2 + x^4 \quad \Rightarrow \quad x^2(b^2 - a^2) = a^4 - 2a^2x^2 \quad \Rightarrow$$

$$x^2(b^2 + a^2) = a^4 \quad \Rightarrow \quad x = \frac{a^2}{\sqrt{a^2 + b^2}}.$$

Now we must check the value of A at this point as well as at the endpoints of the domain to see which gives the maximum value. $A(0) = a\sqrt{b^2 - a^2}$, $A(a) = 0$ and

$$A\left(\frac{a^2}{\sqrt{a^2 + b^2}}\right) = \sqrt{a^2 - \left(\frac{a^2}{\sqrt{a^2 + b^2}}\right)^2}\left[\frac{a^2}{\sqrt{a^2 + b^2}} + \sqrt{\left(\frac{a^2}{\sqrt{a^2 + b^2}}\right)^2 + b^2 - a^2}\right]$$

$$= \frac{ab}{\sqrt{a^2 + b^2}}\left[\frac{a^2}{\sqrt{a^2 + b^2}} + \frac{b^2}{\sqrt{a^2 + b^2}}\right] = \frac{ab(a^2 + b^2)}{a^2 + b^2} = ab$$

Since $b \geq \sqrt{b^2 - a^2}$, $A\left(a^2/\sqrt{a^2 + b^2}\right) \geq A(0)$. So there is an absolute maximum when $x = \dfrac{a^2}{\sqrt{a^2 + b^2}}$. In this case the

horizontal piece should be $\dfrac{2ab}{\sqrt{a^2 + b^2}}$ and the vertical piece should be $\dfrac{a^2 + b^2}{\sqrt{a^2 + b^2}} = \sqrt{a^2 + b^2}$.

61. Note that $|AD| = |AP| + |PD| \;\; \Rightarrow \;\; 5 = x + |PD| \;\; \Rightarrow \;\; |PD| = 5 - x$.

Using the Pythagorean Theorem for $\triangle PDB$ and $\triangle PDC$ gives us

$$L(x) = |AP| + |BP| + |CP| = x + \sqrt{(5 - x)^2 + 2^2} + \sqrt{(5 - x)^2 + 3^2}$$

$$= x + \sqrt{x^2 - 10x + 29} + \sqrt{x^2 - 10x + 34} \quad \Rightarrow$$

$$L'(x) = 1 + \frac{x - 5}{\sqrt{x^2 - 10x + 29}} + \frac{x - 5}{\sqrt{x^2 - 10x + 34}}. \text{ From the graphs of } L$$

and L', it seems that the minimum value of L is about $L(3.59) = 9.35$ m.

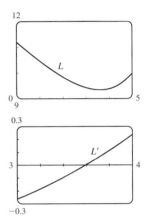

62. We note that since c is the consumption in gallons per hour, and v is the velocity in miles per hour, then

$$\frac{c}{v} = \frac{\text{gallons/hour}}{\text{miles/hour}} = \frac{\text{gallons}}{\text{mile}} \text{ gives us the consumption in gallons per mile, that is, the quantity } G. \text{ To find the minimum,}$$

we calculate $\dfrac{dG}{dv} = \dfrac{d}{dv}\left(\dfrac{c}{v}\right) = \dfrac{v\dfrac{dc}{dv} - c\dfrac{dv}{dv}}{v^2} = \dfrac{v\dfrac{dc}{dv} - c}{v^2}$.

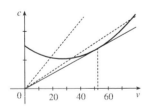

This is 0 when $v\dfrac{dc}{dv} - c = 0 \;\; \Leftrightarrow \;\; \dfrac{dc}{dv} = \dfrac{c}{v}$. This implies that the tangent

line of $c(v)$ passes through the origin, and this occurs when $v \approx 53$ mi/h.

Note that the slope of the secant line through the origin and a point $(v, c(v))$

on the graph is equal to $G(v)$, and it is intuitively clear that G is minimized in

the case where the secant is in fact a tangent.

63.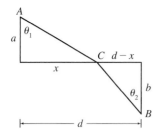

The total time is

$$T(x) = (\text{time from } A \text{ to } C) + (\text{time from } C \text{ to } B)$$

$$= \frac{\sqrt{a^2 + x^2}}{v_1} + \frac{\sqrt{b^2 + (d - x)^2}}{v_2}, \ 0 < x < d$$

$$T'(x) = \frac{x}{v_1 \sqrt{a^2 + x^2}} - \frac{d - x}{v_2 \sqrt{b^2 + (d - x)^2}} = \frac{\sin \theta_1}{v_1} - \frac{\sin \theta_2}{v_2}$$

The minimum occurs when $T'(x) = 0 \ \Rightarrow \ \dfrac{\sin \theta_1}{v_1} = \dfrac{\sin \theta_2}{v_2}$.

[*Note:* $T''(x) > 0$]

64.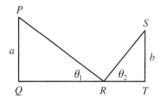

If $d = |QT|$, we minimize $f(\theta_1) = |PR| + |RS| = a \csc \theta_1 + b \csc \theta_2$.

Differentiating with respect to θ_1, and setting $\dfrac{df}{d\theta_1}$ equal to 0, we get

$$\frac{df}{d\theta_1} = 0 = -a \csc \theta_1 \cot \theta_1 - b \csc \theta_2 \cot \theta_2 \frac{d\theta_2}{d\theta_1}.$$

So we need to find an expression for $\dfrac{d\theta_2}{d\theta_1}$. We can do this by observing that $|QT| = \text{constant} = a \cot \theta_1 + b \cot \theta_2$.

Differentiating this equation implicitly with respect to θ_1, we get $-a \csc^2 \theta_1 - b \csc^2 \theta_2 \dfrac{d\theta_2}{d\theta_1} = 0 \ \Rightarrow$

$\dfrac{d\theta_2}{d\theta_1} = -\dfrac{a \csc^2 \theta_1}{b \csc^2 \theta_2}$. We substitute this into the expression for $\dfrac{df}{d\theta_1}$ to get

$$-a \csc \theta_1 \cot \theta_1 - b \csc \theta_2 \cot \theta_2 \left(-\frac{a \csc^2 \theta_1}{b \csc^2 \theta_2} \right) = 0 \ \Leftrightarrow \ -a \csc \theta_1 \cot \theta_1 + a \frac{\csc^2 \theta_1 \cot \theta_2}{\csc \theta_2} = 0 \ \Leftrightarrow$$

$\cot \theta_1 \csc \theta_2 = \csc \theta_1 \cot \theta_2 \ \Leftrightarrow \ \dfrac{\cot \theta_1}{\csc \theta_1} = \dfrac{\cot \theta_2}{\csc \theta_2} \ \Leftrightarrow \ \cos \theta_1 = \cos \theta_2$. Since θ_1 and θ_2 are both acute, we

have $\theta_1 = \theta_2$.

65.

$y^2 = x^2 + z^2$, but triangles CDE and BCA are similar, so

$z/8 = x/\left(4 \sqrt{x - 4} \right) \ \Rightarrow \ z = 2x/\sqrt{x - 4}$. Thus, we minimize

$f(x) = y^2 = x^2 + 4x^2/(x - 4) = x^3/(x - 4), \ 4 < x \le 8$.

$f'(x) = \dfrac{(x - 4)(3x^2) - x^3}{(x - 4)^2} = \dfrac{x^2 [3(x - 4) - x]}{(x - 4)^2} = \dfrac{2x^2 (x - 6)}{(x - 4)^2} = 0$

when $x = 6$. $f'(x) < 0$ when $x < 6$, $f'(x) > 0$ when $x > 6$, so the minimum

occurs when $x = 6$ in.

66.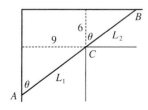

Paradoxically, we solve this maximum problem by solving a minimum problem.

Let L be the length of the line ACB going from wall to wall touching the inner

corner C. As $\theta \to 0$ or $\theta \to \frac{\pi}{2}$, we have $L \to \infty$ and there will be an angle that

makes L a minimum. A pipe of this length will just fit around the corner.

From the diagram, $L = L_1 + L_2 = 9 \csc \theta + 6 \sec \theta \ \Rightarrow \ dL/d\theta = -9 \csc \theta \cot \theta + 6 \sec \theta \tan \theta = 0$ when

$6 \sec \theta \tan \theta = 9 \csc \theta \cot \theta \quad \Leftrightarrow \quad \tan^3 \theta = \frac{9}{6} = 1.5 \quad \Leftrightarrow \quad \tan \theta = \sqrt[3]{1.5}$. Then $\sec^2 \theta = 1 + \left(\frac{3}{2}\right)^{2/3}$ and

$\csc^2 \theta = 1 + \left(\frac{3}{2}\right)^{-2/3}$, so the longest pipe has length $L = 9\left[1 + \left(\frac{3}{2}\right)^{-2/3}\right]^{1/2} + 6\left[1 + \left(\frac{3}{2}\right)^{2/3}\right]^{1/2} \approx 21.07$ ft.

Or, use $\theta = \tan^{-1}\left(\sqrt[3]{1.5}\right) \approx 0.853 \quad \Rightarrow \quad L = 9 \csc \theta + 6 \sec \theta \approx 21.07$ ft.

67.

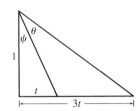

It suffices to maximize $\tan \theta$. Now

$$\frac{3t}{1} = \tan(\psi + \theta) = \frac{\tan \psi + \tan \theta}{1 - \tan \psi \tan \theta} = \frac{t + \tan \theta}{1 - t \tan \theta}. \text{ So}$$

$$3t(1 - t \tan \theta) = t + \tan \theta \quad \Rightarrow \quad 2t = (1 + 3t^2) \tan \theta \quad \Rightarrow \quad \tan \theta = \frac{2t}{1 + 3t^2}.$$

Let $f(t) = \tan \theta = \frac{2t}{1 + 3t^2} \quad \Rightarrow \quad f'(t) = \frac{2(1 + 3t^2) - 2t(6t)}{(1 + 3t^2)^2} = \frac{2(1 - 3t^2)}{(1 + 3t^2)^2} = 0 \quad \Leftrightarrow \quad 1 - 3t^2 = 0 \quad \Leftrightarrow$

$t = \frac{1}{\sqrt{3}}$ since $t \geq 0$. Now $f'(t) > 0$ for $0 \leq t < \frac{1}{\sqrt{3}}$ and $f'(t) < 0$ for $t > \frac{1}{\sqrt{3}}$, so f has an absolute maximum when $t = \frac{1}{\sqrt{3}}$

and $\tan \theta = \frac{2(1/\sqrt{3})}{1 + 3(1/\sqrt{3})^2} = \frac{1}{\sqrt{3}} \quad \Rightarrow \quad \theta = \frac{\pi}{6}$. Substituting for t and θ in $3t = \tan(\psi + \theta)$ gives us

$\sqrt{3} = \tan\left(\psi + \frac{\pi}{6}\right) \quad \Rightarrow \quad \psi = \frac{\pi}{6}$.

68. We maximize the cross-sectional area

$A(\theta) = 10h + 2\left(\frac{1}{2}dh\right) = 10h + dh = 10(10 \sin \theta) + (10 \cos \theta)(10 \sin \theta)$

$\quad = 100(\sin \theta + \sin \theta \cos \theta), \ 0 \leq \theta \leq \frac{\pi}{2}$

$A'(\theta) = 100(\cos \theta + \cos^2 \theta - \sin^2 \theta) = 100(\cos \theta + 2 \cos^2 \theta - 1)$

$\quad = 100(2 \cos \theta - 1)(\cos \theta + 1) = 0$ when $\cos \theta = \frac{1}{2} \quad \Leftrightarrow \quad \theta = \frac{\pi}{3} \quad [\cos \theta \neq -1 \text{ since } 0 \leq \theta \leq \frac{\pi}{2}.]$

Now $A(0) = 0$, $A\left(\frac{\pi}{2}\right) = 100$ and $A\left(\frac{\pi}{3}\right) = 75\sqrt{3} \approx 129.9$, so the maximum occurs when $\theta = \frac{\pi}{3}$.

69.

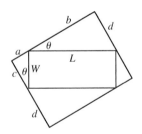

In the small triangle with sides a and c and hypotenuse W, $\sin \theta = \frac{a}{W}$ and

$\cos \theta = \frac{c}{W}$. In the triangle with sides b and d and hypotenuse L, $\sin \theta = \frac{d}{L}$ and

$\cos \theta = \frac{b}{L}$. Thus, $a = W \sin \theta$, $c = W \cos \theta$, $d = L \sin \theta$, and $b = L \cos \theta$, so the

area of the circumscribed rectangle is

$A(\theta) = (a + b)(c + d) = (W \sin \theta + L \cos \theta)(W \cos \theta + L \sin \theta)$

$\quad = W^2 \sin \theta \cos \theta + WL \sin^2 \theta + LW \cos^2 \theta + L^2 \sin \theta \cos \theta$

$\quad = LW \sin^2 \theta + LW \cos^2 \theta + (L^2 + W^2) \sin \theta \cos \theta$

$\quad = LW(\sin^2 \theta + \cos^2 \theta) + (L^2 + W^2) \cdot \frac{1}{2} \cdot 2 \sin \theta \cos \theta = LW + \frac{1}{2}(L^2 + W^2) \sin 2\theta, \ 0 \leq \theta \leq \frac{\pi}{2}$

This expression shows, without calculus, that the maximum value of $A(\theta)$ occurs when $\sin 2\theta = 1 \quad \Leftrightarrow \quad 2\theta = \frac{\pi}{2} \quad \Rightarrow$

$\theta = \frac{\pi}{4}$. So the maximum area is $A\left(\frac{\pi}{4}\right) = LW + \frac{1}{2}(L^2 + W^2) = \frac{1}{2}(L^2 + 2LW + W^2) = \frac{1}{2}(L + W)^2$.

70. (a) Let D be the point such that $a = |AD|$. From the figure, $\sin \theta = \dfrac{b}{|BC|}$ $\Rightarrow$ $|BC| = b \csc \theta$ and

$$\cos \theta = \frac{|BD|}{|BC|} = \frac{a - |AB|}{|BC|} \quad \Rightarrow \quad |BC| = (a - |AB|) \sec \theta. \text{ Eliminating } |BC| \text{ gives}$$

$$(a - |AB|) \sec \theta = b \csc \theta \quad \Rightarrow \quad b \cot \theta = a - |AB| \quad \Rightarrow \quad |AB| = a - b \cot \theta. \text{ The total resistance is}$$

$$R(\theta) = C \frac{|AB|}{r_1^4} + C \frac{|BC|}{r_2^4} = C \left(\frac{a - b \cot \theta}{r_1^4} + \frac{b \csc \theta}{r_2^4} \right).$$

(b) $R'(\theta) = C \left(\dfrac{b \csc^2 \theta}{r_1^4} - \dfrac{b \csc \theta \, \cot \theta}{r_2^4} \right) = bC \csc \theta \left(\dfrac{\csc \theta}{r_1^4} - \dfrac{\cot \theta}{r_2^4} \right).$

$$R'(\theta) = 0 \quad \Leftrightarrow \quad \frac{\csc \theta}{r_1^4} = \frac{\cot \theta}{r_2^4} \quad \Leftrightarrow \quad \frac{r_2^4}{r_1^4} = \frac{\cot \theta}{\csc \theta} = \cos \theta.$$

$R'(\theta) > 0 \quad \Leftrightarrow \quad \dfrac{\csc \theta}{r_1^4} > \dfrac{\cot \theta}{r_2^4} \quad \Rightarrow \quad \cos \theta < \dfrac{r_2^4}{r_1^4}$ and $R'(\theta) < 0$ when $\cos \theta > \dfrac{r_2^4}{r_1^4}$, so there is an absolute minimum

when $\cos \theta = r_2^4 / r_1^4$.

(c) When $r_2 = \frac{2}{3} r_1$, we have $\cos \theta = \left(\frac{2}{3} \right)^4$, so $\theta = \cos^{-1} \left(\frac{2}{3} \right)^4 \approx 79°$.

71. (a)

If $k = $ energy/km over land, then energy/km over water $= 1.4k$.

So the total energy is $E = 1.4k \sqrt{25 + x^2} + k(13 - x)$, $0 \le x \le 13$,

and so $\dfrac{dE}{dx} = \dfrac{1.4kx}{(25 + x^2)^{1/2}} - k$.

Set $\dfrac{dE}{dx} = 0$: $1.4kx = k(25 + x^2)^{1/2}$ $\Rightarrow$ $1.96x^2 = x^2 + 25$ $\Rightarrow$ $0.96x^2 = 25$ $\Rightarrow$ $x = \dfrac{5}{\sqrt{0.96}} \approx 5.1$.

Testing against the value of E at the endpoints: $E(0) = 1.4k(5) + 13k = 20k$, $E(5.1) \approx 17.9k$, $E(13) \approx 19.5k$.

Thus, to minimize energy, the bird should fly to a point about 5.1 km from B.

(b) If W/L is large, the bird would fly to a point C that is closer to B than to D to minimize the energy used flying over water.

If W/L is small, the bird would fly to a point C that is closer to D than to B to minimize the distance of the flight.

$$E = W \sqrt{25 + x^2} + L(13 - x) \quad \Rightarrow \quad \frac{dE}{dx} = \frac{Wx}{\sqrt{25 + x^2}} - L = 0 \text{ when } \frac{W}{L} = \frac{\sqrt{25 + x^2}}{x}. \text{ By the same sort of}$$

argument as in part (a), this ratio will give the minimal expenditure of energy if the bird heads for the point x km from B.

(c) For flight direct to D, $x = 13$, so from part (b), $W/L = \dfrac{\sqrt{25 + 13^2}}{13} \approx 1.07$. There is no value of W/L for which the bird

should fly directly to B. But note that $\lim\limits_{x \to 0^+} (W/L) = \infty$, so if the point at which E is a minimum is close to B, then

W/L is large.

(d) Assuming that the birds instinctively choose the path that minimizes the energy expenditure, we can use the equation for

$dE/dx = 0$ from part (a) with $1.4k = c$, $x = 4$, and $k = 1$: $c(4) = 1 \cdot (25 + 4^2)^{1/2}$ $\Rightarrow$ $c = \sqrt{41}/4 \approx 1.6$.

72. (a) $I(x) \propto \dfrac{\text{strength of source}}{(\text{distance from source})^2}$. Adding the intensities from the left and right lightbulbs,

$$I(x) = \frac{k}{x^2 + d^2} + \frac{k}{(10 - x)^2 + d^2} = \frac{k}{x^2 + d^2} + \frac{k}{x^2 - 20x + 100 + d^2}.$$

(b) The magnitude of the constant k won't affect the location of the point of maximum intensity, so for convenience we take

$k = 1$. $I'(x) = -\dfrac{2x}{(x^2 + d^2)^2} - \dfrac{2(x - 10)}{(x^2 - 20x + 100 + d^2)^2}$.

Substituting $d = 5$ into the equations for $I(x)$ and $I'(x)$, we get

$$I_5(x) = \frac{1}{x^2 + 25} + \frac{1}{x^2 - 20x + 125} \quad \text{and} \quad I_5'(x) = -\frac{2x}{(x^2 + 25)^2} - \frac{2(x - 10)}{(x^2 - 20x + 125)^2}$$

 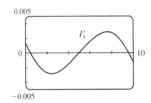

From the graphs, it appears that $I_5(x)$ has a minimum at $x = 5$ m.

(c) Substituting $d = 10$ into the equations for $I(x)$ and $I'(x)$ gives

$$I_{10}(x) = \frac{1}{x^2 + 100} + \frac{1}{x^2 - 20x + 200} \quad \text{and} \quad I_{10}'(x) = -\frac{2x}{\left(x^2 + 100\right)^2} - \frac{2\left(x - 10\right)}{\left(x^2 - 20x + 200\right)^2}$$

From the graphs, it seems that for $d = 10$, the intensity is minimized at the endpoints, that is, $x = 0$ and $x = 10$. The midpoint is now the most brightly lit point!

(d) From the first figures in parts (b) and (c), we see that the minimal illumination changes from the midpoint ($x = 5$ with $d = 5$) to the endpoints ($x = 0$ and $x = 10$ with $d = 10$).

So we try $d = 6$ (see the first figure) and we see that the minimum value still occurs at $x = 5$. Next, we let $d = 8$ (see the second figure) and we see that the minimum value occurs at the endpoints. It appears that for some value of d between 6 and 8, we must have minima at both the midpoint and the endpoints, that is, $I(5)$ must equal $I(0)$. To find this value of d, we solve $I(0) = I(5)$ (with $k = 1$):

$\dfrac{1}{d^2} + \dfrac{1}{100 + d^2} = \dfrac{1}{25 + d^2} + \dfrac{1}{25 + d^2} = \dfrac{2}{25 + d^2} \quad \Rightarrow \quad (25 + d^2)(100 + d^2) + d^2(25 + d^2) = 2d^2(100 + d^2) \quad \Rightarrow$

$2500 + 125d^2 + d^4 + 25d^2 + d^4 = 200d^2 + 2d^4 \quad \Rightarrow \quad 2500 = 50d^2 \quad \Rightarrow \quad d^2 = 50 \quad \Rightarrow \quad d = 5\sqrt{2} \approx 7.071$ [for $0 \le d \le 10$]. The third figure, a graph of $I(0) - I(5)$ with d independent, confirms that $I(0) - I(5) = 0$, that is, $I(0) = I(5)$, when $d = 5\sqrt{2}$. Thus, the point of minimal illumination changes abruptly from the midpoint to the endpoints when $d = 5\sqrt{2}$.

APPLIED PROJECT   The Shape of a Can

1. In this case, the amount of metal used in the making of each top or bottom is $(2r)^2 = 4r^2$. So the quantity we want to minimize is $A = 2\pi rh + 2(4r^2)$. But $V = \pi r^2 h \iff h = V/\pi r^2$. Substituting this expression for h in A gives $A = 2V/r + 8r^2$. Differentiating A with respect to r, we get $dA/dr = -2V/r^2 + 16r = 0 \implies$ $16r^3 = 2V = 2\pi r^2 h \iff \dfrac{h}{r} = \dfrac{8}{\pi} \approx 2.55$. This gives a minimum because $\dfrac{d^2 A}{dr^2} = 16 + \dfrac{4V}{r^3} > 0.$

2.

We need to find the area of metal used up by each end, that is, the area of each hexagon. We subdivide the hexagon into six congruent triangles, each sharing one side (s in the diagram) with the hexagon. We calculate the length of $s = 2r \tan \dfrac{\pi}{6} = \dfrac{2}{\sqrt{3}} r$, so the area of each triangle is $\frac{1}{2} sr = \dfrac{1}{\sqrt{3}} r^2$, and the total area of the hexagon is $6 \cdot \dfrac{1}{\sqrt{3}} r^2 = 2\sqrt{3}\, r^2$. So the quantity we want to minimize is $A = 2\pi rh + 2 \cdot 2\sqrt{3}\, r^2$. Substituting for h as in Problem 1 and differentiating, we get $\dfrac{dA}{dr} = -\dfrac{2V}{r^2} + 8\sqrt{3}\, r.$

Setting this equal to 0, we get $8\sqrt{3}\, r^3 = 2V = 2\pi r^2 h \implies \dfrac{h}{r} = \dfrac{4\sqrt{3}}{\pi} \approx 2.21$. Again this minimizes A because $\dfrac{d^2 A}{dr^2} = 8\sqrt{3} + \dfrac{4V}{r^3} > 0.$

3. Let $C = 4\sqrt{3}\, r^2 + 2\pi rh + k(4\pi r + h) = 4\sqrt{3}\, r^2 + 2\pi r\left(\dfrac{V}{\pi r^2}\right) + k\left(4\pi r + \dfrac{V}{\pi r^2}\right)$. Then $\dfrac{dC}{dr} = 8\sqrt{3}\, r - \dfrac{2V}{r^2} + 4k\pi - \dfrac{2kV}{\pi r^3}$. Setting this equal to 0, dividing by 2 and substituting $\dfrac{V}{r^2} = \pi h$ and $\dfrac{V}{\pi r^3} = \dfrac{h}{r}$ in the second and fourth terms respectively, we get $0 = 4\sqrt{3}\, r - \pi h + 2k\pi - \dfrac{kh}{r} \iff$ $k\left(2\pi - \dfrac{h}{r}\right) = \pi h - 4\sqrt{3}\, r \implies \dfrac{k}{r}\dfrac{2\pi - h/r}{\pi h/r - 4\sqrt{3}} = 1$. We now multiply by $\dfrac{\sqrt[3]{V}}{k}$, noting that $\dfrac{\sqrt[3]{V}}{k}\dfrac{k}{r} = \sqrt[3]{\dfrac{V}{r^3}} = \sqrt[3]{\dfrac{\pi h}{r}},$ and get $\dfrac{\sqrt[3]{V}}{k} = \sqrt[3]{\dfrac{\pi h}{r}} \cdot \dfrac{2\pi - h/r}{\pi h/r - 4\sqrt{3}}.$

4.

Let $\sqrt[3]{V}/k = T$ and $h/r = x$ so that $T(x) = \sqrt[3]{\pi x} \cdot \dfrac{2\pi - x}{\pi x - 4\sqrt{3}}$. We see from the graph of T that when the ratio $\sqrt[3]{V}/k$ is large; that is, either the volume of the can is large or the cost of joining (proportional to k) is small, the optimum value of h/r is about 2.21, but when $\sqrt[3]{V}/k$ is small, indicating small volume or expensive joining, the optimum value of h/r is larger. (The part of the graph for $\sqrt[3]{V}/k < 0$ has no physical meaning, but confirms the location of the asymptote.)

5. Our conclusion is usually true in practice. But there are exceptions, such as cans of tuna, which may have to do with the shape of a reasonable slice of tuna. And for a comfortable grip on a soda or beer can, the geometry of the human hand is a restriction on the radius. Other possible considerations are packaging, transportation and stocking constraints, aesthetic appeal and other marketing concerns. Also, there may be better models than ours which prescribe a differently shaped can in special circumstances.

4.8 Newton's Method

1. (a)

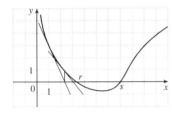

The tangent line at $x = 1$ intersects the x-axis at $x \approx 2.3$, so $x_2 \approx 2.3$. The tangent line at $x = 2.3$ intersects the x-axis at $x \approx 3$, so $x_3 \approx 3.0$.

(b) $x_1 = 5$ would *not* be a better first approximation than $x_1 = 1$ since the tangent line is nearly horizontal. In fact, the second approximation for $x_1 = 5$ appears to be to the left of $x = 1$.

2.

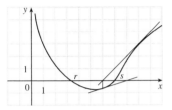

The tangent line at $x = 9$ intersects the x-axis at $x \approx 6.0$, so $x_2 \approx 6.0$. The tangent line at $x = 6.0$ intersects the x-axis at $x \approx 8.0$, so $x_3 \approx 8.0$.

3. Since $x_1 = 3$ and $y = 5x - 4$ is tangent to $y = f(x)$ at $x = 3$, we simply need to find where the tangent line intersects the x-axis. $y = 0 \quad \Rightarrow \quad 5x_2 - 4 = 0 \quad \Rightarrow \quad x_2 = \frac{4}{5}$.

4. (a)

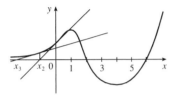

If $x_1 = 0$, then x_2 is negative, and x_3 is even more negative. The sequence of approximations does not converge, that is, Newton's method fails.

(b)

If $x_1 = 1$, the tangent line is horizontal and Newton's method fails.

(c)

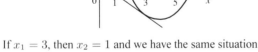

If $x_1 = 3$, then $x_2 = 1$ and we have the same situation as in part (b). Newton's method fails again.

(d)

If $x_1 = 4$, the tangent line is horizontal and Newton's method fails.

(e)

If $x_1 = 5$, then x_2 is greater than 6, x_3 gets closer to 6, and the sequence of approximations converges to 6. Newton's method succeeds!

5. $f(x) = x^3 + 2x - 4 \Rightarrow f'(x) = 3x^2 + 2$, so $x_{n+1} = x_n - \dfrac{x_n^3 + 2x_n - 4}{3x_n^2 + 2}$. Now $x_1 = 1 \Rightarrow$

$x_2 = 1 - \dfrac{1 + 2 - 4}{3 \cdot 1^2 + 2} = 1 - \dfrac{-1}{5} = 1.2 \Rightarrow x_3 = 1.2 - \dfrac{(1.2)^3 + 2(1.2) - 4}{3(1.2)^2 + 2} \approx 1.1797.$

6. $f(x) = \frac{1}{3}x^3 + \frac{1}{2}x^2 + 3 \Rightarrow f'(x) = x^2 + x$, so $x_{n+1} = x_n - \dfrac{\frac{1}{3}x_n^3 + \frac{1}{2}x_n^2 + 3}{x_n^2 + x_n}$. Now $x_1 = -3 \Rightarrow$

$x_2 = -3 - \dfrac{-9 + \frac{9}{2} + 3}{9 - 3} = -3 - \left(-\frac{1}{4}\right) = -2.75 \Rightarrow x_3 = -2.75 - \dfrac{\frac{1}{3}(-2.75)^3 + \frac{1}{2}(-2.75)^2 + 3}{(-2.75)^2 + (-2.75)} \approx -2.7186.$

7. $f(x) = x^5 - x - 1 \Rightarrow f'(x) = 5x^4 - 1$, so $x_{n+1} = x_n - \dfrac{x_n^5 - x_n - 1}{5x_n^4 - 1}$. Now $x_1 = 1 \Rightarrow$

$x_2 = 1 - \dfrac{1 - 1 - 1}{5 - 1} = 1 - \left(-\frac{1}{4}\right) = 1.25 \Rightarrow x_3 = 1.25 - \dfrac{(1.25)^5 - 1.25 - 1}{5(1.25)^4 - 1} \approx 1.1785.$

8. $f(x) = x^5 + 2 \Rightarrow f'(x) = 5x^4$, so $x_{n+1} = x_n - \dfrac{x_n^5 + 2}{5x_n^4}$. Now $x_1 = -1 \Rightarrow$

$x_2 = -1 - \dfrac{(-1)^5 + 2}{5 \cdot (-1)^4} = -1 - \dfrac{1}{5} = -1.2 \Rightarrow x_3 = -1.2 - \dfrac{(-1.2)^5 + 2}{5(-1.2)^4} \approx -1.1529.$

9. $f(x) = x^3 + x + 3 \Rightarrow f'(x) = 3x^2 + 1$, so $x_{n+1} = x_n - \dfrac{x_n^3 + x_n + 3}{3x_n^2 + 1}$.

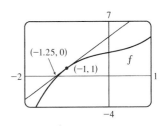

Now $x_1 = -1 \Rightarrow$

$x_2 = -1 - \dfrac{(-1)^3 + (-1) + 3}{3(-1)^2 + 1} = -1 - \dfrac{-1 - 1 + 3}{3 + 1} = -1 - \dfrac{1}{4} = -1.25.$

Newton's method follows the tangent line at $(-1, 1)$ down to its intersection with

the x-axis at $(-1.25, 0)$, giving the second approximation $x_2 = -1.25$.

10. $f(x) = x^4 - x - 1 \Rightarrow f'(x) = 4x^3 - 1$, so $x_{n+1} = x_n - \dfrac{x_n^4 - x_n - 1}{4x_n^3 - 1}$.

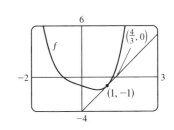

Now $x_1 = 1 \Rightarrow x_2 = 1 - \dfrac{1^4 - 1 - 1}{4 \cdot 1^3 - 1} = 1 - \dfrac{-1}{3} = \dfrac{4}{3}$. Newton's method

follows the tangent line at $(1, -1)$ up to its intersection with the x-axis at $\left(\frac{4}{3}, 0\right)$,

giving the second approximation $x_2 = \frac{4}{3}$.

11. To approximate $x = \sqrt[5]{20}$ (so that $x^5 = 20$), we can take $f(x) = x^5 - 20$. So $f'(x) = 5x^4$, and thus,

$x_{n+1} = x_n - \dfrac{x_n^5 - 20}{5x_n^4}$. Since $\sqrt[5]{32} = 2$ and 32 is reasonably close to 20, we'll use $x_1 = 2$. We need to find approximations

until they agree to eight decimal places. $x_1 = 2 \Rightarrow x_2 = 1.85, x_3 \approx 1.82148614, x_4 \approx 1.82056514,$

$x_5 \approx 1.82056420 \approx x_6$. So $\sqrt[5]{20} \approx 1.82056420$, to eight decimal places.

Here is a quick and easy method for finding the iterations for Newton's method on a programmable calculator.

(The screens shown are from the TI-84 Plus, but the method is similar on other calculators.) Assign $f(x) = x^5 - 20$

to Y_1, and $f'(x) = 5x^4$ to Y_2. Now store $x_1 = 2$ in X and then enter $X - Y_1/Y_2 \to X$ to get $x_2 = 1.85$. By successively pressing the ENTER key, you get the approximations x_3, x_4, ...

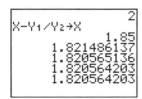

In Derive, load the utility file `SOLVE`. Enter `NEWTON(x^5-20,x,2)` and then `APPROXIMATE` to get $[2, 1.85, 1.82148614, 1.82056514, 1.82056420]$. You can request a specific iteration by adding a fourth argument. For example, `NEWTON(x^5-20,x,2,2)` gives $[2, 1.85, 1.82148614]$.

In Maple, make the assignments $f := x \to x\char`\^5 - 20;$, $g := x \to x - f(x)/D(f)(x);$, and $x := 2.;$. Repeatedly execute the command $x := g(x);$ to generate successive approximations.

In Mathematica, make the assignments $f[x\_] := x\char`\^5 - 20$, $g[x\_] := x - f[x]/f'[x]$, and $x = 2$. Repeatedly execute the command $x = g[x]$ to generate successive approximations.

12. $f(x) = x^{100} - 100 \implies f'(x) = 100x^{99}$, so $x_{n+1} = x_n - \dfrac{x_n^{100} - 100}{100x_n^{99}}$. We need to find approximations until they agree to eight decimal places. $x_1 = 1.05 \implies x_2 \approx 1.04748471$, $x_3 \approx 1.04713448$, $x_4 \approx 1.04712855 \approx x_5$. So $\sqrt[100]{100} \approx 1.04712855$, to eight decimal places.

13. $f(x) = x^4 - 2x^3 + 5x^2 - 6 \implies f'(x) = 4x^3 - 6x^2 + 10x \implies x_{n+1} = x_n - \dfrac{x_n^4 - 2x_n^3 + 5x_n^2 - 6}{4x_n^3 - 6x_n^2 + 10x_n}$. We need to find approximations until they agree to six decimal places. We'll let x_1 equal the midpoint of the given interval, $[1, 2]$. $x_1 = 1.5 \implies x_2 = 1.2625$, $x_3 \approx 1.218808$, $x_4 \approx 1.217563$, $x_5 \approx 1.217562 \approx x_6$. So the root is 1.217562 to six decimal places.

14. $f(x) = 2.2x^5 - 4.4x^3 + 1.3x^2 - 0.9x - 4.0 \implies f'(x) = 11x^4 - 13.2x^2 + 2.6x - 0.9 \implies$
$x_{n+1} = x_n - \dfrac{2.2x_n^5 - 4.4x_n^3 + 1.3x_n^2 - 0.9x_n - 4.0}{11x_n^4 - 13.2x_n^2 + 2.6x_n - 0.9}$. $x_1 = -1.5 \implies x_2 \approx -1.425369$, $x_3 \approx -1.405499$, $x_4 \approx -1.404124$, $x_5 \approx -1.404118 \approx x_6$. So the root is -1.404118 to six decimal places.

15. $\sin x = x^2$, so $f(x) = \sin x - x^2 \implies f'(x) = \cos x - 2x \implies$

$x_{n+1} = x_n - \dfrac{\sin x_n - x_n^2}{\cos x_n - 2x_n}$. From the figure, the positive root of $\sin x = x^2$ is near 1. $x_1 = 1 \implies x_2 \approx 0.891396$, $x_3 \approx 0.876985$, $x_4 \approx 0.876726 \approx x_5$. So the positive root is 0.876726, to six decimal places.

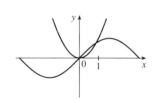

16. $2\cos x = x^4$, so $f(x) = 2\cos x - x^4 \;\Rightarrow\; f'(x) = -2\sin x - 4x^3 \;\Rightarrow\;$

$x_{n+1} = x_n - \dfrac{2\cos x_n - x_n^4}{-2\sin x_n - 4x_n^3}$. From the figure, the positive root

of $2\cos x = x^4$ is near 1. $x_1 = 1 \;\Rightarrow\; x_2 \approx 1.014184$, $x_3 \approx 1.013958 \approx x_4$.

So the positive root is 1.013958, to six decimal places.

17.

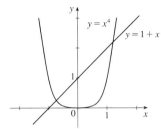

From the graph, we see that there appear to be points of intersection near

$x = -0.7$ and $x = 1.2$. Solving $x^4 = 1 + x$ is the same as solving

$f(x) = x^4 - x - 1 = 0$. $f(x) = x^4 - x - 1 \;\Rightarrow\; f'(x) = 4x^3 - 1$,

so $x_{n+1} = x_n - \dfrac{x_n^4 - x_n - 1}{4x_n^3 - 1}$.

$x_1 = -0.7$	$x_1 = 1.2$
$x_2 \approx -0.725253$	$x_2 \approx 1.221380$
$x_3 \approx -0.724493$	$x_3 \approx 1.220745$
$x_4 \approx -0.724492 \approx x_5$	$x_4 \approx 1.220744 \approx x_5$

To six decimal places, the roots of the equation are -0.724492 and 1.220744.

18.

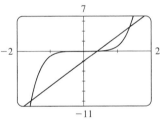

From the graph, we see that reasonable first approximations are $x = 0.5$

and $x = \pm 1.5$. $f(x) = x^5 - 5x + 2 \;\Rightarrow\; f'(x) = 5x^4 - 5$, so

$x_{n+1} = x_n - \dfrac{x_n^5 - 5x_n + 2}{5x_n^4 - 5}$.

$x_1 = -1.5$	$x_1 = 0.5$	$x_1 = 1.5$
$x_2 \approx -1.593846$	$x_2 = 0.4$	$x_2 \approx 1.396923$
$x_3 \approx -1.582241$	$x_3 \approx 0.402102 \approx x_4$	$x_3 \approx 1.373078$
$x_4 \approx -1.582036 \approx x_5$		$x_4 \approx 1.371885$
		$x_5 \approx 1.371882 \approx x_6$

To six decimal places, the roots are -1.582036, 0.402102, and 1.371882.

19.

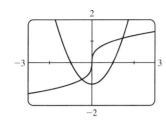

From the graph, we see that there appear to be points of intersection near

$x = -0.5$ and $x = 1.5$. Solving $\sqrt[3]{x} = x^2 - 1$ is the same as solving

$f(x) = \sqrt[3]{x} - x^2 + 1 = 0.$ $f(x) = \sqrt[3]{x} - x^2 + 1$ $\Rightarrow$

$f'(x) = \frac{1}{3}x^{-2/3} - 2x$, so $x_{n+1} = x_n - \dfrac{\sqrt[3]{x_n} - x_n^2 + 1}{\frac{1}{3}x_n^{-2/3} - 2x_n}.$

$x_1 = -0.5$	$x_1 = 1.5$
$x_2 \approx -0.471421$	$x_2 \approx .461653$
$x_3 \approx -0.471074 \approx x_4$	$x_3 \approx 1.461070 \approx x_4$

To six decimal places, the roots are -0.471074 and 1.461070.

20.

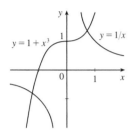

From the graph, we see that there appear to be points of intersection near

$x = -1.2$ and $x = 0.8$. Solving $\dfrac{1}{x} = 1 + x^3$ is the same as solving

$f(x) = \dfrac{1}{x} - 1 - x^3 = 0.$ $f(x) = \dfrac{1}{x} - 1 - x^3$ $\Rightarrow$ $f'(x) = -\dfrac{1}{x^2} - 3x^2$, so

$x_{n+1} = x_n - \dfrac{1/x_n - 1 - x_n^3}{-1/x_n^2 - 3x_n^2}.$

$x_1 = -1.2$	$x_1 = 0.8$
$x_2 \approx -1.221006$	$x_2 \approx 0.724767$
$x_3 \approx -1.220744 \approx x_4$	$x_3 \approx 0.724492 \approx x_4$

To six decimal places, the roots of the equation are -1.220744 and 0.724492.

21. From the graph, there appears to be a point of intersection near $x = 0.6$.

Solving $\cos x = \sqrt{x}$ is the same as solving $f(x) = \cos x - \sqrt{x} = 0$.

$f(x) = \cos x - \sqrt{x}$ $\Rightarrow$ $f'(x) = -\sin x - 1/\left(2\sqrt{x}\right)$, so

$x_{n+1} = x_n - \dfrac{\cos x_n - \sqrt{x_n}}{-\sin x_n - 1/\left(2\sqrt{x}\right)}.$ Now $x_1 = 0.6$ $\Rightarrow$ $x_2 \approx 0.641928$,

$x_3 \approx 0.641714 \approx x_4$. To six decimal places, the root of the equation is 0.641714.

22. From the graph, there appears to be a point of intersection near $x = 0.7$.

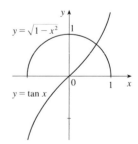

Solving $\tan x = \sqrt{1 - x^2}$ is the same as solving

$f(x) = \tan x - \sqrt{1 - x^2} = 0.$ $f(x) = \tan x - \sqrt{1 - x^2}$ $\Rightarrow$

$f'(x) = \sec^2 x + x/\sqrt{1 - x^2}$, so $x_{n+1} = x_n - \dfrac{\tan x_n - \sqrt{1 - x_n^2}}{\sec^2 x_n + x_n/\sqrt{1 - x_n^2}}.$

$x_1 = 0.7$ $\Rightarrow$ $x_2 \approx 0.652356$, $x_3 \approx 0.649895$, $x_4 \approx 0.649889 \approx x_5$. To six

decimal places, the root of the equation is 0.649889.

23.

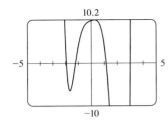

$f(x) = x^6 - x^5 - 6x^4 - x^2 + x + 10 \Rightarrow$

$f'(x) = 6x^5 - 5x^4 - 24x^3 - 2x + 1 \Rightarrow$

$x_{n+1} = x_n - \dfrac{x_n^6 - x_n^5 - 6x_n^4 - x_n^2 + x_n + 10}{6x_n^5 - 5x_n^4 - 24x_n^3 - 2x_n + 1}.$

From the graph of f, there appear to be roots near -1.9, -1.2, 1.1, and 3.

$x_1 = -1.9$	$x_1 = -1.2$	$x_1 = 1.1$	$x_1 = 3$
$x_2 \approx -1.94278290$	$x_2 \approx -1.22006245$	$x_2 \approx 1.14111662$	$x_2 \approx 2.99$
$x_3 \approx -1.93828380$	$x_3 \approx -1.21997997 \approx x_4$	$x_3 \approx 1.13929741$	$x_3 \approx 2.98984106$
$x_4 \approx -1.93822884$		$x_4 \approx 1.13929375 \approx x_5$	$x_4 \approx 2.98984102 \approx x_5$
$x_5 \approx -1.93822883 \approx x_6$			

To eight decimal places, the roots of the equation are -1.93822883, -1.21997997, 1.13929375, and 2.98984102.

24.

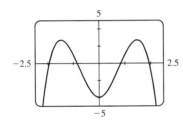

Solving $x^2\left(4 - x^2\right) = \dfrac{4}{x^2 + 1}$ is the same as solving

$f(x) = 4x^2 - x^4 - \dfrac{4}{x^2 + 1} = 0$. $f'(x) = 8x - 4x^3 + \dfrac{8x}{\left(x^2 + 1\right)^2} \Rightarrow$

$x_{n+1} = x_n - \dfrac{4x_n^2 - x_n^4 - 4/\left(x_n^2 + 1\right)}{8x_n - 4x_n^3 + 8x_n/\left(x_n^2 + 1\right)^2}$. From the graph of $f(x)$, there

appear to be roots near $x = \pm 1.9$ and $x = \pm 0.8$. Since f is even, we only need to find the positive roots.

$x_1 = 0.8$	$x_1 = 1.9$
$x_2 \approx 0.84287645$	$x_2 \approx 1.94689103$
$x_3 \approx 0.84310820$	$x_3 \approx 1.94383891$
$x_4 \approx 0.84310821 \approx x_5$	$x_4 \approx 1.94382538 \approx x_5$

To eight decimal places, the roots of the equation are ± 0.84310821 and ± 1.94382538.

25.

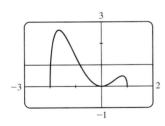

From the graph, $y = x^2\sqrt{2 - x - x^2}$ and $y = 1$ intersect twice, at $x \approx -2$ and

at $x \approx -1$. $f(x) = x^2\sqrt{2 - x - x^2} - 1 \Rightarrow$

$f'(x) = x^2 \cdot \frac{1}{2}(2 - x - x^2)^{-1/2}(-1 - 2x) + (2 - x - x^2)^{1/2} \cdot 2x$

$\quad = \frac{1}{2}x(2 - x - x^2)^{-1/2}[x(-1 - 2x) + 4(2 - x - x^2)]$

$\quad = \dfrac{x(8 - 5x - 6x^2)}{2\sqrt{(2 + x)(1 - x)}},$

so $x_{n+1} = x_n - \dfrac{x_n^2\sqrt{2 - x_n - x_n^2} - 1}{\dfrac{x_n(8 - 5x_n - 6x_n^2)}{2\sqrt{(2 + x_n)(1 - x_n)}}}$. Trying $x_1 = -2$ won't work because $f'(-2)$ is undefined, so we'll try $x_1 = -1.95$.

$$x_1 = -1.95 \qquad\qquad x_1 = -0.8$$
$$x_2 \approx -1.98580357 \qquad\qquad x_2 \approx -0.82674444$$
$$x_3 \approx -1.97899778 \qquad\qquad x_3 \approx -0.82646236$$
$$x_4 \approx -1.97807848 \qquad\qquad x_4 \approx -0.82646233 \approx x_5$$
$$x_5 \approx -1.97806682$$
$$x_6 \approx -1.97806681 \approx x_7$$

To eight decimal places, the roots of the equation are -1.97806681 and -0.82646233.

26.

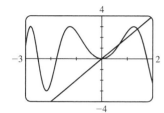

From the equations $y = 3\sin(x^2)$ and $y = 2x$ and the graph, we deduce that one root of the equation $3\sin(x^2) = 2x$ is $x = 0$. We also see that the graphs intersect at approximately $x = 0.7$ and $x = 1.4$. $f(x) = 3\sin(x^2) - 2x$ $\Rightarrow$ $f'(x) = 3\cos(x^2) \cdot 2x - 2$, so $x_{n+1} = x_n - \dfrac{3\sin(x_n^2) - 2x_n}{6x_n\cos(x_n^2) - 2}$.

$$x_1 = 0.7 \qquad\qquad x_1 = 1.4$$
$$x_2 \approx 0.69303689 \qquad\qquad x_2 \approx 1.39530295$$
$$x_3 \approx 0.69299996 \approx x_4 \qquad\qquad x_3 \approx 1.39525078$$
$$\qquad\qquad x_4 \approx 1.39525077 \approx x_5$$

To eight decimal places, the nonzero roots of the equation are 0.69299996 and 1.39525077.

27. (a) $f(x) = x^2 - a$ $\Rightarrow$ $f'(x) = 2x$, so Newton's method gives

$$x_{n+1} = x_n - \frac{x_n^2 - a}{2x_n} = x_n - \frac{1}{2}x_n + \frac{a}{2x_n} = \frac{1}{2}x_n + \frac{a}{2x_n} = \frac{1}{2}\left(x_n + \frac{a}{x_n}\right).$$

(b) Using (a) with $a = 1000$ and $x_1 = \sqrt{900} = 30$, we get $x_2 \approx 31.666667$, $x_3 \approx 31.622807$, and $x_4 \approx 31.622777 \approx x_5$. So $\sqrt{1000} \approx 31.622777$.

28. (a) $f(x) = \dfrac{1}{x} - a$ $\Rightarrow$ $f'(x) = -\dfrac{1}{x^2}$, so $x_{n+1} = x_n - \dfrac{1/x_n - a}{-1/x_n^2} = x_n + x_n - ax_n^2 = 2x_n - ax_n^2$.

(b) Using (a) with $a = 1.6894$ and $x_1 = \frac{1}{2} = 0.5$, we get $x_2 = 0.5754$, $x_3 \approx 0.588485$, and $x_4 \approx 0.588789 \approx x_5$. So $1/1.6984 \approx 0.588789$.

29. $f(x) = x^3 - 3x + 6$ $\Rightarrow$ $f'(x) = 3x^2 - 3$. If $x_1 = 1$, then $f'(x_1) = 0$ and the tangent line used for approximating x_2 is horizontal. Attempting to find x_2 results in trying to divide by zero.

30. $x^3 - x = 1$ $\Leftrightarrow$ $x^3 - x - 1 = 0$. $f(x) = x^3 - x - 1$ $\Rightarrow$ $f'(x) = 3x^2 - 1$, so $x_{n+1} = x_n - \dfrac{x_n^3 - x_n - 1}{3x_n^2 - 1}$.

(a) $x_1 = 1$, $x_2 = 1.5$, $x_3 \approx 1.347826$, $x_4 \approx 1.325200$, $x_5 \approx 1.324718 \approx x_6$

(b) $x_1 = 0.6$, $x_2 = 17.9$, $x_3 \approx 11.946802$, $x_4 \approx 7.985520$, $x_5 \approx 5.356909$, $x_6 \approx 3.624996$, $x_7 \approx 2.505589$, $x_8 \approx 1.820129$, $x_9 \approx 1.461044$, $x_{10} \approx 1.339323$, $x_{11} \approx 1.324913$, $x_{12} \approx 1.324718 \approx x_{13}$

(c) $x_1 = 0.57$, $x_2 \approx -54.165455$, $x_3 \approx -36.114293$, $x_4 \approx -24.082094$, $x_5 \approx -16.063387$, $x_6 \approx -10.721483$,

$x_7 \approx -7.165534$, $x_8 \approx -4.801704$, $x_9 \approx -3.233425$, $x_{10} \approx -2.193674$, $x_{11} \approx -1.496867$, $x_{12} \approx -0.997546$,

$x_{13} \approx -0.496305$, $x_{14} \approx -2.894162$, $x_{15} \approx -1.967962$, $x_{16} \approx -1.341355$, $x_{17} \approx -0.870187$, $x_{18} \approx -0.249949$,

$x_{19} \approx -1.192219$, $x_{20} \approx -0.731952$, $x_{21} \approx 0.355213$, $x_{22} \approx -1.753322$, $x_{23} \approx -1.189420$, $x_{24} \approx -0.729123$,

$x_{25} \approx 0.377844$, $x_{26} \approx -1.937872$, $x_{27} \approx -1.320350$, $x_{28} \approx -0.851919$, $x_{29} \approx -0.200959$, $x_{30} \approx -1.119386$,

$x_{31} \approx -0.654291$, $x_{32} \approx 1.547010$, $x_{33} \approx 1.360051$, $x_{34} \approx 1.325828$, $x_{35} \approx 1.324719$, $x_{36} \approx 1.324718 \approx x_{37}$.

(d)

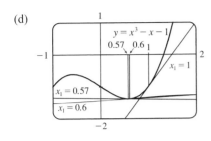

From the figure, we see that the tangent line corresponding to $x_1 = 1$ results in a sequence of approximations that converges quite quickly ($x_5 \approx x_6$). The tangent line corresponding to $x_1 = 0.6$ is close to being horizontal, so x_2 is quite far from the root. But the sequence still converges — just a little more slowly ($x_{12} \approx x_{13}$). Lastly, the tangent line corresponding to $x_1 = 0.57$ is very nearly horizontal, x_2 is farther away from the root, and the sequence takes more iterations to converge ($x_{36} \approx x_{37}$).

31. For $f(x) = x^{1/3}$, $f'(x) = \frac{1}{3}x^{-2/3}$ and

$$x_{n+1} = x_n - \frac{f(x_n)}{f'(x_n)} = x_n - \frac{x_n^{1/3}}{\frac{1}{3}x_n^{-2/3}} = x_n - 3x_n = -2x_n.$$

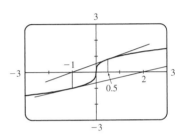

Therefore, each successive approximation becomes twice as large as the previous one in absolute value, so the sequence of approximations fails to converge to the root, which is 0. In the figure, we have $x_1 = 0.5$, $x_2 = -2(0.5) = -1$, and $x_3 = -2(-1) = 2$.

32. According to Newton's Method, for $x_n > 0$,

$$x_{n+1} = x_n - \frac{\sqrt{x_n}}{1/\left(2\sqrt{x_n}\right)} = x_n - 2x_n = -x_n \text{ and for } x_n < 0,$$

$$x_{n+1} = x_n - \frac{-\sqrt{-x_n}}{1/\left(2\sqrt{-x_n}\right)} = x_n - [-2(-x_n)] = -x_n. \text{ So we can see that}$$

after choosing any value x_1 the subsequent values will alternate between $-x_1$ and x_1 and never approach the root.

33. (a) $f(x) = x^6 - x^4 + 3x^3 - 2x$ $\Rightarrow$ $f'(x) = 6x^5 - 4x^3 + 9x^2 - 2$ $\Rightarrow$

$f''(x) = 30x^4 - 12x^2 + 18x$. To find the critical numbers of f, we'll find the zeros of f'. From the graph of f', it appears there are zeros at approximately $x = -1.3$, -0.4, and 0.5. Try $x_1 = -1.3$ $\Rightarrow$

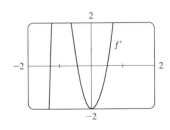

$$x_2 = x_1 - \frac{f'(x_1)}{f''(x_1)} \approx -1.293344 \quad \Rightarrow \quad x_3 \approx -1.293227 \approx x_4.$$

Now try $x_1 = -0.4$ $\Rightarrow$ $x_2 \approx -0.443755$ $\Rightarrow$ $x_3 \approx -0.441735$ $\Rightarrow$ $x_4 \approx -0.441731 \approx x_5$. Finally try $x_1 = 0.5$ $\Rightarrow$ $x_2 \approx 0.507937$ $\Rightarrow$ $x_3 \approx 0.507854 \approx x_4$. Therefore, $x = -1.293227$, -0.441731, and 0.507854 are all the critical numbers correct to six decimal places.

(b) There are two critical numbers where f' changes from negative to positive, so f changes from decreasing to increasing.

$f(-1.293227) \approx -2.0212$ and $f(0.507854) \approx -0.6721$, so -2.0212 is the absolute minimum value of f correct to four decimal places.

34. $f(x) = x \cos x \Rightarrow f'(x) = \cos x - x \sin x$. $f'(x)$ exists for all x, so to find the maximum of f, we can examine the zeros of f'. From the graph of f', we see that a good choice for x_1 is $x_1 = 0.9$. Use $g(x) = \cos x - x \sin x$ and $g'(x) = -2 \sin x - x \cos x$ to obtain $x_2 \approx 0.860781$, $x_3 \approx 0.860334 \approx x_4$. Now we have $f(0) = 0$, $f(\pi) = -\pi$, and $f(0.860334) \approx 0.561096$, so

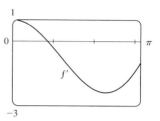

0.561096 is the absolute maximum value of f correct to six decimal places.

35. $y = x^3 + \cos x \Rightarrow y' = 3x^2 - \sin x \Rightarrow y'' = 6x - \cos x \Rightarrow y''' = 6 + \sin x$. Now to solve $y'' = 0$,

try $x_1 = 0$, and then $x_2 = x_1 - \dfrac{y''(x_1)}{y'''(x_1)} = \dfrac{1}{6} \Rightarrow x_3 \approx 0.164419 \approx x_4$. For $x < 0.164419$, $y'' < 0$, and for

$x > 0.164419$, $y'' > 0$. Therefore, the point of inflection, correct to six decimal places, is

$(0.164419, y(0.164419)) \approx (0.164419, 0.990958)$.

36.

$f(x) = -\sin x \Rightarrow f'(x) = -\cos x$. At $x = a$, the slope of the tangent line is $f'(a) = -\cos a$. The line through the origin and $(a, f(a))$ is

$y = \dfrac{-\sin a - 0}{a - 0} x$. If this line is to be tangent to f at $x = a$, then its slope

must equal $f'(a)$. Thus, $\dfrac{-\sin a}{a} = -\cos a \Rightarrow \tan a = a$.

To solve this equation using Newton's method, let $g(x) = \tan x - x$, $g'(x) = \sec^2 x - 1$, and $x_{n+1} = x_n - \dfrac{\tan x_n - x_n}{\sec^2 x_n - 1}$

with $x_1 = 4.5$ (estimated from the figure). $x_2 \approx 4.493614$, $x_3 \approx 4.493410$, $x_4 \approx 4.493409 \approx x_5$. Thus, the slope of the line that has the largest slope is $f'(x_5) \approx 0.217234$.

37. We need to minimize the distance from $(0, 0)$ to an arbitrary point (x, y) on the

curve $y = (x - 1)^2$. $d = \sqrt{x^2 + y^2} \Rightarrow$

$d(x) = \sqrt{x^2 + [(x - 1)^2]^2} = \sqrt{x^2 + (x - 1)^4}$. When $d' = 0$, d will be minimized and equivalently, $s = d^2$ will be minimized, so we will use Newton's method with $f = s'$ and $f' = s''$.

$f(x) = 2x + 4(x - 1)^3 \Rightarrow f'(x) = 2 + 12(x - 1)^2$, so $x_{n+1} = x_n - \dfrac{2x_n + 4(x_n - 1)^3}{2 + 12(x_n - 1)^2}$. Try $x_1 = 0.5 \Rightarrow$

$x_2 = 0.4$, $x_3 \approx 0.410127$, $x_4 \approx 0.410245 \approx x_5$. Now $d(0.410245) \approx 0.537841$ is the minimum distance and the point on the parabola is $(0.410245, 0.347810)$, correct to six decimal places.

38. Let the radius of the circle be r. Using $s = r\theta$, we have $5 = r\theta$ and so $r = 5/\theta$. From the Law of Cosines we get

$4^2 = r^2 + r^2 - 2 \cdot r \cdot r \cdot \cos\theta \Leftrightarrow 16 = 2r^2(1 - \cos\theta) = 2(5/\theta)^2 (1 - \cos\theta)$. Multiplying by θ^2 gives

$16\theta^2 = 50(1 - \cos\theta)$, so we take $f(\theta) = 16\theta^2 + 50\cos\theta - 50$ and

$f'(\theta) = 32\theta - 50\sin\theta$. The formula for Newton's method is

$\theta_{n+1} = \theta_n - \dfrac{16\theta_n^2 + 50\cos\theta_n - 50}{32\theta_n - 50\sin\theta_n}$. From the graph of f, we can use

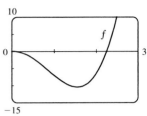

$\theta_1 = 2.2$, giving us $\theta_2 \approx 2.2662$, $\theta_3 \approx 2.2622 \approx \theta_4$. So correct to four

decimal places, the angle is 2.2622 radians $\approx 130°$.

39. In this case, $A = 18{,}000$, $R = 375$, and $n = 5(12) = 60$. So the formula $A = \dfrac{R}{i}[1 - (1 + i)^{-n}]$ becomes

$18{,}000 = \dfrac{375}{x}[1 - (1 + x)^{-60}] \Leftrightarrow 48x = 1 - (1 + x)^{-60}$ [multiply each term by $(1 + x)^{60}$] $\Leftrightarrow$

$48x(1 + x)^{60} - (1 + x)^{60} + 1 = 0$. Let the LHS be called $f(x)$, so that

$$f'(x) = 48x(60)(1 + x)^{59} + 48(1 + x)^{60} - 60(1 + x)^{59}$$
$$= 12(1 + x)^{59}[4x(60) + 4(1 + x) - 5] = 12(1 + x)^{59}(244x - 1)$$

$x_{n+1} = x_n - \dfrac{48x_n(1 + x_n)^{60} - (1 + x_n)^{60} + 1}{12(1 + x_n)^{59}(244x_n - 1)}$. An interest rate of 1% per month seems like a reasonable estimate for

$x = i$. So let $x_1 = 1\% = 0.01$, and we get $x_2 \approx 0.0082202$, $x_3 \approx 0.0076802$, $x_4 \approx 0.0076291$, $x_5 \approx 0.0076286 \approx x_6$.

Thus, the dealer is charging a monthly interest rate of 0.76286% (or 9.55% per year, compounded monthly).

40. (a) $p(x) = x^5 - (2 + r)x^4 + (1 + 2r)x^3 - (1 - r)x^2 + 2(1 - r)x + r - 1 \Rightarrow$

$p'(x) = 5x^4 - 4(2 + r)x^3 + 3(1 + 2r)x^2 - 2(1 - r)x + 2(1 - r)$. So we use

$x_{n+1} = x_n - \dfrac{x_n^5 - (2 + r)x_n^4 + (1 + 2r)x_n^3 - (1 - r)x_n^2 + 2(1 - r)x_n + r - 1}{5x_n^4 - 4(2 + r)x_n^3 + 3(1 + 2r)x_n^2 - 2(1 - r)x_n + 2(1 - r)}$.

We substitute in the value $r \approx 3.04042 \times 10^{-6}$ in order to evaluate the approximations numerically. The libration point

L_1 is slightly less than 1 AU from the sun, so we take $x_1 = 0.95$ as our first approximation, and get $x_2 \approx 0.96682$,

$x_3 \approx 0.97770$, $x_4 \approx 0.98451$, $x_5 \approx 0.98830$, $x_6 \approx 0.98976$, $x_7 \approx 0.98998$, $x_8 \approx 0.98999 \approx x_9$. So, to five decimal

places, L_1 is located 0.98999 AU from the sun (or 0.01001 AU from the earth).

(b) In this case we use Newton's method with the function

$p(x) - 2rx^2 = x^5 - (2 + r)x^4 + (1 + 2r)x^3 - (1 + r)x^2 + 2(1 - r)x + r - 1 \Rightarrow$

$[p(x) - 2rx^2]' = 5x^4 - 4(2 + r)x^3 + 3(1 + 2r)x^2 - 2(1 + r)x + 2(1 - r)$. So

$x_{n+1} = x_n - \dfrac{x_n^5 - (2 + r)x_n^4 + (1 + 2r)x_n^3 - (1 + r)x_n^2 + 2(1 - r)x_n + r - 1}{5x_n^4 - 4(2 + r)x_n^3 + 3(1 + 2r)x_n^2 - 2(1 + r)x_n + 2(1 - r)}$. Again, we substitute

$r \approx 3.04042 \times 10^{-6}$. L_2 is slightly more than 1 AU from the sun and, judging from the result of part (a), probably less

than 0.02 AU from earth. So we take $x_1 = 1.02$ and get $x_2 \approx 1.01422$, $x_3 \approx 1.01118$, $x_4 \approx 1.01018$,

$x_5 \approx 1.01008 \approx x_6$. So, to five decimal places, L_2 is located 1.01008 AU from the sun (or 0.01008 AU from the earth).

4.9 Antiderivatives

1. $f(x) = x - 3 = x^1 - 3 \;\Rightarrow\; F(x) = \dfrac{x^{1+1}}{1+1} - 3x + C = \frac{1}{2}x^2 - 3x + C$

  Check: $F'(x) = \frac{1}{2}(2x) - 3 + 0 = x - 3 = f(x)$

2. $f(x) = \frac{1}{2}x^2 - 2x + 6 \;\Rightarrow\; F(x) = \dfrac{1}{2}\dfrac{x^3}{3} - 2\dfrac{x^2}{2} + 6x + C = \frac{1}{6}x^3 - x^2 + 6x + C$

3. $f(x) = \frac{1}{2} + \frac{3}{4}x^2 - \frac{4}{5}x^3 \;\Rightarrow\; F(x) = \frac{1}{2}x + \dfrac{3}{4}\dfrac{x^{2+1}}{2+1} - \dfrac{4}{5}\dfrac{x^{3+1}}{3+1} + C = \frac{1}{2}x + \frac{1}{4}x^3 - \frac{1}{5}x^4 + C$

  Check: $F'(x) = \frac{1}{2} + \frac{1}{4}(3x^2) - \frac{1}{5}(4x^3) + 0 = \frac{1}{2} + \frac{3}{4}x^2 - \frac{4}{5}x^3 = f(x)$

4. $f(x) = 8x^9 - 3x^6 + 12x^3 \;\Rightarrow\; F(x) = 8\left(\frac{1}{10}x^{10}\right) - 3\left(\frac{1}{7}x^7\right) + 12\left(\frac{1}{4}x^4\right) + C = \frac{4}{5}x^{10} - \frac{3}{7}x^7 + 3x^4 + C$

5. $f(x) = (x+1)(2x-1) = 2x^2 + x - 1 \;\Rightarrow\; F(x) = 2\left(\frac{1}{3}x^3\right) + \frac{1}{2}x^2 - x + C = \frac{2}{3}x^3 + \frac{1}{2}x^2 - x + C$

6. $f(x) = x(2-x)^2 = x\left(4 - 4x + x^2\right) = 4x - 4x^2 + x^3 \;\Rightarrow\;$

  $F(x) = 4\left(\frac{1}{2}x^2\right) - 4\left(\frac{1}{3}x^3\right) + \frac{1}{4}x^4 + C = 2x^2 - \frac{4}{3}x^3 + \frac{1}{4}x^4 + C$

7. $f(x) = 5x^{1/4} - 7x^{3/4} \;\Rightarrow\; F(x) = 5\dfrac{x^{1/4+1}}{\frac{1}{4}+1} - 7\dfrac{x^{3/4+1}}{\frac{3}{4}+1} + C = 5\dfrac{x^{5/4}}{5/4} - 7\dfrac{x^{7/4}}{7/4} + C = 4x^{5/4} - 4x^{7/4} + C$

8. $f(x) = 2x + 3x^{1.7} \;\Rightarrow\; F(x) = x^2 + \dfrac{3}{2.7}x^{2.7} + C = x^2 + \frac{10}{9}x^{2.7} + C$

9. $f(x) = 6\sqrt{x} - \sqrt[6]{x} = 6x^{1/2} - x^{1/6} \;\Rightarrow\;$

  $F(x) = 6\dfrac{x^{1/2+1}}{\frac{1}{2}+1} - \dfrac{x^{1/6+1}}{\frac{1}{6}+1} + C = 6\dfrac{x^{3/2}}{3/2} - \dfrac{x^{7/6}}{7/6} + C = 4x^{3/2} - \frac{6}{7}x^{7/6} + C$

10. $f(x) = \sqrt[4]{x^3} + \sqrt[3]{x^4} = x^{3/4} + x^{4/3} \;\Rightarrow\; F(x) = \dfrac{x^{7/4}}{7/4} + \dfrac{x^{7/3}}{7/3} + C = \frac{4}{7}x^{7/4} + \frac{3}{7}x^{7/3} + C$

11. $f(x) = \dfrac{10}{x^9} = 10x^{-9}$ has domain $(-\infty, 0) \cup (0, \infty)$, so $F(x) = \begin{cases} \dfrac{10x^{-8}}{-8} + C_1 = -\dfrac{5}{4x^8} + C_1 & \text{if } x < 0 \\[2mm] -\dfrac{5}{4x^8} + C_2 & \text{if } x > 0 \end{cases}$

  See Example 1(b) for a similar problem.

12. $g(x) = \dfrac{5 - 4x^3 + 2x^6}{x^6} = 5x^{-6} - 4x^{-3} + 2$ has domain $(-\infty, 0) \cup (0, \infty)$, so

  $G(x) = \begin{cases} 5\dfrac{x^{-5}}{-5} - 4\dfrac{x^{-2}}{-2} + 2x + C_1 = -\dfrac{1}{x^5} + \dfrac{2}{x^2} + 2x + C_1 & \text{if } x < 0 \\[2mm] -\dfrac{1}{x^5} + \dfrac{2}{x^2} + 2x + C_2 & \text{if } x > 0 \end{cases}$

13. $f(u) = \dfrac{u^4 + 3\sqrt{u}}{u^2} = \dfrac{u^4}{u^2} + \dfrac{3u^{1/2}}{u^2} = u^2 + 3u^{-3/2} \;\Rightarrow\;$

  $F(u) = \dfrac{u^3}{3} + 3\dfrac{u^{-3/2+1}}{-3/2+1} + C = \frac{1}{3}u^3 + 3\dfrac{u^{-1/2}}{-1/2} + C = \frac{1}{3}u^3 - \dfrac{6}{\sqrt{u}} + C$

14. $f(t) = 3\cos t - 4\sin t \Rightarrow F(t) = 3(\sin t) - 4(-\cos t) + C = 3\sin t + 4\cos t + C$

15. $g(\theta) = \cos\theta - 5\sin\theta \Rightarrow G(\theta) = \sin\theta - 5(-\cos\theta) + C = \sin\theta + 5\cos\theta + C$

16. $f(\theta) = 6\theta^2 - 7\sec^2\theta \Rightarrow F(\theta) = 2\theta^3 - 7\tan\theta + C_n$ on the interval $\left(n\pi - \frac{\pi}{2}, n\pi + \frac{\pi}{2}\right)$.

17. $f(t) = 2\sec t \tan t + \frac{1}{2}t^{-1/2}$ has domain $\left(0, \frac{\pi}{2}\right)$ and $\left(n\pi - \frac{\pi}{2}, n\pi + \frac{\pi}{2}\right)$ for integers $n \geq 1$. The antiderivative is

$F(t) = 2\sec t + t^{1/2} + C_0$ on the interval $\left(0, \frac{\pi}{2}\right)$ or $F(t) = 2\sec t + t^{1/2} + C_n$ on the interval $\left(n\pi - \frac{\pi}{2}, n\pi + \frac{\pi}{2}\right)$ for

integers $n \geq 1$.

18. $f(x) = 2\sqrt{x} + 6\cos x = 2x^{1/2} + 6\cos x \Rightarrow F(x) = 2\left(\dfrac{x^{3/2}}{3/2}\right) + 6\sin x + C = \frac{4}{3}x^{3/2} + 6\sin x + C$

19. $f(x) = 5x^4 - 2x^5 \Rightarrow F(x) = 5 \cdot \dfrac{x^5}{5} - 2 \cdot \dfrac{x^6}{6} + C = x^5 - \frac{1}{3}x^6 + C.$

$F(0) = 4 \Rightarrow 0^5 - \frac{1}{3} \cdot 0^6 + C = 4 \Rightarrow C = 4$, so $F(x) = x^5 - \frac{1}{3}x^6 + 4.$

The graph confirms our answer since $f(x) = 0$ when F has a local maximum, f is

positive when F is increasing, and f is negative when F is decreasing.

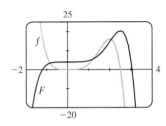

20. $f(x) = x + 2\sin x \Rightarrow F(x) = \frac{1}{2}x^2 - 2\cos x + C.$

$F(0) = -6 \Rightarrow 0 - 2 + C = -6 \Rightarrow C = -4$, so

$F(x) = \frac{1}{2}x^2 - 2\cos x - 4.$

The graph confirms our answer since $f(x) = 0$ when F has a local

minimum, f is positive when F is increasing, and f is negative when F is

decreasing.

21. $f''(x) = 6x + 12x^2 \Rightarrow f'(x) = 6 \cdot \dfrac{x^2}{2} + 12 \cdot \dfrac{x^3}{3} + C = 3x^2 + 4x^3 + C \Rightarrow$

$f(x) = 3 \cdot \dfrac{x^3}{3} + 4 \cdot \dfrac{x^4}{4} + Cx + D = x^3 + x^4 + Cx + D$ 　　[C and D are just arbitrary constants]

22. $f''(x) = 2 + x^3 + x^6 \Rightarrow f'(x) = 2x + \frac{1}{4}x^4 + \frac{1}{7}x^7 + C \Rightarrow f(x) = x^2 + \frac{1}{20}x^5 + \frac{1}{56}x^8 + Cx + D$

23. $f''(x) = \frac{2}{3}x^{2/3} \Rightarrow f'(x) = \dfrac{2}{3}\left(\dfrac{x^{5/3}}{5/3}\right) + C = \frac{2}{5}x^{5/3} + C \Rightarrow f(x) = \dfrac{2}{5}\left(\dfrac{x^{8/3}}{8/3}\right) + Cx + D = \frac{3}{20}x^{8/3} + Cx + D$

24. $f''(x) = 6x + \sin x \Rightarrow f'(x) = 6\left(\dfrac{x^2}{2}\right) - \cos x + C = 3x^2 - \cos x + C \Rightarrow$

$f(x) = 3\left(\dfrac{x^3}{3}\right) - \sin x + Cx + D = x^3 - \sin x + Cx + D$

25. $f'''(t) = 60t^2 \Rightarrow f''(t) = 20t^3 + C \Rightarrow f'(t) = 5t^4 + Ct + D \Rightarrow f(t) = t^5 + \frac{1}{2}Ct^2 + Dt + E$

26. $f'''(t) = t - \sqrt{t} \Rightarrow f''(t) = \frac{1}{2}t^2 - \frac{2}{3}t^{3/2} + C \Rightarrow f'(t) = \frac{1}{6}t^3 - \frac{4}{15}t^{5/2} + Ct + D \Rightarrow$

$f(t) = \frac{1}{24}t^4 - \frac{8}{105}t^{7/2} + \frac{1}{2}Ct^2 + Dt + E$

27. $f'(x) = 1 - 6x \Rightarrow f(x) = x - 3x^2 + C.$ $f(0) = C$ and $f(0) = 8 \Rightarrow C = 8$, so $f(x) = x - 3x^2 + 8.$

28. $f'(x) = 8x^3 + 12x + 3 \Rightarrow f(x) = 2x^4 + 6x^2 + 3x + C.$ $f(1) = 11 + C$ and $f(1) = 6 \Rightarrow$
$11 + C = 6 \Rightarrow C = -5$, so $f(x) = 2x^4 + 6x^2 + 3x - 5$.

29. $f'(x) = \sqrt{x}(6 + 5x) = 6x^{1/2} + 5x^{3/2} \Rightarrow f(x) = 4x^{3/2} + 2x^{5/2} + C.$
$f(1) = 6 + C$ and $f(1) = 10 \Rightarrow C = 4$, so $f(x) = 4x^{3/2} + 2x^{5/2} + 4$.

30. $f'(x) = 2x - 3/x^4 = 2x - 3x^{-4} \Rightarrow f(x) = x^2 + x^{-3} + C$ because we're given that $x > 0$.
$f(1) = 2 + C$ and $f(1) = 3 \Rightarrow C = 1$, so $f(x) = x^2 + 1/x^3 + 1$.

31. $f'(t) = 2\cos t + \sec^2 t \Rightarrow f(t) = 2\sin t + \tan t + C$ because $-\pi/2 < t < \pi/2$.
$f\left(\frac{\pi}{3}\right) = 2(\sqrt{3}/2) + \sqrt{3} + C = 2\sqrt{3} + C$ and $f\left(\frac{\pi}{3}\right) = 4 \Rightarrow C = 4 - 2\sqrt{3}$, so $f(t) = 2\sin t + \tan t + 4 - 2\sqrt{3}$.

32. $f'(x) = x^{-1/3}$ has domain $(-\infty, 0) \cup (0, \infty) \Rightarrow f(x) = \begin{cases} \frac{3}{2}x^{2/3} + C_1 & \text{if } x > 0 \\ \frac{3}{2}x^{2/3} + C_2 & \text{if } x < 0 \end{cases}$

$f(1) = \frac{3}{2} + C_1$ and $f(1) = 1 \Rightarrow C_1 = -\frac{1}{2}$. $f(-1) = \frac{3}{2} + C_2$ and $f(-1) = -1 \Rightarrow C_2 = -\frac{5}{2}$.

Thus, $f(x) = \begin{cases} \frac{3}{2}x^{2/3} - \frac{1}{2} & \text{if } x > 0 \\ \frac{3}{2}x^{2/3} - \frac{5}{2} & \text{if } x < 0 \end{cases}$

33. $f''(x) = 24x^2 + 2x + 10 \Rightarrow f'(x) = 8x^3 + x^2 + 10x + C.$ $f'(1) = 8 + 1 + 10 + C$ and $f'(1) = -3 \Rightarrow$
$19 + C = -3 \Rightarrow C = -22$, so $f'(x) = 8x^3 + x^2 + 10x - 22$ and hence, $f(x) = 2x^4 + \frac{1}{3}x^3 + 5x^2 - 22x + D.$
$f(1) = 2 + \frac{1}{3} + 5 - 22 + D$ and $f(1) = 5 \Rightarrow D = 22 - \frac{7}{3} = \frac{59}{3}$, so $f(x) = 2x^4 + \frac{1}{3}x^3 + 5x^2 - 22x + \frac{59}{3}$.

34. $f''(x) = 4 - 6x - 40x^3 \Rightarrow f'(x) = 4x - 3x^2 - 10x^4 + C.$ $f'(0) = C$ and $f'(0) = 1 \Rightarrow C = 1$, so
$f'(x) = 4x - 3x^2 - 10x^4 + 1$ and hence, $f(x) = 2x^2 - x^3 - 2x^5 + x + D.$ $f(0) = D$ and $f(0) = 2 \Rightarrow D = 2$, so
$f(x) = 2x^2 - x^3 - 2x^5 + x + 2$.

35. $f''(\theta) = \sin\theta + \cos\theta \Rightarrow f'(\theta) = -\cos\theta + \sin\theta + C.$ $f'(0) = -1 + C$ and $f'(0) = 4 \Rightarrow C = 5$, so
$f'(\theta) = -\cos\theta + \sin\theta + 5$ and hence, $f(\theta) = -\sin\theta - \cos\theta + 5\theta + D.$ $f(0) = -1 + D$ and $f(0) = 3 \Rightarrow D = 4$,
so $f(\theta) = -\sin\theta - \cos\theta + 5\theta + 4$.

36. $f''(t) = 3/\sqrt{t} = 3t^{-1/2} \Rightarrow f'(t) = 6t^{1/2} + C.$ $f'(4) = 12 + C$ and $f'(4) = 7 \Rightarrow C = -5$, so $f'(t) = 6t^{1/2} - 5$
and hence, $f(t) = 4t^{3/2} - 5t + D.$ $f(4) = 32 - 20 + D$ and $f(4) = 20 \Rightarrow D = 8$, so $f(t) = 4t^{3/2} - 5t + 8$.

37. $f''(x) = 2 - 12x \Rightarrow f'(x) = 2x - 6x^2 + C \Rightarrow f(x) = x^2 - 2x^3 + Cx + D.$
$f(0) = D$ and $f(0) = 9 \Rightarrow D = 9.$ $f(2) = 4 - 16 + 2C + 9 = 2C - 3$ and $f(2) = 15 \Rightarrow 2C = 18 \Rightarrow$
$C = 9$, so $f(x) = x^2 - 2x^3 + 9x + 9$.

38. $f''(x) = 20x^3 + 12x^2 + 4 \Rightarrow f'(x) = 5x^4 + 4x^3 + 4x + C \Rightarrow f(x) = x^5 + x^4 + 2x^2 + Cx + D.$
$f(0) = D$ and $f(0) = 8 \Rightarrow D = 8.$ $f(1) = 1 + 1 + 2 + C + 8 = C + 12$ and $f(1) = 5 \Rightarrow C = -7$, so
$f(x) = x^5 + x^4 + 2x^2 - 7x + 8$.

39. $f''(x) = 2 + \cos x \;\Rightarrow\; f'(x) = 2x + \sin x + C \;\Rightarrow\; f(x) = x^2 - \cos x + Cx + D.$

$f(0) = -1 + D$ and $f(0) = -1 \;\Rightarrow\; D = 0.$ $f\left(\frac{\pi}{2}\right) = \pi^2/4 + \left(\frac{\pi}{2}\right)C$ and $f\left(\frac{\pi}{2}\right) = 0 \;\Rightarrow\; \left(\frac{\pi}{2}\right)C = -\pi^2/4 \;\Rightarrow\;$

$C = -\frac{\pi}{2}$, so $f(x) = x^2 - \cos x - \left(\frac{\pi}{2}\right)x.$

40. $f'''(x) = \cos x \;\Rightarrow\; f''(x) = \sin x + C.$ $f''(0) = C$ and $f''(0) = 3 \;\Rightarrow\; C = 3.$ $f''(x) = \sin x + 3 \;\Rightarrow\;$

$f'(x) = -\cos x + 3x + D.$ $f'(0) = -1 + D$ and $f'(0) = 2 \;\Rightarrow\; D = 3.$ $f'(x) = -\cos x + 3x + 3 \;\Rightarrow\;$

$f(x) = -\sin x + \frac{3}{2}x^2 + 3x + E.$ $f(0) = E$ and $f(0) = 1 \;\Rightarrow\; E = 1.$ Thus, $f(x) = -\sin x + \frac{3}{2}x^2 + 3x + 1.$

41. Given $f'(x) = 2x + 1$, we have $f(x) = x^2 + x + C.$ Since f passes through $(1, 6)$, $f(1) = 6 \;\Rightarrow\; 1^2 + 1 + C = 6 \;\Rightarrow\;$

$C = 4.$ Therefore, $f(x) = x^2 + x + 4$ and $f(2) = 2^2 + 2 + 4 = 10.$

42. $f'(x) = x^3 \;\Rightarrow\; f(x) = \frac{1}{4}x^4 + C.$ $x + y = 0 \;\Rightarrow\; y = -x \;\Rightarrow\; m = -1.$ Now $m = f'(x) \;\Rightarrow\; -1 = x^3 \;\Rightarrow\;$

$x = -1 \;\Rightarrow\; y = 1$ (from the equation of the tangent line), so $(-1, 1)$ is a point on the graph of $f.$ From f,

$1 = \frac{1}{4}(-1)^4 + C \;\Rightarrow\; C = \frac{3}{4}.$ Therefore, the function is $f(x) = \frac{1}{4}x^4 + \frac{3}{4}.$

43. b is the antiderivative of $f.$ For small x, f is negative, so the graph of its antiderivative must be decreasing. But both a and c are increasing for small x, so only b can be f's antiderivative. Also, f is positive where b is increasing, which supports our conclusion.

44. We know right away that c cannot be f's antiderivative, since the slope of c is not zero at the x-value where $f = 0.$ Now f is positive when a is increasing and negative when a is decreasing, so a is the antiderivative of $f.$

45.

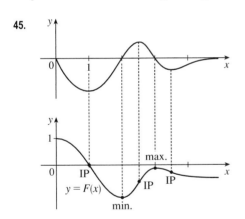

The graph of F must start at $(0, 1).$ Where the given graph, $y = f(x)$, has a local minimum or maximum, the graph of F will have an inflection point.

Where f is negative (positive), F is decreasing (increasing).

Where f changes from negative to positive, F will have a minimum.

Where f changes from positive to negative, F will have a maximum.

Where f is decreasing (increasing), F is concave downward (upward).

46.

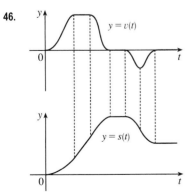

Where v is positive (negative), s is increasing (decreasing).

Where v is increasing (decreasing), s is concave upward (downward).

Where v is horizontal (a steady velocity), s is linear.

47.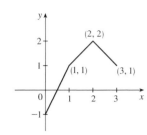

$$f'(x) = \begin{cases} 2 & \text{if } 0 \le x < 1 \\ 1 & \text{if } 1 < x < 2 \\ -1 & \text{if } 2 < x \le 3 \end{cases} \Rightarrow f(x) = \begin{cases} 2x + C & \text{if } 0 \le x < 1 \\ x + D & \text{if } 1 < x < 2 \\ -x + E & \text{if } 2 < x \le 3 \end{cases}$$

$f(0) = -1 \Rightarrow 2(0) + C = -1 \Rightarrow C = -1$. Starting at the point

$(0, -1)$ and moving to the right on a line with slope 2 gets us to the point $(1, 1)$.

The slope for $1 < x < 2$ is 1, so we get to the point $(2, 2)$. Here we have used the fact that f is continuous. We can include the

point $x = 1$ on either the first or the second part of f. The line connecting $(1, 1)$ to $(2, 2)$ is $y = x$, so $D = 0$. The slope for

$2 < x \le 3$ is -1, so we get to $(3, 1)$. $f(3) = 1 \Rightarrow -3 + E = 1 \Rightarrow E = 4$. Thus

$$f(x) = \begin{cases} 2x - 1 & \text{if } 0 \le x \le 1 \\ x & \text{if } 1 < x < 2 \\ -x + 4 & \text{if } 2 \le x \le 3 \end{cases}$$

Note that $f'(x)$ does not exist at $x = 1$ or at $x = 2$.

48. (a)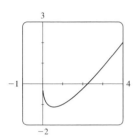

(b) Since $F(0) = 1$, we can start our graph at $(0, 1)$. f has a minimum at about $x = 0.5$, so its derivative is zero there. f is decreasing on $(0, 0.5)$, so its derivative is negative and hence, F is CD on $(0, 0.5)$ and has an IP at $x \approx 0.5$. On $(0.5, 2.2)$, f is negative and increasing (f' is positive), so F is decreasing and CU. On $(2.2, \infty)$, f is positive and increasing, so F is increasing and CU.

(c) $f(x) = 2x - 3\sqrt{x} \Rightarrow$

$F(x) = x^2 - 3 \cdot \frac{2}{3}x^{3/2} + C$.

$F(0) = C$ and $F(0) = 1 \Rightarrow$

$C = 1$, so

$F(x) = x^2 - 2x^{3/2} + 1$.

(d)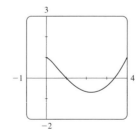

49. $f(x) = \dfrac{\sin x}{1 + x^2}$, $-2\pi \le x \le 2\pi$

Note that the graph of f is one of an odd function, so the graph of F will be one of an even function.

50. $f(x) = \sqrt{x^4 - 2x^2 + 2} - 1$, $-1.5 \le x \le 1.5$

Note that the graph of f is one of an even function, so the graph of F will be one of an odd function.

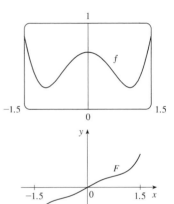

51. $v(t) = s'(t) = \sin t - \cos t \;\Rightarrow\; s(t) = -\cos t - \sin t + C.$ $s(0) = -1 + C$ and $s(0) = 0 \;\Rightarrow\; C = 1$, so $s(t) = -\cos t - \sin t + 1.$

52. $v(t) = s'(t) = 1.5\sqrt{t} \;\Rightarrow\; s(t) = t^{3/2} + C.$ $s(4) = 8 + C$ and $s(4) = 10 \;\Rightarrow\; C = 2$, so $s(t) = t^{3/2} + 2.$

53. $a(t) = v'(t) = t - 2 \;\Rightarrow\; v(t) = \frac{1}{2}t^2 - 2t + C.$ $v(0) = C$ and $v(0) = 3 \;\Rightarrow\; C = 3$, so $v(t) = \frac{1}{2}t^2 - 2t + 3$ and $s(t) = \frac{1}{6}t^3 - t^2 + 3t + D.$ $s(0) = D$ and $s(0) = 1 \;\Rightarrow\; D = 1$, and $s(t) = \frac{1}{6}t^3 - t^2 + 3t + 1.$

54. $a(t) = v'(t) = \cos t + \sin t \;\Rightarrow\; v(t) = \sin t - \cos t + C \;\Rightarrow\; 5 = v(0) = -1 + C \;\Rightarrow\; C = 6$, so $v(t) = \sin t - \cos t + 6 \;\Rightarrow\; s(t) = -\cos t - \sin t + 6t + D \;\Rightarrow\; 0 = s(0) = -1 + D \;\Rightarrow\; D = 1$, so $s(t) = -\cos t - \sin t + 6t + 1.$

55. $a(t) = v'(t) = 10\sin t + 3\cos t \;\Rightarrow\; v(t) = -10\cos t + 3\sin t + C \;\Rightarrow\; s(t) = -10\sin t - 3\cos t + Ct + D.$ $s(0) = -3 + D = 0$ and $s(2\pi) = -3 + 2\pi C + D = 12 \;\Rightarrow\; D = 3$ and $C = \frac{6}{\pi}$. Thus, $s(t) = -10\sin t - 3\cos t + \frac{6}{\pi}t + 3.$

56. $a(t) = t^2 - 4t + 6 \implies v(t) = \frac{1}{3}t^3 - 2t^2 + 6t + C \implies s(t) = \frac{1}{12}t^4 - \frac{2}{3}t^3 + 3t^2 + Ct + D.$ $s(0) = D$ and $s(0) = 0 \implies D = 0.$ $s(1) = \frac{29}{12} + C$ and $s(1) = 20 \implies C = \frac{211}{12}.$ Thus, $s(t) = \frac{1}{12}t^4 - \frac{2}{3}t^3 + 3t^2 + \frac{211}{12}t.$

57. (a) We first observe that since the stone is dropped 450 m above the ground, $v(0) = 0$ and $s(0) = 450.$

$v'(t) = a(t) = -9.8 \implies v(t) = -9.8t + C.$ Now $v(0) = 0 \implies C = 0,$ so $v(t) = -9.8t \implies$
$s(t) = -4.9t^2 + D.$ Last, $s(0) = 450 \implies D = 450 \implies s(t) = 450 - 4.9t^2.$

(b) The stone reaches the ground when $s(t) = 0.$ $450 - 4.9t^2 = 0 \implies t^2 = 450/4.9 \implies t_1 = \sqrt{450/4.9} \approx 9.58$ s.

(c) The velocity with which the stone strikes the ground is $v(t_1) = -9.8\sqrt{450/4.9} \approx -93.9$ m/s.

(d) This is just reworking parts (a) and (b) with $v(0) = -5.$ Using $v(t) = -9.8t + C,$ $v(0) = -5 \implies 0 + C = -5 \implies$
$v(t) = -9.8t - 5.$ So $s(t) = -4.9t^2 - 5t + D$ and $s(0) = 450 \implies D = 450 \implies s(t) = -4.9t^2 - 5t + 450.$
Solving $s(t) = 0$ by using the quadratic formula gives us $t = \left(5 \pm \sqrt{8845}\right)/(-9.8) \implies t_1 \approx 9.09$ s.

58. $v'(t) = a(t) = a \implies v(t) = at + C$ and $v_0 = v(0) = C \implies v(t) = at + v_0 \implies$
$s(t) = \frac{1}{2}at^2 + v_0 t + D \implies s_0 = s(0) = D \implies s(t) = \frac{1}{2}at^2 + v_0 t + s_0$

59. By Exercise 58 with $a = -9.8,$ $s(t) = -4.9t^2 + v_0 t + s_0$ and $v(t) = s'(t) = -9.8t + v_0.$ So
$[v(t)]^2 = (-9.8t + v_0)^2 = (9.8)^2 t^2 - 19.6v_0 t + v_0^2 = v_0^2 + 96.04t^2 - 19.6v_0 t = v_0^2 - 19.6\left(-4.9t^2 + v_0 t\right).$
But $-4.9t^2 + v_0 t$ is just $s(t)$ without the s_0 term; that is, $s(t) - s_0.$ Thus, $[v(t)]^2 = v_0^2 - 19.6[s(t) - s_0].$

60. For the first ball, $s_1(t) = -16t^2 + 48t + 432$ from Example 7. For the second ball, $a(t) = -32 \implies v(t) = -32t + C,$ but
$v(1) = -32(1) + C = 24 \implies C = 56,$ so $v(t) = -32t + 56 \implies s(t) = -16t^2 + 56t + D,$ but
$s(1) = -16(1)^2 + 56(1) + D = 432 \implies D = 392,$ and $s_2(t) = -16t^2 + 56t + 392.$ The balls pass each other
when $s_1(t) = s_2(t) \implies -16t^2 + 48t + 432 = -16t^2 + 56t + 392 \iff 8t = 40 \iff t = 5$ s.

Another solution: From Exercise 58, we have $s_1(t) = -16t^2 + 48t + 432$ and $s_2(t) = -16t^2 + 24t + 432.$
We now want to solve $s_1(t) = s_2(t - 1) \implies -16t^2 + 48t + 432 = -16(t - 1)^2 + 24(t - 1) + 432 \implies$
$48t = 32t - 16 + 24t - 24 \implies 40 = 8t \implies t = 5$ s.

61. Using Exercise 58 with $a = -32,$ $v_0 = 0,$ and $s_0 = h$ (the height of the cliff), we know that the height at time t is
$s(t) = -16t^2 + h.$ $v(t) = s'(t) = -32t$ and $v(t) = -120 \implies -32t = -120 \implies t = 3.75,$ so
$0 = s(3.75) = -16(3.75)^2 + h \implies h = 16(3.75)^2 = 225$ ft.

62. (a) $EIy'' = mg(L - x) + \frac{1}{2}\rho g(L - x)^2 \implies EIy' = -\frac{1}{2}mg(L - x)^2 - \frac{1}{6}\rho g(L - x)^3 + C \implies$

$EIy = \frac{1}{6}mg(L - x)^3 + \frac{1}{24}\rho g(L - x)^4 + Cx + D.$ Since the left end of the board is fixed, we must have $y = y' = 0$
when $x = 0.$ Thus, $0 = -\frac{1}{2}mgL^2 - \frac{1}{6}\rho gL^3 + C$ and $0 = \frac{1}{6}mgL^3 + \frac{1}{24}\rho gL^4 + D.$ It follows that

$EIy = \frac{1}{6}mg(L - x)^3 + \frac{1}{24}\rho g(L - x)^4 + \left(\frac{1}{2}mgL^2 + \frac{1}{6}\rho gL^3\right)x - \left(\frac{1}{6}mgL^3 + \frac{1}{24}\rho gL^4\right)$ and

$f(x) = y = \frac{1}{EI}\left[\frac{1}{6}mg(L - x)^3 + \frac{1}{24}\rho g(L - x)^4 + \left(\frac{1}{2}mgL^2 + \frac{1}{6}\rho gL^3\right)x - \left(\frac{1}{6}mgL^3 + \frac{1}{24}\rho gL^4\right)\right]$

(b) $f(L) < 0$, so the end of the board is a *distance* approximately $-f(L)$ below the horizontal. From our result in (a), we

calculate

$$-f(L) = \frac{-1}{EI}\left[\tfrac{1}{2}mgL^3 + \tfrac{1}{6}\rho gL^4 - \tfrac{1}{6}mgL^3 - \tfrac{1}{24}\rho gL^4\right] = \frac{-1}{EI}\left(\tfrac{1}{3}mgL^3 + \tfrac{1}{8}\rho gL^4\right) = -\frac{gL^3}{EI}\left(\frac{m}{3} + \frac{\rho L}{8}\right)$$

Note: This is positive because g is negative.

63. Marginal cost $= 1.92 - 0.002x = C'(x) \;\Rightarrow\; C(x) = 1.92x - 0.001x^2 + K$. But $C(1) = 1.92 - 0.001 + K = 562 \;\Rightarrow\;$

$K = 560.081$. Therefore, $C(x) = 1.92x - 0.001x^2 + 560.081 \;\Rightarrow\; C(100) = 742.081$, so the cost of producing

100 items is \$742.08.

64. Let the mass, measured from one end, be $m(x)$. Then $m(0) = 0$ and $\rho = \dfrac{dm}{dx} = x^{-1/2} \;\Rightarrow\; m(x) = 2x^{1/2} + C$ and

$m(0) = C = 0$, so $m(x) = 2\sqrt{x}$. Thus, the mass of the 100-centimeter rod is $m(100) = 2\sqrt{100} = 20$ g.

65. Taking the upward direction to be positive we have that for $0 \le t \le 10$ (using the subscript 1 to refer to $0 \le t \le 10$),

$a_1(t) = -(9 - 0.9t) = v_1'(t) \;\Rightarrow\; v_1(t) = -9t + 0.45t^2 + v_0$, but $v_1(0) = v_0 = -10 \;\Rightarrow\;$

$v_1(t) = -9t + 0.45t^2 - 10 = s_1'(t) \;\Rightarrow\; s_1(t) = -\tfrac{9}{2}t^2 + 0.15t^3 - 10t + s_0$. But $s_1(0) = 500 = s_0 \;\Rightarrow\;$

$s_1(t) = -\tfrac{9}{2}t^2 + 0.15t^3 - 10t + 500$. $s_1(10) = -450 + 150 - 100 + 500 = 100$, so it takes

more than 10 seconds for the raindrop to fall. Now for $t > 10$, $a(t) = 0 = v'(t) \;\Rightarrow\;$

$v(t) = \text{constant} = v_1(10) = -9(10) + 0.45(10)^2 - 10 = -55 \;\Rightarrow\; v(t) = -55$.

At 55 m/s, it will take $100/55 \approx 1.8$ s to fall the last 100 m. Hence, the total time is $10 + \frac{100}{55} = \frac{130}{11} \approx 11.8$ s.

66. $v'(t) = a(t) = -22$. The initial velocity is 50 mi/h $= \frac{50 \cdot 5280}{3600} = \frac{220}{3}$ ft/s, so $v(t) = -22t + \frac{220}{3}$.

The car stops when $v(t) = 0 \;\Leftrightarrow\; t = \frac{220}{3 \cdot 22} = \frac{10}{3}$. Since $s(t) = -11t^2 + \frac{220}{3}t$, the distance covered is

$s\left(\frac{10}{3}\right) = -11\left(\frac{10}{3}\right)^2 + \frac{220}{3} \cdot \frac{10}{3} = \frac{1100}{9} = 122.\overline{2}$ ft.

67. $a(t) = k$, the initial velocity is 30 mi/h $= 30 \cdot \frac{5280}{3600} = 44$ ft/s, and the final velocity (after 5 seconds) is

50 mi/h $= 50 \cdot \frac{5280}{3600} = \frac{220}{3}$ ft/s. So $v(t) = kt + C$ and $v(0) = 44 \;\Rightarrow\; C = 44$. Thus, $v(t) = kt + 44 \;\Rightarrow\;$

$v(5) = 5k + 44$. But $v(5) = \frac{220}{3}$, so $5k + 44 = \frac{220}{3} \;\Rightarrow\; 5k = \frac{88}{3} \;\Rightarrow\; k = \frac{88}{15} \approx 5.87$ ft/s^2.

68. $a(t) = -16 \;\Rightarrow\; v(t) = -16t + v_0$ where v_0 is the car's speed (in ft/s) when the brakes were applied. The car stops when

$-16t + v_0 = 0 \;\Leftrightarrow\; t = \frac{1}{16}v_0$. Now $s(t) = \tfrac{1}{2}(-16)t^2 + v_0 t = -8t^2 + v_0 t$. The car travels 200 ft in the time that it takes

to stop, so $s\left(\frac{1}{16}v_0\right) = 200 \;\Rightarrow\; 200 = -8\left(\frac{1}{16}v_0\right)^2 + v_0\left(\frac{1}{16}v_0\right) = \frac{1}{32}v_0^2 \;\Rightarrow\; v_0^2 = 32 \cdot 200 = 6400 \;\Rightarrow\;$

$v_0 = 80$ ft/s $[54.\overline{54}$ mi/h$]$.

69. Let the acceleration be $a(t) = k$ km/h^2. We have $v(0) = 100$ km/h and we can take the initial position $s(0)$ to be 0.

We want the time t_f for which $v(t) = 0$ to satisfy $s(t) < 0.08$ km. In general, $v'(t) = a(t) = k$, so $v(t) = kt + C$, where $C = v(0) = 100$. Now $s'(t) = v(t) = kt + 100$, so $s(t) = \frac{1}{2}kt^2 + 100t + D$, where $D = s(0) = 0$.

Thus, $s(t) = \frac{1}{2}kt^2 + 100t$. Since $v(t_f) = 0$, we have $kt_f + 100 = 0$ or $t_f = -100/k$, so

$s(t_f) = \frac{1}{2}k\left(-\dfrac{100}{k}\right)^2 + 100\left(-\dfrac{100}{k}\right) = 10{,}000\left(\dfrac{1}{2k} - \dfrac{1}{k}\right) = -\dfrac{5{,}000}{k}$. The condition $s(t_f)$ must satisfy is

$-\dfrac{5{,}000}{k} < 0.08 \quad \Rightarrow \quad -\dfrac{5{,}000}{0.08} > k \quad [k \text{ is negative}] \quad \Rightarrow \quad k < -62{,}500 \text{ km/h}^2$, or equivalently,

$k < -\frac{3125}{648} \approx -4.82 \text{ m/s}^2$.

70. (a) For $0 \le t \le 3$ we have $a(t) = 60t \quad \Rightarrow \quad v(t) = 30t^2 + C \quad \Rightarrow \quad v(0) = 0 = C \quad \Rightarrow \quad v(t) = 30t^2$, so

$s(t) = 10t^3 + C \quad \Rightarrow \quad s(0) = 0 = C \quad \Rightarrow \quad s(t) = 10t^3$. Note that $v(3) = 270$ and $s(3) = 270$.

For $3 < t \le 17$: $a(t) = -g = -32 \text{ ft/s} \quad \Rightarrow \quad v(t) = -32(t-3) + C \quad \Rightarrow \quad v(3) = 270 = C \quad \Rightarrow$

$v(t) = -32(t-3) + 270 \quad \Rightarrow \quad s(t) = -16(t-3)^2 + 270(t-3) + C \quad \Rightarrow \quad s(3) = 270 = C \quad \Rightarrow$

$s(t) = -16(t-3)^2 + 270(t-3) + 270$. Note that $v(17) = -178$ and $s(17) = 914$.

For $17 < t \le 22$: The velocity increases linearly from -178 ft/s to -18 ft/s during this period, so

$\dfrac{\Delta v}{\Delta t} = \dfrac{-18 - (-178)}{22 - 17} = \dfrac{160}{5} = 32$. Thus, $v(t) = 32(t-17) - 178 \quad \Rightarrow$

$s(t) = 16(t-17)^2 - 178(t-17) + 914$ and $s(22) = 424$ ft.

For $t > 22$: $v(t) = -18 \quad \Rightarrow \quad s(t) = -18(t-22) + C$. But $s(22) = 424 = C \quad \Rightarrow \quad s(t) = -18(t-22) + 424$.

Therefore, until the rocket lands, we have

$$v(t) = \begin{cases} 30t^2 & \text{if } 0 \le t \le 3 \\ -32(t-3) + 270 & \text{if } 3 < t \le 17 \\ 32(t-17) - 178 & \text{if } 17 < t \le 22 \\ -18 & \text{if } t > 22 \end{cases}$$

and

$$s(t) = \begin{cases} 10t^3 & \text{if } 0 \le t \le 3 \\ -16(t-3)^2 + 270(t-3) + 270 & \text{if } 3 < t \le 17 \\ 16(t-17)^2 - 178(t-17) + 914 & \text{if } 17 < t \le 22 \\ -18(t-22) + 424 & \text{if } t > 22 \end{cases}$$

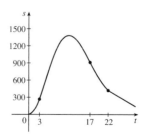

(b) To find the maximum height, set $v(t)$ on $3 < t \le 17$ equal to 0. $-32(t-3) + 270 = 0 \quad \Rightarrow \quad t_1 = 11.4375$ s and the

maximum height is $s(t_1) = -16(t_1 - 3)^2 + 270(t_1 - 3) + 270 = 1409.0625$ ft.

(c) To find the time to land, set $s(t) = -18(t-22) + 424 = 0$. Then $t - 22 = \frac{424}{18} = 23.\overline{5}$, so $t \approx 45.6$ s.

71. (a) First note that $90 \text{ mi/h} = 90 \times \frac{5280}{3600} \text{ ft/s} = 132 \text{ ft/s}$. Then $a(t) = 4 \text{ ft/s}^2 \Rightarrow v(t) = 4t + C$, but $v(0) = 0 \Rightarrow$

$C = 0$. Now $4t = 132$ when $t = \frac{132}{4} = 33$ s, so it takes 33 s to reach 132 ft/s. Therefore, taking $s(0) = 0$, we have

$s(t) = 2t^2$, $0 \le t \le 33$. So $s(33) = 2178$ ft. 15 minutes $= 15(60) = 900$ s, so for $33 < t \le 933$ we have

$v(t) = 132 \text{ ft/s} \Rightarrow s(933) = 132(900) + 2178 = 120{,}978 \text{ ft} = 22.9125 \text{ mi}$.

(b) As in part (a), the train accelerates for 33 s and travels 2178 ft while doing so. Similarly, it decelerates for 33 s and travels

2178 ft at the end of its trip. During the remaining $900 - 66 = 834$ s it travels at 132 ft/s, so the distance traveled is

$132 \cdot 834 = 110{,}088$ ft. Thus, the total distance is $2178 + 110{,}088 + 2178 = 114{,}444 \text{ ft} = 21.675 \text{ mi}$.

(c) $45 \text{ mi} = 45(5280) = 237{,}600$ ft. Subtract $2(2178)$ to take care of the speeding up and slowing down, and we have

233,244 ft at 132 ft/s for a trip of $233{,}244/132 = 1767$ s at 90 mi/h. The total time is

$1767 + 2(33) = 1833 \text{ s} = 30 \text{ min } 33 \text{ s} = 30.55 \text{ min}$.

(d) $37.5(60) = 2250$ s. $2250 - 2(33) = 2184$ s at maximum speed. $2184(132) + 2(2178) = 292{,}644$ total feet or

$292{,}644/5280 = 55.425 \text{ mi}$.

4 Review

CONCEPT CHECK

1. A function f has an **absolute maximum** at $x = c$ if $f(c)$ is the largest function value on the entire domain of f, whereas f has a **local maximum** at c if $f(c)$ is the largest function value when x is near c. See Figure 4 in Section 4.1.

2. (a) See Theorem 4.1.3.

(b) See the Closed Interval Method before Example 8 in Section 4.1.

3. (a) See Theorem 4.1.4.

(b) See Definition 4.1.6.

4. (a) See Rolle's Theorem at the beginning of Section 4.2.

(b) See the Mean Value Theorem in Section 4.2. Geometric interpretation—there is some point P on the graph of a function f [on the interval (a, b)] where the tangent line is parallel to the secant line that connects $(a, f(a))$ and $(b, f(b))$.

5. (a) See the I/D Test before Example 1 in Section 4.3.

(b) If the graph of f lies above all of its tangents on an interval I, then it is called concave upward on I.

(c) See the Concavity Test before Example 4 in Section 4.3.

(d) An inflection point is a point where a curve changes its direction of concavity. They can be found by determining the points at which the second derivative changes sign.

6. (a) See the First Derivative Test after Example 1 in Section 4.3.

(b) See the Second Derivative Test before Example 6 in Section 4.3.

(c) See the note before Example 7 in Section 4.3.

7. (a) See Definitions 4.4.1 and 4.4.5.

(b) See Definitions 4.4.2 and 4.4.6.

(c) See Definition 4.4.7.

(d) See Definition 4.4.3.

8. Without calculus you could get misleading graphs that fail to show the most interesting features of a function. See the discussion at the beginning of Section 4.5 and the first paragraph in Section 4.6.

9. (a) See Figure 2 in Section 4.8.

(b) $x_2 = x_1 - \dfrac{f(x_1)}{f'(x_1)}$

(c) $x_{n+1} = x_n - \dfrac{f(x_n)}{f'(x_n)}$

(d) Newton's method is likely to fail or to work very slowly when $f'(x_1)$ is close to 0. It also fails when $f'(x_i)$ is undefined, such as with $f(x) = 1/x - 2$ and $x_1 = 1$.

10. (a) See the definition at the beginning of Section 4.9.

(b) If F_1 and F_2 are both antiderivatives of f on an interval I, then they differ by a constant.

TRUE-FALSE QUIZ

1. False. For example, take $f(x) = x^3$, then $f'(x) = 3x^2$ and $f'(0) = 0$, but $f(0) = 0$ is not a maximum or minimum; $(0,0)$ is an inflection point.

2. False. For example, $f(x) = |x|$ has an absolute minimum at 0, but $f'(0)$ does not exist.

3. False. For example, $f(x) = x$ is continuous on $(0,1)$ but attains neither a maximum nor a minimum value on $(0,1)$. Don't confuse this with f being continuous on the *closed* interval $[a,b]$, which would make the statement true.

4. True. By the Mean Value Theorem, $f'(c) = \dfrac{f(1) - f(-1)}{1 - (-1)} = \dfrac{0}{2} = 0$. Note that $|c| < 1 \iff c \in (-1, 1)$.

5. True. This is an example of part (b) of the I/D Test.

6. False. For example, the curve $y = f(x) = 1$ has no inflection points but $f''(c) = 0$ for all c.

7. False. $f'(x) = g'(x) \implies f(x) = g(x) + C$. For example, if $f(x) = x + 2$ and $g(x) = x + 1$, then $f'(x) = g'(x) = 1$, but $f(x) \neq g(x)$.

8. False. Assume there is a function f such that $f(1) = -2$ and $f(3) = 0$. Then by the Mean Value Theorem there exists a number $c \in (1, 3)$ such that $f'(c) = \dfrac{f(3) - f(1)}{3 - 1} = \dfrac{0 - (-2)}{2} = 1$. But $f'(x) > 1$ for all x, a contradiction.

9. True. The graph of one such function is sketched.

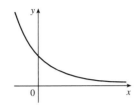

10. False. At any point $(a, f(a))$, we know that $f'(a) < 0$. So since the tangent line at $(a, f(a))$ is not horizontal, it must cross the x-axis—at $x = b$, say. But since $f''(x) > 0$ for all x, the graph of f must lie above all of its tangents; in particular, $f(b) > 0$. But this is a contradiction, since we are given that $f(x) < 0$ for all x.

11. True. Let $x_1 < x_2$ where $x_1, x_2 \in I$. Then $f(x_1) < f(x_2)$ and $g(x_1) < g(x_2)$ [since f and g are increasing on I], so $(f + g)(x_1) = f(x_1) + g(x_1) < f(x_2) + g(x_2) = (f + g)(x_2)$.

12. False. $f(x) = x$ and $g(x) = 2x$ are both increasing on $(0, 1)$, but $f(x) - g(x) = -x$ is not increasing on $(0, 1)$.

13. False. Take $f(x) = x$ and $g(x) = x - 1$. Then both f and g are increasing on $(0, 1)$. But $f(x) g(x) = x(x - 1)$ is not increasing on $(0, 1)$.

14. True. Let $x_1 < x_2$ where $x_1, x_2 \in I$. Then $0 < f(x_1) < f(x_2)$ and $0 < g(x_1) < g(x_2)$ [since f and g are both positive and increasing]. Hence, $f(x_1) g(x_1) < f(x_2) g(x_1) < f(x_2) g(x_2)$. So fg is increasing on I.

15. True. Let $x_1, x_2 \in I$ and $x_1 < x_2$. Then $f(x_1) < f(x_2)$ [f is increasing] $\Rightarrow \dfrac{1}{f(x_1)} > \dfrac{1}{f(x_2)}$ [f is positive] $\Rightarrow$
$g(x_1) > g(x_2) \Rightarrow g(x) = 1/f(x)$ is decreasing on I.

16. False. If f is even, then f' is odd. [See Exercise 83 in Section 3.5.]

17. True. If f is periodic, then there is a number p such that $f(x + p) = f(p)$ for all x. Differentiating gives
$f'(x) = f'(x + p) \cdot (x + p)' = f'(x + p) \cdot 1 = f'(x + p)$, so f' is periodic.

18. False. The most general antiderivative of $f(x) = x^{-2}$ is $F(x) = -1/x + C_1$ for $x < 0$ and $F(x) = -1/x + C_2$ for $x > 0$ [see Example 1(c) in Section 4.9].

19. True. By the Mean Value Theorem, there exists a number c in $(0, 1)$ such that $f(1) - f(0) = f'(c)(1 - 0) = f'(c)$. Since $f'(c)$ is nonzero, $f(1) - f(0) \neq 0$, so $f(1) \neq f(0)$.

EXERCISES

1. $f(x) = x^3 - 6x^2 + 9x + 1$, $[2, 4]$. $f'(x) = 3x^2 - 12x + 9 = 3(x^2 - 4x + 3) = 3(x - 1)(x - 3)$. $f'(x) = 0 \Rightarrow$
$x = 1$ or $x = 3$, but 1 is not in the interval. $f'(x) > 0$ for $3 < x < 4$ and $f'(x) < 0$ for $2 < x < 3$, so $f(3) = 1$ is a local minimum value. Checking the endpoints, we find $f(2) = 3$ and $f(4) = 5$. Thus, $f(3) = 1$ is the absolute minimum value and $f(4) = 5$ is the absolute maximum value.

2. $f(x) = x\sqrt{1-x}$, $[-1, 1]$. $f'(x) = x \cdot \frac{1}{2}(1-x)^{-1/2}(-1) + (1-x)^{1/2}(1) = (1-x)^{-1/2}\left[-\frac{1}{2}x + (1-x)\right] = \dfrac{1 - \frac{3}{2}x}{\sqrt{1-x}}$.

$f'(x) = 0 \Rightarrow x = \frac{2}{3}$. $f'(x)$ does not exist $\Leftrightarrow x = 1$. $f'(x) > 0$ for $-1 < x < \frac{2}{3}$ and $f'(x) < 0$ for $\frac{2}{3} < x < 1$, so

$f\left(\frac{2}{3}\right) = \frac{2}{3}\sqrt{\frac{1}{3}} = \frac{2}{9}\sqrt{3}$ [≈ 0.38] is a local maximum value. Checking the endpoints, we find $f(-1) = -\sqrt{2}$ and $f(1) = 0$.

Thus, $f(-1) = -\sqrt{2}$ is the absolute minimum value and $f\left(\frac{2}{3}\right) = \frac{2}{9}\sqrt{3}$ is the absolute maximum value.

3. $f(x) = \dfrac{3x-4}{x^2+1}$, $[-2, 2]$. $f'(x) = \dfrac{(x^2+1)(3) - (3x-4)(2x)}{(x^2+1)^2} = \dfrac{-(3x^2 - 8x - 3)}{(x^2+1)^2} = \dfrac{-(3x+1)(x-3)}{(x^2+1)^2}$.

$f'(x) = 0 \Rightarrow x = -\frac{1}{3}$ or $x = 3$, but 3 is not in the interval. $f'(x) > 0$ for $-\frac{1}{3} < x < 2$ and $f'(x) < 0$ for

$-2 < x < -\frac{1}{3}$, so $f\left(-\frac{1}{3}\right) = \frac{-5}{10/9} = -\frac{9}{2}$ is a local minimum value. Checking the endpoints, we find $f(-2) = -2$ and

$f(2) = \frac{2}{5}$. Thus, $f\left(-\frac{1}{3}\right) = -\frac{9}{2}$ is the absolute minimum value and $f(2) = \frac{2}{5}$ is the absolute maximum value.

4. $f(x) = (x^2 + 2x)^3$, $[-2, 1]$. $f'(x) = 3(x^2 + 2x)^2(2x+2) = 6(x+1)x^2(x+2)^2$, so the only critical numbers in the

interior of the domain are $x = -1, 0$. $f'(x) < 0$ for $-2 < x < -1$ and $f'(x) > 0$ for $-1 < x < 0$ and $0 < x < 1$, so f is

decreasing on $(-2, -1)$ and increasing on $(-1, 1)$. Thus, $f(-1) = -1$ is a local minimum value. $f(-2) = 0$ and $f(1) = 27$,

so the local minimum value is the absolute minimum value and $f(1) = 27$ is the absolute maximum value.

5. $f(x) = x + \sin 2x$, $[0, \pi]$. $f'(x) = 1 + 2\cos 2x = 0 \Leftrightarrow \cos 2x = -\frac{1}{2} \Leftrightarrow 2x = \frac{2\pi}{3}$ or $\frac{4\pi}{3} \Leftrightarrow x = \frac{\pi}{3}$ or $\frac{2\pi}{3}$.

$f''(x) = -4\sin 2x$, so $f''\left(\frac{\pi}{3}\right) = -4\sin\frac{2\pi}{3} = -2\sqrt{3} < 0$ and $f''\left(\frac{2\pi}{3}\right) = -4\sin\frac{4\pi}{3} = 2\sqrt{3} > 0$, so

$f\left(\frac{\pi}{3}\right) = \frac{\pi}{3} + \frac{\sqrt{3}}{2} \approx 1.91$ is a local maximum value and $f\left(\frac{2\pi}{3}\right) = \frac{2\pi}{3} - \frac{\sqrt{3}}{2} \approx 1.23$ is a local minimum value. Also $f(0) = 0$

and $f(\pi) = \pi$, so $f(0) = 0$ is the absolute minimum value and $f(\pi) = \pi$ is the absolute maximum value.

6. $f(x) = \sin x + \cos^2 x$, $[0, \pi]$. $f'(x) = \cos x - 2\cos x \sin x = \cos x\,(1 - 2\sin x)$, so $f'(x) = 0$ for x in $(0, \pi)$ $\Leftrightarrow$

$\cos x = 0$ or $\sin x = \frac{1}{2} \Leftrightarrow x = \frac{\pi}{6}, \frac{\pi}{2}$, or $\frac{5\pi}{6}$. $f'(x) = \cos x - \sin 2x \Rightarrow f''(x) = -\sin x - 2\cos 2x$, so

$f''\left(\frac{\pi}{6}\right) = -\frac{1}{2} - 2\left(\frac{1}{2}\right) = -\frac{3}{2}$, $f''\left(\frac{\pi}{2}\right) = -1 - 2(-1) = 1$, and $f''\left(\frac{5\pi}{6}\right) = -\frac{1}{2} - 2\left(\frac{1}{2}\right) = -\frac{3}{2}$. Thus, $f\left(\frac{\pi}{6}\right) = \frac{5}{4}$ and

$f\left(\frac{5\pi}{6}\right) = \frac{5}{4}$ are local maxima and $f\left(\frac{\pi}{2}\right) = 1$ is a local minimum. $f(0) = 1$ and $f(\pi) = 1$, so

f has its absolute minimum value of 1 at 0, $\frac{\pi}{2}$, and π. f attains its absolute maximum value of $\frac{5}{4}$ at $\frac{\pi}{6}$ and $\frac{5\pi}{6}$.

7. $\lim\limits_{x\to\infty} \dfrac{3x^4 + x - 5}{6x^4 - 2x^2 + 1} = \lim\limits_{x\to\infty} \dfrac{3 + \frac{1}{x^3} - \frac{5}{x^4}}{6 - \frac{2}{x^2} + \frac{1}{x^4}} = \dfrac{3 + 0 + 0}{6 - 0 + 0} = \dfrac{1}{2}$

8. $\lim\limits_{t\to\infty} \dfrac{t^3 - t + 2}{(2t-1)(t^2 + t + 1)} = \lim\limits_{t\to\infty} \dfrac{1 - 1/t^2 + 2/t^3}{(2 - 1/t)(1 + 1/t + 1/t^2)} = \dfrac{1}{2 \cdot 1} = \dfrac{1}{2}$

9. $\lim\limits_{x\to-\infty} \dfrac{\sqrt{4x^2+1}}{3x-1} = \lim\limits_{x\to-\infty} \dfrac{\sqrt{4x^2+1}/\sqrt{x^2}}{(3x-1)/\sqrt{x^2}} = \lim\limits_{x\to-\infty} \dfrac{\sqrt{4 + 1/x^2}}{-3 + 1/x}$ [since $-x = |x| = \sqrt{x^2}$ for $x < 0$]

$= \dfrac{2}{-3+0} = -\dfrac{2}{3}$

10. $\lim\limits_{x\to-\infty}(x^2 + x^3) = \lim\limits_{x\to-\infty} x^2(1+x) = -\infty$ since $x^2 \to \infty$ and $1 + x \to -\infty$ as $x \to -\infty$.

11. $\displaystyle\lim_{x\to\infty}\left(\sqrt{4x^2+3x}-2x\right)=\lim_{x\to\infty}\frac{\sqrt{4x^2+3x}-2x}{1}\cdot\frac{\sqrt{4x^2+3x}+2x}{\sqrt{4x^2+3x}+2x}=\lim_{x\to\infty}\frac{(4x^2+3x)-4x^2}{\sqrt{4x^2+3x}+2x}$

$$=\lim_{x\to\infty}\frac{3x}{\sqrt{4x^2+3x}+2x}=\lim_{x\to\infty}\frac{3x/\sqrt{x^2}}{\left(\sqrt{4x^2+3x}+2x\right)/\sqrt{x^2}}$$

$$=\lim_{x\to\infty}\frac{3}{\sqrt{4+3/x}+2}\qquad[\text{since }x=|x|=\sqrt{x^2}\text{ for }x>0]$$

$$=\frac{3}{2+2}=\frac{3}{4}$$

12. $0\le\sin^4 x\le 1$, so $0\le\dfrac{\sin^4 x}{\sqrt{x}}\le\dfrac{1}{\sqrt{x}}$. Since $\displaystyle\lim_{x\to\infty}0=0$ and $\displaystyle\lim_{x\to\infty}\frac{1}{\sqrt{x}}=0$, $\displaystyle\lim_{x\to\infty}\frac{\sin^4 x}{\sqrt{x}}=0$ by the Squeeze Theorem.

13. $f(0)=0,\ f'(-2)=f'(1)=f'(9)=0,\ \displaystyle\lim_{x\to\infty}f(x)=0,\ \lim_{x\to 6}f(x)=-\infty,$

$f'(x)<0$ on $(-\infty,-2)$, $(1,6)$, and $(9,\infty)$, $f'(x)>0$ on $(-2,1)$ and $(6,9)$,

$f''(x)>0$ on $(-\infty,0)$ and $(12,\infty)$, $f''(x)<0$ on $(0,6)$ and $(6,12)$

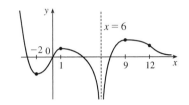

14. For $0<x<1$, $f'(x)=2x$, so $f(x)=x^2+C$. Since $f(0)=0$,

$f(x)=x^2$ on $[0,1]$. For $1<x<3$, $f'(x)=-1$, so $f(x)=-x+D$.

$1=f(1)=-1+D\ \Rightarrow\ D=2$, so $f(x)=2-x$. For $x>3$, $f'(x)=1$,

so $f(x)=x+E$. $-1=f(3)=3+E\ \Rightarrow\ E=-4$, so $f(x)=x-4$.

Since f is even, its graph is symmetric about the y-axis.

15. f is odd, $f'(x)<0$ for $0<x<2$, $f'(x)>0$ for $x>2$,

$f''(x)>0$ for $0<x<3$, $f''(x)<0$ for $x>3$, $\lim_{x\to\infty}f(x)=-2$

16. (a) Using the Test for Monotonic Functions we know that f is increasing on $(-2,0)$ and $(4,\infty)$ because $f'>0$ on $(-2,0)$

and $(4,\infty)$, and that f is decreasing on $(-\infty,-2)$ and $(0,4)$ because $f'<0$ on $(-\infty,-2)$ and $(0,4)$.

(b) Using the First Derivative Test, we know that f has a local maximum at $x=0$ because f' changes from positive to

negative at $x=0$, and that f has a local minimum at $x=4$ because f' changes from negative to positive at $x=4$.

(c)

(d)

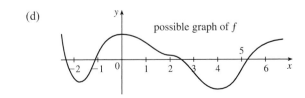

17. $y = f(x) = 2 - 2x - x^3$ **A.** $D = \mathbb{R}$ **B.** y-intercept: $f(0) = 2$.

The x-intercept (approximately 0.770917) can be found using Newton's

Method.   **C.** No symmetry   **D.** No asymptote

H.

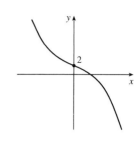

E. $f'(x) = -2 - 3x^2 = -(3x^2 + 2) < 0$, so f is decreasing on $\mathbb{R}$.

F. No extreme value   **G.** $f''(x) = -6x < 0$ on $(0, \infty)$ and $f''(x) > 0$ on

$(-\infty, 0)$, so f is CD on $(0, \infty)$ and CU on $(-\infty, 0)$. There is an IP at $(0, 2)$.

18. $y = f(x) = x^3 - 6x^2 - 15x + 4$ **A.** $D = \mathbb{R}$ **B.** y-intercept: $f(0) = 4$; x-intercepts: $f(x) = 0$ $\Rightarrow$

$x \approx -2.09, 0.24, 7.85$ **C.** No symmetry   **D.** No asymptote

H.

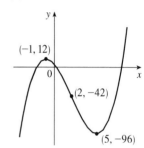

E. $f'(x) = 3x^2 - 12x - 15 = 3(x^2 - 4x - 5) = 3(x+1)(x-5)$,

so f is increasing on $(-\infty, -1)$, decreasing on $(-1, 5)$, and increasing

on $(5, \infty)$.   **F.** Local maximum value $f(-1) = 12$, local minimum

value $f(5) = -96$.   **G.** $f''(x) = 6x - 12 = 6(x-2)$, so f is CD

on $(-\infty, 2)$ and CU on $(2, \infty)$. There is an IP at $(2, -42)$.

19. $y = f(x) = x^4 - 3x^3 + 3x^2 - x = x(x-1)^3$ **A.** $D = \mathbb{R}$ **B.** y-intercept: $f(0) = 0$; x-intercepts: $f(x) = 0$ $\Leftrightarrow$

$x = 0$ or $x = 1$ **C.** No symmetry   **D.** f is a polynomial function and hence, it has no asymptote.

E. $f'(x) = 4x^3 - 9x^2 + 6x - 1$. Since the sum of the coefficients is 0, 1 is a root of f', so

$f'(x) = (x-1)(4x^2 - 5x + 1) = (x-1)^2(4x-1)$. $f'(x) < 0$ $\Rightarrow$ $x < \frac{1}{4}$, so f is decreasing on $\left(-\infty, \frac{1}{4}\right)$

and f is increasing on $\left(\frac{1}{4}, \infty\right)$.   **F.** $f'(x)$ does not change sign at $x = 1$, so

H.

there is not a local extremum there. $f\left(\frac{1}{4}\right) = -\frac{27}{256}$ is a local minimum value.

G. $f''(x) = 12x^2 - 18x + 6 = 6(2x-1)(x-1)$. $f''(x) = 0$ $\Leftrightarrow$ $x = \frac{1}{2}$

or 1. $f''(x) < 0$ $\Leftrightarrow$ $\frac{1}{2} < x < 1$ $\Rightarrow$ f is CD on $\left(\frac{1}{2}, 1\right)$ and CU on

$\left(-\infty, \frac{1}{2}\right)$ and $(1, \infty)$. There are inflection points at $\left(\frac{1}{2}, -\frac{1}{16}\right)$ and $(1, 0)$.

20. $y = f(x) = \dfrac{1}{1 - x^2} = \dfrac{1}{(1+x)(1-x)}$ **A.** $D = \{x \mid x \neq \pm 1\}$ **B.** y-intercept: $f(0) = 1$; no x-intercept

C. $f(-x) = f(x)$, so f is even and the graph of f is symmetric about the y-axis.   **D.** Vertical asymptotes: $x = \pm 1$.

Horizontal asymptote: $y = 0$ **E.** $y' = \dfrac{2x}{(1-x^2)^2} = 0$ $\Leftrightarrow$ $x = 0$, so f is decreasing on $(-\infty, -1)$ and $(-1, 0)$,

and increasing on $(0, 1)$ and $(1, \infty)$.

F. Local minimum value $f(0) = 1$; no local maximum

G. $f''(x) = \dfrac{(1-x^2)^2 \cdot 2 - 2x \cdot 2(1-x^2)(-2x)}{(1-x^2)^4}$

$= \dfrac{2(1-x^2) + 8x^2}{(1-x^2)^3} = \dfrac{6x^2 + 2}{(1-x^2)^3} < 0$ $\Leftrightarrow$ $x^2 > 1$,

H.

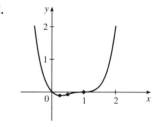

so f is CD on $(-\infty, -1)$ and $(1, \infty)$, and CU on $(-1, 1)$. No IP

21. $y = f(x) = \dfrac{1}{x(x-3)^2}$ **A.** $D = \{x \mid x \neq 0, 3\} = (-\infty, 0) \cup (0, 3) \cup (3, \infty)$ **B.** No intercepts.   **C.** No symmetry.

D. $\displaystyle\lim_{x \to \pm\infty} \frac{1}{x(x-3)^2} = 0$, so $y = 0$ is a HA. $\displaystyle\lim_{x \to 0^+} \frac{1}{x(x-3)^2} = \infty$, $\displaystyle\lim_{x \to 0^-} \frac{1}{x(x-3)^2} = -\infty$, $\displaystyle\lim_{x \to 3} \frac{1}{x(x-3)^2} = \infty$,

so $x = 0$ and $x = 3$ are VA.   **E.** $f'(x) = -\dfrac{(x-3)^2 + 2x(x-3)}{x^2(x-3)^4} = \dfrac{3(1-x)}{x^2(x-3)^3}$ $\Rightarrow$ $f'(x) > 0$ $\Leftrightarrow$ $1 < x < 3$,

so f is increasing on $(1, 3)$ and decreasing on $(-\infty, 0)$, $(0, 1)$, and $(3, \infty)$.   **H.**

F. Local minimum value $f(1) = \frac{1}{4}$ **G.** $f''(x) = \dfrac{6(2x^2 - 4x + 3)}{x^3(x-3)^4}$.

Note that $2x^2 - 4x + 3 > 0$ for all x since it has negative discriminant.

So $f''(x) > 0$ $\Leftrightarrow$ $x > 0$ $\Rightarrow$ f is CU on $(0, 3)$ and $(3, \infty)$ and

CD on $(-\infty, 0)$. No IP

22. $y = f(x) = \dfrac{1}{x^2} - \dfrac{1}{(x-2)^2}$ **A.** $D = \{x \mid x \neq 0, 2\}$ **B.** y-intercept: none; x-intercept: $f(x) = 0$ $\Rightarrow$

$\dfrac{1}{x^2} = \dfrac{1}{(x-2)^2}$ $\Leftrightarrow$ $(x-2)^2 = x^2$ $\Leftrightarrow$ $x^2 - 4x + 4 = x^2$ $\Leftrightarrow$ $4x = 4$ $\Leftrightarrow$ $x = 1$ **C.** No symmetry

D. $\displaystyle\lim_{x \to 0} f(x) = \infty$ and $\displaystyle\lim_{x \to 2} f(x) = -\infty$, so $x = 0$ and $x = 2$ are VA; $\displaystyle\lim_{x \to \pm\infty} f(x) = 0$, so $y = 0$ is a HA

E. $f'(x) = -\dfrac{2}{x^3} + \dfrac{2}{(x-2)^3} > 0$ $\Rightarrow$ $\dfrac{-(x-2)^3 + x^3}{x^3(x-2)^3} > 0$ $\Leftrightarrow$ $\dfrac{-x^3 + 6x^2 - 12x + 8 + x^3}{x^3(x-2)^3} > 0$ $\Leftrightarrow$

$\dfrac{2(3x^2 - 6x + 4)}{x^3(x-2)^3} > 0$. The numerator is positive (the discriminant of the quadratic is negative), so $f'(x) > 0$ if $x < 0$ or

$x > 2$, and hence, f is increasing on $(-\infty, 0)$ and $(2, \infty)$ and decreasing on $(0, 2)$.

F. No local extreme values   **G.** $f''(x) = \dfrac{6}{x^4} - \dfrac{6}{(x-2)^4} > 0$ $\Rightarrow$ **H.**

$\dfrac{(x-2)^4 - x^4}{x^4(x-2)^4} > 0$ $\Leftrightarrow$ $\dfrac{x^4 - 8x^3 + 24x^2 - 32x + 16 - x^4}{x^4(x-2)^4} > 0$ $\Leftrightarrow$

$\dfrac{-8(x^3 - 3x^2 + 4x - 2)}{x^4(x-2)^4} > 0$ $\Leftrightarrow$ $\dfrac{-8(x-1)(x^2 - 2x + 2)}{x^4(x-2)^4} > 0$. So f'' is

positive for $x < 1$ $[x \neq 0]$ and negative for $x > 1$ $[x \neq 2]$. Thus, f is CU on

$(-\infty, 0)$ and $(0, 1)$ and f is CD on $(1, 2)$ and $(2, \infty)$. IP at $(1, 0)$

23. $y = f(x) = \dfrac{x^2}{x+8} = x - 8 + \dfrac{64}{x+8}$ **A.** $D = \{x \mid x \neq -8\}$ **B.** Intercepts are 0   **C.** No symmetry

D. $\displaystyle\lim_{x\to\infty} \dfrac{x^2}{x+8} = \infty$, but $f(x) - (x-8) = \dfrac{64}{x+8} \to 0$ as $x \to \infty$, so $y = x - 8$ is a slant asymptote.

$\displaystyle\lim_{x\to-8^+} \dfrac{x^2}{x+8} = \infty$ and $\displaystyle\lim_{x\to-8^-} \dfrac{x^2}{x+8} = -\infty$, so $x = -8$ is a VA.   **E.** $f'(x) = 1 - \dfrac{64}{(x+8)^2} = \dfrac{x(x+16)}{(x+8)^2} > 0$ $\Leftrightarrow$

$x > 0$ or $x < -16$, so f is increasing on $(-\infty, -16)$ and $(0, \infty)$ and

decreasing on $(-16, -8)$ and $(-8, 0)$.

F. Local maximum value $f(-16) = -32$, local minimum value $f(0) = 0$

G. $f''(x) = 128/(x+8)^3 > 0$ $\Leftrightarrow$ $x > -8$, so f is CU on $(-8, \infty)$ and

CD on $(-\infty, -8)$. No IP

H.

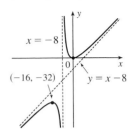

24. $y = f(x) = \sqrt{1-x} + \sqrt{1+x}$ **A.** $1 - x \geq 0$ and $1 + x \geq 0$ $\Rightarrow$ $x \leq 1$ and $x \geq -1$, so $D = [-1, 1]$.

B. y-intercept: $f(0) = 1 + 1 = 2$; no x-intercept because $f(x) > 0$ for all x.

C. $f(-x) = f(x)$, so the curve is symmetric about the y-axis.   **D.** No asymptote

E. $f'(x) = \frac{1}{2}(1-x)^{-1/2}(-1) + \frac{1}{2}(1+x)^{-1/2} = \dfrac{-1}{2\sqrt{1-x}} + \dfrac{1}{2\sqrt{1+x}} = \dfrac{-\sqrt{1+x} + \sqrt{1-x}}{2\sqrt{1-x}\sqrt{1+x}} > 0$ $\Rightarrow$

$-\sqrt{1+x} + \sqrt{1-x} > 0$ $\Rightarrow$ $\sqrt{1-x} > \sqrt{1+x}$ $\Rightarrow$ $1 - x > 1 + x$ $\Rightarrow$ $-2x > 0$ $\Rightarrow$ $x < 0$, so $f'(x) > 0$ for

$-1 < x < 0$ and $f'(x) < 0$ for $0 < x < 1$. Thus, f is increasing on $(-1, 0)$

and decreasing on $(0, 1)$.   **F.** Local maximum value $f(0) = 2$

G. $f''(x) = -\frac{1}{2}\left(-\frac{1}{2}\right)(1-x)^{-3/2}(-1) + \frac{1}{2}\left(-\frac{1}{2}\right)(1+x)^{-3/2}$

$= \dfrac{-1}{4(1-x)^{3/2}} + \dfrac{-1}{4(1+x)^{3/2}} < 0$

for all x in the domain, so f is CD on $(-1, 1)$. No IP

H.

25. $y = f(x) = x\sqrt{2+x}$ **A.** $D = [-2, \infty)$ **B.** y-intercept: $f(0) = 0$; x-intercepts: -2 and 0 **C.** No symmetry

D. No asymptote   **E.** $f'(x) = \dfrac{x}{2\sqrt{2+x}} + \sqrt{2+x} = \dfrac{1}{2\sqrt{2+x}}\left[x + 2(2+x)\right] = \dfrac{3x+4}{2\sqrt{2+x}} = 0$ when $x = -\frac{4}{3}$, so f is

decreasing on $\left(-2, -\frac{4}{3}\right)$ and increasing on $\left(-\frac{4}{3}, \infty\right)$.   **F.** Local minimum value $f\left(-\frac{4}{3}\right) = -\frac{4}{3}\sqrt{\frac{2}{3}} = -\frac{4\sqrt{6}}{9} \approx -1.09$,

no local maximum

G. $f''(x) = \dfrac{2\sqrt{2+x}\cdot 3 - (3x+4)\dfrac{1}{\sqrt{2+x}}}{4(2+x)} = \dfrac{6(2+x) - (3x+4)}{4(2+x)^{3/2}}$

$= \dfrac{3x+8}{4(2+x)^{3/2}}$

$f''(x) > 0$ for $x > -2$, so f is CU on $(-2, \infty)$. No IP

H.

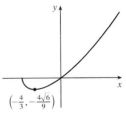

26. $y = f(x) = \sqrt[3]{x^2+1}$ **A.** $D = \mathbb{R}$ **B.** y-intercept: $f(0) = 1$; $f(x) > 0$, so there is no x-intercept   **C.** $f(-x) = f(x)$, so

the curve is symmetric about the y-axis.   **D.** No asymptote   **E.** $f'(x) = \frac{1}{3}(x^2+1)^{-2/3}(2x) = \dfrac{2x}{3(x^2+1)^{2/3}} > 0$ $\Leftrightarrow$

$x > 0$ and $f'(x) < 0$ $\Leftrightarrow$ $x < 0$, so f is increasing on $(0, \infty)$ and decreasing on $(-\infty, 0)$.

F. Local minimum value $f(0) = 1$

G. $f''(x) = \dfrac{3(x^2+1)^{2/3}(2) - 2x(3)\frac{2}{3}(x^2+1)^{-1/3}(2x)}{[3(x^2+1)^{2/3}]^2} = \dfrac{2(x^2+1)^{-1/3}\left[3(x^2+1) - 4x^2\right]}{9(x^2+1)^{4/3}} = \dfrac{2(3-x^2)}{9(x^2+1)^{5/3}}.$

$f''(x) > 0$ $\Leftrightarrow$ $-\sqrt{3} < x < \sqrt{3}$ and $f''(x) < 0$ $\Leftrightarrow$ $x < -\sqrt{3}$ and   **H.**

$x > \sqrt{3}$, so f is CU on $(-\sqrt{3}, \sqrt{3})$ and CD on $(-\infty, -\sqrt{3})$ and $(\sqrt{3}, \infty)$.

There are inflection points at $(\pm\sqrt{3}, f(\pm\sqrt{3})) = (\pm\sqrt{3}, \sqrt[3]{4})$.

27. $y = f(x) = \sin^2 x - 2\cos x$ **A.** $D = \mathbb{R}$ **B.** y-intercept: $f(0) = -2$ **C.** $f(-x) = f(x)$, so f is symmetric with respect

to the y-axis. f has period 2π.   **D.** No asymptote   **E.** $y' = 2\sin x \cos x + 2\sin x = 2\sin x (\cos x + 1)$. $y' = 0$ $\Leftrightarrow$

$\sin x = 0$ or $\cos x = -1$ $\Leftrightarrow$ $x = n\pi$ or $x = (2n+1)\pi$. $y' > 0$ when $\sin x > 0$, since $\cos x + 1 \geq 0$ for all x.

Therefore, $y' > 0$ [and so f is increasing] on $(2n\pi, (2n+1)\pi)$; $y' < 0$ [and so f is decreasing] on $((2n-1)\pi, 2n\pi)$.

F. Local maximum values are $f((2n+1)\pi) = 2$; local minimum values are $f(2n\pi) = -2$.

G. $y' = \sin 2x + 2\sin x$ $\Rightarrow$ $y'' = 2\cos 2x + 2\cos x = 2(2\cos^2 x - 1) + 2\cos x = 4\cos^2 x + 2\cos x - 2$

$$= 2(2\cos^2 x + \cos x - 1) = 2(2\cos x - 1)(\cos x + 1)$$

$y'' = 0$ $\Leftrightarrow$ $\cos x = \frac{1}{2}$ or -1 $\Leftrightarrow$ $x = 2n\pi \pm \frac{\pi}{3}$ or $x = (2n+1)\pi$.   **H.**

$y'' > 0$ [and so f is CU] on $(2n\pi - \frac{\pi}{3}, 2n\pi + \frac{\pi}{3})$; $y'' \leq 0$ [and so f is CD]

on $(2n\pi + \frac{\pi}{3}, 2n\pi + \frac{5\pi}{3})$. There are inflection points at $(2n\pi \pm \frac{\pi}{3}, -\frac{1}{4})$.

28. $y = f(x) = 4x - \tan x$, $-\frac{\pi}{2} < x < \frac{\pi}{2}$ **A.** $D = \left(-\frac{\pi}{2}, \frac{\pi}{2}\right)$.   **B.** y-intercept $= f(0) = 0$ **C.** $f(-x) = -f(x)$, so the

curve is symmetric about $(0,0)$.   **D.** $\displaystyle\lim_{x \to \pi/2^-} (4x - \tan x) = -\infty$, $\displaystyle\lim_{x \to -\pi/2^+} (4x - \tan x) = \infty$, so $x = \frac{\pi}{2}$ and $x = -\frac{\pi}{2}$

are VA.   **E.** $f'(x) = 4 - \sec^2 x > 0$ $\Leftrightarrow$ $\sec x < 2$ $\Leftrightarrow$ $\cos x > \frac{1}{2}$ $\Leftrightarrow$ $-\frac{\pi}{3} < x < \frac{\pi}{3}$, so f is increasing on

$\left(-\frac{\pi}{3}, \frac{\pi}{3}\right)$ and decreasing on $\left(-\frac{\pi}{2}, -\frac{\pi}{3}\right)$ and $\left(\frac{\pi}{3}, \frac{\pi}{2}\right)$.   **F.** $f\left(\frac{\pi}{3}\right) = \frac{4\pi}{3} - \sqrt{3}$ is   **H.**

a local maximum value, $f\left(-\frac{\pi}{3}\right) = \sqrt{3} - \frac{4\pi}{3}$ is a local minimum value.

G. $f''(x) = -2\sec^2 x \tan x > 0$ $\Leftrightarrow$ $\tan x < 0$ $\Leftrightarrow$ $-\frac{\pi}{2} < x < 0$,

so f is CU on $\left(-\frac{\pi}{2}, 0\right)$ and CD on $\left(0, \frac{\pi}{2}\right)$. IP at $(0,0)$

29. $f(x) = \dfrac{x^2 - 1}{x^3}$ $\Rightarrow$ $f'(x) = \dfrac{x^3(2x) - (x^2 - 1)3x^2}{x^6} = \dfrac{3 - x^2}{x^4}$ $\Rightarrow$

$f''(x) = \dfrac{x^4(-2x) - (3 - x^2)4x^3}{x^8} = \dfrac{2x^2 - 12}{x^5}$

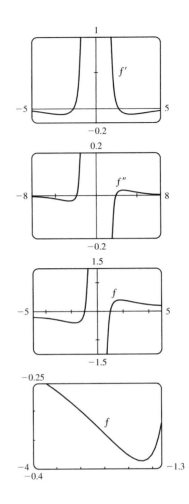

Estimates: From the graphs of f' and f'', it appears that f is increasing on

$(-1.73, 0)$ and $(0, 1.73)$ and decreasing on $(-\infty, -1.73)$ and $(1.73, \infty)$;

f has a local maximum of about $f(1.73) = 0.38$ and a local minimum of about

$f(-1.7) = -0.38$; f is CU on $(-2.45, 0)$ and $(2.45, \infty)$, and CD on

$(-\infty, -2.45)$ and $(0, 2.45)$; and f has inflection points at about

$(-2.45, -0.34)$ and $(2.45, 0.34)$.

Exact: Now $f'(x) = \dfrac{3 - x^2}{x^4}$ is positive for $0 < x^2 < 3$, that is, f is increasing

on $(-\sqrt{3}, 0)$ and $(0, \sqrt{3})$; and $f'(x)$ is negative (and so f is decreasing) on

$(-\infty, -\sqrt{3})$ and $(\sqrt{3}, \infty)$. $f'(x) = 0$ when $x = \pm\sqrt{3}$.

f' goes from positive to negative at $x = \sqrt{3}$, so f has a local maximum of

$f(\sqrt{3}) = \dfrac{(\sqrt{3})^2 - 1}{(\sqrt{3})^3} = \dfrac{2\sqrt{3}}{9}$; and since f is odd, we know that maxima on the

interval $(0, \infty)$ correspond to minima on $(-\infty, 0)$, so f has a local minimum of

$f(-\sqrt{3}) = -\dfrac{2\sqrt{3}}{9}$. Also, $f''(x) = \dfrac{2x^2 - 12}{x^5}$ is positive (so f is CU) on

$(-\sqrt{6}, 0)$ and $(\sqrt{6}, \infty)$, and negative (so f is CD) on $(-\infty, -\sqrt{6})$ and

$(0, \sqrt{6})$. There are IP at $\left(\sqrt{6}, \dfrac{5\sqrt{6}}{36}\right)$ and $\left(-\sqrt{6}, -\dfrac{5\sqrt{6}}{36}\right)$.

30. $f(x) = \dfrac{x^3 - x}{x^2 + x + 3}$ $\Rightarrow$ $f'(x) = \dfrac{x^4 + 2x^3 + 10x^2 - 3}{(x^2 + x + 3)^2}$ $\Rightarrow$ $f''(x) = \dfrac{-6(x^3 - 3x^2 - 12x - 1)}{(x^2 + x + 3)^3}$.

$f(x) = 0$ $\Leftrightarrow$ $x = \pm1$; $f'(x) = 0$ $\Leftrightarrow$ $x \approx -0.57, 0.52$; $f''(x) = 0$ $\Leftrightarrow$ $x \approx -2.21, -0.09, 5.30$.

From the graphs of f' and f'', it appears that f is increasing on $(-\infty, -0.57)$ and $(0.52, \infty)$ and decreasing on

$(-0.57, 0.52)$; f has a local maximum of about $f(-0.57) = 0.14$ and a local minimum of about $f(0.52) = -0.10$;

f is CU on $(-\infty, -2.21)$ and $(-0.09, 5.30)$, and CD on $(-2.21, -0.09)$ and $(5.30, \infty)$; and f has inflection points at about $(-2.21, -1.52)$, $(-0.09, 0.03)$, and $(5.30, 3.95)$.

  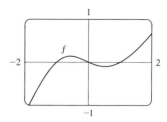

31. $f(x) = 3x^6 - 5x^5 + x^4 - 5x^3 - 2x^2 + 2 \;\Rightarrow\; f'(x) = 18x^5 - 25x^4 + 4x^3 - 15x^2 - 4x \;\Rightarrow$

$f''(x) = 90x^4 - 100x^3 + 12x^2 - 30x - 4$

    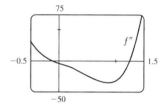

From the graphs of f' and f'', it appears that f is increasing on $(-0.23, 0)$ and $(1.62, \infty)$ and decreasing on $(-\infty, -0.23)$ and $(0, 1.62)$; f has a local maximum of about $f(0) = 2$ and local minima of about $f(-0.23) = 1.96$ and $f(1.62) = -19.2$; f is CU on $(-\infty, -0.12)$ and $(1.24, \infty)$ and CD on $(-0.12, 1.24)$; and f has inflection points at about $(-0.12, 1.98)$ and $(1.24, -12.1)$.

32. $f(x) = x^2 + 6.5 \sin x$, $-5 \le x \le 5 \;\Rightarrow\; f'(x) = 2x + 6.5 \cos x \;\Rightarrow\; f''(x) = 2 - 6.5 \sin x$. $f(x) = 0 \;\Leftrightarrow\;$ $x \approx -2.25$ and $x = 0$; $f'(x) = 0 \;\Leftrightarrow\; x \approx -1.19, 2.40, 3.24$; $f''(x) = 0 \;\Leftrightarrow\; x \approx -3.45, 0.31, 2.83$.

  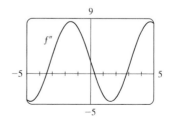

[continued]

From the graphs of f' and f'', it appears that f is decreasing on $(-5, -1.19)$ and $(2.40, 3.24)$ and increasing on $(-1.19, 2.40)$ and $(3.24, 5)$; f has a local maximum of about $f(2.40) = 10.15$ and local minima of about $f(-1.19) = -4.62$ and $f(3.24) = 9.86$; f is CU on $(-3.45, 0.31)$ and $(2.83, 5)$ and CD on $(-5, -3.45)$ and $(0.31, 2.83)$; and f has inflection points at about $(-3.45, 13.93)$, $(0.31, 2.10)$, and $(2.83, 10.00)$.

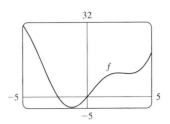

33. Let $f(x) = 3x + 2\cos x + 5$. Then $f(0) = 7 > 0$ and $f(-\pi) = -3\pi - 2 + 5 = -3\pi + 3 = -3(\pi - 1) < 0$, and since f is continuous on $\mathbb{R}$ (hence on $[-\pi, 0]$), the Intermediate Value Theorem assures us that there is at least one zero of f in $[-\pi, 0]$. Now $f'(x) = 3 - 2\sin x > 0$ implies that f is increasing on $\mathbb{R}$, so there is exactly one zero of f, and hence, exactly one real root of the equation $3x + 2\cos x + 5 = 0$.

34. By the Mean Value Theorem, $f'(c) = \dfrac{f(4) - f(0)}{4 - 0}$ $\Leftrightarrow$ $4f'(c) = f(4) - 1$ for some c with $0 < c < 4$. Since $2 \leq f'(c) \leq 5$, we have $4(2) \leq 4f'(c) \leq 4(5)$ $\Leftrightarrow$ $4(2) \leq f(4) - 1 \leq 4(5)$ $\Leftrightarrow$ $8 \leq f(4) - 1 \leq 20$ $\Leftrightarrow$ $9 \leq f(4) \leq 21$.

35. Since f is continuous on $[32, 33]$ and differentiable on $(32, 33)$, then by the Mean Value Theorem there exists a number c in $(32, 33)$ such that $f'(c) = \frac{1}{5}c^{-4/5} = \dfrac{\sqrt[5]{33} - \sqrt[5]{32}}{33 - 32} = \sqrt[5]{33} - 2$, but $\frac{1}{5}c^{-4/5} > 0$ $\Rightarrow$ $\sqrt[5]{33} - 2 > 0$ $\Rightarrow$ $\sqrt[5]{33} > 2$. Also f' is decreasing, so that $f'(c) < f'(32) = \frac{1}{5}(32)^{-4/5} = 0.0125$ $\Rightarrow$ $0.0125 > f'(c) = \sqrt[5]{33} - 2$ $\Rightarrow$ $\sqrt[5]{33} < 2.0125$. Therefore, $2 < \sqrt[5]{33} < 2.0125$.

36. For $(1, 6)$ to be on the curve $y = x^3 + ax^2 + bx + 1$, we have that $6 = 1 + a + b + 1$ $\Rightarrow$ $b = 4 - a$. Now $y' = 3x^2 + 2ax + b$ and $y'' = 6x + 2a$. Also, for $(1, 6)$ to be an inflection point it must be true that $y''(1) = 6(1) + 2a = 0$ $\Rightarrow$ $a = -3$ $\Rightarrow$ $b = 4 - (-3) = 7$. Note that with $a = -3$, we have $y'' = 6x - 6 = 6(x - 1)$, so y'' changes sign at $x = 1$, proving that $(1, 6)$ is a point of inflection. [This does not follow from the fact that $y''(1) = 0$.]

37. (a) $g(x) = f(x^2)$ $\Rightarrow$ $g'(x) = 2xf'(x^2)$ by the Chain Rule. Since $f'(x) > 0$ for all $x \neq 0$, we must have $f'(x^2) > 0$ for $x \neq 0$, so $g'(x) = 0$ $\Leftrightarrow$ $x = 0$. Now $g'(x)$ changes sign (from negative to positive) at $x = 0$, since one of its factors, $f'(x^2)$, is positive for all x, and its other factor, $2x$, changes from negative to positive at this point, so by the First Derivative Test, f has a local and absolute minimum at $x = 0$.

(b) $g'(x) = 2xf'(x^2)$ $\Rightarrow$ $g''(x) = 2[xf''(x^2)(2x) + f'(x^2)] = 4x^2 f''(x^2) + 2f'(x^2)$ by the Product Rule and the Chain Rule. But $x^2 > 0$ for all $x \neq 0$, $f''(x^2) > 0$ [since f is CU for $x > 0$], and $f'(x^2) > 0$ for all $x \neq 0$, so since all of its factors are positive, $g''(x) > 0$ for $x \neq 0$. Whether $g''(0)$ is positive or 0 doesn't matter [since the sign of g'' does not change there]; g is concave upward on $\mathbb{R}$.

38. Call the two integers x and y. Then $x + 4y = 1000$, so $x = 1000 - 4y$. Their product is $P = xy = (1000 - 4y)y$, so our problem is to maximize the function $P(y) = 1000y - 4y^2$, where $0 < y < 250$ and y is an integer. $P'(y) = 1000 - 8y$, so $P'(y) = 0 \iff y = 125$. $P''(y) = -8 < 0$, so $P(125) = 62{,}500$ is an absolute maximum. Since the optimal y turned out to be an integer, we have found the desired pair of numbers, namely $x = 1000 - 4(125) = 500$ and $y = 125$.

39. If $B = 0$, the line is vertical and the distance from $x = -\dfrac{C}{A}$ to (x_1, y_1) is $\left| x_1 + \dfrac{C}{A} \right| = \dfrac{|Ax_1 + By_1 + C|}{\sqrt{A^2 + B^2}}$, so assume

$B \neq 0$. The square of the distance from (x_1, y_1) to the line is $f(x) = (x - x_1)^2 + (y - y_1)^2$ where $Ax + By + C = 0$, so

we minimize $f(x) = (x - x_1)^2 + \left(-\dfrac{A}{B}x - \dfrac{C}{B} - y_1 \right)^2 \Rightarrow f'(x) = 2(x - x_1) + 2\left(-\dfrac{A}{B}x - \dfrac{C}{B} - y_1 \right)\left(-\dfrac{A}{B} \right)$.

$f'(x) = 0 \Rightarrow x = \dfrac{B^2 x_1 - ABy_1 - AC}{A^2 + B^2}$ and this gives a minimum since $f''(x) = 2\left(1 + \dfrac{A^2}{B^2} \right) > 0$. Substituting

this value of x into $f(x)$ and simplifying gives $f(x) = \dfrac{(Ax_1 + By_1 + C)^2}{A^2 + B^2}$, so the minimum distance is

$\sqrt{f(x)} = \dfrac{|Ax_1 + By_1 + C|}{\sqrt{A^2 + B^2}}$.

40. On the hyperbola $xy = 8$, if $d(x)$ is the distance from the point $(x, y) = (x, 8/x)$ to the point $(3, 0)$, then

$[d(x)]^2 = (x - 3)^2 + 64/x^2 = f(x)$. $f'(x) = 2(x - 3) - 128/x^3 = 0 \Rightarrow x^4 - 3x^3 - 64 = 0 \Rightarrow$

$(x - 4)(x^3 + x^2 + 4x + 16) = 0 \Rightarrow x = 4$ since the solution must have $x > 0$. Then $y = \dfrac{8}{4} = 2$, so the point is $(4, 2)$.

41. 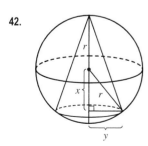 By similar triangles, $\dfrac{y}{x} = \dfrac{r}{\sqrt{x^2 - 2rx}}$, so the area of the triangle is

$A(x) = \frac{1}{2}(2y)x = xy = \dfrac{rx^2}{\sqrt{x^2 - 2rx}} \Rightarrow$

$A'(x) = \dfrac{2rx\sqrt{x^2 - 2rx} - rx^2(x - r)/\sqrt{x^2 - 2rx}}{x^2 - 2rx} = \dfrac{rx^2(x - 3r)}{(x^2 - 2rx)^{3/2}} = 0$

when $x = 3r$.

$A'(x) < 0$ when $2r < x < 3r$, $A'(x) > 0$ when $x > 3r$. So $x = 3r$ gives a minimum and $A(3r) = \dfrac{r(9r^2)}{\sqrt{3}\,r} = 3\sqrt{3}\,r^2$.

42. The volume of the cone is $V = \frac{1}{3}\pi y^2(r + x) = \frac{1}{3}\pi(r^2 - x^2)(r + x)$, $-r \leq x \leq r$.

$V'(x) = \frac{\pi}{3}[(r^2 - x^2)(1) + (r + x)(-2x)] = \frac{\pi}{3}[(r + x)(r - x - 2x)]$

$= \frac{\pi}{3}(r + x)(r - 3x) = 0$ when $x = -r$ or $x = r/3$.

Now $V(r) = 0 = V(-r)$, so the maximum occurs at $x = r/3$ and the volume is

$V\left(\dfrac{r}{3} \right) = \dfrac{\pi}{3}\left(r^2 - \dfrac{r^2}{9} \right)\left(\dfrac{4r}{3} \right) = \dfrac{32\pi r^3}{81}$.

43.

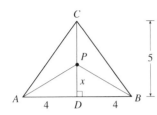

We minimize $L(x) = |PA| + |PB| + |PC| = 2\sqrt{x^2 + 16} + (5 - x)$,

$0 \le x \le 5$. $L'(x) = 2x/\sqrt{x^2 + 16} - 1 = 0 \Leftrightarrow 2x = \sqrt{x^2 + 16} \Leftrightarrow$

$4x^2 = x^2 + 16 \Leftrightarrow x = \frac{4}{\sqrt{3}}$. $L(0) = 13$, $L\left(\frac{4}{\sqrt{3}}\right) \approx 11.9$, $L(5) \approx 12.8$, so the

minimum occurs when $x = \frac{4}{\sqrt{3}} \approx 2.3$.

44.

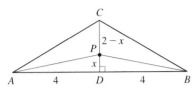

If $|CD| = 2$, the last part of $L(x)$ changes from $(5 - x)$ to $(2 - x)$ with

$0 \le x \le 2$. But we still get $L'(x) = 0 \Leftrightarrow x = \frac{4}{\sqrt{3}}$, which isn't in the interval

$[0, 2]$. Now $L(0) = 10$ and $L(2) = 2\sqrt{20} = 4\sqrt{5} \approx 8.9$. The minimum occurs

when $P = C$.

45. $v = K\sqrt{\dfrac{L}{C} + \dfrac{C}{L}} \Rightarrow \dfrac{dv}{dL} = \dfrac{K}{2\sqrt{(L/C) + (C/L)}}\left(\dfrac{1}{C} - \dfrac{C}{L^2}\right) = 0 \Leftrightarrow \dfrac{1}{C} = \dfrac{C}{L^2} \Leftrightarrow L^2 = C^2 \Leftrightarrow L = C$.

This gives the minimum velocity since $v' < 0$ for $0 < L < C$ and $v' > 0$ for $L > C$.

46.

We minimize the surface area $S = \pi r^2 + 2\pi rh + \frac{1}{2}(4\pi r^2) = 3\pi r^2 + 2\pi rh$.

Solving $V = \pi r^2 h + \frac{2}{3}\pi r^3$ for h, we get $h = \dfrac{V - \frac{2}{3}\pi r^3}{\pi r^2} = \dfrac{V}{\pi r^2} - \frac{2}{3}r$, so

$S(r) = 3\pi r^2 + 2\pi r\left[\dfrac{V}{\pi r^2} - \frac{2}{3}r\right] = \frac{5}{3}\pi r^2 + \dfrac{2V}{r}$.

$S'(r) = -\dfrac{2V}{r^2} + \frac{10}{3}\pi r = \dfrac{\frac{10}{3}\pi r^3 - 2V}{r^2} = 0 \Leftrightarrow \frac{10}{3}\pi r^3 = 2V \Leftrightarrow r^3 = \dfrac{3V}{5\pi} \Leftrightarrow r = \sqrt[3]{\dfrac{3V}{5\pi}}$.

This gives an absolute minimum since $S'(r) < 0$ for $0 < r < \sqrt[3]{\dfrac{3V}{5\pi}}$ and $S'(r) > 0$ for $r > \sqrt[3]{\dfrac{3V}{5\pi}}$. Thus,

$$h = \dfrac{V - \frac{2}{3}\pi \cdot \dfrac{3V}{5\pi}}{\pi\sqrt[3]{\dfrac{(3V)^2}{(5\pi)^2}}} = \dfrac{\left(V - \frac{2}{5}V\right)\sqrt[3]{(5\pi)^2}}{\pi\sqrt[3]{(3V)^2}} = \dfrac{3V\sqrt[3]{(5\pi)^2}}{5\pi\sqrt[3]{(3V)^2}} = \sqrt[3]{\dfrac{3V}{5\pi}} = r$$

47. Let x denote the number of $1 decreases in ticket price. Then the ticket price is $12 - \$1(x)$, and the average attendance is

$11{,}000 + 1000(x)$. Now the revenue per game is

$$R(x) = (\text{price per person}) \times (\text{number of people per game})$$

$$= (12 - x)(11{,}000 + 1000x) = -1000x^2 + 1000x + 132{,}000$$

for $0 \le x \le 4$ [since the seating capacity is 15,000] $\Rightarrow R'(x) = -2000x + 1000 = 0 \Leftrightarrow x = 0.5$. This is a

maximum since $R''(x) = -2000 < 0$ for all x. Now we must check the value of $R(x) = (12 - x)(11{,}000 + 1000x)$ at

$x = 0.5$ and at the endpoints of the domain to see which value of x gives the maximum value of R.

$R(0) = (12)(11{,}000) = 132{,}000$, $R(0.5) = (11.5)(11{,}500) = 132{,}250$, and $R(4) = (8)(15{,}000) = 120{,}000$. Thus, the

maximum revenue of $132,250 per game occurs when the average attendance is 11,500 and the ticket price is $11.50.

48. (a) $C(x) = 1800 + 25x - 0.2x^2 + 0.001x^3$ and

$R(x) = xp(x) = 48.2x - 0.03x^2$.

The profit is maximized when $C'(x) = R'(x)$.

From the figure, we estimate that the tangents are parallel when $x \approx 160$.

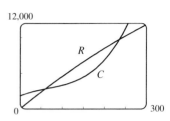

(b) $C'(x) = 25 - 0.4x + 0.003x^2$ and $R'(x) = 48.2 - 0.06x$. $C'(x) = R'(x)$ $\Rightarrow$ $0.003x^2 - 0.34x - 23.2 = 0$ $\Rightarrow$

$x_1 \approx 161.3$ $(x > 0)$. $R''(x) = -0.06$ and $C''(x) = -0.4 + 0.006x$, so $R''(x_1) = -0.06 < C''(x_1) \approx 0.57$ $\Rightarrow$

profit is maximized by producing 161 units.

(c) $c(x) = \dfrac{C(x)}{x} = \dfrac{1800}{x} + 25 - 0.2x + 0.001x^2$ is the average cost. Since

the average cost is minimized when the marginal cost equals the average

cost, we graph $c(x)$ and $C'(x)$ and estimate the point of intersection.

From the figure, $C'(x) = c(x)$ $\Leftrightarrow$ $x \approx 144$.

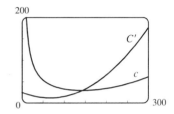

49. $f(x) = x^5 - x^4 + 3x^2 - 3x - 2$ $\Rightarrow$ $f'(x) = 5x^4 - 4x^3 + 6x - 3$, so $x_{n+1} = x_n - \dfrac{x_n^5 - x_n^4 + 3x_n^2 - 3x_n - 2}{5x_n^4 - 4x_n^3 + 6x_n - 3}$.

Now $x_1 = 1$ $\Rightarrow$ $x_2 = 1.5$ $\Rightarrow$ $x_3 \approx 1.343860$ $\Rightarrow$ $x_4 \approx 1.300320$ $\Rightarrow$ $x_5 \approx 1.297396$ $\Rightarrow$

$x_6 \approx 1.297383 \approx x_7$, so the root in $[1, 2]$ is 1.297383, to six decimal places.

50. Graphing $y = \sin x$ and $y = x^2 - 3x + 1$ shows that there are two roots,

one about 0.3 and the other about 2.8. $f(x) = \sin x - x^2 + 3x - 1$ $\Rightarrow$

$f'(x) = \cos x - 2x + 3$ $\Rightarrow$ $x_{n+1} = x_n - \dfrac{\sin x_n - x_n^2 + 3x_n - 1}{\cos x_n - 2x_n + 3}$.

Now $x_1 = 0.3$ $\Rightarrow$ $x_2 \approx 0.268552$ $\Rightarrow$ $x_3 \approx 0.268881 \approx x_4$ and

$x_1 = 2.8$ $\Rightarrow$ $x_2 \approx 2.770354$ $\Rightarrow$ $x_3 \approx 2.770058 \approx x_4$, so to six

decimal places, the roots are 0.268881 and 2.770058.

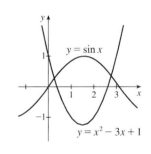

51. $f(t) = \cos t + t - t^2$ $\Rightarrow$ $f'(t) = -\sin t + 1 - 2t$. $f'(t)$ exists for all

t, so to find the maximum of f, we can examine the zeros of f'.

From the graph of f', we see that a good choice for t_1 is $t_1 = 0.3$.

Use $g(t) = -\sin t + 1 - 2t$ and $g'(t) = -\cos t - 2$ to obtain

$t_2 \approx 0.33535293$, $t_3 \approx 0.33541803 \approx t_4$. Since $f''(t) = -\cos t - 2 < 0$

for all t, $f(0.33541803) \approx 1.16718557$ is the absolute maximum.

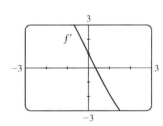

52. $y = f(x) = x \sin x, \ 0 \le x \le 2\pi$. **A.** $D = [0, 2\pi]$ **B.** y-intercept: $f(0) = 0$; x-intercepts: $f(x) = 0 \iff x = 0$ or

$\sin x = 0 \iff x = 0, \pi,$ or 2π. **C.** There is no symmetry on D, but if f is defined for all real numbers x, then f is an even

function. **D.** No asymptote **E.** $f'(x) = x \cos x + \sin x$. To find critical numbers in $(0, 2\pi)$, we graph f' and see that there

are two critical numbers, about 2 and 4.9. To find them more precisely, we use Newton's method, setting

$g(x) = f'(x) = x \cos x + \sin x$, so that $g'(x) = f''(x) = 2 \cos x - x \sin x$ and $x_{n+1} = x_n - \dfrac{x_n \cos x_n + \sin x_n}{2 \cos x_n - x_n \sin x_n}$.

$x_1 = 2 \ \Rightarrow \ x_2 \approx 2.029048, x_3 \approx 2.028758 \approx x_4$ and $x_1 = 4.9 \ \Rightarrow \ x_2 \approx 4.913214, x_3 \approx 4.913180 \approx x_4$, so the

critical numbers, to six decimal places, are $r_1 = 2.028758$ and $r_2 = 4.913180$. By checking sample values of f' in $(0, r_1)$,

(r_1, r_2), and $(r_2, 2\pi)$, we see that f is increasing on $(0, r_1)$, decreasing on (r_1, r_2), and increasing on $(r_2, 2\pi)$. **F.** Local

maximum value $f(r_1) \approx 1.819706$, local minimum value $f(r_2) \approx -4.814470$. **G.** $f''(x) = 2 \cos x - x \sin x$. To find

points where $f''(x) = 0$, we graph f'' and find that $f''(x) = 0$ at about 1 and 3.6. To find the values more precisely, we use

Newton's method. Set $h(x) = f''(x) = 2 \cos x - x \sin x$. Then $h'(x) = -3 \sin x - x \cos x$, so

$x_{n+1} = x_n - \dfrac{2 \cos x_n - x_n \sin x_n}{-3 \sin x_n - x_n \cos x_n}$. $x_1 = 1 \ \Rightarrow \ x_2 \approx 1.078028, x_3 \approx 1.076874 \approx x_4$ and $x_1 = 3.6 \ \Rightarrow$

$x_2 \approx 3.643996, x_3 \approx 3.643597 \approx x_4$, so the zeros of f'', to six decimal places, are $r_3 = 1.076874$ and $r_4 = 3.643597$.

By checking sample values of f'' in $(0, r_3), (r_3, r_4)$, and $(r_4, 2\pi)$, we see that f **H.**

is CU on $(0, r_3)$, CD on (r_3, r_4), and CU on $(r_4, 2\pi)$. f has inflection points at

$(r_3, f(r_3) \approx 0.948166)$ and $(r_4, f(r_4) \approx -1.753240)$.

53. $f'(x) = \sqrt{x^3} + \sqrt[3]{x^2} = x^{3/2} + x^{2/3} \ \Rightarrow \ f(x) = \dfrac{x^{5/2}}{5/2} + \dfrac{x^{5/3}}{5/3} + C = \frac{2}{5}x^{5/2} + \frac{3}{5}x^{5/3} + C$

54. $f'(x) = 8x - 3 \sec^2 x \ \Rightarrow \ f(x) = 8\left(\frac{1}{2}x^2\right) - 3 \tan x + C_n = 4x^2 - 3 \tan x + C_n$ on the interval $\left(n\pi - \frac{\pi}{2}, n\pi + \frac{\pi}{2}\right)$.

55. $f'(t) = 2t - 3 \sin t \ \Rightarrow \ f(t) = t^2 + 3 \cos t + C$.

$f(0) = 3 + C$ and $f(0) = 5 \ \Rightarrow \ C = 2$, so $f(t) = t^2 + 3 \cos t + 2$.

56. $f'(u) = \dfrac{u^2 + \sqrt{u}}{u} = u + u^{-1/2} \ \Rightarrow \ f(u) = \frac{1}{2}u^2 + 2u^{1/2} + C$.

$f(1) = \frac{1}{2} + 2 + C$ and $f(1) = 3 \ \Rightarrow \ C = \frac{1}{2}$, so $f(u) = \frac{1}{2}u^2 + 2\sqrt{u} + \frac{1}{2}$.

57. $f''(x) = 1 - 6x + 48x^2 \quad \Rightarrow \quad f'(x) = x - 3x^2 + 16x^3 + C.$ $f'(0) = C$ and $f'(0) = 2 \quad \Rightarrow \quad C = 2$, so

$f'(x) = x - 3x^2 + 16x^3 + 2$ and hence, $f(x) = \frac{1}{2}x^2 - x^3 + 4x^4 + 2x + D.$

$f(0) = D$ and $f(0) = 1 \quad \Rightarrow \quad D = 1$, so $f(x) = \frac{1}{2}x^2 - x^3 + 4x^4 + 2x + 1.$

58. $f''(x) = 2x^3 + 3x^2 - 4x + 5 \quad \Rightarrow \quad f'(x) = \frac{1}{2}x^4 + x^3 - 2x^2 + 5x + C \quad \Rightarrow$

$f(x) = \frac{1}{10}x^5 + \frac{1}{4}x^4 - \frac{2}{3}x^3 + \frac{5}{2}x^2 + Cx + D.$ $f(0) = D$ and $f(0) = 2 \quad \Rightarrow \quad D = 2.$

$f(1) = \frac{1}{10} + \frac{1}{4} - \frac{2}{3} + \frac{5}{2} + C + 2$ and $f(1) = 0 \quad \Rightarrow \quad C = -\frac{6}{60} - \frac{15}{60} + \frac{40}{60} - \frac{150}{60} - \frac{120}{60} = -\frac{251}{60}$, so

$f(x) = \frac{1}{10}x^5 + \frac{1}{4}x^4 - \frac{2}{3}x^3 + \frac{5}{2}x^2 - \frac{251}{60}x + 2.$

59. $v(t) = s'(t) = 2t - \sin t \quad \Rightarrow \quad s(t) = t^2 + \cos t + C.$

$s(0) = 0 + 1 + C = C + 1$ and $s(0) = 3 \quad \Rightarrow \quad C + 1 = 3 \quad \Rightarrow \quad C = 2$, so $s(t) = t^2 + \cos t + 2.$

60. $a(t) = v'(t) = \sin t + 3\cos t \quad \Rightarrow \quad v(t) = -\cos t + 3\sin t + C.$

$v(0) = -1 + 0 + C$ and $v(0) = 2 \quad \Rightarrow \quad C = 3$, so $v(t) = -\cos t + 3\sin t + 3$ and $s(t) = -\sin t - 3\cos t + 3t + D.$

$s(0) = -3 + D$ and $s(0) = 0 \quad \Rightarrow \quad D = 3$, and $s(t) = -\sin t - 3\cos t + 3t + 3.$

61. $f(x) = x^2 \sin(x^2),\ 0 \le x \le \pi$

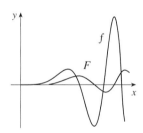

62. $f(x) = x^4 + x^3 + cx^2 \quad \Rightarrow \quad f'(x) = 4x^3 + 3x^2 + 2cx.$ This is 0 when $x(4x^2 + 3x + 2c) = 0 \quad \Leftrightarrow \quad x = 0$

or $4x^2 + 3x + 2c = 0.$ Using the quadratic formula, we find that the roots of this last equation are $x = \dfrac{-3 \pm \sqrt{9 - 32c}}{8}.$

Now if $9 - 32c < 0 \quad \Leftrightarrow \quad c > \frac{9}{32}$, then $(0, 0)$ is the only critical point, a minimum. If $c = \frac{9}{32}$, then there are two critical

points (a minimum at $x = 0$, and a horizontal tangent with no maximum or minimum at $x = -\frac{3}{8}$) and if $c < \frac{9}{32}$, then there are

three critical points except when $c = 0$, in which case the root with the $+$ sign coincides with the critical point at $x = 0.$ For

$0 < c < \frac{9}{32}$, there is a minimum at $x = -\frac{3}{8} - \dfrac{\sqrt{9 - 32c}}{8}$, a maximum at $x = -\frac{3}{8} + \dfrac{\sqrt{9 - 32c}}{8}$, and a minimum at $x = 0.$

For $c = 0$, there is a minimum at $x = -\frac{3}{4}$ and a horizontal tangent with no extremum at $x = 0$, and for $c < 0$, there is a maximum at $x = 0$, and there are minima at $x = -\dfrac{3}{8} \pm \dfrac{\sqrt{9 - 32c}}{8}$. Now we calculate $f''(x) = 12x^2 + 6x + 2c$.

The roots of this equation are $x = \dfrac{-6 \pm \sqrt{36 - 4 \cdot 12 \cdot 2c}}{24}$. So if $36 - 96c \leq 0 \iff c \geq \frac{3}{8}$, then there is no inflection

point. If $c < \frac{3}{8}$, then there are two inflection points at $x = -\dfrac{1}{4} \pm \dfrac{\sqrt{9 - 24c}}{12}$.

Value of c	No. of CP	No. of IP
$c < 0$	3	2
$c = 0$	2	2
$0 < c < \frac{9}{32}$	3	2
$c = \frac{9}{32}$	2	2
$\frac{9}{32} < c < \frac{3}{8}$	1	2
$c \geq \frac{3}{8}$	1	0

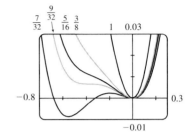

63. Choosing the positive direction to be upward, we have $a(t) = -9.8 \Rightarrow v(t) = -9.8t + v_0$, but $v(0) = 0 = v_0 \Rightarrow$

$v(t) = -9.8t = s'(t) \Rightarrow s(t) = -4.9t^2 + s_0$, but $s(0) = s_0 = 500 \Rightarrow s(t) = -4.9t^2 + 500$. When $s = 0$,

$-4.9t^2 + 500 = 0 \Rightarrow t_1 = \sqrt{\frac{500}{4.9}} \approx 10.1 \Rightarrow v(t_1) = -9.8\sqrt{\frac{500}{4.9}} \approx -98.995$ m/s. Since the canister has been

designed to withstand an impact velocity of 100 m/s, the canister will *not burst*.

64. Let $s_A(t)$ and $s_B(t)$ be the position functions for cars A and B and let $f(t) = s_A(t) - s(t)$. Since A passed B twice, there

must be three values of t such that $f(t) = 0$. Then by three applications of Rolle's Theorem (see Exercise 4.2.22), there is a

number c such that $f''(c) = 0$. So $s_A''(c) = s_B''(c)$; that is, A and B had equal accelerations at $t = c$. We assume that f is

continuous on $[0, T]$ and twice differentiable on $(0, T)$, where T is the total time of the race.

65. (a)

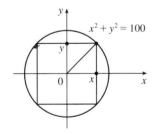

The cross-sectional area of the rectangular beam is

$A = 2x \cdot 2y = 4xy = 4x\sqrt{100 - x^2}, \ 0 \leq x \leq 10$, so

$\dfrac{dA}{dx} = 4x\left(\tfrac{1}{2}\right)(100 - x^2)^{-1/2}(-2x) + (100 - x^2)^{1/2} \cdot 4$

$= \dfrac{-4x^2}{(100 - x^2)^{1/2}} + 4(100 - x^2)^{1/2} = \dfrac{4[-x^2 + (100 - x^2)]}{(100 - x^2)^{1/2}}.$

$\dfrac{dA}{dx} = 0$ when $-x^2 + (100 - x^2) = 0 \Rightarrow x^2 = 50 \Rightarrow x = \sqrt{50} \approx 7.07 \Rightarrow y = \sqrt{100 - (\sqrt{50})^2} = \sqrt{50}$.

Since $A(0) = A(10) = 0$, the rectangle of maximum area is a square.

(b)

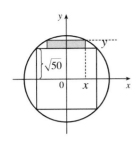

The cross-sectional area of each rectangular plank (shaded in the figure) is

$$A = 2x\left(y - \sqrt{50}\right) = 2x\left[\sqrt{100 - x^2} - \sqrt{50}\right], \ 0 \le x \le \sqrt{50}, \text{ so}$$

$$\frac{dA}{dx} = 2\left(\sqrt{100 - x^2} - \sqrt{50}\right) + 2x\left(\tfrac{1}{2}\right)(100 - x^2)^{-1/2}(-2x)$$

$$= 2(100 - x^2)^{1/2} - 2\sqrt{50} - \frac{2x^2}{(100 - x^2)^{1/2}}$$

Set $\dfrac{dA}{dx} = 0$: $(100 - x^2) - \sqrt{50}\,(100 - x^2)^{1/2} - x^2 = 0 \ \Rightarrow \ 100 - 2x^2 = \sqrt{50}\,(100 - x^2)^{1/2} \ \Rightarrow$

$10{,}000 - 400x^2 + 4x^4 = 50(100 - x^2) \ \Rightarrow \ 4x^4 - 350x^2 + 5000 = 0 \ \Rightarrow \ 2x^4 - 175x^2 + 2500 = 0 \ \Rightarrow$

$x^2 = \dfrac{175 \pm \sqrt{10{,}625}}{4} \approx 69.52 \text{ or } 17.98 \ \Rightarrow \ x \approx 8.34 \text{ or } 4.24.$ But $8.34 > \sqrt{50}$, so $x_1 \approx 4.24 \ \Rightarrow$

$y - \sqrt{50} = \sqrt{100 - x_1^2} - \sqrt{50} \approx 1.99.$ Each plank should have dimensions about $8\frac{1}{2}$ inches by 2 inches.

(c) From the figure in part (a), the width is $2x$ and the depth is $2y$, so the strength is

$$S = k(2x)(2y)^2 = 8kxy^2 = 8kx(100 - x^2) = 800kx - 8kx^3, \ 0 \le x \le 10. \ dS/dx = 800k - 24kx^2 = 0 \text{ when}$$

$24kx^2 = 800k \ \Rightarrow \ x^2 = \frac{100}{3} \ \Rightarrow \ x = \frac{10}{\sqrt{3}} \ \Rightarrow \ y = \sqrt{\frac{200}{3}} = \frac{10\sqrt{2}}{\sqrt{3}} = \sqrt{2}\,x.$ Since $S(0) = S(10) = 0$, the

maximum strength occurs when $x = \frac{10}{\sqrt{3}}$. The dimensions should be $\frac{20}{\sqrt{3}} \approx 11.55$ inches by $\frac{20\sqrt{2}}{\sqrt{3}} \approx 16.33$ inches.

66. (a)

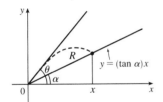

$y = (\tan\theta)x - \dfrac{g}{2v^2\cos^2\theta}x^2.$ The parabola intersects the line when

$$(\tan\alpha)x = (\tan\theta)x - \frac{g}{2v^2\cos^2\theta}x^2 \ \Rightarrow$$

$$x = \frac{(\tan\theta - \tan\alpha)2v^2\cos^2\theta}{g} \ \Rightarrow$$

$$R(\theta) = \frac{x}{\cos\alpha} = \left(\frac{\sin\theta}{\cos\theta} - \frac{\sin\alpha}{\cos\alpha}\right)\frac{2v^2\cos^2\theta}{g\cos\alpha} = \left(\frac{\sin\theta}{\cos\theta} - \frac{\sin\alpha}{\cos\alpha}\right)(\cos\theta\,\cos\alpha)\frac{2v^2\cos\theta}{g\cos^2\alpha}$$

$$= (\sin\theta\,\cos\alpha - \sin\alpha\,\cos\theta)\frac{2v^2\cos\theta}{g\cos^2\alpha} = \sin(\theta - \alpha)\frac{2v^2\cos\theta}{g\cos^2\alpha}$$

(b) $R'(\theta) = \dfrac{2v^2}{g\cos^2\alpha}[\cos\theta\cdot\cos(\theta - \alpha) + \sin(\theta - \alpha)(-\sin\theta)] = \dfrac{2v^2}{g\cos^2\alpha}\cos[\theta + (\theta - \alpha)]$

$$= \frac{2v^2}{g\cos^2\alpha}\cos(2\theta - \alpha) = 0$$

when $\cos(2\theta - \alpha) = 0 \ \Rightarrow \ 2\theta - \alpha = \frac{\pi}{2} \ \Rightarrow \ \theta = \dfrac{\pi/2 + \alpha}{2} = \dfrac{\pi}{4} + \dfrac{\alpha}{2}.$ The First Derivative Test shows that this

gives a maximum value for $R(\theta)$. [This could be done without calculus by applying the formula for $\sin x\,\cos y$ to $R(\theta)$.]

(c)

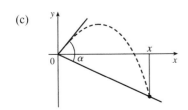

Replacing α by $-\alpha$ in part (a), we get $R(\theta) = \dfrac{2v^2\cos\theta\,\sin(\theta + \alpha)}{g\cos^2\alpha}.$

Proceeding as in part (b), or simply by replacing α by $-\alpha$ in the result of

part (b), we see that $R(\theta)$ is maximized when $\theta = \dfrac{\pi}{4} - \dfrac{\alpha}{2}.$

☐ PROBLEMS PLUS

1. Let $f(x) = \sin x - \cos x$ on $[0, 2\pi]$ since f has period 2π. $f'(x) = \cos x + \sin x = 0 \quad \Leftrightarrow \quad \cos x = -\sin x \quad \Leftrightarrow$

$\tan x = -1 \quad \Leftrightarrow \quad x = \frac{3\pi}{4}$ or $\frac{7\pi}{4}$. Evaluating f at its critical numbers and endpoints, we get $f(0) = -1$, $f\left(\frac{3\pi}{4}\right) = \sqrt{2}$,

$f\left(\frac{7\pi}{4}\right) = -\sqrt{2}$, and $f(2\pi) = -1$. So f has absolute maximum value $\sqrt{2}$ and absolute minimum value $-\sqrt{2}$. Thus,

$-\sqrt{2} \le \sin x - \cos x \le \sqrt{2} \quad \Rightarrow \quad |\sin x - \cos x| \le \sqrt{2}$.

2. $x^2 y^2 (4 - x^2)(4 - y^2) = x^2(4 - x^2)y^2(4 - y^2) = f(x)f(y)$, where $f(t) = t^2(4 - t^2)$.

We will show that $0 \le f(t) \le 4$ for $|t| \le 2$, which gives $0 \le f(x)f(y) \le 16$ for $|x| \le 2$ and $|y| \le 2$.

$f(t) = 4t^2 - t^4 \quad \Rightarrow \quad f'(t) = 8t - 4t^3 = 4t(2 - t^2) = 0 \quad \Rightarrow \quad t = 0$ or $\pm\sqrt{2}$.

$f(0) = 0$, $f(\pm\sqrt{2}) = 2(4 - 2) = 4$, and $f(2) = 0$. So 0 is the absolute minimum value of $f(t)$ on $[-2, 2]$ and 4 is the

absolute maximum value of $f(t)$ on $[-2, 2]$. We conclude that $0 \le f(t) \le 4$ for $|t| \le 2$ and hence, $0 \le f(x)f(y) \le 4^2$ or

$0 \le x^2(4 - x^2)y^2(4 - y^2) \le 16$.

3. First we show that $x(1 - x) \le \frac{1}{4}$ for all x. Let $f(x) = x(1 - x) = x - x^2$. Then $f'(x) = 1 - 2x$. This is 0 when $x = \frac{1}{2}$ and

$f'(x) > 0$ for $x < \frac{1}{2}$, $f'(x) < 0$ for $x > \frac{1}{2}$, so the absolute maximum of f is $f\left(\frac{1}{2}\right) = \frac{1}{4}$. Thus, $x(1 - x) \le \frac{1}{4}$ for all x.

Now suppose that the given assertion is false, that is, $a(1 - b) > \frac{1}{4}$ and $b(1 - a) > \frac{1}{4}$. Multiply these inequalities:

$a(1 - b)b(1 - a) > \frac{1}{16} \quad \Rightarrow \quad [a(1 - a)][b(1 - b)] > \frac{1}{16}$. But we know that $a(1 - a) \le \frac{1}{4}$ and $b(1 - b) \le \frac{1}{4} \quad \Rightarrow$

$[a(1 - a)][b(1 - b)] \le \frac{1}{16}$. Thus, we have a contradiction, so the given assertion is proved.

4. Let $P(a, 1 - a^2)$ be the point of contact. The equation of the tangent line at P is $y - (1 - a^2) = (-2a)(x - a) \quad \Rightarrow$

$y - 1 + a^2 = -2ax + 2a^2 \quad \Rightarrow \quad y = -2ax + a^2 + 1$. To find the x-intercept, put $y = 0$: $2ax = a^2 + 1 \quad \Rightarrow$

$x = \dfrac{a^2 + 1}{2a}$. To find the y-intercept, put $x = 0$: $y = a^2 + 1$. Therefore, the area of the triangle is

$\dfrac{1}{2}\left(\dfrac{a^2 + 1}{2a}\right)(a^2 + 1) = \dfrac{(a^2 + 1)^2}{4a}$. Therefore, we minimize the function $A(a) = \dfrac{(a^2 + 1)^2}{4a}$, $0 < a \le 1$.

$A'(a) = \dfrac{(4a)2(a^2 + 1)(2a) - (a^2 + 1)^2 (4)}{16a^2} = \dfrac{(a^2 + 1)[4a^2 - (a^2 + 1)]}{4a^2} = \dfrac{(a^2 + 1)(3a^2 - 1)}{4a^2}$.

$A'(a) = 0$ when $3a^2 - 1 = 0 \quad \Rightarrow \quad a = \frac{1}{\sqrt{3}}$. $A'(a) < 0$ for $a < \frac{1}{\sqrt{3}}$, $A'(a) > 0$ for $a > \frac{1}{\sqrt{3}}$. So by the First Derivative

Test, there is an absolute minimum when $a = \frac{1}{\sqrt{3}}$. The required point is $\left(\frac{1}{\sqrt{3}}, \frac{2}{3}\right)$ and the corresponding minimum area

is $A\left(\frac{1}{\sqrt{3}}\right) = \frac{4\sqrt{3}}{9}$.

5. Differentiating $x^2 + xy + y^2 = 12$ implicitly with respect to x gives $2x + y + x\dfrac{dy}{dx} + 2y\dfrac{dy}{dx} = 0$, so $\dfrac{dy}{dx} = -\dfrac{2x + y}{x + 2y}$.

At a highest or lowest point, $\dfrac{dy}{dx} = 0 \quad \Leftrightarrow \quad y = -2x$. Substituting $-2x$ for y in the original equation gives

$x^2 + x(-2x) + (-2x)^2 = 12$, so $3x^2 = 12$ and $x = \pm 2$. If $x = 2$, then $y = -2x = -4$, and if $x = -2$ then $y = 4$. Thus, the highest and lowest points are $(-2, 4)$ and $(2, -4)$.

6. (a) $V'(t)$ is the rate of change of the volume of the water with respect to time. $H'(t)$ is the rate of change of the height of the water with respect to time. Since the volume and the height are increasing, $V'(t)$ and $H'(t)$ are positive.

(b) $V'(t)$ is constant, so $V''(t)$ is zero (the slope of a constant function is 0).

(c) At first, the height H of the water increases quickly because the tank is narrow. But as the sphere widens, the rate of increase of the height slows down, reaching a minimum at $t = t_2$. Thus, the height is increasing at a decreasing rate on $(0, t_2)$, so its graph is concave downward and $H''(t_1) < 0$. As the sphere narrows for $t > t_2$, the rate of increase of the height begins to increase, and the graph of H is concave upward. Therefore, $H''(t_2) = 0$ and $H''(t_3) > 0$.

7. $f(x) = \dfrac{1}{1 + |x|} + \dfrac{1}{1 + |x - 2|}$

$$= \begin{cases} \dfrac{1}{1 - x} + \dfrac{1}{1 - (x - 2)} & \text{if } x < 0 \\[2mm] \dfrac{1}{1 + x} + \dfrac{1}{1 - (x - 2)} & \text{if } 0 \le x < 2 \\[2mm] \dfrac{1}{1 + x} + \dfrac{1}{1 + (x - 2)} & \text{if } x \ge 2 \end{cases} \quad \Rightarrow \quad f'(x) = \begin{cases} \dfrac{1}{(1 - x)^2} + \dfrac{1}{(3 - x)^2} & \text{if } x < 0 \\[2mm] \dfrac{-1}{(1 + x)^2} + \dfrac{1}{(3 - x)^2} & \text{if } 0 < x < 2 \\[2mm] \dfrac{-1}{(1 + x)^2} - \dfrac{1}{(x - 1)^2} & \text{if } x > 2 \end{cases}$$

We see that $f'(x) > 0$ for $x < 0$ and $f'(x) < 0$ for $x > 2$. For $0 < x < 2$, we have

$$f'(x) = \frac{1}{(3 - x)^2} - \frac{1}{(x + 1)^2} = \frac{(x^2 + 2x + 1) - (x^2 - 6x + 9)}{(3 - x)^2 (x + 1)^2} = \frac{8(x - 1)}{(3 - x)^2 (x + 1)^2}, \text{ so } f'(x) < 0 \text{ for } 0 < x < 1,$$

$f'(1) = 0$ and $f'(x) > 0$ for $1 < x < 2$. We have shown that $f'(x) > 0$ for $x < 0$; $f'(x) < 0$ for $0 < x < 1$; $f'(x) > 0$ for $1 < x < 2$; and $f'(x) < 0$ for $x > 2$. Therefore, by the First Derivative Test, the local maxima of f are at $x = 0$ and $x = 2$, where f takes the value $\frac{4}{3}$. Therefore, $\frac{4}{3}$ is the absolute maximum value of f.

8. If $f''(x) > 0$ for all x, then f' is increasing on $(-\infty, \infty)$, so $f'(0)$ must be greater than $f'(-1)$. But $f'(0) = 0 < \frac{1}{2} = f'(-1)$, so such a function cannot exist.

9. $A = (x_1, x_1^2)$ and $B = (x_2, x_2^2)$, where x_1 and x_2 are the solutions of the quadratic equation $x^2 = mx + b$. Let $P = (x, x^2)$ and set $A_1 = (x_1, 0)$, $B_1 = (x_2, 0)$, and $P_1 = (x, 0)$. Let $f(x)$ denote the area of triangle PAB. Then $f(x)$ can be expressed in terms of the areas of three trapezoids as follows:

$$f(x) = \text{area}\,(A_1 A B B_1) - \text{area}\,(A_1 A P P_1) - \text{area}\,(B_1 B P P_1)$$

$$= \tfrac{1}{2}(x_1^2 + x_2^2)(x_2 - x_1) - \tfrac{1}{2}(x_1^2 + x^2)(x - x_1) - \tfrac{1}{2}(x^2 + x_2^2)(x_2 - x)$$

After expanding and canceling terms, we get

$$f(x) = \tfrac{1}{2}(x_2 x_1^2 - x_1 x_2^2 - x x_1^2 + x_1 x^2 - x_2 x^2 + x x_2^2) = \tfrac{1}{2}\left[x_1^2 (x_2 - x) + x_2^2 (x - x_1) + x^2 (x_1 - x_2)\right]$$

$$f'(x) = \tfrac{1}{2}\left[-x_1^2 + x_2^2 + 2x(x_1 - x_2)\right]. \quad f''(x) = \tfrac{1}{2}[2(x_1 - x_2)] = x_1 - x_2 < 0 \text{ since } x_2 > x_1.$$

$f'(x) = 0 \;\Rightarrow\; 2x(x_1 - x_2) = x_1^2 - x_2^2 \;\Rightarrow\; x_P = \tfrac{1}{2}(x_1 + x_2).$

$f(x_P) = \tfrac{1}{2}\left(x_1^2\left[\tfrac{1}{2}(x_2 - x_1)\right] + x_2^2\left[\tfrac{1}{2}(x_2 - x_1)\right] + \tfrac{1}{4}(x_1 + x_2)^2(x_1 - x_2)\right)$

$\qquad = \tfrac{1}{2}\left[\tfrac{1}{2}(x_2 - x_1)\left(x_1^2 + x_2^2\right) - \tfrac{1}{4}(x_2 - x_1)(x_1 + x_2)^2\right] = \tfrac{1}{8}(x_2 - x_1)\left[2\left(x_1^2 + x_2^2\right) - \left(x_1^2 + 2x_1x_2 + x_2^2\right)\right]$

$\qquad = \tfrac{1}{8}(x_2 - x_1)\left(x_1^2 - 2x_1x_2 + x_2^2\right) = \tfrac{1}{8}(x_2 - x_1)(x_1 - x_2)^2 = \tfrac{1}{8}(x_2 - x_1)(x_2 - x_1)^2 = \tfrac{1}{8}(x_2 - x_1)^3$

To put this in terms of m and b, we solve the system $y = x_1^2$ and $y = mx_1 + b$, giving us $x_1^2 - mx_1 - b = 0 \;\Rightarrow$

$x_1 = \tfrac{1}{2}\left(m - \sqrt{m^2 + 4b}\right)$. Similarly, $x_2 = \tfrac{1}{2}\left(m + \sqrt{m^2 + 4b}\right)$. The area is then $\tfrac{1}{8}(x_2 - x_1)^3 = \tfrac{1}{8}\left(\sqrt{m^2 + 4b}\right)^3$,

and is attained at the point $P\left(x_P, x_P^2\right) = P\left(\tfrac{1}{2}m, \tfrac{1}{4}m^2\right)$.

Note: Another way to get an expression for $f(x)$ is to use the formula for an area of a triangle in terms of the coordinates of

the vertices: $f(x) = \tfrac{1}{2}\left[\left(x_2x_1^2 - x_1x_2^2\right) + \left(x_1x^2 - xx_1^2\right) + \left(xx_2^2 - x_2x^2\right)\right].$

10. If $f'(x) < 0$ for all x, $f''(x) > 0$ for $|x| > 1$, $f''(x) < 0$ for $|x| < 1$, and

$\displaystyle\lim_{x \to \pm\infty}[f(x) + x] = 0$, then f is decreasing everywhere, concave up on

$(-\infty, -1)$ and $(1, \infty)$, concave down on $(-1, 1)$, and approaches the line

$y = -x$ as $x \to \pm\infty$. An example of such a graph is sketched.

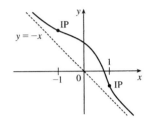

11. $f(x) = \left(a^2 + a - 6\right)\cos 2x + (a - 2)x + \cos 1 \;\Rightarrow\; f'(x) = -\left(a^2 + a - 6\right)\sin 2x\,(2) + (a - 2)$. The derivative exists

for all x, so the only possible critical points will occur where $f'(x) = 0 \;\Leftrightarrow\; 2(a - 2)(a + 3)\sin 2x = a - 2 \;\Leftrightarrow$

either $a = 2$ or $2(a + 3)\sin 2x = 1$, with the latter implying that $\sin 2x = \dfrac{1}{2(a + 3)}$. Since the range of $\sin 2x$ is $[-1, 1]$,

this equation has no solution whenever either $\dfrac{1}{2(a + 3)} < -1$ or $\dfrac{1}{2(a + 3)} > 1$. Solving these inequalities, we get

$-\tfrac{7}{2} < a < -\tfrac{5}{2}.$

12. To sketch the region $\left\{(x, y) \mid 2xy \le |x - y| \le x^2 + y^2\right\}$, we consider two cases.

Case 1: $x \ge y$ This is the case in which (x, y) lies on or below the line $y = x$. The double inequality becomes

$2xy \le x - y \le x^2 + y^2$. The right-hand inequality holds if and only if $x^2 - x + y^2 + y \ge 0 \;\Leftrightarrow$

$\left(x - \tfrac{1}{2}\right)^2 + \left(y + \tfrac{1}{2}\right)^2 \ge \tfrac{1}{2} \;\Leftrightarrow\; (x, y)$ lies on or outside the circle with radius $\tfrac{1}{\sqrt{2}}$ centered at $\left(\tfrac{1}{2}, -\tfrac{1}{2}\right)$.

The left-hand inequality holds if and only if $2xy - x + y \le 0 \;\Leftrightarrow\; xy - \tfrac{1}{2}x + \tfrac{1}{2}y \le 0 \;\Leftrightarrow$

$\left(x + \tfrac{1}{2}\right)\left(y - \tfrac{1}{2}\right) \le -\tfrac{1}{4} \;\Leftrightarrow\; (x, y)$ lies on or below the hyperbola $\left(x + \tfrac{1}{2}\right)\left(y - \tfrac{1}{2}\right) = -\tfrac{1}{4}$, which passes through the

origin and approaches the lines $y = \tfrac{1}{2}$ and $x = -\tfrac{1}{2}$ asymptotically.

Case 2: $y \ge x$ This is the case in which (x, y) lies on or above the line $y = x$. The double inequality becomes

$2xy \le y - x \le x^2 + y^2$. The right-hand inequality holds if and only if $x^2 + x + y^2 - y \ge 0 \;\Leftrightarrow$

$\left(x + \frac{1}{2}\right)^2 + \left(y - \frac{1}{2}\right)^2 \geq \frac{1}{2}$ ⇔ (x, y) lies on or outside the circle of radius $\frac{1}{\sqrt{2}}$ centered at $\left(-\frac{1}{2}, \frac{1}{2}\right)$. The left-hand

inequality holds if and only if $2xy + x - y \leq 0$ ⇔ $xy + \frac{1}{2}x - \frac{1}{2}y \leq 0$ ⇔ $\left(x - \frac{1}{2}\right)\left(y + \frac{1}{2}\right) \leq -\frac{1}{4}$ ⇔ (x, y) lies

on or above the left-hand branch of the hyperbola $\left(x - \frac{1}{2}\right)\left(y + \frac{1}{2}\right) = -\frac{1}{4}$, which passes through the origin and approaches the

lines $y = -\frac{1}{2}$ and $x = \frac{1}{2}$ asymptotically. Therefore, the region of interest consists of the points on or above the left branch of

the hyperbola $\left(x - \frac{1}{2}\right)\left(y + \frac{1}{2}\right) = -\frac{1}{4}$ that are on or outside the circle

$\left(x + \frac{1}{2}\right)^2 + \left(y - \frac{1}{2}\right)^2 = \frac{1}{2}$, together with the points on or below the right

branch of the hyperbola $\left(x + \frac{1}{2}\right)\left(y - \frac{1}{2}\right) = -\frac{1}{4}$ that are on or outside the circle

$\left(x - \frac{1}{2}\right)^2 + \left(y + \frac{1}{2}\right)^2 = \frac{1}{2}$. Note that the inequalities are unchanged when x

and y are interchanged, so the region is symmetric about the line $y = x$. So we

need only have analyzed case 1 and then reflected that region about the line

$y = x$, instead of considering case 2.

13. (a) Let $y = |AD|$, $x = |AB|$, and $1/x = |AC|$, so that $|AB| \cdot |AC| = 1$. We compute

the area $\mathcal{A}$ of $\triangle ABC$ in two ways.

First, $\mathcal{A} = \frac{1}{2}|AB||AC|\sin\frac{2\pi}{3} = \frac{1}{2} \cdot 1 \cdot \frac{\sqrt{3}}{2} = \frac{\sqrt{3}}{4}$.

Second,

$$\mathcal{A} = (\text{area of } \triangle ABD) + (\text{area of } \triangle ACD) = \frac{1}{2}|AB||AD|\sin\frac{\pi}{3} + \frac{1}{2}|AD||AC|\sin\frac{\pi}{3}$$

$$= \frac{1}{2}xy\frac{\sqrt{3}}{2} + \frac{1}{2}y(1/x)\frac{\sqrt{3}}{2} = \frac{\sqrt{3}}{4}y(x + 1/x)$$

Equating the two expressions for the area, we get $\frac{\sqrt{3}}{4}y\left(x + \frac{1}{x}\right) = \frac{\sqrt{3}}{4}$ ⇔ $y = \frac{1}{x + 1/x} = \frac{x}{x^2 + 1}$, $x > 0$.

Another method: Use the Law of Sines on the triangles ABD and ABC. In $\triangle ABD$, we have

$\angle A + \angle B + \angle D = 180°$ ⇔ $60° + \alpha + \angle D = 180°$ ⇔ $\angle D = 120° - \alpha$. Thus,

$\frac{x}{y} = \frac{\sin(120° - \alpha)}{\sin\alpha} = \frac{\sin 120° \cos\alpha - \cos 120° \sin\alpha}{\sin\alpha} = \frac{\frac{\sqrt{3}}{2}\cos\alpha + \frac{1}{2}\sin\alpha}{\sin\alpha}$ ⇒ $\frac{x}{y} = \frac{\sqrt{3}}{2}\cot\alpha + \frac{1}{2}$, and by a

similar argument with $\triangle ABC$, $\frac{\sqrt{3}}{2}\cot\alpha = x^2 + \frac{1}{2}$. Eliminating $\cot\alpha$ gives $\frac{x}{y} = \left(x^2 + \frac{1}{2}\right) + \frac{1}{2}$ ⇒

$y = \frac{x}{x^2 + 1}$, $x > 0$.

(b) We differentiate our expression for y with respect to x to find the maximum:

$\frac{dy}{dx} = \frac{(x^2 + 1) - x(2x)}{(x^2 + 1)^2} = \frac{1 - x^2}{(x^2 + 1)^2} = 0$ when $x = 1$. This indicates a maximum by the First Derivative Test, since

$y'(x) > 0$ for $0 < x < 1$ and $y'(x) < 0$ for $x > 1$, so the maximum value of y is $y(1) = \frac{1}{2}$.

14. (a)

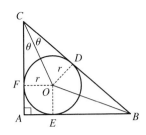

From geometry, two tangents to a circle from a given point have the same length, so $|CF| = |CD|$, $|AE| = |AF|$, and $|BD| = |BE|$. Thus,

$\frac{1}{2}(|BC| + |AC| - |AB|)$

$= \frac{1}{2}[(|BD| + |DC|) + (|AF| + |FC|) - (|AE| + |EB|)]$

$= \frac{1}{2}\left[\left(\underline{|BD|} + |CD|\right) + \left(\underline{|AF|} + |CD|\right) - \left(\underline{|AF|} + \underline{|BD|}\right)\right]$

$= \frac{1}{2}[2\,|CD|] = |CD|$

(b) Using the result from part (a) and the fact that $a = |BC|$, we have $\tan\theta = \dfrac{r}{|CD|}$ $\Rightarrow$

$\dfrac{r}{\tan\theta} = |CD| = \frac{1}{2}(|AC| + |BC| - |AB|) = \frac{1}{2}(a\cos 2\theta + a - a\sin 2\theta)$ $\Leftrightarrow$

$r = \frac{1}{2}a\tan\theta\,(2\cos^2\theta - 1 + 1 - 2\sin\theta\,\cos\theta)$

$= \frac{1}{2}a\,(2\sin\theta\,\cos\theta - 2\sin^2\theta)$ [in terms of θ]

$= \frac{1}{2}a\,(\sin 2\theta + \cos 2\theta - 1)$ [in terms of 2θ]

(c) We differentiate r with respect to θ and set $dr/d\theta$ equal to 0 to find the maximum values:

$dr/d\theta = \frac{1}{2}a\,(2\cos 2\theta - 2\sin 2\theta) = a(\cos 2\theta - \sin 2\theta)$. Since $0 < \theta < \frac{\pi}{4}$, $dr/d\theta = 0$ $\Leftrightarrow$ $\cos 2\theta = \sin 2\theta$ $\Leftrightarrow$

$1 = \tan 2\theta$ $\Leftrightarrow$ $2\theta = \frac{\pi}{4}$ $\Leftrightarrow$ $\theta = \frac{\pi}{8}$. This gives a maximum by the First Derivative Test, since

$dr/d\theta > 0$ for $0 < \theta < \frac{\pi}{8}$, and $dr/d\theta < 0$ for $\frac{\pi}{8} < \theta < \frac{\pi}{4}$. The maximum value is

$r\left(\frac{\pi}{8}\right) = \frac{1}{2}a\left(\sin\frac{\pi}{4} + \cos\frac{\pi}{4} - 1\right) = \frac{1}{2}\left(\sqrt{2} - 1\right)a \approx 0.207a$.

15. (a) $A = \frac{1}{2}bh$ with $\sin\theta = h/c$, so $A = \frac{1}{2}bc\sin\theta$. But A is a constant,

so differentiating this equation with respect to t, we get

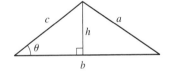

$\dfrac{dA}{dt} = 0 = \dfrac{1}{2}\left[bc\cos\theta\,\dfrac{d\theta}{dt} + b\,\dfrac{dc}{dt}\sin\theta + \dfrac{db}{dt}c\sin\theta\right]$ $\Rightarrow$

$bc\cos\theta\,\dfrac{d\theta}{dt} = -\sin\theta\left[b\,\dfrac{dc}{dt} + c\,\dfrac{db}{dt}\right]$ $\Rightarrow$ $\dfrac{d\theta}{dt} = -\tan\theta\left[\dfrac{1}{c}\dfrac{dc}{dt} + \dfrac{1}{b}\dfrac{db}{dt}\right]$.

(b) We use the Law of Cosines to get the length of side a in terms of those of b and c, and then we differentiate implicitly with

respect to t: $a^2 = b^2 + c^2 - 2bc\cos\theta$ $\Rightarrow$

$2a\,\dfrac{da}{dt} = 2b\,\dfrac{db}{dt} + 2c\,\dfrac{dc}{dt} - 2\left[bc(-\sin\theta)\dfrac{d\theta}{dt} + b\,\dfrac{dc}{dt}\cos\theta + \dfrac{db}{dt}c\cos\theta\right]$ $\Rightarrow$

$\dfrac{da}{dt} = \dfrac{1}{a}\left(b\,\dfrac{db}{dt} + c\,\dfrac{dc}{dt} + bc\sin\theta\,\dfrac{d\theta}{dt} - b\,\dfrac{dc}{dt}\cos\theta - c\,\dfrac{db}{dt}\cos\theta\right)$. Now we substitute our value of a from the Law of

Cosines and the value of $d\theta/dt$ from part (a), and simplify (primes signify differentiation by t):

$\dfrac{da}{dt} = \dfrac{bb' + cc' + bc\sin\theta\left[-\tan\theta(c'/c + b'/b)\right] - (bc' + cb')(\cos\theta)}{\sqrt{b^2 + c^2 - 2bc\cos\theta}}$

$= \dfrac{bb' + cc' - \left[\sin^2\theta\,(bc' + cb') + \cos^2\theta\,(bc' + cb')\right]/\cos\theta}{\sqrt{b^2 + c^2 - 2bc\cos\theta}} = \dfrac{bb' + cc' - (bc' + cb')\sec\theta}{\sqrt{b^2 + c^2 - 2bc\cos\theta}}$

16. Let $x = |AE|$, $y = |AF|$ as shown. The area $\mathcal{A}$ of the $\triangle AEF$ is $\mathcal{A} = \frac{1}{2}xy$. We

need to find a relationship between x and y, so that we can take the derivative

$d\mathcal{A}/dx$ and then find the maximum and minimum areas. Now let A' be the point

on which A ends up after the fold has been performed, and let P be the intersection

of AA' and EF. Note that AA' is perpendicular to EF since we are reflecting A

through the line EF to get to A', and that $|AP| = |PA'|$ for the same reason.

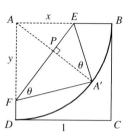

But $|AA'| = 1$, since AA' is a radius of the circle. Since $|AP| + |PA'| = |AA'|$, we have $|AP| = \frac{1}{2}$. Another way to express

the area of the triangle is $\mathcal{A} = \frac{1}{2}|EF||AP| = \frac{1}{2}\sqrt{x^2 + y^2}\left(\frac{1}{2}\right) = \frac{1}{4}\sqrt{x^2 + y^2}$. Equating the two expressions for $\mathcal{A}$, we get

$$\tfrac{1}{2}xy = \tfrac{1}{4}\sqrt{x^2 + y^2} \quad \Rightarrow \quad 4x^2y^2 = x^2 + y^2 \quad \Rightarrow \quad y^2(4x^2 - 1) = x^2 \quad \Rightarrow \quad y = x/\sqrt{4x^2 - 1}.$$

(Note that we could also have derived this result from the similarity of $\triangle A'PE$ and $\triangle A'FE$, that is,

$$\frac{|A'P|}{|PE|} = \frac{|A'F|}{|A'E|} \quad \Rightarrow \quad \frac{\frac{1}{2}}{\sqrt{x^2 - \left(\frac{1}{2}\right)^2}} = \frac{y}{x} \quad \Rightarrow \quad y = \frac{\frac{1}{2}x}{\sqrt{4x^2 - 1}/2} = \frac{x}{\sqrt{4x^2 - 1}}.)$$ Now we can substitute for y and

calculate $\dfrac{d\mathcal{A}}{dx}$: $\mathcal{A} = \dfrac{1}{2}\dfrac{x^2}{\sqrt{4x^2 - 1}} \quad \Rightarrow \quad \dfrac{d\mathcal{A}}{dx} = \dfrac{1}{2}\left[\dfrac{\sqrt{4x^2 - 1}\,(2x) - x^2\left(\frac{1}{2}\right)(4x^2 - 1)^{-1/2}(8x)}{4x^2 - 1}\right]$. This is 0 when

$$2x\sqrt{4x^2 - 1} - 4x^3(4x^2 - 1)^{-1/2} = 0 \quad \Leftrightarrow \quad 2x(4x^2 - 1)^{-1/2}[(4x^2 - 1) - 2x^2] = 0 \quad \Rightarrow \quad (4x^2 - 1) - 2x^2 = 0$$

$[x > 0] \quad \Leftrightarrow \quad 2x^2 = 1 \quad \Rightarrow \quad x = \frac{1}{\sqrt{2}}$. So this is one possible value for an extremum. We must also test the endpoints of the

interval over which x ranges. The largest value that x can attain is 1, and the smallest value of x occurs when $y = 1 \quad \Leftrightarrow$

$1 = x/\sqrt{4x^2 - 1} \quad \Leftrightarrow \quad x^2 = 4x^2 - 1 \quad \Leftrightarrow \quad 3x^2 = 1 \quad \Leftrightarrow \quad x = \frac{1}{\sqrt{3}}$. This will give the same value of $\mathcal{A}$ as will

$x = 1$, since the geometric situation is the same (reflected through the line $y = x$). We calculate

$$\mathcal{A}\left(\tfrac{1}{\sqrt{2}}\right) = \frac{1}{2}\frac{\left(1/\sqrt{2}\right)^2}{\sqrt{4\left(1/\sqrt{2}\right)^2 - 1}} = \frac{1}{4} = 0.25, \text{ and } \mathcal{A}(1) = \frac{1}{2}\frac{1^2}{\sqrt{4(1)^2 - 1}} = \frac{1}{2\sqrt{3}} \approx 0.29. \text{ So the maximum area is}$$

$\mathcal{A}(1) = \mathcal{A}\left(\tfrac{1}{\sqrt{3}}\right) = \frac{1}{2\sqrt{3}}$ and the minimum area is $\mathcal{A}\left(\tfrac{1}{\sqrt{2}}\right) = \frac{1}{4}$.

Another method: Use the angle θ (see diagram above) as a variable:

$$\mathcal{A} = \tfrac{1}{2}xy = \tfrac{1}{2}\left(\tfrac{1}{2}\sec\theta\right)\left(\tfrac{1}{2}\csc\theta\right) = \frac{1}{8\sin\theta\,\cos\theta} = \frac{1}{4\sin 2\theta}. \quad \mathcal{A} \text{ is minimized when } \sin 2\theta \text{ is maximal,}$$

that is, when $\sin 2\theta = 1 \quad \Rightarrow \quad 2\theta = \frac{\pi}{2} \quad \Rightarrow \quad \theta = \frac{\pi}{4}$. Also note that $A'E = x = \frac{1}{2}\sec\theta \le 1 \quad \Rightarrow \quad \sec\theta \le 2 \quad \Rightarrow$

$\cos\theta \ge \frac{1}{2} \quad \Rightarrow \quad \theta \le \frac{\pi}{3}$, and similarly, $A'F = y = \frac{1}{2}\csc\theta \le 1 \quad \Rightarrow \quad \csc\theta \le 2 \quad \Rightarrow \quad \sin\theta \le \frac{1}{2} \quad \Rightarrow \quad \theta \ge \frac{\pi}{6}$.

As above, we find that $\mathcal{A}$ is maximized at these endpoints: $\mathcal{A}\left(\frac{\pi}{6}\right) = \dfrac{1}{4\sin\frac{\pi}{3}} = \dfrac{1}{2\sqrt{3}} = \dfrac{1}{4\sin\frac{2\pi}{3}} = \mathcal{A}\left(\frac{\pi}{3}\right)$; and minimized at

$\theta = \frac{\pi}{4}$: $\mathcal{A}\left(\frac{\pi}{4}\right) = \dfrac{1}{4\sin\frac{\pi}{2}} = \dfrac{1}{4}$.

17. (a) Distance = rate × time, so time = distance/rate. $T_1 = \dfrac{D}{c_1}$, $T_2 = \dfrac{2\,|PR|}{c_1} + \dfrac{|RS|}{c_2} = \dfrac{2h\sec\theta}{c_1} + \dfrac{D - 2h\tan\theta}{c_2}$,

$$T_3 = \frac{2\sqrt{h^2 + D^2/4}}{c_1} = \frac{\sqrt{4h^2 + D^2}}{c_1}.$$

(b) $\dfrac{dT_2}{d\theta} = \dfrac{2h}{c_1}\cdot\sec\theta\tan\theta - \dfrac{2h}{c_2}\sec^2\theta = 0$ when $2h\sec\theta\left(\dfrac{1}{c_1}\tan\theta - \dfrac{1}{c_2}\sec\theta\right) = 0 \;\Rightarrow$

$\dfrac{1}{c_1}\dfrac{\sin\theta}{\cos\theta} - \dfrac{1}{c_2}\dfrac{1}{\cos\theta} = 0 \;\Rightarrow\; \dfrac{\sin\theta}{c_1\cos\theta} = \dfrac{1}{c_2\cos\theta} \;\Rightarrow\; \sin\theta = \dfrac{c_1}{c_2}$. The First Derivative Test shows that this gives

a minimum.

(c) Using part (a) with $D = 1$ and $T_1 = 0.26$, we have $T_1 = \dfrac{D}{c_1} \;\Rightarrow\; c_1 = \dfrac{1}{0.26} \approx 3.85$ km/s. $T_3 = \dfrac{\sqrt{4h^2 + D^2}}{c_1} \;\Rightarrow$

$4h^2 + D^2 = T_3^2 c_1^2 \;\Rightarrow\; h = \tfrac{1}{2}\sqrt{T_3^2 c_1^2 - D^2} = \tfrac{1}{2}\sqrt{(0.34)^2(1/0.26)^2 - 1^2} \approx 0.42$ km. To find c_2, we use $\sin\theta = \dfrac{c_1}{c_2}$

from part (b) and $T_2 = \dfrac{2h\sec\theta}{c_1} + \dfrac{D - 2h\tan\theta}{c_2}$ from part (a). From the figure,

$\sin\theta = \dfrac{c_1}{c_2} \;\Rightarrow\; \sec\theta = \dfrac{c_2}{\sqrt{c_2^2 - c_1^2}}$ and $\tan\theta = \dfrac{c_1}{\sqrt{c_2^2 - c_1^2}}$, so

$T_2 = \dfrac{2hc_2}{c_1\sqrt{c_2^2 - c_1^2}} + \dfrac{D\sqrt{c_2^2 - c_1^2} - 2hc_1}{c_2\sqrt{c_2^2 - c_1^2}}$. Using the values for T_2 [given as 0.32],

h, c_1, and D, we can graph $Y_1 = T_2$ and $Y_2 = \dfrac{2hc_2}{c_1\sqrt{c_2^2 - c_1^2}} + \dfrac{D\sqrt{c_2^2 - c_1^2} - 2hc_1}{c_2\sqrt{c_2^2 - c_1^2}}$ and find their intersection points.

Doing so gives us $c_2 \approx 4.10$ and 7.66, but if $c_2 = 4.10$, then $\theta = \arcsin(c_1/c_2) \approx 69.6°$, which implies that point S is to

the left of point R in the diagram. So $c_2 = 7.66$ km/s.

18. A straight line intersects the curve $y = f(x) = x^4 + cx^3 + 12x^2 - 5x + 2$ in four distinct points if and only if the graph of f

has two inflection points. $f'(x) = 4x^3 + 3cx^2 + 24x - 5$ and $f''(x) = 12x^2 + 6cx + 24$.

$f''(x) = 0 \;\Leftrightarrow\; x = \dfrac{-6c \pm \sqrt{(6c)^2 - 4(12)(24)}}{2(12)}$. There are two distinct roots for $f''(x) = 0$ (and hence two inflection

points) if and only if the discriminant is positive; that is, $36c^2 - 1152 > 0 \;\Leftrightarrow\; c^2 > 32 \;\Leftrightarrow\; |c| > \sqrt{32}$. Thus, the desired

values of c are $c < -4\sqrt{2}$ or $c > 4\sqrt{2}$.

19.

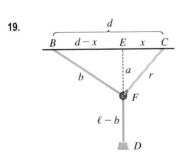

Let $a = |EF|$ and $b = |BF|$ as shown in the figure. Since $\ell = |BF| + |FD|$,

$|FD| = \ell - b$. Now

$|ED| = |EF| + |FD| = a + \ell - b = \sqrt{r^2 - x^2} + \ell - \sqrt{(d - x)^2 + a^2}$

$\quad = \sqrt{r^2 - x^2} + \ell - \sqrt{(d - x)^2 + \left(\sqrt{r^2 - x^2}\right)^2}$

$\quad = \sqrt{r^2 - x^2} + \ell - \sqrt{d^2 - 2dx + x^2 + r^2 - x^2}$

Let $f(x) = \sqrt{r^2 - x^2} + \ell - \sqrt{d^2 + r^2 - 2dx}$.

[continued]

$$f'(x) = \tfrac{1}{2}(r^2 - x^2)^{-1/2}(-2x) - \tfrac{1}{2}(d^2 + r^2 - 2dx)^{-1/2}(-2d) = \frac{-x}{\sqrt{r^2 - x^2}} + \frac{d}{\sqrt{d^2 + r^2 - 2dx}}.$$

$$f'(x) = 0 \;\Rightarrow\; \frac{x}{\sqrt{r^2 - x^2}} = \frac{d}{\sqrt{d^2 + r^2 - 2dx}} \;\Rightarrow\; \frac{x^2}{r^2 - x^2} = \frac{d^2}{d^2 + r^2 - 2dx} \;\Rightarrow$$

$$d^2 x^2 + r^2 x^2 - 2dx^3 = d^2 r^2 - d^2 x^2 \;\Rightarrow\; 0 = 2dx^3 - 2d^2 x^2 - r^2 x^2 + d^2 r^2 \;\Rightarrow$$

$$0 = 2dx^2(x - d) - r^2(x^2 - d^2) \;\Rightarrow\; 0 = 2dx^2(x - d) - r^2(x + d)(x - d) \;\Rightarrow\; 0 = (x - d)[2dx^2 - r^2(x + d)]$$

But $d > r > x$, so $x \neq d$. Thus, we solve $2dx^2 - r^2 x - dr^2 = 0$ for x:

$$x = \frac{-(-r^2) \pm \sqrt{(-r^2)^2 - 4(2d)(-dr^2)}}{2(2d)} = \frac{r^2 \pm \sqrt{r^4 + 8d^2 r^2}}{4d}.$$ Because $\sqrt{r^4 + 8d^2 r^2} > r^2$, the "negative" can be

discarded. Thus, $x = \dfrac{r^2 + \sqrt{r^2}\,\sqrt{r^2 + 8d^2}}{4d} = \dfrac{r^2 + r\,\sqrt{r^2 + 8d^2}}{4d} \;\;[r > 0] = \dfrac{r}{4d}\left(r + \sqrt{r^2 + 8d^2}\right).$ The maximum value

of $|ED|$ occurs at this value of x.

20.

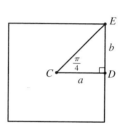

Let $a = \overline{CD}$ denote the distance from the center C of the base to the midpoint D of a side of the base.

Since $\triangle PQR$ is similar to $\triangle DCR$, $\dfrac{a}{h} = \dfrac{r}{\sqrt{h(h - 2r)}} \;\Rightarrow\; a = \dfrac{rh}{\sqrt{h(h - 2r)}} = r\dfrac{\sqrt{h}}{\sqrt{h - 2r}}.$

Let b denote one-half the length of a side of the base. The area A of the base is

$$A = 8(\text{area of } \triangle CDE) = 8\left(\tfrac{1}{2}ab\right) = 4a\left(a \tan \tfrac{\pi}{4}\right) = 4a^2.$$

The volume of the pyramid is $V = \tfrac{1}{3}Ah = \tfrac{1}{3}(4a^2)\,h = \dfrac{4}{3}\left(r\dfrac{\sqrt{h}}{\sqrt{h - 2r}}\right)^2 h = \tfrac{4}{3}r^2 \dfrac{h^2}{h - 2r},$ with domain $h > 2r$.

Now $\dfrac{dV}{dh} = \dfrac{4}{3}r^2 \cdot \dfrac{(h - 2r)(2h) - h^2(1)}{(h - 2r)^2} = \dfrac{4}{3}r^2 \dfrac{h^2 - 4hr}{(h - 2r)^2} = \dfrac{4}{3}r^2 \dfrac{h(h - 4r)}{(h - 2r)^2}$

and $\dfrac{d^2 V}{dh^2} = \dfrac{4}{3}r^2 \cdot \dfrac{(h - 2r)^2(2h - 4r) - (h^2 - 4hr)(2)(h - 2r)(1)}{[(h - 2r)^2]^2}$

$$= \dfrac{4}{3}r^2 \cdot \dfrac{2(h - 2r)\left[(h^2 - 4hr + 4r^2) - (h^2 - 4hr)\right]}{(h - 2r)^2} = \dfrac{8}{3}r^2 \cdot \dfrac{4r^2}{(h - 2r)^3} = \dfrac{32}{3}r^4 \cdot \dfrac{1}{(h - 2r)^3}.$$

The first derivative is equal to zero for $h = 4r$ and the second derivative is positive for $h > 2r$, so the volume of the pyramid is

minimized when $h = 4r$.

To extend our solution to a regular n-gon, we make the following changes:

(1) the number of sides of the base is n

(2) the number of triangles in the base is $2n$

(3) $\angle DCE = \dfrac{\pi}{n}$

(4) $b = a \tan \dfrac{\pi}{n}$

We then obtain the following results: $A = na^2 \tan \dfrac{\pi}{n}$, $V = \dfrac{nr^2}{3} \cdot \tan\left(\dfrac{\pi}{n}\right) \cdot \dfrac{h^2}{h-2r}$, $\dfrac{dV}{dh} = \dfrac{nr^2}{3} \cdot \tan\left(\dfrac{\pi}{n}\right) \cdot \dfrac{h(h-4r)}{(h-2r)^2}$,

and $\dfrac{d^2V}{dh^2} = \dfrac{8nr^4}{3} \cdot \tan\left(\dfrac{\pi}{n}\right) \cdot \dfrac{1}{(h-2r)^3}$. Notice that the answer, $h = 4r$, is independent of the number of sides of the base

of the polygon!

21. $V = \frac{4}{3}\pi r^3 \quad \Rightarrow \quad \dfrac{dV}{dt} = 4\pi r^2 \dfrac{dr}{dt}$. But $\dfrac{dV}{dt}$ is proportional to the surface area, so $\dfrac{dV}{dt} = k \cdot 4\pi r^2$ for some constant k.

Therefore, $4\pi r^2 \dfrac{dr}{dt} = k \cdot 4\pi r^2 \quad \Leftrightarrow \quad \dfrac{dr}{dt} = k = $ constant. An antiderivative of k with respect to t is kt, so $r = kt + C$.

When $t = 0$, the radius r must equal the original radius r_0, so $C = r_0$, and $r = kt + r_0$. To find k we use the fact that when

$t = 3$, $r = 3k + r_0$ and $V = \frac{1}{2}V_0 \quad \Rightarrow \quad \frac{4}{3}\pi(3k+r_0)^3 = \frac{1}{2} \cdot \frac{4}{3}\pi r_0^3 \quad \Rightarrow \quad (3k+r_0)^3 = \frac{1}{2}r_0^3 \quad \Rightarrow \quad 3k + r_0 = \dfrac{1}{\sqrt[3]{2}}r_0 \quad \Rightarrow$

$k = \frac{1}{3}r_0\left(\dfrac{1}{\sqrt[3]{2}} - 1\right)$. Since $r = kt + r_0$, $r = \frac{1}{3}r_0\left(\dfrac{1}{\sqrt[3]{2}} - 1\right)t + r_0$. When the snowball has melted completely we have

$r = 0 \quad \Rightarrow \quad \frac{1}{3}r_0\left(\dfrac{1}{\sqrt[3]{2}} - 1\right)t + r_0 = 0$ which gives $t = \dfrac{3\sqrt[3]{2}}{\sqrt[3]{2}-1}$. Hence, it takes $\dfrac{3\sqrt[3]{2}}{\sqrt[3]{2}-1} - 3 = \dfrac{3}{\sqrt[3]{2}-1} \approx 11\text{ h }33\text{ min}$

longer.

22. By ignoring the bottom hemisphere of the initial spherical bubble, we can rephrase the problem as follows: Prove that the

maximum height of a stack of n hemispherical bubbles is $\sqrt{n}$ if the radius of the bottom hemisphere is 1. We proceed by

induction. The case $n = 1$ is obvious since $\sqrt{1}$ is the height of the first hemisphere. Suppose the assertion is true for $n = k$

and let's suppose we have $k + 1$ hemispherical bubbles forming a stack of maximum height. Suppose the second hemisphere

(counting from the bottom) has radius r. Then by our induction hypothesis (scaled to the setting of a bottom hemisphere of

radius r), the height of the stack formed by the top k bubbles is $\sqrt{k}\,r$. (If it were shorter, then the total stack of $k + 1$ bubbles

wouldn't have maximum height.)

[continued]

The height of the whole stack is $H(r) = \sqrt{k}\, r + \sqrt{1 - r^2}$. (See the figure.)

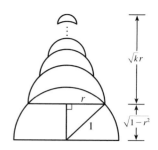

We want to choose r so as to maximize $H(r)$. Note that $0 < r < 1$.

We calculate $H'(r) = \sqrt{k} - \dfrac{r}{\sqrt{1 - r^2}}$ and $H''(r) = \dfrac{-1}{(1 - r^2)^{3/2}}$.

$$H'(r) = 0 \quad \Leftrightarrow \quad r^2 = k(1 - r^2) \quad \Leftrightarrow \quad (k + 1)r^2 = k \quad \Leftrightarrow \quad r = \sqrt{\dfrac{k}{k + 1}}.$$

This is the only critical number in $(0, 1)$ and it represents a local maximum

(hence an absolute maximum) since $H''(r) < 0$ on $(0, 1)$. When $r = \sqrt{\dfrac{k}{k + 1}}$,

$$H(r) = \sqrt{k}\,\dfrac{\sqrt{k}}{\sqrt{k + 1}} + \sqrt{1 - \dfrac{k}{k + 1}} = \dfrac{k}{\sqrt{k + 1}} + \dfrac{1}{\sqrt{k + 1}} = \sqrt{k + 1}.$$ Thus, the assertion is true for $n = k + 1$ when

it is true for $n = k$. By induction, it is true for all positive integers n.

Note: In general, a maximally tall stack of n hemispherical bubbles consists of bubbles with radii

$$1, \ \sqrt{\dfrac{n - 1}{n}}, \ \sqrt{\dfrac{n - 2}{n}}, \ \ldots, \ \sqrt{\dfrac{2}{n}}, \ \sqrt{\dfrac{1}{n}}$$

5 □ INTEGRALS

5.1 Areas and Distances

1. (a) Since f is *increasing*, we can obtain a *lower* estimate by using *left* endpoints. We are instructed to use five rectangles, so $n = 5$.

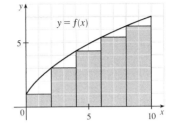

$$L_5 = \sum_{i=1}^{5} f(x_{i-1}) \, \Delta x \qquad [\Delta x = \tfrac{b-a}{n} = \tfrac{10-0}{5} = 2]$$

$$= f(x_0) \cdot 2 + f(x_1) \cdot 2 + f(x_2) \cdot 2 + f(x_3) \cdot 2 + f(x_4) \cdot 2$$

$$= 2\left[f(0) + f(2) + f(4) + f(6) + f(8)\right]$$

$$\approx 2(1 + 3 + 4.3 + 5.4 + 6.3) = 2(20) = 40$$

Since f is *increasing*, we can obtain an *upper* estimate by using *right* endpoints.

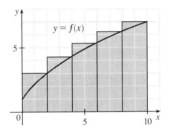

$$R_5 = \sum_{i=1}^{5} f(x_i) \, \Delta x$$

$$= 2\left[f(x_1) + f(x_2) + f(x_3) + f(x_4) + f(x_5)\right]$$

$$= 2\left[f(2) + f(4) + f(6) + f(8) + f(10)\right]$$

$$\approx 2(3 + 4.3 + 5.4 + 6.3 + 7) = 2(26) = 52$$

Comparing R_5 to L_5, we see that we have added the area of the rightmost upper rectangle, $f(10) \cdot 2$, to the sum and subtracted the area of the leftmost lower rectangle, $f(0) \cdot 2$, from the sum.

(b) $L_{10} = \sum_{i=1}^{10} f(x_{i-1}) \, \Delta x \qquad [\Delta x = \tfrac{10-0}{10} = 1]$

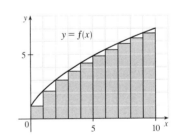

$$= 1\left[f(x_0) + f(x_1) + \cdots + f(x_9)\right]$$

$$= f(0) + f(1) + \cdots + f(9)$$

$$\approx 1 + 2.1 + 3 + 3.7 + 4.3 + 4.9 + 5.4 + 5.8 + 6.3 + 6.7$$

$$= 43.2$$

$$R_{10} = \sum_{i=1}^{10} f(x_i) \, \Delta x = f(1) + f(2) + \cdots + f(10)$$

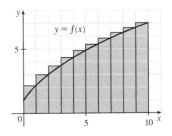

$$= L_{10} + 1 \cdot f(10) - 1 \cdot f(0) \quad \begin{bmatrix} \text{add rightmost upper rectangle,} \\ \text{subtract leftmost lower rectangle} \end{bmatrix}$$

$$= 43.2 + 7 - 1 = 49.2$$

2. (a) (i) $L_6 = \sum\limits_{i=1}^{6} f(x_{i-1})\Delta x \qquad [\Delta x = \frac{12-0}{6} = 2]$

$= 2[f(x_0) + f(x_1) + f(x_2) + f(x_3) + f(x_4) + f(x_5)]$

$= 2[f(0) + f(2) + f(4) + f(6) + f(8) + f(10)]$

$\approx 2(9 + 8.8 + 8.2 + 7.3 + 5.9 + 4.1)$

$= 2(43.3) = 86.6$

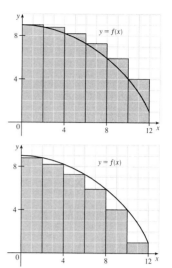

(ii) $R_6 = L_6 + 2 \cdot f(12) - 2 \cdot f(0)$

$\approx 86.6 + 2(1) - 2(9) = 70.6$

[Add area of rightmost lower rectangle
and subtract area of leftmost upper rectangle.]

(iii) $M_6 = \sum\limits_{i=1}^{6} f(x_i^*)\,\Delta x$

$= 2[f(1) + f(3) + f(5) + f(7) + f(9) + f(11)]$

$\approx 2(8.9 + 8.5 + 7.8 + 6.6 + 5.1 + 2.8)$

$= 2(39.7) = 79.4$

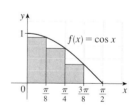

(b) Since f is *decreasing*, we obtain an *overestimate* by using *left* endpoints; that is, L_6.

(c) Since f is *decreasing*, we obtain an *underestimate* by using *right* endpoints; that is, R_6.

(d) M_6 gives the best estimate, since the area of each rectangle appears to be closer to the true area than the overestimates and underestimates in L_6 and R_6.

3. (a) $R_4 = \sum\limits_{i=1}^{4} f(x_i)\,\Delta x \quad \left[\Delta x = \dfrac{\pi/2 - 0}{4} = \dfrac{\pi}{8}\right] \quad = \left[\sum\limits_{i=1}^{4} f(x_i)\right]\Delta x$

$= [f(x_1) + f(x_2) + f(x_3) + f(x_4)]\,\Delta x$

$= \left[\cos\frac{\pi}{8} + \cos\frac{2\pi}{8} + \cos\frac{3\pi}{8} + \cos\frac{4\pi}{8}\right]\frac{\pi}{8}$

$\approx (0.9239 + 0.7071 + 0.3827 + 0)\frac{\pi}{8} \approx 0.7908$

Since f is *decreasing* on $[0, \pi/2]$, an *underestimate* is obtained by using the *right* endpoint approximation, R_4.

(b) $L_4 = \sum\limits_{i=1}^{4} f(x_{i-1})\,\Delta x = \left[\sum\limits_{i=1}^{4} f(x_{i-1})\right]\Delta x$

$= [f(x_0) + f(x_1) + f(x_2) + f(x_3)]\,\Delta x$

$= \left[\cos 0 + \cos\frac{\pi}{8} + \cos\frac{2\pi}{8} + \cos\frac{3\pi}{8}\right]\frac{\pi}{8}$

$\approx (1 + 0.9239 + 0.7071 + 0.3827)\frac{\pi}{8} \approx 1.1835$

L_4 is an overestimate. Alternatively, we could just add the area of the leftmost upper rectangle and subtract the area of the rightmost lower rectangle; that is, $L_4 = R_4 + f(0) \cdot \frac{\pi}{8} - f\left(\frac{\pi}{2}\right) \cdot \frac{\pi}{8}$.

4. (a) $R_4 = \sum_{i=1}^{4} f(x_i)\,\Delta x \qquad \left[\Delta x = \dfrac{4-0}{4} = 1\right]$

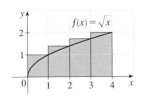

$\qquad = f(x_1)\cdot 1 + f(x_2)\cdot 1 + f(x_3)\cdot 1 + f(x_4)\cdot 1$

$\qquad = f(1) + f(2) + f(3) + f(4)$

$\qquad = \sqrt{1} + \sqrt{2} + \sqrt{3} + \sqrt{4} \approx 6.1463$

Since f is *increasing* on $[0, 4]$, R_4 is an *overestimate*.

(b) $L_4 = \sum_{i=1}^{4} f(x_{i-1})\,\Delta x = f(x_0)\cdot 1 + f(x_1)\cdot 1 + f(x_2)\cdot 1 + f(x_3)\cdot 1$

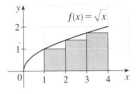

$\qquad = f(0) + f(1) + f(2) + f(3)$

$\qquad = \sqrt{0} + \sqrt{1} + \sqrt{2} + \sqrt{3} \approx 4.1463$

Since f is *increasing* on $[0, 4]$, L_4 is an *underestimate*.

5. (a) $f(x) = 1 + x^2$ and $\Delta x = \dfrac{2-(-1)}{3} = 1 \ \Rightarrow$

$\qquad R_3 = 1\cdot f(0) + 1\cdot f(1) + 1\cdot f(2) = 1\cdot 1 + 1\cdot 2 + 1\cdot 5 = 8.$

$\qquad \Delta x = \dfrac{2-(-1)}{6} = 0.5 \ \Rightarrow$

$\qquad R_6 = 0.5[f(-0.5) + f(0) + f(0.5) + f(1) + f(1.5) + f(2)]$

$\qquad\qquad = 0.5(1.25 + 1 + 1.25 + 2 + 3.25 + 5)$

$\qquad\qquad = 0.5(13.75) = 6.875$

(b) $L_3 = 1\cdot f(-1) + 1\cdot f(0) + 1\cdot f(1) = 1\cdot 2 + 1\cdot 1 + 1\cdot 2 = 5$

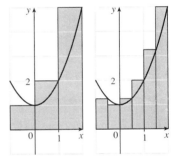

$\qquad L_6 = 0.5[f(-1) + f(-0.5) + f(0) + f(0.5) + f(1) + f(1.5)]$

$\qquad\qquad = 0.5(2 + 1.25 + 1 + 1.25 + 2 + 3.25)$

$\qquad\qquad = 0.5(10.75) = 5.375$

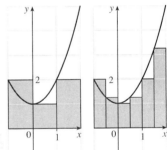

(c) $M_3 = 1\cdot f(-0.5) + 1\cdot f(0.5) + 1\cdot f(1.5)$

$\qquad\quad = 1\cdot 1.25 + 1\cdot 1.25 + 1\cdot 3.25 = 5.75$

$\qquad M_6 = 0.5[f(-0.75) + f(-0.25) + f(0.25)$

$\qquad\qquad\quad + f(0.75) + f(1.25) + f(1.75)]$

$\qquad\qquad = 0.5(1.5625 + 1.0625 + 1.0625 + 1.5625 + 2.5625 + 4.0625)$

$\qquad\qquad = 0.5(11.875) = 5.9375$

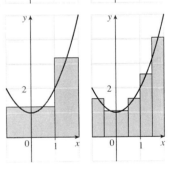

(d) M_6 appears to be the best estimate.

6. (a)

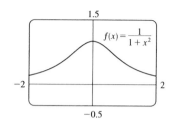

(b) $f(x) = 1/(1 + x^2)$ and $\Delta x = \frac{2-(-2)}{4} = 1 \quad \Rightarrow$

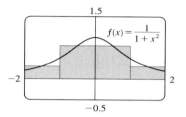

(i) $R_4 = \sum\limits_{i=1}^{4} f(x_i)\,\Delta x$

$\quad = f(-1) \cdot 1 + f(0) \cdot 1 + f(1) \cdot 1 + f(2) \cdot 1$

$\quad = \frac{1}{2} + 1 + \frac{1}{2} + \frac{1}{5} = \frac{11}{5} = 2.2$

(ii) $M_4 = \sum\limits_{i=1}^{4} f(\overline{x}_i)\,\Delta x \qquad \left[\overline{x}_i = \frac{1}{2}(x_{i-1} + x_i)\right]$

$\quad = f(-1.5) \cdot 1 + f(-0.5) \cdot 1 + f(0.5) \cdot 1 + f(1.5) \cdot 1$

$\quad = \frac{4}{13} + \frac{4}{5} + \frac{4}{5} + \frac{4}{13} = \frac{144}{65} \approx 2.2154$

(c) $n = 8$, so $\Delta x = \frac{2-(-2)}{8} = \frac{1}{2}$.

$R_8 = \frac{1}{2}[f(-1.5) + f(-1) + f(-0.5) + f(0) + f(0.5) + f(1) + f(1.5) + f(2)]$

$\quad = \frac{1}{2}\left[\frac{4}{13} + \frac{1}{2} + \frac{4}{5} + 1 + \frac{4}{5} + \frac{1}{2} + \frac{4}{13} + \frac{1}{5}\right] = \frac{287}{130} \approx 2.2077$

$M_8 = \frac{1}{2}[f(-1.75) + f(-1.25) + f(-0.75) + f(-0.25) + f(0.25) + f(0.75) + f(1.25) + f(1.75)]$

$\quad = \frac{1}{2}\left[2\left(\frac{16}{65} + \frac{16}{41} + \frac{16}{25} + \frac{16}{17}\right)\right] \approx 2.2176$

7. Here is one possible algorithm (ordered sequence of operations) for calculating the sums:

1 Let SUM = 0, X_MIN = 0, X_MAX = 1, N = 10 (depending on which sum we are calculating),

DELTA_X = (X_MAX - X_MIN)/N, and RIGHT_ENDPOINT = X_MIN + DELTA_X.

2 Repeat steps 2a, 2b in sequence until RIGHT_ENDPOINT > X_MAX.
2a Add (RIGHT_ENDPOINT)^4 to SUM.
2b Add DELTA_X to RIGHT_ENDPOINT.

At the end of this procedure, (DELTA_X)·(SUM) is equal to the answer we are looking for. We find that

$$R_{10} = \frac{1}{10} \sum_{i=1}^{10} \left(\frac{i}{10}\right)^4 \approx 0.2533, \quad R_{30} = \frac{1}{30} \sum_{i=1}^{30} \left(\frac{i}{30}\right)^4 \approx 0.2170, \quad R_{50} = \frac{1}{50} \sum_{i=1}^{50} \left(\frac{i}{50}\right)^4 \approx 0.2101, \text{ and}$$

$$R_{100} = \frac{1}{100} \sum_{i=1}^{100} \left(\frac{i}{100}\right)^4 \approx 0.2050. \text{ It appears that the exact area is } 0.2.$$

The following display shows the program SUMRIGHT and its output from a TI-83 Plus calculator. To generalize the program, we have input (rather than assign) values for Xmin, Xmax, and N. Also, the function, x^4, is assigned to Y_1, enabling us to evaluate any right sum merely by changing Y_1 and running the program.

8. We can use the algorithm from Exercise 7 with X_MIN = 0, X_MAX = $\pi/2$, and cos(RIGHT_ENDPOINT) instead of (RIGHT_ENDPOINT)^4 in step 2a. We find that $R_{10} = \dfrac{\pi/2}{10}\sum\limits_{i=1}^{10}\cos\left(\dfrac{i\pi}{20}\right) \approx 0.9194$, $R_{30} = \dfrac{\pi/2}{30}\sum\limits_{i=1}^{30}\cos\left(\dfrac{i\pi}{60}\right) \approx 0.9736$,

and $R_{50} = \dfrac{\pi/2}{50}\sum\limits_{i=1}^{50}\cos\left(\dfrac{i\pi}{100}\right) \approx 0.9842$, and $R_{100} = \dfrac{\pi/2}{100}\sum\limits_{i=1}^{100}\cos\left(\dfrac{i\pi}{200}\right) \approx 0.9921$. It appears that the exact area is 1.

9. In Maple, we have to perform a number of steps before getting a numerical answer. After loading the student package [command: `with(student);`] we use the command

   `left_sum:=leftsum(1/(x^2+1),x=0..1,10 [or 30, or 50]);` which gives us the expression in summation notation. To get a numerical approximation to the sum, we use `evalf(left_sum);`. Mathematica does not have a special command for these sums, so we must type them in manually. For example, the first left sum is given by

   `(1/10)*Sum[1/(((i-1)/10)^2+1)],{i,1,10}]`, and we use the `N` command on the resulting output to get a numerical approximation.

   In Derive, we use the `LEFT_RIEMANN` command to get the left sums, but must define the right sums ourselves. (We can define a new function using `LEFT_RIEMANN` with k ranging from 1 to n instead of from 0 to $n-1$.)

   (a) With $f(x) = \dfrac{1}{x^2 + 1}$, $0 \le x \le 1$, the left sums are of the form $L_n = \dfrac{1}{n}\sum\limits_{i=1}^{n}\dfrac{1}{\left(\frac{i-1}{n}\right)^2 + 1}$. Specifically, $L_{10} \approx 0.8100$,

   $L_{30} \approx 0.7937$, and $L_{50} \approx 0.7904$. The right sums are of the form $R_n = \dfrac{1}{n}\sum\limits_{i=1}^{n}\dfrac{1}{\left(\frac{i}{n}\right)^2 + 1}$. Specifically, $R_{10} \approx 0.7600$,

   $R_{30} \approx 0.7770$, and $R_{50} \approx 0.7804$.

(b) In Maple, we use the `leftbox` (with the same arguments as `left_sum`) and `rightbox` commands to generate the graphs.

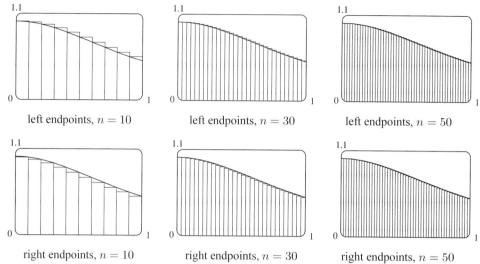

left endpoints, $n = 10$ left endpoints, $n = 30$ left endpoints, $n = 50$

right endpoints, $n = 10$ right endpoints, $n = 30$ right endpoints, $n = 50$

(c) We know that since $y = 1/(x^2 + 1)$ is a decreasing function on $(0, 1)$, all of the left sums are larger than the actual area, and all of the right sums are smaller than the actual area. Since the left sum with $n = 50$ is about $0.7904 < 0.791$ and the right sum with $n = 50$ is about $0.7804 > 0.780$, we conclude that $0.780 < R_{50} <$ exact area $< L_{50} < 0.791$, so the exact area is between 0.780 and 0.791.

10. (a) With $f(x) = x/(x + 2)$, $1 \leq x \leq 4$, and $x_i = 1 + 3i/n$, the left sums are of the form

$$L_n = \sum_{i=1}^{n} f(x_{i-1}) \, \Delta x = \frac{3}{n} \sum_{i=1}^{n} \frac{1 + 3(i - 1)/n}{3 + 3(i - 1)/n}.$$ In particular, $L_{10} \approx 1.5625$, $L_{30} \approx 1.5969$, and $L_{50} \approx 1.6037$.

The right sums are of the form $R_n = \sum_{i=1}^{n} f(x_i) \, \Delta x = \frac{3}{n} \sum_{i=1}^{n} \frac{1 + 3i/n}{3 + 3i/n}.$ In particular, $R_{10} \approx 1.6625$, $R_{30} \approx 1.6302$, and $R_{50} \approx 1.6237$.

(b) In Maple, we use the `leftbox` and `rightbox` commands (with the same arguments as `leftsum` and `rightsum` above) to generate the graphs.

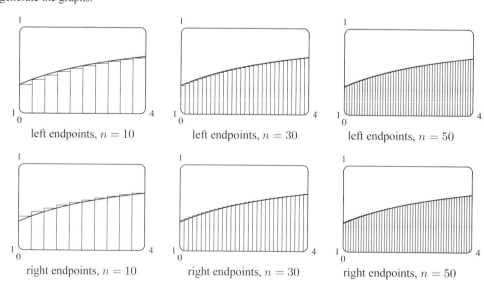

left endpoints, $n = 10$ left endpoints, $n = 30$ left endpoints, $n = 50$

right endpoints, $n = 10$ right endpoints, $n = 30$ right endpoints, $n = 50$

(c) $f'(x) = \dfrac{(x+2)\cdot 1 - x \cdot 1}{(x+2)^2} = \dfrac{2}{(x+2)^2} > 0$, so f is an increasing function. Thus, the left sums are underestimates of the

area A and the right sums are overestimates. The results in part (a) show that $1.603 < L_{50} < A < R_{50} < 1.624$.

11. Since v is an increasing function, L_6 will give us a lower estimate and R_6 will give us an upper estimate.

$$L_6 = (0 \text{ ft/s})(0.5 \text{ s}) + (6.2)(0.5) + (10.8)(0.5) + (14.9)(0.5) + (18.1)(0.5) + (19.4)(0.5) = 0.5(69.4) = 34.7 \text{ ft}$$

$$R_6 = 0.5(6.2 + 10.8 + 14.9 + 18.1 + 19.4 + 20.2) = 0.5(89.6) = 44.8 \text{ ft}$$

12. (a) $d \approx L_5 = (30 \text{ ft/s})(12 \text{ s}) + 28 \cdot 12 + 25 \cdot 12 + 22 \cdot 12 + 24 \cdot 12$

$$= (30 + 28 + 25 + 22 + 24) \cdot 12 = 129 \cdot 12 = 1548 \text{ ft}$$

(b) $d \approx R_5 = (28 + 25 + 22 + 24 + 27) \cdot 12 = 126 \cdot 12 = 1512 \text{ ft}$

(c) The estimates are neither lower nor upper estimates since v is neither an increasing nor a decreasing function of t.

13. Lower estimate for oil leakage: $R_5 = (7.6 + 6.8 + 6.2 + 5.7 + 5.3)(2) = (31.6)(2) = 63.2 \text{ L}$.

Upper estimate for oil leakage: $L_5 = (8.7 + 7.6 + 6.8 + 6.2 + 5.7)(2) = (35)(2) = 70 \text{ L}$.

14. We can find an upper estimate by using the final velocity for each time interval. Thus, the distance d traveled after 62 seconds

can be approximated by

$$d = \sum_{i=1}^{6} v(t_i)\,\Delta t_i = (185 \text{ ft/s})(10 \text{ s}) + 319 \cdot 5 + 447 \cdot 5 + 742 \cdot 12 + 1325 \cdot 27 + 1445 \cdot 3 = 54{,}694 \text{ ft}$$

15. For a decreasing function, using left endpoints gives us an overestimate and using right endpoints results in an underestimate.

We will use M_6 to get an estimate. $\Delta t = 1$, so

$$M_6 = 1[v(0.5) + v(1.5) + v(2.5) + v(3.5) + v(4.5) + v(5.5)] \approx 55 + 40 + 28 + 18 + 10 + 4 = 155 \text{ ft}$$

For a very rough check on the above calculation, we can draw a line from $(0, 70)$ to $(6, 0)$ and calculate the area of the

triangle: $\frac{1}{2}(70)(6) = 210$. This is clearly an overestimate, so our midpoint estimate of 155 is reasonable.

16. For an increasing function, using left endpoints gives us an underestimate and using right endpoints results in an overestimate.

We will use M_6 to get an estimate. $\Delta t = \frac{30 - 0}{6} = 5 \text{ s} = \frac{5}{3600} \text{ h} = \frac{1}{720} \text{ h}$.

$$M_6 = \tfrac{1}{720}[v(2.5) + v(7.5) + v(12.5) + v(17.5) + v(22.5) + v(27.5)]$$

$$= \tfrac{1}{720}(31.25 + 66 + 88 + 103.5 + 113.75 + 119.25) = \tfrac{1}{720}(521.75) \approx 0.725 \text{ km}$$

For a very rough check on the above calculation, we can draw a line from $(0, 0)$ to $(30, 120)$ and calculate the area of the

triangle: $\frac{1}{2}(30)(120) = 1800$. Divide by 3600 to get 0.5, which is clearly an underestimate, making our midpoint estimate of

0.725 seem reasonable. Of course, answers will vary due to different readings of the graph.

17. $f(x) = \sqrt[4]{x}$, $\ 1 \le x \le 16$. $\Delta x = (16 - 1)/n = 15/n$ and $x_i = 1 + i\,\Delta x = 1 + 15i/n$.

$$A = \lim_{n \to \infty} R_n = \lim_{n \to \infty} \sum_{i=1}^{n} f(x_i)\,\Delta x = \lim_{n \to \infty} \sum_{i=1}^{n} \sqrt[4]{1 + \dfrac{15i}{n}} \cdot \dfrac{15}{n}.$$

18. $f(x) = 1 + x^4$, $2 \le x \le 5$. $\Delta x = (5-2)/n = 3/n$ and $x_i = 2 + i\,\Delta x = 2 + 3i/n$.

$$A = \lim_{n \to \infty} R_n = \lim_{n \to \infty} \sum_{i=1}^{n} f(x_i)\,\Delta x = \lim_{n \to \infty} \sum_{i=1}^{n} \left[1 + \left(2 + \frac{3i}{n}\right)^4 \right] \cdot \frac{3}{n}.$$

19. $f(x) = x\cos x$, $0 \le x \le \frac{\pi}{2}$. $\Delta x = (\frac{\pi}{2} - 0)/n = \frac{\pi}{2}/n$ and $x_i = 0 + i\,\Delta x = \frac{\pi}{2}i/n$.

$$A = \lim_{n \to \infty} R_n = \lim_{n \to \infty} \sum_{i=1}^{n} f(x_i)\,\Delta x = \lim_{n \to \infty} \sum_{i=1}^{n} \frac{i\pi}{2n} \cos\left(\frac{i\pi}{2n}\right) \cdot \frac{\pi}{2n}.$$

20. $\displaystyle \lim_{n \to \infty} \sum_{i=1}^{n} \frac{2}{n}\left(5 + \frac{2i}{n}\right)^{10}$ can be interpreted as the area of the region lying under the graph of $y = (5+x)^{10}$ on the interval

$[0,2]$, since for $y = (5+x)^{10}$ on $[0,2]$ with $\Delta x = \dfrac{2-0}{n} = \dfrac{2}{n}$, $x_i = 0 + i\,\Delta x = \dfrac{2i}{n}$, and $x_i^* = x_i$, the expression for the

area is $A = \displaystyle \lim_{n \to \infty} \sum_{i=1}^{n} f(x_i^*)\,\Delta x = \lim_{n \to \infty} \sum_{i=1}^{n} \left(5 + \frac{2i}{n}\right)^{10} \frac{2}{n}$. Note that the answer is not unique. We could use $y = x^{10}$

on $[5,7]$ or, in general, $y = ((5-n)+x)^{10}$ on $[n, n+2]$.

21. $\displaystyle \lim_{n \to \infty} \sum_{i=1}^{n} \frac{\pi}{4n} \tan \frac{i\pi}{4n}$ can be interpreted as the area of the region lying under the graph of $y = \tan x$ on the interval $\left[0, \frac{\pi}{4}\right]$,

since for $y = \tan x$ on $\left[0, \frac{\pi}{4}\right]$ with $\Delta x = \dfrac{\pi/4 - 0}{n} = \dfrac{\pi}{4n}$, $x_i = 0 + i\,\Delta x = \dfrac{i\pi}{4n}$, and $x_i^* = x_i$, the expression for the area is

$A = \displaystyle \lim_{n \to \infty} \sum_{i=1}^{n} f(x_i^*)\,\Delta x = \lim_{n \to \infty} \sum_{i=1}^{n} \tan\left(\frac{i\pi}{4n}\right) \frac{\pi}{4n}$. Note that this answer is not unique, since the expression for the area is

the same for the function $y = \tan(x - k\pi)$ on the interval $\left[k\pi, k\pi + \frac{\pi}{4}\right]$, where k is any integer.

22. (a) $\Delta x = \dfrac{1-0}{n} = \dfrac{1}{n}$ and $x_i = 0 + i\,\Delta x = \dfrac{i}{n}$. $A = \displaystyle \lim_{n \to \infty} R_n = \lim_{n \to \infty} \sum_{i=1}^{n} f(x_i)\,\Delta x = \lim_{n \to \infty} \sum_{i=1}^{n} \left(\frac{i}{n}\right)^3 \cdot \frac{1}{n}$.

(b) $\displaystyle \lim_{n \to \infty} \sum_{i=1}^{n} \frac{i^3}{n^3} \cdot \frac{1}{n} = \lim_{n \to \infty} \frac{1}{n^4} \sum_{i=1}^{n} i^3 = \lim_{n \to \infty} \frac{1}{n^4} \left[\frac{n(n+1)}{2}\right]^2 = \lim_{n \to \infty} \frac{(n+1)^2}{4n^2} = \frac{1}{4} \lim_{n \to \infty} \left(1 + \frac{1}{n}\right)^2 = \frac{1}{4}$

23. (a) $y = f(x) = x^5$. $\Delta x = \dfrac{2-0}{n} = \dfrac{2}{n}$ and $x_i = 0 + i\,\Delta x = \dfrac{2i}{n}$.

$$A = \lim_{n \to \infty} R_n = \lim_{n \to \infty} \sum_{i=1}^{n} f(x_i)\,\Delta x = \lim_{n \to \infty} \sum_{i=1}^{n} \left(\frac{2i}{n}\right)^5 \cdot \frac{2}{n} = \lim_{n \to \infty} \sum_{i=1}^{n} \frac{32i^5}{n^5} \cdot \frac{2}{n} = \lim_{n \to \infty} \frac{64}{n^6} \sum_{i=1}^{n} i^5.$$

(b) $\displaystyle \sum_{i=1}^{n} i^5 \overset{\text{CAS}}{=} \frac{n^2(n+1)^2(2n^2 + 2n - 1)}{12}$

(c) $\displaystyle \lim_{n \to \infty} \frac{64}{n^6} \cdot \frac{n^2(n+1)^2(2n^2 + 2n - 1)}{12} = \frac{64}{12} \lim_{n \to \infty} \frac{(n^2 + 2n + 1)(2n^2 + 2n - 1)}{n^2 \cdot n^2}$

$$= \frac{16}{3} \lim_{n \to \infty} \left(1 + \frac{2}{n} + \frac{1}{n^2}\right)\left(2 + \frac{2}{n} - \frac{1}{n^2}\right) = \frac{16}{3} \cdot 1 \cdot 2 = \frac{32}{3}$$

24. (a) $y = f(x) = x^4 + 5x^2 + x$, $2 \le x \le 7$ $\Rightarrow$ $\Delta x = \dfrac{7-2}{n} = \dfrac{5}{n}$, $x_i = 2 + i\,\Delta x = 2 + \dfrac{5i}{n}$ $\Rightarrow$

$$A = \lim_{n \to \infty} R_n = \lim_{n \to \infty} \frac{5}{n} \sum_{i=1}^{n} \left[\left(2 + \frac{5i}{n}\right)^4 + 5\left(2 + \frac{5i}{n}\right)^2 + \left(2 + \frac{5i}{n}\right)\right]$$

(b) $R_n = \dfrac{5}{n} \cdot \dfrac{4723n^4 + 7845n^3 + 3475n^2 - 125}{6n^3}$

(c) $A = \lim\limits_{n\to\infty} R_n = \dfrac{23{,}615}{6} = 3935.8\overline{3}$

25. $y = f(x) = \cos x.$ $\Delta x = \dfrac{b-0}{n} = \dfrac{b}{n}$ and $x_i = 0 + i\,\Delta x = \dfrac{bi}{n}.$

$$A = \lim_{n\to\infty} R_n = \lim_{n\to\infty} \sum_{i=1}^{n} f(x_i)\,\Delta x = \lim_{n\to\infty} \sum_{i=1}^{n} \cos\left(\frac{bi}{n}\right) \cdot \frac{b}{n} \overset{\text{CAS}}{=} \lim_{n\to\infty} \left[\frac{b\sin\left(b\left(\frac{1}{2n}+1\right)\right)}{2n\sin\left(\frac{b}{2n}\right)} - \frac{b}{2n}\right] \overset{\text{CAS}}{=} \sin b$$

If $b = \frac{\pi}{2}$, then $A = \sin\frac{\pi}{2} = 1.$

26. (a)

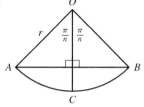

The diagram shows one of the n congruent triangles, $\triangle AOB$, with central angle $2\pi/n$. O is the center of the circle and AB is one of the sides of the polygon.

Radius OC is drawn so as to bisect $\angle AOB$. It follows that OC intersects AB at right angles and bisects AB. Thus, $\triangle AOB$ is divided into two right triangles with legs of length $\frac{1}{2}(AB) = r\sin(\pi/n)$ and $r\cos(\pi/n)$. $\triangle AOB$ has area

$2\cdot\frac{1}{2}[r\sin(\pi/n)][r\cos(\pi/n)] = r^2\sin(\pi/n)\cos(\pi/n) = \frac{1}{2}r^2\sin(2\pi/n),$

so $A_n = n \cdot \text{area}(\triangle AOB) = \frac{1}{2}nr^2\sin(2\pi/n).$

(b) To use Equation 3.4.2, $\lim\limits_{\theta\to 0}\dfrac{\sin\theta}{\theta} = 1$, we need to have the same expression in the denominator as we have in the argument of the sine function—in this case, $2\pi/n$.

$$\lim_{n\to\infty} A_n = \lim_{n\to\infty}\tfrac{1}{2}nr^2\sin(2\pi/n) = \lim_{n\to\infty}\tfrac{1}{2}nr^2\frac{\sin(2\pi/n)}{2\pi/n}\cdot\frac{2\pi}{n} = \lim_{n\to\infty}\frac{\sin(2\pi/n)}{2\pi/n}\pi r^2.\ \text{Let } \theta = \frac{2\pi}{n}.$$

Then as $n\to\infty$, $\theta\to 0$, so $\lim\limits_{n\to\infty}\dfrac{\sin(2\pi/n)}{2\pi/n}\pi r^2 = \lim\limits_{\theta\to 0}\dfrac{\sin\theta}{\theta}\pi r^2 = (1)\,\pi r^2 = \pi r^2.$

5.2 The Definite Integral

1. $f(x) = 3 - \frac{1}{2}x$, $2 \le x \le 14$. $\Delta x = \dfrac{b-a}{n} = \dfrac{14-2}{6} = 2.$

Since we are using left endpoints, $x_i^* = x_{i-1}.$

$L_6 = \sum\limits_{i=1}^{6} f(x_{i-1})\,\Delta x$

$= (\Delta x)\,[f(x_0) + f(x_1) + f(x_2) + f(x_3) + f(x_4) + f(x_5)]$

$= 2[f(2) + f(4) + f(6) + f(8) + f(10) + f(12)]$

$= 2[2 + 1 + 0 + (-1) + (-2) + (-3)] = 2(-3) = -6$

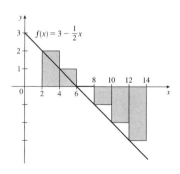

The Riemann sum represents the sum of the areas of the two rectangles above the x-axis minus the sum of the areas of the three rectangles below the x-axis; that is, the *net area* of the rectangles with respect to the x-axis.

2. $f(x) = x^2 - 2x$, $0 \leq x \leq 3$. $\quad \Delta x = \dfrac{b-a}{n} = \dfrac{3-0}{6} = \dfrac{1}{2}$.

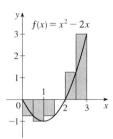

$f(x) = x^2 - 2x$

Since we are using right endpoints, $x_i^* = x_i$.

$R_6 = \displaystyle\sum_{i=1}^{6} f(x_i)\,\Delta x$

$\quad = (\Delta x)\left[f(x_1) + f(x_2) + f(x_3) + f(x_4) + f(x_5) + f(x_6) \right]$

$\quad = \tfrac{1}{2}\left[f\!\left(\tfrac{1}{2}\right) + f(1) + f\!\left(\tfrac{3}{2}\right) + f(2) + f\!\left(\tfrac{5}{2}\right) + f(3) \right]$

$\quad = \tfrac{1}{2}\left(-\tfrac{3}{4} - 1 - \tfrac{3}{4} + 0 + \tfrac{5}{4} + 3\right) = \tfrac{1}{2}\left(\tfrac{7}{4}\right) = \tfrac{7}{8}$

The Riemann sum represents the sum of the areas of the two rectangles above the x-axis minus the sum of the areas of the three rectangles below the x-axis; that is, the *net area* of the rectangles with respect to the x-axis.

3. $M_5 = \displaystyle\sum_{i=1}^{5} f(\overline{x}_i)\,\Delta x \quad [x_i^* = \overline{x}_i = \tfrac{1}{2}(x_{i-1} + x_i)$ is a midpoint and $\Delta x = 1]$

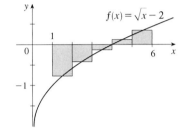

$f(x) = \sqrt{x} - 2$

$\quad = 1\left[f(1.5) + f(2.5) + f(3.5) \right.$

$\qquad\qquad \left. + f(4.5) + f(5.5) \right] \quad [f(x) = \sqrt{x} - 2]$

$\quad \approx -0.856759$

The Riemann sum represents the sum of the areas of the two rectangles above the x-axis minus the sum of the areas of the three rectangles below the x-axis.

4. (a) $f(x) = \sin x$, $0 \leq x \leq \dfrac{3\pi}{2}$. $\quad \Delta x = \dfrac{b-a}{n} = \dfrac{\frac{3\pi}{2} - 0}{6} = \dfrac{\pi}{4}$.

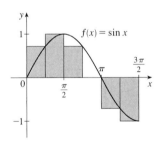

$f(x) = \sin x$

Since we are using right endpoints, $x_i^* = x_i$.

$R_6 = \displaystyle\sum_{i=1}^{6} f(x_i)\,\Delta x$

$\quad = (\Delta x)\left[f(x_1) + f(x_2) + f(x_3) + f(x_4) + f(x_5) + f(x_6) \right]$

$\quad = \dfrac{\pi}{4}\left[f\!\left(\dfrac{\pi}{4}\right) + f\!\left(\dfrac{2\pi}{4}\right) + f\!\left(\dfrac{3\pi}{4}\right) + f\!\left(\dfrac{4\pi}{4}\right) + f\!\left(\dfrac{5\pi}{4}\right) + f\!\left(\dfrac{6\pi}{4}\right) \right]$

$\quad = \dfrac{\pi}{4}\left(\sin\dfrac{\pi}{4} + \sin\dfrac{\pi}{2} + \sin\dfrac{3\pi}{4} + \sin\pi + \sin\dfrac{5\pi}{4} + \sin\dfrac{3\pi}{2} \right)$

$\quad = \dfrac{\pi}{4}\left(\dfrac{\sqrt{2}}{2} + 1 + \dfrac{\sqrt{2}}{2} + 0 - \dfrac{\sqrt{2}}{2} - 1 \right) = \dfrac{\pi\sqrt{2}}{8} \approx 0.555360$

The Riemann sum represents the sum of the areas of the three rectangles above the x-axis minus the sum of the areas of the two rectangles below the x-axis.

(b) Since we are using midpoints, $x_i^* = \overline{x}_i = \frac{1}{2}(x_{i-1} + x_i)$.

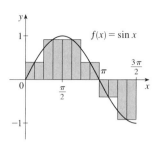

$$M_6 = \sum_{i=1}^{6} f(\overline{x}_i)\,\Delta x$$

$$= (\Delta x)\,[\,f(\overline{x}_1) + f(\overline{x}_2) + f(\overline{x}_3) + f(\overline{x}_4) + f(\overline{x}_5) + f(\overline{x}_6)\,]$$

$$= \frac{\pi}{4}\left[\,f\!\left(\frac{\pi}{8}\right) + f\!\left(\frac{3\pi}{8}\right) + f\!\left(\frac{5\pi}{8}\right) + f\!\left(\frac{7\pi}{8}\right) + f\!\left(\frac{9\pi}{8}\right) + f\!\left(\frac{11\pi}{8}\right)\right]$$

$$= \frac{\pi}{4}\left(\sin\frac{\pi}{8} + \sin\frac{3\pi}{8} + \sin\frac{5\pi}{8} + \sin\frac{7\pi}{8} + \sin\frac{9\pi}{8} + \sin\frac{11\pi}{8}\right)$$

$$\approx \tfrac{\pi}{4}(1.306563) \approx 1.026172$$

The Riemann sum represents the sum of the areas of the four rectangles above the x-axis minus the sum of the areas of the two rectangles below the x-axis. Note that the Riemann sum has the same value as the sum of the areas of the first two rectangles.

5. $\Delta x = (b-a)/n = (8-0)/4 = 8/4 = 2$.

(a) Using the right endpoints to approximate $\int_0^8 f(x)\,dx$, we have

$$\sum_{i=1}^{4} f(x_i)\,\Delta x = 2[f(2) + f(4) + f(6) + f(8)] \approx 2[1 + 2 + (-2) + 1] = 4.$$

(b) Using the left endpoints to approximate $\int_0^8 f(x)\,dx$, we have

$$\sum_{i=1}^{4} f(x_{i-1})\,\Delta x = 2[f(0) + f(2) + f(4) + f(6)] \approx 2[2 + 1 + 2 + (-2)] = 6.$$

(c) Using the midpoint of each subinterval to approximate $\int_0^8 f(x)\,dx$, we have

$$\sum_{i=1}^{4} f(\overline{x}_i)\,\Delta x = 2[f(1) + f(3) + f(5) + f(7)] \approx 2[3 + 2 + 1 + (-1)] = 10.$$

6. (a) Using the right endpoints to approximate $\int_{-3}^{3} g(x)\,dx$, we have

$$\sum_{i=1}^{6} g(x_i)\,\Delta x = 1[g(-2) + g(-1) + g(0) + g(1) + g(2) + g(3)] \approx 1 - 0.5 - 1.5 - 1.5 - 0.5 + 2.5 = -0.5.$$

(b) Using the left endpoints to approximate $\int_{-3}^{3} g(x)\,dx$, we have

$$\sum_{i=1}^{6} g(x_{i-1})\,\Delta x = 1[g(-3) + g(-2) + g(-1) + g(0) + g(1) + g(2)] \approx 2 + 1 - 0.5 - 1.5 - 1.5 - 0.5 = -1.$$

(c) Using the midpoint of each subinterval to approximate $\int_{-3}^{3} g(x)\,dx$, we have

$$\sum_{i=1}^{6} g(\overline{x}_i)\,\Delta x = 1[g(-2.5) + g(-1.5) + g(-0.5) + g(0.5) + g(1.5) + g(2.5)]$$

$$\approx 1.5 + 0 - 1 - 1.75 - 1 + 0.5 = -1.75$$

7. Since f is increasing, $L_5 \le \int_0^{25} f(x)\,dx \le R_5$.

$$\text{Lower estimate} = L_5 = \sum_{i=1}^{5} f(x_{i-1})\,\Delta x = 5[f(0) + f(5) + f(10) + f(15) + f(20)]$$

$$= 5(-42 - 37 - 25 - 6 + 15) = 5(-95) = -475$$

$$\text{Upper estimate} = R_5 = \sum_{i=1}^{5} f(x_i)\,\Delta x = 5[f(5) + f(10) + f(15) + f(20) + f(25)]$$

$$= 5(-37 - 25 - 6 + 15 + 36) = 5(-17) = -85$$

8. (a) Using the right endpoints to approximate $\int_3^9 f(x)\,dx$, we have

$$\sum_{i=1}^{3} f(x_i)\,\Delta x = 2[f(5) + f(7) + f(9)] = 2\,(-0.6 + 0.9 + 1.8) = 4.2.$$

Since f is *increasing*, using *right* endpoints gives an *overestimate*.

(b) Using the left endpoints to approximate $\int_3^9 f(x)\,dx$, we have

$$\sum_{i=1}^{3} f(x_{i-1})\,\Delta x = 2[f(3) + f(5) + f(7)] = 2\,(-3.4 - 0.6 + 0.9) = -6.2.$$

Since f is *increasing*, using *left* endpoints gives an *underestimate*.

(c) Using the midpoint of each interval to approximate $\int_3^9 f(x)\,dx$, we have

$$\sum_{i=1}^{3} f(\overline{x}_i)\,\Delta x = 2[f(4) + f(6) + f(8)] = 2\,(-2.1 + 0.3 + 1.4) = -0.8.$$

We cannot say anything about the midpoint estimate compared to the exact value of the integral.

9. $\Delta x = (10 - 2)/4 = 2$, so the endpoints are 2, 4, 6, 8, and 10, and the midpoints are 3, 5, 7, and 9. The Midpoint Rule

gives $\int_2^{10} \sqrt{x^3 + 1}\,dx \approx \sum_{i=1}^{4} f(\overline{x}_i)\,\Delta x = 2\left(\sqrt{3^3 + 1} + \sqrt{5^3 + 1} + \sqrt{7^3 + 1} + \sqrt{9^3 + 1}\right) \approx 124.1644.$

10. $\Delta x = (\pi/2 - 0)\,/4 = \frac{\pi}{8}$, so the endpoints are 0, $\frac{\pi}{8}$, $\frac{\pi}{4}$, $\frac{3\pi}{8}$, and $\frac{\pi}{2}$, and the midpoints are $\frac{\pi}{16}$, $\frac{3\pi}{16}$, $\frac{5\pi}{16}$, and $\frac{7\pi}{16}$. The Midpoint
Rule gives

$$\int_0^{\pi/2} \cos^4 x\,dx \approx \sum_{i=1}^{4} f(\overline{x}_i)\,\Delta x = \frac{\pi}{8}\left[\cos^4\left(\frac{\pi}{16}\right) + \cos^4\left(\frac{3\pi}{16}\right) + \cos^4\left(\frac{5\pi}{16}\right) + \cos^4\left(\frac{7\pi}{16}\right)\right] = \frac{\pi}{8}\left(\frac{3}{2}\right) \approx 0.5890.$$

11. $\Delta x = (1 - 0)/5 = 0.2$, so the endpoints are 0, 0.2, 0.4, 0.6, 0.8, and 1, and the midpoints are 0.1, 0.3, 0.5, 0.7, and 0.9.
The Midpoint Rule gives

$$\int_0^1 \sin(x^2)\,dx \approx \sum_{i=1}^{5} f(\overline{x}_i)\,\Delta x = 0.2\left[\sin(0.1)^2 + \sin(0.3)^2 + \sin(0.5)^2 + \sin(0.7)^2 + \sin(0.9)^2\right] \approx 0.3084.$$

12. $\Delta x = \dfrac{5 - 1}{4} = 1$, so the endpoints are 1, 2, 3, 4, and 5, and the midpoints are 1.5, 2.5, 3.5, and 4.5. The Midpoint Rule gives

$$\int_1^5 \frac{x-1}{x+1}\,dx \approx \sum_{i=1}^{4} f(\overline{x}_i)\,\Delta x = 1\left[\frac{1.5 - 1}{1.5 + 1} + \frac{2.5 - 1}{2.5 + 1} + \frac{3.5 - 1}{3.5 + 1} + \frac{4.5 - 1}{4.5 + 1}\right] \approx 1.8205.$$

13. In Maple, we use the command `with(student);` to load the sum and box commands, then
`m:=middlesum(sin(x^2),x=0..1,5);` which gives us the sum in summation notation, then `M:=evalf(m);` which
gives $M_5 \approx 0.30843908$, confirming the result of Exercise 11. The command `middlebox(sin(x^2),x=0..1,5)`
generates the graph. Repeating for $n = 10$ and $n = 20$ gives $M_{10} \approx 0.30981629$ and $M_{20} \approx 0.31015563$.

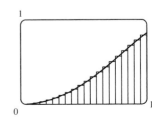

14. See the solution to Exercise 5.1.7 for a possible algorithm to calculate the sums. With $\Delta x = (1-0)/100 = 0.01$ and subinterval endpoints 1, 1.01, 1.02, ..., 1.99, 2, we calculate that the left Riemann sum is

$$L_{100} = \sum_{i=1}^{100} \sin(x_{i-1}^2)\,\Delta x \approx 0.30607, \text{ and the right Riemann sum is } R_{100} = \sum_{i=1}^{100} \sin(x_i^2)\,\Delta x \approx 0.31448.$$

Since $f(x) = \sin(x^2)$ is an increasing function, we must have $L_{100} \le \int_0^1 \sin(x^2)\,dx \le R_{100}$, so

$0.306 < L_{100} \le \int_0^1 \sin(x^2)\,dx \le R_{100} < 0.315$. Therefore, the approximate value $0.3084 \approx 0.31$ in Exercise 11 must be accurate to two decimal places.

15. We'll create the table of values to approximate $\int_0^\pi \sin x\,dx$ by using the

program in the solution to Exercise 5.1.7 with $Y_1 = \sin x$, Xmin $= 0$,

Xmax $= \pi$, and $n = 5, 10, 50$, and 100.

The values of R_n appear to be approaching 2.

n	R_n
5	1.933766
10	1.983524
50	1.999342
100	1.999836

16. $\int_0^2 \sqrt{1+x^4}\,dx$ with $n = 5, 10, 50$, and 100.

n	L_n	R_n
5	3.080614	4.329856
10	3.354110	3.978731
50	3.591540	3.716464
100	3.622383	3.684845

The value of the integral lies between 3.622 and 3.685. Note that $f(x) = \sqrt{1+x^4}$ is increasing on $(0, 2)$. We cannot make a similar statement for $\int_{-1}^2 \sqrt{1+x^4}\,dx$ since f is decreasing on $(-1, 0)$.

17. On $[2, 6]$, $\displaystyle\lim_{n\to\infty} \sum_{i=1}^{n} \frac{1-x_i^2}{4+x_i^2}\,\Delta x = \int_2^6 \frac{1-x^2}{4+x^2}\,dx.$

18. On $[\pi, 2\pi]$, $\displaystyle\lim_{n\to\infty} \sum_{i=1}^{n} \frac{\cos x_i}{x_i}\,\Delta x = \int_\pi^{2\pi} \frac{\cos x}{x}\,dx.$

19. On $[1, 8]$, $\displaystyle\lim_{n\to\infty} \sum_{i=1}^{n} \sqrt{2x_i^* + (x_i^*)^2}\,\Delta x = \int_1^8 \sqrt{2x+x^2}\,dx.$

20. On $[0, 2]$, $\displaystyle\lim_{n\to\infty} \sum_{i=1}^{n} \left[4 - 3(x_i^*)^2 + 6(x_i^*)^5\right]\Delta x = \int_0^2 (4 - 3x^2 + 6x^5)\,dx.$

21. Note that $\Delta x = \dfrac{5-(-1)}{n} = \dfrac{6}{n}$ and $x_i = -1 + i\,\Delta x = -1 + \dfrac{6i}{n}$.

$$\int_{-1}^5 (1+3x)\,dx = \lim_{n\to\infty} \sum_{i=1}^{n} f(x_i)\,\Delta x = \lim_{n\to\infty} \sum_{i=1}^{n}\left[1 + 3\left(-1 + \frac{6i}{n}\right)\right]\frac{6}{n} = \lim_{n\to\infty} \frac{6}{n}\sum_{i=1}^{n}\left[-2 + \frac{18i}{n}\right]$$

$$= \lim_{n\to\infty} \frac{6}{n}\left[\sum_{i=1}^{n}(-2) + \sum_{i=1}^{n}\frac{18i}{n}\right] = \lim_{n\to\infty} \frac{6}{n}\left[-2n + \frac{18}{n}\sum_{i=1}^{n} i\right]$$

$$= \lim_{n\to\infty} \frac{6}{n}\left[-2n + \frac{18}{n}\cdot\frac{n(n+1)}{2}\right] = \lim_{n\to\infty}\left[-12 + \frac{108}{n^2}\cdot\frac{n(n+1)}{2}\right]$$

$$= \lim_{n\to\infty}\left[-12 + 54\frac{n+1}{n}\right] = \lim_{n\to\infty}\left[-12 + 54\left(1 + \frac{1}{n}\right)\right] = -12 + 54\cdot 1 = 42$$

22. $\displaystyle\int_{1}^{4} (x^2 + 2x - 5)\, dx = \lim_{n\to\infty} \sum_{i=1}^{n} f(x_i)\Delta x \qquad [\Delta x = 3/n \text{ and } x_i = 1 + 3i/n]$

$$= \lim_{n\to\infty} \sum_{i=1}^{n} \left[\left(1 + \frac{3i}{n}\right)^2 + 2\left(1 + \frac{3i}{n}\right) - 5\right]\left(\frac{3}{n}\right)$$

$$= \lim_{n\to\infty} \frac{3}{n}\left[\sum_{i=1}^{n}\left(1 + \frac{6i}{n} + \frac{9i^2}{n^2} + 2 + \frac{6i}{n} - 5\right)\right]$$

$$= \lim_{n\to\infty} \frac{3}{n}\left[\sum_{i=1}^{n}\left(\frac{9}{n^2}\cdot i^2 + \frac{12}{n}\cdot i - 2\right)\right] = \lim_{n\to\infty} \frac{3}{n}\left[\frac{9}{n^2}\sum_{i=1}^{n} i^2 + \frac{12}{n}\sum_{i=1}^{n} i - \sum_{i=1}^{n} 2\right]$$

$$= \lim_{n\to\infty} \left(\frac{27}{n^3}\cdot \frac{n(n+1)(2n+1)}{6} + \frac{36}{n^2}\cdot \frac{n(n+1)}{2} - \frac{6}{n}\cdot n\right)$$

$$= \lim_{n\to\infty} \left(\frac{9}{2}\cdot \frac{n+1}{n}\cdot \frac{2n+1}{n} + 18\cdot \frac{n+1}{n} - 6\right)$$

$$= \lim_{n\to\infty} \left[\frac{9}{2}\left(1 + \frac{1}{n}\right)\left(2 + \frac{1}{n}\right) + 18\left(1 + \frac{1}{n}\right) - 6\right] = \tfrac{9}{2}\cdot 1\cdot 2 + 18\cdot 1 - 6 = 21$$

23. Note that $\Delta x = \dfrac{2-0}{n} = \dfrac{2}{n}$ and $x_i = 0 + i\,\Delta x = \dfrac{2i}{n}$.

$$\int_{0}^{2}\left(2 - x^2\right) dx = \lim_{n\to\infty} \sum_{i=1}^{n} f(x_i)\,\Delta x = \lim_{n\to\infty} \sum_{i=1}^{n}\left(2 - \frac{4i^2}{n^2}\right)\left(\frac{2}{n}\right) = \lim_{n\to\infty} \frac{2}{n}\left[\sum_{i=1}^{n} 2 - \frac{4}{n^2}\sum_{i=1}^{n} i^2\right]$$

$$= \lim_{n\to\infty} \frac{2}{n}\left(2n - \frac{4}{n^2}\sum_{i=1}^{n} i^2\right) = \lim_{n\to\infty}\left[4 - \frac{8}{n^3}\cdot \frac{n(n+1)(2n+1)}{6}\right]$$

$$= \lim_{n\to\infty}\left(4 - \frac{4}{3}\cdot \frac{n+1}{n}\cdot \frac{2n+1}{n}\right) = \lim_{n\to\infty}\left[4 - \frac{4}{3}\left(1 + \frac{1}{n}\right)\left(2 + \frac{1}{n}\right)\right] = 4 - \tfrac{4}{3}\cdot 1\cdot 2 = \tfrac{4}{3}$$

24. $\displaystyle\int_{0}^{5}\left(1 + 2x^3\right) dx = \lim_{n\to\infty} \sum_{i=1}^{n} f(x_i)\,\Delta x \qquad [\Delta x = 5/n \text{ and } x_i = 5i/n]$

$$= \lim_{n\to\infty} \sum_{i=1}^{n}\left(1 + 2\cdot \frac{125i^3}{n^3}\right)\left(\frac{5}{n}\right) = \lim_{n\to\infty} \frac{5}{n}\left[\sum_{i=1}^{n} 1 + \frac{250}{n^3}\sum_{i=1}^{n} i^3\right]$$

$$= \lim_{n\to\infty} \frac{5}{n}\left(1\cdot n + \frac{250}{n^3}\sum_{i=1}^{n} i^3\right) = \lim_{n\to\infty}\left[5 + \frac{1250}{n^4}\cdot \frac{n^2(n+1)^2}{4}\right]$$

$$= \lim_{n\to\infty}\left[5 + 312.5\cdot \frac{(n+1)^2}{n^2}\right] = \lim_{n\to\infty}\left[5 + 312.5\left(1 + \frac{1}{n}\right)^2\right]$$

$$= 5 + 312.5 = 317.5$$

25. Note that $\Delta x = \dfrac{2-1}{n} = \dfrac{1}{n}$ and $x_i = 1 + i\,\Delta x = 1 + i(1/n) = 1 + i/n$.

$$\int_1^2 x^3\,dx = \lim_{n\to\infty}\sum_{i=1}^n f(x_i)\,\Delta x = \lim_{n\to\infty}\sum_{i=1}^n\left(1+\frac{i}{n}\right)^3\left(\frac{1}{n}\right) = \lim_{n\to\infty}\frac{1}{n}\sum_{i=1}^n\left(\frac{n+i}{n}\right)^3$$

$$= \lim_{n\to\infty}\frac{1}{n^4}\sum_{i=1}^n\left(n^3+3n^2i+3ni^2+i^3\right) = \lim_{n\to\infty}\frac{1}{n^4}\left[\sum_{i=1}^n n^3 + \sum_{i=1}^n 3n^2i + \sum_{i=1}^n 3ni^2 + \sum_{i=1}^n i^3\right]$$

$$= \lim_{n\to\infty}\frac{1}{n^4}\left[n\cdot n^3 + 3n^2\sum_{i=1}^n i + 3n\sum_{i=1}^n i^2 + \sum_{i=1}^n i^3\right]$$

$$= \lim_{n\to\infty}\left[1 + \frac{3}{n^2}\cdot\frac{n(n+1)}{2} + \frac{3}{n^3}\cdot\frac{n(n+1)(2n+1)}{6} + \frac{1}{n^4}\cdot\frac{n^2(n+1)^2}{4}\right]$$

$$= \lim_{n\to\infty}\left[1 + \frac{3}{2}\cdot\frac{n+1}{n} + \frac{1}{2}\cdot\frac{n+1}{n}\cdot\frac{2n+1}{n} + \frac{1}{4}\cdot\frac{(n+1)^2}{n^2}\right]$$

$$= \lim_{n\to\infty}\left[1 + \frac{3}{2}\left(1+\frac{1}{n}\right) + \frac{1}{2}\left(1+\frac{1}{n}\right)\left(2+\frac{1}{n}\right) + \frac{1}{4}\left(1+\frac{1}{n}\right)^2\right] = 1 + \frac{3}{2} + \frac{1}{2}\cdot 2 + \frac{1}{4} = 3.75$$

26. (a) $\Delta x = (4-0)/8 = 0.5$ and $x_i^* = x_i = 0.5i$.   (b)

$$\int_0^4 (x^2-3x)\,dx \approx \sum_{i=1}^8 f(x_i^*)\,\Delta x$$

$$= 0.5\Big\{\big[0.5^2-3(0.5)\big] + \big[1.0^2-3(1.0)\big] + \cdots$$
$$\qquad + \big[3.5^2-3(3.5)\big] + \big[4.0^2-3(4.0)\big]\Big\}$$

$$= \tfrac{1}{2}\left(-\tfrac{5}{4} - 2 - \tfrac{9}{4} - 2 - \tfrac{5}{4} + 0 + \tfrac{7}{4} + 4\right) = -1.5$$

(c) $$\int_0^4 (x^2-3x)\,dx = \lim_{n\to\infty}\sum_{i=1}^n\left[\left(\frac{4i}{n}\right)^2 - 3\left(\frac{4i}{n}\right)\right]\left(\frac{4}{n}\right)$$

$$= \lim_{n\to\infty}\frac{4}{n}\left[\frac{16}{n^2}\sum_{i=1}^n i^2 - \frac{12}{n}\sum_{i=1}^n i\right]$$

$$= \lim_{n\to\infty}\left[\frac{64}{n^3}\cdot\frac{n(n+1)(2n+1)}{6} - \frac{48}{n^2}\cdot\frac{n(n+1)}{2}\right]$$

$$= \lim_{n\to\infty}\left[\frac{32}{3}\left(1+\frac{1}{n}\right)\left(2+\frac{1}{n}\right) - 24\left(1+\frac{1}{n}\right)\right]$$

$$= \frac{32}{3}\cdot 2 - 24 = -\frac{8}{3}$$

(d) $\int_0^4 (x^2-3x)\,dx = A_1 - A_2$, where A_1 is the area marked $+$ and A_2 is the area marked $-$.

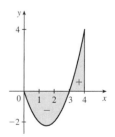

27. $$\int_a^b x\,dx = \lim_{n\to\infty}\frac{b-a}{n}\sum_{i=1}^n\left[a+\frac{b-a}{n}i\right] = \lim_{n\to\infty}\left[\frac{a(b-a)}{n}\sum_{i=1}^n 1 + \frac{(b-a)^2}{n^2}\sum_{i=1}^n i\right]$$

$$= \lim_{n\to\infty}\left[\frac{a(b-a)}{n}n + \frac{(b-a)^2}{n^2}\cdot\frac{n(n+1)}{2}\right] = a(b-a) + \lim_{n\to\infty}\frac{(b-a)^2}{2}\left(1+\frac{1}{n}\right)$$

$$= a(b-a) + \tfrac{1}{2}(b-a)^2 = (b-a)\left(a + \tfrac{1}{2}b - \tfrac{1}{2}a\right) = (b-a)\tfrac{1}{2}(b+a) = \tfrac{1}{2}\left(b^2-a^2\right)$$

28. $\displaystyle\int_a^b x^2\,dx = \lim_{n\to\infty}\frac{b-a}{n}\sum_{i=1}^n\left[a+\frac{b-a}{n}i\right]^2 = \lim_{n\to\infty}\frac{b-a}{n}\sum_{i=1}^n\left[a^2+2a\frac{b-a}{n}i+\frac{(b-a)^2}{n^2}i^2\right]$

$\displaystyle = \lim_{n\to\infty}\left[\frac{(b-a)^3}{n^3}\sum_{i=1}^n i^2 + \frac{2a(b-a)^2}{n^2}\sum_{i=1}^n i + \frac{a^2(b-a)}{n}\sum_{i=1}^n 1\right]$

$\displaystyle = \lim_{n\to\infty}\left[\frac{(b-a)^3}{n^3}\frac{n(n+1)(2n+1)}{6} + \frac{2a(b-a)^2}{n^2}\frac{n(n+1)}{2} + \frac{a^2(b-a)}{n}n\right]$

$\displaystyle = \lim_{n\to\infty}\left[\frac{(b-a)^3}{6}\cdot 1\cdot\left(1+\frac1n\right)\left(2+\frac1n\right) + a(b-a)^2\cdot 1\cdot\left(1+\frac1n\right) + a^2(b-a)\right]$

$\displaystyle = \frac{(b-a)^3}{3} + a(b-a)^2 + a^2(b-a) = \frac{b^3-3ab^2+3a^2b-a^3}{3} + ab^2 - 2a^2b + a^3 + a^2b - a^3$

$\displaystyle = \frac{b^3}{3} - \frac{a^3}{3} - ab^2 + a^2b + ab^2 - a^2b = \frac{b^3-a^3}{3}$

29. $\displaystyle f(x) = \frac{x}{1+x^5}, a=2, b=6$, and $\displaystyle\Delta x = \frac{6-2}{n} = \frac4n$. Using Theorem 4, we get $x_i^* = x_i = 2 + i\,\Delta x = 2 + \frac{4i}{n}$,

so $\displaystyle\int_2^6\frac{x}{1+x^5}\,dx = \lim_{n\to\infty}R_n = \lim_{n\to\infty}\sum_{i=1}^n\frac{2+\frac{4i}{n}}{1+\left(2+\frac{4i}{n}\right)^5}\cdot\frac4n$.

30. $f(x) = x^2\sin x, [a,b] = [0,2\pi], \Delta x = \dfrac{2\pi - 0}{n} = \dfrac{2\pi}{n}$, and $x_i = a + i\,\Delta x = \dfrac{2\pi i}{n}$,

so $\displaystyle\int_0^{2\pi} x^2\sin x\,dx = \lim_{n\to\infty}R_n = \lim_{n\to\infty}\sum_{i=1}^n\left[\left(\frac{2\pi i}{n}\right)^2\sin\left(\frac{2\pi i}{n}\right)\right]\cdot\frac{2\pi}{n}$.

31. $\Delta x = (\pi - 0)/n = \pi/n$ and $x_i^* = x_i = \pi i/n$.

$\displaystyle\int_0^\pi\sin 5x\,dx = \lim_{n\to\infty}\sum_{i=1}^n(\sin 5x_i)\left(\frac\pi n\right) = \lim_{n\to\infty}\sum_{i=1}^n\left(\sin\frac{5\pi i}{n}\right)\frac\pi n \overset{\text{CAS}}{=} \pi\lim_{n\to\infty}\frac1n\cot\left(\frac{5\pi}{2n}\right) \overset{\text{CAS}}{=} \pi\left(\frac{2}{5\pi}\right) = \frac25$

32. $\Delta x = (10-2)/n = 8/n$ and $x_i^* = x_i = 2 + 8i/n$.

$\displaystyle\int_2^{10}x^6\,dx = \lim_{n\to\infty}\sum_{i=1}^n\left(2+\frac{8i}{n}\right)^6\left(\frac8n\right) = 8\lim_{n\to\infty}\frac1n\sum_{i=1}^n\left(2+\frac{8i}{n}\right)^6$

$\displaystyle\overset{\text{CAS}}{=} 8\lim_{n\to\infty}\frac1n\cdot\frac{64(58{,}593n^6 + 164{,}052n^5 + 131{,}208n^4 - 27{,}776n^2 + 2048)}{21n^5}$

$\displaystyle\overset{\text{CAS}}{=} 8\left(\frac{1{,}249{,}984}{7}\right) = \frac{9{,}999{,}872}{7} \approx 1{,}428{,}553.1$

33. (a) Think of $\int_0^2 f(x)\,dx$ as the area of a trapezoid with bases 1 and 3 and height 2. The area of a trapezoid is $A = \frac12(b+B)h$,

so $\int_0^2 f(x)\,dx = \frac12(1+3)2 = 4$.

(b) $\int_0^5 f(x)\,dx = \int_0^2 f(x)\,dx + \int_2^3 f(x)\,dx + \int_3^5 f(x)\,dx$

$\qquad\qquad$ trapezoid $\qquad$ rectangle $\qquad$ triangle

$\qquad = \frac12(1+3)2 + \quad 3\cdot 1 \quad + \quad \frac12\cdot 2\cdot 3 \quad = 4+3+3 = 10$

(c) $\int_5^7 f(x)\,dx$ is the negative of the area of the triangle with base 2 and height 3. $\int_5^7 f(x)\,dx = -\frac12\cdot 2\cdot 3 = -3$.

(d) $\int_7^9 f(x)\,dx$ is the negative of the area of a trapezoid with bases 3 and 2 and height 2, so it equals

$-\frac{1}{2}(B+b)h = -\frac{1}{2}(3+2)2 = -5$. Thus,

$\int_0^9 f(x)\,dx = \int_0^5 f(x)\,dx + \int_5^7 f(x)\,dx + \int_7^9 f(x)\,dx = 10 + (-3) + (-5) = 2$.

34. (a) $\int_0^2 g(x)\,dx = \frac{1}{2}\cdot 4 \cdot 2 = 4$ [area of a triangle]

(b) $\int_2^6 g(x)\,dx = -\frac{1}{2}\pi(2)^2 = -2\pi$ [negative of the area of a semicircle]

(c) $\int_6^7 g(x)\,dx = \frac{1}{2}\cdot 1 \cdot 1 = \frac{1}{2}$ [area of a triangle]

$\int_0^7 g(x)\,dx = \int_0^2 g(x)\,dx + \int_2^6 g(x)\,dx + \int_6^7 g(x)\,dx = 4 - 2\pi + \frac{1}{2} = 4.5 - 2\pi$

35. $\int_0^3 \left(\frac{1}{2}x - 1\right) dx$ can be interpreted as the area of the triangle above the x-axis

minus the area of the triangle below the x-axis; that is,

$\frac{1}{2}(1)\left(\frac{1}{2}\right) - \frac{1}{2}(2)(1) = \frac{1}{4} - 1 = -\frac{3}{4}$.

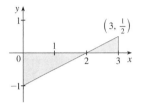

36. $\int_{-2}^2 \sqrt{4 - x^2}\,dx$ can be interpreted as the area under the graph of

$f(x) = \sqrt{4 - x^2}$ between $x = -2$ and $x = 2$. This is equal to half the area of

the circle with radius 2, so $\int_{-2}^2 \sqrt{4 - x^2}\,dx = \frac{1}{2}\pi \cdot 2^2 = 2\pi$.

37. $\int_{-3}^0 \left(1 + \sqrt{9 - x^2}\right) dx$ can be interpreted as the area under the graph of

$f(x) = 1 + \sqrt{9 - x^2}$ between $x = -3$ and $x = 0$. This is equal to one-quarter

the area of the circle with radius 3, plus the area of the rectangle, so

$\int_{-3}^0 \left(1 + \sqrt{9 - x^2}\right) dx = \frac{1}{4}\pi \cdot 3^2 + 1 \cdot 3 = 3 + \frac{9}{4}\pi$.

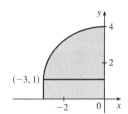

38. $\int_{-1}^3 (3 - 2x)\,dx$ can be interpreted as the area of the triangle above the x-axis

minus the area of the triangle below the x-axis; that is,

$\frac{1}{2}\left(\frac{5}{2}\right)(5) - \frac{1}{2}\left(\frac{3}{2}\right)(3) = \frac{25}{4} - \frac{9}{4} = 4$.

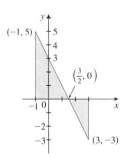

39. $\int_{-1}^2 |x|\,dx$ can be interpreted as the sum of the areas of the two shaded

triangles; that is, $\frac{1}{2}(1)(1) + \frac{1}{2}(2)(2) = \frac{1}{2} + \frac{4}{2} = \frac{5}{2}$.

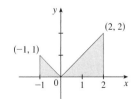

40. $\int_0^{10} |x - 5| \, dx$ can be interpreted as the sum of the areas of the two shaded triangles; that is, $2\left(\frac{1}{2}\right)(5)(5) = 25$.

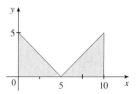

41. $\int_\pi^\pi \sin^2 x \cos^4 x \, dx = 0$ since the limits of intergration are equal.

42. $\int_1^0 3u \sqrt{u^2 + 4} \, du = -\int_0^1 3u \sqrt{u^2 + 4} \, du$ [because we reversed the limits of integration]

$\qquad = -\int_0^1 3x \sqrt{x^2 + 4} \, dx$ [we can use any letter without changing the value of the inegral]

$\qquad = -\left(5\sqrt{5} - 8\right)$ [given value]

$\qquad = 8 - 5\sqrt{5}$

43. $\int_0^1 (5 - 6x^2) \, dx = \int_0^1 5 \, dx - 6 \int_0^1 x^2 \, dx = 5(1 - 0) - 6\left(\frac{1}{3}\right) = 5 - 2 = 3$

44. $\int_2^5 (1 + 3x^4) \, dx = \int_2^5 1 \, dx + \int_2^5 3x^4 \, dx = 1(5 - 2) + 3 \int_2^5 x^4 \, dx = 1(3) + 3(618.6) = 1858.8$

45. $\int_1^4 (2x^2 - 3x + 1) \, dx = 2 \int_1^4 x^2 \, dx - 3 \int_1^4 x \, dx + \int_1^4 1 \, dx$

$\qquad = 2 \cdot \frac{1}{3}(4^3 - 1^3) - 3 \cdot \frac{1}{2}(4^2 - 1^2) + 1(4 - 1) = \frac{45}{2} = 22.5$

46. $\int_0^{\pi/2} (2 \cos x - 5x) \, dx = \int_0^{\pi/2} 2 \cos x \, dx - \int_0^{\pi/2} 5x \, dx = 2 \int_0^{\pi/2} \cos x \, dx - 5 \int_0^{\pi/2} x \, dx$

$\qquad = 2(1) - 5\dfrac{(\pi/2)^2 - 0^2}{2} = 2 - \dfrac{5\pi^2}{8}$

47. $\int_{-2}^2 f(x) \, dx + \int_2^5 f(x) \, dx - \int_{-2}^{-1} f(x) \, dx = \int_{-2}^5 f(x) \, dx + \int_{-1}^{-2} f(x) \, dx$ [by Property 5 and reversing limits]

$\qquad = \int_{-1}^5 f(x) \, dx$ [Property 5]

48. $\int_1^4 f(x) \, dx = \int_1^5 f(x) \, dx - \int_4^5 f(x) \, dx = 12 - 3.6 = 8.4$

49. $\int_0^9 [2f(x) + 3g(x)] \, dx = 2 \int_0^9 f(x) \, dx + 3 \int_0^9 g(x) \, dx = 2(37) + 3(16) = 122$

50. If $f(x) = \begin{cases} 3 & \text{for } x < 3 \\ x & \text{for } x \geq 3 \end{cases}$, then $\int_0^5 f(x) \, dx$ can be interpreted as the area of the shaded region, which consists of a 5-by-3 rectangle surmounted by an isosceles right triangle whose legs have length 2. Thus, $\int_0^5 f(x) \, dx = 5(3) + \frac{1}{2}(2)(2) = 17$.

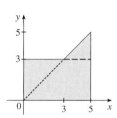

51. Using Integral Comparison Property 8, $m \leq f(x) \leq M \;\Rightarrow\; m(2 - 0) \leq \int_0^2 f(x) \, dx \leq M(2 - 0) \;\Rightarrow\;$ $2m \leq \int_0^2 f(x) \, dx \leq 2M$.

52. $x^2 \leq x$ on $[0, 1]$, so $\sqrt{1 + x^2} \leq \sqrt{1 + x}$ on $[0, 1]$. Hence, $\int_0^1 \sqrt{1 + x^2} \, dx \leq \int_0^1 \sqrt{1 + x} \, dx$ [Property 7].

53. If $-1 \leq x \leq 1$, then $0 \leq x^2 \leq 1$ and $1 \leq 1 + x^2 \leq 2$, so $1 \leq \sqrt{1 + x^2} \leq \sqrt{2}$ and $1[1 - (-1)] \leq \int_{-1}^1 \sqrt{1 + x^2} \, dx \leq \sqrt{2}\,[1 - (-1)]$ [Property 8]; that is, $2 \leq \int_{-1}^1 \sqrt{1 + x^2} \, dx \leq 2\sqrt{2}$.

54. If $\dfrac{\pi}{6} \le x \le \dfrac{\pi}{4}$, then $\cos \dfrac{\pi}{6} \ge \cos x \ge \cos \dfrac{\pi}{4}$ and $\dfrac{\sqrt{2}}{2} \le \cos x \le \dfrac{\sqrt{3}}{2}$, so

$$\dfrac{\sqrt{2}}{2}\left(\dfrac{\pi}{4} - \dfrac{\pi}{6}\right) \le \int_{\pi/6}^{\pi/4} \cos x \, dx \le \dfrac{\sqrt{3}}{2}\left(\dfrac{\pi}{4} - \dfrac{\pi}{6}\right) \quad \text{[Property 8]; that is, } \dfrac{\sqrt{2}\,\pi}{24} \le \int_{\pi/6}^{\pi/4} \cos x \, dx \le \dfrac{\sqrt{3}\,\pi}{24}.$$

55. If $1 \le x \le 4$, then $1 \le \sqrt{x} \le 2$, so $1(4-1) \le \int_1^4 \sqrt{x}\, dx \le 2(4-1)$; that is, $3 \le \int_1^4 \sqrt{x}\, dx \le 6$.

56. If $0 \le x \le 2$, then $1 \le 1 + x^2 \le 5$ and $\dfrac{1}{5} \le \dfrac{1}{1+x^2} \le 1$, so $\dfrac{1}{5}(2-0) \le \int_0^2 \dfrac{1}{1+x^2}\, dx \le 1(2-0)$;

that is, $\dfrac{2}{5} \le \int_0^2 \dfrac{1}{1+x^2}\, dx \le 2$.

57. If $\frac{\pi}{4} \le x \le \frac{\pi}{3}$, then $1 \le \tan x \le \sqrt{3}$, so $1\left(\frac{\pi}{3} - \frac{\pi}{4}\right) \le \int_{\pi/4}^{\pi/3} \tan x\, dx \le \sqrt{3}\left(\frac{\pi}{3} - \frac{\pi}{4}\right)$ or $\frac{\pi}{12} \le \int_{\pi/4}^{\pi/3} \tan x\, dx \le \frac{\pi}{12}\sqrt{3}$.

58. Let $f(x) = x^3 - 3x + 3$ for $0 \le x \le 2$. Then $f'(x) = 3x^2 - 3 = 3(x+1)(x-1)$, so f is decreasing on $(0,1)$ and

increasing on $(1,2)$. f has the absolute minimum value $f(1) = 1$. Since $f(0) = 3$ and $f(2) = 5$, the absolute maximum

value of f is $f(2) = 5$. Thus, $1 \le x^3 - 3x + 3 \le 5$ for x in $[0, 2]$. It follows from Property 8 that

$$1 \cdot (2 - 0) \le \int_0^2 \left(x^3 - 3x + 3\right) dx \le 5 \cdot (2 - 0); \text{ that is, } 2 \le \int_0^2 \left(x^3 - 3x + 3\right) dx \le 10.$$

59. For $-1 \le x \le 1$, $0 \le x^4 \le 1$ and $1 \le \sqrt{1+x^4} \le \sqrt{2}$, so $1[1 - (-1)] \le \int_{-1}^{1} \sqrt{1+x^4}\, dx \le \sqrt{2}\,[1 - (-1)]$

or $2 \le \int_{-1}^{1} \sqrt{1+x^4}\, dx \le 2\sqrt{2}$.

60. Let $f(x) = x - 2\sin x$ for $\pi \le x \le 2\pi$. Then $f'(x) = 1 - 2\cos x$ and $f'(x) = 0 \ \Rightarrow \ \cos x = \frac{1}{2} \ \Rightarrow \ x = \frac{5\pi}{3}$.

f has the absolute maximum value $f\left(\frac{5\pi}{3}\right) = \frac{5\pi}{3} - 2\sin \frac{5\pi}{3} = \frac{5\pi}{3} + \sqrt{3} \approx 6.97$ since $f(\pi) = \pi$ and $f(2\pi) = 2\pi$ are both

smaller than 6.97. Thus, $\pi \le f(x) \le \frac{5\pi}{3} + \sqrt{3} \ \Rightarrow \ \pi(2\pi - \pi) \le \int_\pi^{2\pi} f(x)\, dx \le \left(\frac{5\pi}{3} + \sqrt{3}\right)(2\pi - \pi)$; that is,

$$\pi^2 \le \int_\pi^{2\pi} (x - 2\sin x)\, dx \le \tfrac{5}{3}\pi^2 + \sqrt{3}\,\pi.$$

61. $\sqrt{x^4 + 1} \ge \sqrt{x^4} = x^2$, so $\int_1^3 \sqrt{x^4 + 1}\, dx \ge \int_1^3 x^2\, dx = \frac{1}{3}\left(3^3 - 1^3\right) = \frac{26}{3}$.

62. $0 \le \sin x \le 1$ for $0 \le x \le \frac{\pi}{2}$, so $x \sin x \le x \ \Rightarrow \ \int_0^{\pi/2} x \sin x\, dx \le \int_0^{\pi/2} x\, dx = \frac{1}{2}\left[\left(\frac{\pi}{2}\right)^2 - 0^2\right] = \frac{\pi^2}{8}$.

63. Using right endpoints as in the proof of Property 2, we calculate

$$\int_a^b cf(x)\, dx = \lim_{n\to\infty} \sum_{i=1}^{n} cf(x_i)\, \Delta x = \lim_{n\to\infty} c \sum_{i=1}^{n} f(x_i)\, \Delta x = c \lim_{n\to\infty} \sum_{i=1}^{n} f(x_i)\, \Delta x = c \int_a^b f(x)\, dx.$$

64. As in the proof of Property 2, we write $\int_a^b f(x)\, dx = \lim\limits_{n\to\infty} \sum\limits_{i=1}^{n} f(x_i)\, \Delta x$. Now $f(x_i) \ge 0$ and $\Delta x \ge 0$, so $f(x_i)\, \Delta x \ge 0$ and

therefore $\sum\limits_{i=1}^{n} f(x_i)\, \Delta x \ge 0$. But the limit of nonnegative quantities is nonnegative, so $\int_a^b f(x)\, dx \ge 0$.

65. Since $-|f(x)| \le f(x) \le |f(x)|$, it follows from Property 7 that

$$-\int_a^b |f(x)|\, dx \le \int_a^b f(x)\, dx \le \int_a^b |f(x)|\, dx \quad \Rightarrow \quad \left|\int_a^b f(x)\, dx\right| \le \int_a^b |f(x)|\, dx$$

Note that the definite integral is a real number, and so the following property applies: $-a \le b \le a \ \Rightarrow \ |b| \le a$ for all real

numbers b and nonnegative numbers a.

66. $\left|\int_0^{2\pi} f(x)\sin 2x\,dx\right| \le \int_0^{2\pi} |f(x)\sin 2x|\,dx$ [by Exercise 65] $= \int_0^{2\pi} |f(x)|\,|\sin 2x|\,dx \le \int_0^{2\pi} |f(x)|\,dx$ by Property 7,

since $|\sin 2x| \le 1 \implies |f(x)|\,|\sin 2x| \le |f(x)|$.

67. To show that f is integrable on $[0,1]$, we must show that $\lim\limits_{n\to\infty} \sum\limits_{i=1}^{n} f(x_i^*)\,\Delta x$ exists. Let n denote a positive integer and divide

the interval $[0,1]$ into n equal subintervals $\left[0,\dfrac{1}{n}\right]$, $\left[\dfrac{1}{n},\dfrac{2}{n}\right]$, $\dots$, $\left[\dfrac{n-1}{n},1\right]$. If we choose x_i^* to be a rational number in the ith

subinterval, then we obtain the Riemann sum $\sum\limits_{i=1}^{n} f(x_i^*)\cdot\dfrac{1}{n} = 0$, so $\lim\limits_{n\to\infty} \sum\limits_{i=1}^{n} f(x_i^*)\cdot\dfrac{1}{n} = \lim\limits_{n\to\infty} 0 = 0$. Now suppose we

choose x_i^* to be an irrational number. Then we get $\sum\limits_{i=1}^{n} f(x_i^*)\cdot\dfrac{1}{n} = \sum\limits_{i=1}^{n} 1\cdot\dfrac{1}{n} = n\cdot\dfrac{1}{n} = 1$ for each n, so

$\lim\limits_{n\to\infty} \sum\limits_{i=1}^{n} f(x_i^*)\cdot\dfrac{1}{n} = \lim\limits_{n\to\infty} 1 = 1$. Since the value of $\lim\limits_{n\to\infty} \sum\limits_{i=1}^{n} f(x_i^*)\,\Delta x$ depends on the choice of the sample points x_i^*, the

limit does not exist, and f is not integrable on $[0,1]$.

68. Partition the interval $[0,1]$ into n equal subintervals and choose $x_i^* = \dfrac{1}{n^2}$. Then with $f(x) = \dfrac{1}{x}$,

$\sum\limits_{i=1}^{n} f(x_i^*)\,\Delta x \ge f(x_1^*)\Delta x = \dfrac{1}{1/n^2}\cdot\dfrac{1}{n} = n$. Thus, $\lim\limits_{n\to\infty} \sum\limits_{i=1}^{n} f(x_i^*)\,\Delta x$ can be made arbitrarily large and hence, f is not

integrable on $[0,1]$.

69. $\lim\limits_{n\to\infty} \sum\limits_{i=1}^{n} \dfrac{i^4}{n^5} = \lim\limits_{n\to\infty} \sum\limits_{i=1}^{n} \dfrac{i^4}{n^4}\cdot\dfrac{1}{n} = \lim\limits_{n\to\infty} \sum\limits_{i=1}^{n} \left(\dfrac{i}{n}\right)^4\dfrac{1}{n}$. At this point, we need to recognize the limit as being of the form

$\lim\limits_{n\to\infty} \sum\limits_{i=1}^{n} f(x_i)\,\Delta x$, where $\Delta x = (1-0)/n = 1/n$, $x_i = 0 + i\,\Delta x = i/n$, and $f(x) = x^4$. Thus, the definite integral

is $\int_0^1 x^4\,dx$.

70. $\lim\limits_{n\to\infty} \dfrac{1}{n}\sum\limits_{i=1}^{n} \dfrac{1}{1+(i/n)^2} = \lim\limits_{n\to\infty} \sum\limits_{i=1}^{n} \dfrac{1}{1+(i/n)^2}\cdot\dfrac{1}{n} = \lim\limits_{n\to\infty} \sum\limits_{i=1}^{n} f(x_i)\,\Delta x$, where $\Delta x = (1-0)/n = 1/n$,

$x_i = 0 + i\,\Delta x = i/n$, and $f(x) = \dfrac{1}{1+x^2}$. Thus, the definite integral is $\displaystyle\int_0^1 \dfrac{dx}{1+x^2}$.

71. Choose $x_i = 1 + \dfrac{i}{n}$ and $x_i^* = \sqrt{x_{i-1}x_i} = \sqrt{\left(1+\dfrac{i-1}{n}\right)\left(1+\dfrac{i}{n}\right)}$. Then

$$\int_1^2 x^{-2}\,dx = \lim_{n\to\infty} \dfrac{1}{n}\sum_{i=1}^{n} \dfrac{1}{\left(1+\frac{i-1}{n}\right)\left(1+\frac{i}{n}\right)} = \lim_{n\to\infty} n\sum_{i=1}^{n} \dfrac{1}{(n+i-1)(n+i)}$$

$$= \lim_{n\to\infty} n\sum_{i=1}^{n}\left(\dfrac{1}{n+i-1} - \dfrac{1}{n+i}\right) \quad\text{[by the hint]} \quad = \lim_{n\to\infty} n\left(\sum_{i=0}^{n-1}\dfrac{1}{n+i} - \sum_{i=1}^{n}\dfrac{1}{n+i}\right)$$

$$= \lim_{n\to\infty} n\left(\left[\dfrac{1}{n} + \dfrac{1}{n+1} + \cdots + \dfrac{1}{2n-1}\right] - \left[\dfrac{1}{n+1} + \cdots + \dfrac{1}{2n-1} + \dfrac{1}{2n}\right]\right)$$

$$= \lim_{n\to\infty} n\left(\dfrac{1}{n} - \dfrac{1}{2n}\right) = \lim_{n\to\infty}\left(1 - \tfrac{1}{2}\right) = \tfrac{1}{2}$$

DISCOVERY PROJECT   Area Functions

1. (a)

$y = 2t + 1$

(b)

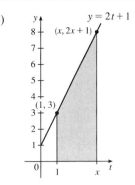

$y = 2t + 1$

$(x, 2x + 1)$

$(1, 3)$

Area of trapezoid $= \frac{1}{2}(b_1 + b_2)h = \frac{1}{2}(3 + 7)2$

$= 10$ square units

Or:

Area of rectangle+area of triangle

$= b_r h_r + \frac{1}{2}b_t h_t = (2)(3) + \frac{1}{2}(2)(4) = 10$ square units

(c) $A'(x) = 2x + 1$. This is the y-coordinate of the point $(x, 2x + 1)$ on the given line.

As in part (a),

$A(x) = \frac{1}{2}[3 + (2x + 1)](x - 1) = \frac{1}{2}(2x + 4)(x - 1)$

$= (x + 2)(x - 1) = x^2 + x - 2$ square units

2. (a)

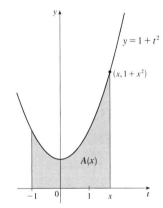

$y = 1 + t^2$

$(x, 1 + x^2)$

$A(x)$

(b) $A(x) = \int_{-1}^{x} (1 + t^2)\, dt = \int_{-1}^{x} 1\, dt + \int_{-1}^{x} t^2\, dt$ [Property 2]

$= 1[x - (-1)] + \dfrac{x^3 - (-1)^3}{3}$ $\begin{bmatrix} \text{Property 1 and} \\ \text{Exercise 5.2.28} \end{bmatrix}$

$= x + 1 + \frac{1}{3}x^3 + \frac{1}{3}$

$= \frac{1}{3}x^3 + x + \frac{4}{3}$

(c) $A'(x) = x^2 + 1$. This is the y-coordinate of the point $\left(x, 1 + x^2\right)$ on the given curve.

(d)

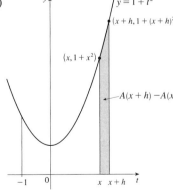

$y = 1 + t^2$

$(x + h, 1 + (x + h)^2)$

$(x, 1 + x^2)$

$A(x + h) - A(x)$

$A(x + h) - A(x)$ is the area under the curve $y = 1 + t^2$ from $t = x$ to $t = x + h$.

(e)

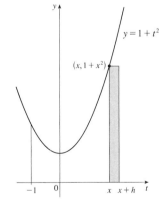

$y = 1 + t^2$

$(x, 1 + x^2)$

An approximating rectangle is shown in the figure. It has height $1 + x^2$, width h, and area $h(1 + x^2)$, so

$A(x + h) - A(x) \approx h(1 + x^2) \quad \Rightarrow \quad \dfrac{A(x + h) - A(x)}{h} \approx 1 + x^2.$

(f) Part (e) says that the average rate of change of A is approximately $1 + x^2$. As h approaches 0, the quotient approaches the instantaneous rate of change—namely, $A'(x)$. So the result of part (c), $A'(x) = x^2 + 1$, is geometrically plausible.

3. (a) $f(x) = \cos(x^2)$

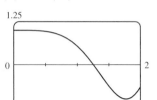

(b) $g(x)$ starts to decrease at that value of x where $\cos(t^2)$ changes from positive to negative; that is, at about $x = 1.25$.

(c) $g(x) = \int_0^x \cos(t^2)\, dt$. Using an integration command, we find that

$g(0) = 0$, $g(0.2) \approx 0.200$, $g(0.4) \approx 0.399$, $g(0.6) \approx 0.592$,

$g(0.8) \approx 0.768$, $g(1.0) \approx 0.905$, $g(1.2) \approx 0.974$, $g(1.4) \approx 0.950$,

$g(1.6) \approx 0.826$, $g(1.8) \approx 0.635$, and $g(2.0) \approx 0.461$.

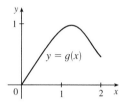

(d) We sketch the graph of g' using the method of Example 1 in Section 3.2.

The graphs of $g'(x)$ and $f(x)$ look alike, so we guess that $g'(x) = f(x)$.

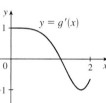

4. In Problems 1 and 2, we showed that if $g(x) = \int_a^x f(t)\, dt$, then $g'(x) = f(x)$, for the functions $f(t) = 2t + 1$ and $f(t) = 1 + t^2$. In Problem 3 we guessed that the same is true for $f(t) = \cos(t^2)$, based on visual evidence. So we conjecture that $g'(x) = f(x)$ for any continuous function f. This turns out to be true and is proved in Section 5.3 (the Fundamental Theorem of Calculus).

5.3 The Fundamental Theorem of Calculus

1. One process undoes what the other one does. The precise version of this statement is given by the Fundamental Theorem of Calculus. See the statement of this theorem and the paragraph that follows it on page 320.

2. (a) $g(x) = \int_0^x f(t)\, dt$, so $g(0) = \int_0^0 f(t)\, dt = 0$.

$g(1) = \int_0^1 f(t)\, dt = \frac{1}{2} \cdot 1 \cdot 1 \quad \text{[area of triangle]} \quad = \frac{1}{2}$.

$g(2) = \int_0^2 f(t)\, dt = \int_0^1 f(t)\, dt + \int_1^2 f(t)\, dt \quad \text{[below the x-axis]}$

$\quad = \frac{1}{2} - \frac{1}{2} \cdot 1 \cdot 1 = 0$.

$g(3) = g(2) + \int_2^3 f(t)\, dt = 0 - \frac{1}{2} \cdot 1 \cdot 1 = -\frac{1}{2}$.

$g(4) = g(3) + \int_3^4 f(t)\, dt = -\frac{1}{2} + \frac{1}{2} \cdot 1 \cdot 1 = 0$.

$g(5) = g(4) + \int_4^5 f(t)\, dt = 0 + 1.5 = 1.5$.

$g(6) = g(5) + \int_5^6 f(t)\, dt = 1.5 + 2.5 = 4$.

(b) $g(7) = g(6) + \int_6^7 f(t)\, dt \approx 4 + 2.2$ [estimate from the graph] $= 6.2$.

(c) The answers from part (a) and part (b) indicate that g has a minimum at

$x = 3$ and a maximum at $x = 7$. This makes sense from the graph of f

since we are subtracting area on $1 < x < 3$ and adding area on $3 < x < 7$.

(d)

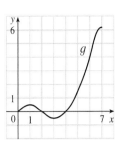

3. (a) $g(x) = \int_0^x f(t)\, dt$.

$g(0) = \int_0^0 f(t)\, dt = 0$

$g(1) = \int_0^1 f(t)\, dt = 1 \cdot 2 = 2$ [rectangle],

$g(2) = \int_0^2 f(t)\, dt = \int_0^1 f(t)\, dt + \int_1^2 f(t)\, dt = g(1) + \int_1^2 f(t)\, dt$

      $= 2 + 1 \cdot 2 + \frac{1}{2} \cdot 1 \cdot 2 = 5$ [rectangle plus triangle],

$g(3) = \int_0^3 f(t)\, dt = g(2) + \int_2^3 f(t)\, dt = 5 + \frac{1}{2} \cdot 1 \cdot 4 = 7$,

$g(6) = g(3) + \int_3^6 f(t)\, dt$ [the integral is negative since f lies under the x-axis]

      $= 7 + \left[- \left(\frac{1}{2} \cdot 2 \cdot 2 + 1 \cdot 2 \right) \right] = 7 - 4 = 3$

(b) g is increasing on $(0, 3)$ because as x increases from 0 to 3, we keep adding more area.

(c) g has a maximum value when we start subtracting area; that is, at $x = 3$.

(d)

4. (a) $g(-3) = \int_{-3}^{-3} f(t)\, dt = 0$, $g(3) = \int_{-3}^{3} f(t)\, dt = \int_{-3}^{0} f(t)\, dt + \int_0^3 f(t)\, dt = 0$ by symmetry, since the area above the

x-axis is the same as the area below the axis.

(b) From the graph, it appears that to the nearest $\frac{1}{2}$, $g(-2) = \int_{-3}^{-2} f(t)\, dt \approx 1$, $g(-1) = \int_{-3}^{-1} f(t)\, dt \approx 3\frac{1}{2}$,

and $g(0) = \int_{-3}^{0} f(t)\, dt \approx 5\frac{1}{2}$.

(c) g is increasing on $(-3, 0)$ because as x increases from -3 to 0, we keep adding more area.

(d) g has a maximum value when we start subtracting area; that is, at $x = 0$.

(e)

(f) The graph of $g'(x)$ is the same as that of $f(x)$, as indicated by FTC1.

5.

 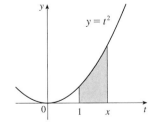

$y = t^2$

(a) By FTC1 with $f(t) = t^2$ and $a = 1$, $g(x) = \int_1^x t^2\, dt$ $\Rightarrow$

$g'(x) = f(x) = x^2$.

(b) Using FTC2, $g(x) = \int_1^x t^2\, dt = \left[\frac{1}{3} t^3 \right]_1^x = \frac{1}{3} x^3 - \frac{1}{3}$ $\Rightarrow$ $g'(x) = x^2$.

6.

 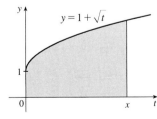

$y = 1 + \sqrt{t}$

(a) By FTC1 with $f(t) = 1 + \sqrt{t}$ and $a = 0$, $g(x) = \int_0^x \left(1 + \sqrt{t} \right) dt$ $\Rightarrow$

$g'(x) = f(x) = 1 + \sqrt{x}$.

(b) Using FTC2, $g(x) = \int_0^x \left(1 + \sqrt{t} \right) dt = \left[t + \frac{2}{3} t^{3/2} \right]_0^x = x + \frac{2}{3} x^{3/2}$ $\Rightarrow$

$g'(x) = 1 + x^{1/2} = 1 + \sqrt{x}$.

7. $f(t) = \dfrac{1}{t^3 + 1}$ and $g(x) = \displaystyle\int_1^x \dfrac{1}{t^3 + 1}\, dt$, so by FTC1, $g'(x) = f(x) = \dfrac{1}{x^3 + 1}$. Note that the lower limit, 1, could be any

real number greater than -1 and not affect this answer.

8. $f(t) = (2 + t^4)^5$ and $g(x) = \displaystyle\int_1^x (2 + t^4)^5\, dt$, so $g'(x) = f(x) = (2 + x^4)^5$.

9. $f(t) = t^2 \sin t$ and $g(y) = \displaystyle\int_2^y t^2 \sin t\, dt$, so by FTC1, $g'(y) = f(y) = y^2 \sin y$.

10. $f(x) = \sqrt{x^2 + 4}$ and $g(r) = \displaystyle\int_0^r \sqrt{x^2 + 4}\, dx$, so by FTC1, $g'(r) = f(r) = \sqrt{r^2 + 4}$.

11. $F(x) = \displaystyle\int_x^\pi \sqrt{1 + \sec t}\, dt = -\int_\pi^x \sqrt{1 + \sec t}\, dt \quad\Rightarrow\quad F'(x) = -\dfrac{d}{dx}\int_\pi^x \sqrt{1 + \sec t}\, dt = -\sqrt{1 + \sec x}$

12. $G(x) = \displaystyle\int_x^1 \cos\sqrt{t}\, dt = -\int_1^x \cos\sqrt{t}\, dt \quad\Rightarrow\quad G'(x) = -\dfrac{d}{dx}\int_1^x \cos\sqrt{t}\, dt = -\cos\sqrt{x}$

13. Let $u = \dfrac{1}{x}$. Then $\dfrac{du}{dx} = -\dfrac{1}{x^2}$. Also, $\dfrac{dh}{dx} = \dfrac{dh}{du}\dfrac{du}{dx}$, so

$$h'(x) = \dfrac{d}{dx}\int_2^{1/x} \sin^4 t\, dt = \dfrac{d}{du}\int_2^u \sin^4 t\, dt \cdot \dfrac{du}{dx} = \sin^4 u\,\dfrac{du}{dx} = \dfrac{-\sin^4(1/x)}{x^2}.$$

14. Let $u = x^2$. Then $\dfrac{du}{dx} = 2x$. Also, $\dfrac{dh}{dx} = \dfrac{dh}{du}\dfrac{du}{dx}$, so

$$h'(x) = \dfrac{d}{dx}\int_0^{x^2} \sqrt{1 + r^3}\, dr = \dfrac{d}{du}\int_0^u \sqrt{1 + r^3}\, dr \cdot \dfrac{du}{dx} = \sqrt{1 + u^3}\,(2x) = 2x\sqrt{1 + (x^2)^3} = 2x\sqrt{1 + x^6}.$$

15. Let $u = \tan x$. Then $\dfrac{du}{dx} = \sec^2 x$. Also, $\dfrac{dy}{dx} = \dfrac{dy}{du}\dfrac{du}{dx}$, so

$$y' = \dfrac{d}{dx}\int_0^{\tan x} \sqrt{t + \sqrt{t}}\, dt = \dfrac{d}{du}\int_0^u \sqrt{t + \sqrt{t}}\, dt \cdot \dfrac{du}{dx} = \sqrt{u + \sqrt{u}}\,\dfrac{du}{dx} = \sqrt{\tan x + \sqrt{\tan x}}\,\sec^2 x.$$

16. Let $u = \cos x$. Then $\dfrac{du}{dx} = -\sin x$. Also, $\dfrac{dy}{dx} = \dfrac{dy}{du}\dfrac{du}{dx}$, so

$$y' = \dfrac{d}{dx}\int_1^{\cos x} (1 + v^2)^{10}\, dv = \dfrac{d}{du}\int_1^u (1 + v^2)^{10}\, dv \cdot \dfrac{du}{dx} = (1 + u^2)^{10}\,\dfrac{du}{dx} = -(1 + \cos^2 x)^{10}\sin x.$$

17. Let $w = 1 - 3x$. Then $\dfrac{dw}{dx} = -3$. Also, $\dfrac{dy}{dx} = \dfrac{dy}{dw}\dfrac{dw}{dx}$, so

$$y' = \dfrac{d}{dx}\int_{1-3x}^1 \dfrac{u^3}{1 + u^2}\, du = \dfrac{d}{dw}\int_w^1 \dfrac{u^3}{1 + u^2}\, du \cdot \dfrac{dw}{dx} = -\dfrac{d}{dw}\int_1^w \dfrac{u^3}{1 + u^2}\, du \cdot \dfrac{dw}{dx} = -\dfrac{w^3}{1 + w^2}(-3) = \dfrac{3(1 - 3x)^3}{1 + (1 - 3x)^2}$$

18. Let $u = \dfrac{1}{x^2}$. Then $\dfrac{du}{dx} = -\dfrac{2}{x^3}$. Also, $\dfrac{dy}{dx} = \dfrac{dy}{du}\dfrac{du}{dx}$, so

$$y' = \dfrac{d}{dx}\int_{1/x^2}^0 \sin^3 t\, dt = \dfrac{d}{du}\int_u^0 \sin^3 t\, dt \cdot \dfrac{du}{dx} = -\dfrac{d}{du}\int_0^u \sin^3 t\, dt \cdot \dfrac{du}{dx} = -\sin^3 u\left(-\dfrac{2}{x^3}\right) = \dfrac{2\sin^3(1/x^2)}{x^3}.$$

19. $\displaystyle\int_{-1}^2 (x^3 - 2x)\, dx = \left[\dfrac{x^4}{4} - x^2\right]_{-1}^2 = \left(\dfrac{2^4}{4} - 2^2\right) - \left(\dfrac{(-1)^4}{4} - (-1)^2\right) = (4 - 4) - \left(\tfrac{1}{4} - 1\right) = 0 - \left(-\tfrac{3}{4}\right) = \tfrac{3}{4}$

20. $\displaystyle\int_{-2}^5 6\, dx = \left[6x\right]_{-2}^5 = 6[5 - (-2)] = 6(7) = 42$

21. $\int_1^4 (5 - 2t + 3t^2)\, dt = \left[5t - t^2 + t^3\right]_1^4 = (20 - 16 + 64) - (5 - 1 + 1) = 68 - 5 = 63$

22. $\int_0^1 \left(1 + \frac{1}{2}u^4 - \frac{2}{5}u^9\right) du = \left[u + \frac{1}{10}u^5 - \frac{1}{25}u^{10}\right]_0^1 = \left(1 + \frac{1}{10} - \frac{1}{25}\right) - 0 = \frac{53}{50}$

23. $\int_0^1 x^{4/5}\, dx = \left[\frac{5}{9}x^{9/5}\right]_0^1 = \frac{5}{9} - 0 = \frac{5}{9}$

24. $\int_1^8 \sqrt[3]{x}\, dx = \int_1^8 x^{1/3}\, dx = \left[\frac{3}{4}x^{4/3}\right]_1^8 = \frac{3}{4}(8^{4/3} - 1^{4/3}) = \frac{3}{4}(2^4 - 1) = \frac{3}{4}(16 - 1) = \frac{3}{4}(15) = \frac{45}{4}$

25. $\int_1^2 \frac{3}{t^4}\, dt = 3\int_1^2 t^{-4}\, dt = 3\left[\frac{t^{-3}}{-3}\right]_1^2 = \frac{3}{-3}\left[\frac{1}{t^3}\right]_1^2 = -1\left(\frac{1}{8} - 1\right) = \frac{7}{8}$

26. $\int_\pi^{2\pi} \cos\theta\, d\theta = \left[\sin\theta\right]_\pi^{2\pi} = \sin 2\pi - \sin \pi = 0 - 0 = 0$

27. $\int_0^2 x(2 + x^5)\, dx = \int_0^2 (2x + x^6)\, dx = \left[x^2 + \frac{1}{7}x^7\right]_0^2 = \left(4 + \frac{128}{7}\right) - (0 + 0) = \frac{156}{7}$

28. $\int_0^1 (3 + x\sqrt{x})\, dx = \int_0^1 (3 + x^{3/2})\, dx = \left[3x + \frac{2}{5}x^{5/2}\right]_0^1 = \left[\left(3 + \frac{2}{5}\right) - 0\right] = \frac{17}{5}$

29. $\int_1^9 \frac{x-1}{\sqrt{x}}\, dx = \int_1^9 \left(\frac{x}{\sqrt{x}} - \frac{1}{\sqrt{x}}\right) dx = \int_1^9 (x^{1/2} - x^{-1/2})\, dx = \left[\frac{2}{3}x^{3/2} - 2x^{1/2}\right]_1^9$

$$= \left(\frac{2}{3}\cdot 27 - 2\cdot 3\right) - \left(\frac{2}{3} - 2\right) = 12 - \left(-\frac{4}{3}\right) = \frac{40}{3}$$

30. $\int_0^2 (y-1)(2y+1)\, dy = \int_0^2 (2y^2 - y - 1)\, dy = \left[\frac{2}{3}y^3 - \frac{1}{2}y^2 - y\right]_0^2 = \left(\frac{16}{3} - 2 - 2\right) - 0 = \frac{4}{3}$

31. $\int_0^{\pi/4} \sec^2 t\, dt = \left[\tan t\right]_0^{\pi/4} = \tan\frac{\pi}{4} - \tan 0 = 1 - 0 = 1$

32. $\int_0^{\pi/4} \sec\theta \tan\theta\, d\theta = \left[\sec\theta\right]_0^{\pi/4} = \sec\frac{\pi}{4} - \sec 0 = \sqrt{2} - 1$

33. $\int_1^2 (1 + 2y)^2\, dy = \int_1^2 (1 + 4y + 4y^2)\, dy = \left[y + 2y^2 + \frac{4}{3}y^3\right]_1^2 = \left(2 + 8 + \frac{32}{3}\right) - \left(1 + 2 + \frac{4}{3}\right) = \frac{62}{3} - \frac{13}{3} = \frac{49}{3}$

34. $\int_1^2 \frac{s^4 + 1}{s^2}\, ds = \int_1^2 (s^2 + s^{-2})\, ds = \left[\frac{1}{3}s^3 - \frac{1}{s}\right]_1^2 = \left(\frac{8}{3} - \frac{1}{2}\right) - \left(\frac{1}{3} - 1\right) = \frac{7}{3} + \frac{1}{2} = \frac{17}{6}$

35. If $f(x) = \begin{cases} \sin x & \text{if } 0 \le x < \pi/2 \\ \cos x & \text{if } \pi/2 \le x \le \pi \end{cases}$ then

$\int_0^\pi f(x)\, dx = \int_0^{\pi/2} \sin x\, dx + \int_{\pi/2}^\pi \cos x\, dx = \left[-\cos x\right]_0^{\pi/2} + \left[\sin x\right]_{\pi/2}^\pi = -\cos\frac{\pi}{2} + \cos 0 + \sin\pi - \sin\frac{\pi}{2}$

$$= -0 + 1 + 0 - 1 = 0$$

Note that f is integrable by Theorem 3 in Section 5.2.

36. If $f(x) = \begin{cases} 2 & \text{if } -2 \le x \le 0 \\ 4 - x^2 & \text{if } 0 < x \le 2 \end{cases}$ then

$\int_{-2}^2 f(x)\, dx = \int_{-2}^0 2\, dx + \int_0^2 (4 - x^2)\, dx = \left[2x\right]_{-2}^0 + \left[4x - \frac{1}{3}x^3\right]_0^2 = [0 - (-4)] + \left(\frac{16}{3} - 0\right) = \frac{28}{3}$

Note that f is integrable by Theorem 3 in Section 5.2.

37. $f(x) = x^{-4}$ is not continuous on the interval $[-2, 1]$, so FTC2 cannot be applied. In fact, f has an infinite discontinuity at $x = 0$, so $\int_{-2}^1 x^{-4}\, dx$ does not exist.

38. $f(x) = \dfrac{4}{x^3}$ is not continuous on the interval $[-1, 2]$, so FTC2 cannot be applied. In fact, f has an infinite discontinuity at

$x = 0$, so $\displaystyle\int_{-1}^{2} \dfrac{4}{x^3}\, dx$ does not exist.

39. $f(\theta) = \sec\theta\,\tan\theta$ is not continuous on the interval $[\pi/3, \pi]$, so FTC2 cannot be applied. In fact, f has an infinite

discontinuity at $x = \pi/2$, so $\int_{\pi/3}^{\pi} \sec\theta\,\tan\theta\,d\theta$ does not exist.

40. $f(x) = \sec^2 x$ is not continuous on the interval $[0, \pi]$, so FTC2 cannot be applied. In fact, f has an infinite discontinuity at

$x = \pi/2$, so $\int_0^{\pi} \sec^2 x\, dx$ does not exist.

41. From the graph, it appears that the area is about 60. The actual area is

$\int_0^{27} x^{1/3}\,dx = \left[\frac{3}{4}x^{4/3}\right]_0^{27} = \frac{3}{4}\cdot 81 - 0 = \frac{243}{4} = 60.75$. This is $\frac{3}{4}$ of the

area of the viewing rectangle.

42. From the graph, it appears that the area is about $\frac{1}{3}$. The actual area is

$\displaystyle\int_1^6 x^{-4}\,dx = \left[\dfrac{x^{-3}}{-3}\right]_1^6 = \left[\dfrac{-1}{3x^3}\right]_1^6 = -\dfrac{1}{3\cdot 216} + \dfrac{1}{3} = \dfrac{215}{648} \approx 0.3318.$

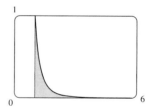

43. It appears that the area under the graph is about $\frac{2}{3}$ of the area of the viewing

rectangle, or about $\frac{2}{3}\pi \approx 2.1$. The actual area is

$\int_0^{\pi} \sin x\, dx = [-\cos x]_0^{\pi} = (-\cos\pi) - (-\cos 0) = -(-1) + 1 = 2.$

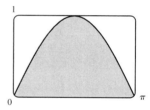

44. Splitting up the region as shown, we estimate that the area under the graph

is $\frac{\pi}{3} + \frac{1}{4}\left(3\cdot\frac{\pi}{3}\right) \approx 1.8$. The actual area is

$\int_0^{\pi/3} \sec^2 x\, dx = [\tan x]_0^{\pi/3} = \sqrt{3} - 0 = \sqrt{3} \approx 1.73.$

45. $\int_{-1}^{2} x^3\, dx = \left[\frac{1}{4}x^4\right]_{-1}^{2} = 4 - \frac{1}{4} = \frac{15}{4} = 3.75$

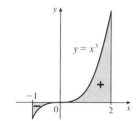

46. $\int_{\pi/4}^{5\pi/2} \sin x\, dx = [-\cos x]_{\pi/4}^{5\pi/2} = 0 + \frac{\sqrt{2}}{2} = \frac{\sqrt{2}}{2}$

47. $g(x) = \displaystyle\int_{2x}^{3x} \frac{u^2 - 1}{u^2 + 1}\, du = \int_{2x}^{0} \frac{u^2 - 1}{u^2 + 1}\, du + \int_{0}^{3x} \frac{u^2 - 1}{u^2 + 1}\, du = -\int_{0}^{2x} \frac{u^2 - 1}{u^2 + 1}\, du + \int_{0}^{3x} \frac{u^2 - 1}{u^2 + 1}\, du \quad \Rightarrow$

$g'(x) = -\dfrac{(2x)^2 - 1}{(2x)^2 + 1} \cdot \dfrac{d}{dx}(2x) + \dfrac{(3x)^2 - 1}{(3x)^2 + 1} \cdot \dfrac{d}{dx}(3x) = -2 \cdot \dfrac{4x^2 - 1}{4x^2 + 1} + 3 \cdot \dfrac{9x^2 - 1}{9x^2 + 1}$

48. $g(x) = \displaystyle\int_{\tan x}^{x^2} \frac{1}{\sqrt{2 + t^4}}\, dt = \int_{\tan x}^{1} \frac{dt}{\sqrt{2 + t^4}} + \int_{1}^{x^2} \frac{dt}{\sqrt{2 + t^4}} = -\int_{1}^{\tan x} \frac{dt}{\sqrt{2 + t^4}} + \int_{1}^{x^2} \frac{dt}{\sqrt{2 + t^4}} \quad \Rightarrow$

$g'(x) = \dfrac{-1}{\sqrt{2 + \tan^4 x}} \dfrac{d}{dx}(\tan x) + \dfrac{1}{\sqrt{2 + x^8}} \dfrac{d}{dx}(x^2) = -\dfrac{\sec^2 x}{\sqrt{2 + \tan^4 x}} + \dfrac{2x}{\sqrt{2 + x^8}}$

49. $y = \displaystyle\int_{\sqrt{x}}^{x^3} \sqrt{t}\, \sin t\, dt = \int_{\sqrt{x}}^{1} \sqrt{t}\, \sin t\, dt + \int_{1}^{x^3} \sqrt{t}\, \sin t\, dt = -\int_{1}^{\sqrt{x}} \sqrt{t}\, \sin t\, dt + \int_{1}^{x^3} \sqrt{t}\, \sin t\, dt \quad \Rightarrow$

$y' = -\sqrt[4]{x}\,(\sin\sqrt{x}) \cdot \dfrac{d}{dx}(\sqrt{x}) + x^{3/2}\sin(x^3) \cdot \dfrac{d}{dx}(x^3) = -\dfrac{\sqrt[4]{x}\,\sin\sqrt{x}}{2\sqrt{x}} + x^{3/2}\sin(x^3)(3x^2)$

$\qquad = 3x^{7/2}\sin(x^3) - \dfrac{\sin\sqrt{x}}{2\sqrt[4]{x}}$

50. $y = \displaystyle\int_{\cos x}^{5x} \cos(u^2)\, du = \int_{0}^{5x} \cos(u^2)\, du - \int_{0}^{\cos x} \cos(u^2)\, du \quad \Rightarrow$

$y' = \cos(25x^2) \cdot \dfrac{d}{dx}(5x) - \cos(\cos^2 x) \cdot \dfrac{d}{dx}(\cos x) = \cos(25x^2) \cdot 5 - \cos(\cos^2 x) \cdot (-\sin x)$

$\qquad = 5\cos(25x^2) + \sin x \cos(\cos^2 x)$

51. $F(x) = \displaystyle\int_{1}^{x} f(t)\, dt \quad \Rightarrow \quad F'(x) = f(x) = \int_{1}^{x^2} \frac{\sqrt{1 + u^4}}{u}\, du \quad \left[\text{since } f(t) = \int_{1}^{t^2} \frac{\sqrt{1 + u^4}}{u}\, du\right] \quad \Rightarrow$

$F''(x) = f'(x) = \dfrac{\sqrt{1 + (x^2)^4}}{x^2} \cdot \dfrac{d}{dx}(x^2) = \dfrac{\sqrt{1 + x^8}}{x^2} \cdot 2x = \dfrac{2\sqrt{1 + x^8}}{x}.$ So $F''(2) = \sqrt{1 + 2^8} = \sqrt{257}.$

52. For the curve to be concave upward, we must have $y'' > 0$.

$y = \displaystyle\int_{0}^{x} \frac{1}{1 + t + t^2}\, dt \quad \Rightarrow \quad y' = \dfrac{1}{1 + x + x^2} \quad \Rightarrow \quad y'' = \dfrac{-(1 + 2x)}{(1 + x + x^2)^2}.$ For this expression to be positive, we must have

$(1 + 2x) < 0$, since $(1 + x + x^2)^2 > 0$ for all x. $(1 + 2x) < 0 \quad \Leftrightarrow \quad x < -\frac{1}{2}$. Thus, the curve is concave upward

on $\left(-\infty, -\frac{1}{2}\right)$.

53. (a) The Fresnel function $S(x) = \int_{0}^{x} \sin\left(\frac{\pi}{2}t^2\right) dt$ has local maximum values where $0 = S'(x) = \sin\left(\frac{\pi}{2}t^2\right)$ and

S' changes from positive to negative. For $x > 0$, this happens when $\frac{\pi}{2}x^2 = (2n - 1)\pi$ [odd multiples of π] $\quad \Leftrightarrow$

$x^2 = 2(2n - 1) \quad \Leftrightarrow \quad x = \sqrt{4n - 2}$, n any positive integer. For $x < 0$, S' changes from positive to negative where

$\frac{\pi}{2}x^2 = 2n\pi$ [even multiples of π] $\quad \Leftrightarrow \quad x^2 = 4n \quad \Leftrightarrow \quad x = -2\sqrt{n}$. S' does not change sign at $x = 0$.

(b) S is concave upward on those intervals where $S''(x) > 0$. Differentiating our expression for $S'(x)$, we get

$S''(x) = \cos\left(\frac{\pi}{2}x^2\right)\left(2\frac{\pi}{2}x\right) = \pi x \cos\left(\frac{\pi}{2}x^2\right)$. For $x > 0$, $S''(x) > 0$ where $\cos\left(\frac{\pi}{2}x^2\right) > 0 \quad \Leftrightarrow \quad 0 < \frac{\pi}{2}x^2 < \frac{\pi}{2}$ or

$\left(2n - \frac{1}{2}\right)\pi < \frac{\pi}{2}x^2 < \left(2n + \frac{1}{2}\right)\pi$, n any integer $\quad \Leftrightarrow \quad 0 < x < 1$ or $\sqrt{4n - 1} < x < \sqrt{4n + 1}$, n any positive integer.

For $x < 0$, $S''(x) > 0$ where $\cos\left(\frac{\pi}{2}x^2\right) < 0 \quad \Leftrightarrow \quad \left(2n - \frac{3}{2}\right)\pi < \frac{\pi}{2}x^2 < \left(2n - \frac{1}{2}\right)\pi$, n any integer $\quad \Leftrightarrow$

$4n - 3 < x^2 < 4n - 1 \quad \Leftrightarrow \quad \sqrt{4n - 3} < |x| < \sqrt{4n - 1} \quad \Rightarrow \quad \sqrt{4n - 3} < -x < \sqrt{4n - 1} \quad \Rightarrow$

$-\sqrt{4n - 3} > x > -\sqrt{4n - 1}$, so the intervals of upward concavity for $x < 0$ are $\left(-\sqrt{4n - 1}, -\sqrt{4n - 3}\right)$, n any positive integer. To summarize: S is concave upward on the intervals $(0, 1)$, $(-\sqrt{3}, -1)$, $(\sqrt{3}, \sqrt{5})$, $(-\sqrt{7}, -\sqrt{5})$, $(\sqrt{7}, 3), \ldots$.

(c) In Maple, we use `plot({int(sin(Pi*t^2/2),t=0..x),0.2},x=0..2);`. Note that Maple recognizes the Fresnel function, calling it `FresnelS(x)`. In Mathematica, we use `Plot[{Integrate[Sin[Pi*t^2/2],{t,0,x}],0.2},{x,0,2}]`. In Derive, we load the utility file FRESNEL and plot `FRESNEL_SIN(x)`. From the graphs, we see that $\int_0^x \sin\left(\frac{\pi}{2}t^2\right) dt = 0.2$ at $x \approx 0.74$.

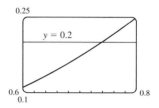

54. (a) In Maple, we should start by setting `si:=int(sin(t)/t,t=0..x);`. In Mathematica, the command is `si=Integrate[Sin[t]/t,{t,0,x}]`. Note that both systems recognize this function; Maple calls it `Si(x)` and Mathematica calls it `SinIntegral[x]`. In Maple, the command to generate the graph is `plot(si,x=-4*Pi..4*Pi);`. In Mathematica, it is `Plot[si,{x,-4*Pi,4*Pi}]`. In Derive, we load the utility file EXP_INT and plot `SI(x)`.

(b) $\text{Si}(x)$ has local maximum values where $\text{Si}'(x)$ changes from positive to negative, passing through 0. From the Fundamental Theorem we know that $\text{Si}'(x) = \dfrac{d}{dx} \displaystyle\int_0^x \dfrac{\sin t}{t}\, dt = \dfrac{\sin x}{x}$, so we must have $\sin x = 0$ for a maximum, and for $x > 0$ we must have $x = (2n - 1)\pi$, n any positive integer, for Si' to be changing from positive to negative at x. For $x < 0$, we must have $x = 2n\pi$, n any positive integer, for a maximum, since the denominator of $\text{Si}'(x)$ is negative for $x < 0$. Thus, the local maxima occur at $x = \pi, -2\pi, 3\pi, -4\pi, 5\pi, -6\pi, \ldots$.

(c) To find the first inflection point, we solve $\text{Si}''(x) = \dfrac{\cos x}{x} - \dfrac{\sin x}{x^2} = 0$. We can see from the graph that the first inflection point lies somewhere between $x = 3$ and $x = 5$. Using a rootfinder gives the value $x \approx 4.4934$. To find the y-coordinate of the inflection point, we evaluate $\text{Si}(4.4934) \approx 1.6556$. So the coordinates of the first inflection point to the right of the origin are about $(4.4934, 1.6556)$. Alternatively, we could graph $S''(x)$ and estimate the first positive x-value at which it changes sign.

(d) It seems from the graph that the function has horizontal asymptotes at $y \approx 1.5$, with $\displaystyle\lim_{x \to \pm\infty} \text{Si}(x) \approx \pm 1.5$ respectively. Using the limit command, we get $\displaystyle\lim_{x \to \infty} \text{Si}(x) = \frac{\pi}{2}$. Since $\text{Si}(x)$ is an odd function, $\displaystyle\lim_{x \to -\infty} \text{Si}(x) = -\frac{\pi}{2}$. So $\text{Si}(x)$ has the horizontal asymptotes $y = \pm\frac{\pi}{2}$.

(e) We use the `fsolve` command in Maple (or `FindRoot` in Mathematica) to find that the solution is $x \approx 1.1$. Or, as in Exercise 53(c), we graph $y = \text{Si}(x)$ and $y = 1$ on the same screen to see where they intersect.

55. (a) By FTC1, $g'(x) = f(x)$. So $g'(x) = f(x) = 0$ at $x = 1, 3, 5, 7$, and 9. g has local maxima at $x = 1$ and 5 (since $f = g'$ changes from positive to negative there) and local minima at $x = 3$ and 7. There is no local maximum or minimum at $x = 9$, since f is not defined for $x > 9$.

(b) We can see from the graph that $\left| \int_0^1 f \, dt \right| < \left| \int_1^3 f \, dt \right| < \left| \int_3^5 f \, dt \right| < \left| \int_5^7 f \, dt \right| < \left| \int_7^9 f \, dt \right|$. So $g(1) = \left| \int_0^1 f \, dt \right|$,

$g(5) = \int_0^5 f \, dt = g(1) - \left| \int_1^3 f \, dt \right| + \left| \int_3^5 f \, dt \right|$, and $g(9) = \int_0^9 f \, dt = g(5) - \left| \int_5^7 f \, dt \right| + \left| \int_7^9 f \, dt \right|$. Thus,

$g(1) < g(5) < g(9)$, and so the absolute maximum of $g(x)$ occurs at $x = 9$.

(c) g is concave downward on those intervals where $g'' < 0$. But $g'(x) = f(x)$, so $g''(x) = f'(x)$, which is negative on (approximately) $\left(\frac{1}{2}, 2 \right)$, $(4, 6)$ and $(8, 9)$. So g is concave downward on these intervals.

(d)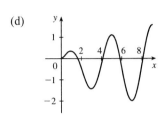

56. (a) By FTC1, $g'(x) = f(x)$. So $g'(x) = f(x) = 0$ at $x = 2, 4, 6, 8$, and 10. g has local maxima at $x = 2$ and 6 (since $f = g'$ changes from positive to negative there) and local minima at $x = 4$ and 8. There is no local maximum or minimum at $x = 10$, since f is not defined for $x > 10$.

(b) We can see from the graph that $\left| \int_0^2 f \, dt \right| > \left| \int_2^4 f \, dt \right| > \left| \int_4^6 f \, dt \right| > \left| \int_6^8 f \, dt \right| > \left| \int_8^{10} f \, dt \right|$. So $g(2) = \left| \int_0^2 f \, dt \right|$,

$g(6) = \int_0^6 f \, dt = g(2) - \left| \int_2^4 f \, dt \right| + \left| \int_4^6 f \, dt \right|$, and $g(10) = \int_0^{10} f \, dt = g(6) - \left| \int_6^8 f \, dt \right| + \left| \int_8^{10} f \, dt \right|$. Thus,

$g(2) > g(6) > g(10)$, and so the absolute maximum of $g(x)$ occurs at $x = 2$.

(c) g is concave downward on those intervals where $g'' < 0$. But $g'(x) = f(x)$, so $g''(x) = f'(x)$, which is negative on $(1, 3)$, $(5, 7)$ and $(9, 10)$. So g is concave downward on these intervals.

(d)

57. $\displaystyle \lim_{n \to \infty} \sum_{i=1}^{n} \frac{i^3}{n^4} = \lim_{n \to \infty} \frac{1 - 0}{n} \sum_{i=1}^{n} \left(\frac{i}{n} \right)^3 = \int_0^1 x^3 \, dx = \left[\frac{x^4}{4} \right]_0^1 = \frac{1}{4}$

58. $\displaystyle \lim_{n \to \infty} \frac{1}{n} \left(\sqrt{\frac{1}{n}} + \sqrt{\frac{2}{n}} + \cdots + \sqrt{\frac{n}{n}} \right) = \lim_{n \to \infty} \frac{1 - 0}{n} \sum_{i=1}^{n} \sqrt{\frac{i}{n}} = \int_0^1 \sqrt{x} \, dx = \left[\frac{2x^{3/2}}{3} \right]_0^1 = \frac{2}{3} - 0 = \frac{2}{3}$

59. Suppose $h < 0$. Since f is continuous on $[x + h, x]$, the Extreme Value Theorem says that there are numbers u and v in $[x + h, x]$ such that $f(u) = m$ and $f(v) = M$, where m and M are the absolute minimum and maximum values of f on $[x + h, x]$. By Property 8 of integrals, $m(-h) \le \int_{x+h}^{x} f(t) \, dt \le M(-h)$; that is, $f(u)(-h) \le -\int_{x}^{x+h} f(t) \, dt \le f(v)(-h)$.

Since $-h > 0$, we can divide this inequality by $-h$: $f(u) \le \dfrac{1}{h} \displaystyle\int_{x}^{x+h} f(t) \, dt \le f(v)$. By Equation 2,

$\dfrac{g(x + h) - g(x)}{h} = \dfrac{1}{h} \displaystyle\int_{x}^{x+h} f(t) \, dt$ for $h \ne 0$, and hence $f(u) \le \dfrac{g(x + h) - g(x)}{h} \le f(v)$, which is Equation 3 in the case where $h < 0$.

60. $\dfrac{d}{dx} \displaystyle\int_{g(x)}^{h(x)} f(t)\, dt = \dfrac{d}{dx}\left[\displaystyle\int_{g(x)}^{a} f(t)\, dt + \displaystyle\int_{a}^{h(x)} f(t)\, dt\right]$ [where a is in the domain of f]

$$= \dfrac{d}{dx}\left[-\int_{a}^{g(x)} f(t)\, dt\right] + \dfrac{d}{dx}\left[\int_{a}^{h(x)} f(t)\, dt\right] = -f(g(x))\, g'(x) + f(h(x))\, h'(x)$$

$$= f(h(x))\, h'(x) - f(g(x))\, g'(x)$$

61. (a) Let $f(x) = \sqrt{x}$ $\Rightarrow$ $f'(x) = 1/(2\sqrt{x}) > 0$ for $x > 0$ $\Rightarrow$ f is increasing on $(0, \infty)$. If $x \geq 0$, then $x^3 \geq 0$, so

$1 + x^3 \geq 1$ and since f is increasing, this means that $f(1 + x^3) \geq f(1)$ $\Rightarrow$ $\sqrt{1 + x^3} \geq 1$ for $x \geq 0$. Next let

$g(t) = t^2 - t$ $\Rightarrow$ $g'(t) = 2t - 1$ $\Rightarrow$ $g'(t) > 0$ when $t \geq 1$. Thus, g is increasing on $(1, \infty)$. And since $g(1) = 0$,

$g(t) \geq 0$ when $t \geq 1$. Now let $t = \sqrt{1 + x^3}$, where $x \geq 0$. $\sqrt{1 + x^3} \geq 1$ (from above) $\Rightarrow$ $t \geq 1$ $\Rightarrow$ $g(t) \geq 0$ $\Rightarrow$

$(1 + x^3) - \sqrt{1 + x^3} \geq 0$ for $x \geq 0$. Therefore, $1 \leq \sqrt{1 + x^3} \leq 1 + x^3$ for $x \geq 0$.

(b) From part (a) and Property 7: $\int_0^1 1\, dx \leq \int_0^1 \sqrt{1 + x^3}\, dx \leq \int_0^1 (1 + x^3)\, dx$ $\Leftrightarrow$

$\big[x\big]_0^1 \leq \int_0^1 \sqrt{1 + x^3}\, dx \leq \big[x + \tfrac{1}{4}x^4\big]_0^1$ $\Leftrightarrow$ $1 \leq \int_0^1 \sqrt{1 + x^3}\, dx \leq 1 + \tfrac{1}{4} = 1.25$.

62. (a) For $0 \leq x \leq 1$, we have $x^2 \leq x$. Since $f(x) = \cos x$ is a decreasing function on $[0, 1]$, $\cos(x^2) \geq \cos x$.

(b) $\pi/6 < 1$, so by part (a), $\cos(x^2) \geq \cos x$ on $[0, \pi/6]$. Thus,

$$\int_0^{\pi/6} \cos(x^2)\, dx \geq \int_0^{\pi/6} \cos x\, dx = \big[\sin x\big]_0^{\pi/6} = \sin(\pi/6) - \sin 0 = \tfrac{1}{2} - 0 = \tfrac{1}{2}.$$

63. $0 < \dfrac{x^2}{x^4 + x^2 + 1} < \dfrac{x^2}{x^4} = \dfrac{1}{x^2}$ on $[5, 10]$, so

$$0 \leq \int_5^{10} \dfrac{x^2}{x^4 + x^2 + 1}\, dx < \int_5^{10} \dfrac{1}{x^2}\, dx = \left[-\dfrac{1}{x}\right]_5^{10} = -\dfrac{1}{10} - \left(-\dfrac{1}{5}\right) = \dfrac{1}{10} = 0.1.$$

64. (a) If $x < 0$, then $g(x) = \int_0^x f(t)\, dt = \int_0^x 0\, dt = 0$.

(b)

If $0 \leq x \leq 1$, then $g(x) = \int_0^x f(t)\, dt = \int_0^x t\, dt = \big[\tfrac{1}{2}t^2\big]_0^x = \tfrac{1}{2}x^2$.

If $1 < x \leq 2$, then

$g(x) = \int_0^x f(t)\, dt = \int_0^1 f(t)\, dt + \int_1^x f(t)\, dt = g(1) + \int_1^x (2 - t)\, dt$

$= \tfrac{1}{2}(1)^2 + \big[2t - \tfrac{1}{2}t^2\big]_1^x = \tfrac{1}{2} + (2x - \tfrac{1}{2}x^2) - (2 - \tfrac{1}{2}) = 2x - \tfrac{1}{2}x^2 - 1$.

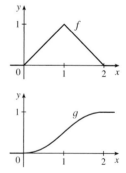

If $x > 2$, then $g(x) = \int_0^x f(t)\, dt = g(2) + \int_2^x 0\, dt = 1 + 0 = 1$. So

$$g(x) = \begin{cases} 0 & \text{if } x < 0 \\ \tfrac{1}{2}x^2 & \text{if } 0 \leq x \leq 1 \\ 2x - \tfrac{1}{2}x^2 - 1 & \text{if } 1 < x \leq 2 \\ 1 & \text{if } x > 2 \end{cases}$$

(c) f is not differentiable at its corners at $x = 0$, 1, and 2. f is differentiable on $(-\infty, 0)$, $(0, 1)$, $(1, 2)$ and $(2, \infty)$.

g is differentiable on $(-\infty, \infty)$.

65. Using FTC1, we differentiate both sides of $6 + \displaystyle\int_a^x \frac{f(t)}{t^2}\,dt = 2\sqrt{x}$ to get $\dfrac{f(x)}{x^2} = 2\,\dfrac{1}{2\sqrt{x}} \;\Rightarrow\; f(x) = x^{3/2}$.

To find a, we substitute $x = a$ in the original equation to obtain $6 + \displaystyle\int_a^a \frac{f(t)}{t^2}\,dt = 2\sqrt{a} \;\Rightarrow\; 6 + 0 = 2\sqrt{a} \;\Rightarrow$

$3 = \sqrt{a} \;\Rightarrow\; a = 9$.

66. The second derivative is the derivative of the first derivative, so we'll apply the Net Change Theorem with $F = h'$.

$\int_1^2 h''(u)\,du = \int_1^2 (h')'(u)\,du = h'(2) - h'(1) = 5 - 2 = 3$. The other information is unnecessary.

67. (a) Let $F(t) = \int_0^t f(s)\,ds$. Then, by FTC1, $F'(t) = f(t) = $ rate of depreciation, so $F(t)$ represents the loss in value over the

interval $[0, t]$.

(b) $C(t) = \dfrac{1}{t}\left[A + \displaystyle\int_0^t f(s)\,ds\right] = \dfrac{A + F(t)}{t}$ represents the average expenditure per unit of t during the interval $[0, t]$,

assuming that there has been only one overhaul during that time period. The company wants to minimize average

expenditure.

(c) $C(t) = \dfrac{1}{t}\left[A + \displaystyle\int_0^t f(s)\,ds\right]$. Using FTC1, we have $C'(t) = -\dfrac{1}{t^2}\left[A + \displaystyle\int_0^t f(s)\,ds\right] + \dfrac{1}{t}f(t)$.

$C'(t) = 0 \;\Rightarrow\; t\,f(t) = A + \displaystyle\int_0^t f(s)\,ds \;\Rightarrow\; f(t) = \dfrac{1}{t}\left[A + \displaystyle\int_0^t f(s)\,ds\right] = C(t)$.

68. (a) $C(t) = \dfrac{1}{t}\displaystyle\int_0^t [f(s) + g(s)]\,ds$. Using FTC1 and the Product Rule, we have

$C'(t) = \dfrac{1}{t}[f(t) + g(t)] - \dfrac{1}{t^2}\displaystyle\int_0^t [f(s) + g(s)]\,ds$. Set $C'(t) = 0$: $\dfrac{1}{t}[f(t) + g(t)] - \dfrac{1}{t^2}\displaystyle\int_0^t [f(s) + g(s)]\,ds = 0 \;\Rightarrow$

$[f(t) + g(t)] - \dfrac{1}{t}\displaystyle\int_0^t [f(s) + g(s)]\,ds = 0 \;\Rightarrow\; [f(t) + g(t)] - C(t) = 0 \;\Rightarrow\; C(t) = f(t) + g(t)$.

(b) For $0 \le t \le 30$, we have $D(t) = \displaystyle\int_0^t \left(\dfrac{V}{15} - \dfrac{V}{450}s\right)ds = \left[\dfrac{V}{15}s - \dfrac{V}{900}s^2\right]_0^t = \dfrac{V}{15}t - \dfrac{V}{900}t^2$.

So $D(t) = V \;\Rightarrow\; \dfrac{V}{15}t - \dfrac{V}{900}t^2 = V \;\Rightarrow\; 60t - t^2 = 900 \;\Rightarrow\; t^2 - 60t + 900 = 0 \;\Rightarrow$

$(t - 30)^2 = 0 \;\Rightarrow\; t = 30$. So the length of time T is 30 months.

(c) $C(t) = \dfrac{1}{t}\displaystyle\int_0^t \left(\dfrac{V}{15} - \dfrac{V}{450}s + \dfrac{V}{12{,}900}s^2\right)ds = \dfrac{1}{t}\left[\dfrac{V}{15}s - \dfrac{V}{900}s^2 + \dfrac{V}{38{,}700}s^3\right]_0^t$

$= \dfrac{1}{t}\left(\dfrac{V}{15}t - \dfrac{V}{900}t^2 + \dfrac{V}{38{,}700}t^3\right) = \dfrac{V}{15} - \dfrac{V}{900}t + \dfrac{V}{38{,}700}t^2 \;\Rightarrow$

$C'(t) = -\dfrac{V}{900} + \dfrac{V}{19{,}350}t = 0$ when $\dfrac{1}{19{,}350}t = \dfrac{1}{900} \;\Rightarrow\; t = 21.5$.

$C(21.5) = \dfrac{V}{15} - \dfrac{V}{900}(21.5) + \dfrac{V}{38{,}700}(21.5)^2 \approx 0.05472V$, $C(0) = \dfrac{V}{15} \approx 0.06667V$, and

$C(30) = \dfrac{V}{15} - \dfrac{V}{900}(30) + \dfrac{V}{38{,}700}(30)^2 \approx 0.05659V$, so the absolute minimum is $C(21.5) \approx 0.05472V$.

(d) As in part (c), we have $C(t) = \dfrac{V}{15} - \dfrac{V}{900}t + \dfrac{V}{38,700}t^2$, so $C(t) = f(t) + g(t)$ ⇔

$$\frac{V}{15} - \frac{V}{900}t + \frac{V}{38,700}t^2 = \frac{V}{15} - \frac{V}{450}t + \frac{V}{12,900}t^2 \quad\Leftrightarrow$$

$$t^2\left(\frac{1}{12,900} - \frac{1}{38,700}\right) = t\left(\frac{1}{450} - \frac{1}{900}\right) \quad\Leftrightarrow\quad t = \frac{1/900}{2/38,700} = \frac{43}{2} = 21.5.$$

This is the value of t that we obtained as the critical number of C in part (c), so we have verified the result of (a) in this case.

69. $\displaystyle\int_1^9 \frac{1}{2x}\,dx = \frac{1}{2}\int_1^9 \frac{1}{x}\,dx = \frac{1}{2}\Big[\ln|x|\Big]_1^9 = \frac{1}{2}(\ln 9 - \ln 1) = \frac{1}{2}\ln 9 - 0 = \ln 9^{1/2} = \ln 3$

70. $\displaystyle\int_0^1 10^x\,dx = \left[\frac{10^x}{\ln 10}\right]_0^1 = \frac{10}{\ln 10} - \frac{1}{\ln 10} = \frac{9}{\ln 10}$

71. $\displaystyle\int_{1/2}^{\sqrt{3}/2} \frac{6}{\sqrt{1-t^2}}\,dt = 6\int_{1/2}^{\sqrt{3}/2} \frac{1}{\sqrt{1-t^2}}\,dt = 6\Big[\sin^{-1} t\Big]_{1/2}^{\sqrt{3}/2} = 6\Big[\sin^{-1}\left(\frac{\sqrt{3}}{2}\right) - \sin^{-1}\left(\frac{1}{2}\right)\Big] = 6\left(\frac{\pi}{3} - \frac{\pi}{6}\right) = 6\left(\frac{\pi}{6}\right) = \pi$

72. $\displaystyle\int_0^1 \frac{4}{t^2+1}\,dt = 4\int_0^1 \frac{1}{1+t^2}\,dt = 4\Big[\tan^{-1} t\Big]_0^1 = 4\left(\tan^{-1} 1 - \tan^{-1} 0\right) = 4\left(\frac{\pi}{4} - 0\right) = \pi$

73. $\displaystyle\int_{-1}^1 e^{u+1}\,du = \Big[e^{u+1}\Big]_{-1}^1 = e^2 - e^0 = e^2 - 1$ [or start with $e^{u+1} = e^u e^1$]

74. $\displaystyle\int_1^2 \frac{4+u^2}{u^3}\,du = \int_1^2 (4u^{-3} + u^{-1})\,du = \left[\frac{4}{-2}u^{-2} + \ln|u|\right]_1^2 = \left[\frac{-2}{u^2} + \ln u\right]_1^2 = \left(-\frac{1}{2} + \ln 2\right) - (-2 + \ln 1) = \frac{3}{2} + \ln 2$

5.4   Indefinite Integrals and the Net Change Theorem

1. $\dfrac{d}{dx}\left[\sqrt{x^2+1} + C\right] = \dfrac{d}{dx}\left[(x^2+1)^{1/2} + C\right] = \frac{1}{2}(x^2+1)^{-1/2} \cdot 2x + 0 = \dfrac{x}{\sqrt{x^2+1}}$

2. $\dfrac{d}{dx}\left[x\sin x + \cos x + C\right] = x\cos x + (\sin x)\cdot 1 - \sin x + 0 = x\cos x$

3. $\dfrac{d}{dx}\left[\sin x - \frac{1}{3}\sin^3 x + C\right] = \dfrac{d}{dx}\left[\sin x - \frac{1}{3}(\sin x)^3 + C\right] = \cos x - \frac{1}{3}\cdot 3(\sin x)^2(\cos x) + 0$

$$= \cos x(1 - \sin^2 x) = \cos x(\cos^2 x) = \cos^3 x$$

4. $\dfrac{d}{dx}\left[\dfrac{2}{3b^2}(bx - 2a)\sqrt{a+bx} + C\right] = \dfrac{d}{dx}\left[\dfrac{2}{3b^2}(bx - 2a)(a + bx)^{1/2} + C\right]$

$$= \frac{2}{3b^2}\left[(bx - 2a)\cdot\frac{1}{2}(a + bx)^{-1/2}(b) + (a + bx)^{1/2}(b)\right] + 0$$

$$= \frac{2}{3b^2}\cdot\frac{1}{2}b(a + bx)^{-1/2}\left[(bx - 2a) + 2(a + bx)\right] = \frac{1}{3b\sqrt{a+bx}}\,[3bx] = \frac{x}{\sqrt{a+bx}}$$

5. $\displaystyle\int (x^2 + x^{-2})\,dx = \frac{x^3}{3} + \frac{x^{-1}}{-1} + C = \frac{1}{3}x^3 - \frac{1}{x} + C$

6. $\displaystyle\int \left(\sqrt{x^3} + \sqrt[3]{x^2}\right)dx = \int (x^{3/2} + x^{2/3})\,dx = \frac{x^{5/2}}{5/2} + \frac{x^{5/3}}{5/3} + C = \frac{2}{5}x^{5/2} + \frac{3}{5}x^{5/3} + C$

7. $\int \left(x^4 - \frac{1}{2}x^3 + \frac{1}{4}x - 2\right) dx = \frac{x^5}{5} - \frac{1}{2}\frac{x^4}{4} + \frac{1}{4}\frac{x^2}{2} - 2x + C = \frac{1}{5}x^5 - \frac{1}{8}x^4 + \frac{1}{8}x^2 - 2x + C$

8. $\int (y^3 + 1.8y^2 - 2.4y) \, dy = \frac{y^4}{4} + 1.8\frac{y^3}{3} - 2.4\frac{y^2}{2} + C = \frac{1}{4}y^4 + 0.6y^3 - 1.2y^2 + C$

9. $\int (1-t)(2+t^2) \, dt = \int (2 - 2t + t^2 - t^3) \, dt = 2t - 2\frac{t^2}{2} + \frac{t^3}{3} - \frac{t^4}{4} + C = 2t - t^2 + \frac{1}{3}t^3 - \frac{1}{4}t^4 + C$

10. $\int v(v^2 + 2)^2 \, dv = \int v(v^4 + 4v^2 + 4) \, dv = \int (v^5 + 4v^3 + 4v) \, dv = \frac{v^6}{6} + 4\frac{v^4}{4} + 4\frac{v^2}{2} + C = \frac{1}{6}v^6 + v^4 + 2v^2 + C$

11. $\int \frac{x^3 - 2\sqrt{x}}{x} \, dx = \int \left(\frac{x^3}{x} - \frac{2x^{1/2}}{x}\right) dx = \int (x^2 - 2x^{-1/2}) \, dx = \frac{x^3}{3} - 2\frac{x^{1/2}}{1/2} + C = \frac{1}{3}x^3 - 4\sqrt{x} + C$

12. $\int \left(u^2 + 1 + \frac{1}{u^2}\right) du = \int (u^2 + 1 + u^{-2}) \, du = \frac{u^3}{3} + u + \frac{u^{-1}}{-1} + C = \frac{1}{3}u^3 + u - \frac{1}{u} + C$

13. $\int (\theta - \csc\theta \cot\theta) \, d\theta = \frac{1}{2}\theta^2 + \csc\theta + C$

14. $\int \sec t \, (\sec t + \tan t) \, dt = \int (\sec^2 t + \sec t \tan t) \, dt = \tan t + \sec t + C$

15. $\int (1 + \tan^2\alpha) \, d\alpha = \int \sec^2\alpha \, d\alpha = \tan\alpha + C$

16. $\int \frac{\sin 2x}{\sin x} \, dx = \int \frac{2\sin x \cos x}{\sin x} \, dx = \int 2\cos x \, dx = 2\sin x + C$

17. $\int \left(\cos x + \frac{1}{2}x\right) dx = \sin x + \frac{1}{4}x^2 + C$. The members of the family in the figure correspond to $C = -5, 0, 5,$ and 10.

18. $\int (1 - x^2)^2 \, dx = \int (1 - 2x^2 + x^4) \, dx = x - \frac{2}{3}x^3 + \frac{1}{5}x^5 + C$

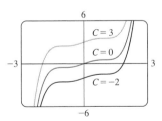

19. $\int_0^2 (6x^2 - 4x + 5) \, dx = \left[6 \cdot \frac{1}{3}x^3 - 4 \cdot \frac{1}{2}x^2 + 5x\right]_0^2 = \left[2x^3 - 2x^2 + 5x\right]_0^2 = (16 - 8 + 10) - 0 = 18$

20. $\int_1^3 (1 + 2x - 4x^3) \, dx = \left[x + 2 \cdot \frac{1}{2}x^2 - 4 \cdot \frac{1}{4}x^4\right]_1^3 = \left[x + x^2 - x^4\right]_1^3 = (3 + 9 - 81) - (1 + 1 - 1) = -69 - 1 = -70$

21. $\int_{-3}^0 (5y^4 - 6y^2 + 14) \, dy = \left[5\left(\frac{1}{5}y^5\right) - 6\left(\frac{1}{3}y^3\right) + 14y\right]_{-3}^0 = \left[y^5 - 2y^3 + 14y\right]_{-3}^0 = 0 - (-243 + 54 - 42) = 231$

22. $\int_{-2}^0 (u^5 - u^3 + u^2) \, du = \left[\frac{1}{6}u^6 - \frac{1}{4}u^4 + \frac{1}{3}u^3\right]_{-2}^0 = 0 - \left(\frac{32}{3} - 4 - \frac{8}{3}\right) = -4$

23. $\int_{-2}^{2}(3u+1)^2\,du = \int_{-2}^{2}\left(9u^2+6u+1\right)\,du = \left[9\cdot\frac{1}{3}u^3+6\cdot\frac{1}{2}u^2+u\right]_{-2}^{2} = \left[3u^3+3u^2+u\right]_{-2}^{2}$

$$= (24+12+2)-(-24+12-2) = 38-(-14) = 52$$

24. $\int_{0}^{4}(2v+5)(3v-1)\,dv = \int_{0}^{4}\left(6v^2+13v-5\right)\,dv = \left[6\cdot\frac{1}{3}v^3+13\cdot\frac{1}{2}v^2-5v\right]_{0}^{4} = \left[2v^3+\frac{13}{2}v^2-5v\right]_{0}^{4}$

$$= (128+104-20)-0 = 212$$

25. $\int_{1}^{4}\sqrt{t}\,(1+t)\,dt = \int_{1}^{4}\left(t^{1/2}+t^{3/2}\right)\,dt = \left[\frac{2}{3}t^{3/2}+\frac{2}{5}t^{5/2}\right]_{1}^{4} = \left(\frac{16}{3}+\frac{64}{5}\right)-\left(\frac{2}{3}+\frac{2}{5}\right) = \frac{14}{3}+\frac{62}{5} = \frac{256}{15}$

26. $\int_{0}^{9}\sqrt{2t}\,dt = \int_{0}^{9}\sqrt{2}\,t^{1/2}\,dt = \left[\sqrt{2}\cdot\frac{2}{3}t^{3/2}\right]_{0}^{9} = \sqrt{2}\cdot\frac{2}{3}\cdot 27-0 = 18\sqrt{2}$

27. $\int_{-2}^{-1}\left(4y^3+\frac{2}{y^3}\right)\,dy = \left[4\cdot\frac{1}{4}y^4+2\cdot\frac{1}{-2}y^{-2}\right]_{-2}^{-1} = \left[y^4-\frac{1}{y^2}\right]_{-2}^{-1} = (1-1)-\left(16-\frac{1}{4}\right) = -\frac{63}{4}$

28. $\int_{1}^{2}\frac{y+5y^7}{y^3}\,dy = \int_{1}^{2}\left(y^{-2}+5y^4\right)\,dy = \left[-y^{-1}+5\cdot\frac{1}{5}y^5\right]_{1}^{2} = \left[-\frac{1}{y}+y^5\right]_{1}^{2} = \left(-\frac{1}{2}+32\right)-(-1+1) = \frac{63}{2}$

29. $\int_{0}^{1}x\left(\sqrt[3]{x}+\sqrt[4]{x}\right)\,dx = \int_{0}^{1}\left(x^{4/3}+x^{5/4}\right)\,dx = \left[\frac{3}{7}x^{7/3}+\frac{4}{9}x^{9/4}\right]_{0}^{1} = \left(\frac{3}{7}+\frac{4}{9}\right)-0 = \frac{55}{63}$

30. $\int_{1}^{2}\left(x+\frac{1}{x}\right)^2\,dx = \int_{1}^{2}\left(x^2+2+x^{-2}\right)\,dx = \left[\frac{x^3}{3}+2x+\frac{x^{-1}}{-1}\right]_{1}^{2} = \left[\frac{x^3}{3}+2x-\frac{1}{x}\right]_{1}^{2}$

$$= \left(\frac{8}{3}+4-\frac{1}{2}\right)-\left(\frac{1}{3}+2-1\right) = \frac{29}{6}$$

31. $\int_{1}^{4}\sqrt{5/x}\,dx = \sqrt{5}\int_{1}^{4}x^{-1/2}\,dx = \sqrt{5}\left[2\sqrt{x}\right]_{1}^{4} = \sqrt{5}\left(2\cdot 2-2\cdot 1\right) = 2\sqrt{5}$

32. $\int_{1}^{9}\frac{3x-2}{\sqrt{x}}\,dx = \int_{1}^{9}\left(3x^{1/2}-2x^{-1/2}\right)\,dx = \left[3\cdot\frac{2}{3}x^{3/2}-2\cdot 2x^{1/2}\right]_{1}^{9} = \left[2x^{3/2}-4x^{1/2}\right]_{1}^{9}$

$$= (54-12)-(2-4) = 44$$

33. $\int_{0}^{\pi}\left(4\sin\theta-3\cos\theta\right)\,d\theta = \left[-4\cos\theta-3\sin\theta\right]_{0}^{\pi} = (4-0)-(-4-0) = 8$

34. $\int_{\pi/4}^{\pi/3}\sec\theta\tan\theta\,d\theta = \left[\sec\theta\right]_{\pi/4}^{\pi/3} = \sec\frac{\pi}{3}-\sec\frac{\pi}{4} = 2-\sqrt{2}$

35. $\int_{0}^{\pi/4}\frac{1+\cos^2\theta}{\cos^2\theta}\,d\theta = \int_{0}^{\pi/4}\left(\frac{1}{\cos^2\theta}+\frac{\cos^2\theta}{\cos^2\theta}\right)\,d\theta = \int_{0}^{\pi/4}\left(\sec^2\theta+1\right)\,d\theta$

$$= \left[\tan\theta+\theta\right]_{0}^{\pi/4} = \left(\tan\frac{\pi}{4}+\frac{\pi}{4}\right)-(0+0) = 1+\frac{\pi}{4}$$

36. $\int_{0}^{\pi/3}\frac{\sin\theta+\sin\theta\tan^2\theta}{\sec^2\theta}\,d\theta = \int_{0}^{\pi/3}\frac{\sin\theta\left(1+\tan^2\theta\right)}{\sec^2\theta}\,d\theta = \int_{0}^{\pi/3}\frac{\sin\theta\sec^2\theta}{\sec^2\theta}\,d\theta = \int_{0}^{\pi/3}\sin\theta\,d\theta$

$$= \left[-\cos\theta\right]_{0}^{\pi/3} = -\frac{1}{2}-(-1) = \frac{1}{2}$$

37. $\int_{1}^{64}\frac{1+\sqrt[3]{x}}{\sqrt{x}}\,dx = \int_{1}^{64}\left(\frac{1}{x^{1/2}}+\frac{x^{1/3}}{x^{1/2}}\right)\,dx = \int_{1}^{64}\left(x^{-1/2}+x^{(1/3)-(1/2)}\right)\,dx = \int_{1}^{64}\left(x^{-1/2}+x^{-1/6}\right)\,dx$

$$= \left[2x^{1/2}+\frac{6}{5}x^{5/6}\right]_{1}^{64} = \left(16+\frac{192}{5}\right)-\left(2+\frac{6}{5}\right) = 14+\frac{186}{5} = \frac{256}{5}$$

38. $\int_0^1 (1+x^2)^3 \, dx = \int_0^1 (1 + 3x^2 + 3x^4 + x^6) \, dx = \left[x + x^3 + \frac{3}{5}x^5 + \frac{1}{7}x^7 \right]_0^1 = \left(1 + 1 + \frac{3}{5} + \frac{1}{7} \right) - 0 = \frac{96}{35}$

39. $\int_0^1 \left(\sqrt[4]{x^5} + \sqrt[5]{x^4} \right) dx = \int_0^1 (x^{5/4} + x^{4/5}) \, dx = \left[\frac{x^{9/4}}{9/4} + \frac{x^{9/5}}{9/5} \right]_0^1 = \left[\frac{4}{9}x^{9/4} + \frac{5}{9}x^{9/5} \right]_0^1 = \frac{4}{9} + \frac{5}{9} - 0 = 1$

40. $\int_1^8 \frac{x-1}{\sqrt[3]{x^2}} \, dx = \int_1^8 (x^{1/3} - x^{-2/3}) \, dx = \left[\frac{x^{4/3}}{4/3} - \frac{x^{1/3}}{1/3} \right]_1^8 = \left[\frac{3}{4}x^{4/3} - 3x^{1/3} \right]_1^8 = \left(\frac{3}{4} \cdot 16 - 3 \cdot 2 \right) - \left(\frac{3}{4} - 3 \right) = \frac{33}{4}$

41. $\int_{-1}^2 (x - 2\,|x|) \, dx = \int_{-1}^0 [x - 2(-x)] \, dx + \int_0^2 [x - 2(x)] \, dx = \int_{-1}^0 3x \, dx + \int_0^2 (-x) \, dx = 3\left[\frac{1}{2}x^2 \right]_{-1}^0 - \left[\frac{1}{2}x^2 \right]_0^2$

$$= 3\left(0 - \frac{1}{2} \right) - (2 - 0) = -\frac{7}{2} = -3.5$$

42. $\int_0^{3\pi/2} |\sin x| \, dx = \int_0^\pi \sin x \, dx + \int_\pi^{3\pi/2} (-\sin x) \, dx = \left[-\cos x \right]_0^\pi + \left[\cos x \right]_\pi^{3\pi/2} = [1 - (-1)] + [0 - (-1)] = 2 + 1 = 3$

43. The graph shows that $y = x + x^2 - x^4$ has x-intercepts at $x = 0$ and at

$x = a \approx 1.32$. So the area of the region that lies under the curve and above the

x-axis is

$\int_0^a (x + x^2 - x^4) \, dx = \left[\frac{1}{2}x^2 + \frac{1}{3}x^3 - \frac{1}{5}x^5 \right]_0^a$

$\qquad = \left(\frac{1}{2}a^2 + \frac{1}{3}a^3 - \frac{1}{5}a^5 \right) - 0 \approx 0.84$

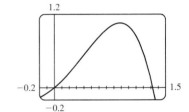

44. The graph shows that $y = 2x + 3x^4 - 2x^6$ has x-intercepts at $x = 0$ and at

$x = a \approx 1.37$. So the area of the region that lies under the curve and above the

x-axis is

$\int_0^a (2x + 3x^4 - 2x^6) \, dx = \left[x^2 + \frac{3}{5}x^5 - \frac{2}{7}x^7 \right]_0^a$

$\qquad = \left(a^2 + \frac{3}{5}a^5 - \frac{2}{7}a^7 \right) - 0 \approx 2.18$

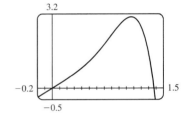

45. $A = \int_0^2 (2y - y^2) \, dy = \left[y^2 - \frac{1}{3}y^3 \right]_0^2 = \left(4 - \frac{8}{3} \right) - 0 = \frac{4}{3}$

46. $y = \sqrt[4]{x} \quad \Rightarrow \quad x = y^4$, so $A = \int_0^1 y^4 \, dy = \left[\frac{1}{5}y^5 \right]_0^1 = \frac{1}{5}$.

47. If $w'(t)$ is the rate of change of weight in pounds per year, then $w(t)$ represents the weight in pounds of the child at age t. We

know from the Net Change Theorem that $\int_5^{10} w'(t) \, dt = w(10) - w(5)$, so the integral represents the increase in the child's

weight (in pounds) between the ages of 5 and 10.

48. $\int_a^b I(t) \, dt = \int_a^b Q'(t) \, dt = Q(b) - Q(a)$ by the Net Change Theorem, so it represents the change in the charge Q from time

$t = a$ to $t = b$.

49. Since $r(t)$ is the rate at which oil leaks, we can write $r(t) = -V'(t)$, where $V(t)$ is the volume of oil at time t. [Note that the

minus sign is needed because V is decreasing, so $V'(t)$ is negative, but $r(t)$ is positive.] Thus, by the Net Change Theorem,

$\int_0^{120} r(t) \, dt = -\int_0^{120} V'(t) \, dt = -[V(120) - V(0)] = V(0) - V(120)$, which is the number of gallons of oil that leaked

from the tank in the first two hours (120 minutes).

50. By the Net Change Theorem, $\int_0^{15} n'(t)\,dt = n(15) - n(0) = n(15) - 100$ represents the increase in the bee population in 15 weeks. So $100 + \int_0^{15} n'(t)\,dt = n(15)$ represents the total bee population after 15 weeks.

51. By the Net Change Theorem, $\int_{1000}^{5000} R'(x)\,dx = R(5000) - R(1000)$, so it represents the increase in revenue when production is increased from 1000 units to 5000 units.

52. The slope of the trail is the rate of change of the elevation E, so $f(x) = E'(x)$. By the Net Change Theorem, $\int_3^5 f(x)\,dx = \int_3^5 E'(x)\,dx = E(5) - E(3)$ is the change in the elevation E between $x = 3$ miles and $x = 5$ miles from the start of the trail.

53. In general, the unit of measurement for $\int_a^b f(x)\,dx$ is the product of the unit for $f(x)$ and the unit for x. Since $f(x)$ is measured in newtons and x is measured in meters, the units for $\int_0^{100} f(x)\,dx$ are newton-meters. (A newton-meter is abbreviated N·m and is called a joule.)

54. The units for $a(x)$ are pounds per foot and the units for x are feet, so the units for da/dx are pounds per foot per foot, denoted (lb/ft)/ft. The unit of measurement for $\int_2^8 a(x)\,dx$ is the product of pounds per foot and feet; that is, pounds.

55. (a) Displacement $= \int_0^3 (3t - 5)\,dt = \left[\frac{3}{2}t^2 - 5t\right]_0^3 = \frac{27}{2} - 15 = -\frac{3}{2}$ m

(b) Distance traveled $= \int_0^3 |3t - 5|\,dt = \int_0^{5/3}(5 - 3t)\,dt + \int_{5/3}^3(3t - 5)\,dt$
$$= \left[5t - \frac{3}{2}t^2\right]_0^{5/3} + \left[\frac{3}{2}t^2 - 5t\right]_{5/3}^3 = \frac{25}{3} - \frac{3}{2}\cdot\frac{25}{9} + \frac{27}{2} - 15 - \left(\frac{3}{2}\cdot\frac{25}{9} - \frac{25}{3}\right) = \frac{41}{6} \text{ m}$$

56. (a) Displacement $= \int_1^6 (t^2 - 2t - 8)\,dt = \left[\frac{1}{3}t^3 - t^2 - 8t\right]_1^6 = (72 - 36 - 48) - \left(\frac{1}{3} - 1 - 8\right) = -\frac{10}{3}$ m

(b) Distance traveled $= \int_1^6 |t^2 - 2t - 8|\,dt = \int_1^6 |(t - 4)(t + 2)|\,dt$
$$= \int_1^4 (-t^2 + 2t + 8)\,dt + \int_4^6 (t^2 - 2t - 8)\,dt = \left[-\frac{1}{3}t^3 + t^2 + 8t\right]_1^4 + \left[\frac{1}{3}t^3 - t^2 - 8t\right]_4^6$$
$$= \left(-\frac{64}{3} + 16 + 32\right) - \left(-\frac{1}{3} + 1 + 8\right) + (72 - 36 - 48) - \left(\frac{64}{3} - 16 - 32\right) = \frac{98}{3} \text{ m}$$

57. (a) $v'(t) = a(t) = t + 4 \Rightarrow v(t) = \frac{1}{2}t^2 + 4t + C \Rightarrow v(0) = C = 5 \Rightarrow v(t) = \frac{1}{2}t^2 + 4t + 5$ m/s

(b) Distance traveled $= \int_0^{10} |v(t)|\,dt = \int_0^{10} \left|\frac{1}{2}t^2 + 4t + 5\right|\,dt = \int_0^{10}\left(\frac{1}{2}t^2 + 4t + 5\right)dt = \left[\frac{1}{6}t^3 + 2t^2 + 5t\right]_0^{10}$
$$= \frac{500}{3} + 200 + 50 = 416\frac{2}{3} \text{ m}$$

58. (a) $v'(t) = a(t) = 2t + 3 \Rightarrow v(t) = t^2 + 3t + C \Rightarrow v(0) = C = -4 \Rightarrow v(t) = t^2 + 3t - 4$

(b) Distance traveled $= \int_0^3 |t^2 + 3t - 4|\,dt = \int_0^3 |(t + 4)(t - 1)|\,dt = \int_0^1(-t^2 - 3t + 4)\,dt + \int_1^3(t^2 + 3t - 4)\,dt$
$$= \left[-\frac{1}{3}t^3 - \frac{3}{2}t^2 + 4t\right]_0^1 + \left[\frac{1}{3}t^3 + \frac{3}{2}t^2 - 4t\right]_1^3$$
$$= \left(-\frac{1}{3} - \frac{3}{2} + 4\right) + \left(9 + \frac{27}{2} - 12\right) - \left(\frac{1}{3} + \frac{3}{2} - 4\right) = \frac{89}{6} \text{ m}$$

59. Since $m'(x) = \rho(x)$, $m = \int_0^4 \rho(x)\,dx = \int_0^4 \left(9 + 2\sqrt{x}\right)dx = \left[9x + \frac{4}{3}x^{3/2}\right]_0^4 = 36 + \frac{32}{3} - 0 = \frac{140}{3} = 46\frac{2}{3}$ kg.

60. By the Net Change Theorem, the amount of water that flows from the tank during the first 10 minutes is
$$\int_0^{10} r(t)\,dt = \int_0^{10}(200 - 4t)\,dt = \left[200t - 2t^2\right]_0^{10} = (2000 - 200) - 0 = 1800 \text{ liters.}$$

61. Let s be the position of the car. We know from Equation 2 that $s(100) - s(0) = \int_0^{100} v(t)\,dt$. We use the Midpoint Rule for

$0 \le t \le 100$ with $n = 5$. Note that the length of each of the five time intervals is 20 seconds $= \frac{20}{3600}$ hour $= \frac{1}{180}$ hour.

So the distance traveled is

$\int_0^{100} v(t)\,dt \approx \frac{1}{180}[v(10) + v(30) + v(50) + v(70) + v(90)] = \frac{1}{180}(38 + 58 + 51 + 53 + 47) = \frac{247}{180} \approx 1.4$ miles.

62. (a) By the Net Change Theorem, the total amount spewed into the atmosphere is $Q(6) - Q(0) = \int_0^6 r(t)\,dt = Q(6)$ since

$Q(0) = 0$. The rate $r(t)$ is positive, so Q is an increasing function. Thus, an upper estimate for $Q(6)$ is R_6 and a lower

estimate for $Q(6)$ is L_6. $\Delta t = \dfrac{b - a}{n} = \dfrac{6 - 0}{6} = 1.$

$R_6 = \displaystyle\sum_{i=1}^{6} r(t_i)\,\Delta t = 10 + 24 + 36 + 46 + 54 + 60 = 230$ tonnes.

$L_6 = \displaystyle\sum_{i=1}^{6} r(t_{i-1})\,\Delta t = R_6 + r(0) - r(6) = 230 + 2 - 60 = 172$ tonnes.

(b) $\Delta t = \dfrac{b - a}{n} = \dfrac{6 - 0}{3} = 2.$ $Q(6) \approx M_3 = 2[r(1) + r(3) + r(5)] = 2(10 + 36 + 54) = 2(100) = 200$ tonnes.

63. From the Net Change Theorem, the increase in cost if the production level is raised

from 2000 yards to 4000 yards is $C(4000) - C(2000) = \int_{2000}^{4000} C'(x)\,dx$.

$\int_{2000}^{4000} C'(x)\,dx = \int_{2000}^{4000} \left(3 - 0.01x + 0.000006x^2\right)dx = \left[3x - 0.005x^2 + 0.000002x^3\right]_{2000}^{4000} = 60{,}000 - 2{,}000 = \$58{,}000$

64. By the Net Change Theorem, the amount of water after four days is

$$25{,}000 + \int_0^4 r(t)\,dt \approx 25{,}000 + M_4 = 25{,}000 + \frac{4 - 0}{4}\left[r(0.5) + r(1.5) + r(2.5) + r(3.5)\right]$$

$$\approx 25{,}000 + [1500 + 1770 + 740 + (-690)] = 28{,}320 \text{ liters}$$

65. (a) We can find the area between the Lorenz curve and the line $y = x$ by subtracting the area under $y = L(x)$ from the area

under $y = x$. Thus,

$$\text{coefficient of inequality} = \frac{\text{area between Lorenz curve and line } y = x}{\text{area under line } y = x} = \frac{\int_0^1 [x - L(x)]\,dx}{\int_0^1 x\,dx}$$

$$= \frac{\int_0^1 [x - L(x)]\,dx}{[x^2/2]_0^1} = \frac{\int_0^1 [x - L(x)]\,dx}{1/2} = 2\int_0^1 [x - L(x)]\,dx$$

(b) $L(x) = \frac{5}{12}x^2 + \frac{7}{12}x$ $\Rightarrow$ $L(50\%) = L\left(\frac{1}{2}\right) = \frac{5}{48} + \frac{7}{24} = \frac{19}{48} = 0.3958\overline{3}$, so the bottom 50% of the households receive

at most about 40% of the income. Using the result in part (a),

$$\text{coefficient of inequality} = 2\int_0^1 [x - L(x)]\,dx = 2\int_0^1 \left(x - \frac{5}{12}x^2 - \frac{7}{12}x\right)dx = 2\int_0^1 \left(\frac{5}{12}x - \frac{5}{12}x^2\right)dx$$

$$= 2\int_0^1 \frac{5}{12}(x - x^2)\,dx = \frac{5}{6}\left[\frac{1}{2}x^2 - \frac{1}{3}x^3\right]_0^1 = \frac{5}{6}\left(\frac{1}{2} - \frac{1}{3}\right) = \frac{5}{6}\left(\frac{1}{6}\right) = \frac{5}{36}$$

66. (a) From Exercise 4.1.66(a), $v(t) = 0.00146t^3 - 0.11553t^2 + 24.98169t - 21.26872$.

(b) $h(125) - h(0) = \int_0^{125} v(t)\,dt = \left[0.000365t^4 - 0.03851t^3 + 12.490845t^2 - 21.26872t\right]_0^{125} \approx 206{,}407$ ft

67. $\int (\sin x + \sinh x)\,dx = -\cos x + \cosh x + C$

68. $\displaystyle\int_{-10}^{10} \frac{2e^x}{\sinh x + \cosh x}\, dx = \int_{-10}^{10} \frac{2e^x}{\dfrac{e^x - e^{-x}}{2} + \dfrac{e^x + e^{-x}}{2}}\, dx = \int_{-10}^{10} \frac{2e^x}{e^x}\, dx = \int_{-10}^{10} 2\, dx = \big[2x\big]_{-10}^{10} = 20 - (-20) = 40$

69. $\displaystyle\int \left(x^2 + 1 + \frac{1}{x^2 + 1}\right) dx = \frac{x^3}{3} + x + \tan^{-1} x + C$

70. $\displaystyle\int_1^2 \frac{(x-1)^3}{x^2}\, dx = \int_1^2 \frac{x^3 - 3x^2 + 3x - 1}{x^2}\, dx = \int_1^2 \left(x - 3 + \frac{3}{x} - \frac{1}{x^2}\right) dx = \left[\tfrac{1}{2}x^2 - 3x + 3\ln|x| + \frac{1}{x}\right]_1^2$

$\qquad = \left(2 - 6 + 3\ln 2 + \tfrac{1}{2}\right) - \left(\tfrac{1}{2} - 3 + 0 + 1\right) = 3\ln 2 - 2$

71. $\displaystyle\int_0^{1/\sqrt{3}} \frac{t^2 - 1}{t^4 - 1}\, dt = \int_0^{1/\sqrt{3}} \frac{t^2 - 1}{(t^2 + 1)(t^2 - 1)}\, dt = \int_0^{1/\sqrt{3}} \frac{1}{t^2 + 1}\, dt = \big[\arctan t\big]_0^{1/\sqrt{3}} = \arctan\left(1/\sqrt{3}\right) - \arctan 0$

$\qquad = \tfrac{\pi}{6} - 0 = \tfrac{\pi}{6}$

72. $B = 3A \;\Rightarrow\; \int_0^b e^x\, dx = 3\int_0^a e^x\, dx \;\Rightarrow\; [e^x]_0^b = 3\,[e^x]_0^a \;\Rightarrow\; e^b - 1 = 3(e^a - 1) \;\Rightarrow\; e^b = 3e^a - 2 \;\Rightarrow$

$b = \ln(3e^a - 2)$

5.5   The Substitution Rule

1. Let $u = 3x$. Then $du = 3\, dx$, so $dx = \tfrac{1}{3}\, du$. Thus,

$\int \cos 3x\, dx = \int \cos u \left(\tfrac{1}{3}\, du\right) = \tfrac{1}{3}\int \cos u\, du = \tfrac{1}{3}\sin u + C = \tfrac{1}{3}\sin 3x + C$. Don't forget that it is often very easy to check

an indefinite integration by differentiating your answer. In this case, $\dfrac{d}{dx}\left(\tfrac{1}{3}\sin 3x + C\right) = \tfrac{1}{3}(\cos 3x)\cdot 3 = \cos 3x$, the

desired result.

2. Let $u = 2 + x^4$. Then $du = 4x^3\, dx$ and $x^3\, dx = \tfrac{1}{4}\, du$,

so $\displaystyle\int x^3 (2 + x^4)^5\, dx = \int u^5 \left(\tfrac{1}{4}\, du\right) = \frac{1}{4}\frac{u^6}{6} + C = \tfrac{1}{24}(2 + x^4)^6 + C$.

3. Let $u = x^3 + 1$. Then $du = 3x^2\, dx$ and $x^2\, dx = \tfrac{1}{3}\, du$, so

$\displaystyle\int x^2 \sqrt{x^3 + 1}\, dx = \int \sqrt{u}\left(\tfrac{1}{3}\, du\right) = \frac{1}{3}\frac{u^{3/2}}{3/2} + C = \frac{1}{3}\cdot\frac{2}{3}u^{3/2} + C = \tfrac{2}{9}(x^3 + 1)^{3/2} + C$.

4. Let $u = 1 - 6t$. Then $du = -6\, dt$ and $dt = -\tfrac{1}{6}\, du$, so

$\displaystyle\int \frac{dt}{(1 - 6t)^4} = \int \frac{-\tfrac{1}{6}\, du}{u^4} = -\tfrac{1}{6}\int u^{-4}\, du = -\frac{1}{6}\frac{u^{-3}}{-3} + C = \frac{1}{18u^3} + C = \frac{1}{18(1 - 6t)^3} + C$.

5. Let $u = \cos\theta$. Then $du = -\sin\theta\, d\theta$ and $\sin\theta\, d\theta = -du$, so

$\displaystyle\int \cos^3\theta \sin\theta\, d\theta = \int u^3 (-du) = -\frac{u^4}{4} + C = -\tfrac{1}{4}\cos^4\theta + C$.

6. Let $u = 1/x$. Then $du = -1/x^2 \, dx$ and $1/x^2 \, dx = -du$, so

$$\int \frac{\sec^2 (1/x)}{x^2} \, dx = \int \sec^2 u \, (-du) = -\tan u + C = -\tan (1/x) + C.$$

7. Let $u = x^2$. Then $du = 2x \, dx$ and $x \, dx = \frac{1}{2} \, du$, so $\int x \sin(x^2) \, dx = \int \sin u \left(\frac{1}{2} \, du\right) = -\frac{1}{2} \cos u + C = -\frac{1}{2} \cos(x^2) + C.$

8. Let $u = x^3 + 5$. Then $du = 3x^2 \, dx$ and $x^2 \, dx = \frac{1}{3} \, du$, so

$$\int x^2 (x^3 + 5)^9 \, dx = \int u^9 \left(\frac{1}{3} \, du\right) = \frac{1}{3} \cdot \frac{1}{10} u^{10} + C = \frac{1}{30}(x^3 + 5)^{10} + C.$$

9. Let $u = 3x - 2$. Then $du = 3 \, dx$ and $dx = \frac{1}{3} \, du$, so $\int (3x - 2)^{20} \, dx = \int u^{20} \left(\frac{1}{3} \, du\right) = \frac{1}{3} \cdot \frac{1}{21} u^{21} + C = \frac{1}{63}(3x - 2)^{21} + C.$

10. Let $u = 3t + 2$. Then $du = 3 \, dt$ and $dt = \frac{1}{3} \, du$, so

$$\int (3t + 2)^{2.4} \, dt = \int u^{2.4} \left(\frac{1}{3} \, du\right) = \frac{1}{3} \frac{u^{3.4}}{3.4} + C = \frac{1}{10.2}(3t + 2)^{3.4} + C.$$

11. Let $u = 2x + x^2$. Then $du = (2 + 2x) \, dx = 2(1 + x) \, dx$ and $(x + 1) \, dx = \frac{1}{2} \, du$, so

$$\int (x + 1)\sqrt{2x + x^2} \, dx = \int \sqrt{u} \left(\frac{1}{2} \, du\right) = \frac{1}{2} \frac{u^{3/2}}{3/2} + C = \frac{1}{3}\left(2x + x^2\right)^{3/2} + C.$$

Or: Let $u = \sqrt{2x + x^2}$. Then $u^2 = 2x + x^2 \;\Rightarrow\; 2u \, du = (2 + 2x) \, dx \;\Rightarrow\; u \, du = (1 + x) \, dx$, so

$\int (x + 1)\sqrt{2x + x^2} \, dx = \int u \cdot u \, du = \int u^2 \, du = \frac{1}{3} u^3 + C = \frac{1}{3}(2x + x^2)^{3/2} + C.$

12. Let $u = x^2 + 1$. Then $du = 2x \, dx$ and $x \, dx = \frac{1}{2} \, du$, so

$$\int \frac{x}{(x^2 + 1)^2} \, dx = \int u^{-2} \left(\frac{1}{2} \, du\right) = \frac{1}{2} \cdot \frac{-1}{u} + C = \frac{-1}{2u} + C = \frac{-1}{2(x^2 + 1)} + C.$$

13. Let $u = \pi t$. Then $du = \pi \, dt$ and $dt = \frac{1}{\pi} \, du$, so $\int \sin \pi t \, dt = \int \sin u \left(\frac{1}{\pi} \, du\right) = \frac{1}{\pi}(-\cos u) + C = -\frac{1}{\pi} \cos \pi t + C.$

14. Let $u = 5t + 4$. Then $du = 5 \, dt$ and $dt = \frac{1}{5} \, du$, so

$$\int \frac{1}{(5t + 4)^{2.7}} \, dt = \int u^{-2.7} \left(\frac{1}{5} \, du\right) = \frac{1}{5} \cdot \frac{1}{-1.7} u^{-1.7} + C = \frac{-1}{8.5} u^{-1.7} + C = \frac{-2}{17(5t + 4)^{1.7}} + C.$$

15. Let $u = 3ax + bx^3$. Then $du = (3a + 3bx^2) \, dx = 3(a + bx^2) \, dx$, so

$$\int \frac{a + bx^2}{\sqrt{3ax + bx^3}} \, dx = \int \frac{\frac{1}{3} \, du}{u^{1/2}} = \frac{1}{3} \int u^{-1/2} \, du = \frac{1}{3} \cdot 2u^2 + C = \frac{2}{3}\sqrt{3ax + bx^3} + C.$$

16. Let $u = 2\theta$. Then $du = 2 \, d\theta$ and $d\theta = \frac{1}{2} \, du$, so $\int \sec 2\theta \tan 2\theta \, d\theta = \int \sec u \tan u \left(\frac{1}{2} \, du\right) = \frac{1}{2} \sec u + C = \frac{1}{2} \sec 2\theta + C.$

17. Let $u = \sqrt{t}$. Then $du = \dfrac{dt}{2\sqrt{t}}$ and $\dfrac{1}{\sqrt{t}} \, dt = 2 \, du$, so $\displaystyle\int \frac{\cos \sqrt{t}}{\sqrt{t}} \, dt = \int \cos u \, (2 \, du) = 2 \sin u + C = 2 \sin \sqrt{t} + C.$

18. Let $u = 1 + x^{3/2}$. Then $du = \frac{3}{2} x^{1/2} \, dx$ and $\sqrt{x} \, dx = \frac{2}{3} \, du$, so

$\int \sqrt{x} \sin(1 + x^{3/2}) \, dx = \int \sin u \left(\frac{2}{3} \, du\right) = \frac{2}{3} \cdot (-\cos u) + C = -\frac{2}{3} \cos(1 + x^{3/2}) + C.$

19. Let $u = \sin\theta$. Then $du = \cos\theta\,d\theta$, so $\int \cos\theta\,\sin^6\theta\,d\theta = \int u^6\,du = \frac{1}{7}u^7 + C = \frac{1}{7}\sin^7\theta + C$.

20. Let $u = 1 + \tan\theta$. Then $du = \sec^2\theta\,d\theta$, so $\int(1+\tan\theta)^5\sec^2\theta\,d\theta = \int u^5\,du = \frac{1}{6}u^6 + C = \frac{1}{6}(1+\tan\theta)^6 + C$.

21. Let $u = 1 + z^3$. Then $du = 3z^2\,dz$ and $z^2\,dz = \frac{1}{3}\,du$, so

$$\int \frac{z^2}{\sqrt[3]{1+z^3}}\,dz = \int u^{-1/3}\left(\frac{1}{3}\,du\right) = \frac{1}{3}\cdot\frac{3}{2}u^{2/3} + C = \frac{1}{2}(1+z^3)^{2/3} + C.$$

22. Let $u = \dfrac{\pi}{x}$. Then $du = -\dfrac{\pi}{x^2}\,dx$ and $\dfrac{1}{x^2}\,dx = -\dfrac{1}{\pi}\,du$, so

$$\int \frac{\cos(\pi/x)}{x^2}\,dx = \int \cos u\left(-\frac{1}{\pi}\,du\right) = -\frac{1}{\pi}\sin u + C = -\frac{1}{\pi}\sin\frac{\pi}{x} + C.$$

23. Let $u = \cot x$. Then $du = -\csc^2 x\,dx$ and $\csc^2 x\,dx = -du$, so

$$\int \sqrt{\cot x}\,\csc^2 x\,dx = \int \sqrt{u}\,(-du) = -\frac{u^{3/2}}{3/2} + C = -\frac{2}{3}(\cot x)^{3/2} + C.$$

24. Let $u = 1 + \tan t$. Then $du = \sec^2 t\,dt$, so

$$\int \frac{dt}{\cos^2 t\,\sqrt{1+\tan t}} = \int \frac{\sec^2 t\,dt}{\sqrt{1+\tan t}} = \int \frac{du}{\sqrt{u}} = \int u^{-1/2}\,du = \frac{u^{1/2}}{1/2} + C = 2\sqrt{1+\tan t} + C.$$

25. Let $u = \sec x$. Then $du = \sec x\,\tan x\,dx$, so

$\int \sec^3 x\,\tan x\,dx = \int \sec^2 x\,(\sec x\,\tan x)\,dx = \int u^2\,du = \frac{1}{3}u^3 + C = \frac{1}{3}\sec^3 x + C$.

26. Let $u = \cos t$. Then $du = -\sin t\,dt$ and $\sin t\,dt = -du$, so

$\int \sin t\,\sec^2(\cos t)\,dt = \int \sec^2 u\cdot(-du) = -\tan u + C = -\tan(\cos t) + C$.

27. Let $u = \sin x$. Then $du = \cos x\,dx$, so $\displaystyle\int \frac{\cos x}{\sin^2 x}\,dx = \int \frac{1}{u^2}\,du = \int u^{-2}\,du = \frac{u^{-1}}{-1} + C = -\frac{1}{u} + C = -\frac{1}{\sin x} + C$

[or $-\csc x + C$].

28. Let $u = 1 - x$. Then $x = 1 - u$ and $dx = -du$, so

$$\int \frac{x^2}{\sqrt{1-x}}\,dx = \int \frac{(1-u)^2}{\sqrt{u}}\,(-du) = -\int \frac{1-2u+u^2}{\sqrt{u}}\,du = -\int\left(u^{-1/2} - 2u^{1/2} + u^{3/2}\right)du$$

$$= -\left(2u^{1/2} - 2\cdot\frac{2}{3}u^{3/2} + \frac{2}{5}u^{5/2}\right) + C = -2\sqrt{1-x} + \frac{4}{3}(1-x)^{3/2} - \frac{2}{5}(1-x)^{5/2} + C$$

29. Let $u = x + 2$. Then $du = dx$, so

$$\int \frac{x}{\sqrt[4]{x+2}}\,dx = \int \frac{u-2}{\sqrt[4]{u}}\,du = \int(u^{3/4} - 2u^{-1/4})\,du = \frac{4}{7}u^{7/4} - 2\cdot\frac{4}{3}u^{3/4} + C$$

$$= \frac{4}{7}(x+2)^{7/4} - \frac{8}{3}(x+2)^{3/4} + C$$

30. Let $u = x^2 + 1$ [so $x^2 = u - 1$]. Then $du = 2x\,dx$ and $x\,dx = \frac{1}{2}\,du$, so

$$\int x^3\sqrt{x^2+1}\,dx = \int x^2\sqrt{x^2+1}\,x\,dx = \int (u-1)\sqrt{u}\left(\tfrac{1}{2}\,du\right) = \tfrac{1}{2}\int(u^{3/2} - u^{1/2})\,du$$

$$= \tfrac{1}{2}\left(\tfrac{2}{5}u^{5/2} - \tfrac{2}{3}u^{3/2}\right) + C = \tfrac{1}{5}(x^2+1)^{5/2} - \tfrac{1}{3}(x^2+1)^{3/2} + C.$$

Or: Let $u = \sqrt{x^2+1}$. Then $u^2 = x^2 + 1 \;\Rightarrow\; 2u\,du = 2x\,dx \;\Rightarrow\; u\,du = x\,dx$, so

$$\int x^3\sqrt{x^2+1}\,dx = \int x^2\sqrt{x^2+1}\,x\,dx = \int(u^2-1)\,u\cdot u\,du = \int(u^4 - u^2)\,du$$

$$= \tfrac{1}{5}u^5 - \tfrac{1}{3}u^3 + C = \tfrac{1}{5}(x^2+1)^{5/2} - \tfrac{1}{3}(x^2+1)^{3/2} + C.$$

Note: This answer can be written as $\frac{1}{15}\sqrt{x^2+1}(3x^4 + x^2 - 2) + C$.

In Exercises 31–34, let $f(x)$ denote the integrand and $F(x)$ its antiderivative (with $C = 0$).

31. $f(x) = x(x^2-1)^3$. $u = x^2 - 1 \;\Rightarrow\; du = 2x\,dx$, so

$$\int x(x^2-1)^3\,dx = \int u^3\left(\tfrac{1}{2}\,du\right) = \tfrac{1}{8}u^4 + C = \tfrac{1}{8}(x^2-1)^4 + C$$

Where f is positive (negative), F is increasing (decreasing). Where f changes from negative to positive (positive to negative), F has a local minimum (maximum).

32. $f(x) = \dfrac{\sin\sqrt{x}}{\sqrt{x}}$. $u = \sqrt{x} \;\Rightarrow\; du = \tfrac{1}{2}x^{-1/2}\,dx = \dfrac{1}{2\sqrt{x}}\,dx$, so

$$\int \frac{\sin\sqrt{x}}{\sqrt{x}}\,dx = \int \sin u\,(2\,du) = -2\cos\sqrt{x} + C$$

Where f is positive (negative), F is increasing (decreasing). Where f changes from negative to positive (positive to negative), F has a local minimum (maximum).

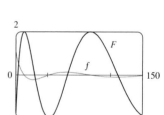

33. $f(x) = \sin^3 x\cos x$. $u = \sin x \;\Rightarrow\; du = \cos x\,dx$, so

$$\int \sin^3 x\cos x\,dx = \int u^3\,du = \tfrac{1}{4}u^4 + C = \tfrac{1}{4}\sin^4 x + C$$

Note that at $x = \frac{\pi}{2}$, f changes from positive to negative and F has a local maximum. Also, both f and F are periodic with period π, so at $x = 0$ and at $x = \pi$, f changes from negative to positive and F has local minima.

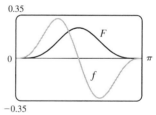

34. $f(\theta) = \tan^2\theta\sec^2\theta$. $u = \tan\theta \;\Rightarrow\; du = \sec^2\theta\,d\theta$, so

$$\int \tan^2\theta\sec^2\theta\,d\theta = \int u^2\,du = \tfrac{1}{3}u^3 + C = \tfrac{1}{3}\tan^3\theta + C$$

Note that f is positive and F is increasing. At $x = 0$, $f = 0$ and F has a horizontal tangent.

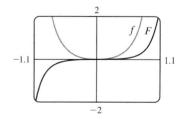

35. Let $u = x - 1$, so $du = dx$. When $x = 0$, $u = -1$; when $x = 2$, $u = 1$. Thus, $\int_0^2 (x-1)^{25}\,dx = \int_{-1}^1 u^{25}\,du = 0$ by Theorem 6(b), since $f(u) = u^{25}$ is an odd function.

36. Let $u = 4 + 3x$, so $du = 3\,dx$. When $x = 0$, $u = 4$; when $x = 7$, $u = 25$. Thus,

$$\int_0^7 \sqrt{4+3x}\,dx = \int_4^{25} \sqrt{u}\left(\tfrac{1}{3}\,du\right) = \frac{1}{3}\left[\frac{u^{3/2}}{3/2}\right]_4^{25} = \tfrac{2}{9}(25^{3/2} - 4^{3/2}) = \tfrac{2}{9}(125 - 8) = \tfrac{234}{9} = 26.$$

37. Let $u = 1 + 2x^3$, so $du = 6x^2\,dx$. When $x = 0$, $u = 1$; when $x = 1$, $u = 3$. Thus,

$$\int_0^1 x^2\left(1+2x^3\right)^5 dx = \int_1^3 u^5\left(\tfrac{1}{6}\,du\right) = \tfrac{1}{6}\left[\tfrac{1}{6}u^6\right]_1^3 = \tfrac{1}{36}(3^6 - 1^6) = \tfrac{1}{36}(729 - 1) = \tfrac{728}{36} = \tfrac{182}{9}.$$

38. Let $u = x^2$, so $du = 2x\,dx$. When $x = 0$, $u = 0$; when $x = \sqrt{\pi}$, $u = \pi$. Thus,

$$\int_0^{\sqrt{\pi}} x\cos(x^2)\,dx = \int_0^{\pi}\cos u\left(\tfrac{1}{2}\,du\right) = \tfrac{1}{2}\left[\sin u\right]_0^{\pi} = \tfrac{1}{2}(\sin\pi - \sin 0) = \tfrac{1}{2}(0-0) = 0.$$

39. Let $u = t/4$, so $du = \tfrac{1}{4}\,dt$. When $t = 0$, $u = 0$; when $t = \pi$, $u = \pi/4$. Thus,

$$\int_0^{\pi}\sec^2(t/4)\,dt = \int_0^{\pi/4}\sec^2 u\,(4\,du) = 4\left[\tan u\right]_0^{\pi/4} = 4\left(\tan\tfrac{\pi}{4} - \tan 0\right) = 4(1-0) = 4.$$

40. Let $u = \pi t$, so $du = \pi\,dt$. When $t = \tfrac{1}{6}$, $u = \tfrac{\pi}{6}$; when $t = \tfrac{1}{2}$, $u = \tfrac{\pi}{2}$. Thus,

$$\int_{1/6}^{1/2}\csc\pi t\,\cot\pi t\,dt = \int_{\pi/6}^{\pi/2}\csc u\,\cot u\left(\tfrac{1}{\pi}\,du\right) = \tfrac{1}{\pi}\left[-\csc u\right]_{\pi/6}^{\pi/2} = -\tfrac{1}{\pi}(1-2) = \tfrac{1}{\pi}.$$

41. $\int_{-\pi/6}^{\pi/6}\tan^3\theta\,d\theta = 0$ by Theorem 6(b), since $f(\theta) = \tan^3\theta$ is an odd function.

42. $\displaystyle\int_{-\pi/2}^{\pi/2}\frac{x^2\sin x}{1+x^6}\,dx = 0$ by Theorem 6(b), since $f(x) = \dfrac{x^2\sin x}{1+x^6}$ is an odd function.

43. Let $u = 1 + 2x$, so $du = 2\,dx$. When $x = 0$, $u = 1$; when $x = 13$, $u = 27$. Thus,

$$\int_0^{13}\frac{dx}{\sqrt[3]{(1+2x)^2}} = \int_1^{27} u^{-2/3}\left(\tfrac{1}{2}\,du\right) = \left[\tfrac{1}{2}\cdot 3u^{1/3}\right]_1^{27} = \tfrac{3}{2}(3-1) = 3.$$

44. Let $u = \sin x$, so $du = \cos x\,dx$. When $x = 0$, $u = 0$; when $x = \tfrac{\pi}{2}$, $u = 1$. Thus,

$$\int_0^{\pi/2}\cos x\,\sin(\sin x)\,dx = \int_0^1\sin u\,du = \left[-\cos u\right]_0^1 = -(\cos 1 - 1) = 1 - \cos 1.$$

45. Let $u = x^2 + a^2$, so $du = 2x\,dx$ and $x\,dx = \tfrac{1}{2}\,du$. When $x = 0$, $u = a^2$; when $x = a$, $u = 2a^2$. Thus,

$$\int_0^a x\sqrt{x^2+a^2}\,dx = \int_{a^2}^{2a^2} u^{1/2}\left(\tfrac{1}{2}\,du\right) = \tfrac{1}{2}\left[\tfrac{2}{3}u^{3/2}\right]_{a^2}^{2a^2} = \left[\tfrac{1}{3}u^{3/2}\right]_{a^2}^{2a^2} = \tfrac{1}{3}\left[(2a^2)^{3/2} - (a^2)^{3/2}\right] = \tfrac{1}{3}\left(2\sqrt{2}-1\right)a^3$$

46. Assume $a > 0$. Let $u = a^2 - x^2$, so $du = -2x\,dx$. When $x = 0$, $u = a^2$; when $x = a$, $u = 0$. Thus,

$$\int_0^a x\sqrt{a^2-x^2}\,dx = \int_{a^2}^0 u^{1/2}\left(-\tfrac{1}{2}\,du\right) = \tfrac{1}{2}\int_0^{a^2} u^{1/2}\,du = \tfrac{1}{2}\cdot\left[\tfrac{2}{3}u^{3/2}\right]_0^{a^2} = \tfrac{1}{3}a^3.$$

47. Let $u = x - 1$, so $u + 1 = x$ and $du = dx$. When $x = 1$, $u = 0$; when $x = 2$, $u = 1$. Thus,

$$\int_1^2 x\sqrt{x-1}\,dx = \int_0^1 (u+1)\sqrt{u}\,du = \int_0^1 (u^{3/2} + u^{1/2})\,du = \left[\tfrac{2}{5}u^{5/2} + \tfrac{2}{3}u^{3/2}\right]_0^1 = \tfrac{2}{5} + \tfrac{2}{3} = \tfrac{16}{15}.$$

48. Let $u = 1 + 2x$, so $x = \tfrac{1}{2}(u-1)$ and $du = 2\,dx$. When $x = 0$, $u = 1$; when $x = 4$, $u = 9$. Thus,

$$\int_0^4 \frac{x\,dx}{\sqrt{1+2x}} = \int_1^9 \frac{\tfrac{1}{2}(u-1)}{\sqrt{u}}\frac{du}{2} = \tfrac{1}{4}\int_1^9 (u^{1/2} - u^{-1/2})\,du = \tfrac{1}{4}\left[\tfrac{2}{3}u^{3/2} - 2u^{1/2}\right]_1^9 = \tfrac{1}{4}\cdot\tfrac{2}{3}\left[u^{3/2} - 3u^{1/2}\right]_1^9$$
$$= \tfrac{1}{6}[(27-9) - (1-3)] = \tfrac{20}{6} = \tfrac{10}{3}$$

49. Let $u = x^{-2}$, so $du = -2x^{-3}\,dx$. When $x = \frac{1}{2}$, $u = 4$; when $x = 1$, $u = 1$. Thus,

$$\int_{1/2}^{1} \frac{\cos(x^{-2})}{x^3}\,dx = \int_{4}^{1} \cos u \left(\frac{du}{-2}\right) = \frac{1}{2}\int_{1}^{4} \cos u\,du = \frac{1}{2}\big[\sin u\big]_{1}^{4} = \frac{1}{2}(\sin 4 - \sin 1).$$

50. Let $u = \dfrac{2\pi t}{T} - \alpha$, so $du = \dfrac{2\pi}{T}\,dt$. When $t = 0$, $u = -\alpha$; when $t = \dfrac{T}{2}$, $u = \pi - \alpha$. Thus,

$$\int_{0}^{\pi/2} \sin\left(\frac{2\pi t}{T} - \alpha\right) dt = \int_{-\alpha}^{\pi-\alpha} \sin u \left(\frac{T}{2\pi}\,du\right) = \frac{T}{2\pi}\big[-\cos u\big]_{-\alpha}^{\pi-\alpha} = -\frac{T}{2\pi}\big[\cos(\pi - \alpha) - \cos(-\alpha)\big]$$

$$= -\frac{T}{2\pi}(-\cos\alpha - \cos\alpha) = -\frac{T}{2\pi}(-2\cos\alpha) = \frac{T}{\pi}\cos\alpha$$

51. From the graph, it appears that the area under the curve is about

$1 + \left(\text{a little more than } \frac{1}{2}\cdot 1\cdot 0.7\right)$, or about 1.4. The exact area is given by

$A = \int_{0}^{1} \sqrt{2x+1}\,dx$. Let $u = 2x + 1$, so $du = 2\,dx$. The limits change to

$2\cdot 0 + 1 = 1$ and $2\cdot 1 + 1 = 3$, and

$A = \int_{1}^{3} \sqrt{u}\left(\frac{1}{2}\,du\right) = \frac{1}{2}\left[\frac{2}{3}u^{3/2}\right]_{1}^{3} = \frac{1}{3}\left(3\sqrt{3} - 1\right) = \sqrt{3} - \frac{1}{3} \approx 1.399.$

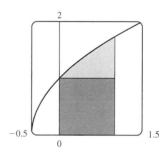

52. From the graph, it appears that the area under the curve is almost $\frac{1}{2}\cdot\pi\cdot 2.6$,

or about 4. The exact area is given by

$$A = \int_{0}^{\pi}(2\sin x - \sin 2x)\,dx = -2\big[\cos x\big]_{0}^{\pi} - \int_{0}^{\pi}\sin 2x\,dx$$

$$= -2(-1-1) - 0 = 4$$

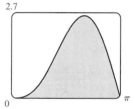

Note: $\int_{0}^{\pi}\sin 2x\,dx = 0$ since it is clear from the graph of $y = \sin 2x$ that $\int_{\pi/2}^{\pi}\sin 2x\,dx = -\int_{0}^{\pi/2}\sin 2x\,dx$.

53. First write the integral as a sum of two integrals:

$I = \int_{-2}^{2}(x+3)\sqrt{4-x^2}\,dx = I_1 + I_2 = \int_{-2}^{2} x\sqrt{4-x^2}\,dx + \int_{-2}^{2} 3\sqrt{4-x^2}\,dx$. $I_1 = 0$ by Theorem 6(b), since

$f(x) = x\sqrt{4-x^2}$ is an odd function and we are integrating from $x = -2$ to $x = 2$. We interpret I_2 as three times the area of

a semicircle with radius 2, so $I = 0 + 3\cdot\frac{1}{2}\left(\pi\cdot 2^2\right) = 6\pi$.

54. Let $u = x^2$. Then $du = 2x\,dx$ and the limits are unchanged ($0^2 = 0$ and $1^2 = 1$), so

$I = \int_{0}^{1} x\sqrt{1-x^4}\,dx = \frac{1}{2}\int_{0}^{1}\sqrt{1-u^2}\,du$. But this integral can be interpreted as the area of a quarter-circle with radius 1.

So $I = \frac{1}{2}\cdot\frac{1}{4}\left(\pi\cdot 1^2\right) = \frac{1}{8}\pi.$

55. The volume of inhaled air in the lungs at time t is

$$V(t) = \int_{0}^{t} f(u)\,du = \int_{0}^{t} \frac{1}{2}\sin\left(\frac{2\pi}{5}u\right)du = \int_{0}^{2\pi t/5} \frac{1}{2}\sin v\left(\frac{5}{2\pi}\,dv\right) \qquad \left[\text{substitute } v = \frac{2\pi}{5}u,\ dv = \frac{2\pi}{5}\,du\right]$$

$$= \frac{5}{4\pi}\big[-\cos v\big]_{0}^{2\pi t/5} = \frac{5}{4\pi}\big[-\cos\left(\frac{2\pi}{5}t\right) + 1\big] = \frac{5}{4\pi}\big[1 - \cos\left(\frac{2\pi}{5}t\right)\big] \text{ liters}$$

56. Let $u = \dfrac{\pi t}{12}$. Then $du = \dfrac{\pi}{12}\, dt$ and

$$\int_0^{24} R(t)\, dt = \int_0^{24} \left[85 - 0.18 \cos\left(\frac{\pi t}{12} \right) \right] dt = \int_0^{2\pi} (85 - 0.18 \cos u)\left(\frac{12}{\pi}\, du \right) = \frac{12}{\pi} \left[85u - 0.18 \sin u \right]_0^{2\pi}$$

$$= \frac{12}{\pi} \left[(85 \cdot 2\pi - 0) - (0 - 0) \right] = 2040 \text{ kcal}$$

57. Let $u = 2x$. Then $du = 2\, dx$, so $\int_0^2 f(2x)\, dx = \int_0^4 f(u)\left(\tfrac{1}{2}\, du \right) = \tfrac{1}{2} \int_0^4 f(u)\, du = \tfrac{1}{2}(10) = 5.$

58. Let $u = x^2$. Then $du = 2x\, dx$, so $\int_0^3 x f(x^2)\, dx = \int_0^9 f(u)\left(\tfrac{1}{2}\, du \right) = \tfrac{1}{2} \int_0^9 f(u)\, du = \tfrac{1}{2}(4) = 2.$

59. Let $u = -x$. Then $du = -dx$, so

$$\int_a^b f(-x)\, dx = \int_{-a}^{-b} f(u)(-du) = \int_{-b}^{-a} f(u)\, du = \int_{-b}^{-a} f(x)\, dx$$

From the diagram, we see that the equality follows from the fact that we are
reflecting the graph of f, and the limits of integration, about the y-axis.

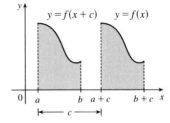

60. Let $u = x + c$. Then $du = dx$, so

$$\int_a^b f(x + c)\, dx = \int_{a+c}^{b+c} f(u)\, du = \int_{a+c}^{b+c} f(x)\, dx$$

From the diagram, we see that the equality follows from the fact that we are
translating the graph of f, and the limits of integration, by a distance c.

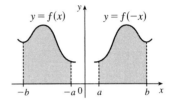

61. Let $u = 1 - x$. Then $x = 1 - u$ and $dx = -du$, so

$$\int_0^1 x^a (1-x)^b\, dx = \int_1^0 (1-u)^a u^b (-du) = \int_0^1 u^b (1-u)^a\, du = \int_0^1 x^b (1-x)^a\, dx.$$

62. Let $u = \pi - x$. Then $du = -dx$. When $x = \pi$, $u = 0$ and when $x = 0$, $u = \pi$. So

$$\int_0^\pi x f(\sin x)\, dx = -\int_\pi^0 (\pi - u) f(\sin(\pi - u))\, du = \int_0^\pi (\pi - u) f(\sin u)\, du$$

$$= \pi \int_0^\pi f(\sin u)\, du - \int_0^\pi u f(\sin u)\, du = \pi \int_0^\pi f(\sin x)\, dx - \int_0^\pi x f(\sin x)\, dx \quad \Rightarrow$$

$$2 \int_0^\pi x f(\sin x)\, dx = \pi \int_0^\pi f(\sin x)\, dx \quad \Rightarrow \quad \int_0^\pi x f(\sin x)\, dx = \frac{\pi}{2} \int_0^\pi f(\sin x)\, dx.$$

63. $\int_0^{\pi/2} f(\cos x)\, dx = \int_0^{\pi/2} f\left[\sin\left(\frac{\pi}{2} - x \right) \right] dx \qquad \left[u = \frac{\pi}{2} - x,\ du = -dx \right]$

$$= -\int_{\pi/2}^0 f(\sin u)(-du) = \int_0^{\pi/2} f(\sin u)\, du = \int_0^{\pi/2} f(\sin x)\, dx$$

Continuity of f is needed in order to apply the substitution rule for definite integrals.

64. In Exercise 63, take $f(x) = x^2$, so $\int_0^{\pi/2} \cos^2 x\, dx = \int_0^{\pi/2} \sin^2 x\, dx$. Now

$$\int_0^{\pi/2} \cos^2 x\, dx + \int_0^{\pi/2} \sin^2 x\, dx = \int_0^{\pi/2} (\cos^2 x + \sin^2 x)\, dx = \int_0^{\pi/2} 1\, dx = \left[x \right]_0^{\pi/2} = \frac{\pi}{2},$$

so $2 \int_0^{\pi/2} \cos^2 x\, dx = \frac{\pi}{2} \quad \Rightarrow \quad \int_0^{\pi/2} \cos^2 x\, dx = \frac{\pi}{4} \quad \left[= \int_0^{\pi/2} \sin^2 x\, dx \right].$

65. Let $u = 5 - 3x$. Then $du = -3\, dx$ and $dx = -\tfrac{1}{3}\, du$, so

$$\int \frac{dx}{5 - 3x} = \int \frac{1}{u}\left(-\tfrac{1}{3}\, du \right) = -\tfrac{1}{3} \ln|u| + C = -\tfrac{1}{3} \ln|5 - 3x| + C.$$

66. Let $u = e^x$. Then $du = e^x\, dx$, so $\int e^x \sin(e^x)\, dx = \int \sin u\, du = -\cos u + C = -\cos(e^x) + C$.

67. Let $u = \ln x$. Then $du = \dfrac{dx}{x}$, so $\displaystyle\int \dfrac{(\ln x)^2}{x}\, dx = \int u^2\, du = \frac{1}{3}u^3 + C = \frac{1}{3}(\ln x)^3 + C$.

68. Let $u = ax + b$. Then $du = a\, dx$ and $dx = (1/a)\, du$, so

$$\int \frac{dx}{ax + b} = \int \frac{(1/a)\, du}{u} = \frac{1}{a}\int \frac{1}{u}\, du = \frac{1}{a}\ln|u| + C = \frac{1}{a}\ln|ax + b| + C.$$

69. Let $u = 1 + e^x$. Then $du = e^x\, dx$, so $\int e^x \sqrt{1 + e^x}\, dx = \int \sqrt{u}\, du = \frac{2}{3}u^{3/2} + C = \frac{2}{3}(1 + e^x)^{3/2} + C$.

Or: Let $u = \sqrt{1 + e^x}$. Then $u^2 = 1 + e^x$ and $2u\, du = e^x\, dx$, so

$\int e^x \sqrt{1 + e^x}\, dx = \int u \cdot 2u\, du = \frac{2}{3}u^3 + C = \frac{2}{3}(1 + e^x)^{3/2} + C$.

70. Let $u = \cos t$. Then $du = -\sin t\, dt$ and $\sin t\, dt = -du$, so $\int e^{\cos t} \sin t\, dt = \int e^u\, (-du) = -e^u + C = -e^{\cos t} + C$.

71. Let $u = \tan x$. Then $du = \sec^2 x\, dx$, so $\int e^{\tan x} \sec^2 x\, dx = \int e^u\, du = e^u + C = e^{\tan x} + C$.

72. Let $u = \tan^{-1} x$. Then $du = \dfrac{dx}{1 + x^2}$, so $\displaystyle\int \dfrac{\tan^{-1} x}{1 + x^2}\, dx = \int u\, du = \dfrac{u^2}{2} + C = \dfrac{(\tan^{-1} x)^2}{2} + C$.

73. Let $u = 1 + x^2$. Then $du = 2x\, dx$, so

$$\int \frac{1 + x}{1 + x^2}\, dx = \int \frac{1}{1 + x^2}\, dx + \int \frac{x}{1 + x^2}\, dx = \tan^{-1} x + \int \frac{\frac{1}{2}\, du}{u} = \tan^{-1} x + \frac{1}{2}\ln|u| + C$$

$$= \tan^{-1} x + \frac{1}{2}\ln\left|1 + x^2\right| + C = \tan^{-1} x + \frac{1}{2}\ln\left(1 + x^2\right) + C \quad [\text{since } 1 + x^2 > 0].$$

74. Let $u = \ln x$. Then $du = (1/x)\, dx$, so $\displaystyle\int \dfrac{\sin(\ln x)}{x}\, dx = \int \sin u\, du = -\cos u + C = -\cos(\ln x) + C$.

75. $\displaystyle\int \dfrac{\sin 2x}{1 + \cos^2 x}\, dx = 2\int \dfrac{\sin x \cos x}{1 + \cos^2 x}\, dx = 2I$. Let $u = \cos x$. Then $du = -\sin x\, dx$, so

$2I = -2\displaystyle\int \dfrac{u\, du}{1 + u^2} = -2 \cdot \frac{1}{2}\ln(1 + u^2) + C = -\ln(1 + u^2) + C = -\ln(1 + \cos^2 x) + C$.

Or: Let $u = 1 + \cos^2 x$.

76. Let $u = \cos x$. Then $du = -\sin x\, dx$ and $\sin x\, dx = -du$, so

$$\int \frac{\sin x}{1 + \cos^2 x}\, dx = \int \frac{-du}{1 + u^2} = -\tan^{-1} u + C = -\tan^{-1}(\cos x) + C.$$

77. $\displaystyle\int \cot x\, dx = \int \dfrac{\cos x}{\sin x}\, dx$. Let $u = \sin x$. Then $du = \cos x\, dx$, so $\displaystyle\int \cot x\, dx = \int \dfrac{1}{u}\, du = \ln|u| + C = \ln|\sin x| + C$.

78. Let $u = x^2$. Then $du = 2x\, dx$, so $\displaystyle\int \dfrac{x}{1 + x^4}\, dx = \int \dfrac{\frac{1}{2}\, du}{1 + u^2} = \frac{1}{2}\tan^{-1} u + C = \frac{1}{2}\tan^{-1}(x^2) + C$.

79. Let $u = \ln x$, so $du = \dfrac{dx}{x}$. When $x = e$, $u = 1$; when $x = e^4$; $u = 4$. Thus,

$$\int_e^{e^4} \frac{dx}{x \sqrt{\ln x}} = \int_1^4 u^{-1/2}\, du = 2\left[u^{1/2}\right]_1^4 = 2(2 - 1) = 2.$$

80. Let $u = -x^2$, so $du = -2x\,dx$. When $x = 0$, $u = 0$; when $x = 1$, $u = -1$. Thus,

$\int_0^1 xe^{-x^2}\,dx = \int_0^{-1} e^u\left(-\frac{1}{2}\,du\right) = -\frac{1}{2}\left[e^u\right]_0^{-1} = -\frac{1}{2}\left(e^{-1} - e^0\right) = \frac{1}{2}(1 - 1/e)$.

81. Let $u = e^z + z$, so $du = (e^z + 1)\,dz$. When $z = 0$, $u = 1$; when $z = 1$, $u = e + 1$. Thus,

$$\int_0^1 \frac{e^z + 1}{e^z + z}\,dz = \int_1^{e+1} \frac{1}{u}\,du = \Big[\ln|u|\Big]_1^{e+1} = \ln|e + 1| - \ln|1| = \ln(e + 1).$$

82. Let $u = \sin^{-1} x$, so $du = \dfrac{dx}{\sqrt{1 - x^2}}$. When $x = 0$, $u = 0$; when $x = \frac{1}{2}$, $u = \frac{\pi}{6}$. Thus,

$$\int_0^{1/2} \frac{\sin^{-1} x}{\sqrt{1 - x^2}}\,dx = \int_0^{\pi/6} u\,du = \left[\frac{u^2}{2}\right]_0^{\pi/6} = \frac{\pi^2}{72}.$$

83. $\dfrac{x \sin x}{1 + \cos^2 x} = x \cdot \dfrac{\sin x}{2 - \sin^2 x} = x f(\sin x)$, where $f(t) = \dfrac{t}{2 - t^2}$. By Exercise 62,

$$\int_0^\pi \frac{x \sin x}{1 + \cos^2 x}\,dx = \int_0^\pi x f(\sin x)\,dx = \frac{\pi}{2}\int_0^\pi f(\sin x)\,dx = \frac{\pi}{2}\int_0^\pi \frac{\sin x}{1 + \cos^2 x}\,dx$$

Let $u = \cos x$. Then $du = -\sin x\,dx$. When $x = \pi$, $u = -1$ and when $x = 0$, $u = 1$. So

$$\frac{\pi}{2}\int_0^\pi \frac{\sin x}{1 + \cos^2 x}\,dx = -\frac{\pi}{2}\int_1^{-1} \frac{du}{1 + u^2} = \frac{\pi}{2}\int_{-1}^1 \frac{du}{1 + u^2} = \frac{\pi}{2}\Big[\tan^{-1} u\Big]_{-1}^1$$

$$= \frac{\pi}{2}[\tan^{-1} 1 - \tan^{-1}(-1)] = \frac{\pi}{2}\left[\frac{\pi}{4} - \left(-\frac{\pi}{4}\right)\right] = \frac{\pi^2}{4}$$

5 Review

CONCEPT CHECK

1. (a) $\sum_{i=1}^n f(x_i^*)\,\Delta x$ is an expression for a Riemann sum of a function f.

   x_i^* is a point in the ith subinterval $[x_{i-1}, x_i]$ and Δx is the length of the subintervals.

   (b) See Figure 1 in Section 5.2.

   (c) In Section 5.2, see Figure 3 and the paragraph beside it.

2. (a) See Definition 5.2.2.

   (b) See Figure 2 in Section 5.2.

   (c) In Section 5.2, see Figure 4 and the paragraph by it (contains "**net area**").

3. See the Fundamental Theorem of Calculus after Example 8 in Section 5.3.

4. (a) See the Net Change Theorem after Example 5 in Section 5.4.

   (b) $\int_{t_1}^{t_2} r(t)\,dt$ represents the change in the amount of water in the reservoir between time t_1 and time t_2.

5. (a) $\int_{60}^{120} v(t)\,dt$ represents the change in position of the particle from $t = 60$ to $t = 120$ seconds.

   (b) $\int_{60}^{120} |v(t)|\,dt$ represents the total distance traveled by the particle from $t = 60$ to 120 seconds.

   (c) $\int_{60}^{120} a(t)\,dt$ represents the change in the velocity of the particle from $t = 60$ to $t = 120$ seconds.

6. (a) $\int f(x)\,dx$ is the family of functions $\{F \mid F' = f\}$. Any two such functions differ by a constant.

(b) The connection is given by the Net Change Theorem: $\int_a^b f(x)\,dx = \left[\int f(x)\,dx\right]_a^b$ if f is continuous.

7. The precise version of this statement is given by the Fundamental Theorem of Calculus. See the statement of this theorem and the paragraph that follows it at the end of Section 5.3.

8. See the Substitution Rule (5.5.4). This says that it is permissible to operate with the dx after an integral sign as if it were a differential.

TRUE-FALSE QUIZ

1. True by Property 2 of the Integral in Section 5.2.

2. False.   Try $a = 0$, $b = 2$, $f(x) = g(x) = 1$ as a counterexample.

3. True by Property 3 of the Integral in Section 5.2.

4. False.   You can't take a variable outside the integral sign. For example, using $f(x) = 1$ on $[0, 1]$,
$$\int_0^1 x\,f(x)\,dx = \int_0^1 x\,dx = \left[\tfrac{1}{2}x^2\right]_0^1 = \tfrac{1}{2} \text{ (a constant) while } x\int_0^1 1\,dx = x\,[x]_0^1 = x \cdot 1 = x \text{ (a variable).}$$

5. False.   For example, let $f(x) = x^2$. Then $\int_0^1 \sqrt{x^2}\,dx = \int_0^1 x\,dx = \tfrac{1}{2}$, but $\sqrt{\int_0^1 x^2\,dx} = \sqrt{\tfrac{1}{3}} = \tfrac{1}{\sqrt{3}}$.

6. True by the Net Change Theorem.

7. True by Comparison Property 7 of the Integral in Section 5.2.

8. False.   For example, let $a = 0$, $b = 1$, $f(x) = 3$, $g(x) = x$. $f(x) > g(x)$ for each x in $(0, 1)$, but $f'(x) = 0 < 1 = g'(x)$ for $x \in (0, 1)$.

9. True.   The integrand is an odd function that is continuous on $[-1, 1]$, so the result follows from Theorem 5.5.6(b).

10. True. $\int_{-5}^5 \left(ax^2 + bx + c\right)dx = \int_{-5}^5 \left(ax^2 + c\right)dx + \int_{-5}^5 bx\,dx$
$$= 2\int_0^5 \left(ax^2 + c\right)dx \text{ [by 5.5.6(a)]} \;+ 0 \text{ [by 5.5.6(b)]}$$

11. False.   The function $f(x) = 1/x^4$ is not bounded on the interval $[-2, 1]$. It has an infinite discontinuity at $x = 0$, so it is not integrable on the interval. (If the integral were to exist, a positive value would be expected, by Comparison Property 6 of Integrals.)

12. False.   See the paragraph before Note 4 and Figure 4 in Section 5.2, and notice that $y = x - x^3 < 0$ for $1 < x \le 2$.

13. False.   For example, the function $y = |x|$ is continuous on $\mathbb{R}$, but has no derivative at $x = 0$.

14. True by FTC1.

15. False. $\int_a^b f(x)\,dx$ is a constant, so $\dfrac{d}{dx}\left(\int_a^b f(x)\,dx\right) = 0$, not $f(x)$ [unless $f(x) = 0$]. Compare the given statement carefully with FTC1, in which the upper limit in the integral is x.

EXERCISES

1. (a)

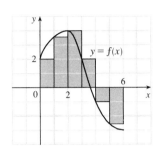

$$L_6 = \sum_{i=1}^{6} f(x_{i-1})\,\Delta x \quad [\Delta x = \tfrac{6-0}{6} = 1]$$

$$= f(x_0) \cdot 1 + f(x_1) \cdot 1 + f(x_2) \cdot 1 + f(x_3) \cdot 1 + f(x_4) \cdot 1 + f(x_5) \cdot 1$$

$$\approx 2 + 3.5 + 4 + 2 + (-1) + (-2.5) = 8$$

The Riemann sum represents the sum of the areas of the four rectangles above the x-axis minus the sum of the areas of the two rectangles below the x-axis.

(b)

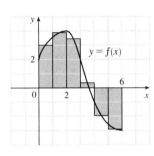

$$M_6 = \sum_{i=1}^{6} f(\overline{x}_i)\,\Delta x \quad [\Delta x = \tfrac{6-0}{6} = 1]$$

$$= f(\overline{x}_1) \cdot 1 + f(\overline{x}_2) \cdot 1 + f(\overline{x}_3) \cdot 1 + f(\overline{x}_4) \cdot 1 + f(\overline{x}_5) \cdot 1 + f(\overline{x}_6) \cdot 1$$

$$= f(0.5) + f(1.5) + f(2.5) + f(3.5) + f(4.5) + f(5.5)$$

$$\approx 3 + 3.9 + 3.4 + 0.3 + (-2) + (-2.9) = 5.7$$

2. (a)

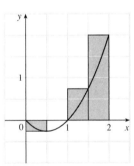

$$f(x) = x^2 - x \text{ and } \Delta x = \tfrac{2-0}{4} = 0.5 \quad \Rightarrow$$

$$R_4 = 0.5f(0.5) + 0.5f(1) + 0.5f(1.5) + 0.5f(2)$$

$$= 0.5(-0.25 + 0 + 0.75 + 2) = 1.25$$

The Riemann sum represents the sum of the areas of the two rectangles above the x-axis minus the area of the rectangle below the x-axis. (The second rectangle vanishes.)

(b) $\displaystyle \int_0^2 (x^2 - x)\,dx = \lim_{n \to \infty} \sum_{i=1}^{n} f(x_i)\,\Delta x \qquad [\Delta x = 2/n \text{ and } x_i = 2i/n]$

$$= \lim_{n \to \infty} \sum_{i=1}^{n} \left(\frac{4i^2}{n^2} - \frac{2i}{n} \right)\left(\frac{2}{n} \right) = \lim_{n \to \infty} \frac{2}{n} \left[\frac{4}{n^2} \sum_{i=1}^{n} i^2 - \frac{2}{n} \sum_{i=1}^{n} i \right]$$

$$= \lim_{n \to \infty} \left[\frac{8}{n^3} \cdot \frac{n(n+1)(2n+1)}{6} - \frac{4}{n^2} \cdot \frac{n(n+1)}{2} \right] = \lim_{n \to \infty} \left[\frac{4}{3} \cdot \frac{n+1}{n} \cdot \frac{2n+1}{n} - 2 \cdot \frac{n+1}{n} \right]$$

$$= \lim_{n \to \infty} \left[\frac{4}{3}\left(1 + \frac{1}{n} \right)\left(2 + \frac{1}{n} \right) - 2\left(1 + \frac{1}{n} \right) \right] = \tfrac{4}{3} \cdot 1 \cdot 2 - 2 \cdot 1 = \tfrac{2}{3}$$

(c) $\displaystyle \int_0^2 (x^2 - x)\,dx = \left[\tfrac{1}{3}x^3 - \tfrac{1}{2}x^2 \right]_0^2 = \left(\tfrac{8}{3} - 2 \right) = \tfrac{2}{3}$

(d)

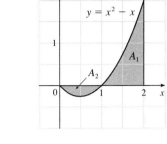

$$\int_0^2 (x^2 - x)\, dx = A_1 - A_2,\ \text{where } A_1 \text{ and } A_2 \text{ are the areas shown in the diagram.}$$

3. $\int_0^1 \left(x + \sqrt{1 - x^2}\right) dx = \int_0^1 x\, dx + \int_0^1 \sqrt{1 - x^2}\, dx = I_1 + I_2.$

I_1 can be interpreted as the area of the triangle shown in the figure

and I_2 can be interpreted as the area of the quarter-circle.

Area $= \frac{1}{2}(1)(1) + \frac{1}{4}(\pi)(1)^2 = \frac{1}{2} + \frac{\pi}{4}.$

4. On $[0, \pi]$, $\displaystyle\lim_{n \to \infty} \sum_{i=1}^{n} \sin x_i\, \Delta x = \int_0^\pi \sin x\, dx = [-\cos x]_0^\pi = -(-1) - (-1) = 2.$

5. $\int_0^6 f(x)\, dx = \int_0^4 f(x)\, dx + \int_4^6 f(x)\, dx \ \Rightarrow\ 10 = 7 + \int_4^6 f(x)\, dx \ \Rightarrow\ \int_4^6 f(x)\, dx = 10 - 7 = 3$

6. (a) $\int_1^5 (x + 2x^5)\, dx = \displaystyle\lim_{n \to \infty} \sum_{i=1}^{n} f(x_i)\, \Delta x \qquad \left[\Delta x = \dfrac{5 - 1}{n} = \dfrac{4}{n},\ x_i = 1 + \dfrac{4i}{n} \right]$

$$= \lim_{n \to \infty} \sum_{i=1}^{n} \left[\left(1 + \frac{4i}{n}\right) + 2\left(1 + \frac{4i}{n}\right)^5 \right] \cdot \frac{4}{n} \stackrel{\text{CAS}}{=} \lim_{n \to \infty} \frac{1305n^4 + 3126n^3 + 2080n^2 - 256}{n^3} \cdot \frac{4}{n}$$

$$= 5220$$

(b) $\int_1^5 (x + 2x^5)\, dx = \left[\frac{1}{2}x^2 + \frac{2}{6}x^6 \right]_1^5 = \left(\frac{25}{2} + \frac{15{,}625}{3} \right) - \left(\frac{1}{2} + \frac{1}{3} \right) = 12 + 5208 = 5220$

7. First note that either a or b must be the graph of $\int_0^x f(t)\, dt$, since $\int_0^0 f(t)\, dt = 0$, and $c(0) \neq 0$. Now notice that $b > 0$ when c is increasing, and that $c > 0$ when a is increasing. It follows that c is the graph of $f(x)$, b is the graph of $f'(x)$, and a is the graph of $\int_0^x f(t)\, dt$.

8. (a) By FTC2, we have $\displaystyle\int_0^{\pi/2} \frac{d}{dx}\left(\sin \frac{x}{2} \cos \frac{x}{3} \right) dx = \left[\sin \frac{x}{2} \cos \frac{x}{3} \right]_0^{\pi/2} = \frac{1}{\sqrt{2}} \cdot \frac{\sqrt{3}}{2} - 0 \cdot 1 = \frac{\sqrt{6}}{4}.$

(b) $\dfrac{d}{dx} \displaystyle\int_0^{\pi/2} \sin \frac{x}{2} \cos \frac{x}{3}\, dx = 0$, since the definite integral is a constant.

(c) $\dfrac{d}{dx} \displaystyle\int_x^{\pi/2} \sin \frac{t}{2} \cos \frac{t}{3}\, dt = \dfrac{d}{dx} \left(-\int_{\pi/2}^x \sin \frac{t}{2} \cos \frac{t}{3}\, dt \right) = -\dfrac{d}{dx} \int_{\pi/2}^x \sin \frac{t}{2} \cos \frac{t}{2}\, dt = -\sin \frac{x}{2} \cos \frac{x}{3}$, by FTC1.

9. $\int_1^2 (8x^3 + 3x^2)\, dx = \left[8 \cdot \frac{1}{4}x^4 + 3 \cdot \frac{1}{3}x^3 \right]_1^2 = \left[2x^4 + x^3 \right]_1^2 = (2 \cdot 2^4 + 2^3) - (2 + 1) = 40 - 3 = 37$

10. $\displaystyle\int_0^T (x^4 - 8x + 7)\, dx = \left[\frac{1}{5}x^5 - 4x^2 + 7x \right]_0^T = \left(\frac{1}{5}T^5 - 4T^2 + 7T \right) - 0 = \frac{1}{5}T^5 - 4T^2 + 7T$

11. $\int_0^1 (1 - x^9)\, dx = \left[x - \frac{1}{10}x^{10} \right]_0^1 = \left(1 - \frac{1}{10} \right) - 0 = \frac{9}{10}$

12. Let $u = 1 - x$, so $du = -dx$ and $dx = -du$. When $x = 0$, $u = 1$; when $x = 1$, $u = 0$. Thus,

$\int_0^1 (1 - x)^9 \, dx = \int_1^0 u^9(-du) = \int_0^1 u^9 \, du = \frac{1}{10} \left[u^{10} \right]_0^1 = \frac{1}{10}(1 - 0) = \frac{1}{10}$.

13. $\displaystyle\int_1^9 \frac{\sqrt{u} - 2u^2}{u} \, du = \int_1^9 (u^{-1/2} - 2u) \, du = \left[2u^{1/2} - u^2 \right]_1^9 = (6 - 81) - (2 - 1) = -76$

14. $\int_0^1 \left(\sqrt[4]{u} + 1 \right)^2 \, du = \int_0^1 (u^{1/2} + 2u^{1/4} + 1) \, du = \left[\frac{2}{3} u^{3/2} + \frac{8}{5} u^{5/4} + u \right]_0^1 = \left(\frac{2}{3} + \frac{8}{5} + 1 \right) - 0 = \frac{49}{15}$

15. Let $u = y^2 + 1$, so $du = 2y \, dy$ and $y \, dy = \frac{1}{2} \, du$. When $y = 0$, $u = 1$; when $y = 1$, $u = 2$. Thus,

$\int_0^1 y(y^2 + 1)^5 \, dy = \int_1^2 u^5 \left(\frac{1}{2} \, du \right) = \frac{1}{2} \left[\frac{1}{6} u^6 \right]_1^2 = \frac{1}{12}(64 - 1) = \frac{63}{12} = \frac{21}{4}$.

16. Let $u = 1 + y^3$, so $du = 3y^2 \, dy$ and $y^2 \, dy = \frac{1}{3} \, du$. When $y = 0$, $u = 1$; when $y = 2$, $u = 9$. Thus,

$\int_0^2 y^2 \sqrt{1 + y^3} \, dy = \int_1^9 u^{1/2} \left(\frac{1}{3} \, du \right) = \frac{1}{3} \left[\frac{2}{3} u^{3/2} \right]_1^9 = \frac{2}{9}(27 - 1) = \frac{52}{9}$.

17. $\displaystyle\int_1^5 \frac{dt}{(t - 4)^2}$ does not exist because the function $f(t) = \dfrac{1}{(t - 4)^2}$ has an infinite discontinuity at $t = 4$;

that is, f is discontinuous on the interval $[1, 5]$.

18. Let $u = 3\pi t$, so $du = 3\pi \, dt$. When $t = 0$, $u = 1$; when $t = 1$, $u = 3\pi$. Thus,

$\displaystyle\int_0^1 \sin(3\pi t) \, dt = \int_0^{3\pi} \sin u \left(\frac{1}{3\pi} \, du \right) = \frac{1}{3\pi} \left[-\cos u \right]_0^{3\pi} = -\frac{1}{3\pi}(-1 - 1) = \frac{2}{3\pi}$.

19. Let $u = v^3$, so $du = 3v^2 \, dv$. When $v = 0$, $u = 0$; when $v = 1$, $u = 1$. Thus,

$\int_0^1 v^2 \cos(v^3) \, dv = \int_0^1 \cos u \left(\frac{1}{3} \, du \right) = \frac{1}{3} \left[\sin u \right]_0^1 = \frac{1}{3}(\sin 1 - 0) = \frac{1}{3} \sin 1$.

20. $\displaystyle\int_{-1}^1 \frac{\sin x}{1 + x^2} \, dx = 0$ by Theorem 5.5.6(b), since $f(x) = \dfrac{\sin x}{1 + x^2}$ is an odd function.

21. $\displaystyle\int_{-\pi/4}^{\pi/4} \frac{t^4 \tan t}{2 + \cos t} \, dt = 0$ by Theorem 5.5.6(b), since $f(t) = \dfrac{t^4 \tan t}{2 + \cos t}$ is an odd function.

22. Let $u = x^2 + 4x$. Then $du = (2x + 4) \, dx = 2(x + 2) \, dx$, so

$\displaystyle\int \frac{x + 2}{\sqrt{x^2 + 4x}} \, dx = \int u^{-1/2} \left(\frac{1}{2} \, du \right) = \frac{1}{2} \cdot 2u^{1/2} + C = \sqrt{u} + C = \sqrt{x^2 + 4x} + C$.

23. Let $u = \sin \pi t$. Then $du = \pi \cos \pi t \, dt$, so $\int \sin \pi t \cos \pi t \, dt = \int u \left(\frac{1}{\pi} \, du \right) = \frac{1}{\pi} \cdot \frac{1}{2} u^2 + C = \frac{1}{2\pi} (\sin \pi t)^2 + C$.

24. Let $u = \cos x$. Then $du = -\sin x \, dx$, so $\int \sin x \cos(\cos x) \, dx = -\int \cos u \, du = -\sin u + C = -\sin(\cos x) + C$.

25. Let $u = 2\theta$. Then $du = 2 \, d\theta$, so

$\int_0^{\pi/8} \sec 2\theta \tan 2\theta \, d\theta = \int_0^{\pi/4} \sec u \tan u \left(\frac{1}{2} \, du \right) = \frac{1}{2} \left[\sec u \right]_0^{\pi/4} = \frac{1}{2} \left(\sec \frac{\pi}{4} - \sec 0 \right) = \frac{1}{2} (\sqrt{2} - 1) = \frac{1}{2} \sqrt{2} - \frac{1}{2}$.

26. Let $u = 1 + \tan t$, so $du = \sec^2 t \, dt$. When $t = 0$, $u = 1$; when $t = \frac{\pi}{4}$, $u = 2$. Thus,

$\int_0^{\pi/4} (1 + \tan t)^3 \sec^2 t \, dt = \int_1^2 u^3 \, du = \left[\frac{1}{4} u^4 \right]_1^2 = \frac{1}{4} (2^4 - 1^4) = \frac{1}{4}(16 - 1) = \frac{15}{4}$.

27. Since $x^2 - 4 < 0$ for $0 \le x < 2$ and $x^2 - 4 > 0$ for $2 < x \le 3$, we have $\left|x^2 - 4\right| = -(x^2 - 4) = 4 - x^2$ for $0 \le x < 2$ and

$\left|x^2 - 4\right| = x^2 - 4$ for $2 < x \le 3$. Thus,

$$\int_0^3 \left|x^2 - 4\right| dx = \int_0^2 (4 - x^2) \, dx + \int_2^3 (x^2 - 4) \, dx = \left[4x - \frac{x^3}{3}\right]_0^2 + \left[\frac{x^3}{3} - 4x\right]_2^3$$

$$= \left(8 - \tfrac{8}{3}\right) - 0 + (9 - 12) - \left(\tfrac{8}{3} - 8\right) = \tfrac{16}{3} - 3 + \tfrac{16}{3} = \tfrac{32}{3} - \tfrac{9}{3} = \tfrac{23}{3}$$

28. Since $\sqrt{x} - 1 < 0$ for $0 \le x < 1$ and $\sqrt{x} - 1 > 0$ for $1 < x \le 4$, we have $\left|\sqrt{x} - 1\right| = -\left(\sqrt{x} - 1\right) = 1 - \sqrt{x}$

for $0 \le x < 1$ and $\left|\sqrt{x} - 1\right| = \sqrt{x} - 1$ for $1 < x \le 4$. Thus,

$$\int_0^4 \left|\sqrt{x} - 1\right| dx = \int_0^1 \left(1 - \sqrt{x}\right) dx + \int_1^4 \left(\sqrt{x} - 1\right) dx = \left[x - \tfrac{2}{3}x^{3/2}\right]_0^1 + \left[\tfrac{2}{3}x^{3/2} - x\right]_1^4$$

$$= \left(1 - \tfrac{2}{3}\right) - 0 + \left(\tfrac{16}{3} - 4\right) - \left(\tfrac{2}{3} - 1\right) = \tfrac{1}{3} + \tfrac{16}{3} - 4 + \tfrac{1}{3} = 6 - 4 = 2$$

In Exercises 29 and 30, let $f(x)$ denote the integrand and $F(x)$ its antiderivative (with $C = 0$).

29. Let $u = 1 + \sin x$. Then $du = \cos x \, dx$, so

$$\int \frac{\cos x \, dx}{\sqrt{1 + \sin x}} = \int u^{-1/2} \, du = 2u^{1/2} + C = 2\sqrt{1 + \sin x} + C.$$

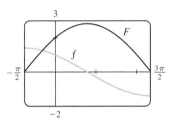

30. Let $u = x^2 + 1$. Then $x^2 = u - 1$ and $x \, dx = \frac{1}{2} \, du$, so

$$\int \frac{x^3}{\sqrt{x^2 + 1}} \, dx = \int \frac{(u - 1)}{\sqrt{u}} \left(\tfrac{1}{2} \, du\right) = \frac{1}{2} \int \left(u^{1/2} - u^{-1/2}\right) du$$

$$= \tfrac{1}{2} \left(\tfrac{2}{3} u^{3/2} - 2u^{1/2}\right) + C = \tfrac{1}{3}(x^2 + 1)^{3/2} - (x^2 + 1)^{1/2} + C$$

$$= \tfrac{1}{3}(x^2 + 1)^{1/2} \left[(x^2 + 1) - 3\right] + C = \tfrac{1}{3}\sqrt{x^2 + 1} \, (x^2 - 2) + C$$

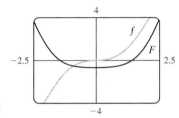

31. From the graph, it appears that the area under the curve $y = x\sqrt{x}$ between $x = 0$ and $x = 4$ is somewhat less than half the area of an 8×4 rectangle, so perhaps about 13 or 14. To find the exact value, we evaluate

$$\int_0^4 x\sqrt{x} \, dx = \int_0^4 x^{3/2} \, dx = \left[\tfrac{2}{5}x^{5/2}\right]_0^4 = \tfrac{2}{5}(4)^{5/2} = \tfrac{64}{5} = 12.8.$$

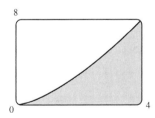

32. From the graph, it seems as though $\int_0^{2\pi} \cos^2 x \sin^3 x \, dx$ is equal to 0. To evaluate

the integral, we write the integral as $I = \int_0^{2\pi} \cos^2 x \left(1 - \cos^2 x\right) \sin x \, dx$ and let

$u = \cos x \Rightarrow du = -\sin x \, dx$. Thus, $I = \int_1^1 u^2 \left(1 - u^2\right)(-du) = 0.$

33. $F(x) = \displaystyle\int_0^x \frac{t^2}{1 + t^3} \, dt \Rightarrow F'(x) = \frac{d}{dx} \int_0^x \frac{t^2}{1 + t^3} \, dt = \frac{x^2}{1 + x^3}$

34. $F(x) = \int_x^1 \sqrt{t + \sin t}\, dt = -\int_1^x \sqrt{t + \sin t}\, dt \quad \Rightarrow \quad F'(x) = -\dfrac{d}{dx} \int_1^x \sqrt{t + \sin t}\, dt = -\sqrt{x + \sin x}$

35. Let $u = x^4$. Then $\dfrac{du}{dx} = 4x^3$. Also, $\dfrac{dg}{dx} = \dfrac{dg}{du}\dfrac{du}{dx}$, so

$$g'(x) = \dfrac{d}{dx} \int_0^{x^4} \cos(t^2)\, dt = \dfrac{d}{du} \int_0^u \cos(t^2)\, dt \cdot \dfrac{du}{dx} = \cos(u^2)\dfrac{du}{dx} = 4x^3 \cos(x^8).$$

36. Let $u = \sin x$. Then $\dfrac{du}{dx} = \cos x$. Also, $\dfrac{dg}{dx} = \dfrac{dg}{du}\dfrac{du}{dx}$, so

$$g'(x) = \dfrac{d}{dx} \int_1^{\sin x} \dfrac{1 - t^2}{1 + t^4}\, dt = \dfrac{d}{du} \int_1^u \dfrac{1 - t^2}{1 + t^4}\, dt \cdot \dfrac{du}{dx} = \dfrac{1 - u^2}{1 + u^4} \cdot \dfrac{du}{dx} = \dfrac{1 - \sin^2 x}{1 + \sin^4 x} \cdot \cos x = \dfrac{\cos^3 x}{1 + \sin^4 x}$$

37. $y = \int_{\sqrt{x}}^x \dfrac{\cos \theta}{\theta}\, d\theta = \int_1^x \dfrac{\cos \theta}{\theta}\, d\theta + \int_{\sqrt{x}}^1 \dfrac{\cos \theta}{\theta}\, d\theta = \int_1^x \dfrac{\cos \theta}{\theta}\, d\theta - \int_1^{\sqrt{x}} \dfrac{\cos \theta}{\theta}\, d\theta \quad \Rightarrow$

$$y' = \dfrac{\cos x}{x} - \dfrac{\cos \sqrt{x}}{\sqrt{x}}\dfrac{1}{2\sqrt{x}} = \dfrac{2\cos x - \cos \sqrt{x}}{2x}$$

38. $y = \int_{2x}^{3x+1} \sin(t^4)\,dt = \int_{2x}^0 \sin(t^4)\, dt + \int_0^{3x+1} \sin(t^4)\, dt = \int_0^{3x+1} \sin(t^4)\, dt - \int_0^{2x} \sin(t^4)\, dt \quad \Rightarrow$

$$y' = \sin[(3x + 1)^4] \cdot \dfrac{d}{dx}(3x + 1) - \sin[(2x)^4] \cdot \dfrac{d}{dx}(2x) = 3\sin[(3x + 1)^4] - 2\sin[(2x)^4]$$

39. If $1 \le x \le 3$, then $\sqrt{1^2 + 3} \le \sqrt{x^2 + 3} \le \sqrt{3^2 + 3} \quad \Rightarrow \quad 2 \le \sqrt{x^2 + 3} \le 2\sqrt{3}$, so

$2(3 - 1) \le \int_1^3 \sqrt{x^2 + 3}\, dx \le 2\sqrt{3}(3 - 1)$; that is, $4 \le \int_1^3 \sqrt{x^2 + 3}\, dx \le 4\sqrt{3}$.

40. If $3 \le x \le 5$, then $4 \le x + 1 \le 6$ and $\dfrac{1}{6} \le \dfrac{1}{x + 1} \le \dfrac{1}{4}$, so $\dfrac{1}{6}(5 - 3) \le \int_3^5 \dfrac{1}{x + 1}\, dx \le \dfrac{1}{4}(5 - 3)$;

that is, $\dfrac{1}{3} \le \int_3^5 \dfrac{1}{x + 1}\, dx \le \dfrac{1}{2}$.

41. $0 \le x \le 1 \quad \Rightarrow \quad 0 \le \cos x \le 1 \quad \Rightarrow \quad x^2 \cos x \le x^2 \quad \Rightarrow \quad \int_0^1 x^2 \cos x\, dx \le \int_0^1 x^2\, dx = \frac{1}{3}[x^3]_0^1 = \frac{1}{3}$ [Property 7].

42. On the interval $\left[\dfrac{\pi}{4}, \dfrac{\pi}{2}\right]$, x is increasing and $\sin x$ is decreasing, so $\dfrac{\sin x}{x}$ is decreasing. Therefore, the largest value of $\dfrac{\sin x}{x}$ on

$\left[\dfrac{\pi}{4}, \dfrac{\pi}{2}\right]$ is $\dfrac{\sin(\pi/4)}{\pi/4} = \dfrac{\sqrt{2}/2}{\pi/4} = \dfrac{2\sqrt{2}}{\pi}$. By Property 8 with $M = \dfrac{2\sqrt{2}}{\pi}$ we get $\displaystyle\int_{\pi/4}^{\pi/2} \dfrac{\sin x}{x}\, dx \le \dfrac{2\sqrt{2}}{\pi}\left(\dfrac{\pi}{2} - \dfrac{\pi}{4}\right) = \dfrac{\sqrt{2}}{2}$.

43. $\Delta x = (3 - 0)/6 = \frac{1}{2}$, so the endpoints are 0, $\frac{1}{2}$, 1, $\frac{3}{2}$, 2, $\frac{5}{2}$, and 3, and the midpoints are $\frac{1}{4}$, $\frac{3}{4}$, $\frac{5}{4}$, $\frac{7}{4}$, $\frac{9}{4}$, and $\frac{11}{4}$.

The Midpoint Rule gives

$$\int_0^3 \sin(x^3)\, dx \approx \sum_{i=1}^6 f(\overline{x}_i)\, \Delta x = \frac{1}{2}\left[\sin\left(\tfrac{1}{4}\right)^3 + \sin\left(\tfrac{3}{4}\right)^3 + \sin\left(\tfrac{5}{4}\right)^3 + \sin\left(\tfrac{7}{4}\right)^3 + \sin\left(\tfrac{9}{4}\right)^3 + \sin\left(\tfrac{11}{4}\right)^3\right] \approx 0.280981.$$

44. (a) Displacement $= \int_0^5 (t^2 - t)\, dt = \left[\frac{1}{3}t^3 - \frac{1}{2}t^2\right]_0^5 = \frac{125}{3} - \frac{25}{2} = \frac{175}{6} = 29.1\overline{6}$ meters

(b) Distance traveled $= \int_0^5 |t^2 - t|\, dt = \int_0^5 |t(t - 1)|\, dt = \int_0^1 (t - t^2)\, dt + \int_1^5 (t^2 - t)\, dt$

$$= \left[\frac{1}{2}t^2 - \frac{1}{3}t^3\right]_0^1 + \left[\frac{1}{3}t^3 - \frac{1}{2}t^2\right]_1^5 = \frac{1}{2} - \frac{1}{3} - 0 + \left(\frac{125}{3} - \frac{25}{2}\right) - \left(\frac{1}{3} - \frac{1}{2}\right) = \frac{177}{6} = 29.5 \text{ meters}$$

45. Note that $r(t) = b'(t)$, where $b(t) =$ the number of barrels of oil consumed up to time t. So, by the Net Change Theorem,

$\int_0^8 r(t)\,dt = b(8) - b(0)$ represents the number of barrels of oil consumed from Jan. 1, 2000, through Jan. 1, 2008.

46. Distance covered $= \int_0^{5.0} v(t)\,dt \approx M_5 = \frac{5.0-0}{5}[v(0.5) + v(1.5) + v(2.5) + v(3.5) + v(4.5)]$

$= 1(4.67 + 8.86 + 10.22 + 10.67 + 10.81) = 45.23$ m

47. We use the Midpoint Rule with $n = 6$ and $\Delta t = \frac{24-0}{6} = 4$. The increase in the bee population was

$\int_0^{24} r(t)\,dt \approx M_6 = 4[r(2) + r(6) + r(10) + r(14) + r(18) + r(22)]$

$\approx 4[50 + 1000 + 7000 + 8550 + 1350 + 150] = 4(18{,}100) = 72{,}400$

48. $A_1 = \frac{1}{2}bh = \frac{1}{2}(2)(2) = 2$, $A_2 = \frac{1}{2}bh = \frac{1}{2}(1)(1) = \frac{1}{2}$, and since

$y = -\sqrt{1 - x^2}$ for $0 \le x \le 1$ represents a quarter-circle with radius 1,

$A_3 = \frac{1}{4}\pi r^2 = \frac{1}{4}\pi(1)^2 = \frac{\pi}{4}$. So

$\int_{-3}^{1} f(x)\,dx = A_1 - A_2 - A_3 = 2 - \frac{1}{2} - \frac{\pi}{4} = \frac{1}{4}(6 - \pi)$

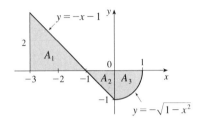

49. Let $u = 2\sin\theta$. Then $du = 2\cos\theta\,d\theta$ and when $\theta = 0$, $u = 0$; when $\theta = \frac{\pi}{2}$, $u = 2$. Thus,

$\int_0^{\pi/2} f(2\sin\theta)\cos\theta\,d\theta = \int_0^2 f(u)\left(\frac{1}{2}\,du\right) = \frac{1}{2}\int_0^2 f(u)\,du = \frac{1}{2}\int_0^2 f(x)\,dx = \frac{1}{2}(6) = 3.$

50. (a) C is increasing on those intervals where C' is positive. By the Fundamental Theorem of Calculus,

$C'(x) = \frac{d}{dx}\left[\int_0^x \cos\left(\frac{\pi}{2}t^2\right)dt\right] = \cos\left(\frac{\pi}{2}x^2\right)$. This is positive when $\frac{\pi}{2}x^2$ is in the interval $\left((2n - \frac{1}{2})\pi, (2n + \frac{1}{2})\pi\right)$,

n any integer. This implies that $(2n - \frac{1}{2})\pi < \frac{\pi}{2}x^2 < (2n + \frac{1}{2})\pi \iff 0 \le |x| \le 1$ or $\sqrt{4n-1} < |x| < \sqrt{4n+1}$, n

any positive integer. So C is increasing on the intervals $[-1, 1]$, $[\sqrt{3}, \sqrt{5}]$, $[-\sqrt{5}, -\sqrt{3}]$, $[\sqrt{7}, 3]$, $[-3, -\sqrt{7}]$, ...

(b) C is concave upward on those intervals where $C'' > 0$. We differentiate C' to find C'': $C'(x) = \cos\left(\frac{\pi}{2}x^2\right) \Rightarrow$

$C''(x) = -\sin\left(\frac{\pi}{2}x^2\right)\left(\frac{\pi}{2}\cdot 2x\right) = -\pi x\sin\left(\frac{\pi}{2}x^2\right)$. For $x > 0$, this is positive where $(2n - 1)\pi < \frac{\pi}{2}x^2 < 2n\pi$, n any

positive integer $\iff \sqrt{2(2n-1)} < x < 2\sqrt{n}$, n any positive integer. Since there is a factor of $-x$ in C'', the intervals

of upward concavity for $x < 0$ are $\left(-\sqrt{2(2n+1)}, -2\sqrt{n}\right)$, n any nonnegative integer. That is, C is concave upward on

$(-\sqrt{2}, 0)$, $(\sqrt{2}, 2)$, $(-\sqrt{6}, -2)$, $(\sqrt{6}, 2\sqrt{2})$, ...

(c)

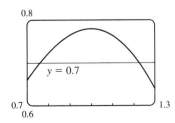

From the graphs, we can determine

that $\int_0^x \cos\left(\frac{\pi}{2}t^2\right)dt = 0.7$ at

$x \approx 0.76$ and $x \approx 1.22$.

(d)

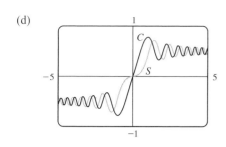

The graphs of $S(x)$ and $C(x)$ have similar shapes, except that S's flattens out near the origin, while C's does not. Note that for $x > 0$, C is increasing where S is concave up, and C is decreasing where S is concave down. Similarly, S is increasing where C is concave down, and S is decreasing where C is concave up. For $x < 0$, these relationships are reversed; that is, C is increasing where S is concave down, and S is increasing where C is concave up. See Example 5.3.3 and Exercise 5.3.53 for a discussion of $S(x)$.

51. $\displaystyle\int_0^x f(t)\,dt = x\sin x + \int_0^x \frac{f(t)}{1+t^2}\,dt \quad\Rightarrow\quad f(x) = x\cos x + \sin x + \frac{f(x)}{1+x^2} \quad$ [by differentiation] $\quad\Rightarrow$

$\displaystyle f(x)\left(1 - \frac{1}{1+x^2}\right) = x\cos x + \sin x \quad\Rightarrow\quad f(x)\left(\frac{x^2}{1+x^2}\right) = x\cos x + \sin x \quad\Rightarrow\quad f(x) = \frac{1+x^2}{x^2}(x\cos x + \sin x)$

52. $2\int_a^x f(t)\,dt = 2\sin x - 1 \quad\Rightarrow\quad \int_a^x f(t)\,dt = \sin x - \frac{1}{2}$. Differentiating both sides using FTC1 gives $f(x) = \cos x$.

We put $x = a$ into the first equation to get $0 = \sin a - \frac{1}{2}$, so $a = \frac{\pi}{6}$ satisfies the given equation.

53. Let $u = f(x)$ and $du = f'(x)\,dx$. So $2\int_a^b f(x)f'(x)\,dx = 2\int_{f(a)}^{f(b)} u\,du = \left[u^2\right]_{f(a)}^{f(b)} = [f(b)]^2 - [f(a)]^2$.

54. Let $F(x) = \displaystyle\int_2^x \sqrt{1+t^3}\,dt$. Then $F'(2) = \lim\limits_{h\to 0} \dfrac{F(2+h) - F(2)}{h} = \lim\limits_{h\to 0}\dfrac{1}{h}\displaystyle\int_2^{2+h}\sqrt{1+t^3}\,dt$, and $F'(x) = \sqrt{1+x^3}$, so

$\lim\limits_{h\to 0}\dfrac{1}{h}\displaystyle\int_2^{2+h}\sqrt{1+t^3}\,dt = F'(2) = \sqrt{1+2^3} = \sqrt{9} = 3$.

55. Let $u = 1 - x$. Then $du = -dx$, so $\int_0^1 f(1-x)\,dx = \int_1^0 f(u)(-du) = \int_0^1 f(u)\,du = \int_0^1 f(x)\,dx$.

56. $\lim\limits_{n\to\infty}\dfrac{1}{n}\left[\left(\dfrac{1}{n}\right)^9 + \left(\dfrac{2}{n}\right)^9 + \left(\dfrac{3}{n}\right)^9 + \cdots + \left(\dfrac{n}{n}\right)^9\right] = \lim\limits_{n\to\infty}\dfrac{1-0}{n}\sum\limits_{i=1}^{n}\left(\dfrac{i}{n}\right)^9 = \displaystyle\int_0^1 x^9\,dx = \left[\dfrac{x^{10}}{10}\right]_0^1 = \dfrac{1}{10}$

The limit is based on Riemann sums using right endpoints and subintervals of equal length.

◻ PROBLEMS PLUS

1. Differentiating both sides of the equation $x \sin \pi x = \int_0^{x^2} f(t)\, dt$ (using FTC1 and the Chain Rule for the right side) gives

   $\sin \pi x + \pi x \cos \pi x = 2x f(x^2)$. Letting $x = 2$ so that $f(x^2) = f(4)$, we obtain $\sin 2\pi + 2\pi \cos 2\pi = 4f(4)$, so

   $f(4) = \frac{1}{4}(0 + 2\pi \cdot 1) = \frac{\pi}{2}$.

2. From the figure, it is clear that the value of a must be between 0 and 1 to obtain the maximum value of the area R.

   $R(a) = \int_a^{a+1}(4x - x^3)\, dx$

   $\qquad = \int_0^{a+1}(4x - x^3)\, dx - \int_0^a(4x - x^3)\, dx$

   By FTC1,

   $R'(a) = 4(a+1) - (a+1)^3 - (4a - a^3)$

   $\qquad = 4a + 4 - (a^3 + 3a^2 + 3a + 1) - 4a + a^3$

   $\qquad = -3a^2 - 3a + 3$

   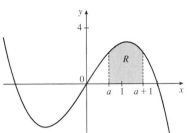

   $R'(a) = 0 \iff a^2 + a - 1 = 0 \implies a = \dfrac{-1 + \sqrt{5}}{2} \approx 0.618$ for $a > 0$. To find the maximum value of R, show that

   $R(a) = \left[2x^2 - \frac{1}{4}x^4 \right]_a^{a+1} = -a^3 - \frac{3}{2}a^2 + 3a + \frac{7}{4}$ and then $R\!\left(\frac{1}{2}\left(-1 + \sqrt{5}\right)\right) = \frac{5}{4}\sqrt{5} \approx 2.795$.

3. Differentiating the given equation, $\int_0^x f(t)\, dt = [f(x)]^2$, using FTC1 gives $f(x) = 2f(x) f'(x) \implies$

   $f(x)[2f'(x) - 1] = 0$, so $f(x) = 0$ or $f'(x) = \frac{1}{2}$. Since $f(x)$ is never 0, we must have $f'(x) = \frac{1}{2}$ and $f'(x) = \frac{1}{2} \implies$

   $f(x) = \frac{1}{2}x + C$. To find C, we substitute into the given equation to get $\int_0^x \left(\frac{1}{2}t + C\right) dt = \left(\frac{1}{2}x + C\right)^2 \iff$

   $\frac{1}{4}x^2 + Cx = \frac{1}{4}x^2 + Cx + C^2$. It follows that $C^2 = 0$, so $C = 0$, and $f(x) = \frac{1}{2}x$.

4. (a)

   From the graph of $f(x) = \dfrac{2cx - x^2}{c^3}$, it appears that the areas

   are equal; that is, the area enclosed is independent of c.

   (b) We first find the x-intercepts of the curve, to determine the limits of integration: $y = 0 \iff 2cx - x^2 = 0 \iff x = 0$

   or $x = 2c$. Now we integrate the function between these limits to find the enclosed area:

   $A = \displaystyle\int_0^{2c} \frac{2cx - x^2}{c^3}\, dx = \frac{1}{c^3}\left[cx^2 - \frac{1}{3}x^3 \right]_0^{2c} = \frac{1}{c^3}\left[c(2c)^2 - \frac{1}{3}(2c)^3 \right] = \frac{1}{c^3}\left[4c^3 - \frac{8}{3}c^3 \right] = \frac{4}{3}$, a constant.

   (c)

   The vertices of the family of parabolas seem to determine a

   branch of a hyperbola.

(d) For a particular c, the vertex is the point where the maximum occurs. We have seen that the x-intercepts are 0 and $2c$, so by

symmetry, the maximum occurs at $x = c$, and its value is $\dfrac{2c(c) - c^2}{c^3} = \dfrac{1}{c}$. So we are interested in the curve consisting of

all points of the form $\left(c, \dfrac{1}{c}\right)$, $c > 0$. This is the part of the hyperbola $y = 1/x$ lying in the first quadrant.

5. $f(x) = \displaystyle\int_0^{g(x)} \dfrac{1}{\sqrt{1+t^3}}\, dt$, where $g(x) = \displaystyle\int_0^{\cos x} [1 + \sin(t^2)]\, dt$. Using FTC1 and the Chain Rule (twice) we have

$f'(x) = \dfrac{1}{\sqrt{1+[g(x)]^3}}\, g'(x) = \dfrac{1}{\sqrt{1+[g(x)]^3}}\, [1 + \sin(\cos^2 x)](-\sin x)$. Now $g\!\left(\tfrac{\pi}{2}\right) = \displaystyle\int_0^0 [1 + \sin(t^2)]\, dt = 0$, so

$f'\!\left(\tfrac{\pi}{2}\right) = \dfrac{1}{\sqrt{1+0}}(1 + \sin 0)(-1) = 1 \cdot 1 \cdot (-1) = -1$.

6. If $f(x) = \int_0^x x^2 \sin(t^2)\, dt = x^2 \int_0^x \sin(t^2)\, dt$, then $f'(x) = x^2 \sin(x^2) + 2x \int_0^x \sin(t^2)\, dt$, by the Product Rule and FTC1.

7. $f(x) = 2 + x - x^2 = (-x + 2)(x + 1) = 0 \quad \Leftrightarrow \quad x = 2$ or $x = -1$. $\quad f(x) \geq 0$ for $x \in [-1, 2]$ and $f(x) < 0$ everywhere

else. The integral $\int_a^b (2 + x - x^2)\, dx$ has a maximum on the interval where the integrand is positive, which is $[-1, 2]$. So

$a = -1$, $b = 2$. (Any larger interval gives a smaller integral since $f(x) < 0$ outside $[-1, 2]$. Any smaller interval also gives a

smaller integral since $f(x) \geq 0$ in $[-1, 2]$.)

8. This sum can be interpreted as a Riemann sum, with the right endpoints of the subintervals as sample

points and with $a = 0$, $b = 10{,}000$, and $f(x) = \sqrt{x}$. So we approximate

$$\sum_{i=1}^{10{,}000} \sqrt{i} \approx \lim_{n \to \infty} \dfrac{10{,}000}{n} \sum_{i=1}^{n} \sqrt{\dfrac{10{,}000i}{n}} = \int_0^{10{,}000} \sqrt{x}\, dx = \left[\tfrac{2}{3} x^{3/2}\right]_0^{10{,}000} = \tfrac{2}{3}(1{,}000{,}000) \approx 666{,}667.$$

Alternate method: We can use graphical methods as follows:

From the figure we see that $\int_{i-1}^{i} \sqrt{x}\, dx < \sqrt{i} < \int_i^{i+1} \sqrt{x}\, dx$, so

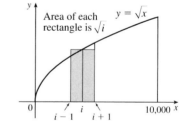

$\int_0^{10{,}000} \sqrt{x}\, dx < \displaystyle\sum_{i=1}^{10{,}000} \sqrt{i} < \int_1^{10{,}001} \sqrt{x}\, dx$. Since

$\int \sqrt{x}\, dx = \tfrac{2}{3} x^{3/2} + C$, we get $\int_0^{10{,}000} \sqrt{x}\, dx = 666{,}666.\overline{6}$ and

$\int_1^{10{,}001} \sqrt{x}\, dx = \tfrac{2}{3}[(10{,}001)^{3/2} - 1] \approx 666{,}766$.

Hence, $666{,}666.\overline{6} < \displaystyle\sum_{i=1}^{10{,}000} \sqrt{i} < 666{,}766$. We can estimate the sum by averaging these bounds:

$\displaystyle\sum_{i=1}^{10{,}000} \approx \dfrac{666{,}666.\overline{6} + 666{,}766}{2} \approx 666{,}716$. The actual value is about $666{,}716.46$.

9. (a) We can split the integral $\int_0^n [\![x]\!]\, dx$ into the sum $\displaystyle\sum_{i=1}^{n} \left[\int_{i-1}^{i} [\![x]\!]\, dx\right]$. But on each of the intervals $[i - 1, i)$ of integration,

$[\![x]\!]$ is a constant function, namely $i - 1$. So the ith integral in the sum is equal to $(i - 1)[i - (i - 1)] = (i - 1)$. So the

original integral is equal to $\displaystyle\sum_{i=1}^{n} (i - 1) = \sum_{i=1}^{n-1} i = \dfrac{(n - 1)n}{2}$.

(b) We can write $\int_a^b [\![x]\!]\, dx = \int_0^b [\![x]\!]\, dx - \int_0^a [\![x]\!]\, dx$.

Now $\int_0^b [\![x]\!]\, dx = \int_0^{[\![b]\!]} [\![x]\!]\, dx + \int_{[\![b]\!]}^b [\![x]\!]\, dx$. The first of these integrals is equal to $\frac{1}{2}([\![b]\!]-1)[\![b]\!]$,

by part (a), and since $[\![x]\!] = [\![b]\!]$ on $[[\![b]\!], b]$, the second integral is just $[\![b]\!](b - [\![b]\!])$. So

$\int_0^b [\![x]\!]\, dx = \frac{1}{2}([\![b]\!]-1)[\![b]\!] + [\![b]\!](b - [\![b]\!]) = \frac{1}{2}[\![b]\!](2b - [\![b]\!] - 1)$ and similarly $\int_0^a [\![x]\!]\, dx = \frac{1}{2}[\![a]\!](2a - [\![a]\!] - 1)$.

Therefore, $\int_a^b [\![x]\!]\, dx = \frac{1}{2}[\![b]\!](2b - [\![b]\!] - 1) - \frac{1}{2}[\![a]\!](2a - [\![a]\!] - 1)$.

10. By FTC1, $\dfrac{d}{dx}\int_0^x \left(\int_1^{\sin t} \sqrt{1+u^4}\, du\right) dt = \int_1^{\sin x}\sqrt{1+u^4}\, du$. Again using FTC1,

$\dfrac{d^2}{dx^2}\int_0^x \left(\int_1^{\sin t}\sqrt{1+u^4}\, du\right) dt = \dfrac{d}{dx}\int_1^{\sin x}\sqrt{1+u^4}\, du = \sqrt{1+\sin^4 x}\,\cos x$.

11. Let $Q(x) = \displaystyle\int_0^x P(t)\, dt = \left[at + \dfrac{b}{2}t^2 + \dfrac{c}{3}t^3 + \dfrac{d}{4}t^4\right]_0^x = ax + \dfrac{b}{2}x^2 + \dfrac{c}{3}x^3 + \dfrac{d}{4}x^4$. Then $Q(0) = 0$, and $Q(1) = 0$ by the

given condition, $a + \dfrac{b}{2} + \dfrac{c}{3} + \dfrac{d}{4} = 0$. Also, $Q'(x) = P(x) = a + bx + cx^2 + dx^3$ by FTC1. By Rolle's Theorem, applied to

Q on $[0, 1]$, there is a number r in $(0, 1)$ such that $Q'(r) = 0$, that is, such that $P(r) = 0$. Thus, the equation $P(x) = 0$ has a

root between 0 and 1.

More generally, if $P(x) = a_0 + a_1 x + a_2 x^2 + \cdots + a_n x^n$ and if $a_0 + \dfrac{a_1}{2} + \dfrac{a_2}{3} + \cdots + \dfrac{a_n}{n+1} = 0$, then the equation

$P(x) = 0$ has a root between 0 and 1. The proof is the same as before:

Let $Q(x) = \displaystyle\int_0^x P(t)\, dt = a_0 x + \dfrac{a_1}{2}x^2 + \dfrac{a_2}{3}x^3 + \cdots + \dfrac{a_n}{n+1}x^n$. Then $Q(0) = Q(1) = 0$ and $Q'(x) = P(x)$. By

Rolle's Theorem applied to Q on $[0, 1]$, there is a number r in $(0, 1)$ such that $Q'(r) = 0$, that is, such that $P(r) = 0$.

12.

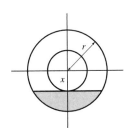

Let x be the distance between the center of the disk and the surface of the liquid.
The wetted circular region has area $\pi r^2 - \pi x^2$ while the unexposed wetted region
(shaded in the diagram) has area $2\int_x^r \sqrt{r^2 - t^2}\, dt$, so the exposed wetted region
has area $A(x) = \pi r^2 - \pi x^2 - 2\int_x^r \sqrt{r^2 - t^2}\, dt$, $0 \le x \le r$. By FTC1, we have
$A'(x) = -2\pi x + 2\sqrt{r^2 - x^2}$.

Now $A'(x) > 0 \;\Rightarrow\; -2\pi x + 2\sqrt{r^2 - x^2} > 0 \;\Rightarrow\; \sqrt{r^2 - x^2} > \pi x \;\Rightarrow\; r^2 - x^2 > \pi^2 x^2 \;\Rightarrow\;$

$r^2 > \pi^2 x^2 + x^2 \;\Rightarrow\; r^2 > x^2(\pi^2 + 1) \;\Rightarrow\; x^2 < \dfrac{r^2}{\pi^2 + 1} \;\Rightarrow\; x < \dfrac{r}{\sqrt{\pi^2 + 1}}$, and we'll call this value x^*.

Since $A'(x) > 0$ for $0 < x < x^*$ and $A'(x) < 0$ for $x^* < x < r$, we have an absolute maximum when $x = x^*$.

13. Note that $\dfrac{d}{dx}\left(\displaystyle\int_0^x \left[\int_0^u f(t)\,dt\right]du\right) = \int_0^x f(t)\,dt$ by FTC1, while

$$\frac{d}{dx}\left[\int_0^x f(u)(x-u)\,du\right] = \frac{d}{dx}\left[x\int_0^x f(u)\,du\right] - \frac{d}{dx}\left[\int_0^x f(u)u\,du\right]$$

$$= \int_0^x f(u)\,du + xf(x) - f(x)x = \int_0^x f(u)\,du$$

Hence, $\int_0^x f(u)(x-u)\,du = \int_0^x\left[\int_0^u f(t)\,dt\right]du + C$. Setting $x = 0$ gives $C = 0$.

14.

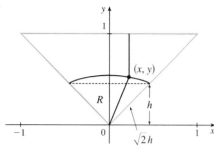

We restrict our attention to the triangle shown. A point in this triangle is closer to the side shown than to any other side, so if we find the area of the region R consisting of all points in the triangle that are closer to the center than to that side, we can multiply this area by 4 to find the total area. We find the equation of the set of points which are equidistant from the center and the side: the distance of the point (x, y) from the side is $1 - y$, and its distance from the center is $\sqrt{x^2 + y^2}$.

So the distances are equal if $\sqrt{x^2 + y^2} = 1 - y \;\Leftrightarrow\; x^2 + y^2 = 1 - 2y + y^2 \;\Leftrightarrow\; y = \frac{1}{2}(1 - x^2)$. Note that the area we are interested in is equal to the area of a triangle plus a crescent-shaped area. To find these areas, we have to find the y-coordinate h of the horizontal line separating them. From the diagram, $1 - h = \sqrt{2}\,h \;\Leftrightarrow\; h = \frac{1}{1+\sqrt{2}} = \sqrt{2} - 1$. We calculate the areas in terms of h, and substitute afterward.

The area of the triangle is $\frac{1}{2}(2h)(h) = h^2$, and the area of the crescent-shaped section is

$$\int_{-h}^h \left[\tfrac{1}{2}(1 - x^2) - h\right]dx = 2\int_0^h\left(\tfrac{1}{2} - h - \tfrac{1}{2}x^2\right)dx = 2\left[\left(\tfrac{1}{2} - h\right)x - \tfrac{1}{6}x^3\right]_0^h = h - 2h^2 - \tfrac{1}{3}h^3.$$

So the area of the whole region is

$$4\left[\left(h - 2h^2 - \tfrac{1}{3}h^3\right) + h^2\right] = 4h\left(1 - h - \tfrac{1}{3}h^2\right) = 4(\sqrt{2} - 1)\left[1 - (\sqrt{2} - 1) - \tfrac{1}{3}(\sqrt{2} - 1)^2\right]$$

$$= 4(\sqrt{2} - 1)\left(1 - \tfrac{1}{3}\sqrt{2}\right) = \tfrac{4}{3}(4\sqrt{2} - 5)$$

15. $\displaystyle\lim_{n\to\infty}\left(\frac{1}{\sqrt{n}\sqrt{n+1}} + \frac{1}{\sqrt{n}\sqrt{n+2}} + \cdots + \frac{1}{\sqrt{n}\sqrt{n+n}}\right)$

$$= \lim_{n\to\infty}\frac{1}{n}\left(\sqrt{\frac{n}{n+1}} + \sqrt{\frac{n}{n+2}} + \cdots + \sqrt{\frac{n}{n+n}}\right)$$

$$= \lim_{n\to\infty}\frac{1}{n}\left(\frac{1}{\sqrt{1 + 1/n}} + \frac{1}{\sqrt{1 + 2/n}} + \cdots + \frac{1}{\sqrt{1 + 1}}\right)$$

$$= \lim_{n\to\infty}\frac{1}{n}\sum_{i=1}^n f\!\left(\frac{i}{n}\right)\qquad\left[\text{where } f(x) = \frac{1}{\sqrt{1 + x}}\right]$$

$$= \int_0^1 \frac{1}{\sqrt{1 + x}}\,dx = \left[2\sqrt{1 + x}\right]_0^1 = 2(\sqrt{2} - 1)$$

16. Note that the graphs of $(x - c)^2$ and $[(x - c) - 2]^2$ intersect when $|x - c| = |x - c - 2|$ ⇔

$c - x = x - c - 2$ ⇔ $x = c + 1$. The integration will proceed differently depending on the value of c.

Case 1: $-2 \leq c < -1$

In this case, $f_c(x) = (x - c - 2)^2$ for $x \in [0, 1]$, so

$$g(c) = \int_0^1 (x - c - 2)^2\, dx = \tfrac{1}{3}\big[(x - c - 2)^3\big]_0^1 = \tfrac{1}{3}\big[(-c - 1)^3 - (-c - 2)^3\big]$$

$$= \tfrac{1}{3}(3c^2 + 9c + 7) = c^2 + 3c + \tfrac{7}{3} = \big(c + \tfrac{3}{2}\big)^2 + \tfrac{1}{12}$$

This is a parabola; its maximum value for $-2 \leq c < -1$ is $g(-2) = \tfrac{1}{3}$, and its minimum value is $g\big(-\tfrac{3}{2}\big) = \tfrac{1}{12}$.

Case 2: $-1 \leq c < 0$

In this case, $f_c(x) = \begin{cases} (x - c)^2 & \text{if } 0 \leq x \leq c + 1 \\ (x - c - 2)^2 & \text{if } c + 1 < x \leq 1 \end{cases}$

Therefore,

$$g(c) = \int_0^1 f_c(x)\, dx = \int_0^{c+1} (x - c)^2\, dx + \int_{c+1}^1 (x - c - 2)^2\, dx$$

$$= \tfrac{1}{3}\big[(x - c)^3\big]_0^{c+1} + \tfrac{1}{3}\big[(x - c - 2)^3\big]_{c+1}^1 = \tfrac{1}{3}\big[1 + c^3 + (-c - 1)^3 - (-1)\big]$$

$$= -c^2 - c + \tfrac{1}{3} = -\big(c + \tfrac{1}{2}\big)^2 + \tfrac{7}{12}$$

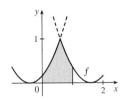

Again, this is a parabola, whose maximum value for $-1 \leq c < 0$ is $g\big(-\tfrac{1}{2}\big) = \tfrac{7}{12}$, and whose minimum value on this c-interval is $g(-1) = \tfrac{1}{3}$.

Case 3: $0 \leq c \leq 2$

In this case, $f_c(x) = (x - c)^2$ for $x \in [0, 1]$, so

$$g(c) = \int_0^1 (x - c)^2\, dx = \tfrac{1}{3}\big[(x - c)^3\big]_0^1 = \tfrac{1}{3}\big[(1 - c)^3 - (-c)^3\big]$$

$$= c^2 - c + \tfrac{1}{3} = \big(c - \tfrac{1}{2}\big)^2 + \tfrac{1}{12}$$

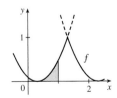

This parabola has a maximum value of $g(2) = \tfrac{7}{3}$ and a minimum value of $g\big(\tfrac{1}{2}\big) = \tfrac{1}{12}$.

We conclude that $g(c)$ has an absolute maximum value of $g(2) = \tfrac{7}{3}$, and absolute minimum values of $g\big(-\tfrac{3}{2}\big) = g\big(\tfrac{1}{2}\big) = \tfrac{1}{12}$.

6 ☐ APPLICATIONS OF INTEGRATION

6.1 Areas Between Curves

1. $A = \int_{x=0}^{x=4} (y_T - y_B)\, dx = \int_0^4 \left[(5x - x^2) - x \right] dx = \int_0^4 (4x - x^2)\, dx = \left[2x^2 - \frac{1}{3}x^3 \right]_0^4 = \left(32 - \frac{64}{3} \right) - (0) = \frac{32}{3}$

2. $A = \int_0^6 \left[2x - (x^2 - 4x) \right] dx = \int_0^6 (6x - x^2)\, dx = \left[3x^2 - \frac{1}{3}x^3 \right]_0^6 = 108 - 72 = 36$

3. $A = \int_{y=0}^{y=1} (x_R - x_L)\, dy = \int_0^1 \left[\sqrt{y} - (y^2 - 1) \right] dy = \int_0^1 (y^{1/2} - y^2 + 1)\, dy = \left[\frac{2}{3}y^{3/2} - \frac{1}{3}y^3 + y \right]_0^1$

$= \left(\frac{2}{3} - \frac{1}{3} + 1 \right) - (0) = \frac{4}{3}$

4. $A = \int_0^3 \left[(2y - y^2) - (y^2 - 4y) \right] dy = \int_0^3 (-2y^2 + 6y)\, dy = \left[-\frac{2}{3}y^3 + 3y^2 \right]_0^3 = (-18 + 27) - 0 = 9$

5. $A = \int_{-1}^2 \left[(9 - x^2) - (x + 1) \right] dx$

$= \int_{-1}^2 (8 - x - x^2)\, dx$

$= \left[8x - \frac{x^2}{2} - \frac{x^3}{3} \right]_{-1}^2$

$= \left(16 - 2 - \frac{8}{3} \right) - \left(-8 - \frac{1}{2} + \frac{1}{3} \right)$

$= 22 - 3 + \frac{1}{2} = \frac{39}{2}$

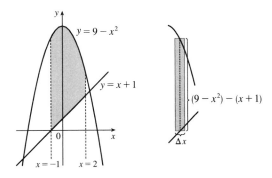

6. $A = \int_{\pi/2}^{\pi} (x - \sin x)\, dx = \left[\frac{x^2}{2} + \cos x \right]_{\pi/2}^{\pi}$

$= \left(\frac{\pi^2}{2} - 1 \right) - \left(\frac{\pi^2}{8} + 0 \right)$

$= \frac{3\pi^2}{8} - 1$

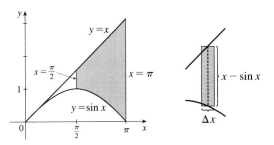

7. The curves intersect when $x = x^2 \iff x^2 - x = 0 \iff$

$x(x - 1) = 0 \iff x = 0$ or 1.

$A = \int_0^1 (x - x^2)\, dx = \left[\frac{1}{2}x^2 - \frac{1}{3}x^3 \right]_0^1 = \frac{1}{2} - \frac{1}{3} = \frac{1}{6}$

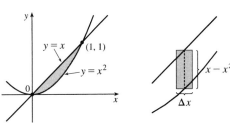

8. The curves intersect when $x^2 - 2x = x + 4 \Leftrightarrow x^2 - 3x - 4 = 0 \Leftrightarrow (x+1)(x-4) = 0 \Leftrightarrow x = -1$ or 4.

$$A = \int_{-1}^{4} \left[x + 4 - (x^2 - 2x) \right] dx$$

$$= \int_{-1}^{4} (-x^2 + 3x + 4) \, dx$$

$$= \left[-\tfrac{1}{3}x^3 + \tfrac{3}{2}x^2 + 4x \right]_{-1}^{4}$$

$$= \left(-\tfrac{64}{3} + 24 + 16 \right) - \left(\tfrac{1}{3} + \tfrac{3}{2} - 4 \right)$$

$$= \tfrac{125}{6}$$

9. First find the points of intersection: $\sqrt{x+3} = \dfrac{x+3}{2} \Rightarrow \left(\sqrt{x+3}\right)^2 = \left(\dfrac{x+3}{2}\right)^2 \Rightarrow x + 3 = \tfrac{1}{4}(x+3)^2 \Rightarrow$

$$4(x+3) - (x+3)^2 = 0 \Rightarrow (x+3)[4 - (x+3)] = 0 \Rightarrow (x+3)(1-x) = 0 \Rightarrow x = -3 \text{ or } 1. \text{ So}$$

$$A = \int_{-3}^{1} \left(\sqrt{x+3} - \frac{x+3}{2} \right) dx$$

$$= \left[\tfrac{2}{3}(x+3)^{3/2} - \frac{(x+3)^2}{4} \right]_{-3}^{1}$$

$$= \left(\tfrac{16}{3} - 4 \right) - (0 - 0) = \tfrac{4}{3}$$

10. $1 + \sqrt{x} = \dfrac{3+x}{3} = 1 + \dfrac{x}{3} \Rightarrow \sqrt{x} = \dfrac{x}{3} \Rightarrow x = \dfrac{x^2}{9} \Rightarrow 9x - x^2 = 0 \Rightarrow x(9-x) = 0 \Rightarrow x = 0 \text{ or } 9, \text{ so}$

$$A = \int_{0}^{9} \left[\left(1 + \sqrt{x}\right) - \left(\frac{3+x}{3}\right) \right] dx = \int_{0}^{9} \left[\left(1 + \sqrt{x}\right) - \left(1 + \frac{x}{3}\right) \right] dx$$

$$= \int_{0}^{9} \left(\sqrt{x} - \tfrac{1}{3}x \right) dx = \left[\tfrac{2}{3}x^{3/2} - \tfrac{1}{6}x^2 \right]_{0}^{9} = 18 - \tfrac{27}{2} = \tfrac{9}{2}$$

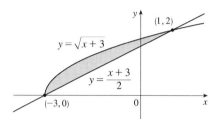

11. $A = \displaystyle\int_{0}^{1} \left(\sqrt{x} - x^2 \right) dx$

$$= \left[\tfrac{2}{3}x^{3/2} - \tfrac{1}{3}x^3 \right]_{0}^{1}$$

$$= \tfrac{2}{3} - \tfrac{1}{3} = \tfrac{1}{3}$$

12. $x^2 = 4x - x^2 \iff 2x^2 - 4x = 0 \iff 2x(x-2) = 0 \iff x = 0$ or 2, so

$$A = \int_0^2 \left[(4x - x^2) - x^2 \right] dx$$

$$= \int_0^2 (4x - 2x^2)\, dx$$

$$= \left[2x^2 - \frac{2}{3}x^3 \right]_0^2$$

$$= 8 - \frac{16}{3} = \frac{8}{3}$$

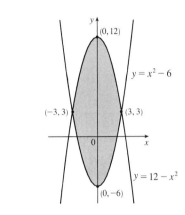

13. $12 - x^2 = x^2 - 6 \iff 2x^2 = 18 \iff$

$x^2 = 9 \iff x = \pm 3$, so

$$A = \int_{-3}^{3} \left[(12 - x^2) - (x^2 - 6) \right] dx$$

$$= 2 \int_0^3 \left(18 - 2x^2 \right) dx \qquad \text{[by symmetry]}$$

$$= 2 \left[18x - \frac{2}{3}x^3 \right]_0^3 = 2\left[(54 - 18) - 0 \right]$$

$$= 2(36) = 72$$

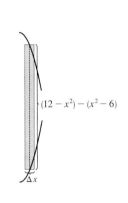

14. $A = \int_0^{2\pi} \left[(2 - \cos x) - \cos x \right] dx$

$$= \int_0^{2\pi} (2 - 2\cos x)\, dx$$

$$= \left[2x - 2\sin x \right]_0^{2\pi}$$

$$= (4\pi - 0) - 0 = 4\pi$$

15. The curves intersect when $8\cos x = \sec^2 x \implies 8\cos^3 x = 1 \implies \cos^3 x = \frac{1}{8} \implies \cos x = \frac{1}{2} \implies$

$x = \frac{\pi}{3}$ for $0 < x < \frac{\pi}{2}$. By symmetry,

$$A = 2\int_0^{\pi/3} (8\cos x - \sec^2 x)\, dx$$

$$= 2 \left[8\sin x - \tan x \right]_0^{\pi/3}$$

$$= 2\left(8 \cdot \frac{\sqrt{3}}{2} - \sqrt{3} \right) = 2\left(3\sqrt{3} \right)$$

$$= 6\sqrt{3}$$

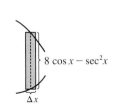

16. $x^3 - x = 3x \iff x^3 - 4x = 0 \iff$

$x(x^2 - 4) = 0 \iff x(x+2)(x-2) = 0 \iff$

$x = 0, -2, \text{ or } 2.$

$A = \int_{-2}^{2} |3x - (x^3 - x)| \, dx$

$\quad = 2\int_{0}^{2} \left[3x - (x^3 - x)\right] dx \qquad \text{[by symmetry]}$

$\quad = 2\int_{0}^{2} (4x - x^3) \, dx = 2\left[2x^2 - \tfrac{1}{4}x^4\right]_0^2 = 2(8 - 4) = 8$

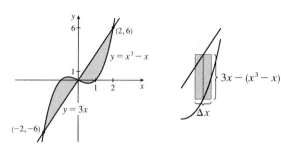

17. $\tfrac{1}{2}x = \sqrt{x} \implies \tfrac{1}{4}x^2 = x \implies x^2 - 4x = 0 \implies x(x-4) = 0 \implies x = 0 \text{ or } 4,$ so

$$A = \int_{0}^{4} \left(\sqrt{x} - \tfrac{1}{2}x\right) dx + \int_{4}^{9} \left(\tfrac{1}{2}x - \sqrt{x}\right) dx = \left[\tfrac{2}{3}x^{3/2} - \tfrac{1}{4}x^2\right]_0^4 + \left[\tfrac{1}{4}x^2 - \tfrac{2}{3}x^{3/2}\right]_4^9$$

$$= \left[\left(\tfrac{16}{3} - 4\right) - 0\right] + \left[\left(\tfrac{81}{4} - 18\right) - \left(4 - \tfrac{16}{3}\right)\right] = \tfrac{81}{4} + \tfrac{32}{3} - 26 = \tfrac{59}{12}$$

18. $A = \int_{-3}^{3} \left|(8 - x^2) - x^2\right| dx = 2\int_{0}^{3} \left|8 - 2x^2\right| dx$

$\quad = 2\int_{0}^{2} (8 - 2x^2) \, dx + 2\int_{2}^{3} (2x^2 - 8) \, dx$

$\quad = 2\left[8x - \tfrac{2}{3}x^3\right]_0^2 + 2\left[\tfrac{2}{3}x^3 - 8x\right]_2^3$

$\quad = 2\left[\left(16 - \tfrac{16}{3}\right) - 0\right] + 2\left[(18 - 24) - \left(\tfrac{16}{3} - 16\right)\right]$

$\quad = 32 - \tfrac{32}{3} + 20 - \tfrac{32}{3} = 52 - \tfrac{64}{3} = \tfrac{92}{3}$

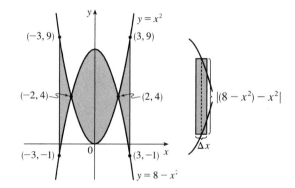

19. $2y^2 = 4 + y^2 \iff y^2 = 4 \iff y = \pm 2,$ so

$A = \int_{-2}^{2} \left[(4 + y^2) - 2y^2\right] dy$

$\quad = 2\int_{0}^{2} (4 - y^2) \, dy \qquad \text{[by symmetry]}$

$\quad = 2\left[4y - \tfrac{1}{3}y^3\right]_0^2 = 2\left(8 - \tfrac{8}{3}\right) = \tfrac{32}{3}$

20. $4x + x^2 = 12 \iff (x+6)(x-2) = 0 \iff x = -6$ or $x = 2$, so $y = -6$ or $y = 2$ and

$$A = \int_{-6}^{2} \left[\left(-\tfrac{1}{4}y^2 + 3\right) - y\right] dy = \left[-\tfrac{1}{12}y^3 - \tfrac{1}{2}y^2 + 3y\right]_{-6}^{2} = \left(-\tfrac{2}{3} - 2 + 6\right) - (18 - 18 - 18) = 22 - \tfrac{2}{3} = \tfrac{64}{3}.$$

21. The curves intersect when $1 - y^2 = y^2 - 1 \iff 2 = 2y^2 \iff y^2 = 1 \iff y = \pm 1$.

$$A = \int_{-1}^{1} \left[(1 - y^2) - (y^2 - 1)\right] dy$$

$$= \int_{-1}^{1} 2(1 - y^2)\, dy$$

$$= 2 \cdot 2 \int_{0}^{1} (1 - y^2)\, dy$$

$$= 4\left[y - \tfrac{1}{3}y^3\right]_0^1 = 4\left(1 - \tfrac{1}{3}\right) = \tfrac{8}{3}$$

22. $A = 2 \int_{0}^{1} \left[\sin\left(\dfrac{\pi x}{2}\right) - x\right] dx$

$= 2\left[-\dfrac{2}{\pi} \cos\left(\dfrac{\pi x}{2}\right) - \dfrac{x^2}{2}\right]_0^1$

$= 2\left[\left(0 - \dfrac{1}{2}\right) - \left(-\dfrac{2}{\pi} - 0\right)\right]$

$= \dfrac{4}{\pi} - 1$

23. Notice that $\cos x = \sin 2x = 2 \sin x \cos x \iff$

$2 \sin x \cos x - \cos x = 0 \iff \cos x\,(2\sin x - 1) = 0 \iff$

$2 \sin x = 1$ or $\cos x = 0 \iff x = \tfrac{\pi}{6}$ or $\tfrac{\pi}{2}$.

$$A = \int_{0}^{\pi/6} (\cos x - \sin 2x)\, dx + \int_{\pi/6}^{\pi/2} (\sin 2x - \cos x)\, dx$$

$$= \left[\sin x + \tfrac{1}{2}\cos 2x\right]_0^{\pi/6} + \left[-\tfrac{1}{2}\cos 2x - \sin x\right]_{\pi/6}^{\pi/2}$$

$$= \tfrac{1}{2} + \tfrac{1}{2}\cdot\tfrac{1}{2} - \left(0 + \tfrac{1}{2}\cdot 1\right) + \left(\tfrac{1}{2} - 1\right) - \left(-\tfrac{1}{2}\cdot\tfrac{1}{2} - \tfrac{1}{2}\right) = \tfrac{1}{2}$$

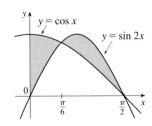

24. The curves intersect when $\cos x = 1 - \cos x$ (on $[0, \pi]$) $\Leftrightarrow$ $2 \cos x = 1$ $\Leftrightarrow$ $\cos x = \frac{1}{2}$ $\Leftrightarrow$ $x = \frac{\pi}{3}$.

$$A = \int_0^{\pi/3} [\cos x - (1 - \cos x)]\, dx + \int_{\pi/3}^{\pi} [(1 - \cos x) - \cos x]\, dx = \int_0^{\pi/3} (2 \cos x - 1)\, dx + \int_{\pi/3}^{\pi} (1 - 2 \cos x)\, dx$$

$$= \Big[2 \sin x - x\Big]_0^{\pi/3} + \Big[x - 2 \sin x\Big]_{\pi/3}^{\pi} = \Big(\sqrt{3} - \frac{\pi}{3}\Big) - 0 + (\pi - 0) - \Big(\frac{\pi}{3} - \sqrt{3}\Big) = 2\sqrt{3} + \frac{\pi}{3}$$

25. From the graph, we see that the curves intersect at $x = 0$, $x = \frac{\pi}{2}$, and $x = \pi$. By symmetry,

$$A = \int_0^{\pi} \left|\cos x - \Big(1 - \frac{2x}{\pi}\Big)\right| dx = 2 \int_0^{\pi/2} \left[\cos x - \Big(1 - \frac{2x}{\pi}\Big)\right] dx = 2 \int_0^{\pi/2} \Big(\cos x - 1 + \frac{2x}{\pi}\Big) dx$$

$$= 2\Big[\sin x - x + \frac{1}{\pi} x^2\Big]_0^{\pi/2} = 2\Big[\Big(1 - \frac{\pi}{2} + \frac{1}{\pi} \cdot \frac{\pi^2}{4}\Big) - 0\Big] = 2\Big(1 - \frac{\pi}{2} + \frac{\pi}{4}\Big) = 2 - \frac{\pi}{2}$$

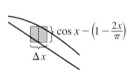

26. For $x > 0$, $x = x^2 - 2$ $\Rightarrow$ $0 = x^2 - x - 2$ $\Rightarrow$ $0 = (x - 2)(x + 1)$ $\Rightarrow$ $x = 2$. By symmetry,

$$A = \int_{-2}^{2} \big[|x| - (x^2 - 2)\big]\, dx$$

$$= 2 \int_0^2 \big[x - (x^2 - 2)\big]\, dx$$

$$= 2 \int_0^2 (x - x^2 + 2)\, dx$$

$$= 2\big[\tfrac{1}{2} x^2 - \tfrac{1}{3} x^3 + 2x\big]_0^2$$

$$= 2\big(2 - \tfrac{8}{3} + 4\big) = \tfrac{20}{3}$$

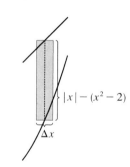

27. Graph the three functions $y = 1/x^2$, $y = x$, and $y = \frac{1}{8}x$; then determine the points of intersection: $(0, 0)$, $(1, 1)$, and $\left(2, \frac{1}{4}\right)$.

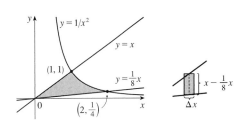

$$A = \int_0^1 \left(x - \frac{1}{8}x\right) dx + \int_1^2 \left(\frac{1}{x^2} - \frac{1}{8}x\right) dx$$

$$= \int_0^1 \frac{7}{8}x \, dx + \int_1^2 \left(x^{-2} - \frac{1}{8}x\right) dx$$

$$= \left[\frac{7}{16}x^2\right]_0^1 + \left[-\frac{1}{x} - \frac{1}{16}x^2\right]_1^2$$

$$= \frac{7}{16} + \left(-\frac{1}{2} - \frac{1}{4}\right) - \left(-1 - \frac{1}{16}\right) = \frac{3}{4}$$

28. The curves $y = 3x^2$ and $y = -4x + 4$ intersect

when $3x^2 = -4x + 4$ [for $x \geq 0$] $\Leftrightarrow$

$3x^2 + 4x - 4 = 0$ $\Leftrightarrow$ $(3x - 2)(x + 2) = 0$ $\Rightarrow$

$x = \frac{2}{3}$. The curves $y = 8x^2$ and $y = -4x + 4$

intersect when $8x^2 = -4x + 4$ [for $x \geq 0$] $\Leftrightarrow$

$8x^2 + 4x - 4 = 0$ $\Leftrightarrow$ $2x^2 + x - 1 = 0$ $\Leftrightarrow$

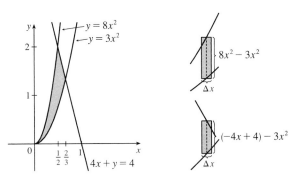

$(2x - 1)(x + 1) = 0$ $\Rightarrow$ $x = \frac{1}{2}$.

$$A = \int_0^{1/2} (8x^2 - 3x^2) \, dx + \int_{1/2}^{2/3} \left[(-4x + 4) - 3x^2\right] dx$$

$$= \int_0^{1/2} 5x^2 \, dx + \int_{1/2}^{2/3} (-3x^2 - 4x + 4) \, dx = \left[\frac{5}{3}x^3\right]_0^{1/2} + \left[-x^3 - 2x^2 + 4x\right]_{1/2}^{2/3}$$

$$= \frac{5}{3}\left(\frac{1}{2}\right)^3 - 0 + \left[-\left(\frac{2}{3}\right)^3 - 2\left(\frac{2}{3}\right)^2 + 4\left(\frac{2}{3}\right)\right] - \left[-\left(\frac{1}{2}\right)^3 - 2\left(\frac{1}{2}\right)^2 + 4\left(\frac{1}{2}\right)\right] = \frac{5}{24} - \frac{8}{27} - \frac{8}{9} + \frac{8}{3} + \frac{1}{8} + \frac{1}{2} - 2$$

$$= \frac{45}{216} - \frac{64}{216} - \frac{192}{216} + \frac{576}{216} + \frac{27}{216} + \frac{108}{216} - \frac{432}{216} = \frac{68}{216} = \frac{17}{54}   \left[\approx 0.315\right]$$

29. An equation of the line through $(0, 0)$ and $(2, 1)$ is $y = \frac{1}{2}x$; through $(0, 0)$

and $(-1, 6)$ is $y = -6x$; through $(2, 1)$ and $(-1, 6)$ is $y = -\frac{5}{3}x + \frac{13}{3}$.

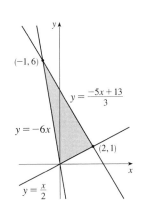

$$A = \int_{-1}^0 \left[\left(-\frac{5}{3}x + \frac{13}{3}\right) - (-6x)\right] dx + \int_0^2 \left[\left(-\frac{5}{3}x + \frac{13}{3}\right) - \frac{1}{2}x\right] dx$$

$$= \int_{-1}^0 \left(\frac{13}{3}x + \frac{13}{3}\right) dx + \int_0^2 \left(-\frac{13}{6}x + \frac{13}{3}\right) dx$$

$$= \frac{13}{3} \int_{-1}^0 (x + 1) \, dx + \frac{13}{3} \int_0^2 \left(-\frac{1}{2}x + 1\right) dx$$

$$= \frac{13}{3}\left[\frac{1}{2}x^2 + x\right]_{-1}^0 + \frac{13}{3}\left[-\frac{1}{4}x^2 + x\right]_0^2$$

$$= \frac{13}{3}\left[0 - \left(\frac{1}{2} - 1\right)\right] + \frac{13}{3}\left[(-1 + 2) - 0\right] = \frac{13}{3} \cdot \frac{1}{2} + \frac{13}{3} \cdot 1 = \frac{13}{2}$$

30. $A = \displaystyle\int_0^2 \left[\left(-\tfrac{4}{5}x + 5\right) - \left(-\tfrac{7}{2}x + 5\right) \right] dx + \int_2^5 \left[\left(-\tfrac{4}{5}x + 5\right) - (x - 4) \right] dx$

$\quad = \displaystyle\int_0^2 \tfrac{27}{10}x \, dx + \int_2^5 \left(-\tfrac{9}{5}x + 9\right) dx$

$\quad = \left[\tfrac{27}{20}x^2\right]_0^2 + \left[-\tfrac{9}{10}x^2 + 9x\right]_2^5$

$\quad = \left(\tfrac{27}{5} - 0\right) + \left(-\tfrac{45}{2} + 45\right) - \left(-\tfrac{18}{5} + 18\right) = \tfrac{27}{2}$

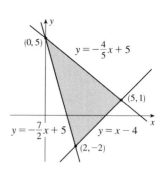

31. The curves intersect when $\sin x = \cos 2x$ (on $[0, \pi/2]$) $\Leftrightarrow$ $\sin x = 1 - 2\sin^2 x$ $\Leftrightarrow$ $2\sin^2 x + \sin x - 1 = 0$ $\Leftrightarrow$

$\quad (2\sin x - 1)(\sin x + 1) = 0 \Rightarrow \sin x = \tfrac{1}{2} \Rightarrow x = \tfrac{\pi}{6}.$

$\quad A = \displaystyle\int_0^{\pi/2} |\sin x - \cos 2x| \, dx$

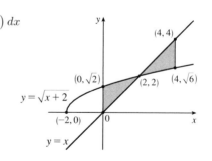

$\quad = \displaystyle\int_0^{\pi/6} (\cos 2x - \sin x) \, dx + \int_{\pi/6}^{\pi/2} (\sin x - \cos 2x) \, dx$

$\quad = \left[\tfrac{1}{2}\sin 2x + \cos x\right]_0^{\pi/6} + \left[-\cos x - \tfrac{1}{2}\sin 2x\right]_{\pi/6}^{\pi/2}$

$\quad = \left(\tfrac{1}{4}\sqrt{3} + \tfrac{1}{2}\sqrt{3}\right) - (0 + 1) + (0 - 0) - \left(-\tfrac{1}{2}\sqrt{3} - \tfrac{1}{4}\sqrt{3}\right)$

$\quad = \tfrac{3}{2}\sqrt{3} - 1$

32. The curves intersect when $\sqrt{x + 2} = x \Rightarrow x + 2 = x^2 \Rightarrow x^2 - x - 2 = 0 \Rightarrow (x - 2)(x + 1) = 0 \Rightarrow$

$\quad x = -1$ or 2. [-1 is extraneous]

$\quad A = \displaystyle\int_0^4 \left|\sqrt{x + 2} - x\right| dx = \int_0^2 \left(\sqrt{x + 2} - x\right) dx + \int_2^4 \left(x - \sqrt{x + 2}\right) dx$

$\quad = \left[\tfrac{2}{3}(x + 2)^{3/2} - \tfrac{1}{2}x^2\right]_0^2 + \left[\tfrac{1}{2}x^2 - \tfrac{2}{3}(x + 2)^{3/2}\right]_2^4$

$\quad = \left(\tfrac{16}{3} - 2\right) - \left[\tfrac{2}{3}\left(2\sqrt{2}\right) - 0\right] + \left[8 - \tfrac{2}{3}\left(6\sqrt{6}\right)\right] - \left(2 - \tfrac{16}{3}\right)$

$\quad = 4 + \tfrac{32}{3} - \tfrac{4}{3}\sqrt{2} - 4\sqrt{6} = \tfrac{44}{3} - 4\sqrt{6} - \tfrac{4}{3}\sqrt{2}$

33. Let $f(x) = \cos^2\left(\dfrac{\pi x}{4}\right) - \sin^2\left(\dfrac{\pi x}{4}\right)$ and $\Delta x = \dfrac{1 - 0}{4}$.

The shaded area is given by

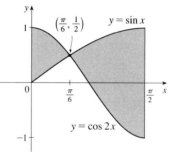

$A = \displaystyle\int_0^1 f(x) \, dx \approx M_4$

$\quad = \tfrac{1}{4}\left[f\left(\tfrac{1}{8}\right) + f\left(\tfrac{3}{8}\right) + f\left(\tfrac{5}{8}\right) + f\left(\tfrac{7}{8}\right)\right]$

$\quad \approx 0.6407$

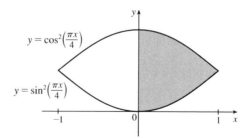

34. The curves intersect when $\sqrt[3]{16 - x^3} = x$ $\Rightarrow$

$16 - x^3 = x^3$ $\Rightarrow$ $2x^3 = 16$ $\Rightarrow$ $x^3 = 8$ $\Rightarrow$ $x = 2$.

Let $f(x) = \sqrt[3]{16 - x^3} - x$ and $\Delta x = \frac{2-0}{4}$.

The shaded area is given by

$A = \int_0^2 f(x)\,dx \approx M_4$

$\qquad = \frac{2}{4}\left[f\left(\frac{1}{4}\right) + f\left(\frac{3}{4}\right) + f\left(\frac{5}{4}\right) + f\left(\frac{7}{4}\right)\right]$

$\qquad \approx 2.8144$

35.

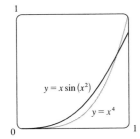

From the graph, we see that the curves intersect at $x = 0$ and $x = a \approx 0.896$, with

$x\sin(x^2) > x^4$ on $(0, a)$. So the area A of the region bounded by the curves is

$$A = \int_0^a \left[x\sin(x^2) - x^4\right]dx = \left[-\tfrac{1}{2}\cos(x^2) - \tfrac{1}{5}x^5\right]_0^a$$

$$= -\tfrac{1}{2}\cos(a^2) - \tfrac{1}{5}a^5 + \tfrac{1}{2} \approx 0.037$$

36.

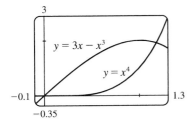

From the graph, we see that the curves intersect at $x = 0$ and at

$x = a \approx 1.17$, with $3x - x^3 > x^4$ on $(0, a)$. So the area of the region

bounded by the curves is

$$A = \int_0^a \left[(3x - x^3) - x^4\right]dx = \left[\tfrac{3}{2}x^2 - \tfrac{1}{4}x^4 - \tfrac{1}{5}x^5\right]_0^a \approx 1.15.$$

37.

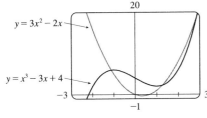

From the graph, we see that the curves intersect at

$x = a \approx -1.11$, $x = b \approx 1.25$, and $x = c \approx 2.86$, with

$x^3 - 3x + 4 > 3x^2 - 2x$ on (a, b) and $3x^2 - 2x > x^3 - 3x + 4$

on (b, c). So the area of the region bounded by the curves is

$$A = \int_a^b \left[(x^3 - 3x + 4) - (3x^2 - 2x)\right]dx + \int_b^c \left[(3x^2 - 2x) - (x^3 - 3x + 4)\right]dx$$

$$= \int_a^b (x^3 - 3x^2 - x + 4)\,dx + \int_b^c (-x^3 + 3x^2 + x - 4)\,dx$$

$$= \left[\tfrac{1}{4}x^4 - x^3 - \tfrac{1}{2}x^2 + 4x\right]_a^b + \left[-\tfrac{1}{4}x^4 + x^3 + \tfrac{1}{2}x^2 - 4x\right]_b^c \approx 8.38$$

38.

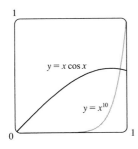

From the graph, we see that the curves intersect at $x = 0$ and $x = a \approx 0.94$, with

$x\cos x > x^{10}$ on $(0, a)$. So the area A of the region bounded by the curves is

$$A = \int_0^a (x\cos x - x^{10})\,dx$$

$$= \left[x\sin x + \cos x - \tfrac{1}{11}x^{11}\right]_0^a \qquad \begin{bmatrix} u = x, & dv = \cos x\,dx \\ du = dx, & v = \sin x \end{bmatrix}$$

$$\approx 0.30$$

39. As the figure illustrates, the curves $y = x$ and $y = x^5 - 6x^3 + 4x$

enclose a four-part region symmetric about the origin (since

$x^5 - 6x^3 + 4x$ and x are odd functions of x). The curves intersect

at values of x where $x^5 - 6x^3 + 4x = x$; that is, where

$x(x^4 - 6x^2 + 3) = 0$. That happens at $x = 0$ and where

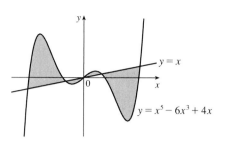

$x^2 = \dfrac{6 \pm \sqrt{36 - 12}}{2} = 3 \pm \sqrt{6}$; that is, at $x = -\sqrt{3 + \sqrt{6}}, -\sqrt{3 - \sqrt{6}}, 0, \sqrt{3 - \sqrt{6}},$ and $\sqrt{3 + \sqrt{6}}$. The exact area is

$$2\int_0^{\sqrt{3+\sqrt{6}}} \left|(x^5 - 6x^3 + 4x) - x\right| dx = 2\int_0^{\sqrt{3+\sqrt{6}}} \left|x^5 - 6x^3 + 3x\right| dx$$

$$= 2\int_0^{\sqrt{3-\sqrt{6}}} (x^5 - 6x^3 + 3x)\, dx + 2\int_{\sqrt{3-\sqrt{6}}}^{\sqrt{3+\sqrt{6}}} (-x^5 + 6x^3 - 3x)\, dx$$

$$\overset{\text{CAS}}{=} 12\sqrt{6} - 9$$

40. The inequality $x \geq 2y^2$ describes the region that lies on, or to the right of,

the parabola $x = 2y^2$. The inequality $x \leq 1 - |y|$ describes the region

that lies on, or to the left of, the curve $x = 1 - |y| = \begin{cases} 1 - y & \text{if } y \geq 0 \\ 1 + y & \text{if } y < 0 \end{cases}$.

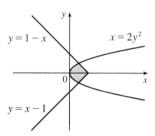

So the given region is the shaded region that lies between the curves.

The graphs of $x = 1 - y$ and $x = 2y^2$ intersect when $1 - y = 2y^2 \iff$

$2y^2 + y - 1 = 0 \iff (2y - 1)(y + 1) = 0 \implies y = \frac{1}{2}$ [for $y \geq 0$]. By symmetry,

$$A = 2\int_0^{1/2} \left[(1 - y) - 2y^2\right] dy = 2\left[-\tfrac{2}{3}y^3 - \tfrac{1}{2}y^2 + y\right]_0^{1/2} = 2\left[\left(-\tfrac{1}{12} - \tfrac{1}{8} + \tfrac{1}{2}\right) - 0\right] = 2\left(\tfrac{7}{24}\right) = \tfrac{7}{12}.$$

41. 1 second $= \frac{1}{3600}$ hour, so 10 s $= \frac{1}{360}$ h. With the given data, we can take $n = 5$ to use the Midpoint Rule.

$\Delta t = \frac{1/360 - 0}{5} = \frac{1}{1800}$, so

$$\text{distance}_{\text{Kelly}} - \text{distance}_{\text{Chris}} = \int_0^{1/360} v_K\, dt - \int_0^{1/360} v_C\, dt = \int_0^{1/360} (v_K - v_C)\, dt$$

$$\approx M_5 = \tfrac{1}{1800}\left[(v_K - v_C)(1) + (v_K - v_C)(3) + (v_K - v_C)(5)\right.$$

$$\left. + (v_K - v_C)(7) + (v_K - v_C)(9)\right]$$

$$= \tfrac{1}{1800}\left[(22 - 20) + (52 - 46) + (71 - 62) + (86 - 75) + (98 - 86)\right]$$

$$= \tfrac{1}{1800}(2 + 6 + 9 + 11 + 12) = \tfrac{1}{1800}(40) = \tfrac{1}{45} \text{ mile, or } 117\tfrac{1}{3} \text{ feet}$$

42. If $x =$ distance from left end of pool and $w = w(x) =$ width at x, then the Midpoint Rule with $n = 4$ and

$\Delta x = \dfrac{b - a}{n} = \dfrac{8 \cdot 2 - 0}{4} = 4$ gives Area $= \int_0^{16} w\, dx \approx 4(6.2 + 6.8 + 5.0 + 4.8) = 4(22.8) = 91.2 \text{ m}^2$.

43. Let $h(x)$ denote the height of the wing at x cm from the left end.

$$A \approx M_5 = \frac{200 - 0}{5}\left[h(20) + h(60) + h(100) + h(140) + h(180)\right]$$

$$= 40(20.3 + 29.0 + 27.3 + 20.5 + 8.7) = 40(105.8) = 4232 \text{ cm}^2$$

44. $A = \int_0^{10} [b(t) - d(t)]\, dt = \int_0^{10} [(2200 + 52.3t + 0.74t^2) - (1460 + 28.8t)]\, dt$

$$= \int_0^{10} (740 + 23.5t + 0.74t^2)\, dt = \left[740t + \frac{23.5}{2}\, t^2 + \frac{0.74}{3}\, t^3 \right]_0^{10}$$

$$= 7400 + 1175 + \tfrac{740}{3} = 8821\tfrac{2}{3} \approx 8822$$

The area A represents the increase in population over a ten-year period.

45. We know that the area under curve A between $t = 0$ and $t = x$ is $\int_0^x v_A(t)\, dt = s_A(x)$, where $v_A(t)$ is the velocity of car A and s_A is its displacement. Similarly, the area under curve B between $t = 0$ and $t = x$ is $\int_0^x v_B(t)\, dt = s_B(x)$.

  (a) After one minute, the area under curve A is greater than the area under curve B. So car A is ahead after one minute.

  (b) The area of the shaded region has numerical value $s_A(1) - s_B(1)$, which is the distance by which A is ahead of B after 1 minute.

  (c) After two minutes, car B is traveling faster than car A and has gained some ground, but the area under curve A from $t = 0$ to $t = 2$ is still greater than the corresponding area for curve B, so car A is still ahead.

  (d) From the graph, it appears that the area between curves A and B for $0 \le t \le 1$ (when car A is going faster), which corresponds to the distance by which car A is ahead, seems to be about 3 squares. Therefore, the cars will be side by side at the time x where the area between the curves for $1 \le t \le x$ (when car B is going faster) is the same as the area for $0 \le t \le 1$. From the graph, it appears that this time is $x \approx 2.2$. So the cars are side by side when $t \approx 2.2$ minutes.

46. The area under $R'(x)$ from $x = 50$ to $x = 100$ represents the change in revenue, and the area under $C'(x)$ from $x = 50$ to $x = 100$ represents the change in cost. The shaded region represents the difference between these two values; that is, the increase in profit as the production level increases from 50 units to 100 units. We use the Midpoint Rule with $n = 5$ and $\Delta x = 10$:

$$M_5 = \Delta x \{ [R'(55) - C'(55)] + [R'(65) - C'(65)] + [R'(75) - C'(75)] + [R'(85) - C'(85)] + [R'(95) - C'(95)] \}$$

$$\approx 10(2.40 - 0.85 + 2.20 - 0.90 + 2.00 - 1.00 + 1.80 - 1.10 + 1.70 - 1.20)$$

$$= 10(5.05) = 50.5 \text{ thousand dollars}$$

Using M_1 would give us $50(2 - 1) = 50$ thousand dollars.

47.

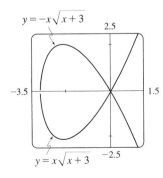

$y = -x\sqrt{x+3}$

$y = x\sqrt{x+3}$

To graph this function, we must first express it as a combination of explicit functions of y; namely, $y = \pm x\sqrt{x+3}$. We can see from the graph that the loop extends from $x = -3$ to $x = 0$, and that by symmetry, the area we seek is just twice the area under the top half of the curve on this interval, the equation of the top half being $y = -x\sqrt{x+3}$. So the area is $A = 2\int_{-3}^0 \left(-x\sqrt{x+3}\right) dx$. We substitute $u = x + 3$, so $du = dx$ and the limits change to 0 and 3, and we get

$$A = -2\int_0^3 [(u-3)\sqrt{u}]\, du = -2\int_0^3 (u^{3/2} - 3u^{1/2})\, du$$

$$= -2\left[\tfrac{2}{5} u^{5/2} - 2u^{3/2}\right]_0^3 = -2\left[\tfrac{2}{5}(3^2\sqrt{3}) - 2(3\sqrt{3})\right] = \tfrac{24}{5}\sqrt{3}$$

48.

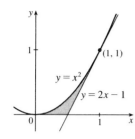

We start by finding the equation of the tangent line to $y = x^2$ at the point $(1, 1)$:

$y' = 2x$, so the slope of the tangent is $2(1) = 2$, and its equation is

$y - 1 = 2(x - 1)$, or $y = 2x - 1$. We would need two integrals to integrate with

respect to x, but only one to integrate with respect to y.

$$A = \int_0^1 \left[\tfrac{1}{2}(y + 1) - \sqrt{y}\right] dy = \left[\tfrac{1}{4}y^2 + \tfrac{1}{2}y - \tfrac{2}{3}y^{3/2}\right]_0^1$$

$$= \tfrac{1}{4} + \tfrac{1}{2} - \tfrac{2}{3} = \tfrac{1}{12}$$

49.

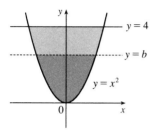

By the symmetry of the problem, we consider only the first quadrant, where

$y = x^2 \ \Rightarrow \ x = \sqrt{y}$. We are looking for a number b such that

$$\int_0^b \sqrt{y}\,dy = \int_b^4 \sqrt{y}\,dy \ \Rightarrow \ \tfrac{2}{3}\left[y^{3/2}\right]_0^b = \tfrac{2}{3}\left[y^{3/2}\right]_b^4 \ \Rightarrow$$

$$b^{3/2} = 4^{3/2} - b^{3/2} \ \Rightarrow \ 2b^{3/2} = 8 \ \Rightarrow \ b^{3/2} = 4 \ \Rightarrow \ b = 4^{2/3} \approx 2.52.$$

50. (a) We want to choose a so that

$$\int_1^a \frac{1}{x^2}\,dx = \int_a^4 \frac{1}{x^2}\,dx \ \Rightarrow \ \left[\frac{-1}{x}\right]_1^a = \left[\frac{-1}{x}\right]_a^4 \ \Rightarrow \ -\frac{1}{a} + 1 = -\frac{1}{4} + \frac{1}{a} \ \Rightarrow \ \frac{5}{4} = \frac{2}{a} \ \Rightarrow \ a = \frac{8}{5}.$$

(b) The area under the curve $y = 1/x^2$ from $x = 1$ to $x = 4$ is $\tfrac{3}{4}$ [take $a = 4$ in the first integral in part (a)]. Now the line

$y = b$ must intersect the curve $x = 1/\sqrt{y}$ and not the line $x = 4$, since the area under the line $y = 1/4^2$ from $x = 1$ to

$x = 4$ is only $\tfrac{3}{16}$, which is less than half of $\tfrac{3}{4}$. We want to choose b so that the upper area in the diagram is half of the total

area under the curve $y = 1/x^2$ from $x = 1$ to $x = 4$. This implies that

$\int_b^1 \left(1/\sqrt{y} - 1\right) dy = \tfrac{1}{2} \cdot \tfrac{3}{4} \ \Rightarrow \ \left[2\sqrt{y} - y\right]_b^1 = \tfrac{3}{8} \ \Rightarrow \ 1 - 2\sqrt{b} + b = \tfrac{3}{8} \ \Rightarrow$

$b - 2\sqrt{b} + \tfrac{5}{8} = 0$. Letting $c = \sqrt{b}$, we get $c^2 - 2c + \tfrac{5}{8} = 0 \ \Rightarrow$

$8c^2 - 16c + 5 = 0$. Thus, $c = \frac{16 \pm \sqrt{256 - 160}}{16} = 1 \pm \frac{\sqrt{6}}{4}$. But $c = \sqrt{b} < 1 \ \Rightarrow$

$c = 1 - \frac{\sqrt{6}}{4} \ \Rightarrow \ b = c^2 = 1 + \tfrac{3}{8} - \frac{\sqrt{6}}{2} = \tfrac{1}{8}\left(11 - 4\sqrt{6}\right) \approx 0.1503.$

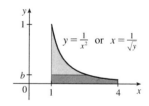

51. We first assume that $c > 0$, since c can be replaced by $-c$ in both equations without changing the graphs, and if $c = 0$ the

curves do not enclose a region. We see from the graph that the enclosed area A lies between $x = -c$ and $x = c$, and by

symmetry, it is equal to four times the area in the first quadrant. The enclosed area is

$$A = 4\int_0^c (c^2 - x^2)\,dx = 4\left[c^2 x - \tfrac{1}{3}x^3\right]_0^c = 4\left(c^3 - \tfrac{1}{3}c^3\right) = 4\left(\tfrac{2}{3}c^3\right) = \tfrac{8}{3}c^3$$

So $A = 576 \ \Leftrightarrow \ \tfrac{8}{3}c^3 = 576 \ \Leftrightarrow \ c^3 = 216 \ \Leftrightarrow \ c = \sqrt[3]{216} = 6.$

Note that $c = -6$ is another solution, since the graphs are the same.

52.

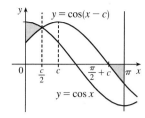

It appears from the diagram that the curves $y = \cos x$ and $y = \cos(x - c)$ intersect halfway between 0 and c, namely, when $x = c/2$. We can verify that this is indeed true by noting that $\cos(c/2 - c) = \cos(-c/2) = \cos(c/2)$. The point where $\cos(x - c)$ crosses the x-axis is $x = \frac{\pi}{2} + c$. So we require that

$\int_0^{c/2} [\cos x - \cos(x - c)] \, dx = - \int_{\pi/2+c}^{\pi} \cos(x - c) \, dx$ [the negative sign on

the RHS is needed since the second area is beneath the x-axis] $\Leftrightarrow$ $[\sin x - \sin(x - c)]_0^{c/2} = -[\sin(x - c)]_{\pi/2+c}^{\pi}$ $\Rightarrow$

$[\sin(c/2) - \sin(-c/2)] - [-\sin(-c)] = -\sin(\pi - c) + \sin[(\frac{\pi}{2} + c) - c]$ $\Leftrightarrow$ $2\sin(c/2) - \sin c = -\sin c + 1$.

[Here we have used the oddness of the sine function, and the fact that $\sin(\pi - c) = \sin c$]. So $2\sin(c/2) = 1$ $\Leftrightarrow$

$\sin(c/2) = \frac{1}{2}$ $\Leftrightarrow$ $c/2 = \frac{\pi}{6}$ $\Leftrightarrow$ $c = \frac{\pi}{3}$.

53. $A = \int_1^2 \left(\frac{1}{x} - \frac{1}{x^2} \right) dx = \left[\ln x + \frac{1}{x} \right]_1^2$

$= \left(\ln 2 + \frac{1}{2} \right) - (\ln 1 + 1)$

$= \ln 2 - \frac{1}{2} \approx 0.19$

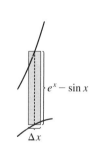

54. $A = \int_0^{\pi/2} (e^x - \sin x) \, dx = \left[e^x + \cos x \right]_0^{\pi/2}$

$= (e^{\pi/2} + 0) - (1 + 1) = e^{\pi/2} - 2$

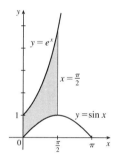

55. The curves intersect when $\tan x = 2 \sin x$ (on $[-\pi/3, \pi/3]$) $\Leftrightarrow$ $\sin x = 2 \sin x \cos x$ $\Leftrightarrow$

$2 \sin x \cos x - \sin x = 0$ $\Leftrightarrow$ $\sin x (2 \cos x - 1) = 0$ $\Leftrightarrow$ $\sin x = 0$ or $\cos x = \frac{1}{2}$ $\Leftrightarrow$ $x = 0$ or $x = \pm\frac{\pi}{3}$.

$A = \int_{-\pi/3}^{\pi/3} (2 \sin x - \tan x) \, dx$

$= 2 \int_0^{\pi/3} (2 \sin x - \tan x) \, dx$ [by symmetry]

$= 2 \left[-2 \cos x - \ln |\sec x| \right]_0^{\pi/3}$

$= 2 [(-1 - \ln 2) - (-2 - 0)]$

$= 2(1 - \ln 2) = 2 - 2 \ln 2$

56. The curve and the line will determine a region when they intersect at two or
more points. So we solve the equation $x/(x^2+1) = mx$ ⇒

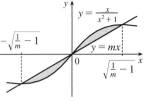

$x = x(mx^2+m)$ ⇒ $x(mx^2+m) - x = 0$ ⇒

$x(mx^2+m-1) = 0$ ⇒ $x = 0$ or $mx^2+m-1 = 0$ ⇒

$x = 0$ or $x^2 = \dfrac{1-m}{m}$ ⇒ $x = 0$ or $x = \pm\sqrt{\dfrac{1}{m}-1}$. Note that if $m = 1$, this has only the solution $x = 0$, and no region

is determined. But if $1/m - 1 > 0$ ⇔ $1/m > 1$ ⇔ $0 < m < 1$, then there are two solutions. [Another way of seeing

this is to observe that the slope of the tangent to $y = x/(x^2+1)$ at the origin is $y'(0) = 1$ and therefore we must have

$0 < m < 1$.] Note that we cannot just integrate between the positive and negative roots, since the curve and the line cross at

the origin. Since mx and $x/(x^2+1)$ are both odd functions, the total area is twice the area between the curves on the interval

$\left[0, \sqrt{1/m-1}\right]$. So the total area enclosed is

$$2\int_0^{\sqrt{1/m-1}} \left[\frac{x}{x^2+1} - mx\right] dx = 2\left[\tfrac{1}{2}\ln(x^2+1) - \tfrac{1}{2}mx^2\right]_0^{\sqrt{1/m-1}} = [\ln(1/m-1+1) - m(1/m-1)] - (\ln 1 - 0)$$

$$= \ln(1/m) - 1 + m = m - \ln m - 1$$

6.2 Volumes

1. A cross-section is a disk with radius $2 - \tfrac{1}{2}x$, so its area is $A(x) = \pi\left(2 - \tfrac{1}{2}x\right)^2$.

$$V = \int_1^2 A(x)\,dx = \int_1^2 \pi\left(2 - \tfrac{1}{2}x\right)^2 dx$$

$$= \pi\int_1^2 \left(4 - 2x + \tfrac{1}{4}x^2\right) dx$$

$$= \pi\left[4x - x^2 + \tfrac{1}{12}x^3\right]_1^2$$

$$= \pi\left[\left(8 - 4 + \tfrac{8}{12}\right) - \left(4 - 1 + \tfrac{1}{12}\right)\right]$$

$$= \pi\left(1 + \tfrac{7}{12}\right) = \tfrac{19}{12}\pi$$

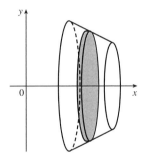

2. A cross-section is a disk with radius $1 - x^2$, so its area is

$A(x) = \pi(1-x^2)^2$.

$$V = \int_{-1}^1 A(x)\,dx = \int_{-1}^1 \pi(1-x^2)^2\,dx$$

$$= 2\pi\int_0^1 \left(1 - 2x^2 + x^4\right) dx = 2\pi\left[x - \tfrac{2}{3}x^3 + \tfrac{1}{5}x^5\right]_0^1$$

$$= 2\pi\left(1 - \tfrac{2}{3} + \tfrac{1}{5}\right) = 2\pi\left(\tfrac{8}{15}\right) = \tfrac{16}{15}\pi$$

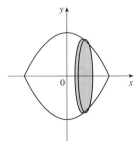

3. A cross-section is a disk with radius $1/x$, so its area is

$$A(x) = \pi(1/x)^2.$$

$$V = \int_1^2 A(x)\,dx = \int_1^2 \pi\left(\frac{1}{x}\right)^2 dx$$

$$= \pi \int_1^2 \frac{1}{x^2}\,dx = \pi\left[-\frac{1}{x}\right]_1^2$$

$$= \pi\left[-\tfrac{1}{2} - (-1)\right] = \tfrac{\pi}{2}$$

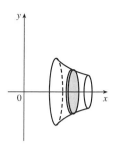

4. A cross-section is a disk with radius $\sqrt{25 - x^2}$, so its area is $A(x) = \pi\left(\sqrt{25 - x^2}\right)^2$.

$$V = \int_2^4 A(x)\,dx = \int_2^4 \pi\left(\sqrt{25 - x^2}\right)^2 dx$$

$$= \pi \int_2^4 (25 - x^2)\,dx = \pi\left[25x - \tfrac{1}{3}x^3\right]_2^4$$

$$= \pi\left[\left(100 - \tfrac{64}{3}\right) - \left(50 - \tfrac{8}{3}\right)\right] = \tfrac{94}{3}\pi$$

5. A cross-section is a disk with radius $2\sqrt{y}$, so its area is

$$A(y) = \pi\left(2\sqrt{y}\right)^2.$$

$$V = \int_0^9 A(y)\,dy = \int_0^9 \pi\left(2\sqrt{y}\right)^2 dy = 4\pi \int_0^9 y\,dy$$

$$= 4\pi\left[\tfrac{1}{2}y^2\right]_0^9 = 2\pi(81) = 162\pi$$

6. A cross-section is a disk with radius $y - y^2$, so its area is $A(y) = \pi(y - y^2)^2$.

$$V = \int_0^1 A(y)\,dy = \int_0^1 \pi(y - y^2)^2 dy$$

$$= \pi(y^4 - 2y^3 + y^2)\,dy = \pi\left[\tfrac{1}{5}y^5 - \tfrac{1}{2}y^4 + \tfrac{1}{3}y^3\right]_0^1$$

$$= \pi\left(\tfrac{1}{5} - \tfrac{1}{2} + \tfrac{1}{3}\right) = \tfrac{\pi}{30}$$

7. A cross-section is a washer (annulus) with inner

radius x^3 and outer radius x, so its area is

$$A(x) = \pi(x)^2 - \pi(x^3)^2 = \pi(x^2 - x^6).$$

$$V = \int_0^1 A(x)\,dx = \int_0^1 \pi(x^2 - x^6)\,dx$$

$$= \pi\left[\tfrac{1}{3}x^3 - \tfrac{1}{7}x^7\right]_0^1 = \pi\left(\tfrac{1}{3} - \tfrac{1}{7}\right) = \tfrac{4}{21}\pi$$

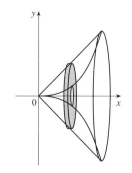

8. A cross-section is a washer with inner radius $\frac{1}{4}x^2$

and outer radius $5 - x^2$, so its area is

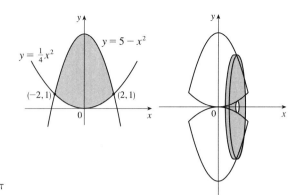

$A(x) = \pi(5 - x^2)^2 - \pi\left(\frac{1}{4}x^2\right)^2$

$\qquad = \pi\left(25 - 10x^2 + x^4 - \frac{1}{16}x^4\right).$

$V = \int_{-2}^{2} A(x)\,dx = \int_{-2}^{2} \pi\left(25 - 10x^2 + \frac{15}{16}x^4\right)dx$

$\qquad = 2\pi \int_0^2 \left(25 - 10x^2 + \frac{15}{16}x^4\right)dx$

$\qquad = 2\pi\left[25x - \frac{10}{3}x^3 + \frac{3}{16}x^5\right]_0^2 = 2\pi\left(50 - \frac{80}{3} + 6\right) = \frac{176}{3}\pi$

9. A cross-section is a washer with inner radius y^2

and outer radius $2y$, so its area is

$A(y) = \pi(2y)^2 - \pi(y^2)^2 = \pi(4y^2 - y^4).$

$V = \int_0^2 A(y)\,dy = \pi\int_0^2 (4y^2 - y^4)\,dy$

$\qquad = \pi\left[\frac{4}{3}y^3 - \frac{1}{5}y^5\right]_0^2 = \pi\left(\frac{32}{3} - \frac{32}{5}\right) = \frac{64}{15}\pi$

10. A cross-section is a washer with inner radius $x = 2\sqrt{y}$ and outer radius 2, so its area is

$A(y) = \pi\left[(2)^2 - \left(2\sqrt{y}\right)^2\right]$

$\qquad = \pi(4 - 4y) = 4\pi(1 - y).$

$V = \int_0^1 A(y)\,dy = \int_0^1 4\pi(1 - y)\,dy$

$\qquad = 4\pi\left[y - \frac{1}{2}y^2\right]_0^1 = 4\pi\left[\left(1 - \frac{1}{2}\right) - 0\right] = 2\pi$

11. A cross-section is a washer with inner radius $1 - \sqrt{x}$ and outer radius $1 - x$, so its area is

$A(x) = \pi(1 - x)^2 - \pi\left(1 - \sqrt{x}\right)^2$

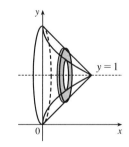

$\qquad = \pi\left[(1 - 2x + x^2) - \left(1 - 2\sqrt{x} + x\right)\right]$

$\qquad = \pi\left(-3x + x^2 + 2\sqrt{x}\right).$

$V = \int_0^1 A(x)\,dx = \pi\int_0^1 \left(-3x + x^2 + 2\sqrt{x}\right)dx$

$\qquad = \pi\left[-\frac{3}{2}x^2 + \frac{1}{3}x^3 + \frac{4}{3}x^{3/2}\right]_0^1 = \pi\left(-\frac{3}{2} + \frac{5}{3}\right) = \frac{\pi}{6}$

12. A cross-section is circular with radius $4 - x^2$, so its area is $A(x) = \pi(4 - x^2)^2 = \pi(16 - 8x^2 + x^4)$.

$$V = \int_{-2}^{2} A(x)\, dx = 2\int_{0}^{2} A(x)\, dx$$

$$= 2\pi \int_{0}^{2} (16 - 8x^2 + x^4)\, dx$$

$$= 2\pi \left[16x - \tfrac{8}{3}x^3 + \tfrac{1}{5}x^5\right]_0^2$$

$$= 2\pi \left(32 - \tfrac{64}{3} + \tfrac{32}{5}\right) = 64\pi \left(1 - \tfrac{2}{3} + \tfrac{1}{5}\right)$$

$$= 64\pi \cdot \tfrac{8}{15} = \tfrac{512\pi}{15}$$

13. A cross-section is a washer with inner radius $(1 + \sec x) - 1 = \sec x$ and outer radius $3 - 1 = 2$, so its area is

$$A(x) = \pi\left[2^2 - (\sec x)^2\right] = \pi(4 - \sec^2 x).$$

$$V = \int_{-\pi/3}^{\pi/3} A(x)\, dx = \int_{-\pi/3}^{\pi/3} \pi(4 - \sec^2 x)\, dx$$

$$= 2\pi \int_{0}^{\pi/3} (4 - \sec^2 x)\, dx \qquad \text{[by symmetry]}$$

$$= 2\pi \left[4x - \tan x\right]_0^{\pi/3} = 2\pi\left[\left(\tfrac{4\pi}{3} - \sqrt{3}\right) - 0\right]$$

$$= 2\pi\left(\tfrac{4\pi}{3} - \sqrt{3}\right)$$

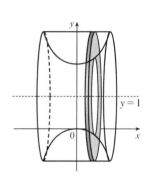

14.
$$V = \int_{1}^{3} \pi \left\{\left[\frac{1}{x^2} - (-1)\right]^2 - \left[0 - (-1)\right]^2\right\} dx$$

$$= \pi \int_{1}^{3} \left[\left(\frac{1}{x^2} + 1\right)^2 - 1^2\right] dx$$

$$= \pi \int_{1}^{3} \left(\frac{1}{x^4} + \frac{2}{x^2}\right) dx = \pi \left[-\frac{1}{3x^3} - \frac{2}{x}\right]_1^3$$

$$= \pi \left[\left(-\tfrac{1}{81} - \tfrac{2}{3}\right) - \left(-\tfrac{1}{3} - 2\right)\right] = \frac{134\pi}{81}$$

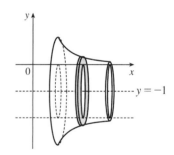

15.
$$V = \int_{-1}^{1} \pi(1 - y^2)^2\, dy = 2\int_{0}^{1} \pi(1 - y^2)^2\, dy$$

$$= 2\pi \int_{0}^{1} (1 - 2y^2 + y^4)\, dy$$

$$= 2\pi \left[y - \tfrac{2}{3}y^3 + \tfrac{1}{5}y^5\right]_0^1$$

$$= 2\pi \cdot \tfrac{8}{15} = \tfrac{16}{15}\pi$$

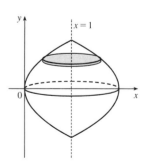

16. $y = \sqrt{x} \quad \Rightarrow \quad x = y^2$, so the outer radius is $2 - y^2$.

$$V = \int_0^1 \pi \left[(2 - y^2)^2 - (2 - y)^2 \right] dy$$

$$= \pi \int_0^1 \left[(4 - 4y^2 + y^4) - (4 - 4y + y^2) \right] dy$$

$$= \pi \int_0^1 (y^4 - 5y^2 + 4y) \, dy$$

$$= \pi \left[\tfrac{1}{5} y^5 - \tfrac{5}{3} y^3 + 2y^2 \right]_0^1$$

$$= \pi \left(\tfrac{1}{5} - \tfrac{5}{3} + 2 \right) = \tfrac{8}{15} \pi$$

17. $y = x^2 \quad \Rightarrow \quad x = \sqrt{y}$ for $x \geq 0$. The outer radius is the distance from $x = -1$ to $x = \sqrt{y}$ and the inner radius is the

distance from $x = -1$ to $x = y^2$.

$$V = \int_0^1 \pi \left\{ \left[\sqrt{y} - (-1) \right]^2 - \left[y^2 - (-1) \right]^2 \right\} dy = \pi \int_0^1 \left[\left(\sqrt{y} + 1 \right)^2 - (y^2 + 1)^2 \right] dy$$

$$= \pi \int_0^1 \left(y + 2\sqrt{y} + 1 - y^4 - 2y^2 - 1 \right) dy = \pi \int_0^1 \left(y + 2\sqrt{y} - y^4 - 2y^2 \right) dy$$

$$= \pi \left[\tfrac{1}{2} y^2 + \tfrac{4}{3} y^{3/2} - \tfrac{1}{5} y^5 - \tfrac{2}{3} y^3 \right]_0^1 = \pi \left(\tfrac{1}{2} + \tfrac{4}{3} - \tfrac{1}{5} - \tfrac{2}{3} \right) = \tfrac{29}{30} \pi$$

18. For $0 \leq y < 2$, a cross-section is an annulus with inner radius $2 - 1$ and outer radius $4 - 1$, the area of which is

$A_1(y) = \pi (4 - 1)^2 - \pi (2 - 1)^2$. For $2 \leq y \leq 4$, a cross-section is an annulus with inner radius $y - 1$ and outer

radius $4 - 1$, the area of which is $A_2(y) = \pi (4 - 1)^2 - \pi (y - 1)^2$.

$$V = \int_0^4 A(y) \, dy = \pi \int_0^2 \left[(4 - 1)^2 - (2 - 1)^2 \right] dy + \pi \int_2^4 \left[(4 - 1)^2 - (y - 1)^2 \right] dy$$

$$= \pi \left[8y \right]_0^2 + \pi \int_2^4 (8 + 2y - y^2) \, dy$$

$$= 16\pi + \pi \left[8y + y^2 - \tfrac{1}{3} y^3 \right]_2^4$$

$$= 16\pi + \pi \left[\left(32 + 16 - \tfrac{64}{3} \right) - \left(16 + 4 - \tfrac{8}{3} \right) \right]$$

$$= \tfrac{76}{3} \pi$$

19. $\mathscr{R}_1$ about OA (the line $y = 0$): $\displaystyle V = \int_0^1 A(x) \, dx = \int_0^1 \pi (x^3)^2 \, dx = \pi \int_0^1 x^6 \, dx = \pi \left[\tfrac{1}{7} x^7 \right]_0^1 = \tfrac{\pi}{7}$

20. $\mathscr{R}_1$ about OC (the line $x = 0$):

$$V = \int_0^1 A(y) \, dy = \int_0^1 \left[\pi (1)^2 - \pi \left(\sqrt[3]{y} \right)^2 \right] dy = \pi \int_0^1 (1 - y^{2/3}) \, dy = \pi \left[y - \tfrac{3}{5} y^{5/3} \right]_0^1 = \pi \left(1 - \tfrac{3}{5} \right) = \tfrac{2}{5} \pi$$

21. $\mathcal{R}_1$ about AB (the line $x = 1$):

$$V = \int_0^1 A(y)\, dy = \int_0^1 \pi\left(1 - \sqrt[3]{y}\right)^2 dy = \pi\int_0^1 (1 - 2y^{1/3} + y^{2/3})\, dy = \pi\left[y - \tfrac{3}{2}y^{4/3} + \tfrac{3}{5}y^{5/3}\right]_0^1$$
$$= \pi\left(1 - \tfrac{3}{2} + \tfrac{3}{5}\right) = \tfrac{\pi}{10}$$

22. $\mathcal{R}_1$ about BC (the line $y = 1$):

$$V = \int_0^1 A(x)\, dx = \int_0^1 \left[\pi(1)^2 - \pi(1 - x^3)^2\right] dx = \pi\int_0^1 \left[1 - (1 - 2x^3 + x^6)\right] dx$$
$$= \pi\int_0^1 (2x^3 - x^6)\, dx = \pi\left[\tfrac{1}{2}x^4 - \tfrac{1}{7}x^7\right]_0^1 = \pi\left(\tfrac{1}{2} - \tfrac{1}{7}\right) = \tfrac{5}{14}\pi$$

23. $\mathcal{R}_2$ about OA (the line $y = 0$):

$$V = \int_0^1 A(x)\, dx = \int_0^1 \left[\pi(1)^2 - \pi\left(\sqrt{x}\right)^2\right] dx = \pi\int_0^1 (1 - x)\, dx = \pi\left[x - \tfrac{1}{2}x^2\right]_0^1 = \pi\left(1 - \tfrac{1}{2}\right) = \tfrac{\pi}{2}$$

24. $\mathcal{R}_2$ about OC (the line $x = 0$): $\quad V = \int_0^1 A(y)\, dy = \int_0^1 \pi(y^2)^2\, dy = \pi\int_0^1 y^4\, dy = \pi\left[\tfrac{1}{5}y^5\right]_0^1 = \tfrac{\pi}{5}$

25. $\mathcal{R}_2$ about AB (the line $x = 1$):

$$V = \int_0^1 A(y)\, dy = \int_0^1 \left[\pi(1)^2 - \pi(1 - y^2)^2\right] dy = \pi\int_0^1 \left[1 - (1 - 2y^2 + y^4)\right] dy = \pi\int_0^1 (2y^2 - y^4)\, dy$$
$$= \pi\left[\tfrac{2}{3}y^3 - \tfrac{1}{5}y^5\right]_0^1 = \pi\left(\tfrac{2}{3} - \tfrac{1}{5}\right) = \tfrac{7}{15}\pi$$

26. $\mathcal{R}_2$ about BC (the line $y = 1$):

$$V = \int_0^1 A(x)\, dx = \int_0^1 \pi\left(1 - \sqrt{x}\right)^2 dx = \pi\int_0^1 (1 - 2x^{1/2} + x)\, dx = \pi\left[x - \tfrac{4}{3}x^{3/2} + \tfrac{1}{2}x^2\right]_0^1 = \pi\left(1 - \tfrac{4}{3} + \tfrac{1}{2}\right) = \tfrac{\pi}{6}$$

27. $\mathcal{R}_3$ about OA (the line $y = 0$):

$$V = \int_0^1 A(x)\, dx = \int_0^1 \left[\pi\left(\sqrt{x}\right)^2 - \pi(x^3)^2\right] dx = \pi\int_0^1 (x - x^6)\, dx = \pi\left[\tfrac{1}{2}x^2 - \tfrac{1}{7}x^7\right]_0^1 = \pi\left(\tfrac{1}{2} - \tfrac{1}{7}\right) = \tfrac{5}{14}\pi.$$

Note: Let $\mathcal{R} = \mathcal{R}_1 + \mathcal{R}_2 + \mathcal{R}_3$. If we rotate $\mathcal{R}$ about any of the segments OA, OC, AB, or BC, we obtain a right circular cylinder of height 1 and radius 1. Its volume is $\pi r^2 h = \pi(1)^2 \cdot 1 = \pi$. As a check for Exercises 19, 23, and 27, we can add the answers, and that sum must equal π. Thus, $\tfrac{\pi}{7} + \tfrac{\pi}{2} + \tfrac{5\pi}{14} = \left(\tfrac{2+7+5}{14}\right)\pi = \pi$.

28. $\mathcal{R}_3$ about OC (the line $x = 0$):

$$V = \int_0^1 A(y)\, dy = \int_0^1 \left[\pi\left(\sqrt[3]{y}\right)^2 - \pi(y^2)^2\right] dy = \pi\int_0^1 (y^{2/3} - y^4)\, dy = \pi\left[\tfrac{3}{5}y^{5/3} - \tfrac{1}{5}y^5\right]_0^1 = \pi\left(\tfrac{3}{5} - \tfrac{1}{5}\right) = \tfrac{2}{5}\pi$$

Note: See the note in Exercise 27. For Exercises 20, 24, and 28, we have $\tfrac{2\pi}{5} + \tfrac{\pi}{5} + \tfrac{2\pi}{5} = \pi$.

29. $\mathcal{R}_3$ about AB (the line $x = 1$):

$$V = \int_0^1 A(y)\, dy = \int_0^1 \left[\pi(1 - y^2)^2 - \pi\left(1 - \sqrt[3]{y}\right)^2\right] dy = \pi\int_0^1 \left[(1 - 2y^2 + y^4) - (1 - 2y^{1/3} + y^{2/3})\right] dy$$
$$= \pi\int_0^1 \left(-2y^2 + y^4 + 2y^{1/3} - y^{2/3}\right) dy = \pi\left[-\tfrac{2}{3}y^3 + \tfrac{1}{5}y^5 + \tfrac{3}{2}y^{4/3} - \tfrac{3}{5}y^{5/3}\right]_0^1 = \pi\left(-\tfrac{2}{3} + \tfrac{1}{5} + \tfrac{3}{2} - \tfrac{3}{5}\right) = \tfrac{13}{30}\pi$$

Note: See the note in Exercise 27. For Exercises 21, 25, and 29, we have $\tfrac{\pi}{10} + \tfrac{7\pi}{15} + \tfrac{13\pi}{30} = \left(\tfrac{3+14+13}{30}\right)\pi = \pi$.

30. $\mathcal{R}_3$ about BC (the line $y = 1$):

$$V = \int_0^1 A(x)\,dx = \int_0^1 \left[\pi(1-x^3)^2 - \pi\left(1-\sqrt{x}\right)^2\right]dx = \pi\int_0^1 \left[(1-2x^3+x^6)-(1-2x^{1/2}+x)\right]dx$$

$$= \pi\int_0^1 (-2x^3+x^6+2x^{1/2}-x)\,dx = \pi\left[-\tfrac{1}{2}x^4 + \tfrac{1}{7}x^7 + \tfrac{4}{3}x^{3/2} - \tfrac{1}{2}x^2\right]_0^1 = \pi\left(-\tfrac{1}{2}+\tfrac{1}{7}+\tfrac{4}{3}-\tfrac{1}{2}\right) = \tfrac{10}{21}\pi$$

Note: See the note in Exercise 27. For Exercises 22, 26, and 30, we have $\frac{5\pi}{14} + \frac{\pi}{6} + \frac{10\pi}{21} = \left(\frac{15+7+20}{42}\right)\pi = \pi$.

31. $V = \pi\displaystyle\int_0^{\pi/4} (1-\tan^3 x)^2\,dx$

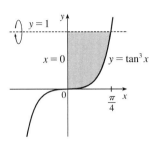

32. $y = (x-2)^4$ and $8x - y = 16$ intersect when

$(x-2)^4 = 8x - 16 = 8(x-2)$ ⇔

$(x-2)^4 - 8(x-2) = 0$ ⇔ $(x-2)[(x-2)^3 - 8] = 0$ ⇔

$x - 2 = 0$ or $x - 2 = 2$ ⇔ $x = 2$ or 4.

$y = (x-2)^4 \Rightarrow x - 2 = \pm\sqrt[4]{y} \Rightarrow$

$x = 2 + \sqrt[4]{y}$ [since $x \geq 2$].

$8x - y = 16 \Rightarrow 8x = y + 16 \Rightarrow x = \tfrac{1}{8}y + 2$.

$V = \pi\displaystyle\int_0^{16} \left\{\left[10 - \left(\tfrac{1}{8}y+2\right)\right]^2 - \left[10 - (2+\sqrt[4]{y})\right]^2\right\}dy$

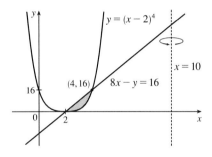

33. $V = \pi\displaystyle\int_0^{\pi} \left[(1-0)^2 - (1-\sin x)^2\right]dx$

$= \pi\displaystyle\int_0^{\pi} \left[1^2 - (1-\sin x)^2\right]dx$

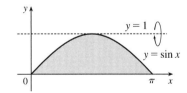

34. $V = \pi\displaystyle\int_0^{\pi} \left\{[\sin x - (-2)]^2 - [0-(-2)]^2\right\}dx$

$= \pi\displaystyle\int_0^{\pi} \left[(\sin x + 2)^2 - 2^2\right]dx$

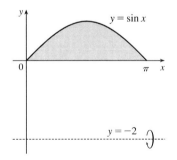

35. $V = \pi \displaystyle\int_{-\sqrt{8}}^{\sqrt{8}} \left\{ [3 - (-2)]^2 - \left[\sqrt{y^2 + 1} - (-2) \right]^2 \right\} dy$

$= \pi \displaystyle\int_{-2\sqrt{2}}^{2\sqrt{2}} \left[5^2 - \left(\sqrt{1 + y^2} + 2 \right)^2 \right] dy$

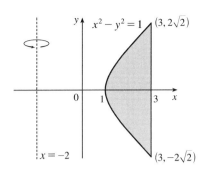

36. $V = \pi \displaystyle\int_{0}^{2\pi} \left\{ (4 - \cos x)^2 - [4 - (2 - \cos x)]^2 \right\} dx$

$= \pi \displaystyle\int_{0}^{2\pi} \left[(4 - \cos x)^2 - (2 + \cos x)^2 \right] dx$

37.

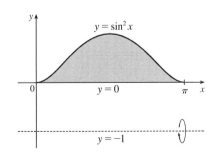

$y = 2 + x^2 \cos x$ and $y = x^4 + x + 1$ intersect at

$x = a \approx -1.288$ and $x = b \approx 0.884$.

$V = \pi \displaystyle\int_{a}^{b} [(2 + x^2 \cos x)^2 - (x^4 + x + 1)^2] dx \approx 23.780$

38. We see from the graph in Exercise 6.1.36 that the x-coordinates of the points of intersection are $x = 0$ and $x = a \approx 1.17$, with

$3x - x^3 > x^4$ on the interval $(0, a)$, so the volume of revolution is

$\pi \displaystyle\int_{0}^{a} [(3x - x^3)^2 - (x^4)^2] dx = \pi \displaystyle\int_{0}^{a} (9x^2 - 6x^4 + x^6 - x^8) dx = \pi \left[3x^3 - \tfrac{6}{5}x^5 + \tfrac{1}{7}x^7 - \tfrac{1}{9}x^9 \right]_{0}^{a} \approx 6.74$

39. $V = \pi \displaystyle\int_{0}^{\pi} \left\{ \left[\sin^2 x - (-1) \right]^2 - [0 - (-1)]^2 \right\} dx$

$\overset{\text{CAS}}{=} \tfrac{11}{8} \pi^2$

40. $V = \pi \int_0^2 \left\{ [2 - (x^2 - 2x)]^2 - [2 - x\cos(\pi x/4)]^2 \right\} dx$

$\overset{\text{CAS}}{=} \dfrac{4(19\pi^2 + 120\pi - 210)}{15\pi}$

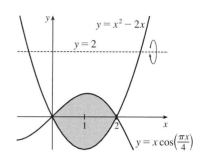

41. $\pi \int_0^{\pi/2} \cos^2 x \, dx$ describes the volume of the solid obtained by rotating the region $\mathcal{R} = \left\{ (x, y) \mid 0 \le x \le \frac{\pi}{2}, 0 \le y \le \cos x \right\}$ of the xy-plane about the x-axis.

42. $\pi \int_2^5 y \, dy = \pi \int_2^5 \left(\sqrt{y} \right)^2 dy$ describes the volume of the solid obtained by rotating the region

$\mathcal{R} = \left\{ (x, y) \mid 2 \le y \le 5, 0 \le x \le \sqrt{y} \right\}$ of the xy-plane about the y-axis.

43. $\pi \int_0^1 (y^4 - y^8) \, dy = \pi \int_0^1 \left[(y^2)^2 - (y^4)^2 \right] dy$ describes the volume of the solid obtained by rotating the region

$\mathcal{R} = \left\{ (x, y) \mid 0 \le y \le 1, y^4 \le x \le y^2 \right\}$ of the xy-plane about the y-axis.

44. $\pi \int_0^{\pi/2} \left[(1 + \cos x)^2 - 1^2 \right] dx$ describes the volume of the solid obtained by rotating the region

$\mathcal{R} = \left\{ (x, y) \mid 0 \le x \le \frac{\pi}{2}, 1 \le y \le 1 + \cos x \right\}$ of the xy-plane about the x-axis.

Or: The solid could be obtained by rotating the region $\mathcal{R}' = \left\{ (x, y) \mid 0 \le x \le \frac{\pi}{2}, 0 \le y \le \cos x \right\}$ about the line $y = -1$.

45. There are 10 subintervals over the 15-cm length, so we'll use $n = 10/2 = 5$ for the Midpoint Rule.

$V = \int_0^{15} A(x) \, dx \approx M_5 = \frac{15 - 0}{5} [A(1.5) + A(4.5) + A(7.5) + A(10.5) + A(13.5)]$

$= 3(18 + 79 + 106 + 128 + 39) = 3 \cdot 370 = 1110 \text{ cm}^3$

46. $V = \int_0^{10} A(x) \, dx \approx M_5 = \frac{10 - 0}{5} [A(1) + A(3) + A(5) + A(7) + A(9)]$

$= 2(0.65 + 0.61 + 0.59 + 0.55 + 0.50) = 2(2.90) = 5.80 \text{ m}^3$

47. (a) $V = \int_2^{10} \pi [f(x)]^2 \, dx \approx \pi \frac{10 - 2}{4} \left\{ [f(3)]^2 + [f(5)]^2 + [f(7)]^2 + [f(9)]^2 \right\}$

$\approx 2\pi \left[(1.5)^2 + (2.2)^2 + (3.8)^2 + (3.1)^2 \right] \approx 196 \text{ units}^3$

(b) $V = \int_0^4 \pi \left[(\text{outer radius})^2 - (\text{inner radius})^2 \right] dy$

$\approx \pi \frac{4 - 0}{4} \left\{ [(9.9)^2 - (2.2)^2] + [(9.7)^2 - (3.0)^2] + [(9.3)^2 - (5.6)^2] + [(8.7)^2 - (6.5)^2] \right\}$

$\approx 838 \text{ units}^3$

48. (a) $V = \int_{-1}^1 \pi \left[\left(ax^3 + bx^2 + cx + d \right) \sqrt{1 - x^2} \right]^2 dx \overset{\text{CAS}}{=} \dfrac{4 \left\{ 5a^2 + 18ac + 3 \left[3b^2 + 14bd + 7(c^2 + 5d^2) \right] \right\} \pi}{315}$

(b) $y = (-0.06x^3 + 0.04x^2 + 0.1x + 0.54)\sqrt{1 - x^2}$ is graphed in the

figure. Substitute $a = -0.06$, $b = 0.04$, $c = 0.1$, and $d = 0.54$ in the

answer for part (a) to get $V \overset{\text{CAS}}{=} \dfrac{3769\pi}{9375} \approx 1.263$.

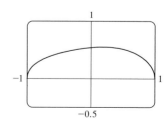

49. We'll form a right circular cone with height h and base radius r by revolving the line $y = \frac{r}{h}x$ about the x-axis.

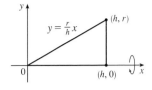

$$V = \pi \int_0^h \left(\frac{r}{h}x\right)^2 dx = \pi \int_0^h \frac{r^2}{h^2}x^2 dx = \pi \frac{r^2}{h^2}\left[\frac{1}{3}x^3\right]_0^h$$

$$= \pi \frac{r^2}{h^2}\left(\frac{1}{3}h^3\right) = \frac{1}{3}\pi r^2 h$$

Another solution: Revolve $x = -\frac{r}{h}y + r$ about the y-axis.

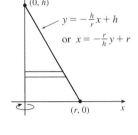

$$V = \pi \int_0^h \left(-\frac{r}{h}y + r\right)^2 dy \overset{*}{=} \pi \int_0^h \left[\frac{r^2}{h^2}y^2 - \frac{2r^2}{h}y + r^2\right] dy$$

$$= \pi \left[\frac{r^2}{3h^2}y^3 - \frac{r^2}{h}y^2 + r^2 y\right]_0^h = \pi\left(\tfrac{1}{3}r^2 h - r^2 h + r^2 h\right) = \tfrac{1}{3}\pi r^2 h$$

* Or use substitution with $u = r - \frac{r}{h}y$ and $du = -\frac{r}{h}dy$ to get

$$\pi \int_r^0 u^2\left(-\frac{h}{r}du\right) = -\pi \frac{h}{r}\left[\frac{1}{3}u^3\right]_r^0 = -\pi\frac{h}{r}\left(-\frac{1}{3}r^3\right) = \frac{1}{3}\pi r^2 h.$$

50. $$V = \pi \int_0^h \left(R - \frac{R-r}{h}y\right)^2 dy$$

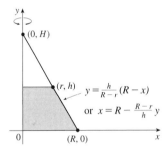

$$= \pi \int_0^h \left[R^2 - \frac{2R(R-r)}{h}y + \left(\frac{R-r}{h}\right)^2 y^2\right] dy$$

$$= \pi \left[R^2 y - \frac{R(R-r)}{h}y^2 + \frac{1}{3}\left(\frac{R-r}{h}\right)^2 y^3\right]_0^h$$

$$= \pi\left[R^2 h - R(R-r)h + \tfrac{1}{3}(R-r)^2 h\right]$$

$$= \tfrac{1}{3}\pi h\left[3Rr + (R^2 - 2Rr + r^2)\right] = \tfrac{1}{3}\pi h(R^2 + Rr + r^2)$$

Another solution: $\dfrac{H}{R} = \dfrac{H-h}{r}$ by similar triangles. Therefore,

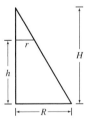

$$Hr = HR - hR \;\Rightarrow\; hR = H(R-r) \;\Rightarrow\; H = \frac{hR}{R-r}. \text{ Now}$$

$$V = \tfrac{1}{3}\pi R^2 H - \tfrac{1}{3}\pi r^2(H-h) \qquad \text{[by Exercise 49]}$$

$$= \tfrac{1}{3}\pi R^2 \frac{hR}{R-r} - \tfrac{1}{3}\pi r^2 \frac{rh}{R-r} \qquad \left[H-h = \frac{rH}{R} = \frac{rhR}{R(R-r)}\right]$$

$$= \frac{1}{3}\pi h \frac{R^3 - r^3}{R-r} = \tfrac{1}{3}\pi h(R^2 + Rr + r^2)$$

$$= \tfrac{1}{3}\left[\pi R^2 + \pi r^2 + \sqrt{(\pi R^2)(\pi r^2)}\right]h = \tfrac{1}{3}\left(A_1 + A_2 + \sqrt{A_1 A_2}\right)h$$

where A_1 and A_2 are the areas of the bases of the frustum. (See Exercise 52 for a related result.)

51. $x^2 + y^2 = r^2 \quad \Leftrightarrow \quad x^2 = r^2 - y^2$

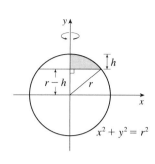

$$V = \pi \int_{r-h}^{r} \left(r^2 - y^2\right) dy = \pi \left[r^2 y - \frac{y^3}{3}\right]_{r-h}^{r}$$

$$= \pi \left\{ \left[r^3 - \frac{r^3}{3}\right] - \left[r^2(r-h) - \frac{(r-h)^3}{3}\right] \right\}$$

$$= \pi \left\{ \tfrac{2}{3}r^3 - \tfrac{1}{3}(r-h)\left[3r^2 - (r-h)^2\right] \right\}$$

$$= \tfrac{1}{3}\pi \left\{ 2r^3 - (r-h)\left[3r^2 - \left(r^2 - 2rh + h^2\right)\right] \right\}$$

$$= \tfrac{1}{3}\pi \left\{ 2r^3 - (r-h)\left[2r^2 + 2rh - h^2\right] \right\}$$

$$= \tfrac{1}{3}\pi \left(2r^3 - 2r^3 - 2r^2 h + rh^2 + 2r^2 h + 2rh^2 - h^3\right)$$

$$= \tfrac{1}{3}\pi \left(3rh^2 - h^3\right) = \tfrac{1}{3}\pi h^2 (3r - h), \text{ or, equivalently, } \pi h^2 \left(r - \frac{h}{3}\right)$$

52. An equation of the line is $x = \dfrac{\Delta x}{\Delta y} y + (x\text{-intercept}) = \dfrac{a/2 - b/2}{h - 0} y + \dfrac{b}{2} = \dfrac{a - b}{2h} y + \dfrac{b}{2}$.

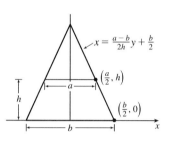

$$V = \int_0^h A(y)\,dy = \int_0^h (2x)^2\,dy$$

$$= \int_0^h \left[2\left(\frac{a-b}{2h}y + \frac{b}{2}\right)\right]^2 dy = \int_0^h \left[\frac{a-b}{h}y + b\right]^2 dy$$

$$= \int_0^h \left[\frac{(a-b)^2}{h^2}y^2 + \frac{2b(a-b)}{h}y + b^2\right] dy$$

$$= \left[\frac{(a-b)^2}{3h^2}y^3 + \frac{b(a-b)}{h}y^2 + b^2 y\right]_0^h$$

$$= \tfrac{1}{3}(a-b)^2 h + b(a-b)h + b^2 h = \tfrac{1}{3}\left(a^2 - 2ab + b^2 + 3ab\right)h$$

$$= \tfrac{1}{3}\left(a^2 + ab + b^2\right)h$$

[Note that this can be written as $\tfrac{1}{3}\left(A_1 + A_2 + \sqrt{A_1 A_2}\right)h$, as in Exercise 50.]

If $a = b$, we get a rectangular solid with volume $b^2 h$. If $a = 0$, we get a square pyramid with volume $\tfrac{1}{3}b^2 h$.

53. For a cross-section at height y, we see from similar triangles that $\dfrac{\alpha/2}{b/2} = \dfrac{h-y}{h}$, so $\alpha = b\left(1 - \dfrac{y}{h}\right)$.

Similarly, for cross-sections having $2b$ as their base and β replacing α, $\beta = 2b\left(1 - \dfrac{y}{h}\right)$. So

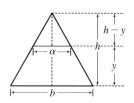

$$V = \int_0^h A(y)\,dy = \int_0^h \left[b\left(1 - \frac{y}{h}\right)\right]\left[2b\left(1 - \frac{y}{h}\right)\right] dy$$

$$= \int_0^h 2b^2 \left(1 - \frac{y}{h}\right)^2 dy = 2b^2 \int_0^h \left(1 - \frac{2y}{h} + \frac{y^2}{h^2}\right) dy$$

$$= 2b^2 \left[y - \frac{y^2}{h} + \frac{y^3}{3h^2}\right]_0^h = 2b^2 \left[h - h + \tfrac{1}{3}h\right]$$

$$= \tfrac{2}{3}b^2 h \quad \left[= \tfrac{1}{3}Bh \text{ where } B \text{ is the area of the base, as with any pyramid.}\right]$$

54. Consider the triangle consisting of two vertices of the base and the center of the base. This triangle is similar to the corresponding triangle at a height y, so $a/b = \alpha/\beta \Rightarrow \alpha = a\beta/b$. Also by similar triangles, $b/h = \beta/(h-y) \Rightarrow \beta = b(h-y)/h$. These two equations imply that $\alpha = a(1 - y/h)$, and since the cross-section is an equilateral triangle, it has area

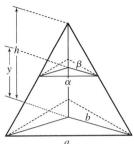

$$A(y) = \frac{1}{2} \cdot \alpha \cdot \frac{\sqrt{3}}{2}\alpha = \frac{a^2(1-y/h)^2}{4}\sqrt{3}, \text{ so}$$

$$V = \int_0^h A(y)\,dy = \frac{a^2\sqrt{3}}{4}\int_0^h \left(1 - \frac{y}{h}\right)^2 dy$$

$$= \frac{a^2\sqrt{3}}{4}\left[-\frac{h}{3}\left(1 - \frac{y}{h}\right)^3\right]_0^h = -\frac{\sqrt{3}}{12}a^2 h(-1) = \frac{\sqrt{3}}{12}a^2 h$$

55. A cross-section at height z is a triangle similar to the base, so we'll multiply the legs of the base triangle, 3 and 4, by a proportionality factor of $(5-z)/5$. Thus, the triangle at height z has area

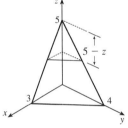

$$A(z) = \frac{1}{2} \cdot 3\left(\frac{5-z}{5}\right) \cdot 4\left(\frac{5-z}{5}\right) = 6\left(1 - \frac{z}{5}\right)^2, \text{ so}$$

$$V = \int_0^5 A(z)\,dz = 6\int_0^5 \left(1 - \frac{z}{5}\right)^2 dz = 6\int_1^0 u^2(-5\,du) \qquad \begin{bmatrix} u = 1 - z/5, \\ du = -\frac{1}{5}\,dz \end{bmatrix}$$

$$= -30\left[\frac{1}{3}u^3\right]_1^0 = -30\left(-\frac{1}{3}\right) = 10 \text{ cm}^3$$

56. A cross-section is shaded in the diagram.

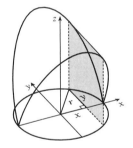

$$A(x) = (2y)^2 = \left(2\sqrt{r^2 - x^2}\right)^2, \text{ so}$$

$$V = \int_{-r}^r A(x)\,dx = 2\int_0^r 4(r^2 - x^2)\,dx$$

$$= 8\left[r^2 x - \frac{1}{3}x^3\right]_0^r = 8\left(\frac{2}{3}r^3\right) = \frac{16}{3}r^3$$

57. If l is a leg of the isosceles right triangle and $2y$ is the hypotenuse,

typical cross-section of length $2y = \sqrt{36 - 9x^2}$

then $l^2 + l^2 = (2y)^2 \Rightarrow 2l^2 = 4y^2 \Rightarrow l^2 = 2y^2$.

$$V = \int_{-2}^2 A(x)\,dx = 2\int_0^2 A(x)\,dx = 2\int_0^2 \frac{1}{2}(l)(l)\,dx = 2\int_0^2 y^2\,dx$$

$$= 2\int_0^2 \frac{1}{4}(36 - 9x^2)\,dx = \frac{9}{2}\int_0^2 (4 - x^2)\,dx$$

$$= \frac{9}{2}\left[4x - \frac{1}{3}x^3\right]_0^2 = \frac{9}{2}\left(8 - \frac{8}{3}\right) = 24$$

58. The cross-section of the base corresponding to the coordinate y has length $x = 1 - y$. The corresponding equilateral triangle

with side s has area $A(y) = s^2 \left(\dfrac{\sqrt{3}}{4} \right) = (1-y)^2 \left(\dfrac{\sqrt{3}}{4} \right)$. Therefore,

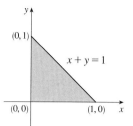

$$V = \int_0^1 A(y)\, dy = \int_0^1 (1-y)^2 \left(\frac{\sqrt{3}}{4} \right) dy$$

$$= \frac{\sqrt{3}}{4} \int_0^1 (1 - 2y + y^2)\, dy = \frac{\sqrt{3}}{4} \left[y - y^2 + \tfrac{1}{3}y^3 \right]_0^1$$

$$= \frac{\sqrt{3}}{4} \left(\frac{1}{3} \right) = \frac{\sqrt{3}}{12}$$

Or: $\displaystyle\int_0^1 (1-y)^2 \left(\frac{\sqrt{3}}{4} \right) dy = \frac{\sqrt{3}}{4} \int_1^0 u^2(-du) \quad [u = 1 - y] = \frac{\sqrt{3}}{4} \left[\frac{1}{3}u^3 \right]_0^1 = \frac{\sqrt{3}}{12}$

59. The cross-section of the base corresponding to the coordinate x has length

$y = 1 - x$. The corresponding square with side s has area

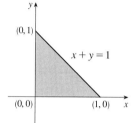

$A(x) = s^2 = (1-x)^2 = 1 - 2x + x^2$. Therefore,

$$V = \int_0^1 A(x)\, dx = \int_0^1 (1 - 2x + x^2)\, dx$$

$$= \left[x - x^2 + \tfrac{1}{3}x^3 \right]_0^1 = \left(1 - 1 + \tfrac{1}{3} \right) - 0 = \tfrac{1}{3}$$

Or: $\displaystyle\int_0^1 (1-x)^2\, dx = \int_1^0 u^2(-du) \quad [u = 1 - x] = \left[\tfrac{1}{3}u^3 \right]_0^1 = \tfrac{1}{3}$

60. The cross-section of the base corresponding to the coordinate y has length

$2x = 2\sqrt{1-y}$. $\left[y = 1 - x^2 \;\Leftrightarrow\; x = \pm\sqrt{1-y} \right]$ The corresponding square

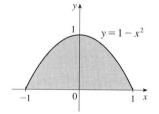

with side s has area $A(x) = s^2 = \left(2\sqrt{1-y} \right)^2 = 4(1-y)$. Therefore,

$V = \int_0^1 A(y)\, dy = \int_0^1 4(1-y)\, dy = 4 \left[y - \tfrac{1}{2}y^2 \right]_0^1 = 4 \left[\left(1 - \tfrac{1}{2} \right) - 0 \right] = 2.$

61. The cross-section of the base b corresponding to the coordinate x has length $1 - x^2$. The height h also has length $1 - x^2$,

so the corresponding isosceles triangle has area $A(x) = \tfrac{1}{2}bh = \tfrac{1}{2}(1-x^2)^2$. Therefore,

$$V = \int_{-1}^1 \tfrac{1}{2}(1-x^2)^2\, dx$$

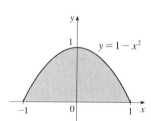

$$= 2 \cdot \tfrac{1}{2} \int_0^1 (1 - 2x^2 + x^4)\, dx \qquad \text{[by symmetry]}$$

$$= \left[x - \tfrac{2}{3}x^3 + \tfrac{1}{5}x^5 \right]_0^1 = \left(1 - \tfrac{2}{3} + \tfrac{1}{5} \right) - 0 = \tfrac{8}{15}$$

62. (a) $V = \int_{-r}^r A(x)\, dx = 2 \int_0^r A(x)\, dx = 2 \int_0^r \tfrac{1}{2}h\left(2\sqrt{r^2 - x^2} \right) dx = 2h \int_0^r \sqrt{r^2 - x^2}\, dx$

(b) Observe that the integral represents one quarter of the area of a circle of radius r, so $V = 2h \cdot \tfrac{1}{4}\pi r^2 = \tfrac{1}{2}\pi h r^2$.

63. (a) The torus is obtained by rotating the circle $(x - R)^2 + y^2 = r^2$ about

the y-axis. Solving for x, we see that the right half of the circle is given by

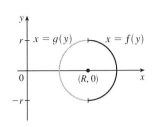

$x = R + \sqrt{r^2 - y^2} = f(y)$ and the left half by $x = R - \sqrt{r^2 - y^2} = g(y)$. So

$V = \pi \int_{-r}^{r} \left\{ [f(y)]^2 - [g(y)]^2 \right\} dy$

$= 2\pi \int_0^r \left[\left(R^2 + 2R\sqrt{r^2 - y^2} + r^2 - y^2 \right) - \left(R^2 - 2R\sqrt{r^2 - y^2} + r^2 - y^2 \right) \right] dy$

$= 2\pi \int_0^r 4R\sqrt{r^2 - y^2} \, dy = 8\pi R \int_0^r \sqrt{r^2 - y^2} \, dy$

(b) Observe that the integral represents a quarter of the area of a circle with radius r, so

$8\pi R \int_0^r \sqrt{r^2 - y^2} \, dy = 8\pi R \cdot \frac{1}{4}\pi r^2 = 2\pi^2 r^2 R.$

64. The cross-sections perpendicular to the y-axis in Figure 17 are rectangles. The rectangle corresponding to the coordinate y has

a base of length $2\sqrt{16 - y^2}$ in the xy-plane and a height of $\frac{1}{\sqrt{3}}y$, since $\angle BAC = 30°$ and $|BC| = \frac{1}{\sqrt{3}}|AB|$. Thus,

$A(y) = \frac{2}{\sqrt{3}}y\sqrt{16 - y^2}$ and

$V = \int_0^4 A(y) \, dy = \frac{2}{\sqrt{3}} \int_0^4 \sqrt{16 - y^2}\, y \, dy = \frac{2}{\sqrt{3}} \int_{16}^0 u^{1/2} \left(-\frac{1}{2}\, du \right)$ [Put $u = 16 - y^2$, so $du = -2y\, dy$]

$= \frac{1}{\sqrt{3}} \int_0^{16} u^{1/2} \, du = \frac{1}{\sqrt{3}} \frac{2}{3} \left[u^{3/2} \right]_0^{16} = \frac{2}{3\sqrt{3}}(64) = \frac{128}{3\sqrt{3}}$

65. (a) Volume$(S_1) = \int_0^h A(z) \, dz =$ Volume(S_2) since the cross-sectional area $A(z)$ at height z is the same for both solids.

(b) By Cavalieri's Principle, the volume of the cylinder in the figure is the same as that of a right circular cylinder with radius r

and height h, that is, $\pi r^2 h$.

66. Each cross-section of the solid S in a plane perpendicular to the

x-axis is a square (since the edges of the cut lie on the

cylinders, which are perpendicular). One-quarter of this square

and one-eighth of S are shown. The area of this quarter-square

is $|PQ|^2 = r^2 - x^2$. Therefore, $A(x) = 4(r^2 - x^2)$ and the

volume of S is

$V = \int_{-r}^{r} A(x) \, dx = 4 \int_{-r}^{r} (r^2 - x^2) \, dx$

$= 8(r^2 - x^2) \, dx = 8\left[r^2 x - \frac{1}{3}x^3 \right]_0^r = \frac{16}{3}r^3$

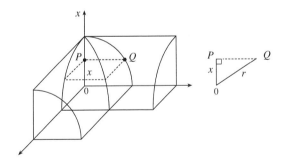

67. The volume is obtained by rotating the area common to two circles of radius r, as

shown. The volume of the right half is

$V_{\text{right}} = \pi \int_0^{r/2} y^2 \, dx = \pi \int_0^{r/2} \left[r^2 - \left(\frac{1}{2}r + x \right)^2 \right] dx$

$= \pi \left[r^2 x - \frac{1}{3}\left(\frac{1}{2}r + x \right)^3 \right]_0^{r/2} = \pi \left[\left(\frac{1}{2}r^3 - \frac{1}{3}r^3 \right) - \left(0 - \frac{1}{24}r^3 \right) \right] = \frac{5}{24}\pi r^3$

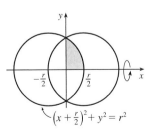

So by symmetry, the total volume is twice this, or $\frac{5}{12}\pi r^3$.

Another solution: We observe that the volume is the twice the volume of a cap of a sphere, so we can use the formula from

Exercise 51 with $h = \frac{1}{2}r$: $V = 2 \cdot \frac{1}{3}\pi h^2 (3r - h) = \frac{2}{3}\pi \left(\frac{1}{2}r \right)^2 \left(3r - \frac{1}{2}r \right) = \frac{5}{12}\pi r^3$.

68. We consider two cases: one in which the ball is not completely submerged and the other in which it is.

Case 1: $0 \le h \le 10$ The ball will not be completely submerged, and so a cross-section of the water parallel to the surface will be the shaded area shown in the first diagram. We can find the area of the cross-section at height x above the bottom of the bowl by using the Pythagorean Theorem: $R^2 = 15^2 - (15 - x)^2$ and $r^2 = 5^2 - (x - 5)^2$, so $A(x) = \pi(R^2 - r^2) = 20\pi x$.

The volume of water when it has depth h is then $V(h) = \int_0^h A(x)\,dx = \int_0^h 20\pi x\,dx = \left[10\pi x^2\right]_0^h = 10\pi h^2$ cm^3, $0 \le h \le 10$.

Case 2: $10 < h \le 15$ In this case we can find the volume by simply subtracting the volume displaced by the ball from the total volume inside the bowl underneath the surface of the water. The total volume underneath the surface is just the volume of a cap of the bowl, so we use the formula from

Exercise 51: $V_{\text{cap}}(h) = \frac{1}{3}\pi h^2(45 - h)$. The volume of

the small sphere is $V_{\text{ball}} = \frac{4}{3}\pi(5)^3 = \frac{500}{3}\pi$, so the total

volume is $V_{\text{cap}} - V_{\text{ball}} = \frac{1}{3}\pi(45h^2 - h^3 - 500)$ cm^3.

69. Take the x-axis to be the axis of the cylindrical hole of radius r.
A quarter of the cross-section through y, perpendicular to the
y-axis, is the rectangle shown. Using the Pythagorean Theorem
twice, we see that the dimensions of this rectangle are

$x = \sqrt{R^2 - y^2}$ and $z = \sqrt{r^2 - y^2}$, so

$\frac{1}{4}A(y) = xz = \sqrt{r^2 - y^2}\,\sqrt{R^2 - y^2}$, and

$$V = \int_{-r}^{r} A(y)\,dy = \int_{-r}^{r} 4\sqrt{r^2 - y^2}\,\sqrt{R^2 - y^2}\,dy = 8\int_{0}^{r} \sqrt{r^2 - y^2}\,\sqrt{R^2 - y^2}\,dy$$

70. The line $y = r$ intersects the semicircle $y = \sqrt{R^2 - x^2}$ when $r = \sqrt{R^2 - x^2} \;\Rightarrow\; r^2 = R^2 - x^2 \;\Rightarrow$
$x^2 = R^2 - r^2 \;\Rightarrow\; x = \pm\sqrt{R^2 - r^2}$. Rotating the shaded region about the x-axis gives us

$$V = \int_{-\sqrt{R^2-r^2}}^{\sqrt{R^2-r^2}} \pi\left[\left(\sqrt{R^2 - x^2}\right)^2 - r^2\right]dx = 2\pi\int_{0}^{\sqrt{R^2-r^2}} (R^2 - x^2 - r^2)\,dx \qquad \text{[by symmetry]}$$

$$= 2\pi\int_{0}^{\sqrt{R^2-r^2}} \left[(R^2 - r^2) - x^2\right]dx = 2\pi\left[(R^2 - r^2)x - \frac{1}{3}x^3\right]_{0}^{\sqrt{R^2-r^2}}$$

$$= 2\pi\left[(R^2 - r^2)^{3/2} - \frac{1}{3}(R^2 - r^2)^{3/2}\right] = 2\pi \cdot \frac{2}{3}(R^2 - r^2)^{3/2} = \frac{4\pi}{3}(R^2 - r^2)^{3/2}$$

Our answer makes sense in limiting cases. As $r \to 0$, $V \to \frac{4}{3}\pi R^3$, which is the volume of the full sphere. As $r \to R$, $V \to 0$, which makes sense because the hole's radius is approaching that of the sphere.

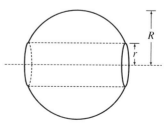

71. (a) The radius of the barrel is the same at each end by symmetry, since the

function $y = R - cx^2$ is even. Since the barrel is obtained by rotating

the graph of the function y about the x-axis, this radius is equal to the

value of y at $x = \frac{1}{2}h$, which is $R - c\left(\frac{1}{2}h\right)^2 = R - d = r$.

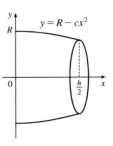

(b) The barrel is symmetric about the y-axis, so its volume is twice the volume of that part of the barrel for $x > 0$. Also, the

barrel is a volume of rotation, so

$$V = 2 \int_0^{h/2} \pi y^2 \, dx = 2\pi \int_0^{h/2} \left(R - cx^2\right)^2 \, dx = 2\pi\left[R^2 x - \tfrac{2}{3}Rcx^3 + \tfrac{1}{5}c^2 x^5\right]_0^{h/2}$$

$$= 2\pi\left(\tfrac{1}{2}R^2 h - \tfrac{1}{12}Rch^3 + \tfrac{1}{160}c^2 h^5\right)$$

Trying to make this look more like the expression we want, we rewrite it as $V = \tfrac{1}{3}\pi h\left[2R^2 + \left(R^2 - \tfrac{1}{2}Rch^2 + \tfrac{3}{80}c^2 h^4\right)\right]$.

But $R^2 - \tfrac{1}{2}Rch^2 + \tfrac{3}{80}c^2 h^4 = \left(R - \tfrac{1}{4}ch^2\right)^2 - \tfrac{1}{40}c^2 h^4 = (R - d)^2 - \tfrac{2}{5}\left(\tfrac{1}{4}ch^2\right)^2 = r^2 - \tfrac{2}{5}d^2$.

Substituting this back into V, we see that $V = \tfrac{1}{3}\pi h\left(2R^2 + r^2 - \tfrac{2}{5}d^2\right)$, as required.

72. It suffices to consider the case where $\mathcal{R}$ is bounded by the curves $y = f(x)$ and $y = g(x)$ for $a \le x \le b$, where $g(x) \le f(x)$

for all x in $[a, b]$, since other regions can be decomposed into subregions of this type. We are concerned with the volume

obtained when $\mathcal{R}$ is rotated about the line $y = -k$, which is equal to

$$V_2 = \pi \int_a^b \left([f(x) + k]^2 - [g(x) + k]^2\right) dx = \pi \int_a^b \left([f(x)]^2 - [g(x)]^2\right) dx + 2\pi k \int_a^b [f(x) - g(x)] \, dx = V_1 + 2\pi k A$$

6.3   Volumes by Cylindrical Shells

1.

If we were to use the "washer" method, we would first have to locate the

local maximum point (a, b) of $y = x(x - 1)^2$ using the methods of

Chapter 4. Then we would have to solve the equation $y = x(x - 1)^2$

for x in terms of y to obtain the functions $x = g_1(y)$ and $x = g_2(y)$

shown in the first figure. This step would be difficult because it involves

the cubic formula. Finally we would find the volume using

$$V = \pi \int_0^b \left\{[g_1(y)]^2 - [g_2(y)]^2\right\} dy.$$

Using shells, we find that a typical approximating shell has radius x, so its circumference is $2\pi x$. Its height is y, that is,

$x(x - 1)^2$. So the total volume is

$$V = \int_0^1 2\pi x\left[x(x - 1)^2\right] dx = 2\pi \int_0^1 \left(x^4 - 2x^3 + x^2\right) dx = 2\pi\left[\frac{x^5}{5} - 2\frac{x^4}{4} + \frac{x^3}{3}\right]_0^1 = \frac{\pi}{15}$$

2.

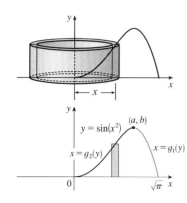

A typical cylindrical shell has circumference $2\pi x$ and height $\sin(x^2)$.

$V = \int_0^{\sqrt{\pi}} 2\pi x \sin(x^2)\, dx$. Let $u = x^2$. Then $du = 2x\, dx$, so

$V = \pi \int_0^{\pi} \sin u\, du = \pi[-\cos u]_0^{\pi} = \pi[1 - (-1)] = 2\pi$. For slicing, we

would first have to locate the local maximum point (a, b) of $y = \sin(x^2)$

using the methods of Chapter 4. Then we would have to solve the equation

$y = \sin(x^2)$ for x in terms of y to obtain the functions $x = g_1(y)$ and

$x = g_2(y)$ shown in the second figure. Finally we would find the volume

using $V = \pi \int_0^{b} \left\{ [g_1(y)]^2 - [g_2(y)]^2 \right\} dy$. Using shells is definitely

preferable to slicing.

3. $V = \displaystyle\int_1^2 2\pi x \cdot \frac{1}{x}\, dx = 2\pi \int_1^2 1\, dx$

$= 2\pi \big[\, x \,\big]_1^2 = 2\pi(2 - 1) = 2\pi$

4. $V = \int_0^1 2\pi x \cdot x^2\, dx = 2\pi \int_0^1 x^3\, dx$

$= 2\pi \big[\tfrac{1}{4}x^4\big]_0^1 = 2\pi \cdot \tfrac{1}{4} = \tfrac{\pi}{2}$

5. $V = \int_0^2 2\pi x(4 - x^2)\, dx = 2\pi \int_0^2 (4x - x^3)\, dx$

$= 2\pi \big[2x^2 - \tfrac{1}{4}x^4\big]_0^2 = 2\pi(8 - 4)$

$= 8\pi$

6. $V = 2\pi \int_0^3 \left\{ x[(3 + 2x - x^2) - (3 - x)] \right\} dx = 2\pi \int_0^3 \left[x(3x - x^2) \right] dx$

$= 2\pi \int_0^3 (3x^2 - x^3)\, dx = 2\pi \big[x^3 - \tfrac{1}{4}x^4\big]_0^3 = 2\pi\big(27 - \tfrac{81}{4}\big) = 2\pi\big(\tfrac{27}{4}\big) = \tfrac{27}{2}\pi$

7. The curves intersect when $4(x-2)^2 = x^2 - 4x + 7 \iff 4x^2 - 16x + 16 = x^2 - 4x + 7 \iff$

$3x^2 - 12x + 9 = 0 \iff 3(x^2 - 4x + 3) = 0 \iff 3(x-1)(x-3) = 0$, so $x = 1$ or 3.

$V = 2\pi \int_1^3 \left\{ x\left[(x^2 - 4x + 7) - 4(x-2)^2\right] \right\} dx = 2\pi \int_1^3 \left[x(x^2 - 4x + 7 - 4x^2 + 16x - 16) \right] dx$

$= 2\pi \int_1^3 \left[x(-3x^2 + 12x - 9) \right] dx = 2\pi(-3) \int_1^3 (x^3 - 4x^2 + 3x)\, dx = -6\pi \left[\frac{1}{4}x^4 - \frac{4}{3}x^3 + \frac{3}{2}x^2 \right]_1^3$

$= -6\pi \left[\left(\frac{81}{4} - 36 + \frac{27}{2} \right) - \left(\frac{1}{4} - \frac{4}{3} + \frac{3}{2} \right) \right] = -6\pi \left(20 - 36 + 12 + \frac{4}{3} \right) = -6\pi \left(-\frac{8}{3} \right) = 16\pi$

8. By slicing:

$V = \int_0^1 \pi \left[\left(\sqrt{y} \right)^2 - (y^2)^2 \right] dy = \pi \int_0^1 (y - y^4)\, dy$

$= \pi \left[\frac{1}{2}y^2 - \frac{1}{5}y^5 \right]_0^1 = \pi \left(\frac{1}{2} - \frac{1}{5} \right) = \frac{3}{10}\pi$

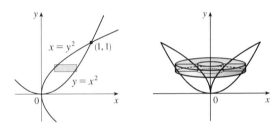

By cylindrical shells:

$V = \int_0^1 2\pi x \left(\sqrt{x} - x^2 \right) dx = 2\pi \int_0^1 (x^{3/2} - x^3)\, dx = 2\pi \left[\frac{2}{5}x^{5/2} - \frac{1}{4}x^4 \right]_0^1$

$= 2\pi \left(\frac{2}{5} - \frac{1}{4} \right) = 2\pi \left(\frac{3}{20} \right) = \frac{3}{10}\pi$

9. $V = \int_1^2 2\pi y(1 + y^2)\, dy = 2\pi \int_1^2 (y + y^3)\, dy = 2\pi \left[\frac{1}{2}y^2 + \frac{1}{4}y^4 \right]_1^2$

$= 2\pi \left[(2 + 4) - \left(\frac{1}{2} + \frac{1}{4} \right) \right] = 2\pi \left(\frac{21}{4} \right) = \frac{21}{2}\pi$

10. $V = \int_0^1 2\pi y \sqrt{y}\,dy = 2\pi \int_0^1 y^{3/2}\,dy$

$\quad = 2\pi \left[\frac{2}{5} y^{5/2}\right]_0^1 = \frac{4}{5}\pi$

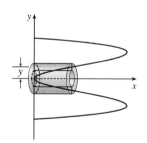

11. $V = 2\pi \int_0^8 \left[y(\sqrt[3]{y} - 0)\right] dy$

$\quad = 2\pi \int_0^8 y^{4/3}\,dy = 2\pi \left[\frac{3}{7} y^{7/3}\right]_0^8$

$\quad = \frac{6\pi}{7}(8^{7/3}) = \frac{6\pi}{7}(2^7) = \frac{768}{7}\pi$

12. $V = 2\pi \int_0^4 \left[y(4y^2 - y^3)\right] dy$

$\quad = 2\pi \int_0^4 (4y^3 - y^4)\,dy$

$\quad = 2\pi \left[y^4 - \frac{1}{5}y^5\right]_0^4 = 2\pi\left(256 - \frac{1024}{5}\right)$

$\quad = 2\pi\left(\frac{256}{5}\right) = \frac{512}{5}\pi$

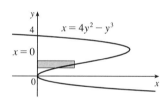

13. The height of the shell is $2 - \left[1 + (y-2)^2\right] = 1 - (y-2)^2 = 1 - (y^2 - 4y + 4) = -y^2 + 4y - 3$.

$\quad V = 2\pi \int_1^3 y(-y^2 + 4y - 3)\,dy$

$\quad = 2\pi \int_1^3 (-y^3 + 4y^2 - 3y)\,dy$

$\quad = 2\pi \left[-\frac{1}{4}y^4 + \frac{4}{3}y^3 - \frac{3}{2}y^2\right]_1^3$

$\quad = 2\pi \left[\left(-\frac{81}{4} + 36 - \frac{27}{2}\right) - \left(-\frac{1}{4} + \frac{4}{3} - \frac{3}{2}\right)\right]$

$\quad = 2\pi\left(\frac{8}{3}\right) = \frac{16}{3}\pi$

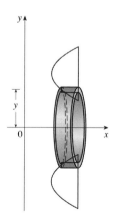

14. $V = \int_0^3 2\pi y \left[4 - (y-1)^2 - (3-y) \right] dy$

$= 2\pi \int_0^3 y(-y^2 + 3y) \, dy$

$= 2\pi \int_0^3 (-y^3 + 3y^2) \, dy = 2\pi \left[-\frac{1}{4}y^4 + y^3 \right]_0^3$

$= 2\pi \left(-\frac{81}{4} + 27 \right) = 2\pi \left(\frac{27}{4} \right) = \frac{27}{2}\pi$

15. The shell has radius $2 - x$, circumference $2\pi(2 - x)$, and height x^4.

$V = \int_0^1 2\pi(2 - x)x^4 \, dx$

$= 2\pi \int_0^1 (2x^4 - x^5) \, dx$

$= 2\pi \left[\frac{2}{5}x^5 - \frac{1}{6}x^6 \right]_0^1$

$= 2\pi \left[\left(\frac{2}{5} - \frac{1}{6} \right) - 0 \right] = 2\pi \left(\frac{7}{30} \right) = \frac{7}{15}\pi$

16. The shell has radius $x - (-1) = x + 1$, circumference $2\pi(x + 1)$, and height $\sqrt{x}$.

$V = \int_0^1 2\pi(x + 1)\sqrt{x} \, dx$

$= 2\pi \int_0^1 (x^{3/2} + x^{1/2}) \, dx$

$= 2\pi \left[\frac{2}{5}x^{5/2} + \frac{2}{3}x^{3/2} \right]_0^1$

$= 2\pi \left[\left(\frac{2}{5} + \frac{2}{3} \right) - 0 \right] = 2\pi \left(\frac{16}{15} \right) = \frac{32}{15}\pi$

17. The shell has radius $x - 1$, circumference $2\pi(x - 1)$, and height $(4x - x^2) - 3 = -x^2 + 4x - 3$.

$V = \int_1^3 2\pi(x - 1)(-x^2 + 4x - 3) \, dx$

$= 2\pi \int_1^3 (-x^3 + 5x^2 - 7x + 3) \, dx$

$= 2\pi \left[-\frac{1}{4}x^4 + \frac{5}{3}x^3 - \frac{7}{2}x^2 + 3x \right]_1^3$

$= 2\pi \left[\left(-\frac{81}{4} + 45 - \frac{63}{2} + 9 \right) - \left(-\frac{1}{4} + \frac{5}{3} - \frac{7}{2} + 3 \right) \right]$

$= 2\pi \left(\frac{4}{3} \right) = \frac{8}{3}\pi$

18. The shell has radius $1 - x$, circumference $2\pi(1 - x)$, and height $(2 - x^2) - x^2 = 2 - 2x^2$.

$V = \int_{-1}^1 2\pi(1 - x)(2 - 2x^2) \, dx$

$= 2\pi(2) \int_{-1}^1 (1 - x)(1 - x^2) \, dx$

$= 4\pi \int_{-1}^1 (1 - x - x^2 + x^3) \, dx$

$= 4\pi(2) \int_0^1 (1 - x^2) \, dx$ [by 5.5.6]

$= 8\pi \left[x - \frac{1}{3}x^3 \right]_0^1 = 8\pi \left[\left(1 - \frac{1}{3} \right) - 0 \right] = 8\pi \left(\frac{2}{3} \right) = \frac{16}{3}\pi$

19. The shell has radius $1 - y$, circumference $2\pi(1 - y)$, and height $1 - \sqrt[3]{y}$ $\quad\left[y = x^3 \quad\Leftrightarrow\quad x = \sqrt[3]{y}\right]$.

$$V = \int_0^1 2\pi(1 - y)(1 - y^{1/3})\,dy$$

$$= 2\pi\int_0^1(1 - y - y^{1/3} + y^{4/3})\,dy$$

$$= 2\pi\left[y - \tfrac{1}{2}y^2 - \tfrac{3}{4}y^{4/3} + \tfrac{3}{7}y^{7/3}\right]_0^1$$

$$= 2\pi\left[\left(1 - \tfrac{1}{2} - \tfrac{3}{4} + \tfrac{3}{7}\right) - 0\right]$$

$$= 2\pi\left(\tfrac{5}{28}\right) = \tfrac{5}{14}\pi$$

20. The shell has radius $y - (-1) = y + 1$, circumference $2\pi(y + 1)$, and height $\sqrt{y} - y^2$.

$$V = \int_0^1 2\pi(y + 1)\left(\sqrt{y} - y^2\right)\,dy$$

$$= 2\pi\int_0^1(y^{3/2} + y^{1/2} - y^3 - y^2)\,dy$$

$$= 2\pi\left[\tfrac{2}{5}y^{5/2} + \tfrac{2}{3}y^{3/2} - \tfrac{1}{4}y^4 - \tfrac{1}{3}y^3\right]_0^1$$

$$= 2\pi\left(\tfrac{2}{5} + \tfrac{2}{3} - \tfrac{1}{4} - \tfrac{1}{3}\right) = 2\pi\left(\tfrac{29}{60}\right) = \tfrac{29}{30}\pi$$

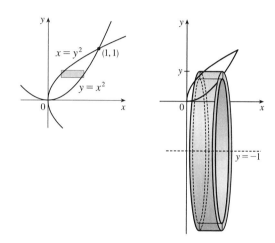

21. $V = \int_{2\pi}^{3\pi} 2\pi x \sin x\,dx$

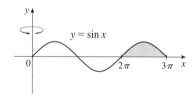

22. $V = \int_0^3 2\pi(7 - x)[(4x - x^2) - x]\,dx$

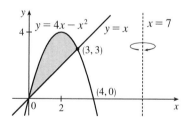

23. $V = \int_0^1 2\pi[x - (-1)]\left(\sin\tfrac{\pi}{2}x - x^4\right)\,dx$

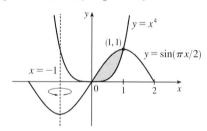

24. $V = \int_0^2 2\pi(2 - x)\left(\dfrac{1}{1 + x^2}\right)\,dx$

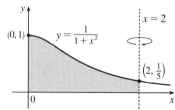

25. $V = \int_0^\pi 2\pi(4-y)\sqrt{\sin y}\, dy$

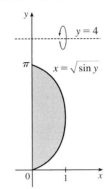

26. $V = \int_{-3}^{3} 2\pi(5-y)\left(4-\sqrt{y^2+7}\right)dy$

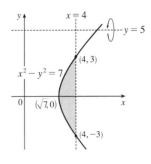

27. $V = \int_0^1 2\pi x\sqrt{1+x^3}\, dx.$ Let $f(x) = x\sqrt{1+x^3}$.

Then the Midpoint Rule with $n=5$ gives

$\int_0^1 f(x)\, dx \approx \frac{1-0}{5}\left[f(0.1)+f(0.3)+f(0.5)+f(0.7)+f(0.9)\right]$

$\approx 0.2(2.9290)$

Multiplying by 2π gives $V \approx 3.68$.

28. $\Delta x = \frac{12-2}{5} = 2$, $n=5$ and $x_i^* = 2+(2i+1)$, where $i = 0, 1, 2, 3, 4$. The values of $f(x)$ are taken directly from the diagram.

$$V = \int_2^{12} 2\pi x f(x)\, dx \approx 2\pi[3f(3)+5f(5)+7f(7)+9f(9)+11f(11)]\cdot 2$$

$$\approx 2\pi[3(2)+5(4)+7(4)+9(2)+11(1)]2 = 332\pi$$

29. $\int_0^3 2\pi x^5\, dx = 2\pi\int_0^3 x(x^4)\, dx$. The solid is obtained by rotating the region $0 \le y \le x^4$, $0 \le x \le 3$ about the y-axis using cylindrical shells.

30. $2\pi\int_0^2 \frac{y}{1+y^2}\, dy = 2\pi\int_0^2 y\left(\frac{1}{1+y^2}\right)dy$. The solid is obtained by rotating the region $0 \le x \le \frac{1}{1+y^2}$, $0 \le y \le 2$ about the x-axis using cylindrical shells.

31. $\int_0^1 2\pi(3-y)(1-y^2)\, dy$. The solid is obtained by rotating the region bounded by (i) $x = 1-y^2$, $x=0$, and $y=0$ or (ii) $x = y^2$, $x=1$, and $y=0$ about the line $y=3$ using cylindrical shells.

32. $\int_0^{\pi/4} 2\pi(\pi-x)(\cos x - \sin x)\, dx$. The solid is obtained by rotating the region bounded by (i) $0 \le y \le \cos x - \sin x$, $0 \le x \le \frac{\pi}{4}$ or (ii) $\sin x \le y \le \cos x$, $0 \le x \le \frac{\pi}{4}$ about the line $x = \pi$ using cylindrical shells.

33.

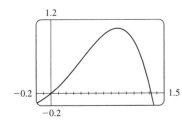

From the graph, the curves intersect at $x=0$ and at $x = a \approx 1.32$, with $x + x^2 - x^4 > 0$ on the interval $(0, a)$. So the volume of the solid obtained by rotating the region about the y-axis is

$$V = 2\pi\int_0^a [x(x+x^2-x^4)]\, dx = 2\pi\int_0^a (x^2+x^3-x^5)\, dx$$

$$= 2\pi\left[\tfrac{1}{3}x^3 + \tfrac{1}{4}x^4 - \tfrac{1}{6}x^6\right]_0^a \approx 4.05$$

34.

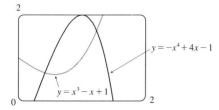

From the graph, the curves intersect at $x = a \approx 0.42$ and $x = b \approx 1.23$, with $-x^4 + 4x - 1 > x^3 - x + 1$ on the interval (a, b). So the volume of the solid obtained by rotating the region about the y-axis is

$$V = 2\pi \int_a^b x \left[(-x^4 + 4x - 1) - (x^3 - x + 1) \right] \, dx$$

$$= 2\pi \int_a^b x(-x^4 - x^3 + 5x - 2) \, dx \approx 3.17$$

35. $V = 2\pi \displaystyle\int_0^{\pi/2} \left[\left(\tfrac{\pi}{2} - x \right) \left(\sin^2 x - \sin^4 x \right) \right] dx$

$\overset{\text{CAS}}{=} \tfrac{1}{32} \pi^3$

36. $V = 2\pi \int_0^\pi \left\{ [x - (-1)](x^3 \sin x) \right\} dx$

$\overset{\text{CAS}}{=} 2\pi(\pi^4 + \pi^3 - 12\pi^2 - 6\pi + 48)$

$= 2\pi^5 + 2\pi^4 - 24\pi^3 - 12\pi^2 + 96\pi$

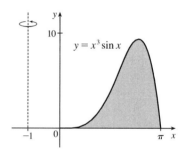

37. Use shells:

$V = \int_2^4 2\pi x(-x^2 + 6x - 8) \, dx = 2\pi \int_2^4 (-x^3 + 6x^2 - 8x) \, dx$

$= 2\pi \left[-\tfrac{1}{4}x^4 + 2x^3 - 4x^2 \right]_2^4$

$= 2\pi[(-64 + 128 - 64) - (-4 + 16 - 16)]$

$= 2\pi(4) = 8\pi$

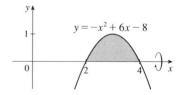

38. Use disks:

$V = \int_2^4 \pi(-x^2 + 6x - 8)^2 \, dx$

$= \pi \int_2^4 (x^4 - 12x^3 + 52x^2 - 96x + 64) \, dx$

$= \pi \left[\tfrac{1}{5}x^5 - 3x^4 + \tfrac{52}{3}x^3 - 48x^2 + 64x \right]_2^4$

$= \pi \left(\tfrac{512}{15} - \tfrac{496}{15} \right) = \tfrac{16}{15}\pi$

39. Use shells:

$$V = \int_1^4 2\pi[x - (-1)][5 - (x^2 - 5x + 9)]\, dx$$

$$= 2\pi \int_1^4 (x+1)(-x^2 + 5x - 4)\, dx$$

$$= 2\pi \int_1^4 (-x^3 + 4x^2 + x - 4)\, dx = 2\pi\left[-\tfrac{1}{4}x^4 + \tfrac{4}{3}x^3 + \tfrac{1}{2}x^2 - 4x\right]_1^4$$

$$= 2\pi\left[\left(-64 + \tfrac{256}{3} + 8 - 16\right) - \left(-\tfrac{1}{4} + \tfrac{4}{3} + \tfrac{1}{2} - 4\right)\right]$$

$$= 2\pi\left(\tfrac{63}{4}\right) = \tfrac{63}{2}\pi$$

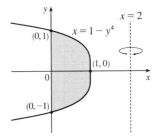

40. Use washers:

$$V = \int_{-1}^1 \pi\left\{[2-0]^2 - [2 - (1 - y^4)]^2\right\} dy$$

$$= 2\pi \int_0^1 \left[4 - (1 + y^4)^2\right] dy \qquad \text{[by symmetry]}$$

$$= 2\pi \int_0^1 \left[4 - (1 + 2y^4 + y^8)\right] dy = 2\pi \int_0^1 \left(3 - 2y^4 - y^8\right) dy$$

$$= 2\pi\left[3y - \tfrac{2}{5}y^5 - \tfrac{1}{9}y^9\right]_0^1 = 2\pi\left(3 - \tfrac{2}{5} - \tfrac{1}{9}\right)$$

$$= 2\pi\left(\tfrac{112}{45}\right) = \tfrac{224}{45}\pi$$

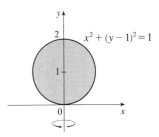

41. Use disks: $x^2 + (y-1)^2 = 1 \;\Leftrightarrow\; x = \pm\sqrt{1 - (y-1)^2}$

$$V = \pi \int_0^2 \left[\sqrt{1 - (y-1)^2}\right]^2 dy = \pi \int_0^2 (2y - y^2)\, dy$$

$$= \pi\left[y^2 - \tfrac{1}{3}y^3\right]_0^2 = \pi\left(4 - \tfrac{8}{3}\right) = \tfrac{4}{3}\pi$$

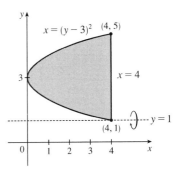

42. Use shells:

$$V = \int_1^5 2\pi(y-1)[4 - (y-3)^2]\, dy$$

$$= 2\pi \int_1^5 (y-1)(-y^2 + 6y - 5)\, dy$$

$$= 2\pi \int_1^5 (-y^3 + 7y^2 - 11y + 5)\, dy$$

$$= 2\pi\left[-\tfrac{1}{4}y^4 + \tfrac{7}{3}y^3 - \tfrac{11}{2}y^2 + 5y\right]_1^5$$

$$= 2\pi\left(\tfrac{275}{12} - \tfrac{19}{12}\right) = \tfrac{128}{3}\pi$$

43. Use shells:

$$V = 2 \int_0^r 2\pi x \sqrt{r^2 - x^2}\, dx$$

$$= -2\pi \int_0^r (r^2 - x^2)^{1/2}(-2x)\, dx$$

$$= \left[-2\pi \cdot \tfrac{2}{3}(r^2 - x^2)^{3/2}\right]_0^r$$

$$= -\tfrac{4}{3}\pi(0 - r^3) = \tfrac{4}{3}\pi r^3$$

44. $V = \int_{R-r}^{R+r} 2\pi x \cdot 2\sqrt{r^2 - (x-R)^2}\, dx$

$\qquad = \int_{-r}^{r} 4\pi(u + R)\sqrt{r^2 - u^2}\, du \qquad$ [let $u = x - R$]

$\qquad = 4\pi R \int_{-r}^{r} \sqrt{r^2 - u^2}\, du + 4\pi \int_{-r}^{r} u\sqrt{r^2 - u^2}\, du$

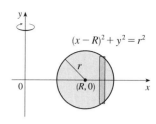

The first integral is the area of a semicircle of radius r, that is, $\frac{1}{2}\pi r^2$, and the second is zero since the integrand is an odd function. Thus, $V = 4\pi R\left(\frac{1}{2}\pi r^2\right) + 4\pi \cdot 0 = 2\pi^2 R r^2$.

45. $V = 2\pi \int_0^r x\left(-\dfrac{h}{r}x + h\right) dx = 2\pi h \int_0^r \left(-\dfrac{x^2}{r} + x\right) dx$

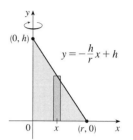

$\qquad = 2\pi h\left[-\dfrac{x^3}{3r} + \dfrac{x^2}{2}\right]_0^r = 2\pi h\,\dfrac{r^2}{6} = \dfrac{\pi r^2 h}{3}$

46. By symmetry, the volume of a napkin ring obtained by drilling a hole of radius r through a sphere with radius R is twice the volume obtained by rotating the area above the x-axis and below the curve $y = \sqrt{R^2 - x^2}$ (the equation of the top half of the cross-section of the sphere), between $x = r$ and $x = R$, about the y-axis. This volume is equal to

$$2\int_{\text{inner radius}}^{\text{outer radius}} 2\pi r h\, dx = 2 \cdot 2\pi \int_r^R x\sqrt{R^2 - x^2}\, dx = 4\pi\left[-\tfrac{1}{3}\left(R^2 - x^2\right)^{3/2}\right]_r^R = \tfrac{4}{3}\pi\left(R^2 - r^2\right)^{3/2}$$

But by the Pythagorean Theorem, $R^2 - r^2 = \left(\frac{1}{2}h\right)^2$, so the volume of the napkin ring is $\frac{4}{3}\pi\left(\frac{1}{2}h\right)^3 = \frac{1}{6}\pi h^3$, which is independent of both R and r; that is, the amount of wood in a napkin ring of height h is the same regardless of the size of the sphere used. Note that most of this calculation has been done already, but with more difficulty, in Exercise 6.2.70.

Another solution: The height of the missing cap is the radius of the sphere minus half the height of the cut-out cylinder, that is, $R - \frac{1}{2}h$. Using Exercise 6.2.51,

$$V_{\text{napkin ring}} = V_{\text{sphere}} - V_{\text{cylinder}} - 2V_{\text{cap}} = \tfrac{4}{3}\pi R^3 - \pi r^2 h - 2 \cdot \tfrac{\pi}{3}\left(R - \tfrac{1}{2}h\right)^2\left[3R - \left(R - \tfrac{1}{2}h\right)\right] = \tfrac{1}{6}\pi h^3$$

6.4 Work

1. $W = Fd = mgd = (40)(9.8)(1.5) = 588$ J

2. $W = Fd = (100)(200) = 20{,}000$ ft-lb

3. $W = \int_a^b f(x)\, dx = \int_0^9 \dfrac{10}{(1+x)^2}\, dx = 10\int_1^{10} \dfrac{1}{u^2}\, du \quad [u = 1 + x,\ du = dx] \ = 10\left[-\dfrac{1}{u}\right]_1^{10} = 10\left(-\tfrac{1}{10} + 1\right) = 9$ ft-lb

4. $W = \int_1^2 \cos\left(\frac{1}{3}\pi x\right) dx = \frac{3}{\pi}\left[\sin\left(\frac{1}{3}\pi x\right)\right]_1^2 = \frac{3}{\pi}\left(\frac{\sqrt{3}}{2} - \frac{\sqrt{3}}{2}\right) = 0$ N·m $= 0$ J.

Interpretation: From $x = 1$ to $x = \frac{3}{2}$, the force does work equal to $\int_1^{3/2} \cos\left(\frac{1}{3}\pi x\right) dx = \frac{3}{\pi}\left(1 - \frac{\sqrt{3}}{2}\right)$ J in accelerating the particle and increasing its kinetic energy. From $x = \frac{3}{2}$ to $x = 2$, the force opposes the motion of the particle, decreasing its kinetic energy. This is negative work, equal in magnitude but opposite in sign to the work done from $x = 1$ to $x = \frac{3}{2}$.

5. The force function is given by $F(x)$ (in newtons) and the work (in joules) is the area under the curve, given by

$$\int_0^8 F(x)\,dx = \int_0^4 F(x)\,dx + \int_4^8 F(x)\,dx = \frac{1}{2}(4)(30) + (4)(30) = 180 \text{ J.}$$

6. $W = \int_4^{20} f(x)\,dx \approx M_4 = \Delta x[f(6) + f(10) + f(14) + f(18)] = \frac{20-4}{4}[5.8 + 8.8 + 8.2 + 5.2] = 4(28) = 112 \text{ J}$

7. $10 = f(x) = kx = \frac{1}{3}k$ [4 inches $= \frac{1}{3}$ foot], so $k = 30$ lb/ft and $f(x) = 30x$. Now 6 inches $= \frac{1}{2}$ foot, so

$W = \int_0^{1/2} 30x\,dx = \left[15x^2\right]_0^{1/2} = \frac{15}{4}$ ft-lb.

8. $25 = f(x) = kx = k(0.1)$ [10 cm $= 0.1$ m], so $k = 250$ N/m and $f(x) = 250x$. Now 5 cm $= 0.05$ m, so

$W = \int_0^{0.05} 250x\,dx = \left[125x^2\right]_0^{0.05} = 125(0.0025) = 0.3125 \approx 0.31$ J.

9. (a) If $\int_0^{0.12} kx\,dx = 2$ J, then $2 = \left[\frac{1}{2}kx^2\right]_0^{0.12} = \frac{1}{2}k(0.0144) = 0.0072k$ and $k = \frac{2}{0.0072} = \frac{2500}{9} \approx 277.78$ N/m.

Thus, the work needed to stretch the spring from 35 cm to 40 cm is

$\int_{0.05}^{0.10} \frac{2500}{9}x\,dx = \left[\frac{1250}{9}x^2\right]_{1/20}^{1/10} = \frac{1250}{9}\left(\frac{1}{100} - \frac{1}{400}\right) = \frac{25}{24} \approx 1.04$ J.

(b) $f(x) = kx$, so $30 = \frac{2500}{9}x$ and $x = \frac{270}{2500}$ m $= 10.8$ cm

10. If $12 = \int_0^1 kx\,dx = \left[\frac{1}{2}kx^2\right]_0^1 = \frac{1}{2}k$, then $k = 24$ lb/ft and the work required is

$\int_0^{3/4} 24x\,dx = \left[12x^2\right]_0^{3/4} = 12 \cdot \frac{9}{16} = \frac{27}{4} = 6.75$ ft-lb.

11. The distance from 20 cm to 30 cm is 0.1 m, so with $f(x) = kx$, we get $W_1 = \int_0^{0.1} kx\,dx = k\left[\frac{1}{2}x^2\right]_0^{0.1} = \frac{1}{200}k$.

Now $W_2 = \int_{0.1}^{0.2} kx\,dx = k\left[\frac{1}{2}x^2\right]_{0.1}^{0.2} = k\left(\frac{4}{200} - \frac{1}{200}\right) = \frac{3}{200}k$. Thus, $W_2 = 3W_1$.

12. Let L be the natural length of the spring in meters. Then

$6 = \int_{0.10-L}^{0.12-L} kx\,dx = \left[\frac{1}{2}kx^2\right]_{0.10-L}^{0.12-L} = \frac{1}{2}k\left[(0.12-L)^2 - (0.10-L)^2\right]$ and

$10 = \int_{0.12-L}^{0.14-L} kx\,dx = \left[\frac{1}{2}kx^2\right]_{0.12-L}^{0.14-L} = \frac{1}{2}k\left[(0.14-L)^2 - (0.12-L)^2\right]$.

Simplifying gives us $12 = k(0.0044 - 0.04L)$ and $20 = k(0.0052 - 0.04L)$. Subtracting the first equation from the second gives $8 = 0.0008k$, so $k = 10,000$. Now the second equation becomes $20 = 52 - 400L$, so $L = \frac{32}{400}$ m $= 8$ cm.

In Exercises 13–20, n is the number of subintervals of length Δx, and x_i^* is a sample point in the ith subinterval $[x_{i-1}, x_i]$.

13. (a) The portion of the rope from x ft to $(x + \Delta x)$ ft below the top of the building weighs $\frac{1}{2}\Delta x$ lb and must be lifted x_i^* ft, so its contribution to the total work is $\frac{1}{2}x_i^*\Delta x$ ft-lb. The total work is

$$W = \lim_{n \to \infty} \sum_{i=1}^n \frac{1}{2}x_i^*\,\Delta x = \int_0^{50} \frac{1}{2}x\,dx = \left[\frac{1}{4}x^2\right]_0^{50} = \frac{2500}{4} = 625 \text{ ft-lb}$$

Notice that the exact height of the building does not matter (as long as it is more than 50 ft).

(b) When half the rope is pulled to the top of the building, the work to lift the top half of the rope is

$W_1 = \int_0^{25} \frac{1}{2}x\,dx = \left[\frac{1}{4}x^2\right]_0^{25} = \frac{625}{4}$ ft-lb. The bottom half of the rope is lifted 25 ft and the work needed to accomplish that is $W_2 = \int_{25}^{50} \frac{1}{2} \cdot 25\,dx = \frac{25}{2}\big[x\big]_{25}^{50} = \frac{625}{2}$ ft-lb. The total work done in pulling half the rope to the top of the building

is $W = W_1 + W_2 = \frac{625}{2} + \frac{625}{4} = \frac{3}{4} \cdot 625 = \frac{1875}{4}$ ft-lb.

14. *Assumptions*:

1. After lifting, the chain is L-shaped, with 4 m of the chain lying along the ground.

2. The chain slides effortlessly and without friction along the ground while its end is lifted.

3. The weight density of the chain is constant throughout its length and therefore equals $(8\text{ kg/m})(9.8\text{ m/s}^2) = 78.4$ N/m.

The part of the chain x m from the lifted end is raised $6 - x$ m if $0 \le x \le 6$ m, and it is lifted 0 m if $x > 6$ m.

Thus, the work needed is

$$W = \lim_{n\to\infty} \sum_{i=1}^{n} (6 - x_i^*) \cdot 78.4\,\Delta x = \int_0^6 (6-x)78.4\,dx = 78.4\Big[6x - \tfrac{1}{2}x^2\Big]_0^6 = (78.4)(18) = 1411.2 \text{ J}$$

15. The work needed to lift the cable is $\lim_{n\to\infty}\sum_{i=1}^{n} 2x_i^*\,\Delta x = \int_0^{500} 2x\,dx = \big[x^2\big]_0^{500} = 250{,}000$ ft-lb. The work needed to lift

the coal is $800\text{ lb} \cdot 500\text{ ft} = 400{,}000$ ft-lb. Thus, the total work required is $250{,}000 + 400{,}000 = 650{,}000$ ft-lb.

16. The work needed to lift the bucket itself is $4\text{ lb} \cdot 80\text{ ft} = 320$ ft-lb. At time t (in seconds) the bucket is $x_i^* = 2t$ ft above its

original 80 ft depth, but it now holds only $(40 - 0.2t)$ lb of water. In terms of distance, the bucket holds $\Big[40 - 0.2\big(\tfrac{1}{2}x_i^*\big)\Big]$ lb

of water when it is x_i^* ft above its original 80 ft depth. Moving this amount of water a distance Δx requires

$\big(40 - \tfrac{1}{10}x_i^*\big)\,\Delta x$ ft-lb of work. Thus, the work needed to lift the water is

$$W = \lim_{n\to\infty}\sum_{i=1}^{n}\Big(40 - \tfrac{1}{10}x_i^*\Big)\,\Delta x = \int_0^{80}\Big(40 - \tfrac{1}{10}x\Big)dx = \Big[40x - \tfrac{1}{20}x^2\Big]_0^{80} = (3200 - 320)\text{ ft-lb}$$

Adding the work of lifting the bucket gives a total of 3200 ft-lb of work.

17. At a height of x meters $(0 \le x \le 12)$, the mass of the rope is $(0.8\text{ kg/m})(12 - x\text{ m}) = (9.6 - 0.8x)$ kg and the mass of the

water is $\big(\tfrac{36}{12}\text{ kg/m}\big)(12 - x\text{ m}) = (36 - 3x)$ kg. The mass of the bucket is 10 kg, so the total mass is

$(9.6 - 0.8x) + (36 - 3x) + 10 = (55.6 - 3.8x)$ kg, and hence, the total force is $9.8(55.6 - 3.8x)$ N. The work needed to lift

the bucket Δx m through the ith subinterval of $[0, 12]$ is $9.8(55.6 - 3.8x_i^*)\Delta x$, so the total work is

$$W = \lim_{n\to\infty}\sum_{i=1}^{n} 9.8(55.6 - 3.8x_i^*)\,\Delta x = \int_0^{12}(9.8)(55.6 - 3.8x)\,dx = 9.8\Big[55.6x - 1.9x^2\Big]_0^{12} = 9.8(393.6) \approx 3857 \text{ J}$$

18. The chain's weight density is $\dfrac{25\text{ lb}}{10\text{ ft}} = 2.5$ lb/ft. The part of the chain x ft below the ceiling (for $5 \le x \le 10$) has to be lifted

$2(x-5)$ ft, so the work needed to lift the ith subinterval of the chain is $2(x_i^* - 5)(2.5\,\Delta x)$. The total work needed is

$$W = \lim_{n\to\infty}\sum_{i=1}^{n} 2(x_i^* - 5)(2.5)\,\Delta x = \int_5^{10}[2(x-5)(2.5)]\,dx = 5\int_5^{10}(x-5)\,dx$$
$$= 5\Big[\tfrac{1}{2}x^2 - 5x\Big]_5^{10} = 5\Big[(50 - 50) - \big(\tfrac{25}{2} - 25\big)\Big] = 5\big(\tfrac{25}{2}\big) = 62.5 \text{ ft-lb}$$

19. A "slice" of water Δx m thick and lying at a depth of x_i^* m (where $0 \le x_i^* \le \tfrac{1}{2}$) has volume $(2 \times 1 \times \Delta x)$ m^3, a mass of

$2000\,\Delta x$ kg, weighs about $(9.8)(2000\,\Delta x) = 19{,}600\,\Delta x$ N, and thus requires about $19{,}600x_i^*\,\Delta x$ J of work for its removal.

So $W = \lim_{n\to\infty}\sum_{i=1}^{n} 19{,}600x_i^*\,\Delta x = \int_0^{1/2} 19{,}600x\,dx = \big[9800x^2\big]_0^{1/2} = 2450$ J.

20. A horizontal cylindrical slice of water Δx ft thick has a volume of $\pi r^2 h = \pi \cdot 12^2 \cdot \Delta x$ ft^3 and weighs about $\left(62.5 \text{ lb/ft}^3\right)\left(144\pi \, \Delta x \text{ ft}^3\right) = 9000\pi \, \Delta x$ lb. If the slice lies x_i^* ft below the edge of the pool (where $1 \le x_i^* \le 5$), then the work needed to pump it out is about $9000\pi x_i^* \, \Delta x$. Thus,

$$W = \lim_{n \to \infty} \sum_{i=1}^{n} 9000\pi x_i^* \, \Delta x = \int_1^5 9000\pi x \, dx = \left[4500\pi x^2\right]_1^5 = 4500\pi(25 - 1) = 108{,}000\pi \text{ ft-lb}$$

21. A rectangular "slice" of water Δx m thick and lying x m above the bottom has width x m and volume $8x \, \Delta x$ m^3. It weighs about $(9.8 \times 1000)(8x \, \Delta x)$ N, and must be lifted $(5 - x)$ m by the pump, so the work needed is about $(9.8 \times 10^3)(5 - x)(8x \, \Delta x)$ J. The total work required is

$$W \approx \int_0^3 (9.8 \times 10^3)(5 - x)8x \, dx = (9.8 \times 10^3) \int_0^3 (40x - 8x^2) \, dx = (9.8 \times 10^3)\left[20x^2 - \tfrac{8}{3}x^3\right]_0^3$$

$$= (9.8 \times 10^3)(180 - 72) = (9.8 \times 10^3)(108) = 1058.4 \times 10^3 \approx 1.06 \times 10^6 \text{ J}$$

22. Let y measure depth (in meters) below the center of the spherical tank, so that $y = -3$ at the top of the tank and $y = -4$ at the spigot. A horizontal disk-shaped "slice" of water Δy m thick and lying at coordinate y has radius $\sqrt{9 - y^2}$ m and volume $\pi r^2 \Delta y = \pi(9 - y^2) \, \Delta y$ m^3. It weighs about $(9.8 \times 1000)\pi(9 - y^2) \, \Delta y$ N and must be lifted $(y + 4)$ m by the pump, so the work needed to pump it out is about $(9.8 \times 10^3)(y + 4)\pi(9 - y^2) \, \Delta y$ J. The total work required is

$$W \approx \int_{-3}^3 (9.8 \times 10^3)(y + 4)\pi(9 - y^2) \, dy = (9.8 \times 10^3)\pi \int_{-3}^3 (9y - y^3 + 36 - 4y^2) \, dy$$

$$= (9.8 \times 10^3)\pi(2)(4) \int_0^3 (9 - y^2) \, dy \qquad \text{[by Theorem 5.5.6]}$$

$$= (78.4 \times 10^3)\pi\left[9y - \tfrac{1}{3}y^3\right]_0^3 = (78.4 \times 10^3)\pi(18) = 1{,}411{,}200\pi \approx 4.43 \times 10^6 \text{ J}$$

23. Let x measure depth (in feet) below the spout at the top of the tank. A horizontal disk-shaped "slice" of water Δx ft thick and lying at coordinate x has radius $\frac{3}{8}(16 - x)$ ft $(\star)$ and volume $\pi r^2 \Delta x = \pi \cdot \frac{9}{64}(16 - x)^2 \, \Delta x$ ft^3. It weighs about $(62.5)\frac{9\pi}{64}(16 - x)^2 \, \Delta x$ lb and must be lifted x ft by the pump, so the work needed to pump it out is about $(62.5)x\frac{9\pi}{64}(16 - x)^2 \, \Delta x$ ft-lb. The total work required is

$$W \approx \int_0^8 (62.5)x\frac{9\pi}{64}(16 - x)^2 \, dx = (62.5)\frac{9\pi}{64} \int_0^8 x(256 - 32x + x^2) \, dx$$

$$= (62.5)\frac{9\pi}{64} \int_0^8 (256x - 32x^2 + x^3) \, dx = (62.5)\frac{9\pi}{64}\left[128x^2 - \tfrac{32}{3}x^3 + \tfrac{1}{4}x^4\right]_0^8$$

$$= (62.5)\frac{9\pi}{64}\left(\frac{11,264}{3}\right) = 33{,}000\pi \approx 1.04 \times 10^5 \text{ ft-lb}$$

$(\star)$ From similar triangles, $\dfrac{d}{8 - x} = \dfrac{3}{8}$.

So $r = 3 + d = 3 + \tfrac{3}{8}(8 - x)$

$\qquad = \dfrac{3(8)}{8} + \dfrac{3}{8}(8 - x)$

$\qquad = \tfrac{3}{8}(16 - x)$

24. Let x measure the distance (in feet) above the bottom of the tank. A horizontal "slice" of water Δx ft thick and lying at coordinate x has volume $10(2x) \, \Delta x$ ft^3. It weighs about $(62.5)20x \, \Delta x$ lb and must be lifted $(6 - x)$ ft by the pump, so the work needed to pump it out is about $(62.5)(6 - x)20x \, \Delta x$ ft-lb. The total work required is

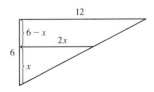

$$W \approx \int_0^6 (62.5)(6 - x)20x \, dx = 1250 \int_0^6 (6x - x^2) \, dx = 1250\left[3x^2 - \tfrac{1}{3}x^3\right]_0^6 = 1250(36) = 45{,}000 \text{ ft-lb.}$$

25. If only 4.7×10^5 J of work is done, then only the water above a certain level (call it h) will be pumped out. So we use the same formula as in Exercise 21, except that the work is fixed, and we are trying to find the lower limit of integration:

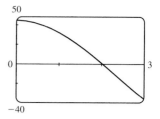

$4.7 \times 10^5 \approx \int_h^3 (9.8 \times 10^3)(5 - x)8x \, dx = (9.8 \times 10^3) \left[20x^2 - \frac{8}{3}x^3 \right]_h^3 \quad \Leftrightarrow$

$\frac{4.7}{9.8} \times 10^2 \approx 48 = \left(20 \cdot 3^2 - \frac{8}{3} \cdot 3^3 \right) - \left(20h^2 - \frac{8}{3}h^3 \right) \quad \Leftrightarrow$

$2h^3 - 15h^2 + 45 = 0$. To find the solution of this equation, we plot $2h^3 - 15h^2 + 45$ between $h = 0$ and $h = 3$. We see that the equation is satisfied for $h \approx 2.0$. So the depth of water remaining in the tank is about 2.0 m.

26. The only changes needed in the solution for Exercise 22 are: (1) change the lower limit from -3 to 0 and (2) change 1000 to 900.

$$W \approx \int_0^3 (9.8 \times 900)(y + 4)\pi(9 - y^2) \, dy = (9.8 \times 900) \, \pi \int_0^3 (9y - y^3 + 36 - 4y^2) \, dy$$

$$= (9.8 \times 900)\pi \left[\frac{9}{2}y^2 - \frac{1}{4}y^4 + 36y - \frac{4}{3}y^3 \right]_0^3 = (9.8 \times 900)\pi(92.25) = 813{,}645\pi$$

$$\approx 2.56 \times 10^6 \text{ J} \quad \text{[about 58\% of the work in Exercise 22]}$$

27. $V = \pi r^2 x$, so V is a function of x and P can also be regarded as a function of x. If $V_1 = \pi r^2 x_1$ and $V_2 = \pi r^2 x_2$, then

$$W = \int_{x_1}^{x_2} F(x) \, dx = \int_{x_1}^{x_2} \pi r^2 P(V(x)) \, dx = \int_{x_1}^{x_2} P(V(x)) \, dV(x) \qquad \text{[Let } V(x) = \pi r^2 x, \text{ so } dV(x) = \pi r^2 \, dx.\text{]}$$

$$= \int_{V_1}^{V_2} P(V) \, dV \quad \text{by the Substitution Rule.}$$

28. $160 \text{ lb/in}^2 = 160 \cdot 144 \text{ lb/ft}^2$, $100 \text{ in}^3 = \frac{100}{1728} \text{ ft}^3$, and $800 \text{ in}^3 = \frac{800}{1728} \text{ ft}^3$.

$k = PV^{1.4} = (160 \cdot 144)\left(\frac{100}{1728} \right)^{1.4} = 23{,}040\left(\frac{25}{432} \right)^{1.4} \approx 426.5$. Therefore, $P \approx 426.5V^{-1.4}$ and

$$W = \int_{100/1728}^{800/1728} 426.5V^{-1.4} \, dV = 426.5 \left[\frac{1}{-0.4}V^{-0.4} \right]_{25/432}^{25/54} = (426.5)(2.5)\left[\left(\frac{432}{25} \right)^{0.4} - \left(\frac{54}{25} \right)^{0.4} \right] \approx 1.88 \times 10^3 \text{ ft-lb.}$$

29. $W = \int_a^b F(r) \, dr = \int_a^b G\frac{m_1 m_2}{r^2} \, dr = Gm_1 m_2 \left[\frac{-1}{r} \right]_a^b = Gm_1 m_2 \left(\frac{1}{a} - \frac{1}{b} \right)$

30. By Exercise 29, $W = GMm\left(\frac{1}{R} - \frac{1}{R + 1{,}000{,}000} \right)$ where M = mass of the earth in kg, R = radius of the earth in m, and m = mass of satellite in kg. (Note that 1000 km = 1,000,000 m.) Thus,

$$W = (6.67 \times 10^{-11})(5.98 \times 10^{24})(1000) \times \left(\frac{1}{6.37 \times 10^6} - \frac{1}{7.37 \times 10^6} \right) \approx 8.50 \times 10^9 \text{ J}$$

6.5 Average Value of a Function

1. $f_{\text{ave}} = \frac{1}{b-a} \int_a^b f(x) \, dx = \frac{1}{4-0} \int_0^4 (4x - x^2) \, dx = \frac{1}{4} \left[2x^2 - \frac{1}{3}x^3 \right]_0^4 = \frac{1}{4}\left[\left(32 - \frac{64}{3} \right) - 0 \right] = \frac{1}{4}\left(\frac{32}{3} \right) = \frac{8}{3}$

2. $f_{\text{ave}} = \frac{1}{b-a} \int_a^b f(x) \, dx = \frac{1}{\pi - (-\pi)} \int_{-\pi}^{\pi} \sin 4x \, dx = 0 \quad$ [by Theorem 5.5.6(b)]

3. $g_{ave} = \frac{1}{b-a} \int_a^b g(x)\,dx = \frac{1}{8-1} \int_1^8 \sqrt[3]{x}\,dx = \frac{1}{7} \left[\frac{3}{4} x^{4/3} \right]_1^8 = \frac{3}{28}(16-1) = \frac{45}{28}$

4. $g_{ave} = \frac{1}{2-0} \int_0^2 x^2 \sqrt{1+x^3}\,dx = \frac{1}{2} \int_1^9 \sqrt{u} \cdot \frac{1}{3}\,du$ $[u = 1+x^3,\ du = 3x^2\,dx]$ $= \frac{1}{6} \left[\frac{2}{3} u^{3/2} \right]_1^9 = \frac{1}{9}(27-1) = \frac{26}{9}$

5. $f_{ave} = \frac{1}{5-0} \int_0^5 t\sqrt{1+t^2}\,dt = \frac{1}{5} \int_1^{26} \sqrt{u} \left(\frac{1}{2}\,du \right)$ $[u = 1+t^2,\ du = 2t\,dt]$

$= \frac{1}{10} \int_1^{26} u^{1/2}\,du = \frac{1}{10} \cdot \frac{2}{3} \left[u^{3/2} \right]_1^{26} = \frac{1}{15}(26^{3/2} - 1)$

6. $f_{ave} = \frac{1}{b-a} \int_a^b f(x)\,dx = \frac{1}{\pi/2 - 0} \int_0^{\pi/2} \sec^2(\theta/2)\,d\theta = \frac{2}{\pi} \left[2\tan(\theta/2) \right]_0^{\pi/2} = \frac{2}{\pi}[2(1) - 0] = \frac{4}{\pi}$

7. $h_{ave} = \frac{1}{\pi - 0} \int_0^\pi \cos^4 x \sin x\,dx = \frac{1}{\pi} \int_1^{-1} u^4(-du)$ $[u = \cos x,\ du = -\sin x\,dx]$

$= \frac{1}{\pi} \int_{-1}^1 u^4\,du = \frac{1}{\pi} \cdot 2 \int_0^1 u^4\,du$ [by Theorem 5.5.6(a)] $= \frac{2}{\pi} \left[\frac{1}{5} u^5 \right]_0^1 = \frac{2}{5\pi}$

8. $h_{ave} = \frac{1}{6-1} \int_1^6 \frac{3}{(1+r)^2}\,dr = \frac{1}{5} \int_2^7 3u^{-2}\,du$ $[u = 1+r,\ du = dr]$

$= -\frac{3}{5} \left[u^{-1} \right]_2^7 = -\frac{3}{5} \left(\frac{1}{7} - \frac{1}{2} \right) = \frac{3}{5} \left(\frac{1}{2} - \frac{1}{7} \right) = \frac{3}{5} \cdot \frac{5}{14} = \frac{3}{14}$

9. (a) $f_{ave} = \frac{1}{5-2} \int_2^5 (x-3)^2\,dx = \frac{1}{3} \left[\frac{1}{3}(x-3)^3 \right]_2^5$ **(c)**

$= \frac{1}{9} \left[2^3 - (-1)^3 \right] = \frac{1}{9}(8+1) = 1$

(b) $f(c) = f_{ave} \iff (c-3)^2 = 1 \iff$

$c - 3 = \pm 1 \iff c = 2 \text{ or } 4$

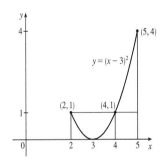

10. (a) $f_{ave} = \frac{1}{4-0} \int_0^4 \sqrt{x}\,dx = \frac{1}{4} \left[\frac{2}{3} x^{3/2} \right]_0^4$ **(c)**

$= \frac{1}{6} \left[x^{3/2} \right]_0^4 = \frac{1}{6}[8-0] = \frac{4}{3}$

(b) $f(c) = f_{ave} \iff \sqrt{c} = \frac{4}{3} \iff c = \frac{16}{9}$

11. (a) $f_{ave} = \frac{1}{\pi - 0} \int_0^\pi (2\sin x - \sin 2x)\,dx$ **(c)**

$= \frac{1}{\pi} \left[-2\cos x + \frac{1}{2}\cos 2x \right]_0^\pi$

$= \frac{1}{\pi} \left[\left(2 + \frac{1}{2} \right) - \left(-2 + \frac{1}{2} \right) \right] = \frac{4}{\pi}$

(b) $f(c) = f_{ave} \iff 2\sin c - \sin 2c = \frac{4}{\pi} \iff$

$c_1 \approx 1.238 \text{ or } c_2 \approx 2.808$

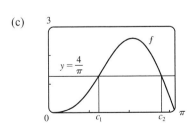

12. (a) $f_{ave} = \dfrac{1}{2-0} \displaystyle\int_0^2 \dfrac{2x}{(1+x^2)^2}\, dx$

$\qquad = \dfrac{1}{2} \displaystyle\int_1^5 \dfrac{1}{u^2}\, du \qquad [u = 1 + x^2,\ du = 2x\, dx]$

$\qquad = \dfrac{1}{2}\left[-\dfrac{1}{u}\right]_1^5 = -\dfrac{1}{2}\left(\dfrac{1}{5} - 1\right) = \dfrac{2}{5}$

(c)

(b) $f(c) = f_{ave} \quad\Leftrightarrow\quad \dfrac{2c}{(1+c^2)^2} = \dfrac{2}{5} \quad\Leftrightarrow\quad 5c = (1+c^2)^2 \quad\Leftrightarrow$

$\qquad c_1 \approx 0.220 \ \text{ or } \ c_2 \approx 1.207$

13. f is continuous on $[1,3]$, so by the Mean Value Theorem for Integrals there exists a number c in $[1,3]$ such that

$\displaystyle\int_1^3 f(x)\, dx = f(c)(3-1) \quad\Rightarrow\quad 8 = 2f(c)$; that is, there is a number c such that $f(c) = \dfrac{8}{2} = 4$.

14. The requirement is that $\dfrac{1}{b-0} \displaystyle\int_0^b f(x)\, dx = 3$. The LHS of this equation is equal to

$\dfrac{1}{b} \displaystyle\int_0^b (2 + 6x - 3x^2)\, dx = \dfrac{1}{b}\left[2x + 3x^2 - x^3\right]_0^b = 2 + 3b - b^2$, so we solve the equation $2 + 3b - b^2 = 3 \quad\Leftrightarrow$

$b^2 - 3b + 1 = 0 \quad\Leftrightarrow\quad b = \dfrac{3 \pm \sqrt{(-3)^2 - 4\cdot 1 \cdot 1}}{2\cdot 1} = \dfrac{3 \pm \sqrt{5}}{2}$. Both roots are valid since they are positive.

15. $f_{ave} = \dfrac{1}{50-20} \displaystyle\int_{20}^{50} f(x)\, dx \approx \dfrac{1}{30} M_3 = \dfrac{1}{30} \cdot \dfrac{50-20}{3}[f(25) + f(35) + f(45)] = \dfrac{1}{3}(38 + 29 + 48) = \dfrac{115}{3} = 38\frac{1}{3}$

16. (a) $v_{ave} = \dfrac{1}{12-0}\displaystyle\int_0^{12} v(t)\, dt = \dfrac{1}{12} I$. Use the Midpoint Rule with $n = 3$ and $\Delta t = \dfrac{12-0}{3} = 4$ to estimate I.

$\qquad I \approx M_3 = 4[v(2) + v(6) + v(10)] = 4[21 + 50 + 66] = 4(137) = 548$. Thus, $v_{ave} \approx \dfrac{1}{12}(548) = 45\frac{2}{3}$ km/h.

(b) Estimating from the graph, $v(t) = 45\frac{2}{3}$ when $t \approx 5.2$ s.

17. Let $t = 0$ and $t = 12$ correspond to 9 AM and 9 PM, respectively.

$$T_{ave} = \dfrac{1}{12-0}\int_0^{12}\left[50 + 14\sin\tfrac{1}{12}\pi t\right] dt = \dfrac{1}{12}\left[50t - 14\cdot\dfrac{12}{\pi}\cos\tfrac{1}{12}\pi t\right]_0^{12}$$

$$= \dfrac{1}{12}\left[50\cdot 12 + 14\cdot\dfrac{12}{\pi} + 14\cdot\dfrac{12}{\pi}\right] = \left(50 + \dfrac{28}{\pi}\right)\,^\circ\text{F} \approx 59\,^\circ\text{F}$$

18. $T_{ave} = \dfrac{1}{5}\displaystyle\int_0^5 4x\, dx = \dfrac{1}{5}\left[2x^2\right]_0^5 = 10\,^\circ\text{C}$

19. $\rho_{ave} = \dfrac{1}{8}\displaystyle\int_0^8 \dfrac{12}{\sqrt{x+1}}\, dx = \dfrac{3}{2}\displaystyle\int_0^8 (x+1)^{-1/2}\, dx = \left[3\sqrt{x+1}\right]_0^8 = 9 - 3 = 6$ kg/m

20. $s = \tfrac{1}{2}gt^2 \quad\Rightarrow\quad t = \sqrt{2s/g}$ [since $t \geq 0$]. Now $v = ds/dt = gt = g\sqrt{2s/g} = \sqrt{2gs} \quad\Rightarrow\quad v^2 = 2gs \quad\Rightarrow\quad s = \dfrac{v^2}{2g}$.

We see that v can be regarded as a function of t or of s: $v = F(t) = gt$ and $v = G(s) = \sqrt{2gs}$. Note that $v_T = F(T) = gT$.

Displacement can be viewed as a function of t: $s = s(t) = \tfrac{1}{2}gt^2$; also $s(t) = \dfrac{v^2}{2g} = \dfrac{[F(t)]^2}{2g}$. When $t = T$, these two

formulas for $s(t)$ imply that

$$\sqrt{2gs(T)} = F(T) = v_T = gT = 2\left(\tfrac{1}{2}gT^2\right)/T = 2s(T)/T \qquad\qquad (\star)$$

The average of the velocities with respect to time t during the interval $[0, T]$ is

$$v_{t\text{-ave}} = F_{\text{ave}} = \frac{1}{T-0} \int_0^T F(t)\, dt = \frac{1}{T}[s(T) - s(0)] \quad \text{[by FTC]} \quad = \frac{s(T)}{T} \quad \text{[since } s(0) = 0\text{]} \quad = \frac{1}{2}v_T \quad \text{[by (⋆)]}$$

But the average of the velocities with respect to displacement s during the corresponding displacement interval $[s(0), s(T)] = [0, s(T)]$ is

$$v_{s\text{-ave}} = G_{\text{ave}} = \frac{1}{s(T) - 0} \int_0^{s(T)} G(s)\, ds = \frac{1}{s(T)} \int_0^{s(T)} \sqrt{2gs}\, ds = \frac{\sqrt{2g}}{s(T)} \int_0^{s(T)} s^{1/2}\, ds$$

$$= \frac{\sqrt{2g}}{s(T)} \cdot \frac{2}{3}\left[s^{3/2}\right]_0^{s(T)} = \frac{2}{3} \cdot \frac{\sqrt{2g}}{s(T)} \cdot \left[s(T)\right]^{3/2} = \frac{2}{3}\sqrt{2gs(T)} = \frac{2}{3}v_T \quad \text{[by (⋆)]}$$

21. $V_{\text{ave}} = \frac{1}{5}\int_0^5 V(t)\, dt = \frac{1}{5}\int_0^5 \frac{5}{4\pi}\left[1 - \cos\left(\frac{2}{5}\pi t\right)\right] dt = \frac{1}{4\pi}\int_0^5 \left[1 - \cos\left(\frac{2}{5}\pi t\right)\right] dt$

$= \frac{1}{4\pi}\left[t - \frac{5}{2\pi}\sin\left(\frac{2}{5}\pi t\right)\right]_0^5 = \frac{1}{4\pi}[(5-0) - 0] = \frac{5}{4\pi} \approx 0.4\text{ L}$

22. $v_{\text{ave}} = \frac{1}{R-0}\int_0^R v(r)\, dr = \frac{1}{R}\int_0^R \frac{P}{4\eta l}(R^2 - r^2)\, dr = \frac{P}{4\eta l R}\left[R^2 r - \frac{1}{3}r^3\right]_0^R = \frac{P}{4\eta l R}\left(\frac{2}{3}\right)R^3 = \frac{PR^2}{6\eta l}.$

Since $v(r)$ is decreasing on $(0, R]$, $v_{\max} = v(0) = \dfrac{PR^2}{4\eta l}$. Thus, $v_{\text{ave}} = \frac{2}{3}v_{\max}$.

23. Let $F(x) = \int_a^x f(t)\, dt$ for x in $[a, b]$. Then F is continuous on $[a, b]$ and differentiable on (a, b), so by the Mean Value Theorem there is a number c in (a, b) such that $F(b) - F(a) = F'(c)(b - a)$. But $F'(x) = f(x)$ by the Fundamental Theorem of Calculus. Therefore, $\int_a^b f(t)\, dt - 0 = f(c)(b - a)$.

24. $f_{\text{ave}}[a, b] = \dfrac{1}{b-a}\int_a^b f(x)\, dx = \dfrac{1}{b-a}\int_a^c f(x)\, dx + \dfrac{1}{b-a}\int_c^b f(x)\, dx$

$= \dfrac{c-a}{b-a}\left[\dfrac{1}{c-a}\int_a^c f(x)\, dx\right] + \dfrac{b-c}{b-a}\left[\dfrac{1}{b-c}\int_c^b f(x)\, dx\right] = \dfrac{c-a}{b-a}f_{\text{ave}}[a, c] + \dfrac{b-c}{b-a}f_{\text{ave}}[c, b]$

6 Review

CONCEPT CHECK

1. (a) See Section 6.1, Figure 2 and Equations 6.1.1 and 6.1.2.

(b) Instead of using "top minus bottom" and integrating from left to right, we use "right minus left" and integrate from bottom to top. See Figures 11 and 12 in Section 6.1.

2. The numerical value of the area represents the number of meters by which Sue is ahead of Kathy after 1 minute.

3. (a) See the discussion in Section 6.2, near Figures 2 and 3, ending in the Definition of Volume.

(b) See the discussion between Examples 5 and 6 in Section 6.2. If the cross-section is a disk, find the radius in terms of x or y and use $A = \pi(\text{radius})^2$. If the cross-section is a washer, find the inner radius r_{in} and outer radius r_{out} and use $A = \pi(r_{\text{out}}^2) - \pi(r_{\text{in}}^2)$.

4. (a) $V = 2\pi r h \,\Delta r = $ (circumference)(height)(thickness)

(b) For a typical shell, find the circumference and height in terms of x or y and calculate

$V = \int_a^b$ (circumference)(height)$(dx$ or $dy)$, where a and b are the limits on x or y.

(c) Sometimes slicing produces washers or disks whose radii are difficult (or impossible) to find explicitly. On other occasions, the cylindrical shell method leads to an easier integral than slicing does.

5. $\int_0^6 f(x)\,dx$ represents the amount of work done. Its units are newton-meters, or joules.

6. (a) The average value of a function f on an interval $[a, b]$ is $f_{\text{ave}} = \dfrac{1}{b - a} \displaystyle\int_a^b f(x)\,dx$.

(b) The Mean Value Theorem for Integrals says that there is a number c at which the value of f is exactly equal to the average value of the function, that is, $f(c) = f_{\text{ave}}$. For a geometric interpretation of the Mean Value Theorem for Integrals, see Figure 2 in Section 6.5 and the discussion that accompanies it.

EXERCISES

1. The curves intersect when $x^2 = 4x - x^2 \iff 2x^2 - 4x = 0 \iff$

$2x(x - 2) = 0 \iff x = 0$ or 2.

$A = \int_0^2 \left[(4x - x^2) - x^2\right] dx = \int_0^2 (4x - 2x^2)\,dx$

$= \left[2x^2 - \frac{2}{3}x^3\right]_0^2 = \left[(8 - \frac{16}{3}) - 0\right] = \frac{8}{3}$

2. $20 - x^2 = x^2 - 12 \iff 32 = 2x^2 \iff x^2 = 16 \iff x = \pm 4$. So

$A = \int_{-4}^4 [(20 - x^2) - (x^2 - 12)]\,dx = \int_{-4}^4 (32 - 2x^2)\,dx$

$= 2\int_0^4 (32 - 2x^2)\,dx$ [even function]

$= 2\left[32x - \frac{2}{3}x^3\right]_0^4 = 2\left(128 - \frac{128}{3}\right) = \frac{512}{3}$

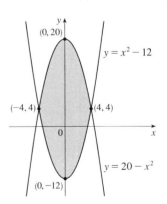

3. If $x \ge 0$, then $|x| = x$, and the graphs intersect when $x = 1 - 2x^2 \iff 2x^2 + x - 1 = 0 \iff (2x - 1)(x + 1) = 0 \iff$

$x = \frac{1}{2}$ or -1, but $-1 < 0$. By symmetry, we can double the area from $x = 0$ to $x = \frac{1}{2}$.

$A = 2\int_0^{1/2} \left[(1 - 2x^2) - x\right] dx = 2\int_0^{1/2} (-2x^2 - x + 1)\,dx$

$= 2\left[-\frac{2}{3}x^3 - \frac{1}{2}x^2 + x\right]_0^{1/2} = 2\left[\left(-\frac{1}{12} - \frac{1}{8} + \frac{1}{2}\right) - 0\right]$

$= 2\left(\frac{7}{24}\right) = \frac{7}{12}$

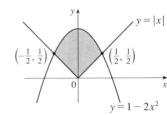

4. $y^2 + 3y = -y \iff y^2 + 4y = 0 \iff y(y+4) = 0 \iff$
$y = 0$ or -4.

$A = \int_{-4}^{0} \left[-y - (y^2 + 3y) \right] dy = \int_{-4}^{0} (-y^2 - 4y) \, dy$

$= \left[-\frac{1}{3} y^3 - 2y^2 \right]_{-4}^{0} = 0 - \left(\frac{64}{3} - 32 \right) = \frac{32}{3}$

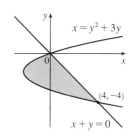

5. $A = \int_{0}^{2} \left[\sin\left(\frac{\pi x}{2} \right) - (x^2 - 2x) \right] dx$

$= \left[-\frac{2}{\pi} \cos\left(\frac{\pi x}{2} \right) - \frac{1}{3} x^3 + x^2 \right]_{0}^{2}$

$= \left(\frac{2}{\pi} - \frac{8}{3} + 4 \right) - \left(-\frac{2}{\pi} - 0 + 0 \right) = \frac{4}{3} + \frac{4}{\pi}$

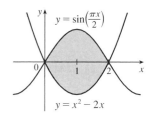

6. $A = \int_{0}^{1} \left(\sqrt{x} - x^2 \right) dx + \int_{1}^{2} \left(x^2 - \sqrt{x} \right) dx$

$= \left[\frac{2}{3} x^{3/2} - \frac{1}{3} x^3 \right]_{0}^{1} + \left[\frac{1}{3} x^3 - \frac{2}{3} x^{3/2} \right]_{1}^{2}$

$= \left[\left(\frac{2}{3} - \frac{1}{3} \right) - 0 \right] + \left[\left(\frac{8}{3} - \frac{4}{3} \sqrt{2} \right) - \left(\frac{1}{3} - \frac{2}{3} \right) \right]$

$= \frac{10}{3} - \frac{4}{3} \sqrt{2}$

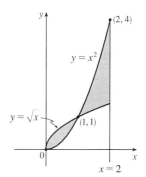

7. Using washers with inner radius x^2 and outer radius $2x$, we have

$V = \pi \int_{0}^{2} \left[(2x)^2 - (x^2)^2 \right] dx = \pi \int_{0}^{2} (4x^2 - x^4) \, dx$

$= \pi \left[\frac{4}{3} x^3 - \frac{1}{5} x^5 \right]_{0}^{2} = \pi \left(\frac{32}{3} - \frac{32}{5} \right)$

$= 32\pi \cdot \frac{2}{15} = \frac{64}{15} \pi$

8. $1 + y^2 = y + 3 \iff y^2 - y - 2 = 0 \iff (y-2)(y+1) = 0 \iff$
$y = 2$ or -1.

$V = \pi \int_{-1}^{2} \left[(y+3)^2 - (1+y^2)^2 \right] dy = \pi \int_{-1}^{2} (y^2 + 6y + 9 - 1 - 2y^2 - y^4) \, dy$

$= \pi \int_{-1}^{2} (8 + 6y - y^2 - y^4) \, dy = \pi \left[8y + 3y^2 - \frac{1}{3} y^3 - \frac{1}{5} y^5 \right]_{-1}^{2}$

$= \pi \left[\left(16 + 12 - \frac{8}{3} - \frac{32}{5} \right) - \left(-8 + 3 + \frac{1}{3} + \frac{1}{5} \right) \right] = \pi \left(33 - \frac{9}{3} - \frac{33}{5} \right) = \frac{117}{5} \pi$

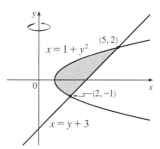

9. $V = \pi \int_{-3}^{3} \left\{ \left[(9 - y^2) - (-1) \right]^2 - \left[0 - (-1) \right]^2 \right\} dy$

$= 2\pi \int_{0}^{3} \left[(10 - y^2)^2 - 1 \right] dy = 2\pi \int_{0}^{3} (100 - 20y^2 + y^4 - 1) \, dy$

$= 2\pi \int_{0}^{3} (99 - 20y^2 + y^4) \, dy = 2\pi \left[99y - \frac{20}{3}y^3 + \frac{1}{5}y^5 \right]_{0}^{3}$

$= 2\pi \left(297 - 180 + \frac{243}{5} \right) = \frac{1656}{5}\pi$

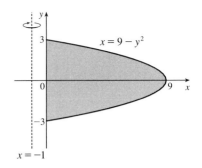

10. $V = \pi \int_{-2}^{2} \left\{ \left[(9 - x^2) - (-1) \right]^2 - \left[(x^2 + 1) - (-1) \right]^2 \right\} dx$

$= \pi \int_{-2}^{2} \left[(10 - x^2)^2 - (x^2 + 2)^2 \right] dx$

$= 2\pi \int_{0}^{2} (96 - 24x^2) \, dx = 48\pi \int_{0}^{2} (4 - x^2) \, dx$

$= 48\pi \left[4x - \frac{1}{3}x^3 \right]_{0}^{2} = 48\pi \left(8 - \frac{8}{3} \right) = 256\pi$

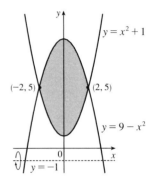

11. The graph of $x^2 - y^2 = a^2$ is a hyperbola with right and left branches.

Solving for y gives us $y^2 = x^2 - a^2 \;\Rightarrow\; y = \pm\sqrt{x^2 - a^2}$.

We'll use shells and the height of each shell is

$\sqrt{x^2 - a^2} - \left(-\sqrt{x^2 - a^2} \right) = 2\sqrt{x^2 - a^2}$.

The volume is $V = \int_{a}^{a+h} 2\pi x \cdot 2\sqrt{x^2 - a^2} \, dx$. To evaluate, let $u = x^2 - a^2$,

so $du = 2x \, dx$ and $x \, dx = \frac{1}{2} \, du$. When $x = a$, $u = 0$, and when $x = a + h$,

$u = (a + h)^2 - a^2 = a^2 + 2ah + h^2 - a^2 = 2ah + h^2$.

Thus, $V = 4\pi \int_{0}^{2ah+h^2} \sqrt{u} \left(\frac{1}{2} \, du \right) = 2\pi \left[\frac{2}{3}u^{3/2} \right]_{0}^{2ah+h^2} = \frac{4}{3}\pi \left(2ah + h^2 \right)^{3/2}$.

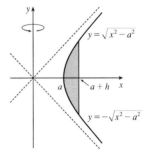

12. A shell has radius x, circumference $2\pi x$, and height $\tan x - x$.

$V = \int_{0}^{\pi/3} 2\pi x \left(\tan x - x \right) dx$

13. A shell has radius $\frac{\pi}{2} - x$, circumference $2\pi \left(\frac{\pi}{2} - x \right)$, and height $\cos^2 x - \frac{1}{4}$.

$y = \cos^2 x$ intersects $y = \frac{1}{4}$ when $\cos^2 x = \frac{1}{4} \;\Leftrightarrow\;$

$\cos x = \pm\frac{1}{2}$ $[\, |x| \le \pi/2 \,] \;\Leftrightarrow\; x = \pm\frac{\pi}{3}$.

$V = \int_{-\pi/3}^{\pi/3} 2\pi \left(\frac{\pi}{2} - x \right) \left(\cos^2 x - \frac{1}{4} \right) dx$

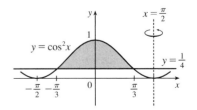

14. A washer has outer radius $2 - x^2$ and inner radius $2 - \sqrt{x}$.

$$V = \int_0^1 \pi \left[(2 - x^2)^2 - \left(2 - \sqrt{x}\right)^2 \right] dx$$

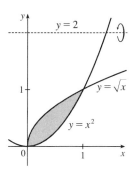

15. (a) A cross-section is a washer with inner radius x^2 and outer radius x.

$$V = \int_0^1 \pi \left[(x)^2 - (x^2)^2 \right] dx = \int_0^1 \pi (x^2 - x^4) \, dx = \pi \left[\tfrac{1}{3}x^3 - \tfrac{1}{5}x^5 \right]_0^1 = \pi \left[\tfrac{1}{3} - \tfrac{1}{5} \right] = \tfrac{2}{15}\pi$$

(b) A cross-section is a washer with inner radius y and outer radius $\sqrt{y}$.

$$V = \int_0^1 \pi \left[\left(\sqrt{y}\right)^2 - y^2 \right] dy = \int_0^1 \pi (y - y^2) \, dy = \pi \left[\tfrac{1}{2}y^2 - \tfrac{1}{3}y^3 \right]_0^1 = \pi \left[\tfrac{1}{2} - \tfrac{1}{3} \right] = \tfrac{\pi}{6}$$

(c) A cross-section is a washer with inner radius $2 - x$ and outer radius $2 - x^2$.

$$V = \int_0^1 \pi \left[(2 - x^2)^2 - (2 - x)^2 \right] dx = \int_0^1 \pi (x^4 - 5x^2 + 4x) \, dx = \pi \left[\tfrac{1}{5}x^5 - \tfrac{5}{3}x^3 + 2x^2 \right]_0^1 = \pi \left[\tfrac{1}{5} - \tfrac{5}{3} + 2 \right] = \tfrac{8}{15}\pi$$

16. (a) $A = \int_0^1 (2x - x^2 - x^3) \, dx = \left[x^2 - \tfrac{1}{3}x^3 - \tfrac{1}{4}x^4 \right]_0^1 = 1 - \tfrac{1}{3} - \tfrac{1}{4} = \tfrac{5}{12}$

(b) A cross-section is a washer with inner radius x^3 and outer radius $2x - x^2$, so its area is $\pi (2x - x^2)^2 - \pi (x^3)^2$.

$$V = \int_0^1 A(x) \, dx = \int_0^1 \pi [(2x - x^2)^2 - (x^3)^2] \, dx = \int_0^1 \pi (4x^2 - 4x^3 + x^4 - x^6) \, dx$$

$$= \pi \left[\tfrac{4}{3}x^3 - x^4 + \tfrac{1}{5}x^5 - \tfrac{1}{7}x^7 \right]_0^1 = \pi \left(\tfrac{4}{3} - 1 + \tfrac{1}{5} - \tfrac{1}{7} \right) = \tfrac{41}{105}\pi$$

(c) Using the method of cylindrical shells,

$$V = \int_0^1 2\pi x (2x - x^2 - x^3) \, dx = \int_0^1 2\pi (2x^2 - x^3 - x^4) \, dx = 2\pi \left[\tfrac{2}{3}x^3 - \tfrac{1}{4}x^4 - \tfrac{1}{5}x^5 \right]_0^1 = 2\pi \left(\tfrac{2}{3} - \tfrac{1}{4} - \tfrac{1}{5} \right) = \tfrac{13}{30}\pi.$$

17. (a) Using the Midpoint Rule on $[0, 1]$ with $f(x) = \tan(x^2)$ and $n = 4$, we estimate

$$A = \int_0^1 \tan(x^2) \, dx \approx \tfrac{1}{4} \left[\tan\left(\left(\tfrac{1}{8}\right)^2 \right) + \tan\left(\left(\tfrac{3}{8}\right)^2 \right) + \tan\left(\left(\tfrac{5}{8}\right)^2 \right) + \tan\left(\left(\tfrac{7}{8}\right)^2 \right) \right] \approx \tfrac{1}{4}(1.53) \approx 0.38$$

(b) Using the Midpoint Rule on $[0, 1]$ with $f(x) = \pi \tan^2(x^2)$ (for disks) and $n = 4$, we estimate

$$V = \int_0^1 f(x) \, dx \approx \tfrac{1}{4}\pi \left[\tan^2\left(\left(\tfrac{1}{8}\right)^2 \right) + \tan^2\left(\left(\tfrac{3}{8}\right)^2 \right) + \tan^2\left(\left(\tfrac{5}{8}\right)^2 \right) + \tan^2\left(\left(\tfrac{7}{8}\right)^2 \right) \right] \approx \tfrac{\pi}{4}(1.114) \approx 0.87$$

18. (a)

From the graph, we see that the curves intersect at $x = 0$ and at $x = a \approx 0.75$, with $1 - x^2 > x^6 - x + 1$ on $(0, a)$.

(b) The area of $\mathcal{R}$ is $A = \int_0^a \left[(1 - x^2) - (x^6 - x + 1) \right] dx = \left[-\tfrac{1}{3}x^3 - \tfrac{1}{7}x^7 + \tfrac{1}{2}x^2 \right]_0^a \approx 0.12.$

(c) Using washers, the volume generated when $\mathcal{R}$ is rotated about the x-axis is

$$V = \pi \int_0^a [(1-x^2)^2 - (x^6 - x + 1)^2]\, dx = \pi \int_0^a (-x^{12} + 2x^7 - 2x^6 + x^4 - 3x^2 + 2x)\, dx$$

$$= \pi \left[-\tfrac{1}{13}x^{13} + \tfrac{1}{4}x^8 - \tfrac{2}{7}x^7 + \tfrac{1}{5}x^5 - x^3 + x^2\right]_0^a \approx 0.54$$

(d) Using shells, the volume generated when $\mathcal{R}$ is rotated about the y-axis is

$$V = \int_0^a 2\pi x[(1-x^2) - (x^6 - x + 1)]\, dx = 2\pi \int_0^a (-x^3 - x^7 + x^2)\, dx = 2\pi \left[-\tfrac{1}{4}x^4 - \tfrac{1}{8}x^8 + \tfrac{1}{3}x^3\right]_0^a \approx 0.31.$$

19. $\int_0^{\pi/2} 2\pi x \cos x\, dx = \int_0^{\pi/2} (2\pi x) \cos x\, dx$

The solid is obtained by rotating the region $\mathcal{R} = \left\{(x,y) \mid 0 \le x \le \tfrac{\pi}{2}, 0 \le y \le \cos x\right\}$ about the y-axis.

20. $\int_0^{\pi/2} 2\pi \cos^2 x\, dx = \int_0^{\pi/2} \pi \left(\sqrt{2}\cos x\right)^2 dx$

The solid is obtained by rotating the region $\mathcal{R} = \left\{(x,y) \mid 0 \le x \le \tfrac{\pi}{2}, 0 \le y \le \sqrt{2}\cos x\right\}$ about the x-axis.

21. $\int_0^{\pi} \pi(2 - \sin x)^2\, dx$

The solid is obtained by rotating the region $\mathcal{R} = \left\{(x,y) \mid 0 \le x \le \pi, 0 \le y \le 2 - \sin x\right\}$ about the x-axis.

22. $\int_0^4 2\pi(6 - y)(4y - y^2)\, dy$

The solid is obtained by rotating the region $\mathcal{R} = \left\{(x,y) \mid 0 \le x \le 4y - y^2, 0 \le y \le 4\right\}$ about the line $y = 6$.

23. Take the base to be the disk $x^2 + y^2 \le 9$. Then $V = \int_{-3}^3 A(x)\, dx$, where $A(x_0)$ is the area of the isosceles right triangle

whose hypotenuse lies along the line $x = x_0$ in the xy-plane. The length of the hypotenuse is $2\sqrt{9 - x^2}$ and the length of

each leg is $\sqrt{2}\sqrt{9 - x^2}$. $A(x) = \tfrac{1}{2}\left(\sqrt{2}\sqrt{9 - x^2}\right)^2 = 9 - x^2$, so

$$V = 2\int_0^3 A(x)\, dx = 2\int_0^3 (9 - x^2)\, dx = 2\left[9x - \tfrac{1}{3}x^3\right]_0^3 = 2(27 - 9) = 36$$

24. $V = \int_{-1}^1 A(x)\, dx = 2\int_0^1 A(x)\, dx = 2\int_0^1 \left[(2 - x^2) - x^2\right]^2 dx = 2\int_0^1 \left[2(1 - x^2)\right]^2 dx$

$$= 8\int_0^1 (1 - 2x^2 + x^4)\, dx = 8\left[x - \tfrac{2}{3}x^3 + \tfrac{1}{5}x^5\right]_0^1 = 8\left(1 - \tfrac{2}{3} + \tfrac{1}{5}\right) = \tfrac{64}{15}$$

25. Equilateral triangles with sides measuring $\tfrac{1}{4}x$ meters have height $\tfrac{1}{4}x \sin 60° = \tfrac{\sqrt{3}}{8}x$. Therefore,

$$A(x) = \tfrac{1}{2} \cdot \tfrac{1}{4}x \cdot \tfrac{\sqrt{3}}{8}x = \tfrac{\sqrt{3}}{64}x^2. \quad V = \int_0^{20} A(x)\, dx = \tfrac{\sqrt{3}}{64}\int_0^{20} x^2\, dx = \tfrac{\sqrt{3}}{64}\left[\tfrac{1}{3}x^3\right]_0^{20} = \tfrac{8000\sqrt{3}}{64 \cdot 3} = \tfrac{125\sqrt{3}}{3}\ \text{m}^3.$$

26. (a) By the symmetry of the problem, we consider only the solid to the right of the origin. The semicircular cross-sections

perpendicular to the x-axis have radius $1 - x$, so $A(x) = \tfrac{1}{2}\pi(1 - x)^2$. Now we can calculate

$$V = 2\int_0^1 A(x)\, dx = 2\int_0^1 \tfrac{1}{2}\pi(1 - x)^2\, dx = \int_0^1 \pi(1 - x)^2\, dx = -\tfrac{\pi}{3}\left[(1 - x)^3\right]_0^1 = \tfrac{\pi}{3}.$$

(b) Cut the solid with a plane perpendicular to the x-axis and passing through the y-axis. Fold the half of the solid in the

region $x \le 0$ under the xy-plane so that the point $(-1, 0)$ comes around and touches the point $(1, 0)$. The resulting solid is

a right circular cone of radius 1 with vertex at $(x, y, z) = (1, 0, 0)$ and with its base in the yz-plane, centered at the origin.

The volume of this cone is $\tfrac{1}{3}\pi r^2 h = \tfrac{1}{3}\pi \cdot 1^2 \cdot 1 = \tfrac{\pi}{3}$.

27. $f(x) = kx$ $\Rightarrow$ $30\text{ N} = k(15 - 12)\text{ cm}$ $\Rightarrow$ $k = 10\text{ N/cm} = 1000\text{ N/m}$. $20\text{ cm} - 12\text{ cm} = 0.08\text{ m}$ $\Rightarrow$

$W = \int_0^{0.08} kx\,dx = 1000 \int_0^{0.08} x\,dx = 500\big[x^2\big]_0^{0.08} = 500(0.08)^2 = 3.2\text{ N·m} = 3.2\text{ J}.$

28. The work needed to raise the elevator alone is $1600\text{ lb} \times 30\text{ ft} = 48{,}000$ ft-lb. The work needed to raise the bottom

170 ft of cable is $170\text{ ft} \times 10\text{ lb/ft} \times 30\text{ ft} = 51{,}000$ ft-lb. The work needed to raise the top 30 ft of cable is

$\int_0^{30} 10x\,dx = \big[5x^2\big]_0^{30} = 5 \cdot 900 = 4500$ ft-lb. Adding these, we see that the total work needed is

$48{,}000 + 51{,}000 + 4{,}500 = 103{,}500$ ft-lb.

29. (a) The parabola has equation $y = ax^2$ with vertex at the origin and passing through

$(4, 4)$. $4 = a \cdot 4^2$ $\Rightarrow$ $a = \frac{1}{4}$ $\Rightarrow$ $y = \frac{1}{4}x^2$ $\Rightarrow$ $x^2 = 4y$ $\Rightarrow$

$x = 2\sqrt{y}$. Each circular disk has radius $2\sqrt{y}$ and is moved $4 - y$ ft.

$$W = \int_0^4 \pi \left(2\sqrt{y}\right)^2 62.5(4 - y)\,dy = 250\pi \int_0^4 y(4 - y)\,dy$$

$$= 250\pi \big[2y^2 - \tfrac{1}{3}y^3\big]_0^4 = 250\pi\left(32 - \tfrac{64}{3}\right) = \tfrac{8000\pi}{3} \approx 8378\text{ ft-lb}$$

(b) In part (a) we knew the final water level (0) but not the amount of work done. Here

we use the same equation, except with the work fixed, and the lower limit of

integration (that is, the final water level — call it h) unknown: $W = 4000$ $\Leftrightarrow$

$250\pi \big[2y^2 - \tfrac{1}{3}y^3\big]_h^4 = 4000$ $\Leftrightarrow$ $\tfrac{16}{\pi} = \big[\left(32 - \tfrac{64}{3}\right) - \left(2h^2 - \tfrac{1}{3}h^3\right)\big]$ $\Leftrightarrow$

$h^3 - 6h^2 + 32 - \tfrac{48}{\pi} = 0$. We graph the function $f(h) = h^3 - 6h^2 + 32 - \tfrac{48}{\pi}$

on the interval $[0, 4]$ to see where it is 0. From the graph, $f(h) = 0$ for $h \approx 2.1$.

So the depth of water remaining is about 2.1 ft.

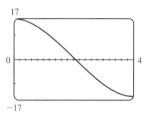

30. $f_{\text{ave}} = \frac{1}{10 - 0} \int_0^{10} t\,\sin(t^2)\,dt = \frac{1}{10} \int_0^{100} \sin u \left(\tfrac{1}{2}\,du\right)$ $[u = t^2,\ du = 2t\,dt]$

$= \frac{1}{20}\Big[-\cos u\Big]_0^{100} = \frac{1}{20}(-\cos 100 + \cos 0) = \frac{1}{20}(1 - \cos 100) \approx 0.007$

31. $\displaystyle\lim_{h \to 0} f_{\text{ave}} = \lim_{h \to 0} \frac{1}{(x + h) - x} \int_x^{x+h} f(t)\,dt = \lim_{h \to 0} \frac{F(x + h) - F(x)}{h}$, where $F(x) = \int_a^x f(t)\,dt$. But we recognize this

limit as being $F'(x)$ by the definition of a derivative. Therefore, $\displaystyle\lim_{h \to 0} f_{\text{ave}} = F'(x) = f(x)$ by FTC1.

32. (a) $\mathcal{R}_1$ is the region below the graph of $y = x^2$ and above the x-axis between $x = 0$ and $x = b$, and $\mathcal{R}_2$ is the region

to the left of the graph of $x = \sqrt{y}$ and to the right of the y-axis between $y = 0$ and $y = b^2$. So the area of $\mathcal{R}_1$ is

$A_1 = \int_0^b x^2\,dx = \big[\tfrac{1}{3}x^3\big]_0^b = \tfrac{1}{3}b^3$, and the area of $\mathcal{R}_2$ is $A_2 = \int_0^{b^2} \sqrt{y}\,dy = \big[\tfrac{2}{3}y^{3/2}\big]_0^{b^2} = \tfrac{2}{3}b^3$. So there is no

solution to $A_1 = A_2$ for $b \neq 0$.

(b) Using disks, we calculate the volume of rotation of $\mathcal{R}_1$ about the x-axis to be $V_{1,x} = \pi \int_0^b (x^2)^2\,dx = \tfrac{1}{5}\pi b^5$.

Using cylindrical shells, we calculate the volume of rotation of $\mathcal{R}_1$ about the y-axis to be

$V_{1,y} = 2\pi \int_0^b x(x^2)\,dx = 2\pi \big[\tfrac{1}{4}x^4\big]_0^b = \tfrac{1}{2}\pi b^4$. So $V_{1,x} = V_{1,y}$ $\Leftrightarrow$ $\tfrac{1}{5}\pi b^5 = \tfrac{1}{2}\pi b^4$ $\Leftrightarrow$ $2b = 5$ $\Leftrightarrow$ $b = \tfrac{5}{2}$.

So the volumes of rotation about the x- and y-axes are the same for $b = \tfrac{5}{2}$.

(c) We use cylindrical shells to calculate the volume of rotation of $\mathcal{R}_2$ about the x-axis:

$$\mathcal{R}_{2,x} = 2\pi \int_0^{b^2} y\left(\sqrt{y}\right) dy = 2\pi \left[\tfrac{2}{5}y^{5/2}\right]_0^{b^2} = \tfrac{4}{5}\pi b^5.$$ We already know the volume of rotation of $\mathcal{R}_1$ about the x-axis

from part (b), and $\mathcal{R}_{1,x} = \mathcal{R}_{2,x} \iff \tfrac{1}{5}\pi b^5 = \tfrac{4}{5}\pi b^5$, which has no solution for $b \neq 0$.

(d) We use disks to calculate the volume of rotation of $\mathcal{R}_2$ about the y-axis: $\mathcal{R}_{2,y} = \pi \int_0^{b^2} \left(\sqrt{y}\right)^2 dy = \pi\left[\tfrac{1}{2}y^2\right]_0^{b^2} = \tfrac{1}{2}\pi b^4.$

We know the volume of rotation of $\mathcal{R}_1$ about the y-axis from part (b), and $\mathcal{R}_{1,y} = \mathcal{R}_{2,y} \iff \tfrac{1}{2}\pi b^4 = \tfrac{1}{2}\pi b^4$. But this

equation is true for all b, so the volumes of rotation about the y-axis are equal for all values of b.

□ PROBLEMS PLUS

1. (a) The area under the graph of f from 0 to t is equal to $\int_0^t f(x)\,dx$, so the requirement is that $\int_0^t f(x)\,dx = t^3$ for all t. We differentiate both sides of this equation with respect to t (with the help of FTC1) to get $f(t) = 3t^2$. This function is positive and continuous, as required.

(b) The volume generated from $x = 0$ to $x = b$ is $\int_0^b \pi [f(x)]^2\,dx$. Hence, we are given that $b^2 = \int_0^b \pi [f(x)]^2\,dx$ for all $b > 0$. Differentiating both sides of this equation with respect to b using the Fundamental Theorem of Calculus gives $2b = \pi [f(b)]^2 \;\Rightarrow\; f(b) = \sqrt{2b/\pi}$, since f is positive. Therefore, $f(x) = \sqrt{2x/\pi}$.

2. The total area of the region bounded by the parabola $y = x - x^2 = x(1 - x)$

and the x-axis is $\int_0^1 (x - x^2)\,dx = \left[\frac{1}{2}x^2 - \frac{1}{3}x^3\right]_0^1 = \frac{1}{6}$. Let the slope of the

line we are looking for be m. Then the area above this line but below the

parabola is $\int_0^a \left[(x - x^2) - mx\right] dx$, where a is the x-coordinate of the point

of intersection of the line and the parabola. We find the point of intersection

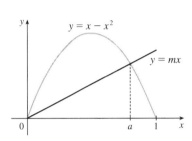

by solving the equation $x - x^2 = mx \;\Leftrightarrow\; 1 - x = m \;\Leftrightarrow\; x = 1 - m$. So the value of a is $1 - m$, and

$$\int_0^{1-m} \left[(x - x^2) - mx\right] dx = \int_0^{1-m} \left[(1 - m)x - x^2\right] dx = \left[\tfrac{1}{2}(1 - m)x^2 - \tfrac{1}{3}x^3\right]_0^{1-m}$$

$$= \tfrac{1}{2}(1 - m)(1 - m)^2 - \tfrac{1}{3}(1 - m)^3 = \tfrac{1}{6}(1 - m)^3$$

We want this to be half of $\frac{1}{6}$, so $\frac{1}{6}(1 - m)^3 = \frac{1}{12} \;\Leftrightarrow\; (1 - m)^3 = \frac{6}{12} \;\Leftrightarrow\; 1 - m = \sqrt[3]{\frac{1}{2}} \;\Leftrightarrow\; m = 1 - \frac{1}{\sqrt[3]{2}}$. So the

slope of the required line is $1 - \frac{1}{\sqrt[3]{2}} \approx 0.206$.

3. Let a and b be the x-coordinates of the points where the line intersects the

curve. From the figure, $R_1 = R_2 \;\Rightarrow$

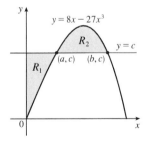

$$\int_0^a \left[c - (8x - 27x^3)\right] dx = \int_a^b \left[(8x - 27x^3) - c\right] dx$$

$$\left[cx - 4x^2 + \tfrac{27}{4}x^4\right]_0^a = \left[4x^2 - \tfrac{27}{4}x^4 - cx\right]_a^b$$

$$ac - 4a^2 + \tfrac{27}{4}a^4 = \left(4b^2 - \tfrac{27}{4}b^4 - bc\right) - \left(4a^2 - \tfrac{27}{4}a^4 - ac\right)$$

$$0 = 4b^2 - \tfrac{27}{4}b^4 - bc = 4b^2 - \tfrac{27}{4}b^4 - b(8b - 27b^3)$$

$$= 4b^2 - \tfrac{27}{4}b^4 - 8b^2 + 27b^4 = \tfrac{81}{4}b^4 - 4b^2$$

$$= b^2\left(\tfrac{81}{4}b^2 - 4\right)$$

So for $b > 0$, $b^2 = \frac{16}{81} \;\Rightarrow\; b = \frac{4}{9}$. Thus, $c = 8b - 27b^3 = 8\left(\frac{4}{9}\right) - 27\left(\frac{64}{729}\right) = \frac{32}{9} - \frac{64}{27} = \frac{32}{27}$.

4. (a) Take slices perpendicular to the line through the center C of the bottom of the glass and the point P where the top surface of the water meets the bottom of the glass.

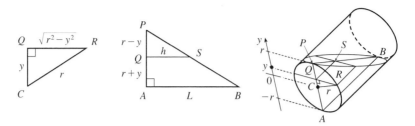

A typical rectangular cross-section y units above the axis of the glass has width $2\,|QR| = 2\,\sqrt{r^2 - y^2}$ and length

$$h = |QS| = \frac{L}{2r}\,(r - y). \quad \left[\text{Triangles } PQS \text{ and } PAB \text{ are similar, so } \frac{h}{L} = \frac{|PQ|}{|PA|} = \frac{r - y}{2r}.\right] \quad \text{Thus,}$$

$$V = \int_{-r}^{r} 2\,\sqrt{r^2 - y^2} \cdot \frac{L}{2r}\,(r - y)\,dy = L \int_{-r}^{r} \left(1 - \frac{y}{r}\right)\sqrt{r^2 - y^2}\,dy$$

$$= L \int_{-r}^{r} \sqrt{r^2 - y^2}\,dy - \frac{L}{r} \int_{-r}^{r} y\,\sqrt{r^2 - y^2}\,dy$$

$$= L \cdot \frac{\pi r^2}{2} - \frac{L}{r} \cdot 0 \quad \begin{bmatrix} \text{the first integral is the area of a semicircle of radius } r, \\ \text{and the second has an odd integrand} \end{bmatrix} \quad = \frac{\pi r^2 L}{2}$$

(b) Slice parallel to the plane through the axis of the glass and the point of contact P. (This is the plane determined by P, B, and C in the figure.) $STUV$ is a typical trapezoidal slice. With respect to an x-axis with origin at C as shown, if S and V have x-coordinate x, then $|SV| = 2\,\sqrt{r^2 - x^2}$. Projecting the trapezoid $STUV$ onto the plane of the triangle PAB (call the projection $S'T'U'V'$), we see that $|AP| = 2r$, $|SV| = 2\,\sqrt{r^2 - x^2}$, and $|S'P| = |V'A| = \frac{1}{2}(|AP| - |SV|) = r - \sqrt{r^2 - x^2}$.

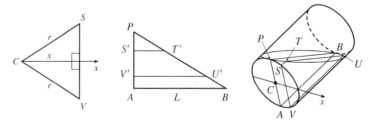

By similar triangles, $\dfrac{|ST|}{|S'P|} = \dfrac{|AB|}{|AP|}$, so $|ST| = \left(r - \sqrt{r^2 - x^2}\right) \cdot \dfrac{L}{2r}$. In the same way, we find that

$\dfrac{|VU|}{|V'P|} = \dfrac{|AB|}{|AP|}$, so $|VU| = |V'P| \cdot \dfrac{L}{2r} = (|AP| - |V'A|) \cdot \dfrac{L}{2r} = \left(r + \sqrt{r^2 - x^2}\right) \cdot \dfrac{L}{2r}$. The

area $A(x)$ of the trapezoid $STUV$ is $\frac{1}{2}|SV| \cdot (|ST| + |VU|)$; that is,

$$A(x) = \frac{1}{2} \cdot 2\,\sqrt{r^2 - x^2} \cdot \left[\left(r - \sqrt{r^2 - x^2}\right) \cdot \frac{L}{2r} + \left(r + \sqrt{r^2 - x^2}\right) \cdot \frac{L}{2r}\right] = L\,\sqrt{r^2 - x^2}. \quad \text{Thus,}$$

$$V = \int_{-r}^{r} A(x)\,dx = L \int_{-r}^{r} \sqrt{r^2 - x^2}\,dx = L \cdot \frac{\pi r^2}{2} = \frac{\pi r^2 L}{2}.$$

(c) See the computation of V in part (a) or part (b).

(d) The volume of the water is exactly half the volume of the cylindrical glass, so $V = \frac{1}{2}\pi r^2 L$.

(e) Choose x-, y-, and z-axes as shown in the figure. Then slices perpendicular to the x-axis are triangular, slices perpendicular to the y-axis are rectangular, and slices perpendicular to the z-axis are segments of circles.

 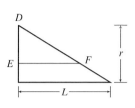

Using triangular slices, we find that the area $A(x)$ of a typical slice DEF, where D has x-coordinate x, is given by

$$A(x) = \tfrac{1}{2}|DE| \cdot |EF| = \tfrac{1}{2}|DE| \cdot \left(\frac{L}{r}|DE|\right) = \frac{L}{2r}|DE|^2 = \frac{L}{2r}(r^2 - x^2).\text{ Thus,}$$

$$V = \int_{-r}^{r} A(x)\,dx = \frac{L}{2r}\int_{-r}^{r}(r^2 - x^2)\,dx = \frac{L}{r}\int_{-r}^{r}(r^2 - x^2)\,dx = \frac{L}{r}\left[r^2 x - \frac{x^3}{3}\right]_0^r$$

$$= \frac{L}{r}\left(r^3 - \frac{r^3}{3}\right) = \frac{L}{r}\cdot\frac{2}{3}r^3 = \tfrac{2}{3}r^2 L \qquad \text{[This is } 2/(3\pi) \approx 0.21 \text{ of the volume of the glass.]}$$

5. (a) $V = \pi h^2(r - h/3) = \frac{1}{3}\pi h^2(3r - h)$. See the solution to Exercise 6.2.51.

(b) The smaller segment has height $h = 1 - x$ and so by part (a) its volume is

$V = \frac{1}{3}\pi(1 - x)^2\,[3(1) - (1 - x)] = \frac{1}{3}\pi(x - 1)^2(x + 2)$. This volume must be $\frac{1}{3}$ of the total volume of the sphere, which is $\frac{4}{3}\pi(1)^3$. So $\frac{1}{3}\pi(x - 1)^2(x + 2) = \frac{1}{3}\left(\frac{4}{3}\pi\right)$ $\Rightarrow$ $(x^2 - 2x + 1)(x + 2) = \frac{4}{3}$ $\Rightarrow$ $x^3 - 3x + 2 = \frac{4}{3}$ $\Rightarrow$ $3x^3 - 9x + 2 = 0$. Using Newton's method with $f(x) = 3x^3 - 9x + 2$, $f'(x) = 9x^2 - 9$, we get

$x_{n+1} = x_n - \dfrac{3x_n^3 - 9x_n + 2}{9x_n^2 - 9}$. Taking $x_1 = 0$, we get $x_2 \approx 0.2222$, and $x_3 \approx 0.2261 \approx x_4$, so, correct to four decimal places, $x \approx 0.2261$.

(c) With $r = 0.5$ and $s = 0.75$, the equation $x^3 - 3rx^2 + 4r^3 s = 0$ becomes $x^3 - 3(0.5)x^2 + 4(0.5)^3(0.75) = 0$ $\Rightarrow$ $x^3 - \frac{3}{2}x^2 + 4\left(\frac{1}{8}\right)\frac{3}{4} = 0$ $\Rightarrow$ $8x^3 - 12x^2 + 3 = 0$. We use Newton's method with $f(x) = 8x^3 - 12x^2 + 3$,

$f'(x) = 24x^2 - 24x$, so $x_{n+1} = x_n - \dfrac{8x_n^3 - 12x_n^2 + 3}{24x_n^2 - 24x_n}$. Take $x_1 = 0.5$. Then $x_2 \approx 0.6667$, and $x_3 \approx 0.6736 \approx x_4$.

So to four decimal places the depth is 0.6736 m.

(d) (i) From part (a) with $r = 5$ in., the volume of water in the bowl is

$V = \frac{1}{3}\pi h^2(3r - h) = \frac{1}{3}\pi h^2(15 - h) = 5\pi h^2 - \frac{1}{3}\pi h^3$. We are given that $\dfrac{dV}{dt} = 0.2 \text{ in}^3/\text{s}$ and we want to find $\dfrac{dh}{dt}$

when $h = 3$. Now $\dfrac{dV}{dt} = 10\pi h \dfrac{dh}{dt} - \pi h^2 \dfrac{dh}{dt}$, so $\dfrac{dh}{dt} = \dfrac{0.2}{\pi(10h - h^2)}$. When $h = 3$, we have

$\dfrac{dh}{dt} = \dfrac{0.2}{\pi(10 \cdot 3 - 3^2)} = \dfrac{1}{105\pi} \approx 0.003 \text{ in/s}$.

(ii) From part (a), the volume of water required to fill the bowl from the instant that the water is 4 in. deep is

$V = \frac{1}{2} \cdot \frac{4}{3}\pi(5)^3 - \frac{1}{3}\pi(4)^2(15 - 4) = \frac{2}{3} \cdot 125\pi - \frac{16}{3} \cdot 11\pi = \frac{74}{3}\pi$. To find the time required to fill the bowl we divide

this volume by the rate: Time $= \dfrac{74\pi/3}{0.2} = \dfrac{370\pi}{3} \approx 387 \text{ s} \approx 6.5 \text{ min}$.

6. (a) The volume above the surface is $\int_0^{L-h} A(y)\,dy = \int_{-h}^{L-h} A(y)\,dy - \int_{-h}^0 A(y)\,dy$. So the proportion of volume above the

surface is $\dfrac{\int_0^{L-h} A(y)\,dy}{\int_{-h}^{L-h} A(y)\,dy} = \dfrac{\int_{-h}^{L-h} A(y)\,dy - \int_{-h}^0 A(y)\,dy}{\int_{-h}^{L-h} A(y)\,dy}$. Now by Archimedes' Principle, we have $F = W \;\Rightarrow$

$\rho_f g \int_{-h}^0 A(y)\,dy = \rho_0 g \int_{-h}^{L-h} A(y)\,dy$, so $\int_{-h}^0 A(y)\,dy = (\rho_0/\rho_f)\int_{-h}^{L-h} A(y)\,dy$. Therefore,

$\dfrac{\int_0^{L-h} A(y)\,dy}{\int_{-h}^{L-h} A(y)\,dy} = \dfrac{\int_{-h}^{L-h} A(y)\,dy - (\rho_0/\rho_f)\int_{-h}^{L-h} A(y)\,dy}{\int_{-h}^{L-h} A(y)\,dy} = \dfrac{\rho_f - \rho_0}{\rho_f}$, so the percentage of volume above the surface

is $100\left(\dfrac{\rho_f - \rho_0}{\rho_f}\right)\%$.

(b) For an iceberg, the percentage of volume above the surface is $100\left(\frac{1030 - 917}{1030}\right)\% \approx 11\%$.

(c) No, the water does not overflow. Let V_i be the volume of the ice cube, and let V_w be the volume of the water which results

from the melting. Then by the formula derived in part (a), the volume of ice above the surface of the water is

$[(\rho_f - \rho_0)/\rho_f]V_i$, so the volume below the surface is $V_i - [(\rho_f - \rho_0)/\rho_f]\,V_i = (\rho_0/\rho_f)V_i$. Now the mass of the ice

cube is the same as the mass of the water which is created when it melts, namely $m = \rho_0 V_i = \rho_f V_w \;\Rightarrow$

$V_w = (\rho_0/\rho_f)V_i$. So when the ice cube melts, the volume of the resulting water is the same as the underwater volume of

the ice cube, and so the water does not overflow.

(d) The figure shows the instant when the height of the exposed part of the ball is y.

Using the formula in Problem 5(a) with $r = 0.4$ and $h = 0.8 - y$, we see that the

volume of the submerged part of the sphere is $\frac{1}{3}\pi(0.8 - y)^2[1.2 - (0.8 - y)]$, so

its weight is $1000g \cdot \frac{1}{3}\pi s^2(1.2 - s)$, where $s = 0.8 - y$. Then the work done to

submerge the sphere is

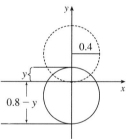

$$W = \int_0^{0.8} g\frac{1000}{3}\pi s^2(1.2 - s)\,ds = g\frac{1000}{3}\pi \int_0^{0.8}(1.2s^2 - s^3)\,ds$$

$$= g\frac{1000}{3}\pi\left[0.4s^3 - \tfrac{1}{4}s^4\right]_0^{0.8} = g\frac{1000}{3}\pi(0.2048 - 0.1024) = 9.8\,\frac{1000}{3}\,\pi(0.1024) \approx 1.05 \times 10^3 \text{ J}$$

7. We are given that the rate of change of the volume of water is $\dfrac{dV}{dt} = -kA(x)$, where k is some positive constant and $A(x)$ is

the area of the surface when the water has depth x. Now we are concerned with the rate of change of the depth of the water

with respect to time, that is, $\dfrac{dx}{dt}$. But by the Chain Rule, $\dfrac{dV}{dt} = \dfrac{dV}{dx}\dfrac{dx}{dt}$, so the first equation can be written

$\dfrac{dV}{dx}\dfrac{dx}{dt} = -kA(x)$ (⋆). Also, we know that the total volume of water up to a depth x is $V(x) = \int_0^x A(s)\,ds$, where $A(s)$ is

the area of a cross-section of the water at a depth s. Differentiating this equation with respect to x, we get $dV/dx = A(x)$.

Substituting this into equation ⋆, we get $A(x)(dx/dt) = -kA(x) \;\Rightarrow\; dx/dt = -k$, a constant.

8. A typical sphere of radius r is shown in the figure. We wish to maximize the shaded volume V, which can be thought of as the volume of a hemisphere of radius r minus the volume of the spherical cap with height $h = 1 - \sqrt{1 - r^2}$ and radius 1.

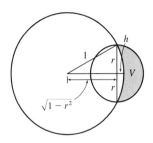

$V = \frac{1}{2} \cdot \frac{4}{3}\pi r^3 - \frac{1}{3}\pi \left(1 - \sqrt{1 - r^2}\right)^2 \left[3(1) - \left(1 - \sqrt{1 - r^2}\right)\right]$ [by Problem 5(a)]

$= \frac{1}{3}\pi \left[2r^3 - \left(2 - 2\sqrt{1 - r^2} - r^2\right)\left(2 + \sqrt{1 - r^2}\right)\right]$

$= \frac{1}{3}\pi \left[2r^3 - 2 + \left(r^2 + 2\right)\sqrt{1 - r^2}\right]$

$V' = \frac{1}{3}\pi \left[6r^2 + \frac{(r^2 + 2)(-r)}{\sqrt{1 - r^2}} + \sqrt{1 - r^2}(2r)\right] = \frac{1}{3}\pi \left[\frac{6r^2\sqrt{1 - r^2} - r(r^2 + 2) + 2r(1 - r^2)}{\sqrt{1 - r^2}}\right]$

$= \frac{1}{3}\pi \left(\frac{6r^2\sqrt{1 - r^2} - 3r^3}{\sqrt{1 - r^2}}\right) = \frac{\pi r^2 \left(2\sqrt{1 - r^2} - r\right)}{\sqrt{1 - r^2}}$

$V'(r) = 0 \quad \Leftrightarrow \quad 2\sqrt{1 - r^2} = r \quad \Leftrightarrow \quad 4 - 4r^2 = r^2 \quad \Leftrightarrow \quad r^2 = \frac{4}{5} \quad \Leftrightarrow \quad r = \frac{2}{\sqrt{5}} \approx 0.89.$

Since $V'(r) > 0$ for $0 < r < \frac{2}{\sqrt{5}}$ and $V'(r) < 0$ for $\frac{2}{\sqrt{5}} < r < 1$, we know that V attains a maximum at $r = \frac{2}{\sqrt{5}}$.

9. We must find expressions for the areas A and B, and then set them equal and see what this says about the curve C. If $P = (a, 2a^2)$, then area A is just $\int_0^a (2x^2 - x^2)\, dx = \int_0^a x^2\, dx = \frac{1}{3}a^3$. To find area B, we use y as the variable of integration. So we find the equation of the middle curve as a function of y: $y = 2x^2 \quad \Leftrightarrow \quad x = \sqrt{y/2}$, since we are concerned with the first quadrant only. We can express area B as

$$\int_0^{2a^2} \left[\sqrt{y/2} - C(y)\right] dy = \left[\frac{4}{3}(y/2)^{3/2}\right]_0^{2a^2} - \int_0^{2a^2} C(y)\, dy = \frac{4}{3}a^3 - \int_0^{2a^2} C(y)\, dy$$

where $C(y)$ is the function with graph C. Setting $A = B$, we get $\frac{1}{3}a^3 = \frac{4}{3}a^3 - \int_0^{2a^2} C(y)\, dy \quad \Leftrightarrow \quad \int_0^{2a^2} C(y)\, dy = a^3$.

Now we differentiate this equation with respect to a using the Chain Rule and the Fundamental Theorem:

$C(2a^2)(4a) = 3a^2 \quad \Rightarrow \quad C(y) = \frac{3}{4}\sqrt{y/2}$, where $y = 2a^2$. Now we can solve for y: $x = \frac{3}{4}\sqrt{y/2} \quad \Rightarrow$

$x^2 = \frac{9}{16}(y/2) \quad \Rightarrow \quad y = \frac{32}{9}x^2.$

10. We want to find the volume of that part of the sphere which is below the surface of the water. As we can see from the diagram, this region is a cap of a sphere with radius r and height $r + d$. If we can find an expression for d in terms of h, r and θ, then we can determine the volume of the region [see Problem 5(a)], and then differentiate with respect to r to find the maximum. We see that

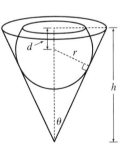

$\sin\theta = \frac{r}{h - d} \quad \Leftrightarrow \quad h - d = \frac{r}{\sin\theta} \quad \Leftrightarrow \quad d = h - r\csc\theta.$

Now we can use the formula from Problem 5(a) to find the volume of water displaced:

$$V = \frac{1}{3}\pi h^2 (3r - h) = \frac{1}{3}\pi (r + d)^2 \left[3r - (r + d)\right] = \frac{1}{3}\pi (r + h - r\csc\theta)^2 (2r - h + r\csc\theta)$$

$$= \frac{\pi}{3}[r(1 - \csc\theta) + h]^2 [r(2 + \csc\theta) - h]$$

[continued]

Now we differentiate with respect to r:

$$dV/dr = \tfrac{\pi}{3}\left([r(1-\csc\theta)+h]^2(2+\csc\theta)+2[r(1-\csc\theta)+h](1-\csc\theta)[r(2+\csc\theta)-h]\right)$$

$$= \tfrac{\pi}{3}[r(1-\csc\theta)+h]([r(1-\csc\theta)+h](2+\csc\theta)+2(1-\csc\theta)[r(2+\csc\theta)-h])$$

$$= \tfrac{\pi}{3}[r(1-\csc\theta)+h](3(2+\csc\theta)(1-\csc\theta)r+[(2+\csc\theta)-2(1-\csc\theta)]h)$$

$$= \tfrac{\pi}{3}[r(1-\csc\theta)+h][3(2+\csc\theta)(1-\csc\theta)r+3h\csc\theta]$$

This is 0 when $r = \dfrac{h}{\csc\theta-1}$ and when $r = \dfrac{h\csc\theta}{(\csc\theta+2)(\csc\theta-1)}$. Now since $V\left(\dfrac{h}{\csc\theta-1}\right) = 0$ (the first factor

vanishes; this corresponds to $d = -r$), the maximum volume of water is displaced when $r = \dfrac{h\csc\theta}{(\csc\theta-1)(\csc\theta+2)}$.

(Our intuition tells us that a maximum value does exist, and it must occur at a critical number.) Multiplying numerator and

denominator by $\sin^2\theta$, we get an alternative form of the answer: $r = \dfrac{h\sin\theta}{\sin\theta+\cos 2\theta}$.

11. (a) Stacking disks along the y-axis gives us $V = \int_0^h \pi\,[f(y)]^2\,dy$.

(b) Using the Chain Rule, $\dfrac{dV}{dt} = \dfrac{dV}{dh}\cdot\dfrac{dh}{dt} = \pi\,[f(h)]^2\,\dfrac{dh}{dt}$.

(c) $kA\sqrt{h} = \pi[f(h)]^2\,\dfrac{dh}{dt}$. Set $\dfrac{dh}{dt} = C$: $\pi[f(h)]^2\,C = kA\sqrt{h} \;\Rightarrow\; [f(h)]^2 = \dfrac{kA}{\pi C}\sqrt{h} \;\Rightarrow\; f(h) = \sqrt{\dfrac{kA}{\pi C}}\,h^{1/4}$; that

is, $f(y) = \sqrt{\dfrac{kA}{\pi C}}\,y^{1/4}$. The advantage of having $\dfrac{dh}{dt} = C$ is that the markings on the container are equally spaced.

12. (a) We first use the cylindrical shell method to express the volume V in terms of h, r, and ω:

$$V = \int_0^r 2\pi xy\,dx = \int_0^r 2\pi x\left[h+\dfrac{\omega^2 x^2}{2g}\right]dx = 2\pi\int_0^r\left(hx+\dfrac{\omega^2 x^3}{2g}\right)dx$$

$$= 2\pi\left[\dfrac{hx^2}{2}+\dfrac{\omega^2 x^4}{8g}\right]_0^r = 2\pi\left[\dfrac{hr^2}{2}+\dfrac{\omega^2 r^4}{8g}\right] = \pi hr^2 + \dfrac{\pi\omega^2 r^4}{4g} \quad\Rightarrow$$

$$h = \dfrac{V-\left(\pi\omega^2 r^4\right)/(4g)}{\pi r^2} = \dfrac{4gV-\pi\omega^2 r^4}{4\pi gr^2}.$$

(b) The surface touches the bottom when $h = 0 \;\Rightarrow\; 4gV - \pi\omega^2 r^4 = 0 \;\Rightarrow\; \omega^2 = \dfrac{4gV}{\pi r^4} \;\Rightarrow\; \omega = \dfrac{2\sqrt{gV}}{\sqrt{\pi}\,r^2}$.

To spill over the top, $y(r) > L \;\Leftrightarrow$

$$L < h + \dfrac{\omega^2 r^2}{2g} = \dfrac{4gV-\pi\omega^2 r^4}{4\pi gr^2} + \dfrac{\omega^2 r^2}{2g} = \dfrac{4gV}{4\pi gr^2} - \dfrac{\pi\omega^2 r^2}{4\pi gr^2} + \dfrac{\omega^2 r^2}{2g}$$

$$= \dfrac{V}{\pi r^2} - \dfrac{\omega^2 r^2}{4g} + \dfrac{\omega^2 r^2}{2g} = \dfrac{V}{\pi r^2} + \dfrac{\omega^2 r^2}{4g} \quad\Leftrightarrow$$

$\dfrac{\omega^2 r^2}{4g} > L - \dfrac{V}{\pi r^2} = \dfrac{\pi r^2 L - V}{\pi r^2} \;\Leftrightarrow\; \omega^2 > \dfrac{4g(\pi r^2 L - V)}{\pi r^4}$. So for spillage, the angular speed should

be $\omega > \dfrac{2\sqrt{g(\pi r^2 L - V)}}{r^2\sqrt{\pi}}$.

(c) (i) Here we have $r = 2$, $L = 7$, $h = 7 - 5 = 2$. When $x = 1$, $y = 7 - 4 = 3$. Therefore, $3 = 2 + \dfrac{\omega^2\cdot 1^2}{2\cdot 32} \;\Rightarrow$

$1 = \dfrac{\omega^2}{2\cdot 32} \;\Rightarrow\; \omega^2 = 64 \;\Rightarrow\; \omega = 8$ rad/s. $V = \pi(2)(2)^2 + \dfrac{\pi\cdot 8^2\cdot 2^4}{4g} = 8\pi + 8\pi = 16\pi$ ft^2.

(ii) At the wall, $x = 2$, so $y = 2 + \dfrac{8^2\cdot 2^2}{2\cdot 32} = 6$ and the surface is $7 - 6 = 1$ ft below the top of the tank.

13. The cubic polynomial passes through the origin, so let its equation be

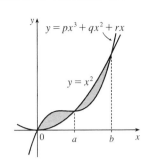

$y = px^3 + qx^2 + rx$. The curves intersect when $px^3 + qx^2 + rx = x^2$ ⇔

$px^3 + (q-1)x^2 + rx = 0$. Call the left side $f(x)$. Since $f(a) = f(b) = 0$,

another form of f is

$$f(x) = px(x-a)(x-b) = px[x^2 - (a+b)x + ab]$$
$$= p[x^3 - (a+b)x^2 + abx]$$

Since the two areas are equal, we must have $\int_0^a f(x)\,dx = -\int_a^b f(x)\,dx$ ⇒

$[F(x)]_0^a = [F(x)]_b^a$ ⇒ $F(a) - F(0) = F(a) - F(b)$ ⇒ $F(0) = F(b)$, where F is an antiderivative of f.

Now $F(x) = \int f(x)\,dx = \int p[x^3 - (a+b)x^2 + abx]\,dx = p\left[\frac{1}{4}x^4 - \frac{1}{3}(a+b)x^3 + \frac{1}{2}abx^2\right] + C$, so

$F(0) = F(b)$ ⇒ $C = p\left[\frac{1}{4}b^4 - \frac{1}{3}(a+b)b^3 + \frac{1}{2}ab^3\right] + C$ ⇒ $0 = p\left[\frac{1}{4}b^4 - \frac{1}{3}(a+b)b^3 + \frac{1}{2}ab^3\right]$ ⇒

$0 = 3b - 4(a+b) + 6a$ [multiply by $12/(pb^3)$, $b \neq 0$] ⇒ $0 = 3b - 4a - 4b + 6a$ ⇒ $b = 2a$.

Hence, b is twice the value of a.

14. (a) Place the round flat tortilla on an xy-coordinate system as shown in

the first figure. An equation of the circle is $x^2 + y^2 = 4^2$ and the

height of a cross-section is $2\sqrt{16 - x^2}$.

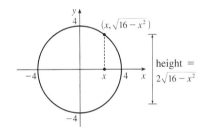

Now look at a cross-section with central angle θ_x as shown in the

second figure (r is the radius of the circular cylinder). The filled area

$A(x)$ is equal to the area $A_1(x)$ of the sector minus the area $A_2(x)$

of the triangle.

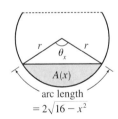

$$A(x) = A_1(x) - A_2(x) = \tfrac{1}{2}r^2\theta_x - \tfrac{1}{2}r^2\sin\theta_x \quad \text{[area formulas from trigonometry]}$$

$$= \tfrac{1}{2}r(r\theta_x) - \tfrac{1}{2}r^2\sin\left(\frac{s}{r}\right) \quad \text{[arc length } s = r\theta_x \Rightarrow \theta_x = s/r]$$

$$= \tfrac{1}{2}r \cdot 2\sqrt{16 - x^2} - \tfrac{1}{2}r^2\sin\left(\frac{2\sqrt{16 - x^2}}{r}\right) \quad [s = 2\sqrt{16 - x^2}]$$

$$= r\sqrt{16 - x^2} - \tfrac{1}{2}r^2\sin\left(\frac{2}{r}\sqrt{16 - x^2}\right) \quad (\star)$$

Note that the central angle θ_x will be small near the ends of the tortilla; that is, when $|x| \approx 4$. But near the center of the

tortilla (when $|x| \approx 0$), the central angle θ_x may exceed $180°$. Thus, the sine of θ_x will be negative and the second term in

$(\star)$ will be positive (actually adding area to the area of the sector). The volume of the taco can be found by integrating the cross-sectional areas from $x = -4$ to $x = 4$. Thus,

$$V(x) = \int_{-4}^{4} A(x)\, dx = \int_{-4}^{4} \left[r\sqrt{16 - x^2} - \tfrac{1}{2}r^2 \sin\left(\frac{2}{r}\sqrt{16 - x^2}\right) \right] dx$$

(b) To find the value of r that maximizes the volume of the taco, we can define

the function

$$V(r) = \int_{-4}^{4} \left[r\sqrt{16 - x^2} - \tfrac{1}{2}r^2 \sin\left(\frac{2}{r}\sqrt{16 - x^2}\right) \right] dx$$

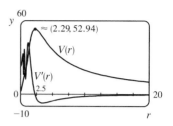

The figure shows a graph of $y = V(r)$ and $y = V'(r)$. The maximum

volume of about 52.94 occurs when $r \approx 2.2912$.

15. We assume that P lies in the region of positive x. Since $y = x^3$ is an odd

function, this assumption will not affect the result of the calculation. Let

$P = (a, a^3)$. The slope of the tangent to the curve $y = x^3$ at P is $3a^2$, and so

the equation of the tangent is $y - a^3 = 3a^2(x - a) \iff y = 3a^2 x - 2a^3$.

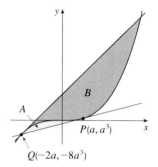

We solve this simultaneously with $y = x^3$ to find the other point of intersection:

$x^3 = 3a^2 x - 2a^3 \iff (x - a)^2(x + 2a) = 0$. So $Q = (-2a, -8a^3)$ is

the other point of intersection. The equation of the tangent at Q is

$y - (-8a^3) = 12a^2[x - (-2a)] \iff y = 12a^2 x + 16a^3$. By symmetry,

this tangent will intersect the curve again at $x = -2(-2a) = 4a$. The curve lies above the first tangent, and

below the second, so we are looking for a relationship between $A = \int_{-2a}^{a} \left[x^3 - (3a^2 x - 2a^3) \right] dx$ and

$B = \int_{-2a}^{4a} \left[(12a^2 x + 16a^3) - x^3 \right] dx$. We calculate $A = \left[\tfrac{1}{4}x^4 - \tfrac{3}{2}a^2 x^2 + 2a^3 x \right]_{-2a}^{a} = \tfrac{3}{4}a^4 - (-6a^4) = \tfrac{27}{4}a^4$, and

$B = \left[6a^2 x^2 + 16a^3 x - \tfrac{1}{4}x^4 \right]_{-2a}^{4a} = 96a^4 - (-12a^4) = 108a^4$. We see that $B = 16A = 2^4 A$. This is because our

calculation of area B was essentially the same as that of area A, with a replaced by $-2a$, so if we replace a with $-2a$ in our

expression for A, we get $\tfrac{27}{4}(-2a)^4 = 108a^4 = B$.

7 ☐ INVERSE FUNCTIONS:
Exponential, Logarithmic, and Inverse Trigonometric Functions

7.1 Inverse Functions

1. (a) See Definition 1.

   (b) It must pass the Horizontal Line Test.

2. (a) $f^{-1}(y) = x \iff f(x) = y$ for any y in B. The domain of f^{-1} is B and the range of f^{-1} is A.

   (b) See the steps in (5).

   (c) Reflect the graph of f about the line $y = x$.

3. f is not one-to-one because $2 \neq 6$, but $f(2) = 2.0 = f(6)$.

4. f is one-to-one since for any two different domain values, there are different range values.

5. No horizontal line intersects the graph of f more than once. Thus, by the Horizontal Line Test, f is one-to-one.

6. The horizontal line $y = 0$ (the x-axis) intersects the graph of f in more than one point. Thus, by the Horizontal Line Test, f is not one-to-one.

7. The horizontal line $y = 0$ (the x-axis) intersects the graph of f in more than one point. Thus, by the Horizontal Line Test, f is not one-to-one.

8. No horizontal line intersects the graph of f more than once. Thus, by the Horizontal Line Test, f is one-to-one.

9. The graph of $f(x) = x^2 - 2x$ is a parabola with axis of symmetry $x = -\dfrac{b}{2a} = -\dfrac{-2}{2(1)} = 1$. Pick any x-values equidistant from 1 to find two equal function values. For example, $f(0) = 0$ and $f(2) = 0$, so f is not one-to-one.

10. The graph of $f(x) = 10 - 3x$ is a line with slope -3. It passes the Horizontal Line Test, so f is one-to-one.

    Algebraic solution: If $x_1 \neq x_2$, then $-3x_1 \neq -3x_2 \Rightarrow 10 - 3x_1 \neq 10 - 3x_2 \Rightarrow f(x_1) \neq f(x_2)$, so f is one-to-one.

11. $g(x) = 1/x$. $x_1 \neq x_2 \Rightarrow 1/x_1 \neq 1/x_2 \Rightarrow g(x_1) \neq g(x_2)$, so g is one-to-one.

    Geometric solution: The graph of g is the hyperbola shown in Figure 14 in Section 1.2. It passes the Horizontal Line Test, so g is one-to-one.

12. $g(x) = |x| \Rightarrow g(-1) = 1 = g(1)$, so g is not one-to-one.

13. $h(x) = 1 + \cos x$ is not one-to-one since $h(a + 2\pi n) = h(a)$ for any real number a and any integer n.

14. $h(x) = 1 + \cos x$, $0 \le x \le \pi \Rightarrow h'(x) = -\sin x \le 0$ on $[0, \pi]$ with equality only at the endpoints, so h is decreasing and hence one-to-one on $[0, \pi]$.

15. A football will attain every height h up to its maximum height twice: once on the way up, and again on the way down. Thus, even if t_1 does not equal t_2, $f(t_1)$ may equal $f(t_2)$, so f is not 1-1.

16. f is not 1-1 because eventually we all stop growing and therefore, there are two times at which we have the same height.

17. Since $f(2) = 9$ and f is 1-1, we know that $f^{-1}(9) = 2$. Remember, if the point $(2, 9)$ is on the graph of f, then the point $(9, 2)$ is on the graph of f^{-1}.

18. $f(x) = x + \cos x \Rightarrow f'(x) = 1 - \sin x \geq 0$, with equality only if $x = \frac{\pi}{2} + 2n\pi$. So f is increasing on $\mathbb{R}$, and hence, 1-1. By inspection, $f(0) = 0 + \cos 0 = 1$, so $f^{-1}(1) = 0$.

19. $h(x) = x + \sqrt{x} \Rightarrow h'(x) = 1 + 1/(2\sqrt{x}) > 0$ on $(0, \infty)$. So h is increasing and hence, 1-1. By inspection, $h(4) = 4 + \sqrt{4} = 6$, so $h^{-1}(6) = 4$.

20. (a) f is 1-1 because it passes the Horizontal Line Test.

(b) Domain of $f = [-3, 3] =$ Range of f^{-1}. Range of $f = [-1, 3] =$ Domain of f^{-1}.

(c) Since $f(0) = 2$, $f^{-1}(2) = 0$.

(d) Since $f(-1.7) \approx 0$, $f^{-1}(0) = -1.7$.

21. We solve $C = \frac{5}{9}(F - 32)$ for F: $\frac{9}{5}C = F - 32 \Rightarrow F = \frac{9}{5}C + 32$. This gives us a formula for the inverse function, that is, the Fahrenheit temperature F as a function of the Celsius temperature C. $F \geq -459.67 \Rightarrow \frac{9}{5}C + 32 \geq -459.67 \Rightarrow \frac{9}{5}C \geq -491.67 \Rightarrow C \geq -273.15$, the domain of the inverse function.

22. $m = \dfrac{m_0}{\sqrt{1 - v^2/c^2}} \Rightarrow 1 - \dfrac{v^2}{c^2} = \dfrac{m_0^2}{m^2} \Rightarrow \dfrac{v^2}{c^2} = 1 - \dfrac{m_0^2}{m^2} \Rightarrow v^2 = c^2\left(1 - \dfrac{m_0^2}{m^2}\right) \Rightarrow v = c\sqrt{1 - \dfrac{m_0^2}{m^2}}$.

This formula gives us the speed v of the particle in terms of its mass m, that is, $v = f^{-1}(m)$.

23. $y = f(x) = 3 - 2x \Rightarrow 2x = 3 - y \Rightarrow x = \dfrac{3 - y}{2}$. Interchange x and y: $y = \dfrac{3 - x}{2}$. So $f^{-1}(x) = \dfrac{3 - x}{2}$.

24. $y = f(x) = \dfrac{4x - 1}{2x + 3} \Rightarrow y(2x + 3) = 4x - 1 \Rightarrow 2xy + 3y = 4x - 1 \Rightarrow 3y + 1 = 4x - 2xy \Rightarrow$

$3y + 1 = (4 - 2y)x \Rightarrow x = \dfrac{3y + 1}{4 - 2y}$. Interchange x and y: $y = \dfrac{3x + 1}{4 - 2x}$. So $f^{-1}(x) = \dfrac{3x + 1}{4 - 2x}$.

25. $f(x) = \sqrt{10 - 3x} \Rightarrow y = \sqrt{10 - 3x} \ (y \geq 0) \Rightarrow y^2 = 10 - 3x \Rightarrow 3x = 10 - y^2 \Rightarrow x = -\frac{1}{3}y^2 + \frac{10}{3}$.

Interchange x and y: $y = -\frac{1}{3}x^2 + \frac{10}{3}$. So $f^{-1}(x) = -\frac{1}{3}x^2 + \frac{10}{3}$. Note that the domain of f^{-1} is $x \geq 0$.

26. $y = f(x) = 2x^3 + 3 \Rightarrow y - 3 = 2x^3 \Rightarrow \dfrac{y - 3}{2} = x^3 \Rightarrow x = \sqrt[3]{\dfrac{y - 3}{2}}$. Interchange x and y: $y = \sqrt[3]{\dfrac{x - 3}{2}}$.

So $f^{-1}(x) = \sqrt[3]{\dfrac{x - 3}{2}}$.

27. For $f(x) = \dfrac{1 - \sqrt{x}}{1 + \sqrt{x}}$, the domain is $x \geq 0$. $f(0) = 1$ and as x increases, y decreases. As $x \to \infty$,

$\dfrac{1 - \sqrt{x}}{1 + \sqrt{x}} \cdot \dfrac{1/\sqrt{x}}{1/\sqrt{x}} = \dfrac{1/\sqrt{x} - 1}{1/\sqrt{x} + 1} \to \dfrac{-1}{1} = -1$, so the range of f is $-1 < y \leq 1$. Thus, the domain of f^{-1} is $-1 < x \leq 1$.

$y = \dfrac{1 - \sqrt{x}}{1 + \sqrt{x}} \Rightarrow y(1 + \sqrt{x}) = 1 - \sqrt{x} \Rightarrow y + y\sqrt{x} = 1 - \sqrt{x} \Rightarrow \sqrt{x} + y\sqrt{x} = 1 - y \Rightarrow$

$$\sqrt{x}(1+y) = 1 - y \quad \Rightarrow \quad \sqrt{x} = \frac{1-y}{1+y} \quad \Rightarrow \quad x = \left(\frac{1-y}{1+y}\right)^2. \text{ Interchange } x \text{ and } y: \quad y = \left(\frac{1-x}{1+x}\right)^2. \text{ So}$$

$$f^{-1}(x) = \left(\frac{1-x}{1+x}\right)^2 \text{ with } -1 < x \le 1.$$

28. $y = f(x) = 2x^2 - 8x, \ x \ge 2 \quad \Rightarrow \quad 2x^2 - 8x - y = 0, \ x \ge 2 \quad \Rightarrow$

$$x = \frac{8 + \sqrt{64 + 8y}}{4} \quad \left[\begin{array}{c} \text{quadratic formula with} \\ \text{parts with } a = 2, \ b = -8, \text{ and } c = -y \end{array}\right] = \frac{8 + 2\sqrt{16 + 2y}}{4} = 2 + \tfrac{1}{2}\sqrt{16 + 2y}. \text{ Interchange } x \text{ and } y:$$

$y = 2 + \tfrac{1}{2}\sqrt{16 + 2x}.$ So $f^{-1}(x) = 2 + \tfrac{1}{2}\sqrt{16 + 2x}.$

Alternate solution (by completing the square): $y = 2x^2 - 8x, \ x \ge 2 \quad \Rightarrow \quad x^2 - 4x = y/2, \ x \ge 2 \quad \Rightarrow$

$$(x-2)^2 = x^2 - 4x + 4 = \frac{y}{2} + 4 = \frac{y+8}{2} = \frac{2y+16}{4}, x \ge 2 \quad \Rightarrow \quad x - 2 = +\sqrt{\frac{2y+16}{4}} \quad \Rightarrow \quad x = 2 + \tfrac{1}{2}\sqrt{2y + 16}.$$

Interchange x and y: $\ y = 2 + \tfrac{1}{2}\sqrt{2x + 16}.$ So $f^{-1}(x) = 2 + \tfrac{1}{2}\sqrt{2x + 16}.$

29. $y = f(x) = x^4 + 1 \quad \Rightarrow \quad y - 1 = x^4 \quad \Rightarrow \quad x = \sqrt[4]{y-1} \ (\text{not } \pm \text{ since}$

$x \ge 0$). Interchange x and y: $y = \sqrt[4]{x-1}.$ So $f^{-1}(x) = \sqrt[4]{x-1}.$ The

graph of $y = \sqrt[4]{x-1}$ is just the graph of $y = \sqrt[4]{x}$ shifted right one unit.

From the graph, we see that f and f^{-1} are reflections about the line $y = x.$

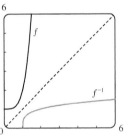

30. $y = f(x) = \sqrt{x^2 + 2x}, \ x > 0 \quad \Rightarrow \quad y > 0 \text{ and } y^2 = x^2 + 2x \quad \Rightarrow$

$x^2 + 2x - y^2 = 0.$ Now we use the quadratic formula:

$$x = \frac{-2 \pm \sqrt{2^2 - 4 \cdot 1 \cdot (-y^2)}}{2 \cdot 1} = -1 \pm \sqrt{1 + y^2}. \text{ But } x > 0, \text{ so the negative}$$

root is inadmissible. Interchange x and y: $y = -1 + \sqrt{1 + x^2}.$

So $f^{-1}(x) = -1 + \sqrt{1 + x^2}, \ x > 0.$

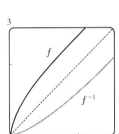

31. Reflect the graph of f about the line $y = x.$ The points $(-1, -2), \ (1, -1),$

$(2, 2),$ and $(3, 3)$ on f are reflected to $(-2, -1), \ (-1, 1), \ (2, 2),$ and $(3, 3)$

on $f^{-1}.$

32. Reflect the graph of f about the line $y = x.$

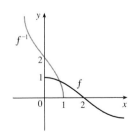

33. (a) $x_1 \neq x_2 \;\Rightarrow\; x_1^3 \neq x_2^3 \;\Rightarrow\; f(x_1) \neq f(x_2)$, so f is one-to-one.

(b) $f'(x) = 3x^2$ and $f(2) = 8 \;\Rightarrow\; f^{-1}(8) = 2$, so $(f^{-1})'(8) = 1/f'(f^{-1}(8)) = 1/f'(2) = \frac{1}{12}$.

(c) $y = x^3 \;\Rightarrow\; x = y^{1/3}$. Interchanging x and y gives $y = x^{1/3}$,

(e)

so $f^{-1}(x) = x^{1/3}$. Domain$\left(f^{-1}\right)$ = range$(f) = \mathbb{R}$.

Range$\left(f^{-1}\right)$ = domain$(f) = \mathbb{R}$.

(d) $f^{-1}(x) = x^{1/3} \;\Rightarrow\; (f^{-1})'(x) = \frac{1}{3}x^{-2/3} \;\Rightarrow$

$(f^{-1})'(8) = \frac{1}{3}\left(\frac{1}{4}\right) = \frac{1}{12}$ as in part (b).

34. (a) $x_1 \neq x_2 \;\Rightarrow\; x_1 - 2 \neq x_2 - 2 \;\Rightarrow\; \sqrt{x_1 - 2} \neq \sqrt{x_2 - 2} \;\Rightarrow\; f(x_1) \neq f(x_2)$, so f is 1-1.

(b) $f(6) = 2$, so $f^{-1}(2) = 6$. Also $f'(x) = \dfrac{1}{2\sqrt{x-2}}$, so $(f^{-1})'(2) = \dfrac{1}{f'(f^{-1}(2))} = \dfrac{1}{f'(6)} = \dfrac{1}{1/4} = 4$.

(c) $y = \sqrt{x - 2} \;\Rightarrow\; y^2 = x - 2 \;\Rightarrow\; x = y^2 + 2$.

(e)

Interchange x and y: $y = x^2 + 2$. So $f^{-1}(x) = x^2 + 2$.

Domain $= [0, \infty)$, range $= [2, \infty)$.

(d) $f^{-1}(x) = x^2 + 2 \;\Rightarrow\; (f^{-1})'(x) = 2x \;\Rightarrow\; (f^{-1})'(2) = 4$.

35. (a) Since $x \geq 0$, $x_1 \neq x_2 \;\Rightarrow\; x_1^2 \neq x_2^2 \;\Rightarrow\; 9 - x_1^2 \neq 9 - x_2^2 \;\Rightarrow\; f(x_1) \neq f(x_2)$, so f is 1-1.

(b) $f'(x) = -2x$ and $f(1) = 8 \;\Rightarrow\; f^{-1}(8) = 1$, so $(f^{-1})'(8) = \dfrac{1}{f'(f^{-1}(8))} = \dfrac{1}{f'(1)} = \dfrac{1}{-2} = -\dfrac{1}{2}$.

(c) $y = 9 - x^2 \;\Rightarrow\; x^2 = 9 - y \;\Rightarrow\; x = \sqrt{9 - y}$.

(e)

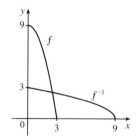

Interchange x and y: $y = \sqrt{9 - x}$, so $f^{-1}(x) = \sqrt{9 - x}$.

Domain(f^{-1}) = range $(f) = [0, 9]$.

Range(f^{-1}) = domain $(f) = [0, 3]$.

(d) $(f^{-1})'(x) = -1/(2\sqrt{9-x}) \;\Rightarrow\; (f^{-1})'(8) = -\frac{1}{2}$ as in part (b).

36. (a) $x_1 \neq x_2 \;\Rightarrow\; x_1 - 1 \neq x_2 - 1 \;\Rightarrow\; \dfrac{1}{x_1 - 1} \neq \dfrac{1}{x_2 - 1} \;\Rightarrow\; f(x_1) \neq f(x_2)$, so f is 1-1.

(b) $f^{-1}(2) = \frac{3}{2}$ since $f\left(\frac{3}{2}\right) = 2$. Also $f'(x) = -1/(x-1)^2$, so $(f^{-1})'(2) = 1/f'\left(\frac{3}{2}\right) = \frac{1}{-4} = -\frac{1}{4}$.

(c) $y = 1/(x-1) \;\Rightarrow\; x - 1 = 1/y \;\Rightarrow\; x = 1 + 1/y$. Interchange

(e)

x and y: $y = 1 + 1/x$. So $f^{-1}(x) = 1 + 1/x$, $x > 0$ (since $y > 1$).

Domain $= (0, \infty)$, range $= (1, \infty)$.

(d) $(f^{-1})'(x) = -1/x^2$, so $(f^{-1})'(2) = -\frac{1}{4}$.

37. $f(0) = 4 \;\Rightarrow\; f^{-1}(4) = 0$, and $f(x) = 2x^3 + 3x^2 + 7x + 4 \;\Rightarrow\; f'(x) = 6x^2 + 6x + 7$ and $f'(0) = 7$.

Thus, $(f^{-1})'(4) = \dfrac{1}{f'(f^{-1}(4))} = \dfrac{1}{f'(0)} = \dfrac{1}{7}$.

38. $f(0) = 2 \Rightarrow f^{-1}(2) = 0$, and $f(x) = x^3 + 3\sin x + 2\cos x \Rightarrow f'(x) = 3x^2 + 3\cos x - 2\sin x$ and $f'(0) = 3$.

Thus, $(f^{-1})'(2) = \dfrac{1}{f'(f^{-1}(2))} = \dfrac{1}{f'(0)} = \dfrac{1}{3}$.

39. $f(0) = 3 \Rightarrow f^{-1}(3) = 0$, and $f(x) = 3 + x^2 + \tan(\pi x/2) \Rightarrow f'(x) = 2x + \frac{\pi}{2}\sec^2(\pi x/2)$ and

$f'(0) = \frac{\pi}{2} \cdot 1 = \frac{\pi}{2}$. Thus, $(f^{-1})'(3) = 1/f'(f^{-1}(3)) = 1/f'(0) = 2/\pi$.

40. $f(1) = 2 \Rightarrow f^{-1}(2) = 1$, and $f(x) = \sqrt{x^3 + x^2 + x + 1} \Rightarrow f'(x) = \dfrac{3x^2 + 2x + 1}{2\sqrt{x^3 + x^2 + x + 1}}$ and

$f'(1) = \dfrac{3 + 2 + 1}{2\sqrt{1 + 1 + 1 + 1}} = \dfrac{3}{2}$. Thus, $(f^{-1})'(2) = 1/f'(f^{-1}(2)) = 1/f'(1) = \frac{2}{3}$.

41. $f(4) = 5 \Rightarrow f^{-1}(5) = 4$. Thus, $(f^{-1})'(5) = \dfrac{1}{f'(f^{-1}(5))} = \dfrac{1}{f'(4)} = \dfrac{1}{2/3} = \dfrac{3}{2}$.

42. $f(3) = 2 \Rightarrow f^{-1}(2) = 3$. Thus, $(f^{-1})'(2) = \dfrac{1}{f'(f^{-1}(2))} = \dfrac{1}{f'(3)} = 9$. Hence, $G(x) = \dfrac{1}{f^{-1}(x)} \Rightarrow$

$G'(x) = -\dfrac{(f^{-1})'(x)}{[f^{-1}(x)]^2} \Rightarrow G'(2) = -\dfrac{(f^{-1})'(2)}{[f^{-1}(2)]^2} = -\dfrac{9}{(3)^2} = -1$.

43. We see that the graph of $y = f(x) = \sqrt{x^3 + x^2 + x + 1}$ is increasing, so f is 1-1.

Enter $x = \sqrt{y^3 + y^2 + y + 1}$ and use your CAS to solve the equation for y.

Using Derive, we get two (irrelevant) solutions involving imaginary expressions,

as well as one which can be simplified to the following:

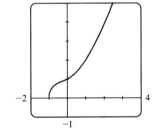

$y = f^{-1}(x) = -\dfrac{\sqrt[3]{4}}{6}\left(\sqrt[3]{D - 27x^2 + 20} - \sqrt[3]{D + 27x^2 - 20} + \sqrt[3]{2}\right)$

where $D = 3\sqrt{3}\sqrt{27x^4 - 40x^2 + 16}$.

Maple and Mathematica each give two complex expressions and one real expression, and the real expression is equivalent

to that given by Derive. For example, Maple's expression simplifies to $\dfrac{1}{6}\dfrac{M^{2/3} - 8 - 2M^{1/3}}{2M^{1/3}}$, where

$M = 108x^2 + 12\sqrt{48 - 120x^2 + 81x^4} - 80$.

44. Since $\sin(2n\pi) = 0$, $h(x) = \sin x$ is not one-to-one. $h'(x) = \cos x > 0$ on $\left(-\frac{\pi}{2}, \frac{\pi}{2}\right)$, so h is increasing and hence 1-1 on

$\left[-\frac{\pi}{2}, \frac{\pi}{2}\right]$. Let $y = f^{-1}(x) = \sin^{-1}x$ so that $\sin y = x$. Differentiating $\sin y = x$ implicitly with respect to x gives us

$\cos y\, \dfrac{dy}{dx} = 1 \Rightarrow \dfrac{dy}{dx} = \dfrac{1}{\cos y}$. Now $\cos^2 y + \sin^2 y = 1 \Rightarrow \cos y = \pm\sqrt{1 - \sin^2 y}$, but since $\cos y > 0$ on $\left(-\frac{\pi}{2}, \frac{\pi}{2}\right)$,

we have $\dfrac{dy}{dx} = \dfrac{1}{\sqrt{1 - \sin^2 y}} = \dfrac{1}{\sqrt{1 - x^2}}$.

45. (a) If the point (x, y) is on the graph of $y = f(x)$, then the point $(x - c, y)$ is that point shifted c units to the left. Since f is

1-1, the point (y, x) is on the graph of $y = f^{-1}(x)$ and the point corresponding to $(x - c, y)$ on the graph of f is

$(y, x - c)$ on the graph of f^{-1}. Thus, the curve's reflection is shifted *down* the same number of units as the curve itself is

shifted to the left. So an expression for the inverse function is $g^{-1}(x) = f^{-1}(x) - c$.

(b) If we compress (or stretch) a curve horizontally, the curve's reflection in the line $y = x$ is compressed (or stretched) *vertically* by the same factor. Using this geometric principle, we see that the inverse of $h(x) = f(cx)$ can be expressed as $h^{-1}(x) = (1/c)\, f^{-1}(x)$.

46. (a) We know that $g'(x) = \dfrac{1}{f'(g(x))}$. Thus,

$$g''(x) = -\frac{f''(g(x)) \cdot g'(x)}{[f'(g(x))]^2} = -\frac{f''(g(x)) \cdot [1/f'(g'(x))]}{[f'(g(x))]^2} = -\frac{f''(g(x))}{f'(g(x))[f'(g(x))]^2} = -\frac{f''(g(x))}{[f'(g(x))]^3}.$$

(b) f is increasing $\Rightarrow$ $f'(g(x)) > 0$ $\Rightarrow$ $[f'(g(x))]^3 > 0$. f is concave upward $\Rightarrow$ $f''(g(x)) > 0$.

So $g''(x) = -\dfrac{f''(g(x))}{[f'(g(x))]^3} < 0$, which implies that g [f's inverse] is concave downward.

7.2 Exponential Functions and Their Derivatives

1. (a) $f(x) = a^x$, $a > 0$ 　　(b) $\mathbb{R}$ 　　(c) $(0, \infty)$ 　　(d) See Figures 6(c), 6(b), and 6(a), respectively.

2. (a) The number e is the value of a such that the slope of the tangent line at $x = 0$ on the graph of $y = a^x$ is exactly 1.

(b) $e \approx 2.71828$ 　　(c) $f(x) = e^x$

3. All of these graphs approach 0 as $x \to -\infty$, all of them pass through the point $(0, 1)$, and all of them are increasing and approach ∞ as $x \to \infty$. The larger the base, the faster the function increases for $x > 0$, and the faster it approaches 0 as $x \to -\infty$.

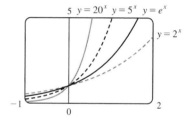

4. The graph of e^{-x} is the reflection of the graph of e^x about the y-axis, and the graph of 8^{-x} is the reflection of that of 8^x about the y-axis. The graph of 8^x increases more quickly than that of e^x for $x > 0$, and approaches 0 faster as $x \to -\infty$.

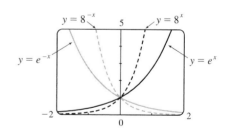

5. The functions with bases greater than 1 (3^x and 10^x) are increasing, while those with bases less than 1 $\left[\left(\frac{1}{3}\right)^x$ and $\left(\frac{1}{10}\right)^x\right]$ are decreasing. The graph of $\left(\frac{1}{3}\right)^x$ is the reflection of that of 3^x about the y-axis, and the graph of $\left(\frac{1}{10}\right)^x$ is the reflection of that of 10^x about the y-axis. The graph of 10^x increases more quickly than that of 3^x for $x > 0$, and approaches 0 faster as $x \to -\infty$.

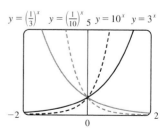

6. Each of the graphs approaches ∞ as $x \to -\infty$, and each approaches 0 as $x \to \infty$. The smaller the base, the faster the function grows as $x \to -\infty$, and the faster it approaches 0 as $x \to \infty$.

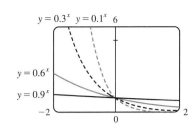

7. We start with the graph of $y = 4^x$ (Figure 3) and then shift 3 units downward. This shift doesn't affect the domain, but the range of $y = 4^x - 3$ is $(-3, \infty)$. There is a horizontal asymptote of $y = -3$.

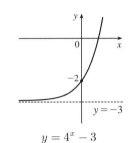

$y = 4^x$ $y = 4^x - 3$

8. We start with the graph of $y = 4^x$ (Figure 3) and then shift 3 units to the right. There is a horizontal asymptote of $y = 0$.

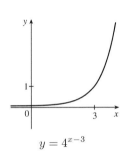

$y = 4^x$ $y = 4^{x-3}$

9. We start with the graph of $y = 2^x$ (Figure 3), reflect it about the y-axis, and then about the x-axis (or just rotate $180°$ to handle both reflections) to obtain the graph of $y = -2^{-x}$. In each graph, $y = 0$ is the horizontal asymptote.

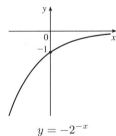

$y = 2^x$ $y = 2^{-x}$ $y = -2^{-x}$

10. We start with the graph of $y = e^x$ (Figure 12), vertically stretch by a factor of 2, and then shift 1 unit upward. There is a horizontal asymptote of $y = 1$.

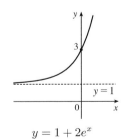

$y = 2e^x$ $y = 1 + 2e^x$

11. We start with the graph of $y = e^x$ (Figure 12) and reflect about the y-axis to get the graph of $y = e^{-x}$. Then we compress the graph vertically by a factor of 2 to obtain the graph of $y = \frac{1}{2}e^{-x}$ and then reflect about the x-axis to get the graph of $y = -\frac{1}{2}e^{-x}$. Finally, we shift the graph upward one unit to get the graph of $y = 1 - \frac{1}{2}e^{-x}$.

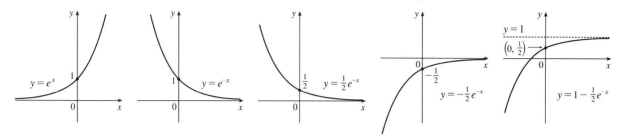

12. We start with the graph of $y = e^x$ (Figure 12) and reflect about the x-axis to get the graph of $y = -e^x$. Then shift the graph upward one unit to get the graph of $y = 1 - e^x$. Finally, we stretch the graph vertically by a factor of 2 to obtain the graph of $y = 2(1 - e^x)$.

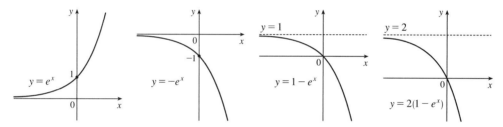

13. (a) To find the equation of the graph that results from shifting the graph of $y = e^x$ 2 units downward, we subtract 2 from the original function to get $y = e^x - 2$.

(b) To find the equation of the graph that results from shifting the graph of $y = e^x$ 2 units to the right, we replace x with $x - 2$ in the original function to get $y = e^{(x-2)}$.

(c) To find the equation of the graph that results from reflecting the graph of $y = e^x$ about the x-axis, we multiply the original function by -1 to get $y = -e^x$.

(d) To find the equation of the graph that results from reflecting the graph of $y = e^x$ about the y-axis, we replace x with $-x$ in the original function to get $y = e^{-x}$.

(e) To find the equation of the graph that results from reflecting the graph of $y = e^x$ about the x-axis and then about the y-axis, we first multiply the original function by -1 (to get $y = -e^x$) and then replace x with $-x$ in this equation to get $y = -e^{-x}$.

14. (a) This reflection consists of first reflecting the graph about the x-axis (giving the graph with equation $y = -e^x$) and then shifting this graph $2 \cdot 4 = 8$ units upward. So the equation is $y = -e^x + 8$.

(b) This reflection consists of first reflecting the graph about the y-axis (giving the graph with equation $y = e^{-x}$) and then shifting this graph $2 \cdot 2 = 4$ units to the right. So the equation is $y = e^{-(x-4)}$.

15. (a) The denominator $1 + e^x$ is never equal to zero because $e^x > 0$, so the domain of $f(x) = 1/(1 + e^x)$ is $\mathbb{R}$.

(b) $1 - e^x = 0 \iff e^x = 1 \iff x = 0$, so the domain of $f(x) = 1/(1 - e^x)$ is $(-\infty, 0) \cup (0, \infty)$.

16. (a) The sine and exponential functions have domain $\mathbb{R}$, so $g(t) = \sin(e^{-t})$ also has domain $\mathbb{R}$.

(b) The function $g(t) = \sqrt{1 - 2^t}$ has domain $\{t \mid 1 - 2^t \geq 0\} = \{t \mid 2^t \leq 1\} = \{t \mid t \leq 0\} = (-\infty, 0]$.

17. Use $y = Ca^x$ with the points $(1, 6)$ and $(3, 24)$. $\quad 6 = Ca^1 \quad \left[C = \frac{6}{a}\right] \quad$ and $24 = Ca^3 \quad \Rightarrow \quad 24 = \left(\dfrac{6}{a}\right)a^3 \quad \Rightarrow$

$4 = a^2 \quad \Rightarrow \quad a = 2 \quad$ [since $a > 0$]   and $C = \frac{6}{2} = 3$. The function is $f(x) = 3 \cdot 2^x$.

18. Given the y-intercept $(0, 2)$, we have $y = Ca^x = 2a^x$. Using the point $\left(2, \frac{2}{9}\right)$ gives us $\frac{2}{9} = 2a^2 \quad \Rightarrow \quad \frac{1}{9} = a^2 \quad \Rightarrow \quad a = \frac{1}{3}$

[since $a > 0$].   The function is $f(x) = 2\left(\frac{1}{3}\right)^x$ or $f(x) = 2(3)^{-x}$.

19. $2 \text{ ft} = 24 \text{ in}$, $f(24) = 24^2 \text{ in} = 576 \text{ in} = 48 \text{ ft}$. $g(24) = 2^{24} \text{ in} = 2^{24}/(12 \cdot 5280) \text{ mi} \approx 265 \text{ mi}$

20. We see from the graphs that for x less than about 1.8, $g(x) = 5^x > f(x) = x^5$, and then near the point $(1.8, 17.1)$ the curves intersect. Then $f(x) > g(x)$ from $x \approx 1.8$ until $x = 5$. At $(5, 3125)$ there is another point of intersection, and for $x > 5$ we see that $g(x) > f(x)$. In fact, g increases much more rapidly than f beyond that point.

21. The graph of g finally surpasses that of f at $x \approx 35.8$.

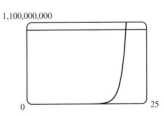

22. We graph $y = e^x$ and $y = 1,000,000,000$ and determine where

$e^x = 1 \times 10^9$. This seems to be true at $x \approx 20.723$, so $e^x > 1 \times 10^9$

for $x > 20.723$.

23. $\lim\limits_{x \to \infty} (1.001)^x = \infty$ by (3), since $1.001 > 1$.

24. By (3), if $a > 1$, $\lim\limits_{x \to -\infty} a^x = 0$, so $\lim\limits_{x \to -\infty} (1.001)^x = 0$.

25. Divide numerator and denominator by e^{3x}: $\lim\limits_{x \to \infty} \dfrac{e^{3x} - e^{-3x}}{e^{3x} + e^{-3x}} = \lim\limits_{x \to \infty} \dfrac{1 - e^{-6x}}{1 + e^{-6x}} = \dfrac{1-0}{1+0} = 1$

26. Let $t = -x^2$. As $x \to \infty$, $t \to -\infty$. So $\lim\limits_{x \to \infty} e^{-x^2} = \lim\limits_{t \to -\infty} e^t = 0$ by (10).

27. Let $t = 3/(2 - x)$. As $x \to 2^+$, $t \to -\infty$. So $\lim\limits_{x \to 2^+} e^{3/(2-x)} = \lim\limits_{t \to -\infty} e^t = 0$ by (10).

28. Let $t = 3/(2 - x)$. As $x \to 2^-$, $t \to \infty$. So $\lim\limits_{x \to 2^-} e^{3/(2-x)} = \lim\limits_{t \to \infty} e^t = \infty$ by (10).

29. Since $-1 \le \cos x \le 1$ and $e^{-2x} > 0$, we have $-e^{-2x} \le e^{-2x} \cos x \le e^{-2x}$. We know that $\lim\limits_{x \to \infty} (-e^{-2x}) = 0$ and

$\lim\limits_{x \to \infty} (e^{-2x}) = 0$, so by the Squeeze Theorem, $\lim\limits_{x \to \infty} (e^{-2x} \cos x) = 0$.

30. If we let $t = \tan x$, then as $x \to (\pi/2)^+$, $t \to -\infty$. Thus, $\lim\limits_{x \to (\pi/2)^+} e^{\tan x} = \lim\limits_{t \to -\infty} e^t = 0$.

31. By the Product Rule, $f(x) = (x^3 + 2x)e^x \quad \Rightarrow$

$$f'(x) = (x^3 + 2x)(e^x)' + e^x(x^3 + 2x)' = (x^3 + 2x)e^x + e^x(3x^2 + 2)$$

$$= e^x[(x^3 + 2x) + (3x^2 + 2)] = e^x(x^3 + 3x^2 + 2x + 2)$$

32. By the Quotient Rule, $y = \dfrac{e^x}{1 + x} \quad \Rightarrow \quad y' = \dfrac{(1+x)\,e^x - e^x(1)}{(1+x)^2} = \dfrac{e^x + xe^x - e^x}{(x+1)^2} = \dfrac{xe^x}{(x+1)^2}$.

33. By (9), $y = e^{ax^3} \quad \Rightarrow \quad y' = e^{ax^3} \dfrac{d}{dx}(ax^3) = 3ax^2 e^{ax^3}$.

34. $y = e^u(\cos u + cu) \quad \Rightarrow \quad y' = e^u(-\sin u + c) + (\cos u + cu)e^u = e^u(\cos u - \sin u + cu + c)$

35. $f(u) = e^{1/u} \quad \Rightarrow \quad f'(u) = e^{1/u} \cdot \dfrac{d}{du}\left(\dfrac{1}{u}\right) = e^{1/u}\left(\dfrac{-1}{u^2}\right) = \left(\dfrac{-1}{u^2}\right) e^{1/u}$

36. By the Product Rule, $g(x) = \sqrt{x}\, e^x = x^{1/2} e^x \quad \Rightarrow \quad g'(x) = x^{1/2}(e^x) + e^x\left(\tfrac{1}{2}x^{-1/2}\right) = \tfrac{1}{2}x^{-1/2}e^x(2x + 1)$.

37. By (9), $F(t) = e^{t \sin 2t} \quad \Rightarrow \quad F'(t) = e^{t \sin 2t}(t \sin 2t)' = e^{t \sin 2t}(t \cdot 2\cos 2t + \sin 2t \cdot 1) = e^{t \sin 2t}(2t \cos 2t + \sin 2t)$

38. $f(t) = \sin(e^t) + e^{\sin t} \quad \Rightarrow \quad f'(t) = \cos(e^t) \cdot e^t + e^{\sin t} \cdot \cos t = e^t \cos(e^t) + e^{\sin t} \cos t$

39. $y = \sqrt{1 + 2e^{3x}} \quad \Rightarrow \quad y' = \dfrac{1}{2}(1 + 2e^{3x})^{-1/2}\dfrac{d}{dx}(1 + 2e^{3x}) = \dfrac{1}{2\sqrt{1 + 2e^{3x}}}(2e^{3x} \cdot 3) = \dfrac{3e^{3x}}{\sqrt{1 + 2e^{3x}}}$

40. $y = e^{k \tan \sqrt{x}} \quad \Rightarrow \quad y' = e^{k \tan \sqrt{x}} \cdot \dfrac{d}{dx}\left(k \tan \sqrt{x}\right) = e^{k \tan \sqrt{x}}\left(k \sec^2 \sqrt{x} \cdot \tfrac{1}{2}x^{-1/2}\right) = \dfrac{k \sec^2 \sqrt{x}}{2\sqrt{x}} e^{k \tan \sqrt{x}}$

41. $y = e^{e^x} \quad \Rightarrow \quad y' = e^{e^x} \cdot \dfrac{d}{dx}(e^x) = e^{e^x} \cdot e^x$ or $e^{e^x + x}$

42. $y = \dfrac{e^u - e^{-u}}{e^u + e^{-u}}$ $\Rightarrow$

$$y' = \frac{(e^u + e^{-u})(e^u - (-e^{-u})) - (e^u - e^{-u})(e^u + (-e^{-u}))}{(e^u + e^{-u})^2} = \frac{e^{2u} + e^0 + e^0 + e^{-2u} - (e^{2u} - e^0 - e^0 + e^{-2u})}{(e^u + e^{-u})^2}$$

$$= \frac{4e^0}{(e^u + e^{-u})^2} = \frac{4}{(e^u + e^{-u})^2}$$

43. By the Quotient Rule, $y = \dfrac{ae^x + b}{ce^x + d}$ $\Rightarrow$

$$y' = \frac{(ce^x + d)(ae^x) - (ae^x + b)(ce^x)}{(ce^x + d)^2} = \frac{(ace^x + ad - ace^x - bc)e^x}{(ce^x + d)^2} = \frac{(ad - bc)e^x}{(ce^x + d)^2}.$$

44. $y = \sqrt{1 + xe^{-2x}}$ $\Rightarrow$ $y' = \frac{1}{2}(1 + xe^{-2x})^{-1/2}\left[x(-2e^{-2x}) + e^{-2x}\right] = \dfrac{e^{-2x}(-2x + 1)}{2\sqrt{1 + xe^{-2x}}}$

45. $y = \cos\left(\dfrac{1 - e^{2x}}{1 + e^{2x}}\right)$ $\Rightarrow$

$$y' = -\sin\left(\frac{1 - e^{2x}}{1 + e^{2x}}\right) \cdot \frac{d}{dx}\left(\frac{1 - e^{2x}}{1 + e^{2x}}\right) = -\sin\left(\frac{1 - e^{2x}}{1 + e^{2x}}\right) \cdot \frac{(1 + e^{2x})(-2e^{2x}) - (1 - e^{2x})(2e^{2x})}{(1 + e^{2x})^2}$$

$$= -\sin\left(\frac{1 - e^{2x}}{1 + e^{2x}}\right) \cdot \frac{-2e^{2x}\left[(1 + e^{2x}) + (1 - e^{2x})\right]}{(1 + e^{2x})^2} = -\sin\left(\frac{1 - e^{2x}}{1 + e^{2x}}\right) \cdot \frac{-2e^{2x}(2)}{(1 + e^{2x})^2} = \frac{4e^{2x}}{(1 + e^{2x})^2} \cdot \sin\left(\frac{1 - e^{2x}}{1 + e^{2x}}\right)$$

46. $f(t) = \sin^2\left(e^{\sin^2 t}\right) = \left[\sin\left(e^{\sin^2 t}\right)\right]^2$ $\Rightarrow$

$$f'(t) = 2\left[\sin\left(e^{\sin^2 t}\right)\right] \cdot \frac{d}{dt}\sin\left(e^{\sin^2 t}\right) = 2\sin\left(e^{\sin^2 t}\right) \cdot \cos\left(e^{\sin^2 t}\right) \cdot \frac{d}{dt}e^{\sin^2 t}$$

$$= 2\sin\left(e^{\sin^2 t}\right)\cos\left(e^{\sin^2 t}\right) \cdot e^{\sin^2 t} \cdot \frac{d}{dt}\sin^2 t = 2\sin\left(e^{\sin^2 t}\right)\cos\left(e^{\sin^2 t}\right)e^{\sin^2 t} \cdot 2\sin t \cos t$$

$$= 4\sin\left(e^{\sin^2 t}\right)\cos\left(e^{\sin^2 t}\right)e^{\sin^2 t}\sin t \cos t$$

47. $y = e^{2x}\cos \pi x$ $\Rightarrow$ $y' = e^{2x}(-\pi \sin \pi x) + (\cos \pi x)(2e^{2x}) = e^{2x}(2\cos \pi x - \pi \sin \pi x)$.

At $(0, 1)$, $y' = 1(2 - 0) = 2$, so an equation of the tangent line is $y - 1 = 2(x - 0)$, or $y = 2x + 1$.

48. $y = \dfrac{e^x}{x}$ $\Rightarrow$ $y' = \dfrac{x \cdot e^x - e^x \cdot 1}{x^2} = \dfrac{e^x(x - 1)}{x^2}$.

At $(1, e)$, $y' = 0$, and an equation of the tangent line is $y - e = 0(x - 1)$, or $y = e$.

49. $\dfrac{d}{dx}\left(e^{x^2 y}\right) = \dfrac{d}{dx}(x + y)$ $\Rightarrow$ $e^{x^2 y}(x^2 y' + y \cdot 2x) = 1 + y'$ $\Rightarrow$ $x^2 e^{x^2 y}y' + 2xye^{x^2 y} = 1 + y'$ $\Rightarrow$

$$x^2 e^{x^2 y}y' - y' = 1 - 2xye^{x^2 y} \Rightarrow y'(x^2 e^{x^2 y} - 1) = 1 - 2xye^{x^2 y} \Rightarrow y' = \frac{1 - 2xye^{x^2 y}}{x^2 e^{x^2 y} - 1}$$

50. $xe^y + ye^x = 1$ $\Rightarrow$ $xe^y y' + e^y \cdot 1 + ye^x + e^x y' = 0$ $\Rightarrow$ $y'(xe^y + e^x) = -e^y - ye^x$ $\Rightarrow$ $y' = -\dfrac{e^y + ye^x}{xe^y + e^x}$. At

$(0, 1)$, $y' = -\dfrac{e + 1 \cdot 1}{0 + 1} = -(e + 1)$, so an equation for the tangent line is $y - 1 = -(e + 1)(x - 0)$, or $y = -(e + 1)x + 1$.

51. $y = e^x + e^{-x/2}$ $\Rightarrow$ $y' = e^x - \frac{1}{2}e^{-x/2}$ $\Rightarrow$ $y'' = e^x + \frac{1}{4}e^{-x/2}$, so

$2y'' - y' - y = 2\left(e^x + \frac{1}{4}e^{-x/2}\right) - \left(e^x - \frac{1}{2}e^{-x/2}\right) - \left(e^x + e^{-x/2}\right) = 0.$

52. $y = Ae^{-x} + Bxe^{-x}$ $\Rightarrow$ $y' = -Ae^{-x} + Be^{-x} - Bxe^{-x} = (B - A)e^{-x} - Bxe^{-x}$ $\Rightarrow$

$y'' = (A - B)e^{-x} - Be^{-x} + Bxe^{-x} = (A - 2B)e^{-x} + Bxe^{-x}$,

so $y'' + 2y' + y = (A - 2B)e^{-x} + Bxe^{-x} + 2\left[(B - A)e^{-x} - Bxe^{-x}\right] + Ae^{-x} + Bxe^{-x} = 0.$

53. $y = e^{rx}$ $\Rightarrow$ $y' = re^{rx}$ $\Rightarrow$ $y'' = r^2 e^{rx}$, so if $y = e^{rx}$ satisfies the differential equation $y'' + 6y' + 8y = 0$,

then $r^2 e^{rx} + 6re^{rx} + 8e^{rx} = 0$; that is, $e^{rx}(r^2 + 6r + 8) = 0$. Since $e^{rx} > 0$ for all x, we must have $r^2 + 6r + 8 = 0$,

or $(r + 2)(r + 4) = 0$, so $r = -2$ or -4.

54. $y = e^{\lambda x}$ $\Rightarrow$ $y' = \lambda e^{\lambda x}$ $\Rightarrow$ $y'' = \lambda^2 e^{\lambda x}$. Thus, $y + y' = y''$ $\Leftrightarrow$ $e^{\lambda x} + \lambda e^{\lambda x} = \lambda^2 e^{\lambda x}$ $\Leftrightarrow$

$e^{\lambda x}(\lambda^2 - \lambda - 1) = 0$ $\Leftrightarrow$ $\lambda = \frac{1 \pm \sqrt{5}}{2}$, since $e^{\lambda x} \neq 0$.

55. $f(x) = e^{2x}$ $\Rightarrow$ $f'(x) = 2e^{2x}$ $\Rightarrow$ $f''(x) = 2 \cdot 2e^{2x} = 2^2 e^{2x}$ $\Rightarrow$

$f'''(x) = 2^2 \cdot 2e^{2x} = 2^3 e^{2x}$ $\Rightarrow$ $\cdots$ $\Rightarrow$ $f^{(n)}(x) = 2^n e^{2x}$

56. $f(x) = xe^{-x}$ $\Rightarrow$ $f'(x) = x(-e^{-x}) + e^{-x} = (1 - x)e^{-x}$ $\Rightarrow$

$f''(x) = (1 - x)(-e^{-x}) + e^{-x}(-1) = (x - 2)e^{-x}$ $\Rightarrow$ $f'''(x) = (x - 2)(-e^{-x}) + e^{-x} = (3 - x)e^{-x}$ $\Rightarrow$

$f^{(4)}(x) = (3 - x)(-e^{-x}) + e^{-x}(-1) = (x - 4)e^{-x}$ $\Rightarrow$ $\cdots$ $\Rightarrow$ $f^{(n)}(x) = (-1)^n (x - n)e^{-x}$.

So $D^{1000} xe^{-x} = (x - 1000)e^{-x}$.

57. (a) $f(x) = e^x + x$ is continuous on $\mathbb{R}$ and $f(-1) = e^{-1} - 1 < 0 < 1 = f(0)$, so by the Intermediate Value Theorem,

$e^x + x = 0$ has a root in $(-1, 0)$.

(b) $f(x) = e^x + x$ $\Rightarrow$ $f'(x) = e^x + 1$, so $x_{n+1} = x_n - \dfrac{e^{x_n} + x_n}{e^{x_n} + 1}$. Using $x_1 = -0.5$, we get $x_2 \approx -0.566311$,

$x_3 \approx -0.567143 \approx x_4$, so the root is -0.567143 to six decimal places.

58.

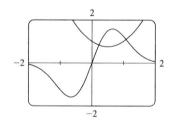

Solving $4e^{-x^2} \sin x = x^2 - x + 1$ is the same as solving

$f(x) = 4e^{-x^2} \sin x - x^2 + x - 1 = 0.$

$f'(x) = 4e^{-x^2}(\cos x - 2x \sin x) - 2x + 1$ $\Rightarrow$

$x_{n+1} = x_n - \dfrac{4e^{-x_n^2} \sin x_n - x_n^2 + x_n - 1}{4e^{-x_n^2}(\cos x_n - 2x_n \sin x_n) - 2x_n + 1}.$

From the graph of f, there appear to be roots near 0.2 and 1.1.

$x_1 = 0.2$	$x_1 = 1.1$
$x_2 \approx 0.21883273$	$x_2 \approx 1.08432830$
$x_3 \approx 0.21916357$	$x_3 \approx 1.08422462 \approx x_4$
$x_4 \approx 0.21916368 \approx x_5$	

To eight decimal places, the roots of the equation are 0.21916368 and 1.08422462.

59. (a) $\lim\limits_{t\to\infty} p(t) = \lim\limits_{t\to\infty} \dfrac{1}{1 + ae^{-kt}} = \dfrac{1}{1 + a \cdot 0} = 1$, since $k > 0 \;\Rightarrow\; -kt \to -\infty \;\Rightarrow\; e^{-kt} \to 0$.

(b) $p(t) = (1 + ae^{-kt})^{-1} \;\Rightarrow\; \dfrac{dp}{dt} = -(1 + ae^{-kt})^{-2}(-kae^{-kt}) = \dfrac{kae^{-kt}}{(1 + ae^{-kt})^2}$

(c)

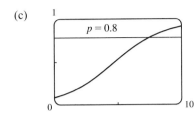

From the graph of $p(t) = (1 + 10e^{-0.5t})^{-1}$, it seems that $p(t) = 0.8$

(indicating that 80% of the population has heard the rumor) when

$t \approx 7.4$ hours.

60. (a)

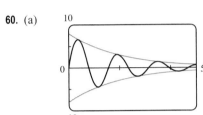

The displacement function is squeezed between the other two functions. This

is because $-1 \le \sin 4t \le 1 \;\Rightarrow\; -8e^{-t/2} \le 8e^{-t/2}\sin 4t \le 8e^{-t/2}$.

(b) The maximum value of the displacement is about 6.6 cm, occurring at $t \approx 0.36$ s. It occurs just before the graph of the

displacement function touches the graph of $8e^{-t/2}$ (when $t = \frac{\pi}{8} \approx 0.39$).

(c) The velocity of the object is the derivative of its displacement function, that is,

$$\frac{d}{dt}\left(8e^{-t/2}\sin 4t\right) = 8\left[e^{-t/2}\cos 4t(4) + \sin 4t\left(-\tfrac{1}{2}\right)e^{-t/2}\right]$$

If the displacement is zero, then we must have $\sin 4t = 0$ (since the exponential term in the displacement function is

always positive). The first time that $\sin 4t = 0$ after $t = 0$ occurs at $t = \frac{\pi}{4}$. Substituting this into our expression for the

velocity, and noting that the second term vanishes, we get $v\left(\frac{\pi}{4}\right) = 8e^{-\pi/8}\cos\left(4 \cdot \frac{\pi}{4}\right) \cdot 4 = -32e^{-\pi/8} \approx -21.6$ cm/s.

(d)

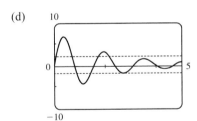

The graph indicates that the displacement is less than 2 cm from equilibrium

whenever t is larger than about 2.8.

61. $f(x) = x - e^x \;\Rightarrow\; f'(x) = 1 - e^x = 0 \;\Leftrightarrow\; e^x = 1 \;\Leftrightarrow\; x = 0$. Now $f'(x) > 0$ for all $x < 0$ and $f'(x) < 0$ for all

$x > 0$, so the absolute maximum value is $f(0) = 0 - 1 = -1$.

62. $g(x) = \dfrac{e^x}{x} \;\Rightarrow\; g'(x) = \dfrac{xe^x - e^x}{x^2} = 0 \;\Leftrightarrow\; e^x(x-1) = 0 \;\Rightarrow\; x = 1$. Now $g'(x) > 0 \;\Leftrightarrow\; \dfrac{xe^x - e^x}{x^2} > 0 \;\Leftrightarrow\;$

$x - 1 > 0 \;\Leftrightarrow\; x > 1$ and $g'(x) < 0 \;\Leftrightarrow\; \dfrac{xe^x - e^x}{x^2} < 0 \;\Leftrightarrow\; x - 1 < 0 \;\Leftrightarrow\; x < 1$. Thus there is an absolute

minimum value of $g(1) = e$ at $x = 1$.

63. $f(x) = xe^{-x^2/8}$, $[-1, 4]$. $f'(x) = x \cdot e^{-x^2/8} \cdot \left(-\frac{x}{4}\right) + e^{-x^2/8} \cdot 1 = e^{-x^2/8}\left(-\frac{x^2}{4} + 1\right)$. Since $e^{-x^2/8}$ is never 0,

$f'(x) = 0 \;\Rightarrow\; -x^2/4 + 1 = 0 \;\Rightarrow\; 1 = x^2/4 \;\Rightarrow\; x^2 = 4 \;\Rightarrow\; x = \pm 2$, but -2 is not in the given interval, $[-1, 4]$.

$f(-1) = -e^{-1/8} \approx -0.88$, $f(2) = 2e^{-1/2} \approx 1.21$, and $f(4) = 4e^{-2} \approx 0.54$. So $f(2) = 2e^{-1/2}$ is the absolute maximum

value and $f(-1) = -e^{-1/8}$ is the absolute minimum value.

64. $f(x) = x^2 e^{-x/2}$, $[-1, 6]$ $\Rightarrow$ $f'(x) = x^2 e^{-x/2}\left(-\frac{1}{2}\right) + e^{-x/2}(2x) = xe^{-x/2}\left(-\frac{1}{2}x + 2\right)$. $f'(x) = 0$ $\Rightarrow$

$x = 0$ or 4. $f(-1) = e^{1/2} \approx 1.65$, $f(0) = 0$, $f(4) = 16e^{-2} \approx 2.17$, and $f(6) = 36e^{-3} \approx 1.79$. Thus, on $[-1, 6]$, the

absolute maximum value of f is $f(4) = 16e^{-2}$ and the absolute minimum value is $f(0) = 0$.

65. (a) $f(x) = (1 - x)e^{-x}$ $\Rightarrow$ $f'(x) = (1 - x)(-e^{-x}) + e^{-x}(-1) = e^{-x}(x - 2) > 0$ $\Rightarrow$ $x > 2$, so f is increasing on

   $(2, \infty)$ and decreasing on $(-\infty, 2)$.

(b) $f''(x) = e^{-x}(1) + (x - 2)(-e^{-x}) = e^{-x}(3 - x) > 0$ $\Leftrightarrow$ $x < 3$, so f is CU on $(-\infty, 3)$ and CD on $(3, \infty)$.

(c) f'' changes sign at $x = 3$, so there is an IP at $\left(3, -2e^{-3}\right)$.

66. (a) $f(x) = \dfrac{e^x}{x^2}$ $\Rightarrow$ $f'(x) = \dfrac{x^2 e^x - e^x (2x)}{(x^2)^2} = \dfrac{xe^x(x - 2)}{x^4} = \dfrac{e^x(x - 2)}{x^3}$. $f'(x) > 0$ $\Leftrightarrow$ $x < 0$ or $x > 2$, so f is

   increasing on $(-\infty, 0)$ and $(2, \infty)$. $f'(x) < 0$ $\Leftrightarrow$ $0 < x < 2$, so f is decreasing on $(0, 2)$.

(b) $f''(x) = \dfrac{x^3[e^x \cdot 1 + (x - 2)e^x] - e^x(x - 2) \cdot 3x^2}{(x^3)^2} = \dfrac{x^2 e^x[x(x - 1) - 3(x - 2)]}{x^6} = \dfrac{e^x(x^2 - 4x + 6)}{x^4}$.

   $x^2 - 4x + 6 = (x^2 - 4x + 4) + 2 = (x - 2)^2 + 2 > 0$, so $f''(x) > 0$ and f is CU on $(-\infty, 0)$ and $(0, \infty)$.

(c) There are no changes in concavity and, hence, there are no points of inflection.

67. $y = f(x) = e^{-1/(x+1)}$ **A.** $D = \{x \mid x \neq -1\} = (-\infty, -1) \cup (-1, \infty)$ **B.** No x-intercept; y-intercept $= f(0) = e^{-1}$

C. No symmetry   **D.** $\displaystyle\lim_{x \to \pm\infty} e^{-1/(x+1)} = 1$ since $-1/(x + 1) \to 0$, so $y = 1$ is a HA. $\displaystyle\lim_{x \to -1^+} e^{-1/(x+1)} = 0$ since

$-1/(x + 1) \to -\infty$, $\displaystyle\lim_{x \to -1^-} e^{-1/(x+1)} = \infty$ since $-1/(x + 1) \to \infty$, so $x = -1$ is a VA.

E. $f'(x) = e^{-1/(x+1)}/(x + 1)^2$ $\Rightarrow$ $f'(x) > 0$ for all x except 1, so

f is increasing on $(-\infty, -1)$ and $(-1, \infty)$.   **F.** No extreme values

G. $f''(x) = \dfrac{e^{-1/(x+1)}}{(x + 1)^4} + \dfrac{e^{-1/(x+1)}(-2)}{(x + 1)^3} = -\dfrac{e^{-1/(x+1)}(2x + 1)}{(x + 1)^4}$ $\Rightarrow$

$f''(x) > 0$ $\Leftrightarrow$ $2x + 1 < 0$ $\Leftrightarrow$ $x < -\frac{1}{2}$, so f is CU on $(-\infty, -1)$

and $\left(-1, -\frac{1}{2}\right)$, and CD on $\left(-\frac{1}{2}, \infty\right)$. f has an IP at $\left(-\frac{1}{2}, e^{-2}\right)$.

H.

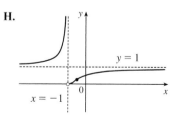

68. $y = f(x) = e^{-x} \sin x$, $0 \le x \le 2\pi$ **A.** $D = \mathbb{R}$ **B.** y-intercept: $f(0) = 0$; x-intercepts: $f(x) = 0 \iff \sin x = 0 \iff$

$x = 0, \pi$, and 2π. **C.** No symmetry **D.** No asymptote **E.** $f'(x) = e^{-x} \cos x + \sin x \,(-e^{-x}) = e^{-x}(\cos x - \sin x)$.

$f'(x) = 0 \iff \cos x = \sin x \iff x = \frac{\pi}{4}, \frac{5\pi}{4}$. $f'(x) > 0$ if x is in $\left(0, \frac{\pi}{4}\right)$ or $\left(\frac{5\pi}{4}, 2\pi\right)$ [f is increasing] and

$f'(x) < 0$ if x is in $\left(\frac{\pi}{4}, \frac{5\pi}{4}\right)$ [f is decreasing]. **F.** Local maximum value $f\left(\frac{\pi}{4}\right)$ and local minimum value $f\left(\frac{5\pi}{4}\right)$

G. $f''(x) = e^{-x}(-\sin x - \cos x) + (\cos x - \sin x)(-e^{-x}) = e^{-x}(-2\cos x)$. $f''(x) > 0 \iff -2\cos x > 0 \iff$

$\cos x < 0 \Rightarrow x$ is in $\left(\frac{\pi}{2}, \frac{3\pi}{2}\right)$ [f is CU] and $f''(x) < 0 \iff$ **H.**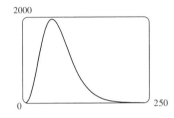

$\cos x > 0 \Rightarrow x$ is in $\left(0, \frac{\pi}{2}\right)$ or $\left(\frac{3\pi}{2}, 2\pi\right)$ [f is CD].

IP at $\left(\frac{\pi}{2} + n\pi, f\left(\frac{\pi}{2} + n\pi\right)\right)$

69. $S(t) = At^p e^{-kt}$ with $A = 0.01$, $p = 4$, and $k = 0.07$. We will find the

zeros of f'' for $f(t) = t^p e^{-kt}$.

$f'(t) = t^p(-ke^{-kt}) + e^{-kt}(pt^{p-1}) = e^{-kt}(-kt^p + pt^{p-1})$

$f''(t) = e^{-kt}(-kpt^{p-1} + p(p-1)t^{p-2}) + (-kt^p + pt^{p-1})(-ke^{-kt})$

$\quad = t^{p-2}e^{-kt}[-kpt + p(p-1) + k^2t^2 - kpt]$

$\quad = t^{p-2}e^{-kt}(k^2t^2 - 2kpt + p^2 - p)$

Using the given values of p and k gives us $f''(t) = t^2 e^{-0.07t}(0.0049t^2 - 0.56t + 12)$. So $S''(t) = 0.01 f''(t)$ and its zeros

are $t = 0$ and the solutions of $0.0049t^2 - 0.56t + 12 = 0$, which are $t_1 = \frac{200}{7} \approx 28.57$ and $t_2 = \frac{600}{7} \approx 85.71$.

At t_1 minutes, the rate of increase of the level of medication in the bloodstream is at its greatest and at t_2 minutes, the rate of

decrease is the greatest.

70. The function $f(x) = e^{\cos x}$ is periodic with

period 2π, so we consider it only on the interval

$[0, 2\pi]$. We see that it has local maxima of about

$f(0) \approx 2.72$ and $f(2\pi) \approx 2.72$, and a local

minimum of about $f(3.14) \approx 0.37$. To find the

exact values, we calculate $f'(x) = -\sin x\, e^{\cos x}$. This is 0 when $-\sin x = 0 \iff x = 0, \pi$ or 2π (since we are only

considering $x \in [0, 2\pi]$). Also $f'(x) > 0 \iff \sin x < 0 \iff 0 < x < \pi$. So $f(0) = f(2\pi) = e$

(both maxima) and $f(\pi) = e^{\cos \pi} = 1/e$ (minimum). To find the inflection points, we calculate and graph

$f''(x) = \frac{d}{dx}\left(-\sin x\, e^{\cos x}\right) = -\cos x\, e^{\cos x} - \sin x(e^{\cos x})(-\sin x) = e^{\cos x}\left(\sin^2 x - \cos x\right)$. From the graph of $f''(x)$,

we see that f has inflection points at $x \approx 0.90$ and at $x \approx 5.38$. These x-coordinates correspond to inflection points

$(0.90, 1.86)$ and $(5.38, 1.86)$.

71. $f(x) = e^{x^3 - x} \to 0$ as $x \to -\infty$, and

$f(x) \to \infty$ as $x \to \infty$. From the graph,

it appears that f has a local minimum of

about $f(0.58) = 0.68$, and a local

maximum of about $f(-0.58) = 1.47$.

To find the exact values, we calculate

$f'(x) = (3x^2 - 1)e^{x^3 - x}$, which is 0 when $3x^2 - 1 = 0$ $\Leftrightarrow$ $x = \pm\frac{1}{\sqrt{3}}$. The negative root corresponds to the local

maximum $f\left(-\frac{1}{\sqrt{3}}\right) = e^{(-1/\sqrt{3})^3 - (-1/\sqrt{3})} = e^{2\sqrt{3}/9}$, and the positive root corresponds to the local minimum

$f\left(\frac{1}{\sqrt{3}}\right) = e^{(1/\sqrt{3})^3 - (1/\sqrt{3})} = e^{-2\sqrt{3}/9}$. To estimate the inflection points, we calculate and graph

$f''(x) = \dfrac{d}{dx}\left[(3x^2 - 1)e^{x^3 - x}\right] = (3x^2 - 1)e^{x^3 - x}(3x^2 - 1) + e^{x^3 - x}(6x) = e^{x^3 - x}(9x^4 - 6x^2 + 6x + 1).$

From the graph, it appears that $f''(x)$ changes sign (and thus f has inflection points) at $x \approx -0.15$ and $x \approx -1.09$. From the

graph of f, we see that these x-values correspond to inflection points at about $(-0.15, 1.15)$ and $(-1.09, 0.82)$.

72. (a) As $|x| \to \infty$, $t = -x^2/(2\sigma^2) \to -\infty$, and $e^t \to 0$. The HA is $y = 0$. Since t takes on its maximum value at $x = 0$, so

does e^t. Showing this result using derivatives, we have $f(x) = e^{-x^2/(2\sigma^2)}$ $\Rightarrow$ $f'(x) = e^{-x^2/(2\sigma^2)}(-x/\sigma^2).$

$f'(x) = 0$ $\Leftrightarrow$ $x = 0$. Because f' changes from positive to negative at $x = 0$, $f(0) = 1$ is a local maximum. For

inflection points, we find $f''(x) = -\dfrac{1}{\sigma^2}\left[e^{-x^2/(2\sigma^2)} \cdot 1 + xe^{-x^2/(2\sigma^2)}(-x/\sigma^2)\right] = \dfrac{-1}{\sigma^2}e^{-x^2/(2\sigma^2)}(1 - x^2/\sigma^2).$

$f''(x) = 0$ $\Leftrightarrow$ $x^2 = \sigma^2$ $\Leftrightarrow$ $x = \pm\sigma$. $f''(x) < 0$ $\Leftrightarrow$ $x^2 < \sigma^2$ $\Leftrightarrow$ $-\sigma < x < \sigma$.

So f is CD on $(-\sigma, \sigma)$ and CU on $(-\infty, -\sigma)$ and (σ, ∞). IP at $(\pm\sigma, e^{-1/2})$.

(b) Since we have IP at $x = \pm\sigma$, the inflection points move away from the y-axis as σ increases.

(c)

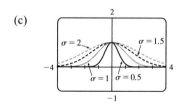

From the graph, we see that as σ increases, the graph tends to spread out and

there is more area between the curve and the x-axis.

73. Let $u = -3x$. Then $du = -3\,dx$, so $\int_0^5 e^{-3x}\,dx = -\frac{1}{3}\int_0^{-15} e^u\,du = -\frac{1}{3}[e^u]_0^{-15} = -\frac{1}{3}\left(e^{-15} - e^0\right) = \frac{1}{3}\left(1 - e^{-15}\right).$

74. Let $u = -x^2$, so $du = -2x\,dx$. When $x = 0$, $u = 0$; when $x = 1$, $u = -1$. Thus,

$\int_0^1 xe^{-x^2}\,dx = \int_0^{-1} e^u\left(-\frac{1}{2}\,du\right) = -\frac{1}{2}[e^u]_0^{-1} = -\frac{1}{2}\left(e^{-1} - e^0\right) = \frac{1}{2}(1 - 1/e).$

75. Let $u = 1 + e^x$. Then $du = e^x\,dx$, so $\int e^x\sqrt{1 + e^x}\,dx = \int \sqrt{u}\,du = \frac{2}{3}u^{3/2} + C = \frac{2}{3}(1 + e^x)^{3/2} + C.$

76. $\displaystyle\int \dfrac{(1 + e^x)^2}{e^x}\,dx = \int \dfrac{1 + 2e^x + e^{2x}}{e^x}\,dx = \int (e^{-x} + 2 + e^x)\,dx = -e^{-x} + 2x + e^x + C$

77. $\int (e^x + e^{-x})^2\,dx = \int (e^{2x} + 2 + e^{-2x})\,dx = \frac{1}{2}e^{2x} + 2x - \frac{1}{2}e^{-2x} + C$

78. $\int e^x (4+e^x)^5 \, dx$ $\begin{bmatrix} u = 4+e^x, \\ du = e^x \, dx \end{bmatrix}$ $= \int u^5 \, du = \frac{1}{6}u^6 + C = \frac{1}{6}(4+e^x)^6 + C$

79. $\int \sin x \, e^{\cos x} \, dx$ $\begin{bmatrix} u = \cos x, \\ du = -\sin x \, dx \end{bmatrix}$ $= \int e^u \, (-du) = -e^u + C = -e^{\cos x} + C$

80. Let $u = \dfrac{1}{x}$. Then $du = -\dfrac{1}{x^2} \, dx$, so $\displaystyle\int \frac{e^{1/x}}{x^2} \, dx = -\int e^u \, du = -e^u + C = -e^{1/x} + C.$

81. Let $u = \sqrt{x}$. Then $du = \dfrac{1}{2\sqrt{x}} \, dx$, so $\displaystyle\int \frac{e^{\sqrt{x}}}{\sqrt{x}} \, dx = 2 \int e^u \, du = 2e^u + C = 2e^{\sqrt{x}} + C.$

82. Let $u = e^x$. Then $du = e^x \, dx$, so $\int e^x \sin(e^x) \, dx = \int \sin u \, du = -\cos u + C = -\cos(e^x) + C.$

83. Area $= \int_0^1 \left(e^{3x} - e^x\right) dx = \left[\frac{1}{3}e^{3x} - e^x\right]_0^1 = \left(\frac{1}{3}e^3 - e\right) - \left(\frac{1}{3} - 1\right) = \frac{1}{3}e^3 - e + \frac{2}{3} \approx 4.644$

84. $f''(x) = 3e^x + 5\sin x$ $\Rightarrow$ $f'(x) = 3e^x - 5\cos x + C$ $\Rightarrow$ $2 = f'(0) = 3 - 5 + C$ $\Rightarrow$ $C = 4$, so
$f'(x) = 3e^x - 5\cos x + 4$ $\Rightarrow$ $f(x) = 3e^x - 5\sin x + 4x + D$ $\Rightarrow$ $1 = f(0) = 3 + D$ $\Rightarrow$ $D = -2$,
so $f(x) = 3e^x - 5\sin x + 4x - 2.$

85. $V = \int_0^1 \pi (e^x)^2 \, dx = \pi \int_0^1 e^{2x} \, dx = \frac{1}{2}\pi \left[e^{2x}\right]_0^1 = \frac{\pi}{2}\left(e^2 - 1\right)$

86. $V = \int_0^1 2\pi x e^{-x^2} \, dx$. Let $u = x^2$.

Thus, $du = 2x \, dx$, so

$V = \pi \int_0^1 e^{-u} \, du = \pi \left[-e^{-u}\right]_0^1 = \pi(1 - 1/e).$

87. (a) $\operatorname{erf}(x) = \dfrac{2}{\sqrt{\pi}} \displaystyle\int_0^x e^{-t^2} \, dt$ $\Rightarrow$ $\displaystyle\int_0^x e^{-t^2} \, dt = \frac{\sqrt{\pi}}{2} \operatorname{erf}(x)$. By Property 5 of definite integrals in Section 5.2,

$\int_0^b e^{-t^2} \, dt = \int_0^a e^{-t^2} \, dt + \int_a^b e^{-t^2} \, dt$, so

$\displaystyle\int_a^b e^{-t^2} \, dt = \int_0^b e^{-t^2} \, dt - \int_0^a e^{-t^2} \, dt = \frac{\sqrt{\pi}}{2} \operatorname{erf}(b) - \frac{\sqrt{\pi}}{2} \operatorname{erf}(a) = \frac{1}{2} \sqrt{\pi} \left[\operatorname{erf}(b) - \operatorname{erf}(a)\right].$

(b) $y = e^{x^2} \operatorname{erf}(x)$ $\Rightarrow$ $y' = 2xe^{x^2} \operatorname{erf}(x) + e^{x^2} \operatorname{erf}'(x) = 2xy + e^{x^2} \cdot \dfrac{2}{\sqrt{\pi}} e^{-x^2}$ [by FTC1] $= 2xy + \dfrac{2}{\sqrt{\pi}}.$

88. Let $r(t) = ae^{bt}$ with $a = 450.268$ and $b = 1.12567$, and $n(t) = $ population after t hours. Since $r(t) = n'(t)$,

$\int_0^3 r(t) \, dt = n(3) - n(0)$ is the total change in the population after three hours. Since we start with 400 bacteria, the population will be

$$n(3) = 400 + \int_0^3 r(t) \, dt = 400 + \int_0^3 ae^{bt} \, dt = 400 + \frac{a}{b}\left[e^{bt}\right]_0^3 = 400 + \frac{a}{b}\left(e^{3b} - 1\right)$$

$$\approx 400 + 11{,}313 = 11{,}713 \text{ bacteria}$$

89. We use Theorem 7.1.7. Note that $f(0) = 3 + 0 + e^0 = 4$, so $f^{-1}(4) = 0$. Also $f'(x) = 1 + e^x$. Therefore,

$$\left(f^{-1}\right)'(4) = \frac{1}{f'(f^{-1}(4))} = \frac{1}{f'(0)} = \frac{1}{1 + e^0} = \frac{1}{2}.$$

90. We recognize this limit as the definition of the derivative of the function $f(x) = e^{\sin x}$ at $x = \pi$, since it is of the form

$$\lim_{x \to \pi} \frac{f(x) - f(\pi)}{x - \pi}.$$ Therefore, the limit is equal to $f'(\pi) = (\cos \pi)e^{\sin \pi} = -1 \cdot e^0 = -1$.

91.

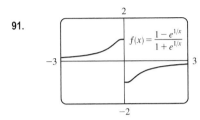

From the graph, it appears that f is an odd function (f is undefined for $x = 0$).

To prove this, we must show that $f(-x) = -f(x)$.

$$f(-x) = \frac{1 - e^{1/(-x)}}{1 + e^{1/(-x)}} = \frac{1 - e^{(-1/x)}}{1 + e^{(-1/x)}} = \frac{1 - \dfrac{1}{e^{1/x}}}{1 + \dfrac{1}{e^{1/x}}} \cdot \frac{e^{1/x}}{e^{1/x}} = \frac{e^{1/x} - 1}{e^{1/x} + 1}$$

$$= -\frac{1 - e^{1/x}}{1 + e^{1/x}} = -f(x)$$

so f is an odd function.

92. We'll start with $b = -1$ and graph $f(x) = \dfrac{1}{1 + ae^{bx}}$ for $a = 0.1$, 1, and 5.

From the graph, we see that there is a horizontal asymptote $y = 0$ as $x \to -\infty$

and a horizontal asymptote $y = 1$ as $x \to \infty$. If $a = 1$, the y-intercept is $\left(0, \frac{1}{2}\right)$.

As a gets smaller (close to 0), the graph of f moves left. As a gets larger, the graph

of f moves right.

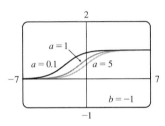

As b changes from -1 to 0, the graph of f is stretched horizontally. As b

changes through large negative values, the graph of f is compressed horizontally.

(This takes care of negatives values of b.)

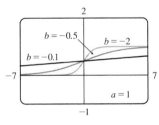

If b is positive, the graph of f is reflected through the y-axis.

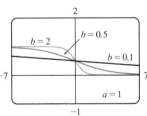

Last, if $b = 0$, the graph of f is the horizontal line $y = 1/(1 + a)$.

93. (a) Let $f(x) = e^x - 1 - x$. Now $f(0) = e^0 - 1 = 0$, and for $x \ge 0$, we have $f'(x) = e^x - 1 \ge 0$. Now, since $f(0) = 0$ and

f is increasing on $[0, \infty)$, $f(x) \ge 0$ for $x \ge 0$ $\Rightarrow$ $e^x - 1 - x \ge 0$ $\Rightarrow$ $e^x \ge 1 + x$.

(b) For $0 \le x \le 1$, $x^2 \le x$, so $e^{x^2} \le e^x$ [since e^x is increasing]. Hence [from (a)] $1 + x^2 \le e^{x^2} \le e^x$.

So $\frac{4}{3} = \int_0^1 \left(1 + x^2\right) dx \le \int_0^1 e^{x^2} dx \le \int_0^1 e^x dx = e - 1 < e$ $\Rightarrow$ $\frac{4}{3} \le \int_0^1 e^{x^2} dx \le e$.

94. (a) Let $f(x) = e^x - 1 - x - \frac{1}{2}x^2$. Thus, $f'(x) = e^x - 1 - x$, which is positive for $x \geq 0$ by Exercise 93(a). Thus $f(x)$ is

increasing on $(0, \infty)$, so on that interval, $0 = f(0) \leq f(x) = e^x - 1 - x - \frac{1}{2}x^2 \Rightarrow e^x \geq 1 + x + \frac{1}{2}x^2$.

(b) Using the same argument as in Exercise 93(b), from part (a) we have $1 + x^2 + \frac{1}{2}x^4 \leq e^{x^2} \leq e^x$

[for $0 \leq x \leq 1$] $\Rightarrow \int_0^1 (1 + x^2 + \frac{1}{2}x^4)\,dx \leq \int_0^1 e^{x^2}\,dx \leq \int_0^1 e^x\,dx \Rightarrow \frac{43}{30} \leq \int_0^1 e^{x^2}\,dx \leq e - 1$.

95. (a) By Exercise 93(a), the result holds for $n = 1$. Suppose that $e^x \geq 1 + x + \frac{x^2}{2!} + \cdots + \frac{x^k}{k!}$ for $x \geq 0$.

Let $f(x) = e^x - 1 - x - \frac{x^2}{2!} - \cdots - \frac{x^k}{k!} - \frac{x^{k+1}}{(k+1)!}$. Then $f'(x) = e^x - 1 - x - \cdots - \frac{x^k}{k!} \geq 0$ by assumption. Hence

$f(x)$ is increasing on $(0, \infty)$. So $0 \leq x$ implies that $0 = f(0) \leq f(x) = e^x - 1 - x - \cdots - \frac{x^k}{k!} - \frac{x^{k+1}}{(k+1)!}$, and hence

$e^x \geq 1 + x + \cdots + \frac{x^k}{k!} + \frac{x^{k+1}}{(k+1)!}$ for $x \geq 0$. Therefore, for $x \geq 0$, $e^x \geq 1 + x + \frac{x^2}{2!} + \cdots + \frac{x^n}{n!}$ for every positive

integer n, by mathematical induction.

(b) Taking $n = 4$ and $x = 1$ in (a), we have $e = e^1 \geq 1 + \frac{1}{2} + \frac{1}{6} + \frac{1}{24} = 2.708\overline{3} > 2.7$.

(c) $e^x \geq 1 + x + \cdots + \frac{x^k}{k!} + \frac{x^{k+1}}{(k+1)!} \Rightarrow \frac{e^x}{x^k} \geq \frac{1}{x^k} + \frac{1}{x^{k-1}} + \cdots + \frac{1}{k!} + \frac{x}{(k+1)!} \geq \frac{x}{(k+1)!}$.

But $\lim\limits_{x \to \infty} \frac{x}{(k+1)!} = \infty$, so $\lim\limits_{x \to \infty} \frac{e^x}{x^k} = \infty$.

7.3 Logarithmic Functions

1. (a) It is defined as the inverse of the exponential function with base a, that is, $\log_a x = y \Leftrightarrow a^y = x$.

(b) $(0, \infty)$ (c) $\mathbb{R}$ (d) See Figure 1.

2. (a) The natural logarithm is the logarithm with base e, denoted $\ln x$.

(b) The common logarithm is the logarithm with base 10, denoted $\log x$.

(c) See Figure 3.

3. (a) $\log_5 125 = 3$ since $5^3 = 125$.   (b) $\log_3 \frac{1}{27} = -3$ since $3^{-3} = \frac{1}{3^3} = \frac{1}{27}$.

4. (a) $\ln(1/e) = \ln 1 - \ln e = 0 - 1 = -1$ (b) $\log_{10}\sqrt{10} = \log_{10} 10^{1/2} = \frac{1}{2}$ by (2).

5. (a) $\log_5 \frac{1}{25} = \log_5 5^{-2} = -2$ by (2).   (b) $e^{\ln 15} = 15$ by (8).

6. (a) $\log_{10} 0.1 = -1$ since $10^{-1} = 0.1$.

(b) $\log_8 320 - \log_8 5 = \log_8 \frac{320}{5} = \log_8 64 = 2$ since $8^2 = 64$.

7. (a) $\log_2 6 - \log_2 15 + \log_2 20 = \log_2\left(\frac{6}{15}\right) + \log_2 20$ [by Law 2]

$= \log_2\left(\frac{6}{15} \cdot 20\right)$ [by Law 1]

$= \log_2 8$, and $\log_2 8 = 3$ since $2^3 = 8$.

(b) $\log_3 100 - \log_3 18 - \log_3 50 = \log_3\left(\frac{100}{18}\right) - \log_3 50 = \log_3\left(\frac{100}{18\cdot 50}\right)$

$$= \log_3\left(\tfrac{1}{9}\right), \text{ and } \log_3\left(\tfrac{1}{9}\right) = -2 \text{ since } 3^{-2} = \tfrac{1}{9}.$$

8. (a) $e^{-2\ln 5} = \left(e^{\ln 5}\right)^{-2} \overset{(6)}{=} 5^{-2} = \dfrac{1}{5^2} = \dfrac{1}{25}$ (b) $\ln\left(\ln e^{e^{10}}\right) \overset{(6)}{=} \ln(e^{10}) \overset{(6)}{=} 10$

9. $\log_2\left(\dfrac{x^3 y}{z^2}\right) = \log_2(x^3 y) - \log_2 z^2 = \log_2 x^3 + \log_2 y - \log_2 z^2 = 3\log_2 x + \log_2 y - 2\log_2 z$

[assuming that the variables are positive]

10. $\ln \sqrt{a(b^2 + c^2)} = \ln(a(b^2 + c^2))^{1/2} = \tfrac{1}{2}\ln(a(b^2 + c^2)) = \tfrac{1}{2}[\ln a + \ln(b^2 + c^2)] = \tfrac{1}{2}\ln a + \tfrac{1}{2}\ln(b^2 + c^2)$

11. $\ln(uv)^{10} = 10\ln(uv) = 10(\ln u + \ln v) = 10\ln u + 10\ln v$

12. $\ln \dfrac{3x^2}{(x+1)^5} = \ln 3x^2 - \ln(x+1)^5 = \ln 3 + \ln x^2 - 5\ln(x+1) = \ln 3 + 2\ln x - 5\ln(x+1)$

13. $\log_{10} a - \log_{10} b + \log_{10} c = \log_{10} \dfrac{a}{b} + \log_{10} c = \log_{10}\left(\dfrac{a}{b}\cdot c\right) = \log_{10} \dfrac{ac}{b}$

14. $\ln(x+y) + \ln(x-y) - 2\ln z = \ln((x+y)(x-y)) - \ln z^2 = \ln(x^2 - y^2) - \ln z^2 = \ln \dfrac{x^2 - y^2}{z^2}$

15. $\ln 5 + 5\ln 3 = \ln 5 + \ln 3^5$ [by Law 3]

$$= \ln(5\cdot 3^5) \qquad \text{[by Law 1]}$$

$$= \ln 1215$$

16. $\ln 3 + \tfrac{1}{3}\ln 8 = \ln 3 + \ln 8^{1/3} = \ln 3 + \ln 2 = \ln(3\cdot 2) = \ln 6$

17. $\ln(1+x^2) + \tfrac{1}{2}\ln x - \ln\sin x = \ln(1+x^2) + \ln x^{1/2} - \ln\sin x = \ln[(1+x^2)\sqrt{x}] - \ln\sin x = \ln \dfrac{(1+x^2)\sqrt{x}}{\sin x}$

18. $\ln(a+b) + \ln(a-b) - 2\ln c = \ln[(a+b)(a-b)] - \ln c^2$ [by Laws 1, 3]

$$= \ln \dfrac{(a+b)(a-b)}{c^2} \qquad \text{[by Law 2]}$$

$$\text{or } \ln \dfrac{a^2 - b^2}{c^2}$$

19. (a) $\log_{12} e = \dfrac{\ln e}{\ln 12} = \dfrac{1}{\ln 12} \approx 0.402430$

(b) $\log_6 13.54 = \dfrac{\ln 13.54}{\ln 6} \approx 1.454240$

(c) $\log_2 \pi = \dfrac{\ln \pi}{\ln 2} \approx 1.651496$

20. To graph the functions, we use $\log_2 x = \dfrac{\ln x}{\ln 2}$, $\log_4 x = \dfrac{\ln x}{\ln 4}$, etc. These

graphs all approach $-\infty$ as $x \to 0^+$, and they all pass through the point

$(1, 0)$. Also, they are all increasing, and all approach ∞ as $x \to \infty$. The

smaller the base, the larger the rate of increase of the function (for $x > 1$)

and the closer the approach to the y-axis (as $x \to 0^+$).

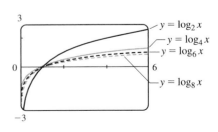

21. To graph these functions, we use $\log_{1.5} x = \dfrac{\ln x}{\ln 1.5}$ and $\log_{50} x = \dfrac{\ln x}{\ln 50}$.

These graphs all approach $-\infty$ as $x \to 0^+$, and they all pass through the

point $(1, 0)$. Also, they are all increasing, and all approach ∞ as $x \to \infty$.

The functions with larger bases increase extremely slowly, and the ones with

smaller bases do so somewhat more quickly. The functions with large bases

approach the y-axis more closely as $x \to 0^+$.

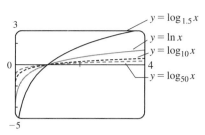

22. We see that the graph of $\ln x$ is the reflection of the graph of e^x about the

line $y = x$, and that the graph of $\log_{10} x$ is the reflection of the graph of 10^x

about the same line. The graph of 10^x increases more quickly than that

of e^x. Also note that $\log_{10} x \to \infty$ as $x \to \infty$ more slowly than $\ln x$.

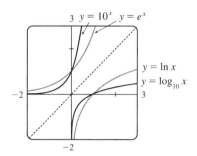

23. (a) Shift the graph of $y = \log_{10} x$ five units to the left to

obtain the graph of $y = \log_{10}(x + 5)$. Note the vertical

asymptote of $x = -5$.

(b) Reflect the graph of $y = \ln x$ about the x-axis to obtain

the graph of $y = -\ln x$.

$$y = \log_{10} x \qquad\qquad y = \log_{10}(x + 5)$$

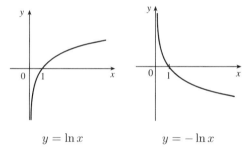

$$y = \ln x \qquad\qquad y = -\ln x$$

24. (a) Reflect the graph of $y = \ln x$ about the y-axis to obtain

the graph of $y = \ln(-x)$.

(b) Reflect the portion of the graph of $y = \ln x$ to the right

of the y-axis about the y-axis. The graph of $y = \ln|x|$

is that reflection in addition to the original portion.

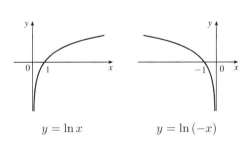

$$y = \ln x \qquad\qquad y = \ln(-x)$$

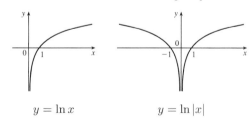

$$y = \ln x \qquad\qquad y = \ln|x|$$

25. (a) $2 \ln x = 1 \;\Rightarrow\; \ln x = \frac{1}{2} \;\Rightarrow\; x = e^{1/2} = \sqrt{e}$

(b) $e^{-x} = 5 \;\Rightarrow\; -x = \ln 5 \;\Rightarrow\; x = -\ln 5$

26. (a) $e^{2x+3} - 7 = 0 \;\Rightarrow\; e^{2x+3} = 7 \;\Rightarrow\; 2x + 3 = \ln 7 \;\Rightarrow\; 2x = \ln 7 - 3 \;\Rightarrow\; x = \frac{1}{2}(\ln 7 - 3)$

(b) $\ln(5 - 2x) = -3 \;\Rightarrow\; 5 - 2x = e^{-3} \;\Rightarrow\; 2x = 5 - e^{-3} \;\Rightarrow\; x = \frac{1}{2}(5 - e^{-3})$

27. (a) $2^{x-5} = 3 \;\Leftrightarrow\; \log_2 3 = x - 5 \;\Leftrightarrow\; x = 5 + \log_2 3$.

$Or: 2^{x-5} = 3 \;\Leftrightarrow\; \ln(2^{x-5}) = \ln 3 \;\Leftrightarrow\; (x - 5)\ln 2 = \ln 3 \;\Leftrightarrow\; x - 5 = \dfrac{\ln 3}{\ln 2} \;\Leftrightarrow\; x = 5 + \dfrac{\ln 3}{\ln 2}$

(b) $\ln x + \ln(x-1) = \ln(x(x-1)) = 1 \iff x(x-1) = e^1 \iff x^2 - x - e = 0$. The quadratic formula (with $a = 1$,

$b = -1$, and $c = -e$) gives $x = \frac{1}{2}\left(1 \pm \sqrt{1+4e}\right)$, but we reject the negative root since the natural logarithm is not

defined for $x < 0$. So $x = \frac{1}{2}\left(1 + \sqrt{1+4e}\right)$.

28. (a) $e^{3x+1} = k \iff 3x + 1 = \ln k \iff x = \frac{1}{3}(\ln k - 1)$

(b) $\log_2(mx) = c \iff mx = 2^c \iff x = 2^c/m$

29. $3xe^x + x^2 e^x = 0 \iff xe^x(3+x) = 0 \iff x = 0 \text{ or } -3$

30. $10(1+e^{-x})^{-1} = 3 \iff (1+e^{-x})^{-1} = \frac{3}{10} \iff 1 + e^{-x} = \frac{10}{3} \iff e^{-x} = \frac{7}{3} \iff -x = \ln\frac{7}{3} \Rightarrow$

$x = -\ln\frac{7}{3} = \ln\left(\frac{7}{3}\right)^{-1} = \ln\frac{3}{7}$

31. $\ln(\ln x) = 1 \iff e^{\ln(\ln x)} = e^1 \iff \ln x = e^1 = e \iff e^{\ln x} = e^e \iff x = e^e$

32. $e^{e^x} = 10 \iff \ln\left(e^{e^x}\right) = \ln 10 \iff e^x \ln e = e^x = \ln 10 \iff \ln e^x = \ln(\ln 10) \iff x = \ln(\ln 10)$

33. $e^{2x} - e^x - 6 = 0 \iff (e^x - 3)(e^x + 2) = 0 \iff e^x = 3 \text{ or } -2 \Rightarrow x = \ln 3 \text{ since } e^x > 0.$

34. $\ln(2x+1) = 2 - \ln x \Rightarrow \ln x + \ln(2x+1) = \ln e^2 \Rightarrow \ln[x(2x+1)] = \ln e^2 \Rightarrow 2x^2 + x = e^2 \Rightarrow$

$2x^2 + x - e^2 = 0 \Rightarrow x = \dfrac{-1 + \sqrt{1+8e^2}}{4}$ [since $x > 0$].

35. (a) $e^{2+5x} = 100 \Rightarrow \ln\left(e^{2+5x}\right) = \ln 100 \Rightarrow 2 + 5x = \ln 100 \Rightarrow 5x = \ln 100 - 2 \Rightarrow$

$x = \frac{1}{5}(\ln 100 - 2) \approx 0.5210$

(b) $\ln(e^x - 2) = 3 \Rightarrow e^x - 2 = e^3 \Rightarrow e^x = e^3 + 2 \Rightarrow x = \ln(e^3 + 2) \approx 3.0949$

36. (a) $\ln\left(1+\sqrt{x}\right) = 2 \Rightarrow 1 + \sqrt{x} = e^2 \Rightarrow \sqrt{x} = e^2 - 1 \Rightarrow x = (e^2 - 1)^2 \approx 40.8200$

(b) $3^{1/(x-4)} = 7 \Rightarrow \ln 3^{1/(x-4)} = \ln 7 \Rightarrow \dfrac{1}{x-4}\ln 3 = \ln 7 \Rightarrow x - 4 = \dfrac{\ln 3}{\ln 7} \Rightarrow x = 4 + \dfrac{\ln 3}{\ln 7} \approx 4.5646$

37. (a) $e^x < 10 \Rightarrow \ln e^x < \ln 10 \Rightarrow x < \ln 10 \Rightarrow x \in (-\infty, \ln 10)$

(b) $\ln x > -1 \Rightarrow e^{\ln x} > e^{-1} \Rightarrow x > e^{-1} \Rightarrow x \in (1/e, \infty)$

38. (a) $2 < \ln x < 9 \Rightarrow e^2 < e^{\ln x} < e^9 \Rightarrow e^2 < x < e^9 \Rightarrow x \in \left(e^2, e^9\right)$

(b) $e^{2-3x} > 4 \Rightarrow \ln e^{2-3x} > \ln 4 \Rightarrow 2 - 3x > \ln 4 \Rightarrow -3x > \ln 4 - 2 \Rightarrow x < -\frac{1}{3}(\ln 4 - 2) \Rightarrow$

$x \in \left(-\infty, \frac{1}{3}(2 - \ln 4)\right)$

39. 3 ft $= 36$ in, so we need x such that $\log_2 x = 36 \iff x = 2^{36} = 68{,}719{,}476{,}736$. In miles, this is

$68{,}719{,}476{,}736$ in $\cdot \dfrac{1 \text{ ft}}{12 \text{ in}} \cdot \dfrac{1 \text{ mi}}{5280 \text{ ft}} \approx 1{,}084{,}587.7$ mi.

40. (a) $v(t) = ce^{-kt} \Rightarrow a(t) = v'(t) = -kce^{-kt} = -kv(t)$

(b) $v(0) = ce^0 = c$, so c is the initial velocity.

(c) $v(t) = ce^{-kt} = c/2 \Rightarrow e^{-kt} = \frac{1}{2} \Rightarrow -kt = \ln\frac{1}{2} = -\ln 2 \Rightarrow t = (\ln 2)/k$

41. If I is the intensity of the 1989 San Francisco earthquake, then $\log_{10}(I/S) = 7.1 \Rightarrow$

$\log_{10}(16I/S) = \log_{10} 16 + \log_{10}(I/S) = \log_{10} 16 + 7.1 \approx 8.3.$

42. Let I_1 and I_2 be the intensities of the music and the mower. Then $10\log_{10}\left(\dfrac{I_1}{I_0}\right)=120$ and $10\log_{10}\left(\dfrac{I_2}{I_0}\right)=106$, so

$$\log_{10}\left(\frac{I_1}{I_2}\right)=\log_{10}\left(\frac{I_1/I_0}{I_2/I_0}\right)=\log_{10}\left(\frac{I_1}{I_0}\right)-\log_{10}\left(\frac{I_2}{I_0}\right)=12-10.6=1.4\ \Rightarrow\ \frac{I_1}{I_2}=10^{1.4}\approx25.$$

43. (a) $n=100\cdot2^{t/3}\ \Rightarrow\ \dfrac{n}{100}=2^{t/3}\ \Rightarrow\ \log_2\left(\dfrac{n}{100}\right)=\dfrac{t}{3}\ \Rightarrow\ t=3\log_2\left(\dfrac{n}{100}\right)$. Using formula (7), we can write this

as $t=f^{-1}(n)=3\cdot\dfrac{\ln(n/100)}{\ln 2}$. This function tells us how long it will take to obtain n bacteria (given the number n).

(b) $n=50{,}000\ \Rightarrow\ t=f^{-1}(50{,}000)=3\cdot\dfrac{\ln\left(\frac{50{,}000}{100}\right)}{\ln 2}=3\left(\dfrac{\ln 500}{\ln 2}\right)\approx26.9$ hours

44. (a) $Q=Q_0(1-e^{-t/a})\ \Rightarrow\ \dfrac{Q}{Q_0}=1-e^{-t/a}\ \Rightarrow\ e^{-t/a}=1-\dfrac{Q}{Q_0}\ \Rightarrow\ -\dfrac{t}{a}=\ln\left(1-\dfrac{Q}{Q_0}\right)\ \Rightarrow$

$t=-a\ln(1-Q/Q_0)$. This gives us the time t necessary to obtain a given charge Q.

(b) $Q=0.9Q_0$ and $a=2\ \Rightarrow\ t=-2\ln(1-0.9(Q_0/Q_0))=-2\ln 0.1\approx4.6$ seconds.

45. Let $t=x^2-9$. Then as $x\to3^+$, $t\to0^+$, and $\lim\limits_{x\to3^+}\ln(x^2-9)=\lim\limits_{t\to0^+}\ln t=-\infty$ by (8).

46. As $x\to2^-$, $8x-x^4=x(8-x^3)\to0^+$ since x is positive and $8-x^3\to0^+$. Thus, $\lim\limits_{x\to2^-}\log_5(8x-x^4)=-\infty$.

47. $\lim\limits_{x\to0}\ln(\cos x)=\ln 1=0$. [$\ln(\cos x)$ is continuous at $x=0$ since it is the composite of two continuous functions.]

48. $\lim\limits_{x\to0^+}\ln(\sin x)=-\infty$ since $\sin x\to0^+$ as $x\to0^+$.

49. $\lim\limits_{x\to\infty}[\ln(1+x^2)-\ln(1+x)]=\lim\limits_{x\to\infty}\ln\dfrac{1+x^2}{1+x}=\ln\left(\lim\limits_{x\to\infty}\dfrac{1+x^2}{1+x}\right)=\ln\left(\lim\limits_{x\to\infty}\dfrac{\frac1x+x}{\frac1x+1}\right)=\infty$, since the limit in

parentheses is ∞.

50. $\lim\limits_{x\to\infty}[\ln(2+x)-\ln(1+x)]=\lim\limits_{x\to\infty}\ln\left(\dfrac{2+x}{1+x}\right)=\lim\limits_{x\to\infty}\ln\left(\dfrac{2/x+1}{1/x+1}\right)=\ln\dfrac11=\ln 1=0$

51. $f(x)=\log_{10}(x^2-9)$. $D_f=\left\{x\mid x^2-9>0\right\}=\left\{x\mid |x|>3\right\}=(-\infty,-3)\cup(3,\infty)$

52. $f(x)=\ln x+\ln(2-x)$. $D_f=\{x\mid x>0\text{ and }2-x>0\}=\{x\mid x>0\text{ and }x<2\}=(0,2)$

53. (a) For $f(x)=\sqrt{3-e^{2x}}$, we must have $3-e^{2x}\geq0\ \Rightarrow\ e^{2x}\leq3\ \Rightarrow\ 2x\leq\ln 3\ \Rightarrow\ x\leq\frac12\ln 3$.

Thus, the domain of f is $(-\infty,\frac12\ln 3]$.

(b) $y=f(x)=\sqrt{3-e^{2x}}$ [note that $y\geq0$] $\Rightarrow\ y^2=3-e^{2x}\ \Rightarrow\ e^{2x}=3-y^2\ \Rightarrow\ 2x=\ln(3-y^2)\ \Rightarrow$

$x=\frac12\ln(3-y^2)$. Interchange x and y: $y=\frac12\ln(3-x^2)$. So $f^{-1}(x)=\frac12\ln(3-x^2)$. For the domain of f^{-1}, we must

have $3-x^2>0\ \Rightarrow\ x^2<3\ \Rightarrow\ |x|<\sqrt3\ \Rightarrow\ -\sqrt3<x<\sqrt3\ \Rightarrow\ 0\leq x<\sqrt3$ since $x\geq0$. Note that the

domain of f^{-1}, $[0,\sqrt3)$, equals the range of f.

54. (a) For $f(x)=\ln(2+\ln x)$, we must have $2+\ln x>0\ \Rightarrow\ \ln x>-2\ \Rightarrow\ x>e^{-2}$. Thus, the domain of f

is (e^{-2},∞).

(6) $y = f(x) = \ln(2 + \ln x)$ ⇒ $e^y = 2 + \ln x$ ⇒ $\ln x = e^y - 2$ ⇒ $x = e^{e^y - 2}$. Interchange x and y: $y = e^{e^x - 2}$.

So $f^{-1}(x) = e^{e^x - 2}$. The domain of f^{-1}, as well as the range of f, is $\mathbb{R}$.

55. $y = \ln(x + 3)$ ⇒ $e^y = e^{\ln(x+3)} = x + 3$ ⇒ $x = e^y - 3$.

Interchange x and y: the inverse function is $y = e^x - 3$.

56. $y = 2^{10^x}$ ⇒ $\log_2 y = 10^x$ ⇒ $\log_{10}(\log_2 y) = x$.

Interchange x and y: $y = \log_{10}(\log_2 x)$ is the inverse function.

57. $y = f(x) = e^{x^3}$ ⇒ $\ln y = x^3$ ⇒ $x = \sqrt[3]{\ln y}$. Interchange x and y: $y = \sqrt[3]{\ln x}$. So $f^{-1}(x) = \sqrt[3]{\ln x}$.

58. $y = (\ln x)^2$, $x \geq 1$, $\ln x = \sqrt{y}$ ⇒ $x = e^{\sqrt{y}}$. Interchange x and y: $y = e^{\sqrt{x}}$ is the inverse function.

59. $y = \log_{10}\left(1 + \dfrac{1}{x}\right)$ ⇒ $10^y = 1 + \dfrac{1}{x}$ ⇒ $\dfrac{1}{x} = 10^y - 1$ ⇒ $x = \dfrac{1}{10^y - 1}$.

Interchange x and y: $y = \dfrac{1}{10^x - 1}$ is the inverse function.

60. $y = f(x) = \dfrac{e^x}{1 + 2e^x}$ ⇒ $y + 2ye^x = e^x$ ⇒ $y = e^x - 2ye^x$ ⇒ $y = e^x(1 - 2y)$ ⇒ $e^x = \dfrac{y}{1 - 2y}$ ⇒

$x = \ln\left(\dfrac{y}{1 - 2y}\right)$. Interchange x and y: $y = \ln\left(\dfrac{x}{1 - 2x}\right)$. So $f^{-1}(x) = \ln\left(\dfrac{x}{1 - 2x}\right)$. Note that the range of f and the

domain of f^{-1} is $(0, \frac{1}{2})$.

61. $f(x) = e^{3x} - e^x$ ⇒ $f'(x) = 3e^{3x} - e^x$. Thus, $f'(x) > 0$ ⇔ $3e^{3x} > e^x$ ⇔ $\dfrac{3e^{3x}}{e^x} > \dfrac{e^x}{e^x}$ ⇔ $3e^{2x} > 1$ ⇔

$e^{2x} > \frac{1}{3}$ ⇔ $2x > \ln\left(\frac{1}{3}\right) = -\ln 3$ ⇔ $x > -\frac{1}{2}\ln 3$, so f is increasing on $\left(-\frac{1}{2}\ln 3, \infty\right)$.

62. $y = 2e^x - e^{-3x}$ ⇒ $y' = 2e^x + 3e^{-3x}$ ⇒ $y'' = 2e^x - 9e^{-3x}$. Thus, $y'' < 0$ ⇔ $2e^x < 9e^{-3x}$ ⇔

$e^{4x} < \frac{9}{2}$ ⇔ $4x < \ln\frac{9}{2}$ ⇔ $x < \frac{1}{4}\ln\frac{9}{2}$, so f is concave downward on $\left(-\infty, \frac{1}{4}\ln\frac{9}{2}\right)$.

63. $y = f(x) = (x^2 - 2)e^{-x}$ ⇒ $y' = (x^2 - 2)(-e^{-x}) + e^{-x}(2x) = e^{-x}(-x^2 + 2x + 2)$ ⇒

$y'' = e^{-x}(-2x + 2) + (-x^2 + 2x + 2)(-e^{-x}) = e^{-x}(x^2 - 4x) = xe^{-x}(x - 4)$.

$y'' > 0$ ⇒ $x < 0$ or $x > 4$ ⇒ f is CU on $(-\infty, 0)$ and $(4, \infty)$.

64.

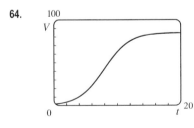

From the graph, we estimate that the most rapid increase in the percentage of households in the United States with at least one VCR occurs at about $t = 8$. To maximize the first derivative, we need to determine the values for which the second derivative is 0. We'll use $V(t) = \dfrac{a}{1 + be^{ct}}$, and substitute $a = 85$, $b = 53$, and $c = -0.5$ later.

$V'(t) = -\dfrac{a(bce^{ct})}{(1 + be^{ct})^2}$ [by the Reciprocal Rule] and

$$V''(t) = -abc \cdot \frac{\left(1 + be^{ct}\right)^2 \cdot ce^{ct} - e^{ct} \cdot 2(1 + be^{ct}) \cdot bce^{ct}}{\left[(1 + be^{ct})^2\right]^2} = \frac{-abc \cdot ce^{ct}(1 + be^{ct})[(1 + be^{ct}) - 2be^{ct}]}{(1 + be^{ct})^4}$$

$$= \frac{-abc^2 e^{ct}(1 - be^{ct})}{(1 + be^{ct})^3}$$

So $V''(t) = 0 \iff 1 = be^{ct} \iff e^{ct} = 1/b \iff ct = \ln(1/b) \iff t = (1/c)\ln(1/b) = -2\ln\frac{1}{53} \approx 7.94$ years, which corresponds to roughly midyear 1988.

65. (a) We have to show that $-f(x) = f(-x)$.

$$-f(x) = -\ln\left(x + \sqrt{x^2 + 1}\right) = \ln\left(\left(x + \sqrt{x^2 + 1}\right)^{-1}\right) = \ln\frac{1}{x + \sqrt{x^2 + 1}}$$

$$= \ln\left(\frac{1}{x + \sqrt{x^2 + 1}} \cdot \frac{x - \sqrt{x^2 + 1}}{x - \sqrt{x^2 + 1}}\right) = \ln\frac{x - \sqrt{x^2 + 1}}{x^2 - x^2 - 1} = \ln\left(\sqrt{x^2 + 1} - x\right) = f(-x)$$

Thus, f is an odd function.

(b) Let $y = \ln\left(x + \sqrt{x^2 + 1}\right)$. Then $e^y = x + \sqrt{x^2 + 1} \iff (e^y - x)^2 = x^2 + 1 \iff e^{2y} - 2xe^y + x^2 = x^2 + 1 \iff$

$2xe^y = e^{2y} - 1 \iff x = \dfrac{e^{2y} - 1}{2e^y} = \frac{1}{2}(e^y - e^{-y})$. Thus, the inverse function is $f^{-1}(x) = \frac{1}{2}(e^x - e^{-x})$.

66. Let (a, e^{-a}) be the point where the tangent meets the curve. The tangent has slope $-e^{-a}$ and is perpendicular to the line $2x - y = 8$, which has slope 2. So $-e^{-a} = -\frac{1}{2} \Rightarrow e^{-a} = \frac{1}{2} \Rightarrow e^a = 2 \Rightarrow a = \ln(e^a) = \ln 2$. Thus, the point on the curve is $\left(\ln 2, \frac{1}{2}\right)$ and the equation of the tangent is $y - \frac{1}{2} = -\frac{1}{2}(x - \ln 2)$ or $x + 2y = 1 + \ln 2$.

67. $x^{1/\ln x} = 2 \Rightarrow \ln(x^{1/\ln x}) = \ln(2) \Rightarrow \dfrac{1}{\ln x} \cdot \ln x = \ln 2 \Rightarrow 1 = \ln 2$, a contradiction, so the given equation has no

solution. The function $f(x) = x^{1/\ln x} = (e^{\ln x})^{1/\ln x} = e^1 = e$ for all $x > 0$, so the function $f(x) = x^{1/\ln x}$ is the constant function $f(x) = e$.

68. (a) $\lim\limits_{x \to \infty} x^{\ln x} = \lim\limits_{x \to \infty} \left(e^{\ln x}\right)^{\ln x} = \lim\limits_{x \to \infty} e^{(\ln x)^2} = \infty$ since $(\ln x)^2 \to \infty$ as $x \to \infty$.

(b) $\lim\limits_{x \to 0^+} x^{-\ln x} = \lim\limits_{x \to 0^+} \left(e^{\ln x}\right)^{-\ln x} = \lim\limits_{x \to 0^+} e^{-(\ln x)^2} = 0$ since $-(\ln x)^2 \to -\infty$ as $x \to 0^+$.

(c) $\lim\limits_{x \to 0^+} x^{1/x} = \lim\limits_{x \to 0^+} \left(e^{\ln x}\right)^{1/x} = \lim\limits_{x \to 0^+} e^{(\ln x/x)} = 0$ since $\dfrac{\ln x}{x} \to -\infty$ as $x \to 0^+$. Note that as $x \to 0^+$, $\ln x$ is a large negative number and x is a small positive number, so $(\ln x)/x \to -\infty$.

(d) $\lim\limits_{x \to \infty} (\ln 2x)^{-\ln x} = \lim\limits_{x \to \infty} \left[e^{\ln(\ln 2x)}\right]^{-\ln x} = \lim\limits_{x \to \infty} e^{-\ln x \ln(\ln 2x)} = 0$ since $-\ln x \ln(\ln 2x) \to -\infty$ as $x \to \infty$.

69. (a) Let $\varepsilon > 0$ be given. We need N such that $|a^x - 0| < \varepsilon$ when $x < N$. But $a^x < \varepsilon \iff x < \log_a \varepsilon$. Let $N = \log_a \varepsilon$.
Then $x < N \Rightarrow x < \log_a \varepsilon \Rightarrow |a^x - 0| = a^x < \varepsilon$, so $\lim\limits_{x \to -\infty} a^x = 0$.

(b) Let $M > 0$ be given. We need N such that $a^x > M$ when $x > N$. But $a^x > M \iff x > \log_a M$. Let $N = \log_a M$.
Then $x > N \Rightarrow x > \log_a M \Rightarrow a^x > M$, so $\lim\limits_{x \to \infty} a^x = \infty$.

70. (a)

 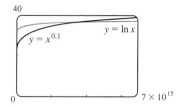

From the graphs, we see that $f(x) = x^{0.1} > g(x) = \ln x$ for approximately $0 < x < 3.06$, and then $g(x) > f(x)$ for

$3.06 < x < 3.43 \times 10^{15}$ (approximately). At that point, the graph of f finally surpasses the graph of g for good.

(b)

(c) From the graph at left, it seems that $\dfrac{\ln x}{x^{0.1}} < 0.1$ whenever

$x > 1.3 \times 10^{28}$ (approximately). So we can take $N = 1.3 \times 10^{28}$,

or any larger number.

71. $\ln(x^2 - 2x - 2) \le 0 \Rightarrow 0 < x^2 - 2x - 2 \le 1$. Now $x^2 - 2x - 2 \le 1$ gives $x^2 - 2x - 3 \le 0$ and hence

$(x - 3)(x + 1) \le 0$. So $-1 \le x \le 3$. Now $0 < x^2 - 2x - 2 \Rightarrow x < 1 - \sqrt{3}$ or $x > 1 + \sqrt{3}$. Therefore,

$\ln(x^2 - 2x - 2) \le 0 \Leftrightarrow -1 \le x < 1 - \sqrt{3}$ or $1 + \sqrt{3} < x \le 3$.

72. (a) The primes less than 25 are 2, 3, 5, 7, 11, 13, 17, 19, and 23. There

are 9 of them, so $\pi(25) = 9$. We use the sieve of Eratosthenes, and

arrive at the figure at right. There are 25 numbers left over, so

$\pi(100) = 25$.

(b) Let $f(n) = \dfrac{\pi(n)}{n/\ln n}$. We compute $f(100) = \dfrac{25}{100/\ln 100} \approx 1.15$,

$f(1000) \approx 1.16$, $f(10^4) \approx 1.13$, $f(10^5) \approx 1.10$, $f(10^6) \approx 1.08$,

and $f(10^7) \approx 1.07$.

(c) By the Prime Number Theorem, the number of primes less than a billion, that is, $\pi(10^9)$, should be close to

$10^9/\ln 10^9 \approx 48{,}254{,}942$. In fact, $\pi(10^9) = 50{,}847{,}543$, so our estimate is off by about 5.1%. Do not attempt this

calculation at home.

7.4 Derivatives of Logarithmic Functions

1. The differentiation formula for logarithmic functions, $\dfrac{d}{dx}(\log_a x) = \dfrac{1}{x \ln a}$, is simplest when $a = e$ because $\ln e = 1$.

2. $f(x) = \ln(x^2 + 10) \Rightarrow f'(x) = \dfrac{1}{x^2 + 10}\dfrac{d}{dx}(x^2 + 10) = \dfrac{2x}{x^2 + 10}$

3. $f(x) = \sin(\ln x) \Rightarrow f'(x) = \cos(\ln x) \cdot \dfrac{d}{dx}\ln x = \cos(\ln x) \cdot \dfrac{1}{x} = \dfrac{\cos(\ln x)}{x}$

4. $f(x) = \ln(\sin^2 x) = \ln(\sin x)^2 = 2\ln|\sin x| \Rightarrow f'(x) = 2 \cdot \dfrac{1}{\sin x} \cdot \cos x = 2\cot x$

5. $f(x) = \log_2(1 - 3x) \;\Rightarrow\; f'(x) = \dfrac{1}{(1 - 3x)\ln 2}\dfrac{d}{dx}(1 - 3x) = \dfrac{-3}{(1 - 3x)\ln 2}$ or $\dfrac{3}{(3x - 1)\ln 2}$

6. $f(x) = \log_5(xe^x) \;\Rightarrow\; f'(x) = \dfrac{1}{xe^x\ln 5}\dfrac{d}{dx}(xe^x) = \dfrac{1}{xe^x\ln 5}(xe^x + e^x \cdot 1) = \dfrac{e^x(x + 1)}{xe^x\ln 5} = \dfrac{x + 1}{x\ln 5}$

Another solution: We can change the form of the function by first using logarithm properties.

$f(x) = \log_5(xe^x) = \log_5 x + \log_5 e^x \;\Rightarrow\; f'(x) = \dfrac{1}{x\ln 5} + \dfrac{1}{e^x\ln 5}\cdot e^x = \dfrac{1}{x\ln 5} + \dfrac{1}{\ln 5}$ or $\dfrac{1 + x}{x\ln 5}$

7. $f(x) = \sqrt[5]{\ln x} = (\ln x)^{1/5} \;\Rightarrow\; f'(x) = \tfrac{1}{5}(\ln x)^{-4/5}\dfrac{d}{dx}(\ln x) = \dfrac{1}{5(\ln x)^{4/5}}\cdot\dfrac{1}{x} = \dfrac{1}{5x\sqrt[5]{(\ln x)^4}}$

8. $f(x) = \ln\sqrt[5]{x} = \ln x^{1/5} = \tfrac{1}{5}\ln x \;\Rightarrow\; f'(x) = \dfrac{1}{5}\cdot\dfrac{1}{x} = \dfrac{1}{5x}$

9. $f(x) = \sin x \ln(5x) \;\Rightarrow\; f'(x) = \sin x \cdot \dfrac{1}{5x}\cdot\dfrac{d}{dx}(5x) + \ln(5x)\cdot\cos x = \dfrac{\sin x \cdot 5}{5x} + \cos x \ln(5x) = \dfrac{\sin x}{x} + \cos x \ln(5x)$

10. $f(t) = \dfrac{1 + \ln t}{1 - \ln t} \;\Rightarrow\; f'(t) = \dfrac{(1 - \ln t)(1/t) - (1 + \ln t)(-1/t)}{(1 - \ln t)^2} = \dfrac{(1/t)[(1 - \ln t) + (1 + \ln t)]}{(1 - \ln t)^2} = \dfrac{2}{t(1 - \ln t)^2}$

11. $F(t) = \ln\dfrac{(2t + 1)^3}{(3t - 1)^4} = \ln(2t + 1)^3 - \ln(3t - 1)^4 = 3\ln(2t + 1) - 4\ln(3t - 1) \;\Rightarrow$

$F'(t) = 3\cdot\dfrac{1}{2t + 1}\cdot 2 - 4\cdot\dfrac{1}{3t - 1}\cdot 3 = \dfrac{6}{2t + 1} - \dfrac{12}{3t - 1}$, or combined, $\dfrac{-6(t + 3)}{(2t + 1)(3t - 1)}$.

12. $h(x) = \ln\big(x + \sqrt{x^2 - 1}\big) \;\Rightarrow\; h'(x) = \dfrac{1}{x + \sqrt{x^2 - 1}}\left(1 + \dfrac{x}{\sqrt{x^2 - 1}}\right) = \dfrac{1}{x + \sqrt{x^2 - 1}}\cdot\dfrac{\sqrt{x^2 - 1} + x}{\sqrt{x^2 - 1}} = \dfrac{1}{\sqrt{x^2 - 1}}$

13. $g(x) = \ln\big(x\sqrt{x^2 - 1}\big) = \ln x + \ln(x^2 - 1)^{1/2} = \ln x + \tfrac{1}{2}\ln(x^2 - 1) \;\Rightarrow$

$g'(x) = \dfrac{1}{x} + \dfrac{1}{2}\cdot\dfrac{1}{x^2 - 1}\cdot 2x = \dfrac{1}{x} + \dfrac{x}{x^2 - 1} = \dfrac{x^2 - 1 + x\cdot x}{x(x^2 - 1)} = \dfrac{2x^2 - 1}{x(x^2 - 1)}$

14. $F(y) = y\ln(1 + e^y) \;\Rightarrow\; F'(y) = y\cdot\dfrac{1}{1 + e^y}\cdot e^y + \ln(1 + e^y)\cdot 1 = \dfrac{ye^y}{1 + e^y} + \ln(1 + e^y)$

15. $f(u) = \dfrac{\ln u}{1 + \ln(2u)} \;\Rightarrow$

$f'(u) = \dfrac{[1 + \ln(2u)]\cdot\frac{1}{u} - \ln u\cdot\frac{1}{2u}\cdot 2}{[1 + \ln(2u)]^2} = \dfrac{\frac{1}{u}[1 + \ln(2u) - \ln u]}{[1 + \ln(2u)]^2} = \dfrac{1 + (\ln 2 + \ln u) - \ln u}{u[1 + \ln(2u)]^2} = \dfrac{1 + \ln 2}{u[1 + \ln(2u)]^2}$

16. $y = \ln(x^4\sin^2 x) = \ln x^4 + \ln(\sin x)^2 = 4\ln x + 2\ln\sin x \;\Rightarrow\; y' = 4\cdot\dfrac{1}{x} + 2\cdot\dfrac{1}{\sin x}\cdot\cos x = \dfrac{4}{x} + 2\cot x$

17. $h(t) = t^3 - 3^t \;\Rightarrow\; h'(t) = 3t^2 - 3^t\ln 3$

18. $y = 10^{\tan\theta} \;\Rightarrow\; y' = 10^{\tan\theta}(\ln 10)(\sec^2\theta)$

19. $y = \ln\big|2 - x - 5x^2\big| \;\Rightarrow\; y' = \dfrac{1}{2 - x - 5x^2}\cdot(-1 - 10x) = \dfrac{-10x - 1}{2 - x - 5x^2}$ or $\dfrac{10x + 1}{5x^2 + x - 2}$

20. $H(z) = \ln \sqrt{\dfrac{a^2 - z^2}{a^2 + z^2}} = \ln \left(\dfrac{a^2 - z^2}{a^2 + z^2}\right)^{1/2} = \dfrac{1}{2} \ln\left(\dfrac{a^2 - z^2}{a^2 + z^2}\right) = \frac{1}{2}\ln(a^2 - z^2) - \frac{1}{2}\ln(a^2 + z^2) \quad \Rightarrow$

$H'(z) = \dfrac{1}{2} \cdot \dfrac{1}{a^2 - z^2} \cdot (-2z) - \dfrac{1}{2} \cdot \dfrac{1}{a^2 + z^2} \cdot (2z) = \dfrac{z}{z^2 - a^2} - \dfrac{z}{z^2 + a^2} = \dfrac{z(z^2 + a^2) - z(z^2 - a^2)}{(z^2 - a^2)(z^2 + a^2)}$

$= \dfrac{z^3 + za^2 - z^3 + za^2}{(z^2 - a^2)(z^2 + a^2)} = \dfrac{2a^2 z}{z^4 - a^4}$

21. $y = \ln(e^{-x} + xe^{-x}) = \ln(e^{-x}(1 + x)) = \ln(e^{-x}) + \ln(1 + x) = -x + \ln(1 + x) \quad \Rightarrow$

$y' = -1 + \dfrac{1}{1 + x} = \dfrac{-1 - x + 1}{1 + x} = -\dfrac{x}{1 + x}$

22. $y = [\ln(1 + e^x)]^2 \quad \Rightarrow \quad y' = 2[\ln(1 + e^x)] \cdot \dfrac{1}{1 + e^x} \cdot e^x = \dfrac{2e^x \ln(1 + e^x)}{1 + e^x}$

23. $y = 2x \log_{10} \sqrt{x} = 2x \log_{10} x^{1/2} = 2x \cdot \frac{1}{2} \log_{10} x = x \log_{10} x \quad \Rightarrow \quad y' = x \cdot \dfrac{1}{x \ln 10} + \log_{10} x \cdot 1 = \dfrac{1}{\ln 10} + \log_{10} x$

Note: $\dfrac{1}{\ln 10} = \dfrac{\ln e}{\ln 10} = \log_{10} e$, so the answer could be written as $\dfrac{1}{\ln 10} + \log_{10} x = \log_{10} e + \log_{10} x = \log_{10} ex.$

24. $y = \log_2(e^{-x} \cos \pi x) = \log_2 e^{-x} + \log_2 \cos \pi x = -x \log_2 e + \log_2 \cos \pi x \quad \Rightarrow$

$y' = -\log_2 e + \dfrac{1}{\cos \pi x \, (\ln 2)} \dfrac{d}{dx}(\cos \pi x) = -\log_2 e + \dfrac{-\pi \sin \pi x}{\cos \pi x \, (\ln 2)} = -\log_2 e - \dfrac{\pi}{\ln 2} \tan \pi x$

Note: $\dfrac{1}{\ln 2} = \dfrac{\ln e}{\ln 2} = \log_2 e$, so the answer could be written as $-\log_2 e - \pi \log_2 e \tan \pi x = (-\log_2 e)(1 + \pi \tan \pi x).$

25. Using Formula 7 and the Chain Rule, $y = 5^{-1/x} \quad \Rightarrow \quad y' = 5^{-1/x}(\ln 5)\left[-1 \cdot (-x^{-2})\right] = 5^{-1/x}(\ln 5)/x^2$

26. $y = 2^{3^{x^2}} \quad \Rightarrow \quad y' = 2^{3^{x^2}}(\ln 2) \dfrac{d}{dx}\left(3^{x^2}\right) = 2^{3^{x^2}}(\ln 2)3^{x^2}(\ln 3)(2x)$

27. $y = x^2 \ln(2x) \quad \Rightarrow \quad y' = x^2 \cdot \dfrac{1}{2x} \cdot 2 + \ln(2x) \cdot (2x) = x + 2x \ln(2x) \quad \Rightarrow$

$y'' = 1 + 2x \cdot \dfrac{1}{2x} \cdot 2 + \ln(2x) \cdot 2 = 1 + 2 + 2\ln(2x) = 3 + 2\ln(2x)$

28. $y = \dfrac{\ln x}{x^2} \quad \Rightarrow \quad y' = \dfrac{x^2(1/x) - (\ln x)(2x)}{(x^2)^2} = \dfrac{x(1 - 2\ln x)}{x^4} = \dfrac{1 - 2\ln x}{x^3} \quad \Rightarrow$

$y'' = \dfrac{x^3(-2/x) - (1 - 2\ln x)(3x^2)}{(x^3)^2} = \dfrac{x^2(-2 - 3 + 6\ln x)}{x^6} = \dfrac{6\ln x - 5}{x^4}$

29. $y = \ln\left(x + \sqrt{1 + x^2}\right) \quad \Rightarrow$

$y' = \dfrac{1}{x + \sqrt{1 + x^2}} \dfrac{d}{dx}\left(x + \sqrt{1 + x^2}\right) = \dfrac{1}{x + \sqrt{1 + x^2}}\left[1 + \tfrac{1}{2}(1 + x^2)^{-1/2}(2x)\right]$

$= \dfrac{1}{x + \sqrt{1 + x^2}}\left(1 + \dfrac{x}{\sqrt{1 + x^2}}\right) = \dfrac{1}{x + \sqrt{1 + x^2}} \cdot \dfrac{\sqrt{1 + x^2} + x}{\sqrt{1 + x^2}} = \dfrac{1}{\sqrt{1 + x^2}} \quad \Rightarrow$

$y'' = -\tfrac{1}{2}(1 + x^2)^{-3/2}(2x) = \dfrac{-x}{(1 + x^2)^{3/2}}$

30. $y = \ln(\sec x + \tan x) \;\Rightarrow\; y' = \dfrac{\sec x \tan x + \sec^2 x}{\sec x + \tan x} = \sec x \;\Rightarrow\; y'' = \sec x \tan x$

31. $f(x) = \dfrac{x}{1 - \ln(x - 1)} \;\Rightarrow$

$$f'(x) = \frac{[1 - \ln(x - 1)] \cdot 1 - x \cdot \dfrac{-1}{x - 1}}{[1 - \ln(x - 1)]^2} = \frac{\dfrac{(x - 1)[1 - \ln(x - 1)] + x}{x - 1}}{[1 - \ln(x - 1)]^2} = \frac{x - 1 - (x - 1)\ln(x - 1) + x}{(x - 1)[1 - \ln(x - 1)]^2}$$

$$= \frac{2x - 1 - (x - 1)\ln(x - 1)}{(x - 1)[1 - \ln(x - 1)]^2}$$

$\text{Dom}(f) = \{x \mid x - 1 > 0 \quad\text{and}\quad 1 - \ln(x - 1) \neq 0\} = \{x \mid x > 1 \quad\text{and}\quad \ln(x - 1) \neq 1\}$

$\qquad = \{x \mid x > 1 \quad\text{and}\quad x - 1 \neq e^1\} = \{x \mid x > 1 \quad\text{and}\quad x \neq 1 + e\} = (1, 1 + e) \cup (1 + e, \infty)$

32. $f(x) = \dfrac{1}{1 + \ln x} \;\Rightarrow\; f'(x) = -\dfrac{1/x}{(1 + \ln x)^2}$ [Reciprocal Rule] $= -\dfrac{1}{x(1 + \ln x)^2}.$

$\text{Dom}(f) = \{x \mid x > 0 \quad\text{and}\quad \ln x \neq -1\} = \{x \mid x > 0 \quad\text{and}\quad x \neq 1/e\} = (0, 1/e) \cup (1/e, \infty).$

33. $f(x) = \ln(x^2 - 2x) \;\Rightarrow\; f'(x) = \dfrac{1}{x^2 - 2x}(2x - 2) = \dfrac{2(x - 1)}{x(x - 2)}.$

$\text{Dom}(f) = \{x \mid x(x - 2) > 0\} = (-\infty, 0) \cup (2, \infty).$

34. $f(x) = \ln \ln \ln x \;\Rightarrow\; f'(x) = \dfrac{1}{\ln \ln x} \cdot \dfrac{1}{\ln x} \cdot \dfrac{1}{x}.$

$\text{Dom}(f) = \{x \mid \ln \ln x > 0\} = \{x \mid \ln x > 1\} = \{x \mid x > e\} = (e, \infty).$

35. $f(x) = \dfrac{\ln x}{1 + x^2} \;\Rightarrow\; f'(x) = \dfrac{(1 + x^2)\left(\dfrac{1}{x}\right) - (\ln x)(2x)}{(1 + x^2)^2},$ so $f'(1) = \dfrac{2(1) - 0(2)}{2^2} = \dfrac{2}{4} = \dfrac{1}{2}.$

36. $f(x) = \ln(1 + e^{2x}) \;\Rightarrow\; f'(x) = \dfrac{1}{1 + e^{2x}}(2e^{2x}) = \dfrac{2e^{2x}}{1 + e^{2x}},$ so $f'(0) = \dfrac{2e^0}{1 + e^0} = \dfrac{2(1)}{1 + 1} = 1.$

37. $y = \ln\left(xe^{x^2}\right) = \ln x + \ln e^{x^2} = \ln x + x^2 \;\Rightarrow\; y' = \dfrac{1}{x} + 2x.$ At $(1, 1),$ the slope of the tangent line is

$y'(1) = 1 + 2 = 3,$ and an equation of the tangent line is $y - 1 = 3(x - 1),$ or $y = 3x - 2.$

38. $y = \ln(x^3 - 7) \;\Rightarrow\; y' = \dfrac{1}{x^3 - 7} \cdot 3x^2 \;\Rightarrow\; y'(2) = \dfrac{12}{8 - 7} = 12,$ so an equation of a tangent line at $(2, 0)$ is

$y - 0 = 12(x - 2)$ or $y = 12x - 24.$

39. $f(x) = \sin x + \ln x \;\Rightarrow\; f'(x) = \cos x + 1/x.$

This is reasonable, because the graph shows that f increases when f' is

positive, and $f'(x) = 0$ when f has a horizontal tangent.

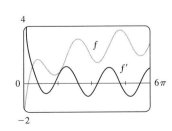

40. $y = \dfrac{\ln x}{x}$ $\Rightarrow$ $y' = \dfrac{x(1/x) - \ln x}{x^2} = \dfrac{1 - \ln x}{x^2}$.

$y'(1) = \dfrac{1 - 0}{1^2} = 1$ and $y'(e) = \dfrac{1 - 1}{e^2} = 0$ $\Rightarrow$ equations of tangent

lines are $y - 0 = 1(x - 1)$ or $y = x - 1$ and $y - 1/e = 0(x - e)$

or $y = 1/e$.

41. $y = (2x + 1)^5 (x^4 - 3)^6$ $\Rightarrow$ $\ln y = \ln\big((2x + 1)^5 (x^4 - 3)^6\big)$ $\Rightarrow$ $\ln y = 5 \ln(2x + 1) + 6 \ln(x^4 - 3)$ $\Rightarrow$

$\dfrac{1}{y} y' = 5 \cdot \dfrac{1}{2x + 1} \cdot 2 + 6 \cdot \dfrac{1}{x^4 - 3} \cdot 4x^3$ $\Rightarrow$

$y' = y\left(\dfrac{10}{2x + 1} + \dfrac{24x^3}{x^4 - 3}\right) = (2x + 1)^5 (x^4 - 3)^6 \left(\dfrac{10}{2x + 1} + \dfrac{24x^3}{x^4 - 3}\right)$.

[The answer could be simplified to $y' = 2(2x + 1)^4 (x^4 - 3)^5 (29x^4 + 12x^3 - 15)$, but this is unnecessary.]

42. $y = \sqrt{x}\, e^{x^2} (x^2 + 1)^{10}$ $\Rightarrow$ $\ln y = \ln \sqrt{x} + \ln e^{x^2} + \ln(x^2 + 1)^{10}$ $\Rightarrow$ $\ln y = \tfrac{1}{2} \ln x + x^2 + 10 \ln(x^2 + 1)$ $\Rightarrow$

$\dfrac{1}{y} y' = \dfrac{1}{2} \cdot \dfrac{1}{x} + 2x + 10 \cdot \dfrac{1}{x^2 + 1} \cdot 2x$ $\Rightarrow$ $y' = \sqrt{x}\, e^{x^2} (x^2 + 1)^{10} \left(\dfrac{1}{2x} + 2x + \dfrac{20x}{x^2 + 1}\right)$

43. $y = \dfrac{\sin^2 x \tan^4 x}{(x^2 + 1)^2}$ $\Rightarrow$ $\ln y = \ln(\sin^2 x \tan^4 x) - \ln(x^2 + 1)^2$ $\Rightarrow$

$\ln y = \ln(\sin x)^2 + \ln(\tan x)^4 - \ln(x^2 + 1)^2$ $\Rightarrow$ $\ln y = 2 \ln|\sin x| + 4 \ln|\tan x| - 2 \ln(x^2 + 1)$ $\Rightarrow$

$\dfrac{1}{y} y' = 2 \cdot \dfrac{1}{\sin x} \cdot \cos x + 4 \cdot \dfrac{1}{\tan x} \cdot \sec^2 x - 2 \cdot \dfrac{1}{x^2 + 1} \cdot 2x$ $\Rightarrow$ $y' = \dfrac{\sin^2 x \tan^4 x}{(x^2 + 1)^2} \left(2 \cot x + \dfrac{4 \sec^2 x}{\tan x} - \dfrac{4x}{x^2 + 1}\right)$

44. $y = \sqrt[4]{\dfrac{x^2 + 1}{x^2 - 1}}$ $\Rightarrow$ $\ln y = \tfrac{1}{4} \ln(x^2 + 1) - \tfrac{1}{4} \ln(x^2 - 1)$ $\Rightarrow$ $\dfrac{1}{y} y' = \dfrac{1}{4} \cdot \dfrac{1}{x^2 + 1} \cdot 2x - \dfrac{1}{4} \cdot \dfrac{1}{x^2 - 1} \cdot 2x$ $\Rightarrow$

$y' = \sqrt[4]{\dfrac{x^2 + 1}{x^2 - 1}} \cdot \dfrac{1}{2}\left(\dfrac{x}{x^2 + 1} - \dfrac{x}{x^2 - 1}\right) = \dfrac{1}{2} \sqrt[4]{\dfrac{x^2 + 1}{x^2 - 1}} \left(\dfrac{-2x}{x^4 - 1}\right) = \dfrac{x}{1 - x^4} \sqrt[4]{\dfrac{x^2 + 1}{x^2 - 1}}$

45. $y = x^x$ $\Rightarrow$ $\ln y = \ln x^x$ $\Rightarrow$ $\ln y = x \ln x$ $\Rightarrow$ $y'/y = x(1/x) + (\ln x) \cdot 1$ $\Rightarrow$ $y' = y(1 + \ln x)$ $\Rightarrow$

$y' = x^x (1 + \ln x)$

46. $y = x^{\cos x}$ $\Rightarrow$ $\ln y = \ln x^{\cos x}$ $\Rightarrow$ $\ln y = \cos x \ln x$ $\Rightarrow$ $\dfrac{1}{y} y' = \cos x \cdot \dfrac{1}{x} + \ln x \cdot (-\sin x)$ $\Rightarrow$

$y' = y\left(\dfrac{\cos x}{x} - \ln x \sin x\right)$ $\Rightarrow$ $y' = x^{\cos x} \left(\dfrac{\cos x}{x} - \ln x \sin x\right)$

47. $y = x^{\sin x}$ $\Rightarrow$ $\ln y = \ln x^{\sin x}$ $\Rightarrow$ $\ln y = \sin x \ln x$ $\Rightarrow$ $\dfrac{y'}{y} = (\sin x) \cdot \dfrac{1}{x} + (\ln x)(\cos x)$ $\Rightarrow$

$y' = y\left(\dfrac{\sin x}{x} + \ln x \cos x\right)$ $\Rightarrow$ $y' = x^{\sin x} \left(\dfrac{\sin x}{x} + \ln x \cos x\right)$

48. $y = \sqrt{x}^{\,x}$ $\Rightarrow$ $\ln y = \ln \sqrt{x}^{\,x}$ $\Rightarrow$ $\ln y = x \ln x^{1/2}$ $\Rightarrow$ $\ln y = \tfrac{1}{2} x \ln x$ $\Rightarrow$ $\dfrac{1}{y} y' = \dfrac{1}{2} x \cdot \dfrac{1}{x} + \ln x \cdot \dfrac{1}{2}$ $\Rightarrow$

$y' = y\big(\tfrac{1}{2} + \tfrac{1}{2} \ln x\big)$ $\Rightarrow$ $y' = \tfrac{1}{2} \sqrt{x}^{\,x} (1 + \ln x)$

49. $y = (\cos x)^x$ $\Rightarrow$ $\ln y = \ln(\cos x)^x$ $\Rightarrow$ $\ln y = x \ln \cos x$ $\Rightarrow$ $\dfrac{1}{y}\, y' = x \cdot \dfrac{1}{\cos x} \cdot (-\sin x) + \ln \cos x \cdot 1$ $\Rightarrow$

$y' = y\left(\ln \cos x - \dfrac{x \sin x}{\cos x}\right)$ $\Rightarrow$ $y' = (\cos x)^x (\ln \cos x - x \tan x)$

50. $y = (\sin x)^{\ln x}$ $\Rightarrow$ $\ln y = \ln(\sin x)^{\ln x}$ $\Rightarrow$ $\ln y = \ln x \cdot \ln \sin x$ $\Rightarrow$ $\dfrac{1}{y}\, y' = \ln x \cdot \dfrac{1}{\sin x} \cdot \cos x + \ln \sin x \cdot \dfrac{1}{x}$ $\Rightarrow$

$y' = y\left(\ln x \cdot \dfrac{\cos x}{\sin x} + \dfrac{\ln \sin x}{x}\right)$ $\Rightarrow$ $y' = (\sin x)^{\ln x}\left(\ln x \cot x + \dfrac{\ln \sin x}{x}\right)$

51. $y = (\tan x)^{1/x}$ $\Rightarrow$ $\ln y = \ln(\tan x)^{1/x}$ $\Rightarrow$ $\ln y = \dfrac{1}{x} \ln \tan x$ $\Rightarrow$

$\dfrac{1}{y}\, y' = \dfrac{1}{x} \cdot \dfrac{1}{\tan x} \cdot \sec^2 x + \ln \tan x \cdot \left(-\dfrac{1}{x^2}\right)$ $\Rightarrow$ $y' = y\left(\dfrac{\sec^2 x}{x \tan x} - \dfrac{\ln \tan x}{x^2}\right)$ $\Rightarrow$

$y' = (\tan x)^{1/x}\left(\dfrac{\sec^2 x}{x \tan x} - \dfrac{\ln \tan x}{x^2}\right)$ or $y' = (\tan x)^{1/x} \cdot \dfrac{1}{x}\left(\csc x \sec x - \dfrac{\ln \tan x}{x}\right)$

52. $y = (\ln x)^{\cos x}$ $\Rightarrow$ $\ln y = \cos x \ln(\ln x)$ $\Rightarrow$ $\dfrac{y'}{y} = \cos x \cdot \dfrac{1}{\ln x} \cdot \dfrac{1}{x} + (\ln \ln x)(-\sin x)$ $\Rightarrow$

$y' = (\ln x)^{\cos x}\left(\dfrac{\cos x}{x \ln x} - \sin x \ln \ln x\right)$

53. $y = \ln(x^2 + y^2)$ $\Rightarrow$ $y' = \dfrac{1}{x^2 + y^2}\dfrac{d}{dx}(x^2 + y^2)$ $\Rightarrow$ $y' = \dfrac{2x + 2yy'}{x^2 + y^2}$ $\Rightarrow$ $x^2 y' + y^2 y' = 2x + 2yy'$ $\Rightarrow$

$x^2 y' + y^2 y' - 2yy' = 2x$ $\Rightarrow$ $(x^2 + y^2 - 2y)y' = 2x$ $\Rightarrow$ $y' = \dfrac{2x}{x^2 + y^2 - 2y}$

54. $x^y = y^x$ $\Rightarrow$ $y \ln x = x \ln y$ $\Rightarrow$ $y \cdot \dfrac{1}{x} + (\ln x) \cdot y' = x \cdot \dfrac{1}{y} \cdot y' + \ln y$ $\Rightarrow$ $y' \ln x - \dfrac{x}{y}\, y' = \ln y - \dfrac{y}{x}$ $\Rightarrow$

$y' = \dfrac{\ln y - y/x}{\ln x - x/y}$

55. $f(x) = \ln(x - 1)$ $\Rightarrow$ $f'(x) = \dfrac{1}{(x-1)} = (x-1)^{-1}$ $\Rightarrow$ $f''(x) = -(x-1)^{-2}$ $\Rightarrow$ $f'''(x) = 2(x-1)^{-3}$ $\Rightarrow$

$f^{(4)}(x) = -2 \cdot 3(x-1)^{-4}$ $\Rightarrow$ $\cdots$ $\Rightarrow$ $f^{(n)}(x) = (-1)^{n-1} \cdot 2 \cdot 3 \cdot 4 \cdot \cdots \cdot (n-1)(x-1)^{-n} = (-1)^{n-1}\dfrac{(n-1)!}{(x-1)^n}$

56. $y = x^8 \ln x$, so $D^9 y = D^8 y' = D^8(8x^7 \ln x + x^7)$. But the eighth derivative of x^7 is 0, so we now have

$D^8(8x^7 \ln x) = D^7(8 \cdot 7x^6 \ln x + 8x^6) = D^7(8 \cdot 7x^6 \ln x) = D^6(8 \cdot 7 \cdot 6x^5 \ln x) = \cdots = D(8!\, x^0 \ln x) = 8!/x.$

57.

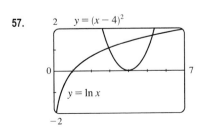

From the graph, it appears that the curves $y = (x-4)^2$ and $y = \ln x$ intersect just to the left of $x = 3$ and to the right of $x = 5$, at about $x = 5.3$. Let $f(x) = \ln x - (x-4)^2$. Then $f'(x) = 1/x - 2(x-4)$, so Newton's Method says that $x_{n+1} = x_n - f(x_n)/f'(x_n) = x_n - \dfrac{\ln x_n - (x_n - 4)^2}{1/x_n - 2(x_n - 4)}$. Taking

$x_0 = 3$, we get $x_1 \approx 2.957738$, $x_2 \approx 2.958516 \approx x_3$, so the first root is 2.958516, to six decimal places. Taking $x_0 = 5$, we get $x_1 \approx 5.290755$, $x_2 \approx 5.290718 \approx x_3$, so the second (and final) root is 5.290718, to six decimal places.

58. $y = \ln(4 - x^2)$

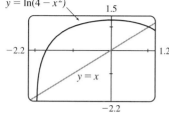

We use Newton's Method with $f(x) = \ln(4 - x^2) - x$ and

$$f'(x) = \frac{1}{4 - x^2}(-2x) - 1 = -1 - \frac{2x}{4 - x^2}.$$ The formula is

$x_{n+1} = x_n - f(x_n)/f'(x_n)$. From the graphs it seems that the roots occur at approximately $x = -1.9$ and $x = 1.1$. However, if we use $x_1 = -1.9$ as an initial approximation to the first root, we get $x_2 \approx -2.009611$, and

$f(x) = \ln(x - 2)^2 - x$ is undefined at this point, making it impossible to calculate x_3. We must use a more accurate first estimate, such as $x_1 = -1.95$. With this approximation, we get $x_1 = -1.95$, $x_2 \approx -1.1967495$, $x_3 \approx -1.964760$, $x_4 \approx x_5 \approx -1.964636$. Calculating the second root gives $x_1 = 1.1$, $x_2 \approx 1.058649$, $x_3 \approx 1.058007$, $x_4 \approx x_5 \approx 1.058006$. So, correct to six decimal places, the two roots of the equation $\ln(4 - x^2) = x$ are $x = -1.964636$ and $x = 1.058006$.

59. $f(x) = \dfrac{\ln x}{\sqrt{x}} \;\Rightarrow\; f'(x) = \dfrac{\sqrt{x}\,(1/x) - (\ln x)[1/(2\sqrt{x}\,)]}{x} = \dfrac{2 - \ln x}{2x^{3/2}} \;\Rightarrow$

$f''(x) = \dfrac{2x^{3/2}(-1/x) - (2 - \ln x)(3x^{1/2})}{4x^3} = \dfrac{3\ln x - 8}{4x^{5/2}} > 0 \;\Leftrightarrow\; \ln x > \tfrac{8}{3} \;\Leftrightarrow\; x > e^{8/3}$, so f is CU on $(e^{8/3}, \infty)$

and CD on $(0, e^{8/3})$. The inflection point is $\left(e^{8/3}, \tfrac{8}{3}e^{-4/3}\right)$.

60. $f(x) = x\ln x$, $f'(x) = \ln x + 1 = 0$ when

$\ln x = -1 \;\Leftrightarrow\; x = e^{-1}$. $f'(x) > 0 \;\Leftrightarrow\; \ln x + 1 > 0 \;\Leftrightarrow\; \ln x > -1 \;\Leftrightarrow\; x > 1/e$. $f'(x) < 0 \;\Leftrightarrow$

$\ln x + 1 < 0 \;\Leftrightarrow\; x < 1/e$. Therefore, there is an absolute minimum value of $f(1/e) = (1/e)\ln(1/e) = -1/e$.

61. $y = f(x) = \ln(\sin x)$

A. $D = \{x \text{ in } \mathbb{R} \mid \sin x > 0\} = \displaystyle\bigcup_{n=-\infty}^{\infty} (2n\pi, (2n+1)\pi) = \cdots \cup (-4\pi, -3\pi) \cup (-2\pi, -\pi) \cup (0, \pi) \cup (2\pi, 3\pi) \cup \cdots$

B. No y-intercept; x-intercepts: $f(x) = 0 \;\Leftrightarrow\; \ln(\sin x) = 0 \;\Leftrightarrow\; \sin x = e^0 = 1 \;\Leftrightarrow\; x = 2n\pi + \frac{\pi}{2}$ for each integer n.   **C.** f is periodic with period 2π.   **D.** $\displaystyle\lim_{x \to (2n\pi)^+} f(x) = -\infty$ and $\displaystyle\lim_{x \to [(2n+1)\pi]^-} f(x) = -\infty$, so the lines $x = n\pi$ are VAs for all integers n.   **E.** $f'(x) = \dfrac{\cos x}{\sin x} = \cot x$, so $f'(x) > 0$ when $2n\pi < x < 2n\pi + \frac{\pi}{2}$ for each integer n, and $f'(x) < 0$ when $2n\pi + \frac{\pi}{2} < x < (2n+1)\pi$. Thus, f is increasing on $\left(2n\pi, 2n\pi + \frac{\pi}{2}\right)$ and decreasing on $\left(2n\pi + \frac{\pi}{2}, (2n+1)\pi\right)$ for each integer n.

F. Local maximum values $f\left(2n\pi + \frac{\pi}{2}\right) = 0$, no local minimum.

G. $f''(x) = -\csc^2 x < 0$, so f is CD on $(2n\pi, (2n+1)\pi)$ for each integer n. No IP

H.

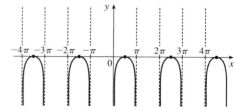

62. $y = \ln(\tan^2 x)$ **A.** $D = \{x \mid x \neq n\pi/2\}$ **B.** x-intercepts $n\pi + \frac{\pi}{4}$, no y-intercept. **C.** $f(-x) = f(x)$, so the curve is

symmetric about the y-axis. Also $f(x + \pi) = f(x)$, so f is periodic with period π, and we consider parts D–G only for

$-\frac{\pi}{2} < x < \frac{\pi}{2}$. **D.** $\displaystyle\lim_{x \to 0} \ln(\tan^2 x) = -\infty$ and $\displaystyle\lim_{x \to (\pi/2)^-} \ln(\tan^2 x) = \infty$, $\displaystyle\lim_{x \to -(\pi/2)^+} \ln(\tan^2 x) = \infty$, so $x = 0$,

$x = \pm\frac{\pi}{2}$ are VA. **E.** $f'(x) = \dfrac{2\tan x \sec^2 x}{\tan^2 x} = 2\,\dfrac{\sec^2 x}{\tan x} > 0$ ⇔

$\tan x > 0$ ⇔ $0 < x < \frac{\pi}{2}$, so f is increasing on $\left(0, \frac{\pi}{2}\right)$ and

decreasing on $\left(-\frac{\pi}{2}, 0\right)$. **F.** No maximum or minimum

G. $f'(x) = \dfrac{2}{\sin x \cos x} = \dfrac{4}{\sin 2x}$ ⇒ $f''(x) = \dfrac{-8\cos 2x}{\sin^2 2x} < 0$

⇔ $\cos 2x > 0$ ⇔ $-\frac{\pi}{4} < x < \frac{\pi}{4}$, so f is CD on $\left(-\frac{\pi}{4}, 0\right)$ and

$\left(0, \frac{\pi}{4}\right)$ and CU on $\left(-\frac{\pi}{2}, -\frac{\pi}{4}\right)$ and $\left(\frac{\pi}{4}, \frac{\pi}{2}\right)$. IP are $\left(\pm\frac{\pi}{4}, 0\right)$.

H.
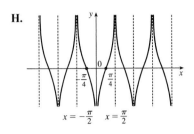

63. $y = f(x) = \ln(1 + x^2)$ **A.** $D = \mathbb{R}$ **B.** Both intercepts are 0. **C.** $f(-x) = f(x)$, so the curve is symmetric about the

y-axis. **D.** $\displaystyle\lim_{x \to \pm\infty} \ln(1 + x^2) = \infty$, no asymptotes. **E.** $f'(x) = \dfrac{2x}{1 + x^2} > 0$ ⇔

$x > 0$, so f is increasing on $(0, \infty)$ and decreasing on $(-\infty, 0)$.

F. $f(0) = 0$ is a local and absolute minimum.

G. $f''(x) = \dfrac{2(1 + x^2) - 2x(2x)}{(1 + x^2)^2} = \dfrac{2(1 - x^2)}{(1 + x^2)^2} > 0$ ⇔

$|x| < 1$, so f is CU on $(-1, 1)$, CD on $(-\infty, -1)$ and $(1, \infty)$. IP

$(1, \ln 2)$ and $(-1, \ln 2)$.

H.

64. $y = f(x) = \ln(x^2 - 3x + 2) = \ln\left[(x - 1)(x - 2)\right]$ **A.** $D = \left\{x \text{ in } \mathbb{R} \mid x^2 - 3x + 2 > 0\right\} = (-\infty, 1) \cup (2, \infty)$.

B. y-intercept: $f(0) = \ln 2$; x-intercepts: $f(x) = 0$ ⇔ $x^2 - 3x + 2 = e^0$ ⇔ $x^2 - 3x + 1 = 0$ ⇔

$x = \frac{3 \pm \sqrt{5}}{2}$ ⇒ $x \approx 0.38, 2.62$ **C.** No symmetry **D.** $\displaystyle\lim_{x \to 1^-} f(x) = \lim_{x \to 2^+} f(x) = -\infty$, so $x = 1$ and $x = 2$ are VAs.

No HA **E.** $f'(x) = \dfrac{2x - 3}{x^2 - 3x + 2} = \dfrac{2(x - 3/2)}{(x - 1)(x - 2)}$, so $f'(x) < 0$ for $x < 1$ and $f'(x) > 0$ for $x > 2$. Thus, f is

decreasing on $(-\infty, 1)$ and increasing on $(2, \infty)$. **F.** No extreme values

G. $f''(x) = \dfrac{(x^2 - 3x + 2) \cdot 2 - (2x - 3)^2}{(x^2 - 3x + 2)^2}$

$= \dfrac{2x^2 - 6x + 4 - 4x^2 + 12x - 9}{(x^2 - 3x + 2)^2} = \dfrac{-2x^2 + 6x - 5}{(x^2 - 3x + 2)^2}$

The numerator is negative for all x and the denominator is positive, so

$f''(x) < 0$ for all x in the domain of f. Thus, f is CD on $(-\infty, 1)$ and $(2, \infty)$.

No IP

H.
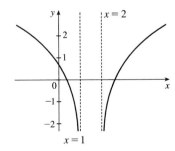

65. We use the CAS to calculate $f'(x) = \dfrac{2 + \sin x + x \cos x}{2x + x \sin x}$ and

$f''(x) = \dfrac{2x^2 \sin x + 4 \sin x - \cos^2 x + x^2 + 5}{x^2 (\cos^2 x - 4 \sin x - 5)}$. From the graphs, it

seems that $f' > 0$ (and so f is increasing) on approximately the intervals

$(0, 2.7)$, $(4.5, 8.2)$ and $(10.9, 14.3)$. It seems that f'' changes sign

(indicating inflection points) at $x \approx 3.8, 5.7, 10.0$ and 12.0.

Looking back at the graph of $f(x) = \ln(2x + x \sin x)$, this implies that the inflection points have approximate coordinates

$(3.8, 1.7)$, $(5.7, 2.1)$, $(10.0, 2.7)$, and $(12.0, 2.9)$.

66. We see that if $c \le 0$, $f(x) = \ln(x^2 + c)$ is only defined for $x^2 > -c \Rightarrow |x| > \sqrt{-c}$, and

$\lim\limits_{x \to \sqrt{-c}^+} f(x) = \lim\limits_{x \to -\sqrt{-c}^-} f(x) = -\infty$, since $\ln y \to -\infty$ as $y \to 0$. Thus, for $c < 0$, there are vertical asymptotes at

$x = \pm\sqrt{c}$, and as c decreases (that is, $|c|$ increases), the asymptotes get further apart. For $c = 0$,

$\lim\limits_{x \to 0} f(x) = -\infty$, so there is a vertical asymptote at $x = 0$. If $c > 0$, there is no asymptote. To find the maxima, minima, and

inflection points, we differentiate: $f(x) = \ln(x^2 + c) \Rightarrow f'(x) = \dfrac{1}{x^2 + c}(2x)$, so by the First Derivative Test there is a

local and absolute minimum at $x = 0$. Differentiating again, we get

$f''(x) = \dfrac{1}{x^2 + c}(2) + 2x\left[-(x^2 + c)^{-2}(2x)\right] = \dfrac{2(c - x^2)}{(x^2 + c)^2}$. Now if

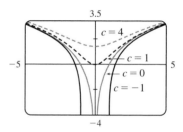

$c \le 0$, this is always negative, so f is concave down on both of the intervals

on which it is defined. If $c > 0$, then f'' changes sign when $c = x^2 \Leftrightarrow$

$x = \pm\sqrt{c}$. So for $c > 0$ there are inflection points at $\pm\sqrt{c}$, and as c

increases, the inflection points get further apart.

67. (a) Using a calculator or CAS, we obtain the model $Q = ab^t$ with $a \approx 100.0124369$ and $b \approx 0.000045145933$.

(b) Use $Q'(t) = ab^t \ln b$ (from Formula 7) with the values of a and b from part (a) to get $Q'(0.04) \approx -670.63$ μA.

The result of Example 2 in Section 2.1 was -670 μA.

68. (a) $P = ab^t$ with $a = 4.502714 \times 10^{-20}$ and $b = 1.029953851$,

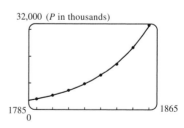

where P is measured in thousands of people. The fit appears to be very good.

(b) **For 1800:** $m_1 = \dfrac{5308 - 3929}{1800 - 1790} = 137.9$, $m_2 = \dfrac{7240 - 5308}{1810 - 1800} = 193.2$.

So $P'(1800) \approx (m_1 + m_2)/2 = 165.55$ thousand people/year.

For 1850: $m_1 = \dfrac{23{,}192 - 17{,}063}{1850 - 1840} = 612.9$, $m_2 = \dfrac{31{,}443 - 23{,}192}{1860 - 1850} = 825.1$.

So $P'(1850) \approx (m_1 + m_2)/2 = 719$ thousand people/year.

(c) Using $P'(t) = ab^t \ln b$ (from Formula 7) with the values of a and b from part (a), we get $P'(1800) \approx 156.85$ and $P'(1850) \approx 686.07$. These estimates are somewhat less than the ones in part (b).

(d) $P(1870) \approx 41{,}946.56$. The difference of 3.4 million people is most likely due to the Civil War (1861–1865).

69. $\displaystyle\int_2^4 \frac{3}{x}\,dx = 3\int_2^4 \frac{1}{x}\,dx = 3\Big[\ln|x|\Big]_2^4 = 3(\ln 4 - \ln 2) = 3\ln\frac{4}{2} = 3\ln 2$

70. $\displaystyle\int_1^2 \frac{4+u^2}{u^3}\,du = \int_1^2 (4u^{-3} + u^{-1})\,du = \left[\frac{4}{-2}u^{-2} + \ln|u|\right]_1^2 = \left[\frac{-2}{u^2} + \ln u\right]_1^2 = \left(-\frac{1}{2} + \ln 2\right) - (-2 + \ln 1) = \frac{3}{2} + \ln 2$

71. $\displaystyle\int_1^2 \frac{dt}{8-3t} = \left[-\frac{1}{3}\ln|8-3t|\right]_1^2 = -\frac{1}{3}\ln 2 - \left(-\frac{1}{3}\ln 5\right) = \frac{1}{3}(\ln 5 - \ln 2) = \frac{1}{3}\ln\frac{5}{2}$

Or: Let $u = 8 - 3t$. Then $du = -3\,dt$, so

$\displaystyle\int_1^2 \frac{dt}{8-3t} = \int_5^2 \frac{-\frac{1}{3}\,du}{u} = \left[-\frac{1}{3}\ln|u|\right]_5^2 = -\frac{1}{3}\ln 2 - \left(-\frac{1}{3}\ln 5\right) = \frac{1}{3}(\ln 5 - \ln 2) = \frac{1}{3}\ln\frac{5}{2}.$

72. $\displaystyle\int_4^9 \left(\sqrt{x} + \frac{1}{\sqrt{x}}\right)^2 dx = \int_4^9 \left(x + 2 + \frac{1}{x}\right)dx = \left[\frac{1}{2}x^2 + 2x + \ln x\right]_4^9 = \frac{81}{2} + 18 + \ln 9 - (8 + 8 + \ln 4)$

$\qquad\qquad = \frac{85}{2} + \ln\frac{9}{4}$

73. $\displaystyle\int_1^e \frac{x^2 + x + 1}{x}\,dx = \int_1^e \left(x + 1 + \frac{1}{x}\right)dx = \left[\frac{1}{2}x^2 + x + \ln x\right]_1^e = \left(\frac{1}{2}e^2 + e + 1\right) - \left(\frac{1}{2} + 1 + 0\right)$

$\qquad\qquad = \frac{1}{2}e^2 + e - \frac{1}{2}$

74. Let $u = \ln x$. Then $du = (1/x)\,dx$, so $\displaystyle\int \frac{\sin(\ln x)}{x}\,dx = \int \sin u\,du = -\cos u + C = -\cos(\ln x) + C.$

75. Let $u = \ln x$. Then $du = \dfrac{dx}{x}$ $\Rightarrow$ $\displaystyle\int \frac{(\ln x)^2}{x}\,dx = \int u^2\,du = \frac{1}{3}u^3 + C = \frac{1}{3}(\ln x)^3 + C.$

76. Let $u = 2 + \sin x$. Then $du = \cos x\,dx$, so

$\displaystyle\int \frac{\cos x}{2 + \sin x}\,dx = \int \frac{1}{u}\,du = \ln|u| + C = \ln|2 + \sin x| + C = \ln(2 + \sin x) + C$ [since $2 + \sin x > 0$].

77. $\displaystyle\int \frac{\sin 2x}{1 + \cos^2 x}\,dx = 2\int \frac{\sin x \cos x}{1 + \cos^2 x}\,dx = 2I.$ Let $u = \cos x$. Then $du = -\sin x\,dx$, so

$2I = -2\displaystyle\int \frac{u\,du}{1 + u^2} = -2 \cdot \frac{1}{2}\ln(1 + u^2) + C = -\ln(1 + u^2) + C = -\ln(1 + \cos^2 x) + C.$

Or: Let $u = 1 + \cos^2 x$.

78. Let $u = e^x + 1$. Then $du = e^x\,dx$, so $\displaystyle\int \frac{e^x}{e^x + 1}\,dx = \int \frac{du}{u} = \ln|u| + C = \ln(e^x + 1) + C.$

79. $\displaystyle\int_1^2 10^t\,dt = \left[\frac{10^t}{\ln 10}\right]_1^2 = \frac{10^2}{\ln 10} - \frac{10^1}{\ln 10} = \frac{100 - 10}{\ln 10} = \frac{90}{\ln 10}$

80. Let $u = x^2$. Then $du = 2x\,dx$, so $\displaystyle\int x2^{x^2}\,dx = \frac{1}{2}\int 2^u\,du = \frac{1}{2}\frac{2^u}{\ln 2} + C = \frac{1}{2\ln 2}2^{x^2} + C.$

81. (a) $\dfrac{d}{dx}\left(\ln|\sin x| + C\right) = \dfrac{1}{\sin x}\cos x = \cot x$

(b) Let $u = \sin x$. Then $du = \cos x\,dx$, so $\displaystyle\int \cot x\,dx = \int \frac{\cos x}{\sin x}\,dx = \int \frac{du}{u} = \ln|u| + C = \ln|\sin x| + C.$

82. Let $u = x - 2$. Then the area is

$$A = -\int_{-4}^{-1}\frac{2}{x-2}\,dx = -2\int_{-6}^{-3}\frac{du}{u} = \left[-2\ln|u|\right]_{-6}^{-3} = -2\ln 3 + 2\ln 6 = 2\ln 2 \approx 1.386.$$

83. The cross-sectional area is $\pi\left(1/\sqrt{x+1}\right)^2 = \pi/(x+1)$. Therefore, the volume is

$$\int_0^1 \frac{\pi}{x+1}\,dx = \pi[\ln(x+1)]_0^1 = \pi(\ln 2 - \ln 1) = \pi\ln 2.$$

84. Using cylindrical shells, we get $V = \displaystyle\int_0^3 \frac{2\pi x}{x^2+1}\,dx = \pi\left[\ln(1+x^2)\right]_0^3 = \pi\ln 10.$

85. $W = \displaystyle\int_{V_1}^{V_2} P\,dV = \int_{600}^{1000}\frac{C}{V}\,dV = C\int_{600}^{1000}\frac{1}{V}\,dV = C\left[\ln|V|\right]_{600}^{1000} = C(\ln 1000 - \ln 600) = C\ln\frac{1000}{600} = C\ln\frac{5}{3}.$

Initially, $PV = C$, where $P = 150$ kPa and $V = 600$ cm^3, so $C = (150)(600) = 90{,}000$. Thus,

$$W = 90{,}000\ln\tfrac{5}{3} \approx 45{,}974 \text{ kPa}\cdot\text{cm}^3 = 45{,}974(10^3\text{ Pa})(10^{-6}\text{ m}^3) = 45{,}974 \text{ Pa}\cdot\text{m}^3 = 45{,}974 \text{ N}\cdot\text{m} \quad [\text{Pa} = \text{N}/\text{m}^2]$$
$$= 45{,}974 \text{ J}$$

86. $f''(x) = x^{-2}$, $x > 0$ $\Rightarrow$ $f'(x) = -1/x + C$ $\Rightarrow$ $f(x) = -\ln x + Cx + D$. $0 = f(1) = C + D$ and

$0 = f(2) = -\ln 2 + 2C + D = -\ln 2 + 2C - C = -\ln 2 + C$ $\Rightarrow$ $C = \ln 2$ and $D = -\ln 2$. So

$f(x) = -\ln x + (\ln 2)x - \ln 2.$

87. $f(x) = 2x + \ln x$ $\Rightarrow$ $f'(x) = 2 + 1/x$. If $g = f^{-1}$, then $f(1) = 2$ $\Rightarrow$ $g(2) = 1$, so

$g'(2) = 1/f'(g(2)) = 1/f'(1) = \frac{1}{3}.$

88. $f(x) = e^x + \ln x$ $\Rightarrow$ $f'(x) = e^x + 1/x$. $h = f^{-1}$ and $f(1) = e$ $\Rightarrow$ $h(e) = 1$, so $h'(e) = 1/f'(1) = 1/(e+1).$

89. The curve and the line will determine a region when they intersect at two or

more points. So we solve the equation $x/(x^2+1) = mx$ $\Rightarrow$ $x = 0$ or

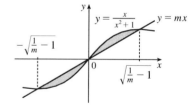

$mx^2 + m - 1 = 0$ $\Rightarrow$ $x = 0$ or $x = \dfrac{\pm\sqrt{-4(m)(m-1)}}{2m} = \pm\sqrt{\dfrac{1}{m} - 1}.$

Note that if $m = 1$, this has only the solution $x = 0$, and no region is

determined. But if $1/m - 1 > 0$ $\Leftrightarrow$ $1/m > 1$ $\Leftrightarrow$ $0 < m < 1$, then there are two solutions. [Another way of seeing this

is to observe that the slope of the tangent to $y = x/(x^2+1)$ at the origin is $y' = 1$ and therefore we must have $0 < m < 1$.]

Note that we cannot just integrate between the positive and negative roots, since the curve and the line cross at the origin.

Since mx and $x/(x^2+1)$ are both odd functions, the total area is twice the area between the curves on the interval

$\left[0, \sqrt{1/m - 1}\right]$. So the total area enclosed is

$$2\int_0^{\sqrt{1/m-1}} \left[\frac{x}{x^2+1} - mx\right] dx = 2\left[\frac{1}{2}\ln(x^2+1) - \frac{1}{2}mx^2\right]_0^{\sqrt{1/m-1}}$$

$$= \left[\ln\left(\frac{1}{m} - 1 + 1\right) - m\left(\frac{1}{m} - 1\right)\right] - (\ln 1 - 0)$$

$$= \ln\left(\frac{1}{m}\right) + m - 1 = m - \ln m - 1$$

90. (a) Let $f(x) = \ln x \;\Rightarrow\; f'(x) = 1/x \;\Rightarrow\; f''(x) = -1/x^2$. The linear approximation to $\ln x$ near 1 is
$\ln x \approx f(1) + f'(1)(x-1) = \ln 1 + \frac{1}{1}(x-1) = x - 1$.

(b)

(c)

From the graph, it appears that the linear approximation is accurate to within 0.1 for x between about 0.62 and 1.51.

91. If $f(x) = \ln(1+x)$, then $f'(x) = \dfrac{1}{1+x}$, so $f'(0) = 1$.

Thus, $\displaystyle\lim_{x\to 0}\frac{\ln(1+x)}{x} = \lim_{x\to 0}\frac{f(x)}{x} = \lim_{x\to 0}\frac{f(x) - f(0)}{x - 0} = f'(0) = 1$.

92. Let $m = n/x$. Then $n = xm$, and as $n \to \infty$, $m \to \infty$.

Therefore, $\displaystyle\lim_{n\to\infty}\left(1 + \frac{x}{n}\right)^n = \lim_{m\to\infty}\left(1 + \frac{1}{m}\right)^{mx} = \left[\lim_{m\to\infty}\left(1 + \frac{1}{m}\right)^m\right]^x = e^x$ by Equation 9.

7.2*   The Natural Logarithmic Function

1. $\ln\dfrac{r^2}{3\sqrt{s}} = \ln r^2 - \ln 3\sqrt{s} = 2\ln r - (\ln 3 + \ln s^{1/2}) = 2\ln r - \ln 3 - \frac{1}{2}\ln s$

2. $\ln\sqrt{a(b^2 + c^2)} = \ln(a(b^2 + c^2))^{1/2} = \frac{1}{2}\ln(a(b^2 + c^2)) = \frac{1}{2}[\ln a + \ln(b^2 + c^2)] = \frac{1}{2}\ln a + \frac{1}{2}\ln(b^2 + c^2)$

3. $\ln(uv)^{10} = 10\ln(uv) = 10(\ln u + \ln v) = 10\ln u + 10\ln v$

4. $\ln\dfrac{3x^2}{(x+1)^5} = \ln 3x^2 - \ln(x+1)^5 = \ln 3 + \ln x^2 - 5\ln(x+1) = \ln 3 + 2\ln x - 5\ln(x+1)$

5. $\ln 5 + 5\ln 3 = \ln 5 + \ln 3^5$ [by Law 3]

          $= \ln(5 \cdot 3^5)$ [by Law 1]

          $= \ln 1215$

6. $\ln 3 + \frac{1}{3}\ln 8 = \ln 3 + \ln 8^{1/3} = \ln 3 + \ln 2 = \ln(3 \cdot 2) = \ln 6$

7. $\ln(1 + x^2) + \frac{1}{2}\ln x - \ln \sin x = \ln(1 + x^2) + \ln x^{1/2} - \ln \sin x = \ln[(1 + x^2)\sqrt{x}] - \ln \sin x = \ln \dfrac{(1 + x^2)\sqrt{x}}{\sin x}$

8. $\ln(a + b) + \ln(a - b) - 2 \ln c = \ln[(a + b)(a - b)] - \ln c^2$ [by Laws 1, 3]

$$= \ln \frac{(a + b)(a - b)}{c^2} \qquad \text{[by Law 2]}$$

$$\text{or } \ln \frac{a^2 - b^2}{c^2}$$

9. Reflect the graph of $y = \ln x$ about the x-axis to obtain the graph of $y = -\ln x$.

10. Reflect the portion of the graph of $y = \ln x$ to the right of the y-axis about the y-axis. The graph of $y = \ln|x|$ is that reflection in addition to the original portion.

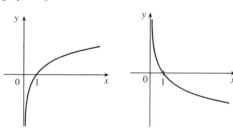

$$y = \ln x \qquad\qquad y = -\ln x$$

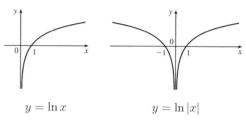

$$y = \ln x \qquad\qquad y = \ln|x|$$

11.

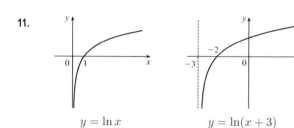

$$y = \ln x \qquad\qquad y = \ln(x + 3)$$

12.

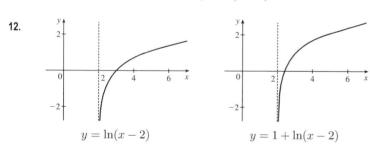

$$y = \ln(x - 2) \qquad\qquad y = 1 + \ln(x - 2)$$

13. Let $t = x^2 - 9$. Then as $x \to 3^+$, $t \to 0^+$, and $\displaystyle\lim_{x \to 3^+} \ln(x^2 - 9) = \lim_{t \to 0^+} \ln t = -\infty$ by (4).

14. $\displaystyle\lim_{x \to \infty} [\ln(2 + x) - \ln(1 + x)] = \lim_{x \to \infty} \ln\left(\frac{2 + x}{1 + x}\right) = \lim_{x \to \infty} \ln\left(\frac{2/x + 1}{1/x + 1}\right) = \ln \frac{1}{1} = \ln 1 = 0$

15. $f(x) = \sqrt{x}\ln x \quad\Rightarrow\quad f'(x) = \dfrac{1}{2\sqrt{x}}\ln x + \sqrt{x}\left(\dfrac{1}{x}\right) = \dfrac{\ln x + 2}{2\sqrt{x}}$

16. $f(x) = \ln(x^2 + 10) \quad\Rightarrow\quad f'(x) = \dfrac{1}{x^2 + 10}\dfrac{d}{dx}(x^2 + 10) = \dfrac{2x}{x^2 + 10}$

17. $f(x) = \sin(\ln x) \quad\Rightarrow\quad f'(x) = \cos(\ln x) \cdot \dfrac{d}{dx}\ln x = \cos(\ln x) \cdot \dfrac{1}{x} = \dfrac{\cos(\ln x)}{x}$

18. $f(x) = \ln(\sin^2 x) = \ln(\sin x)^2 = 2\ln|\sin x| \quad\Rightarrow\quad f'(x) = 2 \cdot \dfrac{1}{\sin x} \cdot \cos x = 2\cot x$

19. $f(x) = \sqrt[5]{\ln x} = (\ln x)^{1/5} \quad\Rightarrow\quad f'(x) = \frac{1}{5}(\ln x)^{-4/5} \dfrac{d}{dx}(\ln x) = \dfrac{1}{5(\ln x)^{4/5}} \cdot \dfrac{1}{x} = \dfrac{1}{5x\sqrt[5]{(\ln x)^4}}$

20. $f(x) = \ln\sqrt[5]{x} = \ln x^{1/5} = \frac{1}{5}\ln x \quad\Rightarrow\quad f'(x) = \dfrac{1}{5} \cdot \dfrac{1}{x} = \dfrac{1}{5x}$

21. $f(x) = \sin x\,\ln(5x) \quad\Rightarrow\quad f'(x) = \sin x \cdot \dfrac{1}{5x} \cdot \dfrac{d}{dx}(5x) + \ln(5x) \cdot \cos x = \dfrac{\sin x \cdot 5}{5x} + \cos x\,\ln(5x) = \dfrac{\sin x}{x} + \cos x\,\ln(5x)$

22. $h(x) = \ln\!\left(x + \sqrt{x^2 - 1}\right) \quad\Rightarrow\quad h'(x) = \dfrac{1}{x + \sqrt{x^2 - 1}}\left(1 + \dfrac{x}{\sqrt{x^2 - 1}}\right) = \dfrac{1}{x + \sqrt{x^2 - 1}} \cdot \dfrac{\sqrt{x^2 - 1} + x}{\sqrt{x^2 - 1}} = \dfrac{1}{\sqrt{x^2 - 1}}$

23. $g(x) = \ln\dfrac{a - x}{a + x} = \ln(a - x) - \ln(a + x) \quad\Rightarrow$

$g'(x) = \dfrac{1}{a - x}(-1) - \dfrac{1}{a + x} = \dfrac{-(a + x) - (a - x)}{(a - x)(a + x)} = \dfrac{-2a}{a^2 - x^2}$

24. $f(t) = \dfrac{1 + \ln t}{1 - \ln t} \quad\Rightarrow\quad f'(t) = \dfrac{(1 - \ln t)(1/t) - (1 + \ln t)(-1/t)}{(1 - \ln t)^2} = \dfrac{(1/t)[(1 - \ln t) + (1 + \ln t)]}{(1 - \ln t)^2} = \dfrac{2}{t(1 - \ln t)^2}$

25. $F(t) = \ln\dfrac{(2t + 1)^3}{(3t - 1)^4} = \ln(2t + 1)^3 - \ln(3t - 1)^4 = 3\ln(2t + 1) - 4\ln(3t - 1) \quad\Rightarrow$

$F'(t) = 3 \cdot \dfrac{1}{2t + 1} \cdot 2 - 4 \cdot \dfrac{1}{3t - 1} \cdot 3 = \dfrac{6}{2t + 1} - \dfrac{12}{3t - 1}$, or combined, $\dfrac{-6(t + 3)}{(2t + 1)(3t - 1)}$.

26. $H(z) = \ln\sqrt{\dfrac{a^2 - z^2}{a^2 + z^2}} = \ln\!\left(\dfrac{a^2 - z^2}{a^2 + z^2}\right)^{1/2} = \dfrac{1}{2}\ln\!\left(\dfrac{a^2 - z^2}{a^2 + z^2}\right) = \frac{1}{2}\ln(a^2 - z^2) - \frac{1}{2}\ln(a^2 + z^2) \quad\Rightarrow$

$H'(z) = \dfrac{1}{2} \cdot \dfrac{1}{a^2 - z^2} \cdot (-2z) - \dfrac{1}{2} \cdot \dfrac{1}{a^2 + z^2} \cdot (2z) = \dfrac{z}{z^2 - a^2} - \dfrac{z}{z^2 + a^2} = \dfrac{z(z^2 + a^2) - z(z^2 - a^2)}{(z^2 - a^2)(z^2 + a^2)}$

$\qquad = \dfrac{z^3 + za^2 - z^3 + za^2}{(z^2 - a^2)(z^2 + a^2)} = \dfrac{2a^2 z}{z^4 - a^4}$

27. $g(x) = \ln\!\left(x\sqrt{x^2 - 1}\right) = \ln x + \ln(x^2 - 1)^{1/2} = \ln x + \frac{1}{2}\ln(x^2 - 1) \quad\Rightarrow$

$g'(x) = \dfrac{1}{x} + \dfrac{1}{2} \cdot \dfrac{1}{x^2 - 1} \cdot 2x = \dfrac{1}{x} + \dfrac{x}{x^2 - 1} = \dfrac{x^2 - 1 + x \cdot x}{x(x^2 - 1)} = \dfrac{2x^2 - 1}{x(x^2 - 1)}$

28. $y = \ln(x^4\sin^2 x) = \ln x^4 + \ln(\sin x)^2 = 4\ln x + 2\ln\sin x \quad\Rightarrow\quad y' = 4 \cdot \dfrac{1}{x} + 2 \cdot \dfrac{1}{\sin x} \cdot \cos x = \dfrac{4}{x} + 2\cot x$

29. $f(u) = \dfrac{\ln u}{1 + \ln(2u)} \quad\Rightarrow$

$f'(u) = \dfrac{[1 + \ln(2u)] \cdot \frac{1}{u} - \ln u \cdot \frac{1}{2u} \cdot 2}{[1 + \ln(2u)]^2} = \dfrac{\frac{1}{u}[1 + \ln(2u) - \ln u]}{[1 + \ln(2u)]^2} = \dfrac{1 + (\ln 2 + \ln u) - \ln u}{u[1 + \ln(2u)]^2} = \dfrac{1 + \ln 2}{u[1 + \ln(2u)]^2}$

30. $y = (\ln\tan x)^2 \quad\Rightarrow\quad y' = 2(\ln\tan x) \cdot \dfrac{1}{\tan x} \cdot \sec^2 x = \dfrac{2(\ln\tan x)\sec^2 x}{\tan x}$

31. $y = \ln\left|2 - x - 5x^2\right| \quad\Rightarrow\quad y' = \dfrac{1}{2 - x - 5x^2} \cdot (-1 - 10x) = \dfrac{-10x - 1}{2 - x - 5x^2}$ or $\dfrac{10x + 1}{5x^2 + x - 2}$

32. $y = \ln \tan^2 x = \ln(\tan x)^2 = 2 \ln \tan x \quad \Rightarrow \quad y' = 2 \dfrac{1}{\tan x} \sec^2 x = 2 \dfrac{\cos x}{\sin x} \dfrac{1}{\cos^2 x} = \dfrac{2}{\sin x \cos x}$ [or $2 \csc x \sec x$]

33. $y = \tan\left[\ln(ax + b)\right] \quad \Rightarrow \quad y' = \sec^2[\ln(ax + b)] \cdot \dfrac{1}{ax + b} \cdot a = \sec^2[\ln(ax + b)] \dfrac{a}{ax + b}$

34. $y = \ln|\cos(\ln x)| \quad \Rightarrow \quad y' = \dfrac{1}{\cos(\ln x)} \left(-\sin(\ln x)\right) \dfrac{1}{x} = -\dfrac{\tan(\ln x)}{x}$

35. $y = x^2 \ln(2x) \quad \Rightarrow \quad y' = x^2 \cdot \dfrac{1}{2x} \cdot 2 + \ln(2x) \cdot (2x) = x + 2x \ln(2x) \quad \Rightarrow$

$y'' = 1 + 2x \cdot \dfrac{1}{2x} \cdot 2 + \ln(2x) \cdot 2 = 1 + 2 + 2\ln(2x) = 3 + 2\ln(2x)$

36. $y = \ln(\sec x + \tan x) \quad \Rightarrow \quad y' = \dfrac{\sec x \tan x + \sec^2 x}{\sec x + \tan x} = \sec x \quad \Rightarrow \quad y'' = \sec x \tan x$

37. $f(x) = \dfrac{x}{1 - \ln(x - 1)} \quad \Rightarrow$

$f'(x) = \dfrac{[1 - \ln(x-1)] \cdot 1 - x \cdot \dfrac{-1}{x-1}}{[1 - \ln(x-1)]^2} = \dfrac{\dfrac{(x-1)[1 - \ln(x-1)] + x}{x-1}}{[1 - \ln(x-1)]^2} = \dfrac{x - 1 - (x-1)\ln(x-1) + x}{(x-1)[1 - \ln(x-1)]^2}$

$= \dfrac{2x - 1 - (x-1)\ln(x-1)}{(x-1)[1 - \ln(x-1)]^2}$

$\mathrm{Dom}(f) = \{x \mid x - 1 > 0 \quad \text{and} \quad 1 - \ln(x-1) \neq 0\} = \{x \mid x > 1 \quad \text{and} \quad \ln(x-1) \neq 1\}$

$= \{x \mid x > 1 \quad \text{and} \quad x - 1 \neq e^1\} = \{x \mid x > 1 \quad \text{and} \quad x \neq 1 + e\} = (1, 1 + e) \cup (1 + e, \infty)$

38. $f(x) = \ln(x^2 - 2x) \quad \Rightarrow \quad f'(x) = \dfrac{1}{x^2 - 2x}(2x - 2) = \dfrac{2(x-1)}{x(x-2)}.$

$\mathrm{Dom}(f) = \{x \mid x(x - 2) > 0\} = (-\infty, 0) \cup (2, \infty).$

39. $f(x) = \sqrt{1 - \ln x}$ is defined $\Leftrightarrow x > 0$ [so that $\ln x$ is defined] and $1 - \ln x \geq 0 \Leftrightarrow$

$x > 0$ and $\ln x \leq 1 \Leftrightarrow 0 < x \leq e$, so the domain of f is $(0, e]$. Now

$f'(x) = \dfrac{1}{2}(1 - \ln x)^{-1/2} \cdot \dfrac{d}{dx}(1 - \ln x) = \dfrac{1}{2\sqrt{1 - \ln x}} \cdot \left(-\dfrac{1}{x}\right) = \dfrac{-1}{2x\sqrt{1 - \ln x}}.$

40. $f(x) = \ln \ln \ln x \quad \Rightarrow \quad f'(x) = \dfrac{1}{\ln \ln x} \cdot \dfrac{1}{\ln x} \cdot \dfrac{1}{x}.$

$\mathrm{Dom}(f) = \{x \mid \ln \ln x > 0\} = \{x \mid \ln x > 1\} = \{x \mid x > e\} = (e, \infty).$

41. $f(x) = \dfrac{\ln x}{1 + x^2} \quad \Rightarrow \quad f'(x) = \dfrac{(1 + x^2)\left(\dfrac{1}{x}\right) - (\ln x)(2x)}{(1 + x^2)^2}$, so $f'(1) = \dfrac{2(1) - 0(2)}{2^2} = \dfrac{2}{4} = \dfrac{1}{2}.$

42. $f(x) = \dfrac{\ln x}{x}$ $\Rightarrow$ $f'(x) = \dfrac{x\left(\dfrac{1}{x}\right) - (\ln x)(1)}{x^2} = \dfrac{1 - \ln x}{x^2}$ $\Rightarrow$ $f''(x) = \dfrac{x^2\left(-\dfrac{1}{x}\right) - (1 - \ln x)(2x)}{(x^2)^2}$, so

$f''(e) = \dfrac{-e - 0}{e^4} = -\dfrac{1}{e^3}$.

43. $f(x) = \sin x + \ln x$ $\Rightarrow$ $f'(x) = \cos x + 1/x$.

This is reasonable, because the graph shows that f increases when f' is

positive, and $f'(x) = 0$ when f has a horizontal tangent.

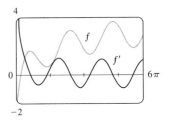

44. $f(x) = \ln(x^2 + x + 1)$ $\Rightarrow$ $f'(x) = \dfrac{1}{x^2 + x + 1}(2x + 1)$. Notice from

the graph that f is increasing when $f'(x)$ is positive.

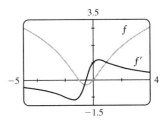

45. $y = \sin(2\ln x)$ $\Rightarrow$ $y' = \cos(2\ln x) \cdot \dfrac{2}{x}$. At $(1, 0)$, $y' = \cos 0 \cdot \dfrac{2}{1} = 2$, so an equation of the tangent line is

$y - 0 = 2 \cdot (x - 1)$, or $y = 2x - 2$.

46. $y = \ln(x^3 - 7)$ $\Rightarrow$ $y' = \dfrac{1}{x^3 - 7} \cdot 3x^2$ $\Rightarrow$ $y'(2) = \dfrac{12}{8 - 7} = 12$, so an equation of a tangent line at $(2, 0)$ is

$y - 0 = 12(x - 2)$ or $y = 12x - 24$.

47. $y = \ln(x^2 + y^2)$ $\Rightarrow$ $y' = \dfrac{1}{x^2 + y^2} \dfrac{d}{dx}(x^2 + y^2)$ $\Rightarrow$ $y' = \dfrac{2x + 2yy'}{x^2 + y^2}$ $\Rightarrow$ $x^2 y' + y^2 y' = 2x + 2yy'$ $\Rightarrow$

$x^2 y' + y^2 y' - 2yy' = 2x$ $\Rightarrow$ $(x^2 + y^2 - 2y)y' = 2x$ $\Rightarrow$ $y' = \dfrac{2x}{x^2 + y^2 - 2y}$

48. $\ln xy = \ln x + \ln y = y \sin x$ $\Rightarrow$ $1/x + y'/y = y\cos x + y'\sin x$ $\Rightarrow$ $y'(1/y - \sin x) = y\cos x - 1/x$ $\Rightarrow$

$y' = \dfrac{y\cos x - 1/x}{1/y - \sin x} = \left(\dfrac{y}{x}\right)\dfrac{xy\cos x - 1}{1 - y\sin x}$

49. $f(x) = \ln(x - 1)$ $\Rightarrow$ $f'(x) = \dfrac{1}{(x - 1)} = (x - 1)^{-1}$ $\Rightarrow$ $f''(x) = -(x - 1)^{-2}$ $\Rightarrow$ $f'''(x) = 2(x - 1)^{-3}$ $\Rightarrow$

$f^{(4)}(x) = -2 \cdot 3(x - 1)^{-4}$ $\Rightarrow$ $\cdots$ $\Rightarrow$ $f^{(n)}(x) = (-1)^{n-1} \cdot 2 \cdot 3 \cdot 4 \cdot \cdots \cdot (n - 1)(x - 1)^{-n} = (-1)^{n-1}\dfrac{(n-1)!}{(x-1)^n}$

50. $y = x^8 \ln x$, so $D^9 y = D^8 y' = D^8(8x^7 \ln x + x^7)$. But the eighth derivative of x^7 is 0, so we now have

$D^8(8x^7 \ln x) = D^7(8 \cdot 7x^6 \ln x + 8x^6) = D^7(8 \cdot 7x^6 \ln x) = D^6(8 \cdot 7 \cdot 6x^5 \ln x) = \cdots = D(8! \, x^0 \ln x) = 8!/x$.

51.

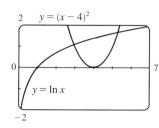

From the graph, it appears that the curves $y = (x - 4)^2$ and $y = \ln x$ intersect just to the left of $x = 3$ and to the right of $x = 5$, at about $x = 5.3$. Let $f(x) = \ln x - (x - 4)^2$. Then $f'(x) = 1/x - 2(x - 4)$, so Newton's Method says that $x_{n+1} = x_n - f(x_n)/f'(x_n) = x_n - \dfrac{\ln x_n - (x_n - 4)^2}{1/x_n - 2(x_n - 4)}$. Taking

$x_0 = 3$, we get $x_1 \approx 2.957738$, $x_2 \approx 2.958516 \approx x_3$, so the first root is 2.958516, to six decimal places. Taking $x_0 = 5$, we get $x_1 \approx 5.290755$, $x_2 \approx 5.290718 \approx x_3$, so the second (and final) root is 5.290718, to six decimal places.

52.

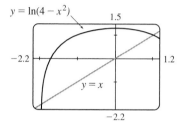

We use Newton's Method with $f(x) = \ln(4 - x^2) - x$ and $f'(x) = \dfrac{1}{4 - x^2}(-2x) - 1 = -1 - \dfrac{2x}{4 - x^2}$. The formula is $x_{n+1} = x_n - f(x_n)/f'(x_n)$. From the graphs it seems that the roots occur at approximately $x = -1.9$ and $x = 1.1$. However, if we use $x_1 = -1.9$ as an initial approximation to the first root, we get $x_2 \approx -2.009611$, and

$f(x) = \ln(x - 2)^2 - x$ is undefined at this point, making it impossible to calculate x_3. We must use a more accurate first estimate, such as $x_1 = -1.95$. With this approximation, we get $x_1 = -1.95$, $x_2 \approx -1.1967495$, $x_3 \approx -1.964760$, $x_4 \approx x_5 \approx -1.964636$. Calculating the second root gives $x_1 = 1.1$, $x_2 \approx 1.058649$, $x_3 \approx 1.058007$, $x_4 \approx x_5 \approx 1.058006$. So, correct to six decimal places, the two roots of the equation $\ln(4 - x^2) = x$ are $x = -1.964636$ and $x = 1.058006$.

53. $y = f(x) = \ln(\sin x)$

A. $D = \{x \text{ in } \mathbb{R} \mid \sin x > 0\} = \bigcup\limits_{n=-\infty}^{\infty} (2n\pi, (2n+1)\pi) = \cdots \cup (-4\pi, -3\pi) \cup (-2\pi, -\pi) \cup (0, \pi) \cup (2\pi, 3\pi) \cup \cdots$

B. No y-intercept; x-intercepts: $f(x) = 0$ $\Leftrightarrow$ $\ln(\sin x) = 0$ $\Leftrightarrow$ $\sin x = e^0 = 1$ $\Leftrightarrow$ $x = 2n\pi + \frac{\pi}{2}$ for each integer n. **C.** f is periodic with period 2π. **D.** $\lim\limits_{x \to (2n\pi)^+} f(x) = -\infty$ and $\lim\limits_{x \to [(2n+1)\pi]^-} f(x) = -\infty$, so the lines

$x = n\pi$ are VAs for all integers n. **E.** $f'(x) = \dfrac{\cos x}{\sin x} = \cot x$, so $f'(x) > 0$ when $2n\pi < x < 2n\pi + \frac{\pi}{2}$ for each

integer n, and $f'(x) < 0$ when $2n\pi + \frac{\pi}{2} < x < (2n+1)\pi$. Thus, f is increasing on $\left(2n\pi, 2n\pi + \frac{\pi}{2}\right)$ and

decreasing on $\left(2n\pi + \frac{\pi}{2}, (2n+1)\pi\right)$ for each integer n. **H.**

F. Local maximum values $f\left(2n\pi + \frac{\pi}{2}\right) = 0$, no local minimum.

G. $f''(x) = -\csc^2 x < 0$, so f is CD on $(2n\pi, (2n+1)\pi)$ for each integer n. No IP

54. $y = \ln(\tan^2 x)$ **A.** $D = \{x \mid x \neq n\pi/2\}$ **B.** x-intercepts $n\pi + \frac{\pi}{4}$, no y-intercept. **C.** $f(-x) = f(x)$, so the curve is

symmetric about the y-axis. Also $f(x + \pi) = f(x)$, so f is periodic with period π, and we consider parts D–G only for

$-\frac{\pi}{2} < x < \frac{\pi}{2}$. **D.** $\displaystyle\lim_{x \to 0} \ln(\tan^2 x) = -\infty$ and $\displaystyle\lim_{x \to (\pi/2)^-} \ln(\tan^2 x) = \infty$, $\displaystyle\lim_{x \to -(-\pi/2)^+} \ln(\tan^2 x) = \infty$, so $x = 0$,

$x = \pm\frac{\pi}{2}$ are VA. **E.** $f'(x) = \dfrac{2\tan x \sec^2 x}{\tan^2 x} = 2\dfrac{\sec^2 x}{\tan x} > 0 \Leftrightarrow$

$\tan x > 0 \Leftrightarrow 0 < x < \frac{\pi}{2}$, so f is increasing on $\left(0, \frac{\pi}{2}\right)$ and

decreasing on $\left(-\frac{\pi}{2}, 0\right)$. **F.** No maximum or minimum

G. $f'(x) = \dfrac{2}{\sin x \cos x} = \dfrac{4}{\sin 2x} \Rightarrow f''(x) = \dfrac{-8\cos 2x}{\sin^2 2x} < 0$

$\Leftrightarrow \cos 2x > 0 \Leftrightarrow -\frac{\pi}{4} < x < \frac{\pi}{4}$, so f is CD on $\left(-\frac{\pi}{4}, 0\right)$ and

$\left(0, \frac{\pi}{4}\right)$ and CU on $\left(-\frac{\pi}{2}, -\frac{\pi}{4}\right)$ and $\left(\frac{\pi}{4}, \frac{\pi}{2}\right)$. IP are $\left(\pm\frac{\pi}{4}, 0\right)$.

H.

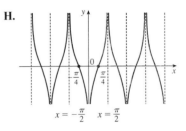

55. $y = f(x) = \ln(1 + x^2)$ **A.** $D = \mathbb{R}$ **B.** Both intercepts are 0. **C.** $f(-x) = f(x)$, so the curve is symmetric about the

y-axis. **D.** $\displaystyle\lim_{x \to \pm\infty} \ln(1 + x^2) = \infty$, no asymptotes. **E.** $f'(x) = \dfrac{2x}{1 + x^2} > 0 \Leftrightarrow$

$x > 0$, so f is increasing on $(0, \infty)$ and decreasing on $(-\infty, 0)$.

F. $f(0) = 0$ is a local and absolute minimum.

H.

G. $f''(x) = \dfrac{2(1 + x^2) - 2x(2x)}{(1 + x^2)^2} = \dfrac{2(1 - x^2)}{(1 + x^2)^2} > 0 \Leftrightarrow$

$|x| < 1$, so f is CU on $(-1, 1)$, CD on $(-\infty, -1)$ and $(1, \infty)$. IP

$(1, \ln 2)$ and $(-1, \ln 2)$.

56. $y = f(x) = \ln(x^2 - 3x + 2) = \ln\left[(x - 1)(x - 2)\right]$ **A.** $D = \{x \text{ in } \mathbb{R} \mid x^2 - 3x + 2 > 0\} = (-\infty, 1) \cup (2, \infty)$.

B. y-intercept: $f(0) = \ln 2$; x-intercepts: $f(x) = 0 \Leftrightarrow x^2 - 3x + 2 = e^0 \Leftrightarrow x^2 - 3x + 1 = 0 \Leftrightarrow$

$x = \frac{3 \pm \sqrt{5}}{2} \Rightarrow x \approx 0.38, 2.62$ **C.** No symmetry **D.** $\displaystyle\lim_{x \to 1^-} f(x) = \lim_{x \to 2^+} f(x) = -\infty$, so $x = 1$ and $x = 2$ are VAs.

No HA **E.** $f'(x) = \dfrac{2x - 3}{x^2 - 3x + 2} = \dfrac{2(x - 3/2)}{(x - 1)(x - 2)}$, so $f'(x) < 0$ for $x < 1$ and $f'(x) > 0$ for $x > 2$. Thus, f is

decreasing on $(-\infty, 1)$ and increasing on $(2, \infty)$. **F.** No extreme values

G. $f''(x) = \dfrac{(x^2 - 3x + 2) \cdot 2 - (2x - 3)^2}{(x^2 - 3x + 2)^2}$

$= \dfrac{2x^2 - 6x + 4 - 4x^2 + 12x - 9}{(x^2 - 3x + 2)^2} = \dfrac{-2x^2 + 6x - 5}{(x^2 - 3x + 2)^2}$

The numerator is negative for all x and the denominator is positive, so

$f''(x) < 0$ for all x in the domain of f. Thus, f is CD on $(-\infty, 1)$ and $(2, \infty)$.

No IP

H.

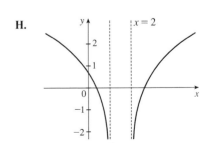

57. We use the CAS to calculate $f'(x) = \dfrac{2 + \sin x + x \cos x}{2x + x \sin x}$ and

$f''(x) = \dfrac{2x^2 \sin x + 4 \sin x - \cos^2 x + x^2 + 5}{x^2(\cos^2 x - 4 \sin x - 5)}$. From the graphs, it

seems that $f' > 0$ (and so f is increasing) on approximately the intervals

$(0, 2.7)$, $(4.5, 8.2)$ and $(10.9, 14.3)$. It seems that f'' changes sign

(indicating inflection points) at $x \approx 3.8$, 5.7, 10.0 and 12.0.

Looking back at the graph of $f(x) = \ln(2x + x \sin x)$, this implies that the inflection points have approximate coordinates

$(3.8, 1.7)$, $(5.7, 2.1)$, $(10.0, 2.7)$, and $(12.0, 2.9)$.

58. We see that if $c \leq 0$, $f(x) = \ln(x^2 + c)$ is only defined for $x^2 > -c \Rightarrow |x| > \sqrt{-c}$, and

$\lim\limits_{x \to \sqrt{-c}^+} f(x) = \lim\limits_{x \to -\sqrt{-c}^-} f(x) = -\infty$, since $\ln y \to -\infty$ as $y \to 0$. Thus, for $c < 0$, there are vertical asymptotes at

$x = \pm\sqrt{c}$, and as c decreases (that is, $|c|$ increases), the asymptotes get further apart. For $c = 0$,

$\lim\limits_{x \to 0} f(x) = -\infty$, so there is a vertical asymptote at $x = 0$. If $c > 0$, there is no asymptote. To find the maxima, minima, and

inflection points, we differentiate: $f(x) = \ln(x^2 + c) \Rightarrow f'(x) = \dfrac{1}{x^2 + c}(2x)$, so by the First Derivative Test there is a

local and absolute minimum at $x = 0$. Differentiating again, we get

$f''(x) = \dfrac{1}{x^2 + c}(2) + 2x\left[-(x^2 + c)^{-2}(2x)\right] = \dfrac{2(c - x^2)}{(x^2 + c)^2}$. Now if

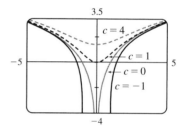

$c \leq 0$, this is always negative, so f is concave down on both of the intervals

on which it is defined. If $c > 0$, then f'' changes sign when $c = x^2 \Leftrightarrow$

$x = \pm\sqrt{c}$. So for $c > 0$ there are inflection points at $\pm\sqrt{c}$, and as c

increases, the inflection points get further apart.

59. $y = (2x + 1)^5(x^4 - 3)^6 \Rightarrow \ln y = \ln\big((2x + 1)^5(x^4 - 3)^6\big) \Rightarrow \ln y = 5\ln(2x + 1) + 6\ln(x^4 - 3) \Rightarrow$

$\dfrac{1}{y}y' = 5 \cdot \dfrac{1}{2x + 1} \cdot 2 + 6 \cdot \dfrac{1}{x^4 - 3} \cdot 4x^3 \Rightarrow$

$y' = y\left(\dfrac{10}{2x + 1} + \dfrac{24x^3}{x^4 - 3}\right) = (2x + 1)^5(x^4 - 3)^6\left(\dfrac{10}{2x + 1} + \dfrac{24x^3}{x^4 - 3}\right)$.

[The answer could be simplified to $y' = 2(2x + 1)^4(x^4 - 3)^5(29x^4 + 12x^3 - 15)$, but this is unnecessary.]

60. $y = \dfrac{(x^3 + 1)^4 \sin^2 x}{x^{1/3}} \Rightarrow \ln|y| = 4\ln|x^3 + 1| + 2\ln|\sin x| - \tfrac{1}{3}\ln|x|$.

So $\dfrac{y'}{y} = 4\dfrac{3x^2}{x^3 + 1} + 2\dfrac{\cos x}{\sin x} - \dfrac{1}{3x} \Rightarrow y' = \dfrac{(x^3 + 1)^4 \sin^2 x}{x^{1/3}}\left(\dfrac{12x^2}{x^3 + 1} + 2\cot x - \dfrac{1}{3x}\right)$.

61. $y = \dfrac{\sin^2 x \tan^4 x}{(x^2+1)^2}$ $\Rightarrow$ $\ln y = \ln(\sin^2 x \tan^4 x) - \ln(x^2+1)^2$ $\Rightarrow$

$\ln y = \ln(\sin x)^2 + \ln(\tan x)^4 - \ln(x^2+1)^2$ $\Rightarrow$ $\ln y = 2\ln|\sin x| + 4\ln|\tan x| - 2\ln(x^2+1)$ $\Rightarrow$

$\dfrac{1}{y}\,y' = 2\cdot\dfrac{1}{\sin x}\cdot\cos x + 4\cdot\dfrac{1}{\tan x}\cdot\sec^2 x - 2\cdot\dfrac{1}{x^2+1}\cdot 2x$ $\Rightarrow$ $y' = \dfrac{\sin^2 x \tan^4 x}{(x^2+1)^2}\left(2\cot x + \dfrac{4\sec^2 x}{\tan x} - \dfrac{4x}{x^2+1}\right)$

62. $y = \sqrt[4]{\dfrac{x^2+1}{x^2-1}}$ $\Rightarrow$ $\ln y = \frac{1}{4}\ln(x^2+1) - \frac{1}{4}\ln(x^2-1)$ $\Rightarrow$ $\dfrac{1}{y}\,y' = \dfrac{1}{4}\cdot\dfrac{1}{x^2+1}\cdot 2x - \dfrac{1}{4}\cdot\dfrac{1}{x^2-1}\cdot 2x$ $\Rightarrow$

$y' = \sqrt[4]{\dfrac{x^2+1}{x^2-1}}\cdot\dfrac{1}{2}\left(\dfrac{x}{x^2+1} - \dfrac{x}{x^2-1}\right) = \dfrac{1}{2}\sqrt[4]{\dfrac{x^2+1}{x^2-1}}\left(\dfrac{-2x}{x^4-1}\right) = \dfrac{x}{1-x^4}\sqrt[4]{\dfrac{x^2+1}{x^2-1}}$

63. $\displaystyle\int_2^4 \frac{3}{x}\,dx = 3\int_2^4 \frac{1}{x}\,dx = 3\Big[\ln|x|\Big]_2^4 = 3(\ln 4 - \ln 2) = 3\ln\frac{4}{2} = 3\ln 2$

64. $\displaystyle\int_1^2 \frac{4+u^2}{u^3}\,du = \int_1^2 (4u^{-3} + u^{-1})\,du = \left[\frac{4}{-2}u^{-2} + \ln|u|\right]_1^2 = \left[\frac{-2}{u^2} + \ln u\right]_1^2 = \left(-\frac{1}{2} + \ln 2\right) - (-2 + \ln 1) = \frac{3}{2} + \ln 2$

65. $\displaystyle\int_1^2 \frac{dt}{8-3t} = \left[-\frac{1}{3}\ln|8-3t|\right]_1^2 = -\frac{1}{3}\ln 2 - \left(-\frac{1}{3}\ln 5\right) = \frac{1}{3}(\ln 5 - \ln 2) = \frac{1}{3}\ln\frac{5}{2}$

Or: Let $u = 8 - 3t$. Then $du = -3\,dt$, so

$\displaystyle\int_1^2 \frac{dt}{8-3t} = \int_5^2 \frac{-\frac{1}{3}\,du}{u} = \left[-\frac{1}{3}\ln|u|\right]_5^2 = -\frac{1}{3}\ln 2 - \left(-\frac{1}{3}\ln 5\right) = \frac{1}{3}(\ln 5 - \ln 2) = \frac{1}{3}\ln\frac{5}{2}.$

66. $\displaystyle\int_4^9 \left(\sqrt{x} + \frac{1}{\sqrt{x}}\right)^2 dx = \int_4^9 \left(x + 2 + \frac{1}{x}\right) dx = \left[\frac{1}{2}x^2 + 2x + \ln x\right]_4^9 = \frac{81}{2} + 18 + \ln 9 - (8 + 8 + \ln 4)$

$= \frac{85}{2} + \ln\frac{9}{4}$

67. $\displaystyle\int_1^e \frac{x^2+x+1}{x}\,dx = \int_1^e \left(x + 1 + \frac{1}{x}\right) dx = \left[\frac{1}{2}x^2 + x + \ln x\right]_1^e = \left(\frac{1}{2}e^2 + e + 1\right) - \left(\frac{1}{2} + 1 + 0\right)$

$= \frac{1}{2}e^2 + e - \frac{1}{2}$

68. Let $u = \ln x$. Then $du = \dfrac{1}{x}\,dx$, so $\displaystyle\int_e^6 \frac{dx}{x\ln x} = \int_1^{\ln 6} \frac{1}{u}\,du = \Big[\ln|u|\Big]_1^{\ln 6} = \ln\ln 6 - \ln 1 = \ln\ln 6$

69. Let $u = \ln x$. Then $du = \dfrac{dx}{x}$ $\Rightarrow$ $\displaystyle\int \frac{(\ln x)^2}{x}\,dx = \int u^2\,du = \frac{1}{3}u^3 + C = \frac{1}{3}(\ln x)^3 + C.$

70. Let $u = 2 + \sin x$. Then $du = \cos x\,dx$, so

$\displaystyle\int \frac{\cos x}{2 + \sin x}\,dx = \int \frac{1}{u}\,du = \ln|u| + C = \ln|2 + \sin x| + C = \ln(2 + \sin x) + C$ [since $2 + \sin x > 0$].

71. $\displaystyle\int \frac{\sin 2x}{1 + \cos^2 x}\,dx = 2\int \frac{\sin x \cos x}{1 + \cos^2 x}\,dx = 2I.$ Let $u = \cos x$. Then $du = -\sin x\,dx$, so

$2I = -2\displaystyle\int \frac{u\,du}{1 + u^2} = -2\cdot\frac{1}{2}\ln(1 + u^2) + C = -\ln(1 + u^2) + C = -\ln(1 + \cos^2 x) + C.$

Or: Let $u = 1 + \cos^2 x$.

72. Let $u = \ln x$. Then $du = (1/x)\,dx$, so $\displaystyle\int \frac{\sin(\ln x)}{x}\,dx = \int \sin u\,du = -\cos u + C = -\cos(\ln x) + C.$

73. (a) $\displaystyle\frac{d}{dx}\left(\ln|\sin x| + C\right) = \frac{1}{\sin x}\cos x = \cot x$

(b) Let $u = \sin x$. Then $du = \cos x\,dx$, so $\displaystyle\int \cot x\,dx = \int \frac{\cos x}{\sin x}\,dx = \int \frac{du}{u} = \ln|u| + C = \ln|\sin x| + C.$

74. Let $u = x - 2$. Then the area is

$$A = -\int_{-4}^{-1} \frac{2}{x-2}\,dx = -2\int_{-6}^{-3} \frac{du}{u} = \left[-2\ln|u|\right]_{-6}^{-3} = -2\ln 3 + 2\ln 6 = 2\ln 2 \approx 1.386.$$

75. The cross-sectional area is $\pi\left(1/\sqrt{x+1}\right)^2 = \pi/(x+1)$. Therefore, the volume is

$$\int_0^1 \frac{\pi}{x+1}\,dx = \pi[\ln(x+1)]_0^1 = \pi(\ln 2 - \ln 1) = \pi\ln 2.$$

76. Using cylindrical shells, we get $\displaystyle V = \int_0^3 \frac{2\pi x}{x^2+1}\,dx = \pi\left[\ln(1+x^2)\right]_0^3 = \pi\ln 10.$

77. $\displaystyle W = \int_{V_1}^{V_2} P\,dV = \int_{600}^{1000} \frac{C}{V}\,dV = C\int_{600}^{1000} \frac{1}{V}\,dV = C\Big[\ln|V|\Big]_{600}^{1000} = C(\ln 1000 - \ln 600) = C\ln\frac{1000}{600} = C\ln\frac{5}{3}.$

Initially, $PV = C$, where $P = 150$ kPa and $V = 600$ cm^3, so $C = (150)(600) = 90{,}000$. Thus,

$$W = 90{,}000\ln\tfrac{5}{3} \approx 45{,}974 \text{ kPa}\cdot\text{cm}^3 = 45{,}974(10^3\text{ Pa})(10^{-6}\text{ m}^3) = 45{,}974 \text{ Pa}\cdot\text{m}^3 = 45{,}974 \text{ N}\cdot\text{m} \quad [\text{Pa} = \text{N/m}^2]$$
$$= 45{,}974 \text{ J}$$

78. $f''(x) = x^{-2}$, $x > 0$ $\Rightarrow$ $f'(x) = -1/x + C$ $\Rightarrow$ $f(x) = -\ln x + Cx + D$. $0 = f(1) = C + D$ and

$0 = f(2) = -\ln 2 + 2C + D = -\ln 2 + 2C - C = -\ln 2 + C$ $\Rightarrow$ $C = \ln 2$ and $D = -\ln 2$. So

$f(x) = -\ln x + (\ln 2)x - \ln 2.$

79. $f(x) = 2x + \ln x$ $\Rightarrow$ $f'(x) = 2 + 1/x$. If $g = f^{-1}$, then $f(1) = 2$ $\Rightarrow$ $g(2) = 1$, so

$g'(2) = 1/f'(g(2)) = 1/f'(1) = \frac{1}{3}.$

80. (a) Let $f(x) = \ln x$ $\Rightarrow$ $f'(x) = 1/x$ $\Rightarrow$ $f''(x) = -1/x^2$. The linear approximation to $\ln x$ near 1 is

$\ln x \approx f(1) + f'(1)(x-1) = \ln 1 + \frac{1}{1}(x-1) = x - 1.$

(b)

(c)

From the graph, it appears that the linear approximation is accurate to within 0.1 for x between about 0.62 and 1.51.

81. (a)

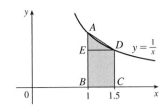

We interpret $\ln 1.5$ as the area under the curve $y = 1/x$ from $x = 1$ to $x = 1.5$. The area of the rectangle $BCDE$ is $\frac{1}{2} \cdot \frac{2}{3} = \frac{1}{3}$. The area of the trapezoid $ABCD$ is $\frac{1}{2} \cdot \frac{1}{2}\left(1 + \frac{2}{3}\right) = \frac{5}{12}$. Thus, by comparing areas, we observe that $\frac{1}{3} < \ln 1.5 < \frac{5}{12}$.

(b) With $f(t) = 1/t$, $n = 10$, and $\Delta t = 0.05$, we have

$$\ln 1.5 = \int_1^{1.5} (1/t)\, dt \approx (0.05)[f(1.025) + f(1.075) + \cdots + f(1.475)]$$

$$= (0.05)\left[\tfrac{1}{1.025} + \tfrac{1}{1.075} + \cdots + \tfrac{1}{1.475}\right] \approx 0.4054$$

82. (a) $y = \dfrac{1}{t}$, $y' = -\dfrac{1}{t^2}$. The slope of AD is $\dfrac{1/2 - 1}{2 - 1} = -\dfrac{1}{2}$. Let c be the t-coordinate of the point on $y = \dfrac{1}{t}$ with slope $-\dfrac{1}{2}$.

Then $-\dfrac{1}{c^2} = -\dfrac{1}{2} \ \Rightarrow \ c^2 = 2 \ \Rightarrow \ c = \sqrt{2}$ since $c > 0$. Therefore the tangent line is given by

$y - \frac{1}{\sqrt{2}} = -\frac{1}{2}\left(t - \sqrt{2}\right) \ \Rightarrow \ y = -\frac{1}{2}t + \sqrt{2}.$

(b)

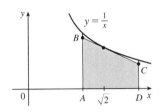

Since the graph of $y = 1/t$ is concave upward, the graph lies above the tangent line, that is, above the line segment BC. Now $|AB| = -\frac{1}{2} + \sqrt{2}$ and $|CD| = -1 + \sqrt{2}$. So the area of the trapezoid $ABCD$ is

$\frac{1}{2}\left[\left(-\frac{1}{2} + \sqrt{2}\right) + \left(-1 + \sqrt{2}\right)1\right] = -\frac{3}{4} + \sqrt{2} \approx 0.6642$. So $\ln 2 >$ area of trapezoid $ABCD > 0.66$.

83.

The area of R_i is $\dfrac{1}{i+1}$ and so $\dfrac{1}{2} + \dfrac{1}{3} + \cdots + \dfrac{1}{n} < \displaystyle\int_1^n \dfrac{1}{t}\, dt = \ln n.$

The area of S_i is $\dfrac{1}{i}$ and so $1 + \dfrac{1}{2} + \cdots + \dfrac{1}{n-1} > \displaystyle\int_1^n \dfrac{1}{t}\, dt = \ln n.$

84. If $f(x) = \ln(x^r)$, then $f'(x) = (1/x^r)(rx^{r-1}) = r/x$. But if $g(x) = r \ln x$, then $g'(x) = r/x$. So f and g must differ by a constant: $\ln(x^r) = r \ln x + C$. Put $x = 1$: $\ln(1^r) = r \ln 1 + C \ \Rightarrow \ C = 0$, so $\ln(x^r) = r \ln x$.

85. The curve and the line will determine a region when they intersect at two or

more points. So we solve the equation $x/(x^2+1) = mx \Rightarrow x = 0$ or

$mx^2 + m - 1 = 0 \Rightarrow x = 0$ or $x = \dfrac{\pm\sqrt{-4(m)(m-1)}}{2m} = \pm\sqrt{\dfrac{1}{m} - 1}$.

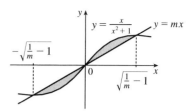

Note that if $m = 1$, this has only the solution $x = 0$, and no region is

determined. But if $1/m - 1 > 0 \Leftrightarrow 1/m > 1 \Leftrightarrow 0 < m < 1$, then there are two solutions. [Another way of seeing this

is to observe that the slope of the tangent to $y = x/(x^2+1)$ at the origin is $y' = 1$ and therefore we must have $0 < m < 1$.]

Note that we cannot just integrate between the positive and negative roots, since the curve and the line cross at the origin.

Since mx and $x/(x^2+1)$ are both odd functions, the total area is twice the area between the curves on the interval

$\left[0, \sqrt{1/m - 1}\right]$. So the total area enclosed is

$$2\int_0^{\sqrt{1/m-1}} \left[\frac{x}{x^2+1} - mx\right] dx = 2\left[\tfrac{1}{2}\ln(x^2+1) - \tfrac{1}{2}mx^2\right]_0^{\sqrt{1/m-1}}$$

$$= \left[\ln\left(\frac{1}{m} - 1 + 1\right) - m\left(\frac{1}{m} - 1\right)\right] - (\ln 1 - 0)$$

$$= \ln\left(\frac{1}{m}\right) + m - 1 = m - \ln m - 1$$

86. (a)

From the graphs, we see that $f(x) = x^{0.1} > g(x) = \ln x$ for approximately $0 < x < 3.06$, and then $g(x) > f(x)$ for

$3.06 < x < 3.43 \times 10^{15}$ (approximately). At that point, the graph of f finally surpasses the graph of g for good.

(b)

(c) From the graph at left, it seems that $\dfrac{\ln x}{x^{0.1}} < 0.1$ whenever

$x > 1.3 \times 10^{28}$ (approximately). So we can take $N = 1.3 \times 10^{28}$,

or any larger number.

87. If $f(x) = \ln(1+x)$, then $f'(x) = \dfrac{1}{1+x}$, so $f'(0) = 1$.

Thus, $\displaystyle\lim_{x\to 0}\frac{\ln(1+x)}{x} = \lim_{x\to 0}\frac{f(x)}{x} = \lim_{x\to 0}\frac{f(x) - f(0)}{x - 0} = f'(0) = 1$.

7.3*   The Natural Exponential Function

1.

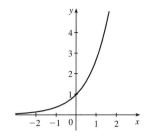

The function value at $x = 0$ is 1 and the slope at $x = 0$ is 1.

2. (a) $e^{\ln 15} = 15$ by (4).

(b) $\ln(1/e) = \ln 1 - \ln e = 0 - 1 = -1$

3. (a) $e^{-2\ln 5} = \left(e^{\ln 5}\right)^{-2} \overset{(4)}{=} 5^{-2} = \dfrac{1}{5^2} = \dfrac{1}{25}$

(b) $\ln\left(\ln e^{e^{10}}\right) \overset{(5)}{=} \ln(e^{10}) \overset{(5)}{=} 10$

4. (a) $\ln e^{\sin x} = \sin x$

(b) $e^{x+\ln x} = e^x e^{\ln x} = xe^x$

5. (a) $2\ln x = 1 \;\Rightarrow\; \ln x = \tfrac{1}{2} \;\Rightarrow\; x = e^{1/2} = \sqrt{e}$

(b) $e^{-x} = 5 \;\Rightarrow\; -x = \ln 5 \;\Rightarrow\; x = -\ln 5$

6. (a) $e^{2x+3} - 7 = 0 \;\Rightarrow\; e^{2x+3} = 7 \;\Rightarrow\; 2x + 3 = \ln 7 \;\Rightarrow\; 2x = \ln 7 - 3 \;\Rightarrow\; x = \tfrac{1}{2}(\ln 7 - 3)$

(b) $\ln(5 - 2x) = -3 \;\Rightarrow\; 5 - 2x = e^{-3} \;\Rightarrow\; 2x = 5 - e^{-3} \;\Rightarrow\; x = \tfrac{1}{2}(5 - e^{-3})$

7. (a) $e^{3x+1} = k \;\Leftrightarrow\; 3x + 1 = \ln k \;\Leftrightarrow\; x = \tfrac{1}{3}(\ln k - 1)$

(b) $\ln x + \ln(x - 1) = \ln(x(x - 1)) = 1 \;\Leftrightarrow\; x(x - 1) = e^1 \;\Leftrightarrow\; x^2 - x - e = 0$. The quadratic formula (with $a = 1$, $b = -1$, and $c = -e$) gives $x = \tfrac{1}{2}\left(1 \pm \sqrt{1 + 4e}\right)$, but we reject the negative root since the natural logarithm is not defined for $x < 0$. So $x = \tfrac{1}{2}\left(1 + \sqrt{1 + 4e}\right)$.

8. (a) $\ln(\ln x) = 1 \;\Leftrightarrow\; e^{\ln(\ln x)} = e^1 \;\Leftrightarrow\; \ln x = e^1 = e \;\Leftrightarrow\; e^{\ln x} = e^e \;\Leftrightarrow\; x = e^e$

(b) $e^{e^x} = 10 \;\Leftrightarrow\; \ln\left(e^{e^x}\right) = \ln 10 \;\Leftrightarrow\; e^x \ln e = e^x = \ln 10 \;\Leftrightarrow\; \ln e^x = \ln(\ln 10) \;\Leftrightarrow\; x = \ln(\ln 10)$

9. $3xe^x + x^2 e^x = 0 \;\Leftrightarrow\; xe^x(3 + x) = 0 \;\Leftrightarrow\; x = 0 \text{ or } -3$

10. $10(1 + e^{-x})^{-1} = 3 \;\Leftrightarrow\; (1 + e^{-x})^{-1} = \tfrac{3}{10} \;\Leftrightarrow\; 1 + e^{-x} = \tfrac{10}{3} \;\Leftrightarrow\; e^{-x} = \tfrac{7}{3} \;\Leftrightarrow\; -x = \ln\tfrac{7}{3} \;\Rightarrow\;$
$x = -\ln\tfrac{7}{3} = \ln\left(\tfrac{7}{3}\right)^{-1} = \ln\tfrac{3}{7}$

11. $e^{2x} - e^x - 6 = 0 \;\Leftrightarrow\; (e^x - 3)(e^x + 2) = 0 \;\Leftrightarrow\; e^x = 3 \text{ or } -2 \;\Rightarrow\; x = \ln 3 \text{ since } e^x > 0.$

12. $\ln(2x + 1) = 2 - \ln x \;\Rightarrow\; \ln x + \ln(2x + 1) = \ln e^2 \;\Rightarrow\; \ln[x(2x + 1)] = \ln e^2 \;\Rightarrow\; 2x^2 + x = e^2 \;\Rightarrow\;$
$2x^2 + x - e^2 = 0 \;\Rightarrow\; x = \dfrac{-1 + \sqrt{1 + 8e^2}}{4} \quad [\text{since } x > 0].$

13. (a) $e^{2+5x} = 100 \;\Rightarrow\; \ln\left(e^{2+5x}\right) = \ln 100 \;\Rightarrow\; 2 + 5x = \ln 100 \;\Rightarrow\; 5x = \ln 100 - 2 \;\Rightarrow\;$
$x = \tfrac{1}{5}(\ln 100 - 2) \approx 0.5210$

(b) $\ln(e^x - 2) = 3 \;\Rightarrow\; e^x - 2 = e^3 \;\Rightarrow\; e^x = e^3 + 2 \;\Rightarrow\; x = \ln(e^3 + 2) \approx 3.0949$

14. (a) $\ln(1 + \sqrt{x}) = 2 \;\Rightarrow\; 1 + \sqrt{x} = e^2 \;\Rightarrow\; \sqrt{x} = e^2 - 1 \;\Rightarrow\; x = (e^2 - 1)^2 \approx 40.8200$

(b) $e^{1/(x-4)} = 7 \;\Rightarrow\; \ln e^{1/(x-4)} = \ln 7 \;\Rightarrow\; \dfrac{1}{x-4} = \ln 7 \;\Rightarrow\; \dfrac{1}{\ln 7} = x - 4 \;\Rightarrow\; x = 4 + \dfrac{1}{\ln 7} \approx 4.5139$

15. (a) $e^x < 10 \;\Rightarrow\; \ln e^x < \ln 10 \;\Rightarrow\; x < \ln 10 \;\Rightarrow\; x \in (-\infty, \ln 10)$

(b) $\ln x > -1 \;\Rightarrow\; e^{\ln x} > e^{-1} \;\Rightarrow\; x > e^{-1} \;\Rightarrow\; x \in (1/e, \infty)$

16. (a) $2 < \ln x < 9 \;\Rightarrow\; e^2 < e^{\ln x} < e^9 \;\Rightarrow\; e^2 < x < e^9 \;\Rightarrow\; x \in (e^2, e^9)$

(b) $e^{2-3x} > 4 \;\Rightarrow\; \ln e^{2-3x} > \ln 4 \;\Rightarrow\; 2 - 3x > \ln 4 \;\Rightarrow\; -3x > \ln 4 - 2 \;\Rightarrow\; x < -\frac{1}{3}(\ln 4 - 2) \;\Rightarrow\;$
$x \in \left(-\infty, \frac{1}{3}(2 - \ln 4)\right)$

17.

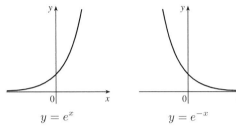

18. We start with the graph of $y = e^x$ (Figure 2), vertically stretch by a factor of 2, and then shift 1 unit upward. There is a horizontal asymptote of $y = 1$.

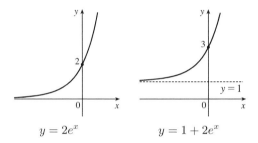

19. We start with the graph of $y = e^x$ (Figure 2) and reflect about the y-axis to get the graph of $y = e^{-x}$. Then we compress the graph vertically by a factor of 2 to obtain the graph of $y = \frac{1}{2}e^{-x}$ and then reflect about the x-axis to get the graph of $y = -\frac{1}{2}e^{-x}$. Finally, we shift the graph upward one unit to get the graph of $y = 1 - \frac{1}{2}e^{-x}$.

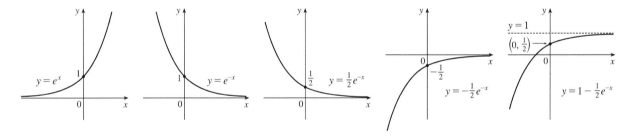

20. We start with the graph of $y = e^x$ (Figure 2) and reflect about the x-axis to get the graph of $y = -e^x$. Then shift the graph upward one unit to get the graph of $y = 1 - e^x$. Finally, we stretch the graph vertically by a factor of 2 to obtain the graph of $y = 2(1 - e^x)$.

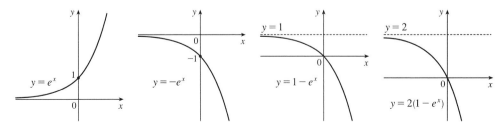

21. (a) For $f(x) = \sqrt{3 - e^{2x}}$, we must have $3 - e^{2x} \geq 0 \Rightarrow e^{2x} \leq 3 \Rightarrow 2x \leq \ln 3 \Rightarrow x \leq \frac{1}{2}\ln 3$.

Thus, the domain of f is $(-\infty, \frac{1}{2}\ln 3]$.

(b) $y = f(x) = \sqrt{3 - e^{2x}}$ [note that $y \geq 0$] $\Rightarrow y^2 = 3 - e^{2x} \Rightarrow e^{2x} = 3 - y^2 \Rightarrow 2x = \ln(3 - y^2) \Rightarrow$

$x = \frac{1}{2}\ln(3 - y^2)$. Interchange x and y: $y = \frac{1}{2}\ln(3 - x^2)$. So $f^{-1}(x) = \frac{1}{2}\ln(3 - x^2)$. For the domain of f^{-1}, we must

have $3 - x^2 > 0 \Rightarrow x^2 < 3 \Rightarrow |x| < \sqrt{3} \Rightarrow -\sqrt{3} < x < \sqrt{3} \Rightarrow 0 \leq x < \sqrt{3}$ since $x \geq 0$. Note that the

domain of f^{-1}, $[0, \sqrt{3})$, equals the range of f.

22. (a) For $f(x) = \ln(2 + \ln x)$, we must have $2 + \ln x > 0 \Rightarrow \ln x > -2 \Rightarrow x > e^{-2}$. Thus, the domain of f

is (e^{-2}, ∞).

(b) $y = f(x) = \ln(2 + \ln x) \Rightarrow e^y = 2 + \ln x \Rightarrow \ln x = e^y - 2 \Rightarrow x = e^{e^y - 2}$. Interchange x and y: $y = e^{e^x - 2}$.

So $f^{-1}(x) = e^{e^x - 2}$. The domain of f^{-1}, as well as the range of f, is $\mathbb{R}$.

23. $y = \ln(x + 3) \Rightarrow e^y = e^{\ln(x+3)} = x + 3 \Rightarrow x = e^y - 3$.

Interchange x and y: the inverse function is $y = e^x - 3$.

24. $y = (\ln x)^2$, $x \geq 1$, $\ln x = \sqrt{y} \Rightarrow x = e^{\sqrt{y}}$. Interchange x and y: $y = e^{\sqrt{x}}$ is the inverse function.

25. $y = f(x) = e^{x^3} \Rightarrow \ln y = x^3 \Rightarrow x = \sqrt[3]{\ln y}$. Interchange x and y: $y = \sqrt[3]{\ln x}$. So $f^{-1}(x) = \sqrt[3]{\ln x}$.

26. $y = f(x) = \dfrac{e^x}{1 + 2e^x} \Rightarrow y + 2ye^x = e^x \Rightarrow y = e^x - 2ye^x \Rightarrow y = e^x(1 - 2y) \Rightarrow e^x = \dfrac{y}{1 - 2y} \Rightarrow$

$x = \ln\left(\dfrac{y}{1 - 2y}\right)$. Interchange x and y: $y = \ln\left(\dfrac{x}{1 - 2x}\right)$. So $f^{-1}(x) = \ln\left(\dfrac{x}{1 - 2x}\right)$. Note that the range of f and the

domain of f^{-1} is $(0, \frac{1}{2})$.

27. Divide numerator and denominator by e^{3x}: $\displaystyle\lim_{x \to \infty} \frac{e^{3x} - e^{-3x}}{e^{3x} + e^{-3x}} = \lim_{x \to \infty} \frac{1 - e^{-6x}}{1 + e^{-6x}} = \frac{1 - 0}{1 + 0} = 1$

28. Let $t = -x^2$. As $x \to \infty$, $t \to -\infty$. So $\displaystyle\lim_{x \to \infty} e^{-x^2} = \lim_{t \to -\infty} e^t = 0$ by (6).

29. Let $t = 3/(2 - x)$. As $x \to 2^+$, $t \to -\infty$. So $\displaystyle\lim_{x \to 2^+} e^{3/(2-x)} = \lim_{t \to -\infty} e^t = 0$ by (6).

30. Let $t = 3/(2 - x)$. As $x \to 2^-$, $t \to \infty$. So $\displaystyle\lim_{x \to 2^-} e^{3/(2-x)} = \lim_{t \to \infty} e^t = \infty$ by (6).

31. Since $-1 \leq \cos x \leq 1$ and $e^{-2x} > 0$, we have $-e^{-2x} \leq e^{-2x}\cos x \leq e^{-2x}$. We know that $\displaystyle\lim_{x \to \infty}(-e^{-2x}) = 0$ and

$\displaystyle\lim_{x \to \infty}(e^{-2x}) = 0$, so by the Squeeze Theorem, $\displaystyle\lim_{x \to \infty}(e^{-2x}\cos x) = 0$.

32. If we let $t = \tan x$, then as $x \to (\pi/2)^+$, $t \to -\infty$. Thus, $\displaystyle\lim_{x \to (\pi/2)^+} e^{\tan x} = \lim_{t \to -\infty} e^t = 0$.

33. By the Product Rule, $f(x) = (x^3 + 2x)e^x \Rightarrow$

$$f'(x) = (x^3 + 2x)(e^x)' + e^x(x^3 + 2x)' = (x^3 + 2x)e^x + e^x(3x^2 + 2)$$

$$= e^x[(x^3 + 2x) + (3x^2 + 2)] = e^x(x^3 + 3x^2 + 2x + 2)$$

34. By the Quotient Rule, $y = \dfrac{e^x}{1+x} \quad \Rightarrow \quad y' = \dfrac{(1+x)\,e^x - e^x(1)}{(1+x)^2} = \dfrac{e^x + xe^x - e^x}{(x+1)^2} = \dfrac{xe^x}{(x+1)^2}.$

35. By (9), $y = e^{ax^3} \quad \Rightarrow \quad y' = e^{ax^3}\dfrac{d}{dx}\left(ax^3\right) = 3ax^2 e^{ax^3}.$

36. $y = e^u(\cos u + cu) \quad \Rightarrow \quad y' = e^u(-\sin u + c) + (\cos u + cu)e^u = e^u(\cos u - \sin u + cu + c)$

37. $f(u) = e^{1/u} \quad \Rightarrow \quad f'(u) = e^{1/u} \cdot \dfrac{d}{du}\left(\dfrac{1}{u}\right) = e^{1/u}\left(\dfrac{-1}{u^2}\right) = \left(\dfrac{-1}{u^2}\right)e^{1/u}$

38. By the Product Rule, $g(x) = \sqrt{x}\,e^x = x^{1/2}e^x \quad \Rightarrow \quad g'(x) = x^{1/2}(e^x) + e^x\left(\tfrac{1}{2}x^{-1/2}\right) = \tfrac{1}{2}x^{-1/2}e^x(2x+1).$

39. By (9), $F(t) = e^{t\sin 2t} \quad \Rightarrow \quad F'(t) = e^{t\sin 2t}(t\sin 2t)' = e^{t\sin 2t}(t\cdot 2\cos 2t + \sin 2t \cdot 1) = e^{t\sin 2t}(2t\cos 2t + \sin 2t)$

40. $f(t) = \sin(e^t) + e^{\sin t} \quad \Rightarrow \quad f'(t) = \cos(e^t)\cdot e^t + e^{\sin t}\cdot \cos t = e^t\cos(e^t) + e^{\sin t}\cos t$

41. $y = \sqrt{1 + 2e^{3x}} \quad \Rightarrow \quad y' = \dfrac{1}{2}(1 + 2e^{3x})^{-1/2}\dfrac{d}{dx}(1 + 2e^{3x}) = \dfrac{1}{2\sqrt{1 + 2e^{3x}}}(2e^{3x}\cdot 3) = \dfrac{3e^{3x}}{\sqrt{1 + 2e^{3x}}}$

42. $y = e^{k\tan\sqrt{x}} \quad \Rightarrow \quad y' = e^{k\tan\sqrt{x}}\cdot \dfrac{d}{dx}\left(k\tan\sqrt{x}\right) = e^{k\tan\sqrt{x}}\left(k\sec^2\sqrt{x}\cdot\tfrac{1}{2}x^{-1/2}\right) = \dfrac{k\sec^2\sqrt{x}}{2\sqrt{x}}e^{k\tan\sqrt{x}}$

43. $y = e^{e^x} \quad \Rightarrow \quad y' = e^{e^x}\cdot \dfrac{d}{dx}\left(e^x\right) = e^{e^x}\cdot e^x \text{ or } e^{e^x+x}$

44. $y = \dfrac{e^u - e^{-u}}{e^u + e^{-u}} \quad \Rightarrow$

$$y' = \dfrac{(e^u + e^{-u})(e^u - (-e^{-u})) - (e^u - e^{-u})(e^u + (-e^{-u}))}{(e^u + e^{-u})^2} = \dfrac{e^{2u} + e^0 + e^0 + e^{-2u} - (e^{2u} - e^0 - e^0 + e^{-2u})}{(e^u + e^{-u})^2}$$

$$= \dfrac{4e^0}{(e^u + e^{-u})^2} = \dfrac{4}{(e^u + e^{-u})^2}$$

45. By the Quotient Rule, $y = \dfrac{ae^x + b}{ce^x + d} \quad \Rightarrow$

$$y' = \dfrac{(ce^x + d)(ae^x) - (ae^x + b)(ce^x)}{(ce^x + d)^2} = \dfrac{(ace^x + ad - ace^x - bc)e^x}{(ce^x + d)^2} = \dfrac{(ad - bc)e^x}{(ce^x + d)^2}.$$

46. $y = \sqrt{1 + xe^{-2x}} \quad \Rightarrow \quad y' = \tfrac{1}{2}\left(1 + xe^{-2x}\right)^{-1/2}\left[x\left(-2e^{-2x}\right) + e^{-2x}\right] = \dfrac{e^{-2x}(-2x + 1)}{2\sqrt{1 + xe^{-2x}}}$

47. $y = \cos\left(\dfrac{1 - e^{2x}}{1 + e^{2x}}\right) \quad \Rightarrow$

$$y' = -\sin\left(\dfrac{1 - e^{2x}}{1 + e^{2x}}\right)\cdot \dfrac{d}{dx}\left(\dfrac{1 - e^{2x}}{1 + e^{2x}}\right) = -\sin\left(\dfrac{1 - e^{2x}}{1 + e^{2x}}\right)\cdot \dfrac{(1 + e^{2x})(-2e^{2x}) - (1 - e^{2x})(2e^{2x})}{(1 + e^{2x})^2}$$

$$= -\sin\left(\dfrac{1 - e^{2x}}{1 + e^{2x}}\right)\cdot \dfrac{-2e^{2x}\left[(1 + e^{2x}) + (1 - e^{2x})\right]}{(1 + e^{2x})^2} = -\sin\left(\dfrac{1 - e^{2x}}{1 + e^{2x}}\right)\cdot \dfrac{-2e^{2x}(2)}{(1 + e^{2x})^2} = \dfrac{4e^{2x}}{(1 + e^{2x})^2}\cdot \sin\left(\dfrac{1 - e^{2x}}{1 + e^{2x}}\right)$$

48. $f(t) = \sin^2\left(e^{\sin^2 t}\right) = \left[\sin\left(e^{\sin^2 t}\right)\right]^2 \quad \Rightarrow$

$$f'(t) = 2\left[\sin\left(e^{\sin^2 t}\right)\right] \cdot \frac{d}{dt} \sin\left(e^{\sin^2 t}\right) = 2\sin\left(e^{\sin^2 t}\right) \cdot \cos\left(e^{\sin^2 t}\right) \cdot \frac{d}{dt} e^{\sin^2 t}$$

$$= 2\sin\left(e^{\sin^2 t}\right)\cos\left(e^{\sin^2 t}\right) \cdot e^{\sin^2 t} \cdot \frac{d}{dt}\sin^2 t = 2\sin\left(e^{\sin^2 t}\right)\cos\left(e^{\sin^2 t}\right) e^{\sin^2 t} \cdot 2\sin t \cos t$$

$$= 4\sin\left(e^{\sin^2 t}\right)\cos\left(e^{\sin^2 t}\right) e^{\sin^2 t} \sin t \cos t$$

49. $y = e^{2x}\cos \pi x \quad \Rightarrow \quad y' = e^{2x}(-\pi \sin \pi x) + (\cos \pi x)(2e^{2x}) = e^{2x}(2\cos \pi x - \pi \sin \pi x)$.

At $(0, 1)$, $y' = 1(2 - 0) = 2$, so an equation of the tangent line is $y - 1 = 2(x - 0)$, or $y = 2x + 1$.

50. $y = \dfrac{e^x}{x} \quad \Rightarrow \quad y' = \dfrac{x \cdot e^x - e^x \cdot 1}{x^2} = \dfrac{e^x(x-1)}{x^2}$.

At $(1, e)$, $y' = 0$, and an equation of the tangent line is $y - e = 0(x - 1)$, or $y = e$.

51. $\dfrac{d}{dx}\left(e^{x^2 y}\right) = \dfrac{d}{dx}(x + y) \quad \Rightarrow \quad e^{x^2 y}(x^2 y' + y \cdot 2x) = 1 + y' \quad \Rightarrow \quad x^2 e^{x^2 y} y' + 2xy e^{x^2 y} = 1 + y' \quad \Rightarrow$

$$x^2 e^{x^2 y} y' - y' = 1 - 2xy e^{x^2 y} \quad \Rightarrow \quad y'(x^2 e^{x^2 y} - 1) = 1 - 2xy e^{x^2 y} \quad \Rightarrow \quad y' = \dfrac{1 - 2xy e^{x^2 y}}{x^2 e^{x^2 y} - 1}$$

52. $xe^y + ye^x = 1 \quad \Rightarrow \quad xe^y y' + e^y \cdot 1 + ye^x + e^x y' = 0 \quad \Rightarrow \quad y'(xe^y + e^x) = -e^y - ye^x \quad \Rightarrow \quad y' = -\dfrac{e^y + ye^x}{xe^y + e^x}$. At

$(0, 1)$, $y' = -\dfrac{e + 1 \cdot 1}{0 + 1} = -(e + 1)$, so an equation for the tangent line is $y - 1 = -(e + 1)(x - 0)$, or $y = -(e + 1)x + 1$.

53. $y = e^x + e^{-x/2} \quad \Rightarrow \quad y' = e^x - \frac{1}{2}e^{-x/2} \quad \Rightarrow \quad y'' = e^x + \frac{1}{4}e^{-x/2}$, so

$$2y'' - y' - y = 2\left(e^x + \tfrac{1}{4}e^{-x/2}\right) - \left(e^x - \tfrac{1}{2}e^{-x/2}\right) - \left(e^x + e^{-x/2}\right) = 0.$$

54. $y = Ae^{-x} + Bxe^{-x} \quad \Rightarrow \quad y' = -Ae^{-x} + Be^{-x} - Bxe^{-x} = (B - A)e^{-x} - Bxe^{-x} \quad \Rightarrow$

$y'' = (A - B)e^{-x} - Be^{-x} + Bxe^{-x} = (A - 2B)e^{-x} + Bxe^{-x}$,

so $y'' + 2y' + y = (A - 2B)e^{-x} + Bxe^{-x} + 2\left[(B - A)e^{-x} - Bxe^{-x}\right] + Ae^{-x} + Bxe^{-x} = 0$.

55. $y = e^{rx} \quad \Rightarrow \quad y' = re^{rx} \quad \Rightarrow \quad y'' = r^2 e^{rx}$, so if $y = e^{rx}$ satisfies the differential equation $y'' + 6y' + 8y = 0$,

then $r^2 e^{rx} + 6re^{rx} + 8e^{rx} = 0$; that is, $e^{rx}(r^2 + 6r + 8) = 0$. Since $e^{rx} > 0$ for all x, we must have $r^2 + 6r + 8 = 0$,

or $(r + 2)(r + 4) = 0$, so $r = -2$ or -4.

56. $y = e^{\lambda x} \quad \Rightarrow \quad y' = \lambda e^{\lambda x} \quad \Rightarrow \quad y'' = \lambda^2 e^{\lambda x}$. Thus, $y + y' = y'' \quad \Leftrightarrow \quad e^{\lambda x} + \lambda e^{\lambda x} = \lambda^2 e^{\lambda x} \quad \Leftrightarrow$

$e^{\lambda x}(\lambda^2 - \lambda - 1) = 0 \quad \Leftrightarrow \quad \lambda = \frac{1 \pm \sqrt{5}}{2}$, since $e^{\lambda x} \neq 0$.

57. $f(x) = e^{2x} \quad \Rightarrow \quad f'(x) = 2e^{2x} \quad \Rightarrow \quad f''(x) = 2 \cdot 2e^{2x} = 2^2 e^{2x} \quad \Rightarrow$

$f'''(x) = 2^2 \cdot 2e^{2x} = 2^3 e^{2x} \quad \Rightarrow \quad \cdots \quad \Rightarrow \quad f^{(n)}(x) = 2^n e^{2x}$

58. $f(x) = xe^{-x} \quad \Rightarrow \quad f'(x) = x(-e^{-x}) + e^{-x} = (1 - x)e^{-x} \quad \Rightarrow$

$f''(x) = (1 - x)(-e^{-x}) + e^{-x}(-1) = (x - 2)e^{-x} \quad \Rightarrow \quad f'''(x) = (x - 2)(-e^{-x}) + e^{-x} = (3 - x)e^{-x} \quad \Rightarrow$

$f^{(4)}(x) = (3 - x)(-e^{-x}) + e^{-x}(-1) = (x - 4)e^{-x} \quad \Rightarrow \quad \cdots \quad \Rightarrow \quad f^{(n)}(x) = (-1)^n(x - n)e^{-x}$.

So $D^{1000}xe^{-x} = (x - 1000)e^{-x}$.

59. (a) $f(x) = e^x + x$ is continuous on $\mathbb{R}$ and $f(-1) = e^{-1} - 1 < 0 < 1 = f(0)$, so by the Intermediate Value Theorem,

$e^x + x = 0$ has a root in $(-1, 0)$.

(b) $f(x) = e^x + x \implies f'(x) = e^x + 1$, so $x_{n+1} = x_n - \dfrac{e^{x_n} + x_n}{e^{x_n} + 1}$. Using $x_1 = -0.5$, we get $x_2 \approx -0.566311$,

$x_3 \approx -0.567143 \approx x_4$, so the root is -0.567143 to six decimal places.

60.

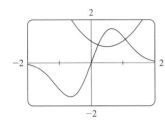

Solving $4e^{-x^2} \sin x = x^2 - x + 1$ is the same as solving

$f(x) = 4e^{-x^2} \sin x - x^2 + x - 1 = 0$.

$f'(x) = 4e^{-x^2}(\cos x - 2x \sin x) - 2x + 1 \implies$

$$x_{n+1} = x_n - \frac{4e^{-x_n^2} \sin x_n - x_n^2 + x_n - 1}{4e^{-x_n^2}(\cos x_n - 2x_n \sin x_n) - 2x_n + 1}.$$

From the figure, we see that the graphs intersect at approximately $x = 0.2$ and $x = 1.1$.

$x_1 = 0.2$	$x_1 = 1.1$
$x_2 \approx 0.21883273$	$x_2 \approx 1.08432830$
$x_3 \approx 0.21916357$	$x_3 \approx 1.08422462 \approx x_4$
$x_4 \approx 0.21916368 \approx x_5$	

To eight decimal places, the roots of the equation are 0.21916368 and 1.08422462.

61. (a) $\displaystyle\lim_{t \to \infty} p(t) = \lim_{t \to \infty} \frac{1}{1 + ae^{-kt}} = \frac{1}{1 + a \cdot 0} = 1$, since $k > 0 \implies -kt \to -\infty \implies e^{-kt} \to 0$.

(b) $p(t) = (1 + ae^{-kt})^{-1} \implies \dfrac{dp}{dt} = -(1 + ae^{-kt})^{-2}(-kae^{-kt}) = \dfrac{kae^{-kt}}{(1 + ae^{-kt})^2}$

(c)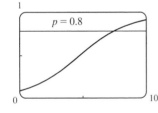

From the graph of $p(t) = (1 + 10e^{-0.5t})^{-1}$, it seems that $p(t) = 0.8$

(indicating that 80% of the population has heard the rumor) when

$t \approx 7.4$ hours.

62.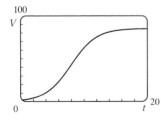

From the graph, we estimate that the most rapid increase in the percentage of

households in the United States with at least one VCR occurs at about $t = 8$.

To maximize the first derivative, we need to determine the values for which the

second derivative is 0. We'll use $V(t) = \dfrac{a}{1 + be^{ct}}$, and substitute $a = 85$,

$b = 53$, and $c = -0.5$ later.

$V'(t) = -\dfrac{a(bce^{ct})}{(1 + be^{ct})^2}$ [by the Reciprocal Rule] and

$V''(t) = -abc \cdot \dfrac{(1 + be^{ct})^2 \cdot ce^{ct} - e^{ct} \cdot 2(1 + be^{ct}) \cdot bce^{ct}}{[(1 + be^{ct})^2]^2} = \dfrac{-abc \cdot ce^{ct}(1 + be^{ct})[(1 + be^{ct}) - 2be^{ct}]}{(1 + be^{ct})^4}$

$= \dfrac{-abc^2 e^{ct}(1 - be^{ct})}{(1 + be^{ct})^3}$

So $V''(t) = 0 \iff 1 = be^{ct} \iff e^{ct} = 1/b \iff ct = \ln(1/b) \iff t = (1/c)\ln(1/b) = -2\ln\frac{1}{53} \approx 7.94$ years,

which corresponds to roughly midyear 1988.

63. $f(x) = x - e^x \Rightarrow f'(x) = 1 - e^x = 0 \Leftrightarrow e^x = 1 \Leftrightarrow x = 0$. Now $f'(x) > 0$ for all $x < 0$ and $f'(x) < 0$ for all $x > 0$, so the absolute maximum value is $f(0) = 0 - 1 = -1$.

64. $g(x) = \dfrac{e^x}{x} \Rightarrow g'(x) = \dfrac{xe^x - e^x}{x^2} = 0 \Leftrightarrow e^x(x-1) = 0 \Rightarrow x = 1$. Now $g'(x) > 0 \Leftrightarrow \dfrac{xe^x - e^x}{x^2} > 0 \Leftrightarrow$

$x - 1 > 0 \Leftrightarrow x > 1$ and $g'(x) < 0 \Leftrightarrow \dfrac{xe^x - e^x}{x^2} < 0 \Leftrightarrow x - 1 < 0 \Leftrightarrow x < 1$. Thus there is an absolute

minimum value of $g(1) = e$ at $x = 1$.

65. $f(x) = xe^{-x^2/8}$, $[-1, 4]$. $f'(x) = x \cdot e^{-x^2/8} \cdot \left(-\frac{x}{4}\right) + e^{-x^2/8} \cdot 1 = e^{-x^2/8}\left(-\frac{x^2}{4} + 1\right)$. Since $e^{-x^2/8}$ is never 0,

$f'(x) = 0 \Rightarrow -x^2/4 + 1 = 0 \Rightarrow 1 = x^2/4 \Rightarrow x^2 = 4 \Rightarrow x = \pm 2$, but -2 is not in the given interval, $[-1, 4]$.

$f(-1) = -e^{-1/8} \approx -0.88$, $f(2) = 2e^{-1/2} \approx 1.21$, and $f(4) = 4e^{-2} \approx 0.54$. So $f(2) = 2e^{-1/2}$ is the absolute maximum

value and $f(-1) = -e^{-1/8}$ is the absolute minimum value.

66. $f(x) = x^2 e^{-x/2}$, $[-1, 6] \Rightarrow f'(x) = x^2 e^{-x/2}\left(-\frac{1}{2}\right) + e^{-x/2}(2x) = xe^{-x/2}\left(-\frac{1}{2}x + 2\right)$. $f'(x) = 0 \Rightarrow$

$x = 0$ or 4. $f(-1) = e^{1/2} \approx 1.65$, $f(0) = 0$, $f(4) = 16e^{-2} \approx 2.17$, and $f(6) = 36e^{-3} \approx 1.79$. Thus, on $[-1, 6]$, the

absolute maximum value of f is $f(4) = 16e^{-2}$ and the absolute minimum value is $f(0) = 0$.

67. (a) $f(x) = (1 - x)e^{-x} \Rightarrow f'(x) = (1-x)(-e^{-x}) + e^{-x}(-1) = e^{-x}(x-2) > 0 \Rightarrow x > 2$, so f is increasing on $(2, \infty)$ and decreasing on $(-\infty, 2)$.

(b) $f''(x) = e^{-x}(1) + (x-2)(-e^{-x}) = e^{-x}(3-x) > 0 \Leftrightarrow x < 3$, so f is CU on $(-\infty, 3)$ and CD on $(3, \infty)$.

(c) f'' changes sign at $x = 3$, so there is an IP at $\left(3, -2e^{-3}\right)$.

68. (a) $f(x) = \dfrac{e^x}{x^2} \Rightarrow f'(x) = \dfrac{x^2 e^x - e^x(2x)}{(x^2)^2} = \dfrac{xe^x(x-2)}{x^4} = \dfrac{e^x(x-2)}{x^3}$. $f'(x) > 0 \Leftrightarrow x < 0$ or $x > 2$, so f is

increasing on $(-\infty, 0)$ and $(2, \infty)$. $f'(x) < 0 \Leftrightarrow 0 < x < 2$, so f is decreasing on $(0, 2)$.

(b) $f''(x) = \dfrac{x^3[e^x \cdot 1 + (x-2)e^x] - e^x(x-2) \cdot 3x^2}{(x^3)^2} = \dfrac{x^2 e^x[x(x-1) - 3(x-2)]}{x^6} = \dfrac{e^x(x^2 - 4x + 6)}{x^4}$.

$x^2 - 4x + 6 = (x^2 - 4x + 4) + 2 = (x-2)^2 + 2 > 0$, so $f''(x) > 0$ and f is CU on $(-\infty, 0)$ and $(0, \infty)$.

(c) There are no changes in concavity and, hence, there are no points of inflection.

69. $y = f(x) = e^{-1/(x+1)}$ **A.** $D = \{x \mid x \neq -1\} = (-\infty, -1) \cup (-1, \infty)$ **B.** No x-intercept; y-intercept $= f(0) = e^{-1}$

C. No symmetry   **D.** $\lim\limits_{x \to \pm\infty} e^{-1/(x+1)} = 1$ since $-1/(x+1) \to 0$, so $y = 1$ is a HA. $\lim\limits_{x \to -1^+} e^{-1/(x+1)} = 0$ since

$-1/(x+1) \to -\infty$, $\lim\limits_{x \to -1^-} e^{-1/(x+1)} = \infty$ since $-1/(x+1) \to \infty$, so $x = -1$ is a VA.

E. $f'(x) = e^{-1/(x+1)}/(x+1)^2 \Rightarrow f'(x) > 0$ for all x except 1, so

f is increasing on $(-\infty, -1)$ and $(-1, \infty)$.   **F.** No extreme values

G. $f''(x) = \dfrac{e^{-1/(x+1)}}{(x+1)^4} + \dfrac{e^{-1/(x+1)}(-2)}{(x+1)^3} = -\dfrac{e^{-1/(x+1)}(2x+1)}{(x+1)^4} \Rightarrow$

$f''(x) > 0 \Leftrightarrow 2x + 1 < 0 \Leftrightarrow x < -\frac{1}{2}$, so f is CU on $(-\infty, -1)$

and $\left(-1, -\frac{1}{2}\right)$, and CD on $\left(-\frac{1}{2}, \infty\right)$. f has an IP at $\left(-\frac{1}{2}, e^{-2}\right)$.

H.

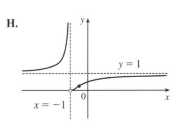

70. $y = f(x) = e^{2x} - e^x$ **A.** $D = \mathbb{R}$ **B.** y-intercept: $f(0) = 0$; x-intercepts: $f(x) = 0 \Rightarrow e^{2x} = e^x \Rightarrow e^x = 1 \Rightarrow x = 0$. **C.** No symmetry **D.** $\lim\limits_{x \to -\infty} e^{2x} - e^x = 0$, so $y = 0$ is a HA. No VA. **E.** $f'(x) = 2e^{2x} - e^x = e^x(2e^x - 1)$,

so $f'(x) > 0 \Leftrightarrow e^x > \frac{1}{2} \Leftrightarrow x > \ln \frac{1}{2} = -\ln 2$ and $f'(x) < 0 \Leftrightarrow$

$e^x < \frac{1}{2} \Leftrightarrow x < \ln \frac{1}{2}$, so f is decreasing on $\left(-\infty, \ln \frac{1}{2}\right)$

and increasing on $\left(\ln \frac{1}{2}, \infty\right)$.

F. Local minimum value $f\left(\ln \frac{1}{2}\right) = e^{2\ln(1/2)} - e^{\ln(1/2)} = \left(\frac{1}{2}\right)^2 - \frac{1}{2} = -\frac{1}{4}$

G. $f''(x) = 4e^{2x} - e^x = e^x(4e^x - 1)$, so $f''(x) > 0 \Leftrightarrow e^x > \frac{1}{4} \Leftrightarrow$

$x > \ln \frac{1}{4}$ and $f''(x) < 0 \Leftrightarrow x < \ln \frac{1}{4}$. Thus, f is CD on $\left(-\infty, \ln \frac{1}{4}\right)$ and

CU on $\left(\ln \frac{1}{4}, \infty\right)$. IP at $\left(\ln \frac{1}{4}, \left(\frac{1}{4}\right)^2 - \frac{1}{4}\right) = \left(\ln \frac{1}{4}, -\frac{3}{16}\right)$

H.
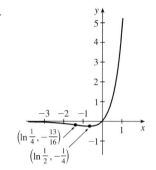

71. $S(t) = At^p e^{-kt}$ with $A = 0.01$, $p = 4$, and $k = 0.07$. We will find the zeros of f'' for $f(t) = t^p e^{-kt}$.

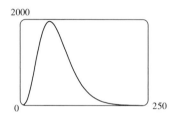

$f'(t) = t^p(-ke^{-kt}) + e^{-kt}(pt^{p-1}) = e^{-kt}(-kt^p + pt^{p-1})$

$f''(t) = e^{-kt}(-kpt^{p-1} + p(p-1)t^{p-2}) + (-kt^p + pt^{p-1})(-ke^{-kt})$

$\qquad = t^{p-2}e^{-kt}[-kpt + p(p-1) + k^2t^2 - kpt]$

$\qquad = t^{p-2}e^{-kt}(k^2t^2 - 2kpt + p^2 - p)$

Using the given values of p and k gives us $f''(t) = t^2e^{-0.07t}(0.0049t^2 - 0.56t + 12)$. So $S''(t) = 0.01f''(t)$ and its zeros are $t = 0$ and the solutions of $0.0049t^2 - 0.56t + 12 = 0$, which are $t_1 = \frac{200}{7} \approx 28.57$ and $t_2 = \frac{600}{7} \approx 85.71$.

At t_1 minutes, the rate of increase of the level of medication in the bloodstream is at its greatest and at t_2 minutes, the rate of decrease is the greatest.

72. The function $f(x) = e^{\cos x}$ is periodic with period 2π, so we consider it only on the interval $[0, 2\pi]$. We see that it has local maxima of about $f(0) \approx 2.72$ and $f(2\pi) \approx 2.72$, and a local minimum of about $f(3.14) \approx 0.37$. To find the

exact values, we calculate $f'(x) = -\sin x \, e^{\cos x}$. This is 0 when $-\sin x = 0 \Leftrightarrow x = 0$, π or 2π (since we are only considering $x \in [0, 2\pi]$). Also $f'(x) > 0 \Leftrightarrow \sin x < 0 \Leftrightarrow 0 < x < \pi$. So $f(0) = f(2\pi) = e$ (both maxima) and $f(\pi) = e^{\cos \pi} = 1/e$ (minimum). To find the inflection points, we calculate and graph

$f''(x) = \dfrac{d}{dx}\left(-\sin x \, e^{\cos x}\right) = -\cos x \, e^{\cos x} - \sin x(e^{\cos x})(-\sin x) = e^{\cos x}\left(\sin^2 x - \cos x\right)$. From the graph of $f''(x)$,

we see that f has inflection points at $x \approx 0.90$ and at $x \approx 5.38$. These x-coordinates correspond to inflection points $(0.90, 1.86)$ and $(5.38, 1.86)$.

73. $f(x) = e^{x^3 - x} \to 0$ as $x \to -\infty$, and $f(x) \to \infty$ as $x \to \infty$. From the graph, it appears that f has a local minimum of about $f(0.58) = 0.68$, and a local maximum of about $f(-0.58) = 1.47$. To find the exact values, we calculate

$f'(x) = (3x^2 - 1)e^{x^3 - x}$, which is 0 when $3x^2 - 1 = 0 \Leftrightarrow x = \pm\frac{1}{\sqrt{3}}$. The negative root corresponds to the local

maximum $f\left(-\frac{1}{\sqrt{3}}\right) = e^{(-1/\sqrt{3})^3 - (-1/\sqrt{3})} = e^{2\sqrt{3}/9}$, and the positive root corresponds to the local minimum

$f\left(\frac{1}{\sqrt{3}}\right) = e^{(1/\sqrt{3})^3 - (1/\sqrt{3})} = e^{-2\sqrt{3}/9}$. To estimate the inflection points, we calculate and graph

$$f''(x) = \frac{d}{dx}\left[(3x^2 - 1)e^{x^3 - x}\right] = (3x^2 - 1)e^{x^3 - x}(3x^2 - 1) + e^{x^3 - x}(6x) = e^{x^3 - x}(9x^4 - 6x^2 + 6x + 1).$$

From the graph, it appears that $f''(x)$ changes sign (and thus f has inflection points) at $x \approx -0.15$ and $x \approx -1.09$. From the graph of f, we see that these x-values correspond to inflection points at about $(-0.15, 1.15)$ and $(-1.09, 0.82)$.

74. (a) As $|x| \to \infty$, $t = -x^2/(2\sigma^2) \to -\infty$, and $e^t \to 0$. The HA is $y = 0$. Since t takes on its maximum value at $x = 0$, so

does e^t. Showing this result using derivatives, we have $f(x) = e^{-x^2/(2\sigma^2)} \Rightarrow f'(x) = e^{-x^2/(2\sigma^2)}(-x/\sigma^2)$.

$f'(x) = 0 \Leftrightarrow x = 0$. Because f' changes from positive to negative at $x = 0$, $f(0) = 1$ is a local maximum. For

inflection points, we find $f''(x) = -\frac{1}{\sigma^2}\left[e^{-x^2/(2\sigma^2)} \cdot 1 + xe^{-x^2/(2\sigma^2)}(-x/\sigma^2)\right] = \frac{-1}{\sigma^2}e^{-x^2/(2\sigma^2)}(1 - x^2/\sigma^2)$.

$f''(x) = 0 \Leftrightarrow x^2 = \sigma^2 \Leftrightarrow x = \pm\sigma$. $f''(x) < 0 \Leftrightarrow x^2 < \sigma^2 \Leftrightarrow -\sigma < x < \sigma$.

So f is CD on $(-\sigma, \sigma)$ and CU on $(-\infty, -\sigma)$ and (σ, ∞). IP at $(\pm\sigma, e^{-1/2})$.

(b) Since we have IP at $x = \pm\sigma$, the inflection points move away from the y-axis as σ increases.

(c)

From the graph, we see that as σ increases, the graph tends to spread out and there is more area between the curve and the x-axis.

75. Let $u = -3x$. Then $du = -3\,dx$, so $\int_0^5 e^{-3x}\,dx = -\frac{1}{3}\int_0^{-15} e^u\,du = -\frac{1}{3}[e^u]_0^{-15} = -\frac{1}{3}\left(e^{-15} - e^0\right) = \frac{1}{3}\left(1 - e^{-15}\right).$

76. Let $u = -x^2$, so $du = -2x\,dx$. When $x = 0$, $u = 0$; when $x = 1$, $u = -1$. Thus,

$$\int_0^1 xe^{-x^2}\,dx = \int_0^{-1} e^u\left(-\frac{1}{2}\,du\right) = -\frac{1}{2}[e^u]_0^{-1} = -\frac{1}{2}\left(e^{-1} - e^0\right) = \frac{1}{2}(1 - 1/e).$$

77. Let $u = 1 + e^x$. Then $du = e^x\,dx$, so $\int e^x\sqrt{1 + e^x}\,dx = \int \sqrt{u}\,du = \frac{2}{3}u^{3/2} + C = \frac{2}{3}(1 + e^x)^{3/2} + C.$

78. $\displaystyle\int \frac{(1 + e^x)^2}{e^x}\,dx = \int \frac{1 + 2e^x + e^{2x}}{e^x}\,dx = \int (e^{-x} + 2 + e^x)\,dx = -e^{-x} + 2x + e^x + C$

79. $\int (e^x + e^{-x})^2\,dx = \int (e^{2x} + 2 + e^{-2x})\,dx = \frac{1}{2}e^{2x} + 2x - \frac{1}{2}e^{-2x} + C$

80. $\int e^x (4 + e^x)^5 \, dx$ $\begin{bmatrix} u = 4 + e^x, \\ du = e^x \, dx \end{bmatrix}$ $= \int u^5 \, du = \frac{1}{6} u^6 + C = \frac{1}{6} (4 + e^x)^6 + C$

81. $\int \sin x \, e^{\cos x} \, dx$ $\begin{bmatrix} u = \cos x, \\ du = -\sin x \, dx \end{bmatrix}$ $= \int e^u \, (-du) = -e^u + C = -e^{\cos x} + C$

82. Let $u = \dfrac{1}{x}$. Then $du = -\dfrac{1}{x^2} \, dx$, so $\displaystyle\int \frac{e^{1/x}}{x^2} \, dx = -\int e^u \, du = -e^u + C = -e^{1/x} + C$.

83. Let $u = \sqrt{x}$. Then $du = \dfrac{1}{2\sqrt{x}} \, dx$, so $\displaystyle\int \frac{e^{\sqrt{x}}}{\sqrt{x}} \, dx = 2 \int e^u \, du = 2e^u + C = 2e^{\sqrt{x}} + C$.

84. Let $u = e^x$. Then $du = e^x \, dx$, so $\int e^x \sin(e^x) \, dx = \int \sin u \, du = -\cos u + C = -\cos(e^x) + C$.

85. Area $= \int_0^1 \left(e^{3x} - e^x \right) dx = \left[\frac{1}{3} e^{3x} - e^x \right]_0^1 = \left(\frac{1}{3} e^3 - e \right) - \left(\frac{1}{3} - 1 \right) = \frac{1}{3} e^3 - e + \frac{2}{3} \approx 4.644$

86. $f''(x) = 3e^x + 5 \sin x$ $\Rightarrow$ $f'(x) = 3e^x - 5 \cos x + C$ $\Rightarrow$ $2 = f'(0) = 3 - 5 + C$ $\Rightarrow$ $C = 4$, so

$f'(x) = 3e^x - 5 \cos x + 4$ $\Rightarrow$ $f(x) = 3e^x - 5 \sin x + 4x + D$ $\Rightarrow$ $1 = f(0) = 3 + D$ $\Rightarrow$ $D = -2$,

so $f(x) = 3e^x - 5 \sin x + 4x - 2$.

87. $V = \int_0^1 \pi (e^x)^2 \, dx = \pi \int_0^1 e^{2x} \, dx = \frac{1}{2} \pi \left[e^{2x} \right]_0^1 = \frac{\pi}{2} \left(e^2 - 1 \right)$

88. $V = \int_0^1 2\pi x e^{-x^2} \, dx$. Let $u = x^2$.

Thus, $du = 2x \, dx$, so

$V = \pi \int_0^1 e^{-u} \, du = \pi \left[-e^{-u} \right]_0^1 = \pi (1 - 1/e)$.

89. (a) $\operatorname{erf}(x) = \dfrac{2}{\sqrt{\pi}} \displaystyle\int_0^x e^{-t^2} \, dt$ $\Rightarrow$ $\displaystyle\int_0^x e^{-t^2} \, dt = \dfrac{\sqrt{\pi}}{2} \operatorname{erf}(x)$. By Property 5 of definite integrals in Section 5.2,

$\int_0^b e^{-t^2} \, dt = \int_0^a e^{-t^2} \, dt + \int_a^b e^{-t^2} \, dt$, so

$\displaystyle\int_a^b e^{-t^2} \, dt = \int_0^b e^{-t^2} \, dt - \int_0^a e^{-t^2} \, dt = \dfrac{\sqrt{\pi}}{2} \operatorname{erf}(b) - \dfrac{\sqrt{\pi}}{2} \operatorname{erf}(a) = \frac{1}{2} \sqrt{\pi} \, [\operatorname{erf}(b) - \operatorname{erf}(a)]$.

(b) $y = e^{x^2} \operatorname{erf}(x)$ $\Rightarrow$ $y' = 2x e^{x^2} \operatorname{erf}(x) + e^{x^2} \operatorname{erf}'(x) = 2xy + e^{x^2} \cdot \dfrac{2}{\sqrt{\pi}} e^{-x^2}$ [by FTC1] $= 2xy + \dfrac{2}{\sqrt{\pi}}$.

90. Let $r(t) = a e^{bt}$ with $a = 450.268$ and $b = 1.12567$, and $n(t) =$ population after t hours. Since $r(t) = n'(t)$,

$\int_0^3 r(t) \, dt = n(3) - n(0)$ is the total change in the population after three hours. Since we start with 400 bacteria, the

population will be

$$n(3) = 400 + \int_0^3 r(t) \, dt = 400 + \int_0^3 a e^{bt} \, dt = 400 + \frac{a}{b} \left[e^{bt} \right]_0^3 = 400 + \frac{a}{b} \left(e^{3b} - 1 \right)$$

$$\approx 400 + 11{,}313 = 11{,}713 \text{ bacteria}$$

91. We use Theorem 7.1.7. Note that $f(0) = 3 + 0 + e^0 = 4$, so $f^{-1}(4) = 0$. Also $f'(x) = 1 + e^x$. Therefore,

$$\left(f^{-1}\right)'(4) = \frac{1}{f'(f^{-1}(4))} = \frac{1}{f'(0)} = \frac{1}{1 + e^0} = \frac{1}{2}.$$

92. We recognize this limit as the definition of the derivative of the function $f(x) = e^{\sin x}$ at $x = \pi$, since it is of the form

$$\lim_{x \to \pi} \frac{f(x) - f(\pi)}{x - \pi}.$$ Therefore, the limit is equal to $f'(\pi) = (\cos \pi)e^{\sin \pi} = -1 \cdot e^0 = -1.$

93.

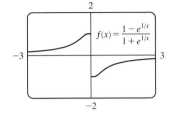

From the graph, it appears that f is an odd function (f is undefined for $x = 0$).
To prove this, we must show that $f(-x) = -f(x)$.

$$f(-x) = \frac{1 - e^{1/(-x)}}{1 + e^{1/(-x)}} = \frac{1 - e^{(-1/x)}}{1 + e^{(-1/x)}} = \frac{1 - \dfrac{1}{e^{1/x}}}{1 + \dfrac{1}{e^{1/x}}} \cdot \frac{e^{1/x}}{e^{1/x}} = \frac{e^{1/x} - 1}{e^{1/x} + 1}$$

$$= -\frac{1 - e^{1/x}}{1 + e^{1/x}} = -f(x)$$

so f is an odd function.

94. We'll start with $b = -1$ and graph $f(x) = \dfrac{1}{1 + ae^{bx}}$ for $a = 0.1$, 1, and 5.

From the graph, we see that there is a horizontal asymptote $y = 0$ as $x \to -\infty$
and a horizontal asymptote $y = 1$ as $x \to \infty$. If $a = 1$, the y-intercept is $\left(0, \frac{1}{2}\right)$.
As a gets smaller (close to 0), the graph of f moves left. As a gets larger, the graph
of f moves right.

As b changes from -1 to 0, the graph of f is stretched horizontally. As b
changes through large negative values, the graph of f is compressed horizontally.
(This takes care of negatives values of b.)

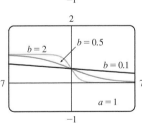

If b is positive, the graph of f is reflected through the y-axis.

Last, if $b = 0$, the graph of f is the horizontal line $y = 1/(1 + a)$.

95. Using the second law of logarithms and Equation 5, we have $\ln(e^x/e^y) = \ln e^x - \ln e^y = x - y = \ln(e^{x-y})$. Since $\ln$ is a
one-to-one function, it follows that $e^x/e^y = e^{x-y}$.

96. Using the third law of logarithms and Equation 5, we have $\ln e^{rx} = rx = r \ln e^x = \ln (e^x)^r$. Since $\ln$ is a one-to-one
function, it follows that $e^{rx} = (e^x)^r$.

97. (a) Let $f(x) = e^x - 1 - x$. Now $f(0) = e^0 - 1 = 0$, and for $x \geq 0$, we have $f'(x) = e^x - 1 \geq 0$. Now, since $f(0) = 0$ and

f is increasing on $[0, \infty)$, $f(x) \geq 0$ for $x \geq 0$ $\Rightarrow$ $e^x - 1 - x \geq 0$ $\Rightarrow$ $e^x \geq 1 + x$.

(b) For $0 \leq x \leq 1$, $x^2 \leq x$, so $e^{x^2} \leq e^x$ [since e^x is increasing]. Hence [from (a)] $1 + x^2 \leq e^{x^2} \leq e^x$.

So $\frac{4}{3} = \int_0^1 \left(1 + x^2\right) dx \leq \int_0^1 e^{x^2} dx \leq \int_0^1 e^x dx = e - 1 < e$ $\Rightarrow$ $\frac{4}{3} \leq \int_0^1 e^{x^2} dx \leq e$.

98. (a) Let $f(x) = e^x - 1 - x - \frac{1}{2}x^2$. Thus, $f'(x) = e^x - 1 - x$, which is positive for $x \geq 0$ by Exercise 97(a). Thus $f(x)$ is

increasing on $(0, \infty)$, so on that interval, $0 = f(0) \leq f(x) = e^x - 1 - x - \frac{1}{2}x^2$ $\Rightarrow$ $e^x \geq 1 + x + \frac{1}{2}x^2$.

(b) Using the same argument as in Exercise 97(b), from part (a) we have $1 + x^2 + \frac{1}{2}x^4 \leq e^{x^2} \leq e^x$

[for $0 \leq x \leq 1$] $\Rightarrow$ $\int_0^1 \left(1 + x^2 + \frac{1}{2}x^4\right) dx \leq \int_0^1 e^{x^2} dx \leq \int_0^1 e^x dx$ $\Rightarrow$ $\frac{43}{30} \leq \int_0^1 e^{x^2} dx \leq e - 1$.

99. (a) By Exercise 97(a), the result holds for $n = 1$. Suppose that $e^x \geq 1 + x + \dfrac{x^2}{2!} + \cdots + \dfrac{x^k}{k!}$ for $x \geq 0$.

Let $f(x) = e^x - 1 - x - \dfrac{x^2}{2!} - \cdots - \dfrac{x^k}{k!} - \dfrac{x^{k+1}}{(k+1)!}$. Then $f'(x) = e^x - 1 - x - \cdots - \dfrac{x^k}{k!} \geq 0$ by assumption. Hence

$f(x)$ is increasing on $(0, \infty)$. So $0 \leq x$ implies that $0 = f(0) \leq f(x) = e^x - 1 - x - \cdots - \dfrac{x^k}{k!} - \dfrac{x^{k+1}}{(k+1)!}$, and hence

$e^x \geq 1 + x + \cdots + \dfrac{x^k}{k!} + \dfrac{x^{k+1}}{(k+1)!}$ for $x \geq 0$. Therefore, for $x \geq 0$, $e^x \geq 1 + x + \dfrac{x^2}{2!} + \cdots + \dfrac{x^n}{n!}$ for every positive

integer n, by mathematical induction.

(b) Taking $n = 4$ and $x = 1$ in (a), we have $e = e^1 \geq 1 + \frac{1}{2} + \frac{1}{6} + \frac{1}{24} = 2.708\overline{3} > 2.7$.

(c) $e^x \geq 1 + x + \cdots + \dfrac{x^k}{k!} + \dfrac{x^{k+1}}{(k+1)!}$ $\Rightarrow$ $\dfrac{e^x}{x^k} \geq \dfrac{1}{x^k} + \dfrac{1}{x^{k-1}} + \cdots + \dfrac{1}{k!} + \dfrac{x}{(k+1)!} \geq \dfrac{x}{(k+1)!}$.

But $\lim\limits_{x \to \infty} \dfrac{x}{(k+1)!} = \infty$, so $\lim\limits_{x \to \infty} \dfrac{e^x}{x^k} = \infty$.

100. (a) The graph of g finally surpasses that of f at $x \approx 35.8$. (b)

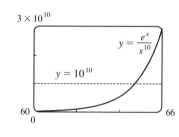

(c) From the graph in part (b), it seems that $e^x/x^{10} > 10^{10}$ whenever $x > 65$, approximately. So we can take $N \geq 65$.

7.4* General Logarithmic and Exponential Functions

1. (a) $a^x = e^{x \ln a}$

(b) The domain of $f(x) = a^x$ is $\mathbb{R}$.

(c) The range of $f(x) = a^x$ $[a \neq 1]$ is $(0, \infty)$.

(d) (i) See Figure 1.   (ii) See Figure 3.   (iii) See Figure 2.

2. (a) $\log_a x$ is the number y such that $a^y = x$.   (b) The domain of $f(x) = \log_a x$ is $(0, \infty)$.

(c) The range of $f(x) = \log_a x$ is $\mathbb{R}$.   (d) See Figure 9.

3. $5^{\sqrt{7}} = \left(e^{\ln 5}\right)^{\sqrt{7}} = e^{\sqrt{7}\ln 5}$ **4.** $10^{x^2} = \left(e^{\ln 10}\right)^{x^2} = e^{x^2 \ln 10}$

5. $(\cos x)^x = \left(e^{\ln \cos x}\right)^x = e^{x \ln(\cos x)}$.   **6.** $x^{\cos x} = \left(e^{\ln x}\right)^{\cos x} = e^{(\cos x)(\ln x)}$

7. (a) $\log_5 125 = 3$ since $5^3 = 125$.   (b) $\log_3 \dfrac{1}{27} = -3$ since $3^{-3} = \dfrac{1}{3^3} = \dfrac{1}{27}$.

8. (a) $\log_{10} \sqrt{10} = \log_{10} 10^{1/2} = \frac{1}{2}$ [analogous to (5) in Section 7.3*].

(b) $\log_8 320 - \log_8 5 = \log_8 \frac{320}{5} = \log_8 64 = 2$ since $8^2 = 64$.

9. (a) $\log_2 6 - \log_2 15 + \log_2 20 = \log_2\left(\frac{6}{15}\right) + \log_2 20$ [by Law 2]

$= \log_2\left(\frac{6}{15} \cdot 20\right)$ [by Law 1]

$= \log_2 8$, and $\log_2 8 = 3$ since $2^3 = 8$.

(b) $\log_3 100 - \log_3 18 - \log_3 50 = \log_3\left(\frac{100}{18}\right) - \log_3 50 = \log_3\left(\frac{100}{18 \cdot 50}\right)$

$= \log_3\left(\frac{1}{9}\right)$, and $\log_3\left(\frac{1}{9}\right) = -2$ since $3^{-2} = \frac{1}{9}$.

10. (a) $\log_a \dfrac{1}{a} = -1$ since $a^{-1} = \dfrac{1}{a}$. $\left[\textit{Or: } \log_a \dfrac{1}{a} = \log_a a^{-1} = -1\right]$

(b) $10^{(\log_{10} 4 + \log_{10} 7)} = 10^{\log_{10} 4} \cdot 10^{\log_{10} 7} = 4 \cdot 7 = 28$

$\left[\textit{Or: } 10^{(\log_{10} 4 + \log_{10} 7)} = 10^{\log_{10}(4 \cdot 7)} = 10^{\log_{10} 28} = 28\right]$

11. All of these graphs approach 0 as $x \to -\infty$, all of them pass through the point $(0, 1)$, and all of them are increasing and approach ∞ as $x \to \infty$. The larger the base, the faster the function increases for $x > 0$, and the faster it approaches 0 as $x \to -\infty$.

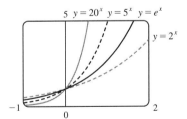

12. The functions with bases greater than 1 (3^x and 10^x) are increasing, while those with bases less than 1 $\left[\left(\frac{1}{3}\right)^x$ and $\left(\frac{1}{10}\right)^x\right]$ are decreasing. The graph of $\left(\frac{1}{3}\right)^x$ is the reflection of that of 3^x about the y-axis, and the graph of $\left(\frac{1}{10}\right)^x$ is the reflection of that of 10^x about the y-axis. The graph of 10^x increases more quickly than that of 3^x for $x > 0$, and approaches 0 faster as $x \to -\infty$.

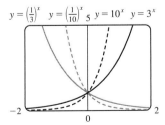

13. (a) $\log_{12} e = \dfrac{\ln e}{\ln 12} = \dfrac{1}{\ln 12} \approx 0.402430$

(b) $\log_6 13.54 = \dfrac{\ln 13.54}{\ln 6} \approx 1.454240$

(c) $\log_2 \pi = \dfrac{\ln \pi}{\ln 2} \approx 1.651496$

14. To graph the functions, we use $\log_2 x = \dfrac{\ln x}{\ln 2}$, $\log_4 x = \dfrac{\ln x}{\ln 4}$, etc. These

graphs all approach $-\infty$ as $x \to 0^+$, and they all pass through the point

$(1, 0)$. Also, they are all increasing, and all approach ∞ as $x \to \infty$. The

smaller the base, the larger the rate of increase of the function (for $x > 1$)

and the closer the approach to the y-axis (as $x \to 0^+$).

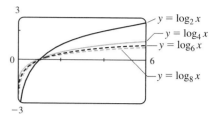

15. To graph these functions, we use $\log_{1.5} x = \dfrac{\ln x}{\ln 1.5}$ and $\log_{50} x = \dfrac{\ln x}{\ln 50}$.

These graphs all approach $-\infty$ as $x \to 0^+$, and they all pass through the

point $(1, 0)$. Also, they are all increasing, and all approach ∞ as $x \to \infty$.

The functions with larger bases increase extremely slowly, and the ones with

smaller bases do so somewhat more quickly. The functions with large bases

approach the y-axis more closely as $x \to 0^+$.

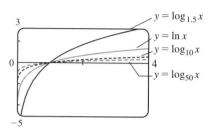

16. We see that the graph of $\ln x$ is the reflection of the graph of e^x about the

line $y = x$, and that the graph of $\log_{10} x$ is the reflection of the graph of 10^x

about the same line. The graph of 10^x increases more quickly than that

of e^x. Also note that $\log_{10} x \to \infty$ as $x \to \infty$ more slowly than $\ln x$.

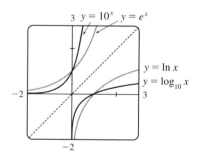

17. Use $y = Ca^x$ with the points $(1, 6)$ and $(3, 24)$. $6 = Ca^1$ $\left[C = \dfrac{6}{a} \right]$ and $24 = Ca^3$ $\Rightarrow$ $24 = \left(\dfrac{6}{a} \right) a^3$ $\Rightarrow$

$4 = a^2$ $\Rightarrow$ $a = 2$ [since $a > 0$]   and $C = \dfrac{6}{2} = 3$. The function is $f(x) = 3 \cdot 2^x$.

18. Given the y-intercept $(0, 2)$, we have $y = Ca^x = 2a^x$. Using the point $\left(2, \dfrac{2}{9} \right)$ gives us $\dfrac{2}{9} = 2a^2$ $\Rightarrow$ $\dfrac{1}{9} = a^2$ $\Rightarrow$ $a = \dfrac{1}{3}$

[since $a > 0$].   The function is $f(x) = 2\left(\dfrac{1}{3} \right)^x$ or $f(x) = 2(3)^{-x}$.

19. (a) 2 ft = 24 in, $f(24) = 24^2$ in = 576 in = 48 ft. $g(24) = 2^{24}$ in = $2^{24} / (12 \cdot 5280)$ mi ≈ 265 mi

(b) 3 ft = 36 in, so we need x such that $\log_2 x = 36$ $\Leftrightarrow$ $x = 2^{36} = 68{,}719{,}476{,}736$. In miles, this is

$68{,}719{,}476{,}736$ in $\cdot \dfrac{1 \text{ ft}}{12 \text{ in}} \cdot \dfrac{1 \text{ mi}}{5280 \text{ ft}} \approx 1{,}084{,}587.7$ mi.

20. We see from the graphs that for x less than about 1.8, $g(x) = 5^x > f(x) = x^5$, and then near the point $(1.8, 17.1)$ the curves intersect. Then $f(x) > g(x)$ from $x \approx 1.8$ until $x = 5$. At $(5, 3125)$ there is another point of intersection, and for $x > 5$ we see that $g(x) > f(x)$. In fact, g increases much more rapidly than f beyond that point.

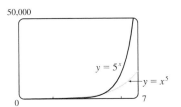

21. $\lim\limits_{x \to \infty} (1.001)^x = \infty$ by Figure 1, since $1.001 > 1$.

22. By Figure 1, if $a > 1$, $\lim\limits_{x \to -\infty} a^x = 0$, so $\lim\limits_{x \to -\infty} (1.001)^x = 0$.

23. $\lim\limits_{t \to \infty} 2^{-t^2} = \lim\limits_{u \to -\infty} 2^u$ [where $u = -t^2$] $= 0$

24. Let $t = x^2 - 5x + 6$. As $x \to 3^+$, $t = (x - 2)(x - 3) \to 0^+$. $\lim\limits_{x \to 3^+} \log_{10}(x^2 - 5x + 6) = \lim\limits_{t \to 0^+} \log_{10} t = -\infty$

[analogous to (4) in Section 7.2*].

25. $h(t) = t^3 - 3^t \Rightarrow h'(t) = 3t^2 - 3^t \ln 3$

26. $g(x) = x^4 4^x \Rightarrow g'(x) = x^4 4^x \ln 4 + 4^x \cdot 4x^3 = x^3 4^x (x \ln 4 + 4)$

27. Using Formula 4 and the Chain Rule, $y = 5^{-1/x} \Rightarrow y' = 5^{-1/x} (\ln 5) \left[-1 \cdot (-x^{-2}) \right] = 5^{-1/x} (\ln 5)/x^2$

28. $y = 10^{\tan \theta} \Rightarrow y' = 10^{\tan \theta} (\ln 10)(\sec^2 \theta)$

29. $f(u) = (2^u + 2^{-u})^{10} \Rightarrow$

$$f'(u) = 10(2^u + 2^{-u})^9 \frac{d}{du} (2^u + 2^{-u}) = 10(2^u + 2^{-u})^9 \left[2^u \ln 2 + 2^{-u} \ln 2 \cdot (-1) \right]$$
$$= 10 \ln 2 (2^u + 2^{-u})^9 (2^u - 2^{-u})$$

30. $y = 2^{3^{x^2}} \Rightarrow y' = 2^{3^{x^2}} (\ln 2) \frac{d}{dx} \left(3^{x^2} \right) = 2^{3^{x^2}} (\ln 2) 3^{x^2} (\ln 3)(2x)$

31. $f(x) = \log_2(1 - 3x) \Rightarrow f'(x) = \dfrac{1}{(1 - 3x) \ln 2} \dfrac{d}{dx} (1 - 3x) = \dfrac{-3}{(1 - 3x) \ln 2}$ or $\dfrac{3}{(3x - 1) \ln 2}$

32. $f(x) = \log_5(xe^x) \Rightarrow f'(x) = \dfrac{1}{xe^x \ln 5} \dfrac{d}{dx} (xe^x) = \dfrac{1}{xe^x \ln 5} (xe^x + e^x \cdot 1) = \dfrac{e^x(x + 1)}{xe^x \ln 5} = \dfrac{x + 1}{x \ln 5}$

Another solution: We can change the form of the function by first using logarithm properties.

$$f(x) = \log_5(xe^x) = \log_5 x + \log_5 e^x \Rightarrow f'(x) = \frac{1}{x \ln 5} + \frac{1}{e^x \ln 5} \cdot e^x = \frac{1}{x \ln 5} + \frac{1}{\ln 5} \text{ or } \frac{1 + x}{x \ln 5}$$

33. $y = 2x \log_{10} \sqrt{x} = 2x \log_{10} x^{1/2} = 2x \cdot \frac{1}{2} \log_{10} x = x \log_{10} x \;\;\Rightarrow\;\; y' = x \cdot \frac{1}{x \ln 10} + \log_{10} x \cdot 1 = \frac{1}{\ln 10} + \log_{10} x$

Note: $\frac{1}{\ln 10} = \frac{\ln e}{\ln 10} = \log_{10} e$, so the answer could be written as $\frac{1}{\ln 10} + \log_{10} x = \log_{10} e + \log_{10} x = \log_{10} ex$.

34. $y = \log_2(e^{-x} \cos \pi x) = \log_2 e^{-x} + \log_2 \cos \pi x = -x \log_2 e + \log_2 \cos \pi x \;\;\Rightarrow\;\;$

$y' = -\log_2 e + \frac{1}{\cos \pi x \, (\ln 2)} \frac{d}{dx}(\cos \pi x) = -\log_2 e + \frac{-\pi \sin \pi x}{\cos \pi x \, (\ln 2)} = -\log_2 e - \frac{\pi}{\ln 2} \tan \pi x$

Note: $\frac{1}{\ln 2} = \frac{\ln e}{\ln 2} = \log_2 e$, so the answer could be written as $-\log_2 e - \pi \log_2 e \tan \pi x = (-\log_2 e)(1 + \pi \tan \pi x)$.

35. $y = x^x \;\;\Rightarrow\;\; \ln y = \ln x^x \;\;\Rightarrow\;\; \ln y = x \ln x \;\;\Rightarrow\;\; y'/y = x(1/x) + (\ln x) \cdot 1 \;\;\Rightarrow\;\; y' = y(1 + \ln x) \;\;\Rightarrow\;\;$
$y' = x^x(1 + \ln x)$

36. $y = x^{\cos x} \;\;\Rightarrow\;\; \ln y = \ln x^{\cos x} \;\;\Rightarrow\;\; \ln y = \cos x \ln x \;\;\Rightarrow\;\; \frac{1}{y} y' = \cos x \cdot \frac{1}{x} + \ln x \cdot (-\sin x) \;\;\Rightarrow\;\;$

$y' = y\left(\frac{\cos x}{x} - \ln x \sin x\right) \;\;\Rightarrow\;\; y' = x^{\cos x}\left(\frac{\cos x}{x} - \ln x \sin x\right)$

37. $y = x^{\sin x} \;\;\Rightarrow\;\; \ln y = \ln x^{\sin x} \;\;\Rightarrow\;\; \ln y = \sin x \ln x \;\;\Rightarrow\;\; \frac{y'}{y} = (\sin x) \cdot \frac{1}{x} + (\ln x)(\cos x) \;\;\Rightarrow\;\;$

$y' = y\left(\frac{\sin x}{x} + \ln x \cos x\right) \;\;\Rightarrow\;\; y' = x^{\sin x}\left(\frac{\sin x}{x} + \ln x \cos x\right)$

38. $y = \sqrt{x}^{\,x} \;\;\Rightarrow\;\; \ln y = \ln \sqrt{x}^{\,x} \;\;\Rightarrow\;\; \ln y = x \ln x^{1/2} \;\;\Rightarrow\;\; \ln y = \frac{1}{2} x \ln x \;\;\Rightarrow\;\; \frac{1}{y} y' = \frac{1}{2} x \cdot \frac{1}{x} + \ln x \cdot \frac{1}{2} \;\;\Rightarrow\;\;$

$y' = y\left(\frac{1}{2} + \frac{1}{2} \ln x\right) \;\;\Rightarrow\;\; y' = \frac{1}{2} \sqrt{x}^{\,x}(1 + \ln x)$

39. $y = (\cos x)^x \;\;\Rightarrow\;\; \ln y = \ln(\cos x)^x \;\;\Rightarrow\;\; \ln y = x \ln \cos x \;\;\Rightarrow\;\; \frac{1}{y} y' = x \cdot \frac{1}{\cos x} \cdot (-\sin x) + \ln \cos x \cdot 1 \;\;\Rightarrow\;\;$

$y' = y\left(\ln \cos x - \frac{x \sin x}{\cos x}\right) \;\;\Rightarrow\;\; y' = (\cos x)^x (\ln \cos x - x \tan x)$

40. $y = (\sin x)^{\ln x} \;\;\Rightarrow\;\; \ln y = \ln(\sin x)^{\ln x} \;\;\Rightarrow\;\; \ln y = \ln x \cdot \ln \sin x \;\;\Rightarrow\;\; \frac{1}{y} y' = \ln x \cdot \frac{1}{\sin x} \cdot \cos x + \ln \sin x \cdot \frac{1}{x} \;\;\Rightarrow\;\;$

$y' = y\left(\ln x \cdot \frac{\cos x}{\sin x} + \frac{\ln \sin x}{x}\right) \;\;\Rightarrow\;\; y' = (\sin x)^{\ln x}\left(\ln x \cot x + \frac{\ln \sin x}{x}\right)$

41. $y = (\tan x)^{1/x} \;\;\Rightarrow\;\; \ln y = \ln(\tan x)^{1/x} \;\;\Rightarrow\;\; \ln y = \frac{1}{x} \ln \tan x \;\;\Rightarrow\;\;$

$\frac{1}{y} y' = \frac{1}{x} \cdot \frac{1}{\tan x} \cdot \sec^2 x + \ln \tan x \cdot \left(-\frac{1}{x^2}\right) \;\;\Rightarrow\;\; y' = y\left(\frac{\sec^2 x}{x \tan x} - \frac{\ln \tan x}{x^2}\right) \;\;\Rightarrow\;\;$

$y' = (\tan x)^{1/x}\left(\frac{\sec^2 x}{x \tan x} - \frac{\ln \tan x}{x^2}\right)$ or $y' = (\tan x)^{1/x} \cdot \frac{1}{x}\left(\csc x \sec x - \frac{\ln \tan x}{x}\right)$

42. $y = (\ln x)^{\cos x} \;\;\Rightarrow\;\; \ln y = \cos x \ln(\ln x) \;\;\Rightarrow\;\; \frac{y'}{y} = \cos x \cdot \frac{1}{\ln x} \cdot \frac{1}{x} + (\ln \ln x)(-\sin x) \;\;\Rightarrow\;\;$

$y' = (\ln x)^{\cos x}\left(\frac{\cos x}{x \ln x} - \sin x \ln \ln x\right)$

43. $y = 10^x \;\;\Rightarrow\;\; y' = 10^x \ln 10$, so at $(1, 10)$, the slope of the tangent line is $10^1 \ln 10 = 10 \ln 10$, and its equation is
$y - 10 = 10 \ln 10 (x - 1)$, or $y = (10 \ln 10)x + 10(1 - \ln 10)$.

44. $f(x) = x^{\cos x} = e^{\ln x \cos x} \quad \Rightarrow$

$$f'(x) = e^{\ln x \cos x}\left[\ln x(-\sin x) + \cos x\left(\frac{1}{x}\right)\right]$$

$$= x^{\cos x}\left[\frac{\cos x}{x} - \sin x \ln x\right]$$

This is reasonable, because the graph shows that f increases when $f'(x)$ is positive.

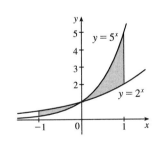

45. $\displaystyle\int_1^2 10^t \, dt = \left[\frac{10^t}{\ln 10}\right]_1^2 = \frac{10^2}{\ln 10} - \frac{10^1}{\ln 10} = \frac{100 - 10}{\ln 10} = \frac{90}{\ln 10}$

46. $\int \left(x^5 + 5^x\right) dx = \frac{1}{6}x^6 + \frac{1}{\ln 5}5^x + C$

47. $\displaystyle\int \frac{\log_{10} x}{x} \, dx = \int \frac{(\ln x)/(\ln 10)}{x} \, dx = \frac{1}{\ln 10}\int \frac{\ln x}{x} \, dx$. Now put $u = \ln x$, so $du = \dfrac{1}{x}\,dx$, and the expression becomes

$$\frac{1}{\ln 10}\int u \, du = \frac{1}{\ln 10}\left(\tfrac{1}{2}u^2 + C_1\right) = \frac{1}{2 \ln 10}(\ln x)^2 + C.$$

Or: The substitution $u = \log_{10} x$ gives $du = \dfrac{dx}{x \ln 10}$ and we get $\displaystyle\int \frac{\log_{10} x}{x} \, dx = \tfrac{1}{2}\ln 10(\log_{10} x)^2 + C.$

48. Let $u = x^2$. Then $du = 2x\,dx$, so $\displaystyle\int x2^{x^2} \, dx = \tfrac{1}{2}\int 2^u \, du = \frac{1}{2}\frac{2^u}{\ln 2} + C = \frac{1}{2\ln 2}2^{x^2} + C.$

49. Let $u = \sin\theta$. Then $du = \cos\theta\,d\theta$ and $\displaystyle\int 3^{\sin\theta}\,\cos\theta\,d\theta = \int 3^u\,du = \frac{3^u}{\ln 3} + C = \frac{1}{\ln 3}3^{\sin\theta} + C.$

50. Let $u = 2^x + 1$. Then $du = 2^x \ln 2\,dx$, so $\displaystyle\int \frac{2^x}{2^x + 1}\,dx = \int \frac{1}{u}\frac{du}{\ln 2} = \frac{1}{\ln 2}\ln|u| + C = \frac{1}{\ln 2}\ln(2^x + 1) + C.$

51. $A = \displaystyle\int_{-1}^0 (2^x - 5^x)\,dx + \int_0^1 (5^x - 2^x)\,dx = \left[\frac{2^x}{\ln 2} - \frac{5^x}{\ln 5}\right]_{-1}^0 + \left[\frac{5^x}{\ln 5} - \frac{2^x}{\ln 2}\right]_0^1$

$$= \left(\frac{1}{\ln 2} - \frac{1}{\ln 5}\right) - \left(\frac{1/2}{\ln 2} - \frac{1/5}{\ln 5}\right) + \left(\frac{5}{\ln 5} - \frac{2}{\ln 2}\right) - \left(\frac{1}{\ln 5} - \frac{1}{\ln 2}\right)$$

$$= \frac{16}{5\ln 5} - \frac{1}{2\ln 2}$$

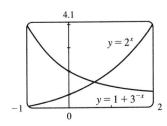

52. Using disks, the volume is $V = \int_0^1 \pi[10^{-x}]^2\,dx = \pi\int_0^1 10^{-2x}\,dx$. To evaluate the integral, we let $u = -2x \quad \Rightarrow$

$du = -2\,dx$, $x = 0 \quad \Rightarrow \quad u = 0$, and $x = 1 \quad \Rightarrow \quad u = -2$, so we have

$$V = -\frac{\pi}{2}\int_0^{-2} 10^u\,du = -\frac{\pi}{2}\left[\frac{1}{\ln 10}10^u\right]_0^{-2} = -\frac{\pi}{2\ln 10}(10^{-2} - 1) = \frac{99\pi}{200\ln 10}$$

53. We see that the graphs of $y = 2^x$ and $y = 1 + 3^{-x}$ intersect at $x \approx 0.6$. We

let $f(x) = 2^x - 1 - 3^{-x}$ and calculate $f'(x) = 2^x \ln 2 + 3^{-x} \ln 3$, and

using the formula $x_{n+1} = x_n - f(x_n)/f'(x_n)$ (Newton's Method), we get

$x_1 = 0.6$, $x_2 \approx x_3 \approx 0.600967$. So, correct to six decimal places, the root

occurs at $x = 0.600967$.

54. $x^y = y^x$ $\Rightarrow$ $y \ln x = x \ln y$ $\Rightarrow$ $y \cdot \dfrac{1}{x} + (\ln x) \cdot y' = x \cdot \dfrac{1}{y} \cdot y' + \ln y$ $\Rightarrow$ $y' \ln x - \dfrac{x}{y} y' = \ln y - \dfrac{y}{x}$ $\Rightarrow$

$y' = \dfrac{\ln y - y/x}{\ln x - x/y}$

55. $y = \log_{10}\left(1 + \dfrac{1}{x}\right)$ $\Rightarrow$ $10^y = 1 + \dfrac{1}{x}$ $\Rightarrow$ $\dfrac{1}{x} = 10^y - 1$ $\Rightarrow$ $x = \dfrac{1}{10^y - 1}$.

Interchange x and y: $y = \dfrac{1}{10^x - 1}$ is the inverse function.

56. $\displaystyle\lim_{x \to 0^+} x^{-\ln x} = \lim_{x \to 0^+} \left(e^{\ln x}\right)^{-\ln x} = \lim_{x \to 0^+} e^{-(\ln x)^2} = 0$ since $-(\ln x)^2 \to -\infty$ as $x \to 0^+$.

57. If I is the intensity of the 1989 San Francisco earthquake, then $\log_{10}(I/S) = 7.1$ $\Rightarrow$

$\log_{10}(16I/S) = \log_{10} 16 + \log_{10}(I/S) = \log_{10} 16 + 7.1 \approx 8.3$.

58. Let I_1 and I_2 be the intensities of the music and the mower. Then $10 \log_{10}\left(\dfrac{I_1}{I_0}\right) = 120$ and $10 \log_{10}\left(\dfrac{I_2}{I_0}\right) = 106$, so

$\log_{10}\left(\dfrac{I_1}{I_2}\right) = \log_{10}\left(\dfrac{I_1/I_0}{I_2/I_0}\right) = \log_{10}\left(\dfrac{I_1}{I_0}\right) - \log_{10}\left(\dfrac{I_2}{I_0}\right) = 12 - 10.6 = 1.4$ $\Rightarrow$ $\dfrac{I_1}{I_2} = 10^{1.4} \approx 25$.

59. We find I with the loudness formula from Exercise 58, substituting $I_0 = 10^{-12}$ and $L = 50$:

$50 = 10 \log_{10} \dfrac{I}{10^{-12}}$ $\Leftrightarrow$ $5 = \log_{10} \dfrac{I}{10^{-12}}$ $\Leftrightarrow$ $10^5 = \dfrac{I}{10^{-12}}$ $\Leftrightarrow$ $I = 10^{-7}$ watt/m². Now we differentiate L with

respect to I: $L = 10 \log_{10} \dfrac{I}{I_0}$ $\Rightarrow$ $\dfrac{dL}{dI} = 10 \dfrac{1}{(I/I_0) \ln 10}\left(\dfrac{1}{I_0}\right) = \dfrac{10}{\ln 10}\left(\dfrac{1}{I}\right)$. Substituting $I = 10^{-7}$, we get

$L'(50) = \dfrac{10}{\ln 10}\left(\dfrac{1}{10^{-7}}\right) = \dfrac{10^8}{\ln 10} \approx 4.34 \times 10^7 \dfrac{\text{dB}}{\text{watt/m}^2}$.

60. (a) $I(x) = I_0 a^x$ $\Rightarrow$ $I'(x) = I_0(\ln a)a^x = (I_0 a^x)\ln a = I(x)\ln a$

(b) We substitute $I_0 = 8$, $a = 0.38$ and $x = 20$ into the first expression for $I'(x)$ above:

$I'(20) = 8(\ln 0.38)(0.38)^{20} \approx -3.05 \times 10^{-8}$.

(c) The average value of the function $I(x)$ between $x = 0$ and $x = 20$ is

$\dfrac{\int_0^{20} I(x)\,dx}{20 - 0} = \dfrac{1}{20}\int_0^{20} 8(0.38)^x\,dx = \dfrac{2}{5}\left[\dfrac{(0.38)^x}{\ln 0.38}\right]_0^{20} = \dfrac{2(0.38^{20} - 1)}{5 \ln 0.38} \approx 0.41$.

61. (a) Using a calculator or CAS, we obtain the model $Q = ab^t$ with $a \approx 100.0124369$ and $b \approx 0.000045145933$.

(b) Use $Q'(t) = ab^t \ln b$ (from Formula 4) with the values of a and b from part (a) to get $Q'(0.04) \approx -670.63\ \mu A$.

The result of Example 2 in Section 2.1 was $-670\ \mu A$.

62. (a) $P = ab^t$ with $a = 4.502714 \times 10^{-20}$ and $b = 1.029953851$,

where P is measured in thousands of people. The fit appears to be very good.

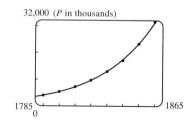

(b) **For 1800:** $m_1 = \dfrac{5308 - 3929}{1800 - 1790} = 137.9$, $m_2 = \dfrac{7240 - 5308}{1810 - 1800} = 193.2$.

So $P'(1800) \approx (m_1 + m_2)/2 = 165.55$ thousand people/year.

For 1850: $m_1 = \dfrac{23{,}192 - 17{,}063}{1850 - 1840} = 612.9$, $m_2 = \dfrac{31{,}443 - 23{,}192}{1860 - 1850} = 825.1$.

So $P'(1850) \approx (m_1 + m_2)/2 = 719$ thousand people/year.

(c) Using $P'(t) = ab^t \ln b$ (from Formula 4) with the values of a and b from part (a), we get $P'(1800) \approx 156.85$ and $P'(1850) \approx 686.07$. These estimates are somewhat less than the ones in part (b).

(d) $P(1870) \approx 41{,}946.56$. The difference of 3.4 million people is most likely due to the Civil War (1861–1865).

63. Using Definition 1 and the second law of exponents for e^x, we have $a^{x-y} = e^{(x-y)\ln a} = e^{x \ln a - y \ln a} = \dfrac{e^{x \ln a}}{e^{y \ln a}} = \dfrac{a^x}{a^y}$.

64. Using Definition 1, the first law of logarithms, and the first law of exponents for e^x, we have

$(ab)^x = e^{x \ln(ab)} = e^{x(\ln a + \ln b)} = e^{x \ln a + x \ln b} = e^{x \ln a} e^{x \ln b} = a^x b^x$.

65. Let $\log_a x = r$ and $\log_a y = s$. Then $a^r = x$ and $a^s = y$.

(a) $xy = a^r a^s = a^{r+s} \quad \Rightarrow \quad \log_a (xy) = r + s = \log_a x + \log_a y$

(b) $\dfrac{x}{y} = \dfrac{a^r}{a^s} = a^{r-s} \quad \Rightarrow \quad \log_a \dfrac{x}{y} = r - s = \log_a x - \log_a y$

(c) $x^y = (a^r)^y = a^{ry} \quad \Rightarrow \quad \log_a (x^y) = ry = y \log_a x$

66. Let $m = n/x$. Then $n = xm$, and as $n \to \infty$, $m \to \infty$.

Therefore, $\displaystyle\lim_{n \to \infty} \left(1 + \frac{x}{n}\right)^n = \lim_{m \to \infty} \left(1 + \frac{1}{m}\right)^{mx} = \left[\lim_{m \to \infty} \left(1 + \frac{1}{m}\right)^{m}\right]^x = e^x$ by Equation 9.

7.5   Exponential Growth and Decay

1. The relative growth rate is $\dfrac{1}{P}\dfrac{dP}{dt} = 0.7944$, so $\dfrac{dP}{dt} = 0.7944P$ and, by Theorem 2, $P(t) = P(0)e^{0.7944t} = 2e^{0.7944t}$.

Thus, $P(6) = 2e^{0.7944(6)} \approx 234.99$ or about 235 members.

2. (a) By Theorem 2, $P(t) = P(0)e^{kt} = 60e^{kt}$. In 20 minutes ($\tfrac{1}{3}$ hour), there are 120 cells, so $P\!\left(\tfrac{1}{3}\right) = 60e^{k/3} = 120 \quad \Rightarrow$

$e^{k/3} = 2 \quad \Rightarrow \quad k/3 = \ln 2 \quad \Rightarrow \quad k = 3 \ln 2 = \ln(2^3) = \ln 8$.

(b) $P(t) = 60e^{(\ln 8)t} = 60 \cdot 8^t$

(c) $P(8) = 60 \cdot 8^8 = 60 \cdot 2^{24} = 1{,}006{,}632{,}960$

(d) $dP/dt = kP \quad \Rightarrow \quad P'(8) = kP(8) = (\ln 8)P(8) \approx 2.093$ billion cells/h

(e) $P(t) = 20{,}000 \quad \Rightarrow \quad 60 \cdot 8^t = 20{,}000 \quad \Rightarrow \quad 8^t = 1000/3 \quad \Rightarrow \quad t \ln 8 = \ln(1000/3) \quad \Rightarrow$

$t = \dfrac{\ln(1000/3)}{\ln 8} \approx 2.79$ h

3. (a) By Theorem 2, $P(t) = P(0)e^{kt} = 100e^{kt}$. Now $P(1) = 100e^{k(1)} = 420 \quad \Rightarrow \quad e^k = \frac{420}{100} \quad \Rightarrow \quad k = \ln 4.2$.

So $P(t) = 100e^{(\ln 4.2)t} = 100(4.2)^t$.

(b) $P(3) = 100(4.2)^3 = 7408.8 \approx 7409$ bacteria

(c) $dP/dt = kP$ $\Rightarrow$ $P'(3) = k \cdot P(3) = (\ln 4.2)\big(100(4.2)^3\big)$ [from part (a)] $\approx 10{,}632$ bacteria/hour

(d) $P(t) = 100(4.2)^t = 10{,}000$ $\Rightarrow$ $(4.2)^t = 100$ $\Rightarrow$ $t = (\ln 100)/(\ln 4.2) \approx 3.2$ hours

4. (a) $y(t) = y(0)e^{kt}$ $\Rightarrow$ $y(2) = y(0)e^{2k} = 600$, $y(8) = y(0)e^{8k} = 75{,}000$. Dividing these equations, we get

$e^{8k}/e^{2k} = 75{,}000/600$ $\Rightarrow$ $e^{6k} = 125$ $\Rightarrow$ $6k = \ln 125 = \ln 5^3 = 3\ln 5$ $\Rightarrow$ $k = \frac{3}{6}\ln 5 = \frac{1}{2}\ln 5$.

Thus, $y(0) = 600/e^{2k} = 600/e^{\ln 5} = \frac{600}{5} = 120$.

(b) $y(t) = y(0)e^{kt} = 120e^{(\ln 5)t/2}$ or $y = 120 \cdot 5^{t/2}$

(c) $y(5) = 120 \cdot 5^{5/2} = 120 \cdot 25\sqrt{5} = 3000\sqrt{5} \approx 6708$ bacteria.

(d) $y(t) = 120 \cdot 5^{t/2}$ $\Rightarrow$ $y'(t) = 120 \cdot 5^{t/2} \cdot \ln 5 \cdot \frac{1}{2} = 60 \cdot \ln 5 \cdot 5^{t/2}$.

$y'(5) = 60 \cdot \ln 5 \cdot 5^{5/2} = 60 \cdot \ln 5 \cdot 25\sqrt{5} \approx 5398$ bacteria/hour.

(e) $y(t) = 200{,}000$ $\Leftrightarrow$ $120e^{(\ln 5)t/2} = 200{,}000$ $\Leftrightarrow$ $e^{(\ln 5)t/2} = \frac{5000}{3}$ $\Leftrightarrow$ $(\ln 5)t/2 = \ln\frac{5000}{3}$ $\Leftrightarrow$

$t = \big(2\ln\frac{5000}{3}\big)\big/\ln 5 \approx 9.2$ h.

5. (a) Let the population (in millions) in the year t be $P(t)$. Since the initial time is the year 1750, we substitute $t - 1750$ for t in

Theorem 2, so the exponential model gives $P(t) = P(1750)e^{k(t-1750)}$. Then $P(1800) = 980 = 790e^{k(1800-1750)}$ $\Rightarrow$

$\frac{980}{790} = e^{k(50)}$ $\Rightarrow$ $\ln\frac{980}{790} = 50k$ $\Rightarrow$ $k = \frac{1}{50}\ln\frac{980}{790} \approx 0.0043104$. So with this model, we have

$P(1900) = 790e^{k(1900-1750)} \approx 1508$ million, and $P(1950) = 790e^{k(1950-1750)} \approx 1871$ million. Both of these

estimates are much too low.

(b) In this case, the exponential model gives $P(t) = P(1850)e^{k(t-1850)}$ $\Rightarrow$ $P(1900) = 1650 = 1260e^{k(1900-1850)}$ $\Rightarrow$

$\ln\frac{1650}{1260} = k(50)$ $\Rightarrow$ $k = \frac{1}{50}\ln\frac{1650}{1260} \approx 0.005393$. So with this model, we estimate

$P(1950) = 1260e^{k(1950-1850)} \approx 2161$ million. This is still too low, but closer than the estimate of $P(1950)$ in part (a).

(c) The exponential model gives $P(t) = P(1900)e^{k(t-1900)}$ $\Rightarrow$ $P(1950) = 2560 = 1650e^{k(1950-1900)}$ $\Rightarrow$

$\ln\frac{2560}{1650} = k(50)$ $\Rightarrow$ $k = \frac{1}{50}\ln\frac{2560}{1650} \approx 0.008785$. With this model, we estimate

$P(2000) = 1650e^{k(2000-1900)} \approx 3972$ million. This is much too low. The discrepancy is explained by the fact that the

world birth rate (average yearly number of births per person) is about the same as always, whereas the mortality rate

(especially the infant mortality rate) is much lower, owing mostly to advances in medical science and to the wars in the first

part of the twentieth century. The exponential model assumes, among other things, that the birth and mortality rates will

remain constant.

6. (a) Let $P(t)$ be the population (in millions) in the year t. Since the initial time is the year 1900, we substitute $t - 1900$ for t in

Theorem 2, and find that the exponential model gives $P(t) = P(1900)e^{k(t-1900)}$ $\Rightarrow$

$P(1910) = 92 = 76e^{k(1910-1900)}$ $\Rightarrow$ $k = \frac{1}{10}\ln\frac{92}{76} \approx 0.0191$. With this model, we estimate

$P(2000) = 76e^{k(2000-1900)} \approx 514$ million. This estimate is much too high. The discrepancy is explained by the fact that, between the years 1900 and 1910, an enormous number of immigrants (compared to the total population) came to the United States. Since that time, immigration (as a proportion of total population) has been much lower. Also, the birth rate in the United States has declined since the turn of the century. So our calculation of the constant k was based partly on factors which no longer exist.

(b) Substituting $t - 1980$ for t in Theorem 2, we find that the exponential model gives $P(t) = P(1980)e^{k(t-1980)}$ $\Rightarrow$

$P(1990) = 250 = 227e^{k(1990-1980)}$ $\Rightarrow$ $k = \frac{1}{10} \ln \frac{250}{227} \approx 0.00965$. With this model, we estimate

$P(2000) = 227e^{k(2000-1980)} \approx 275.3$ million. This is quite accurate. The further estimates are

$P(2010) = 227e^{30k} \approx 303$ million and $P(2020) = 227e^{40k} \approx 334$ million.

(c)

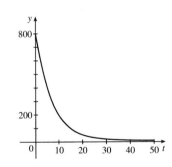

The model in part (a) is quite inaccurate after 1910 (off by 5 million in 1920 and 12 million in 1930). The model in part (b) is more accurate (which is not surprising, since it is based on more recent information).

7. (a) If $y = [\text{N}_2\text{O}_5]$ then by Theorem 2, $\dfrac{dy}{dt} = -0.0005y$ $\Rightarrow$ $y(t) = y(0)e^{-0.0005t} = Ce^{-0.0005t}$.

(b) $y(t) = Ce^{-0.0005t} = 0.9C$ $\Rightarrow$ $e^{-0.0005t} = 0.9$ $\Rightarrow$ $-0.0005t = \ln 0.9$ $\Rightarrow$ $t = -2000 \ln 0.9 \approx 211$ s

8. (a) The mass remaining after t days is $y(t) = y(0)e^{kt} = 800e^{kt}$. Since the half-life is 5.0 days, $y(5) = 800e^{5k} = 400$ $\Rightarrow$

$e^{5k} = \frac{1}{2}$ $\Rightarrow$ $5k = \ln \frac{1}{2}$ $\Rightarrow$ $k = -(\ln 2)/5$, so $y(t) = 800e^{-(\ln 2)t/5} = 800 \cdot 2^{-t/5}$.

(b) $y(30) = 800 \cdot 2^{-30/5} = 12.5$ mg

(c) $800e^{-(\ln 2)t/5} = 1$ $\Leftrightarrow$ $-(\ln 2)\frac{t}{5} = \ln \frac{1}{800} = -\ln 800$ $\Leftrightarrow$

$t = 5 \frac{\ln 800}{\ln 2} \approx 48$ days

(d)

9. (a) If $y(t)$ is the mass (in mg) remaining after t years, then $y(t) = y(0)e^{kt} = 100e^{kt}$.

$y(30) = 100e^{30k} = \frac{1}{2}(100)$ $\Rightarrow$ $e^{30k} = \frac{1}{2}$ $\Rightarrow$ $k = -(\ln 2)/30$ $\Rightarrow$ $y(t) = 100e^{-(\ln 2)t/30} = 100 \cdot 2^{-t/30}$

(b) $y(100) = 100 \cdot 2^{-100/30} \approx 9.92$ mg

(c) $100e^{-(\ln 2)t/30} = 1$ $\Rightarrow$ $-(\ln 2)t/30 = \ln \frac{1}{100}$ $\Rightarrow$ $t = -30 \frac{\ln 0.01}{\ln 2} \approx 199.3$ years

10. (a) If $y(t)$ is the mass after t days and $y(0) = A$, then $y(t) = Ae^{kt}$.

$y(1) = Ae^{k} = 0.945A$ $\Rightarrow$ $e^{k} = 0.945$ $\Rightarrow$ $k = \ln 0.945$.

Then $Ae^{(\ln 0.945)t} = \frac{1}{2}A$ $\Leftrightarrow$ $\ln e^{(\ln 0.945)t} = \ln \frac{1}{2}$ $\Leftrightarrow$ $(\ln 0.945)t = \ln \frac{1}{2}$ $\Leftrightarrow$ $t = -\frac{\ln 2}{\ln 0.945} \approx 12.25$ years.

(b) $Ae^{(\ln 0.945)t} = 0.20A$ $\Leftrightarrow$ $(\ln 0.945)t = \ln \frac{1}{5}$ $\Leftrightarrow$ $t = -\frac{\ln 5}{\ln 0.945} \approx 28.45$ years

11. Let $y(t)$ be the level of radioactivity. Thus, $y(t) = y(0)e^{-kt}$ and k is determined by using the half-life:

$$y(5730) = \tfrac{1}{2}y(0) \quad\Rightarrow\quad y(0)e^{-k(5730)} = \tfrac{1}{2}y(0) \quad\Rightarrow\quad e^{-5730k} = \tfrac{1}{2} \quad\Rightarrow\quad -5730k = \ln\tfrac{1}{2} \quad\Rightarrow\quad k = -\frac{\ln\frac{1}{2}}{5730} = \frac{\ln 2}{5730}.$$

If 74% of the ^{14}C remains, then we know that $y(t) = 0.74y(0)$ $\Rightarrow$ $0.74 = e^{-t(\ln 2)/5730}$ $\Rightarrow$ $\ln 0.74 = -\dfrac{t \ln 2}{5730}$ $\Rightarrow$

$t = -\dfrac{5730(\ln 0.74)}{\ln 2} \approx 2489 \approx 2500$ years.

12. From the information given, we know that $\dfrac{dy}{dx} = 2y$ $\Rightarrow$ $y = Ce^{2x}$ by Theorem 2. To calculate C we use the point $(0, 5)$:

$5 = Ce^{2(0)}$ $\Rightarrow$ $C = 5$. Thus, the equation of the curve is $y = 5e^{2x}$.

13. (a) Using Newton's Law of Cooling, $\dfrac{dT}{dt} = k(T - T_s)$, we have $\dfrac{dT}{dt} = k(T - 75)$. Now let $y = T - 75$, so

$y(0) = T(0) - 75 = 185 - 75 = 110$, so y is a solution of the initial-value problem $dy/dt = ky$ with $y(0) = 110$ and by

Theorem 2 we have $y(t) = y(0)e^{kt} = 110e^{kt}$.

$y(30) = 110e^{30k} = 150 - 75$ $\Rightarrow$ $e^{30k} = \frac{75}{110} = \frac{15}{22}$ $\Rightarrow$ $k = \frac{1}{30}\ln\frac{15}{22}$, so $y(t) = 110e^{\frac{1}{30}t\ln\left(\frac{15}{22}\right)}$ and

$y(45) = 110e^{\frac{45}{30}\ln\left(\frac{15}{22}\right)} \approx 62°$F. Thus, $T(45) \approx 62 + 75 = 137°$F.

(b) $T(t) = 100$ $\Rightarrow$ $y(t) = 25$. $y(t) = 110e^{\frac{1}{30}t\ln\left(\frac{15}{22}\right)} = 25$ $\Rightarrow$ $e^{\frac{1}{30}t\ln\left(\frac{15}{22}\right)} = \frac{25}{110}$ $\Rightarrow$ $\frac{1}{30}t\ln\frac{15}{22} = \ln\frac{25}{110}$ $\Rightarrow$

$t = \dfrac{30\ln\frac{25}{110}}{\ln\frac{15}{22}} \approx 116$ min.

14. (a) Let $T(t) = $ temperature after t minutes. Newton's Law of Cooling implies that $\dfrac{dT}{dt} = k(T - 5)$. Let $y(t) = T(t) - 5$.

Then $\dfrac{dy}{dt} = ky$, so $y(t) = y(0)e^{kt} = 15e^{kt}$ $\Rightarrow$ $T(t) = 5 + 15e^{kt}$ $\Rightarrow$ $T(1) = 5 + 15e^{k} = 12$ $\Rightarrow$ $e^{k} = \frac{7}{15}$ $\Rightarrow$

$k = \ln\frac{7}{15}$, so $T(t) = 5 + 15e^{\ln(7/15)t}$ and $T(2) = 5 + 15e^{2\ln(7/15)} \approx 8.3°$C.

(b) $5 + 15e^{\ln(7/15)t} = 6$ when $e^{\ln(7/15)t} = \frac{1}{15}$ $\Rightarrow$ $\ln\left(\frac{7}{15}\right)t = \ln\frac{1}{15}$ $\Rightarrow$ $t = \dfrac{\ln\frac{1}{15}}{\ln\frac{7}{15}} \approx 3.6$ min.

15. $\dfrac{dT}{dt} = k(T - 20)$. Letting $y = T - 20$, we get $\dfrac{dy}{dt} = ky$, so $y(t) = y(0)e^{kt}$. $y(0) = T(0) - 20 = 5 - 20 = -15$, so

$y(25) = y(0)e^{25k} = -15e^{25k}$, and $y(25) = T(25) - 20 = 10 - 20 = -10$, so $-15e^{25k} = -10$ $\Rightarrow$ $e^{25k} = \frac{2}{3}$. Thus,

$25k = \ln\left(\frac{2}{3}\right)$ and $k = \frac{1}{25}\ln\left(\frac{2}{3}\right)$, so $y(t) = y(0)e^{kt} = -15e^{(1/25)\ln(2/3)t}$. More simply, $e^{25k} = \frac{2}{3}$ $\Rightarrow$ $e^{k} = \left(\frac{2}{3}\right)^{1/25}$ $\Rightarrow$

$e^{kt} = \left(\frac{2}{3}\right)^{t/25}$ $\Rightarrow$ $y(t) = -15 \cdot \left(\frac{2}{3}\right)^{t/25}$.

(a) $T(50) = 20 + y(50) = 20 - 15 \cdot \left(\frac{2}{3}\right)^{50/25} = 20 - 15 \cdot \left(\frac{2}{3}\right)^{2} = 20 - \frac{20}{3} = 13.\overline{3}°$C

(b) $15 = T(t) = 20 + y(t) = 20 - 15 \cdot \left(\frac{2}{3}\right)^{t/25}$ $\Rightarrow$ $15 \cdot \left(\frac{2}{3}\right)^{t/25} = 5$ $\Rightarrow$ $\left(\frac{2}{3}\right)^{t/25} = \frac{1}{3}$ $\Rightarrow$

$(t/25)\ln\left(\frac{2}{3}\right) = \ln\left(\frac{1}{3}\right)$ $\Rightarrow$ $t = 25\ln\left(\frac{1}{3}\right)/\ln\left(\frac{2}{3}\right) \approx 67.74$ min.

16. (a) Similar to Example 3, we have $T_s = 20°\text{C}$ and hence $\dfrac{dT}{dt} = c(T - 20)$. Let $y = T - 20$, so that

$y(0) = T(0) - 20 = 95 - 20 = 75$. Now y satisfies (2), so $y = 75e^{ct}$. We are given that $T(30) = 61$, so

$y(30) = 61 - 20 = 41$ and $41 = 75e^{c(30)}$ $\Rightarrow$ $\frac{41}{75} = e^{30c}$ $\Rightarrow$ $30c = \ln\frac{41}{75}$ $\Rightarrow$ $c = \frac{1}{30}\ln\frac{41}{75} \approx -0.020131$.

Thus, $T(t) = 20 + 75e^{-kt}$, where $k = -c \approx 0.02$.

(b) $T_{\text{ave}} = \frac{1}{30-0}\int_0^{30} T(t)\,dt = \frac{1}{30}\int_0^{30}(20 + 75e^{-kt})\,dt = \frac{1}{30}\left[20t - \frac{75}{k}e^{-kt}\right]_0^{30} = \frac{1}{30}\left[\left(600 - \frac{75}{k}e^{-30k}\right) - \left(0 - \frac{75}{k}\right)\right]$

$= \frac{1}{30}\left(600 - \frac{75}{k}\cdot\frac{41}{75} + \frac{75}{k}\right) = \frac{1}{30}\left(600 + \frac{34}{k}\right) = 20 + \frac{34}{30k} \approx 76.3°\text{C}$

17. (a) Let $P(h)$ be the pressure at altitude h. Then $dP/dh = kP$ $\Rightarrow$ $P(h) = P(0)e^{kh} = 101.3e^{kh}$.

$P(1000) = 101.3e^{1000k} = 87.14$ $\Rightarrow$ $1000k = \ln\left(\frac{87.14}{101.3}\right)$ $\Rightarrow$ $k = \frac{1}{1000}\ln\left(\frac{87.14}{101.3}\right)$ $\Rightarrow$

$P(h) = 101.3\,e^{\frac{1}{1000}h\ln\left(\frac{87.14}{101.3}\right)}$, so $P(3000) = 101.3e^{3\ln\left(\frac{87.14}{101.3}\right)} \approx 64.5$ kPa.

(b) $P(6187) = 101.3\,e^{\frac{6187}{1000}\ln\left(\frac{87.14}{101.3}\right)} \approx 39.9$ kPa

18. (a) Using $A = A_0\left(1 + \dfrac{r}{n}\right)^{nt}$ with $A_0 = 1000$, $r = 0.08$, and $t = 3$, we have:

(b)

(i) Annually: $n = 1$; $A = 1000\left(1 + \frac{0.08}{1}\right)^{1\cdot3} = \$1259.71$

(ii) Quarterly: $n = 4$; $A = 1000\left(1 + \frac{0.08}{4}\right)^{4\cdot3} = \$1268.24$

(iii) Monthly: $n = 12$; $A = 1000\left(1 + \frac{0.08}{12}\right)^{12\cdot3} = \$1270.24$

(iv) Weekly: $n = 52$ $A = 1000\left(1 + \frac{0.08}{52}\right)^{52\cdot3} = \$1271.01$

(v) Daily: $n = 365$; $A = 1000\left(1 + \frac{0.08}{365}\right)^{365\cdot3} = \$1271.22$

(vi) Hourly: $n = 365\cdot24$; $A = 1000\left(1 + \frac{0.08}{365\cdot24}\right)^{365\cdot24\cdot3} = \$1271.25$

(vii) Continuously: $A = 1000e^{(0.08)3} = \$1271.25$

$A_{0.10}(3) = \$1349.86,$

$A_{0.08}(3) = \$1271.25,$ and

$A_{0.06}(3) = \$1197.22.$

19. (a) Using $A = A_0\left(1 + \dfrac{r}{n}\right)^{nt}$ with $A_0 = 3000$, $r = 0.05$, and $t = 5$, we have:

(i) Annually: $n = 1$; $A = 3000\left(1 + \frac{0.05}{1}\right)^{1\cdot5} = \$3828.84$

(ii) Semiannually: $n = 2$; $A = 3000\left(1 + \frac{0.05}{2}\right)^{2\cdot5} = \$3840.25$

(iii) Monthly: $n = 12$; $A = 3000\left(1 + \frac{0.05}{12}\right)^{12\cdot5} = \$3850.08$

(iv) Weekly: $n = 52$; $A = 3000\left(1 + \frac{0.05}{52}\right)^{52\cdot5} = \$3851.61$

(v) Daily: $n = 365$; $A = 3000\left(1 + \frac{0.05}{365}\right)^{365\cdot5} = \$3852.01$

(vi) Continuously: $A = 3000e^{(0.05)5} = \$3852.08$

(b) $dA/dt = 0.05A$ and $A(0) = 3000$.

20. (a) $A_0e^{0.06t} = 2A_0$ $\Leftrightarrow$ $e^{0.06t} = 2$ $\Leftrightarrow$ $0.06t = \ln 2$ $\Leftrightarrow$ $t = \frac{50}{3}\ln 2 \approx 11.55$, so the investment will

double in about 11.55 years.

(b) The annual interest rate in $A = A_0(1 + r)^t$ is r. From part (a), we have $A = A_0e^{0.06t}$. These amounts must be equal,

so $(1 + r)^t = e^{0.06t}$ $\Rightarrow$ $1 + r = e^{0.06}$ $\Rightarrow$ $r = e^{0.06} - 1 \approx 0.0618 = 6.18\%$, which is the equivalent annual

interest rate.

7.6 Inverse Trigonometric Functions

1. (a) $\sin^{-1}\left(\frac{\sqrt{3}}{2}\right) = \frac{\pi}{3}$ since $\sin\frac{\pi}{3} = \frac{\sqrt{3}}{2}$ and $\frac{\pi}{3}$ is in $\left[-\frac{\pi}{2}, \frac{\pi}{2}\right]$.

   (b) $\cos^{-1}(-1) = \pi$ since $\cos\pi = -1$ and π is in $[0, \pi]$.

2. (a) $\tan^{-1}\left(\frac{1}{\sqrt{3}}\right) = \frac{\pi}{6}$ since $\tan\frac{\pi}{6} = \frac{1}{\sqrt{3}}$ and $\frac{\pi}{6}$ is in $\left(-\frac{\pi}{2}, \frac{\pi}{2}\right)$.

   (b) $\sec^{-1} 2 = \frac{\pi}{3}$ since $\sec\frac{\pi}{3} = 2$ and $\frac{\pi}{3}$ is in $\left[0, \frac{\pi}{2}\right) \cup \left[\pi, \frac{3\pi}{2}\right)$.

3. (a) $\arctan 1 = \frac{\pi}{4}$ since $\tan\frac{\pi}{4} = 1$ and $\frac{\pi}{4}$ is in $\left(-\frac{\pi}{2}, \frac{\pi}{2}\right)$.

   (b) $\sin^{-1}\frac{1}{\sqrt{2}} = \frac{\pi}{4}$ since $\sin\frac{\pi}{4} = \frac{1}{\sqrt{2}}$ and $\frac{\pi}{4}$ is in $\left[-\frac{\pi}{2}, \frac{\pi}{2}\right]$.

4. (a) $\cot^{-1}\left(-\sqrt{3}\right) = \frac{5\pi}{6}$ since $\cot\frac{5\pi}{6} = -\sqrt{3}$ and $\frac{5\pi}{6}$ is in $(0, \pi)$.

   (b) $\arccos(-\frac{1}{2}) = \frac{2\pi}{3}$ since $\cos\frac{2\pi}{3} = -\frac{1}{2}$ and $\frac{2\pi}{3}$ is in $[0, \pi]$.

5. (a) In general, $\tan(\arctan x) = x$ for any real number x. Thus, $\tan(\arctan 10) = 10$.

   (b) $\sin^{-1}\left(\sin\frac{7\pi}{3}\right) = \sin^{-1}\left(\sin\frac{\pi}{3}\right) = \sin^{-1}\frac{\sqrt{3}}{2} = \frac{\pi}{3}$ since $\sin\frac{\pi}{3} = \frac{\sqrt{3}}{2}$ and $\frac{\pi}{3}$ is in $\left[-\frac{\pi}{2}, \frac{\pi}{2}\right]$.

   [Recall that $\frac{7\pi}{3} = \frac{\pi}{3} + 2\pi$ and the sine function is periodic with period 2π.]

6. (a) $\tan^{-1}\left(\tan\frac{3\pi}{4}\right) = \tan^{-1}(-1) = -\frac{\pi}{4}$

   (b) $\cos\left(\arcsin\frac{1}{2}\right) = \cos\left(\frac{\pi}{6}\right) = \frac{\sqrt{3}}{2}$

7. Let $\theta = \sin^{-1}\left(\frac{2}{3}\right)$.

   Then $\tan\left(\sin^{-1}\left(\frac{2}{3}\right)\right) = \tan\theta = \frac{2}{\sqrt{5}}$.

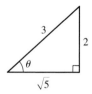

8. Let $\theta = \arccos\frac{3}{5}$.

   Then $\csc\left(\arccos\left(\frac{3}{5}\right)\right) = \csc\theta = \frac{5}{4}$.

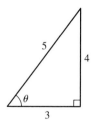

9. Let $\theta = \tan^{-1}\sqrt{2}$. Then

$$\sin\left(2\tan^{-1}\sqrt{2}\right) = \sin(2\theta) = 2\sin\theta\cos\theta$$

$$= 2\left(\frac{\sqrt{2}}{\sqrt{3}}\right)\left(\frac{1}{\sqrt{3}}\right) = \frac{2\sqrt{2}}{3}$$

10. Let $x = \tan^{-1} 2$ and $y = \tan^{-1} 3$. Then

$$\cos\left(\tan^{-1} 2 + \tan^{-1} 3\right) = \cos(x + y) = \cos x \, \cos y - \sin x \, \sin y$$

$$= \frac{1}{\sqrt{5}} \frac{1}{\sqrt{10}} - \frac{2}{\sqrt{5}} \frac{3}{\sqrt{10}}$$

$$= \frac{-5}{\sqrt{50}} = \frac{-5}{5\sqrt{2}} = \frac{-1}{\sqrt{2}}$$

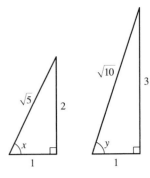

11. Let $y = \sin^{-1} x$. Then $-\frac{\pi}{2} \le y \le \frac{\pi}{2} \;\; \Rightarrow \;\; \cos y \ge 0$, so $\cos(\sin^{-1} x) = \cos y = \sqrt{1 - \sin^2 y} = \sqrt{1 - x^2}$.

12. Let $y = \sin^{-1} x$. Then $\sin y = x$, so from the triangle we see that

$$\tan(\sin^{-1} x) = \tan y = \frac{x}{\sqrt{1 - x^2}}.$$

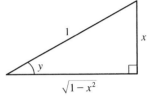

13. Let $y = \tan^{-1} x$. Then $\tan y = x$, so from the triangle we see that

$$\sin(\tan^{-1} x) = \sin y = \frac{x}{\sqrt{1 + x^2}}.$$

14. Let $y = \tan^{-1} x$. Then $\tan y = x$, so from the triangle we see that

$$\cos(2\tan^{-1} x) = \cos 2y = \cos^2 y - \sin^2 y$$

$$= \left(\frac{1}{\sqrt{1 + x^2}}\right)^2 - \left(\frac{x}{\sqrt{1 + x^2}}\right)^2 = \frac{1 - x^2}{1 + x^2}$$

15.

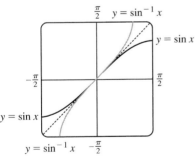

The graph of $\sin^{-1} x$ is the reflection of the graph of $\sin x$ about the line $y = x$.

16.

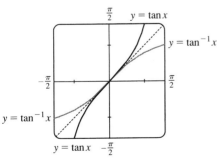

The graph of $\tan^{-1} x$ is the reflection of the graph of $\tan x$ about the line $y = x$.

17. Let $y = \cos^{-1} x$. Then $\cos y = x$ and $0 \le y \le \pi$ $\Rightarrow$ $-\sin y \dfrac{dy}{dx} = 1$ $\Rightarrow$

$$\dfrac{dy}{dx} = -\dfrac{1}{\sin y} = -\dfrac{1}{\sqrt{1 - \cos^2 y}} = -\dfrac{1}{\sqrt{1 - x^2}}. \quad \text{[Note that } \sin y \ge 0 \text{ for } 0 \le y \le \pi.\text{]}$$

18. (a) Let $a = \sin^{-1} x$ and $b = \cos^{-1} x$. Then $\cos a = \sqrt{1 - \sin^2 a} = \sqrt{1 - x^2}$ since $\cos a \ge 0$ for $-\frac{\pi}{2} \le a \le \frac{\pi}{2}$.

Similarly, $\sin b = \sqrt{1 - x^2}$. So

$$\sin(\sin^{-1} x + \cos^{-1} x) = \sin(a + b) = \sin a \cos b + \cos a \sin b = x \cdot x + \sqrt{1 - x^2}\,\sqrt{1 - x^2} = x^2 + (1 - x^2) = 1$$

But $-\frac{\pi}{2} \le \sin^{-1} x + \cos^{-1} x \le \frac{3\pi}{2}$, and so $\sin^{-1} x + \cos^{-1} x = \frac{\pi}{2}$.

(b) We differentiate $\sin^{-1} x + \cos^{-1} x = \frac{\pi}{2}$ with respect to x, and get

$$\dfrac{1}{\sqrt{1 - x^2}} + \dfrac{d}{dx}\left(\cos^{-1} x\right) = 0 \quad \Rightarrow \quad \dfrac{d}{dx}\left(\cos^{-1} x\right) = -\dfrac{1}{\sqrt{1 - x^2}}.$$

19. Let $y = \cot^{-1} x$. Then $\cot y = x$ $\Rightarrow$ $-\csc^2 y \dfrac{dy}{dx} = 1$ $\Rightarrow$ $\dfrac{dy}{dx} = -\dfrac{1}{\csc^2 y} = -\dfrac{1}{1 + \cot^2 y} = -\dfrac{1}{1 + x^2}$.

20. Let $y = \sec^{-1} x$. Then $\sec y = x$ and $y \in \left(0, \frac{\pi}{2}\right] \cup \left[\pi, \frac{3\pi}{2}\right)$. Differentiate with respect to x:

$$\sec y \tan y \left(\dfrac{dy}{dx}\right) = 1 \quad \Rightarrow \quad \dfrac{dy}{dx} = \dfrac{1}{\sec y \tan y} = \dfrac{1}{\sec y \sqrt{\sec^2 y - 1}} = \dfrac{1}{x\sqrt{x^2 - 1}}. \text{ Note that } \tan^2 y = \sec^2 y - 1 \quad \Rightarrow$$

$\tan y = \sqrt{\sec^2 y - 1}$ since $\tan y > 0$ when $0 < y < \frac{\pi}{2}$ or $\pi < y < \frac{3\pi}{2}$.

21. Let $y = \csc^{-1} x$. Then $\csc y = x$ $\Rightarrow$ $-\csc y \cot y \dfrac{dy}{dx} = 1$ $\Rightarrow$

$$\dfrac{dy}{dx} = -\dfrac{1}{\csc y \cot y} = -\dfrac{1}{\csc y \sqrt{\csc^2 y - 1}} = -\dfrac{1}{x\sqrt{x^2 - 1}}. \text{ Note that } \cot y \ge 0 \text{ on the domain of } \csc^{-1} x.$$

22. $y = \sqrt{\tan^{-1} x} = (\tan^{-1} x)^{1/2}$ $\Rightarrow$

$$y' = \tfrac{1}{2}(\tan^{-1} x)^{-1/2} \cdot \dfrac{d}{dx}(\tan^{-1} x) = \dfrac{1}{2\sqrt{\tan^{-1} x}} \cdot \dfrac{1}{1 + x^2} = \dfrac{1}{2\sqrt{\tan^{-1} x}\,(1 + x^2)}$$

23. $y = \tan^{-1} \sqrt{x}$ $\Rightarrow$ $y' = \dfrac{1}{1 + \left(\sqrt{x}\right)^2} \cdot \dfrac{d}{dx}\left(\sqrt{x}\right) = \dfrac{1}{1 + x}\left(\tfrac{1}{2}x^{-1/2}\right) = \dfrac{1}{2\sqrt{x}\,(1 + x)}$

24. $f(x) = x \ln(\arctan x)$ $\Rightarrow$ $f'(x) = x \cdot \dfrac{1}{\arctan x} \cdot \dfrac{1}{1 + x^2} + \ln(\arctan x) \cdot 1 = \dfrac{x}{(1 + x^2)\arctan x} + \ln(\arctan x)$

25. $y = \sin^{-1}(2x + 1)$ $\Rightarrow$

$$y' = \dfrac{1}{\sqrt{1 - (2x + 1)^2}} \cdot \dfrac{d}{dx}(2x + 1) = \dfrac{1}{\sqrt{1 - (4x^2 + 4x + 1)}} \cdot 2 = \dfrac{2}{\sqrt{-4x^2 - 4x}} = \dfrac{1}{\sqrt{-x^2 - x}}$$

26. $g(x) = \sqrt{x^2 - 1}\,\sec^{-1} x$ $\Rightarrow$ $g'(x) = \sqrt{x^2 - 1} \cdot \dfrac{1}{x\sqrt{x^2 - 1}} + \sec^{-1} x \cdot \dfrac{1}{2}\left(x^2 - 1\right)^{-1/2}(2x) = \dfrac{1}{x} + \dfrac{x\sec^{-1} x}{\sqrt{x^2 - 1}}$

$$\left[\text{or } \dfrac{\sqrt{x^2 - 1} + x^2 \sec^{-1} x}{x\sqrt{x^2 - 1}}\right]$$

27. $G(x) = \sqrt{1-x^2} \arccos x \Rightarrow G'(x) = \sqrt{1-x^2} \cdot \dfrac{-1}{\sqrt{1-x^2}} + \arccos x \cdot \dfrac{1}{2}(1-x^2)^{-1/2}(-2x) = -1 - \dfrac{x \arccos x}{\sqrt{1-x^2}}$

28. $F(\theta) = \arcsin\sqrt{\sin\theta} = \arcsin(\sin\theta)^{1/2} \Rightarrow$

$$F'(\theta) = \frac{1}{\sqrt{1-\left(\sqrt{\sin\theta}\right)^2}} \cdot \frac{d}{d\theta}(\sin\theta)^{1/2} = \frac{1}{\sqrt{1-\sin\theta}} \cdot \frac{1}{2}(\sin\theta)^{-1/2}\cdot\cos\theta = \frac{\cos\theta}{2\sqrt{1-\sin\theta}\sqrt{\sin\theta}}$$

29. $y = \cos^{-1}(e^{2x}) \Rightarrow y' = -\dfrac{1}{\sqrt{1-(e^{2x})^2}} \cdot \dfrac{d}{dx}(e^{2x}) = -\dfrac{2e^{2x}}{\sqrt{1-e^{4x}}}$

30. $y = \arctan\sqrt{\dfrac{1-x}{1+x}} = \arctan\left(\dfrac{1-x}{1+x}\right)^{1/2} \Rightarrow$

$$y' = \frac{1}{1+\left(\sqrt{\frac{1-x}{1+x}}\right)^2} \cdot \frac{d}{dx}\left(\frac{1-x}{1+x}\right)^{1/2} = \frac{1}{1+\frac{1-x}{1+x}} \cdot \frac{1}{2}\left(\frac{1-x}{1+x}\right)^{-1/2} \cdot \frac{(1+x)(-1)-(1-x)(1)}{(1+x)^2}$$

$$= \frac{1}{\frac{1+x}{1+x}+\frac{1-x}{1+x}} \cdot \frac{1}{2}\left(\frac{1+x}{1-x}\right)^{1/2} \cdot \frac{-2}{(1+x)^2} = \frac{1+x}{2}\cdot\frac{1}{2}\cdot\frac{(1+x)^{1/2}}{(1-x)^{1/2}}\cdot\frac{-2}{(1+x)^2}$$

$$= \frac{-1}{2(1-x)^{1/2}(1+x)^{1/2}} = \frac{-1}{2\sqrt{1-x^2}}$$

31. $y = \arctan(\cos\theta) \Rightarrow y' = \dfrac{1}{1+(\cos\theta)^2}(-\sin\theta) = -\dfrac{\sin\theta}{1+\cos^2\theta}$

32. $y = \tan^{-1}\left(x-\sqrt{x^2+1}\right) \Rightarrow$

$$y' = \frac{1}{1+\left(x-\sqrt{x^2+1}\right)^2}\left(1-\frac{x}{\sqrt{x^2+1}}\right) = \frac{1}{1+x^2-2x\sqrt{x^2+1}+x^2+1}\left(\frac{\sqrt{x^2+1}-x}{\sqrt{x^2+1}}\right)$$

$$= \frac{\sqrt{x^2+1}-x}{2\left(1+x^2-x\sqrt{x^2+1}\right)\sqrt{x^2+1}} = \frac{\sqrt{x^2+1}-x}{2\left[\sqrt{x^2+1}(1+x^2)-x(x^2+1)\right]} = \frac{\sqrt{x^2+1}-x}{2\left[(1+x^2)\left(\sqrt{x^2+1}-x\right)\right]}$$

$$= \frac{1}{2(1+x^2)}$$

33. $h(t) = \cot^{-1}(t)+\cot^{-1}(1/t) \Rightarrow$

$$h'(t) = -\frac{1}{1+t^2}-\frac{1}{1+(1/t)^2}\cdot\frac{d}{dt}\frac{1}{t} = -\frac{1}{1+t^2}-\frac{t^2}{t^2+1}\cdot\left(-\frac{1}{t^2}\right) = -\frac{1}{1+t^2}+\frac{1}{t^2+1} = 0.$$

Note that this makes sense because $h(t) = \dfrac{\pi}{2}$ for $t>0$ and $h(t) = \dfrac{3\pi}{2}$ for $t<0$.

34. $y = \tan^{-1}\left(\dfrac{x}{a}\right)+\ln\sqrt{\dfrac{x-a}{x+a}} = \tan^{-1}\left(\dfrac{x}{a}\right)+\frac{1}{2}\ln(x-a)-\frac{1}{2}\ln(x+a) \Rightarrow$

$$y' = \frac{a}{x^2+a^2}+\frac{1/2}{x-a}-\frac{1/2}{x+a} = \frac{a}{x^2+a^2}+\frac{a}{x^2-a^2} = \frac{2ax^2}{x^4-a^4}$$

35. $y = \arccos\left(\dfrac{b + a\cos x}{a + b\cos x}\right) \quad \Rightarrow$

$$y' = -\frac{1}{\sqrt{1 - \left(\dfrac{b + a\cos x}{a + b\cos x}\right)^2}} \frac{(a + b\cos x)(-a\sin x) - (b + a\cos x)(-b\sin x)}{(a + b\cos x)^2}$$

$$= \frac{1}{\sqrt{a^2 + b^2\cos^2 x - b^2 - a^2\cos^2 x}} \frac{(a^2 - b^2)\sin x}{|a + b\cos x|} = \frac{1}{\sqrt{a^2 - b^2}\sqrt{1 - \cos^2 x}} \frac{(a^2 - b^2)\sin x}{|a + b\cos x|} = \frac{\sqrt{a^2 - b^2}}{|a + b\cos x|}\frac{\sin x}{|\sin x|}$$

But $0 \le x \le \pi$, so $|\sin x| = \sin x$. Also $a > b > 0 \quad \Rightarrow \quad b\cos x \ge -b > -a$, so $a + b\cos x > 0$. Thus $y' = \dfrac{\sqrt{a^2 - b^2}}{a + b\cos x}$.

36. $f(x) = \arcsin(e^x) \quad \Rightarrow \quad f'(x) = \dfrac{1}{\sqrt{1 - (e^x)^2}} \cdot e^x = \dfrac{e^x}{\sqrt{1 - e^{2x}}}$.

Domain$(f) = \{x \mid -1 \le e^x \le 1\} = \{x \mid 0 < e^x \le 1\} = (-\infty, 0]$.

Domain$(f') = \{x \mid 1 - e^{2x} > 0\} = \{x \mid e^{2x} < 1\} = \{x \mid 2x < 0\} = (-\infty, 0)$.

37. $g(x) = \cos^{-1}(3 - 2x) \quad \Rightarrow \quad g'(x) = -\dfrac{1}{\sqrt{1 - (3 - 2x)^2}}(-2) = \dfrac{2}{\sqrt{1 - (3 - 2x)^2}}$.

Domain$(g) = \{x \mid -1 \le 3 - 2x \le 1\} = \{x \mid -4 \le -2x \le -2\} = \{x \mid 2 \ge x \ge 1\} = [1, 2]$.

Domain$(g') = \{x \mid 1 - (3 - 2x)^2 > 0\} = \{x \mid (3 - 2x)^2 < 1\} = \{x \mid |3 - 2x| < 1\}$

$\qquad = \{x \mid -1 < 3 - 2x < 1\} = \{x \mid -4 < -2x < -2\} = \{x \mid 2 > x > 1\} = (1, 2)$

38. $\tan^{-1}(xy) = 1 + x^2 y \quad \Rightarrow \quad \dfrac{1}{1 + x^2 y^2}(xy' + y \cdot 1) = 0 + x^2 y' + 2xy \quad \Rightarrow$

$$y'\left(\frac{x}{1 + x^2 y^2} - x^2\right) = 2xy - \frac{y}{1 + x^2 y^2} \quad \Rightarrow \quad y' = \frac{2xy - \dfrac{y}{1 + x^2 y^2}}{\dfrac{x}{1 + x^2 y^2} - x^2} = \frac{2xy(1 + x^2 y^2) - y}{x - x^2(1 + x^2 y^2)} = \frac{y(-1 + 2x + 2x^3 y^2)}{x(1 - x - x^3 y^2)}$$

39. $g(x) = x\sin^{-1}\left(\dfrac{x}{4}\right) + \sqrt{16 - x^2} \quad \Rightarrow \quad g'(x) = \sin^{-1}\left(\dfrac{x}{4}\right) + \dfrac{x}{4\sqrt{1 - (x/4)^2}} - \dfrac{x}{\sqrt{16 - x^2}} = \sin^{-1}\left(\dfrac{x}{4}\right) \quad \Rightarrow$

$g'(2) = \sin^{-1}\left(\dfrac{1}{2}\right) = \dfrac{\pi}{6}$

40. $y = 3\arccos\dfrac{x}{2} \quad \Rightarrow \quad y' = 3\left[-\dfrac{1}{\sqrt{1 - (x/2)^2}}\right]\left(\dfrac{1}{2}\right)$, so at $(1, \pi)$, $y' = -\dfrac{3}{2\sqrt{1 - \frac{1}{4}}} = -\sqrt{3}$. An equation of the tangent

line is $y - \pi = -\sqrt{3}\,(x - 1)$, or $y = -\sqrt{3}\,x + \pi + \sqrt{3}$.

41. $f(x) = \sqrt{1 - x^2}\arcsin x \quad \Rightarrow \quad f'(x) = \sqrt{1 - x^2} \cdot \dfrac{1}{\sqrt{1 - x^2}} + \arcsin x \cdot \dfrac{1}{2}\left(1 - x^2\right)^{-1/2}(-2x) = 1 - \dfrac{x\arcsin x}{\sqrt{1 - x^2}}$

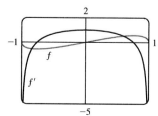

Note that $f' = 0$ where the graph of f has a horizontal tangent. Also note that f' is negative when f is decreasing and f' is positive when f is increasing.

42. $f(x) = \arctan(x^2 - x) \Rightarrow f'(x) = \dfrac{1}{1 + (x^2 - x)^2} \cdot \dfrac{d}{dx}(x^2 - x) = \dfrac{2x - 1}{1 + (x^2 - x)^2}$

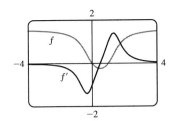

Note that $f' = 0$ where the graph of f has a horizontal tangent. Also note that f' is negative when f is decreasing and f' is positive when f is increasing.

43. $\displaystyle\lim_{x \to -1^+} \sin^{-1} x = \sin^{-1}(-1) = -\frac{\pi}{2}$

44. Let $t = \dfrac{1 + x^2}{1 + 2x^2}$. As $x \to \infty$, $t = \dfrac{1 + x^2}{1 + 2x^2} = \dfrac{1/x^2 + 1}{1/x^2 + 2} \to \dfrac{1}{2}$.

$\displaystyle\lim_{x \to \infty} \arccos\left(\dfrac{1 + x^2}{1 + 2x^2}\right) = \lim_{t \to 1/2} \arccos t = \arccos \frac{1}{2} = \frac{\pi}{3}$.

45. Let $t = e^x$. As $x \to \infty$, $t \to \infty$. $\displaystyle\lim_{x \to \infty} \arctan(e^x) = \lim_{t \to \infty} \arctan t = \frac{\pi}{2}$ by (8).

46. Let $t = \ln x$. As $x \to 0^+$, $t \to -\infty$. $\displaystyle\lim_{x \to 0^+} \tan^{-1}(\ln x) = \lim_{t \to -\infty} \tan^{-1} t = -\frac{\pi}{2}$ by (8).

47.

From the figure, $\tan \alpha = \dfrac{5}{x}$ and $\tan \beta = \dfrac{2}{3 - x}$. Since

$$\alpha + \beta + \theta = 180° = \pi, \ \theta = \pi - \tan^{-1}\left(\frac{5}{x}\right) - \tan^{-1}\left(\frac{2}{3 - x}\right) \Rightarrow$$

$$\frac{d\theta}{dx} = -\frac{1}{1 + \left(\dfrac{5}{x}\right)^2}\left(-\frac{5}{x^2}\right) - \frac{1}{1 + \left(\dfrac{2}{3 - x}\right)^2}\left[\frac{2}{(3 - x)^2}\right]$$

$$= \frac{x^2}{x^2 + 25} \cdot \frac{5}{x^2} - \frac{(3 - x)^2}{(3 - x)^2 + 4} \cdot \frac{2}{(3 - x)^2}.$$

Now $\dfrac{d\theta}{dx} = 0 \Rightarrow \dfrac{5}{x^2 + 25} = \dfrac{2}{x^2 - 6x + 13} \Rightarrow 2x^2 + 50 = 5x^2 - 30x + 65 \Rightarrow$

$3x^2 - 30x + 15 = 0 \Rightarrow x^2 - 10x + 5 = 0 \Rightarrow x = 5 \pm 2\sqrt{5}$. We reject the root with the $+$ sign, since it is larger than 3. $d\theta/dx > 0$ for $x < 5 - 2\sqrt{5}$ and $d\theta/dx < 0$ for $x > 5 - 2\sqrt{5}$, so θ is maximized when

$|AP| = x = 5 - 2\sqrt{5} \approx 0.53$.

48. Let x be the distance from the observer to the wall. Then, from the given figure,

$$\theta = \tan^{-1}\left(\frac{h + d}{x}\right) - \tan^{-1}\left(\frac{d}{x}\right), \ x > 0 \Rightarrow$$

$$\frac{d\theta}{dx} = \frac{1}{1 + [(h + d)/x]^2}\left[-\frac{h + d}{x^2}\right] - \frac{1}{1 + (d/x)^2}\left[-\frac{d}{x^2}\right] = -\frac{h + d}{x^2 + (h + d)^2} + \frac{d}{x^2 + d^2}$$

$$= \frac{d[x^2 + (h + d)^2] - (h + d)(x^2 + d^2)}{[x^2 + (h + d)^2](x^2 + d^2)} = \frac{h^2 d + hd^2 - hx^2}{[x^2 + (h + d)^2](x^2 + d^2)} = 0 \Leftrightarrow$$

$hx^2 = h^2 d + hd^2 \Leftrightarrow x^2 = hd + d^2 \Leftrightarrow x = \sqrt{d(h + d)}$. Since $d\theta/dx > 0$ for all $x < \sqrt{d(h + d)}$ and $d\theta/dx < 0$ for all $x > \sqrt{d(h + d)}$, the absolute maximum occurs when $x = \sqrt{d(h + d)}$.

49.

$$\frac{dx}{dt} = 2 \text{ ft/s}, \sin\theta = \frac{x}{10} \quad \Rightarrow \quad \theta = \sin^{-1}\left(\frac{x}{10}\right), \frac{d\theta}{dx} = \frac{1/10}{\sqrt{1 - (x/10)^2}},$$

$$\frac{d\theta}{dt} = \frac{d\theta}{dx}\frac{dx}{dt} = \frac{1/10}{\sqrt{1 - (x/10)^2}}(2) \text{ rad/s}, \left.\frac{d\theta}{dt}\right]_{x=6} = \frac{2/10}{\sqrt{1 - (6/10)^2}} \text{ rad/s} = \frac{1}{4} \text{ rad/s}$$

50.

$$\frac{d\theta}{dt} = 4 \text{ rev/min} = 8\pi \cdot 60 \text{ rad/h. From the diagram, we see that } \tan\theta = \frac{x}{3} \quad \Rightarrow \quad \theta = \tan^{-1}\left(\frac{x}{3}\right).$$

$$\text{Thus, } 8\pi \cdot 60 = \frac{d\theta}{dt} = \frac{d\theta}{dx}\frac{dx}{dt} = \frac{1/3}{1 + (x/3)^2}\frac{dx}{dt}. \text{ So } \frac{dx}{dt} = 8\pi \cdot 60 \cdot 3\left[1 + \left(\frac{x}{3}\right)^2\right] \text{ km/h, and}$$

$$\text{at } x = 1, \frac{dx}{dt} = 8\pi \cdot 60 \cdot 3\left[1 + \frac{1}{9}\right] \text{ km/h} = 1600\pi \text{ km/h}.$$

51. $y = f(x) = \sin^{-1}(x/(x+1))$ **A.** $D = \{x \mid -1 \le x/(x+1) \le 1\}$. For $x > -1$ we have $-x - 1 \le x \le x + 1 \quad \Leftrightarrow$

$2x \ge -1 \quad \Leftrightarrow \quad x \ge -\frac{1}{2}$, so $D = \left[-\frac{1}{2}, \infty\right)$. **B.** Intercepts are 0 **C.** No symmetry

D. $\lim\limits_{x \to \infty} \sin^{-1}\left(\dfrac{x}{x+1}\right) = \lim\limits_{x \to \infty} \sin^{-1}\left(\dfrac{1}{1+1/x}\right) = \sin^{-1} 1 = \frac{\pi}{2}$, so $y = \frac{\pi}{2}$ is a HA.

E. $f'(x) = \dfrac{1}{\sqrt{1 - [x/(x+1)]^2}}\dfrac{(x+1) - x}{(x+1)^2} = \dfrac{1}{(x+1)\sqrt{2x+1}} > 0,$

so f is increasing on $\left(-\frac{1}{2}, \infty\right)$. **F.** No local maximum or minimum,

$f\left(-\frac{1}{2}\right) = \sin^{-1}(-1) = -\frac{\pi}{2}$ is an absolute minimum

G. $f''(x) = -\dfrac{\sqrt{2x+1} + (x+1)/\sqrt{2x+1}}{(x+1)^2(2x+1)}$

$= -\dfrac{3x+2}{(x+1)^2(2x+1)^{3/2}} < 0$ on D, so f is CD on $\left(-\frac{1}{2}, \infty\right)$.

H.

52. $y = f(x) = \tan^{-1}\left(\dfrac{x-1}{x+1}\right)$ **A.** $D = \{x \mid x \ne -1\}$ **B.** x-intercept $= 1$, y-intercept $= f(0) = \tan^{-1}(-1) = -\frac{\pi}{4}$

C. No symmetry **D.** $\lim\limits_{x \to \pm\infty} \tan^{-1}\left(\dfrac{x-1}{x+1}\right) = \lim\limits_{x \to \pm\infty} \tan^{-1}\left(\dfrac{1 - 1/x}{1 + 1/x}\right) = \tan^{-1} 1 = \frac{\pi}{4}$, so $y = \frac{\pi}{4}$ is a HA.

Also $\lim\limits_{x \to -1^+} \tan^{-1}\left(\dfrac{x-1}{x+1}\right) = -\dfrac{\pi}{2}$ and $\lim\limits_{x \to -1^-} \tan^{-1}\left(\dfrac{x-1}{x+1}\right) = \dfrac{\pi}{2}$.

E. $f'(x) = \dfrac{1}{1 + [(x-1)/(x+1)]^2}\dfrac{(x+1) - (x-1)}{(x+1)^2} = \dfrac{2}{(x+1)^2 + (x-1)^2} = \dfrac{1}{x^2 + 1} > 0,$

so f is increasing on $(-\infty, -1)$ and $(-1, \infty)$. **F.** No extreme values

G. $f''(x) = -2x/\left(x^2 + 1\right)^2 > 0 \quad \Leftrightarrow \quad x < 0$, so f is CU on $(-\infty, -1)$

and $(-1, 0)$, and CD on $(0, \infty)$. IP at $\left(0, -\frac{\pi}{4}\right)$

H.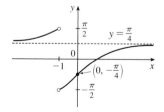

53. $y = f(x) = x - \tan^{-1} x$ **A.** $D = \mathbb{R}$ **B.** Intercepts are 0 **C.** $f(-x) = -f(x)$, so the curve is symmetric about the

origin. **D.** $\lim\limits_{x \to \infty} (x - \tan^{-1} x) = \infty$ and $\lim\limits_{x \to -\infty} (x - \tan^{-1} x) = -\infty$, no HA.

But $f(x) - \left(x - \frac{\pi}{2}\right) = -\tan^{-1} x + \frac{\pi}{2} \to 0$ as $x \to \infty$, and

$f(x) - \left(x + \frac{\pi}{2}\right) = -\tan^{-1} x - \frac{\pi}{2} \to 0$ as $x \to -\infty$, so $y = x \pm \frac{\pi}{2}$ are

slant asymptotes. **E.** $f'(x) = 1 - \dfrac{1}{x^2 + 1} = \dfrac{x^2}{x^2 + 1} > 0$, so f is

increasing on $\mathbb{R}$. **F.** No extrema

H.

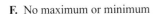

G. $f''(x) = \dfrac{(1 + x^2)(2x) - x^2(2x)}{(1 + x^2)^2} = \dfrac{2x}{(1 + x^2)^2} > 0 \Leftrightarrow x > 0$, so

f is CU on $(0, \infty)$, CD on $(-\infty, 0)$. IP at $(0, 0)$.

54. $y = \tan^{-1}(\ln x)$ **A.** $D = (0, \infty)$ **B.** No y-intercept, x-intercept when $\tan^{-1}(\ln x) = 0 \Leftrightarrow \ln x = 0 \Leftrightarrow x = 1$.

C. No symmetry **D.** $\lim\limits_{x \to \infty} \tan^{-1}(\ln x) = \frac{\pi}{2}$,

so $y = \frac{\pi}{2}$ is a HA. Also $\lim\limits_{x \to 0^+} \tan^{-1}(\ln x) = -\frac{\pi}{2}$.

H.

E. $f'(x) = \dfrac{1}{x[1 + (\ln x)^2]} > 0$, so f is increasing on $(0, \infty)$.

F. No maximum or minimum

G. $f''(x) = \dfrac{-[1 + (\ln x)^2 + x(2\ln x/x)]}{x^2[1 + (\ln x)^2]^2} = -\dfrac{(1 + \ln x)^2}{x^2[1 + (\ln x)^2]^2} < 0$,

so f is CD on $(0, \infty)$.

55. $f(x) = \arctan(\cos(3\arcsin x))$. We use a CAS to compute f' and f'', and to graph f, f', and f'':

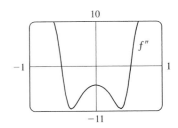

From the graph of f', it appears that the only maximum occurs at $x = 0$ and there are minima at $x = \pm 0.87$. From the graph

of f'', it appears that there are inflection points at $x = \pm 0.52$.

56. First note that the function $f(x) = x - c\sin^{-1} x$ is only defined on the interval $[-1, 1]$, since $\sin^{-1}$ is only defined on that

interval. We differentiate to get $f'(x) = 1 - c/\sqrt{1 - x^2}$. Now if $c \le 0$, then $f'(x) \ge 1$, so there is no extremum and f is

increasing on its domain. If $c > 1$, then $f'(x) < 0$, so there is no local extremum and f is decreasing on its domain, and if

$c = 1$, then there is still no extremum, since $f'(x)$ does not change sign at $x = 0$. So we can only have local extrema if

$0 < c < 1$. In this case, f is increasing where $f'(x) > 0$ ⟺ $\sqrt{1 - x^2} > c$ ⟺ $|x| < \sqrt{1 - c^2}$, and decreasing where $\sqrt{1 - c^2} < |x| \leq 1$. f has a maximum at $x = \sqrt{1 - c^2}$ and a minimum at $x = -\sqrt{1 - c^2}$.

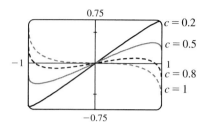

57. $f(x) = \dfrac{2 + x^2}{1 + x^2} = \dfrac{1 + (1 + x^2)}{1 + x^2} = \dfrac{1}{1 + x^2} + 1$ ⟹ $F(x) = \tan^{-1} x + x + C$

58. $f'(x) = 4/\sqrt{1 - x^2}$ ⟹ $f(x) = 4\sin^{-1} x + C$. $f\left(\frac{1}{2}\right) = 4\sin^{-1}\left(\frac{1}{2}\right) + C = 4 \cdot \frac{\pi}{6} + C$ and $f\left(\frac{1}{2}\right) = 1$ ⟹
$\frac{2\pi}{3} + C = 1$ ⟹ $C = 1 - \frac{2\pi}{3}$, so $f(x) = 4\sin^{-1} x + 1 - \frac{2\pi}{3}$.

59. $\displaystyle\int_{1/2}^{\sqrt{3}/2} \dfrac{6}{\sqrt{1 - t^2}}\, dt = 6\int_{1/2}^{\sqrt{3}/2} \dfrac{1}{\sqrt{1 - t^2}}\, dt = 6\Big[\sin^{-1} t\Big]_{1/2}^{\sqrt{3}/2} = 6\Big[\sin^{-1}\left(\frac{\sqrt{3}}{2}\right) - \sin^{-1}\left(\frac{1}{2}\right)\Big] = 6\left(\frac{\pi}{3} - \frac{\pi}{6}\right) = 6\left(\frac{\pi}{6}\right) = \pi$

60. Let $u = \tan^{-1} x$. Then $du = dx/(1 + x^2)$, so $\displaystyle\int \dfrac{\tan^{-1} x}{1 + x^2}\, dx = \int u\, du = \frac{1}{2}u^2 + C = \frac{1}{2}(\tan^{-1} x)^2 + C$.

61. Let $u = 4x$. Then $du = 4\, dx$, so
$\displaystyle\int_0^{\sqrt{3}/4} \dfrac{dx}{1 + 16x^2} = \frac{1}{4}\int_0^{\sqrt{3}} \dfrac{1}{1 + u^2}\, du = \frac{1}{4}\Big[\tan^{-1} u\Big]_0^{\sqrt{3}} = \frac{1}{4}\left(\tan^{-1}\sqrt{3} - \tan^{-1} 0\right) = \frac{1}{4}\left(\frac{\pi}{3} - 0\right) = \frac{\pi}{12}$.

62. Let $u = 2t$. Then $\sqrt{1 - 4t^2} = \sqrt{1 - u^2}$ and $du = 2\, dt$, so
$\displaystyle\int \dfrac{dt}{\sqrt{1 - 4t^2}} = \int \dfrac{\frac{1}{2}\, du}{\sqrt{1 - u^2}} = \frac{1}{2}\sin^{-1} u + C = \frac{1}{2}\sin^{-1}(2t) + C$.

63. Let $u = 1 + x^2$. Then $du = 2x\, dx$, so
$\displaystyle\int \dfrac{1 + x}{1 + x^2}\, dx = \int \dfrac{1}{1 + x^2}\, dx + \int \dfrac{x}{1 + x^2}\, dx = \tan^{-1} x + \int \dfrac{\frac{1}{2}\, du}{u} = \tan^{-1} x + \frac{1}{2}\ln|u| + C$
$\qquad = \tan^{-1} x + \frac{1}{2}\ln|1 + x^2| + C = \tan^{-1} x + \frac{1}{2}\ln(1 + x^2) + C$ [since $1 + x^2 > 0$].

64. Let $u = -\cos x$. Then $du = \sin x\, dx$, so
$\displaystyle\int_0^{\pi/2} \dfrac{\sin x}{1 + \cos^2 x}\, dx = \int_{-1}^0 \dfrac{1}{1 + u^2}\, du = \Big[\tan^{-1} u\Big]_{-1}^0 = \tan^{-1} 0 - \tan^{-1}(-1) = 0 - \left(-\frac{\pi}{4}\right) = \frac{\pi}{4}$.

65. Let $u = \sin^{-1} x$. Then $du = \dfrac{1}{\sqrt{1 - x^2}}\, dx$, so $\displaystyle\int \dfrac{dx}{\sqrt{1 - x^2}\,\sin^{-1} x} = \int \dfrac{1}{u}\, du = \ln|u| + C = \ln|\sin^{-1} x| + C$.

66. Let $u = \frac{1}{2}x$. Then $du = \frac{1}{2}\, dx$ ⟹
$\displaystyle\int \dfrac{dx}{x\sqrt{x^2 - 4}} = \int \dfrac{dx}{2x\sqrt{(x/2)^2 - 1}} = \int \dfrac{2\, du}{4u\sqrt{u^2 - 1}} = \frac{1}{2}\int \dfrac{du}{u\sqrt{u^2 - 1}} = \frac{1}{2}\sec^{-1} u + C = \frac{1}{2}\sec^{-1}\left(\frac{1}{2}x\right) + C$.

67. Let $u = t^3$. Then $du = 3t^2\, dt$ and $\displaystyle\int \dfrac{t^2}{\sqrt{1 - t^6}}\, dt = \int \dfrac{\frac{1}{3}\, du}{\sqrt{1 - u^2}} = \frac{1}{3}\sin^{-1} u + C = \frac{1}{3}\sin^{-1}(t^3) + C$.

68. Let $u = e^{2x}$. Then $du = 2e^{2x}\,dx$ $\Rightarrow$ $\displaystyle\int \frac{e^{2x}\,dx}{\sqrt{1-e^{4x}}} = \frac{1}{2}\int \frac{du}{\sqrt{1-u^2}} = \frac{1}{2}\sin^{-1}u + C = \frac{1}{2}\sin^{-1}(e^{2x}) + C$

69. Let $u = \sqrt{x}$. Then $du = \dfrac{dx}{2\sqrt{x}}$ and $\displaystyle\int \frac{dx}{\sqrt{x}\,(1+x)} = \int \frac{2\,du}{1+u^2} = 2\tan^{-1}u + C = 2\tan^{-1}\sqrt{x} + C.$

70. Let $u = x^2$. Then $du = 2x\,dx$, so $\displaystyle\int \frac{x}{1+x^4}\,dx = \int \frac{\frac{1}{2}\,du}{1+u^2} = \frac{1}{2}\tan^{-1}u + C = \frac{1}{2}\tan^{-1}(x^2) + C.$

71. Let $u = x/a$. Then $du = dx/a$, so $\displaystyle\int \frac{dx}{\sqrt{a^2-x^2}} = \int \frac{dx}{a\sqrt{1-(x/a)^2}} = \int \frac{du}{\sqrt{1-u^2}} = \sin^{-1}u + C = \sin^{-1}\frac{x}{a} + C.$

72. We use the disk method: $A = \displaystyle\int_0^2 \pi\left[\frac{1}{\sqrt{x^2+4}}\right]^2 dx = \pi\int_0^2 \frac{1}{x^2+4}\,dx.$ By Formula 14, this is equal to

$\pi\left[\frac{1}{2}\tan^{-1}(x/2)\right]_0^2 = \frac{\pi}{2}\left(\frac{\pi}{4} - 0\right) = \frac{\pi^2}{8}.$

73.

The integral represents the area below the curve $y = \sin^{-1}x$ on the interval $x \in [0, 1]$. The bounding curves are $y = \sin^{-1}x$ $\Leftrightarrow$ $x = \sin y$, $y = 0$ and $x = 1$. We see that y ranges between $\sin^{-1}0 = 0$ and $\sin^{-1}1 = \frac{\pi}{2}$. So we have to integrate the function $x = 1 - \sin y$ between $y = 0$ and $y = \frac{\pi}{2}$:

$\int_0^1 \sin^{-1}x\,dx = \int_0^{\pi/2}(1 - \sin y)\,dy = \left(\frac{\pi}{2} + \cos\frac{\pi}{2}\right) - (0 + \cos 0) = \frac{\pi}{2} - 1.$

74. Let $a = \arctan x$ and $b = \arctan y$. Then by the addition formula for the tangent (see Reference Page 2 in the textbook),

$\tan(a+b) = \dfrac{\tan a + \tan b}{1 - (\tan a)(\tan b)} = \dfrac{\tan(\arctan x) + \tan(\arctan y)}{1 - \tan(\arctan x)\,\tan(\arctan y)}$ $\Rightarrow$ $\tan(a+b) = \dfrac{x+y}{1-xy}$ $\Rightarrow$

$\arctan x + \arctan y = a + b = \arctan\left(\dfrac{x+y}{1-xy}\right)$, since $-\frac{\pi}{2} < \arctan x + \arctan y < \frac{\pi}{2}.$

75. (a) $\arctan\frac{1}{2} + \arctan\frac{1}{3} = \arctan\left(\dfrac{\frac{1}{2}+\frac{1}{3}}{1-\frac{1}{2}\cdot\frac{1}{3}}\right) = \arctan 1 = \dfrac{\pi}{4}$

(b) $2\arctan\frac{1}{3} + \arctan\frac{1}{7} = \left(\arctan\frac{1}{3} + \arctan\frac{1}{3}\right) + \arctan\frac{1}{7} = \arctan\left(\dfrac{\frac{1}{3}+\frac{1}{3}}{1-\frac{1}{3}\cdot\frac{1}{3}}\right) + \arctan\frac{1}{7}$

$= \arctan\frac{3}{4} + \arctan\frac{1}{7} = \arctan\left(\dfrac{\frac{3}{4}+\frac{1}{7}}{1-\frac{3}{4}\cdot\frac{1}{7}}\right) = \arctan 1 = \dfrac{\pi}{4}$

76. (a) $f(x) = \sin(\sin^{-1}x)$ $\qquad\qquad$ (b) $g(x) = \sin^{-1}(\sin x)$

(c) $g'(x) = \dfrac{d}{dx}\sin^{-1}(\sin x) = \dfrac{1}{\sqrt{1-\sin^2 x}}\cos x = \dfrac{\cos x}{\sqrt{\cos^2 x}} = \dfrac{\cos x}{|\cos x|}$

(d) $h(x) = \cos^{-1}(\sin x)$, so

$$h'(x) = -\frac{\cos x}{\sqrt{1 - \sin^2 x}} = -\frac{\cos x}{|\cos x|}.$$

Notice that $h(x) = \frac{\pi}{2} - g(x)$ because

$\sin^{-1} t + \cos^{-1} t = \frac{\pi}{2}$ for all t.

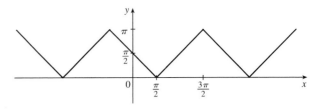

77. Let $f(x) = 2\sin^{-1} x - \cos^{-1}(1 - 2x^2)$. Then $f'(x) = \dfrac{2}{\sqrt{1 - x^2}} - \dfrac{4x}{\sqrt{1 - (1 - 2x^2)^2}} = \dfrac{2}{\sqrt{1 - x^2}} - \dfrac{4x}{2x\sqrt{1 - x^2}} = 0$

[since $x \geq 0$]. Thus $f'(x) = 0$ for all $x \in [0, 1)$. Thus $f(x) = C$. To find C let $x = 0$. Thus

$2\sin^{-1}(0) - \cos^{-1}(1) = 0 = C$. Therefore we see that $f(x) = 2\sin^{-1} x - \cos^{-1}(1 - 2x^2) = 0 \Rightarrow$

$2\sin^{-1} x = \cos^{-1}(1 - 2x^2)$.

78. Let $f(x) = \sin^{-1}\left(\dfrac{x - 1}{x + 1}\right) - 2\tan^{-1}\sqrt{x} + \dfrac{\pi}{2}$. Note that the domain of f is $[0, \infty)$. Thus

$$f'(x) = \frac{1}{\sqrt{1 - \left(\dfrac{x - 1}{x + 1}\right)^2}} \cdot \frac{(x + 1) - (x - 1)}{(x + 1)^2} - \frac{2}{1 + x} \cdot \frac{1}{2\sqrt{x}} = \frac{1}{\sqrt{x}\,(x + 1)} - \frac{1}{\sqrt{x}\,(x + 1)} = 0. \text{ Then } f(x) = C.$$

To find C, we let $x = 0 \Rightarrow \sin^{-1}(-1) - 2\tan^{-1}(0) + \dfrac{\pi}{2} = C \Rightarrow -\dfrac{\pi}{2} - 0 + \dfrac{\pi}{2} = 0 = C$. Thus, $f(x) = 0 \Rightarrow$

$\sin^{-1}\left(\dfrac{x - 1}{x + 1}\right) = 2\tan^{-1}\sqrt{x} - \dfrac{\pi}{2}$.

79. $y = \sec^{-1} x \Rightarrow \sec y = x \Rightarrow \sec y \tan y \dfrac{dy}{dx} = 1 \Rightarrow \dfrac{dy}{dx} = \dfrac{1}{\sec y \tan y}$. Now $\tan^2 y = \sec^2 y - 1 = x^2 - 1$, so

$\tan y = \pm\sqrt{x^2 - 1}$. For $y \in \left[0, \dfrac{\pi}{2}\right)$, $x \geq 1$, so $\sec y = x = |x|$ and $\tan y \geq 0 \Rightarrow \dfrac{dy}{dx} = \dfrac{1}{x\sqrt{x^2 - 1}} = \dfrac{1}{|x|\sqrt{x^2 - 1}}$.

For $y \in \left(\dfrac{\pi}{2}, \pi\right]$, $x \leq -1$, so $|x| = -x$ and $\tan y = -\sqrt{x^2 - 1} \Rightarrow$

$$\frac{dy}{dx} = \frac{1}{\sec y \tan y} = \frac{1}{x\left(-\sqrt{x^2 - 1}\right)} = \frac{1}{(-x)\sqrt{x^2 - 1}} = \frac{1}{|x|\sqrt{x^2 - 1}}$$

80. (a) Since $|\arctan(1/x)| < \dfrac{\pi}{2}$, we have $0 \leq |x\arctan(1/x)| \leq \dfrac{\pi}{2}|x| \to 0$ as $x \to 0$. So, by the Squeeze Theorem,

$\lim\limits_{x \to 0} f(x) = 0 = f(0)$, so f is continuous at 0.

(b) Here $\dfrac{f(x) - f(0)}{x - 0} = \dfrac{x\arctan(1/x) - 0}{x} = \arctan\left(\dfrac{1}{x}\right)$. So (see Exercise 52 in Section 3.2 for a discussion of left- and

right-hand derivatives) $f'_-(0) = \lim\limits_{x \to 0^-} \dfrac{f(x) - f(0)}{x - 0} = \lim\limits_{x \to 0^-} \arctan\left(\dfrac{1}{x}\right) = \lim\limits_{y \to -\infty} \arctan y = -\dfrac{\pi}{2}$, while

$f'_+(0) = \lim\limits_{x \to 0^+} \dfrac{f(x) - f(0)}{x - 0} = \lim\limits_{x \to 0^+} \arctan\left(\dfrac{1}{x}\right) = \lim\limits_{y \to \infty} \arctan y = \dfrac{\pi}{2}$. So $f'(0)$ does not exist.

APPLIED PROJECT   Where To Sit at the Movies

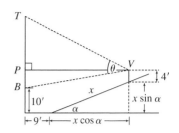

1. $|VP| = 9 + x\cos\alpha$, $|PT| = 35 - (4 + x\sin\alpha) = 31 - x\sin\alpha$, and

$|PB| = (4 + x\sin\alpha) - 10 = x\sin\alpha - 6$. So using the Pythagorean Theorem,

we have $|VT| = \sqrt{|VP|^2 + |PT|^2} = \sqrt{(9 + x\cos\alpha)^2 + (31 - x\sin\alpha)^2} = a$,

and $|VB| = \sqrt{|VP|^2 + |PB|^2} = \sqrt{(9 + x\cos\alpha)^2 + (x\sin\alpha - 6)^2} = b$.

Using the Law of Cosines on $\triangle VBT$, we get $25^2 = a^2 + b^2 - 2ab\cos\theta$ $\Leftrightarrow$

$\cos\theta = \dfrac{a^2 + b^2 - 625}{2ab}$ $\Leftrightarrow$ $\theta = \arccos\left(\dfrac{a^2 + b^2 - 625}{2ab}\right)$, as required.

2. From the graph of θ, it appears that the value of x which maximizes θ is

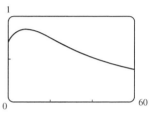

$x \approx 8.25$ ft. Assuming that the first row is at $x = 0$, the row closest to this

value of x is the fourth row, at $x = 9$ ft, and from the graph, the viewing

angle in this row seems to be about 0.85 radians, or about $49°$.

3. With a CAS, we type in the definition of θ, substitute in the proper values of a and b in terms of x and $\alpha = 20° = \frac{\pi}{9}$ radians, and then use the differentiation command to find the derivative. We use a numerical rootfinder and find that the root of the equation $d\theta/dx = 0$ is $x \approx 8.253062$, as approximated in Problem 2.

4. From the graph in Problem 2, it seems that the average value of the function on the interval $[0, 60]$ is about 0.6. We can use a

CAS to approximate $\frac{1}{60}\int_0^{60}\theta(x)\,dx \approx 0.625 \approx 36°$. (The calculation is much faster if we reduce the number of digits of

accuracy required.) The minimum value is $\theta(60) \approx 0.38$ and, from Problem 2, the maximum value is about 0.85.

7.7   Hyperbolic Functions

1. (a) $\sinh 0 = \frac{1}{2}(e^0 - e^0) = 0$ (b) $\cosh 0 = \frac{1}{2}(e^0 + e^0) = \frac{1}{2}(1 + 1) = 1$

2. (a) $\tanh 0 = \dfrac{(e^0 - e^{-0})/2}{(e^0 + e^{-0})/2} = 0$ (b) $\tanh 1 = \dfrac{e^1 - e^{-1}}{e^1 + e^{-1}} = \dfrac{e^2 - 1}{e^2 + 1} \approx 0.76159$

3. (a) $\sinh(\ln 2) = \dfrac{e^{\ln 2} - e^{-\ln 2}}{2} = \dfrac{e^{\ln 2} - (e^{\ln 2})^{-1}}{2} = \dfrac{2 - 2^{-1}}{2} = \dfrac{2 - \frac{1}{2}}{2} = \dfrac{3}{4}$

(b) $\sinh 2 = \frac{1}{2}(e^2 - e^{-2}) \approx 3.62686$

4. (a) $\cosh 3 = \frac{1}{2}(e^3 + e^{-3}) \approx 10.06766$ (b) $\cosh(\ln 3) = \dfrac{e^{\ln 3} + e^{-\ln 3}}{2} = \dfrac{3 + \frac{1}{3}}{2} = \dfrac{5}{3}$

5. (a) $\operatorname{sech} 0 = \dfrac{1}{\cosh 0} = \dfrac{1}{1} = 1$ (b) $\cosh^{-1} 1 = 0$ because $\cosh 0 = 1$.

6. (a) $\sinh 1 = \frac{1}{2}(e^1 - e^{-1}) \approx 1.17520$

(b) Using Equation 3, we have $\sinh^{-1} 1 = \ln\left(1 + \sqrt{1^2 + 1}\right) = \ln\left(1 + \sqrt{2}\right) \approx 0.88137$.

7. $\sinh(-x) = \frac{1}{2}[e^{-x} - e^{-(-x)}] = \frac{1}{2}(e^{-x} - e^x) = -\frac{1}{2}(e^{-x} - e^x) = -\sinh x$

8. $\cosh(-x) = \frac{1}{2}[e^{-x} + e^{-(-x)}] = \frac{1}{2}(e^{-x} + e^x) = \frac{1}{2}(e^x + e^{-x}) = \cosh x$

9. $\cosh x + \sinh x = \frac{1}{2}(e^x + e^{-x}) + \frac{1}{2}(e^x - e^{-x}) = \frac{1}{2}(2e^x) = e^x$

10. $\cosh x - \sinh x = \frac{1}{2}(e^x + e^{-x}) - \frac{1}{2}(e^x - e^{-x}) = \frac{1}{2}(2e^{-x}) = e^{-x}$

11. $\sinh x \cosh y + \cosh x \sinh y = \left[\frac{1}{2}(e^x - e^{-x})\right]\left[\frac{1}{2}(e^y + e^{-y})\right] + \left[\frac{1}{2}(e^x + e^{-x})\right]\left[\frac{1}{2}(e^y - e^{-y})\right]$

$$= \frac{1}{4}[(e^{x+y} + e^{x-y} - e^{-x+y} - e^{-x-y}) + (e^{x+y} - e^{x-y} + e^{-x+y} - e^{-x-y})]$$

$$= \frac{1}{4}(2e^{x+y} - 2e^{-x-y}) = \frac{1}{2}[e^{x+y} - e^{-(x+y)}] = \sinh(x+y)$$

12. $\cosh x \cosh y + \sinh x \sinh y = \left[\frac{1}{2}(e^x + e^{-x})\right]\left[\frac{1}{2}(e^y + e^{-y})\right] + \left[\frac{1}{2}(e^x - e^{-x})\right]\left[\frac{1}{2}(e^y - e^{-y})\right]$

$$= \frac{1}{4}[(e^{x+y} + e^{x-y} + e^{-x+y} + e^{-x-y}) + (e^{x+y} - e^{x-y} - e^{-x+y} + e^{-x-y})]$$

$$= \frac{1}{4}(2e^{x+y} + 2e^{-x-y}) = \frac{1}{2}\left[e^{x+y} + e^{-(x+y)}\right] = \cosh(x+y)$$

13. Divide both sides of the identity $\cosh^2 x - \sinh^2 x = 1$ by $\sinh^2 x$:

$$\frac{\cosh^2 x}{\sinh^2 x} - \frac{\sinh^2 x}{\sinh^2 x} = \frac{1}{\sinh^2 x} \quad \Leftrightarrow \quad \coth^2 x - 1 = \operatorname{csch}^2 x.$$

14. $\tanh(x+y) = \dfrac{\sinh(x+y)}{\cosh(x+y)} = \dfrac{\sinh x \cosh y + \cosh x \sinh y}{\cosh x \cosh y + \sinh x \sinh y} = \dfrac{\dfrac{\sinh x \cosh y}{\cosh x \cosh y} + \dfrac{\cosh x \sinh y}{\cosh x \cosh y}}{\dfrac{\cosh x \cosh y}{\cosh x \cosh y} + \dfrac{\sinh x \sinh y}{\cosh x \cosh y}}$

$$= \frac{\tanh x + \tanh y}{1 + \tanh x \tanh y}$$

15. Putting $y = x$ in the result from Exercise 11, we have

$$\sinh 2x = \sinh(x+x) = \sinh x \cosh x + \cosh x \sinh x = 2\sinh x \cosh x.$$

16. Putting $y = x$ in the result from Exercise 12, we have

$$\cosh 2x = \cosh(x+x) = \cosh x \cosh x + \sinh x \sinh x = \cosh^2 x + \sinh^2 x.$$

17. $\tanh(\ln x) = \dfrac{\sinh(\ln x)}{\cosh(\ln x)} = \dfrac{(e^{\ln x} - e^{-\ln x})/2}{(e^{\ln x} + e^{-\ln x})/2} = \dfrac{x - (e^{\ln x})^{-1}}{x + (e^{\ln x})^{-1}} = \dfrac{x - x^{-1}}{x + x^{-1}} = \dfrac{x - 1/x}{x + 1/x} = \dfrac{(x^2 - 1)/x}{(x^2 + 1)/x} = \dfrac{x^2 - 1}{x^2 + 1}$

18. $\dfrac{1 + \tanh x}{1 - \tanh x} = \dfrac{1 + (\sinh x)/\cosh x}{1 - (\sinh x)/\cosh x} = \dfrac{\cosh x + \sinh x}{\cosh x - \sinh x} = \dfrac{\frac{1}{2}(e^x + e^{-x}) + \frac{1}{2}(e^x - e^{-x})}{\frac{1}{2}(e^x + e^{-x}) - \frac{1}{2}(e^x - e^{-x})} = \dfrac{e^x}{e^{-x}} = e^{2x}$

Or: Using the results of Exercises 9 and 10, $\dfrac{\cosh x + \sinh x}{\cosh x - \sinh x} = \dfrac{e^x}{e^{-x}} = e^{2x}$

19. By Exercise 9, $(\cosh x + \sinh x)^n = (e^x)^n = e^{nx} = \cosh nx + \sinh nx.$

20. $\coth x = \dfrac{1}{\tanh x} \quad \Rightarrow \quad \coth x = \dfrac{1}{\tanh x} = \dfrac{1}{12/13} = \dfrac{13}{12}.$

$\operatorname{sech}^2 x = 1 - \tanh^2 x = 1 - \left(\frac{12}{13}\right)^2 = \frac{25}{169} \quad \Rightarrow \quad \operatorname{sech} x = \frac{5}{13}$ [sech, like cosh, is positive].

$\cosh x = \dfrac{1}{\operatorname{sech} x} \quad \Rightarrow \quad \cosh x = \dfrac{1}{5/13} = \dfrac{13}{5}.$

$$\tanh x = \frac{\sinh x}{\cosh x} \quad \Rightarrow \quad \sinh x = \tanh x \cosh x \quad \Rightarrow \quad \sinh x = \frac{12}{13} \cdot \frac{13}{5} = \frac{12}{5}.$$

$$\operatorname{csch} x = \frac{1}{\sinh x} \quad \Rightarrow \quad \operatorname{csch} x = \frac{1}{12/5} = \frac{5}{12}.$$

21. $\operatorname{sech} x = \dfrac{1}{\cosh x} \quad \Rightarrow \quad \operatorname{sech} x = \dfrac{1}{5/3} = \dfrac{3}{5}.$

$$\cosh^2 x - \sinh^2 x = 1 \quad \Rightarrow \quad \sinh^2 x = \cosh^2 x - 1 = \left(\tfrac{5}{3}\right)^2 - 1 = \tfrac{16}{9} \quad \Rightarrow \quad \sinh x = \tfrac{4}{3} \quad [\text{because } x > 0].$$

$$\operatorname{csch} x = \frac{1}{\sinh x} \quad \Rightarrow \quad \operatorname{csch} x = \frac{1}{4/3} = \frac{3}{4}.$$

$$\tanh x = \frac{\sinh x}{\cosh x} \quad \Rightarrow \quad \tanh x = \frac{4/3}{5/3} = \frac{4}{5}.$$

$$\coth x = \frac{1}{\tanh x} \quad \Rightarrow \quad \coth x = \frac{1}{4/5} = \frac{5}{4}.$$

22. (a)

  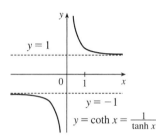

$y = \operatorname{csch} x = \dfrac{1}{\sinh x}$

$y = \operatorname{sech} x = \dfrac{1}{\cosh x}$

$y = \coth x = \dfrac{1}{\tanh x}$

23. (a) $\displaystyle \lim_{x \to \infty} \tanh x = \lim_{x \to \infty} \frac{e^x - e^{-x}}{e^x + e^{-x}} \cdot \frac{e^{-x}}{e^{-x}} = \lim_{x \to \infty} \frac{1 - e^{-2x}}{1 + e^{-2x}} = \frac{1 - 0}{1 + 0} = 1$

(b) $\displaystyle \lim_{x \to -\infty} \tanh x = \lim_{x \to -\infty} \frac{e^x - e^{-x}}{e^x + e^{-x}} \cdot \frac{e^x}{e^x} = \lim_{x \to -\infty} \frac{e^{2x} - 1}{e^{2x} + 1} = \frac{0 - 1}{0 + 1} = -1$

(c) $\displaystyle \lim_{x \to \infty} \sinh x = \lim_{x \to \infty} \frac{e^x - e^{-x}}{2} = \infty$

(d) $\displaystyle \lim_{x \to -\infty} \sinh x = \lim_{x \to -\infty} \frac{e^x - e^{-x}}{2} = -\infty$

(e) $\displaystyle \lim_{x \to \infty} \operatorname{sech} x = \lim_{x \to \infty} \frac{2}{e^x + e^{-x}} = 0$

(f) $\displaystyle \lim_{x \to \infty} \coth x = \lim_{x \to \infty} \frac{e^x + e^{-x}}{e^x - e^{-x}} \cdot \frac{e^{-x}}{e^{-x}} = \lim_{x \to \infty} \frac{1 + e^{-2x}}{1 - e^{-2x}} = \frac{1 + 0}{1 - 0} = 1 \quad [\textit{Or:} \text{ Use part (a)}]$

(g) $\displaystyle \lim_{x \to 0^+} \coth x = \lim_{x \to 0^+} \frac{\cosh x}{\sinh x} = \infty$, since $\sinh x \to 0$ through positive values and $\cosh x \to 1$.

(h) $\displaystyle \lim_{x \to 0^-} \coth x = \lim_{x \to 0^-} \frac{\cosh x}{\sinh x} = -\infty$, since $\sinh x \to 0$ through negative values and $\cosh x \to 1$.

(i) $\displaystyle \lim_{x \to -\infty} \operatorname{csch} x = \lim_{x \to -\infty} \frac{2}{e^x - e^{-x}} = 0$

24. (a) $\dfrac{d}{dx}(\cosh x) = \dfrac{d}{dx}\left[\tfrac{1}{2}(e^x + e^{-x})\right] = \tfrac{1}{2}(e^x - e^{-x}) = \sinh x$

(b) $\dfrac{d}{dx}(\tanh x) = \dfrac{d}{dx}\left(\dfrac{\sinh x}{\cosh x}\right) = \dfrac{\cosh x \cosh x - \sinh x \sinh x}{\cosh^2 x} = \dfrac{\cosh^2 x - \sinh^2 x}{\cosh^2 x} = \dfrac{1}{\cosh^2 x} = \operatorname{sech}^2 x$

(c) $\dfrac{d}{dx}(\operatorname{csch} x) = \dfrac{d}{dx}\left(\dfrac{1}{\sinh x}\right) = -\dfrac{\cosh x}{\sinh^2 x} = -\dfrac{1}{\sinh x}\cdot\dfrac{\cosh x}{\sinh x} = -\operatorname{csch} x \coth x$

(d) $\dfrac{d}{dx}(\operatorname{sech} x) = \dfrac{d}{dx}\left(\dfrac{1}{\cosh x}\right) = -\dfrac{\sinh x}{\cosh^2 x} = -\dfrac{1}{\cosh x}\cdot\dfrac{\sinh x}{\cosh x} = -\operatorname{sech} x \tanh x$

(e) $\dfrac{d}{dx}(\coth x) = \dfrac{d}{dx}\left(\dfrac{\cosh x}{\sinh x}\right) = \dfrac{\sinh x \sinh x - \cosh x \cosh x}{\sinh^2 x} = \dfrac{\sinh^2 x - \cosh^2 x}{\sinh^2 x} = -\dfrac{1}{\sinh^2 x} = -\operatorname{csch}^2 x$

25. Let $y = \sinh^{-1} x$. Then $\sinh y = x$ and, by Example 1(a), $\cosh^2 y - \sinh^2 y = 1 \;\Rightarrow\;$ [with $\cosh y > 0$]

$\cosh y = \sqrt{1 + \sinh^2 y} = \sqrt{1 + x^2}$. So by Exercise 9, $e^y = \sinh y + \cosh y = x + \sqrt{1+x^2} \;\Rightarrow\; y = \ln\left(x + \sqrt{1+x^2}\,\right)$.

26. Let $y = \cosh^{-1} x$. Then $\cosh y = x$ and $y \geq 0$, so $\sinh y = \sqrt{\cosh^2 y - 1} = \sqrt{x^2 - 1}$. So, by Exercise 9,

$e^y = \cosh y + \sinh y = x + \sqrt{x^2 - 1} \;\Rightarrow\; y = \ln\left(x + \sqrt{x^2 - 1}\,\right)$.

Another method: Write $x = \cosh y = \frac{1}{2}\left(e^y + e^{-y}\right)$ and solve a quadratic, as in Example 3.

27. (a) Let $y = \tanh^{-1} x$. Then $x = \tanh y = \dfrac{\sinh y}{\cosh y} = \dfrac{(e^y - e^{-y})/2}{(e^y + e^{-y})/2}\cdot\dfrac{e^y}{e^y} = \dfrac{e^{2y} - 1}{e^{2y} + 1} \;\Rightarrow\; xe^{2y} + x = e^{2y} - 1 \;\Rightarrow\;$

$\quad 1 + x = e^{2y} - xe^{2y} \;\Rightarrow\; 1 + x = e^{2y}(1 - x) \;\Rightarrow\; e^{2y} = \dfrac{1+x}{1-x} \;\Rightarrow\; 2y = \ln\left(\dfrac{1+x}{1-x}\right) \;\Rightarrow\; y = \tfrac{1}{2}\ln\left(\dfrac{1+x}{1-x}\right)$.

(b) Let $y = \tanh^{-1} x$. Then $x = \tanh y$, so from Exercise 18 we have

$\quad e^{2y} = \dfrac{1 + \tanh y}{1 - \tanh y} = \dfrac{1+x}{1-x} \;\Rightarrow\; 2y = \ln\left(\dfrac{1+x}{1-x}\right) \;\Rightarrow\; y = \tfrac{1}{2}\ln\left(\dfrac{1+x}{1-x}\right)$.

28. (a) (i) $y = \operatorname{csch}^{-1} x \;\Leftrightarrow\; \operatorname{csch} y = x \quad (x \neq 0)$

(ii) We sketch the graph of csch^{-1} by reflecting the graph of csch (see Exercise 22) about the line $y = x$.

(iii) Let $y = \operatorname{csch}^{-1} x$. Then $x = \operatorname{csch} y = \dfrac{2}{e^y - e^{-y}} \;\Rightarrow\; xe^y - xe^{-y} = 2 \;\Rightarrow\;$

$\quad x(e^y)^2 - 2e^y - x = 0 \;\Rightarrow\; e^y = \dfrac{1 \pm \sqrt{x^2 + 1}}{x}$. But $e^y > 0$, so for $x > 0$,

$\quad e^y = \dfrac{1 + \sqrt{x^2 + 1}}{x}$ and for $x < 0$, $e^y = \dfrac{1 - \sqrt{x^2 + 1}}{x}$. Thus, $\operatorname{csch}^{-1} x = \ln\left(\dfrac{1}{x} + \dfrac{\sqrt{x^2+1}}{|x|}\right)$.

(b) (i) $y = \operatorname{sech}^{-1} x \;\Leftrightarrow\; \operatorname{sech} y = x$ and $y > 0$.

(ii) We sketch the graph of sech^{-1} by reflecting the graph of sech (see Exercise 22) about the line $y = x$.

(iii) Let $y = \operatorname{sech}^{-1} x$, so $x = \operatorname{sech} y = \dfrac{2}{e^y + e^{-y}} \;\Rightarrow\; xe^y + xe^{-y} = 2 \;\Rightarrow\;$

$\quad x(e^y)^2 - 2e^y + x = 0 \;\Leftrightarrow\; e^y = \dfrac{1 \pm \sqrt{1 - x^2}}{x}$. But $y > 0 \;\Rightarrow\; e^y > 1$.

This rules out the minus sign because $\dfrac{1 - \sqrt{1 - x^2}}{x} > 1 \;\Leftrightarrow\; 1 - \sqrt{1-x^2} > x \;\Leftrightarrow\; 1 - x > \sqrt{1 - x^2} \;\Leftrightarrow\;$

$1 - 2x + x^2 > 1 - x^2 \;\Leftrightarrow\; x^2 > x \;\Leftrightarrow\; x > 1$, but $x = \operatorname{sech} y \leq 1$.

Thus, $e^y = \dfrac{1 + \sqrt{1 - x^2}}{x} \;\Rightarrow\; \operatorname{sech}^{-1} x = \ln\left(\dfrac{1 + \sqrt{1 - x^2}}{x}\right)$.

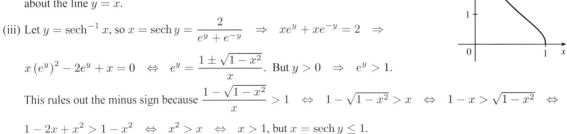

(c) (i) $y = \coth^{-1} x$ $\Leftrightarrow$ $\coth y = x$

   (ii) We sketch the graph of $\coth^{-1}$ by reflecting the graph of $\coth$ (see Exercise 22) about the line $y = x$.

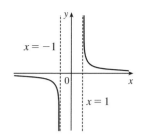

   (iii) Let $y = \coth^{-1} x$. Then $x = \coth y = \dfrac{e^y + e^{-y}}{e^y - e^{-y}}$ $\Rightarrow$

   $xe^y - xe^{-y} = e^y + e^{-y}$ $\Rightarrow$ $(x-1)e^y = (x+1)e^{-y}$ $\Rightarrow$ $e^{2y} = \dfrac{x+1}{x-1}$ $\Rightarrow$

   $2y = \ln \dfrac{x+1}{x-1}$ $\Rightarrow$ $\coth^{-1} x = \dfrac{1}{2} \ln \dfrac{x+1}{x-1}$

29. (a) Let $y = \cosh^{-1} x$. Then $\cosh y = x$ and $y \geq 0$ $\Rightarrow$ $\sinh y \dfrac{dy}{dx} = 1$ $\Rightarrow$

   $\dfrac{dy}{dx} = \dfrac{1}{\sinh y} = \dfrac{1}{\sqrt{\cosh^2 y - 1}} = \dfrac{1}{\sqrt{x^2 - 1}}$ [since $\sinh y \geq 0$ for $y \geq 0$]. *Or:* Use Formula 4.

(b) Let $y = \tanh^{-1} x$. Then $\tanh y = x$ $\Rightarrow$ $\operatorname{sech}^2 y \dfrac{dy}{dx} = 1$ $\Rightarrow$ $\dfrac{dy}{dx} = \dfrac{1}{\operatorname{sech}^2 y} = \dfrac{1}{1 - \tanh^2 y} = \dfrac{1}{1 - x^2}$.

   Or: Use Formula 5.

(c) Let $y = \operatorname{csch}^{-1} x$. Then $\operatorname{csch} y = x$ $\Rightarrow$ $-\operatorname{csch} y \coth y \dfrac{dy}{dx} = 1$ $\Rightarrow$ $\dfrac{dy}{dx} = -\dfrac{1}{\operatorname{csch} y \coth y}$. By Exercise 13,

   $\coth y = \pm\sqrt{\operatorname{csch}^2 y + 1} = \pm\sqrt{x^2 + 1}$. If $x > 0$, then $\coth y > 0$, so $\coth y = \sqrt{x^2 + 1}$. If $x < 0$, then $\coth y < 0$,

   so $\coth y = -\sqrt{x^2 + 1}$. In either case we have $\dfrac{dy}{dx} = -\dfrac{1}{\operatorname{csch} y \coth y} = -\dfrac{1}{|x|\sqrt{x^2 + 1}}$.

(d) Let $y = \operatorname{sech}^{-1} x$. Then $\operatorname{sech} y = x$ $\Rightarrow$ $-\operatorname{sech} y \tanh y \dfrac{dy}{dx} = 1$ $\Rightarrow$

   $\dfrac{dy}{dx} = -\dfrac{1}{\operatorname{sech} y \tanh y} = -\dfrac{1}{\operatorname{sech} y \sqrt{1 - \operatorname{sech}^2 y}} = -\dfrac{1}{x\sqrt{1 - x^2}}$. [Note that $y > 0$ and so $\tanh y > 0$.]

(e) Let $y = \coth^{-1} x$. Then $\coth y = x$ $\Rightarrow$ $-\operatorname{csch}^2 y \dfrac{dy}{dx} = 1$ $\Rightarrow$ $\dfrac{dy}{dx} = -\dfrac{1}{\operatorname{csch}^2 y} = \dfrac{1}{1 - \coth^2 y} = \dfrac{1}{1 - x^2}$

   by Exercise 13.

30. $f(x) = \tanh(1 + e^{2x})$ $\Rightarrow$ $f'(x) = \operatorname{sech}^2(1 + e^{2x}) \dfrac{d}{dx}(1 + e^{2x}) = 2e^{2x} \operatorname{sech}^2(1 + e^{2x})$

31. $f(x) = x \sinh x - \cosh x$ $\Rightarrow$ $f'(x) = x(\sinh x)' + \sinh x \cdot 1 - \sinh x = x \cosh x$

32. $g(x) = \cosh(\ln x)$ $\Rightarrow$ $g'(x) = \sinh(\ln x) \cdot (\ln x)' = \dfrac{1}{x} \sinh(\ln x)$

   Or: $g(x) = \cosh(\ln x) = \dfrac{1}{2}(e^{\ln x} + e^{-\ln x}) = \dfrac{1}{2}(x + x^{-1})$ $\Rightarrow$ $g'(x) = \dfrac{1}{2}(1 - x^{-2}) = \dfrac{1}{2} - 1/(2x^2)$

33. $h(x) = \ln(\cosh x)$ $\Rightarrow$ $h'(x) = \dfrac{1}{\cosh x}(\cosh x)' = \dfrac{\sinh x}{\cosh x} = \tanh x$

34. $y = x \coth(1 + x^2)$ $\Rightarrow$ $y' = x[-\operatorname{csch}^2(1 + x^2) \cdot 2x] + \coth(1 + x^2) \cdot 1 = -2x^2 \operatorname{csch}^2(1 + x^2) + \coth(1 + x^2)$

35. $y = e^{\cosh 3x}$ $\Rightarrow$ $y' = e^{\cosh 3x} \cdot \sinh 3x \cdot 3 = 3e^{\cosh 3x} \sinh 3x$

36. $f(t) = \operatorname{csch} t \, (1 - \ln \operatorname{csch} t) \quad \Rightarrow$

$$f'(t) = \operatorname{csch} t \left[-\frac{1}{\operatorname{csch} t} (-\operatorname{csch} t \, \coth t) \right] + (1 - \ln \operatorname{csch} t)(-\operatorname{csch} t \, \coth t)$$

$$= \operatorname{csch} t \, \coth t - (1 - \ln \operatorname{csch} t) \operatorname{csch} t \, \coth t = \operatorname{csch} t \, \coth t \left[1 - (1 - \ln \operatorname{csch} t) \right] = \operatorname{csch} t \, \coth t \, \ln \operatorname{csch} t$$

37. $f(t) = \operatorname{sech}^2(e^t) = [\operatorname{sech}(e^t)]^2 \quad \Rightarrow$

$$f'(t) = 2[\operatorname{sech}(e^t)] \, [\operatorname{sech}(e^t)]' = 2 \operatorname{sech}(e^t) \left[-\operatorname{sech}(e^t) \tanh(e^t) \cdot e^t \right] = -2e^t \operatorname{sech}^2(e^t) \tanh(e^t)$$

38. $y = \sinh(\cosh x) \quad \Rightarrow \quad y' = \cosh(\cosh x) \cdot \sinh x$

39. $y = \arctan(\tanh x) \quad \Rightarrow \quad y' = \dfrac{1}{1 + (\tanh x)^2} (\tanh x)' = \dfrac{\operatorname{sech}^2 x}{1 + \tanh^2 x}$

40. $y = \sqrt[4]{\dfrac{1 + \tanh x}{1 - \tanh x}} = \left(\dfrac{1 + \tanh x}{1 - \tanh x} \right)^{1/4} \quad \Rightarrow$

$$y' = \frac{1}{4} \left(\frac{1 + \tanh x}{1 - \tanh x} \right)^{-3/4} \frac{(1 - \tanh x)(\operatorname{sech}^2 x) - (1 + \tanh x)(-\operatorname{sech}^2 x)}{(1 - \tanh x)^2}$$

$$= \frac{1 (1 - \tanh x)^{3/4}}{4 (1 + \tanh x)^{3/4}} \frac{\operatorname{sech}^2 x \left[(1 - \tanh x) + (1 + \tanh x) \right]}{(1 - \tanh x)^2} = \frac{2 \operatorname{sech}^2 x \, (1 - \tanh x)^{3/4}}{4(1 + \tanh x)^{3/4}(1 - \tanh x)^2}$$

$$= \frac{\operatorname{sech}^2 x}{2(1 + \tanh x)^{3/4}(1 - \tanh x)^{5/4}}$$

Or: From Exercise 18, $\dfrac{1 + \tanh x}{1 - \tanh x} = e^{2x}$, so $y = \sqrt[4]{\dfrac{1 + \tanh x}{1 - \tanh x}} = \sqrt[4]{e^{2x}} = e^{x/2}$ and $y' = \tfrac{1}{2} e^{x/2}$.

41. $G(x) = \dfrac{1 - \cosh x}{1 + \cosh x} \quad \Rightarrow$

$$G'(x) = \frac{(1 + \cosh x)(-\sinh x) - (1 - \cosh x)(\sinh x)}{(1 + \cosh x)^2} = \frac{-\sinh x - \sinh x \, \cosh x - \sinh x + \sinh x \, \cosh x}{(1 + \cosh x)^2}$$

$$= \frac{-2 \sinh x}{(1 + \cosh x)^2}$$

42. $y = x^2 \sinh^{-1}(2x) \quad \Rightarrow \quad y' = x^2 \cdot \dfrac{1}{\sqrt{1 + (2x)^2}} \cdot 2 + \sinh^{-1}(2x) \cdot 2x = 2x \left[\dfrac{x}{\sqrt{1 + 4x^2}} + \sinh^{-1}(2x) \right]$

43. $y = \tanh^{-1} \sqrt{x} \quad \Rightarrow \quad y' = \dfrac{1}{1 - \left(\sqrt{x} \right)^2} \cdot \dfrac{1}{2} x^{-1/2} = \dfrac{1}{2 \sqrt{x} \, (1 - x)}$

44. $y = x \tanh^{-1} x + \ln \sqrt{1 - x^2} = x \tanh^{-1} x + \tfrac{1}{2} \ln(1 - x^2) \quad \Rightarrow$

$$y' = \tanh^{-1} x + \frac{x}{1 - x^2} + \frac{1}{2} \left(\frac{1}{1 - x^2} \right)(-2x) = \tanh^{-1} x$$

45. $y = x \sinh^{-1}(x/3) - \sqrt{9 + x^2} \quad \Rightarrow$

$$y' = \sinh^{-1} \left(\frac{x}{3} \right) + x \frac{1/3}{\sqrt{1 + (x/3)^2}} - \frac{2x}{2 \sqrt{9 + x^2}} = \sinh^{-1} \left(\frac{x}{3} \right) + \frac{x}{\sqrt{9 + x^2}} - \frac{x}{\sqrt{9 + x^2}} = \sinh^{-1} \left(\frac{x}{3} \right)$$

46. $y = \operatorname{sech}^{-1} \sqrt{1 - x^2} \quad \Rightarrow \quad y' = -\dfrac{1}{\sqrt{1 - x^2} \sqrt{1 - (1 - x^2)}} \dfrac{-2x}{2 \sqrt{1 - x^2}} = \dfrac{x}{(1 - x^2) |x|} = \dfrac{1}{1 - x^2}$ since $x > 0$.

47. $y = \coth^{-1}\sqrt{x^2+1}$ $\Rightarrow$ $y' = \dfrac{1}{1-(x^2+1)}\dfrac{2x}{2\sqrt{x^2+1}} = -\dfrac{1}{x\sqrt{x^2+1}}$

48. (a) Let $a = 0.03291765$. A graph of the central curve,

$y = f(x) = 211.49 - 20.96\cosh ax$, is shown.

(b) $f(0) = 211.49 - 20.96\cosh 0 = 211.49 - 20.96(1) = 190.53$ m.

(c) $y = 100$ $\Rightarrow$ $100 = 211.49 - 20.96\cosh ax$ $\Rightarrow$

$20.96\cosh ax = 111.49$ $\Rightarrow$ $\cosh ax = \dfrac{111.49}{20.96}$ $\Rightarrow$

$ax = \pm\cosh^{-1}\dfrac{111.49}{20.96}$ $\Rightarrow$ $x = \pm\dfrac{1}{a}\cosh^{-1}\dfrac{111.49}{20.96} \approx \pm 71.56$ m. The points are approximately $(\pm 71.56, 100)$.

(d) $f(x) = 211.49 - 20.96\cosh ax$ $\Rightarrow$ $f'(x) = -20.96\sinh ax \cdot a$.

$f'\left(\pm\dfrac{1}{a}\cosh^{-1}\dfrac{111.49}{20.96}\right) = -20.96a\sinh\left[a\left(\pm\dfrac{1}{a}\cosh^{-1}\dfrac{111.49}{20.96}\right)\right] = -20.96a\sinh\left(\pm\cosh^{-1}\dfrac{111.49}{20.96}\right) \approx \mp 3.6.$

So the slope at $(71.56, 100)$ is about -3.6 and the slope at $(-71.56, 100)$ is about 3.6.

49. As the depth d of the water gets large, the fraction $\dfrac{2\pi d}{L}$ gets large, and from Figure 3 or Exercise 23(a), $\tanh\left(\dfrac{2\pi d}{L}\right)$

approaches 1. Thus, $v = \sqrt{\dfrac{gL}{2\pi}\tanh\left(\dfrac{2\pi d}{L}\right)} \approx \sqrt{\dfrac{gL}{2\pi}(1)} = \sqrt{\dfrac{gL}{2\pi}}.$

50.

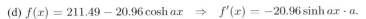

For $y = a\cosh(x/a)$ with $a > 0$, we have the y-intercept equal to a.

As a increases, the graph flattens.

51. (a) $y = 20\cosh(x/20) - 15$ $\Rightarrow$ $y' = 20\sinh(x/20)\cdot\frac{1}{20} = \sinh(x/20)$. Since the right pole is positioned at $x = 7$,

we have $y'(7) = \sinh\frac{7}{20} \approx 0.3572$.

(b) If α is the angle between the tangent line and the x-axis, then $\tan\alpha =$ slope of the line $= \sinh\frac{7}{20}$, so

$\alpha = \tan^{-1}\left(\sinh\frac{7}{20}\right) \approx 0.343$ rad $\approx 19.66°$. Thus, the angle between the line and the pole is $\theta = 90° - \alpha \approx 70.34°$.

52. We differentiate the function twice, then substitute into the differential equation: $y = \dfrac{T}{\rho g}\cosh\dfrac{\rho g x}{T}$ $\Rightarrow$

$\dfrac{dy}{dx} = \dfrac{T}{\rho g}\sinh\left(\dfrac{\rho g x}{T}\right)\dfrac{\rho g}{T} = \sinh\dfrac{\rho g x}{T}$ $\Rightarrow$ $\dfrac{d^2 y}{dx^2} = \cosh\left(\dfrac{\rho g x}{T}\right)\dfrac{\rho g}{T} = \dfrac{\rho g}{T}\cosh\dfrac{\rho g x}{T}$. We evaluate the two sides

separately: LHS $= \dfrac{d^2 y}{dx^2} = \dfrac{\rho g}{T}\cosh\dfrac{\rho g x}{T}$ and RHS $= \dfrac{\rho g}{T}\sqrt{1+\left(\dfrac{dy}{dx}\right)^2} = \dfrac{\rho g}{T}\sqrt{1+\sinh^2\dfrac{\rho g x}{T}} = \dfrac{\rho g}{T}\cosh\dfrac{\rho g x}{T}$,

by the identity proved in Example 1(a).

53. (a) $y = A \sinh mx + B \cosh mx \quad \Rightarrow \quad y' = mA \cosh mx + mB \sinh mx \quad \Rightarrow$

$y'' = m^2 A \sinh mx + m^2 B \cosh mx = m^2 (A \sinh mx + B \cosh mx) = m^2 y$

(b) From part (a), a solution of $y'' = 9y$ is $y(x) = A \sinh 3x + B \cosh 3x$. So $-4 = y(0) = A \sinh 0 + B \cosh 0 = B$, so

$B = -4$. Now $y'(x) = 3A \cosh 3x - 12 \sinh 3x \quad \Rightarrow \quad 6 = y'(0) = 3A \quad \Rightarrow \quad A = 2$, so $y = 2 \sinh 3x - 4 \cosh 3x$.

54. $\lim\limits_{x \to \infty} \dfrac{\sinh x}{e^x} = \lim\limits_{x \to \infty} \dfrac{e^x - e^{-x}}{2e^x} = \lim\limits_{x \to \infty} \dfrac{1 - e^{-2x}}{2} = \dfrac{1 - 0}{2} = \dfrac{1}{2}$

55. The tangent to $y = \cosh x$ has slope 1 when $y' = \sinh x = 1 \quad \Rightarrow \quad x = \sinh^{-1} 1 = \ln(1 + \sqrt{2})$, by Equation 3.

Since $\sinh x = 1$ and $y = \cosh x = \sqrt{1 + \sinh^2 x}$, we have $\cosh x = \sqrt{2}$. The point is $\left(\ln(1 + \sqrt{2}), \sqrt{2}\right)$.

56. $\cosh x = \cosh[\ln(\sec\theta + \tan\theta)] = \dfrac{1}{2}\left[e^{\ln(\sec\theta + \tan\theta)} + e^{-\ln(\sec\theta + \tan\theta)}\right] = \dfrac{1}{2}\left[\sec\theta + \tan\theta + \dfrac{1}{\sec\theta + \tan\theta}\right]$

$= \dfrac{1}{2}\left[\sec\theta + \tan\theta + \dfrac{\sec\theta - \tan\theta}{(\sec\theta + \tan\theta)(\sec\theta - \tan\theta)}\right] = \dfrac{1}{2}\left[\sec\theta + \tan\theta + \dfrac{\sec\theta - \tan\theta}{\sec^2\theta - \tan^2\theta}\right]$

$= \dfrac{1}{2}(\sec\theta + \tan\theta + \sec\theta - \tan\theta) = \sec\theta$

57. Let $u = \cosh x$. Then $du = \sinh x \, dx$, so $\int \sinh x \cosh^2 x \, dx = \int u^2 \, du = \frac{1}{3}u^3 + C = \frac{1}{3}\cosh^3 x + C$.

58. Let $u = 1 + 4x$. Then $du = 4 \, dx$, so $\int \sinh(1 + 4x) \, dx = \frac{1}{4}\int \sinh u \, du = \frac{1}{4}\cosh u + C = \frac{1}{4}\cosh(1 + 4x) + C$.

59. Let $u = \sqrt{x}$. Then $du = \dfrac{dx}{2\sqrt{x}}$ and $\int \dfrac{\sinh\sqrt{x}}{\sqrt{x}} \, dx = \int \sinh u \cdot 2 \, du = 2\cosh u + C = 2\cosh\sqrt{x} + C$.

60. Let $u = \cosh x$. Then $du = \sinh x \, dx$, and $\int \tanh x \, dx = \int \dfrac{\sinh x}{\cosh x} \, dx = \int \dfrac{du}{u} = \ln|u| + C = \ln(\cosh x) + C$.

61. $\int \dfrac{\cosh x}{\cosh^2 x - 1} \, dx = \int \dfrac{\cosh x}{\sinh^2 x} \, dx = \int \dfrac{\cosh x}{\sinh x} \cdot \dfrac{1}{\sinh x} \, dx = \int \coth x \, \operatorname{csch} x \, dx = -\operatorname{csch} x + C$

62. Let $u = 2 + \tanh x$. Then $du = \operatorname{sech}^2 x \, dx$, so

$\int \dfrac{\operatorname{sech}^2 x}{2 + \tanh x} \, dx = \int \dfrac{1}{u} \, du = \ln|u| + C = \ln|2 + \tanh x| + C = \ln(2 + \tanh x) + C$ [since $2 + \tanh x > 1$].

63. Let $t = 3u$. Then $dt = 3 \, du$ and

$\displaystyle\int_4^6 \dfrac{1}{\sqrt{t^2 - 9}} \, dt = \int_{4/3}^2 \dfrac{1}{\sqrt{9u^2 - 9}} \, 3 \, du = \int_{4/3}^2 \dfrac{du}{\sqrt{u^2 - 1}} = \left[\cosh^{-1} u\right]_{4/3}^2 = \cosh^{-1} 2 - \cosh^{-1}\left(\tfrac{4}{3}\right)$ or

$= \left[\cosh^{-1} u\right]_{4/3}^2 = \left[\ln\left(u + \sqrt{u^2 - 1}\right)\right]_{4/3}^2 = \ln(2 + \sqrt{3}) - \ln\left(\dfrac{4 + \sqrt{7}}{3}\right) = \ln\left(\dfrac{6 + 3\sqrt{3}}{4 + \sqrt{7}}\right)$

64. Let $u = 4t$. Then $du = 4 \, dt$ and

$\displaystyle\int_0^1 \dfrac{dt}{\sqrt{16t^2 + 1}} = \int_0^4 \dfrac{\frac{1}{4} \, du}{\sqrt{u^2 + 1}} = \dfrac{1}{4}\left[\sinh^{-1} u\right]_0^4 = \dfrac{1}{4}\left[\ln\left(u + \sqrt{u^2 + 1}\right)\right]_0^4 = \dfrac{1}{4}\left[\ln\left(4 + \sqrt{17}\right) - \ln 1\right] =$

$\frac{1}{4}\ln\left(4 + \sqrt{17}\right)$

65. Let $u = e^x$. Then $du = e^x \, dx$ and $\displaystyle\int \frac{e^x}{1 - e^{2x}} \, dx = \int \frac{du}{1 - u^2} = \tanh^{-1} u + C = \tanh^{-1}(e^x) + C$

$\left[\text{or } \frac{1}{2} \ln\left(\frac{1 + e^x}{1 - e^x}\right) + C \right]$.

66. We want $\int_0^1 \sinh cx \, dx = 1$. To calculate the integral, we put $u = cx$,

so $du = c \, dx$, the upper limit becomes c, and the equation becomes

$\dfrac{1}{c} \displaystyle\int_0^c \sinh u \, du = 1 \iff \dfrac{1}{c}[\cosh c - 1] = 1 \iff \cosh c - 1 = c$.

We plot the function $f(c) = \cosh c - c - 1$, and see that its positive root

lies at approximately $c = 1.62$. So the equation $\int_0^1 \sinh cx \, dx = 1$ holds

for $c \approx 1.62$.

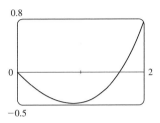

67. (a) From the graphs, we estimate

that the two curves $y = \cosh 2x$

and $y = 1 + \sinh x$ intersect at

$x = 0$ and at $x = a \approx 0.481$.

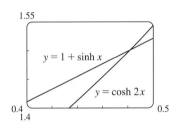

(b) We have found the two roots of the equation $\cosh 2x = 1 + \sinh x$ to be $x = 0$ and $x = a \approx 0.481$. Note from the first

graph that $1 + \sinh x > \cosh 2x$ on the interval $(0, a)$, so the area between the two curves is

$$A = \int_0^a (1 + \sinh x - \cosh 2x) \, dx = \left[x + \cosh x - \tfrac{1}{2} \sinh 2x \right]_0^a$$

$$= [a + \cosh a - \tfrac{1}{2} \sinh 2a] - [0 + \cosh 0 - \tfrac{1}{2} \sinh 0] \approx 0.0402$$

68. The area of the triangle with vertices O, P, and $(\cosh t, 0)$ is $\tfrac{1}{2} \sinh t \cosh t$, and the area under the curve $x^2 - y^2 = 1$, from

$x = 1$ to $x = \cosh t$, is $\int_1^{\cosh t} \sqrt{x^2 - 1} \, dx$. Therefore, the area of the shaded region is

$A(t) = \tfrac{1}{2} \sinh t \cosh t - \int_1^{\cosh t} \sqrt{x^2 - 1} \, dx$. So, by FTC1,

$$A'(t) = \tfrac{1}{2}\left(\cosh^2 t + \sinh^2 t\right) - \sqrt{\cosh^2 t - 1} \sinh t = \tfrac{1}{2}\left(\cosh^2 t + \sinh^2 t\right) - \sqrt{\sinh^2 t} \sinh t$$

$$= \tfrac{1}{2}\left(\cosh^2 t + \sinh^2 t\right) - \sinh^2 t = \tfrac{1}{2}\left(\cosh^2 t - \sinh^2 t\right) = \tfrac{1}{2}(1) = \tfrac{1}{2}$$

Thus $A(t) = \tfrac{1}{2}t + C$, since $A'(t) = \tfrac{1}{2}$. To calculate C, we let $t = 0$. Thus,

$A(0) = \tfrac{1}{2} \sinh 0 \cosh 0 - \int_1^{\cosh 0} \sqrt{x^2 - 1} \, dx = \tfrac{1}{2}(0) + C \Rightarrow C = 0$. Thus $A(t) = \tfrac{1}{2}t$.

69. If $ae^x + be^{-x} = \alpha \cosh(x + \beta)$ [or $\alpha \sinh(x + \beta)$],   then

$ae^x + be^{-x} = \dfrac{\alpha}{2}\left(e^{x+\beta} \pm e^{-x-\beta}\right) = \dfrac{\alpha}{2}\left(e^x e^\beta \pm e^{-x} e^{-\beta}\right) = \left(\dfrac{\alpha}{2} e^\beta\right) e^x \pm \left(\dfrac{\alpha}{2} e^{-\beta}\right) e^{-x}$. Comparing coefficients of e^x

and e^{-x}, we have $a = \dfrac{\alpha}{2} e^\beta$ **(1)**   and $b = \pm \dfrac{\alpha}{2} e^{-\beta}$ **(2)**. We need to find α and β. Dividing equation **(1)** by equation **(2)** gives us $\frac{a}{b} = \pm e^{2\beta}$ $\Rightarrow$ $(\star)$ $2\beta = \ln\left(\pm\frac{a}{b}\right)$ $\Rightarrow$ $\beta = \frac{1}{2}\ln\left(\pm\frac{a}{b}\right)$. Solving equations **(1)** and **(2)** for e^β gives us

$e^\beta = \dfrac{2a}{\alpha}$ and $e^\beta = \pm\dfrac{\alpha}{2b}$, so $\dfrac{2a}{\alpha} = \pm\dfrac{\alpha}{2b}$ $\Rightarrow$ $\alpha^2 = \pm4ab$ $\Rightarrow$ $\alpha = 2\sqrt{\pm ab}$.

$(\star)$ If $\frac{a}{b} > 0$, we use the $+$ sign and obtain a cosh function, whereas if $\frac{a}{b} < 0$, we use the $-$ sign and obtain a sinh

function.

In summary, if a and b have the same sign, we have $ae^x + be^{-x} = 2\sqrt{ab}\cosh\left(x + \frac{1}{2}\ln\frac{a}{b}\right)$, whereas, if a and b have the

opposite sign, then $ae^x + be^{-x} = 2\sqrt{-ab}\sinh\left(x + \frac{1}{2}\ln\left(-\frac{a}{b}\right)\right)$.

7.8   Indeterminate Forms and L'Hospital's Rule

Note: The use of l'Hospital's Rule is indicated by an H above the equal sign: $\overset{\text{H}}{=\!=}$

1. (a) $\displaystyle\lim_{x\to a} \frac{f(x)}{g(x)}$ is an indeterminate form of type $\dfrac{0}{0}$.

(b) $\displaystyle\lim_{x\to a} \frac{f(x)}{p(x)} = 0$ because the numerator approaches 0 while the denominator becomes large.

(c) $\displaystyle\lim_{x\to a} \frac{h(x)}{p(x)} = 0$ because the numerator approaches a finite number while the denominator becomes large.

(d) If $\displaystyle\lim_{x\to a} p(x) = \infty$ and $f(x) \to 0$ through positive values, then $\displaystyle\lim_{x\to a} \frac{p(x)}{f(x)} = \infty$. [For example, take $a = 0$, $p(x) = 1/x^2$,

and $f(x) = x^2$.] If $f(x) \to 0$ through negative values, then $\displaystyle\lim_{x\to a} \frac{p(x)}{f(x)} = -\infty$. [For example, take $a = 0$, $p(x) = 1/x^2$,

and $f(x) = -x^2$.] If $f(x) \to 0$ through both positive and negative values, then the limit might not exist. [For example,

take $a = 0$, $p(x) = 1/x^2$, and $f(x) = x$.]

(e) $\displaystyle\lim_{x\to a} \frac{p(x)}{q(x)}$ is an indeterminate form of type $\dfrac{\infty}{\infty}$.

2. (a) $\displaystyle\lim_{x\to a} [f(x)p(x)]$ is an indeterminate form of type $0 \cdot \infty$.

(b) When x is near a, $p(x)$ is large and $h(x)$ is near 1, so $h(x)p(x)$ is large. Thus, $\displaystyle\lim_{x\to a} [h(x)p(x)] = \infty$.

(c) When x is near a, $p(x)$ and $q(x)$ are both large, so $p(x)q(x)$ is large. Thus, $\displaystyle\lim_{x\to a} [p(x)q(x)] = \infty$.

3. (a) When x is near a, $f(x)$ is near 0 and $p(x)$ is large, so $f(x) - p(x)$ is large negative. Thus, $\displaystyle\lim_{x\to a} [f(x) - p(x)] = -\infty$.

(b) $\displaystyle\lim_{x\to a} [p(x) - q(x)]$ is an indeterminate form of type $\infty - \infty$.

(c) When x is near a, $p(x)$ and $q(x)$ are both large, so $p(x) + q(x)$ is large. Thus, $\displaystyle\lim_{x\to a} [p(x) + q(x)] = \infty$.

4. (a) $\displaystyle\lim_{x\to a} [f(x)]^{g(x)}$ is an indeterminate form of type 0^0.

(b) If $y = [f(x)]^{p(x)}$, then $\ln y = p(x)\ln f(x)$. When x is near a, $p(x) \to \infty$ and $\ln f(x) \to -\infty$, so $\ln y \to -\infty$.

Therefore, $\displaystyle\lim_{x\to a} [f(x)]^{p(x)} = \lim_{x\to a} y = \lim_{x\to a} e^{\ln y} = 0$, provided f^p is defined.

(c) $\lim\limits_{x \to a} [h(x)]^{p(x)}$ is an indeterminate form of type 1^∞.

(d) $\lim\limits_{x \to a} [p(x)]^{f(x)}$ is an indeterminate form of type ∞^0.

(e) If $y = [p(x)]^{q(x)}$, then $\ln y = q(x) \ln p(x)$. When x is near a, $q(x) \to \infty$ and $\ln p(x) \to \infty$, so $\ln y \to \infty$. Therefore,

$$\lim\limits_{x \to a} [p(x)]^{q(x)} = \lim\limits_{x \to a} y = \lim\limits_{x \to a} e^{\ln y} = \infty.$$

(f) $\lim\limits_{x \to a} {}^{q(x)}\!\sqrt{p(x)} = \lim\limits_{x \to a} [p(x)]^{1/q(x)}$ is an indeterminate form of type ∞^0.

5. This limit has the form $\frac{0}{0}$. We can simply factor and simplify to evaluate the limit.

$$\lim_{x \to 1} \frac{x^2 - 1}{x^2 - x} = \lim_{x \to 1} \frac{(x+1)(x-1)}{x(x-1)} = \lim_{x \to 1} \frac{x+1}{x} = \frac{1+1}{1} = 2$$

6. This limit has the form $\frac{0}{0}$. $\lim\limits_{x \to 2} \frac{x^2 + x - 6}{x - 2} = \lim\limits_{x \to 2} \frac{(x+3)(x-2)}{x-2} = \lim\limits_{x \to 2} (x + 3) = 2 + 3 = 5$

7. This limit has the form $\frac{0}{0}$. $\lim\limits_{x \to 1} \frac{x^9 - 1}{x^5 - 1} \overset{H}{=} \lim\limits_{x \to 1} \frac{9x^8}{5x^4} = \frac{9}{5} \lim\limits_{x \to 1} x^4 = \frac{9}{5}(1) = \frac{9}{5}$

8. This limit has the form $\frac{0}{0}$. $\lim\limits_{x \to 1} \frac{x^a - 1}{x^b - 1} \overset{H}{=} \lim\limits_{x \to 1} \frac{ax^{a-1}}{bx^{b-1}} = \frac{a}{b}$

9. This limit has the form $\frac{0}{0}$. $\lim\limits_{x \to (\pi/2)^+} \frac{\cos x}{1 - \sin x} \overset{H}{=} \lim\limits_{x \to (\pi/2)^+} \frac{-\sin x}{-\cos x} = \lim\limits_{x \to (\pi/2)^+} \tan x = -\infty.$

10. This limit has the form $\frac{0}{0}$. $\lim\limits_{x \to 0} \frac{\sin 4x}{\tan 5x} \overset{H}{=} \lim\limits_{x \to 0} \frac{4 \cos 4x}{5 \sec^2(5x)} = \frac{4(1)}{5(1)^2} = \frac{4}{5}$

11. This limit has the form $\frac{0}{0}$. $\lim\limits_{t \to 0} \frac{e^t - 1}{t^3} \overset{H}{=} \lim\limits_{t \to 0} \frac{e^t}{3t^2} = \infty$ since $e^t \to 1$ and $3t^2 \to 0^+$ as $t \to 0$.

12. This limit has the form $\frac{0}{0}$. $\lim\limits_{t \to 0} \frac{e^{3t} - 1}{t} \overset{H}{=} \lim\limits_{t \to 0} \frac{3e^{3t}}{1} = 3$

13. This limit has the form $\frac{0}{0}$. $\lim\limits_{x \to 0} \frac{\tan px}{\tan qx} \overset{H}{=} \lim\limits_{x \to 0} \frac{p \sec^2 px}{q \sec^2 qx} = \frac{p(1)^2}{q(1)^2} = \frac{p}{q}$

14. $\lim\limits_{\theta \to \pi/2} \frac{1 - \sin \theta}{\csc \theta} = \frac{0}{1} = 0$. L'Hospital's Rule does not apply.

15. This limit has the form $\frac{\infty}{\infty}$. $\lim\limits_{x \to \infty} \frac{\ln x}{\sqrt{x}} \overset{H}{=} \lim\limits_{x \to \infty} \frac{1/x}{\frac{1}{2} x^{-1/2}} = \lim\limits_{x \to \infty} \frac{2}{\sqrt{x}} = 0$

16. This limit has the form $\frac{\infty}{\infty}$. $\lim\limits_{x \to \infty} \frac{x + x^2}{1 - 2x^2} \overset{H}{=} \lim\limits_{x \to \infty} \frac{1 + 2x}{-4x} \overset{H}{=} \lim\limits_{x \to \infty} \frac{2}{-4} = -\frac{1}{2}.$

A better method is to divide the numerator and the denominator by x^2: $\lim\limits_{x \to \infty} \frac{x + x^2}{1 - 2x^2} = \lim\limits_{x \to \infty} \frac{\frac{1}{x} + 1}{\frac{1}{x^2} - 2} = \frac{0 + 1}{0 - 2} = -\frac{1}{2}.$

17. $\lim\limits_{x \to 0^+} [(\ln x)/x] = -\infty$ since $\ln x \to -\infty$ as $x \to 0^+$ and dividing by small values of x just increases the magnitude of the quotient $(\ln x)/x$. L'Hospital's Rule does not apply.

18. This limit has the form $\frac{\infty}{\infty}$. $\displaystyle\lim_{x\to\infty}\frac{\ln\ln x}{x}\overset{\text{H}}{=}\lim_{x\to\infty}\frac{\frac{1}{\ln x}\cdot\frac{1}{x}}{1}=\lim_{x\to\infty}\frac{1}{x\ln x}=0$

19. This limit has the form $\frac{\infty}{\infty}$. $\displaystyle\lim_{x\to\infty}\frac{e^x}{x^3}\overset{\text{H}}{=}\lim_{x\to\infty}\frac{e^x}{3x^2}\overset{\text{H}}{=}\lim_{x\to\infty}\frac{e^x}{6x}\overset{\text{H}}{=}\lim_{x\to\infty}\frac{e^x}{6}=\infty$

20. This limit has the form $\frac{0}{0}$. $\displaystyle\lim_{x\to1}\frac{\ln x}{\sin\pi x}\overset{\text{H}}{=}\lim_{x\to1}\frac{1/x}{\pi\cos\pi x}=\frac{1}{\pi(-1)}=-\frac{1}{\pi}$

21. This limit has the form $\frac{0}{0}$. $\displaystyle\lim_{x\to0}\frac{e^x-1-x}{x^2}\overset{\text{H}}{=}\lim_{x\to0}\frac{e^x-1}{2x}\overset{\text{H}}{=}\lim_{x\to0}\frac{e^x}{2}=\frac{1}{2}$

22. This limit has the form $\frac{0}{0}$. $\displaystyle\lim_{x\to0}\frac{e^x-1-x-\frac{1}{2}x^2}{x^3}\overset{\text{H}}{=}\lim_{x\to0}\frac{e^x-1-x}{3x^2}\overset{\text{H}}{=}\lim_{x\to0}\frac{e^x-1}{6x}\overset{\text{H}}{=}\lim_{x\to0}\frac{e^x}{6}=\frac{1}{6}$

23. This limit has the form $\frac{0}{0}$. $\displaystyle\lim_{x\to0}\frac{\tanh x}{\tan x}\overset{\text{H}}{=}\lim_{x\to0}\frac{\operatorname{sech}^2x}{\sec^2x}=\frac{\operatorname{sech}^2 0}{\sec^2 0}=\frac{1}{1}=1$

24. This limit has the form $\frac{0}{0}$.

$$\lim_{x\to0}\frac{x-\sin x}{x-\tan x}\overset{\text{H}}{=}\lim_{x\to0}\frac{1-\cos x}{1-\sec^2 x}\overset{\text{H}}{=}\lim_{x\to0}\frac{-(-\sin x)}{-2\sec x\,(\sec x\,\tan x)}=-\frac{1}{2}\lim_{x\to0}\frac{\sin x\left(\frac{\cos x}{\sin x}\right)}{\sec^2 x}$$

$$=-\tfrac{1}{2}\lim_{x\to0}\cos^3 x=-\tfrac{1}{2}(1)^3=-\tfrac{1}{2}$$

Another method is to write the limit as $\displaystyle\lim_{x\to0}\frac{1-\dfrac{\sin x}{x}}{1-\dfrac{\tan x}{x}}$.

25. This limit has the form $\frac{0}{0}$. $\displaystyle\lim_{t\to0}\frac{5^t-3^t}{t}\overset{\text{H}}{=}\lim_{t\to0}\frac{5^t\ln 5-3^t\ln 3}{1}=\ln 5-\ln 3=\ln\frac{5}{3}$

26. This limit has the form $\frac{0}{0}$. $\displaystyle\lim_{x\to0}\frac{\sin x-x}{x^3}\overset{\text{H}}{=}\lim_{x\to0}\frac{\cos x-1}{3x^2}\overset{\text{H}}{=}\lim_{x\to0}\frac{-\sin x}{6x}\overset{\text{H}}{=}\lim_{x\to0}\frac{-\cos x}{6}=-\frac{1}{6}$

27. This limit has the form $\frac{0}{0}$. $\displaystyle\lim_{x\to0}\frac{\sin^{-1}x}{x}\overset{\text{H}}{=}\lim_{x\to0}\frac{1/\sqrt{1-x^2}}{1}=\lim_{x\to0}\frac{1}{\sqrt{1-x^2}}=\frac{1}{1}=1$

28. This limit has the form $\frac{\infty}{\infty}$. $\displaystyle\lim_{x\to\infty}\frac{(\ln x)^2}{x}\overset{\text{H}}{=}\lim_{x\to\infty}\frac{2(\ln x)(1/x)}{1}=2\lim_{x\to\infty}\frac{\ln x}{x}\overset{\text{H}}{=}2\lim_{x\to\infty}\frac{1/x}{1}=2(0)=0$

29. This limit has the form $\frac{0}{0}$. $\displaystyle\lim_{x\to0}\frac{1-\cos x}{x^2}\overset{\text{H}}{=}\lim_{x\to0}\frac{\sin x}{2x}\overset{\text{H}}{=}\lim_{x\to0}\frac{\cos x}{2}=\frac{1}{2}$

30. This limit has the form $\frac{0}{0}$.

$$\lim_{x\to0}\frac{\cos mx-\cos nx}{x^2}\overset{\text{H}}{=}\lim_{x\to0}\frac{-m\sin mx+n\sin nx}{2x}\overset{\text{H}}{=}\lim_{x\to0}\frac{-m^2\cos mx+n^2\cos nx}{2}=\tfrac{1}{2}\left(n^2-m^2\right)$$

31. $\displaystyle\lim_{x\to0}\frac{x+\sin x}{x+\cos x}=\frac{0+0}{0+1}=\frac{0}{1}=0$. L'Hospital's Rule does not apply.

32. This limit has the form $\frac{0}{0}$. $\displaystyle\lim_{x\to0}\frac{x}{\tan^{-1}(4x)}\overset{\text{H}}{=}\lim_{x\to0}\frac{1}{\dfrac{1}{1+(4x)^2}\cdot 4}=\lim_{x\to0}\frac{1+16x^2}{4}=\frac{1}{4}$

33. This limit has the form $\frac{0}{0}$. $\lim\limits_{x\to 1}\dfrac{1-x+\ln x}{1+\cos\pi x}\overset{\mathrm{H}}{=}\lim\limits_{x\to 1}\dfrac{-1+1/x}{-\pi\sin\pi x}\overset{\mathrm{H}}{=}\lim\limits_{x\to 1}\dfrac{-1/x^2}{-\pi^2\cos\pi x}=\dfrac{-1}{-\pi^2\,(-1)}=-\dfrac{1}{\pi^2}$

34. $\lim\limits_{x\to\infty}\dfrac{\sqrt{x^2+2}}{\sqrt{2x^2+1}}=\lim\limits_{x\to\infty}\sqrt{\dfrac{x^2+2}{2x^2+1}}=\sqrt{\lim\limits_{x\to\infty}\dfrac{x^2+2}{2x^2+1}}=\sqrt{\lim\limits_{x\to\infty}\dfrac{1+2/x^2}{2+1/x^2}}=\sqrt{\dfrac{1}{2}}$

35. This limit has the form $\frac{0}{0}$. $\lim\limits_{x\to 1}\dfrac{x^a-ax+a-1}{(x-1)^2}\overset{\mathrm{H}}{=}\lim\limits_{x\to 1}\dfrac{ax^{a-1}-a}{2(x-1)}\overset{\mathrm{H}}{=}\lim\limits_{x\to 1}\dfrac{a(a-1)x^{a-2}}{2}=\dfrac{a(a-1)}{2}$

36. This limit has the form $\frac{0}{0}$. $\lim\limits_{x\to 0}\dfrac{e^x-e^{-x}-2x}{x-\sin x}\overset{\mathrm{H}}{=}\lim\limits_{x\to 0}\dfrac{e^x+e^{-x}-2}{1-\cos x}\overset{\mathrm{H}}{=}\lim\limits_{x\to 0}\dfrac{e^x-e^{-x}}{\sin x}\overset{\mathrm{H}}{=}\lim\limits_{x\to 0}\dfrac{e^x+e^{-x}}{\cos x}=\dfrac{1+1}{1}=2$

37. This limit has the form $\frac{0}{0}$. $\lim\limits_{x\to 0}\dfrac{\cos x-1+\frac{1}{2}x^2}{x^4}\overset{\mathrm{H}}{=}\lim\limits_{x\to 0}\dfrac{-\sin x+x}{4x^3}\overset{\mathrm{H}}{=}\lim\limits_{x\to 0}\dfrac{-\cos x+1}{12x^2}\overset{\mathrm{H}}{=}\lim\limits_{x\to 0}\dfrac{\sin x}{24x}\overset{\mathrm{H}}{=}\lim\limits_{x\to 0}\dfrac{\cos x}{24}=\dfrac{1}{24}$

38. This limit has the form $\frac{\infty}{\infty}$.

$$\lim_{x\to a^+}\dfrac{\cos x\ln(x-a)}{\ln(e^x-e^a)}=\lim_{x\to a^+}\cos x\lim_{x\to a^+}\dfrac{\ln(x-a)}{\ln(e^x-e^a)}\overset{\mathrm{H}}{=}\cos a\lim_{x\to a^+}\dfrac{\dfrac{1}{x-a}}{\dfrac{1}{e^x-e^a}\cdot e^x}$$

$$=\cos a\lim_{x\to a^+}\dfrac{1}{e^x}\cdot\lim_{x\to a^+}\dfrac{e^x-e^a}{x-a}\overset{\mathrm{H}}{=}\cos a\cdot\dfrac{1}{e^a}\lim_{x\to a^+}\dfrac{e^x}{1}=\cos a\cdot\dfrac{1}{e^a}\cdot e^a=\cos a$$

39. This limit has the form $\infty\cdot 0$.

$$\lim_{x\to\infty}x\sin(\pi/x)=\lim_{x\to\infty}\dfrac{\sin(\pi/x)}{1/x}\overset{\mathrm{H}}{=}\lim_{x\to\infty}\dfrac{\cos(\pi/x)(-\pi/x^2)}{-1/x^2}=\pi\lim_{x\to\infty}\cos(\pi/x)=\pi(1)=\pi$$

40. This limit has the form $\infty\cdot 0$. $\lim\limits_{x\to-\infty}x^2e^x=\lim\limits_{x\to-\infty}\dfrac{x^2}{e^{-x}}\overset{\mathrm{H}}{=}\lim\limits_{x\to-\infty}\dfrac{2x}{-e^{-x}}\overset{\mathrm{H}}{=}\lim\limits_{x\to-\infty}\dfrac{2}{e^{-x}}=\lim\limits_{x\to-\infty}2e^x=0$

41. This limit has the form $\infty\cdot 0$. We'll change it to the form $\frac{0}{0}$.

$$\lim_{x\to 0}\cot 2x\sin 6x=\lim_{x\to 0}\dfrac{\sin 6x}{\tan 2x}\overset{\mathrm{H}}{=}\lim_{x\to 0}\dfrac{6\cos 6x}{2\sec^2 2x}=\dfrac{6(1)}{2(1)^2}=3$$

42. This limit has the form $0\cdot(-\infty)$.

$$\lim_{x\to 0^+}\sin x\ln x=\lim_{x\to 0^+}\dfrac{\ln x}{\csc x}\overset{\mathrm{H}}{=}\lim_{x\to 0^+}\dfrac{1/x}{-\csc x\cot x}=-\lim_{x\to 0^+}\left(\dfrac{\sin x}{x}\cdot\tan x\right)=-\left(\lim_{x\to 0^+}\dfrac{\sin x}{x}\right)\left(\lim_{x\to 0^+}\tan x\right)$$

$$=-1\cdot 0=0$$

43. This limit has the form $\infty\cdot 0$. $\lim\limits_{x\to\infty}x^3e^{-x^2}=\lim\limits_{x\to\infty}\dfrac{x^3}{e^{x^2}}\overset{\mathrm{H}}{=}\lim\limits_{x\to\infty}\dfrac{3x^2}{2xe^{x^2}}=\lim\limits_{x\to\infty}\dfrac{3x}{2e^{x^2}}\overset{\mathrm{H}}{=}\lim\limits_{x\to\infty}\dfrac{3}{4xe^{x^2}}=0$

44. $\lim\limits_{x\to\pi/4}(1-\tan x)\sec x=(1-1)\sqrt{2}=0$. L'Hospital's Rule does not apply.

45. This limit has the form $0\cdot(-\infty)$.

$$\lim_{x\to 1^+}\ln x\tan(\pi x/2)=\lim_{x\to 1^+}\dfrac{\ln x}{\cot(\pi x/2)}\overset{\mathrm{H}}{=}\lim_{x\to 1^+}\dfrac{1/x}{(-\pi/2)\csc^2(\pi x/2)}=\dfrac{1}{(-\pi/2)(1)^2}=-\dfrac{2}{\pi}$$

46. This limit has the form $\infty\cdot 0$.

$$\lim_{x\to\infty}x\tan(1/x)=\lim_{x\to\infty}\dfrac{\tan(1/x)}{1/x}\overset{\mathrm{H}}{=}\lim_{x\to\infty}\dfrac{\sec^2(1/x)(-1/x^2)}{-1/x^2}=\lim_{x\to\infty}\sec^2(1/x)=1^2=1$$

47. This limit has the form $\infty - \infty$.

$$\lim_{x \to 1} \left(\frac{x}{x - 1} - \frac{1}{\ln x} \right) = \lim_{x \to 1} \frac{x \ln x - (x - 1)}{(x - 1) \ln x} \overset{\text{H}}{=} \lim_{x \to 1} \frac{x(1/x) + \ln x - 1}{(x - 1)(1/x) + \ln x} = \lim_{x \to 1} \frac{\ln x}{1 - (1/x) + \ln x}$$

$$\overset{\text{H}}{=} \lim_{x \to 1} \frac{1/x}{1/x^2 + 1/x} \cdot \frac{x^2}{x^2} = \lim_{x \to 1} \frac{x}{1 + x} = \frac{1}{1 + 1} = \frac{1}{2}$$

48. This limit has the form $\infty - \infty$. $\lim_{x \to 0} (\csc x - \cot x) = \lim_{x \to 0} \left(\frac{1}{\sin x} - \frac{\cos x}{\sin x} \right) = \lim_{x \to 0} \frac{1 - \cos x}{\sin x} \overset{\text{H}}{=} \lim_{x \to 0} \frac{\sin x}{\cos x} = 0$

49. We will multiply and divide by the conjugate of the expression to change the form of the expression.

$$\lim_{x \to \infty} \left(\sqrt{x^2 + x} - x \right) = \lim_{x \to \infty} \left(\frac{\sqrt{x^2 + x} - x}{1} \cdot \frac{\sqrt{x^2 + x} + x}{\sqrt{x^2 + x} + x} \right) = \lim_{x \to \infty} \frac{(x^2 + x) - x^2}{\sqrt{x^2 + x} + x}$$

$$= \lim_{x \to \infty} \frac{x}{\sqrt{x^2 + x} + x} = \lim_{x \to \infty} \frac{1}{\sqrt{1 + 1/x} + 1} = \frac{1}{\sqrt{1} + 1} = \frac{1}{2}$$

As an alternate solution, write $\sqrt{x^2 + x} - x$ as $\sqrt{x^2 + x} - \sqrt{x^2}$, factor out $\sqrt{x^2}$, rewrite as $\left(\sqrt{1 + 1/x} - 1 \right) / (1/x)$, and apply l'Hospital's Rule.

50. This limit has the form $\infty - \infty$.

$$\lim_{x \to 0} \left(\cot x - \frac{1}{x} \right) = \lim_{x \to 0} \left(\frac{\cos x}{\sin x} - \frac{1}{x} \right) = \lim_{x \to 0} \frac{x \cos x - \sin x}{x \sin x} \overset{\text{H}}{=} \lim_{x \to 0} \frac{x(-\sin x) + \cos x - \cos x}{x \cos x + \sin x}$$

$$= -\lim_{x \to 0} \frac{x \sin x}{x \cos x + \sin x} \overset{\text{H}}{=} -\lim_{x \to 0} \frac{x \cos x + \sin x}{x(-\sin x) + \cos x + \cos x} = -\frac{0 + 0}{0 + 1 + 1} = 0$$

51. The limit has the form $\infty - \infty$ and we will change the form to a product by factoring out x.

$$\lim_{x \to \infty} (x - \ln x) = \lim_{x \to \infty} x \left(1 - \frac{\ln x}{x} \right) = \infty \text{ since } \lim_{x \to \infty} \frac{\ln x}{x} \overset{\text{H}}{=} \lim_{x \to \infty} \frac{1/x}{1} = 0.$$

52. As $x \to \infty$, $1/x \to 0$, and $e^{1/x} \to 1$. So the limit has the form $\infty - \infty$ and we will change the form to a product by factoring out x.

$$\lim_{x \to \infty} \left(x e^{1/x} - x \right) = \lim_{x \to \infty} x \left(e^{1/x} - 1 \right) = \lim_{x \to \infty} \frac{e^{1/x} - 1}{1/x} \overset{\text{H}}{=} \lim_{x \to \infty} \frac{e^{1/x}(-1/x^2)}{-1/x^2} = \lim_{x \to \infty} e^{1/x} = e^0 = 1$$

53. $y = x^{x^2} \quad \Rightarrow \quad \ln y = x^2 \ln x$, so $\lim_{x \to 0^+} \ln y = \lim_{x \to 0^+} x^2 \ln x = \lim_{x \to 0^+} \frac{\ln x}{1/x^2} \overset{\text{H}}{=} \lim_{x \to 0^+} \frac{1/x}{-2/x^3} = \lim_{x \to 0^+} \left(-\frac{1}{2} x^2 \right) = 0 \quad \Rightarrow$

$\lim_{x \to 0^+} x^{x^2} = \lim_{x \to 0^+} e^{\ln y} = e^0 = 1.$

54. $y = (\tan 2x)^x \quad \Rightarrow \quad \ln y = x \cdot \ln \tan 2x$, so

$$\lim_{x \to 0^+} \ln y = \lim_{x \to 0^+} x \cdot \ln \tan 2x = \lim_{x \to 0^+} \frac{\ln \tan 2x}{1/x} \overset{\text{H}}{=} \lim_{x \to 0^+} \frac{(1/\tan 2x)(2 \sec^2 2x)}{-1/x^2} = \lim_{x \to 0^+} \frac{-2x^2 \cos 2x}{\sin 2x \, \cos^2 2x}$$

$$= \lim_{x \to 0^+} \frac{2x}{\sin 2x} \cdot \lim_{x \to 0^+} \frac{-x}{\cos 2x} = 1 \cdot 0 = 0 \quad \Rightarrow$$

$\lim_{x \to 0^+} (\tan 2x)^x = \lim_{x \to 0^+} e^{\ln y} = e^0 = 1.$

55. $y = (1 - 2x)^{1/x}$ ⇒ $\ln y = \dfrac{1}{x}\ln(1 - 2x)$, so $\lim\limits_{x \to 0} \ln y = \lim\limits_{x \to 0} \dfrac{\ln(1 - 2x)}{x} \overset{\text{H}}{=} \lim\limits_{x \to 0} \dfrac{-2/(1 - 2x)}{1} = -2$ ⇒

$\lim\limits_{x \to 0}(1 - 2x)^{1/x} = \lim\limits_{x \to 0} e^{\ln y} = e^{-2}$.

56. $y = \left(1 + \dfrac{a}{x}\right)^{bx}$ ⇒ $\ln y = bx \ln\left(1 + \dfrac{a}{x}\right)$, so

$\lim\limits_{x \to \infty} \ln y = \lim\limits_{x \to \infty} \dfrac{b\ln(1 + a/x)}{1/x} \overset{\text{H}}{=} \lim\limits_{x \to \infty} \dfrac{b\left(\dfrac{1}{1 + a/x}\right)\left(-\dfrac{a}{x^2}\right)}{-1/x^2} = \lim\limits_{x \to \infty} \dfrac{ab}{1 + a/x} = ab$ ⇒

$\lim\limits_{x \to \infty}\left(1 + \dfrac{a}{x}\right)^{bx} = \lim\limits_{x \to \infty} e^{\ln y} = e^{ab}$.

57. $y = \left(1 + \dfrac{3}{x} + \dfrac{5}{x^2}\right)^x$ ⇒ $\ln y = x \ln\left(1 + \dfrac{3}{x} + \dfrac{5}{x^2}\right)$ ⇒

$\lim\limits_{x \to \infty} \ln y = \lim\limits_{x \to \infty} \dfrac{\ln\left(1 + \dfrac{3}{x} + \dfrac{5}{x^2}\right)}{1/x} \overset{\text{H}}{=} \lim\limits_{x \to \infty} \dfrac{\left(-\dfrac{3}{x^2} - \dfrac{10}{x^3}\right)\Big/\left(1 + \dfrac{3}{x} + \dfrac{5}{x^2}\right)}{-1/x^2} = \lim\limits_{x \to \infty} \dfrac{3 + \dfrac{10}{x}}{1 + \dfrac{3}{x} + \dfrac{5}{x^2}} = 3$,

so $\lim\limits_{x \to \infty}\left(1 + \dfrac{3}{x} + \dfrac{5}{x^2}\right)^x = \lim\limits_{x \to \infty} e^{\ln y} = e^3$.

58. $y = x^{(\ln 2)/(1 + \ln x)}$ ⇒ $\ln y = \dfrac{\ln 2}{1 + \ln x}\ln x$ ⇒

$\lim\limits_{x \to \infty} \ln y = \lim\limits_{x \to \infty} \dfrac{(\ln 2)(\ln x)}{1 + \ln x} \overset{\text{H}}{=} \lim\limits_{x \to \infty} \dfrac{(\ln 2)(1/x)}{1/x} = \lim\limits_{x \to \infty} \ln 2 = \ln 2$, so $\lim\limits_{x \to \infty} x^{(\ln 2)/(1 + \ln x)} = \lim\limits_{x \to \infty} e^{\ln y} = e^{\ln 2} = 2$.

59. $y = x^{1/x}$ ⇒ $\ln y = (1/x)\ln x$ ⇒ $\lim\limits_{x \to \infty} \ln y = \lim\limits_{x \to \infty} \dfrac{\ln x}{x} \overset{\text{H}}{=} \lim\limits_{x \to \infty} \dfrac{1/x}{1} = 0$ ⇒

$\lim\limits_{x \to \infty} x^{1/x} = \lim\limits_{x \to \infty} e^{\ln y} = e^0 = 1$

60. $y = (e^x + x)^{1/x}$ ⇒ $\ln y = \dfrac{1}{x}\ln(e^x + x)$,

so $\lim\limits_{x \to \infty} \ln y = \lim\limits_{x \to \infty} \dfrac{\ln(e^x + x)}{x} \overset{\text{H}}{=} \lim\limits_{x \to \infty} \dfrac{e^x + 1}{e^x + x} \overset{\text{H}}{=} \lim\limits_{x \to \infty} \dfrac{e^x}{e^x + 1} \overset{\text{H}}{=} \lim\limits_{x \to \infty} \dfrac{e^x}{e^x} = 1$ ⇒

$\lim\limits_{x \to \infty}(e^x + x)^{1/x} = \lim\limits_{x \to \infty} e^{\ln y} = e^1 = e$.

61. $y = (4x + 1)^{\cot x}$ ⇒ $\ln y = \cot x \ln(4x + 1)$, so $\lim\limits_{x \to 0^+} \ln y = \lim\limits_{x \to 0^+} \dfrac{\ln(4x + 1)}{\tan x} \overset{\text{H}}{=} \lim\limits_{x \to 0^+} \dfrac{\dfrac{4}{4x + 1}}{\sec^2 x} = 4$ ⇒

$\lim\limits_{x \to 0^+}(4x + 1)^{\cot x} = \lim\limits_{x \to 0^+} e^{\ln y} = e^4$.

62. $y = (2 - x)^{\tan(\pi x/2)}$ ⇒ $\ln y = \tan\left(\dfrac{\pi x}{2}\right)\ln(2 - x)$ ⇒

$\lim\limits_{x \to 1} \ln y = \lim\limits_{x \to 1}\left[\tan\left(\dfrac{\pi x}{2}\right)\ln(2 - x)\right] = \lim\limits_{x \to 1} \dfrac{\ln(2 - x)}{\cot\left(\dfrac{\pi x}{2}\right)} \overset{\text{H}}{=} \lim\limits_{x \to 1} \dfrac{\dfrac{1}{2 - x}(-1)}{-\csc^2\left(\dfrac{\pi x}{2}\right)\cdot\dfrac{\pi}{2}} = \dfrac{2}{\pi}\lim\limits_{x \to 1} \dfrac{\sin^2\left(\dfrac{\pi x}{2}\right)}{2 - x}$

$= \dfrac{2}{\pi}\cdot\dfrac{1^2}{1} = \dfrac{2}{\pi}$ ⇒ $\lim\limits_{x \to 1}(2 - x)^{\tan(\pi x/2)} = \lim\limits_{x \to 1} e^{\ln y} = e^{(2/\pi)}$

63. $y = (\cos x)^{1/x^2}$ $\Rightarrow$ $\ln y = \dfrac{1}{x^2} \ln \cos x$ $\Rightarrow$

$$\lim_{x \to 0^+} \ln y = \lim_{x \to 0^+} \frac{\ln \cos x}{x^2} \overset{\text{H}}{=} \lim_{x \to 0^+} \frac{-\tan x}{2x} \overset{\text{H}}{=} \lim_{x \to 0^+} \frac{-\sec^2 x}{2} = -\frac{1}{2} \quad \Rightarrow$$

$$\lim_{x \to 0^+} (\cos x)^{1/x^2} = \lim_{x \to 0^+} e^{\ln y} = e^{-1/2} = 1/\sqrt{e}$$

64. $y = \left(\dfrac{2x-3}{2x+5}\right)^{2x+1}$ $\Rightarrow$ $\ln y = (2x+1)\ln\left(\dfrac{2x-3}{2x+5}\right)$ $\Rightarrow$

$$\lim_{x \to \infty} \ln y = \lim_{x \to \infty} \frac{\ln(2x-3) - \ln(2x+5)}{1/(2x+1)} \overset{\text{H}}{=} \lim_{x \to \infty} \frac{2/(2x-3) - 2/(2x+5)}{-2/(2x+1)^2} = \lim_{x \to \infty} \frac{-8(2x+1)^2}{(2x-3)(2x+5)}$$

$$= \lim_{x \to \infty} \frac{-8(2+1/x)^2}{(2-3/x)(2+5/x)} = -8 \quad \Rightarrow \quad \lim_{x \to \infty} \left(\frac{2x-3}{2x+5}\right)^{2x+1} = e^{-8}$$

65.

From the graph, if $x = 500$, $y \approx 7.36$. The limit has the form 1^∞.

Now $y = \left(1 + \dfrac{2}{x}\right)^x$ $\Rightarrow$ $\ln y = x \ln\left(1 + \dfrac{2}{x}\right)$ $\Rightarrow$

$$\lim_{x \to \infty} \ln y = \lim_{x \to \infty} \frac{\ln(1 + 2/x)}{1/x} \overset{\text{H}}{=} \lim_{x \to \infty} \frac{\frac{1}{1+2/x}\left(-\frac{2}{x^2}\right)}{-1/x^2}$$

$$= 2 \lim_{x \to \infty} \frac{1}{1 + 2/x} = 2(1) = 2 \quad \Rightarrow$$

$$\lim_{x \to \infty} \left(1 + \frac{2}{x}\right)^x = \lim_{x \to \infty} e^{\ln y} = e^2 \ [\approx 7.39]$$

66.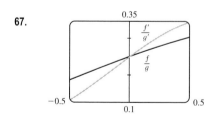

From the graph, as $x \to 0$, $y \approx 0.55$. The limit has the form $\frac{0}{0}$.

$$\lim_{x \to 0} \frac{5^x - 4^x}{3^x - 2^x} \overset{\text{H}}{=} \lim_{x \to 0} \frac{5^x \ln 5 - 4^x \ln 4}{3^x \ln 3 - 2^x \ln 2} = \frac{\ln 5 - \ln 4}{\ln 3 - \ln 2} = \frac{\ln \frac{5}{4}}{\ln \frac{3}{2}} \ [\approx 0.55]$$

67.

From the graph, it appears that $\lim\limits_{x \to 0} \dfrac{f(x)}{g(x)} = \lim\limits_{x \to 0} \dfrac{f'(x)}{g'(x)} = 0.25$.

We calculate $\lim\limits_{x \to 0} \dfrac{f(x)}{g(x)} = \lim\limits_{x \to 0} \dfrac{e^x - 1}{x^3 + 4x} \overset{\text{H}}{=} \lim\limits_{x \to 0} \dfrac{e^x}{3x^2 + 4} = \dfrac{1}{4}$.

68.

From the graph, it appears that $\lim\limits_{x \to 0} \dfrac{f(x)}{g(x)} = \lim\limits_{x \to 0} \dfrac{f'(x)}{g'(x)} = 4$. We calculate

$$\lim_{x \to 0} \frac{f(x)}{g(x)} = \lim_{x \to 0} \frac{2x \sin x}{\sec x - 1} \overset{\text{H}}{=} \lim_{x \to 0} \frac{2(x \cos x + \sin x)}{\sec x \tan x}$$

$$\overset{\text{H}}{=} \lim_{x \to 0} \frac{2(-x \sin x + \cos x + \cos x)}{\sec x(\sec^2 x) + \tan x(\sec x \tan x)} = \frac{4}{1} = 4$$

69. $y = f(x) = xe^{-x}$ **A.** $D = \mathbb{R}$ **B.** Intercepts are 0   **C.** No symmetry

D. $\lim_{x \to \infty} xe^{-x} = \lim_{x \to \infty} \dfrac{x}{e^x} \overset{H}{=} \lim_{x \to \infty} \dfrac{1}{e^x} = 0$, so $y = 0$ is a HA. $\lim_{x \to -\infty} xe^{-x} = -\infty$

E. $f'(x) = e^{-x} - xe^{-x} = e^{-x}(1 - x) > 0$ $\Leftrightarrow$ $x < 1$, so f is increasing

on $(-\infty, 1)$ and decreasing on $(1, \infty)$.   **F.** Absolute and local maximum

value $f(1) = 1/e$.   **G.** $f''(x) = e^{-x}(x - 2) > 0$ $\Leftrightarrow$ $x > 2$, so f is CU

on $(2, \infty)$ and CD on $(-\infty, 2)$.   IP at $\left(2, 2/e^2\right)$

H.

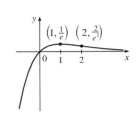

70. $y = f(x) = \dfrac{\ln x}{x^2}$ **A.** $D = (0, \infty)$ **B.** y-intercept: none; x-intercept: $f(x) = 0$ $\Leftrightarrow$ $\ln x = 0$ $\Leftrightarrow$ $x = 1$

C. No symmetry   **D.** $\lim_{x \to 0^+} f(x) = -\infty$, so $x = 0$ is a VA; $\lim_{x \to \infty} \dfrac{\ln x}{x^2} \overset{H}{=} \lim_{x \to \infty} \dfrac{1/x}{2x} = 0$, so $y = 0$ is a HA.

E. $f'(x) = \dfrac{x^2(1/x) - (\ln x)(2x)}{(x^2)^2} = \dfrac{x(1 - 2\ln x)}{x^4} = \dfrac{1 - 2\ln x}{x^3}$. $f'(x) > 0$ $\Leftrightarrow$ $1 - 2\ln x > 0$ $\Leftrightarrow$ $\ln x < \dfrac{1}{2}$ $\Rightarrow$

$0 < x < e^{1/2}$ and $f'(x) < 0$ $\Rightarrow$ $x > e^{1/2}$, so f is increasing on $\left(0, \sqrt{e}\right)$ and decreasing on $\left(\sqrt{e}, \infty\right)$.

F. Local maximum value $f(e^{1/2}) = \dfrac{1/2}{e} = \dfrac{1}{2e}$

G. $f''(x) = \dfrac{x^3(-2/x) - (1 - 2\ln x)(3x^2)}{(x^3)^2}$

$= \dfrac{x^2[-2 - 3(1 - 2\ln x)]}{x^6} = \dfrac{-5 + 6\ln x}{x^4}$

$f''(x) > 0$ $\Leftrightarrow$ $-5 + 6\ln x > 0$ $\Leftrightarrow$ $\ln x > \dfrac{5}{6}$ $\Rightarrow$ $x > e^{5/6}$ [f is CU]

and $f''(x) < 0$ $\Leftrightarrow$ $0 < x < e^{5/6}$ [f is CD]. IP at $(e^{5/6}, 5/(6e^{5/3}))$

H.

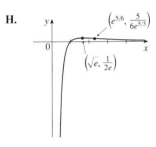

71. $y = f(x) = xe^{-x^2}$ **A.** $D = \mathbb{R}$ **B.** Intercepts are 0   **C.** $f(-x) = -f(x)$, so the curve is symmetric

about the origin.   **D.** $\lim_{x \to \pm\infty} xe^{-x^2} = \lim_{x \to \pm\infty} \dfrac{x}{e^{x^2}} \overset{H}{=} \lim_{x \to \pm\infty} \dfrac{1}{2xe^{x^2}} = 0$, so $y = 0$ is a HA.

E. $f'(x) = e^{-x^2} - 2x^2 e^{-x^2} = e^{-x^2}(1 - 2x^2) > 0$ $\Leftrightarrow$ $x^2 < \dfrac{1}{2}$ $\Leftrightarrow$ $|x| < \dfrac{1}{\sqrt{2}}$, so f is increasing on $\left(-\dfrac{1}{\sqrt{2}}, \dfrac{1}{\sqrt{2}}\right)$

and decreasing on $\left(-\infty, -\dfrac{1}{\sqrt{2}}\right)$ and $\left(\dfrac{1}{\sqrt{2}}, \infty\right)$.   **F.** Local maximum value $f\left(\dfrac{1}{\sqrt{2}}\right) = 1/\sqrt{2e}$, local minimum

value $f\left(-\dfrac{1}{\sqrt{2}}\right) = -1/\sqrt{2e}$ **G.** $f''(x) = -2xe^{-x^2}(1 - 2x^2) - 4xe^{-x^2} = 2xe^{-x^2}(2x^2 - 3) > 0$ $\Leftrightarrow$

$x > \sqrt{\dfrac{3}{2}}$ or $-\sqrt{\dfrac{3}{2}} < x < 0$, so f is CU on $\left(\sqrt{\dfrac{3}{2}}, \infty\right)$ and   **H.**

$\left(-\sqrt{\dfrac{3}{2}}, 0\right)$ and CD on $\left(-\infty, -\sqrt{\dfrac{3}{2}}\right)$ and $\left(0, \sqrt{\dfrac{3}{2}}\right)$.

IP are $(0, 0)$ and $\left(\pm\sqrt{\dfrac{3}{2}}, \pm\sqrt{\dfrac{3}{2}}\,e^{-3/2}\right)$.

72. $y = f(x) = e^x/x$ **A.** $D = \{x \mid x \neq 0\}$ **B.** No intercept   **C.** No symmetry

D. $\lim\limits_{x \to \infty} \dfrac{e^x}{x} \overset{\text{H}}{=} \lim\limits_{x \to \infty} \dfrac{e^x}{1} = \infty, = \lim\limits_{x \to -\infty} \dfrac{e^x}{x} = 0$, so $y = 0$ is a HA. $\lim\limits_{x \to 0^+} \dfrac{e^x}{x} = \infty$, $\lim\limits_{x \to 0^-} \dfrac{e^x}{x} = -\infty$, so $x = 0$ is a VA.

E. $f'(x) = \dfrac{xe^x - e^x}{x^2} > 0 \iff (x-1)e^x > 0 \iff x > 1$, so f is increasing on $(1, \infty)$, and decreasing

on $(-\infty, 0)$ and $(0, 1)$.   **F.** $f(1) = e$ is a local minimum value.

H.

G. $f''(x) = \dfrac{x^2(xe^x) - 2x(xe^x - e^x)}{x^4} = \dfrac{e^x x^2 - 2x + 2)}{x^3} > 0 \iff x > 0$

since $x^2 - 2x + 2 > 0$ for all x. So f is CU on $(0, \infty)$ and CD on $(-\infty, 0)$.

No IP

73. $y = f(x) = x - \ln(1+x)$ **A.** $D = \{x \mid x > -1\} = (-1, \infty)$ **B.** Intercepts are 0   **C.** No symmetry

D. $\lim\limits_{x \to -1^+} [x - \ln(1+x)] = \infty$, so $x = -1$ is a VA. $\lim\limits_{x \to \infty} [x - \ln(1+x)] = \lim\limits_{x \to \infty} x \left[1 - \dfrac{\ln(1+x)}{x}\right] = \infty$,

since $\lim\limits_{x \to \infty} \dfrac{\ln(1+x)}{x} \overset{\text{H}}{=} \lim\limits_{x \to \infty} \dfrac{1/(1+x)}{1} = 0$.

H.

E. $f'(x) = 1 - \dfrac{1}{1+x} = \dfrac{x}{1+x} > 0 \iff x > 0$ since $x + 1 > 0$.

So f is increasing on $(0, \infty)$ and decreasing on $(-1, 0)$.

F. $f(0) = 0$ is an absolute minimum.

G. $f''(x) = 1/(1+x)^2 > 0$, so f is CU on $(-1, \infty)$.

74. $y = f(x) = (x^2 - 3)e^{-x}$ **A.** $D = \mathbb{R}$ **B.** y-intercept: $f(0) = -3$; x-intercepts: $f(x) = 0 \iff x^2 - 3 = 0 \iff$

$x = \pm\sqrt{3}$ **C.** No symmetry   **D.** $\lim\limits_{x \to \infty} (x^2 - 3)e^{-x} = \lim\limits_{x \to \infty} \dfrac{x^2 - 3}{e^x} \overset{\text{H}}{=} \lim\limits_{x \to \infty} \dfrac{2x}{e^x} \overset{\text{H}}{=} \lim\limits_{x \to \infty} \dfrac{2}{e^x} = 0$, so $y = 0$ is a HA.

E. $f'(x) = (x^2 - 3)(-e^{-x}) + e^{-x}(2x) = -e^{-x}[(x^2 - 3) - 2x] = -e^{-x}(x-3)(x+1)$. $f'(x) > 0 \iff -1 < x < 3$

and $f'(x) < 0 \iff x < -1$ or $x > 3$, so f is increasing on $(-1, 3)$ and decreasing on $(-\infty, -1)$ and $(3, \infty)$.

F. Local maximum value $f(3) = 6e^{-3}$; local minimum value $f(-1) = -2e$

G. $f''(x) = (-e^{-x})(2x - 2) + (x^2 - 2x - 3)(e^{-x}) = e^{-x}[-(2x - 2) + (x^2 - 2x - 3)] = e^{-x}(x^2 - 4x - 1)$.

$f''(x) = 0 \iff x = \dfrac{4 \pm \sqrt{20}}{2} = 2 \pm \sqrt{5}$, so $f''(x) > 0 \iff x < 2 - \sqrt{5}$ or   **H.**

$x > 2 + \sqrt{5}$ and $f''(x) < 0 \iff 2 - \sqrt{5} < x < 2 + \sqrt{5}$, so f is CU on

$(-\infty, 2 - \sqrt{5})$ and $(2 + \sqrt{5}, \infty)$ and f is CD on $(2 - \sqrt{5}, 2 + \sqrt{5})$.

IP at $(2 - \sqrt{5}, f(2 - \sqrt{5})) \approx (-0.24, -3.73)$ and

$(2 + \sqrt{5}, f(2 + \sqrt{5})) \approx (4.24, 0.22)$

75. (a) $f(x) = x^{-x}$

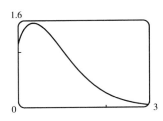

(b) $y = f(x) = x^{-x}$. We note that $\ln f(x) = \ln x^{-x} = -x \ln x = -\dfrac{\ln x}{1/x}$, so

$$\lim_{x \to 0^+} \ln f(x) \overset{H}{=} \lim_{x \to 0^+} -\frac{1/x}{-x^{-2}} = \lim_{x \to 0^+} x = 0. \text{ Thus } \lim_{x \to 0^+} f(x) = \lim_{x \to 0^+} e^{\ln f(x)} = e^0 = 1.$$

(c) From the graph, it appears that there is a local and absolute maximum of about $f(0.37) \approx 1.44$. To find the exact value, we

differentiate: $f(x) = x^{-x} = e^{-x \ln x} \;\Rightarrow\; f'(x) = e^{-x \ln x}\left[-x\left(\dfrac{1}{x}\right) + \ln x(-1)\right] = -x^{-x}(1 + \ln x)$. This is 0 only

when $1 + \ln x = 0 \;\Leftrightarrow\; x = e^{-1}$. Also $f'(x)$ changes from positive to negative at e^{-1}. So the maximum value is

$f(1/e) = (1/e)^{-1/e} = e^{1/e}$.

(d)

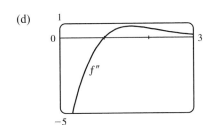

We differentiate again to get

$$f''(x) = -x^{-x}(1/x) + (1 + \ln x)^2(x^{-x}) = x^{-x}\left[(1 + \ln x)^2 - 1/x\right]$$

From the graph of $f''(x)$, it seems that $f''(x)$ changes from negative to positive at $x = 1$, so we estimate that f has an IP at $x = 1$.

76. (a) $f(x) = (\sin x)^{\sin x}$ is continuous where $\sin x > 0$, that is, on intervals

of the form $(2n\pi, (2n+1)\pi)$, so we have graphed f on $(0, \pi)$.

(b) $y = (\sin x)^{\sin x} \;\Rightarrow\; \ln y = \sin x \ln \sin x$, so

$$\lim_{x \to 0^+} \ln y = \lim_{x \to 0^+} \sin x \ln \sin x = \lim_{x \to 0^+} \frac{\ln \sin x}{\csc x} \overset{H}{=} \lim_{x \to 0^+} \frac{\cot x}{-\csc x \cot x} = \lim_{x \to 0^+} (-\sin x) = 0 \;\Rightarrow$$

$$\lim_{x \to 0^+} y = e^0 = 1.$$

(c) It appears that we have a local maximum at $(1.57, 1)$ and local minima at $(0.38, 0.69)$ and $(2.76, 0.69)$.

$$y = (\sin x)^{\sin x} \;\Rightarrow\; \ln y = \sin x \ln \sin x \;\Rightarrow\; \frac{y'}{y} = (\sin x)\left(\frac{\cos x}{\sin x}\right) + (\ln \sin x)\cos x = \cos x\,(1 + \ln \sin x) \;\Rightarrow$$

$y' = (\sin x)^{\sin x}(\cos x)(1 + \ln \sin x).$ $y' = 0 \;\Rightarrow\; \cos x = 0$ or $\ln \sin x = -1 \;\Rightarrow\; x_2 = \frac{\pi}{2}$ or $\sin x = e^{-1}$.

On $(0, \pi)$, $\sin x = e^{-1} \;\Rightarrow\; x_1 = \sin^{-1}(e^{-1})$ and $x_3 = \pi - \sin^{-1}(e^{-1})$. Approximating these points gives us

$(x_1, f(x_1)) \approx (0.3767, 0.6922)$, $(x_2, f(x_2)) \approx (1.5708, 1)$, and $(x_3, f(x_3)) \approx (2.7649, 0.6922)$. The approximations

confirm our estimates.

(d)

From the graph, we see that $f''(x) = 0$ at $x \approx 0.94$ and $x \approx 2.20$.

Since f'' changes sign at these values, they are x-coordinates of inflection points.

77. (a) $f(x) = x^{1/x}$

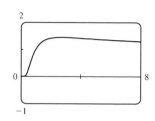

(b) Recall that $a^b = e^{b \ln a}$. $\displaystyle\lim_{x \to 0^+} x^{1/x} = \lim_{x \to 0^+} e^{(1/x) \ln x}$. As $x \to 0^+$, $\dfrac{\ln x}{x} \to -\infty$, so $x^{1/x} = e^{(1/x) \ln x} \to 0$. This

indicates that there is a hole at $(0, 0)$. As $x \to \infty$, we have the indeterminate form ∞^0. $\displaystyle\lim_{x \to \infty} x^{1/x} = \lim_{x \to \infty} e^{(1/x) \ln x}$,

but $\displaystyle\lim_{x \to \infty} \frac{\ln x}{x} \overset{\text{H}}{=} \lim_{x \to \infty} \frac{1/x}{1} = 0$, so $\displaystyle\lim_{x \to \infty} x^{1/x} = e^0 = 1$. This indicates that $y = 1$ is a HA.

(c) Estimated maximum: $(2.72, 1.45)$. No estimated minimum. We use logarithmic differentiation to find any critical

numbers. $y = x^{1/x} \Rightarrow \ln y = \dfrac{1}{x} \ln x \Rightarrow \dfrac{y'}{y} = \dfrac{1}{x} \cdot \dfrac{1}{x} + (\ln x)\left(-\dfrac{1}{x^2}\right) \Rightarrow y' = x^{1/x}\left(\dfrac{1 - \ln x}{x^2}\right) = 0 \Rightarrow$

$\ln x = 1 \Rightarrow x = e$. For $0 < x < e$, $y' > 0$ and for $x > e$, $y' < 0$, so $f(e) = e^{1/e}$ is a local maximum value. This

point is approximately $(2.7183, 1.4447)$, which agrees with our estimate.

(d)

From the graph, we see that $f''(x) = 0$ at $x \approx 0.58$ and $x \approx 4.37$. Since f''

changes sign at these values, they are x-coordinates of inflection points.

78.

The first figure shows representative examples of $f(x) = x^n e^{-x}$ with n odd. n is even in the second figure. All curves pass through the origin and approach $y = 0$ as $x \to \infty$.

$f'(x) = \dfrac{x^n(n - x)}{xe^x} = 0 \Leftrightarrow x = n$ or $x = 0$ (the latter for $n > 1$). At $x = 0$, we have a local minimum for n even. At

$x = n$, we have a local maximum for all n. As n increases, $(n, f(n))$ gets farther away from the origin.

$f''(x) = \dfrac{x^n(x^2 - 2nx + n^2 - n)}{x^2 e^x} = 0 \Leftrightarrow x = n \pm \sqrt{n}$ or $x = 0$ (the latter for $n > 2$). As n increases, the IP move

farther away from the origin—they are symmetric about the line $x = n$.

79. If $c < 0$, then $\displaystyle\lim_{x\to-\infty} f(x) = \lim_{x\to-\infty} xe^{-cx} = \lim_{x\to-\infty} \frac{x}{e^{cx}} \overset{\text{H}}{=} \lim_{x\to-\infty} \frac{1}{ce^{cx}} = 0$, and $\displaystyle\lim_{x\to\infty} f(x) = \infty$.

If $c > 0$, then $\displaystyle\lim_{x\to-\infty} f(x) = -\infty$, and $\displaystyle\lim_{x\to\infty} f(x) \overset{\text{H}}{=} \lim_{x\to\infty} \frac{1}{ce^{cx}} = 0$.

If $c = 0$, then $f(x) = x$, so $\displaystyle\lim_{x\to\pm\infty} f(x) = \pm\infty$, respectively.

So we see that $c = 0$ is a transitional value. We now exclude the case $c = 0$, since we know how the function behaves in that case. To find the maxima and minima of f, we differentiate: $f(x) = xe^{-cx}$ $\Rightarrow$

$f'(x) = x(-ce^{-cx}) + e^{-cx} = (1 - cx)e^{-cx}$. This is 0 when $1 - cx = 0$ $\Leftrightarrow$ $x = 1/c$. If $c < 0$ then this represents a minimum value of $f(1/c) = 1/(ce)$, since $f'(x)$ changes from negative to positive at $x = 1/c$; and if $c > 0$, it represents a maximum value. As $|c|$ increases, the maximum or minimum point gets closer to the origin. To find the inflection points, we differentiate again: $f'(x) = e^{-cx}(1 - cx)$ $\Rightarrow$

$f''(x) = e^{-cx}(-c) + (1 - cx)(-ce^{-cx}) = (cx - 2)ce^{-cx}$. This changes sign when $cx - 2 = 0$ $\Leftrightarrow$ $x = 2/c$. So as $|c|$ increases, the points of inflection get closer to the origin.

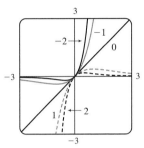

80. We see that both numerator and denominator approach 0, so we can use l'Hospital's Rule:

$$\lim_{x\to a} \frac{\sqrt{2a^3 x - x^4} - a\sqrt[3]{aax}}{a - \sqrt[4]{ax^3}} \overset{\text{H}}{=} \lim_{x\to a} \frac{\frac{1}{2}(2a^3 x - x^4)^{-1/2}(2a^3 - 4x^3) - a\left(\frac{1}{3}\right)(aax)^{-2/3}a^2}{-\frac{1}{4}(ax^3)^{-3/4}(3ax^2)}$$

$$= \frac{\frac{1}{2}(2a^3 a - a^4)^{-1/2}(2a^3 - 4a^3) - \frac{1}{3}a^3(a^2 a)^{-2/3}}{-\frac{1}{4}(aa^3)^{-3/4}(3aa^2)}$$

$$= \frac{(a^4)^{-1/2}(-a^3) - \frac{1}{3}a^3(a^3)^{-2/3}}{-\frac{3}{4}a^3(a^4)^{-3/4}} = \frac{-a - \frac{1}{3}a}{-\frac{3}{4}} = \frac{4}{3}\left(\frac{4}{3}a\right) = \frac{16}{9}a$$

81. $\displaystyle\lim_{x\to\infty} \frac{x}{\sqrt{x^2 + 1}} \overset{\text{H}}{=} \lim_{x\to\infty} \frac{1}{\frac{1}{2}(x^2 + 1)^{-1/2}(2x)} = \lim_{x\to\infty} \frac{\sqrt{x^2 + 1}}{x}$. Repeated applications of l'Hospital's Rule result in the

original limit or the limit of the reciprocal of the function. Another method is to try dividing the numerator and denominator

by x: $\displaystyle\lim_{x\to\infty} \frac{x}{\sqrt{x^2 + 1}} = \lim_{x\to\infty} \frac{x/x}{\sqrt{x^2/x^2 + 1/x^2}} = \lim_{x\to\infty} \frac{1}{\sqrt{1 + 1/x^2}} = \frac{1}{1} = 1$

82. $\displaystyle\lim_{c\to0^+} s(t) = \lim_{c\to0^+} \left(\frac{m}{c}\ln\cosh\sqrt{\frac{gc}{mt}}\right) = m\lim_{c\to0^+} \frac{\ln\cosh\sqrt{ac}}{c}$ [let $a = g/(mt)$]

$$\overset{\text{H}}{=} m\lim_{c\to0^+} \frac{\frac{1}{\cosh\sqrt{ac}}(\sinh\sqrt{ac})\left(\frac{\sqrt{a}}{2\sqrt{c}}\right)}{1} = \frac{m\sqrt{a}}{2}\lim_{c\to0^+} \frac{\tanh\sqrt{ac}}{\sqrt{c}}$$

$$\overset{\text{H}}{=} \frac{m\sqrt{a}}{2}\lim_{c\to0^+} \frac{\text{sech}^2\sqrt{ac}\left[\sqrt{a}/\left(2\sqrt{c}\right)\right]}{1/\left(2\sqrt{c}\right)} = \frac{ma}{2}\lim_{c\to0^+} \text{sech}^2\sqrt{ac} = \frac{ma}{2}(1)^2 = \frac{mg}{2mt} = \frac{g}{2t}$$

83. $\displaystyle\lim_{E\to0^+} P(E) = \lim_{E\to0^+}\left(\frac{e^E+e^{-E}}{e^E-e^{-E}}-\frac{1}{E}\right)$

$\displaystyle = \lim_{E\to0^+}\frac{E\left(e^E+e^{-E}\right)-1\left(e^E-e^{-E}\right)}{\left(e^E-e^{-E}\right)E} = \lim_{E\to0^+}\frac{Ee^E+Ee^{-E}-e^E+e^{-E}}{Ee^E-Ee^{-E}}$ $\left[\text{form is } \frac{0}{0}\right]$

$\displaystyle \overset{\text{H}}{=} \lim_{E\to0^+}\frac{Ee^E+e^E\cdot1+E(-e^{-E})+e^{-E}\cdot1-e^E+(-e^{-E})}{Ee^E+e^E\cdot1-[E(-e^{-E})+e^{-E}\cdot1]}$

$\displaystyle = \lim_{E\to0^+}\frac{Ee^E-Ee^{-E}}{Ee^E+e^E+Ee^{-E}-e^{-E}} = \lim_{E\to0^+}\frac{e^E-e^{-E}}{e^E+\dfrac{e^E}{E}+e^{-E}-\dfrac{e^{-E}}{E}}$ [divide by E]

$\displaystyle = \frac{0}{2+L}, \quad\text{where } L=\lim_{E\to0^+}\frac{e^E-e^{-E}}{E}$ $\left[\text{form is } \frac{0}{0}\right]$ $\displaystyle\overset{\text{H}}{=}\lim_{E\to0^+}\frac{e^E+e^{-E}}{1}=\frac{1+1}{1}=2$

Thus, $\displaystyle\lim_{E\to0^+}P(E)=\frac{0}{2+2}=0.$

84. (a) $\displaystyle\lim_{R\to r^+} v = \lim_{R\to r^+}\left[-c\left(\frac{r}{R}\right)^2\ln\left(\frac{r}{R}\right)\right] = -cr^2\lim_{R\to r^+}\left[\left(\frac{1}{R}\right)^2\ln\left(\frac{r}{R}\right)\right] = -cr^2\cdot\frac{1}{r^2}\cdot\ln1 = -c\cdot0 = 0$

As the insulation of a metal cable becomes thinner, the velocity of an electrical impulse in the cable approaches zero.

(b) $\displaystyle\lim_{r\to0^+} v = \lim_{r\to0^+}\left[-c\left(\frac{r}{R}\right)^2\ln\left(\frac{r}{R}\right)\right] = -\frac{c}{R^2}\lim_{r\to0^+}\left[r^2\ln\left(\frac{r}{R}\right)\right]$ [form is $0\cdot\infty$]

$\displaystyle = -\frac{c}{R^2}\lim_{r\to0^+}\frac{\ln\left(\dfrac{r}{R}\right)}{\dfrac{1}{r^2}}$ [form is ∞/∞] $\displaystyle\overset{\text{H}}{=}-\frac{c}{R^2}\lim_{r\to0^+}\frac{\dfrac{R}{r}\cdot\dfrac{1}{R}}{\dfrac{-2}{r^3}} = -\frac{c}{R^2}\lim_{r\to0^+}\left(-\frac{r^2}{2}\right)=0$

As the radius of the metal cable approaches zero, the velocity of an electrical impulse in the cable approaches zero.

85. First we will find $\displaystyle\lim_{n\to\infty}\left(1+\frac{r}{n}\right)^{nt}$, which is of the form 1^∞. $y=\left(1+\dfrac{r}{n}\right)^{nt}$ $\Rightarrow$ $\ln y = nt\ln\left(1+\dfrac{r}{n}\right)$, so

$\displaystyle\lim_{n\to\infty}\ln y = \lim_{n\to\infty}nt\ln\left(1+\frac{r}{n}\right) = t\lim_{n\to\infty}\frac{\ln(1+r/n)}{1/n} \overset{\text{H}}{=} t\lim_{n\to\infty}\frac{(-r/n^2)}{(1+r/n)(-1/n^2)} = t\lim_{n\to\infty}\frac{r}{1+i/n} = tr$ $\Rightarrow$

$\displaystyle\lim_{n\to\infty}y = e^{rt}.$ Thus, as $n\to\infty$, $A=A_0\left(1+\dfrac{r}{n}\right)^{nt}\to A_0e^{rt}.$

86. (a) $\displaystyle\lim_{t\to\infty} v = \lim_{t\to\infty}\frac{mg}{c}\left(1-e^{-ct/m}\right) = \frac{mg}{c}\lim_{t\to\infty}\left(1-e^{-ct/m}\right) = \frac{mg}{c}(1-0)$ [because $-ct/m\to-\infty$ as $t\to\infty$]

$\displaystyle = \frac{mg}{c}$, which is the speed the object approaches as time goes on, the so-called limiting velocity.

(b) $\displaystyle\lim_{c\to0^+} v = \lim_{c\to0^+}\frac{mg}{c}\left(1-e^{-ct/m}\right) = mg\lim_{c\to0^+}\frac{1-e^{-ct/m}}{c}$ $\left[\text{form is }\frac{0}{0}\right]$

$\displaystyle\overset{\text{H}}{=}mg\lim_{c\to0^+}\frac{(-e^{-ct/m})\cdot(-t/m)}{1} = \frac{mgt}{m}\lim_{c\to0^+}e^{-ct/m} = gt(1) = gt$

The velocity of a falling object in a vacuum is directly proportional to the amount of time it falls.

87. Both numerator and denominator approach 0 as $x\to0$, so we use l'Hospital's Rule (and FTC1):

$$\lim_{x\to0}\frac{S(x)}{x^3} = \lim_{x\to0}\frac{\int_0^x\sin(\pi t^2/2)\,dt}{x^3} \overset{\text{H}}{=} \lim_{x\to0}\frac{\sin(\pi x^2/2)}{3x^2} \overset{\text{H}}{=} \lim_{x\to0}\frac{\pi x\cos(\pi x^2/2)}{6x} = \frac{\pi}{6}\cdot\cos0 = \frac{\pi}{6}$$

88. Both numerator and denominator approach 0 as $a \to 0$, so we use l'Hospital's Rule. (Note that we are differentiating with respect to a, since that is the quantity which is changing.) We also use the Fundamental Theorem of Calculus, Part 1:

$$\lim_{a \to 0} T(x,t) = \lim_{a \to 0} \frac{C \int_0^a e^{-(x-u)^2/(4kt)}\, du}{a\sqrt{4\pi kt}} \overset{\text{H}}{=} \lim_{a \to 0} \frac{Ce^{-(x-a)^2/(4kt)}}{\sqrt{4\pi kt}} = \frac{Ce^{-x^2/(4kt)}}{\sqrt{4\pi kt}}$$

89. Since $f(2) = 0$, the given limit has the form $\frac{0}{0}$.

$$\lim_{x \to 0} \frac{f(2+3x) + f(2+5x)}{x} \overset{\text{H}}{=} \lim_{x \to 0} \frac{f'(2+3x)\cdot 3 + f'(2+5x)\cdot 5}{1} = f'(2)\cdot 3 + f'(2)\cdot 5 = 8f'(2) = 8\cdot 7 = 56$$

90. $L = \lim_{x \to 0}\left(\frac{\sin 2x}{x^3} + a + \frac{b}{x^2}\right) = \lim_{x \to 0}\frac{\sin 2x + ax^3 + bx}{x^3} \overset{\text{H}}{=} \lim_{x \to 0}\frac{2\cos 2x + 3ax^2 + b}{3x^2}$. As $x \to 0$, $3x^2 \to 0$, and

$(2\cos 2x + 3ax^2 + b) \to b + 2$, so the last limit exists only if $b + 2 = 0$, that is, $b = -2$. Thus,

$$\lim_{x \to 0}\frac{2\cos 2x + 3ax^2 - 2}{3x^2} \overset{\text{H}}{=} \lim_{x \to 0}\frac{-4\sin 2x + 6ax}{6x} \overset{\text{H}}{=} \lim_{x \to 0}\frac{-8\cos 2x + 6a}{6} = \frac{6a - 8}{6}, \text{ which is equal to 0 if and only}$$

if $a = \frac{4}{3}$. Hence, $L = 0$ if and only if $b = -2$ and $a = \frac{4}{3}$.

91. Since $\lim_{h \to 0}[f(x+h) - f(x-h)] = f(x) - f(x) = 0$ (f is differentiable and hence continuous) and $\lim_{h \to 0} 2h = 0$, we use

l'Hospital's Rule:

$$\lim_{h \to 0}\frac{f(x+h) - f(x-h)}{2h} \overset{\text{H}}{=} \lim_{h \to 0}\frac{f'(x+h)(1) - f'(x-h)(-1)}{2} = \frac{f'(x) + f'(x)}{2} = \frac{2f'(x)}{2} = f'(x)$$

$\dfrac{f(x+h) - f(x-h)}{2h}$ is the slope of the secant line between

$(x-h, f(x-h))$ and $(x+h, f(x+h))$. As $h \to 0$, this line gets closer

to the tangent line and its slope approaches $f'(x)$.

92. Since $\lim_{h \to 0}[f(x+h) - 2f(x) + f(x-h)] = f(x) - 2f(x) + f(x) = 0$ [f is differentiable and hence continuous]

and $\lim_{h \to 0} h^2 = 0$, we can apply l'Hospital's Rule:

$$\lim_{h \to 0}\frac{f(x+h) - 2f(x) + f(x-h)}{h^2} \overset{\text{H}}{=} \lim_{h \to 0}\frac{f'(x+h) - f'(x-h)}{2h} = f''(x)$$

At the last step, we have applied the result of Exercise 91 to $f'(x)$.

93. $\displaystyle\lim_{x \to \infty}\frac{e^x}{x^n} \overset{\text{H}}{=} \lim_{x \to \infty}\frac{e^x}{nx^{n-1}} \overset{\text{H}}{=} \lim_{x \to \infty}\frac{e^x}{n(n-1)x^{n-2}} \overset{\text{H}}{=} \cdots \overset{\text{H}}{=} \lim_{x \to \infty}\frac{e^x}{n!} = \infty$

94. This limit has the form $\frac{\infty}{\infty}$. $\displaystyle\lim_{x \to \infty}\frac{\ln x}{x^p} \overset{\text{H}}{=} \lim_{x \to \infty}\frac{1/x}{px^{p-1}} = \lim_{x \to \infty}\frac{1}{px^p} = 0$ since $p > 0$.

95. $\displaystyle\lim_{x \to 0^+} x^{\alpha}\ln x = \lim_{x \to 0^+}\frac{\ln x}{x^{-\alpha}} \overset{\text{H}}{=} \lim_{x \to 0^+}\frac{1/x}{-\alpha x^{-\alpha - 1}} = \lim_{x \to 0^+}\frac{x^\alpha}{-\alpha} = 0$ since $\alpha > 0$.

96. Using l'Hospital's Rule and FTC1, we have

$$\lim_{x \to 0}\frac{\int_0^x \sin(t^2)\, dt}{x^3} = \lim_{x \to 0}\frac{\frac{d}{dx}\int_0^x \sin(t^2)\, dt}{3x^2} = \lim_{x \to 0}\frac{\sin(x^2)}{3x^2} = \frac{1}{3}\lim_{x \to 0}\frac{\sin(x^2)}{x^2} = \frac{1}{3}.$$

97. Let the radius of the circle be r. We see that $A(\theta)$ is the area of the whole figure (a sector of the circle with radius 1), minus the area of $\triangle OPR$. But the area of the sector of the circle is $\frac{1}{2}r^2\theta$ (see Reference Page 1), and the area of the triangle is $\frac{1}{2}r\,|PQ| = \frac{1}{2}r(r\sin\theta) = \frac{1}{2}r^2\sin\theta$. So we have $A(\theta) = \frac{1}{2}r^2\theta - \frac{1}{2}r^2\sin\theta = \frac{1}{2}r^2(\theta - \sin\theta)$. Now by elementary trigonometry, $B(\theta) = \frac{1}{2}\,|QR|\,|PQ| = \frac{1}{2}(r - |OQ|)\,|PQ| = \frac{1}{2}(r - r\cos\theta)(r\sin\theta) = \frac{1}{2}r^2(1 - \cos\theta)\sin\theta$. So the limit we want is

$$\lim_{\theta\to0^+}\frac{A(\theta)}{B(\theta)} = \lim_{\theta\to0^+}\frac{\frac{1}{2}r^2(\theta-\sin\theta)}{\frac{1}{2}r^2(1-\cos\theta)\sin\theta} \overset{\text{H}}{=} \lim_{\theta\to0^+}\frac{1-\cos\theta}{(1-\cos\theta)\cos\theta + \sin\theta\,(\sin\theta)}$$

$$= \lim_{\theta\to0^+}\frac{1-\cos\theta}{\cos\theta - \cos^2\theta + \sin^2\theta} \overset{\text{H}}{=} \lim_{\theta\to0^+}\frac{\sin\theta}{-\sin\theta - 2\cos\theta\,(-\sin\theta) + 2\sin\theta\,(\cos\theta)}$$

$$= \lim_{\theta\to0^+}\frac{\sin\theta}{-\sin\theta + 4\sin\theta\,\cos\theta} = \lim_{\theta\to0^+}\frac{1}{-1+4\cos\theta} = \frac{1}{-1+4\cos0} = \frac{1}{3}$$

98. The area $A(t) = \int_0^t \sin(x^2)\,dx$, and the area $B(t) = \frac{1}{2}t\sin(t^2)$. Since $\lim_{t\to0^+}A(t) = 0 = \lim_{t\to0^+}B(t)$, we can use l'Hospital's Rule:

$$\lim_{t\to0^+}\frac{A(t)}{B(t)} \overset{\text{H}}{=} \lim_{t\to0^+}\frac{\sin(t^2)}{\frac{1}{2}\sin(t^2) + \frac{1}{2}t[2t\cos(t^2)]} \qquad \text{[by FTC1 and the Product Rule]}$$

$$\overset{\text{H}}{=} \lim_{t\to0^+}\frac{2t\cos(t^2)}{t\cos(t^2) - 2t^3\sin(t^2) + 2t\cos(t^2)} = \lim_{t\to0^+}\frac{2\cos(t^2)}{3\cos(t^2) - 2t^2\sin(t^2)} = \frac{2}{3-0} = \frac{2}{3}$$

99. (a) We show that $\lim_{x\to0}\dfrac{f(x)}{x^n} = 0$ for every integer $n \ge 0$. Let $y = \dfrac{1}{x^2}$. Then

$$\lim_{x\to0}\frac{f(x)}{x^{2n}} = \lim_{x\to0}\frac{e^{-1/x^2}}{(x^2)^n} = \lim_{y\to\infty}\frac{y^n}{e^y} \overset{\text{H}}{=} \lim_{y\to\infty}\frac{ny^{n-1}}{e^y} \overset{\text{H}}{=} \cdots \overset{\text{H}}{=} \lim_{y\to\infty}\frac{n!}{e^y} = 0 \;\Rightarrow$$

$$\lim_{x\to0}\frac{f(x)}{x^n} = \lim_{x\to0}x^n\frac{f(x)}{x^{2n}} = \lim_{x\to0}x^n\lim_{x\to0}\frac{f(x)}{x^{2n}} = 0. \text{ Thus, } f'(0) = \lim_{x\to0}\frac{f(x)-f(0)}{x-0} = \lim_{x\to0}\frac{f(x)}{x} = 0.$$

(b) Using the Chain Rule and the Quotient Rule we see that $f^{(n)}(x)$ exists for $x \ne 0$. In fact, we prove by induction that for each $n \ge 0$, there is a polynomial p_n and a non-negative integer k_n with $f^{(n)}(x) = p_n(x)f(x)/x^{k_n}$ for $x \ne 0$. This is true for $n = 0$; suppose it is true for the nth derivative. Then $f'(x) = f(x)(2/x^3)$, so

$$f^{(n+1)}(x) = \left[x^{k_n}\left[p_n'(x)\,f(x) + p_n(x)\,f'(x)\right] - k_n x^{k_n-1}p_n(x)\,f(x)\right]x^{-2k_n}$$

$$= \left[x^{k_n}p_n'(x) + p_n(x)(2/x^3) - k_n x^{k_n-1}p_n(x)\right]f(x)x^{-2k_n}$$

$$= \left[x^{k_n+3}p_n'(x) + 2p_n(x) - k_n x^{k_n+2}\,p_n(x)\right]f(x)x^{-(2k_n+3)}$$

which has the desired form.

Now we show by induction that $f^{(n)}(0) = 0$ for all n. By part (a), $f'(0) = 0$. Suppose that $f^{(n)}(0) = 0$. Then

$$f^{(n+1)}(0) = \lim_{x\to0}\frac{f^{(n)}(x) - f^{(n)}(0)}{x-0} = \lim_{x\to0}\frac{f^{(n)}(x)}{x} = \lim_{x\to0}\frac{p_n(x)\,f(x)/x^{k_n}}{x} = \lim_{x\to0}\frac{p_n(x)\,f(x)}{x^{k_n+1}}$$

$$= \lim_{x\to0}p_n(x)\lim_{x\to0}\frac{f(x)}{x^{k_n+1}} = p_n(0)\cdot0 = 0$$

100. (a) For f to be continuous, we need $\lim\limits_{x\to 0} f(x) = f(0) = 1$. We note that for $x \neq 0$, $\ln f(x) = \ln |x|^x = x \ln |x|$.

So $\lim\limits_{x\to 0} \ln f(x) = \lim\limits_{x\to 0} x \ln |x| = \lim\limits_{x\to 0} \dfrac{\ln |x|}{1/x} \overset{\text{H}}{=} \lim\limits_{x\to 0} \dfrac{1/x}{-1/x^2} = 0$. Therefore, $\lim\limits_{x\to 0} f(x) = \lim\limits_{x\to 0} e^{\ln f(x)} = e^0 = 1$.

So f is continuous at 0.

(b) From the graphs, it appears that f is differentiable at 0.

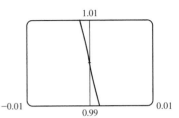

(c) To find f', we use logarithmic differentiation: $\ln f(x) = x \ln |x| \;\Rightarrow\; \dfrac{f'(x)}{f(x)} = x\left(\dfrac{1}{x}\right) + \ln |x| \;\Rightarrow$

$f'(x) = f(x)(1 + \ln |x|) = |x|^x(1 + \ln |x|)$, $x \neq 0$. Now $f'(x) \to -\infty$ as $x \to 0$ [since $|x|^x \to 1$ and

$(1 + \ln |x|) \to -\infty$], so the curve has a vertical tangent at $(0, 1)$ and is therefore not differentiable there.

The fact cannot be seen in the graphs in part (b) because $\ln |x| \to -\infty$ very slowly as $x \to 0$.

7 Review

CONCEPT CHECK

1. (a) A function f is called a *one-to-one function* if it never takes on the same value twice; that is, if $f(x_1) \neq f(x_2)$ whenever

$x_1 \neq x_2$. (Or, f is 1-1 if each output corresponds to only one input.)

   Use the Horizontal Line Test: A function is one-to-one if and only if no horizontal line intersects its graph more than

   once.

(b) If f is a one-to-one function with domain A and range B, then its *inverse function* f^{-1} has domain B and range A and is

   defined by

$$f^{-1}(y) = x \;\Leftrightarrow\; f(x) = y$$

   for any y in B. The graph of f^{-1} is obtained by reflecting the graph of f about the line $y = x$.

(c) $(f^{-1})'(a) = \dfrac{1}{f'(f^{-1}(a))}$

2. (a) The function $f(x) = e^x$ has domain $\mathbb{R}$ and range $(0, \infty)$.

(b) The function $f(x) = \ln x$ has domain $(0, \infty)$ and range $\mathbb{R}$.

(c) The graphs are reflections of one another about the line $y = x$. See Figure 7.3.3 or Figure 7.3*.1.

(d) $\log_a x = \dfrac{\ln x}{\ln a}$

3. (a) The inverse sine function $f(x) = \sin^{-1} x$ is defined as follows:

$$\sin^{-1} x = y \quad \Leftrightarrow \quad \sin y = x \quad \text{and} \quad -\frac{\pi}{2} \le y \le \frac{\pi}{2}$$

Its domain is $-1 \le x \le 1$ and its range is $-\frac{\pi}{2} \le y \le \frac{\pi}{2}$.

(b) The inverse cosine function $f(x) = \cos^{-1} x$ is defined as follows:

$$\cos^{-1} x = y \quad \Leftrightarrow \quad \cos y = x \quad \text{and} \quad 0 \le y \le \pi$$

Its domain is $-1 \le x \le 1$ and its range is $0 \le y \le \pi$.

(c) The inverse tangent function $f(x) = \tan^{-1} x$ is defined as follows:

$$\tan^{-1} x = y \quad \Leftrightarrow \quad \tan y = x \quad \text{and} \quad -\frac{\pi}{2} < y < \frac{\pi}{2}$$

Its domain is $\mathbb{R}$ and its range is $-\frac{\pi}{2} < y < \frac{\pi}{2}$.

4. $\sinh x = \dfrac{e^x - e^{-x}}{2}$, $\cosh x = \dfrac{e^x + e^{-x}}{2}$, $\tanh x = \dfrac{\sinh x}{\cosh x} = \dfrac{e^x - e^{-x}}{e^x + e^{-x}}$

5. (a) $y = e^x \Rightarrow y' = e^x$ (b) $y = a^x \Rightarrow y' = a^x \ln a$

(c) $y = \ln x \Rightarrow y' = 1/x$ (d) $y = \log_a x \Rightarrow y' = 1/(x \ln a)$

(e) $y = \sin^{-1} x \Rightarrow y' = 1/\sqrt{1 - x^2}$ (f) $y = \cos^{-1} x \Rightarrow y' = -1/\sqrt{1 - x^2}$

(g) $y = \tan^{-1} x \Rightarrow y' = 1/(1 + x^2)$ (h) $y = \sinh x \Rightarrow y' = \cosh x$

(i) $y = \cosh x \Rightarrow y' = \sinh x$ (j) $y = \tanh x \Rightarrow y' = \text{sech}^2 x$

(k) $y = \sinh^{-1} x \Rightarrow y' = 1/\sqrt{1 + x^2}$ (l) $y = \cosh^{-1} x \Rightarrow y' = 1/\sqrt{x^2 - 1}$

(m) $y = \tanh^{-1} x \Rightarrow y' = 1/(1 - x^2)$

6. (a) See Definition 7.2.7 (or 7.2*.5).

(b) $e = \lim\limits_{x \to 0} (1 + x)^{1/x}$

(c) The differentiation formula for $y = a^x$ $[y' = a^x \ln a]$ is simplest when $a = e$ because $\ln e = 1$.

(d) The differentiation formula for $y = \log_a x$ $[y' = 1/(x \ln a)]$ is simplest when $a = e$ because $\ln e = 1$.

7. (a) $\dfrac{dy}{dt} = ky$

(b) The equation in part (a) is an appropriate model for population growth, assuming that there is enough room and nutrition to support the growth.

(c) If $y(0) = y_0$, then the solution is $y(t) = y_0 e^{kt}$.

8. (a) See l'Hospital's Rule and the three notes that follow it in Section 7.8.

(b) Write fg as $\dfrac{f}{1/g}$ or $\dfrac{g}{1/f}$.

(c) Convert the difference into a quotient using a common denominator, rationalizing, factoring, or some other method.

(d) Convert the power to a product by taking the natural logarithm of both sides of $y = f^g$ or by writing f^g as $e^{g \ln f}$.

TRUE-FALSE QUIZ

1. True.　If f is one-to-one, with domain $\mathbb{R}$, then $f^{-1}(f(6)) = 6$ by the first cancellation equation [see (4) in Section 7.1].

2. False.　By Theorem 7 in Section 7.1, $(f^{-1})'(6) = \dfrac{1}{f'(f^{-1}(6))}$, *not* $\dfrac{1}{f'(6)}$ unless $f^{-1}(6) = 6$.

3. False.　For example, $\cos \frac{\pi}{2} = \cos\left(-\frac{\pi}{2}\right)$, so $\cos x$ is not 1-1.

4. False.　It is true that $\tan \frac{3\pi}{4} = -1$, but since the range of $\tan^{-1}$ is $\left(-\frac{\pi}{2}, \frac{\pi}{2}\right)$, we must have $\tan^{-1}(-1) = -\frac{\pi}{4}$.

5. True, since $\ln x$ is an increasing function on $(0, \infty)$.

6. True, by Equation 7.4*.1 $[a^x = e^{x \ln a}]$.　*Or:* From (7.3.6), $e^{\sqrt{5} \ln \pi} = (e^{\ln \pi})^{\sqrt{5}} = \pi^{\sqrt{5}}$.

7. True.　We can divide by e^x since $e^x \neq 0$ for every x.

8. False.　For example, $\ln(1 + 1) = \ln 2$, but $\ln 1 + \ln 1 = 0$. In fact $\ln a + \ln b = \ln(ab)$.

9. False.　Let $x = e$. Then $(\ln x)^6 = (\ln e)^6 = 1^6 = 1$, but $6 \ln x = 6 \ln e = 6 \cdot 1 = 6 \neq 1 = (\ln x)^6$.

10. False.　$\dfrac{d}{dx} 10^x = 10^x \ln 10$

11. False.　$\ln 10$ is a constant, so its derivative is 0.

12. True.　$y = e^{3x} \;\Rightarrow\; \ln y = 3x \;\Rightarrow\; x = \frac{1}{3} \ln y \;\Rightarrow\;$ the inverse function is $y = \frac{1}{3} \ln x$.

13. False.　The "-1" is not an exponent; it is an indication of an inverse function.

14. False.　For example, $\tan^{-1} 20$ is defined; $\sin^{-1} 20$ and $\cos^{-1} 20$ are not.

15. True.　See Figure 2 in Section 7.7.

16. True.　$\ln \frac{1}{10} = -\ln 10 = -\int_1^{10} (1/x)\, dx$, by Equation 7.4.4 or by Definition 7.2*.1.

17. True.　$\int_2^{16} (1/x)\, dx = \ln x \big]_2^{16} = \ln 16 - \ln 2 = \ln \frac{16}{2} = \ln 8 = \ln 2^3 = 3 \ln 2$

18. False.　L'Hospital's Rule does not apply since $\displaystyle\lim_{x \to \pi^-} \frac{\tan x}{1 - \cos x} = \frac{0}{2} = 0$.

EXERCISES

1. No. f is not 1-1 because the graph of f fails the Horizontal Line Test.

2. (a) g is one-to-one because it passes the Horizontal Line Test.

(b) When $y = 2$, $x \approx 0.2$. So $g^{-1}(2) \approx 0.2$.

(c) The range of g is $[-1, 3.5]$, which is the same as the domain of g^{-1}.

(d) We reflect the graph of g through the line $y = x$ to obtain the graph of g^{-1}.

3. (a) $f^{-1}(3) = 7$ since $f(7) = 3$.   (b) $(f^{-1})'(3) = \dfrac{1}{f'(f^{-1}(3))} = \dfrac{1}{f'(7)} = \dfrac{1}{8}$

4. $y = \dfrac{x+1}{2x+1}$.   Interchanging x and y gives us $x = \dfrac{y+1}{2y+1}$ $\Rightarrow$ $2xy + x = y + 1$ $\Rightarrow$ $2xy - y = 1 - x$ $\Rightarrow$

$y(2x - 1) = 1 - x$ $\Rightarrow$ $y = \dfrac{1-x}{2x-1} = f^{-1}(x)$.

5.

$y = 5^x - 1$

6.

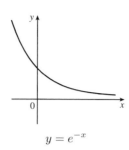

$y = e^{-x}$

$y = -e^{-x}$

7. Reflect the graph of $y = \ln x$ about the x-axis to obtain the graph of $y = -\ln x$.

$y = \ln x$ $y = -\ln x$

8.

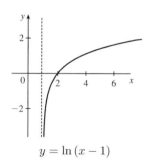

$y = \ln(x - 1)$

9.

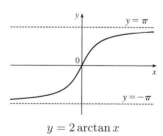

$y = 2\arctan x$

10. We have seen that if $a > 1$, then $a^x > x^a$ for sufficiently large x. (See Exercise 7.2.20.) In general, we could show that
$\lim\limits_{x \to \infty} (a^x/x^a) = \infty$ by using l'Hospital's Rule repeatedly. Also, $\log_a x$ increases much more slowly than either x^a or a^x.
[Compare the graph of $\log_a x$ with those of x^a and a^x, or use l'Hospital's Rule to show that $\lim\limits_{x \to \infty} [(\log_a x)/x^a] = 0$.]
So for large x, $\log_a x < x^a < a^x$.

11. (a) $e^{2\ln 3} = (e^{\ln 3})^2 = 3^2 = 9$ (b) $\log_{10} 25 + \log_{10} 4 = \log_{10}(25 \cdot 4) = \log_{10} 100 = \log_{10} 10^2 = 2$

12. (a) $\ln e^{\pi} = \pi$ (b) $\tan(\arcsin \tfrac{1}{2}) = \tan \tfrac{\pi}{6} = \tfrac{1}{\sqrt{3}}$

13. $\ln x = \tfrac{1}{3}$ $\Leftrightarrow$ $\log_e x = \tfrac{1}{3}$ $\Rightarrow$ $x = e^{1/3}$

14. $e^x = \tfrac{1}{3}$ $\Rightarrow$ $x = \ln \tfrac{1}{3} = \ln 1 - \ln 3 = -\ln 3$

15. $e^{e^x} = 17$ $\Rightarrow$ $\ln e^{e^x} = \ln 17$ $\Rightarrow$ $e^x = \ln 17$ $\Rightarrow$ $\ln e^x = \ln(\ln 17)$ $\Rightarrow$ $x = \ln \ln 17$

16. $\ln(1 + e^{-x}) = 3 \quad \Rightarrow \quad 1 + e^{-x} = e^3 \quad \Rightarrow \quad e^{-x} = e^3 - 1 \quad \Rightarrow \quad \ln e^{-x} = \ln(e^3 - 1) \quad \Rightarrow \quad -x = \ln(e^3 - 1) \quad \Rightarrow$

$x = -\ln(e^3 - 1)$

17. $\ln(x + 1) + \ln(x - 1) = 1 \quad \Rightarrow \quad \ln[(x + 1)(x - 1)] = 1 \quad \Rightarrow \quad \ln(x^2 - 1) = \ln e \quad \Rightarrow \quad x^2 - 1 = e \quad \Rightarrow$

$x^2 = e + 1 \quad \Rightarrow \quad x = \sqrt{e + 1}$ since $\ln(x - 1)$ is defined only when $x > 1$.

18. $\log_5(c^x) = d \quad \Rightarrow \quad x \log_5 c = d \quad \Rightarrow \quad x = \dfrac{d}{\log_5 c}$.

Or: $\log_5(c^x) = d \quad \Rightarrow \quad 5^d = c^x \quad \Rightarrow \quad \ln 5^d = \ln c^x \quad \Rightarrow \quad d \ln 5 = x \ln c \quad \Rightarrow \quad x = \dfrac{d \ln 5}{\ln c}$.

19. $\tan^{-1} x = 1 \quad \Rightarrow \quad \tan \tan^{-1} x = \tan 1 \quad \Rightarrow \quad x = \tan 1 \ (\approx 1.5574)$

20. $\sin x = 0.3 \quad \Rightarrow \quad x = \sin^{-1} 0.3 = \alpha$ for $-\frac{\pi}{2} \le x \le \frac{\pi}{2}$. The reference angle for α is $\pi - \alpha$, so all solutions are

$x = \alpha + 2n\pi$ and $x = \pi - \alpha + 2n\pi$ [or $(2n + 1)\pi - \alpha$]

21. $f(t) = t^2 \ln t \quad \Rightarrow \quad f'(t) = t^2 \cdot \dfrac{1}{t} + (\ln t)(2t) = t + 2t \ln t$ or $t(1 + 2 \ln t)$

22. $g(t) = \dfrac{e^t}{1 + e^t} \quad \Rightarrow \quad g'(t) = \dfrac{(1 + e^t)e^t - e^t(e^t)}{(1 + e^t)^2} = \dfrac{e^t}{(1 + e^t)^2}$

23. $h(\theta) = e^{\tan 2\theta} \quad \Rightarrow \quad h'(\theta) = e^{\tan 2\theta} \cdot \sec^2 2\theta \cdot 2 = 2 \sec^2(2\theta)\, e^{\tan 2\theta}$

24. $h(u) = 10^{\sqrt{u}} \quad \Rightarrow \quad h'(u) = 10^{\sqrt{u}} \cdot \ln 10 \cdot \dfrac{1}{2\sqrt{u}} = \dfrac{(\ln 10)10^{\sqrt{u}}}{2\sqrt{u}}$

25. $y = \ln|\sec 5x + \tan 5x| \quad \Rightarrow$

$y' = \dfrac{1}{\sec 5x + \tan 5x}(\sec 5x \tan 5x \cdot 5 + \sec^2 5x \cdot 5) = \dfrac{5 \sec 5x\, (\tan 5x + \sec 5x)}{\sec 5x + \tan 5x} = 5 \sec 5x$

26. $y = e^{-t}(t^2 - 2t + 2) \quad \Rightarrow$

$y' = e^{-t}(2t - 2) + (t^2 - 2t + 2)(-e^{-t}) = e^{-t}(2t - 2 - t^2 + 2t - 2) = e^{-t}(-t^2 + 4t - 4)$

27. $y = e^{cx}(c \sin x - \cos x) \quad \Rightarrow \quad y' = ce^{cx}(c \sin x - \cos x) + e^{cx}(c \cos x + \sin x) = (c^2 + 1)e^{cx} \sin x$

28. $y = e^{mx} \cos nx \quad \Rightarrow$

$y' = e^{mx}(\cos nx)' + \cos nx\,(e^{mx})' = e^{mx}(-\sin nx \cdot n) + \cos nx\,(e^{mx} \cdot m) = e^{mx}(m \cos nx - n \sin nx)$

29. $y = \ln(\sec^2 x) = 2 \ln|\sec x| \quad \Rightarrow \quad y' = (2/\sec x)(\sec x \tan x) = 2 \tan x$

30. $y = \ln(x^2 e^x) = 2 \ln|x| + x \quad \Rightarrow \quad y' = 2/x + 1$

31. $y = \dfrac{e^{1/x}}{x^2} \quad \Rightarrow \quad y' = \dfrac{x^2(e^{1/x})' - e^{1/x}(x^2)'}{(x^2)^2} = \dfrac{x^2(e^{1/x})(-1/x^2) - e^{1/x}(2x)}{x^4} = \dfrac{-e^{1/x}(1 + 2x)}{x^4}$

32. $y = (\arcsin 2x)^2 \quad \Rightarrow \quad y' = 2(\arcsin 2x) \cdot (\arcsin 2x)' = 2 \arcsin 2x \cdot \dfrac{1}{\sqrt{1 - (2x)^2}} \cdot 2 = \dfrac{4 \arcsin 2x}{\sqrt{1 - 4x^2}}$

33. $y = 3^{x \ln x} \quad \Rightarrow \quad y' = 3^{x \ln x}(\ln 3)\dfrac{d}{dx}(x \ln x) = 3^{x \ln x}(\ln 3)\left(x \cdot \dfrac{1}{x} + \ln x \cdot 1\right) = 3^{x \ln x}(\ln 3)(1 + \ln x)$

34. $y = e^{\cos x} + \cos(e^x)$ $\Rightarrow$ $y' = -\sin x\, e^{\cos x} - e^x \sin(e^x)$

35. $H(v) = v \tan^{-1} v$ $\Rightarrow$ $H'(v) = v \cdot \dfrac{1}{1 + v^2} + \tan^{-1} v \cdot 1 = \dfrac{v}{1 + v^2} + \tan^{-1} v$

36. $F(z) = \log_{10}(1 + z^2)$ $\Rightarrow$ $F'(z) = \dfrac{1}{(\ln 10)(1 + z^2)} \cdot 2z = \dfrac{2z}{(\ln 10)(1 + z^2)}$

37. $y = x \sinh(x^2)$ $\Rightarrow$ $y' = x \cosh(x^2) \cdot 2x + \sinh(x^2) \cdot 1 = 2x^2 \cosh(x^2) + \sinh(x^2)$

38. $y = (\cos x)^x$ $\Rightarrow$ $\ln y = \ln(\cos x)^x = x \ln \cos x$ $\Rightarrow$ $\dfrac{y'}{y} = x \cdot \dfrac{1}{\cos x} \cdot (-\sin x) + \ln \cos x \cdot 1$ $\Rightarrow$

$y' = (\cos x)^x (\ln \cos x - x \tan x)$

39. $y = \ln \sin x - \frac{1}{2} \sin^2 x$ $\Rightarrow$ $y' = \dfrac{1}{\sin x} \cdot \cos x - \frac{1}{2} \cdot 2 \sin x \cdot \cos x = \cot x - \sin x \cos x$

40. $y = \arctan\left(\arcsin \sqrt{x}\,\right)$ $\Rightarrow$ $y' = \dfrac{1}{1 + \left(\arcsin \sqrt{x}\,\right)^2} \cdot \dfrac{1}{\sqrt{1 - x}} \cdot \dfrac{1}{2\sqrt{x}}$

41. $y = \ln\left(\dfrac{1}{x}\right) + \dfrac{1}{\ln x} = \ln x^{-1} + (\ln x)^{-1} = -\ln x + (\ln x)^{-1}$ $\Rightarrow$ $y' = -1 \cdot \dfrac{1}{x} + (-1)(\ln x)^{-2} \cdot \dfrac{1}{x} = -\dfrac{1}{x} - \dfrac{1}{x\,(\ln x)^2}$

42. $xe^y = y - 1$ $\Rightarrow$ $e^y + xe^y y' = y'$ $\Rightarrow$ $y' = e^y / (1 - xe^y)$

43. $y = \ln(\cosh 3x)$ $\Rightarrow$ $y' = (1/\cosh 3x)(\sinh 3x)(3) = 3 \tanh 3x$

44. $y = \dfrac{(x^2 + 1)^4}{(2x + 1)^3 (3x - 1)^5}$ $\Rightarrow$

$\ln y = \ln \dfrac{(x^2 + 1)^4}{(2x + 1)^3 (3x - 1)^5} = \ln(x^2 + 1)^4 - \ln[(2x + 1)^3 (3x - 1)^5] = 4 \ln(x^2 + 1) - [\ln(2x + 1)^3 + \ln(3x - 1)^5]$

$\qquad = 4 \ln(x^2 + 1) - 3 \ln(2x + 1) - 5 \ln(3x - 1)$ $\Rightarrow$

$\dfrac{y'}{y} = 4 \cdot \dfrac{1}{x^2 + 1} \cdot 2x - 3 \cdot \dfrac{1}{2x + 1} \cdot 2 - 5 \cdot \dfrac{1}{3x - 1} \cdot 3$ $\Rightarrow$ $y' = \dfrac{(x^2 + 1)^4}{(2x + 1)^3 (3x - 1)^5}\left(\dfrac{8x}{x^2 + 1} - \dfrac{6}{2x + 1} - \dfrac{15}{3x - 1}\right)$.

[The answer could be simplified to $y' = -\dfrac{(x^2 + 56x + 9)(x^2 + 1)^3}{(2x + 1)^4 (3x - 1)^6}$, but this is unnecessary.]

45. $y = \cosh^{-1}(\sinh x)$ $\Rightarrow$ $y' = (\cosh x) / \sqrt{\sinh^2 x - 1}$

46. $y = x \tanh^{-1} \sqrt{x}$ $\Rightarrow$ $y' = \tanh^{-1} \sqrt{x} + x \dfrac{1}{1 - (\sqrt{x})^2} \dfrac{1}{2\sqrt{x}} = \tanh^{-1} \sqrt{x} + \dfrac{\sqrt{x}}{2(1 - x)}$

47. $y = \cos\left(e^{\sqrt{\tan 3x}}\right)$ $\Rightarrow$

$y' = -\sin\left(e^{\sqrt{\tan 3x}}\right) \cdot \left(e^{\sqrt{\tan 3x}}\right)' = -\sin\left(e^{\sqrt{\tan 3x}}\right) e^{\sqrt{\tan 3x}} \cdot \frac{1}{2}(\tan 3x)^{-1/2} \cdot \sec^2(3x) \cdot 3$

$\qquad = \dfrac{-3 \sin\left(e^{\sqrt{\tan 3x}}\right) e^{\sqrt{\tan 3x}} \sec^2(3x)}{2\sqrt{\tan 3x}}$

48. $\dfrac{d}{dx}\left(\dfrac{1}{2}\tan^{-1}x+\dfrac{1}{4}\ln\dfrac{(x+1)^2}{x^2+1}\right)=\dfrac{d}{dx}\left(\dfrac{1}{2}\tan^{-1}x+\dfrac{1}{2}\ln|x+1|-\dfrac{1}{4}\ln(x^2+1)\right)$

$$=\dfrac{1}{2}\dfrac{1}{x^2+1}+\dfrac{1}{2}\dfrac{1}{x+1}-\dfrac{1}{4}\dfrac{2x}{x^2+1}=\dfrac{1}{2}\left(\dfrac{1}{x^2+1}-\dfrac{x}{x^2+1}+\dfrac{1}{x+1}\right)$$

$$=\dfrac{1}{2}\left(\dfrac{1-x}{x^2+1}+\dfrac{1}{x+1}\right)=\dfrac{1}{2}\left(\dfrac{1-x^2}{(x^2+1)(1+x)}+\dfrac{x^2+1}{(x^2+1)(1+x)}\right)$$

$$=\dfrac{1}{2}\dfrac{2}{(x^2+1)(1+x)}=\dfrac{1}{(1+x)(x^2+1)}$$

49. $f(x)=e^{g(x)}\ \Rightarrow\ f'(x)=e^{g(x)}g'(x)$

50. $f(x)=g(e^x)\ \Rightarrow\ f'(x)=g'(e^x)\,e^x$

51. $f(x)=\ln|g(x)|\ \Rightarrow\ f'(x)=\dfrac{1}{g(x)}g'(x)=\dfrac{g'(x)}{g(x)}$

52. $f(x)=g(\ln x)\ \Rightarrow\ f'(x)=g'(\ln x)\cdot\dfrac{1}{x}=\dfrac{g'(\ln x)}{x}$

53. $f(x)=2^x\ \Rightarrow\ f'(x)=2^x\ln 2\ \Rightarrow\ f''(x)=2^x(\ln 2)^2\ \Rightarrow\ \cdots\ \Rightarrow\ f^{(n)}(x)=2^x(\ln 2)^n$

54. $f(x)=\ln(2x)=\ln 2+\ln x\ \Rightarrow\ f'(x)=x^{-1},\ f''(x)=-x^{-2},\ f'''(x)=2x^{-3},\ f^{(4)}(x)=-2\cdot 3x^{-4},\ldots,$

$f^{(n)}(x)=(-1)^{n-1}(n-1)!\,x^{-n}$

55. We first show it is true for $n=1$: $f'(x)=e^x+xe^x=(x+1)e^x$. We now assume it is true for $n=k$:

$f^{(k)}(x)=(x+k)e^x$. With this assumption, we must show it is true for $n=k+1$:

$f^{(k+1)}(x)=\dfrac{d}{dx}\left[f^{(k)}(x)\right]=\dfrac{d}{dx}[(x+k)e^x]=e^x+(x+k)e^x=[x+(k+1)]\,e^x.$

Therefore, $f^{(n)}(x)=(x+n)e^x$ by mathematical induction.

56. Using implicit differentiation, $y=x+\arctan y\ \Rightarrow\ y'=1+\dfrac{1}{1+y^2}y'\ \Rightarrow\ y'\left(1-\dfrac{1}{1+y^2}\right)=1\ \Rightarrow$

$y'\left(\dfrac{y^2}{1+y^2}\right)=1\ \Rightarrow\ y'=\dfrac{1+y^2}{y^2}=\dfrac{1}{y^2}+1.$

57. $y=(2+x)e^{-x}\ \Rightarrow\ y'=(2+x)(-e^{-x})+e^{-x}\cdot 1=e^{-x}[-(2+x)+1]=e^{-x}(-x-1)$. At $(0,2)$, $y'=1(-1)=-1$,

so an equation of the tangent line is $y-2=-1(x-0)$, or $y=-x+2$.

58. $y=f(x)=x\ln x\ \Rightarrow\ f'(x)=\ln x+1$, so the slope of the tangent at (e,e) is $f'(e)=2$ and an equation is

$y-e=2(x-e)$ or $y=2x-e$.

59. $y=[\ln(x+4)]^2\ \Rightarrow\ y'=2[\ln(x+4)]^1\cdot\dfrac{1}{x+4}\cdot 1=2\dfrac{\ln(x+4)}{x+4}$ and $y'=0\ \Leftrightarrow\ \ln(x+4)=0\ \Leftrightarrow$

$x+4=e^0\ \Rightarrow\ x+4=1\ \Leftrightarrow\ x=-3$, so the tangent is horizontal at the point $(-3,0)$.

60. $f(x) = xe^{\sin x}$ $\Rightarrow$ $f'(x) = x[e^{\sin x}(\cos x)] + e^{\sin x}(1) = e^{\sin x}(x\cos x + 1)$. As a check on our work, we notice from the

graphs that $f'(x) > 0$ when f is increasing. Also, we see in the larger viewing rectangle a certain similarity in the graphs of f

and f': the sizes of the oscillations of f and f' are linked.

61. (a) The line $x - 4y = 1$ has slope $\frac{1}{4}$. A tangent to $y = e^x$ has slope $\frac{1}{4}$ when $y' = e^x = \frac{1}{4}$ $\Rightarrow$ $x = \ln\frac{1}{4} = -\ln 4$.

Since $y = e^x$, the y-coordinate is $\frac{1}{4}$ and the point of tangency is $\left(-\ln 4, \frac{1}{4}\right)$. Thus, an equation of the tangent line

is $y - \frac{1}{4} = \frac{1}{4}(x + \ln 4)$ or $y = \frac{1}{4}x + \frac{1}{4}(\ln 4 + 1)$.

(b) The slope of the tangent at the point (a, e^a) is $\dfrac{d}{dx}e^x\bigg|_{x=a} = e^a$. Thus, an equation of the tangent line is

$y - e^a = e^a(x - a)$. We substitute $x = 0$, $y = 0$ into this equation, since we want the line to pass through the origin:

$0 - e^a = e^a(0 - a)$ $\Leftrightarrow$ $-e^a = e^a(-a)$ $\Leftrightarrow$ $a = 1$. So an equation of the tangent line at the point $(a, e^a) = (1, e)$

is $y - e = e(x - 1)$ or $y = ex$.

62. (a) $\displaystyle\lim_{t\to\infty} C(t) = \lim_{t\to\infty}\left[K(e^{-at} - e^{-bt})\right] = K\lim_{t\to\infty}(e^{-at} - e^{-bt}) = K(0 - 0) = 0$ because $-at \to -\infty$ and $-bt \to -\infty$

as $t \to \infty$.

(b) $C(t) = K(e^{-at} - e^{-bt})$ $\Rightarrow$ $C'(t) = K(e^{-at}(-a) - e^{-bt}(-b)) = K(-ae^{-at} + be^{-bt})$

(c) $C'(t) = 0$ $\Leftrightarrow$ $be^{-bt} = ae^{-at}$ $\Leftrightarrow$ $\dfrac{b}{a} = e^{(-a+b)t}$ $\Leftrightarrow$ $\ln\dfrac{b}{a} = (b-a)t$ $\Leftrightarrow$ $t = \dfrac{\ln(b/a)}{b - a}$

63. $\displaystyle\lim_{x\to\infty} e^{-3x} = 0$ since $-3x \to -\infty$ as $x \to \infty$ and $\displaystyle\lim_{t\to-\infty} e^t = 0$.

64. $\displaystyle\lim_{x\to 10^-} \ln(100 - x^2) = -\infty$ since as $x \to 10^-$, $(100 - x^2) \to 0^+$.

65. Let $t = 2/(x - 3)$. As $x \to 3^-$, $t \to -\infty$. $\displaystyle\lim_{x\to 3^-} e^{2/(x-3)} = \lim_{t\to-\infty} e^t = 0$

66. If $y = x^3 - x = x(x^2 - 1)$, then as $x \to \infty$, $y \to \infty$. $\displaystyle\lim_{x\to\infty} \arctan(x^3 - x) = \lim_{y\to\infty} \arctan y = \frac{\pi}{2}$ by (7.6.8).

67. Let $t = \sinh x$. As $x \to 0^+$, $t \to 0^+$. $\displaystyle\lim_{x\to 0^+} \ln(\sinh x) = \lim_{t\to 0^+} \ln t = -\infty$

68. $-1 \le \sin x \le 1$ $\Rightarrow$ $-e^{-x} \le e^{-x}\sin x \le e^{-x}$. Now $\displaystyle\lim_{x\to\infty}(\pm e^{-x}) = 0$, so by the Squeeze Theorem,

$\displaystyle\lim_{x\to\infty} e^{-x}\sin x = 0$.

69. $\lim\limits_{x\to\infty} \dfrac{(1+2^x)/2^x}{(1-2^x)/2^x} = \lim\limits_{x\to\infty} \dfrac{1/2^x+1}{1/2^x-1} = \dfrac{0+1}{0-1} = -1$

70. Let $t = x/4$, so $x = 4t$. As $x\to\infty$, $t\to\infty$. $\lim\limits_{x\to\infty}\left(1+\dfrac{4}{x}\right)^x = \lim\limits_{t\to\infty}\left(1+\dfrac{1}{t}\right)^{4t} = \left[\lim\limits_{t\to\infty}\left(1+\dfrac{1}{t}\right)^t\right]^4 = e^4$

71. This limit has the form $\frac{0}{0}$. $\lim\limits_{x\to0} \dfrac{\tan\pi x}{\ln(1+x)} \overset{\mathrm{H}}{=} \lim\limits_{x\to0} \dfrac{\pi\sec^2\pi x}{1/(1+x)} = \dfrac{\pi\cdot1^2}{1/1} = \pi$

72. This limit has the form $\frac{0}{0}$. $\lim\limits_{x\to0} \dfrac{1-\cos x}{x^2+x} \overset{\mathrm{H}}{=} \lim\limits_{x\to0} \dfrac{\sin x}{2x+1} = \dfrac{0}{1} = 0$

73. This limit has the form $\frac{0}{0}$. $\lim\limits_{x\to0} \dfrac{e^{4x}-1-4x}{x^2} \overset{\mathrm{H}}{=} \lim\limits_{x\to0} \dfrac{4e^{4x}-4}{2x} \overset{\mathrm{H}}{=} \lim\limits_{x\to0} \dfrac{16e^{4x}}{2} = \lim\limits_{x\to0} 8e^{4x} = 8\cdot1 = 8$

74. This limit has the form $\frac{\infty}{\infty}$. $\lim\limits_{x\to\infty} \dfrac{e^{4x}-1-4x}{x^2} \overset{\mathrm{H}}{=} \lim\limits_{x\to\infty} \dfrac{4e^{4x}-4}{2x} \overset{\mathrm{H}}{=} \lim\limits_{x\to\infty} \dfrac{16e^{4x}}{2} = \lim\limits_{x\to\infty} 8e^{4x} = \infty$

75. This limit has the form $\infty\cdot0$. $\lim\limits_{x\to\infty} x^3e^{-x} = \lim\limits_{x\to\infty} \dfrac{x^3}{e^x} \overset{\mathrm{H}}{=} \lim\limits_{x\to\infty} \dfrac{3x^2}{e^x} \overset{\mathrm{H}}{=} \lim\limits_{x\to\infty} \dfrac{6x}{e^x} \overset{\mathrm{H}}{=} \lim\limits_{x\to\infty} \dfrac{6}{e^x} = 0$

76. This limit has the form $0\cdot(-\infty)$. $\lim\limits_{x\to0^+} x^2\ln x = \lim\limits_{x\to0^+} \dfrac{\ln x}{1/x^2} \overset{\mathrm{H}}{=} \lim\limits_{x\to0^+} \dfrac{1/x}{-2/x^3} = \lim\limits_{x\to0^+} \left(-\tfrac{1}{2}x^2\right) = 0$

77. This limit has the form $\infty-\infty$.

$$\lim\limits_{x\to1^+}\left(\dfrac{x}{x-1}-\dfrac{1}{\ln x}\right) = \lim\limits_{x\to1^+}\left(\dfrac{x\ln x-x+1}{(x-1)\ln x}\right) \overset{\mathrm{H}}{=} \lim\limits_{x\to1^+} \dfrac{x\cdot(1/x)+\ln x-1}{(x-1)\cdot(1/x)+\ln x} = \lim\limits_{x\to1^+} \dfrac{\ln x}{1-1/x+\ln x}$$

$$\overset{\mathrm{H}}{=} \lim\limits_{x\to1^+} \dfrac{1/x}{1/x^2+1/x} = \dfrac{1}{1+1} = \dfrac{1}{2}$$

78. $y = (\tan x)^{\cos x} \;\Rightarrow\; \ln y = \cos x\,\ln\tan x$, so

$$\lim\limits_{x\to(\pi/2)^-}\ln y = \lim\limits_{x\to(\pi/2)^-} \dfrac{\ln\tan x}{\sec x} \overset{\mathrm{H}}{=} \lim\limits_{x\to(\pi/2)^-} \dfrac{(1/\tan x)\sec^2 x}{\sec x\tan x} = \lim\limits_{x\to(\pi/2)^-} \dfrac{\sec x}{\tan^2 x} = \lim\limits_{x\to(\pi/2)^-} \dfrac{\cos x}{\sin^2 x} = \dfrac{0}{1^2} = 0,$$

so $\lim\limits_{x\to(\pi/2)^-} (\tan x)^{\cos x} = \lim\limits_{x\to(\pi/2)^-} e^{\ln y} = e^0 = 1$.

79. $y = f(x) = \tan^{-1}(1/x)$ **A.** $D = \{x \mid x\neq0\}$ **B.** No intercept   **C.** $f(-x) = -f(x)$, so the curve is symmetric

about the origin.   **D.** $\lim\limits_{x\to\pm\infty}\tan^{-1}(1/x) = \tan^{-1}0 = 0$, so $y = 0$ is a HA. $\lim\limits_{x\to0^+}\tan^{-1}(1/x) = \dfrac{\pi}{2}$ and

$\lim\limits_{x\to0^-}\tan^{-1}(1/x) = -\dfrac{\pi}{2}$ since $\dfrac{1}{x}\to\pm\infty$ as $x\to0^\pm$.

H.

E. $f'(x) = \dfrac{1}{1+(1/x)^2}\left(-1/x^2\right) = \dfrac{-1}{x^2+1} \;\Rightarrow\; f'(x) < 0$, so f is

decreasing on $(-\infty,0)$ and $(0,\infty)$.   **F.** No maximum nor minimum

G. $f''(x) = \dfrac{2x}{(x^2+1)^2} > 0 \;\Leftrightarrow\; x > 0$, so f is CU on $(0,\infty)$ and CD on $(-\infty,0)$.

80. $y = f(x) = \sin^{-1}(1/x)$ **A.** $D = \{x \mid -1 \le 1/x \le 1\} = (-\infty, -1] \cup [1, \infty)$. **B.** No intercept

C. $f(-x) = -f(x)$, symmetric about the origin **D.** $\lim\limits_{x \to \pm\infty} \sin^{-1}(1/x) = \sin^{-1}(0) = 0$, so $y = 0$ is a HA.

E. $f'(x) = \dfrac{1}{\sqrt{1 - (1/x)^2}}\left(-\dfrac{1}{x^2}\right) = \dfrac{-1}{\sqrt{x^4 - x^2}} < 0$, so f is decreasing on $(-\infty, -1)$ and $(1, \infty)$.

F. No local extreme value, but $f(1) = \frac{\pi}{2}$ is the absolute maximum value

and $f(-1) = -\frac{\pi}{2}$ is the absolute minimum value.

H.

G. $f''(x) = \dfrac{4x^3 - 2x}{2(x^4 - x^2)^{3/2}} = \dfrac{x(2x^2 - 1)}{(x^4 - x^2)^{3/2}} > 0$ for $x > 1$ and $f''(x) < 0$

for $x < -1$, so f is CU on $(1, \infty)$ and CD on $(-\infty, -1)$. No IP

81. $y = f(x) = x \ln x$ **A.** $D = (0, \infty)$ **B.** No y-intercept; x-intercept 1. **C.** No symmetry **D.** No asymptote

[Note that the graph approaches the point $(0, 0)$ as $x \to 0^+$.]

H.

E. $f'(x) = x(1/x) + (\ln x)(1) = 1 + \ln x$, so $f'(x) \to -\infty$ as $x \to 0^+$ and

$f'(x) \to \infty$ as $x \to \infty$. $f'(x) = 0 \iff \ln x = -1 \iff x = e^{-1} = 1/e$.

$f'(x) > 0$ for $x > 1/e$, so f is decreasing on $(0, 1/e)$ and increasing on

$(1/e, \infty)$. **F.** Local minimum: $f(1/e) = -1/e$. No local maximum.

G. $f''(x) = 1/x$, so $f''(x) > 0$ for $x > 0$. The graph is CU on $(0, \infty)$ and

there is no IP.

82. $y = f(x) = e^{2x - x^2}$ **A.** $D = \mathbb{R}$ **B.** y-intercept 1; no x-intercept **C.** No symmetry **D.** $\lim\limits_{x \to \pm\infty} e^{2x - x^2} = 0$, so $y = 0$

is a HA. **E.** $y = f(x) = e^{2x - x^2} \implies f'(x) = 2(1 - x)e^{2x - x^2} > 0 \iff x < 1$, so f is increasing on $(-\infty, 1)$ and

decreasing on $(1, \infty)$. **F.** $f(1) = e$ is a local and absolute maximum value.

G. $f''(x) = 2(2x^2 - 4x + 1)e^{2x - x^2} = 0 \iff x = 1 \pm \frac{\sqrt{2}}{2}$. **H.**

$f''(x) > 0 \iff x < 1 - \frac{\sqrt{2}}{2}$ or $x > 1 + \frac{\sqrt{2}}{2}$, so f is CU on $\left(-\infty, 1 - \frac{\sqrt{2}}{2}\right)$

and $\left(1 + \frac{\sqrt{2}}{2}, \infty\right)$, and CD on $\left(1 - \frac{\sqrt{2}}{2}, 1 + \frac{\sqrt{2}}{2}\right)$. IP at $\left(1 \pm \frac{\sqrt{2}}{2}, \sqrt{e}\right)$

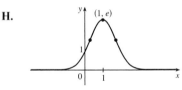

83. $y = f(x) = xe^{-2x}$ **A.** $D = \mathbb{R}$ **B.** y-intercept: $f(0) = 0$; x-intercept: $f(x) = 0 \iff x = 0$

C. No symmetry **D.** $\lim\limits_{x \to \infty} xe^{-2x} = \lim\limits_{x \to \infty} \dfrac{x}{e^{2x}} \overset{\mathrm{H}}{=} \lim\limits_{x \to \infty} \dfrac{1}{2e^{2x}} = 0$, so $y = 0$ is a HA.

E. $f'(x) = x(-2e^{-2x}) + e^{-2x}(1) = e^{-2x}(-2x + 1) > 0 \iff -2x + 1 > 0 \iff x < \frac{1}{2}$ and $f'(x) < 0 \iff x > \frac{1}{2}$,

so f is increasing on $\left(-\infty, \frac{1}{2}\right)$ and decreasing on $\left(\frac{1}{2}, \infty\right)$. **F.** Local maximum value $f\left(\frac{1}{2}\right) = \frac{1}{2}e^{-1} = 1/(2e)$;

no local minimum value

G. $f''(x) = e^{-2x}(-2) + (-2x + 1)(-2e^{-2x})$ **H.**

$= 2e^{-2x}[-1 - (-2x + 1)] = 4(x - 1)e^{-2x}$.

$f''(x) > 0 \iff x > 1$ and $f''(x) < 0 \iff x < 1$, so f is

CU on $(1, \infty)$ and CD on $(-\infty, 1)$. IP at $(1, f(1)) = (1, e^{-2})$

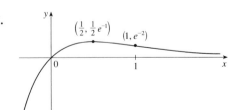

84. $y = f(x) = x + \ln(x^2 + 1)$ **A.** $D = \mathbb{R}$ **B.** y-intercept: $f(0) = 0 + \ln 1 = 0$; x-intercept: $f(x) = 0$ $\Leftrightarrow$

$\ln(x^2 + 1) = -x$ $\Leftrightarrow$ $x^2 + 1 = e^{-x}$ $\Rightarrow$ $x = 0$ since the graphs of $y = x^2 + 1$ and $y = e^{-x}$ intersect only at $x = 0$.

C. No symmetry   **D.** No asymptote   **E.** $f'(x) = 1 + \dfrac{2x}{x^2 + 1} = \dfrac{x^2 + 2x + 1}{x^2 + 1} = \dfrac{(x + 1)^2}{x^2 + 1}$. $f'(x) > 0$ if $x \neq -1$ and

f is increasing on $\mathbb{R}$.   **F.** No local extreme values

G. $f''(x) = \dfrac{(x^2 + 1)2 - 2x(2x)}{(x^2 + 1)^2} = \dfrac{2[(x^2 + 1) - 2x^2]}{(x^2 + 1)^2} = \dfrac{2(1 - x^2)}{(x^2 + 1)^2}$.   **H.**

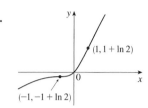

$f''(x) > 0$ $\Leftrightarrow$ $-1 < x < 1$ and $f''(x) < 0$ $\Leftrightarrow$ $x < -1$ or $x > 1$, so f is

CU on $(-1, 1)$ and f is CD on $(-\infty, -1)$ and $(1, \infty)$. IP at $(-1, -1 + \ln 2)$

and $(1, 1 + \ln 2)$

85.

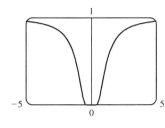

From the graph, we estimate the points of inflection to be about $(\pm 0.82, 0.22)$.

$f(x) = e^{-1/x^2}$ $\Rightarrow$ $f'(x) = 2x^{-3}e^{-1/x^2}$ $\Rightarrow$

$f''(x) = 2[x^{-3}(2x^{-3})e^{-1/x^2} + e^{-1/x^2}(-3x^{-4})] = 2x^{-6}e^{-1/x^2}(2 - 3x^2)$.

This is 0 when $2 - 3x^2 = 0$ $\Leftrightarrow$ $x = \pm\sqrt{\frac{2}{3}}$, so the inflection points

are $\left(\pm\sqrt{\frac{2}{3}}, e^{-3/2}\right)$.

86. We exclude the case $c = 0$, since in that case $f(x) = 0$ for all x. To find the maxima and minima, we differentiate:

$$f(x) = cxe^{-cx^2} \quad \Rightarrow \quad f'(x) = c\left[xe^{-cx^2}(-2cx) + e^{-cx^2}(1)\right] = ce^{-cx^2}(-2cx^2 + 1)$$

This is 0 where $-2cx^2 + 1 = 0$ $\Leftrightarrow$ $x = \pm 1/\sqrt{2c}$. So if $c > 0$, there are two maxima or minima, whose x-coordinates

approach 0 as c increases. The negative root gives a minimum and the positive root gives a maximum, by the First Derivative

Test. By substituting back into the equation, we see that $f\left(\pm 1/\sqrt{2c}\right) = c\left(\pm 1/\sqrt{2c}\right)e^{-c\left(\pm 1/\sqrt{2c}\right)^2} = \pm\sqrt{c/2e}$. So as c

increases, the extreme points become more pronounced. Note that if $c > 0$, then $\lim\limits_{x \to \pm\infty} f(x) = 0$. If $c < 0$, then there are no

extreme values, and $\lim\limits_{x \to \pm\infty} f(x) = \mp\infty$.

To find the points of inflection, we differentiate again: $f'(x) = ce^{-cx^2}(-2cx^2 + 1)$ $\Rightarrow$

$f''(x) = c\left[e^{-cx^2}(-4cx) + (-2cx^2 + 1)(-2cxe^{-cx^2})\right] = -2c^2xe^{-cx^2}(3 - 2cx^2)$. This is 0 at $x = 0$ and where

$3 - 2cx^2 = 0$ $\Leftrightarrow$ $x = \pm\sqrt{3/(2c)}$ $\Rightarrow$ IP at $\left(\pm\sqrt{3/(2c)}, \pm\sqrt{3c/2}\,e^{-3/2}\right)$. If $c > 0$ there are three inflection points,

and as c increases, the x-coordinates of the nonzero inflection points approach 0. If $c < 0$, there is only one inflection point,

the origin.

87. $s(t) = Ae^{-ct}\cos(\omega t + \delta)$ $\Rightarrow$

$v(t) = s'(t) = A\{e^{-ct}[-\omega\sin(\omega t + \delta)] + \cos(\omega t + \delta)(-ce^{-ct})\} = -Ae^{-ct}[\omega\sin(\omega t + \delta) + c\cos(\omega t + \delta)]$ $\Rightarrow$

$a(t) = v'(t) = -A\{e^{-ct}[\omega^2\cos(\omega t + \delta) - c\omega\sin(\omega t + \delta)] + [\omega\sin(\omega t + \delta) + c\cos(\omega t + \delta)](-ce^{-ct})\}$

$\quad = -Ae^{-ct}[\omega^2\cos(\omega t + \delta) - c\omega\sin(\omega t + \delta) - c\omega\sin(\omega t + \delta) - c^2\cos(\omega t + \delta)]$

$\quad = -Ae^{-ct}[(\omega^2 - c^2)\cos(\omega t + \delta) - 2c\omega\sin(\omega t + \delta)] = Ae^{-ct}[(c^2 - \omega^2)\cos(\omega t + \delta) + 2c\omega\sin(\omega t + \delta)]$

88. (a) Let $f(x) = \ln x + x - 3$. Then $f'(x) = 1/x + 1 >$ [for $x > 0$)] and $f(2) \approx -0.307$ and $f(e) \approx 0.718$.

f is differentiable on $(2, e)$, continuous on $[2, e]$ and $f(2) < 0$, $f(e) > 0$. Therefore, by the Intermediate Value Theorem

there exists a number c in $(2, e)$ such that $f(c) = 0$. Thus, there is one root. But $f'(x) > 0$ for $x \in (2, e)$, so f is

increasing on $(2, e)$, which means that there is exactly one root.

(b) We use Newton's Method with $f(x) = \ln x + x - 3$, $f'(x) = 1/x + 1$, and $x_1 = 2$.

$x_2 = x_1 - \dfrac{\ln x_1 + x_1 - 3}{1/x_1 + 1} = 2 - \dfrac{\ln 2 + 2 - 3}{1/2 + 1} \approx 2.20457$. Similarly, $x_3 \approx 2.20794$, $x_4 = 2.20794$. Thus, the root of

the equation, correct to four decimal places, is 2.2079.

89. (a) $y(t) = y(0)e^{kt} = 200e^{kt}$ $\Rightarrow$ $y(0.5) = 200e^{0.5k} = 360$ $\Rightarrow$ $e^{0.5k} = 1.8$ $\Rightarrow$ $0.5k = \ln 1.8$ $\Rightarrow$

$k = 2\ln 1.8 = \ln(1.8)^2 = \ln 3.24$ $\Rightarrow$ $y(t) = 200e^{(\ln 3.24)t} = 200(3.24)^t$

(b) $y(4) = 200(3.24)^4 \approx 22{,}040$ bacteria

(c) $y'(t) = 200(3.24)^t \cdot \ln 3.24$, so $y'(4) = 200(3.24)^4 \cdot \ln 3.24 \approx 25{,}910$ bacteria per hour

(d) $200(3.24)^t = 10{,}000$ $\Rightarrow$ $(3.24)^t = 50$ $\Rightarrow$ $t\ln 3.24 = \ln 50$ $\Rightarrow$ $t = \ln 50/\ln 3.24 \approx 3.33$ hours

90. (a) If $y(t)$ is the mass remaining after t years, then $y(t) = y(0)e^{kt} = 100e^{kt}$. $y(5.24) = 100e^{5.24k} = \frac{1}{2} \cdot 100$ $\Rightarrow$

$e^{5.24k} = \frac{1}{2}$ $\Rightarrow$ $5.24k = -\ln 2$ $\Rightarrow$ $k = -\frac{1}{5.24}\ln 2$ $\Rightarrow$ $y(t) = 100e^{-(\ln 2)t/5.24} = 100 \cdot 2^{-t/5.24}$. Thus,

$y(20) = 100 \cdot 2^{-20/5.24} \approx 7.1$ mg.

(b) $100 \cdot 2^{-t/5.24} = 1$ $\Rightarrow$ $2^{-t/5.24} = \dfrac{1}{100}$ $\Rightarrow$ $-\dfrac{t}{5.24}\ln 2 = \ln\dfrac{1}{100}$ $\Rightarrow$ $t = 5.24\,\dfrac{\ln 100}{\ln 2} \approx 34.8$ years

91. Let $P(t) = \dfrac{64}{1 + 31e^{-0.7944t}} = \dfrac{A}{1 + Be^{ct}} = A(1 + Be^{ct})^{-1}$, where $A = 64$, $B = 31$, and $c = -0.7944$.

$P'(t) = -A(1 + Be^{ct})^{-2}(Bce^{ct}) = -ABce^{ct}(1 + Be^{ct})^{-2}$

$P''(t) = -ABce^{ct}[-2(1 + Be^{ct})^{-3}(Bce^{ct})] + (1 + Be^{ct})^{-2}(-ABc^2e^{ct})$

$\quad = -ABc^2e^{ct}(1 + Be^{ct})^{-3}[-2Be^{ct} + (1 + Be^{ct})] = -\dfrac{ABc^2e^{ct}(1 - Be^{ct})}{(1 + Be^{ct})^3}$

The population is increasing most rapidly when its graph changes from CU to CD; that is, when $P''(t) = 0$ in this case.

$P''(t) = 0$ $\Rightarrow$ $Be^{ct} = 1$ $\Rightarrow$ $e^{ct} = \dfrac{1}{B}$ $\Rightarrow$ $ct = \ln\dfrac{1}{B}$ $\Rightarrow$ $t = \dfrac{\ln(1/B)}{c} = \dfrac{\ln(1/31)}{-0.7944} \approx 4.32$ days. Note that

$P\left(\dfrac{1}{c}\ln\dfrac{1}{B}\right) = \dfrac{A}{1 + Be^{c(1/c)\ln(1/B)}} = \dfrac{A}{1 + Be^{\ln(1/B)}} = \dfrac{A}{1 + B(1/B)} = \dfrac{A}{1+1} = \dfrac{A}{2}$, one-half the limit of P as $t \to \infty$.

92. Let $t = 4u$. Then $dt = 4\,du$ and

$$\int_0^4 \frac{1}{16 + t^2}\,dt = \int_0^1 \frac{1}{16 + 16u^2}\cdot 4\,du = \frac{1}{4}\int_0^1 \frac{du}{1 + u^2} = \frac{1}{4}\Big[\tan^{-1}u\Big]_0^1 = \tfrac{1}{4}(\tan^{-1}1 - \tan^{-1}0) = \tfrac{1}{4}\left(\tfrac{\pi}{4} - 0\right) = \tfrac{\pi}{16}.$$

93. Let $u = -2y^2$. Then $du = -4y\,dy$ and $\int_0^1 ye^{-2y^2}\,dy = \int_0^{-2} e^u\left(-\tfrac{1}{4}\,du\right) = -\tfrac{1}{4}\big[e^u\big]_0^{-2} = -\tfrac{1}{4}(e^{-2}-1) = \tfrac{1}{4}(1-e^{-2})$.

94. $\displaystyle\int_2^5 \frac{dr}{1+2r} = \tfrac{1}{2}\Big[\ln|1+2r|\Big]_2^5 = \tfrac{1}{2}(\ln 11 - \ln 5) = \tfrac{1}{2}\ln\tfrac{11}{5}$

95. Let $u = e^x$, so $du = e^x\,dx$. When $x = 0, u = 1$; when $x = 1, u = e$. Thus,

$$\int_0^1 \frac{e^x}{1 + e^{2x}}\,dx = \int_1^e \frac{1}{1+u^2}\,du = \big[\arctan u\big]_1^e = \arctan e - \arctan 1 = \arctan e - \tfrac{\pi}{4}.$$

96. Let $u = \sin x$. Then $du = \cos x\,dx$, so

$$\int_0^{\pi/2} \frac{\cos x}{1 + \sin^2 x}\,dx = \int_0^1 \frac{1}{1+u^2}\,du = \big[\tan^{-1}u\big]_0^1 = \tan^{-1}1 - \tan^{-1}0 = \tfrac{\pi}{4} - 0 = \tfrac{\pi}{4}.$$

97. Let $u = \sqrt{x}$. Then $du = \dfrac{dx}{2\sqrt{x}} \;\Rightarrow\; \displaystyle\int \frac{e^{\sqrt{x}}}{\sqrt{x}}\,dx = 2\int e^u\,du = 2e^u + C = 2e^{\sqrt{x}} + C.$

98. Let $u = \ln x$. Then $du = \dfrac{dx}{x} \;\Rightarrow\; \displaystyle\int \frac{\cos(\ln x)}{x}\,dx = \int \cos u\,du = \sin u + C = \sin(\ln x) + C.$

99. Let $u = x^2 + 2x$. Then $du = (2x+2)\,dx = 2(x+1)\,dx$ and

$$\int \frac{x+1}{x^2+2x}\,dx = \int \frac{\tfrac{1}{2}\,du}{u} = \tfrac{1}{2}\ln|u| + C = \tfrac{1}{2}\ln|x^2 + 2x| + C.$$

100. Let $u = 1 + \cot x$. Then $du = -\csc^2 x\,dx$, so $\displaystyle\int \frac{\csc^2 x}{1 + \cot x}\,dx = \int \frac{1}{u}(-du) = -\ln|u| + C = -\ln|1+\cot x| + C.$

101. Let $u = \ln(\cos x)$. Then $du = \dfrac{-\sin x}{\cos x}\,dx = -\tan x\,dx \;\Rightarrow\;$

$$\int \tan x\,\ln(\cos x)\,dx = -\int u\,du = -\tfrac{1}{2}u^2 + C = -\tfrac{1}{2}[\ln(\cos x)]^2 + C.$$

102. Let $u = x^2$. Then $du = 2x\,dx$, so $\displaystyle\int \frac{x}{\sqrt{1-x^4}}\,dx = \tfrac{1}{2}\int \frac{du}{\sqrt{1-u^2}} = \tfrac{1}{2}\sin^{-1}u + C = \tfrac{1}{2}\sin^{-1}(x^2) + C.$

103. Let $u = \tan\theta$. Then $du = \sec^2\theta\,d\theta$ and $\displaystyle\int 2^{\tan\theta}\sec^2\theta\,d\theta = \int 2^u\,du = \frac{2^u}{\ln 2} + C = \frac{2^{\tan\theta}}{\ln 2} + C.$

104. $\displaystyle\int \sinh au\,du = \frac{1}{a}\cosh au + C$

105. $\displaystyle\int \left(\frac{1-x}{x}\right)^2 dx = \int \left(\frac{1}{x} - 1\right)^2 dx = \int \left(\frac{1}{x^2} - \frac{2}{x} + 1\right) dx = -\frac{1}{x} - 2\ln|x| + x + C$

106. $1 + e^{2x} > e^{2x} \;\Rightarrow\; \sqrt{1+e^{2x}} > \sqrt{e^{2x}} = e^x \;\Rightarrow\; \int_0^1 \sqrt{1+e^{2x}}\,dx \geq \int_0^1 e^x\,dx = e^x\big]_0^1 = e - 1$

107. $\cos x \leq 1 \;\Rightarrow\; e^x\cos x \leq e^x \;\Rightarrow\; \int_0^1 e^x\cos x\,dx \leq \int_0^1 e^x\,dx = e^x\big]_0^1 = e - 1$

108. For $0 \leq x \leq 1, 0 \leq \sin^{-1} x \leq \frac{\pi}{2}$, so $\int_0^1 x \sin^{-1} x \, dx \leq \int_0^1 x \left(\frac{\pi}{2}\right) dx = \frac{\pi}{4} x^2 \Big]_0^1 = \frac{\pi}{4}$.

109. $f(x) = \int_1^{\sqrt{x}} \frac{e^s}{s} \, ds \quad \Rightarrow \quad f'(x) = \frac{d}{dx} \int_1^{\sqrt{x}} \frac{e^s}{s} \, ds = \frac{e^{\sqrt{x}}}{\sqrt{x}} \frac{d}{dx} \sqrt{x} = \frac{e^{\sqrt{x}}}{\sqrt{x}} \frac{1}{2\sqrt{x}} = \frac{e^{\sqrt{x}}}{2x}$

110. $f(x) = \int_{\ln x}^{2x} e^{-t^2} \, dt \quad \Rightarrow$

$f'(x) = \frac{d}{dx} \int_{\ln x}^{2x} e^{-t^2} \, dt = -\frac{d}{dx} \int_0^{\ln x} e^{-t^2} \, dt + \frac{d}{dx} \int_0^{2x} e^{-t^2} \, dt = -e^{-(\ln x)^2} \left(\frac{1}{x}\right) + e^{-(2x)^2} (2) = -\frac{e^{-(\ln x)^2}}{x} + 2e^{-4x^2}$

111. $f_{\text{ave}} = \frac{1}{4-1} \int_1^4 \frac{1}{x} \, dx = \frac{1}{3} \Big[\ln |x| \Big]_1^4 = \frac{1}{3} [\ln 4 - \ln 1] = \frac{1}{3} \ln 4$

112. $A = \int_{-2}^0 (e^{-x} - e^x) \, dx + \int_0^1 (e^x - e^{-x}) \, dx = \left[-e^{-x} - e^x \right]_{-2}^0 + \left[e^x + e^{-x} \right]_0^1$

$= [(-1 - 1) - (-e^2 - e^{-2})] + [(e + e^{-1}) - (1 + 1)] = e^2 + e + e^{-1} + e^{-2} - 4$

113. $V = \int_0^1 \frac{2\pi x}{1 + x^4} \, dx$ by cylindrical shells. Let $u = x^2 \quad \Rightarrow \quad du = 2x \, dx$. Then

$V = \int_0^1 \frac{\pi}{1 + u^2} \, du = \pi \left[\tan^{-1} u \right]_0^1 = \pi \left(\tan^{-1} 1 - \tan^{-1} 0 \right) = \pi \left(\frac{\pi}{4}\right) = \frac{\pi^2}{4}$.

114. $f(x) = x + x^2 + e^x \quad \Rightarrow \quad f'(x) = 1 + 2x + e^x$ and $f(0) = 1 \quad \Rightarrow \quad g(1) = 0$ [where $g = f^{-1}$],

so $g'(1) = \frac{1}{f'(g(1))} = \frac{1}{f'(0)} = \frac{1}{2}$.

115. $f(x) = \ln x + \tan^{-1} x \quad \Rightarrow \quad f(1) = \ln 1 + \tan^{-1} 1 = \frac{\pi}{4} \quad \Rightarrow \quad g\left(\frac{\pi}{4}\right) = 1$ [where $g = f^{-1}$].

$f'(x) = \frac{1}{x} + \frac{1}{1 + x^2}$, so $g'\left(\frac{\pi}{4}\right) = \frac{1}{f'(1)} = \frac{1}{3/2} = \frac{2}{3}$.

116.

The area of such a rectangle is just the product of its sides, that is, $A(x) = x \cdot e^{-x}$.

We want to find the maximum of this function, so we differentiate:

$A'(x) = x(-e^{-x}) + e^{-x}(1) = e^{-x}(1 - x)$. This is 0 only at $x = 1$, and changes from positive to negative there, so by the First Derivative Test this gives a local maximum. So the largest area is $A(1) = 1/e$.

117.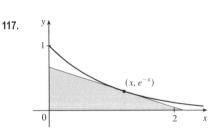

We find the equation of a tangent to the curve $y = e^{-x}$, so that we can find the x- and y-intercepts of this tangent, and then we can find the area of the triangle.

The slope of the tangent at the point (a, e^{-a}) is given by $\frac{d}{dx} e^{-x} \Big]_{x=a} = -e^{-a}$,

and so the equation of the tangent is $y - e^{-a} = -e^{-a}(x - a) \quad \Leftrightarrow$

$y = e^{-a}(a - x + 1)$.

The y-intercept of this line is $y = e^{-a}(a - 0 + 1) = e^{-a}(a + 1)$. To find the x-intercept we set $y = 0 \quad \Rightarrow$

$e^{-a}(a - x + 1) = 0 \quad \Rightarrow \quad x = a + 1$. So the area of the triangle is $A(a) = \frac{1}{2} \left[e^{-a}(a + 1) \right](a + 1) = \frac{1}{2} e^{-a}(a + 1)^2$. We differentiate this with respect to a: $A'(a) = \frac{1}{2}\left[e^{-a}(2)(a+1) + (a+1)^2 e^{-a}(-1)\right] = \frac{1}{2}e^{-a}\left(1 - a^2\right)$. This is 0

at $a = \pm 1$, and the root $a = 1$ gives a maximum, by the First Derivative Test. So the maximum area of the triangle is

$A(1) = \frac{1}{2}e^{-1}(1+1)^2 = 2e^{-1} = 2/e$.

118. Using Theorem 5.2.4 with $a = 0$ and $b = 1$, we have $\int_0^1 e^x\, dx = \lim\limits_{n \to \infty} \dfrac{1}{n} \sum\limits_{i=1}^{n} e^{i/n}$. This series is a geometric series

with $a = r = e^{1/n}$, so $\sum\limits_{i=1}^{n} e^{i/n} = e^{1/n} \dfrac{e^{n/n} - 1}{e^{1/n} - 1} = e^{1/n} \dfrac{e - 1}{e^{1/n} - 1} \quad\Rightarrow$

$\displaystyle\int_0^1 e^x\, dx = \lim\limits_{n \to \infty} \dfrac{1}{n} \sum\limits_{i=1}^{n} e^{i/n} = \lim\limits_{n \to \infty} (e-1)e^{1/n} \dfrac{1/n}{e^{1/n} - 1}$. As $n \to \infty$, $1/n \to 0^+$, so $e^{1/n} \to e^0 = 1$.

Let $t = 1/n$. Then $e^{1/n} - 1 = e^t - 1 \to 0^+$, so l'Hospital's Rule gives $\lim\limits_{t \to 0} \dfrac{t}{e^t - 1} = \lim\limits_{t \to 0} \dfrac{1}{e^t} = 1$ and we have

$\displaystyle\int_0^1 e^x\, dx = \left[\lim\limits_{t \to 0^+} (e-1)e^t\right]\left[\lim\limits_{t \to 0^+} \dfrac{t}{e^t - 1}\right] = e - 1$.

119. $\lim\limits_{x \to -1} F(x) = \lim\limits_{x \to -1} \dfrac{b^{x+1} - a^{x+1}}{x+1} \overset{\text{H}}{=} \lim\limits_{x \to -1} \dfrac{b^{x+1}\ln b - a^{x+1}\ln a}{1} = \ln b - \ln a = F(-1)$, so F is continuous at -1.

120. Let $\theta_1 = \operatorname{arccot} x$, so $\cot\theta_1 = x = x/1$.

So $\sin(\operatorname{arccot} x) = \sin\theta_1 = \dfrac{1}{\sqrt{x^2+1}}$.

Let $\theta_2 = \arctan\left[\dfrac{1}{\sqrt{x^2+1}}\right]$, so $\tan\theta_2 = \dfrac{1}{\sqrt{x^2+1}}$.

Hence, $\cos\{\arctan[\sin(\operatorname{arccot} x)]\} = \cos\theta_2 = \dfrac{\sqrt{x^2+1}}{\sqrt{x^2+2}} = \sqrt{\dfrac{x^2+1}{x^2+2}}$.

121. Using FTC1, we differentiate both sides of the given equation, $\int_0^x f(t)\, dt = xe^{2x} + \int_0^x e^{-t} f(t)\, dt$, and get

$$f(x) = e^{2x} + 2xe^{2x} + e^{-x} f(x) \quad\Rightarrow\quad f(x)\left(1 - e^{-x}\right) = e^{2x} + 2xe^{2x} \quad\Rightarrow\quad f(x) = \dfrac{e^{2x}(1 + 2x)}{1 - e^{-x}}.$$

122. (a) Let $f(x) = x - \ln x - 1$, so $f'(x) = 1 - \dfrac{1}{x} = \dfrac{x-1}{x}$. Since $x > 0$, $f'(x) < 0$ for $0 < x < 1$ and $f'(x) > 0$ for $x > 1$.

So there is an absolute minimum at $x = 1$ with $f(1) = 0$. So for $x > 0$, $x \neq 1$, $x - \ln x - 1 = f(x) > f(1) = 0$, and

hence $\ln x < x - 1$.

(b) Here let $f(x) = \ln x - \dfrac{x-1}{x} = \ln x - 1 + \dfrac{1}{x}$. So $f'(x) = \dfrac{1}{x} - \dfrac{1}{x^2} = \dfrac{x-1}{x^2}$. As in (a), we see that there is an absolute

minimum value at $x = 1$ and that $f(1) = 0$. So for $x > 0$, $x \neq 1$, $\ln x - \dfrac{x-1}{x} = f(x) > f(1) = 0$ and hence

$\dfrac{x-1}{x} < \ln x$.

(c) Let $b > a > 0$, so $b/a > 1$. Letting $x = b/a$ in the inequalities in (a) and (b) gives

$$\frac{b-a}{b} = \frac{b/a - 1}{b/a} < \ln \frac{b}{a} < \frac{b}{a} - 1 = \frac{b-a}{a}.$$ Noting that $\ln \frac{b}{a} = \ln b - \ln a$, the result follows after dividing through

by $b - a$.

(d) Let $f(x) = \ln x$. From the given diagram, we see that

(slope of tangent at $x = b$) < (slope of secant line) < (slope of tangent at $x = a$). Since $f'(x) = \frac{1}{x}$, we therefore have

$\frac{1}{b} < \frac{\ln b - \ln a}{b - a} < \frac{1}{a}$. To make this geometric argument more rigorous, we could use the Mean Value Theorem:

For any a and b with $0 < a < b$, there exists some $c \in (a, b)$ for which $f'(c) = \frac{1}{c} = \frac{\ln b - \ln a}{b - a}$. But $\frac{1}{x}$ is a decreasing

function on $(0, \infty)$, so $\frac{1}{b} < \frac{1}{c} = \frac{\ln b - \ln a}{b - a} < \ln \frac{1}{a}$.

(e) Since $\frac{1}{b} < \frac{1}{x} < \frac{1}{a}$ for $a < x < b$, Property 8 says that $\frac{1}{b}(b-a) < \int_a^b \frac{1}{x}\,dx < \frac{1}{a}(b-a) \quad \Rightarrow$

$\frac{1}{b}(b-a) < \ln b - \ln a < \frac{1}{a}(b-a) \quad \Rightarrow \quad \frac{1}{b} < \frac{\ln b - \ln a}{b - a} < \frac{1}{a}$. (Note from the proof of Property 8 that we are justified

in making all of the inequalities strict.)

☐ PROBLEMS PLUS

1. Let $y = f(x) = e^{-x^2}$. The area of the rectangle under the curve from $-x$ to x is $A(x) = 2xe^{-x^2}$ where $x \geq 0$. We maximize

$A(x)$: $A'(x) = 2e^{-x^2} - 4x^2 e^{-x^2} = 2e^{-x^2}\left(1 - 2x^2\right) = 0 \Rightarrow x = \frac{1}{\sqrt{2}}$. This gives a maximum since $A'(x) > 0$

for $0 \leq x < \frac{1}{\sqrt{2}}$ and $A'(x) < 0$ for $x > \frac{1}{\sqrt{2}}$. We next determine the points of inflection of $f(x)$. Notice that

$f'(x) = -2xe^{-x^2} = -A(x)$. So $f''(x) = -A'(x)$ and hence, $f''(x) < 0$ for $-\frac{1}{\sqrt{2}} < x < \frac{1}{\sqrt{2}}$ and $f''(x) > 0$ for $x < -\frac{1}{\sqrt{2}}$

and $x > \frac{1}{\sqrt{2}}$. So $f(x)$ changes concavity at $x = \pm\frac{1}{\sqrt{2}}$, and the two vertices of the rectangle of largest area are at the inflection

points.

2. We use proof by contradiction. Suppose that $\log_2 5$ is a rational number. Then $\log_2 5 = m/n$ where m and n are positive

integers $\Rightarrow 2^{m/n} = 5 \Rightarrow 2^m = 5^n$. But this is impossible since 2^m is even and 5^n is odd. So $\log_2 5$ is irrational.

3. Consider the statement that $\dfrac{d^n}{dx^n}(e^{ax}\sin bx) = r^n e^{ax}\sin(bx + n\theta)$. For $n = 1$,

$\dfrac{d}{dx}(e^{ax}\sin bx) = ae^{ax}\sin bx + be^{ax}\cos bx$, and

$re^{ax}\sin(bx + \theta) = re^{ax}[\sin bx \cos\theta + \cos bx \sin\theta] = re^{ax}\left(\dfrac{a}{r}\sin bx + \dfrac{b}{r}\cos bx\right) = ae^{ax}\sin bx + be^{ax}\cos bx$

since $\tan\theta = \dfrac{b}{a} \Rightarrow \sin\theta = \dfrac{b}{r}$ and $\cos\theta = \dfrac{a}{r}$. So the statement is true for $n = 1$.

Assume it is true for $n = k$. Then

$$\dfrac{d^{k+1}}{dx^{k+1}}(e^{ax}\sin bx) = \dfrac{d}{dx}\left[r^k e^{ax}\sin(bx + k\theta)\right] = r^k ae^{ax}\sin(bx + k\theta) + r^k e^{ax}b\cos(bx + k\theta)$$

$$= r^k e^{ax}[a\sin(bx + k\theta) + b\cos(bx + k\theta)]$$

But

$\sin[bx + (k+1)\theta] = \sin[(bx + k\theta) + \theta] = \sin(bx + k\theta)\cos\theta + \sin\theta\cos(bx + k\theta) = \frac{a}{r}\sin(bx + k\theta) + \frac{b}{r}\cos(bx + k\theta)$.

Hence, $a\sin(bx + k\theta) + b\cos(bx + k\theta) = r\sin[bx + (k+1)\theta]$. So

$\dfrac{d^{k+1}}{dx^{k+1}}(e^{ax}\sin bx) = r^k e^{ax}[a\sin(bx+k\theta)+b\cos(bx+k\theta)] = r^k e^{ax}[r\sin(bx+(k+1)\theta)] = r^{k+1}e^{ax}[\sin(bx+(k+1)\theta)]$.

Therefore, the statement is true for all n by mathematical induction.

4. Let $y = \tan^{-1}x$. Then $\tan y = x$, so from the triangle we see that

$\sin(\tan^{-1}x) = \sin y = \dfrac{x}{\sqrt{1 + x^2}}$. Using this fact we have that

$\sin(\tan^{-1}(\sinh x)) = \dfrac{\sinh x}{\sqrt{1 + \sinh^2 x}} = \dfrac{\sinh x}{\cosh x} = \tanh x$.

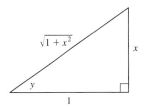

Hence, $\sin^{-1}(\tanh x) = \sin^{-1}(\sin(\tan^{-1}(\sinh x))) = \tan^{-1}(\sinh x)$.

5. We first show that $\dfrac{x}{1 + x^2} < \tan^{-1}x$ for $x > 0$. Let $f(x) = \tan^{-1}x - \dfrac{x}{1 + x^2}$. Then

$f'(x) = \dfrac{1}{1 + x^2} - \dfrac{1(1 + x^2) - x(2x)}{(1 + x^2)^2} = \dfrac{(1 + x^2) - (1 - x^2)}{(1 + x^2)^2} = \dfrac{2x^2}{(1 + x^2)^2} > 0$ for $x > 0$. So $f(x)$ is increasing

607

on $(0, \infty)$. Hence, $0 < x \ \Rightarrow \ 0 = f(0) < f(x) = \tan^{-1} x - \dfrac{x}{1 + x^2}$. So $\dfrac{x}{1 + x^2} < \tan^{-1} x$ for $0 < x$. We next show

that $\tan^{-1} x < x$ for $x > 0$. Let $h(x) = x - \tan^{-1} x$. Then $h'(x) = 1 - \dfrac{1}{1 + x^2} = \dfrac{x^2}{1 + x^2} > 0$. Hence, $h(x)$ is increasing

on $(0, \infty)$. So for $0 < x$, $0 = h(0) < h(x) = x - \tan^{-1} x$. Hence, $\tan^{-1} x < x$ for $x > 0$, and we conclude that

$\dfrac{x}{1 + x^2} < \tan^{-1} x < x$ for $x > 0$.

6. The shaded region has area $\int_0^1 f(x)\, dx = \frac{1}{3}$. The integral $\int_0^1 f^{-1}(y)\, dy$

gives the area of the unshaded region, which we know to be $1 - \frac{1}{3} = \frac{2}{3}$.

So $\int_0^1 f^{-1}(y)\, dy = \frac{2}{3}$.

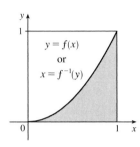

7. By the Fundamental Theorem of Calculus, $f(x) = \int_1^x \sqrt{1 + t^3}\, dt \ \Rightarrow \ f'(x) = \sqrt{1 + x^3} > 0$ for $x > -1$.

So f is increasing on $(-1, \infty)$ and hence is one-to-one. Note that $f(1) = 0$, so $f^{-1}(1) = 0 \ \Rightarrow$

$(f^{-1})'(0) = 1/f'(1) = \frac{1}{\sqrt{2}}$.

8. $y = \dfrac{x}{\sqrt{a^2 - 1}} - \dfrac{2}{\sqrt{a^2 - 1}} \arctan \dfrac{\sin x}{a + \sqrt{a^2 - 1} + \cos x}$. Let $k = a + \sqrt{a^2 - 1}$. Then

$$y' = \dfrac{1}{\sqrt{a^2 - 1}} - \dfrac{2}{\sqrt{a^2 - 1}} \cdot \dfrac{1}{1 + \sin^2 x / (k + \cos x)^2} \cdot \dfrac{\cos x(k + \cos x) + \sin^2 x}{(k + \cos x)^2}$$

$$= \dfrac{1}{\sqrt{a^2 - 1}} - \dfrac{2}{\sqrt{a^2 - 1}} \cdot \dfrac{k \cos x + \cos^2 x + \sin^2 x}{(k + \cos x)^2 + \sin^2 x} = \dfrac{1}{\sqrt{a^2 - 1}} - \dfrac{2}{\sqrt{a^2 - 1}} \cdot \dfrac{k \cos x + 1}{k^2 + 2k \cos x + 1}$$

$$= \dfrac{k^2 + 2k \cos x + 1 - 2k \cos x - 2}{\sqrt{a^2 - 1}\,(k^2 + 2k \cos x + 1)} = \dfrac{k^2 - 1}{\sqrt{a^2 - 1}\,(k^2 + 2k \cos x + 1)}$$

But $k^2 = 2a^2 + 2a\sqrt{a^2 - 1} - 1 = 2a\big(a + \sqrt{a^2 - 1}\big) - 1 = 2ak - 1$, so $k^2 + 1 = 2ak$, and $k^2 - 1 = 2(ak - 1)$.

So $y' = \dfrac{2(ak - 1)}{\sqrt{a^2 - 1}\,(2ak + 2k \cos x)} = \dfrac{ak - 1}{\sqrt{a^2 - 1}\,k\,(a + \cos x)}$. But $ak - 1 = a^2 + a\sqrt{a^2 - 1} - 1 = k\sqrt{a^2 - 1}$,

so $y' = 1/(a + \cos x)$.

9. If $L = \lim\limits_{x \to \infty} \left(\dfrac{x + a}{x - a} \right)^x$, then L has the indeterminate form 1^∞, so

$$\ln L = \lim_{x \to \infty} \ln\left(\dfrac{x + a}{x - a} \right)^x = \lim_{x \to \infty} x \ln\left(\dfrac{x + a}{x - a} \right) = \lim_{x \to \infty} \dfrac{\ln(x + a) - \ln(x - a)}{1/x} \overset{\text{H}}{=} \lim_{x \to \infty} \dfrac{\dfrac{1}{x + a} - \dfrac{1}{x - a}}{-1/x^2}$$

$$= \lim_{x \to \infty} \left[\dfrac{(x - a) - (x + a)}{(x + a)(x - a)} \cdot \dfrac{-x^2}{1} \right] = \lim_{x \to \infty} \dfrac{2ax^2}{x^2 - a^2} = \lim_{x \to \infty} \dfrac{2a}{1 - a^2/x^2} = 2a$$

Hence, $\ln L = 2a$, so $L = e^{2a}$. From the original equation, we want $L = e^1 \ \Rightarrow \ 2a = 1 \ \Rightarrow \ a = \frac{1}{2}$.

10. *Case (i) (first graph):* For $x + y \geq 0$, that is, $y \geq -x$, $|x + y| = x + y \leq e^x \Rightarrow y \leq e^x - x$.

Note that $y = e^x - x$ is always above the line $y = -x$ and that $y = -x$ is a slant asymptote.

Case (ii) (second graph): For $x + y < 0$, that is, $y < -x$, $|x + y| = -x - y \leq e^x \Rightarrow y \geq -x - e^x$.

Note that $-x - e^x$ is always below the line $y = -x$ and $y = -x$ is a slant asymptote.

Putting the two pieces together gives the third graph.

11. Both sides of the inequality are positive, so $\cosh(\sinh x) < \sinh(\cosh x)$

$\Leftrightarrow \quad \cosh^2(\sinh x) < \sinh^2(\cosh x) \quad \Leftrightarrow \quad \sinh^2(\sinh x) + 1 < \sinh^2(\cosh x)$

$\Leftrightarrow \quad 1 < [\sinh(\cosh x) - \sinh(\sinh x)][\sinh(\cosh x) + \sinh(\sinh x)]$

$\Leftrightarrow \quad 1 < \left[\sinh\left(\dfrac{e^x + e^{-x}}{2}\right) - \sinh\left(\dfrac{e^x - e^{-x}}{2}\right)\right]\left[\sinh\left(\dfrac{e^x + e^{-x}}{2}\right) + \sinh\left(\dfrac{e^x - e^{-x}}{2}\right)\right]$

$\Leftrightarrow \quad 1 < [2\cosh(e^x/2)\sinh(e^{-x}/2)][2\sinh(e^x/2)\cosh(e^{-x}/2)]$ [use the addition formulas and cancel]

$\Leftrightarrow \quad 1 < [2\sinh(e^x/2)\cosh(e^x/2)][2\sinh(e^x/2)\cosh(e^x/2)] \quad \Leftrightarrow \quad 1 < \sinh e^x \sinh e^{-x}$,

by the half-angle formula. Now both e^x and e^{-x} are positive, and $\sinh y > y$ for $y > 0$, since $\sinh 0 = 0$ and

$(\sinh y - y)' = \cosh y - 1 > 0$ for $x > 0$, so $1 = e^x e^{-x} < \sinh e^x \sinh e^{-x}$. So, following this chain of reasoning

backward, we arrive at the desired result.

12. First, we recognize some symmetry in the inequality: $\dfrac{e^{x+y}}{xy} \geq e^2 \quad \Leftrightarrow \quad \dfrac{e^x}{x} \cdot \dfrac{e^y}{y} \geq e \cdot e$. This suggests that we need to show

that $\dfrac{e^x}{x} \geq e$ for $x > 0$. If we can do this, then the inequality $\dfrac{e^y}{y} \geq e$ is true, and the given inequality follows. $f(x) = \dfrac{e^x}{x} \Rightarrow$

$f'(x) = \dfrac{xe^x - e^x}{x^2} = \dfrac{e^x(x - 1)}{x^2} = 0 \quad \Rightarrow \quad x = 1$. By the First Derivative Test, we have a minimum of $f(1) = e$, so

$e^x/x \geq e$ for all x.

13.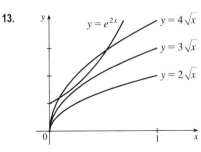

Let $f(x) = e^{2x}$ and $g(x) = k\sqrt{x}$ $[k > 0]$. From the graphs of f and g,

we see that f will intersect g exactly once when f and g share a tangent

line. Thus, we must have $f = g$ and $f' = g'$ at $x = a$.

$$f(a) = g(a) \quad \Rightarrow \quad e^{2a} = k\sqrt{a} \quad (\star)$$

and $\qquad f'(a) = g'(a) \quad \Rightarrow \quad 2e^{2a} = \dfrac{k}{2\sqrt{a}} \quad \Rightarrow \quad e^{2a} = \dfrac{k}{4\sqrt{a}}.$

So we must have $k\sqrt{a} = \dfrac{k}{4\sqrt{a}} \quad \Rightarrow \quad \left(\sqrt{a}\right)^2 = \dfrac{k}{4k} \quad \Rightarrow \quad a = \frac{1}{4}$. From $(\star)$, $e^{2(1/4)} = k\sqrt{1/4} \quad \Rightarrow$

$k = 2e^{1/2} = 2\sqrt{e} \approx 3.297$.

14. We see that at $x = 0$, $f(x) = a^x = 1 + x = 1$, so if $y = a^x$ is to lie above $y = 1 + x$,

the two curves must just touch at $(0, 1)$, that is, we must have $f'(0) = 1$. [To see this

analytically, note that $a^x \geq 1 + x \;\;\Rightarrow\;\; a^x - 1 \geq x \;\;\Rightarrow\;\; \dfrac{a^x - 1}{x} \geq 1$ for $x > 0$, so

$f'(0) = \lim\limits_{x \to 0^+} \dfrac{a^x - 1}{x} \geq 1$. Similarly, for $x < 0$, $a^x - 1 \geq x \;\;\Rightarrow\;\; \dfrac{a^x - 1}{x} \leq 1$, so

$f'(0) = \lim\limits_{x \to 0^-} \dfrac{a^x - 1}{x} \leq 1.$

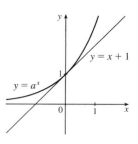

Since $1 \leq f'(0) \leq 1$, we must have $f'(0) = 1$.] But $f'(x) = a^x \ln a \;\;\Rightarrow\;\; f'(0) = \ln a$, so we have $\ln a = 1 \;\;\Leftrightarrow\;\; a = e$.

Another method: The inequality certainly holds for $x \leq -1$, so consider $x > -1$, $x \neq 0$. Then $a^x \geq 1 + x \;\;\Rightarrow\;\;$

$a \geq (1 + x)^{1/x}$ for $x > 0 \;\;\Rightarrow\;\; a \geq \lim\limits_{x \to 0^+} (1 + x)^{1/x} = e$, by Equation 7.4.8 (or Equation 7.4*.8). Also, $a^x \geq 1 + x \;\;\Rightarrow\;\;$

$a \leq (1 + x)^{1/x}$ for $x < 0 \;\;\Rightarrow\;\; a \leq \lim\limits_{x \to 0^-} (1 + x)^{1/x} = e$. So since $e \leq a \leq e$, we must have $a = e$.

15. Suppose that the curve $y = a^x$ intersects the line $y = x$. Then $a^{x_0} = x_0$ for some $x_0 > 0$, and hence $a = x_0^{1/x_0}$. We find the

maximum value of $g(x) = x^{1/x}$, $x > 0$, because if a is larger than the maximum value of this function, then the curve $y = a^x$

does not intersect the line $y = x$. $g'(x) = e^{(1/x)\ln x}\left(-\dfrac{1}{x^2}\ln x + \dfrac{1}{x}\cdot\dfrac{1}{x}\right) = x^{1/x}\left(\dfrac{1}{x^2}\right)(1 - \ln x)$. This is 0 only where

$x = e$, and for $0 < x < e$, $f'(x) > 0$, while for $x > e$, $f'(x) < 0$, so g has an absolute maximum of $g(e) = e^{1/e}$. So if

$y = a^x$ intersects $y = x$, we must have $0 < a \leq e^{1/e}$. Conversely, suppose that $0 < a \leq e^{1/e}$. Then $a^e \leq e$, so the graph of

$y = a^x$ lies below or touches the graph of $y = x$ at $x = e$. Also $a^0 = 1 > 0$, so the graph of $y = a^x$ lies above that of $y = x$

at $x = 0$. Therefore, by the Intermediate Value Theorem, the graphs of $y = a^x$ and $y = x$ must intersect somewhere between

$x = 0$ and $x = e$.

16.

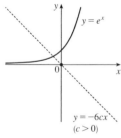

$y = cx^3 + e^x \;\;\Rightarrow\;\; y' = 3cx^2 + e^x \;\;\Rightarrow\;\; y'' = 6cx + e^x$. The curve will have

inflection points when y'' changes sign. $y'' = 0 \;\;\Rightarrow\;\; -6cx = e^x$, so y'' will change

sign when the line $y = -6cx$ intersects the curve $y = e^x$ (but is not tangent to it).

Note that if $c = 0$, the curve is just $y = e^x$, which has no inflection point.

The first figure shows that for $c > 0$, $y = -6cx$ will intersect $y = e^x$ once, so

$y = cx^3 + e^x$ will have one inflection point.

The second figure shows that for $c < 0$, the line $y = -6cx$ can intersect the curve

$y = e^x$ in two points (two inflection points), be tangent to it (no inflection point), or not

intersect it (no inflection point). The tangent line at (a, e^a) has slope e^a, but from the

diagram we see that the slope is $\dfrac{e^a}{a}$. So $\dfrac{e^a}{a} = e^a \;\;\Rightarrow\;\; a = 1$. Thus, the slope is e.

The line $y = -6cx$ must have slope greater than e, so $-6c > e \;\;\Rightarrow\;\; c < -e/6$.

Therefore, the curve $y = cx^3 + e^x$ will have one inflection point if $c > 0$ and two inflection points if $c < -e/6$.

8 ☐ TECHNIQUES OF INTEGRATION

8.1 Integration by Parts

1. Let $u = \ln x$, $dv = x^2\,dx$ $\Rightarrow$ $du = \frac{1}{x}\,dx$, $v = \frac{1}{3}x^3$. Then by Equation 2,

$$\int x^2 \ln x\,dx = (\ln x)\left(\tfrac{1}{3}x^3\right) - \int \left(\tfrac{1}{3}x^3\right)\left(\tfrac{1}{x}\right)\,dx = \tfrac{1}{3}x^3 \ln x - \tfrac{1}{3}\int x^2\,dx = \tfrac{1}{3}x^3 \ln x - \tfrac{1}{3}\left(\tfrac{1}{3}x^3\right) + C$$

$$= \tfrac{1}{3}x^3 \ln x - \tfrac{1}{9}x^3 + C \quad \left[\text{or } \tfrac{1}{3}x^3\left(\ln x - \tfrac{1}{3}\right) + C\right]$$

2. Let $u = \theta$, $dv = \cos\theta\,d\theta$ $\Rightarrow$ $du = d\theta$, $v = \sin\theta$. Then by Equation 2,

$$\int \theta \cos\theta\,d\theta = \theta \sin\theta - \int \sin\theta\,d\theta = \theta \sin\theta + \cos\theta + C.$$

Note: A mnemonic device which is helpful for selecting u when using integration by parts is the LIATE principle of precedence for u:

> $\underline{\text{L}}$ogarithmic
>
> $\underline{\text{I}}$nverse trigonometric
>
> $\underline{\text{A}}$lgebraic
>
> $\underline{\text{T}}$rigonometric
>
> $\underline{\text{E}}$xponential

If the integrand has several factors, then we try to choose among them a u which appears as high as possible on the list. For example, in $\int xe^{2x}\,dx$ the integrand is xe^{2x}, which is the product of an algebraic function (x) and an exponential function (e^{2x}). Since $\underline{\text{A}}$lgebraic appears before $\underline{\text{E}}$xponential, we choose $u = x$. Sometimes the integration turns out to be similar regardless of the selection of u and dv, but it is advisable to refer to LIATE when in doubt.

3. Let $u = x$, $dv = \cos 5x\,dx$ $\Rightarrow$ $du = dx$, $v = \frac{1}{5}\sin 5x$. Then by Equation 2,

$$\int x \cos 5x\,dx = \tfrac{1}{5}x \sin 5x - \int \tfrac{1}{5}\sin 5x\,dx = \tfrac{1}{5}x \sin 5x + \tfrac{1}{25}\cos 5x + C.$$

4. Let $u = x$, $dv = e^{-x}\,dx$ $\Rightarrow$ $du = dx$, $v = -e^{-x}$. Then $\int xe^{-x}\,dx = -xe^{-x} + \int e^{-x}\,dx = -xe^{-x} - e^{-x} + C.$

5. Let $u = r$, $dv = e^{r/2}\,dr$ $\Rightarrow$ $du = dr$, $v = 2e^{r/2}$. Then $\int re^{r/2}\,dr = 2re^{r/2} - \int 2e^{r/2}\,dr = 2re^{r/2} - 4e^{r/2} + C.$

6. Let $u = t$, $dv = \sin 2t\,dt$ $\Rightarrow$ $du = dt$, $v = -\frac{1}{2}\cos 2t$. Then

$$\int t \sin 2t\,dt = -\tfrac{1}{2}t \cos 2t + \tfrac{1}{2}\int \cos 2t\,dt = -\tfrac{1}{2}t \cos 2t + \tfrac{1}{4}\sin 2t + C.$$

7. Let $u = x^2$, $dv = \sin\pi x\,dx$ $\Rightarrow$ $du = 2x\,dx$ and $v = -\frac{1}{\pi}\cos\pi x$. Then

$$I = \int x^2 \sin\pi x\,dx = -\tfrac{1}{\pi}x^2 \cos\pi x + \tfrac{2}{\pi}\int x \cos\pi x\,dx \; (\star). \text{ Next let } U = x, dV = \cos\pi x\,dx \; \Rightarrow \; dU = dx,$$

$$V = \tfrac{1}{\pi}\sin\pi x, \text{ so } \int x \cos\pi x\,dx = \tfrac{1}{\pi}x \sin\pi x - \tfrac{1}{\pi}\int \sin\pi x\,dx = \tfrac{1}{\pi}x \sin\pi x + \tfrac{1}{\pi^2}\cos\pi x + C_1.$$

Substituting for $\int x \cos\pi x\,dx$ in $(\star)$, we get

$$I = -\tfrac{1}{\pi}x^2 \cos\pi x + \tfrac{2}{\pi}\left(\tfrac{1}{\pi}x \sin\pi x + \tfrac{1}{\pi^2}\cos\pi x + C_1\right) = -\tfrac{1}{\pi}x^2 \cos\pi x + \tfrac{2}{\pi^2}x \sin\pi x + \tfrac{2}{\pi^3}\cos\pi x + C, \text{ where } C = \tfrac{2}{\pi}C_1.$$

8. Let $u = x^2$, $dv = \cos mx\, dx$ $\Rightarrow$ $du = 2x\, dx$, $v = \dfrac{1}{m}\sin mx$. Then

$$I = \int x^2 \cos mx\, dx = \frac{1}{m}x^2 \sin mx - \frac{2}{m}\int x \sin mx\, dx \ (\star). \text{ Next let } U = x, dV = \sin mx\, dx \Rightarrow dU = dx,$$

$$V = -\frac{1}{m}\cos mx, \text{ so } \int x \sin mx\, dx = -\frac{1}{m}x\cos mx + \frac{1}{m}\int \cos mx\, dx = -\frac{1}{m}x\cos mx + \frac{1}{m^2}\sin mx + C_1.$$

Substituting for $\int x \sin mx\, dx$ in $(\star)$, we get

$$I = \frac{1}{m}x^2 \sin mx - \frac{2}{m}\left(-\frac{1}{m}x\cos mx + \frac{1}{m^2}\sin mx + C_1\right) = \frac{1}{m}x^2 \sin mx + \frac{2}{m^2}x\cos mx - \frac{2}{m^3}\sin mx + C,$$

where $C = -\dfrac{2}{m}C_1$.

9. Let $u = \ln(2x+1)$, $dv = dx$ $\Rightarrow$ $du = \dfrac{2}{2x+1}dx$, $v = x$. Then

$$\int \ln(2x+1)\, dx = x\ln(2x+1) - \int \frac{2x}{2x+1}\, dx = x\ln(2x+1) - \int \frac{(2x+1)-1}{2x+1}\, dx$$

$$= x\ln(2x+1) - \int\left(1 - \frac{1}{2x+1}\right)dx = x\ln(2x+1) - x + \tfrac{1}{2}\ln(2x+1) + C$$

$$= \tfrac{1}{2}(2x+1)\ln(2x+1) - x + C$$

10. Let $u = \sin^{-1}x$, $dv = dx$ $\Rightarrow$ $du = \dfrac{dx}{\sqrt{1-x^2}}$, $v = x$. Then $\int \sin^{-1}x\, dx = x\sin^{-1}x - \int \dfrac{x}{\sqrt{1-x^2}}\, dx$. Setting

$t = 1 - x^2$, we get $dt = -2x\, dx$, so $-\int \dfrac{x\, dx}{\sqrt{1-x^2}} = -\int t^{-1/2}\left(-\tfrac{1}{2}dt\right) = \tfrac{1}{2}(2t^{1/2}) + C = t^{1/2} + C = \sqrt{1-x^2} + C.$

Hence, $\int \sin^{-1}x\, dx = x\sin^{-1}x + \sqrt{1-x^2} + C.$

11. Let $u = \arctan 4t$, $dv = dt$ $\Rightarrow$ $du = \dfrac{4}{1+(4t)^2}dt = \dfrac{4}{1+16t^2}dt$, $v = t$. Then

$$\int \arctan 4t\, dt = t\arctan 4t - \int \frac{4t}{1+16t^2}\, dt = t\arctan 4t - \frac{1}{8}\int \frac{32t}{1+16t^2}\, dt = t\arctan 4t - \tfrac{1}{8}\ln(1+16t^2) + C.$$

12. Let $u = \ln p$, $dv = p^5 dp$ $\Rightarrow$ $du = \dfrac{1}{p}dp$, $v = \tfrac{1}{6}p^6$. Then $\int p^5 \ln p\, dp = \tfrac{1}{6}p^6 \ln p - \tfrac{1}{6}\int p^5 dp = \tfrac{1}{6}p^6 \ln p - \tfrac{1}{36}p^6 + C.$

13. Let $u = t$, $dv = \sec^2 2t\, dt$ $\Rightarrow$ $du = dt$, $v = \tfrac{1}{2}\tan 2t$. Then

$\int t\sec^2 2t\, dt = \tfrac{1}{2}t\tan 2t - \tfrac{1}{2}\int \tan 2t\, dt = \tfrac{1}{2}t\tan 2t - \tfrac{1}{4}\ln|\sec 2t| + C.$

14. Let $u = s$, $dv = 2^s ds$ $\Rightarrow$ $du = ds$, $v = \dfrac{1}{\ln 2}2^s$. Then

$$\int s\, 2^s ds = \frac{1}{\ln 2}s\, 2^s - \frac{1}{\ln 2}\int 2^s ds = \frac{1}{\ln 2}s\, 2^s - \frac{1}{(\ln 2)^2}2^s + C \left[\text{or } \frac{2^s}{(\ln 2)^2}(s\ln 2 - 1) + C\right].$$

15. First let $u = (\ln x)^2$, $dv = dx$ $\Rightarrow$ $du = 2\ln x \cdot \tfrac{1}{x}dx$, $v = x$. Then by Equation 2,

$I = \int (\ln x)^2 dx = x(\ln x)^2 - 2\int x\ln x \cdot \tfrac{1}{x}dx = x(\ln x)^2 - 2\int \ln x\, dx.$ Next let $U = \ln x$, $dV = dx$ $\Rightarrow$

$dU = 1/x\, dx$, $V = x$ to get $\int \ln x\, dx = x\ln x - \int x\cdot(1/x)\, dx = x\ln x - \int dx = x\ln x - x + C_1.$ Thus,

$I = x(\ln x)^2 - 2(x\ln x - x + C_1) = x(\ln x)^2 - 2x\ln x + 2x + C$, where $C = -2C_1.$

16. Let $u = t, dv = \sinh mt\, dt \Rightarrow du = dt, v = \frac{1}{m}\cosh mt$. Then

$$\int t \sinh mt\, dt = \frac{1}{m}t\cosh mt - \int \frac{1}{m}\cosh mt\, dt = \frac{1}{m}t\cosh mt - \frac{1}{m^2}\sinh mt + C \ [m \neq 0].$$

17. First let $u = \sin 3\theta, dv = e^{2\theta}\, d\theta \Rightarrow du = 3\cos 3\theta\, d\theta, v = \frac{1}{2}e^{2\theta}$. Then

$I = \int e^{2\theta}\sin 3\theta\, d\theta = \frac{1}{2}e^{2\theta}\sin 3\theta - \frac{3}{2}\int e^{2\theta}\cos 3\theta\, d\theta$. Next let $U = \cos 3\theta, dV = e^{2\theta}\, d\theta \Rightarrow dU = -3\sin 3\theta\, d\theta,$

$V = \frac{1}{2}e^{2\theta}$ to get $\int e^{2\theta}\cos 3\theta\, d\theta = \frac{1}{2}e^{2\theta}\cos 3\theta + \frac{3}{2}\int e^{2\theta}\sin 3\theta\, d\theta$. Substituting in the previous formula gives

$I = \frac{1}{2}e^{2\theta}\sin 3\theta - \frac{3}{4}e^{2\theta}\cos 3\theta - \frac{9}{4}\int e^{2\theta}\sin 3\theta\, d\theta = \frac{1}{2}e^{2\theta}\sin 3\theta - \frac{3}{4}e^{2\theta}\cos 3\theta - \frac{9}{4}I \Rightarrow$

$\frac{13}{4}I = \frac{1}{2}e^{2\theta}\sin 3\theta - \frac{3}{4}e^{2\theta}\cos 3\theta + C_1$. Hence, $I = \frac{1}{13}e^{2\theta}(2\sin 3\theta - 3\cos 3\theta) + C$, where $C = \frac{4}{13}C_1$.

18. First let $u = e^{-\theta}, dv = \cos 2\theta\, d\theta \Rightarrow du = -e^{-\theta}\, d\theta, v = \frac{1}{2}\sin 2\theta$. Then

$I = \int e^{-\theta}\cos 2\theta\, d\theta = \frac{1}{2}e^{-\theta}\sin 2\theta - \int \frac{1}{2}\sin 2\theta\left(-e^{-\theta}\, d\theta\right) = \frac{1}{2}e^{-\theta}\sin 2\theta + \frac{1}{2}\int e^{-\theta}\sin 2\theta\, d\theta.$

Next let $U = e^{-\theta}, dV = \sin 2\theta\, d\theta \Rightarrow dU = -e^{-\theta}\, d\theta, V = -\frac{1}{2}\cos 2\theta$, so

$\int e^{-\theta}\sin 2\theta\, d\theta = -\frac{1}{2}e^{-\theta}\cos 2\theta - \int\left(-\frac{1}{2}\right)\cos 2\theta\left(-e^{-\theta}\, d\theta\right) = -\frac{1}{2}e^{-\theta}\cos 2\theta - \frac{1}{2}\int e^{-\theta}\cos 2\theta\, d\theta.$

So $I = \frac{1}{2}e^{-\theta}\sin 2\theta + \frac{1}{2}\left[\left(-\frac{1}{2}e^{-\theta}\cos 2\theta\right) - \frac{1}{2}I\right] = \frac{1}{2}e^{-\theta}\sin 2\theta - \frac{1}{4}e^{-\theta}\cos 2\theta - \frac{1}{4}I \Rightarrow$

$\frac{5}{4}I = \frac{1}{2}e^{-\theta}\sin 2\theta - \frac{1}{4}e^{-\theta}\cos 2\theta + C_1 \Rightarrow I = \frac{4}{5}\left(\frac{1}{2}e^{-\theta}\sin 2\theta - \frac{1}{4}e^{-\theta}\cos 2\theta + C_1\right) = \frac{2}{5}e^{-\theta}\sin 2\theta - \frac{1}{5}e^{-\theta}\cos 2\theta + C.$

19. Let $u = t, dv = \sin 3t\, dt \Rightarrow du = dt, v = -\frac{1}{3}\cos 3t$. Then

$\int_0^\pi t\sin 3t\, dt = \left[-\frac{1}{3}t\cos 3t\right]_0^\pi + \frac{1}{3}\int_0^\pi \cos 3t\, dt = \left(\frac{1}{3}\pi - 0\right) + \frac{1}{9}\left[\sin 3t\right]_0^\pi = \frac{\pi}{3}.$

20. First let $u = x^2 + 1, dv = e^{-x}\, dx \Rightarrow du = 2x\, dx, v = -e^{-x}$. By (6),

$\int_0^1 (x^2+1)e^{-x}\, dx = \left[-(x^2+1)e^{-x}\right]_0^1 + \int_0^1 2xe^{-x}\, dx = -2e^{-1} + 1 + 2\int_0^1 xe^{-x}\, dx.$

Next let $U = x, dV = e^{-x}\, dx \Rightarrow dU = dx, V = -e^{-x}$. By (6) again,

$\int_0^1 xe^{-x}\, dx = \left[-xe^{-x}\right]_0^1 + \int_0^1 e^{-x}\, dx = -e^{-1} + \left[-e^{-x}\right]_0^1 = -e^{-1} - e^{-1} + 1 = -2e^{-1} + 1.$ So

$\int_0^1 (x^2+1)e^{-x}\, dx = -2e^{-1} + 1 + 2(-2e^{-1}+1) = -2e^{-1} + 1 - 4e^{-1} + 2 = -6e^{-1} + 3.$

21. Let $u = t, dv = \cosh t\, dt \Rightarrow du = dt, v = \sinh t$. Then

$\int_0^1 t\cosh t\, dt = \left[t\sinh t\right]_0^1 - \int_0^1 \sinh t\, dt = (\sinh 1 - \sinh 0) - \left[\cosh t\right]_0^1 = \sinh 1 - (\cosh 1 - \cosh 0)$

$= \sinh 1 - \cosh 1 + 1.$

We can use the definitions of sinh and cosh to write the answer in terms of e:

$\sinh 1 - \cosh 1 + 1 = \frac{1}{2}(e^1 - e^{-1}) - \frac{1}{2}(e^1 + e^{-1}) + 1 = -e^{-1} + 1 = 1 - 1/e.$

22. Let $u = \ln y, dv = \frac{1}{\sqrt{y}}\, dy = y^{-1/2}\, dy \Rightarrow du = \frac{1}{y}\, dy, v = 2y^{1/2}$. Then

$\int_4^9 \frac{\ln y}{\sqrt{y}}\, dy = \left[2\sqrt{y}\ln y\right]_4^9 - \int_4^9 2y^{-1/2}\, dy = (6\ln 9 - 4\ln 4) - \left[4\sqrt{y}\right]_4^9 = 6\ln 9 - 4\ln 4 - (12 - 8)$

$= 6\ln 9 - 4\ln 4 - 4$

23. Let $u = \ln x$, $dv = x^{-2}\,dx$ $\Rightarrow$ $du = \dfrac{1}{x}\,dx$, $v = -x^{-1}$. By (6),

$$\int_1^2 \frac{\ln x}{x^2}\,dx = \left[-\frac{\ln x}{x}\right]_1^2 + \int_1^2 x^{-2}\,dx = -\tfrac{1}{2}\ln 2 + \ln 1 + \left[-\frac{1}{x}\right]_1^2 = -\tfrac{1}{2}\ln 2 + 0 - \tfrac{1}{2} + 1 = \tfrac{1}{2} - \tfrac{1}{2}\ln 2.$$

24. First let $u = x^3$, $dv = \cos x\,dx$ $\Rightarrow$ $du = 3x^2\,dx$, $v = \sin x$. Then $I_1 = \int x^3 \cos x\,dx = x^3 \sin x - 3\int x^2 \sin x\,dx$. Next

let $u_1 = x^2$, $dv_1 = \sin x\,dx$ $\Rightarrow$ $du_1 = 2x\,dx$, $v_1 = -\cos x$. Then $I_2 = \int x^2 \sin x\,dx = -x^2 \cos x + 2\int x \cos x\,dx$.

Finally, let $u_2 = x$, $dv_2 = \cos x\,dx$ $\Rightarrow$ $du_2 = dx$, $v_2 = \sin x$. Then

$\int x \cos x\,dx = x \sin x - \int \sin x\,dx = x \sin x + \cos x + C$. Substituting in the expression for I_2, we get

$I_2 = -x^2 \cos x + 2(x \sin x + \cos x + C) = -x^2 \cos x + 2x \sin x + 2\cos x + 2C$. Substituting the last expression for I_2 into

I_1 gives $I_1 = x^3 \sin x - 3(-x^2 \cos x + 2x \sin x + 2\cos x + 2C) = x^3 \sin x + 3x^2 \cos x - 6x \sin x - 6\cos x - 6C$. Thus,

$$\int_0^\pi x^3 \cos x\,dx = \left[x^3 \sin x + 3x^2 \cos x - 6x \sin x - 6\cos x - 6C\right]_0^\pi$$

$$= (0 - 3\pi^2 - 0 + 6 - 6C) - (0 + 0 - 0 - 6 - 6C) = 12 - 3\pi^2$$

25. Let $u = y$, $dv = \dfrac{dy}{e^{2y}} = e^{-2y}\,dy$ $\Rightarrow$ $du = dy$, $v = -\tfrac{1}{2}e^{-2y}$. Then

$$\int_0^1 \frac{y}{e^{2y}}\,dy = \left[-\tfrac{1}{2}ye^{-2y}\right]_0^1 + \tfrac{1}{2}\int_0^1 e^{-2y}\,dy = \left(-\tfrac{1}{2}e^{-2} + 0\right) - \tfrac{1}{4}\left[e^{-2y}\right]_0^1 = -\tfrac{1}{2}e^{-2} - \tfrac{1}{4}e^{-2} + \tfrac{1}{4} = \tfrac{1}{4} - \tfrac{3}{4}e^{-2}.$$

26. Let $u = \arctan(1/x)$, $dv = dx$ $\Rightarrow$ $du = \dfrac{1}{1 + (1/x)^2} \cdot \dfrac{-1}{x^2}\,dx = \dfrac{-dx}{x^2 + 1}$, $v = x$. Then

$$\int_1^{\sqrt{3}} \arctan\left(\frac{1}{x}\right)dx = \left[x\arctan\left(\frac{1}{x}\right)\right]_1^{\sqrt{3}} + \int_1^{\sqrt{3}} \frac{x\,dx}{x^2 + 1} = \sqrt{3}\,\frac{\pi}{6} - 1 \cdot \frac{\pi}{4} + \frac{1}{2}\left[\ln(x^2 + 1)\right]_1^{\sqrt{3}}$$

$$= \frac{\pi\sqrt{3}}{6} - \frac{\pi}{4} + \frac{1}{2}(\ln 4 - \ln 2) = \frac{\pi\sqrt{3}}{6} - \frac{\pi}{4} + \frac{1}{2}\ln\frac{4}{2} = \frac{\pi\sqrt{3}}{6} - \frac{\pi}{4} + \frac{1}{2}\ln 2$$

27. Let $u = \cos^{-1} x$, $dv = dx$ $\Rightarrow$ $du = -\dfrac{dx}{\sqrt{1 - x^2}}$, $v = x$. Then

$$I = \int_0^{1/2} \cos^{-1} x\,dx = \left[x\cos^{-1} x\right]_0^{1/2} + \int_0^{1/2} \frac{x\,dx}{\sqrt{1 - x^2}} = \tfrac{1}{2} \cdot \frac{\pi}{3} + \int_1^{3/4} t^{-1/2}\left[-\tfrac{1}{2}dt\right], \text{ where } t = 1 - x^2 \ \Rightarrow$$

$dt = -2x\,dx$. Thus, $I = \frac{\pi}{6} + \frac{1}{2}\int_{3/4}^1 t^{-1/2}\,dt = \frac{\pi}{6} + \left[\sqrt{t}\right]_{3/4}^1 = \frac{\pi}{6} + 1 - \frac{\sqrt{3}}{2} = \frac{1}{6}\left(\pi + 6 - 3\sqrt{3}\right)$.

28. Let $u = (\ln x)^2$, $dv = x^{-3}\,dx$ $\Rightarrow$ $du = \dfrac{2\ln x}{x}\,dx$, $v = -\tfrac{1}{2}x^{-2}$. Then

$$I = \int_1^2 \frac{(\ln x)^2}{x^3}\,dx = \left[-\frac{(\ln x)^2}{2x^2}\right]_1^2 + \int_1^2 \frac{\ln x}{x^3}\,dx. \text{ Now let } U = \ln x, \ dV = x^{-3}\,dx \ \Rightarrow \ dU = \frac{1}{x}\,dx, \ V = -\tfrac{1}{2}x^{-2}.$$

Then

$$\int_1^2 \frac{\ln x}{x^3}\,dx = \left[-\frac{\ln x}{2x^2}\right]_1^2 + \frac{1}{2}\int_1^2 x^{-3}\,dx = -\frac{1}{8}\ln 2 + 0 + \frac{1}{2}\left[-\frac{1}{2x^2}\right]_1^2 = -\frac{1}{8}\ln 2 + \frac{1}{2}\left(-\frac{1}{8} + \frac{1}{2}\right) = \frac{3}{16} - \frac{1}{8}\ln 2.$$

Thus $I = \left(-\tfrac{1}{8}(\ln 2)^2 + 0\right) + \left(\tfrac{3}{16} - \tfrac{1}{8}\ln 2\right) = -\tfrac{1}{8}(\ln 2)^2 - \tfrac{1}{8}\ln 2 + \tfrac{3}{16}$.

29. Let $u = \ln(\sin x)$, $dv = \cos x \, dx$ $\Rightarrow$ $du = \dfrac{\cos x}{\sin x} \, dx$, $v = \sin x$. Then

$I = \int \cos x \ln(\sin x) \, dx = \sin x \ln(\sin x) - \int \cos x \, dx = \sin x \ln(\sin x) - \sin x + C$.

Another method: Substitute $t = \sin x$, so $dt = \cos x \, dx$. Then $I = \int \ln t \, dt = t \ln t - t + C$ (see Example 2) and so

$I = \sin x \, (\ln \sin x - 1) + C$.

30. Let $u = r^2$, $dv = \dfrac{r}{\sqrt{4+r^2}} \, dr$ $\Rightarrow$ $du = 2r \, dr$, $v = \sqrt{4+r^2}$. By (6),

$$\int_0^1 \frac{r^3}{\sqrt{4+r^2}} \, dr = \left[r^2 \sqrt{4+r^2} \right]_0^1 - 2 \int_0^1 r \sqrt{4+r^2} \, dr = \sqrt{5} - \tfrac{2}{3} \left[(4+r^2)^{3/2} \right]_0^1$$

$$= \sqrt{5} - \tfrac{2}{3} (5)^{3/2} + \tfrac{2}{3}(8) = \sqrt{5} \left(1 - \tfrac{10}{3} \right) + \tfrac{16}{3} = \tfrac{16}{3} - \tfrac{7}{3}\sqrt{5}$$

31. Let $u = (\ln x)^2$, $dv = x^4 \, dx$ $\Rightarrow$ $du = 2 \dfrac{\ln x}{x} \, dx$, $v = \dfrac{x^5}{5}$. By (6),

$$\int_1^2 x^4 (\ln x)^2 \, dx = \left[\frac{x^5}{5} (\ln x)^2 \right]_1^2 - 2 \int_1^2 \frac{x^4}{5} \ln x \, dx = \tfrac{32}{5} (\ln 2)^2 - 0 - 2 \int_1^2 \frac{x^4}{5} \ln x \, dx.$$

Let $U = \ln x$, $dV = \dfrac{x^4}{5} \, dx$ $\Rightarrow$ $dU = \dfrac{1}{x} \, dx$, $V = \dfrac{x^5}{25}$.

Then $\displaystyle \int_1^2 \frac{x^4}{5} \ln x \, dx = \left[\frac{x^5}{25} \ln x \right]_1^2 - \int_1^2 \frac{x^4}{25} \, dx = \tfrac{32}{25} \ln 2 - 0 - \left[\frac{x^5}{125} \right]_1^2 = \tfrac{32}{25} \ln 2 - \left(\tfrac{32}{125} - \tfrac{1}{125} \right)$.

So $\int_1^2 x^4 (\ln x)^2 \, dx = \tfrac{32}{5} (\ln 2)^2 - 2 \left(\tfrac{32}{25} \ln 2 - \tfrac{31}{125} \right) = \tfrac{32}{5} (\ln 2)^2 - \tfrac{64}{25} \ln 2 + \tfrac{62}{125}$.

32. Let $u = \sin(t-s)$, $dv = e^s \, ds$ $\Rightarrow$ $du = -\cos(t-s) \, ds$, $v = e^s$. Then

$I = \int_0^t e^s \sin(t-s) \, ds = \left[e^s \sin(t-s) \right]_0^t + \int_0^t e^s \cos(t-s) \, ds = e^t \sin 0 - e^0 \sin t + I_1$. For I_1, let $U = \cos(t-s)$,

$dV = e^s \, ds$ $\Rightarrow$ $dU = \sin(t-s) \, ds$, $V = e^s$. So $I_1 = \left[e^s \cos(t-s) \right]_0^t - \int_0^t e^s \sin(t-s) \, ds = e^t \cos 0 - e^0 \cos t - I$.

Thus, $I = -\sin t + e^t - \cos t - I$ $\Rightarrow$ $2I = e^t - \cos t - \sin t$ $\Rightarrow$ $I = \tfrac{1}{2}(e^t - \cos t - \sin t)$.

33. Let $y = \sqrt{x}$, so that $dy = \tfrac{1}{2} x^{-1/2} \, dx = \dfrac{1}{2\sqrt{x}} \, dx = \dfrac{1}{2y} \, dx$. Thus, $\int \cos \sqrt{x} \, dx = \int \cos y \, (2y \, dy) = 2 \int y \cos y \, dy$. Now

use parts with $u = y$, $dv = \cos y \, dy$, $du = dy$, $v = \sin y$ to get $\int y \cos y \, dy = y \sin y - \int \sin y \, dy = y \sin y + \cos y + C_1$,

so $\int \cos \sqrt{x} \, dx = 2y \sin y + 2 \cos y + C = 2\sqrt{x} \sin \sqrt{x} + 2 \cos \sqrt{x} + C$.

34. Let $x = -t^2$, so that $dx = -2t \, dt$. Thus, $\int t^3 e^{-t^2} \, dt = \int (-t^2) e^{-t^2} \left(\tfrac{1}{2} \right) (-2t \, dt) = \tfrac{1}{2} \int x e^x \, dx$. Now use parts with

$u = x$, $dv = e^x \, dx$, $du = dx$, $v = e^x$ to get

$\tfrac{1}{2} \int x e^x \, dx = \tfrac{1}{2} \left(x e^x - \int e^x \, dx \right) = \tfrac{1}{2} x e^x - \tfrac{1}{2} e^x + C = -\tfrac{1}{2}(1-x)e^x + C = -\tfrac{1}{2}(1+t^2)e^{-t^2} + C$.

35. Let $x = \theta^2$, so that $dx = 2\theta \, d\theta$. Thus, $\displaystyle \int_{\sqrt{\pi/2}}^{\sqrt{\pi}} \theta^3 \cos(\theta^2) \, d\theta = \int_{\sqrt{\pi/2}}^{\sqrt{\pi}} \theta^2 \cos(\theta^2) \cdot \tfrac{1}{2}(2\theta \, d\theta) = \tfrac{1}{2} \int_{\pi/2}^{\pi} x \cos x \, dx$. Now use

parts with $u = x$, $dv = \cos x \, dx$, $du = dx$, $v = \sin x$ to get

$$\tfrac{1}{2} \int_{\pi/2}^{\pi} x \cos x \, dx = \tfrac{1}{2} \left(\left[x \sin x \right]_{\pi/2}^{\pi} - \int_{\pi/2}^{\pi} \sin x \, dx \right) = \tfrac{1}{2} \left[x \sin x + \cos x \right]_{\pi/2}^{\pi}$$

$$= \tfrac{1}{2}(\pi \sin \pi + \cos \pi) - \tfrac{1}{2} \left(\tfrac{\pi}{2} \sin \tfrac{\pi}{2} + \cos \tfrac{\pi}{2} \right) = \tfrac{1}{2}(\pi \cdot 0 - 1) - \tfrac{1}{2} \left(\tfrac{\pi}{2} \cdot 1 + 0 \right) = -\tfrac{1}{2} - \tfrac{\pi}{4}$$

36. Let $x = \cos t$, so that $dx = -\sin t\, dt$. Thus,

$\int_0^\pi e^{\cos t} \sin 2t\, dt = \int_0^\pi e^{\cos t}(2\sin t \cos t)\, dt = \int_1^{-1} e^x \cdot 2x\,(-dx) = 2\int_{-1}^1 xe^x\, dx$. Now use parts with $u = x$,

$dv = e^x\, dx$, $du = dx$, $v = e^x$ to get

$2\int_{-1}^1 xe^x\, dx = 2\left([xe^x]_{-1}^1 - \int_{-1}^1 e^x\, dx\right) = 2\left(e^1 + e^{-1} - [e^x]_{-1}^1\right) = 2(e + e^{-1} - [e^1 - e^{-1}]) = 2(2e^{-1}) = 4/e.$

37. Let $y = 1 + x$, so that $dy = dx$. Thus, $\int x \ln(1+x)\, dx = \int(y-1)\ln y\, dy$. Now use parts with $u = \ln y$, $dv = (y-1)\, dy$,

$du = \frac{1}{y}\, dy$, $v = \frac{1}{2}y^2 - y$ to get

$$\int(y-1)\ln y\, dy = \left(\tfrac{1}{2}y^2 - y\right)\ln y - \int\left(\tfrac{1}{2}y - 1\right)dy = \tfrac{1}{2}y(y-2)\ln y - \tfrac{1}{4}y^2 + y + C$$
$$= \tfrac{1}{2}(1+x)(x-1)\ln(1+x) - \tfrac{1}{4}(1+x)^2 + 1 + x + C,$$

which can be written as $\frac{1}{2}(x^2 - 1)\ln(1+x) - \frac{1}{4}x^2 + \frac{1}{2}x + \frac{3}{4} + C$.

38. Let $y = \ln x$, so that $dy = \frac{1}{x}\, dx$ $\Rightarrow$ $dx = x\, dy = e^y\, dy$. Thus,

$\int \sin(\ln x)\, dx = \int \sin y\, e^y\, dy = \frac{1}{2}e^y(\sin y - \cos y) + C$ [by Example 4] $= \frac{1}{2}x[\sin(\ln x) - \cos(\ln x)] + C$.

In Exercises 39–42, let $f(x)$ denote the integrand and $F(x)$ its antiderivative (with $C = 0$).

39. Let $u = 2x + 3$, $dv = e^x\, dx$ $\Rightarrow$ $du = 2\, dx$, $v = e^x$. Then

$\int(2x+3)e^x\, dx = (2x+3)e^x - 2\int e^x\, dx = (2x+3)e^x - 2e^x + C$

$= (2x+1)e^x + C$

We see from the graph that this is reasonable, since F has a minimum where f changes from negative to positive.

40. Let $u = \ln x$, $dv = x^{3/2}\, dx$ $\Rightarrow$ $du = \frac{1}{x}\, dx$, $v = \frac{2}{5}x^{5/2}$. Then

$\int x^{3/2}\ln x\, dx = \frac{2}{5}x^{5/2}\ln x - \frac{2}{5}\int x^{3/2}\, dx = \frac{2}{5}x^{5/2}\ln x - \left(\frac{2}{5}\right)^2 x^{5/2} + C$

$= \frac{2}{5}x^{5/2}\ln x - \frac{4}{25}x^{5/2} + C$

We see from the graph that this is reasonable, since F has a minimum where f changes from negative to positive.

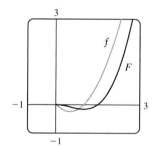

41. Let $u = \frac{1}{2}x^2$, $dv = 2x\sqrt{1+x^2}\, dx$ $\Rightarrow$ $du = x\, dx$, $v = \frac{2}{3}(1+x^2)^{3/2}$.

Then

$\int x^3 \sqrt{1+x^2}\, dx = \frac{1}{2}x^2\left[\frac{2}{3}(1+x^2)^{3/2}\right] - \frac{2}{3}\int x(1+x^2)^{3/2}dx$

$= \frac{1}{3}x^2(1+x^2)^{3/2} - \frac{2}{3}\cdot\frac{2}{5}\cdot\frac{1}{2}(1+x^2)^{5/2} + C$

$= \frac{1}{3}x^2(1+x^2)^{3/2} - \frac{2}{15}(1+x^2)^{5/2} + C$

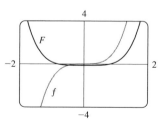

Another method: Use substitution with $u = 1 + x^2$ to get $\frac{1}{5}(1+x^2)^{5/2} - \frac{1}{3}(1+x^2)^{3/2} + C$.

42. First let $u = x^2$, $dv = \sin 2x\, dx$ $\Rightarrow$ $du = 2x\, dx$, $v = -\frac{1}{2}\cos 2x$.

Then $I = \int x^2 \sin 2x\, dx = -\frac{1}{2}x^2 \cos 2x + \int x \cos 2x\, dx$.

Next let $U = x$, $dV = \cos 2x\, dx$ $\Rightarrow$ $dU = dx$, $V = \frac{1}{2}\sin 2x$, so

$\int x \cos 2x\, dx = \frac{1}{2}x \sin 2x - \int \frac{1}{2}\sin 2x\, dx = \frac{1}{2}x \sin 2x + \frac{1}{4}\cos 2x + C$.

Thus, $I = -\frac{1}{2}x^2 \cos 2x + \frac{1}{2}x \sin 2x + \frac{1}{4}\cos 2x + C$.

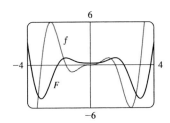

We see from the graph that this is reasonable, since F increases where f is positive and F decreases where f is negative.

Note also that f is an odd function and F is an even function.

43. (a) Take $n = 2$ in Example 6 to get $\displaystyle\int \sin^2 x\, dx = -\frac{1}{2}\cos x \sin x + \frac{1}{2}\int 1\, dx = \frac{x}{2} - \frac{\sin 2x}{4} + C.$

(b) $\int \sin^4 x\, dx = -\frac{1}{4}\cos x \sin^3 x + \frac{3}{4}\int \sin^2 x\, dx = -\frac{1}{4}\cos x \sin^3 x + \frac{3}{8}x - \frac{3}{16}\sin 2x + C$.

44. (a) Let $u = \cos^{n-1} x$, $dv = \cos x\, dx$ $\Rightarrow$ $du = -(n-1)\cos^{n-2} x \sin x\, dx$, $v = \sin x$ in (2):

$$\int \cos^n x\, dx = \cos^{n-1} x \sin x + (n-1)\int \cos^{n-2} x \sin^2 x\, dx$$
$$= \cos^{n-1} x \sin x + (n-1)\int \cos^{n-2} x (1 - \cos^2 x)\, dx$$
$$= \cos^{n-1} x \sin x + (n-1)\int \cos^{n-2} x\, dx - (n-1)\int \cos^n x\, dx$$

Rearranging terms gives $n\int \cos^n x\, dx = \cos^{n-1} x \sin x + (n-1)\int \cos^{n-2} x\, dx$ or

$$\int \cos^n x\, dx = \frac{1}{n}\cos^{n-1} x \sin x + \frac{n-1}{n}\int \cos^{n-2} x\, dx$$

(b) Take $n = 2$ in part (a) to get $\int \cos^2 x\, dx = \frac{1}{2}\cos x \sin x + \frac{1}{2}\int 1\, dx = \frac{x}{2} + \frac{\sin 2x}{4} + C$.

(c) $\int \cos^4 x\, dx = \frac{1}{4}\cos^3 x \sin x + \frac{3}{4}\int \cos^2 x\, dx = \frac{1}{4}\cos^3 x \sin x + \frac{3}{8}x + \frac{3}{16}\sin 2x + C$

45. (a) From Example 6, $\displaystyle\int \sin^n x\, dx = -\frac{1}{n}\cos x \sin^{n-1} x + \frac{n-1}{n}\int \sin^{n-2} x\, dx$. Using (6),

$$\int_0^{\pi/2} \sin^n x\, dx = \left[-\frac{\cos x \sin^{n-1} x}{n}\right]_0^{\pi/2} + \frac{n-1}{n}\int_0^{\pi/2} \sin^{n-2} x\, dx$$

$$= (0 - 0) + \frac{n-1}{n}\int_0^{\pi/2} \sin^{n-2} x\, dx = \frac{n-1}{n}\int_0^{\pi/2} \sin^{n-2} x\, dx$$

(b) Using $n = 3$ in part (a), we have $\int_0^{\pi/2} \sin^3 x\, dx = \frac{2}{3}\int_0^{\pi/2} \sin x\, dx = \left[-\frac{2}{3}\cos x\right]_0^{\pi/2} = \frac{2}{3}$.

Using $n = 5$ in part (a), we have $\int_0^{\pi/2} \sin^5 x\, dx = \frac{4}{5}\int_0^{\pi/2} \sin^3 x\, dx = \frac{4}{5}\cdot\frac{2}{3} = \frac{8}{15}$.

(c) The formula holds for $n = 1$ (that is, $2n + 1 = 3$) by (b). Assume it holds for some $k \geq 1$. Then

$$\int_0^{\pi/2} \sin^{2k+1} x\, dx = \frac{2 \cdot 4 \cdot 6 \cdots (2k)}{3 \cdot 5 \cdot 7 \cdots (2k+1)}.$$ By Example 6,

$$\int_0^{\pi/2} \sin^{2k+3} x\, dx = \frac{2k+2}{2k+3}\int_0^{\pi/2} \sin^{2k+1} x\, dx = \frac{2k+2}{2k+3}\cdot\frac{2 \cdot 4 \cdot 6 \cdots (2k)}{3 \cdot 5 \cdot 7 \cdots (2k+1)}$$

$$= \frac{2 \cdot 4 \cdot 6 \cdots (2k)[2(k+1)]}{3 \cdot 5 \cdot 7 \cdots (2k+1)[2(k+1)+1]},$$

so the formula holds for $n = k + 1$. By induction, the formula holds for all $n \geq 1$.

46. Using Exercise 45(a), we see that the formula holds for $n = 1$, because $\int_0^{\pi/2} \sin^2 x \, dx = \frac{1}{2} \int_0^{\pi/2} 1 \, dx = \frac{1}{2} \left[x \right]_0^{\pi/2} = \frac{1}{2} \cdot \frac{\pi}{2}$.

Now assume it holds for some $k \geq 1$. Then $\displaystyle\int_0^{\pi/2} \sin^{2k} x \, dx = \frac{1 \cdot 3 \cdot 5 \cdot \cdots \cdot (2k-1)}{2 \cdot 4 \cdot 6 \cdot \cdots \cdot (2k)} \frac{\pi}{2}$. By Exercise 45(a),

$$\int_0^{\pi/2} \sin^{2(k+1)} x \, dx = \frac{2k+1}{2k+2} \int_0^{\pi/2} \sin^{2k} x \, dx = \frac{2k+1}{2k+2} \cdot \frac{1 \cdot 3 \cdot 5 \cdot \cdots \cdot (2k-1)}{2 \cdot 4 \cdot 6 \cdot \cdots \cdot (2k)} \frac{\pi}{2}$$

$$= \frac{1 \cdot 3 \cdot 5 \cdot \cdots \cdot (2k-1)(2k+1)}{2 \cdot 4 \cdot 6 \cdot \cdots \cdot (2k)(2k+2)} \cdot \frac{\pi}{2},$$

so the formula holds for $n = k + 1$. By induction, the formula holds for all $n \geq 1$.

47. Let $u = (\ln x)^n$, $dv = dx$ $\Rightarrow$ $du = n(\ln x)^{n-1}(dx/x)$, $v = x$. By Equation 2,

$\int (\ln x)^n \, dx = x(\ln x)^n - \int nx(\ln x)^{n-1}(dx/x) = x(\ln x)^n - n \int (\ln x)^{n-1} \, dx$.

48. Let $u = x^n$, $dv = e^x \, dx$ $\Rightarrow$ $du = nx^{n-1} \, dx$, $v = e^x$. By Equation 2, $\int x^n e^x \, dx = x^n e^x - n \int x^{n-1} e^x \, dx$.

49. $\int \tan^n x \, dx = \int \tan^{n-2} x \tan^2 x \, dx = \int \tan^{n-2} x \, (\sec^2 x - 1) \, dx = \int \tan^{n-2} x \sec^2 x \, dx - \int \tan^{n-2} x \, dx$

$= I - \int \tan^{n-2} x \, dx$.

Let $u = \tan^{n-2} x$, $dv = \sec^2 x \, dx$ $\Rightarrow$ $du = (n-2) \tan^{n-3} x \sec^2 x \, dx$, $v = \tan x$. Then, by Equation 2,

$$I = \tan^{n-1} x - (n-2) \int \tan^{n-2} x \sec^2 x \, dx$$

$$1I = \tan^{n-1} x - (n-2)I$$

$$(n-1)I = \tan^{n-1} x$$

$$I = \frac{\tan^{n-1} x}{n-1}$$

Returning to the original integral, $\int \tan^n x \, dx = \dfrac{\tan^{n-1} x}{n-1} - \int \tan^{n-2} x \, dx$.

50. Let $u = \sec^{n-2} x$, $dv = \sec^2 x \, dx$ $\Rightarrow$ $du = (n-2) \sec^{n-3} x \sec x \tan x \, dx$, $v = \tan x$. Then, by Equation 2,

$$\int \sec^n x \, dx = \tan x \sec^{n-2} x - (n-2) \int \sec^{n-2} x \tan^2 x \, dx$$

$$= \tan x \sec^{n-2} x - (n-2) \int \sec^{n-2} x \, (\sec^2 x - 1) \, dx$$

$$= \tan x \sec^{n-2} x - (n-2) \int \sec^n x \, dx + (n-2) \int \sec^{n-2} x \, dx$$

so $(n-1) \int \sec^n x \, dx = \tan x \sec^{n-2} x + (n-2) \int \sec^{n-2} x \, dx$. If $n - 1 \neq 0$, then

$$\int \sec^n x \, dx = \frac{\tan x \sec^{n-2} x}{n-1} + \frac{n-2}{n-1} \int \sec^{n-2} x \, dx.$$

51. By repeated applications of the reduction formula in Exercise 47,

$$\int (\ln x)^3 \, dx = x (\ln x)^3 - 3 \int (\ln x)^2 \, dx = x(\ln x)^3 - 3 \left[x(\ln x)^2 - 2 \int (\ln x)^1 \, dx \right]$$

$$= x (\ln x)^3 - 3x(\ln x)^2 + 6 \left[x(\ln x)^1 - 1 \int (\ln x)^0 \, dx \right]$$

$$= x (\ln x)^3 - 3x(\ln x)^2 + 6x \ln x - 6 \int 1 \, dx = x (\ln x)^3 - 3x(\ln x)^2 + 6x \ln x - 6x + C$$

52. By repeated applications of the reduction formula in Exercise 48,

$$\int x^4 e^x \, dx = x^4 e^x - 4 \int x^3 e^x \, dx = x^4 e^x - 4\left(x^3 e^x - 3 \int x^2 e^x \, dx\right)$$

$$= x^4 e^x - 4x^3 e^x + 12\left(x^2 e^x - 2 \int x^1 e^x \, dx\right) = x^4 e^x - 4x^3 e^x + 12x^2 e^x - 24\left(x^1 e^x - \int x^0 e^x \, dx\right)$$

$$= x^4 e^x - 4x^3 e^x + 12x^2 e^x - 24x e^x + 24 e^x + C \quad \left[\text{or } e^x\left(x^4 - 4x^3 + 12x^2 - 24x + 24\right) + C\right]$$

53. Area $= \int_0^5 x e^{-0.4x} \, dx$. Let $u = x$, $dv = e^{-0.4x} \, dx \quad \Rightarrow$

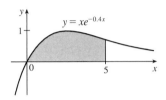

$du = dx$, $v = -2.5 e^{-0.4x}$. Then

$\text{area} = \left[-2.5 x e^{-0.4x}\right]_0^5 + 2.5 \int_0^5 e^{-0.4x} \, dx$

$\phantom{\text{area}} = -12.5 e^{-2} + 0 + 2.5\left[-2.5 e^{-0.4x}\right]_0^5$

$\phantom{\text{area}} = -12.5 e^{-2} - 6.25(e^{-2} - 1) = 6.25 - 18.75 e^{-2} \ \text{or} \ \frac{25}{4} - \frac{75}{4} e^{-2}$

54. The curves $y = x \ln x$ and $y = 5 \ln x$ intersect when $x \ln x = 5 \ln x \ \Leftrightarrow \ x \ln x - 5 \ln x = 0 \ \Leftrightarrow \ (x - 5) \ln x = 0$;

that is, when $x = 1$ or $x = 5$. For $1 < x < 5$, we have $5 \ln x > x \ln x$ since $\ln x > 0$. Thus,

$\text{area} = \int_1^5 (5 \ln x - x \ln x) \, dx = \int_1^5 \left[(5 - x) \ln x\right] dx$. Let $u = \ln x$, $dv = (5 - x) \, dx \quad \Rightarrow \quad du = dx/x$, $v = 5x - \frac{1}{2} x^2$.

Then

$$\text{area} = \left[(\ln x)\left(5x - \tfrac{1}{2} x^2\right)\right]_1^5 - \int_1^5 \left[\left(5x - \tfrac{1}{2} x^2\right)\tfrac{1}{x}\right] dx = (\ln 5)\left(\tfrac{25}{2}\right) - 0 - \int_1^5 \left(5 - \tfrac{1}{2} x\right) dx$$

$$= \tfrac{25}{2} \ln 5 - \left[5x - \tfrac{1}{4} x^2\right]_1^5 = \tfrac{25}{2} \ln 5 - \left[\left(25 - \tfrac{25}{4}\right) - \left(5 - \tfrac{1}{4}\right)\right] = \tfrac{25}{2} \ln 5 - 14$$

55. The curves $y = x \sin x$ and $y = (x - 2)^2$ intersect at $a \approx 1.04748$ and

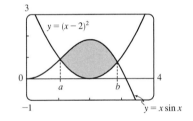

$b \approx 2.87307$, so

$\text{area} = \int_a^b \left[x \sin x - (x - 2)^2\right] dx$

$\phantom{\text{area}} = \left[-x \cos x + \sin x - \tfrac{1}{3}(x - 2)^3\right]_a^b \quad \text{[by Example 1]}$

$\phantom{\text{area}} \approx 2.81358 - 0.63075 = 2.18283$

56. The curves $y = \arctan 3x$ and $y = \frac{1}{2} x$ intersect at $x = \pm a \approx \pm 2.91379$,

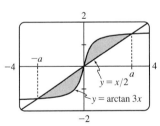

so

$\text{area} = \int_{-a}^a \left|\arctan 3x - \tfrac{1}{2} x\right| dx = 2 \int_0^a \left(\arctan 3x - \tfrac{1}{2} x\right) dx$

$\phantom{\text{area}} = 2\left[x \arctan 3x - \tfrac{1}{6} \ln(1 + 9x^2) - \tfrac{1}{4} x^2\right]_0^a \quad \text{[see Example 5]}$

$\phantom{\text{area}} \approx 2(1.39768) = 2.79536$

57. $V = \int_0^1 2\pi x \cos(\pi x/2) \, dx$. Let $u = x$, $dv = \cos(\pi x/2) \, dx \quad \Rightarrow \quad du = dx$, $v = \frac{2}{\pi} \sin(\pi x/2)$.

$$V = 2\pi \left[\frac{2}{\pi} x \sin\left(\frac{\pi x}{2}\right)\right]_0^1 - 2\pi \cdot \frac{2}{\pi} \int_0^1 \sin\left(\frac{\pi x}{2}\right) dx = 2\pi\left(\frac{2}{\pi} - 0\right) - 4\left[-\frac{2}{\pi} \cos\left(\frac{\pi x}{2}\right)\right]_0^1 = 4 + \frac{8}{\pi}(0 - 1) = 4 - \frac{8}{\pi}.$$

58. Volume $= \int_0^1 2\pi x(e^x - e^{-x}) \, dx = 2\pi \int_0^1 (x e^x - x e^{-x}) \, dx = 2\pi\left[\int_0^1 x e^x \, dx - \int_0^1 x e^{-x} \, dx\right]$ [both integrals by parts]

$$= 2\pi\left[(x e^x - e^x) - (-x e^{-x} - e^{-x})\right]_0^1 = 2\pi[2/e - 0] = 4\pi/e$$

59. Volume $= \int_{-1}^{0} 2\pi(1-x)e^{-x}\,dx$. Let $u = 1-x$, $dv = e^{-x}\,dx$ $\Rightarrow$ $du = -dx$, $v = -e^{-x}$.

$V = 2\pi\left[(1-x)(-e^{-x})\right]_{-1}^{0} - 2\pi\int_{-1}^{0} e^{-x}\,dx = 2\pi\left[(x-1)(e^{-x}) + e^{-x}\right]_{-1}^{0} = 2\pi\left[xe^{-x}\right]_{-1}^{0} = 2\pi(0+e) = 2\pi e$

60. Volume $= \int_{1}^{\pi} 2\pi y \cdot \ln y\,dy = 2\pi\left[\frac{1}{2}y^2\ln y - \frac{1}{4}y^2\right]_{1}^{\pi}$ [by parts with $u = \ln y$ and $dv = y\,dy$]

$= 2\pi\left[\frac{1}{4}y^2(2\ln y - 1)\right]_{1}^{\pi} = 2\pi\left[\frac{\pi^2(2\ln\pi - 1)}{4} - \frac{(0-1)}{4}\right] = \pi^3\ln\pi - \frac{\pi^3}{2} + \frac{\pi}{2}$

61. The average value of $f(x) = x^2\ln x$ on the interval $[1, 3]$ is $f_{\text{ave}} = \dfrac{1}{3-1}\displaystyle\int_{1}^{3} x^2\ln x\,dx = \frac{1}{2}I$.

Let $u = \ln x$, $dv = x^2\,dx$ $\Rightarrow$ $du = (1/x)\,dx$, $v = \frac{1}{3}x^3$.

So $I = \left[\frac{1}{3}x^3\ln x\right]_{1}^{3} - \int_{1}^{3}\frac{1}{3}x^2\,dx = (9\ln 3 - 0) - \left[\frac{1}{9}x^3\right]_{1}^{3} = 9\ln 3 - \left(3 - \frac{1}{9}\right) = 9\ln 3 - \frac{26}{9}$.

Thus, $f_{\text{ave}} = \frac{1}{2}I = \frac{1}{2}\left(9\ln 3 - \frac{26}{9}\right) = \frac{9}{2}\ln 3 - \frac{13}{9}$.

62. The rocket will have height $H = \int_{0}^{60} v(t)\,dt$ after 60 seconds.

$$H = \int_{0}^{60}\left[-gt - v_e\ln\left(\frac{m-rt}{m}\right)\right]dt = -g\left[\frac{1}{2}t^2\right]_{0}^{60} - v_e\left[\int_{0}^{60}\ln(m-rt)\,dt - \int_{0}^{60}\ln m\,dt\right]$$

$$= -g(1800) + v_e(\ln m)(60) - v_e\int_{0}^{60}\ln(m-rt)\,dt$$

Let $u = \ln(m-rt)$, $dv = dt$ $\Rightarrow$ $du = \dfrac{1}{m-rt}(-r)\,dt$, $v = t$. Then

$$\int_{0}^{60}\ln(m-rt)\,dt = \left[t\ln(m-rt)\right]_{0}^{60} + \int_{0}^{60}\frac{rt}{m-rt}\,dt = 60\ln(m-60r) + \int_{0}^{60}\left(-1 + \frac{m}{m-rt}\right)dt$$

$$= 60\ln(m-60r) + \left[-t - \frac{m}{r}\ln(m-rt)\right]_{0}^{60} = 60\ln(m-60r) - 60 - \frac{m}{r}\ln(m-60r) + \frac{m}{r}\ln m$$

So $H = -1800g + 60v_e\ln m - 60v_e\ln(m-60r) + 60v_e + \dfrac{m}{r}v_e\ln(m-60r) - \dfrac{m}{r}v_e\ln m$. Substituting $g = 9.8$,

$m = 30{,}000$, $r = 160$, and $v_e = 3000$ gives us $H \approx 14{,}844$ m.

63. Since $v(t) > 0$ for all t, the desired distance is $s(t) = \int_{0}^{t} v(w)\,dw = \int_{0}^{t} w^2 e^{-w}\,dw$.

First let $u = w^2$, $dv = e^{-w}\,dw$ $\Rightarrow$ $du = 2w\,dw$, $v = -e^{-w}$. Then $s(t) = \left[-w^2 e^{-w}\right]_{0}^{t} + 2\int_{0}^{t} we^{-w}\,dw$.

Next let $U = w$, $dV = e^{-w}\,dw$ $\Rightarrow$ $dU = dw$, $V = -e^{-w}$. Then

$$s(t) = -t^2 e^{-t} + 2\left(\left[-we^{-w}\right]_{0}^{t} + \int_{0}^{t} e^{-w}\,dw\right) = -t^2 e^{-t} + 2\left(-te^{-t} + 0 + \left[-e^{-w}\right]_{0}^{t}\right)$$

$$= -t^2 e^{-t} + 2(-te^{-t} - e^{-t} + 1) = -t^2 e^{-t} - 2te^{-t} - 2e^{-t} + 2 = 2 - e^{-t}(t^2 + 2t + 2) \text{ meters}$$

64. Suppose $f(0) = g(0) = 0$ and let $u = f(x)$, $dv = g''(x)\,dx$ $\Rightarrow$ $du = f'(x)\,dx$, $v = g'(x)$.

Then $\int_{0}^{a} f(x)\,g''(x)\,dx = \left[f(x)\,g'(x)\right]_{0}^{a} - \int_{0}^{a} f'(x)\,g'(x)\,dx = f(a)\,g'(a) - \int_{0}^{a} f'(x)\,g'(x)\,dx$.

Now let $U = f'(x)$, $dV = g'(x)\,dx$ $\Rightarrow$ $dU = f''(x)\,dx$ and $V = g(x)$, so

$\int_{0}^{a} f'(x)\,g'(x)\,dx = \left[f'(x)\,g(x)\right]_{0}^{a} - \int_{0}^{a} f''(x)\,g(x)\,dx = f'(a)\,g(a) - \int_{0}^{a} f''(x)\,g(x)\,dx$.

Combining the two results, we get $\int_{0}^{a} f(x)\,g''(x)\,dx = f(a)\,g'(a) - f'(a)\,g(a) + \int_{0}^{a} f''(x)\,g(x)\,dx$.

65. For $I = \int_1^4 x f''(x)\, dx$, let $u = x$, $dv = f''(x)\, dx$ $\Rightarrow$ $du = dx$, $v = f'(x)$. Then

$$I = \left[x f'(x) \right]_1^4 - \int_1^4 f'(x)\, dx = 4 f'(4) - 1 \cdot f'(1) - [f(4) - f(1)] = 4 \cdot 3 - 1 \cdot 5 - (7 - 2) = 12 - 5 - 5 = 2.$$

We used the fact that f'' is continuous to guarantee that I exists.

66. (a) Take $g(x) = x$ and $g'(x) = 1$ in Equation 1.

(b) By part (a), $\int_a^b f(x)\, dx = bf(b) - a\,f(a) - \int_a^b x\,f'(x)\, dx$. Now let $y = f(x)$, so that $x = g(y)$ and $dy = f'(x)\, dx$.

Then $\int_a^b x\,f'(x)\, dx = \int_{f(a)}^{f(b)} g(y)\, dy$. The result follows.

(c) Part (b) says that the area of region $ABFC$ is

$$= \qquad bf(b) \qquad - \qquad af(a) \qquad - \quad \int_{f(a)}^{f(b)} g(y)\, dy$$

$$= (\text{area of rectangle } OBFE) - (\text{area of rectangle } OACD) - (\text{area of region } DCFE)$$

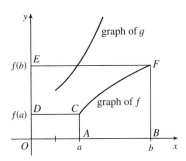

(d) We have $f(x) = \ln x$, so $f^{-1}(x) = e^x$, and since $g = f^{-1}$, we have $g(y) = e^y$. By part (b),

$$\int_1^e \ln x\, dx = e \ln e - 1 \ln 1 - \int_{\ln 1}^{\ln e} e^y\, dy = e - \int_0^1 e^y\, dy = e - \left[e^y \right]_0^1 = e - (e - 1) = 1.$$

67. Using the formula for volumes of rotation and the figure, we see that

Volume $= \int_0^d \pi b^2\, dy - \int_0^c \pi a^2\, dy - \int_c^d \pi [g(y)]^2\, dy = \pi b^2 d - \pi a^2 c - \int_c^d \pi [g(y)]^2\, dy$. Let $y = f(x)$,

which gives $dy = f'(x)\, dx$ and $g(y) = x$, so that $V = \pi b^2 d - \pi a^2 c - \pi \int_a^b x^2 f'(x)\, dx$.

Now integrate by parts with $u = x^2$, and $dv = f'(x)\, dx$ $\Rightarrow$ $du = 2x\, dx$, $v = f(x)$, and

$\int_a^b x^2 f'(x)\, dx = \left[x^2 f(x) \right]_a^b - \int_a^b 2x\,f(x)\, dx = b^2 f(b) - a^2 f(a) - \int_a^b 2x\,f(x)\, dx$, but $f(a) = c$ and $f(b) = d$ $\Rightarrow$

$V = \pi b^2 d - \pi a^2 c - \pi \left[b^2 d - a^2 c - \int_a^b 2x\,f(x)\, dx \right] = \int_a^b 2\pi x\,f(x)\, dx$.

68. (a) We note that for $0 \le x \le \frac{\pi}{2}$, $0 \le \sin x \le 1$, so $\sin^{2n+2} x \le \sin^{2n+1} x \le \sin^{2n} x$. So by the second Comparison Property

of the Integral, $I_{2n+2} \le I_{2n+1} \le I_{2n}$.

(b) Substituting directly into the result from Exercise 46, we get

$$\frac{I_{2n+2}}{I_{2n}} = \frac{\dfrac{1 \cdot 3 \cdot 5 \cdots\cdots [2(n+1)-1]}{2 \cdot 4 \cdot 6 \cdots\cdots [2(n+1)]}\, \dfrac{\pi}{2}}{\dfrac{1 \cdot 3 \cdot 5 \cdots\cdots (2n-1)}{2 \cdot 4 \cdot 6 \cdots\cdots (2n)}\, \dfrac{\pi}{2}} = \frac{2(n+1)-1}{2(n+1)} = \frac{2n+1}{2n+2}$$

(c) We divide the result from part (a) by I_{2n}. The inequalities are preserved since I_{2n} is positive: $\dfrac{I_{2n+2}}{I_{2n}} \le \dfrac{I_{2n+1}}{I_{2n}} \le \dfrac{I_{2n}}{I_{2n}}$.

Now from part (b), the left term is equal to $\dfrac{2n+1}{2n+2}$, so the expression becomes $\dfrac{2n+1}{2n+2} \le \dfrac{I_{2n+1}}{I_{2n}} \le 1$. Now

$\displaystyle\lim_{n\to\infty} \dfrac{2n+1}{2n+2} = \lim_{n\to\infty} 1 = 1$, so by the Squeeze Theorem, $\displaystyle\lim_{n\to\infty} \dfrac{I_{2n+1}}{I_{2n}} = 1$.

(d) We substitute the results from Exercises 45 and 46 into the result from part (c):

$$1 = \lim_{n\to\infty} \frac{I_{2n+1}}{I_{2n}} = \lim_{n\to\infty} \frac{\dfrac{2\cdot4\cdot6\cdots(2n)}{3\cdot5\cdot7\cdots(2n+1)}}{\dfrac{1\cdot3\cdot5\cdots(2n-1)}{2\cdot4\cdot6\cdots(2n)}\dfrac{\pi}{2}} = \lim_{n\to\infty}\left[\frac{2\cdot4\cdot6\cdots(2n)}{3\cdot5\cdot7\cdots(2n+1)}\right]\left[\frac{2\cdot4\cdot6\cdots(2n)}{1\cdot3\cdot5\cdots(2n-1)}\left(\frac{2}{\pi}\right)\right]$$

$$= \lim_{n\to\infty} \frac{2}{1}\cdot\frac{2}{3}\cdot\frac{4}{3}\cdot\frac{4}{5}\cdot\frac{6}{5}\cdot\frac{6}{7}\cdots\frac{2n}{2n-1}\cdot\frac{2n}{2n+1}\cdot\frac{2}{\pi} \qquad \text{[rearrange terms]}$$

Multiplying both sides by $\frac{\pi}{2}$ gives us the *Wallis product*:

$$\frac{\pi}{2} = \frac{2}{1}\cdot\frac{2}{3}\cdot\frac{4}{3}\cdot\frac{4}{5}\cdot\frac{6}{5}\cdot\frac{6}{7}\cdots$$

(e) The area of the kth rectangle is k. At the $2n$th step, the area is increased from $2n-1$ to $2n$ by multiplying the width by $\dfrac{2n}{2n-1}$, and at the $(2n+1)$th step, the area is increased from $2n$ to $2n+1$ by multiplying the height by $\dfrac{2n+1}{2n}$. These two steps multiply the ratio of width to height by $\dfrac{2n}{2n-1}$ and $\dfrac{1}{(2n+1)/(2n)} = \dfrac{2n}{2n+1}$ respectively. So, by part (d), the

limiting ratio is $\dfrac{2}{1}\cdot\dfrac{2}{3}\cdot\dfrac{4}{3}\cdot\dfrac{4}{5}\cdot\dfrac{6}{5}\cdot\dfrac{6}{7}\cdots = \dfrac{\pi}{2}$.

8.2 Trigonometric Integrals

The symbols $\overset{s}{=}$ and $\overset{c}{=}$ indicate the use of the substitutions $\{u = \sin x, du = \cos x\,dx\}$ and $\{u = \cos x, du = -\sin x\,dx\}$, respectively.

1. $\int \sin^3 x \cos^2 x\,dx = \int \sin^2 x \cos^2 x \sin x\,dx = \int(1-\cos^2 x)\cos^2 x \sin x\,dx \overset{c}{=} \int(1-u^2)u^2(-du)$

$= \int(u^2-1)u^2\,du = \int(u^4-u^2)\,du = \frac{1}{5}u^5 - \frac{1}{3}u^3 + C = \frac{1}{5}\cos^5 x - \frac{1}{3}\cos^3 x + C$

2. $\int \sin^6 x \cos^3 x\,dx = \int \sin^6 x \cos^2 x \cos x\,dx = \int \sin^6 x\,(1-\sin^2 x)\cos x\,dx \overset{s}{=} \int u^6(1-u^2)\,du$

$= \int(u^6-u^8)\,du = \frac{1}{7}u^7 - \frac{1}{9}u^9 + C = \frac{1}{7}\sin^7 x - \frac{1}{9}\sin^9 x + C$

3. $\int_{\pi/2}^{3\pi/4} \sin^5 x \cos^3 x\,dx = \int_{\pi/2}^{3\pi/4}\sin^5 x \cos^2 x \cos x\,dx = \int_{\pi/2}^{3\pi/4}\sin^5 x\,(1-\sin^2 x)\cos x\,dx \overset{s}{=} \int_1^{\sqrt{2}/2} u^5(1-u^2)\,du$

$= \int_1^{\sqrt{2}/2}(u^5-u^7)\,du = \left[\frac{1}{6}u^6 - \frac{1}{8}u^8\right]_1^{\sqrt{2}/2} = \left(\frac{1/8}{6} - \frac{1/16}{8}\right) - \left(\frac{1}{6}-\frac{1}{8}\right) = -\frac{11}{384}$

4. $\int_0^{\pi/2}\cos^5 x\,dx = \int_0^{\pi/2}(\cos^2 x)^2\cos x\,dx = \int_0^{\pi/2}(1-\sin^2 x)^2\cos x\,dx \overset{s}{=} \int_0^1(1-u^2)^2\,du$

$= \int_0^1(1-2u^2+u^4)\,du = \left[u-\frac{2}{3}u^3+\frac{1}{5}u^5\right]_0^1 = \left(1-\frac{2}{3}+\frac{1}{5}\right) - 0 = \frac{8}{15}$

5. Let $y = \pi x$, so $dy = \pi \, dx$ and

$$\int \sin^2(\pi x) \cos^5(\pi x)\, dx = \frac{1}{\pi}\int \sin^2 y \, \cos^5 y \, dy = \frac{1}{\pi}\int \sin^2 y \, \cos^4 y \, \cos y \, dy$$

$$= \frac{1}{\pi}\int \sin^2 y \, (1 - \sin^2 y)^2 \, \cos y \, dy \overset{s}{=} \frac{1}{\pi}\int u^2(1 - u^2)^2 \, du = \frac{1}{\pi}\int (u^2 - 2u^4 + u^6)\, du$$

$$= \frac{1}{\pi}\left(\frac{1}{3}u^3 - \frac{2}{5}u^5 + \frac{1}{7}u^7\right) + C = \frac{1}{3\pi}\sin^3 y - \frac{2}{5\pi}\sin^5 y + \frac{1}{7\pi}\sin^7 y + C$$

$$= \frac{1}{3\pi}\sin^3(\pi x) - \frac{2}{5\pi}\sin^5(\pi x) + \frac{1}{7\pi}\sin^7(\pi x) + C$$

6. Let $y = \sqrt{x}$, so that $dy = \dfrac{1}{2\sqrt{x}}\, dx$ and $dx = 2y\, dy$. Then

$$\int \frac{\sin^3(\sqrt{x})}{\sqrt{x}}\, dx = \int \frac{\sin^3 y}{y}\,(2y\, dy) = 2\int \sin^3 y \, dy = 2\int \sin^2 y \, \sin y \, dy = 2\int (1 - \cos^2 y)\,\sin y \, dy$$

$$\overset{c}{=} 2\int (1 - u^2)(-du) = 2\int (u^2 - 1)\, du = 2\left(\frac{1}{3}u^3 - u\right) + C = \frac{2}{3}\cos^3 y - 2\cos y + C$$

$$= \frac{2}{3}\cos^3(\sqrt{x}) - 2\cos\sqrt{x} + C$$

7. $\int_0^{\pi/2} \cos^2\theta \, d\theta = \int_0^{\pi/2} \frac{1}{2}(1 + \cos 2\theta)\, d\theta$ [half-angle identity]

$$= \frac{1}{2}\left[\theta + \frac{1}{2}\sin 2\theta\right]_0^{\pi/2} = \frac{1}{2}\left[\left(\frac{\pi}{2} + 0\right) - (0 + 0)\right] = \frac{\pi}{4}$$

8. $\int_0^{\pi/2} \sin^2(2\theta)\, d\theta = \int_0^{\pi/2} \frac{1}{2}(1 - \cos 4\theta)\, d\theta = \frac{1}{2}\left[\theta - \frac{1}{4}\sin 4\theta\right]_0^{\pi/2} = \frac{1}{2}\left[\left(\frac{\pi}{2} - 0\right) - (0 - 0)\right] = \frac{\pi}{4}$

9. $\int_0^\pi \sin^4(3t)\, dt = \int_0^\pi \left[\sin^2(3t)\right]^2 dt = \int_0^\pi \left[\frac{1}{2}(1 - \cos 6t)\right]^2 dt = \frac{1}{4}\int_0^\pi (1 - 2\cos 6t + \cos^2 6t)\, dt$

$$= \frac{1}{4}\int_0^\pi \left[1 - 2\cos 6t + \frac{1}{2}(1 + \cos 12t)\right] dt = \frac{1}{4}\int_0^\pi \left(\frac{3}{2} - 2\cos 6t + \frac{1}{2}\cos 12t\right) dt$$

$$= \frac{1}{4}\left[\frac{3}{2}t - \frac{1}{3}\sin 6t + \frac{1}{24}\sin 12t\right]_0^\pi = \frac{1}{4}\left[\left(\frac{3\pi}{2} - 0 + 0\right) - (0 - 0 + 0)\right] = \frac{3\pi}{8}$$

10. $\int_0^\pi \cos^6\theta \, d\theta = \int_0^\pi (\cos^2\theta)^3\, d\theta = \int_0^\pi \left[\frac{1}{2}(1 + \cos 2\theta)\right]^3 d\theta = \frac{1}{8}\int_0^\pi (1 + 3\cos 2\theta + 3\cos^2 2\theta + \cos^3 2\theta)\, d\theta$

$$= \frac{1}{8}\left[\theta + \frac{3}{2}\sin 2\theta\right]_0^\pi + \frac{1}{8}\int_0^\pi \left[\frac{3}{2}(1 + \cos 4\theta)\right] d\theta + \frac{1}{8}\int_0^\pi \left[(1 - \sin^2 2\theta)\cos 2\theta\right] d\theta$$

$$= \frac{1}{8}\pi + \frac{3}{16}\left[\theta + \frac{1}{4}\sin 4\theta\right]_0^\pi + \frac{1}{8}\int_0^0 (1 - u^2)\left(\frac{1}{2}du\right) \qquad [\, u = \sin 2\theta,\ du = 2\cos 2\theta \, d\theta\,]$$

$$= \frac{\pi}{8} + \frac{3\pi}{16} + 0 = \frac{5\pi}{16}$$

11. $\int (1 + \cos\theta)^2\, d\theta = \int (1 + 2\cos\theta + \cos^2\theta)\, d\theta = \theta + 2\sin\theta + \frac{1}{2}\int (1 + \cos 2\theta)\, d\theta$

$$= \theta + 2\sin\theta + \frac{1}{2}\theta + \frac{1}{4}\sin 2\theta + C = \frac{3}{2}\theta + 2\sin\theta + \frac{1}{4}\sin 2\theta + C$$

12. Let $u = x$, $dv = \cos^2 x\, dx \ \Rightarrow\ du = dx$, $v = \int \cos^2 x\, dx = \int \frac{1}{2}(1 + \cos 2x)\, dx = \frac{1}{2}x + \frac{1}{4}\sin 2x$, so

$$\int x\cos^2 x\, dx = x\left(\frac{1}{2}x + \frac{1}{4}\sin 2x\right) - \int \left(\frac{1}{2}x + \frac{1}{4}\sin 2x\right) dx = \frac{1}{2}x^2 + \frac{1}{4}x\sin 2x - \frac{1}{4}x^2 + \frac{1}{8}\cos 2x + C$$

$$= \frac{1}{4}x^2 + \frac{1}{4}x\sin 2x + \frac{1}{8}\cos 2x + C$$

13. $\int_0^{\pi/2} \sin^2 x \, \cos^2 x \, dx = \int_0^{\pi/2} \frac{1}{4}(4\sin^2 x \, \cos^2 x)\, dx = \int_0^{\pi/2} \frac{1}{4}(2\sin x \, \cos x)^2 dx = \frac{1}{4}\int_0^{\pi/2} \sin^2 2x \, dx$

$$= \frac{1}{4}\int_0^{\pi/2} \frac{1}{2}(1 - \cos 4x)\, dx = \frac{1}{8}\int_0^{\pi/2}(1 - \cos 4x)\, dx = \frac{1}{8}\left[x - \frac{1}{4}\sin 4x\right]_0^{\pi/2} = \frac{1}{8}\left(\frac{\pi}{2}\right) = \frac{\pi}{16}$$

14. $\int_0^\pi \sin^2 t \, \cos^4 t \, dt = \frac{1}{4} \int_0^\pi (4 \sin^2 t \, \cos^2 t) \cos^2 t \, dt = \frac{1}{4} \int_0^\pi (2 \sin t \, \cos t)^2 \frac{1}{2}(1 + \cos 2t) \, dt$

$\qquad = \frac{1}{8} \int_0^\pi (\sin 2t)^2 (1 + \cos 2t) \, dt = \frac{1}{8} \int_0^\pi (\sin^2 2t + \sin^2 2t \, \cos 2t) \, dt$

$\qquad = \frac{1}{8} \int_0^\pi \sin^2 2t \, dt + \frac{1}{8} \int_0^\pi \sin^2 2t \, \cos 2t \, dt = \frac{1}{8} \int_0^\pi \frac{1}{2}(1 - \cos 4t) \, dt + \frac{1}{8} \left[\frac{1}{3} \cdot \frac{1}{2} \sin^3 2t \right]_0^\pi$

$\qquad = \frac{1}{16} \left[t - \frac{1}{4} \sin 4t \right]_0^\pi + \frac{1}{8}(0 - 0) = \frac{1}{16} \left[(\pi - 0) - 0 \right] = \frac{\pi}{16}$

15. $\displaystyle \int \frac{\cos^5 \alpha}{\sqrt{\sin \alpha}} \, d\alpha = \int \frac{\cos^4 \alpha}{\sqrt{\sin \alpha}} \cos \alpha \, d\alpha = \int \frac{\left(1 - \sin^2 \alpha\right)^2}{\sqrt{\sin \alpha}} \cos \alpha \, d\alpha \overset{s}{=} \int \frac{(1 - u^2)^2}{\sqrt{u}} \, du$

$\qquad = \int \frac{1 - 2u^2 + u^4}{u^{1/2}} \, du = \int \left(u^{-1/2} - 2u^{3/2} + u^{7/2}\right) du = 2u^{1/2} - \frac{4}{5} u^{5/2} + \frac{2}{9} u^{9/2} + C$

$\qquad = \frac{2}{45} u^{1/2}(45 - 18u^2 + 5u^4) + C = \frac{2}{45} \sqrt{\sin \alpha} \left(45 - 18 \sin^2 \alpha + 5 \sin^4 \alpha\right) + C$

16. Let $u = \sin \theta$. Then $du = \cos \theta \, d\theta$ and

$$\int \cos \theta \, \cos^5(\sin \theta) \, d\theta = \int \cos^5 u \, du = \int (\cos^2 u)^2 \cos u \, du = \int (1 - \sin^2 u)^2 \cos u \, du$$

$$= \int (1 - 2\sin^2 u + \sin^4 u) \cos u \, du = I$$

Now let $x = \sin u$. Then $dx = \cos u \, du$ and

$$I = \int (1 - 2x^2 + x^4) \, dx = x - \frac{2}{3}x^3 + \frac{1}{5}x^5 + C = \sin u - \frac{2}{3} \sin^3 u + \frac{1}{5} \sin^5 u + C$$

$$= \sin(\sin \theta) - \frac{2}{3} \sin^3(\sin \theta) + \frac{1}{5} \sin^5(\sin \theta) + C$$

17. $\displaystyle \int \cos^2 x \, \tan^3 x \, dx = \int \frac{\sin^3 x}{\cos x} \, dx \overset{c}{=} \int \frac{(1 - u^2)(-du)}{u} = \int \left[\frac{-1}{u} + u \right] du$

$\qquad = -\ln|u| + \frac{1}{2}u^2 + C = \frac{1}{2} \cos^2 x - \ln|\cos x| + C$

18. $\displaystyle \int \cot^5 \theta \, \sin^4 \theta \, d\theta = \int \frac{\cos^5 \theta}{\sin^5 \theta} \sin^4 \theta \, d\theta = \int \frac{\cos^5 \theta}{\sin \theta} \, d\theta = \int \frac{\cos^4 \theta}{\sin \theta} \cos \theta \, d\theta = \int \frac{\left(1 - \sin^2 \theta\right)^2}{\sin \theta} \cos \theta \, d\theta$

$\qquad \overset{s}{=} \int \frac{(1 - u^2)^2}{u} \, du = \int \frac{1 - 2u^2 + u^4}{u} \, du = \int \left(\frac{1}{u} - 2u + u^3 \right) du$

$\qquad = \ln|u| - u^2 + \frac{1}{4}u^4 + C = \ln|\sin \theta| - \sin^2 \theta + \frac{1}{4} \sin^4 \theta + C$

19. $\displaystyle \int \frac{\cos x + \sin 2x}{\sin x} \, dx = \int \frac{\cos x + 2 \sin x \, \cos x}{\sin x} \, dx = \int \frac{\cos x}{\sin x} \, dx + \int 2 \cos x \, dx \overset{s}{=} \int \frac{1}{u} \, du + 2 \sin x$

$\qquad = \ln|u| + 2 \sin x + C = \ln|\sin x| + 2 \sin x + C$

Or: Use the formula $\int \cot x \, dx = \ln|\sin x| + C$.

20. $\int \cos^2 x \, \sin 2x \, dx = 2 \int \cos^3 x \, \sin x \, dx \overset{c}{=} -2 \int u^3 \, du = -\frac{1}{2}u^4 + C = -\frac{1}{2} \cos^4 x + C$

21. Let $u = \tan x$, $du = \sec^2 x \, dx$. Then $\int \sec^2 x \, \tan x \, dx = \int u \, du = \frac{1}{2}u^2 + C = \frac{1}{2} \tan^2 x + C$.

Or: Let $v = \sec x$, $dv = \sec x \, \tan x \, dx$. Then $\int \sec^2 x \, \tan x \, dx = \int v \, dv = \frac{1}{2}v^2 + C = \frac{1}{2} \sec^2 x + C$.

22. $\int_0^{\pi/2} \sec^4(t/2) \, dt = \int_0^{\pi/4} \sec^4 x \, (2 \, dx) \quad \left[x = t/2, \, dx = \frac{1}{2} \, dt \right] \quad = 2 \int_0^{\pi/4} \sec^2 x \, (1 + \tan^2 x) \, dx$

$\qquad = 2 \int_0^1 (1 + u^2) \, du \quad [u = \tan x, \, du = \sec^2 x \, dx] \quad = 2 \left[u + \frac{1}{3}u^3 \right]_0^1 = 2\left(1 + \frac{1}{3}\right) = \frac{8}{3}$

23. $\int \tan^2 x \, dx = \int (\sec^2 x - 1) \, dx = \tan x - x + C$

24. $\int (\tan^2 x + \tan^4 x) \, dx = \int \tan^2 x \, (1 + \tan^2 x) \, dx = \int \tan^2 x \sec^2 x \, dx = \int u^2 \, du \quad [u = \tan x, \, du = \sec^2 x \, dx]$

$= \frac{1}{3} u^3 + C = \frac{1}{3} \tan^3 x + C$

25. $\int \sec^6 t \, dt = \int \sec^4 t \cdot \sec^2 t \, dt = \int (\tan^2 t + 1)^2 \sec^2 t \, dt = \int (u^2 + 1)^2 \, du \quad [u = \tan t, \, du = \sec^2 t \, dt]$

$= \int (u^4 + 2u^2 + 1) \, du = \frac{1}{5} u^5 + \frac{2}{3} u^3 + u + C = \frac{1}{5} \tan^5 t + \frac{2}{3} \tan^3 t + \tan t + C$

26. $\int_0^{\pi/4} \sec^4 \theta \tan^4 \theta \, d\theta = \int_0^{\pi/4} (\tan^2 \theta + 1) \tan^4 \theta \sec^2 \theta \, d\theta = \int_0^1 (u^2 + 1) u^4 \, du \quad [u = \tan \theta, \, du = \sec^2 \theta \, d\theta]$

$= \int_0^1 (u^6 + u^4) \, du = \left[\frac{1}{7} u^7 + \frac{1}{5} u^5 \right]_0^1 = \frac{1}{7} + \frac{1}{5} = \frac{12}{35}$

27. $\int_0^{\pi/3} \tan^5 x \sec^4 x \, dx = \int_0^{\pi/3} \tan^5 x \, (\tan^2 x + 1) \sec^2 x \, dx = \int_0^{\sqrt{3}} u^5 (u^2 + 1) \, du \quad [u = \tan x, \, du = \sec^2 x \, dx]$

$= \int_0^{\sqrt{3}} (u^7 + u^5) \, du = \left[\frac{1}{8} u^8 + \frac{1}{6} u^6 \right]_0^{\sqrt{3}} = \frac{81}{8} + \frac{27}{6} = \frac{81}{8} + \frac{9}{2} = \frac{81}{8} + \frac{36}{8} = \frac{117}{8}$

Alternate solution:

$\int_0^{\pi/3} \tan^5 x \sec^4 x \, dx = \int_0^{\pi/3} \tan^4 x \sec^3 x \sec x \tan x \, dx = \int_0^{\pi/3} (\sec^2 x - 1)^2 \sec^3 x \sec x \tan x \, dx$

$= \int_1^2 (u^2 - 1)^2 u^3 \, du \quad [u = \sec x, \, du = \sec x \tan x \, dx] \quad = \int_1^2 (u^4 - 2u^2 + 1) u^3 \, du$

$= \int_1^2 (u^7 - 2u^5 + u^3) \, du = \left[\frac{1}{8} u^8 - \frac{1}{3} u^6 + \frac{1}{4} u^4 \right]_1^2 = (32 - \frac{64}{3} + 4) - (\frac{1}{8} - \frac{1}{3} + \frac{1}{4}) = \frac{117}{8}$

28. $\int \tan^3 (2x) \sec^5 (2x) \, dx = \int \tan^2 (2x) \sec^4 (2x) \cdot \sec(2x) \tan(2x) \, dx$

$= \int (u^2 - 1) u^4 (\frac{1}{2} \, du) \quad [u = \sec(2x), \, du = 2\sec(2x)\tan(2x)\,dx]$

$= \frac{1}{2} \int (u^6 - u^4) \, du = \frac{1}{14} u^7 - \frac{1}{10} u^5 + C = \frac{1}{14} \sec^7 (2x) - \frac{1}{10} \sec^5 (2x) + C$

29. $\int \tan^3 x \sec x \, dx = \int \tan^2 x \sec x \tan x \, dx = \int (\sec^2 x - 1) \sec x \tan x \, dx$

$= \int (u^2 - 1) \, du \quad [u = \sec x, \, du = \sec x \tan x \, dx] \quad = \frac{1}{3} u^3 - u + C = \frac{1}{3} \sec^3 x - \sec x + C$

30. $\int_0^{\pi/3} \tan^5 x \sec^6 x \, dx = \int_0^{\pi/3} \tan^5 x \sec^4 x \sec^2 x \, dx = \int_0^{\pi/3} \tan^5 x \, (1 + \tan^2 x)^2 \sec^2 x \, dx$

$= \int_0^{\sqrt{3}} u^5 (1 + u^2)^2 \, du \quad [u = \tan x, \, du = \sec^2 x \, dx] \quad = \int_0^{\sqrt{3}} u^5 (1 + 2u^2 + u^4) \, du$

$= \int_0^{\sqrt{3}} (u^5 + 2u^7 + u^9) \, du = \left[\frac{1}{6} u^6 + \frac{1}{4} u^8 + \frac{1}{10} u^{10} \right]_0^{\sqrt{3}} = \frac{27}{6} + \frac{81}{4} + \frac{243}{10} = \frac{981}{20}$

Alternate solution:

$\int_0^{\pi/3} \tan^5 x \sec^6 x \, dx = \int_0^{\pi/3} \tan^4 x \sec^5 x \sec x \tan x \, dx = \int_0^{\pi/3} (\sec^2 x - 1)^2 \sec^5 x \sec x \tan x \, dx$

$= \int_1^2 (u^2 - 1)^2 u^5 \, du \quad [u = \sec x, \, du = \sec x \tan x \, dx]$

$= \int_1^2 (u^4 - 2u^2 + 1) u^5 \, du = \int_1^2 (u^9 - 2u^7 + u^5) \, du$

$= \left[\frac{1}{10} u^{10} - \frac{1}{4} u^8 + \frac{1}{6} u^6 \right]_1^2 = (\frac{512}{5} - 64 + \frac{32}{3}) - (\frac{1}{10} - \frac{1}{4} + \frac{1}{6}) = \frac{981}{20}$

31. $\int \tan^5 x \, dx = \int (\sec^2 x - 1)^2 \tan x \, dx = \int \sec^4 x \tan x \, dx - 2 \int \sec^2 x \tan x \, dx + \int \tan x \, dx$

$= \int \sec^3 x \sec x \tan x \, dx - 2 \int \tan x \sec^2 x \, dx + \int \tan x \, dx$

$= \frac{1}{4} \sec^4 x - \tan^2 x + \ln |\sec x| + C \quad [\text{or } \frac{1}{4} \sec^4 x - \sec^2 x + \ln |\sec x| + C]$

32. $\int \tan^6 ay \, dy = \int \tan^4 ay \, (\sec^2 ay - 1) \, dy = \int \tan^4 ay \sec^2 ay \, dy - \int \tan^4 ay \, dy$

$\qquad = \frac{1}{5a} \tan^5 ay - \int \tan^2 ay \, (\sec^2 ay - 1) \, dy = \frac{1}{5a} \tan^5 ay - \int \tan^2 ay \sec^2 ay \, dy + \int (\sec^2 ay - 1) \, dy$

$\qquad = \frac{1}{5a} \tan^5 ay - \frac{1}{3a} \tan^3 ay + \frac{1}{a} \tan ay - y + C$

33. $\displaystyle\int \frac{\tan^3 \theta}{\cos^4 \theta} \, d\theta = \int \tan^3 \theta \sec^4 \theta \, d\theta = \int \tan^3 \theta \cdot (\tan^2 \theta + 1) \cdot \sec^2 \theta \, d\theta$

$\qquad = \int u^3 (u^2 + 1) \, du \qquad [u = \tan \theta, \, du = \sec^2 \theta \, d\theta]$

$\qquad = \int (u^5 + u^3) \, du = \frac{1}{6} u^6 + \frac{1}{4} u^4 + C = \frac{1}{6} \tan^6 \theta + \frac{1}{4} \tan^4 \theta + C$

34. $\int \tan^2 x \sec x \, dx = \int (\sec^2 x - 1) \sec x \, dx = \int \sec^3 x \, dx - \int \sec x \, dx$

$\qquad = \frac{1}{2} (\sec x \, \tan x + \ln |\sec x + \tan x|) - \ln |\sec x + \tan x| + C \qquad$ [by Example 8 and (1)]

$\qquad = \frac{1}{2} (\sec x \, \tan x - \ln |\sec x + \tan x|) + C$

35. Let $u = x, \, dv = \sec x \, \tan x \, dx \quad \Rightarrow \quad du = dx, v = \sec x$. Then

$\int x \sec x \, \tan x \, dx = x \sec x - \int \sec x \, dx = x \sec x - \ln |\sec x + \tan x| + C$.

36. $\displaystyle\int \frac{\sin \phi}{\cos^3 \phi} \, d\phi = \int \frac{\sin \phi}{\cos \phi} \cdot \frac{1}{\cos^2 \phi} \, d\phi = \int \tan \phi \sec^2 \phi \, d\phi = \int u \, du \qquad [u = \tan \phi, \, du = \sec^2 \phi \, d\phi]$

$\qquad = \frac{1}{2} u^2 + C = \frac{1}{2} \tan^2 \phi + C$

Alternate solution: Let $u = \cos \phi$ to get $\frac{1}{2} \sec^2 \phi + C$.

37. $\int_{\pi/6}^{\pi/2} \cot^2 x \, dx = \int_{\pi/6}^{\pi/2} (\csc^2 x - 1) \, dx = \left[- \cot x - x \right]_{\pi/6}^{\pi/2} = \left(0 - \frac{\pi}{2} \right) - \left(-\sqrt{3} - \frac{\pi}{6} \right) = \sqrt{3} - \frac{\pi}{3}$

38. $\displaystyle\int_{\pi/4}^{\pi/2} \cot^3 x \, dx = \int_{\pi/4}^{\pi/2} \cot x \, (\csc^2 x - 1) \, dx = \int_{\pi/4}^{\pi/2} \cot x \csc^2 x \, dx - \int_{\pi/4}^{\pi/2} \frac{\cos x}{\sin x} \, dx$

$\qquad = \left[-\frac{1}{2} \cot^2 x - \ln |\sin x| \right]_{\pi/4}^{\pi/2} = (0 - \ln 1) - \left[-\frac{1}{2} - \ln \frac{1}{\sqrt{2}} \right] = \frac{1}{2} + \ln \frac{1}{\sqrt{2}} = \frac{1}{2} (1 - \ln 2)$

39. $\int \cot^3 \alpha \csc^3 \alpha \, d\alpha = \int \cot^2 \alpha \csc^2 \alpha \cdot \csc \alpha \, \cot \alpha \, d\alpha = \int (\csc^2 \alpha - 1) \csc^2 \alpha \cdot \csc \alpha \, \cot \alpha \, d\alpha$

$\qquad = \int (u^2 - 1) u^2 \cdot (-du) \qquad [u = \csc \alpha, \, du = -\csc \alpha \, \cot \alpha \, d\alpha]$

$\qquad = \int (u^2 - u^4) \, du = \frac{1}{3} u^3 - \frac{1}{5} u^5 + C = \frac{1}{3} \csc^3 \alpha - \frac{1}{5} \csc^5 \alpha + C$

40. $\int \csc^4 x \cot^6 x \, dx = \int \cot^6 x \, (\cot^2 x + 1) \csc^2 x \, dx = \int u^6 (u^2 + 1) \cdot (-du) \qquad [u = \cot x, \, du = -\csc^2 x \, dx]$

$\qquad = \int u^6 (u^2 + 1) \cdot (-du) \qquad [u = \cot x, \, du = -\csc^2 x \, dx]$

$\qquad = \int (-u^8 - u^6) \, du = -\frac{1}{9} u^9 - \frac{1}{7} u^7 + C = -\frac{1}{9} \cot^9 x - \frac{1}{7} \cot^7 x + C$

41. $I = \displaystyle\int \csc x \, dx = \int \frac{\csc x \, (\csc x - \cot x)}{\csc x - \cot x} \, dx = \int \frac{-\csc x \, \cot x + \csc^2 x}{\csc x - \cot x} \, dx$. Let $u = \csc x - \cot x \quad \Rightarrow$

$du = (-\csc x \, \cot x + \csc^2 x) \, dx$. Then $I = \int du/u = \ln |u| = \ln |\csc x - \cot x| + C$.

42. Let $u = \csc x$, $dv = \csc^2 x \, dx$. Then $du = -\csc x \cot x \, dx$, $v = -\cot x$ $\Rightarrow$

$$\int \csc^3 x \, dx = -\csc x \cot x - \int \csc x \cot^2 x \, dx = -\csc x \cot x - \int \csc x \left(\csc^2 x - 1\right) dx$$

$$= -\csc x \cot x + \int \csc x \, dx - \int \csc^3 x \, dx$$

Solving for $\int \csc^3 x \, dx$ and using Exercise 41, we get

$$\int \csc^3 x \, dx = -\tfrac{1}{2} \csc x \cot x + \tfrac{1}{2} \int \csc x \, dx = -\tfrac{1}{2} \csc x \cot x + \tfrac{1}{2} \ln |\csc x - \cot x| + C. \text{ Thus,}$$

$$\int_{\pi/6}^{\pi/3} \csc^3 x \, dx = \left[-\tfrac{1}{2} \csc x \cot x + \tfrac{1}{2} \ln |\csc x - \cot x| \right]_{\pi/6}^{\pi/3}$$

$$= -\tfrac{1}{2} \cdot \tfrac{2}{\sqrt{3}} \cdot \tfrac{1}{\sqrt{3}} + \tfrac{1}{2} \ln \left| \tfrac{2}{\sqrt{3}} - \tfrac{1}{\sqrt{3}} \right| + \tfrac{1}{2} \cdot 2 \cdot \sqrt{3} - \tfrac{1}{2} \ln \left| 2 - \sqrt{3} \right|$$

$$= -\tfrac{1}{3} + \sqrt{3} + \tfrac{1}{2} \ln \tfrac{1}{\sqrt{3}} - \tfrac{1}{2} \ln \left(2 - \sqrt{3} \right) \approx 1.7825$$

43. $\int \sin 8x \cos 5x \, dx \overset{2a}{=} \int \tfrac{1}{2} [\sin(8x - 5x) + \sin(8x + 5x)] \, dx = \tfrac{1}{2} \int \sin 3x \, dx + \tfrac{1}{2} \int \sin 13x \, dx$

$$= -\tfrac{1}{6} \cos 3x - \tfrac{1}{26} \cos 13x + C$$

44. $\int \cos \pi x \cos 4\pi x \, dx \overset{2c}{=} \int \tfrac{1}{2} [\cos(\pi x - 4\pi x) + \cos(\pi x + 4\pi x)] \, dx = \tfrac{1}{2} \int \cos(-3\pi x) \, dx + \tfrac{1}{2} \int \cos(5\pi x) \, dx$

$$= \tfrac{1}{2} \int \cos 3\pi x \, dx + \tfrac{1}{2} \int \cos 5\pi x \, dx = \tfrac{1}{6\pi} \sin 3\pi x + \tfrac{1}{10\pi} \sin 5\pi x + C$$

45. $\int \sin 5\theta \sin \theta \, d\theta \overset{2b}{=} \int \tfrac{1}{2} [\cos(5\theta - \theta) - \cos(5\theta + \theta)] \, d\theta = \tfrac{1}{2} \int \cos 4\theta \, d\theta - \tfrac{1}{2} \int \cos 6\theta \, d\theta = \tfrac{1}{8} \sin 4\theta - \tfrac{1}{12} \sin 6\theta + C$

46. $\displaystyle \int \frac{\cos x + \sin x}{\sin 2x} \, dx = \frac{1}{2} \int \frac{\cos x + \sin x}{\sin x \cos x} \, dx = \frac{1}{2} \int (\csc x + \sec x) \, dx$

$$= \tfrac{1}{2} \left(\ln |\csc x - \cot x| + \ln |\sec x + \tan x| \right) + C \qquad \text{[by Exercise 41 and (1)]}$$

47. $\displaystyle \int \frac{1 - \tan^2 x}{\sec^2 x} \, dx = \int \left(\cos^2 x - \sin^2 x \right) dx = \int \cos 2x \, dx = \frac{1}{2} \sin 2x + C$

48. $\displaystyle \int \frac{dx}{\cos x - 1} = \int \frac{1}{\cos x - 1} \cdot \frac{\cos x + 1}{\cos x + 1} \, dx = \int \frac{\cos x + 1}{\cos^2 x - 1} \, dx = \int \frac{\cos x + 1}{-\sin^2 x} \, dx$

$$= \int \left(-\cot x \csc x - \csc^2 x \right) dx = \csc x + \cot x + C$$

49. Let $u = \tan(t^2)$ $\Rightarrow$ $du = 2t \sec^2(t^2) \, dt$. Then $\int t \sec^2(t^2) \tan^4(t^2) \, dt = \int u^4 \left(\tfrac{1}{2} \, du \right) = \tfrac{1}{10} u^5 + C = \tfrac{1}{10} \tan^5(t^2) + C$.

50. Let $u = \tan^7 x$, $dv = \sec x \tan x \, dx$ $\Rightarrow$ $du = 7 \tan^6 x \sec^2 x \, dx$, $v = \sec x$. Then

$$\int \tan^8 x \sec x \, dx = \int \tan^7 x \cdot \sec x \tan x \, dx = \tan^7 x \sec x - \int 7 \tan^6 x \sec^2 x \sec x \, dx$$

$$= \tan^7 x \sec x - 7 \int \tan^6 x \left(\tan^2 x + 1 \right) \sec x \, dx$$

$$= \tan^7 x \sec x - 7 \int \tan^8 x \sec x \, dx - 7 \int \tan^6 x \sec x \, dx.$$

Thus, $8 \int \tan^8 x \sec x \, dx = \tan^7 x \sec x - 7 \int \tan^6 x \sec x \, dx$ and

$$\int_0^{\pi/4} \tan^8 x \sec x \, dx = \frac{1}{8} \left[\tan^7 x \sec x \right]_0^{\pi/4} - \frac{7}{8} \int_0^{\pi/4} \tan^6 x \sec x \, dx = \frac{\sqrt{2}}{8} - \frac{7}{8} I.$$

In Exercises 51–54, let $f(x)$ denote the integrand and $F(x)$ its antiderivative (with $C = 0$).

51. Let $u = x^2$, so that $du = 2x\,dx$. Then

$$\int x\sin^2(x^2)\,dx = \int \sin^2 u\left(\tfrac{1}{2}\,du\right) = \tfrac{1}{2}\int \tfrac{1}{2}(1 - \cos 2u)\,du$$

$$= \tfrac{1}{4}\left(u - \tfrac{1}{2}\sin 2u\right) + C = \tfrac{1}{4}u - \tfrac{1}{4}\left(\tfrac{1}{2}\cdot 2\sin u\,\cos u\right) + C$$

$$= \tfrac{1}{4}x^2 - \tfrac{1}{4}\sin(x^2)\,\cos(x^2) + C$$

We see from the graph that this is reasonable, since F increases where f is positive and F decreases where f is negative. Note also that f is an odd function and F is an even function.

52. $\int \sin^3 x \cos^4 x\,dx = \int \cos^4 x\,(1 - \cos^2 x)\sin x\,dx$

$$\overset{c}{=} \int u^4(1 - u^2)(-du) = \int (u^6 - u^4)\,du$$

$$= \tfrac{1}{7}u^7 - \tfrac{1}{5}u^5 + C = \tfrac{1}{7}\cos^7 x - \tfrac{1}{5}\cos^5 x + C$$

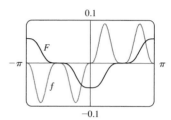

We see from the graph that this is reasonable, since F increases where f is positive and F decreases where f is negative. Note also that f is an odd function and F is an even function.

53. $\int \sin 3x \sin 6x\,dx = \int \tfrac{1}{2}[\cos(3x - 6x) - \cos(3x + 6x)]\,dx$

$$= \tfrac{1}{2}\int (\cos 3x - \cos 9x)\,dx$$

$$= \tfrac{1}{6}\sin 3x - \tfrac{1}{18}\sin 9x + C$$

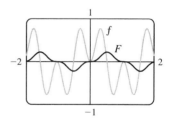

Notice that $f(x) = 0$ whenever F has a horizontal tangent.

54. $\int \sec^4 \tfrac{x}{2}\,dx = \int \left(\tan^2 \tfrac{x}{2} + 1\right)\sec^2 \tfrac{x}{2}\,dx$

$$= \int (u^2 + 1)\,2\,du \quad \left[u = \tan \tfrac{x}{2},\ du = \tfrac{1}{2}\sec^2 \tfrac{x}{2}\,dx\right]$$

$$= \tfrac{2}{3}u^3 + 2u + C = \tfrac{2}{3}\tan^3 \tfrac{x}{2} + 2\tan \tfrac{x}{2} + C$$

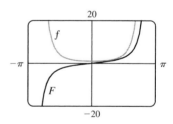

Notice that F is increasing and f is positive on the intervals on which they are defined. Also, F has no horizontal tangent and f is never zero.

55. $f_{\text{ave}} = \tfrac{1}{2\pi}\int_{-\pi}^{\pi} \sin^2 x \cos^3 x\,dx = \tfrac{1}{2\pi}\int_{-\pi}^{\pi} \sin^2 x\,(1 - \sin^2 x)\cos x\,dx$

$$= \tfrac{1}{2\pi}\int_0^0 u^2(1 - u^2)\,du \qquad [\text{where } u = \sin x]$$

$$= 0$$

56. (a) Let $u = \cos x$. Then $du = -\sin x\,dx \ \Rightarrow\ \int \sin x\,\cos x\,dx = \int u(-du) = -\tfrac{1}{2}u^2 + C = -\tfrac{1}{2}\cos^2 x + C_1$.

(b) Let $u = \sin x$. Then $du = \cos x\,dx \ \Rightarrow\ \int \sin x\,\cos x\,dx = \int u\,du = \tfrac{1}{2}u^2 + C = \tfrac{1}{2}\sin^2 x + C_2$.

(c) $\int \sin x\,\cos x\,dx = \int \tfrac{1}{2}\sin 2x\,dx = -\tfrac{1}{4}\cos 2x + C_3$

(d) Let $u = \sin x$, $dv = \cos x\,dx$. Then $du = \cos x\,dx$, $v = \sin x$, so $\int \sin x\,\cos x\,dx = \sin^2 x - \int \sin x\,\cos x\,dx$, by Equation 8.1.2, so $\int \sin x\,\cos x\,dx = \tfrac{1}{2}\sin^2 x + C_4$.

Using $\cos^2 x = 1 - \sin^2 x$ and $\cos 2x = 1 - 2\sin^2 x$, we see that the answers differ only by a constant.

57. $A = \int_{-\pi/4}^{\pi/4} (\cos^2 x - \sin^2 x)\, dx = \int_{-\pi/4}^{\pi/4} \cos 2x\, dx$

$= 2\int_0^{\pi/4} \cos 2x\, dx = 2\left[\frac{1}{2}\sin 2x\right]_0^{\pi/4} = \left[\sin 2x\right]_0^{\pi/4}$

$= 1 - 0 = 1$

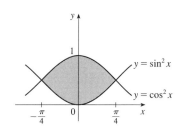

58. $A = \int_{\pi/4}^{5\pi/4} (\sin^3 x - \cos^3 x)\, dx = \int_{\pi/4}^{5\pi/4} \sin^3 x\, dx - \int_{\pi/4}^{5\pi/4} \cos^3 x\, dx$

$= \int_{\pi/4}^{5\pi/4} (1 - \cos^2 x)\sin x\, dx - \int_{\pi/4}^{5\pi/4} (1 - \sin^2 x)\cos x\, dx$

$\overset{c,s}{=} \int_{\sqrt{2}/2}^{-\sqrt{2}/2} (1 - u^2)(-du) - \int_{\sqrt{2}/2}^{-\sqrt{2}/2} (1 - u^2)\, du$

$= 2\int_0^{\sqrt{2}/2} (1 - u^2)\, du + 2\int_0^{\sqrt{2}/2} (1 - u^2)\, du = 4\left[u - \frac{1}{3}u^3\right]_0^{\sqrt{2}/2}$

$= 4\left[\left(\frac{\sqrt{2}}{2} - \frac{1}{3}\cdot\frac{\sqrt{2}}{4}\right) - 0\right] = 2\sqrt{2} - \frac{1}{3}\sqrt{2} = \frac{5}{3}\sqrt{2}$

59.

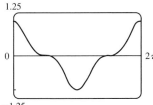

It seems from the graph that $\int_0^{2\pi} \cos^3 x\, dx = 0$, since the area below the x-axis and above the graph looks about equal to the area above the axis and below the graph. By Example 1, the integral is $\left[\sin x - \frac{1}{3}\sin^3 x\right]_0^{2\pi} = 0$. Note that due to symmetry, the integral of any odd power of $\sin x$ or $\cos x$ between limits which differ by $2n\pi$ (n any integer) is 0.

60.

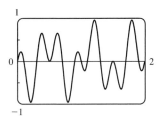

It seems from the graph that $\int_0^2 \sin 2\pi x\, \cos 5\pi x\, dx = 0$, since each bulge above the x-axis seems to have a corresponding depression below the x-axis. To evaluate the integral, we use a trigonometric identity:

$\int_0^1 \sin 2\pi x\, \cos 5\pi x\, dx = \frac{1}{2}\int_0^2 [\sin(2\pi x - 5\pi x) + \sin(2\pi x + 5\pi x)]\, dx$

$= \frac{1}{2}\int_0^2 [\sin(-3\pi x) + \sin 7\pi x]\, dx$

$= \frac{1}{2}\left[\frac{1}{3\pi}\cos(-3\pi x) - \frac{1}{7\pi}\cos 7\pi x\right]_0^2$

$= \frac{1}{2}\left[\frac{1}{3\pi}(1 - 1) - \frac{1}{7\pi}(1 - 1)\right] = 0$

61. Using disks, $V = \int_{\pi/2}^{\pi} \pi \sin^2 x\, dx = \pi\int_{\pi/2}^{\pi} \frac{1}{2}(1 - \cos 2x)\, dx = \pi\left[\frac{1}{2}x - \frac{1}{4}\sin 2x\right]_{\pi/2}^{\pi} = \pi\left(\frac{\pi}{2} - 0 - \frac{\pi}{4} + 0\right) = \frac{\pi^2}{4}$

62. Using disks,

$V = \int_0^{\pi} \pi(\sin^2 x)^2\, dx = 2\pi\int_0^{\pi/2} \left[\frac{1}{2}(1 - \cos 2x)\right]^2 dx$

$= \frac{\pi}{2}\int_0^{\pi/2} (1 - 2\cos 2x + \cos^2 2x)\, dx$

$= \frac{\pi}{2}\int_0^{\pi/2} \left[1 - 2\cos 2x + \frac{1}{2}(1 - \cos 4x)\right] dx$

$= \frac{\pi}{2}\int_0^{\pi/2} \left(\frac{3}{2} - 2\cos 2x - \frac{1}{2}\cos 4x\right) dx = \frac{\pi}{2}\left[\frac{3}{2}x - \sin 2x + \frac{1}{8}\sin 4x\right]_0^{\pi/2}$

$= \frac{\pi}{2}\left[\left(\frac{3\pi}{4} - 0 + 0\right) - (0 - 0 + 0)\right] = \frac{3}{8}\pi^2$

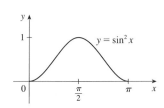

63. Using washers,

$$V = \int_0^{\pi/4} \pi \left[(1 - \sin x)^2 - (1 - \cos x)^2\right] dx$$

$$= \pi \int_0^{\pi/4} \left[(1 - 2\sin x + \sin^2 x) - (1 - 2\cos x + \cos^2 x)\right] dx$$

$$= \pi \int_0^{\pi/4} (2\cos x - 2\sin x + \sin^2 x - \cos^2 x) \, dx$$

$$= \pi \int_0^{\pi/4} (2\cos x - 2\sin x - \cos 2x) \, dx = \pi \left[2\sin x + 2\cos x - \tfrac{1}{2}\sin 2x\right]_0^{\pi/4}$$

$$= \pi\left[\left(\sqrt{2} + \sqrt{2} - \tfrac{1}{2}\right) - (0 + 2 - 0)\right] = \pi\left(2\sqrt{2} - \tfrac{5}{2}\right)$$

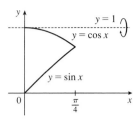

64. Using washers,

$$V = \int_0^{\pi/3} \pi\left\{[\sec x - (-1)]^2 - [\cos x - (-1)]^2\right\} dx$$

$$= \pi \int_0^{\pi/3} \left[(\sec^2 x + 2\sec x + 1) - (\cos^2 x + 2\cos x + 1)\right] dx$$

$$= \pi \int_0^{\pi/3} \left[\sec^2 x + 2\sec x - \tfrac{1}{2}(1 + \cos 2x) - 2\cos x\right] dx$$

$$= \pi \left[\tan x + 2\ln|\sec x + \tan x| - \tfrac{1}{2}x - \tfrac{1}{4}\sin 2x - 2\sin x\right]_0^{\pi/3}$$

$$= \pi\left[\left(\sqrt{3} + 2\ln\left(2 + \sqrt{3}\right) - \tfrac{\pi}{6} - \tfrac{1}{8}\sqrt{3} - \sqrt{3}\right) - 0\right]$$

$$= 2\pi \ln\left(2 + \sqrt{3}\right) - \tfrac{1}{6}\pi^2 - \tfrac{1}{8}\pi\sqrt{3}$$

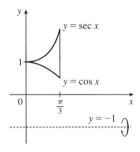

65. $s = f(t) = \int_0^t \sin \omega u \, \cos^2 \omega u \, du$. Let $y = \cos \omega u \ \Rightarrow \ dy = -\omega \sin \omega u \, du$. Then

$$s = -\tfrac{1}{\omega} \int_1^{\cos \omega t} y^2 \, dy = -\tfrac{1}{\omega}\left[\tfrac{1}{3}y^3\right]_1^{\cos \omega t} = \tfrac{1}{3\omega}\left(1 - \cos^3 \omega t\right).$$

66. (a) We want to calculate the square root of the average value of $[E(t)]^2 = [155\sin(120\pi t)]^2 = 155^2 \sin^2(120\pi t)$. First, we calculate the average value itself, by integrating $[E(t)]^2$ over one cycle (between $t = 0$ and $t = \tfrac{1}{60}$, since there are 60 cycles per second) and dividing by $\left(\tfrac{1}{60} - 0\right)$:

$$[E(t)]_{\text{ave}}^2 = \tfrac{1}{1/60} \int_0^{1/60} [155^2 \sin^2(120\pi t)] \, dt = 60 \cdot 155^2 \int_0^{1/60} \tfrac{1}{2}[1 - \cos(240\pi t)] \, dt$$

$$= 60 \cdot 155^2 \left(\tfrac{1}{2}\right)\left[t - \tfrac{1}{240\pi}\sin(240\pi t)\right]_0^{1/60} = 60 \cdot 155^2 \left(\tfrac{1}{2}\right)\left[\left(\tfrac{1}{60} - 0\right) - (0 - 0)\right] = \tfrac{155^2}{2}$$

The RMS value is just the square root of this quantity, which is $\tfrac{155}{\sqrt{2}} \approx 110$ V.

(b) $220 = \sqrt{[E(t)]_{\text{ave}}^2} \ \Rightarrow$

$$220^2 = [E(t)]_{\text{ave}}^2 = \tfrac{1}{1/60} \int_0^{1/60} A^2 \sin^2(120\pi t) \, dt = 60 A^2 \int_0^{1/60} \tfrac{1}{2}[1 - \cos(240\pi t)] \, dt$$

$$= 30 A^2 \left[t - \tfrac{1}{240\pi}\sin(240\pi t)\right]_0^{1/60} = 30 A^2 \left[\left(\tfrac{1}{60} - 0\right) - (0 - 0)\right] = \tfrac{1}{2}A^2$$

Thus, $220^2 = \tfrac{1}{2}A^2 \ \Rightarrow \ A = 220\sqrt{2} \approx 311$ V.

67. Just note that the integrand is odd $[f(-x) = -f(x)]$.

Or: If $m \neq n$, calculate

$$\int_{-\pi}^{\pi} \sin mx \, \cos nx \, dx = \int_{-\pi}^{\pi} \tfrac{1}{2}[\sin(m - n)x + \sin(m + n)x] \, dx = \tfrac{1}{2}\left[-\frac{\cos(m - n)x}{m - n} - \frac{\cos(m + n)x}{m + n}\right]_{-\pi}^{\pi} = 0$$

If $m = n$, then the first term in each set of brackets is zero.

68. $\int_{-\pi}^{\pi} \sin mx \sin nx \, dx = \int_{-\pi}^{\pi} \frac{1}{2}[\cos(m-n)x - \cos(m+n)x] \, dx.$

If $m \neq n$, this is equal to $\frac{1}{2}\left[\dfrac{\sin(m-n)x}{m-n} - \dfrac{\sin(m+n)x}{m+n}\right]_{-\pi}^{\pi} = 0.$

If $m = n$, we get $\int_{-\pi}^{\pi} \frac{1}{2}[1 - \cos(m+n)x] \, dx = \left[\frac{1}{2}x\right]_{-\pi}^{\pi} - \left[\dfrac{\sin(m+n)x}{2(m+n)}\right]_{-\pi}^{\pi} = \pi - 0 = \pi.$

69. $\int_{-\pi}^{\pi} \cos mx \cos nx \, dx = \int_{-\pi}^{\pi} \frac{1}{2}[\cos(m-n)x + \cos(m+n)x] \, dx.$

If $m \neq n$, this is equal to $\frac{1}{2}\left[\dfrac{\sin(m-n)x}{m-n} + \dfrac{\sin(m+n)x}{m+n}\right]_{-\pi}^{\pi} = 0.$

If $m = n$, we get $\int_{-\pi}^{\pi} \frac{1}{2}[1 + \cos(m+n)x] \, dx = \left[\frac{1}{2}x\right]_{-\pi}^{\pi} + \left[\dfrac{\sin(m+n)x}{2(m+n)}\right]_{-\pi}^{\pi} = \pi + 0 = \pi.$

70. $\dfrac{1}{\pi}\int_{-\pi}^{\pi} f(x) \sin mx \, dx = \dfrac{1}{\pi}\int_{-\pi}^{\pi}\left[\left(\displaystyle\sum_{n=1}^{m} a_n \sin nx\right)\sin mx\right] dx = \displaystyle\sum_{n=1}^{m} \dfrac{a_n}{\pi}\int_{-\pi}^{\pi} \sin mx \sin nx \, dx.$ By Exercise 68, every

term is zero except the mth one, and that term is $\dfrac{a_m}{\pi} \cdot \pi = a_m.$

8.3   Trigonometric Substitution

1. Let $x = 3\sec\theta$, where $0 \leq \theta < \frac{\pi}{2}$ or $\pi \leq \theta < \frac{3\pi}{2}$. Then

$dx = 3\sec\theta \tan\theta \, d\theta$ and

$$\sqrt{x^2-9} = \sqrt{9\sec^2\theta - 9} = \sqrt{9(\sec^2\theta - 1)} = \sqrt{9\tan^2\theta}$$
$$= 3\,|\tan\theta| = 3\tan\theta \text{ for the relevant values of } \theta.$$

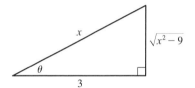

$$\int \frac{1}{x^2\sqrt{x^2-9}} \, dx = \int \frac{1}{9\sec^2\theta \cdot 3\tan\theta} \, 3\sec\theta \tan\theta \, d\theta = \frac{1}{9}\int \cos\theta \, d\theta = \frac{1}{9}\sin\theta + C = \frac{1}{9}\frac{\sqrt{x^2-9}}{x} + C$$

Note that $-\sec(\theta + \pi) = \sec\theta$, so the figure is sufficient for the case $\pi \leq \theta < \frac{3\pi}{2}$.

2. Let $x = 3\sin\theta$, where $-\frac{\pi}{2} \leq \theta \leq \frac{\pi}{2}$. Then $dx = 3\cos\theta \, d\theta$ and

$$\sqrt{9-x^2} = \sqrt{9 - 9\sin^2\theta} = \sqrt{9(1-\sin^2\theta)} = \sqrt{9\cos^2\theta}$$
$$= 3\,|\cos\theta| = 3\cos\theta \text{ for the relevant values of } \theta.$$

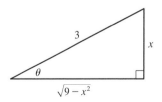

$\int x^3\sqrt{9-x^2} \, dx = \int 3^3\sin^3\theta \cdot 3\cos\theta \cdot 3\cos\theta \, d\theta = 3^5\int \sin^3\theta\cos^2\theta \, d\theta = 3^5\int \sin^2\theta\cos^2\theta \sin\theta \, d\theta$

$\qquad = 3^5\int (1-\cos^2\theta)\cos^2\theta \sin\theta \, d\theta = 3^5\int (1-u^2)u^2\,(-du) \qquad [u = \cos\theta, \, du = -\sin\theta \, d\theta]$

$\qquad = 3^5\int (u^4 - u^2)\, du = 3^5\left(\frac{1}{5}u^5 - \frac{1}{3}u^3\right) + C = 3^5\left(\frac{1}{5}\cos^5\theta - \frac{1}{3}\cos^3\theta\right) + C$

$\qquad = 3^5\left[\frac{1}{5}\dfrac{(9-x^2)^{5/2}}{3^5} - \frac{1}{3}\dfrac{(9-x^2)^{3/2}}{3^3}\right] + C$

$\qquad = \frac{1}{5}(9-x^2)^{5/2} - 3(9-x^2)^{3/2} + C \quad \text{or} \quad -\frac{1}{5}(x^2+6)(9-x^2)^{3/2} + C$

3. Let $x = 3 \tan \theta$, where $-\frac{\pi}{2} < \theta < \frac{\pi}{2}$. Then $dx = 3 \sec^2 \theta \, d\theta$ and

$$\sqrt{x^2 + 9} = \sqrt{9 \tan^2 \theta + 9} = \sqrt{9(\tan^2 \theta + 1)} = \sqrt{9 \sec^2 \theta}$$

$$= 3 |\sec \theta| = 3 \sec \theta \text{ for the relevant values of } \theta.$$

$$\int \frac{x^3}{\sqrt{x^2 + 9}} \, dx = \int \frac{3^3 \tan^3 \theta}{3 \sec \theta} 3 \sec^2 \theta \, d\theta = 3^3 \int \tan^3 \theta \sec \theta d\theta = 3^3 \int \tan^2 \theta \tan \theta \sec \theta \, d\theta$$

$$= 3^3 \int (\sec^2 \theta - 1) \tan \theta \sec \theta \, d\theta = 3^3 \int (u^2 - 1) \, du \qquad [u = \sec \theta, \, du = \sec \theta \tan \theta \, d\theta]$$

$$= 3^3 \left(\tfrac{1}{3} u^3 - u \right) + C = 3^3 \left(\tfrac{1}{3} \sec^3 \theta - \sec \theta \right) + C = 3^3 \left[\frac{1}{3} \frac{\left(x^2 + 9\right)^{3/2}}{3^3} - \frac{\sqrt{x^2 + 9}}{3} \right] + C$$

$$= \tfrac{1}{3}(x^2 + 9)^{3/2} - 9\sqrt{x^2 + 9} + C \quad \text{or} \quad \tfrac{1}{3}(x^2 - 18)\sqrt{x^2 + 9} + C$$

4. Let $x = 4 \sin \theta$, where $-\pi/2 \leq \theta \leq \pi/2$. Then $dx = 4 \cos \theta \, d\theta$ and

$$\sqrt{16 - x^2} = \sqrt{16 - 16 \sin^2 \theta} = \sqrt{16 \cos^2 \theta} = 4 |\cos \theta| = 4 \cos \theta. \text{ When } x = 0, \, 4 \sin \theta = 0 \quad \Rightarrow \quad \theta = 0,$$

and when $x = 2\sqrt{3}$, $4 \sin \theta = 2\sqrt{3} \quad \Rightarrow \quad \sin \theta = \frac{\sqrt{3}}{2} \quad \Rightarrow \quad \theta = \frac{\pi}{3}$. Thus, substitution gives

$$\int_0^{2\sqrt{3}} \frac{x^3}{\sqrt{16 - x^2}} \, dx = \int_0^{\pi/3} \frac{4^3 \sin^3 \theta}{4 \cos \theta} 4 \cos \theta \, d\theta = 4^3 \int_0^{\pi/3} \sin^3 \theta \, d\theta = 4^3 \int_0^{\pi/3} (1 - \cos^2 \theta) \sin \theta \, d\theta$$

$$\overset{c}{=} -4^3 \int_1^{1/2} (1 - u^2) \, du = -64 \left[u - \tfrac{1}{3} u^3 \right]_1^{1/2}$$

$$= -64 \left[\left(\tfrac{1}{2} - \tfrac{1}{24} \right) - \left(1 - \tfrac{1}{3} \right) \right] = -64 \left(-\tfrac{5}{24} \right) = \tfrac{40}{3}$$

Or: Let $u = 16 - x^2$, $x^2 = 16 - u$, $du = -2x \, dx$.

5. Let $t = \sec \theta$, so $dt = \sec \theta \tan \theta \, d\theta$, $t = \sqrt{2} \quad \Rightarrow \quad \theta = \frac{\pi}{4}$, and $t = 2 \quad \Rightarrow \quad \theta = \frac{\pi}{3}$. Then

$$\int_{\sqrt{2}}^2 \frac{1}{t^3 \sqrt{t^2 - 1}} \, dt = \int_{\pi/4}^{\pi/3} \frac{1}{\sec^3 \theta \tan \theta} \sec \theta \tan \theta \, d\theta = \int_{\pi/4}^{\pi/3} \frac{1}{\sec^2 \theta} \, d\theta = \int_{\pi/4}^{\pi/3} \cos^2 \theta \, d\theta$$

$$= \int_{\pi/4}^{\pi/3} \tfrac{1}{2}(1 + \cos 2\theta) \, d\theta = \tfrac{1}{2} \left[\theta + \tfrac{1}{2} \sin 2\theta \right]_{\pi/4}^{\pi/3}$$

$$= \tfrac{1}{2} \left[\left(\tfrac{\pi}{3} + \tfrac{1}{2} \tfrac{\sqrt{3}}{2} \right) - \left(\tfrac{\pi}{4} + \tfrac{1}{2} \cdot 1 \right) \right] = \tfrac{1}{2} \left(\tfrac{\pi}{12} + \tfrac{\sqrt{3}}{4} - \tfrac{1}{2} \right) = \tfrac{\pi}{24} + \tfrac{\sqrt{3}}{8} - \tfrac{1}{4}$$

6. Let $x = \sec \theta$, so $dx = \sec \theta \tan \theta \, d\theta$, $x = 1 \quad \Rightarrow \quad \theta = 0$, and $x = 2 \quad \Rightarrow \quad \theta = \frac{\pi}{3}$. Then

$$\int_1^2 \frac{\sqrt{x^2 - 1}}{x} \, dx = \int_0^{\pi/3} \frac{\tan \theta}{\sec \theta} \sec \theta \tan \theta \, d\theta = \int_0^{\pi/3} \tan^2 \theta \, d\theta = \int_0^{\pi/3} (\sec^2 \theta - 1) \, d\theta$$

$$= \left[\tan \theta - \theta \right]_0^{\pi/3} = \left(\sqrt{3} - \tfrac{\pi}{3} \right) - 0 = \sqrt{3} - \tfrac{\pi}{3}$$

7. Let $x = 5 \sin \theta$, so $dx = 5 \cos \theta \, d\theta$. Then

$$\int \frac{1}{x^2 \sqrt{25 - x^2}} \, dx = \int \frac{1}{5^2 \sin^2 \theta \cdot 5 \cos \theta} 5 \cos \theta \, d\theta = \frac{1}{25} \int \csc^2 \theta \, d\theta$$

$$= -\frac{1}{25} \cot \theta + C = -\frac{1}{25} \frac{\sqrt{25 - x^2}}{x} + C$$

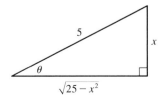

8. Let $x = 10\tan\theta$, so $dx = 10\sec^2\theta\,d\theta$. Then

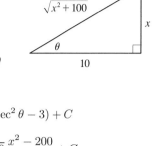

$$\int \frac{x^3}{\sqrt{x^2+100}}\,dx = \int \frac{1000\tan^3\theta}{10\sec\theta}\,10\sec^2\theta\,d\theta$$

$$= 1000\int\tan^3\theta\,\sec\theta\,d\theta = 1000\int(\sec^2\theta-1)\,\sec\theta\,\tan\theta\,d\theta$$

$$= 1000\int(u^2-1)\,du \qquad [u=\sec\theta,\,du=\sec\theta\,\tan\theta\,d\theta]$$

$$= 1000\left(\tfrac{1}{3}u^3-u\right)+C = \tfrac{1000}{3}u(u^2-3)+C = \tfrac{1000}{3}\sec\theta\,(\sec^2\theta-3)+C$$

$$= \frac{1000}{3}\frac{\sqrt{x^2+100}}{10}\left(\frac{x^2+100}{100}-3\right)+C = \frac{100}{3}\sqrt{x^2+100}\,\frac{x^2-200}{100}+C$$

$$= \tfrac{1}{3}(x^2-200)\sqrt{x^2+100}+C$$

9. Let $x = 4\tan\theta$, where $-\tfrac{\pi}{2}<\theta<\tfrac{\pi}{2}$. Then $dx = 4\sec^2\theta\,d\theta$ and

$$\sqrt{x^2+16} = \sqrt{16\tan^2\theta+16} = \sqrt{16(\tan^2\theta+1)}$$

$$= \sqrt{16\sec^2\theta} = 4\,|\sec\theta|$$

$$= 4\sec\theta \text{ for the relevant values of } \theta.$$

$$\int \frac{dx}{\sqrt{x^2+16}} = \int \frac{4\sec^2\theta\,d\theta}{4\sec\theta} = \int \sec\theta\,d\theta = \ln|\sec\theta+\tan\theta|+C_1 = \ln\left|\frac{\sqrt{x^2+16}}{4}+\frac{x}{4}\right|+C_1$$

$$= \ln\left|\sqrt{x^2+16}+x\right|-\ln|4|+C_1 = \ln\left(\sqrt{x^2+16}+x\right)+C, \text{ where } C=C_1-\ln 4.$$

(Since $\sqrt{x^2+16}+x>0$, we don't need the absolute value.)

10. Let $t = \sqrt{2}\tan\theta$, where $-\tfrac{\pi}{2}<\theta<\tfrac{\pi}{2}$. Then $dt = \sqrt{2}\sec^2\theta\,d\theta$ and

$$\sqrt{t^2+2} = \sqrt{2\tan^2\theta+2} = \sqrt{2(\tan^2\theta+1)} = \sqrt{2\sec^2\theta}$$

$$= \sqrt{2}\,|\sec\theta| = \sqrt{2}\sec\theta \text{ for the relevant values of } \theta.$$

$$\int \frac{t^5}{\sqrt{t^2+2}}\,dt = \int \frac{4\sqrt{2}\tan^5\theta}{\sqrt{2}\sec\theta}\sqrt{2}\sec^2\theta\,d\theta = 4\sqrt{2}\int \tan^5\theta\,\sec\theta\,d\theta$$

$$= 4\sqrt{2}\int(\sec^2\theta-1)^2\sec\theta\,\tan\theta\,d\theta = 4\sqrt{2}\int(u^2-1)^2\,du \quad [u=\sec\theta,\,du=\sec\theta\,\tan\theta\,d\theta]$$

$$= 4\sqrt{2}\int(u^4-2u^2+1)\,du = 4\sqrt{2}\left(\tfrac{1}{5}u^5-\tfrac{2}{3}u^3+u\right)+C$$

$$= \frac{4\sqrt{2}}{15}u(3u^4-10u^2+15)+C = \frac{4\sqrt{2}}{15}\cdot\frac{\sqrt{t^2+2}}{\sqrt{2}}\left[3\cdot\frac{(t^2+2)^2}{2^2}-10\frac{t^2+2}{2}+15\right]+C$$

$$= \tfrac{4}{15}\sqrt{t^2+2}\cdot\tfrac{1}{4}\left[3(t^4+4t^2+4)-20(t^2+2)+60\right]+C = \tfrac{1}{15}\sqrt{t^2+2}\,(3t^4-8t^2+32)+C$$

11. Let $2x = \sin\theta$, where $-\tfrac{\pi}{2}\le\theta\le\tfrac{\pi}{2}$. Then $x = \tfrac{1}{2}\sin\theta$, $dx = \tfrac{1}{2}\cos\theta\,d\theta$,

and $\sqrt{1-4x^2} = \sqrt{1-(2x)^2} = \cos\theta$.

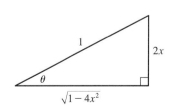

$$\int \sqrt{1-4x^2}\,dx = \int \cos\theta\left(\tfrac{1}{2}\cos\theta\right)d\theta = \tfrac{1}{4}\int(1+\cos 2\theta)\,d\theta$$

$$= \tfrac{1}{4}\left(\theta+\tfrac{1}{2}\sin 2\theta\right)+C = \tfrac{1}{4}(\theta+\sin\theta\,\cos\theta)+C$$

$$= \tfrac{1}{4}\left[\sin^{-1}(2x)+2x\sqrt{1-4x^2}\right]+C$$

12. $\int_0^1 x\sqrt{x^2+4}\,dx = \int_4^5 \sqrt{u}\left(\frac{1}{2}\,du\right)$ $\quad[u=x^2+4,\, du=2x\,dx]$ $\quad = \frac{1}{2}\cdot\frac{2}{3}\left[u^{3/2}\right]_4^5 = \frac{1}{3}\left(5\sqrt{5}-8\right)$

13. Let $x=3\sec\theta$, where $0\le\theta<\frac{\pi}{2}$ or $\pi\le\theta<\frac{3\pi}{2}$. Then

$dx = 3\sec\theta\,\tan\theta\,d\theta$ and $\sqrt{x^2-9}=3\tan\theta$, so

$$\int \frac{\sqrt{x^2-9}}{x^3}\,dx = \int \frac{3\tan\theta}{27\sec^3\theta}\,3\sec\theta\,\tan\theta\,d\theta = \frac{1}{3}\int \frac{\tan^2\theta}{\sec^2\theta}\,d\theta$$

$$= \frac{1}{3}\int \sin^2\theta\,d\theta = \frac{1}{3}\int \frac{1}{2}(1-\cos 2\theta)\,d\theta = \frac{1}{6}\theta - \frac{1}{12}\sin 2\theta + C = \frac{1}{6}\theta - \frac{1}{6}\sin\theta\,\cos\theta + C$$

$$= \frac{1}{6}\sec^{-1}\left(\frac{x}{3}\right) - \frac{1}{6}\frac{\sqrt{x^2-9}}{x}\frac{3}{x} + C = \frac{1}{6}\sec^{-1}\left(\frac{x}{3}\right) - \frac{\sqrt{x^2-9}}{2x^2} + C$$

14. Let $u=\sqrt{5}\,\sin\theta$, so $du=\sqrt{5}\cos\theta\,d\theta$. Then

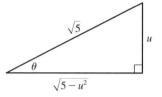

$$\int \frac{du}{u\sqrt{5-u^2}} = \int \frac{1}{\sqrt{5}\,\sin\theta\cdot\sqrt{5}\,\cos\theta}\sqrt{5}\,\cos\theta\,d\theta = \frac{1}{\sqrt{5}}\int \csc\theta\,d\theta$$

$$= \frac{1}{\sqrt{5}}\ln|\csc\theta - \cot\theta| + C \qquad \text{[by Exercise 8.2.41]}$$

$$= \frac{1}{\sqrt{5}}\ln\left|\frac{\sqrt{5}}{u}-\frac{\sqrt{5-u^2}}{u}\right| + C = \frac{1}{\sqrt{5}}\ln\left|\frac{\sqrt{5}-\sqrt{5-u^2}}{u}\right| + C$$

15. Let $x=a\sin\theta$, $dx=a\cos\theta\,d\theta$, $x=0 \Rightarrow \theta=0$ and $x=a \Rightarrow \theta=\frac{\pi}{2}$. Then

$$\int_0^a x^2\sqrt{a^2-x^2}\,dx = \int_0^{\pi/2} a^2\sin^2\theta\,(a\cos\theta)\,a\cos\theta\,d\theta = a^4\int_0^{\pi/2}\sin^2\theta\,\cos^2\theta\,d\theta = a^4\int_0^{\pi/2}\left[\frac{1}{2}(2\sin\theta\,\cos\theta)\right]^2 d\theta$$

$$= \frac{a^4}{4}\int_0^{\pi/2}\sin^2 2\theta\,d\theta = \frac{a^4}{4}\int_0^{\pi/2}\frac{1}{2}(1-\cos 4\theta)\,d\theta = \frac{a^4}{8}\left[\theta - \frac{1}{4}\sin 4\theta\right]_0^{\pi/2}$$

$$= \frac{a^4}{8}\left[\left(\frac{\pi}{2}-0\right)-0\right] = \frac{\pi}{16}a^4$$

16. Let $x=\frac{1}{3}\sec\theta$, so $dx=\frac{1}{3}\sec\theta\,\tan\theta\,d\theta$, $x=\sqrt{2}/3 \Rightarrow \theta=\frac{\pi}{4}$, $x=\frac{2}{3} \Rightarrow \theta=\frac{\pi}{3}$. Then

$$\int_{\sqrt{2}/3}^{2/3}\frac{dx}{x^5\sqrt{9x^2-1}} = \int_{\pi/4}^{\pi/3}\frac{\frac{1}{3}\sec\theta\,\tan\theta\,d\theta}{\left(\frac{1}{3}\right)^5\sec^5\theta\,\tan\theta} = 3^4\int_{\pi/4}^{\pi/3}\cos^4\theta\,d\theta = 81\int_{\pi/4}^{\pi/3}\left[\frac{1}{2}(1+\cos 2\theta)\right]^2 d\theta$$

$$= \frac{81}{4}\int_{\pi/4}^{\pi/3}(1+2\cos 2\theta + \cos^2 2\theta)\,d\theta = \frac{81}{4}\int_{\pi/4}^{\pi/3}\left[1+2\cos 2\theta + \frac{1}{2}(1+\cos 4\theta)\right]d\theta$$

$$= \frac{81}{4}\int_{\pi/4}^{\pi/3}\left(\frac{3}{2}+2\cos 2\theta + \frac{1}{2}\cos 4\theta\right)d\theta = \frac{81}{4}\left[\frac{3}{2}\theta + \sin 2\theta + \frac{1}{8}\sin 4\theta\right]_{\pi/4}^{\pi/3}$$

$$= \frac{81}{4}\left[\left(\frac{\pi}{2}+\frac{\sqrt{3}}{2}-\frac{\sqrt{3}}{16}\right)-\left(\frac{3\pi}{8}+1+0\right)\right] = \frac{81}{4}\left(\frac{\pi}{8}+\frac{7}{16}\sqrt{3}-1\right)$$

17. Let $u=x^2-7$, so $du=2x\,dx$. Then $\int \frac{x}{\sqrt{x^2-7}}\,dx = \frac{1}{2}\int \frac{1}{\sqrt{u}}\,du = \frac{1}{2}\cdot 2\sqrt{u}+C = \sqrt{x^2-7}+C.$

18. Let $ax = b \sec \theta$, so $(ax)^2 = b^2 \sec^2 \theta \Rightarrow$

$(ax)^2 - b^2 = b^2 \sec^2 \theta - b^2 = b^2(\sec^2 \theta - 1) = b^2 \tan^2 \theta$.

So $\sqrt{(ax)^2 - b^2} = b \tan \theta$, $dx = \dfrac{b}{a} \sec \theta \tan \theta \, d\theta$, and

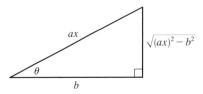

$\displaystyle \int \frac{dx}{[(ax)^2 - b^2]^{3/2}} = \int \frac{\frac{b}{a} \sec \theta \tan \theta}{b^3 \tan^3 \theta} \, d\theta = \frac{1}{ab^2} \int \frac{\sec \theta}{\tan^2 \theta} \, d\theta = \frac{1}{ab^2} \int \frac{\cos \theta}{\sin^2 \theta} \, d\theta = \frac{1}{ab^2} \int \csc \theta \cot \theta \, d\theta$

$\qquad\qquad = -\dfrac{1}{ab^2} \csc \theta + C = -\dfrac{1}{ab^2} \dfrac{ax}{\sqrt{(ax)^2 - b^2}} + C = -\dfrac{x}{b^2 \sqrt{(ax)^2 - b^2}} + C$

19. Let $x = \tan \theta$, where $-\dfrac{\pi}{2} < \theta < \dfrac{\pi}{2}$. Then $dx = \sec^2 \theta \, d\theta$

and $\sqrt{1 + x^2} = \sec \theta$, so

$\displaystyle \int \frac{\sqrt{1 + x^2}}{x} \, dx = \int \frac{\sec \theta}{\tan \theta} \sec^2 \theta \, d\theta = \int \frac{\sec \theta}{\tan \theta} (1 + \tan^2 \theta) \, d\theta$

$\qquad = \int (\csc \theta + \sec \theta \tan \theta) \, d\theta$

$\qquad = \ln |\csc \theta - \cot \theta| + \sec \theta + C \qquad$ [by Exercise 8.2.41]

$\qquad = \ln \left| \dfrac{\sqrt{1 + x^2}}{x} - \dfrac{1}{x} \right| + \dfrac{\sqrt{1 + x^2}}{1} + C = \ln \left| \dfrac{\sqrt{1 + x^2} - 1}{x} \right| + \sqrt{1 + x^2} + C$

20. Let $t = 5 \sin \theta$, where $-\dfrac{\pi}{2} \le \theta \le \dfrac{\pi}{2}$. Then $dt = 5 \cos \theta \, d\theta$

and $\sqrt{25 - t^2} = 5 \cos \theta$, so

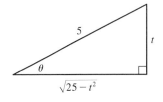

$\displaystyle \int \frac{t}{\sqrt{25 - t^2}} \, dt = \int \frac{5 \sin \theta}{5 \cos \theta} 5 \cos \theta \, d\theta = 5 \int \sin \theta \, d\theta$

$\qquad\qquad = -5 \cos \theta + C = -5 \cdot \dfrac{\sqrt{25 - t^2}}{5} + C = -\sqrt{25 - t^2} + C$

Or: Let $u = 25 - t^2$, so $du = -2t \, dt$.

21. Let $x = \dfrac{3}{5} \sin \theta$, so $dx = \dfrac{3}{5} \cos \theta \, d\theta$, $x = 0 \Rightarrow \theta = 0$, and $x = 0.6 \Rightarrow \theta = \dfrac{\pi}{2}$. Then

$\displaystyle \int_0^{0.6} \frac{x^2}{\sqrt{9 - 25x^2}} \, dx = \int_0^{\pi/2} \frac{\left(\frac{3}{5}\right)^2 \sin^2 \theta}{3 \cos \theta} \left(\frac{3}{5} \cos \theta \, d\theta \right) = \frac{9}{125} \int_0^{\pi/2} \sin^2 \theta \, d\theta$

$\qquad = \dfrac{9}{125} \int_0^{\pi/2} \dfrac{1}{2} (1 - \cos 2\theta) \, d\theta = \dfrac{9}{250} \left[\theta - \dfrac{1}{2} \sin 2\theta \right]_0^{\pi/2}$

$\qquad = \dfrac{9}{250} \left[\left(\dfrac{\pi}{2} - 0 \right) - 0 \right] = \dfrac{9}{500} \pi$

22. Let $x = \tan \theta$, where $-\dfrac{\pi}{2} < \theta < \dfrac{\pi}{2}$. Then $dx = \sec^2 \theta \, d\theta$,

$\sqrt{x^2 + 1} = \sec \theta$ and $x = 0 \Rightarrow \theta = 0$, $x = 1 \Rightarrow \theta = \dfrac{\pi}{4}$, so

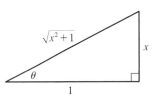

$\displaystyle \int_0^1 \sqrt{x^2 + 1} \, dx = \int_0^{\pi/4} \sec \theta \sec^2 \theta \, d\theta = \int_0^{\pi/4} \sec^3 \theta \, d\theta$

$\qquad = \dfrac{1}{2} \Big[\sec \theta \tan \theta + \ln |\sec \theta + \tan \theta| \Big]_0^{\pi/4} \qquad$ [by Example 8.2.8]

$\qquad = \dfrac{1}{2} \left[\sqrt{2} \cdot 1 + \ln\left(1 + \sqrt{2}\right) - 0 - \ln(1 + 0) \right] = \dfrac{1}{2} \left[\sqrt{2} + \ln\left(1 + \sqrt{2}\right) \right]$

23. $5 + 4x - x^2 = -(x^2 - 4x + 4) + 9 = -(x-2)^2 + 9$. Let

$x - 2 = 3\sin\theta$, $-\frac{\pi}{2} \le \theta \le \frac{\pi}{2}$, so $dx = 3\cos\theta\, d\theta$. Then

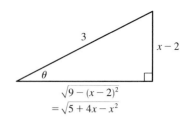

$\int \sqrt{5 + 4x - x^2}\, dx = \int \sqrt{9 - (x-2)^2}\, dx = \int \sqrt{9 - 9\sin^2\theta}\, 3\cos\theta\, d\theta$

$= \int \sqrt{9\cos^2\theta}\, 3\cos\theta\, d\theta = \int 9\cos^2\theta\, d\theta$

$= \frac{9}{2}\int (1 + \cos 2\theta)\, d\theta = \frac{9}{2}\left(\theta + \frac{1}{2}\sin 2\theta\right) + C$

$= \frac{9}{2}\theta + \frac{9}{4}\sin 2\theta + C = \frac{9}{2}\theta + \frac{9}{4}(2\sin\theta\,\cos\theta) + C$

$= \frac{9}{2}\sin^{-1}\left(\frac{x-2}{3}\right) + \frac{9}{2} \cdot \frac{x-2}{3} \cdot \frac{\sqrt{5 + 4x - x^2}}{3} + C$

$= \frac{9}{2}\sin^{-1}\left(\frac{x-2}{3}\right) + \frac{1}{2}(x-2)\sqrt{5 + 4x - x^2} + C$

24. $t^2 - 6t + 13 = (t^2 - 6t + 9) + 4 = (t-3)^2 + 2^2$.

Let $t - 3 = 2\tan\theta$, so $dt = 2\sec^2\theta\, d\theta$. Then

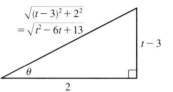

$\int \frac{dt}{\sqrt{t^2 - 6t + 13}} = \int \frac{1}{\sqrt{(2\tan\theta)^2 + 2^2}} 2\sec^2\theta\, d\theta$

$= \int \frac{2\sec^2\theta}{2\sec\theta}\, d\theta = \int \sec\theta\, d\theta = \ln|\sec\theta + \tan\theta| + C_1$ [by Formula 8.2.1]

$= \ln\left|\frac{\sqrt{t^2 - 6t + 13}}{2} + \frac{t-3}{2}\right| + C_1$

$= \ln\left|\sqrt{t^2 - 6t + 13} + t - 3\right| + C$ where $C = C_1 - \ln 2$

25. $x^2 + x + 1 = \left(x^2 + x + \frac{1}{4}\right) + \frac{3}{4} = \left(x + \frac{1}{2}\right)^2 + \left(\frac{\sqrt{3}}{2}\right)^2$. Let

$x + \frac{1}{2} = \frac{\sqrt{3}}{2}\tan\theta$, so $dx = \frac{\sqrt{3}}{2}\sec^2\theta\, d\theta$ and $\sqrt{x^2 + x + 1} = \frac{\sqrt{3}}{2}\sec\theta$.

Then

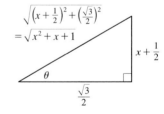

$\int \frac{x}{\sqrt{x^2 + x + 1}}\, dx = \int \frac{\frac{\sqrt{3}}{2}\tan\theta - \frac{1}{2}}{\frac{\sqrt{3}}{2}\sec\theta} \frac{\sqrt{3}}{2}\sec^2\theta\, d\theta$

$= \int \left(\frac{\sqrt{3}}{2}\tan\theta - \frac{1}{2}\right)\sec\theta\, d\theta = \int \frac{\sqrt{3}}{2}\tan\theta\,\sec\theta\, d\theta - \int \frac{1}{2}\sec\theta\, d\theta$

$= \frac{\sqrt{3}}{2}\sec\theta - \frac{1}{2}\ln|\sec\theta + \tan\theta| + C_1$

$= \sqrt{x^2 + x + 1} - \frac{1}{2}\ln\left|\frac{2}{\sqrt{3}}\sqrt{x^2 + x + 1} + \frac{2}{\sqrt{3}}\left(x + \frac{1}{2}\right)\right| + C_1$

$= \sqrt{x^2 + x + 1} - \frac{1}{2}\ln\left|\frac{2}{\sqrt{3}}\left[\sqrt{x^2 + x + 1} + \left(x + \frac{1}{2}\right)\right]\right| + C_1$

$= \sqrt{x^2 + x + 1} - \frac{1}{2}\ln\frac{2}{\sqrt{3}} - \frac{1}{2}\ln\left(\sqrt{x^2 + x + 1} + x + \frac{1}{2}\right) + C_1$

$= \sqrt{x^2 + x + 1} - \frac{1}{2}\ln\left(\sqrt{x^2 + x + 1} + x + \frac{1}{2}\right) + C$, where $C = C_1 - \frac{1}{2}\ln\frac{2}{\sqrt{3}}$

26. $3 + 4x - 4x^2 = -(4x^2 - 4x + 1) + 4 = 2^2 - (2x - 1)^2.$

Let $2x - 1 = 2\sin\theta$, so $2\,dx = 2\cos\theta\,d\theta$ and $\sqrt{3 + 4x - 4x^2} = 2\cos\theta.$

Then

$$\int \frac{x^2}{(3 + 4x - 4x^2)^{3/2}}\,dx = \int \frac{\left[\frac{1}{2}(1 + 2\sin\theta)\right]^2}{(2\cos\theta)^3}\cos\theta\,d\theta$$

$$= \frac{1}{32}\int \frac{1 + 4\sin\theta + 4\sin^2\theta}{\cos^2\theta}\,d\theta = \frac{1}{32}\int (\sec^2\theta + 4\tan\theta\,\sec\theta + 4\tan^2\theta)\,d\theta$$

$$= \tfrac{1}{32}\int [\sec^2\theta + 4\tan\theta\,\sec\theta + 4(\sec^2\theta - 1)]\,d\theta$$

$$= \tfrac{1}{32}\int (5\sec^2\theta + 4\tan\theta\,\sec\theta - 4)\,d\theta = \tfrac{1}{32}(5\tan\theta + 4\sec\theta - 4\theta) + C$$

$$= \frac{1}{32}\left[5\cdot\frac{2x - 1}{\sqrt{3 + 4x - 4x^2}} + 4\cdot\frac{2}{\sqrt{3 + 4x - 4x^2}} - 4\cdot\sin^{-1}\left(\frac{2x - 1}{2}\right)\right] + C$$

$$= \frac{10x + 3}{32\sqrt{3 + 4x - 4x^2}} - \frac{1}{8}\sin^{-1}\left(\frac{2x - 1}{2}\right) + C$$

27. $x^2 + 2x = (x^2 + 2x + 1) - 1 = (x + 1)^2 - 1.$ Let $x + 1 = 1\sec\theta$,

so $dx = \sec\theta\,\tan\theta\,d\theta$ and $\sqrt{x^2 + 2x} = \tan\theta.$ Then

$$\int \sqrt{x^2 + 2x}\,dx = \int \tan\theta\,(\sec\theta\,\tan\theta\,d\theta) = \int \tan^2\theta\,\sec\theta\,d\theta$$

$$= \int (\sec^2\theta - 1)\sec\theta\,d\theta = \int \sec^3\theta\,d\theta - \int \sec\theta\,d\theta$$

$$= \tfrac{1}{2}\sec\theta\,\tan\theta + \tfrac{1}{2}\ln|\sec\theta + \tan\theta| - \ln|\sec\theta + \tan\theta| + C$$

$$= \tfrac{1}{2}\sec\theta\,\tan\theta - \tfrac{1}{2}\ln|\sec\theta + \tan\theta| + C = \tfrac{1}{2}(x + 1)\sqrt{x^2 + 2x} - \tfrac{1}{2}\ln\left|x + 1 + \sqrt{x^2 + 2x}\right| + C$$

28. $x^2 - 2x + 2 = (x^2 - 2x + 1) + 1 = (x - 1)^2 + 1.$ Let $x - 1 = 1\tan\theta$,

so $dx = \sec^2\theta\,d\theta$ and $\sqrt{x^2 - 2x + 2} = \sec\theta.$ Then

$$\int \frac{x^2 + 1}{(x^2 - 2x + 2)^2}\,dx = \int \frac{(\tan\theta + 1)^2 + 1}{\sec^4\theta}\sec^2\theta\,d\theta$$

$$= \int \frac{\tan^2\theta + 2\tan\theta + 2}{\sec^2\theta}\,d\theta$$

$$= \int (\sin^2\theta + 2\sin\theta\,\cos\theta + 2\cos^2\theta)\,d\theta = \int (1 + 2\sin\theta\,\cos\theta + \cos^2\theta)\,d\theta$$

$$= \int \left[1 + 2\sin\theta\,\cos\theta + \tfrac{1}{2}(1 + \cos 2\theta)\right]\,d\theta = \int \left(\tfrac{3}{2} + 2\sin\theta\,\cos\theta + \tfrac{1}{2}\cos 2\theta\right)\,d\theta$$

$$= \tfrac{3}{2}\theta + \sin^2\theta + \tfrac{1}{4}\sin 2\theta + C = \tfrac{3}{2}\theta + \sin^2\theta + \tfrac{1}{2}\sin\theta\,\cos\theta + C$$

$$= \frac{3}{2}\tan^{-1}\left(\frac{x - 1}{1}\right) + \frac{(x - 1)^2}{x^2 - 2x + 2} + \frac{1}{2}\frac{x - 1}{\sqrt{x^2 - 2x + 2}}\frac{1}{\sqrt{x^2 - 2x + 2}} + C$$

$$= \frac{3}{2}\tan^{-1}(x - 1) + \frac{2(x^2 - 2x + 1) + x - 1}{2(x^2 - 2x + 2)} + C = \frac{3}{2}\tan^{-1}(x - 1) + \frac{2x^2 - 3x + 1}{2(x^2 - 2x + 2)} + C$$

[continued]

We can write the answer as

$$\frac{3}{2}\tan^{-1}(x-1) + \frac{(2x^2 - 4x + 4) + x - 3}{2(x^2 - 2x + 2)} + C = \frac{3}{2}\tan^{-1}(x-1) + 1 + \frac{x-3}{2(x^2 - 2x + 2)} + C$$

$$= \frac{3}{2}\tan^{-1}(x-1) + \frac{x-3}{2(x^2 - 2x + 2)} + C_1, \text{ where } C_1 = 1 + C$$

29. Let $u = x^2$, $du = 2x\, dx$. Then

$$\int x\sqrt{1-x^4}\, dx = \int \sqrt{1-u^2}\left(\tfrac{1}{2}\, du\right) = \tfrac{1}{2}\int \cos\theta \cdot \cos\theta\, d\theta \qquad \left[\begin{array}{c}\text{where } u = \sin\theta,\, du = \cos\theta\, d\theta, \\ \text{and } \sqrt{1-u^2} = \cos\theta\end{array}\right]$$

$$= \tfrac{1}{2}\int \tfrac{1}{2}(1 + \cos 2\theta)\, d\theta = \tfrac{1}{4}\theta + \tfrac{1}{8}\sin 2\theta + C = \tfrac{1}{4}\theta + \tfrac{1}{4}\sin\theta\,\cos\theta + C$$

$$= \tfrac{1}{4}\sin^{-1} u + \tfrac{1}{4}u\sqrt{1-u^2} + C = \tfrac{1}{4}\sin^{-1}(x^2) + \tfrac{1}{4}x^2\sqrt{1-x^4} + C$$

30. Let $u = \sin t$, $du = \cos t\, dt$. Then

$$\int_0^{\pi/2}\frac{\cos t}{\sqrt{1 + \sin^2 t}}\, dt = \int_0^1 \frac{1}{\sqrt{1+u^2}}\, du = \int_0^{\pi/4}\frac{1}{\sec\theta}\sec^2\theta\, d\theta \qquad \left[\begin{array}{c}\text{where } u = \tan\theta,\, du = \sec^2\theta\, d\theta, \\ \text{and } \sqrt{1+u^2} = \sec\theta\end{array}\right]$$

$$= \int_0^{\pi/4}\sec\theta\, d\theta = \Big[\ln|\sec\theta + \tan\theta|\Big]_0^{\pi/4} \qquad \text{[by (1) in Section 8.2]}$$

$$= \ln\big(\sqrt{2}+1\big) - \ln(1+0) = \ln\big(\sqrt{2}+1\big)$$

31. (a) Let $x = a\tan\theta$, where $-\frac{\pi}{2} < \theta < \frac{\pi}{2}$. Then $\sqrt{x^2 + a^2} = a\sec\theta$ and

$$\int \frac{dx}{\sqrt{x^2 + a^2}} = \int \frac{a\sec^2\theta\, d\theta}{a\sec\theta} = \int \sec\theta\, d\theta = \ln|\sec\theta + \tan\theta| + C_1 = \ln\left|\frac{\sqrt{x^2+a^2}}{a} + \frac{x}{a}\right| + C_1$$

$$= \ln\big(x + \sqrt{x^2 + a^2}\,\big) + C \quad \text{where } C = C_1 - \ln|a|$$

(b) Let $x = a\sinh t$, so that $dx = a\cosh t\, dt$ and $\sqrt{x^2 + a^2} = a\cosh t$. Then

$$\int \frac{dx}{\sqrt{x^2 + a^2}} = \int \frac{a\cosh t\, dt}{a\cosh t} = t + C = \sinh^{-1}\frac{x}{a} + C.$$

32. (a) Let $x = a\tan\theta$, $-\frac{\pi}{2} < \theta < \frac{\pi}{2}$. Then

$$I = \int \frac{x^2}{(x^2 + a^2)^{3/2}}\, dx = \int \frac{a^2\tan^2\theta}{a^3\sec^3\theta}\,a\sec^2\theta\, d\theta = \int \frac{\tan^2\theta}{\sec\theta}\, d\theta = \int \frac{\sec^2\theta - 1}{\sec\theta}\, d\theta$$

$$= \int (\sec\theta - \cos\theta)\, d\theta = \ln|\sec\theta + \tan\theta| - \sin\theta + C$$

$$= \ln\left|\frac{\sqrt{x^2+a^2}}{a} + \frac{x}{a}\right| - \frac{x}{\sqrt{x^2+a^2}} + C = \ln\big(x + \sqrt{x^2+a^2}\,\big) - \frac{x}{\sqrt{x^2+a^2}} + C_1$$

(b) Let $x = a\sinh t$. Then

$$I = \int \frac{a^2\sinh^2 t}{a^3\cosh^3 t}\,a\cosh t\, dt = \int \tanh^2 t\, dt = \int (1 - \text{sech}^2 t)\, dt = t - \tanh t + C$$

$$= \sinh^{-1}\frac{x}{a} - \frac{x}{\sqrt{a^2 + x^2}} + C$$

33. The average value of $f(x) = \sqrt{x^2 - 1}/x$ on the interval $[1, 7]$ is

$$\frac{1}{7-1} \int_1^7 \frac{\sqrt{x^2-1}}{x}\,dx = \frac{1}{6} \int_0^\alpha \frac{\tan\theta}{\sec\theta} \cdot \sec\theta\tan\theta\,d\theta \quad \begin{bmatrix} \text{where } x = \sec\theta, dx = \sec\theta\tan\theta\,d\theta, \\ \sqrt{x^2-1} = \tan\theta, \text{ and } \alpha = \sec^{-1}7 \end{bmatrix}$$

$$= \tfrac{1}{6} \int_0^\alpha \tan^2\theta\,d\theta = \tfrac{1}{6}\int_0^\alpha (\sec^2\theta - 1)\,d\theta = \tfrac{1}{6}\big[\tan\theta - \theta\big]_0^\alpha$$

$$= \tfrac{1}{6}(\tan\alpha - \alpha) = \tfrac{1}{6}\big(\sqrt{48} - \sec^{-1}7\big)$$

34. $9x^2 - 4y^2 = 36 \;\Rightarrow\; y = \pm\frac{3}{2}\sqrt{x^2-4} \;\Rightarrow\;$

$$\text{area} = 2\int_2^3 \tfrac{3}{2}\sqrt{x^2-4}\,dx = 3\int_2^3 \sqrt{x^2-4}\,dx$$

$$= 3\int_0^\alpha 2\tan\theta\, 2\sec\theta\,\tan\theta\,d\theta \quad \begin{bmatrix} \text{where } x = 2\sec\theta, \\ dx = 2\sec\theta\,\tan\theta\,d\theta, \\ \alpha = \sec^{-1}\left(\frac{3}{2}\right) \end{bmatrix}$$

$$= 12\int_0^\alpha \left(\sec^2\theta - 1\right)\sec\theta\,d\theta = 12\int_0^\alpha \left(\sec^3\theta - \sec\theta\right)d\theta$$

$$= 12\big[\tfrac{1}{2}(\sec\theta\,\tan\theta + \ln|\sec\theta + \tan\theta|) - \ln|\sec\theta + \tan\theta|\big]_0^\alpha$$

$$= 6\Big[\sec\theta\,\tan\theta - \ln|\sec\theta + \tan\theta|\Big]_0^\alpha$$

$$= 6\left[\tfrac{3\sqrt5}{4} - \ln\left(\tfrac{3}{2} + \tfrac{\sqrt5}{2}\right)\right] = \tfrac{9\sqrt5}{2} - 6\ln\left(\tfrac{3+\sqrt5}{2}\right)$$

35. Area of $\triangle POQ = \frac{1}{2}(r\cos\theta)(r\sin\theta) = \frac{1}{2}r^2\sin\theta\,\cos\theta$. Area of region $PQR = \int_{r\cos\theta}^r \sqrt{r^2-x^2}\,dx$.

Let $x = r\cos u \;\Rightarrow\; dx = -r\sin u\,du$ for $\theta \le u \le \frac{\pi}{2}$. Then we obtain

$$\int \sqrt{r^2-x^2}\,dx = \int r\sin u\,(-r\sin u)\,du = -r^2\int \sin^2 u\,du = -\tfrac{1}{2}r^2(u - \sin u\,\cos u) + C$$

$$= -\tfrac{1}{2}r^2\cos^{-1}(x/r) + \tfrac{1}{2}x\sqrt{r^2-x^2} + C$$

so $\quad$ area of region $PQR = \tfrac{1}{2}\big[-r^2\cos^{-1}(x/r) + x\sqrt{r^2-x^2}\big]_{r\cos\theta}^r$

$$= \tfrac{1}{2}\big[0 - (-r^2\theta + r\cos\theta\, r\sin\theta)\big] = \tfrac{1}{2}r^2\theta - \tfrac{1}{2}r^2\sin\theta\,\cos\theta$$

and thus, (area of sector POR) = (area of $\triangle POQ$) + (area of region PQR) = $\frac{1}{2}r^2\theta$.

36. Let $x = \sqrt{2}\sec\theta$, where $0 \le \theta < \frac{\pi}{2}$ or $\pi \le \theta < \frac{3\pi}{2}$, so $dx = \sqrt{2}\sec\theta\tan\theta\,d\theta$. Then

$$\int \frac{dx}{x^4\sqrt{x^2-2}} = \int \frac{\sqrt{2}\sec\theta\,\tan\theta\,d\theta}{4\sec^4\theta\,\sqrt{2}\,\tan\theta}$$

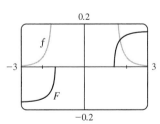

$$= \tfrac{1}{4}\int \cos^3\theta\,d\theta = \tfrac{1}{4}\int \left(1 - \sin^2\theta\right)\cos\theta\,d\theta$$

$$= \tfrac{1}{4}\big[\sin\theta - \tfrac{1}{3}\sin^3\theta\big] + C \qquad [\text{substitute } u = \sin\theta]$$

$$= \frac{1}{4}\left[\frac{\sqrt{x^2-2}}{x} - \frac{(x^2-2)^{3/2}}{3x^3}\right] + C$$

From the graph, it appears that our answer is reasonable. [Notice that $f(x)$ is large when F increases rapidly and small when F levels out.]

37. From the graph, it appears that the curve $y = x^2\sqrt{4 - x^2}$ and the

line $y = 2 - x$ intersect at about $x = a \approx 0.81$ and $x = 2$, with

$x^2\sqrt{4 - x^2} > 2 - x$ on $(a, 2)$. So the area bounded by the curve and the line is

$A \approx \int_a^2 \left[x^2\sqrt{4 - x^2} - (2 - x) \right] dx = \int_a^2 x^2\sqrt{4 - x^2}\, dx - \left[2x - \frac{1}{2}x^2 \right]_a^2$.

To evaluate the integral, we put $x = 2\sin\theta$, where $-\frac{\pi}{2} \le \theta \le \frac{\pi}{2}$. Then

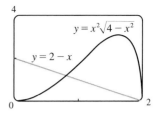

$dx = 2\cos\theta\, d\theta$, $x = 2 \Rightarrow \theta = \sin^{-1} 1 = \frac{\pi}{2}$, and $x = a \Rightarrow \theta = \alpha = \sin^{-1}(a/2) \approx 0.416$. So

$$\int_a^2 x^2\sqrt{4 - x^2}\, dx \approx \int_\alpha^{\pi/2} 4\sin^2\theta\,(2\cos\theta)(2\cos\theta\, d\theta) = 4\int_\alpha^{\pi/2} \sin^2 2\theta\, d\theta = 4\int_\alpha^{\pi/2} \tfrac{1}{2}(1 - \cos 4\theta)\, d\theta$$

$$= 2\left[\theta - \tfrac{1}{4}\sin 4\theta \right]_\alpha^{\pi/2} = 2\left[\left(\tfrac{\pi}{2} - 0 \right) - \left(\alpha - \tfrac{1}{4}(0.996) \right) \right] \approx 2.81$$

Thus, $A \approx 2.81 - \left[\left(2 \cdot 2 - \tfrac{1}{2} \cdot 2^2 \right) - \left(2a - \tfrac{1}{2}a^2 \right) \right] \approx 2.10$.

38. Let $x = b\tan\theta$, so that $dx = b\sec^2\theta\, d\theta$ and $\sqrt{x^2 + b^2} = b\sec\theta$.

$$E(P) = \int_{-a}^{L-a} \frac{\lambda b}{4\pi\varepsilon_0(x^2 + b^2)^{3/2}}\, dx = \frac{\lambda b}{4\pi\varepsilon_0} \int_{\theta_1}^{\theta_2} \frac{1}{(b\sec\theta)^3} b\sec^2\theta\, d\theta$$

$$= \frac{\lambda}{4\pi\varepsilon_0 b} \int_{\theta_1}^{\theta_2} \frac{1}{\sec\theta}\, d\theta = \frac{\lambda}{4\pi\varepsilon_0 b} \int_{\theta_1}^{\theta_2} \cos\theta\, d\theta = \frac{\lambda}{4\pi\varepsilon_0 b} \left[\sin\theta \right]_{\theta_1}^{\theta_2}$$

$$= \frac{\lambda}{4\pi\varepsilon_0 b} \left[\frac{x}{\sqrt{x^2 + b^2}} \right]_{-a}^{L-a} = \frac{\lambda}{4\pi\varepsilon_0 b} \left(\frac{L - a}{\sqrt{(L - a)^2 + b^2}} + \frac{a}{\sqrt{a^2 + b^2}} \right)$$

39. (a) Let $t = a\sin\theta$, $dt = a\cos\theta\, d\theta$, $t = 0 \Rightarrow \theta = 0$ and $t = x \Rightarrow \theta = \sin^{-1}(x/a)$. Then

$$\int_0^x \sqrt{a^2 - t^2}\, dt = \int_0^{\sin^{-1}(x/a)} a\cos\theta\,(a\cos\theta\, d\theta)$$

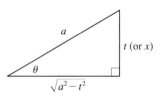

$$= a^2 \int_0^{\sin^{-1}(x/a)} \cos^2\theta\, d\theta = \frac{a^2}{2} \int_0^{\sin^{-1}(x/a)} (1 + \cos 2\theta)\, d\theta$$

$$= \frac{a^2}{2} \left[\theta + \tfrac{1}{2}\sin 2\theta \right]_0^{\sin^{-1}(x/a)} = \frac{a^2}{2} \left[\theta + \sin\theta\,\cos\theta \right]_0^{\sin^{-1}(x/a)}$$

$$= \frac{a^2}{2} \left[\left(\sin^{-1}\left(\frac{x}{a} \right) + \frac{x}{a} \cdot \frac{\sqrt{a^2 - x^2}}{a} \right) - 0 \right]$$

$$= \tfrac{1}{2}a^2\sin^{-1}(x/a) + \tfrac{1}{2}x\sqrt{a^2 - x^2}$$

(b) The integral $\int_0^x \sqrt{a^2 - t^2}\, dt$ represents the area under the curve $y = \sqrt{a^2 - t^2}$ between the vertical lines $t = 0$ and $t = x$.

The figure shows that this area consists of a triangular region and a sector of the circle $t^2 + y^2 = a^2$. The triangular region

has base x and height $\sqrt{a^2 - x^2}$, so its area is $\tfrac{1}{2}x\sqrt{a^2 - x^2}$. The sector has area $\tfrac{1}{2}a^2\theta = \tfrac{1}{2}a^2\sin^{-1}(x/a)$.

40. The curves intersect when $x^2 + \left(\tfrac{1}{2}x^2 \right)^2 = 8 \Leftrightarrow x^2 + \tfrac{1}{4}x^4 = 8 \Leftrightarrow x^4 + 4x^2 - 32 = 0 \Leftrightarrow$

$(x^2 + 8)(x^2 - 4) = 0 \Leftrightarrow x = \pm 2$. The area inside the circle and above the parabola is given by

$A_1 = \int_{-2}^{2} \left(\sqrt{8 - x^2} - \frac{1}{2}x^2 \right) dx = 2 \int_0^2 \sqrt{8 - x^2} \, dx - 2 \int_0^2 \frac{1}{2}x^2 \, dx$

$\qquad = 2 \left[\frac{1}{2}(8) \sin^{-1} \left(\frac{2}{\sqrt{8}} \right) + \frac{1}{2}(2) \sqrt{8 - 2^2} - \frac{1}{2} \left[\frac{1}{3}x^3 \right]_0^2 \right]$ [by Exercise 39]

$\qquad = 8 \sin^{-1} \left(\frac{1}{\sqrt{2}} \right) + 2\sqrt{4} - \frac{8}{3} = 8 \left(\frac{\pi}{4} \right) + 4 - \frac{8}{3} = 2\pi + \frac{4}{3}$

Since the area of the disk is $\pi \left(\sqrt{8} \right)^2 = 8\pi$, the area inside the circle and

below the parabola ia $A_2 = 8\pi - \left(2\pi + \frac{4}{3} \right) = 6\pi - \frac{4}{3}$.

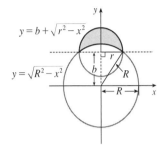

41. Let the equation of the large circle be $x^2 + y^2 = R^2$. Then the equation of

the small circle is $x^2 + (y - b)^2 = r^2$, where $b = \sqrt{R^2 - r^2}$ is the distance

between the centers of the circles. The desired area is

$A = \int_{-r}^{r} \left[\left(b + \sqrt{r^2 - x^2} \right) - \sqrt{R^2 - x^2} \right] dx$

$\qquad = 2 \int_0^r \left(b + \sqrt{r^2 - x^2} - \sqrt{R^2 - x^2} \right) dx$

$\qquad = 2 \int_0^r b \, dx + 2 \int_0^r \sqrt{r^2 - x^2} \, dx - 2 \int_0^r \sqrt{R^2 - x^2} \, dx$

The first integral is just $2br = 2r \sqrt{R^2 - r^2}$. The second integral represents the area of a quarter-circle of radius r, so its value

is $\frac{1}{4}\pi r^2$. To evaluate the other integral, note that

$$\int \sqrt{a^2 - x^2} \, dx = \int a^2 \cos^2 \theta \, d\theta \quad [x = a \sin \theta, \, dx = a \cos \theta \, d\theta] \quad = \left(\frac{1}{2}a^2 \right) \int (1 + \cos 2\theta) \, d\theta$$

$$= \frac{1}{2}a^2 \left(\theta + \frac{1}{2} \sin 2\theta \right) + C = \frac{1}{2}a^2 (\theta + \sin \theta \, \cos \theta) + C$$

$$= \frac{a^2}{2} \arcsin \left(\frac{x}{a} \right) + \frac{a^2}{2} \left(\frac{x}{a} \right) \frac{\sqrt{a^2 - x^2}}{a} + C = \frac{a^2}{2} \arcsin \left(\frac{x}{a} \right) + \frac{x}{2} \sqrt{a^2 - x^2} + C$$

Thus, the desired area is

$$A = 2r \sqrt{R^2 - r^2} + 2 \left(\frac{1}{4}\pi r^2 \right) - \left[R^2 \arcsin(x/R) + x \sqrt{R^2 - x^2} \right]_0^r$$

$$= 2r \sqrt{R^2 - r^2} + \frac{1}{2}\pi r^2 - \left[R^2 \arcsin(r/R) + r \sqrt{R^2 - r^2} \right] = r \sqrt{R^2 - r^2} + \frac{\pi}{2}r^2 - R^2 \arcsin(r/R)$$

42. Note that the circular cross-sections of the tank are the same everywhere, so the

percentage of the total capacity that is being used is equal to the percentage of any

cross-section that is under water. The underwater area is

$A = 2 \int_{-5}^{2} \sqrt{25 - y^2} \, dy$

$\qquad = \left[25 \arcsin(y/5) + y \sqrt{25 - y^2} \right]_{-5}^{2}$ [substitute $y = 5 \sin \theta$]

$\qquad = 25 \arcsin \frac{2}{5} + 2 \sqrt{21} + \frac{25}{2}\pi \approx 58.72 \text{ ft}^2$

so the fraction of the total capacity in use is $\dfrac{A}{\pi (5)^2} \approx \dfrac{58.72}{25\pi} \approx 0.748$ or 74.8%.

43. We use cylindrical shells and assume that $R > r$. $x^2 = r^2 - (y - R)^2 \Rightarrow x = \pm\sqrt{r^2 - (y - R)^2}$,

so $g(y) = 2\sqrt{r^2 - (y - R)^2}$ and

$$V = \int_{R-r}^{R+r} 2\pi y \cdot 2\sqrt{r^2 - (y - R)^2}\, dy = \int_{-r}^{r} 4\pi(u + R)\sqrt{r^2 - u^2}\, du \qquad [\text{where } u = y - R]$$

$$= 4\pi \int_{-r}^{r} u\sqrt{r^2 - u^2}\, du + 4\pi R \int_{-r}^{r} \sqrt{r^2 - u^2}\, du \qquad \begin{bmatrix} \text{where } u = r\sin\theta,\, du = r\cos\theta\, d\theta \\ \text{in the second integral} \end{bmatrix}$$

$$= 4\pi\left[-\tfrac{1}{3}(r^2 - u^2)^{3/2}\right]_{-r}^{r} + 4\pi R \int_{-\pi/2}^{\pi/2} r^2\cos^2\theta\, d\theta = -\tfrac{4\pi}{3}(0 - 0) + 4\pi R r^2 \int_{-\pi/2}^{\pi/2} \cos^2\theta\, d\theta$$

$$= 2\pi R r^2 \int_{-\pi/2}^{\pi/2}(1 + \cos 2\theta)\, d\theta = 2\pi R r^2 \left[\theta + \tfrac{1}{2}\sin 2\theta\right]_{-\pi/2}^{\pi/2} = 2\pi^2 R r^2$$

Another method: Use washers instead of shells, so $V = 8\pi R \int_0^r \sqrt{r^2 - y^2}\, dy$ as in Exercise 6.2.63(a), but evaluate the integral using $y = r\sin\theta$.

8.4 Integration of Rational Functions by Partial Fractions

1. (a) $\dfrac{2x}{(x + 3)(3x + 1)} = \dfrac{A}{x + 3} + \dfrac{B}{3x + 1}$

(b) $\dfrac{1}{x^3 + 2x^2 + x} = \dfrac{1}{x(x^2 + 2x + 1)} = \dfrac{1}{x(x + 1)^2} = \dfrac{A}{x} + \dfrac{B}{x + 1} + \dfrac{C}{(x + 1)^2}$

2. (a) $\dfrac{x}{x^2 + x - 2} = \dfrac{x}{(x + 2)(x - 1)} = \dfrac{A}{x + 2} + \dfrac{B}{x - 1}$

(b) $\dfrac{x^2}{x^2 + x + 2} = \dfrac{(x^2 + x + 2) - (x + 2)}{x^2 + x + 2} = 1 - \dfrac{x + 2}{x^2 + x + 2}$

Notice that $x^2 + x + 2$ can't be factored because its discriminant is $b^2 - 4ac = -7 < 0$.

3. (a) $\dfrac{x^4 + 1}{x^5 + 4x^3} = \dfrac{x^4 + 1}{x^3(x^2 + 4)} = \dfrac{A}{x} + \dfrac{B}{x^2} + \dfrac{C}{x^3} + \dfrac{Dx + E}{x^2 + 4}$

(b) $\dfrac{1}{(x^2 - 9)^2} = \dfrac{1}{[(x + 3)(x - 3)]^2} = \dfrac{1}{(x + 3)^2(x - 3)^2} = \dfrac{A}{x + 3} + \dfrac{B}{(x + 3)^2} + \dfrac{C}{x - 3} + \dfrac{D}{(x - 3)^2}$

4. (a) $\dfrac{x^3}{x^2 + 4x + 3} = x - 4 + \dfrac{13x + 12}{x^2 + 4x + 3} = x - 4 + \dfrac{13x + 12}{(x + 1)(x + 3)} = x - 4 + \dfrac{A}{x + 1} + \dfrac{B}{x + 3}$

(b) $\dfrac{2x + 1}{(x + 1)^3(x^2 + 4)^2} = \dfrac{A}{x + 1} + \dfrac{B}{(x + 1)^2} + \dfrac{C}{(x + 1)^3} + \dfrac{Dx + E}{x^2 + 4} + \dfrac{Fx + G}{(x^2 + 4)^2}$

5. (a) $\dfrac{x^4}{x^4 - 1} = \dfrac{(x^4 - 1) + 1}{x^4 - 1} = 1 + \dfrac{1}{x^4 - 1}$ [or use long division] $= 1 + \dfrac{1}{(x^2 - 1)(x^2 + 1)}$

$$= 1 + \dfrac{1}{(x - 1)(x + 1)(x^2 + 1)} = 1 + \dfrac{A}{x - 1} + \dfrac{B}{x + 1} + \dfrac{Cx + D}{x^2 + 1}$$

(b) $\dfrac{t^4 + t^2 + 1}{(t^2 + 1)(t^2 + 4)^2} = \dfrac{At + B}{t^2 + 1} + \dfrac{Ct + D}{t^2 + 4} + \dfrac{Et + F}{(t^2 + 4)^2}$

6. (a) $\dfrac{x^4}{(x^3+x)(x^2-x+3)} = \dfrac{x^4}{x(x^2+1)(x^2-x+3)} = \dfrac{x^3}{(x^2+1)(x^2-x+3)} = \dfrac{Ax+B}{x^2+1} + \dfrac{Cx+D}{x^2-x+3}$

(b) $\dfrac{1}{x^6-x^3} = \dfrac{1}{x^3(x^3-1)} = \dfrac{1}{x^3(x-1)(x^2+x+1)} = \dfrac{A}{x} + \dfrac{B}{x^2} + \dfrac{C}{x^3} + \dfrac{D}{x-1} + \dfrac{Ex+F}{x^2+x+1}$

7. $\displaystyle\int \frac{x}{x-6}\,dx = \int \frac{(x-6)+6}{x-6}\,dx = \int \left(1 + \frac{6}{x-6}\right)dx = x + 6\ln|x-6| + C$

8. $\displaystyle\int \frac{r^2}{r+4}\,dr = \int \left(\frac{r^2-16}{r+4} + \frac{16}{r+4}\right)dr = \int \left(r-4+\frac{16}{r+4}\right)dr$ [or use long division]

$\qquad = \frac{1}{2}r^2 - 4r + 16\ln|r+4| + C$

9. $\dfrac{x-9}{(x+5)(x-2)} = \dfrac{A}{x+5} + \dfrac{B}{x-2}$. Multiply both sides by $(x+5)(x-2)$ to get $x - 9 = A(x-2) + B(x+5)(*)$, or

equivalently, $x - 9 = (A+B)x - 2A + 5B$. Equating coefficients of x on each side of the equation gives us $1 = A + B$ **(1)**

and equating constants gives us $-9 = -2A + 5B$ **(2)**. Adding two times **(1)** to **(2)** gives us $-7 = 7B$ ⇔ $B = -1$ and

hence, $A = 2$. [Alternatively, to find the coefficients A and B, we may use substitution as follows: substitute 2 for x in $(*)$ to

get $-7 = 7B$ ⇔ $B = -1$, then substitute -5 for x in $(*)$ to get $-14 = -7A$ ⇔ $A = 2$.] Thus,

$\displaystyle\int \frac{x-9}{(x+5)(x-2)}\,dx = \int \left(\frac{2}{x+5} + \frac{-1}{x-2}\right)dx = 2\ln|x+5| - \ln|x-2| + C.$

10. $\dfrac{1}{(t+4)(t-1)} = \dfrac{A}{t+4} + \dfrac{B}{t-1}$ ⇒ $1 = A(t-1) + B(t+4)$.

$t = 1$ ⇒ $1 = 5B$ ⇒ $B = \frac{1}{5}$. $t = -4$ ⇒ $1 = -5A$ ⇒ $A = -\frac{1}{5}$. Thus,

$\displaystyle\int \frac{1}{(t+4)(t-1)}\,dt = \int \left(\frac{-1/5}{t+4} + \frac{1/5}{t-1}\right)dt = -\frac{1}{5}\ln|t+4| + \frac{1}{5}\ln|t-1| + C$ or $\dfrac{1}{5}\ln\left|\dfrac{t-1}{t+4}\right| + C$

11. $\dfrac{1}{x^2-1} = \dfrac{1}{(x+1)(x-1)} = \dfrac{A}{x+1} + \dfrac{B}{x-1}$. Multiply both sides by $(x+1)(x-1)$ to get $1 = A(x-1) + B(x+1)$.

Substituting 1 for x gives $1 = 2B$ ⇔ $B = \frac{1}{2}$. Substituting -1 for x gives $1 = -2A$ ⇔ $A = -\frac{1}{2}$. Thus,

$\displaystyle\int_2^3 \frac{1}{x^2-1}\,dx = \int_2^3 \left(\frac{-1/2}{x+1} + \frac{1/2}{x-1}\right)dx = \left[-\frac{1}{2}\ln|x+1| + \frac{1}{2}\ln|x-1|\right]_2^3$

$\qquad = \left(-\frac{1}{2}\ln 4 + \frac{1}{2}\ln 2\right) - \left(-\frac{1}{2}\ln 3 + \frac{1}{2}\ln 1\right) = \frac{1}{2}(\ln 2 + \ln 3 - \ln 4)$ [or $\frac{1}{2}\ln\frac{3}{2}$]

12. $\dfrac{x-1}{x^2+3x+2} = \dfrac{A}{x+1} + \dfrac{B}{x+2}$. Multiply both sides by $(x+1)(x+2)$ to get $x - 1 = A(x+2) + B(x+1)$. Substituting

-2 for x gives $-3 = -B$ ⇔ $B = 3$. Substituting -1 for x gives $-2 = A$. Thus,

$\displaystyle\int_0^1 \frac{x-1}{x^2+3x+2}\,dx = \int_0^1 \left(\frac{-2}{x+1} + \frac{3}{x+2}\right)dx = \left[-2\ln|x+1| + 3\ln|x+2|\right]_0^1$

$\qquad = (-2\ln 2 + 3\ln 3) - (-2\ln 1 + 3\ln 2) = 3\ln 3 - 5\ln 2$ [or $\ln\frac{27}{32}$]

13. $\displaystyle\int \frac{ax}{x^2-bx}\,dx = \int \frac{ax}{x(x-b)}\,dx = \int \frac{a}{x-b}\,dx = a\ln|x-b| + C$

14. If $a \neq b$, $\dfrac{1}{(x+a)(x+b)} = \dfrac{1}{b-a}\left(\dfrac{1}{x+a} - \dfrac{1}{x+b}\right)$, so if $a \neq b$, then

$$\int \frac{dx}{(x+a)(x+b)} = \frac{1}{b-a}\left(\ln|x+a| - \ln|x+b|\right) + C = \frac{1}{b-a}\ln\left|\frac{x+a}{x+b}\right| + C$$

If $a = b$, then $\displaystyle\int \frac{dx}{(x+a)^2} = -\frac{1}{x+a} + C$.

15. $\dfrac{x^3 - 2x^2 - 4}{x^3 - 2x^2} = 1 + \dfrac{-4}{x^2(x-2)}$. Write $\dfrac{-4}{x^2(x-2)} = \dfrac{A}{x} + \dfrac{B}{x^2} + \dfrac{C}{x-2}$. Multiplying both sides by $x^2(x-2)$ gives

$-4 = Ax(x-2) + B(x-2) + Cx^2$. Substituting 0 for x gives $-4 = -2B \iff B = 2$. Substituting 2 for x gives

$-4 = 4C \iff C = -1$. Equating coefficients of x^2, we get $0 = A + C$, so $A = 1$. Thus,

$$\int_3^4 \frac{x^3 - 2x^2 - 4}{x^3 - 2x^2}\, dx = \int_3^4 \left(1 + \frac{1}{x} + \frac{2}{x^2} - \frac{1}{x-2}\right) dx = \left[x + \ln|x| - \frac{2}{x} - \ln|x-2|\right]_3^4$$

$$= \left[\left(4 + \ln 4 - \tfrac{1}{2} - \ln 2\right) - \left(3 + \ln 3 - \tfrac{2}{3} - 0\right)\right] = \tfrac{7}{6} + \ln \tfrac{2}{3}$$

16. $\dfrac{x^3 - 4x - 10}{x^2 - x - 6} = x + 1 + \dfrac{3x - 4}{(x-3)(x+2)}$. Write $\dfrac{3x-4}{(x-3)(x+2)} = \dfrac{A}{x-3} + \dfrac{B}{x+2}$. Then

$3x - 4 = A(x+2) + B(x-3)$. Taking $x = 3$ and $x = -2$, we get $5 = 5A \iff A = 1$ and $-10 = -5B \iff B = 2$,

so

$$\int_0^1 \frac{x^3 - 4x - 10}{x^2 - x - 6}\, dx = \int_0^1 \left(x + 1 + \frac{1}{x-3} + \frac{2}{x+2}\right) dx = \left[\frac{1}{2}x^2 + x + \ln|x-3| + 2\ln(x+2)\right]_0^1$$

$$= \left(\tfrac{1}{2} + 1 + \ln 2 + 2\ln 3\right) - (0 + 0 + \ln 3 + 2\ln 2) = \tfrac{3}{2} + \ln 3 - \ln 2 = \tfrac{3}{2} + \ln \tfrac{3}{2}$$

17. $\dfrac{4y^2 - 7y - 12}{y(y+2)(y-3)} = \dfrac{A}{y} + \dfrac{B}{y+2} + \dfrac{C}{y-3} \Rightarrow 4y^2 - 7y - 12 = A(y+2)(y-3) + By(y-3) + Cy(y+2)$. Setting

$y = 0$ gives $-12 = -6A$, so $A = 2$. Setting $y = -2$ gives $18 = 10B$, so $B = \tfrac{9}{5}$. Setting $y = 3$ gives $3 = 15C$, so $C = \tfrac{1}{5}$.

Now

$$\int_1^2 \frac{4y^2 - 7y - 12}{y(y+2)(y-3)}\, dy = \int_1^2 \left(\frac{2}{y} + \frac{9/5}{y+2} + \frac{1/5}{y-3}\right) dy = \left[2\ln|y| + \tfrac{9}{5}\ln|y+2| + \tfrac{1}{5}\ln|y-3|\right]_1^2$$

$$= 2\ln 2 + \tfrac{9}{5}\ln 4 + \tfrac{1}{5}\ln 1 - 2\ln 1 - \tfrac{9}{5}\ln 3 - \tfrac{1}{5}\ln 2$$

$$= 2\ln 2 + \tfrac{18}{5}\ln 2 - \tfrac{1}{5}\ln 2 - \tfrac{9}{5}\ln 3 = \tfrac{27}{5}\ln 2 - \tfrac{9}{5}\ln 3 = \tfrac{9}{5}(3\ln 2 - \ln 3) = \tfrac{9}{5}\ln \tfrac{8}{3}$$

18. $\dfrac{x^2 + 2x - 1}{x^3 - x} = \dfrac{x^2 + 2x - 1}{x(x+1)(x-1)} = \dfrac{A}{x} + \dfrac{B}{x+1} + \dfrac{C}{x-1}$. Multiply both sides by $x(x+1)(x-1)$ to get

$x^2 + 2x - 1 = A(x+1)(x-1) + Bx(x-1) + Cx(x+1)$. Substituting 0 for x gives $-1 = -A \iff A = 1$.

Substituting -1 for x gives $-2 = 2B \iff B = -1$. Substituting 1 for x gives $2 = 2C \iff C = 1$. Thus,

$$\int \frac{x^2 + 2x - 1}{x^3 - x}\, dx = \int \left(\frac{1}{x} - \frac{1}{x+1} + \frac{1}{x-1}\right) dx = \ln|x| - \ln|x+1| + \ln|x-1| + C = \ln\left|\frac{x(x-1)}{x+1}\right| + C.$$

19. $\dfrac{1}{(x+5)^2(x-1)} = \dfrac{A}{x+5} + \dfrac{B}{(x+5)^2} + \dfrac{C}{x-1} \Rightarrow 1 = A(x+5)(x-1) + B(x-1) + C(x+5)^2.$

Setting $x = -5$ gives $1 = -6B$, so $B = -\frac{1}{6}$. Setting $x = 1$ gives $1 = 36C$, so $C = \frac{1}{36}$. Setting $x = -2$ gives

$1 = A(3)(-3) + B(-3) + C(3^2) = -9A - 3B + 9C = -9A + \frac{1}{2} + \frac{1}{4} = -9A + \frac{3}{4}$, so $9A = -\frac{1}{4}$ and $A = -\frac{1}{36}$. Now

$\displaystyle\int \frac{1}{(x+5)^2(x-1)}\,dx = \int \left[\frac{-1/36}{x+5} - \frac{1/6}{(x+5)^2} + \frac{1/36}{x-1}\right]dx = -\frac{1}{36}\ln|x+5| + \frac{1}{6(x+5)} + \frac{1}{36}\ln|x-1| + C.$

20. $\dfrac{x^2 - 5x + 16}{(2x+1)(x-2)^2} = \dfrac{A}{2x+1} + \dfrac{B}{x-2} + \dfrac{C}{(x-2)^2} \Rightarrow x^2 - 5x + 16 = A(x-2)^2 + B(x-2)(2x+1) + C(2x+1).$

Setting $x = 2$ gives $10 = 5C$, so $C = 2$. Setting $x = -\frac{1}{2}$ gives $\frac{75}{4} = \frac{25}{4}A$, so $A = 3$. Equating coefficients of x^2, we get

$1 = A + 2B$, so $-2 = 2B$ and $B = -1$. Thus,

$\displaystyle\int \frac{x^2 - 5x + 16}{(2x+1)(x-2)^2}\,dx = \int \left(\frac{3}{2x+1} - \frac{1}{x-2} + \frac{2}{(x-2)^2}\right)dx = \frac{3}{2}\ln|2x+1| - \ln|x-2| - \frac{2}{x-2} + C$

21.
$$x^2 + 4\ \overline{\smash{\big)}\ x^3\ +\ 0x^2\ +\ 0x\ +\ 4}$$
$$\underline{x^3\ +\ 4x}$$
$$-4x\ +\ 4$$

By long division, $\dfrac{x^3 + 4}{x^2 + 4} = x + \dfrac{-4x + 4}{x^2 + 4}$. Thus,

$\displaystyle\int \frac{x^3 + 4}{x^2 + 4}\,dx = \int \left(x + \frac{-4x+4}{x^2+4}\right)dx = \int \left(x - \frac{4x}{x^2+4} + \frac{4}{x^2 + 2^2}\right)dx$

$\displaystyle = \frac{1}{2}x^2 - 4 \cdot \frac{1}{2}\ln|x^2 + 4| + 4 \cdot \frac{1}{2}\tan^{-1}\left(\frac{x}{2}\right) + C = \frac{1}{2}x^2 - 2\ln(x^2 + 4) + 2\tan^{-1}\left(\frac{x}{2}\right) + C$

22. $\dfrac{1}{s^2(s-1)^2} = \dfrac{A}{s} + \dfrac{B}{s^2} + \dfrac{C}{s-1} + \dfrac{D}{(s-1)^2} \Rightarrow 1 = As(s-1)^2 + B(s-1)^2 + Cs^2(s-1) + Ds^2.$

Set $s = 0$, giving $B = 1$. Then set $s = 1$ to get $D = 1$. Equate the coefficients of s^3 to get

$0 = A + C$ or $A = -C$, and finally set $s = 2$ to get $1 = 2A + 1 - 4A + 4$ or $A = 2$. Now

$\displaystyle\int \frac{ds}{s^2(s-1)^2} = \int \left[\frac{2}{s} + \frac{1}{s^2} - \frac{2}{s-1} + \frac{1}{(s-1)^2}\right]ds = 2\ln|s| - \frac{1}{s} - 2\ln|s-1| - \frac{1}{s-1} + C.$

23. $\dfrac{5x^2 + 3x - 2}{x^3 + 2x^2} = \dfrac{5x^2 + 3x - 2}{x^2(x+2)} = \dfrac{A}{x} + \dfrac{B}{x^2} + \dfrac{C}{x+2}$. Multiply by $x^2(x+2)$ to

get $5x^2 + 3x - 2 = Ax(x+2) + B(x+2) + Cx^2$. Set $x = -2$ to get $C = 3$, and take

$x = 0$ to get $B = -1$. Equating the coefficients of x^2 gives $5 = A + C \Rightarrow A = 2$. So

$\displaystyle\int \frac{5x^2 + 3x - 2}{x^3 + 2x^2}\,dx = \int \left(\frac{2}{x} - \frac{1}{x^2} + \frac{3}{x+2}\right)dx = 2\ln|x| + \frac{1}{x} + 3\ln|x+2| + C.$

24. $\dfrac{x^2 - x + 6}{x^3 + 3x} = \dfrac{x^2 - x + 6}{x(x^2 + 3)} = \dfrac{A}{x} + \dfrac{Bx + C}{x^2 + 3}$. Multiply by $x(x^2 + 3)$ to get $x^2 - x + 6 = A(x^2 + 3) + (Bx + C)x$.

Substituting 0 for x gives $6 = 3A \iff A = 2$. The coefficients of the x^2-terms must be equal, so $1 = A + B \Rightarrow$

$B = 1 - 2 = -1$. The coefficients of the x-terms must be equal, so $-1 = C$. Thus,

$$\int \frac{x^2 - x + 6}{x^3 + 3x}\,dx = \int \left(\frac{2}{x} + \frac{-x - 1}{x^2 + 3}\right)dx = \int \left(\frac{2}{x} - \frac{x}{x^2 + 3} - \frac{1}{x^2 + 3}\right)dx$$

$$= 2\ln|x| - \frac{1}{2}\ln(x^2 + 3) - \frac{1}{\sqrt{3}}\tan^{-1}\frac{x}{\sqrt{3}} + C$$

25. $\dfrac{10}{(x - 1)(x^2 + 9)} = \dfrac{A}{x - 1} + \dfrac{Bx + C}{x^2 + 9}$. Multiply both sides by $(x - 1)(x^2 + 9)$ to get

$10 = A(x^2 + 9) + (Bx + C)(x - 1)$ $(\star)$. Substituting 1 for x gives $10 = 10A \iff A = 1$. Substituting 0 for x gives

$10 = 9A - C \Rightarrow C = 9(1) - 10 = -1$. The coefficients of the x^2-terms in $(\star)$ must be equal, so $0 = A + B \Rightarrow$

$B = -1$. Thus,

$$\int \frac{10}{(x - 1)(x^2 + 9)}\,dx = \int \left(\frac{1}{x - 1} + \frac{-x - 1}{x^2 + 9}\right)dx = \int \left(\frac{1}{x - 1} - \frac{x}{x^2 + 9} - \frac{1}{x^2 + 9}\right)dx$$

$$= \ln|x - 1| - \tfrac{1}{2}\ln(x^2 + 9) - \tfrac{1}{3}\tan^{-1}\left(\tfrac{x}{3}\right) + C$$

In the second term we used the substitution $u = x^2 + 9$ and in the last term we used Formula 10.

26. $\displaystyle\int \frac{x^2 + x + 1}{(x^2 + 1)^2}\,dx = \int \frac{x^2 + 1}{(x^2 + 1)^2}\,dx + \int \frac{x}{(x^2 + 1)^2}\,dx = \int \frac{1}{x^2 + 1}\,dx + \frac{1}{2}\int \frac{1}{u^2}\,du \quad [u = x^2 + 1,\, du = 2x\,dx]$

$$= \tan^{-1}x + \frac{1}{2}\left(-\frac{1}{u}\right) + C = \tan^{-1}x - \frac{1}{2(x^2 + 1)} + C$$

27. $\dfrac{x^3 + x^2 + 2x + 1}{(x^2 + 1)(x^2 + 2)} = \dfrac{Ax + B}{x^2 + 1} + \dfrac{Cx + D}{x^2 + 2}$. Multiply both sides by $(x^2 + 1)(x^2 + 2)$ to get

$x^3 + x^2 + 2x + 1 = (Ax + B)(x^2 + 2) + (Cx + D)(x^2 + 1) \iff$

$x^3 + x^2 + 2x + 1 = (Ax^3 + Bx^2 + 2Ax + 2B) + (Cx^3 + Dx^2 + Cx + D) \iff$

$x^3 + x^2 + 2x + 1 = (A + C)x^3 + (B + D)x^2 + (2A + C)x + (2B + D)$. Comparing coefficients gives us the following

system of equations:

$$A + C = 1 \quad \textbf{(1)} \qquad\qquad B + D = 1 \quad \textbf{(2)}$$
$$2A + C = 2 \quad \textbf{(3)} \qquad\qquad 2B + D = 1 \quad \textbf{(4)}$$

Subtracting equation **(1)** from equation **(3)** gives us $A = 1$, so $C = 0$. Subtracting equation **(2)** from equation **(4)** gives us

$B = 0$, so $D = 1$. Thus, $I = \displaystyle\int \frac{x^3 + x^2 + 2x + 1}{(x^2 + 1)(x^2 + 2)}\,dx = \int \left(\frac{x}{x^2 + 1} + \frac{1}{x^2 + 2}\right)dx$. For $\displaystyle\int \frac{x}{x^2 + 1}\,dx$, let $u = x^2 + 1$

so $du = 2x\,dx$ and then $\displaystyle\int \frac{x}{x^2 + 1}\,dx = \frac{1}{2}\int \frac{1}{u}\,du = \frac{1}{2}\ln|u| + C = \frac{1}{2}\ln(x^2 + 1) + C$. For $\displaystyle\int \frac{1}{x^2 + 2}\,dx$, use

Formula 10 with $a = \sqrt{2}$. So $\displaystyle\int \frac{1}{x^2 + 2}\,dx = \int \frac{1}{x^2 + \left(\sqrt{2}\right)^2}\,dx = \frac{1}{\sqrt{2}}\tan^{-1}\frac{x}{\sqrt{2}} + C.$

Thus, $I = \dfrac{1}{2}\ln\left(x^2 + 1\right) + \dfrac{1}{\sqrt{2}}\tan^{-1}\dfrac{x}{\sqrt{2}} + C.$

28. $\dfrac{x^2 - 2x - 1}{(x-1)^2(x^2+1)} = \dfrac{A}{x-1} + \dfrac{B}{(x-1)^2} + \dfrac{Cx+D}{x^2+1} \quad \Rightarrow$

$x^2 - 2x - 1 = A(x-1)\left(x^2+1\right) + B\left(x^2+1\right) + (Cx+D)(x-1)^2$. Setting $x = 1$ gives $B = -1$. Equating the

coefficients of x^3 gives $A = -C$. Equating the constant terms gives $-1 = -A - 1 + D$, so $D = A$,

and setting $x = 2$ gives $-1 = 5A - 5 - 2A + A$ or $A = 1$. We have

$\displaystyle\int \frac{x^2 - 2x - 1}{(x-1)^2(x^2+1)}\,dx = \int \left[\frac{1}{x-1} - \frac{1}{(x-1)^2} - \frac{x-1}{x^2+1}\right]dx = \ln|x-1| + \frac{1}{x-1} - \frac{1}{2}\ln\left(x^2+1\right) + \tan^{-1}x + C.$

29. $\displaystyle\int \frac{x+4}{x^2+2x+5}\,dx = \int \frac{x+1}{x^2+2x+5}\,dx + \int \frac{3}{x^2+2x+5}\,dx = \frac{1}{2}\int \frac{(2x+2)\,dx}{x^2+2x+5} + \int \frac{3\,dx}{(x+1)^2+4}$

$\displaystyle\qquad = \frac{1}{2}\ln\left|x^2+2x+5\right| + 3\int \frac{2\,du}{4(u^2+1)} \qquad \begin{bmatrix}\text{where } x+1 = 2u, \\ \text{and } dx = 2\,du\end{bmatrix}$

$\displaystyle\qquad = \frac{1}{2}\ln(x^2+2x+5) + \frac{3}{2}\tan^{-1}u + C = \frac{1}{2}\ln(x^2+2x+5) + \frac{3}{2}\tan^{-1}\left(\frac{x+1}{2}\right) + C$

30. $\dfrac{3x^2 + x + 4}{x^4 + 3x^2 + 2} = \dfrac{3x^2 + x + 4}{(x^2+1)(x^2+2)} = \dfrac{Ax+B}{x^2+1} + \dfrac{Cx+D}{x^2+2}$. Multiply both sides by $(x^2+1)(x^2+2)$ to get

$3x^2 + x + 4 = (Ax+B)(x^2+2) + (Cx+D)(x^2+1) \quad \Leftrightarrow$

$3x^2 + x + 4 = (Ax^3 + Bx^2 + 2Ax + 2B) + (Cx^3 + Dx^2 + Cx + D) \quad \Leftrightarrow$

$3x^2 + x + 4 = (A+C)x^3 + (B+D)x^2 + (2A+C)x + (2B+D)$. Comparing coefficients gives us the following system

of equations:

$$A + C = 0 \quad \textbf{(1)} \qquad\qquad B + D = 3 \quad \textbf{(2)}$$
$$2A + C = 1 \quad \textbf{(3)} \qquad\qquad 2B + D = 4 \quad \textbf{(4)}$$

Subtracting equation **(1)** from equation **(3)** gives us $A = 1$, so $C = -1$. Subtracting equation **(2)** from equation **(4)** gives us

$B = 1$, so $D = 2$. Thus,

$\displaystyle I = \int \frac{3x^2 + x + 4}{x^4 + 3x^2 + 2}\,dx = \int \frac{x+1}{x^2+1}\,dx + \int \frac{-x+2}{x^2+2}\,dx$

$\displaystyle\qquad = \frac{1}{2}\int \frac{2x}{x^2+1}\,dx + \int \frac{1}{x^2+1}\,dx - \frac{1}{2}\int \frac{2x}{x^2+2}\,dx + 2\int \frac{1}{x^2+\left(\sqrt{2}\right)^2}\,dx$

$\displaystyle\qquad = \tfrac{1}{2}\ln\left|x^2+1\right| + \tan^{-1}x - \tfrac{1}{2}\ln\left|x^2+2\right| + 2\cdot\frac{1}{\sqrt{2}}\tan^{-1}\left(\frac{x}{\sqrt{2}}\right) + C$

$\displaystyle\qquad = \tfrac{1}{2}\ln(x^2+1) - \tfrac{1}{2}\ln(x^2+2) + \tan^{-1}x + \sqrt{2}\tan^{-1}\left(x/\sqrt{2}\right) + C$

31. $\dfrac{1}{x^3 - 1} = \dfrac{1}{(x-1)(x^2+x+1)} = \dfrac{A}{x-1} + \dfrac{Bx+C}{x^2+x+1} \quad \Rightarrow \quad 1 = A(x^2+x+1) + (Bx+C)(x-1).$

Take $x = 1$ to get $A = \frac{1}{3}$. Equating coefficients of x^2 and then comparing the constant terms, we get $0 = \frac{1}{3} + B$, $1 = \frac{1}{3} - C$,

so $B = -\frac{1}{3}$, $C = -\frac{2}{3} \quad \Rightarrow$

$$\int \frac{1}{x^3-1}\,dx = \int \frac{\frac{1}{3}}{x-1}\,dx + \int \frac{-\frac{1}{3}x - \frac{2}{3}}{x^2+x+1}\,dx = \frac{1}{3}\ln|x-1| - \frac{1}{3}\int \frac{x+2}{x^2+x+1}\,dx$$

$$= \frac{1}{3}\ln|x-1| - \frac{1}{3}\int \frac{x+1/2}{x^2+x+1}\,dx - \frac{1}{3}\int \frac{(3/2)\,dx}{(x+1/2)^2 + 3/4}$$

$$= \frac{1}{3}\ln|x-1| - \frac{1}{6}\ln(x^2+x+1) - \frac{1}{2}\left(\frac{2}{\sqrt{3}}\right)\tan^{-1}\left(\frac{x+\frac{1}{2}}{\sqrt{3}/2}\right) + K$$

$$= \frac{1}{3}\ln|x-1| - \frac{1}{6}\ln(x^2+x+1) - \frac{1}{\sqrt{3}}\tan^{-1}\left(\frac{1}{\sqrt{3}}(2x+1)\right) + K$$

32. $\displaystyle\int_0^1 \frac{x}{x^2+4x+13}\,dx = \int_0^1 \frac{\frac{1}{2}(2x+4)}{x^2+4x+13}\,dx - 2\int_0^1 \frac{dx}{(x+2)^2+9}$

$$= \frac{1}{2}\int_{13}^{18} \frac{dy}{y} - 2\int_{2/3}^1 \frac{3\,du}{9u^2+9} \qquad \begin{bmatrix} \text{where } y = x^2+4x+13,\ dy = (2x+4)\,dx, \\ x+2 = 3u, \text{ and } dx = 3\,du \end{bmatrix}$$

$$= \frac{1}{2}\left[\ln y\right]_{13}^{18} - \frac{2}{3}\left[\tan^{-1}u\right]_{2/3}^1 = \frac{1}{2}\ln\frac{18}{13} - \frac{2}{3}\left(\frac{\pi}{4} - \tan^{-1}\left(\frac{2}{3}\right)\right)$$

$$= \frac{1}{2}\ln\frac{18}{13} - \frac{\pi}{6} + \frac{2}{3}\tan^{-1}\left(\frac{2}{3}\right)$$

33. Let $u = x^4 + 4x^2 + 3$, so that $du = (4x^3 + 8x)\,dx = 4(x^3+2x)\,dx$, $x = 0 \Rightarrow u = 3$, and $x = 1 \Rightarrow u = 8$.

Then $\displaystyle\int_0^1 \frac{x^3+2x}{x^4+4x^2+3}\,dx = \int_3^8 \frac{1}{u}\left(\frac{1}{4}\,du\right) = \frac{1}{4}\left[\ln|u|\right]_3^8 = \frac{1}{4}(\ln 8 - \ln 3) = \frac{1}{4}\ln\frac{8}{3}.$

34. $\dfrac{x^3}{x^3+1} = \dfrac{(x^3+1)-1}{x^3+1} = 1 - \dfrac{1}{x^3+1} = 1 - \left(\dfrac{A}{x+1} + \dfrac{Bx+C}{x^2-x+1}\right) \quad \Rightarrow \quad 1 = A(x^2-x+1) + (Bx+C)(x+1).$

Equate the terms of degree 2, 1 and 0 to get $0 = A + B$, $0 = -A + B + C$, $1 = A + C$. Solve the three equations to get

$A = \frac{1}{3}$, $B = -\frac{1}{3}$, and $C = \frac{2}{3}$. So

$$\int \frac{x^3}{x^3+1}\,dx = \int\left[1 - \frac{\frac{1}{3}}{x+1} + \frac{\frac{1}{3}x - \frac{2}{3}}{x^2-x+1}\right]dx = x - \frac{1}{3}\ln|x+1| + \frac{1}{6}\int \frac{2x-1}{x^2-x+1}\,dx - \frac{1}{2}\int \frac{dx}{\left(x-\frac{1}{2}\right)^2 + \frac{3}{4}}$$

$$= x - \frac{1}{3}\ln|x+1| + \frac{1}{6}\ln(x^2-x+1) - \frac{1}{\sqrt{3}}\tan^{-1}\left(\frac{1}{\sqrt{3}}(2x-1)\right) + K$$

35. $\dfrac{1}{x(x^2+4)^2} = \dfrac{A}{x} + \dfrac{Bx+C}{x^2+4} + \dfrac{Dx+E}{(x^2+4)^2} \quad \Rightarrow \quad 1 = A(x^2+4)^2 + (Bx+C)x(x^2+4) + (Dx+E)x.$ Setting $x = 0$

gives $1 = 16A$, so $A = \frac{1}{16}$. Now compare coefficients.

$$1 = \tfrac{1}{16}(x^4+8x^2+16) + (Bx^2+Cx)(x^2+4) + Dx^2 + Ex$$

$$1 = \tfrac{1}{16}x^4 + \tfrac{1}{2}x^2 + 1 + Bx^4 + Cx^3 + 4Bx^2 + 4Cx + Dx^2 + Ex$$

$$1 = \left(\tfrac{1}{16} + B\right)x^4 + Cx^3 + \left(\tfrac{1}{2} + 4B + D\right)x^2 + (4C + E)x + 1$$

So $B + \frac{1}{16} = 0 \ \Rightarrow \ B = -\frac{1}{16}, C = 0, \frac{1}{2} + 4B + D = 0 \ \Rightarrow \ D = -\frac{1}{4}$, and $4C + E = 0 \ \Rightarrow \ E = 0$. Thus,

$$\int \frac{dx}{x(x^2+4)^2} = \int \left(\frac{\frac{1}{16}}{x} + \frac{-\frac{1}{16}x}{x^2+4} + \frac{-\frac{1}{4}x}{(x^2+4)^2} \right) dx = \frac{1}{16} \ln|x| - \frac{1}{16} \cdot \frac{1}{2} \ln|x^2+4| - \frac{1}{4} \left(-\frac{1}{2} \right) \frac{1}{x^2+4} + C$$

$$= \frac{1}{16} \ln|x| - \frac{1}{32} \ln(x^2+4) + \frac{1}{8(x^2+4)} + C$$

36. Let $u = x^5 + 5x^3 + 5x$, so that $du = (5x^4 + 15x^2 + 5)dx = 5(x^4 + 3x^2 + 1)dx$. Then

$$\int \frac{x^4+3x^2+1}{x^5+5x^3+5x} dx = \int \frac{1}{u} \left(\frac{1}{5} du \right) = \frac{1}{5} \ln|u| + C = \frac{1}{5} \ln|x^5 + 5x^3 + 5x| + C$$

37. $\dfrac{x^2-3x+7}{(x^2-4x+6)^2} = \dfrac{Ax+B}{x^2-4x+6} + \dfrac{Cx+D}{(x^2-4x+6)^2} \ \Rightarrow \ x^2-3x+7 = (Ax+B)(x^2-4x+6) + Cx + D \ \Rightarrow$

$x^2 - 3x + 7 = Ax^3 + (-4A+B)x^2 + (6A-4B+C)x + (6B+D)$. So $A = 0, -4A+B = 1 \ \Rightarrow \ B = 1$,

$6A - 4B + C = -3 \ \Rightarrow \ C = 1, 6B + D = 7 \ \Rightarrow \ D = 1$. Thus,

$$I = \int \frac{x^2-3x+7}{(x^2-4x+6)^2} dx = \int \left(\frac{1}{x^2-4x+6} + \frac{x+1}{(x^2-4x+6)^2} \right) dx$$

$$= \int \frac{1}{(x-2)^2+2} dx + \int \frac{x-2}{(x^2-4x+6)^2} dx + \int \frac{3}{(x^2-4x+6)^2} dx$$

$$= I_1 + I_2 + I_3.$$

$$I_1 = \int \frac{1}{(x-2)^2 + \left(\sqrt{2}\right)^2} dx = \frac{1}{\sqrt{2}} \tan^{-1}\left(\frac{x-2}{\sqrt{2}} \right) + C_1$$

$$I_2 = \frac{1}{2} \int \frac{2x-4}{(x^2-4x+6)^2} dx = \frac{1}{2} \int \frac{1}{u^2} du = \frac{1}{2} \left(-\frac{1}{u} \right) + C_2 = -\frac{1}{2(x^2-4x+6)} + C_2$$

$$I_3 = 3 \int \frac{1}{\left[(x-2)^2 + \left(\sqrt{2}\right)^2\right]^2} dx = 3 \int \frac{1}{[2(\tan^2\theta+1)]^2} \sqrt{2} \sec^2\theta \, d\theta \quad \begin{bmatrix} x-2 = \sqrt{2}\tan\theta, \\ dx = \sqrt{2}\sec^2\theta \, d\theta \end{bmatrix}$$

$$= \frac{3\sqrt{2}}{4} \int \frac{\sec^2\theta}{\sec^4\theta} d\theta = \frac{3\sqrt{2}}{4} \int \cos^2\theta \, d\theta = \frac{3\sqrt{2}}{4} \int \frac{1}{2}(1+\cos 2\theta)\, d\theta$$

$$= \frac{3\sqrt{2}}{8} \left(\theta + \frac{1}{2}\sin 2\theta\right) + C_3 = \frac{3\sqrt{2}}{8} \tan^{-1}\left(\frac{x-2}{\sqrt{2}}\right) + \frac{3\sqrt{2}}{8}\left(\frac{1}{2} \cdot 2\sin\theta \, \cos\theta\right) + C_3$$

$$= \frac{3\sqrt{2}}{8} \tan^{-1}\left(\frac{x-2}{\sqrt{2}}\right) + \frac{3\sqrt{2}}{8} \cdot \frac{x-2}{\sqrt{x^2-4x+6}} \cdot \frac{\sqrt{2}}{\sqrt{x^2-4x+6}} + C_3$$

$$= \frac{3\sqrt{2}}{8} \tan^{-1}\left(\frac{x-2}{\sqrt{2}}\right) + \frac{3(x-2)}{4(x^2-4x+6)} + C_3$$

So $I = I_1 + I_2 + I_3 \quad [C = C_1 + C_2 + C_3]$

$$= \frac{1}{\sqrt{2}} \tan^{-1}\left(\frac{x-2}{\sqrt{2}}\right) + \frac{-1}{2(x^2-4x+6)} + \frac{3\sqrt{2}}{8} \tan^{-1}\left(\frac{x-2}{\sqrt{2}}\right) + \frac{3(x-2)}{4(x^2-4x+6)} + C$$

$$= \left(\frac{4\sqrt{2}}{8} + \frac{3\sqrt{2}}{8}\right) \tan^{-1}\left(\frac{x-2}{\sqrt{2}}\right) + \frac{3(x-2)-2}{4(x^2-4x+6)} + C = \frac{7\sqrt{2}}{8} \tan^{-1}\left(\frac{x-2}{\sqrt{2}}\right) + \frac{3x-8}{4(x^2-4x+6)} + C$$

38. $\dfrac{x^3 + 2x^2 + 3x - 2}{(x^2 + 2x + 2)^2} = \dfrac{Ax + B}{x^2 + 2x + 2} + \dfrac{Cx + D}{(x^2 + 2x + 2)^2} \quad \Rightarrow$

$x^3 + 2x^2 + 3x - 2 = (Ax + B)(x^2 + 2x + 2) + Cx + D \quad \Rightarrow$

$x^3 + 2x^2 + 3x - 2 = Ax^3 + (2A + B)x^2 + (2A + 2B + C)x + 2B + D.$

So $A = 1, 2A + B = 2 \quad \Rightarrow \quad B = 0, 2A + 2B + C = 3 \quad \Rightarrow \quad C = 1,$ and $2B + D = -2 \quad \Rightarrow \quad D = -2.$ Thus,

$I = \displaystyle\int \dfrac{x^3 + 2x^2 + 3x - 2}{(x^2 + 2x + 2)^2}\, dx = \int \left(\dfrac{x}{x^2 + 2x + 2} + \dfrac{x - 2}{(x^2 + 2x + 2)^2} \right) dx$

$= \displaystyle\int \dfrac{x + 1}{x^2 + 2x + 2}\, dx + \int \dfrac{-1}{x^2 + 2x + 2}\, dx + \int \dfrac{x + 1}{(x^2 + 2x + 2)^2}\, dx + \int \dfrac{-3}{(x^2 + 2x + 2)^2}\, dx$

$= I_1 + I_2 + I_3 + I_4.$

$I_1 = \displaystyle\int \dfrac{x + 1}{x^2 + 2x + 2}\, dx = \int \dfrac{1}{u} \left(\dfrac{1}{2}\, du \right) \quad \begin{bmatrix} u = x^2 + 2x + 2, \\ du = 2(x + 1)\, dx \end{bmatrix} = \dfrac{1}{2} \ln \left| x^2 + 2x + 2 \right| + C_1$

$I_2 = -\displaystyle\int \dfrac{1}{(x + 1)^2 + 1}\, dx = -\dfrac{1}{1} \tan^{-1} \left(\dfrac{x + 1}{1} \right) + C_2 = -\tan^{-1}(x + 1) + C_2$

$I_3 = \displaystyle\int \dfrac{x + 1}{(x^2 + 2x + 2)^2}\, dx = \int \dfrac{1}{u^2} \left(\dfrac{1}{2}\, du \right) = -\dfrac{1}{2u} + C_3 = -\dfrac{1}{2(x^2 + 2x + 2)} + C_3$

$I_4 = -3\displaystyle\int \dfrac{1}{[(x + 1)^2 + 1]^2}\, dx = -3\int \dfrac{1}{(\tan^2\theta + 1)^2} \sec^2\theta\, d\theta \quad \begin{bmatrix} x + 1 = 1\tan\theta, \\ dx = \sec^2\theta\, d\theta \end{bmatrix}$

$= -3\displaystyle\int \dfrac{1}{\sec^2\theta}\, d\theta = -3\int \cos^2\theta\, d\theta = -\dfrac{3}{2}\int (1 + \cos 2\theta)\, d\theta$

$= -\dfrac{3}{2}\left(\theta + \dfrac{1}{2}\sin 2\theta \right) + C_4 = -\dfrac{3}{2}\theta - \dfrac{3}{2}\left(\dfrac{1}{2} \cdot 2\sin\theta\, \cos\theta \right) + C_4$

$= -\dfrac{3}{2}\tan^{-1}\left(\dfrac{x + 1}{1} \right) - \dfrac{3}{2} \cdot \dfrac{x + 1}{\sqrt{x^2 + 2x + 2}} \cdot \dfrac{1}{\sqrt{x^2 + 2x + 2}} + C_4$

$= -\dfrac{3}{2}\tan^{-1}(x + 1) - \dfrac{3(x + 1)}{2(x^2 + 2x + 2)} + C_4$

So $\quad I = I_1 + I_2 + I_3 + I_4 \qquad [C = C_1 + C_2 + C_3 + C_4]$

$= \dfrac{1}{2}\ln(x^2 + 2x + 2) - \tan^{-1}(x + 1) - \dfrac{1}{2(x^2 + 2x + 2)} - \dfrac{3}{2}\tan^{-1}(x + 1) - \dfrac{3(x + 1)}{2(x^2 + 2x + 2)} + C$

$= \dfrac{1}{2}\ln(x^2 + 2x + 2) - \dfrac{5}{2}\tan^{-1}(x + 1) - \dfrac{3x + 4}{2(x^2 + 2x + 2)} + C$

39. Let $u = \sqrt{x + 1}$. Then $x = u^2 - 1, dx = 2u\, du \quad \Rightarrow$

$\displaystyle\int \dfrac{dx}{x\sqrt{x + 1}} = \int \dfrac{2u\, du}{(u^2 - 1)\, u} = 2\int \dfrac{du}{u^2 - 1} = \ln\left| \dfrac{u - 1}{u + 1} \right| + C = \ln\left| \dfrac{\sqrt{x + 1} - 1}{\sqrt{x + 1} + 1} \right| + C.$

40. Let $u = \sqrt{x + 3}$, so $u^2 = x + 3$ and $2u\,du = dx$. Then

$$\int \frac{dx}{2\sqrt{x+3} + x} = \int \frac{2u\,du}{2u + (u^2 - 3)} = \int \frac{2u}{u^2 + 2u - 3}\,du = \int \frac{2u}{(u+3)(u-1)}\,du. \text{ Now}$$

$$\frac{2u}{(u+3)(u-1)} = \frac{A}{u+3} + \frac{B}{u-1} \quad \Rightarrow \quad 2u = A(u-1) + B(u+3). \text{ Setting } u = 1 \text{ gives } 2 = 4B, \text{ so } B = \tfrac{1}{2}.$$

Setting $u = -3$ gives $-6 = -4A$, so $A = \tfrac{3}{2}$. Thus,

$$\int \frac{2u}{(u+3)(u-1)}\,du = \int \left(\frac{\frac{3}{2}}{u+3} + \frac{\frac{1}{2}}{u-1}\,du \right)$$

$$= \tfrac{3}{2} \ln|u+3| + \tfrac{1}{2} \ln|u-1| + C = \tfrac{3}{2} \ln\!\left(\sqrt{x+3} + 3 \right) + \tfrac{1}{2} \ln\left| \sqrt{x+3} - 1 \right| + C$$

41. Let $u = \sqrt{x}$, so $u^2 = x$ and $dx = 2u\,du$. Thus,

$$\int_9^{16} \frac{\sqrt{x}}{x - 4}\,dx = \int_3^4 \frac{u}{u^2 - 4}\,2u\,du = 2\int_3^4 \frac{u^2}{u^2 - 4}\,du = 2\int_3^4 \left(1 + \frac{4}{u^2 - 4}\right)du \qquad \text{[by long division]}$$

$$= 2 + 8\int_3^4 \frac{du}{(u+2)(u-2)} \quad (\star)$$

Multiply $\dfrac{1}{(u+2)(u-2)} = \dfrac{A}{u+2} + \dfrac{B}{u-2}$ by $(u+2)(u-2)$ to get $1 = A(u-2) + B(u+2)$. Equating coefficients we

get $A + B = 0$ and $-2A + 2B = 1$. Solving gives us $B = \tfrac{1}{4}$ and $A = -\tfrac{1}{4}$, so $\dfrac{1}{(u+2)(u-2)} = \dfrac{-1/4}{u+2} + \dfrac{1/4}{u-2}$ and $(\star)$ is

$$2 + 8\int_3^4 \left(\frac{-1/4}{u+2} + \frac{1/4}{u-2} \right)du = 2 + 8\left[-\tfrac{1}{4}\ln|u+2| + \tfrac{1}{4}\ln|u-2| \right]_3^4 = 2 + \left[2\ln|u-2| - 2\ln|u+2| \right]_3^4$$

$$= 2 + 2\left[\ln\left| \frac{u-2}{u+2} \right| \right]_3^4 = 2 + 2\left(\ln\tfrac{2}{6} - \ln\tfrac{1}{5} \right) = 2 + 2\ln\tfrac{2/6}{1/5}$$

$$= 2 + 2\ln\tfrac{5}{3} \ \text{ or } \ 2 + \ln\!\left(\tfrac{5}{3}\right)^2 = 2 + \ln\tfrac{25}{9}$$

42. Let $u = \sqrt[3]{x}$. Then $x = u^3$, $dx = 3u^2\,du \ \Rightarrow$

$$\int_0^1 \frac{1}{1 + \sqrt[3]{x}}\,dx = \int_0^1 \frac{3u^2\,du}{1 + u} = \int_0^1 \left(3u - 3 + \frac{3}{1+u} \right)du = \left[\tfrac{3}{2}u^2 - 3u + 3\ln(1+u) \right]_0^1 = 3\left(\ln 2 - \tfrac{1}{2} \right).$$

43. Let $u = \sqrt[3]{x^2 + 1}$. Then $x^2 = u^3 - 1$, $2x\,dx = 3u^2\,du \ \Rightarrow$

$$\int \frac{x^3\,dx}{\sqrt[3]{x^2 + 1}} = \int \frac{(u^3 - 1)\frac{3}{2}u^2\,du}{u} = \frac{3}{2}\int (u^4 - u)\,du$$

$$= \tfrac{3}{10}u^5 - \tfrac{3}{4}u^2 + C = \tfrac{3}{10}(x^2 + 1)^{5/3} - \tfrac{3}{4}(x^2 + 1)^{2/3} + C$$

44. Let $u = \sqrt{x}$. Then $x = u^2$, $dx = 2u\,du \ \Rightarrow$

$$\int_{1/3}^3 \frac{\sqrt{x}}{x^2 + x}\,dx = \int_{1/\sqrt{3}}^{\sqrt{3}} \frac{u \cdot 2u\,du}{u^4 + u^2} = 2\int_{1/\sqrt{3}}^{\sqrt{3}} \frac{du}{u^2 + 1} = 2\left[\tan^{-1} u \right]_{1/\sqrt{3}}^{\sqrt{3}} = 2\left(\tfrac{\pi}{3} - \tfrac{\pi}{6} \right) = \tfrac{\pi}{3}.$$

45. If we were to substitute $u = \sqrt{x}$, then the square root would disappear but a cube root would remain. On the other hand, the

substitution $u = \sqrt[3]{x}$ would eliminate the cube root but leave a square root. We can eliminate both roots by means of the

substitution $u = \sqrt[6]{x}$. (Note that 6 is the least common multiple of 2 and 3.)

Let $u = \sqrt[6]{x}$. Then $x = u^6$, so $dx = 6u^5\, du$ and $\sqrt{x} = u^3$, $\sqrt[3]{x} = u^2$. Thus,

$$\int \frac{dx}{\sqrt{x} - \sqrt[3]{x}} = \int \frac{6u^5\, du}{u^3 - u^2} = 6\int \frac{u^5}{u^2\,(u-1)}\, du = 6\int \frac{u^3}{u-1}\, du$$

$$= 6\int \left(u^2 + u + 1 + \frac{1}{u-1} \right) du \qquad \text{[by long division]}$$

$$= 6\left(\tfrac{1}{3}u^3 + \tfrac{1}{2}u^2 + u + \ln|u-1| \right) + C = 2\sqrt{x} + 3\sqrt[3]{x} + 6\sqrt[6]{x} + 6\ln\left|\sqrt[6]{x} - 1\right| + C$$

46. Let $u = \sqrt{1 + \sqrt{x}}$, so that $u^2 = 1 + \sqrt{x}$, $x = (u^2 - 1)^2$, and $dx = 2(u^2 - 1)\cdot 2u\, du = 4u(u^2 - 1)\, du$. Then

$$\int \frac{\sqrt{1 + \sqrt{x}}}{x}\, dx = \int \frac{u}{(u^2 - 1)^2} \cdot 4u(u^2 - 1)\, du = \int \frac{4u^2}{u^2 - 1}\, du = \int \left(4 + \frac{4}{u^2 - 1} \right) du. \text{ Now}$$

$$\frac{4}{u^2 - 1} = \frac{A}{u+1} + \frac{B}{u-1} \quad \Rightarrow \quad 4 = A(u-1) + B(u+1). \text{ Setting } u = 1 \text{ gives } 4 = 2B, \text{ so } B = 2. \text{ Setting } u = -1 \text{ gives}$$

$4 = -2A$, so $A = -2$. Thus,

$$\int \left(4 + \frac{4}{u^2 - 1} \right) du = \int \left(4 - \frac{2}{u+1} + \frac{2}{u-1} \right) du = 4u - 2\ln|u+1| + 2\ln|u-1| + C$$

$$= 4\sqrt{1 + \sqrt{x}} - 2\ln\left(\sqrt{1 + \sqrt{x}} + 1\right) + 2\ln\left(\sqrt{1 + \sqrt{x}} - 1\right) + C$$

47. Let $u = e^x$. Then $x = \ln u$, $dx = \dfrac{du}{u} \quad \Rightarrow$

$$\int \frac{e^{2x}\, dx}{e^{2x} + 3e^x + 2} = \int \frac{u^2\,(du/u)}{u^2 + 3u + 2} = \int \frac{u\, du}{(u+1)(u+2)} = \int \left[\frac{-1}{u+1} + \frac{2}{u+2} \right] du$$

$$= 2\ln|u+2| - \ln|u+1| + C = \ln\frac{(e^x + 2)^2}{e^x + 1} + C$$

48. Let $u = \sin x$. Then $du = \cos x\, dx \quad \Rightarrow$

$$\int \frac{\cos x\, dx}{\sin^2 x + \sin x} = \int \frac{du}{u^2 + u} = \int \frac{du}{u(u+1)} = \int \left[\frac{1}{u} - \frac{1}{u+1} \right] du = \ln\left| \frac{u}{u+1} \right| + C = \ln\left| \frac{\sin x}{1 + \sin x} \right| + C.$$

49. Let $u = \tan t$, so that $du = \sec^2 t\, dt$. Then $\displaystyle\int \frac{\sec^2 t}{\tan^2 t + 3\tan t + 2}\, dt = \int \frac{1}{u^2 + 3u + 2}\, du = \int \frac{1}{(u+1)(u+2)}\, du.$

Now $\dfrac{1}{(u+1)(u+2)} = \dfrac{A}{u+1} + \dfrac{B}{u+2} \quad \Rightarrow \quad 1 = A(u+2) + B(u+1).$

Setting $u = -2$ gives $1 = -B$, so $B = -1$. Setting $u = -1$ gives $1 = A$.

Thus, $\displaystyle\int \frac{1}{(u+1)(u+2)}\, du = \int \left(\frac{1}{u+1} - \frac{1}{u+2} \right) du = \ln|u+1| - \ln|u+2| + C = \ln|\tan t + 1| - \ln|\tan t + 2| + C.$

50. Let $u = e^x$, so that $du = e^x\, dx$. Then $\displaystyle\int \frac{e^x}{(e^x - 2)(e^{2x} + 1)}\, dx = \int \frac{1}{(u-2)(u^2+1)}\, du$. Now

$$\frac{1}{(u-2)(u^2+1)} = \frac{A}{u-2} + \frac{Bu+C}{u^2+1} \quad\Rightarrow\quad 1 = A(u^2+1) + (Bu+C)(u-2).$$ Setting $u = 2$ gives $1 = 5A$, so $A = \tfrac{1}{5}$.

Setting $u = 0$ gives $1 = \tfrac{1}{5} - 2C$, so $C = -\tfrac{2}{5}$. Comparing coefficients of u^2 gives $0 = \tfrac{1}{5} + B$, so $B = -\tfrac{1}{5}$. Thus,

$$\int \frac{1}{(u-2)(u^2+1)}\, du = \int \left(\frac{\tfrac{1}{5}}{u-2} + \frac{-\tfrac{1}{5}u - \tfrac{2}{5}}{u^2+1} \right) du = \frac{1}{5}\int \frac{1}{u-2}\, du - \frac{1}{5}\int \frac{u}{u^2+1}\, du - \frac{2}{5}\int \frac{1}{u^2+1}\, du$$

$$= \tfrac{1}{5}\ln|u-2| - \tfrac{1}{5}\cdot\tfrac{1}{2}\ln|u^2+1| - \tfrac{2}{5}\tan^{-1} u + C$$

$$= \tfrac{1}{5}\ln|e^x - 2| - \tfrac{1}{10}\ln(e^{2x}+1) - \tfrac{2}{5}\tan^{-1} e^x + C$$

51. Let $u = \ln(x^2 - x + 2)$, $dv = dx$. Then $du = \dfrac{2x-1}{x^2 - x + 2}\, dx$, $v = x$, and (by integration by parts)

$$\int \ln(x^2 - x + 2)\, dx = x\ln(x^2 - x + 2) - \int \frac{2x^2 - x}{x^2 - x + 2}\, dx = x\ln(x^2 - x + 2) - \int \left(2 + \frac{x-4}{x^2 - x + 2}\right) dx$$

$$= x\ln(x^2 - x + 2) - 2x - \int \frac{\tfrac{1}{2}(2x-1)}{x^2 - x + 2}\, dx + \frac{7}{2}\int \frac{dx}{(x-\tfrac{1}{2})^2 + \tfrac{7}{4}}$$

$$= x\ln(x^2 - x + 2) - 2x - \frac{1}{2}\ln(x^2 - x + 2) + \frac{7}{2}\int \frac{\tfrac{\sqrt{7}}{2} du}{\tfrac{7}{4}(u^2+1)} \quad \begin{bmatrix} \text{where } x - \tfrac{1}{2} = \tfrac{\sqrt{7}}{2}u, \\ dx = \tfrac{\sqrt{7}}{2}\, du, \\ (x-\tfrac{1}{2})^2 + \tfrac{7}{4} = \tfrac{7}{4}(u^2+1) \end{bmatrix}$$

$$= (x - \tfrac{1}{2})\ln(x^2 - x + 2) - 2x + \sqrt{7}\tan^{-1} u + C$$

$$= (x - \tfrac{1}{2})\ln(x^2 - x + 2) - 2x + \sqrt{7}\tan^{-1} \frac{2x-1}{\sqrt{7}} + C$$

52. Let $u = \tan^{-1} x$, $dv = x\, dx \Rightarrow du = dx/(1+x^2)$, $v = \tfrac{1}{2}x^2$.

Then $\displaystyle\int x\tan^{-1} x\, dx = \frac{1}{2}x^2\tan^{-1} x - \frac{1}{2}\int \frac{x^2}{1+x^2}\, dx$. To evaluate the last integral, use long division or observe that

$$\int \frac{x^2}{1+x^2}\, dx = \int \frac{(1+x^2)-1}{1+x^2}\, dx = \int 1\, dx - \int \frac{1}{1+x^2}\, dx = x - \tan^{-1} x + C_1. \text{ So}$$

$$\int x\tan^{-1} x\, dx = \tfrac{1}{2}x^2\tan^{-1} x - \tfrac{1}{2}(x - \tan^{-1} x + C_1) = \tfrac{1}{2}(x^2\tan^{-1} x + \tan^{-1} x - x) + C.$$

53.

From the graph, we see that the integral will be negative, and we guess that the area is about the same as that of a rectangle with width 2 and height 0.3, so we estimate the integral to be $-(2\cdot 0.3) = -0.6$. Now

$$\frac{1}{x^2 - 2x - 3} = \frac{1}{(x-3)(x+1)} = \frac{A}{x-3} + \frac{B}{x+1} \quad\Leftrightarrow\quad$$

$$1 = (A+B)x + A - 3B, \text{ so } A = -B \text{ and } A - 3B = 1 \quad\Leftrightarrow\quad A = \tfrac{1}{4}$$

and $B = -\tfrac{1}{4}$, so the integral becomes

$$\int_0^2 \frac{dx}{x^2 - 2x - 3} = \frac{1}{4}\int_0^2 \frac{dx}{x-3} - \frac{1}{4}\int_0^2 \frac{dx}{x+1} = \frac{1}{4}\Big[\ln|x-3| - \ln|x+1|\Big]_0^2 = \frac{1}{4}\left[\ln\left|\frac{x-3}{x+1}\right|\right]_0^2$$

$$= \tfrac{1}{4}\left(\ln\tfrac{1}{3} - \ln 3\right) = -\tfrac{1}{2}\ln 3 \approx -0.55$$

54. $\dfrac{1}{x^3 - 2x^2} = \dfrac{1}{x^2(x-2)} = \dfrac{A}{x} + \dfrac{B}{x^2} + \dfrac{C}{x-2}$ $\Rightarrow$ $1 = (A+C)x^2 + (B-2A)x - 2B$, so $A+C = B-2A = 0$ and

$-2B = 1$ $\Rightarrow$ $B = -\frac{1}{2}$, $A = -\frac{1}{4}$, and $C = \frac{1}{4}$. So the general antiderivative of $\dfrac{1}{x^3 - 2x^2}$ is

$$\int \frac{dx}{x^3 - 2x^2} = -\frac{1}{4}\int \frac{dx}{x} - \frac{1}{2}\int \frac{dx}{x^2} + \frac{1}{4}\int \frac{dx}{x-2}$$

$$= -\tfrac{1}{4}\ln|x| - \tfrac{1}{2}(-1/x) + \tfrac{1}{4}\ln|x-2| + C$$

$$= \frac{1}{4}\ln\left|\frac{x-2}{x}\right| + \frac{1}{2x} + C$$

We plot this function with $C = 0$ on the same screen as $y = \dfrac{1}{x^3 - 2x^2}$.

55. $\displaystyle\int \frac{dx}{x^2 - 2x} = \int \frac{dx}{(x-1)^2 - 1} = \int \frac{du}{u^2 - 1}$ [put $u = x - 1$]

$$= \frac{1}{2}\ln\left|\frac{u-1}{u+1}\right| + C \quad \text{[by Equation 6]} \quad = \frac{1}{2}\ln\left|\frac{x-2}{x}\right| + C$$

56. $\displaystyle\int \frac{(2x+1)\,dx}{4x^2 + 12x - 7} = \frac{1}{4}\int \frac{(8x+12)\,dx}{4x^2 + 12x - 7} - \int \frac{2\,dx}{(2x+3)^2 - 16}$

$$= \tfrac{1}{4}\ln\left|4x^2 + 12x - 7\right| - \int \frac{du}{u^2 - 16} \qquad \text{[put } u = 2x + 3]$$

$$= \tfrac{1}{4}\ln\left|4x^2 + 12x - 7\right| - \tfrac{1}{8}\ln\left|(u-4)/(u+4)\right| + C \qquad \text{[by Equation 6]}$$

$$= \tfrac{1}{4}\ln\left|4x^2 + 12x - 7\right| - \tfrac{1}{8}\ln\left|(2x-1)/(2x+7)\right| + C$$

57. (a) If $t = \tan\left(\dfrac{x}{2}\right)$, then $\dfrac{x}{2} = \tan^{-1} t$. The figure gives

$$\cos\left(\frac{x}{2}\right) = \frac{1}{\sqrt{1+t^2}} \text{ and } \sin\left(\frac{x}{2}\right) = \frac{t}{\sqrt{1+t^2}}.$$

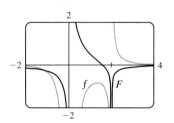

(b) $\cos x = \cos\left(2 \cdot \dfrac{x}{2}\right) = 2\cos^2\left(\dfrac{x}{2}\right) - 1$

$$= 2\left(\frac{1}{\sqrt{1+t^2}}\right)^2 - 1 = \frac{2}{1+t^2} - 1 = \frac{1-t^2}{1+t^2}$$

(c) $\dfrac{x}{2} = \arctan t$ $\Rightarrow$ $x = 2\arctan t$ $\Rightarrow$ $dx = \dfrac{2}{1+t^2}\,dt$

58. Let $t = \tan(x/2)$. Then, using Exercise 57, $dx = \dfrac{2}{1+t^2}\,dt$, $\sin x = \dfrac{2t}{1+t^2}$ $\Rightarrow$

$$\int \frac{dx}{3 - 5\sin x} = \int \frac{2\,dt/(1+t^2)}{3 - 10t/(1+t^2)} = \int \frac{2\,dt}{3(1+t^2) - 10t} = 2\int \frac{dt}{3t^2 - 10t + 3}$$

$$= \frac{1}{4}\int \left[\frac{1}{t-3} - \frac{3}{3t-1}\right] dt = \tfrac{1}{4}(\ln|t-3| - \ln|3t-1|) + C = \frac{1}{4}\ln\left|\frac{\tan(x/2) - 3}{3\tan(x/2) - 1}\right| + C$$

59. Let $t = \tan(x/2)$. Then, using the expressions in Exercise 57, we have

$$\int \frac{1}{3\sin x - 4\cos x}\,dx = \int \frac{1}{3\left(\dfrac{2t}{1+t^2}\right) - 4\left(\dfrac{1-t^2}{1+t^2}\right)}\,\frac{2\,dt}{1+t^2} = 2\int \frac{dt}{3(2t) - 4(1-t^2)} = \int \frac{dt}{2t^2 + 3t - 2}$$

$$= \int \frac{dt}{(2t-1)(t+2)} = \int \left[\frac{2}{5}\frac{1}{2t-1} - \frac{1}{5}\frac{1}{t+2}\right]dt \quad \text{[using partial fractions]}$$

$$= \tfrac{1}{5}\left[\ln|2t-1| - \ln|t+2|\right] + C = \frac{1}{5}\ln\left|\frac{2t-1}{t+2}\right| + C = \frac{1}{5}\ln\left|\frac{2\tan(x/2)-1}{\tan(x/2)+2}\right| + C$$

60. Let $t = \tan(x/2)$. Then, by Exercise 57,

$$\int_{\pi/3}^{\pi/2} \frac{dx}{1+\sin x - \cos x} = \int_{1/\sqrt{3}}^{1} \frac{2\,dt/(1+t^2)}{1 + 2t/(1+t^2) - (1-t^2)/(1+t^2)} = \int_{1/\sqrt{3}}^{1} \frac{2\,dt}{1+t^2+2t-1+t^2}$$

$$= \int_{1/\sqrt{3}}^{1} \left[\frac{1}{t} - \frac{1}{t+1}\right]dt = \Big[\ln t - \ln(t+1)\Big]_{1/\sqrt{3}}^{1} = \ln\frac{1}{2} - \ln\frac{1}{\sqrt{3}+1} = \ln\frac{\sqrt{3}+1}{2}$$

61. Let $t = \tan(x/2)$. Then, by Exercise 57,

$$\int_{0}^{\pi/2} \frac{\sin 2x}{2+\cos x}\,dx = \int_{0}^{\pi/2} \frac{2\sin x\cos x}{2+\cos x}\,dx = \int_{0}^{1} \frac{2\cdot\dfrac{2t}{1+t^2}\cdot\dfrac{1-t^2}{1+t^2}}{2+\dfrac{1-t^2}{1+t^2}}\,\frac{2}{1+t^2}\,dt = \int_{0}^{1} \frac{\dfrac{8t(1-t^2)}{(1+t^2)^2}}{2(1+t^2)+(1-t^2)}\,dt$$

$$= \int_{0}^{1} 8t \cdot \frac{1-t^2}{(t^2+3)(t^2+1)^2}\,dt = I$$

If we now let $u = t^2$, then $\dfrac{1-t^2}{(t^2+3)(t^2+1)^2} = \dfrac{1-u}{(u+3)(u+1)^2} = \dfrac{A}{u+3} + \dfrac{B}{u+1} + \dfrac{C}{(u+1)^2} \quad \Rightarrow$

$1 - u = A(u+1)^2 + B(u+3)(u+1) + C(u+3)$. Set $u = -1$ to get $2 = 2C$, so $C = 1$. Set $u = -3$ to get $4 = 4A$, so $A = 1$. Set $u = 0$ to get $1 = 1 + 3B + 3$, so $B = -1$. So

$$I = \int_{0}^{1} \left[\frac{8t}{t^2+3} - \frac{8t}{t^2+1} + \frac{8t}{(t^2+1)^2}\right]dt = \left[4\ln(t^2+3) - 4\ln(t^2+1) - \frac{4}{t^2+1}\right]_{0}^{1}$$

$$= (4\ln 4 - 4\ln 2 - 2) - (4\ln 3 - 0 - 4) = 8\ln 2 - 4\ln 2 - 4\ln 3 + 2 = 4\ln\tfrac{2}{3} + 2$$

62. $\dfrac{1}{x^3+x} = \dfrac{1}{x(x^2+1)} = \dfrac{A}{x} + \dfrac{Bx+C}{x^2+1} \quad \Rightarrow \quad 1 = A(x^2+1) + (Bx+C)x$. Set $x = 0$ to get $1 = A$. So

$1 = (1+B)x^2 + Cx + 1 \quad \Rightarrow \quad B+1 = 0 \quad [B = -1]$ and $C = 0$. Thus, the area is

$$\int_{1}^{2} \frac{1}{x^3+x}\,dx = \int_{1}^{2} \left(\frac{1}{x} - \frac{x}{x^2+1}\right)dx = \Big[\ln|x| - \tfrac{1}{2}\ln|x^2+1|\Big]_{1}^{2} = \left(\ln 2 - \tfrac{1}{2}\ln 5\right) - \left(0 - \tfrac{1}{2}\ln 2\right)$$

$$= \tfrac{3}{2}\ln 2 - \tfrac{1}{2}\ln 5 \quad \left[\text{or } \tfrac{1}{2}\ln\tfrac{8}{5}\right]$$

63. By long division, $\dfrac{x^2+1}{3x-x^2} = -1 + \dfrac{3x+1}{3x-x^2}$. Now

$\dfrac{3x+1}{3x-x^2} = \dfrac{3x+1}{x(3-x)} = \dfrac{A}{x} + \dfrac{B}{3-x}$ $\Rightarrow$ $3x+1 = A(3-x) + Bx$. Set $x=3$ to get $10 = 3B$, so $B = \frac{10}{3}$. Set $x=0$ to

get $1 = 3A$, so $A = \frac{1}{3}$. Thus, the area is

$$\int_1^2 \frac{x^2+1}{3x-x^2}\,dx = \int_1^2 \left(-1 + \frac{\frac{1}{3}}{x} + \frac{\frac{10}{3}}{3-x}\right) dx = \left[-x + \tfrac{1}{3}\ln|x| - \tfrac{10}{3}\ln|3-x|\right]_1^2$$
$$= \left(-2 + \tfrac{1}{3}\ln 2 - 0\right) - \left(-1 + 0 - \tfrac{10}{3}\ln 2\right) = -1 + \tfrac{11}{3}\ln 2$$

64. (a) We use disks, so the volume is $V = \pi \displaystyle\int_0^1 \left[\dfrac{1}{x^2+3x+2}\right]^2 dx = \pi \int_0^1 \dfrac{dx}{(x+1)^2(x+2)^2}$. To evaluate the integral,

we use partial fractions: $\dfrac{1}{(x+1)^2(x+2)^2} = \dfrac{A}{x+1} + \dfrac{B}{(x+1)^2} + \dfrac{C}{x+2} + \dfrac{D}{(x+2)^2}$ $\Rightarrow$

$1 = A(x+1)(x+2)^2 + B(x+2)^2 + C(x+1)^2(x+2) + D(x+1)^2$. We set $x=-1$, giving $B=1$, then set

$x=-2$, giving $D=1$. Now equating coefficients of x^3 gives $A = -C$, and then equating constants gives

$1 = 4A + 4 + 2(-A) + 1$ $\Rightarrow$ $A = -2$ $\Rightarrow$ $C = 2$. So the expression becomes

$$V = \pi \int_0^1 \left[\frac{-2}{x+1} + \frac{1}{(x+1)^2} + \frac{2}{(x+2)} + \frac{1}{(x+2)^2}\right] dx = \pi\left[2\ln\left|\frac{x+2}{x+1}\right| - \frac{1}{x+1} - \frac{1}{x+2}\right]_0^1$$
$$= \pi\left[\left(2\ln\tfrac{3}{2} - \tfrac{1}{2} - \tfrac{1}{3}\right) - \left(2\ln 2 - 1 - \tfrac{1}{2}\right)\right] = \pi\left(2\ln\tfrac{3/2}{2} + \tfrac{2}{3}\right) = \pi\left(\tfrac{2}{3} + \ln\tfrac{9}{16}\right)$$

(b) In this case, we use cylindrical shells, so the volume is $V = 2\pi \displaystyle\int_0^1 \dfrac{x\,dx}{x^2+3x+2} = 2\pi\int_0^1 \dfrac{x\,dx}{(x+1)(x+2)}$. We use

partial fractions to simplify the integrand: $\dfrac{x}{(x+1)(x+2)} = \dfrac{A}{x+1} + \dfrac{B}{x+2}$ $\Rightarrow$ $x = (A+B)x + 2A + B$. So

$A+B=1$ and $2A+B=0$ $\Rightarrow$ $A=-1$ and $B=2$. So the volume is

$$2\pi\int_0^1 \left[\frac{-1}{x+1} + \frac{2}{x+2}\right] dx = 2\pi\left[-\ln|x+1| + 2\ln|x+2|\right]_0^1$$
$$= 2\pi(-\ln 2 + 2\ln 3 + \ln 1 - 2\ln 2) = 2\pi(2\ln 3 - 3\ln 2) = 2\pi\ln\tfrac{9}{8}$$

65. $\dfrac{P+S}{P[(r-1)P-S]} = \dfrac{A}{P} + \dfrac{B}{(r-1)P-S}$ $\Rightarrow$ $P+S = A[(r-1)P-S] + BP = [(r-1)A+B]P - AS$ $\Rightarrow$

$(r-1)A + B = 1, -A = 1$ $\Rightarrow$ $A = -1, B = r$. Now

$$t = \int \frac{P+S}{P[(r-1)P-S]}\,dP = \int\left[\frac{-1}{P} + \frac{r}{(r-1)P-S}\right]dP = -\int \frac{dP}{P} + \frac{r}{r-1}\int\frac{r-1}{(r-1)P-S}\,dP$$

so $t = -\ln P + \dfrac{r}{r-1}\ln|(r-1)P-S| + C$. Here $r = 0.10$ and $S = 900$, so

$t = -\ln P + \dfrac{0.1}{-0.9}\ln|-0.9P-900| + C = -\ln P - \tfrac{1}{9}\ln(|-1||0.9P+900|) = -\ln P - \tfrac{1}{9}\ln(0.9P+900) + C$.

When $t=0$, $P = 10{,}000$, so $0 = -\ln 10{,}000 - \tfrac{1}{9}\ln(9900) + C$. Thus, $C = \ln 10{,}000 + \tfrac{1}{9}\ln 9900$ [≈ 10.2326], so our equation becomes

$$t = \ln 10{,}000 - \ln P + \tfrac{1}{9} \ln 9900 - \tfrac{1}{9} \ln(0.9P + 900) = \ln \frac{10{,}000}{P} + \frac{1}{9} \ln \frac{9900}{0.9P + 900}$$

$$= \ln \frac{10{,}000}{P} + \frac{1}{9} \ln \frac{1100}{0.1P + 100} = \ln \frac{10{,}000}{P} + \frac{1}{9} \ln \frac{11{,}000}{P + 1000}$$

66. If we subtract and add $2x^2$, we get

$$x^4 + 1 = x^4 + 2x^2 + 1 - 2x^2 = \left(x^2 + 1\right)^2 - 2x^2 = \left(x^2 + 1\right)^2 - \left(\sqrt{2}\,x\right)^2$$

$$= \left[(x^2 + 1) - \sqrt{2}\,x\right]\left[(x^2 + 1) + \sqrt{2}\,x\right] = \left(x^2 - \sqrt{2}\,x + 1\right)\left(x^2 + \sqrt{2}\,x + 1\right)$$

So we can decompose $\dfrac{1}{x^4 + 1} = \dfrac{Ax + B}{x^2 + \sqrt{2}\,x + 1} + \dfrac{Cx + D}{x^2 - \sqrt{2}\,x + 1} \quad \Rightarrow$

$1 = (Ax + B)\left(x^2 - \sqrt{2}\,x + 1\right) + (Cx + D)\left(x^2 + \sqrt{2}\,x + 1\right)$. Setting the constant terms equal gives $B + D = 1$, then

from the coefficients of x^3 we get $A + C = 0$. Now from the coefficients of x we get $A + C + (B - D)\sqrt{2} = 0 \quad \Leftrightarrow$

$[(1 - D) - D]\sqrt{2} = 0 \quad \Rightarrow \quad D = \tfrac{1}{2} \quad \Rightarrow \quad B = \tfrac{1}{2}$, and finally, from the coefficients of x^2 we get

$\sqrt{2}\,(C - A) + B + D = 0 \quad \Rightarrow \quad C - A = -\tfrac{1}{\sqrt{2}} \quad \Rightarrow \quad C = -\tfrac{\sqrt{2}}{4}$ and $A = \tfrac{\sqrt{2}}{4}$. So we rewrite the integrand, splitting the

terms into forms which we know how to integrate:

$$\frac{1}{x^4 + 1} = \frac{\tfrac{\sqrt{2}}{4}x + \tfrac{1}{2}}{x^2 + \sqrt{2}\,x + 1} + \frac{-\tfrac{\sqrt{2}}{4}x + \tfrac{1}{2}}{x^2 - \sqrt{2}\,x + 1} = \frac{1}{4\sqrt{2}}\left[\frac{2x + 2\sqrt{2}}{x^2 + \sqrt{2}\,x + 1} - \frac{2x - 2\sqrt{2}}{x^2 - \sqrt{2}\,x + 1}\right]$$

$$= \frac{\sqrt{2}}{8}\left[\frac{2x + \sqrt{2}}{x^2 + \sqrt{2}\,x + 1} - \frac{2x - \sqrt{2}}{x^2 - \sqrt{2}\,x + 1}\right] + \frac{1}{4}\left[\frac{1}{\left(x + \tfrac{1}{\sqrt{2}}\right)^2 + \tfrac{1}{2}} + \frac{1}{\left(x - \tfrac{1}{\sqrt{2}}\right)^2 + \tfrac{1}{2}}\right]$$

Now we integrate: $\displaystyle \int \frac{dx}{x^4 + 1} = \frac{\sqrt{2}}{8}\ln\left(\frac{x^2 + \sqrt{2}\,x + 1}{x^2 - \sqrt{2}\,x + 1}\right) + \frac{\sqrt{2}}{4}\left[\tan^{-1}\left(\sqrt{2}\,x + 1\right) + \tan^{-1}\left(\sqrt{2}\,x - 1\right)\right] + C.$

67. (a) In Maple, we define $f(x)$, and then use `convert(f,parfrac,x);` to obtain

$$f(x) = \frac{24{,}110/4879}{5x + 2} - \frac{668/323}{2x + 1} - \frac{9438/80{,}155}{3x - 7} + \frac{(22{,}098x + 48{,}935)/260{,}015}{x^2 + x + 5}$$

In Mathematica, we use the command `Apart`, and in Derive, we use `Expand`.

(b) $\displaystyle \int f(x)\,dx = \frac{24{,}110}{4879} \cdot \frac{1}{5}\ln|5x + 2| - \frac{668}{323} \cdot \frac{1}{2}\ln|2x + 1| - \frac{9438}{80{,}155} \cdot \frac{1}{3}\ln|3x - 7|$

$$+ \frac{1}{260{,}015}\int \frac{22{,}098\left(x + \tfrac{1}{2}\right) + 37{,}886}{\left(x + \tfrac{1}{2}\right)^2 + \tfrac{19}{4}}\,dx + C$$

$$= \frac{24{,}110}{4879} \cdot \frac{1}{5}\ln|5x + 2| - \frac{668}{323} \cdot \frac{1}{2}\ln|2x + 1| - \frac{9438}{80{,}155} \cdot \frac{1}{3}\ln|3x - 7|$$

$$+ \frac{1}{260{,}015}\left[22{,}098 \cdot \frac{1}{2}\ln\left(x^2 + x + 5\right) + 37{,}886 \cdot \sqrt{\tfrac{4}{19}}\tan^{-1}\left(\frac{1}{\sqrt{19/4}}\left(x + \tfrac{1}{2}\right)\right)\right] + C$$

$$= \frac{4822}{4879}\ln|5x + 2| - \frac{334}{323}\ln|2x + 1| - \frac{3146}{80{,}155}\ln|3x - 7| + \frac{11{,}049}{260{,}015}\ln\left(x^2 + x + 5\right)$$

$$+ \frac{75{,}772}{260{,}015\sqrt{19}}\tan^{-1}\left[\frac{1}{\sqrt{19}}\left(2x + 1\right)\right] + C$$

[continued]

Using a CAS, we get

$$\frac{4822\ln(5x+2)}{4879} - \frac{334\ln(2x+1)}{323} - \frac{3146\ln(3x-7)}{80{,}155}$$

$$+ \frac{11{,}049\ln(x^2+x+5)}{260{,}015} + \frac{3988\sqrt{19}}{260{,}015}\tan^{-1}\left[\frac{\sqrt{19}}{19}(2x+1)\right]$$

The main difference in this answer is that the absolute value signs and the constant of integration have been omitted. Also, the fractions have been reduced and the denominators rationalized.

68. (a) In Maple, we define $f(x)$, and then use `convert(f,parfrac,x);` to get

$$f(x) = \frac{5828/1815}{(5x-2)^2} - \frac{59{,}096/19{,}965}{5x-2} + \frac{2(2843x+816)/3993}{2x^2+1} + \frac{(313x-251)/363}{(2x^2+1)^2}.$$

In Mathematica, we use the command `Apart`, and in Derive, we use `Expand`.

(b) As we saw in Exercise 67, computer algebra systems omit the absolute value signs in $\int (1/y)\,dy = \ln|y|$. So we use the CAS to integrate the expression in part (a) and add the necessary absolute value signs and constant of integration to get

$$\int f(x)\,dx = -\frac{5828}{9075(5x-2)} - \frac{59{,}096\ln|5x-2|}{99{,}825} + \frac{2843\ln(2x^2+1)}{7986}$$

$$+ \frac{503}{15{,}972}\sqrt{2}\tan^{-1}\left(\sqrt{2}\,x\right) - \frac{1}{2904}\frac{1004x+626}{2x^2+1} + C$$

(c) From the graph, we see that f goes from negative to positive at $x \approx -0.78$, then back to negative at $x \approx 0.8$, and finally back to positive at $x = 1$. Also, $\lim_{x\to 0.4} f(x) = \infty$. So we see (by the First Derivative Test) that $\int f(x)\,dx$ has minima at $x \approx -0.78$ and $x = 1$, and a maximum at $x \approx 0.80$, and that $\int f(x)\,dx$ is unbounded as $x \to 0.4$. Note also that just to the right of $x = 0.4$, f has large values, so $\int f(x)\,dx$ increases rapidly, but slows down as f drops toward 0.

$\int f(x)\,dx$ decreases from about 0.8 to 1, then increases slowly since f stays small and positive.

69. There are only finitely many values of x where $Q(x) = 0$ (assuming that Q is not the zero polynomial). At all other values of x, $F(x)/Q(x) = G(x)/Q(x)$, so $F(x) = G(x)$. In other words, the values of F and G agree at all except perhaps finitely many values of x. By continuity of F and G, the polynomials F and G must agree at those values of x too.

More explicitly: if a is a value of x such that $Q(a) = 0$, then $Q(x) \neq 0$ for all x sufficiently close to a. Thus,

$$F(a) = \lim_{x\to a} F(x) \qquad \text{[by continuity of } F]$$

$$= \lim_{x\to a} G(x) \qquad \text{[whenever } Q(x) \neq 0]$$

$$= G(a) \qquad \text{[by continuity of } G]$$

70. Let $f(x) = ax^2 + bx + c$. We calculate the partial fraction decomposition of $\dfrac{f(x)}{x^2(x+1)^3}$. Since $f(0) = 1$, we must have

$c = 1$, so $\dfrac{f(x)}{x^2(x+1)^3} = \dfrac{ax^2 + bx + 1}{x^2(x+1)^3} = \dfrac{A}{x} + \dfrac{B}{x^2} + \dfrac{C}{x+1} + \dfrac{D}{(x+1)^2} + \dfrac{E}{(x+1)^3}$. Now in order for the integral not to

contain any logarithms (that is, in order for it to be a rational function), we must have $A = C = 0$, so

$ax^2 + bx + 1 = B(x+1)^3 + Dx^2(x+1) + Ex^2$. Equating constant terms gives $B = 1$, then equating coefficients of x

gives $3B = b \implies b = 3$. This is the quantity we are looking for, since $f'(0) = b$.

8.5   Strategy for Integration

1. Let $u = \sin x$, so that $du = \cos x\, dx$. Then $\int \cos x(1 + \sin^2 x)\, dx = \int (1 + u^2)\, du = u + \frac{1}{3}u^3 + C = \sin x + \frac{1}{3}\sin^3 x + C$.

2. $\displaystyle \int \frac{\sin^3 x}{\cos x}\, dx = \int \frac{\sin^2 x \sin x}{\cos x}\, dx = \int \frac{(1 - \cos^2 x)\sin x}{\cos x}\, dx = \int \frac{1 - u^2}{u}\,(-du) \qquad \begin{bmatrix} u = \cos x \\ du = -\sin x\, dx \end{bmatrix}$

$\qquad\qquad = \int \left(u - \frac{1}{u}\right) du = \frac{1}{2}u^2 - \ln|u| + C = \frac{1}{2}\cos^2 x - \ln|\cos x| + C$

3. $\displaystyle \int \frac{\sin x + \sec x}{\tan x}\, dx = \int \left(\frac{\sin x}{\tan x} + \frac{\sec x}{\tan x}\right) dx = \int (\cos x + \csc x)\, dx = \sin x + \ln|\csc x - \cot x| + C$

4. $\displaystyle \int \tan^3 \theta\, d\theta = \int (\sec^2 \theta - 1)\tan \theta\, d\theta = \int \tan \theta \sec^2 \theta\, d\theta - \int \frac{\sin \theta}{\cos \theta}\, d\theta$

$\qquad = \int u\, du + \int \frac{dv}{v} \qquad \begin{bmatrix} u = \tan \theta, & v = \cos \theta, \\ du = \sec^2 \theta\, d\theta & dv = -\sin \theta\, d\theta \end{bmatrix}$

$\qquad = \frac{1}{2}u^2 + \ln|v| + C = \frac{1}{2}\tan^2 \theta + \ln|\cos \theta| + C$

5. $\displaystyle \int_0^2 \frac{2t}{(t-3)^2}\, dt = \int_{-3}^{-1} \frac{2(u+3)}{u^2}\, du \quad \begin{bmatrix} u = t - 3, \\ du = dt \end{bmatrix} = \int_{-3}^{-1} \left(\frac{2}{u} + \frac{6}{u^2}\right) du = \left[2\ln|u| - \frac{6}{u}\right]_{-3}^{-1}$

$\qquad\qquad = (2\ln 1 + 6) - (2\ln 3 + 2) = 4 - 2\ln 3 \text{ or } 4 - \ln 9$

6. Let $u = x^2$. Then $du = 2x\, dx \implies \displaystyle \int \frac{x\, dx}{\sqrt{3 - x^4}} = \frac{1}{2}\int \frac{du}{\sqrt{3 - u^2}} = \frac{1}{2}\sin^{-1}\frac{u}{\sqrt{3}} + C = \frac{1}{2}\sin^{-1}\frac{x^2}{\sqrt{3}} + C.$

7. Let $u = \arctan y$. Then $du = \dfrac{dy}{1 + y^2} \implies \displaystyle \int_{-1}^1 \frac{e^{\arctan y}}{1 + y^2}\, dy = \int_{-\pi/4}^{\pi/4} e^u\, du = \left[e^u\right]_{-\pi/4}^{\pi/4} = e^{\pi/4} - e^{-\pi/4}.$

8. $\displaystyle \int x \csc x \cot x\, dx \quad \begin{bmatrix} u = x, & dv = \csc x \cot x\, dx, \\ du = dx & v = -\csc x \end{bmatrix} = -x\csc x - \int (-\csc x)\, dx = -x\csc x + \ln|\csc x - \cot x| + C$

9. $\displaystyle \int_1^3 r^4 \ln r\, dr \quad \begin{bmatrix} u = \ln r, & dv = r^4\, dr, \\ du = \dfrac{dr}{r} & v = \frac{1}{5}r^5 \end{bmatrix} = \left[\frac{1}{5}r^5 \ln r\right]_1^3 - \int_1^3 \frac{1}{5}r^4\, dr = \frac{243}{5}\ln 3 - 0 - \left[\frac{1}{25}r^5\right]_1^3$

$\qquad\qquad = \frac{243}{5}\ln 3 - \left(\frac{243}{25} - \frac{1}{25}\right) = \frac{243}{5}\ln 3 - \frac{242}{25}$

10. $\dfrac{x-1}{x^2-4x-5} = \dfrac{x-1}{(x-5)(x+1)} = \dfrac{A}{x-5} + \dfrac{B}{x+1}$ $\Rightarrow$ $x - 1 = A(x+1) + B(x-5)$. Setting $x = -1$ gives

$-2 = -6B$, so $B = \frac{1}{3}$. Setting $x = 5$ gives $4 = 6A$, so $A = \frac{2}{3}$. Now

$$\int_0^4 \frac{x-1}{x^2-4x-5}\,dx = \int_0^4 \left(\frac{2/3}{x-5} + \frac{1/3}{x+1}\right)dx = \left[\frac{2}{3}\ln|x-5| + \frac{1}{3}\ln|x+1|\right]_0^4$$

$$= \tfrac{2}{3}\ln 1 + \tfrac{1}{3}\ln 5 - \tfrac{2}{3}\ln 5 - \tfrac{1}{3}\ln 1 = -\tfrac{1}{3}\ln 5$$

11. $\displaystyle\int \frac{x-1}{x^2-4x+5}\,dx = \int \frac{(x-2)+1}{(x-2)^2+1}\,dx = \int \left(\frac{u}{u^2+1} + \frac{1}{u^2+1}\right)du$ $\qquad [u = x-2,\ du = dx]$

$$= \tfrac{1}{2}\ln(u^2+1) + \tan^{-1}u + C = \tfrac{1}{2}\ln(x^2-4x+5) + \tan^{-1}(x-2) + C$$

12. $\displaystyle\int \frac{x}{x^4+x^2+1}\,dx = \int \frac{\frac{1}{2}\,du}{u^2+u+1}\quad \begin{bmatrix} u = x^2, \\ du = 2x\,dx\end{bmatrix} \quad = \frac{1}{2}\int \frac{du}{\left(u+\frac{1}{2}\right)^2 + \frac{3}{4}}$

$$= \frac{1}{2}\int \frac{\frac{\sqrt{3}}{2}\,dv}{\frac{3}{4}(v^2+1)}\quad \begin{bmatrix} u + \frac{1}{2} = \frac{\sqrt{3}}{2}v, \\ du = \frac{\sqrt{3}}{2}\,dv\end{bmatrix} \quad = \frac{\sqrt{3}}{4}\cdot\frac{4}{3}\int \frac{dv}{v^2+1}$$

$$= \tfrac{1}{\sqrt{3}}\tan^{-1}v + C = \tfrac{1}{\sqrt{3}}\tan^{-1}\left(\tfrac{2}{\sqrt{3}}\left(x^2+\tfrac{1}{2}\right)\right) + C$$

13. $\int \sin^3\theta\,\cos^5\theta\,d\theta = \int \cos^5\theta\,\sin^2\theta\,\sin\theta\,d\theta = -\int \cos^5\theta\,(1-\cos^2\theta)(-\sin\theta)\,d\theta$

$$= -\int u^5(1-u^2)\,du \quad \begin{bmatrix} u = \cos\theta, \\ du = -\sin\theta\,d\theta\end{bmatrix}$$

$$= \int(u^7 - u^5)\,du = \tfrac{1}{8}u^8 - \tfrac{1}{6}u^6 + C = \tfrac{1}{8}\cos^8\theta - \tfrac{1}{6}\cos^6\theta + C$$

Another solution:

$\int \sin^3\theta\,\cos^5\theta\,d\theta = \int \sin^3\theta\,(\cos^2\theta)^2\cos\theta\,d\theta = \int \sin^3\theta\,(1-\sin^2\theta)^2\cos\theta\,d\theta$

$$= \int u^3(1-u^2)^2\,du \quad \begin{bmatrix} u = \sin\theta, \\ du = \cos\theta\,d\theta\end{bmatrix} \quad = \int u^3(1-2u^2+u^4)\,du$$

$$= \int(u^3 - 2u^5 + u^7)\,du = \tfrac{1}{4}u^4 - \tfrac{1}{3}u^6 + \tfrac{1}{8}u^8 + C = \tfrac{1}{4}\sin^4\theta - \tfrac{1}{3}\sin^6\theta + \tfrac{1}{8}\sin^8\theta + C$$

14. Let $u = 1 + x^2$, so that $du = 2x\,dx$. Then

$$\int \frac{x^3}{\sqrt{1+x^2}}\,dx = \int \frac{x^2}{\sqrt{1+x^2}}\,(x\,dx) = \int \frac{u-1}{u^{1/2}}\left(\tfrac{1}{2}\,du\right) = \frac{1}{2}\int(u^{1/2} - u^{-1/2})\,du = \frac{1}{2}\left(\frac{2}{3}u^{3/2} - 2u^{1/2}\right) + C$$

$$= \tfrac{1}{3}(1+x^2)^{3/2} - (1+x^2)^{1/2} + C \quad \left[\text{or } \tfrac{1}{3}(x^2-2)\sqrt{1+x^2} + C\right]$$

15. Let $x = \sin\theta$, where $-\frac{\pi}{2} \le \theta \le \frac{\pi}{2}$. Then $dx = \cos\theta\,d\theta$ and $(1-x^2)^{1/2} = \cos\theta$,

so

$$\int \frac{dx}{(1-x^2)^{3/2}} = \int \frac{\cos\theta\,d\theta}{(\cos\theta)^3} = \int \sec^2\theta\,d\theta = \tan\theta + C = \frac{x}{\sqrt{1-x^2}} + C.$$

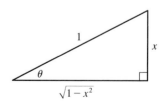

16. $\displaystyle\int_0^{\sqrt{2}/2} \frac{x^2}{\sqrt{1-x^2}}\,dx = \int_0^{\pi/4} \frac{\sin^2\theta}{\cos\theta}\cos\theta\,d\theta \qquad \begin{bmatrix} u = \sin\theta, \\ du = \cos\theta\,d\theta \end{bmatrix}$

$\qquad = \int_0^{\pi/4} \frac{1}{2}(1 - \cos 2\theta)\,d\theta = \frac{1}{2}\left[\theta - \frac{1}{2}\sin 2\theta\right]_0^{\pi/4} = \frac{1}{2}\left[\left(\frac{\pi}{4} - \frac{1}{2}\right) - (0 - 0)\right] = \frac{\pi}{8} - \frac{1}{4}$

17. $\displaystyle\int x\sin^2 x\,dx \qquad \begin{bmatrix} u = x, & dv = \sin^2 x\,dx, \\ du = dx & v = \int \sin^2 x\,dx = \int \frac{1}{2}(1 - \cos 2x)\,dx = \frac{1}{2}x - \frac{1}{2}\sin x\cos x \end{bmatrix}$

$\qquad = \frac{1}{2}x^2 - \frac{1}{2}x\sin x\cos x - \int \left(\frac{1}{2}x - \frac{1}{2}\sin x\cos x\right)dx$

$\qquad = \frac{1}{2}x^2 - \frac{1}{2}x\sin x\cos x - \frac{1}{4}x^2 + \frac{1}{4}\sin^2 x + C = \frac{1}{4}x^2 - \frac{1}{2}x\sin x\cos x + \frac{1}{4}\sin^2 x + C$

Note: $\int \sin x\cos x\,dx = \int s\,ds = \frac{1}{2}s^2 + C$ [where $s = \sin x$, $ds = \cos x\,dx$].

A slightly different method is to write $\int x\sin^2 x\,dx = \int x\cdot\frac{1}{2}(1 - \cos 2x)\,dx = \frac{1}{2}\int x\,dx - \frac{1}{2}\int x\cos 2x\,dx$. If we evaluate

the second integral by parts, we arrive at the equivalent answer $\frac{1}{4}x^2 - \frac{1}{4}x\sin 2x - \frac{1}{8}\cos 2x + C$.

18. Let $u = e^{2t}$, $du = 2e^{2t}\,dt$. Then $\displaystyle\int \frac{e^{2t}}{1 + e^{4t}}\,dt = \int \frac{\frac{1}{2}(2e^{2t})\,dt}{1 + (e^{2t})^2} = \int \frac{\frac{1}{2}\,du}{1 + u^2} = \frac{1}{2}\tan^{-1}u + C = \frac{1}{2}\tan^{-1}(e^{2t}) + C.$

19. Let $u = e^x$. Then $\int e^{x + e^x}\,dx = \int e^{e^x}e^x\,dx = \int e^u\,du = e^u + C = e^{e^x} + C.$

20. Since e^2 is a constant, $\int e^2\,dx = e^2 x + C.$

21. Let $t = \sqrt{x}$, so that $t^2 = x$ and $2t\,dt = dx$. Then $\int \arctan\sqrt{x}\,dx = \int \arctan t\,(2t\,dt) = I$. Now use parts with

$u = \arctan t$, $dv = 2t\,dt \quad\Rightarrow\quad du = \dfrac{1}{1 + t^2}\,dt$, $v = t^2$. Thus,

$I = t^2\arctan t - \displaystyle\int \frac{t^2}{1 + t^2}\,dt = t^2\arctan t - \int\left(1 - \frac{1}{1 + t^2}\right)dt = t^2\arctan t - t + \arctan t + C$

$\qquad = x\arctan\sqrt{x} - \sqrt{x} + \arctan\sqrt{x} + C \quad \left[\text{or } (x + 1)\arctan\sqrt{x} - \sqrt{x} + C\right]$

22. Let $u = 1 + (\ln x)^2$, so that $du = \dfrac{2\ln x}{x}\,dx$. Then

$\displaystyle\int \frac{\ln x}{x\sqrt{1 + (\ln x)^2}}\,dx = \frac{1}{2}\int \frac{1}{\sqrt{u}}\,du = \frac{1}{2}\left(2\sqrt{u}\right) + C = \sqrt{1 + (\ln x)^2} + C.$

23. Let $u = 1 + \sqrt{x}$. Then $x = (u - 1)^2$, $dx = 2(u - 1)\,du \quad\Rightarrow$

$\displaystyle\int_0^1 \left(1 + \sqrt{x}\right)^8 dx = \int_1^2 u^8\cdot 2(u - 1)\,du = 2\int_1^2 (u^9 - u^8)\,du = \left[\frac{1}{5}u^{10} - 2\cdot\frac{1}{9}u^9\right]_1^2 = \frac{1024}{5} - \frac{1024}{9} - \frac{1}{5} + \frac{2}{9} = \frac{4097}{45}.$

24. Let $u = \ln(x^2 - 1)$, $dv = dx \quad\Leftrightarrow\quad du = \dfrac{2x}{x^2 - 1}$, $v = x$. Then

$\displaystyle\int \ln(x^2 - 1)\,dx = x\ln(x^2 - 1) - \int \frac{2x^2}{x^2 - 1}\,dx = x\ln(x^2 - 1) - \int\left[2 + \frac{2}{(x - 1)(x + 1)}\right]dx$

$\qquad = x\ln(x^2 - 1) - \displaystyle\int\left[2 + \frac{1}{x - 1} - \frac{1}{x + 1}\right]dx = x\ln(x^2 - 1) - 2x - \ln|x - 1| + \ln|x + 1| + C$

25. $\dfrac{3x^2 - 2}{x^2 - 2x - 8} = 3 + \dfrac{6x + 22}{(x-4)(x+2)} = 3 + \dfrac{A}{x-4} + \dfrac{B}{x+2}$ $\Rightarrow$ $6x + 22 = A(x+2) + B(x-4)$. Setting

$x = 4$ gives $46 = 6A$, so $A = \frac{23}{3}$. Setting $x = -2$ gives $10 = -6B$, so $B = -\frac{5}{3}$. Now

$\displaystyle\int \dfrac{3x^2 - 2}{x^2 - 2x - 8}\, dx = \int \left(3 + \dfrac{23/3}{x-4} - \dfrac{5/3}{x+2}\right) dx = 3x + \frac{23}{3}\ln|x-4| - \frac{5}{3}\ln|x+2| + C.$

26. $\displaystyle\int \dfrac{3x^2 - 2}{x^3 - 2x - 8}\, dx = \int \dfrac{du}{u} \quad \begin{bmatrix} u = x^3 - 2x - 8, \\ du = (3x^2 - 2)\, dx \end{bmatrix} = \ln|u| + C = \ln|x^3 - 2x - 8| + C$

27. Let $u = 1 + e^x$, so that $du = e^x\, dx = (u-1)\, dx$. Then $\displaystyle\int \dfrac{1}{1 + e^x}\, dx = \int \dfrac{1}{u} \cdot \dfrac{du}{u-1} = \int \dfrac{1}{u(u-1)}\, du = I$. Now

$\dfrac{1}{u(u-1)} = \dfrac{A}{u} + \dfrac{B}{u-1}$ $\Rightarrow$ $1 = A(u-1) + Bu$. Set $u = 1$ to get $1 = B$. Set $u = 0$ to get $1 = -A$, so $A = -1$.

Thus, $I = \displaystyle\int \left(\dfrac{-1}{u} + \dfrac{1}{u-1}\right) du = -\ln|u| + \ln|u-1| + C = -\ln(1 + e^x) + \ln e^x + C = x - \ln(1 + e^x) + C.$

Another method: Multiply numerator and denominator by e^{-x} and let $u = e^{-x} + 1$. This gives the answer in the

form $-\ln(e^{-x} + 1) + C$.

28. $\int \sin\sqrt{at}\, dt = \int \sin u \cdot \frac{2}{a}u\, du$ $[u = \sqrt{at},\ u^2 = at,\ 2u\, du = a\, dt]$ $= \frac{2}{a}\int u \sin u\, du$

$= \frac{2}{a}[-u\cos u + \sin u] + C$ [integration by parts] $= -\frac{2}{a}\sqrt{at}\,\cos\sqrt{at} + \frac{2}{a}\sin\sqrt{at} + C$

$= -2\sqrt{\frac{t}{a}}\,\cos\sqrt{at} + \frac{2}{a}\sin\sqrt{at} + C$

29. $\displaystyle\int_0^5 \dfrac{3w - 1}{w + 2}\, dw = \int_0^5 \left(3 - \dfrac{7}{w+2}\right) dw = \Big[3w - 7\ln|w+2|\Big]_0^5 = 15 - 7\ln 7 + 7\ln 2$

$= 15 + 7(\ln 2 - \ln 7) = 15 + 7\ln\frac{2}{7}$

30. $x^2 - 4x < 0$ on $[0, 4]$, so

$\displaystyle\int_{-2}^2 \left|x^2 - 4x\right|\, dx = \int_{-2}^0 (x^2 - 4x)\, dx + \int_0^2 (4x - x^2)\, dx = \Big[\frac{1}{3}x^3 - 2x^2\Big]_{-2}^0 + \Big[2x^2 - \frac{1}{3}x^3\Big]_0^2$

$= 0 - \left(-\frac{8}{3} - 8\right) + \left(8 - \frac{8}{3}\right) - 0 = 16$

31. As in Example 5,

$\displaystyle\int \sqrt{\dfrac{1+x}{1-x}}\, dx = \int \dfrac{\sqrt{1+x}}{\sqrt{1-x}} \cdot \dfrac{\sqrt{1+x}}{\sqrt{1+x}}\, dx = \int \dfrac{1+x}{\sqrt{1-x^2}}\, dx = \int \dfrac{dx}{\sqrt{1-x^2}} + \int \dfrac{x\, dx}{\sqrt{1-x^2}} = \sin^{-1} x - \sqrt{1 - x^2} + C.$

Another method: Substitute $u = \sqrt{(1+x)/(1-x)}$.

32. $\displaystyle\int \dfrac{\sqrt{2x-1}}{2x+3}\, dx = \int \dfrac{u \cdot u\, du}{u^2 + 4} \quad \begin{bmatrix} u = \sqrt{2x-1},\ 2x + 3 = u^2 + 4, \\ u^2 = 2x - 1,\ u\, du = dx \end{bmatrix} = \int \left(1 - \dfrac{4}{u^2 + 4}\right) du$

$= u - 4 \cdot \frac{1}{2}\tan^{-1}\left(\frac{1}{2}u\right) + C = \sqrt{2x-1} - 2\tan^{-1}\left(\frac{1}{2}\sqrt{2x-1}\right) + C$

33. $3 - 2x - x^2 = -(x^2 + 2x + 1) + 4 = 4 - (x + 1)^2$. Let $x + 1 = 2\sin\theta$,

where $-\frac{\pi}{2} \le \theta \le \frac{\pi}{2}$. Then $dx = 2\cos\theta\,d\theta$ and

$$\int \sqrt{3 - 2x - x^2}\,dx = \int \sqrt{4 - (x+1)^2}\,dx = \int \sqrt{4 - 4\sin^2\theta}\,2\cos\theta\,d\theta$$

$$= 4\int \cos^2\theta\,d\theta = 2\int (1 + \cos 2\theta)\,d\theta$$

$$= 2\theta + \sin 2\theta + C = 2\theta + 2\sin\theta\cos\theta + C$$

$$= 2\sin^{-1}\left(\frac{x+1}{2}\right) + 2 \cdot \frac{x+1}{2} \cdot \frac{\sqrt{3 - 2x - x^2}}{2} + C$$

$$= 2\sin^{-1}\left(\frac{x+1}{2}\right) + \frac{x+1}{2}\sqrt{3 - 2x - x^2} + C$$

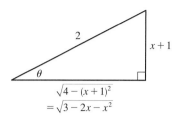

34. $\displaystyle\int_{\pi/4}^{\pi/2} \frac{1 + 4\cot x}{4 - \cot x}\,dx = \int_{\pi/4}^{\pi/2} \left[\frac{(1 + 4\cos x/\sin x)}{(4 - \cos x/\sin x)} \cdot \frac{\sin x}{\sin x}\right] dx = \int_{\pi/4}^{\pi/2} \frac{\sin x + 4\cos x}{4\sin x - \cos x}\,dx$

$$= \int_{3/\sqrt{2}}^{4} \frac{1}{u}\,du \qquad \begin{bmatrix} u = 4\sin x - \cos x, \\ du = (4\cos x + \sin x)\,dx \end{bmatrix}$$

$$= \left[\ln|u|\right]_{3/\sqrt{2}}^{4} = \ln 4 - \ln\frac{3}{\sqrt{2}} = \ln\frac{4}{3/\sqrt{2}} = \ln\left(\frac{4}{3}\sqrt{2}\right)$$

35. Because $f(x) = x^8 \sin x$ is the product of an even function and an odd function, it is odd.

Therefore, $\int_{-1}^{1} x^8 \sin x\,dx = 0$ [by (5.5.6)(b)].

36. $\sin 4x \cos 3x = \frac{1}{2}(\sin x + \sin 7x)$ by Formula 8.2.2(a), so

$$\int \sin 4x \cos 3x\,dx = \frac{1}{2}\int (\sin x + \sin 7x)\,dx = \frac{1}{2}\left[-\cos x - \frac{1}{7}\cos 7x\right] + C = -\frac{1}{2}\cos x - \frac{1}{14}\cos 7x + C.$$

37. $\int_0^{\pi/4} \cos^2\theta \tan^2\theta\,d\theta = \int_0^{\pi/4} \sin^2\theta\,d\theta = \int_0^{\pi/4} \frac{1}{2}(1 - \cos 2\theta)\,d\theta = \left[\frac{1}{2}\theta - \frac{1}{4}\sin 2\theta\right]_0^{\pi/4} = \left(\frac{\pi}{8} - \frac{1}{4}\right) - (0 - 0) = \frac{\pi}{8} - \frac{1}{4}$

38. $\int_0^{\pi/4} \tan^5\theta \sec^3\theta\,d\theta = \int_0^{\pi/4} \left(\tan^2\theta\right)^2 \sec^2\theta \cdot \sec\theta\tan\theta\,d\theta = \int_1^{\sqrt{2}} (u^2 - 1)^2 u^2\,du \qquad \begin{bmatrix} u = \sec\theta, \\ du = \sec\theta\tan\theta\,d\theta \end{bmatrix}$

$$= \int_1^{\sqrt{2}} (u^6 - 2u^4 + u^2)\,du = \left[\frac{1}{7}u^7 - \frac{2}{5}u^5 + \frac{1}{3}u^3\right]_1^{\sqrt{2}}$$

$$= \left(\frac{8}{7}\sqrt{2} - \frac{8}{5}\sqrt{2} + \frac{2}{3}\sqrt{2}\right) - \left(\frac{1}{7} - \frac{2}{5} + \frac{1}{3}\right) = \frac{22}{105}\sqrt{2} - \frac{8}{105} = \frac{2}{105}\left(11\sqrt{2} - 4\right)$$

39. Let $u = \sec\theta$, so that $du = \sec\theta\tan\theta\,d\theta$. Then $\displaystyle\int \frac{\sec\theta\tan\theta}{\sec^2\theta - \sec\theta}\,d\theta = \int \frac{1}{u^2 - u}\,du = \int \frac{1}{u(u - 1)}\,du = I$. Now

$$\frac{1}{u(u-1)} = \frac{A}{u} + \frac{B}{u-1} \quad \Rightarrow \quad 1 = A(u - 1) + Bu.$$ Set $u = 1$ to get $1 = B$. Set $u = 0$ to get $1 = -A$, so $A = -1$.

Thus, $I = \displaystyle\int \left(\frac{-1}{u} + \frac{1}{u-1}\right) du = -\ln|u| + \ln|u - 1| + C = \ln|\sec\theta - 1| - \ln|\sec\theta| + C$ [or $\ln|1 - \cos\theta| + C$].

40. $4y^2 - 4y - 3 = (2y - 1)^2 - 2^2$, so let $u = 2y - 1 \quad \Rightarrow \quad du = 2\,dy$. Thus,

$$\int \frac{dy}{\sqrt{4y^2 - 4y - 3}} = \int \frac{dy}{\sqrt{(2y - 1)^2 - 2^2}} = \frac{1}{2}\int \frac{du}{\sqrt{u^2 - 2^2}}$$

$$= \frac{1}{2}\ln\left|u + \sqrt{u^2 - 2^2}\right| \qquad \text{[by Formula 20 in the table in this section]}$$

$$= \frac{1}{2}\ln\left|2y - 1 + \sqrt{4y^2 - 4y - 3}\right| + C$$

41. Let $u = \theta$, $dv = \tan^2 \theta \, d\theta = (\sec^2 \theta - 1) \, d\theta \;\Rightarrow\; du = d\theta$ and $v = \tan \theta - \theta$. So

$$\int \theta \tan^2 \theta \, d\theta = \theta(\tan \theta - \theta) - \int (\tan \theta - \theta) \, d\theta = \theta \tan \theta - \theta^2 - \ln|\sec \theta| + \tfrac{1}{2}\theta^2 + C$$

$$= \theta \tan \theta - \tfrac{1}{2}\theta^2 - \ln|\sec \theta| + C$$

42. Let $u = \tan^{-1} x$, $dv = \dfrac{1}{x^2} \, dx \;\Rightarrow\; du = \dfrac{1}{1 + x^2} \, dx$, $v = -\dfrac{1}{x}$. Then

$$I = \int \frac{\tan^{-1} x}{x^2} \, dx = -\frac{1}{x} \tan^{-1} x - \int \left(-\frac{1}{x(1 + x^2)}\right) dx = -\frac{1}{x} \tan^{-1} x + \int \left(\frac{A}{x} + \frac{Bx + C}{1 + x^2}\right) dx$$

$$\frac{1}{x(1 + x^2)} = \frac{A}{x} + \frac{Bx + C}{1 + x^2} \;\Rightarrow\; 1 = A(1 + x^2) + (Bx + C)x \;\Rightarrow\; 1 = (A + B)x^2 + Cx + A, \text{ so } C = 0, A = 1,$$

and $A + B = 0 \;\Rightarrow\; B = -1$. Thus,

$$I = -\frac{1}{x} \tan^{-1} x + \int \left(\frac{1}{x} - \frac{x}{1 + x^2}\right) dx = -\frac{1}{x} \tan^{-1} x + \ln|x| - \frac{1}{2}\ln\left|1 + x^2\right| + C$$

$$= -\frac{\tan^{-1} x}{x} + \ln\left|\frac{x}{\sqrt{x^2 + 1}}\right| + C$$

Or: Let $x = \tan \theta$, so that $dx = \sec^2 \theta \, d\theta$. Then $\displaystyle\int \frac{\tan^{-1} x}{x^2} \, dx = \int \frac{\theta}{\tan^2 \theta} \sec^2 \theta \, d\theta = \int \theta \csc^2 \theta \, d\theta = I$. Now use parts

with $u = \theta$, $dv = \csc^2 \theta \, d\theta \;\Rightarrow\; du = d\theta$, $v = -\cot \theta$. Thus,

$$I = -\theta \cot \theta - \int(-\cot \theta) \, d\theta = -\theta \cot \theta + \ln|\sin \theta| + C$$

$$= -\tan^{-1} x \cdot \frac{1}{x} + \ln\left|\frac{x}{\sqrt{x^2 + 1}}\right| + C = -\frac{\tan^{-1} x}{x} + \ln\left|\frac{x}{\sqrt{x^2 + 1}}\right| + C$$

43. Let $u = 1 + e^x$, so that $du = e^x \, dx$. Then $\int e^x \sqrt{1 + e^x} \, dx = \int u^{1/2} \, du = \tfrac{2}{3}u^{3/2} + C = \tfrac{2}{3}(1 + e^x)^{3/2} + C$.

Or: Let $u = \sqrt{1 + e^x}$, so that $u^2 = 1 + e^x$ and $2u \, du = e^x \, dx$. Then

$\int e^x \sqrt{1 + e^x} \, dx = \int u \cdot 2u \, du = \int 2u^2 \, du = \tfrac{2}{3}u^3 + C = \tfrac{2}{3}(1 + e^x)^{3/2} + C$.

44. Let $u = \sqrt{1 + e^x}$. Then $u^2 = 1 + e^x$, $2u \, du = e^x \, dx = (u^2 - 1) \, dx$, and $dx = \dfrac{2u}{u^2 - 1} \, du$, so

$$\int \sqrt{1 + e^x} \, dx = \int u \cdot \frac{2u}{u^2 - 1} \, du = \int \frac{2u^2}{u^2 - 1} \, du = \int \left(2 + \frac{2}{u^2 - 1}\right) du = \int \left(2 + \frac{1}{u - 1} - \frac{1}{u + 1}\right) du$$

$$= 2u + \ln|u - 1| - \ln|u + 1| + C = 2\sqrt{1 + e^x} + \ln\left(\sqrt{1 + e^x} - 1\right) - \ln\left(\sqrt{1 + e^x} + 1\right) + C$$

45. Let $t = x^3$. Then $dt = 3x^2 \, dx \;\Rightarrow\; I = \int x^5 e^{-x^3} \, dx = \tfrac{1}{3} \int te^{-t} \, dt$. Now integrate by parts with $u = t$, $dv = e^{-t} \, dt$:

$I = -\tfrac{1}{3}te^{-t} + \tfrac{1}{3} \int e^{-t} \, dt = -\tfrac{1}{3}te^{-t} - \tfrac{1}{3}e^{-t} + C = -\tfrac{1}{3}e^{-x^3}(x^3 + 1) + C$.

46. $\dfrac{1+\sin x}{1-\sin x} = \dfrac{1+\sin x}{1-\sin x} \cdot \dfrac{1+\sin x}{1+\sin x} = \dfrac{1+2\sin x+\sin^2 x}{1-\sin^2 x} = \dfrac{1+2\sin x+\sin^2 x}{\cos^2 x} = \dfrac{1}{\cos^2 x} + \dfrac{2\sin x}{\cos^2 x} + \dfrac{\sin^2 x}{\cos^2 x}$

$= \sec^2 x + 2\sec x \tan x + \tan^2 x = \sec^2 x + 2\sec x \tan x + \sec^2 x - 1 = 2\sec^2 x + 2\sec x \tan x - 1$

Thus, $\displaystyle\int \dfrac{1+\sin x}{1-\sin x}\,dx = \int (2\sec^2 x + 2\sec x \tan x - 1)\,dx = 2\tan x + 2\sec x - x + C$

47. Let $u = x - 1$, so that $du = dx$. Then

$\int x^3 (x-1)^{-4}\,dx = \int (u+1)^3 u^{-4}\,du = \int (u^3 + 3u^2 + 3u + 1)u^{-4}\,du = \int (u^{-1} + 3u^{-2} + 3u^{-3} + u^{-4})\,du$

$= \ln|u| - 3u^{-1} - \tfrac{3}{2}u^{-2} - \tfrac{1}{3}u^{-3} + C = \ln|x-1| - 3(x-1)^{-1} - \tfrac{3}{2}(x-1)^{-2} - \tfrac{1}{3}(x-1)^{-3} + C$

48. Let $u = x^2$. Then $du = 2x\,dx \ \Rightarrow \ \displaystyle\int \dfrac{x\,dx}{x^4 - a^4} = \int \dfrac{\frac{1}{2}\,du}{u^2 - (a^2)^2} = \dfrac{1}{4a^2}\ln\left|\dfrac{u-a^2}{u+a^2}\right| + C = \dfrac{1}{4a^2}\ln\left|\dfrac{x^2-a^2}{x^2+a^2}\right| + C.$

49. Let $u = \sqrt{4x+1} \ \Rightarrow \ u^2 = 4x + 1 \ \Rightarrow \ 2u\,du = 4\,dx \ \Rightarrow \ dx = \tfrac{1}{2}u\,du.$ So

$\displaystyle\int \dfrac{1}{x\sqrt{4x+1}}\,dx = \int \dfrac{\frac{1}{2}u\,du}{\frac{1}{4}(u^2-1)\,u} = 2\int \dfrac{du}{u^2-1} = 2\left(\tfrac{1}{2}\right)\ln\left|\dfrac{u-1}{u+1}\right| + C \qquad \text{[by Formula 19]}$

$= \ln\left|\dfrac{\sqrt{4x+1}-1}{\sqrt{4x+1}+1}\right| + C$

50. As in Exercise 49, let $u = \sqrt{4x+1}$. Then $\displaystyle\int \dfrac{dx}{x^2\sqrt{4x+1}} = \int \dfrac{\frac{1}{2}u\,du}{\left[\frac{1}{4}(u^2-1)\right]^2 u} = 8\int \dfrac{du}{(u^2-1)^2}.$

Now $\dfrac{1}{(u^2-1)^2} = \dfrac{1}{(u+1)^2(u-1)^2} = \dfrac{A}{u+1} + \dfrac{B}{(u+1)^2} + \dfrac{C}{u-1} + \dfrac{D}{(u-1)^2} \ \Rightarrow$

$1 = A(u+1)(u-1)^2 + B(u-1)^2 + C(u-1)(u+1)^2 + D(u+1)^2. \ u=1 \ \Rightarrow \ D = \tfrac{1}{4}, u = -1 \ \Rightarrow \ B = \tfrac{1}{4}.$

Equating coefficients of u^3 gives $A + C = 0$, and equating coefficients of 1 gives $1 = A + B - C + D \ \Rightarrow$

$1 = A + \tfrac{1}{4} - C + \tfrac{1}{4} \ \Rightarrow \ \tfrac{1}{2} = A - C.$ So $A = \tfrac{1}{4}$ and $C = -\tfrac{1}{4}.$ Therefore,

$\displaystyle\int \dfrac{dx}{x^2\sqrt{4x+1}} = 8\int \left[\dfrac{1/4}{u+1} + \dfrac{1/4}{(u+1)^2} + \dfrac{-1/4}{u-1} + \dfrac{1/4}{(u-1)^2}\right]du$

$= \int \left[\dfrac{2}{u+1} + 2(u+1)^{-2} - \dfrac{2}{u-1} + 2(u-1)^{-2}\right]du$

$= 2\ln|u+1| - \dfrac{2}{u+1} - 2\ln|u-1| - \dfrac{2}{u-1} + C$

$= 2\ln\left(\sqrt{4x+1}+1\right) - \dfrac{2}{\sqrt{4x+1}+1} - 2\ln\left|\sqrt{4x+1}-1\right| - \dfrac{2}{\sqrt{4x+1}-1} + C$

51. Let $2x = \tan\theta \ \Rightarrow \ x = \tfrac{1}{2}\tan\theta,\ dx = \tfrac{1}{2}\sec^2\theta\,d\theta,\ \sqrt{4x^2+1} = \sec\theta,$ so

$\displaystyle\int \dfrac{dx}{x\sqrt{4x^2+1}} = \int \dfrac{\frac{1}{2}\sec^2\theta\,d\theta}{\frac{1}{2}\tan\theta\,\sec\theta} = \int \dfrac{\sec\theta}{\tan\theta}\,d\theta = \int \csc\theta\,d\theta$

$= -\ln|\csc\theta + \cot\theta| + C \qquad [\text{or } \ln|\csc\theta - \cot\theta| + C]$

$= -\ln\left|\dfrac{\sqrt{4x^2+1}}{2x} + \dfrac{1}{2x}\right| + C \qquad \left[\text{or } \ln\left|\dfrac{\sqrt{4x^2+1}}{2x} - \dfrac{1}{2x}\right| + C\right]$

52. Let $u = x^2$. Then $du = 2x\,dx \Rightarrow$

$$\int \frac{dx}{x(x^4+1)} = \int \frac{x\,dx}{x^2(x^4+1)} = \frac{1}{2}\int \frac{du}{u(u^2+1)} = \frac{1}{2}\int \left[\frac{1}{u} - \frac{u}{u^2+1}\right] du = \frac{1}{2}\ln|u| - \frac{1}{4}\ln(u^2+1) + C$$

$$= \frac{1}{2}\ln(x^2) - \frac{1}{4}\ln(x^4+1) + C = \frac{1}{4}\left[\ln(x^4) - \ln(x^4+1)\right] + C = \frac{1}{4}\ln\left(\frac{x^4}{x^4+1}\right) + C$$

Or: Write $I = \displaystyle\int \frac{x^3\,dx}{x^4(x^4+1)}$ and let $u = x^4$.

53. $\displaystyle\int x^2 \sinh(mx)\,dx = \frac{1}{m}x^2\cosh(mx) - \frac{2}{m}\int x\cosh(mx)\,dx \qquad \begin{bmatrix} u = x^2, & dv = \sinh(mx)\,dx, \\ du = 2x\,dx & v = \frac{1}{m}\cosh(mx) \end{bmatrix}$

$$= \frac{1}{m}x^2\cosh(mx) - \frac{2}{m}\left(\frac{1}{m}x\sinh(mx) - \frac{1}{m}\int \sinh(mx)\,dx\right) \qquad \begin{bmatrix} U = x, & dV = \cosh(mx)\,dx, \\ dU = dx & V = \frac{1}{m}\sinh(mx) \end{bmatrix}$$

$$= \frac{1}{m}x^2\cosh(mx) - \frac{2}{m^2}x\sinh(mx) + \frac{2}{m^3}\cosh(mx) + C$$

54. $\int (x+\sin x)^2\,dx = \int (x^2 + 2x\sin x + \sin^2 x)\,dx = \frac{1}{3}x^3 + 2(\sin x - x\cos x) + \frac{1}{2}(x - \sin x\cos x) + C$

$$= \frac{1}{3}x^3 + \frac{1}{2}x + 2\sin x - \frac{1}{2}\sin x\cos x - 2x\cos x + C$$

55. Let $u = \sqrt{x}$, so that $x = u^2$ and $dx = 2u\,du$. Then $\displaystyle\int \frac{dx}{x + x\sqrt{x}} = \int \frac{2u\,du}{u^2 + u^2 \cdot u} = \int \frac{2}{u(1+u)}\,du = I$.

Now $\dfrac{2}{u(1+u)} = \dfrac{A}{u} + \dfrac{B}{1+u} \Rightarrow 2 = A(1+u) + Bu$. Set $u = -1$ to get $2 = -B$, so $B = -2$. Set $u = 0$ to get $2 = A$.

Thus, $I = \displaystyle\int \left(\frac{2}{u} - \frac{2}{1+u}\right) du = 2\ln|u| - 2\ln|1+u| + C = 2\ln\sqrt{x} - 2\ln\left(1 + \sqrt{x}\right) + C$.

56. Let $u = \sqrt{x}$, so that $x = u^2$ and $dx = 2u\,du$. Then

$$\int \frac{dx}{\sqrt{x} + x\sqrt{x}} = \int \frac{2u\,du}{u + u^2 \cdot u} = \int \frac{2}{1+u^2}\,du = 2\tan^{-1}u + C = 2\tan^{-1}\sqrt{x} + C.$$

57. Let $u = \sqrt[3]{x+c}$. Then $x = u^3 - c \Rightarrow$

$$\int x\sqrt[3]{x+c}\,dx = \int (u^3 - c)u \cdot 3u^2\,du = 3\int (u^6 - cu^3)\,du = \frac{3}{7}u^7 - \frac{3}{4}cu^4 + C = \frac{3}{7}(x+c)^{7/3} - \frac{3}{4}c(x+c)^{4/3} + C$$

58. Let $t = \sqrt{x^2 - 1}$. Then $dt = \left(x/\sqrt{x^2-1}\right)dx$, $x^2 - 1 = t^2$, $x = \sqrt{t^2+1}$, so

$$I = \int \frac{x\ln x}{\sqrt{x^2-1}}\,dx = \int \ln\sqrt{t^2+1}\,dt = \frac{1}{2}\int \ln(t^2+1)\,dt.$$ Now use parts with $u = \ln(t^2+1)$, $dv = dt$:

$$I = \frac{1}{2}t\ln(t^2+1) - \int \frac{t^2}{t^2+1}\,dt = \frac{1}{2}t\ln(t^2+1) - \int \left[1 - \frac{1}{t^2+1}\right] dt$$

$$= \frac{1}{2}t\ln(t^2+1) - t + \tan^{-1}t + C = \sqrt{x^2-1}\,\ln x - \sqrt{x^2-1} + \tan^{-1}\sqrt{x^2-1} + C$$

Another method: First integrate by parts with $u = \ln x$, $dv = \left(x/\sqrt{x^2-1}\right)dx$ and then use substitution

$\left(x = \sec\theta \text{ or } u = \sqrt{x^2-1}\,\right)$.

59. Let $u = \sin x$, so that $du = \cos x \, dx$. Then

$$\int \cos x \, \cos^3(\sin x) \, dx = \int \cos^3 u \, du = \int \cos^2 u \, \cos u \, du = \int (1 - \sin^2 u) \cos u \, du$$

$$= \int (\cos u - \sin^2 u \, \cos u) \, du = \sin u - \tfrac{1}{3} \sin^3 u + C = \sin(\sin x) - \tfrac{1}{3} \sin^3(\sin x) + C$$

60. Let $2x = \sec \theta$, so that $2 \, dx = \sec \theta \, \tan \theta \, d\theta$. Then

$$\int \frac{dx}{x^2 \sqrt{4x^2 - 1}} = \int \frac{\tfrac{1}{2} \sec \theta \, \tan \theta \, d\theta}{\tfrac{1}{4} \sec^2 \theta \, \sqrt{\sec^2 \theta - 1}} = \int \frac{2 \tan \theta \, d\theta}{\sec \theta \, \tan \theta}$$

$$= 2 \int \cos \theta \, d\theta = 2 \sin \theta + C$$

$$= 2 \cdot \frac{\sqrt{4x^2 - 1}}{2x} + C = \frac{\sqrt{4x^2 - 1}}{x} + C$$

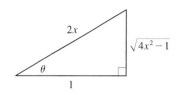

61. Let $y = \sqrt{x}$ so that $dy = \dfrac{1}{2\sqrt{x}} \, dx \;\; \Rightarrow \;\; dx = 2\sqrt{x} \, dy = 2y \, dy$. Then

$$\int \sqrt{x} \, e^{\sqrt{x}} \, dx = \int y e^y (2y \, dy) = \int 2y^2 e^y \, dy \qquad \begin{bmatrix} u = 2y^2, & dv = e^y \, dy, \\ du = 4y \, dy & v = e^y \end{bmatrix}$$

$$= 2y^2 e^y - \int 4y e^y \, dy \qquad \begin{bmatrix} U = 4y, & dV = e^y \, dy, \\ dU = 4 \, dy & V = e^y \end{bmatrix}$$

$$= 2y^2 e^y - (4y e^y - \int 4e^y \, dy) = 2y^2 e^y - 4y e^y + 4e^y + C$$

$$= 2(y^2 - 2y + 2)e^y + C = 2\left(x - 2\sqrt{x} + 2\right) e^{\sqrt{x}} + C$$

62. Let $u = \sqrt[3]{x}$. Then $x = u^3$, $dx = 3u^2 \, du \;\; \Rightarrow$

$$\int \frac{dx}{x + \sqrt[3]{x}} = \int \frac{3u^2 \, du}{u^3 + u} = \frac{3}{2} \int \frac{2u \, du}{u^2 + 1} = \tfrac{3}{2} \ln(u^2 + 1) + C = \tfrac{3}{2} \ln(x^{2/3} + 1) + C.$$

63. Let $u = \cos^2 x$, so that $du = 2 \cos x \, (-\sin x) \, dx$. Then

$$\int \frac{\sin 2x}{1 + \cos^4 x} \, dx = \int \frac{2 \sin x \, \cos x}{1 + (\cos^2 x)^2} \, dx = \int \frac{1}{1 + u^2} (-du) = -\tan^{-1} u + C = -\tan^{-1}(\cos^2 x) + C.$$

64. Let $u = \tan x$. Then

$$\int_{\pi/4}^{\pi/3} \frac{\ln(\tan x) \, dx}{\sin x \, \cos x} = \int_{\pi/4}^{\pi/3} \frac{\ln(\tan x)}{\tan x} \sec^2 x \, dx = \int_1^{\sqrt{3}} \frac{\ln u}{u} \, du = \left[\tfrac{1}{2}(\ln u)^2 \right]_1^{\sqrt{3}} = \tfrac{1}{2}\left(\ln \sqrt{3} \right)^2 = \tfrac{1}{8}(\ln 3)^2.$$

65. $\displaystyle \int \frac{dx}{\sqrt{x + 1} + \sqrt{x}} = \int \left(\frac{1}{\sqrt{x + 1} + \sqrt{x}} \cdot \frac{\sqrt{x + 1} - \sqrt{x}\sqrt{x}}{\sqrt{x + 1} - \sqrt{x}} \right) dx = \int \left(\sqrt{x + 1} - \sqrt{x} \right) dx$

$$= \tfrac{2}{3} \left[(x + 1)^{3/2} - x^{3/2} \right] + C$$

66. $\displaystyle \int \frac{u^3 + 1}{u^3 - u^2} \, du = \int \left[1 + \frac{u^2 + 1}{(u - 1)u^2} \right] du = u + \int \left[\frac{2}{u - 1} - \frac{1}{u} - \frac{1}{u^2} \right] du = u + 2 \ln|u - 1| - \ln|u| + \frac{1}{u} + C.$ Thus,

$$\int_2^3 \frac{u^3 + 1}{u^3 - u^2} \, du = \left[u + 2\ln(u - 1) - \ln u + \frac{1}{u} \right]_2^3 = \left(3 + 2\ln 2 - \ln 3 + \tfrac{1}{3} \right) - \left(2 + 2\ln 1 - \ln 2 + \tfrac{1}{2} \right)$$

$$= 1 + 3 \ln 2 - \ln 3 - \tfrac{1}{6} = \tfrac{5}{6} + \ln \tfrac{8}{3}$$

67. Let $x = \tan\theta$, so that $dx = \sec^2\theta\,d\theta$, $x = \sqrt{3}$ $\Rightarrow$ $\theta = \frac{\pi}{3}$, and $x = 1$ $\Rightarrow$ $\theta = \frac{\pi}{4}$. Then

$$\int_1^{\sqrt{3}} \frac{\sqrt{1+x^2}}{x^2}\,dx = \int_{\pi/4}^{\pi/3} \frac{\sec\theta}{\tan^2\theta}\sec^2\theta\,d\theta = \int_{\pi/4}^{\pi/3} \frac{\sec\theta\,(\tan^2\theta + 1)}{\tan^2\theta}\,d\theta = \int_{\pi/4}^{\pi/3}\left(\frac{\sec\theta\,\tan^2\theta}{\tan^2\theta} + \frac{\sec\theta}{\tan^2\theta}\right)d\theta$$

$$= \int_{\pi/4}^{\pi/3}(\sec\theta + \csc\theta\,\cot\theta)\,d\theta = \Big[\ln|\sec\theta + \tan\theta| - \csc\theta\Big]_{\pi/4}^{\pi/3}$$

$$= \left(\ln\left|2 + \sqrt{3}\right| - \tfrac{2}{\sqrt{3}}\right) - \left(\ln\left|\sqrt{2}+1\right| - \sqrt{2}\right) = \sqrt{2} - \tfrac{2}{\sqrt{3}} + \ln\left(2 + \sqrt{3}\right) - \ln\left(1 + \sqrt{2}\right)$$

68. Let $u = e^x$. Then $x = \ln u$, $dx = du/u$ $\Rightarrow$

$$\int \frac{dx}{1 + 2e^x - e^{-x}} = \int \frac{du/u}{1 + 2u - 1/u} = \int \frac{du}{2u^2 + u - 1} = \int\left[\frac{2/3}{2u-1} - \frac{1/3}{u+1}\right]du$$

$$= \tfrac{1}{3}\ln|2u - 1| - \tfrac{1}{3}\ln|u + 1| + C = \tfrac{1}{3}\ln|(2e^x - 1)/(e^x + 1)| + C$$

69. Let $u = e^x$. Then $x = \ln u$, $dx = du/u$ $\Rightarrow$

$$\int \frac{e^{2x}}{1 + e^x}\,dx = \int \frac{u^2}{1 + u}\frac{du}{u} = \int \frac{u}{1+u}\,du = \int\left(1 - \frac{1}{1+u}\right)du = u - \ln|1 + u| + C = e^x - \ln(1 + e^x) + C.$$

70. Use parts with $u = \ln(x + 1)$, $dv = dx/x^2$:

$$\int \frac{\ln(x+1)}{x^2}\,dx = -\frac{1}{x}\ln(x+1) + \int \frac{dx}{x(x+1)} = -\frac{1}{x}\ln(x+1) + \int\left[\frac{1}{x} - \frac{1}{x+1}\right]dx$$

$$= -\frac{1}{x}\ln(x+1) + \ln|x| - \ln(x+1) + C = -\left(1 + \frac{1}{x}\right)\ln(x+1) + \ln|x| + C$$

71. Let $\theta = \arcsin x$, so that $d\theta = \dfrac{1}{\sqrt{1-x^2}}\,dx$ and $x = \sin\theta$. Then

$$\int \frac{x + \arcsin x}{\sqrt{1-x^2}}\,dx = \int (\sin\theta + \theta)\,d\theta = -\cos\theta + \tfrac{1}{2}\theta^2 + C$$

$$= -\sqrt{1-x^2} + \tfrac{1}{2}(\arcsin x)^2 + C$$

72. $\displaystyle \int \frac{4^x + 10^x}{2^x}\,dx = \int\left(\frac{4^x}{2^x} + \frac{10^x}{2^x}\right)dx = \int (2^x + 5^x)\,dx = \frac{2^x}{\ln 2} + \frac{5^x}{\ln 5} + C$

73. $\displaystyle \frac{1}{(x-2)(x^2+4)} = \frac{A}{x-2} + \frac{Bx+C}{x^2+4}$ $\Rightarrow$ $1 = A(x^2+4) + (Bx+C)(x-2) = (A+B)x^2 + (C-2B)x + (4A - 2C)$.

So $0 = A + B = C - 2B$, $1 = 4A - 2C$. Setting $x = 2$ gives $A = \tfrac{1}{8}$ $\Rightarrow$ $B = -\tfrac{1}{8}$ and $C = -\tfrac{1}{4}$. So

$$\int \frac{1}{(x-2)(x^2+4)}\,dx = \int\left(\frac{\tfrac{1}{8}}{x-2} + \frac{-\tfrac{1}{8}x - \tfrac{1}{4}}{x^2+4}\right)dx = \frac{1}{8}\int \frac{dx}{x-2} - \frac{1}{16}\int \frac{2x\,dx}{x^2+4} - \frac{1}{4}\int \frac{dx}{x^2+4}$$

$$= \tfrac{1}{8}\ln|x - 2| - \tfrac{1}{16}\ln(x^2 + 4) - \tfrac{1}{8}\tan^{-1}(x/2) + C$$

74. Let $u = 2 + \sqrt{x}$, so that $du = \dfrac{1}{2\sqrt{x}}\,dx$. Then

$$\int \frac{dx}{\sqrt{x}\,(2+\sqrt{x}\,)^4} = \int \frac{2\,du}{u^4} = 2\int u^{-4}\,du = -\frac{2}{3}u^{-3} + C = -\frac{2}{3(2+\sqrt{x}\,)^3} + C.$$

75. Let $y = \sqrt{1+e^x}$, so that $y^2 = 1 + e^x$, $2y\,dy = e^x\,dx$, $e^x = y^2 - 1$, and $x = \ln(y^2 - 1)$. Then

$$\int \frac{xe^x}{\sqrt{1+e^x}}\,dx = \int \frac{\ln(y^2-1)}{y}\,(2y\,dy) = 2\int [\ln(y+1) + \ln(y-1)]\,dy$$

$$= 2[(y+1)\ln(y+1) - (y+1) + (y-1)\ln(y-1) - (y-1)] + C \qquad \text{[by Example 8.1.2]}$$

$$= 2[y\ln(y+1) + \ln(y+1) - y - 1 + y\ln(y-1) - \ln(y-1) - y + 1] + C$$

$$= 2[y(\ln(y+1) + \ln(y-1)) + \ln(y+1) - \ln(y-1) - 2y] + C$$

$$= 2\left[y\ln(y^2-1) + \ln\frac{y+1}{y-1} - 2y\right] + C = 2\left[\sqrt{1+e^x}\,\ln(e^x) + \ln\frac{\sqrt{1+e^x}+1}{\sqrt{1+e^x}-1} - 2\sqrt{1+e^x}\right] + C$$

$$= 2x\sqrt{1+e^x} + 2\ln\frac{\sqrt{1+e^x}+1}{\sqrt{1+e^x}-1} - 4\sqrt{1+e^x} + C = 2(x-2)\sqrt{1+e^x} + 2\ln\frac{\sqrt{1+e^x}+1}{\sqrt{1+e^x}-1} + C$$

76. $\int (x^2 - bx)\sin 2x\,dx = -\frac{1}{2}(x^2 - bx)\cos 2x + \frac{1}{2}\int (2x-b)\cos 2x\,dx$

$$\left[u = x^2 - bx,\, dv = \sin 2x\,dx,\, du = (2x-b)\,dx,\, v = -\tfrac{1}{2}\cos 2x\right]$$

$$= -\tfrac{1}{2}(x^2 - bx)\cos 2x + \tfrac{1}{2}\left[\tfrac{1}{2}(2x-b)\sin 2x - \int \sin 2x\,dx\right]$$

$$\left[U = 2x - b,\, dV = \cos 2x\,dx,\, dU = 2\,dx,\, V = \tfrac{1}{2}\sin 2x\right]$$

$$= -\tfrac{1}{2}(x^2 - bx)\cos 2x + \tfrac{1}{4}(2x-b)\sin 2x + \tfrac{1}{4}\cos 2x + C$$

77. Let $u = x^{3/2}$ so that $u^2 = x^3$ and $du = \frac{3}{2}x^{1/2}\,dx \;\Rightarrow\; \sqrt{x}\,dx = \frac{2}{3}\,du$. Then

$$\int \frac{\sqrt{x}}{1+x^3}\,dx = \int \frac{\frac{2}{3}}{1+u^2}\,du = \frac{2}{3}\tan^{-1}u + C = \tfrac{2}{3}\tan^{-1}(x^{3/2}) + C.$$

78. $\displaystyle \int \frac{\sec x\,\cos 2x}{\sin x + \sec x}\,dx = \int \frac{\sec x\,\cos 2x}{\sin x + \sec x}\cdot\frac{2\cos x}{2\cos x}\,dx = \int \frac{2\cos 2x}{2\sin x\cos x + 2}\,dx$

$$= \int \frac{2\cos 2x}{\sin 2x + 2}\,dx = \int \frac{1}{u}\,du \qquad \begin{bmatrix} u = \sin 2x + 2, \\ du = 2\cos 2x\,dx \end{bmatrix}$$

$$= \ln|u| + C = \ln|\sin 2x + 2| + C = \ln(\sin 2x + 2) + C$$

79. Let $u = x$, $dv = \sin^2 x\cos x\,dx \;\Rightarrow\; du = dx$, $v = \frac{1}{3}\sin^3 x$. Then

$$\int x\sin^2 x\,\cos x\,dx = \tfrac{1}{3}x\sin^3 x - \int \tfrac{1}{3}\sin^3 x\,dx = \tfrac{1}{3}x\sin^3 x - \tfrac{1}{3}\int (1-\cos^2 x)\sin x\,dx$$

$$= \frac{1}{3}x\sin^3 x + \frac{1}{3}\int (1-y^2)\,dy \qquad \begin{bmatrix} u = \cos x, \\ du = -\sin x\,dx \end{bmatrix}$$

$$= \tfrac{1}{3}x\sin^3 x + \tfrac{1}{3}y - \tfrac{1}{9}y^3 + C = \tfrac{1}{3}x\sin^3 x + \tfrac{1}{3}\cos x - \tfrac{1}{9}\cos^3 x + C$$

80. $\displaystyle\int \frac{\sin x \cos x}{\sin^4 x + \cos^4 x} \, dx = \int \frac{\sin x \cos x}{(\sin^2 x)^2 + (\cos^2 x)^2} \, dx = \int \frac{\sin x \cos x}{(\sin^2 x)^2 + (1 - \sin^2 x)^2} \, dx$

$$= \int \frac{1}{u^2 + (1-u)^2} \left(\tfrac{1}{2} \, du \right) \qquad \begin{bmatrix} u = \sin^2 x, \\ du = 2 \sin x \cos x \, dx \end{bmatrix}$$

$$= \int \frac{1}{4u^2 - 4u + 2} \, du = \int \frac{1}{(4u^2 - 4u + 1) + 1} \, du$$

$$= \int \frac{1}{(2u-1)^2 + 1} \, du = \frac{1}{2} \int \frac{1}{y^2 + 1} \, dy \qquad \begin{bmatrix} y = 2u - 1, \\ dy = 2 \, du \end{bmatrix}$$

$$= \tfrac{1}{2} \tan^{-1} y + C = \tfrac{1}{2} \tan^{-1}(2u - 1) + C = \tfrac{1}{2} \tan^{-1}(2 \sin^2 x - 1) + C$$

Another solution: $\displaystyle\int \frac{\sin x \cos x}{\sin^4 x + \cos^4 x} \, dx = \int \frac{(\sin x \cos x)/\cos^4 x}{(\sin^4 x + \cos^4 x)/\cos^4 x} \, dx = \int \frac{\tan x \sec^2 x}{\tan^4 x + 1} \, dx$

$$= \int \frac{1}{u^2 + 1} \left(\tfrac{1}{2} \, du \right) \qquad \begin{bmatrix} u = \tan^2 x, \\ du = 2 \tan x \sec^2 x \, dx \end{bmatrix}$$

$$= \tfrac{1}{2} \tan^{-1} u + C = \tfrac{1}{2} \tan^{-1}(\tan^2 x) + C$$

81. The function $y = 2xe^{x^2}$ *does* have an elementary antiderivative, so we'll use this fact to help evaluate the integral.

$$\int (2x^2 + 1)e^{x^2} \, dx = \int 2x^2 e^{x^2} \, dx + \int e^{x^2} \, dx = \int x \left(2xe^{x^2} \right) dx + \int e^{x^2} \, dx$$

$$= xe^{x^2} - \int e^{x^2} \, dx + \int e^{x^2} \, dx \qquad \begin{bmatrix} u = x, & dv = 2xe^{x^2} \, dx, \\ du = dx & v = e^{x^2} \end{bmatrix} \quad = xe^{x^2} + C$$

8.6 Integration Using Tables and Computer Algebra Systems

Keep in mind that there are several ways to approach many of these exercises, and different methods can lead to different forms of the answer.

1. We could make the substitution $u = \sqrt{2}\,x$ to obtain the radical $\sqrt{7 - u^2}$ and then use Formula 33 with $a = \sqrt{7}$.

Alternatively, we will factor $\sqrt{2}$ out of the radical and use $a = \sqrt{\frac{7}{2}}$.

$$\int \frac{\sqrt{7 - 2x^2}}{x^2} \, dx = \sqrt{2} \int \frac{\sqrt{\frac{7}{2} - x^2}}{x^2} \, dx \overset{33}{=} \sqrt{2} \left[-\frac{1}{x} \sqrt{\frac{7}{2} - x^2} - \sin^{-1} \frac{x}{\sqrt{\frac{7}{2}}} \right] + C$$

$$= -\frac{1}{x} \sqrt{7 - 2x^2} - \sqrt{2} \sin^{-1} \left(\sqrt{\tfrac{2}{7}}\, x \right) + C$$

2. $\displaystyle\int \frac{3x}{\sqrt{3 - 2x}} \, dx = 3 \int \frac{x}{\sqrt{3 + (-2)x}} \, dx \overset{55}{=} 3 \left[\frac{2}{3(-2)^2} (-2x - 2 \cdot 3) \sqrt{3 + (-2)x} \right] + C$

$$= \tfrac{1}{2}(-2x - 6) \sqrt{3 - 2x} + C = -(x + 3) \sqrt{3 - 2x} + C$$

3. Let $u = \pi x \;\Rightarrow\; du = \pi \, dx$, so

$$\int \sec^3(\pi x) \, dx = \tfrac{1}{\pi} \int \sec^3 u \, du \overset{71}{=} \tfrac{1}{\pi} \left(\tfrac{1}{2} \sec u \tan u + \tfrac{1}{2} \ln |\sec u + \tan u| \right) + C$$

$$= \tfrac{1}{2\pi} \sec \pi x \tan \pi x + \tfrac{1}{2\pi} \ln |\sec \pi x + \tan \pi x| + C$$

4. $\int e^{2\theta}\sin 3\theta\, d\theta \overset{98}{=} \dfrac{e^{2\theta}}{2^2+3^2}(2\sin 3\theta - 3\cos 3\theta) + C = \frac{2}{13}e^{2\theta}\sin 3\theta - \frac{3}{13}e^{2\theta}\cos 3\theta + C$

5. $\displaystyle\int_0^1 2x\cos^{-1}x\, dx \overset{91}{=} 2\left[\dfrac{2x^2-1}{4}\cos^{-1}x - \dfrac{x\sqrt{1-x^2}}{4}\right]_0^1 = 2\left[\left(\frac{1}{4}\cdot 0 - 0\right) - \left(-\frac{1}{4}\cdot\frac{\pi}{2} - 0\right)\right] = 2\left(\frac{\pi}{8}\right) = \frac{\pi}{4}$

6. $\displaystyle\int_2^3 \dfrac{1}{x^2\sqrt{4x^2-7}}\, dx = \int_4^6 \dfrac{1}{\left(\frac{1}{2}u\right)^2\sqrt{u^2-7}}\left(\dfrac{1}{2}\,du\right)$ $[u=2x,\, du=2\,dx]$

$$= 2\int_4^6 \dfrac{du}{u^2\sqrt{u^2-7}} \overset{45}{=} 2\left[\dfrac{\sqrt{u^2-7}}{7u}\right]_4^6 = 2\left(\dfrac{\sqrt{29}}{42} - \dfrac{3}{28}\right) = \dfrac{\sqrt{29}}{21} - \dfrac{3}{14}$$

7. Let $u=\pi x$, so that $du = \pi\, dx$. Then

$$\int \tan^3(\pi x)\, dx = \int \tan^3 u\left(\frac{1}{\pi}\,du\right) = \frac{1}{\pi}\int \tan^3 u\, du \overset{69}{=} \frac{1}{\pi}\left[\frac{1}{2}\tan^2 u + \ln|\cos u|\right] + C$$

$$= \frac{1}{2\pi}\tan^2(\pi x) + \frac{1}{\pi}\ln|\cos(\pi x)| + C$$

8. Let $u = 1+\sqrt{x}$, so that $du = \dfrac{1}{2\sqrt{x}}\, dx$. Then

$$\int \dfrac{\ln(1+\sqrt{x})}{\sqrt{x}}\, dx = \int \ln u\,(2\,du) = 2\int \ln u\, du \overset{100}{=} 2(u\ln u - u) + C = 2u(\ln u - 1) + C$$

$$= 2(1+\sqrt{x})[\ln(1+\sqrt{x}) - 1] + C$$

9. Let $u=2x$ and $a=3$. Then $du = 2\,dx$ and

$$\int \dfrac{dx}{x^2\sqrt{4x^2+9}} = \int \dfrac{\frac{1}{2}\,du}{\dfrac{u^2}{4}\sqrt{u^2+a^2}} = 2\int \dfrac{du}{u^2\sqrt{a^2+u^2}} \overset{28}{=} -2\dfrac{\sqrt{a^2+u^2}}{a^2u} + C$$

$$= -2\dfrac{\sqrt{4x^2+9}}{9\cdot 2x} + C = -\dfrac{\sqrt{4x^2+9}}{9x} + C$$

10. Let $u=\sqrt{2}\,y$ and $a=\sqrt{3}$. Then $du = \sqrt{2}\,dy$ and

$$\int \dfrac{\sqrt{2y^2-3}}{y^2}\, dy = \int \dfrac{\sqrt{u^2-a^2}}{\frac{1}{2}u^2}\dfrac{du}{\sqrt{2}} = \sqrt{2}\int \dfrac{\sqrt{u^2-a^2}}{u^2}\, du \overset{42}{=} \sqrt{2}\left(-\dfrac{\sqrt{u^2-a^2}}{u} + \ln\left|u + \sqrt{u^2-a^2}\right|\right) + C$$

$$= \sqrt{2}\left(-\dfrac{\sqrt{2y^2-3}}{\sqrt{2}\,y} + \ln\left|\sqrt{2}\,y + \sqrt{2y^2-3}\right|\right) + C$$

$$= -\dfrac{\sqrt{2y^2-3}}{y} + \sqrt{2}\ln\left|\sqrt{2}\,y + \sqrt{2y^2-3}\right| + C$$

11. $\displaystyle\int_{-1}^0 t^2 e^{-t}\, dt \overset{97}{=} \left[\dfrac{1}{-1}t^2 e^{-t}\right]_{-1}^0 - \dfrac{2}{-1}\int_{-1}^0 te^{-t}\, dt = e + 2\int_{-1}^0 te^{-t}\, dt \overset{96}{=} e + 2\left[\dfrac{1}{(-1)^2}(-t-1)e^{-t}\right]_{-1}^0$

$$= e + 2\left[-e^0 + 0\right] = e - 2$$

12. Let $u = x^3 + 1$, so that $du = 3x^2\, dx$. Then

$$\int x^2\operatorname{csch}(x^3+1)\, dx = \int \operatorname{csch}u\left(\frac{1}{3}\,du\right) = \frac{1}{3}\int \operatorname{csch}u\, du \overset{108}{=} \frac{1}{3}\ln\left|\tanh\frac{1}{2}u\right| + C = \frac{1}{3}\ln\left|\tanh\frac{1}{2}(x^3+1)\right| + C$$

13. $\displaystyle\int \frac{\tan^3(1/z)}{z^2}\,dz \quad \begin{bmatrix} u = 1/z, \\ du = -dz/z^2 \end{bmatrix} = -\int \tan^3 u\,du \overset{69}{=} -\tfrac{1}{2}\tan^2 u - \ln|\cos u| + C$

$\qquad\qquad\qquad\qquad\qquad\qquad\qquad = -\tfrac{1}{2}\tan^2\left(\tfrac{1}{z}\right) - \ln\left|\cos\left(\tfrac{1}{z}\right)\right| + C$

14. Let $u = \sqrt{x}$. Then $u^2 = x$ and $2u\,du = dx$, so

$$\int \sin^{-1}\sqrt{x}\,dx = 2\int u\sin^{-1} u\,du \overset{90}{=} \frac{2u^2-1}{2}\sin^{-1} u + \frac{u\sqrt{1-u^2}}{2} + C = \frac{2x-1}{2}\sin^{-1}\sqrt{x} + \frac{\sqrt{x(1-x)}}{2} + C.$$

15. Let $u = e^x$, so that $du = e^x\,dx$ and $e^{2x} = u^2$. Then

$$\int e^{2x}\arctan(e^x)\,dx = \int u^2\arctan u\left(\frac{du}{u}\right) = \int u\arctan u\,du$$

$$\overset{92}{=} \frac{u^2+1}{2}\arctan u - \frac{u}{2} + C = \frac{1}{2}(e^{2x}+1)\arctan(e^x) - \frac{1}{2}e^x + C$$

16. Let $u = x^2$, so that $du = 2x\,dx$. Then

$$\int x\sin(x^2)\cos(3x^2)\,dx = \frac{1}{2}\int \sin u\cos 3u\,du \overset{81}{=} -\frac{1}{2}\frac{\cos(1-3)u}{2(1-3)} - \frac{1}{2}\frac{\cos(1+3)u}{2(1+3)} + C$$

$$= \tfrac{1}{8}\cos 2u - \tfrac{1}{16}\cos 4u + C = \tfrac{1}{8}\cos(2x^2) - \tfrac{1}{16}\cos(4x^2) + C$$

17. Let $z = 6 + 4y - 4y^2 = 6 - (4y^2 - 4y + 1) + 1 = 7 - (2y-1)^2$, $u = 2y - 1$, and $a = \sqrt{7}$. Then $z = a^2 - u^2$, $du = 2\,dy$,
and

$$\int y\sqrt{6+4y-4y^2}\,dy = \int y\sqrt{z}\,dy = \int \tfrac{1}{2}(u+1)\sqrt{a^2-u^2}\,\tfrac{1}{2}\,du = \tfrac{1}{4}\int u\sqrt{a^2-u^2}\,du + \tfrac{1}{4}\int \sqrt{a^2-u^2}\,du$$

$$= \tfrac{1}{4}\int \sqrt{a^2-u^2}\,du - \tfrac{1}{8}\int(-2u)\sqrt{a^2-u^2}\,du$$

$$\overset{30}{=} \frac{u}{8}\sqrt{a^2-u^2} + \frac{a^2}{8}\sin^{-1}\left(\frac{u}{a}\right) - \frac{1}{8}\int \sqrt{w}\,dw \quad \begin{bmatrix} w = a^2 - u^2, \\ dw = -2u\,du \end{bmatrix}$$

$$= \frac{2y-1}{8}\sqrt{6+4y-4y^2} + \frac{7}{8}\sin^{-1}\frac{2y-1}{\sqrt{7}} - \frac{1}{8}\cdot\frac{2}{3}w^{3/2} + C$$

$$= \frac{2y-1}{8}\sqrt{6+4y-4y^2} + \frac{7}{8}\sin^{-1}\frac{2y-1}{\sqrt{7}} - \frac{1}{12}(6+4y-4y^2)^{3/2} + C$$

This can be rewritten as

$$\sqrt{6+4y-4y^2}\left[\frac{1}{8}(2y-1) - \frac{1}{12}(6+4y-4y^2)\right] + \frac{7}{8}\sin^{-1}\frac{2y-1}{\sqrt{7}} + C$$

$$= \left(\frac{1}{3}y^2 - \frac{1}{12}y - \frac{5}{8}\right)\sqrt{6+4y-4y^2} + \frac{7}{8}\sin^{-1}\left(\frac{2y-1}{\sqrt{7}}\right) + C$$

$$= \frac{1}{24}(8y^2 - 2y - 15)\sqrt{6+4y-4y^2} + \frac{7}{8}\sin^{-1}\left(\frac{2y-1}{\sqrt{7}}\right) + C$$

18. $\displaystyle\int \frac{dx}{2x^3 - 3x^2} = \int \frac{dx}{x^2(-3+2x)} \overset{50}{=} -\frac{1}{-3x} + \frac{2}{(-3)^2}\ln\left|\frac{-3+2x}{x}\right| + C = \frac{1}{3x} + \frac{2}{9}\ln\left|\frac{2x-3}{x}\right| + C$

19. Let $u = \sin x$. Then $du = \cos x\,dx$, so

$$\int \sin^2 x\cos x\ln(\sin x)\,dx = \int u^2\ln u\,du \overset{101}{=} \frac{u^{2+1}}{(2+1)^2}\left[(2+1)\ln u - 1\right] + C = \tfrac{1}{9}u^3(3\ln u - 1) + C$$

$$= \tfrac{1}{9}\sin^3 x\left[3\ln(\sin x) - 1\right] + C$$

20. Let $u = \sin\theta$, so that $du = \cos\theta\,d\theta$. Then

$$\int \frac{\sin 2\theta}{\sqrt{5 - \sin\theta}}\,d\theta = \int \frac{2\sin\theta\,\cos\theta}{\sqrt{5 - \sin\theta}}\,d\theta = 2\int \frac{u}{\sqrt{5 - u}}\,du \stackrel{55}{=} 2\cdot\frac{2}{3(-1)^2}\left[-1u - 2(5)\right]\sqrt{5 - u} + C$$

$$= \tfrac{4}{3}(-u - 10)\sqrt{5 - u} + C = -\tfrac{4}{3}(\sin\theta + 10)\sqrt{5 - \sin\theta} + C$$

21. Let $u = e^x$ and $a = \sqrt{3}$. Then $du = e^x\,dx$ and

$$\int \frac{e^x}{3 - e^{2x}}\,dx = \int \frac{du}{a^2 - u^2} \stackrel{19}{=} \frac{1}{2a}\ln\left|\frac{u + a}{u - a}\right| + C = \frac{1}{2\sqrt{3}}\ln\left|\frac{e^x + \sqrt{3}}{e^x - \sqrt{3}}\right| + C.$$

22. Let $u = x^2$ and $a = 2$. Then $du = 2x\,dx$ and

$$\int_0^2 x^3\sqrt{4x^2 - x^4}\,dx = \tfrac{1}{2}\int_0^2 x^2\sqrt{2\cdot 2\cdot x^2 - (x^2)^2}\cdot 2x\,dx = \tfrac{1}{2}\int_0^4 u\sqrt{2au - u^2}\,du$$

$$\stackrel{114}{=} \left[\frac{2u^2 - au - 3a^2}{12}\sqrt{2au - u^2} + \frac{a^3}{4}\cos^{-1}\left(\frac{a - u}{a}\right)\right]_0^4$$

$$= \left[\frac{2u^2 - 2u - 12}{12}\sqrt{4u - u^2} + \frac{8}{4}\cos^{-1}\left(\frac{2 - u}{2}\right)\right]_0^4$$

$$= \left[\frac{u^2 - u - 6}{6}\sqrt{4u - u^2} + 2\cos^{-1}\left(\frac{2 - u}{2}\right)\right]_0^4$$

$$= [0 + 2\cos^{-1}(-1)] - (0 + 2\cos^{-1}1) = 2\cdot\pi - 2\cdot 0 = 2\pi$$

23. $\int \sec^5 x\,dx \stackrel{77}{=} \tfrac{1}{4}\tan x\,\sec^3 x + \tfrac{3}{4}\int \sec^3 x\,dx \stackrel{77}{=} \tfrac{1}{4}\tan x\,\sec^3 x + \tfrac{3}{4}\left(\tfrac{1}{2}\tan x\,\sec x + \tfrac{1}{2}\int \sec x\,dx\right)$

$\stackrel{14}{=} \tfrac{1}{4}\tan x\,\sec^3 x + \tfrac{3}{8}\tan x\,\sec x + \tfrac{3}{8}\ln|\sec x + \tan x| + C$

24. Let $u = 2x$. Then $du = 2\,dx$, so

$$\int \sin^6 2x\,dx = \tfrac{1}{2}\int \sin^6 u\,du \stackrel{73}{=} \tfrac{1}{2}\left(-\tfrac{1}{6}\sin^5 u\,\cos u + \tfrac{5}{6}\int \sin^4 u\,du\right)$$

$$\stackrel{73}{=} -\tfrac{1}{12}\sin^5 u\,\cos u + \tfrac{5}{12}\left(-\tfrac{1}{4}\sin^3 u\,\cos u + \tfrac{3}{4}\int \sin^2 u\,du\right)$$

$$\stackrel{63}{=} -\tfrac{1}{12}\sin^5 u\,\cos u - \tfrac{5}{48}\sin^3 u\,\cos u + \tfrac{5}{16}\left(\tfrac{1}{2}u - \tfrac{1}{4}\sin 2u\right) + C$$

$$= -\tfrac{1}{12}\sin^5 2x\,\cos 2x - \tfrac{5}{48}\sin^3 2x\,\cos 2x - \tfrac{5}{64}\sin 4x + \tfrac{5}{16}x + C$$

25. Let $u = \ln x$ and $a = 2$. Then $du = dx/x$ and

$$\int \frac{\sqrt{4 + (\ln x)^2}}{x}\,dx = \int \sqrt{a^2 + u^2}\,du \stackrel{21}{=} \frac{u}{2}\sqrt{a^2 + u^2} + \frac{a^2}{2}\ln\left(u + \sqrt{a^2 + u^2}\right) + C$$

$$= \tfrac{1}{2}(\ln x)\sqrt{4 + (\ln x)^2} + 2\ln\left[\ln x + \sqrt{4 + (\ln x)^2}\right] + C$$

26. $\int x^4 e^{-x}\,dx \stackrel{97}{=} -x^4 e^{-x} + 4\int x^3 e^{-x}\,dx \stackrel{97}{=} -x^4 e^{-x} + 4\left(-x^3 e^{-x} + 3\int x^2 e^{-x}\,dx\right)$

$\stackrel{97}{=} -(x^4 + 4x^3)e^{-x} + 12\left(-x^2 e^{-x} + 2\int xe^{-x}\,dx\right)$

$\stackrel{96}{=} -(x^4 + 4x^3 + 12x^2)e^{-x} + 24[(-x - 1)e^{-x}] + C = -(x^4 + 4x^3 + 12x^2 + 24x + 24)e^{-x} + C$

So $\int_0^1 x^4 e^{-x}\,dx = \left[-(x^4 + 4x^3 + 12x^2 + 24x + 24)e^{-x}\right]_0^1 = -(1 + 4 + 12 + 24 + 24)e^{-1} + 24e^0 = 24 - 65e^{-1}$.

27. Let $u = e^x$. Then $x = \ln u$, $dx = du/u$, so

$$\int \sqrt{e^{2x} - 1}\,dx = \int \frac{\sqrt{u^2 - 1}}{u}\,du \overset{41}{=} \sqrt{u^2 - 1} - \cos^{-1}(1/u) + C = \sqrt{e^{2x} - 1} - \cos^{-1}(e^{-x}) + C.$$

28. Let $u = \alpha t - 3$ and assume that $\alpha \neq 0$. Then $du = \alpha\,dt$ and

$$\int e^t \sin(\alpha t - 3)\,dt = \frac{1}{\alpha}\int e^{(u+3)/\alpha} \sin u\,du = \frac{1}{\alpha}e^{3/\alpha}\int e^{(1/\alpha)u} \sin u\,du$$

$$\overset{98}{=} \frac{1}{\alpha}e^{3/\alpha}\frac{e^{(1/\alpha)u}}{(1/\alpha)^2 + 1^2}\left(\frac{1}{\alpha}\sin u - \cos u\right) + C = \frac{1}{\alpha}e^{3/\alpha}e^{(1/\alpha)u}\frac{\alpha^2}{1 + \alpha^2}\left(\frac{1}{\alpha}\sin u - \cos u\right) + C$$

$$= \frac{1}{1 + \alpha^2}e^{(u+3)/\alpha}(\sin u - \alpha \cos u) + C = \frac{1}{1 + \alpha^2}e^t\left[\sin(\alpha t - 3) - \alpha \cos(\alpha t - 3)\right] + C$$

29. $\displaystyle\int \frac{x^4\,dx}{\sqrt{x^{10} - 2}} = \int \frac{x^4\,dx}{\sqrt{(x^5)^2 - 2}} = \frac{1}{5}\int \frac{du}{\sqrt{u^2 - 2}}$ $\quad\begin{bmatrix} u = x^5, \\ du = 5x^4\,dx \end{bmatrix}$

$$\overset{43}{=} \tfrac{1}{5}\ln\left|u + \sqrt{u^2 - 2}\right| + C = \tfrac{1}{5}\ln\left|x^5 + \sqrt{x^{10} - 2}\right| + C$$

30. Let $u = \tan\theta$ and $a = 3$. Then $du = \sec^2\theta\,d\theta$ and

$$\int \frac{\sec^2\theta \tan^2\theta}{\sqrt{9 - \tan^2\theta}}\,d\theta = \int \frac{u^2}{\sqrt{a^2 - u^2}}\,du \overset{34}{=} -\frac{u}{2}\sqrt{a^2 - u^2} + \frac{a^2}{2}\sin^{-1}\left(\frac{u}{a}\right) + C$$

$$= -\frac{1}{2}\tan\theta\sqrt{9 - \tan^2\theta} + \frac{9}{2}\sin^{-1}\left(\frac{\tan\theta}{3}\right) + C$$

31. Using cylindrical shells, we get

$$V = 2\pi\int_0^2 x \cdot x\sqrt{4 - x^2}\,dx = 2\pi\int_0^2 x^2\sqrt{4 - x^2}\,dx \overset{31}{=} 2\pi\left[\frac{x}{8}(2x^2 - 4)\sqrt{4 - x^2} + \frac{16}{8}\sin^{-1}\frac{x}{2}\right]_0^2$$

$$= 2\pi[(0 + 2\sin^{-1} 1) - (0 + 2\sin^{-1} 0)] = 2\pi\left(2 \cdot \frac{\pi}{2}\right) = 2\pi^2$$

32. Using disks, we get

$$\text{Volume} = \int_0^{\pi/4} \pi \tan^4 x\,dx \overset{75}{=} \pi\left(\left[\tfrac{1}{3}\tan^3 x\right]_0^{\pi/4} - \int_0^{\pi/4}\tan^2 x\,dx\right) \overset{65}{=} \pi\left[\tfrac{1}{3}\tan^3 x - \tan x + x\right]_0^{\pi/4}$$

$$= \pi\left(\tfrac{1}{3} - 1 + \tfrac{\pi}{4}\right) = \pi\left(\tfrac{\pi}{4} - \tfrac{2}{3}\right)$$

33. (a) $\displaystyle\frac{d}{du}\left[\frac{1}{b^3}\left(a + bu - \frac{a^2}{a + bu} - 2a\ln|a + bu|\right) + C\right] = \frac{1}{b^3}\left[b + \frac{ba^2}{(a + bu)^2} - \frac{2ab}{(a + bu)}\right]$

$$= \frac{1}{b^3}\left[\frac{b(a + bu)^2 + ba^2 - (a + bu)2ab}{(a + bu)^2}\right] = \frac{1}{b^3}\left[\frac{b^3 u^2}{(a + bu)^2}\right] = \frac{u^2}{(a + bu)^2}$$

(b) Let $t = a + bu \Rightarrow dt = b\,du$. Note that $u = \dfrac{t - a}{b}$ and $du = \dfrac{1}{b}\,dt$.

$$\int \frac{u^2\,du}{(a + bu)^2} = \frac{1}{b^3}\int \frac{(t - a)^2}{t^2}\,dt = \frac{1}{b^3}\int \frac{t^2 - 2at + a^2}{t^2}\,dt = \frac{1}{b^3}\int\left(1 - \frac{2a}{t} + \frac{a^2}{t^2}\right)dt$$

$$= \frac{1}{b^3}\left(t - 2a\ln|t| - \frac{a^2}{t}\right) + C = \frac{1}{b^3}\left(a + bu - \frac{a^2}{a + bu} - 2a\ln|a + bu|\right) + C$$

34. (a) $\dfrac{d}{du}\left[\dfrac{u}{8}(2u^2 - a^2)\sqrt{a^2 - u^2} + \dfrac{a^4}{8}\sin^{-1}\dfrac{u}{a} + C\right]$

$$= \dfrac{u}{8}(2u^2 - a^2)\dfrac{-u}{\sqrt{a^2 - u^2}} + \sqrt{a^2 - u^2}\left[\dfrac{u}{8}(4u) + (2u^2 - a^2)\tfrac{1}{8}\right] + \dfrac{a^4}{8}\dfrac{1/a}{\sqrt{1 - u^2/a^2}}$$

$$= -\dfrac{u^2(2u^2 - a^2)}{8\sqrt{a^2 - u^2}} + \sqrt{a^2 - u^2}\left[\dfrac{u^2}{2} + \dfrac{2u^2 - a^2}{8}\right] + \dfrac{a^4}{8\sqrt{a^2 - u^2}}$$

$$= \tfrac{1}{2}(a^2 - u^2)^{-1/2}\left[-\dfrac{u^2}{4}(2u^2 - a^2) + u^2(a^2 - u^2) + \tfrac{1}{4}(a^2 - u^2)(2u^2 - a^2) + \dfrac{a^4}{4}\right]$$

$$= \tfrac{1}{2}(a^2 - u^2)^{-1/2}[2u^2 a^2 - 2u^4] = \dfrac{u^2(a^2 - u^2)}{\sqrt{a^2 - u^2}} = u^2\sqrt{a^2 - u^2}$$

(b) Let $u = a\sin\theta \;\Rightarrow\; du = a\cos\theta\,d\theta$. Then

$$\int u^2\sqrt{a^2 - u^2}\,du = \int a^2\sin^2\theta\, a\sqrt{1 - \sin^2\theta}\,a\cos\theta\,d\theta = a^4\int \sin^2\theta\,\cos^2\theta\,d\theta$$

$$= a^4\int \tfrac{1}{2}(1 + \cos 2\theta)\tfrac{1}{2}(1 - \cos 2\theta)\,d\theta = \tfrac{1}{4}a^4\int(1 - \cos^2 2\theta)\,d\theta$$

$$= \tfrac{1}{4}a^4\int\left[1 - \tfrac{1}{2}(1 + \cos 4\theta)\right]d\theta = \tfrac{1}{4}a^4\left(\tfrac{1}{2}\theta - \tfrac{1}{8}\sin 4\theta\right) + C$$

$$= \tfrac{1}{4}a^4\left(\tfrac{1}{2}\theta - \tfrac{1}{8}\cdot 2\sin 2\theta\,\cos 2\theta\right) + C = \tfrac{1}{4}a^4\left[\tfrac{1}{2}\theta - \tfrac{1}{2}\sin\theta\,\cos\theta(1 - 2\sin^2\theta)\right] + C$$

$$= \dfrac{a^4}{8}\left[\sin^{-1}\dfrac{u}{a} - \dfrac{u}{a}\dfrac{\sqrt{a^2 - u^2}}{a}\left(1 - \dfrac{2u^2}{a^2}\right)\right] + C = \dfrac{a^4}{8}\left[\sin^{-1}\dfrac{u}{a} - \dfrac{u}{a}\dfrac{\sqrt{a^2 - u^2}}{a}\dfrac{a^2 - 2u^2}{a^2}\right] + C$$

$$= \dfrac{u}{8}(2u^2 - a^2)\sqrt{a^2 - u^2} + \dfrac{a^4}{8}\sin^{-1}\dfrac{u}{a} + C$$

35. Maple and Mathematica both give $\int \sec^4 x\,dx = \tfrac{2}{3}\tan x + \tfrac{1}{3}\tan x\,\sec^2 x$, while Derive gives the second

term as $\dfrac{\sin x}{3\cos^3 x} = \dfrac{1}{3}\dfrac{\sin x}{\cos x}\dfrac{1}{\cos^2 x} = \dfrac{1}{3}\tan x\,\sec^2 x$. Using Formula 77, we get

$\int \sec^4 x\,dx = \tfrac{1}{3}\tan x\,\sec^2 x + \tfrac{2}{3}\int \sec^2 x\,dx = \tfrac{1}{3}\tan x\,\sec^2 x + \tfrac{2}{3}\tan x + C$.

36. Derive gives $\int \csc^5 x\,dx = \dfrac{3}{8}\ln\left(\tan\left(\dfrac{x}{2}\right)\right) - \cos x\left(\dfrac{3}{8\sin^2 x} + \dfrac{1}{4\sin^4 x}\right)$ and Maple gives

$-\dfrac{1}{4}\dfrac{\cos x}{\sin^4 x} - \dfrac{3}{8}\dfrac{\cos x}{\sin^2 x} + \dfrac{3}{8}\ln(\csc x - \cot x)$. Using a half-angle identity for tangent, $\tan\dfrac{x}{2} = \dfrac{1 - \cos x}{\sin x}$, we have

$\ln\tan\dfrac{x}{2} = \ln\dfrac{1 - \cos x}{\sin x} = \ln\left(\dfrac{1}{\sin x} - \dfrac{\cos x}{\sin x}\right) = \ln(\csc x - \cot x)$, so those two answers are equivalent.

Mathematica gives

$$I = -\dfrac{3}{32}\csc^2\dfrac{x}{2} - \dfrac{1}{64}\csc^4\dfrac{x}{2} - \dfrac{3}{8}\log\cos\dfrac{x}{2} + \dfrac{3}{8}\log\sin\dfrac{x}{2} + \dfrac{3}{32}\sec^2\dfrac{x}{2} + \dfrac{1}{64}\sec^4\dfrac{x}{2}$$

$$= \dfrac{3}{8}\left(\log\sin\dfrac{x}{2} - \log\cos\dfrac{x}{2}\right) + \dfrac{3}{32}\left(\sec^2\dfrac{x}{2} - \csc^2\dfrac{x}{2}\right) + \dfrac{1}{64}\left(\sec^4\dfrac{x}{2} - \csc^4\dfrac{x}{2}\right)$$

$$= \dfrac{3}{8}\log\dfrac{\sin(x/2)}{\cos(x/2)} + \dfrac{3}{32}\left[\dfrac{1}{\cos^2(x/2)} - \dfrac{1}{\sin^2(x/2)}\right] + \dfrac{1}{64}\left[\dfrac{1}{\cos^4(x/2)} - \dfrac{1}{\sin^4(x/2)}\right]$$

$$= \dfrac{3}{8}\log\tan\dfrac{x}{2} + \dfrac{3}{32}\left[\dfrac{\sin^2(x/2) - \cos^2(x/2)}{\cos^2(x/2)\sin^2(x/2)}\right] + \dfrac{1}{64}\left[\dfrac{\sin^4(x/2) - \cos^4(x/2)}{\cos^4(x/2)\sin^4(x/2)}\right]$$

[continued]

Now
$$\frac{\sin^2(x/2) - \cos^2(x/2)}{\cos^2(x/2)\,\sin^2(x/2)} = \frac{\dfrac{1-\cos x}{2} - \dfrac{1+\cos x}{2}}{\dfrac{1+\cos x}{2} \cdot \dfrac{1-\cos x}{2}} = \frac{-\dfrac{2\cos x}{2}}{\dfrac{1-\cos^2 x}{4}} = \frac{-4\cos x}{\sin^2 x}$$

and
$$\frac{\sin^4(x/2) - \cos^4(x/2)}{\cos^4(x/2)\,\sin^4(x/2)} = \frac{\sin^2(x/2) - \cos^2(x/2)}{\cos^2(x/2)\,\sin^2(x/2)} \, \frac{\sin^2(x/2) + \cos^2(x/2)}{\cos^2(x/2)\,\sin^2(x/2)}$$

$$= \frac{-4\cos x}{\sin^2 x}\, \frac{1}{\dfrac{1+\cos x}{2} \cdot \dfrac{1-\cos x}{2}} = -\frac{4\cos x}{\sin^2 x}\, \frac{4}{1-\cos^2 x} = -\frac{16\cos x}{\sin^4 x}$$

Returning to the expression for I, we get

$$I = \frac{3}{8}\log\tan\frac{x}{2} + \frac{3}{32}\left(\frac{-4\cos x}{\sin^2 x}\right) + \frac{1}{64}\left(\frac{-16\cos x}{\sin^4 x}\right) = \frac{3}{8}\log\tan\frac{x}{2} - \frac{3}{8}\frac{\cos x}{\sin^2 x} - \frac{1}{4}\frac{\cos x}{\sin^4 x},$$

so all are equivalent.

Now use Formula 78 to get

$$\int \csc^5 x\,dx = \frac{-1}{4}\cot x\,\csc^3 x + \frac{3}{4}\int \csc^3 x\,dx = -\frac{1}{4}\frac{\cos x}{\sin x}\frac{1}{\sin^3 x} + \frac{3}{4}\left(\frac{-1}{2}\cot x\,\csc x + \frac{1}{2}\int \csc x\,dx\right)$$

$$= -\frac{1}{4}\frac{\cos x}{\sin^4 x} - \frac{3}{8}\frac{\cos x}{\sin x}\frac{1}{\sin x} + \frac{3}{8}\int \csc x\,dx = -\frac{1}{4}\frac{\cos x}{\sin^4 x} - \frac{3}{8}\frac{\cos x}{\sin^2 x} + \frac{3}{8}\ln|\csc x - \cot x| + C$$

37. Derive gives $\int x^2\sqrt{x^2+4}\,dx = \frac{1}{4}x(x^2+2)\sqrt{x^2+4} - 2\ln\left(\sqrt{x^2+4}+x\right)$. Maple gives

$\frac{1}{4}x(x^2+4)^{3/2} - \frac{1}{2}x\sqrt{x^2+4} - 2\operatorname{arcsinh}\left(\frac{1}{2}x\right)$. Applying the command `convert(%,ln);` yields

$$\frac{1}{4}x(x^2+4)^{3/2} - \frac{1}{2}x\sqrt{x^2+4} - 2\ln\left(\frac{1}{2}x + \frac{1}{2}\sqrt{x^2+4}\right) = \frac{1}{4}x(x^2+4)^{1/2}\left[(x^2+4)-2\right] - 2\ln\left[\left(x+\sqrt{x^2+4}\right)/2\right]$$

$$= \frac{1}{4}x(x^2+2)\sqrt{x^2+4} - 2\ln\left(\sqrt{x^2+4}+x\right) + 2\ln 2$$

Mathematica gives $\frac{1}{4}x(2+x^2)\sqrt{3+x^2} - 2\operatorname{arcsinh}(x/2)$. Applying the `TrigToExp` and `Simplify` commands gives

$\frac{1}{4}\left[x(2+x^2)\sqrt{4+x^2} - 8\log\left(\frac{1}{2}\left(x+\sqrt{4+x^2}\right)\right)\right] = \frac{1}{4}x(x^2+2)\sqrt{x^2+4} - 2\ln\left(x+\sqrt{4+x^2}\right) + 2\ln 2$, so all are

equivalent (without constant).

Now use Formula 22 to get

$$\int x^2\sqrt{2^2+x^2}\,dx = \frac{x}{8}(2^2+2x^2)\sqrt{2^2+x^2} - \frac{2^4}{8}\ln\left(x+\sqrt{2^2+x^2}\right) + C$$

$$= \frac{x}{8}(2)(2+x^2)\sqrt{4+x^2} - 2\ln\left(x+\sqrt{4+x^2}\right) + C$$

$$= \frac{1}{4}x(x^2+2)\sqrt{x^2+4} - 2\ln\left(\sqrt{x^2+4}+x\right) + C$$

38. Derive gives $\displaystyle\int \frac{dx}{e^x(3e^x+2)} = -\frac{e^{-x}}{2} + \frac{3\ln(3e^x+2)}{4} - \frac{3x}{4}$, Maple gives $\frac{3}{4}\ln(3e^x+2) - \frac{1}{2e^x} - \frac{3}{4}\ln(e^x)$, and

Mathematica gives

$$-\frac{e^{-x}}{2} + \frac{3}{4}\log(3+2e^{-x}) = -\frac{e^{-x}}{2} + \frac{3}{4}\log\left(\frac{3e^x+2}{e^x}\right) = -\frac{e^{-x}}{2} + \frac{3}{4}\frac{\ln(3e^x+2)}{\ln e^x} = -\frac{e^{-x}}{2} + \frac{3}{4}\ln(3e^x+2) - \frac{3}{4}x,$$

so all are equivalent. Now let $u = e^x$, so $du = e^x\,dx$ and $dx = du/u$. Then

$$\int \frac{1}{e^x(3e^x+2)}\,dx = \int \frac{1}{u(3u+2)}\,\frac{du}{u} = \int \frac{1}{u^2(2+3u)}\,du \overset{50}{=} -\frac{1}{2u} + \frac{3}{2^2}\ln\left|\frac{2+3u}{u}\right| + C$$

$$= -\frac{1}{2e^x} + \frac{3}{4}\ln(2+3e^x) - \frac{3}{4}\ln e^x + C = -\frac{1}{2e^x} + \frac{3}{4}\ln(3e^x+2) - \frac{3}{4}x + C$$

39. Maple gives $\int x\sqrt{1+2x}\,dx = \frac{1}{10}(1+2x)^{5/2} - \frac{1}{6}(1+2x)^{3/2}$, Mathematica gives $\sqrt{1+2x}\left(\frac{2}{5}x^2 + \frac{1}{15}x - \frac{1}{15}\right)$, and Derive

gives $\frac{1}{15}(1+2x)^{3/2}(3x-1)$. The first two expressions can be simplified to Derive's result. If we use Formula 54, we get

$$\int x\sqrt{1+2x}\,dx = \frac{2}{15(2)^2}(3\cdot 2x - 2\cdot 1)(1+2x)^{3/2} + C = \frac{1}{30}(6x-2)(1+2x)^{3/2} + C = \frac{1}{15}(3x-1)(1+2x)^{3/2}.$$

40. Maple and Derive both give $\int \sin^4 x\,dx = -\frac{1}{4}\sin^3 x\,\cos x - \frac{3}{8}\cos x\,\sin x + \frac{3}{8}x$, while Mathematica gives

$\frac{1}{32}(12x - 8\sin 2x + \sin 4x)$, which can be expanded and simplified to give the other expression. Now

$$\int \sin^4 x\,dx \overset{73}{=} -\frac{1}{4}\sin^3 x\,\cos x + \frac{3}{4}\int \sin^2 x\,dx \overset{63}{=} -\frac{1}{4}\sin^3 x\,\cos x + \frac{3}{4}\left(\frac{1}{2}x - \frac{1}{4}\sin 2x\right) + C$$

$$= -\frac{1}{4}\sin^3 x\,\cos x - \frac{3}{8}\sin x\,\cos x + \frac{3}{8}x + C \text{ since } \sin 2x = 2\sin x\,\cos x$$

41. Maple gives $\int \tan^5 x\,dx = \frac{1}{4}\tan^4 x - \frac{1}{2}\tan^2 x + \frac{1}{2}\ln(1+\tan^2 x)$, Mathematica gives

$\int \tan^5 x\,dx = \frac{1}{4}[-1 - 2\cos(2x)]\sec^4 x - \ln(\cos x)$, and Derive gives $\int \tan^5 x\,dx = \frac{1}{4}\tan^4 x - \frac{1}{2}\tan^2 x - \ln(\cos x)$.

These expressions are equivalent, and none includes absolute value bars or a constant of integration. Note that Mathematica's

and Derive's expressions suggest that the integral is undefined where $\cos x < 0$, which is not the case. Using Formula 75,

$\int \tan^5 x\,dx = \frac{1}{5-1}\tan^{5-1} x - \int \tan^{5-2} x\,dx = \frac{1}{4}\tan^4 x - \int \tan^3 x\,dx$. Using Formula 69,

$\int \tan^3 x\,dx = \frac{1}{2}\tan^2 x + \ln|\cos x| + C$, so $\int \tan^5 x\,dx = \frac{1}{4}\tan^4 x - \frac{1}{2}\tan^2 x - \ln|\cos x| + C$.

42. Derive, Maple, and Mathematica all give $\displaystyle\int \frac{1}{\sqrt{1+\sqrt[3]{x}}}\,dx = \frac{2}{5}\sqrt{\sqrt[3]{x}+1}\left(3\sqrt[3]{x^2} - 4\sqrt[3]{x} + 8\right).$ [Maple adds a

constant of $-\frac{16}{5}$.] We'll change the form of the integral by letting $u = \sqrt[3]{x}$, so that $u^3 = x$ and $3u^2\,du = dx$. Then

$$\int \frac{1}{\sqrt{1+\sqrt[3]{x}}}\,dx = \int \frac{3u^2\,du}{\sqrt{1+u}} \overset{56}{=} 3\left[\frac{2}{15(1)^3}\left(8(1)^2 + 3(1)^2 u^2 - 4(1)(1)u\right)\sqrt{1+u}\right] + C$$

$$= \frac{2}{5}(8 + 3u^2 - 4u)\sqrt{1+u} + C = \frac{2}{5}\left(8 + 3\sqrt[3]{x^2} - 4\sqrt[3]{x}\right)\sqrt{1+\sqrt[3]{x}} + C$$

43. (a) $F(x) = \displaystyle\int f(x)\,dx = \int \frac{1}{x\sqrt{1-x^2}}\,dx \overset{35}{=} -\frac{1}{1}\ln\left|\frac{1+\sqrt{1-x^2}}{x}\right| + C = -\ln\left|\frac{1+\sqrt{1-x^2}}{x}\right| + C.$

f has domain $\{x \mid x \neq 0, 1 - x^2 > 0\} = \{x \mid x \neq 0, |x| < 1\} = (-1, 0) \cup (0, 1)$. F has the same domain.

(b) Derive gives $F(x) = \ln(\sqrt{1-x^2} - 1) - \ln x$ and Mathematica gives $F(x) = \ln x - \ln(1 + \sqrt{1-x^2})$.

Both are correct if you take absolute values of the logarithm arguments, and both would then have the

same domain. Maple gives $F(x) = -\operatorname{arctanh}(1/\sqrt{1-x^2})$. This function has domain

$\{x \mid |x| < 1, -1 < 1/\sqrt{1-x^2} < 1\} = \{x \mid |x| < 1, 1/\sqrt{1-x^2} < 1\} = \{x \mid |x| < 1, \sqrt{1-x^2} > 1\} = \emptyset,$

the empty set! If we apply the command $\mathtt{convert(\%,ln)}$; to Maple's answer, we get

$-\frac{1}{2}\ln\left(\frac{1}{\sqrt{1-x^2}} + 1\right) + \frac{1}{2}\ln\left(1 - \frac{1}{\sqrt{1-x^2}}\right)$, which has the same domain, $\emptyset$.

44. None of Maple, Mathematica and Derive is able to evaluate $\int (1 + \ln x)\sqrt{1 + (x\ln x)^2}\, dx$. However, if we let $u = x\ln x$, then $du = (1 + \ln x)\, dx$ and the integral is simply $\int \sqrt{1 + u^2}\, du$, which any CAS can evaluate. The antiderivative is

$$\tfrac{1}{2}\ln\!\left(x\ln x + \sqrt{1 + (x\ln x)^2}\right) + \tfrac{1}{2}x\ln x\sqrt{1 + (x\ln x)^2} + C.$$

45. Maple gives the antiderivative

$$F(x) = \int \frac{x^2 - 1}{x^4 + x^2 + 1}\, dx = -\tfrac{1}{2}\ln(x^2 + x + 1) + \tfrac{1}{2}\ln(x^2 - x + 1).$$

We can see that at 0, this antiderivative is 0. From the graphs, it appears that F has

a maximum at $x = -1$ and a minimum at $x = 1$ [since $F'(x) = f(x)$ changes

sign at these x-values], and that F has inflection points at $x \approx -1.7$, $x = 0$, and

$x \approx 1.7$ [since $f(x)$ has extrema at these x-values].

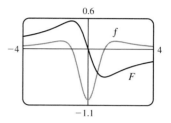

46. Maple gives the antiderivative which, after we use the `simplify` command, becomes

$\int xe^{-x}\sin x\, dx = -\tfrac{1}{2}e^{-x}(\cos x + x\cos x + x\sin x)$. At $x = 0$, this antiderivative has the value $-\tfrac{1}{2}$, so we use

$F(x) = -\tfrac{1}{2}e^{-x}(\cos x + x\cos x + x\sin x) + \tfrac{1}{2}$ to make $F(0) = 0$.

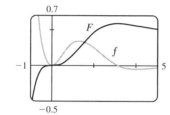

From the graphs, it appears that F has a

minimum at $x \approx -3.1$ and a maximum

at $x \approx 3.1$ [note that $f(x) = 0$

at $x = \pm\pi$], and that F has inflection

points where f' changes sign, at

$x \approx -2.5$, $x = 0$, $x \approx 1.3$ and $x \approx 4.1$.

47. Since $f(x) = \sin^4 x \cos^6 x$ is everywhere positive, we know that its antiderivative F is increasing. Maple gives

$\int f(x)\, dx = -\tfrac{1}{10}\sin^3 x \cos^7 x - \tfrac{3}{80}\sin x \cos^7 x + \tfrac{1}{160}\cos^5 x \sin x + \tfrac{1}{128}\cos^3 x \sin x + \tfrac{3}{256}\cos x \sin x + \tfrac{3}{256}x$

and this expression is 0 at $x = 0$.

F has a minimum at $x = 0$ and a maximum at $x = \pi$.

F has inflection points where f' changes sign, that is, at $x \approx 0.7$, $x = \pi/2$,

and $x \approx 2.5$.

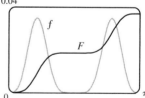

48. From the graph of $f(x) = \dfrac{x^3 - x}{x^6 + 1}$, we can see that F has a maximum

at $x = 0$, and minima at $x \approx \pm 1$. The antiderivative given by Maple is

$F(x) = -\tfrac{1}{3}\ln(x^2 + 1) + \tfrac{1}{6}\ln(x^4 - x^2 + 1)$, and $F(0) = 0$.

Note that f is odd, and its antiderivative F is even.

F has inflection points where f' changes sign, that is, at $x \approx \pm 0.5$ and $x \approx \pm 1.4$.

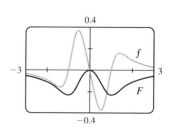

DISCOVERY PROJECT   Patterns in Integrals

1. (a) The CAS results are listed. Note that the absolute value symbols are missing, as is the familiar " $+ C$ ".

(i) $\displaystyle\int \frac{1}{(x+2)(x+3)}\,dx = \ln(x+2) - \ln(x+3)$ (ii) $\displaystyle\int \frac{1}{(x+1)(x+5)}\,dx = \frac{\ln(x+1)}{4} - \frac{\ln(x+5)}{4}$

(iii) $\displaystyle\int \frac{1}{(x+2)(x-5)}\,dx = \frac{\ln(x-5)}{7} - \frac{\ln(x+2)}{7}$ (iv) $\displaystyle\int \frac{1}{(x+2)^2}\,dx = -\frac{1}{x+2}$

(b) If $a \neq b$, it appears that $\ln(x+a)$ is divided by $b - a$ and $\ln(x+b)$ is divided by $a - b$, so we guess that

$$\int \frac{1}{(x+a)(x+b)}\,dx = \frac{\ln(x+a)}{b-a} + \frac{\ln(x+b)}{a-b} + C. \text{ If } a = b, \text{ as in part (a)(iv), it appears that}$$

$$\int \frac{1}{(x+a)^2}\,dx = -\frac{1}{x+a} + C.$$

(c) The CAS verifies our guesses. Now $\dfrac{1}{(x+a)(x+b)} = \dfrac{A}{x+a} + \dfrac{B}{x+b}$ $\Rightarrow$ $1 = A(x+b) + B(x+a)$.

Setting $x = -b$ gives $B = 1/(a-b)$ and setting $x = -a$ gives $A = 1/(b-a)$. So

$$\int \frac{1}{(x+a)(x+b)}\,dx = \int \left[\frac{1/(b-a)}{x+a} + \frac{1/(a-b)}{x+b}\right]dx = \frac{\ln|x+a|}{b-a} + \frac{\ln|x+b|}{a-b} + C$$

and our guess for $a \neq b$ is correct. If $a = b$, then $\dfrac{1}{(x+a)(x+b)} = \dfrac{1}{(x+a)^2} = (x+a)^{-2}$. Letting $u = x + a$ $\Rightarrow$

$du = dx$, we have $\displaystyle\int (x+a)^{-2}\,dx = \int u^{-2}\,du = -\frac{1}{u} + C = -\frac{1}{x+a} + C$, and our guess for $a = b$ is also correct.

2. (a) (i) $\displaystyle\int \sin x \cos 2x\,dx = \frac{\cos x}{2} - \frac{\cos 3x}{6}$ (ii) $\displaystyle\int \sin 3x \cos 7x\,dx = \frac{\cos 4x}{8} - \frac{\cos 10x}{20}$

(iii) $\displaystyle\int \sin 8x \cos 3x\,dx = -\frac{\cos 11x}{22} - \frac{\cos 5x}{10}$

(b) Looking at the sums and differences of a and b in part (a), we guess that

$$\int \sin ax \cos bx\,dx = \frac{\cos((a-b)x)}{2(b-a)} - \frac{\cos((a+b)x)}{2(a+b)} + C$$

Note that $\cos((a-b)x) = \cos((b-a)x)$.

(c) The CAS verifies our guess. Again, we can prove that the guess is correct by differentiating:

$$\frac{d}{dx}\left[\frac{\cos((a-b)x)}{2(b-a)} - \frac{\cos((a+b)x)}{2(a+b)}\right] = \frac{1}{2(b-a)}[-\sin((a-b)x)](a-b) - \frac{1}{2(a+b)}[-\sin((a+b)x)](a+b)$$

$$= \tfrac{1}{2}\sin(ax - bx) + \tfrac{1}{2}\sin(ax + bx)$$

$$= \tfrac{1}{2}(\sin ax \cos bx - \cos ax \sin bx) + \tfrac{1}{2}(\sin ax \cos bx + \cos ax \sin bx)$$

$$= \sin ax \cos bx$$

Our formula is valid for $a \neq b$.

3. (a) (i) $\int \ln x\,dx = x\ln x - x$ (ii) $\int x\ln x\,dx = \frac{1}{2}x^2 \ln x - \frac{1}{4}x^2$

(iii) $\int x^2 \ln x\,dx = \frac{1}{3}x^3 \ln x - \frac{1}{9}x^3$ (iv) $\int x^3 \ln x\,dx = \frac{1}{4}x^4 \ln x - \frac{1}{16}x^4$

(v) $\int x^7 \ln x\,dx = \frac{1}{8}x^8 \ln x - \frac{1}{64}x^8$

(b) We guess that $\int x^n \ln x \, dx = \dfrac{1}{n+1}x^{n+1} \ln x - \dfrac{1}{(n+1)^2}x^{n+1}$.

(c) Let $u = \ln x$, $dv = x^n \, dx$ $\Rightarrow$ $du = \dfrac{dx}{x}$, $v = \dfrac{1}{n+1}x^{n+1}$. Then

$$\int x^n \ln x \, dx = \frac{1}{n+1}x^{n+1} \ln x - \frac{1}{n+1}\int x^n \, dx = \frac{1}{n+1}x^{n+1} \ln x - \frac{1}{n+1} \cdot \frac{1}{n+1}x^{n+1},$$

which verifies our guess. We must have $n + 1 \neq 0$ $\Leftrightarrow$ $n \neq -1$.

4. (a) (i) $\int xe^x \, dx = e^x(x-1)$ **(ii)** $\int x^2 e^x \, dx = e^x(x^2 - 2x + 2)$

(iii) $\int x^3 e^x \, dx = e^x(x^3 - 3x^2 + 6x - 6)$ **(iv)** $\int x^4 e^x \, dx = e^x(x^4 - 4x^3 + 12x^2 - 24x + 24)$

(v) $\int x^5 e^x \, dx = e^x\left(x^5 - 5x^4 + 20x^3 - 60x^2 + 120x - 120\right)$

(b) Notice from part (a) that we can write

$$\int x^4 e^x \, dx = e^x(x^4 - 4x^3 + 4 \cdot 3x^2 - 4 \cdot 3 \cdot 2x + 4 \cdot 3 \cdot 2 \cdot 1)$$

and $$\int x^5 e^x \, dx = e^x(x^5 - 5x^4 + 5 \cdot 4x^3 - 5 \cdot 4 \cdot 3x^2 + 5 \cdot 4 \cdot 3 \cdot 2x - 5 \cdot 4 \cdot 3 \cdot 2 \cdot 1)$$

So we guess that

$$\int x^6 e^x \, dx = e^x(x^6 - 6x^5 + 6 \cdot 5x^4 - 6 \cdot 5 \cdot 4x^3 + 6 \cdot 5 \cdot 4 \cdot 3x^2 - 6 \cdot 5 \cdot 4 \cdot 3 \cdot 2x + 6 \cdot 5 \cdot 4 \cdot 3 \cdot 2 \cdot 1)$$

$$= e^x(x^6 - 6x^5 + 30x^4 - 120x^3 + 360x^2 - 720x + 720)$$

The CAS verifies our guess.

(c) From the results in part (a), as well as our prediction in part (b), we speculate that

$$\int x^n e^x \, dx = e^x\left[x^n - nx^{n-1} + n(n-1)x^{n-2} - n(n-1)(n-2)x^{n-3} + \cdots \pm n!x \mp n!\right] = e^x \sum_{i=0}^{n}(-1)^{n-i}\frac{n!}{i!}\,x^i.$$

(We have reversed the order of the polynomial's terms.)

(d) Let S_n be the statement that $\int x^n e^x \, dx = e^x \sum_{i=0}^{n}(-1)^{n-i}\dfrac{n!}{i!}\,x^i$.

S_1 is true by part (a)(i). Suppose S_k is true for some k, and consider S_{k+1}. Integrating by parts with $u = x^{k+1}$, $dv = e^x \, dx$ $\Rightarrow$ $du = (k+1)x^k \, dx$, $v = e^x$, we get

$$\int x^{k+1} e^x \, dx = x^{k+1}e^x - (k+1)\int x^k e^x \, dx = x^{k+1}e^x - (k+1)\left[e^x \sum_{i=0}^{k}(-1)^{k-i}\frac{k!}{i!}\,x^i\right]$$

$$= e^x\left[x^{k+1} - (k+1)\sum_{i=0}^{k}(-1)^{k-i}\frac{k!}{i!}\,x^i\right] = e^x\left[x^{k+1} + \sum_{i=0}^{k}(-1)^{k-i+1}\frac{(k+1)k!}{i!}\,x^i\right]$$

$$= e^x \sum_{i=0}^{k+1}(-1)^{(k+1)-i}\frac{(k+1)!}{i!}\,x^i$$

This verifies S_n for $n = k + 1$. Thus, by mathematical induction, S_n is true for all n, where n is a positive integer.

8.7 Approximate Integration

1. (a) $\Delta x = (b-a)/n = (4-0)/2 = 2$

$$L_2 = \sum_{i=1}^{2} f(x_{i-1})\,\Delta x = f(x_0)\cdot 2 + f(x_1)\cdot 2 = 2\,[f(0) + f(2)] = 2(0.5 + 2.5) = 6$$

$$R_2 = \sum_{i=1}^{2} f(x_i)\,\Delta x = f(x_1)\cdot 2 + f(x_2)\cdot 2 = 2\,[f(2) + f(4)] = 2(2.5 + 3.5) = 12$$

$$M_2 = \sum_{i=1}^{2} f(\overline{x}_i)\Delta x = f(\overline{x}_1)\cdot 2 + f(\overline{x}_2)\cdot 2 = 2\,[f(1) + f(3)] \approx 2(1.6 + 3.2) = 9.6$$

(b)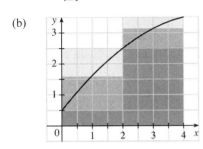

L_2 is an underestimate, since the area under the small rectangles is less than the area under the curve, and R_2 is an overestimate, since the area under the large rectangles is greater than the area under the curve. It appears that M_2 is an overestimate, though it is fairly close to I. See the solution to Exercise 45 for a proof of the fact that if f is concave down on $[a, b]$, then the Midpoint Rule is an overestimate of $\int_a^b f(x)\,dx$.

(c) $T_2 = \left(\frac{1}{2}\,\Delta x\right)[f(x_0) + 2f(x_1) + f(x_2)] = \frac{2}{2}[f(0) + 2f(2) + f(4)] = 0.5 + 2(2.5) + 3.5 = 9.$

This approximation is an underestimate, since the graph is concave down. Thus, $T_2 = 9 < I$. See the solution to Exercise 45 for a general proof of this conclusion.

(d) For any n, we will have $L_n < T_n < I < M_n < R_n$.

2.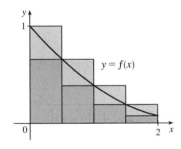

The diagram shows that $L_4 > T_4 > \int_0^2 f(x)\,dx > R_4$, and it appears that M_4 is a bit less than $\int_0^2 f(x)\,dx$. In fact, for any function that is concave upward, it can be shown that $L_n > T_n > \int_0^2 f(x)\,dx > M_n > R_n$.

(a) Since $0.9540 > 0.8675 > 0.8632 > 0.7811$, it follows that $L_n = 0.9540$, $T_n = 0.8675$, $M_n = 0.8632$, and $R_n = 0.7811$.

(b) Since $M_n < \int_0^2 f(x)\,dx < T_n$, we have $0.8632 < \int_0^2 f(x)\,dx < 0.8675$.

3. $f(x) = \cos(x^2)$, $\Delta x = \frac{1-0}{4} = \frac{1}{4}$

(a) $T_4 = \frac{1}{4\cdot 2}\left[f(0) + 2f\left(\frac{1}{4}\right) + 2f\left(\frac{2}{4}\right) + 2f\left(\frac{3}{4}\right) + f(1)\right] \approx 0.895759$

(b) $M_4 = \frac{1}{4}\left[f\left(\frac{1}{8}\right) + f\left(\frac{3}{8}\right) + f\left(\frac{5}{8}\right) + f\left(\frac{7}{8}\right)\right] \approx 0.908907$

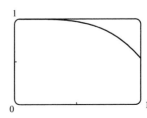

The graph shows that f is concave down on $[0, 1]$. So T_4 is an underestimate and M_4 is an overestimate. We can conclude that $0.895759 < \int_0^1 \cos(x^2)\,dx < 0.908907$.

4.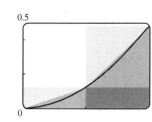

(a) Since f is increasing on $[0, 1]$, L_2 will underestimate I (since the area of the darkest rectangle is less than the area under the curve), and R_2 will overestimate I. Since f is concave upward on $[0, 1]$, M_2 will underestimate I and T_2 will overestimate I (the area under the straight line segments is greater than the area under the curve).

(b) For any n, we will have $L_n < M_n < I < T_n < R_n$.

(c) $L_5 = \sum\limits_{i=1}^{5} f(x_{i-1})\, \Delta x = \frac{1}{5}[f(0.0) + f(0.2) + f(0.4) + f(0.6) + f(0.8)] \approx 0.1187$

$R_5 = \sum\limits_{i=1}^{5} f(x_i)\, \Delta x = \frac{1}{5}[f(0.2) + f(0.4) + f(0.6) + f(0.8) + f(1)] \approx 0.2146$

$M_5 = \sum\limits_{i=1}^{5} f(\overline{x}_i)\, \Delta x = \frac{1}{5}[f(0.1) + f(0.3) + f(0.5) + f(0.7) + f(0.9)] \approx 0.1622$

$T_5 = \left(\frac{1}{2}\,\Delta x\right)[f(0) + 2f(0.2) + 2f(0.4) + 2f(0.6) + 2f(0.8) + f(1)] \approx 0.1666$

From the graph, it appears that the Midpoint Rule gives the best approximation. (This is in fact the case, since $I \approx 0.16371405$.)

5. $f(x) = x^2 \sin x$, $\Delta x = \dfrac{b-a}{n} = \dfrac{\pi - 0}{8} = \dfrac{\pi}{8}$

(a) $M_8 = \frac{\pi}{8}\left[f\left(\frac{\pi}{16}\right) + f\left(\frac{3\pi}{16}\right) + f\left(\frac{5\pi}{16}\right) + \cdots + f\left(\frac{15\pi}{16}\right)\right] \approx 5.932957$

(b) $S_8 = \frac{\pi}{8 \cdot 3}\left[f(0) + 4f\left(\frac{\pi}{8}\right) + 2f\left(\frac{2\pi}{8}\right) + 4f\left(\frac{3\pi}{8}\right) + 2f\left(\frac{4\pi}{8}\right) + 4f\left(\frac{5\pi}{8}\right) + 2f\left(\frac{6\pi}{8}\right) + 4f\left(\frac{7\pi}{8}\right) + f(\pi)\right]$

≈ 5.869247

Actual: $\int_0^\pi x^2 \sin x\, dx \overset{84}{=} \left[-x^2 \cos x\right]_0^\pi + 2\int_0^\pi x\cos x\, dx \overset{83}{=} \left[-\pi^2(-1) - 0\right] + 2\left[\cos x + x\sin x\right]_0^\pi$

$= \pi^2 + 2[(-1 + 0) - (1 + 0)] = \pi^2 - 4 \approx 5.869604$

Errors: $E_M = $ actual $- M_8 = \int_0^\pi x^2 \sin x\, dx - M_8 \approx -0.063353$

$E_S = $ actual $- S_8 = \int_0^\pi x^2 \sin x\, dx - S_8 \approx 0.000357$

6. $f(x) = e^{-\sqrt{x}}$, $\Delta x = \dfrac{b-a}{n} = \dfrac{1-0}{6} = \dfrac{1}{6}$

(a) $M_6 = \frac{1}{6}\left[f\left(\frac{1}{12}\right) + f\left(\frac{3}{12}\right) + f\left(\frac{5}{12}\right) + f\left(\frac{7}{12}\right) + f\left(\frac{9}{12}\right) + f\left(\frac{11}{12}\right)\right] \approx 0.525100$

(b) $S_6 = \frac{1}{6 \cdot 3}\left[f(0) + 4f\left(\frac{1}{6}\right) + 2f\left(\frac{2}{6}\right) + 4f\left(\frac{3}{6}\right) + 2f\left(\frac{4}{6}\right) + 4f\left(\frac{5}{6}\right) + f(1)\right] \approx 0.533979$

Actual: $\int_0^1 e^{-\sqrt{x}}\, dx = \int_0^{-1} e^u 2u\, du \qquad [u = -\sqrt{x},\ u^2 = x,\ 2u\, du = dx]$

$\overset{96}{=} 2[(u-1)e^u]_0^{-1} = 2\left[-2e^{-1} - (-1e^0)\right] = 2 - 4e^{-1} \approx 0.528482$

Errors: $E_M = $ actual $- M_6 = \int_0^1 e^{-\sqrt{x}}\, dx - M_6 \approx 0.003382$

$E_S = $ actual $- S_6 = \int_0^1 e^{-\sqrt{x}}\, dx - S_6 \approx -0.005497$

7. $f(x) = \sqrt[4]{1 + x^2}$, $\Delta x = \dfrac{2 - 0}{8} = \dfrac{1}{4}$

(a) $T_8 = \frac{1}{4 \cdot 2}\left[f(0) + 2f\left(\frac{1}{4}\right) + 2f\left(\frac{1}{2}\right) + \cdots + 2f\left(\frac{3}{2}\right) + 2f\left(\frac{7}{4}\right) + f(2)\right] \approx 2.413790$

(b) $M_8 = \frac{1}{4}\left[f\left(\frac{1}{8}\right) + f\left(\frac{3}{8}\right) + \cdots + f\left(\frac{13}{8}\right) + f\left(\frac{15}{8}\right)\right] \approx 2.411453$

(c) $S_8 = \frac{1}{4 \cdot 3}\left[f(0) + 4f\left(\frac{1}{4}\right) + 2f\left(\frac{1}{2}\right) + 4f\left(\frac{3}{4}\right) + 2f(1) + 4f\left(\frac{5}{4}\right) + 2f\left(\frac{3}{2}\right) + 4f\left(\frac{7}{4}\right) + f(2)\right] \approx 2.412232$

8. $f(x) = \sin\left(x^2\right)$, $\Delta x = \dfrac{\frac{1}{2} - 0}{4} = \dfrac{1}{8}$

(a) $T_4 = \frac{1}{8 \cdot 2}\left[f(0) + 2f\left(\frac{1}{8}\right) + 2f\left(\frac{2}{8}\right) + 2f\left(\frac{3}{8}\right) + f\left(\frac{1}{2}\right)\right] \approx 0.042743$

(b) $M_4 = \frac{1}{8}\left[f\left(\frac{1}{16}\right) + f\left(\frac{3}{16}\right) + f\left(\frac{5}{16}\right) + f\left(\frac{7}{16}\right)\right] \approx 0.040850$

(c) $S_4 = \frac{1}{8 \cdot 3}\left[f(0) + 4f\left(\frac{1}{8}\right) + 2f\left(\frac{2}{8}\right) + 4f\left(\frac{3}{8}\right) + f\left(\frac{1}{2}\right)\right] \approx 0.041478$

9. $f(x) = \dfrac{\ln x}{1 + x}$, $\Delta x = \dfrac{2 - 1}{10} = \dfrac{1}{10}$

(a) $T_{10} = \frac{1}{10 \cdot 2}[f(1) + 2f(1.1) + 2f(1.2) + \cdots + 2f(1.8) + 2f(1.9) + f(2)] \approx 0.146879$

(b) $M_{10} = \frac{1}{10}[f(1.05) + f(1.15) + \cdots + f(1.85) + f(1.95)] \approx 0.147391$

(c) $S_{10} = \frac{1}{10 \cdot 3}[f(1) + 4f(1.1) + 2f(1.2) + 4f(1.3) + 2f(1.4) + 4f(1.5) + 2f(1.6) + 4f(1.7)$
$\qquad\qquad + 2f(1.8) + 4f(1.9) + f(2)]$

$\qquad \approx 0.147219$

10. $f(t) = \dfrac{1}{1 + t^2 + t^4}$, $\Delta t = \dfrac{3 - 0}{6} = \dfrac{1}{2}$

(a) $T_6 = \frac{1}{2 \cdot 2}\left[f(0) + 2f\left(\frac{1}{2}\right) + 2f(1) + 2f\left(\frac{3}{2}\right) + 2f(2) + 2f\left(\frac{5}{2}\right) + f(3)\right] \approx 0.895122$

(b) $M_6 = \frac{1}{2}\left[f\left(\frac{1}{4}\right) + f\left(\frac{3}{4}\right) + f\left(\frac{5}{4}\right) + f\left(\frac{7}{4}\right) + f\left(\frac{9}{4}\right) + f\left(\frac{11}{4}\right)\right] \approx 0.895478$

(c) $S_6 = \frac{1}{2 \cdot 3}\left[f(0) + 4f\left(\frac{1}{2}\right) + 2f(1) + 4f\left(\frac{3}{2}\right) + 2f(2) + 4f\left(\frac{5}{2}\right) + f(3)\right] \approx 0.898014$

11. $f(t) = \sin(e^{t/2})$, $\Delta t = \dfrac{\frac{1}{2} - 0}{8} = \dfrac{1}{16}$

(a) $T_8 = \frac{1}{16 \cdot 2}\left[f(0) + 2f\left(\frac{1}{16}\right) + 2f\left(\frac{2}{16}\right) + \cdots + 2f\left(\frac{7}{16}\right) + f\left(\frac{1}{2}\right)\right] \approx 0.451948$

(b) $M_8 = \frac{1}{16}\left[f\left(\frac{1}{32}\right) + f\left(\frac{3}{32}\right) + f\left(\frac{5}{32}\right) + \cdots + f\left(\frac{13}{32}\right) + f\left(\frac{15}{32}\right)\right] \approx 0.451991$

(c) $S_8 = \frac{1}{16 \cdot 3}\left[f(0) + 4f\left(\frac{1}{16}\right) + 2f\left(\frac{2}{16}\right) + \cdots + 4f\left(\frac{7}{16}\right) + f\left(\frac{1}{2}\right)\right] \approx 0.451976$

12. $f(x) = \sqrt{1 + \sqrt{x}}$, $\Delta x = \dfrac{4 - 0}{8} = \dfrac{1}{2}$

(a) $T_8 = \frac{1}{2 \cdot 2}\left[f(0) + 2f\left(\frac{1}{2}\right) + 2f(1) + \cdots + 2f(3) + 2f\left(\frac{7}{2}\right) + f(4)\right] \approx 6.042985$

(b) $M_8 = \frac{1}{2}\left[f\left(\frac{1}{4}\right) + f\left(\frac{3}{4}\right) + \cdots + f\left(\frac{13}{4}\right) + f\left(\frac{15}{4}\right)\right] \approx 6.084778$

(c) $S_8 = \frac{1}{2 \cdot 3}\left[f(0) + 4f\left(\frac{1}{2}\right) + 2f(1) + 4f\left(\frac{3}{2}\right) + 2f(2) + 4f\left(\frac{5}{2}\right) + 2f(3) + 4f\left(\frac{7}{2}\right) + f(4)\right] \approx 6.061678$

13. $f(t) = e^{\sqrt{t}} \sin t$, $\Delta t = \dfrac{4 - 0}{8} = \dfrac{1}{2}$

(a) $T_8 = \frac{1}{2 \cdot 2}\left[f(0) + 2f\left(\frac{1}{2}\right) + 2f(1) + 2f\left(\frac{3}{2}\right) + 2f(2) + 2f\left(\frac{5}{2}\right) + 2f(3) + 2f\left(\frac{7}{2}\right) + f(4)\right] \approx 4.513618$

(b) $M_8 = \frac{1}{2}\left[f\left(\frac{1}{4}\right) + f\left(\frac{3}{4}\right) + f\left(\frac{5}{4}\right) + f\left(\frac{7}{4}\right) + f\left(\frac{9}{4}\right) + f\left(\frac{11}{4}\right) + f\left(\frac{13}{4}\right) + f\left(\frac{15}{4}\right)\right] \approx 4.748256$

(c) $S_8 = \frac{1}{2 \cdot 3}\left[f(0) + 4f\left(\frac{1}{2}\right) + 2f(1) + 4f\left(\frac{3}{2}\right) + 2f(2) + 4f\left(\frac{5}{2}\right) + 2f(3) + 4f\left(\frac{7}{2}\right) + f(4)\right] \approx 4.675111$

14. $f(z) = \sqrt{z}e^{-z}$, $\Delta z = \frac{1-0}{10} = \frac{1}{10}$

(a) $T_{10} = \frac{1}{10 \cdot 2} \{f(0) + 2[f(0.1) + f(0.2) + \cdots + f(0.9)] + f(1)\} \approx 0.372299$

(b) $M_{10} = \frac{1}{10}[f(0.05) + f(0.15) + f(0.25) + \cdots + f(0.95)] \approx 0.380894$

(c) $S_{10} = \frac{1}{10 \cdot 3}[f(0) + 4f(0.1) + 2f(0.2) + 4f(0.3) + 2f(0.4) + 4f(0.5) + 2f(0.6)$
$\qquad\qquad + 4f(0.7) + 2f(0.8) + 4f(0.9) + f(1)]$
$\qquad \approx 0.376330$

15. $f(x) = \dfrac{\cos x}{x}$, $\Delta x = \dfrac{5-1}{8} = \dfrac{1}{2}$

(a) $T_8 = \frac{1}{2 \cdot 2}\left[f(1) + 2f\left(\frac{3}{2}\right) + 2f(2) + \cdots + 2f(4) + 2f\left(\frac{9}{2}\right) + f(5)\right] \approx -0.495333$

(b) $M_8 = \frac{1}{2}\left[f\left(\frac{5}{4}\right) + f\left(\frac{7}{4}\right) + f\left(\frac{9}{4}\right) + f\left(\frac{11}{4}\right) + f\left(\frac{13}{4}\right) + f\left(\frac{15}{4}\right) + f\left(\frac{17}{4}\right) + f\left(\frac{19}{4}\right)\right] \approx -0.543321$

(c) $S_8 = \frac{1}{2 \cdot 3}\left[f(1) + 4f\left(\frac{3}{2}\right) + 2f(2) + 4f\left(\frac{5}{2}\right) + 2f(3) + 4f\left(\frac{7}{2}\right) + 2f(4) + 4f\left(\frac{9}{2}\right) + f(5)\right] \approx -0.526123$

16. $f(x) = \ln(x^3 + 2)$, $\Delta x = \dfrac{6-4}{10} = \dfrac{1}{5}$

(a) $T_{10} = \frac{1}{5 \cdot 2}[f(4) + 2f(4.2) + 2f(4.4) + \cdots + 2f(5.6) + 2f(5.8) + f(6)] \approx 9.649753$

(b) $M_{10} = \frac{1}{5}[f(4.1) + f(4.3) + \cdots + f(5.7) + f(5.9)] \approx 9.650912$

(c) $S_{10} = \frac{1}{5 \cdot 3}[f(4) + 4f(4.2) + 2f(4.4) + 4f(4.6) + 2f(4.8) + 4f(5) + 2f(5.2) + 4f(5.4)$
$\qquad\qquad + 2f(5.6) + 4f(5.8) + f(6)]$
$\qquad \approx 9.650526$

17. $f(y) = \dfrac{1}{1+y^5}$, $\Delta y = \dfrac{3-0}{6} = \dfrac{1}{2}$

(a) $T_6 = \frac{1}{2 \cdot 2}\left[f(0) + 2f\left(\frac{1}{2}\right) + 2f\left(\frac{2}{2}\right) + 2f\left(\frac{3}{2}\right) + 2f\left(\frac{4}{2}\right) + 2f\left(\frac{5}{2}\right) + f(3)\right] \approx 1.064275$

(b) $M_6 = \frac{1}{2}\left[f\left(\frac{1}{4}\right) + f\left(\frac{3}{4}\right) + f\left(\frac{5}{4}\right) + f\left(\frac{7}{4}\right) + f\left(\frac{9}{4}\right) + f\left(\frac{11}{4}\right)\right] \approx 1.067416$

(c) $S_6 = \frac{1}{2 \cdot 3}\left[f(0) + 4f\left(\frac{1}{2}\right) + 2f\left(\frac{2}{2}\right) + 4f\left(\frac{3}{2}\right) + 2f\left(\frac{4}{2}\right) + 4f\left(\frac{5}{2}\right) + f(3)\right] \approx 1.074915$

18. $f(x) = \cos\sqrt{x}$, $\Delta x = \frac{4-0}{10} = \frac{2}{5} = 0.4$

(a) $T_{10} = \frac{2}{5 \cdot 2}[f(0) + 2f(0.4) + 2f(0.8) + \cdots + 2f(3.2) + 2f(3.6) + f(4)] \approx 0.808532$

(b) $M_{10} = \frac{2}{5}[f(0.2) + f(0.6) + f(1) + \cdots + f(3.4) + f(3.8)] \approx 0.803078$

(c) $S_{10} = \frac{2}{5 \cdot 3}[f(0) + 4f(0.4) + 2f(0.8) + 4f(1.2) + 2f(1.6) + 4f(2) + 2f(2.4)$
$\qquad\qquad + 4f(2.8) + 2f(3.2) + 4f(3.6) + f(4)]$
$\qquad \approx 0.804896$

19. $f(x) = \cos(x^2)$, $\Delta x = \frac{1-0}{8} = \frac{1}{8}$

(a) $T_8 = \frac{1}{8 \cdot 2}\left\{f(0) + 2\left[f\left(\frac{1}{8}\right) + f\left(\frac{2}{8}\right) + \cdots + f\left(\frac{7}{8}\right)\right] + f(1)\right\} \approx 0.902333$

$\ M_8 = \frac{1}{8}\left[f\left(\frac{1}{16}\right) + f\left(\frac{3}{16}\right) + f\left(\frac{5}{16}\right) + \cdots + f\left(\frac{15}{16}\right)\right] = 0.905620$

(b) $f(x) = \cos(x^2)$, $f'(x) = -2x\sin(x^2)$, $f''(x) = -2\sin(x^2) - 4x^2\cos(x^2)$. For $0 \le x \le 1$, sin and cos are positive,

so $|f''(x)| = 2\sin(x^2) + 4x^2\cos(x^2) \le 2 \cdot 1 + 4 \cdot 1 \cdot 1 = 6$ since $\sin(x^2) \le 1$ and $\cos(x^2) \le 1$ for all x,

and $x^2 \le 1$ for $0 \le x \le 1$. So for $n = 8$, we take $K = 6$, $a = 0$, and $b = 1$ in Theorem 3, to get

$|E_T| \le 6 \cdot 1^3/(12 \cdot 8^2) = \frac{1}{128} = 0.0078125$ and $|E_M| \le \frac{1}{256} = 0.00390625$. [A better estimate is obtained by noting

from a graph of f'' that $|f''(x)| \le 4$ for $0 \le x \le 1$.]

(c) Take $K = 6$ [as in part (b)] in Theorem 3. $\quad |E_T| \le \dfrac{K(b-a)^3}{12n^2} \le 0.0001 \quad \Leftrightarrow \quad \dfrac{6(1-0)^3}{12n^2} \le 10^{-4} \quad \Leftrightarrow$

$\dfrac{1}{2n^2} \le \dfrac{1}{10^4} \quad \Leftrightarrow \quad 2n^2 \ge 10^4 \quad \Leftrightarrow \quad n^2 \ge 5000 \quad \Leftrightarrow \quad n \ge 71$. Take $n = 71$ for T_n. For E_M, again take $K = 6$ in

Theorem 3 to get $|E_M| \le 10^{-4} \quad \Leftrightarrow \quad 4n^2 \ge 10^4 \quad \Leftrightarrow \quad n^2 \ge 2500 \quad \Leftrightarrow \quad n \ge 50$. Take $n = 50$ for M_n.

20. $f(x) = e^{1/x}$, $\Delta x = \frac{2-1}{10} = \frac{1}{10}$

(a) $T_{10} = \frac{1}{10 \cdot 2}[f(1) + 2f(1.1) + 2f(1.2) + \cdots + 2f(1.9) + f(2)] \approx 2.021976$

$M_{10} = \frac{1}{10}[f(1.05) + f(1.15) + f(1.25) + \cdots + f(1.95)] \approx 2.019102$

(b) $f(x) = e^{1/x}$, $f'(x) = -\dfrac{1}{x^2}e^{1/x}$, $f''(x) = \dfrac{2x+1}{x^4}e^{1/x}$. Now f'' is decreasing on $[1, 2]$, so let $x = 1$ to take $K = 3e$.

$|E_T| \le \dfrac{3e(2-1)^3}{12(10)^2} = \dfrac{e}{400} \approx 0.006796$. $\quad |E_M| \le \dfrac{|E_T|}{2} = \dfrac{e}{800} \approx 0.003398$.

(c) Take $K = 3e$ [as in part (b)] in Theorem 3. $\quad |E_T| \le \dfrac{K(b-a)^3}{12n^2} \le 0.0001 \quad \Leftrightarrow \quad \dfrac{3e(2-1)^3}{12n^2} \le 10^{-4} \quad \Leftrightarrow$

$\dfrac{e}{4n^2} \le \dfrac{1}{10^4} \quad \Leftrightarrow \quad n^2 \ge \dfrac{10^4 e}{4} \quad \Leftrightarrow \quad n \ge 83$. Take $n = 83$ for T_n. For E_M, again take $K = 3e$ in Theorem 3 to get

$|E_M| \le 10^{-4} \quad \Leftrightarrow \quad n^2 \ge \dfrac{10^4 e}{8} \quad \Leftrightarrow \quad n \ge 59$. Take $n = 59$ for M_n.

21. $f(x) = \sin x$, $\Delta x = \frac{\pi - 0}{10} = \frac{\pi}{10}$

(a) $T_{10} = \frac{\pi}{10 \cdot 2}\left[f(0) + 2f\left(\frac{\pi}{10}\right) + 2f\left(\frac{2\pi}{10}\right) + \cdots + 2f\left(\frac{9\pi}{10}\right) + f(\pi)\right] \approx 1.983524$

$M_{10} = \frac{\pi}{10}\left[f\left(\frac{\pi}{20}\right) + f\left(\frac{3\pi}{20}\right) + f\left(\frac{5\pi}{20}\right) + \cdots + f\left(\frac{19\pi}{20}\right)\right] \approx 2.008248$

$S_{10} = \frac{\pi}{10 \cdot 3}\left[f(0) + 4f\left(\frac{\pi}{10}\right) + 2f\left(\frac{2\pi}{10}\right) + 4f\left(\frac{3\pi}{10}\right) + \cdots + 4f\left(\frac{9\pi}{10}\right) + f(\pi)\right] \approx 2.000110$

Since $I = \int_0^\pi \sin x \, dx = \left[-\cos x\right]_0^\pi = 1 - (-1) = 2$, $E_T = I - T_{10} \approx 0.016476$, $E_M = I - M_{10} \approx -0.008248$,

and $E_S = I - S_{10} \approx -0.000110$.

(b) $f(x) = \sin x \quad \Rightarrow \quad \left|f^{(n)}(x)\right| \le 1$, so take $K = 1$ for all error estimates.

$|E_T| \le \dfrac{K(b-a)^3}{12n^2} = \dfrac{1(\pi - 0)^3}{12(10)^2} = \dfrac{\pi^3}{1200} \approx 0.025839. \quad |E_M| \le \dfrac{|E_T|}{2} = \dfrac{\pi^3}{2400} \approx 0.012919.$

$|E_S| \le \dfrac{K(b-a)^5}{180n^4} = \dfrac{1(\pi - 0)^5}{180(10)^4} = \dfrac{\pi^5}{1,800,000} \approx 0.000170.$

The actual error is about 64% of the error estimate in all three cases.

(c) $|E_T| \le 0.00001 \iff \dfrac{\pi^3}{12n^2} \le \dfrac{1}{10^5} \iff n^2 \ge \dfrac{10^5 \pi^3}{12} \implies n \ge 508.3.$ Take $n = 509$ for T_n.

$|E_M| \le 0.00001 \iff \dfrac{\pi^3}{24n^2} \le \dfrac{1}{10^5} \iff n^2 \ge \dfrac{10^5 \pi^3}{24} \implies n \ge 359.4.$ Take $n = 360$ for M_n.

$|E_S| \le 0.00001 \iff \dfrac{\pi^5}{180n^4} \le \dfrac{1}{10^5} \iff n^4 \ge \dfrac{10^5 \pi^5}{180} \implies n \ge 20.3.$

Take $n = 22$ for S_n (since n must be even).

22. From Example 7(b), we take $K = 76e$ to get $|E_S| \le \dfrac{76e(1)^5}{180n^4} \le 0.00001 \implies n^4 \ge \dfrac{76e}{180(0.00001)} \implies n \ge 18.4.$

Take $n = 20$ (since n must be even).

23. (a) Using a CAS, we differentiate $f(x) = e^{\cos x}$ twice, and find that

$f''(x) = e^{\cos x}(\sin^2 x - \cos x)$. From the graph, we see that the maximum

value of $|f''(x)|$ occurs at the endpoints of the interval $[0, 2\pi]$.

Since $f''(0) = -e$, we can use $K = e$ or $K = 2.8$.

(b) A CAS gives $M_{10} \approx 7.954926518$. (In Maple, use `student[middlesum]`.)

(c) Using Theorem 3 for the Midpoint Rule, with $K = e$, we get $|E_M| \le \dfrac{e(2\pi - 0)^3}{24 \cdot 10^2} \approx 0.280945995.$

With $K = 2.8$, we get $|E_M| \le \dfrac{2.8(2\pi - 0)^3}{24 \cdot 10^2} = 0.289391916.$

(d) A CAS gives $I \approx 7.954926521$.

(e) The actual error is only about 3×10^{-9}, much less than the estimate in part (c).

(f) We use the CAS to differentiate twice more, and then graph

$f^{(4)}(x) = e^{\cos x}(\sin^4 x - 6\sin^2 x \cos x + 3 - 7\sin^2 x + \cos x).$

From the graph, we see that the maximum value of $\left|f^{(4)}(x)\right|$ occurs at the

endpoints of the interval $[0, 2\pi]$. Since $f^{(4)}(0) = 4e$, we can use $K = 4e$

or $K = 10.9$.

(g) A CAS gives $S_{10} \approx 7.953789422$. (In Maple, use `student[simpson]`.)

(h) Using Theorem 4 with $K = 4e$, we get $|E_S| \le \dfrac{4e(2\pi - 0)^5}{180 \cdot 10^4} \approx 0.059153618.$

With $K = 10.9$, we get $|E_S| \le \dfrac{10.9(2\pi - 0)^5}{180 \cdot 10^4} \approx 0.059299814.$

(i) The actual error is about $7.954926521 - 7.953789422 \approx 0.00114$. This is quite a bit smaller than the estimate in part (h), though the difference is not nearly as great as it was in the case of the Midpoint Rule.

(j) To ensure that $|E_S| \leq 0.0001$, we use Theorem 4: $|E_S| \leq \dfrac{4e(2\pi)^5}{180 \cdot n^4} \leq 0.0001 \quad \Rightarrow \quad \dfrac{4e(2\pi)^5}{180 \cdot 0.0001} \leq n^4 \quad \Rightarrow$

$n^4 \geq 5{,}915{,}362 \quad \Leftrightarrow \quad n \geq 49.3$. So we must take $n \geq 50$ to ensure that $|I - S_n| \leq 0.0001$.
($K = 10.9$ leads to the same value of n.)

24. (a) Using the CAS, we differentiate $f(x) = \sqrt{4 - x^3}$ twice, and find

that $f''(x) = -\dfrac{9x^4}{4(4 - x^3)^{3/2}} - \dfrac{3x}{(4 - x^3)^{1/2}}$.

From the graph, we see that $|f''(x)| < 2.2$ on $[-1, 1]$.

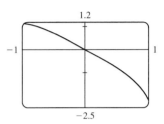

(b) A CAS gives $M_{10} \approx 3.995804152$. (In Maple, use `student[middlesum]`.)

(c) Using Theorem 3 for the Midpoint Rule, with $K = 2.2$, we get $|E_M| \leq \dfrac{2.2\,[1 - (-1)]^3}{24 \cdot 10^2} \approx 0.00733$.

(d) A CAS gives $I \approx 3.995487677$.

(e) The actual error is about -0.0003165, much less than the estimate in part (c).

(f) We use the CAS to differentiate twice more, and then graph

$f^{(4)}(x) = \dfrac{9}{16}\,\dfrac{x^2(x^6 - 224x^3 - 1280)}{(4 - x^3)^{7/2}}$.

From the graph, we see that $\left|f^{(4)}(x)\right| < 18.1$ on $[-1, 1]$.

(g) A CAS gives $S_{10} \approx 3.995449790$. (In Maple, use `student[simpson]`.)

(h) Using Theorem 4 with $K = 18.1$, we get $|E_S| \leq \dfrac{18.1\,[1 - (-1)]^5}{180 \cdot 10^4} \approx 0.000322$.

(i) The actual error is about $3.995487677 - 3.995449790 \approx 0.0000379$. This is quite a bit smaller than the estimate in part (h).

(j) To ensure that $|E_S| \leq 0.0001$, we use Theorem 4: $|E_S| \leq \dfrac{18.1(2)^5}{180 \cdot n^4} \leq 0.0001 \quad \Rightarrow \quad \dfrac{18.1(2)^5}{180 \cdot 0.0001} \leq n^4 \quad \Rightarrow$

$n^4 \geq 32{,}178 \quad \Rightarrow n \geq 13.4$. So we must take $n \geq 14$ to ensure that $|I - S_n| \leq 0.0001$.

25. $I = \int_0^1 xe^x \, dx = [(x-1)e^x]_0^1$ [parts or Formula 96] $= 0 - (-1) = 1$, $f(x) = xe^x$, $\Delta x = 1/n$

$n = 5$: $L_5 = \frac{1}{5}[f(0) + f(0.2) + f(0.4) + f(0.6) + f(0.8)] \approx 0.742943$

$R_5 = \frac{1}{5}[f(0.2) + f(0.4) + f(0.6) + f(0.8) + f(1)] \approx 1.286599$

$T_5 = \frac{1}{5 \cdot 2}[f(0) + 2f(0.2) + 2f(0.4) + 2f(0.6) + 2f(0.8) + f(1)] \approx 1.014771$

$M_5 = \frac{1}{5}[f(0.1) + f(0.3) + f(0.5) + f(0.7) + f(0.9)] \approx 0.992621$

$E_L = I - L_5 \approx 1 - 0.742943 = 0.257057$

$E_R \approx 1 - 1.286599 = -0.286599$

$E_T \approx 1 - 1.014771 = -0.014771$

$E_M \approx 1 - 0.992621 = 0.007379$

$n = 10$: $L_{10} = \frac{1}{10}[f(0) + f(0.1) + f(0.2) + \cdots + f(0.9)] \approx 0.867782$

$R_{10} = \frac{1}{10}[f(0.1) + f(0.2) + \cdots + f(0.9) + f(1)] \approx 1.139610$

$T_{10} = \frac{1}{10 \cdot 2}\{f(0) + 2[f(0.1) + f(0.2) + \cdots + f(0.9)] + f(1)\} \approx 1.003696$

$M_{10} = \frac{1}{10}[f(0.05) + f(0.15) + \cdots + f(0.85) + f(0.95)] \approx 0.998152$

$E_L = I - L_{10} \approx 1 - 0.867782 = 0.132218$

$E_R \approx 1 - 1.139610 = -0.139610$

$E_T \approx 1 - 1.003696 = -0.003696$

$E_M \approx 1 - 0.998152 = 0.001848$

$n = 20$: $L_{20} = \frac{1}{20}[f(0) + f(0.05) + f(0.10) + \cdots + f(0.95)] \approx 0.932967$

$R_{20} = \frac{1}{20}[f(0.05) + f(0.10) + \cdots + f(0.95) + f(1)] \approx 1.068881$

$T_{20} = \frac{1}{20 \cdot 2}\{f(0) + 2[f(0.05) + f(0.10) + \cdots + f(0.95)] + f(1)\} \approx 1.000924$

$M_{20} = \frac{1}{20}[f(0.025) + f(0.075) + f(0.125) + \cdots + f(0.975)] \approx 0.999538$

$E_L = I - L_{20} \approx 1 - 0.932967 = 0.067033$

$E_R \approx 1 - 1.068881 = -0.068881$

$E_T \approx 1 - 1.000924 = -0.000924$

$E_M \approx 1 - 0.999538 = 0.000462$

n	L_n	R_n	T_n	M_n
5	0.742943	1.286599	1.014771	0.992621
10	0.867782	1.139610	1.003696	0.998152
20	0.932967	1.068881	1.000924	0.999538

n	E_L	E_R	E_T	E_M
5	0.257057	−0.286599	−0.014771	0.007379
10	0.132218	−0.139610	−0.003696	0.001848
20	0.067033	−0.068881	−0.000924	0.000462

Observations:

1. E_L and E_R are always opposite in sign, as are E_T and E_M.

2. As n is doubled, E_L and E_R are decreased by about a factor of 2, and E_T and E_M are decreased by a factor of about 4.

3. The Midpoint approximation is about twice as accurate as the Trapezoidal approximation.

4. All the approximations become more accurate as the value of n increases.

5. The Midpoint and Trapezoidal approximations are much more accurate than the endpoint approximations.

26. $I = \int_1^2 \dfrac{1}{x^2}\, dx = \left[-\dfrac{1}{x}\right]_1^2 = -\dfrac{1}{2} - (-1) = \dfrac{1}{2},\, f(x) = \dfrac{1}{x^2},\, \Delta x = \dfrac{1}{n}$

$n = 5$: $L_5 = \frac{1}{5}[f(1) + f(1.2) + f(1.4) + f(1.6) + f(1.8)] \approx 0.580783$

$R_5 = \frac{1}{5}[f(1.2) + f(1.4) + f(1.6) + f(1.8) + f(2)] \approx 0.430783$

$T_5 = \frac{1}{5 \cdot 2}[f(1) + 2f(1.2) + 2f(1.4) + 2f(1.6) + 2f(1.8) + f(2)] \approx 0.505783$

$M_5 = \frac{1}{5}[f(1.1) + f(1.3) + f(1.5) + f(1.7) + f(1.9)] \approx 0.497127$

$E_L = I - L_5 \approx \frac{1}{2} - 0.580783 = -0.080783$

$E_R \approx \frac{1}{2} - 0.430783 = 0.069217$

$E_T \approx \frac{1}{2} - 0.505783 = -0.005783$

$E_M \approx \frac{1}{2} - 0.497127 = 0.002873$

$n = 10$: $L_{10} = \frac{1}{10}[f(1) + f(1.1) + f(1.2) + \cdots + f(1.9)] \approx 0.538955$

$R_{10} = \frac{1}{10}[f(1.1) + f(1.2) + \cdots + f(1.9) + f(2)] \approx 0.463955$

$T_{10} = \frac{1}{10 \cdot 2}\{f(1) + 2[f(1.1) + f(1.2) + \cdots + f(1.9)] + f(2)\} \approx 0.501455$

$M_{10} = \frac{1}{10}[f(1.05) + f(1.15) + \cdots + f(1.85) + f(1.95)] \approx 0.499274$

$E_L = I - L_{10} \approx \frac{1}{2} - 0.538955 = -0.038955$

$E_R \approx \frac{1}{2} - 0.463955 = 0.036049$

$E_T \approx \frac{1}{2} - 0.501455 = -0.001455$

$E_M \approx \frac{1}{2} - 0.499274 = 0.000726$

$n = 20$: $L_{20} = \frac{1}{20}[f(1) + f(1.05) + f(1.10) + \cdots + f(1.95)] \approx 0.519114$

$R_{20} = \frac{1}{20}[f(1.05) + f(1.10) + \cdots + f(1.95) + f(2)] \approx 0.481614$

$T_{20} = \frac{1}{20 \cdot 2}\{f(1) + 2[f(1.05) + f(1.10) + \cdots + f(1.95)] + f(2)\} \approx 0.500364$

$M_{20} = \frac{1}{20}[f(1.025) + f(1.075) + f(1.125) + \cdots + f(1.975)] \approx 0.499818$

$E_L = I - L_{20} \approx \frac{1}{2} - 0.519114 = -0.019114$

$E_R \approx \frac{1}{2} - 0.481614 = 0.018386$

$E_T \approx \frac{1}{2} - 0.500364 = -0.000364$

$E_M \approx \frac{1}{2} - 0.499818 = 0.000182$

n	L_n	R_n	T_n	M_n
5	0.580783	0.430783	0.505783	0.497127
10	0.538955	0.463955	0.501455	0.499274
20	0.519114	0.481614	0.500364	0.499818

n	E_L	E_R	E_T	E_M
5	−0.080783	0.069217	−0.005783	0.002873
10	−0.038955	0.036049	−0.001455	0.000726
20	−0.019114	0.018386	−0.000364	0.000182

Observations:

1. E_L and E_R are always opposite in sign, as are E_T and E_M.

2. As n is doubled, E_L and E_R are decreased by about a factor of 2, and E_T and E_M are decreased by a factor of about 4.

3. The Midpoint approximation is about twice as accurate as the Trapezoidal approximation.

4. All the approximations become more accurate as the value of n increases.

5. The Midpoint and Trapezoidal approximations are much more accurate than the endpoint approximations.

27. $I = \int_0^2 x^4 \, dx = \left[\frac{1}{5}x^5\right]_0^2 = \frac{32}{5} - 0 = 6.4$, $f(x) = x^4$, $\Delta x = \frac{2-0}{n} = \frac{2}{n}$

$n = 6$: $\quad T_6 = \frac{2}{6 \cdot 2}\left\{f(0) + 2\left[f\left(\frac{1}{3}\right) + f\left(\frac{2}{3}\right) + f\left(\frac{3}{3}\right) + f\left(\frac{4}{3}\right) + f\left(\frac{5}{3}\right)\right] + f(2)\right\} \approx 6.695473$

$\qquad\quad M_6 = \frac{2}{6}\left[f\left(\frac{1}{6}\right) + f\left(\frac{3}{6}\right) + f\left(\frac{5}{6}\right) + f\left(\frac{7}{6}\right) + f\left(\frac{9}{6}\right) + f\left(\frac{11}{6}\right)\right] \approx 6.252572$

$\qquad\quad S_6 = \frac{2}{6 \cdot 3}\left[f(0) + 4f\left(\frac{1}{3}\right) + 2f\left(\frac{2}{3}\right) + 4f\left(\frac{3}{3}\right) + 2f\left(\frac{4}{3}\right) + 4f\left(\frac{5}{3}\right) + f(2)\right] \approx 6.403292$

$\qquad\quad E_T = I - T_6 \approx 6.4 - 6.695473 = -0.295473$

$\qquad\quad E_M \approx 6.4 - 6.252572 = 0.147428$

$\qquad\quad E_S \approx 6.4 - 6.403292 = -0.003292$

$n = 12$: $\quad T_{12} = \frac{2}{12 \cdot 2}\left\{f(0) + 2\left[f\left(\frac{1}{6}\right) + f\left(\frac{2}{6}\right) + f\left(\frac{3}{6}\right) + \cdots + f\left(\frac{11}{6}\right)\right] + f(2)\right\} \approx 6.474023$

$\qquad\quad M_6 = \frac{2}{12}\left[f\left(\frac{1}{12}\right) + f\left(\frac{3}{12}\right) + f\left(\frac{5}{12}\right) + \cdots + f\left(\frac{23}{12}\right)\right] \approx 6.363008$

$\qquad\quad S_6 = \frac{2}{12 \cdot 3}\left[f(0) + 4f\left(\frac{1}{6}\right) + 2f\left(\frac{2}{6}\right) + 4f\left(\frac{3}{6}\right) + 2f\left(\frac{4}{6}\right) + \cdots + 4f\left(\frac{11}{6}\right) + f(2)\right] \approx 6.400206$

$\qquad\quad E_T = I - T_{12} \approx 6.4 - 6.474023 = -0.074023$

$\qquad\quad E_M \approx 6.4 - 6.363008 = 0.036992$

$\qquad\quad E_S \approx 6.4 - 6.400206 = -0.000206$

n	T_n	M_n	S_n
6	6.695473	6.252572	6.403292
12	6.474023	6.363008	6.400206

n	E_T	E_M	E_S
6	−0.295473	0.147428	−0.003292
12	−0.074023	0.036992	−0.000206

Observations:

1. E_T and E_M are opposite in sign and decrease by a factor of about 4 as n is doubled.

2. The Simpson's approximation is much more accurate than the Midpoint and Trapezoidal approximations, and E_S seems to decrease by a factor of about 16 as n is doubled.

28. $I = \int_1^4 \frac{1}{\sqrt{x}} \, dx = \left[2\sqrt{x}\right]_1^4 = 4 - 2 = 2$, $f(x) = \frac{1}{\sqrt{x}}$, $\Delta x = \frac{4-1}{n} = \frac{3}{n}$

$n = 6$: $\quad T_6 = \frac{3}{6 \cdot 2}\left\{f(1) + 2\left[f\left(\frac{3}{2}\right) + f\left(\frac{4}{2}\right) + f\left(\frac{5}{2}\right) + f\left(\frac{6}{2}\right) + f\left(\frac{7}{2}\right)\right] + f(4)\right\} \approx 2.008966$

$\qquad\quad M_6 = \frac{3}{6}\left[f\left(\frac{5}{4}\right) + f\left(\frac{7}{4}\right) + f\left(\frac{9}{4}\right) + f\left(\frac{11}{4}\right) + f\left(\frac{13}{4}\right) + f\left(\frac{15}{4}\right)\right] \approx 1.995572$

$\qquad\quad S_6 = \frac{3}{6 \cdot 3}\left[f(1) + 4f\left(\frac{3}{2}\right) + 2f\left(\frac{4}{2}\right) + 4f\left(\frac{5}{2}\right) + 2f\left(\frac{6}{2}\right) + 4f\left(\frac{7}{2}\right) + f(4)\right] \approx 2.000469$

$\qquad\quad E_T = I - T_6 \approx 2 - 2.008966 = -0.008966,$

$\qquad\quad E_M \approx 2 - 1.995572 = 0.004428,$

$\qquad\quad E_S \approx 2 - 2.000469 = -0.000469$

$n = 12$: $\quad T_{12} = \frac{3}{12 \cdot 2}\left\{f(1) + 2\left[f\left(\frac{5}{4}\right) + f\left(\frac{6}{4}\right) + f\left(\frac{7}{4}\right) + \cdots + f\left(\frac{15}{4}\right)\right] + f(4)\right\} \approx 2.002269$

$\qquad\quad M_{12} = \frac{3}{12}\left[f\left(\frac{9}{8}\right) + f\left(\frac{11}{8}\right) + f\left(\frac{13}{8}\right) + \cdots + f\left(\frac{31}{8}\right)\right] \approx 1.998869$

$\qquad\quad S_{12} = \frac{3}{12 \cdot 3}\left[f(1) + 4f\left(\frac{5}{4}\right) + 2f\left(\frac{6}{4}\right) + 4f\left(\frac{7}{4}\right) + 2f\left(\frac{8}{4}\right) + \cdots + 4f\left(\frac{15}{4}\right) + f(4)\right] \approx 2.000036$

$\qquad\quad E_T = I - T_{12} \approx 2 - 2.002269 = -0.002269$

$\qquad\quad E_M \approx 2 - 1.998869 = 0.001131$

$\qquad\quad E_S \approx 2 - 2.000036 = -0.000036$

n	T_n	M_n	S_n
6	2.008966	1.995572	2.000469
12	2.002269	1.998869	2.000036

n	E_T	E_M	E_S
6	−0.008966	0.004428	−0.000469
12	−0.002269	0.001131	−0.000036

Observations:

1. E_T and E_M are opposite in sign and decrease by a factor of about 4 as n is doubled.

2. The Simpson's approximation is much more accurate than the Midpoint and Trapezoidal approximations, and E_S seems to decrease by a factor of about 16 as n is doubled.

29. $\Delta x = (b-a)/n = (6-0)/6 = 1$

(a) $T_6 = \frac{\Delta x}{2}[f(0) + 2f(1) + 2f(2) + 2f(3) + 2f(4) + 2f(5) + f(6)]$

$\approx \frac{1}{2}[3 + 2(5) + 2(4) + 2(2) + 2(2.8) + 2(4) + 1]$

$= \frac{1}{2}(39.6) = 19.8$

(b) $M_6 = \Delta x[f(0.5) + f(1.5) + f(2.5) + f(3.5) + f(4.5) + f(5.5)]$

$\approx 1[4.5 + 4.7 + 2.6 + 2.2 + 3.4 + 3.2]$

$= 20.6$

(c) $S_6 = \frac{\Delta x}{3}[f(0) + 4f(1) + 2f(2) + 4f(3) + 2f(4) + 4f(5) + f(6)]$

$\approx \frac{1}{3}[3 + 4(5) + 2(4) + 4(2) + 2(2.8) + 4(4) + 1]$

$= \frac{1}{3}(61.6) = 20.5\overline{3}$

30. If x = distance from left end of pool and $w = w(x)$ = width at x, then Simpson's Rule with $n = 8$ and $\Delta x = 2$ gives

Area $= \int_0^{16} w \, dx \approx \frac{2}{3}[0 + 4(6.2) + 2(7.2) + 4(6.8) + 2(5.6) + 4(5.0) + 2(4.8) + 4(4.8) + 0] \approx 84 \text{ m}^2$.

31. (a) We are given the function values at the endpoints of 8 intervals of length 0.4, so we'll use the Midpoint Rule with $n = 8/2 = 4$ and $\Delta x = (3.2 - 0)/4 = 0.8$.

$\int_0^{3.2} f(x) \, dx \approx M_4 = 0.8[f(0.4) + f(1.2) + f(2.0) + f(2.8)] = 0.8[6.5 + 6.4 + 7.6 + 8.8]$

$= 0.8(29.3) = 23.44$

(b) $-4 \le f''(x) \le 1 \Rightarrow |f''(x)| \le 4$, so use $K = 4$, $a = 0$, $b = 3.2$, and $n = 4$ in Theorem 3.

So $|E_M| \le \dfrac{4(3.2-0)^3}{24(4)^2} = \dfrac{128}{375} = 0.341\overline{3}$.

32. We use Simpson's Rule with $n = 10$ and $\Delta x = \frac{1}{2}$:

distance $= \int_0^5 v(t) \, dt \approx S_{10} = \frac{1}{2 \cdot 3}[f(0) + 4f(0.5) + 2f(1) + \cdots + 4f(4.5) + f(5)]$

$= \frac{1}{6}[0 + 4(4.67) + 2(7.34) + 4(8.86) + 2(9.73) + 4(10.22)$

$+ 2(10.51) + 4(10.67) + 2(10.76) + 4(10.81) + 10.81]$

$= \frac{1}{6}(268.41) = 44.735 \text{ m}$

33. By the Net Change Theorem, the increase in velocity is equal to $\int_0^6 a(t)\,dt$. We use Simpson's Rule with $n = 6$ and

$\Delta t = (6 - 0)/6 = 1$ to estimate this integral:

$$\int_0^6 a(t)\,dt \approx S_6 = \tfrac{1}{3}[a(0) + 4a(1) + 2a(2) + 4a(3) + 2a(4) + 4a(5) + a(6)]$$

$$\approx \tfrac{1}{3}[0 + 4(0.5) + 2(4.1) + 4(9.8) + 2(12.9) + 4(9.5) + 0] = \tfrac{1}{3}(113.2) = 37.7\overline{3} \text{ ft/s}$$

34. By the Net Change Theorem, the total amount of water that leaked out during the first six hours is equal to $\int_0^6 r(t)\,dt$.

We use Simpson's Rule with $n = 6$ and $\Delta t = \frac{6-0}{6} = 1$ to estimate this integral:

$$\int_0^6 r(t)\,dt \approx S_6 = \tfrac{1}{3}[r(0) + 4r(1) + 2r(2) + 4r(3) + 2r(4) + 4r(5) + r(6)]$$

$$\approx \tfrac{1}{3}[4 + 4(3) + 2(2.4) + 4(1.9) + 2(1.4) + 4(1.1) + 1] = \tfrac{1}{3}(36.6) = 12.2 \text{ liters}$$

The function values were obtained from a high-resolution graph.

35. By the Net Change Theorem, the energy used is equal to $\int_0^6 P(t)\,dt$. We use Simpson's Rule with $n = 12$ and

$\Delta t = \frac{6-0}{12} = \frac{1}{2}$ to estimate this integral:

$$\int_0^6 P(t)\,dt \approx S_{12} = \tfrac{1/2}{3}[P(0) + 4P(0.5) + 2P(1) + 4P(1.5) + 2P(2) + 4P(2.5) + 2P(3)$$

$$+ 4P(3.5) + 2P(4) + 4P(4.5) + 2P(5) + 4P(5.5) + P(6)]$$

$$= \tfrac{1}{6}[1814 + 4(1735) + 2(1686) + 4(1646) + 2(1637) + 4(1609) + 2(1604)$$

$$+ 4(1611) + 2(1621) + 4(1666) + 2(1745) + 4(1886) + 2052]$$

$$= \tfrac{1}{6}(61{,}064) = 10{,}177.\overline{3} \text{ megawatt-hours}$$

36. By the Net Change Theorem, the total amount of data transmitted is equal to $\int_0^8 D(t)\,dt \times 3600$ [since $D(t)$ is measured in

megabits per second and t is in hours]. We use Simpson's Rule with $n = 8$ and $\Delta t = (8 - 0)/8 = 1$ to estimate this integral:

$$\int_0^8 D(t)\,dt \approx S_8 = \tfrac{1}{3}[D(0) + 4D(1) + 2D(2) + 4D(3) + 2D(4) + 4D(5) + 2D(6) + 4D(7) + D(8)]$$

$$\approx \tfrac{1}{3}[0.35 + 4(0.32) + 2(0.41) + 4(0.50) + 2(0.51) + 4(0.56) + 2(0.56) + 4(0.83) + 0.88]$$

$$= \tfrac{1}{3}(13.03) = 4.34\overline{3}$$

Now multiply by 3600 to obtain 15,636 megabits.

37. Let $y = f(x)$ denote the curve. Using cylindrical shells, $V = \int_2^{10} 2\pi x f(x)\,dx = 2\pi \int_2^{10} x f(x)\,dx = 2\pi I_1$.

Now use Simpson's Rule to approximate I_1:

$$I_1 \approx S_8 = \tfrac{10-2}{3(8)}[2f(2) + 4 \cdot 3f(3) + 2 \cdot 4f(4) + 4 \cdot 5f(5) + 2 \cdot 6f(6) + 4 \cdot 7f(7) + 2 \cdot 8f(8) + 4 \cdot 9f(9) + 10f(10)]$$

$$\approx \tfrac{1}{3}[2(0) + 12(1.5) + 8(1.9) + 20(2.2) + 12(3.0) + 28(3.8) + 16(4.0) + 36(3.1) + 10(0)]$$

$$= \tfrac{1}{3}(395.2)$$

Thus, $V \approx 2\pi \cdot \tfrac{1}{3}(395.2) \approx 827.7$ or 828 cubic units.

38. Work $= \int_0^{18} f(x)\,dx \approx S_6 = \tfrac{18-0}{6 \cdot 3}[f(0) + 4f(3) + 2f(6) + 4f(9) + 2f(12) + 4f(15) + f(18)]$

$$= 1 \cdot [9.8 + 4(9.1) + 2(8.5) + 4(8.0) + 2(7.7) + 4(7.5) + 7.4] = 148 \text{ joules}$$

39. Using disks, $V = \int_1^5 \pi(e^{-1/x})^2\, dx = \pi \int_1^5 e^{-2/x}\, dx = \pi I_1$. Now use Simpson's Rule with $f(x) = e^{-2/x}$ to approximate

I_1. $I_1 \approx S_8 = \frac{5-1}{3(8)}\left[f(1) + 4f(1.5) + 2f(2) + 4f(2.5) + 2f(3) + 4f(3.5) + 2f(4) + 4f(4.5) + f(5)\right] \approx \frac{1}{6}(11.4566)$

Thus, $V \approx \pi \cdot \frac{1}{6}(11.4566) \approx 6.0$ cubic units.

40. Using Simpson's Rule with $n = 10$, $\Delta x = \frac{\pi/2}{10}$, $L = 1$, $\theta_0 = \frac{42\pi}{180}$ radians, $g = 9.8 \text{ m/s}^2$, $k^2 = \sin^2\left(\frac{1}{2}\theta_0\right)$, and

$f(x) = 1/\sqrt{1 - k^2 \sin^2 x}$, we get

$$T = 4\sqrt{\frac{L}{g}}\int_0^{\pi/2}\frac{dx}{\sqrt{1 - k^2 \sin^2 x}} \approx 4\sqrt{\frac{L}{g}}\, S_{10}$$

$$= 4\sqrt{\frac{1}{9.8}}\left(\frac{\pi/2}{10 \cdot 3}\right)\left[f(0) + 4f\left(\frac{\pi}{20}\right) + 2f\left(\frac{2\pi}{20}\right) + \cdots + 4f\left(\frac{9\pi}{20}\right) + f\left(\frac{\pi}{2}\right)\right] \approx 2.07665$$

41. $I(\theta) = \dfrac{N^2 \sin^2 k}{k^2}$, where $k = \dfrac{\pi N d \sin\theta}{\lambda}$, $N = 10{,}000$, $d = 10^{-4}$, and $\lambda = 632.8 \times 10^{-9}$. So $I(\theta) = \dfrac{(10^4)^2 \sin^2 k}{k^2}$,

where $k = \dfrac{\pi(10^4)(10^{-4})\sin\theta}{632.8 \times 10^{-9}}$. Now $n = 10$ and $\Delta\theta = \dfrac{10^{-6} - (-10^{-6})}{10} = 2 \times 10^{-7}$, so

$M_{10} = 2 \times 10^{-7}[I(-0.0000009) + I(-0.0000007) + \cdots + I(0.0000009)] \approx 59.4$.

42. $f(x) = \cos(\pi x)$, $\Delta x = \frac{20-0}{10} = 2$ $\Rightarrow$

$T_{10} = \frac{2}{2}\{f(0) + 2[f(2) + f(4) + \cdots + f(18)] + f(20)\} = 1[\cos 0 + 2(\cos 2\pi + \cos 4\pi + \cdots + \cos 18\pi) + \cos 20\pi]$

$= 1 + 2(1 + 1 + 1 + 1 + 1 + 1 + 1 + 1 + 1) + 1 = 20$

The actual value is $\int_0^{20}\cos(\pi x)\, dx = \frac{1}{\pi}\left[\sin\pi x\right]_0^{20} = \frac{1}{\pi}(\sin 20\pi - \sin 0) = 0$. The discrepancy is due to the fact that the

function is sampled only at points of the form $2n$, where its value is $f(2n) = \cos(2n\pi) = 1$.

43. Consider the function f whose graph is shown. The area $\int_0^2 f(x)\, dx$

is close to 2. The Trapezoidal Rule gives

$T_2 = \frac{2-0}{2\cdot 2}\left[f(0) + 2f(1) + f(2)\right] = \frac{1}{2}\left[1 + 2\cdot 1 + 1\right] = 2$.

The Midpoint Rule gives $M_2 = \frac{2-0}{2}\left[f(0.5) + f(1.5)\right] = 1[0 + 0] = 0$,

so the Trapezoidal Rule is more accurate.

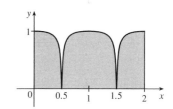

44. Consider the function $f(x) = |x - 1|$, $0 \le x \le 2$. The area $\int_0^2 f(x)\, dx$

is exactly 1. So is the right endpoint approximation:

$R_2 = f(1)\,\Delta x + f(2)\,\Delta x = 0 \cdot 1 + 1 \cdot 1 = 1$. But Simpson's Rule

approximates f with the parabola $y = (x - 1)^2$, shown dashed, and

$S_2 = \dfrac{\Delta x}{3}\left[f(0) + 4f(1) + f(2)\right] = \dfrac{1}{3}\left[1 + 4 \cdot 0 + 1\right] = \dfrac{2}{3}$.

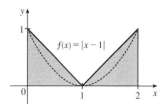

45. Since the Trapezoidal and Midpoint approximations on the interval $[a, b]$ are the sums of the Trapezoidal and Midpoint

approximations on the subintervals $[x_{i-1}, x_i]$, $i = 1, 2, \ldots, n$, we can focus our attention on one such interval. The condition

$f''(x) < 0$ for $a \le x \le b$ means that the graph of f is concave down as in Figure 5. In that figure, T_n is the area of the trapezoid $AQRD$, $\int_a^b f(x)\,dx$ is the area of the region $AQPRD$, and M_n is the area of the trapezoid $ABCD$, so $T_n < \int_a^b f(x)\,dx < M_n$. In general, the condition $f'' < 0$ implies that the graph of f on $[a,b]$ lies above the chord joining the points $(a, f(a))$ and $(b, f(b))$. Thus, $\int_a^b f(x)\,dx > T_n$. Since M_n is the area under a tangent to the graph, and since $f'' < 0$ implies that the tangent lies above the graph, we also have $M_n > \int_a^b f(x)\,dx$. Thus, $T_n < \int_a^b f(x)\,dx < M_n$.

46. Let f be a polynomial of degree ≤ 3; say $f(x) = Ax^3 + Bx^2 + Cx + D$. It will suffice to show that Simpson's estimate is exact when there are two subintervals ($n = 2$), because for a larger even number of subintervals the sum of exact estimates is exact. As in the derivation of Simpson's Rule, we can assume that $x_0 = -h$, $x_1 = 0$, and $x_2 = h$. Then Simpson's approximation is

$$\int_{-h}^h f(x)\,dx \approx \tfrac{1}{3}h[f(-h) + 4f(0) + f(h)] = \tfrac{1}{3}h\big[\big(-Ah^3 + Bh^2 - Ch + D\big) + 4D + \big(Ah^3 + Bh^2 + Ch + D\big)\big]$$
$$= \tfrac{1}{3}h[2Bh^2 + 6D] = \tfrac{2}{3}Bh^3 + 2Dh$$

The exact value of the integral is

$$\int_{-h}^h (Ax^3 + Bx^2 + Cx + D)\,dx = 2\int_0^h (Bx^2 + D)\,dx \qquad \text{[by Theorem 5.5.6(a) and (b)]}$$
$$= 2\big[\tfrac{1}{3}Bx^3 + Dx\big]_0^h = \tfrac{2}{3}Bh^3 + 2Dh$$

Thus, Simpson's Rule is exact.

47. $T_n = \tfrac{1}{2}\Delta x\,[f(x_0) + 2f(x_1) + \cdots + 2f(x_{n-1}) + f(x_n)]$ and

$M_n = \Delta x\,[f(\overline{x}_1) + f(\overline{x}_2) + \cdots + f(\overline{x}_{n-1}) + f(\overline{x}_n)]$, where $\overline{x}_i = \tfrac{1}{2}(x_{i-1} + x_i)$. Now

$T_{2n} = \tfrac{1}{2}\big(\tfrac{1}{2}\Delta x\big)[f(x_0) + 2f(\overline{x}_1) + 2f(x_1) + 2f(\overline{x}_2) + 2f(x_2) + \cdots + 2f(\overline{x}_{n-1}) + 2f(x_{n-1}) + 2f(\overline{x}_n) + f(x_n)]$

so

$$\tfrac{1}{2}(T_n + M_n) = \tfrac{1}{2}T_n + \tfrac{1}{2}M_n$$
$$= \tfrac{1}{4}\Delta x[f(x_0) + 2f(x_1) + \cdots + 2f(x_{n-1}) + f(x_n)] + \tfrac{1}{4}\Delta x[2f(\overline{x}_1) + 2f(\overline{x}_2) + \cdots + 2f(\overline{x}_{n-1}) + 2f(\overline{x}_n)]$$
$$= T_{2n}$$

48. $T_n = \dfrac{\Delta x}{2}\left[f(x_0) + 2\sum_{i=1}^{n-1} f(x_i) + f(x_n)\right]$ and $M_n = \Delta x \sum_{i=1}^{n} f\left(x_i - \dfrac{\Delta x}{2}\right)$, so

$$\tfrac{1}{3}T_n + \tfrac{2}{3}M_n = \tfrac{1}{3}(T_n + 2M_n) = \dfrac{\Delta x}{3\cdot 2}\left[f(x_0) + 2\sum_{i=1}^{n-1} f(x_i) + f(x_n) + 4\sum_{i=1}^{n} f\left(x_i - \dfrac{\Delta x}{2}\right)\right]$$

where $\Delta x = \dfrac{b-a}{n}$. Let $\delta x = \dfrac{b-a}{2n}$. Then $\Delta x = 2\delta x$, so

$$\tfrac{1}{3}T_n + \tfrac{2}{3}M_n = \dfrac{\delta x}{3}\left[f(x_0) + 2\sum_{i=1}^{n-1} f(x_i) + f(x_n) + 4\sum_{i=1}^{n} f(x_i - \delta x)\right]$$
$$= \tfrac{1}{3}\delta x[f(x_0) + 4f(x_1 - \delta x) + 2f(x_1) + 4f(x_2 - \delta x)$$
$$+ 2f(x_2) + \cdots + 2f(x_{n-1}) + 4f(x_n - \delta x) + f(x_n)]$$

Since $x_0, x_1 - \delta x, x_1, x_2 - \delta x, x_2, \ldots, x_{n-1}, x_n - \delta x, x_n$ are the subinterval endpoints for S_{2n}, and since $\delta x = \dfrac{b-a}{2n}$ is the width of the subintervals for S_{2n}, the last expression for $\tfrac{1}{3}T_n + \tfrac{2}{3}M_n$ is the usual expression for S_{2n}. Therefore, $\tfrac{1}{3}T_n + \tfrac{2}{3}M_n = S_{2n}$.

8.8 Improper Integrals

1. (a) Since $\int_1^\infty x^4 e^{-x^4}\,dx$ has an infinite interval of integration, it is an improper integral of Type I.

(b) Since $y = \sec x$ has an infinite discontinuity at $x = \frac{\pi}{2}$, $\int_0^{\pi/2} \sec x\,dx$ is a Type II improper integral.

(c) Since $y = \dfrac{x}{(x-2)(x-3)}$ has an infinite discontinuity at $x = 2$, $\int_0^2 \dfrac{x}{x^2 - 5x + 6}\,dx$ is a Type II improper integral.

(d) Since $\int_{-\infty}^0 \dfrac{1}{x^2 + 5}\,dx$ has an infinite interval of integration, it is an improper integral of Type I.

2. (a) Since $y = \dfrac{1}{2x - 1}$ is defined and continuous on $[1, 2]$, $\int_1^2 \dfrac{1}{2x - 1}\,dx$ is proper.

(b) Since $y = \dfrac{1}{2x - 1}$ has an infinite discontinuity at $x = \frac{1}{2}$, $\int_0^1 \dfrac{1}{2x - 1}\,dx$ is a Type II improper integral.

(c) Since $\int_{-\infty}^\infty \dfrac{\sin x}{1 + x^2}\,dx$ has an infinite interval of integration, it is an improper integral of Type I.

(d) Since $y = \ln(x - 1)$ has an infinite discontinuity at $x = 1$, $\int_1^2 \ln(x - 1)\,dx$ is a Type II improper integral.

3. The area under the graph of $y = 1/x^3 = x^{-3}$ between $x = 1$ and $x = t$ is

$A(t) = \int_1^t x^{-3}\,dx = \left[-\frac{1}{2}x^{-2}\right]_1^t = -\frac{1}{2}t^{-2} - \left(-\frac{1}{2}\right) = \frac{1}{2} - 1/(2t^2)$. So the area for $1 \le x \le 10$ is

$A(10) = 0.5 - 0.005 = 0.495$, the area for $1 \le x \le 100$ is $A(100) = 0.5 - 0.00005 = 0.49995$, and the area for

$1 \le x \le 1000$ is $A(1000) = 0.5 - 0.0000005 = 0.4999995$. The total area under the curve for $x \ge 1$ is

$\displaystyle \lim_{t\to\infty} A(t) = \lim_{t\to\infty} \left[\frac{1}{2} - 1/(2t^2)\right] = \frac{1}{2}$.

4. (a)

(b) The area under the graph of f from $x = 1$ to $x = t$ is

$$F(t) = \int_1^t f(x)\,dx = \int_1^t x^{-1.1}\,dx = \left[-\frac{1}{0.1}x^{-0.1}\right]_1^t$$
$$= -10(t^{-0.1} - 1) = 10(1 - t^{-0.1})$$

and the area under the graph of g is

$$G(t) = \int_1^t g(x)\,dx = \int_1^t x^{-0.9}\,dx = \left[\frac{1}{0.1}x^{0.1}\right]_1^t = 10(t^{0.1} - 1).$$

t	$F(t)$	$G(t)$
10	2.06	2.59
100	3.69	5.85
10^4	6.02	15.12
10^6	7.49	29.81
10^{10}	9	90
10^{20}	9.9	990

(c) The total area under the graph of f is $\lim_{t \to \infty} F(t) = \lim_{t \to \infty} 10(1 - t^{-0.1}) = 10$.

The total area under the graph of g does not exist, since $\lim_{t \to \infty} G(t) = \lim_{t \to \infty} 10(t^{0.1} - 1) = \infty$.

5. $I = \int_1^\infty \dfrac{1}{(3x+1)^2}\, dx = \lim_{t \to \infty} \int_1^t \dfrac{1}{(3x+1)^2}\, dx.$ Now

$\displaystyle \int \dfrac{1}{(3x+1)^2}\, dx = \dfrac{1}{3} \int \dfrac{1}{u^2}\, du \quad [u = 3x+1,\, du = 3\, dx] \quad = -\dfrac{1}{3u} + C = -\dfrac{1}{3(3x+1)} + C,$

so $I = \lim_{t \to \infty} \left[-\dfrac{1}{3(3x+1)} \right]_1^t = \lim_{t \to \infty} \left[-\dfrac{1}{3(3t+1)} + \dfrac{1}{12} \right] = 0 + \dfrac{1}{12} = \dfrac{1}{12}.$ Convergent

6. $\displaystyle \int_{-\infty}^0 \dfrac{1}{2x-5}\, dx = \lim_{t \to -\infty} \int_t^0 \dfrac{1}{2x-5}\, dx = \lim_{t \to -\infty} \left[\tfrac{1}{2} \ln|2x-5| \right]_t^0 = \lim_{t \to \infty} \left[\tfrac{1}{2} \ln 5 - \tfrac{1}{2} \ln|2t-5| \right] = -\infty.$

Divergent

7. $\displaystyle \int_{-\infty}^{-1} \dfrac{1}{\sqrt{2-w}}\, dw = \lim_{t \to -\infty} \int_t^{-1} \dfrac{1}{\sqrt{2-w}}\, dw = \lim_{t \to -\infty} \left[-2\sqrt{2-w} \right]_t^{-1} \quad [u = 2-w,\, du = -dw]$

$= \lim_{t \to -\infty} \left[-2\sqrt{3} + 2\sqrt{2-t} \right] = \infty.$ Divergent

8. $\displaystyle \int_0^\infty \dfrac{x}{(x^2+2)^2}\, dx = \lim_{t \to \infty} \int_0^t \dfrac{x}{(x^2+2)^2}\, dx = \lim_{t \to \infty} \dfrac{1}{2} \left[\dfrac{-1}{x^2+2} \right]_0^t = \dfrac{1}{2} \lim_{t \to \infty} \left(\dfrac{-1}{t^2+2} + \dfrac{1}{2} \right)$

$= \tfrac{1}{2}\left(0 + \tfrac{1}{2} \right) = \tfrac{1}{4}.$ Convergent

9. $\int_4^\infty e^{-y/2}\, dy = \lim_{t \to \infty} \int_4^t e^{-y/2}\, dy = \lim_{t \to \infty} \left[-2e^{-y/2} \right]_4^t = \lim_{t \to \infty} \left(-2e^{-t/2} + 2e^{-2} \right) = 0 + 2e^{-2} = 2e^{-2}.$

Convergent

10. $\int_{-\infty}^{-1} e^{-2t}\, dt = \lim_{x \to -\infty} \int_x^{-1} e^{-2t}\, dt = \lim_{x \to -\infty} \left[-\tfrac{1}{2} e^{-2t} \right]_x^{-1} = \lim_{x \to -\infty} \left[-\tfrac{1}{2} e^2 + \tfrac{1}{2} e^{-2x} \right] = \infty.$ Divergent

11. $\displaystyle \int_{-\infty}^\infty \dfrac{x\, dx}{1+x^2} = \int_{-\infty}^0 \dfrac{x\, dx}{1+x^2} + \int_0^\infty \dfrac{x\, dx}{1+x^2}$ and

$\displaystyle \int_{-\infty}^0 \dfrac{x\, dx}{1+x^2} = \lim_{t \to -\infty} \left[\tfrac{1}{2} \ln\left(1+x^2\right) \right]_t^0 = \lim_{t \to -\infty} \left[0 - \tfrac{1}{2} \ln\left(1+t^2\right) \right] = -\infty.$ Divergent

12. $I = \int_{-\infty}^\infty \left(2 - v^4\right) dv = I_1 + I_2 = \int_{-\infty}^0 \left(2 - v^4\right) dv + \int_0^\infty \left(2 - v^4\right) dv,$ but

$I_1 = \lim_{t \to -\infty} \left[2v - \tfrac{1}{5} v^5 \right]_t^0 = \lim_{t \to -\infty} \left(-2t + \tfrac{1}{5} t^5 \right) = -\infty.$ Since I_1 is divergent, I is divergent, and there is no need to

evaluate I_2.   Divergent

13. $\int_{-\infty}^\infty x e^{-x^2}\, dx = \int_{-\infty}^0 x e^{-x^2}\, dx + \int_0^\infty x e^{-x^2}\, dx.$

$\int_{-\infty}^0 x e^{-x^2}\, dx = \lim_{t \to -\infty} \left(-\tfrac{1}{2} \right) \left[e^{-x^2} \right]_t^0 = \lim_{t \to -\infty} \left(-\tfrac{1}{2} \right) \left(1 - e^{-t^2} \right) = -\tfrac{1}{2} \cdot 1 = -\tfrac{1}{2},$ and

$\int_0^\infty x e^{-x^2}\, dx = \lim_{t \to \infty} \left(-\tfrac{1}{2} \right) \left[e^{-x^2} \right]_0^t = \lim_{t \to \infty} \left(-\tfrac{1}{2} \right) \left(e^{-t^2} - 1 \right) = -\tfrac{1}{2} \cdot (-1) = \tfrac{1}{2}.$

Therefore, $\int_{-\infty}^\infty x e^{-x^2}\, dx = -\tfrac{1}{2} + \tfrac{1}{2} = 0.$ Convergent

14. $\int_1^\infty \dfrac{e^{-\sqrt{x}}}{\sqrt{x}}\,dx = \lim_{t\to\infty}\int_1^t \dfrac{e^{-\sqrt{x}}}{\sqrt{x}}\,dx = \lim_{t\to\infty}\int_1^{\sqrt{t}} e^{-u}(2\,du)$ $\left[\begin{array}{l}u=\sqrt{x},\\ du=dx/(2\sqrt{x})\end{array}\right]$

$\qquad = 2\lim_{t\to\infty}\Big[-e^{-u}\Big]_1^{\sqrt{t}} = 2\lim_{t\to\infty}\left(-e^{-\sqrt{t}}+e^{-1}\right) = 2(0+e^{-1}) = 2e^{-1}.$ Convergent

15. $\int_{2\pi}^\infty \sin\theta\,d\theta = \lim_{t\to\infty}\int_{2\pi}^t \sin\theta\,d\theta = \lim_{t\to\infty}\big[-\cos\theta\big]_{2\pi}^t = \lim_{t\to\infty}(-\cos t+1).$ This limit does not exist, so the integral is

divergent.   Divergent

16. $I = \int_{-\infty}^\infty \cos\pi t\,dt = I_1+I_2 = \int_{-\infty}^0 \cos\pi t\,dt + \int_0^\infty \cos\pi t\,dt$, but $I_1 = \lim_{s\to-\infty}\left[\dfrac{1}{\pi}\sin\pi t\right]_s^0 = \lim_{s\to-\infty}\left(-\dfrac{1}{\pi}\sin\pi t\right)$ and

this limit does not exist. Since I_1 is divergent, I is divergent, and there is no need to evaluate I_2.   Divergent

17. $\int_1^\infty \dfrac{x+1}{x^2+2x}\,dx = \lim_{t\to\infty}\int_1^t \dfrac{\frac{1}{2}(2x+2)}{x^2+2x}\,dx = \tfrac{1}{2}\lim_{t\to\infty}\Big[\ln(x^2+2x)\Big]_1^t = \tfrac{1}{2}\lim_{t\to\infty}\Big[\ln(t^2+2t)-\ln 3\Big] = \infty.$

Divergent

18. $\int_0^\infty \dfrac{dz}{z^2+3z+2} = \lim_{t\to\infty}\int_0^t \left[\dfrac{1}{z+1}-\dfrac{1}{z+2}\right]dz = \lim_{t\to\infty}\left[\ln\!\left(\dfrac{z+1}{z+2}\right)\right]_0^t$

$\qquad = \lim_{t\to\infty}\left[\ln\!\left(\dfrac{t+1}{t+2}\right)-\ln\!\left(\dfrac{1}{2}\right)\right] = \ln 1+\ln 2 = \ln 2.$ Convergent

19. $\int_0^\infty se^{-5s}\,ds = \lim_{t\to\infty}\int_0^t se^{-5s}\,ds = \lim_{t\to\infty}\big[-\tfrac{1}{5}se^{-5s}-\tfrac{1}{25}e^{-5s}\big]$ $\left[\begin{array}{l}\text{by integration by}\\ \text{parts with }u=s\end{array}\right]$

$\qquad = \lim_{t\to\infty}\left(-\tfrac{1}{5}te^{-5t}-\tfrac{1}{25}e^{-5t}+\tfrac{1}{25}\right) = 0-0+\tfrac{1}{25}$ [by l'Hospital's Rule]

$\qquad = \tfrac{1}{25}.$ Convergent

20. $\int_{-\infty}^6 re^{r/3}\,dr = \lim_{t\to-\infty}\int_t^6 re^{r/3}\,dr = \lim_{t\to-\infty}\Big[3re^{r/3}-9e^{r/3}\Big]_t^6$ $\left[\begin{array}{l}\text{by integration by}\\ \text{parts with }u=r\end{array}\right]$

$\qquad = \lim_{t\to-\infty}\left(18e^2-9e^2-3te^{t/3}+9e^{t/3}\right) = 9e^2-0+0$ [by l'Hospital's Rule]

$\qquad = 9e^2.$ Convergent

21. $\int_1^\infty \dfrac{\ln x}{x}\,dx = \lim_{t\to\infty}\left[\dfrac{(\ln x)^2}{2}\right]_1^t$ $\left[\begin{array}{l}\text{by substitution with}\\ u=\ln x,\, du = dx/x\end{array}\right]$ $= \lim_{t\to\infty}\dfrac{(\ln t)^2}{2} = \infty.$ Divergent

22. $I = \int_{-\infty}^\infty x^3 e^{-x^4}\,dx = I_1+I_2 = \int_{-\infty}^0 x^3 e^{-x^4}\,dx + \int_0^\infty x^3 e^{-x^4}\,dx.$ Now

$\qquad\qquad I_2 = \lim_{t\to\infty}\int_0^t x^3 e^{-x^4}\,dx = \lim_{t\to\infty}\int_0^{t^4} e^{-u}\left(\tfrac{1}{4}\,du\right)$ $\left[\begin{array}{l}u=x^4,\\ du = 4x^3\,dx\end{array}\right]$

$\qquad\qquad = \tfrac{1}{4}\lim_{t\to\infty}\Big[-e^{-u}\Big]_0^{t^4} = \tfrac{1}{4}\lim_{t\to\infty}\left(-e^{-t^4}+1\right) = \tfrac{1}{4}(0+1) = \tfrac{1}{4}.$

Since $f(x) = x^3 e^{-x^4}$ is an odd function, $I_1 = -\tfrac{1}{4}$, and hence, $I = 0.$ Convergent

23. $\displaystyle\int_{-\infty}^{\infty} \frac{x^2}{9+x^6}\,dx = \int_{-\infty}^{0} \frac{x^2}{9+x^6}\,dx + \int_{0}^{\infty} \frac{x^2}{9+x^6}\,dx = 2\int_{0}^{\infty} \frac{x^2}{9+x^6}\,dx$ [since the integrand is even].

Now $\displaystyle\int \frac{x^2\,dx}{9+x^6}\quad \begin{bmatrix} u=x^3 \\ du=3x^2dx \end{bmatrix} = \int \frac{\frac{1}{3}\,du}{9+u^2}\quad \begin{bmatrix} u=3v \\ du=3\,dv \end{bmatrix} = \int \frac{\frac{1}{3}(3\,dv)}{9+9v^2} = \frac{1}{9}\int \frac{dv}{1+v^2}$

$$= \frac{1}{9}\tan^{-1} v + C = \frac{1}{9}\tan^{-1}\left(\frac{u}{3}\right) + C = \frac{1}{9}\tan^{-1}\left(\frac{x^3}{3}\right) + C,$$

so $\displaystyle 2\int_{0}^{\infty} \frac{x^2}{9+x^6}\,dx = 2\lim_{t\to\infty}\int_{0}^{t} \frac{x^2}{9+x^6}\,dx = 2\lim_{t\to\infty}\left[\frac{1}{9}\tan^{-1}\left(\frac{x^3}{3}\right)\right]_{0}^{t} = 2\lim_{t\to\infty}\frac{1}{9}\tan^{-1}\left(\frac{t^3}{3}\right) = \frac{2}{9}\cdot\frac{\pi}{2} = \frac{\pi}{9}.$

Convergent

24. $\displaystyle\int_{0}^{\infty} \frac{e^x}{e^{2x}+3}\,dx = \lim_{t\to\infty}\int_{0}^{t} \frac{e^x}{(e^x)^2+(\sqrt{3})^2}\,dx = \lim_{t\to\infty}\left[\frac{1}{\sqrt{3}}\arctan\frac{e^x}{\sqrt{3}}\right]_{0}^{t} = \frac{1}{\sqrt{3}}\lim_{t\to\infty}\left(\arctan\frac{e^t}{\sqrt{3}} - \arctan\frac{1}{\sqrt{3}}\right)$

$$= \frac{1}{\sqrt{3}}\left(\frac{\pi}{2} - \frac{\pi}{6}\right) = \frac{1}{\sqrt{3}}\left(\frac{\pi}{3}\right) = \frac{\pi\sqrt{3}}{9}.\qquad \text{Convergent}$$

25. $\displaystyle\int_{e}^{\infty} \frac{1}{x(\ln x)^3}\,dx = \lim_{t\to\infty}\int_{e}^{t} \frac{1}{x(\ln x)^3}\,dx = \lim_{t\to\infty}\int_{1}^{\ln t} u^{-3}\,du \quad \begin{bmatrix} u=\ln x, \\ du=dx/x \end{bmatrix} = \lim_{t\to\infty}\left[-\frac{1}{2u^2}\right]_{1}^{\ln t}$

$$= \lim_{t\to\infty}\left[-\frac{1}{2(\ln t)^2} + \frac{1}{2}\right] = 0 + \frac{1}{2} = \frac{1}{2}.\qquad \text{Convergent}$$

26. $\displaystyle\int_{0}^{\infty} \frac{x\arctan x}{(1+x^2)^2}\,dx = \lim_{t\to\infty}\int_{0}^{t} \frac{x\arctan x}{(1+x^2)^2}\,dx.$ Let $u = \arctan x$, $dv = \dfrac{x\,dx}{(1+x^2)^2}$. Then $du = \dfrac{dx}{1+x^2}$,

$v = \dfrac{1}{2}\displaystyle\int \frac{2x\,dx}{(1+x^2)^2} = \frac{-1/2}{1+x^2}$, and

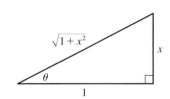

$\displaystyle\int \frac{x\arctan x}{(1+x^2)^2}\,dx = -\frac{1}{2}\frac{\arctan x}{1+x^2} + \frac{1}{2}\int \frac{dx}{(1+x^2)^2}\quad \begin{bmatrix} x=\tan\theta, \\ dx=\sec^2\theta\,d\theta \end{bmatrix}$

$$= -\frac{1}{2}\frac{\arctan x}{1+x^2} + \frac{1}{2}\int \frac{\sec^2\theta\,d\theta}{(\sec^2\theta)^2}$$

$$= -\frac{1}{2}\frac{\arctan x}{1+x^2} + \frac{1}{2}\int \cos^2\theta\,d\theta$$

$$= -\frac{1}{2}\frac{\arctan x}{1+x^2} + \frac{\theta}{4} + \frac{\sin\theta\,\cos\theta}{4} + C$$

$$= -\frac{1}{2}\frac{\arctan x}{1+x^2} + \frac{1}{4}\arctan x + \frac{1}{4}\frac{x}{1+x^2} + C$$

It follows that

$\displaystyle\int_{0}^{\infty} \frac{x\arctan x}{(1+x^2)^2}\,dx = \lim_{t\to\infty}\left[-\frac{1}{2}\frac{\arctan x}{1+x^2} + \frac{1}{4}\arctan x + \frac{1}{4}\frac{x}{1+x^2}\right]_{0}^{t}$

$$= \lim_{t\to\infty}\left(-\frac{1}{2}\frac{\arctan t}{1+t^2} + \frac{1}{4}\arctan t + \frac{1}{4}\frac{t}{1+t^2}\right) = 0 + \frac{1}{4}\cdot\frac{\pi}{2} + 0 = \frac{\pi}{8}.\qquad \text{Convergent}$$

27. $\displaystyle\int_{0}^{1} \frac{3}{x^5}\,dx = \lim_{t\to 0^+}\int_{t}^{1} 3x^{-5}\,dx = \lim_{t\to 0^+}\left[-\frac{3}{4x^4}\right]_{t}^{1} = -\frac{3}{4}\lim_{t\to 0^+}\left(1 - \frac{1}{t^4}\right) = \infty.\qquad \text{Divergent}$

28. $\int_2^3 \dfrac{1}{\sqrt{3-x}}\,dx = \lim\limits_{t\to 3^-}\int_2^t (3-x)^{-1/2}\,dx = \lim\limits_{t\to 3^-}\left[-2(3-x)^{1/2}\right]_2^t = -2\lim\limits_{t\to 3^-}\left(\sqrt{3-t}-\sqrt{1}\right) = -2(0-1) = 2.$

Convergent

29. $\int_{-2}^{14}\dfrac{dx}{\sqrt[4]{x+2}} = \lim\limits_{t\to -2^+}\int_t^{14}(x+2)^{-1/4}\,dx = \lim\limits_{t\to -2^+}\left[\dfrac{4}{3}(x+2)^{3/4}\right]_t^{14} = \dfrac{4}{3}\lim\limits_{t\to -2^+}\left[16^{3/4}-(t+2)^{3/4}\right]$

$= \frac{4}{3}(8-0) = \frac{32}{3}.$ Convergent

30. $\int_6^8 \dfrac{4}{(x-6)^3}\,dx = \lim\limits_{t\to 6^+}\int_t^8 4(x-6)^{-3}\,dx = \lim\limits_{t\to 6^+}\left[-2(x-6)^{-2}\right]_t^8 = -2\lim\limits_{t\to 6^+}\left[\dfrac{1}{2^2}-\dfrac{1}{(t-6)^2}\right] = \infty.$ Divergent

31. $\int_{-2}^3 \dfrac{dx}{x^4} = \int_{-2}^0 \dfrac{dx}{x^4} + \int_0^3 \dfrac{dx}{x^4}$, but $\int_{-2}^0 \dfrac{dx}{x^4} = \lim\limits_{t\to 0^-}\left[-\dfrac{x^{-3}}{3}\right]_{-2}^t = \lim\limits_{t\to 0^-}\left[-\dfrac{1}{3t^3}-\dfrac{1}{24}\right] = \infty.$ Divergent

32. $\int_0^1 \dfrac{dx}{\sqrt{1-x^2}} = \lim\limits_{t\to 1^-}\int_0^t \dfrac{dx}{\sqrt{1-x^2}} = \lim\limits_{t\to 1^-}\left[\sin^{-1}x\right]_0^t = \lim\limits_{t\to 1^-}\sin^{-1}t = \dfrac{\pi}{2}.$ Convergent

33. There is an infinite discontinuity at $x=1$. $\int_0^{33}(x-1)^{-1/5}\,dx = \int_0^1(x-1)^{-1/5}\,dx + \int_1^{33}(x-1)^{-1/5}\,dx.$ Here

$\int_0^1(x-1)^{-1/5}\,dx = \lim\limits_{t\to 1^-}\int_0^t(x-1)^{-1/5}\,dx = \lim\limits_{t\to 1^-}\left[\frac{5}{4}(x-1)^{4/5}\right]_0^t = \lim\limits_{t\to 1^-}\left[\frac{5}{4}(t-1)^{4/5}-\frac{5}{4}\right] = -\frac{5}{4}$ and

$\int_1^{33}(x-1)^{-1/5}\,dx = \lim\limits_{t\to 1^+}\int_t^{33}(x-1)^{-1/5}\,dx = \lim\limits_{t\to 1^+}\left[\frac{5}{4}(x-1)^{4/5}\right]_t^{33} = \lim\limits_{t\to 1^+}\left[\frac{5}{4}\cdot 16 - \frac{5}{4}(t-1)^{4/5}\right] = 20.$

Thus, $\int_0^{33}(x-1)^{-1/5}\,dx = -\frac{5}{4} + 20 = \frac{75}{4}.$ Convergent

34. $f(y) = 1/(4y-1)$ has an infinite discontinuity at $y = \frac{1}{4}$.

$\int_{1/4}^1 \dfrac{1}{4y-1}\,dy = \lim\limits_{t\to(1/4)^+}\int_t^1 \dfrac{1}{4y-1}\,dy = \lim\limits_{t\to(1/4)^+}\left[\frac{1}{4}\ln|4y-1|\right]_t^1 = \lim\limits_{t\to(1/4)^+}\left[\frac{1}{4}\ln 3 - \frac{1}{4}\ln(4t-1)\right] = \infty,$

so $\int_{1/4}^1 \dfrac{1}{4y-1}\,dy$ diverges, and hence, $\int_0^1 \dfrac{1}{4y-1}\,dy$ diverges.   Divergent

35. $I = \int_0^3 \dfrac{dx}{x^2-6x+5} = \int_0^3 \dfrac{dx}{(x-1)(x-5)} = I_1 + I_2 = \int_0^1 \dfrac{dx}{(x-1)(x-5)} + \int_1^3 \dfrac{dx}{(x-1)(x-5)}.$

Now $\dfrac{1}{(x-1)(x-5)} = \dfrac{A}{x-1} + \dfrac{B}{x-5} \Rightarrow 1 = A(x-5) + B(x-1).$

Set $x=5$ to get $1=4B$, so $B=\frac{1}{4}$. Set $x=1$ to get $1=-4A$, so $A=-\frac{1}{4}$. Thus

$$I_1 = \lim\limits_{t\to 1^-}\int_0^t \left(\dfrac{-\frac{1}{4}}{x-1} + \dfrac{\frac{1}{4}}{x-5}\right)dx = \lim\limits_{t\to 1^-}\left[-\frac{1}{4}\ln|x-1| + \frac{1}{4}\ln|x-5|\right]_0^t$$

$$= \lim\limits_{t\to 1^-}\left[\left(-\frac{1}{4}\ln|t-1| + \frac{1}{4}\ln|t-5|\right) - \left(-\frac{1}{4}\ln|-1| + \frac{1}{4}\ln|-5|\right)\right]$$

$$= \infty, \quad\text{since } \lim\limits_{t\to 1^-}\left(-\frac{1}{4}\ln|t-1|\right) = \infty.$$

Since I_1 is divergent, I is divergent.

36. $\int_{\pi/2}^\pi \csc x\,dx = \lim\limits_{t\to \pi^-}\int_{\pi/2}^t \csc x\,dx = \lim\limits_{t\to \pi^-}\left[\ln|\csc x - \cot x|\right]_{\pi/2}^t = \lim\limits_{t\to \pi^-}\left[\ln(\csc t - \cot t) - \ln(1-0)\right]$

$= \lim\limits_{t\to \pi^-}\ln\left(\dfrac{1-\cos t}{\sin t}\right) = \infty.$ Divergent

37. $\displaystyle\int_{-1}^{0} \frac{e^{1/x}}{x^3}\,dx = \lim_{t\to 0^-}\int_{-1}^{t}\frac{1}{x}e^{1/x}\cdot\frac{1}{x^2}\,dx = \lim_{t\to 0^-}\int_{-1}^{1/t} ue^u\,(-du)$ $\qquad\begin{bmatrix} u = 1/x, \\ du = -dx/x^2 \end{bmatrix}$

$\qquad\qquad = \lim_{t\to 0^-}\big[(u-1)e^u\big]_{1/t}^{-1}$ $\quad\begin{bmatrix}\text{use parts} \\ \text{or Formula 96}\end{bmatrix}$ $\quad = \lim_{t\to 0^-}\left[-2e^{-1} - \left(\frac{1}{t}-1\right)e^{1/t}\right]$

$\qquad\qquad = -\dfrac{2}{e} - \lim_{s\to -\infty}(s-1)e^s$ $\quad [s = 1/t]$ $\quad = -\dfrac{2}{e} - \lim_{s\to -\infty}\dfrac{s-1}{e^{-s}} \overset{\text{H}}{=} -\dfrac{2}{e} - \lim_{s\to -\infty}\dfrac{1}{-e^{-s}}$

$\qquad\qquad = -\dfrac{2}{e} - 0 = -\dfrac{2}{e}.$ $\quad$ Convergent

38. $\displaystyle\int_{0}^{1}\frac{e^{1/x}}{x^3}\,dx = \lim_{t\to 0^+}\int_{t}^{1}\frac{1}{x}e^{1/x}\cdot\frac{1}{x^2}\,dx = \lim_{t\to 0^+}\int_{1/t}^{1} ue^u\,(-du)$ $\qquad\begin{bmatrix} u = 1/x, \\ du = -dx/x^2 \end{bmatrix}$

$\qquad\qquad = \lim_{t\to 0^+}\big[(u-1)e^u\big]_{1}^{1/t}$ $\quad\begin{bmatrix}\text{use parts} \\ \text{or Formula 96}\end{bmatrix}$ $\quad = \lim_{t\to 0^+}\left[\left(\frac{1}{t}-1\right)e^{1/t} - 0\right]$

$\qquad\qquad = \lim_{s\to\infty}(s-1)e^s$ $\quad [s = 1/t]$ $\quad = \infty.$ $\quad$ Divergent

39. $I = \displaystyle\int_{0}^{2} z^2\ln z\,dz = \lim_{t\to 0^+}\int_{t}^{2} z^2\ln z\,dz = \lim_{t\to 0^+}\left[\dfrac{z^3}{3^2}(3\ln z - 1)\right]_{t}^{2}$ $\qquad\begin{bmatrix}\text{integrate by parts} \\ \text{or use Formula 101}\end{bmatrix}$

$\qquad = \lim_{t\to 0^+}\left[\tfrac{8}{9}(3\ln 2 - 1) - \tfrac{1}{9}t^3(3\ln t - 1)\right] = \tfrac{8}{3}\ln 2 - \tfrac{8}{9} - \tfrac{1}{9}\lim_{t\to 0^+}\left[t^3(3\ln t - 1)\right] = \tfrac{8}{3}\ln 2 - \tfrac{8}{9} - \tfrac{1}{9}L.$

Now $L = \displaystyle\lim_{t\to 0^+}\left[t^3(3\ln t - 1)\right] = \lim_{t\to 0^+}\dfrac{3\ln t - 1}{t^{-3}} \overset{\text{H}}{=} \lim_{t\to 0^+}\dfrac{3/t}{-3/t^4} = \lim_{t\to 0^+}(-t^3) = 0.$

Thus, $L = 0$ and $I = \tfrac{8}{3}\ln 2 - \tfrac{8}{9}.$ $\qquad$ Convergent

40. Integrate by parts with $u = \ln x$, $dv = dx/\sqrt{x}$ $\quad\Rightarrow\quad du = dx/x$, $v = 2\sqrt{x}$.

$\qquad\displaystyle\int_{0}^{1}\frac{\ln x}{\sqrt{x}}\,dx = \lim_{t\to 0^+}\int_{t}^{1}\frac{\ln x}{\sqrt{x}}\,dx = \lim_{t\to 0^+}\left(\left[2\sqrt{x}\ln x\right]_{t}^{1} - 2\int_{t}^{1}\frac{dx}{\sqrt{x}}\right) = \lim_{t\to 0^+}\left(-2\sqrt{t}\ln t - 4\left[\sqrt{x}\right]_{t}^{1}\right)$

$\qquad\qquad = \lim_{t\to 0^+}\left(-2\sqrt{t}\ln t - 4 + 4\sqrt{t}\right) = -4$

since $\displaystyle\lim_{t\to 0^+}\sqrt{t}\ln t = \lim_{t\to 0^+}\frac{\ln t}{t^{-1/2}} \overset{\text{H}}{=} \lim_{t\to 0^+}\frac{1/t}{-t^{-3/2}/2} = \lim_{t\to 0^+}(-2\sqrt{t}) = 0.$ $\qquad$ Convergent

41.

$\qquad\qquad\text{Area} = \displaystyle\int_{-\infty}^{1} e^x\,dx = \lim_{t\to -\infty}\big[e^x\big]_{t}^{1} = e - \lim_{t\to -\infty} e^t = e$

42.

$\qquad\qquad\text{Area} = \displaystyle\int_{-2}^{\infty} e^{-x/2}\,dx = -2\lim_{t\to\infty}\big[e^{-x/2}\big]_{-2}^{t} = -2\lim_{t\to\infty} e^{-t/2} + 2e = 2e$

43.

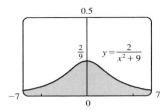

$$\text{Area} = \int_{-\infty}^{\infty} \frac{2}{x^2 + 9}\, dx = 2 \cdot 2 \int_{0}^{\infty} \frac{1}{x^2 + 9}\, dx = 4 \lim_{t \to \infty} \int_{0}^{t} \frac{1}{x^2 + 9}\, dx$$

$$= 4 \lim_{t \to \infty} \left[\frac{1}{3} \tan^{-1} \frac{x}{3} \right]_{0}^{t} = \frac{4}{3} \lim_{t \to \infty} \left[\tan^{-1} \frac{t}{3} - 0 \right] = \frac{4}{3} \cdot \frac{\pi}{2} = \frac{2\pi}{3}$$

44.

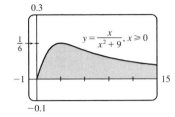

$$\text{Area} = \int_{0}^{\infty} \frac{x}{x^2 + 9}\, dx = \lim_{t \to \infty} \int_{0}^{t} \frac{x}{x^2 + 9}\, dx = \lim_{t \to \infty} \left[\tfrac{1}{2} \ln\left(x^2 + 9\right) \right]_{0}^{t}$$

$$= \tfrac{1}{2} \lim_{t \to \infty} \left[\ln\left(t^2 + 9\right) - \ln 9 \right] = \infty$$

Infinite area

45.

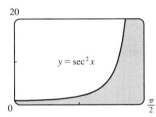

$$\text{Area} = \int_{0}^{\pi/2} \sec^2 x\, dx = \lim_{t \to (\pi/2)^-} \int_{0}^{t} \sec^2 x\, dx = \lim_{t \to (\pi/2)^-} \left[\tan x \right]_{0}^{t}$$

$$= \lim_{t \to (\pi/2)^-} \left(\tan t - 0 \right) = \infty$$

Infinite area

46.

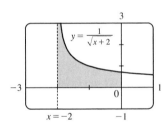

$$\text{Area} = \int_{-2}^{0} \frac{1}{\sqrt{x + 2}}\, dx = \lim_{t \to -2^+} \int_{t}^{0} \frac{1}{\sqrt{x + 2}}\, dx = \lim_{t \to -2^+} \left[2\sqrt{x + 2} \right]_{t}^{0}$$

$$= \lim_{t \to -2^+} \left(2\sqrt{2} - 2\sqrt{t + 2} \right) = 2\sqrt{2} - 0 = 2\sqrt{2}$$

47. (a)

t	$\int_1^t g(x)\, dx$
2	0.447453
5	0.577101
10	0.621306
100	0.668479
1000	0.672957
10,000	0.673407

$g(x) = \dfrac{\sin^2 x}{x^2}$.

It appears that the integral is convergent.

(b) $-1 \leq \sin x \leq 1 \ \Rightarrow \ 0 \leq \sin^2 x \leq 1 \ \Rightarrow \ 0 \leq \dfrac{\sin^2 x}{x^2} \leq \dfrac{1}{x^2}$. Since $\displaystyle\int_{1}^{\infty} \frac{1}{x^2}\, dx$ is convergent

[Equation 2 with $p = 2 > 1$], $\displaystyle\int_{1}^{\infty} \frac{\sin^2 x}{x^2}\, dx$ is convergent by the Comparison Theorem.

(c)

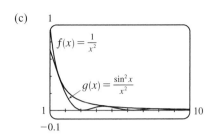

Since $\int_1^\infty f(x)\,dx$ is finite and the area under $g(x)$ is less than the area under $f(x)$ on any interval $[1, t]$, $\int_1^\infty g(x)\,dx$ must be finite; that is, the integral is convergent.

48. (a)

t	$\int_2^t g(x)\,dx$
5	3.830327
10	6.801200
100	23.328769
1000	69.023361
10,000	208.124560

$$g(x) = \frac{1}{\sqrt{x} - 1}.$$

It appears that the integral is divergent.

(b) For $x \geq 2$, $\sqrt{x} > \sqrt{x} - 1 \;\Rightarrow\; \dfrac{1}{\sqrt{x}} < \dfrac{1}{\sqrt{x} - 1}$. Since $\displaystyle\int_2^\infty \frac{1}{\sqrt{x}}\,dx$ is divergent $\;$ [Equation 2 with $p = \frac{1}{2} \leq 1$],

$\displaystyle\int_2^\infty \frac{1}{\sqrt{x} - 1}\,dx$ is divergent by the Comparison Theorem.

(c)

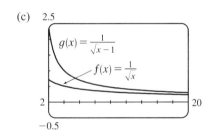

Since $\int_2^\infty f(x)\,dx$ is infinite and the area under $g(x)$ is greater than the area under $f(x)$ on any interval $[2, t]$, $\int_2^\infty g(x)\,dx$ must be infinite; that is, the integral is divergent.

49. For $x > 0$, $\dfrac{x}{x^3 + 1} < \dfrac{x}{x^3} = \dfrac{1}{x^2}$. $\displaystyle\int_1^\infty \frac{1}{x^2}\,dx$ is convergent by Equation 2 with $p = 2 > 1$, so $\displaystyle\int_1^\infty \frac{x}{x^3 + 1}\,dx$ is convergent by the Comparison Theorem. $\displaystyle\int_0^1 \frac{x}{x^3 + 1}\,dx$ is a constant, so $\displaystyle\int_0^\infty \frac{x}{x^3 + 1}\,dx = \int_0^1 \frac{x}{x^3 + 1}\,dx + \int_1^\infty \frac{x}{x^3 + 1}\,dx$ is also convergent.

50. For $x \geq 1$, $\dfrac{2 + e^{-x}}{x} > \dfrac{2}{x}$ $\;$ [since $e^{-x} > 0$] $\;> \dfrac{1}{x}$. $\displaystyle\int_1^\infty \frac{1}{x}\,dx$ is divergent by Equation 2 with $p = 1 \leq 1$, so $\displaystyle\int_1^\infty \frac{2 + e^{-x}}{x}\,dx$ is divergent by the Comparison Theorem.

51. For $x > 1$, $f(x) = \dfrac{x + 1}{\sqrt{x^4 - x}} > \dfrac{x + 1}{\sqrt{x^4}} > \dfrac{x}{x^2} = \dfrac{1}{x}$, so $\displaystyle\int_2^\infty f(x)\,dx$ diverges by comparison with $\displaystyle\int_2^\infty \frac{1}{x}\,dx$, which diverges by Equation 2 with $p = 1 \leq 1$. Thus, $\int_1^\infty f(x)\,dx = \int_1^2 f(x)\,dx + \int_2^\infty f(x)\,dx$ also diverges.

52. For $x \geq 0$, $\arctan x < \dfrac{\pi}{2} < 2$, so $\dfrac{\arctan x}{2 + e^x} < \dfrac{2}{2 + e^x} < \dfrac{2}{e^x} = 2e^{-x}$. Now

$$I = \int_0^\infty 2e^{-x}\, dx = \lim_{t\to\infty} \int_0^t 2e^{-x}\, dx = \lim_{t\to\infty}\left[-2e^{-x}\right]_0^t = \lim_{t\to\infty}\left(-\frac{2}{e^t} + 2\right) = 2,\ \text{so } I \text{ is convergent, and by comparison,}$$

$\displaystyle\int_0^\infty \dfrac{\arctan x}{2 + e^x}\, dx$ is convergent.

53. For $0 < x \leq 1$, $\dfrac{\sec^2 x}{x\sqrt{x}} > \dfrac{1}{x^{3/2}}$. Now

$$I = \int_0^1 x^{-3/2}\, dx = \lim_{t\to 0^+} \int_t^1 x^{-3/2}\, dx = \lim_{t\to 0^+}\left[-2x^{-1/2}\right]_t^1 = \lim_{t\to 0^+}\left(-2 + \frac{2}{\sqrt{t}}\right) = \infty,\ \text{so } I \text{ is divergent, and by}$$

comparison, $\displaystyle\int_0^1 \dfrac{\sec^2 x}{x\sqrt{x}}$ is divergent.

54. For $0 < x \leq 1$, $\dfrac{\sin^2 x}{\sqrt{x}} \leq \dfrac{1}{\sqrt{x}}$. Now

$$I = \int_0^\pi \frac{1}{\sqrt{x}}\, dx = \lim_{t\to 0^+} \int_t^\pi x^{-1/2}\, dx = \lim_{t\to 0^+}\left[2x^{1/2}\right]_t^\pi = \lim_{t\to 0^+}\left(2\pi - 2\sqrt{t}\right) = 2\pi - 0 = 2\pi,\ \text{so } I \text{ is convergent, and by}$$

comparison, $\displaystyle\int_0^\pi \dfrac{\sin^2 x}{\sqrt{x}}\, dx$ is convergent.

55. $\displaystyle\int_0^\infty \frac{dx}{\sqrt{x}\,(1+x)} = \int_0^1 \frac{dx}{\sqrt{x}\,(1+x)} + \int_1^\infty \frac{dx}{\sqrt{x}\,(1+x)} = \lim_{t\to 0^+} \int_t^1 \frac{dx}{\sqrt{x}\,(1+x)} + \lim_{t\to\infty} \int_1^t \frac{dx}{\sqrt{x}\,(1+x)}$. Now

$$\int \frac{dx}{\sqrt{x}\,(1+x)} = \int \frac{2u\, du}{u(1+u^2)} \quad \left[\begin{matrix} u = \sqrt{x},\ x = u^2, \\ dx = 2u\, du \end{matrix}\right] = 2\int \frac{du}{1+u^2} = 2\tan^{-1} u + C = 2\tan^{-1}\sqrt{x} + C,\ \text{so}$$

$$\int_0^\infty \frac{dx}{\sqrt{x}\,(1+x)} = \lim_{t\to 0^+}\left[2\tan^{-1}\sqrt{x}\right]_t^1 + \lim_{t\to\infty}\left[2\tan^{-1}\sqrt{x}\right]_1^t$$

$$= \lim_{t\to 0^+}\left[2\left(\tfrac{\pi}{4}\right) - 2\tan^{-1}\sqrt{t}\right] + \lim_{t\to\infty}\left[2\tan^{-1}\sqrt{t} - 2\left(\tfrac{\pi}{4}\right)\right] = \tfrac{\pi}{2} - 0 + 2\left(\tfrac{\pi}{2}\right) - \tfrac{\pi}{2} = \pi.$$

56. $\displaystyle\int_2^\infty \frac{dx}{x\sqrt{x^2-4}} = \int_2^3 \frac{dx}{x\sqrt{x^2-4}} + \int_3^\infty \frac{dx}{x\sqrt{x^2-4}} = \lim_{t\to 2^+} \int_t^3 \frac{dx}{x\sqrt{x^2-4}} + \lim_{t\to\infty} \int_3^t \frac{dx}{x\sqrt{x^2-4}}$. Now

$$\int \frac{dx}{x\sqrt{x^2-4}} = \int \frac{2\sec\theta\tan\theta\, d\theta}{2\sec\theta\, 2\tan\theta} \quad \left[\begin{matrix} x = 2\sec\theta,\ \text{where} \\ 0 \leq \theta < \pi/2\ \text{or}\ \pi \leq \theta < 3\pi/2 \end{matrix}\right] = \tfrac{1}{2}\theta + C = \tfrac{1}{2}\sec^{-1}\left(\tfrac{1}{2}x\right) + C,\ \text{so}$$

$$\int_2^\infty \frac{dx}{x\sqrt{x^2-4}} = \lim_{t\to 2^+}\left[\tfrac{1}{2}\sec^{-1}\left(\tfrac{1}{2}x\right)\right]_t^3 + \lim_{t\to\infty}\left[\tfrac{1}{2}\sec^{-1}\left(\tfrac{1}{2}x\right)\right]_3^t = \tfrac{1}{2}\sec^{-1}\left(\tfrac{3}{2}\right) - 0 + \tfrac{1}{2}\left(\tfrac{\pi}{2}\right) - \tfrac{1}{2}\sec^{-1}\left(\tfrac{3}{2}\right) = \tfrac{\pi}{4}.$$

57. If $p = 1$, then $\displaystyle\int_0^1 \frac{dx}{x^p} = \lim_{t \to 0^+} \int_t^1 \frac{dx}{x} = \lim_{t \to 0^+} [\ln x]_t^1 = \infty.$ Divergent.

If $p \neq 1$, then $\displaystyle\int_0^1 \frac{dx}{x^p} = \lim_{t \to 0^+} \int_t^1 \frac{dx}{x^p}$ [note that the integral is not improper if $p < 0$]

$$= \lim_{t \to 0^+} \left[\frac{x^{-p+1}}{-p+1} \right]_t^1 = \lim_{t \to 0^+} \frac{1}{1-p} \left[1 - \frac{1}{t^{p-1}} \right]$$

If $p > 1$, then $p - 1 > 0$, so $\dfrac{1}{t^{p-1}} \to \infty$ as $t \to 0^+$, and the integral diverges.

If $p < 1$, then $p - 1 < 0$, so $\dfrac{1}{t^{p-1}} \to 0$ as $t \to 0^+$ and $\displaystyle\int_0^1 \frac{dx}{x^p} = \frac{1}{1-p} \left[\lim_{t \to 0^+} \left(1 - t^{1-p} \right) \right] = \frac{1}{1-p}.$

Thus, the integral converges if and only if $p < 1$, and in that case its value is $\dfrac{1}{1-p}$.

58. Let $u = \ln x$. Then $du = dx/x \;\Rightarrow\; \displaystyle\int_e^\infty \frac{dx}{x\,(\ln x)^p} = \int_1^\infty \frac{du}{u^p}$. By Example 4, this converges to $\dfrac{1}{p-1}$ if $p > 1$

and diverges otherwise.

59. First suppose $p = -1$. Then

$$\int_0^1 x^p \ln x \, dx = \int_0^1 \frac{\ln x}{x} \, dx = \lim_{t \to 0^+} \int_t^1 \frac{\ln x}{x} \, dx = \lim_{t \to 0^+} \left[\tfrac{1}{2}(\ln x)^2 \right]_t^1 = -\tfrac{1}{2} \lim_{t \to 0^+} (\ln t)^2 = -\infty, \text{ so the}$$

integral diverges. Now suppose $p \neq -1$. Then integration by parts gives

$$\int x^p \ln x \, dx = \frac{x^{p+1}}{p+1} \ln x - \int \frac{x^p}{p+1} \, dx = \frac{x^{p+1}}{p+1} \ln x - \frac{x^{p+1}}{(p+1)^2} + C. \text{ If } p < -1, \text{ then } p + 1 < 0, \text{ so}$$

$$\int_0^1 x^p \ln x \, dx = \lim_{t \to 0^+} \left[\frac{x^{p+1}}{p+1} \ln x - \frac{x^{p+1}}{(p+1)^2} \right]_t^1 = \frac{-1}{(p+1)^2} - \left(\frac{1}{p+1} \right) \lim_{t \to 0^+} \left[t^{p+1} \left(\ln t - \frac{1}{p+1} \right) \right] = \infty.$$

If $p > -1$, then $p + 1 > 0$ and

$$\int_0^1 x^p \ln x \, dx = \frac{-1}{(p+1)^2} - \left(\frac{1}{p+1} \right) \lim_{t \to 0^+} \frac{\ln t - 1/(p+1)}{t^{-(p+1)}} \overset{\text{H}}{=} \frac{-1}{(p+1)^2} - \left(\frac{1}{p+1} \right) \lim_{t \to 0^+} \frac{1/t}{-(p+1)t^{-(p+2)}}$$

$$= \frac{-1}{(p+1)^2} + \frac{1}{(p+1)^2} \lim_{t \to 0^+} t^{p+1} = \frac{-1}{(p+1)^2}$$

Thus, the integral converges to $-\dfrac{1}{(p+1)^2}$ if $p > -1$ and diverges otherwise.

60. (a) $\boldsymbol{n = 0}$: $\displaystyle\int_0^\infty x^n e^{-x} \, dx = \lim_{t \to \infty} \int_0^t e^{-x} \, dx = \lim_{t \to \infty} \left[-e^{-x} \right]_0^t = \lim_{t \to \infty} \left[-e^{-t} + 1 \right] = 0 + 1 = 1$

$\boldsymbol{n = 1}$: $\displaystyle\int_0^\infty x^n e^{-x} \, dx = \lim_{t \to \infty} \int_0^t x e^{-x} \, dx.$ To evaluate $\displaystyle\int x e^{-x} \, dx$, we'll use integration by parts

with $u = x, \, dv = e^{-x} \, dx \;\Rightarrow\; du = dx, \, v = -e^{-x}.$

So $\displaystyle\int x e^{-x} \, dx = -x e^{-x} - \int -e^{-x} \, dx = -x e^{-x} - e^{-x} + C = (-x - 1)e^{-x} + C$ and

$\displaystyle\lim_{t \to \infty} \int_0^t x e^{-x} \, dx = \lim_{t \to \infty} \left[(-x - 1)e^{-x} \right]_0^t = \lim_{t \to \infty} \left[(-t - 1)e^{-t} + 1 \right] = \lim_{t \to \infty} \left[-t e^{-t} - e^{-t} + 1 \right]$

$= 0 - 0 + 1$ [use l'Hospital's Rule] $= 1$

$n = 2$: $\int_0^\infty x^n e^{-x}\,dx = \lim\limits_{t\to\infty}\int_0^t x^2 e^{-x}\,dx$. To evaluate $\int x^2 e^{-x}\,dx$, we could use integration by parts

again or Formula 97. Thus,

$$\lim_{t\to\infty}\int_0^t x^2 e^{-x}\,dx = \lim_{t\to\infty}\left[-x^2 e^{-x}\right]_0^t + 2\lim_{t\to\infty}\int_0^t x e^{-x}\,dx$$

$$= 0 + 0 + 2(1) \quad [\text{use l'Hospital's Rule and the result for } n = 1] \quad = 2$$

$n = 3$: $\int_0^\infty x^n e^{-x}\,dx = \lim\limits_{t\to\infty}\int_0^t x^3 e^{-x}\,dx \overset{97}{=} \lim\limits_{t\to\infty}\left[-x^3 e^{-x}\right]_0^t + 3\lim\limits_{t\to\infty}\int_0^t x^2 e^{-x}\,dx$

$$= 0 + 0 + 3(2) \quad [\text{use l'Hospital's Rule and the result for } n = 2] \quad = 6$$

(b) For $n = 1$, 2, and 3, we have $\int_0^\infty x^n e^{-x}\,dx = 1$, 2, and 6. The values for the integral are equal to the factorials for n, so

we guess $\int_0^\infty x^n e^{-x}\,dx = n!$.

(c) Suppose that $\int_0^\infty x^k e^{-x}\,dx = k!$ for some positive integer k. Then $\int_0^\infty x^{k+1} e^{-x}\,dx = \lim\limits_{t\to\infty}\int_0^t x^{k+1} e^{-x}\,dx$.

To evaluate $\int x^{k+1} e^{-x}\,dx$, we use parts with $u = x^{k+1}$, $dv = e^{-x}\,dx \quad \Rightarrow \quad du = (k+1)x^k\,dx$, $v = -e^{-x}$.

So $\int x^{k+1} e^{-x}\,dx = -x^{k+1} e^{-x} - \int -(k+1)x^k e^{-x}\,dx = -x^{k+1} e^{-x} + (k+1)\int x^k e^{-x}\,dx$ and

$$\lim_{t\to\infty}\int_0^t x^{k+1} e^{-x}\,dx = \lim_{t\to\infty}\left[-x^{k+1} e^{-x}\right]_0^t + (k+1)\lim_{t\to\infty}\int_0^t x^k e^{-x}\,dx$$

$$= \lim_{t\to\infty}\left[-t^{k+1} e^{-t} + 0\right] + (k+1)k! = 0 + 0 + (k+1)! = (k+1)!,$$

so the formula holds for $k + 1$. By induction, the formula holds for all positive integers. (Since $0! = 1$, the formula holds

for $n = 0$, too.)

61. (a) $I = \int_{-\infty}^\infty x\,dx = \int_{-\infty}^0 x\,dx + \int_0^\infty x\,dx$, and $\int_0^\infty x\,dx = \lim\limits_{t\to\infty}\int_0^t x\,dx = \lim\limits_{t\to\infty}\left[\frac12 x^2\right]_0^t = \lim\limits_{t\to\infty}\left[\frac12 t^2 - 0\right] = \infty$,

so I is divergent.

(b) $\int_{-t}^t x\,dx = \left[\frac12 x^2\right]_{-t}^t = \frac12 t^2 - \frac12 t^2 = 0$, so $\lim\limits_{t\to\infty}\int_{-t}^t x\,dx = 0$. Therefore, $\int_{-\infty}^\infty x\,dx \neq \lim\limits_{t\to\infty}\int_{-t}^t x\,dx$.

62. Let $k = \dfrac{M}{2RT}$ so that $\overline{v} = \dfrac{4}{\sqrt{\pi}} k^{3/2}\int_0^\infty v^3 e^{-kv^2}\,dv$. Let I denote the integral and use parts to integrate I. Let $\alpha = v^2$,

$d\beta = v e^{-kv^2}\,dv \quad \Rightarrow \quad d\alpha = 2v\,dv$, $\beta = -\dfrac{1}{2k}e^{-kv^2}$:

$$I = \lim_{t\to\infty}\left[-\frac{1}{2k}v^2 e^{-kv^2}\right]_0^t + \frac{1}{k}\int_0^\infty v e^{-kv^2}\,dv_0^t = -\frac{1}{2k}\lim_{t\to\infty}\left(t^2 e^{-kt^2}\right) + \frac{1}{k}\lim_{t\to\infty}\left[-\frac{1}{2k}e^{-kv^2}\right]$$

$$\overset{\text{H}}{=} -\frac{1}{2k}\cdot 0 - \frac{1}{2k^2}(0 - 1) = \frac{1}{2k^2}$$

Thus, $\overline{v} = \dfrac{4}{\sqrt{\pi}} k^{3/2}\cdot\dfrac{1}{2k^2} = \dfrac{2}{(k\pi)^{1/2}} = \dfrac{2}{[\pi M/(2RT)]^{1/2}} = \dfrac{2\sqrt{2}\,\sqrt{RT}}{\sqrt{\pi M}} = \sqrt{\dfrac{8RT}{\pi M}}$.

63. Volume $= \int_1^\infty \pi\left(\dfrac{1}{x}\right)^2\,dx = \pi\lim\limits_{t\to\infty}\int_1^t \dfrac{dx}{x^2} = \pi\lim\limits_{t\to\infty}\left[-\dfrac{1}{x}\right]_1^t = \pi\lim\limits_{t\to\infty}\left(1 - \dfrac{1}{t}\right) = \pi < \infty$.

64. Work $= \int_R^\infty \frac{GMm}{r^2}\,dr = \lim_{t\to\infty} \int_R^t \frac{GMm}{r^2}\,dr = \lim_{t\to\infty} GMm\left[\frac{-1}{r}\right]_R^t = GMm\lim_{t\to\infty}\left(\frac{-1}{t}+\frac{1}{R}\right) = \frac{GMm}{R}$, where

M = mass of the earth $= 5.98\times10^{24}$ kg, m = mass of satellite $= 10^3$ kg, R = radius of the earth $= 6.37\times10^6$ m, and

G = gravitational constant $= 6.67\times10^{-11}$ N·m²/kg.

Therefore, Work $= \dfrac{6.67\times10^{-11}\cdot5.98\times10^{24}\cdot10^3}{6.37\times10^6} \approx 6.26\times10^{10}$ J.

65. Work $= \int_R^\infty F\,dr = \lim_{t\to\infty}\int_R^t \frac{GmM}{r^2}\,dr = \lim_{t\to\infty} GmM\left(\frac{1}{R}-\frac{1}{t}\right) = \frac{GmM}{R}$. The initial kinetic energy provides the work,

so $\frac12 mv_0^2 = \dfrac{GmM}{R} \;\;\Rightarrow\;\; v_0 = \sqrt{\dfrac{2GM}{R}}$.

66. $y(s) = \displaystyle\int_s^R \frac{2r}{\sqrt{r^2-s^2}}\,x(r)\,dr$ and $x(r) = \frac12(R-r)^2 \;\;\Rightarrow$

$$y(s) = \lim_{t\to s^+}\int_t^R \frac{r(R-r)^2}{\sqrt{r^2-s^2}}\,dr = \lim_{t\to s^+}\int_t^R \frac{r^3-2Rr^2+R^2r}{\sqrt{r^2-s^2}}\,dr$$

$$= \lim_{t\to s^+}\left[\int_t^R \frac{r^3\,dr}{\sqrt{r^2-s^2}} - 2R\int_t^R \frac{r^2\,dr}{\sqrt{r^2-s^2}} + R^2\int_t^R \frac{r\,dr}{\sqrt{r^2-s^2}}\right] = \lim_{t\to s^+}\left(I_1 - 2RI_2 + R^2I_3\right) = L$$

For I_1: Let $u = \sqrt{r^2-s^2} \;\;\Rightarrow\;\; u^2 = r^2-s^2$, $r^2 = u^2+s^2$, $2r\,dr = 2u\,du$, so, omitting limits and constant of integration,

$$I_1 = \int \frac{(u^2+s^2)u}{u}\,du = \int(u^2+s^2)\,du = \tfrac13 u^3 + s^2 u = \tfrac13 u(u^2+3s^2)$$
$$= \tfrac13\sqrt{r^2-s^2}\,(r^2-s^2+3s^2) = \tfrac13\sqrt{r^2-s^2}\,(r^2+2s^2)$$

For I_2: Using Formula 44, $I_2 = \dfrac{r}{2}\sqrt{r^2-s^2} + \dfrac{s^2}{2}\ln\left|r+\sqrt{r^2-s^2}\right|$.

For I_3: Let $u = r^2-s^2 \;\;\Rightarrow\;\; du = 2r\,dr$. Then $I_3 = \dfrac12\displaystyle\int\frac{du}{\sqrt{u}} = \tfrac12\cdot2\sqrt{u} = \sqrt{r^2-s^2}$.

Thus,

$$L = \lim_{t\to s^+}\left[\tfrac13\sqrt{r^2-s^2}\,(r^2+2s^2) - 2R\left(\frac{r}{2}\sqrt{r^2-s^2} + \frac{s^2}{2}\ln\left|r+\sqrt{r^2-s^2}\right|\right) + R^2\sqrt{r^2-s^2}\right]_t^R$$

$$= \lim_{t\to s^+}\left[\tfrac13\sqrt{R^2-s^2}(R^2+2s^2) - 2R\left(\frac{R}{2}\sqrt{R^2-s^2} + \frac{s^2}{2}\ln\left|R+\sqrt{R^2-s^2}\right|\right) + R^2\sqrt{R^2-s^2}\right]$$

$$\qquad - \lim_{t\to s^+}\left[\tfrac13\sqrt{t^2-s^2}\,(t^2+2s^2) - 2R\left(\frac{t}{2}\sqrt{t^2-s^2} + \frac{s^2}{2}\ln\left|t+\sqrt{t^2-s^2}\right|\right) + R^2\sqrt{t^2-s^2}\right]$$

$$= \left[\tfrac13\sqrt{R^2-s^2}\,(R^2+2s^2) - Rs^2\ln\left|R+\sqrt{R^2-s^2}\right|\right] - \left[-Rs^2\ln|s|\right]$$

$$= \tfrac13\sqrt{R^2-s^2}\,(R^2+2s^2) - Rs^2\ln\left(\frac{R+\sqrt{R^2-s^2}}{s}\right)$$

67. We would expect a small percentage of bulbs to burn out in the first few hundred hours, most of the bulbs to burn out after close to 700 hours, and a few overachievers to burn on and on.

(a)

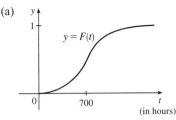

(b) $r(t) = F'(t)$ is the rate at which the fraction $F(t)$ of burnt-out bulbs increases as t increases. This could be interpreted as a fractional burnout rate.

(c) $\int_0^\infty r(t)\, dt = \lim_{x \to \infty} F(x) = 1$, since all of the bulbs will eventually burn out.

68. $I = \int_0^\infty t e^{kt}\, dt = \lim_{s \to \infty} \left[\dfrac{1}{k^2}(kt-1) e^{kt} \right]_0^s$ [Formula 96, or parts] $= \lim_{s \to \infty} \left[\left(\dfrac{1}{k} s e^{ks} - \dfrac{1}{k^2} e^{ks} \right) - \left(-\dfrac{1}{k^2} \right) \right].$

Since $k < 0$ the first two terms approach 0 (you can verify that the first term does so with l'Hospital's Rule), so the limit is equal to $1/k^2$. Thus, $M = -kI = -k(1/k^2) = -1/k = -1/(-0.000121) \approx 8264.5$ years.

69. $I = \int_a^\infty \dfrac{1}{x^2+1}\, dx = \lim_{t \to \infty} \int_a^t \dfrac{1}{x^2+1}\, dx = \lim_{t \to \infty} \left[\tan^{-1} x \right]_a^t = \lim_{t \to \infty} \left(\tan^{-1} t - \tan^{-1} a \right) = \dfrac{\pi}{2} - \tan^{-1} a.$

$I < 0.001 \ \Rightarrow\ \dfrac{\pi}{2} - \tan^{-1} a < 0.001 \ \Rightarrow\ \tan^{-1} a > \dfrac{\pi}{2} - 0.001 \ \Rightarrow\ a > \tan\left(\dfrac{\pi}{2} - 0.001 \right) \approx 1000.$

70. $f(x) = e^{-x^2}$ and $\Delta x = \dfrac{4-0}{8} = \dfrac{1}{2}.$

$\int_0^4 f(x)\, dx \approx S_8 = \dfrac{1}{2 \cdot 3}[f(0) + 4f(0.5) + 2f(1) + \cdots + 2f(3) + 4f(3.5) + f(4)] \approx \dfrac{1}{6}(5.31717808) \approx 0.8862$

Now $x > 4 \ \Rightarrow\ -x \cdot x < -x \cdot 4 \ \Rightarrow\ e^{-x^2} < e^{-4x} \ \Rightarrow\ \int_4^\infty e^{-x^2}\, dx < \int_4^\infty e^{-4x}\, dx.$

$\int_4^\infty e^{-4x}\, dx = \lim_{t \to \infty} \left[-\dfrac{1}{4} e^{-4x} \right]_4^t = -\dfrac{1}{4}\left(0 - e^{-16} \right) = 1/(4e^{16}) \approx 0.0000000281 < 0.0000001$, as desired.

71. (a) $F(s) = \int_0^\infty f(t) e^{-st}\, dt = \int_0^\infty e^{-st}\, dt = \lim_{n \to \infty} \left[-\dfrac{e^{-st}}{s} \right]_0^n = \lim_{n \to \infty} \left(\dfrac{e^{-sn}}{-s} + \dfrac{1}{s} \right).$ This converges to $\dfrac{1}{s}$ only if $s > 0$.

Therefore $F(s) = \dfrac{1}{s}$ with domain $\{ s \mid s > 0 \}.$

(b) $F(s) = \int_0^\infty f(t) e^{-st}\, dt = \int_0^\infty e^t e^{-st}\, dt = \lim_{n \to \infty} \int_0^n e^{t(1-s)}\, dt = \lim_{n \to \infty} \left[\dfrac{1}{1-s} e^{t(1-s)} \right]_0^n$

$= \lim_{n \to \infty} \left(\dfrac{e^{(1-s)n}}{1-s} - \dfrac{1}{1-s} \right)$

This converges only if $1 - s < 0 \ \Rightarrow\ s > 1$, in which case $F(s) = \dfrac{1}{s-1}$ with domain $\{ s \mid s > 1 \}.$

(c) $F(s) = \int_0^\infty f(t) e^{-st}\, dt = \lim_{n \to \infty} \int_0^n t e^{-st}\, dt.$ Use integration by parts: let $u = t$, $dv = e^{-st}\, dt \ \Rightarrow\ du = dt,$

$v = -\dfrac{e^{-st}}{s}$. Then $F(s) = \lim_{n \to \infty} \left[-\dfrac{t}{s} e^{-st} - \dfrac{1}{s^2} e^{-st} \right]_0^n = \lim_{n \to \infty} \left(\dfrac{-n}{se^{sn}} - \dfrac{1}{s^2 e^{sn}} + 0 + \dfrac{1}{s^2} \right) = \dfrac{1}{s^2}$ only if $s > 0$.

Therefore, $F(s) = \dfrac{1}{s^2}$ and the domain of F is $\{ s \mid s > 0 \}.$

72. $0 \le f(t) \le Me^{at}$ $\Rightarrow$ $0 \le f(t)e^{-st} \le Me^{at}e^{-st}$ for $t \ge 0$. Now use the Comparison Theorem:

$$\int_0^\infty Me^{at}e^{-st}\, dt = \lim_{n \to \infty} M \int_0^n e^{t(a-s)}\, dt = M \cdot \lim_{n \to \infty} \left[\frac{1}{a-s}e^{t(a-s)}\right]_0^n = M \cdot \lim_{n \to \infty} \frac{1}{a-s}\left[e^{n(a-s)} - 1\right]$$

This is convergent only when $a - s < 0$ $\Rightarrow$ $s > a$. Therefore, by the Comparison Theorem, $F(s) = \int_0^\infty f(t)\,e^{-st}\, dt$ is also convergent for $s > a$.

73. $G(s) = \int_0^\infty f'(t)e^{-st}\, dt$. Integrate by parts with $u = e^{-st}$, $dv = f'(t)\, dt$ $\Rightarrow$ $du = -se^{-st}$, $v = f(t)$:

$$G(s) = \lim_{n \to \infty} \left[f(t)e^{-st}\right]_0^n + s \int_0^\infty f(t)e^{-st}\, dt = \lim_{n \to \infty} f(n)e^{-sn} - f(0) + sF(s)$$

But $0 \le f(t) \le Me^{at}$ $\Rightarrow$ $0 \le f(t)e^{-st} \le Me^{at}e^{-st}$ and $\lim_{t \to \infty} Me^{t(a-s)} = 0$ for $s > a$. So by the Squeeze Theorem,

$\lim_{t \to \infty} f(t)e^{-st} = 0$ for $s > a$ $\Rightarrow$ $G(s) = 0 - f(0) + sF(s) = sF(s) - f(0)$ for $s > a$.

74. Assume without loss of generality that $a < b$. Then

$$\int_{-\infty}^a f(x)\, dx + \int_a^\infty f(x)\, dx = \lim_{t \to -\infty} \int_t^a f(x)\, dx + \lim_{u \to \infty} \int_a^u f(x)\, dx$$

$$= \lim_{t \to -\infty} \int_t^a f(x)\, dx + \lim_{u \to \infty} \left[\int_a^b f(x)\, dx + \int_b^u f(x)\, dx\right]$$

$$= \lim_{t \to -\infty} \int_t^a f(x)\, dx + \int_a^b f(x)\, dx + \lim_{u \to \infty} \int_b^u f(x)\, dx$$

$$= \lim_{t \to -\infty} \left[\int_t^a f(x)\, dx + \int_a^b f(x)\, dx\right] + \int_b^\infty f(x)\, dx$$

$$= \lim_{t \to -\infty} \int_t^b f(x)\, dx + \int_b^\infty f(x)\, dx = \int_{-\infty}^b f(x)\, dx + \int_b^\infty f(x)\, dx$$

75. We use integration by parts: let $u = x$, $dv = xe^{-x^2}\, dx$ $\Rightarrow$ $du = dx$, $v = -\frac{1}{2}e^{-x^2}$. So

$$\int_0^\infty x^2 e^{-x^2}\, dx = \lim_{t \to \infty} \left[-\frac{1}{2}xe^{-x^2}\right]_0^t + \frac{1}{2}\int_0^\infty e^{-x^2}\, dx = \lim_{t \to \infty} \left[-\frac{t}{2e^{t^2}}\right] + \frac{1}{2}\int_0^\infty e^{-x^2}\, dx = \frac{1}{2}\int_0^\infty e^{-x^2}\, dx$$

(The limit is 0 by l'Hospital's Rule.)

76. $\int_0^\infty e^{-x^2}\, dx$ is the area under the curve $y = e^{-x^2}$ for $0 \le x < \infty$ and $0 < y \le 1$. Solving $y = e^{-x^2}$ for x, we get

$y = e^{-x^2}$ $\Rightarrow$ $\ln y = -x^2$ $\Rightarrow$ $-\ln y = x^2$ $\Rightarrow$ $x = \pm\sqrt{-\ln y}$. Since x is positive, choose $x = \sqrt{-\ln y}$, and

the area is represented by $\int_0^1 \sqrt{-\ln y}\, dy$. Therefore, each integral represents the same area, so the integrals are equal.

77. For the first part of the integral, let $x = 2\tan\theta$ $\Rightarrow$ $dx = 2\sec^2\theta\, d\theta$.

$$\int \frac{1}{\sqrt{x^2+4}}\, dx = \int \frac{2\sec^2\theta}{2\sec\theta}\, d\theta = \int \sec\theta\, d\theta = \ln|\sec\theta + \tan\theta|.$$

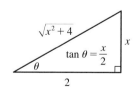

From the figure, $\tan\theta = \frac{x}{2}$, and $\sec\theta = \frac{\sqrt{x^2+4}}{2}$. So

$$I = \int_0^\infty \left(\frac{1}{\sqrt{x^2+4}} - \frac{C}{x+2} \right) dx = \lim_{t\to\infty} \left[\ln \left| \frac{\sqrt{x^2+4}}{2} + \frac{x}{2} \right| - C\ln|x+2| \right]_0^t$$

$$= \lim_{t\to\infty} \left[\ln \frac{\sqrt{t^2+4}+t}{2} - C\ln(t+2) - (\ln 1 - C\ln 2) \right]$$

$$= \lim_{t\to\infty} \left[\ln \left(\frac{\sqrt{t^2+4}+t}{2(t+2)^C} \right) + \ln 2^C \right] = \ln \left(\lim_{t\to\infty} \frac{t+\sqrt{t^2+4}}{(t+2)^C} \right) + \ln 2^{C-1}$$

Now $L = \lim\limits_{t\to\infty} \dfrac{t+\sqrt{t^2+4}}{(t+2)^C} \overset{\text{H}}{=} \lim\limits_{t\to\infty} \dfrac{1+t/\sqrt{t^2+4}}{C(t+2)^{C-1}} = \dfrac{2}{C\lim\limits_{t\to\infty}(t+2)^{C-1}}$.

If $C < 1$, $L = \infty$ and I diverges.

If $C = 1$, $L = 2$ and I converges to $\ln 2 + \ln 2^0 = \ln 2$.

If $C > 1$, $L = 0$ and I diverges to $-\infty$.

78. $I = \int_0^\infty \left(\dfrac{x}{x^2+1} - \dfrac{C}{3x+1} \right) dx = \lim\limits_{t\to\infty} \left[\frac{1}{2}\ln(x^2+1) - \frac{1}{3}C\ln(3x+1) \right]_0^t = \lim\limits_{t\to\infty} \left[\ln(t^2+1)^{1/2} - \ln(3t+1)^{C/3} \right]$

$$= \lim_{t\to\infty} \left(\ln \frac{(t^2+1)^{1/2}}{(3t+1)^{C/3}} \right) = \ln \left(\lim_{t\to\infty} \frac{\sqrt{t^2+1}}{(3t+1)^{C/3}} \right)$$

For $C \le 0$, the integral diverges. For $C > 0$, we have

$$L = \lim_{t\to\infty} \frac{\sqrt{t^2+1}}{(3t+1)^{C/3}} \overset{\text{H}}{=} \lim_{t\to\infty} \frac{t/\sqrt{t^2+1}}{C(3t+1)^{(C/3)-1}} = \frac{1}{C}\lim_{t\to\infty} \frac{1}{(3t+1)^{(C/3)-1}}$$

For $C/3 < 1 \;\Leftrightarrow\; C < 3$, $L = \infty$ and I diverges.

For $C = 3$, $L = \frac{1}{3}$ and $I = \ln\frac{1}{3}$.

For $C > 3$, $L = 0$ and I diverges to $-\infty$.

79. No, $I = \int_0^\infty f(x)\,dx$ must be *divergent*. Since $\lim\limits_{x\to\infty} f(x) = 1$, there must exist an N such that if $x \ge N$, then $f(x) \ge \frac{1}{2}$.

Thus, $I = I_1 + I_2 = \int_0^N f(x)\,dx + \int_N^\infty f(x)\,dx$, where I_1 is an ordinary definite integral that has a finite value, and I_2 is improper and diverges by comparison with the divergent integral $\int_N^\infty \frac{1}{2}\,dx$.

80. As in Exercise 55, we let $I = \int_0^\infty \dfrac{x^a}{1+x^b}\,dx = I_1 + I_2$, where $I_1 = \int_0^1 \dfrac{x^a}{1+x^b}\,dx$ and $I_2 = \int_1^\infty \dfrac{x^a}{1+x^b}\,dx$. We will

show that I_1 converges for $a > -1$ and I_2 converges for $b > a+1$, so that I converges when $a > -1$ and $b > a+1$.

I_1 is improper only when $a < 0$. When $0 \le x \le 1$, we have $\dfrac{1}{1+x^b} \le 1 \;\Rightarrow\; \dfrac{1}{x^{-a}(1+x^b)} \le \dfrac{1}{x^{-a}}$. The integral

$\int_0^1 \dfrac{1}{x^{-a}}\,dx$ converges for $-a < 1$ [or $a > -1$] by Exercise 57, so by the Comparison Theorem, $\int_0^1 \dfrac{1}{x^{-a}(1+x^b)}\,dx$

converges for $-1 < a < 0$. I_1 is not improper when $a \ge 0$, so it has a finite real value in that case. Therefore, I_1 has a finite

real value (converges) when $a > -1$.

I_2 is always improper. When $x \ge 1$, $\dfrac{x^a}{1+x^b} = \dfrac{1}{x^{-a}(1+x^b)} = \dfrac{1}{x^{-a}+x^{b-a}} < \dfrac{1}{x^{b-a}}$. By (2), $\int_1^\infty \dfrac{1}{x^{b-a}}\,dx$ converges

for $b - a > 1$ (or $b > a+1$), so by the Comparison Theorem, $\int_1^\infty \dfrac{x^a}{1+x^b}\,dx$ converges for $b > a+1$.

Thus, I converges if $a > -1$ and $b > a+1$.

8 Review

1. See Formula 8.1.1 or 8.1.2. We try to choose $u = f(x)$ to be a function that becomes simpler when differentiated (or at least not more complicated) as long as $dv = g'(x)\,dx$ can be readily integrated to give v.

2. See the Strategy for Evaluating $\int \sin^m x \cos^n x\,dx$ on page 498.

3. If $\sqrt{a^2 - x^2}$ occurs, try $x = a\sin\theta$; if $\sqrt{a^2 + x^2}$ occurs, try $x = a\tan\theta$, and if $\sqrt{x^2 - a^2}$ occurs, try $x = a\sec\theta$. See the Table of Trigonometric Substitutions on page 503.

4. See Equation 2 and Expressions 7, 9, and 11 in Section 8.4.

5. See the Midpoint Rule, the Trapezoidal Rule, and Simpson's Rule, as well as their associated error bounds, all in Section 8.7. We would expect the best estimate to be given by Simpson's Rule.

6. See Definitions 1(a), (b), and (c) in Section 8.8.

7. See Definitions 3(b), (a), and (c) in Section 8.8.

8. See the Comparison Theorem after Example 8 in Section 8.8.

1. **False.** Since the numerator has a higher degree than the denominator, $\dfrac{x(x^2 + 4)}{x^2 - 4} = x + \dfrac{8x}{x^2 - 4} = x + \dfrac{A}{x + 2} + \dfrac{B}{x - 2}$.

2. **True.** In fact, $A = -1$, $B = C = 1$.

3. **False.** It can be put in the form $\dfrac{A}{x} + \dfrac{B}{x^2} + \dfrac{C}{x - 4}$.

4. **False.** The form is $\dfrac{A}{x} + \dfrac{Bx + C}{x^2 + 4}$.

5. **False.** This is an improper integral, since the denominator vanishes at $x = 1$.

   $$\int_0^4 \frac{x}{x^2 - 1}\,dx = \int_0^1 \frac{x}{x^2 - 1}\,dx + \int_1^4 \frac{x}{x^2 - 1}\,dx \text{ and}$$

   $$\int_0^1 \frac{x}{x^2 - 1}\,dx = \lim_{t \to 1^-} \int_0^t \frac{x}{x^2 - 1}\,dx = \lim_{t \to 1^-} \left[\tfrac{1}{2}\ln|x^2 - 1|\right]_0^t = \lim_{t \to 1^-} \tfrac{1}{2}\ln|t^2 - 1| = \infty$$

   So the integral diverges.

6. **True** by Theorem 8.8.2 with $p = \sqrt{2} > 1$.

7. **False.** See Exercise 61 in Section 8.8.

8. **False.** For example, with $n = 1$ the Trapezoidal Rule is much more accurate than the Midpoint Rule for the function in the diagram.

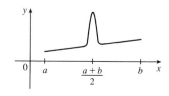

9. (a) True. See the end of Section 8.5.

(b) False. Examples include the functions $f(x) = e^{x^2}$, $g(x) = \sin(x^2)$, and $h(x) = \dfrac{\sin x}{x}$.

10. True. If f is continuous on $[0, \infty)$, then $\int_0^1 f(x)\,dx$ is finite. Since $\int_1^\infty f(x)\,dx$ is finite, so is

$\int_0^\infty f(x)\,dx = \int_0^1 f(x)\,dx + \int_1^\infty f(x)\,dx$.

11. False. If $f(x) = 1/x$, then f is continuous and decreasing on $[1, \infty)$ with $\lim\limits_{x\to\infty} f(x) = 0$, but $\int_1^\infty f(x)\,dx$ is divergent.

12. True. $\int_a^\infty [f(x) + g(x)]\,dx = \lim\limits_{t\to\infty} \int_a^t [f(x) + g(x)]\,dx = \lim\limits_{t\to\infty} \left(\int_a^t f(x)\,dx + \int_a^t g(x)\,dx \right)$

$= \lim\limits_{t\to\infty} \int_a^t f(x)\,dx + \lim\limits_{t\to\infty} \int_a^t g(x)\,dx$ $\begin{bmatrix}\text{since both limits} \\ \text{in the sum exist}\end{bmatrix}$

$= \int_a^\infty f(x)\,dx + \int_a^\infty g(x)\,dx$

Since the two integrals are finite, so is their sum.

13. False. Take $f(x) = 1$ for all x and $g(x) = -1$ for all x. Then $\int_a^\infty f(x)\,dx = \infty$ [divergent]

and $\int_a^\infty g(x)\,dx = -\infty$ [divergent], but $\int_a^\infty [f(x) + g(x)]\,dx = 0$ [convergent].

14. False. $\int_0^\infty f(x)\,dx$ could converge or diverge. For example, if $g(x) = 1$, then $\int_0^\infty f(x)\,dx$ diverges if $f(x) = 1$ and converges if $f(x) = 0$.

EXERCISES

1. $\displaystyle\int_0^5 \frac{x}{x+10}\,dx = \int_0^5 \left(1 - \frac{10}{x+10}\right) dx = \left[x - 10\ln(x+10)\right]_0^5 = 5 - 10\ln 15 + 10\ln 10$

$= 5 + 10\ln\frac{10}{15} = 5 + 10\ln\frac{2}{3}$

2. $\displaystyle\int_0^5 ye^{-0.6y}\,dy$ $\begin{bmatrix}u = y, & dv = e^{-0.6y}\,dy, \\ du = dy & v = -\frac{5}{3}e^{-0.6y}\end{bmatrix} = \left[-\frac{5}{3}ye^{-0.6y}\right]_0^5 - \int_0^5 \left(-\frac{5}{3}e^{-0.6y}\right) dy = -\frac{25}{3}e^{-3} - \frac{25}{9}\left[e^{-0.6y}\right]_0^5$

$= -\frac{25}{3}e^{-3} - \frac{25}{9}(e^{-3} - 1) = -\frac{25}{3}e^{-3} - \frac{25}{9}e^{-3} + \frac{25}{9} = \frac{25}{9} - \frac{100}{9}e^{-3}$

3. $\displaystyle\int_0^{\pi/2} \frac{\cos\theta}{1+\sin\theta}\,d\theta = \left[\ln(1+\sin\theta)\right]_0^{\pi/2} = \ln 2 - \ln 1 = \ln 2$

4. $\displaystyle\int_1^4 \frac{dt}{(2t+1)^3}$ $\begin{bmatrix}u = 2t+1, \\ du = 2\,dt\end{bmatrix} = \int_3^9 \frac{\frac{1}{2}\,du}{u^3} = \frac{-1}{4}\left[\frac{1}{u^2}\right]_3^9 = -\frac{1}{4}\left(\frac{1}{81} - \frac{1}{9}\right) = -\frac{1}{4}\left(-\frac{8}{81}\right) = \frac{2}{81}$

5. $\int_0^{\pi/2} \sin^3\theta \cos^2\theta\,d\theta = \int_0^{\pi/2}(1 - \cos^2\theta)\cos^2\theta \sin\theta\,d\theta = \int_1^0 (1-u^2)u^2\,(-du)$ $\begin{bmatrix}u = \cos\theta, \\ du = -\sin\theta\,d\theta\end{bmatrix}$

$= \int_0^1 (u^2 - u^4)\,du = \left[\frac{1}{3}u^3 - \frac{1}{5}u^5\right]_0^1 = \left(\frac{1}{3} - \frac{1}{5}\right) - 0 = \frac{2}{15}$

6. $\dfrac{1}{y^2 - 4y - 12} = \dfrac{1}{(y-6)(y+2)} = \dfrac{A}{y-6} + \dfrac{B}{y+2} \Rightarrow 1 = A(y+2) + B(y-6)$. Letting $y = -2 \Rightarrow B = -\frac{1}{8}$ and

letting $y = 6 \Rightarrow A = \frac{1}{8}$. So $\displaystyle\int \dfrac{1}{y^2 - 4y - 12}\, dy = \int \left(\dfrac{1/8}{y-6} + \dfrac{-1/8}{y+2}\right) dy = \frac{1}{8}\ln|y-6| - \frac{1}{8}\ln|y+2| + C$.

7. Let $u = \ln t$, $du = dt/t$. Then $\displaystyle\int \dfrac{\sin(\ln t)}{t}\, dt = \int \sin u\, du = -\cos u + C = -\cos(\ln t) + C$.

8. Let $u = \sqrt{e^x - 1}$, so that $u^2 = e^x - 1$, $2u\, du = e^x\, dx$, and $e^x = u^2 + 1$. Then

$$\int \dfrac{1}{\sqrt{e^x - 1}}\, dx = \int \dfrac{1}{u}\dfrac{2u\, du}{u^2 + 1} = 2\int \dfrac{1}{u^2 + 1}\, du = 2\tan^{-1} u + C = 2\tan^{-1}\sqrt{e^x - 1} + C.$$

9. $\displaystyle\int_1^4 x^{3/2}\ln x\, dx \quad \begin{bmatrix} u = \ln x, & dv = x^{3/2}\, dx, \\ du = dx/x & v = \frac{2}{5}x^{5/2} \end{bmatrix} = \frac{2}{5}\Big[x^{5/2}\ln x\Big]_1^4 - \frac{2}{5}\int_1^4 x^{3/2}\, dx = \frac{2}{5}(32\ln 4 - \ln 1) - \frac{2}{5}\Big[\frac{2}{5}x^{5/2}\Big]_1^4$

$$= \tfrac{2}{5}(64\ln 2) - \tfrac{4}{25}(32 - 1) = \tfrac{128}{5}\ln 2 - \tfrac{124}{25} \quad \Big[\text{or } \tfrac{64}{5}\ln 4 - \tfrac{124}{25}\Big]$$

10. Let $u = \arctan x$, $du = dx/(1 + x^2)$. Then

$$\int_0^1 \dfrac{\sqrt{\arctan x}}{1 + x^2}\, dx = \int_0^{\pi/4} \sqrt{u}\, du = \frac{2}{3}\Big[u^{3/2}\Big]_0^{\pi/4} = \frac{2}{3}\left[\dfrac{\pi^{3/2}}{4^{3/2}} - 0\right] = \frac{2}{3}\cdot\frac{1}{8}\pi^{3/2} = \frac{1}{12}\pi^{3/2}.$$

11. Let $x = \sec\theta$. Then

$$\int_1^2 \dfrac{\sqrt{x^2 - 1}}{x}\, dx = \int_0^{\pi/3} \dfrac{\tan\theta}{\sec\theta}\sec\theta\tan\theta\, d\theta = \int_0^{\pi/3} \tan^2\theta\, d\theta = \int_0^{\pi/3}(\sec^2\theta - 1)\, d\theta = \Big[\tan\theta - \theta\Big]_0^{\pi/3} = \sqrt{3} - \frac{\pi}{3}.$$

12. $\displaystyle\int_{-1}^1 \dfrac{\sin x}{1 + x^2}\, dx = 0$ by Theorem 5.5.6(b), since $f(x) = \dfrac{\sin x}{1 + x^2}$ is an odd function.

13. Let $t = \sqrt[3]{x}$. Then $t^3 = x$ and $3t^2\, dt = dx$, so $\int e^{\sqrt[3]{x}}\, dx = \int e^t \cdot 3t^2\, dt = 3I$. To evaluate I, let $u = t^2$,

$dv = e^t\, dt \Rightarrow du = 2t\, dt$, $v = e^t$, so $I = \int t^2 e^t\, dt = t^2 e^t - \int 2t e^t\, dt$. Now let $U = t$, $dV = e^t\, dt \Rightarrow$

$dU = dt$, $V = e^t$. Thus, $I = t^2 e^t - 2\big[te^t - \int e^t\, dt\big] = t^2 e^t - 2te^t + 2e^t + C_1$, and hence

$3I = 3e^t(t^2 - 2t + 2) + C = 3e^{\sqrt[3]{x}}(x^{2/3} - 2x^{1/3} + 2) + C$.

14. $\displaystyle\int \dfrac{x^2 + 2}{x + 2}\, dx = \int\left(x - 2 + \dfrac{6}{x + 2}\right) dx = \frac{1}{2}x^2 - 2x + 6\ln|x + 2| + C$

15. $\dfrac{x - 1}{x^2 + 2x} = \dfrac{x - 1}{x(x + 2)} = \dfrac{A}{x} + \dfrac{B}{x + 2} \Rightarrow x - 1 = A(x + 2) + Bx$. Set $x = -2$ to get $-3 = -2B$, so $B = \frac{3}{2}$. Set $x = 0$

to get $-1 = 2A$, so $A = -\frac{1}{2}$. Thus, $\displaystyle\int \dfrac{x - 1}{x^2 + 2x}\, dx = \int\left(\dfrac{-\frac{1}{2}}{x} + \dfrac{\frac{3}{2}}{x + 2}\right) dx = -\frac{1}{2}\ln|x| + \frac{3}{2}\ln|x + 2| + C$.

16. $\displaystyle\int \dfrac{\sec^6\theta}{\tan^2\theta}\, d\theta = \int \dfrac{(\tan^2\theta + 1)^2\sec^2\theta}{\tan^2\theta}\, d\theta \quad \begin{bmatrix} u = \tan\theta, \\ du = -\sec^2\theta\, d\theta \end{bmatrix} = \int \dfrac{(u^2 + 1)^2}{u^2}\, du = \int \dfrac{u^4 + 2u^2 + 1}{u^2}\, du$

$$= \int\left(u^2 + 2 + \dfrac{1}{u^2}\right) du = \dfrac{u^3}{3} + 2u - \dfrac{1}{u} + C = \frac{1}{3}\tan^3\theta + 2\tan\theta - \cot\theta + C$$

17. Integrate by parts with $u = x$, $dv = \sec x \tan x\,dx$ $\Rightarrow$ $du = dx$, $v = \sec x$:

$\int x \sec x \tan x\,dx = x \sec x - \int \sec x\,dx \overset{14}{=} x \sec x - \ln|\sec x + \tan x| + C.$

18. $\dfrac{x^2 + 8x - 3}{x^3 + 3x^2} = \dfrac{x^2 + 8x - 3}{x^2(x + 3)} = \dfrac{A}{x} + \dfrac{B}{x^2} + \dfrac{C}{x + 3}$ $\Rightarrow$ $x^2 + 8x - 3 = Ax(x + 3) + B(x + 3) + Cx^2.$

Taking $x = 0$, we get $-3 = 3B$, so $B = -1$. Taking $x = -3$, we get $-18 = 9C$, so $C = -2$.

Taking $x = 1$, we get $6 = 4A + 4B + C = 4A - 4 - 2$, so $4A = 12$ and $A = 3$. Now

$\displaystyle\int \dfrac{x^2 + 8x - 3}{x^3 + 3x^2}\,dx = \int \left(\dfrac{3}{x} - \dfrac{1}{x^2} - \dfrac{2}{x + 3} \right) dx = 3\ln|x| + \dfrac{1}{x} - 2\ln|x + 3| + C.$

19. $\displaystyle\int \dfrac{x + 1}{9x^2 + 6x + 5}\,dx = \int \dfrac{x + 1}{(9x^2 + 6x + 1) + 4}\,dx = \int \dfrac{x + 1}{(3x + 1)^2 + 4}\,dx \qquad \begin{bmatrix} u = 3x + 1, \\ du = 3\,dx \end{bmatrix}$

$\qquad = \displaystyle\int \dfrac{\left[\frac{1}{3}(u - 1)\right] + 1}{u^2 + 4}\left(\dfrac{1}{3}\,du\right) = \dfrac{1}{3} \cdot \dfrac{1}{3} \int \dfrac{(u - 1) + 3}{u^2 + 4}\,du$

$\qquad = \dfrac{1}{9}\displaystyle\int \dfrac{u}{u^2 + 4}\,du + \dfrac{1}{9}\int \dfrac{2}{u^2 + 2^2}\,du = \dfrac{1}{9} \cdot \dfrac{1}{2}\ln(u^2 + 4) + \dfrac{2}{9} \cdot \dfrac{1}{2}\tan^{-1}\left(\dfrac{1}{2}u\right) + C$

$\qquad = \tfrac{1}{18}\ln(9x^2 + 6x + 5) + \tfrac{1}{9}\tan^{-1}\left[\tfrac{1}{2}(3x + 1)\right] + C$

20. $\displaystyle\int \tan^5\theta \sec^3\theta\,d\theta = \int \tan^4\theta \sec^2\theta \sec\theta \tan\theta\,d\theta = \int (\sec^2\theta - 1)^2 \sec^2\theta \sec\theta \tan\theta\,d\theta \qquad \begin{bmatrix} u = \sec\theta, \\ du = \sec\theta \tan\theta\,d\theta \end{bmatrix}$

$\qquad = \displaystyle\int (u^2 - 1)^2 u^2\,du = \int (u^6 - 2u^4 + u^2)\,du$

$\qquad = \tfrac{1}{7}u^7 - \tfrac{2}{5}u^5 + \tfrac{1}{3}u^3 + C = \tfrac{1}{7}\sec^7\theta - \tfrac{2}{5}\sec^5\theta + \tfrac{1}{3}\sec^3\theta + C$

21. $\displaystyle\int \dfrac{dx}{\sqrt{x^2 - 4x}} = \int \dfrac{dx}{\sqrt{(x^2 - 4x + 4) - 4}} = \int \dfrac{dx}{\sqrt{(x - 2)^2 - 2^2}}$

$\qquad = \displaystyle\int \dfrac{2\sec\theta \tan\theta\,d\theta}{2\tan\theta} \qquad \begin{bmatrix} x - 2 = 2\sec\theta, \\ dx = 2\sec\theta \tan\theta\,d\theta \end{bmatrix}$

$\qquad = \displaystyle\int \sec\theta\,d\theta = \ln|\sec\theta + \tan\theta| + C_1$

$\qquad = \ln\left| \dfrac{x - 2}{2} + \dfrac{\sqrt{x^2 - 4x}}{2} \right| + C_1$

$\qquad = \ln\left| x - 2 + \sqrt{x^2 - 4x} \right| + C$, where $C = C_1 - \ln 2$

22. Let $x = \sqrt{t}$, so that $x^2 = t$ and $2x\,dx = dt$. Then

$\qquad\qquad \displaystyle\int t e^{\sqrt{t}}\,dt = \int x^2 e^x (2x\,dx) = \int 2x^3 e^x\,dx \qquad \begin{bmatrix} u_1 = 2x^3, & dv_1 = e^x\,dx, \\ du_1 = 6x^2\,dx & v_1 = e^x \end{bmatrix}$

$\qquad\qquad\qquad = 2x^3 e^x - \displaystyle\int 6x^2 e^x\,dx \qquad \begin{bmatrix} u_2 = 6x^2, & dv_2 = e^x\,dx, \\ du_2 = 12x\,dx & v_2 = e^x \end{bmatrix}$

$\qquad\qquad\qquad = 2x^3 e^x - \left(6x^2 e^x - \displaystyle\int 12x e^x\,dx\right) \qquad \begin{bmatrix} u_3 = 12x, & dv_3 = e^x\,dx, \\ du_3 = 12\,dx & v_3 = e^x \end{bmatrix}$

$\qquad\qquad\qquad = 2x^3 e^x - 6x^2 e^x + \left(12x e^x - \displaystyle\int 12 e^x\,dx\right) = 2x^3 e^x - 6x^2 e^x + 12x e^x - 12 e^x + C$

$\qquad\qquad\qquad = 2e^x(x^3 - 3x^2 + 6x - 6) + C = 2e^{\sqrt{t}}\left(t\sqrt{t} - 3t + 6\sqrt{t} - 6\right) + C$

23. Let $x = \tan\theta$, so that $dx = \sec^2\theta\,d\theta$. Then

$$\int \frac{dx}{x\sqrt{x^2+1}} = \int \frac{\sec^2\theta\,d\theta}{\tan\theta\,\sec\theta} = \int \frac{\sec\theta}{\tan\theta}\,d\theta$$

$$= \int \csc\theta\,d\theta = \ln|\csc\theta - \cot\theta| + C$$

$$= \ln\left|\frac{\sqrt{x^2+1}}{x} - \frac{1}{x}\right| + C = \ln\left|\frac{\sqrt{x^2+1}-1}{x}\right| + C$$

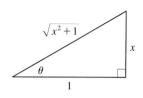

24. Let $u = \cos x$, $dv = e^x\,dx$ $\Rightarrow$ $du = -\sin x\,dx$, $v = e^x$: $(*)$ $I = \int e^x\cos x\,dx = e^x\cos x + \int e^x\sin x\,dx$.

To integrate $\int e^x\sin x\,dx$, let $U = \sin x$, $dV = e^x\,dx$ $\Rightarrow$ $dU = \cos x\,dx$, $V = e^x$. Then

$\int e^x\sin x\,dx = e^x\sin x - \int e^x\cos x\,dx = e^x\sin x - I$. By substitution in $(*)$, $I = e^x\cos x + e^x\sin x - I$ $\Rightarrow$

$2I = e^x(\cos x + \sin x)$ $\Rightarrow$ $I = \frac{1}{2}e^x(\cos x + \sin x) + C$.

25. $\dfrac{3x^3 - x^2 + 6x - 4}{(x^2+1)(x^2+2)} = \dfrac{Ax+B}{x^2+1} + \dfrac{Cx+D}{x^2+2}$ $\Rightarrow$ $3x^3 - x^2 + 6x - 4 = (Ax+B)(x^2+2) + (Cx+D)(x^2+1)$.

Equating the coefficients gives $A + C = 3$, $B + D = -1$, $2A + C = 6$, and $2B + D = -4$ $\Rightarrow$

$A = 3$, $C = 0$, $B = -3$, and $D = 2$. Now

$$\int \frac{3x^3 - x^2 + 6x - 4}{(x^2+1)(x^2+2)}\,dx = 3\int \frac{x-1}{x^2+1}\,dx + 2\int \frac{dx}{x^2+2} = \frac{3}{2}\ln(x^2+1) - 3\tan^{-1}x + \sqrt{2}\tan^{-1}\left(\frac{x}{\sqrt{2}}\right) + C.$$

26. $\int x\sin x\,\cos x\,dx = \int \frac{1}{2}x\sin 2x\,dx$ $\begin{bmatrix} u = \frac{1}{2}x, & dv = \sin 2x\,dx, \\ du = \frac{1}{2}\,dx & v = -\frac{1}{2}\cos 2x \end{bmatrix}$

$$= -\frac{1}{4}x\cos 2x + \int \frac{1}{4}\cos 2x\,dx = -\frac{1}{4}x\cos 2x + \frac{1}{8}\sin 2x + C$$

27. $\int_0^{\pi/2} \cos^3 x\,\sin 2x\,dx = \int_0^{\pi/2} \cos^3 x\,(2\sin x\,\cos x)\,dx = \int_0^{\pi/2} 2\cos^4 x\,\sin x\,dx = \left[-\frac{2}{5}\cos^5 x\right]_0^{\pi/2} = \frac{2}{5}$

28. Let $u = \sqrt[3]{x}$. Then $x = u^3$, $dx = 3u^2\,du$ $\Rightarrow$

$$\int \frac{\sqrt[3]{x}+1}{\sqrt[3]{x}-1}\,dx = \int \frac{u+1}{u-1}\,3u^2\,du = 3\int\left(u^2 + 2u + 2 + \frac{2}{u-1}\right)du$$

$$= u^3 + 3u^2 + 6u + 6\ln|u-1| + C = x + 3x^{2/3} + 6\sqrt[3]{x} + 6\ln|\sqrt[3]{x}-1| + C$$

29. The product of an odd function and an even function is an odd function, so $f(x) = x^5\sec x$ is an odd function.

By Theorem 5.5.6(b), $\int_{-1}^{1} x^5\sec x\,dx = 0$.

30. Let $u = e^{-x}$, $du = -e^{-x}\,dx$. Then

$$\int \frac{dx}{e^x\sqrt{1-e^{-2x}}} = \int \frac{e^{-x}\,dx}{\sqrt{1-(e^{-x})^2}} = \int \frac{-du}{\sqrt{1-u^2}} = -\sin^{-1}u + C = -\sin^{-1}(e^{-x}) + C.$$

31. Let $u = \sqrt{e^x - 1}$. Then $u^2 = e^x - 1$ and $2u\,du = e^x\,dx$. Also, $e^x + 8 = u^2 + 9$. Thus,

$$\int_0^{\ln 10} \frac{e^x\sqrt{e^x-1}}{e^x+8}\,dx = \int_0^3 \frac{u\cdot 2u\,du}{u^2+9} = 2\int_0^3 \frac{u^2}{u^2+9}\,du = 2\int_0^3 \left(1 - \frac{9}{u^2+9}\right)du$$

$$= 2\left[u - \frac{9}{3}\tan^{-1}\left(\frac{u}{3}\right)\right]_0^3 = 2[(3 - 3\tan^{-1}1) - 0] = 2\left(3 - 3\cdot\frac{\pi}{4}\right) = 6 - \frac{3\pi}{2}$$

32. $\displaystyle\int_0^{\pi/4} \frac{x\sin x}{\cos^3 x}\,dx = \int_0^{\pi/4} x\tan x \sec^2 x\,dx \qquad \begin{bmatrix} u = x, & dv = \tan x \sec^2 x\,dx, \\ du = dx & v = \frac{1}{2}\tan^2 x \end{bmatrix}$

$$= \left[\frac{x}{2}\tan^2 x\right]_0^{\pi/4} - \frac{1}{2}\int_0^{\pi/4}\tan^2 x\,dx = \frac{\pi}{8}\cdot 1^2 - 0 - \frac{1}{2}\int_0^{\pi/4}(\sec^2 x - 1)\,dx$$

$$= \frac{\pi}{8} - \frac{1}{2}\Big[\tan x - x\Big]_0^{\pi/4} = \frac{\pi}{8} - \frac{1}{2}\left(1 - \frac{\pi}{4}\right) = \frac{\pi}{4} - \frac{1}{2}$$

33. Let $x = 2\sin\theta \ \Rightarrow\ (4 - x^2)^{3/2} = (2\cos\theta)^3,\ dx = 2\cos\theta\,d\theta$, so

$$\int \frac{x^2}{(4 - x^2)^{3/2}}\,dx = \int \frac{4\sin^2\theta}{8\cos^3\theta}\,2\cos\theta\,d\theta = \int \tan^2\theta\,d\theta = \int (\sec^2\theta - 1)\,d\theta$$

$$= \tan\theta - \theta + C = \frac{x}{\sqrt{4 - x^2}} - \sin^{-1}\!\left(\frac{x}{2}\right) + C$$

34. Integrate by parts twice, first with $u = (\arcsin x)^2,\ dv = dx$:

$$I = \int (\arcsin x)^2\,dx = x(\arcsin x)^2 - \int 2x\arcsin x \left(\frac{dx}{\sqrt{1 - x^2}}\right)$$

Now let $U = \arcsin x,\ dV = \dfrac{x}{\sqrt{1 - x^2}}\,dx \ \Rightarrow\ dU = \dfrac{1}{\sqrt{1 - x^2}}\,dx,\ V = -\sqrt{1 - x^2}$. So

$$I = x(\arcsin x)^2 - 2\left[\arcsin x\left(-\sqrt{1 - x^2}\right) + \int dx\right] = x(\arcsin x)^2 + 2\sqrt{1 - x^2}\arcsin x - 2x + C$$

35. $\displaystyle\int \frac{1}{\sqrt{x + x^{3/2}}}\,dx = \int \frac{dx}{\sqrt{x(1 + \sqrt{x})}} = \int \frac{dx}{\sqrt{x}\sqrt{1 + \sqrt{x}}} \qquad \begin{bmatrix} u = 1 + \sqrt{x}, \\ du = \frac{dx}{2\sqrt{x}} \end{bmatrix} = \int \frac{2\,du}{\sqrt{u}} = \int 2u^{-1/2}\,du$

$$= 4\sqrt{u} + C = 4\sqrt{1 + \sqrt{x}} + C$$

36. $\displaystyle\int \frac{1 - \tan\theta}{1 + \tan\theta}\,d\theta = \int \frac{\dfrac{\cos\theta}{\cos\theta} - \dfrac{\sin\theta}{\cos\theta}}{\dfrac{\cos\theta}{\cos\theta} + \dfrac{\sin\theta}{\cos\theta}}\,d\theta = \int \frac{\cos\theta - \sin\theta}{\cos\theta + \sin\theta}\,d\theta = \ln|\cos\theta + \sin\theta| + C$

37. $\int (\cos x + \sin x)^2 \cos 2x\,dx = \int (\cos^2 x + 2\sin x\cos x + \sin^2 x)\cos 2x\,dx = \int (1 + \sin 2x)\cos 2x\,dx$

$$= \int \cos 2x\,dx + \tfrac{1}{2}\int \sin 4x\,dx = \tfrac{1}{2}\sin 2x - \tfrac{1}{8}\cos 4x + C$$

Or: $\int (\cos x + \sin x)^2 \cos 2x\,dx = \int (\cos x + \sin x)^2(\cos^2 x - \sin^2 x)\,dx$

$$= \int (\cos x + \sin x)^3(\cos x - \sin x)\,dx = \tfrac{1}{4}(\cos x + \sin x)^4 + C_1$$

38. Let $u = x + 2$, so that $du = dx$ and $x = u - 2$. Then

$$\int \frac{x^2}{(x + 2)^3}\,dx = \int \frac{(u - 2)^2}{u^3}\,du = \int \frac{u^2 - 4u + 4}{u^3}\,du = \int \left(\frac{1}{u} - 4u^{-2} + 4u^{-3}\right)du$$

$$= \ln|u| + 4u^{-1} - 2u^{-2} + C = \ln|x + 2| + \frac{4}{x + 2} - \frac{2}{(x + 2)^2} + C$$

39. We'll integrate $I = \displaystyle\int \frac{xe^{2x}}{(1+2x)^2}\,dx$ by parts with $u = xe^{2x}$ and $dv = \dfrac{dx}{(1+2x)^2}$. Then $du = (x \cdot 2e^{2x} + e^{2x} \cdot 1)\,dx$

and $v = -\dfrac{1}{2} \cdot \dfrac{1}{1+2x}$, so

$$I = -\frac{1}{2} \cdot \frac{xe^{2x}}{1+2x} - \int \left[-\frac{1}{2} \cdot \frac{e^{2x}(2x+1)}{1+2x} \right] dx = -\frac{xe^{2x}}{4x+2} + \frac{1}{2} \cdot \frac{1}{2}e^{2x} + C = e^{2x}\left(\frac{1}{4} - \frac{x}{4x+2} \right) + C$$

Thus, $\displaystyle\int_0^{1/2} \frac{xe^{2x}}{(1+2x)^2}\,dx = \left[e^{2x}\left(\frac{1}{4} - \frac{x}{4x+2} \right) \right]_0^{1/2} = e\left(\frac{1}{4} - \frac{1}{8} \right) - 1\left(\frac{1}{4} - 0 \right) = \frac{1}{8}e - \frac{1}{4}.$

40. $\displaystyle\int_{\pi/4}^{\pi/3} \frac{\sqrt{\tan\theta}}{\sin 2\theta}\,d\theta = \int_{\pi/4}^{\pi/3} \frac{\sqrt{\frac{\sin\theta}{\cos\theta}}}{2\sin\theta\cos\theta}\,d\theta = \int_{\pi/4}^{\pi/3} \frac{1}{2}(\sin\theta)^{-1/2}(\cos\theta)^{-3/2}\,d\theta = \int_{\pi/4}^{\pi/3} \frac{1}{2}\left(\frac{\sin\theta}{\cos\theta} \right)^{-1/2}(\cos\theta)^{-2}d\theta$

$$= \int_{\pi/4}^{\pi/3} \frac{1}{2}(\tan\theta)^{-1/2}\sec^2\theta\,d\theta = \left[\sqrt{\tan\theta} \right]_{\pi/4}^{\pi/3} = \sqrt{\sqrt{3}} - \sqrt{1} = \sqrt[4]{3} - 1$$

41. $\displaystyle\int_1^\infty \frac{1}{(2x+1)^3}\,dx = \lim_{t\to\infty} \int_1^t \frac{1}{(2x+1)^3}\,dx = \lim_{t\to\infty} \int_1^t \tfrac{1}{2}(2x+1)^{-3}\,2\,dx = \lim_{t\to\infty} \left[-\frac{1}{4(2x+1)^2} \right]_1^t$

$$= -\frac{1}{4} \lim_{t\to\infty} \left[\frac{1}{(2t+1)^2} - \frac{1}{9} \right] = -\frac{1}{4}\left(0 - \frac{1}{9} \right) = \frac{1}{36}$$

42. $\displaystyle\int_1^\infty \frac{\ln x}{x^4}\,dx = \lim_{t\to\infty} \int_1^t \frac{\ln x}{x^4}\,dx \qquad \left[\begin{array}{ll} u = \ln x, & dv = dx/x^4, \\ du = dx/x & v = -1/(3x^3) \end{array} \right]$

$$= \lim_{t\to\infty} \left[-\frac{\ln x}{3x^3} \right]_1^t + \int_1^t \frac{1}{3x^4}\,dx = \lim_{t\to\infty} \left(-\frac{\ln t}{3t^3} + 0 + \left[\frac{-1}{9x^3} \right]_1^t \right) \overset{\text{H}}{=} \lim_{t\to\infty} \left(-\frac{1}{9t^3} + \left[\frac{-1}{9t^3} + \frac{1}{9} \right] \right)$$

$$= 0 + 0 + \tfrac{1}{9} = \tfrac{1}{9}$$

43. $\displaystyle\int \frac{dx}{x\ln x} \qquad \left[\begin{array}{l} u = \ln x, \\ du = dx/x \end{array} \right] = \int \frac{du}{u} = \ln|u| + C = \ln|\ln x| + C$, so

$\displaystyle\int_2^\infty \frac{dx}{x\ln x} = \lim_{t\to\infty} \int_2^t \frac{dx}{x\ln x} = \lim_{t\to\infty} \left[\ln|\ln x| \right]_2^t = \lim_{t\to\infty} [\ln(\ln t) - \ln(\ln 2)] = \infty$, so the integral is divergent.

44. Let $u = \sqrt{y-2}$. Then $y = u^2 + 2$ and $dy = 2u\,du$, so

$$\int \frac{y\,dy}{\sqrt{y-2}} = \int \frac{(u^2+2)2u\,du}{u} = 2\int (u^2+2)\,du = 2\left[\tfrac{1}{3}u^3 + 2u \right] + C$$

Thus, $\displaystyle\int_2^6 \frac{y\,dy}{\sqrt{y-2}} = \lim_{t\to 2^+} \int_t^6 \frac{y\,dy}{\sqrt{y-2}} = \lim_{t\to 2^+} \left[\tfrac{2}{3}(y-2)^{3/2} + 4\sqrt{y-2} \right]_t^6$

$$= \lim_{t\to 2^+} \left[\tfrac{16}{3} + 8 - \tfrac{2}{3}(t-2)^{3/2} - 4\sqrt{t-2} \right] = \tfrac{40}{3}.$$

45. $\displaystyle\int_0^4 \frac{\ln x}{\sqrt{x}}\,dx = \lim_{t\to 0^+}\int_t^4 \frac{\ln x}{\sqrt{x}}\,dx \overset{*}{=} \lim_{t\to 0^+}\left[2\sqrt{x}\,\ln x - 4\sqrt{x}\,\right]_t^4$

$$= \lim_{t\to 0^+}\left[(2\cdot 2\ln 4 - 4\cdot 2) - (2\sqrt{t}\,\ln t - 4\sqrt{t}\,)\right]\overset{**}{=}(4\ln 4 - 8) - (0 - 0) = 4\ln 4 - 8$$

(*) Let $u = \ln x$, $dv = \dfrac{1}{\sqrt{x}}\,dx \;\Rightarrow\; du = \dfrac{1}{x}\,dx$, $v = 2\sqrt{x}$. Then

$$\int \frac{\ln x}{\sqrt{x}}\,dx = 2\sqrt{x}\,\ln x - 2\int \frac{dx}{\sqrt{x}} = 2\sqrt{x}\,\ln x - 4\sqrt{x} + C$$

(**) $\displaystyle\lim_{t\to 0^+}\left(2\sqrt{t}\,\ln t\right) = \lim_{t\to 0^+}\frac{2\ln t}{t^{-1/2}} \overset{\mathrm{H}}{=} \lim_{t\to 0^+}\frac{2/t}{-\frac{1}{2}t^{-3/2}} = \lim_{t\to 0^+}\left(-4\sqrt{t}\,\right) = 0$

46. Note that $f(x) = 1/(2 - 3x)$ has an infinite discontinuity at $x = \frac{2}{3}$. Now

$$\int_0^{2/3}\frac{1}{2-3x}\,dx = \lim_{t\to (2/3)^-}\int_0^t \frac{1}{2-3x}\,dx = \lim_{t\to(2/3)^-}\left[-\tfrac{1}{3}\ln|2-3x|\right]_0^t = -\tfrac{1}{3}\lim_{t\to(2/3)^-}\left[\ln|2-3t| - \ln 2\right] = \infty$$

Since $\displaystyle\int_0^{2/3}\frac{1}{2-3x}\,dx$ diverges, so does $\displaystyle\int_0^1 \frac{1}{2-3x}\,dx$.

47. $\displaystyle\int_0^1 \frac{x-1}{\sqrt{x}}\,dx = \lim_{t\to 0^+}\int_t^1\left(\frac{x}{\sqrt{x}} - \frac{1}{\sqrt{x}}\right)dx = \lim_{t\to 0^+}\int_t^1 \left(x^{1/2} - x^{-1/2}\right)dx = \lim_{t\to 0^+}\left[\tfrac{2}{3}x^{3/2} - 2x^{1/2}\right]_t^1$

$$= \lim_{t\to 0^+}\left[\left(\tfrac{2}{3} - 2\right) - \left(\tfrac{2}{3}t^{3/2} - 2t^{1/2}\right)\right] = -\tfrac{4}{3} - 0 = -\tfrac{4}{3}$$

48. $I = \displaystyle\int_{-1}^1 \frac{dx}{x^2 - 2x} = \int_{-1}^1 \frac{dx}{x(x-2)} = \int_{-1}^0 \frac{dx}{x(x-2)} + \int_0^1 \frac{dx}{x(x-2)} = I_1 + I_2$. Now

$$\frac{1}{x(x-2)} = \frac{A}{x} + \frac{B}{x-2} \quad\Rightarrow\quad 1 = A(x-2) + Bx.\ \text{Set } x = 2 \text{ to get } 1 = 2B, \text{ so } B = \tfrac{1}{2}.\ \text{Set } x = 0 \text{ to get } 1 = -2A,$$

$A = -\tfrac{1}{2}$. Thus,

$$I_2 = \lim_{t\to 0^+}\int_t^1\left(\frac{-\tfrac{1}{2}}{x} + \frac{\tfrac{1}{2}}{x-2}\right)dx = \lim_{t\to 0^+}\left[-\tfrac{1}{2}\ln|x| + \tfrac{1}{2}\ln|x-2|\right]_t^1 = \lim_{t\to 0^+}\left[(0 + 0) - \left(-\tfrac{1}{2}\ln t + \tfrac{1}{2}\ln|t-2|\right)\right]$$

$$= -\tfrac{1}{2}\ln 2 + \tfrac{1}{2}\lim_{t\to 0^+}\ln t = -\infty.$$

Since I_2 diverges, I is divergent.

49. Let $u = 2x + 1$. Then

$$\int_{-\infty}^\infty \frac{dx}{4x^2 + 4x + 5} = \int_{-\infty}^\infty \frac{\tfrac{1}{2}\,du}{u^2 + 4} = \frac{1}{2}\int_{-\infty}^0 \frac{du}{u^2 + 4} + \frac{1}{2}\int_0^\infty \frac{du}{u^2 + 4}$$

$$= \tfrac{1}{2}\lim_{t\to -\infty}\left[\tfrac{1}{2}\tan^{-1}\left(\tfrac{1}{2}u\right)\right]_t^0 + \tfrac{1}{2}\lim_{t\to\infty}\left[\tfrac{1}{2}\tan^{-1}\left(\tfrac{1}{2}u\right)\right]_0^t = \tfrac{1}{4}\left[0 - \left(-\tfrac{\pi}{2}\right)\right] + \tfrac{1}{4}\left[\tfrac{\pi}{2} - 0\right] = \tfrac{\pi}{4}.$$

50. $\int_1^\infty \dfrac{\tan^{-1}x}{x^2}\,dx = \lim\limits_{t\to\infty} \int_1^t \dfrac{\tan^{-1}x}{x^2}\,dx$. Integrate by parts:

$$\int \frac{\tan^{-1}x}{x^2}\,dx = \frac{-\tan^{-1}x}{x} + \int \frac{1}{x}\frac{dx}{1+x^2} = \frac{-\tan^{-1}x}{x} + \int \left[\frac{1}{x} - \frac{x}{x^2+1}\right]dx$$

$$= \frac{-\tan^{-1}x}{x} + \ln|x| - \tfrac{1}{2}\ln(x^2+1) + C = \frac{-\tan^{-1}x}{x} + \frac{1}{2}\ln\frac{x^2}{x^2+1} + C$$

Thus,

$$\int_1^\infty \frac{\tan^{-1}x}{x^2}\,dx = \lim\limits_{t\to\infty}\left[-\frac{\tan^{-1}x}{x} + \frac{1}{2}\ln\frac{x^2}{x^2+1}\right]_1^t = \lim\limits_{t\to\infty}\left[-\frac{\tan^{-1}t}{t} + \frac{1}{2}\ln\frac{t^2}{t^2+1} + \frac{\pi}{4} - \frac{1}{2}\ln\frac{1}{2}\right]$$

$$= 0 + \tfrac{1}{2}\ln 1 + \tfrac{\pi}{4} + \tfrac{1}{2}\ln 2 = \tfrac{\pi}{4} + \tfrac{1}{2}\ln 2$$

51. We first make the substitution $t = x + 1$, so $\ln(x^2 + 2x + 2) = \ln\left[(x+1)^2 + 1\right] = \ln(t^2 + 1)$. Then we use parts with $u = \ln(t^2 + 1)$, $dv = dt$:

$$\int \ln(t^2+1)\,dt = t\,\ln(t^2+1) - \int \frac{t(2t)\,dt}{t^2+1} = t\,\ln(t^2+1) - 2\int \frac{t^2\,dt}{t^2+1} = t\,\ln(t^2+1) - 2\int \left(1 - \frac{1}{t^2+1}\right)dt$$

$$= t\,\ln(t^2+1) - 2t + 2\arctan t + C$$

$$= (x+1)\ln(x^2+2x+2) - 2x + 2\arctan(x+1) + K, \ \text{where } K = C - 2$$

[Alternatively, we could have integrated by parts immediately with $u = \ln(x^2 + 2x + 2)$.] Notice from the graph that $f = 0$ where F has a horizontal tangent. Also, F is always increasing, and $f \geq 0$.

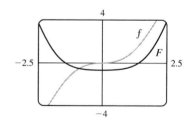

52. Let $u = x^2 + 1$. Then $x^2 = u - 1$ and $x\,dx = \tfrac{1}{2}\,du$, so

$$\int \frac{x^3}{\sqrt{x^2+1}}\,dx = \int \frac{(u-1)}{\sqrt{u}}\left(\tfrac{1}{2}\,du\right) = \frac{1}{2}\int \left(u^{1/2} - u^{-1/2}\right)du$$

$$= \tfrac{1}{2}\left(\tfrac{2}{3}u^{3/2} - 2u^{1/2}\right) + C = \tfrac{1}{3}(x^2+1)^{3/2} - (x^2+1)^{1/2} + C$$

$$= \tfrac{1}{3}(x^2+1)^{1/2}\left[(x^2+1) - 3\right] + C = \tfrac{1}{3}\sqrt{x^2+1}\,(x^2-2) + C$$

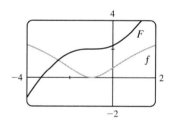

53. From the graph, it seems as though $\int_0^{2\pi} \cos^2 x\,\sin^3 x\,dx$ is equal to 0.

To evaluate the integral, we write the integral as

$I = \int_0^{2\pi} \cos^2 x\,(1 - \cos^2 x)\,\sin x\,dx$ and let $u = \cos x \ \Rightarrow$

$du = -\sin x\,dx$. Thus, $I = \int_1^1 u^2(1 - u^2)(-du) = 0$.

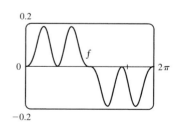

54. (a) To evaluate $\int x^5 e^{-2x}\,dx$ by hand, we would integrate by parts repeatedly, always taking $dv = e^{-2x}$ and starting with

$u = x^5$. Each time we would reduce the degree of the x-factor by 1.

(b) To evaluate the integral using tables, we would use Formula 97 (which is

proved using integration by parts) until the exponent of x was reduced to 1,

and then we would use Formula 96.

(c) $\int x^5 e^{-2x}\,dx = -\frac{1}{8}e^{-2x}(4x^5 + 10x^4 + 20x^3 + 30x^2 + 30x + 15) + C$

(d)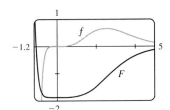

55. $\int \sqrt{4x^2 - 4x - 3}\,dx = \int \sqrt{(2x-1)^2 - 4}\,dx \quad \left[\begin{matrix} u = 2x - 1, \\ du = 2\,dx \end{matrix}\right] = \int \sqrt{u^2 - 2^2}\left(\frac{1}{2}\,du\right)$

$\overset{39}{=} \frac{1}{2}\left(\frac{u}{2}\sqrt{u^2 - 2^2} - \frac{2^2}{2}\ln\left|u + \sqrt{u^2 - 2^2}\right|\right) + C = \frac{1}{4}u\sqrt{u^2 - 4} - \ln\left|u + \sqrt{u^2 - 4}\right| + C$

$= \frac{1}{4}(2x - 1)\sqrt{4x^2 - 4x - 3} - \ln\left|2x - 1 + \sqrt{4x^2 - 4x - 3}\right| + C$

56. $\int \csc^5 t\,dt \overset{78}{=} -\frac{1}{4}\cot t\,\csc^3 t + \frac{3}{4}\int \csc^3 t\,dt \overset{72}{=} -\frac{1}{4}\cot t\,\csc^3 t + \frac{3}{4}\left[-\frac{1}{2}\csc t\,\cot t + \frac{1}{2}\ln|\csc t - \cot t|\right] + C$

$= -\frac{1}{4}\cot t\,\csc^3 t - \frac{3}{8}\csc t\,\cot t + \frac{3}{8}\ln|\csc t - \cot t| + C$

57. Let $u = \sin x$, so that $du = \cos x\,dx$. Then

$\int \cos x\sqrt{4 + \sin^2 x}\,dx = \int \sqrt{2^2 + u^2}\,du \overset{21}{=} \frac{u}{2}\sqrt{2^2 + u^2} + \frac{2^2}{2}\ln\left(u + \sqrt{2^2 + u^2}\right) + C$

$= \frac{1}{2}\sin x\sqrt{4 + \sin^2 x} + 2\ln\left(\sin x + \sqrt{4 + \sin^2 x}\right) + C$

58. Let $u = \sin x$. Then $du = \cos x\,dx$, so

$\displaystyle\int \frac{\cot x\,dx}{\sqrt{1 + 2\sin x}} = \int \frac{du}{u\sqrt{1 + 2u}} \overset{\substack{57\ \text{with}\\ a=1,\,b=2}}{=} \ln\left|\frac{\sqrt{1 + 2u} - 1}{\sqrt{1 + 2u} + 1}\right| + C = \ln\left|\frac{\sqrt{1 + 2\sin x} - 1}{\sqrt{1 + 2\sin x} + 1}\right| + C$

59. (a) $\displaystyle\frac{d}{du}\left[-\frac{1}{u}\sqrt{a^2 - u^2} - \sin^{-1}\left(\frac{u}{a}\right) + C\right] = \frac{1}{u^2}\sqrt{a^2 - u^2} + \frac{1}{\sqrt{a^2 - u^2}} - \frac{1}{\sqrt{1 - u^2/a^2}}\cdot\frac{1}{a}$

$= (a^2 - u^2)^{-1/2}\left[\frac{1}{u^2}(a^2 - u^2) + 1 - 1\right] = \frac{\sqrt{a^2 - u^2}}{u^2}$

(b) Let $u = a\sin\theta \ \Rightarrow\ du = a\cos\theta\,d\theta,\ a^2 - u^2 = a^2(1 - \sin^2\theta) = a^2\cos^2\theta.$

$\displaystyle\int \frac{\sqrt{a^2 - u^2}}{u^2}\,du = \int \frac{a^2\cos^2\theta}{a^2\sin^2\theta}\,d\theta = \int \frac{1 - \sin^2\theta}{\sin^2\theta}\,d\theta = \int(\csc^2\theta - 1)\,d\theta = -\cot\theta - \theta + C$

$= -\frac{\sqrt{a^2 - u^2}}{u} - \sin^{-1}\left(\frac{u}{a}\right) + C$

60. Work backward, and use integration by parts with $U = u^{-(n-1)}$ and $dV = (a + bu)^{-1/2}\, du$ $\Rightarrow$

$$dU = \frac{-(n-1)\,du}{u^n} \text{ and } V = \frac{2}{b}\sqrt{a+bu}, \text{ to get}$$

$$\int \frac{du}{u^{n-1}\sqrt{a+bu}} = \int U\,dV = UV - \int V\,dU = \frac{2\sqrt{a+bu}}{bu^{n-1}} + \frac{2(n-1)}{b}\int \frac{\sqrt{a+bu}}{u^n}\,du$$

$$= \frac{2\sqrt{a+bu}}{bu^{n-1}} + \frac{2(n-1)}{b}\int \frac{a+bu}{u^n\sqrt{a+bu}}\,du$$

$$= \frac{2\sqrt{a+bu}}{bu^{n-1}} + 2(n-1)\int \frac{du}{u^{n-1}\sqrt{a+bu}} + \frac{2a(n-1)}{b}\int \frac{du}{u^n\sqrt{a+bu}}$$

Rearranging the equation gives $\dfrac{2a(n-1)}{b}\displaystyle\int \frac{du}{u^n\sqrt{a+bu}} = -\frac{2\sqrt{a+bu}}{bu^{n-1}} - (2n-3)\displaystyle\int \frac{du}{u^{n-1}\sqrt{a+bu}}$ $\Rightarrow$

$$\int \frac{du}{u^n\sqrt{a+bu}} = \frac{-\sqrt{a+bu}}{a(n-1)u^{n-1}} - \frac{b(2n-3)}{2a(n-1)}\int \frac{du}{u^{n-1}\sqrt{a+bu}}$$

61. For $n \geq 0$, $\int_0^\infty x^n\,dx = \lim\limits_{t\to\infty} \left[x^{n+1}/(n+1)\right]_0^t = \infty$. For $n < 0$, $\int_0^\infty x^n\,dx = \int_0^1 x^n\,dx + \int_1^\infty x^n\,dx$. Both integrals are improper. By (8.8.2), the second integral diverges if $-1 \leq n < 0$. By Exercise 8.8.57, the first integral diverges if $n \leq -1$. Thus, $\int_0^\infty x^n\,dx$ is divergent for all values of n.

62. $I = \displaystyle\int_0^\infty e^{ax}\cos x\,dx = \lim_{t\to\infty}\int_0^t e^{ax}\cos x\,dx \overset{\substack{99\text{ with}\\ b=1}}{=} \lim_{t\to\infty}\left[\frac{e^{ax}}{a^2+1}(a\cos x + \sin x)\right]_0^t$

$$= \lim_{t\to\infty}\left[\frac{e^{at}}{a^2+1}(a\cos t + \sin t) - \frac{1}{a^2+1}(a)\right] = \frac{1}{a^2+1}\lim_{t\to\infty}\left[e^{at}(a\cos t + \sin t) - a\right].$$

For $a \geq 0$, the limit does not exist due to oscillation. For $a < 0$, $\lim\limits_{t\to\infty}\left[e^{at}(a\cos t + \sin t)\right] = 0$ by the Squeeze Theorem,

because $\left|e^{at}(a\cos t + \sin t)\right| \leq e^{at}(|a| + 1)$, so $I = \dfrac{1}{a^2+1}(-a) = -\dfrac{a}{a^2+1}$.

63. $f(x) = \dfrac{1}{\ln x}$, $\Delta x = \dfrac{b-a}{n} = \dfrac{4-2}{10} = \dfrac{1}{5}$

   (a) $T_{10} = \frac{1}{5\cdot 2}\{f(2) + 2[f(2.2) + f(2.4) + \cdots + f(3.8)] + f(4)\} \approx 1.925444$

   (b) $M_{10} = \frac{1}{5}[f(2.1) + f(2.3) + f(2.5) + \cdots + f(3.9)] \approx 1.920915$

   (c) $S_{10} = \frac{1}{5\cdot 3}[f(2) + 4f(2.2) + 2f(2.4) + \cdots + 2f(3.6) + 4f(3.8) + f(4)] \approx 1.922470$

64. $f(x) = \sqrt{x}\cos x$, $\Delta x = \frac{b-a}{n} = \frac{4-1}{10} = \frac{3}{10}$

   (a) $T_{10} = \frac{3}{10\cdot 2}\{f(1) + 2[f(1.3) + f(1.6) + \cdots + f(3.7)] + f(4)\} \approx -2.835151$

   (b) $M_{10} = \frac{3}{10}[f(1.15) + f(1.45) + f(1.75) + \cdots + f(3.85)] \approx -2.856809$

   (c) $S_{10} = \frac{3}{10\cdot 3}[f(1) + 4f(1.3) + 2f(1.6) + \cdots + 2f(3.4) + 4f(3.7) + f(4)] \approx -2.849672$

65. $f(x) = \dfrac{1}{\ln x}$ $\Rightarrow$ $f'(x) = -\dfrac{1}{x(\ln x)^2}$ $\Rightarrow$ $f''(x) = \dfrac{2 + \ln x}{x^2(\ln x)^3} = \dfrac{2}{x^2(\ln x)^3} + \dfrac{1}{x^2(\ln x)^2}$. Note that each term of

$f''(x)$ decreases on $[2, 4]$, so we'll take $K = f''(2) \approx 2.022$. $|E_T| \le \dfrac{K(b-a)^3}{12n^2} \approx \dfrac{2.022(4-2)^3}{12(10)^2} = 0.01348$ and

$|E_M| \le \dfrac{K(b-a)^3}{24n^2} = 0.00674$. $|E_T| \le 0.00001$ $\Leftrightarrow$ $\dfrac{2.022(8)}{12n^2} \le \dfrac{1}{10^5}$ $\Leftrightarrow$ $n^2 \ge \dfrac{10^5(2.022)(8)}{12}$ $\Rightarrow$ $n \ge 367.2$.

Take $n = 368$ for T_n. $|E_M| \le 0.00001$ $\Leftrightarrow$ $n^2 \ge \dfrac{10^5(2.022)(8)}{24}$ $\Rightarrow$ $n \ge 259.6$. Take $n = 260$ for M_n.

66. $\displaystyle\int_1^4 \dfrac{e^x}{x}\,dx \approx S_6 = \dfrac{(4-1)/6}{3}[f(1) + 4f(1.5) + 2f(2) + 4f(2.5) + 2f(3) + 4f(3.5) + f(4)] \approx 17.739438$

67. $\Delta t = \left(\dfrac{10}{60} - 0\right)/10 = \dfrac{1}{60}$.

Distance traveled $= \displaystyle\int_0^{10} v\,dt \approx S_{10}$

$= \dfrac{1}{60 \cdot 3}[40 + 4(42) + 2(45) + 4(49) + 2(52) + 4(54) + 2(56) + 4(57) + 2(57) + 4(55) + 56]$

$= \dfrac{1}{180}(1544) = 8.5\overline{7}$ mi

68. We use Simpson's Rule with $n = 6$ and $\Delta t = \dfrac{24 - 0}{6} = 4$:

Increase in bee population $= \displaystyle\int_0^{24} r(t)\,dt \approx S_6$

$= \dfrac{4}{3}[r(0) + 4r(4) + 2r(8) + 4r(12) + 2r(16) + 4r(20) + r(24)]$

$= \dfrac{4}{3}[0 + 4(300) + 2(3000) + 4(11{,}000) + 2(4000) + 4(400) + 0]$

$= \dfrac{4}{3}(60{,}800) \approx 81{,}067$ bees

69. (a) $f(x) = \sin(\sin x)$. A CAS gives

$f^{(4)}(x) = \sin(\sin x)[\cos^4 x + 7\cos^2 x - 3]$
$\quad\quad + \cos(\sin x)\left[6\cos^2 x \sin x + \sin x\right]$

From the graph, we see that $\left|f^{(4)}(x)\right| < 3.8$ for $x \in [0, \pi]$.

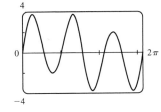

(b) We use Simpson's Rule with $f(x) = \sin(\sin x)$ and $\Delta x = \dfrac{\pi}{10}$:

$\displaystyle\int_0^\pi f(x)\,dx \approx \dfrac{\pi}{10 \cdot 3}\left[f(0) + 4f\!\left(\dfrac{\pi}{10}\right) + 2f\!\left(\dfrac{2\pi}{10}\right) + \cdots + 4f\!\left(\dfrac{9\pi}{10}\right) + f(\pi)\right] \approx 1.786721$

From part (a), we know that $\left|f^{(4)}(x)\right| < 3.8$ on $[0, \pi]$, so we use Theorem 8.7.4 with $K = 3.8$, and estimate the error

as $|E_S| \le \dfrac{3.8(\pi - 0)^5}{180(10)^4} \approx 0.000646$.

(c) If we want the error to be less than 0.00001, we must have $|E_S| \le \dfrac{3.8\pi^5}{180n^4} \le 0.00001$,

so $n^4 \ge \dfrac{3.8\pi^5}{180(0.00001)} \approx 646{,}041.6$ $\Rightarrow$ $n \ge 28.35$. Since n must be even for Simpson's Rule, we must have $n \ge 30$

to ensure the desired accuracy.

70. With an x-axis in the normal position, at $x = 7$ we have $C = 2\pi r = 45 \ \Rightarrow \ r(7) = \frac{2\pi}{45}$.

Using Simpson's Rule with $n = 4$ and $\Delta x = 7$, we have

$$V = \int_0^{28} \pi [r(x)]^2 \ dx \approx S_4 = \frac{7}{3}\left[0 + 4\pi\left(\frac{45}{2\pi}\right)^2 + 2\pi\left(\frac{53}{2\pi}\right)^2 + 4\pi\left(\frac{45}{2\pi}\right)^2 + 0\right] = \frac{7}{3}\left(\frac{21{,}818}{4\pi}\right) \approx 4051 \text{ cm}^3.$$

71. $\dfrac{x^3}{x^5 + 2} \le \dfrac{x^3}{x^5} = \dfrac{1}{x^2}$ for x in $[1, \infty)$. $\displaystyle\int_1^\infty \frac{1}{x^2}\,dx$ is convergent by (8.8.2) with $p = 2 > 1$. Therefore, $\displaystyle\int_1^\infty \frac{x^3}{x^5 + 2}\,dx$ is

convergent by the Comparison Theorem.

72. The line $y = 3$ intersects the hyperbola $y^2 - x^2 = 1$ at two points on its upper branch, namely $\left(-2\sqrt{2}, 3\right)$ and $\left(2\sqrt{2}, 3\right)$.

The desired area is

$$A = \int_{-2\sqrt{2}}^{2\sqrt{2}} \left(3 - \sqrt{x^2 + 1}\right) dx = 2\int_0^{2\sqrt{2}} \left(3 - \sqrt{x^2 + 1}\right) dx \overset{21}{=} 2\left[3x - \tfrac{1}{2}x\sqrt{x^2 + 1} - \tfrac{1}{2}\ln\left(x + \sqrt{x^2 + 1}\right)\right]_0^{2\sqrt{2}}$$

$$= \left[6x - x\sqrt{x^2 + 1} - \ln\left(x + \sqrt{x^2 + 1}\right)\right]_0^{2\sqrt{2}} = 12\sqrt{2} - 2\sqrt{2}\cdot 3 - \ln\left(2\sqrt{2} + 3\right) = 6\sqrt{2} - \ln\left(3 + 2\sqrt{2}\right)$$

Another method: $A = 2\int_1^3 \sqrt{y^2 - 1}\,dy$ and use Formula 39.

73. For x in $\left[0, \frac{\pi}{2}\right]$, $0 \le \cos^2 x \le \cos x$. For x in $\left[\frac{\pi}{2}, \pi\right]$, $\cos x \le 0 \le \cos^2 x$. Thus,

$$\text{area} = \int_0^{\pi/2} (\cos x - \cos^2 x)\,dx + \int_{\pi/2}^\pi (\cos^2 x - \cos x)\,dx$$

$$= \left[\sin x - \tfrac{1}{2}x - \tfrac{1}{4}\sin 2x\right]_0^{\pi/2} + \left[\tfrac{1}{2}x + \tfrac{1}{4}\sin 2x - \sin x\right]_{\pi/2}^\pi = \left[\left(1 - \tfrac{\pi}{4}\right) - 0\right] + \left[\tfrac{\pi}{2} - \left(\tfrac{\pi}{4} - 1\right)\right] = 2$$

74. The curves $y = \dfrac{1}{2 \pm \sqrt{x}}$ are defined for $x \ge 0$. For $x > 0$, $\dfrac{1}{2 - \sqrt{x}} > \dfrac{1}{2 + \sqrt{x}}$. Thus, the required area is

$$\int_0^1 \left(\frac{1}{2 - \sqrt{x}} - \frac{1}{2 + \sqrt{x}}\right) dx = \int_0^1 \left(\frac{1}{2 - u} - \frac{1}{2 + u}\right) 2u\,du \quad \left[u = \sqrt{x}\,\right] \quad = 2\int_0^1 \left(-\frac{u}{u - 2} - \frac{u}{u + 2}\right) du$$

$$= 2\int_0^1 \left(-1 - \frac{2}{u - 2} - 1 + \frac{2}{u + 2}\right) du = 2\left[2\ln\left|\frac{u + 2}{u - 2}\right| - 2u\right]_0^1 = 4\ln 3 - 4.$$

75. Using the formula for disks, the volume is

$$V = \int_0^{\pi/2} \pi \left[f(x)\right]^2 dx = \pi \int_0^{\pi/2} (\cos^2 x)^2 \, dx = \pi \int_0^{\pi/2} \left[\tfrac{1}{2}(1 + \cos 2x)\right]^2 dx$$

$$= \tfrac{\pi}{4} \int_0^{\pi/2} (1 + \cos^2 2x + 2\cos 2x)\,dx = \tfrac{\pi}{4}\int_0^{\pi/2}\left[1 + \tfrac{1}{2}(1 + \cos 4x) + 2\cos 2x\right] dx$$

$$= \tfrac{\pi}{4}\left[\tfrac{3}{2}x + \tfrac{1}{2}\left(\tfrac{1}{4}\sin 4x\right) + 2\left(\tfrac{1}{2}\sin 2x\right)\right]_0^{\pi/2} = \tfrac{\pi}{4}\left[\left(\tfrac{3\pi}{4} + \tfrac{1}{8}\cdot 0 + 0\right) - 0\right] = \tfrac{3\pi^2}{16}$$

76. Using the formula for cylindrical shells, the volume is

$$V = \int_0^{\pi/2} 2\pi x f(x)\,dx = 2\pi \int_0^{\pi/2} x\cos^2 x\,dx = 2\pi \int_0^{\pi/2} x\left[\tfrac{1}{2}(1 + \cos 2x)\right] dx = 2\left(\tfrac{1}{2}\right)\pi\int_0^{\pi/2}(x + x\cos 2x)\,dx$$

$$= \pi\left(\left[\tfrac{1}{2}x^2\right]_0^{\pi/2} + \left[x\left(\tfrac{1}{2}\sin 2x\right)\right]_0^{\pi/2} - \int_0^{\pi/2}\tfrac{1}{2}\sin 2x\,dx\right) \quad \begin{bmatrix} \text{parts with } u = x, \\ dv = \cos 2x\,dx \end{bmatrix}$$

$$= \pi\left[\tfrac{1}{2}\left(\tfrac{\pi}{2}\right)^2 + 0 - \tfrac{1}{2}\left[-\tfrac{1}{2}\cos 2x\right]_0^{\pi/2}\right] = \tfrac{\pi^3}{8} + \tfrac{\pi}{4}(-1 - 1) = \tfrac{1}{8}\left(\pi^3 - 4\pi\right)$$

77. By the Fundamental Theorem of Calculus,

$$\int_0^\infty f'(x)\,dx = \lim_{t\to\infty}\int_0^t f'(x)\,dx = \lim_{t\to\infty}[f(t) - f(0)] = \lim_{t\to\infty}f(t) - f(0) = 0 - f(0) = -f(0).$$

78. (a) $(\tan^{-1}x)_{\text{ave}} = \lim_{t\to\infty}\dfrac{1}{t-0}\int_0^t \tan^{-1}x\,dx \overset{89}{=} \lim_{t\to\infty}\left\{\dfrac{1}{t}\left[x\tan^{-1}x - \tfrac{1}{2}\ln(1+x^2)\right]_0^t\right\}$

$$= \lim_{t\to\infty}\left[\dfrac{1}{t}\left(t\tan^{-1}t - \tfrac{1}{2}\ln(1+t^2)\right)\right] = \lim_{t\to\infty}\left[\tan^{-1}t - \dfrac{\ln(1+t^2)}{2t}\right]$$

$$\overset{\text{H}}{=} \dfrac{\pi}{2} - \lim_{t\to\infty}\dfrac{2t/(1+t^2)}{2} = \dfrac{\pi}{2} - 0 = \dfrac{\pi}{2}$$

(b) $f(x) \geq 0$ and $\int_a^\infty f(x)\,dx$ is divergent $\Rightarrow$ $\lim_{t\to\infty}\int_a^t f(x)\,dx = \infty$.

$$f_{\text{ave}} = \lim_{t\to\infty}\dfrac{\int_a^t f(x)\,dx}{t-a}\,dx \overset{\text{H}}{=} \lim_{t\to\infty}\dfrac{f(t)}{1} \quad [\text{by FTC1}] \quad = \lim_{x\to\infty}f(x), \text{ if this limit exists.}$$

(c) Suppose $\int_a^\infty f(x)\,dx$ converges; that is, $\lim_{t\to\infty}\int_a^t f(x)\,dx = L < \infty$. Then

$$f_{\text{ave}} = \lim_{t\to\infty}\left[\dfrac{1}{t-a}\int_a^t f(x)\,dx\right] = \lim_{t\to\infty}\dfrac{1}{t-a}\cdot\lim_{t\to\infty}\int_a^t f(x)\,dx = 0\cdot L = 0.$$

(d) $(\sin x)_{\text{ave}} = \lim_{t\to\infty}\dfrac{1}{t}\int_0^t \sin x\,dx = \lim_{t\to\infty}\left(\dfrac{1}{t}\left[-\cos x\right]_0^t\right) = \lim_{t\to\infty}\left(-\dfrac{\cos t}{t} + \dfrac{1}{t}\right) = \lim_{t\to\infty}\dfrac{1 - \cos t}{t} = 0$

79. Let $u = 1/x$ $\Rightarrow$ $x = 1/u$ $\Rightarrow$ $dx = -(1/u^2)\,du$.

$$\int_0^\infty \dfrac{\ln x}{1+x^2}\,dx = \int_\infty^0 \dfrac{\ln(1/u)}{1+1/u^2}\left(-\dfrac{du}{u^2}\right) = \int_\infty^0 \dfrac{-\ln u}{u^2+1}(-du) = \int_\infty^0 \dfrac{\ln u}{1+u^2}\,du = -\int_0^\infty \dfrac{\ln u}{1+u^2}\,du$$

Therefore, $\displaystyle\int_0^\infty \dfrac{\ln x}{1+x^2}\,dx = -\int_0^\infty \dfrac{\ln x}{1+x^2}\,dx = 0.$

80. If the distance between P and the point charge is d, then the potential V at P is

$$V = W = \int_\infty^d F\,dr = \int_\infty^d \dfrac{q}{4\pi\varepsilon_0 r^2}\,dr = \lim_{t\to\infty}\dfrac{q}{4\pi\varepsilon_0}\left[-\dfrac{1}{r}\right]_t^d = \dfrac{q}{4\pi\varepsilon_0}\lim_{t\to\infty}\left(-\dfrac{1}{d} + \dfrac{1}{t}\right) = -\dfrac{q}{4\pi\varepsilon_0 d}.$$

☐ PROBLEMS PLUS

1.

 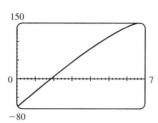

By symmetry, the problem can be reduced to finding the line $x = c$ such that the shaded area is one-third of the area of the

quarter-circle. An equation of the semicircle is $y = \sqrt{49 - x^2}$, so we require that $\int_0^c \sqrt{49 - x^2}\, dx = \frac{1}{3} \cdot \frac{1}{4}\pi(7)^2 \quad \Leftrightarrow$

$\left[\frac{1}{2}x\sqrt{49 - x^2} + \frac{49}{2}\sin^{-1}(x/7)\right]_0^c = \frac{49}{12}\pi$ [by Formula 30] $\Leftrightarrow$ $\frac{1}{2}c\sqrt{49 - c^2} + \frac{49}{2}\sin^{-1}(c/7) = \frac{49}{12}\pi$.

This equation would be difficult to solve exactly, so we plot the left-hand side as a function of c, and find that the equation

holds for $c \approx 1.85$. So the cuts should be made at distances of about 1.85 inches from the center of the pizza.

2. $\displaystyle\int \frac{1}{x^7 - x}\, dx = \int \frac{dx}{x(x^6 - 1)} = \int \frac{x^5}{x^6(x^6 - 1)}\, dx = \frac{1}{6}\int \frac{1}{u(u - 1)}\, du \quad \begin{bmatrix} u = x^6, \\ du = 6x^5\, dx \end{bmatrix}$

$\qquad = \frac{1}{6}\int \left(\frac{1}{u - 1} - \frac{1}{u}\right) du = \frac{1}{6}(\ln|u - 1| - \ln|u|) + C$

$\qquad = \frac{1}{6}\ln\left|\frac{u - 1}{u}\right| + C = \frac{1}{6}\ln\left|\frac{x^6 - 1}{x^6}\right| + C$

Alternate method:

$\qquad\displaystyle\int \frac{1}{x^7 - x}\, dx = \int \frac{x^{-7}}{1 - x^{-6}}\, dx \quad \begin{bmatrix} u = 1 - x^{-6}, \\ du = 6x^{-7}\, dx \end{bmatrix} = \frac{1}{6}\int du/u = \frac{1}{6}\ln|u| + C = \frac{1}{6}\ln\left|1 - x^{-6}\right| + C$

Other methods: Substitute $u = x^3$ or $x^3 = \sec\theta$.

3. The given integral represents the difference of the shaded areas, which appears to

be 0. It can be calculated by integrating with respect to either x or y, so we find x

in terms of y for each curve: $y = \sqrt[3]{1 - x^7} \quad \Rightarrow \quad x = \sqrt[7]{1 - y^3}$ and

$y = \sqrt[7]{1 - x^3} \quad \Rightarrow \quad x = \sqrt[3]{1 - y^7}$, so

$\int_0^1 \left(\sqrt[3]{1 - y^7} - \sqrt[7]{1 - y^3}\right) dy = \int_0^1 \left(\sqrt[7]{1 - x^3} - \sqrt[3]{1 - x^7}\right) dx$. But this

equation is of the form $z = -z$. So $\int_0^1 \left(\sqrt[3]{1 - x^7} - \sqrt[7]{1 - x^3}\right) dx = 0$.

4. The area of each circle is $\pi(1)^2 = \pi$. By symmetry, the area of the union of the two disks is $A = \pi + \pi - 4I$.

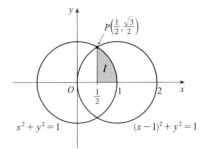

$$I = \int_{1/2}^{1} \sqrt{1 - x^2}\, dx$$

$$\overset{30}{=} \left[\frac{x}{2}\sqrt{1 - x^2} + \frac{1}{2}\sin^{-1}\left(\frac{x}{1}\right) \right]_{1/2}^{1} \qquad \text{[or substitute } x = \sin\theta]$$

$$= \left(0 + \frac{\pi}{4}\right) - \left(\frac{1}{4}\frac{\sqrt{3}}{2} + \frac{1}{2}\frac{\pi}{6}\right) = \frac{\pi}{4} - \frac{\sqrt{3}}{8} - \frac{\pi}{12} = \frac{\pi}{6} - \frac{\sqrt{3}}{8}$$

Thus, $A = 2\pi - 4\left(\frac{\pi}{6} - \frac{\sqrt{3}}{8}\right) = 2\pi - \frac{2\pi}{3} + \frac{\sqrt{3}}{2} = \frac{4\pi}{3} + \frac{\sqrt{3}}{2}$.

Alternate solution (no calculus): The area of the sector, with central angle at the origin, containing I is

$\frac{1}{2}r^2\theta = \frac{1}{2}(1)^2\left(\frac{\pi}{3}\right) = \frac{\pi}{6}$. The area of the triangle with hypotenuse OP is $\frac{1}{2}\left(\frac{1}{2}\right)\left(\frac{\sqrt{3}}{2}\right) = \frac{\sqrt{3}}{8}$.

Thus, the area of I is $\frac{\pi}{6} - \frac{\sqrt{3}}{8}$, as calculated above.

5. The area A of the remaining part of the circle is given by

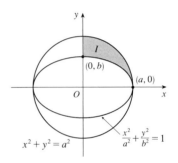

$$A = 4I = 4\int_{0}^{a}\left(\sqrt{a^2 - x^2} - \frac{b}{a}\sqrt{a^2 - x^2}\right)dx = 4\left(1 - \frac{b}{a}\right)\int_{0}^{a}\sqrt{a^2 - x^2}\,dx$$

$$\overset{30}{=} \frac{4}{a}(a - b)\left[\frac{x}{2}\sqrt{a^2 - x^2} + \frac{a^2}{2}\sin^{-1}\frac{x}{a}\right]_{0}^{a}$$

$$= \frac{4}{a}(a - b)\left[\left(0 + \frac{a^2}{2}\frac{\pi}{2}\right) - 0\right] = \frac{4}{a}(a - b)\left(\frac{a^2\pi}{4}\right) = \pi a(a - b),$$

which is the area of an ellipse with semiaxes a and $a - b$.

Alternate solution: Subtracting the area of the ellipse from the area of the circle gives us $\pi a^2 - \pi ab = \pi a\,(a - b)$, as calculated above. (The formula for the area of an ellipse was derived in Example 2 in Section 8.3.)

6. (a) The tangent to the curve $y = f(x)$ at $x = x_0$ has the equation $y - f(x_0) = f'(x_0)(x - x_0)$. The y-intercept of this tangent line is $f(x_0) - f'(x_0)x_0$. Thus, L is the distance from the point $(0, f(x_0) - f'(x_0)x_0)$ to

the point $(x_0, f(x_0))$; that is, $L^2 = x_0^2 + [f'(x_0)]^2 x_0^2$, so $[f'(x_0)]^2 = \dfrac{L^2 - x_0^2}{x_0^2}$ and $f'(x_0) = -\dfrac{\sqrt{L^2 - x_0^2}}{x_0}$

for $0 < x_0 < L$.

(b) $\dfrac{dy}{dx} = -\dfrac{\sqrt{L^2 - x^2}}{x} \quad \Rightarrow \quad y = \int\left(-\dfrac{\sqrt{L^2 - x^2}}{x}\right)dx.$

Let $x = L\sin\theta$. Then $dx = L\cos\theta\,d\theta$ and

$$y = \int \frac{-L\cos\theta\,L\cos\theta\,d\theta}{L\sin\theta} = L\int\frac{\sin^2\theta - 1}{\sin\theta}\,d\theta = L\int(\sin\theta - \csc\theta)\,d\theta$$

$$= -L\cos\theta - L\ln|\csc\theta - \cot\theta| + C = -\sqrt{L^2 - x^2} - L\ln\left(\frac{L}{x} - \frac{\sqrt{L^2 - x^2}}{x}\right) + C$$

When $x = L$, $y = 0$, and $0 = -0 - L\ln(1 - 0) + C$, so $C = 0$. Therefore, $y = -\sqrt{L^2 - x^2} - L\ln\left(\dfrac{L - \sqrt{L^2 - x^2}}{x}\right)$.

7. Recall that $\cos A \cos B = \frac{1}{2}[\cos(A + B) + \cos(A - B)]$. So

$$f(x) = \int_0^\pi \cos t \cos(x - t)\, dt = \frac{1}{2}\int_0^\pi [\cos(t + x - t) + \cos(t - x + t)]\, dt = \frac{1}{2}\int_0^\pi [\cos x + \cos(2t - x)]\, dt$$

$$= \frac{1}{2}\left[t\cos x + \frac{1}{2}\sin(2t - x)\right]_0^\pi = \frac{\pi}{2}\cos x + \frac{1}{4}\sin(2\pi - x) - \frac{1}{4}\sin(-x)$$

$$= \frac{\pi}{2}\cos x + \frac{1}{4}\sin(-x) - \frac{1}{4}\sin(-x) = \frac{\pi}{2}\cos x$$

The minimum of $\cos x$ on this domain is -1, so the minimum value of $f(x)$ is $f(\pi) = -\frac{\pi}{2}$.

8. n is a positive integer, so

$$\int (\ln x)^n\, dx = x(\ln x)^n - \int x \cdot n(\ln x)^{n-1}\,(dx/x) \quad \text{[by parts]} \;=\; x(\ln x)^n - n\int(\ln x)^{n-1}\, dx$$

Thus,
$$\int_0^1 (\ln x)^n\, dx = \lim_{t \to 0^+} \int_t^1 (\ln x)^n\, dx = \lim_{t \to 0^+} [x(\ln x)^n]_t^1 - n \lim_{t \to 0^+} \int_t^1 (\ln x)^{n-1}\, dx$$

$$= -\lim_{t \to 0^+} \frac{(\ln t)^n}{1/t} - n\int_0^1 (\ln x)^{n-1}\, dx = -n\int_0^1 (\ln x)^{n-1}\, dx$$

by repeated application of l'Hospital's Rule. We want to prove that $\int_0^1 (\ln x)^n\, dx = (-1)^n n!$ for every positive integer n. For $n = 1$, we have

$$\int_0^1 (\ln x)^1\, dx = (-1)\int_0^1 (\ln x)^0\, dx = -\int_0^1 dx = -1 \qquad \left[\text{or } \int_0^1 \ln x\, dx = \lim_{t \to 0^+} [x\ln x - x]_t^1 = -1\right]$$

Assuming that the formula holds for n, we find that

$$\int_0^1 (\ln x)^{n+1}\, dx = -(n + 1)\int_0^1 (\ln x)^n\, dx = -(n + 1)(-1)^n n! = (-1)^{n+1}(n + 1)!$$

This is the formula for $n + 1$. Thus, the formula holds for all positive integers n by induction.

9. In accordance with the hint, we let $I_k = \int_0^1 (1 - x^2)^k\, dx$, and we find an expression for I_{k+1} in terms of I_k. We integrate I_{k+1} by parts with $u = (1 - x^2)^{k+1} \;\Rightarrow\; du = (k + 1)(1 - x^2)^k(-2x),\; dv = dx \;\Rightarrow\; v = x$, and then split the remaining integral into identifiable quantities:

$$I_{k+1} = x(1 - x^2)^{k+1}\Big|_0^1 + 2(k + 1)\int_0^1 x^2(1 - x^2)^k\, dx = (2k + 2)\int_0^1 (1 - x^2)^k[1 - (1 - x^2)]\, dx$$

$$= (2k + 2)(I_k - I_{k+1})$$

So $I_{k+1}[1 + (2k + 2)] = (2k + 2)I_k \;\Rightarrow\; I_{k+1} = \dfrac{2k + 2}{2k + 3}I_k$. Now to complete the proof, we use induction:

$I_0 = 1 = \dfrac{2^0 (0!)^2}{1!}$, so the formula holds for $n = 0$. Now suppose it holds for $n = k$. Then

$$I_{k+1} = \frac{2k + 2}{2k + 3}I_k = \frac{2k + 2}{2k + 3}\left[\frac{2^{2k}(k!)^2}{(2k + 1)!}\right] = \frac{2(k + 1)2^{2k}(k!)^2}{(2k + 3)(2k + 1)!} = \frac{2(k + 1)}{2k + 2} \cdot \frac{2(k + 1)2^{2k}(k!)^2}{(2k + 3)(2k + 1)!}$$

$$= \frac{[2(k + 1)]^2\, 2^{2k}(k!)^2}{(2k + 3)(2k + 2)(2k + 1)!} = \frac{2^{2(k+1)}\,[(k + 1)!]^2}{[2(k + 1) + 1]!}$$

So by induction, the formula holds for all integers $n \geq 0$.

10. (a) Since $-1 \le \sin \le 1$, we have $-f(x) \le f(x) \sin nx \le f(x)$, and

the graph of $y = f(x) \sin nx$ oscillates between $f(x)$ and $-f(x)$.

(The diagram shows the case $f(x) = e^x$ and $n = 10$.) As $n \to \infty$,

the graph oscillates more and more frequently; see the graphs in

part (b).

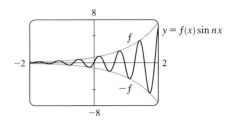

(b) From the graphs of the integrand, it seems that

$\lim_{n \to \infty} \int_0^1 f(x) \sin nx \, dx = 0$, since as n increases,

the integrand oscillates more and more rapidly, and

thus (since f' is continuous) it makes sense that the

areas above the x-axis and below it during each

oscillation approach equality.

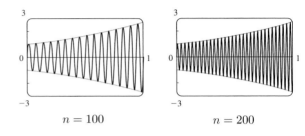

$n = 100$ $\qquad\qquad n = 200$

(c) We integrate by parts with $u = f(x) \;\Rightarrow\; du = f'(x)\,dx$, $dv = \sin nx\,dx \;\Rightarrow\; v = -\dfrac{\cos nx}{n}$:

$$\int_0^1 f(x) \sin nx \, dx = \left[-\frac{f(x) \cos nx}{n} \right]_0^1 + \int_0^1 \frac{\cos nx}{n} f'(x) \, dx = \frac{1}{n} \left(\int_0^1 \cos nx f'(x) \, dx - \left[f(x) \cos nx \right]_0^1 \right)$$

$$= \frac{1}{n} \left[\int_0^1 \cos nx f'(x) \, dx + f(0) - f(1) \cos n \right]$$

Taking absolute values of the first and last terms in this equality, and using the facts that $|\alpha \pm \beta| \le |\alpha| + |\beta|$,

$\int_0^1 f(x)\,dx \le \int_0^1 |f(x)|\,dx$, $|f(0)| = f(0)$ [f is positive], $|f'(x)| \le M$ for $0 \le x \le 1$, and $|\cos nx| \le 1$,

$$\left| \int_0^1 f(x) \sin nx \, dx \right| \le \frac{1}{n} \left[\left| \int_0^1 M \, dx \right| + |f(0)| + |f(1)| \right] = \frac{1}{n} \left[M + |f(0)| + |f(1)| \right]$$

which approaches 0 as $n \to \infty$. The result follows by the Squeeze Theorem.

11. $0 < a < b$. Now

$$\int_0^1 [bx + a(1-x)]^t \, dx = \int_a^b \frac{u^t}{(b-a)} \, du \quad [u = bx + a(1-x)] \quad = \left[\frac{u^{t+1}}{(t+1)(b-a)} \right]_a^b = \frac{b^{t+1} - a^{t+1}}{(t+1)(b-a)}.$$

Now let $y = \lim_{t \to 0} \left[\dfrac{b^{t+1} - a^{t+1}}{(t+1)(b-a)} \right]^{1/t}$. Then $\ln y = \lim_{t \to 0} \left[\dfrac{1}{t} \ln \dfrac{b^{t+1} - a^{t+1}}{(t+1)(b-a)} \right]$. This limit is of the form $0/0$,

so we can apply l'Hospital's Rule to get

$$\ln y = \lim_{t \to 0} \left[\frac{b^{t+1} \ln b - a^{t+1} \ln a}{b^{t+1} - a^{t+1}} - \frac{1}{t+1} \right] = \frac{b \ln b - a \ln a}{b - a} - 1 = \frac{b \ln b}{b - a} - \frac{a \ln a}{b - a} - \ln e = \ln \frac{b^{b/(b-a)}}{e a^{a/(b-a)}}.$$

Therefore, $y = e^{-1} \left(\dfrac{b^b}{a^a} \right)^{1/(b-a)}$.

12.

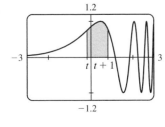

From the graph, it appears that the area under the graph of $f(x) = \sin(e^x)$ on the

interval $[t, t+1]$ is greatest when $t \approx -0.2$. To find the exact value, we write the

integral as $I = \int_t^{t+1} f(x) \, dx = \int_0^{t+1} f(x) \, dx - \int_0^t f(x) \, dx$, and use FTC1 to find

$dI/dt = f(t+1) - f(t) = \sin(e^{t+1}) - \sin(e^t) = 0$ when $\sin(e^{t+1}) = \sin(e^t)$.

Now we have $\sin x = \sin y$ whenever $x - y = 2k\pi$ and also whenever x and y are the same distance from $\left(k + \frac{1}{2}\right)\pi$, k any integer, since $\sin x$ is symmetric about the line $x = \left(k + \frac{1}{2}\right)\pi$. The first possibility is the more obvious one, but if we calculate $e^{t+1} - e^t = 2k\pi$, we get $t = \ln(2k\pi/(e - 1))$, which is about 1.3 for $k = 1$ (the least possible value of k). From the graph, this looks unlikely to give the maximum we are looking for. So instead we set $e^{t+1} - \left(k + \frac{1}{2}\right)\pi = \left(k + \frac{1}{2}\right)\pi - e^t$ ⇔ $e^{t+1} + e^t = (2k+1)\pi$ ⇔ $e^t(e + 1) = (2k+1)\pi$ ⇔ $t = \ln((2k+1)\pi/(e+1))$. Now $k = 0$ ⇒ $t = \ln(\pi/(e+1)) \approx -0.16853$, which does give the maximum value, as we have seen from the graph of f.

13.

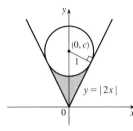

An equation of the circle with center $(0, c)$ and radius 1 is $x^2 + (y - c)^2 = 1^2$, so an equation of the lower semicircle is $y = c - \sqrt{1 - x^2}$. At the points of tangency, the slopes of the line and semicircle must be equal. For $x \geq 0$, we must have

$$y' = 2 \;\Rightarrow\; \frac{x}{\sqrt{1 - x^2}} = 2 \;\Rightarrow\; x = 2\sqrt{1 - x^2} \;\Rightarrow\; x^2 = 4(1 - x^2) \;\Rightarrow$$

$$5x^2 = 4 \;\Rightarrow\; x^2 = \tfrac{4}{5} \;\Rightarrow\; x = \tfrac{2}{5}\sqrt{5} \text{ and so } y = 2\left(\tfrac{2}{5}\sqrt{5}\right) = \tfrac{4}{5}\sqrt{5}.$$

The slope of the perpendicular line segment is $-\frac{1}{2}$, so an equation of the line segment is $y - \frac{4}{5}\sqrt{5} = -\frac{1}{2}\left(x - \frac{2}{5}\sqrt{5}\right)$ ⇔ $y = -\frac{1}{2}x + \frac{1}{5}\sqrt{5} + \frac{4}{5}\sqrt{5}$ ⇔ $y = -\frac{1}{2}x + \sqrt{5}$, so $c = \sqrt{5}$ and an equation of the lower semicircle is $y = \sqrt{5} - \sqrt{1 - x^2}$. Thus, the shaded area is

$$2\int_0^{(2/5)\sqrt{5}} \left[\left(\sqrt{5} - \sqrt{1 - x^2}\right) - 2x\right] dx \overset{30}{=} 2\left[\sqrt{5}\,x - \frac{x}{2}\sqrt{1 - x^2} - \frac{1}{2}\sin^{-1} x - x^2\right]_0^{(2/5)\sqrt{5}}$$

$$= 2\left[2 - \frac{\sqrt{5}}{5}\cdot\frac{1}{\sqrt{5}} - \frac{1}{2}\sin^{-1}\left(\frac{2}{\sqrt{5}}\right) - \frac{4}{5}\right] - 2(0)$$

$$= 2\left[1 - \frac{1}{2}\sin^{-1}\left(\frac{2}{\sqrt{5}}\right)\right] = 2 - \sin^{-1}\left(\frac{2}{\sqrt{5}}\right)$$

14. (a) $M\dfrac{dv}{dt} - ub = -Mg$ ⇒ $(M_0 - bt)\dfrac{dv}{dt} = ub - (M_0 - bt)g$ ⇒ $\dfrac{dv}{dt} = \dfrac{ub}{M_0 - bt} - g$ ⇒

$v(t) = -u\ln(M_0 - bt) - gt + C$. Now $0 = v(0) = -u\ln M_0 + C$, so $C = u\ln M_0$. Thus

$$v(t) = u\ln M_0 - u\ln(M_0 - bt) - gt = u\ln\frac{M_0}{M_0 - bt} - gt.$$

(b) Burnout velocity $= v\left(\dfrac{M_2}{b}\right) = u\ln\dfrac{M_0}{M_0 - M_2} - g\dfrac{M_2}{b} = u\ln\dfrac{M_0}{M_1} - g\dfrac{M_2}{b}$.

Note: The reason for the term "burnout velocity" is that M_2 kilograms of fuel is used in M_2/b seconds, so $v(M_2/b)$ is the rocket's velocity when the fuel is used up.

(c) Height at burnout time $= y\left(\dfrac{M_2}{b}\right)$. Now $\dfrac{dy}{dt} = v(t) = u\ln M_0 - gt - u\ln(M_0 - bt)$, so

$$y(t) = (u\ln M_0)t - \frac{gt^2}{2} - \frac{u}{b}(M_0 - bt)\ln(M_0 - bt) + ut + C.\ \text{Since } 0 = y(0) = \frac{u}{b}M_0\ln M_0 + C,\ \text{we get}$$

$$C = -\frac{u}{b}M_0\ln M_0 \text{ and } y(t) = u(1 + \ln M_0)t - \frac{gt^2}{2} + \frac{u}{b}(M_0 - bt)\ln(M_0 - bt) - \frac{u}{b}M_0\ln M_0.$$

Therefore, the height at burnout is

$$y\left(\frac{M_2}{b}\right) = u(1 + \ln M_0)\frac{M_2}{b} - \frac{g}{2}\left(\frac{M_2}{b}\right)^2 + \frac{u}{b}M_1 \ln M_1 - \frac{u}{b}M_0 \ln M_0$$

$$= \frac{u}{b}M_2 - \frac{u}{b}M_1 \ln M_0 + \frac{u}{b}M_1 \ln M_1 - \frac{g}{2}\left(\frac{M_2}{b}\right)^2 = \frac{u}{b}M_2 + \frac{u}{b}M_1 \ln \frac{M_1}{M_0} - \frac{g}{2}\left(\frac{M_2}{b}\right)^2$$

[In the calculation of $y(M_2/b)$, repeated use was made of the relation $M_0 = M_1 + M_2$. In particular,

$t = M_2/b \ \Rightarrow \ M_0 - bt = M_1$.]

(d) The formula for $y(t)$ in part (c) holds while there is still fuel. Once the fuel is used up, gravity is the only force

acting on the rocket. $-M_1g = M_1\dfrac{dv}{dt} \ \Rightarrow \ \dfrac{dv}{dt} = -g \ \Rightarrow \ v(t) = -gt + c_1$, where $c_1 = v\left(\dfrac{M_2}{b}\right) + \dfrac{gM_2}{b} \ \Rightarrow$

$v(t) = v\left(\dfrac{M_2}{b}\right) - g\left(t - \dfrac{M_2}{b}\right) \ \Rightarrow \ y(t) = v\left(\dfrac{M_2}{b}\right)\left(t - \dfrac{M_2}{b}\right) - \dfrac{g}{2}\left(t - \dfrac{M_2}{b}\right)^2 + c_2$, where $c_2 = y\left(\dfrac{M_2}{b}\right)$,

so $y(t) = y\left(\dfrac{M_2}{b}\right) + v\left(\dfrac{M_2}{b}\right)\left(t - \dfrac{M_2}{b}\right) - \dfrac{g}{2}\left(t - \dfrac{M_2}{b}\right)^2, t \geq \dfrac{M_2}{b}$.

To summarize: For $0 \leq t \leq \dfrac{M_2}{b}$, $y(t) = u(1 + \ln M_0)t - \dfrac{gt^2}{2} + \dfrac{u}{b}(M_0 - bt)\ln(M_0 - bt) - \dfrac{u}{b}M_0 \ln M_0$

[from part (c)], and for $t \geq \dfrac{M_2}{b}$, $y(t) = y\left(\dfrac{M_2}{b}\right) + v\left(\dfrac{M_2}{b}\right)\left(t - \dfrac{M_2}{b}\right) - \dfrac{g}{2}\left(t - \dfrac{M_2}{b}\right)^2$ [from above].

$y\left(\dfrac{M_2}{b}\right)$ and $v\left(\dfrac{M_2}{b}\right)$ are given in parts (c) and (b), respectively.

15. We integrate by parts with $u = \dfrac{1}{\ln(1 + x + t)}$, $dv = \sin t\,dt$, so $du = \dfrac{-1}{(1 + x + t)[\ln(1 + x + t)]^2}$ and $v = -\cos t$. The

integral becomes

$$I = \int_0^\infty \frac{\sin t\,dt}{\ln(1 + x + t)} = \lim_{b \to \infty}\left(\left[\frac{-\cos t}{\ln(1 + x + t)}\right]_0^b - \int_0^b \frac{\cos t\,dt}{(1 + x + t)[\ln(1 + x + t)]^2}\right)$$

$$= \lim_{b \to \infty}\frac{-\cos b}{\ln(1 + x + b)} + \frac{1}{\ln(1 + x)} + \int_0^\infty \frac{-\cos t\,dt}{(1 + x + t)[\ln(1 + x + t)]^2} = \frac{1}{\ln(1 + x)} + J$$

where $J = \int_0^\infty \dfrac{-\cos t\,dt}{(1 + x + t)[\ln(1 + x + t)]^2}$. Now $-1 \leq -\cos t \leq 1$ for all t; in fact, the inequality is strict except

at isolated points. So $-\int_0^\infty \dfrac{dt}{(1 + x + t)[\ln(1 + x + t)]^2} < J < \int_0^\infty \dfrac{dt}{(1 + x + t)[\ln(1 + x + t)]^2} \ \Leftrightarrow$

$-\dfrac{1}{\ln(1 + x)} < J < \dfrac{1}{\ln(1 + x)} \ \Leftrightarrow \ 0 < I < \dfrac{2}{\ln(1 + x)}$.

16. Integration by parts twice, first with $u = f(x)$ and $dv = dx$, then with $u = f'(x)$ and $dv = x\,dx$, gives

$\int_0^1 f(x)\,dx = \frac{1}{2}\int_0^1 x^2 f''(x)\,dx$. So

$$\left|\int_0^1 f(x)\,dx\right| = \frac{1}{2}\left|\int_0^1 x^2 f''(x)\,dx\right| \leq \frac{1}{2}\int_0^1 x^2 |f''(x)|\,dx \leq \frac{3}{2}\int_0^1 x^2\,dx = \frac{1}{2}$$

9 □ FURTHER APPLICATIONS OF INTEGRATION

9.1 Arc Length

1. $y = 2x - 5 \implies L = \int_{-1}^{3} \sqrt{1 + (dy/dx)^2}\, dx = \int_{-1}^{3} \sqrt{1 + (2)^2}\, dx = \sqrt{5}\,[3 - (-1)] = 4\sqrt{5}.$

The arc length can be calculated using the distance formula, since the curve is a line segment, so

$L = [\text{distance from } (-1, -7) \text{ to } (3, 1)] = \sqrt{[3 - (-1)]^2 + [1 - (-7)]^2} = \sqrt{80} = 4\sqrt{5}$

2. Using the arc length formula with $y = \sqrt{2 - x^2} \implies \dfrac{dy}{dx} = -\dfrac{x}{\sqrt{2 - x^2}}$, we get

$$L = \int_0^1 \sqrt{1 + \left(\frac{dy}{dx}\right)^2}\, dx = \int_0^1 \sqrt{1 + \frac{x^2}{2 - x^2}}\, dx = \int_0^1 \frac{\sqrt{2}\, dx}{\sqrt{2 - x^2}} = \sqrt{2} \int_0^1 \frac{dx}{\sqrt{\left(\sqrt{2}\right)^2 - x^2}}$$

$$= \sqrt{2}\left[\sin^{-1}\left(\frac{x}{\sqrt{2}}\right)\right]_0^1 = \sqrt{2}\left[\sin^{-1}\left(\frac{1}{\sqrt{2}}\right) - \sin^{-1} 0\right] = \sqrt{2}\left[\tfrac{\pi}{4} - 0\right] = \sqrt{2}\,\tfrac{\pi}{4}$$

The curve is a one-eighth of a circle with radius $\sqrt{2}$, so the length of the arc is $\frac{1}{8}\left(2\pi \cdot \sqrt{2}\right) = \sqrt{2}\,\frac{\pi}{4}$, as above.

3. $y = \cos x \implies dy/dx = -\sin x \implies 1 + (dy/dx)^2 = 1 + \sin^2 x$. So $L = \int_0^{2\pi} \sqrt{1 + \sin^2 x}\, dx$.

4. $y = xe^{-x^2} \implies dy/dx = xe^{-x^2}(-2x) + e^{-x^2} \cdot 1 = e^{-x^2}(1 - 2x^2) \implies 1 + (dy/dx)^2 = 1 + (1 - 2x^2)^2 e^{-2x^2}.$

So $L = \int_0^1 \sqrt{1 + (1 - 2x^2)^2\, e^{-2x^2}}\, dx.$

5. $x = y + y^3 \implies dx/dy = 1 + 3y^2 \implies 1 + (dx/dy)^2 = 1 + (1 + 3y^2)^2 = 9y^4 + 6y^2 + 2.$

So $L = \int_1^4 \sqrt{9y^4 + 6y^2 + 2}\, dy.$

6. $\dfrac{x^2}{a^2} + \dfrac{y^2}{b^2} = 1$, $y = \pm b\sqrt{1 - x^2/a^2} = \pm\dfrac{b}{a}\sqrt{a^2 - x^2}$ [assume $a > 0$]. $y = \dfrac{b}{a}\sqrt{a^2 - x^2} \implies \dfrac{dy}{dx} = \dfrac{-bx}{a\sqrt{a^2 - x^2}} \implies$

$$\left(\frac{dy}{dx}\right)^2 = \frac{b^2 x^2}{a^2(a^2 - x^2)}.\ \text{So}\ L = 2\int_{-a}^{a}\left[1 + \frac{b^2 x^2}{a^2(a^2 - x^2)}\right]^{1/2} dx = \frac{4}{a}\int_0^a \left[\frac{(b^2 - a^2)x^2 + a^4}{a^2 - x^2}\right]^{1/2} dx.$$

7. $y = 1 + 6x^{3/2} \implies dy/dx = 9x^{1/2} \implies 1 + (dy/dx)^2 = 1 + 81x$. So

$L = \int_0^1 \sqrt{1 + 81x}\, dx = \int_1^{82} u^{1/2}\left(\frac{1}{81}\, du\right) \begin{bmatrix} u = 1 + 81x, \\ du = 81\, dx \end{bmatrix} = \frac{1}{81} \cdot \frac{2}{3}\left[u^{3/2}\right]_1^{82} = \frac{2}{243}\left(82\sqrt{82} - 1\right)$

8. $y^2 = 4(x + 4)^3$, $y > 0 \implies y = 2(x + 4)^{3/2} \implies dy/dx = 3(x + 4)^{1/2} \implies$

$1 + (dy/dx)^2 = 1 + 9(x + 4) = 9x + 37$. So

$L = \int_0^2 \sqrt{9x + 37}\, dx \begin{bmatrix} u = 9x + 37, \\ du = 9\, dx \end{bmatrix} = \int_{37}^{55} u^{1/2}\left(\frac{1}{9}\, du\right) = \frac{1}{9} \cdot \frac{2}{3}\left[u^{3/2}\right]_{37}^{55} = \frac{2}{27}\left(55\sqrt{55} - 37\sqrt{37}\right).$

9. $y = \dfrac{x^5}{6} + \dfrac{1}{10x^3} \quad \Rightarrow \quad \dfrac{dy}{dx} = \dfrac{5}{6}x^4 - \dfrac{3}{10}x^{-4} \quad \Rightarrow$

$1 + (dy/dx)^2 = 1 + \frac{25}{36}x^8 - \frac{1}{2} + \frac{9}{100}x^{-8} = \frac{25}{36}x^8 + \frac{1}{2} + \frac{9}{100}x^{-8} = \left(\frac{5}{6}x^4 + \frac{3}{10}x^{-4}\right)^2$. So

$L = \int_1^2 \sqrt{\left(\frac{5}{6}x^4 + \frac{3}{10}x^{-4}\right)^2}\,dx = \int_1^2 \left(\frac{5}{6}x^4 + \frac{3}{10}x^{-4}\right)dx = \left[\frac{1}{6}x^5 - \frac{1}{10}x^{-3}\right]_1^2 = \left(\frac{32}{6} - \frac{1}{80}\right) - \left(\frac{1}{6} - \frac{1}{10}\right)$

$\qquad = \frac{31}{6} + \frac{7}{80} = \frac{1261}{240}$

10. $x = \dfrac{y^4}{8} + \dfrac{1}{4y^2} \quad \Rightarrow \quad \dfrac{dx}{dy} = \frac{1}{2}y^3 - \frac{1}{2}y^{-3} \quad \Rightarrow$

$1 + (dx/dy)^2 = 1 + \frac{1}{4}y^6 - \frac{1}{2} + \frac{1}{4}y^{-6} = \frac{1}{4}y^6 + \frac{1}{2} + \frac{1}{4}y^{-6} = \left(\frac{1}{2}y^3 + \frac{1}{2}y^{-3}\right)^2$. So

$L = \int_1^2 \sqrt{\left(\frac{1}{2}y^3 + \frac{1}{2}y^{-3}\right)^2}\,dy = \int_1^2 \left(\frac{1}{2}y^3 + \frac{1}{2}y^{-3}\right)dy = \left[\frac{1}{8}y^4 - \frac{1}{4}y^{-2}\right]_1^2 = \left(2 - \frac{1}{16}\right) - \left(\frac{1}{8} - \frac{1}{4}\right)$

$\qquad = 2 + \frac{1}{16} = \frac{33}{16}$.

11. $x = \frac{1}{3}\sqrt{y}\,(y-3) = \frac{1}{3}y^{3/2} - y^{1/2} \quad \Rightarrow \quad dx/dy = \frac{1}{2}y^{1/2} - \frac{1}{2}y^{-1/2} \quad \Rightarrow$

$1 + (dx/dy)^2 = 1 + \frac{1}{4}y - \frac{1}{2} + \frac{1}{4}y^{-1} = \frac{1}{4}y + \frac{1}{2} + \frac{1}{4}y^{-1} = \left(\frac{1}{2}y^{1/2} + \frac{1}{2}y^{-1/2}\right)^2$. So

$L = \int_1^9 \left(\frac{1}{2}y^{1/2} + \frac{1}{2}y^{-1/2}\right)dy = \frac{1}{2}\left[\frac{2}{3}y^{3/2} + 2y^{1/2}\right]_1^9 = \frac{1}{2}\left[\left(\frac{2}{3}\cdot 27 + 2\cdot 3\right) - \left(\frac{2}{3}\cdot 1 + 2\cdot 1\right)\right]$

$\qquad = \frac{1}{2}\left(24 - \frac{8}{3}\right) = \frac{1}{2}\left(\frac{64}{3}\right) = \frac{32}{3}$.

12. $y = \ln(\cos x) \quad \Rightarrow \quad dy/dx = -\tan x \quad \Rightarrow \quad 1 + (dy/dx)^2 = 1 + \tan^2 x = \sec^2 x$. So

$L = \int_0^{\pi/3} \sqrt{\sec^2 x}\,dx = \int_0^{\pi/3} \sec x\,dx = \Big[\ln|\sec x + \tan x|\Big]_0^{\pi/3} = \ln\left(2 + \sqrt{3}\right) - \ln(1+0) = \ln\left(2 + \sqrt{3}\right)$.

13. $y = \ln(\sec x) \quad \Rightarrow \quad \dfrac{dy}{dx} = \dfrac{\sec x \tan x}{\sec x} = \tan x \quad \Rightarrow \quad 1 + \left(\dfrac{dy}{dx}\right)^2 = 1 + \tan^2 x = \sec^2 x$, so

$L = \int_0^{\pi/4} \sqrt{\sec^2 x}\,dx = \int_0^{\pi/4} |\sec x|\,dx = \int_0^{\pi/4} \sec x\,dx = \Big[\ln(\sec x + \tan x)\Big]_0^{\pi/4}$

$\qquad = \ln\left(\sqrt{2} + 1\right) - \ln(1+0) = \ln\left(\sqrt{2} + 1\right)$

14. $y = 3 + \frac{1}{2}\cosh 2x \quad \Rightarrow \quad y' = \sinh 2x \quad \Rightarrow \quad 1 + (dy/dx)^2 = 1 + \sinh^2(2x) = \cosh^2(2x)$. So

$L = \int_0^1 \sqrt{\cosh^2(2x)}\,dx = \int_0^1 \cosh 2x\,dx = \left[\frac{1}{2}\sinh 2x\right]_0^1 = \frac{1}{2}\sinh 2 - 0 = \frac{1}{2}\sinh 2$.

15. $y = \ln(1 - x^2) \quad \Rightarrow \quad y' = \dfrac{1}{1-x^2}\cdot(-2x) \quad \Rightarrow$

$1 + \left(\dfrac{dy}{dx}\right)^2 = 1 + \dfrac{4x^2}{(1-x^2)^2} = \dfrac{1 - 2x^2 + x^4 + 4x^2}{(1-x^2)^2} = \dfrac{1 + 2x^2 + x^4}{(1-x^2)^2} = \dfrac{(1+x^2)^2}{(1-x^2)^2} \quad \Rightarrow$

$$\sqrt{1+\left(\frac{dy}{dx}\right)^2}=\sqrt{\left(\frac{1+x^2}{1-x^2}\right)^2}=\frac{1+x^2}{1-x^2}=-1+\frac{2}{1-x^2}\quad[\text{by division}]\quad=-1+\frac{1}{1+x}+\frac{1}{1-x}\quad[\text{partial fractions}].$$

So $L=\displaystyle\int_0^{1/2}\left(-1+\frac{1}{1+x}+\frac{1}{1-x}\right)dx=\Big[-x+\ln|1+x|-\ln|1-x|\Big]_0^{1/2}=\left(-\tfrac12+\ln\tfrac32-\ln\tfrac12\right)-0=\ln3-\tfrac12.$

16. $y=\sqrt{x-x^2}+\sin^{-1}\left(\sqrt{x}\right)\ \Rightarrow\ \dfrac{dy}{dx}=\dfrac{1-2x}{2\sqrt{x-x^2}}+\dfrac{1}{2\sqrt{x}\sqrt{1-x}}=\dfrac{2-2x}{2\sqrt{x}\sqrt{1-x}}=\sqrt{\dfrac{1-x}{x}}\ \Rightarrow$

$1+\left(\dfrac{dy}{dx}\right)^2=1+\dfrac{1-x}{x}=\dfrac{1}{x}.$ The curve has endpoints $(0,0)$ and $\left(1,\tfrac\pi2\right)$, so $L=\displaystyle\int_0^1\sqrt{\dfrac1x}\,dx=\Big[2\sqrt{x}\Big]_0^1=2.$

17. $y=e^x\ \Rightarrow\ y'=e^x\ \Rightarrow\ 1+(y')^2=1+e^{2x}.$ So

$$L=\int_0^1\sqrt{1+e^{2x}}\,dx=\int_1^e\sqrt{1+u^2}\,\frac{du}{u}\quad\left[\begin{array}{l}u=e^x,\text{ so}\\x=\ln u,dx=du/u\end{array}\right]=\int_1^e\frac{\sqrt{1+u^2}}{u^2}\,u\,du$$

$$=\int_{\sqrt2}^{\sqrt{1+e^2}}\frac{v}{v^2-1}\,v\,dv\quad\left[\begin{array}{l}v=\sqrt{1+u^2},\text{ so}\\v^2=1+u^2,v\,dv=u\,du\end{array}\right]=\int_{\sqrt2}^{\sqrt{1+e^2}}\left(1+\frac{1/2}{v-1}-\frac{1/2}{v+1}\right)dv$$

$$=\left[v+\tfrac12\ln\frac{v-1}{v+1}\right]_{\sqrt2}^{\sqrt{1+e^2}}=\sqrt{1+e^2}+\tfrac12\ln\frac{\sqrt{1+e^2}-1}{\sqrt{1+e^2}+1}-\sqrt2-\tfrac12\ln\frac{\sqrt2-1}{\sqrt2+1}$$

$$=\sqrt{1+e^2}-\sqrt2+\ln\left(\sqrt{1+e^2}-1\right)-1-\ln\left(\sqrt2-1\right)$$

Or: Use Formula 23 for $\int\left(\sqrt{1+u^2}/u\right)du$, or substitute $u=\tan\theta$.

18. $y=\ln\left(\dfrac{e^x+1}{e^x-1}\right)=\ln(e^x+1)-\ln(e^x-1)\ \Rightarrow\ y'=\dfrac{e^x}{e^x+1}-\dfrac{e^x}{e^x-1}=\dfrac{-2e^x}{e^{2x}-1}\ \Rightarrow$

$1+(y')^2=1+\dfrac{4e^{2x}}{(e^{2x}-1)^2}=\dfrac{(e^{2x}+1)^2}{(e^{2x}-1)^2}\ \Rightarrow\ \sqrt{1+(y')^2}=\dfrac{e^{2x}+1}{e^{2x}-1}=\dfrac{e^x+e^{-x}}{e^x-e^{-x}}=\dfrac{\cosh x}{\sinh x}.$

So $L=\displaystyle\int_a^b\frac{\cosh x}{\sinh x}\,dx=\Big[\ln\sinh x\Big]_a^b=\ln\sinh b-\ln\sinh a=\ln\left(\frac{\sinh b}{\sinh a}\right)=\ln\left(\frac{e^b-e^{-b}}{e^a-e^{-a}}\right).$

19. $y=\tfrac12x^2\ \Rightarrow\ dy/dx=x\ \Rightarrow\ 1+(dy/dx)^2=1+x^2.$ So

$$L=\int_{-1}^1\sqrt{1+x^2}\,dx=2\int_0^1\sqrt{1+x^2}\,dx\quad[\text{by symmetry}]\overset{21}{=}2\Big[\tfrac{x}{2}\sqrt{1+x^2}+\tfrac12\ln\left(x+\sqrt{1+x^2}\right)\Big]_0^1\quad\left[\begin{array}{l}\text{or substitute}\\x=\tan\theta\end{array}\right]$$

$$=2\Big[\left(\tfrac12\sqrt2+\tfrac12\ln\left(1+\sqrt2\right)\right)-\left(0+\tfrac12\ln1\right)\Big]=\sqrt2+\ln\left(1+\sqrt2\right)$$

20. $x^2=(y-4)^3\ \Rightarrow\ x=(y-4)^{3/2}\ [\text{for }x>0]\ \Rightarrow\ dx/dy=\tfrac32(y-4)^{1/2}\ \Rightarrow$

$1+(dx/dy)^2=1+\tfrac94(y-4)=\tfrac94y-8.$ So

$$L=\int_5^8\sqrt{\tfrac94y-8}\,dy=\int_{13/4}^{10}\sqrt{u}\left(\tfrac49\,du\right)\quad\left[\begin{array}{l}u=\tfrac94y-8,\\du=\tfrac94\,dy\end{array}\right]=\tfrac49\Big[\tfrac23u^{3/2}\Big]_{13/4}^{10}$$

$$=\tfrac{8}{27}\Big[10^{3/2}-\left(\tfrac{13}{4}\right)^{3/2}\Big]\quad\Big[\text{or }\tfrac{1}{27}\left(80\sqrt{10}-13\sqrt{13}\right)\Big]$$

21.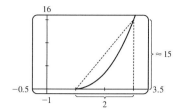

From the figure, the length of the curve is slightly larger than the hypotenuse of the triangle formed by the points $(1, 0)$, $(3, 0)$, and $(3, f(3)) \approx (3, 15)$, where $y = f(x) = \frac{2}{3}(x^2 - 1)^{3/2}$. This length is about $\sqrt{15^2 + 2^2} \approx 15$, so we might estimate the length to be 15.5.

$y = \frac{2}{3}(x^2 - 1)^{3/2}$ $\Rightarrow$ $y' = (x^2 - 1)^{1/2}(2x)$ $\Rightarrow$ $1 + (y')^2 = 1 + 4x^2(x^2 - 1) = 4x^4 - 4x^2 + 1 = (2x^2 - 1)^2$,

so, using the fact that $2x^2 - 1 > 0$ for $1 \le x \le 3$,

$L = \int_1^3 \sqrt{(2x^2 - 1)^2}\, dx = \int_1^3 |2x^2 - 1|\, dx = \int_1^3 (2x^2 - 1)\, dx = \left[\frac{2}{3}x^3 - x\right]_1^3 = (18 - 3) - \left(\frac{2}{3} - 1\right) = \frac{46}{3} = 15.\overline{3}.$

22.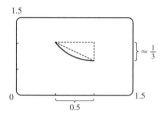

From the figure, the length of the curve is slightly larger than the hypotenuse of the triangle formed by the points $(0.5, f(0.5) \approx 1)$, $(1, f(0.5) \approx 1)$ and $\left(1, \frac{2}{3}\right)$, where $y = f(x) = x^3/6 + 1/(2x)$. This length is about

$\sqrt{\left(\frac{1}{2}\right)^2 + \left(\frac{1}{3}\right)^2} \approx 0.6$, so we might estimate the length to be 0.65.

$y = \frac{x^3}{6} + \frac{1}{2x}$ $\Rightarrow$ $y' = \frac{x^2}{2} - \frac{x^{-2}}{2}$ $\Rightarrow$ $1 + (y')^2 = 1 + \frac{x^4}{4} - \frac{1}{2} + \frac{x^{-4}}{4} = \frac{x^4}{4} + \frac{1}{2} + \frac{x^{-4}}{4} = \left(\frac{x^2}{2} + \frac{x^{-2}}{2}\right)^2$

so, using the fact that the parenthetical expression is positive,

$L = \int_{1/2}^1 \sqrt{\left(\frac{x^2}{2} + \frac{x^{-2}}{2}\right)^2}\, dx = \int_{1/2}^1 \left(\frac{x^2}{2} + \frac{x^{-2}}{2}\right) dx = \left[\frac{x^3}{6} - \frac{1}{2x}\right]_{1/2}^1 = \left(\frac{1}{6} - \frac{1}{2}\right) - \left(\frac{1}{48} - 1\right) = \frac{31}{48} = 0.6458\overline{3}$

23. $y = xe^{-x}$ $\Rightarrow$ $dy/dx = e^{-x} - xe^{-x} = e^{-x}(1 - x)$ $\Rightarrow$ $1 + (dy/dx)^2 = 1 + e^{-2x}(1 - x)^2$. Let

$f(x) = \sqrt{1 + (dy/dx)^2} = \sqrt{1 + e^{-2x}(1 - x)^2}$. Then $L = \int_0^5 f(x)\, dx$. Since $n = 10$, $\Delta x = \frac{5 - 0}{10} = \frac{1}{2}$. Now

$L \approx S_{10} = \frac{1/2}{3}[f(0) + 4f\left(\frac{1}{2}\right) + 2f(1) + 4f\left(\frac{3}{2}\right) + 2f(2) + 4f\left(\frac{5}{2}\right) + 2f(3) + 4f\left(\frac{7}{2}\right) + 2f(4) + 4f\left(\frac{9}{2}\right) + f(5)]$

≈ 5.115840

The value of the integral produced by a calculator is 5.113568 (to six decimal places).

24. $x = y + \sqrt{y}$ $\Rightarrow$ $dx/dy = 1 + \frac{1}{2\sqrt{y}}$ $\Rightarrow$ $1 + (dx/dy)^2 = 1 + \left(1 + \frac{1}{2\sqrt{y}}\right)^2 = 2 + \frac{1}{\sqrt{y}} + \frac{1}{4y}.$

Let $g(y) = \sqrt{1 + (dx/dy)^2}$. Then $L = \int_1^2 g(y)\, dy$. Since $n = 10$, $\Delta y = \frac{2 - 1}{10} = \frac{1}{10}$. Now

$L \approx S_{10} = \frac{1/10}{3}[g(1) + 4g(1.1) + 2g(1.2) + 4g(1.3) + 2g(1.4)$

$+ 4g(1.5) + 2g(1.6) + 4g(1.7) + 2g(1.8) + 4g(1.9) + g(2)] \approx 1.732215,$

which is the same value of the integral produced by a calculator to six decimal places.

25. $y = \sec x \;\Rightarrow\; dy/dx = \sec x \tan x \;\Rightarrow\; L = \int_0^{\pi/3} f(x)\,dx$, where $f(x) = \sqrt{1 + \sec^2 x \tan^2 x}$.

Since $n = 10$, $\Delta x = \dfrac{\pi/3 - 0}{10} = \dfrac{\pi}{30}$. Now

$$L \approx S_{10} = \frac{\pi/30}{3}\left[f(0) + 4f\left(\frac{\pi}{30}\right) + 2f\left(\frac{2\pi}{30}\right) + 4f\left(\frac{3\pi}{30}\right) + 2f\left(\frac{4\pi}{30}\right) + 4f\left(\frac{5\pi}{30}\right)\right.$$

$$\left. + 2f\left(\frac{6\pi}{30}\right) + 4f\left(\frac{7\pi}{30}\right) + 2f\left(\frac{8\pi}{30}\right) + 4f\left(\frac{9\pi}{30}\right) + f\left(\frac{\pi}{3}\right)\right] \approx 1.569619.$$

The value of the integral produced by a calculator is 1.569259 (to six decimal places).

26. $y = x \ln x \;\Rightarrow\; dy/dx = 1 + \ln x$. Let $f(x) = \sqrt{1 + (dy/dx)^2} = \sqrt{1 + (1 + \ln x)^2}$. Then $L = \int_1^3 f(x)\,dx$.

Since $n = 10$, $\Delta x = \frac{3-1}{10} = \frac{1}{5}$. Now

$$L \approx S_{10} = \tfrac{1/5}{3}[f(1) + 4f(1.2) + 2f(1.4) + 4f(1.6) + 2f(1.8) + 4f(2)$$

$$+ 2f(2.2) + 4f(2.4) + 2f(2.6) + 4f(2.8) + f(3)] \approx 3.869618.$$

The value of the integral produced by a calculator is 3.869617 (to six decimal places).

27. (a)

(b)

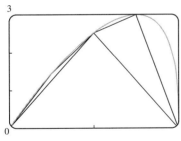

Let $f(x) = y = x\sqrt[3]{4 - x}$. The polygon with one side is just the line segment joining the points $(0, f(0)) = (0, 0)$ and $(4, f(4)) = (4, 0)$, and its length $L_1 = 4$.

The polygon with two sides joins the points $(0, 0)$, $(2, f(2)) = (2, 2\sqrt[3]{2})$ and $(4, 0)$. Its length

$$L_2 = \sqrt{(2 - 0)^2 + \left(2\sqrt[3]{2} - 0\right)^2} + \sqrt{(4 - 2)^2 + \left(0 - 2\sqrt[3]{2}\right)^2} = 2\sqrt{4 + 2^{8/3}} \approx 6.43$$

Similarly, the inscribed polygon with four sides joins the points $(0, 0)$, $\left(1, \sqrt[3]{3}\right)$, $(2, 2\sqrt[3]{2})$, $(3, 3)$, and $(4, 0)$, so its length

$$L_3 = \sqrt{1 + \left(\sqrt[3]{3}\right)^2} + \sqrt{1 + \left(2\sqrt[3]{2} - \sqrt[3]{3}\right)^2} + \sqrt{1 + \left(3 - 2\sqrt[3]{2}\right)^2} + \sqrt{1 + 9} \approx 7.50$$

(c) Using the arc length formula with $\dfrac{dy}{dx} = x\left[\tfrac{1}{3}(4 - x)^{-2/3}(-1)\right] + \sqrt[3]{4 - x} = \dfrac{12 - 4x}{3(4 - x)^{2/3}}$, the length of the curve is

$$L = \int_0^4 \sqrt{1 + \left(\frac{dy}{dx}\right)^2}\,dx = \int_0^4 \sqrt{1 + \left[\frac{12 - 4x}{3(4 - x)^{2/3}}\right]^2}\,dx.$$

(d) According to a CAS, the length of the curve is $L \approx 7.7988$. The actual value is larger than any of the approximations in part (b). This is always true, since any approximating straight line between two points on the curve is shorter than the length of the curve between the two points.

28. (a) Let $f(x) = y = x + \sin x$ with $0 \le x \le 2\pi$.

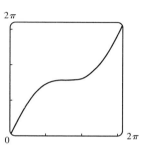

(b) The polygon with one side is just the line segment joining the points $(0, f(0)) = (0, 0)$ and $(2\pi, f(2\pi)) = (2\pi, 2\pi)$, and

its length is $\sqrt{(2\pi - 0)^2 + (2\pi - 0)^2} = 2\sqrt{2}\,\pi \approx 8.9$.

The polygon with two sides joins the points $(0, 0)$, $(\pi, f(\pi)) = (\pi, \pi)$, and

$(2\pi, 2\pi)$. Its length is

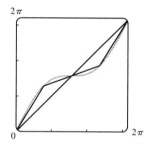

$$\sqrt{(\pi - 0)^2 + (\pi - 0)^2} + \sqrt{(2\pi - \pi)^2 + (2\pi - \pi)^2} = \sqrt{2}\,\pi + \sqrt{2}\,\pi$$
$$= 2\sqrt{2}\,\pi \approx 8.9$$

Note from the diagram that the two approximations are the same because the sides

of the two-sided polygon are in fact on the same line, since $f(\pi) = \pi = \frac{1}{2}f(2\pi)$.

The four-sided polygon joins the points $(0, 0)$, $\left(\frac{\pi}{2}, \frac{\pi}{2} + 1\right)$, (π, π), $\left(\frac{3\pi}{2}, \frac{3\pi}{2} - 1\right)$, and $(2\pi, 2\pi)$, so its length is

$$\sqrt{\left(\frac{\pi}{2}\right)^2 + \left(\frac{\pi}{2} + 1\right)^2} + \sqrt{\left(\frac{\pi}{2}\right)^2 + \left(\frac{\pi}{2} - 1\right)^2} + \sqrt{\left(\frac{\pi}{2}\right)^2 + \left(\frac{\pi}{2} - 1\right)^2} + \sqrt{\left(\frac{\pi}{2}\right)^2 + \left(\frac{\pi}{2} + 1\right)^2} \approx 9.4$$

(c) Using the arc length formula with $dy/dx = 1 + \cos x$, the length of the curve is

$$L = \int_0^{2\pi} \sqrt{1 + (1 + \cos x)^2}\,dx = \int_0^{2\pi} \sqrt{2 + 2\cos x + \cos^2 x}\,dx$$

(d) The CAS approximates the integral as 9.5076. The actual length is larger than the approximations in part (b).

29. $y = \ln x \;\Rightarrow\; dy/dx = 1/x \;\Rightarrow\; 1 + (dy/dx)^2 = 1 + 1/x^2 = (x^2 + 1)/x^2 \;\Rightarrow$

$$L = \int_1^2 \sqrt{\frac{x^2 + 1}{x^2}}\,dx = \int_1^2 \frac{\sqrt{1 + x^2}}{x}\,dx \overset{23}{=} \left[\sqrt{1 + x^2} - \ln\left|\frac{1 + \sqrt{1 + x^2}}{x}\right|\right]_1^2$$

$$= \sqrt{5} - \ln\left(\frac{1 + \sqrt{5}}{2}\right) - \sqrt{2} + \ln\left(1 + \sqrt{2}\right)$$

30. $y = x^{4/3} \;\Rightarrow\; dy/dx = \frac{4}{3}x^{1/3} \;\Rightarrow\; 1 + (dy/dx)^2 = 1 + \frac{16}{9}x^{2/3} \;\Rightarrow$

$$L = \int_0^1 \sqrt{1 + \frac{16}{9}x^{2/3}}\,dx = \int_0^{4/3} \sqrt{1 + u^2}\,\frac{81}{64}u^2\,du \qquad \begin{bmatrix} u = \frac{4}{3}x^{1/3},\; du = \frac{4}{9}x^{-2/3}\,dx, \\ dx = \frac{9}{4}x^{2/3}\,du = \frac{9}{4} \cdot \frac{9}{16}u^2\,du = \frac{81}{64}u^2\,du \end{bmatrix}$$

$$\overset{22}{=} \frac{81}{64}\left[\frac{1}{8}u(1 + 2u^2)\sqrt{1 + u^2} - \frac{1}{8}\ln\left(u + \sqrt{1 + u^2}\right)\right]_0^{4/3} = \frac{81}{64}\left[\frac{1}{6}\left(1 + \frac{32}{9}\right)\sqrt{\frac{25}{9}} - \frac{1}{8}\ln\left(\frac{4}{3} + \sqrt{\frac{25}{9}}\right)\right]$$

$$= \frac{81}{64}\left(\frac{1}{6} \cdot \frac{41}{9} \cdot \frac{5}{3} - \frac{1}{8}\ln 3\right) = \frac{205}{128} - \frac{81}{512}\ln 3 \approx 1.4277586$$

31. $y^{2/3} = 1 - x^{2/3}$ $\Rightarrow$ $y = (1 - x^{2/3})^{3/2}$ $\Rightarrow$

$$\frac{dy}{dx} = \frac{3}{2}(1 - x^{2/3})^{1/2}\left(-\frac{2}{3}x^{-1/3}\right) = -x^{-1/3}(1 - x^{2/3})^{1/2} \quad \Rightarrow$$

$$\left(\frac{dy}{dx}\right)^2 = x^{-2/3}(1 - x^{2/3}) = x^{-2/3} - 1. \text{ Thus}$$

$$L = 4\int_0^1 \sqrt{1 + (x^{-2/3} - 1)}\ dx = 4\int_0^1 x^{-1/3}\ dx = 4\lim_{t\to 0^+}\left[\frac{3}{2}x^{2/3}\right]_t^1 = 6.$$

32. (a)

(b) $y = x^{2/3}$ $\Rightarrow$ $1 + \left(\dfrac{dy}{dx}\right)^2 = 1 + \left(\dfrac{2}{3}x^{-1/3}\right)^2 = 1 + \dfrac{4}{9}x^{-2/3}$. So $L = \int_0^1 \sqrt{1 + \frac{4}{9}x^{-2/3}}\ dx$ [an improper integral].

$x = y^{3/2}$ $\Rightarrow$ $1 + \left(\dfrac{dx}{dy}\right)^2 = 1 + \left(\dfrac{3}{2}y^{1/2}\right)^2 = 1 + \dfrac{9}{4}y$. So $L = \int_0^1 \sqrt{1 + \frac{9}{4}y}\ dy$.

The second integral equals $\dfrac{4}{9}\cdot\dfrac{2}{3}\left[(1 + \frac{9}{4}y)^{3/2}\right]_0^1 = \dfrac{8}{27}\left(\dfrac{13\sqrt{13}}{8} - 1\right) = \dfrac{13\sqrt{13} - 8}{27}$.

The first integral can be evaluated as follows:

$$\int_0^1 \sqrt{1 + \tfrac{4}{9}x^{-2/3}}\ dx = \lim_{t\to 0^+}\int_t^1 \frac{\sqrt{9x^{2/3} + 4}}{3x^{1/3}}\ dx = \lim_{t\to 0^+}\int_{9t^{2/3}}^9 \frac{\sqrt{u + 4}}{18}\ du \qquad \left[\begin{array}{l} u = 9x^{2/3}, \\ du = 6x^{-1/3}\ dx \end{array}\right]$$

$$= \int_0^9 \frac{\sqrt{u + 4}}{18}\ du = \frac{1}{18}\cdot\left[\frac{2}{3}(u + 4)^{3/2}\right]_0^9 = \frac{1}{27}(13^{3/2} - 4^{3/2}) = \frac{13\sqrt{13} - 8}{27}$$

(c) L = length of the arc of this curve from $(-1, 1)$ to $(8, 4)$

$$= \int_0^1 \sqrt{1 + \frac{9}{4}y}\ dy + \int_0^4 \sqrt{1 + \frac{9}{4}y}\ dy = \frac{13\sqrt{13} - 8}{27} + \frac{8}{27}\left[\left(1 + \frac{9}{4}y\right)^{3/2}\right]_0^4 \qquad \text{[from part (b)]}$$

$$= \frac{13\sqrt{13} - 8}{27} + \frac{8}{27}(10\sqrt{10} - 1) = \frac{13\sqrt{13} + 80\sqrt{10} - 16}{27}$$

33. $y = 2x^{3/2}$ $\Rightarrow$ $y' = 3x^{1/2}$ $\Rightarrow$ $1 + (y')^2 = 1 + 9x$. The arc length function with starting point $P_0(1, 2)$ is

$$s(x) = \int_1^x \sqrt{1 + 9t}\ dt = \left[\frac{2}{27}(1 + 9t)^{3/2}\right]_1^x = \frac{2}{27}\left[(1 + 9x)^{3/2} - 10\sqrt{10}\right].$$

34. (a)

(c)

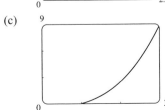

(b) $f(x) = \dfrac{1}{3}x^3 + \dfrac{1}{4x}$, $x > 0$ $\Rightarrow$ $f'(x) = x^2 - \dfrac{1}{4x^2}$. Then

$$1 + [f'(x)]^2 = 1 + \left(x^4 - \frac{1}{2} + \frac{1}{16x^4}\right) = x^4 + \frac{1}{2} + \frac{1}{16x^4} = \left(x^2 + \frac{1}{4x^2}\right)^2,$$

so $\sqrt{1 + [f'(x)]^2} = x^2 + \dfrac{1}{4x^2}$. Thus,

$$s(x) = \int_1^x \sqrt{1 + [f'(t)]^2}\ dt = \int_1^x \left(t^2 + \frac{1}{4t^2}\right)dt = \left[\frac{1}{3}t^3 - \frac{1}{4t}\right]_1^x$$

$$= \left(\frac{1}{3}x^3 - \frac{1}{4x}\right) - \left(\frac{1}{3} - \frac{1}{4}\right) = \frac{1}{3}x^3 - \frac{1}{4x} - \frac{1}{12} \quad \text{for } x \geq 1$$

35. $y = \sin^{-1} x + \sqrt{1 - x^2}$ $\Rightarrow$ $y' = \dfrac{1}{\sqrt{1 - x^2}} - \dfrac{x}{\sqrt{1 - x^2}} = \dfrac{1 - x}{\sqrt{1 - x^2}}$ $\Rightarrow$

$$1 + (y')^2 = 1 + \frac{(1 - x)^2}{1 - x^2} = \frac{1 - x^2 + 1 - 2x + x^2}{1 - x^2} = \frac{2 - 2x}{1 - x^2} = \frac{2(1 - x)}{(1 + x)(1 - x)} = \frac{2}{1 + x} \quad \Rightarrow$$

$\sqrt{1 + (y')^2} = \sqrt{\dfrac{2}{1 + x}}$. Thus, the arc length function with starting point $(0, 1)$ is given by

$$s(x) = \int_0^x \sqrt{1 + [f'(t)]^2}\, dt = \int_0^x \sqrt{\frac{2}{1 + t}}\, dt = \sqrt{2}\left[2\sqrt{1 + t}\right]_0^x = 2\sqrt{2}\left(\sqrt{1 + x} - 1\right).$$

36. $y = 150 - \frac{1}{40}(x - 50)^2$ $\Rightarrow$ $y' = -\frac{1}{20}(x - 50)$ $\Rightarrow$ $1 + (y')^2 = 1 + \frac{1}{20^2}(x - 50)^2$, so the distance traveled by

the kite is

$$L = \int_0^{80} \sqrt{1 + \frac{1}{20^2}(x - 50)^2}\, dx = \int_{-5/2}^{3/2} \sqrt{1 + u^2}\,(20\, du) \qquad \begin{bmatrix} u = \frac{1}{20}(x - 50), \\ du = \frac{1}{20}\, dx \end{bmatrix}$$

$$\overset{21}{=} 20\left[\frac{1}{2} u\sqrt{1 + u^2} + \frac{1}{2}\ln\left(u + \sqrt{1 + u^2}\right)\right]_{-5/2}^{3/2} = 10\left[\frac{3}{2}\sqrt{\frac{13}{4}} + \ln\left(\frac{3}{2} + \sqrt{\frac{13}{4}}\right) + \frac{5}{2}\sqrt{\frac{29}{4}} - \ln\left(-\frac{5}{2} + \sqrt{\frac{29}{4}}\right)\right]$$

$$= \frac{15}{2}\sqrt{13} + \frac{25}{2}\sqrt{29} + 10\ln\left(\frac{3 + \sqrt{13}}{-5 + \sqrt{29}}\right) \approx 122.8 \text{ ft}$$

37. The prey hits the ground when $y = 0$ $\Leftrightarrow$ $180 - \frac{1}{45}x^2 = 0$ $\Leftrightarrow$ $x^2 = 45 \cdot 180$ $\Rightarrow$ $x = \sqrt{8100} = 90$,

since x must be positive. $y' = -\frac{2}{45}x$ $\Rightarrow$ $1 + (y')^2 = 1 + \frac{4}{45^2}x^2$, so the distance traveled by the prey is

$$L = \int_0^{90} \sqrt{1 + \frac{4}{45^2}x^2}\, dx = \int_0^4 \sqrt{1 + u^2}\left(\frac{45}{2}\, du\right) \qquad \begin{bmatrix} u = \frac{2}{45}x, \\ du = \frac{2}{45}\, dx \end{bmatrix}$$

$$\overset{21}{=} \frac{45}{2}\left[\frac{1}{2} u\sqrt{1 + u^2} + \frac{1}{2}\ln\left(u + \sqrt{1 + u^2}\right)\right]_0^4 = \frac{45}{2}\left[2\sqrt{17} + \frac{1}{2}\ln\left(4 + \sqrt{17}\right)\right] = 45\sqrt{17} + \frac{45}{4}\ln\left(4 + \sqrt{17}\right) \approx 209.1 \text{ m}$$

38. Let $y = a - b\cosh cx$, where $a = 211.49$, $b = 20.96$, and $c = 0.03291765$. Then $y' = -bc\sinh cx$ $\Rightarrow$

$1 + (y')^2 = 1 + b^2 c^2 \sinh^2(cx)$. So $L = \int_{-91.2}^{91.2} \sqrt{1 + b^2 c^2 \sinh^2(cx)}\, dx \approx 451.137 \approx 451$, to the nearest meter.

39. The sine wave has amplitude 1 and period 14, since it goes through two periods in a distance of 28 in., so its equation is

$y = 1\sin\left(\frac{2\pi}{14}x\right) = \sin\left(\frac{\pi}{7}x\right)$. The width w of the flat metal sheet needed to make the panel is the arc length of the sine curve

from $x = 0$ to $x = 28$. We set up the integral to evaluate w using the arc length formula with $\frac{dy}{dx} = \frac{\pi}{7}\cos\left(\frac{\pi}{7}x\right)$:

$L = \int_0^{28} \sqrt{1 + \left[\frac{\pi}{7}\cos\left(\frac{\pi}{7}x\right)\right]^2}\, dx = 2\int_0^{14} \sqrt{1 + \left[\frac{\pi}{7}\cos\left(\frac{\pi}{7}x\right)\right]^2}\, dx$. This integral would be very difficult to evaluate exactly,

so we use a CAS, and find that $L \approx 29.36$ inches.

40. (a) $y = c + a\cosh\left(\frac{x}{a}\right)$ $\Rightarrow$ $y' = \sinh\left(\frac{x}{a}\right)$ $\Rightarrow$ $1 + (y')^2 = 1 + \sinh^2\left(\frac{x}{a}\right) = \cosh^2\left(\frac{x}{a}\right)$. So

$L = \int_{-b}^{b} \sqrt{\cosh^2\left(\frac{x}{a}\right)}\, dx = 2\int_0^b \cosh\left(\frac{x}{a}\right) dx = 2\left[a\sinh\left(\frac{x}{a}\right)\right]_0^b = 2a\sinh\left(\frac{b}{a}\right)$.

(b) At $x = 0$, $y = c + a$, so $c + a = 20$. The poles are 50 ft apart, so $b = 25$, and

$L = 51 \quad\Rightarrow\quad 51 = 2a \sinh(b/a)$ [from part (a)]. From the figure, we see

that $y = 51$ intersects $y = 2x \sinh(25/x)$ at $x \approx 72.3843$ for $x > 0$.

So $a \approx 72.3843$ and the wire should be attached at a distance of

$y = c + a \cosh(25/a) = 20 - a + a \cosh(25/a) \approx 24.36$ ft above the

ground.

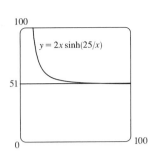

41. $y = \int_1^x \sqrt{t^3 - 1}\, dt \quad\Rightarrow\quad dy/dx = \sqrt{x^3 - 1}$ [by FTC1] $\quad\Rightarrow\quad 1 + (dy/dx)^2 = 1 + \left(\sqrt{x^3 - 1}\right)^2 = x^3 \quad\Rightarrow$

$L = \int_1^4 \sqrt{x^3}\, dx = \int_1^4 x^{3/2}\, dx = \frac{2}{5}\left[x^{5/2}\right]_1^4 = \frac{2}{5}(32 - 1) = \frac{62}{5} = 12.4$

42. By symmetry, the length of the curve in each quadrant is the same,

so we'll find the length in the first quadrant and multiply by 4.

$x^{2k} + y^{2k} = 1 \quad\Rightarrow\quad y^{2k} = 1 - x^{2k} \quad\Rightarrow\quad y = (1 - x^{2k})^{1/(2k)}$

(in the first quadrant), so we use the arc length formula with

$\dfrac{dy}{dx} = \dfrac{1}{2k}(1 - x^{2k})^{1/(2k)-1}(-2kx^{2k-1}) = -x^{2k-1}(1 - x^{2k})^{1/(2k)-1}$

The total length is therefore

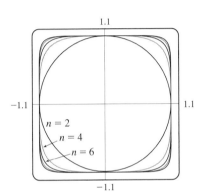

$$L_{2k} = 4\int_0^1 \sqrt{1 + [-x^{2k-1}(1-x^{2k})^{1/(2k)-1}]^2}\, dx = 4\int_0^1 \sqrt{1 + x^{2(2k-1)}(1-x^{2k})^{1/k-2}}\, dx$$

Now from the graph, we see that as k increases, the "corners" of these fat circles get closer to the points $(\pm 1, \pm 1)$ and

$(\pm 1, \mp 1)$, and the "edges" of the fat circles approach the lines joining these four points. It seems plausible that as $k \to \infty$, the

total length of the fat circle with $n = 2k$ will approach the length of the perimeter of the square with sides of length 2. This is

supported by taking the limit as $k \to \infty$ of the equation of the fat circle in the first quadrant: $\displaystyle\lim_{k \to \infty} (1 - x^{2k})^{1/(2k)} = 1$

for $0 \le x < 1$. So we guess that $\displaystyle\lim_{k \to \infty} L_{2k} = 4 \cdot 2 = 8$.

DISCOVERY PROJECT   Arc Length Contest

For advice on how to run the contest and a list of student entries, see the article "Arc Length Contest" by Larry Riddle in

The College Mathematics Journal, Volume 29, No. 4, September 1998, pages 314–320.

9.2 Area of a Surface of Revolution

1. $y = x^4 \quad \Rightarrow \quad dy/dx = 4x^3 \quad \Rightarrow \quad ds = \sqrt{1 + (dy/dx)^2}\, dx = \sqrt{1 + 16x^6}\, dx$

(a) By (7), an integral for the area of the surface obtained by rotating the curve about the x-axis is

$S = \int 2\pi y\, ds = \int_0^1 2\pi x^4 \sqrt{1 + 16x^6}\, dx.$

(b) By (8), an integral for the area of the surface obtained by rotating the curve about the y-axis is

$S = \int 2\pi x\, ds = \int_0^1 2\pi x \sqrt{1 + 16x^6}\, dx.$

2. $y = xe^{-x} \quad \Rightarrow \quad dy/dx = x(-e^{-x}) + e^{-x} = e^{-x}(1 - x) \quad \Rightarrow \quad ds = \sqrt{1 + (dy/dx)^2}\, dx = \sqrt{1 + e^{-2x}(1 - x)^2}\, dx$

(a) By (7), $S = \int 2\pi y\, ds = \int_1^3 2\pi x e^{-x} \sqrt{1 + e^{-2x}(1 - x)^2}\, dx.$

(b) By (8), $S = \int 2\pi x\, ds = \int_1^3 2\pi x \sqrt{1 + e^{-2x}(1 - x)^2}\, dx.$

3. $y = \tan^{-1} x \quad \Rightarrow \quad \dfrac{dy}{dx} = \dfrac{1}{1 + x^2} \quad \Rightarrow \quad ds = \sqrt{1 + \left(\dfrac{dy}{dx}\right)^2}\, dx = \sqrt{1 + \dfrac{1}{(1 + x^2)^2}}\, dx.$

(a) By (7), $S = \int 2\pi y\, ds = \int_0^1 2\pi \tan^{-1} x \sqrt{1 + \dfrac{1}{(1 + x^2)^2}}\, dx.$

(b) By (8), $S = \int 2\pi x\, ds = \int_0^1 2\pi x \sqrt{1 + \dfrac{1}{(1 + x^2)^2}}\, dx.$

4. $x = \sqrt{y - y^2}$ [defined for $0 \le y \le 1$] $\Rightarrow$

$\dfrac{dx}{dy} = \dfrac{1 - 2y}{2\sqrt{y - y^2}} \quad \Rightarrow \quad ds = \sqrt{1 + \left(\dfrac{dx}{dy}\right)^2}\, dy = \sqrt{\dfrac{4(y - y^2) + 1 - 4y + 4y^2}{4(y - y^2)}}\, dy = \sqrt{\dfrac{1}{4y(1 - y)}}\, dy.$

(a) By (7), $S = \int 2\pi y\, ds = \int_0^1 2\pi y \sqrt{\dfrac{1}{4y(1 - y)}}\, dy.$

(b) By (8), $S = \int 2\pi x\, ds = \int_0^1 2\pi \sqrt{y - y^2} \sqrt{\dfrac{1}{4y(1 - y)}}\, dy.$

5. $y = x^3 \quad \Rightarrow \quad y' = 3x^2.$ So

$$S = \int_0^2 2\pi y \sqrt{1 + (y')^2}\, dx = 2\pi \int_0^2 x^3 \sqrt{1 + 9x^4}\, dx \qquad [u = 1 + 9x^4,\, du = 36x^3\, dx]$$

$$= \tfrac{2\pi}{36} \int_1^{145} \sqrt{u}\, du = \tfrac{\pi}{18} \left[\tfrac{2}{3} u^{3/2}\right]_1^{145} = \tfrac{\pi}{27} \left(145\sqrt{145} - 1\right)$$

6. The curve $9x = y^2 + 18$ is symmetric about the x-axis, so we only use its top half, given by

$y = 3\sqrt{x - 2}.$ $\dfrac{dy}{dx} = \dfrac{3}{2\sqrt{x - 2}},$ so $1 + \left(\dfrac{dy}{dx}\right)^2 = 1 + \dfrac{9}{4(x - 2)}.$ Thus,

$$S = \int_2^6 2\pi \cdot 3\sqrt{x - 2} \sqrt{1 + \dfrac{9}{4(x - 2)}}\, dx = 6\pi \int_2^6 \sqrt{x - 2 + \tfrac{9}{4}}\, dx = 6\pi \int_2^6 \left(x + \tfrac{1}{4}\right)^{1/2}\, dx$$

$$= 6\pi \cdot \tfrac{2}{3} \left[\left(x + \tfrac{1}{4}\right)^{3/2}\right]_2^6 = 4\pi \left[\left(\tfrac{25}{4}\right)^{3/2} - \left(\tfrac{9}{4}\right)^{3/2}\right] = 4\pi \left(\tfrac{125}{8} - \tfrac{27}{8}\right) = 4\pi \cdot \tfrac{98}{8} = 49\pi$$

7. $y = \sqrt{1 + 4x}$ $\Rightarrow$ $y' = \frac{1}{2}(1 + 4x)^{-1/2}(4) = \dfrac{2}{\sqrt{1+4x}}$ $\Rightarrow$ $\sqrt{1 + (y')^2} = \sqrt{1 + \dfrac{4}{1+4x}} = \sqrt{\dfrac{5+4x}{1+4x}}$. So

$$S = \int_1^5 2\pi y\, \sqrt{1 + (y')^2}\, dx = 2\pi \int_1^5 \sqrt{1+4x}\, \sqrt{\dfrac{5+4x}{1+4x}}\, dx = 2\pi \int_1^5 \sqrt{4x + 5}\, dx$$

$$= 2\pi \int_9^{25} \sqrt{u}\, \left(\tfrac{1}{4}\, du\right) \quad \begin{bmatrix} u = 4x + 5, \\ du = 4\, dx \end{bmatrix} = \dfrac{2\pi}{4} \left[\tfrac{2}{3} u^{3/2}\right]_9^{25} = \dfrac{\pi}{3}(25^{3/2} - 9^{3/2}) = \dfrac{\pi}{3}(125 - 27) = \dfrac{98}{3}\pi$$

8. $y = c + a\cosh(x/a)$ $\Rightarrow$ $y' = \sinh(x/a)$ $\Rightarrow$ $1 + (y')^2 = 1 + \sinh^2(x/a) = \cosh^2(x/a)$ $\Rightarrow$

$\sqrt{1 + (y')^2} = \cosh(x/a)$. So

$$S = \int_0^a 2\pi y\, \sqrt{1 + (y')^2}\, dx = 2\pi \int_0^a \left[c + a\cosh\left(\dfrac{x}{a}\right)\right]\cosh\left(\dfrac{x}{a}\right) dx = 2\pi \int_0^a \left[c\cosh\left(\dfrac{x}{a}\right) + a\cosh^2\left(\dfrac{x}{a}\right)\right] dx$$

$$= 2\pi \int_0^a \left[c\cosh\left(\dfrac{x}{a}\right) + \dfrac{a}{2}\left(1 + \cosh\left(\dfrac{2x}{a}\right)\right)\right] dx \qquad \left[\cosh^2 x = \tfrac{1}{2}(1 + \cosh 2x)\right]$$

$$= 2\pi \left[ac\sinh\left(\dfrac{x}{a}\right) + \dfrac{ax}{2} + \dfrac{a^2}{4}\sinh\left(\dfrac{2x}{a}\right)\right]_0^a = 2\pi\left(ac\sinh 1 + \tfrac{1}{2}a^2 + \tfrac{1}{4}a^2 \sinh 2\right)$$

9. $y = \sin \pi x$ $\Rightarrow$ $y' = \pi \cos \pi x$ $\Rightarrow$ $1 + (y')^2 = 1 + \pi^2 \cos^2(\pi x)$. So

$$S = \int_0^1 2\pi y\, \sqrt{1 + (y')^2}\, dx = 2\pi \int_0^1 \sin \pi x\, \sqrt{1 + \pi^2 \cos^2(\pi x)}\, dx \qquad \begin{bmatrix} u = \pi \cos \pi x, \\ du = -\pi^2 \sin \pi x\, dx \end{bmatrix}$$

$$= 2\pi \int_\pi^{-\pi} \sqrt{1 + u^2}\, \left(-\dfrac{1}{\pi^2}\, du\right) = \dfrac{2}{\pi} \int_{-\pi}^\pi \sqrt{1 + u^2}\, du$$

$$= \dfrac{4}{\pi} \int_0^\pi \sqrt{1 + u^2}\, du \overset{21}{=} \dfrac{4}{\pi}\left[\dfrac{u}{2}\sqrt{1 + u^2} + \tfrac{1}{2}\ln\left(u + \sqrt{1 + u^2}\right)\right]_0^\pi$$

$$= \dfrac{4}{\pi}\left[\left(\dfrac{\pi}{2}\sqrt{1 + \pi^2} + \tfrac{1}{2}\ln\left(\pi + \sqrt{1 + \pi^2}\right)\right) - 0\right] = 2\sqrt{1 + \pi^2} + \dfrac{2}{\pi}\ln\left(\pi + \sqrt{1 + \pi^2}\right)$$

10. $y = \dfrac{x^3}{6} + \dfrac{1}{2x}$ $\Rightarrow$ $\dfrac{dy}{dx} = \dfrac{x^2}{2} - \dfrac{1}{2x^2}$ $\Rightarrow$ $\sqrt{1 + \left(\dfrac{dy}{dx}\right)^2} = \sqrt{\dfrac{x^4}{4} + \dfrac{1}{2} + \dfrac{1}{4x^4}} = \sqrt{\left(\dfrac{x^2}{2} + \dfrac{1}{2x^2}\right)^2} = \dfrac{x^2}{2} + \dfrac{1}{2x^2}$ $\Rightarrow$

$$S = \int_{1/2}^1 2\pi\left(\dfrac{x^3}{6} + \dfrac{1}{2x}\right)\left(\dfrac{x^2}{2} + \dfrac{1}{2x^2}\right) dx = 2\pi \int_{1/2}^1 \left(\dfrac{x^5}{12} + \dfrac{x}{12} + \dfrac{x}{4} + \dfrac{1}{4x^3}\right) dx$$

$$= 2\pi \int_{1/2}^1 \left(\dfrac{x^5}{12} + \dfrac{x}{3} + \dfrac{x^{-3}}{4}\right) dx = 2\pi\left[\dfrac{x^6}{72} + \dfrac{x^2}{6} - \dfrac{x^{-2}}{8}\right]_{1/2}^1$$

$$= 2\pi\left[\left(\tfrac{1}{72} + \tfrac{1}{6} - \tfrac{1}{8}\right) - \left(\tfrac{1}{64 \cdot 72} + \tfrac{1}{24} - \tfrac{1}{2}\right)\right] = 2\pi\left(\tfrac{263}{512}\right) = \dfrac{263}{256}\pi$$

11. $x = \tfrac{1}{3}(y^2 + 2)^{3/2}$ $\Rightarrow$ $dx/dy = \tfrac{1}{2}(y^2 + 2)^{1/2}(2y) = y\sqrt{y^2 + 2}$ $\Rightarrow$ $1 + (dx/dy)^2 = 1 + y^2(y^2 + 2) = (y^2 + 1)^2$.

So $S = 2\pi \int_1^2 y(y^2 + 1)\, dy = 2\pi\left[\tfrac{1}{4}y^4 + \tfrac{1}{2}y^2\right]_1^2 = 2\pi\left(4 + 2 - \tfrac{1}{4} - \tfrac{1}{2}\right) = \dfrac{21\pi}{2}$.

12. $x = 1 + 2y^2$ $\Rightarrow$ $1 + (dx/dy)^2 = 1 + (4y)^2 = 1 + 16y^2$.

So $S = 2\pi \int_1^2 y\, \sqrt{1 + 16y^2}\, dy = \dfrac{\pi}{16} \int_1^2 (16y^2 + 1)^{1/2} 32y\, dy = \dfrac{\pi}{16}\left[\tfrac{2}{3}(16y^2 + 1)^{3/2}\right]_1^2 = \dfrac{\pi}{24}\left(65\sqrt{65} - 17\sqrt{17}\right)$.

13. $y = \sqrt[3]{x} \quad \Rightarrow \quad x = y^3 \quad \Rightarrow \quad 1 + (dx/dy)^2 = 1 + 9y^4$. So

$$S = 2\pi \int_1^2 x \sqrt{1 + (dx/dy)^2}\, dy = 2\pi \int_1^2 y^3 \sqrt{1 + 9y^4}\, dy = \frac{2\pi}{36} \int_1^2 \sqrt{1 + 9y^4}\, 36y^3\, dy = \frac{\pi}{18} \left[\frac{2}{3} \left(1 + 9y^4 \right)^{3/2} \right]_1^2$$

$$= \frac{\pi}{27} \left(145\sqrt{145} - 10\sqrt{10} \right)$$

14. $y = 1 - x^2 \quad \Rightarrow \quad 1 + (dy/dx)^2 = 1 + 4x^2 \quad \Rightarrow$

$$S = 2\pi \int_0^1 x \sqrt{1 + 4x^2}\, dx = \frac{\pi}{4} \int_0^1 8x \sqrt{4x^2 + 1}\, dx = \frac{\pi}{4} \left[\frac{2}{3} \left(4x^2 + 1 \right)^{3/2} \right]_0^1 = \frac{\pi}{6} \left(5\sqrt{5} - 1 \right)$$

15. $x = \sqrt{a^2 - y^2} \quad \Rightarrow \quad dx/dy = \frac{1}{2} (a^2 - y^2)^{-1/2}(-2y) = -y/\sqrt{a^2 - y^2} \quad \Rightarrow$

$$1 + \left(\frac{dx}{dy} \right)^2 = 1 + \frac{y^2}{a^2 - y^2} = \frac{a^2 - y^2}{a^2 - y^2} + \frac{y^2}{a^2 - y^2} = \frac{a^2}{a^2 - y^2} \quad \Rightarrow$$

$$S = \int_0^{a/2} 2\pi \sqrt{a^2 - y^2} \frac{a}{\sqrt{a^2 - y^2}}\, dy = 2\pi \int_0^{a/2} a\, dy = 2\pi a \left[y \right]_0^{a/2} = 2\pi a \left(\frac{a}{2} - 0 \right) = \pi a^2.$$

Note that this is $\frac{1}{4}$ the surface area of a sphere of radius a, and the length of the interval $y = 0$ to $y = a/2$ is $\frac{1}{4}$ the length of the interval $y = -a$ to $y = a$.

16. $y = \frac{1}{4} x^2 - \frac{1}{2} \ln x \quad \Rightarrow \quad \frac{dy}{dx} = \frac{x}{2} - \frac{1}{2x} \quad \Rightarrow \quad 1 + \left(\frac{dy}{dx} \right)^2 = 1 + \frac{x^2}{4} - \frac{1}{2} + \frac{1}{4x^2} = \frac{x^2}{4} + \frac{1}{2} + \frac{1}{4x^2} = \left(\frac{x}{2} + \frac{1}{2x} \right)^2$. So

$$S = \int_1^2 2\pi x \sqrt{\left(\frac{x}{2} + \frac{1}{2x} \right)^2}\, dx = 2\pi \int_1^2 x \left(\frac{x}{2} + \frac{1}{2x} \right) dx = \pi \int_1^2 (x^2 + 1)\, dx = \pi \left[\frac{1}{3} x^3 + x \right]_1^2$$

$$= \pi \left[\left(\frac{8}{3} + 2 \right) - \left(\frac{1}{3} + 1 \right) \right] = \frac{10}{3} \pi$$

17. $y = \ln x \quad \Rightarrow \quad dy/dx = 1/x \quad \Rightarrow \quad 1 + (dy/dx)^2 = 1 + 1/x^2 \quad \Rightarrow \quad S = \int_1^3 2\pi \ln x \sqrt{1 + 1/x^2}\, dx$.

Let $f(x) = \ln x \sqrt{1 + 1/x^2}$. Since $n = 10$, $\Delta x = \frac{3-1}{10} = \frac{1}{5}$. Then

$S \approx S_{10} = 2\pi \cdot \frac{1/5}{3} \left[f(1) + 4f(1.2) + 2f(1.4) + \cdots + 2f(2.6) + 4f(2.8) + f(3) \right] \approx 9.023754$.

The value of the integral produced by a calculator is 9.024262 (to six decimal places).

18. $y = x + \sqrt{x} \quad \Rightarrow \quad dy/dx = 1 + \frac{1}{2} x^{-1/2} \quad \Rightarrow \quad 1 + (dy/dx)^2 = 2 + x^{-1/2} + \frac{1}{4} x^{-1} \quad \Rightarrow$

$$S = \int_1^2 2\pi \left(x + \sqrt{x} \right) \sqrt{2 + \frac{1}{\sqrt{x}} + \frac{1}{4x}}\, dx.$$ Let $f(x) = (x + \sqrt{x}) \sqrt{2 + \frac{1}{\sqrt{x}} + \frac{1}{4x}}$. Since $n = 10$, $\Delta x = \frac{2-1}{10} = \frac{1}{10}$.

Then $S \approx S_{10} = 2\pi \cdot \frac{1/10}{3} \left[f(1) + 4f(1.1) + 2f(1.2) + \cdots + 2f(1.8) + 4f(1.9) + f(2) \right] \approx 29.506566$.

The value of the integral produced by a calculator is 29.506568 (to six decimal places).

19. $y = \sec x \quad \Rightarrow \quad dy/dx = \sec x \tan x \quad \Rightarrow \quad 1 + (dy/dx)^2 = 1 + \sec^2 x \tan^2 x \quad \Rightarrow$

$S = \int_0^{\pi/3} 2\pi \sec x \sqrt{1 + \sec^2 x \tan^2 x}\, dx$. Let $f(x) = \sec x \sqrt{1 + \sec^2 x \tan^2 x}$. Since $n = 10$, $\Delta x = \frac{\pi/3 - 0}{10} = \frac{\pi}{30}$.

Then $S \approx S_{10} = 2\pi \cdot \frac{\pi/30}{3} \left[f(0) + 4f\left(\frac{\pi}{30} \right) + 2f\left(\frac{2\pi}{30} \right) + \cdots + 2f\left(\frac{8\pi}{30} \right) + 4f\left(\frac{9\pi}{30} \right) + f\left(\frac{\pi}{3} \right) \right] \approx 13.527296$.

The value of the integral produced by a calculator is 13.516987 (to six decimal places).

20. $y = e^{-x^2}$ $\Rightarrow$ $dy/dx = -2xe^{-x^2}$ $\Rightarrow$ $1 + (dy/dx)^2 = 1 + 4x^2 e^{-2x^2}$ $\Rightarrow$

$S = \int_0^1 2\pi e^{-x^2} \sqrt{1 + 4x^2 e^{-2x^2}}\, dx$. Let $f(x) = e^{-x^2} \sqrt{1 + 4x^2 e^{-2x^2}}$. Since $n = 10$, $\Delta x = \frac{1-0}{10} = \frac{1}{10}$.

Then $S \approx S_{10} = 2\pi \cdot \frac{1/10}{3} \left[f(0) + 4f\left(\frac{1}{10}\right) + 2f\left(\frac{2}{10}\right) + \cdots + 2f\left(\frac{8}{10}\right) + 4f\left(\frac{9}{10}\right) + f(1) \right] \approx 5.537658$.

The value of the integral produced by a calculator is 5.537643 (to six decimal places).

21. $y = 1/x$ $\Rightarrow$ $ds = \sqrt{1 + (dy/dx)^2}\, dx = \sqrt{1 + (-1/x^2)^2}\, dx = \sqrt{1 + 1/x^4}\, dx$ $\Rightarrow$

$S = \int_1^2 2\pi \cdot \frac{1}{x} \sqrt{1 + \frac{1}{x^4}}\, dx = 2\pi \int_1^2 \frac{\sqrt{x^4 + 1}}{x^3}\, dx = 2\pi \int_1^4 \frac{\sqrt{u^2 + 1}}{u^2} \left(\frac{1}{2}\, du\right)$ $\qquad [u = x^2, du = 2x\, dx]$

$= \pi \int_1^4 \frac{\sqrt{1 + u^2}}{u^2}\, du \overset{24}{=} \pi \left[-\frac{\sqrt{1+u^2}}{u} + \ln\left(u + \sqrt{1 + u^2}\right) \right]_1^4$

$= \pi \left[-\frac{\sqrt{17}}{4} + \ln\left(4 + \sqrt{17}\right) + \frac{\sqrt{2}}{1} - \ln\left(1 + \sqrt{2}\right) \right] = \frac{\pi}{4} \left[4\ln\left(\sqrt{17} + 4\right) - 4\ln\left(\sqrt{2} + 1\right) - \sqrt{17} + 4\sqrt{2} \right]$

22. $y = \sqrt{x^2 + 1}$ $\Rightarrow$ $\dfrac{dy}{dx} = \dfrac{x}{\sqrt{x^2 + 1}}$ $\Rightarrow$ $ds = \sqrt{1 + \left(\dfrac{dy}{dx}\right)^2}\, dx = \sqrt{1 + \dfrac{x^2}{x^2 + 1}}\, dx$ $\Rightarrow$

$S = \int_0^3 2\pi \sqrt{x^2 + 1} \sqrt{1 + \frac{x^2}{x^2 + 1}}\, dx = 2\pi \int_0^3 \sqrt{2x^2 + 1}\, dx = 2\sqrt{2}\,\pi \int_0^3 \sqrt{x^2 + \left(\frac{1}{\sqrt{2}}\right)^2}\, dx$

$\overset{21}{=} 2\sqrt{2}\pi \left[\frac{1}{2} x \sqrt{x^2 + \frac{1}{2}} + \frac{1}{4} \ln\left(x + \sqrt{x^2 + \frac{1}{2}}\right) \right]_0^3 = 2\sqrt{2}\pi \left[\frac{3}{2}\sqrt{9 + \frac{1}{2}} + \frac{1}{4}\ln\left(3 + \sqrt{9 + \frac{1}{2}}\right) - \frac{1}{4}\ln\frac{1}{\sqrt{2}} \right]$

$= 2\sqrt{2}\pi \left[\frac{3}{2}\sqrt{\frac{19}{2}} + \frac{1}{4}\ln\left(3 + \sqrt{\frac{19}{2}}\right) + \frac{1}{4}\ln\sqrt{2} \right] = 2\sqrt{2}\pi \left[\frac{3}{2}\frac{\sqrt{19}}{\sqrt{2}} + \frac{1}{4}\ln\left(3\sqrt{2} + \sqrt{19}\right) \right]$

$= 3\sqrt{19}\pi + \frac{\pi}{\sqrt{2}}\ln\left(3\sqrt{2} + \sqrt{19}\right)$

23. $y = x^3$ and $0 \le y \le 1$ $\Rightarrow$ $y' = 3x^2$ and $0 \le x \le 1$.

$S = \int_0^1 2\pi x \sqrt{1 + (3x^2)^2}\, dx = 2\pi \int_0^3 \sqrt{1 + u^2}\, \frac{1}{6}\, du$ $\quad \begin{bmatrix} u = 3x^2, \\ du = 6x\, dx \end{bmatrix}$ $\quad = \frac{\pi}{3} \int_0^3 \sqrt{1 + u^2}\, du$

$\overset{21}{=}$ [or use CAS] $\frac{\pi}{3} \left[\frac{1}{2} u \sqrt{1 + u^2} + \frac{1}{2} \ln\left(u + \sqrt{1 + u^2}\right) \right]_0^3 = \frac{\pi}{3} \left[\frac{3}{2}\sqrt{10} + \frac{1}{2}\ln\left(3 + \sqrt{10}\right) \right] = \frac{\pi}{6} \left[3\sqrt{10} + \ln\left(3 + \sqrt{10}\right) \right]$

24. $y = \ln(x + 1)$, $0 \le x \le 1$. $ds = \sqrt{1 + \left(\dfrac{dy}{dx}\right)^2}\, dx = \sqrt{1 + \left(\dfrac{1}{x + 1}\right)^2}\, dx$, so

$S = \int_0^1 2\pi x \sqrt{1 + \frac{1}{(x+1)^2}}\, dx = \int_1^2 2\pi (u - 1) \sqrt{1 + \frac{1}{u^2}}\, du$ $\qquad [u = x + 1, du = dx]$

$= 2\pi \int_1^2 u \frac{\sqrt{1 + u^2}}{u}\, du - 2\pi \int_1^2 \frac{\sqrt{1 + u^2}}{u}\, du = 2\pi \int_1^2 \sqrt{1 + u^2}\, du - 2\pi \int_1^2 \frac{\sqrt{1 + u^2}}{u}\, du$

$\overset{21,23}{=}$ [or use CAS] $2\pi \left[\frac{1}{2} u \sqrt{1 + u^2} + \frac{1}{2}\ln\left(u + \sqrt{1 + u^2}\right) \right]_1^2 - 2\pi \left[\sqrt{1 + u^2} - \ln\left(\frac{1 + \sqrt{1 + u^2}}{u}\right) \right]_1^2$

$= 2\pi \left[\sqrt{5} + \frac{1}{2}\ln\left(2 + \sqrt{5}\right) - \frac{1}{2}\sqrt{2} - \frac{1}{2}\ln\left(1 + \sqrt{2}\right) \right] - 2\pi \left[\sqrt{5} - \ln\left(\frac{1 + \sqrt{5}}{2}\right) - \sqrt{2} + \ln\left(1 + \sqrt{2}\right) \right]$

$= 2\pi \left[\frac{1}{2}\ln\left(2 + \sqrt{5}\right) + \ln\left(\frac{1 + \sqrt{5}}{2}\right) + \frac{\sqrt{2}}{2} - \frac{3}{2}\ln\left(1 + \sqrt{2}\right) \right]$

25. $S = 2\pi \int_{1}^{\infty} y \sqrt{1 + \left(\dfrac{dy}{dx}\right)^2}\, dx = 2\pi \int_{1}^{\infty} \dfrac{1}{x}\sqrt{1 + \dfrac{1}{x^4}}\, dx = 2\pi \int_{1}^{\infty} \dfrac{\sqrt{x^4 + 1}}{x^3}\, dx$. Rather than trying to

evaluate this integral, note that $\sqrt{x^4 + 1} > \sqrt{x^4} = x^2$ for $x > 0$. Thus, if the area is finite,

$S = 2\pi \int_{1}^{\infty} \dfrac{\sqrt{x^4 + 1}}{x^3}\, dx > 2\pi \int_{1}^{\infty} \dfrac{x^2}{x^3}\, dx = 2\pi \int_{1}^{\infty} \dfrac{1}{x}\, dx$. But we know that this integral diverges, so the area S is

infinite.

26. $S = \int_{0}^{\infty} 2\pi y \sqrt{1 + (dy/dx)^2}\, dx = 2\pi \int_{0}^{\infty} e^{-x}\sqrt{1 + (-e^{-x})^2}\, dx$ $[y = e^{-x}, y' = -e^{-x}]$.

Evaluate $I = \int e^{-x}\sqrt{1 + (-e^{-x})^2}\, dx$ by using the substitution $u = -e^{-x}$, $du = e^{-x}\, dx$:

$I = \int \sqrt{1 + u^2}\, du \overset{21}{=} \tfrac{1}{2}u\sqrt{1 + u^2} + \tfrac{1}{2}\ln\left(u + \sqrt{1 + u^2}\right) + C = \tfrac{1}{2}(-e^{-x})\sqrt{1 + e^{-2x}} + \tfrac{1}{2}\ln\left(-e^{-x} + \sqrt{1 + e^{-2x}}\right) + C.$

Returning to the surface area integral, we have

$S = 2\pi \lim\limits_{t \to \infty} \int_{0}^{t} e^{-x}\sqrt{1 + (-e^{-x})^2}\, dx = 2\pi \lim\limits_{t \to \infty} \left[\tfrac{1}{2}(-e^{-x})\sqrt{1 + e^{-2x}} + \tfrac{1}{2}\ln\left(-e^{-x} + \sqrt{1 + e^{-2x}}\right)\right]_{0}^{t}$

$= 2\pi \lim\limits_{t \to \infty} \left\{\left[\tfrac{1}{2}(-e^{-t})\sqrt{1 + e^{-2t}} + \tfrac{1}{2}\ln\left(-e^{-t} + \sqrt{1 + e^{-2t}}\right)\right] - \left[\tfrac{1}{2}(-1)\sqrt{1 + 1} + \tfrac{1}{2}\ln\left(-1 + \sqrt{1 + 1}\right)\right]\right\}$

$= 2\pi \left\{\left[\tfrac{1}{2}(0)\sqrt{1} + \tfrac{1}{2}\ln\left(0 + \sqrt{1}\right)\right] - \left[-\tfrac{1}{2}\sqrt{2} + \tfrac{1}{2}\ln\left(-1 + \sqrt{2}\right)\right]\right\}$

$= 2\pi\left\{[0] + \tfrac{1}{2}\left[\sqrt{2} - \ln\left(\sqrt{2} - 1\right)\right]\right\} = \pi\left[\sqrt{2} - \ln\left(\sqrt{2} - 1\right)\right]$

27. Since $a > 0$, the curve $3ay^2 = x(a - x)^2$ only has points with $x \ge 0$.

[$3ay^2 \ge 0 \;\Rightarrow\; x(a - x)^2 \ge 0 \;\Rightarrow\; x \ge 0$.]

The curve is symmetric about the x-axis (since the equation is unchanged

when y is replaced by $-y$). $y = 0$ when $x = 0$ or a, so the curve's loop

extends from $x = 0$ to $x = a$.

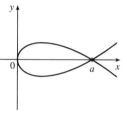

$\dfrac{d}{dx}(3ay^2) = \dfrac{d}{dx}[x(a - x)^2] \;\Rightarrow\; 6ay\dfrac{dy}{dx} = x \cdot 2(a - x)(-1) + (a - x)^2 \;\Rightarrow\; \dfrac{dy}{dx} = \dfrac{(a - x)[-2x + a - x]}{6ay} \;\Rightarrow$

$\left(\dfrac{dy}{dx}\right)^2 = \dfrac{(a - x)^2(a - 3x)^2}{36a^2y^2} = \dfrac{(a - x)^2(a - 3x)^2}{36a^2} \cdot \dfrac{3a}{x(a - x)^2}$ $\begin{bmatrix}\text{the last fraction} \\ \text{is } 1/y^2\end{bmatrix} = \dfrac{(a - 3x)^2}{12ax} \;\Rightarrow$

$1 + \left(\dfrac{dy}{dx}\right)^2 = 1 + \dfrac{a^2 - 6ax + 9x^2}{12ax} = \dfrac{12ax}{12ax} + \dfrac{a^2 - 6ax + 9x^2}{12ax} = \dfrac{a^2 + 6ax + 9x^2}{12ax} = \dfrac{(a + 3x)^2}{12ax}$ for $x \ne 0$.

(a) $S = \int_{x=0}^{a} 2\pi y\, ds = 2\pi \int_{0}^{a} \dfrac{\sqrt{x}\,(a - x)}{\sqrt{3a}} \cdot \dfrac{a + 3x}{\sqrt{12ax}}\, dx = 2\pi \int_{0}^{a} \dfrac{(a - x)(a + 3x)}{6a}\, dx$

$= \dfrac{\pi}{3a} \int_{0}^{a}(a^2 + 2ax - 3x^2)\, dx = \dfrac{\pi}{3a}\left[a^2x + ax^2 - x^3\right]_{0}^{a} = \dfrac{\pi}{3a}(a^3 + a^3 - a^3) = \dfrac{\pi}{3a} \cdot a^3 = \dfrac{\pi a^2}{3}.$

Note that we have rotated the top half of the loop about the x-axis. This generates the full surface.

(b) We must rotate the full loop about the y-axis, so we get double the area obtained by rotating the top half of the loop:

$S = 2 \cdot 2\pi \int_{x=0}^{a} x\, ds = 4\pi \int_{0}^{a} x\,\dfrac{a + 3x}{\sqrt{12ax}}\, dx = \dfrac{4\pi}{2\sqrt{3a}} \int_{0}^{a} x^{1/2}(a + 3x)\, dx = \dfrac{2\pi}{\sqrt{3a}} \int_{0}^{a}(ax^{1/2} + 3x^{3/2})\, dx$

$= \dfrac{2\pi}{\sqrt{3a}}\left[\dfrac{2}{3}ax^{3/2} + \dfrac{6}{5}x^{5/2}\right]_{0}^{a} = \dfrac{2\pi\sqrt{3}}{3\sqrt{a}}\left(\dfrac{2}{3}a^{5/2} + \dfrac{6}{5}a^{5/2}\right) = \dfrac{2\pi\sqrt{3}}{3}\left(\dfrac{2}{3} + \dfrac{6}{5}\right)a^2 = \dfrac{2\pi\sqrt{3}}{3}\left(\dfrac{28}{15}\right)a^2$

$= \dfrac{56\pi\sqrt{3}\,a^2}{45}$

28. In general, if the parabola $y = ax^2$, $-c \le x \le c$, is rotated about the y-axis, the surface area it generates is

$$2\pi \int_0^c x\sqrt{1 + (2ax)^2}\, dx = 2\pi \int_0^{2ac} \frac{u}{2a}\sqrt{1 + u^2}\, \frac{1}{2a}\, du \quad \begin{bmatrix} u = 2ax, \\ du = 2a\, dx \end{bmatrix} = \frac{\pi}{4a^2} \int_0^{2ac} \left(1 + u^2\right)^{1/2} 2u\, du$$

$$= \frac{\pi}{4a^2}\left[\tfrac{2}{3}\left(1 + u^2\right)^{3/2}\right]_0^{2ac} = \frac{\pi}{6a^2}\left[\left(1 + 4a^2c^2\right)^{3/2} - 1\right]$$

Here $2c = 10$ ft and $ac^2 = 2$ ft, so $c = 5$ and $a = \frac{2}{25}$. Thus, the surface area is

$$S = \frac{\pi}{6}\frac{625}{4}\left[\left(1 + 4\cdot\frac{4}{625}\cdot 25\right)^{3/2} - 1\right] = \frac{625\pi}{24}\left[\left(1 + \frac{16}{25}\right)^{3/2} - 1\right] = \frac{625\pi}{24}\left(\frac{41\sqrt{41}}{125} - 1\right) = \frac{5\pi}{24}\left(41\sqrt{41} - 125\right) \approx 90.01 \text{ ft}^2.$$

29. (a) $\dfrac{x^2}{a^2} + \dfrac{y^2}{b^2} = 1 \ \Rightarrow\ \dfrac{y\,(dy/dx)}{b^2} = -\dfrac{x}{a^2} \ \Rightarrow\ \dfrac{dy}{dx} = -\dfrac{b^2 x}{a^2 y} \ \Rightarrow$

$$1 + \left(\frac{dy}{dx}\right)^2 = 1 + \frac{b^4 x^2}{a^4 y^2} = \frac{b^4 x^2 + a^4 y^2}{a^4 y^2} = \frac{b^4 x^2 + a^4 b^2\left(1 - x^2/a^2\right)}{a^4 b^2\left(1 - x^2/a^2\right)} = \frac{a^4 b^2 + b^4 x^2 - a^2 b^2 x^2}{a^4 b^2 - a^2 b^2 x^2}$$

$$= \frac{a^4 + b^2 x^2 - a^2 x^2}{a^4 - a^2 x^2} = \frac{a^4 - \left(a^2 - b^2\right)x^2}{a^2\left(a^2 - x^2\right)}$$

The ellipsoid's surface area is twice the area generated by rotating the first-quadrant portion of the ellipse about the x-axis. Thus,

$$S = 2\int_0^a 2\pi y\sqrt{1 + \left(\frac{dy}{dx}\right)^2}\, dx = 4\pi \int_0^a \frac{b}{a}\sqrt{a^2 - x^2}\, \frac{\sqrt{a^4 - (a^2 - b^2)x^2}}{a\sqrt{a^2 - x^2}}\, dx = \frac{4\pi b}{a^2}\int_0^a \sqrt{a^4 - (a^2 - b^2)x^2}\, dx$$

$$= \frac{4\pi b}{a^2}\int_0^{a\sqrt{a^2 - b^2}} \sqrt{a^4 - u^2}\, \frac{du}{\sqrt{a^2 - b^2}} \quad \left[u = \sqrt{a^2 - b^2}\, x\right] \ \overset{30}{=}\ \frac{4\pi b}{a^2\sqrt{a^2 - b^2}}\left[\frac{u}{2}\sqrt{a^4 - u^2} + \frac{a^4}{2}\sin^{-1}\left(\frac{u}{a^2}\right)\right]_0^{a\sqrt{a^2 - b^2}}$$

$$= \frac{4\pi b}{a^2\sqrt{a^2 - b^2}}\left[\frac{a\sqrt{a^2 - b^2}}{2}\sqrt{a^4 - a^2(a^2 - b^2)} + \frac{a^4}{2}\sin^{-1}\frac{\sqrt{a^2 - b^2}}{a}\right] = 2\pi\left[b^2 + \frac{a^2 b\sin^{-1}\frac{\sqrt{a^2 - b^2}}{a}}{\sqrt{a^2 - b^2}}\right]$$

(b) $\dfrac{x^2}{a^2} + \dfrac{y^2}{b^2} = 1 \ \Rightarrow\ \dfrac{x\,(dx/dy)}{a^2} = -\dfrac{y}{b^2} \ \Rightarrow\ \dfrac{dx}{dy} = -\dfrac{a^2 y}{b^2 x} \ \Rightarrow$

$$1 + \left(\frac{dx}{dy}\right)^2 = 1 + \frac{a^4 y^2}{b^4 x^2} = \frac{b^4 x^2 + a^4 y^2}{b^4 x^2} = \frac{b^4 a^2(1 - y^2/b^2) + a^4 y^2}{b^4 a^2(1 - y^2/b^2)} = \frac{a^2 b^4 - a^2 b^2 y^2 + a^4 y^2}{a^2 b^4 - a^2 b^2 y^2}$$

$$= \frac{b^4 - b^2 y^2 + a^2 y^2}{b^4 - b^2 y^2} = \frac{b^4 - (b^2 - a^2)y^2}{b^2(b^2 - y^2)}$$

The oblate spheroid's surface area is twice the area generated by rotating the first-quadrant portion of the ellipse about the y-axis. Thus,

$$S = 2\int_0^b 2\pi x\sqrt{1 + \left(\frac{dx}{dy}\right)^2}\, dy = 4\pi \int_0^b \frac{a}{b}\sqrt{b^2 - y^2}\, \frac{\sqrt{b^4 - (b^2 - a^2)y^2}}{b\sqrt{b^2 - y^2}}\, dy$$

$$= \frac{4\pi a}{b^2}\int_0^b \sqrt{b^4 - (b^2 - a^2)y^2}\, dy = \frac{4\pi a}{b^2}\int_0^{b\sqrt{b^2 - a^2}} \sqrt{b^4 - u^2}\, \frac{du}{\sqrt{b^2 - a^2}} \quad \left[u = \sqrt{b^2 - a^2}\, y\right]$$

$$\overset{30}{=}\ \frac{4\pi a}{b^2\sqrt{b^2 - a^2}}\left[\frac{u}{2}\sqrt{b^4 - u^2} + \frac{b^4}{2}\sin^{-1}\left(\frac{u}{b^2}\right)\right]_0^{b\sqrt{b^2 - a^2}}$$

$$= \frac{4\pi a}{b^2\sqrt{b^2 - a^2}}\left[\frac{b\sqrt{b^2 - a^2}}{2}\sqrt{b^4 - b^2(b^2 - a^2)} + \frac{b^4}{2}\sin^{-1}\frac{\sqrt{b^2 - a^2}}{b}\right] = 2\pi\left[a^2 + \frac{ab^2\sin^{-1}\frac{\sqrt{b^2 - a^2}}{b}}{\sqrt{b^2 - a^2}}\right]$$

Notice that this result can be obtained from the answer in part (a) by interchanging a and b.

30. The upper half of the torus is generated by rotating the curve $(x - R)^2 + y^2 = r^2$, $y > 0$, about the y-axis.

$$y\frac{dy}{dx} = -(x - R) \quad \Rightarrow \quad 1 + \left(\frac{dy}{dx}\right)^2 = 1 + \frac{(x - R)^2}{y^2} = \frac{y^2 + (x - R)^2}{y^2} = \frac{r^2}{r^2 - (x - R)^2}. \text{ Thus,}$$

$$S = 2\int_{R-r}^{R+r} 2\pi x \sqrt{1 + \left(\frac{dy}{dx}\right)^2}\, dx = 4\pi \int_{R-r}^{R+r} \frac{rx}{\sqrt{r^2 - (x - R)^2}}\, dx = 4\pi r \int_{-r}^{r} \frac{u + R}{\sqrt{r^2 - u^2}}\, du \quad [u = x - R]$$

$$= 4\pi r \int_{-r}^{r} \frac{u\, du}{\sqrt{r^2 - u^2}} + 4\pi Rr \int_{-r}^{r} \frac{du}{\sqrt{r^2 - u^2}} = 4\pi r \cdot 0 + 8\pi Rr \int_{0}^{r} \frac{du}{\sqrt{r^2 - u^2}} \quad \begin{bmatrix} \text{since the first integrand is odd} \\ \text{and the second is even} \end{bmatrix}$$

$$= 8\pi Rr \left[\sin^{-1}(u/r)\right]_{0}^{r} = 8\pi Rr\left(\tfrac{\pi}{2}\right) = 4\pi^2 Rr$$

31. The analogue of $f(x_i^*)$ in the derivation of (4) is now $c - f(x_i^*)$, so

$$S = \lim_{n \to \infty} \sum_{i=1}^{n} 2\pi[c - f(x_i^*)]\sqrt{1 + [f'(x_i^*)]^2}\, \Delta x = \int_a^b 2\pi[c - f(x)]\sqrt{1 + [f'(x)]^2}\, dx.$$

32. $y = x^{1/2} \quad \Rightarrow \quad y' = \tfrac{1}{2}x^{-1/2} \quad \Rightarrow \quad 1 + (y')^2 = 1 + 1/4x$, so by Exercise 31, $S = \int_0^4 2\pi\left(4 - \sqrt{x}\right)\sqrt{1 + 1/(4x)}\, dx$.

Using a CAS, we get $S = 2\pi \ln\left(\sqrt{17} + 4\right) + \tfrac{\pi}{6}\left(31\sqrt{17} + 1\right) \approx 80.6095$.

33. For the upper semicircle, $f(x) = \sqrt{r^2 - x^2}$, $f'(x) = -x/\sqrt{r^2 - x^2}$. The surface area generated is

$$S_1 = \int_{-r}^{r} 2\pi\left(r - \sqrt{r^2 - x^2}\right)\sqrt{1 + \frac{x^2}{r^2 - x^2}}\, dx = 4\pi \int_0^r \left(r - \sqrt{r^2 - x^2}\right)\frac{r}{\sqrt{r^2 - x^2}}\, dx$$

$$= 4\pi \int_0^r \left(\frac{r^2}{\sqrt{r^2 - x^2}} - r\right) dx$$

For the lower semicircle, $f(x) = -\sqrt{r^2 - x^2}$ and $f'(x) = \dfrac{x}{\sqrt{r^2 - x^2}}$, so $S_2 = 4\pi \int_0^r \left(\dfrac{r^2}{\sqrt{r^2 - x^2}} + r\right) dx$.

Thus, the total area is $S = S_1 + S_2 = 8\pi \int_0^r \left(\dfrac{r^2}{\sqrt{r^2 - x^2}}\right) dx = 8\pi\left[r^2 \sin^{-1}\left(\dfrac{x}{r}\right)\right]_0^r = 8\pi r^2\left(\dfrac{\pi}{2}\right) = 4\pi^2 r^2$.

34. Take the sphere $x^2 + y^2 + z^2 = \tfrac{1}{4}d^2$ and let the intersecting planes be

$y = c$ and $y = c + h$, where $-\tfrac{1}{2}d \le c \le \tfrac{1}{2}d - h$. The sphere intersects the

xy-plane in the circle $x^2 + y^2 = \tfrac{1}{4}d^2$. From this equation, we get

$$x\frac{dx}{dy} + y = 0, \text{ so } \frac{dx}{dy} = -\frac{y}{x}. \text{ The desired surface area is}$$

$$S = 2\pi \int x\, ds = 2\pi \int_c^{c+h} x\sqrt{1 + (dx/dy)^2}\, dy = 2\pi \int_c^{c+h} x\sqrt{1 + y^2/x^2}\, dy = 2\pi \int_c^{c+h} \sqrt{x^2 + y^2}\, dy$$

$$= 2\pi \int_c^{c+h} \tfrac{1}{2}d\, dy = \pi d \int_c^{c+h} dy = \pi dh$$

35. In the derivation of (4), we computed a typical contribution to the surface area to be $2\pi\dfrac{y_{i-1} + y_i}{2}|P_{i-1}P_i|$,

the area of a frustum of a cone. When $f(x)$ is not necessarily positive, the approximations $y_i = f(x_i) \approx f(x_i^*)$ and

$y_{i-1} = f(x_{i-1}) \approx f(x_i^*)$ must be replaced by $y_i = |f(x_i)| \approx |f(x_i^*)|$ and $y_{i-1} = |f(x_{i-1})| \approx |f(x_i^*)|$. Thus,

$2\pi\dfrac{y_{i-1} + y_i}{2}|P_{i-1}P_i| \approx 2\pi|f(x_i^*)|\sqrt{1 + [f'(x_i^*)]^2}\, \Delta x$. Continuing with the rest of the derivation as before,

we obtain $S = \int_a^b 2\pi|f(x)|\sqrt{1 + [f'(x)]^2}\, dx$.

36. Since $g(x) = f(x) + c$, we have $g'(x) = f'(x)$. Thus,

$$S_g = \int_a^b 2\pi g(x) \sqrt{1 + [g'(x)]^2} \, dx = \int_a^b 2\pi [f(x) + c] \sqrt{1 + [f'(x)]^2} \, dx$$

$$= \int_a^b 2\pi f(x) \sqrt{1 + [f'(x)]^2} \, dx + 2\pi c \int_a^b \sqrt{1 + [f'(x)]^2} \, dx = S_f + 2\pi c L$$

DISCOVERY PROJECT   Rotating on a Slant

1.

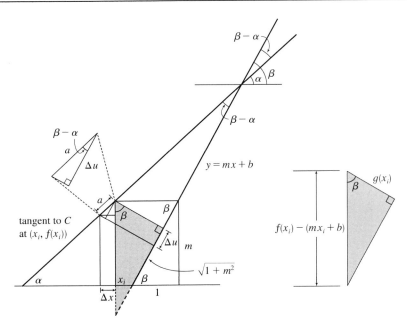

In the figure, the segment a lying above the interval $[x_i - \Delta x, x_i]$ along the tangent to C has length

$\Delta x \sec \alpha = \Delta x \sqrt{1 + \tan^2 \alpha} = \sqrt{1 + [f'(x_i)]^2} \, \Delta x$. The segment from $(x_i, f(x_i))$ drawn perpendicular to the line

$y = mx + b$ has length

$$g(x_i) = [f(x_i) - mx_i - b] \cos \beta = \frac{f(x_i) - mx_i - b}{\sec \beta} = \frac{f(x_i) - mx_i - b}{\sqrt{1 + \tan^2 \beta}} = \frac{f(x_i) - mx_i - b}{\sqrt{1 + m^2}}$$

Also, $\cos(\beta - \alpha) = \dfrac{\Delta u}{\Delta x \sec \alpha} \quad \Rightarrow$

$$\Delta u = \Delta x \sec \alpha \, \cos(\beta - \alpha) = \Delta x \, \frac{\cos \beta \, \cos \alpha + \sin \beta \, \sin \alpha}{\cos \alpha} = \Delta x (\cos \beta + \sin \beta \, \tan \alpha)$$

$$= \Delta x \left[\frac{1}{\sqrt{1 + m^2}} + \frac{m}{\sqrt{1 + m^2}} f'(x_i) \right] = \frac{1 + m f'(x_i)}{\sqrt{1 + m^2}} \Delta x$$

Thus,

$$\text{Area}(\mathcal{R}) = \lim_{n \to \infty} \sum_{i=1}^{n} g(x_i) \, \Delta u = \lim_{n \to \infty} \sum_{i=1}^{n} \frac{f(x_i) - mx_i - b}{\sqrt{1 + m^2}} \cdot \frac{1 + m f'(x_i)}{\sqrt{1 + m^2}} \Delta x$$

$$= \frac{1}{1 + m^2} \int_p^q [f(x) - mx - b][1 + m f'(x)] \, dx$$

2. From Problem 1 with $m = 1$, $f(x) = x + \sin x$, $mx + b = x - 2$, $p = 0$, and $q = 2\pi$,

$$\text{Area} = \frac{1}{1 + 1^2} \int_0^{2\pi} [x + \sin x - (x - 2)][1 + 1(1 + \cos x)]\, dx = \frac{1}{2} \int_0^{2\pi} (\sin x + 2)(2 + \cos x)\, dx$$

$$= \frac{1}{2} \int_0^{2\pi} (2 \sin x + \sin x \cos x + 4 + 2 \cos x)\, dx = \frac{1}{2} \left[-2 \cos x + \frac{1}{2} \sin^2 x + 4x + 2 \sin x \right]_0^{2\pi}$$

$$= \frac{1}{2} [(-2 + 0 + 8\pi + 0) - (-2 + 0 + 0 + 0)] = \frac{1}{2}(8\pi) = 4\pi$$

3. $V = \lim\limits_{n \to \infty} \sum\limits_{i=1}^{n} \pi [g(x_i)]^2 \, \Delta u = \lim\limits_{n \to \infty} \sum\limits_{i=1}^{n} \pi \left[\dfrac{f(x_i) - mx_i - b}{\sqrt{1 + m^2}} \right]^2 \dfrac{1 + mf'(x_i)}{\sqrt{1 + m^2}} \, \Delta x$

$$= \frac{\pi}{(1 + m^2)^{3/2}} \int_p^q [f(x) - mx - b]^2 [1 + mf'(x)]\, dx$$

4. $V = \dfrac{\pi}{(1 + 1^2)^{3/2}} \displaystyle\int_0^{2\pi} (x + \sin x - x + 2)^2 (1 + 1 + \cos x)\, dx$

$$= \frac{\pi}{2\sqrt{2}} \int_0^{2\pi} (\sin x + 2)^2 (\cos x + 2)\, dx = \frac{\pi}{2\sqrt{2}} \int_0^{2\pi} (\sin^2 x + 4 \sin x + 4)(\cos x + 2)\, dx$$

$$= \frac{\pi}{2\sqrt{2}} \int_0^{2\pi} (\sin^2 x \cos x + 4 \sin x \cos x + 4 \cos x + 2 \sin^2 x + 8 \sin x + 8)\, dx$$

$$= \frac{\pi}{2\sqrt{2}} \left[\frac{1}{3} \sin^3 x + 2 \sin^2 x + 4 \sin x + x - \frac{1}{2} \sin 2x - 8 \cos x + 8x \right]_0^{2\pi} \quad \text{[since } 2 \sin^2 x = 1 - \cos 2x\text{]}$$

$$= \frac{\pi}{2\sqrt{2}} [(2\pi - 8 + 16\pi) - (-8)] = \frac{9\sqrt{2}}{2} \pi^2$$

5. $S = \displaystyle\int_p^q 2\pi g(x) \sqrt{1 + [f'(x)]^2} \, dx = \dfrac{2\pi}{\sqrt{1 + m^2}} \int_p^q [f(x) - mx - b] \sqrt{1 + [f'(x)]^2} \, dx$

6. From Problem 5 with $f(x) = \sqrt{x}$, $p = 0$, $q = 4$, $m = \frac{1}{2}$, and $b = 0$,

$$S = \frac{2\pi}{\sqrt{1 + \left(\frac{1}{2}\right)^2}} \int_0^4 \left(\sqrt{x} - \frac{1}{2}x \right) \sqrt{1 + \left(\frac{1}{2\sqrt{x}} \right)^2} \, dx \overset{\text{CAS}}{=} \frac{\pi}{\sqrt{5}} \left[\frac{\ln(\sqrt{17} + 4)}{32} + \frac{37\sqrt{17}}{24} - \frac{1}{3} \right] \approx 8.554$$

9.3 Applications to Physics and Engineering

1. The weight density of water is $\delta = 62.5$ lb/ft^3.

(a) $P = \delta d \approx (62.5 \text{ lb/ft}^3)(3 \text{ ft}) = 187.5 \text{ lb/ft}^2$

(b) $F = PA \approx (187.5 \text{ lb/ft}^2)(5 \text{ ft})(2 \text{ ft}) = 1875$ lb. (*A* is the area of the bottom of the tank.)

(c) As in Example 1, the area of the ith strip is $2\,(\Delta x)$ and the pressure is $\delta d = \delta x_i$. Thus,

$\quad F = \int_0^3 \delta x \cdot 2\, dx \approx (62.5)(2) \int_0^3 x\, dx = 125 \left[\frac{1}{2}x^2 \right]_0^3 = 125\left(\frac{9}{2}\right) = 562.5$ lb.

2. (a) $P = \rho g d = (820 \text{ kg/m}^3)(9.8 \text{ m/s}^2)(1.5 \text{ m}) = 12{,}054 \text{ Pa} \approx 12$ kPa

(b) $F = PA = (12{,}054 \text{ Pa})(8 \text{ m})(4 \text{ m}) \approx 3.86 \times 10^5$ N   (*A* is the area at the bottom of the tank.)

(c) The area of the ith strip is $4(\Delta x)$ and the pressure is $\rho g d = \rho g\, x_i$. Thus,

$\quad F = \int_0^{1.5} \rho g x \cdot 4\, dx = (820)(9.8) \cdot 4 \int_0^{1.5} x\, dx = 32{,}144 \left[\frac{1}{2}x^2 \right]_0^{3/2} = 16{,}072 \left(\frac{9}{4}\right) \approx 3.62 \times 10^4$ N.

In Exercises 3–9, n is the number of subintervals of length Δx and x_i^* is a sample point in the ith subinterval $[x_{i-1}, x_i]$.

3. Set up a vertical x-axis as shown, with $x = 0$ at the water's surface and x increasing in the downward direction. Then the area of the ith rectangular strip is $6\,\Delta x$ and the pressure on the strip is δx_i^* (where $\delta \approx 62.5 \text{ lb/ft}^3$). Thus, the hydrostatic force on the strip is $\delta x_i^* \cdot 6\,\Delta x$ and the total hydrostatic force $\approx \sum_{i=1}^{n} \delta x_i^* \cdot 6\,\Delta x$. The total force

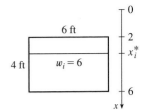

$$F = \lim_{n \to \infty} \sum_{i=1}^{n} \delta x_i^* \cdot 6\,\Delta x = \int_2^6 \delta x \cdot 6\,dx = 6\delta \int_2^6 x\,dx = 6\delta\left[\tfrac{1}{2}x^2\right]_2^6 = 6\delta(18 - 2) = 96\delta \approx 6000 \text{ lb}$$

4. Set up a vertical x-axis as shown. Then the area of the ith rectangular strip is

$$\tfrac{4}{3}(4 - x_i^*)\,\Delta x. \quad \left[\text{By similar triangles, } \frac{w_i}{4 - x_i^*} = \frac{4}{3}, \text{ so } w_i = \tfrac{4}{3}(4 - x_i^*).\right]$$

The pressure on the strip is δx_i^*, so the hydrostatic force on the strip is

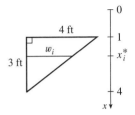

$\delta x_i^* \cdot \tfrac{4}{3}(4 - x_i^*)\,\Delta x$ and the total force on the plate $\approx \sum_{i=1}^{n} \delta x_i^* \cdot \tfrac{4}{3}(4 - x_i^*)\,\Delta x$. The total

force

$$F = \lim_{n \to \infty} \sum_{i=1}^{n} \delta x_i^* \cdot \tfrac{4}{3}(4 - x_i^*)\,\Delta x = \int_1^4 \delta x \cdot \tfrac{4}{3}(4 - x)\,dx = \tfrac{4}{3}\delta \int_1^4 (4x - x^2)\,dx$$

$$= \tfrac{4}{3}\delta\left[2x^2 - \tfrac{1}{3}x^3\right]_1^4 = \tfrac{4}{3}\delta\left[\left(32 - \tfrac{64}{3}\right) - \left(2 - \tfrac{1}{3}\right)\right] = \tfrac{4}{3}\delta(9) = 12\delta \approx 750 \text{ lb}$$

5. Set up a vertical x-axis as shown. The base of the triangle shown in the figure has length $\sqrt{3^2 - (x_i^*)^2}$, so $w_i = 2\sqrt{9 - (x_i^*)^2}$, and the area of the ith rectangular strip is $2\sqrt{9 - (x_i^*)^2}\,\Delta x$. The ith rectangular strip is $(x_i^* - 1)$ m below the surface level of the water, so the pressure on the strip is $\rho g(x_i^* - 1)$. The hydrostatic force on the strip is $\rho g(x_i^* - 1) \cdot 2\sqrt{9 - (x_i^*)^2}\,\Delta x$ and the total

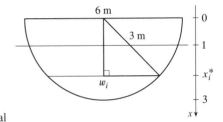

force on the plate $\approx \sum_{i=1}^{n} \rho g(x_i^* - 1) \cdot 2\sqrt{9 - (x_i^*)^2}\,\Delta x$. The total force

$$F = \lim \sum_{i=1}^{n} \rho g(x_i^* - 1) \cdot 2\sqrt{9 - (x_i^*)^2}\,\Delta x = 2\rho g \int_1^3 (x - 1)\sqrt{9 - x^2}\,dx$$

$$= 2\rho g \int_1^3 x\sqrt{9 - x^2}\,dx - 2\rho g \int_1^3 \sqrt{9 - x^2}\,dx \overset{30}{=} 2\rho g\left[-\tfrac{1}{3}(9 - x^2)^{3/2}\right]_1^3 - 2\rho g\left[\frac{x}{2}\sqrt{9 - x^2} + \tfrac{9}{2}\sin^{-1}\left(\frac{x}{3}\right)\right]_1^3$$

$$= 2\rho g\left[0 + \tfrac{1}{3}(8\sqrt{8})\right] - 2\rho g\left[\left(0 + \tfrac{9}{2} \cdot \tfrac{\pi}{2}\right) - \left(\tfrac{1}{2}\sqrt{8} + \tfrac{9}{2}\sin^{-1}\left(\tfrac{1}{3}\right)\right)\right]$$

$$= \tfrac{32}{3}\sqrt{2}\,\rho g - \tfrac{9\pi}{2}\rho g + 2\sqrt{2}\,\rho g + 9\left[\sin^{-1}\left(\tfrac{1}{3}\right)\right]\rho g = \left(\tfrac{38}{3}\sqrt{2} - \tfrac{9\pi}{2} + 9\sin^{-1}\left(\tfrac{1}{3}\right)\right)\rho g$$

$$\approx 6.835 \cdot 1000 \cdot 9.8 \approx 6.7 \times 10^4 \text{ N}$$

Note: If you set up a typical coordinate system with the water level at $y = -1$, then $F = \int_{-3}^{-1} \rho g(-1 - y)2\sqrt{9 - y^2}\,dy$.

6. By similar triangles, $w_i/4 = x_i^*/5$, so $w_i = \frac{4}{5}x_i^*$ and the area of the ith strip is $\frac{4}{5}x_i^* \Delta x$.

The pressure on the strip is $\rho g x_i^*$, so the hydrostatic force on the strip is $\rho g x_i^* \cdot \frac{4}{5}x_i^* \Delta x$

and the total force on the plate $\approx \sum_{i=1}^{n} \rho g x_i^* \cdot \frac{4}{5}x_i^* \Delta x$. The total force

$$F = \lim_{n\to\infty} \sum_{i=1}^{n} \rho g x_i^* \cdot \tfrac{4}{5}x_i^* \Delta x = \int_0^5 \rho g x \cdot \tfrac{4}{5}x \, dx = \tfrac{4}{5}\rho g \left[\tfrac{1}{3}x^3\right]_0^5 = \tfrac{4}{5}\rho g \cdot \tfrac{125}{3} = \tfrac{100}{3}\rho g$$

$$\approx \tfrac{100}{3} \cdot 1000 \cdot 9.8 \approx 3.3 \times 10^5 \text{ N}$$

7. Set up a vertical x-axis as shown. Then the area of the ith rectangular strip is

$$\left(2 - \frac{2}{\sqrt{3}}x_i^*\right)\Delta x. \left[\text{By similar triangles, } \frac{w_i}{2} = \frac{\sqrt{3}-x_i^*}{\sqrt{3}}, \text{ so } w_i = 2 - \frac{2}{\sqrt{3}}x_i^*.\right]$$

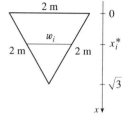

The pressure on the strip is $\rho g x_i^*$, so the hydrostatic force on the strip is

$\rho g x_i^* \left(2 - \frac{2}{\sqrt{3}}x_i^*\right)\Delta x$ and the hydrostatic force on the plate $\approx \sum_{i=1}^{n} \rho g x_i^* \left(2 - \frac{2}{\sqrt{3}}x_i^*\right)\Delta x$.

The total force

$$F = \lim_{n\to\infty} \sum_{i=1}^{n} \rho g x_i^* \left(2 - \frac{2}{\sqrt{3}}x_i^*\right)\Delta x = \int_0^{\sqrt{3}} \rho g x \left(2 - \frac{2}{\sqrt{3}}x\right)dx = \rho g \int_0^{\sqrt{3}}\left(2x - \frac{2}{\sqrt{3}}x^2\right)dx$$

$$= \rho g \left[x^2 - \frac{2}{3\sqrt{3}}x^3\right]_0^{\sqrt{3}} = \rho g \left[(3-2)-0\right] = \rho g \approx 1000 \cdot 9.8 = 9.8 \times 10^3 \text{ N}$$

8. Set up a vertical x-axis as shown. Then the area of the ith rectangular strip is

$$(4 - x_i^*)\Delta x. \left[\text{By similar triangles, } \frac{w_i}{4} = \frac{4-x_i^*}{4}, \text{ so } w_i = 4 - x_i^*.\right] \text{ The } i\text{th}$$

rectangular strip is $(x_i^* - 1)$ m below the surface level of the water, so the pressure on the

strip is $\rho g(x_i^* - 1)$. The hydrostatic force on the strip is $\rho g(x_i^* - 1)(4 - x_i^*)\Delta x$ and the

hydrostatic force on the plate $\approx \sum_{i=1}^{n} \rho g(x_i^* - 1)(4 - x_i^*)\Delta x$. The total force

$$F = \lim_{n\to\infty} \sum_{i=1}^{n} \rho g(x_i^* - 1)(4 - x_i^*)\Delta x = \int_1^2 \rho g(x-1)(4-x)\,dx = \rho g \int_1^2 (-x^2 + 5x - 4)\,dx$$

$$= \rho g \left[-\tfrac{1}{3}x^3 + \tfrac{5}{2}x^2 - 4x\right]_1^2 = \rho g \left[\left(-\tfrac{8}{3} + 10 - 8\right) - \left(-\tfrac{1}{3} + \tfrac{5}{2} - 4\right)\right]$$

$$= \tfrac{7}{6}\rho g \approx \tfrac{7}{6} \cdot 1000 \cdot 9.8 \approx 1.14 \times 10^4 \text{ N}$$

Note: If you let the water level correspond to $x = 0$, then $F = \int_0^1 \rho g x(3-x)\,dx$.

9. Set up coordinate axes as shown in the figure. The length of the ith strip is

$2\sqrt{25-(y_i^*)^2}$ and its area is $2\sqrt{25-(y_i^*)^2}\,\Delta y$. The pressure on this strip is

approximately $\delta d_i = 62.5(7-y_i^*)$ and so the force on the strip is approximately

$62.5(7-y_i^*)2\sqrt{25-(y_i^*)^2}\,\Delta y$. The total force

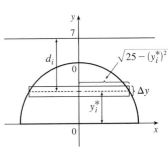

$$F = \lim_{n\to\infty}\sum_{i=1}^{n}62.5(7-y_i^*)2\sqrt{25-(y_i^*)^2}\,\Delta y = 125\int_0^5(7-y)\sqrt{25-y^2}\,dy$$

$$= 125\left\{\int_0^5 7\sqrt{25-y^2}\,dy - \int_0^5 y\sqrt{25-y^2}\,dy\right\} = 125\left\{7\int_0^5\sqrt{25-y^2}\,dy - \left[-\tfrac{1}{3}(25-y^2)^{3/2}\right]_0^5\right\}$$

$$= 125\left\{7\left(\tfrac{1}{4}\pi\cdot5^2\right)+\tfrac{1}{3}(0-125)\right\} = 125\left(\tfrac{175\pi}{4}-\tfrac{125}{3}\right) \approx 11{,}972 \approx 1.2\times10^4\text{ lb}$$

10. Set up coordinate axes as shown in the figure. For the *top half,* the length

of the ith strip is $2\left(a/\sqrt{2}-y_i^*\right)$ and its area is $2\left(a/\sqrt{2}-y_i^*\right)\Delta y$.

The pressure on this strip is approximately $\delta d_i = \delta\left(a/\sqrt{2}-y_i^*\right)$ and so the

force on the strip is approximately $2\delta\left(a/\sqrt{2}-y_i^*\right)^2\Delta y$. The total force

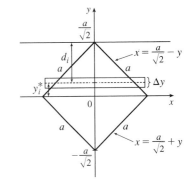

$$F_1 = \lim_{n\to\infty}\sum_{i=1}^{n}2\delta\left(\frac{a}{\sqrt{2}}-y_i^*\right)^2\Delta y = 2\delta\int_0^{a/\sqrt{2}}\left(\frac{a}{\sqrt{2}}-y\right)^2 dy$$

$$= 2\delta\left[-\frac{1}{3}\left(\frac{a}{\sqrt{2}}-y\right)^3\right]_0^{a/\sqrt{2}} = -\frac{2}{3}\delta\left[0-\left(\frac{a}{\sqrt{2}}\right)^3\right] = \frac{2\delta}{3}\frac{a^3}{2\sqrt{2}} = \frac{\sqrt{2}\,a^3\delta}{6}$$

For the *bottom half,* the length is $2\left(a/\sqrt{2}+y_i^*\right)$ and the total force is

$$F_2 = \lim_{n\to\infty}\sum_{i=1}^{n}2\delta\left(\frac{a}{\sqrt{2}}+y_i^*\right)\left(\frac{a}{\sqrt{2}}-y_i^*\right)\Delta y = 2\delta\int_{-a/\sqrt{2}}^{0}\left(\frac{a^2}{2}-y^2\right)dy = 2\delta\left[\tfrac{1}{2}a^2 y-\tfrac{1}{3}y^3\right]_{-a/\sqrt{2}}^{0}$$

$$= 2\delta\left[0-\left(-\frac{\sqrt{2}\,a^3}{4}+\frac{\sqrt{2}\,a^3}{12}\right)\right] = 2\delta\left(\frac{\sqrt{2}\,a^3}{6}\right) = \frac{2\sqrt{2}\,a^3\delta}{6}\qquad[F_2=2F_1]$$

Thus, the total force $F = F_1 + F_2 = \dfrac{3\sqrt{2}\,a^3\delta}{6} = \dfrac{\sqrt{2}\,a^3\delta}{2}$.

11. Set up a vertical x-axis as shown. Then the area of the ith rectangular strip is

$\dfrac{a}{h}(2h-x_i^*)\,\Delta x$. $\left[\text{By similar triangles, }\dfrac{w_i}{2h-x_i^*}=\dfrac{2a}{2h},\text{ so }w_i=\dfrac{a}{h}(2h-x_i^*).\right]$

The pressure on the strip is δx_i^*, so the hydrostatic force on the plate

$\approx \displaystyle\sum_{i=1}^{n}\delta x_i^*\frac{a}{h}(2h-x_i^*)\,\Delta x$. The total force

$$F = \lim_{n\to\infty}\sum_{i=1}^{n}\delta x_i^*\frac{a}{h}(2h-x_i^*)\,\Delta x = \delta\frac{a}{h}\int_0^h x(2h-x)\,dx = \frac{a\delta}{h}\int_0^h\left(2hx-x^2\right)dx$$

$$= \frac{a\delta}{h}\left[hx^2-\tfrac{1}{3}x^3\right]_0^h = \frac{a\delta}{h}\left(h^3-\tfrac{1}{3}h^3\right) = \frac{a\delta}{h}\left(\frac{2h^3}{3}\right) = \tfrac{2}{3}\delta ah^2$$

12. The area of the ith rectangular strip is $2\sqrt{2y_i^*}\,\Delta y$ and the pressure on it is $\delta d_i = \delta(8 - y_i^*)$.

$$F = \int_0^8 \delta(8 - y)\,2\sqrt{2y}\,dy = 42 \cdot 2 \cdot \sqrt{2}\int_0^8 (8-y)y^{1/2}\,dy$$

$$= 84\sqrt{2}\int_0^8 (8y^{1/2} - y^{3/2})\,dy = 84\sqrt{2}\left[8 \cdot \tfrac{2}{3}y^{3/2} - \tfrac{2}{5}y^{5/2}\right]_0^8$$

$$= 84\sqrt{2}\left[8 \cdot \tfrac{2}{3} \cdot 16\sqrt{2} - \tfrac{2}{5} \cdot 128\sqrt{2}\right]$$

$$= 84\sqrt{2} \cdot 256\sqrt{2}\left(\tfrac{1}{3} - \tfrac{1}{5}\right) = 43{,}008 \cdot \tfrac{2}{15} = 5734.4\text{ lb}$$

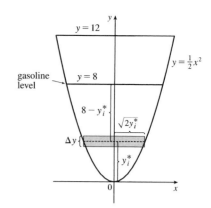

13. By similar triangles, $\dfrac{8}{4\sqrt{3}} = \dfrac{w_i}{x_i^*} \;\Rightarrow\; w_i = \dfrac{2x_i^*}{\sqrt{3}}$. The area of the ith

rectangular strip is $\dfrac{2x_i^*}{\sqrt{3}}\,\Delta x$ and the pressure on it is $\rho g\left(4\sqrt{3} - x_i^*\right)$.

$$F = \int_0^{4\sqrt{3}} \rho g\left(4\sqrt{3} - x\right)\frac{2x}{\sqrt{3}}\,dx = 8\rho g\int_0^{4\sqrt{3}} x\,dx - \frac{2\rho g}{\sqrt{3}}\int_0^{4\sqrt{3}} x^2\,dx$$

$$= 4\rho g\left[x^2\right]_0^{4\sqrt{3}} - \frac{2\rho g}{3\sqrt{3}}\left[x^3\right]_0^{4\sqrt{3}} = 192\rho g - \frac{2\rho g}{3\sqrt{3}}\,64 \cdot 3\sqrt{3} = 192\rho g - 128\rho g = 64\rho g$$

$$\approx 64(840)(9.8) \approx 5.27 \times 10^5\text{ N}$$

14. $F = \int_0^2 \rho g(10 - x)2\sqrt{4 - x^2}\,dx$

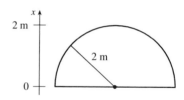

$$= 20\rho g\int_0^2 \sqrt{4 - x^2}\,dx - \rho g\int_0^2 \sqrt{4 - x^2}\,2x\,dx$$

$$= 20\rho g\tfrac{1}{4}\pi(2^2) - \rho g\int_0^4 u^{1/2}\,du \qquad [u = 4 - x^2,\, du = -2x\,dx]$$

$$= 20\pi\rho g - \tfrac{2}{3}\rho g\left[u^{3/2}\right]_0^4 = 20\pi\rho g - \tfrac{16}{3}\rho g = \rho g\left(20\pi - \tfrac{16}{3}\right)$$

$$= (1000)(9.8)\left(20\pi - \tfrac{16}{3}\right) \approx 5.63 \times 10^5\text{ N}$$

15. (a) The top of the cube has depth $d = 1\text{ m} - 20\text{ cm} = 80\text{ cm} = 0.8\text{ m}$.

$$F = \rho g d A \approx (1000)(9.8)(0.8)(0.2)^2 = 313.6 \approx 314\text{ N}$$

(b) The area of a strip is $0.2\,\Delta x$ and the pressure on it is $\rho g x_i^*$.

$$F = \int_{0.8}^1 \rho g x(0.2)\,dx = 0.2\rho g\left[\tfrac{1}{2}x^2\right]_{0.8}^1 = (0.2\rho g)(0.18) = 0.036\rho g = 0.036(1000)(9.8) = 352.8 \approx 353\text{ N}$$

16. The height of the dam is $h = \sqrt{70^2 - 25^2}\cos 30° = 15\sqrt{19}\left(\tfrac{\sqrt{3}}{2}\right)$.

The width of the trapezoid is $w = 50 + 2a$.

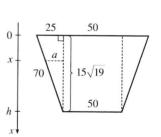

By similar triangles, $\dfrac{25}{h} = \dfrac{a}{h - x} \;\Rightarrow\; a = \dfrac{25}{h}(h - x)$. Thus,

$$w = 50 + 2 \cdot \frac{25}{h}(h - x) = 50 + \frac{50}{h} \cdot h - \frac{50}{h} \cdot x = 50 + 50 - \frac{50x}{h} = 100 - \frac{50x}{h}.$$

From the small triangle in the second figure, $\cos 30° = \dfrac{\Delta x}{z}$ ⇒

$z = \Delta x \sec 30° = 2\,\Delta x /\sqrt{3}.$

$F = \displaystyle\int_0^h \delta x \left(100 - \frac{50x}{h}\right)\frac{2}{\sqrt{3}}\,dx = \frac{200\delta}{\sqrt{3}}\int_0^h x\,dx - \frac{100\delta}{h\sqrt{3}}\int_0^h x^2\,dx$

$= \dfrac{200\delta}{\sqrt{3}}\dfrac{h^2}{2} - \dfrac{100\delta}{h\sqrt{3}}\dfrac{h^3}{3} = \dfrac{200\delta h^2}{3\sqrt{3}} = \dfrac{200(62.5)}{3\sqrt{3}}\cdot\dfrac{12{,}825}{4} \approx 7.71\times 10^6$ lb

17. (a) The area of a strip is $20\,\Delta x$ and the pressure on it is δx_i.

$$F = \int_0^3 \delta x 20\,dx = 20\delta\left[\tfrac{1}{2}x^2\right]_0^3 = 20\delta\cdot\tfrac{9}{2} = 90\delta$$

$$= 90(62.5) = 5625\text{ lb} \approx 5.63\times 10^3\text{ lb}$$

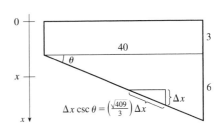

(b) $F = \int_0^9 \delta x 20\,dx = 20\delta\left[\tfrac{1}{2}x^2\right]_0^9 = 20\delta\cdot\tfrac{81}{2} = 810\delta = 810(62.5) = 50{,}625\text{ lb} \approx 5.06\times 10^4\text{ lb.}$

(c) For the first 3 ft, the length of the side is constant at 40 ft. For $3 < x \le 9$, we can use similar triangles to find the length a:

$\dfrac{a}{40} = \dfrac{9-x}{6}$ ⇒ $a = 40\cdot\dfrac{9-x}{6}.$

$F = \int_0^3 \delta x 40\,dx + \int_3^9 \delta x(40)\dfrac{9-x}{6}\,dx = 40\delta\left[\tfrac{1}{2}x^2\right]_0^3 + \tfrac{20}{3}\delta\int_3^9(9x-x^2)\,dx = 180\delta + \tfrac{20}{3}\delta\left[\tfrac{9}{2}x^2 - \tfrac{1}{3}x^3\right]_3^9$

$= 180\delta + \tfrac{20}{3}\delta\left[\left(\tfrac{729}{2} - 243\right) - \left(\tfrac{81}{2} - 9\right)\right] = 180\delta + 600\delta = 780\delta = 780(62.5) = 48{,}750\text{ lb} \approx 4.88\times 10^4\text{ lb}$

(d) For any right triangle with hypotenuse on the bottom,

$\sin\theta = \dfrac{\Delta x}{\text{hypotenuse}}$ ⇒

hypotenuse $= \Delta x \csc\theta = \Delta x\dfrac{\sqrt{40^2+6^2}}{6} = \dfrac{\sqrt{409}}{3}\Delta x.$

$F = \int_3^9 \delta x 20\dfrac{\sqrt{409}}{3}\,dx = \tfrac{1}{3}\left(20\sqrt{409}\right)\delta\left[\tfrac{1}{2}x^2\right]_3^9$

$= \tfrac{1}{3}\cdot 10\sqrt{409}\,\delta(81-9) \approx 303{,}356\text{ lb} \approx 3.03\times 10^5\text{ lb}$

18. Partition the interval $[a,b]$ by points x_i as usual and choose $x_i^* \in [x_{i-1},x_i]$ for each i. The ith horizontal strip of the immersed plate is approximated by a rectangle of height Δx_i and width $w(x_i^*)$, so its area is $A_i \approx w(x_i^*)\,\Delta x_i$. For small Δx_i, the pressure P_i on the ith strip is almost constant and $P_i \approx \rho g x_i^*$ by Equation 1. The hydrostatic force F_i acting on the ith strip is $F_i = P_i A_i \approx \rho g x_i^* w(x_i^*)\,\Delta x_i$. Adding these forces and taking the limit as $n \to \infty$, we obtain the hydrostatic force on the immersed plate:

$$F = \lim_{n\to\infty}\sum_{i=1}^n F_i = \lim_{n\to\infty}\sum_{i=1}^n \rho g x_i^* w(x_i^*)\,\Delta x_i = \int_a^b \rho g x w(x)\,dx$$

19. $F = \int_2^5 \rho g x \cdot w(x)\,dx$, where $w(x)$ is the width of the plate at depth x. Since $n = 6$, $\Delta x = \dfrac{5-2}{6} = \dfrac{1}{2}$, and

$F \approx S_6$

$= \rho g\cdot\dfrac{1/2}{3}[2\cdot w(2) + 4\cdot 2.5\cdot w(2.5) + 2\cdot 3\cdot w(3) + 4\cdot 3.5\cdot w(3.5) + 2\cdot 4\cdot w(4) + 4\cdot 4.5\cdot w(4.5) + 5\cdot w(5)]$

$= \tfrac{1}{6}\rho g(2\cdot 0 + 10\cdot 0.8 + 6\cdot 1.7 + 14\cdot 2.4 + 8\cdot 2.9 + 18\cdot 3.3 + 5\cdot 3.6)$

$= \tfrac{1}{6}(1000)(9.8)(152.4) \approx 2.5\times 10^5\text{ N}$

20. (a) From Equation 8, $\overline{x} = \frac{1}{A}\int_a^b xw(x)\,dx \;\Rightarrow\; A\overline{x} = \int_a^b xw(x)\,dx \;\Rightarrow\; \rho g A\overline{x} = \rho g \int_a^b xw(x)\,dx \;\Rightarrow$

$(\rho g\overline{x})A = \int_a^b \rho g\, xw(x)\,dx = F$ by Exercise 18.

(b) For the figure in Exercise 10, let the coordinates of the centroid $(\overline{x}, \overline{y}) = (a/\sqrt{2}, 0)$.

$$F = (\rho g\overline{x})A = \rho g\,\frac{a}{\sqrt{2}}\,a^2 = \delta\,\frac{\sqrt{2}\,a}{2}\,a^2 = \frac{\sqrt{2}\,a^3\delta}{2}.$$

21. The moment M of the system about the origin is $M = \sum_{i=1}^{2} m_i x_i = m_1 x_1 + m_2 x_2 = 40 \cdot 2 + 30 \cdot 5 = 230.$

The mass m of the system is $m = \sum_{i=1}^{2} m_i = m_1 + m_2 = 40 + 30 = 70.$

The center of mass of the system is $M/m = \frac{230}{70} = \frac{23}{7}.$

22. $M = m_1 x_1 + m_2 x_2 + m_3 x_3 = 25(-2) + 20(3) + 10(7) = 80$; $\overline{x} = M/(m_1 + m_2 + m_3) = \frac{80}{55} = \frac{16}{11}.$

23. $m = \sum_{i=1}^{3} m_i = 6 + 5 + 10 = 21.$

$M_x = \sum_{i=1}^{3} m_i\, y_i = 6(5) + 5(-2) + 10(-1) = 10$; $M_y = \sum_{i=1}^{3} m_i\, x_i = 6(1) + 5(3) + 10(-2) = 1.$

$\overline{x} = \dfrac{M_y}{m} = \dfrac{1}{21}$ and $\overline{y} = \dfrac{M_x}{m} = \dfrac{10}{21}$, so the center of mass of the system is $\left(\frac{1}{21}, \frac{10}{21}\right).$

24. $M_x = \sum_{i=1}^{4} m_i\, y_i = 6(-2) + 5(4) + 1(-7) + 4(-1) = -3,\ M_y = \sum_{i=1}^{4} m_i\, x_i = 6(1) + 5(3) + 1(-3) + 4(6) = 42,$

and $m = \sum_{i=1}^{4} m_i = 16$, so $\overline{x} = \dfrac{M_y}{m} = \dfrac{42}{16} = \dfrac{21}{8}$ and $\overline{y} = \dfrac{M_x}{m} = -\dfrac{3}{16}$; the center of mass is $(\overline{x}, \overline{y}) = \left(\frac{21}{8}, -\frac{3}{16}\right).$

25. Since the region in the figure is symmetric about the y-axis, we know

that $\overline{x} = 0$. The region is "bottom-heavy," so we know that $\overline{y} < 2$,

and we might guess that $\overline{y} = 1.5$.

$A = \int_{-2}^{2}(4 - x^2)\,dx = 2\int_0^2 (4 - x^2)\,dx = 2\left[4x - \frac{1}{3}x^3\right]_0^2$

$= 2\left(8 - \frac{8}{3}\right) = \frac{32}{3}.$

$\overline{x} = \frac{1}{A}\int_{-2}^{2} x(4 - x^2)\,dx = 0$ since $f(x) = x(4 - x^2)$ is an odd

function (or since the region is symmetric about the y-axis).

$\overline{y} = \frac{1}{A}\int_{-2}^{2}\frac{1}{2}(4 - x^2)^2\,dx = \frac{3}{32}\cdot\frac{1}{2}\cdot 2\int_0^2 (16 - 8x^2 + x^4)\,dx = \frac{3}{32}\left[16x - \frac{8}{3}x^3 + \frac{1}{5}x^5\right]_0^2$

$= \frac{3}{32}\left(32 - \frac{64}{3} + \frac{32}{5}\right) = 3\left(1 - \frac{2}{3} + \frac{1}{5}\right) = 3\left(\frac{8}{15}\right) = \frac{8}{5}$

Thus, the centroid is $(\overline{x}, \overline{y}) = \left(0, \frac{8}{5}\right).$

26. The region in the figure is "left-heavy" and "bottom-heavy," so we know $\overline{x} < 1$
and $\overline{y} < 1.5$, and we might guess that $\overline{x} = 0.7$ and $\overline{y} = 1.2$.

$3x + 2y = 6 \iff 2y = 6 - 3x \iff y = 3 - \frac{3}{2}x.$

$A = \int_0^2 \left(3 - \frac{3}{2}x\right) dx = \left[3x - \frac{3}{4}x^2\right]_0^2 = 6 - 3 = 3.$

$\overline{x} = \frac{1}{A}\int_0^2 x\left(3 - \frac{3}{2}x\right) dx = \frac{1}{3}\int_0^2 \left(3x - \frac{3}{2}x^2\right) dx = \frac{1}{3}\left[\frac{3}{2}x^2 - \frac{1}{2}x^3\right]_0^2$

$= \frac{1}{3}(6 - 4) = \frac{2}{3}.$

$\overline{y} = \frac{1}{A}\int_0^2 \frac{1}{2}\left(3 - \frac{3}{2}x\right)^2 dx = \frac{1}{3}\cdot\frac{1}{2}\int_0^2 \left(9 - 9x + \frac{9}{4}x^2\right) dx = \frac{1}{6}\left[9x - \frac{9}{2}x^2 + \frac{3}{4}x^3\right]_0^2 = \frac{1}{6}(18 - 18 + 6) = 1.$

Thus, the centroid is $(\overline{x}, \overline{y}) = \left(\frac{2}{3}, 1\right)$.

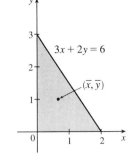

27. The region in the figure is "right-heavy" and "bottom-heavy," so we know
$\overline{x} > 0.5$ and $\overline{y} < 1$, and we might guess that $\overline{x} = 0.6$ and $\overline{y} = 0.9$.

$A = \int_0^1 e^x \, dx = [e^x]_0^1 = e - 1.$

$\overline{x} = \frac{1}{A}\int_0^1 xe^x \, dx = \frac{1}{e-1}[xe^x - e^x]_0^1 \qquad \text{[by parts]}$

$= \frac{1}{e-1}[0 - (-1)] = \frac{1}{e-1}.$

$\overline{y} = \frac{1}{A}\int_0^1 \frac{1}{2}(e^x)^2 \, dx = \frac{1}{e-1}\cdot\frac{1}{4}[e^{2x}]_0^1 = \frac{1}{4(e-1)}(e^2 - 1) = \frac{e+1}{4}.$

Thus, the centroid is $(\overline{x}, \overline{y}) = \left(\frac{1}{e-1}, \frac{e+1}{4}\right) \approx (0.58, 0.93).$

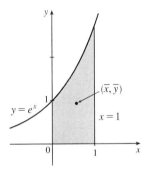

28. The region in the figure is "left-heavy" and "bottom-heavy," so we know
$\overline{x} < 1.5$ and $\overline{y} < 0.5$, and we might guess that $\overline{x} = 1.4$ and $\overline{y} = 0.4$.

$A = \int_1^2 \frac{1}{x} \, dx = [\ln x]_1^2 = \ln 2.$ $\overline{x} = \frac{1}{A}\int_1^2 x\cdot\frac{1}{x} \, dx = \frac{1}{A}[x]_1^2 = \frac{1}{A} = \frac{1}{\ln 2}.$

$\overline{y} = \frac{1}{A}\int_1^2 \frac{1}{2}\left(\frac{1}{x}\right)^2 dx = \frac{1}{2A}\int_1^2 x^{-2} \, dx = \frac{1}{2A}\left[-\frac{1}{x}\right]_1^2$

$= \frac{1}{2\ln 2}\left(-\frac{1}{2} + 1\right) = \frac{1}{4\ln 2}.$

Thus, the centroid is $(\overline{x}, \overline{y}) = \left(\frac{1}{\ln 2}, \frac{1}{4\ln 2}\right) \approx (1.44, 0.36).$

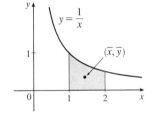

29. $A = \int_0^1 (x^{1/2} - x^2) \, dx = \left[\frac{2}{3}x^{3/2} - \frac{1}{3}x^3\right]_0^1 = \left(\frac{2}{3} - \frac{1}{3}\right) - 0 = \frac{1}{3}.$

$\overline{x} = \frac{1}{A}\int_0^1 x(x^{1/2} - x^2) \, dx = 3\int_0^1 (x^{3/2} - x^3) \, dx$

$= 3\left[\frac{2}{5}x^{5/2} - \frac{1}{4}x^4\right]_0^1 = 3\left(\frac{2}{5} - \frac{1}{4}\right) = 3\left(\frac{3}{20}\right) = \frac{9}{20}.$

$\overline{y} = \frac{1}{A}\int_0^1 \frac{1}{2}\left[(x^{1/2})^2 - (x^2)^2\right] dx = 3\left(\frac{1}{2}\right)\int_0^1 (x - x^4) \, dx$

$= \frac{3}{2}\left[\frac{1}{2}x^2 - \frac{1}{5}x^5\right]_0^1 = \frac{3}{2}\left(\frac{1}{2} - \frac{1}{5}\right) = \frac{3}{2}\left(\frac{3}{10}\right) = \frac{9}{20}.$

Thus, the centroid is $(\overline{x}, \overline{y}) = \left(\frac{9}{20}, \frac{9}{20}\right).$

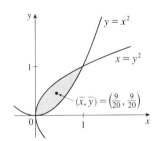

30. $A = \int_{-1}^{2} (x + 2 - x^2)\, dx = \left[\frac{1}{2}x^2 + 2x - \frac{1}{3}x^3\right]_{-1}^{2}$

$= \left(2 + 4 - \frac{8}{3}\right) - \left(\frac{1}{2} - 2 + \frac{1}{3}\right) = \frac{9}{2}.$

$\bar{x} = \frac{1}{A} \int_{-1}^{2} x(x + 2 - x^2)\, dx = \frac{2}{9} \int_{-1}^{2} \left(x^2 + 2x - x^3\right) dx$

$= \frac{2}{9}\left[\frac{1}{3}x^3 + x^2 - \frac{1}{4}x^4\right]_{-1}^{2}$

$= \frac{2}{9}\left[\left(\frac{8}{3} + 4 - 4\right) - \left(-\frac{1}{3} + 1 - \frac{1}{4}\right)\right] = \frac{2}{9} \cdot \frac{9}{4} = \frac{1}{2}.$

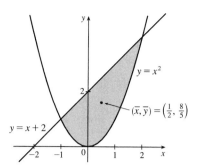

$\bar{y} = \frac{1}{A} \int_{-1}^{2} \frac{1}{2}\left[(x+2)^2 - (x^2)^2\right] dx = \frac{2}{9} \cdot \frac{1}{2} \int_{-1}^{2} \left(x^2 + 4x + 4 - x^4\right) dx = \frac{1}{9}\left[\frac{1}{3}x^3 + 2x^2 + 4x - \frac{1}{5}x^5\right]_{-1}^{2}$

$= \frac{1}{9}\left[\left(\frac{8}{3} + 8 + 8 - \frac{32}{5}\right) - \left(-\frac{1}{3} + 2 - 4 + \frac{1}{5}\right)\right] = \frac{1}{9}\left(18 + \frac{9}{3} - \frac{33}{5}\right) = \frac{1}{9} \cdot \frac{72}{5} = \frac{8}{5}.$

Thus, the centroid is $(\bar{x}, \bar{y}) = \left(\frac{1}{2}, \frac{8}{5}\right).$

31. $A = \int_{0}^{\pi/4} (\cos x - \sin x)\, dx = \left[\sin x + \cos x\right]_{0}^{\pi/4} = \sqrt{2} - 1.$

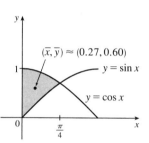

$\bar{x} = A^{-1} \int_{0}^{\pi/4} x(\cos x - \sin x)\, dx$

$= A^{-1}\left[x(\sin x + \cos x) + \cos x - \sin x\right]_{0}^{\pi/4}$ [integration by parts]

$= A^{-1}\left(\frac{\pi}{4}\sqrt{2} - 1\right) = \frac{\frac{1}{4}\pi\sqrt{2} - 1}{\sqrt{2} - 1}.$

$\bar{y} = A^{-1} \int_{0}^{\pi/4} \frac{1}{2}(\cos^2 x - \sin^2 x)\, dx = \frac{1}{2A} \int_{0}^{\pi/4} \cos 2x\, dx = \frac{1}{4A}\left[\sin 2x\right]_{0}^{\pi/4} = \frac{1}{4A} = \frac{1}{4\left(\sqrt{2} - 1\right)}.$

Thus, the centroid is $(\bar{x}, \bar{y}) = \left(\dfrac{\pi\sqrt{2} - 4}{4\left(\sqrt{2} - 1\right)}, \dfrac{1}{4\left(\sqrt{2} - 1\right)}\right) \approx (0.27, 0.60).$

32. $A = \int_{0}^{1} x^3\, dx + \int_{1}^{2} (2 - x)\, dx = \left[\frac{1}{4}x^4\right]_{0}^{1} + \left[2x - \frac{1}{2}x^2\right]_{1}^{2}$

$= \frac{1}{4} + (4 - 2) - \left(2 - \frac{1}{2}\right) = \frac{3}{4}.$

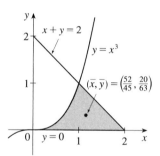

$\bar{x} = \frac{1}{A}\left[\int_{0}^{1} x(x^3)\, dx + \int_{1}^{2} x(2 - x)\, dx\right] = \frac{4}{3}\left[\int_{0}^{1} x^4\, dx + \int_{1}^{2} (2x - x^2)\, dx\right]$

$= \frac{4}{3}\left\{\left[\frac{1}{5}x^5\right]_{0}^{1} + \left[x^2 - \frac{1}{3}x^3\right]_{1}^{2}\right\} = \frac{4}{3}\left[\frac{1}{5} + \left(4 - \frac{8}{3}\right) - \left(1 - \frac{1}{3}\right)\right]$

$= \frac{4}{3}\left(\frac{13}{15}\right) = \frac{52}{45}.$

$\bar{y} = \frac{1}{A}\left[\int_{0}^{1} \frac{1}{2}(x^3)^2\, dx + \int_{1}^{2} \frac{1}{2}(2 - x)^2\, dx\right] = \frac{2}{3}\left[\int_{0}^{1} x^6\, dx + \int_{1}^{2} (x - 2)^2\, dx\right] = \frac{2}{3}\left\{\left[\frac{1}{7}x^7\right]_{0}^{1} + \left[\frac{1}{3}(x - 2)^3\right]_{1}^{2}\right\}$

$= \frac{2}{3}\left(\frac{1}{7} - 0 + 0 + \frac{1}{3}\right) = \frac{2}{3}\left(\frac{10}{21}\right) = \frac{20}{63}.$

Thus, the centroid is $(\bar{x}, \bar{y}) = \left(\frac{52}{45}, \frac{20}{63}\right).$

33. From the figure we see that $\overline{y} = 0$. Now

$$A = \int_0^5 2\sqrt{5-x}\,dx = 2\Big[-\tfrac{2}{3}(5-x)^{3/2}\Big]_0^5 = 2\Big(0 + \tfrac{2}{3}\cdot 5^{3/2}\Big) = \tfrac{20}{3}\sqrt{5},$$

so

$$\overline{x} = \tfrac{1}{A}\int_0^5 x\Big[\sqrt{5-x} - \big(-\sqrt{5-x}\big)\Big]\,dx = \tfrac{1}{A}\int_0^5 2x\sqrt{5-x}\,dx$$

$$= \tfrac{1}{A}\int_{\sqrt{5}}^0 2\big(5-u^2\big)u(-2u)\,du \qquad \begin{bmatrix} u = \sqrt{5-x},\; x = 5-u^2, \\ u^2 = 5-x,\; dx = -2u\,du \end{bmatrix}$$

$$= \tfrac{4}{A}\int_0^{\sqrt{5}} u^2\big(5-u^2\big)\,du = \tfrac{4}{A}\Big[\tfrac{5}{3}u^3 - \tfrac{1}{5}u^5\Big]_0^{\sqrt{5}} = \tfrac{3}{5\sqrt{5}}\Big(\tfrac{25}{3}\sqrt{5} - 5\sqrt{5}\Big) = 5 - 3 = 2.$$

Thus, the centroid is $(\overline{x}, \overline{y}) = (2, 0)$.

34. The quarter-circle has equation $y = \sqrt{1-x^2}$ for $0 \le x \le 1$ and the line has equation $y = x - 1$.

$A = \tfrac{1}{4}\pi(1)^2 + \tfrac{1}{2}$, so $m = \rho A = 3\big(\tfrac{\pi}{4} + \tfrac{1}{2}\big) = \tfrac{3}{4}(\pi + 2)$.

$$M_x = \rho\int_0^1 \tfrac{1}{2}\Big[\big(\sqrt{1-x^2}\big)^2 - (x-1)^2\Big]\,dx = \tfrac{3}{2}\int_0^1\Big[\big(1-x^2\big) - \big(x^2 - 2x + 1\big)\Big]\,dx$$

$$= \tfrac{3}{2}\int_0^1\big(-2x^2 + 2x\big)\,dx = \tfrac{3}{2}\Big[-\tfrac{2}{3}x^3 + x^2\Big]_0^1 = \tfrac{3}{2}\big(-\tfrac{2}{3} + 1\big) = \tfrac{3}{2}\big(\tfrac{1}{3}\big) = \tfrac{1}{2}$$

$$M_y = \rho\int_0^1 x\Big[\sqrt{1-x^2} - (x-1)\Big]\,dx = 3\int_0^1\big(x\sqrt{1-x^2} - x^2 + x\big)\,dx$$

$$= 3\Big[-\tfrac{1}{3}\big(1-x^2\big)^{3/2} - \tfrac{1}{3}x^3 + \tfrac{1}{2}x^2\Big]_0^1 = 3\Big[\big(0 - \tfrac{1}{3} + \tfrac{1}{2}\big) - \big(-\tfrac{1}{3} - 0 + 0\big)\Big] = \tfrac{3}{2}$$

$\overline{x} = \dfrac{M_y}{m} = \dfrac{\tfrac{3}{2}}{\tfrac{3}{4}(\pi+2)} = \dfrac{2}{\pi+2}$ and $\overline{y} = \dfrac{M_x}{m} = \dfrac{\tfrac{1}{2}}{\tfrac{3}{4}(\pi+2)} = \dfrac{2}{3(\pi+2)}$. Thus, the centroid is $\left(\dfrac{2}{\pi+2}, \dfrac{2}{3(\pi+2)}\right)$.

35. The line has equation $y = \tfrac{3}{4}x$. $A = \tfrac{1}{2}(4)(3) = 6$, so $m = \rho A = 10(6) = 60$.

$$M_x = \rho\int_0^4 \tfrac{1}{2}\big(\tfrac{3}{4}x\big)^2\,dx = 10\int_0^4 \tfrac{9}{32}x^2\,dx = \tfrac{45}{16}\Big[\tfrac{1}{3}x^3\Big]_0^4 = \tfrac{45}{16}\big(\tfrac{64}{3}\big) = 60$$

$$M_y = \rho\int_0^4 x\big(\tfrac{3}{4}x\big)\,dx = \tfrac{15}{2}\int_0^4 x^2\,dx = \tfrac{15}{2}\Big[\tfrac{1}{3}x^3\Big]_0^4 = \tfrac{15}{2}\big(\tfrac{64}{3}\big) = 160$$

$\overline{x} = \dfrac{M_y}{m} = \dfrac{160}{60} = \dfrac{8}{3}$ and $\overline{y} = \dfrac{M_x}{m} = \dfrac{60}{60} = 1$. Thus, the centroid is $(\overline{x}, \overline{y}) = \big(\tfrac{8}{3}, 1\big)$.

36. We'll use $n = 8$, so $\Delta x = \dfrac{b-a}{n} = \dfrac{8-0}{8} = 1$.

$$A = \int_0^8 f(x)\,dx \approx S_{10} = \tfrac{1}{3}[f(0) + 4f(1) + 2f(2) + 4f(3) + 2f(4) + 4f(5) + 2f(6) + 4f(7) + f(8)]$$

$$\approx \tfrac{1}{3}[0 + 4(2.0) + 2(2.6) + 4(2.3) + 2(2.2) + 4(3.3) + 2(4.0) + 4(3.2) + 0]$$

$$= \tfrac{1}{3}(60.8) = 20.2\overline{6} \quad \Big[\text{or } \tfrac{304}{15}\Big]$$

Now $\displaystyle\int_0^8 x f(x)\,dx \approx \tfrac{1}{3}[0 \cdot f(0) + 4 \cdot 1 \cdot f(1) + 2 \cdot 2 \cdot f(2) + 4 \cdot 3 \cdot f(3)$

$$+ 2 \cdot 4 \cdot f(4) + 4 \cdot 5 \cdot f(5) + 2 \cdot 6 \cdot f(6) + 4 \cdot 7 \cdot f(7) + 8 \cdot f(8)]$$

$$\approx \tfrac{1}{3}[0 + 8 + 10.4 + 27.6 + 17.6 + 66 + 48 + 89.6 + 0]$$

$$= \tfrac{1}{3}(267.2) = 89.0\overline{6} \quad \Big[\text{or } \tfrac{1336}{15}\Big], \text{ so } \overline{x} = \tfrac{1}{A}\int_0^8 x f(x)\,dx \approx 4.39.$$

Also, $\displaystyle\int_0^8 [f(x)]^2\,dx \approx \tfrac{1}{3}[(0^2 + 4(2.0)^2 + 2(2.6)^2 + 4(2.3)^2 + 2(2.2)^2 + 4(3.3)^2 + 2(4.0)^2 + 4(3.2)^2 + 0^2)]$

$$= \tfrac{1}{3}(176.88) = 58.96, \text{ so } \overline{y} = \tfrac{1}{A}\int_0^8 \tfrac{1}{2}[f(x)]^2\,dx \approx 1.45.$$

Thus, the centroid is $(\overline{x}, \overline{y}) \approx (4.4, 1.5)$.

37. $A = \int_0^2 (2^x - x^2)\, dx = \left[\dfrac{2^x}{\ln 2} - \dfrac{x^3}{3}\right]_0^2$

$= \left(\dfrac{4}{\ln 2} - \dfrac{8}{3}\right) - \dfrac{1}{\ln 2} = \dfrac{3}{\ln 2} - \dfrac{8}{3} \approx 1.661418.$

$\bar{x} = \dfrac{1}{A}\int_0^2 x(2^x - x^2)\, dx = \dfrac{1}{A}\int_0^2 (x2^x - x^3)\, dx$

$= \dfrac{1}{A}\left[\dfrac{x2^x}{\ln 2} - \dfrac{2^x}{(\ln 2)^2} - \dfrac{x^4}{4}\right]_0^2 \quad \text{[use parts]}$

$= \dfrac{1}{A}\left[\dfrac{8}{\ln 2} - \dfrac{4}{(\ln 2)^2} - 4 + \dfrac{1}{(\ln 2)^2}\right] = \dfrac{1}{A}\left[\dfrac{8}{\ln 2} - \dfrac{3}{(\ln 2)^2} - 4\right] \approx \dfrac{1}{A}(1.297453) \approx 0.781$

$\bar{y} = \dfrac{1}{A}\int_0^2 \tfrac{1}{2}\left[(2^x)^2 - (x^2)^2\right] dx = \dfrac{1}{A}\int_0^2 \tfrac{1}{2}(2^{2x} - x^4)\, dx = \dfrac{1}{A}\cdot\dfrac{1}{2}\left[\dfrac{2^{2x}}{2\ln 2} - \dfrac{x^5}{5}\right]_0^2$

$= \dfrac{1}{A}\cdot\dfrac{1}{2}\left(\dfrac{16}{2\ln 2} - \dfrac{32}{5} - \dfrac{1}{2\ln 2}\right) = \dfrac{1}{A}\left(\dfrac{15}{4\ln 2} - \dfrac{16}{5}\right) \approx \dfrac{1}{A}(2.210106) \approx 1.330$

Thus, the centroid is $(\bar{x}, \bar{y}) \approx (0.781, 1.330).$

Since the position of a centroid is independent of density when the density is constant, we will assume for convenience that $\rho = 1$ in Exercises 38 and 39.

38. The curves $y = x + \ln x$ and $y = x^3 - x$ intersect at

$(a, c) \approx (0.447141, -0.357742)$ and $(b, d) \approx (1.507397, 1.917782).$

$A = \int_a^b (x + \ln x - x^3 + x)\, dx = \int_a^b (2x + \ln x - x^3)\, dx$

$\overset{100}{=}\left[x^2 + x\ln x - x - \tfrac{1}{4}x^4\right]_a^b \approx 0.709781$

$\bar{x} = \dfrac{1}{A}\int_a^b x(2x + \ln x - x^3)\, dx = \dfrac{1}{A}\int_a^b (2x^2 + x\ln x - x^4)\, dx$

$\overset{101}{=} \dfrac{1}{A}\left[\tfrac{2}{3}x^3 + \tfrac{1}{4}x^2(2\ln x - 1) - \tfrac{1}{5}x^5\right]_a^b \approx \dfrac{1}{A}(0.699489) \approx 0.985501$

$\bar{y} = \dfrac{1}{A}\int_a^b \tfrac{1}{2}\left[(x + \ln x)^2 - (x^3 - x)^2\right] dx = \dfrac{1}{2A}\int_a^b \left[2x\ln x + (\ln x)^2 - x^6 + 2x^4\right] dx$

$\overset{\substack{101\text{ and}\\ \text{parts}}}{=} \dfrac{1}{2A}\left[x^2\ln x - \tfrac{1}{2}x^2 + x(\ln x)^2 - 2x\ln x + 2x - \tfrac{1}{7}x^7 + \tfrac{2}{5}x^5\right]_a^b \approx \dfrac{1}{2A}(0.765092) \approx 0.538964$

Thus, the centroid is $(\bar{x}, \bar{y}) \approx (0.986, 0.539).$

39. Choose x- and y-axes so that the base (one side of the triangle) lies along the x-axis with the other vertex along the positive y-axis as shown. From geometry, we know the medians intersect at a point $\tfrac{2}{3}$ of the way from each vertex (along the median) to the opposite side. The median from B goes to the midpoint $\left(\tfrac{1}{2}(a + c), 0\right)$ of side AC, so the point of intersection of the medians is $\left(\tfrac{2}{3}\cdot\tfrac{1}{2}(a + c), \tfrac{1}{3}b\right) = \left(\tfrac{1}{3}(a + c), \tfrac{1}{3}b\right).$

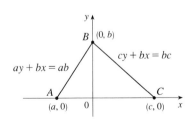

This can also be verified by finding the equations of two medians, and solving them simultaneously to find their point of intersection. Now let us compute the location of the centroid of the triangle. The area is $A = \tfrac{1}{2}(c - a)b.$

$$\bar{x} = \frac{1}{A}\left[\int_a^0 x \cdot \frac{b}{a}(a-x)\,dx + \int_0^c x \cdot \frac{b}{c}(c-x)\,dx\right] = \frac{1}{A}\left[\frac{b}{a}\int_a^0 (ax-x^2)\,dx + \frac{b}{c}\int_0^c (cx-x^2)\,dx\right]$$

$$= \frac{b}{Aa}\left[\frac{1}{2}ax^2 - \frac{1}{3}x^3\right]_a^0 + \frac{b}{Ac}\left[\frac{1}{2}cx^2 - \frac{1}{3}x^3\right]_0^c = \frac{b}{Aa}\left[-\frac{1}{2}a^3 + \frac{1}{3}a^3\right] + \frac{b}{Ac}\left[\frac{1}{2}c^3 - \frac{1}{3}c^3\right]$$

$$= \frac{2}{a(c-a)} \cdot \frac{-a^3}{6} + \frac{2}{c(c-a)} \cdot \frac{c^3}{6} = \frac{1}{3(c-a)}(c^2-a^2) = \frac{a+c}{3}$$

and

$$\bar{y} = \frac{1}{A}\left[\int_a^0 \frac{1}{2}\left(\frac{b}{a}(a-x)\right)^2 dx + \int_0^c \frac{1}{2}\left(\frac{b}{c}(c-x)\right)^2 dx\right]$$

$$= \frac{1}{A}\left[\frac{b^2}{2a^2}\int_a^0 (a^2 - 2ax + x^2)\,dx + \frac{b^2}{2c^2}\int_0^c (c^2 - 2cx + x^2)\,dx\right]$$

$$= \frac{1}{A}\left[\frac{b^2}{2a^2}[a^2 x - ax^2 + \tfrac{1}{3}x^3]_a^0 + \frac{b^2}{2c^2}[c^2 x - cx^2 + \tfrac{1}{3}x^3]_0^c\right]$$

$$= \frac{1}{A}\left[\frac{b^2}{2a^2}(-a^3 + a^3 - \tfrac{1}{3}a^3) + \frac{b^2}{2c^2}(c^3 - c^3 + \tfrac{1}{3}c^3)\right] = \frac{1}{A}\left[\frac{b^2}{6}(-a+c)\right] = \frac{2}{(c-a)b} \cdot \frac{(c-a)b^2}{6} = \frac{b}{3}$$

Thus, the centroid is $(\bar{x}, \bar{y}) = \left(\dfrac{a+c}{3}, \dfrac{b}{3}\right)$, as claimed.

Remarks: Actually the computation of $\bar{y}$ is all that is needed. By considering each side of the triangle in turn to be the base, we see that the centroid is $\frac{1}{3}$ of the way from each side to the opposite vertex and must therefore be the intersection of the medians.

The computation of $\bar{y}$ in this problem (and many others) can be simplified by using horizontal rather than vertical approximating rectangles. If the length of a thin rectangle at coordinate y is $\ell(y)$, then its area is $\ell(y)\,\Delta y$, its mass is $\rho\ell(y)\,\Delta y$, and its moment about the x-axis is $\Delta M_x = \rho y \ell(y)\,\Delta y$. Thus,

$$M_x = \int \rho y \ell(y)\,dy \qquad \text{and} \qquad \bar{y} = \frac{\int \rho y \ell(y)\,dy}{\rho A} = \frac{1}{A}\int y\ell(y)\,dy$$

In this problem, $\ell(y) = \dfrac{c-a}{b}(b-y)$ by similar triangles, so

$$\bar{y} = \frac{1}{A}\int_0^b \frac{c-a}{b} y(b-y)\,dy = \frac{2}{b^2}\int_0^b (by - y^2)\,dy = \frac{2}{b^2}\left[\tfrac{1}{2}by^2 - \tfrac{1}{3}y^3\right]_0^b = \frac{2}{b^2} \cdot \frac{b^3}{6} = \frac{b}{3}$$

Notice that only one integral is needed when this method is used.

40. Divide the lamina into three rectangles with masses 2, 2 and 6, with centroids $\left(-\frac{3}{2}, 1\right)$, $\left(0, \frac{1}{2}\right)$ and $\left(2, \frac{3}{2}\right)$, respectively. The total mass of the lamina is 10. So, using Formulas 5, 6, and 7, we have

$$\bar{x} = \frac{M_y}{m} = \frac{1}{m}\sum_{i=1}^3 m_i x_i = \tfrac{1}{10}\left[2\left(-\tfrac{3}{2}\right) + 2(0) + 6(2)\right] = \tfrac{1}{10}(9)$$

and

$$\bar{y} = \frac{M_x}{m} = \frac{1}{m}\sum_{i=1}^3 m_i y_i = \tfrac{1}{10}\left[2(1) + 2\left(\tfrac{1}{2}\right) + 6\left(\tfrac{3}{2}\right)\right] = \tfrac{1}{10}(12).$$

Thus, the centroid is $(\bar{x}, \bar{y}) = \left(\frac{9}{10}, \frac{6}{5}\right)$.

41. Divide the lamina into two triangles and one rectangle with respective masses of 2, 2 and 4, so that the total mass is 8. Using

the result of Exercise 39, the triangles have centroids $\left(-1, \frac{2}{3}\right)$ and $\left(1, \frac{2}{3}\right)$. The centroid of the rectangle (its center) is $\left(0, -\frac{1}{2}\right)$.

So, using Formulas 5 and 7, we have $\overline{y} = \dfrac{M_x}{m} = \dfrac{1}{m}\sum_{i=1}^{3} m_i\, y_i = \frac{1}{8}\left[2\left(\frac{2}{3}\right) + 2\left(\frac{2}{3}\right) + 4\left(-\frac{1}{2}\right)\right] = \frac{1}{8}\left(\frac{2}{3}\right) = \frac{1}{12}$, and $\overline{x} = 0$,

since the lamina is symmetric about the line $x = 0$. Thus, the centroid is $(\overline{x}, \overline{y}) = \left(0, \frac{1}{12}\right)$.

42. The parabola has equation $y = kx^2$ and passes through (a, b),

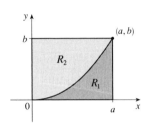

so $b = ka^2 \quad\Rightarrow\quad k = \dfrac{b}{a^2}$ and hence, $y = \dfrac{b}{a^2}x^2$.

R_1 has area $A_1 = \displaystyle\int_0^a \frac{b}{a^2}x^2\, dx = \frac{b}{a^2}\left[\frac{1}{3}x^3\right]_0^a = \frac{b}{a^2}\left(\frac{a^3}{3}\right) = \frac{1}{3}ab.$

Since R has area ab, R_2 has area $A_2 = ab - \frac{1}{3}ab = \frac{2}{3}ab.$

For R_1: $\quad \overline{x}_1 = \dfrac{1}{A_1}\displaystyle\int_0^a x\left(\frac{b}{a^2}x^2\right)dx = \frac{3}{ab}\frac{b}{a^2}\int_0^a x^3\,dx = \frac{3}{a^3}\left[\frac{1}{4}x^4\right]_0^a = \frac{3}{a^3}\left(\frac{1}{4}a^4\right) = \frac{3}{4}a$

$\qquad\qquad \overline{y}_1 = \dfrac{1}{A_1}\displaystyle\int_0^a \frac{1}{2}\left(\frac{b}{a^2}x^2\right)^2 dx = \frac{3}{ab}\frac{b^2}{2a^4}\int_0^a x^4\,dx = \frac{3b}{2a^5}\left[\frac{1}{5}x^5\right]_0^a = \frac{3b}{2a^5}\left(\frac{1}{5}a^5\right) = \frac{3}{10}b$

Thus, the centroid for R_1 is $(\overline{x}_1, \overline{y}_1) = \left(\frac{3}{4}a, \frac{3}{10}b\right).$

For R_2: $\quad \overline{x}_2 = \dfrac{1}{A_2}\displaystyle\int_0^a x\left(b - \frac{b}{a^2}x^2\right)dx = \frac{3}{2ab}\int_0^a b\left(x - \frac{1}{a^2}x^3\right)dx = \frac{3}{2a}\left[\frac{1}{2}x^2 - \frac{1}{4a^2}x^4\right]_0^a$

$\qquad\qquad = \frac{3}{2a}\left(\frac{a^2}{2} - \frac{a^2}{4}\right) = \frac{3}{2a}\left(\frac{a^2}{4}\right) = \frac{3}{8}a$

$\qquad\qquad \overline{y}_2 = \dfrac{1}{A_2}\displaystyle\int_0^a \frac{1}{2}\left[(b)^2 - \left(\frac{b}{a^2}x^2\right)^2\right]dx = \frac{3}{2ab}\frac{1}{2}\int_0^a b^2\left(1 - \frac{1}{a^4}x^4\right)dx = \frac{3b}{4a}\left[x - \frac{1}{5a^4}x^5\right]_0^a$

$\qquad\qquad = \frac{3b}{4a}\left(a - \frac{1}{5}a\right) = \frac{3b}{4a}\left(\frac{4a}{5}\right) = \frac{3}{5}b$

Thus, the centroid for R_2 is $(\overline{x}_2, \overline{y}_2) = \left(\frac{3}{8}a, \frac{3}{5}b\right).$ Note the relationships: $A_2 = 2A_1$, $\overline{x}_1 = 2\,\overline{x}_2$, $\overline{y}_2 = 2\,\overline{y}_1.$

43. $\displaystyle\int_a^b (cx + d)\,f(x)\,dx = \int_a^b cx\,f(x)\,dx + \int_a^b d\,f(x)\,dx = c\int_a^b x\,f(x)\,dx + d\int_a^b f(x)\,dx = c\overline{x}A + d\int_a^b f(x)\,dx$ [by (8)]

$\qquad\qquad = c\overline{x}\displaystyle\int_a^b f(x)\,dx + d\int_a^b f(x)\,dx = (c\overline{x} + d)\int_a^b f(x)\,dx$

44. A sphere can be generated by rotating a semicircle about its diameter. By Example 4, the center of mass travels a distance

$2\pi\overline{y} = 2\pi\left(\dfrac{4r}{3\pi}\right) = \dfrac{8r}{3}$, so by the Theorem of Pappus, the volume of the sphere is $V = Ad = \dfrac{\pi r^2}{2}\cdot\dfrac{8r}{3} = \dfrac{4}{3}\pi r^3.$

45. A cone of height h and radius r can be generated by rotating a right triangle

about one of its legs as shown. By Exercise 39, $\overline{x} = \frac{1}{3}r$, so by the Theorem of

Pappus, the volume of the cone is

$\qquad V = Ad = \left(\frac{1}{2}\cdot\text{base}\cdot\text{height}\right)\cdot(2\pi\overline{x}) = \frac{1}{2}rh\cdot 2\pi\left(\frac{1}{3}r\right) = \frac{1}{3}\pi r^2 h.$

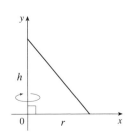

46. From the symmetry in the figure, $\overline{y} = 4$. So the distance traveled by the centroid

when rotating the triangle about the x-axis is $d = 2\pi \cdot 4 = 8\pi$. The area of the

triangle is $A = \frac{1}{2}bh = \frac{1}{2}(2)(3) = 3$. By the Theorem of Pappus, the volume of

the resulting solid is $Ad = 3(8\pi) = 24\pi$.

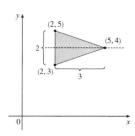

47. Suppose the region lies between two curves $y = f(x)$ and $y = g(x)$ where $f(x) \geq g(x)$, as illustrated in Figure 13.

Choose points x_i with $a = x_0 < x_1 < \cdots < x_n = b$ and choose x_i^* to be the midpoint of the ith subinterval; that is,

$x_i^* = \overline{x}_i = \frac{1}{2}(x_{i-1} + x_i)$. Then the centroid of the ith approximating rectangle R_i is its center $C_i = \left(\overline{x}_i, \frac{1}{2}[f(\overline{x}_i) + g(\overline{x}_i)]\right)$.

Its area is $[f(\overline{x}_i) - g(\overline{x}_i)]\,\Delta x$, so its mass is

$\rho[f(\overline{x}_i) - g(\overline{x}_i)]\,\Delta x$. Thus, $M_y(R_i) = \rho[f(\overline{x}_i) - g(\overline{x}_i)]\,\Delta x \cdot \overline{x}_i = \rho\overline{x}_i\,[f(\overline{x}_i) - g(\overline{x}_i)]\,\Delta x$ and

$M_x(R_i) = \rho[f(\overline{x}_i) - g(\overline{x}_i)]\,\Delta x \cdot \frac{1}{2}[f(\overline{x}_i) + g(\overline{x}_i)] = \rho \cdot \frac{1}{2}\left[f(\overline{x}_i)^2 - g(\overline{x}_i)^2\right]\Delta x$. Summing over i and taking the limit

as $n \to \infty$, we get $M_y = \lim\limits_{n\to\infty} \sum_i \rho\overline{x}_i\,[f(\overline{x}_i) - g(\overline{x}_i)]\,\Delta x = \rho\int_a^b x[f(x) - g(x)]\,dx$ and

$M_x = \lim\limits_{n\to\infty} \sum_i \rho \cdot \frac{1}{2}\left[f(\overline{x}_i)^2 - g(\overline{x}_i)^2\right]\Delta x = \rho\int_a^b \frac{1}{2}\left[f(x)^2 - g(x)^2\right]dx.$

Thus, $\overline{x} = \dfrac{M_y}{m} = \dfrac{M_y}{\rho A} = \dfrac{1}{A}\int_a^b x[f(x) - g(x)]\,dx$ and $\overline{y} = \dfrac{M_x}{m} = \dfrac{M_x}{\rho A} = \dfrac{1}{A}\int_a^b \frac{1}{2}\left[f(x)^2 - g(x)^2\right]dx.$

48. (a) Let $0 \leq x \leq 1$. If $n < m$, then $x^n > x^m$; that is, raising x to a larger

power produces a smaller number.

(b) Using Formulas 9 and the fact that the area of $\mathcal{R}$ is

$A = \int_0^1 (x^n - x^m)\,dx = \dfrac{1}{n+1} - \dfrac{1}{m+1} = \dfrac{m-n}{(n+1)(m+1)}$, we get

$$\overline{x} = \dfrac{(n+1)(m+1)}{m-n}\int_0^1 x[x^n - x^m]\,dx = \dfrac{(n+1)(m+1)}{m-n}\int_0^1 \left(x^{n+1} - x^{m+1}\right)dx$$

$$= \dfrac{(n+1)(m+1)}{m-n}\left[\dfrac{1}{n+2} - \dfrac{1}{m+2}\right] = \dfrac{(n+1)(m+1)}{(n+2)(m+2)}$$

and

$$\overline{y} = \dfrac{(n+1)(m+1)}{m-n}\int_0^1 \frac{1}{2}\left[(x^n)^2 - (x^m)^2\right]dx = \dfrac{(n+1)(m+1)}{2(m-n)}\int_0^1 \left(x^{2n} - x^{2m}\right)dx$$

$$= \dfrac{(n+1)(m+1)}{2(m-n)}\left[\dfrac{1}{2n+1} - \dfrac{1}{2m+1}\right] = \dfrac{(n+1)(m+1)}{(2n+1)(2m+1)}$$

(c) If we take $n = 3$ and $m = 4$, then

$$(\overline{x}, \overline{y}) = \left(\dfrac{4 \cdot 5}{5 \cdot 6}, \dfrac{4 \cdot 5}{7 \cdot 9}\right) = \left(\dfrac{2}{3}, \dfrac{20}{63}\right)$$

which lies outside $\mathcal{R}$ since $\left(\dfrac{2}{3}\right)^3 = \dfrac{8}{27} < \dfrac{20}{63}$. This is the simplest of many

possibilities.

DISCOVERY PROJECT Complementary Coffee Cups

1. Cup A has volume $V_A = \int_0^h \pi [f(y)]^2 \, dy$ and cup B has volume

$$V_B = \int_0^h \pi [k - f(y)]^2 \, dy = \int_0^h \pi \{ k^2 - 2kf(y) + [f(y)]^2 \} \, dy$$

$$= \left[\pi k^2 y \right]_0^h - 2\pi k \int_0^h f(y) \, dy + \int_0^h \pi [f(y)]^2 \, dy = \pi k^2 h - 2\pi k A_1 + V_A$$

Thus, $V_A = V_B \iff \pi k (kh - 2A_1) = 0 \iff k = 2(A_1/h)$; that is, k is twice the average value of f on the interval $[0, h]$.

2. From Problem 1, $V_A = V_B \iff kh = 2A_1 \iff A_1 + A_2 = 2A_1 \iff A_2 = A_1$.

3. Let $\overline{x}_1$ and $\overline{x}_2$ denote the x-coordinates of the centroids of A_1 and A_2, respectively. By Pappus's Theorem,

$$V_A = 2\pi \overline{x}_1 A_1 \text{ and } V_B = 2\pi (k - \overline{x}_2) A_2, \text{ so } V_A = V_B \iff \overline{x}_1 A_1 = k A_2 - \overline{x}_2 A_2 \iff k A_2 = \overline{x}_1 A_1 + \overline{x}_2 A_2 \overset{(\star)}{\iff}$$

$k A_2 = \frac{1}{2} k (A_1 + A_2) \iff \frac{1}{2} k A_2 = \frac{1}{2} k A_1 \iff A_2 = A_1$, as shown in Problem 2. [$(\star)$ The sum of the moments of the regions of areas A_1 and A_2 about the y-axis equals the moment of the entire k-by-h rectangle about the y-axis.]

So, since $A_1 + A_2 = kh$, we have $V_A = V_B \iff A_1 = A_2 \iff A_1 = \frac{1}{2}(A_1 + A_2) \iff A_1 = \frac{1}{2}(kh) \iff$

$k = 2 (A_1/h)$, as shown in Problem 1.

4. We'll use a cup that is $h = 8$ cm high with a diameter of 6 cm on the top and the bottom and symmetrically bulging to a diameter of 8 cm in the middle (all inside dimensions).

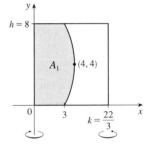

For an equation, we'll use a parabola with a vertex at $(4, 4)$; that is,

$x = a(y - 4)^2 + 4$. To find a, use the point $(3, 0)$:

$3 = a(0 - 4)^2 + 4 \implies -1 = 16a \implies a = -\frac{1}{16}$. To find k, we'll use the

relationship in Problem 1, so we need A_1.

$$A_1 = \int_0^8 \left[-\frac{1}{16}(y - 4)^2 + 4 \right] dy = \int_{-4}^4 \left(-\frac{1}{16} u^2 + 4 \right) du \quad [u = y - 4]$$

$$= 2 \int_0^4 \left(-\frac{1}{16} u^2 + 4 \right) du = 2 \left[-\frac{1}{48} u^3 + 4u \right]_0^4 = 2 \left(-\frac{4}{3} + 16 \right) = \frac{88}{3}.$$

Thus, $k = 2(A_1/h) = 2 \left(\frac{88/3}{8} \right) = \frac{22}{3}$.

So with $h = 8$ and curve $x = -\frac{1}{16}(y - 4)^2 + 4$, we have

$$V_A = \int_0^8 \pi \left[-\frac{1}{16}(y - 4)^2 + 4 \right]^2 dy = \pi \int_{-4}^4 \left(-\frac{1}{16} u^2 + 4 \right)^2 du \quad [u = y - 4] \quad = 2\pi \int_0^4 \left(\frac{1}{256} u^4 - \frac{1}{2} u^2 + 16 \right) du$$

$$= 2\pi \left[\frac{1}{1280} u^5 - \frac{1}{6} u^3 + 16u \right]_0^4 = 2\pi \left(\frac{4}{5} - \frac{32}{3} + 64 \right) = 2\pi \left(\frac{812}{15} \right) = \frac{1624}{15} \pi$$

This is approximately 340 cm³ or 11.5 fl. oz. And with $k = \frac{22}{3}$, we know from Problem 1 that cup B holds the same amount.

9.4 Applications to Economics and Biology

1. By the Net Change Theorem, $C(2000) - C(0) = \int_0^{2000} C'(x)\,dx$ ⇒

$$C(2000) = 20{,}000 + \int_0^{2000} (5 - 0.008x + 0.000009x^2)\,dx = 20{,}000 + \left[5x - 0.004x^2 + 0.000003x^3\right]_0^{2000}$$

$$= 20{,}000 + 10{,}000 - 0.004(4{,}000{,}000) + 0.000003(8{,}000{,}000{,}000) = 30{,}000 - 16{,}000 + 24{,}000$$

$$= \$38{,}000$$

2. By the Net Change Theorem, $R(5000) - R(1000) = \int_{1000}^{5000} R'(x)\,dx$ ⇒

$$R(5000) = 12{,}400 + \int_{1000}^{5000} (12 - 0.0004x)\,dx = 12{,}400 + \left[12x - 0.0002x^2\right]_{1000}^{5000}$$

$$= 12{,}400 + (60{,}000 - 5{,}000) - (12{,}000 - 200) = \$55{,}600$$

3. If the production level is raised from 1200 units to 1600 units, then the increase in cost is

$$C(1600) - C(1200) = \int_{1200}^{1600} C'(x)\,dx = \int_{1200}^{1600} (74 + 1.1x - 0.002x^2 + 0.00004x^3)\,dx$$

$$= \left[74x + 0.55x^2 - \tfrac{0.002}{3}x^3 + 0.00001x^4\right]_{1200}^{1600} = 64{,}331{,}733.33 - 20{,}464{,}800 = \$43{,}866{,}933.33$$

4. Consumer surplus $= \int_0^{300} \left[p(x) - p(300)\right]dx$

$$= \int_0^{300} [20 - 0.05x - (5)]\,dx$$

$$= \int_0^{300} (15 - 0.05x)\,dx = \left[15x - 0.025x^2\right]_0^{300}$$

$$= 4500 - 2250 = \$2250$$

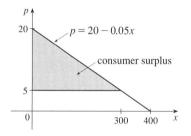

5. $p(x) = 10$ ⇒ $\dfrac{450}{x+8} = 10$ ⇒ $x + 8 = 45$ ⇒ $x = 37$.

$$\text{Consumer surplus} = \int_0^{37} [p(x) - 10]\,dx = \int_0^{37} \left(\frac{450}{x+8} - 10\right)dx$$

$$= \left[450\ln(x+8) - 10x\right]_0^{37} = (450\ln 45 - 370) - 450\ln 8$$

$$= 450\ln\left(\tfrac{45}{8}\right) - 370 \approx \$407.25$$

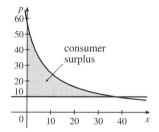

6. $p_S(x) = 3 + 0.01x^2$. $P = p_S(10) = 3 + 1 = 4$.

$$\text{Producer surplus} = \int_0^{10} [P - p_S(x)]\,dx = \int_0^{10} \left[4 - 3 - 0.01x^2\right]dx$$

$$= \left[x - \tfrac{0.01}{3}x^3\right]_0^{10} \approx 10 - 3.33 = \$6.67$$

7. $P = p_S(x)$ ⇒ $400 = 200 + 0.2x^{3/2}$ ⇒ $200 = 0.2x^{3/2}$ ⇒ $1000 = x^{3/2}$ ⇒ $x = 1000^{2/3} = 100$.

$$\text{Producer surplus} = \int_0^{100} [P - p_S(x)]\,dx = \int_0^{100} [400 - (200 + 0.2x^{3/2})]\,dx = \int_0^{100} \left(200 - \tfrac{1}{5}x^{3/2}\right)dx$$

$$= \left[200x - \tfrac{2}{25}x^{5/2}\right]_0^{100} = 20{,}000 - 8{,}000 = \$12{,}000$$

8. $p = 50 - \frac{1}{20}x$ and $p = 20 + \frac{1}{10}x$ intersect at $p = 40$ and $x = 200$.

Consumer surplus $= \int_0^{200} \left(50 - \frac{1}{20}x - 40\right) dx = \left[10x - \frac{1}{40}x^2\right]_0^{200} = \$1000$

Producer surplus $= \int_0^{200} \left(40 - 20 - \frac{1}{10}x\right) dx = \left[20x - \frac{1}{20}x^2\right]_0^{200} = \$2000$

9. $p(x) = \dfrac{800{,}000e^{-x/5000}}{x + 20{,}000} = 16 \quad \Rightarrow \quad x = x_1 \approx 3727.04$.

Consumer surplus $= \int_0^{x_1} [p(x) - 16] \, dx \approx \$37{,}753$

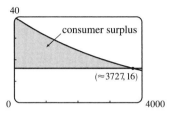

10. The demand function is linear with slope $\frac{-0.5}{35} = -\frac{1}{70}$ and $p(400) = 7.5$, so an equation is $p - 7.5 = -\frac{1}{70}(x - 400)$ or

$p = -\frac{1}{70}x + \frac{185}{14}$. A selling price of \$6 implies that $6 = -\frac{1}{70}x + \frac{185}{14} \quad \Rightarrow \quad \frac{1}{70}x = \frac{185}{14} - \frac{84}{14} = \frac{101}{14} \quad \Rightarrow \quad x = 505$.

Consumer surplus $= \int_0^{505} \left(-\frac{1}{70}x + \frac{185}{14} - 6\right) dx = \left[-\frac{1}{140}x^2 + \frac{101}{14}x\right]_0^{505} \approx \$1821.61$.

11. $f(8) - f(4) = \int_4^8 f'(t) \, dt = \int_4^8 \sqrt{t} \, dt = \left[\frac{2}{3}t^{3/2}\right]_4^8 = \frac{2}{3}\left(16\sqrt{2} - 8\right) \approx \$9.75$ million

12. The total revenue R obtained in the first four years is

$$R = \int_0^4 f(t) \, dt = \int_0^4 9000\sqrt{1 + 2t} \, dt = \int_1^9 9000u^{1/2}\left(\frac{1}{2}\,du\right) \qquad [u = 1 + 2t, du = 2\,dt]$$

$$= 4500\left[\frac{2}{3}u^{3/2}\right]_1^9 = 3000(27 - 1) = \$78{,}000$$

13. $N = \displaystyle\int_a^b Ax^{-k}\, dx = A\left[\dfrac{x^{-k+1}}{-k+1}\right]_a^b = \dfrac{A}{1-k}\left(b^{1-k} - a^{1-k}\right)$.

Similarly, $\displaystyle\int_a^b Ax^{1-k}\, dx = A\left[\dfrac{x^{2-k}}{2-k}\right]_a^b = \dfrac{A}{2-k}\left(b^{2-k} - a^{2-k}\right)$.

Thus, $\bar{x} = \dfrac{1}{N}\displaystyle\int_a^b Ax^{1-k}\, dx = \dfrac{[A/(2-k)](b^{2-k} - a^{2-k})}{[A/(1-k)](b^{1-k} - a^{1-k})} = \dfrac{(1-k)(b^{2-k} - a^{2-k})}{(2-k)(b^{1-k} - a^{1-k})}$.

14. $n(9) - n(5) = \displaystyle\int_5^9 \left(2200 + 10e^{0.8t}\right) dt = \left[2200t + \dfrac{10e^{0.8t}}{0.8}\right]_5^9 = \left[2200t\right]_5^9 + \dfrac{25}{2}\left[e^{0.8t}\right]_5^9$

$$= 2200(9 - 5) + 12.5(e^{7.2} - e^4) \approx 24{,}860$$

15. $F = \dfrac{\pi P R^4}{8\eta l} = \dfrac{\pi(4000)(0.008)^4}{8(0.027)(2)} \approx 1.19 \times 10^{-4} \text{ cm}^3/\text{s}$

16. If the flux remains constant, then $\dfrac{\pi P_0 R_0^4}{8\eta l} = \dfrac{\pi P R^4}{8\eta l} \quad \Rightarrow \quad P_0 R_0^4 = P R^4 \quad \Rightarrow \quad \dfrac{P}{P_0} = \left(\dfrac{R_0}{R}\right)^4$.

$R = \frac{3}{4}R_0 \quad \Rightarrow \quad \dfrac{P}{P_0} = \left(\dfrac{R_0}{\frac{3}{4}R_0}\right)^4 \quad \Rightarrow \quad P = P_0\left(\frac{4}{3}\right)^4 \approx 3.1605P_0 > 3P_0$; that is, the blood pressure is more than tripled.

17. From (3), $F = \dfrac{A}{\int_0^T c(t)\,dt} = \dfrac{6}{20I}$, where

$$I = \int_0^{10} te^{-0.6t}\,dt = \left[\dfrac{1}{(-0.6)^2}(-0.6t-1)e^{-0.6t}\right]_0^{10}\;\begin{bmatrix}\text{integrating}\\\text{by parts}\end{bmatrix} = \tfrac{1}{0.36}(-7e^{-6}+1)$$

Thus, $F = \dfrac{6(0.36)}{20(1-7e^{-6})} = \dfrac{0.108}{1-7e^{-6}} \approx 0.1099$ L/s or 6.594 L/min.

18. As in Example 2, we will estimate the cardiac output using Simpson's Rule with $\Delta t = (20-0)/10 = 2$.

$$\int_0^{20} c(t)\,dt \approx \tfrac{2}{3}[c(0)+4c(2)+2c(4)+4c(6)+2c(8)+4c(10)+2c(12)+4c(14)+2c(16)+4c(18)+c(20)]$$

$$= \tfrac{2}{3}[0+4(2.4)+2(5.1)+4(7.8)+2(7.6)+4(5.4)+2(3.9)+4(2.3)+2(1.6)+4(0.7)+0]$$

$$= \tfrac{2}{3}(110.8) \approx 73.87 \text{ mg}\cdot\text{s/L}$$

Therefore, $F \approx \dfrac{A}{73.87} = \dfrac{8}{73.87} \approx 0.1083$ L/s or 6.498 L/min.

19. As in Example 2, we will estimate the cardiac output using Simpson's Rule with $\Delta t = (16-0)/8 = 2$.

$$\int_0^{16} c(t)\,dt \approx \tfrac{2}{3}[c(0)+4c(2)+2c(4)+4c(6)+2c(8)+4c(10)+2c(12)+4c(14)+c(16)]$$

$$\approx \tfrac{2}{3}[0+4(6.1)+2(7.4)+4(6.7)+2(5.4)+4(4.1)+2(3.0)+4(2.1)+1.5]$$

$$= \tfrac{2}{3}(109.1) = 72.7\overline{3} \text{ mg}\cdot\text{s/L}$$

Therefore, $F \approx \dfrac{A}{72.7\overline{3}} = \dfrac{7}{72.7\overline{3}} \approx 0.0962$ L/s or 5.77 L/min.

9.5 Probability

1. (a) $\int_{30,000}^{40,000} f(x)\,dx$ is the probability that a randomly chosen tire will have a lifetime between 30,000 and 40,000 miles.

(b) $\int_{25,000}^{\infty} f(x)\,dx$ is the probability that a randomly chosen tire will have a lifetime of at least 25,000 miles.

2. (a) The probability that you drive to school in less than 15 minutes is $\int_0^{15} f(t)\,dt$.

(b) The probability that it takes you more than half an hour to get to school is $\int_{30}^{\infty} f(t)\,dt$.

3. (a) In general, we must satisfy the two conditions that are mentioned before Example 1—namely, (1) $f(x) \geq 0$ for all x, and (2) $\int_{-\infty}^{\infty} f(x)\,dx = 1$. For $0 \leq x \leq 4$, we have $f(x) = \tfrac{3}{64}x\sqrt{16-x^2} \geq 0$, so $f(x) \geq 0$ for all x. Also,

$$\int_{-\infty}^{\infty} f(x)\,dx = \int_0^4 \tfrac{3}{64}x\sqrt{16-x^2}\,dx = -\tfrac{3}{128}\int_0^4 (16-x^2)^{1/2}(-2x)\,dx = -\tfrac{3}{128}\left[\tfrac{2}{3}(16-x^2)^{3/2}\right]_0^4$$

$$= -\tfrac{1}{64}\left[(16-x^2)^{3/2}\right]_0^4 = -\tfrac{1}{64}(0-64) = 1.$$

Therefore, f is a probability density function.

(b) $P(X < 2) = \int_{-\infty}^2 f(x)\,dx = \int_0^2 \tfrac{3}{64}x\sqrt{16-x^2}\,dx = -\tfrac{3}{128}\int_0^2 (16-x^2)^{1/2}(-2x)\,dx$

$$= -\tfrac{3}{128}\left[\tfrac{2}{3}(16-x^2)^{3/2}\right]_0^2 = -\tfrac{1}{64}\left[(16-x^2)^{3/2}\right]_0^2 = -\tfrac{1}{64}(12^{3/2}-16^{3/2})$$

$$= \tfrac{1}{64}(64-12\sqrt{12}) = \tfrac{1}{64}(64-24\sqrt{3}) = 1 - \tfrac{3}{8}\sqrt{3} \approx 0.350481$$

4. (a) Since $f(x) = xe^{-x} \geq 0$ if $x \geq 0$ and $f(x) = 0$ if $x < 0$, it follows that $f(x) \geq 0$ for all x. Also,

$$\int_{-\infty}^{\infty} f(x)\,dx = \int_0^{\infty} xe^{-x}\,dx = \lim_{t \to \infty} \int_0^t xe^{-x}\,dx \overset{96}{=} \quad [\text{or by parts}] \quad \lim_{t \to \infty} \left[(-x-1)e^{-x}\right]_0^t$$

$$= \lim_{t \to \infty}\left[(-t-1)e^{-t}+1\right] = 1 - \lim_{t \to \infty} \frac{t+1}{e^t} \overset{\text{H}}{=} 1 - \lim_{t \to \infty} \frac{1}{e^t} = 1 - 0 = 1$$

Thus, f is a probability density function.

(b) $P\left(1 \leq X \leq 2\right) = \int_1^2 xe^{-x}\,dx = \left[(-x-1)e^{-x}\right]_1^2 = -3e^{-2} + 2e^{-1} = 2/e - 3/e^2 \ [\approx 0.33]$

5. (a) In general, we must satisfy the two conditions that are mentioned before Example 1—namely, (1) $f(x) \geq 0$ for all x, and (2) $\int_{-\infty}^{\infty} f(x)\,dx = 1$. If $c \geq 0$, then $f(x) \geq 0$, so condition (1) is satisfied. For condition (2), we see that

$$\int_{-\infty}^{\infty} f(x)\,dx = \int_{-\infty}^{\infty} \frac{c}{1+x^2}\,dx \text{ and}$$

$$\int_0^{\infty} \frac{c}{1+x^2}\,dx = \lim_{t \to \infty} \int_0^t \frac{c}{1+x^2}\,dx = c \lim_{t \to \infty} \left[\tan^{-1} x\right]_0^t = c \lim_{t \to \infty} \tan^{-1} t = c\left(\frac{\pi}{2}\right)$$

Similarly, $\displaystyle\int_{-\infty}^0 \frac{c}{1+x^2}\,dx = c\left(\frac{\pi}{2}\right)$, so $\displaystyle\int_{-\infty}^{\infty} \frac{c}{1+x^2}\,dx = 2c\left(\frac{\pi}{2}\right) = c\pi$.

Since $c\pi$ must equal 1, we must have $c = 1/\pi$ so that f is a probability density function.

(b) $P\left(-1 < X < 1\right) = \displaystyle\int_{-1}^1 \frac{1/\pi}{1+x^2}\,dx = \frac{2}{\pi}\int_0^1 \frac{1}{1+x^2}\,dx = \frac{2}{\pi}\left[\tan^{-1} x\right]_0^1 = \frac{2}{\pi}\left(\frac{\pi}{4} - 0\right) = \frac{1}{2}$

6. (a) For $0 \leq x \leq 1$, we have $f(x) = kx^2(1-x)$, which is nonnegative if and only if $k \geq 0$. Also,

$$\int_{-\infty}^{\infty} f(x)\,dx = \int_0^1 kx^2(1-x)\,dx = k\int_0^1 (x^2 - x^3)\,dx = k\left[\frac{1}{3}x^3 - \frac{1}{4}x^4\right]_0^1 = k/12. \text{ Now } k/12 = 1 \ \Leftrightarrow \ k = 12.$$

Therefore, f is a probability density function if and only if $k = 12$.

(b) Let $k = 12$.

$$P\left(X \geq \tfrac{1}{2}\right) = \int_{1/2}^{\infty} f(x)\,dx = \int_{1/2}^1 12x^2(1-x)\,dx = \int_{1/2}^1 (12x^2 - 12x^3)\,dx = \left[4x^3 - 3x^4\right]_{1/2}^1$$

$$= (4-3) - \left(\tfrac{1}{2} - \tfrac{3}{16}\right) = 1 - \tfrac{5}{16} = \tfrac{11}{16}$$

(c) The mean

$$\mu = \int_{-\infty}^{\infty} xf(x)\,dx = \int_0^1 x \cdot 12x^2(1-x)\,dx = 12\int_0^1 (x^3 - x^4)\,dx = 12\left[\tfrac{1}{4}x^4 - \tfrac{1}{5}x^5\right]_0^1 = 12\left(\tfrac{1}{4} - \tfrac{1}{5}\right) = \tfrac{12}{20} = \tfrac{3}{5}.$$

7. (a) In general, we must satisfy the two conditions that are mentioned before Example 1—namely, (1) $f(x) \geq 0$ for all x, and (2) $\int_{-\infty}^{\infty} f(x)\,dx = 1$. Since $f(x) = 0$ or $f(x) = 0.1$, condition (1) is satisfied. For condition (2), we see that

$$\int_{-\infty}^{\infty} f(x)\,dx = \int_0^{10} 0.1\,dx = \left[\tfrac{1}{10}x\right]_0^{10} = 1. \text{ Thus, } f(x) \text{ is a probability density function for the spinner's values.}$$

(b) Since all the numbers between 0 and 10 are equally likely to be selected, we expect the mean to be halfway between the endpoints of the interval; that is, $x = 5$.

$$\mu = \int_{-\infty}^{\infty} xf(x)\,dx = \int_0^{10} x(0.1)\,dx = \left[\tfrac{1}{20}x^2\right]_0^{10} = \tfrac{100}{20} = 5, \quad \text{as expected.}$$

8. (a) As in the preceding exercise, (1) $f(x) \geq 0$ and (2) $\int_{-\infty}^{\infty} f(x)\,dx = \int_{0}^{10} f(x)\,dx = \frac{1}{2}(10)(0.2)$ [area of a triangle] $= 1$.

So $f(x)$ is a probability density function.

(b) (i) $P(X < 3) = \int_{0}^{3} f(x)\,dx = \frac{1}{2}(3)(0.1) = \frac{3}{20} = 0.15$

(ii) We first compute $P(X > 8)$ and then subtract that value and our answer in (i) from 1 (the total probability).

$P(X > 8) = \int_{8}^{10} f(x)\,dx = \frac{1}{2}(2)(0.1) = \frac{2}{20} = 0.10$. So $P(3 \leq X \leq 8) = 1 - 0.15 - 0.10 = 0.75$.

(c) We find equations of the lines from $(0,0)$ to $(6, 0.2)$ and from $(6, 0.2)$ to $(10, 0)$, and find that

$$f(x) = \begin{cases} \frac{1}{30}x & \text{if } 0 \leq x < 6 \\ -\frac{1}{20}x + \frac{1}{2} & \text{if } 6 \leq x < 10 \\ 0 & \text{otherwise} \end{cases}$$

$$\mu = \int_{-\infty}^{\infty} x f(x)\,dx = \int_{0}^{6} x\left(\frac{1}{30}x\right) dx + \int_{6}^{10} x\left(-\frac{1}{20}x + \frac{1}{2}\right) dx = \left[\frac{1}{90}x^3\right]_{0}^{6} + \left[-\frac{1}{60}x^3 + \frac{1}{4}x^2\right]_{6}^{10}$$

$$= \frac{216}{90} + \left(-\frac{1000}{60} + \frac{100}{4}\right) - \left(-\frac{216}{60} + \frac{36}{4}\right) = \frac{16}{3} = 5.\overline{3}$$

9. We need to find m so that $\int_{m}^{\infty} f(t)\,dt = \frac{1}{2}$ $\Rightarrow$ $\lim_{x \to \infty} \int_{m}^{x} \frac{1}{5}e^{-t/5}\,dt = \frac{1}{2}$ $\Rightarrow$ $\lim_{x \to \infty} \left[\frac{1}{5}(-5)e^{-t/5}\right]_{m}^{x} = \frac{1}{2}$ $\Rightarrow$

$(-1)(0 - e^{-m/5}) = \frac{1}{2}$ $\Rightarrow$ $e^{-m/5} = \frac{1}{2}$ $\Rightarrow$ $-m/5 = \ln\frac{1}{2}$ $\Rightarrow$ $m = -5\ln\frac{1}{2} = 5\ln 2 \approx 3.47$ min.

10. (a) $\mu = 1000$ $\Rightarrow$ $f(t) = \begin{cases} 0 & \text{if } t < 0 \\ \frac{1}{1000}e^{-t/1000} & \text{if } t \geq 0 \end{cases}$

(i) $P(0 \leq X \leq 200) = \int_{0}^{200} \frac{1}{1000}e^{-t/1000}\,dt = \left[-e^{-t/1000}\right]_{0}^{200} = -e^{-1/5} + 1 \approx 0.181$

(ii) $P(X > 800) = \int_{800}^{\infty} \frac{1}{1000}e^{-t/1000}\,dt = \lim_{x \to \infty} \left[-e^{-t/1000}\right]_{800}^{x} = 0 + e^{-4/5} \approx 0.449$

(b) We need to find m so that $\int_{m}^{\infty} f(t)\,dt = \frac{1}{2}$ $\Rightarrow$ $\lim_{x \to \infty} \int_{m}^{x} \frac{1}{1000}e^{-t/1000}\,dt = \frac{1}{2}$ $\Rightarrow$ $\lim_{x \to \infty} \left[-e^{-t/1000}\right]_{m}^{x} = \frac{1}{2}$ $\Rightarrow$

$0 + e^{-m/1000} = \frac{1}{2}$ $\Rightarrow$ $-m/1000 = \ln\frac{1}{2}$ $\Rightarrow$ $m = -1000\ln\frac{1}{2} = 1000\ln 2 \approx 693.1$ h.

11. We use an exponential density function with $\mu = 2.5$ min.

(a) $P(X > 4) = \int_{4}^{\infty} f(t)\,dt = \lim_{x \to \infty} \int_{4}^{x} \frac{1}{2.5}e^{-t/2.5}\,dt = \lim_{x \to \infty} \left[-e^{-t/2.5}\right]_{4}^{x} = 0 + e^{-4/2.5} \approx 0.202$

(b) $P(0 \leq X \leq 2) = \int_{0}^{2} f(t)\,dt = \left[-e^{-t/2.5}\right]_{0}^{2} = -e^{-2/2.5} + 1 \approx 0.551$

(c) We need to find a value a so that $P(X \geq a) = 0.02$, or, equivalently, $P(0 \leq X \leq a) = 0.98$ $\Leftrightarrow$

$\int_{0}^{a} f(t)\,dt = 0.98$ $\Leftrightarrow$ $\left[-e^{-t/2.5}\right]_{0}^{a} = 0.98$ $\Leftrightarrow$ $-e^{-a/2.5} + 1 = 0.98$ $\Leftrightarrow$ $e^{-a/2.5} = 0.02$ $\Leftrightarrow$

$-a/2.5 = \ln 0.02$ $\Leftrightarrow$ $a = -2.5\ln\frac{1}{50} = 2.5\ln 50 \approx 9.78$ min ≈ 10 min. The ad should say that if you aren't served within 10 minutes, you get a free hamburger.

12. (a) With $\mu = 69$ and $\sigma = 2.8$, we have $P(65 \leq X \leq 73) = \int_{65}^{73} \frac{1}{2.8\sqrt{2\pi}} \exp\left(-\frac{(x-69)^2}{2 \cdot 2.8^2}\right) dx \approx 0.847$

(using a calculator or computer to estimate the integral).

(b) $P(X > 6 \text{ feet}) = P(X > 72 \text{ inches}) = 1 - P(0 \leq X \leq 72) \approx 1 - 0.858 = 0.142$, so 14.2% of the adult male population is more than 6 feet tall.

13. $P(X \geq 10) = \int_{10}^{\infty} \frac{1}{4.2\sqrt{2\pi}} \exp\left(-\frac{(x-9.4)^2}{2 \cdot 4.2^2}\right) dx$. To avoid the improper integral we approximate it by the integral from

10 to 100. Thus, $P(X \geq 10) \approx \int_{10}^{100} \frac{1}{4.2\sqrt{2\pi}} \exp\left(-\frac{(x-9.4)^2}{2 \cdot 4.2^2}\right) dx \approx 0.443$ (using a calculator or computer to estimate

the integral), so about 44 percent of the households throw out at least 10 lb of paper a week.

Note: We can't evaluate $1 - P(0 \leq X \leq 10)$ for this problem since a significant amount of area lies to the left of $X = 0$.

14. (a) $P(0 \leq X \leq 480) = \int_{0}^{480} \frac{1}{12\sqrt{2\pi}} \exp\left(-\frac{(x-500)^2}{2 \cdot 12^2}\right) dx \approx 0.0478$ (using a calculator or computer to estimate the

integral), so there is about a 4.78% chance that a particular box contains less than 480 g of cereal.

(b) We need to find μ so that $P(0 \leq X < 500) = 0.05$. Using our calculator or computer to find $P(0 \leq X \leq 500)$ for

various values of μ, we find that if $\mu = 519.73$, $P = 0.05007$; and if $\mu = 519.74$, $P = 0.04998$. So a good target weight

is at least 519.74 g.

15. (a) $P(0 \leq X \leq 100) = \int_{0}^{100} \frac{1}{8\sqrt{2\pi}} \exp\left(-\frac{(x-112)^2}{2 \cdot 8^2}\right) dx \approx 0.0668$ (using a calculator or computer to estimate the

integral), so there is about a 6.68% chance that a randomly chosen vehicle is traveling at a legal speed.

(b) $P(X \geq 125) = \int_{125}^{\infty} \frac{1}{8\sqrt{2\pi}} \exp\left(-\frac{(x-112)^2}{2 \cdot 8^2}\right) dx = \int_{125}^{\infty} f(x)\, dx$. In this case, we could use a calculator or computer

to estimate either $\int_{125}^{300} f(x)\, dx$ or $1 - \int_{0}^{125} f(x)\, dx$. Both are approximately 0.0521, so about 5.21% of the motorists are

targeted.

16. $f(x) = \frac{1}{\sigma\sqrt{2\pi}} e^{-(x-\mu)^2/(2\sigma^2)} \quad \Rightarrow \quad f'(x) = \frac{1}{\sigma\sqrt{2\pi}} e^{-(x-\mu)^2/(2\sigma^2)} \frac{-2(x-\mu)}{2\sigma^2} = \frac{-1}{\sigma^3\sqrt{2\pi}} e^{-(x-\mu)^2/(2\sigma^2)}(x-\mu) \quad \Rightarrow$

$\qquad f''(x) = \frac{-1}{\sigma^3\sqrt{2\pi}}\left[e^{-(x-\mu)^2/(2\sigma^2)} \cdot 1 + (x-\mu)e^{-(x-\mu)^2/(2\sigma^2)}\frac{-2(x-\mu)}{2\sigma^2}\right]$

$\qquad = \frac{-1}{\sigma^3\sqrt{2\pi}}e^{-(x-\mu)^2/(2\sigma^2)}\left[1 - \frac{(x-\mu)^2}{\sigma^2}\right] = \frac{1}{\sigma^5\sqrt{2\pi}}e^{-(x-\mu)^2/(2\sigma^2)}\left[(x-\mu)^2 - \sigma^2\right]$

$f''(x) < 0 \quad \Rightarrow \quad (x-\mu)^2 - \sigma^2 < 0 \quad \Rightarrow \quad |x-\mu| < \sigma \quad \Rightarrow \quad -\sigma < x - \mu < \sigma \quad \Rightarrow \quad \mu - \sigma < x < \mu + \sigma$ and similarly,

$f''(x) > 0 \quad \Rightarrow \quad x < \mu - \sigma$ or $x > \mu + \sigma$. Thus, f changes concavity and has inflection points at $x = \mu \pm \sigma$.

17. $P(\mu - 2\sigma \leq X \leq \mu + 2\sigma) = \int_{\mu-2\sigma}^{\mu+2\sigma} \frac{1}{\sigma\sqrt{2\pi}} \exp\left(-\frac{(x-\mu)^2}{2\sigma^2}\right) dx$. Substituting $t = \frac{x-\mu}{\sigma}$ and $dt = \frac{1}{\sigma} dx$ gives us

$\int_{-2}^{2} \frac{1}{\sigma\sqrt{2\pi}} e^{-t^2/2}(\sigma\, dt) = \frac{1}{\sqrt{2\pi}} \int_{-2}^{2} e^{-t^2/2}\, dt \approx 0.9545$.

18. Let $f(x) = \begin{cases} 0 & \text{if } x < 0 \\ ce^{-cx} & \text{if } x \geq 0 \end{cases}$ where $c = 1/\mu$. By using parts, tables, or a CAS, we find that

(1): $\int xe^{bx}\, dx = (e^{bx}/b^2)(bx - 1)$

(2): $\int x^2 e^{bx}\, dx = (e^{bx}/b^3)(b^2x^2 - 2bx + 2)$

Now

$$\sigma^2 = \int_{-\infty}^{\infty}(x-\mu)^2 f(x)\, dx = \int_{-\infty}^{0}(x-\mu)^2 f(x)\, dx + \int_{0}^{\infty}(x-\mu)^2 f(x)\, dx$$

$$= 0 + \lim_{t\to\infty} c\int_{0}^{t}(x-\mu)^2 e^{-cx}\, dx = c \cdot \lim_{t\to\infty} \int_{0}^{t}\left(x^2 e^{-cx} - 2x\mu e^{-cx} + \mu^2 e^{-cx}\right) dx$$

Next we use (2) and (1) with $b = -c$ to get

$$\sigma^2 = c \lim_{t \to \infty} \left[-\frac{e^{-cx}}{c^3}(c^2 x^2 + 2cx + 2) - 2\mu \frac{e^{-cx}}{c^2}(-cx - 1) + \mu^2 \frac{e^{-cx}}{-c} \right]_0^t$$

Using l'Hospital's Rule several times, along with the fact that $\mu = 1/c$, we get

$$\sigma^2 = c \left[0 - \left(-\frac{2}{c^3} + \frac{2}{c} \cdot \frac{1}{c^2} + \frac{1}{c^2} \cdot \frac{1}{-c} \right) \right] = c \left(\frac{1}{c^3} \right) = \frac{1}{c^2} \quad \Rightarrow \quad \sigma = \frac{1}{c} = \mu$$

19. (a) First $p(r) = \dfrac{4}{a_0^3} r^2 e^{-2r/a_0} \geq 0$ for $r \geq 0$. Next,

$$\int_{-\infty}^{\infty} p(r)\, dr = \int_0^{\infty} \frac{4}{a_0^3} r^2 e^{-2r/a_0}\, dr = \frac{4}{a_0^3} \lim_{t \to \infty} \int_0^t r^2 e^{-2r/a_0}\, dr$$

By using parts, tables, or a CAS [or as in Exercise 18] , we find that $\int x^2 e^{bx}\, dx = (e^{bx}/b^3)(b^2 x^2 - 2bx + 2)$. (⋆)

Next, we use (⋆) (with $b = -2/a_0$) and l'Hospital's Rule to get $\dfrac{4}{a_0^3}\left[\dfrac{a_0^3}{-8}(-2) \right] = 1$. This satisfies the second condition for

a function to be a probability density function.

(b) Using l'Hospital's Rule, $\dfrac{4}{a_0^3} \lim_{r \to \infty} \dfrac{r^2}{e^{2r/a_0}} = \dfrac{4}{a_0^3} \lim_{r \to \infty} \dfrac{2r}{(2/a_0)e^{2r/a_0}} = \dfrac{2}{a_0^2} \lim_{r \to \infty} \dfrac{2}{(2/a_0)e^{2r/a_0}} = 0$.

To find the maximum of p, we differentiate:

$$p'(r) = \frac{4}{a_0^3}\left[r^2 e^{-2r/a_0}\left(-\frac{2}{a_0} \right) + e^{-2r/a_0}(2r) \right] = \frac{4}{a_0^3} e^{-2r/a_0}(2r)\left(-\frac{r}{a_0} + 1 \right)$$

$$p'(r) = 0 \quad \Leftrightarrow \quad r = 0 \text{ or } 1 = \frac{r}{a_0} \quad \Leftrightarrow \quad r = a_0 \quad [a_0 \approx 5.59 \times 10^{-11} \text{ m}].$$

$p'(r)$ changes from positive to negative at $r = a_0$, so $p(r)$ has its maximum value at $r = a_0$.

(c) It is fairly difficult to find a viewing rectangle, but knowing the maximum

value from part (b) helps.

$$p(a_0) = \frac{4}{a_0^3} a_0^2 e^{-2a_0/a_0} = \frac{4}{a_0} e^{-2} \approx 9{,}684{,}098{,}979$$

With a maximum of nearly 10 billion and a total area under the curve of 1,

we know that the "hump" in the graph must be extremely narrow.

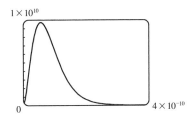

(d) $P(r) = \displaystyle\int_0^r \frac{4}{a_0^3} s^2 e^{-2s/a_0}\, ds \quad \Rightarrow \quad P(4a_0) = \int_0^{4a_0} \frac{4}{a_0^3} s^2 e^{-2s/a_0}\, ds$. Using (⋆) from part (a) [with $b = -2/a_0$],

$$P(4a_0) = \frac{4}{a_0^3}\left[\frac{e^{-2s/a_0}}{-8/a_0^3}\left(\frac{4}{a_0^2}s^2 + \frac{4}{a_0}s + 2 \right) \right]_0^{4a_0} = \frac{4}{a_0^3}\left(\frac{a_0^3}{-8} \right)[e^{-8}(64 + 16 + 2) - 1(2)] = -\frac{1}{2}(82 e^{-8} - 2)$$

$$= 1 - 41 e^{-8} \approx 0.986$$

(e) $\mu = \displaystyle\int_{-\infty}^{\infty} r\, p(r)\, dr = \frac{4}{a_0^3} \lim_{t \to \infty} \int_0^t r^3 e^{-2r/a_0}\, dr$. Integrating by parts three times or using a CAS, we find that

$$\int x^3 e^{bx}\, dx = \frac{e^{bx}}{b^4}(b^3 x^3 - 3b^2 x^2 + 6bx - 6). \text{ So with } b = -\frac{2}{a_0}, \text{ we use l'Hospital's Rule, and get}$$

$$\mu = \frac{4}{a_0^3}\left[-\frac{a_0^4}{16}(-6) \right] = \tfrac{3}{2} a_0.$$

9 Review

---

1. (a) The length of a curve is defined to be the limit of the lengths of the inscribed polygons, as described near Figure 3 in Section 9.1.

(b) See Equation 9.1.2.

(c) See Equation 9.1.4.

2. (a) $S = \int_a^b 2\pi f(x) \sqrt{1 + [f'(x)]^2} \, dx$

(b) If $x = g(y)$, $c \le y \le d$, then $S = \int_c^d 2\pi y \sqrt{1 + [g'(y)]^2} \, dy$.

(c) $S = \int_a^b 2\pi x \sqrt{1 + [f'(x)]^2} \, dx$ or $S = \int_c^d 2\pi g(y) \sqrt{1 + [g'(y)]^2} \, dy$

3. Let $c(x)$ be the cross-sectional length of the wall (measured parallel to the surface of the fluid) at depth x. Then the hydrostatic force against the wall is given by $F = \int_a^b \delta x c(x) \, dx$, where a and b are the lower and upper limits for x at points of the wall and δ is the weight density of the fluid.

4. (a) The center of mass is the point at which the plate balances horizontally.

(b) See Equations 9.3.8.

5. If a plane region $\mathcal{R}$ that lies entirely on one side of a line ℓ in its plane is rotated about ℓ, then the volume of the resulting solid is the product of the area of $\mathcal{R}$ and the distance traveled by the centroid of $\mathcal{R}$.

6. See Figure 3 in Section 9.4, and the discussion which precedes it.

7. (a) See the definition in the first paragraph of the subsection *Cardiac Output* in Section 9.4.

(b) See the discussion in the second paragraph of the subsection *Cardiac Output* in Section 9.4.

8. A probability density function f is a function on the domain of a continuous random variable X such that $\int_a^b f(x) \, dx$ measures the probability that X lies between a and b. Such a function f has nonnegative values and satisfies the relation $\int_D f(x) \, dx = 1$, where D is the domain of the corresponding random variable X. If $D = \mathbb{R}$, or if we define $f(x) = 0$ for real numbers $x \notin D$, then $\int_{-\infty}^{\infty} f(x) \, dx = 1$. (Of course, to work with f in this way, we must assume that the integrals of f exist.)

9. (a) $\int_0^{130} f(x) \, dx$ represents the probability that the weight of a randomly chosen female college student is less than 130 pounds.

(b) $\mu = \int_{-\infty}^{\infty} x f(x) \, dx = \int_0^{\infty} x f(x) \, dx$

(c) The median of f is the number m such that $\int_m^{\infty} f(x) \, dx = \frac{1}{2}$.

10. See the discussion near Equation 3 in Section 9.5.

EXERCISES

1. $y = \frac{1}{6}(x^2 + 4)^{3/2}$ $\Rightarrow$ $dy/dx = \frac{1}{4}(x^2 + 4)^{1/2}(2x)$ $\Rightarrow$

$1 + (dy/dx)^2 = 1 + \left[\frac{1}{2}x(x^2+4)^{1/2}\right]^2 = 1 + \frac{1}{4}x^2(x^2+4) = \frac{1}{4}x^4 + x^2 + 1 = \left(\frac{1}{2}x^2 + 1\right)^2.$

Thus, $L = \int_0^3 \sqrt{\left(\frac{1}{2}x^2 + 1\right)^2}\, dx = \int_0^3 \left(\frac{1}{2}x^2 + 1\right) dx = \left[\frac{1}{6}x^3 + x\right]_0^3 = \frac{15}{2}.$

2. $y = 2\ln\left(\sin\frac{1}{2}x\right)$ $\Rightarrow$ $\dfrac{dy}{dx} = 2\cdot\dfrac{1}{\sin\left(\frac{1}{2}x\right)}\cdot\cos\left(\frac{1}{2}x\right)\cdot\frac{1}{2} = \cot\left(\frac{1}{2}x\right)$ $\Rightarrow$ $1 + \left(\dfrac{dy}{dx}\right)^2 = 1 + \cot^2\left(\frac{1}{2}x\right) = \csc^2\left(\frac{1}{2}x\right).$

Thus,

$$L = \int_{\pi/3}^{\pi} \sqrt{\csc^2\left(\tfrac{1}{2}x\right)}\, dx = \int_{\pi/3}^{\pi} \left|\csc\left(\tfrac{1}{2}x\right)\right| dx = \int_{\pi/3}^{\pi} \csc\left(\tfrac{1}{2}x\right) dx = \int_{\pi/6}^{\pi/2} \csc u\,(2\,du) \quad \begin{bmatrix} u = \frac{1}{2}x, \\ du = \frac{1}{2}\,dx \end{bmatrix}$$

$$= 2\Big[\ln\left|\csc u - \cot u\right|\Big]_{\pi/6}^{\pi/2} = 2\left[\ln\left|\csc\tfrac{\pi}{2} - \cot\tfrac{\pi}{2}\right| - \ln\left|\csc\tfrac{\pi}{6} - \cot\tfrac{\pi}{6}\right|\right]$$

$$= 2\left[\ln|1 - 0| - \ln\left|2 - \sqrt{3}\,\right|\right] = -2\ln\left(2 - \sqrt{3}\right) \approx 2.63$$

3. (a) $y = \dfrac{x^4}{16} + \dfrac{1}{2x^2} = \dfrac{1}{16}x^4 + \dfrac{1}{2}x^{-2}$ $\Rightarrow$ $\dfrac{dy}{dx} = \dfrac{1}{4}x^3 - x^{-3}$ $\Rightarrow$

$1 + (dy/dx)^2 = 1 + \left(\frac{1}{4}x^3 - x^{-3}\right)^2 = 1 + \frac{1}{16}x^6 - \frac{1}{2} + x^{-6} = \frac{1}{16}x^6 + \frac{1}{2} + x^{-6} = \left(\frac{1}{4}x^3 + x^{-3}\right)^2.$

Thus, $L = \int_1^2 \left(\frac{1}{4}x^3 + x^{-3}\right) dx = \left[\frac{1}{16}x^4 - \frac{1}{2}x^{-2}\right]_1^2 = \left(1 - \frac{1}{8}\right) - \left(\frac{1}{16} - \frac{1}{2}\right) = \frac{21}{16}.$

(b) $S = \int_1^2 2\pi x\left(\frac{1}{4}x^3 + x^{-3}\right) dx = 2\pi\int_1^2 \left(\frac{1}{4}x^4 + x^{-2}\right) dx = 2\pi\left[\frac{1}{20}x^5 - \frac{1}{x}\right]_1^2$

$= 2\pi\left[\left(\frac{32}{20} - \frac{1}{2}\right) - \left(\frac{1}{20} - 1\right)\right] = 2\pi\left(\frac{8}{5} - \frac{1}{2} - \frac{1}{20} + 1\right) = 2\pi\left(\frac{41}{20}\right) = \frac{41}{10}\pi$

4. (a) $y = x^2$ $\Rightarrow$ $1 + (y')^2 = 1 + 4x^2$ $\Rightarrow$

$$S = \int_0^1 2\pi x\sqrt{1 + 4x^2}\, dx = \int_1^5 \frac{\pi}{4}\sqrt{u}\, du \quad [u = 1 + 4x^2] = \frac{\pi}{6}\left[u^{3/2}\right]_1^5 = \frac{\pi}{6}(5^{3/2} - 1)$$

(b) $y = x^2$ $\Rightarrow$ $1 + (y')^2 = 1 + 4x^2.$ So

$$S = 2\pi\int_0^1 x^2\sqrt{1 + 4x^2}\, dx = 2\pi\int_0^2 \frac{1}{4}u^2\sqrt{1 + u^2}\,\frac{1}{2}\,du \quad [u = 2x] \quad = \frac{\pi}{4}\int_0^2 u^2\sqrt{1 + u^2}\, du$$

$$= \frac{\pi}{4}\left[\frac{1}{8}u(1 + 2u^2)\sqrt{1 + u^2} - \frac{1}{8}\ln\left|u + \sqrt{1 + u^2}\right|\right]_0^2 \qquad [u = \tan\theta \text{ or use Formula 22}]$$

$$= \frac{\pi}{4}\left[\frac{1}{4}(9)\sqrt{5} - \frac{1}{8}\ln(2 + \sqrt{5}) - 0\right] = \frac{\pi}{32}\left[18\sqrt{5} - \ln(2 + \sqrt{5})\right]$$

5. $y = e^{-x^2}$ $\Rightarrow$ $dy/dx = -2xe^{-x^2}$ $\Rightarrow$ $1 + (dy/dx)^2 = 1 + 4x^2e^{-2x^2}.$ Let $f(x) = \sqrt{1 + 4x^2e^{-2x^2}}.$ Then

$$L = \int_0^3 f(x)\, dx \approx S_6 = \frac{(3-0)/6}{3}\left[f(0) + 4f(0.5) + 2f(1) + 4f(1.5) + 2f(2) + 4f(2.5) + f(3)\right] \approx 3.292287$$

6. $S = \int_0^3 2\pi y\, ds = \int_0^3 2\pi e^{-x^2}\sqrt{1 + 4x^2e^{-2x^2}}\, dx.$ Let $g(x) = 2\pi e^{-x^2}\sqrt{1 + 4x^2e^{-2x^2}}.$ Then

$$S = \int_0^3 g(x)\, dx \approx S_6 = \frac{(3-0)/6}{3}\left[g(0) + 4g(0.5) + 2g(1) + 4g(1.5) + 2g(2) + 4g(2.5) + g(3)\right] \approx 6.648327.$$

7. $y = \int_1^x \sqrt{\sqrt{t} - 1}\, dt \quad \Rightarrow \quad dy/dx = \sqrt{\sqrt{x} - 1} \quad \Rightarrow \quad 1 + (dy/dx)^2 = 1 + \left(\sqrt{x} - 1\right) = \sqrt{x}.$

Thus, $L = \int_1^{16} \sqrt{\sqrt{x}}\, dx = \int_1^{16} x^{1/4}\, dx = \frac{4}{5}\left[x^{5/4}\right]_1^{16} = \frac{4}{5}(32 - 1) = \frac{124}{5}.$

8. $S = \int_1^{16} 2\pi x\, ds = 2\pi \int_1^{16} x \cdot x^{1/4}\, dx = 2\pi \int_1^{16} x^{5/4}\, dx = 2\pi \cdot \frac{4}{9}\left[x^{9/4}\right]_1^{16} = \frac{8\pi}{9}(512 - 1) = \frac{4088}{9}\pi$

9. As in Example 1 of Section 9.3, $\dfrac{a}{2 - x} = \dfrac{1}{2} \quad \Rightarrow \quad 2a = 2 - x$ and $w = 2(1.5 + a) = 3 + 2a = 3 + 2 - x = 5 - x.$

Thus, $F = \int_0^2 \rho g x (5 - x)\, dx = \rho g \left[\frac{5}{2}x^2 - \frac{1}{3}x^3\right]_0^2 = \rho g \left(10 - \frac{8}{3}\right) = \frac{22}{3}\delta \quad [\rho g = \delta] \quad \approx \frac{22}{3} \cdot 62.5 \approx 458$ lb.

10. $F = \int_0^4 \delta(4 - y)2\left(2\sqrt{y}\right) dy = 4\delta \int_0^4 (4y^{1/2} - y^{3/2})\, dy$

$\qquad = 4\delta \left[\frac{8}{3}y^{3/2} - \frac{2}{5}y^{5/2}\right]_0^4 = 4\delta\left(\frac{64}{3} - \frac{64}{5}\right) = 256\delta\left(\frac{1}{3} - \frac{1}{5}\right)$

$\qquad = \frac{512}{15}\delta \approx 2133.3$ lb $\qquad [\delta \approx 62.5 \text{ lb/ft}^3]$

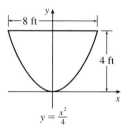

11. $A = \int_0^4 \left(\sqrt{x} - \frac{1}{2}x\right) dx = \left[\frac{2}{3}x^{3/2} - \frac{1}{4}x^2\right]_0^4 = \frac{16}{3} - 4 = \frac{4}{3}$

$\overline{x} = \frac{1}{A}\int_0^4 x\left(\sqrt{x} - \frac{1}{2}x\right) dx = \frac{3}{4}\int_0^4 \left(x^{3/2} - \frac{1}{2}x^2\right) dx$

$\qquad = \frac{3}{4}\left[\frac{2}{5}x^{5/2} - \frac{1}{6}x^3\right]_0^4 = \frac{3}{4}\left(\frac{64}{5} - \frac{64}{6}\right) = \frac{3}{4}\left(\frac{64}{30}\right) = \frac{8}{5}$

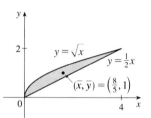

$\overline{y} = \frac{1}{A}\int_0^4 \frac{1}{2}\left[\left(\sqrt{x}\right)^2 - \left(\frac{1}{2}x\right)^2\right] dx = \frac{3}{4}\int_0^4 \frac{1}{2}\left(x - \frac{1}{4}x^2\right) dx = \frac{3}{8}\left[\frac{1}{2}x^2 - \frac{1}{12}x^3\right]_0^4 = \frac{3}{8}\left(8 - \frac{16}{3}\right) = \frac{3}{8}\left(\frac{8}{3}\right) = 1$

Thus, the centroid is $(\overline{x}, \overline{y}) = \left(\frac{8}{5}, 1\right).$

12. From the symmetry of the region, $\overline{x} = \frac{\pi}{2}.$ $\quad A = \int_{\pi/4}^{3\pi/4} \sin x\, dx = \left[-\cos x\right]_{\pi/4}^{3\pi/4} = \frac{1}{\sqrt{2}} - \left(-\frac{1}{\sqrt{2}}\right) = \sqrt{2}$

$\overline{y} = \frac{1}{A}\int_{\pi/4}^{3\pi/4} \frac{1}{2}\sin^2 x\, dx = \frac{1}{A}\int_{\pi/4}^{3\pi/4} \frac{1}{4}(1 - \cos 2x)\, dx$

$\qquad = \frac{1}{4\sqrt{2}}\left[x - \frac{1}{2}\sin 2x\right]_{\pi/4}^{3\pi/4}$

$\qquad = \frac{1}{4\sqrt{2}}\left[\frac{3\pi}{4} - \frac{1}{2}(-1) - \frac{\pi}{4} + \frac{1}{2}\cdot 1\right] = \frac{1}{4\sqrt{2}}\left(\frac{\pi}{2} + 1\right)$

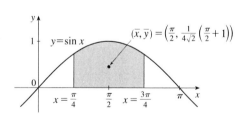

Thus, the centroid is $(\overline{x}, \overline{y}) = \left(\frac{\pi}{2}, \frac{1}{4\sqrt{2}}\left(\frac{\pi}{2} + 1\right)\right) \approx (1.57, 0.45).$

13. An equation of the line passing through $(0, 0)$ and $(3, 2)$ is $y = \frac{2}{3}x.$ $A = \frac{1}{2} \cdot 3 \cdot 2 = 3.$ Therefore, using Equations 9.3.8,

$\overline{x} = \frac{1}{3}\int_0^3 x\left(\frac{2}{3}x\right) dx = \frac{2}{27}\left[x^3\right]_0^3 = 2$ and $\overline{y} = \frac{1}{3}\int_0^3 \frac{1}{2}\left(\frac{2}{3}x\right)^2 dx = \frac{2}{81}\left[x^3\right]_0^3 = \frac{2}{3}.$ Thus, the centroid is $(\overline{x}, \overline{y}) = \left(2, \frac{2}{3}\right).$

14. Suppose first that the large rectangle were complete, so that its mass would be $6 \cdot 3 = 18$. Its centroid would be $\left(1, \frac{3}{2}\right)$. The mass removed from this object to create the one being studied is 3. The centroid of the cut-out piece is $\left(\frac{3}{2}, \frac{3}{2}\right)$. Therefore, for the actual lamina, whose mass is 15, $\overline{x} = \frac{18}{15}(1) - \frac{3}{15}\left(\frac{3}{2}\right) = \frac{9}{10}$, and $\overline{y} = \frac{3}{2}$, since the lamina is symmetric about the line $y = \frac{3}{2}$. Thus, the centroid is $(\overline{x}, \overline{y}) = \left(\frac{9}{10}, \frac{3}{2}\right)$.

15. The centroid of this circle, $(1, 0)$, travels a distance $2\pi(1)$ when the lamina is rotated about the y-axis. The area of the circle is $\pi(1)^2$. So by the Theorem of Pappus, $V = A(2\pi\overline{x}) = \pi(1)^2 2\pi(1) = 2\pi^2$.

16. The semicircular region has an area of $\frac{1}{2}\pi r^2$, and sweeps out a sphere of radius r when rotated about the x-axis.

$\overline{x} = 0$ because of symmetry about the line $x = 0$. And by the Theorem of Pappus, $V = A(2\pi\overline{y}) \Rightarrow$

$\frac{4}{3}\pi r^3 = \frac{1}{2}\pi r^2(2\pi\overline{y}) \Rightarrow \overline{y} = \frac{4}{3\pi}r$. Thus, the centroid is $(\overline{x}, \overline{y}) = \left(0, \frac{4}{3\pi}r\right)$.

17. $x = 100 \Rightarrow P = 2000 - 0.1(100) - 0.01(100)^2 = 1890$

$$\text{Consumer surplus} = \int_0^{100}[p(x) - P]\,dx = \int_0^{100}\left(2000 - 0.1x - 0.01x^2 - 1890\right)dx$$

$$= \left[110x - 0.05x^2 - \frac{0.01}{3}x^3\right]_0^{100} = 11{,}000 - 500 - \frac{10{,}000}{3} \approx \$7166.67$$

18. $\int_0^{24} c(t)\,dt \approx S_{12} = \frac{24 - 0}{12 \cdot 3}[1(0) + 4(1.9) + 2(3.3) + 4(5.1) + 2(7.6) + 4(7.1) + 2(5.8)$

$$+ \ 4(4.7) + 2(3.3) + 4(2.1) + 2(1.1) + 4(0.5) + 1(0)]$$

$$= \tfrac{2}{3}(127.8) = 85.2 \text{ mg} \cdot \text{s/L}$$

Therefore, $F \approx A/85.2 = 6/85.2 \approx 0.0704$ L/s or 4.225 L/min.

19. $f(x) = \begin{cases} \frac{\pi}{20}\sin\left(\frac{\pi}{10}x\right) & \text{if } 0 \le x \le 10 \\ 0 & \text{if } x < 0 \text{ or } x > 10 \end{cases}$

(a) $f(x) \ge 0$ for all real numbers x and

$$\int_{-\infty}^{\infty} f(x)\,dx = \int_0^{10} \frac{\pi}{20}\sin\left(\frac{\pi}{10}x\right)dx = \frac{\pi}{20} \cdot \frac{10}{\pi}\left[-\cos\left(\frac{\pi}{10}x\right)\right]_0^{10} = \tfrac{1}{2}(-\cos\pi + \cos 0) = \tfrac{1}{2}(1 + 1) = 1$$

Therefore, f is a probability density function.

(b) $P(X < 4) = \int_{-\infty}^{4} f(x)\,dx = \int_0^4 \frac{\pi}{20}\sin\left(\frac{\pi}{10}x\right)dx = \tfrac{1}{2}\left[-\cos\left(\frac{\pi}{10}x\right)\right]_0^4 = \tfrac{1}{2}\left(-\cos\frac{2\pi}{5} + \cos 0\right)$

$$\approx \tfrac{1}{2}(-0.309017 + 1) \approx 0.3455$$

(c) $\mu = \int_{-\infty}^{\infty} xf(x)\,dx = \int_0^{10} \frac{\pi}{20}x\sin\left(\frac{\pi}{10}x\right)dx$

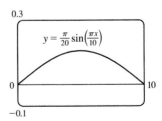

$$= \int_0^{\pi} \frac{\pi}{20} \cdot \frac{10}{\pi}u(\sin u)\left(\frac{10}{\pi}\right)du \qquad \left[u = \tfrac{\pi}{10}x,\ du = \tfrac{\pi}{10}\,dx\right]$$

$$= \frac{5}{\pi}\int_0^{\pi} u\sin u\,du \overset{82}{=} \frac{5}{\pi}\left[\sin u - u\cos u\right]_0^{\pi} = \frac{5}{\pi}[0 - \pi(-1)] = 5$$

This answer is expected because the graph of f is symmetric about the line $x = 5$.

20. $P(250 \leq X \leq 280) = \int_{250}^{280} \frac{1}{15\sqrt{2\pi}} \exp\left(\frac{-(x-268)^2}{2 \cdot 15^2}\right) dx \approx 0.673.$ Thus, the percentage of pregnancies that last

between 250 and 280 days is about 67.3%.

21. (a) The probability density function is $f(t) = \begin{cases} 0 & \text{if } t < 0 \\ \frac{1}{8}e^{-t/8} & \text{if } t \geq 0 \end{cases}$

$$P(0 \leq X \leq 3) = \int_0^3 \frac{1}{8}e^{-t/8}\, dt = \left[-e^{-t/8}\right]_0^3 = -e^{-3/8} + 1 \approx 0.3127$$

(b) $P(X > 10) = \int_{10}^{\infty} \frac{1}{8}e^{-t/8}\, dt = \lim_{x \to \infty}\left[-e^{-t/8}\right]_{10}^{x} = \lim_{x \to \infty}\left(-e^{-x/8} + e^{-10/8}\right) = 0 + e^{-5/4} \approx 0.2865$

(c) We need to find m such that $P(X \geq m) = \frac{1}{2} \;\Rightarrow\; \int_m^{\infty} \frac{1}{8}e^{-t/8}\, dt = \frac{1}{2} \;\Rightarrow\; \lim_{x \to \infty}\left[-e^{-t/8}\right]_m^{x} = \frac{1}{2} \;\Rightarrow\;$

$\lim_{x \to \infty}\left(-e^{-x/8} + e^{-m/8}\right) = \frac{1}{2} \;\Rightarrow\; e^{-m/8} = \frac{1}{2} \;\Rightarrow\; -m/8 = \ln\frac{1}{2} \;\Rightarrow\; m = -8\ln\frac{1}{2} = 8\ln 2 \approx 5.55$ minutes.

PROBLEMS PLUS

1. $x^2 + y^2 \le 4y \iff x^2 + (y-2)^2 \le 4$, so S is part of a circle, as shown
in the diagram. The area of S is

$$\int_0^1 \sqrt{4y - y^2}\, dy \overset{113}{=} \left[\frac{y-2}{2} \sqrt{4y - y^2} + 2\cos^{-1}\left(\frac{2-y}{2}\right) \right]_0^1 \quad [a = 2]$$

$$= -\tfrac{1}{2}\sqrt{3} + 2\cos^{-1}\left(\tfrac{1}{2}\right) - 2\cos^{-1} 1$$

$$= -\frac{\sqrt{3}}{2} + 2\left(\frac{\pi}{3}\right) - 2(0) = \frac{2\pi}{3} - \frac{\sqrt{3}}{2}$$

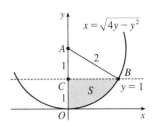

Another method (without calculus): Note that $\theta = \angle CAB = \frac{\pi}{3}$, so the area is

$$\text{(area of sector } OAB) - \text{(area of } \triangle ABC) = \tfrac{1}{2}\left(2^2\right)\tfrac{\pi}{3} - \tfrac{1}{2}(1)\sqrt{3} = \frac{2\pi}{3} - \frac{\sqrt{3}}{2}$$

2. $y = \pm\sqrt{x^3 - x^4} \implies$ The loop of the curve is symmetric about $y = 0$, and therefore $\overline{y} = 0$. At each point x
where $0 \le x \le 1$, the lamina has a vertical length of $\sqrt{x^3 - x^4} - \left(-\sqrt{x^3 - x^4}\right) = 2\sqrt{x^3 - x^4}$. Therefore,

$$\overline{x} = \frac{\int_0^1 x \cdot 2\sqrt{x^3 - x^4}\, dx}{\int_0^1 2\sqrt{x^3 - x^4}\, dx} = \frac{\int_0^1 x\sqrt{x^3 - x^4}\, dx}{\int_0^1 \sqrt{x^3 - x^4}\, dx}. \text{ We evaluate the integrals separately:}$$

$$\int_0^1 x\sqrt{x^3 - x^4}\, dx = \int_0^1 x^{5/2}\sqrt{1 - x}\, dx$$

$$= \int_0^{\pi/2} 2\sin^6\theta \cos\theta \sqrt{1 - \sin^2\theta}\, d\theta \qquad \left[\begin{array}{l} \sin\theta = \sqrt{x},\, \cos\theta\, d\theta = dx/(2\sqrt{x}), \\ 2\sin\theta\, \cos\theta\, d\theta = dx \end{array} \right]$$

$$= \int_0^{\pi/2} 2\sin^6\theta \cos^2\theta\, d\theta = \int_0^{\pi/2} 2\left[\tfrac{1}{2}(1 - \cos 2\theta)\right]^3 \tfrac{1}{2}(1 + \cos 2\theta)\, d\theta$$

$$= \int_0^{\pi/2} \tfrac{1}{8}(1 - 2\cos 2\theta + 2\cos^3 2\theta - \cos^4 2\theta)\, d\theta$$

$$= \int_0^{\pi/2} \tfrac{1}{8}\left[1 - 2\cos 2\theta + 2\cos 2\theta(1 - \sin^2 2\theta) - \tfrac{1}{4}(1 + \cos 4\theta)^2\right] d\theta$$

$$= \tfrac{1}{8}\left[\theta - \tfrac{1}{3}\sin^3 2\theta\right]_0^{\pi/2} - \tfrac{1}{32}\int_0^{\pi/2}\left(1 + 2\cos 4\theta + \cos^2 4\theta\right) d\theta$$

$$= \frac{\pi}{16} - \tfrac{1}{32}\left[\theta + \tfrac{1}{2}\sin 4\theta\right]_0^{\pi/2} - \tfrac{1}{64}\int_0^{\pi/2}(1 + \cos 8\theta)\, d\theta$$

$$= \frac{\pi}{16} - \tfrac{1}{32}\left[\theta + \tfrac{1}{2}\sin 4\theta\right]_0^{\pi/2} - \tfrac{1}{64}\int_0^{\pi/2}(1 + \cos 8\theta)\, d\theta$$

$$= \frac{3\pi}{64} - \tfrac{1}{64}\left[\theta + \tfrac{1}{8}\sin 8\theta\right]_0^{\pi/2} = \frac{5\pi}{128}$$

$$\int_0^1 \sqrt{x^3 - x^4}\, dx = \int_0^1 x^{3/2}\sqrt{1 - x}\, dx = \int_0^{\pi/2} 2\sin^4\theta \cos\theta\sqrt{1 - \sin^2\theta}\, d\theta \qquad [\sin\theta = \sqrt{x}]$$

$$= \int_0^{\pi/2} 2\sin^4\theta \cos^2\theta\, d\theta = \int_0^{\pi/2} 2 \cdot \tfrac{1}{4}(1 - \cos 2\theta)^2 \cdot \tfrac{1}{2}(1 + \cos 2\theta)\, d\theta$$

$$= \int_0^{\pi/2} \tfrac{1}{4}(1 - \cos 2\theta - \cos^2 2\theta + \cos^3 2\theta)\, d\theta$$

$$= \int_0^{\pi/2} \tfrac{1}{4}\left[1 - \cos 2\theta - \tfrac{1}{2}(1 + \cos 4\theta) + \cos 2\theta(1 - \sin^2 2\theta)\right] d\theta$$

$$= \tfrac{1}{4}\left[\tfrac{\theta}{2} - \tfrac{1}{8}\sin 4\theta - \tfrac{1}{6}\sin^3 2\theta\right]_0^{\pi/2} = \frac{\pi}{16}$$

Therefore, $\overline{x} = \dfrac{5\pi/128}{\pi/16} = \dfrac{5}{8}$, and $(\overline{x}, \overline{y}) = \left(\tfrac{5}{8}, 0\right)$.

3. (a) The two spherical zones, whose surface areas we will call S_1 and S_2, are

generated by rotation about the y-axis of circular arcs, as indicated in the figure.

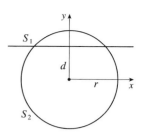

The arcs are the upper and lower portions of the circle $x^2 + y^2 = r^2$ that are

obtained when the circle is cut with the line $y = d$. The portion of the upper arc

in the first quadrant is sufficient to generate the upper spherical zone. That

portion of the arc can be described by the relation $x = \sqrt{r^2 - y^2}$ for

$d \leq y \leq r$. Thus, $dx/dy = -y/\sqrt{r^2 - y^2}$ and

$$ds = \sqrt{1 + \left(\frac{dx}{dy}\right)^2}\, dy = \sqrt{1 + \frac{y^2}{r^2 - y^2}}\, dy = \sqrt{\frac{r^2}{r^2 - y^2}}\, dy = \frac{r\, dy}{\sqrt{r^2 - y^2}}$$

From Formula 9.2.8 we have

$$S_1 = \int_d^r 2\pi x \sqrt{1 + \left(\frac{dx}{dy}\right)^2}\, dy = \int_d^r 2\pi \sqrt{r^2 - y^2}\, \frac{r\, dy}{\sqrt{r^2 - y^2}} = \int_d^r 2\pi r\, dy = 2\pi r(r - d)$$

Similarly, we can compute $S_2 = \int_{-r}^d 2\pi x \sqrt{1 + (dx/dy)^2}\, dy = \int_{-r}^d 2\pi r\, dy = 2\pi r(r + d)$. Note that $S_1 + S_2 = 4\pi r^2$,

the surface area of the entire sphere.

(b) $r = 3960$ mi and $d = r\,(\sin 75°) \approx 3825$ mi,

so the surface area of the Arctic Ocean is about

$2\pi r(r-d) \approx 2\pi(3960)(135) \approx 3.36\times10^6$ mi².

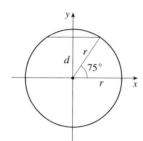

(c) The area on the sphere lies between planes $y = y_1$ and $y = y_2$, where $y_2 - y_1 = h$. Thus, we compute the surface area on

the sphere to be $S = \int_{y_1}^{y_2} 2\pi x \sqrt{1 + \left(\frac{dx}{dy}\right)^2}\, dy = \int_{y_1}^{y_2} 2\pi r\, dy = 2\pi r(y_2 - y_1) = 2\pi rh.$

This equals the lateral area of a cylinder of radius r and height h, since such

a cylinder is obtained by rotating the line $x = r$ about the y-axis, so the

surface area of the cylinder between the planes $y = y_1$ and $y = y_2$ is

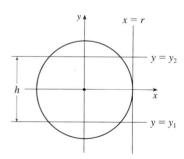

$$A = \int_{y_1}^{y_2} 2\pi x \sqrt{1 + \left(\frac{dx}{dy}\right)^2}\, dy = \int_{y_1}^{y_2} 2\pi r \sqrt{1 + 0^2}\, dy$$

$$= 2\pi ry \Big|_{y=y_1}^{y_2} = 2\pi r(y_2 - y_1) = 2\pi rh$$

(d) $h = 2r \sin 23.45° \approx 3152$ mi, so the surface area of the

Torrid Zone is $2\pi r h \approx 2\pi(3960)(3152) \approx 7.84 \times 10^7$ mi^2.

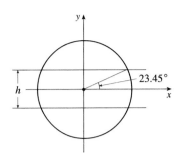

4. (a) Since the right triangles OAT and OTB are similar, we have $\dfrac{r+H}{r} = \dfrac{r}{a}$ $\Rightarrow$

$a = \dfrac{r^2}{r+H}$. The surface area visible from B is $S = \int_a^r 2\pi x \sqrt{1 + (dx/dy)^2}\, dy$.

From $x^2 + y^2 = r^2$, we get $\dfrac{d}{dy}(x^2 + y^2) = \dfrac{d}{dy}(r^2)$ $\Rightarrow$ $2x\dfrac{dx}{dy} + 2y = 0$ $\Rightarrow$

$\dfrac{dx}{dy} = -\dfrac{y}{x}$ and $1 + \left(\dfrac{dx}{dy}\right)^2 = \dfrac{x^2 + y^2}{x^2} = \dfrac{r^2}{x^2}$. Thus,

$$S = \int_a^r 2\pi x \cdot \frac{r}{x}\, dy = 2\pi r(r - a) = 2\pi r\left(r - \frac{r^2}{r+H}\right) = 2\pi r^2\left(1 - \frac{r}{r+H}\right) = 2\pi r^2 \cdot \frac{H}{r+H} = \frac{2\pi r^2 H}{r+H}.$$

(b) Assume $R \geq r$. If a light is placed at point L, at a distance x from

the center of the sphere of radius r, then from part (a) we find that

the total illuminated area A on the two spheres is [with $r + H = x$

and $r + H = d - x$].

$$A(x) = \frac{2\pi r^2(x - r)}{x} + \frac{2\pi R^2(d - x - R)}{d - x} \quad [r \leq x \leq d - R]. \quad \frac{A(x)}{2\pi} = r^2\left(1 - \frac{r}{x}\right) + R^2\left(1 - \frac{R}{d - x}\right),$$

so $A'(x) = 0$ $\Leftrightarrow$ $0 = r^2 \cdot \dfrac{r}{x^2} + R^2 \cdot \dfrac{-R}{(d-x)^2}$ $\Leftrightarrow$ $\dfrac{r^3}{x^2} = \dfrac{R^3}{(d-x)^2}$ $\Leftrightarrow$ $\dfrac{(d-x)^2}{x^2} = \dfrac{R^3}{r^3}$ $\Leftrightarrow$

$\left(\dfrac{d}{x} - 1\right)^2 = \left(\dfrac{R}{r}\right)^3$ $\Rightarrow$ $\dfrac{d}{x} - 1 = \left(\dfrac{R}{r}\right)^{3/2}$ $\Leftrightarrow$ $\dfrac{d}{x} = 1 + \left(\dfrac{R}{r}\right)^{3/2}$ $\Leftrightarrow$ $x = x^* = \dfrac{d}{1 + (R/r)^{3/2}}$.

Now $A'(x) = 2\pi\left(\dfrac{r^3}{x^2} - \dfrac{R^3}{(d-x)^2}\right)$ $\Rightarrow$ $A''(x) = 2\pi\left(-\dfrac{2r^3}{x^3} - \dfrac{2R^3}{(d-x)^3}\right)$ and $A''(x^*) < 0$, so we have a

local maximum at $x = x^*$.

However, x^* may not be an allowable value of x — we must show that x^* is between r and $d - R$.

(1) $x^* \geq r$ $\Leftrightarrow$ $\dfrac{d}{1 + (R/r)^{3/2}} \geq r$ $\Leftrightarrow$ $d \geq r + R\sqrt{R/r}$

(2) $x^* \leq d - R$ $\Leftrightarrow$ $\dfrac{d}{1 + (R/r)^{3/2}} \leq d - R$ $\Leftrightarrow$ $d \leq d - R + d\left(\dfrac{R}{r}\right)^{3/2} - R\left(\dfrac{R}{r}\right)^{3/2}$ $\Leftrightarrow$

$R + R\left(\dfrac{R}{r}\right)^{3/2} \leq d\left(\dfrac{R}{r}\right)^{3/2}$ $\Leftrightarrow$ $d \geq \dfrac{R}{(R/r)^{3/2}} + R = R + r\sqrt{r/R}$, but

$R + r\sqrt{r/R} \leq R + r$, and since $d > r + R$ [given], we conclude that $x^* \leq d - R$.

Thus, from (1) and (2), x^* is not an allowable value of x if $d < r + R\sqrt{R/r}$.

So A may have a maximum at $x = r$, x^*, or $d - R$.

$$A(r) = \frac{2\pi R^2(d - r - R)}{d - r} \qquad \text{and} \qquad A(d - R) = \frac{2\pi r^2(d - r - R)}{d - R}$$

$A(r) > A(d - R) \;\Leftrightarrow\; \dfrac{R^2}{d - r} > \dfrac{r^2}{d - R} \;\Leftrightarrow\; R^2(d - R) > r^2(d - r) \;\Leftrightarrow\; R^2 d - R^3 > r^2 d - r^3 \;\Leftrightarrow\;$

$R^2 d - r^2 d > R^3 - r^3 \;\Leftrightarrow\; d(R - r)(R + r) > (R - r)(R^2 + Rr + r^2) \;\Leftrightarrow\; d > (R^2 + Rr + r^2)/(R + r) \;\Leftrightarrow\;$

$d > [(R + r)^2 - Rr]/(R + r) \;\Leftrightarrow\; d > R + r - Rr/(R + r)$. Now $R + r - Rr/(R + r) < R + r$, and we know that

$d > R + r$, so we conclude that $A(r) > A(d - R)$.

In conclusion, A has an absolute maximum at $x = x^*$ provided $d \geq r + R\sqrt{R/r}$; otherwise, A has its maximum

at $x = r$.

5. (a) Choose a vertical x-axis pointing downward with its origin at the surface. In order to calculate the pressure at depth z,

consider n subintervals of the interval $[0, z]$ by points x_i and choose a point $x_i^* \in [x_{i-1}, x_i]$ for each i. The thin layer of

water lying between depth x_{i-1} and depth x_i has a density of approximately $\rho(x_i^*)$, so the weight of a piece of that layer

with unit cross-sectional area is $\rho(x_i^*)g\,\Delta x$. The total weight of a column of water extending from the surface to depth z

(with unit cross-sectional area) would be approximately $\sum\limits_{i=1}^{n} \rho(x_i^*)g\,\Delta x$. The estimate becomes exact if we take the limit

as $n \to \infty$; weight (or force) per unit area at depth z is $W = \lim\limits_{n\to\infty} \sum\limits_{i=1}^{n} \rho(x_i^*)g\,\Delta x$. In other words, $P(z) = \int_0^z \rho(x)g\,dx$.

More generally, if we make no assumptions about the location of the origin, then $P(z) = P_0 + \int_0^z \rho(x)g\,dx$, where P_0 is

the pressure at $x = 0$. Differentiating, we get $dP/dz = \rho(z)g$.

(b)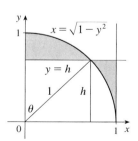

$$F = \int_{-r}^{r} P(L + x) \cdot 2\sqrt{r^2 - x^2}\,dx$$

$$= \int_{-r}^{r} \left(P_0 + \int_0^{L+x} \rho_0 e^{z/H} g\,dz \right) \cdot 2\sqrt{r^2 - x^2}\,dx$$

$$= P_0 \int_{-r}^{r} 2\sqrt{r^2 - x^2}\,dx + \rho_0 g H \int_{-r}^{r} \left(e^{(L+x)/H} - 1 \right) \cdot 2\sqrt{r^2 - x^2}\,dx$$

$$= (P_0 - \rho_0 g H) \int_{-r}^{r} 2\sqrt{r^2 - x^2}\,dx + \rho_0 g H \int_{-r}^{r} e^{(L+x)/H} \cdot 2\sqrt{r^2 - x^2}\,dx$$

$$= (P_0 - \rho_0 g H)(\pi r^2) + \rho_0 g H e^{L/H} \int_{-r}^{r} e^{x/H} \cdot 2\sqrt{r^2 - x^2}\,dx$$

6. The problem can be reduced to finding the line which minimizes the shaded

area in the diagram. An equation of the circle in the first quadrant is

$x = \sqrt{1 - y^2}$. So the shaded area is

$$A(h) = \int_0^h \left(1 - \sqrt{1 - y^2} \right) dy + \int_h^1 \sqrt{1 - y^2}\,dy$$

$$= \int_0^h \left(1 - \sqrt{1 - y^2} \right) dy - \int_1^h \sqrt{1 - y^2}\,dy$$

$$A'(h) = 1 - \sqrt{1 - h^2} - \sqrt{1 - h^2} \quad \text{[by FTC]} \; = 1 - 2\sqrt{1 - h^2}$$

$A' = 0 \iff \sqrt{1 - h^2} = \frac{1}{2} \implies 1 - h^2 = \frac{1}{4} \implies h^2 = \frac{3}{4} \implies h = \frac{\sqrt{3}}{2}.$

$A''(h) = -2 \cdot \frac{1}{2}(1 - h^2)^{-1/2}(-2h) = \dfrac{2h}{\sqrt{1 - h^2}} > 0$, so $h = \dfrac{\sqrt{3}}{2}$ gives a minimum value of A.

Note: Another strategy is to use the angle θ as the variable (see the diagram above) and show that

$A = \theta + \cos\theta - \frac{\pi}{4} - \frac{1}{2}\sin 2\theta$, which is minimized when $\theta = \frac{\pi}{6}$.

7. To find the height of the pyramid, we use similar triangles. The first figure shows a cross-section of the pyramid passing through the top and through two opposite corners of the square base. Now $|BD| = b$, since it is a radius of the sphere, which has diameter $2b$ since it is tangent to the opposite sides of the square base. Also, $|AD| = b$ since $\triangle ADB$ is isosceles. So the height is $|AB| = \sqrt{b^2 + b^2} = \sqrt{2}\,b$.

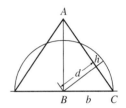

We first observe that the shared volume is equal to half the volume of the sphere, minus the sum of the four equal volumes (caps of the sphere) cut off by the triangular faces of the pyramid. See Exercise 6.2.51 for a derivation of the formula for the volume of a cap of a sphere. To use the formula, we need to find the perpendicular distance h of each triangular face from the surface of the sphere. We first find the distance d from the center of the sphere to one of the triangular faces. The third figure shows a cross-section of the pyramid through the top and through the midpoints of opposite sides of the square base. From similar triangles we find that

$$\frac{d}{b} = \frac{|AB|}{|AC|} = \frac{\sqrt{2}\,b}{\sqrt{b^2 + \left(\sqrt{2}\,b\right)^2}} \implies d = \frac{\sqrt{2}\,b^2}{\sqrt{3b^2}} = \frac{\sqrt{6}}{3}\,b$$

So $h = b - d = b - \frac{\sqrt{6}}{3}b = \frac{3 - \sqrt{6}}{3}b$. So, using the formula $V = \pi h^2(r - h/3)$ from Exercise 6.2.51 with $r = b$, we find that the volume of each of the caps is $\pi\left(\frac{3 - \sqrt{6}}{3}b\right)^2\left(b - \frac{3 - \sqrt{6}}{3 \cdot 3}b\right) = \frac{15 - 6\sqrt{6}}{9} \cdot \frac{6 + \sqrt{6}}{9}\pi b^3 = \left(\frac{2}{3} - \frac{7}{27}\sqrt{6}\right)\pi b^3$. So, using our first observation, the shared volume is $V = \frac{1}{2}\left(\frac{4}{3}\pi b^3\right) - 4\left(\frac{2}{3} - \frac{7}{27}\sqrt{6}\right)\pi b^3 = \left(\frac{28}{27}\sqrt{6} - 2\right)\pi b^3$.

8. Orient the positive x-axis as in the figure.

Suppose that the plate has height h and is symmetric about the x-axis. At depth x below the water ($2 \le x \le 2 + h$), let the width of the plate be $2f(x)$. Now each of the n horizontal strips has height h/n and the ith strip ($1 \le i \le n$) goes from

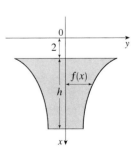

$x = 2 + \left(\dfrac{i - 1}{n}\right)h$ to $x = 2 + \left(\dfrac{i}{n}\right)h$. The hydrostatic force on the ith strip is $F(i) = \displaystyle\int_{2 + [(i-1)/n]h}^{2 + (i/n)h} 62.5x[2f(x)]\,dx.$

If we now let $x[2f(x)] = k$ (a constant) so that $f(x) = k/(2x)$, then

$$F(i) = \int_{2+[(i-1)/n]h}^{2+(i/n)h} 62.5k \, dx = 62.5k \left[x \right]_{2+[(i-1)/n]h}^{2+(i/n)h} = 62.5k \left[\left(2 + \frac{i}{n} h \right) - \left(2 + \frac{i-1}{n} h \right) \right] = 62.5k \left(\frac{h}{n} \right)$$

So the hydrostatic force on the ith strip is independent of i, that is, the force on each strip is the same. So the plate can be shaped as shown in the figure. (In fact, the required condition is satisfied whenever the plate has width C/x at depth x, for some constant C. Many shapes are possible.)

9. We can assume that the cut is made along a vertical line $x = b > 0$, that the disk's boundary is the circle $x^2 + y^2 = 1$, and that the center of mass of the smaller piece (to the right of $x = b$) is $\left(\frac{1}{2}, 0\right)$. We wish to find b to two decimal places. We have $\dfrac{1}{2} = \overline{x} = \dfrac{\int_b^1 x \cdot 2\sqrt{1-x^2}\,dx}{\int_b^1 2\sqrt{1-x^2}\,dx}$. Evaluating the

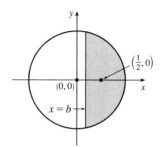

numerator gives us $-\int_b^1 (1-x^2)^{1/2}(-2x)\,dx = -\frac{2}{3}\left[(1-x^2)^{3/2}\right]_b^1 = -\frac{2}{3}\left[0 - (1-b^2)^{3/2}\right] = \frac{2}{3}(1-b^2)^{3/2}$.

Using Formula 30 in the table of integrals, we find that the denominator is

$\left[x\sqrt{1-x^2} + \sin^{-1}x\right]_b^1 = \left(0 + \frac{\pi}{2}\right) - \left(b\sqrt{1-b^2} + \sin^{-1}b\right)$. Thus, we have $\dfrac{1}{2} = \overline{x} = \dfrac{\frac{2}{3}(1-b^2)^{3/2}}{\frac{\pi}{2} - b\sqrt{1-b^2} - \sin^{-1}b}$, or,

equivalently, $\frac{2}{3}(1-b^2)^{3/2} = \frac{\pi}{4} - \frac{1}{2}b\sqrt{1-b^2} - \frac{1}{2}\sin^{-1}b$. Solving this equation numerically with a calculator or CAS, we obtain $b \approx 0.138173$, or $b = 0.14$ m to two decimal places.

10. $A_1 = 30 \ \Rightarrow \ \frac{1}{2}bh = 30 \ \Rightarrow \ bh = 60$.

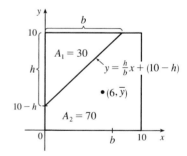

$\overline{x} = 6 \ \Rightarrow \ \dfrac{1}{A_2}\displaystyle\int_0^{10} xf(x)\,dx = 6 \ \Rightarrow$

$\displaystyle\int_0^b x\left(\frac{h}{b}x + 10 - h\right)dx + \int_b^{10} x(10)\,dx = 6(70) \ \Rightarrow$

$\displaystyle\int_0^b \left(\frac{h}{b}x^2 + 10x - hx\right)dx + 10 \cdot \frac{1}{2}\left[x^2\right]_b^{10} = 420 \ \Rightarrow$

$\left[\dfrac{h}{3b}x^3 + 5x^2 - \dfrac{h}{2}x^2\right]_0^b + 5(100 - b^2) = 420 \ \Rightarrow \ \frac{1}{3}hb^2 + 5b^2 - \frac{1}{2}hb^2 + 500 - 5b^2 = 420 \ \Rightarrow \ 80 = \frac{1}{6}hb^2 \ \Rightarrow$

$480 = (hb)b \ \Rightarrow \ 480 = 60b \ \Rightarrow \ b = 8$. So $h = \frac{60}{8} = \frac{15}{2}$ and an equation of the line is

$y = \dfrac{15/2}{8}x + \left(10 - \dfrac{15}{2}\right) = \dfrac{15}{16}x + \dfrac{5}{2}$. Now

$$\overline{y} = \frac{1}{A_2}\int_0^{10} \frac{1}{2}[f(x)]^2\,dx = \frac{1}{70 \cdot 2}\left[\int_0^8 \left(\frac{15}{16}x + \frac{5}{2}\right)^2 dx + \int_8^{10}(10)^2\,dx\right]$$

$$= \frac{1}{140}\left[\int_0^8 \left(\frac{225}{256}x^2 + \frac{75}{16}x + \frac{25}{4}\right)dx + 100(10 - 8)\right] = \frac{1}{140}\left(\left[\frac{225}{768}x^3 + \frac{75}{32}x^2 + \frac{25}{4}x\right]_0^8 + 200\right)$$

$$= \frac{1}{140}(150 + 150 + 50 + 200) = \frac{550}{140} = \frac{55}{14}$$

Another solution: Assume that the right triangle cut from the square has legs a cm and b cm long as shown. The triangle has area 30 cm^2, so $\frac{1}{2}ab = 30$ and $ab = 60$. We place the square in the first quadrant of the xy-plane as shown, and we let T, R, and S denote the triangle, the remaining portion of the square, and the full square, respectively. By symmetry, the centroid of S is $(5, 5)$. By Exercise 9.3.39, the centroid of T is $\left(\dfrac{b}{3}, 10 - \dfrac{a}{3}\right)$.

We are given that the centroid of R is $(6, c)$, where c is to be determined. We take the density of the square to be 1, so that areas can be used as masses. Then T has mass $m_T = 30$, S has mass $m_S = 100$, and R has mass $m_R = m_S - m_T = 70$. As in Exercises 40 and 41 of Section 9.3, we view S as consisting of a mass m_T at the centroid $(\overline{x}_T, \overline{y}_T)$ of T and a mass R at the centroid $(\overline{x}_R, \overline{y}_R)$ of R. Then $\overline{x}_S = \dfrac{m_T \overline{x}_T + m_R \overline{x}_R}{m_T + m_R}$ and $\overline{y}_S = \dfrac{m_T \overline{y}_T + m_R \overline{y}_R}{m_T + m_R}$; that is, $5 = \dfrac{30(b/3) + 70(6)}{100}$

and $5 = \dfrac{30(10 - a/3) + 70c}{100}$.

Solving the first equation for b, we get $b = 8$ cm. Since $ab = 60$ cm^2,

it follows that $a = \frac{60}{8} = 7.5$ cm. Now the second equation says that

$70c = 200 + 10a$, so $7c = 20 + a = \frac{55}{2}$ and $c = \frac{55}{14} = 3.9\overline{285714}$ cm.

The solution is depicted in the figure.

11. If $h = L$, then $P = \dfrac{\text{area under } y = L \sin \theta}{\text{area of rectangle}} = \dfrac{\int_0^{\pi} L \sin \theta \, d\theta}{\pi L} = \dfrac{[-\cos \theta]_0^{\pi}}{\pi} = \dfrac{-(-1)+1}{\pi} = \dfrac{2}{\pi}$.

If $h = L/2$, then $P = \dfrac{\text{area under } y = \frac{1}{2}L \sin \theta}{\text{area of rectangle}} = \dfrac{\int_0^{\pi} \frac{1}{2}L \sin \theta \, d\theta}{\pi L} = \dfrac{[-\cos \theta]_0^{\pi}}{2\pi} = \dfrac{2}{2\pi} = \dfrac{1}{\pi}$.

12. (a) The total set of possibilities can be identified with the rectangular region $\mathscr{R} = \{(\theta, y) \mid 0 \le y < L, 0 \le \theta < \pi\}$. Even when $h > L$, the needle intersects at least one line if and only if $y \le h \sin \theta$. Let $\mathscr{R}_1 = \{(\theta, y) \mid 0 \le y \le h \sin \theta, 0 \le \theta < \pi\}$. When $h \le L$, $\mathscr{R}_1$ is contained in $\mathscr{R}$, but that is no longer true when $h > L$. Thus, the probability that the needle intersects a line becomes

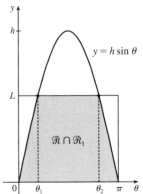

$$P = \dfrac{\text{area}(\mathscr{R} \cap \mathscr{R}_1)}{\text{area}(\mathscr{R})} = \dfrac{\text{area}(\mathscr{R} \cap \mathscr{R}_1)}{\pi L}$$

When $h > L$, the curve $y = h \sin \theta$ intersects the line $y = L$ twice—at $\left(\sin^{-1}(L/h), L\right)$ and at $\left(\pi - \sin^{-1}(L/h), L\right)$. Set $\theta_1 = \sin^{-1}(L/h)$ and $\theta_2 = \pi - \theta_1$. Then

$$\text{area}(\mathcal{R} \cap \mathcal{R}_1) = \int_0^{\theta_1} h \sin\theta \, d\theta + \int_{\theta_1}^{\theta_2} L \, d\theta + \int_{\theta_2}^{\pi} h \sin\theta \, d\theta$$

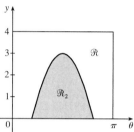

$$= 2\int_0^{\theta_1} h \sin\theta \, d\theta + L(\theta_2 - \theta_1) = 2h\left[-\cos\theta\right]_0^{\theta_1} + L(\pi - 2\theta_1)$$

$$= 2h(1 - \cos\theta_1) + L(\pi - 2\theta_1)$$

$$= 2h\left(1 - \frac{\sqrt{h^2 - L^2}}{h}\right) + L\left[\pi - 2\sin^{-1}\left(\frac{L}{h}\right)\right]$$

$$= 2h - 2\sqrt{h^2 - L^2} + \pi L - 2L\sin^{-1}\left(\frac{L}{h}\right)$$

We are told that $L = 4$ and $h = 7$, so $\text{area}(\mathcal{R} \cap \mathcal{R}_1) = 14 - 2\sqrt{33} + 4\pi - 8\sin^{-1}\left(\frac{4}{7}\right) \approx 10.21128$ and

$P = \frac{1}{4\pi}\text{area}(\mathcal{R} \cap \mathcal{R}_1) \approx 0.812588$. (By comparison, $P = \frac{2}{\pi} \approx 0.636620$ when $h = L$, as shown in the solution to

Problem 11.)

(b) The needle intersects at least two lines when $y + L \le h\sin\theta$; that is, when

$y \le h\sin\theta - L$. Set $\mathcal{R}_2 = \{(\theta, y) \mid 0 \le y \le h\sin\theta - L, 0 \le \theta < \pi\}$.

Then the probability that the needle intersects at least two lines is

$$P_2 = \frac{\text{area}(\mathcal{R} \cap \mathcal{R}_2)}{\text{area}(\mathcal{R})} = \frac{\text{area}(\mathcal{R} \cap \mathcal{R}_2)}{\pi L}$$

When $L = 4$ and $h = 7$, $\mathcal{R}_2$ is contained in $\mathcal{R}$ (see the figure). Thus,

$$P_2 = \frac{1}{4\pi}\text{area}(\mathcal{R}_2) = \frac{1}{4\pi}\int_{\sin^{-1}(4/7)}^{\pi - \sin^{-1}(4/7)}(7\sin\theta - 4)\, d\theta = \frac{1}{4\pi} \cdot 2\int_{\sin^{-1}(4/7)}^{\pi/2}(7\sin\theta - 4)\, d\theta$$

$$= \frac{1}{2\pi}\left[-7\cos\theta - 4\theta\right]_{\sin^{-1}(4/7)}^{\pi/2} = \frac{1}{2\pi}\left[0 - 2\pi + 7\frac{\sqrt{33}}{7} + 4\sin^{-1}\left(\frac{4}{7}\right)\right] = \frac{\sqrt{33} + 4\sin^{-1}\left(\frac{4}{7}\right) - 2\pi}{2\pi}$$

$$\approx 0.301497$$

(c) The needle intersects at least three lines when $y + 2L \le h\sin\theta$: that is, when $y \le h\sin\theta - 2L$. Set

$\mathcal{R}_3 = \{(\theta, y) \mid 0 \le y \le h\sin\theta - 2L, 0 \le \theta < \pi\}$. Then the probability that the needle intersects at least three lines is

$P_3 = \dfrac{\text{area}(\mathcal{R} \cap \mathcal{R}_3)}{\text{area}(\mathcal{R})} = \dfrac{\text{area}(\mathcal{R} \cap \mathcal{R}_3)}{\pi L}$. (At this point, the generalization to P_n, n any positive integer, should be clear.)

Under the given assumption,

$$P_3 = \frac{1}{\pi L}\text{area}(\mathcal{R}_3) = \frac{1}{\pi L}\int_{\sin^{-1}(2L/h)}^{\pi - \sin^{-1}(2L/h)}(h\sin\theta - 2L)\, d\theta = \frac{2}{\pi L}\int_{\sin^{-1}(2L/h)}^{\pi/2}(h\sin\theta - 2L)\, d\theta$$

$$= \frac{2}{\pi L}\left[-h\cos\theta - 2L\theta\right]_{\sin^{-1}(2L/h)}^{\pi/2} = \frac{2}{\pi L}\left[-\pi L + \sqrt{h^2 - 4L^2} + 2L\sin^{-1}(2L/h)\right]$$

Note that the probability that a needle touches exactly one line is $P_1 - P_2$, the probability that it touches exactly two lines

is $P_2 - P_3$, and so on.

10 □ DIFFERENTIAL EQUATIONS

10.1 Modeling with Differential Equations

1. $y = x - x^{-1} \quad \Rightarrow \quad y' = 1 + x^{-2}$. To show that y is a solution of the differential equation, we will substitute the expressions for y and y' in the left-hand side of the equation and show that the left-hand side is equal to the right-hand side.

$$\text{LHS} = xy' + y = x(1 + x^{-2}) + (x - x^{-1}) = x + x^{-1} + x - x^{-1} = 2x = \text{RHS}$$

2. $y = \sin x \, \cos x - \cos x \quad \Rightarrow \quad y' = \sin x \, (-\sin x) + \cos x \, (\cos x) - (-\sin x) = \cos^2 x - \sin^2 x + \sin x$.

$$\text{LHS} = y' + (\tan x)y = \cos^2 x - \sin^2 x + \sin x + (\tan x)(\sin x \, \cos x - \cos x)$$

$$= \cos^2 x - \sin^2 x + \sin x + \sin^2 x - \sin x = \cos^2 x = \text{RHS},$$

so y is a solution of the differential equation. Also, $y(0) = \sin 0 \, \cos 0 - \cos 0 = 0 \cdot 1 - 1 = -1$, so the initial condition is satisfied.

3. (a) $y = e^{rx} \quad \Rightarrow \quad y' = re^{rx} \quad \Rightarrow \quad y'' = r^2 e^{rx}$. Substituting these expressions into the differential equation $2y'' + y' - y = 0$, we get $2r^2 e^{rx} + re^{rx} - e^{rx} = 0 \quad \Rightarrow \quad (2r^2 + r - 1)e^{rx} = 0 \quad \Rightarrow$ $(2r - 1)(r + 1) = 0$ [since e^{rx} is never zero] $\quad \Rightarrow \quad r = \frac{1}{2}$ or -1.

(b) Let $r_1 = \frac{1}{2}$ and $r_2 = -1$, so we need to show that every member of the family of functions $y = ae^{x/2} + be^{-x}$ is a solution of the differential equation $2y'' + y' - y = 0$.

$y = ae^{x/2} + be^{-x} \quad \Rightarrow \quad y' = \frac{1}{2}ae^{x/2} - be^{-x} \quad \Rightarrow \quad y'' = \frac{1}{4}ae^{x/2} + be^{-x}$.

$$\text{LHS} = 2y'' + y' - y = 2\left(\frac{1}{4}ae^{x/2} + be^{-x}\right) + \left(\frac{1}{2}ae^{x/2} - be^{-x}\right) - (ae^{x/2} + be^{-x})$$

$$= \frac{1}{2}ae^{x/2} + 2be^{-x} + \frac{1}{2}ae^{x/2} - be^{-x} - ae^{x/2} - be^{-x}$$

$$= \left(\frac{1}{2}a + \frac{1}{2}a - a\right)e^{x/2} + (2b - b - b)e^{-x}$$

$$= 0 = \text{RHS}$$

4. (a) $y = \cos kt \quad \Rightarrow \quad y' = -k \sin kt \quad \Rightarrow \quad y'' = -k^2 \cos kt$. Substituting these expressions into the differential equation $4y'' = -25y$, we get $4(-k^2 \cos kt) = -25(\cos kt) \quad \Rightarrow \quad (25 - 4k^2)\cos kt = 0$ [for all t] $\quad \Rightarrow \quad 25 - 4k^2 = 0 \quad \Rightarrow$ $k^2 = \frac{25}{4} \quad \Rightarrow \quad k = \pm\frac{5}{2}$.

(b) $y = A \sin kt + B \cos kt \quad \Rightarrow \quad y' = Ak \cos kt - Bk \sin kt \quad \Rightarrow \quad y'' = -Ak^2 \sin kt - Bk^2 \cos kt$.

The given differential equation $4y'' = -25y$ is equivalent to $4y'' + 25y = 0$. Thus,

$$\text{LHS} = 4y'' + 25y = 4(-Ak^2 \sin kt - Bk^2 \cos kt) + 25(A \sin kt + B \cos kt)$$

$$= -4Ak^2 \sin kt - 4Bk^2 \cos kt + 25A \sin kt + 25B \cos kt$$

$$= (25 - 4k^2)A \sin kt + (25 - 4k^2)B \cos kt$$

$$= 0 \quad \text{since } k^2 = \frac{25}{4}.$$

5. (a) $y = \sin x \;\Rightarrow\; y' = \cos x \;\Rightarrow\; y'' = -\sin x$.

LHS $= y'' + y = -\sin x + \sin x = 0 \neq \sin x$, so $y = \sin x$ **is not** a solution of the differential equation.

(b) $y = \cos x \;\Rightarrow\; y' = -\sin x \;\Rightarrow\; y'' = -\cos x$.

LHS $= y'' + y = -\cos x + \cos x = 0 \neq \sin x$, so $y = \cos x$ **is not** a solution of the differential equation.

(c) $y = \frac{1}{2}x \sin x \;\Rightarrow\; y' = \frac{1}{2}(x\cos x + \sin x) \;\Rightarrow\; y'' = \frac{1}{2}(-x\sin x + \cos x + \cos x)$.

LHS $= y'' + y = \frac{1}{2}(-x\sin x + 2\cos x) + \frac{1}{2}x \sin x = \cos x \neq \sin x$, so $y = \frac{1}{2}x \sin x$ **is not** a solution of the differential equation.

(d) $y = -\frac{1}{2}x\cos x \;\Rightarrow\; y' = -\frac{1}{2}(-x\sin x + \cos x) \;\Rightarrow\; y'' = -\frac{1}{2}(-x\cos x - \sin x - \sin x)$.

LHS $= y'' + y = -\frac{1}{2}(-x\cos x - 2\sin x) + \left(-\frac{1}{2}x\cos x\right) = \sin x = $ RHS, so $y = -\frac{1}{2}x\cos x$ **is a** solution of the differential equation.

6. (a) $y = \dfrac{\ln x + C}{x} \;\Rightarrow\; y' = \dfrac{x \cdot (1/x) - (\ln x + C)}{x^2} = \dfrac{1 - \ln x - C}{x^2}$.

$$\text{LHS} = x^2 y' + xy = x^2 \cdot \frac{1 - \ln x - C}{x^2} + x \cdot \frac{\ln x + C}{x}$$

$$= 1 - \ln x - C + \ln x + C = 1 = \text{RHS}, \text{ so } y \text{ is a solution of the differential equation.}$$

(b)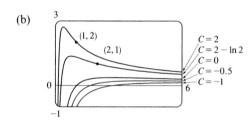

A few notes about the graph of $y = (\ln x + C)/x$:

(1) There is a vertical asymptote of $x = 0$.

(2) There is a horizontal asymptote of $y = 0$.

(3) $y = 0 \;\Rightarrow\; \ln x + C = 0 \;\Rightarrow\; x = e^{-C}$, so there is an x-intercept at e^{-C}.

(4) $y' = 0 \;\Rightarrow\; \ln x = 1 - C \;\Rightarrow\; x = e^{1-C}$, so there is a local maximum at $x = e^{1-C}$.

(c) $y(1) = 2 \;\Rightarrow\; 2 = \dfrac{\ln 1 + C}{1} \;\Rightarrow\; 2 = C$, so the solution is $y = \dfrac{\ln x + 2}{x}$ [shown in part (b)].

(d) $y(2) = 1 \;\Rightarrow\; 1 = \dfrac{\ln 2 + C}{2} \;\Rightarrow\; 2 + \ln 2 + C \;\Rightarrow\; C = 2 - \ln 2$, so the solution is $y = \dfrac{\ln x + 2 - \ln 2}{x}$ [shown in part (b)].

7. (a) Since the derivative $y' = -y^2$ is always negative (or 0 if $y = 0$), the function y must be decreasing (or equal to 0) on any interval on which it is defined.

(b) $y = \dfrac{1}{x + C} \;\Rightarrow\; y' = -\dfrac{1}{(x+C)^2}$. LHS $= y' = -\dfrac{1}{(x+C)^2} = -\left(\dfrac{1}{x+C}\right)^2 = -y^2 = $ RHS

(c) $y = 0$ is a solution of $y' = -y^2$ that is not a member of the family in part (b).

(d) If $y(x) = \dfrac{1}{x+C}$, then $y(0) = \dfrac{1}{0+C} = \dfrac{1}{C}$. Since $y(0) = 0.5$, $\dfrac{1}{C} = \dfrac{1}{2} \;\Rightarrow\; C = 2$, so $y = \dfrac{1}{x+2}$.

8. (a) If x is close to 0, then xy^3 is close to 0, and hence, y' is close to 0. Thus, the graph of y must have a tangent line that is nearly horizontal. If x is large, then xy^3 is large, and the graph of y must have a tangent line that is nearly vertical. (In both cases, we assume reasonable values for y.)

(b) $y = (c - x^2)^{-1/2}$ $\Rightarrow$ $y' = x(c - x^2)^{-3/2}$. RHS $= xy^3 = x[(c - x^2)^{-1/2}]^3 = x(c - x^2)^{-3/2} = y' =$ LHS

(c)

When x is close to 0, y' is also close to 0.

As x gets larger, so does $|y'|$.

(d) $y(0) = (c - 0)^{-1/2} = 1/\sqrt{c}$ and $y(0) = 2$ $\Rightarrow$ $\sqrt{c} = \frac{1}{2}$ $\Rightarrow$ $c = \frac{1}{4}$, so $y = \left(\frac{1}{4} - x^2\right)^{-1/2}$.

9. (a) $\dfrac{dP}{dt} = 1.2P\left(1 - \dfrac{P}{4200}\right)$. Now $\dfrac{dP}{dt} > 0$ $\Rightarrow$ $1 - \dfrac{P}{4200} > 0$ [assuming that $P > 0$] $\Rightarrow$ $\dfrac{P}{4200} < 1$ $\Rightarrow$

$P < 4200$ $\Rightarrow$ the population is increasing for $0 < P < 4200$.

(b) $\dfrac{dP}{dt} < 0$ $\Rightarrow$ $P > 4200$

(c) $\dfrac{dP}{dt} = 0$ $\Rightarrow$ $P = 4200$ or $P = 0$

10. (a) $y = k$ $\Rightarrow$ $y' = 0$, so $\dfrac{dy}{dt} = y^4 - 6y^3 + 5y^2$ $\Leftrightarrow$ $0 = k^4 - 6k^3 + 5k^2$ $\Leftrightarrow$ $k^2(k^2 - 6k + 5) = 0$ $\Leftrightarrow$

$k^2(k - 1)(k - 5) = 0$ $\Leftrightarrow$ $k = 0, 1,$ or 5

(b) y is increasing $\Leftrightarrow$ $\dfrac{dy}{dt} > 0$ $\Leftrightarrow$ $y^2(y - 1)(y - 5) > 0$ $\Leftrightarrow$ $y \in (-\infty, 0) \cup (0, 1) \cup (5, \infty)$

(c) y is decreasing $\Leftrightarrow$ $\dfrac{dy}{dt} < 0$ $\Leftrightarrow$ $y \in (1, 5)$

11. (a) This function is increasing *and* also decreasing. But $dy/dt = e^t(y - 1)^2 \geq 0$ for all t, implying that the graph of the solution of the differential equation cannot be decreasing on any interval.

(b) When $y = 1$, $dy/dt = 0$, but the graph does not have a horizontal tangent line.

12. The graph for this exercise is shown in the figure at the right.

A. $y' = 1 + xy > 1$ for points in the first quadrant, but we can see that $y' < 0$ for some points in the first quadrant.

B. $y' = -2xy = 0$ when $x = 0$, but we can see that $y' > 0$ for $x = 0$.

Thus, equations A and B are incorrect, so the correct equation is C.

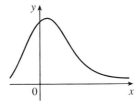

C. $y' = 1 - 2xy$ seems reasonable since:

(1) When $x = 0$, y' could be 1.

(2) When $x < 0$, y' could be greater than 1.

(3) Solving $y' = 1 - 2xy$ for y gives us $y = \dfrac{1 - y'}{2x}$. If y' takes on small negative values, then as $x \to \infty$, $y \to 0^+$, as shown in the figure.

13. (a) P increases most rapidly at the beginning, since there are usually many simple, easily-learned sub-skills associated with learning a skill. As t increases, we would expect dP/dt to remain positive, but decrease. This is because as time progresses, the only points left to learn are the more difficult ones.

(b) $\dfrac{dP}{dt} = k(M - P)$ is always positive, so the level of performance P is increasing. As P gets close to M, dP/dt gets close to 0; that is, the performance levels off, as explained in part (a).

(c)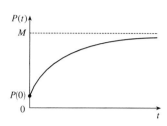

14. (a) The coffee cools most quickly as soon as it is removed from the heat source. The rate of cooling decreases toward 0 since the coffee approaches room temperature.

(b) $\dfrac{dy}{dt} = k(y - R)$, where k is a proportionality constant, y is the temperature of the coffee, and R is the room temperature. The initial condition is $y(0) = 95°\text{C}$. The answer and the model support each other because as y approaches R, dy/dt approaches 0, so the model seems appropriate.

(c)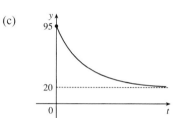

10.2   Direction Fields and Euler's Method

1. (a)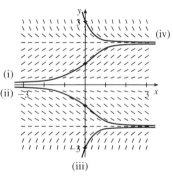

(b) It appears that the constant functions $y = 0$, $y = -2$, and $y = 2$ are equilibrium solutions. Note that these three values of y satisfy the given differential equation $y' = y\left(1 - \frac{1}{4}y^2\right)$.

2. (a)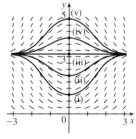

(b) From the figure, it appears that $y = \pi$ is an equilibrium solution. From the equation $y' = x \sin y$, we see that $y = n\pi$ (n an integer) describes all the equilibrium solutions.

3. $y' = 2 - y$. The slopes at each point are independent of x, so the slopes are the same along each line parallel to the x-axis. Thus, III is the direction field for this equation. Note that for $y = 2$, $y' = 0$.

4. $y' = x(2 - y) = 0$ on the lines $x = 0$ and $y = 2$. Direction field I satisfies these conditions.

5. $y' = x + y - 1 = 0$ on the line $y = -x + 1$. Direction field IV satisfies this condition. Notice also that on the line $y = -x$ we have $y' = -1$, which is true in IV.

6. $y' = \sin x \sin y = 0$ on the lines $x = 0$ and $y = 0$, and $y' > 0$ for $0 < x < \pi, 0 < y < \pi$. Direction field II satisfies these conditions.

7. (a) $y(0) = 1$

   (b) $y(0) = 2$

   (c) $y(0) = -1$

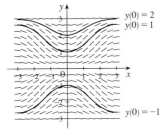

8. (a) $y(0) = -1$

   (b) $y(0) = 0$

   (c) $y(0) = 1$

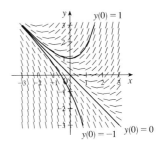

9.

x	y	$y' = 1 + y$
0	0	1
0	1	2
0	2	3
0	-3	-2
0	-2	-1

Note that for $y = -1$, $y' = 0$. The three solution curves sketched go through $(0, 0)$, $(0, -1)$, and $(0, -2)$.

10.

x	y	$y' = x^2 - y^2$
± 1	± 3	-8
± 3	± 1	8
± 1	± 0.5	0.75
± 0.5	± 1	-0.75

Note that $y' = 0$ for $y = \pm x$. If $|x| < |y|$, then $y' < 0$; that is, the slopes are negative for all points in quadrants I and II above both of the lines $y = x$ and $y = -x$, and all points in quadrants III and IV below both of the lines $y = -x$ and $y = x$. A similar statement holds for positive slopes.

11.

x	y	$y' = y - 2x$
-2	-2	2
-2	2	6
2	2	-2
2	-2	-6

Note that $y' = 0$ for any point on the line $y = 2x$. The slopes are positive to the left of the line and negative to the right of the line. The solution curve in the graph passes through $(1, 0)$.

12.

x	y	$y' = 1 - xy$
± 1	± 1	0
± 2	± 2	-3
± 2	∓ 2	5

Note that $y' = 0$ for any point on the hyperbola $xy = 1$ (or $y = 1/x$). The slopes are negative at points "inside" the branches and positive at points everywhere else. The solution curve in the graph passes through $(0, 0)$.

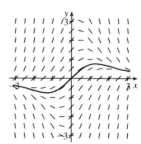

13.

x	y	$y' = y + xy$
0	± 2	± 2
1	± 2	± 4
-3	± 2	∓ 4

Note that $y' = y(x + 1) = 0$ for any point on $y = 0$ or on $x = -1$. The slopes are positive when the factors y and $x + 1$ have the same sign and negative when they have opposite signs. The solution curve in the graph passes through $(0, 1)$.

14.

x	y	$y' = x - xy$
± 2	0	± 2
± 2	3	∓ 4
± 2	-1	± 4

Note that $y' = x(1 - y) = 0$ for any point on $x = 0$ or on $y = 1$. The slopes are positive when the factors x and $1 - y$ have the same sign and negative when they have opposite signs. The solution curve in the graph passes through $(1, 0)$.

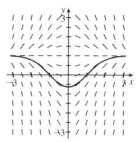

15. In Maple, we can use either `directionfield` (in Maple's share library) or `DEtools[DEplot]` to plot the direction field. To plot the solution, we can either use the initial-value option in `directionfield`, or actually solve the equation.

In Mathematica, we use `PlotVectorField` for the direction field, and the `Plot[Evaluate[...]]` construction to plot the solution, which is

$$y = 2\arctan\left(e^{x^3/3} \cdot \tan\frac{1}{2}\right).$$

In Derive, use `Direction_Field` (in utility file ODE_APPR) to plot the direction field. Then use `DSOLVE1(-x^2*SIN(y),1,x,y,0,1)` (in utility file ODE1) to solve the equation. Simplify each result.

16. See Exercise 15 for specific CAS directions. The exact solution is

$$y = \frac{2\left(3 - e^{2x^2}\right)}{e^{2x^2} + 3}$$

17.

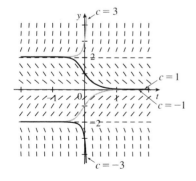

The direction field is for the differential equation $y' = y^3 - 4y$.

$L = \lim_{t \to \infty} y(t)$ exists for $-2 \le c \le 2$;

$L = \pm 2$ for $c = \pm 2$ and $L = 0$ for $-2 < c < 2$.

For other values of c, L does not exist.

18.

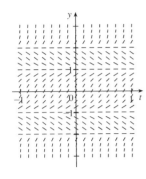

Note that when $f(y) = 0$ on the graph in the text, we have $y' = f(y) = 0$; so we get horizontal segments at $y = \pm 1, \pm 2$. We get segments with negative slopes only for $1 < |y| < 2$. All other segments have positive slope. For the limiting behavior of solutions:

- If $y(0) > 2$, then $\lim\limits_{t \to \infty} y = \infty$ and $\lim\limits_{t \to -\infty} y = 2$.
- If $1 < y(0) < 2$, then $\lim\limits_{t \to \infty} y = 1$ and $\lim\limits_{t \to -\infty} y = 2$.
- If $-1 < y(0) < 1$, then $\lim\limits_{t \to \infty} y = 1$ and $\lim\limits_{t \to -\infty} y = -1$.
- If $-2 < y(0) < -1$, then $\lim\limits_{t \to \infty} y = -2$ and $\lim\limits_{t \to -\infty} y = -1$.
- If $y < -2$, then $\lim\limits_{t \to \infty} y = -2$ and $\lim\limits_{t \to -\infty} y = -\infty$.

19. (a) $y' = F(x, y) = y$ and $y(0) = 1 \quad \Rightarrow \quad x_0 = 0, y_0 = 1$.

(i) $h = 0.4$ and $y_1 = y_0 + hF(x_0, y_0) \quad \Rightarrow \quad y_1 = 1 + 0.4 \cdot 1 = 1.4$. $x_1 = x_0 + h = 0 + 0.4 = 0.4$,

so $y_1 = y(0.4) = 1.4$.

(ii) $h = 0.2 \quad \Rightarrow \quad x_1 = 0.2$ and $x_2 = 0.4$, so we need to find y_2.

$y_1 = y_0 + hF(x_0, y_0) = 1 + 0.2y_0 = 1 + 0.2 \cdot 1 = 1.2$,

$y_2 = y_1 + hF(x_1, y_1) = 1.2 + 0.2y_1 = 1.2 + 0.2 \cdot 1.2 = 1.44$.

(iii) $h = 0.1 \quad \Rightarrow \quad x_4 = 0.4$, so we need to find y_4. $y_1 = y_0 + hF(x_0, y_0) = 1 + 0.1y_0 = 1 + 0.1 \cdot 1 = 1.1$,

$y_2 = y_1 + hF(x_1, y_1) = 1.1 + 0.1y_1 = 1.1 + 0.1 \cdot 1.1 = 1.21$,

$y_3 = y_2 + hF(x_2, y_2) = 1.21 + 0.1y_2 = 1.21 + 0.1 \cdot 1.21 = 1.331$,

$y_4 = y_3 + hF(x_3, y_3) = 1.331 + 0.1y_3 = 1.331 + 0.1 \cdot 1.331 = 1.4641$.

(b)

We see that the estimates are underestimates since they are all below the graph of $y = e^x$.

(c) (i) For $h = 0.4$: (exact value) − (approximate value) $= e^{0.4} - 1.4 \approx 0.0918$

(ii) For $h = 0.2$: (exact value) − (approximate value) $= e^{0.4} - 1.44 \approx 0.0518$

(iii) For $h = 0.1$: (exact value) − (approximate value) $= e^{0.4} - 1.4641 \approx 0.0277$

Each time the step size is halved, the error estimate also appears to be halved (approximately).

20.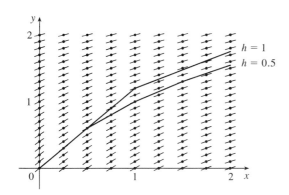

As x increases, the slopes decrease and all of the estimates are above the true values. Thus, all of the estimates are overestimates.

21. $h = 0.5$, $x_0 = 1$, $y_0 = 0$, and $F(x, y) = y - 2x$.

Note that $x_1 = x_0 + h = 1 + 0.5 = 1.5$, $x_2 = 2$, and $x_3 = 2.5$.

$y_1 = y_0 + hF(x_0, y_0) = 0 + 0.5F(1, 0) = 0.5[0 - 2(1)] = -1$.

$y_2 = y_1 + hF(x_1, y_1) = -1 + 0.5F(1.5, -1) = -1 + 0.5[-1 - 2(1.5)] = -3$.

$y_3 = y_2 + hF(x_2, y_2) = -3 + 0.5F(2, -3) = -3 + 0.5[-3 - 2(2)] = -6.5$.

$y_4 = y_3 + hF(x_3, y_3) = -6.5 + 0.5F(2.5, -6.5) = -6.5 + 0.5[-6.5 - 2(2.5)] = -12.25$.

22. $h = 0.2$, $x_0 = 0$, $y_0 = 0$, and $F(x, y) = 1 - xy$.

Note that $x_1 = x_0 + h = 0 + 0.2 = 0.2$, $x_2 = 0.4$, $x_3 = 0.6$, and $x_4 = 0.8$.

$y_1 = y_0 + hF(x_0, y_0) = 0 + 0.2F(0, 0) = 0.2[1 - (0)(0)] = 0.2$.

$y_2 = y_1 + hF(x_1, y_1) = 0.2 + 0.2F(0.2, 0.2) = 0.2 + 0.2[1 - (0.2)(0.2)] = 0.392$.

$y_3 = y_2 + hF(x_2, y_2) = 0.392 + 0.2F(0.4, 0.392) = 0.392 + 0.2[1 - (0.4)(0.392)] = 0.56064$.

$y_4 = y_3 + hF(x_3, y_3) = 0.56064 + 0.2[1 - (0.6)(0.56064)] = 0.6933632$.

$y_5 = y_4 + hF(x_4, y_4) = 0.6933632 + 0.2[1 - (0.8)(0.6933632)] = 0.782425088$.

Thus, $y(1) \approx 0.7824$.

23. $h = 0.1$, $x_0 = 0$, $y_0 = 1$, and $F(x, y) = y + xy$.

Note that $x_1 = x_0 + h = 0 + 0.1 = 0.1$, $x_2 = 0.2$, $x_3 = 0.3$, and $x_4 = 0.4$.

$y_1 = y_0 + hF(x_0, y_0) = 1 + 0.1F(0, 1) = 1 + 0.1[1 + (0)(1)] = 1.1$.

$y_2 = y_1 + hF(x_1, y_1) = 1.1 + 0.1F(0.1, 1.1) = 1.1 + 0.1[1.1 + (0.1)(1.1)] = 1.221$.

$y_3 = y_2 + hF(x_2, y_2) = 1.221 + 0.1F(0.2, 1.221) = 1.221 + 0.1[1.221 + (0.2)(1.221)] = 1.36752$.

$y_4 = y_3 + hF(x_3, y_3) = 1.36752 + 0.1F(0.3, 1.36752) = 1.36752 + 0.1[1.36752 + (0.3)(1.36752)]$
$\qquad = 1.5452976$.

$y_5 = y_4 + hF(x_4, y_4) = 1.5452976 + 0.1F(0.4, 1.5452976)$
$\qquad = 1.5452976 + 0.1[1.5452976 + (0.4)(1.5452976)] = 1.761639264$.

Thus, $y(0.5) \approx 1.7616$.

24. (a) $h = 0.2$, $x_0 = 1$, $y_0 = 0$, and $F(x, y) = x - xy$.

We need to find y_2, because $x_1 = 1.2$ and $x_2 = 1.4$.

$y_1 = y_0 + hF(x_0, y_0) = 0 + 0.2F(1, 0) = 0.2[1 - (1)(0)] = 0.2$.

$y_2 = y_1 + hF(x_1, y_1) = 0.2 + 0.2F(1.2, 0.2) = 0.2 + 0.2[1.2 - (1.2)(0.2)] = 0.392 \approx y(1.4)$.

(b) Now $h = 0.1$, so we need to find y_4.

$y_1 = 0 + 0.1[1 - (1)(0)] = 0.1$,

$y_2 = 0.1 + 0.1[1.1 - (1.1)(0.1)] = 0.199$,

$y_3 = 0.199 + 0.1[1.2 - (1.2)(0.199)] = 0.29512$, and

$y_4 = 0.29512 + 0.1[1.3 - (1.3)(0.29512)] = 0.3867544 \approx y(1.4)$.

25. (a) $dy/dx + 3x^2 y = 6x^2 \Rightarrow y' = 6x^2 - 3x^2 y$. Store this expression in Y_1 and use the following simple program to evaluate $y(1)$ for each part, using $H = h = 1$ and $N = 1$ for part (i), $H = 0.1$ and $N = 10$ for part (ii), and so forth.

$$h \rightarrow H: 0 \rightarrow X: 3 \rightarrow Y:$$
$$\text{For}(I, 1, N): Y + H \times Y_1 \rightarrow Y: X + H \rightarrow X:$$
$$\text{End(loop)}:$$
$$\text{Display Y.} \quad [\text{To see all iterations, include this statement in the loop.}]$$

(i) $H = 1, N = 1 \Rightarrow y(1) = 3$
(ii) $H = 0.1, N = 10 \Rightarrow y(1) \approx 2.3928$
(iii) $H = 0.01, N = 100 \Rightarrow y(1) \approx 2.3701$
(iv) $H = 0.001, N = 1000 \Rightarrow y(1) \approx 2.3681$

(b) $y = 2 + e^{-x^3} \Rightarrow y' = -3x^2 e^{-x^3}$

$$\text{LHS} = y' + 3x^2 y = -3x^2 e^{-x^3} + 3x^2 \left(2 + e^{-x^3}\right) = -3x^2 e^{-x^3} + 6x^2 + 3x^2 e^{-x^3} = 6x^2 = \text{RHS}$$

$y(0) = 2 + e^{-0} = 2 + 1 = 3$

(c) The exact value of $y(1)$ is $2 + e^{-1^3} = 2 + e^{-1}$.
   (i) For $h = 1$: (exact value) $-$ (approximate value) $= 2 + e^{-1} - 3 \approx -0.6321$
   (ii) For $h = 0.1$: (exact value) $-$ (approximate value) $= 2 + e^{-1} - 2.3928 \approx -0.0249$
   (iii) For $h = 0.01$: (exact value) $-$ (approximate value) $= 2 + e^{-1} - 2.3701 \approx -0.0022$
   (iv) For $h = 0.001$: (exact value) $-$ (approximate value) $= 2 + e^{-1} - 2.3681 \approx -0.0002$

   In (ii)–(iv), it seems that when the step size is divided by 10, the error estimate is also divided by 10 (approximately).

26. (a) We use the program from the solution to Exercise 25   (b)
   with $Y_1 = x^3 - y^3$, $H = 0.01$, and $N = \frac{2-0}{0.01} = 200$.
   With $(x_0, y_0) = (0, 1)$, we get $y(2) \approx 1.9000$.

Notice from the graph that $y(2) \approx 1.9$, which serves as
a check on our calculation in part (a).

27. (a) $R\dfrac{dQ}{dt} + \dfrac{1}{C}Q = E(t)$ becomes $5Q' + \dfrac{1}{0.05}Q = 60$

or $Q' + 4Q = 12$.

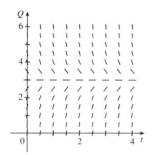

(b) From the graph, it appears that the limiting value of the charge Q is about 3.

(c) If $Q' = 0$, then $4Q = 12$ $\Rightarrow$ $Q = 3$ is an equilibrium solution.

(d)

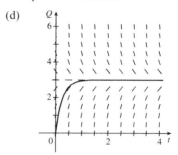

(e) $Q' + 4Q = 12$ $\Rightarrow$ $Q' = 12 - 4Q$. Now $Q(0) = 0$, so $t_0 = 0$ and $Q_0 = 0$.

$$Q_1 = Q_0 + hF(t_0, Q_0) = 0 + 0.1(12 - 4 \cdot 0) = 1.2$$

$$Q_2 = Q_1 + hF(t_1, Q_1) = 1.2 + 0.1(12 - 4 \cdot 1.2) = 1.92$$

$$Q_3 = Q_2 + hF(t_2, Q_2) = 1.92 + 0.1(12 - 4 \cdot 1.92) = 2.352$$

$$Q_4 = Q_3 + hF(t_3, Q_3) = 2.352 + 0.1(12 - 4 \cdot 2.352) = 2.6112$$

$$Q_5 = Q_4 + hF(t_4, Q_4) = 2.6112 + 0.1(12 - 4 \cdot 2.6112) = 2.76672$$

Thus, $Q_5 = Q(0.5) \approx 2.77$ C.

28. (a) From Exercise 10.1.14, we have $dy/dt = k(y - R)$. We are given that $R = 20°$C and $dy/dt = -1°$C/min when

$y = 70°$C. Thus, $-1 = k(70 - 20)$ $\Rightarrow$ $k = -\dfrac{1}{50}$ and the differential equation becomes $dy/dt = -\dfrac{1}{50}(y - 20)$.

(b)

The limiting value of the temperature is $20°$C; that is, the temperature of the room.

(c) From part (a), $dy/dt = -\dfrac{1}{50}(y - 20)$. With $t_0 = 0$, $y_0 = 95$, and $h = 2$ min, we get

$$y_1 = y_0 + hF(t_0, y_0) = 95 + 2\left[-\tfrac{1}{50}(95 - 20)\right] = 92$$

$$y_2 = y_1 + hF(t_1, y_1) = 92 + 2\left[-\tfrac{1}{50}(92 - 20)\right] = 89.12$$

$$y_3 = y_2 + hF(t_2, y_2) = 89.12 + 2\left[-\tfrac{1}{50}(89.12 - 20)\right] = 86.3552$$

$$y_4 = y_3 + hF(t_3, y_3) = 86.3552 + 2\left[-\tfrac{1}{50}(86.3552 - 20)\right] = 83.700992$$

$$y_5 = y_4 + hF(t_4, y_4) = 83.700992 + 2\left[-\tfrac{1}{50}(83.700992 - 20)\right] = 81.15295232$$

Thus, $y(10) \approx 81.15°$C.

10.3 Separable Equations

1. $\dfrac{dy}{dx} = \dfrac{y}{x}$ $\Rightarrow$ $\dfrac{dy}{y} = \dfrac{dx}{x}$ $[y \neq 0]$ $\Rightarrow$ $\displaystyle\int \dfrac{dy}{y} = \int \dfrac{dx}{x}$ $\Rightarrow$ $\ln|y| = \ln|x| + C$ $\Rightarrow$

$|y| = e^{\ln|x|+C} = e^{\ln|x|}e^C = e^C|x|$ $\Rightarrow$ $y = Kx$, where $K = \pm e^C$ is a constant. (In our derivation, K was nonzero,

but we can restore the excluded case $y = 0$ by allowing K to be zero.)

2. $\dfrac{dy}{dx} = \dfrac{\sqrt{x}}{e^y}$ $\Rightarrow$ $e^y\,dy = \sqrt{x}\,dx$ $\Rightarrow$ $\int e^y\,dy = \int x^{1/2}\,dx$ $\Rightarrow$ $e^y = \tfrac{2}{3}x^{3/2} + C$ $\Rightarrow$ $y = \ln\!\left(\tfrac{2}{3}x^{3/2} + C\right)$

3. $(x^2+1)y' = xy$ $\Rightarrow$ $\dfrac{dy}{dx} = \dfrac{xy}{x^2+1}$ $\Rightarrow$ $\dfrac{dy}{y} = \dfrac{x\,dx}{x^2+1}$ $[y \neq 0]$ $\Rightarrow$ $\displaystyle\int \dfrac{dy}{y} = \int \dfrac{x\,dx}{x^2+1}$ $\Rightarrow$

$\ln|y| = \tfrac{1}{2}\ln(x^2+1) + C$ $[u = x^2+1,\, du = 2x\,dx]$ $= \ln(x^2+1)^{1/2} + \ln e^C = \ln\!\left(e^C\sqrt{x^2+1}\right)$ $\Rightarrow$

$|y| = e^C\sqrt{x^2+1}$ $\Rightarrow$ $y = K\sqrt{x^2+1}$, where $K = \pm e^C$ is a constant. (In our derivation, K was nonzero, but we can

restore the excluded case $y = 0$ by allowing K to be zero.)

4. $y' = y^2 \sin x$ $\Rightarrow$ $\dfrac{dy}{dx} = y^2 \sin x$ $\Rightarrow$ $\dfrac{dy}{y^2} = \sin x\,dx$ $[y \neq 0]$ $\Rightarrow$ $\displaystyle\int \dfrac{dy}{y^2} = \int \sin x\,dx$ $\Rightarrow$

$-\dfrac{1}{y} = -\cos x + C$ $\Rightarrow$ $\dfrac{1}{y} = \cos x - C$ $\Rightarrow$ $y = \dfrac{1}{\cos x + K}$, where $K = -C$. $y = 0$ is also a solution.

5. $(1 + \tan y)\,y' = x^2 + 1$ $\Rightarrow$ $(1 + \tan y)\dfrac{dy}{dx} = x^2 + 1$ $\Rightarrow$ $\left(1 + \dfrac{\sin y}{\cos y}\right)dy = (x^2+1)\,dx$ $\Rightarrow$

$\displaystyle\int\left(1 - \dfrac{-\sin y}{\cos y}\right)dy = \int (x^2+1)\,dx$ $\Rightarrow$ $y - \ln|\cos y| = \tfrac{1}{3}x^3 + x + C$.

Note: The left side is equivalent to $y + \ln|\sec y|$.

6. $\dfrac{du}{dr} = \dfrac{1 + \sqrt{r}}{1 + \sqrt{u}}$ $\Rightarrow$ $\left(1 + \sqrt{u}\right)du = \left(1 + \sqrt{r}\right)dr$ $\Rightarrow$ $\int(1 + u^{1/2})\,du = \int(1 + r^{1/2})\,dr$ $\Rightarrow$

$u + \tfrac{2}{3}u^{3/2} = r + \tfrac{2}{3}r^{3/2} + C$

7. $\dfrac{dy}{dt} = \dfrac{te^t}{y\sqrt{1 + y^2}}$ $\Rightarrow$ $y\sqrt{1 + y^2}\,dy = te^t\,dt$ $\Rightarrow$ $\int y\sqrt{1 + y^2}\,dy = \int te^t\,dt$ $\Rightarrow$ $\tfrac{1}{3}\left(1 + y^2\right)^{3/2} = te^t - e^t + C$

[where the first integral is evaluated by substitution and the second by parts] $\Rightarrow$ $1 + y^2 = [3(te^t - e^t + C)]^{2/3}$ $\Rightarrow$

$y = \pm\sqrt{[3(te^t - e^t + C)]^{2/3} - 1}$

8. $\dfrac{dy}{d\theta} = \dfrac{e^y \sin^2\theta}{y \sec\theta}$ $\Rightarrow$ $\dfrac{y}{e^y}\,dy = \dfrac{\sin^2\theta}{\sec\theta}\,d\theta$ $\Rightarrow$ $\int ye^{-y}\,dy = \int \sin^2\theta\cos\theta\,d\theta$. Integrating the left side by parts with

$u = y$, $dv = e^{-y}\,dy$ and the right side by the substitution $u = \sin\theta$, we obtain $-ye^{-y} - e^{-y} = \tfrac{1}{3}\sin^3\theta + C$. We cannot

solve explicitly for y.

9. $\dfrac{du}{dt} = 2 + 2u + t + tu$ $\Rightarrow$ $\dfrac{du}{dt} = (1 + u)(2 + t)$ $\Rightarrow$ $\displaystyle\int \dfrac{du}{1 + u} = \int (2 + t)\,dt$ $[u \neq -1]$ $\Rightarrow$

$\ln|1 + u| = \tfrac{1}{2}t^2 + 2t + C$ $\Rightarrow$ $|1 + u| = e^{t^2/2 + 2t + C} = Ke^{t^2/2 + 2t}$, where $K = e^C$ $\Rightarrow$ $1 + u = \pm Ke^{t^2/2 + 2t}$ $\Rightarrow$

$u = -1 \pm Ke^{t^2/2 + 2t}$ where $K > 0$. $u = -1$ is also a solution, so $u = -1 + Ae^{t^2/2 + 2t}$, where A is an arbitrary constant.

10. $\dfrac{dz}{dt} + e^{t+z} = 0$ $\Rightarrow$ $\dfrac{dz}{dt} = -e^t e^z$ $\Rightarrow$ $\int e^{-z}\,dz = -\int e^t\,dt$ $\Rightarrow$ $-e^{-z} = -e^t + C$ $\Rightarrow$ $e^{-z} = e^t - C$ $\Rightarrow$

$\dfrac{1}{e^z} = e^t - C$ $\Rightarrow$ $e^z = \dfrac{1}{e^t - C}$ $\Rightarrow$ $z = \ln\!\left(\dfrac{1}{e^t - C}\right)$ $\Rightarrow$ $z = -\ln(e^t - C)$

11. $\dfrac{dy}{dx} = \dfrac{x}{y}$ $\Rightarrow$ $y\,dy = x\,dx$ $\Rightarrow$ $\int y\,dy = \int x\,dx$ $\Rightarrow$ $\tfrac{1}{2}y^2 = \tfrac{1}{2}x^2 + C.$ $y(0) = -3$ $\Rightarrow$

$\tfrac{1}{2}(-3)^2 = \tfrac{1}{2}(0)^2 + C$ $\Rightarrow$ $C = \tfrac{9}{2}$, so $\tfrac{1}{2}y^2 = \tfrac{1}{2}x^2 + \tfrac{9}{2}$ $\Rightarrow$ $y^2 = x^2 + 9$ $\Rightarrow$ $y = -\sqrt{x^2 + 9}$ since $y(0) = -3 < 0$.

12. $\dfrac{dy}{dx} = \dfrac{y\cos x}{1+y^2}$, $y(0) = 1$. $(1+y^2)\,dy = y\cos x\,dx$ $\Rightarrow$ $\dfrac{1+y^2}{y}\,dy = \cos x\,dx$ $\Rightarrow$ $\int\!\left(\dfrac{1}{y} + y\right)dy = \int \cos x\,dx$ $\Rightarrow$

$\ln|y| + \tfrac{1}{2}y^2 = \sin x + C.$ $y(0) = 1$ $\Rightarrow$ $\ln 1 + \tfrac{1}{2} = \sin 0 + C$ $\Rightarrow$ $C = \tfrac{1}{2}$, so $\ln|y| + \tfrac{1}{2}y^2 = \sin x + \tfrac{1}{2}$.

We cannot solve explicitly for y.

13. $x\cos x = (2y + e^{3y})\,y'$ $\Rightarrow$ $x\cos x\,dx = (2y + e^{3y})\,dy$ $\Rightarrow$ $\int (2y + e^{3y})\,dy = \int x\cos x\,dx$ $\Rightarrow$

$y^2 + \tfrac{1}{3}e^{3y} = x\sin x + \cos x + C$ [where the second integral is evaluated using integration by parts].

Now $y(0) = 0$ $\Rightarrow$ $0 + \tfrac{1}{3} = 0 + 1 + C$ $\Rightarrow$ $C = -\tfrac{2}{3}$. Thus, a solution is $y^2 + \tfrac{1}{3}e^{3y} = x\sin x + \cos x - \tfrac{2}{3}$.

We cannot solve explicitly for y.

14. $\dfrac{dP}{dt} = \sqrt{Pt}$ $\Rightarrow$ $dP/\sqrt{P} = \sqrt{t}\,dt$ $\Rightarrow$ $\int P^{-1/2}\,dP = \int t^{1/2}\,dt$ $\Rightarrow$ $2P^{1/2} = \tfrac{2}{3}t^{3/2} + C.$

$P(1) = 2$ $\Rightarrow$ $2\sqrt{2} = \tfrac{2}{3} + C$ $\Rightarrow$ $C = 2\sqrt{2} - \tfrac{2}{3}$, so $2P^{1/2} = \tfrac{2}{3}t^{3/2} + 2\sqrt{2} - \tfrac{2}{3}$ $\Rightarrow$ $\sqrt{P} = \tfrac{1}{3}t^{3/2} + \sqrt{2} - \tfrac{1}{3}$ $\Rightarrow$

$P = \left(\tfrac{1}{3}t^{3/2} + \sqrt{2} - \tfrac{1}{3}\right)^2$.

15. $\dfrac{du}{dt} = \dfrac{2t + \sec^2 t}{2u}$, $u(0) = -5$. $\int 2u\,du = \int (2t + \sec^2 t)\,dt$ $\Rightarrow$ $u^2 = t^2 + \tan t + C$,

where $[u(0)]^2 = 0^2 + \tan 0 + C$ $\Rightarrow$ $C = (-5)^2 = 25$. Therefore, $u^2 = t^2 + \tan t + 25$, so $u = \pm\sqrt{t^2 + \tan t + 25}$.

Since $u(0) = -5$, we must have $u = -\sqrt{t^2 + \tan t + 25}$.

16. $xy' + y = y^2$ $\Rightarrow$ $x\dfrac{dy}{dx} = y^2 - y$ $\Rightarrow$ $x\,dy = (y^2 - y)\,dx$ $\Rightarrow$ $\dfrac{dy}{y^2 - y} = \dfrac{dx}{x}$ $\Rightarrow$

$\int \dfrac{dy}{y(y-1)} = \int \dfrac{dx}{x}$ $[y \neq 0, 1]$ $\Rightarrow$ $\int\!\left(\dfrac{1}{y-1} - \dfrac{1}{y}\right)dy = \int \dfrac{dx}{x}$ $\Rightarrow$ $\ln|y-1| - \ln|y| = \ln|x| + C$ $\Rightarrow$

$\ln\left|\dfrac{y-1}{y}\right| = \ln\left(e^C\,|x|\right)$ $\Rightarrow$ $\left|\dfrac{y-1}{y}\right| = e^C\,|x|$ $\Rightarrow$ $\dfrac{y-1}{y} = Kx$, where $K = \pm e^C$ $\Rightarrow$ $1 - \dfrac{1}{y} = Kx$ $\Rightarrow$

$\dfrac{1}{y} = 1 - Kx$ $\Rightarrow$ $y = \dfrac{1}{1 - Kx}$. [The excluded cases, $y = 0$ and $y = 1$, are ruled out by the initial condition $y(1) = -1$.]

Now $y(1) = -1$ $\Rightarrow$ $-1 = \dfrac{1}{1-K}$ $\Rightarrow$ $1 - K = -1$ $\Rightarrow$ $K = 2$, so $y = \dfrac{1}{1 - 2x}$.

17. $y' \tan x = a + y$, $0 < x < \pi/2$ $\Rightarrow$ $\dfrac{dy}{dx} = \dfrac{a+y}{\tan x}$ $\Rightarrow$ $\dfrac{dy}{a+y} = \cot x\, dx$ $[a + y \neq 0]$ $\Rightarrow$

$\displaystyle\int \dfrac{dy}{a+y} = \int \dfrac{\cos x}{\sin x}\, dx$ $\Rightarrow$ $\ln|a+y| = \ln|\sin x| + C$ $\Rightarrow$ $|a+y| = e^{\ln|\sin x|+C} = e^{\ln|\sin x|} \cdot e^C = e^C |\sin x|$ $\Rightarrow$

$a + y = K \sin x$, where $K = \pm e^C$. (In our derivation, K was nonzero, but we can restore the excluded case

$y = -a$ by allowing K to be zero.) $y(\pi/3) = a$ $\Rightarrow$ $a + a = K \sin\left(\dfrac{\pi}{3}\right)$ $\Rightarrow$ $2a = K \dfrac{\sqrt{3}}{2}$ $\Rightarrow$ $K = \dfrac{4a}{\sqrt{3}}$.

Thus, $a + y = \dfrac{4a}{\sqrt{3}} \sin x$ and so $y = \dfrac{4a}{\sqrt{3}} \sin x - a$.

18. $\dfrac{dL}{dt} = kL^2 \ln t$ $\Rightarrow$ $\dfrac{dL}{L^2} = k \ln t\, dt$ $\Rightarrow$ $\displaystyle\int \dfrac{dL}{L^2} = \int k \ln t\, dt$ $\Rightarrow$ $-\dfrac{1}{L} = kt \ln t - \displaystyle\int k\, dt$

[by parts with $u = \ln t$, $dv = k\, dt$] $\Rightarrow$ $-\dfrac{1}{L} = kt \ln t - kt + C$ $\Rightarrow$ $L = \dfrac{1}{kt - kt \ln t - C}$.

$L(1) = -1$ $\Rightarrow$ $-1 = \dfrac{1}{k - k \ln 1 - C}$ $\Rightarrow$ $C - k = 1$ $\Rightarrow$ $C = k + 1$. Thus, $L = \dfrac{1}{kt - kt \ln t - k - 1}$.

19. If the slope at the point (x, y) is xy, then we have $\dfrac{dy}{dx} = xy$ $\Rightarrow$ $\dfrac{dy}{y} = x\, dx$ $[y \neq 0]$ $\Rightarrow$ $\displaystyle\int \dfrac{dy}{y} = \int x\, dx$ $\Rightarrow$

$\ln|y| = \frac{1}{2}x^2 + C$. $y(0) = 1$ $\Rightarrow$ $\ln 1 = 0 + C$ $\Rightarrow$ $C = 0$. Thus, $|y| = e^{x^2/2}$ $\Rightarrow$ $y = \pm e^{x^2/2}$, so $y = e^{x^2/2}$

since $y(0) = 1 > 0$. Note that $y = 0$ is not a solution because it doesn't satisfy the initial condition $y(0) = 1$.

20. $f'(x) = f(x)(1 - f(x))$ $\Rightarrow$ $\dfrac{dy}{dx} = y(1 - y)$ $\Rightarrow$ $\dfrac{dy}{y(1-y)} = dx$ $[y \neq 0, 1]$ $\Rightarrow$ $\displaystyle\int \dfrac{dy}{y(1-y)} = \int dx$ $\Rightarrow$

$\displaystyle\int \left(\dfrac{A}{y} + \dfrac{B}{1-y}\right) dy = \int dx$ $\Rightarrow$ $\displaystyle\int \left(\dfrac{1}{y} + \dfrac{1}{1-y}\right) dy = \int dx$ $\Rightarrow$ $\ln|y| - \ln|1-y| = x + C$ $\Rightarrow$

$\ln\left|\dfrac{y}{1-y}\right| = x + C$ $\Rightarrow$ $\left|\dfrac{y}{1-y}\right| = e^{x+C}$ $\Rightarrow$ $\dfrac{y}{1-y} = Ke^x$, where $K = \pm e^C$ $\Rightarrow$

$y = (1 - y)Ke^x = Ke^x - yKe^x$ $\Rightarrow$ $y + yKe^x = Ke^x$ $\Rightarrow$ $y(1 + Ke^x) = Ke^x$ $\Rightarrow$ $y = \dfrac{Ke^x}{1 + Ke^x}$.

$f(0) = \frac{1}{2}$ $\Rightarrow$ $\dfrac{1}{2} = \dfrac{K}{1 + K}$ $\Rightarrow$ $1 + K = 2K$ $\Rightarrow$ $K = 1$, so $y = \dfrac{e^x}{1 + e^x}$ $\left[\text{or } \dfrac{1}{1 + e^{-x}}\right]$.

Note that $y = 0$ and $y = 1$ are not solutions because they don't satisfy the initial condition $f(0) = \frac{1}{2}$.

21. $u = x + y$ $\Rightarrow$ $\dfrac{d}{dx}(u) = \dfrac{d}{dx}(x + y)$ $\Rightarrow$ $\dfrac{du}{dx} = 1 + \dfrac{dy}{dx}$, but $\dfrac{dy}{dx} = x + y = u$, so $\dfrac{du}{dx} = 1 + u$ $\Rightarrow$

$\dfrac{du}{1+u} = dx$ $[u \neq -1]$ $\Rightarrow$ $\displaystyle\int \dfrac{du}{1+u} = \int dx$ $\Rightarrow$ $\ln|1 + u| = x + C$ $\Rightarrow$ $|1 + u| = e^{x+C}$ $\Rightarrow$

$1 + u = \pm e^C e^x$ $\Rightarrow$ $u = \pm e^C e^x - 1$ $\Rightarrow$ $x + y = \pm e^C e^x - 1$ $\Rightarrow$ $y = Ke^x - x - 1$, where $K = \pm e^C \neq 0$.

If $u = -1$, then $-1 = x + y$ $\Rightarrow$ $y = -x - 1$, which is just $y = Ke^x - x - 1$ with $K = 0$. Thus, the general solution

is $y = Ke^x - x - 1$, where $K \in \mathbb{R}$.

22. $xy' = y + xe^{y/x}$ ⇒ $y' = y/x + e^{y/x}$ ⇒ $\dfrac{dy}{dx} = v + e^v$. Also, $v = y/x$ ⇒ $xv = y$ ⇒ $\dfrac{dy}{dx} = x\dfrac{dv}{dx} + v$,

so $v + e^v = x\dfrac{dv}{dx} + v$ ⇒ $\dfrac{dv}{e^v} = \dfrac{dx}{x}$ $[x \neq 0]$ ⇒ $\displaystyle\int \dfrac{dv}{e^v} = \int \dfrac{dx}{x}$ ⇒ $-e^{-v} = \ln|x| + C$ ⇒

$e^{-v} = -\ln|x| - C$ ⇒ $-v = \ln(-\ln|x| - C)$ ⇒ $y/x = -\ln(-\ln|x| - C)$ ⇒ $y = -x\ln(-\ln|x| - C)$.

23. (a) $y' = 2x\sqrt{1 - y^2}$ ⇒ $\dfrac{dy}{dx} = 2x\sqrt{1 - y^2}$ ⇒ $\dfrac{dy}{\sqrt{1 - y^2}} = 2x\,dx$ ⇒ $\displaystyle\int \dfrac{dy}{\sqrt{1 - y^2}} = \int 2x\,dx$ ⇒

$\sin^{-1} y = x^2 + C$ for $-\dfrac{\pi}{2} \leq x^2 + C \leq \dfrac{\pi}{2}$.

(b) $y(0) = 0$ ⇒ $\sin^{-1} 0 = 0^2 + C$ ⇒ $C = 0$,

so $\sin^{-1} y = x^2$ and $y = \sin(x^2)$ for

$-\sqrt{\pi/2} \leq x \leq \sqrt{\pi/2}$.

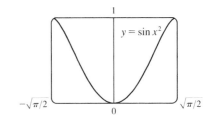

(c) For $\sqrt{1 - y^2}$ to be a real number, we must have $-1 \leq y \leq 1$; that is, $-1 \leq y(0) \leq 1$. Thus, the initial-value problem

$y' = 2x\sqrt{1 - y^2}$, $y(0) = 2$ does *not* have a solution.

24. $e^{-y}y' + \cos x = 0$ ⇔ $\displaystyle\int e^{-y}\,dy = -\int \cos x\,dx$ ⇔ $-e^{-y} = -\sin x + C_1$ ⇔ $y = -\ln(\sin x + C)$. The solution

is periodic, with period 2π. Note that for $C > 1$, the domain of the solution is $\mathbb{R}$, but for $-1 < C \leq 1$ it is only defined on the

intervals where $\sin x + C > 0$, and it is meaningless for $C \leq -1$, since then $\sin x + C \leq 0$, and the logarithm is undefined.

$C = -0.5$

$C = 0.5$

$C = 1$

$C = 1.5$

$C = 3$

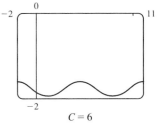
$C = 6$

For $-1 < C < 1$, the solution curve consists of concave-up pieces separated by intervals on which the solution is not defined

(where $\sin x + C \leq 0$). For $C = 1$, the solution curve consists of concave-up pieces separated by vertical asymptotes at the

points where $\sin x + C = 0$ ⇔ $\sin x = -1$. For $C > 1$, the curve is continuous, and as C increases, the graph moves

downward, and the amplitude of the oscillations decreases.

25. $\dfrac{dy}{dx} = \dfrac{\sin x}{\sin y}$, $y(0) = \dfrac{\pi}{2}$. So $\int \sin y \, dy = \int \sin x \, dx$ ⇔

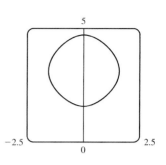

$-\cos y = -\cos x + C$ ⇔ $\cos y = \cos x - C$. From the initial condition,

we need $\cos \frac{\pi}{2} = \cos 0 - C$ ⇒ $0 = 1 - C$ ⇒ $C = 1$, so the solution is

$\cos y = \cos x - 1$. Note that we cannot take $\cos^{-1}$ of both sides, since that would

unnecessarily restrict the solution to the case where $-1 \le \cos x - 1$ ⇔ $0 \le \cos x$,

as $\cos^{-1}$ is defined only on $[-1, 1]$. Instead we plot the graph using Maple's

`plots[implicitplot]` or Mathematica's `Plot[Evaluate[···]]`.

26. $\dfrac{dy}{dx} = \dfrac{x\sqrt{x^2 + 1}}{ye^y}$ ⇔ $\int ye^y \, dy = \int x\sqrt{x^2 + 1} \, dx$. We use parts on the LHS with $u = y$, $dv = e^y \, dy$, and on the RHS

we use the substitution $z = x^2 + 1$, so $dz = 2x \, dx$. The equation becomes $ye^y - \int e^y \, dy = \frac{1}{2}\int \sqrt{z} \, dz$ ⇔

$e^y(y - 1) = \frac{1}{3}(x^2 + 1)^{3/2} + C$, so we see that the curves are symmetric about the y-axis. Every point (x, y) in the plane lies

on one of the curves, namely the one for which $C = (y - 1)e^y - \frac{1}{3}(x^2 + 1)^{3/2}$. For example, along the y-axis,

$C = (y - 1)e^y - \frac{1}{3}$, so the origin lies on the curve with $C = -\frac{4}{3}$. We use Maple's `plots[implicitplot]` command or

`Plot[Evaluate[···]]` in Mathematica to plot the solution curves for various values of C.

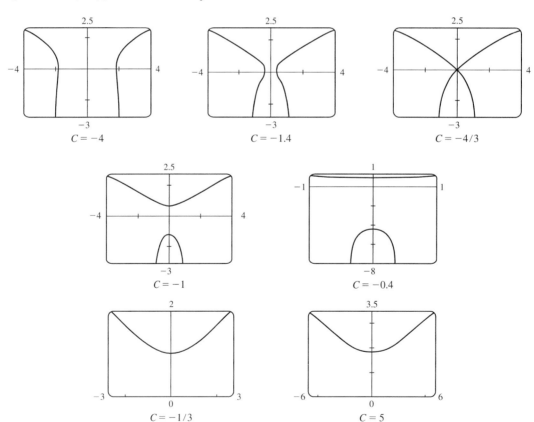

It seems that the transitional values of C are $-\frac{4}{3}$ and $-\frac{1}{3}$. For $C < -\frac{4}{3}$, the graph consists of left and right branches. At

$C = -\frac{4}{3}$, the two branches become connected at the origin, and as C increases, the graph splits into top and bottom branches.

At $C = -\frac{1}{3}$, the bottom half disappears. As C increases further, the graph moves upward, but doesn't change shape much.

27. (a)

x	y	$y' = 1/y$
0	0.5	2
0	−0.5	−2
0	1	1
0	−1	−1
0	2	0.5

x	y	$y' = 1/y$
0	−2	−0.5
0	4	0.25
0	3	$0.\overline{3}$
0	0.25	4
0	$0.\overline{3}$	3

(b) $y' = 1/y \Rightarrow dy/dx = 1/y \Rightarrow$

$y\, dy = dx \Rightarrow \int y\, dy = \int dx \Rightarrow \frac{1}{2}y^2 = x + C \Rightarrow$

$y^2 = 2(x + C)$ or $y = \pm\sqrt{2(x + C)}$.

(c)

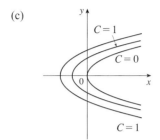

28. (a)

x	y	$y' = x^2/y$
1	1	1
−1	1	1
−1	−1	−1
1	−1	−1
1	2	0.5
2	1	4
2	2	2
1	0.5	2
0.5	1	0.25
2	0.5	8

(b) $y' = x^2/y \Rightarrow y\, dy = x^2\, dx$,

so $\frac{1}{2}y^2 = \frac{1}{3}x^3 + C_1$,

or $y = \pm\left(\frac{2}{3}x^3 + C\right)^{1/2}$.

(c)

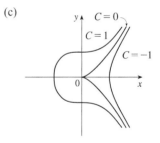

29. The curves $x^2 + 2y^2 = k^2$ form a family of ellipses with major axis on the x-axis. Differentiating gives

$\frac{d}{dx}\left(x^2 + 2y^2\right) = \frac{d}{dx}\left(k^2\right) \Rightarrow 2x + 4yy' = 0 \Rightarrow 4yy' = -2x \Rightarrow y' = \dfrac{-x}{2y}$. Thus, the slope of the tangent line

at any point (x, y) on one of the ellipses is $y' = \dfrac{-x}{2y}$, so the orthogonal trajectories

must satisfy $y' = \dfrac{2y}{x} \Leftrightarrow \dfrac{dy}{dx} = \dfrac{2y}{x} \Leftrightarrow \dfrac{dy}{y} = 2 = \dfrac{dx}{x} \Leftrightarrow$

$\int \dfrac{dy}{y} = 2 \int \dfrac{dx}{x} \Leftrightarrow \ln|y| = 2\ln|x| + C_1 \Leftrightarrow \ln|y| = \ln|x|^2 + C_1 \Leftrightarrow$

$|y| = e^{\ln x^2 + C_1} \Leftrightarrow y = \pm x^2 \cdot e^{C_1} = Cx^2$. This is a family of parabolas.

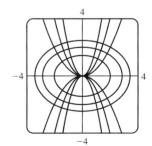

30. The curves $y^2 = kx^3$ form a family of power functions. Differentiating gives $\dfrac{d}{dx}\left(y^2\right) = \dfrac{d}{dx}\left(kx^3\right) \Rightarrow 2yy' = 3kx^2 \Rightarrow$

$y' = \dfrac{3kx^2}{2y} = \dfrac{3(y^2/x^3)x^2}{2y} = \dfrac{3y}{2x}$, the slope of the tangent line at (x, y) on one of the curves. Thus, the orthogonal trajectories must satisfy $y' = -\dfrac{2x}{3y}$ $\Leftrightarrow$ $\dfrac{dy}{dx} = -\dfrac{2x}{3y}$ $\Leftrightarrow$

$3y\,dy = -2x\,dx$ $\Leftrightarrow$ $\int 3y\,dy = \int -2x\,dx$ $\Leftrightarrow$ $\frac{3}{2}y^2 = -x^2 + C_1$ $\Leftrightarrow$

$3y^2 = -2x^2 + C_2$ $\Leftrightarrow$ $2x^2 + 3y^2 = C$. This is a family of ellipses.

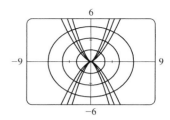

31. The curves $y = k/x$ form a family of hyperbolas with asymptotes $x = 0$ and $y = 0$. Differentiating gives

$\dfrac{d}{dx}(y) = \dfrac{d}{dx}\left(\dfrac{k}{x}\right)$ $\Rightarrow$ $y' = -\dfrac{k}{x^2}$ $\Rightarrow$ $y' = -\dfrac{xy}{x^2}$ [since $y = k/x$ $\Rightarrow$ $xy = k$] $\Rightarrow$ $y' = -\dfrac{y}{x}$. Thus, the slope

of the tangent line at any point (x, y) on one of the hyperbolas is $y' = -y/x$,

so the orthogonal trajectories must satisfy $y' = x/y$ $\Leftrightarrow$ $\dfrac{dy}{dx} = \dfrac{x}{y}$ $\Leftrightarrow$

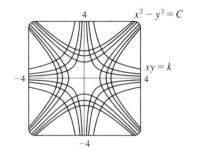

$y\,dy = x\,dx$ $\Leftrightarrow$ $\int y\,dy = \int x\,dx$ $\Leftrightarrow$ $\frac{1}{2}y^2 = \frac{1}{2}x^2 + C_1$ $\Leftrightarrow$

$y^2 = x^2 + C_2$ $\Leftrightarrow$ $x^2 - y^2 = C$. This is a family of hyperbolas with

asymptotes $y = \pm x$.

32. Differentiating $y = \dfrac{x}{1 + kx}$ gives $y' = \dfrac{1}{(1 + kx)^2}$, but $k = \dfrac{x - y}{xy}$, so

$y' = \dfrac{1}{\left(1 + \dfrac{x - y}{y}\right)^2} = \dfrac{y^2}{x^2}$. Thus, the orthogonal trajectories must satisfy

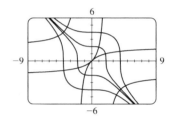

$y' = -x^2/y^2$ $\Leftrightarrow$ $y^2\,dy = -x^2\,dx$ $\Leftrightarrow$ $\int y^2\,dy = -\int x^2\,dx$ $\Leftrightarrow$

$\frac{1}{3}y^3 = -\frac{1}{3}x^3 + C_1$ $\Leftrightarrow$ $y^3 = C - x^3$ $\Leftrightarrow$ $y = \sqrt[3]{C - x^3}$.

33. From Exercise 10.2.27, $\dfrac{dQ}{dt} = 12 - 4Q$ $\Leftrightarrow$ $\displaystyle\int \dfrac{dQ}{12 - 4Q} = \int dt$ $\Leftrightarrow$ $-\frac{1}{4}\ln|12 - 4Q| = t + C$ $\Leftrightarrow$

$\ln|12 - 4Q| = -4t - 4C$ $\Leftrightarrow$ $|12 - 4Q| = e^{-4t - 4C}$ $\Leftrightarrow$ $12 - 4Q = Ke^{-4t}$ $[K = \pm e^{-4C}]$ $\Leftrightarrow$

$4Q = 12 - Ke^{-4t}$ $\Leftrightarrow$ $Q = 3 - Ae^{-4t}$ $[A = K/4]$. $Q(0) = 0$ $\Leftrightarrow$ $0 = 3 - A$ $\Leftrightarrow$ $A = 3$ $\Leftrightarrow$

$Q(t) = 3 - 3e^{-4t}$. As $t \to \infty$, $Q(t) \to 3 - 0 = 3$ (the limiting value).

34. From Exercise 10.2.28, $\dfrac{dy}{dt} = -\frac{1}{50}(y - 20)$ $\Leftrightarrow$ $\displaystyle\int \dfrac{dy}{y - 20} = \int \left(-\frac{1}{50}\right) dt$ $\Leftrightarrow$ $\ln|y - 20| = -\frac{1}{50}t + C$ $\Leftrightarrow$

$y - 20 = Ke^{-t/50}$ $\Leftrightarrow$ $y(t) = Ke^{-t/50} + 20$. $y(0) = 95$ $\Leftrightarrow$ $95 = K + 20$ $\Leftrightarrow$ $K = 75$ $\Leftrightarrow$

$y(t) = 75e^{-t/50} + 20$.

35. $\dfrac{dP}{dt} = k(M - P)$ $\Leftrightarrow$ $\displaystyle\int \dfrac{dP}{P - M} = \int (-k)\,dt$ $\Leftrightarrow$ $\ln|P - M| = -kt + C$ $\Leftrightarrow$ $|P - M| = e^{-kt + C}$ $\Leftrightarrow$

$P - M = Ae^{-kt}$ $[A = \pm e^C]$ $\Leftrightarrow$ $P = M + Ae^{-kt}$. If we assume that performance is at level 0 when $t = 0$, then

$P(0) = 0$ $\Leftrightarrow$ $0 = M + A$ $\Leftrightarrow$ $A = -M$ $\Leftrightarrow$ $P(t) = M - Me^{-kt}$. $\displaystyle\lim_{t \to \infty} P(t) = M - M \cdot 0 = M$.

36. (a) $\dfrac{dx}{dt} = k(a - x)(b - x)$, $a \neq b$. Using partial fractions, $\dfrac{1}{(a - x)(b - x)} = \dfrac{1/(b - a)}{a - x} - \dfrac{1/(b - a)}{b - x}$, so

$$\int \frac{dx}{(a - x)(b - x)} = \int k\,dt \quad \Rightarrow \quad \frac{1}{b - a}\left(-\ln|a - x| + \ln|b - x|\right) = kt + C \quad \Rightarrow \quad \ln\left|\frac{b - x}{a - x}\right| = (b - a)(kt + C).$$

The concentrations $[A] = a - x$ and $[B] = b - x$ cannot be negative, so $\dfrac{b - x}{a - x} \geq 0$ and $\left|\dfrac{b - x}{a - x}\right| = \dfrac{b - x}{a - x}$.

We now have $\ln\left(\dfrac{b - x}{a - x}\right) = (b - a)(kt + C)$. Since $x(0) = 0$, we get $\ln\left(\dfrac{b}{a}\right) = (b - a)C$. Hence,

$$\ln\left(\frac{b - x}{a - x}\right) = (b - a)kt + \ln\left(\frac{b}{a}\right) \quad \Rightarrow \quad \frac{b - x}{a - x} = \frac{b}{a}e^{(b - a)kt} \quad \Rightarrow \quad x = \frac{b[e^{(b - a)kt} - 1]}{be^{(b - a)kt}/a - 1} = \frac{ab[e^{(b - a)kt} - 1]}{be^{(b - a)kt} - a}\;\frac{\text{moles}}{\text{L}}.$$

(b) If $b = a$, then $\dfrac{dx}{dt} = k(a - x)^2$, so $\displaystyle\int \frac{dx}{(a - x)^2} = \int k\,dt$ and $\dfrac{1}{a - x} = kt + C$. Since $x(0) = 0$, we get $C = \dfrac{1}{a}$.

Thus, $a - x = \dfrac{1}{kt + 1/a}$ and $x = a - \dfrac{a}{akt + 1} = \dfrac{a^2 kt}{akt + 1}\;\dfrac{\text{moles}}{\text{L}}$. Suppose $x = [C] = a/2$ when $t = 20$. Then

$$x(20) = a/2 \quad \Rightarrow \quad \frac{a}{2} = \frac{20a^2 k}{20ak + 1} \quad \Rightarrow \quad 40a^2 k = 20a^2 k + a \quad \Rightarrow \quad 20a^2 k = a \quad \Rightarrow \quad k = \frac{1}{20a}, \text{ so}$$

$$x = \frac{a^2 t/(20a)}{1 + at/(20a)} = \frac{at/20}{1 + t/20} = \frac{at}{t + 20}\;\frac{\text{moles}}{\text{L}}.$$

37. (a) If $a = b$, then $\dfrac{dx}{dt} = k(a - x)(b - x)^{1/2}$ becomes $\dfrac{dx}{dt} = k(a - x)^{3/2} \quad \Rightarrow \quad (a - x)^{-3/2}\,dx = k\,dt \quad \Rightarrow$

$$\int (a - x)^{-3/2}\,dx = \int k\,dt \quad \Rightarrow \quad 2(a - x)^{-1/2} = kt + C \quad \text{[by substitution]} \quad \Rightarrow \quad \frac{2}{kt + C} = \sqrt{a - x} \quad \Rightarrow$$

$$\left(\frac{2}{kt + C}\right)^2 = a - x \quad \Rightarrow \quad x(t) = a - \frac{4}{(kt + C)^2}. \text{ The initial concentration of HBr is } 0, \text{ so } x(0) = 0 \quad \Rightarrow$$

$$0 = a - \frac{4}{C^2} \quad \Rightarrow \quad \frac{4}{C^2} = a \quad \Rightarrow \quad C^2 = \frac{4}{a} \quad \Rightarrow \quad C = 2/\sqrt{a} \quad [C \text{ is positive since } kt + C = 2(a - x)^{-1/2} > 0].$$

Thus, $x(t) = a - \dfrac{4}{(kt + 2/\sqrt{a}\,)^2}$.

(b) $\dfrac{dx}{dt} = k(a - x)(b - x)^{1/2} \quad \Rightarrow \quad \dfrac{dx}{(a - x)\sqrt{b - x}} = k\,dt \quad \Rightarrow \quad \displaystyle\int \frac{dx}{(a - x)\sqrt{b - x}} = \int k\,dt \quad (\star).$

From the hint, $u = \sqrt{b - x} \quad \Rightarrow \quad u^2 = b - x \quad \Rightarrow \quad 2u\,du = -dx$, so

$$\int \frac{dx}{(a - x)\sqrt{b - x}} = \int \frac{-2u\,du}{[a - (b - u^2)]u} = -2\int \frac{du}{a - b + u^2} = -2\int \frac{du}{\left(\sqrt{a - b}\right)^2 + u^2}$$

$$\overset{17}{=} -2\left(\frac{1}{\sqrt{a - b}}\tan^{-1}\frac{u}{\sqrt{a - b}}\right)$$

So $(\star)$ becomes $\dfrac{-2}{\sqrt{a - b}}\tan^{-1}\dfrac{\sqrt{b - x}}{\sqrt{a - b}} = kt + C$. Now $x(0) = 0 \quad \Rightarrow \quad C = \dfrac{-2}{\sqrt{a - b}}\tan^{-1}\dfrac{\sqrt{b}}{\sqrt{a - b}}$ and we have

$$\frac{-2}{\sqrt{a - b}}\tan^{-1}\frac{\sqrt{b - x}}{\sqrt{a - b}} = kt - \frac{2}{\sqrt{a - b}}\tan^{-1}\frac{\sqrt{b}}{\sqrt{a - b}} \quad \Rightarrow \quad \frac{2}{\sqrt{a - b}}\left(\tan^{-1}\sqrt{\frac{b}{a - b}} - \tan^{-1}\sqrt{\frac{b - x}{a - b}}\right) = kt \quad \Rightarrow$$

$$t(x) = \frac{2}{k\sqrt{a - b}}\left(\tan^{-1}\sqrt{\frac{b}{a - b}} - \tan^{-1}\sqrt{\frac{b - x}{a - b}}\right).$$

38. If $S = \dfrac{dT}{dr}$, then $\dfrac{dS}{dr} = \dfrac{d^2T}{dr^2}$. The differential equation $\dfrac{d^2T}{dr^2} + \dfrac{2}{r}\dfrac{dT}{dr} = 0$ can be written as $\dfrac{dS}{dr} + \dfrac{2}{r}S = 0$. Thus,

$$\dfrac{dS}{dr} = \dfrac{-2S}{r} \quad \Rightarrow \quad \dfrac{dS}{S} = -\dfrac{2}{r}\,dr \quad \Rightarrow \quad \int \dfrac{1}{S}\,dS = \int -\dfrac{2}{r}\,dr \quad \Rightarrow \quad \ln|S| = -2\ln|r| + C. \text{ Assuming } S = dT/dr > 0$$

and $r > 0$, we have $S = e^{-2\ln r + C} = e^{\ln r^{-2}}e^C = r^{-2}k \;\; [k = e^C] \;\; \Rightarrow \;\; S = \dfrac{1}{r^2}k \;\; \Rightarrow \;\; \dfrac{dT}{dr} = \dfrac{1}{r^2}k \;\; \Rightarrow$

$$dT = \dfrac{1}{r^2}k\,dr \quad \Rightarrow \quad \int dT = \int \dfrac{1}{r^2}k\,dr \quad \Rightarrow \quad T(r) = -\dfrac{k}{r} + A.$$

$T(1) = 15 \;\; \Rightarrow \;\; 15 = -k + A \text{ (1) and } T(2) = 25 \;\; \Rightarrow \;\; 25 = -\tfrac{1}{2}k + A \text{ (2)}.$

Now solve for k and A: $-2(2) + (1) \;\; \Rightarrow \;\; -35 = -A$, so $A = 35$ and $k = 20$, and $T(r) = -20/r + 35$.

39. (a) $\dfrac{dC}{dt} = r - kC \;\; \Rightarrow \;\; \dfrac{dC}{dt} = -(kC - r) \;\; \Rightarrow \;\; \int \dfrac{dC}{kC - r} = \int -dt \;\; \Rightarrow \;\; (1/k)\ln|kC - r| = -t + M_1 \;\; \Rightarrow$

$\ln|kC - r| = -kt + M_2 \;\; \Rightarrow \;\; |kC - r| = e^{-kt + M_2} \;\; \Rightarrow \;\; kC - r = M_3 e^{-kt} \;\; \Rightarrow \;\; kC = M_3 e^{-kt} + r \;\; \Rightarrow$

$C(t) = M_4 e^{-kt} + r/k.\; C(0) = C_0 \;\; \Rightarrow \;\; C_0 = M_4 + r/k \;\; \Rightarrow \;\; M_4 = C_0 - r/k \;\; \Rightarrow$

$C(t) = (C_0 - r/k)e^{-kt} + r/k.$

(b) If $C_0 < r/k$, then $C_0 - r/k < 0$ and the formula for $C(t)$ shows that $C(t)$ increases and $\lim\limits_{t\to\infty} C(t) = r/k$.

As t increases, the formula for $C(t)$ shows how the role of C_0 steadily diminishes as that of r/k increases.

40. (a) Use 1 billion dollars as the x-unit and 1 day as the t-unit. Initially, there is \$10 billion of old currency in circulation, so all of the \$50 million returned to the banks is old. At time t, the amount of new currency is $x(t)$ billion dollars, so $10 - x(t)$ billion dollars of currency is old. The fraction of circulating money that is old is $[10 - x(t)]/10$, and the amount of old currency being returned to the banks each day is $\dfrac{10 - x(t)}{10}\, 0.05$ billion dollars. This amount of new currency per day is introduced into circulation, so $\dfrac{dx}{dt} = \dfrac{10 - x}{10}\cdot 0.05 = 0.005(10 - x)$ billion dollars per day.

(b) $\dfrac{dx}{10 - x} = 0.005\,dt \;\; \Rightarrow \;\; \dfrac{-dx}{10 - x} = -0.005\,dt \;\; \Rightarrow \;\; \ln(10 - x) = -0.005t + c \;\; \Rightarrow \;\; 10 - x = Ce^{-0.005t},$

where $C = e^c \;\; \Rightarrow \;\; x(t) = 10 - Ce^{-0.005t}$. From $x(0) = 0$, we get $C = 10$, so $x(t) = 10(1 - e^{-0.005t})$.

(c) The new bills make up 90% of the circulating currency when $x(t) = 0.9 \cdot 10 = 9$ billion dollars.

$9 = 10(1 - e^{-0.005t}) \;\; \Rightarrow \;\; 0.9 = 1 - e^{-0.005t} \;\; \Rightarrow \;\; e^{-0.005t} = 0.1 \;\; \Rightarrow \;\; -0.005t = -\ln 10 \;\; \Rightarrow$

$t = 200\ln 10 \approx 460.517 \text{ days} \approx 1.26 \text{ years}.$

41. (a) Let $y(t)$ be the amount of salt (in kg) after t minutes. Then $y(0) = 15$. The amount of liquid in the tank is 1000 L at all times, so the concentration at time t (in minutes) is $y(t)/1000$ kg/L and $\dfrac{dy}{dt} = -\left[\dfrac{y(t)}{1000}\dfrac{\text{kg}}{\text{L}}\right]\left(10\dfrac{\text{L}}{\text{min}}\right) = -\dfrac{y(t)}{100}\dfrac{\text{kg}}{\text{min}}.$

$$\int \dfrac{dy}{y} = -\dfrac{1}{100}\int dt \;\; \Rightarrow \;\; \ln y = -\dfrac{t}{100} + C, \text{ and } y(0) = 15 \;\; \Rightarrow \;\; \ln 15 = C, \text{ so } \ln y = \ln 15 - \dfrac{t}{100}.$$

It follows that $\ln\left(\dfrac{y}{15}\right) = -\dfrac{t}{100}$ and $\dfrac{y}{15} = e^{-t/100}$, so $y = 15e^{-t/100}$ kg.

(b) After 20 minutes, $y = 15e^{-20/100} = 15e^{-0.2} \approx 12.3$ kg.

42. Let $y(t)$ be the amount of carbon dioxide in the room after t minutes. Then $y(0) = 0.0015(180) = 0.27 \text{ m}^3$. The amount of air in the room is 180 m^3 at all times, so the percentage at time t (in mimutes) is $y(t)/180 \times 100$, and the change in the amount of carbon dioxide with respect to time is

$$\frac{dy}{dt} = (0.0005)\left(2 \; \frac{\text{m}^3}{\text{min}}\right) - \frac{y(t)}{180}\left(2 \; \frac{\text{m}^3}{\text{min}}\right) = 0.001 - \frac{y}{90} = \frac{9 - 100y}{9000} \; \frac{\text{m}^3}{\text{min}}$$

Hence, $\displaystyle\int \frac{dy}{9 - 100y} = \int \frac{dt}{9000}$ and $-\frac{1}{100}\ln|9 - 100y| = \frac{1}{9000}t + C$. Because $y(0) = 0.27$, we have

$-\frac{1}{100}\ln 18 = C$, so $-\frac{1}{100}\ln|9 - 100y| = \frac{1}{9000}t - \frac{1}{100}\ln 18 \;\Rightarrow\; \ln|9 - 100y| = -\frac{1}{90}t + \ln 18 \;\Rightarrow\;$

$\ln|9 - 100y| = \ln e^{-t/90} + \ln 18 \;\Rightarrow\; \ln|9 - 100y| = \ln(18e^{-t/90})$, and $|9 - 100y| = 18e^{-t/90}$. Since y is continuous, $y(0) = 0.27$, and the right-hand side is never zero, we deduce that $9 - 100y$ is always negative. Thus, $|9 - 100y| = 100y - 9$ and we have $100y - 9 = 18e^{-t/90} \;\Rightarrow\; 100y = 9 + 18e^{-t/90} \;\Rightarrow\; y = 0.09 + 0.18e^{-t/90}$. The percentage of carbon dioxide in the room is

$$p(t) = \frac{y}{180} \times 100 = \frac{0.09 + 0.18e^{-t/90}}{180} \times 100 = (0.0005 + 0.001e^{-t/90}) \times 100 = 0.05 + 0.1e^{-t/90}$$

In the long run, we have $\lim\limits_{t \to \infty} p(t) = 0.05 + 0.1(0) = 0.05$; that is, the amount of carbon dioxide approaches 0.05% as time goes on.

43. Let $y(t)$ be the amount of alcohol in the vat after t minutes. Then $y(0) = 0.04(500) = 20$ gal. The amount of beer in the vat is 500 gallons at all times, so the percentage at time t (in minutes) is $y(t)/500 \times 100$, and the change in the amount of alcohol with respect to time t is $\dfrac{dy}{dt} = \text{rate in} - \text{rate out} = 0.06\left(5 \; \dfrac{\text{gal}}{\text{min}}\right) - \dfrac{y(t)}{500}\left(5 \; \dfrac{\text{gal}}{\text{min}}\right) = 0.3 - \dfrac{y}{100} = \dfrac{30 - y}{100} \; \dfrac{\text{gal}}{\text{min}}.$

Hence, $\displaystyle\int \frac{dy}{30 - y} = \int \frac{dt}{100}$ and $-\ln|30 - y| = \frac{1}{100}t + C$. Because $y(0) = 20$, we have $-\ln 10 = C$, so

$-\ln|30 - y| = \frac{1}{100}t - \ln 10 \;\Rightarrow\; \ln|30 - y| = -t/100 + \ln 10 \;\Rightarrow\; \ln|30 - y| = \ln e^{-t/100} + \ln 10 \;\Rightarrow\;$

$\ln|30 - y| = \ln(10e^{-t/100}) \;\Rightarrow\; |30 - y| = 10e^{-t/100}$. Since y is continuous, $y(0) = 20$, and the right-hand side is never zero, we deduce that $30 - y$ is always positive. Thus, $30 - y = 10e^{-t/100} \;\Rightarrow\; y = 30 - 10e^{-t/100}$. The percentage of alcohol is $p(t) = y(t)/500 \times 100 = y(t)/5 = 6 - 2e^{-t/100}$. The percentage of alcohol after one hour is $p(60) = 6 - 2e^{-60/100} \approx 4.9$.

44. (a) If $y(t)$ is the amount of salt (in kg) after t minutes, then $y(0) = 0$ and the total amount of liquid in the tank remains constant at 1000 L.

$$\frac{dy}{dt} = \left(0.05 \; \frac{\text{kg}}{\text{L}}\right)\left(5 \; \frac{\text{L}}{\text{min}}\right) + \left(0.04 \; \frac{\text{kg}}{\text{L}}\right)\left(10 \; \frac{\text{L}}{\text{min}}\right) - \left(\frac{y(t)}{1000} \; \frac{\text{kg}}{\text{L}}\right)\left(15 \; \frac{\text{L}}{\text{min}}\right)$$

$$= 0.25 + 0.40 - 0.015y = 0.65 - 0.015y = \frac{130 - 3y}{200} \; \frac{\text{kg}}{\text{min}}$$

Hence, $\displaystyle\int \frac{dy}{130 - 3y} = \int \frac{dt}{200}$ and $-\frac{1}{3}\ln|130 - 3y| = \frac{1}{200}t + C$. Because $y(0) = 0$, we have $-\frac{1}{3}\ln 130 = C$,

so $-\frac{1}{3}\ln|130 - 3y| = \frac{1}{200}t - \frac{1}{3}\ln 130 \;\Rightarrow\; \ln|130 - 3y| = -\frac{3}{200}t + \ln 130 = \ln(130e^{-3t/200})$, and

$|130 - 3y| = 130e^{-3t/200}$. Since y is continuous, $y(0) = 0$, and the right-hand side is never zero, we deduce that

$130 - 3y$ is always positive. Thus, $130 - 3y = 130e^{-3t/200}$ and $y = \frac{130}{3}(1 - e^{-3t/200})$ kg.

(b) After one hour, $y = \frac{130}{3}(1 - e^{-3\cdot 60/200}) = \frac{130}{3}(1 - e^{-0.9}) \approx 25.7$ kg.

Note: As $t \to \infty$, $y(t) \to \frac{130}{3} = 43\frac{1}{3}$ kg.

45. Assume that the raindrop begins at rest, so that $v(0) = 0$. $dm/dt = km$ and $(mv)' = gm$ $\Rightarrow$ $mv' + vm' = gm$ $\Rightarrow$

$mv' + v(km) = gm$ $\Rightarrow$ $v' + vk = g$ $\Rightarrow$ $\frac{dv}{dt} = g - kv$ $\Rightarrow$ $\int \frac{dv}{g - kv} = \int dt$ $\Rightarrow$

$-(1/k)\ln|g - kv| = t + C$ $\Rightarrow$ $\ln|g - kv| = -kt - kC$ $\Rightarrow$ $g - kv = Ae^{-kt}$. $v(0) = 0$ $\Rightarrow$ $A = g$.

So $kv = g - ge^{-kt}$ $\Rightarrow$ $v = (g/k)(1 - e^{-kt})$. Since $k > 0$, as $t \to \infty$, $e^{-kt} \to 0$ and therefore, $\lim_{t\to\infty} v(t) = g/k$.

46. (a) $m\frac{dv}{dt} = -kv$ $\Rightarrow$ $\frac{dv}{v} = -\frac{k}{m} dt$ $\Rightarrow$ $\ln|v| = -\frac{k}{m}t + C$. Since $v(0) = v_0$, $\ln|v_0| = C$. Therefore,

$\ln\left|\frac{v}{v_0}\right| = -\frac{k}{m}t$ $\Rightarrow$ $\left|\frac{v}{v_0}\right| = e^{-kt/m}$ $\Rightarrow$ $v(t) = \pm v_0 e^{-kt/m}$. The sign is $+$ when $t = 0$, and we assume

v is continuous, so that the sign is $+$ for all t. Thus, $v(t) = v_0 e^{-kt/m}$. $ds/dt = v_0 e^{-kt/m}$ $\Rightarrow$

$s(t) = -\frac{mv_0}{k}e^{-kt/m} + C'$.

From $s(0) = s_0$, we get $s_0 = -\frac{mv_0}{k} + C'$, so $C' = s_0 + \frac{mv_0}{k}$ and $s(t) = s_0 + \frac{mv_0}{k}(1 - e^{-kt/m})$.

The distance traveled from time 0 to time t is $s(t) - s_0$, so the total distance traveled is $\lim_{t\to\infty}[s(t) - s_0] = \frac{mv_0}{k}$.

Note: In finding the limit, we use the fact that $k > 0$ to conclude that $\lim_{t\to\infty} e^{-kt/m} = 0$.

(b) $m\frac{dv}{dt} = -kv^2$ $\Rightarrow$ $\frac{dv}{v^2} = -\frac{k}{m} dt$ $\Rightarrow$ $\frac{-1}{v} = -\frac{kt}{m} + C$ $\Rightarrow$ $\frac{1}{v} = \frac{kt}{m} - C$. Since $v(0) = v_0$,

$C = -\frac{1}{v_0}$ and $\frac{1}{v} = \frac{kt}{m} + \frac{1}{v_0}$. Therefore, $v(t) = \frac{1}{kt/m + 1/v_0} = \frac{mv_0}{kv_0 t + m}$. $\frac{ds}{dt} = \frac{mv_0}{kv_0 t + m}$ $\Rightarrow$

$s(t) = \frac{m}{k}\int \frac{kv_0 dt}{kv_0 t + m} = \frac{m}{k}\ln|kv_0 t + m| + C'$. Since $s(0) = s_0$, we get $s_0 = \frac{m}{k}\ln m + C'$ $\Rightarrow$

$C' = s_0 - \frac{m}{k}\ln m$ $\Rightarrow$ $s(t) = s_0 + \frac{m}{k}(\ln|kv_0 t + m| - \ln m) = s_0 + \frac{m}{k}\ln\left|\frac{kv_0 t + m}{m}\right|$.

We can rewrite the formulas for $v(t)$ and $s(t)$ as $v(t) = \frac{v_0}{1 + (kv_0/m)t}$ and $s(t) = s_0 + \frac{m}{k}\ln\left|1 + \frac{kv_0}{m}t\right|$.

Remarks: This model of horizontal motion through a resistive medium was designed to handle the case in which $v_0 > 0$. Then the term $-kv^2$ representing the resisting force causes the object to decelerate. The absolute value in the expression for $s(t)$ is unnecessary (since k, v_0, and m are all positive), and $\lim_{t\to\infty} s(t) = \infty$. In other words, the object travels infinitely far. However, $\lim_{t\to\infty} v(t) = 0$. When $v_0 < 0$, the term $-kv^2$ increases the magnitude of the object's negative velocity. According to the formula for $s(t)$, the position of the object approaches $-\infty$ as t approaches $m/k(-v_0)$:

$\lim_{t \to -m/(kv_0)} s(t) = -\infty$. Again the object travels infinitely far, but this time the feat is accomplished in a finite amount of

time. Notice also that $\lim_{t \to -m/(kv_0)} v(t) = -\infty$ when $v_0 < 0$, showing that the speed of the object increases without limit.

47. (a) The rate of growth of the area is jointly proportional to $\sqrt{A(t)}$ and $M - A(t)$; that is, the rate is proportional to the product of those two quantities. So for some constant k, $dA/dt = k\sqrt{A}(M - A)$. We are interested in the maximum of the function dA/dt (when the tissue grows the fastest), so we differentiate, using the Chain Rule and then substituting for dA/dt from the differential equation:

$$\frac{d}{dt}\left(\frac{dA}{dt}\right) = k\left[\sqrt{A}(-1)\frac{dA}{dt} + (M - A)\cdot\tfrac{1}{2}A^{-1/2}\frac{dA}{dt}\right] = \tfrac{1}{2}kA^{-1/2}\frac{dA}{dt}\left[-2A + (M - A)\right]$$

$$= \tfrac{1}{2}kA^{-1/2}\left[k\sqrt{A}(M - A)\right][M - 3A] = \tfrac{1}{2}k^2(M - A)(M - 3A)$$

This is 0 when $M - A = 0$ [this situation never actually occurs, since the graph of $A(t)$ is asymptotic to the line $y = M$, as in the logistic model] and when $M - 3A = 0 \iff A(t) = M/3$. This represents a maximum by the First Derivative Test, since $\dfrac{d}{dt}\left(\dfrac{dA}{dt}\right)$ goes from positive to negative when $A(t) = M/3$.

(b) From the CAS, we get $A(t) = M\left(\dfrac{Ce^{\sqrt{M}kt} - 1}{Ce^{\sqrt{M}kt} + 1}\right)^2$. To get C in terms of the initial area A_0 and the maximum area M,

we substitute $t = 0$ and $A = A_0 = A(0)$: $A_0 = M\left(\dfrac{C - 1}{C + 1}\right)^2 \iff (C + 1)\sqrt{A_0} = (C - 1)\sqrt{M} \iff$

$C\sqrt{A_0} + \sqrt{A_0} = C\sqrt{M} - \sqrt{M} \iff \sqrt{M} + \sqrt{A_0} = C\sqrt{M} - C\sqrt{A_0} \iff$

$\sqrt{M} + \sqrt{A_0} = C\left(\sqrt{M} - \sqrt{A_0}\right) \iff C = \dfrac{\sqrt{M} + \sqrt{A_0}}{\sqrt{M} - \sqrt{A_0}}$. [Notice that if $A_0 = 0$, then $C = 1$.]

48. (a) According to the hint we use the Chain Rule: $m\dfrac{dv}{dt} = m\dfrac{dv}{dx}\cdot\dfrac{dx}{dt} = mv\dfrac{dv}{dx} = -\dfrac{mgR^2}{(x + R)^2} \Rightarrow$

$\displaystyle\int v\, dv = \int \dfrac{-gR^2\, dx}{(x + R)^2} \Rightarrow \dfrac{v^2}{2} = \dfrac{gR^2}{x + R} + C$. When $x = 0$, $v = v_0$, so $\dfrac{v_0^2}{2} = \dfrac{gR^2}{0 + R} + C \Rightarrow$

$C = \tfrac{1}{2}v_0^2 - gR \Rightarrow \tfrac{1}{2}v^2 - \tfrac{1}{2}v_0^2 = \dfrac{gR^2}{x + R} - gR$. Now at the top of its flight, the rocket's velocity will be 0, and its

height will be $x = h$. Solving for v_0: $-\tfrac{1}{2}v_0^2 = \dfrac{gR^2}{h + R} - gR \Rightarrow \dfrac{v_0^2}{2} = g\left[-\dfrac{R^2}{R + h} + \dfrac{R(R + h)}{R + h}\right] = \dfrac{gRh}{R + h} \Rightarrow$

$v_0 = \sqrt{\dfrac{2gRh}{R + h}}$.

(b) $v_e = \lim_{h \to \infty} v_0 = \lim_{h \to \infty}\sqrt{\dfrac{2gRh}{R + h}} = \lim_{h \to \infty}\sqrt{\dfrac{2gR}{(R/h) + 1}} = \sqrt{2gR}$

(c) $v_e = \sqrt{2\cdot 32\text{ ft/s}^2 \cdot 3960\text{ mi}\cdot 5280\text{ ft/mi}} \approx 36{,}581\text{ ft/s} \approx 6.93\text{ mi/s}$

APPLIED PROJECT   How Fast Does a Tank Drain?

1. (a) $V = \pi r^2 h$ $\Rightarrow$ $\dfrac{dV}{dt} = \pi r^2 \dfrac{dh}{dt}$ [implicit differentiation] $\Rightarrow$

$$\frac{dh}{dt} = \frac{1}{\pi r^2}\frac{dV}{dt} = \frac{1}{\pi r^2}\left(-a\sqrt{2gh}\right) = \frac{1}{\pi 2^2}\left[-\pi\left(\tfrac{1}{12}\right)^2\sqrt{2\cdot 32}\sqrt{h}\right] = -\tfrac{1}{72}\sqrt{h}$$

(b) $\dfrac{dh}{dt} = -\tfrac{1}{72}\sqrt{h}$ $\Rightarrow$ $h^{-1/2}\,dh = -\tfrac{1}{72}\,dt$ $\Rightarrow$ $2\sqrt{h} = -\tfrac{1}{72}t + C.$

$h(0) = 6$ $\Rightarrow$ $2\sqrt{6} = 0 + C$ $\Rightarrow$ $C = 2\sqrt{6}$ $\Rightarrow$ $h(t) = \left(-\tfrac{1}{144}t + \sqrt{6}\right)^2.$

(c) We want to find t when $h = 0$, so we set $h = 0 = \left(-\tfrac{1}{144}t + \sqrt{6}\right)^2$ $\Rightarrow$ $t = 144\sqrt{6} \approx 5$ min 53 s.

2. (a) $\dfrac{dh}{dt} = k\sqrt{h}$ $\Rightarrow$ $h^{-1/2}\,dh = k\,dt$ $[h \neq 0]$ $\Rightarrow$ $2\sqrt{h} = kt + C$ $\Rightarrow$

t (in s)	$h(t)$ (in cm)
10	8.7
20	7.5
30	6.4
40	5.4
50	4.5
60	3.6

$h(t) = \tfrac{1}{4}(kt + C)^2$. Since $h(0) = 10$ cm, the relation $2\sqrt{h(t)} = kt + C$

gives us $2\sqrt{10} = C$. Also, $h(68) = 3$ cm, so $2\sqrt{3} = 68k + 2\sqrt{10}$ and

$k = -\dfrac{\sqrt{10} - \sqrt{3}}{34}$. Thus,

$h(t) = \dfrac{1}{4}\left(2\sqrt{10} - \dfrac{\sqrt{10} - \sqrt{3}}{34}t\right)^2 \approx 10 - 0.133t + 0.00044t^2.$

Here is a table of values of $h(t)$ correct to one decimal place.

(b) The answers to this part are to be obtained experimentally. See the article by Tom Farmer and Fred Gass, *Physical Demonstrations in the Calculus Classroom,* College Mathematics Journal 1992, pp. 146–148.

3. $V(t) = \pi r^2 h(t) = 100\pi h(t)$ $\Rightarrow$ $\dfrac{dV}{dh} = 100\pi$ and $\dfrac{dV}{dt} = \dfrac{dV}{dh}\dfrac{dh}{dt} = 100\pi\dfrac{dh}{dt}.$

Diameter $= 2.5$ inches $\Rightarrow$ radius $= 1.25$ inches $= \tfrac{5}{4}\cdot\tfrac{1}{12}$ foot $= \tfrac{5}{48}$ foot. Thus, $\dfrac{dV}{dt} = -a\sqrt{2gh}$ $\Rightarrow$

$100\pi\dfrac{dh}{dt} = -\pi\left(\tfrac{5}{48}\right)^2\sqrt{2\cdot 32h} = -\dfrac{25\pi}{288}\sqrt{h}$ $\Rightarrow$ $\dfrac{dh}{dt} = -\dfrac{\sqrt{h}}{1152}$ $\Rightarrow$ $\int h^{-1/2}\,dh = \int -\dfrac{1}{1152}\,dt$ $\Rightarrow$

$2\sqrt{h} = -\dfrac{1}{1152}t + C$ $\Rightarrow$ $\sqrt{h} = -\dfrac{1}{2304}t + k$ $\Rightarrow$ $h(t) = \left(-\dfrac{1}{2304}t + k\right)^2.$ The water pressure after t seconds is

$62.5h(t)$ lb/ft^2, so the condition that the pressure be at least 2160 lb/ft^2 for 10 minutes (600 seconds) is the condition

$62.5 \cdot h(600) \geq 2160$; that is, $\left(k - \tfrac{600}{2304}\right)^2 \geq \tfrac{2160}{62.5}$ $\Rightarrow$ $\left|k - \tfrac{25}{96}\right| \geq \sqrt{34.56}$ $\Rightarrow$ $k \geq \tfrac{25}{96} + \sqrt{34.56}$. Now $h(0) = k^2$,

so the height of the tank should be at least $\left(\tfrac{25}{96} + \sqrt{34.56}\right)^2 \approx 37.69$ ft.

4. (a) If the radius of the circular cross-section at height h is r, then the Pythagorean Theorem gives $r^2 = 2^2 - (2 - h)^2$ since

the radius of the tank is 2 m. So $A(h) = \pi r^2 = \pi[4 - (2 - h)^2] = \pi(4h - h^2)$. Thus, $A(h)\dfrac{dh}{dt} = -a\sqrt{2gh}$ $\Rightarrow$

$\pi(4h - h^2)\dfrac{dh}{dt} = -\pi(0.01)^2\sqrt{2\cdot 10h}$ $\Rightarrow$ $(4h - h^2)\dfrac{dh}{dt} = -0.0001\sqrt{20h}.$

(b) From part (a) we have $(4h^{1/2} - h^{3/2})\, dh = \left(-0.0001\sqrt{20}\right) dt \;\Rightarrow\; \frac{8}{3}h^{3/2} - \frac{2}{5}h^{5/2} = \left(-0.0001\sqrt{20}\right)t + C.$

$h(0) = 2 \;\Rightarrow\; \frac{8}{3}(2)^{3/2} - \frac{2}{5}(2)^{5/2} = C \;\Rightarrow\; C = \left(\frac{16}{3} - \frac{8}{5}\right)\sqrt{2} = \frac{56}{15}\sqrt{2}.$ To find out how long it will take to drain all

the water we evaluate t when $h = 0$: $0 = \left(-0.0001\sqrt{20}\right)t + C \;\Rightarrow\;$

$$t = \frac{C}{0.0001\sqrt{20}} = \frac{56\sqrt{2}/15}{0.0001\sqrt{20}} = \frac{11{,}200\sqrt{10}}{3} \approx 11{,}806 \text{ s} \approx 3\,\text{h}\,17\,\text{min}$$

APPLIED PROJECT   Which Is Faster, Going Up or Coming Down?

1. $mv' = -pv - mg \;\Rightarrow\; m\dfrac{dv}{dt} = -(pv + mg) \;\Rightarrow\; \displaystyle\int \dfrac{dv}{pv + mg} = \int -\dfrac{1}{m}\, dt \;\Rightarrow\;$

$\dfrac{1}{p}\ln(pv + mg) = -\dfrac{1}{m}t + C$ $[pv + mg > 0]$.   At $t = 0$, $v = v_0$, so $C = \dfrac{1}{p}\ln(pv_0 + mg)$.

Thus, $\dfrac{1}{p}\ln(pv + mg) = -\dfrac{1}{m}t + \dfrac{1}{p}\ln(pv_0 + mg) \;\Rightarrow\; \ln(pv + mg) = -\dfrac{p}{m}t + \ln(pv_0 + mg) \;\Rightarrow\;$

$pv + mg = e^{-pt/m}(pv_0 + mg) \;\Rightarrow\; pv = (pv_0 + mg)e^{-pt/m} - mg \;\Rightarrow\; v(t) = \left(v_0 + \dfrac{mg}{p}\right)e^{-pt/m} - \dfrac{mg}{p}.$

2. $y(t) = \displaystyle\int v(t)\, dt = \int \left[\left(v_0 + \dfrac{mg}{p}\right)e^{-pt/m} - \dfrac{mg}{p}\right] dt = \left(v_0 + \dfrac{mg}{p}\right)e^{-pt/m}\left(-\dfrac{m}{p}\right) - \dfrac{mg}{p}t + C.$

At $t = 0$, $y = 0$, so $C = \left(v_0 + \dfrac{mg}{p}\right)\dfrac{m}{p}$. Thus,

$$y(t) = \left(v_0 + \frac{mg}{p}\right)\frac{m}{p} - \left(v_0 + \frac{mg}{p}\right)\frac{m}{p}e^{-pt/m} - \frac{mgt}{p} = \left(v_0 + \frac{mg}{p}\right)\frac{m}{p}\left(1 - e^{-pt/m}\right) - \frac{mgt}{p}$$

3. $v(t) = 0 \;\Rightarrow\; \dfrac{mg}{p} = \left(v_0 + \dfrac{mg}{p}\right)e^{-pt/m} \;\Rightarrow\; e^{pt/m} = \dfrac{pv_0}{mg} + 1 \;\Rightarrow\; \dfrac{pt}{m} = \ln\left(\dfrac{pv_0}{mg} + 1\right) \;\Rightarrow\;$

$t_1 = \dfrac{m}{p}\ln\left(\dfrac{mg + pv_0}{mg}\right)$. With $m = 1$, $v_0 = 20$, $p = \frac{1}{10}$, and $g = 9.8$, we have $t_1 = 10\ln\left(\frac{11.8}{9.8}\right) \approx 1.86$ s.

4.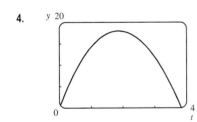

The figure shows the graph of $y = 1180\left(1 - e^{-0.1t}\right) - 98t$. The zeros are

at $t = 0$ and $t_2 \approx 3.84$. Thus, $t_1 - 0 \approx 1.86$ and $t_2 - t_1 \approx 1.98$. So the

time it takes to come down is about 0.12 s longer than the time it takes to go

up; hence, going up is faster.

5. $y(2t_1) = \left(v_0 + \dfrac{mg}{p}\right)\dfrac{m}{p}(1 - e^{-2pt_1/m}) - \dfrac{mg}{p}\cdot 2t_1$

$= \left(\dfrac{pv_0 + mg}{p}\right)\dfrac{m}{p}\left[1 - (e^{pt_1/m})^{-2}\right] - \dfrac{mg}{p}\cdot 2\dfrac{m}{p}\ln\left(\dfrac{pv_0 + mg}{mg}\right)$

Substituting $x = e^{pt_1/m} = \dfrac{pv_0}{mg} + 1 = \dfrac{pv_0 + mg}{mg}$ (from Problem 3), we get

$y(2t_1) = \left(x\cdot\dfrac{mg}{p}\right)\dfrac{m}{p}(1 - x^{-2}) - \dfrac{m^2g}{p^2}\cdot 2\ln x = \dfrac{m^2g}{p^2}\left(x - \dfrac{1}{x} - 2\ln x\right)$. Now $p > 0$, $m > 0$, $t_1 > 0 \;\Rightarrow\;$

$x = e^{pt_1/m} > e^0 = 1.$ $f(x) = x - \dfrac{1}{x} - 2\ln x$ $\Rightarrow$ $f'(x) = 1 + \dfrac{1}{x^2} - \dfrac{2}{x} = \dfrac{x^2 - 2x + 1}{x^2} = \dfrac{(x-1)^2}{x^2} > 0$

for $x > 1$ $\Rightarrow$ $f(x)$ is increasing for $x > 1$. Since $f(1) = 0$, it follows that $f(x) > 0$ for every $x > 1$. Therefore,

$y(2t_1) = \dfrac{m^2 g}{p^2} f(x)$ is positive, which means that the ball has not yet reached the ground at time $2t_1$. This tells us that the

time spent going up is always less than the time spent coming down, so *ascent is faster*.

10.4 Models for Population Growth

1. (a) $dP/dt = 0.05P - 0.0005P^2 = 0.05P(1 - 0.01P) = 0.05P(1 - P/100)$. Comparing to Equation 4,

  $dP/dt = kP(1 - P/K)$, we see that the carrying capacity is $K = 100$ and the value of k is 0.05.

(b) The slopes close to 0 occur where P is near 0 or 100. The largest slopes appear to be on the line $P = 50$. The solutions

  are increasing for $0 < P_0 < 100$ and decreasing for $P_0 > 100$.

(c)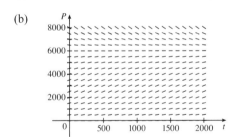
All of the solutions approach $P = 100$ as t increases. As in part (b), the solutions differ since for $0 < P_0 < 100$ they are increasing, and for $P_0 > 100$ they are decreasing. Also, some have an IP and some don't. It appears that the solutions which have $P_0 = 20$ and $P_0 = 40$ have inflection points at $P = 50$.

(d) The equilibrium solutions are $P = 0$ (trivial solution) and $P = 100$. The increasing solutions move away from $P = 0$ and

  all nonzero solutions approach $P = 100$ as $t \to \infty$.

2. (a) $K = 6000$ and $k = 0.0015$ $\Rightarrow$ $dP/dt = 0.0015P(1 - P/6000)$.

(b)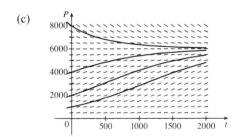
All of the solution curves approach 6000 as $t \to \infty$.

(c)
The curves with $P_0 = 1000$ and $P_0 = 2000$ appear to be concave upward at first and then concave downward. The curve with $P_0 = 4000$ appears to be concave downward everywhere. The curve with $P_0 = 8000$ appears to be concave upward everywhere. The inflection points are where the population grows the fastest.

(d) See the solution to Exercise 10.2.25 for a possible program to calculate $P(50)$. [In this case, we use X = 0, H = 1,

  N = 50, $Y_1 = 0.0015y(1 - y/6000)$, and Y = 1000.] We find that $P(50) \approx 1064$.

(e) Using Equation 7 with $K = 6000$, $k = 0.0015$, and $P_0 = 1000$, we have $P(t) = \dfrac{K}{1 + Ae^{-kt}} = \dfrac{6000}{1 + Ae^{-0.0015t}}$,

where $A = \dfrac{K - P_0}{P_0} = \dfrac{6000 - 1000}{1000} = 5$. Thus, $P(50) = \dfrac{6000}{1 + 5e^{-0.0015(50)}} \approx 1064.1$, which is extremely close to the

estimate obtained in part (d).

(f)

The curves are very similar.

3. (a) $\dfrac{dy}{dt} = ky\left(1 - \dfrac{y}{K}\right)$ $\Rightarrow$ $y(t) = \dfrac{K}{1 + Ae^{-kt}}$ with $A = \dfrac{K - y(0)}{y(0)}$. With $K = 8 \times 10^7$, $k = 0.71$, and

$y(0) = 2 \times 10^7$, we get the model $y(t) = \dfrac{8 \times 10^7}{1 + 3e^{-0.71t}}$, so $y(1) = \dfrac{8 \times 10^7}{1 + 3e^{-0.71}} \approx 3.23 \times 10^7$ kg.

(b) $y(t) = 4 \times 10^7$ $\Rightarrow$ $\dfrac{8 \times 10^7}{1 + 3e^{-0.71t}} = 4 \times 10^7$ $\Rightarrow$ $2 = 1 + 3e^{-0.71t}$ $\Rightarrow$ $e^{-0.71t} = \frac{1}{3}$ $\Rightarrow$

$-0.71t = \ln\frac{1}{3}$ $\Rightarrow$ $t = \dfrac{\ln 3}{0.71} \approx 1.55$ years

4. (a)

From the graph, we estimate the carrying capacity K for the yeast population to be 680.

(b) An estimate of the initial relative growth rate is $\dfrac{1}{P_0} \dfrac{dP}{dt} = \dfrac{1}{18} \cdot \dfrac{39 - 18}{2 - 0} = \dfrac{7}{12} = 0.58\overline{3}$.

(c) An exponential model is $P(t) = 18e^{7t/12}$. A logistic model is $P(t) = \dfrac{680}{1 + Ae^{-7t/12}}$, where $A = \dfrac{680 - 18}{18} = \dfrac{331}{9}$.

(d)

Time in Hours	Observed Values	Exponential Model	Logistic Model
0	18	18	18
2	39	58	55
4	80	186	149
6	171	596	322
8	336	1914	505
10	509	6147	614
12	597	19,739	658
14	640	63,389	673
16	664	203,558	678
18	672	653,679	679

The exponential model is a poor fit for anything beyond the first two observed values. The logistic model varies more for the middle values than it does for the values at either end, but provides a good general fit, as shown in the figure.

(e) $P(7) = \dfrac{680}{1 + \frac{331}{9}e^{-7(7/12)}} \approx 420$ yeast cells

5. (a) We will assume that the difference in the birth and death rates is 20 million/year. Let $t = 0$ correspond to the year 1990 and use a unit of 1 billion for all calculations. $k \approx \dfrac{1}{P}\dfrac{dP}{dt} = \dfrac{1}{5.3}(0.02) = \dfrac{1}{265}$, so

$$\frac{dP}{dt} = kP\left(1 - \frac{P}{K}\right) = \frac{1}{265}P\left(1 - \frac{P}{100}\right), \qquad P \text{ in billions}$$

(b) $A = \dfrac{K - P_0}{P_0} = \dfrac{100 - 5.3}{5.3} = \dfrac{947}{53} \approx 17.8679.$ $P(t) = \dfrac{K}{1 + Ae^{-kt}} = \dfrac{100}{1 + \frac{947}{53}e^{-(1/265)t}}$, so $P(10) \approx 5.49$ billion.

(c) $P(110) \approx 7.81$, and $P(510) \approx 27.72$. The predictions are 7.81 billion in the year 2100 and 27.72 billion in 2500.

(d) If $K = 50$, then $P(t) = \dfrac{50}{1 + \frac{447}{53}e^{-(1/265)t}}$. So $P(10) \approx 5.48$, $P(110) \approx 7.61$, and $P(510) \approx 22.41$. The predictions become 5.48 billion in the year 2000, 7.61 billion in 2100, and 22.41 billion in the year 2500.

6. (a) If we assume that the carrying capacity for the world population is 100 billion, it would seem reasonable that the carrying capacity for the US is 3–5 billion by using current populations and simple proportions. We will use $K = 4$ billion or 4000 million. With $t = 0$ corresponding to 1980, we have $P(t) = \dfrac{4000}{1 + \left(\frac{4000 - 250}{250}\right)e^{-kt}} = \dfrac{4000}{1 + 15e^{-kt}}$.

(b) $P(10) = 275 \;\Rightarrow\; \dfrac{4000}{1 + 15e^{-10k}} = 275 \;\Rightarrow\; 1 + 15e^{-10k} = \dfrac{4000}{275} \;\Rightarrow\; e^{-10k} = \dfrac{\frac{160}{11} - 1}{15} \;\Rightarrow$

$-10k = \ln\dfrac{149}{165} \;\Rightarrow\; k = -\dfrac{1}{10}\ln\dfrac{149}{165} \approx 0.01019992.$

(c) $2100 - 1990 = 110$ and $P(110) \approx 680$ million.

$2200 - 1990 = 210$ and $P(210) \approx 1449$ million, or about 1.4 billion.

(d) $P(t) = 350 \;\Rightarrow\; \dfrac{4000}{1 + 15e^{-kt}} = 350 \;\Rightarrow\; 1 + 15e^{-kt} = \dfrac{80}{7} \;\Rightarrow\; e^{-kt} = \dfrac{73}{7}\cdot\dfrac{1}{15} \;\Rightarrow\; -kt = \ln\dfrac{73}{105} \;\Rightarrow$

$t = 10\,\dfrac{\ln\frac{73}{105}}{\ln\frac{149}{165}} \approx 35.64 \approx 36$. So we predict that the US population will exceed 350 million in the year

$1990 + 36 = 2026.$

7. (a) Our assumption is that $\dfrac{dy}{dt} = ky(1 - y)$, where y is the fraction of the population that has heard the rumor.

(b) Using the logistic equation (4), $\dfrac{dP}{dt} = kP\left(1 - \dfrac{P}{K}\right)$, we substitute $y = \dfrac{P}{K}$, $P = Ky$, and $\dfrac{dP}{dt} = K\dfrac{dy}{dt}$,

to obtain $K\dfrac{dy}{dt} = k(Ky)(1 - y) \;\Leftrightarrow\; \dfrac{dy}{dt} = ky(1 - y)$, our equation in part (a).

Now the solution to (4) is $P(t) = \dfrac{K}{1 + Ae^{-kt}}$, where $A = \dfrac{K - P_0}{P_0}$.

We use the same substitution to obtain $Ky = \dfrac{K}{1 + \frac{K - Ky_0}{Ky_0}e^{-kt}} \;\Rightarrow\; y = \dfrac{y_0}{y_0 + (1 - y_0)e^{-kt}}.$

Alternatively, we could use the same steps as outlined in the solution of Equation 4.

(c) Let t be the number of hours since 8 AM. Then $y_0 = y(0) = \dfrac{80}{1000} = 0.08$ and $y(4) = \dfrac{1}{2}$, so

$\dfrac{1}{2} = y(4) = \dfrac{0.08}{0.08 + 0.92e^{-4k}}$. Thus, $0.08 + 0.92e^{-4k} = 0.16$, $e^{-4k} = \dfrac{0.08}{0.92} = \dfrac{2}{23}$, and $e^{-k} = \left(\dfrac{2}{23}\right)^{1/4}$, so $y = \dfrac{0.08}{0.08 + 0.92(2/23)^{t/4}} = \dfrac{2}{2 + 23(2/23)^{t/4}}$. Solving this equation for t, we get

$$2y + 23y\left(\frac{2}{23}\right)^{t/4} = 2 \;\Rightarrow\; \left(\frac{2}{23}\right)^{t/4} = \frac{2-2y}{23y} \;\Rightarrow\; \left(\frac{2}{23}\right)^{t/4} = \frac{2}{23}\cdot\frac{1-y}{y} \;\Rightarrow\; \left(\frac{2}{23}\right)^{t/4-1} = \frac{1-y}{y}.$$

It follows that $\dfrac{t}{4} - 1 = \dfrac{\ln[(1-y)/y]}{\ln \frac{2}{23}}$, so $t = 4\left[1 + \dfrac{\ln((1-y)/y)}{\ln \frac{2}{23}}\right]$.

When $y = 0.9$, $\dfrac{1-y}{y} = \frac{1}{9}$, so $t = 4\left(1 - \dfrac{\ln 9}{\ln \frac{2}{23}}\right) \approx 7.6$ h or 7 h 36 min. Thus, 90% of the population will have heard

the rumor by 3:36 PM.

8. (a) $P(0) = P_0 = 400$, $P(1) = 1200$ and $K = 10{,}000$. From the solution to the logistic differential equation

$$P(t) = \frac{P_0 K}{P_0 + (K - P_0)e^{-kt}}, \text{ we get } P = \frac{400\,(10{,}000)}{400 + (9600)e^{-kt}} = \frac{10{,}000}{1 + 24e^{-kt}}. \quad P(1) = 1200 \;\Rightarrow$$

$$1 + 24e^{-k} = \tfrac{100}{12} \;\Rightarrow\; e^k = \tfrac{288}{88} \;\Rightarrow\; k = \ln\tfrac{36}{11}. \text{ So } P = \frac{10{,}000}{1 + 24e^{-t\ln(36/11)}} = \frac{10{,}000}{1 + 24\cdot(11/36)^t}.$$

(b) $5000 = \dfrac{10{,}000}{1 + 24(11/36)^t} \;\Rightarrow\; 24\left(\tfrac{11}{36}\right)^t = 1 \;\Rightarrow\; t\ln\tfrac{11}{36} = \ln\tfrac{1}{24} \;\Rightarrow\; t \approx 2.68$ years.

9. (a) $\dfrac{dP}{dt} = kP\left(1 - \dfrac{P}{K}\right) \;\Rightarrow\; \dfrac{d^2P}{dt^2} = k\left[P\left(-\dfrac{1}{K}\dfrac{dP}{dt}\right) + \left(1 - \dfrac{P}{K}\right)\dfrac{dP}{dt}\right] = k\dfrac{dP}{dt}\left(-\dfrac{P}{K} + 1 - \dfrac{P}{K}\right)$

$$= k\left[kP\left(1 - \frac{P}{K}\right)\right]\left(1 - \frac{2P}{K}\right) = k^2 P\left(1 - \frac{P}{K}\right)\left(1 - \frac{2P}{K}\right)$$

(b) P grows fastest when P' has a maximum, that is, when $P'' = 0$. From part (a), $P'' = 0 \;\Leftrightarrow\; P = 0, P = K$,

or $P = K/2$. Since $0 < P < K$, we see that $P'' = 0 \;\Leftrightarrow\; P = K/2$.

10. First we keep k constant (at 0.1, say) and change P_0 in the function

$P = \dfrac{10P_0}{P_0 + (10 - P_0)e^{-0.1t}}$. (Notice that P_0 is the P-intercept.) If $P_0 = 0$,

the function is 0 everywhere. For $0 < P_0 < 5$, the curve has an inflection

point, which moves to the right as P_0 decreases. If $5 < P_0 < 10$, the graph is

concave down everywhere. (We are considering only $t \geq 0$.) If $P_0 = 10$, the

function is the constant function $P = 10$, and if $P_0 > 10$, the function decreases. For all $P_0 \neq 0$, $\lim\limits_{t\to\infty} P = 10$.

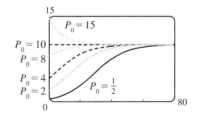

Now we instead keep P_0 constant (at $P_0 = 1$) and change k in the function

$P = \dfrac{10}{1 + 9e^{-kt}}$. It seems that as k increases, the graph approaches the line

$P = 10$ more and more quickly. (Note that the only difference in the shape of

the curves is in the horizontal scaling; if we choose suitable x-scales, the

graphs all look the same.)

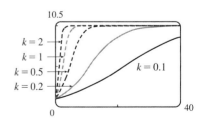

11. Following the hint, we choose $t = 0$ to correspond to 1960 and subtract 94,000 from each of the population figures. We then use a calculator to obtain the models and add 94,000 to get the exponential function

$P_E(t) = 1578.3(1.0933)^t + 94{,}000$ and the logistic function

$P_L(t) = \dfrac{32{,}658.5}{1 + 12.75e^{-0.1706t}} + 94{,}000$. P_L is a reasonably accurate

model, while P_E is not, since an exponential model would only be used for the first few data points.

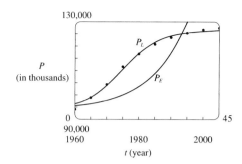

12. Following the hint, we choose $t = 0$ to correspond to 1955 and subtract 29,000 from each of the population figures. We then use a calculator to obtain the models and add 29,000 to get the exponential function

$P_E(t) = 1094(1.0668)^t + 29{,}000$ and the logistic function

$P_L(t) = \dfrac{11{,}103.3}{1 + 12.34e^{-0.1471t}} + 29{,}000$. P_L is a reasonably accurate

model, while P_E is not, since an exponential model would only be used for the first few data points.

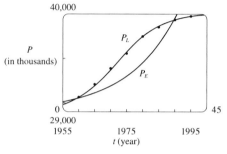

13. (a) $\dfrac{dP}{dt} = kP - m = k\left(P - \dfrac{m}{k}\right)$. Let $y = P - \dfrac{m}{k}$, so $\dfrac{dy}{dt} = \dfrac{dP}{dt}$ and the differential equation becomes $\dfrac{dy}{dt} = ky$.

The solution is $y = y_0 e^{kt} \;\Rightarrow\; P - \dfrac{m}{k} = \left(P_0 - \dfrac{m}{k}\right)e^{kt} \;\Rightarrow\; P(t) = \dfrac{m}{k} + \left(P_0 - \dfrac{m}{k}\right)e^{kt}$.

(b) Since $k > 0$, there will be an exponential expansion $\;\Leftrightarrow\; P_0 - \dfrac{m}{k} > 0 \;\Leftrightarrow\; m < kP_0$.

(c) The population will be constant if $P_0 - \dfrac{m}{k} = 0 \;\Leftrightarrow\; m = kP_0$. It will decline if $P_0 - \dfrac{m}{k} < 0 \;\Leftrightarrow\; m > kP_0$.

(d) $P_0 = 8{,}000{,}000$, $k = \alpha - \beta = 0.016$, $m = 210{,}000 \;\Rightarrow\; m > kP_0 \,(= 128{,}000)$, so by part (c), the population was declining.

14. (a) $\dfrac{dy}{dt} = ky^{1+c} \;\Rightarrow\; y^{-1-c}\,dy = k\,dt \;\Rightarrow\; \dfrac{y^{-c}}{-c} = kt + C$. Since $y(0) = y_0$, we have $C = \dfrac{y_0^{-c}}{-c}$. Thus,

$\dfrac{y^{-c}}{-c} = kt + \dfrac{y_0^{-c}}{-c}$, or $y^{-c} = y_0^{-c} - ckt$. So $y^c = \dfrac{1}{y_0^{-c} - ckt} = \dfrac{y_0^c}{1 - cy_0^c kt}$ and $y(t) = \dfrac{y_0}{(1 - cy_0^c kt)^{1/c}}$.

(b) $y(t) \to \infty$ as $1 - cy_0^c kt \to 0$, that is, as $t \to \dfrac{1}{cy_0^c k}$. Define $T = \dfrac{1}{cy_0^c k}$. Then $\displaystyle\lim_{t \to T^-} y(t) = \infty$.

(c) According to the data given, we have $c = 0.01$, $y(0) = 2$, and $y(3) = 16$, where the time t is given in months. Thus,

$y_0 = 2$ and $16 = y(3) = \dfrac{y_0}{(1 - cy_0^c k \cdot 3)^{1/c}}$. Since $T = \dfrac{1}{cy_0^c k}$, we will solve for $cy_0^c k$. $\;16 = \dfrac{2}{(1 - 3cy_0^c k)^{100}} \;\Rightarrow\;$

$1 - 3cy_0^c k = \left(\frac{1}{8}\right)^{0.01} = 8^{-0.01} \;\Rightarrow\; cy_0^c k = \frac{1}{3}\left(1 - 8^{-0.01}\right)$. Thus, doomsday occurs when

$t = T = \dfrac{1}{cy_0^c k} = \dfrac{3}{1 - 8^{-0.01}} \approx 145.77$ months or 12.15 years.

15. (a) The term -15 represents a harvesting of fish at a constant rate—in this case, 15 fish/week. This is the rate at which fish are caught.

(b)

(c) From the graph in part (b), it appears that $P(t) = 250$ and $P(t) = 750$ are the equilibrium solutions. We confirm this analytically by solving the equation $dP/dt = 0$ as follows: $0.08P(1 - P/1000) - 15 = 0 \Rightarrow$
$0.08P - 0.00008P^2 - 15 = 0 \Rightarrow$
$-0.00008(P^2 - 1000P + 187{,}500) = 0 \Rightarrow$
$(P - 250)(P - 750) = 0 \Rightarrow P = 250$ or 750.

(d)

For $0 < P_0 < 250$, $P(t)$ decreases to 0. For $P_0 = 250$, $P(t)$ remains constant. For $250 < P_0 < 750$, $P(t)$ increases and approaches 750. For $P_0 = 750$, $P(t)$ remains constant. For $P_0 > 750$, $P(t)$ decreases and approaches 750.

(e) $\dfrac{dP}{dt} = 0.08P\left(1 - \dfrac{P}{1000}\right) - 15 \quad \Leftrightarrow \quad -\dfrac{100{,}000}{8} \cdot \dfrac{dP}{dt} = (0.08P - 0.00008P^2 - 15) \cdot \left(-\dfrac{100{,}000}{8}\right) \quad \Leftrightarrow$

$-12{,}500\dfrac{dP}{dt} = P^2 - 1000P + 187{,}500 \quad \Leftrightarrow \quad \dfrac{dP}{(P - 250)(P - 750)} = -\dfrac{1}{12{,}500}\,dt \quad \Leftrightarrow$

$\displaystyle\int\left(\dfrac{-1/500}{P - 250} + \dfrac{1/500}{P - 750}\right)dP = -\dfrac{1}{12{,}500}\,dt \quad \Leftrightarrow \quad \int\left(\dfrac{1}{P - 250} - \dfrac{1}{P - 750}\right)dP = \tfrac{1}{25}\,dt \quad \Leftrightarrow$

$\ln|P - 250| - \ln|P - 750| = \tfrac{1}{25}t + C \quad \Leftrightarrow \quad \ln\left|\dfrac{P - 250}{P - 750}\right| = \tfrac{1}{25}t + C \quad \Leftrightarrow \quad \left|\dfrac{P - 250}{P - 750}\right| = e^{t/25+C} = ke^{t/25} \quad \Leftrightarrow$

$\dfrac{P - 250}{P - 750} = ke^{t/25} \quad \Leftrightarrow \quad P - 250 = Pke^{t/25} - 750ke^{t/25} \quad \Leftrightarrow \quad P - Pke^{t/25} = 250 - 750ke^{t/25} \quad \Leftrightarrow$

$P(t) = \dfrac{250 - 750ke^{t/25}}{1 - ke^{t/25}}$. If $t = 0$ and $P = 200$, then $200 = \dfrac{250 - 750k}{1 - k} \quad \Leftrightarrow \quad 200 - 200k = 250 - 750k \quad \Leftrightarrow$

$550k = 50 \quad \Leftrightarrow \quad k = \tfrac{1}{11}$. Similarly, if $t = 0$ and $P = 300$, then

$k = -\tfrac{1}{9}$. Simplifying P with these two values of k gives us

$P(t) = \dfrac{250(3e^{t/25} - 11)}{e^{t/25} - 11}$ and $P(t) = \dfrac{750(e^{t/25} + 3)}{e^{t/25} + 9}$.

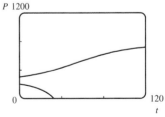

16. (a)

$c = 0$

$c = 10$

$c = 20$

$c = 21$

$c = 25$

$c = 30$

(b) For $0 \leq c \leq 20$, there is at least one equilibrium solution. For $c > 20$, the population always dies out.

(c) $\dfrac{dP}{dt} = 0.08P - 0.00008P^2 - c$. $\dfrac{dP}{dt} = 0 \Leftrightarrow P = \dfrac{-0.08 \pm \sqrt{(0.08)^2 - 4(-0.00008)(-c)}}{2(-0.00008)}$, which has at least

one solution when the discriminant is nonnegative $\Rightarrow 0.0064 - 0.00032c \geq 0 \Leftrightarrow c \leq 20$. For $0 \leq c \leq 20$, there is

at least one value of P such that $dP/dt = 0$ and hence, at least one equilibrium solution. For $c > 20$, $dP/dt < 0$ and the

population always dies out.

(d) The weekly catch should be less than 20 fish per week.

17. (a) $\dfrac{dP}{dt} = (kP)\left(1 - \dfrac{P}{K}\right)\left(1 - \dfrac{m}{P}\right)$. If $m < P < K$, then $dP/dt = (+)(+)(+) = + \Rightarrow P$ is increasing.

If $0 < P < m$, then $dP/dt = (+)(+)(-) = - \Rightarrow P$ is decreasing.

(b)

$k = 0.08$, $K = 1000$, and $m = 200 \Rightarrow$

$$\frac{dP}{dt} = 0.08P\left(1 - \frac{P}{1000}\right)\left(1 - \frac{200}{P}\right)$$

For $0 < P_0 < 200$, the population dies out. For $P_0 = 200$, the population is steady. For $200 < P_0 < 1000$, the population increases and approaches 1000. For $P_0 > 1000$, the population decreases and approaches 1000. The equilibrium solutions are $P(t) = 200$ and $P(t) = 1000$.

(c) $\dfrac{dP}{dt} = kP\left(1 - \dfrac{P}{K}\right)\left(1 - \dfrac{m}{P}\right) = kP\left(\dfrac{K - P}{K}\right)\left(\dfrac{P - m}{P}\right) = \dfrac{k}{K}(K - P)(P - m) \Leftrightarrow$

$\displaystyle\int \dfrac{dP}{(K - P)(P - m)} = \int \dfrac{k}{K}\, dt$. By partial fractions, $\dfrac{1}{(K - P)(P - m)} = \dfrac{A}{K - P} + \dfrac{B}{P - m}$, so

$A(P - m) + B(K - P) = 1$.

If $P = m$, $B = \dfrac{1}{K - m}$; if $P = K$, $A = \dfrac{1}{K - m}$, so $\dfrac{1}{K - m}\displaystyle\int\left(\dfrac{1}{K - P} + \dfrac{1}{P - m}\right) dP = \int \dfrac{k}{K}\, dt \Rightarrow$

$$\frac{1}{K-m}\left(-\ln|K-P|+\ln|P-m|\right)=\frac{k}{K}t+M \quad\Rightarrow\quad \frac{1}{K-m}\ln\left|\frac{P-m}{K-P}\right|=\frac{k}{K}t+M \quad\Rightarrow$$

$$\ln\left|\frac{P-m}{K-P}\right|=(K-m)\frac{k}{K}t+M_1 \quad\Leftrightarrow\quad \frac{P-m}{K-P}=De^{(K-m)(k/K)t} \quad [D=\pm e^{M_1}].$$

Let $t=0$: $\dfrac{P_0-m}{K-P_0}=D$. So $\dfrac{P-m}{K-P}=\dfrac{P_0-m}{K-P_0}e^{(K-m)(k/K)t}$. Solving for P, we get

$$P(t)=\frac{m(K-P_0)+K(P_0-m)e^{(K-m)(k/K)t}}{K-P_0+(P_0-m)e^{(K-m)(k/K)t}}.$$

(d) If $P_0<m$, then $P_0-m<0$. Let $N(t)$ be the numerator of the expression for $P(t)$ in part (c). Then

$$N(0)=P_0(K-m)>0, \text{ and } P_0-m<0 \quad\Leftrightarrow\quad \lim_{t\to\infty}K(P_0-m)e^{(K-m)(k/K)t}=-\infty \quad\Rightarrow\quad \lim_{t\to\infty}N(t)=-\infty.$$

Since N is continuous, there is a number t such that $N(t)=0$ and thus $P(t)=0$. So the species will become extinct.

18. (a) $\dfrac{dP}{dt}=c\ln\left(\dfrac{K}{P}\right)P \quad\Rightarrow\quad \displaystyle\int\frac{dP}{P\ln(K/P)}=\int c\,dt$. Let $u=\ln\left(\dfrac{K}{P}\right)=\ln K-\ln P \quad\Rightarrow\quad du=-\dfrac{dP}{P} \quad\Rightarrow$

$$\int -\frac{du}{u}=ct+D \quad\Rightarrow\quad \ln|u|=-ct-D \quad\Rightarrow\quad |u|=e^{-(ct+D)} \quad\Rightarrow\quad |\ln(K/P)|=e^{-(ct+D)} \quad\Rightarrow$$

$\ln(K/P)=\pm e^{-(ct+D)}$. Letting $t=0$, we get $\ln(K/P_0)=\pm e^{-D}$, so

$$\ln(K/P)=\pm e^{-ct-D}=\pm e^{-ct}e^{-D}=\ln(K/P_0)e^{-ct} \quad\Rightarrow\quad K/P=e^{\ln(K/P_0)e^{-ct}} \quad\Rightarrow$$

$$P(t)=Ke^{-\ln(K/P_0)e^{-ct}}, c\neq 0.$$

(b) $\displaystyle\lim_{t\to\infty}P(t)=\lim_{t\to\infty}Ke^{-\ln(K/P_0)e^{-ct}}=Ke^{-\ln(K/P_0)\cdot 0}=Ke^0=K$

(c)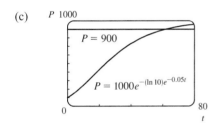

The graphs look very similar. For the Gompertz function, $P(40)\approx 732$, nearly the same as the logistic function. The Gompertz function reaches $P=900$ at $t\approx 61.7$ and its value at $t=80$ is about 959, so it doesn't increase quite as fast as the logistic curve.

(d) $\dfrac{dP}{dt}=c\ln\left(\dfrac{K}{P}\right)P=cP(\ln K-\ln P) \quad\Rightarrow$

$$\frac{d^2P}{dt^2}=c\left[P\left(-\frac{1}{P}\frac{dP}{dt}\right)+(\ln K-\ln P)\frac{dP}{dt}\right]=c\frac{dP}{dt}\left[-1+\ln\left(\frac{K}{P}\right)\right]$$

$$=c[c\ln(K/P)P][\ln(K/P)-1]=c^2P\ln(K/P)\,[\ln(K/P)-1]$$

Since $0<P<K$, $P''=0 \quad\Leftrightarrow\quad \ln(K/P)=1 \quad\Leftrightarrow\quad K/P=e \quad\Leftrightarrow\quad P=K/e$. $P''>0$ for $0<P<K/e$ and $P''<0$ for $K/e<P<K$, so P' is a maximum (and P grows fastest) when $P=K/e$.

Note: If $P>K$, then $\ln(K/P)<0$, so $P''(t)>0$.

19. (a) $dP/dt=kP\cos(rt-\phi) \quad\Rightarrow\quad (dP)/P=k\cos(rt-\phi)\,dt \quad\Rightarrow\quad \int(dP)/P=k\int\cos(rt-\phi)\,dt \quad\Rightarrow$

$\ln P=(k/r)\sin(rt-\phi)+C$. (Since this is a growth model, $P>0$ and we can write $\ln P$ instead of $\ln|P|$.) Since $P(0)=P_0$, we obtain $\ln P_0=(k/r)\sin(-\phi)+C=-(k/r)\sin\phi+C \quad\Rightarrow\quad C=\ln P_0+(k/r)\sin\phi$. Thus,

$\ln P = (k/r)\sin(rt - \phi) + \ln P_0 + (k/r)\sin\phi$, which we can rewrite as $\ln(P/P_0) = (k/r)[\sin(rt - \phi) + \sin\phi]$ or, after exponentiation, $P(t) = P_0 e^{(k/r)[\sin(rt-\phi)+\sin\phi]}$.

(b) As k increases, the amplitude increases, but the minimum value stays the same.

As r increases, the amplitude and the period decrease.

A change in ϕ produces slight adjustments in the phase shift and amplitude.

Comparing values of k with $P_0 = 1$, $r = 2$, and $\phi = \pi/2$

Comparing values of r with $P_0 = 1$, $k = 1$, and $\phi = \pi/2$

Comparing values of ϕ with $P_0 = 1$, $k = 1$, and $r = 2$

$P(t)$ oscillates between $P_0 e^{(k/r)(1+\sin\phi)}$ and $P_0 e^{(k/r)(-1+\sin\phi)}$ (the extreme values are attained when $rt - \phi$ is an odd multiple of $\frac{\pi}{2}$), so $\lim\limits_{t\to\infty} P(t)$ does not exist.

20. (a) $dP/dt = kP\cos^2(rt - \phi) \Rightarrow (dP)/P = k\cos^2(rt - \phi)\,dt \Rightarrow \int(dP)/P = k\int\cos^2(rt - \phi)\,dt \Rightarrow$

$\ln P = k\int \dfrac{1 + \cos(2(rt - \phi))}{2}\,dt = \dfrac{k}{2}t + \dfrac{k}{4r}\sin(2(rt - \phi)) + C$. From $P(0) = P_0$, we get

$\ln P_0 = \dfrac{k}{4r}\sin(-2\phi) + C = C - \dfrac{k}{4r}\sin 2\phi$, so $C = \ln P_0 + \dfrac{k}{4r}\sin 2\phi$ and

$\ln P = \dfrac{k}{2}t + \dfrac{k}{4r}\sin(2(rt - \phi)) + \ln P_0 + \dfrac{k}{4r}\sin 2\phi$. Simplifying, we get

$\ln \dfrac{P}{P_0} = \dfrac{k}{2}t + \dfrac{k}{4r}[\sin(2(rt - \phi)) + \sin 2\phi] = f(t)$, or $P(t) = P_0 e^{f(t)}$.

(b) An increase in k stretches the graph of P vertically while maintaining $P(0) = P_0$.

An increase in r compresses the graph of P horizontally—similar to changing the period in Exercise 19.

As in Exercise 19, a change in ϕ only makes slight adjustments in the growth of P, as shown in the figure.

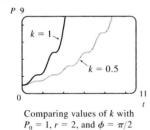

Comparing values of k with $P_0 = 1$, $r = 2$, and $\phi = \pi/2$

Comparing values of r with $P_0 = 1$, $k = 0.5$, and $\phi = \pi/2$

Comparing values of ϕ with $P_0 = 1$, $k = 0.5$, and $r = 2$

$f'(t) = k/2 + [k/(4r)][2r\cos(2(rt - \phi))] = (k/2)[1 + \cos(2(rt - \phi))] \geq 0$. Since $P(t) = P_0 e^{f(t)}$, we have

$P'(t) = P_0 f'(t)e^{f(t)} \geq 0$, with equality only when $\cos(2(rt - \phi)) = -1$; that is, when $rt - \phi$ is an odd multiple of $\frac{\pi}{2}$.

Therefore, $P(t)$ is an increasing function on $(0, \infty)$. P can also be written as $P(t) = P_0 e^{kt/2} e^{(k/4r)[\sin(2(rt-\phi))+\sin 2\phi]}$.

The second exponential oscillates between $e^{(k/4r)(1+\sin 2\phi)}$ and $e^{(k/4r)(-1+\sin 2\phi)}$, while the first one, $e^{kt/2}$, grows without bound. So $\lim\limits_{t\to\infty} P(t) = \infty$.

21. By Equation 7, $P(t) = \dfrac{K}{1 + Ae^{-kt}}$. By comparison, if $c = (\ln A)/k$ and $u = \frac{1}{2}k(t - c)$, then

$$1 + \tanh u = 1 + \frac{e^u - e^{-u}}{e^u + e^{-u}} = \frac{e^u + e^{-u}}{e^u + e^{-u}} + \frac{e^u - e^{-u}}{e^u + e^{-u}} = \frac{2e^u}{e^u + e^{-u}} \cdot \frac{e^{-u}}{e^{-u}} = \frac{2}{1 + e^{-2u}}$$

and $e^{-2u} = e^{-k(t-c)} = e^{kc}e^{-kt} = e^{\ln A}e^{-kt} = Ae^{-kt}$, so

$$\tfrac{1}{2}K\big[1 + \tanh\big(\tfrac{1}{2}k(t - c)\big)\big] = \frac{K}{2}[1 + \tanh u] = \frac{K}{2} \cdot \frac{2}{1 + e^{-2u}} = \frac{K}{1 + e^{-2u}} = \frac{K}{1 + Ae^{-kt}} = P(t).$$

APPLIED PROJECT   Calculus and Baseball

1. (a) $F = ma = m\dfrac{dv}{dt}$, so by the Substitution Rule we have

$$\int_{t_0}^{t_1} F(t)\,dt = \int_{t_0}^{t_1} m\left(\frac{dv}{dt}\right) dt = m\int_{v_0}^{v_1} dv = \big[mv\big]_{v_0}^{v_1} = mv_1 - mv_0 = p(t_1) - p(t_0)$$

   (b) (i) We have $v_1 = 110$ mi/h $= \frac{110(5280)}{3600}$ ft/s $= 161.\overline{3}$ ft/s, $v_0 = -90$ mi/h $= -132$ ft/s, and the mass of the

        baseball is $m = \dfrac{w}{g} = \dfrac{5/16}{32} = \dfrac{5}{512}$. So the change in momentum is

$$p(t_1) - p(t_0) = mv_1 - mv_0 = \tfrac{5}{512}\big[161.\overline{3} - (-132)\big] \approx 2.86 \text{ slug-ft/s}.$$

      (ii) From part (a) and part (b)(i), we have $\int_0^{0.001} F(t)\,dt = p(0.001) - p(0) \approx 2.86$, so the average force over the

        interval $[0, 0.001]$ is $\frac{1}{0.001}\int_0^{0.001} F(t)\,dt \approx \frac{1}{0.001}(2.86) = 2860$ lb.

2. (a) $W = \displaystyle\int_{s_0}^{s_1} F(s)\,ds$, where $F(s) = m\dfrac{dv}{dt} = m\dfrac{dv}{ds}\dfrac{ds}{dt} = mv\dfrac{dv}{ds}$ and so, by the Substitution Rule,

$$W = \int_{s_0}^{s_1} F(s)\,ds = \int_{s_0}^{s_1} mv\frac{dv}{ds}\,ds = \int_{v(s_0)}^{v(s_1)} mv\,dv = \big[\tfrac{1}{2}mv^2\big]_{v_0}^{v_1} = \tfrac{1}{2}mv_1^2 - \tfrac{1}{2}mv_0^2$$

   (b) From part (b)(i), 90 mi/h $= 132$ ft/s. Assume $v_0 = v(s_0) = 0$ and $v_1 = v(s_1) = 132$ ft/s [note that s_1 is the point of

      release of the baseball]. $m = \frac{5}{512}$, so the work done is $W = \tfrac{1}{2}mv_1^2 - \tfrac{1}{2}mv_0^2 = \tfrac{1}{2} \cdot \frac{5}{512} \cdot (132)^2 \approx 85$ ft-lb.

3. (a) Here we have a differential equation of the form $dv/dt = kv$, so by Theorem 10.4.2, the solution is $v(t) = v(0)e^{kt}$.

    In this case $k = -\frac{1}{10}$ and $v(0) = 100$ ft/s, so $v(t) = 100e^{-t/10}$. We are interested in the time t that the ball takes to travel

    280 ft, so we find the distance function

$$s(t) = \int_0^t v(x)\,dx = \int_0^t 100e^{-x/10}\,dx = 100\Big[-10e^{-x/10}\Big]_0^t = -1000(e^{-t/10} - 1) = 1000(1 - e^{-t/10})$$

    Now we set $s(t) = 280$ and solve for t: $280 = 1000(1 - e^{-t/10}) \;\Rightarrow\; 1 - e^{-t/10} = \frac{7}{25} \;\Rightarrow$

    $-\frac{1}{10}t = \ln\big(1 - \frac{7}{25}\big) \;\Rightarrow\; t \approx 3.285$ seconds.

   (b) Let x be the distance of the shortstop from home plate. We calculate the time for the ball to reach home plate as a function

      of x, then differentiate with respect to x to find the value of x which corresponds to the minimum time. The total time that

      it takes the ball to reach home is the sum of the times of the two throws, plus the relay time $\big(\frac{1}{2}$ s$\big)$. The distance from the

      fielder to the shortstop is $280 - x$, so to find the time t_1 taken by the first throw, we solve the equation

$s_1(t_1) = 280 - x \iff 1 - e^{-t_1/10} = \dfrac{280 - x}{1000} \iff t_1 = -10\ln\dfrac{720 + x}{1000}$. We find the time t_2 taken by the second

throw if the shortstop throws with velocity w, since we see that this velocity varies in the rest of the problem. We use

$v = we^{-t/10}$ and isolate t_2 in the equation $s(t_2) = 10w(1 - e^{-t_2/10}) = x \iff e^{-t_2/10} = 1 - \dfrac{x}{10w} \iff$

$t_2 = -10\ln\dfrac{10w - x}{10w}$, so the total time is $t_w(x) = \dfrac{1}{2} - 10\left[\ln\dfrac{720 + x}{1000} + \ln\dfrac{10w - x}{10w}\right]$.

To find the minimum, we differentiate: $\dfrac{dt_w}{dx} = -10\left[\dfrac{1}{720 + x} - \dfrac{1}{10w - x}\right]$, which changes from negative to positive

when $720 + x = 10w - x \iff x = 5w - 360$. By the First Derivative Test, t_w has a minimum at this distance from the

shortstop to home plate. So if the shortstop throws at $w = 105$ ft/s from a point $x = 5(105) - 360 = 165$ ft from home

plate, the minimum time is $t_{105}(165) = \frac{1}{2} - 10\left(\ln\frac{720 + 165}{1000} + \ln\frac{1050 - 165}{1050}\right) \approx 3.431$ seconds. This is longer than the

time taken in part (a), so in this case the manager should encourage a direct throw. If $w = 115$ ft/s, then $x = 215$ ft from

home, and the minimum time is $t_{115}(215) = \frac{1}{2} - 10\left(\ln\frac{720 + 215}{1000} + \ln\frac{1150 - 215}{1150}\right) \approx 3.242$ seconds. This is less than the

time taken in part (a), so in this case, the manager should encourage a relayed throw.

(c) In general, the minimum time is $t_w(5w - 360) = \dfrac{1}{2} - 10\left[\ln\dfrac{360 + 5w}{1000} + \ln\dfrac{360 + 5w}{10w}\right] = \dfrac{1}{2} - 10\ln\dfrac{(w + 72)^2}{400w}$.

We want to find out when this is about 3.285 seconds, the same time as the

direct throw. From the graph, we estimate that this is the case for

$w \approx 112.8$ ft/s. So if the shortstop can throw the ball with this velocity,

then a relayed throw takes the same time as a direct throw.

10.5 Linear Equations

1. $y' + \cos x = y \implies y' + (-1)y = -\cos x$ is linear since it can be put into the standard linear form (1),
$y' + P(x)\,y = Q(x)$.

2. $y' + \cos y = \tan x$ is not linear since it cannot be put into the standard linear form (1), $y' + P(x)\,y = Q(x)$.
$[\cos y$ is not of the form $P(x)\,y.]$

3. $yy' + xy = x^2 \implies y' + x = x^2/y \implies y' - x^2/y = -x$ is not linear since it cannot be put into the standard linear
form (1), $y' + P(x)\,y = Q(x)$.

4. $xy + \sqrt{x} = e^x y' \implies y' = xye^{-x} + \sqrt{x}\,e^{-x} \implies y' + (-xe^{-x})y = \sqrt{x}\,e^{-x}$ is linear since it can be put into the
standard linear form (1), $y' + P(x)\,y = Q(x)$.

5. Comparing the given equation, $y' + 2y = 2e^x$, with the general form, $y' + P(x)y = Q(x)$, we see that $P(x) = 2$ and the

integrating factor is $I(x) = e^{\int P(x)\,dx} = e^{\int 2\,dx} = e^{2x}$. Multiplying the differential equation by $I(x)$ gives

$e^{2x}y' + 2e^{2x}y = 2e^{3x} \implies (e^{2x}y)' = 2e^{3x} \implies e^{2x}y = \int 2e^{3x}\,dx \implies e^{2x}y = \frac{2}{3}e^{3x} + C \implies y = \frac{2}{3}e^x + Ce^{-2x}$.

6. $y' = x + 5y$ $\Rightarrow$ $y' - 5y = x$. $I(x) = e^{\int P(x)\,dx} = e^{\int(-5)\,dx} = e^{-5x}$. Multiplying the differential equation by $I(x)$

gives $e^{-5x}y' - 5e^{-5x}y = xe^{-5x}$ $\Rightarrow$ $(e^{-5x}y)' = xe^{-5x}$ $\Rightarrow$ $e^{-5x}y = \int xe^{-5x}\,dx = -\tfrac{1}{5}xe^{-5x} - \tfrac{1}{25}e^{-5x} + C$

[by parts] $\Rightarrow$ $y = -\tfrac{1}{5}x - \tfrac{1}{25} + Ce^{5x}$.

7. $xy' - 2y = x^2$ [divide by x] $\Rightarrow$ $y' + \left(-\dfrac{2}{x}\right)y = x$ $(\star)$.

$I(x) = e^{\int P(x)\,dx} = e^{\int(-2/x)\,dx} = e^{-2\ln|x|} = e^{\ln|x|^{-2}} = e^{\ln(1/x^2)} = 1/x^2$. Multiplying the differential equation $(\star)$

by $I(x)$ gives $\dfrac{1}{x^2}y' - \dfrac{2}{x^3}y = \dfrac{1}{x}$ $\Rightarrow$ $\left(\dfrac{1}{x^2}y\right)' = \dfrac{1}{x}$ $\Rightarrow$ $\dfrac{1}{x^2}y = \ln|x| + C$ $\Rightarrow$

$y = x^2(\ln|x| + C) = x^2\ln|x| + Cx^2$.

8. $x^2y' + 2xy = \cos^2 x$ $\Rightarrow$ $y' + \dfrac{2}{x}y = \dfrac{\cos^2 x}{x^2}$. $I(x) = e^{\int P(x)\,dx} = e^{\int 2/x\,dx} = e^{2\ln|x|} = e^{\ln(x^2)} = x^2$.

Multiplying by $I(x)$ gives us our original equation back. You may have noticed this immediately, since $P(x)$ is the derivative

of the coefficient of y'. We rewrite it as $(x^2y)' = \cos^2 x$. Thus,

$x^2y = \int \cos^2 x\,dx = \int \tfrac{1}{2}(1 + \cos 2x)\,dx = \tfrac{1}{2}x + \tfrac{1}{4}\sin 2x + C$ $\Rightarrow$

$y = \dfrac{1}{2x} + \dfrac{1}{4x^2}\sin 2x + \dfrac{C}{x^2}$ or $y = \dfrac{1}{2x} + \dfrac{1}{2x^2}\sin x\cos x + \dfrac{C}{x^2}$.

9. Since $P(x)$ is the derivative of the coefficient of y' [$P(x) = 1$ and the coefficient is x], we can write the differential equation

$xy' + y = \sqrt{x}$ in the easily integrable form $(xy)' = \sqrt{x}$ $\Rightarrow$ $xy = \tfrac{2}{3}x^{3/2} + C$ $\Rightarrow$ $y = \tfrac{2}{3}\sqrt{x} + C/x$.

10. $y' + y = \sin(e^x)$, so $I(x) = e^{\int P(x)\,dx} = e^{\int 1\,dx} = e^x$. Multiplying the differential equation by $I(x)$ gives

$e^xy' + e^xy = e^x\sin(e^x)$ $\Rightarrow$ $(e^xy)' = e^x\sin(e^x)$ $\Rightarrow$ $e^xy = \int e^x\sin(e^x)\,dx$ $\Rightarrow$ $e^xy = -\cos(e^x) + C$ $\Rightarrow$

$y = -e^{-x}\cos(e^x) + Ce^{-x}$.

11. $\sin x\dfrac{dy}{dx} + (\cos x)y = \sin(x^2)$ $\Rightarrow$ $[(\sin x)y]' = \sin(x^2)$ $\Rightarrow$ $(\sin x)y = \int\sin(x^2)\,dx$ $\Rightarrow$ $y = \dfrac{\int\sin(x^2)\,dx + C}{\sin x}$.

12. $x\dfrac{dy}{dx} - 4y = x^4e^x$ $\Rightarrow$ $y' - \dfrac{4}{x}y = x^3e^x$.

$I(x) = e^{\int P(x)\,dx} = e^{\int(-4/x)\,dx} = e^{-4\ln|x|} = \left(e^{\ln|x|}\right)^{-4} = |x|^{-4} = x^{-4}$. Multiplying the differential equation by $I(x)$

gives $x^{-4}y' - 4x^{-5}y = x^{-1}e^x$ $\Rightarrow$ $(x^{-4}y)' = e^x/x$ $\Rightarrow$ $x^{-4}y = \int(e^x/x)\,dx$ $\Rightarrow$ $y = \left(\int(e^x/x)\,dx + C\right)x^4$.

13. $(1+t)\dfrac{du}{dt} + u = 1 + t$, $t > 0$ [divide by $1+t$] $\Rightarrow$ $\dfrac{du}{dt} + \dfrac{1}{1+t}u = 1$ $(\star)$, which has the

form $u' + P(t)u = Q(t)$. The integrating factor is $I(t) = e^{\int P(t)\,dt} = e^{\int[1/(1+t)]\,dt} = e^{\ln(1+t)} = 1 + t$.

Multiplying $(\star)$ by $I(t)$ gives us our original equation back. We rewrite it as $[(1+t)u]' = 1 + t$. Thus,

$(1+t)u = \int(1+t)\,dt = t + \tfrac{1}{2}t^2 + C$ $\Rightarrow$ $u = \dfrac{t + \tfrac{1}{2}t^2 + C}{1+t}$ or $u = \dfrac{t^2 + 2t + 2C}{2(t+1)}$.

14. $t \ln t \dfrac{dr}{dt} + r = te^t$ $\Rightarrow$ $\dfrac{dr}{dt} + \dfrac{1}{t \ln t} r = \dfrac{e^t}{\ln t}$. $I(t) = e^{\int dt/(t \ln t)} = e^{\ln(\ln t)} = \ln t$. Multiplying by $\ln t$ gives

$$\ln t \dfrac{dr}{dt} + \dfrac{1}{t} r = e^t \quad \Rightarrow \quad [(\ln t)r]' = e^t \quad \Rightarrow \quad (\ln t)r = e^t + C \quad \Rightarrow \quad r = \dfrac{e^t + C}{\ln t}.$$

15. $y' = x + y$ $\Rightarrow$ $y' + (-1)y = x$. $I(x) = e^{\int(-1)\,dx} = e^{-x}$. Multiplying by e^{-x} gives $e^{-x}y' - e^{-x}y = xe^{-x}$ $\Rightarrow$

$(e^{-x}y)' = xe^{-x}$ $\Rightarrow$ $e^{-x}y = \int xe^{-x}\,dx = -xe^{-x} - e^{-x} + C$ [integration by parts with $u = x$, $dv = e^{-x}\,dx$] $\Rightarrow$

$y = -x - 1 + Ce^x$. $y(0) = 2$ $\Rightarrow$ $-1 + C = 2$ $\Rightarrow$ $C = 3$, so $y = -x - 1 + 3e^x$.

16. $t \dfrac{dy}{dt} + 2y = t^3$, $t > 0$, $y(1) = 0$. Divide by t to get $\dfrac{dy}{dt} + \dfrac{2}{t} y = t^2$, which is linear. $I(t) = e^{\int(2/t)\,dt} = e^{2 \ln t} = t^2$.

Multiplying by t^2 gives $t^2 \dfrac{dy}{dt} + 2ty = t^4$ $\Rightarrow$ $(t^2 y)' = t^4$ $\Rightarrow$ $t^2 y = \frac{1}{5}t^5 + C$ $\Rightarrow$ $y = \dfrac{t^3}{5} + \dfrac{C}{t^2}$. Thus,

$0 = y(1) = \frac{1}{5} + C$ $\Rightarrow$ $C = -\frac{1}{5}$, so $y = \dfrac{t^3}{5} - \dfrac{1}{5t^2}$.

17. $\dfrac{dv}{dt} - 2tv = 3t^2 e^{t^2}$, $v(0) = 5$. $I(t) = e^{\int(-2t)\,dt} = e^{-t^2}$. Multiply the differential equation by $I(t)$ to get

$e^{-t^2} \dfrac{dv}{dt} - 2te^{-t^2}v = 3t^2$ $\Rightarrow$ $\left(e^{-t^2}v\right)' = 3t^2$ $\Rightarrow$ $e^{-t^2}v = \int 3t^2\,dt = t^3 + C$ $\Rightarrow$ $v = t^3 e^{t^2} + Ce^{t^2}$.

$5 = v(0) = 0 \cdot 1 + C \cdot 1 = C$, so $v = t^3 e^{t^2} + 5e^{t^2}$.

18. $2xy' + y = 6x$, $x > 0$ $\Rightarrow$ $y' + \dfrac{1}{2x} y = 3$. $I(x) = e^{\int 1/(2x)\,dx} = e^{(1/2) \ln x} = e^{\ln x^{1/2}} = \sqrt{x}$. Multiplying by $\sqrt{x}$

gives $\sqrt{x}\, y' + \dfrac{1}{2\sqrt{x}} y = 3\sqrt{x}$ $\Rightarrow$ $(\sqrt{x}\, y)' = 3\sqrt{x}$ $\Rightarrow$ $\sqrt{x}\, y = \int 3\sqrt{x}\,dx = 2x^{3/2} + C$ $\Rightarrow$ $y = 2x + \dfrac{C}{\sqrt{x}}$.

$y(4) = 20$ $\Rightarrow$ $8 + \dfrac{C}{2} = 20$ $\Rightarrow$ $C = 24$, so $y = 2x + \dfrac{24}{\sqrt{x}}$.

19. $xy' = y + x^2 \sin x$ $\Rightarrow$ $y' - \dfrac{1}{x} y = x \sin x$. $I(x) = e^{\int(-1/x)\,dx} = e^{-\ln x} = e^{\ln x^{-1}} = \dfrac{1}{x}$.

Multiplying by $\dfrac{1}{x}$ gives $\dfrac{1}{x} y' - \dfrac{1}{x^2} y = \sin x$ $\Rightarrow$ $\left(\dfrac{1}{x} y\right)' = \sin x$ $\Rightarrow$ $\dfrac{1}{x} y = -\cos x + C$ $\Rightarrow$ $y = -x \cos x + Cx$.

$y(\pi) = 0$ $\Rightarrow$ $-\pi \cdot (-1) + C\pi = 0$ $\Rightarrow$ $C = -1$, so $y = -x \cos x - x$.

20. $(x^2 + 1)\dfrac{dy}{dx} + 3x(y - 1) = 0$ $\Rightarrow$ $(x^2 + 1)y' + 3xy = 3x$ $\Rightarrow$ $y' + \dfrac{3x}{x^2 + 1} y = \dfrac{3x}{x^2 + 1}$.

$I(x) = e^{\int 3x/(x^2+1)\,dx} = e^{(3/2) \ln|x^2+1|} = \left(e^{\ln(x^2+1)}\right)^{3/2} = (x^2 + 1)^{3/2}$. Multiplying by $(x^2 + 1)^{3/2}$ gives

$(x^2 + 1)^{3/2} y' + 3x(x^2 + 1)^{1/2} y = 3x(x^2 + 1)^{1/2}$ $\Rightarrow$ $\left[(x^2 + 1)^{3/2}y\right]' = 3x(x^2 + 1)^{1/2}$ $\Rightarrow$

$(x^2 + 1)^{3/2} y = \int 3x(x^2 + 1)^{1/2}\,dx = (x^2 + 1)^{3/2} + C$ $\Rightarrow$ $y = 1 + C(x^2 + 1)^{-3/2}$. Since $y(0) = 2$, we have

$2 = 1 + C(1)$ $\Rightarrow$ $C = 1$ and hence, $y = 1 + (x^2 + 1)^{-3/2}$.

21. $xy' + 2y = e^x$ $\Rightarrow$ $y' + \dfrac{2}{x}\,y = \dfrac{e^x}{x}$.

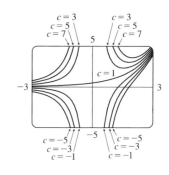

$I(x) = e^{\int (2/x)\,dx} = e^{2\ln|x|} = \left(e^{\ln|x|}\right)^2 = |x|^2 = x^2$.

Multiplying by $I(x)$ gives $x^2\,y' + 2xy = xe^x$ $\Rightarrow$ $(x^2 y)' = xe^x$ $\Rightarrow$

$x^2 y = \int xe^x\,dx = (x-1)e^x + C$ [by parts] $\Rightarrow$

$y = [(x-1)e^x + C]/x^2$. The graphs for $C = -5, -3, -1, 1, 3, 5$, and 7 are

shown. $C = 1$ is a transitional value. For $C < 1$, there is an inflection point and

for $C > 1$, there is a local minimum. As $|C|$ gets larger, the "branches" get

further from the origin.

22. $I(x) = e^{\int \cos x\,dx} = e^{\sin x}$. Multiplying the differential equation by $I(x)$ gives

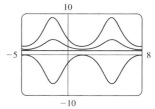

$e^{\sin x} y' + \cos x \cdot e^{\sin x}\,y = \cos x \cdot e^{\sin x}$ $\Rightarrow$ $(e^{\sin x} y)' = \cos x \cdot e^{\sin x}$ $\Rightarrow$

$y = e^{-\sin x}\left[\int \cos x \cdot e^{\sin x}\,dx + C\right] = 1 + Ce^{-\sin x}$. The graphs for $C = -3$,

$0, 1$, and 3 are shown. As the values of C get further from zero the graph is

stretched away from the line $y = 1$, which is the value for $C = 0$. The graphs are

all periodic in x, with a period of 2π.

23. Setting $u = y^{1-n}$, $\dfrac{du}{dx} = (1-n)\,y^{-n}\dfrac{dy}{dx}$ or $\dfrac{dy}{dx} = \dfrac{y^n}{1-n}\dfrac{du}{dx} = \dfrac{u^{n/(1-n)}}{1-n}\dfrac{du}{dx}$. Then the Bernoulli differential equation

becomes $\dfrac{u^{n/(1-n)}}{1-n}\dfrac{du}{dx} + P(x)u^{1/(1-n)} = Q(x)u^{n/(1-n)}$ or $\dfrac{du}{dx} + (1-n)P(x)u = Q(x)(1-n)$.

24. Here $xy' + y = -xy^2$ $\Rightarrow$ $y' + \dfrac{y}{x} = -y^2$, so $n = 2$, $P(x) = \dfrac{1}{x}$ and $Q(x) = -1$. Setting $u = y^{-1}$, u satisfies

$u' - \dfrac{1}{x}u = 1$. Then $I(x) = e^{\int(-1/x)dx} = \dfrac{1}{x}$ (for $x > 0$) and $u = x\left(\int \dfrac{1}{x}\,dx + C\right) = x(\ln|x| + C)$. Thus,

$y = \dfrac{1}{x(C + \ln|x|)}$.

25. Here $y' + \dfrac{2}{x}y = \dfrac{y^3}{x^2}$, so $n = 3$, $P(x) = \dfrac{2}{x}$ and $Q(x) = \dfrac{1}{x^2}$. Setting $u = y^{-2}$, u satisfies $u' - \dfrac{4u}{x} = -\dfrac{2}{x^2}$.

Then $I(x) = e^{\int(-4/x)\,dx} = x^{-4}$ and $u = x^4\left(\int -\dfrac{2}{x^6}\,dx + C\right) = x^4\left(\dfrac{2}{5x^5} + C\right) = Cx^4 + \dfrac{2}{5x}$.

Thus, $y = \pm\left(Cx^4 + \dfrac{2}{5x}\right)^{-1/2}$.

26. $xy'' + 2y' = 12x^2$ and $u = y'$ $\Rightarrow$ $xu' + 2u = 12x^2$ $\Rightarrow$ $u' + \dfrac{2}{x}u = 12x$.

$I(x) = e^{\int(2/x)\,dx} = e^{2\ln|x|} = \left(e^{\ln|x|}\right)^2 = |x|^2 = x^2$. Multiplying the last differential equation by x^2 gives

$x^2 u' + 2xu = 12x^3$ $\Rightarrow$ $(x^2 u)' = 12x^3$ $\Rightarrow$ $x^2 u = \int 12x^3\,dx = 3x^4 + C$ $\Rightarrow$ $u = 3x^2 + C/x^2$ $\Rightarrow$

$y' = 3x^2 + C/x^2$ $\Rightarrow$ $y = x^3 - C/x + D$.

27. (a) $2\dfrac{dI}{dt} + 10I = 40$ or $\dfrac{dI}{dt} + 5I = 20$. Then the integrating factor is $e^{\int 5\,dt} = e^{5t}$. Multiplying the differential equation

by the integrating factor gives $e^{5t}\dfrac{dI}{dt} + 5Ie^{5t} = 20e^{5t} \quad \Rightarrow \quad (e^{5t}I)' = 20e^{5t} \quad \Rightarrow$

$I(t) = e^{-5t}\left[\int 20e^{5t}\,dt + C\right] = 4 + Ce^{-5t}$. But $0 = I(0) = 4 + C$, so $I(t) = 4 - 4e^{-5t}$.

(b) $I(0.1) = 4 - 4e^{-0.5} \approx 1.57$ A

28. (a) $\dfrac{dI}{dt} + 20I = 40\sin 60t$, so the integrating factor is e^{20t}. Multiplying the differential equation by the integrating factor

gives $e^{20t}\dfrac{dI}{dt} + 20Ie^{20t} = 40e^{20t}\sin 60t \quad \Rightarrow \quad (e^{20t}I)' = 40e^{20t}\sin 60t \quad \Rightarrow$

$$I(t) = e^{-20t}\left[\int 40e^{20t}\sin 60t\,dt + C\right] = e^{-20t}\left[40e^{20t}\left(\tfrac{1}{4000}\right)(20\sin 60t - 60\cos 60t)\right] + Ce^{-20t}$$

$$= \dfrac{\sin 60t - 3\cos 60t}{5} + Ce^{-20t}$$

But $1 = I(0) = -\tfrac{3}{5} + C$, so $I(t) = \dfrac{\sin 60t - 3\cos 60t + 8e^{-20t}}{5}$.

(b) $I(0.1) = \dfrac{\sin 6 - 3\cos 6 + 8e^{-2}}{5} \approx -0.42$ A

(c)

29. $5\dfrac{dQ}{dt} + 20Q = 60$ with $Q(0) = 0$ C. Then the integrating factor is $e^{\int 4\,dt} = e^{4t}$, and multiplying the differential

equation by the integrating factor gives $e^{4t}\dfrac{dQ}{dt} + 4e^{4t}Q = 12e^{4t} \quad \Rightarrow \quad (e^{4t}Q)' = 12e^{4t} \quad \Rightarrow$

$Q(t) = e^{-4t}\left[\int 12e^{4t}\,dt + C\right] = 3 + Ce^{-4t}$. But $0 = Q(0) = 3 + C$ so $Q(t) = 3(1 - e^{-4t})$ is the charge at time t

and $I = dQ/dt = 12e^{-4t}$ is the current at time t.

30. $2\dfrac{dQ}{dt} + 100Q = 10\sin 60t$ or $\dfrac{dQ}{dt} + 50Q = 5\sin 60t$. Then the integrating factor is $e^{\int 50\,dt} = e^{50t}$, and multiplying the

differential equation by the integrating factor gives $e^{50t}\dfrac{dQ}{dt} + 50e^{50t}Q = 5e^{50t}\sin 60t \quad \Rightarrow \quad (e^{50t}Q)' = 5e^{50t}\sin 60t \quad \Rightarrow$

$$Q(t) = e^{-50t}\left[\int 5e^{50t}\sin 60t\,dt + C\right] = e^{-50t}\left[5e^{50t}\left(\tfrac{1}{6100}\right)(50\sin 60t - 60\cos 60t)\right] + Ce^{-50t}$$

$$= \tfrac{1}{122}(5\sin 60t - 6\cos 60t) + Ce^{-50t}$$

But $0 = Q(0) = -\tfrac{6}{122} + C$ so $C = \tfrac{3}{61}$ and $Q(t) = \dfrac{5\sin 60t - 6\cos 60t}{122} + \dfrac{3e^{-50t}}{61}$ is the charge at time t, while the current

is $I(t) = \dfrac{dQ}{dt} = \dfrac{150\cos 60t + 180\sin 60t - 150e^{-50t}}{61}$.

31. $\dfrac{dP}{dt} + kP = kM$, so $I(t) = e^{\int k\, dt} = e^{kt}$. Multiplying the differential equation

by $I(t)$ gives $e^{kt}\dfrac{dP}{dt} + kPe^{kt} = kMe^{kt} \;\;\Rightarrow\;\; (e^{kt}P)' = kMe^{kt} \;\;\Rightarrow$

$P(t) = e^{-kt}\left(\int kMe^{kt}\, dt + C\right) = M + Ce^{-kt}, \; k > 0.$ Furthermore, it is

reasonable to assume that $0 \le P(0) \le M$, so $-M \le C \le 0$.

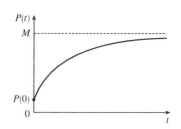

32. Since $P(0) = 0$, we have $P(t) = M(1 - e^{-kt})$. If $P_1(t)$ is Jim's learning curve, then $P_1(1) = 25$ and $P_1(2) = 45$. Hence,

$25 = M_1(1 - e^{-k})$ and $45 = M_1(1 - e^{-2k})$, so $1 - 25/M_1 = e^{-k}$ or $k = -\ln\left(1 - \dfrac{25}{M_1}\right) = \ln\left(\dfrac{M_1}{M_1 - 25}\right)$. But

$45 = M_1(1 - e^{-2k})$ so $45 = M_1\left[1 - \left(\dfrac{M_1 - 25}{M_1}\right)^2\right]$ or $45 = \dfrac{50M_1 - 625}{M_1}$. Thus, $M_1 = 125$ is the maximum number of

units per hour Jim is capable of processing. Similarly, if $P_2(t)$ is Mark's learning curve, then $P_2(1) = 35$ and $P_2(2) = 50$.

So $k = \ln\left(\dfrac{M_2}{M_2 - 35}\right)$ and $50 = M_2\left[1 - \left(\dfrac{M_2 - 35}{M_2}\right)^2\right]$ or $M_2 = 61.25$. Hence the maximum number of units per hour

for Mark is approximately 61. Another approach would be to use the midpoints of the intervals so that $P_1(0.5) = 25$ and

$P_1(1.5) = 45$. Doing so gives us $M_1 \approx 52.6$ and $M_2 \approx 51.8$.

33. $y(0) = 0$ kg.   Salt is added at a rate of $\left(0.4\,\dfrac{\text{kg}}{\text{L}}\right)\left(5\,\dfrac{\text{L}}{\text{min}}\right) = 2\,\dfrac{\text{kg}}{\text{min}}$. Since solution is drained from the tank at a rate of

3 L/min, but salt solution is added at a rate of 5 L/min, the tank, which starts out with 100 L of water, contains $(100 + 2t)$ L

of liquid after t min. Thus, the salt concentration at time t is $\dfrac{y(t)}{100 + 2t}\,\dfrac{\text{kg}}{\text{L}}$. Salt therefore leaves the tank at a rate of

$\left(\dfrac{y(t)}{100 + 2t}\,\dfrac{\text{kg}}{\text{L}}\right)\left(3\,\dfrac{\text{L}}{\text{min}}\right) = \dfrac{3y}{100 + 2t}\,\dfrac{\text{kg}}{\text{min}}$. Combining the rates at which salt enters and leaves the tank, we get

$\dfrac{dy}{dt} = 2 - \dfrac{3y}{100 + 2t}$. Rewriting this equation as $\dfrac{dy}{dt} + \left(\dfrac{3}{100 + 2t}\right)y = 2$, we see that it is linear.

$$I(t) = \exp\left(\int \frac{3\, dt}{100 + 2t}\right) = \exp\left(\tfrac{3}{2}\ln(100 + 2t)\right) = (100 + 2t)^{3/2}$$

Multiplying the differential equation by $I(t)$ gives $(100 + 2t)^{3/2}\dfrac{dy}{dt} + 3(100 + 2t)^{1/2}y = 2(100 + 2t)^{3/2} \;\;\Rightarrow$

$[(100 + 2t)^{3/2}y]' = 2(100 + 2t)^{3/2} \;\;\Rightarrow\;\; (100 + 2t)^{3/2}y = \tfrac{2}{5}(100 + 2t)^{5/2} + C \;\;\Rightarrow$

$y = \tfrac{2}{5}(100 + 2t) + C(100 + 2t)^{-3/2}$. Now $0 = y(0) = \tfrac{2}{5}(100) + C \cdot 100^{-3/2} = 40 + \tfrac{1}{1000}C \;\;\Rightarrow\;\; C = -40{,}000$, so

$y = \left[\tfrac{2}{5}(100 + 2t) - 40{,}000(100 + 2t)^{-3/2}\right]$ kg. From this solution (no pun intended), we calculate the salt concentration

at time t to be $C(t) = \dfrac{y(t)}{100 + 2t} = \left[\dfrac{-40{,}000}{(100 + 2t)^{5/2}} + \dfrac{2}{5}\right]\dfrac{\text{kg}}{\text{L}}$. In particular, $C(20) = \dfrac{-40{,}000}{140^{5/2}} + \dfrac{2}{5} \approx 0.2275\,\dfrac{\text{kg}}{\text{L}}$

and $y(20) = \tfrac{2}{5}(140) - 40{,}000(140)^{-3/2} \approx 31.85$ kg.

34. Let $y(t)$ denote the amount of chlorine in the tank at time t (in seconds). $y(0) = (0.05 \text{ g/L}) (400 \text{ L}) = 20$ g. The amount of

liquid in the tank at time t is $(400 - 6t)$ L since 4 L of water enters the tank each second and 10 L of liquid leaves the tank

each second. Thus, the concentration of chlorine at time t is $\dfrac{y(t)}{400 - 6t} \dfrac{\text{g}}{\text{L}}$. Chlorine doesn't enter the tank, but it leaves at a rate

of $\left[\dfrac{y(t)}{400 - 6t} \dfrac{\text{g}}{\text{L}}\right]\left[10 \dfrac{\text{L}}{\text{s}}\right] = \dfrac{10\, y(t)}{400 - 6t} \dfrac{\text{g}}{\text{s}} = \dfrac{5\, y(t)}{200 - 3t} \dfrac{\text{g}}{\text{s}}$. Therefore, $\dfrac{dy}{dt} = -\dfrac{5y}{200 - 3t} \Rightarrow \displaystyle\int \dfrac{dy}{y} = \int \dfrac{-5\,dt}{200 - 3t} \Rightarrow$

$\ln y = \tfrac{5}{3}\ln(200 - 3t) + C \Rightarrow y = \exp\left(\tfrac{5}{3}\ln(200 - 3t) + C\right) = e^C(200 - 3t)^{5/3}$. Now $20 = y(0) = e^C \cdot 200^{5/3} \Rightarrow$

$e^C = \dfrac{20}{200^{5/3}}$, so $y(t) = 20\dfrac{(200 - 3t)^{5/3}}{200^{5/3}} = 20(1 - 0.015t)^{5/3}$ g for $0 \le t \le 66\tfrac{2}{3}$ s, at which time the tank is empty.

35. (a) $\dfrac{dv}{dt} + \dfrac{c}{m}v = g$ and $I(t) = e^{\int (c/m)\,dt} = e^{(c/m)t}$, and multiplying the differential equation by

$I(t)$ gives $e^{(c/m)t}\dfrac{dv}{dt} + \dfrac{vce^{(c/m)t}}{m} = ge^{(c/m)t} \Rightarrow \left[e^{(c/m)t}v\right]' = ge^{(c/m)t}$. Hence,

$v(t) = e^{-(c/m)t}\left[\displaystyle\int ge^{(c/m)t}\,dt + K\right] = mg/c + Ke^{-(c/m)t}$. But the object is dropped from rest, so $v(0) = 0$ and

$K = -mg/c$. Thus, the velocity at time t is $v(t) = (mg/c)[1 - e^{-(c/m)t}]$.

(b) $\lim\limits_{t \to \infty} v(t) = mg/c$

(c) $s(t) = \int v(t)\,dt = (mg/c)[t + (m/c)e^{-(c/m)t}] + c_1$ where $c_1 = s(0) - m^2g/c^2$.

$s(0)$ is the initial position, so $s(0) = 0$ and $s(t) = (mg/c)[t + (m/c)e^{-(c/m)t}] - m^2g/c^2$.

36. $v = (mg/c)(1 - e^{-ct/m}) \Rightarrow$

$$\dfrac{dv}{dm} = \dfrac{mg}{c}\left(0 - e^{-ct/m} \cdot \dfrac{ct}{m^2}\right) + \dfrac{g}{c}(1 - e^{-ct/m}) \cdot 1 = -\dfrac{gt}{m}e^{-ct/m} + \dfrac{g}{c} - \dfrac{g}{c}e^{-ct/m}$$

$$= \dfrac{g}{c}\left(1 - e^{-ct/m} - \dfrac{ct}{m}e^{-ct/m}\right) \Rightarrow$$

$\dfrac{c}{g}\dfrac{dv}{dm} = 1 - \left(1 + \dfrac{ct}{m}\right)e^{-ct/m} = 1 - \dfrac{1 + ct/m}{e^{ct/m}} = 1 - \dfrac{1 + Q}{e^Q}$, where $Q = \dfrac{ct}{m} \ge 0$. Since $e^Q > 1 + Q$ for all $Q > 0$,

it follows that $dv/dm > 0$ for $t > 0$. In other words, for all $t > 0$, v increases as m increases.

10.6 Predator-Prey Systems

1. (a) $dx/dt = -0.05x + 0.0001xy$. If $y = 0$, we have $dx/dt = -0.05x$, which indicates that in the absence of y, x declines at

a rate proportional to itself. So x represents the predator population and y represents the prey population. The growth of

the prey population, $0.1y$ (from $dy/dt = 0.1y - 0.005xy$), is restricted only by encounters with predators (the term

$-0.005xy$). The predator population increases only through the term $0.0001xy$; that is, by encounters with the prey and

not through additional food sources.

(b) $dy/dt = -0.015y + 0.00008xy$. If $x = 0$, we have $dy/dt = -0.015y$, which indicates that in the absence of x, y would

decline at a rate proportional to itself. So y represents the predator population and x represents the prey population. The

growth of the prey population, $0.2x$ (from $dx/dt = 0.2x - 0.0002x^2 - 0.006xy = 0.2x(1 - 0.001x) - 0.006xy$), is restricted by a carrying capacity of 1000 [from the term $1 - 0.001x = 1 - x/1000$] and by encounters with predators (the term $-0.006xy$). The predator population increases only through the term $0.00008xy$; that is, by encounters with the prey and not through additional food sources.

2. (a) $dx/dt = 0.12x - 0.0006x^2 + 0.00001xy$. $dy/dt = 0.08y + 0.00004xy$.

The xy terms represent encounters between the two species x and y. An increase in y makes dx/dt (the growth rate of x) larger due to the positive term $0.00001xy$. An increase in x makes dy/dt (the growth rate of y) larger due to the positive term $0.00004xy$. Hence, the system describes a cooperation model.

(b) $dx/dt = 0.15x - 0.0002x^2 - 0.0006xy = 0.15x(1 - x/750) - 0.0006xy$.

$dy/dt = 0.2y - 0.00008y^2 - 0.0002xy = 0.2y(1 - y/2500) - 0.0002xy$.

The system shows that x and y have carrying capacities of 750 and 2500. An increase in x reduces the growth rate of y due to the negative term $-0.0002xy$. An increase in y reduces the growth rate of x due to the negative term $-0.0006xy$. Hence, the system describes a competition model.

3. (a) At $t = 0$, there are about 300 rabbits and 100 foxes. At $t = t_1$, the number of foxes reaches a minimum of about 20 while the number of rabbits is about 1000. At $t = t_2$, the number of rabbits reaches a maximum of about 2400, while the number of foxes rebounds to 100. At $t = t_3$, the number of rabbits decreases to about 1000 and the number of foxes reaches a maximum of about 315. As t increases, the number of foxes decreases greatly to 100, and the number of rabbits decreases to 300 (the initial populations), and the cycle starts again.

(b)

4. (a) At $t = 0$, there are about 600 rabbits and 160 foxes. At $t = t_1$, the number of rabbits reaches a minimum of about 80 and the number of foxes is also 80. At $t = t_2$, the number of foxes reaches a minimum of about 25 while the number of rabbits rebounds to 1000. At $t = t_3$, the number of foxes has increased to 40 and the rabbit population has reached a maximum of about 1750. The curve ends at $t = t_4$, where the number of foxes has increased to 65 and the number of rabbits has decreased to about 950.

(b)

5.

6.

7. $\dfrac{dW}{dR} = \dfrac{-0.02W + 0.00002RW}{0.08R - 0.001RW}$ ⟺ $(0.08 - 0.001W)R\,dW = (-0.02 + 0.00002R)W\,dR$ ⟺

$\dfrac{0.08 - 0.001W}{W}\,dW = \dfrac{-0.02 + 0.00002R}{R}\,dR$ ⟺ $\displaystyle\int\left(\dfrac{0.08}{W} - 0.001\right)dW = \int\left(-\dfrac{0.02}{R} + 0.00002\right)dR$ ⟺

$0.08\ln|W| - 0.001W = -0.02\ln|R| + 0.00002R + K$ ⟺ $0.08\ln W + 0.02\ln R = 0.001W + 0.00002R + K$ ⟺

$\ln\left(W^{0.08}R^{0.02}\right) = 0.00002R + 0.001W + K$ ⟺ $W^{0.08}R^{0.02} = e^{0.00002R + 0.001W + K}$ ⟺

$R^{0.02}W^{0.08} = Ce^{0.00002R}e^{0.001W}$ ⟺ $\dfrac{R^{0.02}W^{0.08}}{e^{0.00002R}e^{0.001W}} = C$. In general, if $\dfrac{dy}{dx} = \dfrac{-ry + bxy}{kx - axy}$, then $C = \dfrac{x^r y^k}{e^{bx}e^{ay}}$.

8. (a) A and L are constant ⟹ $A' = 0$ and $L' = 0$ ⟹ $\left\{\begin{array}{l} 0 = 2A - 0.01AL \\ 0 = -0.5L + 0.0001AL \end{array}\right\}$ ⟹ $\left\{\begin{array}{l} 0 = A(2 - 0.01L) \\ 0 = L(-0.5 + 0.0001A) \end{array}\right.$

So either $A = L = 0$ or $L = \frac{2}{0.01} = 200$ and $A = \frac{0.5}{0.0001} = 5000$. The trivial solution $A = L = 0$ just says that if there aren't any aphids or ladybugs, then the populations will not change. The non-trivial solution, $L = 200$ and $A = 5000$, indicates the population sizes needed so that there are no changes in either the number of aphids or the number of ladybugs.

(b) $\dfrac{dL}{dA} = \dfrac{dL/dt}{dA/dt} = \dfrac{-0.5L + 0.0001AL}{2A - 0.01AL}$

(c) The solution curves (phase trajectories) are all closed curves that have the equilibrium point $(5000, 200)$ inside them.

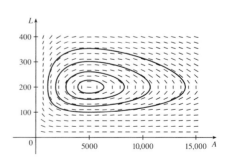

(d)

At $P_0(1000, 200)$, $dA/dt = 0$ and $dL/dt = -80 < 0$, so the number of ladybugs is decreasing and hence, we are proceeding in a counterclockwise direction. At P_0, there aren't enough aphids to support the ladybug population, so the number of ladybugs decreases and the number of aphids begins to increase. The ladybug population reaches a minimum at $P_1(5000, 100)$ while the aphid population increases in a dramatic way, reaching its maximum at $P_2(14, 250, 200)$.

Meanwhile, the ladybug population is increasing from P_1 to $P_3(5000, 355)$, and as we pass through P_2, the increasing number of ladybugs starts to deplete the aphid population. At P_3 the ladybugs reach a maximum population, and start to decrease due to the reduced aphid population. Both populations then decrease until P_0, where the cycle starts over again.

(e) Both graphs have the same period and the graph of L peaks about a quarter of a cycle after the graph of A.

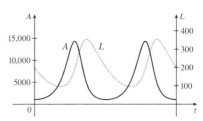

9. (a) Letting $W = 0$ gives us $dR/dt = 0.08R(1 - 0.0002R)$. $dR/dt = 0$ $\Leftrightarrow$ $R = 0$ or 5000. Since $dR/dt > 0$ for $0 < R < 5000$, we would expect the rabbit population to *increase* to 5000 for these values of R. Since $dR/dt < 0$ for $R > 5000$, we would expect the rabbit population to *decrease* to 5000 for these values of R. Hence, in the absence of wolves, we would expect the rabbit population to stabilize at 5000.

(b) R and W are constant $\Rightarrow$ $R' = 0$ and $W' = 0$ $\Rightarrow$

$$\begin{cases} 0 = 0.08R(1 - 0.0002R) - 0.001RW \\ 0 = -0.02W + 0.00002RW \end{cases} \Rightarrow \begin{cases} 0 = R[0.08(1 - 0.0002R) - 0.001W] \\ 0 = W(-0.02 + 0.00002R) \end{cases}$$

The second equation is true if $W = 0$ or $R = \frac{0.02}{0.00002} = 1000$. If $W = 0$ in the first equation, then either $R = 0$ or $R = \frac{1}{0.0002} = 5000$ [as in part (a)]. If $R = 1000$, then $0 = 1000[0.08(1 - 0.0002 \cdot 1000) - 0.001W]$ $\Leftrightarrow$ $0 = 80(1 - 0.2) - W$ $\Leftrightarrow$ $W = 64$.

Case (i): $W = 0$, $R = 0$: both populations are zero

Case (ii): $W = 0$, $R = 5000$: see part (a)

Case (iii): $R = 1000$, $W = 64$: the predator/prey interaction balances and the populations are stable.

(c) The populations of wolves and rabbits fluctuate around 64 and 1000, respectively, and eventually stabilize at those values.

(d)

10. (a) If $L = 0$, $dA/dt = 2A(1 - 0.0001A)$, so $dA/dt = 0 \Leftrightarrow A = 0$ or $A = \frac{1}{0.0001} = 10{,}000$. Since $dA/dt > 0$ for

$0 < A < 10{,}000$, we expect the aphid population to *increase* to 10,000 for these values of A. Since $dA/dt < 0$ for

$A > 10{,}000$, we expect the aphid population to *decrease* to 10,000 for these values of A. Hence, in the absence of

ladybugs we expect the aphid population to stabilize at 10,000.

(b) A and L are constant $\Rightarrow A' = 0$ and $L' = 0 \Rightarrow$

$$\begin{cases} 0 = 2A(1 - 0.0001A) - 0.01AL \\ 0 = -0.5L + 0.0001AL \end{cases} \Rightarrow \begin{cases} 0 = A[2(1 - 0.0001A) - 0.01L] \\ 0 = L(-0.5 + 0.0001A) \end{cases}$$

The second equation is true if $L = 0$ or $A = \frac{0.5}{0.0001} = 5000$. If $L = 0$ in the first equation, then either $A = 0$ or

$A = \frac{1}{0.0001} = 10{,}000$. If $A = 5000$, then $0 = 5000[2(1 - 0.0001 \cdot 5000) - 0.01L] \Leftrightarrow$

$0 = 10{,}000(1 - 0.5) - 50L \Leftrightarrow 50L = 5000 \Leftrightarrow L = 100$.

The equilibrium solutions are: (i) $L = 0, A = 0$ (ii) $L = 0, A = 10{,}000$ (iii) $A = 5000, L = 100$

(c) $\dfrac{dL}{dA} = \dfrac{dL/dt}{dA/dt} = \dfrac{-0.5L + 0.0001AL}{2A(1 - 0.0001A) - 0.01AL}$

(d)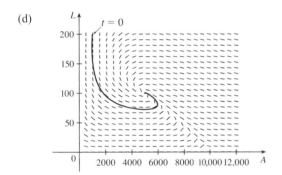

All of the phase trajectories spiral tightly around the equilibrium solution $(5000, 100)$.

(e)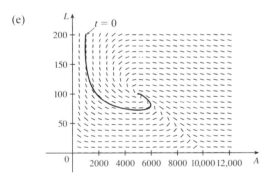

At $t = 0$, the ladybug population decreases rapidly and the aphid population decreases slightly before beginning to increase. As the aphid population continues to increase, the ladybug population reaches a minimum at about $(5000, 75)$. The ladybug population starts to increase and quickly stabilizes at 100, while the aphid population stabilizes at 5000.

(f)

The graph of A peaks just after the graph of L has a minimum.

10 Review

1. (a) A differential equation is an equation that contains an unknown function and one or more of its derivatives.

   (b) The order of a differential equation is the order of the highest derivative that occurs in the equation.

   (c) An initial condition is a condition of the form $y(t_0) = y_0$.

2. $y' = x^2 + y^2 \geq 0$ for all x and y. $y' = 0$ only at the origin, so there is a horizontal tangent at $(0,0)$, but nowhere else. The graph of the solution is increasing on every interval.

3. See the paragraph preceding Example 1 in Section 10.2.

4. See the paragraph next to Figure 14 in Section 10.2.

5. A separable equation is a first-order differential equation in which the expression for dy/dx can be factored as a function of x times a function of y, that is, $dy/dx = g(x)f(y)$. We can solve the equation by integrating both sides of the equation $dy/f(y) = g(x)dx$ and solving for y.

6. A first-order linear differential equation is a differential equation that can be put in the form $\dfrac{dy}{dx} + P(x)\,y = Q(x)$, where P and Q are continuous functions on a given interval. To solve such an equation, multiply it by the integrating factor $I(x) = e^{\int P(x)dx}$ to put it in the form $[I(x)\,y]' = I(x)\,Q(x)$ and then integrate both sides to get $I(x)\,y = \int I(x)\,Q(x)\,dx$, that is, $e^{\int P(x)\,dx}y = \int e^{\int P(x)\,dx}Q(x)\,dx$. Solving for y gives us $y = e^{-\int P(x)\,dx}\int e^{\int P(x)\,dx}Q(x)\,dx$.

7. (a) $\dfrac{dy}{dt} = ky$; the relative growth rate, $\dfrac{1}{y}\dfrac{dy}{dt}$, is constant.

   (b) The equation in part (a) is an appropriate model for population growth, assuming that there is enough room and nutrition to support the growth.

   (c) If $y(0) = y_0$, then the solution is $y(t) = y_0e^{kt}$.

8. (a) $dP/dt = kP(1 - P/K)$, where K is the carrying capacity.

   (b) The equation in part (a) is an appropriate model for population growth, assuming that the population grows at a rate proportional to the size of the population in the beginning, but eventually levels off and approaches its carrying capacity because of limited resources.

9. (a) $dF/dt = kF - aFS$ and $dS/dt = -rS + bFS$.

   (b) In the absence of sharks, an ample food supply would support exponential growth of the fish population, that is, $dF/dt = kF$, where k is a positive constant. In the absence of fish, we assume that the shark population would decline at a rate proportional to itself, that is, $dS/dt = -rS$, where r is a positive constant.

TRUE-FALSE QUIZ

1. True. Since $y^4 \geq 0$, $y' = -1 - y^4 < 0$ and the solutions are decreasing functions.

2. True. $y = \dfrac{\ln x}{x} \quad \Rightarrow \quad y' = \dfrac{1 - \ln x}{x^2}$.

LHS $= x^2 y' + xy = x^2 \cdot \dfrac{1 - \ln x}{x^2} + x \cdot \dfrac{\ln x}{x} = (1 - \ln x) + \ln x = 1 = $ RHS, so $y = \dfrac{\ln x}{x}$ is a solution

of $x^2 y' + xy = 1$.

3. False. $x + y$ cannot be written in the form $g(x) f(y)$.

4. True. $y' = 3y - 2x + 6xy - 1 = 6xy - 2x + 3y - 1 = 2x(3y - 1) + 1(3y - 1) = (2x + 1)(3y - 1)$, so y' can

be written in the form $g(x) f(y)$, and hence, is separable.

5. True. $e^x y' = y \quad \Rightarrow \quad y' = e^{-x} y \quad \Rightarrow \quad y' + (-e^{-x}) y = 0$, which is of the form $y' + P(x) y = Q(x)$, so the

equation is linear.

6. False. $y' + xy = e^y$ cannot be put in the form $y' + P(x) y = Q(x)$, so it is not linear.

7. True. By comparing $\dfrac{dy}{dt} = 2y\left(1 - \dfrac{y}{5}\right)$ with the logistic differential equation (10.4.4), we see that the carrying

capacity is 5; that is, $\lim\limits_{t \to \infty} y = 5$.

EXERCISES

1. (a)

(b) $\lim\limits_{t \to \infty} y(t)$ appears to be finite for $0 \leq c \leq 4$. In fact

$\lim\limits_{t \to \infty} y(t) = 4$ for $c = 4$, $\lim\limits_{t \to \infty} y(t) = 2$ for $0 < c < 4$, and

$\lim\limits_{t \to \infty} y(t) = 0$ for $c = 0$. The equilibrium solutions are

$y(t) = 0$, $y(t) = 2$, and $y(t) = 4$.

2. (a)

We sketch the direction field and four solution curves, as shown.

Note that the slope $y' = x/y$ is not defined on the line $y = 0$.

(b) $y' = x/y \quad \Leftrightarrow \quad y\,dy = x\,dx \quad \Leftrightarrow \quad y^2 = x^2 + C$. For $C = 0$, this is the pair of lines $y = \pm x$. For $C \neq 0$, it is the

hyperbola $x^2 - y^2 = -C$.

3. (a)

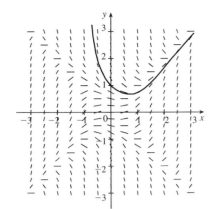

We estimate that when $x = 0.3$, $y = 0.8$, so $y(0.3) \approx 0.8$.

(b) $h = 0.1$, $x_0 = 0$, $y_0 = 1$ and $F(x, y) = x^2 - y^2$. So $y_n = y_{n-1} + 0.1\left(x_{n-1}^2 - y_{n-1}^2\right)$. Thus,

$y_1 = 1 + 0.1\left(0^2 - 1^2\right) = 0.9$, $y_2 = 0.9 + 0.1\left(0.1^2 - 0.9^2\right) = 0.82$, $y_3 = 0.82 + 0.1\left(0.2^2 - 0.82^2\right) = 0.75676$.

This is close to our graphical estimate of $y(0.3) \approx 0.8$.

(c) The centers of the horizontal line segments of the direction field are located on the lines $y = x$ and $y = -x$.

When a solution curve crosses one of these lines, it has a local maximum or minimum.

4. (a) $h = 0.2$, $x_0 = 0$, $y_0 = 1$ and $F(x, y) = 2xy^2$. We need y_2.

$y_1 = 1 + 0.2(2 \cdot 0 \cdot 1^2) = 1$, $y_2 = 1 + 0.2(2 \cdot 0.2 \cdot 1^2) = 1.08 \approx y(0.4)$.

(b) $h = 0.1$ now, so $y_1 = 1 + 0.1(2 \cdot 0 \cdot 1^2) = 1$, $y_2 = 1 + 0.1(2 \cdot 0.1 \cdot 1^2) = 1.02$,

$y_3 = 1.02 + 0.1(2 \cdot 0.2 \cdot 1.02^2) \approx 1.06162$, $y_4 = 1.06162 + 0.1(2 \cdot 0.3 \cdot 1.06162^2) \approx 1.1292 \approx y(0.4)$.

(c) The equation is separable, so we write $\dfrac{dy}{y^2} = 2x\,dx$ $\Rightarrow$ $\displaystyle\int \dfrac{dy}{y^2} = \int 2x\,dx$ $\Leftrightarrow$ $-\dfrac{1}{y} = x^2 + C$, but $y(0) = 1$, so

$C = -1$ and $y(x) = \dfrac{1}{1 - x^2}$ $\Leftrightarrow$ $y(0.4) = \dfrac{1}{1 - 0.16} \approx 1.1905$. From this we see that the approximation was greatly improved by increasing the number of steps, but the approximations were still far off.

5. $y' = xe^{-\sin x} - y\cos x$ $\Rightarrow$ $y' + (\cos x)\,y = xe^{-\sin x}$ $(\star)$. This is a linear equation and the integrating factor is

$I(x) = e^{\int \cos x\,dx} = e^{\sin x}$. Multiplying $(\star)$ by $e^{\sin x}$ gives $e^{\sin x}\,y' + e^{\sin x}(\cos x)\,y = x$ $\Rightarrow$ $\left(e^{\sin x}\,y\right)' = x$ $\Rightarrow$

$e^{\sin x}\,y = \frac{1}{2}x^2 + C$ $\Rightarrow$ $y = \left(\frac{1}{2}x^2 + C\right)e^{-\sin x}$.

6. $\dfrac{dx}{dt} = 1 - t + x - tx = 1(1 - t) + x(1 - t) = (1 + x)(1 - t)$ $\Rightarrow$ $\dfrac{dx}{1 + x} = (1 - t)\,dt$ $\Rightarrow$

$\displaystyle\int \dfrac{dx}{1 + x} = \int (1 - t)\,dt$ $\Rightarrow$ $\ln|1 + x| = t - \frac{1}{2}t^2 + C$ $\Rightarrow$ $|1 + x| = e^{t - t^2/2 + C}$ $\Rightarrow$

$1 + x = \pm e^{t - t^2/2} \cdot e^C$ $\Rightarrow$ $x = -1 + Ke^{t - t^2/2}$, where K is any nonzero constant.

7. $2ye^{y^2}y' = 2x + 3\sqrt{x}$ $\Rightarrow$ $2ye^{y^2}\dfrac{dy}{dx} = 2x + 3\sqrt{x}$ $\Rightarrow$ $2ye^{y^2}\,dy = \left(2x + 3\sqrt{x}\right)dx$ $\Rightarrow$

$\int 2ye^{y^2}\,dy = \int \left(2x + 3\sqrt{x}\right)dx$ $\Rightarrow$ $e^{y^2} = x^2 + 2x^{3/2} + C$ $\Rightarrow$ $y^2 = \ln(x^2 + 2x^{3/2} + C)$ $\Rightarrow$

$y = \pm\sqrt{\ln(x^2 + 2x^{3/2} + C)}$

15. (a) Using (4) and (7) in Section 10.4, we see that for $\dfrac{dP}{dt} = 0.1P\left(1 - \dfrac{P}{2000}\right)$ with $P(0) = 100$, we have $k = 0.1$,

$K = 2000$, $P_0 = 100$, and $A = \dfrac{2000 - 100}{100} = 19$. Thus, the solution of the initial-value problem is

$P(t) = \dfrac{2000}{1 + 19e^{-0.1t}}$ and $P(20) = \dfrac{2000}{1 + 19e^{-2}} \approx 560$.

(b) $P = 1200 \Leftrightarrow 1200 = \dfrac{2000}{1 + 19e^{-0.1t}} \Leftrightarrow 1 + 19e^{-0.1t} = \dfrac{2000}{1200} \Leftrightarrow 19e^{-0.1t} = \dfrac{5}{3} - 1 \Leftrightarrow$

$e^{-0.1t} = \left(\dfrac{2}{3}\right)/19 \Leftrightarrow -0.1t = \ln\dfrac{2}{57} \Leftrightarrow t = -10\ln\dfrac{2}{57} \approx 33.5$.

16. (a) Let $t = 0$ correspond to 1990 so that $P(t) = 5.28e^{kt}$ is a starting point for the model. When $t = 10$, $P = 6.07$.

So $6.07 = 5.28e^{10k} \Rightarrow 10k = \ln\dfrac{6.07}{5.28} \Rightarrow k = \dfrac{1}{10}\ln\dfrac{6.07}{5.28} \approx 0.01394$. For the year 2020, $t = 30$, and

$P(30) = 5.28e^{30k} \approx 8.02$ billion.

(b) $P = 10 \Rightarrow 5.28e^{kt} = 10 \Rightarrow \dfrac{10}{5.28} = e^{kt} \Rightarrow kt = \ln\dfrac{10}{5.28} \Rightarrow t = 10\dfrac{\ln\frac{10}{5.28}}{\ln\frac{6.07}{5.28}} \approx 45.8$ years, that is,

in $1990 + 45 = 2035$.

(c) $P(t) = \dfrac{K}{1 + Ae^{-kt}} = \dfrac{100}{1 + Ae^{-kt}}$, where $A = \dfrac{100 - 5.28}{5.28} \approx 17.94$. Using $k = \dfrac{1}{10}\ln\dfrac{6.07}{5.28}$ from part (a), a model is

$P(t) \approx \dfrac{100}{1 + 17.94e^{-0.01394t}}$ and $P(30) \approx 7.81$ billion, slightly lower than our estimate of 8.02 billion in part (a).

(d) $P = 10 \Rightarrow 1 + Ae^{-kt} = \dfrac{100}{10} \Rightarrow Ae^{-kt} = 9 \Rightarrow e^{-kt} = 9/A \Rightarrow -kt = \ln(9/A) \Rightarrow$

$t = -\dfrac{1}{k}\ln\dfrac{9}{A} \approx 49.47$ years (that is, in 2039), which is later than the prediction of 2035 in part (b).

17. (a) $\dfrac{dL}{dt} \propto L_\infty - L \Rightarrow \dfrac{dL}{dt} = k(L_\infty - L) \Rightarrow \displaystyle\int \dfrac{dL}{L_\infty - L} = \int k\,dt \Rightarrow -\ln|L_\infty - L| = kt + C \Rightarrow$

$\ln|L_\infty - L| = -kt - C \Rightarrow |L_\infty - L| = e^{-kt-C} \Rightarrow L_\infty - L = Ae^{-kt} \Rightarrow L = L_\infty - Ae^{-kt}$.

At $t = 0$, $L = L(0) = L_\infty - A \Rightarrow A = L_\infty - L(0) \Rightarrow L(t) = L_\infty - [L_\infty - L(0)]e^{-kt}$.

(b) $L_\infty = 53$ cm, $L(0) = 10$ cm, and $k = 0.2 \Rightarrow L(t) = 53 - (53 - 10)e^{-0.2t} = 53 - 43e^{-0.2t}$.

18. Denote the amount of salt in the tank (in kg) by y. $y(0) = 0$ since initially there is only water in the tank.

The rate at which y increases is equal to the rate at which salt flows into the tank minus the rate at which it flows out.

That rate is $\dfrac{dy}{dt} = 0.1\dfrac{\text{kg}}{\text{L}} \times 10\dfrac{\text{L}}{\text{min}} - \dfrac{y}{100}\dfrac{\text{kg}}{\text{L}} \times 10\dfrac{\text{L}}{\text{min}} = 1 - \dfrac{y}{10}\dfrac{\text{kg}}{\text{min}} \Rightarrow \displaystyle\int \dfrac{dy}{10 - y} = \int \dfrac{1}{10}\,dt \Rightarrow$

$-\ln|10 - y| = \dfrac{1}{10}t + C \Rightarrow 10 - y = Ae^{-t/10}$. $y(0) = 0 \Rightarrow 10 = A \Rightarrow y = 10(1 - e^{-t/10})$.

At $t = 6$ minutes, $y = 10(1 - e^{-6/10}) \approx 4.512$ kg.

19. Let P represent the population and I the number of infected people. The rate of spread dI/dt is jointly proportional to I and

to $P - I$, so for some constant k, $\dfrac{dI}{dt} = kI(P - I) = (kP)I\left(1 - \dfrac{I}{P}\right)$. From Equation 10.4.7 with $K = P$ and k replaced

by kP, we have $I(t) = \dfrac{P}{1 + Ae^{-kPt}} = \dfrac{I_0 P}{I_0 + (P - I_0)e^{-kPt}}$.

Now, measuring t in days, we substitute $t = 7$, $P = 5000$, $I_0 = 160$ and $I(7) = 1200$ to find k:

$$1200 = \frac{160 \cdot 5000}{160 + (5000 - 160)e^{-5000 \cdot 7 \cdot k}} \quad \Leftrightarrow \quad 3 = \frac{2000}{160 + 4840e^{-35,000k}} \quad \Leftrightarrow \quad 480 + 14{,}520e^{-35,000k} = 2000 \quad \Leftrightarrow$$

$$e^{-35,000k} = \frac{2000 - 480}{14{,}520} \quad \Leftrightarrow \quad -35{,}000k = \ln \frac{38}{363} \quad \Leftrightarrow \quad k = \frac{-1}{35{,}000} \ln \frac{38}{363} \approx 0.00006448. \text{ Next, let}$$

$I = 5000 \times 80\% = 4000$, and solve for t:
$$4000 = \frac{160 \cdot 5000}{160 + (5000 - 160)e^{-k \cdot 5000 \cdot t}} \quad \Leftrightarrow \quad 1 = \frac{200}{160 + 4840e^{-5000kt}} \quad \Leftrightarrow$$

$$160 + 4840e^{-5000kt} = 200 \quad \Leftrightarrow \quad e^{-5000kt} = \frac{200 - 160}{4840} \quad \Leftrightarrow \quad -5000kt = \ln \frac{1}{121} \quad \Leftrightarrow$$

$$t = \frac{-1}{5000k} \ln \frac{1}{121} = \frac{1}{\frac{1}{7} \ln \frac{38}{363}} \cdot \ln \frac{1}{121} = 7 \cdot \frac{\ln 121}{\ln \frac{363}{38}} \approx 14.875. \text{ So it takes about 15 days for 80\% of the population}$$

to be infected.

20. $\dfrac{1}{R} \dfrac{dR}{dt} = \dfrac{k}{S} \dfrac{dS}{dt} \quad \Rightarrow \quad \dfrac{d}{dt}(\ln R) = \dfrac{d}{dt}(k \ln S) \quad \Rightarrow \quad \ln R = k \ln S + C \quad \Rightarrow$

$R = e^{k \ln S + C} = e^C \left(e^{\ln S} \right)^k \quad \Rightarrow \quad R = AS^k$, where $A = e^C$ is a positive constant.

21. $\dfrac{dh}{dt} = -\dfrac{R}{V}\left(\dfrac{h}{k+h} \right) \quad \Rightarrow \quad \displaystyle\int \frac{k+h}{h}\, dh = \int \left(-\frac{R}{V} \right) dt \quad \Rightarrow \quad \int \left(1 + \frac{k}{h} \right) dh = -\frac{R}{V} \int 1\, dt \quad \Rightarrow$

$h + k \ln h = -\dfrac{R}{V}\, t + C$. This equation gives a relationship between h and t, but it is not possible to isolate h and express it in

terms of t.

22. $dx/dt = 0.4x - 0.002xy$, $dy/dt = -0.2y + 0.000008xy$

(a) The xy terms represent encounters between the birds and the insects. Since the y-population increases from these terms
and the x-population decreases, we expect y to represent the birds and x the insects.

(b) x and y are constant $\quad \Rightarrow \quad x' = 0$ and $y' = 0 \quad \Rightarrow$

$$\begin{cases} 0 = 0.4x - 0.002xy \\ 0 = -0.2y + 0.000008xy \end{cases} \quad \Rightarrow \quad \begin{cases} 0 = 0.4x(1 - 0.005y) \\ 0 = -0.2y(1 - 0.00004x) \end{cases} \quad \Rightarrow \quad y = 0 \text{ and } x = 0 \text{ (zero populations)}$$

or $y = \frac{1}{0.005} = 200$ and $x = \frac{1}{0.00004} = 25{,}000$. The non-trivial solution represents the population sizes needed so that
there are no changes in either the number of birds or the number of insects.

(c) $\dfrac{dy}{dx} = \dfrac{dy/dt}{dx/dt} = \dfrac{-0.2y + 0.000008xy}{0.4x - 0.002xy}$

(d)

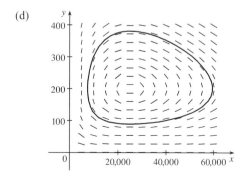

At $(x, y) = (40{,}000, 100)$, $dx/dt = 8000 > 0$, so as t increases we
are proceeding in a counterclockwise direction. The populations
increase to approximately $(59{,}646, 200)$, at which point the insect
population starts to decrease. The birds attain a maximum population
of about 380 when the insect population is 25,000. The populations
decrease to about $(7370, 200)$, at which point the insect population
starts to increase. The birds attain a minimum population of about 88
when the insect population is 25,000, and then the cycle repeats.

(e)

Both graphs have the same period and the bird population peaks about a quarter-cycle after the insect population.

23. (a) $dx/dt = 0.4x(1 - 0.000005x) - 0.002xy$, $dy/dt = -0.2y + 0.000008xy$. If $y = 0$, then $dx/dt = 0.4x(1 - 0.000005x)$, so $dx/dt = 0$ $\Leftrightarrow$ $x = 0$ or $x = 200{,}000$, which shows that the insect population increases logistically with a carrying capacity of 200,000. Since $dx/dt > 0$ for $0 < x < 200{,}000$ and $dx/dt < 0$ for $x > 200{,}000$, we expect the insect population to stabilize at 200,000.

(b) x and y are constant $\Rightarrow$ $x' = 0$ and $y' = 0$ $\Rightarrow$

$$\begin{cases} 0 = 0.4x(1 - 0.000005x) - 0.002xy \\ 0 = -0.2y + 0.000008xy \end{cases} \Rightarrow \begin{cases} 0 = 0.4x[(1 - 0.000005x) - 0.005y] \\ 0 = y(-0.2 + 0.000008x) \end{cases}$$

The second equation is true if $y = 0$ or $x = \frac{0.2}{0.000008} = 25{,}000$. If $y = 0$ in the first equation, then either $x = 0$ or $x = \frac{1}{0.000005} = 200{,}000$. If $x = 25{,}000$, then $0 = 0.4(25{,}000)[(1 - 0.000005 \cdot 25{,}000) - 0.005y]$ $\Rightarrow$ $0 = 10{,}000[(1 - 0.125) - 0.005y]$ $\Rightarrow$ $0 = 8750 - 50y$ $\Rightarrow$ $y = 175$.

Case (i): $y = 0$, $x = 0$: Zero populations

Case (ii): $y = 0$, $x = 200{,}000$: In the absence of birds, the insect population is always 200,000.

Case (iii): $x = 25{,}000$, $y = 175$: The predator/prey interaction balances and the populations are stable.

(c) The populations of the birds and insects fluctuate around 175 and 25,000, respectively, and eventually stabilize at those values.

(d)

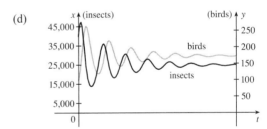

24. First note that, in this question, "weighs" is used in the informal sense, so what we really require is Barbara's mass m in kg as a function of t. Barbara's net intake of calories per day at time t (measured in days) is $c(t) = 1600 - 850 - 15m(t) = 750 - 15m(t)$, where $m(t)$ is her mass at time t. We are given that $m(0) = 60$ kg and $\frac{dm}{dt} = \frac{c(t)}{10{,}000}$, so $\frac{dm}{dt} = \frac{750 - 15m}{10{,}000} = \frac{150 - 3m}{2000} = \frac{-3(m - 50)}{2000}$ with $m(0) = 60$. From $\int \frac{dm}{m - 50} = \int \frac{-3\,dt}{2000}$, we get $\ln|m - 50| = -\frac{3}{2000}t + C$. Since $m(0) = 60$, $C = \ln 10$. Now $\ln \frac{|m - 50|}{10} = -\frac{3t}{2000}$, so $|m - 50| = 10e^{-3t/2000}$. The quantity $m - 50$ is continuous, initially positive, and the right-hand side is never zero. Thus, $m - 50$ is positive for all t, and $m(t) = 50 + 10e^{-3t/2000}$ kg. As $t \to \infty$, $m(t) \to 50$ kg. Thus, Barbara's mass gradually settles down to 50 kg.

25. (a) $\dfrac{d^2y}{dx^2} = k\sqrt{1 + \left(\dfrac{dy}{dx}\right)^2}$. Setting $z = \dfrac{dy}{dx}$, we get $\dfrac{dz}{dx} = k\sqrt{1 + z^2}$ $\Rightarrow$ $\dfrac{dz}{\sqrt{1 + z^2}} = k\,dx$. Using Formula 25 gives

$\ln\left(z + \sqrt{1 + z^2}\,\right) = kx + c$ $\Rightarrow$ $z + \sqrt{1 + z^2} = Ce^{kx}$ [where $C = e^c$] $\Rightarrow$ $\sqrt{1 + z^2} = Ce^{kx} - z$ $\Rightarrow$

$1 + z^2 = C^2 e^{2kx} - 2Ce^{kx}z + z^2$ $\Rightarrow$ $2Ce^{kx}z = C^2 e^{2kx} - 1$ $\Rightarrow$ $z = \dfrac{C}{2}e^{kx} - \dfrac{1}{2C}e^{-kx}$. Now

$\dfrac{dy}{dx} = \dfrac{C}{2}e^{kx} - \dfrac{1}{2C}e^{-kx}$ $\Rightarrow$ $y = \dfrac{C}{2k}e^{kx} + \dfrac{1}{2Ck}e^{-kx} + C'$. From the diagram in the text, we see that $y(0) = a$

and $y(\pm b) = h$. $\quad a = y(0) = \dfrac{C}{2k} + \dfrac{1}{2Ck} + C'$ $\Rightarrow$ $C' = a - \dfrac{C}{2k} - \dfrac{1}{2Ck}$ $\Rightarrow$

$y = \dfrac{C}{2k}(e^{kx} - 1) + \dfrac{1}{2Ck}\left(e^{-kx} - 1\right) + a$. From $h = y(\pm b)$, we find $h = \dfrac{C}{2k}\left(e^{kb} - 1\right) + \dfrac{1}{2Ck}\left(e^{-kb} - 1\right) + a$

and $h = \dfrac{C}{2k}\left(e^{-kb} - 1\right) + \dfrac{1}{2Ck}\left(e^{kb} - 1\right) + a$. Subtracting the second equation from the first, we get

$0 = \dfrac{C}{k}\dfrac{e^{kb} - e^{-kb}}{2} - \dfrac{1}{Ck}\dfrac{e^{kb} - e^{-kb}}{2} = \dfrac{1}{k}\left(C - \dfrac{1}{C}\right)\sinh kb$.

Now $k > 0$ and $b > 0$, so $\sinh kb > 0$ and $C = \pm 1$. If $C = 1$, then

$y = \dfrac{1}{2k}\left(e^{kx} - 1\right) + \dfrac{1}{2k}\left(e^{-kx} - 1\right) + a = \dfrac{1}{k}\dfrac{e^{kx} + e^{-kx}}{2} - \dfrac{1}{k} + a = a + \dfrac{1}{k}(\cosh kx - 1)$. If $C = -1$,

then $y = -\dfrac{1}{2k}\left(e^{kx} - 1\right) - \dfrac{1}{2k}\left(e^{-kx} - 1\right) + a = \dfrac{-1}{k}\dfrac{e^{kx} + e^{-kx}}{2} + \dfrac{1}{k} + a = a - \dfrac{1}{k}(\cosh kx - 1)$.

Since $k > 0$, $\cosh kx \ge 1$, and $y \ge a$, we conclude that $C = 1$ and $y = a + \dfrac{1}{k}(\cosh kx - 1)$, where

$h = y(b) = a + \dfrac{1}{k}(\cosh kb - 1)$. Since $\cosh(kb) = \cosh(-kb)$, there is no further information to extract from the

condition that $y(b) = y(-b)$. However, we could replace a with the expression $h - \dfrac{1}{k}(\cosh kb - 1)$, obtaining

$y = h + \dfrac{1}{k}(\cosh kx - \cosh kb)$. It would be better still to keep a in the expression for y, and use the expression for h to

solve for k in terms of a, b, and h. That would enable us to express y in terms of x and the given parameters a, b, and h.

Sadly, it is not possible to solve for k in closed form. That would have to be done by numerical methods when specific

parameter values are given.

(b) The length of the cable is

$$L = \int_{-b}^{b}\sqrt{1 + (dy/dx)^2}\,dx = \int_{-b}^{b}\sqrt{1 + \sinh^2 kx}\,dx = \int_{-b}^{b}\cosh kx\,dx = 2\int_{0}^{b}\cosh kx\,dx$$

$$= 2\left[(1/k)\sinh kx\right]_{0}^{b} = (2/k)\sinh kb$$

PROBLEMS PLUS

1. We use the Fundamental Theorem of Calculus to differentiate the given equation:

$$[f(x)]^2 = 100 + \int_0^x \left\{ [f(t)]^2 + [f'(t)]^2 \right\} dt \quad \Rightarrow \quad 2f(x)f'(x) = [f(x)]^2 + [f'(x)]^2 \quad \Rightarrow$$

$[f(x)]^2 + [f'(x)]^2 - 2f(x)f'(x) = 0 \quad \Rightarrow \quad [f(x) - f'(x)]^2 = 0 \quad \Leftrightarrow \quad f(x) = f'(x)$. We can solve this as a separable

equation, or else use Theorem 10.4.2 with $k = 1$, which says that the solutions are $f(x) = Ce^x$. Now $[f(0)]^2 = 100$, so

$f(0) = C = \pm 10$, and hence $f(x) = \pm 10e^x$ are the only functions satisfying the given equation.

2. $(fg)' = f'g'$, where $f(x) = e^{x^2} \quad \Rightarrow \quad \left(e^{x^2} g \right)' = 2xe^{x^2} g'$. Since the student's mistake did not affect the answer,

$\left(e^{x^2} g \right)' = e^{x^2} g' + 2xe^{x^2} g = 2xe^{x^2} g'$. So $(2x - 1)g' = 2xg$, or $\dfrac{g'}{g} = \dfrac{2x}{2x - 1} = 1 + \dfrac{1}{2x - 1} \quad \Rightarrow$

$\ln|g(x)| = x + \frac{1}{2} \ln(2x - 1) + C \quad \Rightarrow \quad g(x) = Ae^x \sqrt{2x - 1}$.

3. $f'(x) = \lim\limits_{h \to 0} \dfrac{f(x + h) - f(x)}{h} = \lim\limits_{h \to 0} \dfrac{f(x)\left[f(h) - 1 \right]}{h}$ [since $f(x + h) = f(x)f(h)$]

$= f(x) \lim\limits_{h \to 0} \dfrac{f(h) - 1}{h} = f(x) \lim\limits_{h \to 0} \dfrac{f(h) - f(0)}{h - 0} = f(x)f'(0) = f(x)$

Therefore, $f'(x) = f(x)$ for all x and from Theorem 10.4.2 we get $f(x) = Ae^x$. Now $f(0) = 1 \quad \Rightarrow \quad A = 1 \quad \Rightarrow$

$f(x) = e^x$.

4. $\left(\displaystyle\int f(x)\, dx \right) \left(\displaystyle\int \dfrac{dx}{f(x)} \right) = -1 \quad \Rightarrow \quad \displaystyle\int \dfrac{dx}{f(x)} = \dfrac{-1}{\int f(x)\, dx} \quad \Rightarrow \quad \dfrac{1}{f(x)} = \dfrac{f(x)}{\left[\int f(x)\, dx \right]^2}$ [after differentiating] $\Rightarrow$

$\int f(x)\, dx = \pm f(x)$ [after taking square roots] $\Rightarrow \quad f(x) = \pm f'(x)$ [after differentiating again] $\Rightarrow \quad y = Ae^x$ or

$y = Ae^{-x}$ by Theorem 10.4.2. Therefore, $f(x) = Ae^x$ or $f(x) = Ae^{-x}$, for all nonzero constants A, are the functions

satisfying the original equation.

5. "The area under the graph of f from 0 to x is proportional to the $(n + 1)$st power of $f(x)$" translates to

$\int_0^x f(t)\, dt = k[f(x)]^{n+1}$ for some constant k. By FTC1, $\dfrac{d}{dx} \displaystyle\int_0^x f(t)\, dt = \dfrac{d}{dx} \left\{ k[f(x)]^{n+1} \right\} \quad \Rightarrow$

$f(x) = k(n + 1)[f(x)]^n f'(x) \quad \Rightarrow \quad 1 = k(n + 1)[f(x)]^{n-1} f'(x) \quad \Rightarrow \quad 1 = k(n + 1)y^{n-1} \dfrac{dy}{dx} \quad \Rightarrow$

$k(n + 1)y^{n-1}\, dy = dx \quad \Rightarrow \quad \int k(n + 1)y^{n-1}\, dy = \int dx \quad \Rightarrow \quad k(n + 1)\dfrac{1}{n}y^n = x + C$.

Now $f(0) = 0 \quad \Rightarrow \quad 0 = 0 + C \quad \Rightarrow \quad C = 0$ and then $f(1) = 1 \quad \Rightarrow \quad k(n + 1)\dfrac{1}{n} = 1 \quad \Rightarrow \quad k = \dfrac{n}{n + 1}$,

so $y^n = x$ and $y = f(x) = x^{1/n}$.

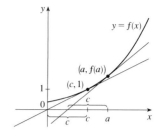

6. Let $y = f(x)$ be a curve that passes through the point $(c, 1)$ and whose subtangents all have length c. The tangent line at $x = a$ has equation $y - f(a) = f'(a)(x - a)$. Assuming $f(a) \neq 0$ and $f'(a) \neq 0$, it has x-intercept $a - \dfrac{f(a)}{f'(a)}$ [let $y = 0$ and solve for x]. Thus, the length of the subtangent is c, so $\left| a - \left(a - \dfrac{f(a)}{f'(a)} \right) \right| = \left| \dfrac{f(a)}{f'(a)} \right| = c \quad \Rightarrow \quad \dfrac{f'(a)}{f(a)} = \pm\dfrac{1}{c}$.

Now $\dfrac{f'(x)}{f(x)} = \pm\dfrac{1}{c} \quad \Rightarrow \quad f'(x) = \pm\dfrac{1}{c} f(x) \quad \Rightarrow \quad \dfrac{dy}{dx} = \pm\dfrac{1}{c} y \quad \Rightarrow \quad \dfrac{dy}{y} = \pm\dfrac{1}{c} dx \quad \Rightarrow \quad \int \dfrac{1}{y}\, dy = \pm\dfrac{1}{c} \int dx \quad \Rightarrow$

$\ln|y| = \pm\dfrac{1}{c} x + K$. Since $f(c) = 1$, $\ln 1 = \pm 1 + K \quad \Rightarrow \quad K = \mp 1$. Thus, $y = e^{\pm x/c \mp 1}$, or $y = e^{\pm(x/c - 1)}$. One curve is an increasing exponential (as shown in the figure) and the other curve is its reflection about the line $x = c$.

7. Let $y(t)$ denote the temperature of the peach pie t minutes after 5:00 PM and R the temperature of the room. Newton's Law of Cooling gives us $dy/dt = k(y - R)$. Solving for y we get $\dfrac{dy}{y - R} = k\, dt \quad \Rightarrow \quad \ln|y - R| = kt + C \quad \Rightarrow$

$|y - R| = e^{kt + C} \quad \Rightarrow \quad y - R = \pm e^{kt} \cdot e^C \quad \Rightarrow \quad y = Me^{kt} + R$, where M is a nonzero constant. We are given temperatures at three times.

$$y(0) = 100 \quad \Rightarrow \quad 100 = M + R \quad \Rightarrow \quad R = 100 - M$$
$$y(10) = 80 \quad \Rightarrow \quad 80 = Me^{10k} + R \quad \textbf{(1)}$$
$$y(20) = 65 \quad \Rightarrow \quad 65 = Me^{20k} + R \quad \textbf{(2)}$$

Substituting $100 - M$ for R in **(1)** and **(2)** gives us

$$-20 = Me^{10k} - M \quad \textbf{(3)} \quad \text{and} \quad -35 = Me^{20k} - M \quad \textbf{(4)}$$

Dividing **(3)** by **(4)** gives us $\dfrac{-20}{-35} = \dfrac{M(e^{10k} - 1)}{M(e^{20k} - 1)} \quad \Rightarrow \quad \dfrac{4}{7} = \dfrac{e^{10k} - 1}{e^{20k} - 1} \quad \Rightarrow \quad 4e^{20k} - 4 = 7e^{10k} - 7 \quad \Rightarrow$

$4e^{20k} - 7e^{10k} + 3 = 0$. This is a quadratic equation in e^{10k}. $(4e^{10k} - 3)(e^{10k} - 1) = 0 \quad \Rightarrow \quad e^{10k} = \tfrac{3}{4}$ or $1 \quad \Rightarrow$

$10k = \ln\tfrac{3}{4}$ or $\ln 1 \quad \Rightarrow \quad k = \tfrac{1}{10}\ln\tfrac{3}{4}$ since k is a nonzero constant of proportionality. Substituting $\tfrac{3}{4}$ for e^{10k} in **(3)** gives us

$-20 = M \cdot \tfrac{3}{4} - M \quad \Rightarrow \quad -20 = -\tfrac{1}{4}M \quad \Rightarrow \quad M = 80$. Now $R = 100 - M$ so $R = 20°\text{C}$.

8. Let b be the number of hours before noon that it began to snow, t the time measured in hours after noon, and $x = x(t) = $ distance traveled by the plow at time t. Then $dx/dt = $ speed of plow. Since the snow falls steadily, the height at time t is $h(t) = k(t + b)$, where k is a constant. We are given that the rate of removal is constant, say R (in m³/h).

If the width of the path is w, then $R = \text{height} \times \text{width} \times \text{speed} = h(t) \times w \times \dfrac{dx}{dt} = k(t + b)w\dfrac{dx}{dt}$. Thus, $\dfrac{dx}{dt} = \dfrac{C}{t + b}$,

where $C = \dfrac{R}{kw}$ is a constant. This is a separable equation. $\displaystyle\int dx = C \int \dfrac{dt}{t + b} \quad \Rightarrow \quad x(t) = C\ln(t + b) + K$.

Put $t = 0$: $0 = C\ln b + K \quad \Rightarrow \quad K = -C\ln b$, so $x(t) = C\ln(t + b) - C\ln b = C\ln(1 + t/b)$.

Put $t = 1$: $6000 = C\ln(1 + 1/b)$ [$x = 6$ km].

Put $t = 2$: $9000 = C\ln(1 + 2/b)$ [$x = (6 + 3)$ km].

Solve for b: $\dfrac{\ln(1+1/b)}{6000} = \dfrac{\ln(1+2/b)}{9000}$ $\Rightarrow$ $3\ln\left(1+\dfrac{1}{b}\right) = 2\ln\left(1+\dfrac{2}{b}\right)$ $\Rightarrow$ $\left(1+\dfrac{1}{b}\right)^3 = \left(1+\dfrac{2}{b}\right)^2$ $\Rightarrow$

$1+\dfrac{3}{b}+\dfrac{3}{b^2}+\dfrac{1}{b^3} = 1+\dfrac{4}{b}+\dfrac{4}{b^2}$ $\Rightarrow$ $\dfrac{1}{b}+\dfrac{1}{b^2}-\dfrac{1}{b^3}=0$ $\Rightarrow$ $b^2+b-1=0$ $\Rightarrow$ $b = \dfrac{-1\pm\sqrt{5}}{2}$.

But $b > 0$, so $b = \dfrac{-1+\sqrt{5}}{2} \approx 0.618\text{ h} \approx 37$ min. The snow began to fall $\dfrac{\sqrt{5}-1}{2}$ hours before noon; that is, at about 11:23 AM.

9. (a) While running from $(L, 0)$ to (x, y), the dog travels a distance

$s = \int_x^L \sqrt{1+(dy/dx)^2}\,dx = -\int_L^x \sqrt{1+(dy/dx)^2}\,dx$, so

$\dfrac{ds}{dx} = -\sqrt{1+(dy/dx)^2}$. The dog and rabbit run at the same speed, so the

rabbit's position when the dog has traveled a distance s is $(0, s)$. Since the

dog runs straight for the rabbit, $\dfrac{dy}{dx} = \dfrac{s-y}{0-x}$ (see the figure).

Thus, $s = y - x\dfrac{dy}{dx}$ $\Rightarrow$ $\dfrac{ds}{dx} = \dfrac{dy}{dx} - \left(x\dfrac{d^2y}{dx^2}+1\dfrac{dy}{dx}\right) = -x\dfrac{d^2y}{dx^2}$. Equating the two expressions for $\dfrac{ds}{dx}$

gives us $x\dfrac{d^2y}{dx^2} = \sqrt{1+\left(\dfrac{dy}{dx}\right)^2}$, as claimed.

(b) Letting $z = \dfrac{dy}{dx}$, we obtain the differential equation $x\dfrac{dz}{dx} = \sqrt{1+z^2}$, or $\dfrac{dz}{\sqrt{1+z^2}} = \dfrac{dx}{x}$. Integrating:

$\ln x = \displaystyle\int \dfrac{dz}{\sqrt{1+z^2}} \overset{25}{=} \ln\left(z+\sqrt{1+z^2}\right) + C$. When $x = L$, $z = dy/dx = 0$, so $\ln L = \ln 1 + C$. Therefore,

$C = \ln L$, so $\ln x = \ln\left(\sqrt{1+z^2}+z\right) + \ln L = \ln\left[L\left(\sqrt{1+z^2}+z\right)\right]$ $\Rightarrow$ $x = L\left(\sqrt{1+z^2}+z\right)$ $\Rightarrow$

$\sqrt{1+z^2} = \dfrac{x}{L} - z$ $\Rightarrow$ $1+z^2 = \left(\dfrac{x}{L}\right)^2 - \dfrac{2xz}{L} + z^2$ $\Rightarrow$ $\left(\dfrac{x}{L}\right)^2 - 2z\left(\dfrac{x}{L}\right) - 1 = 0$ $\Rightarrow$

$z = \dfrac{(x/L)^2-1}{2(x/L)} = \dfrac{x^2-L^2}{2Lx} = \dfrac{x}{2L} - \dfrac{L}{2}\dfrac{1}{x}$ [for $x > 0$]. Since $z = \dfrac{dy}{dx}$, $y = \dfrac{x^2}{4L} - \dfrac{L}{2}\ln x + C_1$.

Since $y = 0$ when $x = L$, $0 = \dfrac{L}{4} - \dfrac{L}{2}\ln L + C_1$ $\Rightarrow$ $C_1 = \dfrac{L}{2}\ln L - \dfrac{L}{4}$. Thus,

$y = \dfrac{x^2}{4L} - \dfrac{L}{2}\ln x + \dfrac{L}{2}\ln L - \dfrac{L}{4} = \dfrac{x^2-L^2}{4L} - \dfrac{L}{2}\ln\left(\dfrac{x}{L}\right)$.

(c) As $x \to 0^+$, $y \to \infty$, so the dog never catches the rabbit.

10. (a) If the dog runs twice as fast as the rabbit, then the rabbit's position when the dog has traveled a distance s is $(0, s/2)$.

Since the dog runs straight toward the rabbit, the tangent line to the dog's path has slope $\dfrac{dy}{dx} = \dfrac{s/2-y}{0-x}$.

Thus, $s = 2y - 2x\dfrac{dy}{dx}$ $\Rightarrow$ $\dfrac{ds}{dx} = 2\dfrac{dy}{dx} - \left(2x\dfrac{d^2y}{dx^2}+2\dfrac{dy}{dx}\right) = -2x\dfrac{d^2y}{dx^2}$.

From Problem 9(a), $\dfrac{ds}{dx} = -\sqrt{1+\left(\dfrac{dy}{dx}\right)^2}$, so $2x\dfrac{d^2y}{dx^2} = \sqrt{1+\left(\dfrac{dy}{dx}\right)^2}$.

Letting $z = \dfrac{dy}{dx}$, we obtain the differential equation $2x\dfrac{dz}{dx} = \sqrt{1+z^2}$, or $\dfrac{2\,dz}{\sqrt{1+z^2}} = \dfrac{dx}{x}$. Integrating, we get

$\ln x = \displaystyle\int \dfrac{2\,dz}{\sqrt{1+z^2}} = 2\ln\left(\sqrt{1+z^2}+z\right) + C$. [See Problem 9(b).]

When $x = L$, $z = dy/dx = 0$, so $\ln L = 2\ln 1 + C = C$. Thus,

$$\ln x = 2\ln\left(\sqrt{1+z^2} + z\right) + \ln L = \ln\left(L\left(\sqrt{1+z^2} + z\right)^2\right) \quad\Rightarrow\quad x = L\left(\sqrt{1+z^2} + z\right)^2 \quad\Rightarrow$$

$$\sqrt{1+z^2} = \sqrt{\frac{x}{L}} - z \quad\Rightarrow\quad 1 + z^2 = \frac{x}{L} - 2\sqrt{\frac{x}{L}}\,z + z^2 \quad\Rightarrow\quad 2\sqrt{\frac{x}{L}}\,z = \frac{x}{L} - 1 \quad\Rightarrow$$

$$\frac{dy}{dx} = z = \frac{1}{2}\sqrt{\frac{x}{L}} - \frac{1}{2\sqrt{x/L}} = \frac{1}{2\sqrt{L}}x^{1/2} - \frac{\sqrt{L}}{2}x^{-1/2} \quad\Rightarrow\quad y = \frac{1}{3\sqrt{L}}x^{3/2} - \sqrt{L}\,x^{1/2} + C_1.$$

When $x = L$, $y = 0$, so $0 = \frac{1}{3\sqrt{L}}L^{3/2} - \sqrt{L}\,L^{1/2} + C_1 = \frac{L}{3} - L + C_1 = C_1 - \frac{2}{3}L$. Therefore, $C_1 = \frac{2}{3}L$ and

$$y = \frac{x^{3/2}}{3\sqrt{L}} - \sqrt{L}\,x^{1/2} + \frac{2}{3}L.$$ As $x \to 0$, $y \to \frac{2}{3}L$, so the dog catches the rabbit when the rabbit is at $\left(0, \frac{2}{3}L\right)$.

(At that point, the dog has traveled a distance of $\frac{4}{3}L$, twice as far as the rabbit has run.)

(b) As in the solutions to part (a) and Problem 9, we get $z = \dfrac{dy}{dx} = \dfrac{x^2}{2L^2} - \dfrac{L^2}{2x^2}$ and hence $y = \dfrac{x^3}{6L^2} + \dfrac{L^2}{2x} - \dfrac{2}{3}L$.

We want to minimize the distance D from the dog at (x, y) to the rabbit at $(0, 2s)$. Now $s = \frac{1}{2}y - \frac{1}{2}x\dfrac{dy}{dx} \quad\Rightarrow$

$$2s = y - xz \quad\Rightarrow\quad y - 2s = xz = x\left(\frac{x^2}{2L^2} - \frac{L^2}{2x^2}\right) = \frac{x^3}{2L^2} - \frac{L^2}{2x}, \text{ so}$$

$$D = \sqrt{(x-0)^2 + (y-2s)^2} = \sqrt{x^2 + \left(\frac{x^3}{2L^2} - \frac{L^2}{2x}\right)^2} = \sqrt{\frac{x^6}{4L^4} + \frac{x^2}{2} + \frac{L^4}{4x^2}} = \sqrt{\left(\frac{x^3}{2L^2} + \frac{L^2}{2x}\right)^2}$$

$$= \frac{x^3}{2L^2} + \frac{L^2}{2x}$$

$$D' = 0 \quad\Leftrightarrow\quad \frac{3x^2}{2L^2} - \frac{L^2}{2x^2} = 0 \quad\Leftrightarrow\quad \frac{3x^2}{2L^2} = \frac{L^2}{2x^2} \quad\Leftrightarrow\quad x^4 = \frac{L^4}{3} \quad\Leftrightarrow\quad x = \frac{L}{\sqrt[4]{3}}, x > 0, L > 0.$$

Since $D''(x) = \dfrac{3x}{L^2} + \dfrac{L^2}{x^3} > 0$ for all $x > 0$, we know that $D\left(\dfrac{L}{\sqrt[4]{3}}\right) = \dfrac{(L\cdot 3^{-1/4})^3}{2L^2} + \dfrac{L^2}{2L\cdot 3^{-1/4}} = \dfrac{2L}{3^{3/4}}$ is the

minimum value of D, that is, the closest the dog gets to the rabbit. The positions at this distance are

Dog: $(x, y) = \left(\dfrac{L}{\sqrt[4]{3}}, \left(\dfrac{5}{3^{7/4}} - \dfrac{2}{3}\right)L\right) = \left(\dfrac{L}{\sqrt[4]{3}}, \dfrac{5\sqrt[4]{3} - 6}{9}L\right)$

Rabbit: $(0, 2s) = \left(0, \dfrac{8\sqrt[4]{3}L}{9} - \dfrac{2L}{3}\right) = \left(0, \dfrac{8\sqrt[4]{3} - 6}{9}L\right)$

11. (a) We are given that $V = \frac{1}{3}\pi r^2 h$, $dV/dt = 60{,}000\pi$ ft^3/h, and $r = 1.5h = \frac{3}{2}h$. So $V = \frac{1}{3}\pi\left(\frac{3}{2}h\right)^2 h = \frac{3}{4}\pi h^3 \quad\Rightarrow$

$$\frac{dV}{dt} = \frac{3}{4}\pi\cdot 3h^2\frac{dh}{dt} = \frac{9}{4}\pi h^2\frac{dh}{dt}. \text{ Therefore, } \frac{dh}{dt} = \frac{4(dV/dt)}{9\pi h^2} = \frac{240{,}000\pi}{9\pi h^2} = \frac{80{,}000}{3h^2} \;(\star) \quad\Rightarrow$$

$\int 3h^2\,dh = \int 80{,}000\,dt \quad\Rightarrow\quad h^3 = 80{,}000t + C$. When $t = 0$, $h = 60$. Thus, $C = 60^3 = 216{,}000$, so

$h^3 = 80{,}000t + 216{,}000$. Let $h = 100$. Then $100^3 = 1{,}000{,}000 = 80{,}000t + 216{,}000 \quad\Rightarrow$

$80{,}000t = 784{,}000 \quad\Rightarrow\quad t = 9.8$, so the time required is 9.8 hours.

(b) The floor area of the silo is $F = \pi\cdot 200^2 = 40{,}000\pi$ ft^2, and the area of the base of the pile is

$A = \pi r^2 = \pi\left(\frac{3}{2}h\right)^2 = \frac{9\pi}{4}h^2$. So the area of the floor which is not covered when $h = 60$ is

$F - A = 40{,}000\pi - 8100\pi = 31{,}900\pi \approx 100{,}217$ ft^2. Now $A = \frac{9\pi}{4}h^2 \quad\Rightarrow\quad dA/dt = \frac{9\pi}{4}\cdot 2h\,(dh/dt)$, and from $(\star)$ in part (a) we know that when $h = 60$, $dh/dt = \frac{80,000}{3(60)^2} = \frac{200}{27}$ ft/h. Therefore,

$dA/dt = \frac{9\pi}{4}(2)(60)\left(\frac{200}{27}\right) = 2000\pi \approx 6283$ ft^2/h.

(c) At $h = 90$ ft, $dV/dt = 60{,}000\pi - 20{,}000\pi = 40{,}000\pi$ ft^3/h. From $(\star)$ in part (a),

$\dfrac{dh}{dt} = \dfrac{4(dV/dt)}{9\pi h^2} = \dfrac{4(40{,}000\pi)}{9\pi h^2} = \dfrac{160{,}000}{9h^2} \;\Rightarrow\; \int 9h^2\,dh = \int 160{,}000\,dt \;\Rightarrow\; 3h^3 = 160{,}000t + C.$ When $t = 0$,

$h = 90$; therefore, $C = 3 \cdot 729{,}000 = 2{,}187{,}000$. So $3h^3 = 160{,}000t + 2{,}187{,}000$. At the top, $h = 100 \;\Rightarrow$

$3(100)^3 = 160{,}000t + 2{,}187{,}000 \;\Rightarrow\; t = \frac{813{,}000}{160{,}000} \approx 5.1$. The pile reaches the top after about 5.1 h.

12. Let $P(a, b)$ be any first-quadrant point on the curve $y = f(x)$. The tangent line at P has equation $y - b = f'(a)(x - a)$, or

equivalently, $y = mx + b - ma$, where $m = f'(a)$. If $Q(0, c)$ is the y-intercept, then $c = b - am$. If $R(k, 0)$ is the

x-intercept, then $k = \dfrac{am - b}{m} = a - \dfrac{b}{m}$. Since the tangent line is bisected at P, we know that $|PQ| = |PR|$; that is,

$\sqrt{(a - 0)^2 + [b - (b - am)]^2} = \sqrt{[a - (a - b/m)]^2 + (b - 0)^2}$. Squaring and simplifying gives us

$a^2 + a^2m^2 = b^2/m^2 + b^2 \;\Rightarrow\; a^2m^2 + a^2m^4 = b^2 + b^2m^2 \;\Rightarrow\; a^2m^4 + (a^2 - b^2)m^2 - b^2 = 0 \;\Rightarrow$

$(a^2m^2 - b^2)(m^2 + 1) = 0 \;\Rightarrow\; m^2 = b^2/a^2$. Since m is the slope of the line from a positive y-intercept to a positive

x-intercept, m must be negative. Since a and b are positive, we have $m = -b/a$, so we will solve the equivalent differential

equation $\dfrac{dy}{dx} = -\dfrac{y}{x} \;\Rightarrow\; \dfrac{dy}{y} = -\dfrac{dx}{x} \;\Rightarrow\; \int \dfrac{dy}{y} = -\int \dfrac{dx}{x} \;\Rightarrow\; \ln y = -\ln x + C \; [x, y > 0] \;\Rightarrow$

$y = e^{-\ln x + C} = e^{\ln x^{-1}} \cdot e^C = x^{-1} \cdot A \;\Rightarrow\; y = A/x$. Since the point $(3, 2)$ is on the curve, $3 = A/2 \;\Rightarrow\; A = 6$

and the curve is $y = 6/x$ with $x > 0$.

13. Let $P(a, b)$ be any point on the curve. If m is the slope of the tangent line at P, then $m = y'(a)$, and an equation of the

normal line at P is $y - b = -\dfrac{1}{m}(x - a)$, or equivalently, $y = -\dfrac{1}{m}x + b + \dfrac{a}{m}$. The y-intercept is always 6, so

$b + \dfrac{a}{m} = 6 \;\Rightarrow\; \dfrac{a}{m} = 6 - b \;\Rightarrow\; m = \dfrac{a}{6 - b}$. We will solve the equivalent differential equation $\dfrac{dy}{dx} = \dfrac{x}{6 - y} \;\Rightarrow$

$(6 - y)\,dy = x\,dx \;\Rightarrow\; \int (6 - y)\,dy = \int x\,dx \;\Rightarrow\; 6y - \frac{1}{2}y^2 = \frac{1}{2}x^2 + C \;\Rightarrow\; 12y - y^2 = x^2 + K.$

Since $(3, 2)$ is on the curve, $12(2) - 2^2 = 3^2 + K \;\Rightarrow\; K = 11$. So the curve is given by $12y - y^2 = x^2 + 11 \;\Rightarrow$

$x^2 + y^2 - 12y + 36 = -11 + 36 \;\Rightarrow\; x^2 + (y - 6)^2 = 25$, a circle with center $(0, 6)$ and radius 5.

14. Suppose C is a curve with the required property and let $P = (x_0, y_0)$ be a point on C. The equation of the normal line to C at

P is $y - y_0 = -\dfrac{1}{y_0'}(x - x_0)$, where y_0' is the value of $\dfrac{dy}{dx}$ at $x = x_0$. This equation makes sense only if $y_0' \neq 0$. If $y_0' = 0$,

then the normal line at P is $x = x_0$, which does not intersect the y-axis at all unless $x_0 = 0$.

So let's assume that $y_0' \neq 0$. Then the normal line to C at P intersects the x-axis at $(x_0 + y_0y_0', 0)$, and it intersects the

y-axis at $(0, y_0 + x_0/y_0')$. The condition on C implies that

$$[\text{distance from } P(x_0, y_0) \text{ to } (0, y_0 + x_0/y_0')] = [\text{distance from } (0, y_0 + x_0/y_0') \text{ to } (x_0 + y_0y_0', 0)]$$
$$\sqrt{(0 - x_0)^2 + (y_0 + x_0/y_0' - y_0)^2} = \sqrt{(x_0 + y_0y_0' - 0)^2 + [0 - (y_0 + x_0/y_0')]^2}$$

Squaring both sides, we get $x_0^2 + x_0^2/(y_0')^2 = (x_0 + y_0y_0')^2 + (y_0 + x_0/y_0')^2$ or

$x_0^2 + \dfrac{x_0^2}{(y_0')^2} = x_0^2 + 2x_0 y_0 y_0' + y_0^2 (y_0')^2 + y_0^2 + 2\dfrac{x_0 y_0}{y_0'} + \dfrac{x_0^2}{(y_0')^2}$. Subtracting $x_0^2 + \dfrac{x_0^2}{(y_0')^2}$ from both sides and multiplying by

y_0', we get

$$0 = y_0^2 y_0' + y_0^{2(y_0')^3} + 2x_0 y_0 [1 + (y_0')^2] = y_0 \{ y_0 y_0' + y_0^3 (y_0') + 2x_0 [1 + (y_0')^2] \}$$

$$= y_0 \{ y_0 y_0' [1 + (y_0')^2] + 2x_0 [1 + (y_0')^2] \} = y_0 (y_0 y_0' + 2x_0)[1 + (y_0')^2]$$

Since $1 + (y_0')^2 \geq 1 > 0$, we conclude that $y_0 (y_0 y_0' + 2x_0) = 0$. Now P is an arbitrary point on C for which $y_0' \neq 0$. Thus, we have shown that $y(yy' + 2x) = 0$ for points (x, y) along C where $y' \neq 0$. One solution of this equation is $y = 0$, but that curve (the x-axis) doesn't satisfy the condition required of C, since its normal lines at points for $x \neq 0$ don't intersect the y-axis. Thus, we can focus our attention on points of C where $y \neq 0$, and conclude that $yy' + 2x = 0$ at points of C where $y \neq 0$ and $y' \neq 0$. Integrating both sides of $yy' + 2x = 0$, we get $\frac{1}{2} y^2 + x^2 = c$. Clearly $c > 0$ (since $y \neq 0$), so we can write $c = a^2$, where $a = \sqrt{c} > 0$. Thus, $\frac{1}{2} y^2 + x^2 = a^2$ and $x^2/a^2 + y^2/(\sqrt{2}a)^2 = 1$. This shows that C is (part of) the ellipse centered at $(0, 0)$ with semimajor axis $\sqrt{2}\,a$ in the y-direction and semiminor axis a in the x-direction.

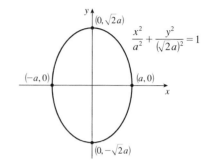

The points of C where $y = 0$ or $y' = 0$ are the vertices $(0, \pm\sqrt{2}\,a)$ and $(\pm a, 0)$. At these points, the condition on C is satisfied in a degenerate way. [When $P = (\pm a, 0)$, the normal line at P *is* the x-axis, so *all* the points of the normal line can be viewed as points of intersection with the x-axis. The intersection with the y-axis at $(0, 0)$ is midway between $(a, 0)$ and $(-a, 0)$; one of these points is P, and the other can be regarded as an intersection of the normal line with the x-axis. Similarly, when $P = (0, \pm\sqrt{2}\,a)$, the normal line is the y-axis, and the point $(0, \pm\sqrt{2}\,a/2)$, which can be regarded as an intersection of the normal line with the y-axis, is midway between P and $(0, 0)$, the intersection with the x-axis.]

Conversely, if C is part of the ellipse $\dfrac{x^2}{a^2} + \dfrac{y^2}{2a^2} = 1$ for some $a > 0$, then the normal line at a point (x_0, y_0) of C (other than the four vertices) has equation $y - y_0 = \dfrac{y_0}{2x_0}(x - x_0)$. Its intersections with the coordinate axes are $\left(0, \dfrac{y_0}{2}\right)$ and $(-x_0, 0)$. $\left[\text{distance from } (x_0, y_0) \text{ to } \left(0, \dfrac{y_0}{2}\right)\right]^2 = x_0^2 + \dfrac{y_0^2}{4}$ and $\left[\text{distance from } \left(0, \dfrac{y_0}{2}\right) \text{ to } (-x_0, 0)\right]^2 = x_0^2 + \dfrac{y_0^2}{4}$, so the required condition is met at points other than the four vertices. As we have explained, if we are willing to interpret the condition broadly, then it can be viewed as holding even at the four vertices.

Another method: Let $P(x_0, y_0)$ be a point on the curve. Since the midpoint of the line segment determined by the normal line from (x_0, y_0) to its intersection with the x-axis has x-coordinate 0, the x-coordinate of the point of intersection with the x-axis must be $-x_0$. Hence, the normal line has slope $\dfrac{y_0 - 0}{x_0 - (-x_0)} = \dfrac{y_0}{2x_0}$. So the tangent line has slope $-\dfrac{2x_0}{y_0}$. This gives the

differential equation $y' = -\dfrac{2x}{y} \;\Rightarrow\; y\,dy = -2x\,dx \;\Rightarrow\; \int y\,dy = \int(-2x)\,dx \;\Rightarrow\; \frac{1}{2}y^2 = -x^2 + C \;\Rightarrow\;$

$x^2 + \frac{1}{2}y^2 = C$ $[C > 0]$.

11 ☐ PARAMETRIC EQUATIONS AND POLAR COORDINATES

11.1 Curves Defined by Parametric Equations

1. $x = 1 + \sqrt{t}, \quad y = t^2 - 4t, \quad 0 \le t \le 5$

t	0	1	2	3	4	5
x	1	2	$1 + \sqrt{2}$	$1 + \sqrt{3}$	3	$1 + \sqrt{5}$
			2.41	2.73		3.24
y	0	-3	-4	-3	0	5

2. $x = 2\cos t, \quad y = t - \cos t, \quad 0 \le t \le 2\pi$

t	0	$\pi/2$	π	$3\pi/2$	2π
x	2	0	-2	0	2
y	-1	$\pi/2$	$\pi + 1$	$3\pi/2$	$2\pi - 1$
		1.57	4.14	4.71	5.28

3. $x = 5\sin t, \quad y = t^2, \quad -\pi \le t \le \pi$

t	$-\pi$	$-\pi/2$	0	$\pi/2$	π
x	0	-5	0	5	0
y	π^2	$\pi^2/4$	0	$\pi^2/4$	π^2
	9.87	2.47		2.47	9.87

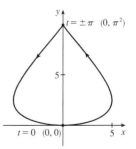

4. $x = e^{-t} + t, \quad y = e^t - t, \quad -2 \le t \le 2$

t	-2	-1	0	1	2
x	$e^2 - 2$	$e - 1$	1	$e^{-1} + 1$	$e^{-2} + 2$
	5.39	1.72		1.37	2.14
y	$e^{-2} + 2$	$e^{-1} + 1$	1	$e - 1$	$e^2 - 2$
	2.14	1.37		1.72	5.39

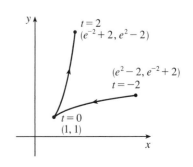

5. $x = 3t - 5$, $\quad y = 2t + 1$

(a)

t	-2	-1	0	1	2	3	4
x	-11	-8	-5	-2	1	4	7
y	-3	-1	1	3	5	7	9

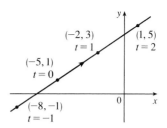

(b) $x = 3t - 5$ $\Rightarrow$ $3t = x + 5$ $\Rightarrow$ $t = \frac{1}{3}(x + 5)$ $\Rightarrow$

$y = 2 \cdot \frac{1}{3}(x + 5) + 1$, so $y = \frac{2}{3}x + \frac{13}{3}$.

6. $x = 1 + t$, $\quad y = 5 - 2t$, $\quad -2 \le t \le 3$

(a)

t	-2	-1	0	1	2	3
x	-1	0	1	2	3	4
y	9	7	5	3	1	-1

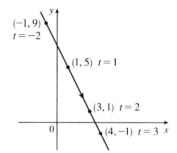

(b) $x = 1 + t$ $\Rightarrow$ $t = x - 1$ $\Rightarrow$ $y = 5 - 2(x - 1)$,

so $y = -2x + 7$, $\quad -1 \le x \le 4$.

7. $x = t^2 - 2$, $\quad y = 5 - 2t$, $\quad -3 \le t \le 4$

(a)

t	-3	-2	-1	0	1	2	3	4
x	7	2	-1	-2	-1	2	7	14
y	11	9	7	5	3	1	-1	-3

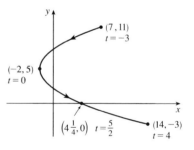

(b) $y = 5 - 2t$ $\Rightarrow$ $2t = 5 - y$ $\Rightarrow$ $t = \frac{1}{2}(5 - y)$ $\Rightarrow$

$x = \left[\frac{1}{2}(5 - y)\right]^2 - 2$, so $x = \frac{1}{4}(5 - y)^2 - 2$, $\quad -3 \le y \le 11$.

8. $x = 1 + 3t$, $\quad y = 2 - t^2$

(a)

t	-3	-2	-1	0	1	2	3
x	-8	-5	-2	1	4	7	10
y	-7	-2	1	2	1	-2	-7

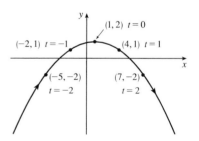

(b) $x = 1 + 3t$ $\Rightarrow$ $t = \frac{1}{3}(x - 1)$ $\Rightarrow$ $y = 2 - \left[\frac{1}{3}(x - 1)\right]^2$,

so $y = -\frac{1}{9}(x - 1)^2 + 2$.

9. $x = \sqrt{t}$, $y = 1 - t$

(a)

t	0	1	2	3	4
x	0	1	1.414	1.732	2
y	1	0	-1	-2	-3

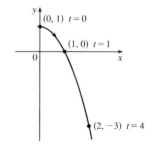

(b) $x = \sqrt{t}$ $\Rightarrow$ $t = x^2$ $\Rightarrow$ $y = 1 - t = 1 - x^2$. Since $t \ge 0$, $x \ge 0$.

So the curve is the right half of the parabola $y = 1 - x^2$.

10. $x = t^2, y = t^3$

(a)

t	-2	-1	0	1	2
x	4	1	0	1	4
y	-8	-1	0	1	8

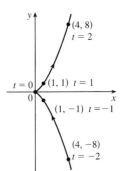

(b) $y = t^3 \;\Rightarrow\; t = \sqrt[3]{y} \;\Rightarrow\; x = t^2 = \left(\sqrt[3]{y}\right)^2 = y^{2/3}. \quad t \in \mathbb{R}, y \in \mathbb{R}, x \geq 0.$

11. (a) $x = \sin\theta, y = \cos\theta, 0 \leq \theta \leq \pi.$

$x^2 + y^2 = \sin^2\theta + \cos^2\theta = 1.$ Since $0 \leq \theta \leq \pi$,

we have $\sin\theta \geq 0$, so $x \geq 0$. Thus, the curve is the

right half of the circle $x^2 + y^2 = 1$.

(b)

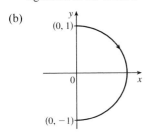

12. (a) $x = 4\cos\theta, y = 5\sin\theta, -\pi/2 \leq \theta \leq \pi/2.$

$\left(\frac{x}{4}\right)^2 + \left(\frac{y}{5}\right)^2 = \cos^2\theta + \sin^2\theta = 1$, which is an

ellipse with x-intercepts $(\pm 4, 0)$ and y-intercepts

$(0, \pm 5)$. We obtain the portion of the ellipse with

$x \geq 0$ since $4\cos\theta \geq 0$ for $-\pi/2 \leq \theta \leq \pi/2$.

(b)

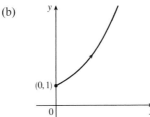

13. (a) $x = \sin t, y = \csc t, 0 < t < \frac{\pi}{2}.$

$y = \csc t = \dfrac{1}{\sin t} = \dfrac{1}{x}$. For $0 < t < \frac{\pi}{2}$, we have

$0 < x < 1$ and $y > 1$. Thus, the curve is the portion

of the hyperbola $y = 1/x$ with $y > 1$.

(b)

14. (a) $x = e^t - 1, y = e^{2t}.$ $y = (e^t)^2 = (x+1)^2$ and

since $x > -1$, we have the right side of the parabola

$y = (x+1)^2$.

(b)

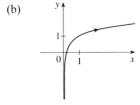

15. (a) $x = e^{2t} \;\Rightarrow\; 2t = \ln x \;\Rightarrow\; t = \frac{1}{2}\ln x.$

$y = t + 1 = \frac{1}{2}\ln x + 1.$

(b)

16. (a) $x = \ln t, y = \sqrt{t}, t \geq 1.$

$x = \ln t \;\Rightarrow\; t = e^x \;\Rightarrow\; y = \sqrt{t} = e^{x/2}, x \geq 0.$

(b)

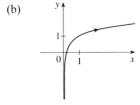

17. (a) $x = \sinh t$, $y = \cosh t$ $\Rightarrow$ $y^2 - x^2 = \cosh^2 t - \sinh^2 t = 1$. Since
$y = \cosh t \geq 1$, we have the upper branch of the hyperbola $y^2 - x^2 = 1$.

(b)

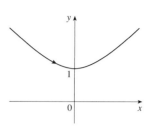

18. (a) $x = 2\cosh t$, $y = 5\sinh t$ $\Rightarrow$ $\dfrac{x}{2} = \cosh t$, $\dfrac{y}{5} = \sinh t$ $\Rightarrow$

$\left(\dfrac{x}{2}\right)^2 = \cosh^2 t$, $\left(\dfrac{y}{5}\right)^2 = \sinh^2 t$. Since $\cosh^2 t - \sinh^2 t = 1$, we have

$\dfrac{x^2}{4} - \dfrac{y^2}{25} = 1$, a hyperbola. Because $x \geq 2$, we have the right branch of the

hyperbola.

(b)

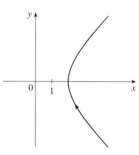

19. $x = 3 + 2\cos t$, $y = 1 + 2\sin t$, $\pi/2 \leq t \leq 3\pi/2$. By Example 4 with $r = 2$, $h = 3$, and $k = 1$, the motion of the particle

takes place on a circle centered at $(3, 1)$ with a radius of 2. As t goes from $\frac{\pi}{2}$ to $\frac{3\pi}{2}$, the particle starts at the point $(3, 3)$ and

moves counterclockwise to $(3, -1)$ [one-half of a circle].

20. $x = 2\sin t$, $y = 4 + \cos t$ $\Rightarrow$ $\sin t = \dfrac{x}{2}$, $\cos t = y - 4$. $\sin^2 t + \cos^2 t = 1$ $\Rightarrow$ $\left(\dfrac{x}{2}\right)^2 + (y-4)^2 = 1$. The motion

of the particle takes place on an ellipse centered at $(0, 4)$. As t goes from 0 to $\frac{3\pi}{2}$, the particle starts at the point $(0, 5)$ and

moves clockwise to $(-2, 4)$ [three-quarters of an ellipse].

21. $x = 5\sin t$, $y = 2\cos t$ $\Rightarrow$ $\sin t = \dfrac{x}{5}$, $\cos t = \dfrac{y}{2}$. $\sin^2 t + \cos^2 t = 1$ $\Rightarrow$ $\left(\dfrac{x}{5}\right)^2 + \left(\dfrac{y}{2}\right)^2 = 1$. The motion of the

particle takes place on an ellipse centered at $(0, 0)$. As t goes from $-\pi$ to 5π, the particle starts at the point $(0, -2)$ and moves

clockwise around the ellipse 3 times.

22. $y = \cos^2 t = 1 - \sin^2 t = 1 - x^2$. The motion of the particle takes place on the parabola $y = 1 - x^2$. As t goes from -2π to

$-\pi$, the particle starts at the point $(0, 1)$, moves to $(1, 0)$, and goes back to $(0, 1)$. As t goes from $-\pi$ to 0, the particle moves

to $(-1, 0)$ and goes back to $(0, 1)$. The particle repeats this motion as t goes from 0 to 2π.

23. We must have $1 \leq x \leq 4$ and $2 \leq y \leq 3$. So the graph of the curve must be contained in the rectangle $[1, 4]$ by $[2, 3]$.

24. (a) From the first graph, we have $1 \leq x \leq 2$. From the second graph, we have $-1 \leq y \leq 1$. The only choice that satisfies

either of those conditions is III.

(b) From the first graph, the values of x cycle through the values from -2 to 2 four times. From the second graph, the values

of y cycle through the values from -2 to 2 six times. Choice I satisfies these conditions.

(c) From the first graph, the values of x cycle through the values from -2 to 2 three times. From the second graph, we have

$0 \leq y \leq 2$. Choice IV satisfies these conditions.

(d) From the first graph, the values of x cycle through the values from -2 to 2 two times. From the second graph, the values of

y do the same thing. Choice II satisfies these conditions.

25. When $t = -1$, $(x, y) = (0, -1)$. As t increases to 0, x decreases to -1 and y

increases to 0. As t increases from 0 to 1, x increases to 0 and y increases to 1.

As t increases beyond 1, both x and y increase. For $t < -1$, x is positive and

decreasing and y is negative and increasing. We could achieve greater accuracy

by estimating x- and y-values for selected values of t from the given graphs and

plotting the corresponding points.

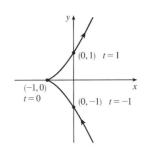

26. For $t < -1$, x is positive and decreasing, while y is negative and increasing (these

points are in Quadrant IV). When $t = -1$, $(x, y) = (0, 0)$ and, as t increases from

-1 to 0, x becomes negative and y increases from 0 to 1. At $t = 0$, $(x, y) = (0, 1)$

and, as t increases from 0 to 1, y decreases from 1 to 0 and x is positive. At

$t = 1$, $(x, y) = (0, 0)$ again, so the loop is completed. For $t > 1$, x and y both

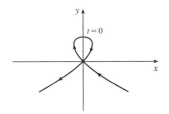

become large negative. This enables us to draw a rough sketch. We could achieve greater accuracy by estimating x- and

y-values for selected values of t from the given graphs and plotting the corresponding points.

27. When $t = 0$ we see that $x = 0$ and $y = 0$, so the curve starts at the origin. As t

increases from 0 to $\frac{1}{2}$, the graphs show that y increases from 0 to 1 while x

increases from 0 to 1, decreases to 0 and to -1, then increases back to 0, so we

arrive at the point $(0, 1)$. Similarly, as t increases from $\frac{1}{2}$ to 1, y decreases from 1

to 0 while x repeats its pattern, and we arrive back at the origin. We could achieve greater accuracy by estimating x- and

y-values for selected values of t from the given graphs and plotting the corresponding points.

28. (a) $x = t^4 - t + 1 = (t^4 + 1) - t > 0$ [think of the graphs of $y = t^4 + 1$ and $y = t$] and $y = t^2 \geq 0$, so these equations

are matched with graph V.

(b) $y = \sqrt{t} \geq 0$. $x = t^2 - 2t = t(t - 2)$ is negative for $0 < t < 2$, so these equations are matched with graph I.

(c) $x = \sin 2t$ has period $2\pi/2 = \pi$. Note that

$y(t + 2\pi) = \sin[t + 2\pi + \sin 2(t + 2\pi)] = \sin(t + 2\pi + \sin 2t) = \sin(t + \sin 2t) = y(t)$, so y has period 2π.

These equations match graph II since x cycles through the values -1 to 1 twice as y cycles through those values once.

(d) $x = \cos 5t$ has period $2\pi/5$ and $y = \sin 2t$ has period π, so x will take on the values -1 to 1, and then 1 to -1, before y

takes on the values -1 to 1. Note that when $t = 0$, $(x, y) = (1, 0)$. These equations are matched with graph VI.

(e) $x = t + \sin 4t$, $y = t^2 + \cos 3t$.   As t becomes large, t and t^2 become the dominant terms in the expressions for x and

y, so the graph will look like the graph of $y = x^2$, but with oscillations. These equations are matched with graph IV.

(f) $x = \dfrac{\sin 2t}{4 + t^2}$, $y = \dfrac{\cos 2t}{4 + t^2}$.   As $t \to \infty$, x and y both approach 0. These equations are matched with graph III.

29. As in Example 6, we let $y = t$ and $x = t - 3t^3 + t^5$ and use a t-interval of $[-3, 3]$.

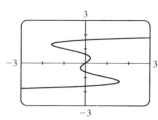

30. We use $x_1 = t$, $y_1 = t^5$ and $x_2 = t(t-1)^2$, $y_2 = t$ with $-3 \le t \le 3$.

There are 3 points of intersection; $(0, 0)$ is fairly obvious. The point in quadrant III

is approximately $(-0.8, -0.4)$ and the point in quadrant I is approximately

$(1.1, 1.8)$.

31. (a) $x = x_1 + (x_2 - x_1)t$, $y = y_1 + (y_2 - y_1)t$, $0 \le t \le 1$. Clearly the curve passes through $P_1(x_1, y_1)$ when $t = 0$ and

through $P_2(x_2, y_2)$ when $t = 1$. For $0 < t < 1$, x is strictly between x_1 and x_2 and y is strictly between y_1 and y_2. For

every value of t, x and y satisfy the relation $y - y_1 = \dfrac{y_2 - y_1}{x_2 - x_1}(x - x_1)$, which is the equation of the line through

$P_1(x_1, y_1)$ and $P_2(x_2, y_2)$.

Finally, any point (x, y) on that line satisfies $\dfrac{y - y_1}{y_2 - y_1} = \dfrac{x - x_1}{x_2 - x_1}$; if we call that common value t, then the given

parametric equations yield the point (x, y); and any (x, y) on the line between $P_1(x_1, y_1)$ and $P_2(x_2, y_2)$ yields a value of

t in $[0, 1]$. So the given parametric equations exactly specify the line segment from $P_1(x_1, y_1)$ to $P_2(x_2, y_2)$.

(b) $x = -2 + [3 - (-2)]t = -2 + 5t$ and $y = 7 + (-1 - 7)t = 7 - 8t$ for $0 \le t \le 1$.

32. For the side of the triangle from A to B, use $(x_1, y_1) = (1, 1)$ and $(x_2, y_2) = (4, 2)$.

Hence, the equations are

$$x = x_1 + (x_2 - x_1)t = 1 + (4 - 1)t = 1 + 3t,$$
$$y = y_1 + (y_2 - y_1)t = 1 + (2 - 1)t = 1 + t.$$

Graphing $x = 1 + 3t$ and $y = 1 + t$ with $0 \le t \le 1$ gives us the side of the

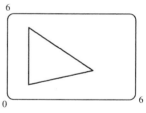

triangle from A to B. Similarly, for the side BC we use $x = 4 - 3t$ and $y = 2 + 3t$, and for the side AC we use $x = 1$

and $y = 1 + 4t$.

33. The circle $x^2 + (y-1)^2 = 4$ has center $(0, 1)$ and radius 2, so by Example 4 it can be represented by $x = 2\cos t$,

$y = 1 + 2\sin t$, $0 \le t \le 2\pi$. This representation gives us the circle with a counterclockwise orientation starting at $(2, 1)$.

(a) To get a clockwise orientation, we could change the equations to $x = 2\cos t$, $y = 1 - 2\sin t$, $0 \le t \le 2\pi$.

(b) To get three times around in the counterclockwise direction, we use the original equations $x = 2\cos t$, $y = 1 + 2\sin t$ with

the domain expanded to $0 \le t \le 6\pi$.

(c) To start at $(0, 3)$ using the original equations, we must have $x_1 = 0$; that is, $2\cos t = 0$. Hence, $t = \frac{\pi}{2}$. So we use

$x = 2\cos t$, $y = 1 + 2\sin t$, $\frac{\pi}{2} \le t \le \frac{3\pi}{2}$.

Alternatively, if we want t to start at 0, we could change the equations of the curve. For example, we could use

$x = -2\sin t$, $y = 1 + 2\cos t$, $0 \le t \le \pi$.

34. (a) Let $x^2/a^2 = \sin^2 t$ and $y^2/b^2 = \cos^2 t$ to obtain $x = a\sin t$ and

$y = b\cos t$ with $0 \le t \le 2\pi$ as possible parametric equations for the ellipse

$x^2/a^2 + y^2/b^2 = 1$.

(b) The equations are $x = 3\sin t$ and $y = b\cos t$ for $b \in \{1, 2, 4, 8\}$.

(c) As b increases, the ellipse stretches vertically.

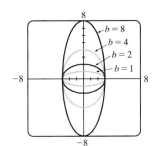

35. *Big circle:* It's centered at $(2, 2)$ with a radius of 2, so by Example 4, parametric equations are

$$x = 2 + 2\cos t, \qquad y = 2 + 2\sin t, \qquad 0 \le t \le 2\pi$$

Small circles: They are centered at $(1, 3)$ and $(3, 3)$ with a radius of 0.1. By Example 4, parametric equations are

and

(left) $x = 1 + 0.1\cos t, \qquad y = 3 + 0.1\sin t, \qquad 0 \le t \le 2\pi$

(right) $x = 3 + 0.1\cos t, \qquad y = 3 + 0.1\sin t, \qquad 0 \le t \le 2\pi$

Semicircle: It's the lower half of a circle centered at $(2, 2)$ with radius 1. By Example 4, parametric equations are

$$x = 2 + 1\cos t, \qquad y = 2 + 1\sin t, \qquad \pi \le t \le 2\pi$$

To get all four graphs on the same screen with a typical graphing calculator, we need to change the last t-interval to $[0, 2\pi]$ in order to match the others. We can do this by changing t to $0.5t$. This change gives us the upper half. There are several ways to get the lower half—one is to change the "+" to a "−" in the y-assignment, giving us

$$x = 2 + 1\cos(0.5t), \qquad y = 2 - 1\sin(0.5t), \qquad 0 \le t \le 2\pi$$

36. If you are using a calculator or computer that can overlay graphs (using multiple t-intervals), the following is appropriate.

Left side: $x = 1$ and y goes from 1.5 to 4, so use

$$x = 1, \qquad y = t, \qquad 1.5 \le t \le 4$$

Right side: $x = 10$ and y goes from 1.5 to 4, so use

$$x = 10, \qquad y = t, \qquad 1.5 \le t \le 4$$

Bottom: x goes from 1 to 10 and $y = 1.5$, so use

$$x = t, \qquad y = 1.5, \qquad 1 \le t \le 10$$

Handle: It starts at $(10, 4)$ and ends at $(13, 7)$, so use

$$x = 10 + t, \qquad y = 4 + t, \qquad 0 \le t \le 3$$

Left wheel: It's centered at $(3, 1)$, has a radius of 1, and appears to go about $30°$ above the horizontal, so use

$$x = 3 + 1\cos t, \qquad y = 1 + 1\sin t, \qquad \tfrac{5\pi}{6} \le t \le \tfrac{13\pi}{6}$$

Right wheel: Similar to the left wheel with center $(8, 1)$, so use

$$x = 8 + 1\cos t, \qquad y = 1 + 1\sin t, \qquad \tfrac{5\pi}{6} \le t \le \tfrac{13\pi}{6}$$

If you are using a calculator or computer that cannot overlay graphs (using one t-interval), the following is appropriate. We'll start by picking the t-interval $[0, 2.5]$ since it easily matches the t-values for the two sides. We now need to find parametric equations for all graphs with $0 \le t \le 2.5$.

Left side: $x = 1$ and y goes from 1.5 to 4, so use

$$x = 1, \qquad y = 1.5 + t, \qquad 0 \le t \le 2.5$$

Right side: $x = 10$ and y goes from 1.5 to 4, so use

$$x = 10, \qquad y = 1.5 + t, \qquad 0 \le t \le 2.5$$

Bottom: x goes from 1 to 10 and $y = 1.5$, so use

$$x = 1 + 3.6t, \qquad y = 1.5, \qquad 0 \le t \le 2.5$$

To get the *x*-assignment, think of creating a linear function such that when $t = 0$, $x = 1$ and when $t = 2.5$, $x = 10$. We can use the point-slope form of a line with $(t_1, x_1) = (0, 1)$ and $(t_2, x_2) = (2.5, 10)$.

$$x - 1 = \frac{10 - 1}{2.5 - 0}(t - 0) \quad \Rightarrow \quad x = 1 + 3.6t.$$

Handle: It starts at $(10, 4)$ and ends at $(13, 7)$, so use

$$x = 10 + 1.2t, \qquad y = 4 + 1.2t, \qquad 0 \le t \le 2.5$$

$(t_1, x_1) = (0, 10)$ and $(t_2, x_2) = (2.5, 13)$ gives us $x - 10 = \dfrac{13 - 10}{2.5 - 0}(t - 0) \quad \Rightarrow \quad x = 10 + 1.2t$.

$(t_1, y_1) = (0, 4)$ and $(t_2, y_2) = (2.5, 7)$ gives us $y - 4 = \dfrac{7 - 4}{2.5 - 0}(t - 0) \quad \Rightarrow \quad y = 4 + 1.2t$.

Left wheel: It's centered at $(3, 1)$, has a radius of 1, and appears to go about $30°$ above the horizontal, so use

$$x = 3 + 1\cos\left(\tfrac{8\pi}{15}t + \tfrac{5\pi}{6}\right), \qquad y = 1 + 1\sin\left(\tfrac{8\pi}{15}t + \tfrac{5\pi}{6}\right), \qquad 0 \le t \le 2.5$$

$(t_1, \theta_1) = \left(0, \tfrac{5\pi}{6}\right)$ and $(t_2, \theta_2) = \left(\tfrac{5}{2}, \tfrac{13\pi}{6}\right)$ gives us $\theta - \tfrac{5\pi}{6} = \dfrac{\tfrac{13\pi}{6} - \tfrac{5\pi}{6}}{\tfrac{5}{2} - 0}(t - 0) \quad \Rightarrow \quad \theta = \tfrac{5\pi}{6} + \tfrac{8\pi}{15}t$.

Right wheel: Similar to the left wheel with center $(8, 1)$, so use

$$x = 8 + 1\cos\left(\tfrac{8\pi}{15}t + \tfrac{5\pi}{6}\right), \qquad y = 1 + 1\sin\left(\tfrac{8\pi}{15}t + \tfrac{5\pi}{6}\right), \qquad 0 \le t \le 2.5$$

37. (a) $x = t^3 \quad \Rightarrow \quad t = x^{1/3}$, so $y = t^2 = x^{2/3}$.

We get the entire curve $y = x^{2/3}$ traversed in a left to right direction.

(b) $x = t^6 \quad \Rightarrow \quad t = x^{1/6}$, so $y = t^4 = x^{4/6} = x^{2/3}$.

Since $x = t^6 \ge 0$, we only get the right half of the curve $y = x^{2/3}$.

(c) $x = e^{-3t} = (e^{-t})^3$ [so $e^{-t} = x^{1/3}$],

$y = e^{-2t} = (e^{-t})^2 = (x^{1/3})^2 = x^{2/3}$.

If $t < 0$, then x and y are both larger than 1. If $t > 0$, then x and y are between 0 and 1. Since $x > 0$ and $y > 0$, the curve never quite reaches the origin.

38. (a) $x = t$, so $y = t^{-2} = x^{-2}$. We get the entire curve $y = 1/x^2$ traversed in a left-to-right direction.

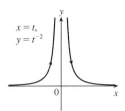

(b) $x = \cos t$, $y = \sec^2 t = \dfrac{1}{\cos^2 t} = \dfrac{1}{x^2}$. Since $\sec t \geq 1$, we only get the parts of the curve $y = 1/x^2$ with $y \geq 1$. We get the first quadrant portion of the curve when $x > 0$, that is, $\cos t > 0$, and we get the second quadrant portion of the curve when $x < 0$, that is, $\cos t < 0$.

(c) $x = e^t$, $y = e^{-2t} = (e^t)^{-2} = x^{-2}$. Since e^t and e^{-2t} are both positive, we only get the first quadrant portion of the curve $y = 1/x^2$.

39. The case $\frac{\pi}{2} < \theta < \pi$ is illustrated. C has coordinates $(r\theta, r)$ as in Example 7, and Q has coordinates $(r\theta, r + r\cos(\pi - \theta)) = (r\theta, r(1 - \cos\theta))$ [since $\cos(\pi - \alpha) = \cos\pi\cos\alpha + \sin\pi\sin\alpha = -\cos\alpha$], so P has coordinates $(r\theta - r\sin(\pi - \theta), r(1 - \cos\theta)) = (r(\theta - \sin\theta), r(1 - \cos\theta))$ [since $\sin(\pi - \alpha) = \sin\pi\cos\alpha - \cos\pi\sin\alpha = \sin\alpha$]. Again we have the parametric equations $x = r(\theta - \sin\theta)$, $y = r(1 - \cos\theta)$.

40. The first two diagrams depict the case $\pi < \theta < \frac{3\pi}{2}$, $d < r$. As in Example 7, C has coordinates $(r\theta, r)$. Now Q (in the second diagram) has coordinates $(r\theta, r + d\cos(\theta - \pi)) = (r\theta, r - d\cos\theta)$, so a typical point P of the trochoid has coordinates $(r\theta + d\sin(\theta - \pi), r - d\cos\theta)$. That is, P has coordinates (x, y), where $x = r\theta - d\sin\theta$ and $y = r - d\cos\theta$. When $d = r$, these equations agree with those of the cycloid.

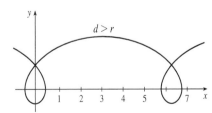

41. It is apparent that $x = |OQ|$ and $y = |QP| = |ST|$. From the diagram, $x = |OQ| = a\cos\theta$ and $y = |ST| = b\sin\theta$. Thus, the parametric equations are $x = a\cos\theta$ and $y = b\sin\theta$. To eliminate θ we rearrange: $\sin\theta = y/b$ $\Rightarrow$ $\sin^2\theta = (y/b)^2$ and $\cos\theta = x/a$ $\Rightarrow$ $\cos^2\theta = (x/a)^2$. Adding the two equations: $\sin^2\theta + \cos^2\theta = 1 = x^2/a^2 + y^2/b^2$. Thus, we have an ellipse.

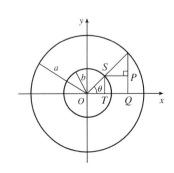

42. A has coordinates $(a\cos\theta, a\sin\theta)$. Since OA is perpendicular to AB, $\triangle OAB$ is a right triangle and B has coordinates

$(a\sec\theta, 0)$. It follows that P has coordinates $(a\sec\theta, b\sin\theta)$. Thus, the parametric equations are $x = a\sec\theta$, $y = b\sin\theta$.

43. $C = (2a\cot\theta, 2a)$, so the x-coordinate of P is $x = 2a\cot\theta$. Let $B = (0, 2a)$.

Then $\angle OAB$ is a right angle and $\angle OBA = \theta$, so $|OA| = 2a\sin\theta$ and

$A = ((2a\sin\theta)\cos\theta, (2a\sin\theta)\sin\theta)$. Thus, the y-coordinate of P

is $y = 2a\sin^2\theta$.

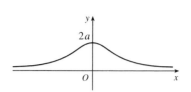

44. (a) Let θ be the angle of inclination of segment OP. Then $|OB| = \dfrac{2a}{\cos\theta}$. Let $C = (2a, 0)$.   (b)

Then by use of right triangle OAC we see that $|OA| = 2a\cos\theta$. Now

$$|OP| = |AB| = |OB| - |OA|$$

$$= 2a\left(\frac{1}{\cos\theta} - \cos\theta\right) = 2a\frac{1-\cos^2\theta}{\cos\theta} = 2a\frac{\sin^2\theta}{\cos\theta} = 2a\sin\theta\tan\theta$$

So P has coordinates $x = 2a\sin\theta\tan\theta \cdot \cos\theta = 2a\sin^2\theta$ and

$y = 2a\sin\theta\tan\theta \cdot \sin\theta = 2a\sin^2\theta\tan\theta$.

45. (a)

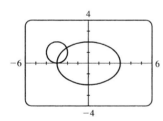

There are 2 points of intersection:

$(-3, 0)$ and approximately $(-2.1, 1.4)$.

(b) A collision point occurs when $x_1 = x_2$ and $y_1 = y_2$ for the same t. So solve the equations:

$$3\sin t = -3 + \cos t \quad \textbf{(1)}$$

$$2\cos t = 1 + \sin t \quad \textbf{(2)}$$

From **(2)**, $\sin t = 2\cos t - 1$. Substituting into **(1)**, we get $3(2\cos t - 1) = -3 + \cos t \;\Rightarrow\; 5\cos t = 0 \;(\star) \;\Rightarrow$

$\cos t = 0 \;\Rightarrow\; t = \frac{\pi}{2}$ or $\frac{3\pi}{2}$. We check that $t = \frac{3\pi}{2}$ satisfies **(1)** and **(2)** but $t = \frac{\pi}{2}$ does not. So the only collision point

occurs when $t = \frac{3\pi}{2}$, and this gives the point $(-3, 0)$. [We could check our work by graphing x_1 and x_2 together as

functions of t and, on another plot, y_1 and y_2 as functions of t. If we do so, we see that the only value of t for which *both*

pairs of graphs intersect is $t = \frac{3\pi}{2}$.]

(c) The circle is centered at $(3, 1)$ instead of $(-3, 1)$. There are still 2 intersection points: $(3, 0)$ and $(2.1, 1.4)$, but there are

no collision points, since $(\star)$ in part (b) becomes $5\cos t = 6 \;\Rightarrow\; \cos t = \frac{6}{5} > 1$.

46. (a) If $\alpha = 30°$ and $v_0 = 500$ m/s, then the equations become $x = (500\cos 30°)t = 250\sqrt{3}\,t$ and

$y = (500\sin 30°)t - \frac{1}{2}(9.8)t^2 = 250t - 4.9t^2$. $y = 0$ when $t = 0$ (when the gun is fired) and again when

$t = \frac{250}{4.9} \approx 51$ s. Then $x = \left(250\sqrt{3}\right)\left(\frac{250}{4.9}\right) \approx 22{,}092$ m, so the bullet hits the ground about 22 km from the gun.

The formula for y is quadratic in t. To find the maximum y-value, we will complete the square:

$$y = -4.9\left(t^2 - \frac{250}{4.9}t\right) = -4.9\left[t^2 - \frac{250}{4.9}t + \left(\frac{125}{4.9}\right)^2\right] + \frac{125^2}{4.9} = -4.9\left(t - \frac{125}{4.9}\right)^2 + \frac{125^2}{4.9} \le \frac{125^2}{4.9}$$

with equality when $t = \frac{125}{4.9}$ s, so the maximum height attained is $\frac{125^2}{4.9} \approx 3189$ m.

(b)

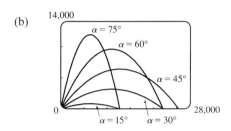

14,000

$\alpha = 75°$

$\alpha = 60°$

$\alpha = 45°$

0

28,000

$\alpha = 15°$ $\alpha = 30°$

As α $(0° < \alpha < 90°)$ increases up to $45°$, the projectile attains a greater height and a greater range. As α increases past $45°$, the projectile attains a greater height, but its range decreases.

(c) $x = (v_0 \cos\alpha)t \quad \Rightarrow \quad t = \dfrac{x}{v_0 \cos\alpha}$.

$$y = (v_0 \sin\alpha)t - \tfrac{1}{2}gt^2 \quad \Rightarrow \quad y = (v_0 \sin\alpha)\,\frac{x}{v_0 \cos\alpha} - \frac{g}{2}\left(\frac{x}{v_0 \cos\alpha}\right)^2 = (\tan\alpha)x - \left(\frac{g}{2v_0^2 \cos^2\alpha}\right)x^2,$$

which is the equation of a parabola (quadratic in x).

47. $x = t^2,\, y = t^3 - ct$. We use a graphing device to produce the graphs for various values of c with $-\pi \le t \le \pi$. Note that all the members of the family are symmetric about the x-axis. For $c < 0$, the graph does not cross itself, but for $c = 0$ it has a cusp at $(0, 0)$ and for $c > 0$ the graph crosses itself at $x = c$, so the loop grows larger as c increases.

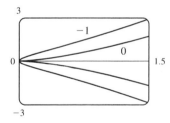

3

-1

0

0

1.5

-3

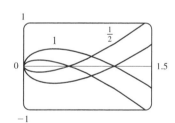

1

1

$\frac{1}{2}$

0

1.5

-1

48. $x = 2ct - 4t^3,\, y = -ct^2 + 3t^4$. We use a graphing device to produce the graphs for various values of c with $-\pi \le t \le \pi$. Note that all the members of the family are symmetric about the y-axis. When $c < 0$, the graph resembles that of a polynomial of even degree, but when $c = 0$ there is a corner at the origin, and when $c > 0$, the graph crosses itself at the origin, and has two cusps below the x-axis. The size of the "swallowtail" increases as c increases.

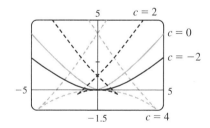

5    $c = 2$

$c = 0$

$c = -2$

-5

5

-1.5

$c = 4$

49. Note that all the Lissajous figures are symmetric about the x-axis. The parameters a and b simply stretch the graph in the x- and y-directions respectively. For $a = b = n = 1$ the graph is simply a circle with radius 1. For $n = 2$ the graph crosses itself at the origin and there are loops above and below the x-axis. In general, the figures have $n-1$ points of intersection, all of which are on the y-axis, and a total of n closed loops.

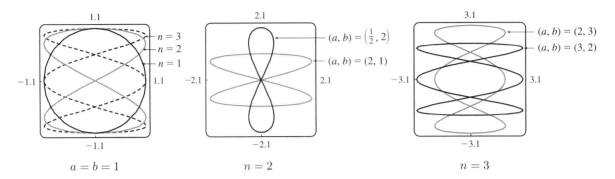

$a = b = 1$ $n = 2$ $n = 3$

50. $x = \cos t$, $y = \sin t - \sin ct$.   If $c = 1$, then $y = 0$, and the curve is simply the line segment from $(-1, 0)$ to $(1, 0)$. The graphs are shown for $c = 2, 3, 4$ and 5.

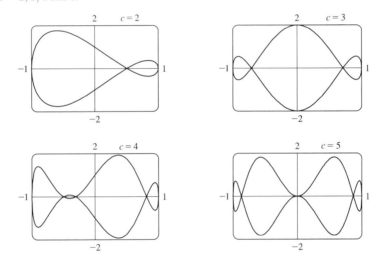

It is easy to see that all the curves lie in the rectangle $[-1, 1]$ by $[-2, 2]$. When c is an integer, $x(t + 2\pi) = x(t)$ and $y(t + 2\pi) = y(t)$, so the curve is closed. When c is a positive integer greater than 1, the curve intersects the x-axis $c + 1$ times and has c loops (one of which degenerates to a tangency at the origin when c is an odd integer of the form $4k + 1$).

As c increases, the curve's loops become thinner, but stay in the region bounded by the semicircles $y = \pm(1 + \sqrt{1 - x^2})$ and the line segments from $(-1, -1)$ to $(-1, 1)$ and from $(1, -1)$ to $(1, 1)$. This is true because $|y| = |\sin t - \sin ct| \le |\sin t| + |\sin ct| \le \sqrt{1 - x^2} + 1$. This curve appears to fill the entire region when c is very large, as shown in the figure for $c = 1000$.

When c is a fraction, we get a variety of shapes with multiple loops, but always within the same region. For some fractional values, such as $c = 2.359$, the curve again appears to fill the region.

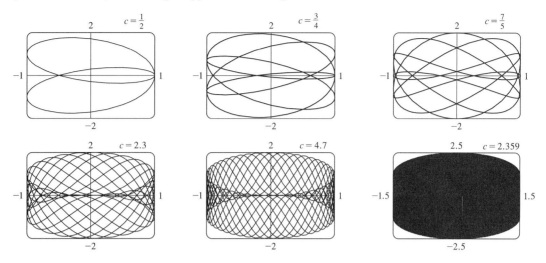

LABORATORY PROJECT   Running Circles Around Circles

1. The center Q of the smaller circle has coordinates $((a-b)\cos\theta, (a-b)\sin\theta)$.

Arc PS on circle C has length $a\theta$ since it is equal in length to arc AS

(the smaller circle rolls without slipping against the larger.)

Thus, $\angle PQS = \dfrac{a}{b}\theta$ and $\angle PQT = \dfrac{a}{b}\theta - \theta$, so P has coordinates

$$x = (a-b)\cos\theta + b\cos(\angle PQT) = (a-b)\cos\theta + b\cos\left(\frac{a-b}{b}\theta\right)$$

and $y = (a-b)\sin\theta - b\sin(\angle PQT) = (a-b)\sin\theta - b\sin\left(\dfrac{a-b}{b}\theta\right).$

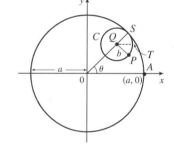

2. With $b = 1$ and a a positive integer greater than 2, we obtain a hypocycloid of a cusps. Shown in the figure is the graph for $a = 4$. Let $a = 4$ and $b = 1$. Using the sum identities to expand $\cos 3\theta$ and $\sin 3\theta$, we obtain

$$x = 3\cos\theta + \cos 3\theta = 3\cos\theta + (4\cos^3\theta - 3\cos\theta) = 4\cos^3\theta$$

and $y = 3\sin\theta - \sin 3\theta = 3\sin\theta - (3\sin\theta - 4\sin^3\theta) = 4\sin^3\theta.$

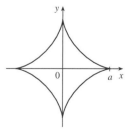

3. The graphs at the right are obtained with $b = 1$ and $a = \frac{1}{2}, \frac{1}{3}, \frac{1}{4}$, and $\frac{1}{10}$ with $-2\pi \le \theta \le 2\pi$. We conclude that as the denominator d increases, the graph gets smaller, but maintains the basic shape shown.

Letting $d = 2$ and $n = 3, 5,$ and 7 with $-2\pi \le \theta \le 2\pi$ gives us the following:

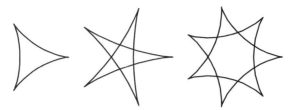

So if d is held constant and n varies, we get a graph with n cusps (assuming n/d is in lowest form). When $n = d + 1$, we obtain a hypocycloid of n cusps. As n increases, we must expand the range of θ in order to get a closed curve. The following graphs have $a = \frac{3}{2}, \frac{5}{4}$, and $\frac{11}{10}$.

4. If $b = 1$, the equations for the hypocycloid are

$$x = (a - 1)\cos\theta + \cos\left((a - 1)\theta\right) \qquad y = (a - 1)\sin\theta - \sin\left((a - 1)\theta\right)$$

which is a hypocycloid of a cusps (from Problem 2). In general, if $a > 1$, we get a figure with cusps on the "outside ring" and if $a < 1$, the cusps are on the "inside ring". In any case, as the values of θ get larger, we get a figure that looks more and more like a washer. If we were to graph the hypocycloid for all values of θ, every point on the washer would eventually be arbitrarily close to a point on the curve.

$$a = \sqrt{2}, \quad -10\pi \le \theta \le 10\pi$$

$$a = e - 2, \quad 0 \le \theta \le 446$$

5. The center Q of the smaller circle has coordinates $((a + b)\cos\theta, (a + b)\sin\theta)$.

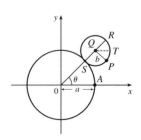

Arc PS has length $a\theta$ (as in Problem 1), so that $\angle PQS = \dfrac{a\theta}{b}$, $\angle PQR = \pi - \dfrac{a\theta}{b}$,

and $\angle PQT = \pi - \dfrac{a\theta}{b} - \theta = \pi - \left(\dfrac{a + b}{b}\right)\theta$ since $\angle RQT = \theta$.

Thus, the coordinates of P are

$$x = (a + b)\cos\theta + b\cos\left(\pi - \frac{a + b}{b}\theta\right) = (a + b)\cos\theta - b\cos\left(\frac{a + b}{b}\theta\right)$$

and $\quad y = (a + b)\sin\theta - b\sin\left(\pi - \dfrac{a + b}{b}\theta\right) = (a + b)\sin\theta - b\sin\left(\dfrac{a + b}{b}\theta\right).$

6. Let $b = 1$ and the equations become

$$x = (a+1)\cos\theta - \cos((a+1)\theta) \qquad y = (a+1)\sin\theta - \sin((a+1)\theta)$$

If $a = 1$, we have a cardioid. If a is a positive integer greater than 1, we get the graph of an "a-leafed clover", with cusps that are a units from the origin. (Some of the pairs of figures are not to scale.)

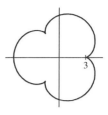

$a = 3, -2\pi \le \theta \le 2\pi$

$a = 10, -2\pi \le \theta \le 2\pi$

If $a = n/d$ with $n = 1$, we obtain a figure that does not increase in size and requires $-d\pi \le \theta \le d\pi$ to be a closed curve traced exactly once.

$a = \frac{1}{4}, -4\pi \le \theta \le 4\pi$

$a = \frac{1}{7}, -7\pi \le \theta \le 7\pi$

Next, we keep d constant and let n vary. As n increases, so does the size of the figure. There is an n-pointed star in the middle.

$a = \frac{2}{5}, -5\pi \le \theta \le 5\pi$

$a = \frac{7}{5}, -5\pi \le \theta \le 5\pi$

Now if $n = d + 1$ we obtain figures similar to the previous ones, but the size of the figure does not increase.

$a = \frac{4}{3}, -3\pi \le \theta \le 3\pi$

$a = \frac{7}{6}, -6\pi \le \theta \le 6\pi$

If a is irrational, we get washers that increase in size as a increases.

$a = \sqrt{2}, 0 \le \theta \le 200$

$a = e - 2, 0 \le \theta \le 446$

11.2 Calculus with Parametric Curves

1. $x = t\sin t$, $y = t^2 + t$ $\Rightarrow$ $\dfrac{dy}{dt} = 2t + 1$, $\dfrac{dx}{dt} = t\cos t + \sin t$, and $\dfrac{dy}{dx} = \dfrac{dy/dt}{dx/dt} = \dfrac{2t+1}{t\cos t + \sin t}$.

2. $x = \dfrac{1}{t}$, $y = \sqrt{t}\,e^{-t}$ $\Rightarrow$ $\dfrac{dy}{dt} = t^{1/2}(-e^{-t}) + e^{-t}\left(\frac{1}{2}t^{-1/2}\right) = \frac{1}{2}t^{-1/2}e^{-t}(-2t + 1) = \dfrac{-2t+1}{2t^{1/2}e^t}$, $\dfrac{dx}{dt} = -\dfrac{1}{t^2}$, and

   $\dfrac{dy}{dx} = \dfrac{dy/dt}{dx/dt} = \dfrac{-2t+1}{2t^{1/2}e^t}\left(-\dfrac{t^2}{1}\right) = \dfrac{(2t-1)t^{3/2}}{2e^t}$.

3. $x = t^4 + 1$, $y = t^3 + t$; $t = -1$. $\dfrac{dy}{dt} = 3t^2 + 1$, $\dfrac{dx}{dt} = 4t^3$, and $\dfrac{dy}{dx} = \dfrac{dy/dt}{dx/dt} = \dfrac{3t^2+1}{4t^3}$. When $t = -1$,

   $(x, y) = (2, -2)$ and $dy/dx = \frac{4}{-4} = -1$, so an equation of the tangent to the curve at the point corresponding to $t = -1$

   is $y - (-2) = (-1)(x - 2)$, or $y = -x$.

4. $x = t - t^{-1}$, $y = 1 + t^2$; $t = 1$. $\dfrac{dy}{dt} = 2t$, $\dfrac{dx}{dt} = 1 + t^{-2} = \dfrac{t^2+1}{t^2}$, and $\dfrac{dy}{dx} = \dfrac{dy/dt}{dx/dt} = 2t\left(\dfrac{t^2}{t^2+1}\right) = \dfrac{2t^3}{t^2+1}$.

   When $t = 1$, $(x, y) = (0, 2)$ and $dy/dx = \frac{2}{2} = 1$, so an equation of the tangent to the curve at the point corresponding to

   $t = 1$ is $y - 2 = 1(x - 0)$, or $y = x + 2$.

5. $x = e^{\sqrt{t}}$, $y = t - \ln t^2$; $t = 1$. $\dfrac{dy}{dt} = 1 - \dfrac{2t}{t^2} = 1 - \dfrac{2}{t}$, $\dfrac{dx}{dt} = \dfrac{e^{\sqrt{t}}}{2\sqrt{t}}$, and $\dfrac{dy}{dx} = \dfrac{dy/dt}{dx/dt} = \dfrac{1 - 2/t}{e^{\sqrt{t}}/(2\sqrt{t})} \cdot \dfrac{2t}{2t} = \dfrac{2t-4}{\sqrt{t}e^{\sqrt{t}}}$.

   When $t = 1$, $(x, y) = (e, 1)$ and $\dfrac{dy}{dx} = -\dfrac{2}{e}$, so an equation of the tangent line is $y - 1 = -\dfrac{2}{e}(x - e)$, or $y = -\dfrac{2}{e}x + 3$.

6. $x = \cos\theta + \sin 2\theta$, $y = \sin\theta + \cos 2\theta$; $\theta = 0$. $\dfrac{dy}{dx} = \dfrac{dy/d\theta}{dx/d\theta} = \dfrac{\cos\theta - 2\sin 2\theta}{-\sin\theta + 2\cos 2\theta}$. When $\theta = 0$, $(x, y) = (1, 1)$ and

   $dy/dx = \frac{1}{2}$, so an equation of the tangent to the curve is $y - 1 = \frac{1}{2}(x - 1)$, or $y = \frac{1}{2}x + \frac{1}{2}$.

7. (a) $x = 1 + \ln t$, $y = t^2 + 2$; $(1, 3)$. $\dfrac{dy}{dt} = 2t$, $\dfrac{dx}{dt} = \dfrac{1}{t}$, and $\dfrac{dy}{dx} = \dfrac{dy/dt}{dx/dt} = \dfrac{2t}{1/t} = 2t^2$.

   At $(1, 3)$, $x = 1 + \ln t = 1$ $\Rightarrow$ $\ln t = 0$ $\Rightarrow$ $t = 1$ and $\dfrac{dy}{dx} = 2$, so an equation of the tangent is $y - 3 = 2(x - 1)$,

   or $y = 2x + 1$.

   (b) $x = 1 + \ln t$ $\Rightarrow$ $x - 1 = \ln t$ $\Rightarrow$ $t = e^{x-1}$, so $y = (e^{x-1})^2 + 2 = e^{2x-2} + 2$ and $\dfrac{dy}{dx} = 2e^{2x-2}$.

   When $x = 1$, $\dfrac{dy}{dx} = 2e^0 = 2$, so an equation of the tangent is $y = 2x + 1$, as in part (a).

8. (a) $x = \tan\theta$, $y = \sec\theta$; $(1, \sqrt{2})$. $\dfrac{dy}{dx} = \dfrac{dy/d\theta}{dx/d\theta} = \dfrac{\sec\theta\tan\theta}{\sec^2\theta} = \dfrac{\tan\theta}{\sec\theta} = \sin\theta$.

   When $(x, y) = (1, \sqrt{2})$, $\theta = \frac{\pi}{4}$ (or $\frac{\pi}{4} + 2\pi n$ for some integer n), so $dy/dx = \sin\frac{\pi}{4} = \sqrt{2}/2$.

   Thus, an equation of the tangent to the curve is $y - \sqrt{2} = (\sqrt{2}/2)(x - 1)$, or $y = (\sqrt{2}/2)x + (\sqrt{2}/2)$.

(b) $\tan^2\theta + 1 = \sec^2\theta \Rightarrow x^2 + 1 = y^2$, so $\dfrac{d}{dx}(x^2+1) = \dfrac{d}{dx}(y^2) \Rightarrow 2x = 2y\dfrac{dy}{dx}$. When $(x,y) = \left(1, \sqrt{2}\right)$,

$\dfrac{dy}{dx} = \dfrac{x}{y} = \dfrac{1}{\sqrt{2}} = \dfrac{\sqrt{2}}{2}$, so an equation of the tangent is $y - \sqrt{2} = \left(\sqrt{2}/2\right)(x-1)$, as in part (a).

9. $x = 6\sin t,\ y = t^2 + t;\ (0,0)$.

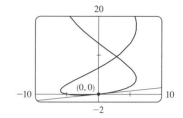

$\dfrac{dy}{dx} = \dfrac{dy/dt}{dx/dt} = \dfrac{2t+1}{6\cos t}$. The point $(0,0)$ corresponds to $t = 0$, so the

slope of the tangent at that point is $\frac{1}{6}$. An equation of the tangent is therefore

$y - 0 = \frac{1}{6}(x-0)$, or $y = \frac{1}{6}x$.

10. $x = \cos t + \cos 2t,\ y = \sin t + \sin 2t;\ (-1,1)$.

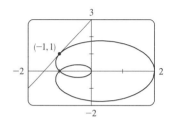

$\dfrac{dy}{dx} = \dfrac{dy/dt}{dx/dt} = \dfrac{\cos t + 2\cos 2t}{-\sin t - 2\sin 2t}$. To find the value of t corresponding to

the point $(-1,1)$, solve $x = -1 \Rightarrow \cos t + \cos 2t = -1 \Rightarrow$

$\cos t + 2\cos^2 t - 1 = -1 \Rightarrow \cos t\,(1 + 2\cos t) = 0 \Rightarrow \cos t = 0$ or

$\cos t = -\frac{1}{2}$. The interval $[0, 2\pi]$ gives the complete curve, so we need only find

the values of t in this interval. Thus, $t = \frac{\pi}{2}$ or $t = \frac{2\pi}{3}$ or $t = \frac{4\pi}{3}$. Checking $t = \frac{\pi}{2}, \frac{3\pi}{2}, \frac{2\pi}{3}$, and $\frac{4\pi}{3}$ in the equation for y,

we find that $t = \frac{\pi}{2}$ corresponds to $(-1,1)$. The slope of the tangent at $(-1,1)$ with $t = \frac{\pi}{2}$ is $\dfrac{0-2}{-1-0} = 2$. An equation

of the tangent is therefore $y - 1 = 2(x+1)$, or $y = 2x + 3$.

11. $x = 4 + t^2,\ y = t^2 + t^3 \Rightarrow \dfrac{dy}{dx} = \dfrac{dy/dt}{dx/dt} = \dfrac{2t+3t^2}{2t} = 1 + \dfrac{3}{2}t \Rightarrow$

$\dfrac{d^2y}{dx^2} = \dfrac{d}{dx}\left(\dfrac{dy}{dx}\right) = \dfrac{d(dy/dx)/dt}{dx/dt} = \dfrac{(d/dt)\left(1+\frac{3}{2}t\right)}{2t} = \dfrac{3/2}{2t} = \dfrac{3}{4t}.$

The curve is CU when $\dfrac{d^2y}{dx^2} > 0$, that is, when $t > 0$.

12. $x = t^3 - 12t,\ y = t^2 - 1 \Rightarrow \dfrac{dy}{dx} = \dfrac{dy/dt}{dx/dt} = \dfrac{2t}{3t^2-12} \Rightarrow$

$\dfrac{d^2y}{dx^2} = \dfrac{\dfrac{d}{dt}\left(\dfrac{dy}{dx}\right)}{dx/dt} = \dfrac{\dfrac{(3t^2-12)\cdot 2 - 2t(6t)}{(3t^2-12)^2}}{3t^2-12} = \dfrac{-6t^2-24}{(3t^2-12)^3} = \dfrac{-6(t^2+4)}{3^3(t^2-4)^3} = \dfrac{-2(t^2+4)}{9(t^2-4)^3}.$

Thus, the curve is CU when $t^2 - 4 < 0 \Rightarrow |t| < 2 \Rightarrow -2 < t < 2$.

13. $x = t - e^t,\ y = t + e^{-t} \Rightarrow$

$\dfrac{dy}{dx} = \dfrac{dy/dt}{dx/dt} = \dfrac{1-e^{-t}}{1-e^t} = \dfrac{1-\dfrac{1}{e^t}}{1-e^t} = \dfrac{\dfrac{e^t-1}{e^t}}{1-e^t} = -e^{-t} \Rightarrow \dfrac{d^2y}{dx^2} = \dfrac{\dfrac{d}{dt}\left(\dfrac{dy}{dx}\right)}{dx/dt} = \dfrac{\dfrac{d}{dt}(-e^{-t})}{dx/dt} = \dfrac{e^{-t}}{1-e^t}.$

The curve is CU when $e^t < 1$ [since $e^{-t} > 0$] $\Rightarrow t < 0$.

14. $x = t + \ln t$, $y = t - \ln t$ $\Rightarrow$ $\dfrac{dy}{dx} = \dfrac{dy/dt}{dx/dt} = \dfrac{1 - 1/t}{1 + 1/t} = \dfrac{t-1}{t+1} = 1 - \dfrac{2}{t+1}$ $\Rightarrow$

$\dfrac{d^2y}{dx^2} = \dfrac{\dfrac{d}{dt}\left(\dfrac{dy}{dx}\right)}{dx/dt} = \dfrac{\dfrac{d}{dt}\left(1 - \dfrac{2}{t+1}\right)}{1 + 1/t} = \dfrac{2/(t+1)^2}{(t+1)/t} = \dfrac{2t}{(t+1)^3}$, so the curve is CU for all t in its domain,

that is, $t > 0$ [$t < -1$ not in domain].

15. $x = 2\sin t$, $y = 3\cos t$, $0 < t < 2\pi$.

$\dfrac{dy}{dx} = \dfrac{dy/dt}{dx/dt} = \dfrac{-3\sin t}{2\cos t} = -\dfrac{3}{2}\tan t$, so $\dfrac{d^2y}{dx^2} = \dfrac{\dfrac{d}{dt}\left(\dfrac{dy}{dx}\right)}{dx/dt} = \dfrac{-\frac{3}{2}\sec^2 t}{2\cos t} = -\dfrac{3}{4}\sec^3 t$.

The curve is CU when $\sec^3 t < 0$ $\Rightarrow$ $\sec t < 0$ $\Rightarrow$ $\cos t < 0$ $\Rightarrow$ $\frac{\pi}{2} < t < \frac{3\pi}{2}$.

16. $x = \cos 2t$, $y = \cos t$, $0 < t < \pi$.

$\dfrac{dy}{dx} = \dfrac{dy/dt}{dx/dt} = \dfrac{-\sin t}{-2\sin 2t} = \dfrac{\sin t}{2 \cdot 2\sin t \cos t} = \dfrac{1}{4\cos t} = \dfrac{1}{4}\sec t$, so $\dfrac{d^2y}{dx^2} = \dfrac{\dfrac{d}{dt}\left(\dfrac{dy}{dx}\right)}{dx/dt} = \dfrac{\frac{1}{4}\sec t \tan t}{-4\sin t \cos t} = -\dfrac{1}{16}\sec^3 t$.

The curve is CU when $\sec^3 t < 0$ $\Rightarrow$ $\sec t < 0$ $\Rightarrow$ $\cos t < 0$ $\Rightarrow$ $\frac{\pi}{2} < t < \pi$.

17. $x = 10 - t^2$, $y = t^3 - 12t$.

$\dfrac{dy}{dt} = 3t^2 - 12 = 3(t+2)(t-2)$, so $\dfrac{dy}{dt} = 0$ $\Leftrightarrow$

$t = \pm 2$ $\Leftrightarrow$ $(x, y) = (6, \mp 16)$.

$\dfrac{dx}{dt} = -2t$, so $\dfrac{dx}{dt} = 0$ $\Leftrightarrow$ $t = 0$ $\Leftrightarrow$ $(x, y) = (10, 0)$.

The curve has horizontal tangents at $(6, \pm 16)$ and a vertical

tangent at $(10, 0)$.

18. $x = 2t^3 + 3t^2 - 12t$, $y = 2t^3 + 3t^2 + 1$.

$\dfrac{dy}{dt} = 6t^2 + 6t = 6t(t+1)$, so $\dfrac{dy}{dt} = 0$ $\Leftrightarrow$

$t = 0$ or -1 $\Leftrightarrow$ $(x, y) = (0, 1)$ or $(13, 2)$.

$\dfrac{dx}{dt} = 6t^2 + 6t - 12 = 6(t+2)(t-1)$, so $\dfrac{dx}{dt} = 0$ $\Leftrightarrow$

$t = -2$ or 1 $\Leftrightarrow$ $(x, y) = (20, -3)$ or $(-7, 6)$.

The curve has horizontal tangents at $(0, 1)$ and $(13, 2)$, and vertical tangents at $(20, -3)$ and $(-7, 6)$.

19. $x = 2\cos\theta$, $y = \sin 2\theta$.

$\dfrac{dy}{d\theta} = 2\cos 2\theta$, so $\dfrac{dy}{d\theta} = 0 \ \Leftrightarrow \ 2\theta = \dfrac{\pi}{2} + n\pi$

[n an integer] $\ \Leftrightarrow \ \theta = \dfrac{\pi}{4} + \dfrac{\pi}{2}n \ \Leftrightarrow$

$(x, y) = \left(\pm\sqrt{2}, \pm 1\right)$. Also, $\dfrac{dx}{d\theta} = -2\sin\theta$, so

$\dfrac{dx}{d\theta} = 0 \ \Leftrightarrow \ \theta = n\pi \ \Leftrightarrow \ (x, y) = (\pm 2, 0)$.

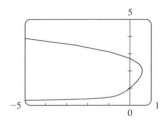

The curve has horizontal tangents at $\left(\pm\sqrt{2}, \pm 1\right)$ (four points), and vertical tangents at $(\pm 2, 0)$.

20. $x = \cos 3\theta$, $y = 2\sin\theta$. $dy/d\theta = 2\cos\theta$, so $dy/d\theta = 0 \ \Leftrightarrow \ \theta = \dfrac{\pi}{2} + n\pi$ (n an integer) $\ \Leftrightarrow \ (x, y) = (0, \pm 2)$.

Also, $dx/d\theta = -3\sin 3\theta$, so $dx/d\theta = 0 \ \Leftrightarrow \ 3\theta = n\pi \ \Leftrightarrow \ \theta = \dfrac{\pi}{3}n \ \Leftrightarrow \ (x, y) = (\pm 1, 0)$ or $\left(\pm 1, \pm\sqrt{3}\right)$.

The curve has horizontal tangents at $(0, \pm 2)$, and vertical tangents at $(\pm 1, 0)$, $\left(\pm 1, -\sqrt{3}\right)$ and $\left(\pm 1, \sqrt{3}\right)$.

21. From the graph, it appears that the rightmost point on the curve $x = t - t^6$, $y = e^t$

is about $(0.6, 2)$. To find the exact coordinates, we find the value of t for which the

graph has a vertical tangent, that is, $0 = dx/dt = 1 - 6t^5 \ \Leftrightarrow \ t = 1/\sqrt[5]{6}$.

Hence, the rightmost point is

$\left(1/\sqrt[5]{6} - 1/\left(6\sqrt[5]{6}\right), e^{1/\sqrt[5]{6}}\right) = \left(5 \cdot 6^{-6/5}, e^{6^{-1/5}}\right) \approx (0.58, 2.01)$.

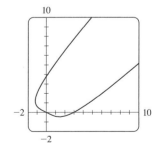

22. From the graph, it appears that the lowest point and the leftmost point on the curve

$x = t^4 - 2t$, $y = t + t^4$ are $(1.5, -0.5)$ and $(-1.2, 1.2)$, respectively. To find the

exact coordinates, we solve $dy/dt = 0$ (horizontal tangents) and $dx/dt = 0$

(vertical tangents).

$\dfrac{dy}{dt} = 0 \ \Leftrightarrow \ 1 + 4t^3 = 0 \ \Leftrightarrow \ t = -\dfrac{1}{\sqrt[3]{4}}$, so the lowest point is

$\left(\dfrac{1}{\sqrt[3]{256}} + \dfrac{2}{\sqrt[3]{4}}, -\dfrac{1}{\sqrt[3]{4}} + \dfrac{1}{\sqrt[3]{256}}\right) = \left(\dfrac{9}{\sqrt[3]{256}}, -\dfrac{3}{\sqrt[3]{256}}\right) \approx (1.42, -0.47)$.

$\dfrac{dx}{dt} = 0 \ \Leftrightarrow \ 4t^3 - 2 = 0 \ \Leftrightarrow \ t = \dfrac{1}{\sqrt[3]{2}}$, so the leftmost point is

$\left(\dfrac{1}{\sqrt[3]{16}} - \dfrac{2}{\sqrt[3]{2}}, \dfrac{1}{\sqrt[3]{2}} + \dfrac{1}{\sqrt[3]{16}}\right) = \left(-\dfrac{3}{\sqrt[3]{16}}, \dfrac{3}{\sqrt[3]{16}}\right) \approx (-1.19, 1.19)$.

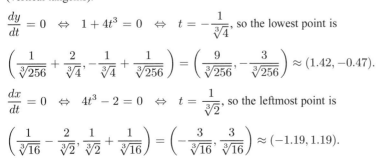

23. We graph the curve $x = t^4 - 2t^3 - 2t^2$, $y = t^3 - t$ in the viewing rectangle $[-2, 1.1]$ by $[-0.5, 0.5]$. This rectangle corresponds approximately to $t \in [-1, 0.8]$.

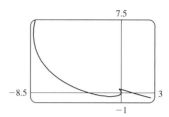

We estimate that the curve has horizontal tangents at about $(-1, -0.4)$ and $(-0.17, 0.39)$ and vertical tangents at about $(0, 0)$ and $(-0.19, 0.37)$. We calculate $\dfrac{dy}{dx} = \dfrac{dy/dt}{dx/dt} = \dfrac{3t^2 - 1}{4t^3 - 6t^2 - 4t}$. The horizontal tangents occur when $dy/dt = 3t^2 - 1 = 0 \iff t = \pm\frac{1}{\sqrt{3}}$, so both horizontal tangents are shown in our graph. The vertical tangents occur when $dx/dt = 2t(2t^2 - 3t - 2) = 0 \iff 2t(2t + 1)(t - 2) = 0 \iff t = 0, -\frac{1}{2}$ or 2. It seems that we have missed one vertical tangent, and indeed if we plot the curve on the t-interval $[-1.2, 2.2]$ we see that there is another vertical tangent at $(-8, 6)$.

24. We graph the curve $x = t^4 + 4t^3 - 8t^2$, $y = 2t^2 - t$ in the viewing rectangle $[-3.7, 0.2]$ by $[-0.2, 1.4]$. It appears that there is a horizontal tangent at about $(-0.4, -0.1)$, and vertical tangents at about $(-3, 1)$ and $(0, 0)$.

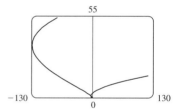

We calculate $\dfrac{dy}{dx} = \dfrac{dy/dt}{dx/dt} = \dfrac{4t - 1}{4t^3 + 12t^2 - 16t}$, so there is a horizontal tangent where $dy/dt = 4t - 1 = 0 \iff t = \frac{1}{4}$.

This point (the lowest point) is shown in the first graph. There are vertical tangents where $dx/dt = 4t^3 + 12t^2 - 16t = 0 \iff 4t(t^2 + 3t - 4) = 0 \iff 4t(t + 4)(t - 1) = 0$. We have missed one vertical tangent corresponding to $t = -4$, and if we plot the graph for $t \in [-5, 3]$, we see that the curve has another vertical tangent line at approximately $(-128, 36)$.

25. $x = \cos t$, $y = \sin t \cos t$. $\quad dx/dt = -\sin t$, $dy/dt = -\sin^2 t + \cos^2 t = \cos 2t$.

$(x, y) = (0, 0) \iff \cos t = 0 \iff t$ is an odd multiple of $\frac{\pi}{2}$. When $t = \frac{\pi}{2}$,

$dx/dt = -1$ and $dy/dt = -1$, so $dy/dx = 1$. When $t = \frac{3\pi}{2}$, $dx/dt = 1$ and

$dy/dt = -1$. So $dy/dx = -1$. Thus, $y = x$ and $y = -x$ are both tangent to the curve at $(0, 0)$.

26.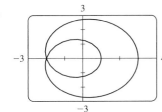

From the graph, we discover that the graph of the curve $x = \cos t + 2\cos 2t$, $y = \sin t + 2\sin 2t$ crosses itself at the point $(-2, 0)$. To find t at $(-2, 0)$, solve $y = 0 \;\Leftrightarrow\; \sin t + 2\sin 2t = 0 \;\Leftrightarrow\; \sin t + 4\sin t \cos t = 0 \;\Leftrightarrow\; \sin t (1 + 4\cos t) = 0 \;\Leftrightarrow\; \sin t = 0$ or $\cos t = -\frac{1}{4}$. We find that $t = \pm\arccos\left(-\frac{1}{4}\right)$ corresponds to $(-2, 0)$.

Now $\dfrac{dy}{dx} = \dfrac{dy/dt}{dx/dt} = \dfrac{\cos t + 4\cos 2t}{-\sin t - 4\sin 2t} = -\dfrac{\cos t + 8\cos^2 t - 4}{\sin t + 8\sin t \cos t}$. When $t = \arccos\left(-\frac{1}{4}\right)$, $\cos t = -\frac{1}{4}$, $\sin t = \dfrac{\sqrt{15}}{4}$,

and $\dfrac{dy}{dx} = -\dfrac{-\frac{1}{4} + \frac{1}{2} - 4}{\frac{\sqrt{15}}{4} - \frac{\sqrt{15}}{2}} = -\dfrac{-\frac{15}{4}}{-\frac{\sqrt{15}}{4}} = -\sqrt{15}$. By symmetry, $t = -\arccos\left(-\frac{1}{4}\right) \;\Rightarrow\; \dfrac{dy}{dx} = \sqrt{15}$.

The tangent lines are $y - 0 = \pm\sqrt{15}\,(x + 2)$, or $y = \sqrt{15}\,x + 2\sqrt{15}$ and $y = -\sqrt{15}\,x - 2\sqrt{15}$.

27. $x = r\theta - d\sin\theta, \; y = r - d\cos\theta$.

(a) $\dfrac{dx}{d\theta} = r - d\cos\theta, \; \dfrac{dy}{d\theta} = d\sin\theta$, so $\dfrac{dy}{dx} = \dfrac{d\sin\theta}{r - d\cos\theta}$.

(b) If $0 < d < r$, then $|d\cos\theta| \le d < r$, so $r - d\cos\theta \ge r - d > 0$. This shows that $dx/d\theta$ never vanishes, so the trochoid can have no vertical tangent if $d < r$.

28. $x = a\cos^3\theta, \; y = a\sin^3\theta$.

(a) $\dfrac{dx}{d\theta} = -3a\cos^2\theta\sin\theta, \; \dfrac{dy}{d\theta} = 3a\sin^2\theta\cos\theta$, so $\dfrac{dy}{dx} = -\dfrac{\sin\theta}{\cos\theta} = -\tan\theta$.

(b) The tangent is horizontal $\;\Leftrightarrow\; dy/dx = 0 \;\Leftrightarrow\; \tan\theta = 0 \;\Leftrightarrow\; \theta = n\pi \;\Leftrightarrow\; (x, y) = (\pm a, 0)$.

The tangent is vertical $\;\Leftrightarrow\; \cos\theta = 0 \;\Leftrightarrow\; \theta$ is an odd multiple of $\frac{\pi}{2} \;\Leftrightarrow\; (x, y) = (0, \pm a)$.

(c) $dy/dx = \pm 1 \;\Leftrightarrow\; \tan\theta = \pm 1 \;\Leftrightarrow\; \theta$ is an odd multiple of $\frac{\pi}{4} \;\Leftrightarrow\; (x, y) = \left(\pm\frac{\sqrt{2}}{4}a, \pm\frac{\sqrt{2}}{4}a\right)$

[All sign choices are valid.]

29. $x = 2t^3, \; y = 1 + 4t - t^2 \;\Rightarrow\; \dfrac{dy}{dx} = \dfrac{dy/dt}{dx/dt} = \dfrac{4 - 2t}{6t^2}$. Now solve $\dfrac{dy}{dx} = 1 \;\Leftrightarrow\; \dfrac{4 - 2t}{6t^2} = 1 \;\Leftrightarrow\;$

$6t^2 + 2t - 4 = 0 \;\Leftrightarrow\; 2(3t - 2)(t + 1) = 0 \;\Leftrightarrow\; t = \frac{2}{3}$ or $t = -1$. If $t = \frac{2}{3}$, the point is $\left(\frac{16}{27}, \frac{29}{9}\right)$, and if $t = -1$, the point is $(-2, -4)$.

30. $x = 3t^2 + 1, \; y = 2t^3 + 1, \; \dfrac{dx}{dt} = 6t, \; \dfrac{dy}{dt} = 6t^2$, so $\dfrac{dy}{dx} = \dfrac{6t^2}{6t} = t$ [even where $t = 0$].

So at the point corresponding to parameter value t, an equation of the tangent line is $y - (2t^3 + 1) = t[x - (3t^2 + 1)]$.

If this line is to pass through $(4, 3)$, we must have $3 - (2t^3 + 1) = t[4 - (3t^2 + 1)] \;\Leftrightarrow\; 2t^3 - 2 = 3t^3 - 3t \;\Leftrightarrow\;$

$t^3 - 3t + 2 = 0 \;\Leftrightarrow\; (t - 1)^2(t + 2) = 0 \;\Leftrightarrow\; t = 1$ or -2. Hence, the desired equations are $y - 3 = x - 4$, or

$y = x - 1$, tangent to the curve at $(4, 3)$, and $y - (-15) = -2(x - 13)$, or $y = -2x + 11$, tangent to the curve at $(13, -15)$.

31. By symmetry of the ellipse about the x- and y-axes,

$$A = 4 \int_0^a y \, dx = 4 \int_{\pi/2}^0 b \sin \theta \, (-a \sin \theta) \, d\theta = 4ab \int_0^{\pi/2} \sin^2 \theta \, d\theta = 4ab \int_0^{\pi/2} \tfrac{1}{2}(1 - \cos 2\theta) \, d\theta$$

$$= 2ab \big[\theta - \tfrac{1}{2} \sin 2\theta \big]_0^{\pi/2} = 2ab \big(\tfrac{\pi}{2}\big) = \pi ab$$

32. The curve $x = t^2 - 2t = t(t - 2)$, $y = \sqrt{t}$ intersects the y-axis when $x = 0$,

that is, when $t = 0$ and $t = 2$. The corresponding values of y are 0 and $\sqrt{2}$.

The shaded area is given by

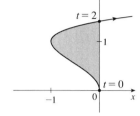

$$\int_{y=0}^{y=\sqrt{2}} (x_R - x_L) \, dy = \int_{t=0}^{t=2} [0 - x(t)] \, y'(t) \, dt = -\int_0^2 (t^2 - 2t)\left(\frac{1}{2\sqrt{t}} \, dt\right)$$

$$= -\int_0^2 \left(\tfrac{1}{2} t^{3/2} - t^{1/2}\right) dt = -\left[\tfrac{1}{5} t^{5/2} - \tfrac{2}{3} t^{3/2}\right]_0^2$$

$$= -\left(\tfrac{1}{5} \cdot 2^{5/2} - \tfrac{2}{3} \cdot 2^{3/2}\right) = -2^{1/2}\left(\tfrac{4}{5} - \tfrac{4}{3}\right) = -\sqrt{2}\left(-\tfrac{8}{15}\right) = \tfrac{8}{15}\sqrt{2}$$

33. The curve $x = 1 + e^t$, $y = t - t^2 = t(1 - t)$ intersects the x-axis when $y = 0$,

that is, when $t = 0$ and $t = 1$. The corresponding values of x are 2 and $1 + e$.

The shaded area is given by

$$\int_{x=2}^{x=1+e} (y_T - y_B) \, dx = \int_{t=0}^{t=1} [y(t) - 0] \, x'(t) \, dt = \int_0^1 (t - t^2) e^t \, dt$$

$$= \int_0^1 t e^t \, dt - \int_0^1 t^2 e^t \, dt = \int_0^1 t e^t \, dt - \left[t^2 e^t\right]_0^1 + 2\int_0^1 t e^t \, dt \qquad \text{[Formula 97 or parts]}$$

$$= 3\int_0^1 t e^t \, dt - (e - 0) = 3\left[(t - 1)e^t\right]_0^1 - e \qquad \text{[Formula 96 or parts]}$$

$$= 3[0 - (-1)] - e = 3 - e$$

34. By symmetry, $A = 4\int_0^a y \, dx = 4\int_{\pi/2}^0 a \sin^3 \theta(-3a \cos^2 \theta \sin \theta) \, d\theta = 12a^2 \int_0^{\pi/2} \sin^4 \theta \, \cos^2 \theta \, d\theta$. Now

$$\int \sin^4 \theta \, \cos^2 \theta \, d\theta = \int \sin^2 \theta \big(\tfrac{1}{4} \sin^2 2\theta\big) \, d\theta = \tfrac{1}{8}\int (1 - \cos 2\theta) \sin^2 2\theta \, d\theta$$

$$= \tfrac{1}{8}\int \left[\tfrac{1}{2}(1 - \cos 4\theta) - \sin^2 2\theta \, \cos 2\theta\right] d\theta = \tfrac{1}{16}\theta - \tfrac{1}{64} \sin 4\theta - \tfrac{1}{48} \sin^3 2\theta + C$$

so $\int_0^{\pi/2} \sin^4 \theta \, \cos^2 \theta \, d\theta = \left[\tfrac{1}{16}\theta - \tfrac{1}{64} \sin 4\theta - \tfrac{1}{48} \sin^3 2\theta\right]_0^{\pi/2} = \tfrac{\pi}{32}$. Thus, $A = 12a^2\big(\tfrac{\pi}{32}\big) = \tfrac{3}{8}\pi a^2$.

35. $x = r\theta - d \sin \theta$, $y = r - d \cos \theta$.

$$A = \int_0^{2\pi r} y \, dx = \int_0^{2\pi} (r - d \cos \theta)(r - d \cos \theta) \, d\theta = \int_0^{2\pi} (r^2 - 2dr \cos \theta + d^2 \cos^2 \theta) \, d\theta$$

$$= \left[r^2 \theta - 2dr \sin \theta + \tfrac{1}{2}d^2\big(\theta + \tfrac{1}{2}\sin 2\theta\big)\right]_0^{2\pi} = 2\pi r^2 + \pi d^2$$

36. (a) By symmetry, the area of $\mathcal{R}$ is twice the area inside $\mathcal{R}$ above the x-axis. The top half of the loop is described by

$$x = t^2, \ y = t^3 - 3t, \ -\sqrt{3} \le t \le 0, \text{ so, using the Substitution Rule with } y = t^3 - 3t \text{ and } dx = 2t \, dt, \text{ we find that}$$

$$\text{area} = 2\int_0^3 y \, dx = 2\int_0^{-\sqrt{3}} (t^3 - 3t) 2t \, dt = 2\int_0^{-\sqrt{3}} (2t^4 - 6t^2) \, dt = 2\left[\tfrac{2}{5}t^5 - 2t^3\right]_0^{-\sqrt{3}}$$

$$= 2\left[\tfrac{2}{5}(-3^{1/2})^5 - 2(-3^{1/2})^3\right] = 2\left[\tfrac{2}{5}(-9\sqrt{3}) - 2(-3\sqrt{3})\right] = \tfrac{24}{5}\sqrt{3}$$

(b) Here we use the formula for disks and use the Substitution Rule as in part (a):

$$\text{volume} = \pi \int_0^3 y^2 \, dx = \pi \int_0^{-\sqrt{3}} (t^3 - 3t)^2 2t \, dt = 2\pi \int_0^{-\sqrt{3}} (t^6 - 6t^4 + 9t^2) t \, dt = 2\pi \left[\tfrac{1}{8} t^8 - t^6 + \tfrac{9}{4} t^4 \right]_0^{-\sqrt{3}}$$

$$= 2\pi \left[\tfrac{1}{8} (-3^{1/2})^8 - (-3^{1/2})^6 + \tfrac{9}{4} (-3^{1/2})^4 \right] = 2\pi \left[\tfrac{81}{8} - 27 + \tfrac{81}{4} \right] = \tfrac{27}{4} \pi$$

(c) By symmetry, the y-coordinate of the centroid is 0. To find the x-coordinate, we note that it is the same as the x-coordinate of the centroid of the top half of $\mathcal{R}$, the area of which is $\tfrac{1}{2} \cdot \tfrac{24}{5} \sqrt{3} = \tfrac{12}{5} \sqrt{3}$. So, using Formula 9.3.8 with $A = \tfrac{12}{5} \sqrt{3}$, we get

$$\overline{x} = \tfrac{5}{12\sqrt{3}} \int_0^3 xy \, dx = \tfrac{5}{12\sqrt{3}} \int_0^{-\sqrt{3}} t^2 (t^3 - 3t) 2t \, dt = \tfrac{5}{6\sqrt{3}} \left[\tfrac{1}{7} t^7 - \tfrac{3}{5} t^5 \right]_0^{-\sqrt{3}}$$

$$= \tfrac{5}{6\sqrt{3}} \left[\tfrac{1}{7} (-3^{1/2})^7 - \tfrac{3}{5} (-3^{1/2})^5 \right] = \tfrac{5}{6\sqrt{3}} \left[-\tfrac{27}{7} \sqrt{3} + \tfrac{27}{5} \sqrt{3} \right] = \tfrac{9}{7}$$

So the coordinates of the centroid of $\mathcal{R}$ are $(x, y) = \left(\tfrac{9}{7}, 0 \right)$.

37. $x = t - t^2$, $y = \tfrac{4}{3} t^{3/2}$, $1 \le t \le 2$. $dx/dt = 1 - 2t$ and $dy/dt = 2t^{1/2}$, so

$$(dx/dt)^2 + (dy/dt)^2 = (1 - 2t)^2 + (2t^{1/2})^2 = 1 - 4t + 4t^2 + 4t = 1 + 4t^2.$$

Thus, $L = \int_a^b \sqrt{(dx/dt)^2 + (dy/dt)^2} \, dt = \int_1^2 \sqrt{1 + 4t^2} \, dt \approx 3.1678.$

38. $x = 1 + e^t$, $y = t^2$, $-3 \le t \le 3$. $dx/dt = e^t$ and $dy/dt = 2t$, so $(dx/dt)^2 + (dy/dt)^2 = e^{2t} + 4t^2.$

Thus, $L = \int_a^b \sqrt{(dx/dt)^2 + (dy/dt)^2} \, dt = \int_{-3}^3 \sqrt{e^{2t} + 4t^2} \, dt \approx 30.5281.$

39. $x = t + \cos t$, $y = t - \sin t$, $0 \le t \le 2\pi$. $dx/dt = 1 - \sin t$ and $dy/dt = 1 - \cos t$, so

$$\left(\tfrac{dx}{dt} \right)^2 + \left(\tfrac{dy}{dt} \right)^2 = (1 - \sin t)^2 + (1 - \cos t)^2 = (1 - 2\sin t + \sin^2 t) + (1 - 2\cos t + \cos^2 t) = 3 - 2\sin t - 2\cos t.$$

Thus, $L = \int_a^b \sqrt{(dx/dt)^2 + (dy/dt)^2} \, dt = \int_0^{2\pi} \sqrt{3 - 2\sin t - 2\cos t} \, dt \approx 10.0367.$

40. $x = \ln t$, $y = \sqrt{t + 1}$, $1 \le t \le 5$. $\dfrac{dx}{dt} = \dfrac{1}{t}$ and $\dfrac{dy}{dt} = \dfrac{1}{2\sqrt{t+1}}$, so $\left(\dfrac{dx}{dt} \right)^2 + \left(\dfrac{dy}{dt} \right)^2 = \dfrac{1}{t^2} + \dfrac{1}{4(t+1)} = \dfrac{t^2 + 4t + 4}{4t^2(t+1)}.$

Thus, $L = \int_a^b \sqrt{\left(\dfrac{dx}{dt} \right)^2 + \left(\dfrac{dy}{dt} \right)^2} \, dt = \int_1^5 \sqrt{\dfrac{t^2 + 4t + 4}{4t^2(t+1)}} \, dt = \int_1^5 \sqrt{\dfrac{(t+2)^2}{(2t)^2(t+1)}} \, dt = \int_1^5 \dfrac{t+2}{2t\sqrt{t+1}} \, dt \approx 1.9310.$

41.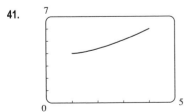

$x = 1 + 3t^2$, $y = 4 + 2t^3$, $0 \le t \le 1$.

$dx/dt = 6t$ and $dy/dt = 6t^2$, so $(dx/dt)^2 + (dy/dt)^2 = 36t^2 + 36t^4.$

Thus, $L = \int_0^1 \sqrt{36t^2 + 36t^4} \, dt = \int_0^1 6t \sqrt{1 + t^2} \, dt$

$$= 6 \int_1^2 \sqrt{u} \left(\tfrac{1}{2} du \right) \quad [u = 1 + t^2, \, du = 2t \, dt]$$

$$= 3 \left[\tfrac{2}{3} u^{3/2} \right]_1^2 = 2(2^{3/2} - 1) = 2 \left(2\sqrt{2} - 1 \right)$$

42.

$x = e^t + e^{-t}$, $y = 5 - 2t$, $0 \le t \le 3$.

$dx/dt = e^t - e^{-t}$ and $dy/dt = -2$, so

$\left(\frac{dx}{dt}\right)^2 + \left(\frac{dy}{dt}\right)^2 = e^{2t} - 2 + e^{-2t} + 4 = e^{2t} + 2 + e^{-2t} = (e^t + e^{-t})^2$ and

$L = \int_0^3 (e^t + e^{-t})\, dt = \left[e^t - e^{-t}\right]_0^3 = e^3 - e^{-3} - (1 - 1) = e^3 - e^{-3}$.

43. $x = \dfrac{t}{1+t}$, $y = \ln(1+t)$, $0 \le t \le 2$.　$\dfrac{dx}{dt} = \dfrac{(1+t)\cdot 1 - t\cdot 1}{(1+t)^2} = \dfrac{1}{(1+t)^2}$ and $\dfrac{dy}{dt} = \dfrac{1}{1+t}$,

so $\left(\dfrac{dx}{dt}\right)^2 + \left(\dfrac{dy}{dt}\right)^2 = \dfrac{1}{(1+t)^4} + \dfrac{1}{(1+t)^2} = \dfrac{1}{(1+t)^4}\left[1 + (1+t)^2\right] = \dfrac{t^2 + 2t + 2}{(1+t)^4}$. Thus,

$$L = \int_0^2 \frac{\sqrt{t^2 + 2t + 2}}{(1+t)^2}\, dt = \int_1^3 \frac{\sqrt{u^2+1}}{u^2}\, du \quad \begin{bmatrix} u = t+1, \\ du = dt \end{bmatrix} \stackrel{24}{=} \left[-\frac{\sqrt{u^2+1}}{u} + \ln\left(u + \sqrt{u^2+1}\right) \right]_1^3$$

$$= -\frac{\sqrt{10}}{3} + \ln\left(3 + \sqrt{10}\right) + \sqrt{2} - \ln\left(1 + \sqrt{2}\right)$$

44. $x = 3\cos t - \cos 3t$, $y = 3\sin t - \sin 3t$, $0 \le t \le \pi$.　$\frac{dx}{dt} = -3\sin t + 3\sin 3t$ and $\frac{dy}{dt} = 3\cos t - 3\cos 3t$, so

$$\left(\tfrac{dx}{dt}\right)^2 + \left(\tfrac{dy}{dt}\right)^2 = 9\sin^2 t - 18\sin t \sin 3t + 9\sin^2(3t) + 9\cos^2 t - 18\cos t \cos 3t + 9\cos^2(3t)$$

$$= 9(\cos^2 t + \sin^2 t) - 18(\cos t \cos 3t + \sin t \sin 3t) + 9[\cos^2(3t) + \sin^2(3t)]$$

$$= 9(1) - 18\cos(t - 3t) + 9(1) = 18 - 18\cos(-2t) = 18(1 - \cos 2t)$$

$$= 18[1 - (1 - 2\sin^2 t)] = 36\sin^2 t.$$

Thus, $L = \int_0^\pi \sqrt{36\sin^2 t}\, dt = 6\int_0^\pi |\sin t|\, dt = 6\int_0^\pi \sin t\, dt = -6\left[\cos t\right]_0^\pi = -6(-1 - 1) = 12$.

45. $x = e^t \cos t$, $y = e^t \sin t$, $0 \le t \le \pi$.

$$\left(\tfrac{dx}{dt}\right)^2 + \left(\tfrac{dy}{dt}\right)^2 = [e^t(\cos t - \sin t)]^2 + [e^t(\sin t + \cos t)]^2$$

$$= (e^t)^2(\cos^2 t - 2\cos t \sin t + \sin^2 t)$$

$$\qquad + (e^t)^2(\sin^2 t + 2\sin t \cos t + \cos^2 t$$

$$= e^{2t}(2\cos^2 t + 2\sin^2 t) = 2e^{2t}$$

Thus, $L = \int_0^\pi \sqrt{2e^{2t}}\, dt = \int_0^\pi \sqrt{2}\, e^t\, dt = \sqrt{2}\left[e^t\right]_0^\pi = \sqrt{2}(e^\pi - 1)$.

46. $x = \cos t + \ln(\tan \tfrac{1}{2}t)$, 　$y = \sin t$, 　$\pi/4 \le t \le 3\pi/4$.

$\dfrac{dx}{dt} = -\sin t + \dfrac{\tfrac{1}{2}\sec^2(t/2)}{\tan(t/2)} = -\sin t + \dfrac{1}{2\sin(t/2)\,\cos(t/2)} = -\sin t + \dfrac{1}{\sin t}$ and $\dfrac{dy}{dt} = \cos t$, so

$\left(\dfrac{dx}{dt}\right)^2 + \left(\dfrac{dy}{dt}\right)^2 = \sin^2 t - 2 + \dfrac{1}{\sin^2 t} + \cos^2 t = 1 - 2 + \csc^2 t = \cot^2 t$. Thus,

$$L = \int_{\pi/4}^{3\pi/4} |\cot t|\, dt = 2\int_{\pi/4}^{\pi/2} \cot t\, dt$$

$$= 2\Big[\ln|\sin t|\Big]_{\pi/4}^{\pi/2} = 2\Big(\ln 1 - \ln \frac{1}{\sqrt{2}}\Big)$$

$$= 2\big(0 + \ln\sqrt{2}\,\big) = 2\big(\tfrac{1}{2}\ln 2\big) = \ln 2.$$

47.

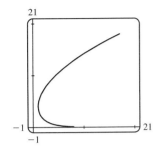

$$x = e^t - t,\ y = 4e^{t/2},\ -8 \le t \le 3$$

$$\Big(\tfrac{dx}{dt}\Big)^2 + \Big(\tfrac{dy}{dt}\Big)^2 = (e^t - 1)^2 + (2e^{t/2})^2 = e^{2t} - 2e^t + 1 + 4e^t$$

$$= e^{2t} + 2e^t + 1 = (e^t + 1)^2$$

$$L = \int_{-8}^{3} \sqrt{(e^t + 1)^2}\, dt = \int_{-8}^{3}(e^t + 1)\, dt = \big[e^t + t\big]_{-8}^{3t}$$

$$= (e^3 + 3) - (e^{-8} - 8) = e^3 - e^{-8} + 11$$

48. $x = 3t - t^3, y = 3t^2.$ $dx/dt = 3 - 3t^2$ and $dy/dt = 6t$, so

$$\Big(\tfrac{dx}{dt}\Big)^2 + \Big(\tfrac{dy}{dt}\Big)^2 = (3 - 3t^2)^2 + (6t)^2 = (3 + 3t^2)^2$$

and the length of the loop is given by

$$L = \int_{-\sqrt{3}}^{\sqrt{3}} (3 + 3t^2)\, dt = 2\int_{0}^{\sqrt{3}} (3 + 3t^2)\, dt = 2[3t + t^3]_0^{\sqrt{3}}$$

$$= 2\big(3\sqrt{3} + 3\sqrt{3}\,\big) = 12\sqrt{3}.$$

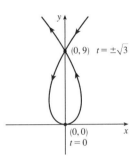

49. $x = t - e^t,\ y = t + e^t,\ -6 \le t \le 6.$

$$\Big(\tfrac{dx}{dt}\Big)^2 + \Big(\tfrac{dy}{dt}\Big)^2 = (1 - e^t)^2 + (1 + e^t)^2 = (1 - 2e^t + e^{2t}) + (1 + 2e^t + e^{2t}) = 2 + 2e^{2t}, \text{ so } L = \int_{-6}^{6} \sqrt{2 + 2e^{2t}}\, dt.$$

Set $f(t) = \sqrt{2 + 2e^{2t}}$. Then by Simpson's Rule with $n = 6$ and $\Delta t = \frac{6-(-6)}{6} = 2$, we get

$$L \approx \tfrac{2}{3}[f(-6) + 4f(-4) + 2f(-2) + 4f(0) + 2f(2) + 4f(4) + f(6)] \approx 612.3053.$$

50. $x = 2a\cot\theta \ \Rightarrow\ dx/dt = -2a\csc^2\theta$ and $y = 2a\sin^2\theta \ \Rightarrow\ dy/dt = 4a\sin\theta\cos\theta = 2a\sin 2\theta.$

So $L = \int_{\pi/4}^{\pi/2}\sqrt{4a^2\csc^4\theta + 4a^2\sin^2 2\theta}\, d\theta = 2a\int_{\pi/4}^{\pi/2}\sqrt{\csc^4\theta + \sin^2 2\theta}\, d\theta.$ Using Simpson's Rule with

$n = 4, \Delta\theta = \frac{\pi/2 - \pi/4}{4} = \frac{\pi}{16},$ and $f(\theta) = \sqrt{\csc^4\theta + \sin^2 2\theta}$, we get

$$L \approx 2a \cdot S_4 = (2a)\tfrac{\pi}{16\cdot 3}\Big[f\big(\tfrac{\pi}{4}\big) + 4f\big(\tfrac{5\pi}{16}\big) + 2f\big(\tfrac{3\pi}{8}\big) + 4f\big(\tfrac{7\pi}{16}\big) + f\big(\tfrac{\pi}{2}\big)\Big] \approx 2.2605a.$$

51. $x = \sin^2 t,\ y = \cos^2 t,\ 0 \le t \le 3\pi.$

$$(dx/dt)^2 + (dy/dt)^2 = (2\sin t\cos t)^2 + (-2\cos t\sin t)^2 = 8\sin^2 t\cos^2 t = 2\sin^2 2t \ \Rightarrow$$

$$\text{Distance} = \int_0^{3\pi}\sqrt{2}\,|\sin 2t|\, dt = 6\sqrt{2}\int_0^{\pi/2}\sin 2t\, dt \ \text{[by symmetry]}\ = -3\sqrt{2}\Big[\cos 2t\Big]_0^{\pi/2} = -3\sqrt{2}\,(-1 - 1) = 6\sqrt{2}.$$

The full curve is traversed as t goes from 0 to $\frac{\pi}{2}$, because the curve is the segment of $x + y = 1$ that lies in the first quadrant

(since $x, y \ge 0$), and this segment is completely traversed as t goes from 0 to $\frac{\pi}{2}$. Thus, $L = \int_0^{\pi/2}\sin 2t\, dt = \sqrt{2}$, as above.

52. $x = \cos^2 t$, $y = \cos t$, $0 \le t \le 4\pi$. $\left(\frac{dx}{dt}\right)^2 + \left(\frac{dy}{dt}\right)^2 = (-2\cos t \sin t)^2 + (-\sin t)^2 = \sin^2 t \, (4\cos^2 t + 1)$

$$\text{Distance} = \int_0^{4\pi} |\sin t| \sqrt{4\cos^2 t + 1}\, dt = 4\int_0^\pi \sin t \sqrt{4\cos^2 t + 1}\, dt$$

$$= -4\int_1^{-1} \sqrt{4u^2 + 1}\, du \quad [u = \cos t,\, du = -\sin t\, dt] \quad = 4\int_{-1}^1 \sqrt{4u^2 + 1}\, du$$

$$= 8\int_0^1 \sqrt{4u^2 + 1}\, du = 8\int_0^{\tan^{-1} 2} \sec\theta \cdot \tfrac{1}{2}\sec^2\theta\, d\theta \quad [2u = \tan\theta,\, 2\, du = \sec^2\theta\, d\theta]$$

$$= 4\int_0^{\tan^{-1} 2} \sec^3\theta\, d\theta \overset{71}{=} \left[2\sec\theta\tan\theta + 2\ln|\sec\theta + \tan\theta|\right]_0^{\tan^{-1} 2} = 4\sqrt{5} + 2\ln\left(\sqrt{5} + 2\right)$$

Thus, $L = \int_0^\pi |\sin t| \sqrt{4\cos^2 t + 1}\, dt = \sqrt{5} + \tfrac{1}{2}\ln\left(\sqrt{5} + 2\right)$.

53. $x = a\sin\theta$, $y = b\cos\theta$, $0 \le \theta \le 2\pi$.

$$\left(\frac{dx}{dt}\right)^2 + \left(\frac{dy}{dt}\right)^2 = (a\cos\theta)^2 + (-b\sin\theta)^2 = a^2\cos^2\theta + b^2\sin^2\theta = a^2(1 - \sin^2\theta) + b^2\sin^2\theta$$

$$= a^2 - (a^2 - b^2)\sin^2\theta = a^2 - c^2\sin^2\theta = a^2\left(1 - \frac{c^2}{a^2}\sin^2\theta\right) = a^2(1 - e^2\sin^2\theta)$$

So $L = 4\int_0^{\pi/2} \sqrt{a^2\left(1 - e^2\sin^2\theta\right)}\, d\theta$ [by symmetry] $= 4a\int_0^{\pi/2} \sqrt{1 - e^2\sin^2\theta}\, d\theta$.

54. $x = a\cos^3\theta$, $y = a\sin^3\theta$.

$$\left(\frac{dx}{dt}\right)^2 + \left(\frac{dy}{dt}\right)^2 = (-3a\cos^2\theta\,\sin\theta)^2 + (3a\sin^2\theta\,\cos\theta)^2$$

$$= 9a^2\cos^4\theta\,\sin^2\theta + 9a^2\sin^4\theta\,\cos^2\theta$$

$$= 9a^2\sin^2\theta\,\cos^2\theta(\cos^2\theta + \sin^2\theta) = 9a^2\sin^2\theta\,\cos^2\theta.$$

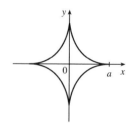

The graph has four-fold symmetry and the curve in the first quadrant corresponds to $0 \le \theta \le \pi/2$. Thus,

$$L = 4\int_0^{\pi/2} 3a\sin\theta\,\cos\theta\, d\theta \qquad \text{[since } a > 0 \text{ and } \sin\theta \text{ and } \cos\theta \text{ are positive for } 0 \le \theta \le \pi/2\text{]}$$

$$= 12a\left[\tfrac{1}{2}\sin^2\theta\right]_0^{\pi/2} = 12a\left(\tfrac{1}{2} - 0\right) = 6a$$

55. (a) $x = 11\cos t - 4\cos(11t/2)$, $y = 11\sin t - 4\sin(11t/2)$.

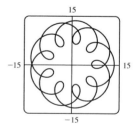

Notice that $0 \le t \le 2\pi$ does not give the complete curve because $x(0) \ne x(2\pi)$. In fact, we must take $t \in [0, 4\pi]$ in order to obtain the complete curve, since the first term in each of the parametric equations has period 2π and the second has period $\frac{2\pi}{11/2} = \frac{4\pi}{11}$, and the least common integer multiple of these two numbers is 4π.

(b) We use the CAS to find the derivatives dx/dt and dy/dt, and then use Theorem 6 to find the arc length. Recent versions of Maple express the integral $\int_0^{4\pi} \sqrt{(dx/dt)^2 + (dy/dt)^2}\, dt$ as $88E\left(2\sqrt{2}\,i\right)$, where $E(x)$ is the elliptic integral

$$\int_0^1 \frac{\sqrt{1 - x^2 t^2}}{\sqrt{1 - t^2}}\, dt$$ and i is the imaginary number $\sqrt{-1}$.

Some earlier versions of Maple (as well as Mathematica) cannot do the integral exactly, so we use the command
`evalf(Int(sqrt(diff(x,t)^2+diff(y,t)^2),t=0..4*Pi));` to estimate the length, and find that the arc length is approximately 294.03. Derive's `Para_arc_length` function in the utility file `Int_apps` simplifies the integral to $11 \int_0^{4\pi} \sqrt{-4\cos t\, \cos\left(\frac{11t}{2}\right) - 4\sin t\, \sin\left(\frac{11t}{2}\right) + 5}\, dt$.

56. (a) It appears that as $t \to \infty$, $(x, y) \to \left(\frac{1}{2}, \frac{1}{2}\right)$, and as $t \to -\infty$, $(x, y) \to \left(-\frac{1}{2}, -\frac{1}{2}\right)$.

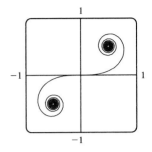

(b) By the Fundamental Theorem of Calculus, $dx/dt = \cos\left(\frac{\pi}{2}t^2\right)$ and $dy/dt = \sin\left(\frac{\pi}{2}t^2\right)$, so by Formula 4, the length of the curve from the origin to the point with parameter value t is

$$L = \int_0^t \sqrt{\left(\frac{dx}{du}\right)^2 + \left(\frac{dy}{du}\right)^2}\, du = \int_0^t \sqrt{\cos^2\left(\frac{\pi}{2}u^2\right) + \sin^2\left(\frac{\pi}{2}u^2\right)}\, du$$
$$= \int_0^t 1\, du = t \qquad [\text{or } -t \text{ if } t < 0]$$

We have used u as the dummy variable so as not to confuse it with the upper limit of integration.

57. $x = 1 + te^t$, $y = (t^2 + 1)e^t$, $0 \le t \le 1$.

$$\left(\frac{dx}{dt}\right)^2 + \left(\frac{dy}{dt}\right)^2 = (te^t + e^t)^2 + [(t^2 + 1)e^t + e^t(2t)]^2 = [e^t(t + 1)]^2 + [e^t(t^2 + 2t + 1)]^2$$
$$= e^{2t}(t + 1)^2 + e^{2t}(t + 1)^4 = e^{2t}(t + 1)^2[1 + (t + 1)^2], \quad \text{so}$$

$$S = \int 2\pi y\, ds = \int_0^1 2\pi(t^2 + 1)e^t \sqrt{e^{2t}(t + 1)^2(t^2 + 2t + 2)}\, dt = \int_0^1 2\pi(t^2 + 1)e^{2t}(t + 1)\sqrt{t^2 + 2t + 2}\, dt \approx 103.5999$$

58. $x = \sin^2 t$, $y = \sin 3t$, $0 \le t \le \frac{\pi}{3}$. $dx/dt = 2\sin t \cos t = \sin 2t$ and $dy/dt = 3\cos 3t$, so

$$(dx/dt)^2 + (dy/dt)^2 = \sin^2 2t + 9\cos^2 3t \text{ and } S = \int 2\pi y\, ds = \int_0^{\pi/3} 2\pi \sin 3t \sqrt{\sin^2 2t + 9\cos^2 3t}\, dt \approx 7.4775.$$

59. $x = t^3$, $y = t^2$, $0 \le t \le 1$. $\left(\frac{dx}{dt}\right)^2 + \left(\frac{dy}{dt}\right)^2 = \left(3t^2\right)^2 + (2t)^2 = 9t^4 + 4t^2$.

$$S = \int_0^1 2\pi y \sqrt{\left(\frac{dx}{dt}\right)^2 + \left(\frac{dy}{dt}\right)^2}\, dt = \int_0^1 2\pi t^2 \sqrt{9t^4 + 4t^2}\, dt = 2\pi \int_0^1 t^2 \sqrt{t^2(9t^2 + 4)}\, dt$$

$$= 2\pi \int_4^{13} \left(\frac{u - 4}{9}\right)\sqrt{u}\left(\frac{1}{18}\, du\right) \quad \begin{bmatrix} u = 9t^2 + 4, \ t^2 = (u - 4)/9, \\ du = 18t\, dt, \text{ so } t\, dt = \frac{1}{18}\, du \end{bmatrix} \quad = \frac{2\pi}{9 \cdot 18} \int_4^{13} (u^{3/2} - 4u^{1/2})\, du$$

$$= \frac{\pi}{81}\left[\frac{2}{5}u^{5/2} - \frac{8}{3}u^{3/2}\right]_4^{13} = \frac{\pi}{81} \cdot \frac{2}{15}\left[3u^{5/2} - 20u^{3/2}\right]_4^{13}$$

$$= \frac{2\pi}{1215}\left[\left(3 \cdot 13^2 \sqrt{13} - 20 \cdot 13 \sqrt{13}\right) - (3 \cdot 32 - 20 \cdot 8)\right] = \frac{2\pi}{1215}\left(247\sqrt{13} + 64\right)$$

60. $x = 3t - t^3$, $y = 3t^2$, $0 \le t \le 1$. $\left(\frac{dx}{dt}\right)^2 + \left(\frac{dy}{dt}\right)^2 = (3 - 3t^2)^2 + (6t)^2 = 9(1 + 2t^2 + t^4) = [3(1 + t^2)]^2$.

$$S = \int_0^1 2\pi \cdot 3t^2 \cdot 3(1 + t^2)\, dt = 18\pi \int_0^1 (t^2 + t^4)\, dt = 18\pi\left[\frac{1}{3}t^3 + \frac{1}{5}t^5\right]_0^1 = \frac{48}{5}\pi$$

61. $x = a\cos^3\theta$, $y = a\sin^3\theta$, $0 \le \theta \le \frac{\pi}{2}$. $\left(\frac{dx}{d\theta}\right)^2 + \left(\frac{dy}{d\theta}\right)^2 = (-3a\cos^2\theta \sin\theta)^2 + (3a\sin^2\theta \cos\theta)^2 = 9a^2\sin^2\theta \cos^2\theta$.

$$S = \int_0^{\pi/2} 2\pi \cdot a\sin^3\theta \cdot 3a\sin\theta \cos\theta\, d\theta = 6\pi a^2 \int_0^{\pi/2} \sin^4\theta \cos\theta\, d\theta = \frac{6}{5}\pi a^2\left[\sin^5\theta\right]_0^{\pi/2} = \frac{6}{5}\pi a^2$$

62. $\left(\frac{dx}{d\theta}\right)^2 + \left(\frac{dy}{d\theta}\right)^2 = (-2\sin\theta + 2\sin 2\theta)^2 + (2\cos\theta - 2\cos 2\theta)^2$

$$= 4[(\sin^2\theta - 2\sin\theta\,\sin 2\theta + \sin^2 2\theta) + (\cos^2\theta - 2\cos\theta\,\cos 2\theta + \cos^2 2\theta)]$$

$$= 4[1 + 1 - 2(\cos 2\theta\cos\theta + \sin 2\theta\sin\theta)] = 8[1 - \cos(2\theta - \theta)] = 8(1 - \cos\theta)$$

We plot the graph with parameter interval $[0, 2\pi]$, and see that we should only

integrate between 0 and π. (If the interval $[0, 2\pi]$ were taken, the surface of

revolution would be generated twice.) Also note that

$y = 2\sin\theta - \sin 2\theta = 2\sin\theta(1 - \cos\theta)$. So

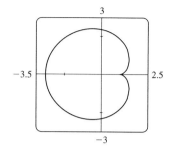

$S = \int_0^\pi 2\pi \cdot 2\sin\theta(1 - \cos\theta)\, 2\sqrt{2}\sqrt{1 - \cos\theta}\, d\theta$

$\quad = 8\sqrt{2}\pi \int_0^\pi (1 - \cos\theta)^{3/2}\sin\theta\, d\theta = 8\sqrt{2}\pi \int_0^2 \sqrt{u^3}\, du \quad \left[\begin{array}{l} u = 1 - \cos\theta, \\ du = \sin\theta\, d\theta \end{array}\right]$

$\quad = 8\sqrt{2}\pi \left[\left(\frac{2}{5}\right)u^{5/2}\right]_0^2 = \frac{16}{5}\sqrt{2}\pi(2^{5/2}) = \frac{128}{5}\pi$

63. $x = t + t^3,\ y = t - \dfrac{1}{t^2},\ 1 \le t \le 2.\quad \frac{dx}{dt} = 1 + 3t^2$ and $\frac{dy}{dt} = 1 + \dfrac{2}{t^3}$, so $\left(\frac{dx}{dt}\right)^2 + \left(\frac{dy}{dt}\right)^2 = (1 + 3t^2)^2 + \left(1 + \dfrac{2}{t^3}\right)^2$

and $S = \displaystyle\int 2\pi y\, ds = \int_1^2 2\pi\left(t - \dfrac{1}{t^2}\right)\sqrt{(1 + 3t^2)^2 + \left(1 + \dfrac{2}{t^3}\right)^2}\, dt \approx 59.101.$

64. $S = \int_{\pi/4}^{\pi/2} 2\pi \cdot 2a\sin^2\theta\sqrt{\csc^4\theta + \sin^2 2\theta}\, dt = 4\pi a \int_{\pi/4}^{\pi/2} \sin^2\theta\sqrt{\csc^4\theta + \sin^2 2\theta}\, d\theta.$

Using Simpson's Rule with $n = 4$, $\Delta\theta = \frac{\pi/2 - \pi/4}{4} = \frac{\pi}{16}$, and $f(\theta) = \sin^2\theta\sqrt{\csc^4\theta + \sin^2 2\theta}$, we get

$S \approx (4\pi a)\frac{\pi}{16\cdot 3}\left[f\left(\frac{\pi}{4}\right) + 4f\left(\frac{5\pi}{16}\right) + 2f\left(\frac{3\pi}{8}\right) + 4f\left(\frac{7\pi}{16}\right) + f\left(\frac{\pi}{2}\right)\right] \approx 11.0893a.$

65. $x = 3t^2,\ y = 2t^3,\ 0 \le t \le 5\ \Rightarrow\ \left(\frac{dx}{dt}\right)^2 + \left(\frac{dy}{dt}\right)^2 = (6t)^2 + (6t^2)^2 = 36t^2(1 + t^2)\ \Rightarrow$

$S = \int_0^5 2\pi x\sqrt{(dx/dt)^2 + (dy/dt)^2}\, dt = \int_0^5 2\pi(3t^2)6t\sqrt{1 + t^2}\, dt = 18\pi \int_0^5 t^2\sqrt{1 + t^2}\, 2t\, dt$

$\quad = 18\pi \int_1^{26} (u - 1)\sqrt{u}\, du \quad \left[\begin{array}{l} u = 1 + t^2, \\ du = 2t\, dt \end{array}\right] \quad = 18\pi \int_1^{26} (u^{3/2} - u^{1/2})\, du = 18\pi\left[\frac{2}{5}u^{5/2} - \frac{2}{3}u^{3/2}\right]_1^{26}$

$\quad = 18\pi\left[\left(\frac{2}{5}\cdot 676\sqrt{26} - \frac{2}{3}\cdot 26\sqrt{26}\right) - \left(\frac{2}{5} - \frac{2}{3}\right)\right] = \frac{24}{5}\pi\left(949\sqrt{26} + 1\right)$

66. $x = e^t - t,\ y = 4e^{t/2},\ 0 \le t \le 1.$ $\left(\frac{dx}{dt}\right)^2 + \left(\frac{dy}{dt}\right)^2 = (e^t - 1)^2 + (2e^{t/2})^2 = e^{2t} + 2e^t + 1 = (e^t + 1)^2.$

$S = \int_0^1 2\pi(e^t - t)\sqrt{(e^t - 1)^2 + (2e^{t/2})^2}\, dt = \int_0^1 2\pi(e^t - t)(e^t + 1)d$

$\quad = 2\pi\left[\frac{1}{2}e^{2t} + e^t - (t - 1)e^t - \frac{1}{2}t^2\right]_0^1 = \pi(e^2 + 2e - 6)$

67. If f' is continuous and $f'(t) \ne 0$ for $a \le t \le b$, then either $f'(t) > 0$ for all t in $[a, b]$ or $f'(t) < 0$ for all t in $[a, b]$. Thus, f

is monotonic (in fact, strictly increasing or strictly decreasing) on $[a, b]$. It follows that f has an inverse. Set $F = g \circ f^{-1}$,

that is, define F by $F(x) = g(f^{-1}(x))$. Then $x = f(t)\ \Rightarrow\ f^{-1}(x) = t$, so $y = g(t) = g(f^{-1}(x)) = F(x)$.

68. By Formula 9.2.5 with $y = F(x)$, $S = \int_a^b 2\pi F(x)\sqrt{1 + [F'(x)]^2}\, dx$. But by Formula 11.2.2,

$$1 + [F'(x)]^2 = 1 + \left(\frac{dy}{dx}\right)^2 = 1 + \left(\frac{dy/dt}{dx/dt}\right)^2 = \frac{(dx/dt)^2 + (dy/dt)^2}{(dx/dt)^2}.$$ Using the Substitution Rule with $x = x(t)$,

where $a = x(\alpha)$ and $b = x(\beta)$, we have $\left[\text{since } dx = \dfrac{dx}{dt}\, dt\right]$

$$S = \int_\alpha^\beta 2\pi\, F(x(t))\sqrt{\frac{(dx/dt)^2 + (dy/dt)^2}{(dx/dt)^2}}\,\frac{dx}{dt}\, dt = \int_\alpha^\beta 2\pi y\sqrt{\left(\frac{dx}{dt}\right)^2 + \left(\frac{dy}{dt}\right)^2}\, dt,$$ which is Formula 11.2.7.

69. (a) $\phi = \tan^{-1}\left(\dfrac{dy}{dx}\right)$ $\Rightarrow$ $\dfrac{d\phi}{dt} = \dfrac{d}{dt}\tan^{-1}\left(\dfrac{dy}{dx}\right) = \dfrac{1}{1 + (dy/dx)^2}\left[\dfrac{d}{dt}\left(\dfrac{dy}{dx}\right)\right]$. But $\dfrac{dy}{dx} = \dfrac{dy/dt}{dx/dt} = \dfrac{\dot{y}}{\dot{x}}$ $\Rightarrow$

$$\frac{d}{dt}\left(\frac{dy}{dx}\right) = \frac{d}{dt}\left(\frac{\dot{y}}{\dot{x}}\right) = \frac{\ddot{y}\dot{x} - \ddot{x}\dot{y}}{\dot{x}^2} \quad\Rightarrow\quad \frac{d\phi}{dt} = \frac{1}{1 + (\dot{y}/\dot{x})^2}\left(\frac{\ddot{y}\dot{x} - \ddot{x}\dot{y}}{\dot{x}^2}\right) = \frac{\dot{x}\ddot{y} - \ddot{x}\dot{y}}{\dot{x}^2 + \dot{y}^2}.$$ Using the Chain Rule, and the

fact that $s = \int_0^t \sqrt{\left(\frac{dx}{dt}\right)^2 + \left(\frac{dy}{dt}\right)^2}\, dt$ $\Rightarrow$ $\frac{ds}{dt} = \sqrt{\left(\frac{dx}{dt}\right)^2 + \left(\frac{dy}{dt}\right)^2} = \left(\dot{x}^2 + \dot{y}^2\right)^{1/2}$, we have that

$$\frac{d\phi}{ds} = \frac{d\phi/dt}{ds/dt} = \left(\frac{\dot{x}\ddot{y} - \ddot{x}\dot{y}}{\dot{x}^2 + \dot{y}^2}\right)\frac{1}{(\dot{x}^2 + \dot{y}^2)^{1/2}} = \frac{\dot{x}\ddot{y} - \ddot{x}\dot{y}}{(\dot{x}^2 + \dot{y}^2)^{3/2}}.$$ So $\kappa = \left|\dfrac{d\phi}{ds}\right| = \left|\dfrac{\dot{x}\ddot{y} - \ddot{x}\dot{y}}{(\dot{x}^2 + \dot{y}^2)^{3/2}}\right| = \dfrac{|\dot{x}\ddot{y} - \ddot{x}\dot{y}|}{(\dot{x}^2 + \dot{y}^2)^{3/2}}.$

(b) $x = x$ and $y = f(x)$ $\Rightarrow$ $\dot{x} = 1$, $\ddot{x} = 0$ and $\dot{y} = \dfrac{dy}{dx}$, $\ddot{y} = \dfrac{d^2y}{dx^2}$.

So $\kappa = \dfrac{\left|1\cdot(d^2y/dx^2) - 0\cdot(dy/dx)\right|}{[1 + (dy/dx)^2]^{3/2}} = \dfrac{\left|d^2y/dx^2\right|}{[1 + (dy/dx)^2]^{3/2}}.$

70. (a) $y = x^2$ $\Rightarrow$ $\dfrac{dy}{dx} = 2x$ $\Rightarrow$ $\dfrac{d^2y}{dx^2} = 2$. So $\kappa = \dfrac{\left|d^2y/dx^2\right|}{[1 + (dy/dx)^2]^{3/2}} = \dfrac{2}{(1 + 4x^2)^{3/2}}$, and at $(1, 1)$,

$$\kappa = \frac{2}{5^{3/2}} = \frac{2}{5\sqrt{5}}.$$

(b) $\kappa' = \dfrac{d\kappa}{dx} = -3(1 + 4x^2)^{-5/2}(8x) = 0$ $\Leftrightarrow$ $x = 0$ $\Rightarrow$ $y = 0$. This is a maximum since $\kappa' > 0$ for $x < 0$ and

$\kappa' < 0$ for $x > 0$. So the parabola $y = x^2$ has maximum curvature at the origin.

71. $x = \theta - \sin\theta$ $\Rightarrow$ $\dot{x} = 1 - \cos\theta$ $\Rightarrow$ $\ddot{x} = \sin\theta$, and $y = 1 - \cos\theta$ $\Rightarrow$ $\dot{y} = \sin\theta$ $\Rightarrow$ $\ddot{y} = \cos\theta$. Therefore,

$$\kappa = \frac{\left|\cos\theta - \cos^2\theta - \sin^2\theta\right|}{[(1 - \cos\theta)^2 + \sin^2\theta]^{3/2}} = \frac{\left|\cos\theta - (\cos^2\theta + \sin^2\theta)\right|}{(1 - 2\cos\theta + \cos^2\theta + \sin^2\theta)^{3/2}} = \frac{|\cos\theta - 1|}{(2 - 2\cos\theta)^{3/2}}.$$ The top of the arch is

characterized by a horizontal tangent, and from Example 2(b) in Section 11.2, the tangent is horizontal when $\theta = (2n - 1)\pi$,

so take $n = 1$ and substitute $\theta = \pi$ into the expression for κ: $\kappa = \dfrac{|\cos\pi - 1|}{(2 - 2\cos\pi)^{3/2}} = \dfrac{|-1 - 1|}{[2 - 2(-1)]^{3/2}} = \dfrac{1}{4}.$

72. (a) Every straight line has parametrizations of the form $x = a + vt$, $y = b + wt$, where a, b are arbitrary and v, $w \neq 0$.

For example, a straight line passing through distinct points (a, b) and (c, d) can be described as the parametrized curve

$x = a + (c - a)t$, $y = b + (d - b)t$. Starting with $x = a + vt$, $y = b + wt$, we compute $\dot{x} = v$, $\dot{y} = w$, $\ddot{x} = \ddot{y} = 0$,

and $\kappa = \dfrac{|v\cdot 0 - w\cdot 0|}{(v^2 + w^2)^{3/2}} = 0.$

(b) Parametric equations for a circle of radius r are $x = r \cos \theta$ and $y = r \sin \theta$. We can take the center to be the origin.

So $\dot{x} = -r \sin \theta \;\Rightarrow\; \ddot{x} = -r \cos \theta$ and $\dot{y} = r \cos \theta \;\Rightarrow\; \ddot{y} = -r \sin \theta$. Therefore,

$$\kappa = \frac{\left| r^2 \sin^2 \theta + r^2 \cos^2 \theta \right|}{(r^2 \sin^2 \theta + r^2 \cos^2 \theta)^{3/2}} = \frac{r^2}{r^3} = \frac{1}{r}. \text{ And so for any } \theta \text{ (and thus any point), } \kappa = \frac{1}{r}.$$

73. The coordinates of T are $(r \cos \theta, r \sin \theta)$. Since TP was unwound from

arc TA, TP has length $r\theta$. Also $\angle PTQ = \angle PTR - \angle QTR = \frac{1}{2}\pi - \theta$,

so P has coordinates $x = r \cos \theta + r\theta \cos\left(\frac{1}{2}\pi - \theta\right) = r(\cos \theta + \theta \sin \theta)$,

$y = r \sin \theta - r\theta \sin\left(\frac{1}{2}\pi - \theta\right) = r(\sin \theta - \theta \cos \theta)$.

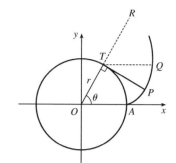

74. If the cow walks with the rope taut, it traces out the portion of the

involute in Exercise 73 corresponding to the range $0 \le \theta \le \pi$, arriving at

the point $(-r, \pi r)$ when $\theta = \pi$. With the rope now fully extended, the

cow walks in a semicircle of radius πr, arriving at $(-r, -\pi r)$. Finally,

the cow traces out another portion of the involute, namely the reflection

about the x-axis of the initial involute path. (This corresponds to the

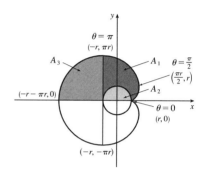

range $-\pi \le \theta \le 0$.) Referring to the figure, we see that the total grazing

area is $2(A_1 + A_3)$. A_3 is one-quarter of the area of a circle of radius πr, so $A_3 = \frac{1}{4}\pi(\pi r)^2 = \frac{1}{4}\pi^3 r^2$. We will compute

$A_1 + A_2$ and then subtract $A_2 = \frac{1}{2}\pi r^2$ to obtain A_1.

To find $A_1 + A_2$, first note that the rightmost point of the involute is $\left(\frac{\pi}{2}r, r\right)$. [To see this, note that $dx/d\theta = 0$ when

$\theta = 0$ or $\frac{\pi}{2}$. $\theta = 0$ corresponds to the cusp at $(r, 0)$ and $\theta = \frac{\pi}{2}$ corresponds to $\left(\frac{\pi}{2}r, r\right)$.] The leftmost point of the involute is

$(-r, \pi r)$. Thus, $A_1 + A_2 = \int_{\theta=\pi}^{\pi/2} y\, dx - \int_{\theta=0}^{\pi/2} y\, dx = \int_{\theta=\pi}^{0} y\, dx$.

Now $y\, dx = r(\sin \theta - \theta \cos \theta)\, r\theta \cos \theta\, d\theta = r^2(\theta \sin \theta \, \cos \theta - \theta^2 \cos^2 \theta)\, d\theta$. Integrate:

$(1/r^2) \int y\, dx = -\theta \cos^2 \theta - \frac{1}{2}(\theta^2 - 1)\sin \theta\, \cos \theta - \frac{1}{6}\theta^3 + \frac{1}{2}\theta + C.$ This enables us to compute

$$A_1 + A_2 = r^2 \left[-\theta \cos^2 \theta - \tfrac{1}{2}(\theta^2 - 1)\sin \theta \, \cos \theta - \tfrac{1}{6}\theta^3 + \tfrac{1}{2}\theta \right]_\pi^0 = r^2 \left[0 - \left(-\pi - \frac{\pi^3}{6} + \frac{\pi}{2} \right) \right] = r^2 \left(\frac{\pi}{2} + \frac{\pi^3}{6} \right)$$

Therefore, $A_1 = (A_1 + A_2) - A_2 = \frac{1}{6}\pi^3 r^2$, so the grazing area is $2(A_1 + A_3) = 2\left(\frac{1}{6}\pi^3 r^2 + \frac{1}{4}\pi^3 r^2\right) = \frac{5}{6}\pi^3 r^2$.

LABORATORY PROJECT   Bézier Curves

1. The parametric equations for a cubic Bézier curve are

$$x = x_0(1-t)^3 + 3x_1 t(1-t)^2 + 3x_2 t^2(1-t) + x_3 t^3$$

$$y = y_0(1-t)^3 + 3y_1 t(1-t)^2 + 3y_2 t^2(1-t) + y_3 t^3$$

where $0 \le t \le 1$. We are given the points $P_0(x_0, y_0) = (4, 1)$, $P_1(x_1, y_1) = (28, 48)$, $P_2(x_2, y_2) = (50, 42)$, and $P_3(x_3, y_3) = (40, 5)$. The curve is then given by

$$x(t) = 4(1-t)^3 + 3 \cdot 28t(1-t)^2 + 3 \cdot 50t^2(1-t) + 40t^3$$

$$y(t) = 1(1-t)^3 + 3 \cdot 48t(1-t)^2 + 3 \cdot 42t^2(1-t) + 5t^3$$

where $0 \le t \le 1$. The line segments are of the form $x = x_0 + (x_1 - x_0)t$, $y = y_0 + (y_1 - y_0)t$:

$P_0 P_1$	$x = 4 + 24t,$	$y = 1 + 47t$
$P_1 P_2$	$x = 28 + 22t,$	$y = 48 - 6t$
$P_2 P_3$	$x = 50 - 10t,$	$y = 42 - 37t$

2. It suffices to show that the slope of the tangent at P_0 is the same as that of line segment $P_0 P_1$, namely $\dfrac{y_1 - y_0}{x_1 - x_0}$.

We calculate the slope of the tangent to the Bézier curve:

$$\frac{dy/dt}{dx/dt} = \frac{-3y_0(1-t)^2 + 3y_1\left[-2t(1-t) + (1-t)^2\right] + 3y_2\left[-t^2 + (2t)(1-t)\right] + 3y_3 t^2}{-3x_0^2(1-t) + 3x_1\left[-2t(1-t) + (1-t)^2\right] + 3x_2[-t^2 + (2t)(1-t)] + 3x_3 t^2}$$

At point P_0, $t = 0$, so the slope of the tangent is $\dfrac{-3y_0 + 3y_1}{-3x_0 + 3x_1} = \dfrac{y_1 - y_0}{x_1 - x_0}$. So the tangent to the curve at P_0 passes through P_1. Similarly, the slope of the tangent at point P_3 [where $t = 1$] is $\dfrac{-3y_2 + 3y_3}{-3x_2 + 3x_3} = \dfrac{y_3 - y_2}{x_3 - x_2}$, which is also the slope of line $P_2 P_3$.

3. It seems that if P_1 were to the right of P_2, a loop would appear.

We try setting $P_1 = (110, 30)$, and the resulting curve does indeed have a loop.

4. Based on the behavior of the Bézier curve in Problems 1–3, we suspect that the four control points should be in an exaggerated C shape. We try $P_0(10, 12)$, $P_1(4, 15)$, $P_2(4, 5)$, and $P_3(10, 8)$, and these produce a decent C. If you are using a CAS, it may be necessary to instruct it to make the x- and y-scales the same so as not to distort the figure (this is called a "constrained projection" in Maple.)

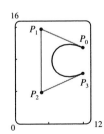

5. We use the same P_0 and P_1 as in Problem 4, and use part of our C as the top of an S. To prevent the center line from slanting up too much, we move P_2 up to $(4, 6)$ and P_3 down and to the left, to $(8, 7)$. In order to have a smooth joint between the top and bottom halves of the S (and a symmetric S), we determine points P_4, P_5, and P_6 by rotating points P_2, P_1, and P_0 about the center of the letter (point P_3). The points are therefore $P_4(12, 8)$, $P_5(12, -1)$, and $P_6(6, 2)$.

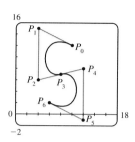

11.3 Polar Coordinates

1. (a) $\left(2, \frac{\pi}{3}\right)$

By adding 2π to $\frac{\pi}{3}$, we obtain the point $\left(2, \frac{7\pi}{3}\right)$. The direction opposite $\frac{\pi}{3}$ is $\frac{4\pi}{3}$, so $\left(-2, \frac{4\pi}{3}\right)$ is a point that satisfies the $r < 0$ requirement.

(b) $\left(1, -\frac{3\pi}{4}\right)$

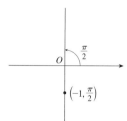

$r > 0$: $\left(1, -\frac{3\pi}{4} + 2\pi\right) = \left(1, \frac{5\pi}{4}\right)$

$r < 0$: $\left(-1, -\frac{3\pi}{4} + \pi\right) = \left(-1, \frac{\pi}{4}\right)$

(c) $\left(-1, \frac{\pi}{2}\right)$

$r > 0$: $\left(-(-1), \frac{\pi}{2} + \pi\right) = \left(1, \frac{3\pi}{2}\right)$

$r < 0$: $\left(-1, \frac{\pi}{2} + 2\pi\right) = \left(-1, \frac{5\pi}{2}\right)$

2. (a) $\left(1, \frac{7\pi}{4}\right)$

$r > 0$: $\left(1, \frac{7\pi}{4} - 2\pi\right) = \left(1, -\frac{\pi}{4}\right)$

$r < 0$: $\left(-1, \frac{7\pi}{4} - \pi\right) = \left(-1, \frac{3\pi}{4}\right)$

(b) $\left(-3, \frac{\pi}{6}\right)$

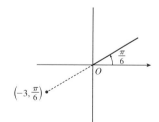

$r > 0$: $\left(-(-3), \frac{\pi}{6} + \pi\right) = \left(3, \frac{7\pi}{6}\right)$

$r < 0$: $\left(-3, \frac{\pi}{6} + 2\pi\right) = \left(-3, \frac{13\pi}{6}\right)$

(c) $(1, -1)$

$\theta = -1$ radian $\approx -57.3°$

$r > 0$: $(1, -1 + 2\pi)$

$r < 0$: $(-1, -1 + \pi)$

3. (a)

$x = 1\cos\pi = 1(-1) = -1$ and

$y = 1\sin\pi = 1(0) = 0$ give us

the Cartesian coordinates $(-1, 0)$.

(b)

$x = 2\cos\left(-\frac{2\pi}{3}\right) = 2\left(-\frac{1}{2}\right) = -1$ and

$y = 2\sin\left(-\frac{2\pi}{3}\right) = 2\left(-\frac{\sqrt{3}}{2}\right) = -\sqrt{3}$

give us $\left(-1, -\sqrt{3}\right)$.

(c)

$x = -2\cos\frac{3\pi}{4} = -2\left(-\frac{\sqrt{2}}{2}\right) = \sqrt{2}$ and

$y = -2\sin\frac{3\pi}{4} = -2\left(\frac{\sqrt{2}}{2}\right) = -\sqrt{2}$

gives us $\left(\sqrt{2}, -\sqrt{2}\right)$.

4. (a)

$x = -\sqrt{2}\cos\frac{5\pi}{4} = -\sqrt{2}\left(-\frac{\sqrt{2}}{2}\right) = 1$ and

$y = -\sqrt{2}\sin\frac{5\pi}{4} = -\sqrt{2}\left(-\frac{\sqrt{2}}{2}\right) = 1$

gives us $(1, 1)$.

(b)

$x = 1\cos\frac{5\pi}{2} = 1(0) = 0$ and

$y = 1\sin\frac{5\pi}{2} = 1(1) = 1$

gives us $(0, 1)$.

(c)

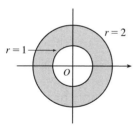

$x = 2\cos\left(-\frac{7\pi}{6}\right) = 2\left(-\frac{\sqrt{3}}{2}\right) = -\sqrt{3}$ and

$y = 2\sin\left(-\frac{7\pi}{6}\right) = 2\left(\frac{1}{2}\right) = 1$

give us $\left(-\sqrt{3}, 1\right)$.

5. (a) $x = 2$ and $y = -2$ $\Rightarrow$ $r = \sqrt{2^2 + (-2)^2} = 2\sqrt{2}$ and $\theta = \tan^{-1}\left(\frac{-2}{2}\right) = -\frac{\pi}{4}$. Since $(2, -2)$ is in the fourth

quadrant, the polar coordinates are (i) $\left(2\sqrt{2}, \frac{7\pi}{4}\right)$ and (ii) $\left(-2\sqrt{2}, \frac{3\pi}{4}\right)$.

(b) $x = -1$ and $y = \sqrt{3}$ $\Rightarrow$ $r = \sqrt{(-1)^2 + \left(\sqrt{3}\right)^2} = 2$ and $\theta = \tan^{-1}\left(\frac{\sqrt{3}}{-1}\right) = \frac{2\pi}{3}$. Since $\left(-1, \sqrt{3}\right)$ is in the second

quadrant, the polar coordinates are (i) $\left(2, \frac{2\pi}{3}\right)$ and (ii) $\left(-2, \frac{5\pi}{3}\right)$.

6. (a) $x = 3\sqrt{3}$ and $y = 3$ $\Rightarrow$ $r = \sqrt{\left(3\sqrt{3}\right)^2 + 3^2} = \sqrt{27 + 9} = 6$ and $\theta = \tan^{-1}\left(\frac{3}{3\sqrt{3}}\right) = \tan^{-1}\left(\frac{1}{\sqrt{3}}\right) = \frac{\pi}{6}$. Since

$\left(3\sqrt{3}, 3\right)$ is in the first quadrant, the polar coordinates are (i) $\left(6, \frac{\pi}{6}\right)$ and (ii) $\left(-6, \frac{7\pi}{6}\right)$.

(b) $x = 1$ and $y = -2$ $\Rightarrow$ $r = \sqrt{1^2 + (-2)^2} = \sqrt{5}$ and $\theta = \tan^{-1}\left(\frac{-2}{1}\right) = -\tan^{-1} 2$. Since $(1, -2)$ is in the fourth

quadrant, the polar coordinates are (i) $\left(\sqrt{5}, 2\pi - \tan^{-1} 2\right)$ and (ii) $\left(-\sqrt{5}, \pi - \tan^{-1} 2\right)$.

7. The curves $r = 1$ and $r = 2$ represent circles with center

O and radii 1 and 2. The region in the plane satisfying

$1 \le r \le 2$ consists of both circles and the shaded region

between them in the figure.

8. $r \ge 0$, $\pi/3 \le \theta \le 2\pi/3$

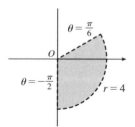

9. The region satisfying $0 \le r < 4$ and $-\pi/2 \le \theta < \pi/6$

does not include the circle $r = 4$ nor the line $\theta = \frac{\pi}{6}$.

10. $2 < r \le 5$, $3\pi/4 < \theta < 5\pi/4$

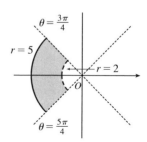

11. $2 < r < 3, \quad \frac{5\pi}{3} \le \theta \le \frac{7\pi}{3}$

12. $r \ge 1, \pi \le \theta \le 2\pi$

13. Converting the polar coordinates $(2, \pi/3)$ and $(4, 2\pi/3)$ to Cartesian coordinates gives us $\left(2\cos\frac{\pi}{3}, 2\sin\frac{\pi}{3}\right) = \left(1, \sqrt{3}\right)$ and $\left(4\cos\frac{2\pi}{3}, 4\sin\frac{2\pi}{3}\right) = \left(-2, 2\sqrt{3}\right)$. Now use the distance formula.

$$d = \sqrt{(x_2 - x_1)^2 + (y_2 - y_1)^2} = \sqrt{(-2 - 1)^2 + \left(2\sqrt{3} - \sqrt{3}\right)^2} = \sqrt{9 + 3} = \sqrt{12} = 2\sqrt{3}$$

14. The points (r_1, θ_1) and (r_2, θ_2) in Cartesian coordinates are $(r_1 \cos\theta_1, r_1 \sin\theta_1)$ and $(r_2 \cos\theta_2, r_2 \sin\theta_2)$, respectively. The *square* of the distance between them is

$$(r_2 \cos\theta_2 - r_1 \cos\theta_1)^2 + (r_2 \sin\theta_2 - r_1 \sin\theta_1)^2$$
$$= \left(r_2^2 \cos^2\theta_2 - 2r_1 r_2 \cos\theta_1 \cos\theta_2 + r_1^2 \cos^2\theta_1\right) + \left(r_2^2 \sin^2\theta_2 - 2r_1 r_2 \sin\theta_1 \sin\theta_2 + r_1^2 \sin^2\theta_1\right)$$
$$= r_1^2\left(\sin^2\theta_1 + \cos^2\theta_1\right) + r_2^2\left(\sin^2\theta_2 + \cos^2\theta_2\right) - 2r_1 r_2(\cos\theta_1 \cos\theta_2 + \sin\theta_1 \sin\theta_2)$$
$$= r_1^2 - 2r_1 r_2 \cos(\theta_1 - \theta_2) + r_2^2,$$

so the distance between them is $\sqrt{r_1^2 - 2r_1 r_2 \cos(\theta_1 - \theta_2) + r_2^2}$.

15. $r = 2 \quad\Leftrightarrow\quad \sqrt{x^2 + y^2} = 2 \quad\Leftrightarrow\quad x^2 + y^2 = 4$, a circle of radius 2 centered at the origin.

16. $r\cos\theta = 1 \quad\Leftrightarrow\quad x = 1$, a vertical line.

17. $r = 3\sin\theta \quad\Rightarrow\quad r^2 = 3r\sin\theta \quad\Leftrightarrow\quad x^2 + y^2 = 3y \quad\Leftrightarrow\quad x^2 + \left(y - \frac{3}{2}\right)^2 = \left(\frac{3}{2}\right)^2$, a circle of radius $\frac{3}{2}$ centered at $\left(0, \frac{3}{2}\right)$. The first two equations are actually equivalent since $r^2 = 3r\sin\theta \quad\Rightarrow\quad r(r - 3\sin\theta) = 0 \quad\Rightarrow\quad r = 0$ or $r = 3\sin\theta$. But $r = 3\sin\theta$ gives the point $r = 0$ (the pole) when $\theta = 0$. Thus, the single equation $r = 3\sin\theta$ is equivalent to the compound condition ($r = 0$ or $r = 3\sin\theta$).

18. $r = 2\sin\theta + 2\cos\theta \quad\Rightarrow\quad r^2 = 2r\sin\theta + 2r\cos\theta \quad\Leftrightarrow\quad x^2 + y^2 = 2y + 2x \quad\Leftrightarrow$
$(x^2 - 2x + 1) + (y^2 - 2y + 1) = 2 \quad\Leftrightarrow\quad (x - 1)^2 + (y - 1)^2 = 2$. The first implication is reversible since $r^2 = 2r\sin\theta + 2r\cos\theta \quad\Rightarrow\quad r = 0$ or $r = 2\sin\theta + 2\cos\theta$, but the curve $r = 2\sin\theta + 2\cos\theta$ passes through the pole $(r = 0)$ when $\theta = -\frac{\pi}{4}$, so $r = 2\sin\theta + 2\cos\theta$ includes the single point of $r = 0$. The curve is a circle of radius $\sqrt{2}$, centered at $(1, 1)$.

19. $r = \csc\theta \quad\Leftrightarrow\quad r = \dfrac{1}{\sin\theta} \quad\Leftrightarrow\quad r\sin\theta = 1 \quad\Leftrightarrow\quad y = 1$, a horizontal line 1 unit above the x-axis.

20. $r = \tan\theta \sec\theta = \dfrac{\sin\theta}{\cos^2\theta} \quad\Rightarrow\quad r\cos^2\theta = \sin\theta \quad\Leftrightarrow\quad (r\cos\theta)^2 = r\sin\theta \quad\Leftrightarrow\quad x^2 = y$, a parabola with vertex at the origin opening upward. The first implication is reversible since $\cos\theta = 0$ would imply $\sin\theta = r\cos^2\theta = 0$, contradicting the fact that $\cos^2\theta + \sin^2\theta = 1$.

21. $x = 3 \iff r\cos\theta = 3 \iff r = 3/\cos\theta \iff r = 3\sec\theta$.

22. $x^2 + y^2 = 9 \iff r^2 = 9 \iff r = 3$. $[r = -3$ gives the same curve.$]$

23. $x = -y^2 \iff r\cos\theta = -r^2\sin^2\theta \iff \cos\theta = -r\sin^2\theta \iff r = -\dfrac{\cos\theta}{\sin^2\theta} = -\cot\theta\csc\theta$.

24. $x + y = 9 \iff r\cos\theta + r\sin\theta = 9 \iff r = 9/(\cos\theta + \sin\theta)$.

25. $x^2 + y^2 = 2cx \iff r^2 = 2cr\cos\theta \iff r^2 - 2cr\cos\theta = 0 \iff r(r - 2c\cos\theta) = 0 \iff r = 0$ or $r = 2c\cos\theta$.
$r = 0$ is included in $r = 2c\cos\theta$ when $\theta = \frac{\pi}{2} + n\pi$, so the curve is represented by the single equation $r = 2c\cos\theta$.

26. $xy = 4 \iff (r\cos\theta)(r\sin\theta) = 4 \iff r^2\left(\frac{1}{2}\cdot 2\sin\theta\cos\theta\right) = 4 \iff r^2\sin 2\theta = 8 \Rightarrow r^2 = 8\csc 2\theta$

27. (a) The description leads immediately to the polar equation $\theta = \frac{\pi}{6}$, and the Cartesian equation $y = \tan\left(\frac{\pi}{6}\right)x = \frac{1}{\sqrt{3}}x$ is slightly more difficult to derive.

(b) The easier description here is the Cartesian equation $x = 3$.

28. (a) Because its center is not at the origin, it is more easily described by its Cartesian equation, $(x - 2)^2 + (y - 3)^2 = 5^2$.

(b) This circle is more easily given in polar coordinates: $r = 4$. The Cartesian equation is also simple: $x^2 + y^2 = 16$.

29. $\theta = -\pi/6$

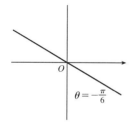

30. $r^2 - 3r + 2 = 0 \iff (r - 1)(r - 2) = 0 \iff$
$r = 1$ or $r = 2$

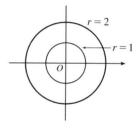

31. $r = \sin\theta \iff r^2 = r\sin\theta \iff x^2 + y^2 = y \iff$
$x^2 + \left(y - \frac{1}{2}\right)^2 = \left(\frac{1}{2}\right)^2$. The reasoning here is the same
as in Exercise 17. This is a circle of radius $\frac{1}{2}$ centered
at $\left(0, \frac{1}{2}\right)$.

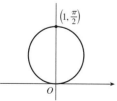

32. $r = -3\cos\theta \iff r^2 = -3r\cos\theta \iff$
$x^2 + y^2 = -3x \iff \left(x + \frac{3}{2}\right)^2 + y^2 = \left(\frac{3}{2}\right)^2$.
This curve is a circle of radius $\frac{3}{2}$ centered at $\left(-\frac{3}{2}, 0\right)$.

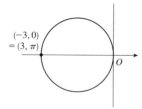

33. $r = 2(1 - \sin\theta)$. This curve is a cardioid.

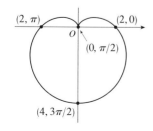

34. $r = 1 - 3\cos\theta$. This is a limaçon.

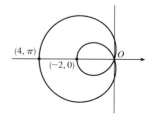

35. $r = \theta, \quad \theta \geq 0$

36. $r = \ln \theta, \ \theta \geq 1$

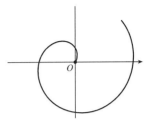

37. $r = 4 \sin 3\theta$

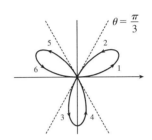

38. $r = \cos 5\theta$

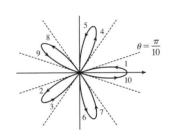

39. $r = 2 \cos 4\theta$

40. $r = 3 \cos 6\theta$

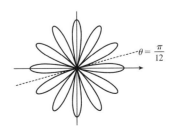

41. $r = 1 - 2\sin\theta$

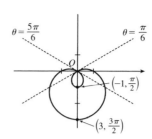

42. $r = 2 + \sin\theta$

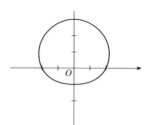

43. $r^2 = 9\sin 2\theta$

44. $r^2 = \cos 4\theta$

45. $r = 2\cos\left(\frac{3}{2}\theta\right)$

46. $r^2\theta = 1 \quad \Leftrightarrow \quad r = \pm 1/\sqrt{\theta}$ for $\theta > 0$

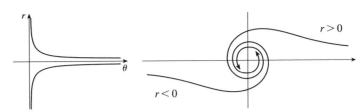

47. $r = 1 + 2\cos 2\theta$

48. $r = 1 + 2\cos(\theta/2)$

49. For $\theta = 0$, π, and 2π, r has its minimum value of about 0.5. For $\theta = \frac{\pi}{2}$ and $\frac{3\pi}{2}$, r attains its maximum value of 2.

We see that the graph has a similar shape for $0 \le \theta \le \pi$ and $\pi \le \theta \le 2\pi$.

50.

  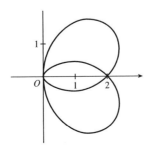

51. $x = (r) \cos \theta = (4 + 2 \sec \theta) \cos \theta = 4 \cos \theta + 2$. Now, $r \to \infty \Rightarrow$

$(4 + 2 \sec \theta) \to \infty \Rightarrow \theta \to \left(\frac{\pi}{2}\right)^-$ or $\theta \to \left(\frac{3\pi}{2}\right)^+$ [since we need only

consider $0 \le \theta < 2\pi$], so $\lim\limits_{r \to \infty} x = \lim\limits_{\theta \to \pi/2^-} (4 \cos \theta + 2) = 2$. Also,

$r \to -\infty \Rightarrow (4 + 2 \sec \theta) \to -\infty \Rightarrow \theta \to \left(\frac{\pi}{2}\right)^+$ or $\theta \to \left(\frac{3\pi}{2}\right)^-$, so

$\lim\limits_{r \to -\infty} x = \lim\limits_{\theta \to \pi/2^+} (4 \cos \theta + 2) = 2$. Therefore, $\lim\limits_{r \to \pm\infty} x = 2 \Rightarrow x = 2$ is a vertical asymptote.

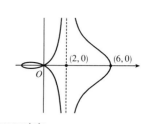

52. $y = r \sin \theta = 2 \sin \theta - \csc \theta \sin \theta = 2 \sin \theta - 1$.

$r \to \infty \Rightarrow (2 - \csc \theta) \to \infty \Rightarrow$

$\csc \theta \to -\infty \Rightarrow \theta \to \pi^+$ [since we need

only consider $0 \le \theta < 2\pi$] and so

$\lim\limits_{r \to \infty} y = \lim\limits_{\theta \to \pi^+} 2 \sin \theta - 1 = -1$.

Also $r \to -\infty \Rightarrow (2 - \csc \theta) \to -\infty \Rightarrow \csc \theta \to \infty \Rightarrow \theta \to \pi^-$ and so $\lim\limits_{r \to -\infty} x = \lim\limits_{\theta \to \pi^-} 2 \sin \theta - 1 = -1$.

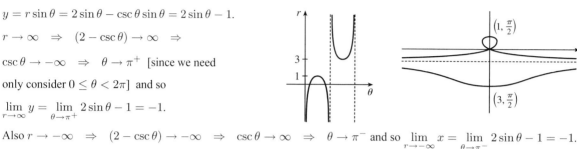

Therefore $\lim\limits_{r \to \pm\infty} y = -1 \Rightarrow y = -1$ is a horizontal asymptote.

53. To show that $x = 1$ is an asymptote we must prove $\lim\limits_{r \to \pm\infty} x = 1$.

$x = (r) \cos \theta = (\sin \theta \tan \theta) \cos \theta = \sin^2 \theta$. Now, $r \to \infty \Rightarrow \sin \theta \tan \theta \to \infty \Rightarrow$

$\theta \to \left(\frac{\pi}{2}\right)^-$, so $\lim\limits_{r \to \infty} x = \lim\limits_{\theta \to \pi/2^-} \sin^2 \theta = 1$. Also, $r \to -\infty \Rightarrow \sin \theta \tan \theta \to -\infty \Rightarrow$

$\theta \to \left(\frac{\pi}{2}\right)^+$, so $\lim\limits_{r \to -\infty} x = \lim\limits_{\theta \to \pi/2^+} \sin^2 \theta = 1$. Therefore, $\lim\limits_{r \to \pm\infty} x = 1 \Rightarrow x = 1$ is

a vertical asymptote. Also notice that $x = \sin^2 \theta \ge 0$ for all θ, and $x = \sin^2 \theta \le 1$ for all θ. And $x \ne 1$, since the curve is not

defined at odd multiples of $\frac{\pi}{2}$. Therefore, the curve lies entirely within the vertical strip $0 \le x < 1$.

54. The equation is $(x^2 + y^2)^3 = 4x^2 y^2$, but using polar coordinates we know that

$x^2 + y^2 = r^2$ and $x = r \cos \theta$ and $y = r \sin \theta$. Substituting into the given

equation: $r^6 = 4r^2 \cos^2 \theta \, r^2 \sin^2 \theta \Rightarrow r^2 = 4 \cos^2 \theta \sin^2 \theta \Rightarrow$

$r = \pm 2 \cos \theta \sin \theta = \pm \sin 2\theta$. $r = \pm \sin 2\theta$ is sketched at right.

55. (a) We see that the curve $r = 1 + c \sin \theta$ crosses itself at the origin, where $r = 0$ (in fact the inner loop corresponds to

negative r-values,) so we solve the equation of the limaçon for $r = 0 \Leftrightarrow c \sin \theta = -1 \Leftrightarrow \sin \theta = -1/c$. Now if

$|c| < 1$, then this equation has no solution and hence there is no inner loop. But if $c < -1$, then on the interval $(0, 2\pi)$

the equation has the two solutions $\theta = \sin^{-1}(-1/c)$ and $\theta = \pi - \sin^{-1}(-1/c)$, and if $c > 1$, the solutions are

$\theta = \pi + \sin^{-1}(1/c)$ and $\theta = 2\pi - \sin^{-1}(1/c)$. In each case, $r < 0$ for θ between the two solutions, indicating a loop.

(b) For $0 < c < 1$, the dimple (if it exists) is characterized by the fact that y has a local maximum at $\theta = \frac{3\pi}{2}$. So we determine

for what c-values $\dfrac{d^2y}{d\theta^2}$ is negative at $\theta = \frac{3\pi}{2}$, since by the Second Derivative Test this indicates a maximum:

$$y = r\sin\theta = \sin\theta + c\sin^2\theta \quad\Rightarrow\quad \frac{dy}{d\theta} = \cos\theta + 2c\sin\theta\,\cos\theta = \cos\theta + c\sin 2\theta \quad\Rightarrow\quad \frac{d^2y}{d\theta^2} = -\sin\theta + 2c\cos 2\theta.$$

At $\theta = \frac{3\pi}{2}$, this is equal to $-(-1) + 2c(-1) = 1 - 2c$, which is negative only for $c > \frac{1}{2}$. A similar argument shows that

for $-1 < c < 0$, y only has a local minimum at $\theta = \frac{\pi}{2}$ (indicating a dimple) for $c < -\frac{1}{2}$.

56. (a) $r = \sqrt{\theta}$, $0 \le \theta \le 16\pi$. r increases as θ increases and there are eight full revolutions. The graph must be either II or V.

When $\theta = 2\pi$, $r = \sqrt{2\pi} \approx 2.5$ and when $\theta = 16\pi$, $r = \sqrt{16\pi} \approx 7$, so the last revolution intersects the polar axis at

approximately 3 times the distance that the first revolution intersects the polar axis, which is depicted in graph V.

(b) $r = \theta^2$, $0 \le \theta \le 16\pi$.   See part (a). This is graph II.

(c) $r = \cos(\theta/3)$. $0 \le \frac{\theta}{3} \le 2\pi \quad\Rightarrow\quad 0 \le \theta \le 6\pi$, so this curve will repeat itself every 6π radians.

$\cos\left(\frac{\theta}{3}\right) = 0 \quad\Rightarrow\quad \frac{\theta}{3} = \frac{\pi}{2} + \pi n \quad\Rightarrow\quad \theta = \frac{3\pi}{2} + 3\pi n$, so there will be two "pole" values, $\frac{3\pi}{2}$ and $\frac{9\pi}{2}$.

This is graph VI.

(d) $r = 1 + 2\cos\theta$ is a limaçon [see Exercise 55(a)] with $c = 2$. This is graph III.

(e) Since $-1 \le \sin 3\theta \le 1$, $1 \le 2 + \sin 3\theta \le 3$, so $r = 2 + \sin 3\theta$ is never 0; that is, the curve never intersects the pole.

This is graph I.

(f) $r = 1 + 2\sin 3\theta$.   Solving $r = 0$ will give us many "pole" values, so this is graph IV.

57. $r = 2\sin\theta \quad\Rightarrow\quad x = r\cos\theta = 2\sin\theta\,\cos\theta = \sin 2\theta$, $y = r\sin\theta = 2\sin^2\theta \quad\Rightarrow$

$$\frac{dy}{dx} = \frac{dy/d\theta}{dx/d\theta} = \frac{2\cdot 2\sin\theta\,\cos\theta}{\cos 2\theta \cdot 2} = \frac{\sin 2\theta}{\cos 2\theta} = \tan 2\theta$$

When $\theta = \dfrac{\pi}{6}$, $\dfrac{dy}{dx} = \tan\left(2 \cdot \dfrac{\pi}{6}\right) = \tan\dfrac{\pi}{3} = \sqrt{3}$.   [*Another method:*   Use Equation 3.]

58. $r = 2 - \sin\theta \quad\Rightarrow\quad x = r\cos\theta = (2 - \sin\theta)\cos\theta$, $y = r\sin\theta = (2 - \sin\theta)\sin\theta \quad\Rightarrow$

$$\frac{dy}{dx} = \frac{dy/d\theta}{dx/d\theta} = \frac{(2 - \sin\theta)\cos\theta + \sin\theta(-\cos\theta)}{(2 - \sin\theta)(-\sin\theta) + \cos\theta(-\cos\theta)} = \frac{2\cos\theta - 2\sin\theta\,\cos\theta}{-2\sin\theta + \sin^2\theta - \cos^2\theta} = \frac{2\cos\theta - \sin 2\theta}{-2\sin\theta - \cos 2\theta}$$

When $\theta = \dfrac{\pi}{3}$, $\dfrac{dy}{dx} = \dfrac{2(1/2) - \left(\sqrt{3}/2\right)}{-2\left(\sqrt{3}/2\right) - (-1/2)} = \dfrac{1 - \sqrt{3}/2}{-\sqrt{3} + 1/2} \cdot \dfrac{2}{2} = \dfrac{2 - \sqrt{3}}{1 - 2\sqrt{3}}$.

59. $r = 1/\theta \quad\Rightarrow\quad x = r\cos\theta = (\cos\theta)/\theta$, $y = r\sin\theta = (\sin\theta)/\theta \quad\Rightarrow$

$$\frac{dy}{dx} = \frac{dy/d\theta}{dx/d\theta} = \frac{\sin\theta(-1/\theta^2) + (1/\theta)\cos\theta}{\cos\theta(-1/\theta^2) - (1/\theta)\sin\theta} \cdot \frac{\theta^2}{\theta^2} = \frac{-\sin\theta + \theta\cos\theta}{-\cos\theta - \theta\sin\theta}$$

When $\theta = \pi$, $\dfrac{dy}{dx} = \dfrac{-0 + \pi(-1)}{-(-1) - \pi(0)} = \dfrac{-\pi}{1} = -\pi$.

60. $r = \cos(\theta/3) \quad \Rightarrow \quad x = r\cos\theta = \cos(\theta/3)\cos\theta,\ y = r\sin\theta = \cos(\theta/3)\sin\theta \quad \Rightarrow$

$$\frac{dy}{dx} = \frac{dy/d\theta}{dx/d\theta} = \frac{\cos(\theta/3)\cos\theta + \sin\theta\ \left(-\frac{1}{3}\sin(\theta/3)\right)}{\cos(\theta/3)\ (-\sin\theta) + \cos\theta\ \left(-\frac{1}{3}\sin(\theta/3)\right)}$$

When $\theta = \pi$, $\dfrac{dy}{dx} = \dfrac{\frac{1}{2}(-1) + (0)\left(-\sqrt{3}/6\right)}{\frac{1}{2}(0) + (-1)\left(-\sqrt{3}/6\right)} = \dfrac{-1/2}{\sqrt{3}/6} = -\dfrac{3}{\sqrt{3}} = -\sqrt{3}.$

61. $r = \cos 2\theta \quad \Rightarrow \quad x = r\cos\theta = \cos 2\theta\ \cos\theta,\ y = r\sin\theta = \cos 2\theta\ \sin\theta \quad \Rightarrow$

$$\frac{dy}{dx} = \frac{dy/d\theta}{dx/d\theta} = \frac{\cos 2\theta\ \cos\theta + \sin\theta\ (-2\sin 2\theta)}{\cos 2\theta\ (-\sin\theta) + \cos\theta\ (-2\sin 2\theta)}$$

When $\theta = \dfrac{\pi}{4}$, $\dfrac{dy}{dx} = \dfrac{0\left(\sqrt{2}/2\right) + \left(\sqrt{2}/2\right)(-2)}{0\left(-\sqrt{2}/2\right) + \left(\sqrt{2}/2\right)(-2)} = \dfrac{-\sqrt{2}}{-\sqrt{2}} = 1.$

62. $r = 1 + 2\cos\theta \quad \Rightarrow \quad x = r\cos\theta = (1 + 2\cos\theta)\cos\theta,\ y = r\sin\theta = (1 + 2\cos\theta)\sin\theta \quad \Rightarrow$

$$\frac{dy}{dx} = \frac{dy/d\theta}{dx/d\theta} = \frac{(1 + 2\cos\theta)\cos\theta + \sin\theta\ (-2\sin\theta)}{(1 + 2\cos\theta)(-\sin\theta) + \cos\theta\ (-2\sin\theta)}$$

When $\theta = \dfrac{\pi}{3}$, $\dfrac{dy}{dx} = \dfrac{2\left(\frac{1}{2}\right) + \left(\sqrt{3}/2\right)\left(-\sqrt{3}\right)}{2\left(-\sqrt{3}/2\right) + \frac{1}{2}\left(-\sqrt{3}\right)} \cdot \dfrac{2}{2} = \dfrac{2 - 3}{-2\sqrt{3} - \sqrt{3}} = \dfrac{-1}{-3\sqrt{3}} = \dfrac{\sqrt{3}}{9}.$

63. $r = 3\cos\theta \quad \Rightarrow \quad x = r\cos\theta = 3\cos\theta\ \cos\theta,\ y = r\sin\theta = 3\cos\theta\ \sin\theta \quad \Rightarrow$

$\dfrac{dy}{d\theta} = -3\sin^2\theta + 3\cos^2\theta = 3\cos 2\theta = 0 \quad \Rightarrow \quad 2\theta = \dfrac{\pi}{2} \text{ or } \dfrac{3\pi}{2} \quad \Leftrightarrow \quad \theta = \dfrac{\pi}{4} \text{ or } \dfrac{3\pi}{4}.$

So the tangent is horizontal at $\left(\dfrac{3}{\sqrt{2}}, \dfrac{\pi}{4}\right)$ and $\left(-\dfrac{3}{\sqrt{2}}, \dfrac{3\pi}{4}\right)\ \left[\text{same as } \left(\dfrac{3}{\sqrt{2}}, -\dfrac{\pi}{4}\right)\right].$

$\dfrac{dx}{d\theta} = -6\sin\theta\cos\theta = -3\sin 2\theta = 0 \quad \Rightarrow \quad 2\theta = 0 \text{ or } \pi \quad \Leftrightarrow \quad \theta = 0 \text{ or } \dfrac{\pi}{2}.$ So the tangent is vertical at $(3, 0)$ and $\left(0, \dfrac{\pi}{2}\right).$

64. $r = 1 - \sin\theta \quad \Rightarrow \quad x = r\cos\theta = \cos\theta\ (1 - \sin\theta),\ y = r\sin\theta = \sin\theta\ (1 - \sin\theta) \quad \Rightarrow$

$\dfrac{dy}{d\theta} = \sin\theta\ (-\cos\theta) + (1 - \sin\theta)\ \cos\theta = \cos\theta\ (1 - 2\sin\theta) = 0 \quad \Rightarrow \quad \cos\theta = 0 \text{ or } \sin\theta = \dfrac{1}{2} \quad \Rightarrow$

$\theta = \dfrac{\pi}{6}, \dfrac{\pi}{2}, \dfrac{5\pi}{6}, \text{ or } \dfrac{3\pi}{2} \quad \Rightarrow \quad$ horizontal tangent at $\left(\dfrac{1}{2}, \dfrac{\pi}{6}\right),\ \left(\dfrac{1}{2}, \dfrac{5\pi}{6}\right),$ and $\left(2, \dfrac{3\pi}{2}\right).$

$$\dfrac{dx}{d\theta} = \cos\theta\ (-\cos\theta) + (1 - \sin\theta)(-\sin\theta) = -\cos^2\theta - \sin\theta + \sin^2\theta = 2\sin^2\theta - \sin\theta - 1$$

$$= (2\sin\theta + 1)(\sin\theta - 1) = 0 \quad \Rightarrow$$

$\sin\theta = -\dfrac{1}{2} \text{ or } 1 \quad \Rightarrow \quad \theta = \dfrac{7\pi}{6}, \dfrac{11\pi}{6}, \text{ or } \dfrac{\pi}{2} \quad \Rightarrow \quad$ vertical tangent at $\left(\dfrac{3}{2}, \dfrac{7\pi}{6}\right), \left(\dfrac{3}{2}, \dfrac{11\pi}{6}\right),$ and $\left(0, \dfrac{\pi}{2}\right).$

Note that the tangent is vertical, not horizontal, when $\theta = \dfrac{\pi}{2}$, since

$$\lim_{\theta \to (\pi/2)^-} \frac{dy/d\theta}{dx/d\theta} = \lim_{\theta \to (\pi/2)^-} \frac{\cos\theta\ (1 - 2\sin\theta)}{(2\sin\theta + 1)(\sin\theta - 1)} = \infty \quad \text{and} \quad \lim_{\theta \to (\pi/2)^+} \frac{dy/d\theta}{dx/d\theta} = -\infty.$$

65. $r = 1 + \cos\theta \quad \Rightarrow \quad x = r\cos\theta = \cos\theta\ (1 + \cos\theta),\ y = r\sin\theta = \sin\theta\ (1 + \cos\theta) \quad \Rightarrow$

$\dfrac{dy}{d\theta} = (1 + \cos\theta)\ \cos\theta - \sin^2\theta = 2\cos^2\theta + \cos\theta - 1 = (2\cos\theta - 1)(\cos\theta + 1) = 0 \quad \Rightarrow \quad \cos\theta = \dfrac{1}{2} \text{ or } -1 \quad \Rightarrow$

$\theta = \dfrac{\pi}{3}, \pi, \text{ or } \dfrac{5\pi}{3} \quad \Rightarrow \quad$ horizontal tangent at $\left(\dfrac{3}{2}, \dfrac{\pi}{3}\right), (0, \pi),$ and $\left(\dfrac{3}{2}, \dfrac{5\pi}{3}\right).$

$\dfrac{dx}{d\theta} = -(1 + \cos\theta)\sin\theta - \cos\theta\sin\theta = -\sin\theta\ (1 + 2\cos\theta) = 0 \quad \Rightarrow \quad \sin\theta = 0 \text{ or } \cos\theta = -\dfrac{1}{2} \quad \Rightarrow$

$\theta = 0, \pi, \dfrac{2\pi}{3}, \text{ or } \dfrac{4\pi}{3} \quad \Rightarrow \quad$ vertical tangent at $(2, 0), \left(\dfrac{1}{2}, \dfrac{2\pi}{3}\right),$ and $\left(\dfrac{1}{2}, \dfrac{4\pi}{3}\right).$

Note that the tangent is horizontal, not vertical when $\theta = \pi$, since $\lim\limits_{\theta \to \pi} \dfrac{dy/d\theta}{dx/d\theta} = 0.$

66. $r = e^\theta \quad \Rightarrow \quad x = r \cos\theta = e^\theta \cos\theta, \; y = r \sin\theta = e^\theta \sin\theta \quad \Rightarrow$

$\frac{dy}{d\theta} = e^\theta \sin\theta + e^\theta \cos\theta = e^\theta (\sin\theta + \cos\theta) = 0 \quad \Rightarrow \quad \sin\theta = -\cos\theta \quad \Rightarrow \quad \tan\theta = -1 \quad \Rightarrow$

$\theta = -\frac{1}{4}\pi + n\pi \; [n \text{ any integer}] \quad \Rightarrow \quad$ horizontal tangents at $\left(e^{\pi(n-1/4)}, \pi\left(n - \frac{1}{4}\right)\right)$.

$\frac{dx}{d\theta} = e^\theta \cos\theta - e^\theta \sin\theta = e^\theta (\cos\theta - \sin\theta) = 0 \quad \Rightarrow \quad \sin\theta = \cos\theta \quad \Rightarrow \quad \tan\theta = 1 \quad \Rightarrow$

$\theta = \frac{1}{4}\pi + n\pi \; [n \text{ any integer}] \quad \Rightarrow \quad$ vertical tangents at $\left(e^{\pi(n+1/4)}, \pi\left(n + \frac{1}{4}\right)\right)$.

67. $r = 2 + \sin\theta \quad \Rightarrow \quad x = r \cos\theta = (2 + \sin\theta)\cos\theta, \; y = r\sin\theta = (2 + \sin\theta)\sin\theta \quad \Rightarrow$

$\frac{dy}{d\theta} = (2 + \sin\theta)\cos\theta + \sin\theta \cos\theta = \cos\theta \cdot 2(1 + \sin\theta) = 0 \quad \Rightarrow \quad \cos\theta = 0 \text{ or } \sin\theta = -1 \quad \Rightarrow$

$\theta = \frac{\pi}{2} \text{ or } \frac{3\pi}{2} \quad \Rightarrow \quad$ horizontal tangent at $\left(3, \frac{\pi}{2}\right)$ and $\left(1, \frac{3\pi}{2}\right)$.

$\frac{dx}{d\theta} = (2 + \sin\theta)(-\sin\theta) + \cos\theta \cos\theta = -2\sin\theta - \sin^2\theta + 1 - \sin^2\theta = -2\sin^2\theta - 2\sin\theta + 1 \quad \Rightarrow$

$\sin\theta = \frac{2 \pm \sqrt{4 + 8}}{-4} = \frac{2 \pm 2\sqrt{3}}{-4} = \frac{1 - \sqrt{3}}{-2} \quad \left[\frac{1 + \sqrt{3}}{-2} < -1\right] \quad \Rightarrow$

$\theta_1 = \sin^{-1}\left(-\frac{1}{2} + \frac{1}{2}\sqrt{3}\right)$ and $\theta_2 = \pi - \theta_1 \quad \Rightarrow \quad$ vertical tangent at $\left(\frac{3}{2} + \frac{1}{2}\sqrt{3}, \theta_1\right)$ and $\left(\frac{3}{2} + \frac{1}{2}\sqrt{3}, \theta_2\right)$.

Note that $r(\theta_1) = 2 + \sin\left[\sin^{-1}\left(-\frac{1}{2} + \frac{1}{2}\sqrt{3}\right)\right] = 2 - \frac{1}{2} + \frac{1}{2}\sqrt{3} = \frac{3}{2} + \frac{1}{2}\sqrt{3}$.

68. By differentiating implicitly, $r^2 = \sin 2\theta \quad \Rightarrow \quad 2r\,(dr/d\theta) = 2\cos 2\theta \quad \Rightarrow$

$dr/d\theta = (1/r)\cos 2\theta$, so $dy/d\theta = (dr/d\theta)\sin\theta + r\cos\theta \quad \Rightarrow$

$$\frac{dy}{d\theta} = \frac{1}{r}\cos 2\theta \, \sin\theta + r\cos\theta = \frac{1}{r}\left(\cos 2\theta \, \sin\theta + r^2 \cos\theta\right)$$

$$= \frac{1}{r}\left(\cos 2\theta \, \sin\theta + \sin 2\theta \, \cos\theta\right) = \frac{1}{r}\sin 3\theta$$

This is 0 when $\sin 3\theta = 0 \quad \Rightarrow \quad \theta = 0, \frac{\pi}{3}$ or $\frac{4\pi}{3}$ (restricting θ to the domain of the lemniscate), so there are horizontal

tangents at $\left(\sqrt[4]{\frac{3}{4}}, \frac{\pi}{3}\right), \left(\sqrt[4]{\frac{3}{4}}, \frac{4\pi}{3}\right)$ and $(0,0)$. Similarly, $dx/d\theta = (1/r)\cos 3\theta = 0$ when $\theta = \frac{\pi}{6}$ or $\frac{7\pi}{6}$, so there are vertical

tangents at $\left(\sqrt[4]{\frac{3}{4}}, \frac{\pi}{6}\right)$ and $\left(\sqrt[4]{\frac{3}{4}}, \frac{7\pi}{6}\right)$ [and $(0,0)$].

69. $r = a\sin\theta + b\cos\theta \quad \Rightarrow \quad r^2 = ar\sin\theta + br\cos\theta \quad \Rightarrow \quad x^2 + y^2 = ay + bx \quad \Rightarrow$

$x^2 - bx + \left(\frac{1}{2}b\right)^2 + y^2 - ay + \left(\frac{1}{2}a\right)^2 = \left(\frac{1}{2}b\right)^2 + \left(\frac{1}{2}a\right)^2 \quad \Rightarrow \quad \left(x - \frac{1}{2}b\right)^2 + \left(y - \frac{1}{2}a\right)^2 = \frac{1}{4}(a^2 + b^2)$, and this is a circle

with center $\left(\frac{1}{2}b, \frac{1}{2}a\right)$ and radius $\frac{1}{2}\sqrt{a^2 + b^2}$.

70. These curves are circles which intersect at the origin and at $\left(\frac{1}{\sqrt{2}}a, \frac{\pi}{4}\right)$. At the origin, the first circle has a horizontal

tangent and the second a vertical one, so the tangents are perpendicular here. For the first circle $[r = a\sin\theta]$,

$dy/d\theta = a\cos\theta \sin\theta + a\sin\theta \cos\theta = a\sin 2\theta = a$ at $\theta = \frac{\pi}{4}$ and $dx/d\theta = a\cos^2\theta - a\sin^2\theta = a\cos 2\theta = 0$

at $\theta = \frac{\pi}{4}$, so the tangent here is vertical. Similarly, for the second circle $[r = a\cos\theta]$, $dy/d\theta = a\cos 2\theta = 0$ and

$dx/d\theta = -a\sin 2\theta = -a$ at $\theta = \frac{\pi}{4}$, so the tangent is horizontal, and again the tangents are perpendicular.

Note for Exercises 71–76: Maple is able to plot polar curves using the `polarplot` command, or using the `coords=polar` option in a regular `plot` command. In Mathematica, use `PolarPlot`. In Derive, change to `Polar` under `Options State`. If your graphing device cannot plot polar equations, you must convert to parametric equations. For example, in Exercise 71, $x = r\cos\theta = [1 + 2\sin(\theta/2)]\cos\theta$, $y = r\sin\theta = [1 + 2\sin(\theta/2)]\sin\theta$.

71. $r = 1 + 2\sin(\theta/2)$. The parameter interval is $[0, 4\pi]$.

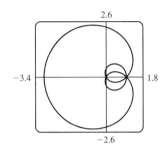

72. $r = \sqrt{1 - 0.8\sin^2\theta}$. The parameter interval is $[0, 2\pi]$.

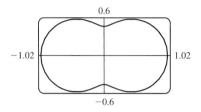

73. $r = e^{\sin\theta} - 2\cos(4\theta)$. The parameter interval is $[0, 2\pi]$.

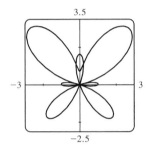

74. $r = \sin^2(4\theta) + \cos(4\theta)$. The parameter interval is $[0, 2\pi]$.

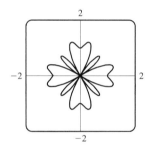

75. $r = 2 - 5\sin(\theta/6)$. The parameter interval is $[-6\pi, 6\pi]$.

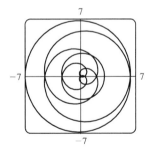

76. $r = \cos(\theta/2) + \cos(\theta/3)$. The parameter interval is $[-6\pi, 6\pi]$.

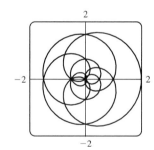

77. It appears that the graph of $r = 1 + \sin\left(\theta - \frac{\pi}{6}\right)$ is the same shape as the graph of $r = 1 + \sin\theta$, but rotated counterclockwise about the origin by $\frac{\pi}{6}$. Similarly, the graph of $r = 1 + \sin\left(\theta - \frac{\pi}{3}\right)$ is rotated by $\frac{\pi}{3}$. In general, the graph of $r = f(\theta - \alpha)$ is the same shape as that of $r = f(\theta)$, but rotated counterclockwise through α about the origin. That is, for any point (r_0, θ_0) on the curve $r = f(\theta)$, the point $(r_0, \theta_0 + \alpha)$ is on the curve $r = f(\theta - \alpha)$, since $r_0 = f(\theta_0) = f((\theta_0 + \alpha) - \alpha)$.

78.

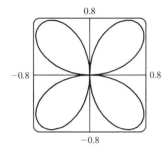

From the graph, the highest points seem to have $y \approx 0.77$. To find the exact value, we solve $dy/d\theta = 0$. $y = r\sin\theta = \sin\theta\,\sin 2\theta \;\Rightarrow$

$$dy/d\theta = 2\sin\theta\,\cos 2\theta + \cos\theta\,\sin 2\theta$$
$$= 2\sin\theta\,(2\cos^2\theta - 1) + \cos\theta\,(2\sin\theta\,\cos\theta)$$
$$= 2\sin\theta\,(3\cos^2\theta - 1)$$

In the first quadrant, this is 0 when $\cos\theta = \frac{1}{\sqrt{3}} \;\Leftrightarrow\; \sin\theta = \sqrt{\frac{2}{3}} \;\Leftrightarrow$

$y = 2\sin^2\theta\,\cos\theta = 2\cdot\frac{2}{3}\cdot\frac{1}{\sqrt{3}} = \frac{4}{9}\sqrt{3} \approx 0.77$.

79. (a) $r = \sin n\theta$.

$n = 2$ $n = 3$ $n = 4$ $n = 5$

From the graphs, it seems that when n is even, the number of loops in the curve (called a rose) is $2n$, and when n is odd, the number of loops is simply n. This is because in the case of n odd, every point on the graph is traversed twice, due to the fact that

$$r(\theta + \pi) = \sin[n(\theta + \pi)] = \sin n\theta\,\cos n\pi + \cos n\theta\,\sin n\pi = \begin{cases} \sin n\theta & \text{if } n \text{ is even} \\ -\sin n\theta & \text{if } n \text{ is odd} \end{cases}$$

(b) The graph of $r = |\sin n\theta|$ has $2n$ loops whether n is odd or even, since $r(\theta + \pi) = r(\theta)$.

$n = 2$ $n = 3$ $n = 4$ $n = 5$

80. $r = 1 + c \sin n\theta$. We vary n while keeping c constant at 2. As n changes, the curves change in the same way as those in Exercise 79: the number of loops increases. Note that if n is even, the smaller loops are outside the larger ones; if n is odd, they are inside.

$c = 2$

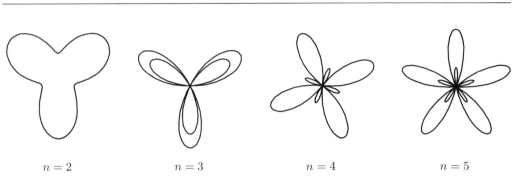

$n = 2$ $n = 3$ $n = 4$ $n = 5$

Now we vary c while keeping $n = 3$. As c increases toward 0, the entire graph gets smaller (the graphs below are not to scale) and the smaller loops shrink in relation to the large ones. At $c = -1$, the small loops disappear entirely, and for $-1 < c < 1$, the graph is a simple, closed curve (at $c = 0$ it is a circle). As c continues to increase, the same changes are seen, but in reverse order, since $1 + (-c)\sin n\theta = 1 + c\sin n(\theta + \pi)$, so the graph for $c = c_0$ is the same as that for $c = -c_0$, with a rotation through π. As $c \to \infty$, the smaller loops get relatively closer in size to the large ones. Note that the distance between the outermost points of corresponding inner and outer loops is always 2. Maple's `animate` command (or Mathematica's `Animate`) is very useful for seeing the changes that occur as c varies.

$n = 3$

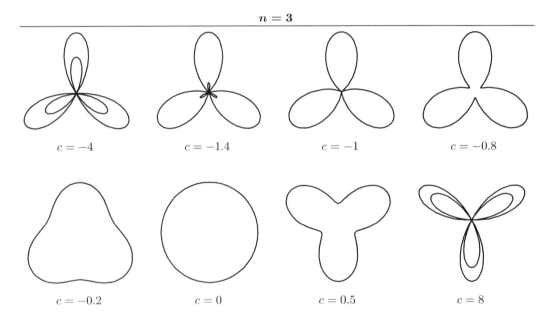

$c = -4$ $c = -1.4$ $c = -1$ $c = -0.8$

$c = -0.2$ $c = 0$ $c = 0.5$ $c = 8$

81. $r = \dfrac{1 - a\cos\theta}{1 + a\cos\theta}$. We start with $a = 0$, since in this case the curve is simply the circle $r = 1$.

As a increases, the graph moves to the left, and its right side becomes flattened. As a increases through about 0.4, the right side seems to grow a dimple, which upon closer investigation (with narrower θ-ranges) seems to appear at $a \approx 0.42$ [the actual value is $\sqrt{2} - 1$]. As $a \to 1$, this dimple becomes more pronounced, and the curve begins to stretch out horizontally, until at $a = 1$ the denominator vanishes at $\theta = \pi$, and the dimple becomes an actual cusp. For $a > 1$ we must choose our parameter interval carefully, since $r \to \infty$ as $1 + a\cos\theta \to 0$ $\Leftrightarrow$ $\theta \to \pm\cos^{-1}(-1/a)$. As a increases from 1, the curve splits into two parts. The left part has a loop, which grows larger as a increases, and the right part grows broader vertically, and its left tip develops a dimple when $a \approx 2.42$ [actually, $\sqrt{2} + 1$]. As a increases, the dimple grows more and more pronounced. If $a < 0$, we get the same graph as we do for the corresponding positive a-value, but with a rotation through π about the pole, as happened when c was replaced with $-c$ in Exercise 80.

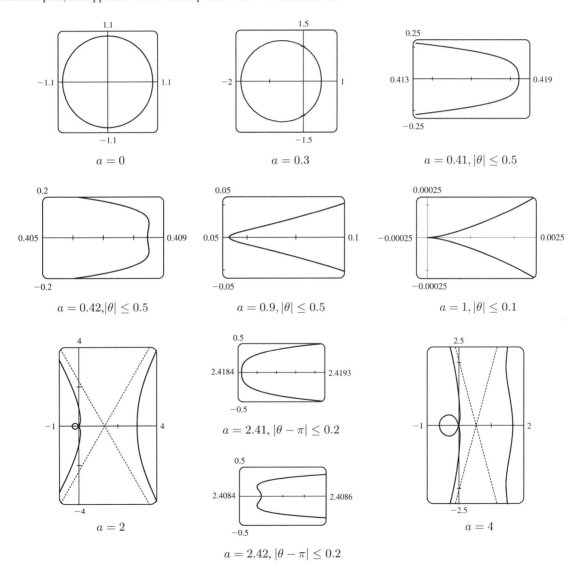

82. Most graphing devices cannot plot implicit polar equations, so we must first find an explicit expression (or expressions) for r in terms of θ, a, and c. We note that the given equation, $r^4 - 2c^2 r^2 \cos 2\theta + c^4 - a^4 = 0$, is a quadratic in r^2, so we use the quadratic formula and find that

$$r^2 = \frac{2c^2 \cos 2\theta \pm \sqrt{4c^4 \cos^2 2\theta - 4(c^4 - a^4)}}{2} = c^2 \cos 2\theta \pm \sqrt{a^4 - c^4 \sin^2 2\theta}$$

so $r = \pm\sqrt{c^2 \cos 2\theta \pm \sqrt{a^4 - c^4 \sin^2 2\theta}}$. So for each graph, we must plot four curves to be sure of plotting all the points which satisfy the given equation. Note that all four functions have period π.

We start with the case $a = c = 1$, and the resulting curve resembles the symbol for infinity. If we let a decrease, the curve splits into two symmetric parts, and as a decreases further, the parts become smaller, further apart, and rounder. If instead we let a increase from 1, the two lobes of the curve join together, and as a increases further they continue to merge, until at $a \approx 1.4$, the graph no longer has dimples, and has an oval shape. As $a \to \infty$, the oval becomes larger and rounder, since the c^2 and c^4 terms lose their significance. Note that the shape of the graph seems to depend only on the ratio c/a, while the size of the graph varies as c and a jointly increase.

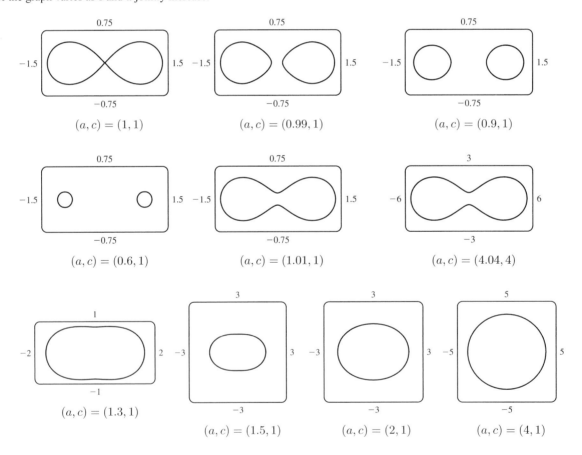

83. $\tan\psi = \tan(\phi - \theta) = \dfrac{\tan\phi - \tan\theta}{1 + \tan\phi\,\tan\theta} = \dfrac{\dfrac{dy}{dx} - \tan\theta}{1 + \dfrac{dy}{dx}\tan\theta} = \dfrac{\dfrac{dy/d\theta}{dx/d\theta} - \tan\theta}{1 + \dfrac{dy/d\theta}{dx/d\theta}\tan\theta}$

$= \dfrac{\dfrac{dy}{d\theta} - \dfrac{dx}{d\theta}\tan\theta}{\dfrac{dx}{d\theta} + \dfrac{dy}{d\theta}\tan\theta} = \dfrac{\left(\dfrac{dr}{d\theta}\sin\theta + r\cos\theta\right) - \tan\theta\left(\dfrac{dr}{d\theta}\cos\theta - r\sin\theta\right)}{\left(\dfrac{dr}{d\theta}\cos\theta - r\sin\theta\right) + \tan\theta\left(\dfrac{dr}{d\theta}\sin\theta + r\cos\theta\right)} = \dfrac{r\cos\theta + r\cdot\dfrac{\sin^2\theta}{\cos\theta}}{\dfrac{dr}{d\theta}\cos\theta + \dfrac{dr}{d\theta}\cdot\dfrac{\sin^2\theta}{\cos\theta}}$

$= \dfrac{r\cos^2\theta + r\sin^2\theta}{\dfrac{dr}{d\theta}\cos^2\theta + \dfrac{dr}{d\theta}\sin^2\theta} = \dfrac{r}{dr/d\theta}$

84. (a) $r = e^\theta \;\Rightarrow\; dr/d\theta = e^\theta$, so by Exercise 83, $\tan\psi = r/e^\theta = 1 \;\Rightarrow$
$\psi = \arctan 1 = \frac{\pi}{4}$.

(b) The Cartesian equation of the tangent line at $(1,0)$ is $y = x - 1$, and that of
the tangent line at $(0, e^{\pi/2})$ is $y = e^{\pi/2} - x$.

(c) Let a be the tangent of the angle between the tangent and radial lines, that
is, $a = \tan\psi$. Then, by Exercise 83, $a = \dfrac{r}{dr/d\theta} \;\Rightarrow\; \dfrac{dr}{d\theta} = \dfrac{1}{a}r \;\Rightarrow$
$r = Ce^{\theta/a}$ [by Theorem 10.4.2].

11.4 Areas and Lengths in Polar Coordinates

1. $r = \theta^2$, $0 \le \theta \le \frac{\pi}{4}$. $\quad A = \displaystyle\int_0^{\pi/4} \tfrac{1}{2}r^2\,d\theta = \int_0^{\pi/4} \tfrac{1}{2}(\theta^2)^2\,d\theta = \int_0^{\pi/4} \tfrac{1}{2}\theta^4\,d\theta = \left[\tfrac{1}{10}\theta^5\right]_0^{\pi/4} = \tfrac{1}{10}\left(\tfrac{\pi}{4}\right)^5 = \tfrac{1}{10{,}240}\pi^5$

2. $r = e^{\theta/2}$, $\pi \le \theta \le 2\pi$. $\quad A = \displaystyle\int_\pi^{2\pi} \tfrac{1}{2}(e^{\theta/2})^2\,d\theta = \int_\pi^{2\pi} \tfrac{1}{2}e^\theta\,d\theta = \tfrac{1}{2}\left[e^\theta\right]_\pi^{2\pi} = \tfrac{1}{2}(e^{2\pi} - e^\pi)$

3. $r = \sin\theta$, $\frac{\pi}{3} \le \theta \le \frac{2\pi}{3}$.

$A = \displaystyle\int_{\pi/3}^{2\pi/3} \tfrac{1}{2}\sin^2\theta\,d\theta = \tfrac{1}{4}\int_{\pi/3}^{2\pi/3}(1 - \cos 2\theta)\,d\theta = \tfrac{1}{4}\left[\theta - \tfrac{1}{2}\sin 2\theta\right]_{\pi/3}^{2\pi/3} = \tfrac{1}{4}\left[\tfrac{2\pi}{3} - \tfrac{1}{2}\sin\tfrac{4\pi}{3} - \tfrac{\pi}{3} + \tfrac{1}{2}\sin\tfrac{2\pi}{3}\right]$

$= \tfrac{1}{4}\left[\tfrac{2\pi}{3} - \tfrac{1}{2}\left(-\tfrac{\sqrt{3}}{2}\right) - \tfrac{\pi}{3} + \tfrac{1}{2}\left(\tfrac{\sqrt{3}}{2}\right)\right] = \tfrac{1}{4}\left(\tfrac{\pi}{3} + \tfrac{\sqrt{3}}{2}\right) = \tfrac{\pi}{12} + \tfrac{\sqrt{3}}{8}$

4. $r = \sqrt{\sin\theta}$, $0 \le \theta \le \pi$. $\quad A = \displaystyle\int_0^\pi \tfrac{1}{2}\left(\sqrt{\sin\theta}\right)^2\,d\theta = \int_0^\pi \tfrac{1}{2}\sin\theta\,d\theta = \left[-\tfrac{1}{2}\cos\theta\right]_0^\pi = \tfrac{1}{2} + \tfrac{1}{2} = 1$

5. $r = \sqrt{\theta}$, $0 \le \theta \le 2\pi$. $\quad A = \displaystyle\int_0^{2\pi} \tfrac{1}{2}r^2\,d\theta = \int_0^{2\pi} \tfrac{1}{2}\left(\sqrt{\theta}\right)^2\,d\theta = \int_0^{2\pi} \tfrac{1}{2}\theta\,d\theta = \left[\tfrac{1}{4}\theta^2\right]_0^{2\pi} = \pi^2$

6. $r = 1 + \cos\theta$, $0 \le \theta \le \pi$.

$A = \displaystyle\int_0^\pi \tfrac{1}{2}(1 + \cos\theta)^2\,d\theta = \tfrac{1}{2}\int_0^\pi (1 + 2\cos\theta + \cos^2\theta)\,d\theta = \tfrac{1}{2}\int_0^\pi \left[1 + 2\cos\theta + \tfrac{1}{2}(1 + \cos 2\theta)\right]d\theta$

$= \tfrac{1}{2}\displaystyle\int_0^\pi \left(\tfrac{3}{2} + 2\cos\theta + \tfrac{1}{2}\cos 2\theta\right)d\theta = \tfrac{1}{2}\left[\tfrac{3}{2}\theta + 2\sin\theta + \tfrac{1}{4}\sin 2\theta\right]_0^\pi = \tfrac{1}{2}\left(\tfrac{3}{2}\pi + 0 + 0\right) - \tfrac{1}{2}(0) = \tfrac{3\pi}{4}$

7. $r = 4 + 3\sin\theta$, $-\frac{\pi}{2} \le \theta \le \frac{\pi}{2}$.

$$A = \int_{-\pi/2}^{\pi/2} \frac{1}{2}((4 + 3\sin\theta)^2 \, d\theta = \frac{1}{2}\int_{-\pi/2}^{\pi/2} (16 + 24\sin\theta + 9\sin^2\theta)\, d\theta = \frac{1}{2}\int_{-\pi/2}^{\pi/2} (16 + 9\sin^2\theta)\, d\theta \quad \text{[by Theorem 5.5.6(b)]}$$

$$= \frac{1}{2} \cdot 2\int_{0}^{\pi/2} \left[16 + 9 \cdot \frac{1}{2}(1 - \cos 2\theta)\right] d\theta \qquad \text{[by Theorem 5.5.6(a)]}$$

$$= \int_{0}^{\pi/2} \left(\frac{41}{2} - \frac{9}{2}\cos 2\theta\right) d\theta = \left[\frac{41}{2}\theta - \frac{9}{4}\sin 2\theta\right]_{0}^{\pi/2} = \left(\frac{41\pi}{4} - 0\right) - (0 - 0) = \frac{41\pi}{4}$$

8. $r = \sin 2\theta$, $0 \le \theta \le \frac{\pi}{2}$.

$$A = \int_{0}^{\pi/2} \frac{1}{2}\sin^2 2\theta \, d\theta = \frac{1}{2}\int_{0}^{\pi/2} \frac{1}{2}(1 - \cos 4\theta)\, d\theta = \frac{1}{4}\left[\theta - \frac{1}{4}\sin 4\theta\right]_{0}^{\pi/2} = \frac{1}{4}\left(\frac{\pi}{2}\right) = \frac{\pi}{8}$$

9. The area above the polar axis is bounded by $r = 3\cos\theta$ for $\theta = 0$

to $\theta = \pi/2$ [*not* π]. By symmetry,

$$A = 2\int_{0}^{\pi/2} \frac{1}{2}r^2 \, d\theta = \int_{0}^{\pi/2}(3\cos\theta)^2 \, d\theta = 3^2 \int_{0}^{\pi/2} \cos^2\theta \, d\theta$$

$$= 9\int_{0}^{\pi/2} \frac{1}{2}(1 + \cos 2\theta)\, d\theta = \frac{9}{2}\left[\theta + \frac{1}{2}\sin 2\theta\right]_{0}^{\pi/2} = \frac{9}{2}\left[\left(\frac{\pi}{2} + 0\right) - (0 + 0)\right] = \frac{9\pi}{4}$$

Also, note that this is a circle with radius $\frac{3}{2}$, so its area is $\pi\left(\frac{3}{2}\right)^2 = \frac{9\pi}{4}$.

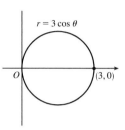

10. $A = \int_{0}^{2\pi} \frac{1}{2}r^2 \, d\theta = \int_{0}^{2\pi} \frac{1}{2}[3(1 + \cos\theta)]^2 \, d\theta$

$$= \frac{9}{2}\int_{0}^{2\pi}(1 + 2\cos\theta + \cos^2\theta)\, d\theta$$

$$= \frac{9}{2}\int_{0}^{2\pi} \left[1 + 2\cos\theta + \frac{1}{2}(1 + \cos 2\theta)\right] d\theta$$

$$= \frac{9}{2}\left[\frac{3}{2}\theta + 2\sin\theta + \frac{1}{4}\sin 2\theta\right]_{0}^{2\pi} = \frac{27}{2}\pi$$

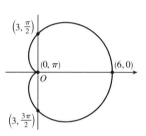

11. The curve goes through the pole when $\theta = \pi/4$, so we'll find the area for

$0 \le \theta \le \pi/4$ and multiply it by 4.

$$A = 4\int_{0}^{\pi/4} \frac{1}{2}r^2 \, d\theta = 2\int_{0}^{\pi/4}(4\cos 2\theta)\, d\theta$$

$$= 8\int_{0}^{\pi/4} \cos 2\theta \, d\theta = 4\left[\sin 2\theta\right]_{0}^{\pi/4} = 4$$

12. To find the area that the curve encloses, we'll double the area to the left of the

vertical axis.

$$A = 2\int_{\pi/2}^{3\pi/2} \frac{1}{2}(2 - \sin\theta)^2 \, d\theta = \int_{\pi/2}^{3\pi/2}(4 - 4\sin\theta + \sin^2\theta)\, d\theta$$

$$= \int_{\pi/2}^{3\pi/2}[4 - 4\sin\theta + \frac{1}{2}(1 - \cos 2\theta)]\, d\theta = \int_{\pi/2}^{3\pi/2} \left(\frac{9}{2} - 4\sin\theta - \frac{1}{2}\cos 2\theta\right) d\theta$$

$$= \left[\frac{9}{2}\theta + 4\cos\theta - \frac{1}{4}\sin 2\theta\right]_{\pi/2}^{3\pi/2} = \left(\frac{27\pi}{4}\right) - \left(\frac{9\pi}{4}\right) = \frac{9\pi}{2}$$

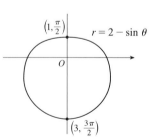

Or: We could have doubled the area to the right of the vertical axis and integrated from $-\pi/2$ to $\pi/2$.

Or: We could have integrated from 0 to 2π [simpler arithmetic].

13. One-sixth of the area lies above the polar axis and is bounded by the curve

$r = 2\cos 3\theta$ for $\theta = 0$ to $\theta = \pi/6$.

$$A = 6 \int_0^{\pi/6} \frac{1}{2}(2\cos 3\theta)^2 \, d\theta = 12 \int_0^{\pi/6} \cos^2 3\theta \, d\theta$$

$$= \frac{12}{2} \int_0^{\pi/6} (1 + \cos 6\theta) \, d\theta$$

$$= 6 \left[\theta + \frac{1}{6}\sin 6\theta \right]_0^{\pi/6} = 6\left(\frac{\pi}{6}\right) = \pi$$

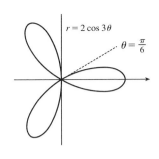

14. $A = \int_0^{2\pi} \frac{1}{2}(2 + \cos 2\theta)^2 \, d\theta = \frac{1}{2}\int_0^{2\pi}(4 + 4\cos 2\theta + \cos^2 2\theta) \, d\theta$

$$= \frac{1}{2}\int_0^{2\pi}\left(4 + 4\cos 2\theta + \frac{1}{2} + \frac{1}{2}\cos 4\theta\right) d\theta$$

$$= \frac{1}{2}\left[\frac{9}{2}\theta + 2\sin 2\theta + \frac{1}{8}\sin 4\theta\right]_0^{2\pi} = \frac{1}{2}(9\pi) = \frac{9\pi}{2}$$

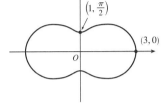

15. $A = \int_0^{2\pi} \frac{1}{2}(1 + 2\sin 6\theta)^2 \, d\theta = \frac{1}{2}\int_0^{2\pi}(1 + 4\sin 6\theta + 4\sin^2 6\theta) \, d\theta$

$$= \frac{1}{2}\int_0^{2\pi}\left[1 + 4\sin 6\theta + 4\cdot\frac{1}{2}(1 - \cos 12\theta)\right] d\theta$$

$$= \frac{1}{2}\int_0^{2\pi}(3 + 4\sin 6\theta - 2\cos 12\theta) \, d\theta$$

$$= \frac{1}{2}\left[3\theta - \frac{2}{3}\cos 6\theta - \frac{1}{6}\sin 12\theta\right]_0^{2\pi}$$

$$= \frac{1}{2}\left[\left(6\pi - \frac{2}{3} - 0\right) - \left(0 - \frac{2}{3} - 0\right)\right] = 3\pi$$

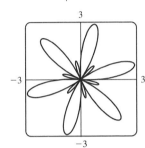

16. $A = \int_0^{\pi} \frac{1}{2}(2\sin\theta + 3\sin 9\theta)^2 \, d\theta = 2\int_0^{\pi/2} \frac{1}{2}(2\sin\theta + 3\sin 9\theta)^2 \, d\theta$

$$= \int_0^{\pi/2}(4\sin^2\theta + 12\sin\theta\sin 9\theta + 9\sin^2 9\theta) \, d\theta$$

$$= \int_0^{\pi/2}\left[2(1 - \cos 2\theta) + 12\cdot\frac{1}{2}(\cos(\theta - 9\theta) - \cos(\theta + 9\theta)) + \frac{9}{2}(1 - \cos 18\theta)\right] d\theta$$

[integration by parts could be used for $\int \sin\theta \sin 9\theta \, d\theta$]

$$= \int_0^{\pi/2}\left(2 - 2\cos 2\theta + 6\cos 8\theta - 6\cos 10\theta + \frac{9}{2} - \frac{9}{2}\cos 18\theta\right) d\theta$$

$$= \left[\frac{13}{2}\theta - \sin 2\theta + \frac{3}{4}\sin 8\theta - \frac{3}{5}\sin 10\theta - \frac{1}{4}\sin 18\theta\right]_0^{\pi/2} = \frac{13\pi}{4}$$

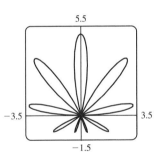

17. The shaded loop is traced out from $\theta = 0$ to $\theta = \pi/2$.

$$A = \int_0^{\pi/2} \frac{1}{2}r^2 \, d\theta = \frac{1}{2}\int_0^{\pi/2}\sin^2 2\theta \, d\theta$$

$$= \frac{1}{2}\int_0^{\pi/2} \frac{1}{2}(1 - \cos 4\theta) \, d\theta = \frac{1}{4}\left[\theta - \frac{1}{4}\sin 4\theta\right]_0^{\pi/2}$$

$$= \frac{1}{4}\left(\frac{\pi}{2}\right) = \frac{\pi}{8}$$

18. $A = \int_0^{\pi/3} \frac{1}{2}(4\sin 3\theta)^2 \, d\theta = 8\int_0^{\pi/3}\sin^2 3\theta \, d\theta$

$$= 4\int_0^{\pi/3}(1 - \cos 6\theta) \, d\theta$$

$$= 4\left[\theta - \frac{1}{6}\sin 6\theta\right]_0^{\pi/3} = \frac{4\pi}{3}$$

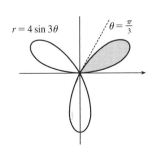

19. $r = 0 \Rightarrow 3\cos 5\theta = 0 \Rightarrow 5\theta = \frac{\pi}{2} \Rightarrow \theta = \frac{\pi}{10}$.

$A = \int_{-\pi/10}^{\pi/10} \frac{1}{2}(3\cos 5\theta)^2 \, d\theta = \int_0^{\pi/10} 9\cos^2 5\theta \, d\theta = \frac{9}{2} \int_0^{\pi/10}(1 + \cos 10\theta) \, d\theta = \frac{9}{2}\left[\theta + \frac{1}{10}\sin 10\theta\right]_0^{\pi/10} = \frac{9\pi}{20}$

20. $r = 0 \Rightarrow 2\sin 6\theta = 0 \Rightarrow 6\theta = 0$ or $\pi \Rightarrow \theta = 0$ or $\frac{\pi}{6}$.

$A = \int_0^{\pi/6} \frac{1}{2}(2\sin 6\theta)^2 \, d\theta = \int_0^{\pi/6} 2\sin^2 6\theta \, d\theta = 2\int_0^{\pi/6} \frac{1}{2}(1 - \cos 12\theta) \, d\theta = \left[\theta - \frac{1}{12}\sin 12\theta\right]_0^{\pi/6} = \frac{\pi}{6}$

21.

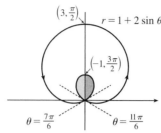

This is a limaçon, with inner loop traced out between $\theta = \frac{7\pi}{6}$ and $\frac{11\pi}{6}$ [found by solving $r = 0$].

$A = 2\int_{7\pi/6}^{3\pi/2} \frac{1}{2}(1 + 2\sin\theta)^2 \, d\theta = \int_{7\pi/6}^{3\pi/2}\left(1 + 4\sin\theta + 4\sin^2\theta\right) d\theta = \int_{7\pi/6}^{3\pi/2}\left[1 + 4\sin\theta + 4\cdot\frac{1}{2}(1 - \cos 2\theta)\right] d\theta$

$= \left[\theta - 4\cos\theta + 2\theta - \sin 2\theta\right]_{7\pi/6}^{3\pi/2} = \left(\frac{9\pi}{2}\right) - \left(\frac{7\pi}{2} + 2\sqrt{3} - \frac{\sqrt{3}}{2}\right) = \pi - \frac{3\sqrt{3}}{2}$

22. To determine when the strophoid $r = 2\cos\theta - \sec\theta$ passes through the pole, we solve

$r = 0 \Rightarrow 2\cos\theta - \frac{1}{\cos\theta} = 0 \Rightarrow 2\cos^2\theta - 1 = 0 \Rightarrow \cos^2\theta = \frac{1}{2} \Rightarrow$

$\cos\theta = \pm\frac{1}{\sqrt{2}} \Rightarrow \theta = \frac{\pi}{4}$ or $\theta = \frac{3\pi}{4}$ for $0 \le \theta \le \pi$ with $\theta \ne \frac{\pi}{2}$.

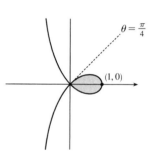

$A = 2\int_0^{\pi/4} \frac{1}{2}(2\cos\theta - \sec\theta)^2 \, d\theta = \int_0^{\pi/4}(4\cos^2\theta - 4 + \sec^2\theta) \, d\theta$

$= \int_0^{\pi/4}\left[4\cdot\frac{1}{2}(1 + \cos 2\theta) - 4 + \sec^2\theta\right] d\theta = \int_0^{\pi/4}(-2 + 2\cos 2\theta + \sec^2\theta) \, d\theta$

$= \left[-2\theta + \sin 2\theta + \tan\theta\right]_0^{\pi/4} = \left(-\frac{\pi}{2} + 1 + 1\right) - 0 = 2 - \frac{\pi}{2}$

23. $2\cos\theta = 1 \Rightarrow \cos\theta = \frac{1}{2} \Rightarrow \theta = \frac{\pi}{3}$ or $\frac{5\pi}{3}$.

$A = 2\int_0^{\pi/3} \frac{1}{2}[(2\cos\theta)^2 - 1^2] \, d\theta = \int_0^{\pi/3}(4\cos^2\theta - 1) \, d\theta$

$= \int_0^{\pi/3}\left\{4\left[\frac{1}{2}(1 + \cos 2\theta)\right] - 1\right\} d\theta = \int_0^{\pi/3}(1 + 2\cos 2\theta) \, d\theta$

$= \left[\theta + \sin 2\theta\right]_0^{\pi/3} = \frac{\pi}{3} + \frac{\sqrt{3}}{2}$

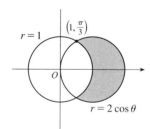

24. $1 - \sin\theta = 1 \Rightarrow \sin\theta = 0 \Rightarrow \theta = 0$ or $\pi \Rightarrow$

$A = \int_\pi^{2\pi} \frac{1}{2}\left[(1 - \sin\theta)^2 - 1\right] d\theta = \frac{1}{2}\int_\pi^{2\pi}(\sin^2\theta - 2\sin\theta) \, d\theta$

$= \frac{1}{4}\int_\pi^{2\pi}(1 - \cos 2\theta - 4\sin\theta) \, d\theta = \frac{1}{4}\left[\theta - \frac{1}{2}\sin 2\theta + 4\cos\theta\right]_\pi^{2\pi}$

$= \frac{1}{4}\pi + 2$

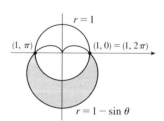

25. To find the area inside the leminiscate $r^2 = 8 \cos 2\theta$ and outside the circle $r = 2$,

we first note that the two curves intersect when $r^2 = 8 \cos 2\theta$ and $r = 2$,

that is, when $\cos 2\theta = \frac{1}{2}$. For $-\pi < \theta \le \pi$, $\cos 2\theta = \frac{1}{2} \iff 2\theta = \pm\pi/3$

or $\pm 5\pi/3 \iff \theta = \pm\pi/6$ or $\pm 5\pi/6$. The figure shows that the desired area is

4 times the area between the curves from 0 to $\pi/6$. Thus,

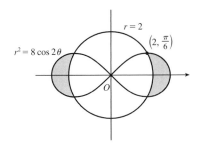

$$A = 4 \int_0^{\pi/6} \left[\tfrac{1}{2}(8 \cos 2\theta) - \tfrac{1}{2}(2)^2\right] d\theta = 8 \int_0^{\pi/6} (2 \cos 2\theta - 1) \, d\theta$$

$$= 8 \left[\sin 2\theta - \theta\right]_0^{\pi/6} = 8\left(\sqrt{3}/2 - \pi/6\right) = 4\sqrt{3} - 4\pi/3$$

26. To find the shaded area A, we'll find the area A_1 inside the curve $r = 2 + \sin \theta$

and subtract $\pi\left(\frac{3}{2}\right)^2$ since $r = 3 \sin \theta$ is a circle with radius $\frac{3}{2}$.

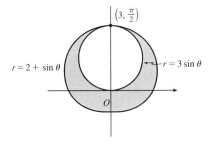

$$A_1 = \int_0^{2\pi} \tfrac{1}{2}(2 + \sin \theta)^2 \, d\theta = \tfrac{1}{2}\int_0^{2\pi}(4 + 4\sin \theta + \sin^2 \theta) \, d\theta$$

$$= \tfrac{1}{2}\int_0^{2\pi}\left[4 + 4\sin \theta + \tfrac{1}{2} \cdot (1 - \cos 2\theta)\right] d\theta$$

$$= \tfrac{1}{2}\int_0^{2\pi}\left(\tfrac{9}{2} + 4\sin \theta - \tfrac{1}{2}\cos 2\theta\right) d\theta$$

$$= \tfrac{1}{2}\left[\tfrac{9}{2}\theta - 4\cos \theta - \tfrac{1}{4}\sin 2\theta\right]_0^{2\pi} = \tfrac{1}{2}[(9\pi - 4) - (-4)] = \tfrac{9\pi}{2}$$

So $A = A_1 - \frac{9\pi}{4} = \frac{9\pi}{2} - \frac{9\pi}{4} = \frac{9\pi}{4}$.

27. $3 \cos \theta = 1 + \cos \theta \iff \cos \theta = \frac{1}{2} \implies \theta = \frac{\pi}{3}$ or $-\frac{\pi}{3}$.

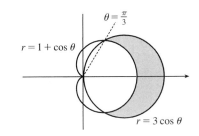

$$A = 2 \int_0^{\pi/3} \tfrac{1}{2}\left[(3 \cos \theta)^2 - (1 + \cos \theta)^2\right] d\theta$$

$$= \int_0^{\pi/3}(8 \cos^2 \theta - 2 \cos \theta - 1) \, d\theta = \int_0^{\pi/3}[4(1 + \cos 2\theta) - 2 \cos \theta - 1] \, d\theta$$

$$= \int_0^{\pi/3}(3 + 4 \cos 2\theta - 2 \cos \theta) \, d\theta = \left[3\theta + 2 \sin 2\theta - 2 \sin \theta\right]_0^{\pi/3}$$

$$= \pi + \sqrt{3} - \sqrt{3} = \pi$$

28. $3 \sin \theta = 2 - \sin \theta \implies 4 \sin \theta = 2 \implies \sin \theta = \frac{1}{2} \implies \theta = \frac{\pi}{6}$ or $\frac{5\pi}{6}$.

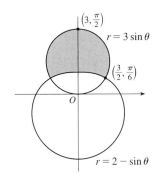

$$A = 2 \int_{\pi/6}^{\pi/2} \tfrac{1}{2}\left[(3 \sin \theta)^2 - (2 - \sin \theta)^2\right] d\theta$$

$$= \int_{\pi/6}^{\pi/2}(9 \sin^2 \theta - 4 + 4 \sin \theta - \sin^2 \theta] \, d\theta$$

$$= \int_{\pi/6}^{\pi/2}(8 \sin^2 \theta + 4 \sin \theta - 4) \, d\theta$$

$$= 4 \int_{\pi/6}^{\pi/2}\left[2 \cdot \tfrac{1}{2}(1 - \cos 2\theta) + \sin \theta - 1\right] d\theta$$

$$= 4 \int_{\pi/6}^{\pi/2}(\sin \theta - \cos 2\theta) \, d\theta = 4\left[-\cos \theta - \tfrac{1}{2}\sin 2\theta\right]_{\pi/6}^{\pi/2}$$

$$= 4\left[(0 - 0) - \left(-\tfrac{\sqrt{3}}{2} - \tfrac{\sqrt{3}}{4}\right)\right] = 4\left(\tfrac{3\sqrt{3}}{4}\right) = 3\sqrt{3}$$

29. $\sqrt{3}\cos\theta = \sin\theta \quad\Rightarrow\quad \sqrt{3} = \dfrac{\sin\theta}{\cos\theta} \quad\Rightarrow\quad \tan\theta = \sqrt{3} \quad\Rightarrow\quad \theta = \frac{\pi}{3}.$

$A = \int_0^{\pi/3} \frac{1}{2}(\sin\theta)^2\,d\theta + \int_{\pi/3}^{\pi/2} \frac{1}{2}\left(\sqrt{3}\cos\theta\right)^2\,d\theta$

$\quad = \int_0^{\pi/3} \frac{1}{2}\cdot\frac{1}{2}(1-\cos 2\theta)\,d\theta + \int_{\pi/3}^{\pi/2} \frac{1}{2}\cdot 3\cdot\frac{1}{2}(1+\cos 2\theta)\,d\theta$

$\quad = \frac{1}{4}\left[\theta - \frac{1}{2}\sin 2\theta\right]_0^{\pi/3} + \frac{3}{4}\left[\theta + \frac{1}{2}\sin 2\theta\right]_{\pi/3}^{\pi/2}$

$\quad = \frac{1}{4}\left[\left(\frac{\pi}{3} - \frac{\sqrt{3}}{4}\right) - 0\right] + \frac{3}{4}\left[\left(\frac{\pi}{2} + 0\right) - \left(\frac{\pi}{3} + \frac{\sqrt{3}}{4}\right)\right]$

$\quad = \frac{\pi}{12} - \frac{\sqrt{3}}{16} + \frac{\pi}{8} - \frac{3\sqrt{3}}{16} = \frac{5\pi}{24} - \frac{\sqrt{3}}{4}$

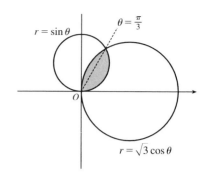

30. $A = 4\int_0^{\pi/2} \frac{1}{2}(1-\cos\theta)^2\,d\theta = 2\int_0^{\pi/2}(1 - 2\cos\theta + \cos^2\theta)\,d\theta$

$\quad = 2\int_0^{\pi/2}\left[1 - 2\cos\theta + \frac{1}{2}(1+\cos 2\theta)\right]d\theta$

$\quad = 2\int_0^{\pi/2}\left(\frac{3}{2} - 2\cos\theta + \frac{1}{2}\cos 2\theta\right)d\theta = \int_0^{\pi/2}(3 - 4\cos\theta + \cos 2\theta)\,d\theta$

$\quad = \left[3\theta - 4\sin\theta + \frac{1}{2}\sin 2\theta\right]_0^{\pi/2} = \frac{3\pi}{2} - 4$

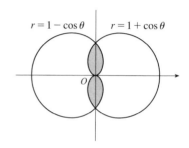

31. $\sin 2\theta = \cos 2\theta \quad\Rightarrow\quad \dfrac{\sin 2\theta}{\cos 2\theta} = 1 \quad\Rightarrow\quad \tan 2\theta = 1 \quad\Rightarrow\quad 2\theta = \frac{\pi}{4} \quad\Rightarrow$

$\theta = \frac{\pi}{8} \quad\Rightarrow$

$\qquad A = 8\cdot 2\int_0^{\pi/8} \frac{1}{2}\sin^2 2\theta\,d\theta = 8\int_0^{\pi/8} \frac{1}{2}(1-\cos 4\theta)\,d\theta$

$\qquad\quad = 4\left[\theta - \frac{1}{4}\sin 4\theta\right]_0^{\pi/8} = 4\left(\frac{\pi}{8} - \frac{1}{4}\cdot 1\right) = \frac{\pi}{2} - 1$

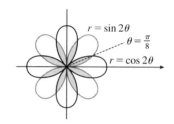

32. $3 + 2\cos\theta = 3 + 2\sin\theta \quad\Rightarrow\quad \cos\theta = \sin\theta \quad\Rightarrow\quad \theta = \frac{\pi}{4}$ or $\frac{5\pi}{4}.$

$A = 2\int_{\pi/4}^{5\pi/4} \frac{1}{2}(3 + 2\cos\theta)^2\,d\theta = \int_{\pi/4}^{5\pi/4}(9 + 12\cos\theta + 4\cos^2\theta)\,d\theta$

$\quad = \int_{\pi/4}^{5\pi/4}\left[9 + 12\cos\theta + 4\cdot\frac{1}{2}(1+\cos 2\theta)\right]d\theta$

$\quad = \int_{\pi/4}^{5\pi/4}(11 + 12\cos\theta + 2\cos 2\theta)\,d\theta = \left[11\theta + 12\sin\theta + \sin 2\theta\right]_{\pi/4}^{5\pi/4}$

$\quad = \left(\frac{55\pi}{4} - 6\sqrt{2} + 1\right) - \left(\frac{11\pi}{4} + 6\sqrt{2} + 1\right) = 11\pi - 12\sqrt{2}$

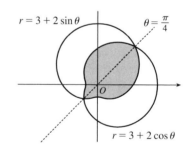

33. $\sin 2\theta = \cos 2\theta \quad\Rightarrow\quad \tan 2\theta = 1 \quad\Rightarrow\quad 2\theta = \frac{\pi}{4} \quad\Rightarrow\quad \theta = \frac{\pi}{8}$

$\qquad A = 4\int_0^{\pi/8} \frac{1}{2}\sin 2\theta\,d\theta \quad$ [since $r^2 = \sin 2\theta$]

$\qquad\quad = \int_0^{\pi/8} 2\sin 2\theta\,d\theta = \left[-\cos 2\theta\right]_0^{\pi/8}$

$\qquad\quad = -\frac{1}{2}\sqrt{2} - (-1) = 1 - \frac{1}{2}\sqrt{2}$

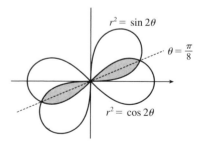

34. Let $\alpha = \tan^{-1}(b/a)$. Then

$$A = \int_0^\alpha \tfrac{1}{2}(a \sin\theta)^2 \, d\theta + \int_\alpha^{\pi/2} \tfrac{1}{2}(b \cos\theta)^2 \, d\theta$$

$$= \tfrac{1}{4}a^2 \left[\theta - \tfrac{1}{2}\sin 2\theta\right]_0^\alpha + \tfrac{1}{4}b^2 \left[\theta + \tfrac{1}{2}\sin 2\theta\right]_\alpha^{\pi/2}$$

$$= \tfrac{1}{4}\alpha(a^2 - b^2) + \tfrac{1}{8}\pi b^2 - \tfrac{1}{4}(a^2 + b^2)(\sin\alpha \, \cos\alpha)$$

$$= \tfrac{1}{4}(a^2 - b^2) \tan^{-1}(b/a) + \tfrac{1}{8}\pi b^2 - \tfrac{1}{4}ab$$

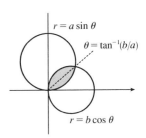

$r = a \sin\theta$

$\theta = \tan^{-1}(b/a)$

$r = b \cos\theta$

35. The darker shaded region (from $\theta = 0$ to $\theta = 2\pi/3$) represents $\frac{1}{2}$ of the desired area plus $\frac{1}{2}$ of the area of the inner loop. From this area, we'll subtract $\frac{1}{2}$ of the area of the inner loop (the lighter shaded region from $\theta = 2\pi/3$ to $\theta = \pi$), and then double that difference to obtain the desired area.

$$A = 2\left[\int_0^{2\pi/3} \tfrac{1}{2}\left(\tfrac{1}{2} + \cos\theta\right)^2 \, d\theta - \int_{2\pi/3}^\pi \tfrac{1}{2}\left(\tfrac{1}{2} + \cos\theta\right)^2 \, d\theta\right]$$

$$= \int_0^{2\pi/3}\left(\tfrac{1}{4} + \cos\theta + \cos^2\theta\right) d\theta - \int_{2\pi/3}^\pi\left(\tfrac{1}{4} + \cos\theta + \cos^2\theta\right) d\theta$$

$$= \int_0^{2\pi/3}\left[\tfrac{1}{4} + \cos\theta + \tfrac{1}{2}(1 + \cos 2\theta)\right] d\theta$$

$$\qquad - \int_{2\pi/3}^\pi\left[\tfrac{1}{4} + \cos\theta + \tfrac{1}{2}(1 + \cos 2\theta)\right] d\theta$$

$$= \left[\frac{\theta}{4} + \sin\theta + \frac{\theta}{2} + \frac{\sin 2\theta}{4}\right]_0^{2\pi/3} - \left[\frac{\theta}{4} + \sin\theta + \frac{\theta}{2} + \frac{\sin 2\theta}{4}\right]_{2\pi/3}^\pi$$

$$= \left(\frac{\pi}{6} + \frac{\sqrt{3}}{2} + \frac{\pi}{3} - \frac{\sqrt{3}}{8}\right) - \left(\frac{\pi}{4} + \frac{\pi}{2}\right) + \left(\frac{\pi}{6} + \frac{\sqrt{3}}{2} + \frac{\pi}{3} - \frac{\sqrt{3}}{8}\right)$$

$$= \frac{\pi}{4} + \frac{3}{4}\sqrt{3} = \tfrac{1}{4}\left(\pi + 3\sqrt{3}\right)$$

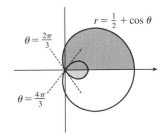

$r = \frac{1}{2} + \cos\theta$

$\theta = \frac{2\pi}{3}$

$\theta = \frac{4\pi}{3}$

36. $r = 0 \Rightarrow 1 + 2\cos 3\theta = 0 \Rightarrow \cos 3\theta = -\frac{1}{2} \Rightarrow 3\theta = \frac{2\pi}{3}, \frac{4\pi}{3}$ [for $0 \leq 3\theta \leq 2\pi$] $\Rightarrow \theta = \frac{2\pi}{9}, \frac{4\pi}{9}$. The darker shaded region (from $\theta = 0$ to $\theta = 2\pi/9$) represents $\frac{1}{2}$ of the desired area plus $\frac{1}{2}$ of the area of the inner loop. From this area, we'll subtract $\frac{1}{2}$ of the area of the inner loop (the lighter shaded region from $\theta = 2\pi/9$ to $\theta = \pi/3$), and then double that difference to obtain the desired area.

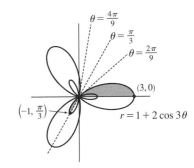

$\theta = \frac{4\pi}{9}$

$\theta = \frac{\pi}{3}$

$\theta = \frac{2\pi}{9}$

$(3, 0)$

$\left(-1, \frac{\pi}{3}\right)$

$r = 1 + 2\cos 3\theta$

$$A = 2\left[\int_0^{2\pi/9} \tfrac{1}{2}(1 + 2\cos 3\theta)^2 \, d\theta - \int_{2\pi/9}^{\pi/3} \tfrac{1}{2}(1 + 2\cos 3\theta)^2 \, d\theta\right]$$

Now

$$r^2 = (1 + 2\cos 3\theta)^2 = 1 + 4\cos 3\theta + 4\cos^2 3\theta = 1 + 4\cos 3\theta + 4 \cdot \tfrac{1}{2}(1 + \cos 6\theta)$$

$$= 1 + 4\cos 3\theta + 2 + 2\cos 6\theta = 3 + 4\cos 3\theta + 2\cos 6\theta$$

and $\int r^2 d\theta = 3\theta + \frac{4}{3}\sin 3\theta + \frac{1}{3}\sin 6\theta + C$, so

$$A = \left[3\theta + \tfrac{4}{3}\sin 3\theta + \tfrac{1}{3}\sin 6\theta\right]_0^{2\pi/9} - \left[3\theta + \tfrac{4}{3}\sin 3\theta + \tfrac{1}{3}\sin 6\theta\right]_{2\pi/9}^{\pi/3}$$

$$= \left[\left(\frac{2\pi}{3} + \frac{4}{3} \cdot \frac{\sqrt{3}}{2} + \frac{1}{3} \cdot \frac{-\sqrt{3}}{2}\right) - 0\right] - \left[(\pi + 0 + 0) - \left(\frac{2\pi}{3} + \frac{4}{3} \cdot \frac{\sqrt{3}}{2} + \frac{1}{3} \cdot \frac{-\sqrt{3}}{2}\right)\right]$$

$$= \frac{4\pi}{3} + \frac{4}{3}\sqrt{3} - \frac{1}{3}\sqrt{3} - \pi = \frac{\pi}{3} + \sqrt{3}$$

37. The pole is a point of intersection.

$$1 + \sin\theta = 3\sin\theta \quad\Rightarrow\quad 1 = 2\sin\theta \quad\Rightarrow\quad \sin\theta = \tfrac{1}{2} \quad\Rightarrow$$

$$\theta = \tfrac{\pi}{6} \text{ or } \tfrac{5\pi}{6}.$$

The other two points of intersection are $\left(\tfrac{3}{2}, \tfrac{\pi}{6}\right)$ and $\left(\tfrac{3}{2}, \tfrac{5\pi}{6}\right)$.

$r = 3\sin\theta$

$r = 1 + \sin\theta$

38. The pole is a point of intersection.

$$1 - \cos\theta = 1 + \sin\theta \quad\Rightarrow\quad -\cos\theta = \sin\theta \quad\Rightarrow\quad -1 = \tan\theta \quad\Rightarrow$$

$$\theta = \tfrac{3\pi}{4} \text{ or } \tfrac{7\pi}{4}.$$

The other two points of intersection are $\left(1 + \tfrac{\sqrt{2}}{2}, \tfrac{3\pi}{4}\right)$ and $\left(1 - \tfrac{\sqrt{2}}{2}, \tfrac{7\pi}{4}\right)$.

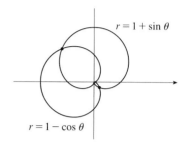

$r = 1 + \sin\theta$

$r = 1 - \cos\theta$

39. $2\sin 2\theta = 1 \quad\Rightarrow\quad \sin 2\theta = \tfrac{1}{2} \quad\Rightarrow\quad 2\theta = \tfrac{\pi}{6}, \tfrac{5\pi}{6}, \tfrac{13\pi}{6}, \text{ or } \tfrac{17\pi}{6}.$

By symmetry, the eight points of intersection are given by

$(1, \theta)$, where $\theta = \tfrac{\pi}{12}, \tfrac{5\pi}{12}, \tfrac{13\pi}{12}, \text{ and } \tfrac{17\pi}{12}$, and

$(-1, \theta)$, where $\theta = \tfrac{7\pi}{12}, \tfrac{11\pi}{12}, \tfrac{19\pi}{12}, \text{ and } \tfrac{23\pi}{12}$.

[There are many ways to describe these points.]

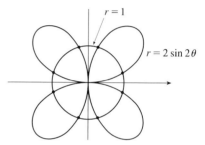

$r = 1$

$r = 2\sin 2\theta$

40. Clearly the pole lies on both curves. $\sin 3\theta = \cos 3\theta \quad\Rightarrow\quad \tan 3\theta = 1 \quad\Rightarrow$

$3\theta = \tfrac{\pi}{4} + n\pi$ [n any integer] $\quad\Rightarrow\quad \theta = \tfrac{\pi}{12} + \tfrac{\pi}{3}n \quad\Rightarrow$

$\theta = \tfrac{\pi}{12}, \tfrac{5\pi}{12}, \text{ or } \tfrac{3\pi}{4}$, so the three remaining intersection points are $\left(\tfrac{1}{\sqrt{2}}, \tfrac{\pi}{12}\right)$,

$\left(-\tfrac{1}{\sqrt{2}}, \tfrac{5\pi}{12}\right)$, and $\left(\tfrac{1}{\sqrt{2}}, \tfrac{3\pi}{4}\right)$.

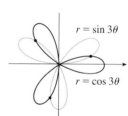

$r = \sin 3\theta$

$r = \cos 3\theta$

41. The pole is a point of intersection. $\sin\theta = \sin 2\theta = 2\sin\theta\,\cos\theta \quad\Leftrightarrow$

$\sin\theta\,(1 - 2\cos\theta) = 0 \quad\Leftrightarrow\quad \sin\theta = 0 \text{ or } \cos\theta = \tfrac{1}{2} \quad\Rightarrow$

$\theta = 0, \pi, \tfrac{\pi}{3}, \text{ or } -\tfrac{\pi}{3} \quad\Rightarrow$ the other intersection points are $\left(\tfrac{\sqrt{3}}{2}, \tfrac{\pi}{3}\right)$

and $\left(\tfrac{\sqrt{3}}{2}, \tfrac{2\pi}{3}\right)$ [by symmetry].

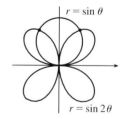

$r = \sin\theta$

$r = \sin 2\theta$

42. Clearly the pole is a point of intersection. $\sin 2\theta = \cos 2\theta \quad\Rightarrow$

$\tan 2\theta = 1 \quad\Rightarrow\quad 2\theta = \tfrac{\pi}{4} + 2n\pi$ [since $\sin 2\theta$ and $\cos 2\theta$ must be

positive in the equations] $\quad\Rightarrow\quad \theta = \tfrac{\pi}{8} + n\pi \quad\Rightarrow\quad \theta = \tfrac{\pi}{8} \text{ or } \tfrac{9\pi}{8}.$

So the curves also intersect at $\left(\tfrac{1}{\sqrt[4]{2}}, \tfrac{\pi}{8}\right)$ and $\left(\tfrac{1}{\sqrt[4]{2}}, \tfrac{9\pi}{8}\right)$.

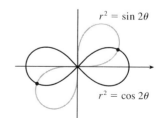

$r^2 = \sin 2\theta$

$r^2 = \cos 2\theta$

43.

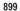

From the first graph, we see that the pole is one point of intersection. By zooming in or using the cursor, we find the θ-values of the intersection points to be $\alpha \approx 0.88786 \approx 0.89$ and $\pi - \alpha \approx 2.25$. (The first of these values may be more easily estimated by plotting $y = 1 + \sin x$ and $y = 2x$ in rectangular coordinates; see the second graph.) By symmetry, the total area contained is twice the area contained in the first quadrant, that is,

$$A = 2\int_0^\alpha \tfrac{1}{2}(2\theta)^2\,d\theta + 2\int_\alpha^{\pi/2} \tfrac{1}{2}(1+\sin\theta)^2\,d\theta = \int_0^\alpha 4\theta^2\,d\theta + \int_\alpha^{\pi/2}\left[1 + 2\sin\theta + \tfrac{1}{2}(1-\cos 2\theta)\right]d\theta$$

$$= \left[\tfrac{4}{3}\theta^3\right]_0^\alpha + \left[\theta - 2\cos\theta + \left(\tfrac{1}{2}\theta - \tfrac{1}{4}\sin 2\theta\right)\right]_\alpha^{\pi/2} = \tfrac{4}{3}\alpha^3 + \left[\left(\tfrac{\pi}{2} + \tfrac{\pi}{4}\right) - \left(\alpha - 2\cos\alpha + \tfrac{1}{2}\alpha - \tfrac{1}{4}\sin 2\alpha\right)\right] \approx 3.4645$$

44.

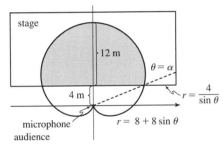

We need to find the shaded area A in the figure. The horizontal line representing the front of the stage has equation $y = 4 \quad \Leftrightarrow$ $r\sin\theta = 4 \quad \Rightarrow \quad r = 4/\sin\theta$. This line intersects the curve $r = 8 + 8\sin\theta$ when $8 + 8\sin\theta = \dfrac{4}{\sin\theta} \quad \Rightarrow$ $8\sin\theta + 8\sin^2\theta = 4 \quad \Rightarrow \quad 2\sin^2\theta + 2\sin\theta - 1 = 0 \quad \Rightarrow$

$$\sin\theta = \frac{-2 \pm \sqrt{4+8}}{4} = \frac{-2 \pm 2\sqrt{3}}{4} = \frac{-1+\sqrt{3}}{2} \quad \text{[the other value is less than } -1] \quad \Rightarrow \quad \theta = \sin^{-1}\left(\frac{\sqrt{3}-1}{2}\right).$$

This angle is about $21.5°$ and is denoted by α in the figure.

$$A = 2\int_\alpha^{\pi/2} \tfrac{1}{2}(8+8\sin\theta)^2\,d\theta - 2\int_\alpha^{\pi/2} \tfrac{1}{2}(4\csc\theta)^2\,d\theta = 64\int_\alpha^{\pi/2}(1 + 2\sin\theta + \sin^2\theta)\,d\theta - 16\int_\alpha^{\pi/2}\csc^2\theta\,d\theta$$

$$= 64\int_\alpha^{\pi/2}\left(1 + 2\sin\theta + \tfrac{1}{2} - \tfrac{1}{2}\cos 2\theta\right)d\theta + 16\int_\alpha^{\pi/2}(-\csc^2\theta)\,d\theta = 64\left[\tfrac{3}{2}\theta - 2\cos\theta - \tfrac{1}{4}\sin 2\theta\right]_\alpha^{\pi/2} + 16\left[\cot\theta\right]_\alpha^{\pi/2}$$

$$= 16\left[6\theta - 8\cos\theta - \sin 2\theta + \cot\theta\right]_\alpha^{\pi/a} = 16[(3\pi - 0 - 0 + 0) - (6\alpha - 8\cos\alpha - \sin 2\alpha + \cot\alpha)]$$

$$= 48\pi - 96\alpha + 128\cos\alpha + 16\sin 2\alpha - 16\cot\alpha$$

From the figure, $x^2 + \left(\sqrt{3}-1\right)^2 = 2^2 \quad \Rightarrow \quad x^2 = 4 - \left(3 - 2\sqrt{3} + 1\right) \quad \Rightarrow$

$x^2 = 2\sqrt{3} = \sqrt{12}$, so $x = \sqrt{2\sqrt{3}} = \sqrt[4]{12}$. Using the trigonometric relationships for a right triangle and the identity $\sin 2\alpha = 2\sin\alpha\cos\alpha$, we continue:

$$A = 48\pi - 96\alpha + 128 \cdot \frac{\sqrt[4]{12}}{2} + 16 \cdot 2 \cdot \frac{\sqrt{3}-1}{2} \cdot \frac{\sqrt[4]{12}}{2} - 16 \cdot \frac{\sqrt[4]{12}}{\sqrt{3}-1} \cdot \frac{\sqrt{3}+1}{\sqrt{3}+1}$$

$$= 48\pi - 96\alpha + 64\sqrt[4]{12} + 8\sqrt[4]{12}\left(\sqrt{3}-1\right) - 8\sqrt[4]{12}\left(\sqrt{3}+1\right) = 48\pi + 48\sqrt[4]{12} - 96\sin^{-1}\left(\frac{\sqrt{3}-1}{2}\right)$$

$$\approx 204.16 \text{ m}^2$$

45. $L = \int_a^b \sqrt{r^2 + (dr/d\theta)^2}\, d\theta = \int_0^{\pi/3} \sqrt{(3\sin\theta)^2 + (3\cos\theta)^2}\, d\theta = \int_0^{\pi/3} \sqrt{9(\sin^2\theta + \cos^2\theta)}\, d\theta$

$= 3\int_0^{\pi/3} d\theta = 3\big[\theta\big]_0^{\pi/3} = 3\big(\frac{\pi}{3}\big) = \pi.$

As a check, note that the circumference of a circle with radius $\frac{3}{2}$ is $2\pi\big(\frac{3}{2}\big) = 3\pi$, and since $\theta = 0$ to $\pi = \frac{\pi}{3}$ traces out $\frac{1}{3}$ of the

circle (from $\theta = 0$ to $\theta = \pi$), $\frac{1}{3}(3\pi) = \pi.$

46. $L = \int_a^b \sqrt{r^2 + (dr/d\theta)^2}\, d\theta = \int_0^{2\pi} \sqrt{(e^{2\theta})^2 + (2e^{2\theta})^2}\, d\theta = \int_0^{2\pi} \sqrt{e^{4\theta} + 4e^{4\theta}}\, d\theta = \int_0^{2\pi} \sqrt{5e^{4\theta}}\, d\theta$

$= \sqrt{5}\int_0^{2\pi} e^{2\theta}\, d\theta = \frac{\sqrt{5}}{2}\big[e^{2\theta}\big]_0^{2\pi} = \frac{\sqrt{5}}{2}(e^{4\pi} - 1)$

47. $L = \int_a^b \sqrt{r^2 + (dr/d\theta)^2}\, d\theta = \int_0^{2\pi} \sqrt{(\theta^2)^2 + (2\theta)^2}\, d\theta = \int_0^{2\pi} \sqrt{\theta^4 + 4\theta^2}\, d\theta$

$= \int_0^{2\pi} \sqrt{\theta^2(\theta^2 + 4)}\, d\theta = \int_0^{2\pi} \theta\sqrt{\theta^2 + 4}\, d\theta$

Now let $u = \theta^2 + 4$, so that $du = 2\theta\, d\theta \quad \big[\theta\, d\theta = \frac{1}{2}\, du\big] \quad$ and

$\int_0^{2\pi} \theta\sqrt{\theta^2 + 4}\, d\theta = \int_4^{4\pi^2 + 4} \frac{1}{2}\sqrt{u}\, du = \frac{1}{2}\cdot\frac{2}{3}\big[u^{3/2}\big]_4^{4(\pi^2+1)} = \frac{1}{3}[4^{3/2}(\pi^2 + 1)^{3/2} - 4^{3/2}] = \frac{8}{3}[(\pi^2 + 1)^{3/2} - 1]$

48. $L = \int_a^b \sqrt{r^2 + (dr/d\theta)^2}\, d\theta = \int_0^{2\pi} \sqrt{\theta^2 + 1}\, d\theta \stackrel{21}{=} \Big[\frac{\theta}{2}\sqrt{\theta^2 + 1} + \frac{1}{2}\ln\big(\theta + \sqrt{\theta^2 + 1}\big)\Big]_0^{2\pi}$

$= \pi\sqrt{4\pi^2 + 1} + \frac{1}{2}\ln\big(2\pi + \sqrt{4\pi^2 + 1}\big)$

49. The curve $r = 3\sin 2\theta$ is completely traced with

$0 \le \theta \le 2\pi$.

$r^2 + \big(\frac{dr}{d\theta}\big)^2 = (3\sin 2\theta)^2 + (6\cos 2\theta)^2 \quad\Rightarrow$

$L = \int_0^{2\pi} \sqrt{9\sin^2 2\theta + 36\cos^2 2\theta}\, d\theta \approx 29.0653$

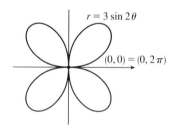

$r = 3\sin 2\theta$

$(0,0) = (0, 2\pi)$

50. The curve $r = 4\sin 3\theta$ is completely traced with

$0 \le \theta \le \pi$.

$r^2 + \big(\frac{dr}{d\theta}\big)^2 = (4\sin 3\theta)^2 + (12\cos 3\theta)^2 \quad\Rightarrow$

$L = \int_0^{\pi} \sqrt{16\sin^2 3\theta + 144\cos^2 3\theta}\, d\theta \approx 26.7298$

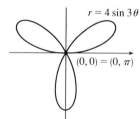

$r = 4\sin 3\theta$

$(0,0) = (0, \pi)$

51. The curve $r = \sin\left(\frac{\theta}{2}\right)$ is completely traced

with $0 \le \theta \le 4\pi$.

$r^2 + \left(\frac{dr}{d\theta}\right)^2 = \sin^2\left(\frac{\theta}{2}\right) + \left[\frac{1}{2}\cos\left(\frac{\theta}{2}\right)\right]^2 \;\Rightarrow$

$L = \displaystyle\int_0^{4\pi} \sqrt{\sin^2\left(\frac{\theta}{2}\right) + \frac{1}{4}\cos^2\left(\frac{\theta}{2}\right)}\, d\theta$

≈ 9.6884

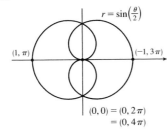

52. The curve $r = 1 + \cos\left(\frac{\theta}{3}\right)$ is completely traced

with $0 \le \theta \le 6\pi$.

$r^2 + \left(\frac{dr}{d\theta}\right)^2 = \left[1 + \cos\left(\frac{\theta}{3}\right)\right]^2 + \left[-\frac{1}{3}\sin\left(\frac{\theta}{3}\right)\right]^2 \;\Rightarrow$

$L = \displaystyle\int_0^{6\pi} \sqrt{\left[1 + \cos\left(\frac{\theta}{3}\right)\right]^2 + \frac{1}{9}\sin^2\left(\frac{\theta}{3}\right)}\, d\theta$

≈ 19.6676

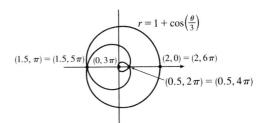

53. The curve $r = \cos^4(\theta/4)$ is completely traced with $0 \le \theta \le 4\pi$.

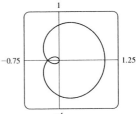

$r^2 + (dr/d\theta)^2 = [\cos^4(\theta/4)]^2 + \left[4\cos^3(\theta/4)\cdot(-\sin(\theta/4))\cdot\frac{1}{4}\right]^2$

$= \cos^8(\theta/4) + \cos^6(\theta/4)\sin^2(\theta/4)$

$= \cos^6(\theta/4)[\cos^2(\theta/4) + \sin^2(\theta/4)] = \cos^6(\theta/4)$

$L = \int_0^{4\pi} \sqrt{\cos^6(\theta/4)}\, d\theta = \int_0^{4\pi} \left|\cos^3(\theta/4)\right|\, d\theta$

$= 2\int_0^{2\pi}\cos^3(\theta/4)\, d\theta$ [since $\cos^3(\theta/4) \ge 0$ for $0 \le \theta \le 2\pi$] $= 8\int_0^{\pi/2}\cos^3 u\, du$ $\left[u = \frac{1}{4}\theta\right]$

$\overset{68}{=} 8\left[\frac{1}{3}(2 + \cos^2 u)\sin u\right]_0^{\pi/2} = \frac{8}{3}[(2\cdot1) - (3\cdot0)] = \frac{16}{3}$

54. The curve $r = \cos^2(\theta/2)$ is completely traced with $0 \le \theta \le 2\pi$.

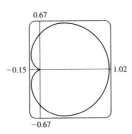

$r^2 + (dr/d\theta)^2 = [\cos^2(\theta/2)]^2 + \left[2\cos(\theta/2)\cdot(-\sin(\theta/2))\cdot\frac{1}{2}\right]^2$

$= \cos^4(\theta/2) + \cos^2(\theta/2)\sin^2(\theta/2)$

$= \cos^2(\theta/2)[\cos^2(\theta/2) + \sin^2(\theta/2)]$

$= \cos^2(\theta/2)$

$L = \int_0^{2\pi} \sqrt{\cos^2(\theta/2)}\, d\theta = \int_0^{2\pi} |\cos(\theta/2)|\, d\theta = 2\int_0^{\pi}\cos(\theta/2)\, d\theta$ [since $\cos(\theta/2) \ge 0$ for $0 \le \theta \le \pi$]

$= 4\int_0^{\pi/2}\cos u\, du$ $\left[u = \frac{1}{2}\theta\right]$ $= 4\big[\sin u\big]_0^{\pi/2} = 4(1 - 0) = 4$

55. (a) From (11.2.7),

$$S = \int_a^b 2\pi y \sqrt{(dx/d\theta)^2 + (dy/d\theta)^2}\, d\theta$$

$$= \int_a^b 2\pi y \sqrt{r^2 + (dr/d\theta)^2}\, d\theta \qquad \text{[from the derivation of Equation 11.4.5]}$$

$$= \int_a^b 2\pi r \sin\theta \sqrt{r^2 + (dr/d\theta)^2}\, d\theta$$

(b) The curve $r^2 = \cos 2\theta$ goes through the pole when $\cos 2\theta = 0 \;\Rightarrow$

$2\theta = \frac{\pi}{2} \;\Rightarrow\; \theta = \frac{\pi}{4}$. We'll rotate the curve from $\theta = 0$ to $\theta = \frac{\pi}{4}$ and double

this value to obtain the total surface area generated.

$r^2 = \cos 2\theta \;\Rightarrow\; 2r\dfrac{dr}{d\theta} = -2\sin 2\theta \;\Rightarrow\; \left(\dfrac{dr}{d\theta}\right)^2 = \dfrac{\sin^2 2\theta}{r^2} = \dfrac{\sin^2 2\theta}{\cos 2\theta}.$

$$S = 2\int_0^{\pi/4} 2\pi\sqrt{\cos 2\theta}\,\sin\theta\,\sqrt{\cos 2\theta + (\sin^2 2\theta)/\cos 2\theta}\,d\theta = 4\pi\int_0^{\pi/4}\sqrt{\cos 2\theta}\,\sin\theta\,\sqrt{\dfrac{\cos^2 2\theta + \sin^2 2\theta}{\cos 2\theta}}\,d\theta$$

$$= 4\pi\int_0^{\pi/4}\sqrt{\cos 2\theta}\,\sin\theta\,\dfrac{1}{\sqrt{\cos 2\theta}}\,d\theta = 4\pi\int_0^{\pi/4}\sin\theta\,d\theta = 4\pi\big[-\cos\theta\big]_0^{\pi/4} = -4\pi\left(\dfrac{\sqrt{2}}{2}-1\right) = 2\pi\left(2-\sqrt{2}\right)$$

56. (a) Rotation around $\theta = \frac{\pi}{2}$ is the same as rotation around the y-axis, that is, $S = \int_a^b 2\pi x\,ds$ where

$ds = \sqrt{(dx/dt)^2 + (dy/dt)^2}\,dt$ for a parametric equation, and for the special case of a polar equation, $x = r\cos\theta$ and

$ds = \sqrt{(dx/d\theta)^2 + (dy/d\theta)^2}\,d\theta = \sqrt{r^2 + (dr/d\theta)^2}\,d\theta$ [see the derivation of Equation 11.4.5]. Therefore, for a polar

equation rotated around $\theta = \frac{\pi}{2}$, $S = \int_a^b 2\pi r\cos\theta\sqrt{r^2 + (dr/d\theta)^2}\,d\theta$.

(b) As in the solution for Exercise 55(b), we can double the surface area generated by rotating the curve from $\theta = 0$ to $\theta = \frac{\pi}{4}$

to obtain the total surface area.

$$S = 2\int_0^{\pi/4} 2\pi\sqrt{\cos 2\theta}\,\cos\theta\,\sqrt{\cos 2\theta + (\sin^2 2\theta)/\cos 2\theta}\,d\theta = 4\pi\int_0^{\pi/4}\sqrt{\cos 2\theta}\,\cos\theta\,\sqrt{\dfrac{\cos^2 2\theta + \sin^2 2\theta}{\cos 2\theta}}\,d\theta$$

$$= 4\pi\int_0^{\pi/4}\sqrt{\cos 2\theta}\,\cos\theta\,\dfrac{1}{\sqrt{\cos 2\theta}}\,d\theta = 4\pi\int_0^{\pi/4}\cos\theta\,d\theta = 4\pi\big[\sin\theta\big]_0^{\pi/4} = 4\pi\left(\dfrac{\sqrt{2}}{2}-0\right) = 2\sqrt{2}\,\pi$$

11.5 Conic Sections

1. $x = 2y^2 \;\Rightarrow\; y^2 = \frac{1}{2}x$. $4p = \frac{1}{2}$, so $p = \frac{1}{8}$. The vertex
is $(0,0)$, the focus is $\left(\frac{1}{8},0\right)$, and the directrix is $x = -\frac{1}{8}$.

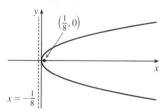

2. $4y + x^2 = 0 \;\Rightarrow\; x^2 = -4y$. $4p = -4$, so $p = -1$.
The vertex is $(0,0)$, the focus is $(0,-1)$, and the directrix
is $y = 1$.

3. $4x^2 = -y \;\Rightarrow\; x^2 = -\frac{1}{4}y$. $4p = -\frac{1}{4}$, so $p = -\frac{1}{16}$.
The vertex is $(0,0)$, the focus is $\left(0,-\frac{1}{16}\right)$, and the
directrix is $y = \frac{1}{16}$.

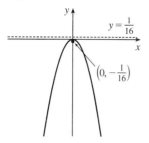

4. $y^2 = 12x$. $4p = 12$, so $p = 3$. The vertex is $(0,0)$, the
focus is $(3,0)$, and the directrix is $x = -3$.

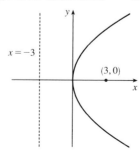

5. $(x + 2)^2 = 8(y - 3)$. $4p = 8$, so $p = 2$. The vertex is $(-2, 3)$, the focus is $(-2, 5)$, and the directrix is $y = 1$.

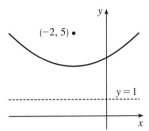

6. $x - 1 = (y + 5)^2$. $4p = 1$, so $p = \frac{1}{4}$. The vertex is $(1, -5)$, the focus is $\left(\frac{5}{4}, -5\right)$, and the directrix is $x = \frac{3}{4}$.

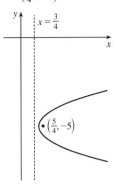

7. $y^2 + 2y + 12x + 25 = 0$ $\Rightarrow$
$y^2 + 2y + 1 = -12x - 24$ $\Rightarrow$
$(y + 1)^2 = -12(x + 2)$. $4p = -12$, so $p = -3$.
The vertex is $(-2, -1)$, the focus is $(-5, -1)$, and the directrix is $x = 1$.

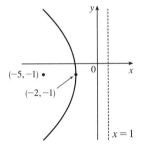

8. $y + 12x - 2x^2 = 16$ $\Rightarrow$ $2x^2 - 12x = y - 16$ $\Rightarrow$
$2(x^2 - 6x + 9) = y - 16 + 18$ $\Rightarrow$
$2(x - 3)^2 = y + 2$ $\Rightarrow$ $(x - 3)^2 = \frac{1}{2}(y + 2)$.
$4p = \frac{1}{2}$, so $p = \frac{1}{8}$. The vertex is $(3, -2)$, the focus is $\left(3, -\frac{15}{8}\right)$, and the directrix is $y = -\frac{17}{8}$.

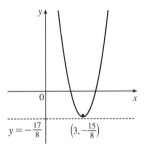

9. The equation has the form $y^2 = 4px$, where $p < 0$.
Since the parabola passes through $(-1, 1)$, we have
$1^2 = 4p(-1)$, so $4p = -1$ and an equation is $y^2 = -x$
or $x = -y^2$. $4p = -1$, so $p = -\frac{1}{4}$ and the focus is
$\left(-\frac{1}{4}, 0\right)$ while the directrix is $x = \frac{1}{4}$.

10. The vertex is $(2, -2)$, so the equation is of the form
$(x - 2)^2 = 4p(y + 2)$, where $p > 0$. The point $(0, 0)$ is
on the parabola, so $4 = 4p(2)$ and $4p = 2$. Thus, an
equation is $(x - 2)^2 = 2(y + 2)$. $4p = 2$, so $p = \frac{1}{2}$ and
the focus is $\left(2, -\frac{3}{2}\right)$ while the directrix is $y = -\frac{5}{2}$.

11. $\dfrac{x^2}{9} + \dfrac{y^2}{5} = 1$ $\Rightarrow$ $a = \sqrt{9} = 3, b = \sqrt{5}$,
$c = \sqrt{a^2 - b^2} = \sqrt{9 - 5} = 2$. The ellipse is centered at
$(0, 0)$, with vertices at $(\pm 3, 0)$. The foci are $(\pm 2, 0)$.

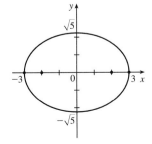

12. $\dfrac{x^2}{64} + \dfrac{y^2}{100} = 1$ $\Rightarrow$ $a = \sqrt{100} = 10, b = \sqrt{64} = 8$,
$c = \sqrt{a^2 - b^2} = \sqrt{100 - 64} = 6$. The ellipse is centered
at $(0, 0)$, with vertices at $(0, \pm 10)$. The foci are $(0, \pm 6)$.

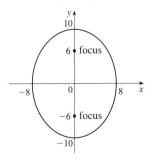

13. $4x^2 + y^2 = 16 \;\Rightarrow\; \dfrac{x^2}{4} + \dfrac{y^2}{16} = 1 \;\Rightarrow$

$a = \sqrt{16} = 4, b = \sqrt{4} = 2,$

$c = \sqrt{a^2 - b^2} = \sqrt{16 - 4} = 2\sqrt{3}$. The ellipse is

centered at $(0,0)$, with vertices at $(0, \pm 4)$. The foci

are $\left(0, \pm 2\sqrt{3}\right)$.

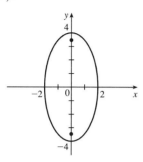

14. $4x^2 + 25y^2 = 25 \;\Rightarrow\; \dfrac{x^2}{25/4} + \dfrac{y^2}{1} = 1 \;\Rightarrow$

$a = \sqrt{\dfrac{25}{4}} = \dfrac{5}{2}, b = \sqrt{1} = 1,$

$c = \sqrt{a^2 - b^2} = \sqrt{\dfrac{25}{4} - 1} = \sqrt{\dfrac{21}{4}} = \dfrac{\sqrt{21}}{2}$. The ellipse

is centered at $(0,0)$, with vertices at $\left(\pm\dfrac{5}{2}, 0\right)$. The foci

are $\left(\pm\dfrac{\sqrt{21}}{2}, 0\right)$.

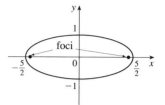

15. $9x^2 - 18x + 4y^2 = 27 \;\Leftrightarrow$

$9(x^2 - 2x + 1) + 4y^2 = 27 + 9 \;\Leftrightarrow$

$9(x - 1)^2 + 4y^2 = 36 \;\Leftrightarrow\; \dfrac{(x-1)^2}{4} + \dfrac{y^2}{9} = 1 \;\Rightarrow$

$a = 3, b = 2, c = \sqrt{5} \;\Rightarrow\;$ center $(1, 0)$,

vertices $(1, \pm 3)$, foci $\left(1, \pm\sqrt{5}\right)$

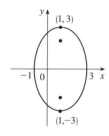

16. $x^2 + 3y^2 + 2x - 12y + 10 = 0 \;\Leftrightarrow$

$x^2 + 2x + 1 + 3(y^2 - 4y + 4) = -10 + 1 + 12 \;\Leftrightarrow$

$(x + 1)^2 + 3(y - 2)^2 = 3 \;\Leftrightarrow$

$\dfrac{(x+1)^2}{3} + \dfrac{(y-2)^2}{1} = 1 \;\Rightarrow\; a = \sqrt{3}, b = 1,$

$c = \sqrt{2} \;\Rightarrow\;$ center $(-1, 2)$, vertices $\left(-1 \pm \sqrt{3}, 2\right)$,

foci $\left(-1 \pm \sqrt{2}, 2\right)$

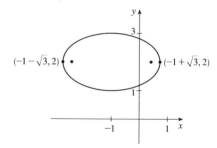

17. The center is $(0,0)$, $a = 3$, and $b = 2$, so an equation is $\dfrac{x^2}{4} + \dfrac{y^2}{9} = 1$. $c = \sqrt{a^2 - b^2} = \sqrt{5}$, so the foci are $\left(0, \pm\sqrt{5}\right)$.

18. The ellipse is centered at $(2, 1)$, with $a = 3$ and $b = 2$. An equation is $\dfrac{(x-2)^2}{9} + \dfrac{(y-1)^2}{4} = 1$. $c = \sqrt{a^2 - b^2} = \sqrt{5}$, so the foci are $\left(2 \pm \sqrt{5}, 1\right)$.

19. $\dfrac{x^2}{144} - \dfrac{y^2}{25} = 1 \;\Rightarrow\; a = 12, b = 5, c = \sqrt{144 + 25} = 13 \;\Rightarrow$

center $(0, 0)$, vertices $(\pm 12, 0)$, foci $(\pm 13, 0)$, asymptotes $y = \pm\dfrac{5}{12}x$.

Note: It is helpful to draw a $2a$-by-$2b$ rectangle whose center is the center

of the hyperbola. The asymptotes are the extended diagonals of the

rectangle.

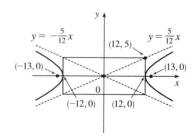

20. $\dfrac{y^2}{16} - \dfrac{x^2}{36} = 1 \;\Rightarrow\; a = 4,\, b = 6,$

$c = \sqrt{a^2 + b^2} = \sqrt{16 + 36} = \sqrt{52} = 2\sqrt{13}.$ The center is $(0, 0)$, the

vertices are $(0, \pm 4)$, the foci are $\left(0, \pm 2\sqrt{13}\right)$, and the asymptotes are the

lines $y = \pm\frac{a}{b}x = \pm\frac{2}{3}x.$

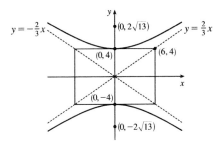

21. $y^2 - x^2 = 4 \;\Leftrightarrow\; \dfrac{y^2}{4} - \dfrac{x^2}{4} = 1 \;\Rightarrow\; a = \sqrt{4} = 2 = b,$

$c = \sqrt{4 + 4} = 2\sqrt{2} \;\Rightarrow\;$ center $(0, 0)$, vertices $(0, \pm 2)$,

foci $\left(0, \pm 2\sqrt{2}\right)$, asymptotes $y = \pm x$

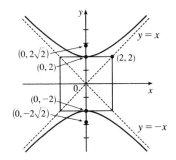

22. $9x^2 - 4y^2 = 36 \;\Leftrightarrow\; \dfrac{x^2}{4} - \dfrac{y^2}{9} = 1 \;\Rightarrow\; a = \sqrt{4} = 2,\, b = \sqrt{9} = 3,$

$c = \sqrt{4 + 9} = \sqrt{13} \;\Rightarrow\;$ center $(0, 0)$, vertices $(\pm 2, 0)$,

foci $\left(\pm\sqrt{13}, 0\right)$, asymptotes $y = \pm\frac{3}{2}x$

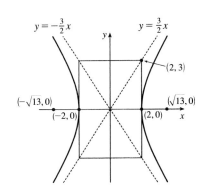

23. $4x^2 - y^2 - 24x - 4y + 28 = 0 \;\Leftrightarrow\;$

$4(x^2 - 6x + 9) - (y^2 + 4y + 4) = -28 + 36 - 4 \;\Leftrightarrow\;$

$4(x - 3)^2 - (y + 2)^2 = 4 \;\Leftrightarrow\; \dfrac{(x - 3)^2}{1} - \dfrac{(y + 2)^2}{4} = 1 \;\Rightarrow\;$

$a = \sqrt{1} = 1,\, b = \sqrt{4} = 2,\, c = \sqrt{1 + 4} = \sqrt{5} \;\Rightarrow\;$

center $(3, -2)$, vertices $(4, -2)$ and $(2, -2)$, foci $\left(3 \pm \sqrt{5}, -2\right)$,

asymptotes $y + 2 = \pm 2(x - 3).$

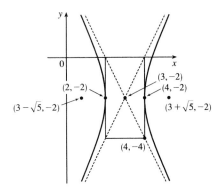

24. $y^2 - 4x^2 - 2y + 16x = 31$ $\Leftrightarrow$

$(y^2 - 2y + 1) - 4(x^2 - 4x + 4) = 31 + 1 - 16$ $\Leftrightarrow$

$(y - 1)^2 - 4(x - 2)^2 = 16$ $\Leftrightarrow$

$\dfrac{(y - 1)^2}{16} - \dfrac{(x - 2)^2}{4} = 1$ $\Rightarrow$ $a = \sqrt{16} = 4, b = \sqrt{4} = 2,$

$c = \sqrt{16 + 4} = \sqrt{20}$ $\Rightarrow$ center $(2, 1)$, vertices $(2, 1 \pm 4)$,

foci $\left(2, 1 \pm \sqrt{20}\right)$, asymptotes $y - 1 = \pm 2(x - 2)$.

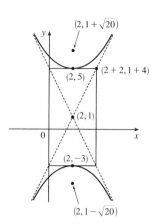

25. $x^2 = y + 1$ $\Leftrightarrow$ $x^2 = 1(y + 1)$. This is an equation of a *parabola* with $4p = 1$, so $p = \frac{1}{4}$. The vertex is $(0, -1)$ and the

focus is $\left(0, -\frac{3}{4}\right)$.

26. $x^2 = y^2 + 1$ $\Leftrightarrow$ $x^2 - y^2 = 1$. This is an equation of a *hyperbola* with vertices $(\pm 1, 0)$. The foci are at

$\left(\pm\sqrt{1 + 1}, 0\right) = \left(\pm\sqrt{2}, 0\right)$.

27. $x^2 = 4y - 2y^2$ $\Leftrightarrow$ $x^2 + 2y^2 - 4y = 0$ $\Leftrightarrow$ $x^2 + 2(y^2 - 2y + 1) = 2$ $\Leftrightarrow$ $x^2 + 2(y - 1)^2 = 2$ $\Leftrightarrow$

$\dfrac{x^2}{2} + \dfrac{(y - 1)^2}{1} = 1$. This is an equation of an *ellipse* with vertices at $\left(\pm\sqrt{2}, 1\right)$. The foci are at $\left(\pm\sqrt{2 - 1}, 1\right) = (\pm 1, 1)$.

28. $y^2 - 8y = 6x - 16$ $\Leftrightarrow$ $y^2 - 8y + 16 = 6x$ $\Leftrightarrow$ $(y - 4)^2 = 6x$. This is an equation of a *parabola* with $4p = 6$,

so $p = \frac{3}{2}$. The vertex is $(0, 4)$ and the focus is $\left(\frac{3}{2}, 4\right)$.

29. $y^2 + 2y = 4x^2 + 3$ $\Leftrightarrow$ $y^2 + 2y + 1 = 4x^2 + 4$ $\Leftrightarrow$ $(y + 1)^2 - 4x^2 = 4$ $\Leftrightarrow$ $\dfrac{(y + 1)^2}{4} - x^2 = 1$. This is an equation

of a *hyperbola* with vertices $(0, -1 \pm 2) = (0, 1)$ and $(0, -3)$. The foci are at $\left(0, -1 \pm \sqrt{4 + 1}\right) = \left(0, -1 \pm \sqrt{5}\right)$.

30. $4x^2 + 4x + y^2 = 0$ $\Leftrightarrow$ $4\left(x^2 + x + \frac{1}{4}\right) + y^2 = 1$ $\Leftrightarrow$ $4\left(x + \frac{1}{2}\right)^2 + y^2 = 1$ $\Leftrightarrow$ $\dfrac{\left(x + \frac{1}{2}\right)^2}{1/4} + y^2 = 1$. This is an

equation of an *ellipse* with vertices $\left(-\frac{1}{2}, 0 \pm 1\right) = \left(-\frac{1}{2}, \pm 1\right)$. The foci are at $\left(-\frac{1}{2}, 0 \pm \sqrt{1 - \frac{1}{4}}\right) = \left(-\frac{1}{2}, \pm\sqrt{3}/2\right)$.

31. The parabola with vertex $(0, 0)$ and focus $(0, -2)$ opens downward and has $p = -2$, so its equation is $x^2 = 4py = -8y$.

32. The parabola with vertex $(1, 0)$ and directrix $x = -5$ opens to the right and has $p = 6$, so its equation is

$y^2 = 4p(x - 1) = 24(x - 1)$.

33. The distance from the focus $(-4, 0)$ to the directrix $x = 2$ is $2 - (-4) = 6$, so the distance from the focus to the vertex is

$\frac{1}{2}(6) = 3$ and the vertex is $(-1, 0)$. Since the focus is to the left of the vertex, $p = -3$. An equation is $y^2 = 4p(x + 1)$ $\Rightarrow$

$y^2 = -12(x + 1)$.

34. The distance from the focus $(3, 6)$ to the vertex $(3, 2)$ is $6 - 2 = 4$. Since the focus is above the vertex, $p = 4$.

An equation is $(x - 3)^2 = 4p(y - 2)$ $\Rightarrow$ $(x - 3)^2 = 16(y - 2)$.

35. A parabola with vertical axis and vertex $(2, 3)$ has equation $y - 3 = a(x - 2)^2$. Since it passes through $(1, 5)$, we have $5 - 3 = a(1 - 2)^2 \ \Rightarrow \ a = 2$, so an equation is $y - 3 = 2(x - 2)^2$.

36. A parabola with horizontal axis has equation $x = ay^2 + by + c$. Since the parabola passes through the point $(-1, 0)$, substitute -1 for x and 0 for y: $-1 = 0 + 0 + c$. Now with $c = -1$, substitute 1 for x and -1 for y: $1 = a - b - 1$ **(1)**; and then 3 for x and 1 for y: $3 = a + b - 1$ **(2)**. Add **(1)** and **(2)** to get $4 = 2a - 2 \ \Rightarrow \ a = 3$ and then $b = 1$. Thus, the equation is $x = 3y^2 + y - 1$.

37. The ellipse with foci $(\pm 2, 0)$ and vertices $(\pm 5, 0)$ has center $(0, 0)$ and a horizontal major axis, with $a = 5$ and $c = 2$, so $b^2 = a^2 - c^2 = 25 - 4 = 21$. An equation is $\dfrac{x^2}{25} + \dfrac{y^2}{21} = 1$.

38. The ellipse with foci $(0, \pm 5)$ and vertices $(0, \pm 13)$ has center $(0, 0)$ and a vertical major axis, with $c = 5$ and $a = 13$, so $b = \sqrt{a^2 - c^2} = 12$. An equation is $\dfrac{x^2}{144} + \dfrac{y^2}{169} = 1$.

39. Since the vertices are $(0, 0)$ and $(0, 8)$, the ellipse has center $(0, 4)$ with a vertical axis and $a = 4$. The foci at $(0, 2)$ and $(0, 6)$ are 2 units from the center, so $c = 2$ and $b = \sqrt{a^2 - c^2} = \sqrt{4^2 - 2^2} = \sqrt{12}$. An equation is $\dfrac{(x - 0)^2}{b^2} + \dfrac{(y - 4)^2}{a^2} = 1 \ \Rightarrow \ \dfrac{x^2}{12} + \dfrac{(y - 4)^2}{16} = 1$.

40. Since the foci are $(0, -1)$ and $(8, -1)$, the ellipse has center $(4, -1)$ with a horizontal axis and $c = 4$. The vertex $(9, -1)$ is 5 units from the center, so $a = 5$ and $b = \sqrt{a^2 - c^2} = \sqrt{5^2 - 4^2} = \sqrt{9}$. An equation is $\dfrac{(x - 4)^2}{a^2} + \dfrac{(y + 1)^2}{b^2} = 1 \ \Rightarrow \ \dfrac{(x - 4)^2}{25} + \dfrac{(y + 1)^2}{9} = 1$.

41. An equation of an ellipse with center $(-1, 4)$ and vertex $(-1, 0)$ is $\dfrac{(x + 1)^2}{b^2} + \dfrac{(y - 4)^2}{4^2} = 1$. The focus $(-1, 6)$ is 2 units from the center, so $c = 2$. Thus, $b^2 + 2^2 = 4^2 \ \Rightarrow \ b^2 = 12$, and the equation is $\dfrac{(x + 1)^2}{12} + \dfrac{(y - 4)^2}{16} = 1$.

42. Foci $F_1(-4, 0)$ and $F_2(4, 0) \ \Rightarrow \ c = 4$ and an equation is $\dfrac{x^2}{a^2} + \dfrac{y^2}{b^2} = 1$. The ellipse passes through $P(-4, 1.8)$, so $2a = |PF_1| + |PF_2| \ \Rightarrow \ 2a = 1.8 + \sqrt{8^2 + (1.8)^2} \ \Rightarrow \ 2a = 1.8 + 8.2 \ \Rightarrow \ a = 5$. $b^2 = a^2 - c^2 = 25 - 16 = 9$ and the equation is $\dfrac{x^2}{25} + \dfrac{y^2}{9} = 1$.

43. An equation of a hyperbola with vertices $(\pm 3, 0)$ is $\dfrac{x^2}{3^2} - \dfrac{y^2}{b^2} = 1$. Foci $(\pm 5, 0) \ \Rightarrow \ c = 5$ and $3^2 + b^2 = 5^2 \ \Rightarrow \ b^2 = 25 - 9 = 16$, so the equation is $\dfrac{x^2}{9} - \dfrac{y^2}{16} = 1$.

44. An equation of a hyperbola with vertices $(0, \pm 2)$ is $\dfrac{y^2}{2^2} - \dfrac{x^2}{b^2} = 1$. Foci $(0, \pm 5) \ \Rightarrow \ c = 5$ and $2^2 + b^2 = 5^2 \ \Rightarrow \ b^2 = 25 - 4 = 21$, so the equation is $\dfrac{y^2}{4} - \dfrac{x^2}{21} = 1$.

45. The center of a hyperbola with vertices $(-3, -4)$ and $(-3, 6)$ is $(-3, 1)$, so $a = 5$ and an equation is

$\dfrac{(y-1)^2}{5^2} - \dfrac{(x+3)^2}{b^2} = 1$. Foci $(-3, -7)$ and $(-3, 9)$ $\Rightarrow$ $c = 8$, so $5^2 + b^2 = 8^2$ $\Rightarrow$ $b^2 = 64 - 25 = 39$ and the

equation is $\dfrac{(y-1)^2}{25} - \dfrac{(x+3)^2}{39} = 1$.

46. The center of a hyperbola with vertices $(-1, 2)$ and $(7, 2)$ is $(3, 2)$, so $a = 4$ and an equation is $\dfrac{(x-3)^2}{4^2} - \dfrac{(y-2)^2}{b^2} = 1$.

Foci $(-2, 2)$ and $(8, 2)$ $\Rightarrow$ $c = 5$, so $4^2 + b^2 = 5^2$ $\Rightarrow$ $b^2 = 25 - 16 = 9$ and the equation is

$\dfrac{(x-3)^2}{16} - \dfrac{(y-2)^2}{9} = 1$.

47. The center of a hyperbola with vertices $(\pm 3, 0)$ is $(0, 0)$, so $a = 3$ and an equation is $\dfrac{x^2}{3^2} - \dfrac{y^2}{b^2} = 1$.

Asymptotes $y = \pm 2x$ $\Rightarrow$ $\dfrac{b}{a} = 2$ $\Rightarrow$ $b = 2(3) = 6$ and the equation is $\dfrac{x^2}{9} - \dfrac{y^2}{36} = 1$.

48. The center of a hyperbola with foci $(2, 0)$ and $(2, 8)$ is $(2, 4)$, so $c = 4$ and an equation is $\dfrac{(y-4)^2}{a^2} - \dfrac{(x-2)^2}{b^2} = 1$.

The asymptote $y = 3 + \frac{1}{2}x$ has slope $\frac{1}{2}$, so $\dfrac{a}{b} = \dfrac{1}{2}$ $\Rightarrow$ $b = 2a$ and $a^2 + b^2 = c^2$ $\Rightarrow$ $a^2 + (2a)^2 = 4^2$ $\Rightarrow$

$5a^2 = 16$ $\Rightarrow$ $a^2 = \frac{16}{5}$ and so $b^2 = 16 - \frac{16}{5} = \frac{64}{5}$. Thus, an equation is $\dfrac{(y-4)^2}{16/5} - \dfrac{(x-2)^2}{64/5} = 1$.

49. In Figure 8, we see that the point on the ellipse closest to a focus is the closer vertex (which is a distance

$a - c$ from it) while the farthest point is the other vertex (at a distance of $a + c$). So for this lunar orbit,

$(a - c) + (a + c) = 2a = (1728 + 110) + (1728 + 314)$, or $a = 1940$; and $(a + c) - (a - c) = 2c = 314 - 110$,

or $c = 102$. Thus, $b^2 = a^2 - c^2 = 3{,}753{,}196$, and the equation is $\dfrac{x^2}{3{,}763{,}600} + \dfrac{y^2}{3{,}753{,}196} = 1$.

50. (a) Choose V to be the origin, with x-axis through V and F. Then F is $(p, 0)$, A is $(p, 5)$, so substituting A into the

equation $y^2 = 4px$ gives $25 = 4p^2$ so $p = \frac{5}{2}$ and $y^2 = 10x$.

(b) $x = 11$ $\Rightarrow$ $y = \sqrt{110}$ $\Rightarrow$ $|CD| = 2\sqrt{110}$

51. (a) Set up the coordinate system so that A is $(-200, 0)$ and B is $(200, 0)$.

$|PA| - |PB| = (1200)(980) = 1{,}176{,}000 \text{ ft} = \frac{2450}{11} \text{ mi} = 2a$ $\Rightarrow$ $a = \frac{1225}{11}$, and $c = 200$ so

$b^2 = c^2 - a^2 = \dfrac{3{,}339{,}375}{121}$ $\Rightarrow$ $\dfrac{121x^2}{1{,}500{,}625} - \dfrac{121y^2}{3{,}339{,}375} = 1$.

(b) Due north of B $\Rightarrow$ $x = 200$ $\Rightarrow$ $\dfrac{(121)(200)^2}{1{,}500{,}625} - \dfrac{121y^2}{3{,}339{,}375} = 1$ $\Rightarrow$ $y = \dfrac{133{,}575}{539} \approx 248 \text{ mi}$

52. $|PF_1| - |PF_2| = \pm 2a$ $\Leftrightarrow$ $\sqrt{(x+c)^2 + y^2} - \sqrt{(x-c)^2 + y^2} = \pm 2a$ $\Leftrightarrow$

$\sqrt{(x+c)^2 + y^2} = \sqrt{(x-c)^2 + y^2} \pm 2a$ $\Leftrightarrow$ $(x+c)^2 + y^2 = (x-c)^2 + y^2 + 4a^2 \pm 4a\sqrt{(x-c)^2 + y^2}$ $\Leftrightarrow$

$$4cx - 4a^2 = \pm 4a \sqrt{(x-c)^2 + y^2} \quad \Leftrightarrow \quad c^2x^2 - 2a^2cx + a^4 = a^2(x^2 - 2cx + c^2 + y^2) \quad \Leftrightarrow$$

$$(c^2 - a^2)x^2 - a^2y^2 = a^2(c^2 - a^2) \quad \Leftrightarrow \quad b^2x^2 - a^2y^2 = a^2b^2 \ [\text{where } b^2 = c^2 - a^2] \quad \Leftrightarrow \quad \frac{x^2}{a^2} - \frac{y^2}{b^2} = 1$$

53. The function whose graph is the upper branch of this hyperbola is concave upward. The function is

$$y = f(x) = a\sqrt{1 + \frac{x^2}{b^2}} = \frac{a}{b}\sqrt{b^2 + x^2}, \text{ so } y' = \frac{a}{b}x(b^2 + x^2)^{-1/2} \text{ and}$$

$$y'' = \frac{a}{b}\left[(b^2 + x^2)^{-1/2} - x^2(b^2 + x^2)^{-3/2}\right] = ab(b^2 + x^2)^{-3/2} > 0 \text{ for all } x, \text{ and so } f \text{ is concave upward.}$$

54. We can follow exactly the same sequence of steps as in the derivation of Formula 4, except we use the points $(1, 1)$ and

$(-1, -1)$ in the distance formula (first equation of that derivation) so $\sqrt{(x-1)^2 + (y-1)^2} + \sqrt{(x+1)^2 + (y+1)^2} = 4$

will lead (after moving the second term to the right, squaring, and simplifying) to $2\sqrt{(x+1)^2 + (y+1)^2} = x + y + 4$,

which, after squaring and simplifying again, leads to $3x^2 - 2xy + 3y^2 = 8$.

55. (a) If $k > 16$, then $k - 16 > 0$, and $\dfrac{x^2}{k} + \dfrac{y^2}{k-16} = 1$ is an *ellipse* since it is the sum of two squares on the left side.

(b) If $0 < k < 16$, then $k - 16 < 0$, and $\dfrac{x^2}{k} + \dfrac{y^2}{k-16} = 1$ is a *hyperbola* since it is the difference of two squares on the

left side.

(c) If $k < 0$, then $k - 16 < 0$, and there is *no curve* since the left side is the sum of two negative terms, which cannot equal 1.

(d) In case (a), $a^2 = k$, $b^2 = k - 16$, and $c^2 = a^2 - b^2 = 16$, so the foci are at $(\pm 4, 0)$. In case (b), $k - 16 < 0$, so $a^2 = k$,

$b^2 = 16 - k$, and $c^2 = a^2 + b^2 = 16$, and so again the foci are at $(\pm 4, 0)$.

56. (a) $y^2 = 4px \quad \Rightarrow \quad 2yy' = 4p \quad \Rightarrow \quad y' = \dfrac{2p}{y}$, so the tangent line is

$$y - y_0 = \frac{2p}{y_0}(x - x_0) \quad \Rightarrow \quad yy_0 - y_0^2 = 2p(x - x_0) \quad \Leftrightarrow$$

$$yy_0 - 4px_0 = 2px - 2px_0 \quad \Rightarrow \quad yy_0 = 2p(x + x_0).$$

(b) The x-intercept is $-x_0$.

57. $x^2 = 4py \quad \Rightarrow \quad 2x = 4py' \quad \Rightarrow \quad y' = \dfrac{x}{2p}$, so the tangent line at (x_0, y_0) is

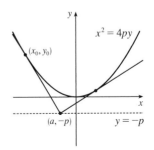

$y - \dfrac{x_0^2}{4p} = \dfrac{x_0}{2p}(x - x_0)$. This line passes through the point $(a, -p)$ on the

directrix, so $-p - \dfrac{x_0^2}{4p} = \dfrac{x_0}{2p}(a - x_0) \quad \Rightarrow \quad -4p^2 - x_0^2 = 2ax_0 - 2x_0^2 \quad \Leftrightarrow$

$x_0^2 - 2ax_0 - 4p^2 = 0 \quad \Leftrightarrow \quad x_0^2 - 2ax_0 + a^2 = a^2 + 4p^2 \quad \Leftrightarrow$

$(x_0 - a)^2 = a^2 + 4p^2 \quad \Leftrightarrow \quad x_0 = a \pm \sqrt{a^2 + 4p^2}$. The slopes of the tangent lines at $x = a \pm \sqrt{a^2 + 4p^2}$

are $\dfrac{a \pm \sqrt{a^2 + 4p^2}}{2p}$, so the product of the two slopes is

$$\frac{a + \sqrt{a^2 + 4p^2}}{2p} \cdot \frac{a - \sqrt{a^2 + 4p^2}}{2p} = \frac{a^2 - (a^2 + 4p^2)}{4p^2} = \frac{-4p^2}{4p^2} = -1,$$

showing that the tangent lines are perpendicular.

58. Without a loss of generality, let the ellipse, hyperbola, and foci be as shown in the figure.

The curves intersect (eliminate y^2) $\Rightarrow$

$$B^2\left(\frac{x^2}{A^2} - \frac{y^2}{B^2}\right) + b^2\left(\frac{x^2}{a^2} + \frac{y^2}{b^2}\right) = B^2 + b^2 \quad \Rightarrow$$

$$\frac{B^2 x^2}{A^2} + \frac{b^2 x^2}{a^2} = B^2 + b^2 \quad \Rightarrow \quad x^2\left(\frac{B^2}{A^2} + \frac{b^2}{a^2}\right) = B^2 + b^2 \quad \Rightarrow$$

$$x^2 = \frac{B^2 + b^2}{\dfrac{a^2 B^2 + b^2 A^2}{A^2 a^2}} = \frac{A^2 a^2 (B^2 + b^2)}{a^2 B^2 + b^2 A^2}.$$

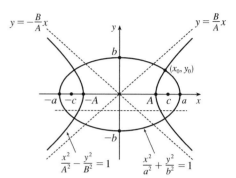

Similarly, $y^2 = \dfrac{B^2 b^2 (a^2 - A^2)}{b^2 A^2 + a^2 B^2}$.

Next we find the slopes of the tangent lines of the curves: $\dfrac{x^2}{a^2} + \dfrac{y^2}{b^2} = 1 \Rightarrow \dfrac{2x}{a^2} + \dfrac{2yy'}{b^2} = 0 \Rightarrow \dfrac{yy'}{b^2} = -\dfrac{x}{a^2} \Rightarrow$

$y'_E = -\dfrac{b^2}{a^2}\dfrac{x}{y}$ and $\dfrac{x^2}{A^2} - \dfrac{y^2}{B^2} = 1 \Rightarrow \dfrac{2x}{A^2} - \dfrac{2yy'}{B^2} = 0 \Rightarrow \dfrac{yy'}{B^2} = \dfrac{x}{A^2} \Rightarrow y'_H = \dfrac{B^2}{A^2}\dfrac{x}{y}$. The product of the slopes

at (x_0, y_0) is $y'_E\, y'_H = -\dfrac{b^2 B^2 x_0^2}{a^2 A^2 y_0^2} = -\dfrac{b^2 B^2 \left[\dfrac{A^2 a^2 (B^2 + b^2)}{a^2 B^2 + b^2 A^2}\right]}{a^2 A^2 \left[\dfrac{B^2 b^2 (a^2 - A^2)}{b^2 A^2 + a^2 B^2}\right]} = -\dfrac{B^2 + b^2}{a^2 - A^2}$. Since $a^2 - b^2 = c^2$ and $A^2 + B^2 = c^2$,

we have $a^2 - b^2 = A^2 + B^2 \Rightarrow a^2 - A^2 = b^2 + B^2$, so the product of the slopes is -1, and hence, the tangent lines at each point of intersection are perpendicular.

59. For $x^2 + 4y^2 = 4$, or $x^2/4 + y^2 = 1$, use the parametrization $x = 2\cos t,\ y = \sin t,\ 0 \le t \le 2\pi$ to get

$$L = 4\int_0^{\pi/2} \sqrt{(dx/dt)^2 + (dy/dt)^2}\, dt = 4\int_0^{\pi/2} \sqrt{4\sin^2 t + \cos^2 t}\, dt = 4\int_0^{\pi/2} \sqrt{3\sin^2 t + 1}\, dt$$

Using Simpson's Rule with $n = 10$, $\Delta t = \frac{\pi/2 - 0}{10} = \frac{\pi}{20}$, and $f(t) = \sqrt{3\sin^2 t + 1}$, we get

$$L \approx \tfrac{4}{3}\left(\tfrac{\pi}{20}\right)\left[f(0) + 4f\left(\tfrac{\pi}{20}\right) + 2f\left(\tfrac{2\pi}{20}\right) + \cdots + 2f\left(\tfrac{8\pi}{20}\right) + 4f\left(\tfrac{9\pi}{20}\right) + f\left(\tfrac{\pi}{2}\right)\right] \approx 9.69$$

60. The length of the major axis is $2a$, so $a = \frac{1}{2}(1.18 \times 10^{10}) = 5.9 \times 10^9$. The length of the minor axis is $2b$, so

$b = \frac{1}{2}(1.14 \times 10^{10}) = 5.7 \times 10^9$. An equation of the ellipse is $\dfrac{x^2}{a^2} + \dfrac{y^2}{b^2} = 1$, or converting into parametric equations,

$x = a\cos\theta$ and $y = b\sin\theta$. So

$$L = 4\int_0^{\pi/2} \sqrt{(dx/d\theta)^2 + (dy/d\theta)^2}\, d\theta = 4\int_0^{\pi/2} \sqrt{a^2\sin^2\theta + b^2\cos^2\theta}\, d\theta$$

Using Simpson's Rule with $n = 10$, $\Delta\theta = \frac{\pi/2 - 0}{10} = \frac{\pi}{20}$, and $f(\theta) = \sqrt{a^2 \sin^2\theta + b^2 \cos^2\theta}$, we get

$$L \approx 4 \cdot S_{10} = 4 \cdot \frac{\pi}{20 \cdot 3}\left[f(0) + 4f\left(\frac{\pi}{20}\right) + 2f\left(\frac{2\pi}{20}\right) + \cdots + 2f\left(\frac{8\pi}{20}\right) + 4f\left(\frac{9\pi}{20}\right) + f\left(\frac{\pi}{2}\right)\right] \approx 3.64 \times 10^{10} \text{ km}$$

61. $\dfrac{x^2}{a^2} - \dfrac{y^2}{b^2} = 1 \;\Rightarrow\; \dfrac{y^2}{b^2} = \dfrac{x^2 - a^2}{a^2} \;\Rightarrow\; y = \pm\dfrac{b}{a}\sqrt{x^2 - a^2}$.

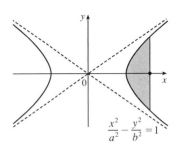

$A = 2\displaystyle\int_a^c \frac{b}{a}\sqrt{x^2 - a^2}\,dx \overset{39}{=} \frac{2b}{a}\left[\frac{x}{2}\sqrt{x^2 - a^2} - \frac{a^2}{2}\ln\left|x + \sqrt{x^2 - a^2}\right|\,\right]_a^c$

$= \dfrac{b}{a}\left[c\sqrt{c^2 - a^2} - a^2\ln\left|c + \sqrt{c^2 - a^2}\,\right| + a^2\ln|a|\,\right]$

Since $a^2 + b^2 = c^2, c^2 - a^2 = b^2$, and $\sqrt{c^2 - a^2} = b$.

$= \dfrac{b}{a}\left[cb - a^2\ln(c+b) + a^2\ln a\right] = \dfrac{b}{a}\left[cb + a^2(\ln a - \ln(b+c))\right]$

$= b^2 c/a + ab\ln[a/(b+c)]$, where $c^2 = a^2 + b^2$.

62. (a) $\dfrac{x^2}{a^2} + \dfrac{y^2}{b^2} = 1 \;\Rightarrow\; \dfrac{y^2}{b^2} = \dfrac{a^2 - x^2}{a^2} \;\Rightarrow\; y = \pm\dfrac{b}{a}\sqrt{a^2 - x^2}$.

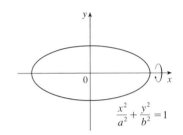

$V = \displaystyle\int_{-a}^{a} \pi\left(\frac{b}{a}\sqrt{a^2 - x^2}\right)^2 dx = 2\pi\frac{b^2}{a^2}\int_0^a (a^2 - x^2)\,dx$

$= \dfrac{2\pi b^2}{a^2}\left[a^2 x - \frac{1}{3}x^3\right]_0^a = \dfrac{2\pi b^2}{a^2}\left(\frac{2a^3}{3}\right) = \dfrac{4}{3}\pi b^2 a$

(b) $\dfrac{x^2}{a^2} + \dfrac{y^2}{b^2} = 1 \;\Rightarrow\; \dfrac{x^2}{a^2} = \dfrac{b^2 - y^2}{b^2} \;\Rightarrow\; x = \pm\dfrac{a}{b}\sqrt{b^2 - y^2}$.

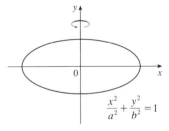

$V = \displaystyle\int_{-b}^{b} \pi\left(\frac{a}{b}\sqrt{b^2 - y^2}\right)^2 dy = 2\pi\frac{a^2}{b^2}\int_0^b (b^2 - y^2)\,dy$

$= \dfrac{2\pi a^2}{b^2}\left[b^2 y - \frac{1}{3}y^3\right]_0^b = \dfrac{2\pi a^2}{b^2}\left(\frac{2b^3}{3}\right) = \dfrac{4}{3}\pi a^2 b$

63. Differentiating implicitly, $\dfrac{x^2}{a^2} + \dfrac{y^2}{b^2} = 1 \;\Rightarrow\; \dfrac{2x}{a^2} + \dfrac{2yy'}{b^2} = 0 \;\Rightarrow\; y' = -\dfrac{b^2 x}{a^2 y}$ $[y \neq 0]$. Thus, the slope of the tangent

line at P is $-\dfrac{b^2 x_1}{a^2 y_1}$. The slope of $F_1 P$ is $\dfrac{y_1}{x_1 + c}$ and of $F_2 P$ is $\dfrac{y_1}{x_1 - c}$. By the formula in Problem 15 on text page 202,

we have

$$\tan\alpha = \frac{\dfrac{y_1}{x_1 + c} + \dfrac{b^2 x_1}{a^2 y_1}}{1 - \dfrac{b^2 x_1 y_1}{a^2 y_1(x_1 + c)}} = \frac{a^2 y_1^2 + b^2 x_1(x_1 + c)}{a^2 y_1(x_1 + c) - b^2 x_1 y_1} = \frac{a^2 b^2 + b^2 c x_1}{c^2 x_1 y_1 + a^2 c y_1}\quad\left[\begin{array}{l}\text{using } b^2 x_1^2 + a^2 y_1^2 = a^2 b^2,\\ \text{and } a^2 - b^2 = c^2\end{array}\right]$$

$$= \frac{b^2\left(cx_1 + a^2\right)}{cy_1\left(cx_1 + a^2\right)} = \frac{b^2}{cy_1}$$

and $\quad \tan\beta = \dfrac{-\dfrac{b^2 x_1}{a^2 y_1} - \dfrac{y_1}{x_1 - c}}{1 - \dfrac{b^2 x_1 y_1}{a^2 y_1(x_1 - c)}} = \dfrac{-a^2 y_1^2 - b^2 x_1(x_1 - c)}{a^2 y_1(x_1 - c) - b^2 x_1 y_1} = \dfrac{-a^2 b^2 + b^2 c x_1}{c^2 x_1 y_1 - a^2 c y_1} = \dfrac{b^2\left(cx_1 - a^2\right)}{cy_1(cx_1 - a^2)} = \dfrac{b^2}{cy_1}$

Thus, $\alpha = \beta$.

64. The slopes of the line segments F_1P and F_2P are $\dfrac{y_1}{x_1+c}$ and $\dfrac{y_1}{x_1-c}$, where P is (x_1, y_1). Differentiating implicitly,

$$\frac{2x}{a^2} - \frac{2yy'}{b^2} = 0 \quad \Rightarrow \quad y' = \frac{b^2x}{a^2y} \quad \Rightarrow \quad \text{the slope of the tangent at } P \text{ is } \frac{b^2x_1}{a^2y_1}, \text{ so by the formula in Problem 15 on text}$$

page 202,

$$\tan\alpha = \frac{\dfrac{b^2x_1}{a^2y_1} - \dfrac{y_1}{x_1+c}}{1 + \dfrac{b^2x_1y_1}{a^2y_1(x_1+c)}} = \frac{b^2x_1(x_1+c) - a^2y_1^2}{a^2y_1(x_1+c) + b^2x_1y_1} = \frac{b^2(cx_1+a^2)}{cy_1(cx_1+a^2)} \quad \left[\begin{array}{c} \text{using } x_1^2/a^2 - y_1^2/b^2 = 1, \\ \text{and } a^2 + b^2 = c^2 \end{array}\right] = \frac{b^2}{cy_1}$$

and

$$\tan\beta = \frac{-\dfrac{b^2x_1}{a^2y_1} + \dfrac{y_1}{x_1-c}}{1 + \dfrac{b^2x_1y_1}{a^2y_1(x_1-c)}} = \frac{-b^2x_1(x_1-c) + a^2y_1^2}{a^2y_1(x_1-c) + b^2x_1y_1} = \frac{b^2(cx_1-a^2)}{cy_1(cx_1-a^2)} = \frac{b^2}{cy_1}$$

So $\alpha = \beta$.

11.6 Conic Sections in Polar Coordinates

1. The directrix $y = 6$ is above the focus at the origin, so we use the form with "$+ e\sin\theta$" in the denominator. [See Theorem 6

and Figure 2(c).] $\quad r = \dfrac{ed}{1 + e\sin\theta} = \dfrac{\frac{7}{4}\cdot 6}{1 + \frac{7}{4}\sin\theta} = \dfrac{42}{4 + 7\sin\theta}$

2. The directrix $x = 4$ is to the right of the focus at the origin, so we use the form with "$+ e\cos\theta$" in the denominator.

$e = 1$ for a parabola, so an equation is $r = \dfrac{ed}{1 + e\cos\theta} = \dfrac{1\cdot 4}{1 + 1\cos\theta} = \dfrac{4}{1 + \cos\theta}$

3. The directrix $x = -5$ is to the left of the focus at the origin, so we use the form with "$- e\cos\theta$" in the denominator.

$r = \dfrac{ed}{1 - e\cos\theta} = \dfrac{\frac{3}{4}\cdot 5}{1 - \frac{3}{4}\cos\theta} = \dfrac{15}{4 - 3\cos\theta}$

4. The directrix $y = -2$ is below the focus at the origin, so we use the form with "$- e\sin\theta$" in the denominator.

$r = \dfrac{ed}{1 - e\sin\theta} = \dfrac{2\cdot 2}{1 - 2\sin\theta} = \dfrac{4}{1 - 2\sin\theta}$

5. The vertex $(4, 3\pi/2)$ is 4 units below the focus at the origin, so the directrix is 8 units below the focus $(d = 8)$, and we use the

form with "$-e\sin\theta$" in the denominator. $e = 1$ for a parabola, so an equation is $r = \dfrac{ed}{1 - e\sin\theta} = \dfrac{1(8)}{1 - 1\sin\theta} = \dfrac{8}{1 - \sin\theta}$.

6. The vertex $P(1, \pi/2)$ is 1 unit above the focus F at the origin, so $|PF| = 1$ and we use the form with "$+e\sin\theta$" in the

denominator. The distance from the focus to the directrix l is d, so

$$e = \frac{|PF|}{|Pl|} \quad \Rightarrow \quad 0.8 = \frac{1}{d-1} \quad \Rightarrow \quad 0.8d - 0.8 = 1 \quad \Rightarrow \quad 0.8d = 1.8 \quad \Rightarrow \quad d = 2.25.$$

An equation is $r = \dfrac{ed}{1 + e\sin\theta} = \dfrac{0.8(2.25)}{1 + 0.8\sin\theta}\cdot\dfrac{5}{5} = \dfrac{9}{5 + 4\sin\theta}$.

7. The directrix $r = 4\sec\theta$ (equivalent to $r\cos\theta = 4$ or $x = 4$) is to the right of the focus at the origin, so we will use the form

with "$+e\cos\theta$" in the denominator. The distance from the focus to the directrix is $d = 4$, so an equation is

$r = \dfrac{ed}{1 + e\cos\theta} = \dfrac{\frac{1}{2}(4)}{1 + \frac{1}{2}\cos\theta}\cdot\dfrac{2}{2} = \dfrac{4}{2 + \cos\theta}$.

8. The directrix $r = -6 \csc \theta$ (equivalent to $r \sin \theta = -6$ or $y = -6$) is below the focus at the origin, so we will use the form
with "$-e \sin \theta$" in the denominator. The distance from the focus to the directrix is $d = 6$, so an equation is

$$r = \frac{ed}{1 - e \sin \theta} = \frac{3(6)}{1 - 3 \sin \theta} = \frac{18}{1 - 3 \sin \theta}.$$

9. $r = \dfrac{1}{1 + \sin \theta} = \dfrac{ed}{1 + e \sin \theta}$, where $d = e = 1$.

(a) Eccentricity $= e = 1$

(b) Since $e = 1$, the conic is a parabola.

(c) Since "$+e \sin \theta$" appears in the denominator, the directrix is above the
focus at the origin. $d = |Fl| = 1$, so an equation of the directrix is $y = 1$.

(d) The vertex is at $\left(\frac{1}{2}, \frac{\pi}{2}\right)$, midway between the focus and the directrix.

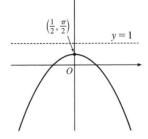

10. $r = \dfrac{12}{3 - 10 \cos \theta} \cdot \dfrac{1/3}{1/3} = \dfrac{4}{1 - \frac{10}{3} \cos \theta}$, where $e = \frac{10}{3}$ and $ed = 4$ $\Rightarrow$ $d = 4\left(\frac{3}{10}\right) = \frac{6}{5}$.

(a) Eccentricity $= e = \frac{10}{3}$

(b) Since $e = \frac{10}{3} > 1$, the conic is a hyperbola.

(c) Since "$-e \cos \theta$" appears in the denominator, the directrix is to the left of the
focus at the origin. $d = |Fl| = \frac{6}{5}$, so an equation of the directrix is $x = -\frac{6}{5}$.

(d) The vertices are $\left(-\frac{12}{7}, 0\right)$ and $\left(\frac{12}{13}, \pi\right)$, so the center is midway between them,
that is, $\left(\frac{120}{91}, \pi\right)$.

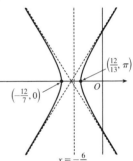

11. $r = \dfrac{12}{4 - \sin \theta} \cdot \dfrac{1/4}{1/4} = \dfrac{3}{1 - \frac{1}{4} \sin \theta}$, where $e = \frac{1}{4}$ and $ed = 3$ $\Rightarrow$ $d = 12$.

(a) Eccentricity $= e = \frac{1}{4}$

(b) Since $e = \frac{1}{4} < 1$, the conic is an ellipse.

(c) Since "$-e \sin \theta$" appears in the denominator, the directrix is below the focus
at the origin. $d = |Fl| = 12$, so an equation of the directrix is $y = -12$.

(d) The vertices are $\left(4, \frac{\pi}{2}\right)$ and $\left(\frac{12}{5}, \frac{3\pi}{2}\right)$, so the center is midway between them,
that is, $\left(\frac{4}{5}, \frac{\pi}{2}\right)$.

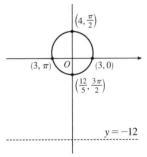

12. $r = \dfrac{3}{2 + 2 \cos \theta} \cdot \dfrac{1/2}{1/2} = \dfrac{3/2}{1 + 1 \cos \theta}$, where $e = 1$ and $ed = \frac{3}{2}$ $\Rightarrow$ $d = \frac{3}{2}$.

(a) Eccentricity $= e = 1$

(b) Since $e = 1$, the conic is a parabola.

(c) Since "$+e \cos \theta$" appears in the denominator, the directrix is to the right of
the focus at the origin. $d = |Fl| = \frac{3}{2}$, so an equation of the directrix is
$x = \frac{3}{2}$.

(d) The vertex is at $\left(\frac{3}{4}, 0\right)$, midway between the focus and directrix.

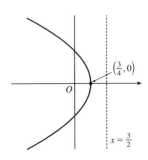

13. $r = \dfrac{9}{6 + 2\cos\theta} \cdot \dfrac{1/6}{1/6} = \dfrac{3/2}{1 + \frac{1}{3}\cos\theta}$, where $e = \frac{1}{3}$ and $ed = \frac{3}{2}$ $\Rightarrow$ $d = \frac{9}{2}$.

(a) Eccentricity $= e = \frac{1}{3}$

(b) Since $e = \frac{1}{3} < 1$, the conic is an ellipse.

(c) Since "$+e\cos\theta$" appears in the denominator, the directrix is to the right of

the focus at the origin. $d = |Fl| = \frac{9}{2}$, so an equation of the directrix is

$x = \frac{9}{2}$.

(d) The vertices are $\left(\frac{9}{8}, 0\right)$ and $\left(\frac{9}{4}, \pi\right)$, so the center is midway between them,

that is, $\left(\frac{9}{16}, \pi\right)$.

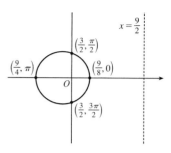

14. $r = \dfrac{8}{4 + 5\sin\theta} \cdot \dfrac{1/4}{1/4} = \dfrac{2}{1 + \frac{5}{4}\sin\theta}$, where $e = \frac{5}{4}$ and $ed = 2$ $\Rightarrow$ $d = 2\left(\frac{4}{5}\right) = \frac{8}{5}$.

(a) Eccentricity $= e = \frac{5}{4}$

(b) Since $e = \frac{5}{4} > 1$, the conic is a hyperbola.

(c) Since "$+e\sin\theta$" appears in the denominator, the directrix is above the

focus at the origin. $d = |Fl| = \frac{8}{5}$, so an equation of the directrix is $y = \frac{8}{5}$.

(d) The vertices are $\left(\frac{8}{9}, \frac{\pi}{2}\right)$ and $\left(-8, \frac{3\pi}{2}\right)$, so the center is midway between them,

that is, $\left(\frac{40}{9}, \frac{\pi}{2}\right)$.

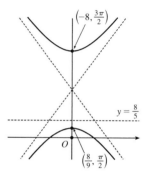

15. $r = \dfrac{3}{4 - 8\cos\theta} \cdot \dfrac{1/4}{1/4} = \dfrac{3/4}{1 - 2\cos\theta}$, where $e = 2$ and $ed = \frac{3}{4}$ $\Rightarrow$ $d = \frac{3}{8}$.

(a) Eccentricity $= e = 2$

(b) Since $e = 2 > 1$, the conic is a hyperbola.

(c) Since "$-e\cos\theta$" appears in the denominator, the directrix is to the left of

the focus at the origin. $d = |Fl| = \frac{3}{8}$, so an equation of the directrix is

$x = -\frac{3}{8}$.

(d) The vertices are $\left(-\frac{3}{4}, 0\right)$ and $\left(\frac{1}{4}, \pi\right)$, so the center is midway between them,

that is, $\left(\frac{1}{2}, \pi\right)$.

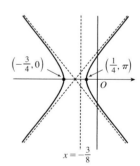

16. $r = \dfrac{10}{5 - 6\sin\theta} \cdot \dfrac{1/5}{1/5} = \dfrac{2}{1 - \frac{6}{5}\sin\theta}$, where $e = \frac{6}{5}$ and $ed = 2$ $\Rightarrow$ $d = 2\left(\frac{5}{6}\right) = \frac{5}{3}$.

(a) Eccentricity $= e = \frac{6}{5}$

(b) Since $e = \frac{6}{5} > 1$, the conic is a hyperbola.

(c) Since "$-e\sin\theta$" appears in the denominator, the directrix is below the focus

at the origin. $d = |Fl| = \frac{5}{3}$, so an equation of the directrix is $y = -\frac{5}{3}$.

(d) The vertices are $\left(-10, \frac{\pi}{2}\right)$ and $\left(\frac{10}{11}, \frac{3\pi}{2}\right)$, so the center is midway between them,

that is, $\left(\frac{60}{11}, \frac{3\pi}{2}\right)$.

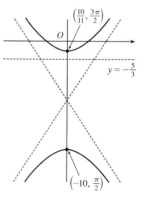

17. (a) $r = \dfrac{1}{1 - 2\sin\theta}$, where $e = 2$ and $ed = 1$ $\Rightarrow$ $d = \frac{1}{2}$. The eccentricity

$e = 2 > 1$, so the conic is a hyperbola. Since "$-e\sin\theta$" appears in the

denominator, the directrix is below the focus at the origin. $d = |Fl| = \frac{1}{2}$,

so an equation of the directrix is $y = -\frac{1}{2}$. The vertices are $\left(-1, \frac{\pi}{2}\right)$ and

$\left(\frac{1}{3}, \frac{3\pi}{2}\right)$, so the center is midway between them, that is, $\left(\frac{2}{3}, \frac{3\pi}{2}\right)$.

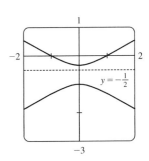

(b) By the discussion that precedes Example 4, the equation

is $r = \dfrac{1}{1 - 2\sin\left(\theta - \frac{3\pi}{4}\right)}$.

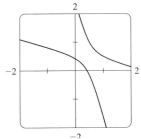

18. $r = \dfrac{4}{5 + 6\cos\theta} = \dfrac{4/5}{1 + \frac{6}{5}\cos\theta}$, so $e = \frac{6}{5}$ and $ed = \frac{4}{5}$ $\Rightarrow$ $d = \frac{2}{3}$.

An equation of the directrix is $x = \frac{2}{3}$ $\Rightarrow$ $r\cos\theta = \frac{2}{3}$ $\Rightarrow$ $r = \dfrac{2}{3\cos\theta}$.

If the hyperbola is rotated about its focus (the origin) through an angle $\pi/3$,

its equation is the same as that of the original, with θ replaced by $\theta - \frac{\pi}{3}$

(see Example 4), so $r = \dfrac{4}{5 + 6\cos\left(\theta - \frac{\pi}{3}\right)}$.

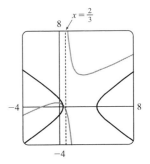

19. For $e < 1$ the curve is an ellipse. It is nearly circular when e is close to 0. As e
increases, the graph is stretched out to the right, and grows larger (that is, its
right-hand focus moves to the right while its left-hand focus remains at the
origin.) At $e = 1$, the curve becomes a parabola with focus at the origin.

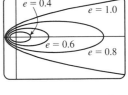

20. (a) The value of d does not seem to affect the shape of the conic (a parabola) at
all, just its size, position, and orientation (for $d < 0$ it opens upward, for
$d > 0$ it opens downward).

(b) We consider only positive values of e. When $0 < e < 1$, the conic is an ellipse. As $e \rightarrow 0^+$, the graph approaches perfect roundness and zero size. As e increases, the ellipse becomes more elongated, until at $e = 1$ it turns into a parabola. For $e > 1$, the conic is a hyperbola, which moves downward and gets broader as e continues to increase.

$e = 0.1$

$e = 0.5$

$e = 0.9$

$e = 1$

$e = 1.1$

$e = 1.5$

$e = 10$

21. $|PF| = e\,|Pl| \Rightarrow r = e[d - r\cos(\pi - \theta)] = e(d + r\cos\theta) \Rightarrow$

$r(1 - e\cos\theta) = ed \Rightarrow r = \dfrac{ed}{1 - e\cos\theta}$

22. $|PF| = e\,|Pl| \Rightarrow r = e[d - r\sin\theta] \Rightarrow r(1 + e\sin\theta) = ed \Rightarrow$

$r = \dfrac{ed}{1 + e\sin\theta}$

23. $|PF| = e\,|Pl| \Rightarrow r = e[d - r\sin(\theta - \pi)] = e(d + r\sin\theta) \Rightarrow$

$r(1 - e\sin\theta) = ed \Rightarrow r = \dfrac{ed}{1 - e\sin\theta}$

24. The parabolas intersect at the two points where $\dfrac{c}{1+\cos\theta}=\dfrac{d}{1-\cos\theta}$ $\Rightarrow$ $\cos\theta=\dfrac{c-d}{c+d}$ $\Rightarrow$ $r=\dfrac{c+d}{2}$.

For the first parabola, $\dfrac{dr}{d\theta}=\dfrac{c\sin\theta}{(1+\cos\theta)^2}$, so

$$\frac{dy}{dx}=\frac{(dr/d\theta)\sin\theta+r\cos\theta}{(dr/d\theta)\cos\theta-r\sin\theta}=\frac{c\sin^2\theta+c\cos\theta(1+\cos\theta)}{c\sin\theta\cos\theta-c\sin\theta(1+\cos\theta)}=\frac{1+\cos\theta}{-\sin\theta}$$

and similarly for the second, $\dfrac{dy}{dx}=\dfrac{1-\cos\theta}{\sin\theta}=\dfrac{\sin\theta}{1+\cos\theta}$. Since the product of these slopes is -1, the parabolas intersect at right angles.

25. We are given $e=0.093$ and $a=2.28\times10^8$. By (7), we have

$$r=\frac{a(1-e^2)}{1+e\cos\theta}=\frac{2.28\times10^8[1-(0.093)^2]}{1+0.093\cos\theta}\approx\frac{2.26\times10^8}{1+0.093\cos\theta}$$

26. We are given $e=0.048$ and $2a=1.56\times10^9$ $\Rightarrow$ $a=7.8\times10^8$. By (7), we have

$$r=\frac{a(1-e^2)}{1+e\cos\theta}=\frac{7.8\times10^8[1-(0.048)^2]}{1+0.048\cos\theta}\approx\frac{7.78\times10^8}{1+0.048\cos\theta}$$

27. Here $2a=$ length of major axis $=36.18$ AU $\Rightarrow$ $a=18.09$ AU and $e=0.97$. By (7), the equation of the orbit is

$r=\dfrac{18.09[1-(0.97)^2]}{1-0.97\cos\theta}\approx\dfrac{1.07}{1-0.97\cos\theta}$. By (8), the maximum distance from the comet to the sun is $18.09(1+0.97)\approx35.64$ AU or about 3.314 billion miles.

28. Here $2a=$ length of major axis $=356.5$ AU $\Rightarrow$ $a=178.25$ AU and $e=0.9951$. By (7), the equation of the orbit is $r=\dfrac{178.25[1-(0.9951)^2]}{1-0.9951\cos\theta}\approx\dfrac{1.7426}{1-0.9951\cos\theta}$. By (8), the minimum distance from the comet to the sun is $178.25(1-0.9951)\approx0.8734$ AU or about 81 million miles.

29. The minimum distance is at perihelion, where $4.6\times10^7=r=a(1-e)=a(1-0.206)=a(0.794)$ $\Rightarrow$ $a=4.6\times10^7/0.794$. So the maximum distance, which is at aphelion, is $r=a(1+e)=(4.6\times10^7/0.794)(1.206)\approx7.0\times10^7$ km.

30. At perihelion, $r=a(1-e)=4.43\times10^9$, and at aphelion, $r=a(1+e)=7.37\times10^9$. Adding, we get $2a=11.80\times10^9$, so $a=5.90\times10^9$ km. Therefore $1+e=a(1+e)/a=\frac{7.37}{5.90}\approx1.249$ and $e\approx0.249$.

31. From Exercise 29, we have $e=0.206$ and $a(1-e)=4.6\times10^7$ km. Thus, $a=4.6\times10^7/0.794$. From (7), we can write the equation of Mercury's orbit as $r=a\dfrac{1-e^2}{1-e\cos\theta}$. So since

$$\frac{dr}{d\theta}=\frac{-a(1-e^2)e\sin\theta}{(1-e\cos\theta)^2}\quad\Rightarrow$$

$$r^2+\left(\frac{dr}{d\theta}\right)^2=\frac{a^2(1-e^2)^2}{(1-e\cos\theta)^2}+\frac{a^2(1-e^2)^2e^2\sin^2\theta}{(1-e\cos\theta)^4}=\frac{a^2(1-e^2)^2}{(1-e\cos\theta)^4}(1-2e\cos\theta+e^2)$$

the length of the orbit is

$$L=\int_0^{2\pi}\sqrt{r^2+(dr/d\theta)^2}\,d\theta=a(1-e^2)\int_0^{2\pi}\frac{\sqrt{1+e^2-2e\cos\theta}}{(1-e\cos\theta)^2}\,d\theta\approx3.6\times10^8\text{ km}$$

This seems reasonable, since Mercury's orbit is nearly circular, and the circumference of a circle of radius a is $2\pi a\approx3.6\times10^8$ km.

11 Review

CONCEPT CHECK

1. (a) A parametric curve is a set of points of the form $(x, y) = (f(t), g(t))$, where f and g are continuous functions of a variable t.

(b) Sketching a parametric curve, like sketching the graph of a function, is difficult to do in general. We can plot points on the curve by finding $f(t)$ and $g(t)$ for various values of t, either by hand or with a calculator or computer. Sometimes, when f and g are given by formulas, we can eliminate t from the equations $x = f(t)$ and $y = g(t)$ to get a Cartesian equation relating x and y. It may be easier to graph that equation than to work with the original formulas for x and y in terms of t.

2. (a) You can find $\dfrac{dy}{dx}$ as a function of t by calculating $\dfrac{dy}{dx} = \dfrac{dy/dt}{dx/dt}$ [if $dx/dt \neq 0$].

(b) Calculate the area as $\int_a^b y\,dx = \int_\alpha^\beta g(t)\,f'(t)dt$ [or $\int_\beta^\alpha g(t)\,f'(t)dt$ if the leftmost point is $(f(\beta), g(\beta))$ rather than $(f(\alpha), g(\alpha))$].

3. (a) $L = \int_\alpha^\beta \sqrt{(dx/dt)^2 + (dy/dt)^2}\,dt = \int_\alpha^\beta \sqrt{[f'(t)]^2 + [g'(t)]^2}\,dt$

(b) $S = \int_\alpha^\beta 2\pi y \sqrt{(dx/dt)^2 + (dy/dt)^2}\,dt = \int_\alpha^\beta 2\pi g(t) \sqrt{[f'(t)]^2 + [g'(t)]^2}\,dt$

4. (a) See Figure 5 in Section 11.3.

(b) $x = r \cos\theta, \; y = r \sin\theta$

(c) To find a polar representation (r, θ) with $r \geq 0$ and $0 \leq \theta < 2\pi$, first calculate $r = \sqrt{x^2 + y^2}$. Then θ is specified by $\cos\theta = x/r$ and $\sin\theta = y/r$.

5. (a) Calculate $\dfrac{dy}{dx} = \dfrac{\dfrac{dy}{d\theta}}{\dfrac{dx}{d\theta}} = \dfrac{\dfrac{d}{d\theta}(y)}{\dfrac{d}{d\theta}(x)} = \dfrac{\dfrac{d}{d\theta}(r\sin\theta)}{\dfrac{d}{d\theta}(r\cos\theta)} = \dfrac{\left(\dfrac{dr}{d\theta}\right)\sin\theta + r\cos\theta}{\left(\dfrac{dr}{d\theta}\right)\cos\theta - r\sin\theta}$, where $r = f(\theta)$.

(b) Calculate $A = \int_a^b \tfrac{1}{2} r^2 \, d\theta = \int_a^b \tfrac{1}{2}[f(\theta)]^2 \, d\theta$

(c) $L = \int_a^b \sqrt{(dx/d\theta)^2 + (dy/d\theta)^2}\,d\theta = \int_a^b \sqrt{r^2 + (dr/d\theta)^2}\,d\theta = \int_a^b \sqrt{[f(\theta)]^2 + [f'(\theta)]^2}\,d\theta$

6. (a) A parabola is a set of points in a plane whose distances from a fixed point F (the focus) and a fixed line l (the directrix) are equal.

(b) $x^2 = 4py$; $y^2 = 4px$

7. (a) An ellipse is a set of points in a plane the sum of whose distances from two fixed points (the foci) is a constant.

(b) $\dfrac{x^2}{a^2} + \dfrac{y^2}{a^2 - c^2} = 1$.

8. (a) A hyperbola is a set of points in a plane the difference of whose distances from two fixed points (the foci) is a constant. This difference should be interpreted as the larger distance minus the smaller distance.

(b) $\dfrac{x^2}{a^2} - \dfrac{y^2}{c^2 - a^2} = 1$

(c) $y = \pm \dfrac{\sqrt{c^2 - a^2}}{a}\,x$

9. (a) If a conic section has focus F and corresponding directrix l, then the eccentricity e is the fixed ratio $|PF|\,/\,|Pl|$ for points P of the conic section.

(b) $e < 1$ for an ellipse; $e > 1$ for a hyperbola; $e = 1$ for a parabola.

(c) $x = d$: $r = \dfrac{ed}{1 + e\cos\theta}$. $x = -d$: $r = \dfrac{ed}{1 - e\cos\theta}$. $y = d$: $r = \dfrac{ed}{1 + e\sin\theta}$. $y = -d$: $r = \dfrac{ed}{1 - e\sin\theta}$.

TRUE-FALSE QUIZ

1. False. Consider the curve defined by $x = f(t) = (t-1)^3$ and $y = g(t) = (t-1)^2$. Then $g'(t) = 2(t-1)$, so $g'(1) = 0$, but its graph has a *vertical* tangent when $t = 1$.   *Note:* The statement is true if $f'(1) \neq 0$ when $g'(1) = 0$.

2. False. If $x = f(t)$ and $y = g(t)$ are twice differentiable, then $\dfrac{d^2y}{dx^2} = \dfrac{d}{dx}\left(\dfrac{dy}{dx}\right) = \dfrac{\dfrac{d}{dt}\left(\dfrac{dy}{dx}\right)}{\dfrac{dx}{dt}}$.

3. False. For example, if $f(t) = \cos t$ and $g(t) = \sin t$ for $0 \le t \le 4\pi$, then the curve is a circle of radius 1, hence its length is 2π, but $\int_0^{4\pi} \sqrt{[f'(t)]^2 + [g'(t)]^2}\, dt = \int_0^{4\pi} \sqrt{(-\sin t)^2 + (\cos t)^2}\, dt = \int_0^{4\pi} 1\, dt = 4\pi$, since as t increases from 0 to 4π, the circle is traversed twice.

4. False. If $(r, \theta) = (1, \pi)$, then $(x, y) = (-1, 0)$, so $\tan^{-1}(y/x) = \tan^{-1} 0 = 0 \neq \theta$. The statement is true for points in quadrants I and IV.

5. True. The curve $r = 1 - \sin 2\theta$ is unchanged if we rotate it through $180°$ about O because $1 - \sin 2(\theta + \pi) = 1 - \sin(2\theta + 2\pi) = 1 - \sin 2\theta$. So it's unchanged if we replace r by $-r$. (See the discussion after Example 8 in Section 11.3.) In other words, it's the same curve as $r = -(1 - \sin 2\theta) = \sin 2\theta - 1$.

6. True. The polar equation $r = 2$, the Cartesian equation $x^2 + y^2 = 4$, and the parametric equations $x = 2\sin 3t$, $y = 2\cos 3t$ $[0 \le t \le 2\pi]$ all describe the circle of radius 2 centered at the origin.

7. False. The first pair of equations gives the portion of the parabola $y = x^2$ with $x \ge 0$, whereas the second pair of equations traces out the whole parabola $y = x^2$.

8. True. $y^2 = 2y + 3x \;\Leftrightarrow\; (y-1)^2 = 3x + 1 = 3\left(x + \frac{1}{3}\right) = 4\left(\frac{3}{4}\right)\left(x + \frac{1}{3}\right)$, which is the equation of a parabola with vertex $\left(-\frac{1}{3}, 1\right)$ and focus $\left(-\frac{1}{3} + \frac{3}{4}, 1\right)$, opening to the right.

9. True. By rotating and translating the parabola, we can assume it has an equation of the form $y = cx^2$, where $c > 0$. The tangent at the point (a, ca^2) is the line $y - ca^2 = 2ca(x - a)$; i.e., $y = 2cax - ca^2$. This tangent meets the parabola at the points (x, cx^2) where $cx^2 = 2cax - ca^2$. This equation is equivalent to $x^2 = 2ax - a^2$ [since $c > 0$]. But $x^2 = 2ax - a^2 \;\Leftrightarrow\; x^2 - 2ax + a^2 = 0 \;\Leftrightarrow\; (x - a)^2 = 0 \;\Leftrightarrow\; x = a \;\Leftrightarrow\; (x, cx^2) = (a, ca^2)$. This shows that each tangent meets the parabola at exactly one point.

10. True. Consider a hyperbola with focus at the origin, oriented so that its polar equation is $r = \dfrac{ed}{1 + e\cos\theta}$, where $e > 1$. The directrix is $x = d$, but along the hyperbola we have $x = r\cos\theta = \dfrac{ed\cos\theta}{1 + e\cos\theta} = d\left(\dfrac{e\cos\theta}{1 + e\cos\theta}\right) \neq d$.

EXERCISES

1. $x = t^2 + 4t$, $y = 2 - t$, $-4 \le t \le 1$. $t = 2 - y$, so

$x = (2-y)^2 + 4(2-y) = 4 - 4y + y^2 + 8 - 4y = y^2 - 8y + 12$ ⟺

$x + 4 = y^2 - 8y + 16 = (y-4)^2$. This is part of a parabola with vertex

$(-4, 4)$, opening to the right.

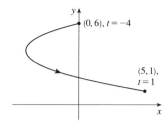

2. $x = 1 + e^{2t}$, $y = e^t$.

$x = 1 + e^{2t} = 1 + (e^t)^2 = 1 + y^2$, $y > 0$.

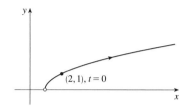

3. $y = \sec\theta = \dfrac{1}{\cos\theta} = \dfrac{1}{x}$. Since $0 \le \theta \le \pi/2$, $0 < x \le 1$ and $y \ge 1$.

This is part of the hyperbola $y = 1/x$.

4. $x = 2\cos\theta$, $y = 1 + \sin\theta$, $\cos^2\theta + \sin^2\theta = 1$ ⟹

$\left(\dfrac{x}{2}\right)^2 + (y-1)^2 = 1$ ⟹ $\dfrac{x^2}{4} + (y-1)^2 = 1$. This is an ellipse,

centered at $(0, 1)$, with semimajor axis of length 2 and semiminor axis of

length 1.

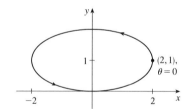

5. Three different sets of parametric equations for the curve $y = \sqrt{x}$ are

   (i) $x = t$, $y = \sqrt{t}$

   (ii) $x = t^4$, $y = t^2$

   (iii) $x = \tan^2 t$, $y = \tan t$, $0 \le t < \pi/2$

There are many other sets of equations that also give this curve.

6. For $t < -1$, $x > 0$ and $y < 0$ with x decreasing and y increasing. When

$t = -1$, $(x, y) = (0, 0)$. When $-1 < t < 0$, we have $-1 < x < 0$ and

$0 < y < 1/2$. When $t = 0$, $(x, y) = (-1, 0)$. When $0 < t < 1$,

$-1 < x < 0$ and $-\frac{1}{2} < y < 0$. When $t = 1$, $(x, y) = (0, 0)$ again.

When $t > 1$, both x and y are positive and increasing.

7. (a)

The Cartesian coordinates are $x = 4\cos\frac{2\pi}{3} = 4\left(-\frac{1}{2}\right) = -2$ and

$y = 4\sin\frac{2\pi}{3} = 4\left(\frac{\sqrt{3}}{2}\right) = 2\sqrt{3}$, that is, the point $\left(-2, 2\sqrt{3}\right)$.

(b) Given $x = -3$ and $y = 3$, we have $r = \sqrt{(-3)^2 + 3^2} = \sqrt{18} = 3\sqrt{2}$. Also, $\tan\theta = \frac{y}{x}$ $\Rightarrow$ $\tan\theta = \frac{3}{-3}$, and since $(-3, 3)$ is in the second quadrant, $\theta = \frac{3\pi}{4}$. Thus, one set of polar coordinates for $(-3, 3)$ is $\left(3\sqrt{2}, \frac{3\pi}{4}\right)$, and two others are $\left(3\sqrt{2}, \frac{11\pi}{4}\right)$ and $\left(-3\sqrt{2}, \frac{7\pi}{4}\right)$.

8. $1 \le r < 2,\ \frac{\pi}{6} \le \theta \le \frac{5\pi}{6}$

9. $r = 1 - \cos\theta$. This cardioid is symmetric about the polar axis.

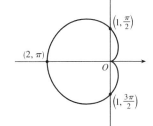

10. $r = \sin 4\theta$. This is an eight-leaved rose.

11. $r = \cos 3\theta$. This is a three-leaved rose. The curve is traced twice.

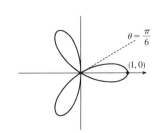

12. $r = 3 + \cos 3\theta$. The curve is symmetric about the horizontal axis.

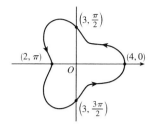

13. $r = 1 + \cos 2\theta$. The curve is symmetric about the pole and both the horizontal and vertical axes.

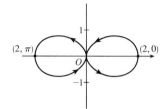

14. $r = 2\cos(\theta/2)$. The curve is symmetric about the pole and both the horizontal and vertical axes.

15. $r = \dfrac{3}{1 + 2\sin\theta}$ $\Rightarrow$ $e = 2 > 1$, so the conic is a hyperbola. $de = 3$ $\Rightarrow$ $d = \frac{3}{2}$ and the form "$+2\sin\theta$" imply that the directrix is above the focus at the origin and has equation $y = \frac{3}{2}$. The vertices are $\left(1, \frac{\pi}{2}\right)$ and $\left(-3, \frac{3\pi}{2}\right)$.

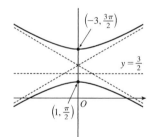

16. $r = \dfrac{3}{2 - 2\cos\theta} \cdot \dfrac{1/2}{1/2} = \dfrac{3/2}{1 - 1\cos\theta}$ $\Rightarrow$ $e = 1$, so the conic is a parabola. $de = \frac{3}{2}$ $\Rightarrow$ $d = \frac{3}{2}$ and the form "$-2\cos\theta$" imply that the directrix is to the left of the focus at the origin and has equation $x = -\frac{3}{2}$. The vertex is $\left(\frac{3}{4}, \pi\right)$.

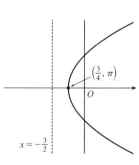

17. $x + y = 2$ $\Leftrightarrow$ $r\cos\theta + r\sin\theta = 2$ $\Leftrightarrow$ $r(\cos\theta + \sin\theta) = 2$ $\Leftrightarrow$ $r = \dfrac{2}{\cos\theta + \sin\theta}$

18. $x^2 + y^2 = 2$ $\Rightarrow$ $r^2 = 2$ $\Rightarrow$ $r = \sqrt{2}$. $[r = -\sqrt{2}$ gives the same curve.$]$

19. $r = (\sin\theta)/\theta$. As $\theta \to \pm\infty$, $r \to 0$.
As $\theta \to 0$, $r \to 1$. In the first figure,
there are an infinite number of
x-intercepts at $x = \pi n$, n a nonzero
integer. These correspond to pole
points in the second figure.

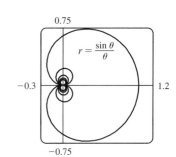

20. $r = \dfrac{2}{4 - 3\cos\theta} = \dfrac{1/2}{1 - \frac{3}{4}\cos\theta} \quad \Rightarrow \quad e = \frac{3}{4}$ and $d = \frac{2}{3}$. The equation of

the directrix is $x = -\frac{2}{3} \quad \Rightarrow \quad r = -2/(3\cos\theta)$. To obtain the equation

of the rotated ellipse, we replace θ in the original equation with $\theta - \frac{2\pi}{3}$,

and get $r = \dfrac{2}{4 - 3\cos\left(\theta - \frac{2\pi}{3}\right)}$.

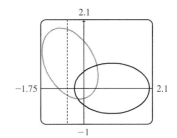

21. $x = \ln t$, $y = 1 + t^2$; $t = 1$. $\dfrac{dy}{dt} = 2t$ and $\dfrac{dx}{dt} = \dfrac{1}{t}$, so $\dfrac{dy}{dx} = \dfrac{dy/dt}{dx/dt} = \dfrac{2t}{1/t} = 2t^2$.

When $t = 1$, $(x, y) = (0, 2)$ and $dy/dx = 2$.

22. $x = t^3 + 6t + 1$, $y = 2t - t^2$; $t = -1$. $\dfrac{dy}{dx} = \dfrac{dy/dt}{dx/dt} = \dfrac{2 - 2t}{3t^2 + 6}$. When $t = -1$, $(x, y) = (-6, -3)$ and $\dfrac{dy}{dx} = \dfrac{4}{9}$.

23. $r = e^{-\theta} \quad \Rightarrow \quad y = r\sin\theta = e^{-\theta}\sin\theta$ and $x = r\cos\theta = e^{-\theta}\cos\theta \quad \Rightarrow$

$\dfrac{dy}{dx} = \dfrac{dy/d\theta}{dx/d\theta} = \dfrac{\frac{dr}{d\theta}\sin\theta + r\cos\theta}{\frac{dr}{d\theta}\cos\theta - r\sin\theta} = \dfrac{-e^{-\theta}\sin\theta + e^{-\theta}\cos\theta}{-e^{-\theta}\cos\theta - e^{-\theta}\sin\theta} \cdot \dfrac{-e^{\theta}}{-e^{\theta}} = \dfrac{\sin\theta - \cos\theta}{\cos\theta + \sin\theta}$.

When $\theta = \pi$, $\dfrac{dy}{dx} = \dfrac{0 - (-1)}{-1 + 0} = \dfrac{1}{-1} = -1$.

24. $r = 3 + \cos 3\theta \quad \Rightarrow \quad \dfrac{dy}{dx} = \dfrac{dy/d\theta}{dx/d\theta} = \dfrac{\frac{dr}{d\theta}\sin\theta + r\cos\theta}{\frac{dr}{d\theta}\cos\theta - r\sin\theta} = \dfrac{-3\sin 3\theta\sin\theta + (3 + \cos 3\theta)\cos\theta}{-3\sin 3\theta\cos\theta - (3 + \cos 3\theta)\sin\theta}$.

When $\theta = \pi/2$, $\dfrac{dy}{dx} = \dfrac{(-3)(-1)(1) + (3 + 0)\cdot 0}{(-3)(-1)(0) - (3 + 0)\cdot 1} = \dfrac{3}{-3} = -1$.

25. $x = t + \sin t$, $y = t - \cos t \quad \Rightarrow \quad \dfrac{dy}{dx} = \dfrac{dy/dt}{dx/dt} = \dfrac{1 + \sin t}{1 + \cos t} \quad \Rightarrow$

$\dfrac{d^2y}{dx^2} = \dfrac{\frac{d}{dt}\left(\frac{dy}{dx}\right)}{dx/dt} = \dfrac{\dfrac{(1 + \cos t)\cos t - (1 + \sin t)(-\sin t)}{(1 + \cos t)^2}}{1 + \cos t} = \dfrac{\cos t + \cos^2 t + \sin t + \sin^2 t}{(1 + \cos t)^3} = \dfrac{1 + \cos t + \sin t}{(1 + \cos t)^3}$

26. $x = 1 + t^2$, $y = t - t^3$. $\dfrac{dy}{dt} = 1 - 3t^2$ and $\dfrac{dx}{dt} = 2t$, so $\dfrac{dy}{dx} = \dfrac{dy/dt}{dx/dt} = \dfrac{1 - 3t^2}{2t} = \frac{1}{2}t^{-1} - \frac{3}{2}t$.

$\dfrac{d^2y}{dx^2} = \dfrac{d(dy/dx)/dt}{dx/dt} = \dfrac{-\frac{1}{2}t^{-2} - \frac{3}{2}}{2t} = -\frac{1}{4}t^{-3} - \frac{3}{4}t^{-1} = -\dfrac{1}{4t^3}\left(1 + 3t^2\right) = -\dfrac{3t^2 + 1}{4t^3}$.

27. We graph the curve $x = t^3 - 3t$, $y = t^2 + t + 1$ for $-2.2 \le t \le 1.2$.

By zooming in or using a cursor, we find that the lowest point is about

$(1.4, 0.75)$. To find the exact values, we find the t-value at which

$dy/dt = 2t + 1 = 0 \iff t = -\frac{1}{2} \iff (x, y) = \left(\frac{11}{8}, \frac{3}{4}\right)$.

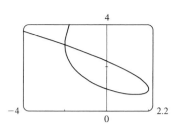

28. We estimate the coordinates of the point of intersection to be $(-2, 3)$. In fact this is exact, since both $t = -2$ and $t = 1$ give

the point $(-2, 3)$. So the area enclosed by the loop is

$$\int_{t=-2}^{t=1} y \, dx = \int_{-2}^{1} (t^2 + t + 1)(3t^2 - 3) \, dt = \int_{-2}^{1} (3t^4 + 3t^3 - 3t - 3) \, dt$$

$$= \left[\frac{3}{5}t^5 + \frac{3}{4}t^4 - \frac{3}{2}t^2 - 3t\right]_{-2}^{1} = \left(\frac{3}{5} + \frac{3}{4} - \frac{3}{2} - 3\right) - \left[-\frac{96}{5} + 12 - 6 - (-6)\right] = \frac{81}{20}$$

29. $x = 2a \cos t - a \cos 2t \implies \dfrac{dx}{dt} = -2a \sin t + 2a \sin 2t = 2a \sin t (2 \cos t - 1) = 0 \iff$

$\sin t = 0$ or $\cos t = \frac{1}{2} \implies t = 0, \frac{\pi}{3}, \pi,$ or $\frac{5\pi}{3}$.

$y = 2a \sin t - a \sin 2t \implies \dfrac{dy}{dt} = 2a \cos t - 2a \cos 2t = 2a(1 + \cos t - 2\cos^2 t) = 2a(1 - \cos t)(1 + 2\cos t) = 0 \implies$

$t = 0, \frac{2\pi}{3},$ or $\frac{4\pi}{3}$.

Thus the graph has vertical tangents where

$t = \frac{\pi}{3}, \pi$ and $\frac{5\pi}{3}$, and horizontal tangents where

$t = \frac{2\pi}{3}$ and $\frac{4\pi}{3}$. To determine what the slope is

where $t = 0$, we use l'Hospital's Rule to evaluate

$\lim\limits_{t \to 0} \dfrac{dy/dt}{dx/dt} = 0$, so there is a horizontal tangent

there.

t	x	y
0	a	0
$\frac{\pi}{3}$	$\frac{3}{2}a$	$\frac{\sqrt{3}}{2}a$
$\frac{2\pi}{3}$	$-\frac{1}{2}a$	$\frac{3\sqrt{3}}{2}a$
π	$-3a$	0
$\frac{4\pi}{3}$	$-\frac{1}{2}a$	$-\frac{3\sqrt{3}}{2}a$
$\frac{5\pi}{3}$	$\frac{3}{2}a$	$-\frac{\sqrt{3}}{2}a$

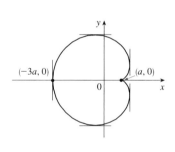

30. From Exercise 29, $x = 2a \cos t - a \cos 2t$, $y = 2a \sin t - a \sin 2t \implies$

$$A = 2 \int_{\pi}^{0} (2a \sin t - a \sin 2t)(-2a \sin t + 2a \sin 2t) \, dt = 4a^2 \int_{0}^{\pi} (2 \sin^2 t + \sin^2 2t - 3 \sin t \sin 2t) \, dt$$

$$= 4a^2 \int_{0}^{\pi} \left[(1 - \cos 2t) + \frac{1}{2}(1 - \cos 4t) - 6 \sin^2 t \cos t\right] dt = 4a^2 \left[t - \frac{1}{2}\sin 2t + \frac{1}{2}t - \frac{1}{8}\sin 4t - 2\sin^3 t\right]_{0}^{\pi}$$

$$= 4a^2 \left(\frac{3}{2}\right)\pi = 6\pi a^2$$

31. The curve $r^2 = 9 \cos 5\theta$ has 10 "petals." For instance, for $-\frac{\pi}{10} \le \theta \le \frac{\pi}{10}$, there are two petals, one with $r > 0$ and one

with $r < 0$.

$$A = 10 \int_{-\pi/10}^{\pi/10} \frac{1}{2}r^2 \, d\theta = 5 \int_{-\pi/10}^{\pi/10} 9 \cos 5\theta \, d\theta = 5 \cdot 9 \cdot 2 \int_{0}^{\pi/10} \cos 5\theta \, d\theta = 18 \left[\sin 5\theta\right]_{0}^{\pi/10} = 18$$

32. $r = 1 - 3 \sin \theta$. The inner loop is traced out as θ goes from $\alpha = \sin^{-1}\left(\frac{1}{3}\right)$ to $\pi - \alpha$, so

$$A = \int_{\alpha}^{\pi-\alpha} \frac{1}{2}r^2 \, d\theta = \int_{\alpha}^{\pi/2} (1 - 3 \sin \theta)^2 \, d\theta = \int_{\alpha}^{\pi/2} \left[1 - 6 \sin \theta + \frac{9}{2}(1 - \cos 2\theta)\right] d\theta$$

$$= \left[\frac{11}{2}\theta + 6 \cos \theta - \frac{9}{4}\sin 2\theta\right]_{\alpha}^{\pi/2} = \frac{11}{4}\pi - \frac{11}{2}\sin^{-1}\left(\frac{1}{3}\right) - 3\sqrt{2}$$

33. The curves intersect when $4\cos\theta = 2 \Rightarrow \cos\theta = \frac{1}{2} \Rightarrow \theta = \pm\frac{\pi}{3}$

for $-\pi \le \theta \le \pi$. The points of intersection are $\left(2, \frac{\pi}{3}\right)$ and $\left(2, -\frac{\pi}{3}\right)$.

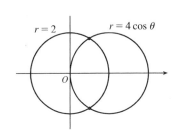

34. The two curves clearly both contain the pole. For other points of intersection, $\cot\theta = 2\cos(\theta + 2n\pi)$ or

$-2\cos(\theta + \pi + 2n\pi)$, both of which reduce to $\cot\theta = 2\cos\theta \Leftrightarrow \cos\theta = 2\sin\theta\cos\theta \Leftrightarrow \cos\theta(1 - 2\sin\theta) = 0 \Rightarrow$

$\cos\theta = 0$ or $\sin\theta = \frac{1}{2} \Rightarrow \theta = \frac{\pi}{6}, \frac{\pi}{2}, \frac{5\pi}{6}$ or $\frac{3\pi}{2} \Rightarrow$ intersection points are $\left(0, \frac{\pi}{2}\right)$, $\left(\sqrt{3}, \frac{\pi}{6}\right)$, and $\left(\sqrt{3}, \frac{11\pi}{6}\right)$.

35. The curves intersect where $2\sin\theta = \sin\theta + \cos\theta \Rightarrow$

$\sin\theta = \cos\theta \Rightarrow \theta = \frac{\pi}{4}$, and also at the origin (at which $\theta = \frac{3\pi}{4}$

on the second curve).

$$A = \int_0^{\pi/4} \tfrac{1}{2}(2\sin\theta)^2\,d\theta + \int_{\pi/4}^{3\pi/4} \tfrac{1}{2}(\sin\theta + \cos\theta)^2\,d\theta$$

$$= \int_0^{\pi/4}(1 - \cos 2\theta)\,d\theta + \tfrac{1}{2}\int_{\pi/4}^{3\pi/4}(1 + \sin 2\theta)\,d\theta$$

$$= \left[\theta - \tfrac{1}{2}\sin 2\theta\right]_0^{\pi/4} + \left[\tfrac{1}{2}\theta - \tfrac{1}{4}\cos 2\theta\right]_{\pi/4}^{3\pi/4} = \tfrac{1}{2}(\pi - 1)$$

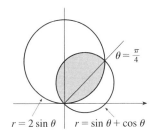

36. $A = 2\int_{-\pi/2}^{\pi/6} \tfrac{1}{2}\left[(2 + \cos 2\theta)^2 - (2 + \sin\theta)^2\right]d\theta$

$$= \int_{-\pi/2}^{\pi/6}\left[4\cos 2\theta + \cos^2 2\theta - 4\sin\theta - \sin^2\theta\right]d\theta$$

$$= \left[2\sin 2\theta + \tfrac{1}{2}\theta + \tfrac{1}{8}\sin 4\theta + 4\cos\theta - \tfrac{1}{2}\theta + \tfrac{1}{4}\sin 2\theta\right]_{-\pi/2}^{\pi/6}$$

$$= \tfrac{51}{16}\sqrt{3}$$

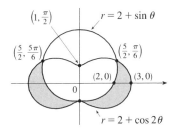

37. $x = 3t^2$, $y = 2t^3$.

$$L = \int_0^2 \sqrt{(dx/dt)^2 + (dy/dt)^2}\,dt = \int_0^2 \sqrt{(6t)^2 + (6t^2)^2}\,dt = \int_0^2 \sqrt{36t^2 + 36t^4}\,dt = \int_0^2 \sqrt{36t^2}\sqrt{1 + t^2}\,dt$$

$$= \int_0^2 6\,|t|\,\sqrt{1 + t^2}\,dt = 6\int_0^2 t\sqrt{1 + t^2}\,dt = 6\int_1^5 u^{1/2}\left(\tfrac{1}{2}du\right) \qquad \left[u = 1 + t^2,\, du = 2t\,dt\right]$$

$$= 6 \cdot \tfrac{1}{2} \cdot \tfrac{2}{3}\left[u^{3/2}\right]_1^5 = 2(5^{3/2} - 1) = 2\left(5\sqrt{5} - 1\right)$$

38. $x = 2 + 3t$, $y = \cosh 3t \Rightarrow (dx/dt)^2 + (dy/dt)^2 = 3^2 + (3\sinh 3t)^2 = 9(1 + \sinh^2 3t) = 9\cosh^2 3t$, so

$$L = \int_0^1 \sqrt{9\cosh^2 3t}\,dt = \int_0^1 |3\cosh 3t|\,dt = \int_0^1 3\cosh 3t\,dt = \left[\sinh 3t\right]_0^1 = \sinh 3 - \sinh 0 = \sinh 3.$$

39. $L = \int_\pi^{2\pi} \sqrt{r^2 + (dr/d\theta)^2}\,d\theta = \int_\pi^{2\pi}\sqrt{(1/\theta)^2 + (-1/\theta^2)^2}\,d\theta = \int_\pi^{2\pi}\dfrac{\sqrt{\theta^2 + 1}}{\theta^2}\,d\theta$

$$\overset{24}{=}\left[-\dfrac{\sqrt{\theta^2 + 1}}{\theta} + \ln\left(\theta + \sqrt{\theta^2 + 1}\right)\right]_\pi^{2\pi} = \dfrac{\sqrt{\pi^2 + 1}}{\pi} - \dfrac{\sqrt{4\pi^2 + 1}}{2\pi} + \ln\left(\dfrac{2\pi + \sqrt{4\pi^2 + 1}}{\pi + \sqrt{\pi^2 + 1}}\right)$$

$$= \dfrac{2\sqrt{\pi^2 + 1} - \sqrt{4\pi^2 + 1}}{2\pi} + \ln\left(\dfrac{2\pi + \sqrt{4\pi^2 + 1}}{\pi + \sqrt{\pi^2 + 1}}\right)$$

40. $L = \int_0^\pi \sqrt{r^2 + (dr/d\theta)^2}\, d\theta = \int_0^\pi \sqrt{\sin^6\left(\frac{1}{3}\theta\right) + \sin^4\left(\frac{1}{3}\theta\right)\cos^2\left(\frac{1}{3}\theta\right)}\, d\theta$

$= \int_0^\pi \sin^2\left(\frac{1}{3}\theta\right) d\theta = \left[\frac{1}{2}\left(\theta - \frac{3}{2}\sin\left(\frac{2}{3}\theta\right)\right)\right]_0^\pi = \frac{1}{2}\pi - \frac{3}{8}\sqrt{3}$

41. $x = 4\sqrt{t},\ y = \dfrac{t^3}{3} + \dfrac{1}{2t^2},\ 1 \le t \le 4 \quad \Rightarrow$

$S = \int_1^4 2\pi y \sqrt{(dx/dt)^2 + (dy/dt)^2}\, dt = \int_1^4 2\pi\left(\frac{1}{3}t^3 + \frac{1}{2}t^{-2}\right)\sqrt{\left(2/\sqrt{t}\right)^2 + (t^2 - t^{-3})^2}\, dt$

$= 2\pi\int_1^4 \left(\frac{1}{3}t^3 + \frac{1}{2}t^{-2}\right)\sqrt{(t^2 + t^{-3})^2}\, dt = 2\pi\int_1^4\left(\frac{1}{3}t^5 + \frac{5}{6} + \frac{1}{2}t^{-5}\right) dt = 2\pi\left[\frac{1}{18}t^6 + \frac{5}{6}t - \frac{1}{8}t^{-4}\right]_1^4 = \frac{471{,}295}{1024}\pi$

42. $x = 2 + 3t,\quad y = \cosh 3t \quad \Rightarrow \quad (dx/dt)^2 + (dy/dt)^2 = 3^2 + (3\sinh 3t)^2 = 9(1 + \sinh^2 3t) = 9\cosh^2 3t$, so

$S = \int_0^1 2\pi y\, ds = \int_0^1 2\pi \cosh 3t\sqrt{9\cosh^2 3t}\, dt = \int_0^1 2\pi \cosh 3t\, |3\cosh 3t|\, dt = \int_0^1 2\pi \cosh 3t \cdot 3\cosh 3t\, dt$

$= 6\pi\int_0^1 \cosh^2 3t\, dt = 6\pi\int_0^1 \frac{1}{2}(1 + \cosh 6t)\, dt = 3\pi\left[t + \frac{1}{6}\sinh 6t\right]_0^1 = 3\pi\left(1 + \frac{1}{6}\sinh 6\right) = 3\pi + \frac{\pi}{2}\sinh 6$

43. For all c except -1, the curve is asymptotic to the line $x = 1$. For $c < -1$, the curve bulges to the right near $y = 0$. As c increases, the bulge becomes smaller, until at $c = -1$ the curve is the straight line $x = 1$. As c continues to increase, the curve bulges to the left, until at $c = 0$ there is a cusp at the origin. For $c > 0$, there is a loop to the left of the origin, whose size and roundness increase as c increases. Note that the x-intercept of the curve is always $-c$.

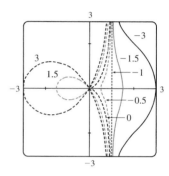

44. For a close to 0, the graph consists of four thin petals. As a increases, the petals get wider, until as $a \to \infty$, each petal occupies almost its entire quarter-circle.

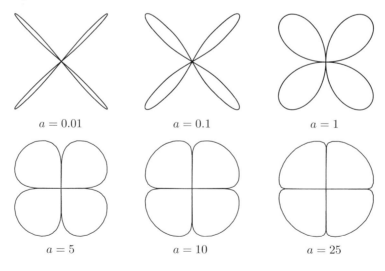

$a = 0.01$ $a = 0.1$ $a = 1$

$a = 5$ $a = 10$ $a = 25$

45. $\dfrac{x^2}{9} + \dfrac{y^2}{8} = 1$ is an ellipse with center $(0,0)$.

$a = 3, b = 2\sqrt{2}, c = 1 \Rightarrow$

foci $(\pm 1, 0)$, vertices $(\pm 3, 0)$.

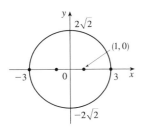

46. $4x^2 - y^2 = 16 \Leftrightarrow \dfrac{x^2}{4} - \dfrac{y^2}{16} = 1$ is a hyperbola

with center $(0,0)$, vertices $(\pm 2, 0)$, $a = 2, b = 4$,

$c = \sqrt{16+4} = 2\sqrt{5}$, foci $\left(\pm 2\sqrt{5}, 0\right)$ and

asymptotes $y = \pm 2x$.

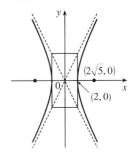

47. $6y^2 + x - 36y + 55 = 0 \Leftrightarrow$

$6(y^2 - 6y + 9) = -(x + 1) \Leftrightarrow$

$(y-3)^2 = -\frac{1}{6}(x+1)$, a parabola with vertex $(-1, 3)$,

opening to the left, $p = -\frac{1}{24} \Rightarrow$ focus $\left(-\frac{25}{24}, 3\right)$ and

directrix $x = -\frac{23}{24}$.

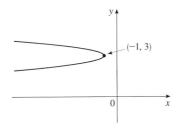

48. $25x^2 + 4y^2 + 50x - 16y = 59 \Leftrightarrow$

$25(x+1)^2 + 4(y-2)^2 = 100 \Leftrightarrow$

$\frac{1}{4}(x+1)^2 + \frac{1}{25}(y-2)^2 = 1$ is an ellipse centered at

$(-1, 2)$ with foci on the line $x = -1$, vertices $(-1, 7)$

and $(-1, -3)$; $a = 5, b = 2 \Rightarrow c = \sqrt{21} \Rightarrow$

foci $\left(-1, 2 \pm \sqrt{21}\right)$.

49. The ellipse with foci $(\pm 4, 0)$ and vertices $(\pm 5, 0)$ has center $(0, 0)$ and a horizontal major axis, with $a = 5$ and $c = 4$,

so $b^2 = a^2 - c^2 = 5^2 - 4^2 = 9$. An equation is $\dfrac{x^2}{25} + \dfrac{y^2}{9} = 1$.

50. The distance from the focus $(2, 1)$ to the directrix $x = -4$ is $2 - (-4) = 6$, so the distance from the focus to the vertex

is $\frac{1}{2}(6) = 3$ and the vertex is $(-1, 1)$. Since the focus is to the right of the vertex, $p = 3$. An equation is

$(y-1)^2 = 4 \cdot 3[x - (-1)]$, or $(y-1)^2 = 12(x+1)$.

51. The center of a hyperbola with foci $(0, \pm 4)$ is $(0, 0)$, so $c = 4$ and an equation is $\dfrac{y^2}{a^2} - \dfrac{x^2}{b^2} = 1$.

The asymptote $y = 3x$ has slope 3, so $\dfrac{a}{b} = \dfrac{3}{1} \Rightarrow a = 3b$ and $a^2 + b^2 = c^2 \Rightarrow (3b)^2 + b^2 = 4^2 \Rightarrow$

$10b^2 = 16 \Rightarrow b^2 = \frac{8}{5}$ and so $a^2 = 16 - \frac{8}{5} = \frac{72}{5}$. Thus, an equation is $\dfrac{y^2}{72/5} - \dfrac{x^2}{8/5} = 1$, or $\dfrac{5y^2}{72} - \dfrac{5x^2}{8} = 1$.

52. Center is $(3, 0)$, and $a = \frac{8}{2} = 4, c = 2 \Leftrightarrow b = \sqrt{4^2 - 2^2} = \sqrt{12} \Rightarrow$

an equation of the ellipse is $\dfrac{(x-3)^2}{12} + \dfrac{y^2}{16} = 1$.

53. $x^2 = -(y - 100)$ has its vertex at $(0, 100)$, so one of the vertices of the ellipse is $(0, 100)$. Another form of the equation of a

parabola is $x^2 = 4p(y - 100)$ so $4p(y - 100) = -(y - 100)$ $\Rightarrow$ $4p = -1$ $\Rightarrow$ $p = -\frac{1}{4}$. Therefore the shared focus is

found at $\left(0, \frac{399}{4}\right)$ so $2c = \frac{399}{4} - 0$ $\Rightarrow$ $c = \frac{399}{8}$ and the center of the ellipse is $\left(0, \frac{399}{8}\right)$. So $a = 100 - \frac{399}{8} = \frac{401}{8}$ and

$b^2 = a^2 - c^2 = \dfrac{401^2 - 399^2}{8^2} = 25$. So the equation of the ellipse is $\dfrac{x^2}{b^2} + \dfrac{\left(y - \frac{399}{8}\right)^2}{a^2} = 1$ $\Rightarrow$ $\dfrac{x^2}{25} + \dfrac{\left(y - \frac{399}{8}\right)^2}{\left(\frac{401}{8}\right)^2} = 1$,

or $\dfrac{x^2}{25} + \dfrac{(8y - 399)^2}{160{,}801} = 1$.

54. $\dfrac{x^2}{a^2} + \dfrac{y^2}{b^2} = 1$ $\Rightarrow$ $\dfrac{2x}{a^2} + \dfrac{2y}{b^2}\dfrac{dy}{dx} = 0$ $\Rightarrow$ $\dfrac{dy}{dx} = -\dfrac{b^2}{a^2}\dfrac{x}{y}$. Therefore $\dfrac{dy}{dx} = m$ $\Leftrightarrow$ $y = -\dfrac{b^2}{a^2}\dfrac{x}{m}$. Combining this

condition with $\dfrac{x^2}{a^2} + \dfrac{y^2}{b^2} = 1$, we find that $x = \pm\dfrac{a^2 m}{\sqrt{a^2 m^2 + b^2}}$. In other words, the two points on the ellipse where the

tangent has slope m are $\left(\pm\dfrac{a^2 m}{\sqrt{a^2 m^2 + b^2}}, \mp\dfrac{b^2}{\sqrt{a^2 m^2 + b^2}}\right)$. The tangent lines at these points have the equations

$y \pm \dfrac{b^2}{\sqrt{a^2 m^2 + b^2}} = m\left(x \mp \dfrac{a^2 m}{\sqrt{a^2 m^2 + b^2}}\right)$ or $y = mx \mp \dfrac{a^2 m^2}{\sqrt{a^2 m^2 + b^2}} \mp \dfrac{b^2}{\sqrt{a^2 m^2 + b^2}} = mx \mp \sqrt{a^2 m^2 + b^2}$.

55. Directrix $x = 4$ $\Rightarrow$ $d = 4$, so $e = \frac{1}{3}$ $\Rightarrow$ $r = \dfrac{ed}{1 + e\cos\theta} = \dfrac{4}{3 + \cos\theta}$.

56. See the end of the proof of Theorem 11.6.1. If $e > 1$, then $1 - e^2 < 0$ and Equations 11.6.4 become $a^2 = \dfrac{e^2 d^2}{(e^2 - 1)^2}$ and

$b^2 = \dfrac{e^2 d^2}{e^2 - 1}$, so $\dfrac{b^2}{a^2} = e^2 - 1$. The asymptotes $y = \pm\dfrac{b}{a}x$ have slopes $\pm\dfrac{b}{a} = \pm\sqrt{e^2 - 1}$, so the angles they make with the

polar axis are $\pm\tan^{-1}\left[\sqrt{e^2 - 1}\right] = \cos^{-1}(\pm 1/e)$.

57. In polar coordinates, an equation for the circle is $r = 2a\sin\theta$. Thus, the coordinates of Q are $x = r\cos\theta = 2a\sin\theta\,\cos\theta$

and $y = r\sin\theta = 2a\sin^2\theta$. The coordinates of R are $x = 2a\cot\theta$ and $y = 2a$. Since P is the midpoint of QR, we use the

midpoint formula to get $x = a(\sin\theta\,\cos\theta + \cot\theta)$ and $y = a(1 + \sin^2\theta)$.

PROBLEMS PLUS

1. $x = \int_1^t \dfrac{\cos u}{u}\, du$, $y = \int_1^t \dfrac{\sin u}{u}\, du$, so by FTC1, we have $\dfrac{dx}{dt} = \dfrac{\cos t}{t}$ and $\dfrac{dy}{dt} = \dfrac{\sin t}{t}$. Vertical tangent lines occur when

$\dfrac{dx}{dt} = 0 \;\Leftrightarrow\; \cos t = 0$. The parameter value corresponding to $(x, y) = (0, 0)$ is $t = 1$, so the nearest vertical tangent

occurs when $t = \frac{\pi}{2}$. Therefore, the arc length between these points is

$$L = \int_1^{\pi/2} \sqrt{\left(\dfrac{dx}{dt}\right)^2 + \left(\dfrac{dy}{dt}\right)^2}\, dt = \int_1^{\pi/2} \sqrt{\dfrac{\cos^2 t}{t^2} + \dfrac{\sin^2 t}{t^2}}\, dt = \int_1^{\pi/2} \dfrac{dt}{t} = \Big[\ln t\Big]_1^{\pi/2} = \ln \frac{\pi}{2}$$

2. (a) The curve $x^4 + y^4 = x^2 + y^2$ is symmetric about both axes and about the line $y = x$ (since interchanging x

and y does not change the equation) so we need only consider $y \geq x \geq 0$ to begin with. Implicit differentiation gives

$4x^3 + 4y^3 y' = 2x + 2yy' \;\Rightarrow\; y' = \dfrac{x(1 - 2x^2)}{y(2y^2 - 1)} \;\Rightarrow\; y' = 0$ when $x = 0$ and when $x = \pm\frac{1}{\sqrt{2}}$. If $x = 0$, then

$y^4 = y^2 \;\Rightarrow\; y^2(y^2 - 1) = 0 \;\Rightarrow\; y = 0$ or ± 1. The point $(0, 0)$ can't be a highest or lowest point because it is

isolated. [If $-1 < x < 1$ and $-1 < y < 1$, then $x^4 < x^2$ and $y^4 < y^2 \;\Rightarrow\; x^4 + y^4 < x^2 + y^2$, except for $(0, 0)$.]

If $x = \frac{1}{\sqrt{2}}$, then $x^2 = \frac{1}{2}$, $x^4 = \frac{1}{4}$, so $\frac{1}{4} + y^4 = \frac{1}{2} + y^2 \;\Rightarrow\; 4y^4 - 4y^2 - 1 = 0 \;\Rightarrow\; y^2 = \dfrac{4 \pm \sqrt{16+16}}{8} = \dfrac{1 \pm \sqrt{2}}{2}$.

But $y^2 > 0$, so $y^2 = \dfrac{1 + \sqrt{2}}{2} \;\Rightarrow\; y = \pm\sqrt{\frac{1}{2}(1 + \sqrt{2})}$. Near the point $(0, 1)$, the denominator of y' is positive and the

numerator changes from negative to positive as x increases through 0, so $(0, 1)$ is a local minimum point. At

$\left(\frac{1}{\sqrt{2}}, \sqrt{\dfrac{1 + \sqrt{2}}{2}}\right)$, y' changes from positive to negative, so that point gives a maximum. By symmetry, the highest points

on the curve are $\left(\pm\frac{1}{\sqrt{2}}, \sqrt{\dfrac{1 + \sqrt{2}}{2}}\right)$ and the lowest points are $\left(\pm\frac{1}{\sqrt{2}}, -\sqrt{\dfrac{1 + \sqrt{2}}{2}}\right)$.

(b) We use the information from part (a), together with symmetry with respect to the
axes and the lines $y = \pm x$, to sketch the curve.

(c) In polar coordinates, $x^4 + y^4 = x^2 + y^2$ becomes $r^4 \cos^4 \theta + r^4 \sin^4 \theta = r^2$ or

$r^2 = \dfrac{1}{\cos^4 \theta + \sin^4 \theta}$. By the symmetry shown in part (b), the area enclosed by

the curve is $A = 8 \int_0^{\pi/4} \frac{1}{2} r^2\, d\theta = 4 \int_0^{\pi/4} \dfrac{d\theta}{\cos^4 \theta + \sin^4 \theta} \stackrel{\text{CAS}}{=} \sqrt{2}\,\pi$.

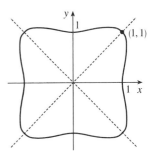

3. In terms of x and y, we have $x = r \cos \theta = (1 + c \sin \theta) \cos \theta = \cos \theta + c \sin \theta \cos \theta = \cos \theta + \frac{1}{2} c \sin 2\theta$ and

$y = r \sin \theta = (1 + c \sin \theta) \sin \theta = \sin \theta + c \sin^2 \theta$. Now $-1 \leq \sin \theta \leq 1 \;\Rightarrow\; -1 \leq \sin \theta + c \sin^2 \theta \leq 1 + c \leq 2$, so

$-1 \leq y \leq 2$. Furthermore, $y = 2$ when $c = 1$ and $\theta = \frac{\pi}{2}$, while $y = -1$ for $c = 0$ and $\theta = \frac{3\pi}{2}$. Therefore, we need a viewing

rectangle with $-1 \leq y \leq 2$.

To find the x-values, look at the equation $x = \cos \theta + \frac{1}{2} c \sin 2\theta$ and use the fact that $\sin 2\theta \geq 0$ for $0 \leq \theta \leq \frac{\pi}{2}$ and

$\sin 2\theta \leq 0$ for $-\frac{\pi}{2} \leq \theta \leq 0$. [Because $r = 1 + c \sin \theta$ is symmetric about the y-axis, we only need to consider

$-\frac{\pi}{2} \le \theta \le \frac{\pi}{2}$.] So for $-\frac{\pi}{2} \le \theta \le 0$, x has a maximum value when $c = 0$ and then $x = \cos\theta$ has a maximum value

of 1 at $\theta = 0$. Thus, the maximum value of x must occur on $\left[0, \frac{\pi}{2}\right]$ with $c = 1$. Then $x = \cos\theta + \frac{1}{2}\sin 2\theta \Rightarrow$

$\frac{dx}{d\theta} = -\sin\theta + \cos 2\theta = -\sin\theta + 1 - 2\sin^2\theta \Rightarrow \frac{dx}{d\theta} = -(2\sin\theta - 1)(\sin\theta + 1) = 0$ when $\sin\theta = -1$ or $\frac{1}{2}$

[but $\sin\theta \ne -1$ for $0 \le \theta \le \frac{\pi}{2}$]. If $\sin\theta = \frac{1}{2}$, then $\theta = \frac{\pi}{6}$ and

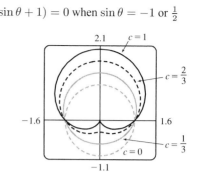

$x = \cos\frac{\pi}{6} + \frac{1}{2}\sin\frac{\pi}{3} = \frac{3}{4}\sqrt{3}$. Thus, the maximum value of x is $\frac{3}{4}\sqrt{3}$, and,

by symmetry, the minimum value is $-\frac{3}{4}\sqrt{3}$. Therefore, the smallest

viewing rectangle that contains every member of the family of polar curves

$r = 1 + c\sin\theta$, where $0 \le c \le 1$, is $\left[-\frac{3}{4}\sqrt{3}, \frac{3}{4}\sqrt{3}\right] \times [-1, 2]$.

4. (a) Let us find the polar equation of the path of the bug that starts in the upper
right corner of the square. If the polar coordinates of this bug, at a
particular moment, are (r, θ), then the polar coordinates of the bug that it is
crawling toward must be $\left(r, \theta + \frac{\pi}{2}\right)$. (The next bug must be the same
distance from the origin and the angle between the lines joining the bugs to
the pole must be $\frac{\pi}{2}$.) The Cartesian coordinates of the first bug are

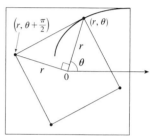

$(r\cos\theta, r\sin\theta)$ and for the second bug we have

$x = r\cos\left(\theta + \frac{\pi}{2}\right) = -r\sin\theta$, $y = r\sin\left(\theta + \frac{\pi}{2}\right) = r\cos\theta$. So the slope of the line joining the bugs is

$\dfrac{r\cos\theta - r\sin\theta}{-r\sin\theta - r\cos\theta} = \dfrac{\sin\theta - \cos\theta}{\sin\theta + \cos\theta}$. This must be equal to the slope of the tangent line at (r, θ), so by

Equation 11.3.3 we have $\dfrac{(dr/d\theta)\sin\theta + r\cos\theta}{(dr/d\theta)\cos\theta - r\sin\theta} = \dfrac{\sin\theta - \cos\theta}{\sin\theta + \cos\theta}$. Solving for $\dfrac{dr}{d\theta}$, we get

$\dfrac{dr}{d\theta}\sin^2\theta + \dfrac{dr}{d\theta}\sin\theta\cos\theta + r\sin\theta\cos\theta + r\cos^2\theta = \dfrac{dr}{d\theta}\sin\theta\cos\theta - \dfrac{dr}{d\theta}\cos^2\theta - r\sin^2\theta + r\sin\theta\cos\theta \Rightarrow$

$\dfrac{dr}{d\theta}(\sin^2\theta + \cos^2\theta) + r(\cos^2\theta + \sin^2\theta) = 0 \Rightarrow \dfrac{dr}{d\theta} = -r$. Solving this differential equation as a separable

equation (as in Section 10.3), or using Theorem 10.4.2 with $k = -1$, we get $r = Ce^{-\theta}$. To determine C we use the fact

that, at its starting position, $\theta = \frac{\pi}{4}$ and $r = \frac{1}{\sqrt{2}}a$, so $\frac{1}{\sqrt{2}}a = Ce^{-\pi/4} \Rightarrow C = \frac{1}{\sqrt{2}}ae^{\pi/4}$. Therefore, a polar equation of

the bug's path is $r = \frac{1}{\sqrt{2}}ae^{\pi/4}e^{-\theta}$ or $r = \frac{1}{\sqrt{2}}ae^{(\pi/4)-\theta}$.

(b) The distance traveled by this bug is $L = \int_{\pi/4}^{\infty}\sqrt{r^2 + (dr/d\theta)^2}\,d\theta$, where $\dfrac{dr}{d\theta} = \dfrac{a}{\sqrt{2}}e^{\pi/4}(-e^{-\theta})$ and so

$r^2 + (dr/d\theta)^2 = \frac{1}{2}a^2e^{\pi/2}e^{-2\theta} + \frac{1}{2}a^2e^{\pi/2}e^{-2\theta} = a^2e^{\pi/2}e^{-2\theta}$. Thus

$$L = \int_{\pi/4}^{\infty}ae^{\pi/4}e^{-\theta}\,d\theta = ae^{\pi/4}\lim_{t\to\infty}\int_{\pi/4}^{t}e^{-\theta}\,d\theta = ae^{\pi/4}\lim_{t\to\infty}\left[-e^{-\theta}\right]_{\pi/4}^{t}$$

$$= ae^{\pi/4}\lim_{t\to\infty}\left[e^{-\pi/4} - e^{-t}\right] = ae^{\pi/4}e^{-\pi/4} = a$$

5. (a) If (a, b) lies on the curve, then there is some parameter value t_1 such that $\dfrac{3t_1}{1 + t_1^3} = a$ and $\dfrac{3t_1^2}{1 + t_1^3} = b$. If $t_1 = 0$,

the point is $(0, 0)$, which lies on the line $y = x$. If $t_1 \neq 0$, then the point corresponding to $t = \dfrac{1}{t_1}$ is given by

$$x = \frac{3(1/t_1)}{1 + (1/t_1)^3} = \frac{3t_1^2}{t_1^3 + 1} = b, \; y = \frac{3(1/t_1)^2}{1 + (1/t_1)^3} = \frac{3t_1}{t_1^3 + 1} = a. \;\; \text{So } (b, a) \text{ also lies on the curve. [Another way to see}$$

this is to do part (e) first; the result is immediate.] The curve intersects the line $y = x$ when $\dfrac{3t}{1 + t^3} = \dfrac{3t^2}{1 + t^3}$ $\Rightarrow$

$t = t^2$ $\Rightarrow$ $t = 0$ or 1, so the points are $(0, 0)$ and $\left(\tfrac{3}{2}, \tfrac{3}{2}\right)$.

(b) $\dfrac{dy}{dt} = \dfrac{(1 + t^3)(6t) - 3t^2(3t^2)}{(1 + t^3)^2} = \dfrac{6t - 3t^4}{(1 + t^3)^2} = 0$ when $6t - 3t^4 = 3t(2 - t^3) = 0$ $\Rightarrow$ $t = 0$ or $t = \sqrt[3]{2}$, so there are

horizontal tangents at $(0, 0)$ and $\left(\sqrt[3]{2},\, \sqrt[3]{4}\right)$. Using the symmetry from part (a), we see that there are vertical tangents at

$(0, 0)$ and $\left(\sqrt[3]{4},\, \sqrt[3]{2}\right)$.

(c) Notice that as $t \to -1^+$, we have $x \to -\infty$ and $y \to \infty$. As $t \to -1^-$, we have $x \to \infty$ and $y \to -\infty$. Also

$$y - (-x - 1) = y + x + 1 = \frac{3t + 3t^2 + (1 + t^3)}{1 + t^3} = \frac{(t + 1)^3}{1 + t^3} = \frac{(t + 1)^2}{t^2 - t + 1} \to 0 \text{ as } t \to -1. \text{ So } y = -x - 1 \text{ is a}$$

slant asymptote.

(d) $\dfrac{dx}{dt} = \dfrac{(1 + t^3)(3) - 3t(3t^2)}{(1 + t^3)^2} = \dfrac{3 - 6t^3}{(1 + t^3)^2}$ and from part (b) we have $\dfrac{dy}{dt} = \dfrac{6t - 3t^4}{(1 + t^3)^2}$. So $\dfrac{dy}{dx} = \dfrac{dy/dt}{dx/dt} = \dfrac{t(2 - t^3)}{1 - 2t^3}$.

Also $\dfrac{d^2y}{dx^2} = \dfrac{\dfrac{d}{dt}\left(\dfrac{dy}{dx}\right)}{dx/dt} = \dfrac{2(1 + t^3)^4}{3(1 - 2t^3)^3} > 0 \;\Leftrightarrow\; t < \dfrac{1}{\sqrt[3]{2}}$.

So the curve is concave upward there and has a minimum point at $(0, 0)$

and a maximum point at $\left(\sqrt[3]{2},\, \sqrt[3]{4}\right)$. Using this together with the

information from parts (a), (b), and (c), we sketch the curve.

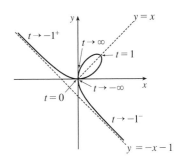

(e) $x^3 + y^3 = \left(\dfrac{3t}{1 + t^3}\right)^3 + \left(\dfrac{3t^2}{1 + t^3}\right)^3 = \dfrac{27t^3 + 27t^6}{(1 + t^3)^3} = \dfrac{27t^3(1 + t^3)}{(1 + t^3)^3} = \dfrac{27t^3}{(1 + t^3)^2}$ and

$3xy = 3\left(\dfrac{3t}{1 + t^3}\right)\left(\dfrac{3t^2}{1 + t^3}\right) = \dfrac{27t^3}{(1 + t^3)^2}$, so $x^3 + y^3 = 3xy$.

(f) We start with the equation from part (e) and substitute $x = r\cos\theta$, $y = r\sin\theta$. Then $x^3 + y^3 = 3xy$ $\Rightarrow$

$r^3 \cos^3\theta + r^3 \sin^3\theta = 3r^2 \cos\theta \sin\theta$. For $r \neq 0$, this gives $r = \dfrac{3\cos\theta\sin\theta}{\cos^3\theta + \sin^3\theta}$. Dividing numerator and denominator

by $\cos^3\theta$, we obtain $r = \dfrac{3\left(\dfrac{1}{\cos\theta}\right)\dfrac{\sin\theta}{\cos\theta}}{1 + \dfrac{\sin^3\theta}{\cos^3\theta}} = \dfrac{3\sec\theta\tan\theta}{1 + \tan^3\theta}$.

(g) The loop corresponds to $\theta \in \left(0, \frac{\pi}{2}\right)$, so its area is

$$A = \int_0^{\pi/2} \frac{r^2}{2} \, d\theta = \frac{1}{2} \int_0^{\pi/2} \left(\frac{3 \sec \theta \tan \theta}{1 + \tan^3 \theta}\right)^2 d\theta = \frac{9}{2} \int_0^{\pi/2} \frac{\sec^2 \theta \tan^2 \theta}{(1 + \tan^3 \theta)^2} \, d\theta = \frac{9}{2} \int_0^{\infty} \frac{u^2 \, du}{(1 + u^3)^2} \quad [\text{let } u = \tan \theta]$$

$$= \lim_{b \to \infty} \frac{9}{2} \left[-\frac{1}{3}(1 + u^3)^{-1}\right]_0^b = \frac{3}{2}$$

(h) By symmetry, the area between the folium and the line $y = -x - 1$ is equal to the enclosed area in the third quadrant, plus twice the enclosed area in the fourth quadrant. The area in the third quadrant is $\frac{1}{2}$, and since $y = -x - 1 \implies$

$r \sin \theta = -r \cos \theta - 1 \implies r = -\dfrac{1}{\sin \theta + \cos \theta}$, the area in the fourth quadrant is

$\dfrac{1}{2} \displaystyle\int_{-\pi/2}^{-\pi/4} \left[\left(-\dfrac{1}{\sin \theta + \cos \theta}\right)^2 - \left(\dfrac{3 \sec \theta \tan \theta}{1 + \tan^3 \theta}\right)^2\right] d\theta \stackrel{\text{CAS}}{=} \dfrac{1}{2}$. Therefore, the total area is $\frac{1}{2} + 2\left(\frac{1}{2}\right) = \frac{3}{2}$.

6. (a) Since the smaller circle rolls without slipping around C, the amount of arc traversed on C ($2r\theta$ in the figure) must equal the amount of arc of the smaller circle that has been in contact with C. Since the smaller circle has radius r, it must have turned through an angle of $2r\theta/r = 2\theta$. In addition to turning through an angle 2θ, the little circle has rolled through an angle θ against C. Thus, P has turned through an angle of 3θ as shown in the figure. (If the little circle had turned through an angle of 2θ with its center pinned to the x-axis,

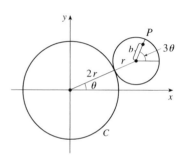

then P would have turned only 2θ instead of 3θ. The movement of the little circle around C adds θ to the angle.) From the figure, we see that the center of the small circle has coordinates $(3r \cos \theta, 3r \sin \theta)$. Thus, P has coordinates (x, y), where $x = b \cos 3\theta + 3r \cos \theta$ and $y = b \sin 3\theta + 3r \sin \theta$.

(b)

$b = \frac{1}{5}r$

$b = \frac{2}{5}r$

$b = \frac{3}{5}r$

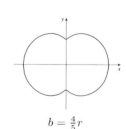

$b = \frac{4}{5}r$

(c) The diagram gives an alternate description of point P on the epitrochoid. Q moves around a circle of radius b, and P rotates one-third as fast with respect to Q at a distance of $3r$. Place an equilateral triangle with sides of length $3\sqrt{3}r$ so that its centroid is at Q and

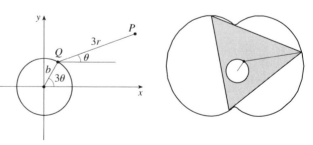

one vertex is at P. (The distance from the centroid to a vertex is $\frac{1}{\sqrt{3}}$ times the length of a side of the equilateral triangle.)

As θ increases by $\frac{2\pi}{3}$, the point Q travels once around the circle of radius b, returning to its original position. At the same time, P (and the rest of the triangle) rotate through an angle of $\frac{2\pi}{3}$ about Q, so P's position is occupied by another vertex. In this way, we see that the epitrochoid traced out by P is simultaneously traced out by the other two vertices as well. The whole equilateral triangle sits inside the epitrochoid (touching it only with its vertices) and each vertex traces out the curve once while the centroid moves around the circle three times.

(d) We view the epitrochoid as being traced out in the same way as in part (c), by a rotor for which the distance from its center to each vertex is $3r$, so it has radius $6r$. To show that the rotor fits inside the epitrochoid, it suffices to show that for any position of the tracing point P, there are no points on the opposite side of the rotor which are outside the epitrochoid. But the most likely case of intersection is when P is on the y-axis, so as long as the diameter of the rotor $\left(\text{which is } 3\sqrt{3}\,r\right)$ is less than the distance between the y-intercepts, the rotor will fit. The y-intercepts occur when $\theta = \frac{\pi}{2}$ or $\theta = \frac{3\pi}{2}$ $\Rightarrow$ $y = -b + 3r$ or $y = b - 3r$, so the distance between the intercepts is $(-b + 3r) - (b - 3r) = 6r - 2b$, and the rotor will fit if $3\sqrt{3}\,r \leq 6r - 2b$ $\Leftrightarrow$ $2b \leq 6r - 3\sqrt{3}\,r$ $\Leftrightarrow$ $b \leq \frac{3}{2}\left(2 - \sqrt{3}\right)r$.

12 □ INFINITE SEQUENCES AND SERIES

12.1 Sequences

1. (a) A sequence is an ordered list of numbers. It can also be defined as a function whose domain is the set of positive integers.

   (b) The terms a_n approach 8 as n becomes large. In fact, we can make a_n as close to 8 as we like by taking n sufficiently large.

   (c) The terms a_n become large as n becomes large. In fact, we can make a_n as large as we like by taking n sufficiently large.

2. (a) From Definition 1, a convergent sequence is a sequence for which $\lim\limits_{n \to \infty} a_n$ exists. Examples: $\{1/n\}$, $\{1/2^n\}$

   (b) A divergent sequence is a sequence for which $\lim\limits_{n \to \infty} a_n$ *does not* exist. Examples: $\{n\}$, $\{\sin n\}$

3. $a_n = 1 - (0.2)^n$, so the sequence is $\{0.8, 0.96, 0.992, 0.9984, 0.99968, \ldots\}$.

4. $a_n = \dfrac{n+1}{3n-1}$, so the sequence is $\left\{ \dfrac{2}{2}, \dfrac{3}{5}, \dfrac{4}{8}, \dfrac{5}{11}, \dfrac{6}{14}, \ldots \right\} = \left\{ 1, \dfrac{3}{5}, \dfrac{1}{2}, \dfrac{5}{11}, \dfrac{3}{7}, \ldots \right\}$.

5. $a_n = \dfrac{3(-1)^n}{n!}$, so the sequence is $\left\{ \dfrac{-3}{1}, \dfrac{3}{2}, \dfrac{-3}{6}, \dfrac{3}{24}, \dfrac{-3}{120}, \ldots \right\} = \left\{ -3, \dfrac{3}{2}, -\dfrac{1}{2}, \dfrac{1}{8}, -\dfrac{1}{40}, \ldots \right\}$.

6. $a_n = 2 \cdot 4 \cdot 6 \cdots \cdot (2n)$, so the sequence is $\{2, 2 \cdot 4, 2 \cdot 4 \cdot 6, 2 \cdot 4 \cdot 6 \cdot 8, 2 \cdot 4 \cdot 6 \cdot 8 \cdot 10, \ldots\} = \{2, 8, 48, 384, 3840, \ldots\}$.

7. $a_1 = 3$, $a_{n+1} = 2a_n - 1$. Each term is defined in terms of the preceding term.

   $a_2 = 2a_1 - 1 = 2(3) - 1 = 5$. $a_3 = 2a_2 - 1 = 2(5) - 1 = 9$. $a_4 = 2a_3 - 1 = 2(9) - 1 = 17$.

   $a_5 = 2a_4 - 1 = 2(17) - 1 = 33$. The sequence is $\{3, 5, 9, 17, 33, \ldots\}$.

8. $a_1 = 4$, $a_{n+1} = \dfrac{a_n}{a_n - 1}$. Each term is defined in terms of the preceding term.

   $a_2 = \dfrac{a_1}{a_1 - 1} = \dfrac{4}{4 - 1} = \dfrac{4}{3}$. $a_3 = \dfrac{a_2}{a_2 - 1} = \dfrac{4/3}{\frac{4}{3} - 1} = \dfrac{4/3}{1/3} = 4$. Since $a_3 = a_1$, we can see that the terms of the sequence

   will alternately equal 4 and $4/3$, so the sequence is $\left\{ 4, \dfrac{4}{3}, 4, \dfrac{4}{3}, 4, \ldots \right\}$.

9. $\left\{ 1, \dfrac{1}{3}, \dfrac{1}{5}, \dfrac{1}{7}, \dfrac{1}{9}, \ldots \right\}$. The denominator of the nth term is the nth positive odd integer, so $a_n = \dfrac{1}{2n-1}$.

10. $\left\{ 1, \dfrac{1}{3}, \dfrac{1}{9}, \dfrac{1}{27}, \dfrac{1}{81}, \ldots \right\}$. The denominator of the nth term is the $(n-1)$st power of 3, so $a_n = \dfrac{1}{3^{n-1}}$.

11. $\{2, 7, 12, 17, \ldots\}$. Each term is larger than the preceding one by 5, so $a_n = a_1 + d(n-1) = 2 + 5(n-1) = 5n - 3$.

12. $\left\{ -\dfrac{1}{4}, \dfrac{2}{9}, -\dfrac{3}{16}, \dfrac{4}{25}, \ldots \right\}$. The numerator of the nth term is n and its denominator is $(n+1)^2$. Including the alternating signs,

    we get $a_n = (-1)^n \dfrac{n}{(n+1)^2}$.

13. $\left\{ 1, -\dfrac{2}{3}, \dfrac{4}{9}, -\dfrac{8}{27}, \ldots \right\}$. Each term is $-\dfrac{2}{3}$ times the preceding one, so $a_n = \left(-\dfrac{2}{3} \right)^{n-1}$.

14. $\{5, 1, 5, 1, 5, 1, \ldots\}$. The average of 5 and 1 is 3, so we can think of the sequence as alternately adding 2 and -2 to 3.

Thus, $a_n = 3 + (-1)^{n+1} \cdot 2$.

15. The first six terms of $a_n = \dfrac{n}{2n+1}$ are $\dfrac{1}{3}, \dfrac{2}{5}, \dfrac{3}{7}, \dfrac{4}{9}, \dfrac{5}{11}, \dfrac{6}{13}$. It appears that the sequence is approaching $\dfrac{1}{2}$.

$$\lim_{n \to \infty} \frac{n}{2n+1} = \lim_{n \to \infty} \frac{1}{2 + 1/n} = \frac{1}{2}$$

16. $\{\cos(n\pi/3)\}_{n=1}^9 = \{\frac{1}{2}, -\frac{1}{2}, -1, -\frac{1}{2}, \frac{1}{2}, 1, \frac{1}{2}, -\frac{1}{2}, -1\}$. The sequence does not appear to have a limit. The values will cycle through the first six numbers in the sequence—never approaching a particular number.

17. $a_n = 1 - (0.2)^n$, so $\lim\limits_{n \to \infty} a_n = 1 - 0 = 1$ by (9).   Converges

18. $a_n = \dfrac{n^3}{n^3 + 1} = \dfrac{n^3/n^3}{(n^3 + 1)/n^3} = \dfrac{1}{1 + 1/n^3}$, so $a_n \to \dfrac{1}{1+0} = 1$ as $n \to \infty$.   Converges

19. $a_n = \dfrac{3 + 5n^2}{n + n^2} = \dfrac{(3 + 5n^2)/n^2}{(n + n^2)/n^2} = \dfrac{5 + 3/n^2}{1 + 1/n}$, so $a_n \to \dfrac{5+0}{1+0} = 5$ as $n \to \infty$.   Converges

20. $a_n = \dfrac{n^3}{n+1} = \dfrac{n^3/n}{(n+1)/n} = \dfrac{n^2}{1 + 1/n^2}$, so $a_n \to \infty$ as $n \to \infty$ since $\lim\limits_{n \to \infty} n^2 = \infty$ and $\lim\limits_{n \to \infty} (1 + 1/n^2) = 1$.   Diverges

21. Because the natural exponential function is continuous at 0, Theorem 7 enables us to write

$$\lim_{n \to \infty} a_n = \lim_{n \to \infty} e^{1/n} = e^{\lim_{n \to \infty}(1/n)} = e^0 = 1.   \text{Converges}$$

22. $a_n = \dfrac{3^{n+2}}{5^n} = \dfrac{3^2 3^n}{5^n} = 9\left(\dfrac{3}{5}\right)^n$, so $\lim\limits_{n \to \infty} a_n = 9 \lim\limits_{n \to \infty} \left(\dfrac{3}{5}\right)^n = 9 \cdot 0 = 0$ by (9) with $r = \dfrac{3}{5}$.   Converges

23. If $b_n = \dfrac{2n\pi}{1 + 8n}$, then $\lim\limits_{n \to \infty} b_n = \lim\limits_{n \to \infty} \dfrac{(2n\pi)/n}{(1 + 8n)/n} = \lim\limits_{n \to \infty} \dfrac{2\pi}{1/n + 8} = \dfrac{2\pi}{8} = \dfrac{\pi}{4}$. Since tan is continuous at $\frac{\pi}{4}$, by

Theorem 7, $\lim\limits_{n \to \infty} \tan\left(\dfrac{2n\pi}{1 + 8n}\right) = \tan\left(\lim\limits_{n \to \infty} \dfrac{2n\pi}{1 + 8n}\right) = \tan\dfrac{\pi}{4} = 1.$ Converges

24. Using the last limit law for sequences and the continuity of the square root function,

$$\lim_{n \to \infty} a_n = \lim_{n \to \infty} \sqrt{\frac{n+1}{9n+1}} = \sqrt{\lim_{n \to \infty} \frac{n+1}{9n+1}} = \sqrt{\lim_{n \to \infty} \frac{1 + 1/n}{9 + 1/n}} = \sqrt{\frac{1}{9}} = \frac{1}{3}.   \text{Converges}$$

25. $a_n = \dfrac{(-1)^{n-1} n}{n^2 + 1} = \dfrac{(-1)^{n-1}}{n + 1/n}$, so $0 \le |a_n| = \dfrac{1}{n + 1/n} \le \dfrac{1}{n} \to 0$ as $n \to \infty$, so $a_n \to 0$ by the Squeeze Theorem and

Theorem 6.   Converges

26. $a_n = \dfrac{(-1)^n n^3}{n^3 + 2n^2 + 1}$. Now $|a_n| = \dfrac{n^3}{n^3 + 2n^2 + 1} = \dfrac{1}{1 + \frac{2}{n} + \frac{1}{n^3}} \to 1$ as $n \to \infty$, but the terms of the sequence $\{a_n\}$

alternate in sign, so the sequence $a_1, a_3, a_5, \ldots$ converges to -1 and the sequence $a_2, a_4, a_6, \ldots$ converges to $+1$.

This shows that the given sequence diverges since its terms don't approach a single real number.

27. $a_n = \cos(n/2)$.   This sequence diverges since the terms don't approach any particular real number as $n \to \infty$. The terms take on values between -1 and 1.

28. $a_n = \cos(2/n)$.   As $n \to \infty$, $2/n \to 0$, so $\cos(2/n) \to \cos 0 = 1$ because cos is continuous.   Converges

29. $a_n = \dfrac{(2n-1)!}{(2n+1)!} = \dfrac{(2n-1)!}{(2n+1)(2n)(2n-1)!} = \dfrac{1}{(2n+1)(2n)} \to 0$ as $n \to \infty$.   Converges

30. $2n \to \infty$ as $n \to \infty$, so since $\lim\limits_{x \to \infty} \arctan x = \frac{\pi}{2}$, we have $\lim\limits_{n \to \infty} \arctan 2n = \frac{\pi}{2}$.   Converges

31. $a_n = \dfrac{e^n + e^{-n}}{e^{2n} - 1} \cdot \dfrac{e^{-n}}{e^{-n}} = \dfrac{1 + e^{-2n}}{e^n - e^{-n}} \to 0$ as $n \to \infty$ because $1 + e^{-2n} \to 1$ and $e^n - e^{-n} \to \infty$.   Converges

32. $a_n = \dfrac{\ln n}{\ln 2n} = \dfrac{\ln n}{\ln 2 + \ln n} = \dfrac{1}{\frac{\ln 2}{\ln n} + 1} \to \dfrac{1}{0 + 1} = 1$ as $n \to \infty$.   Converges

33. $a_n = n^2 e^{-n} = \dfrac{n^2}{e^n}$. Since $\lim\limits_{x \to \infty} \dfrac{x^2}{e^x} \overset{\text{H}}{=} \lim\limits_{x \to \infty} \dfrac{2x}{e^x} \overset{\text{H}}{=} \lim\limits_{x \to \infty} \dfrac{2}{e^x} = 0$, it follows from Theorem 3 that $\lim\limits_{n \to \infty} a_n = 0$.   Converges

34. $a_n = n \cos n\pi = n(-1)^n$.   Since $|a_n| = n \to \infty$ as $n \to \infty$, the given sequence diverges.

35. $0 \le \dfrac{\cos^2 n}{2^n} \le \dfrac{1}{2^n}$ [since $0 \le \cos^2 n \le 1$],   so since $\lim\limits_{n \to \infty} \dfrac{1}{2^n} = 0$, $\left\{ \dfrac{\cos^2 n}{2^n} \right\}$ converges to 0 by the Squeeze Theorem.

36. $a_n = \ln(n+1) - \ln n = \ln\left(\dfrac{n+1}{n}\right) = \ln\left(1 + \dfrac{1}{n}\right) \to \ln(1) = 0$ as $n \to \infty$ because ln is continuous.   Converges

37. $a_n = n \sin(1/n) = \dfrac{\sin(1/n)}{1/n}$.   Since $\lim\limits_{x \to \infty} \dfrac{\sin(1/x)}{1/x} = \lim\limits_{t \to 0^+} \dfrac{\sin t}{t}$ [where $t = 1/x$] $= 1$, it follows from Theorem 3 that $\{a_n\}$ converges to 1.

38. $a_n = \sqrt[n]{2^{1+3n}} = (2^{1+3n})^{1/n} = (2^1 2^{3n})^{1/n} = 2^{1/n} 2^3 = 8 \cdot 2^{1/n}$, so

$\lim\limits_{n \to \infty} a_n = 8 \lim\limits_{n \to \infty} 2^{1/n} = 8 \cdot 2^{\lim_{n \to \infty}(1/n)} = 8 \cdot 2^0 = 8$ by Theorem 7, since the function $f(x) = 2^x$ is continuous at 0.

Convergent

39. $y = \left(1 + \dfrac{2}{x}\right)^x \quad \Rightarrow \quad \ln y = x \ln\left(1 + \dfrac{2}{x}\right)$, so

$\lim\limits_{x \to \infty} \ln y = \lim\limits_{x \to \infty} \dfrac{\ln(1 + 2/x)}{1/x} \overset{\text{H}}{=} \lim\limits_{x \to \infty} \dfrac{\left(\dfrac{1}{1+2/x}\right)\left(-\dfrac{2}{x^2}\right)}{-1/x^2} = \lim\limits_{x \to \infty} \dfrac{2}{1 + 2/x} = 2 \quad \Rightarrow$

$\lim\limits_{x \to \infty} \left(1 + \dfrac{2}{x}\right)^x = \lim\limits_{x \to \infty} e^{\ln y} = e^2$, so by Theorem 3, $\lim\limits_{n \to \infty} \left(1 + \dfrac{2}{n}\right)^n = e^2$.   Convergent

40. $a_n = \dfrac{\sin 2n}{1 + \sqrt{n}}$. $|a_n| \le \dfrac{1}{1 + \sqrt{n}}$ and $\lim\limits_{n \to \infty} \dfrac{1}{1 + \sqrt{n}} = 0$, so $\dfrac{-1}{1 + \sqrt{n}} \le a_n \le \dfrac{1}{1 + \sqrt{n}} \quad \Rightarrow \quad \lim\limits_{n \to \infty} a_n = 0$ by the Squeeze Theorem.   Converges

41. $a_n = \ln(2n^2 + 1) - \ln(n^2 + 1) = \ln\left(\dfrac{2n^2 + 1}{n^2 + 1}\right) = \ln\left(\dfrac{2 + 1/n^2}{1 + 1/n^2}\right) \to \ln 2$ as $n \to \infty$.   Convergent

42. $\lim\limits_{x\to\infty}\dfrac{(\ln x)^2}{x}\overset{\text{H}}{=}\lim\limits_{x\to\infty}\dfrac{2(\ln x)(1/x)}{1}=2\lim\limits_{x\to\infty}\dfrac{\ln x}{x}\overset{\text{H}}{=}2\lim\limits_{x\to\infty}\dfrac{1/x}{1}=0$, so by Theorem 3, $\lim\limits_{n\to\infty}\dfrac{(\ln n)^2}{n}=0$. Convergent

43. $\{0,1,0,0,1,0,0,0,1,\ldots\}$ diverges since the sequence takes on only two values, 0 and 1, and never stays arbitrarily close to either one (or any other value) for n sufficiently large.

44. $\left\{\frac{1}{1},\frac{1}{3},\frac{1}{2},\frac{1}{4},\frac{1}{3},\frac{1}{5},\frac{1}{4},\frac{1}{6},\ldots\right\}$. $a_{2n-1}=\dfrac{1}{n}$ and $a_{2n}=\dfrac{1}{n+2}$ for all positive integers n. $\lim\limits_{n\to\infty}a_n=0$ since

$\lim\limits_{n\to\infty}a_{2n-1}=\lim\limits_{n\to\infty}\dfrac{1}{n}=0$ and $\lim\limits_{n\to\infty}a_{2n}=\lim\limits_{n\to\infty}\dfrac{1}{n+2}=0$. For n sufficiently large, a_n can be made as close to 0

as we like.   Converges

45. $a_n=\dfrac{n!}{2^n}=\dfrac{1}{2}\cdot\dfrac{2}{2}\cdot\dfrac{3}{2}\cdot\ \cdots\ \cdot\dfrac{(n-1)}{2}\cdot\dfrac{n}{2}\geq\dfrac{1}{2}\cdot\dfrac{n}{2}$ [for $n>1$] $=\dfrac{n}{4}\to\infty$ as $n\to\infty$, so $\{a_n\}$ diverges.

46. $0<|a_n|=\dfrac{3^n}{n!}=\dfrac{3}{1}\cdot\dfrac{3}{2}\cdot\dfrac{3}{3}\cdot\ \cdots\ \cdot\dfrac{3}{(n-1)}\cdot\dfrac{3}{n}\leq\dfrac{3}{1}\cdot\dfrac{3}{2}\cdot\dfrac{3}{n}$ [for $n>2$] $=\dfrac{27}{2n}\to0$ as $n\to\infty$, so by the Squeeze

Theorem and Theorem 6, $\{(-3)^n/n!\}$ converges to 0.

47.

From the graph, it appears that the sequence converges to 1.

$\{(-2/e)^n\}$ converges to 0 by (9), and hence $\{1+(-2/e)^n\}$

converges to $1+0=1$.

48.

From the graph, it appears that the sequence converges to a number

greater than 3.

$$\lim\limits_{n\to\infty}a_n=\lim\limits_{n\to\infty}\sqrt{n}\sin\left(\dfrac{\pi}{\sqrt{n}}\right)=\lim\limits_{n\to\infty}\dfrac{\sin\left(\pi/\sqrt{n}\right)}{\pi/\sqrt{n}}\cdot\pi$$

$$=\lim\limits_{x\to0^+}\dfrac{\sin x}{x}\cdot\pi\quad\left[x=\pi/\sqrt{n}\right]\ =1\cdot\pi=\pi.$$

49.

From the graph, it appears that the sequence converges to $\frac{1}{2}$.

As $n\to\infty$,

$$a_n=\sqrt{\dfrac{3+2n^2}{8n^2+n}}=\sqrt{\dfrac{3/n^2+2}{8+1/n}}\quad\Rightarrow\quad\sqrt{\dfrac{0+2}{8+0}}=\sqrt{\dfrac{1}{4}}=\dfrac{1}{2},$$

so $\lim\limits_{n\to\infty}a_n=\frac{1}{2}$.

50.

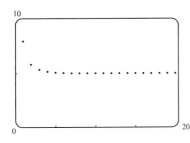

From the graph, it appears that the sequence converges to 5.

$$5 = \sqrt[n]{5^n} \le \sqrt[n]{3^n + 5^n} \le \sqrt[n]{5^n + 5^n} = \sqrt[n]{2}\,\sqrt[n]{5^n}$$

$$= \sqrt[n]{2} \cdot 5 \to 5 \text{ as } n \to \infty \quad \left[\lim_{n \to \infty} 2^{1/n} = 2^0 = 1 \right]$$

Hence, $a_n \to 5$ by the Squeeze Theorem.

Alternate solution: Let $y = (3^x + 5^x)^{1/x}$. Then

$$\lim_{x \to \infty} \ln y = \lim_{x \to \infty} \frac{\ln (3^x + 5^x)}{x} \overset{\text{H}}{=} \lim_{x \to \infty} \frac{3^x \ln 3 + 5^x \ln 5}{3^x + 5^x} = \lim_{x \to \infty} \frac{\left(\frac{3}{5}\right)^x \ln 3 + \ln 5}{\left(\frac{3}{5}\right)^x + 1} = \ln 5,$$

so $\lim_{x \to \infty} y = e^{\ln 5} = 5$, and so $\left\{ \sqrt[n]{3^n + 5^n} \right\}$ converges to 5.

51.

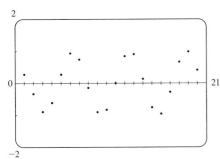

From the graph, it appears that the sequence $\{a_n\} = \left\{ \dfrac{n^2 \cos n}{1 + n^2} \right\}$ is

divergent, since it oscillates between 1 and -1 (approximately). To

prove this, suppose that $\{a_n\}$ converges to L. If $b_n = \dfrac{n^2}{1 + n^2}$, then

$\{b_n\}$ converges to 1, and $\lim_{n \to \infty} \dfrac{a_n}{b_n} = \dfrac{L}{1} = L$. But $\dfrac{a_n}{b_n} = \cos n$, so

$\lim_{n \to \infty} \dfrac{a_n}{b_n}$ does not exist. This contradiction shows that $\{a_n\}$ diverges.

52.

From the graphs, it seems that the sequence diverges. $a_n = \dfrac{1 \cdot 3 \cdot 5 \cdot \cdots \cdot (2n - 1)}{n!}$. We first prove by induction that

$a_n \ge \left(\dfrac{3}{2}\right)^{n-1}$ for all n. This is clearly true for $n = 1$, so let $P(n)$ be the statement that the above is true for n. We must

show it is then true for $n + 1$. $a_{n+1} = a_n \cdot \dfrac{2n + 1}{n + 1} \ge \left(\dfrac{3}{2}\right)^{n-1} \cdot \dfrac{2n + 1}{n + 1}$ (induction hypothesis). But $\dfrac{2n + 1}{n + 1} \ge \dfrac{3}{2}$

[since $2(2n + 1) \ge 3(n + 1) \iff 4n + 2 \ge 3n + 3 \iff n \ge 1$], and so we get that $a_{n+1} \ge \left(\dfrac{3}{2}\right)^{n-1} \cdot \dfrac{3}{2} = \left(\dfrac{3}{2}\right)^n$ which

is $P(n + 1)$. Thus, we have proved our first assertion, so since $\left\{ \left(\dfrac{3}{2}\right)^{n-1} \right\}$ diverges [by (9)], so does the given sequence $\{a_n\}$.

53.

From the graph, it appears that the sequence approaches 0.

$$0 < a_n = \frac{1 \cdot 3 \cdot 5 \cdot \cdots \cdot (2n-1)}{(2n)^n} = \frac{1}{2n} \cdot \frac{3}{2n} \cdot \frac{5}{2n} \cdot \cdots \cdot \frac{2n-1}{2n}$$

$$\leq \frac{1}{2n} \cdot (1) \cdot (1) \cdot \cdots \cdot (1) = \frac{1}{2n} \to 0 \text{ as } n \to \infty$$

So by the Squeeze Theorem, $\left\{ \dfrac{1 \cdot 3 \cdot 5 \cdot \cdots \cdot (2n-1)}{(2n)^n} \right\}$ converges to 0.

54. (a) $a_1 = 1$, $a_{n+1} = 4 - a_n$ for $n \geq 1$. $a_1 = 1$, $a_2 = 4 - a_1 = 4 - 1 = 3$, $a_3 = 4 - a_2 = 4 - 3 = 1$,

$a_4 = 4 - a_3 = 4 - 1 = 3$, $a_5 = 4 - a_4 = 4 - 3 = 1$. Since the terms of the sequence alternate between 1 and 3,

the sequence is divergent.

(b) $a_1 = 2$, $a_2 = 4 - a_1 = 4 - 2 = 2$, $a_3 = 4 - a_2 = 4 - 2 = 2$. Since all of the terms are 2, $\lim\limits_{n\to\infty} a_n = 2$ and hence, the

sequence is convergent.

55. (a) $a_n = 1000(1.06)^n \Rightarrow a_1 = 1060$, $a_2 = 1123.60$, $a_3 = 1191.02$, $a_4 = 1262.48$, and $a_5 = 1338.23$.

(b) $\lim\limits_{n\to\infty} a_n = 1000 \lim\limits_{n\to\infty} (1.06)^n$, so the sequence diverges by (9) with $r = 1.06 > 1$.

56. $a_{n+1} = \begin{cases} \frac{1}{2}a_n & \text{if } a_n \text{ is an even number} \\ 3a_n + 1 & \text{if } a_n \text{ is an odd number} \end{cases}$ When $a_1 = 11$, the first 40 terms are 11, 34, 17, 52, 26, 13, 40, 20, 10, 5,

16, 8, 4, 2, 1, 4, 2, 1, 4, 2, 1, 4, 2, 1, 4, 2, 1, 4, 2, 1, 4, 2, 1, 4, 2, 1, 4, 2, 1, 4. When $a_1 = 25$, the first 40 terms are 25, 76, 38,

19, 58, 29, 88, 44, 22, 11, 34, 17, 52, 26, 13, 40, 20, 10, 5, 16, 8, 4, 2, 1, 4, 2, 1, 4, 2, 1, 4, 2, 1, 4, 2, 1, 4, 2, 1, 4.

The famous Collatz conjecture is that this sequence always reaches 1, regardless of the starting point a_1.

57. If $|r| \geq 1$, then $\{r^n\}$ diverges by (9), so $\{nr^n\}$ diverges also, since $|nr^n| = n \, |r^n| \geq |r^n|$. If $|r| < 1$ then

$$\lim\limits_{x\to\infty} xr^x = \lim\limits_{x\to\infty} \frac{x}{r^{-x}} \overset{\text{H}}{=} \lim\limits_{x\to\infty} \frac{1}{(-\ln r)\,r^{-x}} = \lim\limits_{x\to\infty} \frac{r^x}{-\ln r} = 0, \text{ so } \lim\limits_{n\to\infty} nr^n = 0, \text{ and hence } \{nr^n\} \text{ converges}$$

whenever $|r| < 1$.

58. (a) Let $\lim\limits_{n\to\infty} a_n = L$. By Definition 2, this means that for every $\varepsilon > 0$ there is an integer N such that $|a_n - L| < \varepsilon$

whenever $n > N$. Thus, $|a_{n+1} - L| < \varepsilon$ whenever $n + 1 > N \Leftrightarrow n > N - 1$. It follows that $\lim\limits_{n\to\infty} a_{n+1} = L$ and so

$\lim\limits_{n\to\infty} a_n = \lim\limits_{n\to\infty} a_{n+1}$.

(b) If $L = \lim\limits_{n\to\infty} a_n$ then $\lim\limits_{n\to\infty} a_{n+1} = L$ also, so L must satisfy $L = 1/(1 + L) \Rightarrow L^2 + L - 1 = 0 \Rightarrow L = \frac{-1 + \sqrt{5}}{2}$

(since L has to be nonnegative if it exists).

59. Since $\{a_n\}$ is a decreasing sequence, $a_n > a_{n+1}$ for all $n \geq 1$. Because all of its terms lie between 5 and 8, $\{a_n\}$ is a

bounded sequence. By the Monotonic Sequence Theorem, $\{a_n\}$ is convergent; that is, $\{a_n\}$ has a limit L. L must be less than

8 since $\{a_n\}$ is decreasing, so $5 \leq L < 8$.

60. The terms of $a_n = (-2)^{n+1}$ alternate in sign, so the sequence is not monotonic. The first five terms are $4, -8, 16, -32$, and 64. Since $\lim_{n \to \infty} |a_n| = \lim_{n \to \infty} 2^{n+1} = \infty$, the sequence is not bounded.

61. $a_n = \dfrac{1}{2n + 3}$ is decreasing since $a_{n+1} = \dfrac{1}{2(n + 1) + 3} = \dfrac{1}{2n + 5} < \dfrac{1}{2n + 3} = a_n$ for each $n \geq 1$. The sequence is bounded since $0 < a_n \leq \frac{1}{5}$ for all $n \geq 1$. Note that $a_1 = \frac{1}{5}$.

62. $a_n = \dfrac{2n - 3}{3n + 4}$ defines an increasing sequence since for $f(x) = \dfrac{2x - 3}{3x + 4}$,

$f'(x) = \dfrac{(3x + 4)(2) - (2x - 3)(3)}{(3x + 4)^2} = \dfrac{17}{(3x + 4)^2} > 0$. The sequence is bounded since $a_n \geq a_1 = -\frac{1}{7}$ for $n \geq 1$,

and $a_n < \dfrac{2n - 3}{3n} < \dfrac{2n}{3n} = \dfrac{2}{3}$ for $n \geq 1$.

63. The terms of $a_n = n(-1)^n$ alternate in sign, so the sequence is not monotonic. The first five terms are $-1, 2, -3, 4$, and -5. Since $\lim_{n \to \infty} |a_n| = \lim_{n \to \infty} n = \infty$, the sequence is not bounded.

64. $a_n = ne^{-n}$ defines a positive decreasing sequence since the function $f(x) = xe^{-x}$ is decreasing for $x > 1$.
$[f'(x) = e^{-x} - xe^{-x} = e^{-x}(1 - x) < 0$ for $x > 1.]$ The sequence is bounded above by $a_1 = \frac{1}{e}$ and below by 0.

65. $a_n = \dfrac{n}{n^2 + 1}$ defines a decreasing sequence since for $f(x) = \dfrac{x}{x^2 + 1}$, $f'(x) = \dfrac{(x^2 + 1)(1) - x(2x)}{(x^2 + 1)^2} = \dfrac{1 - x^2}{(x^2 + 1)^2} \leq 0$
for $x \geq 1$. The sequence is bounded since $0 < a_n \leq \frac{1}{2}$ for all $n \geq 1$.

66. $a_n = n + \dfrac{1}{n}$ defines an increasing sequence since the function $g(x) = x + \dfrac{1}{x}$ is increasing for $x > 1$. $[g'(x) = 1 - 1/x^2 > 0$
for $x > 1.]$ The sequence is unbounded since $a_n \to \infty$ as $n \to \infty$. (It is, however, bounded below by $a_1 = 2$.)

67. For $\left\{ \sqrt{2}, \sqrt{2\sqrt{2}}, \sqrt{2\sqrt{2\sqrt{2}}}, \ldots \right\}$, $a_1 = 2^{1/2}$, $a_2 = 2^{3/4}$, $a_3 = 2^{7/8}, \ldots$, so $a_n = 2^{(2^n - 1)/2^n} = 2^{1 - (1/2^n)}$.

$\lim_{n \to \infty} a_n = \lim_{n \to \infty} 2^{1 - (1/2^n)} = 2^1 = 2$.

Alternate solution: Let $L = \lim_{n \to \infty} a_n$. (We could show the limit exists by showing that $\{a_n\}$ is bounded and increasing.)

Then L must satisfy $L = \sqrt{2 \cdot L} \Rightarrow L^2 = 2L \Rightarrow L(L - 2) = 0$. $L \neq 0$ since the sequence increases, so $L = 2$.

68. (a) Let P_n be the statement that $a_{n+1} \geq a_n$ and $a_n \leq 3$. P_1 is obviously true. We will assume that P_n is true and

then show that as a consequence P_{n+1} must also be true. $a_{n+2} \geq a_{n+1} \Leftrightarrow \sqrt{2 + a_{n+1}} \geq \sqrt{2 + a_n} \Leftrightarrow$

$2 + a_{n+1} \geq 2 + a_n \Leftrightarrow a_{n+1} \geq a_n$, which is the induction hypothesis. $a_{n+1} \leq 3 \Leftrightarrow \sqrt{2 + a_n} \leq 3 \Leftrightarrow$

$2 + a_n \leq 9 \Leftrightarrow a_n \leq 7$, which is certainly true because we are assuming that $a_n \leq 3$. So P_n is true for all n, and so

$a_1 \leq a_n \leq 3$ (showing that the sequence is bounded), and hence by the Monotonic Sequence Theorem, $\lim_{n \to \infty} a_n$ exists.

(b) If $L = \lim_{n \to \infty} a_n$, then $\lim_{n \to \infty} a_{n+1} = L$ also, so $L = \sqrt{2 + L} \Rightarrow L^2 = 2 + L \Leftrightarrow L^2 - L - 2 = 0 \Leftrightarrow$

$(L + 1)(L - 2) = 0 \Leftrightarrow L = 2$ [since L can't be negative].

69. $a_1 = 1$, $a_{n+1} = 3 - \dfrac{1}{a_n}$.   We show by induction that $\{a_n\}$ is increasing and bounded above by 3. Let P_n be the proposition

that $a_{n+1} > a_n$ and $0 < a_n < 3$. Clearly P_1 is true. Assume that P_n is true. Then $a_{n+1} > a_n$ $\Rightarrow$ $\dfrac{1}{a_{n+1}} < \dfrac{1}{a_n}$ $\Rightarrow$

$-\dfrac{1}{a_{n+1}} > -\dfrac{1}{a_n}$. Now $a_{n+2} = 3 - \dfrac{1}{a_{n+1}} > 3 - \dfrac{1}{a_n} = a_{n+1}$ $\Leftrightarrow$ P_{n+1}. This proves that $\{a_n\}$ is increasing and bounded

above by 3, so $1 = a_1 < a_n < 3$, that is, $\{a_n\}$ is bounded, and hence convergent by the Monotonic Sequence Theorem.

If $L = \lim\limits_{n \to \infty} a_n$, then $\lim\limits_{n \to \infty} a_{n+1} = L$ also, so L must satisfy $L = 3 - 1/L$ $\Rightarrow$ $L^2 - 3L + 1 = 0$ $\Rightarrow$ $L = \frac{3 \pm \sqrt{5}}{2}$.

But $L > 1$, so $L = \frac{3 + \sqrt{5}}{2}$.

70. $a_1 = 2$, $a_{n+1} = \dfrac{1}{3 - a_n}$.   We use induction. Let P_n be the statement that $0 < a_{n+1} \leq a_n \leq 2$. Clearly P_1 is true, since

$a_2 = 1/(3 - 2) = 1$. Now assume that P_n is true. Then $a_{n+1} \leq a_n$ $\Rightarrow$ $-a_{n+1} \geq -a_n$ $\Rightarrow$ $3 - a_{n+1} \geq 3 - a_n$ $\Rightarrow$

$a_{n+2} = \dfrac{1}{3 - a_{n+1}} \leq \dfrac{1}{3 - a_n} = a_{n+1}$. Also $a_{n+2} > 0$ [since $3 - a_{n+1}$ is positive] and $a_{n+1} \leq 2$ by the induction

hypothesis, so P_{n+1} is true. To find the limit, we use the fact that $\lim\limits_{n \to \infty} a_n = \lim\limits_{n \to \infty} a_{n+1}$ $\Rightarrow$ $L = \frac{1}{3 - L}$ $\Rightarrow$

$L^2 - 3L + 1 = 0$ $\Rightarrow$ $L = \frac{3 \pm \sqrt{5}}{2}$. But $L \leq 2$, so we must have $L = \frac{3 - \sqrt{5}}{2}$.

71. (a) Let a_n be the number of rabbit pairs in the nth month. Clearly $a_1 = 1 = a_2$. In the nth month, each pair that is

2 or more months old (that is, a_{n-2} pairs) will produce a new pair to add to the a_{n-1} pairs already present. Thus,

$a_n = a_{n-1} + a_{n-2}$, so that $\{a_n\} = \{f_n\}$, the Fibonacci sequence.

(b) $a_n = \dfrac{f_{n+1}}{f_n}$ $\Rightarrow$ $a_{n-1} = \dfrac{f_n}{f_{n-1}} = \dfrac{f_{n-1} + f_{n-2}}{f_{n-1}} = 1 + \dfrac{f_{n-2}}{f_{n-1}} = 1 + \dfrac{1}{f_{n-1}/f_{n-2}} = 1 + \dfrac{1}{a_{n-2}}$. If $L = \lim\limits_{n \to \infty} a_n$,

then $L = \lim\limits_{n \to \infty} a_{n-1}$ and $L = \lim\limits_{n \to \infty} a_{n-2}$, so L must satisfy $L = 1 + \dfrac{1}{L}$ $\Rightarrow$ $L^2 - L - 1 = 0$ $\Rightarrow$ $L = \frac{1 + \sqrt{5}}{2}$

[since L must be positive].

72. (a) If f is continuous, then $f(L) = f\left(\lim\limits_{n \to \infty} a_n\right) = \lim\limits_{n \to \infty} f(a_n) = \lim\limits_{n \to \infty} a_{n+1} = \lim\limits_{n \to \infty} a_n = L$ by Exercise 58(a).

(b) By repeatedly pressing the cosine key on the calculator (that is, taking cosine of the previous answer) until the displayed

value stabilizes, we see that $L \approx 0.73909$.

73. (a)

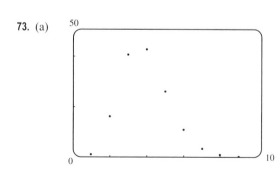

From the graph, it appears that the sequence $\left\{\dfrac{n^5}{n!}\right\}$

converges to 0, that is, $\lim\limits_{n \to \infty} \dfrac{n^5}{n!} = 0$.

(b)

From the first graph, it seems that the smallest possible value of N corresponding to $\varepsilon = 0.1$ is 9, since $n^5/n! < 0.1$ whenever $n \geq 10$, but $9^5/9! > 0.1$. From the second graph, it seems that for $\varepsilon = 0.001$, the smallest possible value for N is 11 since $n^5/n! < 0.001$ whenever $n \geq 12$.

74. Let $\varepsilon > 0$ and let N be any positive integer larger than $\ln(\varepsilon)/\ln|r|$. If $n > N$, then $n > \ln(\varepsilon)/\ln|r| \Rightarrow n\ln|r| < \ln\varepsilon$ [since $|r| < 1 \Rightarrow \ln|r| < 0$] $\Rightarrow \ln(|r|^n) < \ln\varepsilon \Rightarrow |r|^n < \varepsilon \Rightarrow |r^n - 0| < \varepsilon$, and so by Definition 2, $\lim\limits_{n\to\infty} r^n = 0$.

75. Theorem 6: If $\lim\limits_{n\to\infty} |a_n| = 0$ then $\lim\limits_{n\to\infty} -|a_n| = 0$, and since $-|a_n| \leq a_n \leq |a_n|$, we have that $\lim\limits_{n\to\infty} a_n = 0$ by the Squeeze Theorem.

76. Theorem 7: If $\lim\limits_{n\to\infty} a_n = L$ and the function f is continuous at L, then $\lim\limits_{n\to\infty} f(a_n) = f(L)$.

Proof: We must show that, given a number $\varepsilon > 0$, there is an integer N such that $|f(a_n) - f(L)| < \varepsilon$ whenever $n > N$. Suppose $\varepsilon > 0$. Since f is continuous at L, there is a number $\delta > 0$ such that $|f(x) - f(L)| < \varepsilon$ if $|x - L| < \delta$. Since $\lim\limits_{n\to\infty} a_n = L$, there is an integer N such that $|a_n - L| < \delta$ if $n > N$. Suppose $n > N$. Then $0 < |a_n - L| < \delta$, so $|f(a_n) - f(L)| < \varepsilon$.

77. To Prove: If $\lim\limits_{n\to\infty} a_n = 0$ and $\{b_n\}$ is bounded, then $\lim\limits_{n\to\infty} (a_n b_n) = 0$.

Proof: Since $\{b_n\}$ is bounded, there is a positive number M such that $|b_n| \leq M$ and hence, $|a_n|\,|b_n| \leq |a_n|\,M$ for all $n \geq 1$. Let $\varepsilon > 0$ be given. Since $\lim\limits_{n\to\infty} a_n = 0$, there is an integer N such that $|a_n - 0| < \dfrac{\varepsilon}{M}$ if $n > N$. Then

$$|a_n b_n - 0| = |a_n b_n| = |a_n|\,|b_n| \leq |a_n|\,M = |a_n - 0|\,M < \frac{\varepsilon}{M} \cdot M = \varepsilon \text{ for all } n > N. \text{ Since } \varepsilon \text{ was arbitrary,}$$

$\lim\limits_{n\to\infty} (a_n b_n) = 0$.

78. (a) $\dfrac{b^{n+1} - a^{n+1}}{b - a} = b^n + b^{n-1}a + b^{n-2}a^2 + b^{n-3}a^3 + \cdots + ba^{n-1} + a^n$

$< b^n + b^{n-1}b + b^{n-2}b^2 + b^{n-3}b^3 + \cdots + bb^{n-1} + b^n = (n+1)b^n$

(b) Since $b - a > 0$, we have $b^{n+1} - a^{n+1} < (n+1)b^n(b - a) \Rightarrow b^{n+1} - (n+1)b^n(b - a) < a^{n+1} \Rightarrow b^n[(n+1)a - nb] < a^{n+1}$.

(c) With this substitution, $(n+1)a - nb = 1$, and so $b^n = \left(1 + \dfrac{1}{n}\right)^n < a^{n+1} = \left(1 + \dfrac{1}{n+1}\right)^{n+1}$.

(d) With this substitution, we get $\left(1 + \dfrac{1}{2n}\right)^n \left(\dfrac{1}{2}\right) < 1 \;\Rightarrow\; \left(1 + \dfrac{1}{2n}\right)^n < 2 \;\Rightarrow\; \left(1 + \dfrac{1}{2n}\right)^{2n} < 4.$

(e) $a_n < a_{2n}$ since $\{a_n\}$ is increasing, so $a_n < a_{2n} < 4.$

(f) Since $\{a_n\}$ is increasing and bounded above by 4, $a_1 \le a_n \le 4$, and so $\{a_n\}$ is bounded and monotonic, and hence has a limit by the Monotonic Sequence Theorem.

79. (a) First we show that $a > a_1 > b_1 > b$.

$a_1 - b_1 = \dfrac{a+b}{2} - \sqrt{ab} = \tfrac{1}{2}\left(a - 2\sqrt{ab} + b\right) = \tfrac{1}{2}\left(\sqrt{a} - \sqrt{b}\right)^2 > 0 \quad \text{[since } a > b\text{]} \;\Rightarrow\; a_1 > b_1.$ Also

$a - a_1 = a - \tfrac{1}{2}(a + b) = \tfrac{1}{2}(a - b) > 0$ and $b - b_1 = b - \sqrt{ab} = \sqrt{b}\left(\sqrt{b} - \sqrt{a}\right) < 0$, so $a > a_1 > b_1 > b$. In the same

way we can show that $a_1 > a_2 > b_2 > b_1$ and so the given assertion is true for $n = 1$. Suppose it is true for $n = k$, that is,

$a_k > a_{k+1} > b_{k+1} > b_k$. Then

$a_{k+2} - b_{k+2} = \tfrac{1}{2}(a_{k+1} + b_{k+1}) - \sqrt{a_{k+1}b_{k+1}} = \tfrac{1}{2}\left(a_{k+1} - 2\sqrt{a_{k+1}b_{k+1}} + b_{k+1}\right) = \tfrac{1}{2}\left(\sqrt{a_{k+1}} - \sqrt{b_{k+1}}\right)^2 > 0,$

$a_{k+1} - a_{k+2} = a_{k+1} - \tfrac{1}{2}(a_{k+1} + b_{k+1}) = \tfrac{1}{2}(a_{k+1} - b_{k+1}) > 0$, and

$b_{k+1} - b_{k+2} = b_{k+1} - \sqrt{a_{k+1}b_{k+1}} = \sqrt{b_{k+1}}\left(\sqrt{b_{k+1}} - \sqrt{a_{k+1}}\right) < 0 \;\Rightarrow\; a_{k+1} > a_{k+2} > b_{k+2} > b_{k+1},$

so the assertion is true for $n = k + 1$. Thus, it is true for all n by mathematical induction.

(b) From part (a) we have $a > a_n > a_{n+1} > b_{n+1} > b_n > b$, which shows that both sequences, $\{a_n\}$ and $\{b_n\}$, are monotonic and bounded. So they are both convergent by the Monotonic Sequence Theorem.

(c) Let $\lim\limits_{n\to\infty} a_n = \alpha$ and $\lim\limits_{n\to\infty} b_n = \beta$. Then $\lim\limits_{n\to\infty} a_{n+1} = \lim\limits_{n\to\infty} \dfrac{a_n + b_n}{2} \;\Rightarrow\; \alpha = \dfrac{\alpha + \beta}{2} \;\Rightarrow$

$2\alpha = \alpha + \beta \;\Rightarrow\; \alpha = \beta.$

80. (a) Let $\varepsilon > 0$. Since $\lim\limits_{n\to\infty} a_{2n} = L$, there exists N_1 such that $|a_{2n} - L| < \varepsilon$ for $n > N_1$. Since $\lim\limits_{n\to\infty} a_{2n+1} = L$, there exists N_2 such that $|a_{2n+1} - L| < \varepsilon$ for $n > N_2$. Let $N = \max\{2N_1, 2N_2 + 1\}$ and let $n > N$. If n is even, then $n = 2m$ where $m > N_1$, so $|a_n - L| = |a_{2m} - L| < \varepsilon$. If n is odd, then $n = 2m + 1$, where $m > N_2$, so $|a_n - L| = |a_{2m+1} - L| < \varepsilon$. Therefore $\lim\limits_{n\to\infty} a_n = L$.

(b) $a_1 = 1$, $a_2 = 1 + \dfrac{1}{1+1} = \dfrac{3}{2} = 1.5$, $a_3 = 1 + \dfrac{1}{5/2} = \dfrac{7}{5} = 1.4$, $a_4 = 1 + \dfrac{1}{12/5} = \dfrac{17}{12} = 1.41\overline{6}$,

$a_5 = 1 + \dfrac{1}{29/12} = \dfrac{41}{29} \approx 1.413793$, $a_6 = 1 + \dfrac{1}{70/29} = \dfrac{99}{70} \approx 1.414286$, $a_7 = 1 + \dfrac{1}{169/70} = \dfrac{239}{169} \approx 1.414201$,

$a_8 = 1 + \dfrac{1}{408/169} = \dfrac{577}{408} \approx 1.414216$. Notice that $a_1 < a_3 < a_5 < a_7$ and $a_2 > a_4 > a_6 > a_8$. It appears that the odd terms are increasing and the even terms are decreasing. Let's prove that $a_{2n-2} > a_{2n}$ and $a_{2n-1} < a_{2n+1}$ by

mathematical induction. Suppose that $a_{2k-2} > a_{2k}$. Then $1 + a_{2k-2} > 1 + a_{2k} \;\Rightarrow\; \dfrac{1}{1 + a_{2k-2}} < \dfrac{1}{1 + a_{2k}} \;\Rightarrow$

$1 + \dfrac{1}{1 + a_{2k-2}} < 1 + \dfrac{1}{1 + a_{2k}} \;\Rightarrow\; a_{2k-1} < a_{2k+1} \;\Rightarrow\; 1 + a_{2k-1} < 1 + a_{2k+1} \;\Rightarrow$

$\dfrac{1}{1 + a_{2k-1}} > \dfrac{1}{1 + a_{2k+1}} \;\Rightarrow\; 1 + \dfrac{1}{1 + a_{2k-1}} > 1 + \dfrac{1}{1 + a_{2k+1}} \;\Rightarrow\; a_{2k} > a_{2k+2}.$ We have thus shown, by induction, that the odd terms are increasing and the even terms are decreasing. Also all terms lie between 1 and 2, so both $\{a_n\}$ and $\{b_n\}$ are bounded monotonic sequences and are therefore convergent by the Monotonic Sequence Theorem. Let $\lim\limits_{n\to\infty} a_{2n} = L$. Then $\lim\limits_{n\to\infty} a_{2n+2} = L$ also. We have

$$a_{n+2} = 1 + \frac{1}{1 + 1 + 1/(1 + a_n)} = 1 + \frac{1}{(3 + 2a_n)/(1 + a_n)} = \frac{4 + 3a_n}{3 + 2a_n}$$

so $a_{2n+2} = \dfrac{4 + 3a_{2n}}{3 + 2a_{2n}}$. Taking limits of both sides, we get $L = \dfrac{4 + 3L}{3 + 2L}$ $\Rightarrow$ $3L + 2L^2 = 4 + 3L$ $\Rightarrow$ $L^2 = 2$ $\Rightarrow$

$L = \sqrt{2}$ [since $L > 0$]. Thus, $\lim\limits_{n\to\infty} a_{2n} = \sqrt{2}$. Similarly we find that $\lim\limits_{n\to\infty} a_{2n+1} = \sqrt{2}$. So, by part (a),

$\lim\limits_{n\to\infty} a_n = \sqrt{2}$.

81. (a) Suppose $\{p_n\}$ converges to p. Then $p_{n+1} = \dfrac{bp_n}{a + p_n}$ $\Rightarrow$ $\lim\limits_{n\to\infty} p_{n+1} = \dfrac{b \lim\limits_{n\to\infty} p_n}{a + \lim\limits_{n\to\infty} p_n}$ $\Rightarrow$ $p = \dfrac{bp}{a + p}$ $\Rightarrow$

$p^2 + ap = bp$ $\Rightarrow$ $p(p + a - b) = 0$ $\Rightarrow$ $p = 0$ or $p = b - a$.

(b) $p_{n+1} = \dfrac{bp_n}{a + p_n} = \dfrac{\left(\dfrac{b}{a}\right)p_n}{1 + \dfrac{p_n}{a}} < \left(\dfrac{b}{a}\right)p_n$ since $1 + \dfrac{p_n}{a} > 1$.

(c) By part (b), $p_1 < \left(\dfrac{b}{a}\right)p_0$, $p_2 < \left(\dfrac{b}{a}\right)p_1 < \left(\dfrac{b}{a}\right)^2 p_0$, $p_3 < \left(\dfrac{b}{a}\right)p_2 < \left(\dfrac{b}{a}\right)^3 p_0$, etc. In general, $p_n < \left(\dfrac{b}{a}\right)^n p_0$,

so $\lim\limits_{n\to\infty} p_n \le \lim\limits_{n\to\infty} \left(\dfrac{b}{a}\right)^n \cdot p_0 = 0$ since $b < a$. $\left[\text{By result 9, } \lim\limits_{n\to\infty} r^n = 0 \text{ if } -1 < r < 1. \text{ Here } r = \dfrac{b}{a} \in (0, 1).\right]$

(d) Let $a < b$. We first show, by induction, that if $p_0 < b - a$, then $p_n < b - a$ and $p_{n+1} > p_n$.

For $n = 0$, we have $p_1 - p_0 = \dfrac{bp_0}{a + p_0} - p_0 = \dfrac{p_0(b - a - p_0)}{a + p_0} > 0$ since $p_0 < b - a$. So $p_1 > p_0$.

Now we suppose the assertion is true for $n = k$, that is, $p_k < b - a$ and $p_{k+1} > p_k$. Then

$b - a - p_{k+1} = b - a - \dfrac{bp_k}{a + p_k} = \dfrac{a(b - a) + bp_k - ap_k - bp_k}{a + p_k} = \dfrac{a(b - a - p_k)}{a + p_k} > 0$ because $p_k < b - a$. So

$p_{k+1} < b - a$. And $p_{k+2} - p_{k+1} = \dfrac{bp_{k+1}}{a + p_{k+1}} - p_{k+1} = \dfrac{p_{k+1}(b - a - p_{k+1})}{a + p_{k+1}} > 0$ since $p_{k+1} < b - a$. Therefore,

$p_{k+2} > p_{k+1}$. Thus, the assertion is true for $n = k + 1$. It is therefore true for all n by mathematical induction.

A similar proof by induction shows that if $p_0 > b - a$, then $p_n > b - a$ and $\{p_n\}$ is decreasing.

In either case the sequence $\{p_n\}$ is bounded and monotonic, so it is convergent by the Monotonic Sequence Theorem. It then follows from part (a) that $\lim\limits_{n\to\infty} p_n = b - a$.

LABORATORY PROJECT Logistic Sequences

1. To write such a program in Maple it is best to calculate all the points first and then graph them. One possible sequence of commands [taking $p_0 = \frac{1}{2}$ and $k = 1.5$ for the difference equation] is

```
t:='t';p(0):=1/2;k:=1.5;
for j from 1 to 20  do p(j):=k*p(j-1)*(1-p(j-1)) od;
plot([seq([t,p(t)] t=0..20)],t=0..20,p=0..0.5,style=point);
```

In Mathematica, we can use the following program:

```
p[0]=1/2
k=1.5
p[j_]:=k*p[j-1]*(1-p[j-1])
P=Table[p[t],{t,20}]
ListPlot[P]
```

With $p_0 = \frac{1}{2}$ and $k = 1.5$:

n	p_n	n	p_n	n	p_n
0	0.5	7	0.3338465076	14	0.3333373303
1	0.375	8	0.3335895255	15	0.3333353318
2	0.3515625	9	0.3334613309	16	0.3333343326
3	0.3419494629	10	0.3333973076	17	0.3333338329
4	0.3375300416	11	0.3333653143	18	0.3333335831
5	0.3354052689	12	0.3333493223	19	0.3333334582
6	0.3343628617	13	0.3333413274	20	0.3333333958

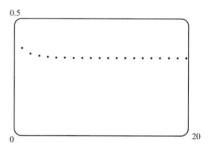

With $p_0 = \frac{1}{2}$ and $k = 2.5$:

n	p_n	n	p_n	n	p_n
0	0.5	7	0.6004164790	14	0.5999967417
1	0.625	8	0.5997913269	15	0.6000016291
2	0.5859375	9	0.6001042277	16	0.5999991854
3	0.6065368651	10	0.5999478590	17	0.6000004073
4	0.5966247409	11	0.6000260637	18	0.5999997964
5	0.6016591486	12	0.5999869664	19	0.6000001018
6	0.5991635437	13	0.6000065164	20	0.5999999491

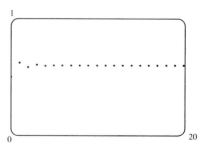

Both of these sequences seem to converge (the first to about $\frac{1}{3}$, the second to about 0.60).

With $p_0 = \frac{7}{8}$ and $k = 1.5$:

n	p_n	n	p_n	n	p_n
0	0.875	7	0.3239166554	14	0.3332554829
1	0.1640625	8	0.3284919837	15	0.3332943990
2	0.2057189941	9	0.3308775005	16	0.3333138639
3	0.2450980344	10	0.3320963702	17	0.3333235980
4	0.2775374819	11	0.3327125567	18	0.3333284655
5	0.3007656421	12	0.3330223670	19	0.3333308994
6	0.3154585059	13	0.3331777051	20	0.3333321164

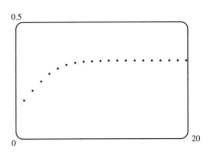

With $p_0 = \frac{7}{8}$ and $k = 2.5$:

n	p_n	n	p_n	n	p_n
0	0.875	7	0.6016572368	14	0.5999869815
1	0.2734375	8	0.5991645155	15	0.6000065088
2	0.4966735840	9	0.6004159972	16	0.5999967455
3	0.6249723374	10	0.5997915688	17	0.6000016272
4	0.5859547872	11	0.6001041070	18	0.5999991864
5	0.6065294364	12	0.5999479194	19	0.6000004068
6	0.5966286980	13	0.6000260335	20	0.5999997966

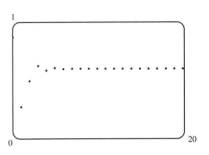

The limit of the sequence seems to depend on k, but not on p_0.

2. With $p_0 = \frac{7}{8}$ and $k = 3.2$:

n	p_n	n	p_n	n	p_n
0	0.875	7	0.5830728495	14	0.7990633827
1	0.35	8	0.7779164854	15	0.5137954979
2	0.728	9	0.5528397669	16	0.7993909896
3	0.6336512	10	0.7910654689	17	0.5131681132
4	0.7428395416	11	0.5288988570	18	0.7994451225
5	0.6112926626	12	0.7973275394	19	0.5130643795
6	0.7603646184	13	0.5171082698	20	0.7994538304

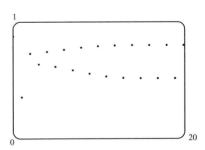

It seems that eventually the terms fluctuate between two values (about 0.5 and 0.8 in this case).

3. With $p_0 = \frac{7}{8}$ and $k = 3.42$:

n	p_n	n	p_n	n	p_n
0	0.875	7	0.4523028596	14	0.8442074951
1	0.3740625	8	0.8472194412	15	0.4498025048
2	0.8007579316	9	0.4426802161	16	0.8463823232
3	0.5456427596	10	0.8437633929	17	0.4446659586
4	0.8478752457	11	0.4508474156	18	0.8445284520
5	0.4411212220	12	0.8467373602	19	0.4490464985
6	0.8431438501	13	0.4438243545	20	0.8461207931

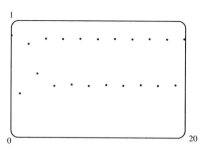

With $p_0 = \frac{7}{8}$ and $k = 3.45$:

n	p_n	n	p_n	n	p_n
0	0.875	7	0.4670259170	14	0.8403376122
1	0.37734375	8	0.8587488490	15	0.4628875685
2	0.8105962830	9	0.4184824586	16	0.8577482026
3	0.5296783241	10	0.8395743720	17	0.4209559716
4	0.8594612299	11	0.4646778983	18	0.8409445432
5	0.4167173034	12	0.8581956045	19	0.4614610237
6	0.8385707740	13	0.4198508858	20	0.8573758782

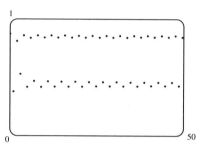

From the graphs above, it seems that for k between 3.4 and 3.5, the terms eventually fluctuate between four values. In the graph below, the pattern followed by the terms is $0.395, 0.832, 0.487, 0.869, 0.395, \ldots$. Note that even for $k = 3.42$ (as in the first graph), there are four distinct "branches"; even after 1000 terms, the first and third terms in the pattern differ by about 2×10^{-9}, while the first and fifth terms differ by only 2×10^{-10}. With $p_0 = \frac{7}{8}$ and $k = 3.48$:

4.

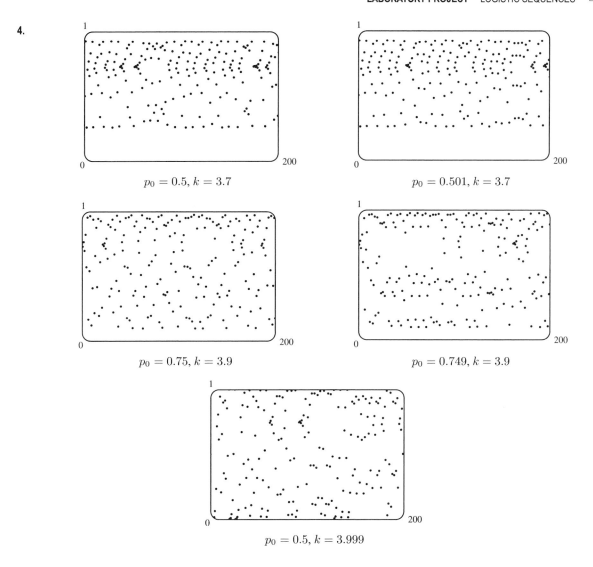

$p_0 = 0.5, k = 3.7$

$p_0 = 0.501, k = 3.7$

$p_0 = 0.75, k = 3.9$

$p_0 = 0.749, k = 3.9$

$p_0 = 0.5, k = 3.999$

From the graphs, it seems that if p_0 is changed by 0.001, the whole graph changes completely. (Note, however, that this might be partially due to accumulated round-off error in the CAS. These graphs were generated by Maple with 100-digit accuracy, and different degrees of accuracy give different graphs.) There seem to be some some fleeting patterns in these graphs, but on the whole they are certainly very chaotic. As k increases, the graph spreads out vertically, with more extreme values close to 0 or 1.

12.2 Series

1. (a) A sequence is an ordered list of numbers whereas a series is the *sum* of a list of numbers.

(b) A series is convergent if the sequence of partial sums is a convergent sequence. A series is divergent if it is not convergent.

2. $\sum_{n=1}^{\infty} a_n = 5$ means that by adding sufficiently many terms of the series we can get as close as we like to the number 5.

In other words, it means that $\lim_{n \to \infty} s_n = 5$, where s_n is the nth partial sum, that is, $\sum_{i=1}^{n} a_i$.

3.

n	s_n
1	-2.40000
2	-1.92000
3	-2.01600
4	-1.99680
5	-2.00064
6	-1.99987
7	-2.00003
8	-1.99999
9	-2.00000
10	-2.00000

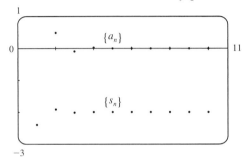

From the graph and the table, it seems that the series converges to -2. In fact, it is a geometric

series with $a = -2.4$ and $r = -\frac{1}{5}$, so its sum is $\sum_{n=1}^{\infty} \dfrac{12}{(-5)^n} = \dfrac{-2.4}{1 - \left(-\frac{1}{5}\right)} = \dfrac{-2.4}{1.2} = -2$.

Note that the dot corresponding to $n = 1$ is part of both $\{a_n\}$ and $\{s_n\}$.

TI-86 Note: To graph $\{a_n\}$ and $\{s_n\}$, set your calculator to Param mode and DrawDot mode. (DrawDot is under

GRAPH, MORE, FORMT (F3).) Now under E(t) = make the assignments: `xt1=t`, `yt1=12/(-5)^t`, `xt2=t`,

`yt2=sum seq(yt1,t,1,t,1)`. (sum and seq are under LIST, OPS (F5), MORE.) Under WIND use

`1,10,1,0,10,1,-3,1,1` to obtain a graph similar to the one above. Then use TRACE (F4) to see the values.

4.

n	s_n
1	0.50000
2	1.90000
3	3.60000
4	5.42353
5	7.30814
6	9.22706
7	11.16706
8	13.12091
9	15.08432
10	17.05462

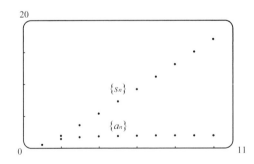

The series $\sum_{n=1}^{\infty} \dfrac{2n^2 - 1}{n^2 + 1}$ diverges, since its terms do not approach 0.

5.

n	s_n
1	1.55741
2	−0.62763
3	−0.77018
4	0.38764
5	−2.99287
6	−3.28388
7	−2.41243
8	−9.21214
9	−9.66446
10	−9.01610

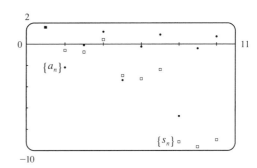

The series $\sum\limits_{n=1}^{\infty} \tan n$ diverges, since its terms do not approach 0.

6.

n	s_n
1	1.00000
2	1.60000
3	1.96000
4	2.17600
5	2.30560
6	2.38336
7	2.43002
8	2.45801
9	2.47481
10	2.48488

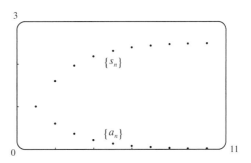

From the graph and the table, it seems that the series converges to 2.5. In fact, it is a geometric

series with $a = 1$ and $r = 0.6$, so its sum is $\sum\limits_{n=1}^{\infty} (0.6)^{n-1} = \dfrac{1}{1 - 0.6} = \dfrac{1}{2/5} = 2.5$.

7.

n	s_n
1	0.29289
2	0.42265
3	0.50000
4	0.55279
5	0.59175
6	0.62204
7	0.64645
8	0.66667
9	0.68377
10	0.69849

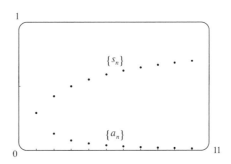

From the graph and the table, it seems that the series converges.

$$\sum_{n=1}^{k} \left(\frac{1}{\sqrt{n}} - \frac{1}{\sqrt{n+1}} \right) = \left(\frac{1}{\sqrt{1}} - \frac{1}{\sqrt{2}} \right) + \left(\frac{1}{\sqrt{2}} - \frac{1}{\sqrt{3}} \right) + \cdots + \left(\frac{1}{\sqrt{k}} - \frac{1}{\sqrt{k+1}} \right)$$

$$= 1 - \frac{1}{\sqrt{k+1}},$$

so $\sum\limits_{n=1}^{\infty} \left(\dfrac{1}{\sqrt{n}} - \dfrac{1}{\sqrt{n+1}} \right) = \lim\limits_{k \to \infty} \left(1 - \dfrac{1}{\sqrt{k+1}} \right) = 1.$

8.

n	s_n
2	0.12500
3	0.19167
4	0.23333
5	0.26190
6	0.28274
7	0.29861
8	0.31111
9	0.32121
10	0.32955
11	0.33654

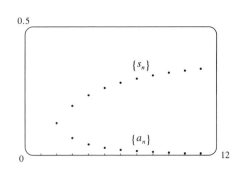

From the graph and the table, it seems that the series converges.

$$\frac{1}{n(n+2)} = \frac{1/2}{n} - \frac{1/2}{n+2} = \frac{1}{2}\left(\frac{1}{n} - \frac{1}{n+2}\right), \text{ so}$$

$$\sum_{n=2}^{k}\frac{1}{n(n+2)} = \frac{1}{2}\left(\frac{1}{2}-\frac{1}{4}\right) + \frac{1}{2}\left(\frac{1}{3}-\frac{1}{5}\right) + \frac{1}{2}\left(\frac{1}{4}-\frac{1}{6}\right) + \cdots + \frac{1}{2}\left(\frac{1}{k}-\frac{1}{k+2}\right)$$

$$= \frac{1}{2}\left(\frac{1}{2}+\frac{1}{3}-\frac{1}{k+1}-\frac{1}{k+2}\right).$$

As $k \to \infty$, this sum approaches $\frac{1}{2}\left(\frac{5}{6}-0\right) = \frac{5}{12}$.

9. (a) $\lim\limits_{n\to\infty} a_n = \lim\limits_{n\to\infty}\frac{2n}{3n+1} = \frac{2}{3}$, so the *sequence* $\{a_n\}$ is convergent by (12.1.1).

(b) Since $\lim\limits_{n\to\infty} a_n = \frac{2}{3} \neq 0$, the *series* $\sum\limits_{n=1}^{\infty} a_n$ is divergent by the Test for Divergence.

10. (a) Both $\sum\limits_{i=1}^{n} a_i$ and $\sum\limits_{j=1}^{n} a_j$ represent the sum of the first n terms of the sequence $\{a_n\}$, that is, the nth partial sum.

(b) $\sum\limits_{i=1}^{n} a_j = \underbrace{a_j + a_j + \cdots + a_j}_{n \text{ terms}} = na_j$, which, in general, is not the same as $\sum\limits_{i=1}^{n} a_i = a_1 + a_2 + \cdots + a_n$.

11. $3 + 2 + \frac{4}{3} + \frac{8}{9} + \cdots$ is a geometric series with first term $a = 3$ and common ratio $r = \frac{2}{3}$. Since $|r| = \frac{2}{3} < 1$, the series converges to $\frac{a}{1-r} = \frac{3}{1-2/3} = \frac{3}{1/3} = 9$.

12. $\frac{1}{8} - \frac{1}{4} + \frac{1}{2} - 1 + \cdots$ is a geometric series with ratio $r = -2$. Since $|r| = 2 > 1$, the series diverges.

13. $3 - 4 + \frac{16}{3} - \frac{64}{9} + \cdots$ is a geometric series with ratio $r = -\frac{4}{3}$. Since $|r| = \frac{4}{3} > 1$, the series diverges.

14. $1 + 0.4 + 0.16 + 0.064 + \cdots$ is a geometric series with ratio $r = 0.4 = \frac{2}{5}$. Since $|r| = \frac{2}{5} < 1$, the series converges to

$$\frac{a}{1-r} = \frac{1}{1-2/5} = \frac{5}{3}.$$

15. $\sum\limits_{n=1}^{\infty} 6(0.9)^{n-1}$ is a geometric series with first term $a = 6$ and ratio $r = 0.9$. Since $|r| = 0.9 < 1$, the series converges to

$$\frac{a}{1-r} = \frac{6}{1-0.9} = \frac{6}{0.1} = 60.$$

16. $\sum_{n=1}^{\infty} \frac{10^n}{(-9)^{n-1}} = \sum_{n=1}^{\infty} \frac{10(10)^{n-1}}{(-9)^{n-1}} = 10 \sum_{n=1}^{\infty} \left(-\frac{10}{9}\right)^{n-1}$. The latter series is geometric with $a = 10$ and ratio $r = -\frac{10}{9}$.

Since $|r| = \frac{10}{9} > 1$, the series diverges.

17. $\sum_{n=1}^{\infty} \frac{(-3)^{n-1}}{4^n} = \frac{1}{4} \sum_{n=1}^{\infty} \left(-\frac{3}{4}\right)^{n-1}$. The latter series is geometric with $a = 1$ and ratio $r = -\frac{3}{4}$. Since $|r| = \frac{3}{4} < 1$, it

converges to $\frac{1}{1 - (-3/4)} = \frac{4}{7}$. Thus, the given series converges to $\left(\frac{1}{4}\right)\left(\frac{4}{7}\right) = \frac{1}{7}$.

18. $\sum_{n=0}^{\infty} \frac{1}{\left(\sqrt{2}\right)^n}$ is a geometric series with ratio $r = \frac{1}{\sqrt{2}}$. Since $|r| = \frac{1}{\sqrt{2}} < 1$, the series converges. Its sum is

$\frac{1}{1 - 1/\sqrt{2}} = \frac{\sqrt{2}}{\sqrt{2} - 1} = \frac{\sqrt{2}}{\sqrt{2} - 1} \cdot \frac{\sqrt{2} + 1}{\sqrt{2} + 1} = \sqrt{2}\left(\sqrt{2} + 1\right) = 2 + \sqrt{2}$.

19. $\sum_{n=0}^{\infty} \frac{\pi^n}{3^{n+1}} = \frac{1}{3} \sum_{n=0}^{\infty} \left(\frac{\pi}{3}\right)^n$ is a geometric series with ratio $r = \frac{\pi}{3}$. Since $|r| > 1$, the series diverges.

20. $\sum_{n=1}^{\infty} \frac{e^n}{3^{n-1}} = 3 \sum_{n=1}^{\infty} \left(\frac{e}{3}\right)^n$ is a geometric series with first term $3(e/3) = e$ and ratio $r = \frac{e}{3}$. Since $|r| < 1$, the series

converges. Its sum is $\frac{e}{1 - e/3} = \frac{3e}{3 - e}$.

21. $\sum_{n=1}^{\infty} \frac{1}{2n} = \frac{1}{2} \sum_{n=1}^{\infty} \frac{1}{n}$ diverges since each of its partial sums is $\frac{1}{2}$ times the corresponding partial sum of the harmonic series

$\sum_{n=1}^{\infty} \frac{1}{n}$, which diverges. $\left[\text{If } \sum_{n=1}^{\infty} \frac{1}{2n} \text{ were to converge, then } \sum_{n=1}^{\infty} \frac{1}{n} \text{ would also have to converge by Theorem 8(i).}\right]$

In general, constant multiples of divergent series are divergent.

22. $\sum_{n=1}^{\infty} \frac{n+1}{2n-3}$ diverges by the Test for Divergence since $\lim_{n \to \infty} a_n = \lim_{n \to \infty} \frac{n+1}{2n-3} = \frac{1}{2} \neq 0$.

23. $\sum_{k=2}^{\infty} \frac{k^2}{k^2 - 1}$ diverges by the Test for Divergence since $\lim_{k \to \infty} a_k = \lim_{k \to \infty} \frac{k^2}{k^2 - 1} = 1 \neq 0$.

24. $\sum_{k=1}^{\infty} \frac{k(k+2)}{(k+3)^2}$ diverges by the Test for Divergence since $\lim_{k \to \infty} a_k = \lim_{k \to \infty} \frac{k(k+2)}{(k+3)^2} = \lim_{k \to \infty} \frac{1 \cdot (1 + 2/k)}{(1 + 3/k)^2} = 1 \neq 0$.

25. Converges.

$$\sum_{n=1}^{\infty} \frac{1 + 2^n}{3^n} = \sum_{n=1}^{\infty} \left(\frac{1}{3^n} + \frac{2^n}{3^n}\right) = \sum_{n=1}^{\infty} \left[\left(\frac{1}{3}\right)^n + \left(\frac{2}{3}\right)^n\right] \qquad \text{[sum of two convergent geometric series]}$$

$$= \frac{1/3}{1 - 1/3} + \frac{2/3}{1 - 2/3} = \frac{1}{2} + 2 = \frac{5}{2}$$

26. $\sum_{n=1}^{\infty} \frac{1 + 3^n}{2^n} = \sum_{n=1}^{\infty} \left(\frac{1}{2^n} + \frac{3^n}{2^n}\right) = \sum_{n=1}^{\infty} \left[\left(\frac{1}{2}\right)^n + \left(\frac{3}{2}\right)^n\right] = \sum_{n=1}^{\infty} \left(\frac{1}{2}\right)^n + \sum_{n=1}^{\infty} \left(\frac{3}{2}\right)^n$. The first series is a convergent

geometric series ($|r| = \frac{1}{2} < 1$), but the second series is a divergent geometric series ($|r| = \frac{3}{2} \geq 1$), so the original series

is divergent.

27. $\sum\limits_{n=1}^{\infty} \sqrt[n]{2} = 2 + \sqrt{2} + \sqrt[3]{2} + \sqrt[4]{2} + \cdots$ diverges by the Test for Divergence since

$$\lim_{n\to\infty} a_n = \lim_{n\to\infty} \sqrt[n]{2} = \lim_{n\to\infty} 2^{1/n} = 2^0 = 1 \neq 0.$$

28. $\sum\limits_{n=1}^{\infty} \left[(0.8)^{n-1} - (0.3)^n\right] = \sum\limits_{n=1}^{\infty} (0.8)^{n-1} - \sum\limits_{n=1}^{\infty} (0.3)^n$ [difference of two convergent geometric series]

$$= \frac{1}{1 - 0.8} - \frac{0.3}{1 - 0.3} = 5 - \frac{3}{7} = \frac{32}{7}$$

29. $\sum\limits_{n=1}^{\infty} \ln\left(\dfrac{n^2 + 1}{2n^2 + 1}\right)$ diverges by the Test for Divergence since

$$\lim_{n\to\infty} a_n = \lim_{n\to\infty} \ln\left(\frac{n^2+1}{2n^2+1}\right) = \ln\left(\lim_{n\to\infty} \frac{n^2+1}{2n^2+1}\right) = \ln\tfrac{1}{2} \neq 0.$$

30. $\sum\limits_{k=1}^{\infty} (\cos 1)^k$ is a geometric series with ratio $r = \cos 1 \approx 0.540302$. It converges because $|r| < 1$. Its sum is

$$\frac{\cos 1}{1 - \cos 1} \approx 1.175343.$$

31. $\sum\limits_{n=1}^{\infty} \arctan n$ diverges by the Test for Divergence since $\lim\limits_{n\to\infty} a_n = \lim\limits_{n\to\infty} \arctan n = \frac{\pi}{2} \neq 0$.

32. $\sum\limits_{n=1}^{\infty} \left(\dfrac{3}{5^n} + \dfrac{2}{n}\right)$ diverges because $\sum\limits_{n=1}^{\infty} \dfrac{2}{n} = 2\sum\limits_{n=1}^{\infty} \dfrac{1}{n}$ diverges. (If it converged, then $\dfrac{1}{2} \cdot 2 \sum\limits_{n=1}^{\infty} \dfrac{1}{n}$ would also converge by

Theorem 8(i), but we know from Example 7 that the harmonic series $\sum\limits_{n=1}^{\infty} \dfrac{1}{n}$ diverges.) If the given series converges, then the

difference $\sum\limits_{n=1}^{\infty} \left(\dfrac{3}{5^n} + \dfrac{2}{n}\right) - \sum\limits_{n=1}^{\infty} \dfrac{3}{5^n}$ must converge (since $\sum\limits_{n=1}^{\infty} \dfrac{3}{5^n}$ is a convergent geometric series) and equal $\sum\limits_{n=1}^{\infty} \dfrac{2}{n}$, but we

have just seen that $\sum\limits_{n=1}^{\infty} \dfrac{2}{n}$ diverges, so the given series must also diverge.

33. $\sum\limits_{n=1}^{\infty} \dfrac{1}{e^n} = \sum\limits_{n=1}^{\infty} \left(\dfrac{1}{e}\right)^n$ is a geometric series with first term $a = \dfrac{1}{e}$ and ratio $r = \dfrac{1}{e}$. Since $|r| = \dfrac{1}{e} < 1$, the series converges

to $\dfrac{1/e}{1 - 1/e} = \dfrac{1/e}{1 - 1/e} \cdot \dfrac{e}{e} = \dfrac{1}{e - 1}$. By Example 6, $\sum\limits_{n=1}^{\infty} \dfrac{1}{n(n+1)} = 1$. Thus, by Theorem 8(ii),

$$\sum_{n=1}^{\infty} \left(\frac{1}{e^n} + \frac{1}{n(n+1)}\right) = \sum_{n=1}^{\infty} \frac{1}{e^n} + \sum_{n=1}^{\infty} \frac{1}{n(n+1)} = \frac{1}{e-1} + 1 = \frac{1}{e-1} + \frac{e-1}{e-1} = \frac{e}{e-1}.$$

34. $\sum\limits_{n=1}^{\infty} \dfrac{e^n}{n^2}$ diverges by the Test for Divergence since $\lim\limits_{n\to\infty} a_n = \lim\limits_{n\to\infty} \dfrac{e^n}{n^2} = \lim\limits_{x\to\infty} \dfrac{e^x}{x^2} \overset{\text{H}}{=} \lim\limits_{x\to\infty} \dfrac{e^x}{2x} \overset{\text{H}}{=} \lim\limits_{x\to\infty} \dfrac{e^x}{2} = \infty \neq 0$.

35. Using partial fractions, the partial sums of the series $\sum\limits_{n=2}^{\infty} \dfrac{2}{n^2 - 1}$ are

$$s_n = \sum_{i=2}^{n} \frac{2}{(i-1)(i+1)} = \sum_{i=2}^{n} \left(\frac{1}{i-1} - \frac{1}{i+1}\right)$$

$$= \left(1 - \frac{1}{3}\right) + \left(\frac{1}{2} - \frac{1}{4}\right) + \left(\frac{1}{3} - \frac{1}{5}\right) + \cdots + \left(\frac{1}{n-3} - \frac{1}{n-1}\right) + \left(\frac{1}{n-2} - \frac{1}{n}\right)$$

This sum is a telescoping series and $s_n = 1 + \frac{1}{2} - \frac{1}{n-1} - \frac{1}{n}$.

Thus, $\sum_{n=2}^{\infty} \frac{2}{n^2 - 1} = \lim_{n \to \infty} s_n = \lim_{n \to \infty} \left(1 + \frac{1}{2} - \frac{1}{n-1} - \frac{1}{n} \right) = \frac{3}{2}$.

36. For the series $\sum_{n=1}^{\infty} \frac{2}{n^2 + 4n + 3}$, $s_n = \sum_{i=1}^{n} \frac{2}{i^2 + 4i + 3} = \sum_{i=1}^{n} \left(\frac{1}{i+1} - \frac{1}{i+3} \right)$ [using partial fractions]. The latter sum is

$\left(\frac{1}{2} - \frac{1}{4} \right) + \left(\frac{1}{3} - \frac{1}{5} \right) + \left(\frac{1}{4} - \frac{1}{6} \right) + \left(\frac{1}{5} - \frac{1}{7} \right) + \cdots + \left(\frac{1}{n} - \frac{1}{n+2} \right) + \left(\frac{1}{n+1} - \frac{1}{n+3} \right) = \frac{1}{2} + \frac{1}{3} - \frac{1}{n+2} - \frac{1}{n+3}$

[telescoping series]

Thus, $\sum_{n=1}^{\infty} \frac{2}{n^2 + 4n + 3} = \lim_{n \to \infty} s_n = \lim_{n \to \infty} \left(\frac{1}{2} + \frac{1}{3} - \frac{1}{n+2} - \frac{1}{n+3} \right) = \frac{1}{2} + \frac{1}{3} = \frac{5}{6}$. Converges

37. For the series $\sum_{n=1}^{\infty} \frac{3}{n(n+3)}$, $s_n = \sum_{i=1}^{n} \frac{3}{i(i+3)} = \sum_{i=1}^{n} \left(\frac{1}{i} - \frac{1}{i+3} \right)$ [using partial fractions]. The latter sum is

$\left(1 - \frac{1}{4} \right) + \left(\frac{1}{2} - \frac{1}{5} \right) + \left(\frac{1}{3} - \frac{1}{6} \right) + \left(\frac{1}{4} - \frac{1}{7} \right) + \cdots + \left(\frac{1}{n-3} - \frac{1}{n} \right) + \left(\frac{1}{n-2} - \frac{1}{n+1} \right) + \left(\frac{1}{n-1} - \frac{1}{n+2} \right) + \left(\frac{1}{n} - \frac{1}{n+3} \right)$

$= 1 + \frac{1}{2} + \frac{1}{3} - \frac{1}{n+1} - \frac{1}{n+2} - \frac{1}{n+3}$ [telescoping series]

Thus, $\sum_{n=1}^{\infty} \frac{3}{n(n+3)} = \lim_{n \to \infty} s_n = \lim_{n \to \infty} \left(1 + \frac{1}{2} + \frac{1}{3} - \frac{1}{n+1} - \frac{1}{n+2} - \frac{1}{n+3} \right) = 1 + \frac{1}{2} + \frac{1}{3} = \frac{11}{6}$. Converges

38. For the series $\sum_{n=1}^{\infty} \ln \frac{n}{n+1}$,

$s_n = (\ln 1 - \ln 2) + (\ln 2 - \ln 3) + (\ln 3 - \ln 4) + \cdots + [\ln n - \ln(n+1)] = \ln 1 - \ln(n+1) = -\ln(n+1)$

[telescoping series]

Thus, $\lim_{n \to \infty} s_n = -\infty$, so the series is divergent.

39. For the series $\sum_{n=1}^{\infty} \left(e^{1/n} - e^{1/(n+1)} \right)$,

$s_n = \sum_{i=1}^{n} \left(e^{1/i} - e^{1/(i+1)} \right) = (e^1 - e^{1/2}) + (e^{1/2} - e^{1/3}) + \cdots + \left(e^{1/n} - e^{1/(n+1)} \right) = e - e^{1/(n+1)}$

[telescoping series]

Thus, $\sum_{n=1}^{\infty} \left(e^{1/n} - e^{1/(n+1)} \right) = \lim_{n \to \infty} s_n = \lim_{n \to \infty} \left(e - e^{1/(n+1)} \right) = e - e^0 = e - 1$.   Converges

40. For the series $\sum_{n=1}^{\infty} \left(\cos \frac{1}{n^2} - \cos \frac{1}{(n+1)^2} \right)$,

$s_n = \sum_{i=1}^{n} \left(\cos \frac{1}{i^2} - \cos \frac{1}{(i+1)^2} \right) = \left(\cos 1 - \cos \frac{1}{4} \right) + \left(\cos \frac{1}{4} - \cos \frac{1}{9} \right) + \cdots + \left(\cos \frac{1}{n^2} - \cos \frac{1}{(n+1)^2} \right)$

$= \cos 1 - \cos \frac{1}{(n+1)^2}$ [telescoping series]

Thus, $\sum_{n=1}^{\infty} \left(\cos \frac{1}{n^2} - \cos \frac{1}{(n+1)^2} \right) = \lim_{n \to \infty} s_n = \lim_{n \to \infty} \left(\cos 1 - \cos \frac{1}{(n+1)^2} \right) = \cos 1 - \cos 0 = \cos 1 - 1$.

Converges

41. $0.\overline{2} = \frac{2}{10} + \frac{2}{10^2} + \cdots$ is a geometric series with $a = \frac{2}{10}$ and $r = \frac{1}{10}$. It converges to $\frac{a}{1-r} = \frac{2/10}{1 - 1/10} = \frac{2}{9}$.

42. $0.\overline{73} = \frac{73}{10^2} + \frac{73}{10^4} + \cdots = \frac{73/10^2}{1 - 1/10^2} = \frac{73/100}{99/100} = \frac{73}{99}$

43. $3.\overline{417} = 3 + \dfrac{417}{10^3} + \dfrac{417}{10^6} + \cdots$. Now $\dfrac{417}{10^3} + \dfrac{417}{10^6} + \cdots$ is a geometric series with $a = \dfrac{417}{10^3}$ and $r = \dfrac{1}{10^3}$.

It converges to $\dfrac{a}{1-r} = \dfrac{417/10^3}{1 - 1/10^3} = \dfrac{417/10^3}{999/10^3} = \dfrac{417}{999}$. Thus, $3.\overline{417} = 3 + \dfrac{417}{999} = \dfrac{3414}{999} = \dfrac{1138}{333}$.

44. $6.2\overline{54} = 6.2 + \dfrac{54}{10^3} + \dfrac{54}{10^5} + \cdots = 6.2 + \dfrac{54/10^3}{1 - 1/10^2} = \dfrac{62}{10} + \dfrac{54}{990} = \dfrac{6192}{990} = \dfrac{344}{55}$

45. $1.53\overline{42} = 1.53 + \dfrac{42}{10^4} + \dfrac{42}{10^6} + \cdots$. Now $\dfrac{42}{10^4} + \dfrac{42}{10^6} + \cdots$ is a geometric series with $a = \dfrac{42}{10^4}$ and $r = \dfrac{1}{10^2}$.

It converges to $\dfrac{a}{1-r} = \dfrac{42/10^4}{1 - 1/10^2} = \dfrac{42/10^4}{99/10^2} = \dfrac{42}{9900}$.

Thus, $1.53\overline{42} = 1.53 + \dfrac{42}{9900} = \dfrac{153}{100} + \dfrac{42}{9900} = \dfrac{15{,}147}{9900} + \dfrac{42}{9900} = \dfrac{15{,}189}{9900}$ or $\dfrac{5063}{3300}$.

46. $7.\overline{12345} = 7 + \dfrac{12{,}345}{10^5} + \dfrac{12{,}345}{10^{10}} + \cdots$. Now $\dfrac{12{,}345}{10^5} + \dfrac{12{,}345}{10^{10}} + \cdots$ is a geometric series with $a = \dfrac{12{,}345}{10^5}$ and $r = \dfrac{1}{10^5}$.

It converges to $\dfrac{a}{1-r} = \dfrac{12{,}345/10^5}{1 - 1/10^5} = \dfrac{12{,}345/10^5}{99{,}999/10^5} = \dfrac{12{,}345}{99{,}999}$.

Thus, $7.\overline{12345} = 7 + \dfrac{12{,}345}{99{,}999} = \dfrac{699{,}993}{99{,}999} + \dfrac{12{,}345}{99{,}999} = \dfrac{712{,}338}{99{,}999}$ or $\dfrac{237{,}446}{33{,}333}$.

47. $\displaystyle\sum_{n=1}^{\infty} \dfrac{x^n}{3^n} = \sum_{n=1}^{\infty} \left(\dfrac{x}{3}\right)^n$ is a geometric series with $r = \dfrac{x}{3}$, so the series converges $\Leftrightarrow |r| < 1 \Leftrightarrow \dfrac{|x|}{3} < 1 \Leftrightarrow |x| < 3$;

that is, $-3 < x < 3$. In that case, the sum of the series is $\dfrac{a}{1-r} = \dfrac{x/3}{1 - x/3} = \dfrac{x/3}{1 - x/3} \cdot \dfrac{3}{3} = \dfrac{x}{3 - x}$.

48. $\displaystyle\sum_{n=1}^{\infty} (x - 4)^n$ is a geometric series with $r = x - 4$, so the series converges $\Leftrightarrow |r| < 1 \Leftrightarrow |x - 4| < 1 \Leftrightarrow$

$3 < x < 5$. In that case, the sum of the series is $\dfrac{x - 4}{1 - (x - 4)} = \dfrac{x - 4}{5 - x}$.

49. $\displaystyle\sum_{n=0}^{\infty} 4^n x^n = \sum_{n=0}^{\infty} (4x)^n$ is a geometric series with $r = 4x$, so the series converges $\Leftrightarrow |r| < 1 \Leftrightarrow 4|x| < 1 \Leftrightarrow$

$|x| < \frac{1}{4}$. In that case, the sum of the series is $\dfrac{1}{1 - 4x}$.

50. $\displaystyle\sum_{n=0}^{\infty} \dfrac{(x + 3)^n}{2^n}$ is a geometric series with $r = \dfrac{x + 3}{2}$, so the series converges $\Leftrightarrow |r| < 1 \Leftrightarrow \dfrac{|x + 3|}{2} < 1 \Leftrightarrow$

$|x + 3| < 2 \Leftrightarrow -5 < x < -1$. For these values of x, the sum of the series is $\dfrac{1}{1 - (x + 3)/2} = \dfrac{2}{2 - (x + 3)} = -\dfrac{2}{x + 1}$.

51. $\displaystyle\sum_{n=0}^{\infty} \dfrac{\cos^n x}{2^n}$ is a geometric series with first term 1 and ratio $r = \dfrac{\cos x}{2}$, so it converges $\Leftrightarrow |r| < 1$. But $|r| = \dfrac{|\cos x|}{2} \le \dfrac{1}{2}$

for all x. Thus, the series converges for all real values of x and the sum of the series is $\dfrac{1}{1 - (\cos x)/2} = \dfrac{2}{2 - \cos x}$.

52. Because $\dfrac{1}{n} \to 0$ and $\ln$ is continuous, we have $\displaystyle\lim_{n\to\infty} \ln\left(1 + \dfrac{1}{n}\right) = \ln 1 = 0$.

We now show that the series $\displaystyle\sum_{n=1}^{\infty} \ln\left(1 + \dfrac{1}{n}\right) = \sum_{n=1}^{\infty} \ln\left(\dfrac{n + 1}{n}\right) = \sum_{n=1}^{\infty} [\ln(n + 1) - \ln n]$ diverges.

$s_n = (\ln 2 - \ln 1) + (\ln 3 - \ln 2) + \cdots + (\ln(n + 1) - \ln n) = \ln(n + 1) - \ln 1 = \ln(n + 1)$.

As $n \to \infty$, $s_n = \ln(n + 1) \to \infty$, so the series diverges.

53. After defining f, We use `convert(f,parfrac);` in Maple, `Apart` in Mathematica, or `Expand Rational` and `Simplify` in Derive to find that the general term is $\dfrac{3n^2 + 3n + 1}{(n^2 + n)^3} = \dfrac{1}{n^3} - \dfrac{1}{(n+1)^3}$. So the nth partial sum is

$$s_n = \sum_{k=1}^{n} \left(\frac{1}{k^3} - \frac{1}{(k+1)^3} \right) = \left(1 - \frac{1}{2^3} \right) + \left(\frac{1}{2^3} - \frac{1}{3^3} \right) + \cdots + \left(\frac{1}{n^3} - \frac{1}{(n+1)^3} \right) = 1 - \frac{1}{(n+1)^3}$$

The series converges to $\lim\limits_{n\to\infty} s_n = 1$. This can be confirmed by directly computing the sum using `sum(f,1..infinity);` (in Maple), `Sum[f,{n,1,Infinity}]` (in Mathematica), or `Calculus Sum` (from 1 to ∞) and `Simplify` (in Derive).

54. See Exercise 53 for specific CAS commands. $\dfrac{1}{n^3 - n} = \dfrac{1/2}{n-1} - \dfrac{1}{n} + \dfrac{1/2}{n+1}$. So the nth partial sum is

$$s_n = \sum_{k=2}^{n} \left(\frac{1/2}{k-1} - \frac{1}{k} + \frac{1/2}{k+1} \right)$$

$$= \left(\frac{1/2}{1} - \frac{1}{2} + \frac{1/2}{3} \right) + \left(\frac{1/2}{2} - \frac{1}{3} + \frac{1/2}{4} \right) + \left(\frac{1/2}{3} - \frac{1}{4} + \frac{1/2}{5} \right)$$

$$+ \left(\frac{1/2}{4} - \frac{1}{5} + \frac{1/2}{6} \right) + \cdots + \left(\frac{1/2}{n-2} - \frac{1}{n-1} + \frac{1/2}{n} \right) + \left(\frac{1/2}{n-1} - \frac{1}{n} + \frac{1/2}{n+1} \right)$$

$$= \frac{1/2}{1} + \left(-\frac{1}{2} + \frac{1/2}{2} \right) + \left(\frac{1/2}{3} - \frac{1}{3} + \frac{1/2}{3} \right) + \left(\frac{1/2}{4} - \frac{1}{4} + \frac{1/2}{4} \right) + \cdots + \left(\frac{1/2}{n} - \frac{1}{n} + \frac{1/2}{n+1} \right)$$

$$= \frac{1}{2} + \left(-\frac{1}{4} \right) + 0 + 0 + \cdots + \frac{1/2}{n} - \frac{1}{n} + \frac{1/2}{n+1}$$

The series converges to $\lim\limits_{n\to\infty} s_n = \frac{1}{4}$.

55. For $n = 1$, $a_1 = 0$ since $s_1 = 0$. For $n > 1$,

$$a_n = s_n - s_{n-1} = \frac{n-1}{n+1} - \frac{(n-1)-1}{(n-1)+1} = \frac{(n-1)n - (n+1)(n-2)}{(n+1)n} = \frac{2}{n(n+1)}$$

Also, $\sum\limits_{n=1}^{\infty} a_n = \lim\limits_{n\to\infty} s_n = \lim\limits_{n\to\infty} \dfrac{1 - 1/n}{1 + 1/n} = 1$.

56. $a_1 = s_1 = 3 - \frac{1}{2} = \frac{5}{2}$. For $n \neq 1$,

$$a_n = s_n - s_{n-1} = \left(3 - n2^{-n} \right) - \left[3 - (n-1)2^{-(n-1)} \right] = -\frac{n}{2^n} + \frac{n-1}{2^{n-1}} \cdot \frac{2}{2} = \frac{2(n-1)}{2^n} - \frac{n}{2^n} = \frac{n-2}{2^n}$$

Also, $\sum\limits_{n=1}^{\infty} a_n = \lim\limits_{n\to\infty} s_n = \lim\limits_{n\to\infty} \left(3 - \dfrac{n}{2^n} \right) = 3$ because $\lim\limits_{x\to\infty} \dfrac{x}{2^x} \overset{\text{H}}{=} \lim\limits_{x\to\infty} \dfrac{1}{2^x \ln 2} = 0$.

57. (a) The first step in the chain occurs when the local government spends D dollars. The people who receive it spend a fraction c of those D dollars, that is, Dc dollars. Those who receive the Dc dollars spend a fraction c of it, that is, Dc^2 dollars. Continuing in this way, we see that the total spending after n transactions is

$$S_n = D + Dc + Dc^2 + \cdots + Dc^{n-1} = \frac{D(1-c^n)}{1-c} \quad \text{by (3)}.$$

(b) $\lim\limits_{n\to\infty} S_n = \lim\limits_{n\to\infty} \dfrac{D(1-c^n)}{1-c} = \dfrac{D}{1-c} \lim\limits_{n\to\infty} (1 - c^n) = \dfrac{D}{1-c} \quad \left[\text{since } 0 < c < 1 \quad \Rightarrow \quad \lim\limits_{n\to\infty} c^n = 0 \right]$

$\qquad = \dfrac{D}{s} \quad [\text{since } c + s = 1] = kD \quad [\text{since } k = 1/s]$

If $c = 0.8$, then $s = 1 - c = 0.2$ and the multiplier is $k = 1/s = 5$.

58. (a) Initially, the ball falls a distance H, then rebounds a distance rH, falls rH, rebounds r^2H, falls r^2H, etc. The total distance it travels is

$$H + 2rH + 2r^2H + 2r^3H + \cdots = H(1 + 2r + 2r^2 + 2r^3 + \cdots) = H[1 + 2r(1 + r + r^2 + \cdots)]$$

$$= H\left[1 + 2r\left(\frac{1}{1-r}\right)\right] = H\left(\frac{1+r}{1-r}\right) \text{ meters}$$

(b) From Example 3 in Section 2.1, we know that a ball falls $\frac{1}{2}gt^2$ meters in t seconds, where g is the gravitational acceleration. Thus, a ball falls h meters in $t = \sqrt{2h/g}$ seconds. The total travel time in seconds is

$$\sqrt{\frac{2H}{g}} + 2\sqrt{\frac{2H}{g}}r + 2\sqrt{\frac{2H}{g}}r^2 + 2\sqrt{\frac{2H}{g}}r^3 + \cdots = \sqrt{\frac{2H}{g}}\left[1 + 2\sqrt{r} + 2\sqrt{r}^2 + 2\sqrt{r}^3 + \cdots\right]$$

$$= \sqrt{\frac{2H}{g}}\left(1 + 2\sqrt{r}\left[1 + \sqrt{r} + \sqrt{r}^2 + \cdots\right]\right)$$

$$= \sqrt{\frac{2H}{g}}\left[1 + 2\sqrt{r}\left(\frac{1}{1-\sqrt{r}}\right)\right] = \sqrt{\frac{2H}{g}}\frac{1+\sqrt{r}}{1-\sqrt{r}}$$

(c) It will help to make a chart of the time for each descent and each rebound of the ball, together with the velocity just before and just after each bounce. Recall that the time in seconds needed to fall h meters is $\sqrt{2h/g}$. The ball hits the ground with velocity $-g\sqrt{2h/g} = -\sqrt{2hg}$ (taking the upward direction to be positive) and rebounds with velocity $kg\sqrt{2h/g} = k\sqrt{2hg}$, taking time $k\sqrt{2h/g}$ to reach the top of its bounce, where its velocity is 0. At that point, its height is k^2h. All these results follow from the formulas for vertical motion with gravitational acceleration $-g$:

$$\frac{d^2y}{dt^2} = -g \quad \Rightarrow \quad v = \frac{dy}{dt} = v_0 - gt \quad \Rightarrow \quad y = y_0 + v_0t - \tfrac{1}{2}gt^2.$$

number of descent	time of descent	speed before bounce	speed after bounce	time of ascent	peak height
1	$\sqrt{2H/g}$	$\sqrt{2Hg}$	$k\sqrt{2Hg}$	$k\sqrt{2H/g}$	k^2H
2	$\sqrt{2k^2H/g}$	$\sqrt{2k^2Hg}$	$k\sqrt{2k^2Hg}$	$k\sqrt{2k^2H/g}$	k^4H
3	$\sqrt{2k^4H/g}$	$\sqrt{2k^4Hg}$	$k\sqrt{2k^4Hg}$	$k\sqrt{2k^4H/g}$	k^6H
$\cdots$	$\cdots$	$\cdots$	$\cdots$	$\cdots$	$\cdots$

The total travel time in seconds is

$$\sqrt{\frac{2H}{g}} + k\sqrt{\frac{2H}{g}} + k\sqrt{\frac{2H}{g}} + k^2\sqrt{\frac{2H}{g}} + k^2\sqrt{\frac{2H}{g}} + \cdots = \sqrt{\frac{2H}{g}}(1 + 2k + 2k^2 + 2k^3 + \cdots)$$

$$= \sqrt{\frac{2H}{g}}\left[1 + 2k(1 + k + k^2 + \cdots)\right]$$

$$= \sqrt{\frac{2H}{g}}\left[1 + 2k\left(\frac{1}{1-k}\right)\right] = \sqrt{\frac{2H}{g}}\frac{1+k}{1-k}$$

Another method: We could use part (b). At the top of the bounce, the height is $k^2h = rh$, so $\sqrt{r} = k$ and the result follows from part (b).

59. $\sum_{n=2}^{\infty} (1+c)^{-n}$ is a geometric series with $a = (1+c)^{-2}$ and $r = (1+c)^{-1}$, so the series converges when

$\left| (1+c)^{-1} \right| < 1 \quad \Leftrightarrow \quad |1+c| > 1 \quad \Leftrightarrow \quad 1+c > 1 \text{ or } 1+c < -1 \quad \Leftrightarrow \quad c > 0 \text{ or } c < -2.$ We calculate the sum of the

series and set it equal to 2: $\dfrac{(1+c)^{-2}}{1-(1+c)^{-1}} = 2 \quad \Leftrightarrow \quad \left(\dfrac{1}{1+c} \right)^2 = 2 - 2\left(\dfrac{1}{1+c} \right) \quad \Leftrightarrow \quad 1 = 2(1+c)^2 - 2(1+c) \quad \Leftrightarrow$

$2c^2 + 2c - 1 = 0 \quad \Leftrightarrow \quad c = \dfrac{-2 \pm \sqrt{12}}{4} = \dfrac{\pm\sqrt{3}-1}{2}.$ However, the negative root is inadmissible because $-2 < \dfrac{-\sqrt{3}-1}{2} < 0.$

So $c = \dfrac{\sqrt{3}-1}{2}.$

60. $\sum_{n=0}^{\infty} e^{nc} = \sum_{n=0}^{\infty} (e^c)^n$ is a geometric series with $a = (e^c)^0 = 1$ and $r = e^c$. If $e^c < 1$, it has sum $\dfrac{1}{1-e^c}$, so $\dfrac{1}{1-e^c} = 10 \quad \Rightarrow$

$\frac{1}{10} = 1 - e^c \quad \Rightarrow \quad e^c = \frac{9}{10} \quad \Rightarrow \quad c = \ln \frac{9}{10}.$

61. $e^{s_n} = e^{1 + \frac{1}{2} + \frac{1}{3} + \cdots + \frac{1}{n}} = e^1 e^{1/2} e^{1/3} \cdots e^{1/n} > (1+1)\left(1+\frac{1}{2}\right)\left(1+\frac{1}{3}\right)\cdots\left(1+\frac{1}{n}\right) \qquad [e^x > 1 + x]$

$\qquad = \dfrac{2}{1} \dfrac{3}{2} \dfrac{4}{3} \cdots \dfrac{n+1}{n} = n + 1$

Thus, $e^{s_n} > n + 1$ and $\lim\limits_{n \to \infty} e^{s_n} = \infty.$ Since $\{s_n\}$ is increasing, $\lim\limits_{n \to \infty} s_n = \infty,$ implying that the harmonic series is

divergent.

62. The area between $y = x^{n-1}$ and $y = x^n$ for $0 \le x \le 1$ is

$$\int_0^1 (x^{n-1} - x^n)\, dx = \left[\frac{x^n}{n} - \frac{x^{n+1}}{n+1} \right]_0^1 = \frac{1}{n} - \frac{1}{n+1}$$

$$= \frac{(n+1) - n}{n(n+1)} = \frac{1}{n(n+1)}$$

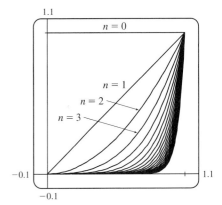

We can see from the diagram that as $n \to \infty$, the sum of the areas

between the successive curves approaches the area of the unit square, that

is, 1. So $\sum\limits_{n=1}^{\infty} \dfrac{1}{n(n+1)} = 1.$

63. Let d_n be the diameter of C_n. We draw lines from the centers of the C_i to

the center of D (or C), and using the Pythagorean Theorem, we can write

$1^2 + \left(1 - \frac{1}{2}d_1\right)^2 = \left(1 + \frac{1}{2}d_1\right)^2 \quad \Leftrightarrow$

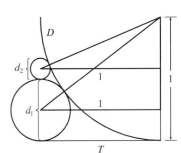

$1 = \left(1 + \frac{1}{2}d_1\right)^2 - \left(1 - \frac{1}{2}d_1\right)^2 = 2d_1$ [difference of squares] $\quad \Rightarrow \quad d_1 = \frac{1}{2}.$

Similarly,

$1 = \left(1 + \frac{1}{2}d_2\right)^2 - \left(1 - d_1 - \frac{1}{2}d_2\right)^2 = 2d_2 + 2d_1 - d_1^2 - d_1 d_2$

$= (2 - d_1)(d_1 + d_2) \quad \Leftrightarrow$

$d_2 = \dfrac{1}{2 - d_1} - d_1 = \dfrac{(1-d_1)^2}{2 - d_1}, \ 1 = \left(1 + \frac{1}{2}d_3\right)^2 - \left(1 - d_1 - d_2 - \frac{1}{2}d_3\right)^2 \quad \Leftrightarrow \quad d_3 = \dfrac{[1 - (d_1 + d_2)]^2}{2 - (d_1 + d_2)},$ and in general,

$d_{n+1} = \dfrac{\left(1 - \sum_{i=1}^{n} d_i\right)^2}{2 - \sum_{i=1}^{n} d_i}$. If we actually calculate d_2 and d_3 from the formulas above, we find that they are $\dfrac{1}{6} = \dfrac{1}{2 \cdot 3}$ and

$\dfrac{1}{12} = \dfrac{1}{3 \cdot 4}$ respectively, so we suspect that in general, $d_n = \dfrac{1}{n(n+1)}$. To prove this, we use induction: Assume that for all

$k \le n,\ d_k = \dfrac{1}{k(k+1)} = \dfrac{1}{k} - \dfrac{1}{k+1}$. Then $\sum_{i=1}^{n} d_i = 1 - \dfrac{1}{n+1} = \dfrac{n}{n+1}$ [telescoping sum]. Substituting this into our

formula for d_{n+1}, we get $d_{n+1} = \dfrac{\left[1 - \dfrac{n}{n+1}\right]^2}{2 - \left(\dfrac{n}{n+1}\right)} = \dfrac{\dfrac{1}{(n+1)^2}}{\dfrac{n+2}{n+1}} = \dfrac{1}{(n+1)(n+2)}$, and the induction is complete.

Now, we observe that the partial sums $\sum_{i=1}^{n} d_i$ of the diameters of the circles approach 1 as $n \to \infty$; that is,

$\displaystyle\sum_{n=1}^{\infty} a_n = \sum_{n=1}^{\infty} \dfrac{1}{n(n+1)} = 1$, which is what we wanted to prove.

64. $|CD| = b \sin\theta$, $|DE| = |CD|\sin\theta = b\sin^2\theta$, $|EF| = |DE|\sin\theta = b\sin^3\theta$, ... Therefore,

$|CD| + |DE| + |EF| + |FG| + \cdots = b \displaystyle\sum_{n=1}^{\infty} \sin^n\theta = b\left(\dfrac{\sin\theta}{1 - \sin\theta}\right)$ since this is a geometric series with $r = \sin\theta$

and $|\sin\theta| < 1$ $\quad$ [because $0 < \theta < \frac{\pi}{2}$].

65. The series $1 - 1 + 1 - 1 + 1 - 1 + \cdots$ diverges (geometric series with $r = -1$) so we cannot say that

$0 = 1 - 1 + 1 - 1 + 1 - 1 + \cdots$.

66. If $\displaystyle\sum_{n=1}^{\infty} a_n$ is convergent, then $\lim\limits_{n\to\infty} a_n = 0$ by Theorem 6, so $\lim\limits_{n\to\infty} \dfrac{1}{a_n} \ne 0$, and so $\displaystyle\sum_{n=1}^{\infty} \dfrac{1}{a_n}$ is divergent by the Test for

Divergence.

67. $\displaystyle\sum_{n=1}^{\infty} ca_n = \lim_{n\to\infty} \sum_{i=1}^{n} ca_i = \lim_{n\to\infty} c\sum_{i=1}^{n} a_i = c\lim_{n\to\infty} \sum_{i=1}^{n} a_i = c\sum_{n=1}^{\infty} a_n$, which exists by hypothesis.

68. If $\sum ca_n$ were convergent, then $\sum(1/c)(ca_n) = \sum a_n$ would be also, by Theorem 8(i). But this is not the case, so $\sum ca_n$

must diverge.

69. Suppose on the contrary that $\sum(a_n + b_n)$ converges. Then $\sum(a_n + b_n)$ and $\sum a_n$ are convergent series. So by

Theorem 8(iii), $\sum[(a_n + b_n) - a_n]$ would also be convergent. But $\sum[(a_n + b_n) - a_n] = \sum b_n$, a contradiction, since

$\sum b_n$ is given to be divergent.

70. No. For example, take $\sum a_n = \sum n$ and $\sum b_n = \sum(-n)$, which both diverge, yet $\sum(a_n + b_n) = \sum 0$, which converges

with sum 0.

71. The partial sums $\{s_n\}$ form an increasing sequence, since $s_n - s_{n-1} = a_n > 0$ for all n. Also, the sequence $\{s_n\}$ is bounded

since $s_n \le 1000$ for all n. So by the Monotonic Sequence Theorem, the sequence of partial sums converges, that is, the series

$\sum a_n$ is convergent.

72. (a) RHS $= \dfrac{1}{f_{n-1} f_n} - \dfrac{1}{f_n f_{n+1}} = \dfrac{f_n f_{n+1} - f_n f_{n-1}}{f_n^2 f_{n-1} f_{n+1}} = \dfrac{f_{n+1} - f_{n-1}}{f_n f_{n-1} f_{n+1}} = \dfrac{(f_{n-1} + f_n) - f_{n-1}}{f_n f_{n-1} f_{n+1}} = \dfrac{1}{f_{n-1} f_{n+1}} =$ LHS

(b) $\displaystyle\sum_{n=2}^{\infty}\frac{1}{f_{n-1}f_{n+1}} = \sum_{n=2}^{\infty}\left(\frac{1}{f_{n-1}f_n}-\frac{1}{f_n f_{n+1}}\right)$ [from part (a)]

$\displaystyle = \lim_{n\to\infty}\left[\left(\frac{1}{f_1 f_2}-\frac{1}{f_2 f_3}\right)+\left(\frac{1}{f_2 f_3}-\frac{1}{f_3 f_4}\right)+\left(\frac{1}{f_3 f_4}-\frac{1}{f_4 f_5}\right)+\cdots+\left(\frac{1}{f_{n-1}f_n}-\frac{1}{f_n f_{n+1}}\right)\right]$

$\displaystyle = \lim_{n\to\infty}\left(\frac{1}{f_1 f_2}-\frac{1}{f_n f_{n+1}}\right)=\frac{1}{f_1 f_2}-0=\frac{1}{1\cdot 1}=1$ because $f_n\to\infty$ as $n\to\infty$.

(c) $\displaystyle\sum_{n=2}^{\infty}\frac{f_n}{f_{n-1}f_{n+1}} = \sum_{n=2}^{\infty}\left(\frac{f_n}{f_{n-1}f_n}-\frac{f_n}{f_n f_{n+1}}\right)$ [as above]

$\displaystyle = \sum_{n=2}^{\infty}\left(\frac{1}{f_{n-1}}-\frac{1}{f_{n+1}}\right)$

$\displaystyle = \lim_{n\to\infty}\left[\left(\frac{1}{f_1}-\frac{1}{f_3}\right)+\left(\frac{1}{f_2}-\frac{1}{f_4}\right)+\left(\frac{1}{f_3}-\frac{1}{f_5}\right)+\left(\frac{1}{f_4}-\frac{1}{f_6}\right)+\cdots+\left(\frac{1}{f_{n-1}}-\frac{1}{f_{n+1}}\right)\right]$

$\displaystyle = \lim_{n\to\infty}\left(\frac{1}{f_1}+\frac{1}{f_2}-\frac{1}{f_n}-\frac{1}{f_{n+1}}\right)=1+1-0-0=2$ because $f_n\to\infty$ as $n\to\infty$.

73. (a) At the first step, only the interval $\left(\frac{1}{3},\frac{2}{3}\right)$ (length $\frac{1}{3}$) is removed. At the second step, we remove the intervals $\left(\frac{1}{9},\frac{2}{9}\right)$ and $\left(\frac{7}{9},\frac{8}{9}\right)$, which have a total length of $2\cdot\left(\frac{1}{3}\right)^2$. At the third step, we remove 2^2 intervals, each of length $\left(\frac{1}{3}\right)^3$. In general, at the nth step we remove 2^{n-1} intervals, each of length $\left(\frac{1}{3}\right)^n$, for a length of $2^{n-1}\cdot\left(\frac{1}{3}\right)^n = \frac{1}{3}\left(\frac{2}{3}\right)^{n-1}$. Thus, the total length of all removed intervals is $\displaystyle\sum_{n=1}^{\infty}\frac{1}{3}\left(\frac{2}{3}\right)^{n-1} = \frac{1/3}{1-2/3}=1$ [geometric series with $a=\frac{1}{3}$ and $r=\frac{2}{3}$]. Notice that at the nth step, the leftmost interval that is removed is $\left(\left(\frac{1}{3}\right)^n,\left(\frac{2}{3}\right)^n\right)$, so we never remove 0, and 0 is in the Cantor set. Also, the rightmost interval removed is $\left(1-\left(\frac{2}{3}\right)^n,1-\left(\frac{1}{3}\right)^n\right)$, so 1 is never removed. Some other numbers in the Cantor set are $\frac{1}{3},\frac{2}{3},\frac{1}{9},\frac{2}{9},\frac{7}{9}$, and $\frac{8}{9}$.

(b) The area removed at the first step is $\frac{1}{9}$; at the second step, $8\cdot\left(\frac{1}{9}\right)^2$; at the third step, $(8)^2\cdot\left(\frac{1}{9}\right)^3$. In general, the area removed at the nth step is $(8)^{n-1}\left(\frac{1}{9}\right)^n = \frac{1}{9}\left(\frac{8}{9}\right)^{n-1}$, so the total area of all removed squares is

$$\sum_{n=1}^{\infty}\frac{1}{9}\left(\frac{8}{9}\right)^{n-1} = \frac{1/9}{1-8/9}=1.$$

74. (a)

a_1	1	2	4	1	1	1000
a_2	2	3	1	4	1000	1
a_3	1.5	2.5	2.5	2.5	500.5	500.5
a_4	1.75	2.75	1.75	3.25	750.25	250.75
a_5	1.625	2.625	2.125	2.875	625.375	375.625
a_6	1.6875	2.6875	1.9375	3.0625	687.813	313.188
a_7	1.65625	2.65625	2.03125	2.96875	656.594	344.406
a_8	1.67188	2.67188	1.98438	3.01563	672.203	328.797
a_9	1.66406	2.66406	2.00781	2.99219	664.398	336.602
a_{10}	1.66797	2.66797	1.99609	3.00391	668.301	332.699
a_{11}	1.66602	2.66602	2.00195	2.99805	666.350	334.650
a_{12}	1.66699	2.66699	1.99902	3.00098	667.325	333.675

The limits seem to be $\frac{5}{3},\frac{8}{3}, 2, 3, 667$, and 334. Note that the limits appear to be "weighted" more toward a_2. In general, we guess that the limit is $\dfrac{a_1+2a_2}{3}$.

(b) $a_{n+1} - a_n = \frac{1}{2}(a_n + a_{n-1}) - a_n = -\frac{1}{2}(a_n - a_{n-1}) = -\frac{1}{2}\left[\frac{1}{2}(a_{n-1} + a_{n-2}) - a_{n-1}\right]$

$\qquad = -\frac{1}{2}\left[-\frac{1}{2}(a_{n-1} - a_{n-2})\right] = \cdots = \left(-\frac{1}{2}\right)^{n-1}(a_2 - a_1)$

Note that we have used the formula $a_k = \frac{1}{2}(a_{k-1} + a_{k-2})$ a total of $n - 1$ times in this calculation, once for each k between 3 and $n + 1$. Now we can write

$$a_n = a_1 + (a_2 - a_1) + (a_3 - a_2) + \cdots + (a_{n-1} - a_{n-2}) + (a_n - a_{n-1})$$

$$= a_1 + \sum_{k=1}^{n-1}(a_{k+1} - a_k) = a_1 + \sum_{k=1}^{n-1}\left(-\frac{1}{2}\right)^{k-1}(a_2 - a_1)$$

and so

$$\lim_{n \to \infty} a_n = a_1 + (a_2 - a_1)\sum_{k=1}^{\infty}\left(-\frac{1}{2}\right)^{k-1} = a_1 + (a_2 - a_1)\left[\frac{1}{1 - (-1/2)}\right] = a_1 + \frac{2}{3}(a_2 - a_1) = \frac{a_1 + 2a_2}{3}.$$

75. (a) For $\sum_{n=1}^{\infty}\dfrac{n}{(n+1)!}$, $s_1 = \dfrac{1}{1 \cdot 2} = \dfrac{1}{2}$, $s_2 = \dfrac{1}{2} + \dfrac{2}{1 \cdot 2 \cdot 3} = \dfrac{5}{6}$, $s_3 = \dfrac{5}{6} + \dfrac{3}{1 \cdot 2 \cdot 3 \cdot 4} = \dfrac{23}{24}$,

$s_4 = \dfrac{23}{24} + \dfrac{4}{1 \cdot 2 \cdot 3 \cdot 4 \cdot 5} = \dfrac{119}{120}$. The denominators are $(n+1)!$, so a guess would be $s_n = \dfrac{(n+1)! - 1}{(n+1)!}$.

(b) For $n = 1$, $s_1 = \dfrac{1}{2} = \dfrac{2! - 1}{2!}$, so the formula holds for $n = 1$. Assume $s_k = \dfrac{(k+1)! - 1}{(k+1)!}$. Then

$$s_{k+1} = \frac{(k+1)! - 1}{(k+1)!} + \frac{k+1}{(k+2)!} = \frac{(k+1)! - 1}{(k+1)!} + \frac{k+1}{(k+1)!(k+2)} = \frac{(k+2)! - (k+2) + k + 1}{(k+2)!}$$

$$= \frac{(k+2)! - 1}{(k+2)!}$$

Thus, the formula is true for $n = k + 1$. So by induction, the guess is correct.

(c) $\lim_{n \to \infty} s_n = \lim_{n \to \infty} \dfrac{(n+1)! - 1}{(n+1)!} = \lim_{n \to \infty}\left[1 - \dfrac{1}{(n+1)!}\right] = 1$ and so $\sum_{n=1}^{\infty}\dfrac{n}{(n+1)!} = 1.$

76.

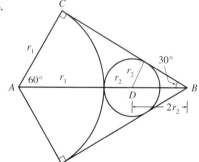

Let $r_1 =$ radius of the large circle, $r_2 =$ radius of next circle, and so on. From the figure we have $\angle BAC = 60°$ and $\cos 60° = r_1/|AB|$, so $|AB| = 2r_1$ and $|DB| = 2r_2$. Therefore, $2r_1 = r_1 + r_2 + 2r_2 \Rightarrow r_1 = 3r_2$. In general, we have $r_{n+1} = \frac{1}{3}r_n$, so the total area is

$$A = \pi r_1^2 + 3\pi r_2^2 + 3\pi r_3^2 + \cdots = \pi r_1^2 + 3\pi r_2^2\left(1 + \frac{1}{3^2} + \frac{1}{3^4} + \frac{1}{3^6} + \cdots\right)$$

$$= \pi r_1^2 + 3\pi r_2^2 \cdot \frac{1}{1 - 1/9} = \pi r_1^2 + \frac{27}{8}\pi r_2^2$$

Since the sides of the triangle have length 1, $|BC| = \frac{1}{2}$ and $\tan 30° = \dfrac{r_1}{1/2}$. Thus, $r_1 = \dfrac{\tan 30°}{2} = \dfrac{1}{2\sqrt{3}} \Rightarrow r_2 = \dfrac{1}{6\sqrt{3}}$,

so $A = \pi\left(\dfrac{1}{2\sqrt{3}}\right)^2 + \dfrac{27\pi}{8}\left(\dfrac{1}{6\sqrt{3}}\right)^2 = \dfrac{\pi}{12} + \dfrac{\pi}{32} = \dfrac{11\pi}{96}$. The area of the triangle is $\dfrac{\sqrt{3}}{4}$, so the circles occupy about 83.1% of the area of the triangle.

12.3 The Integral Test and Estimates of Sums

1. The picture shows that $a_2 = \dfrac{1}{2^{1.3}} < \displaystyle\int_1^2 \dfrac{1}{x^{1.3}}\,dx$,

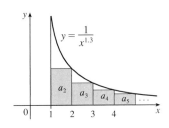

$a_3 = \dfrac{1}{3^{1.3}} < \displaystyle\int_2^3 \dfrac{1}{x^{1.3}}\,dx$, and so on, so $\displaystyle\sum_{n=2}^{\infty} \dfrac{1}{n^{1.3}} < \displaystyle\int_1^{\infty} \dfrac{1}{x^{1.3}}\,dx$. The

integral converges by (8.8.2) with $p = 1.3 > 1$, so the series converges.

2. From the first figure, we see that

$\displaystyle\int_1^6 f(x)\,dx < \sum_{i=1}^{5} a_i$. From the second figure,

we see that $\displaystyle\sum_{i=2}^{6} a_i < \int_1^6 f(x)\,dx$. Thus, we

have $\displaystyle\sum_{i=2}^{6} a_i < \int_1^6 f(x)\,dx < \sum_{i=1}^{5} a_i$.

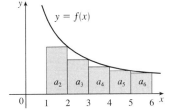

3. The function $f(x) = 1/\sqrt[5]{x} = x^{-1/5}$ is continuous, positive, and decreasing on $[1, \infty)$, so the Integral Test applies.

$\displaystyle\int_1^{\infty} x^{-1/5}\,dx = \lim_{t\to\infty} \int_1^t x^{-1/5}\,dx = \lim_{t\to\infty} \left[\tfrac{5}{4}x^{4/5}\right]_1^t = \lim_{t\to\infty} \left(\tfrac{5}{4}t^{4/5} - \tfrac{5}{4}\right) = \infty$, so $\displaystyle\sum_{n=1}^{\infty} 1/\sqrt[5]{n}$ diverges.

4. The function $f(x) = 1/x^5$ is continuous, positive, and decreasing on $[1, \infty)$, so the Integral Test applies.

$\displaystyle\int_1^{\infty} \dfrac{1}{x^5}\,dx = \lim_{t\to\infty} \int_1^t x^{-5}\,dx = \lim_{t\to\infty} \left[\dfrac{x^{-4}}{-4}\right]_1^t = \lim_{t\to\infty} \left(-\dfrac{1}{4t^4} + \dfrac{1}{4}\right) = \dfrac{1}{4}$.

Since this improper integral is convergent, the series $\displaystyle\sum_{n=1}^{\infty} \dfrac{1}{n^5}$ is also convergent by the Integral Test.

5. The function $f(x) = \dfrac{1}{(2x+1)^3}$ is continuous, positive, and decreasing on $[1, \infty)$, so the Integral Test applies.

$\displaystyle\int_1^{\infty} \dfrac{1}{(2x+1)^3}\,dx = \lim_{t\to\infty} \int_1^t \dfrac{1}{(2x+1)^3}\,dx = \lim_{t\to\infty} \left[-\dfrac{1}{4}\dfrac{1}{(2x+1)^2}\right]_1^t = \lim_{t\to\infty} \left(-\dfrac{1}{4(2t+1)^2} + \dfrac{1}{36}\right) = \dfrac{1}{36}$.

Since this improper integral is convergent, the series $\displaystyle\sum_{n=1}^{\infty} \dfrac{1}{(2n+1)^3}$ is also convergent by the Integral Test.

6. The function $f(x) = 1/\sqrt{x+4} = (x+4)^{-1/2}$ is continuous, positive, and decreasing on $[1, \infty)$, so the Integral Test applies.

$\displaystyle\int_1^{\infty} (x+4)^{-1/2}\,dx = \lim_{t\to\infty} \int_1^t (x+4)^{-1/2}\,dx = \lim_{t\to\infty} \left[2(x+4)^{1/2}\right]_1^t = \lim_{t\to\infty} \left(2\sqrt{t+4} - 2\sqrt{5}\right) = \infty$, so the series

$\displaystyle\sum_{n=1}^{\infty} 1/\sqrt{n+4}$ diverges.

7. $f(x) = xe^{-x}$ is continuous and positive on $[1, \infty)$. $f'(x) = -xe^{-x} + e^{-x} = e^{-x}(1 - x) < 0$ for $x > 1$, so f is decreasing

on $[1, \infty)$. Thus, the Integral Test applies.

$\displaystyle\int_1^{\infty} xe^{-x}\,dx = \lim_{b\to\infty} \int_1^b xe^{-x}\,dx = \lim_{b\to\infty} \left[-xe^{-x} - e^{-x}\right]_1^b$ [by parts] $= \lim_{b\to\infty} [-be^{-b} - e^{-b} + e^{-1} + e^{-1}] = 2/e$

since $\displaystyle\lim_{b\to\infty} be^{-b} = \lim_{b\to\infty} (b/e^b) \overset{\text{H}}{=} \lim_{b\to\infty} (1/e^b) = 0$ and $\displaystyle\lim_{b\to\infty} e^{-b} = 0$. Thus, $\sum_{n=1}^{\infty} ne^{-n}$ converges.

8. The function $f(x) = \dfrac{x+2}{x+1} = 1 + \dfrac{1}{x+1}$ is continuous, positive, and decreasing on $[1, \infty)$, so the Integral Test applies.

$$\int_1^\infty f(x)\,dx = \lim_{t \to \infty} \int_1^t \left(1 + \frac{1}{x+1}\right) dx = \lim_{t \to \infty} \left[x + \ln(x+1)\right]_1^t = \lim_{t \to \infty} (t + \ln(t+1) - 1 - \ln 2) = \infty, \text{ so}$$

$\displaystyle\int_1^\infty \frac{x+2}{x+1}\,dx$ is divergent and the series $\displaystyle\sum_{n=1}^\infty \frac{n+2}{n+1}$ is divergent.

Note: $\displaystyle\lim_{n \to \infty} \frac{n+2}{n+1} = 1$, so the given series diverges by the Test for Divergence.

9. The series $\displaystyle\sum_{n=1}^\infty \frac{1}{n^{0.85}}$ is a p-series with $p = 0.85 \le 1$, so it diverges by (1). Therefore, the series $\displaystyle\sum_{n=1}^\infty \frac{2}{n^{0.85}}$ must also diverge,

for if it converged, then $\displaystyle\sum_{n=1}^\infty \frac{1}{n^{0.85}}$ would have to converge [by Theorem 8(i) in Section 12.2].

10. $\displaystyle\sum_{n=1}^\infty n^{-1.4}$ and $\displaystyle\sum_{n=1}^\infty n^{-1.2}$ are p-series with $p > 1$, so they converge by (1). Thus, $\displaystyle\sum_{n=1}^\infty 3n^{-1.2}$ converges by Theorem 8(i) in

Section 12.2. It follows from Theorem 8(ii) that the given series $\displaystyle\sum_{n=1}^\infty (n^{-1.4} + 3n^{-1.2})$ also converges.

11. $1 + \dfrac{1}{8} + \dfrac{1}{27} + \dfrac{1}{64} + \dfrac{1}{125} + \cdots = \displaystyle\sum_{n=1}^\infty \frac{1}{n^3}$. This is a p-series with $p = 3 > 1$, so it converges by (1).

12. $1 + \dfrac{1}{2\sqrt{2}} + \dfrac{1}{3\sqrt{3}} + \dfrac{1}{4\sqrt{4}} + \dfrac{1}{5\sqrt{5}} + \cdots = \displaystyle\sum_{n=1}^\infty \frac{1}{n\sqrt{n}} = \sum_{n=1}^\infty \frac{1}{n^{3/2}}$. This is a p-series with $p = \frac{3}{2} > 1$, so it converges by (1).

13. $1 + \dfrac{1}{3} + \dfrac{1}{5} + \dfrac{1}{7} + \dfrac{1}{9} + \cdots = \displaystyle\sum_{n=1}^\infty \frac{1}{2n-1}$. The function $f(x) = \dfrac{1}{2x-1}$ is

continuous, positive, and decreasing on $[1, \infty)$, so the Integral Test applies.

$$\int_1^\infty \frac{1}{2x-1}\,dx = \lim_{t \to \infty} \int_1^t \frac{1}{2x-1}\,dx = \lim_{t \to \infty} \left[\tfrac{1}{2} \ln|2x-1|\right]_1^t = \tfrac{1}{2}\lim_{t \to \infty} (\ln(2t-1) - 0) = \infty, \text{ so the series } \sum_{n=1}^\infty \frac{1}{2n-1}$$

diverges.

14. $\dfrac{1}{5} + \dfrac{1}{8} + \dfrac{1}{11} + \dfrac{1}{14} + \dfrac{1}{17} + \cdots = \displaystyle\sum_{n=1}^\infty \frac{1}{3n+2}$. The function $f(x) = \dfrac{1}{3x+2}$ is continuous, positive, and decreasing on

$[1, \infty)$, so the Integral Test applies.

$$\int_1^\infty \frac{1}{3x+2}\,dx = \lim_{t \to \infty} \int_1^t \frac{1}{3x+2}\,dx = \lim_{t \to \infty} \left[\tfrac{1}{3} \ln|3x+2|\right]_1^t = \tfrac{1}{3}\lim_{t \to \infty} (\ln(3t+2) - \ln 5) = \infty, \text{ so the series}$$

$\displaystyle\sum_{n=1}^\infty \frac{1}{3n+2}$ diverges.

15. $\displaystyle\sum_{n=1}^\infty \frac{5 - 2\sqrt{n}}{n^3} = 5 \sum_{n=1}^\infty \frac{1}{n^3} - 2 \sum_{n=1}^\infty \frac{1}{n^{5/2}}$ by Theorem 12.2.8, since $\displaystyle\sum_{n=1}^\infty \frac{1}{n^3}$ and $\displaystyle\sum_{n=1}^\infty \frac{1}{n^{5/2}}$ both converge by (1)

$\left[\text{with } p = 3 > 1 \text{ and } p = \frac{5}{2} > 1\right]$. Thus, $\displaystyle\sum_{n=1}^\infty \frac{5 - 2\sqrt{n}}{n^3}$ converges.

16. $f(x) = \dfrac{x^2}{x^3 + 1}$ is continuous and positive on $[2, \infty)$, and also decreasing since $f'(x) = \dfrac{x(2 - x^3)}{(x^3 + 1)^2} < 0$ for $x \geq 2$,

so we can use the Integral Test [note that f is *not* decreasing on $[1, \infty)$].

$$\int_2^\infty \frac{x^2}{x^3 + 1}\, dx = \lim_{t \to \infty} \left[\tfrac{1}{3} \ln(x^3 + 1)\right]_2^t = \tfrac{1}{3} \lim_{t \to \infty} \left[\ln(t^3 + 1) - \ln 9\right] = \infty, \text{ so the series } \sum_{n=2}^\infty \frac{n^2}{n^3 + 1} \text{ diverges, and so does}$$

the given series, $\displaystyle\sum_{n=1}^\infty \frac{n^2}{n^3 + 1}$.

17. The function $f(x) = \dfrac{1}{x^2 + 4}$ is continuous, positive, and decreasing on $[1, \infty)$, so we can apply the Integral Test.

$$\int_1^\infty \frac{1}{x^2 + 4}\, dx = \lim_{t \to \infty} \int_1^t \frac{1}{x^2 + 4}\, dx = \lim_{t \to \infty} \left[\frac{1}{2} \tan^{-1} \frac{x}{2}\right]_1^t = \frac{1}{2} \lim_{t \to \infty} \left[\tan^{-1}\left(\frac{t}{2}\right) - \tan^{-1}\left(\frac{1}{2}\right)\right]$$
$$= \frac{1}{2}\left[\frac{\pi}{2} - \tan^{-1}\left(\frac{1}{2}\right)\right]$$

Therefore, the series $\displaystyle\sum_{n=1}^\infty \frac{1}{n^2 + 4}$ converges.

18. The function $f(x) = \dfrac{3x + 2}{x(x + 1)} = \dfrac{2}{x} + \dfrac{1}{x + 1}$ [by partial fractions] is continuous, positive, and decreasing on $[1, \infty)$ since it

is the sum of two such functions. Thus, we can apply the Integral Test.

$$\int_1^\infty \frac{3x + 2}{x(x + 1)}\, dx = \lim_{t \to \infty} \int_1^t \left[\frac{2}{x} + \frac{1}{x + 1}\right] dx = \lim_{t \to \infty} \left[2 \ln x + \ln(x + 1)\right]_1^t = \lim_{t \to \infty} [2 \ln t + \ln(t + 1) - \ln 2] = \infty.$$

Thus, the series $\displaystyle\sum_{n=1}^\infty \frac{3n + 2}{n(n + 1)}$ diverges.

19. $\displaystyle\sum_{n=1}^\infty \frac{\ln n}{n^3} = \sum_{n=2}^\infty \frac{\ln n}{n^3}$ since $\dfrac{\ln 1}{1} = 0$. The function $f(x) = \dfrac{\ln x}{x^3}$ is continuous and positive on $[2, \infty)$.

$$f'(x) = \frac{x^3(1/x) - (\ln x)(3x^2)}{(x^3)^2} = \frac{x^2 - 3x^2 \ln x}{x^6} = \frac{1 - 3 \ln x}{x^4} < 0 \quad \Leftrightarrow \quad 1 - 3 \ln x < 0 \quad \Leftrightarrow \quad \ln x > \tfrac{1}{3} \quad \Leftrightarrow$$

$x > e^{1/3} \approx 1.4$, so f is decreasing on $[2, \infty)$, and the Integral Test applies.

$$\int_2^\infty \frac{\ln x}{x^3}\, dx = \lim_{t \to \infty} \int_2^t \frac{\ln x}{x^3}\, dx \overset{(\star)}{=} \lim_{t \to \infty} \left[-\frac{\ln x}{2x^2} - \frac{1}{4x^2}\right]_1^t = \lim_{t \to \infty} \left[-\frac{1}{4t^2}(2 \ln t + 1) + \frac{1}{4}\right] \overset{(\star\star)}{=} \frac{1}{4}, \text{ so the series } \sum_{n=2}^\infty \frac{\ln n}{n^3}$$

converges.

$(\star)$: $u = \ln x$, $dv = x^{-3}\, dx \quad \Rightarrow \quad du = (1/x)\, dx$, $v = -\tfrac{1}{2}x^{-2}$, so

$$\int \frac{\ln x}{x^3}\, dx = -\tfrac{1}{2}x^{-2} \ln x - \int -\tfrac{1}{2}x^{-2}(1/x)\, dx = -\tfrac{1}{2}x^{-2} \ln x + \tfrac{1}{2} \int x^{-3}\, dx = -\tfrac{1}{2}x^{-2} \ln x - \tfrac{1}{4}x^{-2} + C.$$

$(\star\star)$: $\displaystyle\lim_{t \to \infty} \left(-\frac{2 \ln t + 1}{4t^2}\right) \overset{\text{H}}{=} -\lim_{t \to \infty} \frac{2/t}{8t} = -\tfrac{1}{4} \lim_{t \to \infty} \frac{1}{t^2} = 0.$

20. The function $f(x) = \dfrac{1}{x^2 - 4x + 5} = \dfrac{1}{(x-2)^2 + 1}$ is continuous, positive, and decreasing on $[2, \infty)$, so the Integral Test

applies.

$$\int_2^\infty f(x)\,dx = \lim_{t\to\infty} \int_2^t f(x)\,dx = \lim_{t\to\infty} \int_2^t \frac{1}{(x-2)^2+1}\,dx = \lim_{t\to\infty} \big[\tan^{-1}(x-2)\big]_2^t = \lim_{t\to\infty}\big[\tan^{-1}(t-2) - \tan^{-1}0\big]$$
$$= \frac{\pi}{2} - 0 = \frac{\pi}{2}$$

so the series $\sum_{n=2}^\infty \dfrac{1}{n^2 - 4n + 5}$ converges. Of course, this means that $\sum_{n=1}^\infty \dfrac{1}{n^2 - 4n + 5}$ converges too.

21. $f(x) = \dfrac{1}{x \ln x}$ is continuous and positive on $[2, \infty)$, and also decreasing since $f'(x) = -\dfrac{1+\ln x}{x^2(\ln x)^2} < 0$ for $x > 2$, so we can

use the Integral Test. $\int_2^\infty \dfrac{1}{x \ln x}\,dx = \lim_{t\to\infty} [\ln(\ln x)]_2^t = \lim_{t\to\infty} [\ln(\ln t) - \ln(\ln 2)] = \infty$, so the series $\sum_{n=2}^\infty \dfrac{1}{n \ln n}$ diverges.

22. The function $f(x) = \dfrac{1}{x(\ln x)^2}$ is continuous, positive, and decreasing on $[2, \infty)$, so the Integral Test applies.

$$\int_2^\infty f(x)\,dx = \lim_{t\to\infty}\int_2^t \frac{1}{x(\ln x)^2}\,dx = \lim_{t\to\infty}\left[\frac{-1}{\ln x}\right]_2^t \quad \text{[by substitution with } u = \ln x] \quad = -\lim_{t\to\infty}\left(\frac{1}{\ln t} - \frac{1}{\ln 2}\right) = \frac{1}{\ln 2},$$

so the series $\sum_{n=2}^\infty \dfrac{1}{n(\ln n)^2}$ converges.

23. The function $f(x) = e^{1/x}/x^2$ is continuous, positive, and decreasing on $[1, \infty)$, so the Integral Test applies.

[$g(x) = e^{1/x}$ is decreasing and dividing by x^2 doesn't change that fact.]

$$\int_1^\infty f(x)\,dx = \lim_{t\to\infty}\int_1^t \frac{e^{1/x}}{x^2}\,dx = \lim_{t\to\infty}\left[-e^{1/x}\right]_1^t = -\lim_{t\to\infty}(e^{1/t} - e) = -(1-e) = e-1, \text{ so the series } \sum_{n=1}^\infty \frac{e^{1/n}}{n^2}$$

converges.

24. $f(x) = \dfrac{x^2}{e^x} \Rightarrow f'(x) = \dfrac{e^x(2x) - x^2 e^x}{(e^x)^2} = \dfrac{xe^x(2-x)}{(e^x)^2} = \dfrac{x(2-x)}{e^x} < 0$ for $x > 2$, so f is continuous, positive, and

decreasing on $[3, \infty)$ and so the Integral Test applies.

$$\int_3^\infty f(x)\,dx = \lim_{t\to\infty}\int_3^t \frac{x^2}{e^x}\,dx \overset{(\star)}{=} \lim_{t\to\infty}\left[-e^{-x}(x^2 + 2x + 2)\right]_3^t = -\lim_{t\to\infty}\left[e^{-t}(t^2 + 2t + 2) - e^{-3}(17)\right] \overset{(\star\star)}{=} \frac{17}{e^3},$$

so the series $\sum_{n=3}^\infty \dfrac{n^2}{e^n}$ converges.

$(\star)$: $\int x^2 e^{-x}\,dx \overset{97}{=} -x^2 e^{-x} + 2\int xe^{-x}\,dx \overset{97}{=} -x^2 e^{-x} + 2(-xe^{-x} + \int e^{-x}\,dx)$
$$= -x^2 e^{-x} - 2xe^{-x} - 2e^{-x} + C = -e^{-x}(x^2 + 2x + 2) + C.$$

$(\star\star)$: $\lim_{t\to\infty} \dfrac{t^2 + 2t + 2}{e^t} \overset{H}{=} \lim_{t\to\infty} \dfrac{2t+2}{e^t} \overset{H}{=} \lim_{t\to\infty} \dfrac{2}{e^t} = 0$.

25. The function $f(x) = \dfrac{1}{x^3 + x}$ is continuous, positive, and decreasing on $[1, \infty)$, so the Integral Test applies. We use partial

fractions to evaluate the integral:

$$\int_1^\infty \frac{1}{x^3 + x}\, dx = \lim_{t\to\infty} \int_1^t \left[\frac{1}{x} - \frac{x}{1+x^2}\right] dx = \lim_{t\to\infty} \left[\ln x - \frac{1}{2}\ln(1+x^2)\right]_1^t = \lim_{t\to\infty}\left[\ln\frac{x}{\sqrt{1+x^2}}\right]_1^t$$

$$= \lim_{t\to\infty}\left(\ln\frac{t}{\sqrt{1+t^2}} - \ln\frac{1}{\sqrt{2}}\right) = \lim_{t\to\infty}\left(\ln\frac{1}{\sqrt{1+1/t^2}} + \frac{1}{2}\ln 2\right) = \frac{1}{2}\ln 2$$

so the series $\sum\limits_{n=1}^\infty \dfrac{1}{n^3 + n}$ converges.

26. The function $f(x) = \dfrac{x}{x^4 + 1}$ is positive, continuous, and decreasing on $[1, \infty)$. [Note that

$f'(x) = \dfrac{x^4 + 1 - 4x^4}{(x^4+1)^2} = \dfrac{1 - 3x^4}{(x^4+1)^2} < 0$ on $[1, \infty)$.] Thus, we can apply the Integral Test.

$$\int_1^\infty \frac{x}{x^4 + 1}\, dx = \lim_{t\to\infty}\int_1^t \frac{\frac{1}{2}(2x)}{1+(x^2)^2}\, dx = \lim_{t\to\infty}\left[\frac{1}{2}\tan^{-1}(x^2)\right]_1^t = \frac{1}{2}\lim_{t\to\infty}[\tan^{-1}(t^2) - \tan^{-1} 1] = \frac{1}{2}\left(\frac{\pi}{2} - \frac{\pi}{4}\right) = \frac{\pi}{8}$$

so the series $\sum\limits_{n=1}^\infty \dfrac{n}{n^4 + 1}$ converges.

27. We have already shown (in Exercise 21) that when $p = 1$ the series $\sum\limits_{n=2}^\infty \dfrac{1}{n(\ln n)^p}$ diverges, so assume that $p \neq 1$.

$f(x) = \dfrac{1}{x(\ln x)^p}$ is continuous and positive on $[2, \infty)$, and $f'(x) = -\dfrac{p + \ln x}{x^2(\ln x)^{p+1}} < 0$ if $x > e^{-p}$, so that f is eventually

decreasing and we can use the Integral Test.

$$\int_2^\infty \frac{1}{x(\ln x)^p}\, dx = \lim_{t\to\infty}\left[\frac{(\ln x)^{1-p}}{1-p}\right]_2^t \quad \text{[for } p \neq 1\text{]} = \lim_{t\to\infty}\left[\frac{(\ln t)^{1-p}}{1-p} - \frac{(\ln 2)^{1-p}}{1-p}\right]$$

This limit exists whenever $1 - p < 0 \iff p > 1$, so the series converges for $p > 1$.

28. $f(x) = \dfrac{1}{x\ln x\,[\ln(\ln x)]^p}$ is positive and continuous on $[3, \infty)$. For $p \geq 0$, f clearly decreases on $[3, \infty)$; and for $p < 0$,

it can be verified that f is ultimately decreasing. Thus, we can apply the Integral Test.

$$I = \int_3^\infty \frac{dx}{x\ln x\,[\ln(\ln x)]^p} = \lim_{t\to\infty}\int_3^t \frac{[\ln(\ln x)]^{-p}}{x\ln x}\, dx = \lim_{t\to\infty}\left[\frac{[\ln(\ln x)]^{-p+1}}{-p+1}\right]_3^t \quad \text{[for } p \neq 1\text{]}$$

$$= \lim_{t\to\infty}\left[\frac{[\ln(\ln t)]^{-p+1}}{-p+1} - \frac{[\ln(\ln 3)]^{-p+1}}{-p+1}\right],$$

which exists whenever $-p + 1 < 0 \iff p > 1$. If $p = 1$, then $I = \lim_{t\to\infty}\big[\ln(\ln(\ln x))\big]_3^t = \infty$. Therefore,

$\sum\limits_{n=3}^\infty \dfrac{1}{n\ln n\,[\ln(\ln n)]^p}$ converges for $p > 1$.

29. Clearly the series cannot converge if $p \geq -\frac{1}{2}$, because then $\lim\limits_{n\to\infty} n(1+n^2)^p \neq 0$. So assume $p < -\frac{1}{2}$. Then

$f(x) = x(1+x^2)^p$ is continuous, positive, and eventually decreasing on $[1, \infty)$, and we can use the Integral Test.

$$\int_1^\infty x(1+x^2)^p dx = \lim_{t\to\infty} \left[\frac{1}{2} \cdot \frac{(1+x^2)^{p+1}}{p+1} \right]_1^t = \frac{1}{2(p+1)} \lim_{t\to\infty} \left[(1+t^2)^{p+1} - 2^{p+1} \right].$$

This limit exists and is finite $\Leftrightarrow$ $p+1 < 0$ $\Leftrightarrow$ $p < -1$, so the series converges whenever $p < -1$.

30. If $p \leq 0$, $\lim\limits_{n\to\infty} \dfrac{\ln n}{n^p} = \infty$ and the series diverges, so assume $p > 0$. $f(x) = \dfrac{\ln x}{x^p}$ is positive and continuous and $f'(x) < 0$

for $x > e^{1/p}$, so f is eventually decreasing and we can use the Integral Test. Integration by parts gives

$$\int_1^\infty \frac{\ln x}{x^p} dx = \lim_{t\to\infty} \left[\frac{x^{1-p}\left[(1-p)\ln x - 1\right]}{(1-p)^2} \right]_1^t \quad (\text{for } p \neq 1) = \frac{1}{(1-p)^2} \left[\lim_{t\to\infty} t^{1-p}\left[(1-p)\ln t - 1\right] + 1 \right], \text{ which exists}$$

whenever $1 - p < 0$ $\Leftrightarrow$ $p > 1$. Thus, $\sum\limits_{n=1}^{\infty} \dfrac{\ln n}{n^p}$ converges $\Leftrightarrow$ $p > 1$.

31. Since this is a p-series with $p = x$, $\zeta(x)$ is defined when $x > 1$. Unless specified otherwise, the domain of a function f is the set of real numbers x such that the expression for $f(x)$ makes sense and defines a real number. So, in the case of a series, it's the set of real numbers x such that the series is convergent.

32. (a) $f(x) = 1/x^4$ is positive and continuous and $f'(x) = -4/x^5$ is negative for $x > 0$, and so the Integral Test applies.

$$\sum_{n=1}^{\infty} \frac{1}{n^4} \approx s_{10} = \frac{1}{1^4} + \frac{1}{2^4} + \frac{1}{3^4} + \cdots + \frac{1}{10^4} \approx 1.082037.$$

$$R_{10} \leq \int_{10}^{\infty} \frac{1}{x^4} dx = \lim_{t\to\infty} \left[\frac{1}{-3x^3} \right]_{10}^t = \lim_{t\to\infty} \left(-\frac{1}{3t^3} + \frac{1}{3(10)^3} \right) = \frac{1}{3000}, \text{ so the error is at most } 0.000\overline{3}.$$

(b) $s_{10} + \int_{11}^{\infty} \frac{1}{x^4} dx \leq s \leq s_{10} + \int_{10}^{\infty} \frac{1}{x^4} dx \Rightarrow s_{10} + \frac{1}{3(11)^3} \leq s \leq s_{10} + \frac{1}{3(10)^3} \Rightarrow$

$1.082037 + 0.000250 = 1.082287 \leq s \leq 1.082037 + 0.000333 = 1.082370$, so we get $s \approx 1.08233$ with error ≤ 0.00005.

(c) $R_n \leq \int_n^{\infty} \frac{1}{x^4} dx = \frac{1}{3n^3}$. So $R_n < 0.00001 \Rightarrow \frac{1}{3n^3} < \frac{1}{10^5} \Rightarrow 3n^3 > 10^5 \Rightarrow n > \sqrt[3]{(10)^5/3} \approx 32.2$, that is, for $n > 32$.

33. (a) $f(x) = \dfrac{1}{x^2}$ is positive and continuous and $f'(x) = -\dfrac{2}{x^3}$ is negative for $x > 0$, and so the Integral Test applies.

$$\sum_{n=1}^{\infty} \frac{1}{n^2} \approx s_{10} = \frac{1}{1^2} + \frac{1}{2^2} + \frac{1}{3^2} + \cdots + \frac{1}{10^2} \approx 1.549768.$$

$$R_{10} \leq \int_{10}^{\infty} \frac{1}{x^2} dx = \lim_{t\to\infty} \left[\frac{-1}{x} \right]_{10}^t = \lim_{t\to\infty} \left(-\frac{1}{t} + \frac{1}{10} \right) = \frac{1}{10}, \text{ so the error is at most } 0.1.$$

(b) $s_{10} + \int_{11}^{\infty} \frac{1}{x^2} dx \leq s \leq s_{10} + \int_{10}^{\infty} \frac{1}{x^2} dx \Rightarrow s_{10} + \frac{1}{11} \leq s \leq s_{10} + \frac{1}{10} \Rightarrow$

$1.549768 + 0.090909 = 1.640677 \leq s \leq 1.549768 + 0.1 = 1.649768$, so we get $s \approx 1.64522$ (the average of 1.640677 and 1.649768) with error ≤ 0.005 (the maximum of $1.649768 - 1.64522$ and $1.64522 - 1.640677$, rounded up).

(c) $R_n \leq \int_n^{\infty} \frac{1}{x^2} dx = \frac{1}{n}$. So $R_n < 0.001$ if $\frac{1}{n} < \frac{1}{1000} \Leftrightarrow n > 1000$.

34. $f(x) = 1/x^5$ is positive and continuous and $f'(x) = -5/x^6$ is negative for $x > 0$, and so the Integral Test applies. Using (2),

$$R_n \le \int_n^\infty x^{-5}\,dx = \lim_{t\to\infty}\left[\frac{-1}{4x^4}\right]_n^t = \frac{1}{4n^4}.$$ If we take $n = 5$, then $s_5 \approx 1.036662$ and $R_5 \le 0.0004$. So $s \approx s_5 \approx 1.037$.

35. $f(x) = 1/(2x+1)^6$ is continuous, positive, and decreasing on $[1,\infty)$, so the Integral Test applies. Using (2),

$$R_n \le \int_n^\infty (2x+1)^{-6}\,dx = \lim_{t\to\infty}\left[\frac{-1}{10(2x+1)^5}\right]_n^t = \frac{1}{10(2n+1)^5}.$$ To be correct to five decimal places, we want

$$\frac{1}{10(2n+1)^5} \le \frac{5}{10^6} \quad\Leftrightarrow\quad (2n+1)^5 \ge 20{,}000 \quad\Leftrightarrow\quad n \ge \tfrac{1}{2}\big(\sqrt[5]{20{,}000} - 1\big) \approx 3.12,\ \text{so use } n = 4.$$

$$s_4 = \sum_{n=1}^{4} \frac{1}{(2n+1)^6} = \frac{1}{3^6} + \frac{1}{5^6} + \frac{1}{7^6} + \frac{1}{9^6} \approx 0.001\,446 \approx 0.00145.$$

36. $f(x) = \dfrac{1}{x(\ln x)^2}$ is positive and continuous and $f'(x) = -\dfrac{\ln x + 2}{x^2(\ln x)^3}$ is negative for $x > 1$, so the Integral Test applies.

Using (2), we need $0.01 > \displaystyle\int_n^\infty \frac{dx}{x(\ln x)^2} = \lim_{t\to\infty}\left[\frac{-1}{\ln x}\right]_n^t = \frac{1}{\ln n}$. This is true for $n > e^{100}$, so we would have to take this

many terms, which would be problematic because $e^{100} \approx 2.7 \times 10^{43}$.

37. $\displaystyle\sum_{n=1}^\infty n^{-1.001} = \sum_{n=1}^\infty \frac{1}{n^{1.001}}$ is a convergent p-series with $p = 1.001 > 1$. Using (2), we get

$$R_n \le \int_n^\infty x^{-1.001}\,dx = \lim_{t\to\infty}\left[\frac{x^{-0.001}}{-0.001}\right]_n^t = -1000 \lim_{t\to\infty}\left[\frac{1}{x^{0.001}}\right]_n^t = -1000\left(-\frac{1}{n^{0.001}}\right) = \frac{1000}{n^{0.001}}.$$

We want $R_n < 0.000\,000\,005 \quad\Leftrightarrow\quad \dfrac{1000}{n^{0.001}} < 5\times 10^{-9} \quad\Leftrightarrow\quad n^{0.001} > \dfrac{1000}{5\times 10^{-9}} \quad\Leftrightarrow$

$$n > \big(2\times 10^{11}\big)^{1000} = 2^{1000} \times 10^{11{,}000} \approx 1.07 \times 10^{301} \times 10^{11{,}000} = 1.07 \times 10^{11{,}301}.$$

38. (a) $f(x) = \left(\dfrac{\ln x}{x}\right)^2$ is continuous and positive for $x > 1$, and since $f'(x) = \dfrac{2\ln x\,(1 - \ln x)}{x^3} < 0$ for $x > e$, we can apply

the Integral Test. Using a CAS, we get $\displaystyle\int_1^\infty \left(\frac{\ln x}{x}\right)^2 dx = 2$, so the series also converges.

(b) Since the Integral Test applies, the error in $s \approx s_n$ is $R_n \le \displaystyle\int_n^\infty \left(\frac{\ln x}{x}\right)^2 dx = \frac{(\ln n)^2 + 2\ln n + 2}{n}$.

(c) By graphing the functions $y_1 = \dfrac{(\ln x)^2 + 2\ln x + 2}{x}$ and $y_2 = 0.05$, we see that $y_1 < y_2$ for $n \ge 1373$.

(d) Using the CAS to sum the first 1373 terms, we get $s_{1373} \approx 1.94$.

39. (a) From the figure, $a_2 + a_3 + \cdots + a_n \le \int_1^n f(x)\,dx$, so with

$$f(x) = \frac{1}{x},\ \frac{1}{2} + \frac{1}{3} + \frac{1}{4} + \cdots + \frac{1}{n} \le \int_1^n \frac{1}{x}\,dx = \ln n.$$

Thus, $s_n = 1 + \dfrac{1}{2} + \dfrac{1}{3} + \dfrac{1}{4} + \cdots + \dfrac{1}{n} \le 1 + \ln n.$

(b) By part (a), $s_{10^6} \le 1 + \ln 10^6 \approx 14.82 < 15$ and

$$s_{10^9} \le 1 + \ln 10^9 \approx 21.72 < 22.$$

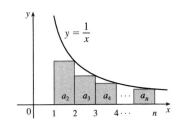

40. (a) The sum of the areas of the n rectangles in the graph to the right is

$1 + \dfrac{1}{2} + \dfrac{1}{3} + \cdots + \dfrac{1}{n}$. Now $\displaystyle\int_1^{n+1} \dfrac{dx}{x}$ is less than this sum because

the rectangles extend above the curve $y = 1/x$, so

$\displaystyle\int_1^{n+1} \dfrac{1}{x}\, dx = \ln(n+1) < 1 + \dfrac{1}{2} + \dfrac{1}{3} + \cdots + \dfrac{1}{n}$, and since

$\ln n < \ln(n+1)$, $0 < 1 + \dfrac{1}{2} + \dfrac{1}{3} + \cdots + \dfrac{1}{n} - \ln n = t_n$.

(b) The area under $f(x) = 1/x$ between $x = n$ and $x = n+1$ is

$\displaystyle\int_n^{n+1} \dfrac{dx}{x} = \ln(n+1) - \ln n$, and this is clearly greater than the area of

the inscribed rectangle in the figure to the right $\left[\text{which is } \dfrac{1}{n+1}\right]$, so

$t_n - t_{n+1} = [\ln(n+1) - \ln n] - \dfrac{1}{n+1} > 0$, and so $t_n > t_{n+1}$, so $\{t_n\}$ is a decreasing sequence.

(c) We have shown that $\{t_n\}$ is decreasing and that $t_n > 0$ for all n. Thus, $0 < t_n \leq t_1 = 1$, so $\{t_n\}$ is a bounded monotonic sequence, and hence converges by the Monotonic Sequence Theorem.

41. $b^{\ln n} = \left(e^{\ln b}\right)^{\ln n} = \left(e^{\ln n}\right)^{\ln b} = n^{\ln b} = \dfrac{1}{n^{-\ln b}}$. This is a p-series, which converges for all b such that $-\ln b > 1$ $\Leftrightarrow$

$\ln b < -1$ $\Leftrightarrow$ $b < e^{-1}$ $\Leftrightarrow$ $b < 1/e$ [with $b > 0$].

42. For the series $\displaystyle\sum_{n=1}^{\infty} \left(\dfrac{c}{n} - \dfrac{1}{n+1}\right)$,

$$s_n = \sum_{i=1}^{n} \left(\dfrac{c}{i} - \dfrac{1}{i+1}\right) = \left(\dfrac{c}{1} - \dfrac{1}{2}\right) + \left(\dfrac{c}{2} - \dfrac{1}{3}\right) + \left(\dfrac{c}{3} - \dfrac{1}{4}\right) + \cdots + \left(\dfrac{c}{n} - \dfrac{1}{n+1}\right)$$

$$= \dfrac{c}{1} + \dfrac{c-1}{2} + \dfrac{c-1}{3} + \dfrac{c-1}{4} + \cdots + \dfrac{c-1}{n} - \dfrac{1}{n+1} = c + (c-1)\left(\dfrac{1}{2} + \dfrac{1}{3} + \dfrac{1}{4} + \cdots + \dfrac{1}{n}\right) - \dfrac{1}{n+1}$$

Thus, $\displaystyle\sum_{n=1}^{\infty} \left(\dfrac{c}{n} - \dfrac{1}{n+1}\right) = \lim_{n\to\infty} s_n = \lim_{n\to\infty}\left[c + (c-1)\sum_{i=2}^{n} \dfrac{1}{i} - \dfrac{1}{n+1}\right]$. Since a constant multiple of a divergent series

is divergent, the last limit exists only if $c - 1 = 0$, so the original series converges only if $c = 1$.

12.4 The Comparison Tests

1. (a) We cannot say anything about $\sum a_n$. If $a_n > b_n$ for all n and $\sum b_n$ is convergent, then $\sum a_n$ could be convergent or divergent. (See the note after Example 2.)

(b) If $a_n < b_n$ for all n, then $\sum a_n$ is convergent. [This is part (i) of the Comparison Test.]

2. (a) If $a_n > b_n$ for all n, then $\sum a_n$ is divergent. [This is part (ii) of the Comparison Test.]

(b) We cannot say anything about $\sum a_n$. If $a_n < b_n$ for all n and $\sum b_n$ is divergent, then $\sum a_n$ could be convergent or divergent.

3. $\dfrac{n}{2n^3 + 1} < \dfrac{n}{2n^3} = \dfrac{1}{2n^2} < \dfrac{1}{n^2}$ for all $n \geq 1$, so $\displaystyle\sum_{n=1}^{\infty} \dfrac{n}{2n^3 + 1}$ converges by comparison with $\displaystyle\sum_{n=1}^{\infty} \dfrac{1}{n^2}$, which converges

because it is a p-series with $p = 2 > 1$.

4. $\dfrac{n^3}{n^4 - 1} > \dfrac{n^3}{n^4} = \dfrac{1}{n}$ for all $n \geq 2$, so $\displaystyle\sum_{n=2}^{\infty} \dfrac{n^3}{n^4 - 1}$ diverges by comparison with $\displaystyle\sum_{n=2}^{\infty} \dfrac{1}{n}$, which diverges because it is a p-series

with $p = 1 \leq 1$ (the harmonic series).

5. $\dfrac{n+1}{n\sqrt{n}} > \dfrac{n}{n\sqrt{n}} = \dfrac{1}{\sqrt{n}}$ for all $n \geq 1$, so $\displaystyle\sum_{n=1}^{\infty} \dfrac{n+1}{n\sqrt{n}}$ diverges by comparison with $\displaystyle\sum_{n=1}^{\infty} \dfrac{1}{\sqrt{n}}$, which diverges because it is a

p-series with $p = \frac{1}{2} \leq 1$.

6. $\dfrac{n-1}{n^2\sqrt{n}} < \dfrac{n}{n^2 n^{1/2}} = \dfrac{1}{n^{3/2}}$ for all $n \geq 1$, so $\displaystyle\sum_{n=1}^{\infty} \dfrac{n-1}{n^2\sqrt{n}}$ converges by comparison with $\displaystyle\sum_{n=1}^{\infty} \dfrac{1}{n^{3/2}}$, which converges because

it is a p-series with $p = \frac{3}{2} > 1$.

7. $\dfrac{9^n}{3 + 10^n} < \dfrac{9^n}{10^n} = \left(\dfrac{9}{10}\right)^n$ for all $n \geq 1$. $\displaystyle\sum_{n=1}^{\infty} \left(\dfrac{9}{10}\right)^n$ is a convergent geometric series $\left(|r| = \frac{9}{10} < 1\right)$, so $\displaystyle\sum_{n=1}^{\infty} \dfrac{9^n}{3 + 10^n}$

converges by the Comparison Test.

8. $\dfrac{4 + 3^n}{2^n} > \dfrac{3^n}{2^n} = \left(\dfrac{3}{2}\right)^n$ for all $n \geq 1$, so $\displaystyle\sum_{n=1}^{\infty} \dfrac{4 + 3^n}{2^n}$ diverges by comparison with the divergent geometric series $\displaystyle\sum_{n=1}^{\infty} \left(\dfrac{3}{2}\right)^n$.

9. $\dfrac{\cos^2 n}{n^2 + 1} \leq \dfrac{1}{n^2 + 1} < \dfrac{1}{n^2}$, so the series $\displaystyle\sum_{n=1}^{\infty} \dfrac{\cos^2 n}{n^2 + 1}$ converges by comparison with the p-series $\displaystyle\sum_{n=1}^{\infty} \dfrac{1}{n^2}$ $[p = 2 > 1]$.

10. $\dfrac{n^2 - 1}{3n^4 + 1} < \dfrac{n^2}{3n^4 + 1} < \dfrac{n^2}{3n^4} = \dfrac{1}{3}\dfrac{1}{n^2}$. $\displaystyle\sum_{n=1}^{\infty} \dfrac{n^2 - 1}{3n^4 + 1}$ converges by comparison with $\displaystyle\sum_{n=1}^{\infty} \dfrac{1}{3n^2}$, which converges because it is

a constant multiple of a convergent p-series $[p = 2 > 1]$. The terms of the given series are positive for $n > 1$, which is good

enough.

11. $\dfrac{n-1}{n\,4^n}$ is positive for $n > 1$ and $\dfrac{n-1}{n\,4^n} < \dfrac{n}{n\,4^n} = \dfrac{1}{4^n} = \left(\dfrac{1}{4}\right)^n$, so $\displaystyle\sum_{n=1}^{\infty} \dfrac{n-1}{n\,4^n}$ converges by comparison with the convergent

geometric series $\displaystyle\sum_{n=1}^{\infty} \left(\dfrac{1}{4}\right)^n$.

12. $\dfrac{1 + \sin n}{10^n} \leq \dfrac{2}{10^n}$ and $\displaystyle\sum_{n=0}^{\infty} \dfrac{2}{10^n} = 2\displaystyle\sum_{n=0}^{\infty} \left(\dfrac{1}{10}\right)^n$, so the given series converges by comparison with a constant multiple of a

convergent geometric series.

13. $\dfrac{\arctan n}{n^{1.2}} < \dfrac{\pi/2}{n^{1.2}}$ for all $n \geq 1$, so $\displaystyle\sum_{n=1}^{\infty} \dfrac{\arctan n}{n^{1.2}}$ converges by comparison with $\dfrac{\pi}{2}\displaystyle\sum_{n=1}^{\infty} \dfrac{1}{n^{1.2}}$, which converges because it is a

constant times a p-series with $p = 1.2 > 1$.

14. $\dfrac{\sqrt{n}}{n-1} > \dfrac{\sqrt{n}}{n} = \dfrac{1}{\sqrt{n}}$, so $\displaystyle\sum_{n=2}^{\infty} \dfrac{\sqrt{n}}{n-1}$ diverges by comparison with the divergent (partial) p-series $\displaystyle\sum_{n=2}^{\infty} \dfrac{1}{\sqrt{n}}$ $\left[p = \frac{1}{2} \leq 1\right]$.

15. $\dfrac{2+(-1)^n}{n\sqrt{n}} \leq \dfrac{3}{n\sqrt{n}}$, and $\sum\limits_{n=1}^{\infty} \dfrac{3}{n\sqrt{n}}$ converges because it is a constant multiple of the convergent p-series $\sum\limits_{n=1}^{\infty} \dfrac{1}{n\sqrt{n}}$

$\left[p = \frac{3}{2} > 1\right]$, so the given series converges by the Comparison Test.

16. $\dfrac{1}{\sqrt{n^3+1}} < \dfrac{1}{\sqrt{n^3}} = \dfrac{1}{n^{3/2}}$, so $\sum\limits_{n=1}^{\infty} \dfrac{1}{\sqrt{n^3+1}}$ converges by comparison with the convergent p-series

$\sum\limits_{n=1}^{\infty} \dfrac{1}{n^{3/2}}$ $\left[p = \frac{3}{2} > 1\right]$.

17. Use the Limit Comparison Test with $a_n = \dfrac{1}{\sqrt{n^2+1}}$ and $b_n = \dfrac{1}{n}$:

$\lim\limits_{n\to\infty} \dfrac{a_n}{b_n} = \lim\limits_{n\to\infty} \dfrac{n}{\sqrt{n^2+1}} = \lim\limits_{n\to\infty} \dfrac{1}{\sqrt{1+(1/n^2)}} = 1 > 0$. Since the harmonic series $\sum\limits_{n=1}^{\infty} \dfrac{1}{n}$ diverges, so does

$\sum\limits_{n=1}^{\infty} \dfrac{1}{\sqrt{n^2+1}}$.

18. Use the Limit Comparison Test with $a_n = \dfrac{1}{2n+3}$ and $b_n = \dfrac{1}{n}$: $\lim\limits_{n\to\infty} \dfrac{a_n}{b_n} = \lim\limits_{n\to\infty} \dfrac{n}{2n+3} = \lim\limits_{n\to\infty} \dfrac{1}{2+(3/n)} = \dfrac{1}{2} > 0$.

Since the harmonic series $\sum\limits_{n=1}^{\infty} \dfrac{1}{n}$ diverges, so does $\sum\limits_{n=1}^{\infty} \dfrac{1}{2n+3}$.

19. Use the Limit Comparison Test with $a_n = \dfrac{1+4^n}{1+3^n}$ and $b_n = \dfrac{4^n}{3^n}$:

$\lim\limits_{n\to\infty} \dfrac{a_n}{b_n} = \lim\limits_{n\to\infty} \dfrac{\dfrac{1+4^n}{1+3^n}}{\dfrac{4^n}{3^n}} = \lim\limits_{n\to\infty} \dfrac{1+4^n}{1+3^n} \cdot \dfrac{3^n}{4^n} = \lim\limits_{n\to\infty} \dfrac{1+4^n}{4^n} \cdot \dfrac{3^n}{1+3^n} = \lim\limits_{n\to\infty} \left(\dfrac{1}{4^n}+1\right) \cdot \dfrac{1}{\dfrac{1}{3^n}+1} = 1 > 0$

Since the geometric series $\sum b_n = \sum \left(\frac{4}{3}\right)^n$ diverges, so does $\sum\limits_{n=1}^{\infty} \dfrac{1+4^n}{1+3^n}$. Alternatively, use the Comparison Test with

$\dfrac{1+4^n}{1+3^n} > \dfrac{1+4^n}{3^n+3^n} > \dfrac{4^n}{2(3^n)} = \dfrac{1}{2}\left(\dfrac{4}{3}\right)^n$ or use the Test for Divergence.

20. $4^n > n$ for all $n \geq 1$ since the function $f(x) = 4^x - x$ satisfies $f(1) = 3$ and $f'(x) = 4^x \ln 4 - 1 > 0$ for $x \geq 1$, so

$\dfrac{n+4^n}{n+6^n} < \dfrac{4^n+4^n}{n+6^n} < \dfrac{2 \cdot 4^n}{6^n} = 2\left(\frac{4}{6}\right)^n$, so the series $\sum\limits_{n=1}^{\infty} \dfrac{n+4^n}{n+6^n}$ converges by comparison with $2 \sum\limits_{n=1}^{\infty} \left(\frac{2}{3}\right)^n$, which is a

constant multiple of a convergent geometric series $\left[|r| = \frac{2}{3} < 1\right]$.

Or: Use the Limit Comparison Test with $a_n = \dfrac{n+4^n}{n+6^n}$ and $b_n = \left(\frac{2}{3}\right)^n$.

21. Use the Limit Comparison Test with $a_n = \dfrac{\sqrt{n+2}}{2n^2+n+1}$ and $b_n = \dfrac{1}{n^{3/2}}$:

$\lim\limits_{n\to\infty} \dfrac{a_n}{b_n} = \lim\limits_{n\to\infty} \dfrac{n^{3/2}\sqrt{n+2}}{2n^2+n+1} = \lim\limits_{n\to\infty} \dfrac{(n^{3/2}\sqrt{n+2})/(n^{3/2}\sqrt{n})}{(2n^2+n+1)/n^2} = \lim\limits_{n\to\infty} \dfrac{\sqrt{1+2/n}}{2+1/n+1/n^2} = \dfrac{\sqrt{1}}{2} = \dfrac{1}{2} > 0$.

Since $\sum\limits_{n=1}^{\infty} \dfrac{1}{n^{3/2}}$ is a convergent p-series $\left[p = \frac{3}{2} > 1\right]$, the series $\sum\limits_{n=1}^{\infty} \dfrac{\sqrt{n+2}}{2n^2+n+1}$ also converges.

22. Use the Limit Comparison Test with $a_n = \dfrac{n+2}{(n+1)^3}$ and $b_n = \dfrac{1}{n^2}$:

$$\lim_{n \to \infty} \frac{a_n}{b_n} = \lim_{n \to \infty} \frac{n^2(n+2)}{(n+1)^3} = \lim_{n \to \infty} \frac{1 + \frac{2}{n}}{\left(1 + \frac{1}{n}\right)^3} = 1 > 0. \text{ Since } \sum_{n=3}^{\infty} \frac{1}{n^2} \text{ is a convergent (partial) } p\text{-series } [p = 2 > 1],$$

the series $\displaystyle\sum_{n=3}^{\infty} \frac{n+2}{(n+1)^3}$ also converges.

23. Use the Limit Comparison Test with $a_n = \dfrac{5+2n}{(1+n^2)^2}$ and $b_n = \dfrac{1}{n^3}$:

$$\lim_{n \to \infty} \frac{a_n}{b_n} = \lim_{n \to \infty} \frac{n^3(5+2n)}{(1+n^2)^2} = \lim_{n \to \infty} \frac{5n^3 + 2n^4}{(1+n^2)^2} \cdot \frac{1/n^4}{1/(n^2)^2} = \lim_{n \to \infty} \frac{\frac{5}{n} + 2}{\left(\frac{1}{n^2} + 1\right)^2} = 2 > 0. \text{ Since } \sum_{n=1}^{\infty} \frac{1}{n^3} \text{ is a convergent}$$

p-series $[p = 3 > 1]$, the series $\displaystyle\sum_{n=1}^{\infty} \frac{5+2n}{(1+n^2)^2}$ also converges.

24. If $a_n = \dfrac{n^2 - 5n}{n^3 + n + 1}$ and $b_n = \dfrac{1}{n}$, then $\displaystyle\lim_{n \to \infty} \frac{a_n}{b_n} = \lim_{n \to \infty} \frac{n^3 - 5n^2}{n^3 + n + 1} = \lim_{n \to \infty} \frac{1 - 5/n}{1 + 1/n^2 + 1/n^3} = 1 > 0,$

so $\displaystyle\sum_{n=1}^{\infty} \frac{n^2 - 5n}{n^3 + n + 1}$ diverges by the Limit Comparison Test with the divergent harmonic series $\displaystyle\sum_{n=1}^{\infty} \frac{1}{n}$.

(Note that $a_n > 0$ for $n \geq 6$.)

25. If $a_n = \dfrac{1 + n + n^2}{\sqrt{1 + n^2 + n^6}}$ and $b_n = \dfrac{1}{n}$, then $\displaystyle\lim_{n \to \infty} \frac{a_n}{b_n} = \lim_{n \to \infty} \frac{n + n^2 + n^3}{\sqrt{1 + n^2 + n^6}} = \lim_{n \to \infty} \frac{1/n^2 + 1/n + 1}{\sqrt{1/n^6 + 1/n^4 + 1}} = 1 > 0,$

so $\displaystyle\sum_{n=1}^{\infty} \frac{1 + n + n^2}{\sqrt{1 + n^2 + n^6}}$ diverges by the Limit Comparison Test with the divergent harmonic series $\displaystyle\sum_{n=1}^{\infty} \frac{1}{n}$.

26. If $a_n = \dfrac{n+5}{\sqrt[3]{n^7 + n^2}}$ and $b_n = \dfrac{n}{\sqrt[3]{n^7}} = \dfrac{n}{n^{7/3}} = \dfrac{1}{n^{4/3}}$, then

$$\lim_{n \to \infty} \frac{a_n}{b_n} = \lim_{n \to \infty} \frac{n^{7/3} + 5n^{4/3}}{(n^7 + n^2)^{1/3}} \cdot \frac{n^{-7/3}}{n^{-7/3}} = \lim_{n \to \infty} \frac{1 + 5/n}{\left[(n^7 + n^2)/n^7\right]^{1/3}} = \lim_{n \to \infty} \frac{1 + 5/n}{(1 + 1/n^5)^{1/3}} = \frac{1+0}{(1+0)^{1/3}} = 1 > 0,$$

so $\displaystyle\sum_{n=1}^{\infty} \frac{n+5}{\sqrt[3]{n^7 + n^2}}$ converges by the Limit Comparison Test with the convergent p-series $\displaystyle\sum_{n=1}^{\infty} \frac{1}{n^{4/3}}$.

27. Use the Limit Comparison Test with $a_n = \left(1 + \dfrac{1}{n}\right)^2 e^{-n}$ and $b_n = e^{-n}$: $\displaystyle\lim_{n \to \infty} \frac{a_n}{b_n} = \lim_{n \to \infty} \left(1 + \frac{1}{n}\right)^2 = 1 > 0.$ Since

$\displaystyle\sum_{n=1}^{\infty} e^{-n} = \sum_{n=1}^{\infty} \frac{1}{e^n}$ is a convergent geometric series $\left[|r| = \frac{1}{e} < 1\right]$, the series $\displaystyle\sum_{n=1}^{\infty} \left(1 + \frac{1}{n}\right)^2 e^{-n}$ also converges.

28. $\dfrac{e^{1/n}}{n} > \dfrac{1}{n}$ for all $n \geq 1$, so $\displaystyle\sum_{n=1}^{\infty} \frac{e^{1/n}}{n}$ diverges by comparison with the harmonic series $\displaystyle\sum_{n=1}^{\infty} \frac{1}{n}$.

29. Clearly $n! = n(n-1)(n-2) \cdots (3)(2) \geq 2 \cdot 2 \cdot 2 \cdots \cdots 2 \cdot 2 = 2^{n-1}$, so $\dfrac{1}{n!} \leq \dfrac{1}{2^{n-1}}$. $\displaystyle\sum_{n=1}^{\infty} \frac{1}{2^{n-1}}$ is a convergent geometric

series $\left[|r| = \frac{1}{2} < 1\right]$, so $\displaystyle\sum_{n=1}^{\infty} \frac{1}{n!}$ converges by the Comparison Test.

30. $\dfrac{n!}{n^n} = \dfrac{1 \cdot 2 \cdot 3 \cdot \cdots \cdot (n-1)n}{n \cdot n \cdot n \cdot \cdots \cdot n \cdot n} \le \dfrac{1}{n} \cdot \dfrac{2}{n} \cdot 1 \cdot 1 \cdot \cdots \cdot 1$ for $n \ge 2$, so since $\displaystyle\sum_{n=1}^{\infty} \dfrac{2}{n^2}$ converges $[p = 2 > 1]$, $\displaystyle\sum_{n=1}^{\infty} \dfrac{n!}{n^n}$ converges

also by the Comparison Test.

31. Use the Limit Comparison Test with $a_n = \sin\left(\dfrac{1}{n}\right)$ and $b_n = \dfrac{1}{n}$. Then $\sum a_n$ and $\sum b_n$ are series with positive terms and

$$\lim_{n \to \infty} \dfrac{a_n}{b_n} = \lim_{n \to \infty} \dfrac{\sin(1/n)}{1/n} = \lim_{\theta \to 0} \dfrac{\sin\theta}{\theta} = 1 > 0. \text{ Since } \sum_{n=1}^{\infty} b_n \text{ is the divergent harmonic series,}$$

$\displaystyle\sum_{n=1}^{\infty} \sin(1/n)$ also diverges.   [Note that we could also use l'Hospital's Rule to evaluate the limit:

$$\lim_{x \to \infty} \dfrac{\sin(1/x)}{1/x} \overset{\text{H}}{=} \lim_{x \to \infty} \dfrac{\cos(1/x) \cdot \left(-1/x^2\right)}{-1/x^2} = \lim_{x \to \infty} \cos \dfrac{1}{x} = \cos 0 = 1.]$$

32. Use the Limit Comparison Test with $a_n = \dfrac{1}{n^{1+1/n}}$ and $b_n = \dfrac{1}{n}$. $\displaystyle\lim_{n \to \infty} \dfrac{a_n}{b_n} = \lim_{n \to \infty} \dfrac{n}{n^{1+1/n}} = \lim_{n \to \infty} \dfrac{1}{n^{1/n}} = 1$

$\left[\text{since } \displaystyle\lim_{x \to \infty} x^{1/x} = 1 \text{ by l'Hospital's Rule}\right]$, so $\displaystyle\sum_{n=1}^{\infty} \dfrac{1}{n}$ diverges   [harmonic series] $\Rightarrow$ $\displaystyle\sum_{n=1}^{\infty} \dfrac{1}{n^{1+1/n}}$ diverges.

33. $\displaystyle\sum_{n=1}^{10} \dfrac{1}{\sqrt{n^4+1}} = \dfrac{1}{\sqrt{2}} + \dfrac{1}{\sqrt{17}} + \dfrac{1}{\sqrt{82}} + \cdots + \dfrac{1}{\sqrt{10{,}001}} \approx 1.24856.$ Now $\dfrac{1}{\sqrt{n^4+1}} < \dfrac{1}{\sqrt{n^4}} = \dfrac{1}{n^2}$, so the error is

$$R_{10} \le T_{10} \le \int_{10}^{\infty} \dfrac{1}{x^2}\,dx = \lim_{t \to \infty} \left[-\dfrac{1}{x}\right]_{10}^{t} = \lim_{t \to \infty} \left(-\dfrac{1}{t} + \dfrac{1}{10}\right) = \dfrac{1}{10} = 0.1.$$

34. $\displaystyle\sum_{n=1}^{10} \dfrac{\sin^2 n}{n^3} = \dfrac{\sin^2 1}{1} + \dfrac{\sin^2 2}{8} + \dfrac{\sin^2 3}{27} + \cdots + \dfrac{\sin^2 10}{1000} \approx 0.83253.$ Now $\dfrac{\sin^2 n}{n^3} \le \dfrac{1}{n^3}$, so the error is

$$R_{10} \le T_{10} \le \int_{10}^{\infty} \dfrac{1}{x^3}\,dx = \lim_{t \to \infty} \left[-\dfrac{1}{2x^2}\right]_{10}^{t} = \lim_{t \to \infty} \left(-\dfrac{1}{2t^2} + \dfrac{1}{200}\right) = \dfrac{1}{200} = 0.005.$$

35. $\displaystyle\sum_{n=1}^{10} \dfrac{1}{1+2^n} = \dfrac{1}{3} + \dfrac{1}{5} + \dfrac{1}{9} + \cdots + \dfrac{1}{1025} \approx 0.76352.$ Now $\dfrac{1}{1+2^n} < \dfrac{1}{2^n}$, so the error is

$$R_{10} \le T_{10} = \sum_{n=11}^{\infty} \dfrac{1}{2^n} = \dfrac{1/2^{11}}{1-1/2} \quad \text{[geometric series]} \approx 0.00098.$$

36. $\displaystyle\sum_{n=1}^{10} \dfrac{n}{(n+1)3^n} = \dfrac{1}{6} + \dfrac{2}{27} + \dfrac{3}{108} + \cdots + \dfrac{10}{649{,}539} \approx 0.283597.$ Now $\dfrac{n}{(n+1)3^n} < \dfrac{n}{n \cdot 3^n} = \dfrac{1}{3^n}$, so the error is

$$R_{10} \le T_{10} = \sum_{n=11}^{\infty} \dfrac{1}{3^n} = \dfrac{1/3^{11}}{1-1/3} \approx 0.0000085.$$

37. Since $\dfrac{d_n}{10^n} \le \dfrac{9}{10^n}$ for each n, and since $\displaystyle\sum_{n=1}^{\infty} \dfrac{9}{10^n}$ is a convergent geometric series $\left(|r| = \tfrac{1}{10} < 1\right)$, $0.d_1 d_2 d_3 \ldots = \displaystyle\sum_{n=1}^{\infty} \dfrac{d_n}{10^n}$

will always converge by the Comparison Test.

38. Clearly, if $p < 0$ then the series diverges, since $\displaystyle\lim_{n \to \infty} \dfrac{1}{n^p \ln n} = \infty.$ If $0 \le p \le 1$, then $n^p \ln n \le n \ln n$ $\Rightarrow$

$\dfrac{1}{n^p \ln n} \ge \dfrac{1}{n \ln n}$ and $\displaystyle\sum_{n=2}^{\infty} \dfrac{1}{n \ln n}$ diverges (Exercise 12.3.21), so $\displaystyle\sum_{n=2}^{\infty} \dfrac{1}{n^p \ln n}$ diverges. If $p > 1$, use the Limit Comparison

Test with $a_n = \dfrac{1}{n^p \ln n}$ and $b_n = \dfrac{1}{n^p}$. $\sum\limits_{n=2}^{\infty} b_n$ converges, and $\lim\limits_{n\to\infty} \dfrac{a_n}{b_n} = \lim\limits_{n\to\infty} \dfrac{1}{\ln n} = 0$, so $\sum\limits_{n=2}^{\infty} \dfrac{1}{n^p \ln n}$ also converges.

(Or use the Comparison Test, since $n^p \ln n > n^p$ for $n > e$.) In summary, the series converges if and only if $p > 1$.

39. Since $\sum a_n$ converges, $\lim\limits_{n\to\infty} a_n = 0$, so there exists N such that $|a_n - 0| < 1$ for all $n > N$ $\Rightarrow$ $0 \le a_n < 1$ for

all $n > N$ $\Rightarrow$ $0 \le a_n^2 \le a_n$. Since $\sum a_n$ converges, so does $\sum a_n^2$ by the Comparison Test.

40. (a) Since $\lim\limits_{n\to\infty} (a_n/b_n) = 0$, there is a number $N > 0$ such that $|a_n/b_n - 0| < 1$ for all $n > N$, and so $a_n < b_n$ since a_n

and b_n are positive. Thus, since $\sum b_n$ converges, so does $\sum a_n$ by the Comparison Test.

(b) (i) If $a_n = \dfrac{\ln n}{n^3}$ and $b_n = \dfrac{1}{n^2}$, then $\lim\limits_{n\to\infty} \dfrac{a_n}{b_n} = \lim\limits_{n\to\infty} \dfrac{\ln n}{n} = \lim\limits_{x\to\infty} \dfrac{\ln x}{x} \overset{\text{H}}{=} \lim\limits_{x\to\infty} \dfrac{1/x}{1} = 0$, so $\sum\limits_{n=1}^{\infty} \dfrac{\ln n}{n^3}$ converges by

part (a).

(ii) If $a_n = \dfrac{\ln n}{\sqrt{n}e^n}$ and $b_n = \dfrac{1}{e^n}$, then $\lim\limits_{n\to\infty} \dfrac{a_n}{b_n} = \lim\limits_{n\to\infty} \dfrac{\ln n}{\sqrt{n}} = \lim\limits_{x\to\infty} \dfrac{\ln x}{\sqrt{x}} \overset{\text{H}}{=} \lim\limits_{x\to\infty} \dfrac{1/x}{1/(2\sqrt{x})} = \lim\limits_{x\to\infty} \dfrac{2}{\sqrt{x}} = 0$. Now

$\sum b_n$ is a convergent geometric series with ratio $r = 1/e$ $[\,|r| < 1\,]$, so $\sum a_n$ converges by part (a).

41. (a) Since $\lim\limits_{n\to\infty} \dfrac{a_n}{b_n} = \infty$, there is an integer N such that $\dfrac{a_n}{b_n} > 1$ whenever $n > N$. (Take $M = 1$ in Definition 12.1.5.)

Then $a_n > b_n$ whenever $n > N$ and since $\sum b_n$ is divergent, $\sum a_n$ is also divergent by the Comparison Test.

(b) (i) If $a_n = \dfrac{1}{\ln n}$ and $b_n = \dfrac{1}{n}$ for $n \ge 2$, then $\lim\limits_{n\to\infty} \dfrac{a_n}{b_n} = \lim\limits_{n\to\infty} \dfrac{n}{\ln n} = \lim\limits_{x\to\infty} \dfrac{x}{\ln x} \overset{\text{H}}{=} \lim\limits_{x\to\infty} \dfrac{1}{1/x} = \lim\limits_{x\to\infty} x = \infty$,

so by part (a), $\sum\limits_{n=2}^{\infty} \dfrac{1}{\ln n}$ is divergent.

(ii) If $a_n = \dfrac{\ln n}{n}$ and $b_n = \dfrac{1}{n}$, then $\sum\limits_{n=1}^{\infty} b_n$ is the divergent harmonic series and $\lim\limits_{n\to\infty} \dfrac{a_n}{b_n} = \lim\limits_{n\to\infty} \ln n = \lim\limits_{x\to\infty} \ln x = \infty$,

so $\sum\limits_{n=1}^{\infty} a_n$ diverges by part (a).

42. Let $a_n = \dfrac{1}{n^2}$ and $b_n = \dfrac{1}{n}$. Then $\lim\limits_{n\to\infty} \dfrac{a_n}{b_n} = \lim\limits_{n\to\infty} \dfrac{1}{n} = 0$, but $\sum b_n$ diverges while $\sum a_n$ converges.

43. $\lim\limits_{n\to\infty} n a_n = \lim\limits_{n\to\infty} \dfrac{a_n}{1/n}$, so we apply the Limit Comparison Test with $b_n = \dfrac{1}{n}$. Since $\lim\limits_{n\to\infty} n a_n > 0$ we know that either both

series converge or both series diverge, and we also know that $\sum\limits_{n=1}^{\infty} \dfrac{1}{n}$ diverges $[p\text{-series with } p = 1]$. Therefore, $\sum a_n$ must be

divergent.

44. First we observe that, by l'Hospital's Rule, $\lim\limits_{x\to 0} \dfrac{\ln(1+x)}{x} = \lim\limits_{x\to 0} \dfrac{1}{1+x} = 1$. Also, if $\sum a_n$ converges, then $\lim\limits_{n\to\infty} a_n = 0$ by

Theorem 12.2.6. Therefore, $\lim\limits_{n\to\infty} \dfrac{\ln(1+a_n)}{a_n} = \lim\limits_{x\to 0} \dfrac{\ln(1+x)}{x} = 1 > 0$. We are given that $\sum a_n$ is convergent and $a_n > 0$.

Thus, $\sum \ln(1+a_n)$ is convergent by the Limit Comparison Test.

45. Yes. Since $\sum a_n$ is a convergent series with positive terms, $\lim\limits_{n\to\infty} a_n = 0$ by Theorem 12.2.6, and $\sum b_n = \sum \sin(a_n)$ is a

series with positive terms (for large enough n). We have $\lim\limits_{n\to\infty} \dfrac{b_n}{a_n} = \lim\limits_{n\to\infty} \dfrac{\sin(a_n)}{a_n} = 1 > 0$ by Theorem 3.4.2. Thus, $\sum b_n$

is also convergent by the Limit Comparison Test.

46. Yes. Since $\sum a_n$ converges, its terms approach 0 as $n \to \infty$, so for some integer N, $a_n \le 1$ for all $n \ge N$. But then

$\sum_{n=1}^{\infty} a_n b_n = \sum_{n=1}^{N-1} a_n b_n + \sum_{n=N}^{\infty} a_n b_n \le \sum_{n=1}^{N-1} a_n b_n + \sum_{n=N}^{\infty} b_n$. The first term is a finite sum, and the second term

converges since $\sum_{n=1}^{\infty} b_n$ converges. So $\sum a_n b_n$ converges by the Comparison Test.

12.5 Alternating Series

1. (a) An alternating series is a series whose terms are alternately positive and negative.

(b) An alternating series $\sum\limits_{n=1}^{\infty} (-1)^{n-1} b_n$ converges if $0 < b_{n+1} \le b_n$ for all n and $\lim\limits_{n\to\infty} b_n = 0$. (This is the Alternating

Series Test.)

(c) The error involved in using the partial sum s_n as an approximation to the total sum s is the remainder $R_n = s - s_n$ and the

size of the error is smaller than b_{n+1}; that is, $|R_n| \le b_{n+1}$. (This is the Alternating Series Estimation Theorem.)

2. $-\dfrac{1}{3} + \dfrac{2}{4} - \dfrac{3}{5} + \dfrac{4}{6} - \dfrac{5}{7} + \cdots = \sum\limits_{n=1}^{\infty} (-1)^n \dfrac{n}{n+2}$. Here $a_n = (-1)^n \dfrac{n}{n+2}$. Since $\lim\limits_{n\to\infty} a_n \ne 0$ (in fact the limit does not

exist), the series diverges by the Test for Divergence.

3. $\dfrac{4}{7} - \dfrac{4}{8} + \dfrac{4}{9} - \dfrac{4}{10} + \dfrac{4}{11} - \cdots = \sum\limits_{n=1}^{\infty} (-1)^{n-1} \dfrac{4}{n+6}$. Now $b_n = \dfrac{4}{n+6} > 0$, $\{b_n\}$ is decreasing, and $\lim\limits_{n\to\infty} b_n = 0$, so the

series converges by the Alternating Series Test.

4. $\dfrac{1}{\sqrt{2}} - \dfrac{1}{\sqrt{3}} + \dfrac{1}{\sqrt{4}} - \dfrac{1}{\sqrt{5}} + \dfrac{1}{\sqrt{6}} - \cdots = \sum\limits_{n=1}^{\infty} (-1)^{n-1} \dfrac{1}{\sqrt{n+1}}$. Now $b_n = \dfrac{1}{\sqrt{n+1}} > 0$, $\{b_n\}$ is decreasing, and

$\lim\limits_{n\to\infty} b_n = 0$, so the series converges by the Alternating Series Test.

5. $\sum\limits_{n=1}^{\infty} a_n = \sum\limits_{n=1}^{\infty} (-1)^{n-1} \dfrac{1}{2n+1} = \sum\limits_{n=1}^{\infty} (-1)^{n-1} b_n$. Now $b_n = \dfrac{1}{2n+1} > 0$, $\{b_n\}$ is decreasing, and $\lim\limits_{n\to\infty} b_n = 0$, so the

series converges by the Alternating Series Test.

6. $\sum\limits_{n=1}^{\infty} a_n = \sum\limits_{n=1}^{\infty} (-1)^{n-1} \dfrac{1}{\ln(n+4)} = \sum\limits_{n=1}^{\infty} (-1)^{n-1} b_n$. Now $b_n = \dfrac{1}{\ln(n+4)} > 0$, $\{b_n\}$ is decreasing, and $\lim\limits_{n\to\infty} b_n = 0$, so

the series converges by the Alternating Series Test.

7. $\sum\limits_{n=1}^{\infty} a_n = \sum\limits_{n=1}^{\infty} (-1)^n \dfrac{3n-1}{2n+1} = \sum\limits_{n=1}^{\infty} (-1)^n b_n$. Now $\lim\limits_{n\to\infty} b_n = \lim\limits_{n\to\infty} \dfrac{3 - 1/n}{2 + 1/n} = \dfrac{3}{2} \ne 0$. Since $\lim\limits_{n\to\infty} a_n \ne 0$

(in fact the limit does not exist), the series diverges by the Test for Divergence.

8. $b_n = \dfrac{n}{\sqrt{n^3+2}} > 0$ for $n \geq 1$. $\{b_n\}$ is decreasing for $n \geq 2$ since

$$\left(\dfrac{x}{\sqrt{x^3+2}}\right)' = \dfrac{(x^3+2)^{1/2}(1) - x \cdot \frac{1}{2}(x^3+2)^{-1/2}(3x^2)}{\left(\sqrt{x^3+2}\right)^2} = \dfrac{\frac{1}{2}(x^3+2)^{-1/2}[2(x^3+2) - 3x^3]}{(x^3+2)^1} = \dfrac{4-x^3}{2(x^3+2)^{3/2}} < 0 \text{ for}$$

$x > \sqrt[3]{4} \approx 1.6$. Also, $\lim\limits_{n\to\infty} b_n = \lim\limits_{n\to\infty} \dfrac{n/n}{\sqrt{n^3+2}/\sqrt{n^2}} = \lim\limits_{n\to\infty} \dfrac{1}{\sqrt{n+2/n^2}} = 0$. Thus, the series $\sum\limits_{n=1}^{\infty} (-1)^n \dfrac{n}{\sqrt{n^3+2}}$

converges by the Alternating Series Test.

9. $b_n = \dfrac{n}{10^n} > 0$ for $n \geq 1$. $\{b_n\}$ is decreasing for $n \geq 1$ since

$$\left(\dfrac{x}{10^x}\right)' = \dfrac{10^x(1) - x \cdot 10^x \ln 10}{(10^x)^2} = \dfrac{10^x(1 - x \ln 10)}{(10^x)^2} = \dfrac{1 - x \ln 10}{10^x} < 0 \text{ for } 1 - x \ln 10 < 0 \ \Rightarrow \ x \ln 10 > 1 \ \Rightarrow$$

$x > \dfrac{1}{\ln 10} \approx 0.4$. Also, $\lim\limits_{n\to\infty} b_n = \lim\limits_{n\to\infty} \dfrac{n}{10^n} = \lim\limits_{x\to\infty} \dfrac{x}{10^x} \overset{H}{=} \lim\limits_{x\to\infty} \dfrac{x}{10^x \ln 10} = 0$. Thus, the series $\sum\limits_{n=1}^{\infty} (-1)^n \dfrac{n}{10^n}$

converges by the Alternating Series Test.

10. $\sum\limits_{n=1}^{\infty} a_n = \sum\limits_{n=1}^{\infty} (-1)^n \dfrac{\sqrt{n}}{1+2\sqrt{n}} = \sum\limits_{n=1}^{\infty} (-1)^n b_n$. Now $\lim\limits_{n\to\infty} b_n = \lim\limits_{n\to\infty} \dfrac{1}{2+1/\sqrt{n}} = \dfrac{1}{2} \neq 0$. Since $\lim\limits_{n\to\infty} a_n \neq 0$

(in fact the limit does not exist), the series diverges by the Test for Divergence.

11. $b_n = \dfrac{n^2}{n^3+4} > 0$ for $n \geq 1$. $\{b_n\}$ is decreasing for $n \geq 2$ since

$$\left(\dfrac{x^2}{x^3+4}\right)' = \dfrac{(x^3+4)(2x) - x^2(3x^2)}{(x^3+4)^2} = \dfrac{x(2x^3+8-3x^3)}{(x^3+4)^2} = \dfrac{x(8-x^3)}{(x^3+4)^2} < 0 \text{ for } x > 2. \text{ Also,}$$

$\lim\limits_{n\to\infty} b_n = \lim\limits_{n\to\infty} \dfrac{1/n}{1+4/n^3} = 0$. Thus, the series $\sum\limits_{n=1}^{\infty} (-1)^{n+1} \dfrac{n^2}{n^3+4}$ converges by the Alternating Series Test.

12. $b_n = \dfrac{e^{1/n}}{n} > 0$ for $n \geq 1$. $\{b_n\}$ is decreasing since $\left(\dfrac{e^{1/x}}{x}\right)' = \dfrac{x \cdot e^{1/x}(-1/x^2) - e^{1/x} \cdot 1}{x^2} = \dfrac{-e^{1/x}(1+x)}{x^3} < 0$ for

$x > 0$. Also, $\lim\limits_{n\to\infty} b_n = 0$ since $\lim\limits_{n\to\infty} e^{1/n} = 1$. Thus, the series $\sum\limits_{n=1}^{\infty} (-1)^{n-1} \dfrac{e^{1/n}}{n}$ converges by the Alternating Series Test.

13. $\sum\limits_{n=2}^{\infty} (-1)^n \dfrac{n}{\ln n}$. $\lim\limits_{n\to\infty} \dfrac{n}{\ln n} = \lim\limits_{x\to\infty} \dfrac{x}{\ln x} \overset{H}{=} \lim\limits_{x\to\infty} \dfrac{1}{1/x} = \infty$, so the series diverges by the Test for Divergence.

14. $\sum\limits_{n=1}^{\infty} (-1)^{n-1}\left(\dfrac{\ln n}{n}\right) = 0 + \sum\limits_{n=2}^{\infty} (-1)^{n-1}\left(\dfrac{\ln n}{n}\right)$. $b_n = \dfrac{\ln n}{n} > 0$ for $n \geq 2$, and if $f(x) = \dfrac{\ln x}{x}$, then

$f'(x) = \dfrac{1-\ln x}{x^2} < 0$ for $x > e$, so $\{b_n\}$ is eventually decreasing. Also,

$\lim\limits_{n\to\infty} b_n = \lim\limits_{n\to\infty} \dfrac{\ln n}{n} = \lim\limits_{x\to\infty} \dfrac{\ln x}{x} \overset{H}{=} \lim\limits_{x\to\infty} \dfrac{1/x}{1} = 0$, so the series converges by the Alternating Series Test.

15. $\sum\limits_{n=1}^{\infty} \dfrac{\cos n\pi}{n^{3/4}} = \sum\limits_{n=1}^{\infty} \dfrac{(-1)^n}{n^{3/4}}$. $b_n = \dfrac{1}{n^{3/4}}$ is decreasing and positive and $\lim\limits_{n\to\infty} \dfrac{1}{n^{3/4}} = 0$, so the series converges by the

Alternating Series Test.

16. $\sin\left(\dfrac{n\pi}{2}\right) = 0$ if n is even and $(-1)^k$ if $n = 2k+1$, so the series $\displaystyle\sum_{n=1}^{\infty} \dfrac{\sin(n\pi/2)}{n!} = \sum_{n=0}^{\infty} \dfrac{(-1)^n}{(2n+1)!}$.

$b_n = \dfrac{1}{(2n+1)!} > 0$, $\{b_n\}$ is decreasing, and $\displaystyle\lim_{n\to\infty} \dfrac{1}{(2n+1)!} = 0$, so the series converges by the Alternating Series Test.

17. $\displaystyle\sum_{n=1}^{\infty} (-1)^n \sin\left(\dfrac{\pi}{n}\right)$. $b_n = \sin\left(\dfrac{\pi}{n}\right) > 0$ for $n \geq 2$ and $\sin\left(\dfrac{\pi}{n}\right) \geq \sin\left(\dfrac{\pi}{n+1}\right)$, and $\displaystyle\lim_{n\to\infty} \sin\left(\dfrac{\pi}{n}\right) = \sin 0 = 0$, so the series converges by the Alternating Series Test.

18. $\displaystyle\sum_{n=1}^{\infty} (-1)^n \cos\left(\dfrac{\pi}{n}\right)$. $\displaystyle\lim_{n\to\infty} \cos\left(\dfrac{\pi}{n}\right) = \cos(0) = 1$, so $\displaystyle\lim_{n\to\infty} (-1)^n \cos\left(\dfrac{\pi}{n}\right)$ does not exist and the series diverges by the Test for Divergence.

19. $\dfrac{n^n}{n!} = \dfrac{n \cdot n \cdot \cdots \cdot n}{1 \cdot 2 \cdot \cdots \cdot n} \geq n \quad \Rightarrow \quad \displaystyle\lim_{n\to\infty} \dfrac{n^n}{n!} = \infty \quad \Rightarrow \quad \displaystyle\lim_{n\to\infty} \dfrac{(-1)^n n^n}{n!}$ does not exist. So the series diverges by the Test for Divergence.

20. $\displaystyle\sum_{n=1}^{\infty} \left(-\dfrac{n}{5}\right)^n$ diverges by the Test for Divergence since $\displaystyle\lim_{n\to\infty} \left(\dfrac{n}{5}\right)^n = \infty \quad \Rightarrow \quad \displaystyle\lim_{n\to\infty} \left(-\dfrac{n}{5}\right)^n$ does not exist.

21.

n	a_n	s_n
1	1	1
2	-0.35355	0.64645
3	0.19245	0.83890
4	-0.125	0.71390
5	0.08944	0.80334
6	-0.06804	0.73530
7	0.05399	0.78929
8	-0.04419	0.74510
9	0.03704	0.78214
10	-0.03162	0.75051

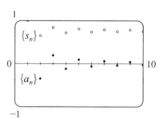

By the Alternating Series Estimation Theorem, the error in the approximation

$\displaystyle\sum_{n=1}^{\infty} \dfrac{(-1)^{n-1}}{n^{3/2}} \approx 0.75051$ is $|s - s_{10}| \leq b_{11} = 1/(11)^{3/2} \approx 0.0275$ (to four decimal places, rounded up).

22.

n	a_n	s_n
1	1	1
2	-0.125	0.875
3	0.03704	0.91204
4	-0.01563	0.89641
5	0.008	0.90441
6	-0.00463	0.89978
7	0.00292	0.90270
8	-0.00195	0.90074
9	0.00137	0.90212
10	-0.001	0.90112

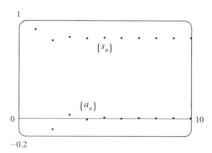

By the Alternating Series Estimation Theorem, the error in the approximation

$\displaystyle\sum_{n=1}^{\infty} \dfrac{(-1)^{n-1}}{n^3} \approx 0.90112$ is $|s - s_{10}| \leq b_{11} = 1/11^3 \approx 0.0007513$.

23. The series $\sum\limits_{n=1}^{\infty} \dfrac{(-1)^{n+1}}{n^6}$ satisfies (i) of the Alternating Series Test because $\dfrac{1}{(n+1)^6} < \dfrac{1}{n^6}$ and (ii) $\lim\limits_{n\to\infty} \dfrac{1}{n^6} = 0$, so the

series is convergent. Now $b_5 = \dfrac{1}{5^6} = 0.000064 > 0.00005$ and $b_6 = \dfrac{1}{6^6} \approx 0.00002 < 0.00005$, so by the Alternating Series

Estimation Theorem, $n = 5$. (That is, since the 6th term is less than the desired error, we need to add the first 5 terms to get the

sum to the desired accuracy.)

24. The series $\sum\limits_{n=1}^{\infty} \dfrac{(-1)^n}{n\,5^n}$ satisfies (i) of the Alternating Series Test because $\dfrac{1}{(n+1)5^{n+1}} < \dfrac{1}{n\,5^n}$ and (ii) $\lim\limits_{n\to\infty} \dfrac{1}{n\,5^n} = 0$, so

the series is convergent. Now $b_4 = \dfrac{1}{4 \cdot 5^4} = 0.0004 > 0.0001$ and $b_5 = \dfrac{1}{5 \cdot 5^5} = 0.000064 < 0.0001$, so by the Alternating

Series Estimation Theorem, $n = 4$. (That is, since the 5th term is less than the desired error, we need to add the first 4 terms to

get the sum to the desired accuracy.)

25. The series $\sum\limits_{n=0}^{\infty} \dfrac{(-1)^n}{10^n\, n!}$ satisfies (i) of the Alternating Series Test because $\dfrac{1}{10^{n+1}(n+1)!} < \dfrac{1}{10^n\, n!}$ and (ii) $\lim\limits_{n\to\infty} \dfrac{1}{10^n\, n!} = 0$,

so the series is convergent. Now $b_3 = \dfrac{1}{10^3\, 3!} \approx 0.000\,167 > 0.000\,005$ and $b_4 = \dfrac{1}{10^4\, 4!} = 0.000\,004 < 0.000\,005$, so by

the Alternating Series Estimation Theorem, $n = 4$ (since the series starts with $n = 0$, not $n = 1$). (That is, since the 5th term

is less than the desired error, we need to add the first 4 terms to get the sum to the desired accuracy.)

26. The series $\sum\limits_{n=1}^{\infty} (-1)^{n-1} n e^{-n} = \sum\limits_{n=1}^{\infty} (-1)^{n-1} \dfrac{n}{e^n}$ satisfies (i) of the Alternating Series Test because

$\left(\dfrac{x}{e^x}\right)' = \dfrac{e^x(1) - x e^x}{(e^x)^2} = \dfrac{e^x(1-x)}{(e^x)^2} = \dfrac{1-x}{e^x} < 0$ for $x > 1$ and (ii) $\lim\limits_{n\to\infty} \dfrac{n}{e^n} = \lim\limits_{x\to\infty} \dfrac{x}{e^x} \overset{\text{H}}{=} \lim\limits_{x\to\infty} \dfrac{1}{e^x} = 0$, so the series is

convergent. Now $b_6 = 6/e^6 \approx 0.015 > 0.01$ and $b_7 = 7/e^7 \approx 0.006 < 0.01$, so by the Alternating Series Estimation

Theorem, $n = 6$. (That is, since the 7th term is less than the desired error, we need to add the first 6 terms to get the sum to the

desired accuracy.)

27. $b_7 = \dfrac{1}{7^5} = \dfrac{1}{16,807} \approx 0.000\,059\,5$, so

$$\sum\limits_{n=1}^{\infty} \dfrac{(-1)^{n+1}}{n^5} \approx s_6 = \sum\limits_{n=1}^{6} \dfrac{(-1)^{n+1}}{n^5} = 1 - \tfrac{1}{32} + \tfrac{1}{243} - \tfrac{1}{1024} + \tfrac{1}{3125} - \tfrac{1}{7776} \approx 0.972\,080.$$ Adding b_7 to s_6 does not change

the fourth decimal place of s_6, so the sum of the series, correct to four decimal places, is 0.9721.

28. $b_6 = \dfrac{6}{8^6} = \dfrac{6}{262,144} \approx 0.000\,023$, so $\sum\limits_{n=1}^{\infty} \dfrac{(-1)^n n}{8^n} \approx s_5 = \sum\limits_{n=1}^{5} \dfrac{(-1)^n n}{8^n} = -\tfrac{1}{8} + \tfrac{2}{64} - \tfrac{3}{512} + \tfrac{4}{4096} - \tfrac{5}{32,768} \approx -0.098\,785$.

Adding b_6 to s_5 does not change the fourth decimal place of s_5, so the sum of the series, correct to four decimal places,

is -0.0988.

29. $b_7 = \dfrac{7^2}{10^7} = 0.000\,004\,9$, so

$$\sum\limits_{n=1}^{\infty} \dfrac{(-1)^{n-1} n^2}{10^n} \approx s_6 = \sum\limits_{n=1}^{6} \dfrac{(-1)^{n-1} n^2}{10^n} = \tfrac{1}{10} - \tfrac{4}{100} + \tfrac{9}{1000} - \tfrac{16}{10,000} + \tfrac{25}{100,000} - \tfrac{36}{1,000,000} = 0.067\,614.$$ Adding b_7 to s_6

does not change the fourth decimal place of s_6, so the sum of the series, correct to four decimal places, is 0.0676.

30. $b_6 = \dfrac{1}{3^6 \cdot 6!} = \dfrac{1}{524{,}880} \approx 0.000\,001\,9$, so

$\displaystyle\sum_{n=1}^{\infty} \dfrac{(-1)^n}{3^n n!} \approx s_5 = \sum_{n=1}^{5} \dfrac{(-1)^n}{3^n n!} = -\dfrac{1}{3} + \dfrac{1}{18} - \dfrac{1}{162} + \dfrac{1}{1944} - \dfrac{1}{29{,}160} \approx -0.283\,471$. Adding b_6 to s_5 does not change the

fourth decimal place of s_5, so the sum of the series, correct to four decimal places, is -0.2835.

31. $\displaystyle\sum_{n=1}^{\infty} \dfrac{(-1)^{n-1}}{n} = 1 - \dfrac{1}{2} + \dfrac{1}{3} - \dfrac{1}{4} + \cdots + \dfrac{1}{49} - \dfrac{1}{50} + \dfrac{1}{51} - \dfrac{1}{52} + \cdots$. The 50th partial sum of this series is an

underestimate, since $\displaystyle\sum_{n=1}^{\infty} \dfrac{(-1)^{n-1}}{n} = s_{50} + \left(\dfrac{1}{51} - \dfrac{1}{52} \right) + \left(\dfrac{1}{53} - \dfrac{1}{54} \right) + \cdots$, and the terms in parentheses are all positive.

The result can be seen geometrically in Figure 1.

32. If $p > 0$, $\dfrac{1}{(n+1)^p} \leq \dfrac{1}{n^p}$ ($\{1/n^p\}$ is decreasing) and $\displaystyle\lim_{n\to\infty} \dfrac{1}{n^p} = 0$, so the series converges by the Alternating Series Test.

If $p \leq 0$, $\displaystyle\lim_{n\to\infty} \dfrac{(-1)^{n-1}}{n^p}$ does not exist, so the series diverges by the Test for Divergence. Thus, $\displaystyle\sum_{n=1}^{\infty} \dfrac{(-1)^{n-1}}{n^p}$

converges $\Leftrightarrow$ $p > 0$.

33. Clearly $b_n = \dfrac{1}{n+p}$ is decreasing and eventually positive and $\displaystyle\lim_{n\to\infty} b_n = 0$ for any p. So the series converges (by the

Alternating Series Test) for any p for which every b_n is defined, that is, $n + p \neq 0$ for $n \geq 1$, or p is not a negative integer.

34. Let $f(x) = \dfrac{(\ln x)^p}{x}$. Then $f'(x) = \dfrac{(\ln x)^{p-1}(p - \ln x)}{x^2} < 0$ if $x > e^p$ so f is eventually decreasing for every p. Clearly

$\displaystyle\lim_{n\to\infty} \dfrac{(\ln n)^p}{n} = 0$ if $p \leq 0$, and if $p > 0$ we can apply l'Hospital's Rule $\llbracket p + 1 \rrbracket$ times to get a limit of 0 as well. So the series

converges for all p (by the Alternating Series Test).

35. $\sum b_{2n} = \sum 1/(2n)^2$ clearly converges (by comparison with the p-series for $p = 2$). So suppose that $\sum (-1)^{n-1} b_n$

converges. Then by Theorem 12.2.8(ii), so does $\sum \left[(-1)^{n-1} b_n + b_n \right] = 2\left(1 + \dfrac{1}{3} + \dfrac{1}{5} + \cdots \right) = 2\sum \dfrac{1}{2n-1}$. But this

diverges by comparison with the harmonic series, a contradiction. Therefore, $\sum (-1)^{n-1} b_n$ must diverge. The Alternating

Series Test does not apply since $\{b_n\}$ is not decreasing.

36. (a) We will prove this by induction. Let $P(n)$ be the proposition that $s_{2n} = h_{2n} - h_n$. $P(1)$ is the statement $s_2 = h_2 - h_1$,

which is true since $1 - \dfrac{1}{2} = \left(1 + \dfrac{1}{2} \right) - 1$. So suppose that $P(n)$ is true. We will show that $P(n+1)$ must be true as a

consequence.

$$h_{2n+2} - h_{n+1} = \left(h_{2n} + \dfrac{1}{2n+1} + \dfrac{1}{2n+2} \right) - \left(h_n + \dfrac{1}{n+1} \right) = (h_{2n} - h_n) + \dfrac{1}{2n+1} - \dfrac{1}{2n+2}$$

$$= s_{2n} + \dfrac{1}{2n+1} - \dfrac{1}{2n+2} = s_{2n+2}$$

which is $P(n+1)$, and proves that $s_{2n} = h_{2n} - h_n$ for all n.

(b) We know that $h_{2n} - \ln(2n) \to \gamma$ and $h_n - \ln n \to \gamma$ as $n \to \infty$. So

$s_{2n} = h_{2n} - h_n = [h_{2n} - \ln(2n)] - (h_n - \ln n) + [\ln(2n) - \ln n]$, and

$\displaystyle\lim_{n\to\infty} s_{2n} = \gamma - \gamma + \lim_{n\to\infty} [\ln(2n) - \ln n] = \lim_{n\to\infty} (\ln 2 + \ln n - \ln n) = \ln 2$.

12.6   Absolute Convergence and the Ratio and Root Tests

1. (a) Since $\lim\limits_{n\to\infty}\left|\dfrac{a_{n+1}}{a_n}\right| = 8 > 1$, part (b) of the Ratio Test tells us that the series $\sum a_n$ is divergent.

(b) Since $\lim\limits_{n\to\infty}\left|\dfrac{a_{n+1}}{a_n}\right| = 0.8 < 1$, part (a) of the Ratio Test tells us that the series $\sum a_n$ is absolutely convergent (and

therefore convergent).

(c) Since $\lim\limits_{n\to\infty}\left|\dfrac{a_{n+1}}{a_n}\right| = 1$, the Ratio Test fails and the series $\sum a_n$ might converge or it might diverge.

2. The series $\sum\limits_{n=1}^{\infty}\dfrac{n^2}{2^n}$ has positive terms and $\lim\limits_{n\to\infty}\dfrac{a_{n+1}}{a_n} = \lim\limits_{n\to\infty}\left[\dfrac{(n+1)^2}{2^{n+1}}\cdot\dfrac{2^n}{n^2}\right] = \lim\limits_{n\to\infty}\left(1+\dfrac{1}{n}\right)^2\cdot\dfrac{1}{2} = \dfrac{1}{2} < 1$, so the

series is absolutely convergent by the Ratio Test.

3. $\sum\limits_{n=0}^{\infty}\dfrac{(-10)^n}{n!}$. Using the Ratio Test, $\lim\limits_{n\to\infty}\left|\dfrac{a_{n+1}}{a_n}\right| = \lim\limits_{n\to\infty}\left|\dfrac{(-10)^{n+1}}{(n+1)!}\cdot\dfrac{n!}{(-10)^n}\right| = \lim\limits_{n\to\infty}\left|\dfrac{-10}{n+1}\right| = 0 < 1$, so the series is

absolutely convergent.

4. $\sum\limits_{n=1}^{\infty}(-1)^{n-1}\dfrac{2^n}{n^4}$ diverges by the Test for Divergence. $\lim\limits_{n\to\infty}\dfrac{2^n}{n^4} = \infty$, so $\lim\limits_{n\to\infty}(-1)^{n-1}\dfrac{2^n}{n^4}$ does not exist.

5. $\sum\limits_{n=1}^{\infty}\dfrac{(-1)^{n+1}}{\sqrt[4]{n}}$ converges by the Alternating Series Test, but $\sum\limits_{n=1}^{\infty}\dfrac{1}{\sqrt[4]{n}}$ is a divergent p-series $\left(p=\frac{1}{4}\leq 1\right)$, so the given series

is conditionally convergent.

6. $\sum\limits_{n=1}^{\infty}\dfrac{1}{n^4}$ is a convergent p-series $(p=4>1)$, so $\sum\limits_{n=1}^{\infty}\dfrac{(-1)^n}{n^4}$ is absolutely convergent.

7. $\lim\limits_{k\to\infty}\left|\dfrac{a_{k+1}}{a_k}\right| = \lim\limits_{k\to\infty}\left[\dfrac{(k+1)\left(\frac{2}{3}\right)^{k+1}}{k\left(\frac{2}{3}\right)^k}\right] = \lim\limits_{k\to\infty}\dfrac{k+1}{k}\left(\dfrac{2}{3}\right)^1 = \dfrac{2}{3}\lim\limits_{k\to\infty}\left(1+\dfrac{1}{k}\right) = \frac{2}{3}(1) = \frac{2}{3} < 1$, so the series

$\sum\limits_{n=1}^{\infty} k\left(\frac{2}{3}\right)^k$ is absolutely convergent by the Ratio Test. Since the terms of this series are positive, absolute convergence is the

same as convergence.

8. $\lim\limits_{n\to\infty}\left|\dfrac{a_{n+1}}{a_n}\right| = \lim\limits_{n\to\infty}\left[\dfrac{(n+1)!}{100^{n+1}}\cdot\dfrac{100^n}{n!}\right] = \lim\limits_{n\to\infty}\dfrac{n+1}{100} = \infty$, so the series $\sum\limits_{n=1}^{\infty}\dfrac{n!}{100^n}$ diverges by the Ratio Test.

9. $\lim\limits_{n\to\infty}\left|\dfrac{a_{n+1}}{a_n}\right| = \lim\limits_{n\to\infty}\left[\dfrac{(1.1)^{n+1}}{(n+1)^4}\cdot\dfrac{n^4}{(1.1)^n}\right] = \lim\limits_{n\to\infty}\dfrac{(1.1)n^4}{(n+1)^4} = (1.1)\lim\limits_{n\to\infty}\dfrac{1}{\dfrac{(n+1)^4}{n^4}} = (1.1)\lim\limits_{n\to\infty}\dfrac{1}{(1+1/n)^4}$

$= (1.1)(1) = 1.1 > 1,$

so the series $\sum\limits_{n=1}^{\infty}(-1)^n\dfrac{(1.1)^n}{n^4}$ diverges by the Ratio Test.

10. $\sum_{n=1}^{\infty} (-1)^n \dfrac{n}{\sqrt{n^3+2}}$ converges by the Alternating Series Test (see Exercise 12.5.8). Let $a_n = \dfrac{1}{\sqrt{n}}$ with $b_n = \dfrac{n}{\sqrt{n^3+2}}$.

Then $\lim\limits_{n\to\infty} \dfrac{a_n}{b_n} = \lim\limits_{n\to\infty} \left(\dfrac{1}{\sqrt{n}} \cdot \dfrac{\sqrt{n^3+2}}{n} \right) = \lim\limits_{n\to\infty} \dfrac{\sqrt{n^3+2}}{\sqrt{n^3}} = \lim\limits_{n\to\infty} \sqrt{1 + \dfrac{2}{n^3}} = 1 > 0$, so $\sum_{n=1}^{\infty} \dfrac{n}{\sqrt{n^3+2}}$ diverges by

limit comparision with the divergent p-series $\sum_{n=1}^{\infty} \dfrac{1}{\sqrt{n}}$ $\left[p = \frac{1}{2} \leq 1 \right]$. Thus, $\sum_{n=1}^{\infty} (-1)^n \dfrac{n}{\sqrt{n^3+2}}$ is conditionally convergent.

11. Since $0 \leq \dfrac{e^{1/n}}{n^3} \leq \dfrac{e}{n^3} = e\left(\dfrac{1}{n^3}\right)$ and $\sum_{n=1}^{\infty} \dfrac{1}{n^3}$ is a convergent p-series $[p = 3 > 1]$, $\sum_{n=1}^{\infty} \dfrac{e^{1/n}}{n^3}$ converges, and so

$\sum_{n=1}^{\infty} \dfrac{(-1)^n e^{1/n}}{n^3}$ is absolutely convergent.

12. $\left| \dfrac{\sin 4n}{4^n} \right| \leq \dfrac{1}{4^n}$, so $\sum_{n=1}^{\infty} \left| \dfrac{\sin 4n}{4^n} \right|$ converges by comparison with the convergent geometric series $\sum_{n=1}^{\infty} \dfrac{1}{4^n}$ $\left[|r| = \frac{1}{4} < 1 \right]$.

Thus, $\sum_{n=1}^{\infty} \dfrac{\sin 4n}{4^n}$ is absolutely convergent.

13. $\lim\limits_{n\to\infty} \left| \dfrac{a_{n+1}}{a_n} \right| = \lim\limits_{n\to\infty} \left[\dfrac{10^{n+1}}{(n+2)\,4^{2n+3}} \cdot \dfrac{(n+1)\,4^{2n+1}}{10^n} \right] = \lim\limits_{n\to\infty} \left(\dfrac{10}{4^2} \cdot \dfrac{n+1}{n+2} \right) = \dfrac{5}{8} < 1$, so the series $\sum_{n=1}^{\infty} \dfrac{10^n}{(n+1)4^{2n+1}}$

is absolutely convergent by the Ratio Test. Since the terms of this series are positive, absolute convergence is the same as convergence.

14. $\lim\limits_{n\to\infty} \left| \dfrac{a_{n+1}}{a_n} \right| = \lim\limits_{n\to\infty} \left[\dfrac{(n+1)^2\,2^{n+1}}{(n+1)!} \cdot \dfrac{n!}{n^2 2^n} \right] = \lim\limits_{n\to\infty} \left[\left(1 + \dfrac{1}{n}\right)^2 \cdot \dfrac{2}{n+1} \right] = 0$, so the series $\sum_{n=1}^{\infty} (-1)^{n+1} \dfrac{n^2 2^n}{n!}$ is

absolutely convergent by the Ratio Test.

15. $\left| \dfrac{(-1)^n \arctan n}{n^2} \right| < \dfrac{\pi/2}{n^2}$, so since $\sum_{n=1}^{\infty} \dfrac{\pi/2}{n^2} = \dfrac{\pi}{2} \sum_{n=1}^{\infty} \dfrac{1}{n^2}$ converges $(p = 2 > 1)$, the given series $\sum_{n=1}^{\infty} \dfrac{(-1)^n \arctan n}{n^2}$

converges absolutely by the Comparison Test.

16. $n^{2/3} - 2 > 0$ for $n \geq 3$, so $\dfrac{3 - \cos n}{n^{2/3} - 2} > \dfrac{1}{n^{2/3} - 2} > \dfrac{1}{n^{2/3}}$ for $n \geq 3$. Since $\sum_{n=1}^{\infty} \dfrac{1}{n^{2/3}}$ diverges $\left[p = \frac{2}{3} \leq 1 \right]$, so does

$\sum_{n=1}^{\infty} \dfrac{3 - \cos n}{n^{2/3} - 2}$ by the Comparison Test.

17. $\sum_{n=2}^{\infty} \dfrac{(-1)^n}{\ln n}$ converges by the Alternating Series Test since $\lim\limits_{n\to\infty} \dfrac{1}{\ln n} = 0$ and $\left\{ \dfrac{1}{\ln n} \right\}$ is decreasing. Now $\ln n < n$, so

$\dfrac{1}{\ln n} > \dfrac{1}{n}$, and since $\sum_{n=2}^{\infty} \dfrac{1}{n}$ is the divergent (partial) harmonic series, $\sum_{n=2}^{\infty} \dfrac{1}{\ln n}$ diverges by the Comparison Test. Thus,

$\sum_{n=2}^{\infty} \dfrac{(-1)^n}{\ln n}$ is conditionally convergent.

18. $\lim\limits_{n\to\infty} \left| \dfrac{a_{n+1}}{a_n} \right| = \lim\limits_{n\to\infty} \dfrac{(n+1)!/(n+1)^{n+1}}{n!/n^n} = \lim\limits_{n\to\infty} \dfrac{n^n}{(n+1)^n} = \lim\limits_{n\to\infty} \dfrac{1}{(1+1/n)^n} = \dfrac{1}{e} < 1$, so the series $\sum_{n=1}^{\infty} \dfrac{n!}{n^n}$

converges absolutely by the Ratio Test.

19. $\dfrac{|\cos{(n\pi/3)}|}{n!} \le \dfrac{1}{n!}$ and $\displaystyle\sum_{n=1}^{\infty} \dfrac{1}{n!}$ converges (use the Ratio Test), so the series $\displaystyle\sum_{n=1}^{\infty} \dfrac{\cos(n\pi/3)}{n!}$ converges absolutely by the

Comparison Test.

20. $\displaystyle\lim_{n\to\infty} \sqrt[n]{|a_n|} = \lim_{n\to\infty} \sqrt[n]{\left|\dfrac{(-2)^n}{n^n}\right|} = \lim_{n\to\infty} \dfrac{2}{n} = 0 < 1$, so the series $\displaystyle\sum_{n=1}^{\infty} \dfrac{(-2)^n}{n^n}$ is absolutely convergent by the Root Test.

21. $\displaystyle\lim_{n\to\infty} \sqrt[n]{|a_n|} = \lim_{n\to\infty} \dfrac{n^2+1}{2n^2+1} = \lim_{n\to\infty} \dfrac{1+1/n^2}{2+1/n^2} = \dfrac{1}{2} < 1$, so the series $\displaystyle\sum_{n=1}^{\infty} \left(\dfrac{n^2+1}{2n^2+1}\right)^n$ is absolutely convergent by the

Root Test.

22. $\displaystyle\lim_{n\to\infty} \sqrt[n]{|a_n|} = \lim_{n\to\infty} \sqrt[n]{\left|\left(\dfrac{-2n}{n+1}\right)^{5n}\right|} = \lim_{n\to\infty} \dfrac{2^5 n^5}{(n+1)^5} = 32 \lim_{n\to\infty} \dfrac{1}{\left(\dfrac{n+1}{n}\right)^5} = 32 \lim_{n\to\infty} \dfrac{1}{(1+1/n)^5}$

$$= 32(1) = 32 > 1,$$

so the series $\displaystyle\sum_{n=2}^{\infty} \left(\dfrac{-2n}{n+1}\right)^{5n}$ diverges by the Root Test.

23. $\displaystyle\lim_{n\to\infty} \sqrt[n]{|a_n|} = \lim_{n\to\infty} \sqrt[n]{\left(1+\dfrac{1}{n}\right)^{n^2}} = \lim_{n\to\infty} \left(1+\dfrac{1}{n}\right)^n = e > 1$ [by Equation 7.4.9 (or 7.4*.9)], so the series

$\displaystyle\sum_{n=1}^{\infty} \left(1+\dfrac{1}{n}\right)^{n^2}$ diverges by the Root Test.

24. $\displaystyle\lim_{n\to\infty} \sqrt[n]{|a_n|} = \lim_{n\to\infty} \sqrt[n]{\left|\dfrac{n}{(\ln n)^n}\right|} = \lim_{n\to\infty} \dfrac{n^{1/n}}{\ln n} \overset{(\ast)}{=} 0 < 1$, so the series $\displaystyle\sum_{n=2}^{\infty} \dfrac{n}{(\ln n)^n}$ is absolutely convergent by the

Root Test.

$(\ast)$ Let $y = x^{1/x}$. Then $\ln y = \dfrac{1}{x} \ln x$, so $\displaystyle\lim_{x\to\infty} \ln y = \lim_{x\to\infty} \dfrac{\ln x}{x} \overset{\text{H}}{=} \lim_{x\to\infty} \dfrac{1/x}{1} = 0 \;\Rightarrow\; \lim_{x\to\infty} y = e^0 = 1$.

25. Use the Ratio Test with the series

$$1 - \dfrac{1\cdot 3}{3!} + \dfrac{1\cdot 3\cdot 5}{5!} - \dfrac{1\cdot 3\cdot 5\cdot 7}{7!} + \cdots + (-1)^{n-1}\dfrac{1\cdot 3\cdot 5\cdots (2n-1)}{(2n-1)!} + \cdots = \sum_{n=1}^{\infty} (-1)^{n-1}\dfrac{1\cdot 3\cdot 5\cdots (2n-1)}{(2n-1)!}.$$

$$\lim_{n\to\infty}\left|\dfrac{a_{n+1}}{a_n}\right| = \lim_{n\to\infty}\left|\dfrac{(-1)^n \cdot 1\cdot 3\cdot 5\cdots (2n-1)[2(n+1)-1]}{[2(n+1)-1]!} \cdot \dfrac{(2n-1)!}{(-1)^{n-1}\cdot 1\cdot 3\cdot 5\cdots (2n-1)}\right|$$

$$= \lim_{n\to\infty}\left|\dfrac{(-1)(2n+1)(2n-1)!}{(2n+1)(2n)(2n-1)!}\right| = \lim_{n\to\infty} \dfrac{1}{2n} = 0 < 1,$$

so the given series is absolutely convergent and therefore convergent.

26. Use the Ratio Test with the series $\dfrac{2}{5} + \dfrac{2\cdot 6}{5\cdot 8} + \dfrac{2\cdot 6\cdot 10}{5\cdot 8\cdot 11} + \dfrac{2\cdot 6\cdot 10\cdot 14}{5\cdot 8\cdot 11\cdot 14} + \cdots = \displaystyle\sum_{n=1}^{\infty} \dfrac{2\cdot 6\cdot 10\cdot 14\cdots (4n-2)}{5\cdot 8\cdot 11\cdot 14\cdots (3n+2)}.$

$$\lim_{n\to\infty}\left|\dfrac{a_{n+1}}{a_n}\right| = \lim_{n\to\infty}\left|\dfrac{2\cdot 6\cdot 10\cdots (4n-2)[4(n+1)-2]}{5\cdot 8\cdot 11\cdots (3n+2)[3(n+1)+2]} \cdot \dfrac{5\cdot 8\cdot 11\cdots (3n+2)}{2\cdot 6\cdot 10\cdots (4n-2)}\right| = \lim_{n\to\infty} \dfrac{4n+2}{3n+5} = \dfrac{4}{3} > 1,$$

so the given series is divergent.

27. $\displaystyle\sum_{n=1}^{\infty} \dfrac{2\cdot 4\cdot 6\cdots (2n)}{n!} = \sum_{n=1}^{\infty} \dfrac{(2\cdot 1)\cdot(2\cdot 2)\cdot(2\cdot 3)\cdots (2\cdot n)}{n!} = \sum_{n=1}^{\infty} \dfrac{2^n n!}{n!} = \sum_{n=1}^{\infty} 2^n$, which diverges by the Test for

Divergence since $\displaystyle\lim_{n\to\infty} 2^n = \infty$.

28. $\lim\limits_{n\to\infty}\left|\dfrac{a_{n+1}}{a_n}\right| = \lim\limits_{n\to\infty}\left|\dfrac{\dfrac{2^{n+1}\,(n+1)!}{5\cdot 8\cdot 11\cdot\,\cdots\,\cdot(3n+5)}}{\dfrac{2^n n!}{5\cdot 8\cdot 11\cdot\,\cdots\,\cdot(3n+2)}}\right| = \lim\limits_{n\to\infty}\dfrac{2(n+1)}{3n+5} = \dfrac{2}{3} < 1$, so the series converges absolutely by the

Ratio Test.

29. By the recursive definition, $\lim\limits_{n\to\infty}\left|\dfrac{a_{n+1}}{a_n}\right| = \lim\limits_{n\to\infty}\left|\dfrac{5n+1}{4n+3}\right| = \dfrac{5}{4} > 1$, so the series diverges by the Ratio Test.

30. By the recursive definition, $\lim\limits_{n\to\infty}\left|\dfrac{a_{n+1}}{a_n}\right| = \lim\limits_{n\to\infty}\left|\dfrac{2+\cos n}{\sqrt{n}}\right| = 0 < 1$, so the series converges absolutely by the Ratio Test.

31. (a) $\lim\limits_{n\to\infty}\left|\dfrac{1/(n+1)^3}{1/n^3}\right| = \lim\limits_{n\to\infty}\dfrac{n^3}{(n+1)^3} = \lim\limits_{n\to\infty}\dfrac{1}{(1+1/n)^3} = 1.$ Inconclusive

(b) $\lim\limits_{n\to\infty}\left|\dfrac{(n+1)}{2^{n+1}}\cdot\dfrac{2^n}{n}\right| = \lim\limits_{n\to\infty}\dfrac{n+1}{2n} = \lim\limits_{n\to\infty}\left(\dfrac{1}{2}+\dfrac{1}{2n}\right) = \dfrac{1}{2}.$ Conclusive (convergent)

(c) $\lim\limits_{n\to\infty}\left|\dfrac{(-3)^n}{\sqrt{n+1}}\cdot\dfrac{\sqrt{n}}{(-3)^{n-1}}\right| = 3\lim\limits_{n\to\infty}\sqrt{\dfrac{n}{n+1}} = 3\lim\limits_{n\to\infty}\sqrt{\dfrac{1}{1+1/n}} = 3.$ Conclusive (divergent)

(d) $\lim\limits_{n\to\infty}\left|\dfrac{\sqrt{n+1}}{1+(n+1)^2}\cdot\dfrac{1+n^2}{\sqrt{n}}\right| = \lim\limits_{n\to\infty}\left[\sqrt{1+\dfrac{1}{n}}\cdot\dfrac{1/n^2+1}{1/n^2+(1+1/n)^2}\right] = 1.$ Inconclusive

32. We use the Ratio Test:

$$\lim\limits_{n\to\infty}\left|\dfrac{a_{n+1}}{a_n}\right| = \lim\limits_{n\to\infty}\left|\dfrac{[(n+1)!]^2/[k(n+1)]!}{(n!)^2/(kn)!}\right| = \lim\limits_{n\to\infty}\left|\dfrac{(n+1)^2}{[k(n+1)]\,[k(n+1)-1]\cdots[kn+1]}\right|$$

Now if $k=1$, then this is equal to $\lim\limits_{n\to\infty}\left|\dfrac{(n+1)^2}{(n+1)}\right| = \infty$, so the series diverges; if $k=2$, the limit is

$\lim\limits_{n\to\infty}\left|\dfrac{(n+1)^2}{(2n+2)(2n+1)}\right| = \dfrac{1}{4} < 1$, so the series converges, and if $k>2$, then the highest power of n in the denominator is

larger than 2, and so the limit is 0, indicating convergence. So the series converges for $k\geq 2$.

33. (a) $\lim\limits_{n\to\infty}\left|\dfrac{a_{n+1}}{a_n}\right| = \lim\limits_{n\to\infty}\left|\dfrac{x^{n+1}}{(n+1)!}\cdot\dfrac{n!}{x^n}\right| = \lim\limits_{n\to\infty}\left|\dfrac{x}{n+1}\right| = |x|\lim\limits_{n\to\infty}\dfrac{1}{n+1} = |x|\cdot 0 = 0 < 1$, so by the Ratio Test the

series $\sum\limits_{n=0}^{\infty}\dfrac{x^n}{n!}$ converges for all x.

(b) Since the series of part (a) always converges, we must have $\lim\limits_{n\to\infty}\dfrac{x^n}{n!} = 0$ by Theorem 12.2.6.

34. (a) $R_n = a_{n+1}+a_{n+2}+a_{n+3}+a_{n+4}+\cdots = a_{n+1}\left(1+\dfrac{a_{n+2}}{a_{n+1}}+\dfrac{a_{n+3}}{a_{n+1}}+\dfrac{a_{n+4}}{a_{n+1}}+\cdots\right)$

$= a_{n+1}\left(1+\dfrac{a_{n+2}}{a_{n+1}}+\dfrac{a_{n+3}}{a_{n+2}}\dfrac{a_{n+2}}{a_{n+1}}+\dfrac{a_{n+4}}{a_{n+3}}\dfrac{a_{n+3}}{a_{n+2}}\dfrac{a_{n+2}}{a_{n+1}}+\cdots\right)$

$= a_{n+1}(1+r_{n+1}+r_{n+2}r_{n+1}+r_{n+3}r_{n+2}r_{n+1}+\cdots)$ $(\star)$

$\leq a_{n+1}(1+r_{n+1}+r_{n+1}^2+r_{n+1}^3+\cdots)$ [since $\{r_n\}$ is decreasing] $= \dfrac{a_{n+1}}{1-r_{n+1}}$

(b) Note that since $\{r_n\}$ is increasing and $r_n\to L$ as $n\to\infty$, we have $r_n < L$ for all n. So, starting with equation $(\star)$,

$R_n = a_{n+1}(1+r_{n+1}+r_{n+2}r_{n+1}+r_{n+3}r_{n+2}r_{n+1}+\cdots)\leq a_{n+1}(1+L+L^2+L^3+\cdots) = \dfrac{a_{n+1}}{1-L}.$

35. (a) $s_5 = \displaystyle\sum_{n=1}^{5} \frac{1}{n2^n} = \frac{1}{2} + \frac{1}{8} + \frac{1}{24} + \frac{1}{64} + \frac{1}{160} = \frac{661}{960} \approx 0.68854$. Now the ratios

$$r_n = \frac{a_{n+1}}{a_n} = \frac{n2^n}{(n+1)2^{n+1}} = \frac{n}{2(n+1)} \text{ form an increasing sequence, since}$$

$$r_{n+1} - r_n = \frac{n+1}{2(n+2)} - \frac{n}{2(n+1)} = \frac{(n+1)^2 - n(n+2)}{2(n+1)(n+2)} = \frac{1}{2(n+1)(n+2)} > 0. \text{ So by Exercise 34(b), the error}$$

in using s_5 is $R_5 \le \dfrac{a_6}{1 - \lim\limits_{n\to\infty} r_n} = \dfrac{1/(6 \cdot 2^6)}{1 - 1/2} = \dfrac{1}{192} \approx 0.00521$.

(b) The error in using s_n as an approximation to the sum is $R_n = \dfrac{a_{n+1}}{1 - \frac{1}{2}} = \dfrac{2}{(n+1)2^{n+1}}$. We want $R_n < 0.00005 \quad \Leftrightarrow$

$$\frac{1}{(n+1)2^n} < 0.00005 \quad \Leftrightarrow \quad (n+1)2^n > 20{,}000. \text{ To find such an } n \text{ we can use trial and error or a graph. We calculate}$$

$(11+1)2^{11} = 24{,}576$, so $s_{11} = \displaystyle\sum_{n=1}^{11} \frac{1}{n2^n} \approx 0.693109$ is within 0.00005 of the actual sum.

36. $s_{10} = \displaystyle\sum_{n=1}^{10} \frac{n}{2^n} = \frac{1}{2} + \frac{2}{4} + \frac{3}{8} + \cdots + \frac{10}{1024} \approx 1.988$. The ratios $r_n = \dfrac{a_{n+1}}{a_n} = \dfrac{n+1}{2^{n+1}} \cdot \dfrac{2^n}{n} = \dfrac{n+1}{2n} = \dfrac{1}{2}\left(1 + \dfrac{1}{n}\right)$ form a

decreasing sequence, and $r_{11} = \dfrac{11+1}{2(11)} = \dfrac{12}{22} = \dfrac{6}{11} < 1$, so by Exercise 34(a), the error in using s_{10} to approximate the sum

of the series $\displaystyle\sum_{n=1}^{\infty} \frac{n}{2^n}$ is $R_{10} \le \dfrac{a_{11}}{1 - r_{11}} = \dfrac{\frac{11}{2048}}{1 - \frac{6}{11}} = \dfrac{121}{10{,}240} \approx 0.0118$.

37. (i) Following the hint, we get that $|a_n| < r^n$ for $n \ge N$, and so since the geometric series $\sum_{n=1}^{\infty} r^n$ converges $[0 < r < 1]$,

the series $\sum_{n=N}^{\infty} |a_n|$ converges as well by the Comparison Test, and hence so does $\sum_{n=1}^{\infty} |a_n|$, so $\sum_{n=1}^{\infty} a_n$ is absolutely

convergent.

(ii) If $\lim\limits_{n\to\infty} \sqrt[n]{|a_n|} = L > 1$, then there is an integer N such that $\sqrt[n]{|a_n|} > 1$ for all $n \ge N$, so $|a_n| > 1$ for $n \ge N$. Thus,

$\lim\limits_{n\to\infty} a_n \ne 0$, so $\sum_{n=1}^{\infty} a_n$ diverges by the Test for Divergence.

(iii) Consider $\displaystyle\sum_{n=1}^{\infty} \frac{1}{n}$ [diverges] and $\displaystyle\sum_{n=1}^{\infty} \frac{1}{n^2}$ [converges]. For each sum, $\lim\limits_{n\to\infty} \sqrt[n]{|a_n|} = 1$, so the Root Test is inconclusive.

38. (a) $\lim\limits_{n\to\infty} \left| \dfrac{a_{n+1}}{a_n} \right| = \lim\limits_{n\to\infty} \left| \dfrac{[4(n+1)]! \, [1103 + 26{,}390(n+1)]}{[(n+1)!]^4 \, 396^{4(n+1)}} \cdot \dfrac{(n!)^4 \, 396^{4n}}{(4n)! \, (1103 + 26{,}390n)} \right|$

$$= \lim_{n\to\infty} \frac{(4n+4)(4n+3)(4n+2)(4n+1)(26{,}390n + 27{,}493)}{(n+1)^4 \, 396^4 \, (26{,}390n + 1103)} = \frac{4^4}{396^4} = \frac{1}{99^4} < 1,$$

so by the Ratio Test, the series $\displaystyle\sum_{n=0}^{\infty} \frac{(4n)! \, (1103 + 26{,}390n)}{(n!)^4 \, 396^{4n}}$ converges.

(b) $\dfrac{1}{\pi} = \dfrac{2\sqrt{2}}{9801} \displaystyle\sum_{n=0}^{\infty} \frac{(4n)! \, (1103 + 26{,}390n)}{(n!)^4 \, 396^{4n}}$

With the first term $(n = 0)$, $\dfrac{1}{\pi} \approx \dfrac{2\sqrt{2}}{9801} \cdot \dfrac{1103}{1} \quad \Rightarrow \quad \pi \approx 3.141\,592\,73$, so we get 6 correct decimal places of π, which is

$3.141\,592\,653\,589\,793\,238$ to 18 decimal places.

With the second term $(n = 1)$, $\dfrac{1}{\pi} \approx \dfrac{2\sqrt{2}}{9801}\left(\dfrac{1103}{1} + \dfrac{4!\,(1103 + 26{,}390)}{396^4}\right)$ $\Rightarrow$ $\pi \approx 3.141\,592\,653\,589\,793\,878$, so

we get 15 correct decimal places of π.

39. (a) Since $\sum a_n$ is absolutely convergent, and since $\left|a_n^+\right| \le |a_n|$ and $\left|a_n^-\right| \le |a_n|$ (because a_n^+ and a_n^- each equal

either a_n or 0), we conclude by the Comparison Test that both $\sum a_n^+$ and $\sum a_n^-$ must be absolutely convergent.

Or: Use Theorem 12.2.8.

(b) We will show by contradiction that both $\sum a_n^+$ and $\sum a_n^-$ must diverge. For suppose that $\sum a_n^+$ converged. Then so

would $\sum\left(a_n^+ - \tfrac{1}{2}a_n\right)$ by Theorem 12.2.8. But $\sum\left(a_n^+ - \tfrac{1}{2}a_n\right) = \sum\left[\tfrac{1}{2}\left(a_n + |a_n|\right) - \tfrac{1}{2}a_n\right] = \tfrac{1}{2}\sum |a_n|$, which

diverges because $\sum a_n$ is only conditionally convergent. Hence, $\sum a_n^+$ can't converge. Similarly, neither can $\sum a_n^-$.

40. Let $\sum b_n$ be the rearranged series constructed in the hint. [This series can be constructed by virtue of the result of

Exercise 39(b).] This series will have partial sums s_n that oscillate in value back and forth across r. Since $\lim\limits_{n\to\infty} a_n = 0$

(by Theorem 12.2.6), and since the size of the oscillations $|s_n - r|$ is always less than $|a_n|$ because of the way $\sum b_n$ was

constructed, we have that $\sum b_n = \lim\limits_{n\to\infty} s_n = r$.

12.7 Strategy for Testing Series

1. $\dfrac{1}{n + 3^n} < \dfrac{1}{3^n} = \left(\dfrac{1}{3}\right)^n$ for all $n \ge 1$. $\sum\limits_{n=1}^{\infty}\left(\dfrac{1}{3}\right)^n$ is a convergent geometric series $\left[|r| = \tfrac{1}{3} < 1\right]$, so $\sum\limits_{n=1}^{\infty}\dfrac{1}{n + 3^n}$

converges by the Comparison Test.

2. $\lim\limits_{n\to\infty} \sqrt[n]{|a_n|} = \lim\limits_{n\to\infty} \sqrt[n]{\left|\dfrac{(2n + 1)^n}{n^{2n}}\right|} = \lim\limits_{n\to\infty}\dfrac{2n + 1}{n^2} = \lim\limits_{n\to\infty}\left(\dfrac{2}{n} + \dfrac{1}{n^2}\right) = 0 < 1$, so the series $\sum\limits_{n=1}^{\infty}\dfrac{(2n + 1)^n}{n^{2n}}$

converges by the Root Test.

3. $\lim\limits_{n\to\infty} |a_n| = \lim\limits_{n\to\infty}\dfrac{n}{n + 2} = 1$, so $\lim\limits_{n\to\infty} a_n = \lim\limits_{n\to\infty} (-1)^n \dfrac{n}{n + 2}$ does not exist. Thus, the series $\sum\limits_{n=1}^{\infty} (-1)^n \dfrac{n}{n + 2}$ diverges by

the Test for Divergence.

4. $b_n = \dfrac{n}{n^2 + 2} > 0$ for $n \ge 1$. $\{b_n\}$ is decreasing for $n \ge 2$ since $\left(\dfrac{x}{x^2 + 2}\right)' = \dfrac{(x^2 + 2)(1) - x(2x)}{(x^2 + 2)^2} = \dfrac{2 - x^2}{(x^2 + 2)^2} < 0$

for $x \ge \sqrt{2}$. Also, $\lim\limits_{n\to\infty} b_n = \lim\limits_{n\to\infty}\dfrac{n}{n^2 + 2} = \lim\limits_{n\to\infty}\dfrac{1/n}{1 + 2/n^2} = 0$. Thus, the series $\sum\limits_{n=1}^{\infty} (-1)^n \dfrac{n}{n^2 + 2}$ converges by the

Alternating Series Test.

5. $\lim\limits_{n\to\infty}\left|\dfrac{a_{n+1}}{a_n}\right| = \lim\limits_{n\to\infty}\left|\dfrac{(n + 1)^2\, 2^n}{(-5)^{n+1}} \cdot \dfrac{(-5)^n}{n^2\, 2^{n-1}}\right| = \lim\limits_{n\to\infty}\dfrac{2(n + 1)^2}{5n^2} = \dfrac{2}{5}\lim\limits_{n\to\infty}\left(1 + \dfrac{1}{n}\right)^2 = \dfrac{2}{5}(1) = \dfrac{2}{5} < 1$, so the series

$\sum\limits_{n=1}^{\infty}\dfrac{n^2\, 2^{n-1}}{(-5)^n}$ converges by the Ratio Test.

6. Use the Limit Comparison Test with $a_n = \dfrac{1}{2n+1}$ and $b_n = \dfrac{1}{n}$: $\displaystyle\lim_{n\to\infty}\frac{a_n}{b_n} = \lim_{n\to\infty}\frac{n}{2n+1} = \lim_{n\to\infty}\frac{1}{2+(1/n)} = \frac{1}{2} > 0.$

Since the harmonic series $\displaystyle\sum_{n=1}^{\infty}\frac{1}{n}$ diverges, so does $\displaystyle\sum_{n=1}^{\infty}\frac{1}{2n+1}$.   [*Or:* Use the Integral Test.]

7. Let $f(x) = \dfrac{1}{x\sqrt{\ln x}}$. Then f is positive, continuous, and decreasing on $[2,\infty)$, so we can apply the Integral Test.

Since $\displaystyle\int \frac{1}{x\sqrt{\ln x}}\,dx$ $\begin{bmatrix} u = \ln x, \\ du = dx/x \end{bmatrix} = \int u^{-1/2}\,du = 2u^{1/2} + C = 2\sqrt{\ln x} + C$, we find

$\displaystyle\int_{2}^{\infty}\frac{dx}{x\sqrt{\ln x}} = \lim_{t\to\infty}\int_{2}^{t}\frac{dx}{x\sqrt{\ln x}} = \lim_{t\to\infty}\Big[2\sqrt{\ln x}\Big]_{2}^{t} = \lim_{t\to\infty}\Big(2\sqrt{\ln t} - 2\sqrt{\ln 2}\Big) = \infty.$ Since the integral diverges, the

given series $\displaystyle\sum_{n=2}^{\infty}\frac{1}{n\sqrt{\ln n}}$ diverges.

8. $\displaystyle\sum_{k=1}^{\infty}\frac{2^k k!}{(k+2)!} = \sum_{k=1}^{\infty}\frac{2^k}{(k+1)(k+2)}$. Using the Ratio Test, we get

$\displaystyle\lim_{k\to\infty}\left|\frac{a_{k+1}}{a_k}\right| = \lim_{k\to\infty}\left|\frac{2^{k+1}}{(k+2)(k+3)}\cdot\frac{(k+1)(k+2)}{2^k}\right| = \lim_{k\to\infty}\left(2\cdot\frac{k+1}{k+3}\right) = 2 > 1,$ so the series diverges.

Or: Use the Test for Divergence.

9. $\displaystyle\sum_{k=1}^{\infty}k^2 e^{-k} = \sum_{k=1}^{\infty}\frac{k^2}{e^k}$. Using the Ratio Test, we get

$\displaystyle\lim_{k\to\infty}\left|\frac{a_{k+1}}{a_k}\right| = \lim_{k\to\infty}\left|\frac{(k+1)^2}{e^{k+1}}\cdot\frac{e^k}{k^2}\right| = \lim_{k\to\infty}\left[\left(\frac{k+1}{k}\right)^2\cdot\frac{1}{e}\right] = 1^2\cdot\frac{1}{e} = \frac{1}{e} < 1,$ so the series converges.

10. Let $f(x) = x^2 e^{-x^3}$. Then f is continuous and positive on $[1,\infty)$, and $f'(x) = \dfrac{x(2-3x^3)}{e^{x^3}} < 0$ for $x \geq 1$, so f is

decreasing on $[1,\infty)$ as well, and we can apply the Integral Test. $\displaystyle\int_{1}^{\infty}x^2 e^{-x^3}\,dx = \lim_{t\to\infty}\left[-\tfrac{1}{3}e^{-x^3}\right]_{1}^{t} = \frac{1}{3e}$, so the integral

converges, and hence, the series converges.

11. $b_n = \dfrac{1}{n\ln n} > 0$ for $n \geq 2$, $\{b_n\}$ is decreasing, and $\displaystyle\lim_{n\to\infty}b_n = 0$, so the given series $\displaystyle\sum_{n=2}^{\infty}\frac{(-1)^{n+1}}{n\ln n}$ converges by the

Alternating Series Test.

12. The series $\displaystyle\sum_{n=1}^{\infty}\sin n$ diverges by the Test for Divergence since $\displaystyle\lim_{n\to\infty}\sin n$ does not exist.

13. $\displaystyle\lim_{n\to\infty}\left|\frac{a_{n+1}}{a_n}\right| = \lim_{n\to\infty}\left|\frac{3^{n+1}(n+1)^2}{(n+1)!}\cdot\frac{n!}{3^n n^2}\right| = \lim_{n\to\infty}\frac{3(n+1)^2}{(n+1)n^2} = 3\lim_{n\to\infty}\frac{n+1}{n^2} = 0 < 1$, so the series $\displaystyle\sum_{n=1}^{\infty}\frac{3^n n^2}{n!}$

converges by the Ratio Test.

14. $\left|\dfrac{\sin 2n}{1+2^n}\right| \leq \dfrac{1}{1+2^n} < \dfrac{1}{2^n} = \left(\dfrac{1}{2}\right)^n$, so the series $\displaystyle\sum_{n=1}^{\infty}\left|\frac{\sin 2n}{1+2^n}\right|$ converges by comparison with the geometric series

$\displaystyle\sum_{n=1}^{\infty}\left(\frac{1}{2}\right)^n$ with $|r| = \tfrac{1}{2} < 1$. Thus, the series $\displaystyle\sum_{n=1}^{\infty}\frac{\sin 2n}{1+2^n}$ converges absolutely, implying convergence.

15. $\lim\limits_{n\to\infty}\left|\dfrac{a_{n+1}}{a_n}\right| = \lim\limits_{n\to\infty}\left|\dfrac{(n+1)!}{2\cdot5\cdot8\cdot\cdots\cdot(3n+2)[3(n+1)+2]}\cdot\dfrac{2\cdot5\cdot8\cdot\cdots\cdot(3n+2)}{n!}\right| = \lim\limits_{n\to\infty}\dfrac{n+1}{3n+5} = \dfrac{1}{3} < 1,$

so the series $\displaystyle\sum_{n=0}^{\infty}\dfrac{n!}{2\cdot5\cdot8\cdot\cdots\cdot(3n+2)}$ converges by the Ratio Test.

16. Using the Limit Comparison Test with $a_n = \dfrac{n^2+1}{n^3+1}$ and $b_n = \dfrac{1}{n}$, we have

$$\lim\limits_{n\to\infty}\dfrac{a_n}{b_n} = \lim\limits_{n\to\infty}\left(\dfrac{n^2+1}{n^3+1}\cdot\dfrac{n}{1}\right) = \lim\limits_{n\to\infty}\dfrac{n^3+n}{n^3+1} = \lim\limits_{n\to\infty}\dfrac{1+1/n^2}{1+1/n^3} = 1 > 0.\ \text{Since}\ \sum_{n=1}^{\infty} b_n\ \text{is the divergent harmonic}$$

series, $\displaystyle\sum_{n=1}^{\infty} a_n$ is also divergent.

17. $\lim\limits_{n\to\infty} 2^{1/n} = 2^0 = 1$, so $\lim\limits_{n\to\infty}(-1)^n\,2^{1/n}$ does not exist and the series $\displaystyle\sum_{n=1}^{\infty}(-1)^n 2^{1/n}$ diverges by the Test for Divergence.

18. $b_n = \dfrac{1}{\sqrt{n}-1}$ for $n\geq2$. $\{b_n\}$ is a decreasing sequence of positive numbers and $\lim\limits_{n\to\infty} b_n = 0$, so $\displaystyle\sum_{n=2}^{\infty}\dfrac{(-1)^{n-1}}{\sqrt{n}-1}$ converges by

the Alternating Series Test.

19. Let $f(x) = \dfrac{\ln x}{\sqrt{x}}$. Then $f'(x) = \dfrac{2-\ln x}{2x^{3/2}} < 0$ when $\ln x > 2$ or $x > e^2$, so $\dfrac{\ln n}{\sqrt{n}}$ is decreasing for $n > e^2$.

By l'Hospital's Rule, $\lim\limits_{n\to\infty}\dfrac{\ln n}{\sqrt{n}} = \lim\limits_{n\to\infty}\dfrac{1/n}{1/\left(2\sqrt{n}\right)} = \lim\limits_{n\to\infty}\dfrac{2}{\sqrt{n}} = 0$, so the series $\displaystyle\sum_{n=1}^{\infty}(-1)^n\dfrac{\ln n}{\sqrt{n}}$ converges by the

Alternating Series Test.

20. $\lim\limits_{k\to\infty}\left|\dfrac{a_{k+1}}{a_k}\right| = \lim\limits_{k\to\infty}\left|\dfrac{k+6}{5^{k+1}}\cdot\dfrac{5^k}{k+5}\right| = \dfrac{1}{5}\lim\limits_{k\to\infty}\dfrac{k+6}{k+5} = \dfrac{1}{5} < 1$, so the series $\displaystyle\sum_{k=1}^{\infty}\dfrac{k+5}{5^k}$ converges by the Ratio Test.

21. $\displaystyle\sum_{n=1}^{\infty}\dfrac{(-2)^{2n}}{n^n} = \sum_{n=1}^{\infty}\left(\dfrac{4}{n}\right)^n$. $\lim\limits_{n\to\infty}\sqrt[n]{|a_n|} = \lim\limits_{n\to\infty}\dfrac{4}{n} = 0 < 1$, so the given series is absolutely convergent by the Root Test.

22. $\dfrac{\sqrt{n^2-1}}{n^3+2n^2+5} < \dfrac{n}{n^3+2n^2+5} < \dfrac{n}{n^3} = \dfrac{1}{n^2}$ for $n\geq1$, so $\displaystyle\sum_{n=1}^{\infty}\dfrac{\sqrt{n^2-1}}{n^3+2n^2+5}$ converges by the Comparison Test with the

convergent p-series $\displaystyle\sum_{n=1}^{\infty} 1/n^2\ [p=2>1]$.

23. Using the Limit Comparison Test with $a_n = \tan\left(\dfrac{1}{n}\right)$ and $b_n = \dfrac{1}{n}$, we have

$$\lim\limits_{n\to\infty}\dfrac{a_n}{b_n} = \lim\limits_{n\to\infty}\dfrac{\tan(1/n)}{1/n} = \lim\limits_{x\to\infty}\dfrac{\tan(1/x)}{1/x} \overset{\text{H}}{=} \lim\limits_{x\to\infty}\dfrac{\sec^2(1/x)\cdot(-1/x^2)}{-1/x^2} = \lim\limits_{x\to\infty}\sec^2(1/x) = 1^2 = 1 > 0.\ \text{Since}$$

$\displaystyle\sum_{n=1}^{\infty} b_n$ is the divergent harmonic series, $\displaystyle\sum_{n=1}^{\infty} a_n$ is also divergent.

24. $\lim\limits_{n\to\infty} a_n = \lim\limits_{n\to\infty}\left(n\sin\dfrac{1}{n}\right) = \lim\limits_{n\to\infty}\dfrac{\sin(1/n)}{1/n} = \lim\limits_{x\to0^+}\dfrac{\sin x}{x} = 1\neq0$, so the series $\displaystyle\sum_{n=1}^{\infty} n\sin(1/n)$ diverges by the

Test for Divergence.

25. Use the Ratio Test. $\lim\limits_{n\to\infty}\left|\dfrac{a_{n+1}}{a_n}\right| = \lim\limits_{n\to\infty}\left|\dfrac{(n+1)!}{e^{(n+1)^2}}\cdot\dfrac{e^{n^2}}{n!}\right| = \lim\limits_{n\to\infty}\dfrac{(n+1)n!\cdot e^{n^2}}{e^{n^2+2n+1}n!} = \lim\limits_{n\to\infty}\dfrac{n+1}{e^{2n+1}} = 0 < 1$, so $\sum\limits_{n=1}^{\infty}\dfrac{n!}{e^{n^2}}$

converges.

26. $\lim\limits_{n\to\infty}\left|\dfrac{a_{n+1}}{a_n}\right| = \lim\limits_{n\to\infty}\dfrac{a_{n+1}}{a_n} = \lim\limits_{n\to\infty}\left(\dfrac{n^2+2n+2}{5^{n+1}}\cdot\dfrac{5^n}{n^2+1}\right) = \lim\limits_{n\to\infty}\left(\dfrac{1+2/n+2/n^2}{1+1/n^2}\cdot\dfrac{1}{5}\right) = \dfrac{1}{5} < 1$, so $\sum\limits_{n=1}^{\infty}\dfrac{n^2+1}{5^n}$

converges by the Ratio Test.

27. $\displaystyle\int_2^{\infty}\dfrac{\ln x}{x^2}\,dx = \lim\limits_{t\to\infty}\left[-\dfrac{\ln x}{x}-\dfrac{1}{x}\right]_1^t$ [using integration by parts] $\overset{\text{H}}{=}1$. So $\sum\limits_{n=1}^{\infty}\dfrac{\ln n}{n^2}$ converges by the Integral Test, and since

$\dfrac{k\ln k}{(k+1)^3} < \dfrac{k\ln k}{k^3} = \dfrac{\ln k}{k^2}$, the given series $\sum\limits_{k=1}^{\infty}\dfrac{k\ln k}{(k+1)^3}$ converges by the Comparison Test.

28. Since $\left\{\dfrac{1}{n}\right\}$ is a decreasing sequence, $e^{1/n}\le e^{1/1}=e$ for all $n\ge1$, and $\sum\limits_{n=1}^{\infty}\dfrac{e}{n^2}$ converges $(p=2>1)$, so $\sum\limits_{n=1}^{\infty}\dfrac{e^{1/n}}{n^2}$

converges by the Comparison Test. (Or use the Integral Test.)

29. $\sum\limits_{n=1}^{\infty}a_n = \sum\limits_{n=1}^{\infty}(-1)^n\dfrac{1}{\cosh n} = \sum\limits_{n=1}^{\infty}(-1)^n b_n$. Now $b_n = \dfrac{1}{\cosh n} > 0$, $\{b_n\}$ is decreasing, and $\lim\limits_{n\to\infty}b_n=0$, so the series

converges by the Alternating Series Test.

Or: Write $\dfrac{1}{\cosh n} = \dfrac{2}{e^n+e^{-n}} < \dfrac{2}{e^n}$ and $\sum\limits_{n=1}^{\infty}\dfrac{1}{e^n}$ is a convergent geometric series, so $\sum\limits_{n=1}^{\infty}\dfrac{1}{\cosh n}$ is convergent by the

Comparison Test. So $\sum\limits_{n=1}^{\infty}(-1)^n\dfrac{1}{\cosh n}$ is absolutely convergent and therefore convergent.

30. Let $f(x) = \dfrac{\sqrt{x}}{x+5}$. Then $f(x)$ is continuous and positive on $[1,\infty)$, and since $f'(x) = \dfrac{5-x}{2\sqrt{x}\,(x+5)^2} < 0$ for $x>5$, $f(x)$ is

eventually decreasing, so we can use the Alternating Series Test. $\lim\limits_{n\to\infty}\dfrac{\sqrt{n}}{n+5} = \lim\limits_{n\to\infty}\dfrac{1}{n^{1/2}+5n^{-1/2}} = 0$, so the series

$\sum\limits_{j=1}^{\infty}(-1)^j\dfrac{\sqrt{j}}{j+5}$ converges.

31. $\lim\limits_{k\to\infty}a_k = \lim\limits_{k\to\infty}\dfrac{5^k}{3^k+4^k} = $ [divide by 4^k] $\lim\limits_{k\to\infty}\dfrac{(5/4)^k}{(3/4)^k+1} = \infty$ since $\lim\limits_{k\to\infty}\left(\dfrac{3}{4}\right)^k = 0$ and $\lim\limits_{k\to\infty}\left(\dfrac{5}{4}\right)^k = \infty$.

Thus, $\sum\limits_{k=1}^{\infty}\dfrac{5^k}{3^k+4^k}$ diverges by the Test for Divergence.

32. $\lim\limits_{n\to\infty}\sqrt[n]{|a_n|} = \lim\limits_{n\to\infty}\sqrt[n]{\left|\dfrac{(n!)^n}{n^{4n}}\right|} = \lim\limits_{n\to\infty}\dfrac{n!}{n^4} = \lim\limits_{n\to\infty}\left[\dfrac{n}{n}\cdot\dfrac{n-1}{n}\cdot\dfrac{n-2}{n}\cdot\dfrac{n-3}{n}\cdot(n-4)!\right]$

$= \lim\limits_{n\to\infty}\left[\left(1-\dfrac{1}{n}\right)\left(1-\dfrac{2}{n}\right)\left(1-\dfrac{3}{n}\right)(n-4)!\right] = \infty$,

so the series $\sum\limits_{n=1}^{\infty}\dfrac{(n!)^n}{n^{4n}}$ diverges by the Root Test.

33. Let $a_n = \dfrac{\sin(1/n)}{\sqrt{n}}$ and $b_n = \dfrac{1}{n\sqrt{n}}$. Then $\lim\limits_{n\to\infty}\dfrac{a_n}{b_n} = \lim\limits_{n\to\infty}\dfrac{\sin(1/n)}{1/n} = 1 > 0$, so $\sum\limits_{n=1}^{\infty}\dfrac{\sin(1/n)}{\sqrt{n}}$ converges by limit

comparison with the convergent p-series $\sum\limits_{n=1}^{\infty}\dfrac{1}{n^{3/2}}$ $[p=3/2>1]$.

34. $0 \le n \cos^2 n \le n$, so $\dfrac{1}{n + n \cos^2 n} \ge \dfrac{1}{n + n} = \dfrac{1}{2n}$. Thus, $\displaystyle\sum_{n=1}^{\infty} \dfrac{1}{n + n \cos^2 n}$ diverges by comparison with $\displaystyle\sum_{n=1}^{\infty} \dfrac{1}{2n}$, which is

a constant multiple of the (divergent) harmonic series.

35. $\displaystyle\lim_{n\to\infty} \sqrt[n]{|a_n|} = \lim_{n\to\infty} \left(\dfrac{n}{n+1}\right)^{n^2/n} = \lim_{n\to\infty} \dfrac{1}{\left[(n+1)/n\right]^n} = \dfrac{1}{\displaystyle\lim_{n\to\infty} (1 + 1/n)^n} = \dfrac{1}{e} < 1$, so the series $\displaystyle\sum_{n=1}^{\infty} \left(\dfrac{n}{n+1}\right)^{n^2}$

converges by the Root Test.

36. Note that $(\ln n)^{\ln n} = \left(e^{\ln \ln n}\right)^{\ln n} = \left(e^{\ln n}\right)^{\ln \ln n} = n^{\ln \ln n}$ and $\ln \ln n \to \infty$ as $n \to \infty$, so $\ln \ln n > 2$ for sufficiently

large n. For these n we have $(\ln n)^{\ln n} > n^2$, so $\dfrac{1}{(\ln n)^{\ln n}} < \dfrac{1}{n^2}$. Since $\displaystyle\sum_{n=2}^{\infty} \dfrac{1}{n^2}$ converges $[p = 2 > 1]$, so does

$\displaystyle\sum_{n=2}^{\infty} \dfrac{1}{(\ln n)^{\ln n}}$ by the Comparison Test.

37. $\displaystyle\lim_{n\to\infty} \sqrt[n]{|a_n|} = \lim_{n\to\infty} (2^{1/n} - 1) = 1 - 1 = 0 < 1$, so the series $\displaystyle\sum_{n=1}^{\infty} \left(\sqrt[n]{2} - 1\right)^n$ converges by the Root Test.

38. Use the Limit Comparison Test with $a_n = \sqrt[n]{2} - 1$ and $b_n = 1/n$. Then

$$\lim_{n\to\infty} \dfrac{a_n}{b_n} = \lim_{n\to\infty} \dfrac{2^{1/n} - 1}{1/n} = \lim_{x\to\infty} \dfrac{2^{1/x} - 1}{1/x} \overset{H}{=} \lim_{x\to\infty} \dfrac{2^{1/x} \cdot \ln 2 \cdot (-1/x^2)}{-1/x^2} = \lim_{x\to\infty} (2^{1/x} \cdot \ln 2) = 1 \cdot \ln 2 = \ln 2 > 0.$$

So since $\displaystyle\sum_{n=1}^{\infty} b_n$ diverges (harmonic series), so does $\displaystyle\sum_{n=1}^{\infty} \left(\sqrt[n]{2} - 1\right)$.

Alternate solution: $\sqrt[n]{2} - 1 = \dfrac{1}{2^{(n-1)/n} + 2^{(n-2)/n} + 2^{(n-3)/n} + \cdots + 2^{1/n} + 1}$ [rationalize the numerator] $\ge \dfrac{1}{2n}$,

and since $\displaystyle\sum_{n=1}^{\infty} \dfrac{1}{2n} = \dfrac{1}{2} \sum_{n=1}^{\infty} \dfrac{1}{n}$ diverges (harmonic series), so does $\displaystyle\sum_{n=1}^{\infty} \left(\sqrt[n]{2} - 1\right)$ by the Comparison Test.

12.8 Power Series

1. A power series is a series of the form $\sum_{n=0}^{\infty} c_n x^n = c_0 + c_1 x + c_2 x^2 + c_3 x^3 + \cdots$, where x is a variable and the c_n's are

constants called the coefficients of the series.

More generally, a series of the form $\sum_{n=0}^{\infty} c_n (x - a)^n = c_0 + c_1 (x - a) + c_2 (x - a)^2 + \cdots$ is called a power series in

$(x - a)$ or a power series centered at a or a power series about a, where a is a constant.

2. (a) Given the power series $\sum_{n=0}^{\infty} c_n (x - a)^n$, the radius of convergence is:

(i) 0 if the series converges only when $x = a$

(ii) ∞ if the series converges for all x, or

(iii) a positive number R such that the series converges if $|x - a| < R$ and diverges if $|x - a| > R$.

In most cases, R can be found by using the Ratio Test.

(b) The interval of convergence of a power series is the interval that consists of all values of x for which the series converges.

Corresponding to the cases in part (a), the interval of convergence is: (i) the single point $\{a\}$, (ii) all real numbers; that is,

the real number line $(-\infty, \infty)$, or (iii) an interval with endpoints $a - R$ and $a + R$ which can contain neither, either, or

both of the endpoints. In this case, we must test the series for convergence at each endpoint to determine the interval of

convergence.

3. If $a_n = \dfrac{x^n}{\sqrt{n}}$, then $\displaystyle\lim_{n\to\infty}\left|\dfrac{a_{n+1}}{a_n}\right| = \lim_{n\to\infty}\left|\dfrac{x^{n+1}}{\sqrt{n+1}}\cdot\dfrac{\sqrt{n}}{x^n}\right| = \lim_{n\to\infty}\left|\dfrac{x}{\sqrt{n+1}/\sqrt{n}}\right| = \lim_{n\to\infty}\dfrac{|x|}{\sqrt{1+1/n}} = |x|.$

By the Ratio Test, the series $\displaystyle\sum_{n=1}^{\infty}\dfrac{x^n}{\sqrt{n}}$ converges when $|x| < 1$, so the radius of convergence $R = 1$. Now we'll check the

endpoints, that is, $x = \pm 1$. When $x = 1$, the series $\displaystyle\sum_{n=1}^{\infty}\dfrac{1}{\sqrt{n}}$ diverges because it is a p-series with $p = \frac{1}{2} \le 1$. When $x = -1$,

the series $\displaystyle\sum_{n=1}^{\infty}\dfrac{(-1)^n}{\sqrt{n}}$ converges by the Alternating Series Test. Thus, the interval of convergence is $I = [-1, 1)$.

4. If $a_n = \dfrac{(-1)^n x^n}{n+1}$, then $\displaystyle\lim_{n\to\infty}\left|\dfrac{a_{n+1}}{a_n}\right| = \lim_{n\to\infty}\left|\dfrac{x^{n+1}}{n+2}\cdot\dfrac{n+1}{x^n}\right| = \lim_{n\to\infty}\dfrac{|x|}{1+1/(n+1)} = |x|.$

By the Ratio Test, the series $\displaystyle\sum_{n=0}^{\infty}\dfrac{(-1)^n x^n}{n+1}$ converges when $|x| < 1$, so $R = 1$. When $x = -1$, the series diverges because it

is the harmonic series; when $x = 1$, it is the alternating harmonic series, which converges by the Alternating Series Test.

Thus, $I = (-1, 1]$.

5. If $a_n = \dfrac{(-1)^{n-1} x^n}{n^3}$, then

$$\lim_{n\to\infty}\left|\dfrac{a_{n+1}}{a_n}\right| = \lim_{n\to\infty}\left|\dfrac{(-1)^n x^{n+1}}{(n+1)^3}\cdot\dfrac{n^3}{(-1)^{n-1}x^n}\right| = \lim_{n\to\infty}\left|\dfrac{(-1)xn^3}{(n+1)^3}\right| = \lim_{n\to\infty}\left[\left(\dfrac{n}{n+1}\right)^3 |x|\right] = 1^3\cdot|x| = |x|.\text{ By the}$$

Ratio Test, the series $\displaystyle\sum_{n=1}^{\infty}\dfrac{(-1)^{n-1} x^n}{n^3}$ converges when $|x| < 1$, so the radius of convergence $R = 1$. Now we'll check the

endpoints, that is, $x = \pm 1$. When $x = 1$, the series $\displaystyle\sum_{n=1}^{\infty}\dfrac{(-1)^{n-1}}{n^3}$ converges by the Alternating Series Test. When $x = -1$,

the series $\displaystyle\sum_{n=1}^{\infty}\dfrac{(-1)^{n-1}(-1)^n}{n^3} = -\sum_{n=1}^{\infty}\dfrac{1}{n^3}$ converges because it is a constant multiple of a convergent p-series $[p = 3 > 1]$.

Thus, the interval of convergence is $I = [-1, 1]$.

6. $a_n = \sqrt{n}\, x^n$, so we need $\displaystyle\lim_{n\to\infty}\left|\dfrac{a_{n+1}}{a_n}\right| = \lim_{n\to\infty}\dfrac{\sqrt{n+1}\,|x|^{n+1}}{\sqrt{n}\,|x|^n} = \lim_{n\to\infty}\sqrt{1+\dfrac{1}{n}}\,|x| = |x| < 1$ for convergence (by the

Ratio Test), so $R = 1$. When $x = \pm 1$, $\displaystyle\lim_{n\to\infty}|a_n| = \lim_{n\to\infty}\sqrt{n} = \infty$, so the series diverges by the Test for Divergence.

Thus, $I = (-1, 1)$.

7. If $a_n = \dfrac{x^n}{n!}$, then $\displaystyle\lim_{n\to\infty}\left|\dfrac{a_{n+1}}{a_n}\right| = \lim_{n\to\infty}\left|\dfrac{x^{n+1}}{(n+1)!}\cdot\dfrac{n!}{x^n}\right| = \lim_{n\to\infty}\left|\dfrac{x}{n+1}\right| = |x|\lim_{n\to\infty}\dfrac{1}{n+1} = |x|\cdot 0 = 0 < 1$ for *all* real x.

So, by the Ratio Test, $R = \infty$ and $I = (-\infty, \infty)$.

8. Here the Root Test is easier. If $a_n = n^n x^n$ then $\displaystyle\lim_{n\to\infty}\sqrt[n]{|a_n|} = \lim_{n\to\infty}n\,|x| = \infty$ if $x \ne 0$, so $R = 0$ and $I = \{0\}$.

9. If $a_n = (-1)^n\dfrac{n^2 x^n}{2^n}$, then

$$\lim_{n\to\infty}\left|\dfrac{a_{n+1}}{a_n}\right| = \lim_{n\to\infty}\left|\dfrac{(n+1)^2\, x^{n+1}}{2^{n+1}}\cdot\dfrac{2^n}{n^2\, x^n}\right| = \lim_{n\to\infty}\left|\dfrac{x(n+1)^2}{2n^2}\right| = \lim_{n\to\infty}\left[\dfrac{|x|}{2}\left(1+\dfrac{1}{n}\right)^2\right] = \dfrac{|x|}{2}(1)^2 = \tfrac{1}{2}|x|.\text{ By the}$$

Ratio Test, the series $\sum_{n=1}^{\infty} (-1)^n \dfrac{n^2 x^n}{2^n}$ converges when $\frac{1}{2}|x| < 1$ $\Leftrightarrow$ $|x| < 2$, so the radius of convergence is $R = 2$.

When $x = \pm 2$, both series $\sum_{n=1}^{\infty} (-1)^n \dfrac{n^2 (\pm 2)^n}{2^n} = \sum_{n=1}^{\infty} (\mp 1)^n n^2$ diverge by the Test for Divergence since

$\lim_{n \to \infty} |(\mp 1)^n n^2| = \infty$. Thus, the interval of convergence is $I = (-2, 2)$.

10. If $a_n = \dfrac{10^n x^n}{n^3}$, then

$$\lim_{n \to \infty} \left| \dfrac{a_{n+1}}{a_n} \right| = \lim_{n \to \infty} \left| \dfrac{10^{n+1} x^{n+1}}{(n+1)^3} \cdot \dfrac{n^3}{10^n x^n} \right| = \lim_{n \to \infty} \left| \dfrac{10 x \, n^3}{(n+1)^3} \right| = \lim_{n \to \infty} \dfrac{10 |x|}{(1 + 1/n)^3} = \dfrac{10 |x|}{1^3} = 10 |x|$$

By the Ratio Test, the series $\sum_{n=1}^{\infty} \dfrac{10^n x^n}{n^3}$ converges when $10 |x| < 1$ $\Leftrightarrow$ $|x| < \frac{1}{10}$, so the radius of convergence is $R = \frac{1}{10}$.

When $x = -\frac{1}{10}$, the series converges by the Alternating Series Test; when $x = \frac{1}{10}$, the series converges because it is a p-series with $p = 3 > 1$. Thus, the interval of convergence is $I = \left[-\frac{1}{10}, \frac{1}{10} \right]$.

11. $a_n = \dfrac{(-2)^n x^n}{\sqrt[4]{n}}$, so $\lim_{n \to \infty} \left| \dfrac{a_{n+1}}{a_n} \right| = \lim_{n \to \infty} \dfrac{2^{n+1} |x|^{n+1}}{\sqrt[4]{n+1}} \cdot \dfrac{\sqrt[4]{n}}{2^n |x|^n} = \lim_{n \to \infty} 2 |x| \sqrt[4]{\dfrac{n}{n+1}} = 2 |x|$, so by the Ratio Test, the

series converges when $2 |x| < 1$ $\Leftrightarrow$ $|x| < \frac{1}{2}$, so $R = \frac{1}{2}$. When $x = -\frac{1}{2}$, we get the divergent p-series $\sum_{n=1}^{\infty} \dfrac{1}{\sqrt[4]{n}}$

$\left[p = \frac{1}{4} \leq 1 \right]$. When $x = \frac{1}{2}$, we get the series $\sum_{n=1}^{\infty} \dfrac{(-1)^n}{\sqrt[4]{n}}$, which converges by the Alternating Series Test.

Thus, $I = \left(-\frac{1}{2}, \frac{1}{2} \right]$.

12. $a_n = \dfrac{x^n}{5^n n^5}$, so $\lim_{n \to \infty} \left| \dfrac{a_{n+1}}{a_n} \right| = \lim_{n \to \infty} \left| \dfrac{x^{n+1}}{5^{n+1} (n+1)^5} \cdot \dfrac{5^n n^5}{x^n} \right| = \lim_{n \to \infty} \dfrac{|x|}{5} \left(\dfrac{n}{n+1} \right)^5 = \dfrac{|x|}{5}$. By the Ratio Test, the series

$\sum_{n=0}^{\infty} \dfrac{x^n}{5^n n^5}$ converges when $\dfrac{|x|}{5} < 1$ $\Leftrightarrow$ $|x| < 5$, so $R = 5$. When $x = -5$, we get the series $\sum_{n=1}^{\infty} \dfrac{(-1)^n}{n^5}$, which

converges by the Alternating Series Test. When $x = 5$, we get the convergent p-series $\sum_{n=1}^{\infty} \dfrac{1}{n^5}$ $[p = 5 > 1]$.

Thus, $I = [-5, 5]$.

13. If $a_n = (-1)^n \dfrac{x^n}{4^n \ln n}$, then $\lim_{n \to \infty} \left| \dfrac{a_{n+1}}{a_n} \right| = \lim_{n \to \infty} \left| \dfrac{x^{n+1}}{4^{n+1} \ln(n+1)} \cdot \dfrac{4^n \ln n}{x^n} \right| = \dfrac{|x|}{4} \lim_{n \to \infty} \dfrac{\ln n}{\ln(n+1)} = \dfrac{|x|}{4} \cdot 1$

[by l'Hospital's Rule] $= \dfrac{|x|}{4}$. By the Ratio Test, the series converges when $\dfrac{|x|}{4} < 1$ $\Leftrightarrow$ $|x| < 4$, so $R = 4$. When

$x = -4$, $\sum_{n=2}^{\infty} (-1)^n \dfrac{x^n}{4^n \ln n} = \sum_{n=2}^{\infty} \dfrac{[(-1)(-4)]^n}{4^n \ln n} = \sum_{n=2}^{\infty} \dfrac{1}{\ln n}$. Since $\ln n < n$ for $n \geq 2$, $\dfrac{1}{\ln n} > \dfrac{1}{n}$ and $\sum_{n=2}^{\infty} \dfrac{1}{n}$ is the

divergent harmonic series (without the $n = 1$ term), $\sum_{n=2}^{\infty} \dfrac{1}{\ln n}$ is divergent by the Comparison Test. When $x = 4$,

$\sum_{n=2}^{\infty} (-1)^n \dfrac{x^n}{4^n \ln n} = \sum_{n=2}^{\infty} (-1)^n \dfrac{1}{\ln n}$, which converges by the Alternating Series Test. Thus, $I = (-4, 4]$.

14. $a_n = (-1)^n \dfrac{x^{2n}}{(2n)!}$, so $\lim_{n \to \infty} \left| \dfrac{a_{n+1}}{a_n} \right| = \lim_{n \to \infty} \dfrac{|x|^{2n+2}}{(2n+2)!} \cdot \dfrac{(2n)!}{|x|^{2n}} = \lim_{n \to \infty} \dfrac{|x|^2}{(2n+1)(2n+2)} = 0 < 1$. Thus, by the Ratio

Test, the series converges for *all* real x and we have $R = \infty$ and $I = (-\infty, \infty)$.

15. If $a_n = \dfrac{(x-2)^n}{n^2+1}$, then $\lim\limits_{n\to\infty}\left|\dfrac{a_{n+1}}{a_n}\right| = \lim\limits_{n\to\infty}\left|\dfrac{(x-2)^{n+1}}{(n+1)^2+1}\cdot\dfrac{n^2+1}{(x-2)^n}\right| = |x-2|\lim\limits_{n\to\infty}\dfrac{n^2+1}{(n+1)^2+1} = |x-2|$. By the

Ratio Test, the series $\sum\limits_{n=0}^{\infty}\dfrac{(x-2)^n}{n^2+1}$ converges when $|x-2|<1$ $[R=1]$ $\Leftrightarrow$ $-1<x-2<1$ $\Leftrightarrow$ $1<x<3$. When

$x=1$, the series $\sum\limits_{n=0}^{\infty}(-1)^n\dfrac{1}{n^2+1}$ converges by the Alternating Series Test; when $x=3$, the series $\sum\limits_{n=0}^{\infty}\dfrac{1}{n^2+1}$ converges by

comparison with the p-series $\sum\limits_{n=1}^{\infty}\dfrac{1}{n^2}$ $[p=2>1]$. Thus, the interval of convergence is $I=[1,3]$.

16. If $a_n = (-1)^n\dfrac{(x-3)^n}{2n+1}$, then $\lim\limits_{n\to\infty}\left|\dfrac{a_{n+1}}{a_n}\right| = \lim\limits_{n\to\infty}\left|\dfrac{(x-3)^{n+1}}{2n+3}\cdot\dfrac{2n+1}{(x-3)^n}\right| = |x-3|\lim\limits_{n\to\infty}\dfrac{2n+1}{2n+3} = |x-3|$. By the

Ratio Test, the series $\sum\limits_{n=0}^{\infty}(-1)^n\dfrac{(x-3)^n}{2n+1}$ converges when $|x-3|<1$ $[R=1]$ $\Leftrightarrow$ $-1<x-3<1$ $\Leftrightarrow$ $2<x<4$.

When $x=2$, the series $\sum\limits_{n=0}^{\infty}\dfrac{1}{2n+1}$ diverges by limit comparison with the harmonic series (or by the Integral Test); when

$x=4$, the series $\sum\limits_{n=0}^{\infty}(-1)^n\dfrac{1}{2n+1}$ converges by the Alternating Series Test. Thus, the interval of convergence is $I=(2,4]$.

17. If $a_n = \dfrac{3^n(x+4)^n}{\sqrt{n}}$, then $\lim\limits_{n\to\infty}\left|\dfrac{a_{n+1}}{a_n}\right| = \lim\limits_{n\to\infty}\left|\dfrac{3^{n+1}(x+4)^{n+1}}{\sqrt{n+1}}\cdot\dfrac{\sqrt{n}}{3^n(x+4)^n}\right| = 3|x+4|\lim\limits_{n\to\infty}\dfrac{\sqrt{n}}{\sqrt{n+1}} = 3|x+4|$.

By the Ratio Test, the series $\sum\limits_{n=1}^{\infty}\dfrac{3^n(x+4)^n}{\sqrt{n}}$ converges when $3|x+4|<1$ $\Leftrightarrow$ $|x+4|<\frac{1}{3}$ $\left[R=\frac{1}{3}\right]$ $\Leftrightarrow$

$-\frac{1}{3}<x+4<\frac{1}{3}$ $\Leftrightarrow$ $-\frac{13}{3}<x<-\frac{11}{3}$. When $x=-\frac{13}{3}$, the series $\sum\limits_{n=1}^{\infty}(-1)^n\dfrac{1}{\sqrt{n}}$ converges by the Alternating Series

Test; when $x=-\frac{11}{3}$, the series $\sum\limits_{n=1}^{\infty}\dfrac{1}{\sqrt{n}}$ diverges $\left[p=\frac{1}{2}\le 1\right]$. Thus, the interval of convergence is $I=\left[-\frac{13}{3},-\frac{11}{3}\right)$.

18. If $a_n = \dfrac{n}{4^n}(x+1)^n$, then $\lim\limits_{n\to\infty}\left|\dfrac{a_{n+1}}{a_n}\right| = \lim\limits_{n\to\infty}\left|\dfrac{(n+1)(x+1)^{n+1}}{4^{n+1}}\cdot\dfrac{4^n}{n(x+1)^n}\right| = \dfrac{|x+1|}{4}\lim\limits_{n\to\infty}\dfrac{n+1}{n} = \dfrac{|x+1|}{4}$.

By the Ratio Test, the series $\sum\limits_{n=1}^{\infty}\dfrac{n}{4^n}(x+1)^n$ converges when $\dfrac{|x+1|}{4}<1$ $\Leftrightarrow$ $|x+1|<4$ $[R=4]$ $\Leftrightarrow$

$-4<x+1<4$ $\Leftrightarrow$ $-5<x<3$. When $x=-5$ or 3, both series $\sum\limits_{n=1}^{\infty}(\mp 1)^n\, n$ diverge by the Test for Divergence since

$\lim\limits_{n\to\infty}|(\mp 1)^n\, n| = \infty$. Thus, the interval of convergence is $I=(-5,3)$.

19. If $a_n = \dfrac{(x-2)^n}{n^n}$, then $\lim\limits_{n\to\infty}\sqrt[n]{|a_n|} = \lim\limits_{n\to\infty}\dfrac{|x-2|}{n} = 0$, so the series converges for all x (by the Root Test).

$R=\infty$ and $I=(-\infty,\infty)$.

20. $\lim\limits_{n\to\infty}\left|\dfrac{a_{n+1}}{a_n}\right| = \lim\limits_{n\to\infty}\left|\dfrac{(3x-2)^{n+1}}{(n+1)3^{n+1}}\cdot\dfrac{n3^n}{(3x-2)^n}\right| = \lim\limits_{n\to\infty}\left(\dfrac{|3x-2|}{3}\cdot\dfrac{1}{1+1/n}\right) = \dfrac{|3x-2|}{3} = \left|x-\frac{2}{3}\right|$, so by the Ratio

Test, the series converges when $\left|x-\frac{2}{3}\right|<1$ $\Leftrightarrow$ $-\frac{1}{3}<x<\frac{5}{3}$. $R=1$. When $x=-\frac{1}{3}$, the series is $\sum\limits_{n=1}^{\infty}\dfrac{(-1)^n}{n}$, the

convergent alternating harmonic series. When $x=\frac{5}{3}$, the series becomes the divergent harmonic series. Thus, $I=\left[-\frac{1}{3},\frac{5}{3}\right)$.

21. $a_n = \dfrac{n}{b^n}(x-a)^n$, where $b > 0$.

$$\lim_{n\to\infty}\left|\frac{a_{n+1}}{a_n}\right| = \lim_{n\to\infty}\frac{(n+1)\,|x-a|^{n+1}}{b^{n+1}}\cdot\frac{b^n}{n\,|x-a|^n} = \lim_{n\to\infty}\left(1+\frac{1}{n}\right)\frac{|x-a|}{b} = \frac{|x-a|}{b}.$$

By the Ratio Test, the series converges when $\dfrac{|x-a|}{b} < 1 \Leftrightarrow |x-a| < b$ [so $R = b$] $\Leftrightarrow -b < x-a < b \Leftrightarrow$

$a - b < x < a + b$. When $|x-a| = b$, $\lim\limits_{n\to\infty}|a_n| = \lim\limits_{n\to\infty} n = \infty$, so the series diverges. Thus, $I = (a-b, a+b)$.

22. $a_n = \dfrac{n(x-4)^n}{n^3+1}$, so

$$\lim_{n\to\infty}\left|\frac{a_{n+1}}{a_n}\right| = \lim_{n\to\infty}\frac{(n+1)\,|x-4|^{n+1}}{(n+1)^3+1}\cdot\frac{n^3+1}{n\,|x-4|^n} = \lim_{n\to\infty}\left(1+\frac{1}{n}\right)\frac{n^3+1}{n^3+3n^2+3n+2}\,|x-4| = |x-4|.$$

By the Ratio Test, the series converges when $|x-4| < 1$ [so $R = 1$] $\Leftrightarrow -1 < x-4 < 1 \Leftrightarrow 3 < x < 5$. When

$|x-4| = 1$, $\sum\limits_{n=1}^{\infty}|a_n| = \sum\limits_{n=1}^{\infty}\dfrac{n}{n^3+1}$, which converges by comparison with the convergent p-series $\sum\limits_{n=1}^{\infty}\dfrac{1}{n^2}$ $[p = 2 > 1]$.

Thus, $I = [3,5]$.

23. If $a_n = n!\,(2x-1)^n$, then $\lim\limits_{n\to\infty}\left|\dfrac{a_{n+1}}{a_n}\right| = \lim\limits_{n\to\infty}\left|\dfrac{(n+1)!\,(2x-1)^{n+1}}{n!(2x-1)^n}\right| = \lim\limits_{n\to\infty}(n+1)\,|2x-1| \to \infty$ as $n\to\infty$

for all $x \neq \frac{1}{2}$. Since the series diverges for all $x \neq \frac{1}{2}$, $R = 0$ and $I = \left\{\frac{1}{2}\right\}$.

24. $a_n = \dfrac{n^2 x^n}{2\cdot4\cdot6\cdots\cdots(2n)} = \dfrac{n^2 x^n}{2^n n!} = \dfrac{nx^n}{2^n(n-1)!}$, so

$$\lim_{n\to\infty}\left|\frac{a_{n+1}}{a_n}\right| = \lim_{n\to\infty}\frac{(n+1)\,|x|^{n+1}}{2^{n+1}n!}\cdot\frac{2^n(n-1)!}{n\,|x|^n} = \lim_{n\to\infty}\frac{n+1}{n^2}\frac{|x|}{2} = 0.$$ Thus, by the Ratio Test, the series converges for

all real x and we have $R = \infty$ and $I = (-\infty, \infty)$.

25. $\lim\limits_{n\to\infty}\left|\dfrac{a_{n+1}}{a_n}\right| = \lim\limits_{n\to\infty}\left[\dfrac{|4x+1|^{n+1}}{(n+1)^2}\cdot\dfrac{n^2}{|4x+1|^n}\right] = \lim\limits_{n\to\infty}\dfrac{|4x+1|}{(1+1/n)^2} = |4x+1|$, so by the Ratio Test, the series

converges when $|4x+1| < 1 \Leftrightarrow -1 < 4x+1 < 1 \Leftrightarrow -2 < 4x < 0 \Leftrightarrow -\frac{1}{2} < x < 0$, so $R = \frac{1}{4}$. When $x = -\frac{1}{2}$,

the series becomes $\sum\limits_{n=1}^{\infty}\dfrac{(-1)^n}{n^2}$, which converges by the Alternating Series Test. When $x = 0$, the series becomes $\sum\limits_{n=1}^{\infty}\dfrac{1}{n^2}$,

a convergent p-series $[p = 2 > 1]$. $\quad I = \left[-\frac{1}{2}, 0\right]$.

26. If $a_n = \dfrac{x^{2n}}{n\,(\ln n)^2}$, then $\lim\limits_{n\to\infty}\left|\dfrac{a_{n+1}}{a_n}\right| = \lim\limits_{n\to\infty}\left|\dfrac{x^{2n+2}}{(n+1)[\ln(n+1)]^2}\cdot\dfrac{n\,(\ln n)^2}{x^{2n}}\right| = |x^2|\lim\limits_{n\to\infty}\dfrac{n\,(\ln n)^2}{(n+1)[\ln(n+1)]^2} = x^2$.

By the Ratio Test, the series $\sum\limits_{n=2}^{\infty}\dfrac{x^{2n}}{n\,(\ln n)^2}$ converges when $x^2 < 1 \Leftrightarrow |x| < 1$, so $R = 1$. When $x = \pm 1$, $x^{2n} = 1$, the

series $\sum\limits_{n=2}^{\infty}\dfrac{1}{n\,(\ln n)^2}$ converges by the Integral Test (see Exercise 12.3.22). Thus, the interval of convergence is $I = [-1, 1]$.

27. If $a_n = \dfrac{x^n}{1 \cdot 3 \cdot 5 \cdot \cdots \cdot (2n-1)}$, then

$$\lim_{n \to \infty} \left| \frac{a_{n+1}}{a_n} \right| = \lim_{n \to \infty} \left| \frac{x^{n+1}}{1 \cdot 3 \cdot 5 \cdot \cdots \cdot (2n-1)(2n+1)} \cdot \frac{1 \cdot 3 \cdot 5 \cdot \cdots \cdot (2n-1)}{x^n} \right| = \lim_{n \to \infty} \frac{|x|}{2n+1} = 0 < 1. \text{ Thus, by the}$$

Ratio Test, the series $\displaystyle\sum_{n=1}^{\infty} \dfrac{x^n}{1 \cdot 3 \cdot 5 \cdot \cdots \cdot (2n-1)}$ converges for *all* real x and we have $R = \infty$ and $I = (-\infty, \infty)$.

28. If $a_n = \dfrac{n!\, x^n}{1 \cdot 3 \cdot 5 \cdot \cdots \cdot (2n-1)}$, then

$$\lim_{n \to \infty} \left| \frac{a_{n+1}}{a_n} \right| = \lim_{n \to \infty} \left| \frac{(n+1)!\, x^{n+1}}{1 \cdot 3 \cdot 5 \cdot \cdots \cdot (2n-1)(2n+1)} \cdot \frac{1 \cdot 3 \cdot 5 \cdot \cdots \cdot (2n-1)}{n!\, x^n} \right| = \lim_{n \to \infty} \frac{(n+1)\,|x|}{2n+1} = \tfrac{1}{2}\,|x|.$$

By the Ratio Test, the series $\displaystyle\sum_{n=1}^{\infty} a_n$ converges when $\tfrac{1}{2}\,|x| < 1 \;\Rightarrow\; |x| < 2$, so $R = 2$. When $x = \pm 2$,

$$|a_n| = \frac{n!\, 2^n}{1 \cdot 3 \cdot 5 \cdot \cdots \cdot (2n-1)} = \frac{[1 \cdot 2 \cdot 3 \cdot \cdots \cdot n]\, 2^n}{[1 \cdot 3 \cdot 5 \cdot \cdots \cdot (2n-1)]} = \frac{2 \cdot 4 \cdot 6 \cdot \cdots \cdot 2n}{1 \cdot 3 \cdot 5 \cdot \cdots \cdot (2n-1)} > 1, \text{ so both endpoint series}$$

diverge by the Test for Divergence. Thus, the interval of convergence is $I = (-2, 2)$.

29. (a) We are given that the power series $\sum_{n=0}^{\infty} c_n x^n$ is convergent for $x = 4$. So by Theorem 3, it must converge for at least

$-4 < x \le 4$. In particular, it converges when $x = -2$; that is, $\sum_{n=0}^{\infty} c_n(-2)^n$ is convergent.

(b) It does not follow that $\sum_{n=0}^{\infty} c_n(-4)^n$ is necessarily convergent. [See the comments after Theorem 3 about convergence at

the endpoint of an interval. An example is $c_n = (-1)^n / (n4^n)$.]

30. We are given that the power series $\sum_{n=0}^{\infty} c_n x^n$ is convergent for $x = -4$ and divergent when $x = 6$. So by Theorem 3 it

converges for at least $-4 \le x < 4$ and diverges for at least $x \ge 6$ and $x < -6$. Therefore:

(a) It converges when $x = 1$; that is, $\sum c_n$ is convergent.

(b) It diverges when $x = 8$; that is, $\sum c_n 8^n$ is divergent.

(c) It converges when $x = -3$; that is, $\sum c_n(-3^n)$ is convergent.

(d) It diverges when $x = -9$; that is, $\sum c_n(-9)^n = \sum (-1)^n c_n 9^n$ is divergent.

31. If $a_n = \dfrac{(n!)^k}{(kn)!}\, x^n$, then

$$\lim_{n \to \infty} \left| \frac{a_{n+1}}{a_n} \right| = \lim_{n \to \infty} \frac{[(n+1)!]^k\, (kn)!}{(n!)^k\, [k(n+1)]!}\, |x| = \lim_{n \to \infty} \frac{(n+1)^k}{(kn+k)(kn+k-1)\cdots(kn+2)(kn+1)}\, |x|$$

$$= \lim_{n \to \infty} \left[\frac{(n+1)}{(kn+1)} \frac{(n+1)}{(kn+2)} \cdots \frac{(n+1)}{(kn+k)} \right] |x|$$

$$= \lim_{n \to \infty} \left[\frac{n+1}{kn+1} \right] \lim_{n \to \infty} \left[\frac{n+1}{kn+2} \right] \cdots \lim_{n \to \infty} \left[\frac{n+1}{kn+k} \right] |x|$$

$$= \left(\frac{1}{k} \right)^k |x| < 1 \quad \Leftrightarrow \quad |x| < k^k \text{ for convergence, and the radius of convergence is } R = k^k.$$

32. (a) Note that the four intervals in parts (a)–(d) have midpoint $m = \frac{1}{2}(p+q)$ and radius of convergence $r = \frac{1}{2}(q-p)$. We also

know that the power series $\sum\limits_{n=0}^{\infty} x^n$ has interval of convergence $(-1, 1)$. To change the radius of convergence to r, we can

change x^n to $\left(\dfrac{x}{r}\right)^n$. To shift the midpoint of the interval of convergence, we can replace x with $x - m$. Thus, a power

series whose interval of convergence is (p, q) is $\sum\limits_{n=0}^{\infty} \left(\dfrac{x-m}{r}\right)^n$, where $m = \frac{1}{2}(p+q)$ and $r = \frac{1}{2}(q-p)$.

(b) Similar to Example 2, we know that $\sum\limits_{n=1}^{\infty} \dfrac{x^n}{n}$ has interval of convergence $[-1, 1)$. By introducing the factor $(-1)^n$

in a_n, the interval of convergence changes to $(-1, 1]$. Now change the midpoint and radius as in part (a) to get

$\sum\limits_{n=1}^{\infty} (-1)^n \dfrac{1}{n} \left(\dfrac{x-m}{r}\right)^n$ as a power series whose interval of convergence is $(p, q]$.

(c) As in part (b), $\sum\limits_{n=1}^{\infty} \dfrac{1}{n} \left(\dfrac{x-m}{r}\right)^n$ is a power series whose interval of convergence is $[p, q)$.

(d) If we increase the exponent on n (to say, $n = 2$), in the power series in part (c), then when $x = q$, the power series

$\sum\limits_{n=1}^{\infty} \dfrac{1}{n^2} \left(\dfrac{x-m}{r}\right)^n$ will converge by comparison to the p-series with $p = 2 > 1$, and the interval of convergence will

be $[p, q]$.

33. No. If a power series is centered at a, its interval of convergence is symmetric about a. If a power series has an infinite radius

of convergence, then its interval of convergence must be $(-\infty, \infty)$, not $[0, \infty)$.

34. The partial sums of the series $\sum_{n=0}^{\infty} x^n$ definitely do not converge

to $f(x) = 1/(1-x)$ for $x \geq 1$, since f is undefined at $x = 1$ and

negative on $(1, \infty)$, while all the partial sums are positive on this

interval. The partial sums also fail to converge to f for $x \leq -1$,

since $0 < f(x) < 1$ on this interval, while the partial sums are

either larger than 1 or less than 0. The partial sums seem to

converge to f on $(-1, 1)$. This graphical evidence is consistent

with what we know about geometric series: convergence for

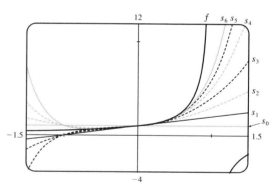

$|x| < 1$, divergence for $|x| \geq 1$ (see Examples 1 and 5 in Section 12.2).

35. (a) If $a_n = \dfrac{(-1)^n \, x^{2n+1}}{n!(n+1)! \, 2^{2n+1}}$, then

$$\lim_{n\to\infty} \left|\dfrac{a_{n+1}}{a_n}\right| = \lim_{n\to\infty} \left|\dfrac{x^{2n+3}}{(n+1)!(n+2)! \, 2^{2n+3}} \cdot \dfrac{n!(n+1)! \, 2^{2n+1}}{x^{2n+1}}\right| = \left(\dfrac{x}{2}\right)^2 \lim_{n\to\infty} \dfrac{1}{(n+1)(n+2)} = 0 \text{ for all } x.$$

So $J_1(x)$ converges for all x and its domain is $(-\infty, \infty)$.

(b), (c) The initial terms of $J_1(x)$ up to $n = 5$ are $a_0 = \dfrac{x}{2}$,

$a_1 = -\dfrac{x^3}{16}, a_2 = \dfrac{x^5}{384}, a_3 = -\dfrac{x^7}{18{,}432}, a_4 = \dfrac{x^9}{1{,}474{,}560}$,

and $a_5 = -\dfrac{x^{11}}{176{,}947{,}200}$. The partial sums seem to

approximate $J_1(x)$ well near the origin, but as $|x|$ increases,

we need to take a large number of terms to get a good

approximation.

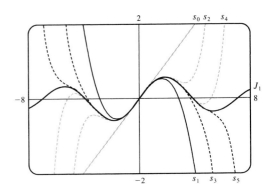

36. (a) $A(x) = 1 + \sum\limits_{n=1}^{\infty} a_n$, where $a_n = \dfrac{x^{3n}}{2 \cdot 3 \cdot 5 \cdot 6 \cdot \cdots \cdot (3n-1)(3n)}$, so $\lim\limits_{n \to \infty} \left| \dfrac{a_{n+1}}{a_n} \right| = |x|^3 \lim\limits_{n \to \infty} \dfrac{1}{(3n+2)(3n+3)} = 0$

for all x, so the domain is $\mathbb{R}$.

(b), (c)

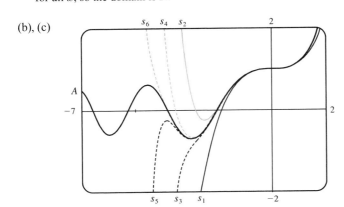

$s_0 = 1$ has been omitted from the graph. The partial sums seem to approximate $A(x)$ well near the origin, but as $|x|$ increases, we need to take a large number of terms to get a good approximation.

To plot A, we must first define $A(x)$ for the CAS. Note that for $n \geq 1$, the denominator of a_n is

$2 \cdot 3 \cdot 5 \cdot 6 \cdot \cdots \cdot (3n-1) \cdot 3n = \dfrac{(3n)!}{1 \cdot 4 \cdot 7 \cdot \cdots \cdot (3n-2)} = \dfrac{(3n)!}{\prod_{k=1}^{n}(3k-2)}$, so $a_n = \dfrac{\prod_{k=1}^{n}(3k-2)}{(3n)!} x^{3n}$ and thus

$A(x) = 1 + \sum\limits_{n=1}^{\infty} \dfrac{\prod_{k=1}^{n}(3k-2)}{(3n)!} x^{3n}$. Both Maple and Mathematica are able to plot A if we define it this way, and Derive

is able to produce a similar graph using a suitable partial sum of $A(x)$.

    Derive, Maple and Mathematica all have two initially known Airy functions, called AI·SERIES(z,m) and

BI·SERIES(z,m) from BESSEL.MTH in Derive and AiryAi and AiryBi in Maple and Mathematica (just Ai and

Bi in older versions of Maple). However, it is very difficult to solve for A in terms of the CAS's Airy functions, although

in fact $A(x) = \dfrac{\sqrt{3}\,\text{AiryAi}(x) + \text{AiryBi}(x)}{\sqrt{3}\,\text{AiryAi}(0) + \text{AiryBi}(0)}$.

37. $s_{2n-1} = 1 + 2x + x^2 + 2x^3 + x^4 + 2x^5 + \cdots + x^{2n-2} + 2x^{2n-1}$

$= 1(1 + 2x) + x^2(1 + 2x) + x^4(1 + 2x) + \cdots + x^{2n-2}(1 + 2x) = (1 + 2x)(1 + x^2 + x^4 + \cdots + x^{2n-2})$

$= (1 + 2x)\dfrac{1 - x^{2n}}{1 - x^2}$ [by (12.2.3)] with $r = x^2$] $\to \dfrac{1 + 2x}{1 - x^2}$ as $n \to \infty$ [by (12.2.4)], when $|x| < 1$.

Also $s_{2n} = s_{2n-1} + x^{2n} \to \dfrac{1 + 2x}{1 - x^2}$ since $x^{2n} \to 0$ for $|x| < 1$. Therefore, $s_n \to \dfrac{1 + 2x}{1 - x^2}$ since s_{2n} and s_{2n-1} both

approach $\dfrac{1 + 2x}{1 - x^2}$ as $n \to \infty$. Thus, the interval of convergence is $(-1, 1)$ and $f(x) = \dfrac{1 + 2x}{1 - x^2}$.

38. $s_{4n-1} = c_0 + c_1 x + c_2 x^2 + c_3 x^3 + c_0 x^4 + c_1 x^5 + c_2 x^6 + c_3 x^7 + \cdots + c_3 x^{4n-1}$

$$= \left(c_0 + c_1 x + c_2 x^2 + c_3 x^3\right)\left(1 + x^4 + x^8 + \cdots + x^{4n-4}\right) \rightarrow \frac{c_0 + c_1 x + c_2 x^2 + c_3 x^3}{1 - x^4} \text{ as } n \to \infty$$

[by (12.2.4) with $r = x^4$] for $\left|x^4\right| < 1 \quad \Leftrightarrow \quad |x| < 1$. Also s_{4n}, s_{4n+1}, s_{4n+2} have the same limits (for example,

$s_{4n} = s_{4n-1} + c_0 x^{4n}$ and $x^{4n} \to 0$ for $|x| < 1$). So if at least one of c_0, c_1, c_2, and c_3 is nonzero, then the interval of

convergence is $(-1, 1)$ and $f(x) = \dfrac{c_0 + c_1 x + c_2 x^2 + c_3 x^3}{1 - x^4}$.

39. We use the Root Test on the series $\sum c_n x^n$. We need $\lim\limits_{n \to \infty} \sqrt[n]{|c_n x^n|} = |x| \lim\limits_{n \to \infty} \sqrt[n]{|c_n|} = c\,|x| < 1$ for convergence, or

$|x| < 1/c$, so $R = 1/c$.

40. Suppose $c_n \neq 0$. Applying the Ratio Test to the series $\sum c_n (x - a)^n$, we find that

$$L = \lim_{n \to \infty}\left|\frac{a_{n+1}}{a_n}\right| = \lim_{n \to \infty}\left|\frac{c_{n+1}(x - a)^{n+1}}{c_n(x - a)^n}\right| = \lim_{n \to \infty}\frac{|x - a|}{|c_n/c_{n+1}|}\,(*) = \frac{|x - a|}{\lim\limits_{n \to \infty}|c_n/c_{n+1}|} \text{ (if } \lim_{n \to \infty}|c_n/c_{n+1}| \neq 0\text{), so the}$$

series converges when $\dfrac{|x - a|}{\lim\limits_{n \to \infty}|c_n/c_{n+1}|} < 1 \quad \Leftrightarrow \quad |x - a| < \lim\limits_{n \to \infty}\left|\dfrac{c_n}{c_{n+1}}\right|$. Thus, $R = \lim\limits_{n \to \infty}\left|\dfrac{c_n}{c_{n+1}}\right|$. If $\lim\limits_{n \to \infty}\left|\dfrac{c_n}{c_{n+1}}\right| = 0$

and $|x - a| \neq 0$, then $(*)$ shows that $L = \infty$ and so the series diverges, and hence, $R = 0$. Thus, in all cases,

$$R = \lim_{n \to \infty}\left|\frac{c_n}{c_{n+1}}\right|.$$

41. For $2 < x < 3$, $\sum c_n x^n$ diverges and $\sum d_n x^n$ converges. By Exercise 12.2.69, $\sum(c_n + d_n)\,x^n$ diverges. Since both series converge for $|x| < 2$, the radius of convergence of $\sum(c_n + d_n)\,x^n$ is 2.

42. Since $\sum c_n x^n$ converges whenever $|x| < R$, $\sum c_n x^{2n} = \sum c_n\left(x^2\right)^n$ converges whenever $\left|x^2\right| < R \quad \Leftrightarrow \quad |x| < \sqrt{R}$, so the second series has radius of convergence $\sqrt{R}$.

12.9 Representations of Functions as Power Series

1. If $f(x) = \sum\limits_{n=0}^{\infty} c_n x^n$ has radius of convergence 10, then $f'(x) = \sum\limits_{n=1}^{\infty} n c_n x^{n-1}$ also has radius of convergence 10 by

Theorem 2.

2. If $f(x) = \sum\limits_{n=0}^{\infty} b_n x^n$ converges on $(-2, 2)$, then $\int f(x)dx = C + \sum\limits_{n=0}^{\infty}\dfrac{b_n}{n + 1}x^{n+1}$ has the same radius of convergence

(by Theorem 2), but may not have the same interval of convergence—it may happen that the integrated series converges at an endpoint (or both endpoints).

3. Our goal is to write the function in the form $\dfrac{1}{1 - r}$, and then use Equation (1) to represent the function as a sum of a power

series. $f(x) = \dfrac{1}{1 + x} = \dfrac{1}{1 - (-x)} = \sum\limits_{n=0}^{\infty}(-x)^n = \sum\limits_{n=0}^{\infty}(-1)^n x^n$ with $|-x| < 1 \quad \Leftrightarrow \quad |x| < 1$, so $R = 1$ and $I = (-1, 1)$.

4. $f(x) = \dfrac{3}{1 - x^4} = 3\left(\dfrac{1}{1 - x^4}\right) = 3(1 + x^4 + x^8 + x^{12} + \cdots) = 3\sum\limits_{n=0}^{\infty}(x^4)^n = \sum\limits_{n=0}^{\infty}3x^{4n}$

with $\left|x^4\right| < 1 \quad \Leftrightarrow \quad |x| < 1$, so $R = 1$ and $I = (-1, 1)$.

$$\left[\text{Note that } 3\sum\limits_{n=0}^{\infty}(x^4)^n \text{ converges} \quad \Leftrightarrow \quad \sum\limits_{n=0}^{\infty}(x^4)^n \text{ converges, so the appropriate condition [from equation (1)] is } \left|x^4\right| < 1.\right]$$

5. $f(x) = \dfrac{2}{3-x} = \dfrac{2}{3}\left(\dfrac{1}{1-x/3}\right) = \dfrac{2}{3}\sum\limits_{n=0}^{\infty}\left(\dfrac{x}{3}\right)^n$ or, equivalently, $2\sum\limits_{n=0}^{\infty}\dfrac{1}{3^{n+1}}x^n$. The series converges when $\left|\dfrac{x}{3}\right| < 1$,

that is, when $|x| < 3$, so $R = 3$ and $I = (-3,3)$.

6. $f(x) = \dfrac{1}{x+10} = \dfrac{1}{10}\left(\dfrac{1}{1-(-x/10)}\right) = \dfrac{1}{10}\sum\limits_{n=0}^{\infty}\left(-\dfrac{x}{10}\right)^n$ or, equivalently, $\sum\limits_{n=0}^{\infty}(-1)^n\dfrac{1}{10^{n+1}}x^n$. The series converges

when $\left|\dfrac{x}{10}\right| < 1$, that is, when $|x| < 10$, so $R = 10$ and $I = (-10,10)$.

7. $f(x) = \dfrac{x}{9+x^2} = \dfrac{x}{9}\left[\dfrac{1}{1+(x/3)^2}\right] = \dfrac{x}{9}\left[\dfrac{1}{1-\{-(x/3)^2\}}\right] = \dfrac{x}{9}\sum\limits_{n=0}^{\infty}\left[-\left(\dfrac{x}{3}\right)^2\right]^n = \dfrac{x}{9}\sum\limits_{n=0}^{\infty}(-1)^n\dfrac{x^{2n}}{9^n} = \sum\limits_{n=0}^{\infty}(-1)^n\dfrac{x^{2n+1}}{9^{n+1}}$

The geometric series $\sum\limits_{n=0}^{\infty}\left[-\left(\dfrac{x}{3}\right)^2\right]^n$ converges when $\left|-\left(\dfrac{x}{3}\right)^2\right| < 1 \Leftrightarrow \dfrac{|x^2|}{9} < 1 \Leftrightarrow |x|^2 < 9 \Leftrightarrow |x| < 3$, so

$R = 3$ and $I = (-3,3)$.

8. $f(x) = \dfrac{x}{2x^2+1} = x\left(\dfrac{1}{1-(-2x^2)}\right) = x\sum\limits_{n=0}^{\infty}(-2x^2)^n$ or, equivalently, $\sum\limits_{n=0}^{\infty}(-1)^n 2^n x^{2n+1}$. The series converges when

$|-2x^2| < 1 \Rightarrow |x^2| < \tfrac{1}{2} \Rightarrow |x| < \dfrac{1}{\sqrt{2}}$, so $R = \dfrac{1}{\sqrt{2}}$ and $I = \left(-\dfrac{1}{\sqrt{2}}, \dfrac{1}{\sqrt{2}}\right)$.

9. $f(x) = \dfrac{1+x}{1-x} = (1+x)\left(\dfrac{1}{1-x}\right) = (1+x)\sum\limits_{n=0}^{\infty}x^n = \sum\limits_{n=0}^{\infty}x^n + \sum\limits_{n=0}^{\infty}x^{n+1} = 1 + \sum\limits_{n=1}^{\infty}x^n + \sum\limits_{n=1}^{\infty}x^n = 1 + 2\sum\limits_{n=1}^{\infty}x^n$.

The series converges when $|x| < 1$, so $R = 1$ and $I = (-1,1)$.

A second approach: $f(x) = \dfrac{1+x}{1-x} = \dfrac{-(1-x)+2}{1-x} = -1 + 2\left(\dfrac{1}{1-x}\right) = -1 + 2\sum\limits_{n=0}^{\infty}x^n = 1 + 2\sum\limits_{n=1}^{\infty}x^n$.

A third approach:

$f(x) = \dfrac{1+x}{1-x} = (1+x)\left(\dfrac{1}{1-x}\right) = (1+x)(1+x+x^2+x^3+\cdots)$

$= (1+x+x^2+x^3+\cdots) + (x+x^2+x^3+x^4+\cdots) = 1 + 2x + 2x^2 + 2x^3 + \cdots = 1 + 2\sum\limits_{n=1}^{\infty}x^n$.

10. $f(x) = \dfrac{x^2}{a^3-x^3} = \dfrac{x^2}{a^3}\cdot\dfrac{1}{1-x^3/a^3} = \dfrac{x^2}{a^3}\sum\limits_{n=0}^{\infty}\left(\dfrac{x^3}{a^3}\right)^n = \sum\limits_{n=0}^{\infty}\dfrac{x^{3n+2}}{a^{3n+3}}$. The series converges when $|x^3/a^3| < 1 \Leftrightarrow$

$|x^3| < |a^3| \Leftrightarrow |x| < |a|$, so $R = |a|$ and $I = (-|a|, |a|)$.

11. $f(x) = \dfrac{3}{x^2-x-2} = \dfrac{3}{(x-2)(x+1)} = \dfrac{A}{x-2} + \dfrac{B}{x+1} \Rightarrow 3 = A(x+1) + B(x-2)$. Let $x = 2$ to get $A = 1$ and

$x = -1$ to get $B = -1$. Thus

$\dfrac{3}{x^2-x-2} = \dfrac{1}{x-2} - \dfrac{1}{x+1} = \dfrac{1}{-2}\left(\dfrac{1}{1-(x/2)}\right) - \dfrac{1}{1-(-x)} = -\dfrac{1}{2}\sum\limits_{n=0}^{\infty}\left(\dfrac{x}{2}\right)^n - \sum\limits_{n=0}^{\infty}(-x)^n$

$= \sum\limits_{n=0}^{\infty}\left[-\dfrac{1}{2}\left(\dfrac{1}{2}\right)^n - 1(-1)^n\right]x^n = \sum\limits_{n=0}^{\infty}\left[(-1)^{n+1} - \dfrac{1}{2^{n+1}}\right]x^n$

We represented f as the sum of two geometric series; the first converges for $x \in (-2,2)$ and the second converges for $(-1,1)$.

Thus, the sum converges for $x \in (-1,1) = I$.

12. $f(x) = \dfrac{x+2}{2x^2 - x - 1} = \dfrac{x+2}{(2x+1)(x-1)} = \dfrac{A}{2x+1} + \dfrac{B}{x-1}$ $\Rightarrow$ $x + 2 = A(x-1) + B(2x+1)$. Let $x = 1$ to get

$3 = 3B$ $\Rightarrow$ $B = 1$ and $x = -\frac{1}{2}$ to get $\frac{3}{2} = -\frac{3}{2}A$ $\Rightarrow$ $A = -1$. Thus,

$$\dfrac{x+2}{2x^2 - x - 1} = \dfrac{-1}{2x+1} + \dfrac{1}{x-1} = -1\left(\dfrac{1}{1-(-2x)}\right) - 1\left(\dfrac{1}{1-x}\right) = -\sum_{n=0}^{\infty}(-2x)^n - \sum_{n=0}^{\infty}x^n$$

$$= -\sum_{n=0}^{\infty}\left[(-2)^n + 1\right]x^n$$

We represented f as the sum of two geometric series; the first converges for $x \in \left(-\frac{1}{2}, \frac{1}{2}\right)$ and the second converges for $(-1, 1)$. Thus, the sum converges for $x \in \left(-\frac{1}{2}, \frac{1}{2}\right) = I$.

13. (a) $f(x) = \dfrac{1}{(1+x)^2} = \dfrac{d}{dx}\left(\dfrac{-1}{1+x}\right) = -\dfrac{d}{dx}\left[\sum_{n=0}^{\infty}(-1)^n x^n\right]$ [from Exercise 3]

$= \sum_{n=1}^{\infty}(-1)^{n+1}nx^{n-1}$ [from Theorem 2(i)] $= \sum_{n=0}^{\infty}(-1)^n(n+1)x^n$ with $R = 1$.

In the last step, note that we *decreased* the initial value of the summation variable n by 1, and then *increased* each occurrence of n in the term by 1 [also note that $(-1)^{n+2} = (-1)^n$].

(b) $f(x) = \dfrac{1}{(1+x)^3} = -\dfrac{1}{2}\dfrac{d}{dx}\left[\dfrac{1}{(1+x)^2}\right] = -\dfrac{1}{2}\dfrac{d}{dx}\left[\sum_{n=0}^{\infty}(-1)^n(n+1)x^n\right]$ [from part (a)]

$= -\dfrac{1}{2}\sum_{n=1}^{\infty}(-1)^n(n+1)nx^{n-1} = \dfrac{1}{2}\sum_{n=0}^{\infty}(-1)^n(n+2)(n+1)x^n$ with $R = 1$.

(c) $f(x) = \dfrac{x^2}{(1+x)^3} = x^2 \cdot \dfrac{1}{(1+x)^3} = x^2 \cdot \dfrac{1}{2}\sum_{n=0}^{\infty}(-1)^n(n+2)(n+1)x^n$ [from part (b)]

$= \dfrac{1}{2}\sum_{n=0}^{\infty}(-1)^n(n+2)(n+1)x^{n+2}$

To write the power series with x^n rather than x^{n+2}, we will *decrease* each occurrence of n in the term by 2 and *increase* the initial value of the summation variable by 2. This gives us $\dfrac{1}{2}\sum_{n=2}^{\infty}(-1)^n(n)(n-1)x^n$ with $R = 1$.

14. (a) $\dfrac{1}{1+x} = \dfrac{1}{1-(-x)} = \sum_{n=0}^{\infty}(-1)^n x^n$ [geometric series with $R = 1$], so

$f(x) = \ln(1+x) = \int \dfrac{dx}{1+x} = \int\left[\sum_{n=0}^{\infty}(-1)^n x^n\right]dx = C + \sum_{n=0}^{\infty}(-1)^n\dfrac{x^{n+1}}{n+1} = \sum_{n=1}^{\infty}\dfrac{(-1)^{n-1}x^n}{n}$

$[C = 0$ since $f(0) = \ln 1 = 0]$, with $R = 1$

(b) $f(x) = x\ln(1+x) = x\left[\sum_{n=1}^{\infty}\dfrac{(-1)^{n-1}x^n}{n}\right]$ [by part (a)] $= \sum_{n=1}^{\infty}\dfrac{(-1)^{n-1}x^{n+1}}{n} = \sum_{n=2}^{\infty}\dfrac{(-1)^n x^n}{n-1}$ with $R = 1$.

(c) $f(x) = \ln(x^2 + 1) = \sum_{n=1}^{\infty}\dfrac{(-1)^{n-1}(x^2)^n}{n}$ [by part (a)] $= \sum_{n=1}^{\infty}\dfrac{(-1)^{n-1}x^{2n}}{n}$ with $R = 1$.

15. $f(x) = \ln(5 - x) = -\int\dfrac{dx}{5-x} = -\dfrac{1}{5}\int\dfrac{dx}{1-x/5} = -\dfrac{1}{5}\int\left[\sum_{n=0}^{\infty}\left(\dfrac{x}{5}\right)^n\right]dx = C - \dfrac{1}{5}\sum_{n=0}^{\infty}\dfrac{x^{n+1}}{5^n(n+1)} = C - \sum_{n=1}^{\infty}\dfrac{x^n}{n\,5^n}$

Putting $x = 0$, we get $C = \ln 5$. The series converges for $|x/5| < 1$ $\Leftrightarrow$ $|x| < 5$, so $R = 5$.

16. We know that $\dfrac{1}{1-2x} = \displaystyle\sum_{n=0}^{\infty}(2x)^n$. Differentiating, we get $\dfrac{2}{(1-2x)^2} = \displaystyle\sum_{n=1}^{\infty} 2^n n x^{n-1} = \displaystyle\sum_{n=0}^{\infty} 2^{n+1}(n+1)x^n$, so

$$f(x) = \frac{x^2}{(1-2x)^2} = \frac{x^2}{2}\cdot\frac{2}{(1-2x)^2} = \frac{x^2}{2}\sum_{n=0}^{\infty}2^{n+1}(n+1)x^n = \sum_{n=0}^{\infty}2^n(n+1)x^{n+2} \text{ or } \sum_{n=2}^{\infty}2^{n-2}(n-1)x^n,$$

with $R = \frac{1}{2}$.

17. $\dfrac{1}{2-x} = \dfrac{1}{2(1-x/2)} = \dfrac{1}{2}\displaystyle\sum_{n=0}^{\infty}\left(\dfrac{x}{2}\right)^n = \displaystyle\sum_{n=0}^{\infty}\dfrac{1}{2^{n+1}}x^n$ for $\left|\dfrac{x}{2}\right| < 1 \iff |x| < 2$. Now

$$\frac{1}{(x-2)^2} = \frac{d}{dx}\left(\frac{1}{2-x}\right) = \frac{d}{dx}\left(\sum_{n=0}^{\infty}\frac{1}{2^{n+1}}x^n\right) = \sum_{n=1}^{\infty}\frac{n}{2^{n+1}}x^{n-1} = \sum_{n=0}^{\infty}\frac{n+1}{2^{n+2}}x^n. \text{ So}$$

$$f(x) = \frac{x^3}{(x-2)^2} = x^3\sum_{n=0}^{\infty}\frac{n+1}{2^{n+2}}x^n = \sum_{n=0}^{\infty}\frac{n+1}{2^{n+2}}x^{n+3} \text{ or } \sum_{n=3}^{\infty}\frac{n-2}{2^{n-1}}x^n \text{ for } |x| < 2. \text{ Thus, } R = 2 \text{ and } I = (-2,2).$$

18. From Example 7, $g(x) = \arctan x = \displaystyle\sum_{n=0}^{\infty}(-1)^n\dfrac{x^{2n+1}}{2n+1}$. Thus,

$$f(x) = \arctan(x/3) = \sum_{n=0}^{\infty}(-1)^n\frac{(x/3)^{2n+1}}{2n+1} = \sum_{n=0}^{\infty}(-1)^n\frac{1}{3^{2n+1}(2n+1)}x^{2n+1} \text{ for } \left|\frac{x}{3}\right| < 1 \iff |x| < 3, \text{ so } R = 3.$$

19. $f(x) = \dfrac{x}{x^2+16} = \dfrac{x}{16}\left(\dfrac{1}{1-(-x^2/16)}\right) = \dfrac{x}{16}\displaystyle\sum_{n=0}^{\infty}\left(-\dfrac{x^2}{16}\right)^n = \dfrac{x}{16}\displaystyle\sum_{n=0}^{\infty}(-1)^n\dfrac{1}{16^n}x^{2n} = \displaystyle\sum_{n=0}^{\infty}(-1)^n\dfrac{1}{16^{n+1}}x^{2n+1}.$

The series converges when $\left|-x^2/16\right| < 1 \iff x^2 < 16 \iff |x| < 4$, so $R = 4$. The partial sums are $s_1 = \dfrac{x}{16}$,

$s_2 = s_1 - \dfrac{x^3}{16^2}, s_3 = s_2 + \dfrac{x^5}{16^3}, s_4 = s_3 - \dfrac{x^7}{16^4}, s_5 = s_4 + \dfrac{x^9}{16^5}, \ldots$. Note that s_1 corresponds to the first term of the infinite

sum, regardless of the value of the summation variable and the value of the exponent.

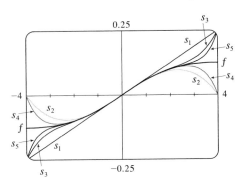

As n increases, $s_n(x)$ approximates f better on the interval of convergence, which is $(-4, 4)$.

20. $f(x) = \ln(x^2 + 4) \implies f'(x) = \dfrac{2x}{x^2+4} = \dfrac{2x}{4}\left(\dfrac{1}{1-(-x^2/4)}\right) = \dfrac{x}{2}\displaystyle\sum_{n=0}^{\infty}\left(-\dfrac{x^2}{4}\right)^n = \displaystyle\sum_{n=0}^{\infty}(-1)^n\dfrac{x^{2n+1}}{2^{2n+1}}$,

so $f(x) = \displaystyle\int\sum_{n=0}^{\infty}(-1)^n\dfrac{x^{2n+1}}{2^{2n+1}}\,dx = C + \displaystyle\sum_{n=0}^{\infty}(-1)^n\dfrac{x^{2n+2}}{2^{2n+1}(2n+2)} = \ln 4 + \displaystyle\sum_{n=0}^{\infty}(-1)^n\dfrac{x^{2n+2}}{(n+1)2^{2n+2}}$

$[f(0) = \ln 4, \text{ so } C = \ln 4]$. The series converges when $\left|-x^2/4\right| < 1 \iff x^2 < 4 \iff |x| < 2, \text{ so } R = 2. \text{ If}$

$x = \pm 2$, then $f(x) = \ln 4 + \sum_{n=0}^{\infty} (-1)^n \dfrac{1}{n+1}$, which converges by the Alternating Series Test. The partial sums

are $s_0 = \ln 4$ $[\approx 1.39]$, $s_1 = s_0 + \dfrac{x^2}{4}$, $s_2 = s_1 - \dfrac{x^4}{2 \cdot 2^4}$, $s_3 = s_2 + \dfrac{x^6}{3 \cdot 2^6}$, $s_4 = s_3 - \dfrac{x^8}{4 \cdot 2^8}$, $\ldots$.

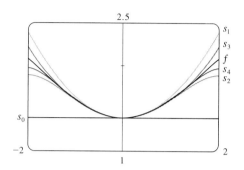

As n increases, $s_n(x)$ approximates f better on the interval of convergence, which is $[-2, 2]$.

21. $f(x) = \ln\left(\dfrac{1+x}{1-x}\right) = \ln(1+x) - \ln(1-x) = \displaystyle\int \dfrac{dx}{1+x} + \int \dfrac{dx}{1-x} = \int \dfrac{dx}{1-(-x)} + \int \dfrac{dx}{1-x}$

$= \displaystyle\int \left[\sum_{n=0}^{\infty} (-1)^n x^n + \sum_{n=0}^{\infty} x^n \right] dx = \int [(1 - x + x^2 - x^3 + x^4 - \cdots) + (1 + x + x^2 + x^3 + x^4 + \cdots)]\, dx$

$= \displaystyle\int (2 + 2x^2 + 2x^4 + \cdots)\, dx = \int \sum_{n=0}^{\infty} 2x^{2n}\, dx = C + \sum_{n=0}^{\infty} \dfrac{2x^{2n+1}}{2n+1}$

But $f(0) = \ln \frac{1}{1} = 0$, so $C = 0$ and we have $f(x) = \sum_{n=0}^{\infty} \dfrac{2x^{2n+1}}{2n+1}$ with $R = 1$. If $x = \pm 1$, then $f(x) = \pm 2 \sum_{n=0}^{\infty} \dfrac{1}{2n+1}$,

which both diverge by the Limit Comparison Test with $b_n = \dfrac{1}{n}$. The partial sums are $s_1 = \dfrac{2x}{1}$, $s_2 = s_1 + \dfrac{2x^3}{3}$,

$s_3 = s_2 + \dfrac{2x^5}{5}$, $\ldots$.

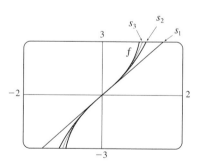

As n increases, $s_n(x)$ approximates f better on the interval of convergence, which is $(-1, 1)$.

22. $f(x) = \tan^{-1}(2x) = 2 \displaystyle\int \dfrac{dx}{1+4x^2} = 2 \int \sum_{n=0}^{\infty} (-1)^n \left(4x^2\right)^n\, dx = 2 \int \sum_{n=0}^{\infty} (-1)^n 4^n x^{2n}\, dx$

$= C + 2 \displaystyle\sum_{n=0}^{\infty} \dfrac{(-1)^n 4^n x^{2n+1}}{2n+1} = \sum_{n=0}^{\infty} \dfrac{(-1)^n 2^{2n+1} x^{2n+1}}{2n+1} \qquad [f(0) = \tan^{-1} 0 = 0, \text{ so } C = 0]$

The series converges when $\left|4x^2\right| < 1 \iff |x| < \frac{1}{2}$, so $R = \frac{1}{2}$. If $x = \pm\frac{1}{2}$, then $f(x) = \sum_{n=0}^{\infty} (-1)^n \dfrac{1}{2n+1}$ and

$f(x) = \sum_{n=0}^{\infty} (-1)^{n+1} \dfrac{1}{2n+1}$, respectively. Both series converge by the Alternating Series Test. The partial sums are

$s_1 = \dfrac{2x}{1}$, $s_2 = s_1 - \dfrac{2^3 x^3}{3}$, $s_3 = s_2 + \dfrac{2^5 x^5}{5}, \ \dots$.

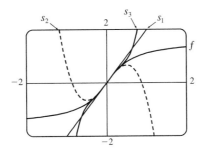

As n increases, $s_n(x)$ approximates f better on the interval of convergence, which is $\left[-\tfrac{1}{2}, \tfrac{1}{2}\right]$.

23. $\dfrac{t}{1-t^8} = t \cdot \dfrac{1}{1-t^8} = t \sum_{n=0}^{\infty} (t^8)^n = \sum_{n=0}^{\infty} t^{8n+1} \quad \Rightarrow \quad \displaystyle\int \dfrac{t}{1-t^8}\,dt = C + \sum_{n=0}^{\infty} \dfrac{t^{8n+2}}{8n+2}$. The series for $\dfrac{1}{1-t^8}$ converges

when $|t^8| < 1 \quad \Leftrightarrow \quad |t| < 1$, so $R = 1$ for that series and also the series for $t/(1-t^8)$. By Theorem 2, the series for

$\displaystyle\int \dfrac{t}{1-t^8}\,dt$ also has $R = 1$.

24. By Example 6, $\ln(1-t) = -\sum_{n=1}^{\infty} \dfrac{t^n}{n}$ for $|t| < 1$, so $\dfrac{\ln(1-t)}{t} = -\sum_{n=1}^{\infty} \dfrac{t^{n-1}}{n}$ and $\displaystyle\int \dfrac{\ln(1-t)}{t}\,dt = C - \sum_{n=1}^{\infty} \dfrac{t^n}{n^2}$.

By Theorem 2, $R = 1$.

25. By Example 7, $\tan^{-1} x = \sum_{n=0}^{\infty} (-1)^n \dfrac{x^{2n+1}}{2n+1}$ with $R = 1$, so

$x - \tan^{-1} x = x - \left(x - \dfrac{x^3}{3} + \dfrac{x^5}{5} - \dfrac{x^7}{7} + \cdots \right) = \dfrac{x^3}{3} - \dfrac{x^5}{5} + \dfrac{x^7}{7} - \cdots = \sum_{n=1}^{\infty} (-1)^{n+1} \dfrac{x^{2n+1}}{2n+1}$ and

$\dfrac{x - \tan^{-1} x}{x^3} = \sum_{n=1}^{\infty} (-1)^{n+1} \dfrac{x^{2n-2}}{2n+1}$, so

$\displaystyle\int \dfrac{x - \tan^{-1} x}{x^3}\,dx = C + \sum_{n=1}^{\infty} (-1)^{n+1} \dfrac{x^{2n-1}}{(2n+1)(2n-1)} = C + \sum_{n=1}^{\infty} (-1)^{n+1} \dfrac{x^{2n-1}}{4n^2-1}$. By Theorem 2, $R = 1$.

26. By Example 7, $\displaystyle\int \tan^{-1}(x^2)\,dx = \int \sum_{n=0}^{\infty} (-1)^n \dfrac{(x^2)^{2n+1}}{2n+1}\,dx = C + \sum_{n=0}^{\infty} (-1)^n \dfrac{x^{4n+3}}{(2n+1)(4n+3)}$ with $R = 1$.

27. $\dfrac{1}{1+x^5} = \dfrac{1}{1-(-x^5)} = \sum_{n=0}^{\infty} (-x^5)^n = \sum_{n=0}^{\infty} (-1)^n x^{5n} \quad \Rightarrow$

$\displaystyle\int \dfrac{1}{1+x^5}\,dx = \int \sum_{n=0}^{\infty} (-1)^n x^{5n}\,dx = C + \sum_{n=0}^{\infty} (-1)^n \dfrac{x^{5n+1}}{5n+1}$. Thus,

$I = \displaystyle\int_0^{0.2} \dfrac{1}{1+x^5}\,dx = \left[x - \dfrac{x^6}{6} + \dfrac{x^{11}}{11} - \cdots \right]_0^{0.2} = 0.2 - \dfrac{(0.2)^6}{6} + \dfrac{(0.2)^{11}}{11} - \cdots$. The series is alternating, so if we use

the first two terms, the error is at most $(0.2)^{11}/11 \approx 1.9 \times 10^{-9}$. So $I \approx 0.2 - (0.2)^6/6 \approx 0.199989$ to six decimal places.

28. From Example 6, we know $\ln(1 - x) = -\sum\limits_{n=1}^{\infty} \dfrac{x^n}{n}$, so

$$\ln\left(1 + x^4\right) = \ln\left[1 - \left(-x^4\right)\right] = -\sum_{n=1}^{\infty} \frac{\left(-x^4\right)^n}{n} = \sum_{n=1}^{\infty} (-1)^{n+1} \frac{x^{4n}}{n} \quad \Rightarrow$$

$$\int \ln\left(1 + x^4\right) dx = \int \sum_{n=1}^{\infty} (-1)^{n+1} \frac{x^{4n}}{n}\, dx = C + \sum_{n=1}^{\infty} (-1)^{n+1} \frac{x^{4n+1}}{n(4n + 1)}. \text{ Thus,}$$

$$I = \int_0^{0.4} \ln\left(1 + x^4\right) dx = \left[\frac{x^5}{5} - \frac{x^9}{18} + \frac{x^{13}}{39} - \frac{x^{17}}{68} + \cdots\right]_0^{0.4} = \frac{(0.4)^5}{5} - \frac{(0.4)^9}{18} + \frac{(0.4)^{13}}{39} - \frac{(0.4)^{17}}{68} + \cdots.$$

The series is alternating, so if we use the first three terms, the error is at most $(0.4)^{17}/68 \approx 2.5 \times 10^{-9}$.

So $I \approx (0.4)^5/5 - (0.4)^9/18 + (0.9)^{13}/39 \approx 0.002034$ to six decimal places.

29. We substitute $3x$ for x in Example 7, and find that

$$\int x \arctan(3x)\, dx = \int x \sum_{n=0}^{\infty} (-1)^n \frac{(3x)^{2n+1}}{2n + 1}\, dx = \int \sum_{n=0}^{\infty} (-1)^n \frac{3^{2n+1} x^{2n+2}}{2n + 1}\, dx = C + \sum_{n=0}^{\infty} (-1)^n \frac{3^{2n+1} x^{2n+3}}{(2n + 1)(2n + 3)}$$

So

$$\int_0^{0.1} x \arctan(3x)\, dx = \left[\frac{3x^3}{1 \cdot 3} - \frac{3^3 x^5}{3 \cdot 5} + \frac{3^5 x^7}{5 \cdot 7} - \frac{3^7 x^9}{7 \cdot 9} + \cdots\right]_0^{0.1}$$

$$= \frac{1}{10^3} - \frac{9}{5 \times 10^5} + \frac{243}{35 \times 10^7} - \frac{2187}{63 \times 10^9} + \cdots.$$

The series is alternating, so if we use three terms, the error is at most $\dfrac{2187}{63 \times 10^9} \approx 3.5 \times 10^{-8}$. So

$$\int_0^{0.1} x \arctan(3x)\, dx \approx \frac{1}{10^3} - \frac{9}{5 \times 10^5} + \frac{243}{35 \times 10^7} \approx 0.000\,983 \text{ to six decimal places.}$$

30. $\displaystyle\int_0^{0.3} \frac{x^2}{1 + x^4}\, dx = \int_0^{0.3} x^2 \sum_{n=0}^{\infty} (-1)^n x^{4n}\, dx = \sum_{n=0}^{\infty} \left[\frac{(-1)^n x^{4n+3}}{4n + 3}\right]_0^{0.3} = \sum_{n=0}^{\infty} \frac{(-1)^n\, 3^{4n+3}}{(4n + 3)10^{4n+3}}$

$$= \frac{3^3}{3 \times 10^3} - \frac{3^7}{7 \times 10^7} + \frac{3^{11}}{11 \times 10^{11}} - \cdots$$

The series is alternating, so if we use only two terms, the error is at most $\dfrac{3^{11}}{11 \times 10^{11}} \approx 0.000\,000\,16$. So, to six decimal

places, $\displaystyle\int_0^{0.3} \frac{x^2}{1 + x^4}\, dx \approx \frac{3^3}{3 \times 10^3} - \frac{3^7}{7 \times 10^7} \approx 0.008\,969$.

31. Using the result of Example 6, $\ln(1 - x) = -\sum\limits_{n=1}^{\infty} \dfrac{x^n}{n}$, with $x = -0.1$, we have

$$\ln 1.1 = \ln[1 - (-0.1)] = 0.1 - \frac{0.01}{2} + \frac{0.001}{3} - \frac{0.0001}{4} + \frac{0.00001}{5} - \cdots. \text{ The series is alternating, so if we use only}$$

the first four terms, the error is at most $\dfrac{0.00001}{5} = 0.000002$. So $\ln 1.1 \approx 0.1 - \dfrac{0.01}{2} + \dfrac{0.001}{3} - \dfrac{0.0001}{4} \approx 0.09531$.

32. $f(x) = \displaystyle\sum_{n=0}^{\infty} \frac{(-1)^n x^{2n}}{(2n)!} \quad \Rightarrow \quad f'(x) = \sum_{n=1}^{\infty} \frac{(-1)^n 2n x^{2n-1}}{(2n)!}$ [the first term disappears], so

$$f''(x) = \sum_{n=1}^{\infty} \frac{(-1)^n (2n)(2n - 1) x^{2n-2}}{(2n)!} = \sum_{n=1}^{\infty} \frac{(-1)^n x^{2(n-1)}}{[2(n - 1)]!} = \sum_{n=0}^{\infty} \frac{(-1)^{n+1} x^{2n}}{(2n)!} \quad \text{[substituting } n + 1 \text{ for } n\text{]}$$

$$= -\sum_{n=0}^{\infty} \frac{(-1)^n x^{2n}}{(2n)!} = -f(x) \quad \Rightarrow \quad f''(x) + f(x) = 0.$$

33. (a) $J_0(x) = \sum\limits_{n=0}^{\infty} \dfrac{(-1)^n \, x^{2n}}{2^{2n} \, (n!)^2}$, $J_0'(x) = \sum\limits_{n=1}^{\infty} \dfrac{(-1)^n \, 2nx^{2n-1}}{2^{2n} \, (n!)^2}$, and $J_0''(x) = \sum\limits_{n=1}^{\infty} \dfrac{(-1)^n \, 2n(2n-1)x^{2n-2}}{2^{2n} \, (n!)^2}$, so

$$x^2 J_0''(x) + x J_0'(x) + x^2 J_0(x) = \sum_{n=1}^{\infty} \frac{(-1)^n \, 2n(2n-1)x^{2n}}{2^{2n} \, (n!)^2} + \sum_{n=1}^{\infty} \frac{(-1)^n \, 2nx^{2n}}{2^{2n} \, (n!)^2} + \sum_{n=0}^{\infty} \frac{(-1)^n \, x^{2n+2}}{2^{2n} \, (n!)^2}$$

$$= \sum_{n=1}^{\infty} \frac{(-1)^n \, 2n(2n-1)x^{2n}}{2^{2n} \, (n!)^2} + \sum_{n=1}^{\infty} \frac{(-1)^n \, 2nx^{2n}}{2^{2n} \, (n!)^2} + \sum_{n=1}^{\infty} \frac{(-1)^{n-1} \, x^{2n}}{2^{2n-2} \, [(n-1)!]^2}$$

$$= \sum_{n=1}^{\infty} \frac{(-1)^n \, 2n(2n-1)x^{2n}}{2^{2n} \, (n!)^2} + \sum_{n=1}^{\infty} \frac{(-1)^n \, 2nx^{2n}}{2^{2n} \, (n!)^2} + \sum_{n=1}^{\infty} \frac{(-1)^n (-1)^{-1} 2^2 n^2 x^{2n}}{2^{2n} \, (n!)^2}$$

$$= \sum_{n=1}^{\infty} (-1)^n \left[\frac{2n(2n-1) + 2n - 2^2 n^2}{2^{2n} \, (n!)^2} \right] x^{2n}$$

$$= \sum_{n=1}^{\infty} (-1)^n \left[\frac{4n^2 - 2n + 2n - 4n^2}{2^{2n} \, (n!)^2} \right] x^{2n} = 0$$

(b) $\displaystyle\int_0^1 J_0(x)\, dx = \int_0^1 \left[\sum_{n=0}^{\infty} \frac{(-1)^n \, x^{2n}}{2^{2n} \, (n!)^2} \right] dx = \int_0^1 \left(1 - \frac{x^2}{4} + \frac{x^4}{64} - \frac{x^6}{2304} + \cdots \right) dx$

$$= \left[x - \frac{x^3}{3 \cdot 4} + \frac{x^5}{5 \cdot 64} - \frac{x^7}{7 \cdot 2304} + \cdots \right]_0^1 = 1 - \frac{1}{12} + \frac{1}{320} - \frac{1}{16{,}128} + \cdots$$

Since $\frac{1}{16{,}128} \approx 0.000062$, it follows from The Alternating Series Estimation Theorem that, correct to three decimal places,

$\int_0^1 J_0(x)\, dx \approx 1 - \frac{1}{12} + \frac{1}{320} \approx 0.920$.

34. (a) $J_1(x) = \sum\limits_{n=0}^{\infty} \dfrac{(-1)^n x^{2n+1}}{n!\,(n+1)!\,2^{2n+1}}$, $J_1'(x) = \sum\limits_{n=0}^{\infty} \dfrac{(-1)^n \, (2n+1)\, x^{2n}}{n!\,(n+1)!\,2^{2n+1}}$, and $J_1''(x) = \sum\limits_{n=1}^{\infty} \dfrac{(-1)^n \, (2n+1)\,(2n)\, x^{2n-1}}{n!\,(n+1)!\,2^{2n+1}}$.

$x^2 J_1''(x) + x J_1'(x) + \left(x^2 - 1 \right) J_1(x)$

$$= \sum_{n=1}^{\infty} \frac{(-1)^n \, (2n+1)(2n)x^{2n+1}}{n!\,(n+1)!\,2^{2n+1}} + \sum_{n=0}^{\infty} \frac{(-1)^n \, (2n+1)x^{2n+1}}{n!\,(n+1)!\,2^{2n+1}}$$

$$+ \sum_{n=0}^{\infty} \frac{(-1)^n \, x^{2n+3}}{n!\,(n+1)!\,2^{2n+1}} - \sum_{n=0}^{\infty} \frac{(-1)^n \, x^{2n+1}}{n!\,(n+1)!\,2^{2n+1}}$$

$$= \sum_{n=1}^{\infty} \frac{(-1)^n \, (2n+1)(2n)x^{2n+1}}{n!\,(n+1)!\,2^{2n+1}} + \sum_{n=0}^{\infty} \frac{(-1)^n \, (2n+1)x^{2n+1}}{n!\,(n+1)!\,2^{2n+1}}$$

$$- \sum_{n=1}^{\infty} \frac{(-1)^n \, x^{2n+1}}{(n-1)!\,n!\,2^{2n-1}} - \sum_{n=0}^{\infty} \frac{(-1)^n \, x^{2n+1}}{n!\,(n+1)!\,2^{2n+1}} \qquad \begin{bmatrix} \text{Replace } n \text{ with } n-1 \\ \text{in the third term} \end{bmatrix}$$

$$= \frac{x}{2} - \frac{x}{2} + \sum_{n=1}^{\infty} (-1)^n \left[\frac{(2n+1)(2n) + (2n+1) - (n)(n+1)2^2 - 1}{n!\,(n+1)!\,2^{2n+1}} \right] x^{2n+1} = 0$$

(b) $J_0(x) = \sum\limits_{n=0}^{\infty} \dfrac{(-1)^n \, x^{2n}}{2^{2n} \, (n!)^2} \quad \Rightarrow$

$$J_0'(x) = \sum_{n=1}^{\infty} \frac{(-1)^n \, (2n)x^{2n-1}}{2^{2n} \, (n!)^2} = \sum_{n=0}^{\infty} \frac{(-1)^{n+1} \, 2(n+1)x^{2n+1}}{2^{2n+2} \, [(n+1)!]^2} \qquad [\text{Replace } n \text{ with } n+1]$$

$$= - \sum_{n=0}^{\infty} \frac{(-1)^n \, x^{2n+1}}{2^{2n+1}(n+1)!\, n!} \qquad [\text{cancel 2 and } n+1; \text{ take } -1 \text{ outside sum}] \quad = -J_1(x)$$

35. (a) $f(x) = \sum\limits_{n=0}^{\infty} \dfrac{x^n}{n!}$ $\Rightarrow$ $f'(x) = \sum\limits_{n=1}^{\infty} \dfrac{nx^{n-1}}{n!} = \sum\limits_{n=1}^{\infty} \dfrac{x^{n-1}}{(n-1)!} = \sum\limits_{n=0}^{\infty} \dfrac{x^n}{n!} = f(x)$

(b) By Theorem 10.4.2, the only solution to the differential equation $df(x)/dx = f(x)$ is $f(x) = Ke^x$, but $f(0) = 1$, so $K = 1$ and $f(x) = e^x$.

Or: We could solve the equation $df(x)/dx = f(x)$ as a separable differential equation.

36. $\dfrac{|\sin nx|}{n^2} \le \dfrac{1}{n^2}$, so $\sum\limits_{n=1}^{\infty} \dfrac{\sin nx}{n^2}$ converges by the Comparison Test. $\dfrac{d}{dx}\left(\dfrac{\sin nx}{n^2}\right) = \dfrac{\cos nx}{n}$, so when $x = 2k\pi$

[k an integer], $\sum\limits_{n=1}^{\infty} f_n'(x) = \sum\limits_{n=1}^{\infty} \dfrac{\cos(2kn\pi)}{n} = \sum\limits_{n=1}^{\infty} \dfrac{1}{n}$, which diverges [harmonic series]. $f_n''(x) = -\sin nx$, so

$\sum\limits_{n=1}^{\infty} f_n''(x) = -\sum\limits_{n=1}^{\infty} \sin nx$, which converges only if $\sin nx = 0$, or $x = k\pi$ [k an integer].

37. If $a_n = \dfrac{x^n}{n^2}$, then by the Ratio Test, $\lim\limits_{n\to\infty}\left|\dfrac{a_{n+1}}{a_n}\right| = \lim\limits_{n\to\infty}\left|\dfrac{x^{n+1}}{(n+1)^2} \cdot \dfrac{n^2}{x^n}\right| = |x|\lim\limits_{n\to\infty}\left(\dfrac{n}{n+1}\right)^2 = |x| < 1$ for

convergence, so $R = 1$. When $x = \pm 1$, $\sum\limits_{n=1}^{\infty}\left|\dfrac{x^n}{n^2}\right| = \sum\limits_{n=1}^{\infty}\dfrac{1}{n^2}$ which is a convergent p-series ($p = 2 > 1$), so the interval of

convergence for f is $[-1, 1]$. By Theorem 2, the radii of convergence of f' and f'' are both 1, so we need only check the

endpoints. $f(x) = \sum\limits_{n=1}^{\infty}\dfrac{x^n}{n^2}$ $\Rightarrow$ $f'(x) = \sum\limits_{n=1}^{\infty}\dfrac{nx^{n-1}}{n^2} = \sum\limits_{n=0}^{\infty}\dfrac{x^n}{n+1}$, and this series diverges for $x = 1$ (harmonic series)

and converges for $x = -1$ (Alternating Series Test), so the interval of convergence is $[-1, 1)$. $f''(x) = \sum\limits_{n=1}^{\infty}\dfrac{nx^{n-1}}{n+1}$ diverges

at both 1 and -1 (Test for Divergence) since $\lim\limits_{n\to\infty}\dfrac{n}{n+1} = 1 \ne 0$, so its interval of convergence is $(-1, 1)$.

38. (a) $\sum\limits_{n=1}^{\infty} nx^{n-1} = \sum\limits_{n=0}^{\infty}\dfrac{d}{dx}x^n = \dfrac{d}{dx}\left[\sum\limits_{n=0}^{\infty}x_n\right] = \dfrac{d}{dx}\left[\dfrac{1}{1-x}\right] = -\dfrac{1}{(1-x)^2}(-1) = \dfrac{1}{(1-x)^2}$, $|x| < 1$.

(b) (i) $\sum\limits_{n=1}^{\infty} nx^n = x\sum\limits_{n=1}^{\infty} nx^{n-1} = x\left[\dfrac{1}{(1-x)^2}\right]$ [from part (a)] $= \dfrac{x}{(1-x)^2}$ for $|x| < 1$.

(ii) Put $x = \tfrac{1}{2}$ in (i): $\sum\limits_{n=1}^{\infty}\dfrac{n}{2^n} = \sum\limits_{n=1}^{\infty}n\left(\tfrac{1}{2}\right)^n = \dfrac{1/2}{(1-1/2)^2} = 2$.

(c) (i) $\sum\limits_{n=2}^{\infty}n(n-1)x^n = x^2\sum\limits_{n=2}^{\infty}n(n-1)x^{n-2} = x^2\dfrac{d}{dx}\left[\sum\limits_{n=1}^{\infty}nx^{n-1}\right] = x^2\dfrac{d}{dx}\dfrac{1}{(1-x)^2}$

$= x^2\dfrac{2}{(1-x)^3} = \dfrac{2x^2}{(1-x)^3}$ for $|x| < 1$.

(ii) Put $x = \tfrac{1}{2}$ in (i): $\sum\limits_{n=2}^{\infty}\dfrac{n^2-n}{2^n} = \sum\limits_{n=2}^{\infty}n(n-1)\left(\tfrac{1}{2}\right)^n = \dfrac{2(1/2)^2}{(1-1/2)^3} = 4$.

(iii) From (b)(ii) and (c)(ii), we have $\sum\limits_{n=1}^{\infty}\dfrac{n^2}{2^n} = \sum\limits_{n=1}^{\infty}\dfrac{n^2-n}{2^n} + \sum\limits_{n=1}^{\infty}\dfrac{n}{2^n} = 4 + 2 = 6$.

39. By Example 7, $\tan^{-1} x = \sum\limits_{n=0}^{\infty} (-1)^n \dfrac{x^{2n+1}}{2n+1}$ for $|x| < 1$. In particular, for $x = \dfrac{1}{\sqrt{3}}$, we

have $\dfrac{\pi}{6} = \tan^{-1}\left(\dfrac{1}{\sqrt{3}}\right) = \sum\limits_{n=0}^{\infty} (-1)^n \dfrac{(1/\sqrt{3})^{2n+1}}{2n+1} = \sum\limits_{n=0}^{\infty} (-1)^n \left(\dfrac{1}{3}\right)^n \dfrac{1}{\sqrt{3}} \dfrac{1}{2n+1}$, so

$$\pi = \dfrac{6}{\sqrt{3}} \sum_{n=0}^{\infty} \dfrac{(-1)^n}{(2n+1)3^n} = 2\sqrt{3} \sum_{n=0}^{\infty} \dfrac{(-1)^n}{(2n+1)3^n}.$$

40. (a) $\displaystyle\int_0^{1/2} \dfrac{dx}{x^2 - x + 1} = \int_0^{1/2} \dfrac{dx}{(x-1/2)^2 + 3/4} \qquad \left[x - \dfrac{1}{2} = \dfrac{\sqrt{3}}{2} u, \; u = \dfrac{2}{\sqrt{3}}\left(x - \dfrac{1}{2}\right), \; dx = \dfrac{\sqrt{3}}{2} du \right]$

$$= \int_{-1/\sqrt{3}}^{0} \dfrac{(\sqrt{3}/2)\, du}{(3/4)(u^2 + 1)} = \dfrac{2\sqrt{3}}{3} \Big[\tan^{-1} u \Big]_{-1/\sqrt{3}}^{0} \quad = \dfrac{2}{\sqrt{3}} \left[0 - \left(-\dfrac{\pi}{6} \right) \right] = \dfrac{\pi}{3\sqrt{3}}$$

(b) $\dfrac{1}{x^3 + 1} = \dfrac{1}{(x+1)(x^2 - x + 1)} \quad \Rightarrow$

$$\dfrac{1}{x^2 - x + 1} = (x+1)\left(\dfrac{1}{1+x^3} \right) = (x+1)\dfrac{1}{1 - (-x^3)} = (x+1) \sum_{n=0}^{\infty} (-1)^n x^{3n}$$

$$= \sum_{n=0}^{\infty} (-1)^n x^{3n+1} + \sum_{n=0}^{\infty} (-1)^n x^{3n} \quad \text{for } |x| < 1 \quad \Rightarrow$$

$$\int \dfrac{dx}{x^2 - x + 1} = C + \sum_{n=0}^{\infty} (-1)^n \dfrac{x^{3n+2}}{3n+2} + \sum_{n=0}^{\infty} (-1)^n \dfrac{x^{3n+1}}{3n+1} \quad \text{for } |x| < 1 \quad \Rightarrow$$

$$\int_0^{1/2} \dfrac{dx}{x^2 - x + 1} = \sum_{n=0}^{\infty} (-1)^n \left[\dfrac{1}{4 \cdot 8^n (3n+2)} + \dfrac{1}{2 \cdot 8^n (3n+1)} \right] = \dfrac{1}{4} \sum_{n=0}^{\infty} \dfrac{(-1)^n}{8^n} \left(\dfrac{2}{3n+1} + \dfrac{1}{3n+2} \right).$$

By part (a), this equals $\dfrac{\pi}{3\sqrt{3}}$, so $\pi = \dfrac{3\sqrt{3}}{4} \sum\limits_{n=0}^{\infty} \dfrac{(-1)^n}{8^n} \left(\dfrac{2}{3n+1} + \dfrac{1}{3n+2} \right)$.

12.10   Taylor and Maclaurin Series

1. Using Theorem 5 with $\sum\limits_{n=0}^{\infty} b_n (x-5)^n$, $b_n = \dfrac{f^{(n)}(a)}{n!}$, so $b_8 = \dfrac{f^{(8)}(5)}{8!}$.

2. (a) Using Equation 6, a power series expansion of f at 1 must have the form $f(1) + f'(1)(x-1) + \cdots$. Comparing to the

given series, $1.6 - 0.8(x-1) + \cdots$, we must have $f'(1) = -0.8$. But from the graph, $f'(1)$ is positive. Hence, the given

series is *not* the Taylor series of f centered at 1.

(b) A power series expansion of f at 2 must have the form $f(2) + f'(2)(x-2) + \frac{1}{2} f''(2)(x-2)^2 + \cdots$. Comparing to the

given series, $2.8 + 0.5(x-2) + 1.5(x-2)^2 - 0.1(x-2)^3 + \cdots$, we must have $\frac{1}{2} f''(2) = 1.5$; that is, $f''(2)$ is positive.

But from the graph, f is concave downward near $x = 2$, so $f''(2)$ must be negative. Hence, the given series is *not* the

Taylor series of f centered at 2.

3. Since $f^{(n)}(0) = (n + 1)!$, Equation 7 gives the Maclaurin series

$$\sum_{n=0}^{\infty} \frac{f^{(n)}(0)}{n!} x^n = \sum_{n=0}^{\infty} \frac{(n + 1)!}{n!} x^n = \sum_{n=0}^{\infty} (n + 1)x^n. \text{ Applying the Ratio Test with } a_n = (n + 1)x^n \text{ gives us}$$

$$\lim_{n \to \infty} \left| \frac{a_{n+1}}{a_n} \right| = \lim_{n \to \infty} \left| \frac{(n + 2)x^{n+1}}{(n + 1)x^n} \right| = |x| \lim_{n \to \infty} \frac{n + 2}{n + 1} = |x| \cdot 1 = |x|. \text{ For convergence, we must have } |x| < 1, \text{ so the}$$

radius of convergence $R = 1$.

4. Since $f^{(n)}(4) = \dfrac{(-1)^n n!}{3^n (n + 1)}$, Equation 6 gives the Taylor series

$$\sum_{n=0}^{\infty} \frac{f^{(n)}(4)}{n!} (x - 4)^n = \sum_{n=0}^{\infty} \frac{(-1)^n n!}{3^n (n + 1) n!} (x - 4)^n = \sum_{n=0}^{\infty} \frac{(-1)^n}{3^n (n + 1)} (x - 4)^n, \text{ which is the Taylor series for } f \text{ centered}$$

at 4. Apply the Ratio Test to find the radius of convergence R.

$$\lim_{n \to \infty} \left| \frac{a_{n+1}}{a_n} \right| = \lim_{n \to \infty} \left| \frac{(-1)^{n+1}(x - 4)^{n+1}}{3^{n+1}(n + 2)} \cdot \frac{3^n(n + 1)}{(-1)^n(x - 4)^n} \right| = \lim_{n \to \infty} \left| \frac{(-1)(x - 4)(n + 1)}{3(n + 2)} \right|$$

$$= \frac{1}{3} |x - 4| \lim_{n \to \infty} \frac{n + 1}{n + 2} = \frac{1}{3} |x - 4|$$

For convergence, $\frac{1}{3} |x - 4| < 1 \quad \Leftrightarrow \quad |x - 4| < 3$, so $R = 3$.

5.

n	$f^{(n)}(x)$	$f^{(n)}(0)$
0	$(1 - x)^{-2}$	1
1	$2(1 - x)^{-3}$	2
2	$6(1 - x)^{-4}$	6
3	$24(1 - x)^{-5}$	24
4	$120(1 - x)^{-6}$	120
⋮	⋮	⋮

$$(1 - x)^{-2} = f(0) + f'(0)x + \frac{f''(0)}{2!} x^2 + \frac{f'''(0)}{3!} x^3 + \frac{f^{(4)}(0)}{4!} x^4 + \cdots$$

$$= 1 + 2x + \frac{6}{2} x^2 + \frac{24}{6} x^3 + \frac{120}{24} x^4 + \cdots$$

$$= 1 + 2x + 3x^2 + 4x^3 + 5x^4 + \cdots = \sum_{n=0}^{\infty} (n + 1)x^n$$

$$\lim_{n \to \infty} \left| \frac{a_{n+1}}{a_n} \right| = \lim_{n \to \infty} \left| \frac{(n + 2)x^{n+1}}{(n + 1)x^n} \right| = |x| \lim_{n \to \infty} \frac{n + 2}{n + 1} = |x| (1) = |x| < 1$$

for convergence, so $R = 1$.

6.

n	$f^{(n)}(x)$	$f^{(n)}(0)$
0	$\ln(1 + x)$	0
1	$(1 + x)^{-1}$	1
2	$-(1 + x)^{-2}$	-1
3	$2(1 + x)^{-3}$	2
4	$-6(1 + x)^{-4}$	-6
5	$24(1 + x)^{-5}$	24
⋮	⋮	⋮

$$\ln(1 + x) = f(0) + f'(0)x + \frac{f''(0)}{2!} x^2$$

$$+ \frac{f'''(0)}{3!} x^3 + \frac{f^{(4)}(0)}{4!} x^4 + \frac{f^{(5)}(0)}{5!} x^5 + \cdots$$

$$= 0 + x - \frac{1}{2} x^2 + \frac{2}{6} x^3 - \frac{6}{24} x^4 + \frac{24}{120} x^5 - \cdots$$

$$= x - \frac{x^2}{2} + \frac{x^3}{3} - \frac{x^4}{4} + \frac{x^5}{5} - \cdots = \sum_{n=1}^{\infty} \frac{(-1)^{n-1}}{n} x^n$$

$$\lim_{n \to \infty} \left| \frac{a_{n+1}}{a_n} \right| = \lim_{n \to \infty} \left| \frac{x^{n+1}}{n + 1} \cdot \frac{n}{x^n} \right| = \lim_{n \to \infty} \frac{|x|}{1 + 1/n} = |x| < 1 \text{ for convergence,}$$

so $R = 1$.

7.

n	$f^{(n)}(x)$	$f^{(n)}(0)$
0	$\sin \pi x$	0
1	$\pi \cos \pi x$	π
2	$-\pi^2 \sin \pi x$	0
3	$-\pi^3 \cos \pi x$	$-\pi^3$
4	$\pi^4 \sin \pi x$	0
5	$\pi^5 \cos \pi x$	π^5
$\vdots$	$\vdots$	$\vdots$

$$\sin \pi x = f(0) + f'(0)x + \frac{f''(0)}{2!}x^2 + \frac{f'''(0)}{3!}x^3$$

$$+ \frac{f^{(4)}(0)}{4!}x^4 + \frac{f^{(5)}(0)}{5!}x^5 + \cdots$$

$$= 0 + \pi x + 0 - \frac{\pi^3}{3!}x^3 + 0 + \frac{\pi^5}{5!}x^5 + \cdots$$

$$= \pi x - \frac{\pi^3}{3!}x^3 + \frac{\pi^5}{5!}x^5 - \frac{\pi^7}{7!}x^7 + \cdots$$

$$= \sum_{n=0}^{\infty} (-1)^n \frac{\pi^{2n+1}}{(2n+1)!} x^{2n+1}$$

$$\lim_{n\to\infty} \left| \frac{a_{n+1}}{a_n} \right| = \lim_{n\to\infty} \left| \frac{\pi^{2n+3} x^{2n+3}}{(2n+3)!} \cdot \frac{(2n+1)!}{\pi^{2n+1} x^{2n+1}} \right| = \lim_{n\to\infty} \frac{\pi^2 x^2}{(2n+3)(2n+2)}$$

$$= 0 < 1 \quad \text{for all } x, \text{ so } R = \infty.$$

8.

n	$f^{(n)}(x)$	$f^{(n)}(0)$
0	$\cos 3x$	1
1	$-3 \sin 3x$	0
2	$-3^2 \cos 3x$	-3^2
3	$3^3 \sin 3x$	0
4	$3^4 \cos 3x$	3^4
$\vdots$	$\vdots$	$\vdots$

$$\cos 3x = \sum_{n=0}^{\infty} \frac{f^{(n)}(0)}{n!} x^n = \sum_{n=0}^{\infty} (-1)^n \frac{3^{2n}}{(2n)!} x^{2n}$$

$$\lim_{n\to\infty} \left| \frac{a_{n+1}}{a_n} \right| = \lim_{n\to\infty} \left| \frac{3^{2n+2} x^{2n+2}}{(2n+2)!} \cdot \frac{(2n)!}{3^{2n} x^{2n}} \right| = \lim_{n\to\infty} \frac{3^2 x^2}{(2n+2)(2n+1)}$$

$$= 0 < 1 \quad \text{for all } x, \text{ so } R = \infty.$$

9.

n	$f^{(n)}(x)$	$f^{(n)}(0)$
0	e^{5x}	1
1	$5e^{5x}$	5
2	$5^2 e^{5x}$	25
3	$5^3 e^{5x}$	125
4	$5^4 e^{5x}$	625
$\vdots$	$\vdots$	$\vdots$

$$e^{5x} = \sum_{n=0}^{\infty} \frac{f^{(n)}(0)}{n!} x^n = \sum_{n=0}^{\infty} \frac{5^n}{n!} x^n.$$

$$\lim_{n\to\infty} \left| \frac{a_{n+1}}{a_n} \right| = \lim_{n\to\infty} \left[\frac{5^{n+1} |x|^{n+1}}{(n+1)!} \cdot \frac{n!}{5^n |x|^n} \right] = \lim_{n\to\infty} \frac{5|x|}{n+1} = 0 < 1$$

for all x, so $R = \infty$.

10.

n	$f^{(n)}(x)$	$f^{(n)}(0)$
0	xe^x	0
1	$(x+1)e^x$	1
2	$(x+2)e^x$	2
3	$(x+3)e^x$	3
$\vdots$	$\vdots$	$\vdots$

$$xe^x = \sum_{n=0}^{\infty} \frac{f^{(n)}(0)}{n!} x^n = \sum_{n=0}^{\infty} \frac{n}{n!} x^n = \sum_{n=1}^{\infty} \frac{n}{n!} x^n = \sum_{n=1}^{\infty} \frac{x^n}{(n-1)!}.$$

$$\lim_{n\to\infty} \left| \frac{a_{n+1}}{a_n} \right| = \lim_{n\to\infty} \left[\frac{|x|^{n+1}}{n!} \cdot \frac{(n-1)!}{|x|^n} \right] = \lim_{n\to\infty} \frac{|x|}{n} = 0 < 1 \text{ for all } x,$$

so $R = \infty$.

11.

n	$f^{(n)}(x)$	$f^{(n)}(0)$
0	$\sinh x$	0
1	$\cosh x$	1
2	$\sinh x$	0
3	$\cosh x$	1
4	$\sinh x$	0
⋮	⋮	⋮

$f^{(n)}(0) = \begin{cases} 0 & \text{if } n \text{ is even} \\ 1 & \text{if } n \text{ is odd} \end{cases}$ so $\sinh x = \sum_{n=0}^{\infty} \frac{x^{2n+1}}{(2n+1)!}$.

Use the Ratio Test to find R. If $a_n = \frac{x^{2n+1}}{(2n+1)!}$, then

$$\lim_{n\to\infty} \left| \frac{a_{n+1}}{a_n} \right| = \lim_{n\to\infty} \left| \frac{x^{2n+3}}{(2n+3)!} \cdot \frac{(2n+1)!}{x^{2n+1}} \right| = x^2 \cdot \lim_{n\to\infty} \frac{1}{(2n+3)(2n+2)}$$

$$= 0 < 1 \quad \text{for all } x, \text{ so } R = \infty.$$

12.

n	$f^{(n)}(x)$	$f^{(n)}(0)$
0	$\cosh x$	1
1	$\sinh x$	0
2	$\cosh x$	1
3	$\sinh x$	0
⋮	⋮	⋮

$f^{(n)}(0) = \begin{cases} 1 & \text{if } n \text{ is even} \\ 0 & \text{if } n \text{ is odd} \end{cases}$ so $\cosh x = \sum_{n=0}^{\infty} \frac{x^{2n}}{(2n)!}$.

Use the Ratio Test to find R. If $a_n = \frac{x^{2n}}{(2n)!}$, then

$$\lim_{n\to\infty} \left| \frac{a_{n+1}}{a_n} \right| = \lim_{n\to\infty} \left| \frac{x^{2n+2}}{(2n+2)!} \cdot \frac{(2n)!}{x^{2n}} \right| = x^2 \cdot \lim_{n\to\infty} \frac{1}{(2n+2)(2n+1)}$$

$$= 0 < 1 \quad \text{for all } x, \text{ so } R = \infty.$$

13.

n	$f^{(n)}(x)$	$f^{(n)}(1)$
0	$x^4 - 3x^2 + 1$	-1
1	$4x^3 - 6x$	-2
2	$12x^2 - 6$	6
3	$24x$	24
4	24	24
5	0	0
6	0	0
⋮	⋮	⋮

$f^{(n)}(x) = 0$ for $n \geq 5$, so f has a finite series expansion about $a = 1$.

$$f(x) = x^4 - 3x^2 + 1 = \sum_{n=0}^{4} \frac{f^{(n)}(1)}{n!}(x-1)^n$$

$$= \frac{-1}{0!}(x-1)^0 + \frac{-2}{1!}(x-1)^1 + \frac{6}{2!}(x-1)^2 + \frac{24}{3!}(x-1)^3 + \frac{24}{4!}(x-1)^4$$

$$= -1 - 2(x-1) + 3(x-1)^2 + 4(x-1)^3 + (x-1)^4$$

A finite series converges for all x, so $R = \infty$.

14.

n	$f^{(n)}(x)$	$f^{(n)}(-2)$
0	$x - x^3$	6
1	$1 - 3x^2$	-11
2	$-6x$	12
3	-6	-6
4	0	0
5	0	0
⋮	⋮	⋮

$f^{(n)}(x) = 0$ for $n \geq 4$, so f has a finite series expansion about $a = -2$.

$$f(x) = x - x^3 = \sum_{n=0}^{3} \frac{f^{(n)}(-2)}{n!}(x+2)^n$$

$$= \frac{6}{0!}(x+2)^0 + \frac{-11}{1!}(x+2)^1 + \frac{12}{2!}(x+2)^2 + \frac{-6}{3!}(x+2)^3$$

$$= 6 - 11(x+2) + 6(x+2)^2 - (x+2)^3$$

A finite series converges for all x, so $R = \infty$.

15. $f(x) = e^x$ $\Rightarrow$ $f^{(n)}(x) = e^x$, so $f^{(n)}(3) = e^3$ and $e^x = \sum_{n=0}^{\infty} \frac{e^3}{n!}(x-3)^n$. If $a_n = \frac{e^3}{n!}(x-3)^n$, then

$$\lim_{n\to\infty} \left| \frac{a_{n+1}}{a_n} \right| = \lim_{n\to\infty} \left| \frac{e^3(x-3)^{n+1}}{(n+1)!} \cdot \frac{n!}{e^3(x-3)^n} \right| = \lim_{n\to\infty} \frac{|x-3|}{n+1} = 0 < 1 \text{ for all } x, \text{ so } R = \infty.$$

16.

n	$f^{(n)}(x)$	$f^{(n)}(-3)$
0	$1/x$	$-1/3$
1	$-1/x^2$	$-1/3^2$
2	$2/x^3$	$-2/3^3$
3	$-6/x^4$	$-6/3^4$
4	$24/x^5$	$-24/3^5$
$\vdots$	$\vdots$	$\vdots$

$$f(x) = \frac{1}{x} = \sum_{n=0}^{\infty} \frac{f^{(n)}(-3)}{n!}(x+3)^n$$

$$= \frac{-1/3}{0!}(x+3)^0 + \frac{-1/3^2}{1!}(x+3)^1 + \frac{-2/3^3}{2!}(x+3)^2$$

$$+ \frac{-6/3^4}{3!}(x+3)^3 + \frac{-24/3^5}{4!}(x+3)^4 + \cdots$$

$$= \sum_{n=0}^{\infty} \frac{-n!/3^{n+1}}{n!}(x+3)^n = -\sum_{n=0}^{\infty} \frac{(x+3)^n}{3^{n+1}}$$

$$\lim_{n\to\infty}\left|\frac{a_{n+1}}{a_n}\right| = \lim_{n\to\infty}\left|\frac{(x+3)^{n+1}}{3^{n+2}} \cdot \frac{3^{n+1}}{(x+3)^n}\right| = \lim_{n\to\infty}\frac{|x+3|}{3} = \frac{|x+3|}{3} < 1$$

for convergence, so $|x+3| < 3$ and $R = 3$.

17.

n	$f^{(n)}(x)$	$f^{(n)}(\pi)$
0	$\cos x$	-1
1	$-\sin x$	0
2	$-\cos x$	1
3	$\sin x$	0
4	$\cos x$	-1
$\vdots$	$\vdots$	$\vdots$

$$\cos x = \sum_{k=0}^{\infty} \frac{f^{(k)}(\pi)}{k!}(x-\pi)^k = -1 + \frac{(x-\pi)^2}{2!} - \frac{(x-\pi)^4}{4!} + \frac{(x-\pi)^6}{6!} - \cdots$$

$$= \sum_{n=0}^{\infty}(-1)^{n+1}\frac{(x-\pi)^{2n}}{(2n)!}$$

$$\lim_{n\to\infty}\left|\frac{a_{n+1}}{a_n}\right| = \lim_{n\to\infty}\left[\frac{|x-\pi|^{2n+2}}{(2n+2)!} \cdot \frac{(2n)!}{|x-\pi|^{2n}}\right] = \lim_{n\to\infty}\frac{|x-\pi|^2}{(2n+2)(2n+1)} = 0 < 1$$

for all x, so $R = \infty$.

18.

n	$f^{(n)}(x)$	$f^{(n)}(\pi/2)$
0	$\sin x$	1
1	$\cos x$	0
2	$-\sin x$	-1
3	$-\cos x$	0
4	$\sin x$	1
$\vdots$	$\vdots$	$\vdots$

$$\sin x = \sum_{k=0}^{\infty}\frac{f^{(k)}(\pi/2)}{k!}\left(x-\frac{\pi}{2}\right)^k$$

$$= 1 - \frac{(x-\pi/2)^2}{2!} + \frac{(x-\pi/2)^4}{4!} - \frac{(x-\pi/2)^6}{6!} + \cdots$$

$$= \sum_{n=0}^{\infty}(-1)^n\frac{(x-\pi/2)^{2n}}{(2n)!}$$

$$\lim_{n\to\infty}\left|\frac{a_{n+1}}{a_n}\right| = \lim_{n\to\infty}\left[\frac{|x-\pi/2|^{2n+2}}{(2n+2)!} \cdot \frac{(2n)!}{|x-\pi/2|^{2n}}\right]$$

$$= \lim_{n\to\infty}\frac{|x-\pi/2|^2}{(2n+2)(2n+1)} = 0 < 1 \quad\text{for all } x\text{, so } R = \infty.$$

19.

n	$f^{(n)}(x)$	$f^{(n)}(9)$
0	$x^{-1/2}$	$\frac{1}{3}$
1	$-\frac{1}{2}x^{-3/2}$	$-\frac{1}{2}\cdot\frac{1}{3^3}$
2	$\frac{3}{4}x^{-5/2}$	$-\frac{1}{2}\cdot\left(-\frac{3}{2}\right)\cdot\frac{1}{3^5}$
3	$-\frac{15}{8}x^{-7/2}$	$-\frac{1}{2}\cdot\left(-\frac{3}{2}\right)\cdot\left(-\frac{5}{2}\right)\cdot\frac{1}{3^7}$
$\vdots$	$\vdots$	$\vdots$

$$\frac{1}{\sqrt{x}} = \frac{1}{3} - \frac{1}{2\cdot3^3}(x-9) + \frac{3}{2^2\cdot3^5}\frac{(x-9)^2}{2!}$$

$$- \frac{3\cdot5}{2^3\cdot3^7}\frac{(x-9)^3}{3!} + \cdots$$

$$= \frac{1}{3} + \sum_{n=1}^{\infty}(-1)^n\frac{1\cdot3\cdot5\cdot\cdots\cdot(2n-1)}{2^n\cdot3^{2n+1}\cdot n!}(x-9)^n.$$

[continued]

$$\lim_{n\to\infty}\left|\frac{a_{n+1}}{a_n}\right| = \lim_{n\to\infty}\left[\frac{1\cdot 3\cdot 5\cdot\,\cdots\,\cdot(2n-1)[2(n+1)-1]\,|x-9|^{n+1}}{2^{n+1}\cdot 3^{[2(n+1)+1]}\cdot(n+1)!}\cdot\frac{2^n\cdot 3^{2n+1}\cdot n!}{1\cdot 3\cdot 5\cdot\,\cdots\,\cdot(2n-1)\,|x-9|^n}\right]$$

$$= \lim_{n\to\infty}\left[\frac{(2n+1)\,|x-9|}{2\cdot 3^2(n+1)}\right] = \frac{1}{9}\,|x-9| < 1$$

for convergence, so $|x-9| < 9$ and $R = 9$.

20.

n	$f^{(n)}(x)$	$f^{(n)}(1)$
0	x^{-2}	1
1	$-2x^{-3}$	-2
2	$6x^{-4}$	6
3	$-24x^{-5}$	-24
4	$120x^{-6}$	120
$\vdots$	$\vdots$	$\vdots$

$$x^{-2} = 1 - 2(x-1) + 6\cdot\frac{(x-1)^2}{2!} - 24\cdot\frac{(x-1)^3}{3!} + 120\cdot\frac{(x-1)^4}{4!} - \cdots$$

$$= 1 - 2(x-1) + 3(x-1)^2 - 4(x-1)^3 + 5(x-1)^4 - \cdots$$

$$= \sum_{n=0}^{\infty}(-1)^n(n+1)(x-1)^n.$$

$$\lim_{n\to\infty}\left|\frac{a_{n+1}}{a_n}\right| = \lim_{n\to\infty}\frac{(n+2)\,|x-1|^{n+1}}{(n+1)\,|x-1|^n} = \lim_{n\to\infty}\left[\frac{n+2}{n+1}\cdot|x-1|\right]$$

$$= |x-1| < 1 \quad\text{for convergence, so } R = 1.$$

21. If $f(x) = \sin \pi x$, then $f^{(n+1)}(x) = \pm\pi^{n+1}\sin\pi x$ or $\pm\pi^{n+1}\cos\pi x$. In each case, $\left|f^{(n+1)}(x)\right| \le \pi^{n+1}$, so by Formula 9

with $a = 0$ and $M = \pi^{n+1}$, $|R_n(x)| \le \dfrac{\pi^{n+1}}{(n+1)!}\,|x|^{n+1} = \dfrac{|\pi x|^{n+1}}{(n+1)!}$. Thus, $|R_n(x)| \to 0$ as $n \to \infty$ by Equation 10.

So $\lim\limits_{n\to\infty} R_n(x) = 0$ and, by Theorem 8, the series in Exercise 7 represents $\sin\pi x$ for all x.

22. If $f(x) = \sin x$, then $f^{(n+1)}(x) = \pm\sin x$ or $\pm\cos x$. In each case, $\left|f^{(n+1)}(x)\right| \le 1$, so by Formula 9 with $a = 0$ and

$M = 1$, $|R_n(x)| \le \dfrac{1}{(n+1)!}\left|x - \dfrac{\pi}{2}\right|^{n+1}$. Thus, $|R_n(x)| \to 0$ as $n \to \infty$ by Equation 10. So $\lim\limits_{n\to\infty} R_n(x) = 0$ and, by

Theorem 8, the series in Exercise 18 represents $\sin x$ for all x.

23. If $f(x) = \sinh x$, then for all n, $f^{(n+1)}(x) = \cosh x$ or $\sinh x$. Since $|\sinh x| < |\cosh x| = \cosh x$ for all x, we have

$\left|f^{(n+1)}(x)\right| \le \cosh x$ for all n. If d is any positive number and $|x| \le d$, then $\left|f^{(n+1)}(x)\right| \le \cosh x \le \cosh d$, so by

Formula 9 with $a = 0$ and $M = \cosh d$, we have $|R_n(x)| \le \dfrac{\cosh d}{(n+1)!}\,|x|^{n+1}$. It follows that $|R_n(x)| \to 0$ as $n \to \infty$ for

$|x| \le d$ (by Equation 10). But d was an arbitrary positive number. So by Theorem 8, the series represents $\sinh x$ for all x.

24. If $f(x) = \cosh x$, then for all n, $f^{(n+1)}(x) = \cosh x$ or $\sinh x$. Since $|\sinh x| < |\cosh x| = \cosh x$ for all x, we have

$\left|f^{(n+1)}(x)\right| \le \cosh x$ for all n. If d is any positive number and $|x| \le d$, then $\left|f^{(n+1)}(x)\right| \le \cosh x \le \cosh d$, so by

Formula 9 with $a = 0$ and $M = \cosh d$, we have $|R_n(x)| \le \dfrac{\cosh d}{(n+1)!}\,|x|^{n+1}$. It follows that $|R_n(x)| \to 0$ as $n \to \infty$ for

$|x| \le d$ (by Equation 10). But d was an arbitrary positive number. So by Theorem 8, the series represents $\cosh x$ for all x.

25. The general binomial series in (17) is

$$(1+x)^k = \sum_{n=0}^{\infty} \binom{k}{n} x^n = 1 + kx + \frac{k(k-1)}{2!}x^2 + \frac{k(k-1)(k-2)}{3!}x^3 + \cdots.$$

$$(1+x)^{1/2} = \sum_{n=0}^{\infty} \binom{\frac{1}{2}}{n} x^n = 1 + \left(\tfrac{1}{2}\right)x + \frac{\left(\frac{1}{2}\right)\left(-\frac{1}{2}\right)}{2!}x^2 + \frac{\left(\frac{1}{2}\right)\left(-\frac{1}{2}\right)\left(-\frac{3}{2}\right)}{3!}x^3 + \cdots$$

$$= 1 + \frac{x}{2} - \frac{x^2}{2^2 \cdot 2!} + \frac{1 \cdot 3 \cdot x^3}{2^3 \cdot 3!} - \frac{1 \cdot 3 \cdot 5 \cdot x^4}{2^4 \cdot 4!} + \cdots$$

$$= 1 + \frac{x}{2} + \sum_{n=2}^{\infty} \frac{(-1)^{n-1} 1 \cdot 3 \cdot 5 \cdot \cdots \cdot (2n-3)x^n}{2^n \cdot n!} \quad \text{for } |x| < 1, \quad \text{so } R = 1.$$

26. $\dfrac{1}{(1+x)^4} = (1+x)^{-4} = \displaystyle\sum_{n=0}^{\infty} \binom{-4}{n} x^n$. The binomial coefficient is

$$\binom{-4}{n} = \frac{(-4)(-5)(-6) \cdot \cdots \cdot (-4-n+1)}{n!} = \frac{(-4)(-5)(-6) \cdot \cdots \cdot [-(n+3)]}{n!}$$

$$= \frac{(-1)^n \cdot 2 \cdot 3 \cdot 4 \cdot 5 \cdot 6 \cdot \cdots \cdot (n+1)(n+2)(n+3)}{2 \cdot 3 \cdot n!} = \frac{(-1)^n (n+1)(n+2)(n+3)}{6}$$

Thus, $\dfrac{1}{(1+x)^4} = \displaystyle\sum_{n=0}^{\infty} \frac{(-1)^n (n+1)(n+2)(n+3)}{6} x^n$ for $|x| < 1$, so $R = 1$.

27. $\dfrac{1}{(2+x)^3} = \dfrac{1}{[2(1+x/2)]^3} = \dfrac{1}{8}\left(1+\dfrac{x}{2}\right)^{-3} = \dfrac{1}{8}\displaystyle\sum_{n=0}^{\infty}\binom{-3}{n}\left(\dfrac{x}{2}\right)^n$. The binomial coefficient is

$$\binom{-3}{n} = \frac{(-3)(-4)(-5) \cdot \cdots \cdot (-3-n+1)}{n!} = \frac{(-3)(-4)(-5) \cdot \cdots \cdot [-(n+2)]}{n!}$$

$$= \frac{(-1)^n \cdot 2 \cdot 3 \cdot 4 \cdot 5 \cdot \cdots \cdot (n+1)(n+2)}{2 \cdot n!} = \frac{(-1)^n (n+1)(n+2)}{2}$$

Thus, $\dfrac{1}{(2+x)^3} = \dfrac{1}{8}\displaystyle\sum_{n=0}^{\infty}\frac{(-1)^n (n+1)(n+2)}{2}\frac{x^n}{2^n} = \displaystyle\sum_{n=0}^{\infty}\frac{(-1)^n (n+1)(n+2)x^n}{2^{n+4}}$ for $\left|\dfrac{x}{2}\right| < 1 \iff |x| < 2$, so $R = 2$.

28. $(1-x)^{2/3} = \displaystyle\sum_{n=0}^{\infty}\binom{\frac{2}{3}}{n}(-x)^n = 1 + \tfrac{2}{3}(-x) + \frac{\frac{2}{3}\left(-\frac{1}{3}\right)}{2!}(-x)^2 + \frac{\frac{2}{3}\left(-\frac{1}{3}\right)\left(-\frac{4}{3}\right)}{3!}(-x)^3 + \cdots$

$$= 1 - \tfrac{2}{3}x + \sum_{n=2}^{\infty}\frac{(-1)^{n-1}(-1)^n \cdot 2 \cdot [1 \cdot 4 \cdot 7 \cdot \cdots \cdot (3n-5)]}{3^n \cdot n!}x^n$$

$$= 1 - \tfrac{2}{3}x - 2\sum_{n=2}^{\infty}\frac{1 \cdot 4 \cdot 7 \cdot \cdots \cdot (3n-5)}{3^n \cdot n!}x^n$$

and $|-x| < 1 \iff |x| < 1$, so $R = 1$.

29. $\sin x = \displaystyle\sum_{n=0}^{\infty}(-1)^n \frac{x^{2n+1}}{(2n+1)!} \Rightarrow f(x) = \sin(\pi x) = \displaystyle\sum_{n=0}^{\infty}(-1)^n \frac{(\pi x)^{2n+1}}{(2n+1)!} = \displaystyle\sum_{n=0}^{\infty}(-1)^n \frac{\pi^{2n+1}}{(2n+1)!}x^{2n+1}, R = \infty.$

30. $\cos x = \sum_{n=0}^{\infty}(-1)^n \dfrac{x^{2n}}{(2n)!}$ $\Rightarrow$ $f(x) = \cos(\pi x/2) = \sum_{n=0}^{\infty}(-1)^n \dfrac{(\pi x/2)^{2n}}{(2n)!} = \sum_{n=0}^{\infty}(-1)^n \dfrac{\pi^{2n}}{2^{2n}\,(2n)!}\,x^{2n},\ R = \infty.$

31. $e^x = \sum_{n=0}^{\infty}\dfrac{x^n}{n!}$ $\Rightarrow$ $e^{2x} = \sum_{n=0}^{\infty}\dfrac{(2x)^n}{n!} = \sum_{n=0}^{\infty}\dfrac{2^n\,x^n}{n!}$, so $f(x) = e^x + e^{2x} = \sum_{n=0}^{\infty}\dfrac{1}{n!}\,x^n + \sum_{n=0}^{\infty}\dfrac{2^n}{n!}\,x^n = \sum_{n=0}^{\infty}\dfrac{2^n+1}{n!}\,x^n,$

$R = \infty.$

32. $e^x = \sum_{n=0}^{\infty}\dfrac{x^n}{n!}$ $\Rightarrow$ $2e^{-x} = 2\sum_{n=0}^{\infty}\dfrac{(-x)^n}{n!} = 2\sum_{n=0}^{\infty}\dfrac{(-1)^n\,x^n}{n!}$, so $f(x) = e^x + 2e^{-x} = \sum_{n=0}^{\infty}\dfrac{[1+2(-1)^n]}{n!}\,x^n,\ R = \infty.$

33. $\cos x = \sum_{n=0}^{\infty}(-1)^n \dfrac{x^{2n}}{(2n)!}$ $\Rightarrow$ $\cos\left(\tfrac{1}{2}x^2\right) = \sum_{n=0}^{\infty}(-1)^n \dfrac{\left(\tfrac{1}{2}x^2\right)^{2n}}{(2n)!} = \sum_{n=0}^{\infty}(-1)^n \dfrac{x^{4n}}{2^{2n}\,(2n)!}$, so

$f(x) = x\cos\left(\tfrac{1}{2}x^2\right) = \sum_{n=0}^{\infty}(-1)^n \dfrac{1}{2^{2n}(2n)!}\,x^{4n+1},\ R = \infty.$

34. $\tan^{-1}x = \sum_{n=0}^{\infty}(-1)^n \dfrac{x^{2n+1}}{2n+1}$ $\Rightarrow$ $\tan^{-1}(x^3) = \sum_{n=0}^{\infty}(-1)^n \dfrac{(x^3)^{2n+1}}{2n+1} = \sum_{n=0}^{\infty}(-1)^n \dfrac{x^{6n+3}}{2n+1}$, so

$x^2\tan^{-1}(x^3) = \sum_{n=0}^{\infty}(-1)^n \dfrac{1}{2n+1}\,x^{6n+5};\ \left|x^3\right| < 1\ \Leftrightarrow\ |x| < 1,$ so $R = 1.$

35. We must write the binomial in the form $(1+$ expression$)$, so we'll factor out a 4.

$$\dfrac{x}{\sqrt{4+x^2}} = \dfrac{x}{\sqrt{4(1+x^2/4)}} = \dfrac{x}{2\sqrt{1+x^2/4}} = \dfrac{x}{2}\left(1+\dfrac{x^2}{4}\right)^{-1/2} = \dfrac{x}{2}\sum_{n=0}^{\infty}\binom{-\tfrac{1}{2}}{n}\left(\dfrac{x^2}{4}\right)^n$$

$$= \dfrac{x}{2}\left[1 + \left(-\tfrac{1}{2}\right)\dfrac{x^2}{4} + \dfrac{\left(-\tfrac{1}{2}\right)\left(-\tfrac{3}{2}\right)}{2!}\left(\dfrac{x^2}{4}\right)^2 + \dfrac{\left(-\tfrac{1}{2}\right)\left(-\tfrac{3}{2}\right)\left(-\tfrac{5}{2}\right)}{3!}\left(\dfrac{x^2}{4}\right)^3 + \cdots\right]$$

$$= \dfrac{x}{2} + \dfrac{x}{2}\sum_{n=1}^{\infty}(-1)^n \dfrac{1\cdot 3\cdot 5\cdot \cdots \cdot(2n-1)}{2^n\cdot 4^n\cdot n!}\,x^{2n}$$

$$= \dfrac{x}{2} + \sum_{n=1}^{\infty}(-1)^n \dfrac{1\cdot 3\cdot 5\cdot \cdots \cdot(2n-1)}{n!\,2^{3n+1}}\,x^{2n+1}\ \text{and}\ \dfrac{x^2}{4} < 1\ \Leftrightarrow\ \dfrac{|x|}{2} < 1\ \Leftrightarrow\ |x| < 2,\ \text{so}\ R = 2.$$

36. $\dfrac{x^2}{\sqrt{2+x}} = \dfrac{x^2}{\sqrt{2(1+x/2)}} = \dfrac{x^2}{\sqrt{2}}\left(1+\dfrac{x}{2}\right)^{-1/2} = \dfrac{x^2}{\sqrt{2}}\sum_{n=0}^{\infty}\binom{-\tfrac{1}{2}}{n}\left(\dfrac{x}{2}\right)^n$

$$= \dfrac{x^2}{\sqrt{2}}\left[1 + \left(-\tfrac{1}{2}\right)\left(\dfrac{x}{2}\right) + \dfrac{\left(-\tfrac{1}{2}\right)\left(-\tfrac{3}{2}\right)}{2!}\left(\dfrac{x}{2}\right)^2 + \dfrac{\left(-\tfrac{1}{2}\right)\left(-\tfrac{3}{2}\right)\left(-\tfrac{5}{2}\right)}{3!}\left(\dfrac{x}{2}\right)^3 + \cdots\right]$$

$$= \dfrac{x^2}{\sqrt{2}} + \dfrac{x^2}{\sqrt{2}}\sum_{n=1}^{\infty}(-1)^n \dfrac{1\cdot 3\cdot 5\cdot \cdots \cdot(2n-1)}{n!\,2^{2n}}\,x^n$$

$$= \dfrac{x^2}{\sqrt{2}} + \sum_{n=1}^{\infty}(-1)^n \dfrac{1\cdot 3\cdot 5\cdot \cdots \cdot(2n-1)}{n!\,2^{2n+1/2}}\,x^{n+2}\ \text{and}\ \left|\dfrac{x}{2}\right| < 1\ \Leftrightarrow\ |x| < 2,\ \text{so}\ R = 2.$$

37. $\sin^2 x = \dfrac{1}{2}(1-\cos 2x) = \dfrac{1}{2}\left[1 - \sum_{n=0}^{\infty}\dfrac{(-1)^n(2x)^{2n}}{(2n)!}\right] = \dfrac{1}{2}\left[1 - 1 - \sum_{n=1}^{\infty}\dfrac{(-1)^n(2x)^{2n}}{(2n)!}\right] = \sum_{n=1}^{\infty}\dfrac{(-1)^{n+1}2^{2n-1}x^{2n}}{(2n)!},$

$R = \infty$

38. $\dfrac{x - \sin x}{x^3} = \dfrac{1}{x^3}\left[x - \sum\limits_{n=0}^{\infty} \dfrac{(-1)^n x^{2n+1}}{(2n+1)!} \right] = \dfrac{1}{x^3}\left[x - x - \sum\limits_{n=1}^{\infty} \dfrac{(-1)^n x^{2n+1}}{(2n+1)!} \right] = \dfrac{1}{x^3}\left[-\sum\limits_{n=0}^{\infty} \dfrac{(-1)^{n+1} x^{2n+3}}{(2n+3)!} \right]$

$= \dfrac{1}{x^3}\sum\limits_{n=0}^{\infty} \dfrac{(-1)^n x^{2n+3}}{(2n+3)!} = \sum\limits_{n=0}^{\infty} \dfrac{(-1)^n x^{2n}}{(2n+3)!}$

and this series also gives the required value at $x = 0$ (namely $1/6$); $R = \infty$.

39. $\cos x = \sum\limits_{n=0}^{\infty} (-1)^n \dfrac{x^{2n}}{(2n)!} \;\;\Rightarrow\;\; f(x) = \cos(x^2) = \sum\limits_{n=0}^{\infty} \dfrac{(-1)^n (x^2)^{2n}}{(2n)!} = \sum\limits_{n=0}^{\infty} \dfrac{(-1)^n x^{4n}}{(2n)!}, R = \infty$

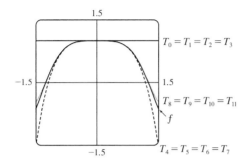

Notice that, as n increases, $T_n(x)$

becomes a better approximation to $f(x)$.

40. $e^x \overset{(11)}{=} \sum\limits_{n=0}^{\infty} \dfrac{x^n}{n!}$, so $e^{-x^2} = \sum\limits_{n=0}^{\infty} \dfrac{(-x^2)^n}{n!} = \sum\limits_{n=0}^{\infty} (-1)^n \dfrac{x^{2n}}{n!}$.

Also, $\cos x \overset{(16)}{=} \sum\limits_{n=0}^{\infty} (-1)^n \dfrac{x^{2n}}{(2n)!}$, so

$f(x) = e^{-x^2} + \cos x = \sum\limits_{n=0}^{\infty} (-1)^n \left(\dfrac{1}{n!} + \dfrac{1}{(2n)!} \right) x^{2n}$

$= 2 - \dfrac{3}{2}x^2 + \dfrac{13}{24}x^4 - \dfrac{121}{720}x^6 + \cdots .$

The series for e^x and $\cos x$ converge for all x, so the same is true of the series

for $f(x)$; that is, $R = \infty$. From the graphs of f and the first few Taylor

polynomials, we see that $T_n(x)$ provides a closer fit to $f(x)$ near 0 as n increases.

41. $e^x \overset{(11)}{=} \sum\limits_{n=0}^{\infty} \dfrac{x^n}{n!}$, so $e^{-x} = \sum\limits_{n=0}^{\infty} \dfrac{(-x)^n}{n!} = \sum\limits_{n=0}^{\infty} (-1)^n \dfrac{x^n}{n!}$, so

$f(x) = xe^{-x} = \sum\limits_{n=0}^{\infty} (-1)^n \dfrac{1}{n!} x^{n+1}$

$= x - x^2 + \tfrac{1}{2}x^3 - \tfrac{1}{6}x^4 + \tfrac{1}{24}x^5 - \tfrac{1}{120}x^6 + \cdots$

$= \sum\limits_{n=1}^{\infty} (-1)^{n-1} \dfrac{x^n}{(n-1)!}$

The series for e^x converges for all x, so the same is true of the series

for $f(x)$; that is, $R = \infty$. From the graphs of f and the first few Taylor

polynomials, we see that $T_n(x)$ provides a closer fit to $f(x)$ near 0 as n increases.

42. From Example 6 in Section 12.9, $\ln(1-x) = -\sum_{n=1}^{\infty} \frac{x^n}{n}$ for $|x| < 1$.

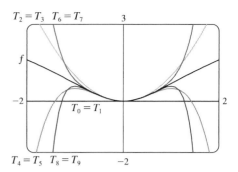

$$\ln(1+x^2) = \ln[1-(-x^2)] = -\sum_{n=1}^{\infty} \frac{(-x^2)^n}{n} = \sum_{n=1}^{\infty} (-1)^{n+1} \frac{1}{n} x^{2n},$$

so $f(x) = \ln(1+x^2) = x^2 - \frac{1}{2}x^4 + \frac{1}{3}x^6 - \frac{1}{4}x^8 + \frac{1}{5}x^{10} - \cdots$. This

series converges for $|x^2| < 1 \iff |x| < 1$, so $R = 1$. From the graphs

of f and the first few Taylor polynomials, we see that $T_n(x)$ provides a

closer fit to $f(x)$ near 0 as n increases.

43. $e^x = \sum_{n=0}^{\infty} \frac{x^n}{n!}$, so $e^{-0.2} = \sum_{n=0}^{\infty} \frac{(-0.2)^n}{n!} = 1 - 0.2 + \frac{1}{2!}(0.2)^2 - \frac{1}{3!}(0.2)^3 + \frac{1}{4!}(0.2)^4 - \frac{1}{5!}(0.2)^5 + \frac{1}{6!}(0.2)^6 - \cdots$.

But $\frac{1}{6!}(0.2)^6 = 8.\overline{8} \times 10^{-8}$, so by the Alternating Series Estimation Theorem, $e^{-0.2} \approx \sum_{n=0}^{5} \frac{(-0.2)^n}{n!} \approx 0.81873$, correct to

five decimal places.

44. $3° = \frac{\pi}{60}$ radians and $\sin x = \sum_{n=0}^{\infty} \frac{(-1)^n x^{2n+1}}{(2n+1)!}$, so

$$\sin \frac{\pi}{60} = \frac{\pi}{60} - \frac{\left(\frac{\pi}{60}\right)^3}{3!} + \frac{\left(\frac{\pi}{60}\right)^5}{5!} - \cdots = \frac{\pi}{60} - \frac{\pi^3}{1{,}296{,}000} + \frac{\pi^5}{93{,}312{,}000{,}000} - \cdots. \text{ But } \frac{\pi^5}{93{,}312{,}000{,}000} < 10^{-8}, \text{ so by}$$

the Alternating Series Estimation Theorem, $\sin \frac{\pi}{60} \approx \frac{\pi}{60} - \frac{\pi^3}{1{,}296{,}000} \approx 0.05234$.

45. (a) $1/\sqrt{1-x^2} = \left[1+(-x^2)\right]^{-1/2} = 1 + \left(-\frac{1}{2}\right)(-x^2) + \frac{\left(-\frac{1}{2}\right)\left(-\frac{3}{2}\right)}{2!}(-x^2)^2 + \frac{\left(-\frac{1}{2}\right)\left(-\frac{3}{2}\right)\left(-\frac{5}{2}\right)}{3!}(-x^2)^3 + \cdots$

$$= 1 + \sum_{n=1}^{\infty} \frac{1 \cdot 3 \cdot 5 \cdot \cdots \cdot (2n-1)}{2^n \cdot n!} x^{2n}$$

(b) $\sin^{-1} x = \int \frac{1}{\sqrt{1-x^2}} \, dx = C + x + \sum_{n=1}^{\infty} \frac{1 \cdot 3 \cdot 5 \cdot \cdots \cdot (2n-1)}{(2n+1)2^n \cdot n!} x^{2n+1}$

$$= x + \sum_{n=1}^{\infty} \frac{1 \cdot 3 \cdot 5 \cdot \cdots \cdot (2n-1)}{(2n+1)2^n \cdot n!} x^{2n+1} \quad \text{since } 0 = \sin^{-1} 0 = C.$$

46. (a) $1/\sqrt[4]{1+x} = (1+x)^{-1/4} = \sum_{n=0}^{\infty} \binom{-\frac{1}{4}}{n} x^n = 1 - \frac{1}{4}x + \frac{\left(-\frac{1}{4}\right)\left(-\frac{5}{4}\right)}{2!}x^2 + \frac{\left(-\frac{1}{4}\right)\left(-\frac{5}{4}\right)\left(-\frac{9}{4}\right)}{3!}x^3 + \cdots$

$$= 1 - \frac{1}{4}x + \sum_{n=2}^{\infty} (-1)^n \frac{1 \cdot 5 \cdot 9 \cdot \cdots \cdot (4n-3)}{4^n \cdot n!} x^n$$

(b) $1/\sqrt[4]{1+x} = 1 - \frac{1}{4}x + \frac{5}{32}x^2 - \frac{15}{128}x^3 + \frac{195}{2048}x^4 - \cdots$. $1/\sqrt[4]{1.1} = 1/\sqrt[4]{1+0.1}$, so let $x = 0.1$. The sum of the first four

terms is then $1 - \frac{1}{4}(0.1) + \frac{5}{32}(0.1)^2 - \frac{15}{128}(0.1)^3 \approx 0.976$. The fifth term is $\frac{195}{2048}(0.1)^4 \approx 0.000\,009\,5$, which does not

affect the third decimal place of the sum, so we have $1/\sqrt[4]{1.1} \approx 0.976$. (Note that the third decimal place of the sum of the

first three terms is affected by the fourth term, so we need to use more than three terms for the sum.)

47. $\cos x \overset{(16)}{=} \sum\limits_{n=0}^{\infty} (-1)^n \dfrac{x^{2n}}{(2n)!} \quad \Rightarrow \quad \cos(x^3) = \sum\limits_{n=0}^{\infty} (-1)^n \dfrac{(x^3)^{2n}}{(2n)!} = \sum\limits_{n=0}^{\infty} (-1)^n \dfrac{x^{6n}}{(2n)!} \quad \Rightarrow$

$x \cos(x^3) = \sum\limits_{n=0}^{\infty} (-1)^n \dfrac{x^{6n+1}}{(2n)!} \quad \Rightarrow \quad \int x \cos(x^3)\, dx = C + \sum\limits_{n=0}^{\infty} (-1)^n \dfrac{x^{6n+2}}{(6n+2)(2n)!}$, with $R = \infty$.

48. $e^x \overset{(11)}{=} \sum\limits_{n=0}^{\infty} \dfrac{x^n}{n!} \quad \Rightarrow \quad e^x - 1 = \sum\limits_{n=1}^{\infty} \dfrac{x^n}{n!} \quad \Rightarrow \quad \dfrac{e^x - 1}{x} = \sum\limits_{n=1}^{\infty} \dfrac{x^{n-1}}{n!} \quad \Rightarrow \quad \int \dfrac{e^x - 1}{x}\, dx = C + \sum\limits_{n=1}^{\infty} \dfrac{x^n}{n \cdot n!}$,

with $R = \infty$.

49. $\cos x \overset{(16)}{=} \sum\limits_{n=0}^{\infty} (-1)^n \dfrac{x^{2n}}{(2n)!} \quad \Rightarrow \quad \cos x - 1 = \sum\limits_{n=1}^{\infty} (-1)^n \dfrac{x^{2n}}{(2n)!} \quad \Rightarrow \quad \dfrac{\cos x - 1}{x} = \sum\limits_{n=1}^{\infty} (-1)^n \dfrac{x^{2n-1}}{(2n)!} \quad \Rightarrow$

$\int \dfrac{\cos x - 1}{x}\, dx = C + \sum\limits_{n=1}^{\infty} (-1)^n \dfrac{x^{2n}}{2n \cdot (2n)!}$, with $R = \infty$.

50. $\arctan x = \sum\limits_{n=0}^{\infty} (-1)^n \dfrac{x^{2n+1}}{2n+1} \quad \Rightarrow \quad \arctan(x^2) = \sum\limits_{n=0}^{\infty} (-1)^n \dfrac{(x^2)^{2n+1}}{2n+1} = \sum\limits_{n=0}^{\infty} (-1)^n \dfrac{x^{4n+2}}{2n+1} \quad \Rightarrow$

$\int \arctan(x^2)\, dx = C + \sum\limits_{n=0}^{\infty} (-1)^n \dfrac{x^{4n+3}}{(2n+1)(4n+3)}$, with $R = 1$.

51. By Exercise 47, $\int x \cos(x^3)\, dx = C + \sum\limits_{n=0}^{\infty} (-1)^n \dfrac{x^{6n+2}}{(6n+2)(2n)!}$, so

$\int_0^1 x \cos(x^3)\, dx = \left[\sum\limits_{n=0}^{\infty} (-1)^n \dfrac{x^{6n+2}}{(6n+2)(2n)!} \right]_0^1 = \sum\limits_{n=0}^{\infty} \dfrac{(-1)^n}{(6n+2)(2n)!} = \dfrac{1}{2} - \dfrac{1}{8 \cdot 2!} + \dfrac{1}{14 \cdot 4!} - \dfrac{1}{20 \cdot 6!} + \cdots$, but

$\dfrac{1}{20 \cdot 6!} = \dfrac{1}{14{,}400} \approx 0.000\,069$, so $\int_0^1 x \cos(x^3)\, dx \approx \dfrac{1}{2} - \dfrac{1}{16} + \dfrac{1}{336} \approx 0.440$ (correct to three decimal places) by the

Alternating Series Estimation Theorem.

52. From the table of Maclaurin series in this section, we see that

$\tan^{-1} x = \sum\limits_{n=0}^{\infty} (-1)^n \dfrac{x^{2n+1}}{2n+1}$ for x in $[-1, 1]$ and $\sin x = \sum\limits_{n=0}^{\infty} (-1)^n \dfrac{x^{2n+1}}{(2n+1)!}$ for all real numbers x, so

$\tan^{-1}(x^3) + \sin(x^3) = \sum\limits_{n=0}^{\infty} (-1)^n \dfrac{x^{6n+3}}{2n+1} + \sum\limits_{n=0}^{\infty} (-1)^n \dfrac{x^{6n+3}}{(2n+1)!}$ for x^3 in $[-1, 1] \quad \Leftrightarrow \quad x$ in $[-1, 1]$. Thus,

$I = \int_0^{0.2} [\tan^{-1}(x^3) + \sin(x^3)]\, dx = \int_0^{0.2} \sum\limits_{n=0}^{\infty} (-1)^n x^{6n+3} \left(\dfrac{1}{2n+1} + \dfrac{1}{(2n+1)!} \right) dx$

$= \left[\sum\limits_{n=0}^{\infty} (-1)^n \dfrac{x^{6n+4}}{6n+4} \left(\dfrac{1}{2n+1} + \dfrac{1}{(2n+1)!} \right) \right]_0^{0.2} = \sum\limits_{n=0}^{\infty} (-1)^n \dfrac{(0.2)^{6n+4}}{6n+4} \left(\dfrac{1}{2n+1} + \dfrac{1}{(2n+1)!} \right)$

$= \dfrac{(0.2)^4}{4}(1+1) - \dfrac{(0.2)^{10}}{10} \left(\dfrac{1}{3} + \dfrac{1}{3!} \right) + \cdots$

But $\dfrac{(0.2)^{10}}{10} \left(\dfrac{1}{3} + \dfrac{1}{3!} \right) = \dfrac{(0.2)^{10}}{20} = 5.12 \times 10^{-9}$, so by the Alternating Series Estimation Theorem,

$I \approx \dfrac{(0.2)^4}{2} = 0.000\,80$ (correct to five decimal places). [Actually, the value is $0.000\,800\,0$, correct to seven decimal places.]

53. $\sqrt{1+x^4} = (1+x^4)^{1/2} = \sum\limits_{n=0}^{\infty} \binom{1/2}{n}(x^4)^n$, so $\int \sqrt{1+x^4}\,dx = C + \sum\limits_{n=0}^{\infty}\binom{1/2}{n}\dfrac{x^{4n+1}}{4n+1}$ and hence, since $0.4 < 1$,

we have

$$I = \int_0^{0.4}\sqrt{1+x^4}\,dx = \sum_{n=0}^{\infty}\binom{1/2}{n}\dfrac{(0.4)^{4n+1}}{4n+1}$$

$$= (1)\dfrac{(0.4)^1}{0!} + \dfrac{\frac{1}{2}}{1!}\dfrac{(0.4)^5}{5} + \dfrac{\frac{1}{2}\left(-\frac{1}{2}\right)}{2!}\dfrac{(0.4)^9}{9} + \dfrac{\frac{1}{2}\left(-\frac{1}{2}\right)\left(-\frac{3}{2}\right)}{3!}\dfrac{(0.4)^{13}}{13} + \dfrac{\frac{1}{2}\left(-\frac{1}{2}\right)\left(-\frac{3}{2}\right)\left(-\frac{5}{2}\right)}{4!}\dfrac{(0.4)^{17}}{17} + \cdots$$

$$= 0.4 + \dfrac{(0.4)^5}{10} - \dfrac{(0.4)^9}{72} + \dfrac{(0.4)^{13}}{208} - \dfrac{5(0.4)^{17}}{2176} + \cdots$$

Now $\dfrac{(0.4)^9}{72} \approx 3.6 \times 10^{-6} < 5 \times 10^{-6}$, so by the Alternating Series Estimation Theorem, $I \approx 0.4 + \dfrac{(0.4)^5}{10} \approx 0.40102$

(correct to five decimal places).

54. $\int_0^{0.5} x^2 e^{-x^2}\,dx = \int_0^{0.5}\sum\limits_{n=0}^{\infty}\dfrac{(-1)^n x^{2n+2}}{n!}\,dx = \sum\limits_{n=0}^{\infty}\left[\dfrac{(-1)^n x^{2n+3}}{n!(2n+3)}\right]_0^{1/2} = \sum\limits_{n=0}^{\infty}\dfrac{(-1)^n}{n!(2n+3)2^{2n+3}}$ and since the term

with $n=2$ is $\dfrac{1}{1792} < 0.001$, we use $\sum\limits_{n=0}^{1}\dfrac{(-1)^n}{n!(2n+3)2^{2n+3}} = \dfrac{1}{24} - \dfrac{1}{160} \approx 0.0354$.

55. $\lim\limits_{x\to0}\dfrac{x - \tan^{-1}x}{x^3} = \lim\limits_{x\to0}\dfrac{x - \left(x - \frac{1}{3}x^3 + \frac{1}{5}x^5 - \frac{1}{7}x^7 + \cdots\right)}{x^3} = \lim\limits_{x\to0}\dfrac{\frac{1}{3}x^3 - \frac{1}{5}x^5 + \frac{1}{7}x^7 - \cdots}{x^3}$

$$= \lim_{x\to0}\left(\tfrac{1}{3} - \tfrac{1}{5}x^2 + \tfrac{1}{7}x^4 - \cdots\right) = \tfrac{1}{3}$$

since power series are continuous functions.

56. $\lim\limits_{x\to0}\dfrac{1 - \cos x}{1 + x - e^x} = \lim\limits_{x\to0}\dfrac{1 - \left(1 - \frac{1}{2!}x^2 + \frac{1}{4!}x^4 - \frac{1}{6!}x^6 + \cdots\right)}{1 + x - \left(1 + x + \frac{1}{2!}x^2 + \frac{1}{3!}x^3 + \frac{1}{4!}x^4 + \frac{1}{5!}x^5 + \frac{1}{6!}x^6 + \cdots\right)}$

$$= \lim_{x\to0}\dfrac{\frac{1}{2!}x^2 - \frac{1}{4!}x^4 + \frac{1}{6!}x^6 - \cdots}{-\frac{1}{2!}x^2 - \frac{1}{3!}x^3 - \frac{1}{4!}x^4 - \frac{1}{5!}x^5 - \frac{1}{6!}x^6 - \cdots}$$

$$= \lim_{x\to0}\dfrac{\frac{1}{2!} - \frac{1}{4!}x^2 + \frac{1}{6!}x^4 - \cdots}{-\frac{1}{2!} - \frac{1}{3!}x - \frac{1}{4!}x^2 - \frac{1}{5!}x^3 - \frac{1}{6!}x^4 - \cdots} = \dfrac{\frac{1}{2} - 0}{-\frac{1}{2} - 0} = -1$$

since power series are continuous functions.

57. $\lim\limits_{x\to0}\dfrac{\sin x - x + \frac{1}{6}x^3}{x^5} = \lim\limits_{x\to0}\dfrac{\left(x - \frac{1}{3!}x^3 + \frac{1}{5!}x^5 - \frac{1}{7!}x^7 + \cdots\right) - x + \frac{1}{6}x^3}{x^5}$

$$= \lim_{x\to0}\dfrac{\frac{1}{5!}x^5 - \frac{1}{7!}x^7 + \cdots}{x^5} = \lim_{x\to0}\left(\dfrac{1}{5!} - \dfrac{x^2}{7!} + \dfrac{x^4}{9!} - \cdots\right) = \dfrac{1}{5!} = \dfrac{1}{120}$$

since power series are continuous functions.

58. $\lim\limits_{x\to0}\dfrac{\tan x - x}{x^3} = \lim\limits_{x\to0}\dfrac{\left(x + \frac{1}{3}x^3 + \frac{2}{15}x^5 + \cdots\right) - x}{x^3} = \lim\limits_{x\to0}\dfrac{\frac{1}{3}x^3 + \frac{2}{15}x^5 + \cdots}{x^3} = \lim\limits_{x\to0}\left(\tfrac{1}{3} + \tfrac{2}{15}x^2 + \cdots\right) = \tfrac{1}{3}$

since power series are continuous functions.

59. From Equation 11, we have $e^{-x^2} = 1 - \dfrac{x^2}{1!} + \dfrac{x^4}{2!} - \dfrac{x^6}{3!} + \cdots$ and we know that $\cos x = 1 - \dfrac{x^2}{2!} + \dfrac{x^4}{4!} - \cdots$ from

Equation 16. Therefore, $e^{-x^2} \cos x = \left(1 - x^2 + \frac{1}{2}x^4 - \cdots\right)\left(1 - \frac{1}{2}x^2 + \frac{1}{24}x^4 - \cdots\right)$. Writing only the terms with

degree ≤ 4, we get $e^{-x^2} \cos x = 1 - \frac{1}{2}x^2 + \frac{1}{24}x^4 - x^2 + \frac{1}{2}x^4 + \frac{1}{2}x^4 + \cdots = 1 - \frac{3}{2}x^2 + \frac{25}{24}x^4 + \cdots$.

60. $\sec x = \dfrac{1}{\cos x} \overset{(16)}{=} \dfrac{1}{1 - \frac{1}{2}x^2 + \frac{1}{24}x^4 - \cdots}$.

$$
\begin{array}{r}
1 + \frac{1}{2}x^2 + \frac{5}{24}x^4 + \cdots \\
\hline
1 - \frac{1}{2}x^2 + \frac{1}{24}x^4 - \cdots \enspace\big|\enspace 1 \\
1 - \frac{1}{2}x^2 + \frac{1}{24}x^4 - \cdots \\
\hline
\frac{1}{2}x^2 - \frac{1}{24}x^4 + \cdots \\
\frac{1}{2}x^2 - \frac{1}{4}x^4 + \cdots \\
\hline
\frac{5}{24}x^4 + \cdots \\
\frac{5}{24}x^4 + \cdots \\
\hline
\cdots
\end{array}
$$

From the long division above, $\sec x = 1 + \frac{1}{2}x^2 + \frac{5}{24}x^4 + \cdots$.

61. $\dfrac{x}{\sin x} \overset{(15)}{=} \dfrac{x}{x - \frac{1}{6}x^3 + \frac{1}{120}x^5 - \cdots}$.

$$
\begin{array}{r}
1 + \frac{1}{6}x^2 + \frac{7}{360}x^4 + \cdots \\
\hline
x - \frac{1}{6}x^3 + \frac{1}{120}x^5 - \cdots \enspace\big|\enspace x \\
x - \frac{1}{6}x^3 + \frac{1}{120}x^5 - \cdots \\
\hline
\frac{1}{6}x^3 - \frac{1}{120}x^5 + \cdots \\
\frac{1}{6}x^3 - \frac{1}{36}x^5 + \cdots \\
\hline
\frac{7}{360}x^5 + \cdots \\
\frac{7}{360}x^5 + \cdots \\
\hline
\cdots
\end{array}
$$

From the long division above, $\dfrac{x}{\sin x} = 1 + \frac{1}{6}x^2 + \frac{7}{360}x^4 + \cdots$.

62. From Example 6 in Section 12.9, we have $\ln(1 - x) = -x - \frac{1}{2}x^2 - \frac{1}{3}x^3 - \cdots$, $|x| < 1$. Therefore,

$$e^x \ln(1 - x) = \left(1 + x + \frac{1}{2}x^2 + \cdots\right)\left(-x - \frac{1}{2}x^2 - \frac{1}{3}x^3 - \cdots\right)$$

$$= -x - \frac{1}{2}x^2 - \frac{1}{3}x^3 - x^2 - \frac{1}{2}x^3 - \frac{1}{2}x^3 - \cdots = -x - \frac{3}{2}x^2 - \frac{4}{3}x^3 - \cdots, \quad |x| < 1$$

63. $\displaystyle\sum_{n=0}^{\infty} (-1)^n \frac{x^{4n}}{n!} = \sum_{n=0}^{\infty} \frac{\left(-x^4\right)^n}{n!} = e^{-x^4}$, by (11).

64. $\displaystyle\sum_{n=0}^{\infty} \frac{(-1)^n \pi^{2n}}{6^{2n}(2n)!} = \sum_{n=0}^{\infty} (-1)^n \frac{\left(\frac{\pi}{6}\right)^{2n}}{(2n)!} = \cos \frac{\pi}{6} = \frac{\sqrt{3}}{2}$, by (16).

65. $\displaystyle\sum_{n=0}^{\infty} \frac{(-1)^n \, \pi^{2n+1}}{4^{2n+1}(2n+1)!} = \sum_{n=0}^{\infty} \frac{(-1)^n \left(\frac{\pi}{4}\right)^{2n+1}}{(2n+1)!} = \sin\frac{\pi}{4} = \frac{1}{\sqrt{2}}$, by (15).

66. $\displaystyle\sum_{n=0}^{\infty} \frac{3^n}{5^n \, n!} = \sum_{n=0}^{\infty} \frac{(3/5)^n}{n!} = e^{3/5}$, by (11).

67. $3 + \dfrac{9}{2!} + \dfrac{27}{3!} + \dfrac{81}{4!} + \cdots = \dfrac{3^1}{1!} + \dfrac{3^2}{2!} + \dfrac{3^3}{3!} + \dfrac{3^4}{4!} + \cdots = \displaystyle\sum_{n=1}^{\infty} \dfrac{3^n}{n!} = \sum_{n=0}^{\infty} \dfrac{3^n}{n!} - 1 = e^3 - 1$, by (11).

68. $1 - \ln 2 + \dfrac{(\ln 2)^2}{2!} - \dfrac{(\ln 2)^3}{3!} + \cdots = \displaystyle\sum_{n=0}^{\infty} \dfrac{(-\ln 2)^n}{n!} = e^{-\ln 2} = \left(e^{\ln 2}\right)^{-1} = 2^{-1} = \frac{1}{2}$, by (11).

69. Assume that $|f'''(x)| \le M$, so $f'''(x) \le M$ for $a \le x \le a+d$. Now $\int_a^x f'''(t)\,dt \le \int_a^x M\,dt \quad\Rightarrow$

$f''(x) - f''(a) \le M(x-a) \quad\Rightarrow\quad f''(x) \le f''(a) + M(x-a)$. Thus, $\int_a^x f''(t)\,dt \le \int_a^x [f''(a) + M(t-a)]\,dt \quad\Rightarrow$

$f'(x) - f'(a) \le f''(a)(x-a) + \frac{1}{2}M(x-a)^2 \quad\Rightarrow\quad f'(x) \le f'(a) + f''(a)(x-a) + \frac{1}{2}M(x-a)^2 \quad\Rightarrow$

$\int_a^x f'(t)\,dt \le \int_a^x \left[f'(a) + f''(a)(t-a) + \frac{1}{2}M(t-a)^2\right] dt \quad\Rightarrow$

$f(x) - f(a) \le f'(a)(x-a) + \frac{1}{2}f''(a)(x-a)^2 + \frac{1}{6}M(x-a)^3$. So

$f(x) - f(a) - f'(a)(x-a) - \frac{1}{2}f''(a)(x-a)^2 \le \frac{1}{6}M(x-a)^3$. But

$R_2(x) = f(x) - T_2(x) = f(x) - f(a) - f'(a)(x-a) - \frac{1}{2}f''(a)(x-a)^2$, so $R_2(x) \le \frac{1}{6}M(x-a)^3$.

A similar argument using $f'''(x) \ge -M$ shows that $R_2(x) \ge -\frac{1}{6}M(x-a)^3$. So $|R_2(x_2)| \le \frac{1}{6}M\,|x-a|^3$.

Although we have assumed that $x > a$, a similar calculation shows that this inequality is also true if $x < a$.

70. (a) $f(x) = \begin{cases} e^{-1/x^2} & \text{if } x \ne 0 \\ 0 & \text{if } x = 0 \end{cases}$ so $f'(0) = \displaystyle\lim_{x \to 0} \frac{f(x) - f(0)}{x - 0} = \lim_{x \to 0} \frac{e^{-1/x^2}}{x} = \lim_{x \to 0} \frac{1/x}{e^{1/x^2}} = \lim_{x \to 0} \frac{x}{2e^{1/x^2}} = 0$

(using l'Hospital's Rule and simplifying in the penultimate step). Similarly, we can use the definition of the derivative and l'Hospital's Rule to show that $f''(0) = 0$, $f^{(3)}(0) = 0$, ..., $f^{(n)}(0) = 0$, so that the Maclaurin series for f consists entirely of zero terms. But since $f(x) \ne 0$ except for $x = 0$, we see that f cannot equal its Maclaurin series except at $x = 0$.

(b)

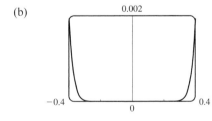

From the graph, it seems that the function is extremely flat at the origin. In fact, it could be said to be "infinitely flat" at $x = 0$, since all of its derivatives are 0 there.

71. (a) $g(x) = \sum\limits_{n=0}^{\infty} \binom{k}{n} x^n \quad\Rightarrow\quad g'(x) = \sum\limits_{n=1}^{\infty} \binom{k}{n} n x^{n-1}$, so

$$(1+x)g'(x) = (1+x)\sum\limits_{n=1}^{\infty} \binom{k}{n} n x^{n-1} = \sum\limits_{n=1}^{\infty} \binom{k}{n} n x^{n-1} + \sum\limits_{n=1}^{\infty} \binom{k}{n} n x^n$$

$$= \sum\limits_{n=0}^{\infty} \binom{k}{n+1} (n+1)x^n + \sum\limits_{n=0}^{\infty} \binom{k}{n} n x^n \qquad \left[\begin{array}{c}\text{Replace } n \text{ with } n+1 \\ \text{in the first series}\end{array}\right]$$

$$= \sum\limits_{n=0}^{\infty} (n+1)\frac{k(k-1)(k-2)\cdots(k-n+1)(k-n)}{(n+1)!} x^n + \sum\limits_{n=0}^{\infty}\left[(n)\frac{k(k-1)(k-2)\cdots(k-n+1)}{n!}\right] x^n$$

$$= \sum\limits_{n=0}^{\infty} \frac{(n+1)k(k-1)(k-2)\cdots(k-n+1)}{(n+1)!}\left[(k-n)+n\right] x^n$$

$$= k\sum\limits_{n=0}^{\infty} \frac{k(k-1)(k-2)\cdots(k-n+1)}{n!} x^n = k\sum\limits_{n=0}^{\infty} \binom{k}{n} x^n = kg(x)$$

Thus, $g'(x) = \dfrac{kg(x)}{1+x}$.

(b) $h(x) = (1+x)^{-k}\, g(x) \quad\Rightarrow$

$$h'(x) = -k(1+x)^{-k-1}g(x) + (1+x)^{-k}g'(x) \qquad \text{[Product Rule]}$$

$$= -k(1+x)^{-k-1}g(x) + (1+x)^{-k}\frac{kg(x)}{1+x} \qquad \text{[from part (a)]}$$

$$= -k(1+x)^{-k-1}g(x) + k(1+x)^{-k-1}g(x) = 0$$

(c) From part (b) we see that $h(x)$ must be constant for $x \in (-1,1)$, so $h(x) = h(0) = 1$ for $x \in (-1,1)$.

Thus, $h(x) = 1 = (1+x)^{-k}\, g(x) \quad\Leftrightarrow\quad g(x) = (1+x)^k$ for $x \in (-1,1)$.

72. By Exercise 25, $\sqrt{1+x} = (1+x)^{1/2} = 1 + \dfrac{x}{2} + \sum\limits_{n=2}^{\infty} \dfrac{(-1)^{n-1} 1 \cdot 3 \cdot 5 \cdots\cdot (2n-3)x^n}{2^n \cdot n!}$, so

$$\left(1-x^2\right)^{1/2} = 1 - \frac{1}{2}x^2 - \sum\limits_{n=2}^{\infty} \frac{1 \cdot 3 \cdot 5 \cdots\cdot (2n-3)}{2^n \cdot n!} x^{2n} \text{ and}$$

$$\sqrt{1 - e^2 \sin^2\theta} = 1 - \frac{1}{2}e^2 \sin^2\theta - \sum\limits_{n=2}^{\infty} \frac{1 \cdot 3 \cdot 5 \cdots\cdot (2n-3)}{2^n \cdot n!} e^{2n} \sin^{2n}\theta. \text{ Thus,}$$

$$L = 4a\int_0^{\pi/2} \sqrt{1 - e^2 \sin^2\theta}\; d\theta = 4a\int_0^{\pi/2}\left(1 - \frac{1}{2}e^2 \sin^2\theta - \sum\limits_{n=2}^{\infty} \frac{1 \cdot 3 \cdot 5 \cdots\cdot (2n-3)}{2^n \cdot n!} e^{2n} \sin^{2n}\theta\right) d\theta$$

$$= 4a\left[\frac{\pi}{2} - \frac{e^2}{2}S_1 - \sum\limits_{n=2}^{\infty} \frac{1 \cdot 3 \cdot 5 \cdots\cdot (2n-3)}{n!}\left(\frac{e^2}{2}\right)^n S_n\right]$$

where $S_n = \displaystyle\int_0^{\pi/2} \sin^{2n}\theta\; d\theta = \dfrac{1 \cdot 3 \cdot 5 \cdots\cdot (2n-1)}{2 \cdot 4 \cdot 6 \cdots\cdot 2n}\dfrac{\pi}{2}$ by Exercise 46 of Section 8.1.

[continued]

$$L = 4a\left(\frac{\pi}{2}\right)\left[1 - \frac{e^2}{2}\cdot\frac{1}{2} - \sum_{n=2}^{\infty} \frac{1\cdot3\cdot5\cdots\cdots(2n-3)}{n!}\left(\frac{e^2}{2}\right)^n \frac{1\cdot3\cdot5\cdots\cdots(2n-1)}{2\cdot4\cdot6\cdots\cdots2n}\right]$$

$$= 2\pi a\left[1 - \frac{e^2}{4} - \sum_{n=2}^{\infty} \frac{e^{2n}}{2^n}\cdot\frac{1^2\cdot3^2\cdot5^2\cdots\cdots(2n-3)^2(2n-1)}{n!\cdot2^n\cdot n!}\right]$$

$$= 2\pi a\left[1 - \frac{e^2}{4} - \sum_{n=2}^{\infty} \frac{e^{2n}}{4^n}\left(\frac{1\cdot3\cdots\cdots(2n-3)}{n!}\right)^2 (2n-1)\right]$$

$$= 2\pi a\left[1 - \frac{e^2}{4} - \frac{3e^4}{64} - \frac{5e^6}{256} - \cdots\right] = \frac{\pi a}{128}(256 - 64e^2 - 12e^4 - 5e^6 - \cdots)$$

LABORATORY PROJECT   An Elusive Limit

1. $f(x) = \dfrac{n(x)}{d(x)} = \dfrac{\sin(\tan x) - \tan(\sin x)}{\arcsin(\arctan x) - \arctan(\arcsin x)}$

The table of function values were obtained using Maple with 10 digits of precision. The results of this project will vary depending on the CAS and precision level. It appears that as $x \to 0^+$, $f(x) \to \frac{10}{3}$. Since f is an even function, we have $f(x) \to \frac{10}{3}$ as $x \to 0$.

x	$f(x)$
1	1.1838
0.1	0.9821
0.01	2.0000
0.001	3.3333
0.0001	3.3333

2. The graph is inconclusive about the limit of f as $x \to 0$.

3. The limit has the indeterminate form $\frac{0}{0}$. Applying l'Hospital's Rule, we obtain the form $\frac{0}{0}$ six times. Finally, on the seventh application we obtain $\lim\limits_{x\to0}\dfrac{n^{(7)}(x)}{d^{(7)}(x)} = \dfrac{-168}{-168} = 1$.

4. $\lim\limits_{x\to0} f(x) = \lim\limits_{x\to0}\dfrac{n(x)}{d(x)} \overset{\text{CAS}}{=} \lim\limits_{x\to0}\dfrac{-\frac{1}{30}x^7 - \frac{29}{756}x^9 + \cdots}{-\frac{1}{30}x^7 + \frac{13}{756}x^9 + \cdots}$

$$= \lim_{x\to0}\frac{\left(-\frac{1}{30}x^7 - \frac{29}{756}x^9 + \cdots\right)/x^7}{\left(-\frac{1}{30}x^7 + \frac{13}{756}x^9 + \cdots\right)/x^7} = \lim_{x\to0}\frac{-\frac{1}{30} - \frac{29}{756}x^2 + \cdots}{-\frac{1}{30} + \frac{13}{756}x^2 + \cdots} = \frac{-\frac{1}{30}}{-\frac{1}{30}} = 1$$

Note that $n^{(7)}(x) = d^{(7)}(x) = -\frac{7!}{30} = -\frac{5040}{30} = -168$, which agrees with the result in Problem 3.

5. The limit command gives the result that $\lim\limits_{x\to0} f(x) = 1$.

6. The strange results (with only 10 digits of precision) must be due to the fact that the terms being subtracted in the numerator and denominator are very close in value when $|x|$ is small. Thus, the differences are imprecise (have few correct digits).

12.11 Applications of Taylor Polynomials

1. (a)

n	$f^{(n)}(x)$	$f^{(n)}(0)$	$T_n(x)$
0	$\cos x$	1	1
1	$-\sin x$	0	1
2	$-\cos x$	-1	$1 - \frac{1}{2}x^2$
3	$\sin x$	0	$1 - \frac{1}{2}x^2$
4	$\cos x$	1	$1 - \frac{1}{2}x^2 + \frac{1}{24}x^4$
5	$-\sin x$	0	$1 - \frac{1}{2}x^2 + \frac{1}{24}x^4$
6	$-\cos x$	-1	$1 - \frac{1}{2}x^2 + \frac{1}{24}x^4 - \frac{1}{720}x^6$

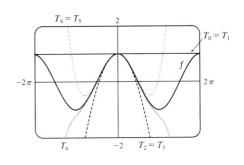

(b)

x	f	$T_0 = T_1$	$T_2 = T_3$	$T_4 = T_5$	T_6
$\frac{\pi}{4}$	0.7071	1	0.6916	0.7074	0.7071
$\frac{\pi}{2}$	0	1	-0.2337	0.0200	-0.0009
π	-1	1	-3.9348	0.1239	-1.2114

(c) As n increases, $T_n(x)$ is a good approximation to $f(x)$ on a larger and larger interval.

2. (a)

n	$f^{(n)}(x)$	$f^{(n)}(1)$	$T_n(x)$
0	x^{-1}	1	1
1	$-x^{-2}$	-1	$1 - (x-1) = 2 - x$
2	$2x^{-3}$	2	$1 - (x-1) + (x-1)^2 = x^2 - 3x + 3$
3	$-6x^{-4}$	-6	$1 - (x-1) + (x-1)^2 - (x-1)^3 = -x^3 + 4x^2 - 6x + 4$

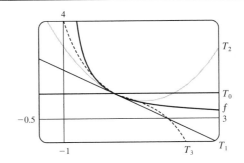

(b)

x	f	T_0	T_1	T_2	T_3
0.9	$1.\overline{1}$	1	1.1	1.11	1.111
1.3	0.7692	1	0.7	0.79	0.763

(c) As n increases, $T_n(x)$ is a good approximation to $f(x)$ on a larger and larger interval.

3.

n	$f^{(n)}(x)$	$f^{(n)}(2)$
0	$1/x$	$\frac{1}{2}$
1	$-1/x^2$	$-\frac{1}{4}$
2	$2/x^3$	$\frac{1}{4}$
3	$-6/x^4$	$-\frac{3}{8}$

$$T_3(x) = \sum_{n=0}^{3} \frac{f^{(n)}(2)}{n!} (x-2)^n$$

$$= \frac{\frac{1}{2}}{0!} - \frac{\frac{1}{4}}{1!}(x-2) + \frac{\frac{1}{4}}{2!}(x-2)^2 - \frac{\frac{3}{8}}{3!}(x-2)^3$$

$$= \frac{1}{2} - \frac{1}{4}(x-2) + \frac{1}{8}(x-2)^2 - \frac{1}{16}(x-2)^3$$

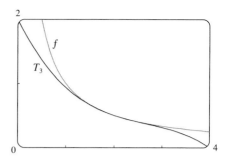

4.

n	$f^{(n)}(x)$	$f^{(n)}(0)$
0	$x + e^{-x}$	1
1	$1 - e^{-x}$	0
2	e^{-x}	1
3	$-e^{-x}$	-1

$$T_3(x) = \sum_{n=0}^{3} \frac{f^{(n)}(0)}{n!} (x-0)^n$$

$$= \frac{1}{0!} + \frac{0}{1!}x + \frac{1}{2!}x^2 - \frac{1}{3!}x^3 = 1 + \frac{1}{2}x^2 - \frac{1}{6}x^3$$

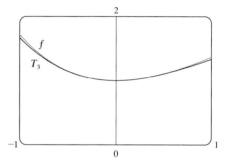

5.

n	$f^{(n)}(x)$	$f^{(n)}(\pi/2)$
0	$\cos x$	0
1	$-\sin x$	-1
2	$-\cos x$	0
3	$\sin x$	1

$$T_3(x) = \sum_{n=0}^{3} \frac{f^{(n)}(\pi/2)}{n!} \left(x - \frac{\pi}{2}\right)^n$$

$$= -\left(x - \frac{\pi}{2}\right) + \frac{1}{6}\left(x - \frac{\pi}{2}\right)^3$$

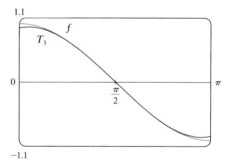

6.

n	$f^{(n)}(x)$	$f^{(n)}(0)$
0	$e^{-x}\sin x$	0
1	$e^{-x}(\cos x - \sin x)$	1
2	$-2e^{-x}\cos x$	-2
3	$2e^{-x}(\cos x + \sin x)$	2

$$T_3(x) = \sum_{n=0}^{3} \frac{f^{(n)}(0)}{n!} x^n = x - x^2 + \frac{1}{3}x^3$$

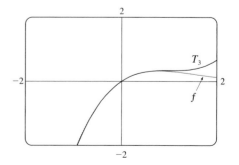

7.

n	$f^{(n)}(x)$	$f^{(n)}(0)$
0	$\arcsin x$	0
1	$\dfrac{1}{\sqrt{1-x^2}}$	1
2	$\dfrac{x}{(1-x^2)^{3/2}}$	0
3	$\dfrac{2x^2+1}{(1-x^2)^{5/2}}$	1

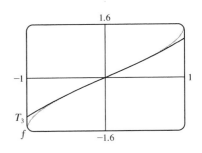

$$T_3(x) = \sum_{n=0}^{3} \frac{f^{(n)}(0)}{n!} x^n = x + \frac{x^3}{6}$$

8.

n	$f^{(n)}(x)$	$f^{(n)}(1)$
0	$\dfrac{\ln x}{x}$	0
1	$\dfrac{1-\ln x}{x^2}$	1
2	$\dfrac{-3+2\ln x}{x^3}$	-3
3	$\dfrac{11-6\ln x}{x^4}$	11

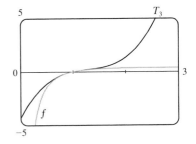

$$T_3(x) = \sum_{n=0}^{3} \frac{f^{(n)}(1)}{n!} (x-1)^n = (x-1) - \tfrac{3}{2}(x-1)^2 + \tfrac{11}{6}(x-1)^3$$

9.

n	$f^{(n)}(x)$	$f^{(n)}(0)$
0	xe^{-2x}	0
1	$(1-2x)e^{-2x}$	1
2	$4(x-1)e^{-2x}$	-4
3	$4(3-2x)e^{-2x}$	12

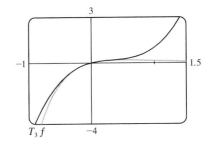

$$T_3(x) = \sum_{n=0}^{3} \frac{f^{(n)}(0)}{n!} x^n = \tfrac{0}{1}\cdot 1 + \tfrac{1}{1}x^1 + \tfrac{-4}{2}x^2 + \tfrac{12}{6}x^3 = x - 2x^2 + 2x^3$$

10.

n	$f^{(n)}(x)$	$f^{(n)}(1)$
0	$\tan^{-1} x$	$\dfrac{\pi}{4}$
1	$\dfrac{1}{1+x^2}$	$\dfrac{1}{2}$
2	$\dfrac{-2x}{(1+x^2)^2}$	$-\dfrac{1}{2}$
3	$\dfrac{6x^2-2}{(1+x^2)^3}$	$\dfrac{1}{2}$

$$T_3(x) = \sum_{n=0}^{3} \frac{f^{(n)}(1)}{n!} (x-1)^n = \frac{\pi}{4} + \frac{1/2}{1}(x-1)^1 + \frac{-1/2}{2}(x-1)^2 + \frac{1/2}{6}(x-1)^3$$

$$= \tfrac{\pi}{4} + \tfrac{1}{2}(x-1) - \tfrac{1}{4}(x-1)^2 + \tfrac{1}{12}(x-1)^3$$

11. You may be able to simply find the Taylor polynomials for

$f(x) = \cot x$ using your CAS. We will list the values of $f^{(n)}(\pi/4)$

for $n = 0$ to $n = 5$.

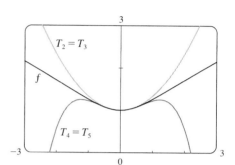

n	0	1	2	3	4	5
$f^{(n)}(\pi/4)$	1	-2	4	-16	80	-512

$$T_5(x) = \sum_{n=0}^{5} \frac{f^{(n)}(\pi/4)}{n!}\left(x - \tfrac{\pi}{4}\right)^n$$

$$= 1 - 2\left(x - \tfrac{\pi}{4}\right) + 2\left(x - \tfrac{\pi}{4}\right)^2 - \tfrac{8}{3}\left(x - \tfrac{\pi}{4}\right)^3 + \tfrac{10}{3}\left(x - \tfrac{\pi}{4}\right)^4 - \tfrac{64}{15}\left(x - \tfrac{\pi}{4}\right)^5$$

For $n = 2$ to $n = 5$, $T_n(x)$ is the polynomial consisting of all the terms up to and including the $\left(x - \tfrac{\pi}{4}\right)^n$ term.

12. You may be able to simply find the Taylor polynomials for

$f(x) = \sqrt[3]{1 + x^2}$ using your CAS. We will list the values of $f^{(n)}(0)$

for $n = 0$ to $n = 5$.

n	0	1	2	3	4	5
$f^{(n)}(0)$	1	0	$\tfrac{2}{3}$	0	$-\tfrac{8}{3}$	0

$$T_5(x) = \sum_{n=0}^{5} \frac{f^{(n)}(0)}{n!} x^n = 1 + \tfrac{1}{3}x^2 - \tfrac{1}{9}x^4$$

For $n = 2$ to $n = 5$, $T_n(x)$ is the polynomial consisting of all the terms up to and including the x^n term. Note that $T_2 = T_3$ and $T_4 = T_5$.

13.

n	$f^{(n)}(x)$	$f^{(n)}(4)$
0	$\sqrt{x}$	2
1	$\tfrac{1}{2}x^{-1/2}$	$\tfrac{1}{4}$
2	$-\tfrac{1}{4}x^{-3/2}$	$-\tfrac{1}{32}$
3	$\tfrac{3}{8}x^{-5/2}$	

(a) $f(x) = \sqrt{x} \approx T_2(x) = 2 + \dfrac{1}{4}(x - 4) - \dfrac{1/32}{2!}(x - 4)^2$

$$= 2 + \tfrac{1}{4}(x - 4) - \tfrac{1}{64}(x - 4)^2$$

(b) $|R_2(x)| \le \dfrac{M}{3!}|x - 4|^3$, where $|f'''(x)| \le M$. Now $4 \le x \le 4.2 \;\Rightarrow$

$|x - 4| \le 0.2 \;\Rightarrow\; |x - 4|^3 \le 0.008$. Since $f'''(x)$ is decreasing

on $[4, 4.2]$, we can take $M = |f'''(4)| = \tfrac{3}{8}4^{-5/2} = \tfrac{3}{256}$, so

$$|R_2(x)| \le \dfrac{3/256}{6}(0.008) = \dfrac{0.008}{512} = 0.000\,015\,625.$$

From the graph of $|R_2(x)| = |\sqrt{x} - T_2(x)|$, it seems that the

error is less than 1.52×10^{-5} on $[4, 4.2]$.

(c)

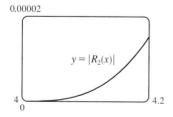

14.

n	$f^{(n)}(x)$	$f^{(n)}(1)$
0	x^{-2}	1
1	$-2x^{-3}$	-2
2	$6x^{-4}$	6
3	$-24x^{-5}$	

(a) $f(x) = x^{-2} \approx T_2(x) = 1 - 2(x-1) + \dfrac{6}{2!}(x-1)^2 = 1 - 2(x-1) + 3(x-1)^2$

(b) $|R_2(x)| \le \dfrac{M}{3!}|x-1|^3$, where $|f'''(x)| \le M$. Now $0.9 \le x \le 1.1 \;\Rightarrow\;$

$|x-1| \le 0.1 \;\Rightarrow\; |x-1|^3 \le 0.001$. Since $f'''(x)$ is decreasing on

$[0.9, 1.1]$, we can take $M = |f'''(0.9)| = \dfrac{24}{(0.9)^5}$, so

$|R_2(x)| \le \dfrac{24/(0.9)^5}{6}(0.001) = \dfrac{0.004}{0.59049} \approx 0.006\,774\,04$.

From the graph of $|R_2(x)| = |x^{-2} - T_2(x)|$, it seems that

the error is less than 0.0046 on $[0.9, 1.1]$.

(c)

15.

n	$f^{(n)}(x)$	$f^{(n)}(1)$
0	$x^{2/3}$	1
1	$\frac{2}{3}x^{-1/3}$	$\frac{2}{3}$
2	$-\frac{2}{9}x^{-4/3}$	$-\frac{2}{9}$
3	$\frac{8}{27}x^{-7/3}$	$\frac{8}{27}$
4	$-\frac{56}{81}x^{-10/3}$	

(a) $f(x) = x^{2/3} \approx T_3(x) = 1 + \frac{2}{3}(x-1) - \dfrac{2/9}{2!}(x-1)^2 + \dfrac{8/27}{3!}(x-1)^3$

$= 1 + \frac{2}{3}(x-1) - \frac{1}{9}(x-1)^2 + \frac{4}{81}(x-1)^3$

(b) $|R_3(x)| \le \dfrac{M}{4!}|x-1|^4$, where $\left|f^{(4)}(x)\right| \le M$. Now $0.8 \le x \le 1.2 \;\Rightarrow\;$

$|x-1| \le 0.2 \;\Rightarrow\; |x-1|^4 \le 0.0016$. Since $\left|f^{(4)}(x)\right|$ is decreasing

on $[0.8, 1.2]$, we can take $M = \left|f^{(4)}(0.8)\right| = \frac{56}{81}(0.8)^{-10/3}$, so

$|R_3(x)| \le \dfrac{\frac{56}{81}(0.8)^{-10/3}}{24}(0.0016) \approx 0.000\,096\,97$.

From the graph of $|R_3(x)| = \left|x^{2/3} - T_3(x)\right|$, it seems that the

error is less than $0.000\,053\,3$ on $[0.8, 1.2]$.

(c)

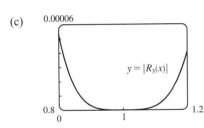

16.

n	$f^{(n)}(x)$	$f^{(n)}(\pi/6)$
0	$\sin x$	$1/2$
1	$\cos x$	$\sqrt{3}/2$
2	$-\sin x$	$-1/2$
3	$-\cos x$	$-\sqrt{3}/2$
4	$\sin x$	$1/2$
5	$\cos x$	

(a) $f(x) = \sin x \approx T_4(x)$

$= \frac{1}{2} + \frac{\sqrt{3}}{2}\left(x - \frac{\pi}{6}\right) - \frac{1}{4}\left(x - \frac{\pi}{6}\right)^2 - \frac{\sqrt{3}}{12}\left(x - \frac{\pi}{6}\right)^3 + \frac{1}{48}\left(x - \frac{\pi}{6}\right)^4$

(b) $|R_4(x)| \le \dfrac{M}{5!}\left|x - \frac{\pi}{6}\right|^5$, where $\left|f^{(5)}(x)\right| \le M$. Now $0 \le x \le \frac{\pi}{3} \;\Rightarrow\;$

$-\frac{\pi}{6} \le x - \frac{\pi}{6} \le \frac{\pi}{6} \;\Rightarrow\; \left|x - \frac{\pi}{6}\right| \le \frac{\pi}{6} \;\Rightarrow\; \left|x - \frac{\pi}{6}\right|^5 \le \left(\frac{\pi}{6}\right)^5$. Since

$\left|f^{(5)}(x)\right|$ is decreasing on $\left[0, \frac{\pi}{3}\right]$, we can take $M = \left|f^{(5)}(0)\right| = \cos 0 = 1$,

so $|R_4(x)| \le \dfrac{1}{5!}\left(\dfrac{\pi}{6}\right)^5 \approx 0.000\,328$.

(c)

0.0004

$y = |R_4(x)|$

0 　 $\dfrac{\pi}{6}$ 　 $\dfrac{\pi}{3}$

From the graph of $|R_4(x)| = |\sin x - T_4(x)|$, it seems that the error is less than $0.000\,297$ on $\left[0, \frac{\pi}{3}\right]$.

17.

n	$f^{(n)}(x)$	$f^{(n)}(0)$
0	$\sec x$	1
1	$\sec x \tan x$	0
2	$\sec x \left(2\sec^2 x - 1\right)$	1
3	$\sec x \tan x \left(6\sec^2 x - 1\right)$	

(a) $f(x) = \sec x \approx T_2(x) = 1 + \frac{1}{2}x^2$

(b) $|R_2(x)| \le \dfrac{M}{3!}\,|x|^3$, where $\left|f^{(3)}(x)\right| \le M$. Now $-0.2 \le x \le 0.2 \;\Rightarrow\; |x| \le 0.2 \;\Rightarrow\; |x|^3 \le (0.2)^3$.

$f^{(3)}(x)$ is an odd function and it is increasing on $[0, 0.2]$ since $\sec x$ and $\tan x$ are increasing on $[0, 0.2]$,

so $\left|f^{(3)}(x)\right| \le f^{(3)}(0.2) \approx 1.085\,158\,892$. Thus, $|R_2(x)| \le \dfrac{f^{(3)}(0.2)}{3!}\,(0.2)^3 \approx 0.001\,447$.

(c)

0.0004

$y = |R_2(x)|$

-0.2 　 0 　 0.2

From the graph of $|R_2(x)| = |\sec x - T_2(x)|$, it seems that the error is less than $0.000\,339$ on $[-0.2, 0.2]$.

18.

n	$f^{(n)}(x)$	$f^{(n)}(1)$
0	$\ln(1 + 2x)$	$\ln 3$
1	$2/(1 + 2x)$	$\frac{2}{3}$
2	$-4/(1 + 2x)^2$	$-\frac{4}{9}$
3	$16/(1 + 2x)^3$	$\frac{16}{27}$
4	$-96/(1 + 2x)^4$	

(a) $f(x) = \ln(1 + 2x) \approx T_3(x)$

$$= \ln 3 + \tfrac{2}{3}(x - 1) - \frac{4/9}{2!}(x - 1)^2 + \frac{16/27}{3!}(x - 1)^3$$

(b) $|R_3(x)| \le \dfrac{M}{4!}\,|x - 1|^4$, where $\left|f^{(4)}(x)\right| \le M$. Now $0.5 \le x \le 1.5 \;\Rightarrow$

$-0.5 \le x - 1 \le 0.5 \;\Rightarrow\; |x - 1| \le 0.5 \;\Rightarrow\; |x - 1|^4 \le \frac{1}{16}$, and

letting $x = 0.5$ gives $M = 6$, so $|R_3(x)| \le \dfrac{6}{4!} \cdot \dfrac{1}{16} = \dfrac{1}{64} = 0.015\,625$.

(c)

0.005

$y = |R_3(x)|$

0.5 　 0 　 1.5

From the graph of $|R_3(x)| = |\ln(1 + 2x) - T_3(x)|$, it seems that the error is less than 0.005 on $[0.5, 1.5]$.

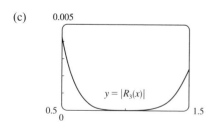

19.

n	$f^{(n)}(x)$	$f^{(n)}(0)$
0	e^{x^2}	1
1	$e^{x^2}(2x)$	0
2	$e^{x^2}(2+4x^2)$	2
3	$e^{x^2}(12x+8x^3)$	0
4	$e^{x^2}(12+48x^2+16x^4)$	

(a) $f(x) = e^{x^2} \approx T_3(x) = 1 + \dfrac{2}{2!}x^2 = 1 + x^2$

(b) $|R_3(x)| \le \dfrac{M}{4!}|x|^4$, where $\left|f^{(4)}(x)\right| \le M$. Now $0 \le x \le 0.1 \Rightarrow$

$x^4 \le (0.1)^4$, and letting $x = 0.1$ gives

$|R_3(x)| \le \dfrac{e^{0.01}(12+0.48+0.0016)}{24}(0.1)^4 \approx 0.00006.$

(c)

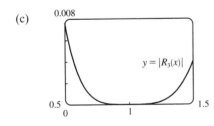

0.00008

$y = |R_3(x)|$

0

0.1

From the graph of $|R_3(x)| = \left|e^{x^2} - T_3(x)\right|$, it appears that the

error is less than $0.000\,051$ on $[0, 0.1]$.

20.

n	$f^{(n)}(x)$	$f^{(n)}(1)$
0	$x \ln x$	0
1	$\ln x + 1$	1
2	$1/x$	1
3	$-1/x^2$	-1
4	$2/x^3$	

(a) $f(x) = x \ln x \approx T_3(x) = (x-1) + \frac{1}{2}(x-1)^2 - \frac{1}{6}(x-1)^3$

(b) $|R_3(x)| \le \dfrac{M}{4!}|x-1|^4$, where $\left|f^{(4)}(x)\right| \le M$. Now $0.5 \le x \le 1.5 \Rightarrow$

$|x-1| \le \frac{1}{2} \Rightarrow |x-1|^4 \le \frac{1}{16}$. Since $\left|f^{(4)}(x)\right|$ is decreasing on

$[0.5, 1.5]$, we can take $M = \left|f^{(4)}(0.5)\right| = 2/(0.5)^3 = 16$, so

$|R_3(x)| \le \frac{16}{24}(1/16) = \frac{1}{24} = 0.041\overline{6}.$

(c)

0.008

$y = |R_3(x)|$

0.5

0

1

1.5

From the graph of $|R_3(x)| = |x \ln x - T_3(x)|$, it seems that the

error is less than 0.0076 on $[0.5, 1.5]$.

21.

n	$f^{(n)}(x)$	$f^{(n)}(0)$
0	$x \sin x$	0
1	$\sin x + x \cos x$	0
2	$2\cos x - x \sin x$	2
3	$-3\sin x - x \cos x$	0
4	$-4\cos x + x \sin x$	-4
5	$5\sin x + x \cos x$	

(a) $f(x) = x \sin x \approx T_4(x) = \dfrac{2}{2!}(x-0)^2 + \dfrac{-4}{4!}(x-0)^4 = x^2 - \dfrac{1}{6}x^4$

(b) $|R_4(x)| \le \dfrac{M}{5!}|x|^5$, where $\left|f^{(5)}(x)\right| \le M$. Now $-1 \le x \le 1 \Rightarrow$

$|x| \le 1$, and a graph of $f^{(5)}(x)$ shows that $\left|f^{(5)}(x)\right| \le 5$ for $-1 \le x \le 1$.

Thus, we can take $M = 5$ and get $|R_4(x)| \le \dfrac{5}{5!} \cdot 1^5 = \dfrac{1}{24} = 0.041\overline{6}.$

(c)

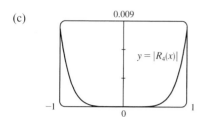

From the graph of $|R_4(x)| = |x \sin x - T_4(x)|$, it seems that the error is less than 0.0082 on $[-1, 1]$.

22.

n	$f^{(n)}(x)$	$f^{(n)}(0)$
0	$\sinh 2x$	0
1	$2\cosh 2x$	2
2	$4\sinh 2x$	0
3	$8\cosh 2x$	8
4	$16\sinh 2x$	0
5	$32\cosh 2x$	32
6	$64\sinh 2x$	

(a) $f(x) = \sinh 2x \approx T_5(x) = 2x + \frac{8}{3!}x^3 + \frac{32}{5!}x^5 = 2x + \frac{4}{3}x^3 + \frac{4}{15}x^5$

(b) $|R_5(x)| \le \frac{M}{6!}|x|^6$, where $\left| f^{(6)}(x) \right| \le M$. For x in $[-1, 1]$, we have $|x| \le 1$. Since $f^{(6)}(x)$ is an increasing odd function on $[-1, 1]$, we see that $\left| f^{(6)}(x) \right| \le f^{(6)}(1) = 64\sinh 2 = 32(e^2 - e^{-2}) \approx 232.119$, so we can take $M = 232.12$ and get $|R_5(x)| \le \frac{232.12}{720} \cdot 1^6 \approx 0.3224$.

(c)

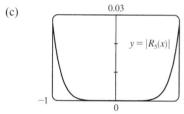

From the graph of $|R_5(x)| = |\sinh 2x - T_5(x)|$, it seems that the error is less than 0.027 on $[-1, 1]$.

23. From Exercise 5, $\cos x = -\left(x - \frac{\pi}{2}\right) + \frac{1}{6}\left(x - \frac{\pi}{2}\right)^3 + R_3(x)$, where $|R_3(x)| \le \frac{M}{4!}\left|x - \frac{\pi}{2}\right|^4$ with $\left| f^{(4)}(x) \right| = |\cos x| \le M = 1$. Now $x = 80° = (90° - 10°) = \left(\frac{\pi}{2} - \frac{\pi}{18}\right) = \frac{4\pi}{9}$ radians, so the error is $\left|R_3\left(\frac{4\pi}{9}\right)\right| \le \frac{1}{24}\left(\frac{\pi}{18}\right)^4 \approx 0.000\,039$, which means our estimate would *not* be accurate to five decimal places. However, $T_3 = T_4$, so we can use $\left|R_4\left(\frac{4\pi}{9}\right)\right| \le \frac{1}{120}\left(\frac{\pi}{18}\right)^5 \approx 0.000\,001$. Therefore, to five decimal places, $\cos 80° \approx -\left(-\frac{\pi}{18}\right) + \frac{1}{6}\left(-\frac{\pi}{18}\right)^3 \approx 0.17365$.

24. From Exercise 16, $\sin x = \frac{1}{2} + \frac{\sqrt{3}}{2}\left(x - \frac{\pi}{6}\right) - \frac{1}{4}\left(x - \frac{\pi}{6}\right)^2 - \frac{\sqrt{3}}{12}\left(x - \frac{\pi}{6}\right)^3 + \frac{1}{48}\left(x - \frac{\pi}{6}\right)^4 + R_4(x)$, where $|R_4(x)| \le \frac{M}{5!}\left|x - \frac{\pi}{6}\right|^5$ with $\left| f^{(5)}(x) \right| = |\cos x| \le M = 1$. Now $x = 38° = (30° + 8°) = \left(\frac{\pi}{6} + \frac{2\pi}{45}\right)$ radians, so the error is $\left|R_4\left(\frac{38\pi}{180}\right)\right| \le \frac{1}{120}\left(\frac{2\pi}{45}\right)^5 \approx 0.000\,000\,44$, which means our estimate will be accurate to five decimal places. Therefore, to five decimal places, $\sin 38° = \frac{1}{2} + \frac{\sqrt{3}}{2}\left(\frac{2\pi}{45}\right) - \frac{1}{4}\left(\frac{2\pi}{45}\right)^2 - \frac{\sqrt{3}}{12}\left(\frac{2\pi}{45}\right)^3 + \frac{1}{48}\left(\frac{2\pi}{45}\right)^4 \approx 0.61566$.

25. All derivatives of e^x are e^x, so $|R_n(x)| \le \frac{e^x}{(n+1)!}|x|^{n+1}$, where $0 < x < 0.1$. Letting $x = 0.1$, $R_n(0.1) \le \frac{e^{0.1}}{(n+1)!}(0.1)^{n+1} < 0.00001$, and by trial and error we find that $n = 3$ satisfies this inequality since $R_3(0.1) < 0.0000046$. Thus, by adding the four terms of the Maclaurin series for e^x corresponding to $n = 0, 1, 2,$ and 3, we can estimate $e^{0.1}$ to within 0.00001. (In fact, this sum is $1.10516\overline{6}$ and $e^{0.1} \approx 1.10517$.)

26. Example 6 in Section 12.9 gives the Maclaurin series for $\ln(1-x)$ as $-\sum\limits_{n=1}^{\infty}\dfrac{x^n}{n}$ for $|x| < 1$. Thus,

$\ln 1.4 = \ln[1-(-0.4)] = -\sum\limits_{n=1}^{\infty}\dfrac{(-0.4)^n}{n} = \sum\limits_{n=1}^{\infty}(-1)^{n+1}\dfrac{(0.4)^n}{n}$. Since this is an alternating series, the error is less than the

first neglected term by the Alternating Series Estimation Theorem, and we find that $|a_6| = (0.4)^6/6 \approx 0.0007 < 0.001$. So

we need the first five (nonzero) terms of the Maclaurin series for the desired accuracy. (In fact, this sum is approximately

0.33698 and $\ln 1.4 \approx 0.33647$.)

27. $\sin x = x - \dfrac{1}{3!}x^3 + \dfrac{1}{5!}x^5 - \cdots$. By the Alternating Series

Estimation Theorem, the error in the approximation

$\sin x = x - \dfrac{1}{3!}x^3$ is less than $\left|\dfrac{1}{5!}x^5\right| < 0.01 \quad \Leftrightarrow$

$\left|x^5\right| < 120(0.01) \quad \Leftrightarrow \quad |x| < (1.2)^{1/5} \approx 1.037$. The curves

$y = x - \frac{1}{6}x^3$ and $y = \sin x - 0.01$ intersect at $x \approx 1.043$, so

the graph confirms our estimate. Since both the sine function

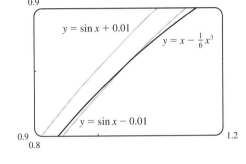

and the given approximation are odd functions, we need to check the estimate only for $x > 0$. Thus, the desired range of

values for x is $-1.037 < x < 1.037$.

28. $\cos x = 1 - \dfrac{1}{2!}x^2 + \dfrac{1}{4!}x^4 - \dfrac{1}{6!}x^6 + \cdots$. By the Alternating Series

Estimation Theorem, the error is less than $\left|-\dfrac{1}{6!}x^6\right| < 0.005 \quad \Leftrightarrow$

$x^6 < 720(0.005) \quad \Leftrightarrow \quad |x| < (3.6)^{1/6} \approx 1.238$. The curves

$y = 1 - \frac{1}{2}x^2 + \frac{1}{24}x^4$ and $y = \cos x + 0.005$ intersect at $x \approx 1.244$,

so the graph confirms our estimate. Since both the cosine function

and the given approximation are even functions, we need to check

the estimate only for $x > 0$. Thus, the desired range of values for x is $-1.238 < x < 1.238$.

29. $\arctan x = x - \dfrac{x^3}{3} + \dfrac{x^5}{5} - \dfrac{x^7}{7} + \cdots$. By the Alternating Series

Estimation Theorem, the error is less than $\left|-\frac{1}{7}x^7\right| < 0.05 \quad \Leftrightarrow$

$\left|x^7\right| < 0.35 \quad \Leftrightarrow \quad |x| < (0.35)^{1/7} \approx 0.8607$. The curves

$y = x - \frac{1}{3}x^3 + \frac{1}{5}x^5$ and $y = \arctan x + 0.05$ intersect at

$x \approx 0.9245$, so the graph confirms our estimate. Since both the

arctangent function and the given approximation are odd functions,

we need to check the estimate only for $x > 0$. Thus, the desired

range of values for x is $-0.86 < x < 0.86$.

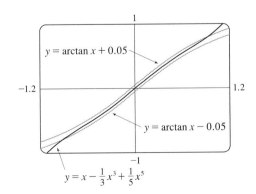

30. $f(x) = \sum\limits_{n=0}^{\infty} \dfrac{f^{(n)}(4)}{n!}(x-4)^n = \sum\limits_{n=0}^{\infty} \dfrac{(-1)^n n!}{3^n(n+1) n!}(x-4)^n = \sum\limits_{n=0}^{\infty} \dfrac{(-1)^n}{3^n(n+1)}(x-4)^n$. Now

$f(5) = \sum\limits_{n=0}^{\infty} \dfrac{(-1)^n}{3^n(n+1)} = \sum\limits_{n=0}^{\infty} (-1)^n b_n$ is the sum of an alternating series that satisfies (i) $b_{n+1} \le b_n$ and

(ii) $\lim\limits_{n \to \infty} b_n = 0$, so by the Alternating Series Estimation Theorem, $|R_5(5)| = |f(5) - T_5(5)| \le b_6$, and

$b_6 = \dfrac{1}{3^6(7)} = \dfrac{1}{5103} \approx 0.000196 < 0.0002$; that is, the fifth-degree Taylor polynomial approximates $f(5)$ with error less

than 0.0002.

31. Let $s(t)$ be the position function of the car, and for convenience set $s(0) = 0$. The velocity of the car is $v(t) = s'(t)$ and the

acceleration is $a(t) = s''(t)$, so the second degree Taylor polynomial is $T_2(t) = s(0) + v(0)t + \dfrac{a(0)}{2}t^2 = 20t + t^2$. We

estimate the distance traveled during the next second to be $s(1) \approx T_2(1) = 20 + 1 = 21$ m. The function $T_2(t)$ would not be

accurate over a full minute, since the car could not possibly maintain an acceleration of 2 m/s^2 for that long (if it did, its final

speed would be 140 m/s ≈ 313 mi/h!).

32. (a)

n	$\rho^{(n)}(t)$	$\rho^{(n)}(20)$
0	$\rho_{20}e^{\alpha(t-20)}$	ρ_{20}
1	$\alpha\rho_{20}e^{\alpha(t-20)}$	$\alpha\rho_{20}$
2	$\alpha^2\rho_{20}e^{\alpha(t-20)}$	$\alpha^2\rho_{20}$

The linear approximation is

$$T_1(t) = \rho(20) + \rho'(20)(t-20) = \rho_{20}[1 + \alpha(t-20)]$$

The quadratic approximation is

$$T_2(t) = \rho(20) + \rho'(20)(t-20) + \dfrac{\rho''(20)}{2}(t-20)^2$$

$$= \rho_{20}\left[1 + \alpha(t-20) + \tfrac{1}{2}\alpha^2(t-20)^2\right]$$

(b)

(c)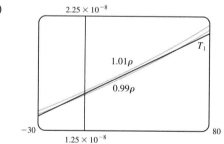

From the graph, it seems that $T_1(t)$ is within 1% of $\rho(t)$, that

is, $0.99\rho(t) \le T_1(t) \le 1.01\rho(t)$, for $-14°C \le t \le 58°C$.

33. $E = \dfrac{q}{D^2} - \dfrac{q}{(D+d)^2} = \dfrac{q}{D^2} - \dfrac{q}{D^2(1+d/D)^2} = \dfrac{q}{D^2}\left[1 - \left(1 + \dfrac{d}{D}\right)^{-2}\right]$.

We use the Binomial Series to expand $(1 + d/D)^{-2}$:

$E = \dfrac{q}{D^2}\left[1 - \left(1 - 2\left(\dfrac{d}{D}\right) + \dfrac{2 \cdot 3}{2!}\left(\dfrac{d}{D}\right)^2 - \dfrac{2 \cdot 3 \cdot 4}{3!}\left(\dfrac{d}{D}\right)^3 + \cdots\right)\right] = \dfrac{q}{D^2}\left[2\left(\dfrac{d}{D}\right) - 3\left(\dfrac{d}{D}\right)^2 + 4\left(\dfrac{d}{D}\right)^3 - \cdots\right]$

$\approx \dfrac{q}{D^2} \cdot 2\left(\dfrac{d}{D}\right) = 2qd \cdot \dfrac{1}{D^3}$

when D is much larger than d; that is, when P is far away from the dipole.

34. (a) $\dfrac{n_1}{\ell_o} + \dfrac{n_2}{\ell_i} = \dfrac{1}{R}\left(\dfrac{n_2 s_i}{\ell_i} - \dfrac{n_1 s_o}{\ell_o}\right)$ [Equation 1] where

$$\ell_o = \sqrt{R^2 + (s_o + R)^2 - 2R(s_o + R)\cos\phi} \quad \text{and} \quad \ell_i = \sqrt{R^2 + (s_i - R)^2 + 2R(s_i - R)\cos\phi} \quad \textbf{(2)}$$

Using $\cos\phi \approx 1$ gives

$$\ell_o = \sqrt{R^2 + (s_o + R)^2 - 2R(s_o + R)} = \sqrt{R^2 + s_o^2 + 2Rs_o + R^2 - 2Rs_o - 2R^2} = \sqrt{s_o^2} = s_o$$

and similarly, $\ell_i = s_i$. Thus, Equation 1 becomes $\dfrac{n_1}{s_o} + \dfrac{n_2}{s_i} = \dfrac{1}{R}\left(\dfrac{n_2 s_i}{s_i} - \dfrac{n_1 s_o}{s_o}\right) \quad \Rightarrow \quad \dfrac{n_1}{s_o} + \dfrac{n_2}{s_i} = \dfrac{n_2 - n_1}{R}.$

(b) Using $\cos\phi \approx 1 - \tfrac{1}{2}\phi^2$ in **(2)** gives us

$$\ell_o = \sqrt{R^2 + (s_o + R)^2 - 2R(s_o + R)\left(1 - \tfrac{1}{2}\phi^2\right)}$$

$$= \sqrt{R^2 + s_o^2 + 2Rs_o + R^2 - 2Rs_o + Rs_o\phi^2 - 2R^2 + R^2\phi^2} = \sqrt{s_o^2 + Rs_o\phi^2 + R^2\phi^2}$$

Anticipating that we will use the binomial series expansion $(1+x)^k \approx 1 + kx$, we can write the last expression for ℓ_o as

$$s_o\sqrt{1 + \phi^2\left(\dfrac{R}{s_o} + \dfrac{R^2}{s_o^2}\right)} \text{ and similarly, } \ell_i = s_i\sqrt{1 - \phi^2\left(\dfrac{R}{s_i} - \dfrac{R^2}{s_i^2}\right)}. \text{ Thus, from Equation 1,}$$

$$\dfrac{n_1}{\ell_o} + \dfrac{n_2}{\ell_i} = \dfrac{1}{R}\left(\dfrac{n_2 s_i}{\ell_i} - \dfrac{n_1 s_o}{\ell_o}\right) \quad \Leftrightarrow \quad n_1\ell_o^{-1} + n_2\ell_i^{-1} = \dfrac{n_2}{R}\cdot\dfrac{s_i}{\ell_i} - \dfrac{n_1}{R}\cdot\dfrac{s_o}{\ell_o} \quad \Leftrightarrow$$

$$\dfrac{n_1}{s_o}\left[1 + \phi^2\left(\dfrac{R}{s_o} + \dfrac{R^2}{s_o^2}\right)\right]^{-1/2} + \dfrac{n_2}{s_i}\left[1 - \phi^2\left(\dfrac{R}{s_i} - \dfrac{R^2}{s_i^2}\right)\right]^{-1/2}$$

$$= \dfrac{n_2}{R}\left[1 - \phi^2\left(\dfrac{R}{s_i} - \dfrac{R^2}{s_i^2}\right)\right]^{-1/2} - \dfrac{n_1}{R}\left[1 + \phi^2\left(\dfrac{R}{s_o} + \dfrac{R^2}{s_o^2}\right)\right]^{-1/2}$$

Approximating the expressions for ℓ_o^{-1} and ℓ_i^{-1} by the first two terms in their binomial series, we get

$$\dfrac{n_1}{s_o}\left[1 - \tfrac{1}{2}\phi^2\left(\dfrac{R}{s_o} + \dfrac{R^2}{s_o^2}\right)\right] + \dfrac{n_2}{s_i}\left[1 + \tfrac{1}{2}\phi^2\left(\dfrac{R}{s_i} - \dfrac{R^2}{s_i^2}\right)\right]$$

$$= \dfrac{n_2}{R}\left[1 + \tfrac{1}{2}\phi^2\left(\dfrac{R}{s_i} - \dfrac{R^2}{s_i^2}\right)\right] - \dfrac{n_1}{R}\left[1 - \tfrac{1}{2}\phi^2\left(\dfrac{R}{s_o} + \dfrac{R^2}{s_o^2}\right)\right] \quad \Leftrightarrow$$

$$\dfrac{n_1}{s_o} - \dfrac{n_1\phi^2}{2s_o}\left(\dfrac{R}{s_o} + \dfrac{R^2}{s_o^2}\right) + \dfrac{n_2}{s_i} + \dfrac{n_2\phi^2}{2s_i}\left(\dfrac{R}{s_i} - \dfrac{R^2}{s_i^2}\right) = \dfrac{n_2}{R} + \dfrac{n_2\phi^2}{2R}\left(\dfrac{R}{s_i} - \dfrac{R^2}{s_i^2}\right) - \dfrac{n_1}{R} + \dfrac{n_1\phi^2}{2R}\left(\dfrac{R}{s_o} + \dfrac{R^2}{s_o^2}\right) \quad \Leftrightarrow$$

$$\dfrac{n_1}{s_o} + \dfrac{n_2}{s_i} = \dfrac{n_2}{R} - \dfrac{n_1}{R} + \dfrac{n_1\phi^2}{2s_o}\left(\dfrac{R}{s_o} + \dfrac{R^2}{s_o^2}\right) + \dfrac{n_1\phi^2}{2R}\left(\dfrac{R}{s_o} + \dfrac{R^2}{s_o^2}\right) + \dfrac{n_2\phi^2}{2R}\left(\dfrac{R}{s_i} - \dfrac{R^2}{s_i^2}\right) - \dfrac{n_2\phi^2}{2s_i}\left(\dfrac{R}{s_i} - \dfrac{R^2}{s_i^2}\right)$$

$$= \dfrac{n_2 - n_1}{R} + \dfrac{n_1\phi^2}{2}\left(\dfrac{R}{s_o} + \dfrac{R^2}{s_o^2}\right)\left(\dfrac{1}{s_o} + \dfrac{1}{R}\right) + \dfrac{n_2\phi^2}{2}\left(\dfrac{R}{s_i} - \dfrac{R^2}{s_i^2}\right)\left(\dfrac{1}{R} - \dfrac{1}{s_i}\right)$$

$$= \dfrac{n_2 - n_1}{R} + \dfrac{n_1\phi^2 R^2}{2s_o}\left(\dfrac{1}{R} + \dfrac{1}{s_o}\right)\left(\dfrac{1}{R} + \dfrac{1}{s_o}\right) + \dfrac{n_2\phi^2 R^2}{2s_i}\left(\dfrac{1}{R} - \dfrac{1}{s_i}\right)\left(\dfrac{1}{R} - \dfrac{1}{s_i}\right)$$

$$= \dfrac{n_2 - n_1}{R} + \phi^2 R^2\left[\dfrac{n_1}{2s_o}\left(\dfrac{1}{R} + \dfrac{1}{s_o}\right)^2 + \dfrac{n_2}{2s_i}\left(\dfrac{1}{R} - \dfrac{1}{s_i}\right)^2\right]$$

From Figure 8, we see that $\sin\phi = h/R$. So if we approximate $\sin\phi$ with ϕ, we get $h = R\phi$ and $h^2 = \phi^2 R^2$ and hence, Equation 4, as desired.

35. (a) If the water is deep, then $2\pi d/L$ is large, and we know that $\tanh x \to 1$ as $x \to \infty$. So we can approximate $\tanh(2\pi d/L) \approx 1$, and so $v^2 \approx gL/(2\pi) \quad \Leftrightarrow \quad v \approx \sqrt{gL/(2\pi)}$.

(b) From the table, the first term in the Maclaurin series of $\tanh x$ is x, so if the water is shallow, we can approximate $\tanh \dfrac{2\pi d}{L} \approx \dfrac{2\pi d}{L}$, and so $v^2 \approx \dfrac{gL}{2\pi} \cdot \dfrac{2\pi d}{L} \quad \Leftrightarrow \quad v \approx \sqrt{gd}$.

n	$f^{(n)}(x)$	$f^{(n)}(0)$
0	$\tanh x$	0
1	$\operatorname{sech}^2 x$	1
2	$-2\operatorname{sech}^2 x \tanh x$	0
3	$2\operatorname{sech}^2 x \,(3\tanh^2 x - 1)$	-2

(c) Since $\tanh x$ is an odd function, its Maclaurin series is alternating, so the error in the approximation $\tanh \dfrac{2\pi d}{L} \approx \dfrac{2\pi d}{L}$ is less than the first neglected term, which is $\dfrac{|f'''(0)|}{3!}\left(\dfrac{2\pi d}{L}\right)^3 = \dfrac{1}{3}\left(\dfrac{2\pi d}{L}\right)^3$.

If $L > 10d$, then $\dfrac{1}{3}\left(\dfrac{2\pi d}{L}\right)^3 < \dfrac{1}{3}\left(2\pi \cdot \dfrac{1}{10}\right)^3 = \dfrac{\pi^3}{375}$, so the error in the approximation $v^2 = gd$ is less than $\dfrac{gL}{2\pi} \cdot \dfrac{\pi^3}{375} \approx 0.0132gL$.

36. (a) $4\sqrt{\dfrac{L}{g}} \displaystyle\int_0^{\pi/2} \dfrac{dx}{\sqrt{1 - k^2\sin^2 x}} = 4\sqrt{\dfrac{L}{g}} \int_0^{\pi/2} \left[1 + (-k^2\sin^2 x)\right]^{-1/2} dx$

$= 4\sqrt{\dfrac{L}{g}} \displaystyle\int_0^{\pi/2} \left[1 - \dfrac{1}{2}(-k^2\sin^2 x) + \dfrac{\frac{1}{2}\cdot\frac{3}{2}}{2!}(-k^2\sin^2 x)^2 - \dfrac{\frac{1}{2}\cdot\frac{3}{2}\cdot\frac{5}{2}}{3!}(-k^2\sin^2 x)^3 + \cdots\right] dx$

$= 4\sqrt{\dfrac{L}{g}} \displaystyle\int_0^{\pi/2} \left[1 + \left(\dfrac{1}{2}\right)k^2\sin^2 x + \left(\dfrac{1\cdot 3}{2\cdot 4}\right)k^4\sin^4 x + \left(\dfrac{1\cdot 3\cdot 5}{2\cdot 4\cdot 6}\right)k^6\sin^6 x + \cdots\right] dx$

$= 4\sqrt{\dfrac{L}{g}} \left[\dfrac{\pi}{2} + \left(\dfrac{1}{2}\right)\left(\dfrac{1}{2}\cdot\dfrac{\pi}{2}\right)k^2 + \left(\dfrac{1\cdot 3}{2\cdot 4}\right)\left(\dfrac{1\cdot 3}{2\cdot 4}\cdot\dfrac{\pi}{2}\right)k^4 + \left(\dfrac{1\cdot 3\cdot 5}{2\cdot 4\cdot 6}\right)\left(\dfrac{1\cdot 3\cdot 5}{2\cdot 4\cdot 6}\cdot\dfrac{\pi}{2}\right)k^6 + \cdots\right]$

[split up the integral and use the result from Exercise 8.1.46]

$= 2\pi\sqrt{\dfrac{L}{g}} \left[1 + \dfrac{1^2}{2^2}k^2 + \dfrac{1^2\cdot 3^2}{2^2\cdot 4^2}k^4 + \dfrac{1^2\cdot 3^2\cdot 5^2}{2^2\cdot 4^2\cdot 6^2}k^6 + \cdots\right]$

(b) The first of the two inequalities is true because all of the terms in the series are positive. For the second,

$$T = 2\pi\sqrt{\dfrac{L}{g}} \left[1 + \dfrac{1^2}{2^2}k^2 + \dfrac{1^2\cdot 3^2}{2^2\cdot 4^2}k^4 + \dfrac{1^2\cdot 3^2\cdot 5^2}{2^2\cdot 4^2\cdot 6^2}k^6 + \dfrac{1^2\cdot 3^2\cdot 5^2\cdot 7^2}{2^2\cdot 4^2\cdot 6^2\cdot 8^2}k^8 + \cdots\right]$$

$$\le 2\pi\sqrt{\dfrac{L}{g}} \left[1 + \tfrac{1}{4}k^2 + \tfrac{1}{4}k^4 + \tfrac{1}{4}k^6 + \tfrac{1}{4}k^8 + \cdots\right]$$

The terms in brackets (after the first) form a geometric series with $a = \tfrac{1}{4}k^2$ and $r = k^2 = \sin^2\left(\tfrac{1}{2}\theta_0\right) < 1$.

So $T \le 2\pi\sqrt{\dfrac{L}{g}} \left[1 + \dfrac{k^2/4}{1 - k^2}\right] = 2\pi\sqrt{\dfrac{L}{g}}\,\dfrac{4 - 3k^2}{4 - 4k^2}$.

(c) We substitute $L = 1$, $g = 9.8$, and $k = \sin(10°/2) \approx 0.08716$, and the inequality from part (b) becomes

$2.01090 \leq T \leq 2.01093$, so $T \approx 2.0109$. The estimate $T \approx 2\pi\sqrt{L/g} \approx 2.0071$ differs by about 0.2%.

If $\theta_0 = 42°$, then $k \approx 0.35837$ and the inequality becomes $2.07153 \leq T \leq 2.08103$, so $T \approx 2.0763$.

The one-term estimate is the same, and the discrepancy between the two estimates increases to about 3.4%.

37. (a) L is the length of the arc subtended by the angle θ, so $L = R\theta$ $\Rightarrow$

$\theta = L/R$. Now $\sec\theta = (R+C)/R$ $\Rightarrow$ $R\sec\theta = R+C$ $\Rightarrow$

$C = R\sec\theta - R = R\sec(L/R) - R$.

(b) Extending the result in Exercise 17, we have $f^{(4)}(x) = \sec x\,(18\sec^2 x\,\tan^2 x + 6\sec^4 x - \sec^2 x - \tan^2 x)$,

so $f^{(4)}(0) = 5$, and $\sec x \approx T_4(x) = 1 + \frac{1}{2}x^2 + \frac{5}{24}x^4$. By part (a),

$$C \approx R\left[1 + \frac{1}{2}\left(\frac{L}{R}\right)^2 + \frac{5}{24}\left(\frac{L}{R}\right)^4\right] - R = R + \frac{1}{2}R\cdot\frac{L^2}{R^2} + \frac{5}{24}R\cdot\frac{L^4}{R^4} - R = \frac{L^2}{2R} + \frac{5L^4}{24R^3}.$$

(c) Taking $L = 100$ km and $R = 6370$ km, the formula in part (a) says that

$C = R\sec(L/R) - R = 6370\sec(100/6370) - 6370 \approx 0.785\,009\,965\,44$ km.

The formula in part (b) says that $C \approx \dfrac{L^2}{2R} + \dfrac{5L^4}{24R^3} = \dfrac{100^2}{2\cdot 6370} + \dfrac{5\cdot 100^4}{24\cdot 6370^3} \approx 0.785\,009\,957\,36$ km.

The difference between these two results is only $0.000\,000\,008\,08$ km, or $0.000\,008\,08$ m!

38. $T_n(x) = f(a) + \dfrac{f'(a)}{1!}(x-a) + \dfrac{f''(a)}{2!}(x-a)^2 + \cdots + \dfrac{f^{(n)}(a)}{n!}(x-a)^n$. Let $0 \leq m \leq n$. Then

$$T_n^{(m)}(x) = m!\,\frac{f^{(m)}(a)}{m!}(x-a)^0 + (m+1)(m)\cdots(2)\frac{f^{(m+1)}(a)}{(m+1)!}(x-a)^1 + \cdots$$

$$+ n(n-1)\cdots(n-m+1)\frac{f^{(n)}(a)}{n!}(x-a)^{n-m}$$

For $x = a$, all terms in this sum except the first one are 0, so $T_n^{(m)}(a) = \dfrac{m!\,f^{(m)}(a)}{m!} = f^{(m)}(a)$.

39. Using $f(x) = T_n(x) + R_n(x)$ with $n = 1$ and $x = r$, we have $f(r) = T_1(r) + R_1(r)$, where T_1 is the first-degree Taylor

polynomial of f at a. Because $a = x_n$, $f(r) = f(x_n) + f'(x_n)(r - x_n) + R_1(r)$. But r is a root of f, so $f(r) = 0$

and we have $0 = f(x_n) + f'(x_n)(r - x_n) + R_1(r)$. Taking the first two terms to the left side gives us

$f'(x_n)(x_n - r) - f(x_n) = R_1(r)$. Dividing by $f'(x_n)$, we get $x_n - r - \dfrac{f(x_n)}{f'(x_n)} = \dfrac{R_1(r)}{f'(x_n)}$. By the formula for Newton's

method, the left side of the preceding equation is $x_{n+1} - r$, so $|x_{n+1} - r| = \left|\dfrac{R_1(r)}{f'(x_n)}\right|$. Taylor's Inequality gives us

$|R_1(r)| \leq \dfrac{|f''(r)|}{2!}|r - x_n|^2$. Combining this inequality with the facts $|f''(x)| \leq M$ and $|f'(x)| \geq K$ gives us

$|x_{n+1} - r| \leq \dfrac{M}{2K}|x_n - r|^2$.

APPLIED PROJECT Radiation from the Stars

1. If we write $f(\lambda) = \dfrac{8\pi hc\lambda^{-5}}{e^{hc/(\lambda kT)} - 1} = \dfrac{a\lambda^{-5}}{e^{b/(\lambda T)} - 1}$, then as $\lambda \to 0^+$, it is of the form ∞/∞, and as $\lambda \to \infty$ it is of the form $0/0$, so in either case we can use l'Hospital's Rule. First of all,

$$\lim_{\lambda \to \infty} f(\lambda) \overset{H}{=} \lim_{\lambda \to \infty} \frac{a\left(-5\lambda^{-6}\right)}{-\dfrac{bT}{(\lambda T)^2}e^{b/(\lambda T)}} = 5\frac{aT}{b}\lim_{\lambda \to \infty}\frac{\lambda^2\lambda^{-6}}{e^{b/(\lambda T)}} = 5\frac{aT}{b}\lim_{\lambda \to \infty}\frac{\lambda^{-4}}{e^{b/(\lambda T)}} = 0$$

Also, $$\lim_{\lambda \to 0^+} f(\lambda) \overset{H}{=} 5\frac{aT}{b}\lim_{\lambda \to 0^+}\frac{\lambda^{-4}}{e^{b/(\lambda T)}} \overset{H}{=} 5\frac{aT}{b}\lim_{\lambda \to 0^+}\frac{-4\lambda^{-5}}{-\dfrac{bT}{(\lambda T)^2}e^{b/(\lambda T)}} = 20\frac{aT^2}{b^2}\lim_{\lambda \to 0^+}\frac{\lambda^{-3}}{e^{b/(\lambda T)}}$$

This is still indeterminate, but note that each time we use l'Hospital's Rule, we gain a factor of λ in the numerator, as well as a constant factor, and the denominator is unchanged. So if we use l'Hospital's Rule three more times, the exponent of λ in the numerator will become 0. That is, for some $\{k_i\}$, all constant,

$$\lim_{\lambda \to 0^+} f(\lambda) \overset{H}{=} k_1\lim_{\lambda \to 0^+}\frac{\lambda^{-3}}{e^{b/(\lambda T)}} \overset{H}{=} k_2\lim_{\lambda \to 0^+}\frac{\lambda^{-2}}{e^{b/(\lambda T)}} \overset{H}{=} k_3\lim_{\lambda \to 0^+}\frac{\lambda^{-1}}{e^{b/(\lambda T)}} \overset{H}{=} k_4\lim_{\lambda \to 0^+}\frac{1}{e^{b/(\lambda T)}} = 0$$

2. We expand the denominator of Planck's Law using the Taylor series $e^x = 1 + x + \dfrac{x^2}{2!} + \dfrac{x^3}{3!} + \cdots$ with $x = \dfrac{hc}{\lambda kT}$, and use the fact that if λ is large, then all subsequent terms in the Taylor expansion are very small compared to the first one, so we can approximate using the Taylor polynomial T_1:

$$f(\lambda) = \frac{8\pi hc\lambda^{-5}}{e^{hc/(\lambda kT)} - 1} = \frac{8\pi hc\lambda^{-5}}{\left[1 + \dfrac{hc}{\lambda kT} + \dfrac{1}{2!}\left(\dfrac{hc}{\lambda kT}\right)^2 + \dfrac{1}{3!}\left(\dfrac{hc}{\lambda kT}\right)^3 + \cdots\right] - 1} \approx \frac{8\pi hc\lambda^{-5}}{\left(1 + \dfrac{hc}{\lambda kT}\right) - 1} = \frac{8\pi kT}{\lambda^4}$$

which is the Rayleigh-Jeans Law.

3. To convert to μm, we substitute $\lambda/10^6$ for λ in both laws. The first figure shows that the two laws are similar for large λ. The second figure shows that the two laws are very different for short wavelengths (Planck's Law gives a maximum at $\lambda \approx 0.51$ μm; the Rayleigh-Jeans Law gives no minimum or maximum.).

 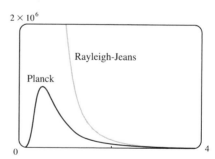

4. From the graph in Problem 3, $f(\lambda)$ has a maximum under Planck's Law at $\lambda \approx 0.51$ μm.

5.

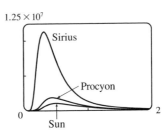

As T gets larger, the total area under the curve increases, as we would expect: the hotter the star, the more energy it emits. Also, as T increases, the λ-value of the maximum decreases, so the higher the temperature, the shorter the peak wavelength (and consequently the average wavelength) of light emitted. This is why Sirius is a blue star and Betelgeuse is a red star: most of Sirius's light is of a fairly short wavelength; that is, a higher frequency, toward the blue end of the spectrum, whereas most of Betelgeuse's light is of a lower frequency, toward the red end of the spectrum.

12   Review

CONCEPT CHECK

1. (a) See Definition 12.1.1.

(b) See Definition 12.2.2.

(c) The terms of the sequence $\{a_n\}$ approach 3 as n becomes large.

(d) By adding sufficiently many terms of the series, we can make the partial sums as close to 3 as we like.

2. (a) See Definition 12.1.11.

(b) A sequence is monotonic if it is either increasing or decreasing.

(c) By Theorem 12.1.12, every bounded, monotonic sequence is convergent.

3. (a) See (4) in Section 12.2.

(b) The p-series $\sum\limits_{n=1}^{\infty} \dfrac{1}{n^p}$ is convergent if $p > 1$.

4. If $\sum a_n = 3$, then $\lim\limits_{n \to \infty} a_n = 0$ and $\lim\limits_{n \to \infty} s_n = 3$.

5. (a) *Test for Divergence:* If $\lim\limits_{n \to \infty} a_n$ does not exist or if $\lim\limits_{n \to \infty} a_n \neq 0$, then the series $\sum_{n=1}^{\infty} a_n$ is divergent.

(b) *Integral Test:* Suppose f is a continuous, positive, decreasing function on $[1, \infty)$ and let $a_n = f(n)$. Then the series $\sum_{n=1}^{\infty} a_n$ is convergent if and only if the improper integral $\int_{1}^{\infty} f(x)\,dx$ is convergent. In other words:

  (i) If $\int_{1}^{\infty} f(x)\,dx$ is convergent, then $\sum_{n=1}^{\infty} a_n$ is convergent.

  (ii) If $\int_{1}^{\infty} f(x)\,dx$ is divergent, then $\sum_{n=1}^{\infty} a_n$ is divergent.

(c) *Comparison Test:* Suppose that $\sum a_n$ and $\sum b_n$ are series with positive terms.

  (i) If $\sum b_n$ is convergent and $a_n \leq b_n$ for all n, then $\sum a_n$ is also convergent.

  (ii) If $\sum b_n$ is divergent and $a_n \geq b_n$ for all n, then $\sum a_n$ is also divergent.

(d) *Limit Comparison Test:* Suppose that $\sum a_n$ and $\sum b_n$ are series with positive terms. If $\lim\limits_{n \to \infty} (a_n/b_n) = c$, where c is a finite number and $c > 0$, then either both series converge or both diverge.

(e) *Alternating Series Test:* If the alternating series $\sum_{n=1}^{\infty}(-1)^{n-1}b_n = b_1 - b_2 + b_3 - b_4 + b_5 - b_6 + \cdots$ $[b_n > 0]$

satisfies (i) $b_{n+1} \le b_n$ for all n and (ii) $\lim\limits_{n\to\infty} b_n = 0$, then the series is convergent.

(f) *Ratio Test:*

(i) If $\lim\limits_{n\to\infty} \left| \dfrac{a_{n+1}}{a_n} \right| = L < 1$, then the series $\sum_{n=1}^{\infty} a_n$ is absolutely convergent (and therefore convergent).

(ii) If $\lim\limits_{n\to\infty} \left| \dfrac{a_{n+1}}{a_n} \right| = L > 1$ or $\lim\limits_{n\to\infty} \left| \dfrac{a_{n+1}}{a_n} \right| = \infty$, then the series $\sum_{n=1}^{\infty} a_n$ is divergent.

(iii) If $\lim\limits_{n\to\infty} \left| \dfrac{a_{n+1}}{a_n} \right| = 1$, the Ratio Test is inconclusive; that is, no conclusion can be drawn about the convergence or

divergence of $\sum a_n.$

(g) *Root Test:*

(i) If $\lim\limits_{n\to\infty} \sqrt[n]{|a_n|} = L < 1$, then the series $\sum_{n=1}^{\infty} a_n$ is absolutely convergent (and therefore convergent).

(ii) If $\lim\limits_{n\to\infty} \sqrt[n]{|a_n|} = L > 1$ or $\lim\limits_{n\to\infty} \sqrt[n]{|a_n|} = \infty$, then the series $\sum_{n=1}^{\infty} a_n$ is divergent.

(iii) If $\lim\limits_{n\to\infty} \sqrt[n]{|a_n|} = 1$, the Root Test is inconclusive..

6. (a) A series $\sum a_n$ is called *absolutely convergent* if the series of absolute values $\sum |a_n|$ is convergent.

(b) If a series $\sum a_n$ is absolutely convergent, then it is convergent.

(c) A series $\sum a_n$ is called *conditionally convergent* if it is convergent but not absolutely convergent.

7. (a) Use (3) in Section 12.3.

(b) See Example 5 in Section 12.4.

(c) By adding terms until you reach the desired accuracy given by the Alternating Series Estimation Theorem on page 748.

8. (a) $\sum_{n=0}^{\infty} c_n(x-a)^n$

(b) Given the power series $\sum_{n=0}^{\infty} c_n(x-a)^n$, the radius of convergence is:

(i) 0 if the series converges only when $x = a$

(ii) ∞ if the series converges for all x, or

(iii) a positive number R such that the series converges if $|x - a| < R$ and diverges if $|x - a| > R$.

(c) The interval of convergence of a power series is the interval that consists of all values of x for which the series converges. Corresponding to the cases in part (b), the interval of convergence is: (i) the single point $\{a\}$, (ii) all real numbers, that is, the real number line $(-\infty, \infty)$, or (iii) an interval with endpoints $a - R$ and $a + R$ which can contain neither, either, or both of the endpoints. In this case, we must test the series for convergence at each endpoint to determine the interval of convergence.

9. (a), (b) See Theorem 12.9.2.

10. (a) $T_n(x) = \sum_{i=0}^{n} \dfrac{f^{(i)}(a)}{i!}(x-a)^i$

(b) $\sum_{n=0}^{\infty} \dfrac{f^{(n)}(a)}{n!}(x-a)^n$

(c) $\displaystyle\sum_{n=0}^{\infty} \frac{f^{(n)}(0)}{n!}\, x^n$ [$a = 0$ in part (b)]

(d) See Theorem 12.10.8.

(e) See Taylor's Inequality (12.10.9).

11. (a)–(e) See Table 1 on page 779.

12. See the binomial series (12.10.17) for the expansion. The radius of convergence for the binomial series is 1.

TRUE-FALSE QUIZ

1. False. See Note 2 after Theorem 12.2.6.

2. False. The series $\displaystyle\sum_{n=1}^{\infty} n^{-\sin 1} = \sum_{n=1}^{\infty} \frac{1}{n^{\sin 1}}$ is a p-series with $p = \sin 1 \approx 0.84 \le 1$, so the series diverges.

3. True. If $\lim_{n\to\infty} a_n = L$, then given any $\varepsilon > 0$, we can find a positive integer N such that $|a_n - L| < \varepsilon$ whenever $n > N$.

        If $n > N$, then $2n + 1 > N$ and $|a_{2n+1} - L| < \varepsilon$. Thus, $\lim_{n\to\infty} a_{2n+1} = L$.

4. True by Theorem 12.8.3.

    Or: Use the Comparison Test to show that $\sum c_n(-2)^n$ converges absolutely.

5. False. For example, take $c_n = (-1)^n/(n6^n)$.

6. True by Theorem 12.8.3.

7. False, since $\displaystyle\lim_{n\to\infty}\left|\frac{a_{n+1}}{a_n}\right| = \lim_{n\to\infty}\left|\frac{1}{(n+1)^3}\cdot\frac{n^3}{1}\right| = \lim_{n\to\infty}\left|\frac{n^3}{(n+1)^3}\cdot\frac{1/n^3}{1/n^3}\right| = \lim_{n\to\infty}\frac{1}{(1+1/n)^3} = 1.$

8. True, since $\displaystyle\lim_{n\to\infty}\left|\frac{a_{n+1}}{a_n}\right| = \lim_{n\to\infty}\left|\frac{1}{(n+1)!}\cdot\frac{n!}{1}\right| = \lim_{n\to\infty}\frac{1}{n+1} = 0 < 1.$

9. False. See the note after Example 2 in Section 12.4.

10. True, since $\displaystyle\frac{1}{e} = e^{-1}$ and $\displaystyle e^x = \sum_{n=0}^{\infty}\frac{x^n}{n!}$, so $\displaystyle e^{-1} = \sum_{n=0}^{\infty}\frac{(-1)^n}{n!}.$

11. True. See (9) in Section 12.1.

12. True, because if $\sum |a_n|$ is convergent, then so is $\sum a_n$ by Theorem 12.6.3.

13. True. By Theorem 12.10.5 the coefficient of x^3 is $\dfrac{f'''(0)}{3!} = \dfrac{1}{3} \;\Rightarrow\; f'''(0) = 2.$

    Or: Use Theorem 12.9.2 to differentiate f three times.

14. False. Let $a_n = n$ and $b_n = -n$. Then $\{a_n\}$ and $\{b_n\}$ are divergent, but $a_n + b_n = 0$, so $\{a_n + b_n\}$ is convergent.

15. False. For example, let $a_n = b_n = (-1)^n$. Then $\{a_n\}$ and $\{b_n\}$ are divergent, but $a_n b_n = 1$, so $\{a_n b_n\}$ is convergent.

16. True by the Monotonic Sequence Theorem, since $\{a_n\}$ is decreasing and $0 < a_n \le a_1$ for all n $\;\Rightarrow\;$ $\{a_n\}$ is bounded.

17. True by Theorem 12.6.3. $\left[\sum (-1)^n \, a_n \text{ is absolutely convergent and hence convergent.}\right]$

18. True. $\lim\limits_{n\to\infty} \dfrac{a_{n+1}}{a_n} < 1 \;\Rightarrow\; \sum a_n$ converges (Ratio Test) $\;\Rightarrow\; \lim\limits_{n\to\infty} a_n = 0$ [Theorem 12.2.6].

19. True. $0.99999\ldots = 0.9 + 0.9(0.1)^1 + 0.9(0.1)^2 + 0.9(0.1)^3 + \cdots = \sum\limits_{n=1}^{\infty} (0.9)(0.1)^{n-1} = \dfrac{0.9}{1 - 0.1} = 1$ by the formula

  for the sum of a geometric series $[S = a_1/(1-r)]$ with ratio r satisfying $|r| < 1$.

20. False.   Let $a_n = (0.1)^n$ and $b_n = (0.2)^n$. Then $\sum\limits_{n=1}^{\infty} a_n = \sum\limits_{n=1}^{\infty} (0.1)^n = \dfrac{0.1}{1 - 0.1} = \dfrac{1}{9} = A$,

  $\sum\limits_{n=1}^{\infty} b_n = \sum\limits_{n=1}^{\infty} (0.2)^n = \dfrac{0.2}{1 - 0.2} = \dfrac{1}{4} = B$, and $\sum\limits_{n=1}^{\infty} a_n b_n = \sum\limits_{n=1}^{\infty} (0.02)^n = \dfrac{0.02}{1 - 0.02} = \dfrac{1}{49}$, but

  $AB = \frac{1}{9} \cdot \frac{1}{4} = \frac{1}{36}$.

EXERCISES

1. $\left\{\dfrac{2+n^3}{1+2n^3}\right\}$ converges since $\lim\limits_{n\to\infty} \dfrac{2+n^3}{1+2n^3} = \lim\limits_{n\to\infty} \dfrac{2/n^3 + 1}{1/n^3 + 2} = \dfrac{1}{2}$.

2. $a_n = \dfrac{9^{n+1}}{10^n} = 9 \cdot \left(\frac{9}{10}\right)^n$, so $\lim\limits_{n\to\infty} a_n = 9 \lim\limits_{n\to\infty} \left(\frac{9}{10}\right)^n = 9 \cdot 0 = 0$ by (12.1.9).

3. $\lim\limits_{n\to\infty} a_n = \lim\limits_{n\to\infty} \dfrac{n^3}{1+n^2} = \lim\limits_{n\to\infty} \dfrac{n}{1/n^2 + 1} = \infty$, so the sequence diverges.

4. $a_n = \cos(n\pi/2)$, so $a_n = 0$ if n is odd and $a_n = \pm 1$ if n is even. As n increases, a_n keeps cycling through the values

  $0, 1, 0, -1$, so the sequence $\{a_n\}$ is divergent.

5. $|a_n| = \left|\dfrac{n \sin n}{n^2 + 1}\right| \leq \dfrac{n}{n^2 + 1} < \dfrac{1}{n}$, so $|a_n| \to 0$ as $n \to \infty$. Thus, $\lim\limits_{n\to\infty} a_n = 0$. The sequence $\{a_n\}$ is convergent.

6. $a_n = \dfrac{\ln n}{\sqrt{n}}$. Let $f(x) = \dfrac{\ln x}{\sqrt{x}}$ for $x > 0$. Then $\lim\limits_{x\to\infty} f(x) = \lim\limits_{x\to\infty} \dfrac{\ln x}{\sqrt{x}} \overset{\text{H}}{=} \lim\limits_{x\to\infty} \dfrac{1/x}{1/(2\sqrt{x})} = \lim\limits_{x\to\infty} \dfrac{2}{\sqrt{x}} = 0$.

  Thus, by Theorem 3 in Section 12.1, $\{a_n\}$ converges and $\lim\limits_{n\to\infty} a_n = 0$.

7. $\left\{\left(1 + \dfrac{3}{n}\right)^{4n}\right\}$ is convergent. Let $y = \left(1 + \dfrac{3}{x}\right)^{4x}$. Then

  $\lim\limits_{x\to\infty} \ln y = \lim\limits_{x\to\infty} 4x \ln(1 + 3/x) = \lim\limits_{x\to\infty} \dfrac{\ln(1 + 3/x)}{1/(4x)} \overset{\text{H}}{=} \lim\limits_{x\to\infty} \dfrac{\dfrac{1}{1 + 3/x}\left(-\dfrac{3}{x^2}\right)}{-1/(4x^2)} = \lim\limits_{x\to\infty} \dfrac{12}{1 + 3/x} = 12$, so

  $\lim\limits_{x\to\infty} y = \lim\limits_{n\to\infty} \left(1 + \dfrac{3}{n}\right)^{4n} = e^{12}$.

8. $\left\{\dfrac{(-10)^n}{n!}\right\}$ converges, since $\dfrac{10^n}{n!} = \dfrac{10 \cdot 10 \cdot 10 \cdots\cdots 10}{1 \cdot 2 \cdot 3 \cdots\cdots 10} \cdot \dfrac{10 \cdot 10 \cdots\cdots 10}{11 \cdot 12 \cdots\cdots n} \leq 10^{10} \left(\dfrac{10}{11}\right)^{n-10} \to 0$ as $n \to \infty$, so

  $\lim\limits_{n\to\infty} \dfrac{(-10)^n}{n!} = 0$ [Squeeze Theorem]. *Or:* Use (12.10.10).

9. We use induction, hypothesizing that $a_{n-1} < a_n < 2$. Note first that $1 < a_2 = \frac{1}{3}(1+4) = \frac{5}{3} < 2$, so the hypothesis holds

for $n = 2$. Now assume that $a_{k-1} < a_k < 2$. Then $a_k = \frac{1}{3}(a_{k-1}+4) < \frac{1}{3}(a_k+4) < \frac{1}{3}(2+4) = 2$. So $a_k < a_{k+1} < 2$,

and the induction is complete. To find the limit of the sequence, we note that $L = \lim\limits_{n\to\infty} a_n = \lim\limits_{n\to\infty} a_{n+1} \Rightarrow$

$L = \frac{1}{3}(L+4) \Rightarrow L = 2$.

10. $\lim\limits_{x\to\infty} \dfrac{x^4}{e^x} \overset{\text{H}}{=} \lim\limits_{x\to\infty} \dfrac{4x^3}{e^x} \overset{\text{H}}{=} \lim\limits_{x\to\infty} \dfrac{12x^2}{e^x} \overset{\text{H}}{=} \lim\limits_{x\to\infty} \dfrac{24x}{e^x} \overset{\text{H}}{=} \lim\limits_{x\to\infty} \dfrac{24}{e^x} = 0$

Then we conclude from Theorem 12.1.3 that $\lim\limits_{n\to\infty} n^4 e^{-n} = 0$.

From the graph, it seems that $12^4 e^{-12} > 0.1$, but $n^4 e^{-n} < 0.1$

whenever $n > 12$. So the smallest value of N corresponding to

$\varepsilon = 0.1$ in the definition of the limit is $N = 12$.

11. $\dfrac{n}{n^3+1} < \dfrac{n}{n^3} = \dfrac{1}{n^2}$, so $\sum\limits_{n=1}^{\infty} \dfrac{n}{n^3+1}$ converges by the Comparison Test with the convergent p-series $\sum\limits_{n=1}^{\infty} \dfrac{1}{n^2}$ $[p = 2 > 1]$.

12. Let $a_n = \dfrac{n^2+1}{n^3+1}$ and $b_n = \dfrac{1}{n}$, so $\lim\limits_{n\to\infty} \dfrac{a_n}{b_n} = \lim\limits_{n\to\infty} \dfrac{n^3+n}{n^3+1} = \lim\limits_{n\to\infty} \dfrac{1+1/n^2}{1+1/n^3} = 1 > 0$.

Since $\sum\limits_{n=1}^{\infty} b_n$ is the divergent harmonic series, $\sum\limits_{n=1}^{\infty} a_n$ also diverges by the Limit Comparison Test.

13. $\lim\limits_{n\to\infty} \left| \dfrac{a_{n+1}}{a_n} \right| = \lim\limits_{n\to\infty} \left[\dfrac{(n+1)^3}{5^{n+1}} \cdot \dfrac{5^n}{n^3} \right] = \lim\limits_{n\to\infty} \left(1 + \dfrac{1}{n}\right)^3 \cdot \dfrac{1}{5} = \dfrac{1}{5} < 1$, so $\sum\limits_{n=1}^{\infty} \dfrac{n^3}{5^n}$ converges by the Ratio Test.

14. Let $b_n = \dfrac{1}{\sqrt{n+1}}$. Then b_n is positive for $n \geq 1$, the sequence $\{b_n\}$ is decreasing, and $\lim\limits_{n\to\infty} b_n = 0$, so the series

$\sum\limits_{n=1}^{\infty} \dfrac{(-1)^n}{\sqrt{n+1}}$ converges by the Alternating Series Test.

15. Let $f(x) = \dfrac{1}{x\sqrt{\ln x}}$. Then f is continuous, positive, and decreasing on $[2, \infty)$, so the Integral Test applies.

$$\int_2^{\infty} f(x)\,dx = \lim\limits_{t\to\infty} \int_2^t \dfrac{1}{x\sqrt{\ln x}}\,dx \quad \left[u = \ln x,\, du = \dfrac{1}{x}\,dx\right] = \lim\limits_{t\to\infty} \int_{\ln 2}^{\ln t} u^{-1/2}\,du = \lim\limits_{t\to\infty} \left[2\sqrt{u}\right]_{\ln 2}^{\ln t}$$

$$= \lim\limits_{t\to\infty} \left(2\sqrt{\ln t} - 2\sqrt{\ln 2}\right) = \infty,$$

so the series $\sum\limits_{n=2}^{\infty} \dfrac{1}{n\sqrt{\ln n}}$ diverges.

16. $\lim\limits_{n\to\infty} \dfrac{n}{3n+1} = \dfrac{1}{3}$, so $\lim\limits_{n\to\infty} \ln\left(\dfrac{n}{3n+1}\right) = \ln\dfrac{1}{3} \neq 0$. Thus, the series $\sum\limits_{n=1}^{\infty} \ln\left(\dfrac{n}{3n+1}\right)$ diverges by the Test for

Divergence.

17. $|a_n| = \left| \dfrac{\cos 3n}{1+(1.2)^n} \right| \leq \dfrac{1}{1+(1.2)^n} < \dfrac{1}{(1.2)^n} = \left(\dfrac{5}{6}\right)^n$, so $\sum\limits_{n=1}^{\infty} |a_n|$ converges by comparison with the convergent geometric

series $\sum\limits_{n=1}^{\infty} \left(\dfrac{5}{6}\right)^n$ $\left[r = \dfrac{5}{6} < 1\right]$. It follows that $\sum\limits_{n=1}^{\infty} a_n$ converges (by Theorem 3 in Section 12.6).

18. $\lim\limits_{n\to\infty} \sqrt[n]{|a_n|} = \lim\limits_{n\to\infty} \sqrt[n]{\left|\dfrac{n^{2n}}{(1+2n^2)^n}\right|} = \lim\limits_{n\to\infty} \dfrac{n^2}{1+2n^2} = \lim\limits_{n\to\infty} \dfrac{1}{1/n^2+2} = \dfrac{1}{2} < 1$, so $\sum\limits_{n=1}^{\infty} \dfrac{n^{2n}}{(1+2n^2)^n}$ converges by the

Root Test.

19. $\lim\limits_{n\to\infty} \left|\dfrac{a_{n+1}}{a_n}\right| = \lim\limits_{n\to\infty} \dfrac{1\cdot 3\cdot 5\cdot\,\cdots\,\cdot(2n-1)(2n+1)}{5^{n+1}\,(n+1)!} \cdot \dfrac{5^n\, n!}{1\cdot 3\cdot 5\cdot\,\cdots\,\cdot(2n-1)} = \lim\limits_{n\to\infty} \dfrac{2n+1}{5(n+1)} = \dfrac{2}{5} < 1$, so the series

converges by the Ratio Test.

20. $\sum\limits_{n=1}^{\infty} \dfrac{(-5)^{2n}}{n^2\,9^n} = \sum\limits_{n=1}^{\infty} \dfrac{1}{n^2}\left(\dfrac{25}{9}\right)^n$. Now $\lim\limits_{n\to\infty} \left|\dfrac{a_{n+1}}{a_n}\right| = \lim\limits_{n\to\infty} \dfrac{25^{n+1}}{(n+1)^2\cdot 9^{n+1}} \cdot \dfrac{n^2\cdot 9^n}{25^n} = \lim\limits_{n\to\infty} \dfrac{25n^2}{9(n+1)^2} = \dfrac{25}{9} > 1$,

so the series diverges by the Ratio Test.

21. $b_n = \dfrac{\sqrt{n}}{n+1} > 0$, $\{b_n\}$ is decreasing, and $\lim\limits_{n\to\infty} b_n = 0$, so the series $\sum\limits_{n=1}^{\infty} (-1)^{n-1} \dfrac{\sqrt{n}}{n+1}$ converges by the Alternating

Series Test.

22. Use the Limit Comparison Test with $a_n = \dfrac{\sqrt{n+1}-\sqrt{n-1}}{n} = \dfrac{2}{n\left(\sqrt{n+1}+\sqrt{n-1}\right)}$ (rationalizing the numerator) and

$b_n = \dfrac{1}{n^{3/2}}$. $\lim\limits_{n\to\infty} \dfrac{a_n}{b_n} = \lim\limits_{n\to\infty} \dfrac{2\sqrt{n}}{\sqrt{n+1}+\sqrt{n-1}} = 1$, so since $\sum\limits_{n=1}^{\infty} b_n$ converges $\left[p = \tfrac{3}{2} > 1\right]$, $\sum\limits_{n=1}^{\infty} a_n$ converges also.

23. Consider the series of absolute values: $\sum\limits_{n=1}^{\infty} n^{-1/3}$ is a p-series with $p = \tfrac{1}{3} \le 1$ and is therefore divergent. But if we apply the

Alternating Series Test, we see that $b_n = \dfrac{1}{\sqrt[3]{n}} > 0$, $\{b_n\}$ is decreasing, and $\lim\limits_{n\to\infty} b_n = 0$, so the series $\sum\limits_{n=1}^{\infty} (-1)^{n-1}\, n^{-1/3}$

converges. Thus, $\sum\limits_{n=1}^{\infty} (-1)^{n-1}\, n^{-1/3}$ is conditionally convergent.

24. $\sum\limits_{n=1}^{\infty} \left|(-1)^{n-1}\, n^{-3}\right| = \sum\limits_{n=1}^{\infty} n^{-3}$ is a convergent p-series $[p = 3 > 1]$. Therefore, $\sum\limits_{n=1}^{\infty} (-1)^{n-1}\, n^{-3}$ is absolutely convergent.

25. $\left|\dfrac{a_{n+1}}{a_n}\right| = \left|\dfrac{(-1)^{n+1}(n+2)3^{n+1}}{2^{2n+3}} \cdot \dfrac{2^{2n+1}}{(-1)^n(n+1)3^n}\right| = \dfrac{n+2}{n+1}\cdot\dfrac{3}{4} = \dfrac{1+(2/n)}{1+(1/n)}\cdot\dfrac{3}{4} \to \dfrac{3}{4} < 1$ as $n \to \infty$, so by the Ratio

Test, $\sum\limits_{n=1}^{\infty} \dfrac{(-1)^n(n+1)3^n}{2^{2n+1}}$ is absolutely convergent.

26. $\lim\limits_{x\to\infty} \dfrac{\sqrt{x}}{\ln x} \overset{\text{H}}{=} \lim\limits_{x\to\infty} \dfrac{1/(2\sqrt{x})}{1/x} = \lim\limits_{x\to\infty} \dfrac{\sqrt{x}}{2} = \infty$. Therefore, $\lim\limits_{n\to\infty} \dfrac{(-1)^n\sqrt{n}}{\ln n} \ne 0$, so the given series is divergent by the

Test for Divergence.

27. $\sum\limits_{n=1}^{\infty} \dfrac{(-3)^{n-1}}{2^{3n}} = \sum\limits_{n=1}^{\infty} \dfrac{(-3)^{n-1}}{(2^3)^n} = \sum\limits_{n=1}^{\infty} \dfrac{(-3)^{n-1}}{8^n} = \dfrac{1}{8}\sum\limits_{n=1}^{\infty} \dfrac{(-3)^{n-1}}{8^{n-1}} = \dfrac{1}{8}\sum\limits_{n=1}^{\infty} \left(-\dfrac{3}{8}\right)^{n-1} = \dfrac{1}{8}\left(\dfrac{1}{1-(-3/8)}\right)$

$= \dfrac{1}{8}\cdot\dfrac{8}{11} = \dfrac{1}{11}$

28. $\displaystyle\sum_{n=1}^{\infty} \frac{1}{n(n+3)} = \sum_{n=1}^{\infty} \left[\frac{1}{3n} - \frac{1}{3(n+3)}\right]$ [partial fractions].

$$s_n = \sum_{i=1}^{n} \left[\frac{1}{3i} - \frac{1}{3(i+3)}\right] = \frac{1}{3} + \frac{1}{6} + \frac{1}{9} - \frac{1}{3(n+1)} - \frac{1}{3(n+2)} - \frac{1}{3(n+3)} \text{ (telescoping sum), so}$$

$$\sum_{n=1}^{\infty} \frac{1}{n(n+3)} = \lim_{n\to\infty} s_n = \frac{1}{3} + \frac{1}{6} + \frac{1}{9} = \frac{11}{18}.$$

29. $\displaystyle\sum_{n=1}^{\infty} [\tan^{-1}(n+1) - \tan^{-1} n] = \lim_{n\to\infty} s_n$

$$= \lim_{n\to\infty} [(\tan^{-1} 2 - \tan^{-1} 1) + (\tan^{-1} 3 - \tan^{-1} 2) + \cdots + (\tan^{-1}(n+1) - \tan^{-1} n)]$$

$$= \lim_{n\to\infty} [\tan^{-1}(n+1) - \tan^{-1} 1] = \frac{\pi}{2} - \frac{\pi}{4} = \frac{\pi}{4}$$

30. $\displaystyle\sum_{n=0}^{\infty} \frac{(-1)^n \pi^n}{3^{2n}(2n)!} = \sum_{n=0}^{\infty} (-1)^n \frac{1}{(2n)!} \cdot \frac{\pi^n}{3^{2n}} = \sum_{n=0}^{\infty} (-1)^n \frac{1}{(2n)!} \cdot \left(\frac{\sqrt{\pi}}{3}\right)^{2n} = \cos\left(\frac{\sqrt{\pi}}{3}\right)$ since $\cos x = \displaystyle\sum_{n=0}^{\infty} (-1)^n \frac{x^{2n}}{(2n)!}$

for all x.

31. $1 - e + \dfrac{e^2}{2!} - \dfrac{e^3}{3!} + \dfrac{e^4}{4!} - \cdots = \displaystyle\sum_{n=0}^{\infty} (-1)^n \frac{e^n}{n!} = \sum_{n=0}^{\infty} \frac{(-e)^n}{n!} = e^{-e}$ since $e^x = \displaystyle\sum_{n=0}^{\infty} \frac{x^n}{n!}$ for all x.

32. $4.17\overline{326} = 4.17 + \dfrac{326}{10^5} + \dfrac{326}{10^8} + \cdots = 4.17 + \dfrac{326/10^5}{1 - 1/10^3} = \dfrac{417}{100} + \dfrac{326}{99,900} = \dfrac{416,909}{99,900}$

33. $\cosh x = \dfrac{1}{2}(e^x + e^{-x}) = \dfrac{1}{2}\left(\displaystyle\sum_{n=0}^{\infty} \frac{x^n}{n!} + \sum_{n=0}^{\infty} \frac{(-x)^n}{n!}\right)$

$$= \frac{1}{2}\left[\left(1 + x + \frac{x^2}{2!} + \frac{x^3}{3!} + \frac{x^4}{4!} + \cdots\right) + \left(1 - x + \frac{x^2}{2!} - \frac{x^3}{3!} + \frac{x^4}{4!} - \cdots\right)\right]$$

$$= \frac{1}{2}\left(2 + 2 \cdot \frac{x^2}{2!} + 2 \cdot \frac{x^4}{4!} + \cdots\right) = 1 + \frac{1}{2}x^2 + \sum_{n=2}^{\infty} \frac{x^{2n}}{(2n)!} \geq 1 + \frac{1}{2}x^2 \quad \text{for all } x$$

34. $\displaystyle\sum_{n=1}^{\infty} (\ln x)^n$ is a geometric series which converges whenever $|\ln x| < 1 \Rightarrow -1 < \ln x < 1 \Rightarrow e^{-1} < x < e.$

35. $\displaystyle\sum_{n=1}^{\infty} \frac{(-1)^{n+1}}{n^5} = 1 - \frac{1}{32} + \frac{1}{243} - \frac{1}{1024} + \frac{1}{3125} - \frac{1}{7776} + \frac{1}{16,807} - \frac{1}{32,768} + \cdots.$

Since $b_8 = \dfrac{1}{8^5} = \dfrac{1}{32,768} < 0.000031$, $\displaystyle\sum_{n=1}^{\infty} \frac{(-1)^{n+1}}{n^5} \approx \sum_{n=1}^{7} \frac{(-1)^{n+1}}{n^5} \approx 0.9721.$

36. (a) $s_5 = \displaystyle\sum_{n=1}^{5} \frac{1}{n^6} = 1 + \frac{1}{2^6} + \cdots + \frac{1}{5^6} \approx 1.017305.$ The series $\displaystyle\sum_{n=1}^{\infty} \frac{1}{n^6}$ converges by the Integral Test, so we estimate the

remainder R_5 with (12.3.2): $R_5 \leq \displaystyle\int_5^{\infty} \frac{dx}{x^6} = \left[-\frac{x^{-5}}{5}\right]_5^{\infty} = \frac{5^{-5}}{5} = 0.000064.$ So the error is at most $0.000064.$

(b) In general, $R_n \leq \displaystyle\int_n^{\infty} \frac{dx}{x^6} = \frac{1}{5n^5}.$ If we take $n = 9$, then $s_9 \approx 1.01734$ and $R_9 \leq \dfrac{1}{5 \cdot 9^5} \approx 3.4 \times 10^{-6}.$

So to five decimal places, $\displaystyle\sum_{n=1}^{\infty} \frac{1}{n^5} \approx \sum_{n=1}^{9} \frac{1}{n^5} \approx 1.01734.$

Another method: Use (12.3.3) instead of (12.3.2).

37. $\sum_{n=1}^{\infty} \dfrac{1}{2+5^n} \approx \sum_{n=1}^{8} \dfrac{1}{2+5^n} \approx 0.18976224.$ To estimate the error, note that $\dfrac{1}{2+5^n} < \dfrac{1}{5^n}$, so the remainder term is

$$R_8 = \sum_{n=9}^{\infty} \dfrac{1}{2+5^n} < \sum_{n=9}^{\infty} \dfrac{1}{5^n} = \dfrac{1/5^9}{1-1/5} = 6.4 \times 10^{-7} \quad \left[\text{geometric series with } a = \tfrac{1}{5^9} \text{ and } r = \tfrac{1}{5}\right].$$

38. (a) $\lim\limits_{n \to \infty} \left| \dfrac{a_{n+1}}{a_n} \right| = \lim\limits_{n \to \infty} \left| \dfrac{(n+1)^{n+1}}{[2(n+1)]!} \cdot \dfrac{(2n)!}{n^n} \right| = \lim\limits_{n \to \infty} \dfrac{(n+1)^n (n+1)^1}{(2n+2)(2n+1)n^n} = \lim\limits_{n \to \infty} \left(\dfrac{n+1}{n} \right)^n \dfrac{1}{2(2n+1)}$

$$= \lim\limits_{n \to \infty} \left(1 + \dfrac{1}{n} \right)^n \dfrac{1}{2(2n+1)} = e \cdot 0 = 0 < 1$$

so the series converges by the Ratio Test.

(b) The series in part (a) is convergent, so $\lim\limits_{n \to \infty} a_n = \lim\limits_{n \to \infty} \dfrac{n^n}{(2n)!} = 0$ by Theorem 12.2.6.

39. Use the Limit Comparison Test. $\lim\limits_{n \to \infty} \left| \dfrac{\left(\frac{n+1}{n}\right) a_n}{a_n} \right| = \lim\limits_{n \to \infty} \dfrac{n+1}{n} = \lim\limits_{n \to \infty} \left(1 + \dfrac{1}{n} \right) = 1 > 0.$

Since $\sum |a_n|$ is convergent, so is $\sum \left| \left(\dfrac{n+1}{n} \right) a_n \right|$, by the Limit Comparison Test.

40. $\lim\limits_{n \to \infty} \left| \dfrac{a_{n+1}}{a_n} \right| = \lim\limits_{n \to \infty} \left| \dfrac{x^{n+1}}{(n+1)^2 \, 5^{n+1}} \cdot \dfrac{n^2 5^n}{x^n} \right| = \lim\limits_{n \to \infty} \dfrac{1}{(1+1/n)^2} \dfrac{|x|}{5} = \dfrac{|x|}{5}$, so by the Ratio Test, $\sum_{n=1}^{\infty} (-1)^n \dfrac{x^n}{n^2 \, 5^n}$

converges when $\dfrac{|x|}{5} < 1 \;\Leftrightarrow\; |x| < 5$, so $R = 5$. When $x = -5$, the series becomes the convergent p-series $\sum_{n=1}^{\infty} \dfrac{1}{n^2}$ with

$p = 2 > 1$. When $x = 5$, the series becomes $\sum_{n=1}^{\infty} \dfrac{(-1)^n}{n^2}$, which converges by the Alternating Series Test. Thus, $I = [-5, 5]$.

41. $\lim\limits_{n \to \infty} \left| \dfrac{a_{n+1}}{a_n} \right| = \lim\limits_{n \to \infty} \left[\dfrac{|x+2|^{n+1}}{(n+1)\,4^{n+1}} \cdot \dfrac{n\,4^n}{|x+2|^n} \right] = \lim\limits_{n \to \infty} \left[\dfrac{n}{n+1} \dfrac{|x+2|}{4} \right] = \dfrac{|x+2|}{4} < 1 \;\Leftrightarrow\; |x+2| < 4$, so $R = 4$.

$|x+2| < 4 \;\Leftrightarrow\; -4 < x+2 < 4 \;\Leftrightarrow\; -6 < x < 2$. If $x = -6$, then the series $\sum_{n=1}^{\infty} \dfrac{(x+2)^n}{n\,4^n}$ becomes

$\sum_{n=1}^{\infty} \dfrac{(-4)^n}{n4^n} = \sum_{n=1}^{\infty} \dfrac{(-1)^n}{n}$, the alternating harmonic series, which converges by the Alternating Series Test. When $x = 2$, the

series becomes the harmonic series $\sum_{n=1}^{\infty} \dfrac{1}{n}$, which diverges. Thus, $I = [-6, 2)$.

42. $\lim\limits_{n \to \infty} \left| \dfrac{a_{n+1}}{a_n} \right| = \lim\limits_{n \to \infty} \left| \dfrac{2^{n+1}\,(x-2)^{n+1}}{(n+3)!} \cdot \dfrac{(n+2)!}{2^n(x-2)^n} \right| = \lim\limits_{n \to \infty} \dfrac{2}{n+3} |x-2| = 0 < 1$, so the series $\sum_{n=1}^{\infty} \dfrac{2^n\,(x-2)^n}{(n+2)!}$

converges for all x. $R = \infty$ and $I = (-\infty, \infty)$.

43. $\lim\limits_{n \to \infty} \left| \dfrac{a_{n+1}}{a_n} \right| = \lim\limits_{n \to \infty} \left| \dfrac{2^{n+1}(x-3)^{n+1}}{\sqrt{n+4}} \cdot \dfrac{\sqrt{n+3}}{2^n(x-3)^n} \right| = 2\,|x-3| \lim\limits_{n \to \infty} \sqrt{\dfrac{n+3}{n+4}} = 2\,|x-3| < 1 \;\Leftrightarrow\; |x-3| < \tfrac{1}{2}$,

so $R = \tfrac{1}{2}$. $|x-3| < \tfrac{1}{2} \;\Leftrightarrow\; -\tfrac{1}{2} < x-3 < \tfrac{1}{2} \;\Leftrightarrow\; \tfrac{5}{2} < x < \tfrac{7}{2}$. For $x = \tfrac{7}{2}$, the series $\sum_{n=1}^{\infty} \dfrac{2^n\,(x-3)^n}{\sqrt{n+3}}$ becomes

$\sum_{n=0}^{\infty} \dfrac{1}{\sqrt{n+3}} = \sum_{n=3}^{\infty} \dfrac{1}{n^{1/2}}$, which diverges $\left[p = \tfrac{1}{2} \le 1 \right]$, but for $x = \tfrac{5}{2}$, we get $\sum_{n=0}^{\infty} \dfrac{(-1)^n}{\sqrt{n+3}}$, which is a convergent

alternating series, so $I = \left[\tfrac{5}{2}, \tfrac{7}{2} \right)$.

44. $\lim\limits_{n\to\infty}\left|\dfrac{a_{n+1}}{a_n}\right| = \lim\limits_{n\to\infty}\left|\dfrac{(2n+2)!\,x^{n+1}}{[(n+1)!]^2}\cdot\dfrac{(n!)^2}{(2n)!\,x^n}\right| = \lim\limits_{n\to\infty}\dfrac{(2n+2)(2n+1)}{(n+1)(n+1)}\,|x| = 4\,|x|.$

To converge, we must have $4\,|x| < 1 \quad\Leftrightarrow\quad |x| < \frac14$, so $R = \frac14$.

45.

n	$f^{(n)}(x)$	$f^{(n)}\left(\frac{\pi}{6}\right)$
0	$\sin x$	$\frac12$
1	$\cos x$	$\frac{\sqrt3}{2}$
2	$-\sin x$	$-\frac12$
3	$-\cos x$	$-\frac{\sqrt3}{2}$
4	$\sin x$	$\frac12$
$\vdots$	$\vdots$	$\vdots$

$\sin x = f\left(\dfrac{\pi}{6}\right) + f'\left(\dfrac{\pi}{6}\right)\left(x - \dfrac{\pi}{6}\right) + \dfrac{f''\left(\frac{\pi}{6}\right)}{2!}\left(x - \dfrac{\pi}{6}\right)^2 + \dfrac{f^{(3)}\left(\frac{\pi}{6}\right)}{3!}\left(x - \dfrac{\pi}{6}\right)^3 + \dfrac{f^{(4)}\left(\frac{\pi}{6}\right)}{4!}\left(x - \dfrac{\pi}{6}\right)^4 + \cdots$

$\qquad = \dfrac12\left[1 - \dfrac{1}{2!}\left(x - \dfrac{\pi}{6}\right)^2 + \dfrac{1}{4!}\left(x - \dfrac{\pi}{6}\right)^4 - \cdots\right] + \dfrac{\sqrt3}{2}\left[\left(x - \dfrac{\pi}{6}\right) - \dfrac{1}{3!}\left(x - \dfrac{\pi}{6}\right)^3 + \cdots\right]$

$\qquad = \dfrac12\displaystyle\sum_{n=0}^{\infty}(-1)^n\dfrac{1}{(2n)!}\left(x - \dfrac{\pi}{6}\right)^{2n} + \dfrac{\sqrt3}{2}\sum_{n=0}^{\infty}(-1)^n\dfrac{1}{(2n+1)!}\left(x - \dfrac{\pi}{6}\right)^{2n+1}$

46.

n	$f^{(n)}(x)$	$f^{(n)}\left(\frac{\pi}{3}\right)$
0	$\cos x$	$\frac12$
1	$-\sin x$	$-\frac{\sqrt3}{2}$
2	$-\cos x$	$-\frac12$
3	$\sin x$	$\frac{\sqrt3}{2}$
4	$\cos x$	$\frac12$
$\vdots$	$\vdots$	$\vdots$

$\cos x = f\left(\dfrac{\pi}{3}\right) + f'\left(\dfrac{\pi}{3}\right)\left(x - \dfrac{\pi}{3}\right) + \dfrac{f''\left(\frac{\pi}{3}\right)}{2!}\left(x - \dfrac{\pi}{3}\right)^2 + \dfrac{f^{(3)}\left(\frac{\pi}{3}\right)}{3!}\left(x - \dfrac{\pi}{3}\right)^3 + \dfrac{f^{(4)}\left(\frac{\pi}{3}\right)}{4!}\left(x - \dfrac{\pi}{3}\right)^4 + \cdots$

$\qquad = \dfrac12\left[1 - \dfrac{1}{2!}\left(x - \dfrac{\pi}{3}\right)^2 + \dfrac{1}{4!}\left(x - \dfrac{\pi}{3}\right)^4 - \cdots\right] + \dfrac{\sqrt3}{2}\left[-\left(x - \dfrac{\pi}{3}\right) + \dfrac{1}{3!}\left(x - \dfrac{\pi}{3}\right)^3 - \cdots\right]$

$\qquad = \dfrac12\displaystyle\sum_{n=0}^{\infty}(-1)^n\dfrac{1}{(2n)!}\left(x - \dfrac{\pi}{3}\right)^{2n} + \dfrac{\sqrt3}{2}\sum_{n=0}^{\infty}(-1)^{n+1}\dfrac{1}{(2n+1)!}\left(x - \dfrac{\pi}{3}\right)^{2n+1}$

47. $\dfrac{1}{1+x} = \dfrac{1}{1-(-x)} = \displaystyle\sum_{n=0}^{\infty}(-x)^n = \sum_{n=0}^{\infty}(-1)^n x^n$ for $|x| < 1 \quad\Rightarrow\quad \dfrac{x^2}{1+x} = \sum_{n=0}^{\infty}(-1)^n x^{n+2}$ with $R = 1$.

48. $\tan^{-1}x = \displaystyle\sum_{n=0}^{\infty}(-1)^n\dfrac{x^{2n+1}}{2n+1}$ with interval of convergence $[-1, 1]$, so

$\tan^{-1}(x^2) = \displaystyle\sum_{n=0}^{\infty}(-1)^n\dfrac{(x^2)^{2n+1}}{2n+1} = \sum_{n=0}^{\infty}(-1)^n\dfrac{x^{4n+2}}{2n+1}$, which converges when $x^2 \in [-1, 1] \quad\Leftrightarrow\quad x \in [-1, 1]$.

Therefore, $R = 1$.

49. $\dfrac{1}{1-x} = \displaystyle\sum_{n=0}^{\infty} x^n$ for $|x| < 1 \quad\Rightarrow\quad \ln(1-x) = -\displaystyle\int \dfrac{dx}{1-x} = -\displaystyle\int \sum_{n=0}^{\infty} x^n\,dx = C - \displaystyle\sum_{n=0}^{\infty} \dfrac{x^{n+1}}{n+1}.$

$\ln(1-0) = C - 0 \quad\Rightarrow\quad C = 0 \quad\Rightarrow\quad \ln(1-x) = -\displaystyle\sum_{n=0}^{\infty}\dfrac{x^{n+1}}{n+1} = \sum_{n=1}^{\infty}\dfrac{-x^n}{n}$ with $R = 1$.

50. $e^x = \displaystyle\sum_{n=0}^{\infty}\dfrac{x^n}{n!} \quad\Rightarrow\quad e^{2x} = \displaystyle\sum_{n=0}^{\infty}\dfrac{(2x)^n}{n!} \quad\Rightarrow\quad xe^{2x} = x\displaystyle\sum_{n=0}^{\infty}\dfrac{2^n\,x^n}{n!} = \displaystyle\sum_{n=0}^{\infty}\dfrac{2^n\,x^{n+1}}{n!},\ R = \infty$

51. $\sin x = \displaystyle\sum_{n=0}^{\infty}\dfrac{(-1)^n\,x^{2n+1}}{(2n+1)!} \quad\Rightarrow\quad \sin(x^4) = \displaystyle\sum_{n=0}^{\infty}\dfrac{(-1)^n\,(x^4)^{2n+1}}{(2n+1)!} = \displaystyle\sum_{n=0}^{\infty}\dfrac{(-1)^n\,x^{8n+4}}{(2n+1)!}$ for all x, so the radius of

convergence is ∞.

52. $e^x = \displaystyle\sum_{n=0}^{\infty}\dfrac{x^n}{n!} \quad\Rightarrow\quad 10^x = e^{(\ln 10)x} = \displaystyle\sum_{n=0}^{\infty}\dfrac{[(\ln 10)x]^n}{n!} = \displaystyle\sum_{n=0}^{\infty}\dfrac{(\ln 10)^n x^n}{n!},\ R = \infty$

53. $f(x) = \dfrac{1}{\sqrt[4]{16-x}} = \dfrac{1}{\sqrt[4]{16(1-x/16)}} = \dfrac{1}{\sqrt[4]{16}\left(1-\frac{1}{16}x\right)^{1/4}} = \tfrac{1}{2}\left(1 - \tfrac{1}{16}x\right)^{-1/4}$

$= \dfrac{1}{2}\left[1 + \left(-\dfrac{1}{4}\right)\left(-\dfrac{x}{16}\right) + \dfrac{\left(-\frac{1}{4}\right)\left(-\frac{5}{4}\right)}{2!}\left(-\dfrac{x}{16}\right)^2 + \dfrac{\left(-\frac{1}{4}\right)\left(-\frac{5}{4}\right)\left(-\frac{9}{4}\right)}{3!}\left(-\dfrac{x}{16}\right)^3 + \cdots\right]$

$= \dfrac{1}{2} + \displaystyle\sum_{n=1}^{\infty}\dfrac{1 \cdot 5 \cdot 9 \cdot\ \cdots\ \cdot (4n-3)}{2 \cdot 4^n \cdot n! \cdot 16^n}\,x^n = \dfrac{1}{2} + \displaystyle\sum_{n=1}^{\infty}\dfrac{1 \cdot 5 \cdot 9 \cdot\ \cdots\ \cdot (4n-3)}{2^{6n+1}\,n!}\,x^n$

for $\left|-\dfrac{x}{16}\right| < 1 \quad\Leftrightarrow\quad |x| < 16$, so $R = 16$.

54. $(1-3x)^{-5} = \displaystyle\sum_{n=0}^{\infty}\binom{-5}{n}(-3x)^n = 1 + (-5)(-3x) + \dfrac{(-5)(-6)}{2!}(-3x)^2 + \dfrac{(-5)(-6)(-7)}{3!}(-3x)^3 + \cdots$

$= 1 + \displaystyle\sum_{n=1}^{\infty}\dfrac{5 \cdot 6 \cdot 7 \cdot\ \cdots\ \cdot (n+4) \cdot 3^n\,x^n}{n!}$ for $|-3x| < 1 \quad\Leftrightarrow\quad |x| < \tfrac{1}{3}$, so $R = \tfrac{1}{3}$.

55. $e^x = \displaystyle\sum_{n=0}^{\infty}\dfrac{x^n}{n!}$, so $\dfrac{e^x}{x} = \dfrac{1}{x}\displaystyle\sum_{n=0}^{\infty}\dfrac{x^n}{n!} = \displaystyle\sum_{n=0}^{\infty}\dfrac{x^{n-1}}{n!} = x^{-1} + \displaystyle\sum_{n=1}^{\infty}\dfrac{x^{n-1}}{n!} = \dfrac{1}{x} + \displaystyle\sum_{n=1}^{\infty}\dfrac{x^{n-1}}{n!}$ and

$\displaystyle\int \dfrac{e^x}{x}\,dx = C + \ln|x| + \displaystyle\sum_{n=1}^{\infty}\dfrac{x^n}{n \cdot n!}.$

56. $(1+x^4)^{1/2} = \displaystyle\sum_{n=0}^{\infty}\binom{\frac{1}{2}}{n}(x^4)^n = 1 + \left(\tfrac{1}{2}\right)x^4 + \dfrac{\left(\frac{1}{2}\right)\left(-\frac{1}{2}\right)}{2!}(x^4)^2 + \dfrac{\left(\frac{1}{2}\right)\left(-\frac{1}{2}\right)\left(-\frac{3}{2}\right)}{3!}(x^4)^3 + \cdots$

$= 1 + \tfrac{1}{2}x^4 - \tfrac{1}{8}x^8 + \tfrac{1}{16}x^{12} - \cdots$

so $\displaystyle\int_0^1 (1+x^4)^{1/2}\,dx = \left[x + \tfrac{1}{10}x^5 - \tfrac{1}{72}x^9 + \tfrac{1}{208}x^{13} - \cdots\right]_0^1 = 1 + \tfrac{1}{10} - \tfrac{1}{72} + \tfrac{1}{208} - \cdots.$

This is an alternating series, so by the Alternating Series Test, the error in the approximation

$\displaystyle\int_0^1 (1+x^4)^{1/2}\,dx \approx 1 + \tfrac{1}{10} - \tfrac{1}{72} \approx 1.086$ is less than $\tfrac{1}{208}$, sufficient for the desired accuracy.

Thus, correct to two decimal places, $\displaystyle\int_0^1 (1+x^4)^{1/2}\,dx \approx 1.09.$

57. (a)

n	$f^{(n)}(x)$	$f^{(n)}(1)$
0	$x^{1/2}$	1
1	$\frac{1}{2}x^{-1/2}$	$\frac{1}{2}$
2	$-\frac{1}{4}x^{-3/2}$	$-\frac{1}{4}$
3	$\frac{3}{8}x^{-5/2}$	$\frac{3}{8}$
4	$-\frac{15}{16}x^{-7/2}$	$-\frac{15}{16}$
$\vdots$	$\vdots$	$\vdots$

$$\sqrt{x} \approx T_3(x) = 1 + \frac{1/2}{1!}(x-1) - \frac{1/4}{2!}(x-1)^2 + \frac{3/8}{3!}(x-1)^3$$

$$= 1 + \tfrac{1}{2}(x-1) - \tfrac{1}{8}(x-1)^2 + \tfrac{1}{16}(x-1)^3$$

(b)

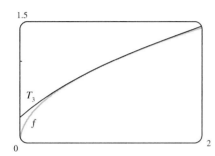

(c) $|R_3(x)| \le \dfrac{M}{4!}|x-1|^4$, where $\left|f^{(4)}(x)\right| \le M$ with

$f^{(4)}(x) = -\frac{15}{16}x^{-7/2}$. Now $0.9 \le x \le 1.1 \;\Rightarrow$

$-0.1 \le x-1 \le 0.1 \;\Rightarrow\; (x-1)^4 \le (0.1)^4$,

and letting $x = 0.9$ gives $M = \dfrac{15}{16(0.9)^{7/2}}$, so

$$|R_3(x)| \le \frac{15}{16(0.9)^{7/2}\,4!}(0.1)^4 \approx 0.000\,005\,648$$

$$\approx 0.000\,006 = 6 \times 10^{-6}$$

(d)

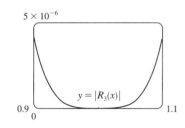

From the graph of $|R_3(x)| = |\sqrt{x} - T_3(x)|$, it appears

that the error is less than 5×10^{-6} on $[0.9, 1.1]$.

58. (a)

n	$f^{(n)}(x)$	$f^{(n)}(0)$
0	$\sec x$	1
1	$\sec x \tan x$	0
2	$\sec x \tan^2 x + \sec^3 x$	1
3	$\sec x \tan^3 x + 5\sec^3 x \tan x$	0
$\vdots$	$\vdots$	$\vdots$

$$\sec x \approx T_2(x) = 1 + \tfrac{1}{2}x^2$$

(b)

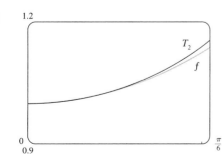

(c) $|R_2(x)| \le \dfrac{M}{3!}|x|^3$, where $\left|f^{(3)}(x)\right| \le M$ with

$f^{(3)}(x) = \sec x \tan^3 x + 5\sec^3 x \tan x$.

Now $0 \le x \le \frac{\pi}{6} \;\Rightarrow\; x^3 \le \left(\frac{\pi}{6}\right)^3$, and letting $x = \frac{\pi}{6}$ gives

$M = \frac{14}{3}$, so $|R_2(x)| \le \dfrac{14}{3 \cdot 6}\left(\dfrac{\pi}{6}\right)^3 \approx 0.111648$.

(d)

0.02

$y = |R_2(x)|$

0

$\frac{\pi}{6}$

From the graph of $|R_2(x)| = |\sec x - T_2(x)|$, it appears that

the error is less than 0.02 on $\left[0, \frac{\pi}{6}\right]$.

59. $\sin x = \displaystyle\sum_{n=0}^{\infty} (-1)^n \dfrac{x^{2n+1}}{(2n+1)!} = x - \dfrac{x^3}{3!} + \dfrac{x^5}{5!} - \dfrac{x^7}{7!} + \cdots$, so $\sin x - x = -\dfrac{x^3}{3!} + \dfrac{x^5}{5!} - \dfrac{x^7}{7!} + \cdots$ and

$\dfrac{\sin x - x}{x^3} = -\dfrac{1}{3!} + \dfrac{x^2}{5!} - \dfrac{x^4}{7!} + \cdots$. Thus, $\displaystyle\lim_{x \to 0} \dfrac{\sin x - x}{x^3} = \lim_{x \to 0} \left(-\dfrac{1}{6} + \dfrac{x^2}{120} - \dfrac{x^4}{5040} + \cdots \right) = -\dfrac{1}{6}$.

60. (a) $F = \dfrac{mgR^2}{(R+h)^2} = \dfrac{mg}{(1+h/R)^2} = mg \displaystyle\sum_{n=0}^{\infty} \binom{-2}{n} \left(\dfrac{h}{R} \right)^n$ [binomial series]

(b) We expand $F = mg \left[1 - 2\,(h/R) + 3\,(h/R)^2 - \cdots \right]$.

This is an alternating series, so by the Alternating Series

Estimation Theorem, the error in the approximation $F = mg$

is less than $2mgh/R$, so for accuracy within 1% we want

$\left| \dfrac{2mgh/R}{mgR^2/(R+h)^2} \right| < 0.01 \quad \Leftrightarrow \quad \dfrac{2h(R+h)^2}{R^3} < 0.01$.

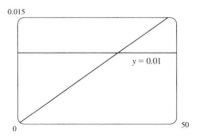

0.015

$y = 0.01$

0

50

This inequality would be difficult to solve for h, so we substitute $R = 6,400$ km and plot both sides of the inequality.

It appears that the approximation is accurate to within 1% for $h < 31$ km.

61. $f(x) = \displaystyle\sum_{n=0}^{\infty} c_n\, x^n \quad \Rightarrow \quad f(-x) = \sum_{n=0}^{\infty} c_n (-x)^n = \sum_{n=0}^{\infty} (-1)^n c_n\, x^n$

(a) If f is an odd function, then $f(-x) = -f(x) \quad \Rightarrow \quad \displaystyle\sum_{n=0}^{\infty} (-1)^n c_n x^n = \sum_{n=0}^{\infty} -c_n x^n$. The coefficients of any power series

are uniquely determined (by Theorem 12.10.5), so $(-1)^n\, c_n = -c_n$.

If n is even, then $(-1)^n = 1$, so $c_n = -c_n \quad \Rightarrow \quad 2c_n = 0 \quad \Rightarrow \quad c_n = 0$. Thus, all even coefficients are 0, that is,

$c_0 = c_2 = c_4 = \cdots = 0$.

(b) If f is even, then $f(-x) = f(x) \quad \Rightarrow \quad \displaystyle\sum_{n=0}^{\infty} (-1)^n c_n\, x^n = \sum_{n=0}^{\infty} c_n\, x^n \quad \Rightarrow \quad (-1)^n\, c_n = c_n$.

If n is odd, then $(-1)^n = -1$, so $-c_n = c_n \quad \Rightarrow \quad 2c_n = 0 \quad \Rightarrow \quad c_n = 0$. Thus, all odd coefficients are 0,

that is, $c_1 = c_3 = c_5 = \cdots = 0$.

62. $e^x = \displaystyle\sum_{n=0}^{\infty} \dfrac{x^n}{n!} \quad \Rightarrow \quad f(x) = e^{x^2} = \sum_{n=0}^{\infty} \dfrac{(x^2)^n}{n!} = \sum_{n=0}^{\infty} \dfrac{x^{2n}}{n!} = \sum_{n=0}^{\infty} \dfrac{1}{n!}\, x^{2n}$. By Theorem 12.10.6 with $a = 0$, we also have

$f(x) = \displaystyle\sum_{k=0}^{\infty} \dfrac{f^{(k)}(0)}{k!}\, x^k$. Comparing coefficients for $k = 2n$, we have $\dfrac{f^{(2n)}(0)}{(2n)!} = \dfrac{1}{n!} \quad \Rightarrow \quad f^{(2n)}(0) = \dfrac{(2n)!}{n!}$.

PROBLEMS PLUS

1. It would be far too much work to compute 15 derivatives of f. The key idea is to remember that $f^{(n)}(0)$ occurs in the coefficient of x^n in the Maclaurin series of f. We start with the Maclaurin series for sin: $\sin x = x - \dfrac{x^3}{3!} + \dfrac{x^5}{5!} - \cdots$.

   Then $\sin(x^3) = x^3 - \dfrac{x^9}{3!} + \dfrac{x^{15}}{5!} - \cdots$, and so the coefficient of x^{15} is $\dfrac{f^{(15)}(0)}{15!} = \dfrac{1}{5!}$. Therefore,

   $$f^{(15)}(0) = \frac{15!}{5!} = 6 \cdot 7 \cdot 8 \cdot 9 \cdot 10 \cdot 11 \cdot 12 \cdot 13 \cdot 14 \cdot 15 = 10{,}897{,}286{,}400.$$

2. We use the problem-solving strategy of taking cases:

   Case (i):   If $|x| < 1$, then $0 \le x^2 < 1$, so $\displaystyle\lim_{n \to \infty} x^{2n} = 0$ [see Example 10 in Section 12.1]

   and $f(x) = \displaystyle\lim_{n \to \infty} \frac{x^{2n} - 1}{x^{2n} + 1} = \frac{0 - 1}{0 + 1} = -1.$

   Case (ii):   If $|x| = 1$, that is, $x = \pm 1$, then $x^2 = 1$, so $f(x) = \displaystyle\lim_{n \to \infty} \frac{x^{2n} - 1}{x^{2n} + 1} = \lim_{n \to \infty} \frac{1 - 1}{1 + 1} = 0.$

   Case (iii):   If $|x| > 1$, then $x^2 > 1$, so $\displaystyle\lim_{n \to \infty} x^{2n} = \infty$ and $f(x) = \displaystyle\lim_{n \to \infty} \frac{x^{2n} - 1}{x^{2n} + 1} = \lim_{n \to \infty} \frac{1 - (1/x^{2n})}{1 + (1/x^{2n})} = \frac{1 - 0}{1 + 0} = 1.$

   Thus, $f(x) = \begin{cases} 1 & \text{if } x < -1 \\ 0 & \text{if } x = -1 \\ -1 & \text{if } -1 < x < 1 \\ 0 & \text{if } x = 1 \\ 1 & \text{if } x > 1 \end{cases}$

   The graph shows that f is continuous everywhere except at $x = \pm 1$.

3. (a) From Formula 14a in Appendix D, with $x = y = \theta$, we get $\tan 2\theta = \dfrac{2 \tan \theta}{1 - \tan^2 \theta}$, so $\cot 2\theta = \dfrac{1 - \tan^2 \theta}{2 \tan \theta}$ $\Rightarrow$

   $$2 \cot 2\theta = \frac{1 - \tan^2 \theta}{\tan \theta} = \cot \theta - \tan \theta.$$ Replacing θ by $\frac{1}{2}x$, we get $2 \cot x = \cot \frac{1}{2}x - \tan \frac{1}{2}x$, or

   $$\tan \tfrac{1}{2}x = \cot \tfrac{1}{2}x - 2 \cot x.$$

   (b) From part (a) with $\dfrac{x}{2^{n-1}}$ in place of x, $\tan \dfrac{x}{2^n} = \cot \dfrac{x}{2^n} - 2 \cot \dfrac{x}{2^{n-1}}$, so the nth partial sum of $\displaystyle\sum_{n=1}^{\infty} \frac{1}{2^n} \tan \frac{x}{2^n}$ is

   $$
   \begin{aligned}
   s_n &= \frac{\tan(x/2)}{2} + \frac{\tan(x/4)}{4} + \frac{\tan(x/8)}{8} + \cdots + \frac{\tan(x/2^n)}{2^n} \\
   &= \left[\frac{\cot(x/2)}{2} - \cot x\right] + \left[\frac{\cot(x/4)}{4} - \frac{\cot(x/2)}{2}\right] + \left[\frac{\cot(x/8)}{8} - \frac{\cot(x/4)}{4}\right] + \cdots \\
   &\quad + \left[\frac{\cot(x/2^n)}{2^n} - \frac{\cot(x/2^{n-1})}{2^{n-1}}\right] = -\cot x + \frac{\cot(x/2^n)}{2^n} \quad \text{[telescoping sum]}
   \end{aligned}
   $$

   Now $\dfrac{\cot(x/2^n)}{2^n} = \dfrac{\cos(x/2^n)}{2^n \sin(x/2^n)} = \dfrac{\cos(x/2^n)}{x} \cdot \dfrac{x/2^n}{\sin(x/2^n)} \to \dfrac{1}{x} \cdot 1 = \dfrac{1}{x}$ as $n \to \infty$ since $x/2^n \to 0$

for $x \neq 0$. Therefore, if $x \neq 0$ and $x \neq k\pi$ where k is any integer, then

$$\sum_{n=1}^{\infty} \frac{1}{2^n} \tan \frac{x}{2^n} = \lim_{n \to \infty} s_n = \lim_{n \to \infty} \left(-\cot x + \frac{1}{2^n} \cot \frac{x}{2^n} \right) = -\cot x + \frac{1}{x}$$

If $x = 0$, then all terms in the series are 0, so the sum is 0.

4. $|AP_2|^2 = 2$, $|AP_3|^2 = 2 + 2^2$, $|AP_4|^2 = 2 + 2^2 + (2^2)^2$, $|AP_5|^2 = 2 + 2^2 + (2^2)^2 + (2^3)^2, \ldots$,

$|AP_n|^2 = 2 + 2^2 + (2^2)^2 + \cdots + (2^{n-2})^2$ [for $n \geq 3$] $= 2 + (4 + 4^2 + 4^3 + \cdots + 4^{n-2})$

$\qquad = 2 + \dfrac{4(4^{n-2} - 1)}{4 - 1}$ [finite geometric sum with $a = 4, r = 4$] $= \dfrac{6}{3} + \dfrac{4^{n-1} - 4}{3} = \dfrac{2}{3} + \dfrac{4^{n-1}}{3}$

So $\tan \angle P_n A P_{n+1} = \dfrac{|P_n P_{n+1}|}{|AP_n|} = \dfrac{2^{n-1}}{\sqrt{\dfrac{2}{3} + \dfrac{4^{n-1}}{3}}} = \dfrac{\sqrt{4^{n-1}}}{\sqrt{\dfrac{2}{3} + \dfrac{4^{n-1}}{3}}} = \dfrac{1}{\sqrt{\dfrac{2}{3 \cdot 4^{n-1}} + \dfrac{1}{3}}} \to \sqrt{3}$ as $n \to \infty$.

Thus, $\angle P_n A P_{n+1} \to \frac{\pi}{3}$ as $n \to \infty$.

5. (a) At each stage, each side is replaced by four shorter sides, each of length $\frac{1}{3}$ of the side length at the preceding stage. Writing s_0 and ℓ_0 for the number of sides and the length of the side of the initial triangle, we generate the table at right. In general, we have $s_n = 3 \cdot 4^n$ and $\ell_n = \left(\frac{1}{3}\right)^n$, so the length of the perimeter at the nth stage of construction is $p_n = s_n \ell_n = 3 \cdot 4^n \cdot \left(\frac{1}{3}\right)^n = 3 \cdot \left(\frac{4}{3}\right)^n$.

$s_0 = 3$	$\ell_0 = 1$
$s_1 = 3 \cdot 4$	$\ell_1 = 1/3$
$s_2 = 3 \cdot 4^2$	$\ell_2 = 1/3^2$
$s_3 = 3 \cdot 4^3$	$\ell_3 = 1/3^3$
$\vdots$	$\vdots$

(b) $p_n = \dfrac{4^n}{3^{n-1}} = 4\left(\dfrac{4}{3}\right)^{n-1}$. Since $\frac{4}{3} > 1$, $p_n \to \infty$ as $n \to \infty$.

(c) The area of each of the small triangles added at a given stage is one-ninth of the area of the triangle added at the preceding stage. Let a be the area of the original triangle. Then the area a_n of each of the small triangles added at stage n is

$a_n = a \cdot \dfrac{1}{9^n} = \dfrac{a}{9^n}$. Since a small triangle is added to each side at every stage, it follows that the total area A_n added to the

figure at the nth stage is $A_n = s_{n-1} \cdot a_n = 3 \cdot 4^{n-1} \cdot \dfrac{a}{9^n} = a \cdot \dfrac{4^{n-1}}{3^{2n-1}}$. Then the total area enclosed by the snowflake

curve is $A = a + A_1 + A_2 + A_3 + \cdots = a + a \cdot \dfrac{1}{3} + a \cdot \dfrac{4}{3^3} + a \cdot \dfrac{4^2}{3^5} + a \cdot \dfrac{4^3}{3^7} + \cdots$. After the first term, this is a

geometric series with common ratio $\dfrac{4}{9}$, so $A = a + \dfrac{a/3}{1 - \frac{4}{9}} = a + \dfrac{a}{3} \cdot \dfrac{9}{5} = \dfrac{8a}{5}$. But the area of the original equilateral

triangle with side 1 is $a = \dfrac{1}{2} \cdot 1 \cdot \sin \dfrac{\pi}{3} = \dfrac{\sqrt{3}}{4}$. So the area enclosed by the snowflake curve is $\dfrac{8}{5} \cdot \dfrac{\sqrt{3}}{4} = \dfrac{2\sqrt{3}}{5}$.

6. Let the series $S = 1 + \frac{1}{2} + \frac{1}{3} + \frac{1}{4} + \frac{1}{6} + \frac{1}{8} + \frac{1}{9} + \frac{1}{12} + \cdots$. Then every term in S is of the form $\dfrac{1}{2^m 3^n}$, $m, n \geq 0$, and

furthermore each term occurs only once. So we can write

$$S = \sum_{m=0}^{\infty} \sum_{n=0}^{\infty} \frac{1}{2^m 3^n} = \sum_{m=0}^{\infty} \sum_{n=0}^{\infty} \frac{1}{2^m} \frac{1}{3^n} = \sum_{m=0}^{\infty} \frac{1}{2^m} \sum_{n=0}^{\infty} \frac{1}{3^n} = \frac{1}{1 - \frac{1}{2}} \cdot \frac{1}{1 - \frac{1}{3}} = 2 \cdot \frac{3}{2} = 3$$

7. (a) Let $a = \arctan x$ and $b = \arctan y$. Then, from Formula 14b in Appendix D,

$$\tan(a - b) = \frac{\tan a - \tan b}{1 + \tan a \, \tan b} = \frac{\tan(\arctan x) - \tan(\arctan y)}{1 + \tan(\arctan x)\tan(\arctan y)} = \frac{x - y}{1 + xy}$$

Now $\arctan x - \arctan y = a - b = \arctan(\tan(a - b)) = \arctan \dfrac{x - y}{1 + xy}$ since $-\dfrac{\pi}{2} < a - b < \dfrac{\pi}{2}$.

(b) From part (a) we have

$$\arctan \tfrac{120}{119} - \arctan \tfrac{1}{239} = \arctan \frac{\frac{120}{119} - \frac{1}{239}}{1 + \frac{120}{119} \cdot \frac{1}{239}} = \arctan \frac{\frac{28{,}561}{28{,}441}}{\frac{28{,}561}{28{,}441}} = \arctan 1 = \tfrac{\pi}{4}$$

(c) Replacing y by $-y$ in the formula of part (a), we get $\arctan x + \arctan y = \arctan \dfrac{x + y}{1 - xy}$. So

$$4 \arctan \tfrac{1}{5} = 2\left(\arctan \tfrac{1}{5} + \arctan \tfrac{1}{5}\right) = 2 \arctan \frac{\frac{1}{5} + \frac{1}{5}}{1 - \frac{1}{5} \cdot \frac{1}{5}} = 2 \arctan \tfrac{5}{12} = \arctan \tfrac{5}{12} + \arctan \tfrac{5}{12}$$

$$= \arctan \frac{\frac{5}{12} + \frac{5}{12}}{1 - \frac{5}{12} \cdot \frac{5}{12}} = \arctan \tfrac{120}{119}$$

Thus, from part (b), we have $4 \arctan \tfrac{1}{5} - \arctan \tfrac{1}{239} = \arctan \tfrac{120}{119} - \arctan \tfrac{1}{239} = \tfrac{\pi}{4}$.

(d) From Example 7 in Section 12.9 we have $\arctan x = x - \dfrac{x^3}{3} + \dfrac{x^5}{5} - \dfrac{x^7}{7} + \dfrac{x^9}{9} - \dfrac{x^{11}}{11} + \cdots$, so

$$\arctan \frac{1}{5} = \frac{1}{5} - \frac{1}{3 \cdot 5^3} + \frac{1}{5 \cdot 5^5} - \frac{1}{7 \cdot 5^7} + \frac{1}{9 \cdot 5^9} - \frac{1}{11 \cdot 5^{11}} + \cdots$$

This is an alternating series and the size of the terms decreases to 0, so by the Alternating Series Estimation Theorem, the sum lies between s_5 and s_6, that is, $0.197395560 < \arctan \tfrac{1}{5} < 0.197395562$.

(e) From the series in part (d) we get $\arctan \dfrac{1}{239} = \dfrac{1}{239} - \dfrac{1}{3 \cdot 239^3} + \dfrac{1}{5 \cdot 239^5} - \cdots$. The third term is less than 2.6×10^{-13}, so by the Alternating Series Estimation Theorem, we have, to nine decimal places, $\arctan \tfrac{1}{239} \approx s_2 \approx 0.004184076$. Thus, $0.004184075 < \arctan \tfrac{1}{239} < 0.004184077$.

(f) From part (c) we have $\pi = 16 \arctan \tfrac{1}{5} - 4 \arctan \tfrac{1}{239}$, so from parts (d) and (e) we have

$$16(0.197395560) - 4(0.004184077) < \pi < 16(0.197395562) - 4(0.004184075) \quad \Rightarrow$$

$3.141592652 < \pi < 3.141592692$. So, to 7 decimal places, $\pi \approx 3.1415927$.

8. (a) Let $a = \operatorname{arccot} x$ and $b = \operatorname{arccot} y$ where $0 < a - b < \pi$. Then

$$\cot(a - b) = \frac{1}{\tan(a - b)} = \frac{1 + \tan a \, \tan b}{\tan a - \tan b} = \frac{\frac{1}{\cot a} \cdot \frac{1}{\cot b} + 1}{\frac{1}{\cot a} - \frac{1}{\cot b}} \cdot \frac{\cot a \, \cot b}{\cot a \, \cot b}$$

$$= \frac{1 + \cot a \, \cot b}{\cot b - \cot a} = \frac{1 + \cot(\operatorname{arccot} x)\cot(\operatorname{arccot} y)}{\cot(\operatorname{arccot} y) - \cot(\operatorname{arccot} x)} = \frac{1 + xy}{y - x}$$

Now $\operatorname{arccot} x - \operatorname{arccot} y = a - b = \operatorname{arccot}(\cot(a - b)) = \operatorname{arccot} \dfrac{1 + xy}{y - x}$ since $0 < a - b < \pi$.

(b) Applying the identity in part (a) with $x = n$ and $y = n + 1$, we have

$$\operatorname{arccot}(n^2 + n + 1) = \operatorname{arccot}(1 + n(n+1)) = \operatorname{arccot}\frac{1 + n(n+1)}{(n+1) - n} = \operatorname{arccot} n - \operatorname{arccot}(n+1)$$

Thus, we have a telescoping series with nth partial sum

$$s_n = [\operatorname{arccot} 0 - \operatorname{arccot} 1] + [\operatorname{arccot} 1 - \operatorname{arccot} 2] + \cdots + [\operatorname{arccot} n - \operatorname{arccot}(n+1)] = \operatorname{arccot} 0 - \operatorname{arccot}(n+1).$$

Thus, $\displaystyle\sum_{n=0}^{\infty} \operatorname{arccot}(n^2 + n + 1) = \lim_{n \to \infty} s_n = \lim_{n \to \infty} [\operatorname{arccot} 0 - \operatorname{arccot}(n+1)] = \frac{\pi}{2} - 0 = \frac{\pi}{2}.$

9. We start with the geometric series $\displaystyle\sum_{n=0}^{\infty} x^n = \frac{1}{1-x}$, $|x| < 1$, and differentiate:

$$\sum_{n=1}^{\infty} nx^{n-1} = \frac{d}{dx}\left(\sum_{n=0}^{\infty} x^n\right) = \frac{d}{dx}\left(\frac{1}{1-x}\right) = \frac{1}{(1-x)^2} \text{ for } |x| < 1 \quad \Rightarrow \quad \sum_{n=1}^{\infty} nx^n = x\sum_{n=1}^{\infty} nx^{n-1} = \frac{x}{(1-x)^2}$$

for $|x| < 1$. Differentiate again:

$$\sum_{n=1}^{\infty} n^2 x^{n-1} = \frac{d}{dx}\frac{x}{(1-x)^2} = \frac{(1-x)^2 - x \cdot 2(1-x)(-1)}{(1-x)^4} = \frac{x+1}{(1-x)^3} \quad \Rightarrow \quad \sum_{n=1}^{\infty} n^2 x^n = \frac{x^2 + x}{(1-x)^3} \quad \Rightarrow$$

$$\sum_{n=1}^{\infty} n^3 x^{n-1} = \frac{d}{dx}\frac{x^2 + x}{(1-x)^3} = \frac{(1-x)^3(2x+1) - (x^2 + x)3(1-x)^2(-1)}{(1-x)^6} = \frac{x^2 + 4x + 1}{(1-x)^4} \quad \Rightarrow$$

$$\sum_{n=1}^{\infty} n^3 x^n = \frac{x^3 + 4x^2 + x}{(1-x)^4}, \; |x| < 1.$$ The radius of convergence is 1 because that is the radius of convergence for the

geometric series we started with. If $x = \pm 1$, the series is $\sum n^3 (\pm 1)^n$, which diverges by the Test For Divergence, so the interval of convergence is $(-1, 1)$.

10. Let's first try the case $k = 1$: $a_0 + a_1 = 0 \quad \Rightarrow \quad a_1 = -a_0 \quad \Rightarrow$

$$\lim_{n \to \infty} \left(a_0\sqrt{n} + a_1\sqrt{n+1}\right) = \lim_{n \to \infty} \left(a_0\sqrt{n} - a_0\sqrt{n+1}\right) = a_0 \lim_{n \to \infty} \left(\sqrt{n} - \sqrt{n+1}\right)\frac{\sqrt{n} + \sqrt{n+1}}{\sqrt{n} + \sqrt{n+1}}$$

$$= a_0 \lim_{n \to \infty} \frac{-1}{\sqrt{n} + \sqrt{n+1}} = 0$$

In general we have $a_0 + a_1 + \cdots + a_k = 0 \quad \Rightarrow \quad a_k = -a_0 - a_1 - \cdots - a_{k-1} \quad \Rightarrow$

$$\lim_{n \to \infty} \left(a_0\sqrt{n} + a_1\sqrt{n+1} + a_2\sqrt{n+2} + \cdots + a_k\sqrt{n+k}\right)$$

$$= \lim_{n \to \infty} \left(a_0\sqrt{n} + a_1\sqrt{n+1} + \cdots + a_{k-1}\sqrt{n+k-1} - a_0\sqrt{n+k} - a_1\sqrt{n+k} - \cdots - a_{k-1}\sqrt{n+k}\right)$$

$$= a_0 \lim_{n \to \infty} \left(\sqrt{n} - \sqrt{n+k}\right) + a_1 \lim_{n \to \infty} \left(\sqrt{n+1} - \sqrt{n+k}\right) + \cdots + a_{k-1} \lim_{n \to \infty} \left(\sqrt{n+k-1} - \sqrt{n+k}\right)$$

Each of these limits is 0 by the same type of simplification as in the case $k = 1$. So we have

$$\lim_{n \to \infty} \left(a_0\sqrt{n} + a_1\sqrt{n+1} + a_2\sqrt{n+2} + \cdots + a_k\sqrt{n+k}\right) = a_0(0) + a_1(0) + \cdots + a_{k-1}(0) = 0$$

11. $\ln\left(1 - \dfrac{1}{n^2}\right) = \ln\left(\dfrac{n^2 - 1}{n^2}\right) = \ln\dfrac{(n+1)(n-1)}{n^2} = \ln[(n+1)(n-1)] - \ln n^2$

$$= \ln(n+1) + \ln(n-1) - 2\ln n = \ln(n-1) - \ln n - \ln n + \ln(n+1)$$

$$= \ln\frac{n-1}{n} - [\ln n - \ln(n+1)] = \ln\frac{n-1}{n} - \ln\frac{n}{n+1}.$$

Let $s_k = \sum_{n=2}^{k} \ln\left(1 - \frac{1}{n^2}\right) = \sum_{n=2}^{k}\left(\ln\frac{n-1}{n} - \ln\frac{n}{n+1}\right)$ for $k \geq 2$. Then

$$s_k = \left(\ln\frac{1}{2} - \ln\frac{2}{3}\right) + \left(\ln\frac{2}{3} - \ln\frac{3}{4}\right) + \cdots + \left(\ln\frac{k-1}{k} - \ln\frac{k}{k+1}\right) = \ln\frac{1}{2} - \ln\frac{k}{k+1}, \text{ so}$$

$$\sum_{n=2}^{\infty} \ln\left(1 - \frac{1}{n^2}\right) = \lim_{k\to\infty} s_k = \lim_{k\to\infty}\left(\ln\frac{1}{2} - \ln\frac{k}{k+1}\right) = \ln\frac{1}{2} - \ln 1 = \ln 1 - \ln 2 - \ln 1 = -\ln 2.$$

12. Place the y-axis as shown and let the length of each book be L. We want to show that the center of mass of the system of n books lies above the table, that is, $\bar{x} < L$. The x-coordinates of the centers of mass of the books are

$$x_1 = \frac{L}{2}, x_2 = \frac{L}{2(n-1)} + \frac{L}{2}, x_3 = \frac{L}{2(n-1)} + \frac{L}{2(n-2)} + \frac{L}{2}, \text{ and so on.}$$

Each book has the same mass m, so if there are n books, then

$$\bar{x} = \frac{mx_1 + mx_2 + \cdots + mx_n}{mn} = \frac{x_1 + x_2 + \cdots + x_n}{n}$$

$$= \frac{1}{n}\left[\frac{L}{2} + \left(\frac{L}{2(n-1)} + \frac{L}{2}\right) + \left(\frac{L}{2(n-1)} + \frac{L}{2(n-2)} + \frac{L}{2}\right) + \cdots \right.$$

$$\left. + \left(\frac{L}{2(n-1)} + \frac{L}{2(n-2)} + \cdots + \frac{L}{4} + \frac{L}{2} + \frac{L}{2}\right)\right]$$

$$= \frac{L}{n}\left[\frac{n-1}{2(n-1)} + \frac{n-2}{2(n-2)} + \cdots + \frac{2}{4} + \frac{1}{2} + \frac{n}{2}\right] = \frac{L}{n}\left[(n-1)\frac{1}{2} + \frac{n}{2}\right] = \frac{2n-1}{2n}L < L$$

This shows that, no matter how many books are added according to the given scheme, the center of mass lies above the table. It remains to observe that the series $\frac{1}{2} + \frac{1}{4} + \frac{1}{6} + \frac{1}{8} + \cdots = \frac{1}{2}\sum(1/n)$ is divergent (harmonic series), so we can make the top book extend as far as we like beyond the edge of the table if we add enough books.

13. (a)

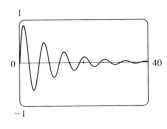

The x-intercepts of the curve occur where $\sin x = 0 \Leftrightarrow x = n\pi$, n an integer. So using the formula for disks (and either a CAS or $\sin^2 x = \frac{1}{2}(1 - \cos 2x)$ and Formula 99 to evaluate the integral), the volume of the nth bead is

$$V_n = \pi \int_{(n-1)\pi}^{n\pi} \left(e^{-x/10}\sin x\right)^2 dx = \pi \int_{(n-1)\pi}^{n\pi} e^{-x/5}\sin^2 x\, dx$$

$$= \frac{250\pi}{101}\left(e^{-(n-1)\pi/5} - e^{-n\pi/5}\right)$$

(b) The total volume is

$$\pi \int_0^\infty e^{-x/5}\sin^2 x\, dx = \sum_{n=1}^{\infty} V_n = \frac{250\pi}{101}\sum_{n=1}^{\infty}\left[e^{-(n-1)\pi/5} - e^{-n\pi/5}\right] = \frac{250\pi}{101} \quad \text{[telescoping sum]}.$$

Another method: If the volume in part (a) has been written as $V_n = \frac{250\pi}{101}e^{-n\pi/5}\left(e^{\pi/5} - 1\right)$, then we recognize $\sum_{n=1}^{\infty} V_n$

as a geometric series with $a = \frac{250\pi}{101}\left(1 - e^{-\pi/5}\right)$ and $r = e^{-\pi/5}$.

14. First notice that both series are absolutely convergent (p-series with $p > 1$.) Let the given expression be called x. Then

$$x = \frac{1 + \frac{1}{2^p} + \frac{1}{3^p} + \frac{1}{4^p} + \cdots}{1 - \frac{1}{2^p} + \frac{1}{3^p} - \frac{1}{4^p} + \cdots} = \frac{1 + \left(2 \cdot \frac{1}{2^p} - \frac{1}{2^p}\right) + \frac{1}{3^p} + \left(2 \cdot \frac{1}{4^p} - \frac{1}{4^p}\right) + \cdots}{1 - \frac{1}{2^p} + \frac{1}{3^p} - \frac{1}{4^p} + \cdots}$$

$$= \frac{\left(1 - \frac{1}{2^p} + \frac{1}{3^p} - \frac{1}{4^p} + \cdots\right) + \left(2 \cdot \frac{1}{2^p} + 2 \cdot \frac{1}{4^p} + 2 \cdot \frac{1}{6^p} + \cdots\right)}{1 - \frac{1}{2^p} + \frac{1}{3^p} - \frac{1}{4^p} + \cdots}$$

$$= 1 + \frac{2\left(\frac{1}{2^p} + \frac{1}{4^p} + \frac{1}{6^p} + \frac{1}{8^p} + \cdots\right)}{1 - \frac{1}{2^p} + \frac{1}{3^p} - \frac{1}{4^p} + \cdots} = 1 + \frac{\frac{1}{2^{p-1}}\left(1 + \frac{1}{2^p} + \frac{1}{3^p} + \frac{1}{4^p} + \cdots\right)}{1 - \frac{1}{2^p} + \frac{1}{3^p} - \frac{1}{4^p} + \cdots} = 1 + 2^{1-p}x$$

Therefore, $x = 1 + 2^{1-p}x \iff x - 2^{1-p}x = 1 \iff x(1 - 2^{1-p}) = 1 \iff x = \dfrac{1}{1 - 2^{1-p}}$.

15. If L is the length of a side of the equilateral triangle, then the area is $A = \frac{1}{2}L \cdot \frac{\sqrt{3}}{2}L = \frac{\sqrt{3}}{4}L^2$ and so $L^2 = \frac{4}{\sqrt{3}}A$.

Let r be the radius of one of the circles. When there are n rows of circles, the figure shows that

$$L = \sqrt{3}\,r + r + (n-2)(2r) + r + \sqrt{3}\,r = r\left(2n - 2 + 2\sqrt{3}\right), \text{ so } r = \frac{L}{2\left(n + \sqrt{3} - 1\right)}.$$

The number of circles is $1 + 2 + \cdots + n = \dfrac{n(n+1)}{2}$, and so the total area of the circles is

$$A_n = \frac{n(n+1)}{2}\pi r^2 = \frac{n(n+1)}{2}\pi \frac{L^2}{4\left(n + \sqrt{3} - 1\right)^2}$$

$$= \frac{n(n+1)}{2}\pi \frac{4A/\sqrt{3}}{4\left(n + \sqrt{3} - 1\right)^2} = \frac{n(n+1)}{\left(n + \sqrt{3} - 1\right)^2}\frac{\pi A}{2\sqrt{3}} \Rightarrow$$

$$\frac{A_n}{A} = \frac{n(n+1)}{\left(n + \sqrt{3} - 1\right)^2}\frac{\pi}{2\sqrt{3}}$$

$$= \frac{1 + 1/n}{\left[1 + (\sqrt{3} - 1)/n\right]^2}\frac{\pi}{2\sqrt{3}} \to \frac{\pi}{2\sqrt{3}} \text{ as } n \to \infty$$

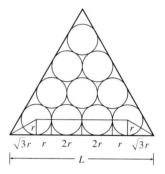

16. Given $a_0 = a_1 = 1$ and $a_n = \dfrac{(n-1)(n-2)a_{n-1} - (n-3)a_{n-2}}{n(n-1)}$, we calculate the next few terms of the sequence:

$$a_2 = \frac{1 \cdot 0 \cdot a_1 - (-1)a_0}{2 \cdot 1} = \frac{1}{2}, \ a_3 = \frac{2 \cdot 1 \cdot a_2 - 0 \cdot a_1}{3 \cdot 2} = \frac{1}{6}, \ a_4 = \frac{3 \cdot 2 \cdot a_3 - 1 \cdot a_2}{4 \cdot 3} = \frac{1}{24}.$$

It seems that $a_n = \dfrac{1}{n!}$, so we try to prove this by induction. The first step is done, so assume $a_k = \dfrac{1}{k!}$ and $a_{k-1} = \dfrac{1}{(k-1)!}$.

Then $a_{k+1} = \dfrac{k(k-1)a_k - (k-2)a_{k-1}}{(k+1)k} = \dfrac{\dfrac{k(k-1)}{k!} - \dfrac{k-2}{(k-1)!}}{(k+1)k} = \dfrac{(k-1) - (k-2)}{[(k+1)(k)](k-1)!} = \dfrac{1}{(k+1)!}$ and the induction

is complete. Therefore, $\displaystyle\sum_{n=0}^{\infty} a_n = \sum_{n=0}^{\infty} \frac{1}{n!} = e$.

17. As in Section 12.9 we have to integrate the function x^x by integrating series. Writing $x^x = (e^{\ln x})^x = e^{x \ln x}$ and using the

Maclaurin series for e^x, we have $x^x = (e^{\ln x})^x = e^{x \ln x} = \sum_{n=0}^{\infty} \dfrac{(x \ln x)^n}{n!} = \sum_{n=0}^{\infty} \dfrac{x^n (\ln x)^n}{n!}$. As with power series, we can

integrate this series term-by-term: $\displaystyle\int_0^1 x^x \, dx = \sum_{n=0}^{\infty} \int_0^1 \dfrac{x^n (\ln x)^n}{n!} \, dx = \sum_{n=0}^{\infty} \dfrac{1}{n!} \int_0^1 x^n (\ln x)^n \, dx$. We integrate by parts

with $u = (\ln x)^n$, $dv = x^n \, dx$, so $du = \dfrac{n(\ln x)^{n-1}}{x} \, dx$ and $v = \dfrac{x^{n+1}}{n+1}$:

$$\int_0^1 x^n (\ln x)^n \, dx = \lim_{t \to 0^+} \int_t^1 x^n (\ln x)^n \, dx = \lim_{t \to 0^+} \left[\dfrac{x^{n+1}}{n+1} (\ln x)^n \right]_t^1 - \lim_{t \to 0^+} \int_t^1 \dfrac{n}{n+1} x^n (\ln x)^{n-1} \, dx$$

$$= 0 - \dfrac{n}{n+1} \int_0^1 x^n (\ln x)^{n-1} \, dx$$

(where l'Hospital's Rule was used to help evaluate the first limit). Further integration by parts gives

$$\int_0^1 x^n (\ln x)^k \, dx = -\dfrac{k}{n+1} \int_0^1 x^n (\ln x)^{k-1} \, dx \text{ and, combining these steps, we get}$$

$$\int_0^1 x^n (\ln x)^n \, dx = \dfrac{(-1)^n \, n!}{(n+1)^n} \int_0^1 x^n \, dx = \dfrac{(-1)^n \, n!}{(n+1)^{n+1}} \quad \Rightarrow$$

$$\int_0^1 x^x \, dx = \sum_{n=0}^{\infty} \dfrac{1}{n!} \int_0^1 x^n (\ln x)^n \, dx = \sum_{n=0}^{\infty} \dfrac{1}{n!} \dfrac{(-1)^n \, n!}{(n+1)^{n+1}} = \sum_{n=0}^{\infty} \dfrac{(-1)^n}{(n+1)^{n+1}} = \sum_{n=1}^{\infty} \dfrac{(-1)^{n-1}}{n^n}.$$

18. **(a)** Since P_n is defined as the midpoint of $P_{n-4}P_{n-3}$, $x_n = \frac{1}{2}(x_{n-4} + x_{n-3})$ for $n \geq 5$. So we prove by induction that

$\frac{1}{2}x_n + x_{n+1} + x_{n+2} + x_{n+3} = 2$. The case $n = 1$ is immediate, since $\frac{1}{2} \cdot 0 + 1 + 1 + 0 = 2$. Assume that the result

holds for $n = k - 1$, that is, $\frac{1}{2}x_{k-1} + x_k + x_{k+1} + x_{k+2} = 2$. Then for $n = k$,

$$\frac{1}{2}x_k + x_{k+1} + x_{k+2} + x_{k+3} = \frac{1}{2}x_k + x_{k+1} + x_{k+2} + \frac{1}{2}(x_{k+3-4} + x_{k+3-3}) \quad \text{[by above]}$$

$$= \frac{1}{2}x_{k-1} + x_k + x_{k+1} + x_{k+2} = 2 \quad \text{[by the induction hypothesis]}$$

Similarly, for $n \geq 5$, $y_n = \frac{1}{2}(y_{n-4} + y_{n-3})$, so the same argument as above holds for y, with 2 replaced by

$\frac{1}{2}y_1 + y_2 + y_3 + y_4 = \frac{1}{2} \cdot 1 + 1 + 0 + 0 = \frac{3}{2}$. So $\frac{1}{2}y_n + y_{n+1} + y_{n+2} + y_{n+3} = \frac{3}{2}$ for all n.

(b) $\displaystyle\lim_{n \to \infty} \left(\frac{1}{2}x_n + x_{n+1} + x_{n+2} + x_{n+3} \right) = \frac{1}{2} \lim_{n \to \infty} x_n + \lim_{n \to \infty} x_{n+1} + \lim_{n \to \infty} x_{n+2} + \lim_{n \to \infty} x_{n+3} = 2$. Since all

the limits on the left hand side are the same, we get $\frac{7}{2} \lim_{n \to \infty} x_n = 2 \quad \Rightarrow \quad \lim_{n \to \infty} x_n = \frac{4}{7}$. In the same way,

$\frac{7}{2} \lim_{n \to \infty} y_n = \frac{3}{2} \quad \Rightarrow \quad \lim_{n \to \infty} y_n = \frac{3}{7}$, so $P = \left(\frac{4}{7}, \frac{3}{7} \right)$.

19. Let $f(x) = \sum_{m=0}^{\infty} c_m x^m$ and $g(x) = e^{f(x)} = \sum_{n=0}^{\infty} d_n x^n$. Then $g'(x) = \sum_{n=0}^{\infty} n d_n x^{n-1}$, so $n d_n$ occurs as the coefficient

of x^{n-1}. But also

$$g'(x) = e^{f(x)} f'(x) = \left(\sum_{n=0}^{\infty} d_n x^n \right) \left(\sum_{m=1}^{\infty} m c_m x^{m-1} \right)$$

$$= \left(d_0 + d_1 x + d_2 x^2 + \cdots + d_{n-1} x^{n-1} + \cdots \right) \left(c_1 + 2c_2 x + 3c_3 x^2 + \cdots + n c_n x^{n-1} + \cdots \right)$$

so the coefficient of x^{n-1} is $c_1 d_{n-1} + 2c_2 d_{n-2} + 3c_3 d_{n-3} + \cdots + n c_n d_0 = \sum_{i=1}^{n} i c_i d_{n-i}$. Therefore, $n d_n = \sum_{i=1}^{n} i c_i d_{n-i}$.

20. Suppose the base of the first right triangle has length a. Then by repeated use of the Pythagorean theorem, we find that the base of the second right triangle has length $\sqrt{1 + a^2}$, the base of the third right triangle has length $\sqrt{2 + a^2}$, and in general, the nth right triangle has base of length $\sqrt{n - 1 + a^2}$ and hypotenuse of length $\sqrt{n + a^2}$. Thus, $\theta_n = \tan^{-1}\left(1/\sqrt{n - 1 + a^2}\right)$ and

$$\sum_{n=1}^{\infty} \theta_n = \sum_{n=1}^{\infty} \tan^{-1}\left(\frac{1}{\sqrt{n - 1 + a^2}}\right) = \sum_{n=0}^{\infty} \tan^{-1}\left(\frac{1}{\sqrt{n + a^2}}\right). \text{ We wish to show that this series diverges.}$$

First notice that the series $\displaystyle\sum_{n=1}^{\infty} \frac{1}{\sqrt{n + a^2}}$ diverges by the Limit Comparison Test with the divergent p-series $\displaystyle\sum_{n=1}^{\infty} \frac{1}{\sqrt{n}}$

$\left[p = \frac{1}{2} \leq 1\right]$ since $\displaystyle\lim_{n \to \infty} \frac{1/\sqrt{n + a^2}}{1/\sqrt{n}} = \lim_{n \to \infty} \frac{\sqrt{n}}{\sqrt{n + a^2}} = \lim_{n \to \infty} \sqrt{\frac{n}{n + a^2}} = \lim_{n \to \infty} \sqrt{\frac{1}{1 + a^2/n}} = 1 > 0.$ Thus,

$\displaystyle\sum_{n=0}^{\infty} \frac{1}{\sqrt{n + a^2}}$ also diverges. Now $\displaystyle\sum_{n=0}^{\infty} \tan^{-1}\left(\frac{1}{\sqrt{n + a^2}}\right)$ diverges by the Limit Comparison Test with $\displaystyle\sum_{n=0}^{\infty} \frac{1}{\sqrt{n + a^2}}$ since

$$\lim_{n \to \infty} \frac{\tan^{-1}\left(1/\sqrt{n + a^2}\right)}{1/\sqrt{n + a^2}} = \lim_{x \to \infty} \frac{\tan^{-1}\left(1/\sqrt{x + a^2}\right)}{1/\sqrt{x + a^2}} = \lim_{y \to \infty} \frac{\tan^{-1}(1/y)}{1/y} \qquad \left[y = \sqrt{x + a^2}\,\right]$$

$$= \lim_{z \to 0^+} \frac{\tan^{-1} z}{z} \quad [z = 1/y] \quad \overset{\text{H}}{=} \quad \lim_{z \to 0^+} \frac{1/(1 + z^2)}{1} = 1 > 0$$

Thus, $\displaystyle\sum_{n=1}^{\infty} \theta_n$ is a divergent series.

21. Call the series S. We group the terms according to the number of digits in their denominators:

$$S = \underbrace{\left(\tfrac{1}{1} + \tfrac{1}{2} + \cdots + \tfrac{1}{8} + \tfrac{1}{9}\right)}_{g_1} + \underbrace{\left(\tfrac{1}{11} + \cdots + \tfrac{1}{99}\right)}_{g_2} + \underbrace{\left(\tfrac{1}{111} + \cdots + \tfrac{1}{999}\right)}_{g_3} + \cdots$$

Now in the group g_n, since we have 9 choices for each of the n digits in the denominator, there are 9^n terms.

Furthermore, each term in g_n is less than $\frac{1}{10^{n-1}}$ [except for the first term in g_1]. So $g_n < 9^n \cdot \frac{1}{10^{n-1}} = 9\left(\frac{9}{10}\right)^{n-1}$.

Now $\displaystyle\sum_{n=1}^{\infty} 9\left(\frac{9}{10}\right)^{n-1}$ is a geometric series with $a = 9$ and $r = \frac{9}{10} < 1$. Therefore, by the Comparison Test,

$$S = \sum_{n=1}^{\infty} g_n < \sum_{n=1}^{\infty} 9\left(\frac{9}{10}\right)^{n-1} = \frac{9}{1 - 9/10} = 90.$$

22. (a) Let $f(x) = \dfrac{x}{1 - x - x^2} = \displaystyle\sum_{n=0}^{\infty} c_n x^n = c_0 + c_1 x + c_2 x^2 + c_3 x^3 + \cdots$. Then

$$x = (1 - x - x^2)(c_0 + c_1 x + c_2 x^2 + c_3 x^3 + \cdots)$$

$$x = c_0 + c_1 x + c_2 x^2 + c_3 x^3 + c_4 x^4 + c_5 x^5 + \cdots$$

$$- c_0 x - c_1 x^2 - c_2 x^3 - c_3 x^4 - c_4 x^5 - \cdots$$

$$- c_0 x^2 - c_1 x^3 - c_2 x^4 - c_3 x^5 - \cdots$$

$$x = c_0 + (c_1 - c_0)x + (c_2 - c_1 - c_0)x^2 + (c_3 - c_2 - c_1)x^3 + \cdots$$

Comparing coefficients of powers of x gives us $c_0 = 0$ and

$$c_1 - c_0 = 1 \qquad \Rightarrow \qquad c_1 = c_0 + 1 = 1$$

$$c_2 - c_1 - c_0 = 0 \qquad \Rightarrow \qquad c_2 = c_1 + c_0 = 1 + 0 = 1$$

$$c_3 - c_2 - c_1 = 0 \qquad \Rightarrow \qquad c_3 = c_2 + c_1 = 1 + 1 = 2$$

In general, we have $c_n = c_{n-1} + c_{n-2}$ for $n \geq 3$. Each c_n is equal to the nth Fibonacci number, that is,

$$\sum_{n=0}^{\infty} c_n x^n = \sum_{n=1}^{\infty} c_n x^n = \sum_{n=1}^{\infty} f_n x^n$$

(b) Completing the square on $x^2 + x - 1$ gives us

$$\left(x^2 + x + \frac{1}{4} \right) - 1 - \frac{1}{4} = \left(x + \frac{1}{2} \right)^2 - \frac{5}{4} = \left(x + \frac{1}{2} \right)^2 - \left(\frac{\sqrt{5}}{2} \right)^2$$

$$= \left(x + \frac{1}{2} + \frac{\sqrt{5}}{2} \right) \left(x + \frac{1}{2} - \frac{\sqrt{5}}{2} \right) = \left(x + \frac{1 + \sqrt{5}}{2} \right) \left(x + \frac{1 - \sqrt{5}}{2} \right)$$

So $\dfrac{x}{1 - x - x^2} = \dfrac{-x}{x^2 + x - 1} = \dfrac{-x}{\left(x + \frac{1 + \sqrt{5}}{2} \right) \left(x + \frac{1 - \sqrt{5}}{2} \right)}$. The factors in the denominator are linear,

so the partial fraction decomposition is

$$\frac{-x}{\left(x + \frac{1 + \sqrt{5}}{2} \right) \left(x + \frac{1 - \sqrt{5}}{2} \right)} = \frac{A}{x + \frac{1 + \sqrt{5}}{2}} + \frac{B}{x + \frac{1 - \sqrt{5}}{2}} - x = A \left(x + \frac{1 - \sqrt{5}}{2} \right) + B \left(x + \frac{1 + \sqrt{5}}{2} \right)$$

If $x = \frac{-1 + \sqrt{5}}{2}$, then $-\frac{-1 + \sqrt{5}}{2} = B \sqrt{5} \Rightarrow B = \frac{1 - \sqrt{5}}{2\sqrt{5}}$.

If $x = \frac{-1 - \sqrt{5}}{2}$, then $-\frac{-1 - \sqrt{5}}{2} = A(-\sqrt{5}) \Rightarrow A = \frac{1 + \sqrt{5}}{-2\sqrt{5}}$. Thus,

$$\frac{x}{1 - x - x^2} = \frac{\frac{1 + \sqrt{5}}{-2\sqrt{5}}}{x + \frac{1 + \sqrt{5}}{2}} + \frac{\frac{1 - \sqrt{5}}{2\sqrt{5}}}{x + \frac{1 - \sqrt{5}}{2}} = \frac{\frac{1 + \sqrt{5}}{-2\sqrt{5}}}{x + \frac{1 + \sqrt{5}}{2}} \cdot \frac{\frac{2}{1 + \sqrt{5}}}{\frac{2}{1 + \sqrt{5}}} + \frac{\frac{1 - \sqrt{5}}{2\sqrt{5}}}{x + \frac{1 - \sqrt{5}}{2}} \cdot \frac{\frac{2}{1 - \sqrt{5}}}{\frac{2}{1 - \sqrt{5}}}$$

$$= \frac{-1/\sqrt{5}}{1 + \frac{2}{1 + \sqrt{5}} x} + \frac{1/\sqrt{5}}{1 + \frac{2}{1 - \sqrt{5}} x} = -\frac{1}{\sqrt{5}} \sum_{n=0}^{\infty} \left(-\frac{2}{1 + \sqrt{5}} x \right)^n + \frac{1}{\sqrt{5}} \sum_{n=0}^{\infty} \left(-\frac{2}{1 - \sqrt{5}} x \right)^n$$

$$= \frac{1}{\sqrt{5}} \sum_{n=0}^{\infty} \left[\left(\frac{-2}{1 - \sqrt{5}} \right)^n - \left(\frac{-2}{1 + \sqrt{5}} \right)^n \right] x^n$$

$$= \frac{1}{\sqrt{5}} \sum_{n=1}^{\infty} \left[\frac{(-2)^n (1 + \sqrt{5})^n - (-2)^n (1 - \sqrt{5})^n}{(1 - \sqrt{5})^n (1 + \sqrt{5})^n} \right] x^n \qquad \text{[the } n = 0 \text{ term is 0]}$$

$$= \frac{1}{\sqrt{5}} \sum_{n=1}^{\infty} \left[\frac{(-2)^n \left((1 + \sqrt{5})^n - (1 - \sqrt{5})^n \right)}{(1 - 5)^n} \right] x^n$$

$$= \frac{1}{\sqrt{5}} \sum_{n=1}^{\infty} \left[\frac{(1 + \sqrt{5})^n - (1 - \sqrt{5})^n}{2^n} \right] x^n \qquad [(-4)^n = (-2)^n \cdot 2^n]$$

From part (a), this series must equal $\sum_{n=1}^{\infty} f_n x^n$, so $f_n = \dfrac{(1 + \sqrt{5})^n - (1 - \sqrt{5})^n}{2^n \sqrt{5}}$, which is an explicit formula for

the nth Fibonacci number.

23. $u = 1 + \dfrac{x^3}{3!} + \dfrac{x^6}{6!} + \dfrac{x^9}{9!} + \cdots, \; v = x + \dfrac{x^4}{4!} + \dfrac{x^7}{7!} + \dfrac{x^{10}}{10!} + \cdots, \; w = \dfrac{x^2}{2!} + \dfrac{x^5}{5!} + \dfrac{x^8}{8!} + \cdots.$

Use the Ratio Test to show that the series for u, v, and w have positive radii of convergence (∞ in each case), so
Theorem 12.9.2 applies, and hence, we may differentiate each of these series:

$$\frac{du}{dx} = \frac{3x^2}{3!} + \frac{6x^5}{6!} + \frac{9x^8}{9!} + \cdots = \frac{x^2}{2!} + \frac{x^5}{5!} + \frac{x^8}{8!} + \cdots = w$$

Similarly, $\dfrac{dv}{dx} = 1 + \dfrac{x^3}{3!} + \dfrac{x^6}{6!} + \dfrac{x^9}{9!} + \cdots = u$, and $\dfrac{dw}{dx} = x + \dfrac{x^4}{4!} + \dfrac{x^7}{7!} + \dfrac{x^{10}}{10!} + \cdots = v.$

So $u' = w$, $v' = u$, and $w' = v$. Now differentiate the left hand side of the desired equation:

$$\frac{d}{dx}(u^3 + v^3 + w^3 - 3uvw) = 3u^2u' + 3v^2v' + 3w^2w' - 3(u'vw + uv'w + uvw')$$

$$= 3u^2w + 3v^2u + 3w^2v - 3(vw^2 + u^2w + uv^2) = 0 \quad \Rightarrow$$

$u^3 + v^3 + w^3 - 3uvw = C$. To find the value of the constant C, we put $x = 0$ in the last equation and get

$1^3 + 0^3 + 0^3 - 3(1 \cdot 0 \cdot 0) = C \quad \Rightarrow \quad C = 1$, so $u^3 + v^3 + w^3 - 3uvw = 1.$

24. To prove: If $n > 1$, then the nth partial sum $s_n = \displaystyle\sum_{i=1}^{n} \frac{1}{i}$ of the harmonic series is not an integer.

Proof: Let 2^k be the largest power of 2 that is less than or equal to n and let M be the product of all the odd positive integers
that are less than or equal to n. Suppose that $s_n = m$, an integer. Then $M2^k s_n = M2^k m$. Since $n \geq 2$, we have $k \geq 1$, and
hence, $M2^k m$ is an even integer. We will show that $M2^k s_n$ is an odd integer, contradicting the equality $M2^k s_n = M2^k m$
and showing that the supposition that s_n is an integer must have been wrong.

$M2^k s_n = M2^k \displaystyle\sum_{i=1}^{n} \frac{1}{i} = \sum_{i=1}^{n} \frac{M2^k}{i}$. If $1 \leq i \leq n$ and i is odd, then $\dfrac{M}{i}$ is an odd integer since i is one of the odd integers

that were multiplied together to form M. Thus, $\dfrac{M2^k}{i}$ is an even integer in this case. If $1 \leq i \leq n$ and i is even, then we can

write $i = 2^r l$, where 2^r is the largest power of 2 dividing i and l is odd. If $r < k$, then $\dfrac{M2^k}{i} = \dfrac{2^k}{2^r} \cdot \dfrac{M}{l} = 2^{k-r}\dfrac{M}{l}$, which is

an even integer, the product of the even integer 2^{k-r} and the odd integer $\dfrac{M}{l}$. If $r = k$, then $l = 1$, since $l > 1 = l \geq 2 \quad \Rightarrow$

$i = 2^k l \geq 2^k \cdot 2 = 2^{k+1}$, contrary to the choice of 2^k as the largest power of 2 that is less than or equal to n. This shows that

$r = k$ only when $i = 2^k$. In that case, $\dfrac{M2^k}{i} = M$, an *odd* integer. Since $\dfrac{M2^k}{i}$ is an even integer for every i except 2^k and

$\dfrac{M2^k}{i}$ is an odd integer when $i = 2^k$, we see that $M2^k s_n$ is an odd integer. This concludes the proof.

☐ APPENDIXES

A Numbers, Inequalities, and Absolute Values

1. $|5 - 23| = |-18| = 18$

2. $|5| - |-23| = 5 - 23 = -18$

3. $|-\pi| = \pi$ because $\pi > 0$.

4. $|\pi - 2| = \pi - 2$ because $\pi - 2 > 0$.

5. $\left|\sqrt{5} - 5\right| = -\left(\sqrt{5} - 5\right) = 5 - \sqrt{5}$ because $\sqrt{5} - 5 < 0$.

6. $\left| |-2| - |-3| \right| = |2 - 3| = |-1| = 1$

7. If $x < 2$, $x - 2 < 0$, so $|x - 2| = -(x - 2) = 2 - x$.

8. If $x > 2$, $x - 2 > 0$, so $|x - 2| = x - 2$.

9. $|x + 1| = \begin{cases} x + 1 & \text{if } x + 1 \geq 0 \\ -(x+1) & \text{if } x + 1 < 0 \end{cases} = \{-x - 1 \quad \text{if } x < -1$

10. $|2x - 1| = \begin{cases} 2x - 1 & \text{if } 2x - 1 \geq 0 \\ -(2x - 1) & \text{if } 2x - 1 < 0 \end{cases} = \begin{cases} 2x - 1 & \text{if } x \geq \frac{1}{2} \\ 1 - 2x & \text{if } x < \frac{1}{2} \end{cases}$

11. $\left|x^2 + 1\right| = x^2 + 1$ [since $x^2 + 1 \geq 0$ for all x].

12. Determine when $1 - 2x^2 < 0 \iff 1 < 2x^2 \iff x^2 > \frac{1}{2} \iff \sqrt{x^2} > \sqrt{\frac{1}{2}} \iff |x| > \sqrt{\frac{1}{2}} \iff$

$x < -\frac{1}{\sqrt{2}}$ or $x > \frac{1}{\sqrt{2}}$. Thus, $\left|1 - 2x^2\right| = \begin{cases} 1 - 2x^2 & \text{if } -\frac{1}{\sqrt{2}} \leq x \leq \frac{1}{\sqrt{2}} \\ 2x^2 - 1 & \text{if } x < -\frac{1}{\sqrt{2}} \text{ or } x > \frac{1}{\sqrt{2}} \end{cases}$

13. $2x + 7 > 3 \iff 2x > -4 \iff x > -2$, so $x \in (-2, \infty)$.

14. $3x - 11 < 4 \iff 3x < 15 \iff x < 5$, so $x \in (-\infty, 5)$.

15. $1 - x \leq 2 \iff -x \leq 1 \iff x \geq -1$, so $x \in [-1, \infty)$.

16. $4 - 3x \geq 6 \iff -3x \geq 2 \iff x \leq -\frac{2}{3}$, so $x \in \left(-\infty, -\frac{2}{3}\right]$.

17. $2x + 1 < 5x - 8 \iff 9 < 3x \iff 3 < x$, so $x \in (3, \infty)$.

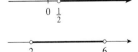

18. $1 + 5x > 5 - 3x \iff 8x > 4 \iff x > \frac{1}{2}$, so $x \in \left(\frac{1}{2}, \infty\right)$.

19. $-1 < 2x - 5 < 7 \iff 4 < 2x < 12 \iff 2 < x < 6$, so $x \in (2, 6)$.

20. $1 < 3x + 4 \le 16$ ⇔ $-3 < 3x \le 12$ ⇔ $-1 < x \le 4$, so $x \in (-1, 4]$.

21. $0 \le 1 - x < 1$ ⇔ $-1 \le -x < 0$ ⇔ $1 \ge x > 0$, so $x \in (0, 1]$.

22. $-5 \le 3 - 2x \le 9$ ⇔ $-8 \le -2x \le 6$ ⇔ $4 \ge x \ge -3$, so $x \in [-3, 4]$.

23. $4x < 2x + 1 \le 3x + 2$. So $4x < 2x + 1$ ⇔ $2x < 1$ ⇔ $x < \frac{1}{2}$, and

$2x + 1 \le 3x + 2$ ⇔ $-1 \le x$. Thus, $x \in \left[-1, \frac{1}{2}\right)$.

24. $2x - 3 < x + 4 < 3x - 2$. So $2x - 3 < x + 4$ ⇔ $x < 7$, and

$x + 4 < 3x - 2$ ⇔ $6 < 2x$ ⇔ $3 < x$, so $x \in (3, 7)$.

25. $(x - 1)(x - 2) > 0$.

    Case 1: (both factors are positive, so their product is positive) $x - 1 > 0$ ⇔ $x > 1$,

        and $x - 2 > 0$ ⇔ $x > 2$, so $x \in (2, \infty)$.

    Case 2: (both factors are negative, so their product is positive) $x - 1 < 0$ ⇔ $x < 1$,

        and $x - 2 < 0$ ⇔ $x < 2$, so $x \in (-\infty, 1)$.

    Thus, the solution set is $(-\infty, 1) \cup (2, \infty)$.

26. $(2x + 3)(x - 1) \ge 0$.

    Case 1: $2x + 3 \ge 0$ ⇔ $x \ge -\frac{3}{2}$, and $x - 1 \ge 0$ ⇔ $x \ge 1$, so $x \in [1, \infty)$.

    Case 2: $2x + 3 \le 0$ ⇔ $x \le -\frac{3}{2}$, and $x - 1 \le 0$ ⇔ $x \le 1$, so $x \in \left(-\infty, -\frac{3}{2}\right]$.

    Thus, the solution set is $\left(-\infty, -\frac{3}{2}\right] \cup [1, \infty)$.

27. $2x^2 + x \le 1$ ⇔ $2x^2 + x - 1 \le 0$ ⇔ $(2x - 1)(x + 1) \le 0$.

    Case 1: $2x - 1 \ge 0$ ⇔ $x \ge \frac{1}{2}$, and $x + 1 \le 0$ ⇔ $x \le -1$,

    which is an impossible combination.

    Case 2: $2x - 1 \le 0$ ⇔ $x \le \frac{1}{2}$, and $x + 1 \ge 0$ ⇔ $x \ge -1$, so $x \in \left[-1, \frac{1}{2}\right]$.

    Thus, the solution set is $\left[-1, \frac{1}{2}\right]$.

28. $x^2 < 2x + 8$ ⇔ $x^2 - 2x - 8 < 0$ ⇔ $(x - 4)(x + 2) < 0$.

    Case 1: $x > 4$ and $x < -2$, which is impossible.

    Case 2: $x < 4$ and $x > -2$.

    Thus, the solution set is $(-2, 4)$.

29. $x^2 + x + 1 > 0$ ⇔ $x^2 + x + \frac{1}{4} + \frac{3}{4} > 0$ ⇔ $\left(x + \frac{1}{2}\right)^2 + \frac{3}{4} > 0$. But since

$\left(x + \frac{1}{2}\right)^2 \ge 0$ for every real x, the original inequality will be true for all real x as well.

Thus, the solution set is $(-\infty, \infty)$.

30. $x^2 + x > 1 \iff x^2 + x - 1 > 0$. Using the quadratic formula, we obtain

$x^2 + x - 1 = \left(x - \frac{-1 - \sqrt{5}}{2} \right) \left(x - \frac{-1 + \sqrt{5}}{2} \right) > 0$.

Case 1: $x - \frac{-1 - \sqrt{5}}{2} > 0$ and $x - \frac{-1 + \sqrt{5}}{2} > 0$, so that $x > \frac{-1 + \sqrt{5}}{2}$.

Case 2: $x - \frac{-1 - \sqrt{5}}{2} < 0$ and $x - \frac{-1 + \sqrt{5}}{2} < 0$, so that $x < \frac{-1 - \sqrt{5}}{2}$.

Thus, the solution set is $\left(-\infty, \frac{-1 - \sqrt{5}}{2} \right) \cup \left(\frac{-1 + \sqrt{5}}{2}, \infty \right)$.

31. $x^2 < 3 \iff x^2 - 3 < 0 \iff \left(x - \sqrt{3} \right)\left(x + \sqrt{3} \right) < 0$.

Case 1: $x > \sqrt{3}$ and $x < -\sqrt{3}$, which is impossible.

Case 2: $x < \sqrt{3}$ and $x > -\sqrt{3}$.

Thus, the solution set is $\left(-\sqrt{3}, \sqrt{3} \right)$.

Another method: $x^2 < 3 \iff |x| < \sqrt{3} \iff -\sqrt{3} < x < \sqrt{3}$.

32. $x^2 \geq 5 \iff x^2 - 5 \geq 0 \iff \left(x - \sqrt{5} \right)\left(x + \sqrt{5} \right) \geq 0$.

Case 1: $x \geq \sqrt{5}$ and $x \geq -\sqrt{5}$, so $x \in [\sqrt{5}, \infty)$.

Case 2: $x \leq \sqrt{5}$ and $x \leq -\sqrt{5}$, so $x \in (-\infty, -\sqrt{5}]$.

Thus, the solution set is $\left(-\infty, -\sqrt{5} \right] \cup \left[\sqrt{5}, \infty \right)$.

Another method: $x^2 \geq 5 \iff |x| \geq \sqrt{5} \iff x \geq \sqrt{5}$ or $x \leq -\sqrt{5}$.

33. $x^3 - x^2 \leq 0 \iff x^2(x - 1) \leq 0$. Since $x^2 \geq 0$ for all x, the inequality is satisfied when $x - 1 \leq 0 \iff x \leq 1$.

Thus, the solution set is $(-\infty, 1]$.

34. $(x + 1)(x - 2)(x + 3) = 0 \iff x = -1, 2,$ or -3. Construct a chart:

Interval	$x + 1$	$x - 2$	$x + 3$	$(x + 1)(x - 2)(x + 3)$
$x < -3$	$-$	$-$	$-$	$-$
$-3 < x < -1$	$-$	$-$	$+$	$+$
$-1 < x < 2$	$+$	$-$	$+$	$-$
$x > 2$	$+$	$+$	$+$	$+$

Thus, $(x + 1)(x - 2)(x + 3) \geq 0$ on $[-3, -1]$ and $[2, \infty)$, and the solution set is $[-3, -1] \cup [2, \infty)$.

35. $x^3 > x \iff x^3 - x > 0 \iff x(x^2 - 1) > 0 \iff x(x - 1)(x + 1) > 0$. Construct a chart:

Interval	x	$x - 1$	$x + 1$	$x(x - 1)(x + 1)$
$x < -1$	$-$	$-$	$-$	$-$
$-1 < x < 0$	$-$	$-$	$+$	$+$
$0 < x < 1$	$+$	$-$	$+$	$-$
$x > 1$	$+$	$+$	$+$	$+$

Since $x^3 > x$ when the last column is positive, the solution set is $(-1, 0) \cup (1, \infty)$.

36. $x^3 + 3x < 4x^2 \quad \Leftrightarrow \quad x^3 - 4x^2 + 3x < 0 \quad \Leftrightarrow \quad x(x^2 - 4x + 3) < 0 \quad \Leftrightarrow \quad x(x-1)(x-3) < 0.$

Interval	x	$x-1$	$x-3$	$x(x-1)(x-3)$
$x < 0$	$-$	$-$	$-$	$-$
$0 < x < 1$	$+$	$-$	$-$	$+$
$1 < x < 3$	$+$	$+$	$-$	$-$
$x > 3$	$+$	$+$	$+$	$+$

Thus, the solution set is $(-\infty, 0) \cup (1, 3)$.

37. $1/x < 4$. This is clearly true for $x < 0$. So suppose $x > 0$. then $1/x < 4 \quad \Leftrightarrow$

$1 < 4x \quad \Leftrightarrow \quad \frac{1}{4} < x$. Thus, the solution set is $(-\infty, 0) \cup \left(\frac{1}{4}, \infty\right)$.

38. $-3 < 1/x \le 1$. We solve the two inequalities separately and take the intersection of the solution sets. First, $-3 < 1/x$ is

clearly true for $x > 0$. So suppose $x < 0$. Then $-3 < 1/x \quad \Leftrightarrow \quad -3x > 1 \quad \Leftrightarrow \quad x < -\frac{1}{3}$, so for this inequality, the solution

set is $\left(-\infty, -\frac{1}{3}\right) \cup (0, \infty)$. Now $1/x \le 1$ is clearly true if $x < 0$. So suppose $x > 0$. Then $1/x \le 1 \quad \Leftrightarrow \quad 1 \le x$, and the

solution set here is $(-\infty, 0) \cup [1, \infty)$.

Taking the intersection of the two solution sets gives the final solution set:

$\left(-\infty, -\frac{1}{3}\right) \cup [1, \infty)$.

39. $C = \frac{5}{9}(F - 32) \quad \Rightarrow \quad F = \frac{9}{5}C + 32$. So $50 \le F \le 95 \quad \Rightarrow \quad 50 \le \frac{9}{5}C + 32 \le 95 \quad \Rightarrow \quad 18 \le \frac{9}{5}C \le 63 \quad \Rightarrow$

$10 \le C \le 35$. So the interval is $[10, 35]$.

40. Since $20 \le C \le 30$ and $C = \frac{5}{9}(F - 32)$, we have $20 \le \frac{5}{9}(F - 32) \le 30 \quad \Rightarrow \quad 36 \le F - 32 \le 54 \quad \Rightarrow \quad 68 \le F \le 86$.

So the interval is $[68, 86]$.

41. (a) Let T represent the temperature in degrees Celsius and h the height in km. $T = 20$ when $h = 0$ and T decreases by $10°C$

for every km ($1°C$ for each 100-m rise). Thus, $T = 20 - 10h$ when $0 \le h \le 12$.

(b) From part (a), $T = 20 - 10h \quad \Rightarrow \quad 10h = 20 - T \quad \Rightarrow \quad h = 2 - T/10$. So $0 \le h \le 5 \quad \Rightarrow \quad 0 \le 2 - T/10 \le 5 \quad \Rightarrow$

$-2 \le -T/10 \le 3 \quad \Rightarrow \quad -20 \le -T \le 30 \quad \Rightarrow \quad 20 \ge T \ge -30 \quad \Rightarrow \quad -30 \le T \le 20$. Thus, the range of

temperatures (in $°C$) to be expected is $[-30, 20]$.

42. The ball will be at least 32 ft above the ground if $h \ge 32 \quad \Leftrightarrow \quad 128 + 16t - 16t^2 \ge 32 \quad \Leftrightarrow \quad 16t^2 - 16t - 96 \le 0 \quad \Leftrightarrow$

$16(t - 3)(t + 2) \le 0$. $t = 3$ and $t = -2$ are endpoints of the interval we're looking for, and constructing a table gives

$-2 \le t \le 3$. But $t \ge 0$, so the ball will be at least 32 ft above the ground in the time interval $[0, 3]$.

43. $|2x| = 3 \quad \Leftrightarrow \quad$ either $2x = 3$ or $2x = -3 \quad \Leftrightarrow \quad x = \frac{3}{2}$ or $x = -\frac{3}{2}$.

44. $|3x + 5| = 1 \quad \Leftrightarrow \quad$ either $3x + 5 = 1$ or -1. In the first case, $3x = -4 \quad \Leftrightarrow \quad x = -\frac{4}{3}$, and in the second case,

$3x = -6 \quad \Leftrightarrow \quad x = -2$. So the solutions are -2 and $-\frac{4}{3}$.

45. $|x + 3| = |2x + 1| \quad \Leftrightarrow \quad$ either $x + 3 = 2x + 1$ or $x + 3 = -(2x + 1)$. In the first case, $x = 2$, and in the second case,

$x + 3 = -2x - 1 \quad \Leftrightarrow \quad 3x = -4 \quad \Leftrightarrow \quad x = -\frac{4}{3}$. So the solutions are $-\frac{4}{3}$ and 2.

46. $\left|\dfrac{2x-1}{x+1}\right| = 3$ ⟺ either $\dfrac{2x-1}{x+1} = 3$ or $\dfrac{2x-1}{x+1} = -3$. In the first case, $2x-1 = 3x+3$ ⟺ $x = -4$, and

in the second case, $2x-1 = -3x-3$ ⟺ $x = -\frac{2}{5}$.

47. By Property 5 of absolute values, $|x| < 3$ ⟺ $-3 < x < 3$, so $x \in (-3,3)$.

48. By Properties 4 and 6 of absolute values, $|x| \geq 3$ ⟺ $x \leq -3$ or $x \geq 3$, so $x \in (-\infty, -3] \cup [3, \infty)$.

49. $|x-4| < 1$ ⟺ $-1 < x-4 < 1$ ⟺ $3 < x < 5$, so $x \in (3,5)$.

50. $|x-6| < 0.1$ ⟺ $-0.1 < x-6 < 0.1$ ⟺ $5.9 < x < 6.1$, so $x \in (5.9, 6.1)$.

51. $|x+5| \geq 2$ ⟺ $x+5 \geq 2$ or $x+5 \leq -2$ ⟺ $x \geq -3$ or $x \leq -7$, so $x \in (-\infty, -7] \cup [-3, \infty)$.

52. $|x+1| \geq 3$ ⟺ $x+1 \geq 3$ or $x+1 \leq -3$ ⟺ $x \geq 2$ or $x \leq -4$, so $x \in (-\infty, -4] \cup [2, \infty)$.

53. $|2x-3| \leq 0.4$ ⟺ $-0.4 \leq 2x-3 \leq 0.4$ ⟺ $2.6 \leq 2x \leq 3.4$ ⟺ $1.3 \leq x \leq 1.7$, so $x \in [1.3, 1.7]$.

54. $|5x-2| < 6$ ⟺ $-6 < 5x-2 < 6$ ⟺ $-4 < 5x < 8$ ⟺ $-\frac{4}{5} < x < \frac{8}{5}$, so $x \in \left(-\frac{4}{5}, \frac{8}{5}\right)$.

55. $1 \leq |x| \leq 4$. So either $1 \leq x \leq 4$ or $1 \leq -x \leq 4$ ⟺ $-1 \geq x \geq -4$. Thus, $x \in [-4, -1] \cup [1, 4]$.

56. $0 < |x-5| < \frac{1}{2}$. Clearly $0 < |x-5|$ for $x \neq 5$. Now $|x-5| < \frac{1}{2}$ ⟺ $-\frac{1}{2} < x-5 < \frac{1}{2}$ ⟺ $4.5 < x < 5.5$. So the
solution set is $(4.5, 5) \cup (5, 5.5)$.

57. $a(bx-c) \geq bc$ ⟺ $bx-c \geq \dfrac{bc}{a}$ ⟺ $bx \geq \dfrac{bc}{a}+c = \dfrac{bc+ac}{a}$ ⟺ $x \geq \dfrac{bc+ac}{ab}$

58. $a \leq bx+c < 2a$ ⟺ $a-c \leq bx < 2a-c$ ⟺ $\dfrac{a-c}{b} \leq x < \dfrac{2a-c}{b}$ (since $b > 0$)

59. $ax+b < c$ ⟺ $ax < c-b$ ⟺ $x > \dfrac{c-b}{a}$ [since $a < 0$]

60. $\dfrac{ax+b}{c} \leq b$ ⟺ $ax+b \geq bc$ [since $c < 0$] ⟺ $ax \geq bc-b$ ⟺ $x \leq \dfrac{b(c-1)}{a}$ [since $a < 0$]

61. $|(x+y)-5| = |(x-2)+(y-3)| \leq |x-2|+|y-3| < 0.01 + 0.04 = 0.05$

62. Use the Triangle Inequality: $|x+3| < \frac{1}{2}$ ⟹

$|4x+13| = |4(x+3)+1| \leq |4(x+3)|+|1| = 4|x+3|+1 < 4\left(\frac{1}{2}\right)+1 = 3$

Another method: $|x+3| < \frac{1}{2}$ ⟹ $-\frac{1}{2} < x+3 < \frac{1}{2}$ ⟹ $-2 < 4x+12 < 2$ ⟹ $-1 < 4x+13 < 3$ ⟹
$|4x+13| < 3$

63. If $a < b$ then $a+a < a+b$ and $a+b < b+b$. So $2a < a+b < 2b$. Dividing by 2, we get $a < \frac{1}{2}(a+b) < b$.

64. If $0 < a < b$, then $\dfrac{1}{ab} > 0$. So $a < b$ ⟹ $\dfrac{1}{ab} \cdot a < \dfrac{1}{ab} \cdot b$ ⟺ $\dfrac{1}{b} < \dfrac{1}{a}$.

65. $|ab| = \sqrt{(ab)^2} = \sqrt{a^2b^2} = \sqrt{a^2}\sqrt{b^2} = |a||b|$

66. $\left|\dfrac{a}{b}\right||b| = \left|\dfrac{a}{b} \cdot b\right| = |a|$ [using the result of Exercise 65]. Dividing the equation through by $|b|$ gives $\left|\dfrac{a}{b}\right| = \dfrac{|a|}{|b|}$.

67. If $0 < a < b$, then $a \cdot a < a \cdot b$ and $a \cdot b < b \cdot b$ [using Rule 3 of Inequalities]. So $a^2 < ab < b^2$ and hence $a^2 < b^2$.

68. Following the hint, the Triangle Inequality becomes $|(x - y) + y| \leq |x - y| + |y|$ $\Leftrightarrow$ $|x| \leq |x - y| + |y|$ $\Leftrightarrow$
$|x - y| \geq |x| - |y|$.

69. Observe that the sum, difference and product of two integers is always an integer. Let the rational numbers be represented
by $r = m/n$ and $s = p/q$ (where m, n, p and q are integers with $n \neq 0$, $q \neq 0$). Now $r + s = \dfrac{m}{n} + \dfrac{p}{q} = \dfrac{mq + pn}{nq}$,
but $mq + pn$ and nq are both integers, so $\dfrac{mq + pn}{nq} = r + s$ is a rational number by definition. Similarly,

$r - s = \dfrac{m}{n} - \dfrac{p}{q} = \dfrac{mq - pn}{nq}$ is a rational number. Finally, $r \cdot s = \dfrac{m}{n} \cdot \dfrac{p}{q} = \dfrac{mp}{nq}$ but mp and nq are both integers, so

$\dfrac{mp}{nq} = r \cdot s$ is a rational number by definition.

70. (a) No. Consider the case of $\sqrt{2}$ and $-\sqrt{2}$. Both are irrational numbers, yet $\sqrt{2} + \left(-\sqrt{2}\right) = 0$ and 0,
being an integer, is not irrational.

(b) No. Consider the case of $\sqrt{2}$ and $\sqrt{2}$. Both are irrational numbers, yet $\sqrt{2} \cdot \sqrt{2} = 2$ is not irrational.

B Coordinate Geometry and Lines

1. Use the distance formula with $P_1(x_1, y_1) = (1, 1)$ and $P_2(x_2, y_2) = (4, 5)$ to get
$$|P_1P_2| = \sqrt{(4 - 1)^2 + (5 - 1)^2} = \sqrt{3^2 + 4^2} = \sqrt{25} = 5$$

2. The distance from $(1, -3)$ to $(5, 7)$ is $\sqrt{(5 - 1)^2 + [7 - (-3)]^2} = \sqrt{4^2 + 10^2} = \sqrt{116} = 2\sqrt{29}$.

3. The distance from $(6, -2)$ to $(-1, 3)$ is $\sqrt{-1 - 6)^2 + [3 - (-2)]^2} = \sqrt{(-7)^2 + 5^2} = \sqrt{74}$.

4. The distance from $(1, -6)$ to $(-1, -3)$ is $\sqrt{(-1 - 1)^2 + [-3 - (-6)]^2} = \sqrt{(-2)^2 + 3^2} = \sqrt{13}$.

5. The distance from $(2, 5)$ to $(4, -7)$ is $\sqrt{(4 - 2)^2 + (-7 - 5)^2} = \sqrt{2^2 + (-12)^2} = \sqrt{148} = 2\sqrt{37}$.

6. The distance from (a, b) to (b, a) is $\sqrt{(b - a)^2 + (a - b)^2} = \sqrt{(a - b)^2 + (a - b)^2} = \sqrt{2(a - b)^2} = \sqrt{2}\,|a - b|$.

7. The slope m of the line through $P(1, 5)$ and $Q(4, 11)$ is $m = \dfrac{11 - 5}{4 - 1} = \dfrac{6}{3} = 2$.

8. The slope m of the line through $P(-1, 6)$ and $Q(4, -3)$ is $m = \dfrac{-3 - 6}{4 - (-1)} = -\dfrac{9}{5}$.

9. The slope m of the line through $P(-3, 3)$ and $Q(-1, -6)$ is $m = \dfrac{-6 - 3}{-1 - (-3)} = -\dfrac{9}{2}$.

10. The slope m of the line through $P(-1, -4)$ and $Q(6, 0)$ is $m = \dfrac{0 - (-4)}{6 - (-1)} = \dfrac{4}{7}$.

11. Using $A(0, 2)$, $B(-3, -1)$, and $C(-4, 3)$, we have $|AC| = \sqrt{(-4 - 0)^2 + (3 - 2)^2} = \sqrt{(-4)^2 + 1^2} = \sqrt{17}$ and
$|BC| = \sqrt{[-4 - (-3)]^2 + [3 - (-1)]^2} = \sqrt{(-1)^2 + 4^2} = \sqrt{17}$, so the triangle has two sides of equal length, and is
isosceles.

12. (a) Using $A(6, -7)$, $B(11, -3)$, and $C(2, -2)$, we have

$$|AB| = \sqrt{(11 - 6)^2 + [-3 - (-7)]^2} = \sqrt{5^2 + 4^2} = \sqrt{41},$$

$$|AC| = \sqrt{(2 - 6)^2 + [-2 - (-7)]^2} = \sqrt{(-4)^2 + 5^2} = \sqrt{41}, \text{ and}$$

$$|BC| = \sqrt{(2 - 11)^2 + [-2 - (-3)]^2} = \sqrt{(-9)^2 + 1^2} = \sqrt{82}.$$

Thus, $|AB|^2 + |AC|^2 = 41 + 41 = 82 = |BC|^2$ and so $\triangle ABC$ is a right triangle.

(b) $m_{AB} = \dfrac{-3 - (-7)}{11 - 6} = \dfrac{4}{5}$ and $m_{AC} = \dfrac{-2 - (-7)}{2 - 6} = -\dfrac{5}{4}$. Thus $m_{AB} \cdot m_{AC} = -1$ and so AB is perpendicular to AC

and $\triangle ABC$ must be a right triangle.

(c) Taking lengths from part (a), the base is $\sqrt{41}$ and the height is $\sqrt{41}$. Thus the area is $\frac{1}{2}bh = \frac{1}{2}\sqrt{41}\sqrt{41} = \frac{41}{2}$.

13. Using $A(-2, 9)$, $B(4, 6)$, $C(1, 0)$, and $D(-5, 3)$, we have

$$|AB| = \sqrt{[4 - (-2)]^2 + (6 - 9)^2} = \sqrt{6^2 + (-3)^2} = \sqrt{45} = \sqrt{9}\sqrt{5} = 3\sqrt{5},$$

$$|BC| = \sqrt{(1 - 4)^2 + (0 - 6)^2} = \sqrt{(-3)^2 + (-6)^2} = \sqrt{45} = \sqrt{9}\sqrt{5} = 3\sqrt{5},$$

$$|CD| = \sqrt{(-5 - 1)^2 + (3 - 0)^2} = \sqrt{(-6)^2 + 3^2} = \sqrt{45} = \sqrt{9}\sqrt{5} = 3\sqrt{5}, \text{ and}$$

$$|DA| = \sqrt{[-2 - (-5)]^2 + (9 - 3)^2} = \sqrt{3^2 + 6^2} = \sqrt{45} = \sqrt{9}\sqrt{5} = 3\sqrt{5}. \text{ So all sides are of equal length and we have a}$$

rhombus. Moreover, $m_{AB} = \dfrac{6 - 9}{4 - (-2)} = -\dfrac{1}{2}$, $m_{BC} = \dfrac{0 - 6}{1 - 4} = 2$, $m_{CD} = \dfrac{3 - 0}{-5 - 1} = -\dfrac{1}{2}$, and

$m_{DA} = \dfrac{9 - 3}{-2 - (-5)} = 2$, so the sides are perpendicular. Thus, A, B, C, and D are vertices of a square.

14. (a) Using $A(-1, 3)$, $B(3, 11)$, and $C(5, 15)$, we have

$$|AB| = \sqrt{[3 - (-1)]^2 + (11 - 3)^2} = \sqrt{4^2 + 8^2} = \sqrt{80} = 4\sqrt{5},$$

$$|BC| = \sqrt{(5 - 3)^2 + (15 - 11)^2} = \sqrt{2^2 + 4^2} = \sqrt{20} = 2\sqrt{5}, \text{ and}$$

$$|AC| = \sqrt{[5 - (-1)]^2 + (15 - 3)^2} = \sqrt{6^2 + 12^2} = \sqrt{180} = 6\sqrt{5}. \text{ Thus, } |AC| = |AB| + |BC|.$$

(b) $m_{AB} = \dfrac{11 - 3}{3 - (-1)} = \dfrac{8}{4} = 2$ and $m_{AC} = \dfrac{15 - 3}{5 - (-1)} = \dfrac{12}{6} = 2$. Since the segments AB and AC have the same slope, A,

B and C must be collinear.

15. For the vertices $A(1, 1)$, $B(7, 4)$, $C(5, 10)$, and $D(-1, 7)$, the slope of the line segment AB is $\dfrac{4 - 1}{7 - 1} = \dfrac{1}{2}$, the slope of CD

is $\dfrac{7 - 10}{-1 - 5} = \dfrac{1}{2}$, the slope of BC is $\dfrac{10 - 4}{5 - 7} = -3$, and the slope of DA is $\dfrac{1 - 7}{1 - (-1)} = -3$. So AB is parallel to CD and

BC is parallel to DA. Hence $ABCD$ is a parallelogram.

16. For the vertices $A(1, 1)$, $B(11, 3)$, $C(10, 8)$, and $D(0, 6)$, the slopes of the four sides are $m_{AB} = \dfrac{3 - 1}{11 - 1} = \dfrac{1}{5}$,

$m_{BC} = \dfrac{8 - 3}{10 - 11} = -5$, $m_{CD} = \dfrac{6 - 8}{0 - 10} = \dfrac{1}{5}$, and $m_{DA} = \dfrac{1 - 6}{1 - 0} = -5$. Hence $AB \parallel CD$, $BC \parallel DA$, $AB \perp BC$,

$BC \perp CD$, $CD \perp DA$, and $DA \perp AB$, and so $ABCD$ is a rectangle.

17. The graph of the equation $x = 3$ is a vertical line with x-intercept 3. The line does not have a slope.

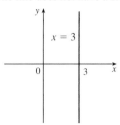

18. The graph of the equation $y = -2$ is a horizontal line with y-intercept -2. The line has slope 0.

19. $xy = 0 \quad \Leftrightarrow \quad x = 0$ or $y = 0$. The graph consists of the coordinate axes.

20. $|y| = 1 \quad \Leftrightarrow \quad y = 1$ or $y = -1$

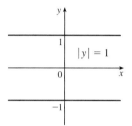

21. By the point-slope form of the equation of a line, an equation of the line through $(2, -3)$ with slope 6 is

$$y - (-3) = 6(x - 2) \text{ or } y = 6x - 15.$$

22. $y - 4 = -3[x - (-1)]$ or $y = -3x + 1$

23. $y - 7 = \frac{2}{3}(x - 1)$ or $y = \frac{2}{3}x + \frac{19}{3}$

24. $y - (-5) = -\frac{7}{2}[x - (-3)]$ or $y = -\frac{7}{2}x - \frac{31}{2}$

25. The slope of the line through $(2, 1)$ and $(1, 6)$ is $m = \dfrac{6 - 1}{1 - 2} = -5$, so an equation of the line is

$$y - 1 = -5(x - 2) \text{ or } y = -5x + 11.$$

26. For $(-1, -2)$ and $(4, 3)$, $m = \dfrac{3 - (-2)}{4 - (-1)} = 1$. An equation of the line is $y - 3 = 1(x - 4)$ or $y = x - 1$.

27. By the slope-intercept form of the equation of a line, an equation of the line is $y = 3x - 2$.

28. By the slope-intercept form of the equation of a line, an equation of the line is $y = \frac{2}{5}x + 4$.

29. Since the line passes through $(1, 0)$ and $(0, -3)$, its slope is $m = \dfrac{-3 - 0}{0 - 1} = 3$, so an equation is $y = 3x - 3$.

Another method: From Exercise 61, $\dfrac{x}{1} + \dfrac{y}{-3} = 1 \quad \Rightarrow \quad -3x + y = -3 \quad \Rightarrow \quad y = 3x - 3.$

30. For $(-8, 0)$ and $(0, 6)$, $m = \dfrac{6 - 0}{0 - (-8)} = \dfrac{3}{4}$. So an equation is $y = \frac{3}{4}x + 6$.

Another method: From Exercise 61, $\dfrac{x}{-8} + \dfrac{y}{6} = 1 \quad \Rightarrow \quad -3x + 4y = 24 \quad \Rightarrow \quad y = \frac{3}{4}x + 6$.

31. The line is parallel to the x-axis, so it is horizontal and must have the form $y = k$. Since it goes through the point $(x, y) = (4, 5)$, the equation is $y = 5$.

32. The line is parallel to the y-axis, so it is vertical and must have the form $x = k$. Since it goes through the point $(x, y) = (4, 5)$, the equation is $x = 4$.

33. Putting the line $x + 2y = 6$ into its slope-intercept form gives us $y = -\frac{1}{2}x + 3$, so we see that this line has slope $-\frac{1}{2}$. Thus, we want the line of slope $-\frac{1}{2}$ that passes through the point $(1, -6)$: $y - (-6) = -\frac{1}{2}(x - 1) \quad \Leftrightarrow \quad y = -\frac{1}{2}x - \frac{11}{2}$.

34. $2x + 3y + 4 = 0 \quad \Leftrightarrow \quad y = -\frac{2}{3}x - \frac{4}{3}$, so $m = -\frac{2}{3}$ and the required line is $y = -\frac{2}{3}x + 6$.

35. $2x + 5y + 8 = 0 \quad \Leftrightarrow \quad y = -\frac{2}{5}x - \frac{8}{5}$. Since this line has slope $-\frac{2}{5}$, a line perpendicular to it would have slope $\frac{5}{2}$, so the required line is $y - (-2) = \frac{5}{2}[x - (-1)] \quad \Leftrightarrow \quad y = \frac{5}{2}x + \frac{1}{2}$.

36. $4x - 8y = 1 \quad \Leftrightarrow \quad y = \frac{1}{2}x - \frac{1}{8}$. Since this line has slope $\frac{1}{2}$, a line perpendicular to it would have slope -2, so the required line is $y - \left(-\frac{2}{3}\right) = -2\left(x - \frac{1}{2}\right) \quad \Leftrightarrow \quad y = -2x + \frac{1}{3}$.

37. $x + 3y = 0 \quad \Leftrightarrow \quad y = -\frac{1}{3}x$, so the slope is $-\frac{1}{3}$ and the y-intercept is 0.

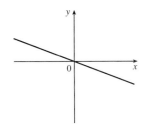

38. $2x - 5y = 0 \quad \Leftrightarrow \quad y = \frac{2}{5}x$, so the slope is $\frac{2}{5}$ and the y-intercept is 0.

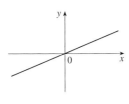

39. $y = -2$ is a horizontal line with slope 0 and y-intercept -2.

40. $2x - 3y + 6 = 0 \quad \Leftrightarrow \quad y = \frac{2}{3}x + 2$, so the slope is $\frac{2}{3}$ and the y-intercept is 2.

41. $3x - 4y = 12 \quad \Leftrightarrow \quad y = \frac{3}{4}x - 3$, so the slope is $\frac{3}{4}$ and the y-intercept is -3.

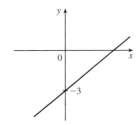

42. $4x + 5y = 10 \quad \Leftrightarrow \quad y = -\frac{4}{5}x + 2$, so the slope is $-\frac{4}{5}$ and the y-intercept is 2.

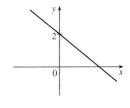

43. $\{(x, y) \mid x < 0\}$

44. $\{(x, y) \mid y > 0\}$

45. $\{(x, y) \mid xy < 0\} =$
$\quad \{(x, y) \mid x < 0 \text{ and } y > 0\}$
$\quad \cup \{(x, y) \mid x > 0 \text{ and } y < 0\}$

46. $\{(x, y) \mid x \geq 1 \text{ and } y < 3\}$

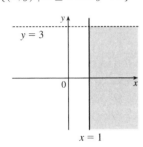

47. $\left\{(x, y) \,\middle|\, |x| \leq 2\right\} =$
$\quad \{(x, y) \mid -2 \leq x \leq 2\}$

48. $\left\{(x, y) \,\middle|\, |x| < 3 \text{ and } |y| < 2\right\}$

49. $\{(x, y) \mid 0 \leq y \leq 4, x \leq 2\}$

50. $\{(x, y) \mid y > 2x - 1\}$

51. $\{(x, y) \mid 1 + x \leq y \leq 1 - 2x\}$

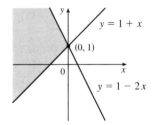

52. $\left\{(x, y) \mid -x \leq y < \frac{1}{2}(x + 3)\right\}$

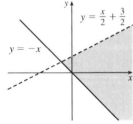

53. Let $P(0, y)$ be a point on the y-axis. The distance from P to $(5, -5)$ is
$$\sqrt{(5 - 0)^2 + (-5 - y)^2} = \sqrt{5^2 + (y + 5)^2}.$$ The distance from P to $(1, 1)$ is
$$\sqrt{(1 - 0)^2 + (1 - y)^2} = \sqrt{1^2 + (y - 1)^2}.$$ We want these distances to be equal:
$$\sqrt{5^2 + (y + 5)^2} = \sqrt{1^2 + (y - 1)^2} \quad \Leftrightarrow \quad 5^2 + (y + 5)^2 = 1^2 + (y - 1)^2 \quad \Leftrightarrow$$
$$25 + (y^2 + 10y + 25) = 1 + (y^2 - 2y + 1) \quad \Leftrightarrow \quad 12y = -48 \quad \Leftrightarrow \quad y = -4.$$
So the desired point is $(0, -4)$.

54. Let M be the point $\left(\dfrac{x_1 + x_2}{2}, \dfrac{y_1 + y_2}{2}\right)$. Then

$$|MP_1|^2 = \left(x_1 - \frac{x_1 + x_2}{2}\right)^2 + \left(y_1 - \frac{y_1 + y_2}{2}\right)^2 = \left(\frac{x_1 - x_2}{2}\right)^2 + \left(\frac{y_1 - y_2}{2}\right)^2$$

$$|MP_2|^2 = \left(x_2 - \frac{x_1 + x_2}{2}\right)^2 + \left(y_2 - \frac{y_1 + y_2}{2}\right)^2 = \left(\frac{x_2 - x_1}{2}\right)^2 + \left(\frac{y_2 - y_1}{2}\right)^2$$

Hence, $|MP_1| = |MP_2|$; that is, M is equidistant from P_1 and P_2.

55. (a) Using the midpoint formula from Exercise 54 with $(1, 3)$ and $(7, 15)$, we get $\left(\frac{1+7}{2}, \frac{3+15}{2}\right) = (4, 9)$.

(b) Using the midpoint formula from Exercise 54 with $(-1, 6)$ and $(8, -12)$, we get $\left(\frac{-1+8}{2}, \frac{6+(-12)}{2}\right) = \left(\frac{7}{2}, -3\right)$.

56. With $A(1, 0)$, $B(3, 6)$, and $C(8, 2)$, the midpoint M_1 of AB is $\left(\frac{1+3}{2}, \frac{0+6}{2}\right) = (2, 3)$, the midpoint M_2 of BC is

$\left(\frac{3+8}{2}, \frac{6+2}{2}\right) = \left(\frac{11}{2}, 4\right)$, and the midpoint M_3 of CA is $\left(\frac{8+1}{2}, \frac{2+0}{2}\right) = \left(\frac{9}{2}, 1\right)$. The lengths of the medians are

$$|AM_2| = \sqrt{\left(\tfrac{11}{2} - 1\right)^2 + (4 - 0)^2} = \sqrt{\left(\tfrac{9}{2}\right)^2 + 4^2} = \sqrt{\tfrac{145}{4}} = \tfrac{\sqrt{145}}{2}$$

$$|BM_3| = \sqrt{\left(\tfrac{9}{2} - 3\right)^2 + (1 - 6)^2} = \sqrt{\left(\tfrac{3}{2}\right)^2 + (-5)^2} = \sqrt{\tfrac{109}{4}} = \tfrac{\sqrt{109}}{2}$$

$$|CM_1| = \sqrt{(2 - 8)^2 + (3 - 2)^2} = \sqrt{(-6)^2 + 1^2} = \sqrt{37}$$

57. $2x - y = 4 \Leftrightarrow y = 2x - 4 \Rightarrow m_1 = 2$ and $6x - 2y = 10 \Leftrightarrow 2y = 6x - 10 \Leftrightarrow y = 3x - 5 \Rightarrow m_2 = 3$.

Since $m_1 \neq m_2$, the two lines are not parallel. To find the point of intersection: $2x - 4 = 3x - 5 \Leftrightarrow x = 1 \Rightarrow$

$y = -2$. Thus, the point of intersection is $(1, -2)$.

58. $3x - 5y + 19 = 0 \Leftrightarrow 5y = 3x + 19 \Leftrightarrow y = \frac{3}{5}x + \frac{19}{5} \Rightarrow m_1 = \frac{3}{5}$ and $10x + 6y - 50 = 0 \Leftrightarrow$

$6y = -10x + 50 \Leftrightarrow y = -\frac{5}{3}x + \frac{25}{3} \Rightarrow m_2 = -\frac{5}{3}$. Since $m_1 m_2 = \frac{3}{5}\left(-\frac{5}{3}\right) = -1$, the two lines are perpendicular.

To find the point of intersection: $\frac{3}{5}x + \frac{19}{5} = -\frac{5}{3}x + \frac{25}{3} \Leftrightarrow 9x + 57 = -25x + 125 \Leftrightarrow 34x = 68 \Leftrightarrow x = 2 \Rightarrow$

$y = \frac{3}{5} \cdot 2 + \frac{19}{5} = \frac{25}{5} = 5$. Thus, the point of intersection is $(2, 5)$.

59. With $A(1, 4)$ and $B(7, -2)$, the slope of segment AB is $\frac{-2 - 4}{7 - 1} = -1$, so its perpendicular bisector has slope 1. The midpoint

of AB is $\left(\frac{1+7}{2}, \frac{4+(-2)}{2}\right) = (4, 1)$, so an equation of the perpendicular bisector is $y - 1 = 1(x - 4)$ or $y = x - 3$.

60. (a) Side PQ has slope $\frac{4-0}{3-1} = 2$, so its equation is $y - 0 = 2(x - 1) \Leftrightarrow y = 2x - 2$. Side QR has slope $\frac{6-4}{-1-3} = -\frac{1}{2}$, so

its equation is $y - 4 = -\frac{1}{2}(x - 3) \Leftrightarrow y = -\frac{1}{2}x + \frac{11}{2}$. Side RP has slope $\frac{0-6}{1-(-1)} = -3$, so its equation is

$y - 0 = -3(x - 1) \Leftrightarrow y = -3x + 3$.

(b) M_1 (the midpoint of PQ) has coordinates $\left(\frac{1+3}{2}, \frac{0+4}{2}\right) = (2, 2)$. M_2 (the midpoint of QR) has coordinates

$\left(\frac{3-1}{2}, \frac{4+6}{2}\right) = (1, 5)$. M_3 (the midpoint of RP) has coordinates $\left(\frac{1-1}{2}, \frac{0+6}{2}\right) = (0, 3)$. RM_1 has slope $\frac{2-6}{2-(-1)} = -\frac{4}{3}$

and hence equation $y - 2 = -\frac{4}{3}(x - 2) \Leftrightarrow y = -\frac{4}{3}x + \frac{14}{3}$. PM_2 is a vertical line with equation $x = 1$. QM_3 has

slope $\frac{3-4}{0-3} = \frac{1}{3}$ and hence equation $y - 3 = \frac{1}{3}(x - 0) \Leftrightarrow y = \frac{1}{3}x + 3$. PM_2 and RM_1 intersect where $x = 1$ and

$y = -\frac{4}{3}(1) + \frac{14}{3} = \frac{10}{3}$, or at $\left(1, \frac{10}{3}\right)$. PM_2 and QM_3 intersect where $x = 1$ and $y = \frac{1}{3}(1) + 3 = \frac{10}{3}$, or at $\left(1, \frac{10}{3}\right)$, so

this is the point where all three medians intersect.

61. (a) Since the x-intercept is a, the point $(a, 0)$ is on the line, and similarly since the y-intercept is b, $(0, b)$ is on the line. Hence, the slope of the line is $m = \dfrac{b - 0}{0 - a} = -\dfrac{b}{a}$. Substituting into $y = mx + b$ gives $y = -\dfrac{b}{a}x + b \iff \dfrac{b}{a}x + y = b \iff \dfrac{x}{a} + \dfrac{y}{b} = 1$.

(b) Letting $a = 6$ and $b = -8$ gives $\dfrac{x}{6} + \dfrac{y}{-8} = 1 \iff -8x + 6y = -48$ [multiply by -48] $\iff 6y = 8x - 48 \iff 3y = 4x - 24 \iff y = \frac{4}{3}x - 8$.

62. (a) Let $d =$ distance traveled (in miles) and $t =$ time elapsed (in hours). At $t = 0$, $d = 0$ and at $t = 50$ minutes $= 50 \cdot \frac{1}{60} = \frac{5}{6}$ h, $d = 40$. Thus, we have two points: $(0, 0)$ and $\left(\frac{5}{6}, 40\right)$, so $m = \dfrac{40 - 0}{5/6 - 0} = 48$ and $d = 48t$.

(b)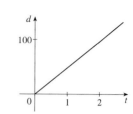

(c) The slope is 48 and represents the car's speed in mi/h.

C   Graphs of Second-Degree Equations

1. An equation of the circle with center $(3, -1)$ and radius 5 is $(x - 3)^2 + (y + 1)^2 = 5^2 = 25$.

2. An equation of the circle with center $(-2, -8)$ and radius 10 is $(x + 2)^2 + (y + 8)^2 = 10^2 = 100$.

3. The equation has the form $x^2 + y^2 = r^2$. Since $(4, 7)$ lies on the circle, we have $4^2 + 7^2 = r^2 \Rightarrow r^2 = 65$. So the required equation is $x^2 + y^2 = 65$.

4. The equation has the form $(x + 1)^2 + (y - 5)^2 = r^2$. Since $(-4, -6)$ lies on the circle, we have $r^2 = (-4 + 1)^2 + (-6 - 5)^2 = 130$. So an equation is $(x + 1)^2 + (y - 5)^2 = 130$.

5. $x^2 + y^2 - 4x + 10y + 13 = 0 \iff x^2 - 4x + y^2 + 10y = -13 \iff (x^2 - 4x + 4) + (y^2 + 10y + 25) = -13 + 4 + 25 = 16 \iff (x - 2)^2 + (y + 5)^2 = 4^2$. Thus, we have a circle with center $(2, -5)$ and radius 4.

6. $x^2 + y^2 + 6y + 2 = 0 \iff x^2 + (y^2 + 6y + 9) = -2 + 9 \iff x^2 + (y + 3)^2 = 7$. Thus, we have a circle with center $(0, -3)$ and radius $\sqrt{7}$.

7. $x^2 + y^2 + x = 0 \iff \left(x^2 + x + \frac{1}{4}\right) + y^2 = \frac{1}{4} \iff \left(x + \frac{1}{2}\right)^2 + y^2 = \left(\frac{1}{2}\right)^2$. Thus, we have a circle with center $\left(-\frac{1}{2}, 0\right)$ and radius $\frac{1}{2}$.

8. $16x^2 + 16y^2 + 8x + 32y + 1 = 0 \iff 16\left(x^2 + \frac{1}{2}x + \frac{1}{16}\right) + 16(y^2 + 2y + 1) = -1 + 1 + 16 \iff 16\left(x + \frac{1}{4}\right)^2 + 16(y + 1)^2 = 16 \iff \left(x + \frac{1}{4}\right)^2 + (y + 1)^2 = 1$. Thus, we have a circle with center $\left(-\frac{1}{4}, -1\right)$ and radius 1.

9. $2x^2 + 2y^2 - x + y = 1 \iff 2\left(x^2 - \frac{1}{2}x + \frac{1}{16}\right) + 2\left(y^2 + \frac{1}{2}y + \frac{1}{16}\right) = 1 + \frac{1}{8} + \frac{1}{8} \iff 2\left(x - \frac{1}{4}\right)^2 + 2\left(y + \frac{1}{4}\right)^2 = \frac{5}{4} \iff \left(x - \frac{1}{4}\right)^2 + \left(y + \frac{1}{4}\right)^2 = \frac{5}{8}$. Thus, we have a circle with center $\left(\frac{1}{4}, -\frac{1}{4}\right)$ and radius $\dfrac{\sqrt{5}}{2\sqrt{2}} = \dfrac{\sqrt{10}}{4}$.

10. $x^2 + y^2 + ax + by + c = 0 \iff \left(x^2 + ax + \frac{1}{4}a^2\right) + \left(y^2 + by + \frac{1}{4}b^2\right) = -c + \frac{1}{4}a^2 + \frac{1}{4}b^2 \iff \left(x + \frac{1}{2}a\right)^2 + \left(y + \frac{1}{2}b\right)^2 = \frac{1}{4}(a^2 + b^2 - 4c)$. For this to represent a nondegenerate circle, $\frac{1}{4}(a^2 + b^2 - 4c) > 0$ or $a^2 + b^2 > 4c$. If this condition is satisfied, the circle has center $\left(-\frac{1}{2}a, -\frac{1}{2}b\right)$ and radius $\frac{1}{2}\sqrt{a^2 + b^2 - 4c}$.

11. $y = -x^2$. Parabola

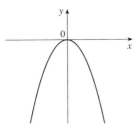

12. $y^2 - x^2 = 1$. Hyperbola

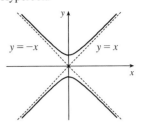

13. $x^2 + 4y^2 = 16$ ⟺ $\dfrac{x^2}{16} + \dfrac{y^2}{4} = 1$. Ellipse

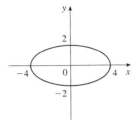

14. $x = -2y^2$. Parabola

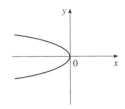

15. $16x^2 - 25y^2 = 400$ ⟺ $\dfrac{x^2}{25} - \dfrac{y^2}{16} = 1$. Hyperbola

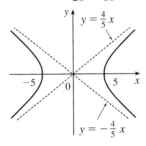

16. $25x^2 + 4y^2 = 100$ ⟺ $\dfrac{x^2}{4} + \dfrac{y^2}{25} = 1$. Ellipse

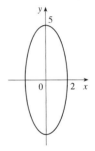

17. $4x^2 + y^2 = 1$ ⟺ $\dfrac{x^2}{1/4} + y^2 = 1$. Ellipse

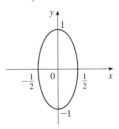

18. $y = x^2 + 2$. Parabola with vertex at $(0, 2)$

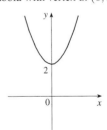

19. $x = y^2 - 1$. Parabola with vertex at $(-1, 0)$

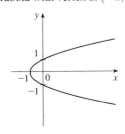

20. $9x^2 - 25y^2 = 225$ ⟺ $\dfrac{x^2}{25} - \dfrac{y^2}{9} = 1$. Hyperbola

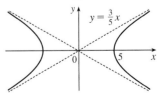

21. $9y^2 - x^2 = 9 \iff y^2 - \dfrac{x^2}{9} = 1$. Hyperbola

22. $2x^2 + 5y^2 = 10 \iff \dfrac{x^2}{5} + \dfrac{y^2}{2} = 1$. Ellipse

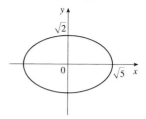

23. $xy = 4$. Hyperbola

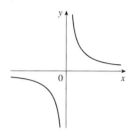

24. $y = x^2 + 2x = \left(x^2 + 2x + 1\right) - 1 = (x+1)^2 - 1$.

Parabola with vertex at $(-1, -1)$

25. $9(x-1)^2 + 4(y-2)^2 = 36 \iff$

$\dfrac{(x-1)^2}{4} + \dfrac{(y-2)^2}{9} = 1$. Ellipse centered at $(1, 2)$

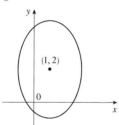

26. $16x^2 + 9y^2 - 36y = 108 \iff$

$16x^2 + 9(y^2 - 4y + 4) = 108 + 36 = 144 \iff$

$\dfrac{x^2}{9} + \dfrac{(y-2)^2}{16} = 1$. Ellipse centered at $(0, 2)$

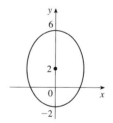

27. $y = x^2 - 6x + 13 = \left(x^2 - 6x + 9\right) + 4 = (x-3)^2 + 4$.

Parabola with vertex at $(3, 4)$

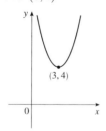

28. $x^2 - y^2 - 4x + 3 = 0 \iff$

$\left(x^2 - 4x + 4\right) - y^2 = -3 + 4 = 1 \iff$

$(x-2)^2 - y^2 = 1$. Hyperbola centered at $(2, 0)$

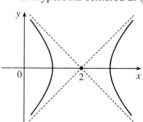

29. $x = 4 - y^2 = -y^2 + 4$. Parabola with vertex at $(4, 0)$

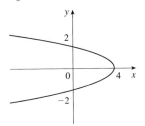

30. $y^2 - 2x + 6y + 5 = 0 \iff y^2 + 6y + 9 = 2x + 4 \iff$
$(y + 3)^2 = 2(x + 2)$. Parabola with vertex $(-2, -3)$

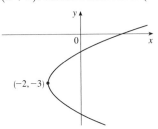

31. $x^2 + 4y^2 - 6x + 5 = 0 \iff$
$(x^2 - 6x + 9) + 4y^2 = -5 + 9 = 4 \iff$
$\dfrac{(x-3)^2}{4} + y^2 = 1$. Ellipse centered at $(3, 0)$

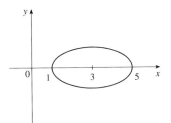

32. $4x^2 + 9y^2 - 16x + 54y + 61 = 0 \iff$
$4(x^2 - 4x + 4) + 9(y^2 + 6y + 9) = -61 + 16 + 81 = 36$
$\iff \dfrac{(x-2)^2}{9} + \dfrac{(y+3)^2}{4} = 1$. Ellipse centered at $(2, -3)$

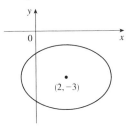

33. $y = 3x$ and $y = x^2$ intersect where $3x = x^2 \iff$
$0 = x^2 - 3x = x(x - 3)$, that is, at $(0, 0)$ and $(3, 9)$.

34. $y = 4 - x^2$, $x - 2y = 2$. Substitute y from the first
equation into the second: $x - 2(4 - x^2) = 2 \iff$
$2x^2 + x - 10 = 0 \iff (2x + 5)(x - 2) = 0 \iff$
$x = -\frac{5}{2}$ or 2. So the points of intersection are $\left(-\frac{5}{2}, -\frac{9}{4}\right)$
and $(2, 0)$.

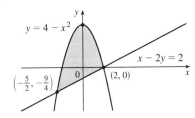

35. The parabola must have an equation of the form $y = a(x - 1)^2 - 1$. Substituting $x = 3$ and $y = 3$ into the equation gives
$3 = a(3 - 1)^2 - 1$, so $a = 1$, and the equation is $y = (x - 1)^2 - 1 = x^2 - 2x$. Note that using the other point $(-1, 3)$ would
have given the same value for a, and hence the same equation.

36. The ellipse has an equation of the form $\dfrac{x^2}{a^2} + \dfrac{y^2}{b^2} = 1$. Substituting $x = 1$ and $y = -\dfrac{10\sqrt{2}}{3}$ gives

$\dfrac{1^2}{a^2} + \dfrac{\left(-10\sqrt{2}/3\right)^2}{b^2} = \dfrac{1}{a^2} + \dfrac{200}{9b^2} = 1$. Substituting $x = -2$ and $y = \dfrac{5\sqrt{5}}{3}$ gives $\dfrac{(-2)^2}{a^2} + \dfrac{\left(5\sqrt{5}/3\right)^2}{b^2} = \dfrac{4}{a^2} + \dfrac{125}{9b^2} = 1$.

From the first equation, $\dfrac{1}{a^2} = 1 - \dfrac{200}{9b^2}$. Putting this into the second equation gives $4\left(1 - \dfrac{200}{9b^2}\right) + \dfrac{125}{9b^2} = 1 \iff$

$3 = \dfrac{675}{9b^2} \iff b^2 = \dfrac{675}{27} = 25$, so $b = 5$. Hence $\dfrac{1}{a^2} = 1 - \dfrac{200}{9(5)^2} = \dfrac{1}{9}$ and so $a = 3$. The equation of the ellipse

is $\dfrac{x^2}{9} + \dfrac{y^2}{25} = 1$.

37. $\left\{(x, y) \mid x^2 + y^2 \le 1\right\}$

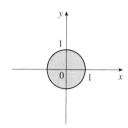

38. $\left\{(x, y) \mid x^2 + y^2 > 4\right\}$

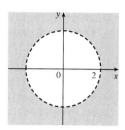

39. $\left\{(x, y) \mid y \ge x^2 - 1\right\}$

40. $\left\{(x, y) \mid x^2 + 4y^2 \le 4\right\}$

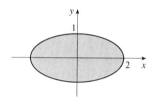

D Trigonometry

1. $210° = 210° \left(\dfrac{\pi}{180°}\right) = \dfrac{7\pi}{6}$ rad

2. $300° = 300° \left(\dfrac{\pi}{180°}\right) = \dfrac{5\pi}{3}$ rad

3. $9° = 9° \left(\dfrac{\pi}{180°}\right) = \dfrac{\pi}{20}$ rad

4. $-315° = -315° \left(\dfrac{\pi}{180°}\right) = -\dfrac{7\pi}{4}$ rad

5. $900° = 900° \left(\dfrac{\pi}{180°}\right) = 5\pi$ rad

6. $36° = 36° \left(\dfrac{\pi}{180°}\right) = \dfrac{\pi}{5}$ rad

7. 4π rad $= 4\pi \left(\dfrac{180°}{\pi}\right) = 720°$

8. $-\dfrac{7\pi}{2}$ rad $= -\dfrac{7\pi}{2} \left(\dfrac{180°}{\pi}\right) = -630°$

9. $\dfrac{5\pi}{12}$ rad $= \dfrac{5\pi}{12} \left(\dfrac{180°}{\pi}\right) = 75°$

10. $\dfrac{8\pi}{3}$ rad $= \dfrac{8\pi}{3} \left(\dfrac{180°}{\pi}\right) = 480°$

11. $-\dfrac{3\pi}{8}$ rad $= -\dfrac{3\pi}{8} \left(\dfrac{180°}{\pi}\right) = -67.5°$

12. 5 rad $= 5 \left(\dfrac{180°}{\pi}\right) = \left(\dfrac{900}{\pi}\right)°$

13. Using Formula 3, $a = r\theta = 36 \cdot \dfrac{\pi}{12} = 3\pi$ cm.

14. Using Formula 3, $a = r\theta = 10 \cdot 72° \left(\dfrac{\pi}{180°}\right) = 4\pi$ cm.

15. Using Formula 3, $\theta = a/r = \dfrac{1}{1.5} = \dfrac{2}{3}$ rad $= \dfrac{2}{3} \left(\dfrac{180°}{\pi}\right) = \left(\dfrac{120}{\pi}\right)° \approx 38.2°$.

16. $a = r\theta \implies r = \dfrac{a}{\theta} = \dfrac{6}{3\pi/4} = \dfrac{8}{\pi}$ cm

17.

315°

18.

−150°

19.

0

$-\frac{3\pi}{4}$

20.

$\frac{7\pi}{3}$

21.

2 rad

22.

−3 rad

23.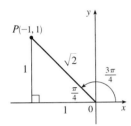

$P(-1,1)$

$\sqrt{2}$

1

$\frac{3\pi}{4}$

$\frac{\pi}{4}$

1

From the diagram we see that a point on the terminal side is $P(-1,1)$. Therefore, taking $x=-1, y=1, r=\sqrt{2}$ in the definitions of the trigonometric ratios, we have $\sin\frac{3\pi}{4}=\frac{1}{\sqrt{2}}$, $\cos\frac{3\pi}{4}=-\frac{1}{\sqrt{2}}$, $\tan\frac{3\pi}{4}=-1$, $\csc\frac{3\pi}{4}=\sqrt{2}$, $\sec\frac{3\pi}{4}=-\sqrt{2}$, and $\cot\frac{3\pi}{4}=-1$.

24.

$\frac{4\pi}{3}$

$\sqrt{3}$ 2 $\frac{\pi}{6}$

$P(-1,-\sqrt{3})$

From the diagram and Figure 8, we see that a point on the terminal side is $P(-1,-\sqrt{3})$. Therefore, taking $x=-1, y=-\sqrt{3}, r=2$ in the definitions of the trigonometric ratios, we have $\sin\frac{4\pi}{3}=-\frac{\sqrt{3}}{2}$, $\cos\frac{4\pi}{3}=-\frac{1}{2}$, $\tan\frac{4\pi}{3}=\sqrt{3}$, $\csc\frac{4\pi}{3}=-\frac{2}{\sqrt{3}}$, $\sec\frac{4\pi}{3}=-2$, and $\cot\frac{4\pi}{3}=\frac{1}{\sqrt{3}}$.

25.

$P(0,1)$

$\frac{9\pi}{2}$

From the diagram we see that a point on the terminal side is $P(0,1)$. Therefore taking $x=0, y=1, r=1$ in the definitions of the trigonometric ratios, we have $\sin\frac{9\pi}{2}=1$, $\cos\frac{9\pi}{2}=0$, $\tan\frac{9\pi}{2}=y/x$ is undefined since $x=0$, $\csc\frac{9\pi}{2}=1$, $\sec\frac{9\pi}{2}=r/x$ is undefined since $x=0$, and $\cot\frac{9\pi}{2}=0$.

26.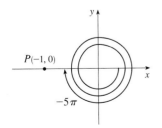

$P(-1,0)$

-5π

From the diagram, we see that a point on the terminal side is $P(-1,0)$. Therefore taking $x=-1, y=0, r=1$ in the definitions of the trigonometric ratios we have $\sin(-5\pi)=0$, $\cos(-5\pi)=-1$, $\tan(-5\pi)=0$, $\csc(-5\pi)$ is undefined, $\sec(-5\pi)=-1$, and $\cot(-5\pi)$ is undefined.

27.

Using Figure 8 we see that a point on the terminal side is $P\left(-\sqrt{3}, 1\right)$.

Therefore taking $x = -\sqrt{3}$, $y = 1$, $r = 2$ in the definitions of the

trigonometric ratios, we have $\sin \frac{5\pi}{6} = \frac{1}{2}$, $\cos \frac{5\pi}{6} = -\frac{\sqrt{3}}{2}$,

$\tan \frac{5\pi}{6} = -\frac{1}{\sqrt{3}}$, $\csc \frac{5\pi}{6} = 2$, $\sec \frac{5\pi}{6} = -\frac{2}{\sqrt{3}}$, and $\cot \frac{5\pi}{6} = -\sqrt{3}$.

28.

From the diagram, we see that a point on the terminal side is $P(-1, 1)$.

Therefore taking $x = -1$, $y = 1$, $r = \sqrt{2}$ in the definitions of the

trigonometric ratios we have $\sin \frac{11\pi}{4} = \frac{1}{\sqrt{2}}$, $\cos \frac{11\pi}{4} = -\frac{1}{\sqrt{2}}$,

$\tan \frac{11\pi}{4} = -1$, $\csc \frac{11\pi}{4} = \sqrt{2}$, $\sec \frac{11\pi}{4} = -\sqrt{2}$, and $\cot \frac{11\pi}{4} = -1$.

29. $\sin \theta = y/r = \frac{3}{5}$ $\Rightarrow$ $y = 3$, $r = 5$, and $x = \sqrt{r^2 - y^2} = 4$ (since $0 < \theta < \frac{\pi}{2}$). Therefore taking $x = 4$, $y = 3$, $r = 5$ in

the definitions of the trigonometric ratios, we have $\cos \theta = \frac{4}{5}$, $\tan \theta = \frac{3}{4}$, $\csc \theta = \frac{5}{3}$, $\sec \theta = \frac{5}{4}$, and $\cot \theta = \frac{4}{3}$.

30. Since $0 < \alpha < \frac{\pi}{2}$, α is in the first quadrant where x and y are both positive. Therefore, $\tan \alpha = y/x = \frac{2}{1}$ $\Rightarrow$ $y = 2$,

$x = 1$, and $r = \sqrt{x^2 + y^2} = \sqrt{5}$. Taking $x = 1$, $y = 2$, $r = \sqrt{5}$ in the definitions of the trigonometric ratios, we have

$\sin \alpha = \frac{2}{\sqrt{5}}$, $\cos \alpha = \frac{1}{\sqrt{5}}$, $\csc \alpha = \frac{\sqrt{5}}{2}$, $\sec \alpha = \sqrt{5}$, and $\cot \alpha = \frac{1}{2}$.

31. $\frac{\pi}{2} < \phi < \pi$ $\Rightarrow$ ϕ is in the second quadrant, where x is negative and y is positive. Therefore

$\sec \phi = r/x = -1.5 = -\frac{3}{2}$ $\Rightarrow$ $r = 3$, $x = -2$, and $y = \sqrt{r^2 - x^2} = \sqrt{5}$. Taking $x = -2$, $y = \sqrt{5}$, and $r = 3$ in the

definitions of the trigonometric ratios, we have $\sin \phi = \frac{\sqrt{5}}{3}$, $\cos \phi = -\frac{2}{3}$, $\tan \phi = -\frac{\sqrt{5}}{2}$, $\csc \phi = \frac{3}{\sqrt{5}}$, and $\cot \theta = -\frac{2}{\sqrt{5}}$.

32. Since $\pi < x < \frac{3\pi}{2}$, x is in the third quadrant where x and y are both negative. Therefore $\cos x = x/r = -\frac{1}{3}$ $\Rightarrow$ $x = -1$,

$r = 3$, and $y = -\sqrt{r^2 - x^2} = -\sqrt{8} = -2\sqrt{2}$. Taking $x = -1$, $r = 3$, $y = -2\sqrt{2}$ in the definitions of the trigonometric

ratios, we have $\sin x = -\frac{2\sqrt{2}}{3}$, $\tan x = 2\sqrt{2}$, $\csc x = -\frac{3}{2\sqrt{2}}$, $\sec x = -3$, and $\cot x = \frac{1}{2\sqrt{2}}$.

33. $\pi < \beta < 2\pi$ means that β is in the third or fourth quadrant where y is negative. Also since $\cot \beta = x/y = 3$ which is

positive, x must also be negative. Therefore $\cot \beta = x/y = \frac{3}{1}$ $\Rightarrow$ $x = -3$, $y = -1$, and $r = \sqrt{x^2 + y^2} = \sqrt{10}$. Taking

$x = -3$, $y = -1$ and $r = \sqrt{10}$ in the definitions of the trigonometric ratios, we have $\sin \beta = -\frac{1}{\sqrt{10}}$, $\cos \beta = -\frac{3}{\sqrt{10}}$,

$\tan \beta = \frac{1}{3}$, $\csc \beta = -\sqrt{10}$, and $\sec \beta = -\frac{\sqrt{10}}{3}$.

34. Since $\frac{3\pi}{2} < \theta < 2\pi$, θ is in the fourth quadrant where x is positive and y is negative. Therefore $\csc \theta = r/y = -\frac{4}{3}$ $\Rightarrow$

$r = 4$, $y = -3$, and $x = \sqrt{r^2 - y^2} = \sqrt{7}$. Taking $x = \sqrt{7}$, $y = -3$, and $r = 4$ in the definitions of the trigonometric ratios,

we have $\sin \theta = -\frac{3}{4}$, $\cos \theta = \frac{\sqrt{7}}{4}$, $\tan \theta = -\frac{3}{\sqrt{7}}$, $\sec \theta = \frac{4}{\sqrt{7}}$, and $\cot \theta = -\frac{\sqrt{7}}{3}$.

35. $\sin 35° = \dfrac{x}{10} \Rightarrow x = 10 \sin 35° \approx 5.73576$ cm

36. $\cos 40° = \dfrac{x}{25} \Rightarrow x = 25 \cos 40° \approx 19.15111$ cm

37. $\tan \dfrac{2\pi}{5} = \dfrac{x}{8} \Rightarrow x = 8 \tan \dfrac{2\pi}{5} \approx 24.62147$ cm

38. $\cos \dfrac{3\pi}{8} = \dfrac{22}{x} \Rightarrow x = \dfrac{22}{\cos \dfrac{3\pi}{8}} \approx 57.48877$ cm

39.

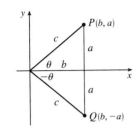

(a) From the diagram we see that $\sin \theta = \dfrac{y}{r} = \dfrac{a}{c}$, and $\sin(-\theta) = \dfrac{-a}{c} = -\dfrac{a}{c} = -\sin \theta$.

(b) Again from the diagram we see that $\cos \theta = \dfrac{x}{r} = \dfrac{b}{c} = \cos(-\theta)$.

40. (a) Using (12a) and (12b), we have

$$\tan(x+y) = \frac{\sin(x+y)}{\cos(x+y)} = \frac{\sin x \, \cos y + \cos x \, \sin y}{\cos x \, \cos y - \sin x \, \sin y} = \frac{\dfrac{\sin x \, \cos y}{\cos x \, \cos y} + \dfrac{\cos x \, \sin y}{\cos x \, \cos y}}{\dfrac{\cos x \, \cos y}{\cos x \, \cos y} - \dfrac{\sin x \, \sin y}{\cos x \, \cos y}} = \frac{\tan x + \tan y}{1 - \tan x \, \tan y}$$

(b) From (10a) and (10b), we have $\tan(-\theta) = -\tan \theta$, so (14a) implies that

$$\tan(x-y) = \tan(x+(-y)) = \frac{\tan x + \tan(-y)}{1 - \tan x \, \tan(-y)} = \frac{\tan x - \tan y}{1 + \tan x \, \tan y}$$

41. (a) Using (12a) and (13a), we have

$$\tfrac{1}{2}[\sin(x+y) + \sin(x-y)] = \tfrac{1}{2}[\sin x \, \cos y + \cos x \, \sin y + \sin x \, \cos y - \cos x \, \sin y] = \tfrac{1}{2}(2 \sin x \, \cos y) = \sin x \, \cos y.$$

(b) This time, using (12b) and (13b), we have

$$\tfrac{1}{2}[\cos(x+y) + \cos(x-y)] = \tfrac{1}{2}[\cos x \, \cos y - \sin x \, \sin y + \cos x \, \cos y + \sin x \, \sin y] = \tfrac{1}{2}(2 \cos x \, \cos y) = \cos x \, \cos y.$$

(c) Again using (12b) and (13b), we have

$$\tfrac{1}{2}[\cos(x-y) - \cos(x+y)] = \tfrac{1}{2}[\cos x \, \cos y + \sin x \, \sin y - \cos x \, \cos y + \sin x \, \sin y]$$
$$= \tfrac{1}{2}(2 \sin x \, \sin y) = \sin x \, \sin y$$

42. Using (13b), $\cos \left(\dfrac{\pi}{2} - x \right) = \cos \dfrac{\pi}{2} \cos x + \sin \dfrac{\pi}{2} \sin x = 0 \cdot \cos x + 1 \cdot \sin x = \sin x.$

43. Using (12a), we have $\sin \left(\dfrac{\pi}{2} + x \right) = \sin \dfrac{\pi}{2} \cos x + \cos \dfrac{\pi}{2} \sin x = 1 \cdot \cos x + 0 \cdot \sin x = \cos x.$

44. Using (13a), we have $\sin(\pi - x) = \sin \pi \cos x - \cos \pi \sin x = 0 \cdot \cos x - (-1) \sin x = \sin x.$

45. Using (6), we have $\sin \theta \cot \theta = \sin \theta \cdot \dfrac{\cos \theta}{\sin \theta} = \cos \theta.$

46. $(\sin x + \cos x)^2 = \sin^2 x + 2 \sin x \, \cos x + \cos^2 x = (\sin^2 x + \cos^2 x) + \sin 2x \ \ [\text{by (15a)}] \ = 1 + \sin 2x \ \ [\text{by (7)}]$

47. $\sec y - \cos y = \dfrac{1}{\cos y} - \cos y \ \ [\text{by (6)}] \ = \dfrac{1 - \cos^2 y}{\cos y} = \dfrac{\sin^2 y}{\cos y} \ \ [\text{by (7)}] \ = \dfrac{\sin y}{\cos y} \sin y = \tan y \, \sin y \ \ [\text{by (6)}]$

48. $\tan^2 \alpha - \sin^2 \alpha = \dfrac{\sin^2 \alpha}{\cos^2 \alpha} - \sin^2 \alpha = \dfrac{\sin^2 \alpha - \sin^2 \alpha \cos^2 \alpha}{\cos^2 \alpha} = \dfrac{\sin^2 \alpha \, (1 - \cos^2 \alpha)}{\cos^2 \alpha} = \tan^2 \alpha \, \sin^2 \alpha \ \ [\text{by (6), (7)}]$

49. $\cot^2\theta + \sec^2\theta = \dfrac{\cos^2\theta}{\sin^2\theta} + \dfrac{1}{\cos^2\theta}$ [by (6)] $= \dfrac{\cos^2\theta \, \cos^2\theta + \sin^2\theta}{\sin^2\theta \, \cos^2\theta}$

$\qquad = \dfrac{(1-\sin^2\theta)(1-\sin^2\theta)+\sin^2\theta}{\sin^2\theta\,\cos^2\theta}$ [by (7)] $= \dfrac{1-\sin^2\theta+\sin^4\theta}{\sin^2\theta\,\cos^2\theta}$

$\qquad = \dfrac{\cos^2\theta+\sin^4\theta}{\sin^2\theta\,\cos^2\theta}$ [by (7)] $= \dfrac{1}{\sin^2\theta} + \dfrac{\sin^2\theta}{\cos^2\theta} = \csc^2\theta + \tan^2\theta$ [by (6)]

50. $2\csc 2t = \dfrac{2}{\sin 2t} = \dfrac{2}{2\sin t\,\cos t}$ [by (15a)] $= \dfrac{1}{\sin t\,\cos t} = \sec t\,\csc t$

51. Using (14a), we have $\tan 2\theta = \tan(\theta+\theta) = \dfrac{\tan\theta+\tan\theta}{1-\tan\theta\,\tan\theta} = \dfrac{2\tan\theta}{1-\tan^2\theta}$.

52. $\dfrac{1}{1-\sin\theta} + \dfrac{1}{1+\sin\theta} = \dfrac{1+\sin\theta+1-\sin\theta}{(1-\sin\theta)(1+\sin\theta)} = \dfrac{2}{1-\sin^2\theta} = \dfrac{2}{\cos^2\theta}$ [by (7)] $= 2\sec^2\theta$

53. Using (15a) and (16a),

$\quad \sin x\,\sin 2x + \cos x\,\cos 2x = \sin x\,(2\sin x\,\cos x) + \cos x\,(2\cos^2 x - 1) = 2\sin^2 x\,\cos x + 2\cos^3 x - \cos x$

$\qquad = 2(1-\cos^2 x)\,\cos x + 2\cos^3 x - \cos x$ [by (7)]

$\qquad = 2\cos x - 2\cos^3 x + 2\cos^3 x - \cos x = \cos x$

$\quad$ *Or:* $\sin x\,\sin 2x + \cos x\,\cos 2x = \cos(2x-x)$ [by 13(b)] $= \cos x$

54. We start with the right side using equations (12a) and (13a):

$\quad \sin(x+y)\sin(x-y) = (\sin x\,\cos y + \cos x\,\sin y)(\sin x\,\cos y - \cos x\,\sin y)$

$\qquad = \sin^2 x\,\cos^2 y - \sin x\,\cos y\,\cos x\,\sin y + \cos x\,\sin y\,\sin x\,\cos y - \cos^2 x\,\sin^2 y$

$\qquad = \sin^2 x\,(1-\sin^2 y) - (1-\sin^2 x)\,\sin^2 y$ [by (7)]

$\qquad = \sin^2 x - \sin^2 x\,\sin^2 y - \sin^2 y + \sin^2 x\,\sin^2 y = \sin^2 x - \sin^2 y$

55. $\dfrac{\sin\phi}{1-\cos\phi} = \dfrac{\sin\phi}{1-\cos\phi}\cdot\dfrac{1+\cos\phi}{1+\cos\phi} = \dfrac{\sin\phi\,(1+\cos\phi)}{1-\cos^2\phi} = \dfrac{\sin\phi\,(1+\cos\phi)}{\sin^2\phi}$ [by (7)]

$\qquad = \dfrac{1+\cos\phi}{\sin\phi} = \dfrac{1}{\sin\phi} + \dfrac{\cos\phi}{\sin\phi} = \csc\phi + \cot\phi$ [by (6)]

56. $\tan x + \tan y = \dfrac{\sin x}{\cos x} + \dfrac{\sin y}{\cos y} = \dfrac{\sin x\,\cos y + \cos x\,\sin y}{\cos x\,\cos y} = \dfrac{\sin(x+y)}{\cos x\,\cos y}$ [by (12a)]

57. Using (12a),

$\quad \sin 3\theta + \sin\theta = \sin(2\theta+\theta) + \sin\theta = \sin 2\theta\,\cos\theta + \cos 2\theta\,\sin\theta + \sin\theta$

$\qquad = \sin 2\theta\,\cos\theta + (2\cos^2\theta - 1)\,\sin\theta + \sin\theta$ [by (16a)]

$\qquad = \sin 2\theta\,\cos\theta + 2\cos^2\theta\,\sin\theta - \sin\theta + \sin\theta = \sin 2\theta\,\cos\theta + \sin 2\theta\,\cos\theta$ [by (15a)]

$\qquad = 2\sin 2\theta\,\cos\theta$

58. We use (12b) with $x = 2\theta$, $y = \theta$ to get

$\quad \cos 3\theta = \cos(2\theta+\theta) = \cos 2\theta\,\cos\theta - \sin 2\theta\,\sin\theta$

$\qquad = (2\cos^2\theta - 1)\,\cos\theta - 2\sin^2\theta\,\cos\theta$ [by (16a) and (15a)]

$\qquad = (2\cos^2\theta - 1)\,\cos\theta - 2(1-\cos^2\theta)\,\cos\theta$ [by (7)]

$\qquad = 2\cos^3\theta - \cos\theta - 2\cos\theta + 2\cos^3\theta = 4\cos^3\theta - 3\cos\theta$

59. Since $\sin x = \frac{1}{3}$ we can label the opposite side as having length 1, the hypotenuse as having length 3, and use the Pythagorean Theorem to get that the adjacent side has length $\sqrt{8}$. Then, from the diagram, $\cos x = \frac{\sqrt{8}}{3}$. Similarly we have that $\sin y = \frac{3}{5}$. Now use (12a):

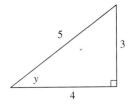

$$\sin(x + y) = \sin x \cos y + \cos x \sin y = \frac{1}{3} \cdot \frac{4}{5} + \frac{\sqrt{8}}{3} \cdot \frac{3}{5} = \frac{4}{15} + \frac{3\sqrt{8}}{15} = \frac{4 + 6\sqrt{2}}{15}.$$

60. Use (12b) and the values for $\sin y$ and $\cos x$ obtained in Exercise 59 to get

$$\cos(x + y) = \cos x \cos y - \sin x \sin y = \frac{\sqrt{8}}{3} \cdot \frac{4}{5} - \frac{1}{3} \cdot \frac{3}{5} = \frac{8\sqrt{2} - 3}{15}$$

61. Using (13b) and the values for $\cos x$ and $\sin y$ obtained in Exercise 59, we have

$$\cos(x - y) = \cos x \cos y + \sin x \sin y = \frac{\sqrt{8}}{3} \cdot \frac{4}{5} + \frac{1}{3} \cdot \frac{3}{5} = \frac{8\sqrt{2} + 3}{15}$$

62. Using (13a) and the values for $\sin y$ and $\cos x$ obtained in Exercise 59, we get

$$\sin(x - y) = \sin x \cos y - \cos x \sin y = \frac{1}{3} \cdot \frac{4}{5} - \frac{\sqrt{8}}{3} \cdot \frac{3}{5} = \frac{4 - 6\sqrt{2}}{15}$$

63. Using (15a) and the values for $\sin y$ and $\cos y$ obtained in Exercise 59, we have $\sin 2y = 2 \sin y \cos y = 2 \cdot \frac{3}{5} \cdot \frac{4}{5} = \frac{24}{25}$.

64. Using (16a) with $\cos y = \frac{4}{5}$, we have $\cos 2y = 2\cos^2 y - 1 = 2\left(\frac{4}{5}\right)^2 - 1 = \frac{32}{25} - 1 = \frac{7}{25}$.

65. $2 \cos x - 1 = 0 \iff \cos x = \frac{1}{2} \implies x = \frac{\pi}{3}, \frac{5\pi}{3}$ for $x \in [0, 2\pi]$.

66. $3 \cot^2 x = 1 \iff 3 = 1/\cot^2 x \iff \tan^2 x = 3 \iff \tan x = \pm\sqrt{3} \implies x = \frac{\pi}{3}, \frac{2\pi}{3}, \frac{4\pi}{3},$ and $\frac{5\pi}{3}$.

67. $2 \sin^2 x = 1 \iff \sin^2 x = \frac{1}{2} \iff \sin x = \pm\frac{1}{\sqrt{2}} \implies x = \frac{\pi}{4}, \frac{3\pi}{4}, \frac{5\pi}{4}, \frac{7\pi}{4}$.

68. $|\tan x| = 1 \iff \tan x = -1$ or $\tan x = 1 \iff x = \frac{3\pi}{4}, \frac{7\pi}{4}$ or $x = \frac{\pi}{4}, \frac{5\pi}{4}$.

69. Using (15a), we have $\sin 2x = \cos x \iff 2 \sin x \cos x - \cos x = 0 \iff \cos x(2 \sin x - 1) = 0 \iff \cos x = 0$ or $2 \sin x - 1 = 0 \implies x = \frac{\pi}{2}, \frac{3\pi}{2}$ or $\sin x = \frac{1}{2} \implies x = \frac{\pi}{6}$ or $\frac{5\pi}{6}$. Therefore, the solutions are $x = \frac{\pi}{6}, \frac{\pi}{2}, \frac{5\pi}{6}, \frac{3\pi}{2}$.

70. By (15a), $2 \cos x + \sin 2x = 0 \iff 2 \cos x + 2 \sin x \cos x = 0 \iff 2 \cos x (1 + \sin x) = 0 \iff \cos x = 0$ or $1 + \sin x = 0 \iff x = \frac{\pi}{2}, \frac{3\pi}{2}$ or $\sin x = -1 \implies x = \frac{3}{2}\pi$. So the solutions are $x = \frac{\pi}{2}, \frac{3\pi}{2}$.

71. $\sin x = \tan x \iff \sin x - \tan x = 0 \iff \sin x - \dfrac{\sin x}{\cos x} = 0 \iff \sin x \left(1 - \dfrac{1}{\cos x}\right) = 0 \iff \sin x = 0$ or $1 - \dfrac{1}{\cos x} = 0 \implies x = 0, \pi, 2\pi$ or $1 = \dfrac{1}{\cos x} \implies \cos x = 1 \implies x = 0, 2\pi$. Therefore the solutions are $x = 0, \pi, 2\pi$.

72. By (16a), $2 + \cos 2x = 3 \cos x \iff 2 + 2\cos^2 x - 1 = 3 \cos x \iff 2 \cos^2 x - 3 \cos x + 1 = 0 \iff (2 \cos x - 1)(\cos x - 1) = 0 \iff \cos x = 1$ or $\cos x = \frac{1}{2} \implies x = 0, 2\pi$ or $x = \frac{\pi}{3}, \frac{5\pi}{3}$.

73. We know that $\sin x = \frac{1}{2}$ when $x = \frac{\pi}{6}$ or $\frac{5\pi}{6}$, and from Figure 13(a), we see that $\sin x \leq \frac{1}{2} \implies 0 \leq x \leq \frac{\pi}{6}$ or $\frac{5\pi}{6} \leq x \leq 2\pi$ for $x \in [0, 2\pi]$.

74. $2 \cos x + 1 > 0 \implies 2 \cos x > -1 \implies \cos x > -\frac{1}{2}$. $\cos x = -\frac{1}{2}$ when $x = \frac{2\pi}{3}, \frac{4\pi}{3}$ and from Figure 13(b), we see that $\cos x > -\frac{1}{2}$ when $0 \leq x < \frac{2\pi}{3}, \frac{4\pi}{3} < x \leq 2\pi$.

75. $\tan x = -1$ when $x = \frac{3\pi}{4}, \frac{7\pi}{4}$, and $\tan x = 1$ when $x = \frac{\pi}{4}$ or $\frac{5\pi}{4}$. From Figure 14(a) we see that $-1 < \tan x < 1 \;\Rightarrow$

$0 \le x < \frac{\pi}{4}, \frac{3\pi}{4} < x < \frac{5\pi}{4}$, and $\frac{7\pi}{4} < x \le 2\pi$.

76. We know that $\sin x = \cos x$ when $x = \frac{\pi}{4}, \frac{5\pi}{4}$, and from the diagram we see that $\sin x > \cos x$ when $\frac{\pi}{4} < x < \frac{5\pi}{4}$.

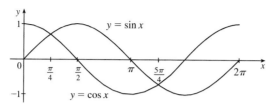

77. $y = \cos\left(x - \frac{\pi}{3}\right)$. We start with the graph of $y = \cos x$ and shift it $\frac{\pi}{3}$ units to the right.

78. $y = \tan 2x$. Start with the graph of $y = \tan x$ with period π and compress it to a period of $\frac{\pi}{2}$.

79. $y = \frac{1}{3}\tan\left(x - \frac{\pi}{2}\right)$. We start with the graph of $y = \tan x$, shift it $\frac{\pi}{2}$ units to the right and compress it to $\frac{1}{3}$ of its original vertical size.

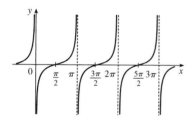

80. $y = 1 + \sec x$. Start with the graph of $y = \sec x$ and raise it by one unit.

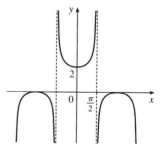

81. $y = |\sin x|$. We start with the graph of $y = \sin x$ and reflect the parts below the x-axis about the x-axis.

82. $y = 2 + \sin\left(x + \frac{\pi}{4}\right)$. Start with the graph of $y = \sin x$, and shift it $\frac{\pi}{4}$ units to the left and 2 units up.

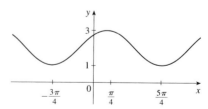

83. From the figure in the text, we see that $x = b\cos\theta$, $y = b\sin\theta$, and from the distance formula we have that the

distance c from (x, y) to $(a, 0)$ is $c = \sqrt{(x - a)^2 + (y - 0)^2} \;\Rightarrow$

$$c^2 = (b\cos\theta - a)^2 + (b\sin\theta)^2 = b^2\cos^2\theta - 2ab\cos\theta + a^2 + b^2\sin^2\theta$$

$$= a^2 + b^2(\cos^2\theta + \sin^2\theta) - 2ab\cos\theta = a^2 + b^2 - 2ab\cos\theta \quad \text{[by (7)]}$$

84. $|AB|^2 = |AC|^2 + |BC|^2 - 2\,|AC|\,|BC|\cos\angle C = (820)^2 + (910)^2 - 2(820)(910)\cos 103° \approx 1{,}836{,}217 \Rightarrow$
$|AB| \approx 1355$ m

85. Using the Law of Cosines, we have $c^2 = 1^2 + 1^2 - 2(1)(1)\cos(\alpha - \beta) = 2\left[1 - \cos(\alpha - \beta)\right]$. Now, using the distance
formula, $c^2 = |AB|^2 = (\cos\alpha - \cos\beta)^2 + (\sin\alpha - \sin\beta)^2$. Equating these two expressions for c^2, we get
$2[1 - \cos(\alpha - \beta)] = \cos^2\alpha + \sin^2\alpha + \cos^2\beta + \sin^2\beta - 2\cos\alpha\,\cos\beta - 2\sin\alpha\,\sin\beta \Rightarrow$
$1 - \cos(\alpha - \beta) = 1 - \cos\alpha\,\cos\beta - \sin\alpha\,\sin\beta \Rightarrow \cos(\alpha - \beta) = \cos\alpha\,\cos\beta + \sin\alpha\,\sin\beta.$

86. $\cos(x + y) = \cos(x - (-y)) = \cos x\,\cos(-y) + \sin x\,\sin(-y)$
$= \cos x\,\cos y - \sin x\,\sin y$ [using Equations (10a) and (10b)]

87. In Exercise 86 we used the subtraction formula for cosine to prove the addition formula for cosine. Using that formula with
$x = \frac{\pi}{2} - \alpha,\, y = \beta$, we get $\cos\left[\left(\frac{\pi}{2} - \alpha\right) + \beta\right] = \cos\left(\frac{\pi}{2} - \alpha\right)\cos\beta - \sin\left(\frac{\pi}{2} - \alpha\right)\sin\beta \Rightarrow$
$\cos\left[\frac{\pi}{2} - (\alpha - \beta)\right] = \cos\left(\frac{\pi}{2} - \alpha\right)\cos\beta - \sin\left(\frac{\pi}{2} - \alpha\right)\sin\beta$. Now we use the identities given in the problem,
$\cos\left(\frac{\pi}{2} - \theta\right) = \sin\theta$ and $\sin\left(\frac{\pi}{2} - \theta\right) = \cos\theta$, to get $\sin(\alpha - \beta) = \sin\alpha\,\cos\beta - \cos\alpha\,\sin\beta.$

88. If $0 < \theta < \frac{\pi}{2}$, we have the case depicted in the first diagram. In this case, we see that the height of the triangle
is $h = a\sin\theta$. If $\frac{\pi}{2} \le \theta < \pi$, we have the case depicted in the second diagram. In this case, the height of the triangle is
$h = a\sin(\pi - \theta) = a\sin\theta$ (by the identity proved in Exercise 44). So in either case, the area of the triangle is
$\frac{1}{2}bh = \frac{1}{2}ab\sin\theta.$

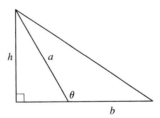

89. Using the formula from Exercise 88, the area of the triangle is $\frac{1}{2}(10)(3)\sin 107° \approx 14.34457$ cm^2.

E   Sigma Notation

1. $\displaystyle\sum_{i=1}^{5} \sqrt{i} = \sqrt{1} + \sqrt{2} + \sqrt{3} + \sqrt{4} + \sqrt{5}$

2. $\displaystyle\sum_{i=1}^{6} \frac{1}{i+1} = \frac{1}{2} + \frac{1}{3} + \frac{1}{4} + \frac{1}{5} + \frac{1}{6} + \frac{1}{7}$

3. $\displaystyle\sum_{i=4}^{6} 3^i = 3^4 + 3^5 + 3^6$

4. $\displaystyle\sum_{i=4}^{6} i^3 = 4^3 + 5^3 + 6^3$

5. $\displaystyle\sum_{k=0}^{4} \frac{2k-1}{2k+1} = -1 + \frac{1}{3} + \frac{3}{5} + \frac{5}{7} + \frac{7}{9}$

6. $\displaystyle\sum_{k=5}^{8} x^k = x^5 + x^6 + x^7 + x^8$

7. $\displaystyle\sum_{i=1}^{n} i^{10} = 1^{10} + 2^{10} + 3^{10} + \cdots + n^{10}$

8. $\displaystyle\sum_{j=n}^{n+3} j^2 = n^2 + (n+1)^2 + (n+2)^2 + (n+3)^2$

9. $\displaystyle\sum_{j=0}^{n-1} (-1)^j = 1 - 1 + 1 - 1 + \cdots + (-1)^{n-1}$

10. $\displaystyle\sum_{i=1}^{n} f(x_i)\,\Delta x_i = f(x_1)\,\Delta x_1 + f(x_2)\,\Delta x_2 + f(x_3)\,\Delta x_3 + \cdots + f(x_n)\,\Delta x_n$

11. $1 + 2 + 3 + 4 + \cdots + 10 = \sum_{i=1}^{10} i$

12. $\sqrt{3} + \sqrt{4} + \sqrt{5} + \sqrt{6} + \sqrt{7} = \sum_{i=3}^{7} \sqrt{i}$

13. $\dfrac{1}{2} + \dfrac{2}{3} + \dfrac{3}{4} + \dfrac{4}{5} + \cdots + \dfrac{19}{20} = \sum_{i=1}^{19} \dfrac{i}{i+1}$

14. $\dfrac{3}{7} + \dfrac{4}{8} + \dfrac{5}{9} + \dfrac{6}{10} + \cdots + \dfrac{23}{27} = \sum_{i=3}^{23} \dfrac{i}{i+4}$

15. $2 + 4 + 6 + 8 + \cdots + 2n = \sum_{i=1}^{n} 2i$

16. $1 + 3 + 5 + 7 + \cdots + (2n-1) = \sum_{i=1}^{n} (2i-1)$

17. $1 + 2 + 4 + 8 + 16 + 32 = \sum_{i=0}^{5} 2^i$

18. $\dfrac{1}{1} + \dfrac{1}{4} + \dfrac{1}{9} + \dfrac{1}{16} + \dfrac{1}{25} + \dfrac{1}{36} = \sum_{i=1}^{6} \dfrac{1}{i^2}$

19. $x + x^2 + x^3 + \cdots + x^n = \sum_{i=1}^{n} x^i$

20. $1 - x + x^2 - x^3 + \cdots + (-1)^n x^n = \sum_{i=0}^{n} (-1)^i x^i$

21. $\sum_{i=4}^{8} (3i-2) = [3(4)-2] + [3(5)-2] + [3(6)-2] + [3(7)-2] + [3(8)-2] = 10 + 13 + 16 + 19 + 22 = 80$

22. $\sum_{i=3}^{6} i(i+2) = 3 \cdot 5 + 4 \cdot 6 + 5 \cdot 7 + 6 \cdot 8 = 15 + 24 + 35 + 48 = 122$

23. $\sum_{j=1}^{6} 3^{j+1} = 3^2 + 3^3 + 3^4 + 3^5 + 3^6 + 3^7 = 9 + 27 + 81 + 243 + 729 + 2187 = 3276$

(For a more general method, see Exercise 47.)

24. $\sum_{k=0}^{8} \cos k\pi = \cos 0 + \cos \pi + \cos 2\pi + \cos 3\pi + \cos 4\pi + \cos 5\pi + \cos 6\pi + \cos 7\pi + \cos 8\pi$

$$= 1 - 1 + 1 - 1 + 1 - 1 + 1 - 1 + 1 = 1$$

25. $\sum_{n=1}^{20} (-1)^n = -1 + 1 - 1 + 1 - 1 + 1 - 1 + 1 - 1 + 1 - 1 + 1 - 1 + 1 - 1 + 1 - 1 + 1 - 1 + 1 = 0$

26. $\sum_{i=1}^{100} 4 = \underbrace{4 + 4 + 4 + \cdots + 4}_{(100 \text{ summands})} = 100 \cdot 4 = 400$

27. $\sum_{i=0}^{4} (2^i + i^2) = (1+0) + (2+1) + (4+4) + (8+9) + (16+16) = 61$

28. $\sum_{i=-2}^{4} 2^{3-i} = 2^5 + 2^4 + 2^3 + 2^2 + 2^1 + 2^0 + 2^{-1} = 63.5$

29. $\sum_{i=1}^{n} 2i = 2 \sum_{i=1}^{n} i = 2 \cdot \dfrac{n(n+1)}{2}$ [by Theorem 3(c)] $= n(n+1)$

30. $\sum_{i=1}^{n} (2 - 5i) = \sum_{i=1}^{n} 2 - \sum_{i=1}^{n} 5i = 2n - 5 \sum_{i=1}^{n} i = 2n - \dfrac{5n(n+1)}{2} = \dfrac{4n}{2} - \dfrac{5n^2 + 5n}{2} = -\dfrac{n(5n+1)}{2}$

31. $\sum_{i=1}^{n} (i^2 + 3i + 4) = \sum_{i=1}^{n} i^2 + 3 \sum_{i=1}^{n} i + \sum_{i=1}^{n} 4 = \dfrac{n(n+1)(2n+1)}{6} + \dfrac{3n(n+1)}{2} + 4n$

$$= \tfrac{1}{6}[(2n^3 + 3n^2 + n) + (9n^2 + 9n) + 24n] = \tfrac{1}{6}(2n^3 + 12n^2 + 34n) = \tfrac{1}{3}n(n^2 + 6n + 17)$$

32. $\displaystyle\sum_{i=1}^{n}(3+2i)^2 = \sum_{i=1}^{n}(9+12i+4i^2) = \sum_{i=1}^{n}9+12\sum_{i=1}^{n}i+4\sum_{i=1}^{n}i^2 = 9n+6n(n+1)+\frac{2n(n+1)(2n+1)}{3}$

$\displaystyle = \frac{27n+18n^2+18n+4n^3+6n^2+2n}{3} = \tfrac13(4n^3+24n^2+47n) = \tfrac13 n(4n^2+24n+47)$

33. $\displaystyle\sum_{i=1}^{n}(i+1)(i+2) = \sum_{i=1}^{n}(i^2+3i+2) = \sum_{i=1}^{n}i^2+3\sum_{i=1}^{n}i+\sum_{i=1}^{n}2 = \frac{n(n+1)(2n+1)}{6}+\frac{3n(n+1)}{2}+2n$

$\displaystyle = \frac{n(n+1)}{6}[(2n+1)+9]+2n = \frac{n(n+1)}{3}(n+5)+2n$

$\displaystyle = \frac{n}{3}[(n+1)(n+5)+6] = \frac{n}{3}(n^2+6n+11)$

34. $\displaystyle\sum_{i=1}^{n}i(i+1)(i+2) = \sum_{i=1}^{n}(i^3+3i^2+2i) = \sum_{i=1}^{n}i^3+3\sum_{i=1}^{n}i^2+2\sum_{i=1}^{n}i$

$\displaystyle = \left[\frac{n(n+1)}{2}\right]^2+\frac{3n(n+1)(2n+1)}{6}+\frac{2n(n+1)}{2}$

$\displaystyle = n(n+1)\left[\frac{n(n+1)}{4}+\frac{2n+1}{2}+1\right] = \frac{n(n+1)}{4}(n^2+n+4n+2+4)$

$\displaystyle = \frac{n(n+1)}{4}(n^2+5n+6) = \frac{n(n+1)(n+2)(n+3)}{4}$

35. $\displaystyle\sum_{i=1}^{n}(i^3-i-2) = \sum_{i=1}^{n}i^3-\sum_{i=1}^{n}i-\sum_{i=1}^{n}2 = \left[\frac{n(n+1)}{2}\right]^2-\frac{n(n+1)}{2}-2n$

$\displaystyle = \tfrac14 n(n+1)[n(n+1)-2]-2n = \tfrac14 n(n+1)(n+2)(n-1)-2n$

$\displaystyle = \tfrac14 n[(n+1)(n-1)(n+2)-8] = \tfrac14 n[(n^2-1)(n+2)-8] = \tfrac14 n(n^3+2n^2-n-10)$

36. By Theorem 3(c) we have that $\displaystyle\sum_{i=1}^{n}i = \frac{n(n+1)}{2} = 78 \Leftrightarrow n(n+1)=156 \Leftrightarrow n^2+n-156=0 \Leftrightarrow$

$(n+13)(n-12)=0 \Leftrightarrow n=12$ or -13. But $n=-13$ produces a negative answer for the sum, so $n=12$.

37. By Theorem 2(a) and Example 3, $\displaystyle\sum_{i=1}^{n}c = c\sum_{i=1}^{n}1 = cn$.

38. Let S_n be the statement that $\displaystyle\sum_{i=1}^{n}i^3 = \left[\frac{n(n+1)}{2}\right]^2$.

1. S_1 is true because $1^3 = \left(\frac{1\cdot2}{2}\right)^2$.

2. Assume S_k is true. Then $\displaystyle\sum_{i=1}^{k}i^3 = \left[\frac{k(k+1)}{2}\right]^2$, so

$\displaystyle\sum_{i=1}^{k+1}i^3 = \left[\frac{k(k+1)}{2}\right]^2+(k+1)^3 = \frac{(k+1)^2}{4}[k^2+4(k+1)] = \frac{(k+1)^2}{4}(k+2)^2 = \left(\frac{(k+1)[(k+1)+1]}{2}\right)^2$

showing that S_{k+1} is true.

Therefore, S_n is true for all n by mathematical induction.

39. $\sum_{i=1}^{n} \left[(i+1)^4 - i^4\right] = (2^4 - 1^4) + (3^4 - 2^4) + (4^4 - 3^4) + \cdots + \left[(n+1)^4 - n^4\right]$

$$= (n+1)^4 - 1^4 = n^4 + 4n^3 + 6n^2 + 4n$$

On the other hand,

$$\sum_{i=1}^{n} \left[(i+1)^4 - i^4\right] = \sum_{i=1}^{n} (4i^3 + 6i^2 + 4i + 1) = 4\sum_{i=1}^{n} i^3 + 6\sum_{i=1}^{n} i^2 + 4\sum_{i=1}^{n} i + \sum_{i=1}^{n} 1$$

$$= 4S + n(n+1)(2n+1) + 2n(n+1) + n \qquad \left[\text{where } S = \sum_{i=1}^{n} i^3\right]$$

$$= 4S + 2n^3 + 3n^2 + n + 2n^2 + 2n + n = 4S + 2n^3 + 5n^2 + 4n$$

Thus, $n^4 + 4n^3 + 6n^2 + 4n = 4S + 2n^3 + 5n^2 + 4n$, from which it follows that

$$4S = n^4 + 2n^3 + n^2 = n^2(n^2 + 2n + 1) = n^2(n+1)^2 \text{ and } S = \left[\frac{n(n+1)}{2}\right]^2.$$

40. The area of G_i is

$$\left(\sum_{k=1}^{i} k\right)^2 - \left(\sum_{k=1}^{i-1} k\right)^2 = \left[\frac{i(i+1)}{2}\right]^2 - \left[\frac{(i-1)i}{2}\right]^2 = \frac{i^2}{4}\left[(i+1)^2 - (i-1)^2\right]$$

$$= \frac{i^2}{4}\left[(i^2 + 2i + 1) - (i^2 - 2i + 1)\right] = \frac{i^2}{4}(4i) = i^3$$

Thus, the area of $ABCD$ is $\sum_{i=1}^{n} i^3 = \left[\frac{n(n+1)}{2}\right]^2.$

41. (a) $\sum_{i=1}^{n} \left[i^4 - (i-1)^4\right] = (1^4 - 0^4) + (2^4 - 1^4) + (3^4 - 2^4) + \cdots + \left[n^4 - (n-1)^4\right] = n^4 - 0 = n^4$

(b) $\sum_{i=1}^{100} \left(5^i - 5^{i-1}\right) = (5^1 - 5^0) + (5^2 - 5^1) + (5^3 - 5^2) + \cdots + \left(5^{100} - 5^{99}\right) = 5^{100} - 5^0 = 5^{100} - 1$

(c) $\sum_{i=3}^{99} \left(\frac{1}{i} - \frac{1}{i+1}\right) = \left(\frac{1}{3} - \frac{1}{4}\right) + \left(\frac{1}{4} - \frac{1}{5}\right) + \left(\frac{1}{5} - \frac{1}{6}\right) + \cdots + \left(\frac{1}{99} - \frac{1}{100}\right) = \frac{1}{3} - \frac{1}{100} = \frac{97}{300}$

(d) $\sum_{i=1}^{n} (a_i - a_{i-1}) = (a_1 - a_0) + (a_2 - a_1) + (a_3 - a_2) + \cdots + (a_n - a_{n-1}) = a_n - a_0$

42. Summing the inequalities $-|a_i| \le a_i \le |a_i|$ for $i = 1, 2, \ldots, n$, we get $-\sum_{i=1}^{n} |a_i| \le \sum_{i=1}^{n} a_i \le \sum_{i=1}^{n} |a_i|$. Since $|x| \le c \Leftrightarrow$

$-c \le x \le c$, we have $\left|\sum_{i=1}^{n} a_i\right| \le \sum_{i=1}^{n} |a_i|$.   *Another method:* Use mathematical induction.

43. $\lim_{n\to\infty} \sum_{i=1}^{n} \frac{1}{n}\left(\frac{i}{n}\right)^2 = \lim_{n\to\infty} \frac{1}{n^3} \sum_{i=1}^{n} i^2 = \lim_{n\to\infty} \frac{1}{n^3} \frac{n(n+1)(2n+1)}{6} = \lim_{n\to\infty} \frac{1}{6}\left(1 + \frac{1}{n}\right)\left(2 + \frac{1}{n}\right) = \frac{1}{6}(1)(2) = \frac{1}{3}$

44. $\lim_{n\to\infty} \sum_{i=1}^{n} \frac{1}{n}\left[\left(\frac{i}{n}\right)^3 + 1\right] = \lim_{n\to\infty} \sum_{i=1}^{n} \left[\frac{i^3}{n^4} + \frac{1}{n}\right] = \lim_{n\to\infty} \left[\frac{1}{n^4} \sum_{i=1}^{n} i^3 + \frac{1}{n} \sum_{i=1}^{n} 1\right] = \lim_{n\to\infty} \left[\frac{1}{n^4}\left(\frac{n(n+1)}{2}\right)^2 + \frac{1}{n}(n)\right]$

$$= \lim_{n\to\infty} \frac{1}{4}\left(1 + \frac{1}{n}\right)^2 + 1 = \frac{1}{4} + 1 = \frac{5}{4}$$

45. $\displaystyle\lim_{n\to\infty}\sum_{i=1}^{n}\frac{2}{n}\left[\left(\frac{2i}{n}\right)^{3}+5\left(\frac{2i}{n}\right)\right]=\lim_{n\to\infty}\sum_{i=1}^{n}\left[\frac{16}{n^{4}}i^{3}+\frac{20}{n^{2}}i\right]=\lim_{n\to\infty}\left[\frac{16}{n^{4}}\sum_{i=1}^{n}i^{3}+\frac{20}{n^{2}}\sum_{i=1}^{n}i\right]$

$\displaystyle=\lim_{n\to\infty}\left[\frac{16}{n^{4}}\frac{n^{2}(n+1)^{2}}{4}+\frac{20}{n^{2}}\frac{n(n+1)}{2}\right]=\lim_{n\to\infty}\left[\frac{4(n+1)^{2}}{n^{2}}+\frac{10n(n+1)}{n^{2}}\right]$

$\displaystyle=\lim_{n\to\infty}\left[4\left(1+\frac{1}{n}\right)^{2}+10\left(1+\frac{1}{n}\right)\right]=4\cdot1+10\cdot1=14$

46. $\displaystyle\lim_{n\to\infty}\sum_{i=1}^{n}\frac{3}{n}\left[\left(1+\frac{3i}{n}\right)^{3}-2\left(1+\frac{3i}{n}\right)\right]=\lim_{n\to\infty}\sum_{i=1}^{n}\frac{3}{n}\left[1+\frac{9i}{n}+\frac{27i^{2}}{n^{2}}+\frac{27i^{3}}{n^{3}}-2-\frac{6i}{n}\right]$

$\displaystyle=\lim_{n\to\infty}\sum_{i=1}^{n}\left[\frac{81}{n^{4}}i^{3}+\frac{81}{n^{3}}i^{2}+\frac{9}{n^{2}}i-\frac{3}{n}\right]$

$\displaystyle=\lim_{n\to\infty}\left[\frac{81}{n^{4}}\frac{n^{2}(n+1)^{2}}{4}+\frac{81}{n^{3}}\frac{n(n+1)(2n+1)}{6}+\frac{9}{n^{2}}\frac{n(n+1)}{2}-\frac{3}{n}n\right]$

$\displaystyle=\lim_{n\to\infty}\left[\frac{81}{4}\left(1+\frac{1}{n}\right)^{2}+\frac{27}{2}\left(1+\frac{1}{n}\right)\left(2+\frac{1}{n}\right)+\frac{9}{2}\left(1+\frac{1}{n}\right)-3\right]$

$\displaystyle=\frac{81}{4}+\frac{54}{2}+\frac{9}{2}-3=\frac{195}{4}$

47. Let $\displaystyle S=\sum_{i=1}^{n}ar^{i-1}=a+ar+ar^{2}+\cdots+ar^{n-1}$. Multiplying both sides by r gives us

$rS=ar+ar^{2}+\cdots+ar^{n-1}+ar^{n}$. Subtracting the first equation from the second, we find

$(r-1)S=ar^{n}-a=a(r^{n}-1)$, so $\displaystyle S=\frac{a(r^{n}-1)}{r-1}$ [since $r\neq1$].

48. $\displaystyle\sum_{i=1}^{n}\frac{3}{2^{i-1}}=3\sum_{i=1}^{n}\left(\frac{1}{2}\right)^{i-1}=\frac{3\left[\left(\frac{1}{2}\right)^{n}-1\right]}{\frac{1}{2}-1}$ [using Exercise 47 with $a=3$ and $r=\frac{1}{2}$] $=6\left[1-\left(\frac{1}{2}\right)^{n}\right]$

49. $\displaystyle\sum_{i=1}^{n}(2i+2^{i})=2\sum_{i=1}^{n}i+\sum_{i=1}^{n}2\cdot2^{i-1}=2\frac{n(n+1)}{2}+\frac{2(2^{n}-1)}{2-1}=2^{n+1}+n^{2}+n-2$.

For the first sum we have used Theorems 2(a) and 3(c), and for the second, Exercise 47 with $a=r=2$.

50. $\displaystyle\sum_{i=1}^{m}\left[\sum_{j=1}^{n}(i+j)\right]=\sum_{i=1}^{m}\left[\sum_{j=1}^{n}i+\sum_{j=1}^{n}j\right]$ [Theorem 2(b)] $=\sum_{i=1}^{m}\left[ni+\frac{n(n+1)}{2}\right]$ [Theorem 3(b) and 3(c)]

$\displaystyle=\sum_{i=1}^{m}ni+\sum_{i=1}^{m}\frac{n(n+1)}{2}=\frac{nm(m+1)}{2}+\frac{nm(n+1)}{2}=\frac{nm}{2}(m+n+2)$

G Complex Numbers

1. $(5-6i)+(3+2i)=(5+3)+(-6+2)i=8+(-4)i=8-4i$

2. $\left(4-\frac{1}{2}i\right)-\left(9+\frac{5}{2}i\right)=(4-9)+\left(-\frac{1}{2}-\frac{5}{2}\right)i=-5+(-3)i=-5-3i$

3. $(2+5i)(4-i)=2(4)+2(-i)+(5i)(4)+(5i)(-i)=8-2i+20i-5i^{2}=8+18i-5(-1)$

$=8+18i+5=13+18i$

4. $(1 - 2i)(8 - 3i) = 8 - 3i - 16i + 6(-1) = 2 - 19i$

5. $\overline{12 + 7i} = 12 - 7i$

6. $2i\left(\frac{1}{2} - i\right) = i - 2(-1) = 2 + i \Rightarrow \overline{2i\left(\frac{1}{2} - i\right)} = \overline{2 + i} = 2 - i$

7. $\dfrac{1 + 4i}{3 + 2i} = \dfrac{1 + 4i}{3 + 2i} \cdot \dfrac{3 - 2i}{3 - 2i} = \dfrac{3 - 2i + 12i - 8(-1)}{3^2 + 2^2} = \dfrac{11 + 10i}{13} = \dfrac{11}{13} + \dfrac{10}{13}i$

8. $\dfrac{3 + 2i}{1 - 4i} = \dfrac{3 + 2i}{1 - 4i} \cdot \dfrac{1 + 4i}{1 + 4i} = \dfrac{3 + 12i + 2i + 8(-1)}{1^2 + 4^2} = \dfrac{-5 + 14i}{17} = -\dfrac{5}{17} + \dfrac{14}{17}i$

9. $\dfrac{1}{1 + i} = \dfrac{1}{1 + i} \cdot \dfrac{1 - i}{1 - i} = \dfrac{1 - i}{1 - (-1)} = \dfrac{1 - i}{2} = \dfrac{1}{2} - \dfrac{1}{2}i$

10. $\dfrac{3}{4 - 3i} = \dfrac{3}{4 - 3i} \cdot \dfrac{4 + 3i}{4 + 3i} = \dfrac{12 + 9i}{16 - 9(-1)} = \dfrac{12}{25} + \dfrac{9}{25}i$

11. $i^3 = i^2 \cdot i = (-1)i = -i$

12. $i^{100} = (i^2)^{50} = (-1)^{50} = 1$

13. $\sqrt{-25} = \sqrt{25}\,i = 5i$

14. $\sqrt{-3}\,\sqrt{-12} = \sqrt{3}\,i\,\sqrt{12}\,i = \sqrt{3 \cdot 12}\,i^2 = \sqrt{36}\,(-1) = -6$

15. $\overline{12 - 5i} = 12 + 15i$ and $|12 - 15i| = \sqrt{12^2 + (-5)^2} = \sqrt{144 + 25} = \sqrt{169} = 13$

16. $\overline{-1 + 2\sqrt{2}\,i} = -1 - 2\sqrt{2}\,i$ and $\left|-1 + 2\sqrt{2}\,i\right| = \sqrt{(-1)^2 + \left(2\sqrt{2}\right)^2} = \sqrt{1 + 8} = \sqrt{9} = 3$

17. $\overline{-4i} = \overline{0 - 4i} = 0 + 4i = 4i$ and $|-4i| = \sqrt{0^2 + (-4)^2} = \sqrt{16} = 4$

18. Let $z = a + bi$ and $w = c + di$.

(a) $\overline{z + w} = \overline{(a + bi) + (c + di)} = \overline{(a + c) + (b + d)i} = (a + c) - (b + d)i = (a - bi) + (c - di) = \overline{z} + \overline{w}$

(b) $\overline{zw} = \overline{(a + bi)(c + di)} = \overline{(ac - bd) + (ad + bc)i} = (ac - bd) - (ad + bc)i.$

On the other hand, $\overline{z}\,\overline{w} = (a - bi)(c - di) = (ac - bd) - (ad + bc)i = \overline{zw}.$

(c) Use mathematical induction and part (b): Let S_n be the statement that $\overline{z^n} = \overline{z}^{\,n}$. S_1 is true because $\overline{z^1} = \overline{z} = \overline{z}^{\,1}$. Assume S_k is true, that is $\overline{z^k} = \overline{z}^{\,k}$. Then $\overline{z^{k+1}} = \overline{z^{1+k}} = \overline{z z^k} = \overline{z}\,\overline{z^k}$ [part (b) with $w = z^k$] $= \overline{z}^{\,1}\overline{z}^{\,k} = \overline{z}^{\,1+k} = \overline{z}^{\,k+1}$, which shows that S_{k+1} is true. Therefore, by mathematical induction, $\overline{z^n} = \overline{z}^{\,n}$ for every positive integer n.

Another proof: Use part (b) with $w = z$, and mathematical induction.

19. $4x^2 + 9 = 0 \Leftrightarrow 4x^2 = -9 \Leftrightarrow x^2 = -\frac{9}{4} \Leftrightarrow x = \pm\sqrt{-\frac{9}{4}} = \pm\sqrt{\frac{9}{4}}\,i = \pm\frac{3}{2}i.$

20. $x^4 = 1 \Leftrightarrow x^4 - 1 = 0 \Leftrightarrow (x^2 - 1)(x^2 + 1) = 0 \Leftrightarrow x^2 - 1 = 0$ or $x^2 + 1 = 0 \Leftrightarrow x = \pm 1$ or $x = \pm i.$

21. By the quadratic formula, $x^2 + 2x + 5 = 0 \Leftrightarrow x = \dfrac{-2 \pm \sqrt{2^2 - 4(1)(5)}}{2(1)} = \dfrac{-2 \pm \sqrt{-16}}{2} = \dfrac{-2 \pm 4i}{2} = -1 \pm 2i.$

22. $2x^2 - 2x + 1 = 0 \iff x = \dfrac{-(-2) \pm \sqrt{(-2)^2 - 4(2)(1)}}{2(2)} = \dfrac{2 \pm \sqrt{-4}}{4} = \dfrac{2 \pm 2i}{4} = \dfrac{1}{2} \pm \dfrac{1}{2} i$

23. By the quadratic formula, $z^2 + z + 2 = 0 \iff z = \dfrac{-1 \pm \sqrt{1^2 - 4(1)(2)}}{2(1)} = \dfrac{-1 \pm \sqrt{-7}}{2} = -\dfrac{1}{2} \pm \dfrac{\sqrt{7}}{2} i.$

24. $z^2 + \frac{1}{2} z + \frac{1}{4} = 0 \iff 4z^2 + 2z + 1 = 0 \iff$

$z = \dfrac{-2 \pm \sqrt{2^2 - 4(4)(1)}}{2(4)} = \dfrac{-2 \pm \sqrt{-12}}{8} = \dfrac{-2 \pm 2\sqrt{3}\, i}{8} = -\dfrac{1}{4} \pm \dfrac{\sqrt{3}}{4} i$

25. For $z = -3 + 3i$, $r = \sqrt{(-3)^2 + 3^2} = 3\sqrt{2}$ and $\tan\theta = \frac{3}{-3} = -1 \implies \theta = \frac{3\pi}{4}$ (since z lies in the second quadrant).

Therefore, $-3 + 3i = 3\sqrt{2}\left(\cos\frac{3\pi}{4} + i\sin\frac{3\pi}{4}\right)$.

26. For $z = 1 - \sqrt{3}\, i$, $r = \sqrt{1^2 + \left(-\sqrt{3}\right)^2} = 2$ and $\tan\theta = \frac{-\sqrt{3}}{1} = -\sqrt{3} \implies \theta = \frac{5\pi}{3}$ (since z lies in the fourth quadrant).

Therefore, $1 - \sqrt{3}\, i = 2\left(\cos\frac{5\pi}{3} + i\sin\frac{5\pi}{3}\right)$.

27. For $z = 3 + 4i$, $r = \sqrt{3^2 + 4^2} = 5$ and $\tan\theta = \frac{4}{3} \implies \theta = \tan^{-1}\left(\frac{4}{3}\right)$ (since z lies in the first quadrant). Therefore,

$3 + 4i = 5\left[\cos\left(\tan^{-1}\frac{4}{3}\right) + i\sin\left(\tan^{-1}\frac{4}{3}\right)\right].$

28. For $z = 8i$, $r = \sqrt{0^2 + 8^2} = 8$ and $\tan\theta = \frac{8}{0}$ is undefined, so $\theta = \frac{\pi}{2}$ (since z lies on the positive imaginary axis). Therefore,

$8i = 8\left(\cos\frac{\pi}{2} + i\sin\frac{\pi}{2}\right)$.

29. For $z = \sqrt{3} + i$, $r = \sqrt{\left(\sqrt{3}\right)^2 + 1^2} = 2$ and $\tan\theta = \frac{1}{\sqrt{3}} \implies \theta = \frac{\pi}{6} \implies z = 2\left(\cos\frac{\pi}{6} + i\sin\frac{\pi}{6}\right)$.

For $w = 1 + \sqrt{3}\, i$, $r = 2$ and $\tan\theta = \sqrt{3} \implies \theta = \frac{\pi}{3} \implies w = 2\left(\cos\frac{\pi}{3} + i\sin\frac{\pi}{3}\right)$.

Therefore, $zw = 2 \cdot 2\left[\cos\left(\frac{\pi}{6} + \frac{\pi}{3}\right) + i\sin\left(\frac{\pi}{6} + \frac{\pi}{3}\right)\right] = 4\left(\cos\frac{\pi}{2} + i\sin\frac{\pi}{2}\right)$,

$z/w = \frac{2}{2}\left[\cos\left(\frac{\pi}{6} - \frac{\pi}{3}\right) + i\sin\left(\frac{\pi}{6} - \frac{\pi}{3}\right)\right] = \cos\left(-\frac{\pi}{6}\right) + i\sin\left(-\frac{\pi}{6}\right)$, and $1 = 1 + 0i = 1(\cos 0 + i\sin 0) \implies$

$1/z = \frac{1}{2}\left[\cos\left(0 - \frac{\pi}{6}\right) + i\sin\left(0 - \frac{\pi}{6}\right)\right] = \frac{1}{2}\left[\cos\left(-\frac{\pi}{6}\right) + i\sin\left(-\frac{\pi}{6}\right)\right]$. For $1/z$, we could also use the formula that precedes

Example 5 to obtain $1/z = \frac{1}{2}\left(\cos\frac{\pi}{6} - i\sin\frac{\pi}{6}\right)$.

30. For $z = 4\sqrt{3} - 4i$, $r = \sqrt{\left(4\sqrt{3}\right)^2 + (-4)^2} = \sqrt{64} = 8$ and $\tan\theta = \frac{-4}{4\sqrt{3}} = -\frac{1}{\sqrt{3}} \implies \theta = \frac{11\pi}{6} \implies$

$z = 8\left(\cos\frac{11\pi}{6} + i\sin\frac{11\pi}{6}\right)$. For $w = 8i$, $r = \sqrt{0^2 + 8^2} = 8$ and $\tan\theta = \frac{8}{0}$ is undefined, so $\theta = \frac{\pi}{2} \implies$

$w = 8\left(\cos\frac{\pi}{2} + i\sin\frac{\pi}{2}\right)$. Therefore, $zw = 8 \cdot 8\left[\cos\left(\frac{11\pi}{6} + \frac{\pi}{2}\right) + i\sin\left(\frac{11\pi}{6} + \frac{\pi}{2}\right)\right] = 64\left(\cos\frac{\pi}{3} + i\sin\frac{\pi}{3}\right)$,

$z/w = \frac{8}{8}\left[\cos\left(\frac{11\pi}{6} - \frac{\pi}{2}\right) + i\sin\left(\frac{11\pi}{6} - \frac{\pi}{2}\right)\right] = \cos\frac{4\pi}{3} + i\sin\frac{4\pi}{3}$, and

$1 = 1 + 0i = 1(\cos 0 + i\sin 0) \implies 1/z = \frac{1}{8}\left[\cos\left(0 - \frac{11\pi}{6}\right) + i\sin\left(0 - \frac{11\pi}{6}\right)\right] = \frac{1}{8}\left[\cos\left(\frac{\pi}{6}\right) + i\sin\left(\frac{\pi}{6}\right)\right]$.

For $1/z$, we could also use the formula that precedes Example 5 to obtain $1/z = \frac{1}{8}\left(\cos\frac{11\pi}{6} - i\sin\frac{11\pi}{6}\right)$.

31. For $z = 2\sqrt{3} - 2i$, $r = \sqrt{\left(2\sqrt{3}\right)^2 + (-2)^2} = 4$ and $\tan\theta = \frac{-2}{2\sqrt{3}} = -\frac{1}{\sqrt{3}} \Rightarrow \theta = -\frac{\pi}{6} \Rightarrow$

$z = 4\left[\cos\left(-\frac{\pi}{6}\right) + i\sin\left(-\frac{\pi}{6}\right)\right]$. For $w = -1 + i$, $r = \sqrt{2}$, $\tan\theta = \frac{1}{-1} = -1 \Rightarrow \theta = \frac{3\pi}{4} \Rightarrow$

$w = \sqrt{2}\left(\cos\frac{3\pi}{4} + i\sin\frac{3\pi}{4}\right)$. Therefore, $zw = 4\sqrt{2}\left[\cos\left(-\frac{\pi}{6} + \frac{3\pi}{4}\right) + i\sin\left(-\frac{\pi}{6} + \frac{3\pi}{4}\right)\right] = 4\sqrt{2}\left(\cos\frac{7\pi}{12} + i\sin\frac{7\pi}{12}\right)$,

$z/w = \frac{4}{\sqrt{2}}\left[\cos\left(-\frac{\pi}{6} - \frac{3\pi}{4}\right) + i\sin\left(-\frac{\pi}{6} - \frac{3\pi}{4}\right)\right] = \frac{4}{\sqrt{2}}\left[\cos\left(-\frac{11\pi}{12}\right) + i\sin\left(-\frac{11\pi}{12}\right)\right] = 2\sqrt{2}\left(\cos\frac{13\pi}{12} + i\sin\frac{13\pi}{12}\right)$, and

$1/z = \frac{1}{4}\left[\cos\left(-\frac{\pi}{6}\right) - i\sin\left(-\frac{\pi}{6}\right)\right] = \frac{1}{4}\left(\cos\frac{\pi}{6} + i\sin\frac{\pi}{6}\right)$.

32. For $z = 4\left(\sqrt{3} + i\right) = 4\sqrt{3} + 4i$, $r = \sqrt{\left(4\sqrt{3}\right)^2 + 4^2} = \sqrt{64} = 8$ and $\tan\theta = \frac{4}{4\sqrt{3}} = \frac{1}{\sqrt{3}} \Rightarrow \theta = \frac{\pi}{6} \Rightarrow$

$z = 8\left(\cos\frac{\pi}{6} + i\sin\frac{\pi}{6}\right)$. For $w = -3 - 3i$, $r = \sqrt{(-3)^2 + (-3)^2} = \sqrt{18} = 3\sqrt{2}$ and $\tan\theta = \frac{-3}{-3} = 1 \Rightarrow \theta = \frac{5\pi}{4} \Rightarrow$

$w = 3\sqrt{2}\left(\cos\frac{5\pi}{4} + i\sin\frac{5\pi}{4}\right)$. Therefore, $zw = 8 \cdot 3\sqrt{2}\left[\cos\left(\frac{\pi}{6} + \frac{5\pi}{4}\right) + i\sin\left(\frac{\pi}{6} + \frac{5\pi}{4}\right)\right] = 24\sqrt{2}\left(\cos\frac{17\pi}{12} + i\sin\frac{17\pi}{12}\right)$,

$z/w = \frac{8}{3\sqrt{2}}\left[\cos\left(\frac{\pi}{6} - \frac{5\pi}{4}\right) + i\sin\left(\frac{\pi}{6} - \frac{5\pi}{4}\right)\right] = \frac{4\sqrt{2}}{3}\left[\cos\left(-\frac{13\pi}{12}\right) + i\sin\left(-\frac{13\pi}{12}\right)\right]$, and $1/z = \frac{1}{8}\left(\cos\frac{\pi}{6} - i\sin\frac{\pi}{6}\right)$.

33. For $z = 1 + i$, $r = \sqrt{2}$ and $\tan\theta = \frac{1}{1} = 1 \Rightarrow \theta = \frac{\pi}{4} \Rightarrow z = \sqrt{2}\left(\cos\frac{\pi}{4} + i\sin\frac{\pi}{4}\right)$. So by De Moivre's Theorem,

$$(1 + i)^{20} = \left[\sqrt{2}\left(\cos\frac{\pi}{4} + i\sin\frac{\pi}{4}\right)\right]^{20} = (2^{1/2})^{20}\left(\cos\frac{20 \cdot \pi}{4} + i\sin\frac{20 \cdot \pi}{4}\right) = 2^{10}(\cos 5\pi + i\sin 5\pi)$$

$$= 2^{10}[-1 + i(0)] = -2^{10} = -1024$$

34. For $z = 1 - \sqrt{3}i$, $r = \sqrt{1^2 + \left(-\sqrt{3}\right)^2} = 2$ and $\tan\theta = \frac{-\sqrt{3}}{1} = -\sqrt{3} \Rightarrow \theta = \frac{5\pi}{3} \Rightarrow z = 2\left(\cos\frac{5\pi}{3} + i\sin\frac{5\pi}{3}\right)$.

So by De Moivre's Theorem,

$$\left(1 - \sqrt{3}i\right)^5 = \left[2\left(\cos\frac{5\pi}{3} + i\sin\frac{5\pi}{3}\right)\right]^5 = 2^5\left(\cos\frac{5 \cdot 5\pi}{3} + i\sin\frac{5 \cdot 5\pi}{3}\right) = 2^5\left(\cos\frac{\pi}{3} + i\sin\frac{\pi}{3}\right)$$

$$= 32\left(\frac{1}{2} + \frac{\sqrt{3}}{2}i\right) = 16 + 16\sqrt{3}i$$

35. For $z = 2\sqrt{3} + 2i$, $r = \sqrt{\left(2\sqrt{3}\right)^2 + 2^2} = \sqrt{16} = 4$ and $\tan\theta = \frac{2}{2\sqrt{3}} = \frac{1}{\sqrt{3}} \Rightarrow \theta = \frac{\pi}{6} \Rightarrow z = 4\left(\cos\frac{\pi}{6} + i\sin\frac{\pi}{6}\right)$.

So by De Moivre's Theorem,

$$\left(2\sqrt{3} + 2i\right)^5 = \left[4\left(\cos\frac{\pi}{6} + i\sin\frac{\pi}{6}\right)\right]^5 = 4^5\left(\cos\frac{5\pi}{6} + i\sin\frac{5\pi}{6}\right) = 1024\left[-\frac{\sqrt{3}}{2} + \frac{1}{2}i\right] = -512\sqrt{3} + 512i.$$

36. For $z = 1 - i$, $r = \sqrt{2}$ and $\tan\theta = \frac{-1}{1} = -1 \Rightarrow \theta = \frac{7\pi}{4} \Rightarrow z = \sqrt{2}\left(\cos\frac{7\pi}{4} + i\sin\frac{7\pi}{4}\right) \Rightarrow$

$$(1 - i)^8 = \left[\sqrt{2}\left(\cos\frac{7\pi}{4} + i\sin\frac{7\pi}{4}\right)\right]^8 = 2^4\left(\cos\frac{8 \cdot 7\pi}{4} + i\sin\frac{8 \cdot 7\pi}{4}\right) = 16(\cos 14\pi + i\sin 14\pi) = 16(1 + 0i) = 16.$$

37. $1 = 1 + 0i = 1(\cos 0 + i\sin 0)$. Using Equation 3 with $r = 1$, $n = 8$, and $\theta = 0$, we have

$$w_k = 1^{1/8}\left[\cos\left(\frac{0 + 2k\pi}{8}\right) + i\sin\left(\frac{0 + 2k\pi}{8}\right)\right] = \cos\frac{k\pi}{4} + i\sin\frac{k\pi}{4}, \text{ where } k = 0, 1, 2, \ldots, 7.$$

$w_0 = 1(\cos 0 + i\sin 0) = 1$, $w_1 = 1\left(\cos\frac{\pi}{4} + i\sin\frac{\pi}{4}\right) = \frac{1}{\sqrt{2}} + \frac{1}{\sqrt{2}}i$,

$w_2 = 1\left(\cos\frac{\pi}{2} + i\sin\frac{\pi}{2}\right) = i$, $w_3 = 1\left(\cos\frac{3\pi}{4} + i\sin\frac{3\pi}{4}\right) = -\frac{1}{\sqrt{2}} + \frac{1}{\sqrt{2}}i$,

$w_4 = 1(\cos\pi + i\sin\pi) = -1$, $w_5 = 1\left(\cos\frac{5\pi}{4} + i\sin\frac{5\pi}{4}\right) = -\frac{1}{\sqrt{2}} - \frac{1}{\sqrt{2}}i$,

$w_6 = 1\left(\cos\frac{3\pi}{2} + i\sin\frac{3\pi}{2}\right) = -i$, $w_7 = 1\left(\cos\frac{7\pi}{4} + i\sin\frac{7\pi}{4}\right) = \frac{1}{\sqrt{2}} - \frac{1}{\sqrt{2}}i$

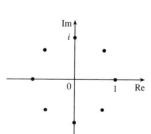

38. $32 = 32 + 0i = 32(\cos 0 + i \sin 0)$. Using Equation 3 with $r = 32$, $n = 5$, and $\theta = 0$, we have

$$w_k = 32^{1/5}\left[\cos\left(\frac{0 + 2k\pi}{5}\right) + i \sin\left(\frac{0 + 2k\pi}{5}\right)\right] = 2\left(\cos \tfrac{2}{5}\pi k + i \sin \tfrac{2}{5}\pi k\right), \text{ where } k = 0, 1, 2, 3, 4.$$

$w_0 = 2(\cos 0 + i \sin 0) = 2$

$w_1 = 2\left(\cos \frac{2\pi}{5} + i \sin \frac{2\pi}{5}\right)$

$w_2 = 2\left(\cos \frac{4\pi}{5} + i \sin \frac{4\pi}{5}\right)$

$w_3 = 2\left(\cos \frac{6\pi}{5} + i \sin \frac{6\pi}{5}\right)$

$w_4 = 2\left(\cos \frac{8\pi}{5} + i \sin \frac{8\pi}{5}\right)$

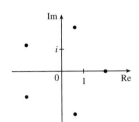

39. $i = 0 + i = 1\left(\cos \frac{\pi}{2} + i \sin \frac{\pi}{2}\right)$. Using Equation 3 with $r = 1$, $n = 3$, and $\theta = \frac{\pi}{2}$, we have

$$w_k = 1^{1/3}\left[\cos\left(\frac{\frac{\pi}{2} + 2k\pi}{3}\right) + i \sin\left(\frac{\frac{\pi}{2} + 2k\pi}{3}\right)\right], \text{ where } k = 0, 1, 2.$$

$w_0 = \left(\cos \frac{\pi}{6} + i \sin \frac{\pi}{6}\right) = \frac{\sqrt{3}}{2} + \frac{1}{2}i$

$w_1 = \left(\cos \frac{5\pi}{6} + i \sin \frac{5\pi}{6}\right) = -\frac{\sqrt{3}}{2} + \frac{1}{2}i$

$w_2 = \left(\cos \frac{9\pi}{6} + i \sin \frac{9\pi}{6}\right) = -i$

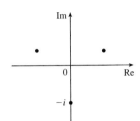

40. $1 + i = \sqrt{2}\left(\cos \frac{\pi}{4} + i \sin \frac{\pi}{4}\right)$. Using Equation 3 with $r = \sqrt{2}$, $n = 3$, and $\theta = \frac{\pi}{4}$, we have

$$w_k = \left(\sqrt{2}\right)^{1/3}\left[\cos\left(\frac{\frac{\pi}{4} + 2k\pi}{3}\right) + i \sin\left(\frac{\frac{\pi}{4} + 2k\pi}{3}\right)\right], \text{ where } k = 0, 1, 2.$$

$w_0 = 2^{1/6}\left(\cos \frac{\pi}{12} + i \sin \frac{\pi}{12}\right)$

$w_1 = 2^{1/6}\left(\cos \frac{3\pi}{4} + i \sin \frac{3\pi}{4}\right) = 2^{1/6}\left(-\frac{1}{\sqrt{2}} + \frac{1}{\sqrt{2}}i\right) = -2^{-1/3} + 2^{-1/3}i$

$w_2 = 2^{1/6}\left(\cos \frac{17\pi}{12} + i \sin \frac{17\pi}{12}\right)$

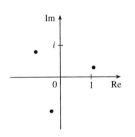

41. Using Euler's formula (6) with $y = \frac{\pi}{2}$, we have $e^{i\pi/2} = \cos \frac{\pi}{2} + i \sin \frac{\pi}{2} = 0 + 1i = i$.

42. Using Euler's formula (6) with $y = 2\pi$, we have $e^{2\pi i} = \cos 2\pi + i \sin 2\pi = 1$.

43. Using Euler's formula (6) with $y = \frac{\pi}{3}$, we have $e^{i\pi/3} = \cos \frac{\pi}{3} + i \sin \frac{\pi}{3} = \frac{1}{2} + \frac{\sqrt{3}}{2}i$.

44. Using Euler's formula (6) with $y = -\pi$, we have $e^{-i\pi} = \cos(-\pi) + i \sin(-\pi) = -1$.

45. Using Equation 7 with $x = 2$ and $y = \pi$, we have $e^{2+i\pi} = e^2 e^{i\pi} = e^2(\cos \pi + i \sin \pi) = e^2(-1 + 0) = -e^2$.

46. Using Equation 7 with $x = \pi$ and $y = 1$, we have $e^{\pi+i} = e^\pi \cdot e^{1i} = e^\pi(\cos 1 + i \sin 1) = e^\pi \cos 1 + (e^\pi \sin 1)i$.

47. Take $r = 1$ and $n = 3$ in De Moivre's Theorem to get

$$[1(\cos\theta + i\sin\theta)]^3 = 1^3(\cos 3\theta + i\sin 3\theta)$$

$$(\cos\theta + i\sin\theta)^3 = \cos 3\theta + i\sin 3\theta$$

$$\cos^3\theta + 3(\cos^2\theta)(i\sin\theta) + 3(\cos\theta)(i\sin\theta)^2 + (i\sin\theta)^3 = \cos 3\theta + i\sin 3\theta$$

$$\cos^3\theta + (3\cos^2\theta\,\sin\theta)i - 3\cos\theta\,\sin^2\theta - (\sin^3\theta)i = \cos 3\theta + i\sin 3\theta$$

$$(\cos^3\theta - 3\sin^2\theta\,\cos\theta) + (3\sin\theta\,\cos^2\theta - \sin^3\theta)i = \cos 3\theta + i\sin 3\theta$$

Equating real and imaginary parts gives $\cos 3\theta = \cos^3\theta - 3\sin^2\theta\,\cos\theta$ and $\sin 3\theta = 3\sin\theta\,\cos^2\theta - \sin^3\theta$.

48. Using Formula 6,

$$e^{ix} + e^{-ix} = (\cos x + i\sin x) + [\cos(-x) + i\sin(-x)] = \cos x + i\sin x + \cos x - i\sin x = 2\cos x$$

Thus, $\cos x = \dfrac{e^{ix} + e^{-ix}}{2}$. Similarly,

$$e^{ix} - e^{-ix} = (\cos x + i\sin x) - [\cos(-x) + i\sin(-x)] = \cos x + i\sin x - \cos x - (-i\sin x) = 2i\sin x$$

Therefore, $\sin x = \dfrac{e^{ix} - e^{-ix}}{2i}$.

49. $F(x) = e^{rx} = e^{(a+bi)x} = e^{ax+bxi} = e^{ax}(\cos bx + i\sin bx) = e^{ax}\cos bx + i(e^{ax}\sin bx) \;\Rightarrow$

$$F'(x) = (e^{ax}\cos bx)' + i(e^{ax}\sin bx)'$$

$$= (ae^{ax}\cos bx - be^{ax}\sin bx) + i(ae^{ax}\sin bx + be^{ax}\cos bx)$$

$$= a[e^{ax}(\cos bx + i\sin bx)] + b[e^{ax}(-\sin bx + i\cos bx)]$$

$$= ae^{rx} + b[e^{ax}(i^2\sin bx + i\cos bx)]$$

$$= ae^{rx} + bi[e^{ax}(\cos bx + i\sin bx)] = ae^{rx} + bie^{rx} = (a+bi)e^{rx} = re^{rx}$$

50. (a) From Exercise 49, $F(x) = e^{(1+i)x} \;\Rightarrow\; F'(x) = (1+i)e^{(1+i)x}$. So

$$\int e^{(1+i)x}\,dx = \frac{1}{1+i}\int F'(x)\,dx = \frac{1}{1+i}F(x) + C = \frac{1-i}{2}F(x) + C = \frac{1-i}{2}e^{(1+i)x} + C$$

(b) $\int e^{(1+i)x}\,dx = \int e^x e^{ix}\,dx = \int e^x(\cos x + i\sin x)\,dx = \int e^x\cos x\,dx + i\int e^x\sin x$ **(1)**.

Also,

$$\frac{1-i}{2}e^{(1+i)x} = \tfrac{1}{2}e^{(1+i)x} - \tfrac{1}{2}ie^{(1+i)x} = \tfrac{1}{2}e^{x+ix} - \tfrac{1}{2}ie^{x+ix}$$

$$= \tfrac{1}{2}e^x(\cos x + i\sin x) - \tfrac{1}{2}ie^x(\cos x + i\sin x)$$

$$= \tfrac{1}{2}e^x\cos x + \tfrac{1}{2}e^x\sin x + \tfrac{1}{2}ie^x\sin x - \tfrac{1}{2}ie^x\cos x$$

$$= \tfrac{1}{2}e^x(\cos x + \sin x) + i\left[\tfrac{1}{2}e^x(\sin x - \cos x)\right] \quad \textbf{(2)}$$

Equating the real and imaginary parts in **(1)** and **(2)**, we see that $\int e^x\cos x\,dx = \tfrac{1}{2}e^x(\cos x + \sin x) + C$ and $\int e^x\sin x\,dx = \tfrac{1}{2}e^x(\sin x - \cos x) + C$.

8. $x^2 y' - y = 2x^3 e^{-1/x}$ $\Rightarrow$ $y' - \dfrac{1}{x^2} y = 2xe^{-1/x}$ $(\star)$. This is a linear equation and the integrating factor is

$I(x) = e^{\int(-1/x^2)\,dx} = e^{1/x}$. Multiplying $(\star)$ by $e^{1/x}$ gives $e^{1/x}\,y' - e^{1/x} \cdot \dfrac{1}{x^2} y = 2x$ $\Rightarrow$ $(e^{1/x}\,y)' = 2x$ $\Rightarrow$

$e^{1/x}\,y = x^2 + C$ $\Rightarrow$ $y = e^{-1/x}(x^2 + C)$.

9. $\dfrac{dr}{dt} + 2tr = r$ $\Rightarrow$ $\dfrac{dr}{dt} = r - 2tr = r(1 - 2t)$ $\Rightarrow$ $\displaystyle\int \dfrac{dr}{r} = \int (1 - 2t)\,dt$ $\Rightarrow$ $\ln|r| = t - t^2 + C$ $\Rightarrow$

$|r| = e^{t - t^2 + C} = ke^{t-t^2}$. Since $r(0) = 5$, $5 = ke^0 = k$. Thus, $r(t) = 5e^{t-t^2}$.

10. $(1 + \cos x)y' = (1 + e^{-y})\sin x$ $\Rightarrow$ $\dfrac{dy}{1 + e^{-y}} = \dfrac{\sin x\,dx}{1 + \cos x}$ $\Rightarrow$ $\displaystyle\int \dfrac{dy}{1 + 1/e^y} = \int \dfrac{\sin x\,dx}{1 + \cos x}$ $\Rightarrow$

$\displaystyle\int \dfrac{e^y\,dy}{1 + e^y} = \int \dfrac{\sin x\,dx}{1 + \cos x}$ $\Rightarrow$ $\ln|1 + e^y| = -\ln|1 + \cos x| + C$ $\Rightarrow$ $\ln(1 + e^y) = -\ln(1 + \cos x) + C$ $\Rightarrow$

$1 + e^y = e^{-\ln(1+\cos x)} \cdot e^C$ $\Rightarrow$ $e^y = ke^{-\ln(1+\cos x)} - 1$ $\Rightarrow$ $y = \ln[ke^{-\ln(1+\cos x)} - 1]$. Since $y(0) = 0$,

$0 = \ln[ke^{-\ln 2} - 1]$ $\Rightarrow$ $e^0 = k\left(\tfrac{1}{2}\right) - 1$ $\Rightarrow$ $k = 4$. Thus, $y(x) = \ln[4e^{-\ln(1+\cos x)} - 1]$. An equivalent form

is $y(x) = \ln\dfrac{3 - \cos x}{1 + \cos x}$.

11. $xy' - y = x\ln x$ $\Rightarrow$ $y' - \dfrac{1}{x} y = \ln x$. $I(x) = e^{\int(-1/x)\,dx} = e^{-\ln|x|} = \left(e^{\ln|x|}\right)^{-1} = |x|^{-1} = 1/x$ since the condition

$y(1) = 2$ implies that we want a solution with $x > 0$. Multiplying the last differential equation by $I(x)$ gives

$\dfrac{1}{x} y' - \dfrac{1}{x^2} y = \dfrac{1}{x}\ln x$ $\Rightarrow$ $\left(\dfrac{1}{x} y\right)' = \dfrac{1}{x}\ln x$ $\Rightarrow$ $\dfrac{1}{x} y = \displaystyle\int \dfrac{\ln x}{x}\,dx$ $\Rightarrow$ $\dfrac{1}{x} y = \tfrac{1}{2}(\ln x)^2 + C$ $\Rightarrow$

$y = \tfrac{1}{2}x(\ln x)^2 + Cx$. Now $y(1) = 2$ $\Rightarrow$ $2 = 0 + C$ $\Rightarrow$ $C = 2$, so $y = \tfrac{1}{2}x(\ln x)^2 + 2x$.

12. $y' = 3x^2 e^y$ $\Rightarrow$ $\dfrac{dy}{dx} = 3x^2 e^y$ $\Rightarrow$ $e^{-y}\,dy = 3x^2\,dx$ $\Rightarrow$

$\int e^{-y}\,dy = \int 3x^2\,dx$ $\Rightarrow$ $-e^{-y} = x^3 + C$. Now $y(0) = 1$ $\Rightarrow$

$-e^{-1} = C$, so $-e^{-y} = x^3 - e^{-1}$ $\Rightarrow$ $e^{-y} = -x^3 + e^{-1}$ $\Rightarrow$

$-y = \ln(-x^3 + e^{-1})$ $\Rightarrow$ $y = -\ln(-x^3 + e^{-1})$. To find the domain,

solve $-x^3 + e^{-1} > 0$ $\Rightarrow$ $x^3 < e^{-1}$ $\Rightarrow$ $x < e^{-1/3}$, so the domain is

$(-\infty, e^{-1/3})$ and $x = e^{-1/3}$ $[\approx 0.72]$ is a vertical asymptote.

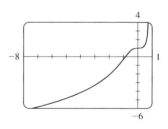

13. $\dfrac{d}{dx}(y) = \dfrac{d}{dx}(ke^x)$ $\Rightarrow$ $y' = ke^x = y$, so the orthogonal trajectories must have $y' = -\dfrac{1}{y}$ $\Rightarrow$ $\dfrac{dy}{dx} = -\dfrac{1}{y}$ $\Rightarrow$

$y\,dy = -dx$ $\Rightarrow$ $\int y\,dy = -\int dx$ $\Rightarrow$ $\tfrac{1}{2}y^2 = -x + C$ $\Rightarrow$ $x = C - \tfrac{1}{2}y^2$, which are parabolas with a horizontal axis.

14. $\dfrac{d}{dx}(y) = \dfrac{d}{dx}(e^{kx})$ $\Rightarrow$ $y' = ke^{kx} = ky = \dfrac{\ln y}{x} \cdot y$, so the orthogonal trajectories must have $y' = -\dfrac{x}{y\ln y}$ $\Rightarrow$

$\dfrac{dy}{dx} = -\dfrac{x}{y\ln y}$ $\Rightarrow$ $y\ln y\,dy = -x\,dx$ $\Rightarrow$ $\int y\ln y\,dy = -\int x\,dx$ $\Rightarrow$ $\tfrac{1}{2}y^2 \ln y - \tfrac{1}{4}y^2$ [parts with $u = \ln y$,

$dv = y\,dy$] $= -\tfrac{1}{2}x^2 + C_1$ $\Rightarrow$ $2y^2 \ln y - y^2 = C - 2x^2$.